30 *Years*

Previous Solved Papers

AF660732

GATE 2025

Civil Engineering

- Fully solved with explanations
- Analysis of previous papers
- Topicwise presentation
- Thoroughly revised & updated

MADE EASY PUBLICATIONS

www.madeeasy.in

MADE EASY Publications Pvt. Ltd.

Corporate Office: 44-A/4, Kalu Sarai (Near Hauz Khas Metro Station), New Delhi-110016 | **Ph. :** 9021300500

Email : infomep@madeeasy.in | **Web :** www.madeeasypublications.org

GATE - 2025

Civil Engineering

Topicwise Previous GATE Solved Papers (1995-2024)

Editions

1st Edition	:	2008
2nd Edition	:	2009
3rd Edition	:	2010
4th Edition	:	2011
5th Edition	:	2012
6th Edition	:	2013
7th Edition	:	2014
8th Edition	:	2015
9th Edition	:	2016
10th Edition	:	2017
11th Edition	:	2018
12th Edition	:	2019
13th Edition	:	2020
14th Edition	:	2021
15th Edition	:	2022
16th Edition	:	2023
17th Edition	**:**	**2024**

Preface

B. Singh (Ex. IES)

Over the period of time the GATE examination has become more challenging due to increasing number of candidates. Though every candidate has ability to succeed but competitive environment, in-depth knowledge, quality guidance and good source of study is required to achieve high level goals.

The new edition of **GATE 2025 Solved Papers : Civil Engineering** has been fully revised, updated and edited. The whole book has been divided into topicwise sections.

At the beginning of each subject, analysis of previous papers are given to improve the understanding of subject.

I have true desire to serve student community by way of providing good source of study and quality guidance. I hope this book will be proved an important tool to succeed in GATE examination. Any suggestions from the readers for the improvement of this book are most welcome.

B. Singh (Ex. IES)
Chairman and Managing Director
MADE EASY Group

GATE-2025

Civil Engineering

CONTENTS

Solid Mechanics

UNIT

I

CONTENTS

Solid Mechanics

Syllabus

Bending moment and shear force in statically determinate beaM; Simple stress and strain relationships; Simple bending theory, flexural and shear stresses, shear centre; Uniform torsion, Transformation of stress; buckling of column, combined and direct bending stresses.

Analysis of Previous GATE Papers

Exam Year	1 Mark Ques.	2 Marks Ques.	Total Marks
1995	2	–	2
1996	3	–	3
1997	2	–	2
1998	–	1	2
1999	4	–	4
2000	7	–	7
2001	1	–	1
2002	2	1	4
2003	2	5	12
2004	1	6	13
2005	2	3	8
2006	3	9	21
2007	3	5	13
2008	–	8	16
2009	2	5	12
2010	5	1	7
2011	1	3	7
2012	4	2	8
2013	4	2	8
2014 Set-1	1	3	7
2014 Set-2	1	5	11

Exam Year	1 Mark Ques.	2 Marks Ques.	Total Marks
2015 Set-1	1	2	5
2015 Set-2	2	3	8
2016 Set-1	–	3	6
2016 Set-2	–	3	6
2017 Set-1	2	2	6
2017 Set 2	1	4	9
2018 Set-1	2	3	8
2018 Set-2	–	1	2
2019 Set-1	2	1	4
2019 Set-2	2	1	4
2020 Set-1	1	2	5
2020 Set-2	2	2	6
2021 Set-1	2	1	4
2021 Set-2	3	1	5
2022 Set-1	2	–	2
2022 Set-2	2	1	4
2023 Set-1	1	2	5
2023 Set-2	2	4	10
2024 Set-1	1	3	7
2024 Set-2	2	1	4

1 Properties of Metals, Stress & Strain

1.1 The material that exhibits the same elastic properties in all direction at a point is said to be
(a) homogeneous (b) orthotropic
(c) viscoelastic (d) isotropic [1995 : 1 M]

1.2 The stress-strain diagram for two materials *A* and *B* is shown below:

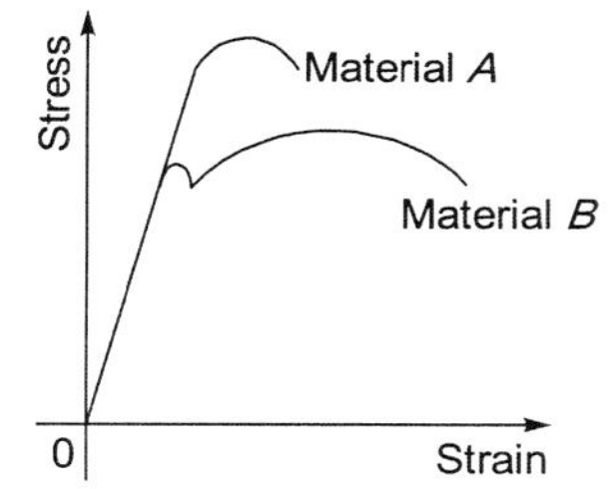

The following statements are made based on this diagram:
I. Material *A* is more brittle than material *B*.
II. The ultimate strength of material *B* is more than that of *A*.
With reference to the above statements, which of the following applies?
(a) Both the statements are false
(b) Both the statements are true
(c) I is true but II is false
(d) I is false but II is true [2000 : 1 M]

1.3 The dimensions for the flexural rigidity of a beam element in mass (*M*), length (*L*) and time (*T*) is given by
(a) MT^{-2} (b) ML^3T^{-2}
(c) $ML^{-1}T^{-2}$ (d) $ML^{-1}T^2$ [2000 : 1 M]

1.4 The shear modulus (*G*), modulus of elasticity (*E*) and the Poisson's ratio (ν) of a material are related as

(a) $G=\frac{E}{[2(1+\nu)]}$ (b) $E=\frac{G}{[2(1+\nu)]}$

(c) $G=\frac{E}{[2(1-\nu)]}$ (d) $G=\frac{E}{[2(\nu-1)]}$

[2002 : 1 M]

1.5 A bar of varying square cross-section is loaded symmetrically as shown in the figure. Loads shown are placed on one of the axes of symmetry of cross-section. Ignoring self weight, the maximum tensile stress in N/mm^2 anywhere is

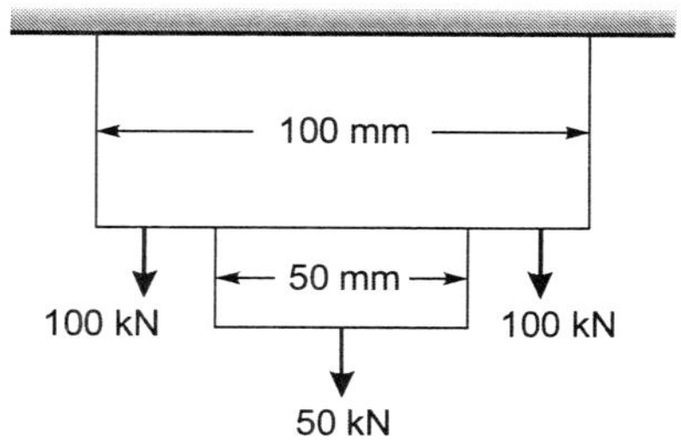

(a) 16.0 (b) 20.0
(c) 25.0 (d) 30.0 [2003 : 1 M]

1.6 For linear elastic systeM, the type of displacement function for the strain energy is
(a) linear (b) quadratic
(c) cubic (d) quartic [2004 : 1 M]

1.7 The symmetry of stress tensor at a point in the body under equilibrium is obtained from
(a) conservation of mass
(b) force equilibrium equations
(c) moment equilibrium equations
(d) conservation of energy [2005 : 1 M]

1.8 For an isotropic material, the relationship between the Young's modulus (*E*), shear modulus (*G*) and Poisson's ratio (μ) is given by

(a) $G=\frac{E}{2(1+\mu)}$ (b) $E=\frac{G}{2(1+\mu)}$

(c) $G=\frac{E}{(1+2\mu)}$ (d) $G=\frac{E}{2(1-2\mu)}$

[2007 : 1 M]

1.9 A metal bar of length 100 mm is inserted between two rigid supports and its temperature is increased by 10°C. If the coefficient of thermal expansion is 12×10^{-6} per °C and the Young's modulus is 2×10^5 MPa, the stress in the bar is
(a) zero (b) 12 MPa
(c) 24 MPa (d) 2400 MPa
[2007 : 2 M]

1.10 A rigid bar is suspended by three rods made of the same material. The area and length of the central rod are 3A and L, respectively while that of the two outer rods are 2A and 2L, respectively. If a downward force of 50 kN is applied to the rigid bar, the forces in the central and each of the outer rods will be

(a) 16.67 kN each (b) 30 kN and 15 kN
(c) 30 kN and 10 kN (d) 21.4 kN and 14.3 kN
[2007 : 2 M]

1.11 U_1 and U_2 are the strain energies stored in a prismatic bar due to axial tensile forces P_1 and P_2, respectively. The strain energy U stored in the same bar due to combined action of P_1 and P_2 will be
(a) $U = U_1 + U_2$ (b) $U = U_1 U_2$
(c) $U < U_1 + U_2$ (d) $U > U_1 + U_2$
[2007 : 2 M]

1.12 A mild steel specimen is under uniaxial tensile stress. Young's modulus and yield stress for mild steel are 2×10^5 MPa and 250 MPa respectively. The maximum amount of strain energy per unit volume that can be stored in this specimen without permanent set is
(a) 156 Nmm/mm^3 (b) 15.6 Nmm/mm^3
(c) 1.56 Nmm/mm^3 (d) 0.156 Nmm/mm^3
[2008 : 1 M]

1.13 A vertical rod *PQ* of length L is fixed at its top end *P* and has a flange fixed to the bottom end *Q*. A weight W is dropped vertically from a height h (<L) on to the flange. The axial stress in the rod can be reduced by
(a) increasing the length of the rod
(b) decreasing the length of the rod
(c) decreasing the area of cross-section of the rod
(d) increasing the modulus of elasticity of the material [2008 : 2 M]

1.14 The number of independent elastic constants for a linear elastic isotropic and homo-geneous material is
(a) 4 (b) 3
(c) 2 (d) 1 [2010 : 1 M]

Linked Answer Questions 1.15 and 1.16:

A rigid beam is hinged at one end and supported on linear elastic springs (both having a stiffness of 'k') at points '1' and '2', and an inclined load acts at '2', as shown.

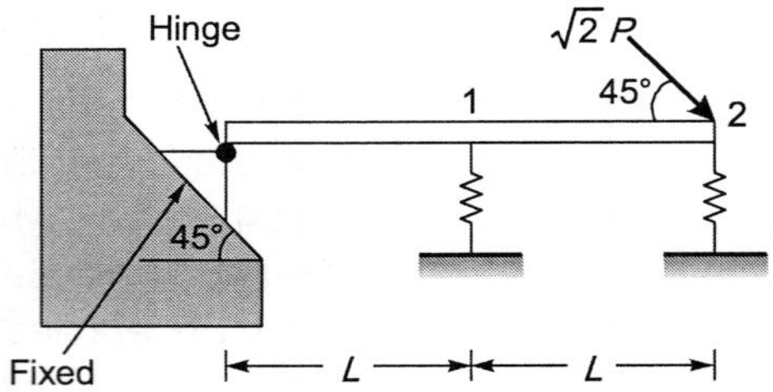

1.15 Which of the following options represents the deflections δ_1 and δ_2 at points '1' and '2'?

(a) $\delta_1 = \frac{2}{5}\left(\frac{2P}{k}\right)$ and $\delta_2 = \frac{4}{5}\left(\frac{2P}{k}\right)$

(b) $\delta_1 = \frac{2}{5}\left(\frac{P}{k}\right)$ and $\delta_2 = \frac{4}{5}\left(\frac{P}{k}\right)$

(c) $\delta_1 = \frac{2}{5}\left(\frac{P}{\sqrt{2}k}\right)$ and $\delta_2 = \frac{4}{5}\left(\frac{P}{\sqrt{2}k}\right)$

(d) $\delta_1 = \frac{2}{5}\left(\frac{\sqrt{2}P}{k}\right)$ and $\delta_2 = \frac{4}{5}\left(\frac{\sqrt{2}P}{k}\right)$

[2011 : 2 M]

1.16 If the load P equals 100 kN, which of the following options represents forces R_1 and R_2 in the springs at points '1' and '2'?
(a) R_1 = 20 kN and R_2 = 40 kN
(b) R_1 = 50 kN and R_2 = 50 kN
(c) R_1 = 30 kN and R_2 = 60 kN
(d) R_1 = 40 kN and R_2 = 80 kN [2011 : 2 M]

1.17 The Poisson's ratio is defined as

(a) $\left|\frac{\text{axial stress}}{\text{lateral stress}}\right|$ (b) $\left|\frac{\text{lateral strain}}{\text{axial strain}}\right|$

(c) $\left|\frac{\text{lateral stress}}{\text{axial stress}}\right|$ (d) $\left|\frac{\text{axial strain}}{\text{lateral strain}}\right|$

[2012 : 1 M]

1.18 Creep strains are
(a) caused due to dead load only
(b) caused due to live load only
(c) caused due to cyclic load only
(d) independent of load [2013 : 1 M]

1.19 A box of weight 100 kN shown in the figure is to be lifted without swinging. If all forces are coplanar, the magnitude and direction (θ) of the force (F) with respect to x-axis should be

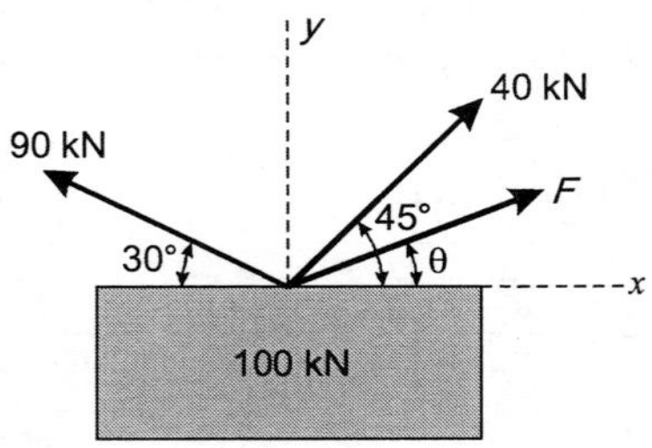

(a) F = 56.389 kN and θ = 28.28°
(b) F = – 56.389 kN and θ = – 28.28°
(c) F = 9.055 kN and θ = 1.414°
(d) F = – 9.055 kN and θ = – 1.414°
[2014 : 2 M, Set-I]

1.20 The values of axial stress (σ) in kN/m^2, bending moment (M) in kNm, and shear force (V) in kN acting at point P for the arrangement shown in the figure are respectively

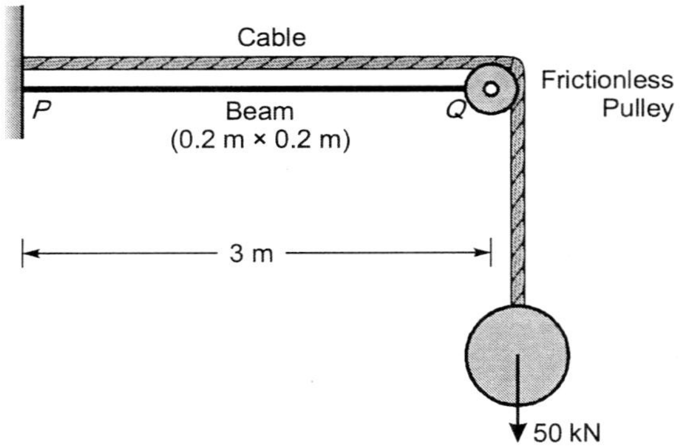

(a) 1000, 75 and 25 (b) 1250, 150 and 50
(c) 1500, 225 and 75 (d) 1750, 300 and 100

[2014 : 2 M, Set-II]

1.21 For the state of stresses (in MPa) shown in the figure below, the maximum shear stress (in MPa) is ______.

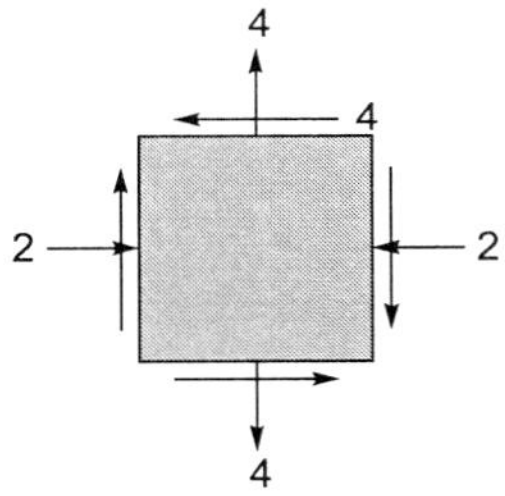

[2014 : 2 M, Set-II]

1.22 A tapered circular rod of diameter varying from 20 mm to 10 mm is connected to another uniform circular rod of diameter 10 mm as shown in the following figure. Both bars are made of same material with the modulus of elasticity, $E = 2 \times 10^5$ MPa. When subjected to a load $P = 30\pi$ kN, the deflection at point A is ____ mm.

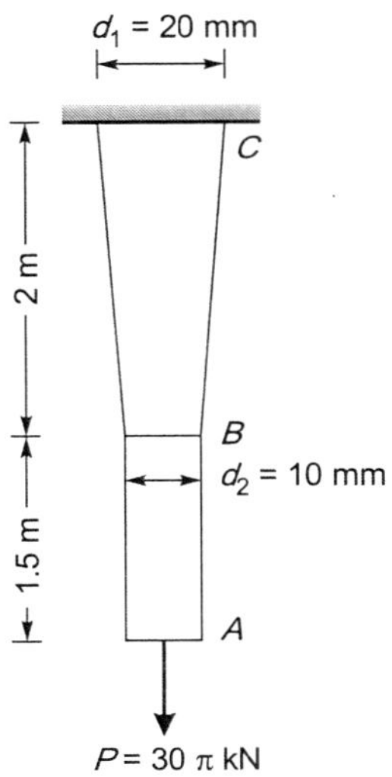

[2015 : 2 M, Set-I]

1.23 An elastic isotropic body is in a hydrostatic state of stress as shown in the figure. For no change in the volume to occur, what should be its Poisson's ratio?

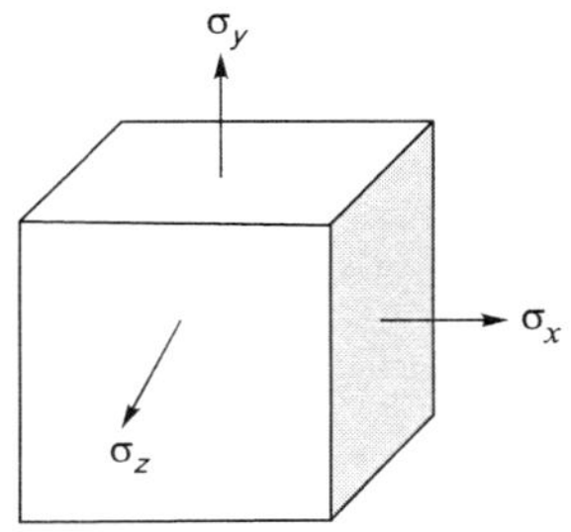

(a) 0.00 (b) 0.25
(c) 0.50 (d) 1.00

[2016 : 2 M, Set-II]

1.24 An elastic bar of length L, uniform cross-sectional area A, coefficient of thermal expansion α, and Young's modulus E is fixed at the two ends. The temperature of the bar is increased by T, resulting in an axial stress σ. Keeping all other parameters unchanged, if the length of the bar is doubled, the axial stress would be

(a) σ (b) 2σ
(c) 0.5σ (d) 0.25σ

[2017 : 1 M, Set-I]

1.25 Consider the stepped bar made with a linear elastic material and subjected to an axial load of 1 kN, as shown in the figure.

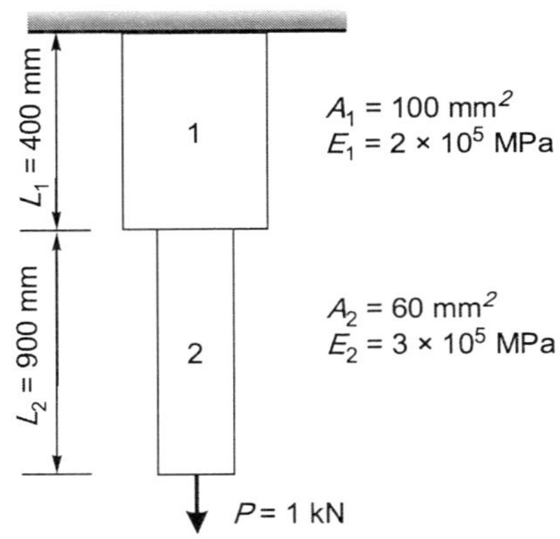

Segments 1 and 2 have cross-sectional area of 100 mm^2 and 60 mm^2. Young's modulus of 2×10^5 MPa and 3×10^5 MPa, and length of 400 mm and 900 mm, respectively. The strain energy (in N-mm upto one decimal place) in the bar due to the axial load is _________.

[2017 : 2 M, Set-I]

1.26 In a material under a state of plane strain, a 10×10 mm square centered at a point gets deformed as shown in the figure.

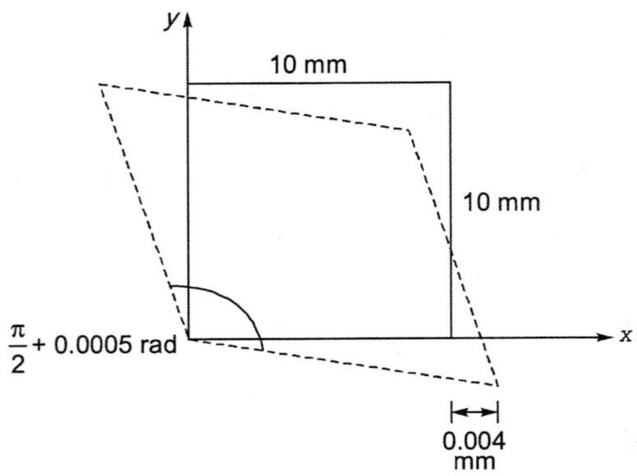

If the shear strain γ_{xy} at this point is expressed as 0.001 k (in rad), the value of k is

(a) 0.50 (b) 0.25
(c) –0.25 (d) –0.50

[2017 : 1 M, Set-II]

1.27 A 2 m long, axially loaded mild steel rod of 8 mm diameter exhibits the load-displacement (P-δ) behavior as shown in the figure.

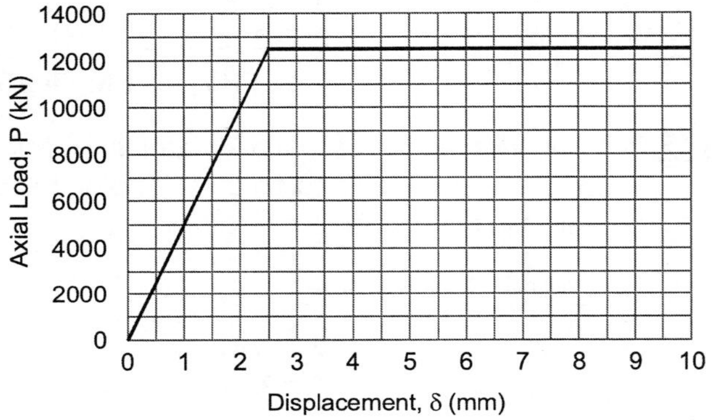

Assume the yield stress of steel as 250 MPa. The complementary strain energy (in N-mm) stored in the bar up to its linear elastic behaviour will be ___________.

[2017 : 2 M, Set-II]

1.28 A plate in equilibrium is subjected to uniform stresses along its edges with magnitude σ_{xx} = 30 MPa and σ_{yy} = 50 MPa as shown in the figure

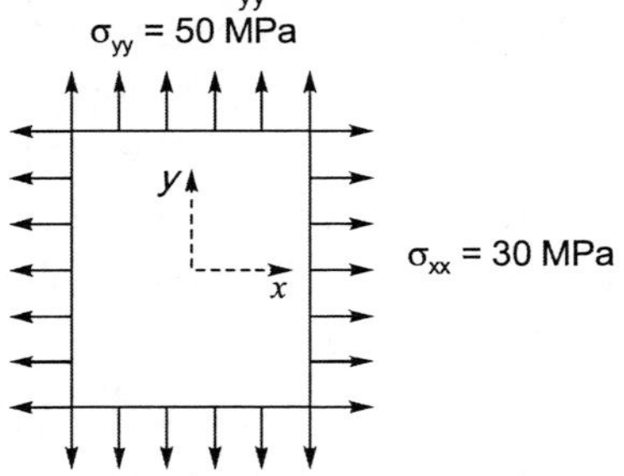

The Young's modulus of the material is 2 × 10^{11} N/m^2 and the Poisson's ratio is 0.3. If σ_{zz} is negligibly small and assumed to be zero, then the strain ε_{zz} is

(a) -120×10^{-6} (b) -60×10^{-6}
(c) 0.0 (d) 120×10^{-6}

[2018 : 2 M, Set-I]

1.29 A rigid, uniform, weightless, horizontal bar is connected to three vertical members P, Q and R as shown in the figure (not drawn to the scale). All three members have identical axial stiffness of 10 kN/mm. The lower ends of bars P and R rest on a rigid horizontal surface. When NO load is applied, a gap of 2 mm exists between the lower end of the bar Q and the rigid horizontal surface. When a vertical load W is placed on the horizontal bar in the downward direction, the bar still remains horizontal and gets displaced by 5 mm in the vertically downward direction.

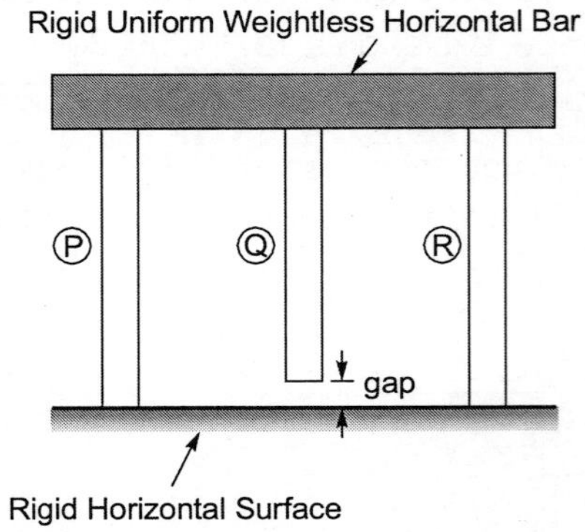

The magnitude of the load W (in kN, round off to the nearest integer), is ______.

[2020 : 2 M, Set-I]

1.30 A prismatic linearly elastic bar of length, L, cross-sectional area A, and made up of a material with Young's modulus E, is subjected to axial tensile force as shown in the figures. When the bar is subjected to axial tensile force P_1 and P_2, the strain energies stored in the bar are U_1 and U_2, respectively.

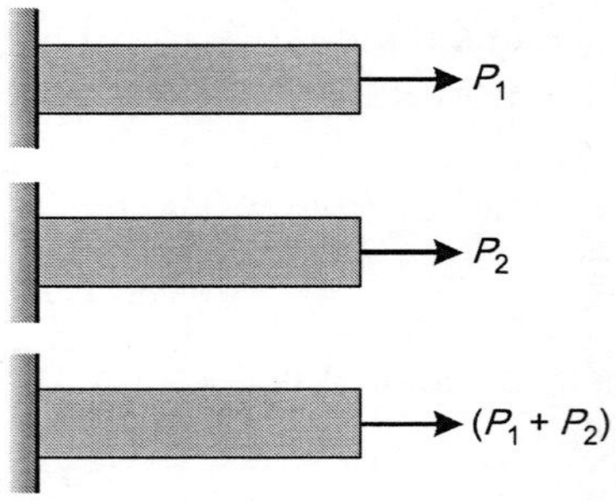

If U is the strain energy stored in the same bar when subjected to an axial tensile force $(P_1 + P_2)$, the correct relationship is

(a) $U = U_1 - U_2$ (b) $U = U_1 + U_2$
(c) $U < U_1 + U_2$ (d) $U > U_1 + U_2$

[2020 : 2 M, Set-II]

1.31 A square plate O-P-Q-R of a linear elastic material with sides 1.0 m is loaded in a state of plane stress. Under a given stress condition, the plate deforM to a new configuration O-P′-Q′-R′ as shown in the figure (not to scale). Under the given deformation, the edges of the plate remain straight.

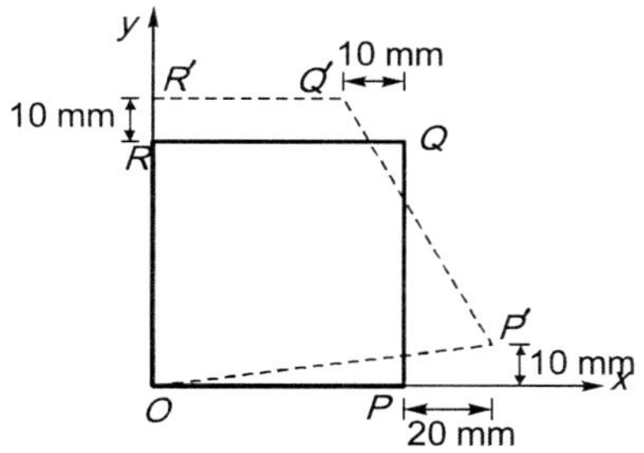

The horizontal displacement of the point (0.5 m, 0.5 m) in the plate O-P-Q-R (in mm, round off to one decimal place) is ________.

[2021 : 2 M, Set-I]

1.32 Strain hardening of structural steel means

(a) strengthening steel member externally for reducing strain experienced

(b) experiencing higher stress than yield stress with increased deformation

(c) strain occurring before plastic flow of steel material

(d) decrease in the stress experienced with increasing strain

[2021 : 1 M, Set-II]

1.33 Stresses acting on an infinitesimal soil element are shown in the figure (with $\sigma_z > \sigma_x$). The major and minor principal stresses are σ_1 and σ_3, respectively. Considering the compressive stresses as positive,which one of the following expressions correctly represents the angle between the major principal stress plane and the horizontal plane?

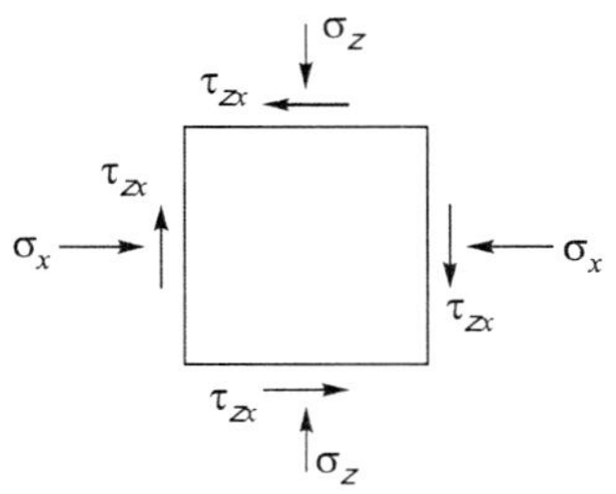

(a) $\tan^{-1}\left(\dfrac{\tau_{zx}}{\sigma_1+\sigma_3}\right)$ (b) $\tan^{-1}\left(\dfrac{\tau_{zx}}{\sigma_3-\sigma_x}\right)$

(c) $\tan^{-1}\left(\dfrac{\tau_{zx}}{\sigma_1-\sigma_x}\right)$ (d) $\tan^{-1}\left(\dfrac{\tau_{zx}}{\sigma_1+\sigma_x}\right)$

[2022 : 1 M, Set-II]

1.34 For a linear elastic and isotropic material, the correct relationship among Young's modulus of elasticity (E), Poisson's ratio (v), and shear modulus (G) is

(a) $G=\dfrac{E}{2(1+v)}$ (b) $G=\dfrac{E}{(1+2v)}$

(c) $E=\dfrac{G}{2(1+v)}$ (d) $E=\dfrac{G}{(1+2v)}$

[2022 : 2 M, Set-II]

1.35 Consider two linearly elastic rods HI and IJ. each of length b, as shown in the figure. The rods are co-linear, and confined between two fixed supports at if and J. Both the rods are initially stress free, The coefficient of linear thermal expansion is a for both the rods. The temperature of the rod IJ is raised by ΔT. whereas the temperature of rod HI remains unchanged. An external horizontal force P is now applied at node I. It is given that $a = 10^{-6}\,°C^{-1}$, $\Delta T = 50°$ C, b = 2m, AE = 10^6N. The axial rigidities of the rods HI and U are 2 AE and AE, respectively. To make the axial force in rod HI equal to zero, the value of the external force P (in N) is ____. (rounded off to the nearest integer).

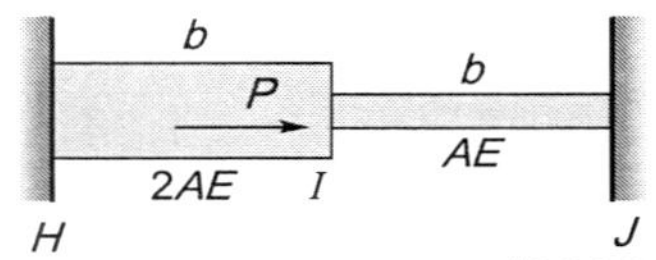

[2022 : 1 M, Set-II]

1.36 A hanger is made of two bars of different sizes. Each bar has a square cross-section. The hanger is loaded by three-point loads in the mid vertical plane as shown in the figure. Ignore the self-weight of the hanger. What is the maximum tensile stress in N/mm^2 anywhere in the hanger without considering stress concentration effects?

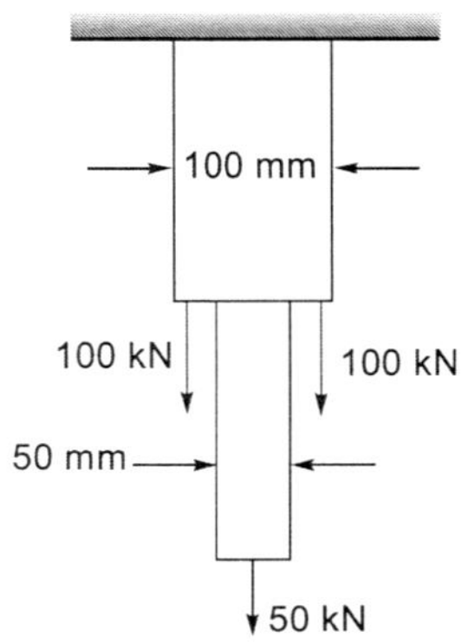

(a) 15.0 (b) 25.0

(c) 35.0 (d) 45.0

[2023 : 1 M, Set-I]

1.37 A 2D thin plate with modulus of elasticity, $E = 1.0$ N/m^2, and Poisson's ratio, $\mu = 0.5$, is in plane stress condition. The displacement field in the plate is given by $u = Cx^2y$. and $v = 0$, where u and v are displacements (in m) along the X and Y directions, respectively, and C is a constant (in m^{-2}). The distances x and y along X and Y, respectively, are in m. The stress in the X direction is $\sigma_{XX} = 40xy$ N/m^2, and the shear stress is $\tau_{XY} = \alpha x^2$ N/m^2. What is the value of α (in N/m^4, in integer)? _______

[2023 : 2 M, Set-II]

1.38 A 3 m long, horizontal, rigid, uniform bean PQ has negligible mass. The beam is subjected to a 3kN concentrated vertically downward force at 1 m from P, as shown in the figure. The beam is resting on vertical linear springs at the ends P and Q. For the spring at the end P, the spring constant $K_P = 100$kN/m.

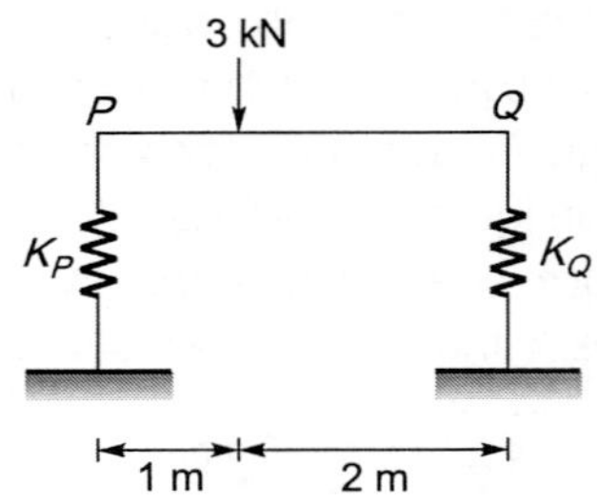

(Figure NOT to scale)

If the beam DOES NOT rotate under the application of the force and displaces only vertically, the value of the spring constant K_Q (in kN/m) for the spring at the end Q

(a) 200 (b) 50

(c) 100 (d) 150

[2024 : 1 M, Set-II]

Answers Properties of Metals, Stress & Strain

1.1 (d)	1.2 (c)	1.3 (b)	1.4 (a)	1.5 (c)	1.6 (b)	1.7 (c)
1.8 (a)	1.9 (c)	1.10 (c)	1.11 (d)	1.12 (d)	1.13 (a)	1.14 (c)
1.15 (b)	1.16 (d)	1.17 (b)	1.18 (a)	1.19 (a)	1.20 (b)	1.21 (5.0)
1.22 (15)	1.23 (c)	1.24 (a)	1.25 (35)	1.26 (d)	1.27 (15625000)	
1.28 (a)	1.29 (130)	1.30 (d)	1.31 (2.5)	1.32 (b)	1.33 (c)	1.34 (a)
1.35 (50)	1.36 (b)	1.37 (5)	1.38 (b)			

Explanations Properties of Metals, Stress & Strain

1.1 (d)

The material that exhibits the same elastic properties in all directions at a point is said to be **isotropic**.

If the material exhibits different properties in different directions, then it is said to be **Anisotropic**.

1.2 (c)

Since strain in material B is more, hence it is more ductile than material A i.e., material A is more brittle than material B. Hence **statement I is true**. Material A can reach upto higher stress level hence ultimate strength of material A is more than that of material B. Hence **statement II is false**.

1.3 (b)

Flexural regidity = $EI = ML^{-1}T^{-2} \times L^4 = ML^3T^{-2}$

1.4 (a)

For isotropic and homogenous materials,

$$E = 2G(1+\mu)$$
$$E = 3k(1-2\mu)$$

1.5 (c)

Method-I

The stress in lower bar

$$= \frac{50 \times 1000}{50 \times 50} = 20 \text{ N/mm}^2$$

The stress in upper bar

$$= \frac{250 \times 1000}{100 \times 100} = 25 \text{ N/mm}^2$$

Thus the maximum tensile stress any where in the bar is 25 N/mm^2.

Method-II

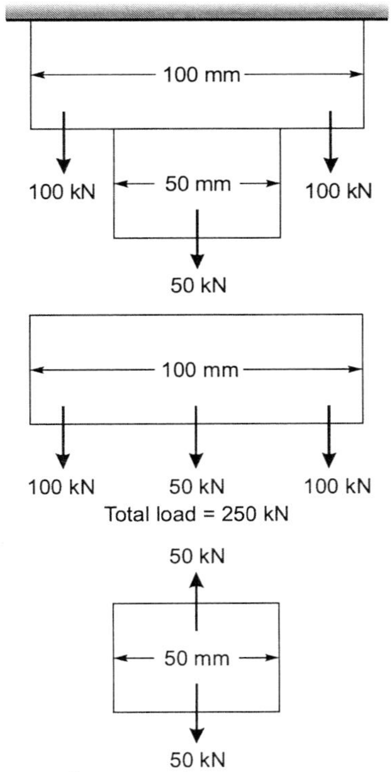

Stress in upper bar

$$= \frac{250 \times 1000}{100 \times 100} = 25 \text{ N/mm}^2$$

$$\text{Stress in lower bar} = \frac{50 \times 10^3}{50 \times 50} = 20 \text{ N/mm}^2$$

So, maximum tensile stress = 25 N/mm^2

1.6 (b)

$$\text{Strain Energy} = \frac{1}{2} \times \sigma \times \varepsilon = \frac{1}{2} E\varepsilon^2$$

Since strain is directly proportional to displacement so strain energy is directly proportional to quadratic equation of displacement.

1.7 (c)

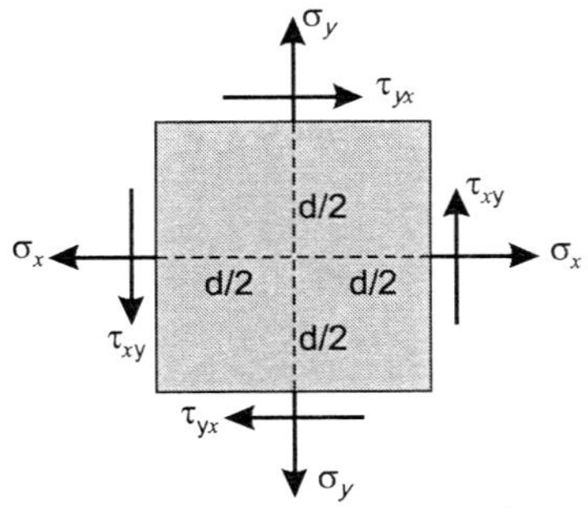

Taking moment equilibrium about the centre, we get,

$$\tau_{yx} \times \frac{d}{2} + \tau_{yx} \times \frac{d}{2} = \tau_{xy} \times \frac{d}{2} + \tau_{xy} \times \frac{d}{2}$$

$$\therefore \qquad \tau_{xy} = \tau_{yx}$$

1.8 (a)

For isotropic and homogenous materials,

$$E = 2G(1 + \mu)$$

$$E = 3k(1 - 2\mu)$$

1.9 (c)

Method-I

Temperature stress

$$= \alpha TE$$

$$= 12 \times 10^{-6} \times 10 \times 2 \times 10^5$$

$$= 24 \text{ MPa}$$

Method-II

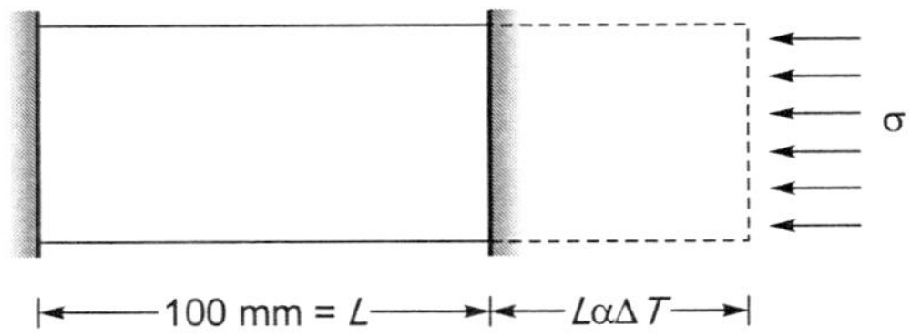

Due to temperature,

$$\Delta L = L\alpha\Delta T$$

But since support in fixed so, expansion is not allowed so stress is developed in the bar which is compressive in nature.

Now, $\text{Expansion due to temperature} = \text{Compression due to stress}$

$$L\alpha\Delta T = \frac{\sigma}{E} \times L$$

$$\sigma = E\alpha\Delta T = 1 \times 10^5 \times 12 \times 10^{-6} \times 10$$

$$\simeq 24 \text{ MPa}$$

1.10 (c)

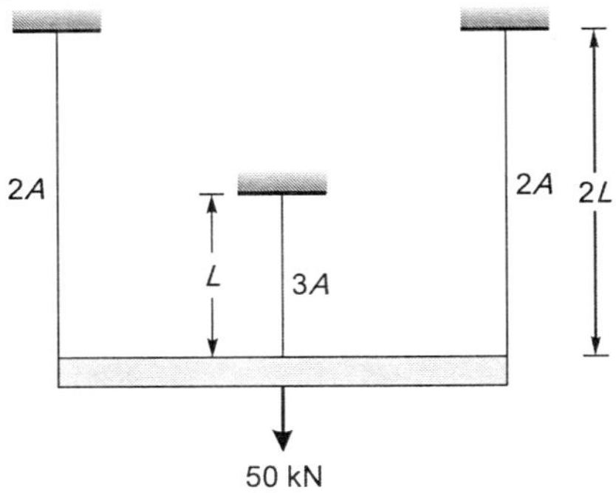

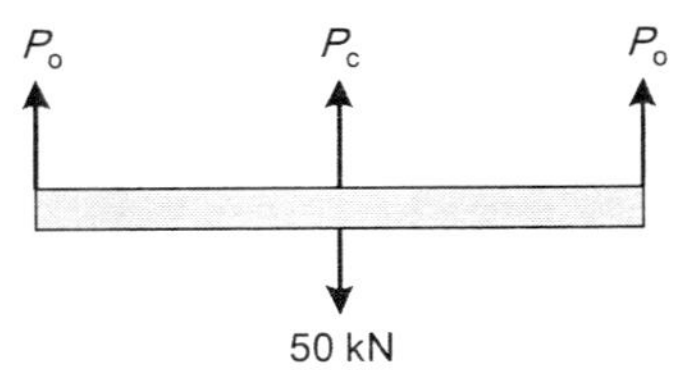

From symmetry, the force in each of outer rods is P_0 and force in the central rod is P_C, then,

$$2P_0 + P_C = 50 \quad \text{...(i)}$$

Also, the elongation of central rod and outer rods is same.

Since all the rods are connected by rigid bar. So, deflection in each rod will be same

$$\therefore \quad \frac{P_0 L_0}{A_0 E} = \frac{P_C L_C}{A_C E}$$

$$\Rightarrow \quad \frac{P_0 \times 2L}{2A} = \frac{P_C \times L}{3A}$$

$$\Rightarrow \quad P_C = 3P_0 \quad \text{...(ii)}$$

Solving (i) and (ii), we get,

$P_C = 30$ kN and $P_0 = 10$ kN

1.11 (d)

Strain energy stored in a prismatic bar due to axial load P_1,

$$U_1 = \frac{P_1^2 L}{2AE}$$

Strain energy stored in a prismatic bar due to axial load, P_2,

$$U_2 = \frac{P_2^2 L}{2AE}$$

Strain energy stored in a prismatic bar due to combined action of P_1 and P_2,

$$U = \frac{(P_1 + P_2)^2 L}{2AE}$$

$$U = \frac{P_1^2 L}{2AE} + \frac{P_2^2 L}{2AE} + \frac{P_1 P_2 L}{AE}$$

$$U = U_1 + U_2 + \frac{P_1 P_2 L}{AE}$$

$$\Rightarrow \quad U > U_1 + U_2$$

1.12 (d)

The strain energy per unit volume may be given as

$$U = \frac{1}{2} \times \text{Stresses} \times \text{Strain}$$

$$U = \frac{1}{2} \times \frac{\sigma_y^2}{E} = \frac{1}{2} \times \frac{(250)^2}{2 \times 10^5} = 0.156 \text{ N-mm/mm}^3$$

1.13 (a)

The kinetic energy of the weight (W) at flange level will be transformed in the form of strain energy in the rod.

We know that,

$U = \sigma^2/2E \times$ Area of cross-section of rod $\times$ length of rod.

Since, energy remains constant, hence to reduce the axial stress in the rod, the length of the rod should be increased or area of cross-section of the rod should be increased or modulus of elasticity should be decreased.

1.14 (c)

Type of material	Number of independent elastic constants
Isotropic	2
Orthotropic	9
Antisotropic	21

1.15 (b)

The free diagram of the beam is shown below,

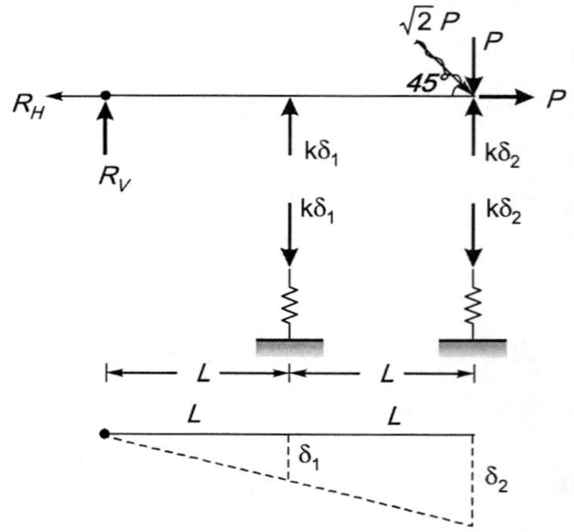

From similar triangles, we get,

$$\frac{L}{\delta_1} = \frac{2L}{\delta_2}$$

$$\Rightarrow \quad \delta_2 = 2\delta_1 \quad \text{...(i)}$$

Taking moments about hinge, we get,

$P \times 2L - k\delta_2 \times 2L - k\delta_1 \times L = 0$

$\Rightarrow 2P - k(2\delta_2 + \delta_1) = 0$ [$\because$ from (i)]

$\Rightarrow 2P - k(4\delta_1 + \delta_1) = 0$

$$\Rightarrow \quad \delta_1 = \frac{2P}{5k}$$

From (i), we get,

$$\delta_2 = 2 \times \frac{2P}{5k} = \frac{4P}{5k}$$

1.16 (d)

$$R_1 = k\delta_1 = k \times \frac{2P}{5k} = \frac{2 \times 100}{5} = 40 \text{ kN}$$

$$R_2 = k\delta_2 = k \times \frac{4P}{5k} = \frac{4 \times 100}{5} = 80 \text{ kN}$$

1.17 (b)

$$\mu = \left| \frac{\text{Lateral strain}}{\text{Axial strain}} \right|$$

1.18 (a)

Creep strains occur with time at a constant level of stress. So they occur due to permanent load i.e., Dead Load.

1.19 (a)

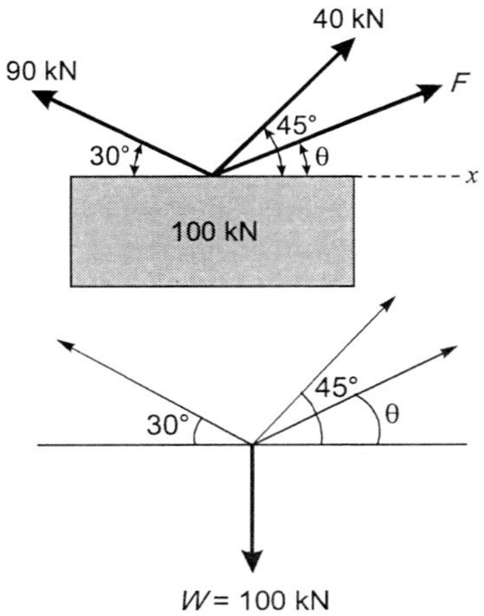

For no swinging force should be balanced i.e.,

$$\Sigma F_H = 0$$
$$\Sigma F_V = 0$$
$$\Sigma F_{horizontal} = 0$$
$$90 \cos 30° = 40 \cos 45° + F \cos \theta$$
$$77.94 = 28.28 + F \cos \theta$$

or $$F \cos\theta = 49.66 \text{ kN} \quad ...(i)$$

$$\Sigma F_{vertical} = 100$$
$$90 \sin 30° + 40 \sin 45° + F \sin \theta = W$$
$$45 + 28.28 + F \sin \theta = 100$$

or $$F \sin\theta = 26.72 \text{ kN} \quad ...(ii)$$

Dividing eq. (ii) by eq. (i), we get

$$\tan \theta = \frac{26.72}{49.66} = 0.5380$$

or $$\theta = \tan^{-1}(0.5380) = 28.28°$$

Substituting $\theta = 28.28$ in eq. (i), we get

$$F \cos 28.28° = 49.66$$

or $$F = \frac{49.66}{\cos 28.28°} = \frac{49.66}{0.88} = 56.43 \text{ kN}$$

1.20 (b)

Loading after removing the cable,

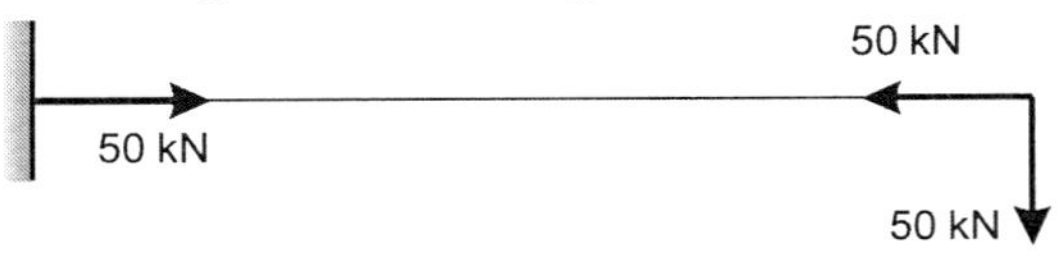

$$\text{Axial stress} = \frac{50}{0.2 \times 0.2} = 1250 \text{ kN/m}^2$$

$$\text{Bending moment} = 50 \times 3 = 150 \text{ kNm}$$

$$\text{Shear force} = 50 \text{ kN}$$

1.21 Sol.

$$\tau_{max} = \frac{\sigma_1 - \sigma_2}{2} = \sqrt{\left(\frac{\sigma_x - \sigma_y}{2}\right)^2 + \tau_{xy}^2} = 5.0 \text{ MPa}$$

NOTE: In this case σ_{P1} = +ve

So it can be found out by Mohr circle, otherwise always check for absolute shear stress.

1.22 Sol.

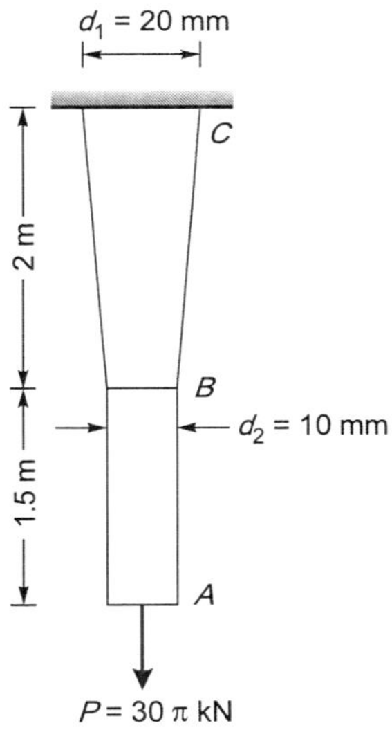

Total elongation,

AB is uniform

So, $$\Delta = \frac{PL}{AE}$$

BC is tapered

$$\Delta = \frac{PL}{\frac{\pi}{4} d_1 d_2 E}$$

$$\Delta = \Delta_{AB} + \Delta_{BC}$$

$$= \frac{PL}{AE} + \frac{4PL}{\pi d_1 d_2 E}$$

$$= \frac{30\pi \times 10^3 \times 1.5 \times 10^3}{\frac{\pi}{4} \times (10)^2 \times 10^5}$$

$$+\frac{30\pi\times10^3\times2\times10^3}{\frac{\pi}{4}\times10\times20\times2\times10^5}$$

$$=(9+6)\text{ mm} = 15\text{ mm}$$

1.23 (c)

Volumetric strain,

$$\epsilon_V = \left(\frac{\sigma_x+\sigma_y+\sigma_z}{3}\right)(1-2\mu)$$

$$\Rightarrow \quad \frac{\delta V}{V} = \left(\frac{\sigma_x+\sigma_y+\sigma_z}{3}\right)(1-2\mu)$$

As $\Delta V = 0$

$\Rightarrow$ Either $\sigma_x + \sigma_y + \sigma_z = 0$ or $1 - 2\mu = 0$

$\Rightarrow \quad 1 - 2\mu = 0$

$\mu = 0.5$

1.24 (a)

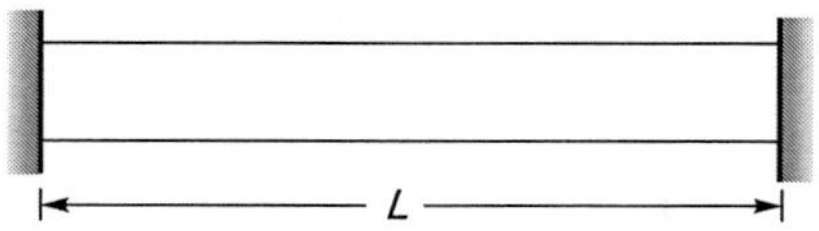

$$\sigma = \alpha TE$$

$\therefore$ Length have no effect on thermal stress.

$\therefore$ Axial stress is only 'σ'.

1.25 Sol.

$A_1 = 100\text{ mm}^2$, $EI = 2\times10^5$ MPa

$A_2 = 60\text{ mm}^2$, $EI = 3\times10^5$ MPa

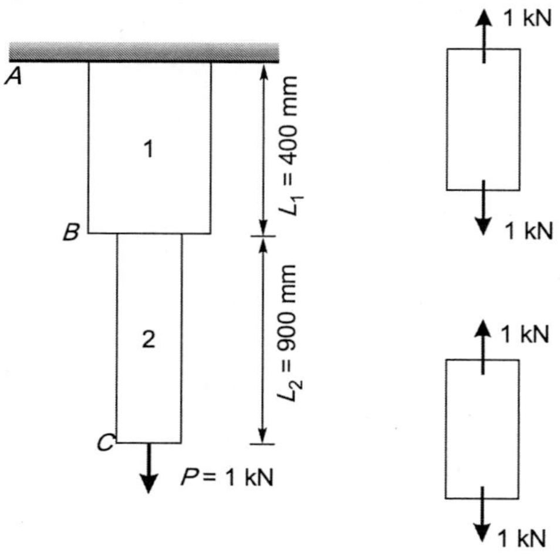

$$\Delta_{AC} = \Delta_{AB} + \Delta_{BC}$$

$$= \frac{1\times10^3\times400}{100\times2\times10^5}+\frac{1\times10^3\times900}{60\times3\times10^5}$$

$$= 0.02 + 0.05 = 0.07\text{ mm}$$

$$U = \frac{1}{2}\times P\times\Delta = \frac{1}{2}\times1\times1000\times0.07 = 35\text{ N-mm}$$

1.26 (d)

According to the sign convention,

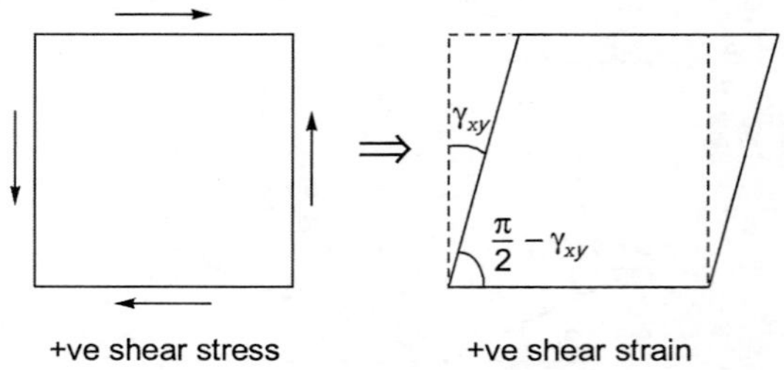

In question since angle has been increase therefore shear strain should be negative.

$$\therefore \quad \gamma_{xy} = -0.0005\text{ rad}$$

$$= 0.001k$$

$$-0.0005 = 0.001k$$

$$\Rightarrow \quad k = -0.50$$

1.27 Sol.

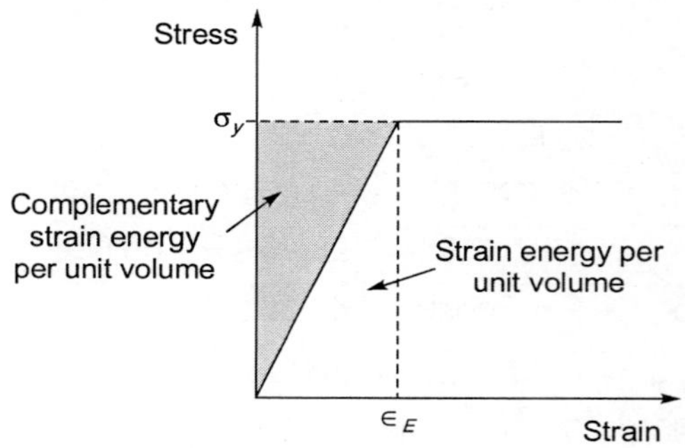

$$\text{Elastic strain}, \epsilon_E = \frac{\Delta L}{L} = \frac{2.5}{2000} = 1.25\times10^{-2}$$

$$\therefore \text{Elastic strain energy} = \frac{1}{2}\sigma_y\,\epsilon_E\,AL$$

$$= \frac{1}{2}\times250\times1.25\times10^{-3}\times\frac{\pi}{4}\times8^2\times2000$$

$$= 15707.96\text{ Nmm}$$

Note: For linear elastic material both complementary energy and strain energy is same.

OR

By considering given graph in question, between Axial Load and Displacement the solution will be as follows:

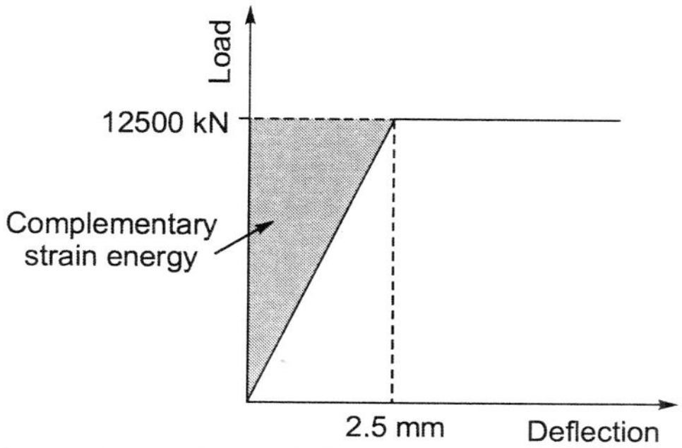

Complementary stain energy,

$$U = \frac{1}{2}P\delta = \frac{1}{2}\left(12500\times10^3\right)\times2.5$$
$$= 15625000 \text{ Nmm}$$

It means there is some error in the given data.

1.28 (a)

$$\sigma_{xx} = 30 \text{ MPa},\ \sigma_{yy} = 50 \text{ MPa},\ \sigma_{zz} = 0$$

$$\varepsilon_{zz} = \frac{\sigma_{ZZ}}{E} - \mu\frac{\sigma_{xx}}{E} - \mu\frac{\sigma_{yy}}{E} = -\frac{\mu}{E}(\sigma_{xx} + \sigma_{yy})$$

$$= -\frac{0.3}{2\times10^5}(30+50) = -120\times10^{-6}$$

1.29 Sol.

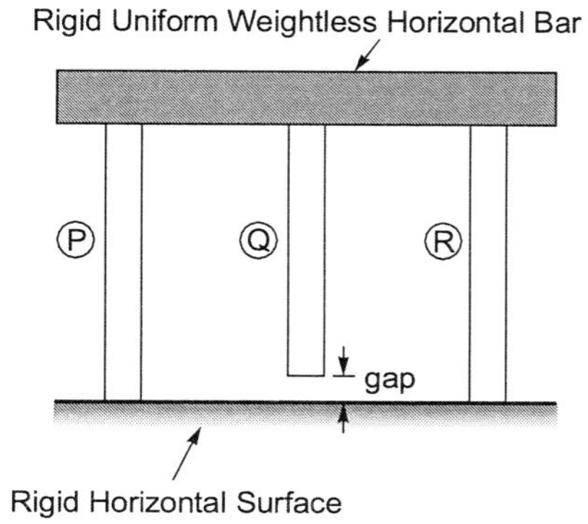

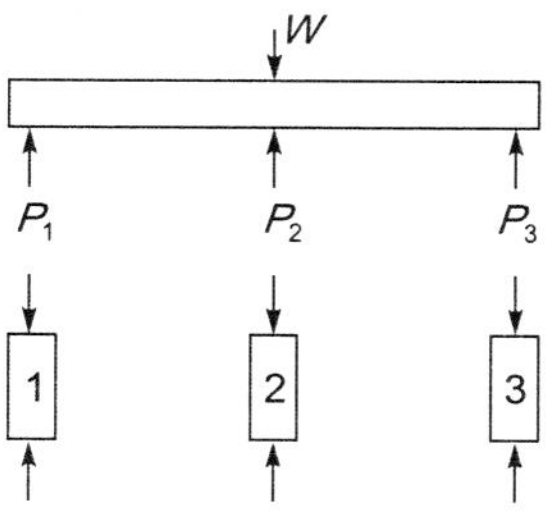

$$P_1 + P_1 + P_2 = W \quad \text{...(i)}$$
$$P_1 = P_3$$
$$\delta_1 = 5 \text{ mm} = \frac{P_1L}{AE} \quad \frac{AE}{L} = 10\ \frac{\text{kN}}{\text{mm}}$$
$$\delta_2 = 3 \text{ mm} = \frac{P_2L}{AE}$$

So, $P_1 = 10 \times 5 = 50$ kN

$$P_2 = 10 \times 3 = 30 \text{ kN}$$
$$W = 2(50) + 30 = 130 \text{ kN}$$

1.30 (d)

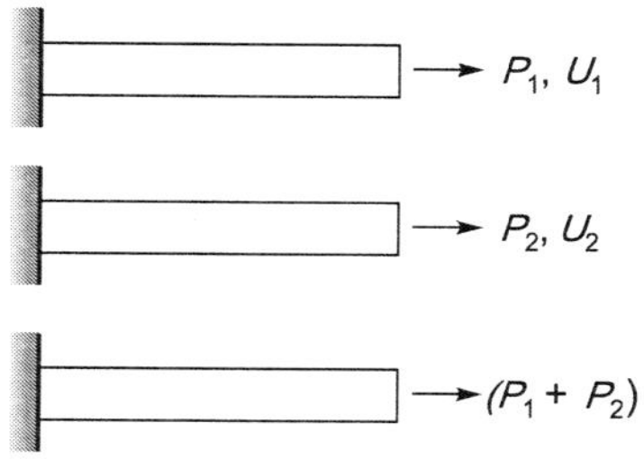

$$U_1 = \frac{P_1^2L}{2AE}$$
$$U_2 = \frac{P_2^2L}{2AE}$$
$$U = \frac{(P_1+P_2)^2L}{2AE}$$

$$(P_1+P_2)^2 > P_1^2 + P_2^2$$
$$U > U_1 + U_2$$

1.31 Sol.

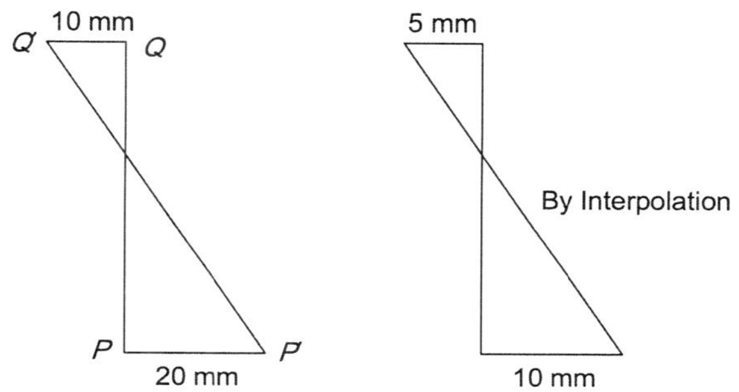

So horizontal displacement of the point (0.5 m, 0.5 m)

$$= -2.5 \text{ mm} + 5 \text{ mm} = 2.5 \text{ mm}$$

1.32 (c)

In strain hardening region, further addition of stress gives additional strain. However, strain inceases with faster rate in this region. The material in this range in this region. The material in this range undergoes changes in its crystalline structure, resulting in increased resistance to further deformation.

1.33 (c)

$$\Sigma F_x = 0$$
$$\Rightarrow -\sigma_x(BC) - \tau_Z \times (AB) + \sigma_1 \sin\theta = 0$$
$$\Rightarrow \sigma_x\left(\frac{AC\sin\theta}{\cos\theta}\right) + \tau_{Zx}\left(\frac{AC\cos\theta}{\cos\theta}\right)$$
$$= \sigma_1\frac{\sin\theta\times AC}{\cos\theta}$$
$$\Rightarrow \sigma_x \tan\theta + \tau_{Zx} = \sigma_1 \tan\theta$$
$$\Rightarrow \tan\theta\,(\sigma_1 - \sigma_x) = \tau_{Zx}$$

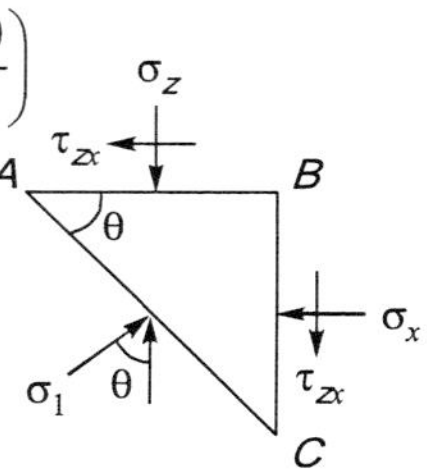

$$\Rightarrow \quad \tan\theta = \left(\frac{\tau_{zx}}{\sigma_1 - \sigma_x}\right)$$

1.34 (a)

$E = 2G(1 + \mu)$; G = Shear modulus

μ = Poission's ratio; E = Young's modulus

1.35 Sol.

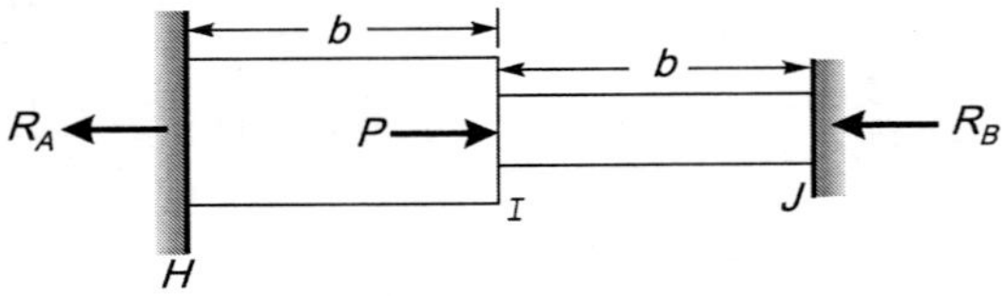

For horizontal force equilibrium,

$$R_A + R_B = P$$

Since there is no axial force in rod HI, thereore $R_A = 0$.

Now check for rod IJ,

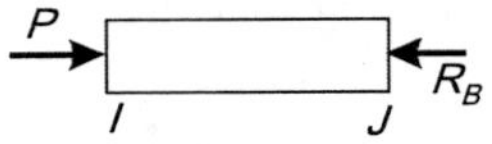

$$\therefore \quad R_B = P$$

Now, as rod IJ is fixed from both end, so net deflection due to increase in temperature will be 0.

$$\therefore b\alpha T + \left[-\frac{Pb}{AE}\right] = 0$$

$$\Rightarrow \quad b\alpha T = \frac{Pb}{AE}$$

$$\Rightarrow \quad P = AE\alpha T$$

$$\Rightarrow \quad P = 10^6 \times 10^{-6} \times 50$$

$$\Rightarrow \quad P = 50 \text{ N}$$

1.36 (b)

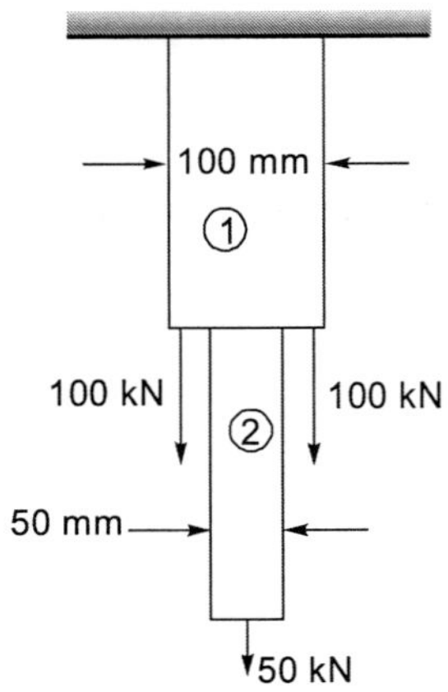

Stress in member (1),

$$(\sigma_1) = \frac{P_1}{A_1} = \frac{250 \times 10^3}{100 \times 100} = 25 \text{ N/mm}^2$$

Stress in member (2),

$$(\sigma_2) = \frac{P_2}{A_2} = \frac{50 \times 10^3}{50 \times 50} = 20 \text{ N/mm}^2$$

$$\therefore \quad \sigma_{max} = 25 \text{ N/mm}^2$$

1.37 Sol.

We know that,

$$\sigma_x = \frac{E}{(1-\mu^2)}(\epsilon_x + \mu \epsilon_y)$$

$$\Rightarrow \quad \sigma_x = \frac{E}{(1-\mu^2)}\left(\frac{\partial u}{\partial x} + \mu\frac{\partial v}{\partial y}\right)$$

$$\Rightarrow \quad \sigma_x = \frac{E}{(1-\mu^2)}(2Cxy + 0)$$

$$\Rightarrow \quad \sigma_x = \frac{2ECxy}{(1-\mu^2)}$$

Given, $\sigma_x = 40xy$ N/m^2

$$\Rightarrow \quad \frac{2EC\cancel{xy}}{(1-\mu^2)} = 40\cancel{xy}$$

$$\Rightarrow \quad C = \frac{40(1-\mu^2)}{2E} = \frac{40(1-0.5^2)}{2\times 1} = 15$$

Now, $$\gamma_{xy} = \frac{\partial u}{\partial y} + \frac{\partial v}{\partial x}$$

$$\gamma_{xy} = Cx^2 + 0$$

Also, $$G = \frac{E}{2(1+\mu)} = \frac{\tau_{xy}}{\gamma_{xy}}$$

$$\Rightarrow \quad \tau_{xy} = \frac{Cx^2 \times 1}{2(1+0.5)} = \left(\frac{15}{2\times 1.5}\right)x^2 = 5x^2$$

On comparing with $\tau_{xy} = \alpha x^2$, the value of α is 5.

1.38 (b)

Let V_P and V_Q be forces in spring and stiffness of spring at Q is k_Q.

From equilibrium equations,

$$\Sigma F_y = 0$$

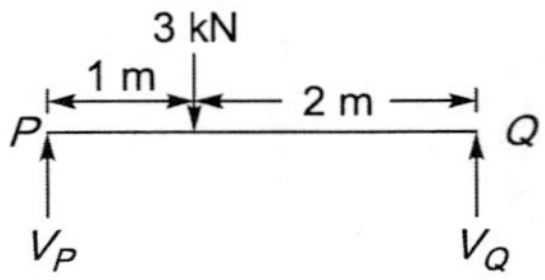

$$\Rightarrow \quad V_P + V_Q = 3$$

Also, $\Sigma M_P = 0$

$$\Rightarrow V_Q \times 3 - 3 \times 1 = 0$$

$$\Rightarrow \quad V_Q = 1 \text{ kN and } V_P = 2 \text{ kN}$$

Now, $\delta_P = \delta_Q$

$$\Rightarrow \quad \frac{2}{100} = \frac{1}{k_Q}$$

$$\Rightarrow \quad k_Q = 50 \text{ kN/m}$$

■■■■

2 Shear Force and Bending Moment

2.1 If the shear force at a section of beam under bending is equal to zero then the bending moment at the section is
(a) zero (b) maximum
(c) minimum (d) constant [1995 : 1 M]

2.2 A propped cantilever of span l, is loaded with uniformly distributed load of intensity w/unit length, all through the span. Bending moment at the fixed end is
(a) $\frac{wl^2}{8}$ (b) $\frac{wl^2}{2}$
(c) $\frac{wl^2}{12}$ (d) $\frac{wl^2}{24}$ [1997 : 1 M]

2.3 A frame $ABCD$ is supported by a roller at A and is on a hinge at C as shown below

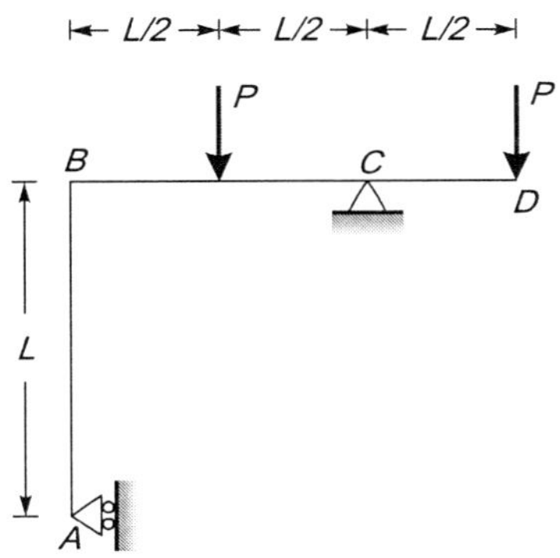

The reaction at the roller end A is given by
(a) P (b) $2P$
(c) $P/2$ (d) Zero [2000 : 1 M]

2.4 The bending moment (in kNm units) at the mid span location X in the beam with overhangs shown below is equal to

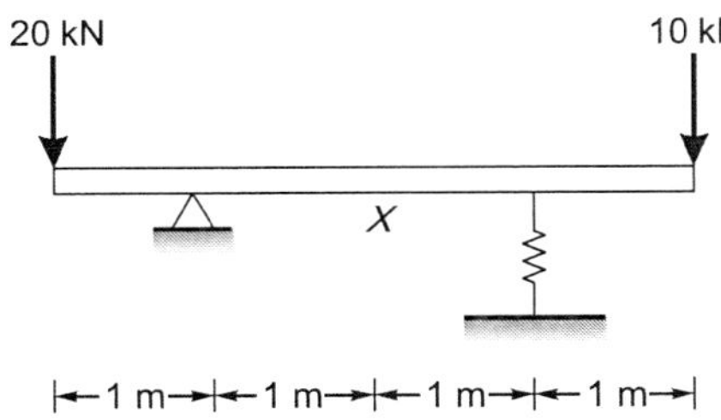

(a) 0 (b) –10
(c) –15 (d) –20 [2001 : 1 M]

2.5 For the loading given in the figure below, two statements (I and II) are made

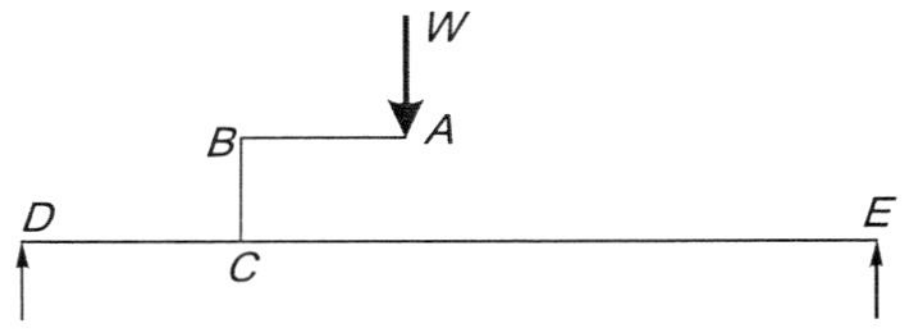

I. Member AB carries shear force and bending moment
II. Member BC carries axial load and shear force
Which of these statement is true?
(a) Statement I True but II is False
(b) Statement I is False but II is True
(c) Both statement I and II are True
(d) Both statement I and II are False
[2002 : 1 M]

2.6 A curved member with a straight vertical leg is carrying a vertical load at Z, as shown in the figure. The stress resultants in the XY segment are

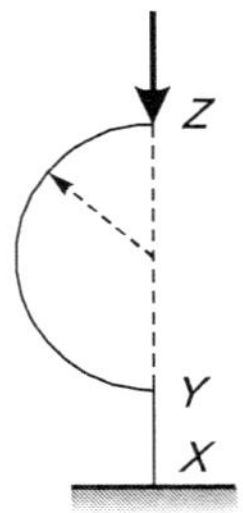

(a) bending moment, shear force and axial force
(b) bending moment and axial force only
(c) bending moment and shear force only
(d) axial force only [2003 : 1 M]

2.7 List-I shows different loads acting on a beam and List-II shows different bending moment distributions. Match the load with the corresponding bending moment diagram.

List-I	List-II
A. 	1. 
B. 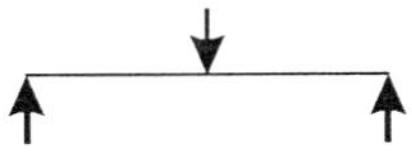	2. 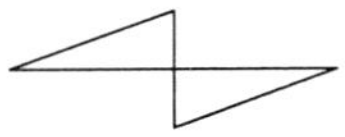

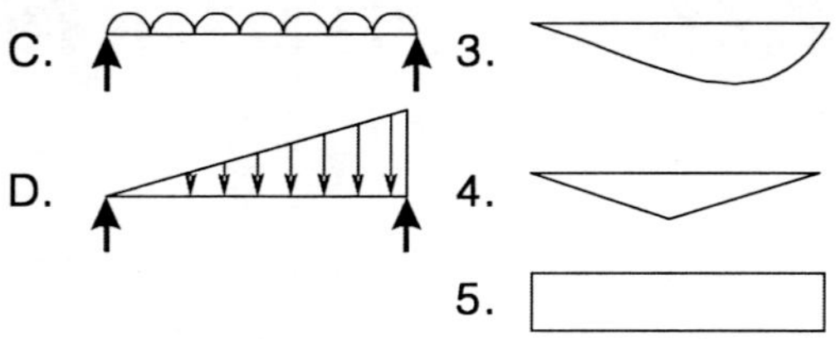

Codes:

	A	B	C	D
(a)	4	2	1	3
(b)	5	4	1	3
(c)	2	5	3	1
(d)	2	4	1	3

[2003 : 2 M]

Common Data for Question 2.8 and 2.9

A three-span continuous beam has an internal hinge at *B*. Section *B* is at the mid-span of *AC*. Section *E* is at the mid-span of *CG*. The 20 kN load is applied at section *B* whereas 10 kN loads are applied at sections *D* and *F* as shown in the figure. Span *GH* is subjected to uniformly distributed load of magnitude 5 kN/m. For the loading shown, shear force immediate to the right of section *E* is 9.84 kN upwards and the hogging moment at section *E* is 10.31 kN-m. [$CD = DE = EF = FG = 1$ m]

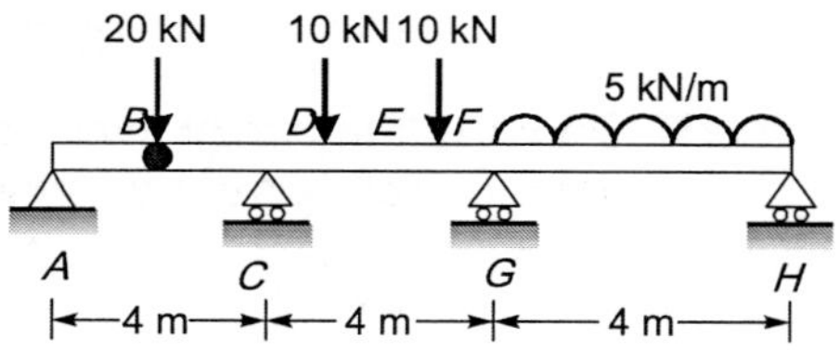

2.8 The magnitude of the shear force immediate to the left and immediate to the right of section *B* are, respectively

(a) 0 and 20 kN
(b) 10 kN and 10 kN
(c) 20 kN and 0
(d) 9.84 kN and 10.16 kN [2004 : 2 M]

2.9 The vertical reaction at support *H* is

(a) 15 kN upward
(b) 9.84 kN upward
(c) 15 kN downward
(d) 9.84 kN downward [2004 : 2 M]

2.10 The bending moment diagram for a beam is given below:

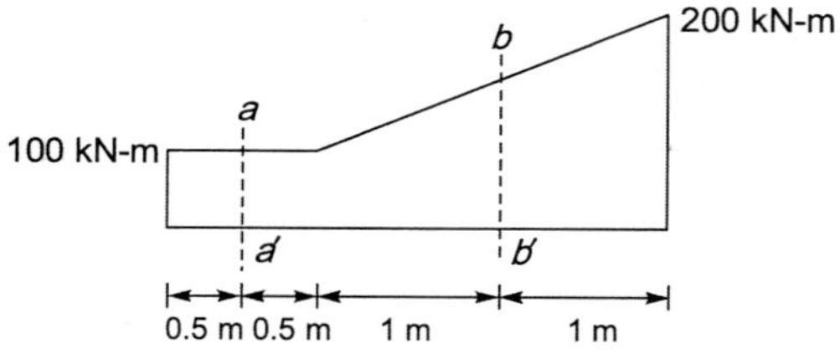

The shear force at sections *aa′* and *bb′* respectively are of the magnitude

(a) 100 kN, 150 kN
(b) zero, 100 kN
(c) zero, 50 kN
(d) 100 kN, 100 kN [2005 : 2 M]

2.11 A simply supported beam *AB* has the bending moment diagram as shown in the following figure:

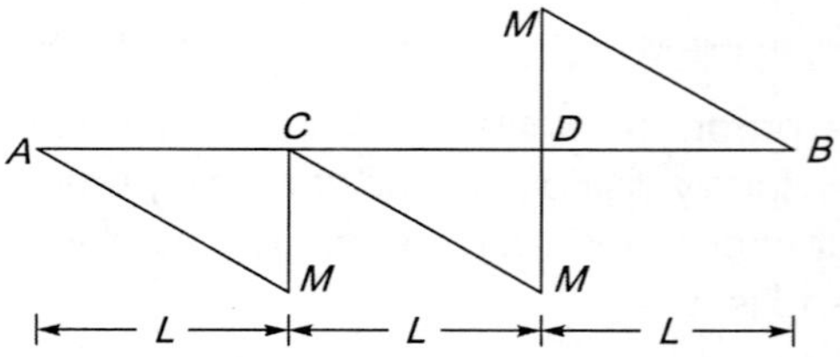

The beam is possibly under the action of following loads

(a) Couples of *M* at *C* and 2*M* at *D*
(b) Couples of 2*M* at *C* and *M* at *D*
(c) Concentrated loads of *M*/*L* at *C* and 2*M*/*L* at *D*
(d) Concentrated load of *M*/*L* at *C* and couple of 2*M* at *D* [2006 : 2 M]

2.12 Match **List-I** (Shear Force DiagraM) beaM with **List-II** (DiagraM of beaM with supports and loading) and select the correct answer by using the codes given below the lists:

List-I

A.

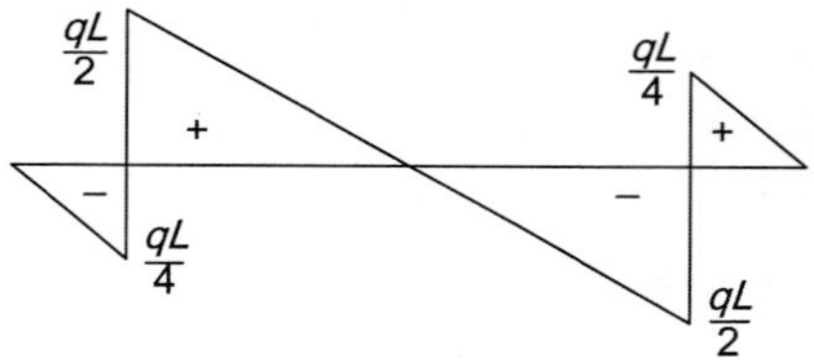

B.

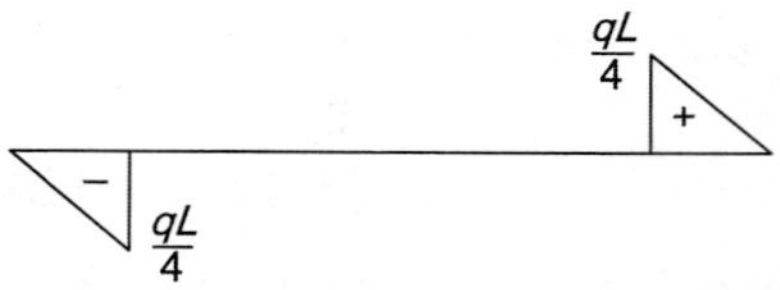

C.

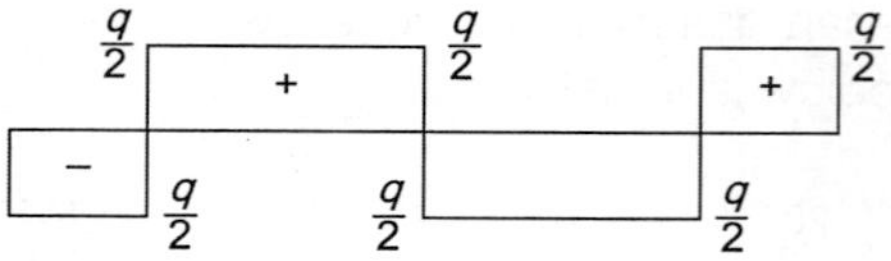

D.

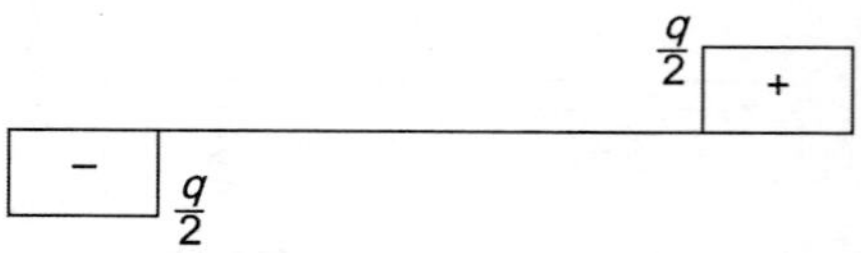

List-II

1. 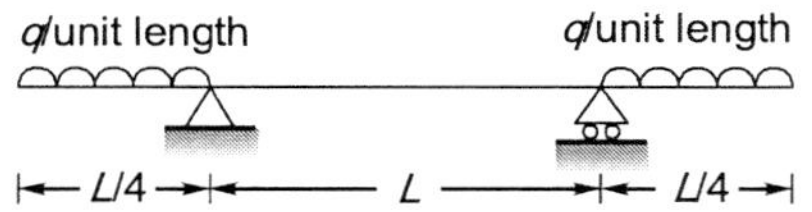

2. 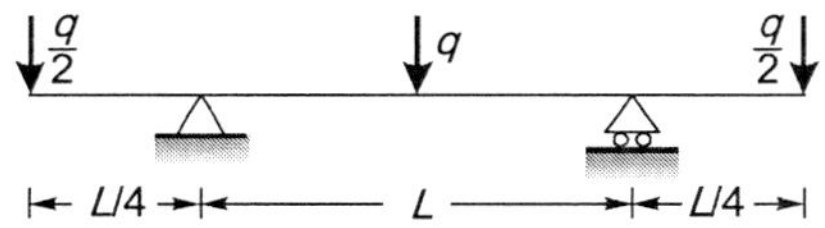

3. 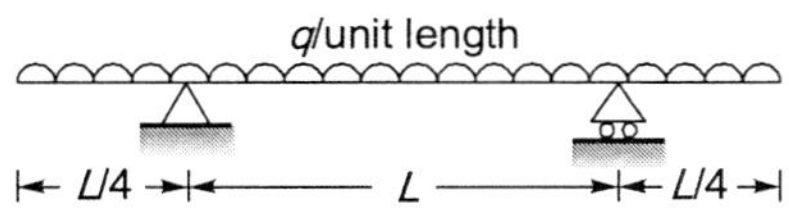

4. 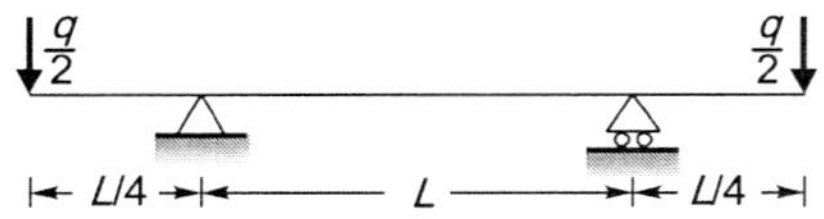

Codes:

	A	B	C	D
(a)	3	1	2	4
(b)	3	4	2	1
(c)	2	1	4	3
(d)	2	4	3	1

[2009 : 2 M]

2.13 Two people weighing W each are sitting on a plank of length L floating on water at $L/4$ from either end. Neglecting the weight of the plank, the bending moment at the centre of the plank is

(a) $\frac{WL}{8}$ (b) $\frac{WL}{16}$

(c) $\frac{WL}{32}$ (d) zero [2010 : 1 M]

2.14 For the simply supported beam of length L, subjected to a uniformly distributed moment M kN-m per unit length as shown in the figure, the bending moment (in kN-m) at the mid-span of the beam is

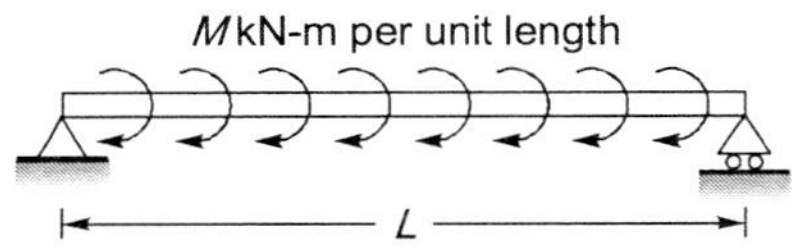

(a) Zero (b) M

(c) ML (d) M/L [2010 : 2 M]

2.15 For the cantilever bracket, $PQRS$, loaded as shown in the adjoining figure ($PQ = RS = L$, and $QR = 2L$), which of the following statements is **FALSE**?

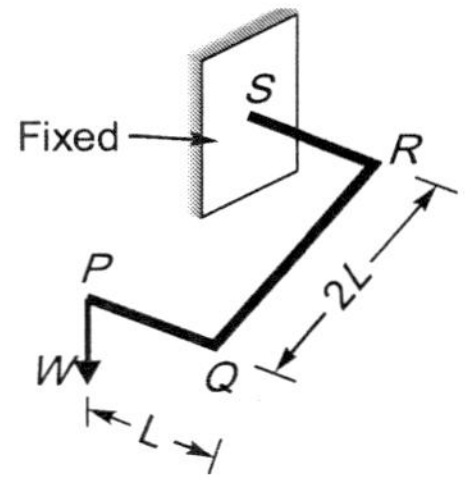

(a) The portion RS has a constant twisting moment with a value of $2WL$.

(b) The portion QR has a varying twisting moment with a maximum value of WL.

(c) The portion PQ has a varying bending moment with a maximum value of WL.

(d) The portion PQ has no twisting moment.

[2011 : 2 M]

2.16 The following statement are related to bending of beaM

I. The slope of the bending moment diagram is equal to the shear force.

II. The slope of the shear force diagram is equal to the load intensity.

III. The slope of the curvature is equal to the flexural rotation.

IV. The second derivative of the deflection is equal to the curvature.

The only FALSE statement is

(a) I (b) II

(c) III (d) IV [2012 : 1 M]

2.17 A rigid member ACB is shown in the figure. The member is supported at A and B by pinned and guided roller supports, respectively. A force P acts at C as shown. Let R_{Ah} and R_{Bh} be the horizontal reactions at supports A and B, respectively, and R_{Av} be the vertical reaction at support A. Self-weight of the member may be ignored.

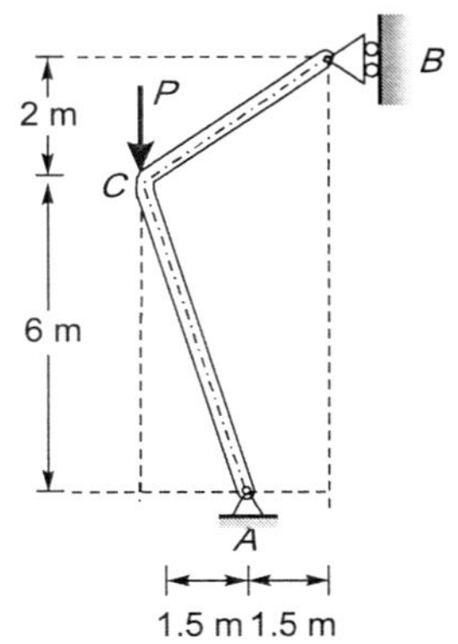

Which one of the following sets gives the correct magnitudes of R_{Av}, R_{Bh} and R_{Av}?

(a) $R_{Av} = 0$; $R_{Bh} = \frac{1}{3}P$; and $R_{Ah} = \frac{2}{3}P$

(b) $R_{Av} = 0$; $R_{Bh} = \frac{2}{3}P$; and $R_{Ah} = \frac{1}{3}P$

(c) $R_{Av} = P$; $R_{Bh} = \frac{3}{8}P$; and $R_{Ah} = \frac{1.5}{8}P$

(d) $R_{Av} = P$; $R_{Bh} = \frac{1.5}{8}P$; and $R_{Ah} = \frac{1.5}{8}P$

[2016 : 2 M, Set-I]

2.18 Consider the three prismatic beaM with the clamped supports P, Q and R as shown in the figures.

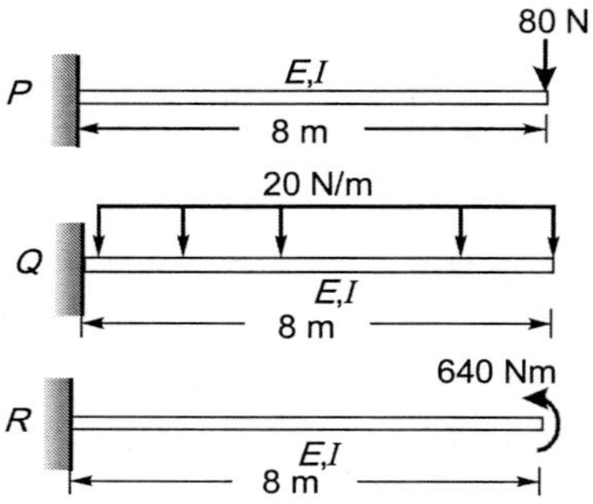

Given that the modulus of elasticity, E is 2.5×10^4 MPa; and the moment of inertia, I is 8×10^8 mm^4, the correct comparison of the magnitudes of the shear force S and the bending moment M developed at the supports is

(a) $S_P < S_Q < S_R$; $M_P = M_Q = M_R$

(b) $S_P = S_Q < S_R$; $M_P = M_Q > M_R$

(c) $S_P < S_Q > S_R$; $M_P = M_Q = M_R$

(d) $S_P < S_Q < S_R$; $M_P < M_Q < M_R$

[2017 : 2 M, Set-II]

2.19 A simply supported beam is subjected to a uniformly distributed load. Which one of the following statements is true?

(a) Maximum or minimum shear force occurs where the curvature is zero.

(b) Maximum or minimum bending moment occurs where the shear force is zero.

(c) Maximum or minimum bending moment occurs where the curvature is zero.

(d) Maximum bending moment and maximum shear force occur at the same section.

[2017 : 1 M, Set-I]

2.20 A weightless cantilever beam of span L is loaded as shown in the figure. For the entire span of the beam, the material properties are identical and the cross-section is rectangular with constant width.

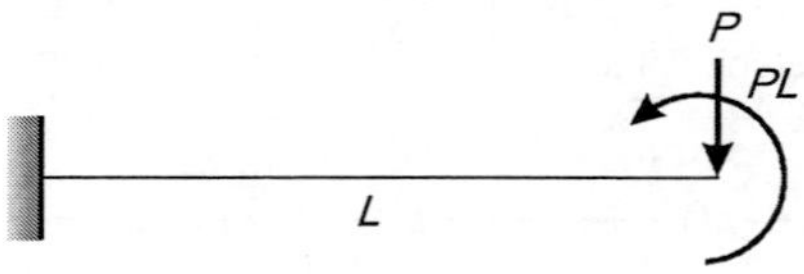

From the flexure-critical perspective, the most economical longitudinal profile of the beam to carry the given loads amongst the options given below, is

(a) 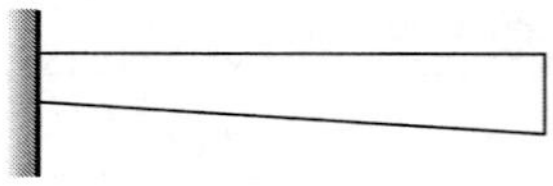

(b) 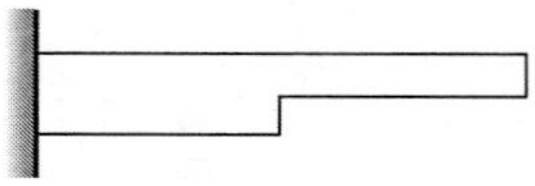

(c) 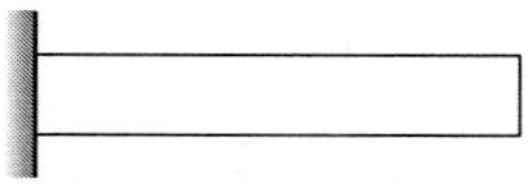

(d) 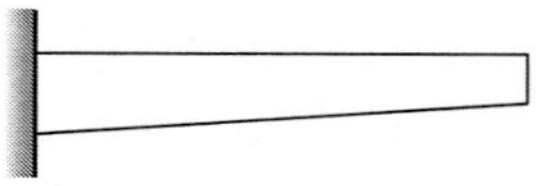

[2020 : 1 M, Set-II]

2.21 A propped cantilever beam XY, with an internal hinge at the middle, is carrying a uniformly distributed load of 10 kN/m, as shown in the figure.

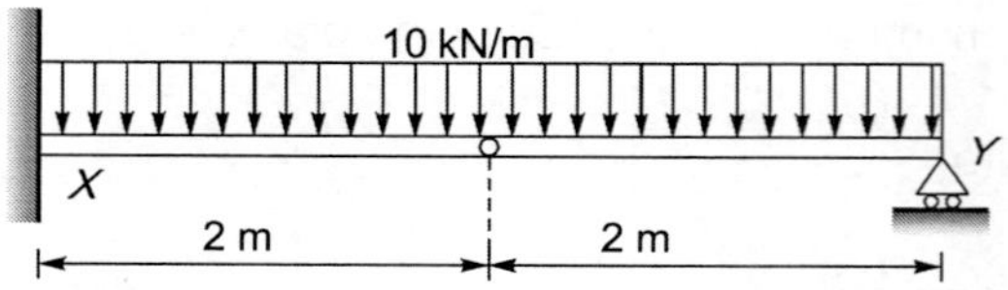

The vertical reaction at support X (in kN, in integer) is ________.

[2021 : 1 M, Set-II]

2.22 A beam is subjected to a system of coplanar forces as shown in the figure. The magnitude of vertical reaction at Support P is ________ N. (round off to one decimal place).

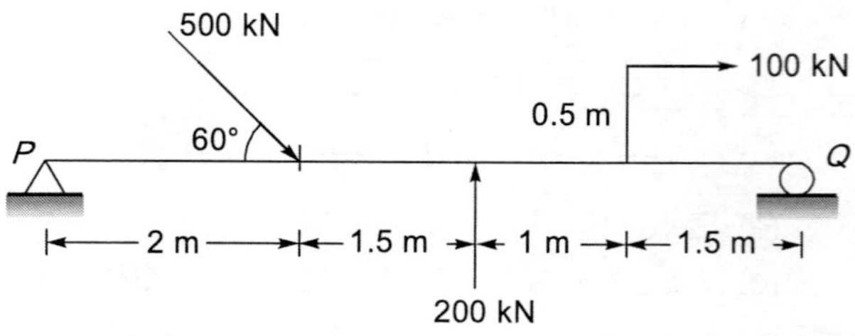

[2023 : 2 M, Set-II]

2.23 An idealised frame supports a load as shown in the figure. The horizontal component of the force transferred from the horizontal member PQ to the vertical member RS at P is ______ N (round off to one decimal place).

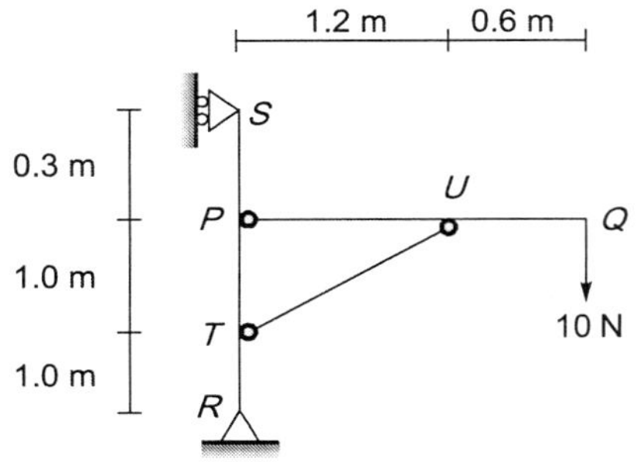

[2023 : 2 M, Set-II]

2.24 The beam shown in the figure is subjected to a uniformly distributed downward load of intensity q between supports A and B.

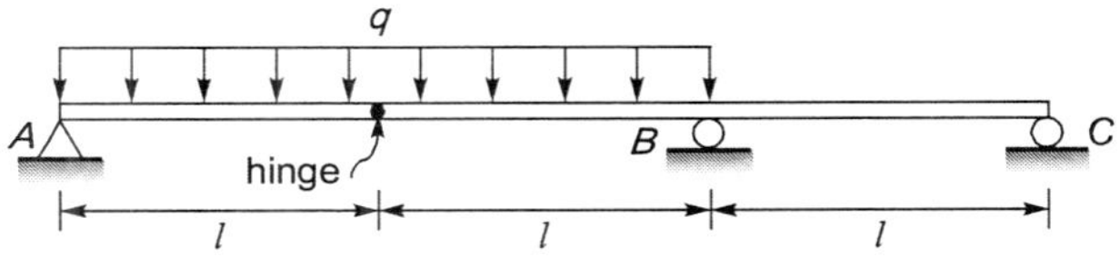

Considering the upward reactions as positive, the support reactions are

(a) $R_A = \frac{ql}{2}; R_B = \frac{5ql}{2}; R_C = 0$

(b) $R_A = \frac{ql}{2}; R_B = \frac{5ql}{2}; R_C = -ql$

(c) $R_A = -ql; R_B = \frac{5ql}{2}; R_C = \frac{ql}{2}$

(d) $R_A = \frac{ql}{2}; R_B = ql; R_C = \frac{ql}{2}$

[2024 : 2 M, Set-I]

Answers Shear Force and Bending Moment

2.1	(d)	2.2	(a)	2.3	(d)	2.4	(c)	2.5	(a)	2.6	(d)	2.7	(d)
2.8	(a)	2.9	(b)	2.10	(c)	2.11	(a)	2.12	(a)	2.13	(d)	2.14	(a)
2.15	(b)	2.16	(c)	2.17	(d)	2.18	(c)	2.19	(b)	2.20	(a)	2.21	(30)
2.22	(197.0)	2.23	(18)	2.24	(b)								

Explanations Shear Force and Bending Moment

2.1 (d)

$$SF = \frac{dM}{dx}$$

$$\frac{dM}{dx} = 0$$

$$M = \text{Constant}$$

If shear force at a section of beam under bending is zero, then the bending moment at that section is constant.

2.2 (a)

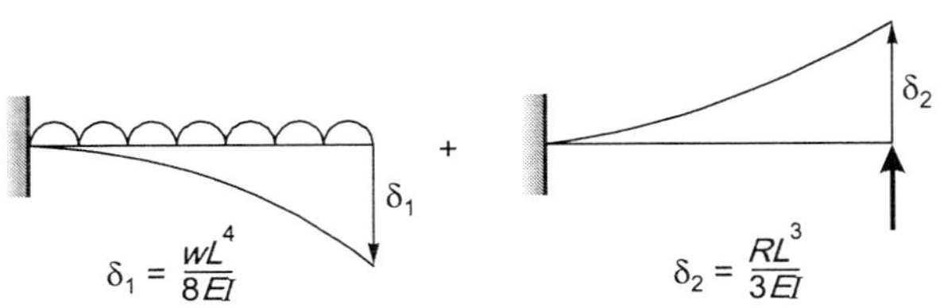

$$\delta_T = 0 = \left(\frac{-wL^4}{8EI}\right) + \left(\frac{RL^3}{3EI}\right) = 0$$

$$R = \frac{3wl}{8}$$

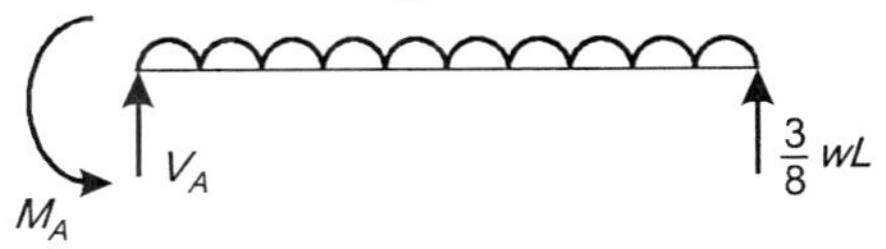

$$\left(-\frac{3}{8}wL.L\right) + \left(wL\frac{L}{2}\right) + (-M_A) = 0$$

$$M_A = \frac{wL^2}{8}$$

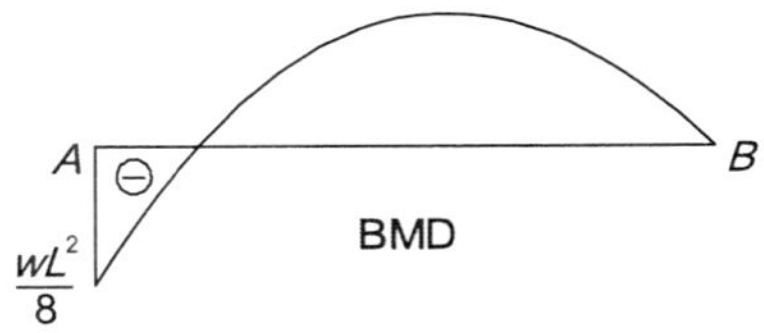

2.3 (d)

The frame is statically determinate.

Taking sum of moment about C, we have

$$\Sigma M_C = 0$$

$$R_A \times L + P\frac{L}{2} - P\frac{L}{2} = 0$$

$$\Rightarrow \quad R_A = 0$$

2.4 (c)

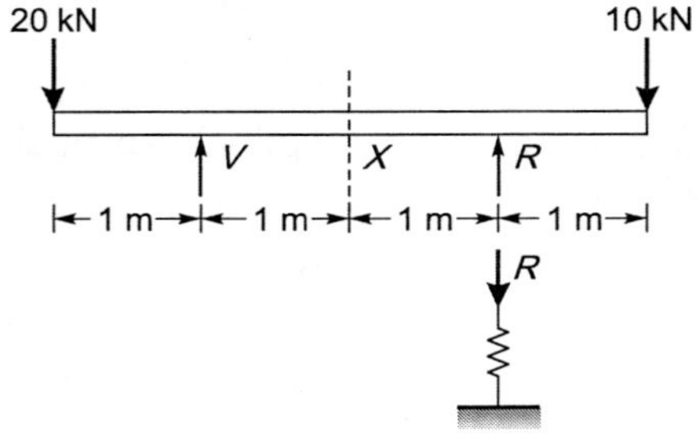

$$R + V = 30 \quad \ldots(i)$$

Now, $V \times 2 - 20 \times 3 + 10 \times 1 = 0$

or $\quad V = \dfrac{60 - 10}{2} = 25\,\text{kN}$

$\therefore \quad R = 5$ kN

BM at $\quad X = 25 \times 1 - 20 \times 2 = -15$ kNm

2.5 (a)

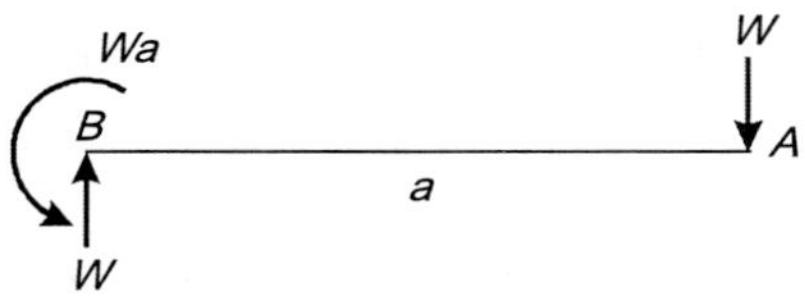

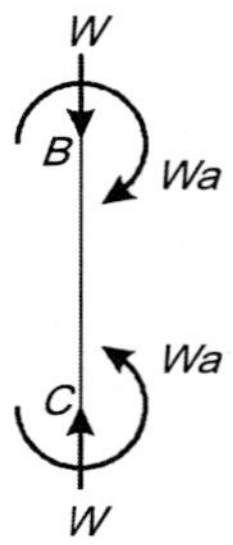

Member *AB* carries shear force and bending moment both. Hence statement-I is true.
Member *BC* carries axial load and bending moment only. Hence statement-II is false.

2.6 (d)

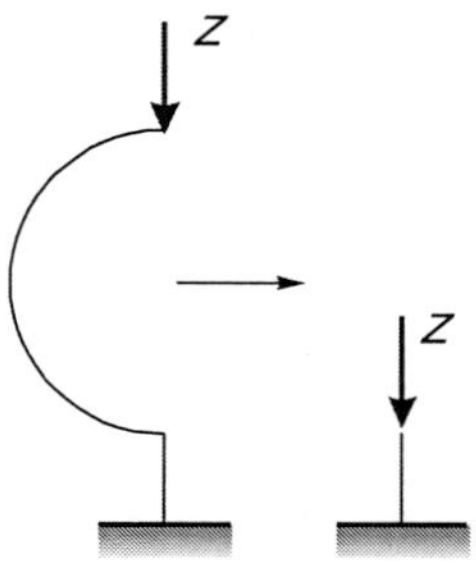

There is no eccentricity between the *XY* segment and the load. So, it is subjected to axial force only. But the curved *YZ* segment is subjected to axial force, shear force and bending moment.

2.7 (d)

A.

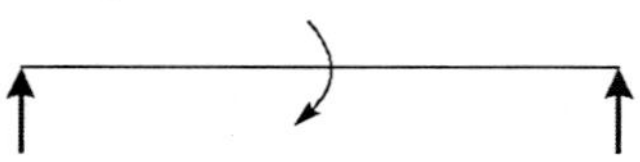

Since couple is acting on beam so sudden change will take place in BM diagram.

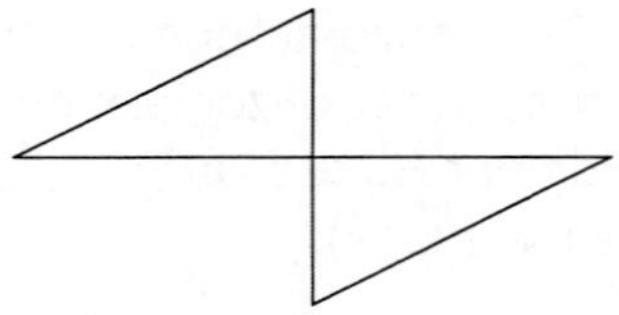

B.

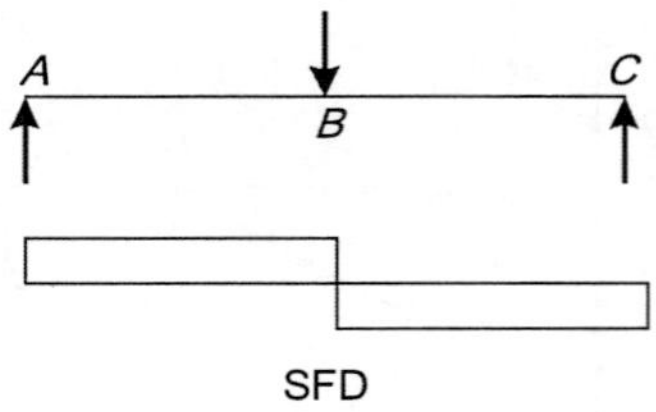

For concentration load, SF is constant between AB & BC so BM is a linear and there is a sudden slope change at B as SF changes, so BMD is

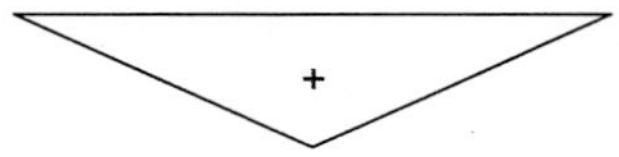

C.

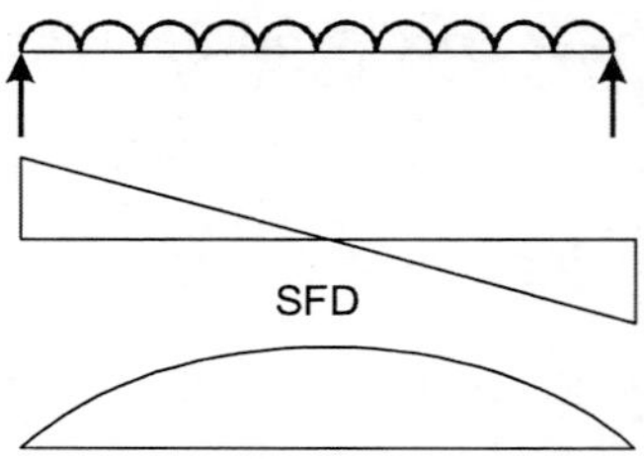

SFD is linear so BMD is parabolic with maxima at centre.

D.

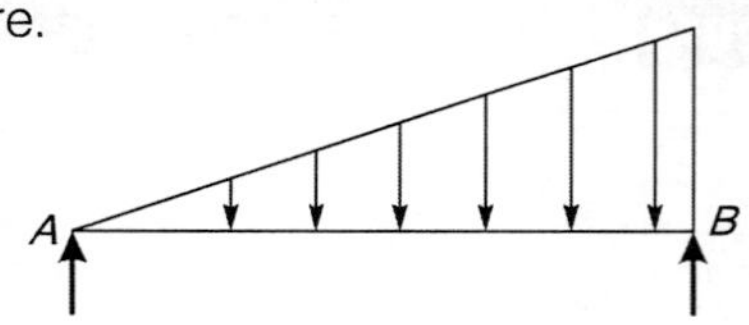

$$L = \frac{-dV}{dx}, \quad V = \frac{dM}{dx}$$

Since load is linearly varying implies shear force is parabolic which implies BMD is cubic. So the only option is (3)

2.8 (a)

The moment about *B* from left = 0
If R_A is the reaction at *A*, then,

$$R_A \times 2 = 0$$

$$\Rightarrow \quad R_A = 0$$

$\therefore$Shear force immediate to the left of *B*,

$$= R_A = 0$$

Shear force immediate to the right of *B*,

$$= 20\text{ kN }(\downarrow)$$

2.9 (b)

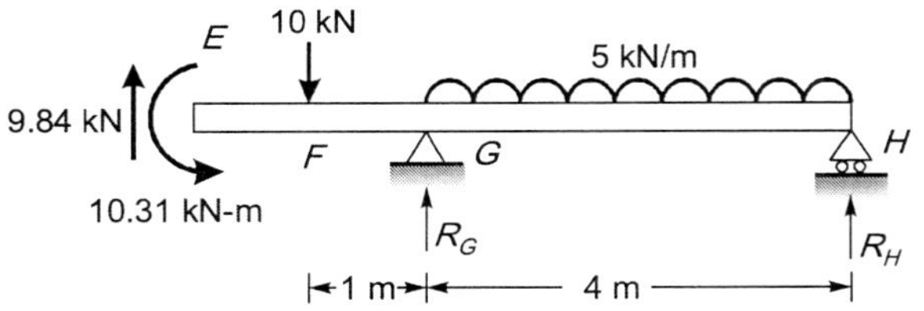

Taking moments about G,

$$R_H \times 4 - 5 \times 4 \times 2 + 10.31 - 9.84 \times 2 + 10 \times 1 = 0$$

$$\Rightarrow \qquad R_H = \frac{39.37}{4} = 9.84 \text{ kN}$$

2.10 (c)

Since $\quad V = \dfrac{dM}{dx}$ from BMD slope at aa'=0,

The bending moment to the left as well as right of section aa' is constant which means shear force is zero at aa'.

Slope at bb'

$$\text{Shear force at bb'} = \frac{200 - 100}{2} = 50 \text{ kN}$$

2.11 (a)

No loading between A and C. At C, there is point moment because BMD is changing suddenly. Similarly a point moment $2M$ is at D.

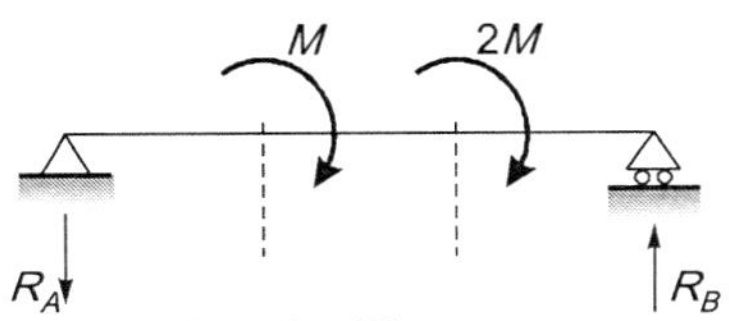

Loading Diagram

$$R_A = R_B = \frac{3M}{3L} = \frac{M}{L}$$

2.12 (a)

∵ Load intensity = Slope os shear force diagram

i.e., $\quad w = \dfrac{dF}{dx}$

A. Figure represents some slope of SFD in the whole span, it means UDL in the whole span i.e., figure (3).

B. Figure shows slope of SFD in the end span and zero slope in the middle span, so it means UDL in the end span and no loading in the middle span i.e., figure (1).

C. Figure shows zero slope of SFD in the whole span. It means UDL is not acting in the whole span, only point load acts, further shear force change sign in the middle span it means some point load is acting in middle span i.e., fig. (2).

D. Figure shows zero slope of SFD in the end span and zero shear force in the middle span, it means UDL is not acting in the whole span i.e., figure (4).

2.13 (d)

The plank will be balanced by the buoyant force acting under its bottom. Let the intensity of buoyant force be w.

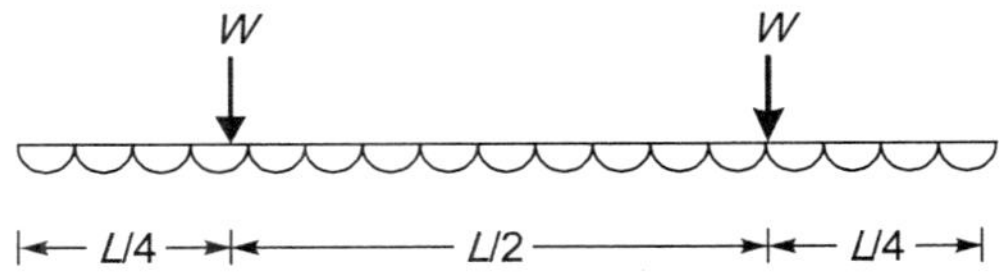

For equilibrium, $\qquad w \times L = W + W$

$$\Rightarrow \qquad w = \frac{2W}{L} \text{ upwards}$$

Thus, the bending moment at the centre of the plank will be,

$$M = \frac{2W}{L} \times \frac{L}{2} \times \frac{L}{4} - W \times \frac{L}{4}$$

$$\Rightarrow \qquad M = \frac{WL}{4} - \frac{WL}{4} = 0$$

2.14 (a)

Let the reaction at the right hand support be V_R upwards. Taking moments about left hand support, we get,

$$V_R \times L - ML = 0 \quad \Rightarrow \quad V_R = M$$

Thus, the reaction at the left hand support V_L will be M downwards.

∴ Moment at the mid-span

$$= -M \times \frac{L}{2} + M \times \frac{L}{2} = 0$$

Infact the bending moment through out the beam is zero.

2.15 (b)

Bending moment at a distance x from P, $M_x = Wx$ Member PQ has varying BM with a maximum value of WL at Q and it has no twisting moment.

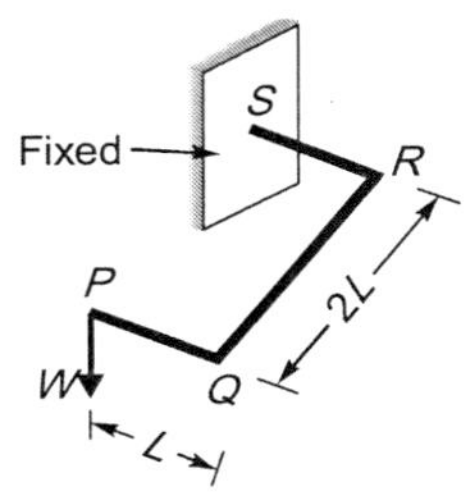

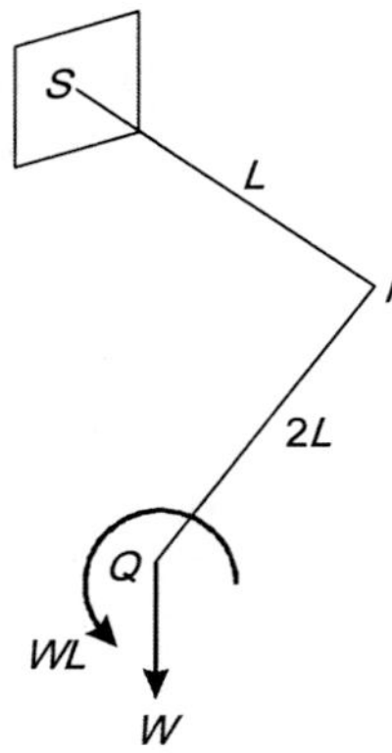

Now moment WL will act as a **twisting moment** for the member QR. Portion QR has varying BM with a maximum value of $2WL$ at R.

Now moment $2WL$ will act as a **twisting moment** for portion RS which is constant in nature.

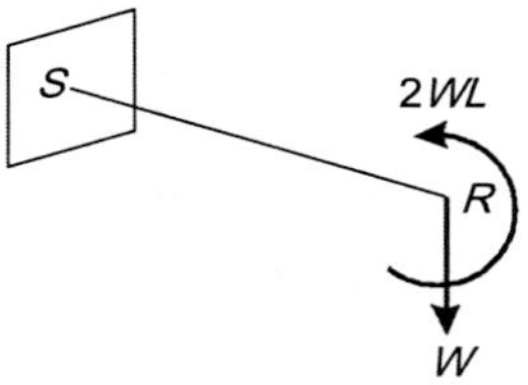

2.16 (c)

We know that,

$$\frac{dS}{dx} = W \qquad \frac{dM}{dX} = S_x$$

$$EI.\frac{d^2y}{dx^2} = M \qquad \therefore \frac{d^2y}{dx^2} = \frac{M}{EI}$$

Also, $$\frac{M}{I} = \frac{f}{y} = \frac{E}{R} \qquad \therefore \frac{M}{EI} = \frac{1}{R}$$

$$\therefore \qquad \frac{d^2y}{dx^2} = \frac{1}{R}$$

2.17 (d)

$$\Sigma F_y = 0$$

$$\Rightarrow \quad R_{av} = P$$

$$\Sigma F_x = 0$$

$$\Rightarrow \quad R_{Bh} = R_{Ah}$$

$$\Sigma M_A = 0$$

$$R_{Bh} \times 8 - P \times 1.5 = 0$$

$$\Rightarrow \quad R_{Bh} = \frac{1.5P}{8}$$

$$\Rightarrow \quad R_{Ah} = \frac{1.5P}{8}$$

$$R_{Av} = P$$

$$R_{Bh} = \frac{1.5P}{8}$$

$$R_{Ah} = \frac{1.5P}{8}$$

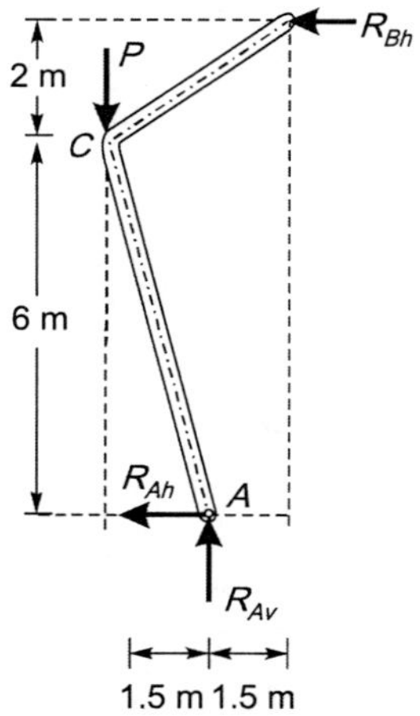

2.18 (c)

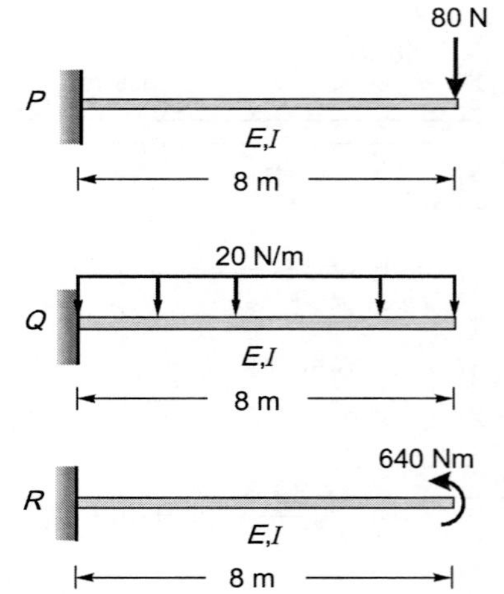

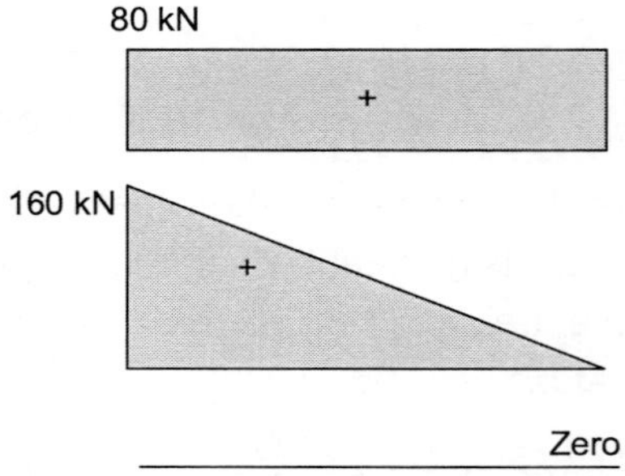

SF Diagram

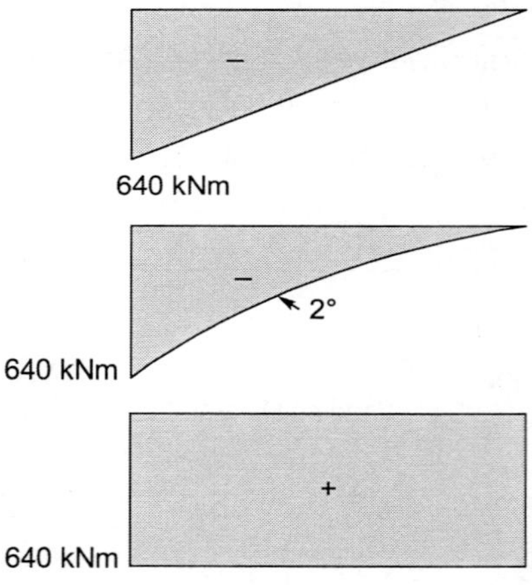

BM Diagram

$$S_P < S_Q > S_R$$

$$M_P = M_Q = M_R$$

Hence correct option is (c).

2.19 (b)

As we know

$$\frac{dM}{dx} = V_x \qquad \text{...(i)}$$

From (i),

For M to be maximum or minimum, V_x should be zero.

2.20 (a)

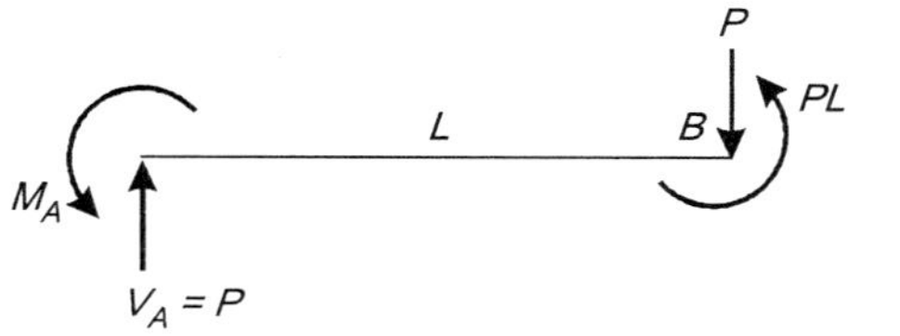

$$(-PL) + (PL) + (-M_A) = 0$$
$$M_A = 0$$

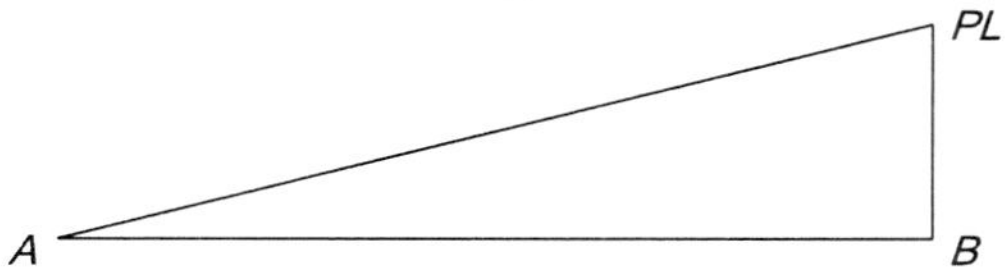

Bending moment diagram

For most economical,
Maximum cross-section is given where maximum bending moment occurs.

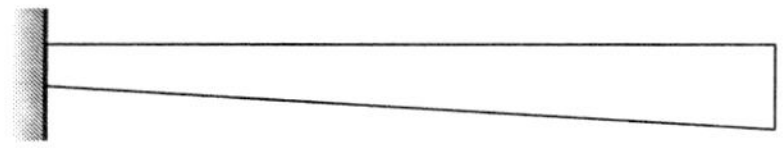

So, option (a) is correct.

2.21 Sol.

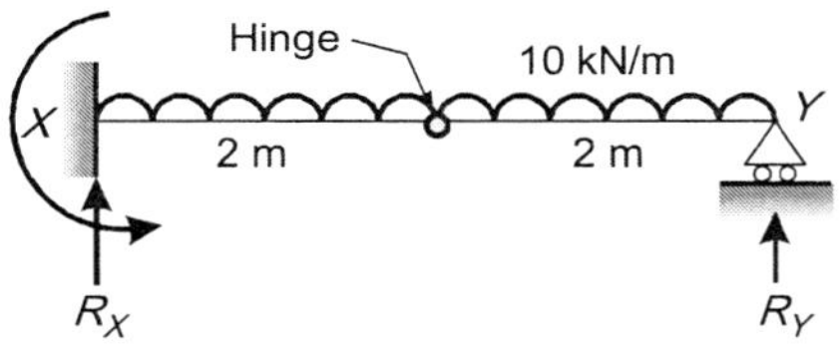

BM = 0 at hinge

$$R_Y \times 2 - 10 \times 2 \times 1 = 0$$
$$R_Y = 10 \text{ kN}$$
$$R_X + R_Y = 10 \times 4$$
$$R_X + 10 = 40$$
$$R_X = 30 \text{ kN}$$

2.22 Sol.

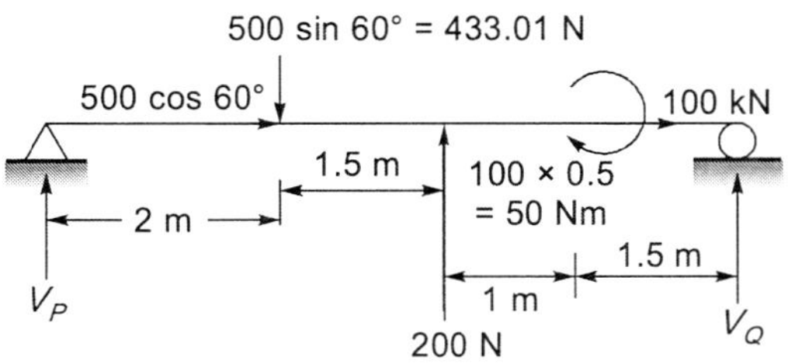

Taking moment at P

$\Sigma M_P = 0$

$$433.01 \times 2 - 200 \times 3.5 + 50 - V_Q \times 6 = 0$$

$\Rightarrow$ $V_Q = 36.00$ N

Also, $\Sigma F_y = 0$

$\Rightarrow$ $V_P + 200 + V_Q = 433.01$

$\Rightarrow$ $V_P = 433.01 - 200 - 36.00$

$\Rightarrow$ $V_P = 197.01$ N

2.23 Sol.

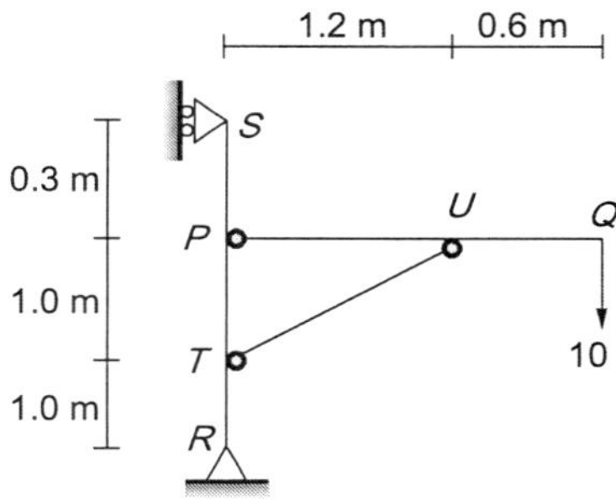

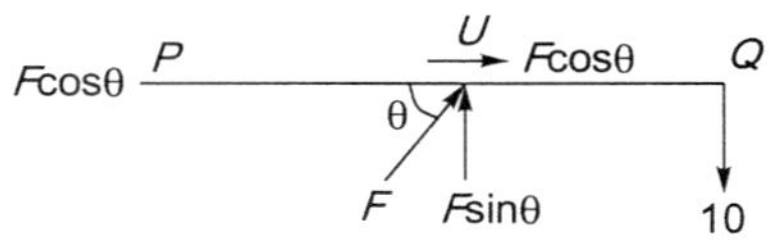

$$\Sigma M_A = 0$$

$\Rightarrow F\sin\theta \times 1.2 = 10 \times 1.8$

$$\Rightarrow \quad F = \frac{18}{1.2\sin\theta} = 23.4307 \text{ N}$$

Horizontal force at $P = F\cos\theta$

$$= 23.43 \times \frac{1.2}{\sqrt{1.2^2 + 1^2}} = 18 \text{ N}$$

2.24 (b)

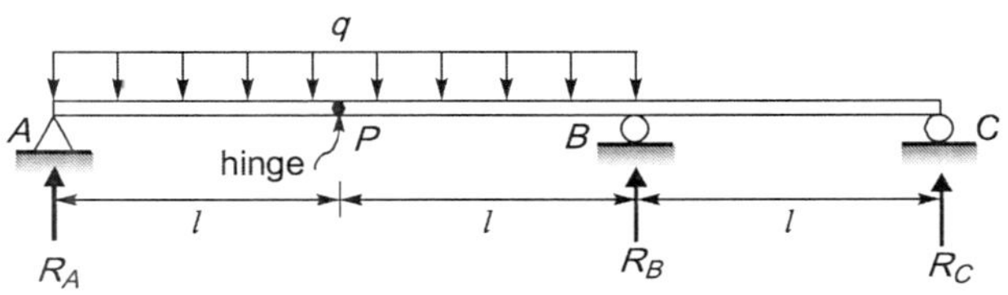

$BM_P = 0$

$$\therefore \quad R_A(l) = q \cdot l \cdot \frac{l}{2} \quad \text{...(1)}$$

$$-R_C(2l) + (-R_B \times l) + \left(q \times l \times \frac{l}{2}\right) = 0 \quad \text{...(2)}$$

$$R_A + R_B + R_C = q\,(2l) \quad \text{...(3)}$$

From (1), (2) and (3)

$$R_A = \frac{ql}{2},\ R_B = \frac{5ql}{2},\ R_C = -ql$$

■■■■

3 Principal Stress and Principal Strain

3.1 If an element of a stressed body is in a state of pure shear with a magnitude of 80 N/mm^2, the magnitude of maximum principal stress at that location is

(a) 80 N/mm^2 (b) 113.14 N/mm^2

(c) 120 N/mm^2 (d) 56.57 N/mm^2

[1999 : 1 M]

3.2 Two perpendicular axes x and y of a section are called principal axes when

(a) Moments of inertial about the axes are equal ($I_x = I_y$)

(b) Product moment of inertial (I_{xy}) is zero

(c) Product of moment of inertia ($I_x . I_y$) is zero

(d) Moment of inertial about one of the axis is greater than the other [1999 : 1 M]

3.3 Pick the incorrect statement from the following four statements

(a) On the plane which carries maximum normal stress, the shear stress is zero

(b) Principal planes are mutually orthogonal

(c) On the plane which carries maximum shear stress, the normal stress is zero

(d) The principal stress axes and principal strain axes coincide for an isotropic material

[2000 : 1 M]

3.4 The state of two dimensional stresses acting on a concrete lamina consists of a direct tensile stress, σ_x = 1.5 N/mm^2, and shear stress, τ = 1.20 N/mm^2, which cause cracking of concrete. Then the tensile strength of the concrete (in N/mm^2) is

(a) 1.50 (b) 2.08

(c) 2.17 (d) 2.29 [2003 : 2 M]

3.5 In a two dimensional stress analysis, the state of stress at a point is shown below. If σ = 120 MPa and τ = 70 MPa, σ_x and σ_y, are respectively,

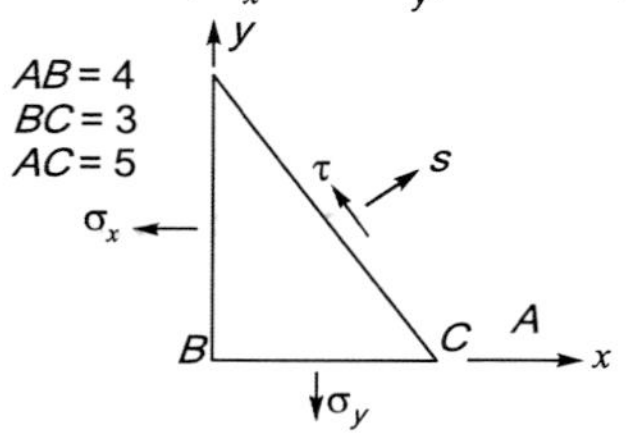

(a) 26.7 MPa and 172.5 MPa

(b) 54 MPa and 128 MPa

(c) 67.5 MPa and 213.3 MPa

(d) 16 MPa and 138 MPa [2004 : 2 M]

3.6 The components of strain tensor at a point in the plane strain case can be obtained by measuring longitudinal strain in following directions

(a) along any two arbitrary directions

(b) along any three arbitrary directions

(c) along two mutually orthogonal directions

(d) along any arbitrary direction [2005 : 1 M]

3.7 If principal stresses in a two-dimensional case are –10 MPa and 20 MPa respectively, then maximum shear stress at the point is

(a) 10 MPa (b) 15 MPa

(c) 20 MPa (d) 30 MPa [2005 : 2 M]

3.8 Mohr's circle for the state of stress defined by $\begin{bmatrix} 30 & 0 \\ 0 & 30 \end{bmatrix}$ MPa is a circle with

(a) center at (0, 0) and radius 30 MPa

(b) center at (0, 0) and radius 60 MPa

(c) center at (30, 0) and radius 30 MPa

(d) center at (30, 0) and zero radius

[2006 : 1 M]

3.9 An axially loaded bar is subjected to a normal stress of 173 MPa. The shear stress in the bar is

(a) 75 MPa (b) 86.5 MPa

(c) 100 MPa (d) 122.3 MPa

[2007 : 1 M]

3.10 Consider the following statements:

1. On a principal plane, only normal stress acts.
2. On a principal plane, both normal and shear stresses act.
3. On a principal plane, only shear stress acts.
4. Isotropic state of stress is independent of frame of reference.

Which of these statements is/are correct?

(a) 1 and 4 (b) 2 only

(c) 2 and 4 (d) 2 and 3 [2009 : 2 M]

3.11 The major and minor principal stresses at a point are 3 MPa and – 3 MPa respectively. The maximum shear stress at the point is
(a) zero (b) 3 MPa
(c) 6 MPa (d) 9 MPa [2010 : 1 M]

3.12 If a small concrete cube is submerged deep in still water in such a way that the pressure exerted on all faces of the cube is *p*, then the maximum shear stress developed inside the cube is
(a) 0 (b) *p*/2
(c) *p* (d) 2*p* [2012 : 1 M]

3.13 The state of 2*D* stress at a point is given by a matrix

$$\begin{bmatrix} \sigma_{xx} & \tau_{xy} \\ \tau_{yx} & \sigma_{yy} \end{bmatrix} = \begin{bmatrix} 100 & 30 \\ 30 & 20 \end{bmatrix} \text{MPa}$$

The maximum shear stress in MPa is
(a) 50 (b) 75
(c) 100 (d) 110 [2013 : 2 M]

3.14 Two triangular wedges are glued together as shown in the following figure. The stress acting normal to the interface, σ_n is __________ MPa.

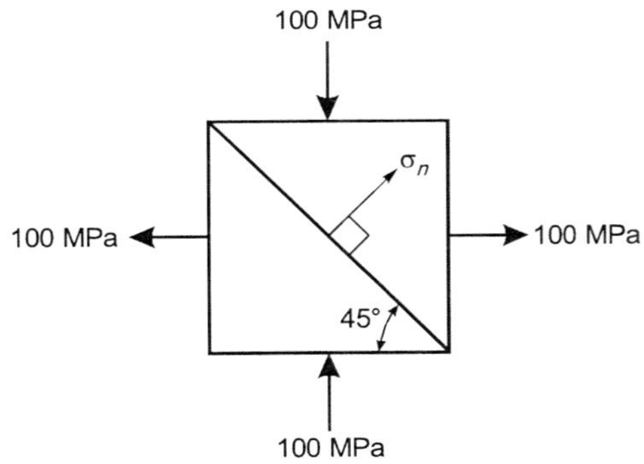

[2015 : 1 M, Set-I]

3.15 For the plane stress situation shown in the figure, the maximum shear stress and the plane on which it acts are

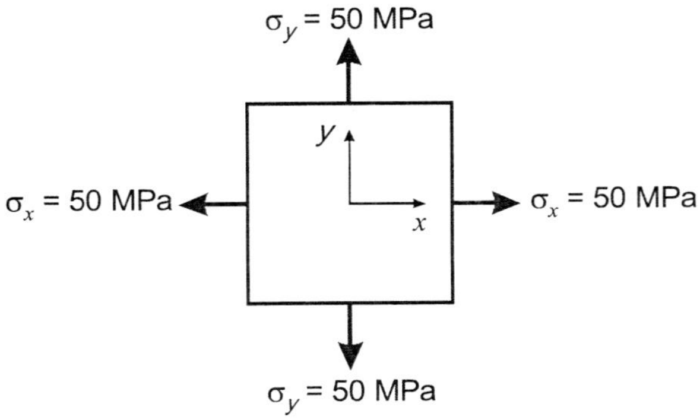

(a) –50 MPa, on a plane 45° clockwise w.r.t. *x*-axis
(b) –50 MPa, on a plane 45° anti-clockwise w.r.t. *x*-axis
(c) 50 MPa, at all orientations
(d) Zero, at all orientations [2015 : 1 M, Set-II]

3.16 For the stress state (in MPa) shown in the figure, the major principal stress is 10 MPa.

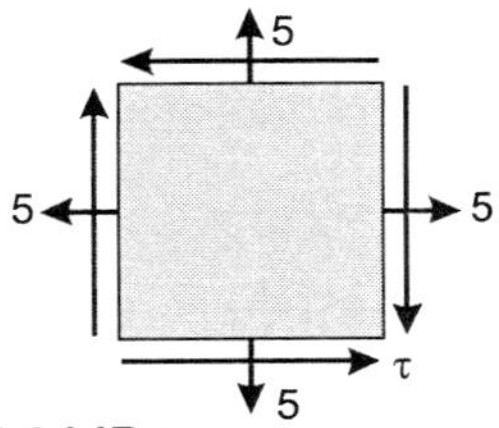

The shear stress τ is
(a) 10.0 MPa (b) 5.0 MPa
(c) 2.5 MPa (d) 0.0 MPa
[2016 : 2 M, Set-II]

3.17 An element is subjected to biaxial normal tensile strains of 0.0030 and 0.0020. The normal strain in the plane of maximum shear strain is
(a) Zero (b) 0.0050
(c) 0.0010 (d) 0.0025
[2019 : 1 M, Set-I]

3.18 For a plane stress problem, the state of stress at a point P is represented by the stress element as shown in figure.

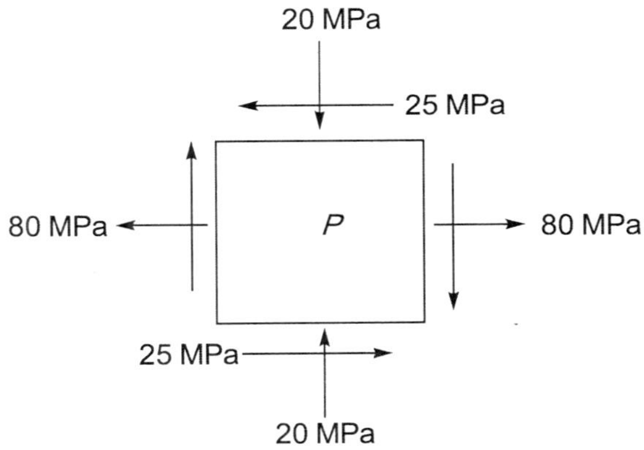

By how much angle (θ) in degrees the stress element should be rotated in order to get the planes of maximum shear stress?

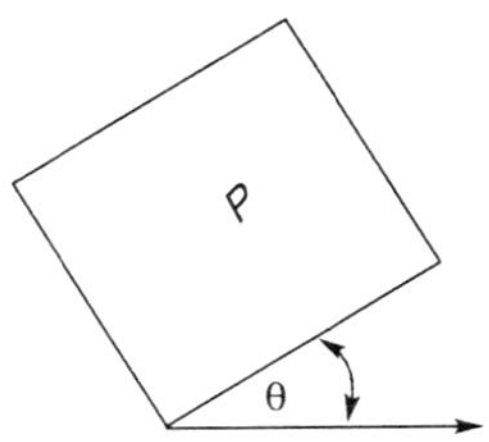

(a) 31.7 (b) 13.3
(c) 26.6 (d) 48.3
[2019 : 2 M, Set-II]

3.19 In a two-dimensional stress analysis, the state of stress at a point P is

$$[\sigma]=\begin{bmatrix}\sigma_{xx} & \tau_{xy}\\ \tau_{xy} & \sigma_{yy}\end{bmatrix}$$

The necessary and sufficient condition for existence of the state of pure shear at the point P, is

(a) $\tau_{xy}=0$

(b) $(\sigma_{xx}-\sigma_{yy})^2+4\tau_{xy}^2=0$

(c) $\sigma_{xx}+\sigma_{yy}=0$

(d) $\sigma_{xx}\sigma_{yy}-\tau_{xy}^2=0$ [2020 : 1 M, Set-I]

3.20 The state of stress represented by Mohr's circle shown in the figure is

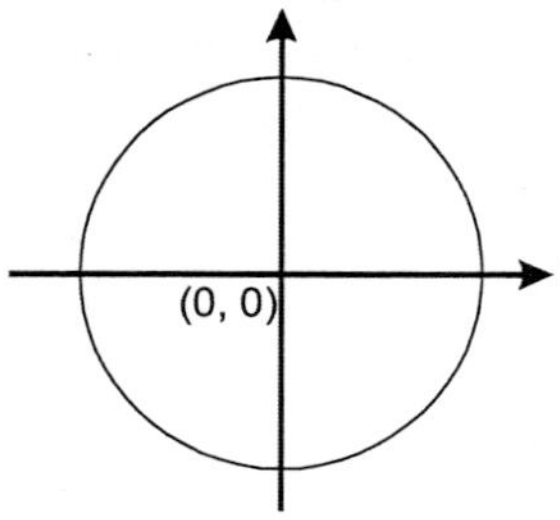

(a) hydrostatic stress
(b) uniaxial tension
(c) biaxial tension of equal magnitude
(d) pure shear [2020 : 1 M, Set-II]

3.21 The state of stress in a deformable body is shown in the figure. Consider transformation of the stress from the x-y coordinates system to the X-Y coordinates system The angle θ, locating the X-axis, is assumed to be positive when measured from the x-axis is counter-clockwise direction.

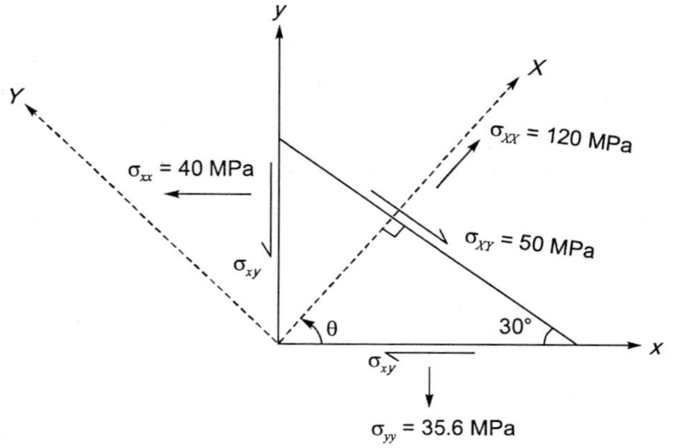

The absolute magnitude of the shear stress components σ_{xy} (in MPa, round off to one decimal place) in x-y coordinate system is ________

[2021 : 1 M, Set-I]

3.22 The infinitesimal element shown in the figure (not to scale) represents the state of stress at a point in a body. What is the magnitude of the maximum principal stress (in N/mm^2, in integer) at the point?

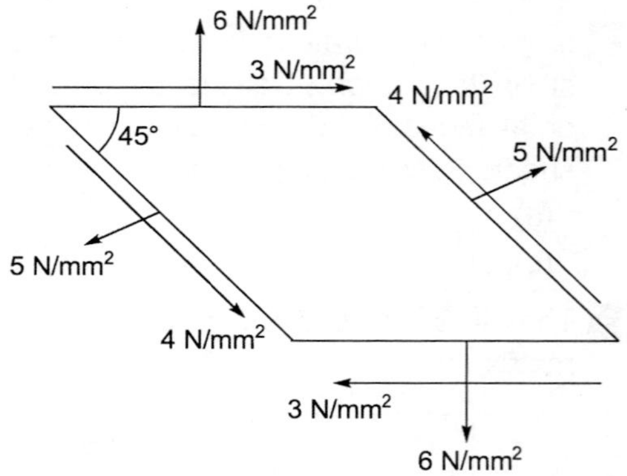

[2023 : 2 M, Set-I]

3.23 In a two-dimensional stress analysis, the state of stress at a point is shown in the figure. The values of length of *PQ*, *QR*, and *RP* are 4, 3 and 5 units, respectively. The principal stresses are ________. (round off to one decimal place)

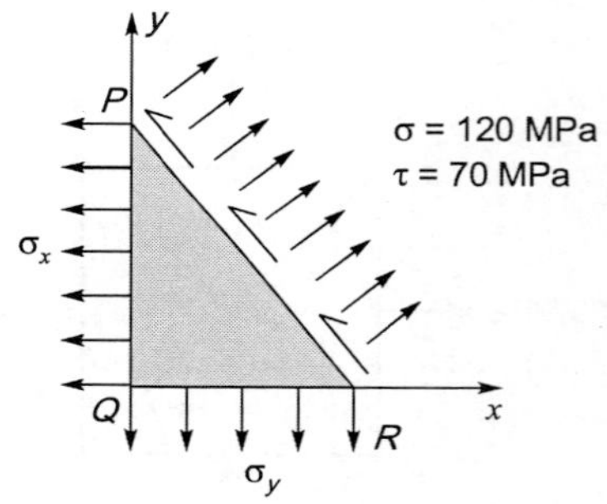

(a) σ_x = 26.7 MPa, σ_y = 172.5 MPa
(b) σ_x = 54.0 MPa, σ_y = 128.5 MPa
(c) σ_x = 67.5 MPa, σ_y = 213.3 MPa
(d) σ_x = 16.0 MPa, σ_y = 138.5 MPa

[2023 : 2 M, Set-II]

3.24 The three-dimensional state of stress at a point is given by

$$\sigma=\begin{pmatrix}10 & 0 & 0\\ 0 & 40 & 0\\ 0 & 0 & 0\end{pmatrix}\text{ MPa.}$$

The maximum shear stress at the point is

(a) 15 MPa (b) 20 MPa
(c) 25 MPa (d) 5 MPa

[2024 : 1 M, Set-I]

3.25 Find the correct match between the plane stress states and the Mohr's circles.

(P) 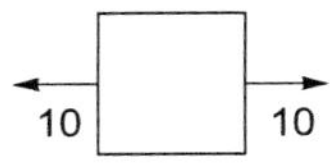

(I) 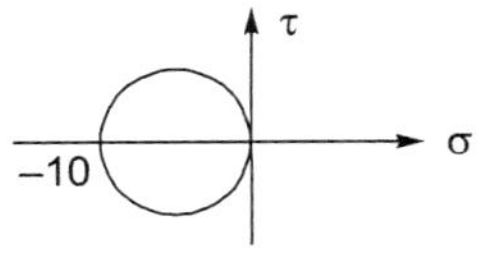

(Q) 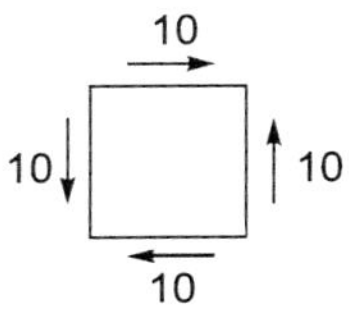

(II) 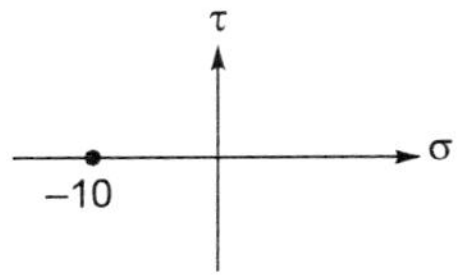

(R) 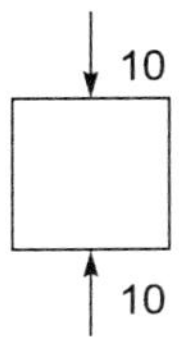

(III) 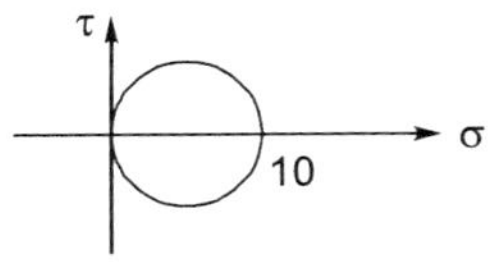

(S) 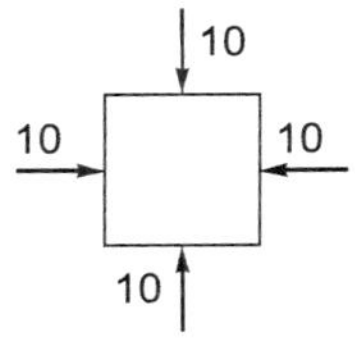

(IV) 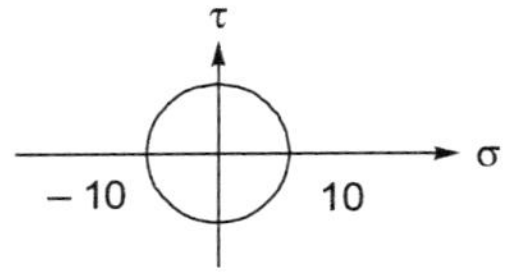

(a) (P)-(III): (Q)-(IV); (R)-(I); (S)-(II)
(b) (P)-(III); (Q)-(II): (R)-(I): (S)-(IV)
(c) (P)-(I): (Q)-(IV): (R)-(III). (S)-(II)
(d) (P)-(I) (Q)-(II). (R)-(III); (S)-(IV)

[2024 : 2 M, Set-I]

■■■■

Answers Principal Stress and Principal Strain

3.1 (a)	3.2 (b)	3.3 (c)	3.4 (c)	3.5 (c)	3.6 (b)	3.7 (b)
3.8 (d)	3.9 (b)	3.10 (a)	3.11 (b)	3.12 (a)	3.13 (a)	3.14 (0)
3.15 (d)	3.16 (b)	3.17 (d)	3.18 (a)	3.19 (c)	3.20 (d)	3.21 (96.2)
3.22 (7)	3.23 (c)	3.24 (b)	3.25 (a)			

Explanations Principal Stress and Principal Strain

3.1 (a)

For the case of pure shear, maximum principal stress occurs at the diagonals and is given by

$$\sigma = \pm\tau = 80 \text{ N/mm}^2$$

3.2 (b)

Principal axes are those axes about which product moment of inertia (I_{xy}) is zero and moment of inertia is either maximum or minimum.

3.3 (c)

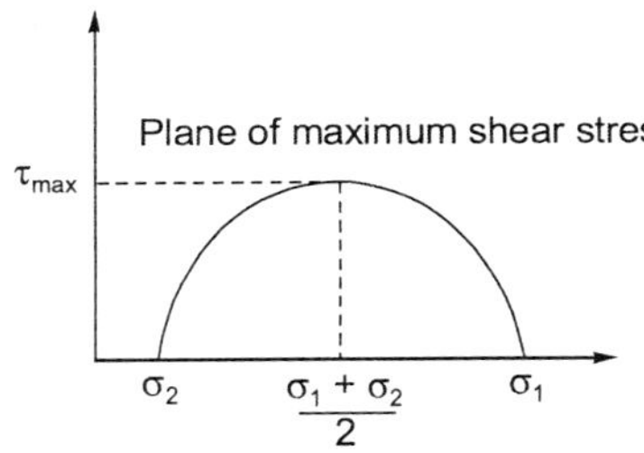

Plane of maximum shaer stress has normal stress of $\dfrac{\sigma_1+\sigma_2}{2}$.

3.4 (c)

Maximum principal stress

$$= \frac{\sigma_x}{2} + \sqrt{\left(\frac{\sigma_x}{2}\right)^2 + \tau^2}$$

$$= \frac{1.5}{2} + \sqrt{\left(\frac{1.5}{2}\right)^2 + (1.20)^2} = 2.17 \text{ N/mm}^2$$

3.5 (c)

Method-I

Let, $\angle CAB = \theta$

$$\therefore \sin\theta = \frac{3}{5};\ \cos\theta = \frac{4}{5};\ \tan\theta = \frac{3}{4}$$

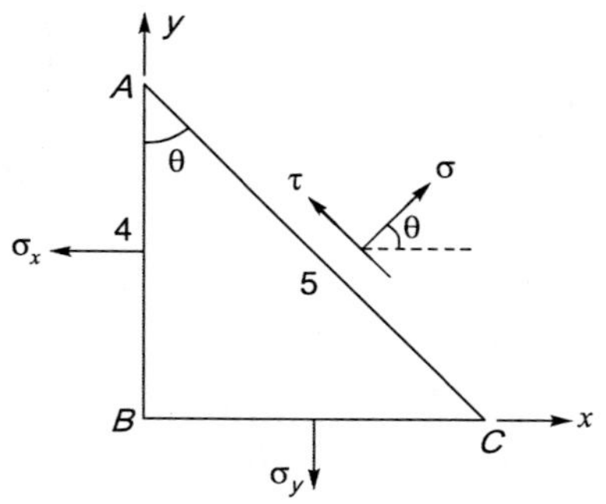

Thus from force equilibrium,

$$\sigma_x \times AB = AC \times (\sigma \cos\theta - \tau \sin\theta)$$

$$\Rightarrow \quad \sigma_x = \frac{5}{4} \times \left(120 \times \frac{4}{5} - 70 \times \frac{3}{5}\right)$$

$$\Rightarrow \quad \sigma_x = 67.5 \text{ MPa}$$

And, $\sigma_y \times BC = AC \times (\sigma \sin\theta + \tau \cos\theta)$

$$\Rightarrow \quad \sigma_y = \frac{5}{3} \times \left(120 \times \frac{3}{5} + 70 \times \frac{4}{5}\right)$$

$$\Rightarrow \quad \sigma_y = 213.3 \text{ MPa}$$

Method-II

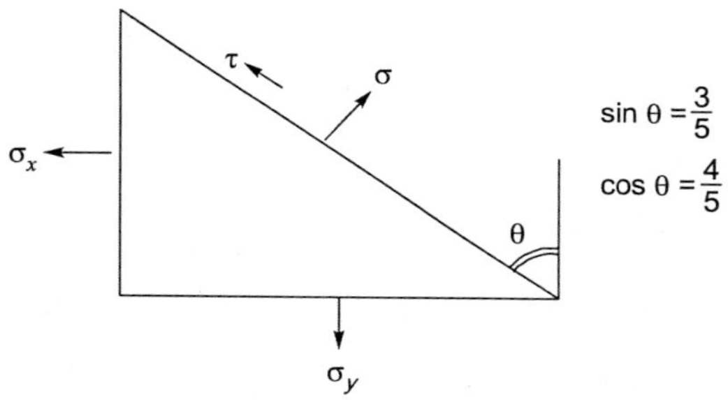

$$\sigma = \sigma_x \cos^2\theta + \sigma_y \sin^2\theta + \cancelto{0}{\tau_{xy}} \sin 2\theta$$

$$\sigma = \sigma_x \times \frac{16}{25} + \sigma_y \times \frac{9}{25}$$

$$120 = \frac{16\sigma_x + 9\sigma_y}{25} \quad \text{...(i)}$$

$$\tau = -\left(\frac{\sigma_x - \sigma_y}{2}\right) \sin 2\theta + \cancelto{0}{\tau_{xy}} \cos 2\theta$$

$$= \frac{\sigma_y - \sigma_x}{2} \times 2 \times \frac{3}{5} \times \frac{4}{5}$$

$$70 = \frac{12}{25}(\sigma_y - \sigma_x) \quad \text{...(ii)}$$

Solving eq. (i) and (ii)

$\sigma_x = 67.5$

$\sigma_y = 213.3$

3.6 (b)

In case of plane strain condition, the components of strain tensor at a point are ϵ_x, ϵ_y and ϕ_{xy}.
Here we have three unknowns so we require 3 equations to find them and these unknowns can be find by the equations of longitudinal strain,

$$\epsilon_1 = \epsilon_x \cos^2\theta_1 + \epsilon_y \sin^2\theta_1 + \frac{\phi_{xy}}{2} \sin 2\theta_1$$

$$\epsilon_2 = \epsilon_x \cos^2\theta_2 + \epsilon_y \sin^2\theta_2 + \frac{\phi_{xy}}{2} \sin 2\theta_2$$

$$\epsilon_3 = \epsilon_x \cos^2\theta_3 + \epsilon_y \sin^2\theta_3 + \frac{\phi_{xy}}{2} \sin 2\theta_3$$

Therefore to find out component of strain tensor in plain strain condition, we measure longitudinal strains (ϵ_1, ϵ_2 and ϵ_3) along any three arbitrary direction.

3.7 (b)

Maximum shear stress

$$= \frac{\sigma_1 - \sigma_2}{2} = \frac{20 - (-10)}{2} = 15 \text{ MPa}$$

3.8 (d)

$$\sigma_{p1} = 30, \quad \sigma_{p2} = 30$$

The maximum and minimum principal stresses are same. So, radius of circle becomes zero and centre is at (30, 0). The circle is represented by a point.

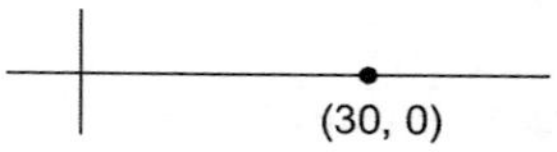

3.9 (b)

$$\text{Shear stress} = \frac{\sigma_1 - \sigma_2}{2} = \frac{173 - 0}{2} = 86.5 \text{ MPa}$$

3.10 (a)

On a principal plane, only normal stresses act. No shear stresses act on the principal plane.

3.11 (b)

Method-I

Maximum shear stress at the point is given by,

$$\tau_{max} = \frac{\sigma_1 - \sigma_2}{2} = \frac{3 - (-3)}{2} = 3 \text{ MPa}$$

Method-II

By Mohr Cicle

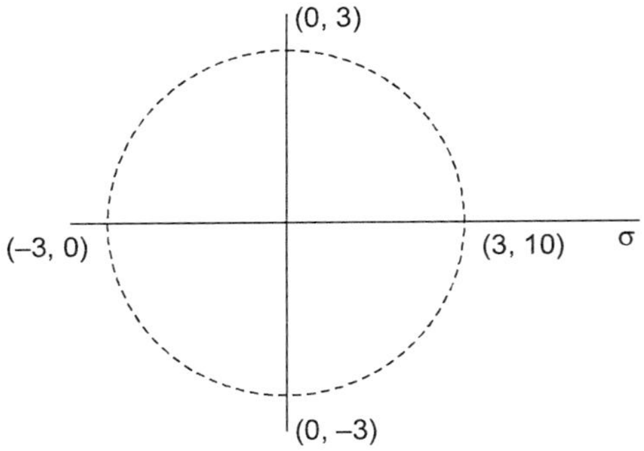

$\tau_{max} = 3$ MPa

3.12 (a)

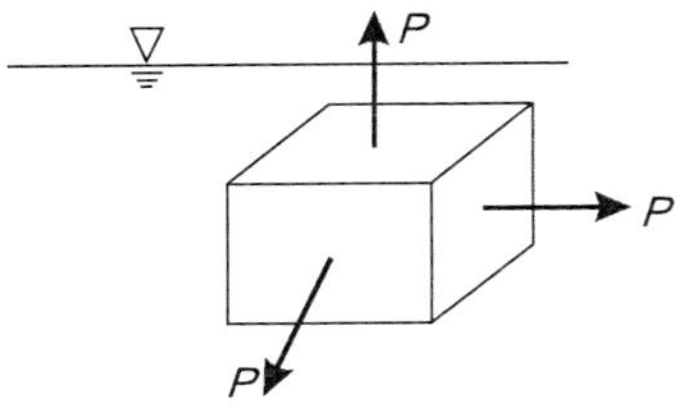

Maximum shear stress,

$$\tau = \frac{\sigma_1 - \sigma_2}{2} = \frac{P-P}{2} = 0$$

Note: It is the case of hdyrostatic force, Mohr circle of stress is a point. Hence, $\tau_{max} = 0$.

3.13 (a)

$$\sigma_{1/2} = \frac{\sigma_x + \sigma_y}{2} \pm \sqrt{\left(\frac{\sigma_x - \sigma_y}{2}\right)^2 + \tau_{xy}^2}$$

$$= \frac{100+20}{2} \pm \sqrt{\left(\frac{100-20}{2}\right)^2 + (30)^2}$$

$$= 60 \pm 50$$

$$\sigma_1 = 110; \quad \sigma_2 = 10$$

$$\therefore \tau_{max} = \frac{\sigma_1 - \sigma_2}{2} = \frac{110-10}{2} = 50 \text{ MPa}$$

Note: τ_{max} absolute $= \frac{110-0}{2} = 55$ MPa

But 55 MPa is not the option so taken $(\tau_{max})_{plane}$ = 50 MPa.

3.14 Sol.

Method-I

As plane *AB* and *BC* are principle planes, therefore Mohr's circle for the given condition is,

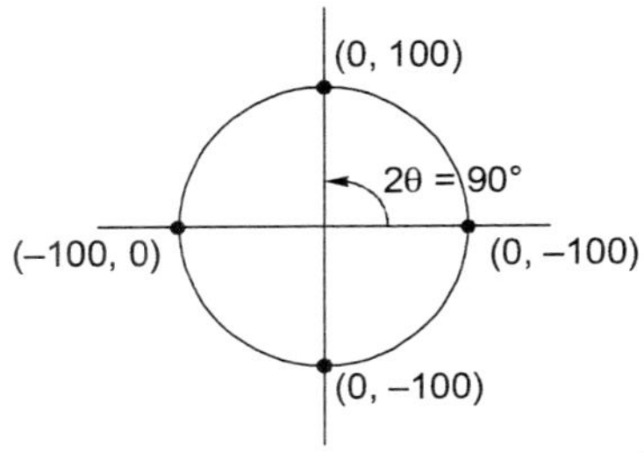

Here, normal stress is zero at 45° to the principle plane.

Method-II

$$\sigma_{P_1} = 100 \text{ MPa}, \; \sigma_{P_2} = -100 \text{ MPa}$$

$\Rightarrow$ at 45°, it is a plane of maximum shear stress

$$\sigma_n = \frac{\sigma_{P_1} + \sigma_{P_2}}{2} = 0$$

3.15 (d)

Under hydrostatic loading condition, stresses at a point in all directions are equal and hence no shear stress.

Alternatively,

$$\tau = \frac{\sigma_1 - \sigma_2}{2} = \frac{50-50}{2} = 0$$

Thus, Mohr's circle reduces to a point.

Hence shear stress at all orientations is zero.

3.16 (b)

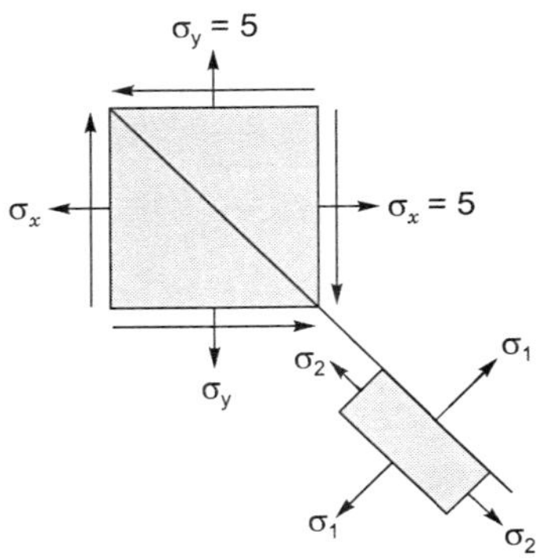

$$\sigma_x + \sigma_y = \sigma_1 + \sigma_2$$

$$\Rightarrow \quad 5 + 5 = 10 + \sigma_2$$

$$\Rightarrow \quad \sigma_2 = 0$$

Now, $$\sigma_{1/2} = \frac{\sigma_x + \sigma_y}{2} \pm \sqrt{\left(\frac{\sigma_x - \sigma_y}{2}\right)^2 + \tau_{xy}^2}$$

$$\therefore \quad \sigma_1 = \frac{5+5}{2} + \sqrt{\left(\frac{5-5}{2}\right)^2 + \tau_{xy}^2}$$

$$\Rightarrow \quad 10 = 5 + \tau_{xy}$$

$$\therefore \quad \tau_{xy} = 5 \text{ MPa}$$

3.17 (d)

$$\epsilon_x = 0.0030,\ \epsilon_y = 0.0020$$

Normal strain in the plane of maximum shear strain

$$= \frac{\epsilon_x + \epsilon_y}{2} = 0.0025$$

3.18 (a)

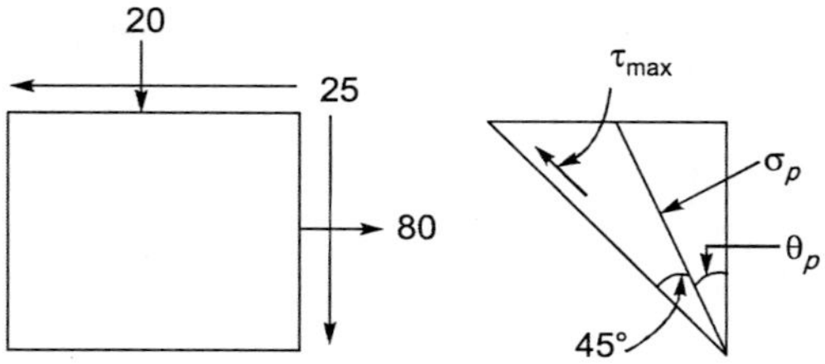

$$\sigma_x = 80$$
$$\sigma_y = -20$$
$$\tau_{xy} = -25$$

Angle of plane of max shear

$$\theta = \theta_p + 45°$$

$$\tan 2\theta_p = \frac{2\tau_{xy}}{\sigma_x - \sigma_y} = \frac{-50}{100}$$

$$\theta_p = -13.28°$$

$\therefore$ $$\theta = 31.71°$$

3.19 (c)

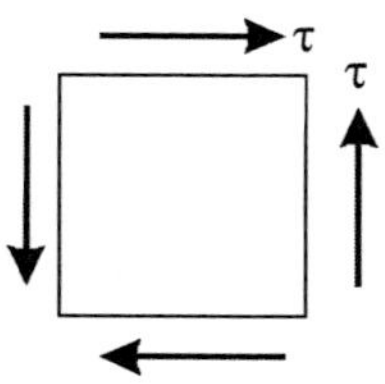

In pure shear condition

$$\sigma_x = 0,\ \sigma_y = 0,\ \tau_{xy} = \tau$$

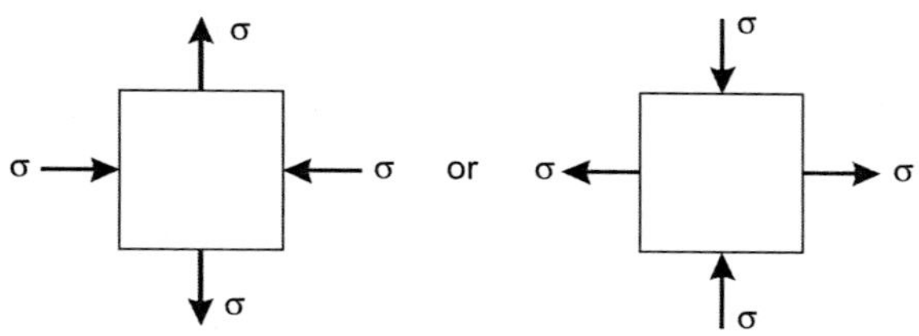

$\sigma_{xx} = -\sigma$, $\sigma_{yy} = \sigma$, $\tau_{xy} = 0$ or $\sigma_{xx} = \sigma$, $\sigma_{yy} = -\sigma$, $\tau_{xy} = 0$

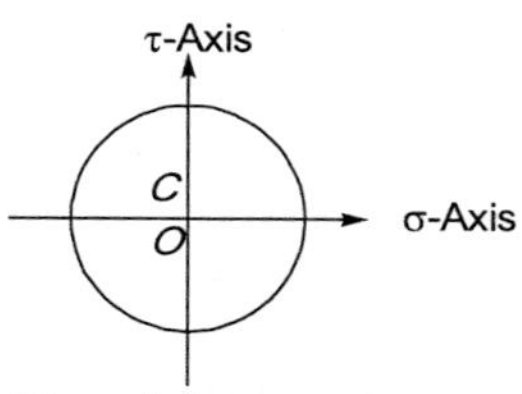

For this condition (c) is correct

$$\sigma_{xx} + \sigma_{yy} = 0$$

3.20 (d)

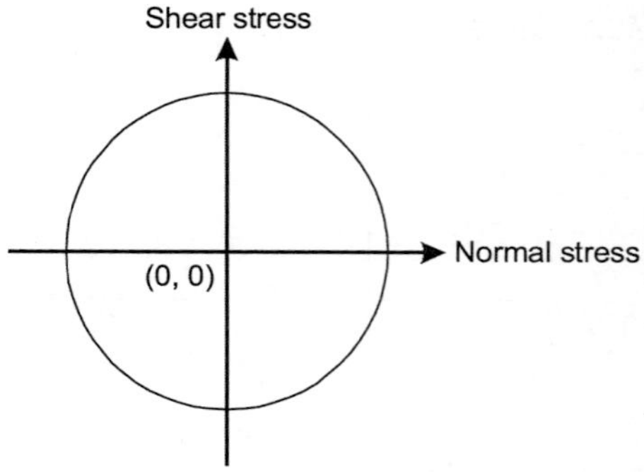

In pure shear condition, Mohr's circle has its center at origin.

3.21 Sol.

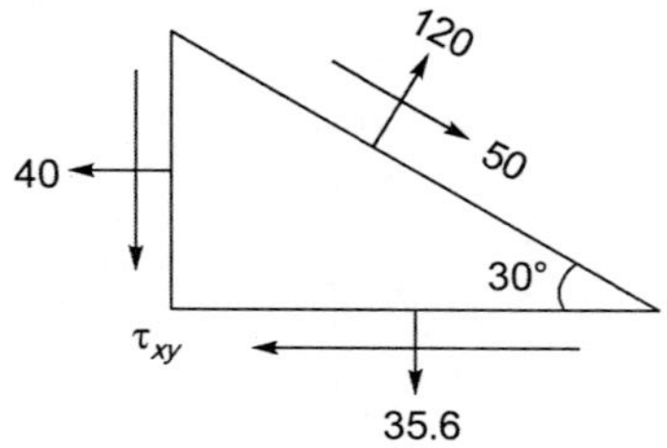

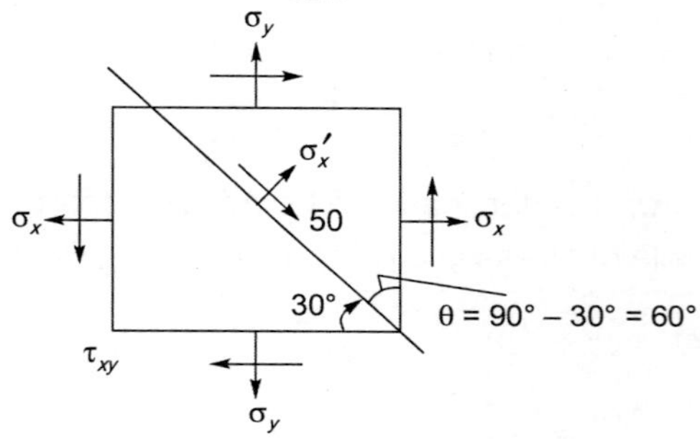

$$\sigma_x' = \frac{\sigma_x + \sigma_y}{2} + \left(\frac{\sigma_x - \sigma_y}{2}\right)\cos 2\theta + \tau_{xy}\sin 2\theta$$

Here $\theta = 60°$

$\sigma_x = 40$ MPa, $\sigma_y = 35.6$, $\sigma_x' = 120$, $\tau_{x'y'} = -50$

Substituting the values in above equation, we get

$$\tau_{xy} = 96.186 \text{ MPa} \simeq 96.2 \text{ MPa}$$

3.22 Sol.

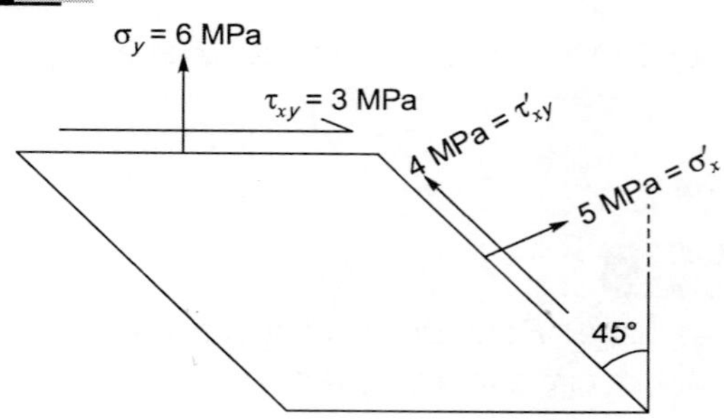

Now, $\sigma'_x = \sigma_x \cos^2\theta + \sigma_y \sin^2\theta + \tau_{xy}\sin 2\theta$ ...(i)

Here $\theta = 45°$

$\sigma'_x = 5$ MPa

$\sigma_y = 6$ MPa

$\tau_{xy} = 3$ MPa

$5 = \sigma_x \cos^2 45° + 6\sin^2 45° + 3\sin^2 \times 45°$

$\frac{\sigma_x}{2} = 5 - 3 - 6 \times \frac{1}{2}$

$\Rightarrow \sigma_x = -2$ MPa

From eq. (i), we get, $\sigma_x = -2$ MPa

Now, $\sigma_{1/2} =$

$$\frac{\sigma_x + \sigma_y}{2} \pm \frac{1}{2}\sqrt{(\sigma_y - \sigma_x)^2 + 4\tau_{xy}^2}$$

$$= \frac{-2+6}{2} \pm \frac{1}{2}\sqrt{(6+2)^2 + 4 \times 3^2}$$

$\therefore \sigma_1 = 7$ MPa and $\sigma_2 = -3$ MPa

3.23 (c)

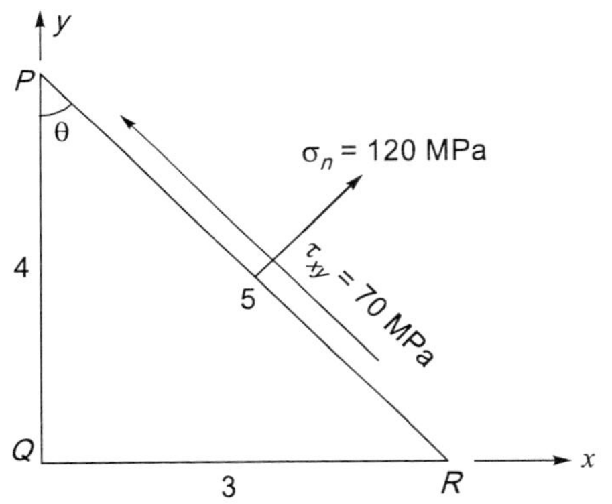

$\cos\theta = \frac{4}{5}$ and $\sin\theta = \frac{3}{5}$; $\theta = 36.87°$

$\because \quad \sigma_n = \sigma_x \cos^2\theta + \sigma_y \sin^2\theta$

$\Rightarrow \quad 120 = \sigma_x \times \left(\frac{4}{5}\right)^2 + \sigma_y \left(\frac{3}{5}\right)^2$

$\Rightarrow \quad 25 \times 120 = 16\sigma_x + 9\sigma_y$ ... (i)

Also, $\tau_{xy} = (\sigma_y - \sigma_x)\sin\theta\cos\theta$

$\Rightarrow \quad 70 = (\sigma_y - \sigma_x)\left(\frac{4}{5}\right)\left(\frac{3}{5}\right)$

$\Rightarrow \quad 70 \times 25 = 12\sigma_y + 12\sigma_x$ ... (ii)

Solving (i) and (ii),

$\sigma_x = 67.5$ MPa and $\sigma_y = 213.3$ MPa

3.24 (b)

In the given matrix, shear stresses are zero. So, the elements on diagonal of this matrix are principal stresses.

So,

$$\tau_{max} = \left|\left(\frac{\sigma_{p1} - \sigma_{p2}}{2}\right), \left(\frac{\sigma_{p2} - \sigma_{p3}}{2}\right), \left(\frac{\sigma_{p3} - \sigma_{p1}}{2}\right)\right|_{max}$$

$$= \left|\left(\frac{10-40}{2}\right), \left(\frac{40-0}{2}\right), \left(\frac{0-10}{2}\right)\right|_{max}$$

$$= |(-15), 20, (-5)|_{max}$$

$$= 20 \text{ MPa}$$

3.25 (a)

■■■■

4 Bending and Shear Stresses

4.1 A homogeneous, simply supported prismatic beam of width B, depth D and span L is subjected to a concentrated load of magnitude P. The load can be placed anywhere along the span of the beam. The maximum flexural stress developed in beam is

(a) $\frac{2}{3}\frac{PL}{BD^2}$ (b) $\frac{3}{4}\frac{PL}{BD^2}$

(c) $\frac{4}{3}\frac{PL}{BD^2}$ (d) $\frac{3}{2}\frac{PL}{BD^2}$ **[2004 : 2 M]**

4.2 A beam with the cross-section given below is subjected to a positive bending moment (causing compression at the top) of 16 kN-m acting around the horizontal axis. The tensile force acting on the hatched area of the cross-section is

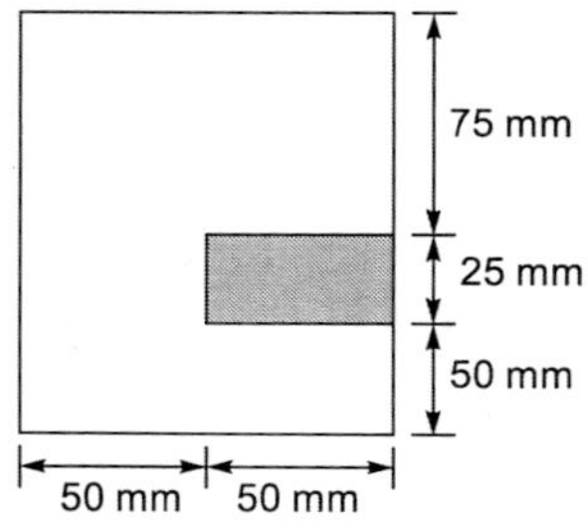

(a) zero (b) 5.9 kN

(c) 8.9 kN (d) 17.8 kN **[2006 : 2 M]**

4.3 If a beam of rectangular cross-section is subjected to a vertical shear force V, the shear force carried by the upper one-third of the cross-section is

(a) zero (b) $\frac{7V}{27}$

(c) $\frac{8V}{27}$ (d) $\frac{V}{3}$ **[2006 : 2 M]**

4.4 I-section of a beam is formed by gluing wooden planks as shown in the figure below. If this beam transmits a constant vertical shear force of 3000 N, the glue at any of the four joints will be subjected to a shear force (in kN per meter length) of

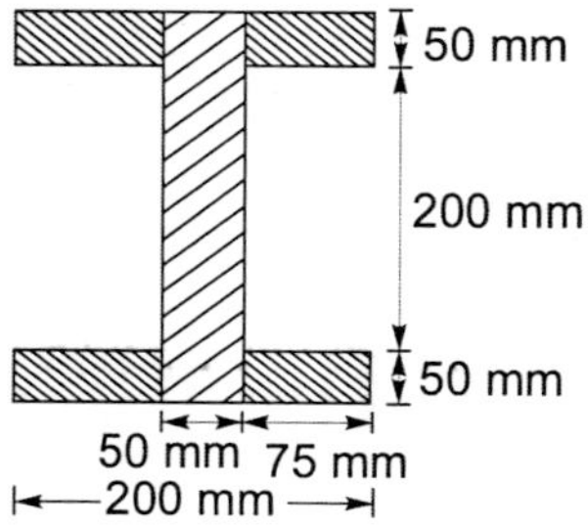

(a) 3.0 (b) 4.0

(c) 8.0 (d) 10.7 **[2006 : 2 M]**

4.5 The shear stress at the neutral axis in a beam of triangular section with a base of 40 mm and height 20 mm, subjected to a shear force of 3 kN is

(a) 3 MPa (b) 6 MPa

(c) 10 MPa (d) 20 MPa **[2007 : 2 M]**

4.6 The maximum tensile stress at the section X-X shown in the figure below is

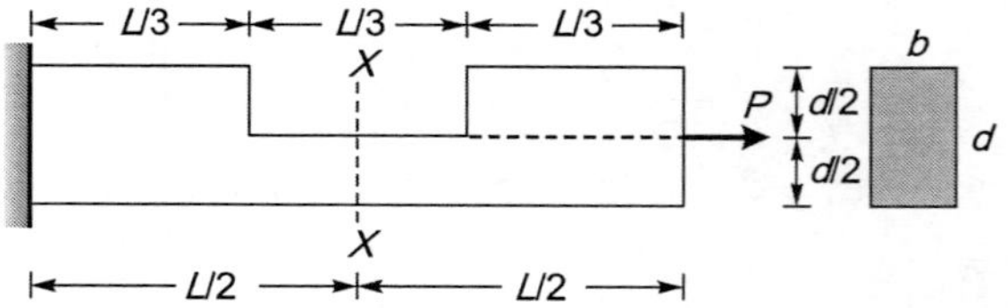

(a) $\frac{8P}{bd}$ (b) $\frac{6P}{bd}$

(c) $\frac{4P}{bd}$ (d) $\frac{2P}{bd}$ **[2008 : 2 M]**

4.7 Consider a simply supported beam with a uniformly distributed load having a neutral axis (NA) as shown. For points P (on the neutral axis) and Q (at the bottom of the beam) the state of stress is best represented by which of the following pairs?

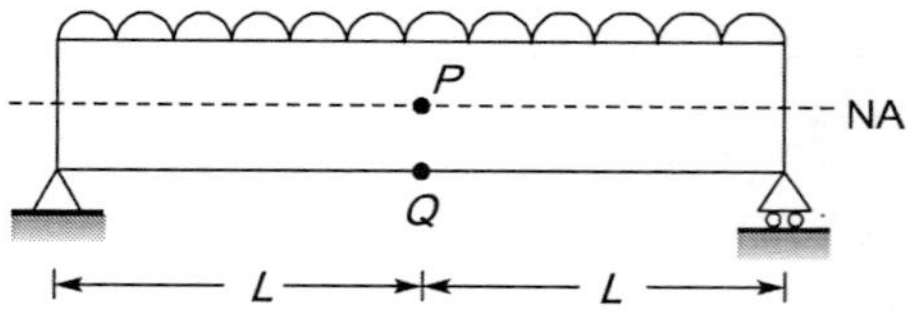

(a) 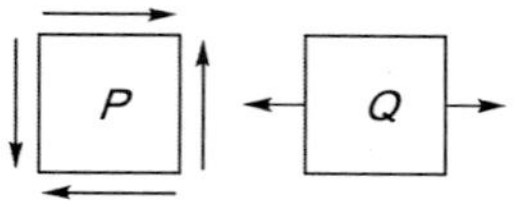

(b) 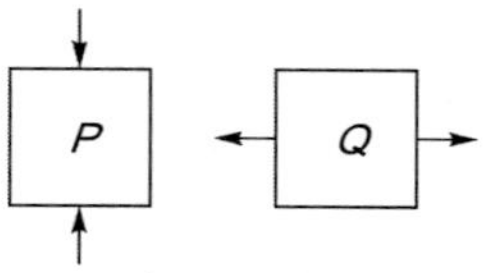

(c) 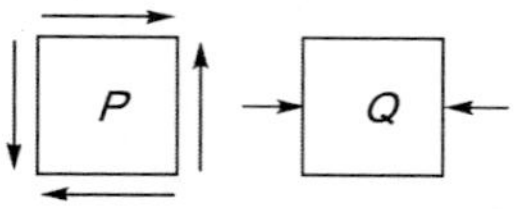

(d) 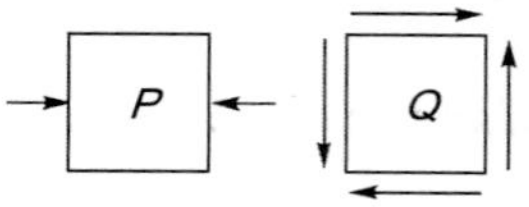

[2011 : 1 M]

4.8 The "Plane section remain plane" assumption in bending theory implies
(a) strain profile is linear
(b) stress profile is linear
(c) both profiles are linear
(d) shear deformation is neglected [2013 : 1 M]

4.9 A symmetric I-section (with width of each flange = 50 mm, thickness of web = 10 mm) of steel is subjected to a shear force of 100 kN. Find the magnitude of the shear stress (in N/mm^2) in the web at its junction with the top flange ______.
[2013 : 1 M]

4.10 The first moment of area about the axis of bending for a beam cross-section is
(a) moment of inertia
(b) section modulus
(c) shape factor
(d) polar moment of inertia [2014 : 1 M, Set-II]

4.11 A simply supported reinforced concrete beam of length 10 m sags while undergoing shrinkage. Assuming a uniform curvature of 0.004 m^{-1} along the span, the maximum deflection (in m) of the beam at mid-span is ______.
[2015 : 2 M, Set-II]

4.12 A 450 mm long plain concrete prism is subjected to the concentrated vertical loads as shown in the figure. Cross section of the prism is given as 150 mm × 150 mm. Considering linear stress distribution across the cross-section, the modulus of rupture (expressed in MPa) is ____.

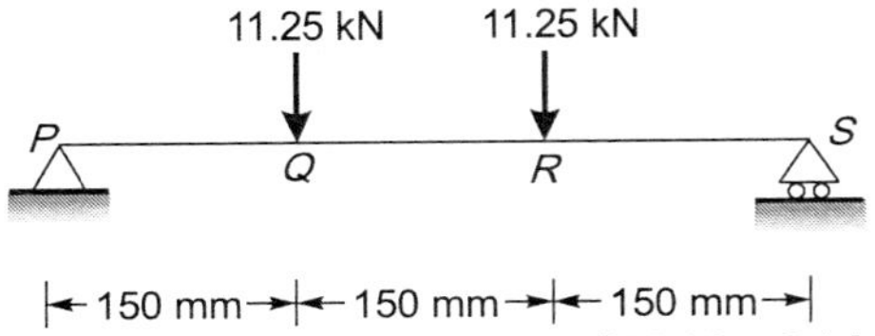

[2016 : 2 M, Set-II]

4.13 A cantilever beam of length 2 m with a square section of side length 0.1 m is loaded vertically at the free end. The vertical displacement at the free end is 5 mm. The beam is made of steel with Young's modulus of 2.0×10^{11} N/m^2. The maximum bending stress at the fixed end of the cantilever is
(a) 20.0 MPa (b) 37.5 MPa
(c) 60.0 MPa (d) 75.0 MPa
[2018 : 2 M, Set-I]

4.14 An 8 m long simply-supported elastic beam of rectangular cross-section (100 mm × 200 mm) is subjected to a uniformly distributed load of 10 kN/m over its entire span. The maximum principal stress (in MPa, up to two decimal places) at a point located at the extreme compression edge of a cross-section and at 2 m from the support is ______. [2018 : 2 M, Set-II]

4.15 For a given loading on a rectangular plain concrete beam with an overall depth of 500 mm, the compressive strain and tensile strain developed at the extreme fibers are of the same magnitude of 2.5×10^{-4}. The curvature in the beam cross-section (in m^{-1}, round off to 3 decimal places), is______.
[2019 : 1 M, Set-I]

4.16 Cross section of a built-up wooden beam as shown in figure (not drawn to scale) is subjected to a vertical shear force of 8 kN. The beam is symmetrical about the neutral axis (NA), shown, and the moment of inertia about N.A. is 1.5×10^9 mm^4. Considering that the nails at the location P are spaced longitudinally (along the length of the beam) at 60 mm, each of the nails at P will be subjected to the shear force of

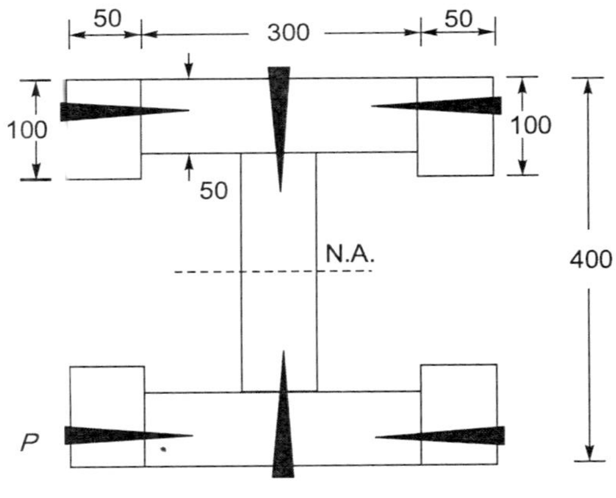

All dimensions are in mm

(a) 240 N (b) 480 N
(c) 60 N (d) 120 N
[2019 : 2 M, Set-I]

4.17 For a channel section subjected to a downward vertical shear force at its centroid, which one of the following represents the correct distribution of shear stress in flange and web?

(a) 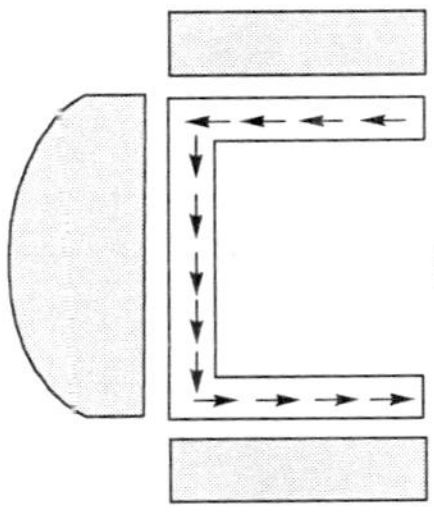

(b) 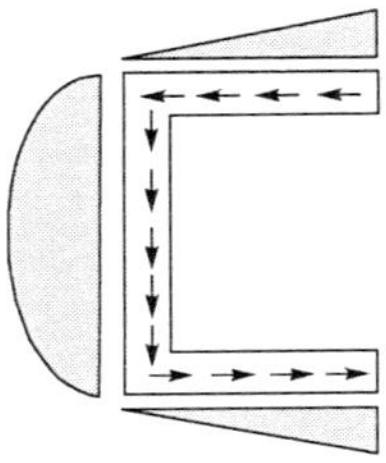

(c) 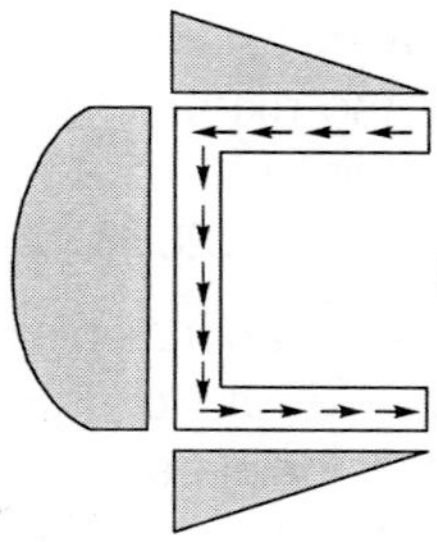 (d) 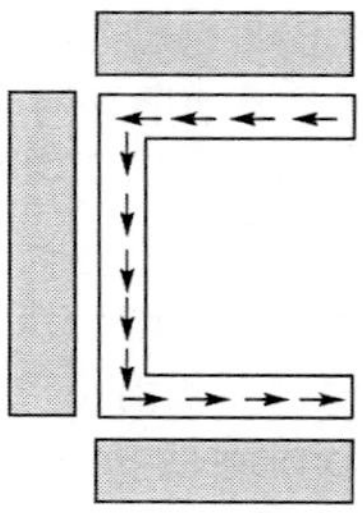

[2019 : 1 M, Set-II]

■■■■

Answers Bending and Shear Stresses

4.1 (d)	4.2 (c)	4.3 (b)	4.4 (b)	4.5 (c)	4.6 (a)	4.7 (*)
4.8 (a)	4.9 (71.1)	4.10 (b)	4.11 (0.05)	4.12 (3)	4.13 (b)	4.14 (90)
4.15 (0.001)	4.16 (a)	4.17 (c)				

Explanations Bending and Shear Stresses

4.1 (d)

When the concentrated load is placed at the midspan, maximum bending moment will develop at the mid span.

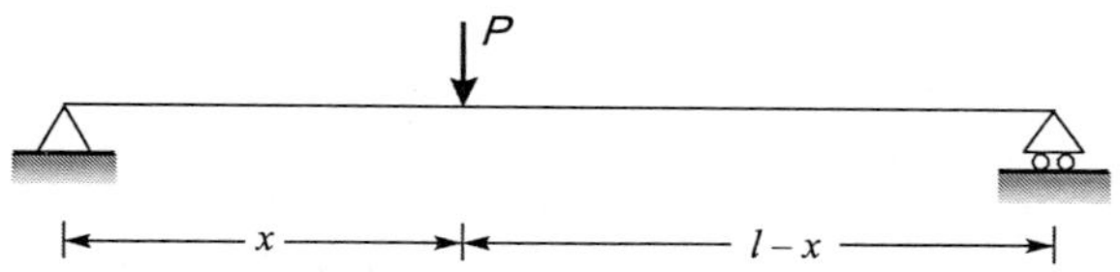

$$M = \frac{Px(l-x)}{l}$$

$$\Rightarrow \quad \frac{dM}{dn} = 0$$

Under load will be maximum

$$n = \frac{l}{2}$$

Now, $$f = \frac{M}{I}y \qquad \left[\because M = \frac{PL}{4}\right]$$

$$= \frac{\frac{PL}{4}\times\frac{D}{2}}{\frac{BD^3}{12}} = \frac{3PL}{2BD^2}$$

4.2 (c)

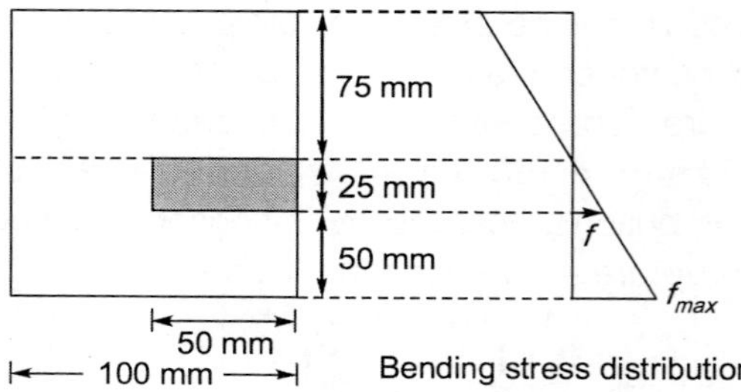

Bending stress distribution

$$f = \frac{M}{I}\times y = \frac{16\times10^6\times12}{100\times150^3}\times25 = 14.22\text{ N/mm}^2$$

From similar triangles, we have,

$$\therefore \text{ Tensile force} = \frac{1}{2}\times25\times14.22\times50\times10^{-3}$$
$$= 8.88 \approx 8.9\text{ kN}$$

4.3 (b)

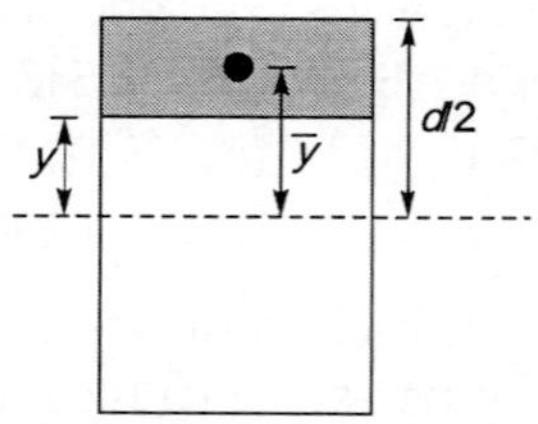

Shear stress at distance y from neutral axis,

$$\tau = \frac{SA\bar{y}}{Ib} = \frac{V \times \left(\frac{d}{2} - y\right) \times b \times \left(\frac{(d/2)+y}{2}\right)}{Ib}$$

$$\Rightarrow \quad \tau = \frac{V \times \left(\frac{d^2}{4} - y^2\right)}{2I}$$

$$\therefore \quad dF = \tau \times b\,dy = \frac{V \times \left(\frac{d^2}{4} - y^2\right)}{2I} \times b\,dy$$

Integrating both sides, we get,

$$\Rightarrow \quad F = \frac{Vb}{2I}\int_{d/6}^{d/2}\left(\frac{d^2}{4} - y^2\right)dy$$

$$= \frac{Vb}{2I}\left[\frac{d^2}{4}y - \frac{y^3}{3}\right]_{d/6}^{d/2}$$

$$= \frac{Vb}{2I}\left[\frac{d^3}{8} - \frac{d^3}{24} - \frac{d^3}{24} + \frac{d^3}{648}\right]$$

$$= \frac{Vb}{2I} \times \frac{d^3}{8} \times \frac{28}{81}$$

$$= \frac{Vb}{2bd^3} \times \frac{d^3}{8} \times \frac{28}{81} \times 12 = \frac{7V}{27}$$

4.4 (b)

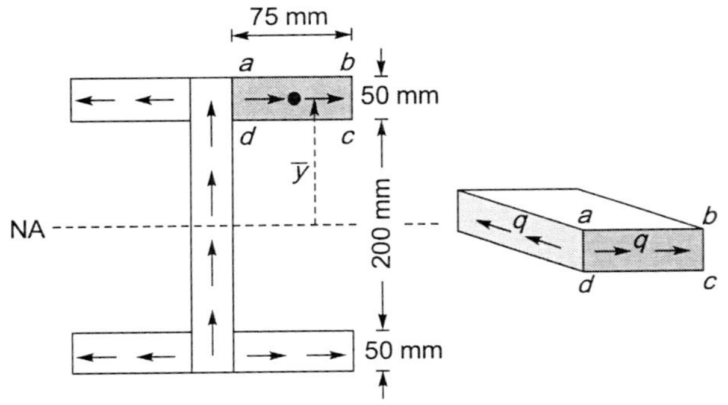

Considering shaded area 'abcd' of plank,

Shear flow, $q = \frac{VQ}{I}$

Shear flow is measured as force per unit length along the longitudinal axis of a beam and over the cross-section is as shown above.

$$Q = A\bar{y}$$

$$Q = (75 \times 50) \times 125 = 468750 \text{ mm}^3$$

Moment of inertia of I-section about NA,

$$I = 350 \times 10^6 \text{ mm}^4$$

$$V = 3000 \text{ N}$$

$$q = \frac{468750 \times 3000}{350 \times 10^6} = 4.018 \text{ N/mm}$$

So, resistance offered by glue to keep I-section intact,

$$q = 4.018 \text{ N/mm} \simeq 4 \text{ kN/m}$$

4.5 (c)

Method-I

Shear stress, $t = \frac{SA\bar{y}}{Ib}$

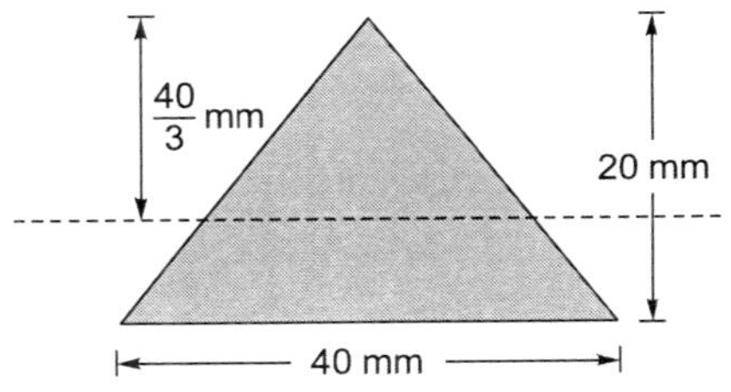

Width at a distance of $\frac{40}{3}$ mm from the top

$$= \frac{40}{20} \times \frac{40}{3} = \frac{80}{3} \text{ mm}$$

$$\therefore \quad \tau = \frac{3 \times 10^3 \times \left(\frac{1}{2} \times \frac{80}{3} \times \frac{40}{3}\right) \times \left(\frac{1}{3} \times \frac{40}{3}\right)}{\left(\frac{40 \times 20^3}{36}\right) \times \frac{80}{3}}$$

$$= \frac{3 \times 10^3 \times 3200 \times 40 \times 36 \times 3}{162 \times 3200 \times 20^3} = 10 \text{ MPa}$$

Method-II

$$\tau_{N.A.} = \frac{4}{3}\tau_{avg} = \frac{4}{3} \times \frac{3000}{\frac{1}{2} \times 40 \times 20} = 10 \text{ MPa}$$

4.6 (a)

The section at X–X may be shown as in the figure below:

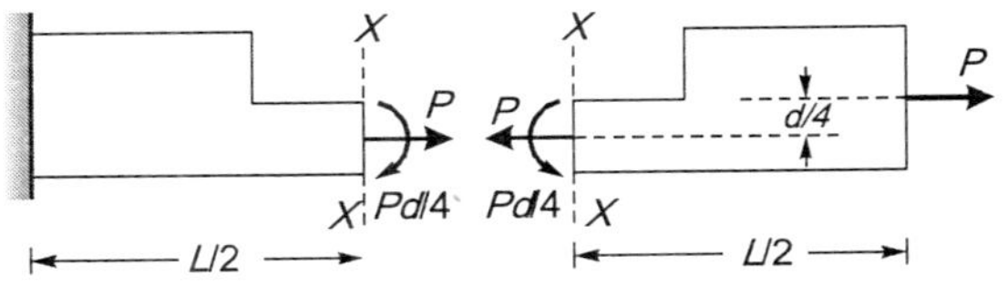

The maximum tensile stress at the section X–X is,

$$\sigma = \frac{P}{A} + \frac{M}{Z}$$

$$= \frac{P}{b \times (d/2)} + \frac{P \times (d/4) \times 6}{b \times (d^2/4)}$$

$$= \frac{2P}{bd} + \frac{6P}{bd} = \frac{8P}{bd}$$

4.7 (*)

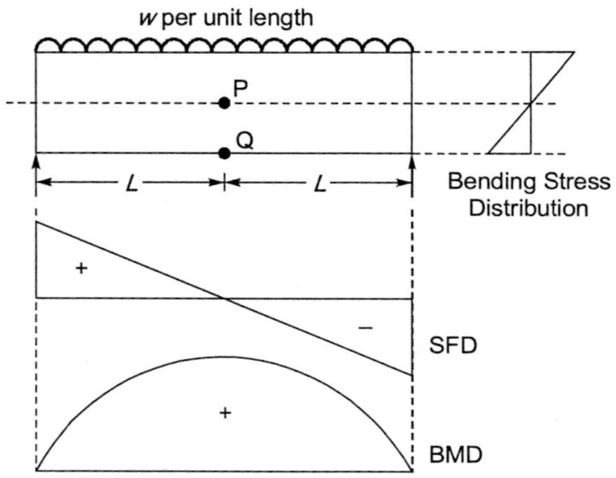

Point *P*: Point *P* lies on NA, hence bending stress is zero at point *P*.

Point *P* also lies at mid span, so shear force, $V = 0$

$\Rightarrow$ Shear stress, $\tau = 0$

$\therefore$ State of stress of point *P* will be,

Point *Q*: At point *Q* flexural stress is maximum and nature of which is tensile due to downward loading. Point *Q* lies at the extreme of beam, therefore, shear stress at point *Q* is zero.

$\therefore$ State of stress of point *Q* will be,

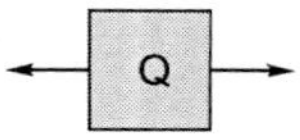

4.8 (a)

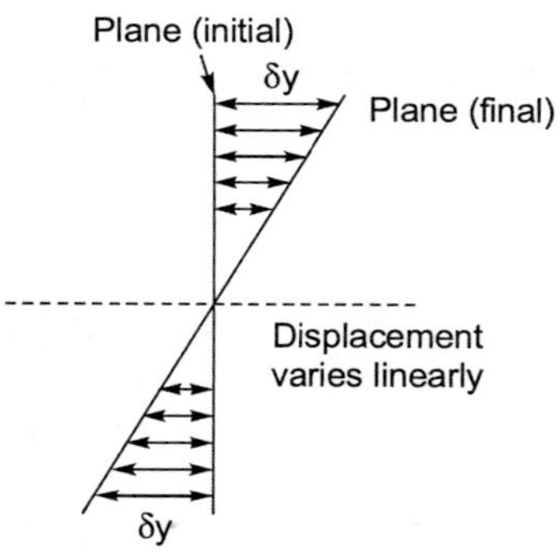

Since, $E \propto \delta y$

So, strain varies linearly.

4.9 Sol.

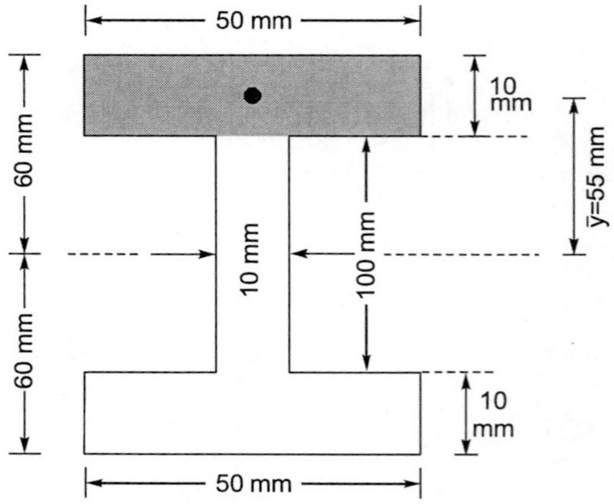

$$I = \frac{50 \times 120^3}{12} - \frac{40 \times 100^3}{12}$$

$$= 3.866 \times 10^6 \text{ mm}^4$$

$$q = \frac{SA\bar{y}}{Ib} = \frac{100 \times 10^3 \times 50 \times 10 \times 55}{3.866 \times 10^6 \times 10}$$

$$= 71.13 \text{ MPa}$$

4.10 (b)

- First moment of area about the axis of bending for a beam cross section is section modulus.
- Second moment of area about the axis of bending for a beam cross section is moment of inertia of section about that axis.

4.11 Sol.

Method-I

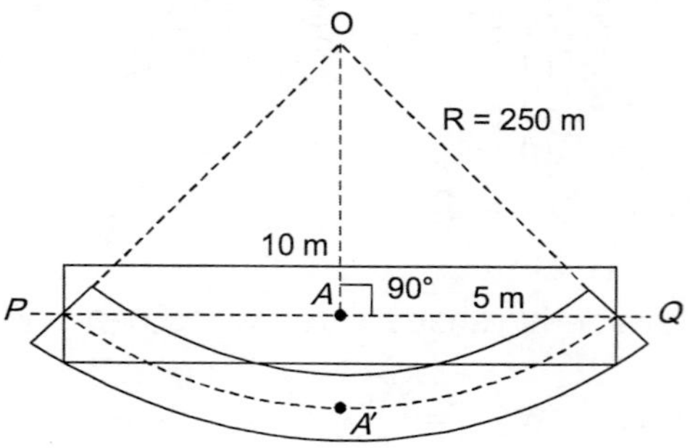

Radius, $R = \dfrac{1}{C} = 250$ m

$$OA = \sqrt{250^2 - 5^2} = 249.95 \text{ m}$$

Deflection $= AA' = 250 - 249.95 = 0.05$ m

Method-II

$$\delta = \frac{l^2}{8R}$$

$$(2R-\delta)\,\delta = \frac{l^2}{4}$$

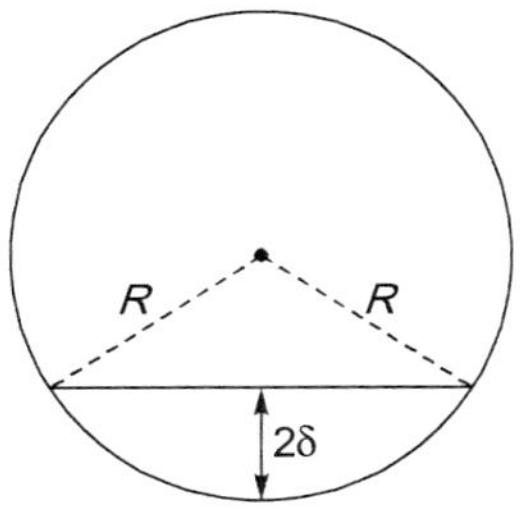

$$\frac{1}{R} = 0.004$$

$$\delta = \frac{l^2}{8R}$$

$$\Rightarrow \quad \delta = \frac{10^2}{8}\times 0.004$$

$$\delta = 0.05\text{ m}$$

4.12 Sol.

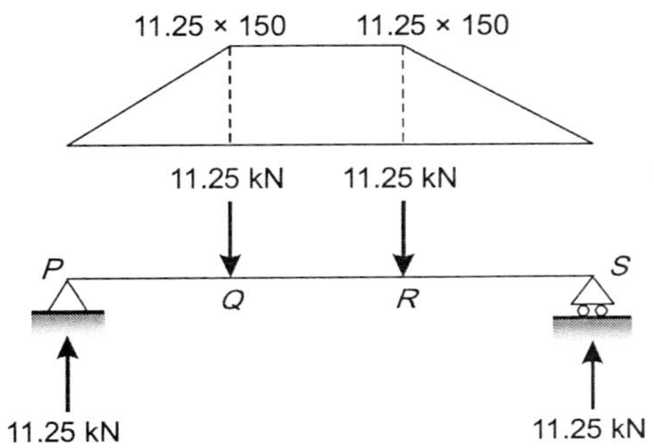

$$BM_Q = 11.25 \times 150$$
$$= 1.6875 \times 10^6 \text{ N-mm}$$

$$\Rightarrow \quad \frac{\sigma}{y} = \frac{M_Q}{I}$$

where, $y = \dfrac{150}{2} = 75$ mm and $I = \dfrac{(150)^4}{12}$

$$\Rightarrow \quad \sigma = \frac{1.6875\times 10^6 \times 75}{(150)^4/12} = 3\text{ MPa}$$

4.13 (b)

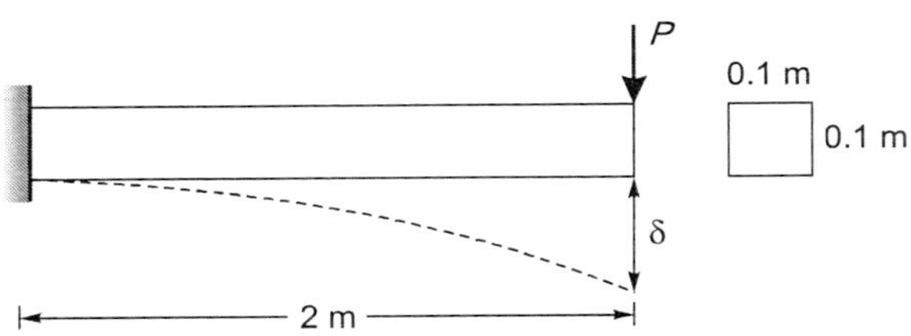

$$I = \frac{(0.1)^4}{12}$$

Deflection $\delta = \dfrac{Pl^3}{3EI}$

$$\Rightarrow \quad 5\times 10^{-3}\text{ m} = \frac{P(2)^3}{3\times 2\times 10^{11}\times\dfrac{(0.1)^4}{12}}$$

$\Rightarrow \quad P = 3125$ N

Now, $\quad M = Pl = 3125 \times 2 = 6250$ Nm

As, $\quad \dfrac{M}{I} = \dfrac{\sigma}{y} = \dfrac{E}{R}$

$$\Rightarrow \quad \sigma_{max} = \frac{M}{Z} = \frac{6250}{\dfrac{(0.1)^3}{6}}$$

$$= 37.5 \times 10^6 \text{ N/m}^2 = 37.5 \text{ MPa}$$

4.14 Sol.

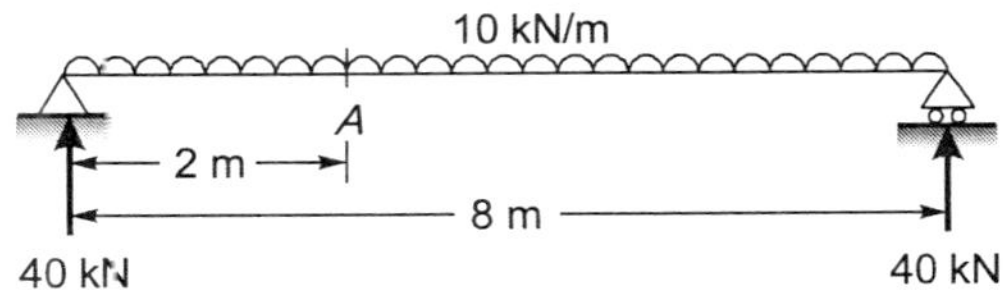

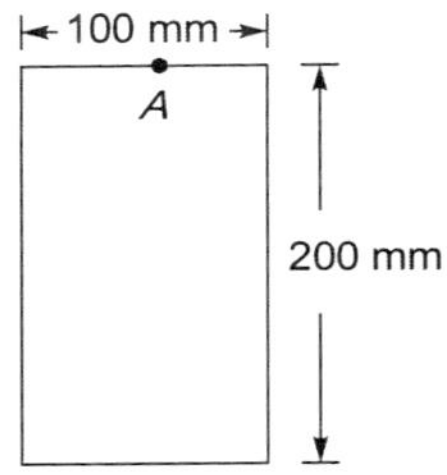

$$V_A + V_B = 80$$
$$V_A = V_B = 40$$

[Due to symmetry]

$$M_A = (-10\times 2\times 1) + 40\times 2$$
$$= 60 \text{ kNm}$$

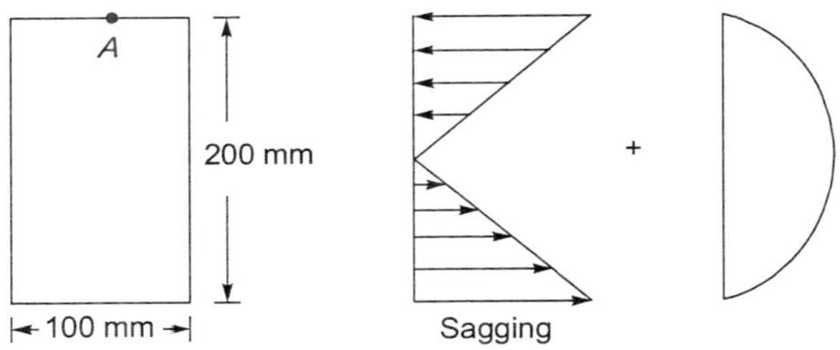

At extreme compression edge,
Bending stress,

$$\sigma = \frac{My}{I}$$

$$= \frac{(60 \times 10^6) \times \frac{200}{2}}{100 \times \frac{200^3}{12}} = 90 \text{ MPa}$$

Direct shear stress = 0

Principal stress,

$$\sigma_{p_1} / \sigma_{p_2} = \frac{90+0}{2} \pm \frac{1}{2}\sqrt{(0-90)^2 + y(0)^2}$$

$$\sigma_{p_1} = 90 \text{ MPa}$$

So principal stress

$$= 90 \text{ N/mm}^2 = 90 \text{ MPa}$$

4.15 Sol.

Given: $D = 500$ mm

$$\text{Strain} = \frac{\text{Stress}}{E} = \frac{f}{E}$$

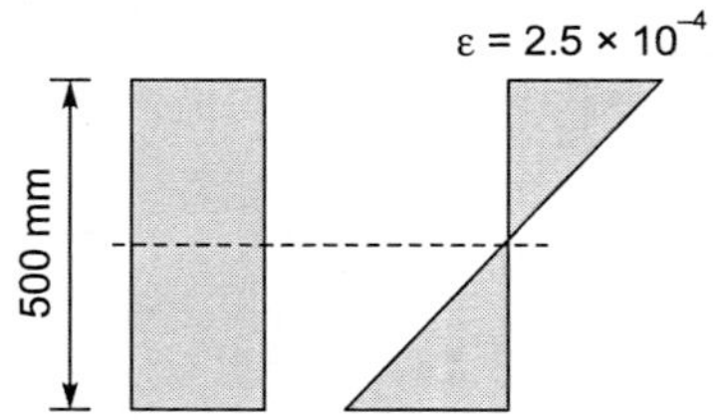

$$\frac{M}{I} = \frac{E}{R} = \frac{f}{y}$$

$$\Rightarrow \quad \frac{E}{R} = \frac{f}{y}$$

$$\Rightarrow \quad \frac{1}{R} = \frac{f/E}{y} = \frac{\epsilon}{y}$$

$$= \frac{2.5 \times 10^{-4}}{250}$$

$$= 1 \times 10^{-6} = 0.001$$

4.16 (a)

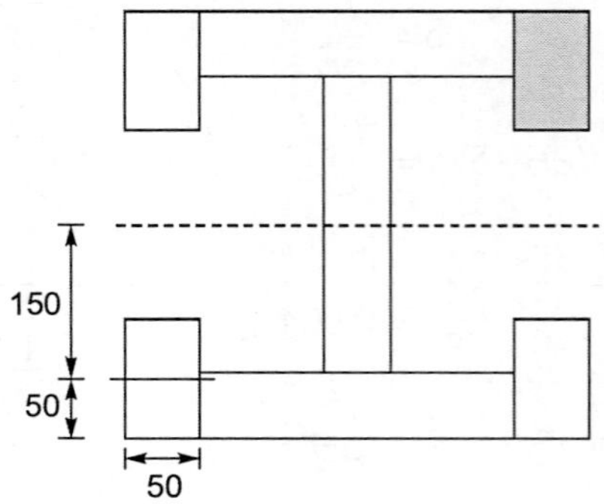

Shear flow, $q = \dfrac{SA\bar{y}}{I}$

$$= \frac{8000 \times 50 \times 100 \times 150}{1.5 \times 10^9}$$

$$= 4 \text{ N/mm}$$

Distance between two nails l = 60 mm

∴S.F. resisted by each nail = $q \times l$ = 240 N

4.17 (c)

Shear flow distribution for channel.

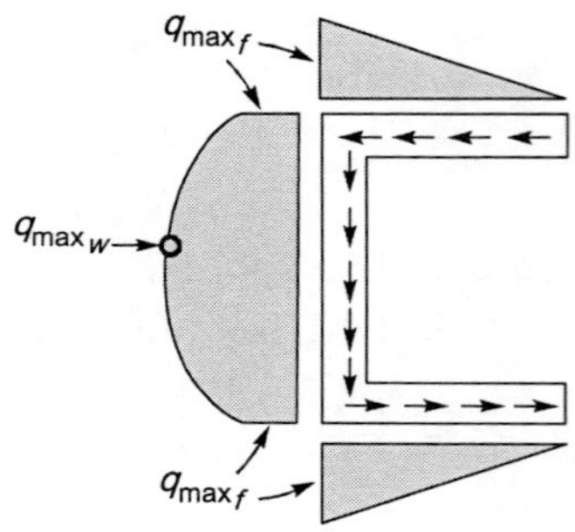

■■■■

5 Deflection of BeaM

5.1 The deflection of cantilever beam at free end B applied with a moment M at the same point is

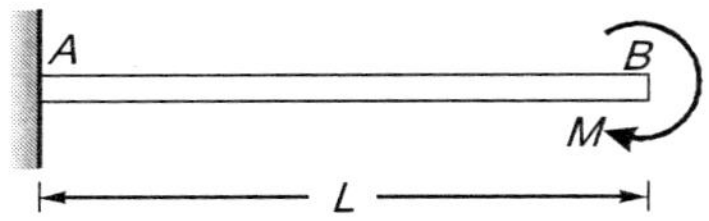

(a) $\dfrac{ML^2}{EI}$ (b) $\dfrac{ML^2}{2EI}$

(c) $\dfrac{ML^2}{3EI}$ (d) $\dfrac{ML^2}{4EI}$ [1996 : 1 M]

5.2 M-θ relationship for a simply supported beam shown below is given by

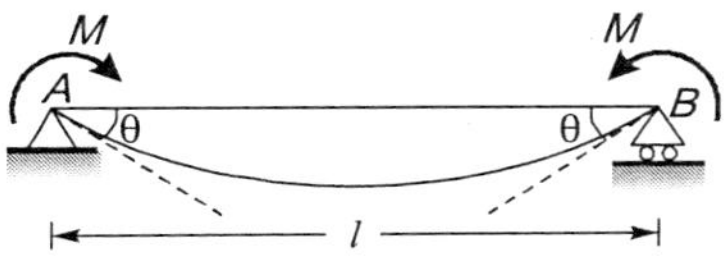

(a) $\dfrac{Ml}{EI} = 2\theta$ (b) $\dfrac{Ml}{EI} = 3\theta$

(c) $\dfrac{Ml}{EI} = 4\theta$ (d) $\dfrac{Ml}{EI} = 6\theta$ [1996 : 1 M]

5.3 A cantilever beam of span L is loaded with a concentrated load P at the free end. Deflection of the beam at the free end is

(a) $\dfrac{PL^3}{48EI}$ (b) $\dfrac{5PL^3}{384EI}$

(c) $\dfrac{PL^3}{3EI}$ (d) $\dfrac{PL^3}{6EI}$ [1997 : 1 M]

5.4 A cantilever beam is as shown in figure. The moment to be applied at free end for zero vertical deflection at that point is

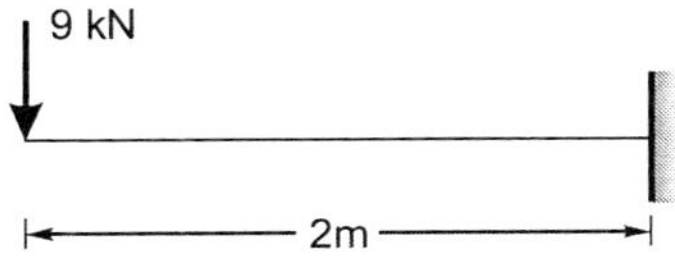

(a) 9 kNm clockwise
(b) 9 kNm anti-clockwise
(c) 12 kNm clockwise
(d) 12 kNm anti-clockwise [1998 : 2 M]

5.5 The slope of the elastic curve at the free end of a cantilever beam of span L, and with flexural rigidity EI, subjected to uniformly distributed load of intensity w is

(a) $\dfrac{wL^3}{6EI}$ (b) $\dfrac{wL^3}{3EI}$

(c) $\dfrac{wL^4}{8EI}$ (d) $\dfrac{wL^3}{2EI}$ [1999 : 1 M]

5.6 A two span beam with an internal hinge is shown below

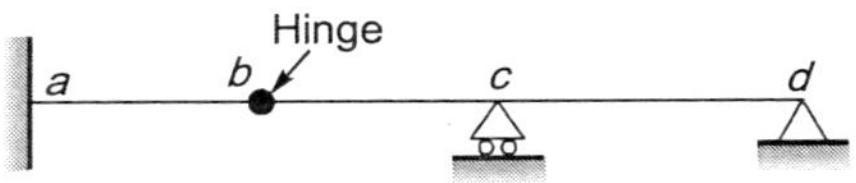

The conjugate beam corresponding to this beam is

(a) 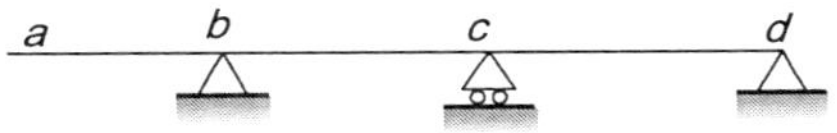

(b) 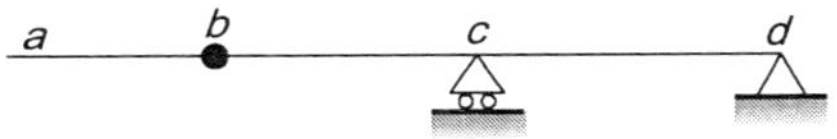

(c) 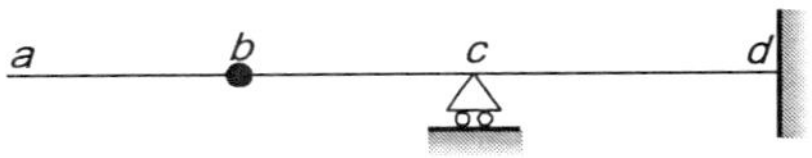

(d) 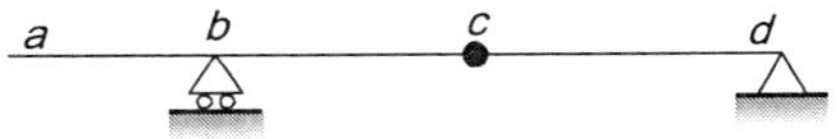

[2000 : 1 M]

5.7 For the structure shown below, the vertical deflection at point A is given by

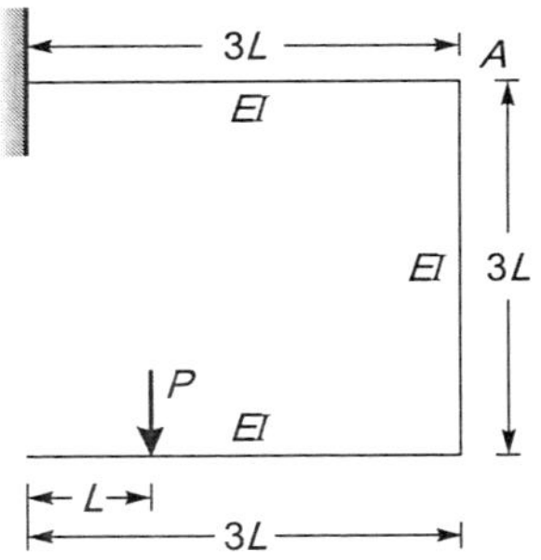

(a) $\dfrac{PL^3}{81EI}$ (b) $\dfrac{2PL^3}{81EI}$

(c) Zero (d) $\dfrac{PL^3}{72EI}$ [2000 : 1 M]

5.8 In the propped cantilever beam carrying a uniformly distributed load of w N/m, shown in the following figure, the reaction at the support B is

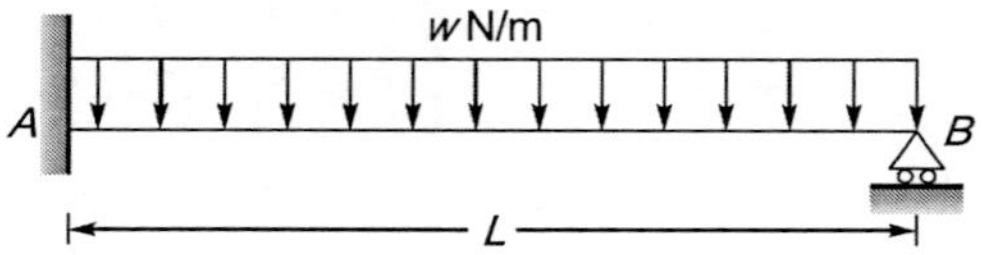

(a) $\frac{5}{8}wL$ (b) $\frac{3}{8}wL$

(c) $\frac{1}{2}wL$ (d) $\frac{3}{4}wL$ [2002 : 2 M]

5.9 A "H" shaped frame of uniform flexural rigidity EI is loaded as shown in the figure. The relative outward displacement between points K and O is

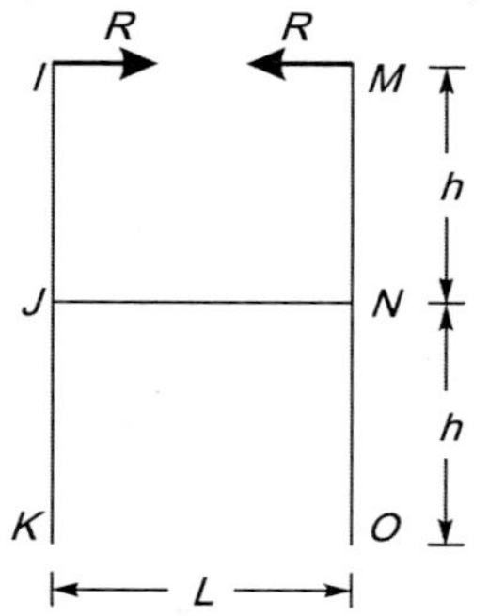

(a) $\frac{RLh^2}{EI}$ (b) $\frac{RL^2h}{EI}$

(c) $\frac{RLh^2}{3EI}$ (d) $\frac{RL^2h}{3EI}$ [2003 : 2 M]

5.10 A simply supported beam of uniform rectangular cross-section of width b and depth h is subjected to linear temperature gradient, 0° at the top and T° at the bottom, as shown in the figure. The coefficient of linear expansion of the beam material is α. The resulting vertical deflection at the mid-span of the beam is

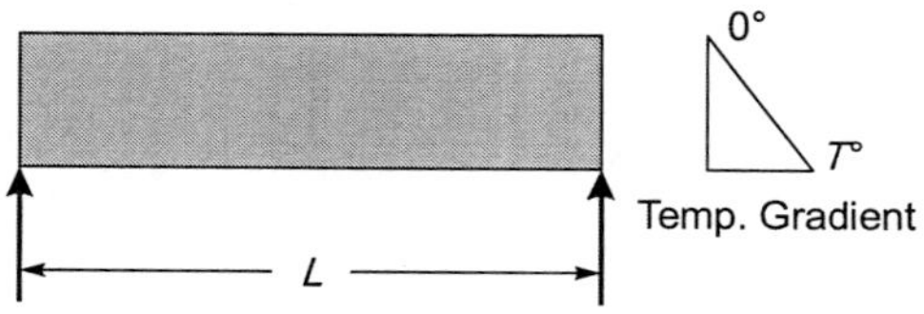

(a) $\frac{\alpha Th^2}{8L}$ upward

(b) $\frac{\alpha TL^2}{8h}$ upward

(c) $\frac{\alpha Th^2}{8L}$ downward

(d) $\frac{\alpha TL^2}{8h}$ downward [2003 : 2 M]

5.11 For the linear elastic beam shown in the figure, the flexural rigidity, EI is 781250 kNm^2. When w = 10 kN/m, the vertical reaction R_A at A is 50 kN. The value of R_A for w=100 kN/m is

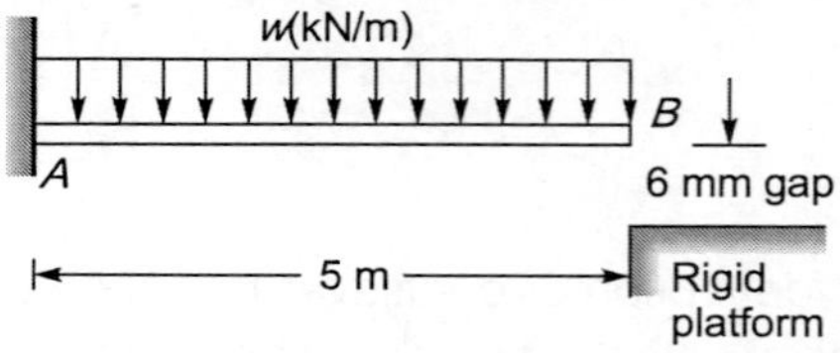

(a) 500 kN (b) 425 kN

(c) 250 kN (d) 75 kN [2004 : 2 M]

5.12 Consider the beam AB shown in the figure below. Part AC of the beam is rigid while Part CB has the flexural rigidity EI. Identify the correct combination of deflection at end B and bending moment at end A, respectively

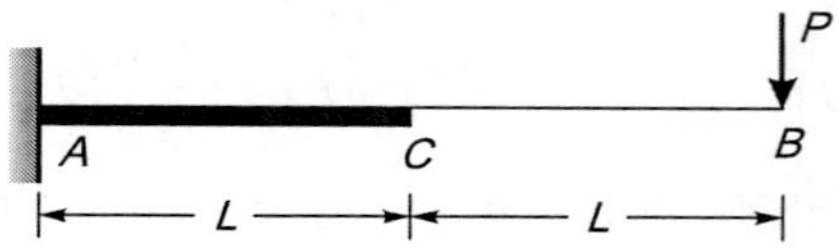

(a) $\frac{PL^3}{3EI}$, $2PL$ (b) $\frac{PL^3}{3EI}$, PL

(c) $\frac{8PL^3}{3EI}$, $2PL$ (d) $\frac{8PL^3}{3EI}$, PL

[2006 : 2 M]

Common Data for Questions 5.13 and 5.14

Consider a propped cantilever beam ABC under two loads of magnitude P each as shown in the figure below. Flexural rigidity of the beam is EI.

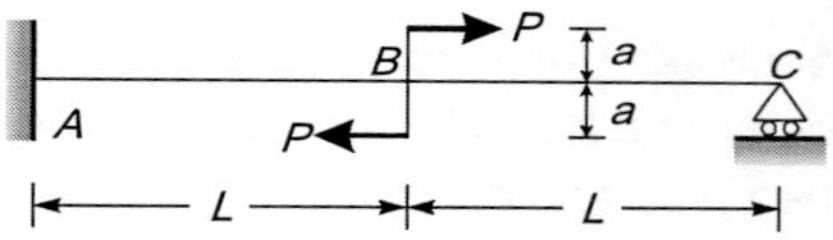

5.13 The reaction at C is

(a) $\frac{9Pa}{16L}$ (upwards) (b) $\frac{9Pa}{16L}$ (downwards)

(c) $\frac{9Pa}{8L}$ (upwards) (d) $\frac{9Pa}{8L}$ (downwards)

[2006 : 2 M]

5.14 The rotation at B is

(a) $\frac{5PLa}{16EI}$ (clockwise)

(b) $\frac{5PLa}{16EI}$ (anticlockwise)

(c) $\frac{59PLa}{16EI}$ (clockwise)

(d) $\frac{59PLa}{16EI}$ (anticlockwise) [2006 : 2 M]

5.15 Vertical reaction developed at *B* in the frame below due to the applied load of 100 kN (with 150,000 mm^2 cross-sectional area and 3.125×10^9 mm^4 moment of inertia for both members) is

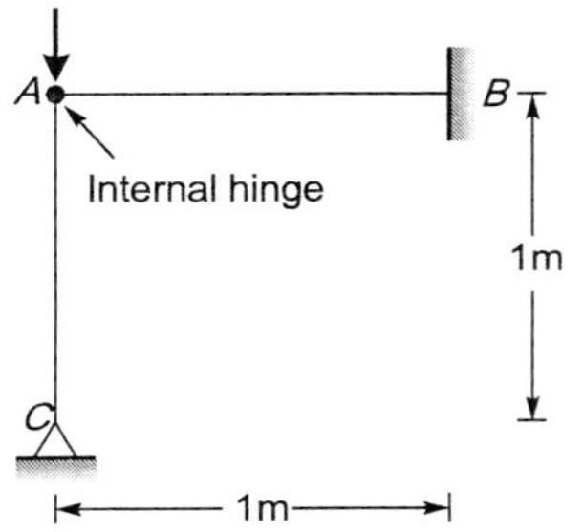

(a) 5.9 kN (b) 30.2 kN
(c) 66.3 kN (d) 94.1 kN [2006 : 2 M]

5.16 The stepped cantilever is subjected to moments *M* as shown in the figure below. The vertical deflection at the free end (neglecting the self weight) is

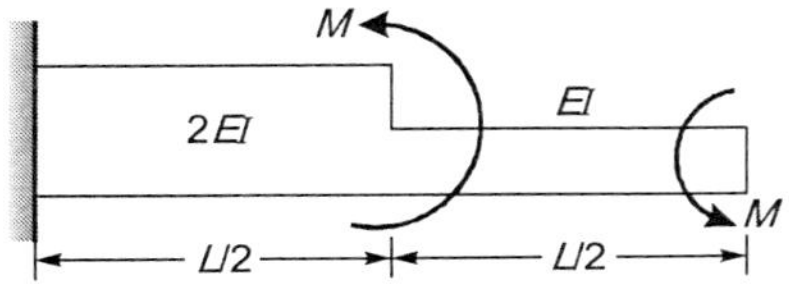

(a) $\frac{ML^2}{8EI}$ (b) $\frac{ML^2}{4EI}$
(c) $\frac{ML^2}{2EI}$ (d) Zero [2008 : 2 M]

Linked Answer Questions 5.17 and 5.18

Beam *GHI* is supported by three pontoons as shown in the figure below. The horizontal cross-sectional area of each pontoon is 8 m^2, the flexural rigidity of the beam is 10000 kN-m^2 and the unit weight of water is 10 kN/m^3.

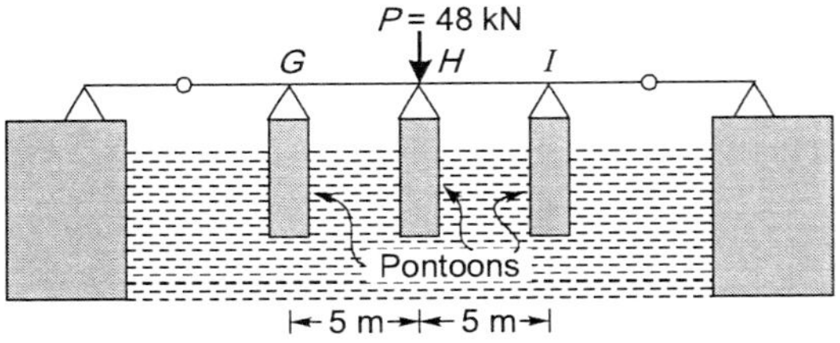

5.17 When the middle pontoon is removed, the deflection at *H* will be
(a) 0.2 m (b) 0.4 m
(c) 0.6 m (d) 0.8 m [2008 : 2 M]

5.18 When the middle pontoon is brought back to its position as shown in the figure above, the reaction at *H* will be
(a) 8.6 kN (b) 15.7 kN
(c) 19.2 kN (d) 24.2 kN [2008 : 2 M]

Linked Answer Questions 5.19 and 5.20.

In the cantilever beam *PQR* shown in figure below, the segment *PQ* has flexural rigidity *EI* and the segment *QR* has infinite flexural rigidity

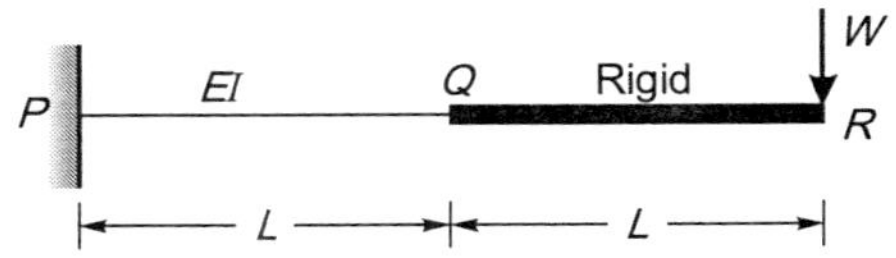

5.19 The deflection and slope of the beam at *Q* are respectively

(a) $\frac{5WL^3}{6EI}$ and $\frac{3WL^2}{2EI}$ (b) $\frac{WL^3}{3EI}$ and $\frac{WL^2}{2EI}$
(c) $\frac{WL^3}{2EI}$ and $\frac{WL^2}{EI}$ (d) $\frac{WL^3}{3EI}$ and $\frac{3WL^2}{2EI}$

[2009 : 2 M]

5.20 The deflection of the beam at *R* is

(a) $\frac{8WL^3}{EI}$ (b) $\frac{5WL^3}{6EI}$
(c) $\frac{7WL^3}{3EI}$ (d) $\frac{8WL^3}{6EI}$ [2009 : 2 M]

5.21 A simply supported beam is subjected to a uniformly distributed load of intensity *w* per unit length, on half of the span from one end. The length of the span and the flexural stiffness are denoted as *l* and *EI* respectively. The deflection at mid-span of the beam is

(a) $\frac{5}{6144}\frac{wl^4}{EI}$ (b) $\frac{5}{768}\frac{wl^4}{EI}$
(c) $\frac{5}{384}\frac{wl^4}{EI}$ (d) $\frac{5}{192}\frac{wl^4}{EI}$

[2012 : 2 M]

5.22 A uniform beam (*EI* = constant) *PQ* in the form of a quarter circle of radius *R* is fixed at end *P* and free at the end *Q*, where a load *W* is applied as shown. The vertical downward displacement δ_Q at the loaded point *Q* is given by $\delta_Q = \beta\left(\frac{WR^3}{EI}\right)$. Find the value of β correct to 4-decimal place.

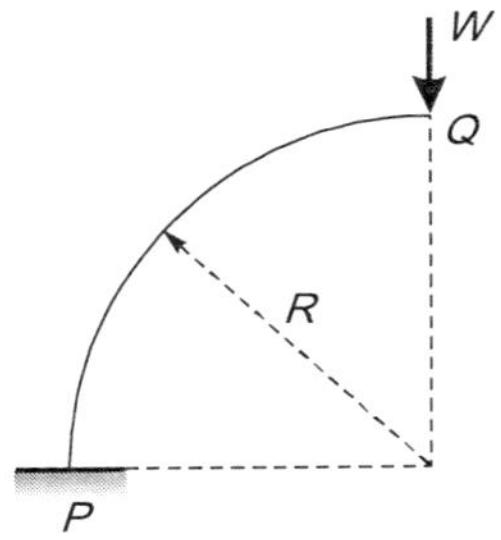

[2013 : 2 M]

5.23 For the cantilever beam of span 3 m (shown below), a concentrated load of 20 kN applied at the free end causes a vertical displacement of 2 mm at a section located at a distance of 1 m from the fixed end. If a concentrated vertically downward load of 10 kN is applied at the section located at a distance of 1 m from the fixed end (with no other load on the beam), the maximum vertical displacement in the same beam (in mm) is ________.

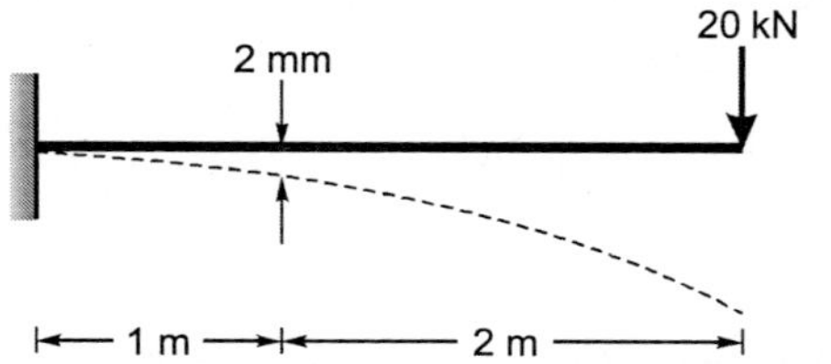

[2014 : 2 M, Set-I]

5.24 The axial load (in kN) in the member PQ for the arrangement/assembly shown in the figure given below is ________.

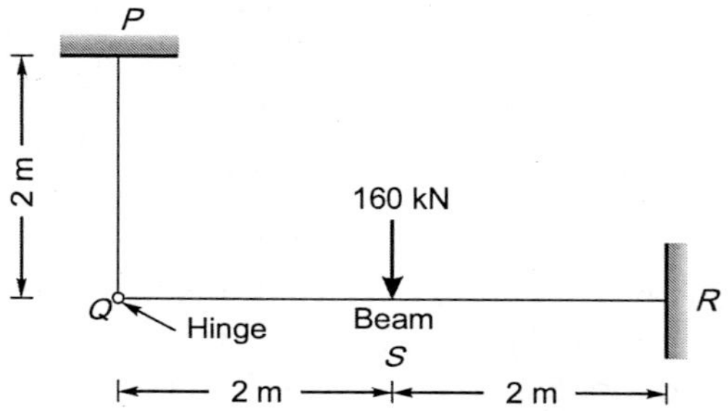

[2014 : 2 M, Set-II]

5.25 The tension (in kN) in a 10 m long cable, shown in the figure, neglecting its self-weight is

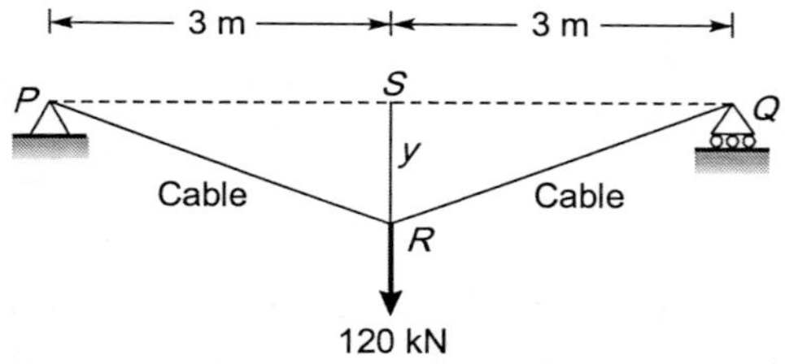

(a) 120 (b) 75
(c) 60 (d) 45 [2014:2 M, Set-II]

5.26 The beam of an overall depth 250 mm (shown below) is used in a building subjected to two different thermal environments. The temperatures at the top and bottom surfaces of the beam are 36°C and 72°C respectively. Considering coefficient of thermal expansion (α) as 1.50×10^{-5} per °C, the vertical deflection of the beam (in mm) at its mid-span due to temperature gradient is ________.

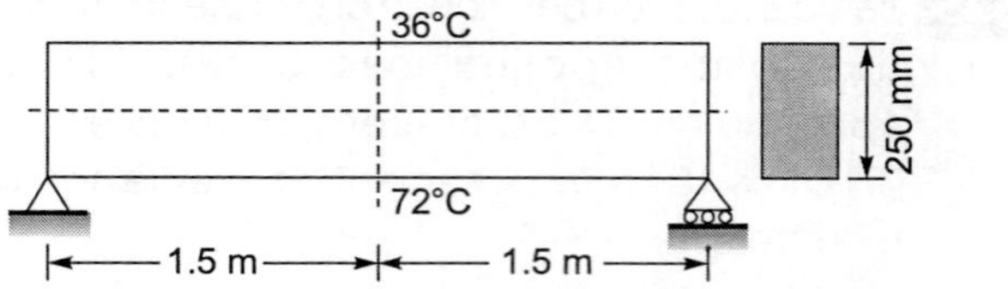

[2014 : 2 M, Set-II]

5.27 A horizontal beam ABC is loaded as shown in the figure below. The distance of the point of contraflexure from end A (in m) is ________.

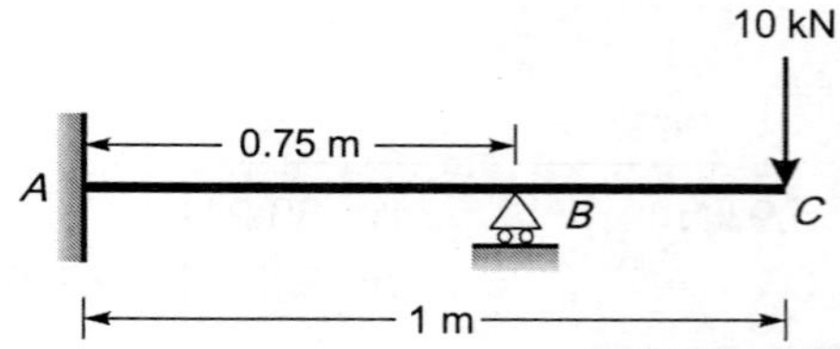

[2015 : 1 M, Set-II]

5.28 A steel strip of length, $L = 200$ mm is fixed at end A and rests at B on a vertical spring of stiffness, $k = 2$ N/mm. The steel strip is 5 mm wide and 10 mm thick. A vertical load, $P = 50$ N is applied at B, as shown in the figure. Considering $E = 200$ GPa, the force (in N) developed in the spring is ________.

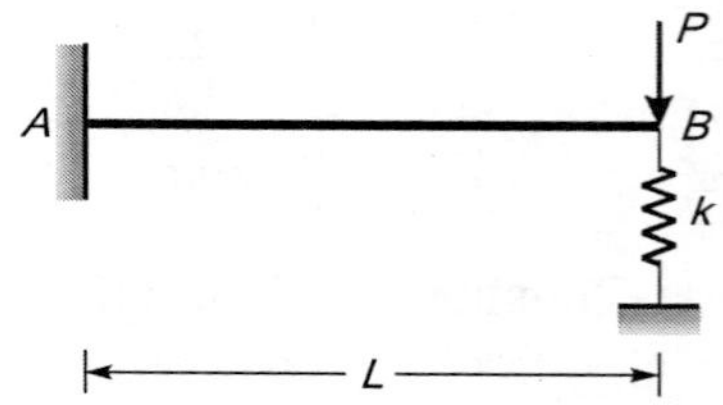

[2015 : 2 M, Set-II]

5.29 Two beaM are connected by a linear spring as shown in the following figure. For a load P as shown in the figure, the percentage of the applied load P carried by the spring is ________.

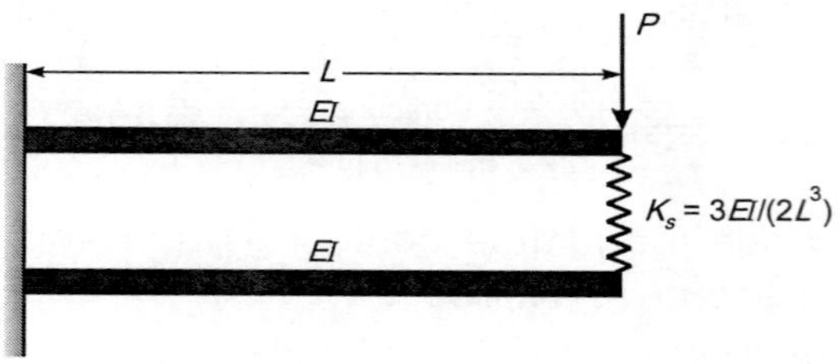

[2015 : 2 M, Set-I]

5.30 A 3 m long simply supported beam of uniform cross-section is subjected to a uniformly distributed load of $w = 20$ kN/m in the central 1 m as shown in the figure

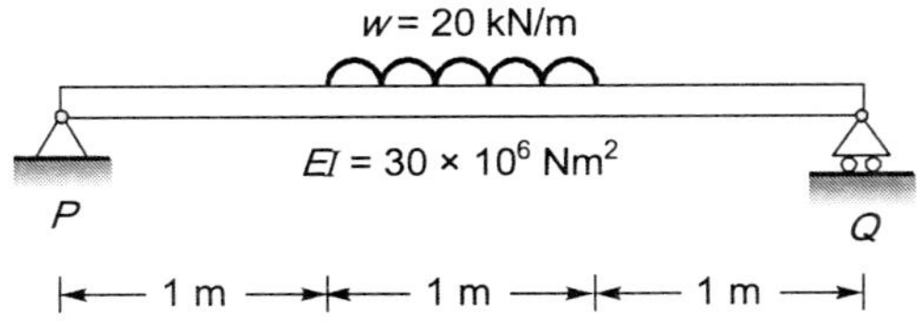

If the flexural rigidity (EI) of the beam is 30×10^6 N-m^2, the maximum slope (expressed in radians) of the deformed beam is

(a) 0.681×10^{-7} (b) 0.943×10^{-7}

(c) 4.310×10^{-7} (d) 5.910×10^{-7}

[2016 : 2 M, Set-I]

5.31 Two beaM PQ (fixed at P with a roller support at Q, as shown in Figure I, which allows vertical movement) and XZ (with a hinge at Y) are shown in the Figures I and II respectively. The spans of PQ and XZ are L and $2L$ respectively. Both the beaM are under the action of uniformly distributed load (w) and have the same flexural stiffness, EI (where, E and I respectively denote modulus of elasticity and moment of inertia about axis of bending). Let the maximum deflection and maximum rotation be δ_{max1} and θ_{max1}, respectively, in the case of beam PQ and the corresponding quantities for the beam XZ be δ_{max2} and θ_{max2}, respectively.

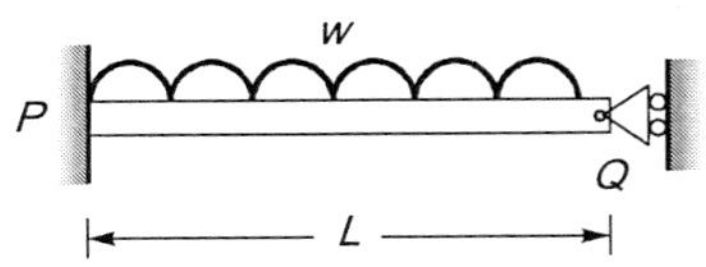

Figure-I

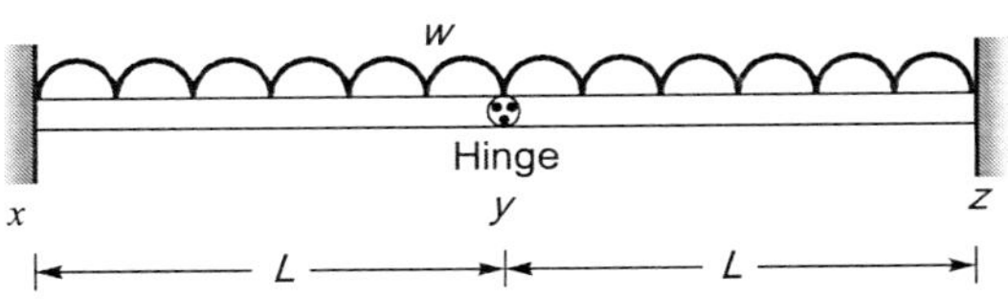

Figure-II

Which one of the following relationship is true?

(a) $\delta_{max1} \neq \delta_{max2}$ and $\theta_{max1} \neq \theta_{max2}$

(b) $\delta_{max1} = \delta_{max2}$ and $\theta_{max1} \neq \theta_{max2}$

(c) $\delta_{max1} \neq \delta_{max2}$ and $\theta_{max1} = \theta_{max2}$

(d) $\delta_{max1} = \delta_{max2}$ and $\theta_{max1} = \theta_{max2}$

[2016 : 2 M, Set-I]

5.32 Two prismatic beaM having the same flexural rigidity of 1000 kN-m^2 are shown in the figures.

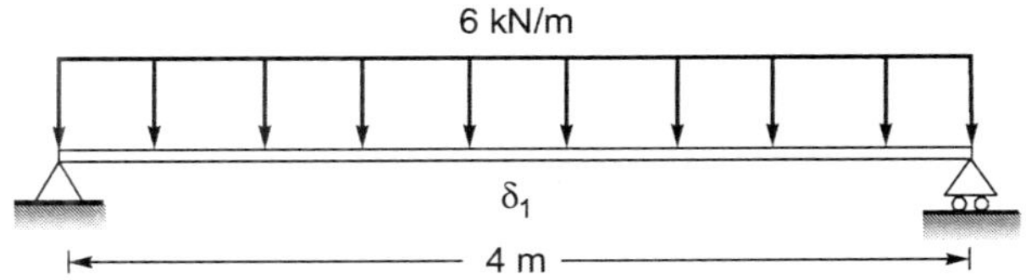

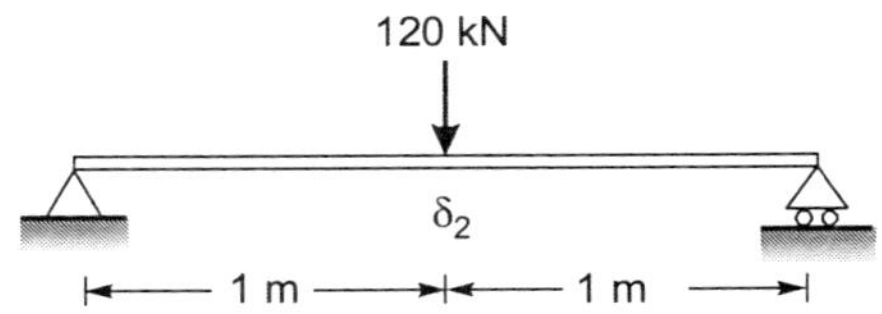

If the mid-span deflections of these beaM are denoted by δ_1 and δ_2 (as indicated in the figures), the correct option is

(a) $\delta_1 = \delta_2$ (b) $\delta_1 < \delta_2$

(c) $\delta_1 > \delta_2$ (d) $\delta_1 >> \delta_2$

[2017 : 2 M, Set-II]

5.33 The figure shows a simply supported beam PQ of uniform flexural rigidity EI carrying two moments M and $2M$

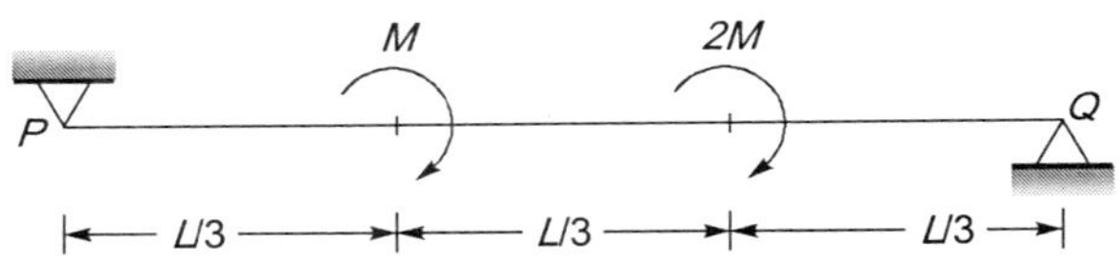

The slope at P will be

(a) 0 (b) $\frac{ML}{(9EI)}$

(c) $\frac{ML}{(6EI)}$ (d) $\frac{ML}{(3EI)}$

[2018 : 2 M, Set-I]

5.34 A cantilever beam PQ of uniform flexural rigidity (EI) is subjected to a concentrated moment M at R as shown in the figure.

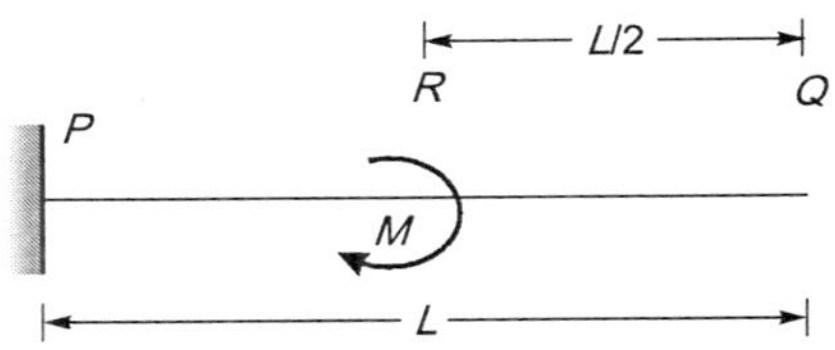

The deflection at the free end Q is

(a) $\frac{ML^2}{6EI}$ (b) $\frac{ML^2}{4EI}$

(c) $\frac{3ML^2}{4EI}$ (d) $\frac{3ML^2}{8EI}$

[2020 : 2 M, Set-I]

5.35 The planar structure RST shown in the figure is roller-supported at S and pin-supported at R. Members RS and ST have uniform flexural rigidity (EI) and S is a rigid joint. Consider only bending deformation and neglect effects of self-weight and axial stiffening.

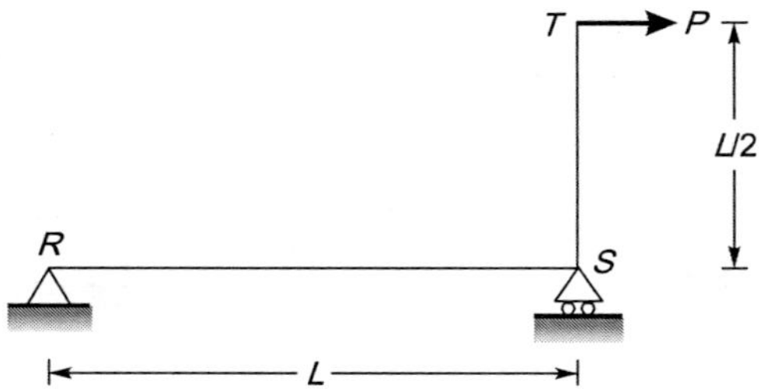

When the structure is subjected to a concentrated horizontal load P at the end T, the magnitude of rotation at the support R, is

(a) $\frac{PL}{6EI}$ (b) $\frac{PL^3}{12EI}$

(c) $\frac{PL^2}{6EI}$ (d) $\frac{PL^2}{12EI}$

[2020 : 2 M, Set-II]

5.36 The equation of deformation is derived to be $y = x^2 - xL$ for a beam shown in the figure.

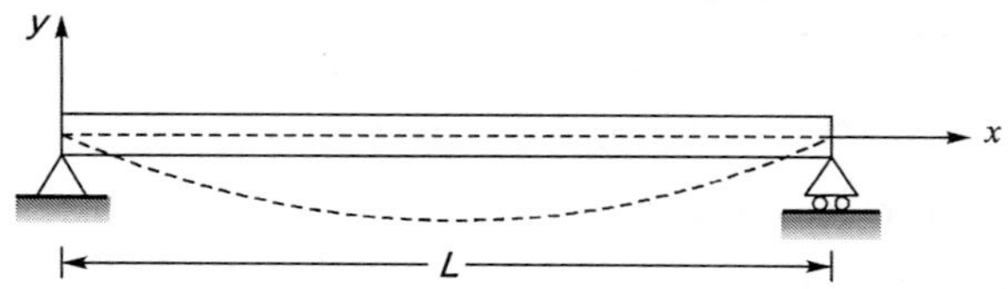

The curvature of the beam at the mid-span (in units, in integer) will be _______.

[2021 : 1 M, Set-I]

5.37 Consider the beam shown in the figure (not to scale), on a hinge support at end A and a roller support at end B. The beam has a constant flexural rigidity, and is subjected to the external moments of magnitude M at one-third spans, as shown in the figure. Which of the following statements is/are TRUE?

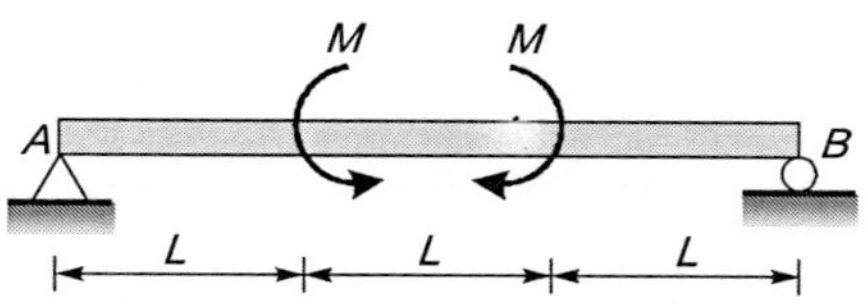

(a) Support reactions are zero
(b) Shear force is zero everywhere
(c) Bending moment is zero everywhere
(d) Deflection is zero everywhere

[2023 : 2 M, Set-I]

5.38 When a simply-supported elastic beam of span L and flexural rigidity EI (E is the modulus of elasticity and I is the moment of inertia of the section) is loaded with a uniformly distributed load w per unit length, the deflection at the mid-span is

$$\Delta_0 = \frac{5}{384}\frac{wL^4}{EI}$$

If the load on one half of the span is now removed, the mid-span deflection

(a) reduces to $\Delta_0/2$
(b) reduces to a value less than $\Delta_0/2$
(c) reduces to a value greater than $\Delta_0/2$
(d) remains unchanged at Δ_0

[2023 : 1 M, Set-II]

5.39 For the 6 m long horizontal cantilever beam PQR shown in the figure, Q is the midpoint. Segment PQ of the beam has flexural rigidity $EI = 2 \times 10^5$ kN.m^2 whereas the segment QR has infinite flexural rigidity. Segment QR is subjected to uniformly distributed, vertically downward load of 5kN/m.

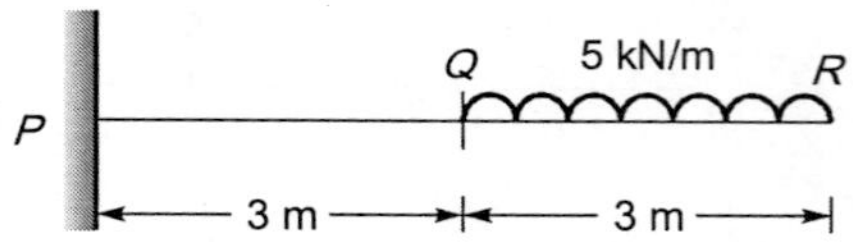

(Figure NOT to scale)

The magnitude of the vertical displacement (in mm) at point Q is ___________ (rounded off to 3 decimal places).

[2024 : 2 M, Set-II]

■■■■

Answers Deflection of Beam

5.1 (b)	5.2 (a)	5.3 (c)	5.4 (c)	5.5 (a)	5.6 (d)	5.7 (c)
5.8 (b)	5.9 (a)	5.10 (d)	5.11 (b)	5.12 (a)	5.13 (c)	5.14 (a)
5.15 (a)	5.16 (c)	5.17 (b)	5.18 (c)	5.19 (a)	5.20 (c)	5.21 (b)
5.22 (0.78)	5.23 (1)	5.24 (50)	5.25 (b)	5.26 (2.43)	5.27 (0.25)	5.28 (3.0)
5.29 (25)	5.30 (*)	5.31 (d)	5.32 (a)	5.33 (c)	5.34 (d)	5.35 (d)
5.36 (2)	5.37 (a, b)	5.38 (a)	5.39 (1.181)			

Explanations Deflection of Beam

5.1 (b)

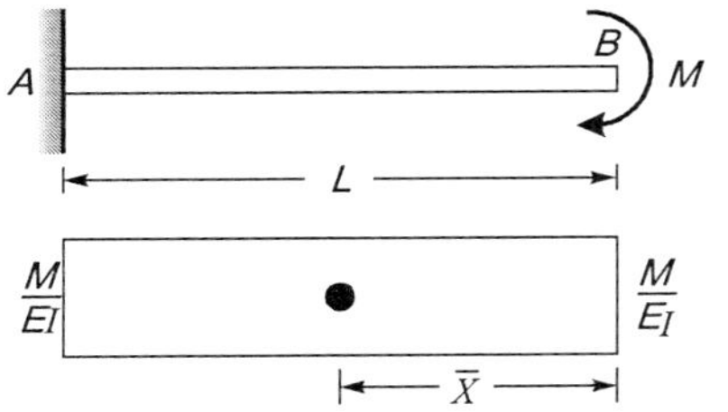

$\frac{M}{EI}$ Diagram

From moment area theorem,
Downward deflection of beam AB at free and 'B',

$$\delta_B = A\bar{X}$$
$$= \frac{M.L}{EI} \cdot \frac{L}{2} = \frac{ML^2}{2EI}$$

5.2 (a)

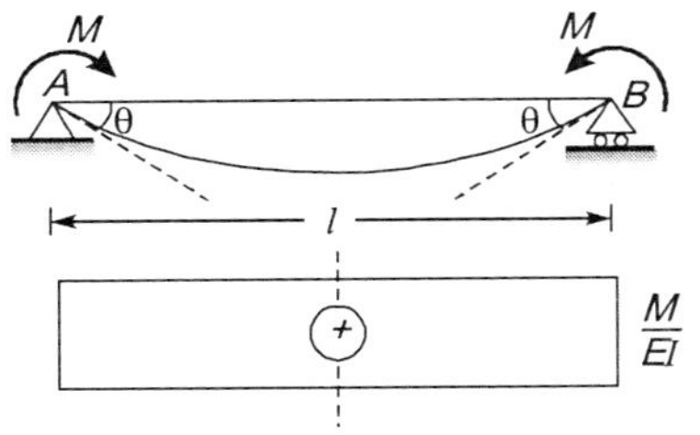

$$\theta_C - \theta_A = \text{Area of } M/EI \text{ diagram}$$
$$\Rightarrow \quad 0 - \theta_A = \frac{M}{EI} \times \frac{l}{2} = \frac{Ml}{2EI}$$
$$\therefore \quad \theta_A = -\frac{Ml}{2EI}$$
$$\therefore \quad \theta_A = \frac{Ml}{2EI} (\text{anticlockwise})$$
$$\Rightarrow \quad \frac{Ml}{EI} = 2\theta$$

5.3 (c)

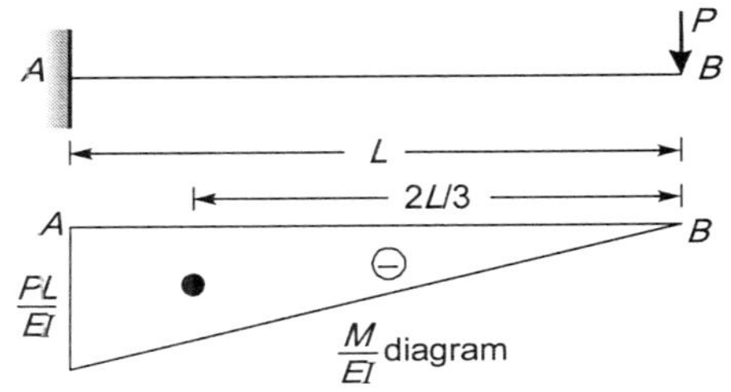

From moment area method,

$$\Delta_{B/A} = A\bar{X}$$
$$= \frac{1}{2} \times L \times \frac{PL}{EI} \times \frac{2L}{3} = \frac{PL^3}{3EI}$$

5.4 (c)

Method-I

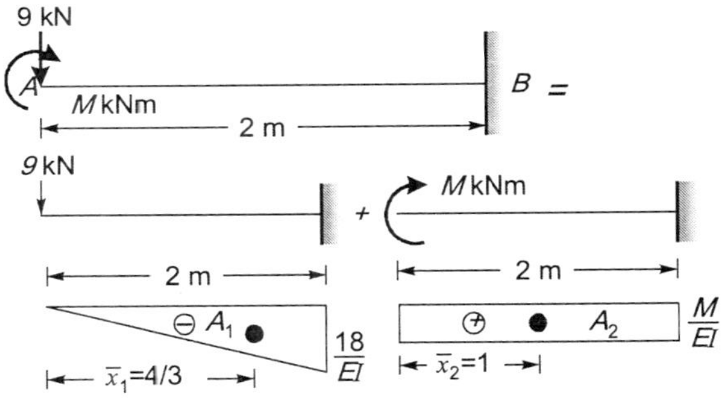

$$\delta_{A/B} = \text{Moment of area of } \frac{M}{EI} \text{ diagram.}$$
$$= A_1\bar{x}_1 + A_2\bar{x}_2$$
$$\Rightarrow \quad 0 = -\frac{1}{2} \times \frac{18}{EI} \times 2\left(\frac{2}{3} \times 2\right) + \frac{M}{EI} \times 2\left(\frac{2}{2}\right)$$
$$\Rightarrow \quad \frac{18}{EI} \times \frac{4}{3} = \frac{2M}{EI}$$
$$\Rightarrow \quad M = 12 \text{ kNm (clockwise)}$$

Method-II

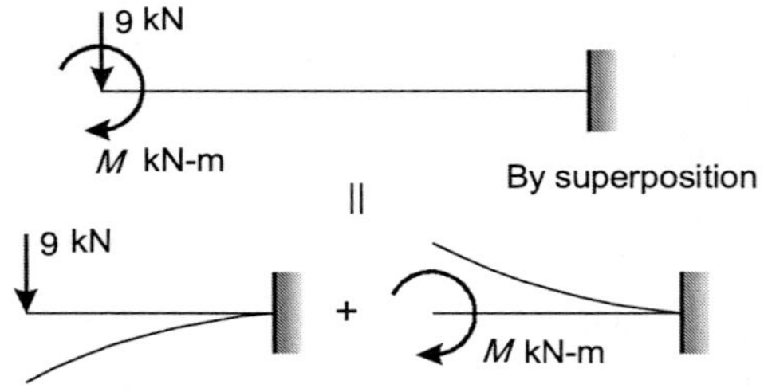

(↓) defl. due to load = (↑) defl. due to moment

$$\frac{Pl^3}{3EI} = \frac{Ml^2}{2EI}$$

$$\frac{9\times(2)^3}{3EI} = \frac{M\times(2)^2}{2EI}$$

$$\frac{9\times 8}{3\times 2} = M$$

$$\Rightarrow \quad 12\text{ kNm} = M\text{ (CW)}$$

5.5 (a)

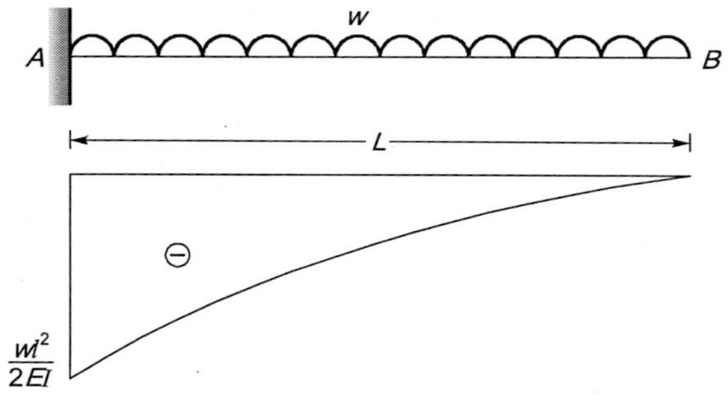

By moment area method

$$\theta_B - \theta_A = \frac{1}{3}\times L\times\frac{wl^2}{2EI} = \frac{wl^3}{6EI}$$

5.6 (d)

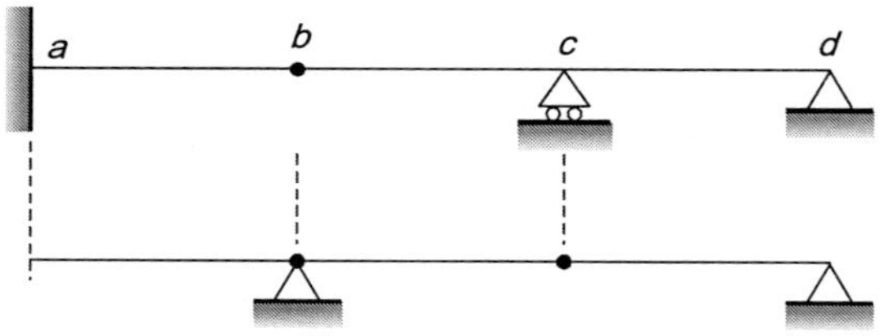

(Conjugate beam)

For a fixed beam, deflection and slope are **zero at fixed end**, so in the conjugate beam bending moment and shear force should be zero. Hence fixed end becomes free end.

Real beam	Conjugate beam
Fixed End	Free End
Internal Hinge	Internal Hinge Support
Internal Hinge Support	Internal Hinge
Hinge Support at Ends	Hinge Support at Ends

5.7 (c)

Consider free body diagram,

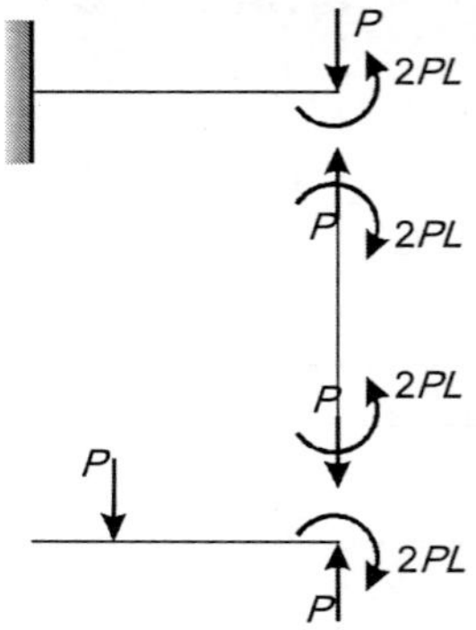

By superposition

$$\text{Deflection at } A = \frac{PL^3}{3EI} - \frac{ML^2}{2EI}$$

$$= \frac{P(3L)^3}{3EI} - \frac{2PL(3L)^2}{2EI} = \frac{P\times 27L^3}{3EI} - \frac{2PL9L^2}{2EI} = 0$$

5.8 (b)

Deflection at proped end $= 0$

$$\frac{w\times L^4}{8EI} - \frac{R_B\times L^3}{3EI} = 0$$

$$\Rightarrow \quad R_B = \frac{3}{8}wL$$

5.9 (a)

The bending moment in the member $JN = R\times h$ (sagging)

$\therefore$ Slope at J or $N = \dfrac{Rh\times L}{2EI}$

Thus, outward displacement between points

$$K \text{ and } O = \frac{RhL}{2EI}\times h + \frac{RhL}{2EI}\times h = \frac{Rh^2L}{EI}$$

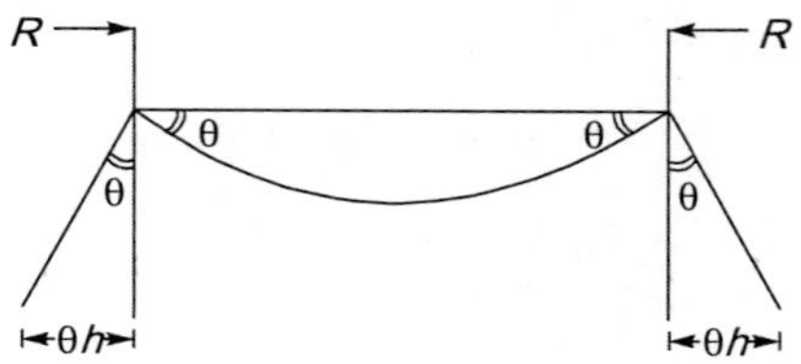

5.10 (d)

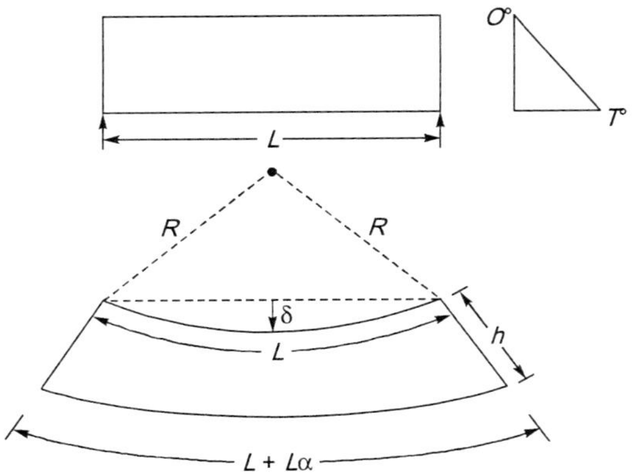

From circle

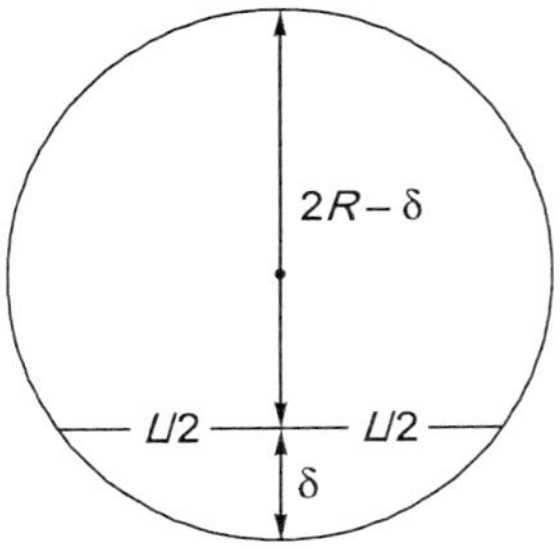

$$(2R-\delta)\times\delta = \frac{L^2}{4}$$

$$2R\delta-\delta^2 = \frac{L^2}{4}$$

$$\delta^2 <<< 2R\delta$$

$$\delta = \frac{L^2}{8R} \quad \text{...(i)}$$

$$\frac{R}{L} = \frac{R+h}{L+\alpha T}$$

$$\Rightarrow \quad 1+\alpha T = 1+\frac{4}{R}$$

Put in eq. (i)

$$R == \frac{h}{\alpha T}$$

$$\delta == \frac{L^2\alpha T}{8\times h}$$

5.11 (b)

The deflection at the free end due to 10 kN/m,

$$\delta = \frac{wL^4}{8EI}$$

$$= \frac{10\times(5)^4\times 1000}{8\times 781250}$$

$$= 1 \text{ mm}$$

The gap between the beam and rigid platform is 6 mm. Hence, no reaction will be developed when w = 10 kN/m.

Now, deflection at the free end for w (100 kN/m) will be = 10 × 1 mm = 10 mm

But, this cannot be possible because margin of deflection is only 6 m.

Thus w = 100 kN/m will induce a reaction R_B at B.

$$\therefore \frac{wL^4}{8EI}-\frac{R_BL^3}{3EI} = \text{Permissible deflection}$$

$$\Rightarrow \frac{100\times(5)^4}{8\times 781250}-\frac{R_B\times(5)^3}{3\times 781250} = \frac{6}{1000}$$

$$\Rightarrow \frac{10}{1000}-\frac{6}{1000} = \frac{R_B\times 125}{3\times 781250}$$

$$\Rightarrow \quad R_B = 75 \text{ kN}$$

$$\therefore \quad R_A = (100\times 5-75) = 425 \text{ kN}$$

5.12 (a)

Part AC of the beam is rigid. Hence C will act as a fixed end. Thus the deflection at B will be given as $\delta_B = \frac{PL^3}{3EI}$

But the bending moment does not depend on the rigidity or flexibility of the beam.

∴ BM at $A = P\times 2L = 2PL$

5.13 (c)

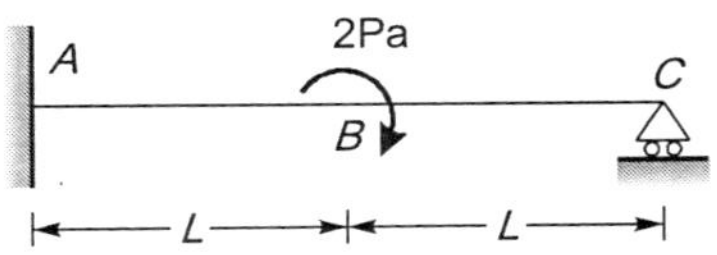

The moment at point $B = 2Pa$

In the cantilever beam ABC, the deflection at C due to moment $2Pa$ will be given as,

$$\delta_c == \delta_B+\theta_B\times L$$

$$\delta_c == \frac{ML^2}{2EI}+\frac{ML}{EI}\times L$$

$$== \frac{2PaL^2}{2EI}+\frac{2PaL}{EI}\times L$$

$$== \frac{3PaL^2}{EI} \text{ (downwards)}$$

∴ The reaction at C will be upwards,

$$\delta_c' == \frac{R(2L)^3}{3EI} == \frac{8RL^3}{3EI} \text{ (upwards)}$$

Thus, $\delta_c = \delta_c'$

$$\frac{3PaL^2}{EI} = \frac{8RL^3}{3EI}$$

$$\Rightarrow \quad R = \frac{9Pa}{8L} \text{ (upwards)}$$

5.14 (a)

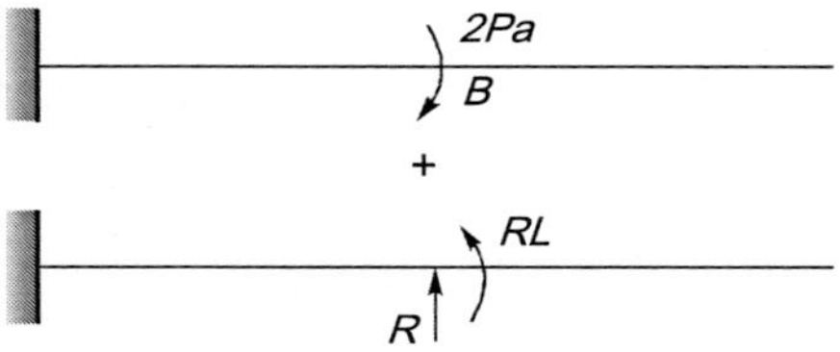

The rotation at B:

(i) Due to moment,

$$\theta_{B_1} = \frac{2Pa \times L}{EI} \text{ (clockwise)}$$

(ii) Due to reaction R,

$$\theta_{B_2} = \frac{RL^2}{2EI} + \frac{RL^2}{EI} = \frac{3RL^2}{2EI} = \frac{27}{16}\frac{PaL}{EI}$$

(anticlockwise)

$$\therefore \quad \theta_B = \theta_{B_1} - \theta_{B_2}$$

$$= \left(2 - \frac{27}{16}\right)\frac{PaL}{EI} = \frac{5}{16}\frac{PaL}{EI} \text{ (clockwise)}$$

5.15 (a)

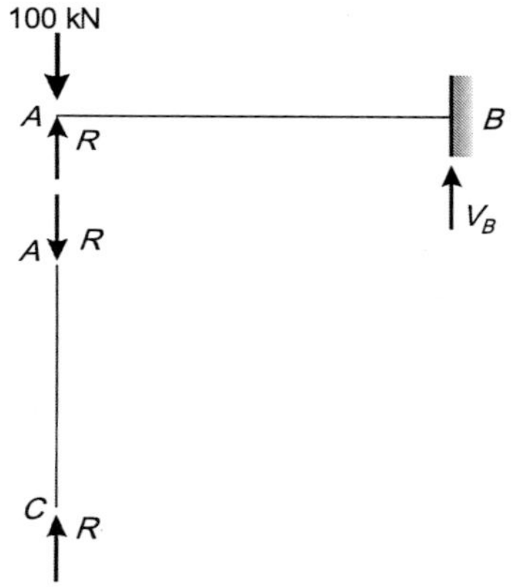

Deflection at A in beam AB = Compression in column AC,

$$\frac{(100 - R)L^3}{3EI} = \frac{RL}{AE}$$

$$\Rightarrow \quad \frac{(100 - R) \times (1 \times 1000)^3}{3 \times 3.125 \times 10^9} = \frac{R}{150000}$$

$$\Rightarrow \quad 100 - R = 0.0625$$

$$\Rightarrow \quad 100 = 1.0625\,R$$

$$\Rightarrow \quad R = 94.1 \text{ kN}$$

$$\therefore \quad V_B = 100 - R = 100 - 94.1 = 5.9 \text{ kN}$$

5.16 (c)

Using **Moment Area Method,**

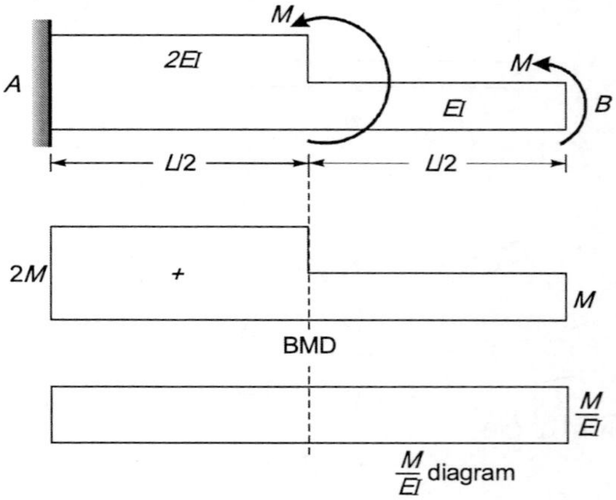

Deflection at B w.r.t. A

= Moment of area of $\frac{M}{EI}$ diagram between A and B about B

$$= \frac{M}{EI} \times L \times \frac{L}{2} = \frac{ML^2}{2EI}$$

5.17 (b)

The reactions at the ends are zero as there are hinges to left of G and right of I. Hence when the middle pontoon is removed, the beam GHI acts as a simply supported beam.

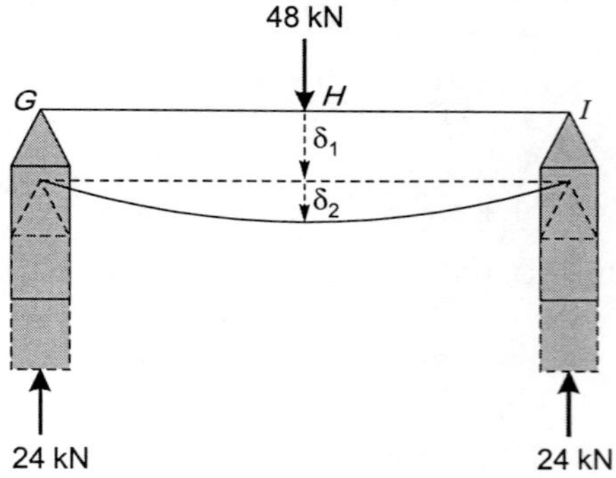

The deflection at H will be due to the load at H as well as due to the downward settlement of pontoons at G and I in water. Since the loading is symmetrical, both the pontoons will be immersed to same height and produce equal reaction (24 kN). Let it be δ_1.

$\therefore$ $\delta_1 \times$ area of cross-section of pontoon $\times$ unit weight of water = 24

$$\Rightarrow \quad \delta_1 \times 8 \times 10 = 24$$

$$\Rightarrow \quad \delta_1 = 0.3 \text{ m}$$

Also, deflection at H due to bending in beam,

$$\delta_2 = \frac{PL^3}{48EI} = \frac{48 \times (10)^3}{48 \times 10^4} = 0.1 \text{ m}$$

∴ Final deflection at *H*,

$$\Delta = \delta_1 + \delta_2 = 0.3 + 0.1 = 0.4 \text{ m}$$

5.18 (c)

Let the elastic deflection at *H* be δ_2,

$$\therefore \quad \delta_2 = \frac{(P-R)L^3}{48EI} \quad \ldots(i)$$

The reactions at *G* and I will be same, as the beam is symmetrically loaded.

Let the reaction at each *G* and I be *Q*.

Using principle of buoyancy, we get,

$\delta_1 \times$ area of cross-section of pontoon $\times \gamma_w = Q$

$$\Rightarrow \quad \delta_1 \times 8 \times 10 = Q$$

$$\Rightarrow \quad \delta_1 = \frac{Q}{80} \quad \ldots(ii)$$

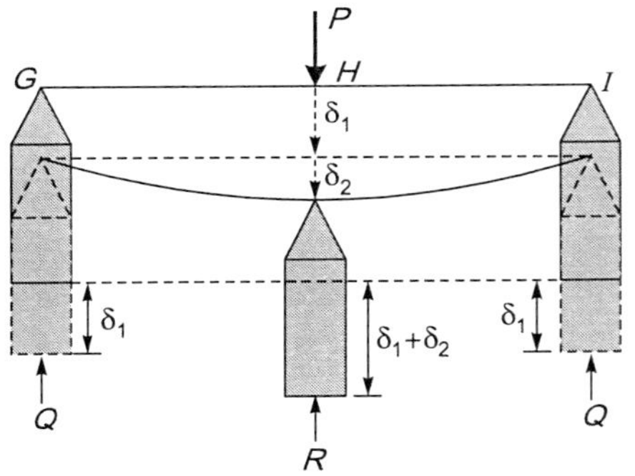

Also, we have,

$$Q + Q + R = P$$

$$\Rightarrow \quad 2Q + R = 48 \quad \ldots(iii)$$

Also, $(\delta_1 + \delta_2) \times$ area of cross-section of Pontoon $\times \gamma_w = R$

$$\Rightarrow (\delta_1 + \delta_2) \times 8 \times 10 = R$$

$$\Rightarrow \quad \delta_1 + \delta_2 = \frac{R}{80}$$

$$\Rightarrow \quad \frac{Q}{80} + \delta_2 = \frac{R}{80} \quad \text{[from (ii)]}$$

$$\Rightarrow \quad \frac{48-R}{2 \times 80} + \delta_2 = \frac{R}{80} \quad \text{[from (iii)]}$$

$$\Rightarrow \quad \delta_2 = \frac{2R - 48 + R}{160}$$

$$\Rightarrow \quad \delta_2 = \frac{3R-48}{160}$$

$$\therefore \quad \frac{(48-R) \times 10^3}{48 \times 10^4} = \frac{3R-48}{160} \quad \text{[from (i)]}$$

$$\Rightarrow \quad R = 19.2 \text{ kN}$$

5.19 (a)

The given cantilever beam can be modified into a beam as shown below,

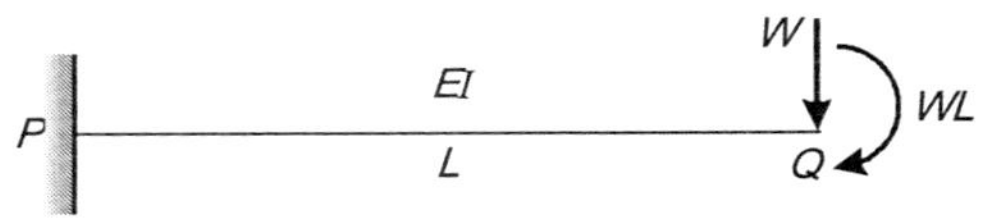

$$\text{Deflection at } Q = \frac{WL^3}{3EI} + \frac{WL \times L^2}{2EI}$$

$$= \frac{2WL^3 + 3WL^3}{6EI} = \frac{5WL^3}{6EI}$$

$$\text{Slope at} \quad Q = \frac{WL^2}{2EI} + \frac{WL \times L}{EI}$$

$$= \frac{WL^2 + 2WL^2}{2EI} = \frac{3WL^2}{2EI}$$

5.20 (c)

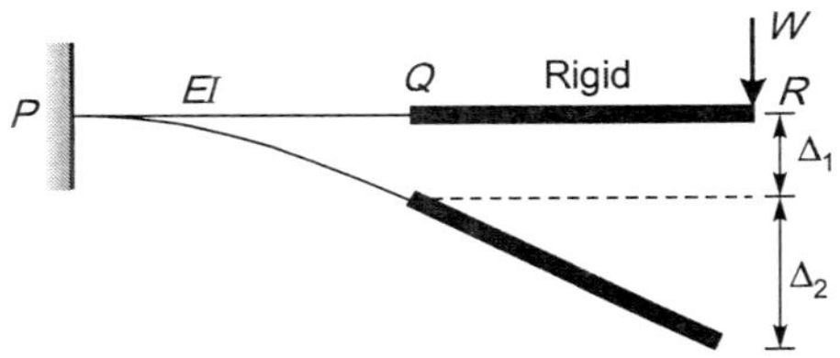

Since the portion *QR* of the beam is rigid, *QR* will remain straight.

Deflection of $R = \Delta_1 + \Delta_2$

= Deflection at *Q* + Slope at $Q \times L$

$$= \frac{5WL^3}{6EI} + \frac{3WL^2}{2EI} \times L = \frac{5WL^3 + 9WL^3}{6EI}$$

$$= \frac{14WL^3}{6EI} = \frac{7WL^3}{3EI}$$

5.21 (b)

Method-I

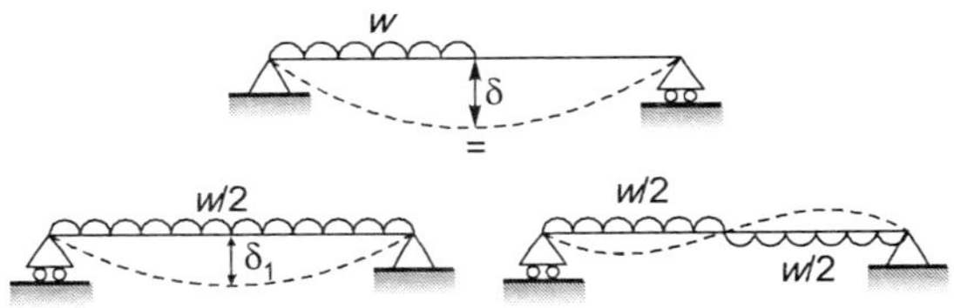

$$\delta = \delta_1 + 0$$

$$= \frac{5}{384} \frac{(w/2)\,l^4}{EI} \qquad \delta = \frac{5}{768} \frac{wl^4}{EI}$$

Method-II

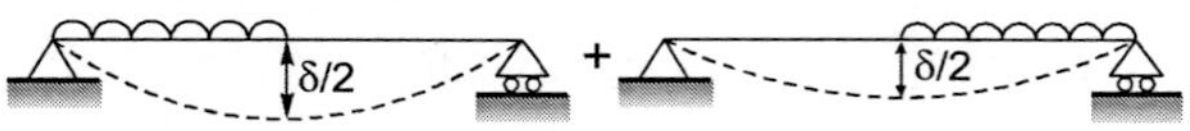

Total deflection,

$$\frac{\delta}{2}+\frac{\delta}{2} = \frac{5}{384}\frac{wl^4}{EI}$$

$$2\left(\frac{\delta}{2}\right) = \frac{5}{768}\frac{wl^4}{EI}$$

$$\frac{\delta}{2} = \frac{5}{768}\frac{wl^4}{EI}$$

5.22 Sol.

$$U = \int \frac{M^2 dx}{2EI} = \int_0^{\pi/2} \frac{(WR\sin\theta)^2 R d\theta}{2EI}$$

$$= \int_0^{\pi/2} \frac{P^2R^3}{2EI}\sin^2\theta d\theta$$

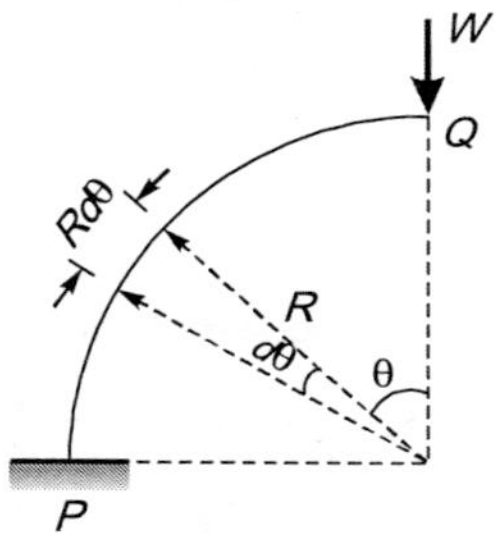

$$= \frac{P^2R^3}{2EI}\int_0^{\pi/2}\frac{(1-\cos 2\theta)d\theta}{2}$$

$$= \frac{P^2R^3}{4EI}\left[\theta - \frac{\sin 2\theta}{2}\right]_0^{\pi/2}$$

$$= \frac{P^2R^3}{4EI}\left[\frac{\pi}{2}\right] = \frac{P^2R^3\pi}{8EI}$$

$$\therefore\ \delta_Q = \frac{\partial U}{\partial W} = \frac{2PR^3\pi}{8EI} = \frac{PR^3}{EI}\frac{\pi}{4}$$

$$\therefore\ \delta_Q = \frac{\pi}{4}\left(\frac{WR^3}{EI}\right)$$

$$\therefore\ \beta = \frac{\pi}{4} = 0.7854$$

5.23 Sol.

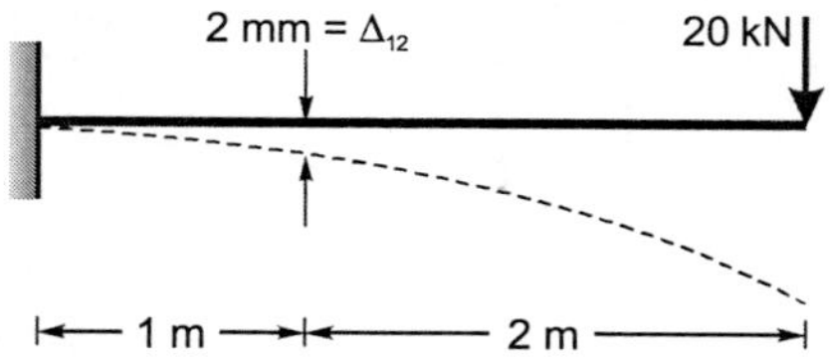

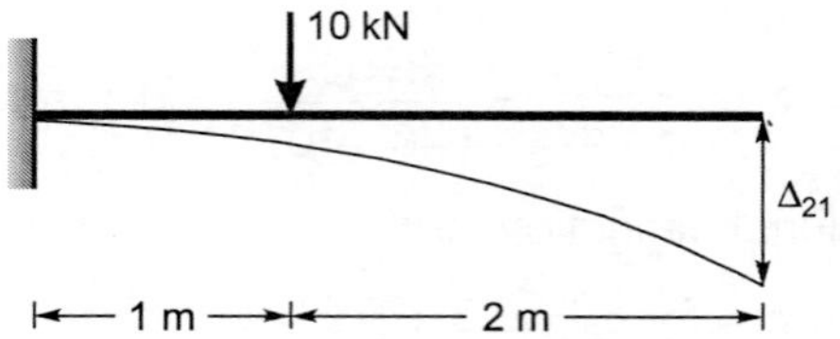

From Betti's law $P_1 \times \Delta_{12} = P_2 \times \Delta_{21}$

$$\Rightarrow\quad 10 \times 2 = 20 \times \Delta_{21}$$

$$\Rightarrow\quad \Delta_{21} = 1 \text{ mm}$$

5.24 Sol.

Free body diagram,

For principle of superposition,

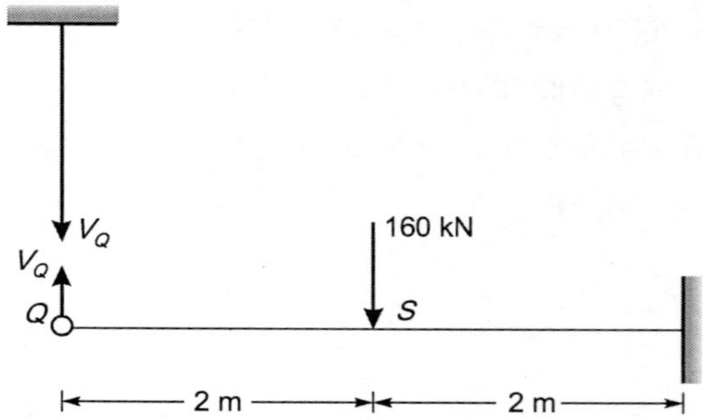

$$\frac{160\times 2^3}{3EI}+\frac{160\times 2^2}{2EI}\times 2-\frac{V_QL^3}{3EI} = \frac{V_QL}{AE} \quad \text{...(i)}$$

Deflections due to axial forces will be very less as compared to bending forces.

So we can neglect the axial deformation.

∴ From equation (i),

$$\frac{160\times 2^3}{3EI}+\frac{160\times 2^2}{2EI}\times 2-\frac{V_Q4^3}{3EI} = 0$$

$$\Rightarrow\quad \frac{160\times 2^3}{3}+\frac{160\times 2^2\times 2}{2} = \frac{V_Q\times 4^3}{3}$$

$$\Rightarrow\quad V_Q = 50 \text{ kN}$$

5.25 (b)

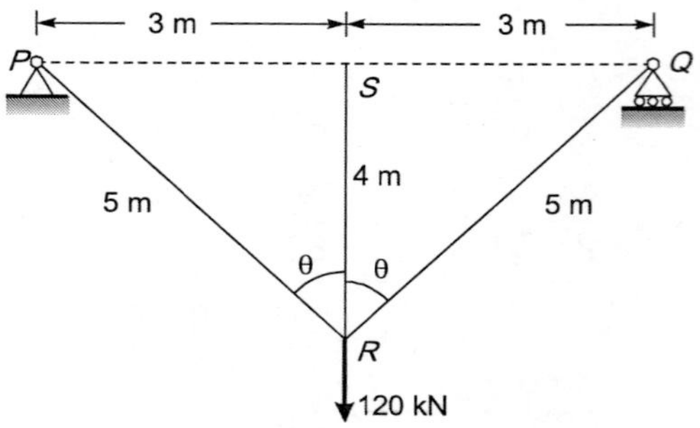

Free body diagram

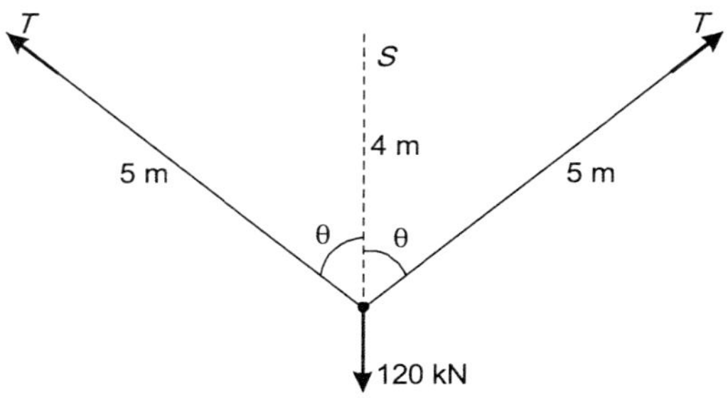

$$\Sigma F_y = 0$$

$$\Rightarrow \quad 2T\cos\theta = 120 \quad \ldots(i)$$

Here, $\cos\theta = \dfrac{4}{5}$

$$2T \times \frac{4}{5} = 120$$

$$\Rightarrow \quad T = \frac{120 \times 5}{2 \times 4} = 75 \text{ kN}$$

5.26 Sol.

Method-I

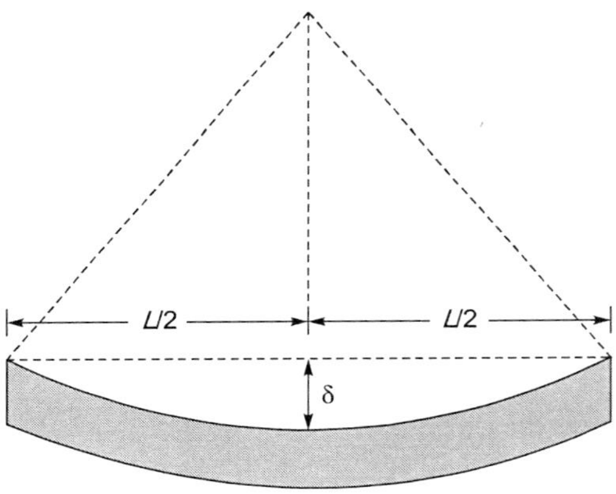

$$R = \frac{h}{\alpha T}$$

From properties of circle,

$$(2R - \delta)\delta = \frac{L}{2} \times \frac{L}{2}$$

(Considering 'δ' very small so neglect δ^2)

$$\Rightarrow \quad \delta = \frac{L^2}{8R} = \frac{L^2 \alpha T}{8h}$$

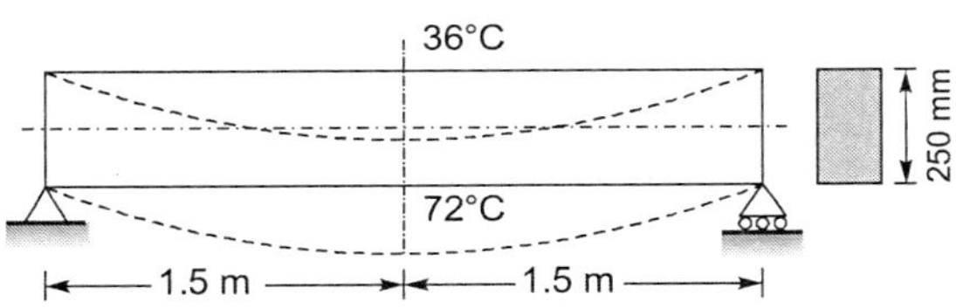

Given, $\alpha = 1.50 \times 10^{-5}$/°C

$$\Rightarrow \quad \delta = \frac{\alpha T L^2}{8h}$$

$$= \frac{1.50 \times 10^{-5} \times (72 - 36) \times 3^2}{8 \times \left(250 \times 10^{-3}\right)}$$

$$= 2.43 \text{ mm}$$

Method-II

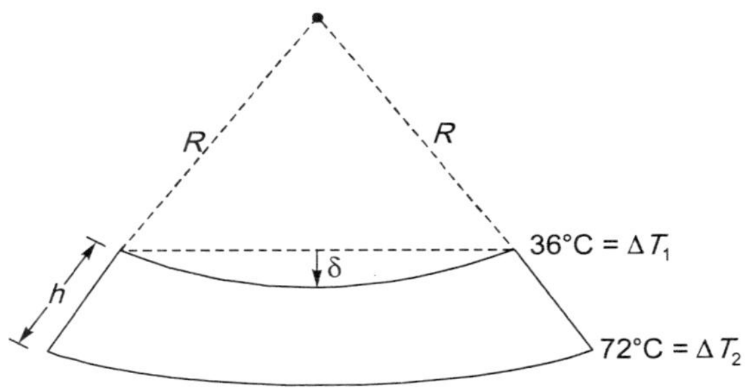

$$(2R - \delta)\delta = \frac{L^2}{4}$$

$$\delta = \frac{L^2}{8R}$$

$$\frac{R}{L + L\alpha\Delta T_1} = \frac{R + h}{L + L\alpha\Delta T_2}$$

$$\frac{\alpha(\Delta T_2 - \Delta T_1)}{1 + \alpha\Delta T_1} = \frac{h}{R}$$

Since $\alpha\Delta T_1 \lll 1$

$$R = \frac{h}{\alpha(\Delta T_2 - \Delta T_1)}$$

$$\delta = \frac{2\alpha(\Delta T_2 - \Delta T_1)}{8h}$$

$$= \frac{3^2 \times 1.5 \times 10^{-5} \times 36}{8 \times 250 \times 10^{-3}}$$

$$= 2.43 \text{ mm}$$

5.27 Sol.

Reaction at *B*,

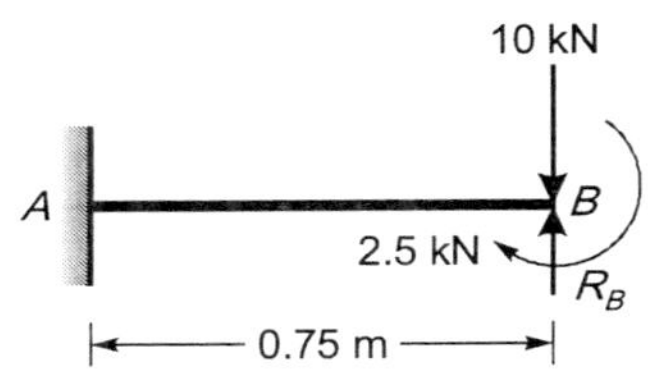

$$\Delta \text{ due to } R_B = \frac{R_B l^3}{3EI}$$

$$\Delta \text{ due to } M = \frac{Ml^2}{2EI}$$

$$\Delta_B = 0 \text{ (Compatibility condition)}$$

$$\frac{10\times(0.75)^3}{3EI} + \frac{2.5\times0.75^2}{2EI} - \frac{R_B\times0.75^3}{3EI} = 0$$

$$\frac{9R_B}{64EI} = \frac{135}{64EI}$$

$$\therefore \quad R_B = 15 \text{ kN}$$

BM at a distance x from free end,

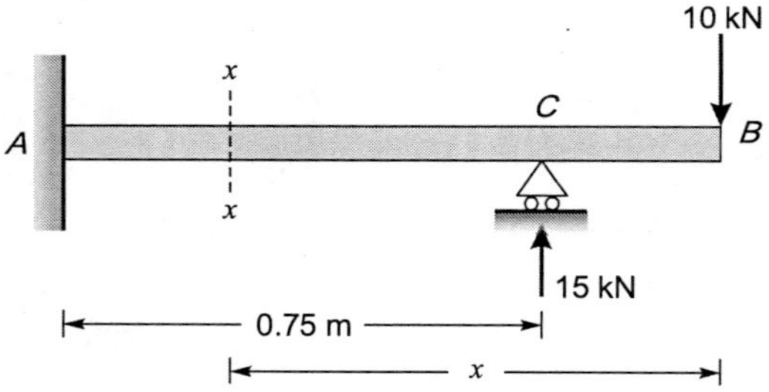

$$\text{BM}_x = 10\times x - 15\times(x-0.25) = 0$$

$$\Rightarrow \quad 10x = 15x - 3.75$$

$$\Rightarrow \quad 5x = 3.75$$

$$\therefore \quad x = 0.75 \text{ m}$$

$\therefore$ From end A, distance is 0.25 m.

5.28 Sol.

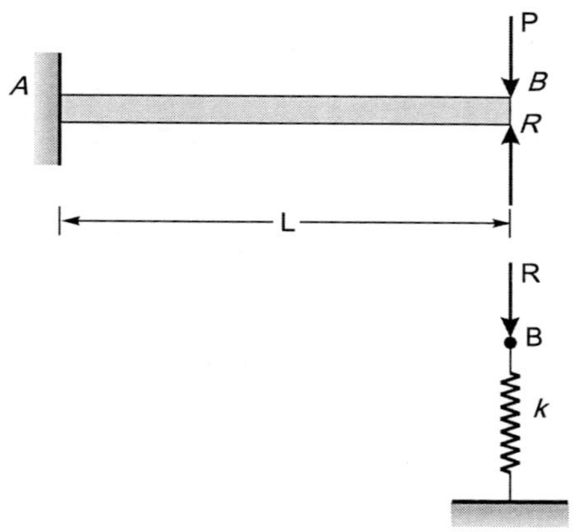

Deflection of point B = Deflection of spring

$$\Delta_B = \frac{(P-R)L^3}{3EI} = \frac{R}{k}$$

Where, R = Force in the spring,

$$\Rightarrow \quad \frac{(50-R)200^3}{3\times200\times10^3\times\dfrac{5\times10^3}{12}} = \frac{R}{2}$$

$$0.064(50-R) = R$$

$$3.2 = R + 0.064R$$

$$R = 3.0075 \text{ N}$$

5.29 Sol.

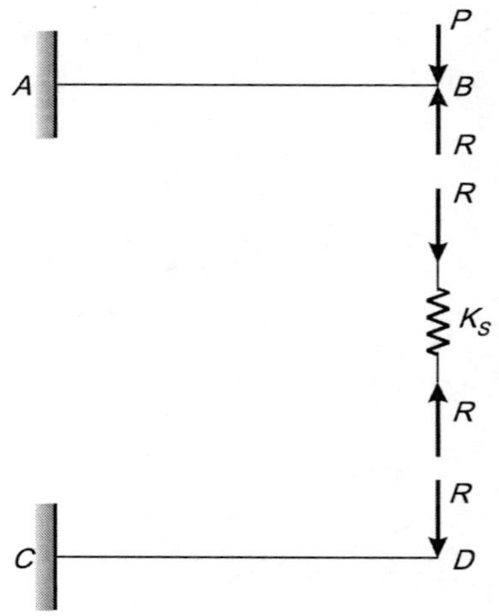

$$\Delta_{\text{spring}} = \Delta_B - \Delta_D$$

Compression of spring

$$= \frac{(P-R)L^3}{3EI} - \frac{RL^3}{3EI} = \frac{R}{K_S}$$

$$\frac{(P-R)L^3}{3EI} - \frac{RL^3}{3EI} = \frac{R\times(2L^3)}{3EI}$$

$$(P-R) - R = 2R$$

$$P = 4R$$

$$R = \frac{P}{4}$$

% force carried by spring = 25%

5.30 (*)

Method-I

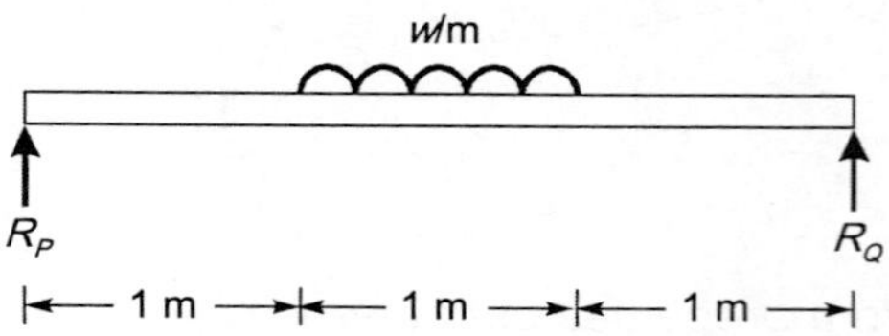

Due to symmetrical loading,

$$R_P = R_Q = \frac{w\times1}{2} = \frac{w}{2}$$

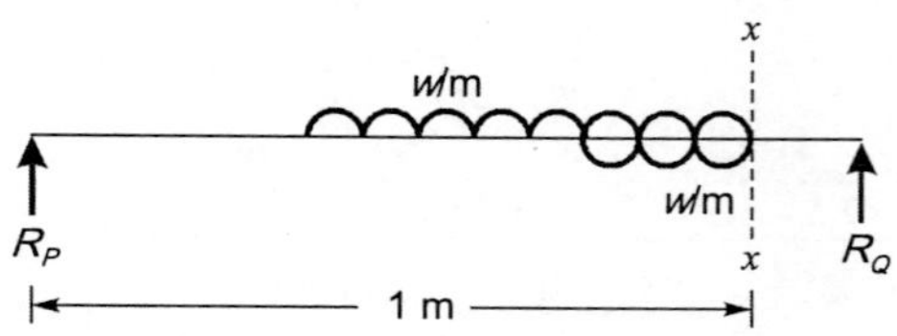

$$M_x = R_P x \,\Big/ -\frac{w}{2}(x-1)^2 \,\Big/ +\frac{w}{2}(x-2)^2$$

According to **Macaulay method,**

$$EI\frac{d^2y}{dx^2} = R_p x \,/\, -\frac{w}{2}(x-1)^2 \,/\, +\frac{w}{2}(x-2)^2$$

After integrating once

$$EI\frac{dy}{dx} = \frac{R_P x^2}{2} + c_1 \,/\, -\frac{w}{6}(x-1)^3 \,/\, +\frac{w}{6}(x-2)^3$$

Due to symmetrical loading slope will be zero at mid section (x = 1.5 m),

$$\therefore \text{ at } x = 15 \text{ m}, \quad \frac{dy}{dx} = 0$$

$$0 = \frac{R_P(1.5)^2}{2} + c_1 - \frac{w}{6}(1.5-1)^3$$

$$= \frac{9R_P}{8} + c_1 - \frac{w}{48} = \frac{9w}{16} + c_1 - \frac{w}{48}$$

$$\therefore \quad c_1 = \frac{w}{48} - \frac{9w}{16} = -\frac{26w}{48} = -\frac{13w}{24}$$

$\therefore$ Equation of slope,

$$EI\cdot\frac{dy}{dx} = \frac{R_P x^2}{2} - \frac{13w}{24} \,/\, -\frac{w}{6}(x-1)^3 \,/\, +\frac{w(x-2)^3}{6}$$

The slope will be maximum at the support,

$\therefore\ \theta_{max(x\,=\,0)}$

$$= \frac{-13w}{24EI} = \frac{-13\times20\times10^3}{24\times30\times10^6} = -0.361\times10^{-3}\text{ radian}$$

Method-II

Moment Area Method,

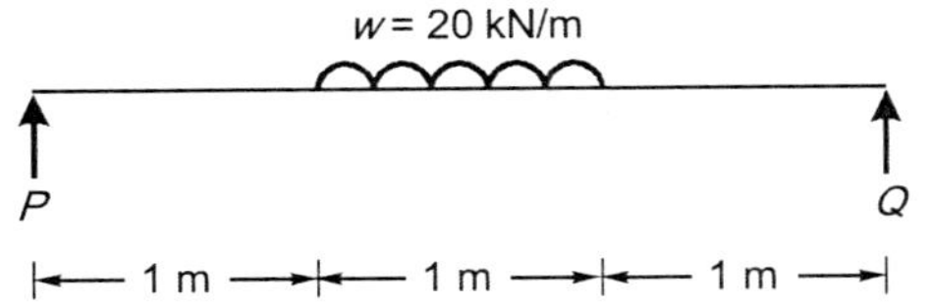

$EI = 30 \times 10^6$ Nm^2

$$M_A = P\times x \qquad (x = 1\text{ m})$$

$$= 10 \times 1 = 10 \text{ kNm}$$

$$M_B = P_x - \frac{W}{2}(x-1)^2 \qquad (x = 1.5\text{m})$$

$$= 10\times1.5 - \frac{20\times0.5^2}{2} = 12.5\text{ kNm}$$

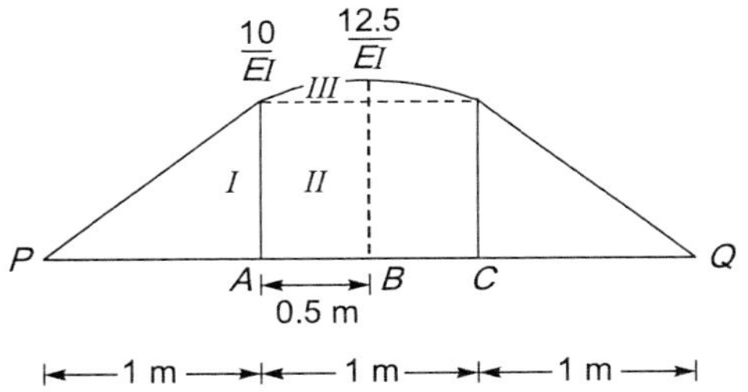

M/EI diagram

$\theta_A - \theta_C$ = Area of M/EI diagram between points P and B.

$$\cancel{\theta_B}^{\,0} - \theta_A = A_I + A_{II} + A_{III}$$

$$= \left(\frac{1}{2}\times1\times\frac{10}{EI}\right) + \left(0.5\times\frac{10}{EI}\right) + \left(\frac{2}{3}\times0.5\times\frac{2.5}{EI}\right)$$

$$= \frac{5}{EI} + \frac{5}{EI} + \frac{0.833}{EI}$$

$$\theta_A = -\frac{10.833}{EI} = -\frac{10.833\times10^3}{30\times10^6}$$

$$= -0.3611 \times 10^{-3} \text{ radians}$$

5.31 (d)

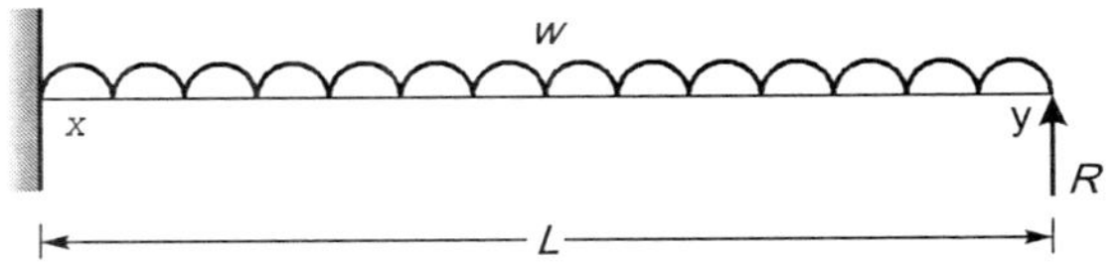

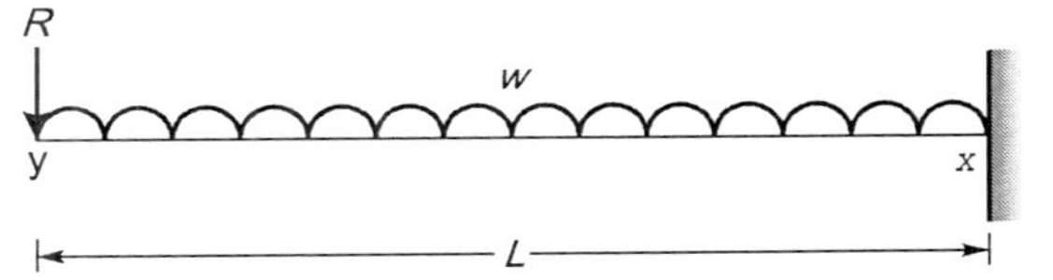

Deflection in beam xy at y = Deflection in beam yz at y,

$$\Rightarrow \quad \frac{wL^4}{8EI} - \frac{RL^3}{3EI} = \frac{RL^3}{3EI} + \frac{wL^4}{8EI}$$

$$\therefore \quad R = 0$$

In beam PQ also at support Q, vertical reaction in zero because of roller support.

So, beam PQ, xy and yz are same.

$$\therefore \quad Q_{max1} = Q_{max2} \text{ and } \delta_{max1} = \delta_{max2}$$

5.32 (a)

$$\delta_1 = \frac{5}{384}\times\frac{wl^4}{EI}$$

$$= \frac{5}{384}\times\frac{6(4)^4}{1000} = 20\times10^{-3}\text{ m}$$

$$\delta_2 = \frac{Pl^3}{48EI} = \frac{120\times2^3}{48\times1000} = 20\times10^{-3}\text{ m}$$

$$\therefore \quad \delta_1 = \delta_2$$

5.33 (c)

Method-I

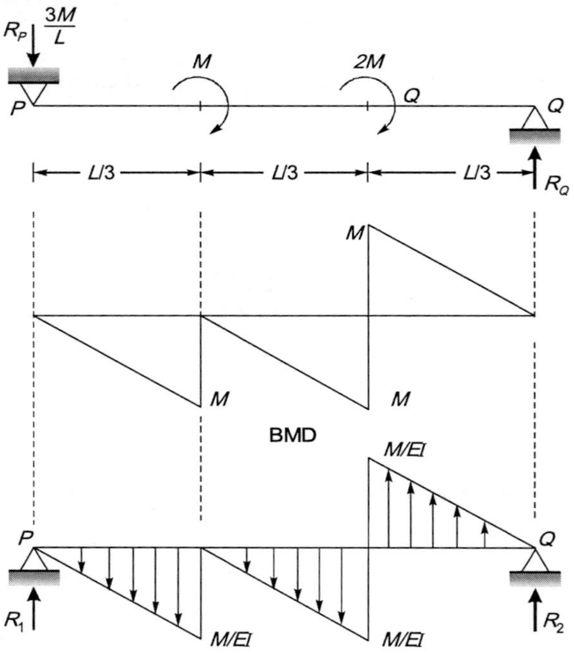

Conjugate Beam

Now $R_1 + R_2 = \frac{1}{2}\times\frac{L}{3}\times\frac{M}{EI} = \frac{ML}{6EI}$

$\Sigma M_Q = 0$

$$R_1L - \frac{1}{2}\times\frac{L}{3}\times\frac{M}{EI}\left(\frac{2L}{3}+\frac{L}{9}\right) - \frac{1}{2}\times\frac{L}{3}\times\frac{M}{EI}\left(\frac{L}{3}+\frac{L}{9}\right) + \frac{1}{2}\times\frac{L}{3}\times\frac{M}{EI}\times\frac{2L}{9} = 0$$

$$R_1 = \frac{7ML}{54EI} + \frac{4ML}{54EI} - \frac{ML}{27EI} = \frac{ML}{6EI}$$

$\therefore$ Slope at P, $\theta_P = SF_P = \frac{ML}{6EI}$

Method-II

Moment area method:

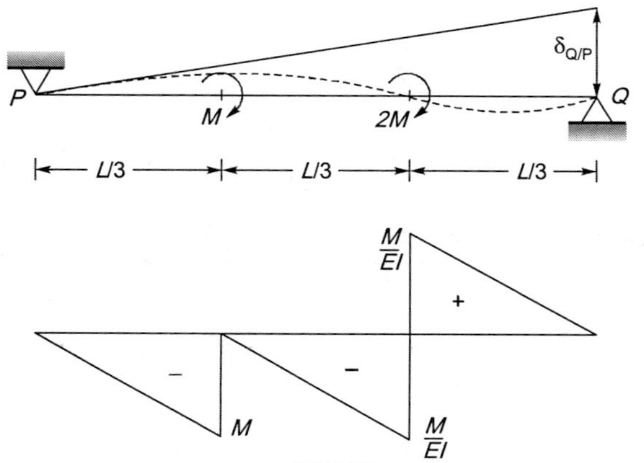

BMD/EI

$\delta_{Q/P}$ = Defleciton of point Q wrt to tangent at point P

$$\theta_Q = \frac{\delta_{Q/P}}{L}$$

$$\delta_{Q/P} = \left(-\frac{1}{2}\times\frac{M}{EI}\times\frac{L}{3}\right)\left(\frac{2L}{3}+\frac{L}{9}\right) + \left(-\frac{1}{2}\times\frac{M}{EI}\times\frac{L}{3}\right)\left(\frac{L}{3}+\frac{L}{9}\right) + \left(\frac{1}{2}\times\frac{M}{EI}\times\frac{L}{3}\right)\left(\frac{2L}{9}\right)$$

$$= \frac{ML^2}{6EI}$$

$$\theta_Q = \frac{\delta_{Q/P}}{L} = \frac{\frac{ML^2}{6EI}}{L} = \frac{ML}{6EI}$$

5.34 (d)

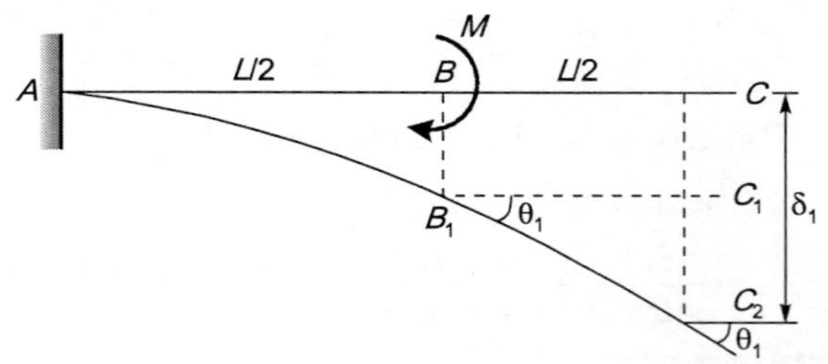

$$\delta_c = \delta_1 = CC_1 + C_1C_2$$

$$= BB_1 + C_1C_2$$

$$= \frac{M\left(\frac{L}{2}\right)^2}{2EI} + \frac{M\left(\frac{L}{2}\right)}{EI}\left(\frac{L}{2}\right)$$

$$\delta_c = \frac{3}{8}\frac{ML^2}{EI}$$

5.35 (d)

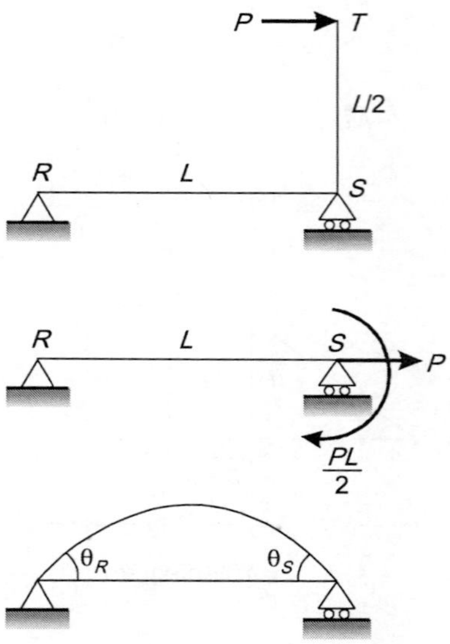

$$\theta_R = \frac{(PL/2)L}{6EI}$$

$$\theta_R = \frac{PL^2}{12EI}$$

5.36 Sol.

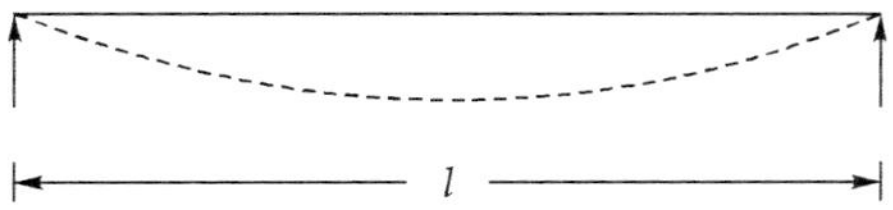

Given, $y = x^2 - xl$

Curvature at mid section is $\frac{1}{R} = \frac{d^2y}{dx^2}$

$$\frac{dy}{dx} = 2x - l$$

$$\frac{d^2y}{dx^2} = 2$$

5.37 (a, b)

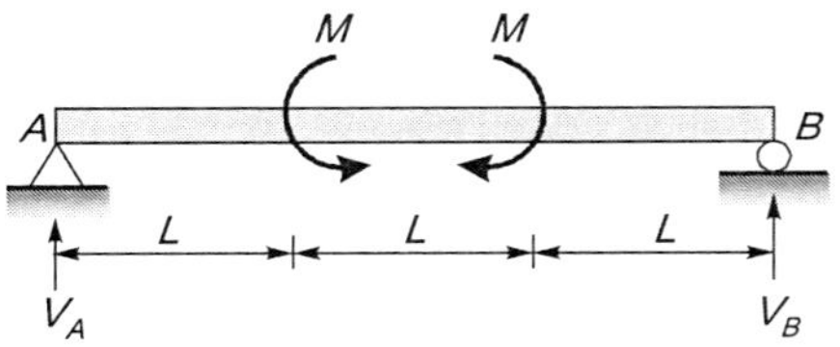

$$\Sigma F_y = 0$$

$\therefore$ $V_A + V_B = 0$ ...(i)

$$\Sigma M_A = 0$$

$\Rightarrow$ $V_B \times L - M + M = 0$

$\therefore$ $V_B = 0$

From equation (i), we get

$$V_A = 0$$

$\therefore$ Shear force is zero everywhere in the beaM, and reactions are also zero.

Bending moment diagram is shown below

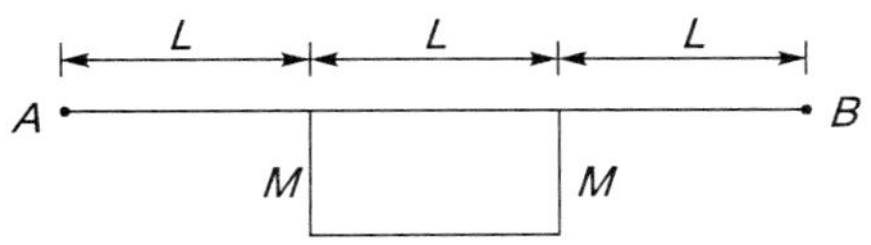

So, bending moment is not zero everywhere, and thus, also deflection is also not zero everywhere.

5.38 (a)

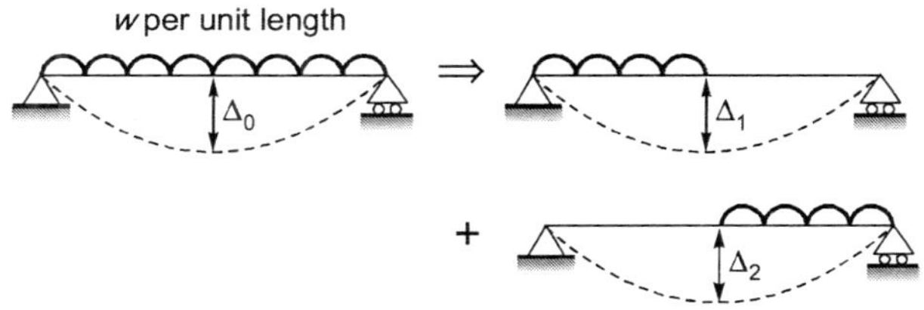

$\Delta_1 = \Delta_2 = \Delta'$ (say)

$\Delta_0 = \Delta_1 + \Delta_2 = 2\Delta'$

$$\Delta_0 = \frac{5}{384}\frac{WL^4}{EI}$$

$$\Delta' = \frac{\Delta_0}{2} = \frac{1}{2}\left(\frac{5}{384}\frac{WL^4}{EI}\right)$$

So, midspan deflection reduced to $\Delta_0/2$.

5.39 Sol.

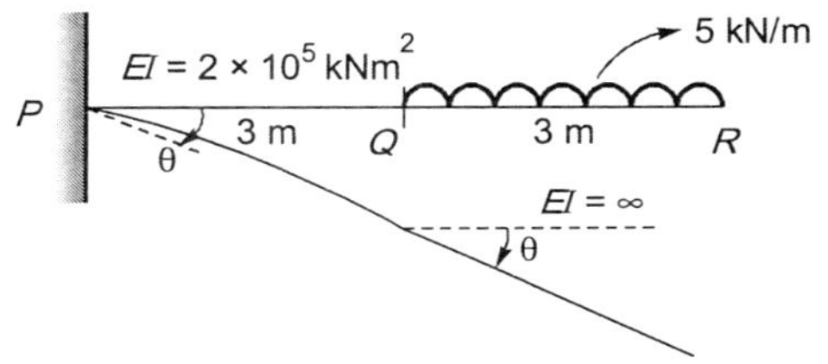

$\therefore$ QR will remain straight

[Transferring UDL to point load and moment at Q]

From principle of superposition,

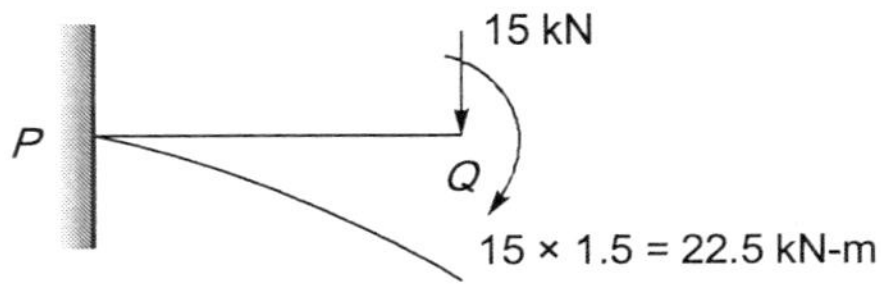

Deflection at Q,

$$\delta_Q = \frac{PL^3}{3EI} + \frac{ML^2}{2EI} = \frac{15 \times 3^3}{3 \times 2 \times 10^5} + \frac{22.5 \times 3^2}{2 \times 2 \times 10^5}$$

$$= 1.181 \times 10^{-3} \text{ m} = 1.181 \text{ mm}$$

■■■■

6 Torsion of Shafts and Pressure Vessels

6.1 A circular solid shaft of span $L = 5$ m is fixed at one end and free at the other end. A twisting moment $T = 100$ kN-m is applied at the free end. The torsional rigidity GJ is 50000 kN-m^2/rad. Following statements are made for this shaft:

1. The maximum rotation is 0.01 rad.
2. The torsional strain energy is 1 kN-m.

With reference to the above statements, which of the following applies?

(a) Both statements are true
(b) Statement 1 is true but 2 is false
(c) Statement 2 is true but 1 is false
(d) Both the statements are false [2004 : 2 M]

6.2 A circular shaft shown in the figure is subjected to torsion T at two points A and B. The torsional rigidity of portions CA and BD is GJ_1 and that of portion AB is GJ_2. The rotations of shaft at points A and B are θ_1 and θ_2. The rotation θ_1 is

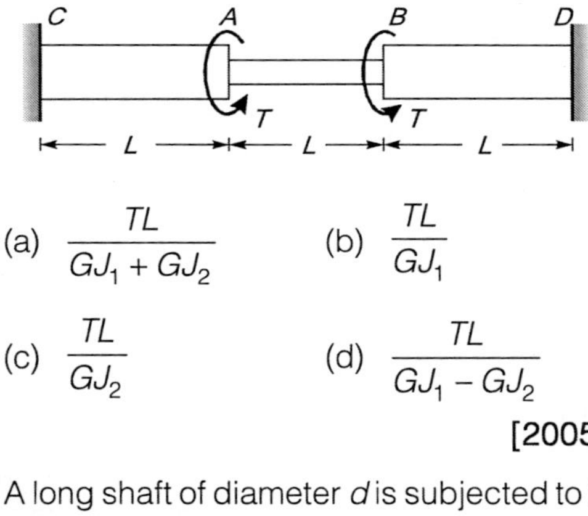

(a) $\dfrac{TL}{GJ_1 + GJ_2}$ (b) $\dfrac{TL}{GJ_1}$

(c) $\dfrac{TL}{GJ_2}$ (d) $\dfrac{TL}{GJ_1 - GJ_2}$

[2005 : 2 M]

6.3 A long shaft of diameter d is subjected to twisting moment T at its ends. The maximum normal stress acting at its cross-section is equal to

(a) Zero (b) $\dfrac{16T}{\pi d^3}$

(c) $\dfrac{32T}{\pi d^3}$ (d) $\dfrac{64T}{\pi d^3}$ [2006 : 1 M]

6.4 A thin-walled long cylindrical tank of inside radius r is subjected simultaneously to internal gas pressure p and axial compressive force F at its ends. In order to produce 'pure shear' state of stress in the wall of the cylinder, F should be equal to

(a) $p\pi r^2$ (b) $2p\pi r^2$
(c) $3p\pi r^2$ (d) $4p\pi r^2$ [2006 : 2 M]

6.5 The maximum and minimum shear stresses in a hollow circular shaft of outer diameter 20 mm and thickness 2 mm, subjected to a torque of 92.7 N-m will be

(a) 59 MPa and 47.2 MPa
(b) 100 MPa and 80 MPa
(c) 118 MPa and 160 MPa
(d) 200 MPa and 160 MPa [2007 : 2 M]

6.6 The maximum shear stress in a solid shaft of circular cross-section having diameter d subjected to a torque T is τ. If the torque is increased by four times and the diameter of the shaft is increased by two times, the maximum shear stress in the shaft will be

(a) 2τ (b) τ
(c) $\tau/2$ (d) $\tau/4$ [2008 : 2 M]

6.7 A thin walled cylindrical pressure vessel having a radius of 0.5 m and wall thickness of 25 mm is subjected to an internal pressure of 700 kPa. The hoop stress developed is

(a) 14 MPa (b) 1.4 MPa
(c) 0.14 MPa (d) 0.014 MPa

[2009 : 1 M]

6.8 A hollow circular shaft has an outer diameter of 100 mm and a wall thickness of 25 mm. The allowable shear stress in the shaft is 125 MPa. The maximum torque the shaft can transmit is

(a) 46 kN-m (b) 24.5 kN-m
(c) 23 kN-m (d) 11.5 kN-m

[2009 : 2 M]

6.9 A solid circular shaft of diameter d and length L is fixed at one end and free at the other end. A torque T is applied at the free end. The shear modulus of the material is G. The angle of twist at the free end is

(a) $\dfrac{16\,TL}{\pi d^4 G}$ (b) $\dfrac{32\,TL}{\pi d^4 G}$

(c) $\dfrac{64TL}{\pi d^4 G}$ (d) $\dfrac{128TL}{\pi d^4 G}$ [2010 : 1 M]

6.10 A hollow circular shaft has an outer diameter of 100 mm and inner diameter of 50 mm. If the allowable shear stress is 125 MPa, the maximum torque (in kN-m) that the shaft can resist is ________.

[2017 : 2 M, Set-II]

6.11 A solid circular beam with radius of 0.25 m and length of 2 m is subjected to a twisting moment of 20 kNm about the z-axis at the free end, which is the only load acting as shown in the figure. The stress component τ_{xy} at Point 'M' in the cross-section of the beam at a distance of 1 m from the fixed end is

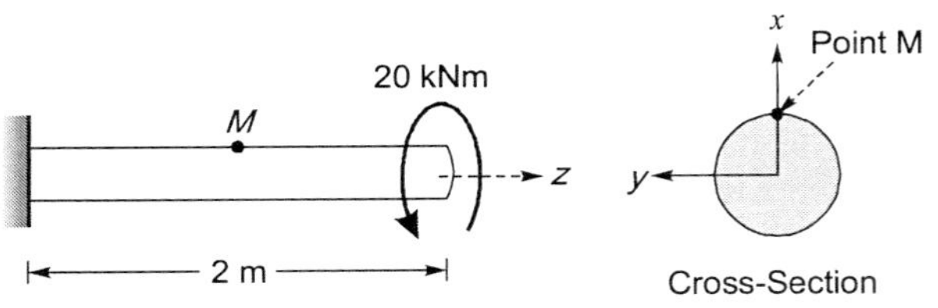

(a) 0.0 MPa (b) 0.51 MPa
(c) 0.815 MPa (d) 2.0 MPa

[2018 : 1 M, Set-I]

6.12 A closed thin-walled tube has thickness, t, mean enclosed area within the boundary of the centerline of tube's thickness, A_m and shear stress, τ. Torsional moment of resistance T, of the section would be

(a) $2\tau A_m t$ (b) $4\tau A_m t$
(c) $\tau A_m t$ (d) $0.5\tau A_m t$

[2019 : 1 M, Set-II]

6.13 A solid circular torsional member OPQ is subjected to torsional moments as shown in the figure (not to scale). The yield shear strength of the constituent material is 160 MPa.

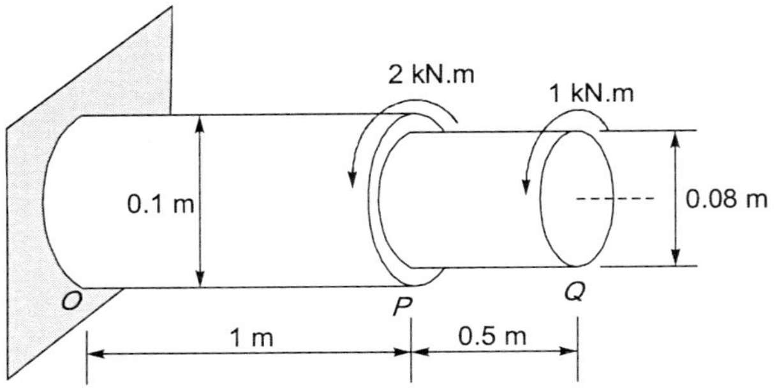

The absolute maximum shear stress in the member (in MPa, round off to one decimal place) is ________.

[2021 : 1 M, Set-II]

6.14 The hoop stress at a point on the surface of a thin cylindrical pressure vessel is computed to be 30.0 MPa. The value of maximum shear stress at this point is

(a) 7.5 MPa (b) 15.0 MPa
(c) 30 MPa (d) 22.5 MPa

[2022 : 1 M, Set-I]

6.15 A circular solid shaft of span L = 5 m is fixed at one end and free at the other end. A torque T = 100 kN.m is applied at the free end. The shear modulus and polar moment of inertia of the section are denoted as G and J, respectively. The torsional rigidity GJ is 50,000 kN.m^2/rad. The following are reported for this shaft:

Statement i) The rotation at the free end is 0.01 rad
Statement ii) The torsional strain energy is 1.0 kN.m

With reference to the above statements, which of the following is true?

(a) Both the statements are correct
(b) Statement i) is correct, but Statement ii) is wrong
(c) Statement i) is wrong, but Statement ii) is correct
(d) Both the statements are wrong

[2023 : 1 M, Set-II]

6.16 A homogeneous shaft PQR with fixed supports at both ends is subjected to a torsional moment T at point Q, as shown in the figure. The polar moments of inertia of the portions PQ and QR of the shaft with circular cross-sections are J_1 and J_2 respectively. The torsional moment reactions at the supports P and R are T_P and T_R, respectively.

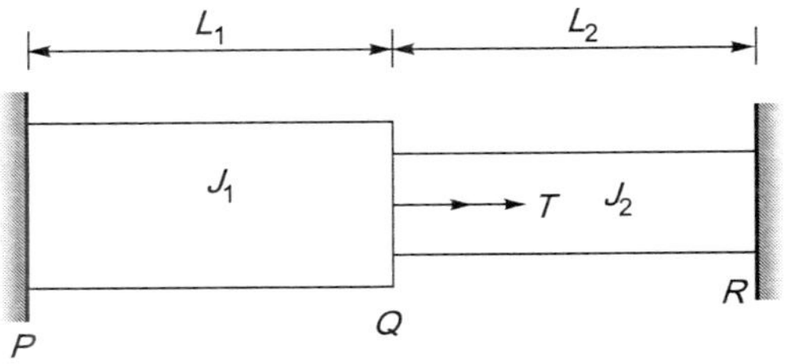

(Figure NOT to scale)

If $T_P/T_R = 4$ and $J_1/J_2 = 2$. The ratio of the lengths L_1/L_2 is

(a) 4.00 (b) 2.00
(c) 0.50 (d) 0.25

[2024 : 2 M, Set-I]

Answers Torsion of Shafts and Pressure Vessels

6.1 (b)	6.2 (b)	6.3 (a)	6.4 (c)	6.5 (b)	6.6 (c)	6.7 (a)
6.8 (c)	6.9 (b)	6.10 (23)	6.11 (a)	6.12 (a)	6.13 (15.3)	6.14 (a, b)
6.15 (b)	6.16 (c)					

Explanations Torsion of Shafts and Pressure Vessels

6.1 (b)

$$\therefore \quad \frac{G\theta}{L} = \frac{T}{J}$$

$$\therefore \quad \theta = \frac{TL}{GJ} = \frac{100 \times 5}{50000}$$

$$= 0.01 \text{ rad}$$

Also, **torsional strain energy** $= \frac{1}{2} T\theta$

$$= \frac{1}{2} \times 100 \times 0.01$$

$$= 0.5 \text{ kN-m}$$

6.2 (b)

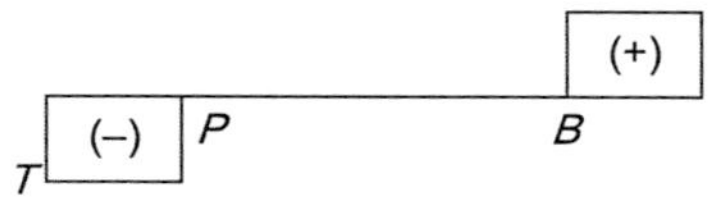

The symmetry of the shaft shows that there is no torsion on section AB,

$$\therefore \text{ Rotation, } \theta_1 = \frac{TL}{GJ_1}$$

6.3 (a)

Cross-section of the shaft is the section perpendicular to the longitudinal axis of shaft. In case of twisting moment acting at its ends, normal stress is zero on the cross-section only shear stress acts.

6.4 (c)

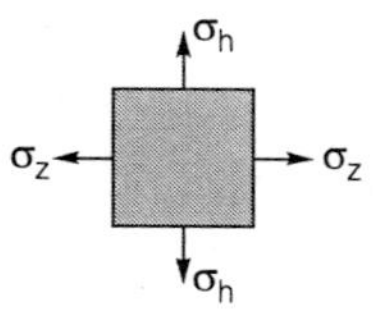

Hoop stress, $\sigma_h = \frac{pr}{t}$

Longitudinal stress,

$$\sigma_z = \frac{pr}{2t} - \frac{F}{2\pi rt}$$

Now, for pure shear state, σ_z should be compressive and is equal to σ_h.

$$\sigma_h + \sigma_z = 0$$

$$\frac{pr}{t} + \frac{pr}{2t} - \frac{F}{2\pi rt} = 0$$

$$\therefore \quad F = 3\pi r p^2$$

6.5 (b)

$$\because \quad \frac{f_s}{R} = \frac{T}{J}$$

Here, $J = \frac{\pi}{32}(20^4 - 16^4) \text{ mm}^4$;

$$R_1 = \frac{20}{2} = 10 \text{ mm};$$

$$T = 92.7 \text{ N-m};$$

$$R_2 = \frac{16}{2} = 8 \text{ mm}$$

$$f_{s_1} = \frac{TR_1}{J} = \frac{92.7 \times 10^3 \times 10}{(\pi/32) \times (20^4 - 16^4)}$$

$$= 99.96 \text{ MPa} \approx 100 \text{ MPa}$$

and

$$f_{s_2} = \frac{TR_2}{J} = \frac{92.7 \times 10^3 \times 8}{(\pi/32) \times (20^4 - 16^4)}$$

$$= 79.96 \text{ MPa} \simeq 80 \text{ MPa}$$

6.6 (c)

Method-I

$$\therefore \quad \frac{\tau}{R} = \frac{T}{I_P}$$

$$\Rightarrow \quad \frac{\tau_1}{\tau_2} = \frac{R_1}{R_2} \times \frac{T_1}{T_2} \times \frac{I_{P_2}}{I_{P_1}}$$

$$\Rightarrow \quad \frac{\tau_1}{\tau_2} = \frac{R_1}{2R_1} \times \frac{T_1}{4T_1} \times \left(\frac{2R_1}{R_1}\right)^4$$

$$\Rightarrow \quad \frac{\tau_1}{\tau_2} = \frac{1}{2} \times \frac{1}{4} \times 16$$

$$\Rightarrow \quad \tau_2 = \frac{\tau_1}{2}$$

$$\Rightarrow \quad \tau_2 = \frac{\tau}{2}$$

Method-II

$$\tau_{max} = \frac{16T}{\pi D^3} = \tau$$

$$\tau_{max}' = \frac{16 \times 4 \times T}{8\pi D^3} = \frac{\tau}{2}$$

6.7 (a)

$$\text{Hoop stress} = \frac{pd}{2t} = \frac{700\times10^3\times2\times0.5}{2\times25\times10^{-3}}$$

$$= 14\times10^6 \text{ Pa} = 14 \text{ MPa}$$

6.8 (c)

Method-I

$$\therefore \quad \frac{T}{I_P} = \frac{\tau}{R}$$

$$\Rightarrow \quad T = \frac{\tau}{R}\times I_P$$

$$= 125\times\frac{\pi}{32}\times(100^4-50^4)\times\frac{2}{100}\times10^{-6}$$

$$= 23.00 \text{ kN-m}$$

Method-II

$$\tau_{max} = \frac{16T}{\pi D^3\left(1-\left(\frac{d}{D}\right)^4\right)}$$

$$T = \frac{125\times\pi\times100^3\left(1-\left(\frac{50}{100}\right)^4\right)}{16}$$

$$= 23 \text{ kNm}$$

6.9 (b)

Angle of twist is given by,

$$\theta = \frac{TL}{GI_P} \quad \text{but } I_P = \frac{\pi d^4}{32}$$

$$\Rightarrow \quad \theta = \frac{TL}{G\times\frac{\pi d^4}{32}} = \frac{32\,TL}{\pi d^4 G}$$

6.10 Sol.

$$\therefore \quad \frac{T}{I_p} = \frac{\tau}{r}$$

$$\Rightarrow \quad \tau = \frac{T}{Z_p} = \frac{T}{\frac{\pi(D^4-d^4)}{16\times D}} = \frac{16T}{\frac{\pi}{D}(D^4-d^4)}$$

$$\Rightarrow \quad \tau = \frac{16T}{\pi D^3\left[1-\left(\frac{d}{D}\right)^4\right]}$$

$$\Rightarrow \quad 125 = \frac{16T}{\pi(100)^3\left[1-\left(\frac{50}{100}\right)^4\right]}$$

$$\Rightarrow \quad T = 23 \text{ kNm}$$

6.11 (a)

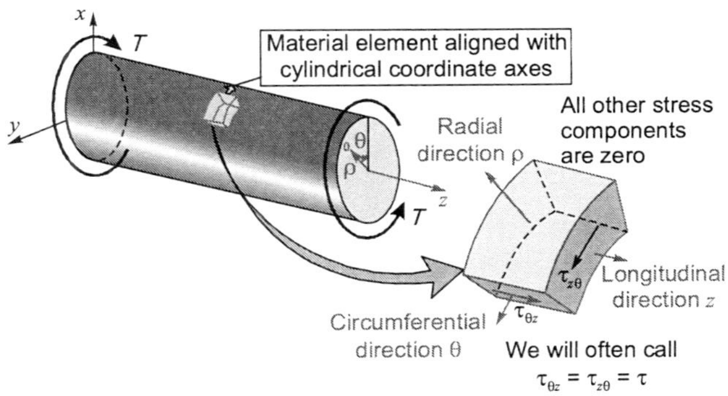

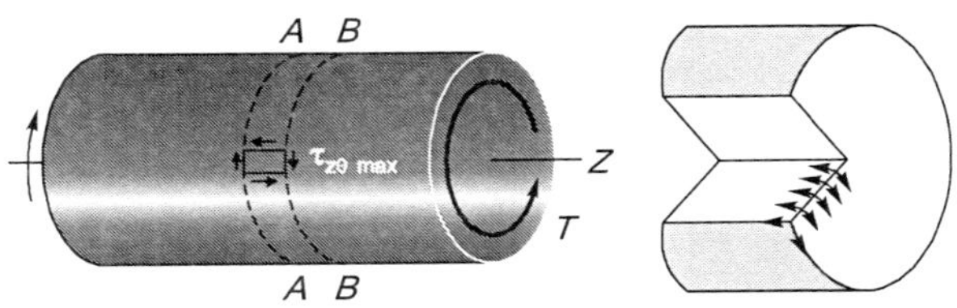

The only non-zero stresses are $\tau_{\theta z} = \tau_{z\theta} = \tau$, if θ is 90° then $\theta = y$

Hence $\quad \tau_{zy} = \tau_{yz} = \tau_{max}$

$$= 16T/\pi d^3 = 0.815 \text{ MPa}$$

But in rest of the planes shear stresses are zero, hence, $\tau_{xy} = \tau_{yx} = 0$

6.12 (a)

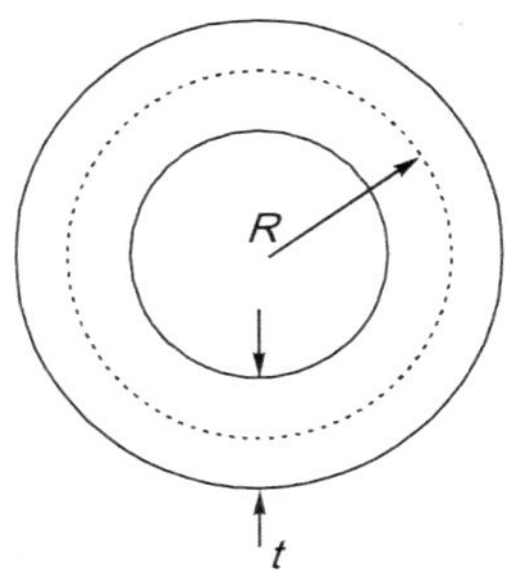

$$\text{Shear stress,} \quad \tau = \frac{T}{J}R = \frac{T}{2\pi R^3 t}R$$

$$\tau = \frac{T}{2\pi R^2 t} = \frac{T}{2A_m\cdot t}$$

$$\therefore \quad T = 2\tau A_m t$$

6.13 Sol.

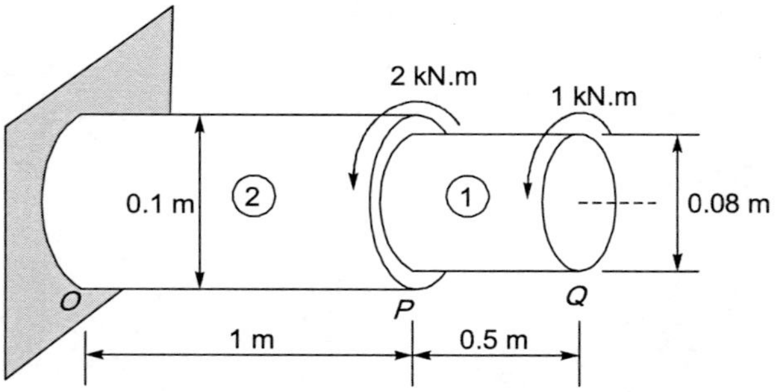

$$\tau_{maxOP} = \frac{16T_{OP}}{\pi d_{OP}^3} = \frac{16\times3\times10^3}{\pi\times100^3}$$

$$= 15.27\ \text{N/mm}^2$$

$$\tau_{maxPQ} = \frac{16T_{PQ}}{\pi d_{PQ}^3} = \frac{16\times1\times10^6}{\pi\times80^3}$$

$$= 9.94\ \text{N/mm}^2$$

$$\tau_{max} = 9.94\ \text{N/mm}^2$$

Absolute max shear stress

$$= 15.27\ \text{N/mm}^2$$

$$\simeq 15.3\ \text{N/mm}^2$$

6.14 (a, b)

We know that,

Hoop stress,

$$\sigma_h = \frac{Pd}{2t} = 30\ \text{N/mm}^2\ \text{(given)}$$

Also, longitudinal stress,

$$\sigma_l = \frac{Pd}{4t} = \frac{30}{2} = 15\ \text{N/mm}^2$$

Radial stress,

$$\sigma_R = 0$$

So the principal stress are,

$$\sigma_1 = \sigma_h = 30\ \text{N/mm}^2$$
$$\sigma_2 = \sigma_l = 15\ \text{N/mm}^2$$
$$\sigma_3 = \sigma_R = 0\ \text{N/mm}^2$$

$$(\tau_{max})_{abs} = \max.\left\{\left|\frac{\sigma_1-\sigma_2}{2}\right|, \left|\frac{\sigma_2-\sigma_3}{2}\right|, \left|\frac{\sigma_3-\sigma_1}{2}\right|\right\}$$

$$= \max.\left\{\left|\frac{30-15}{2}\right|, \left|\frac{15-0}{2}\right|, \left|\frac{0-30}{2}\right|\right\}$$

$$= \max.\{7.5,\ 7.5,\ 15\}$$
$$= 15\ \text{N/mm}^2$$

Note: $\tau_{max} = \left|\frac{\sigma_1-\sigma_2}{2}\right| = \left|\frac{30-15}{2}\right| = 7.5\ \text{N/mm}^2$

M were awarded for both 15 MPa and 7.5 MPa.

6.15 (b)

Rotation at free end (θ)

$$\frac{\tau L}{GJ} = \frac{100\times5(\text{kNm}^2)}{50000(\text{kNm}^2/\text{rad})}$$

$$\frac{\tau L}{GJ} = \frac{100\times5(\text{kNm}^2)}{50000(\text{kNm}^2/\text{rad})} = 0.01\ \text{rad}$$

$$\text{Torsional strain energy} = \frac{1}{2}\times T\times\theta$$

$$= \frac{1}{2}\times100\times0.01 = 0.5\ \text{kN-m}$$

6.16 (c)

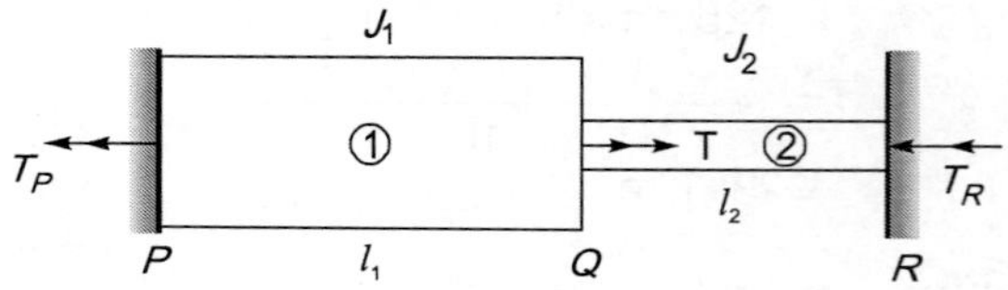

Let T_P and T_R be the resistive torques at support P and R respectively

From equilibrium condition,

$$T_P + T_R = T$$

$\because$ Given $T_P = 4\ T_R$

$$T_R = \frac{T}{5} \text{ and } T_P = \frac{4T}{5}$$

$$\theta_{PQ} + \theta_{QR} = 0$$

$$\Rightarrow \frac{T_PL_1}{GJ_1} + \frac{T_RL_2}{GJ_2} = 0$$

$$\Rightarrow \frac{\frac{4TL_1}{1}}{J_1} + \frac{\frac{TL_2}{5}}{J_2} = 0$$

For $J_1 = 2\ J_2$, $\frac{L_1}{L_2} = \frac{1}{2}$

■■■■

7 Theory of Columns & Shear Centre

7.1 Shear centre for an angle purlin is located at

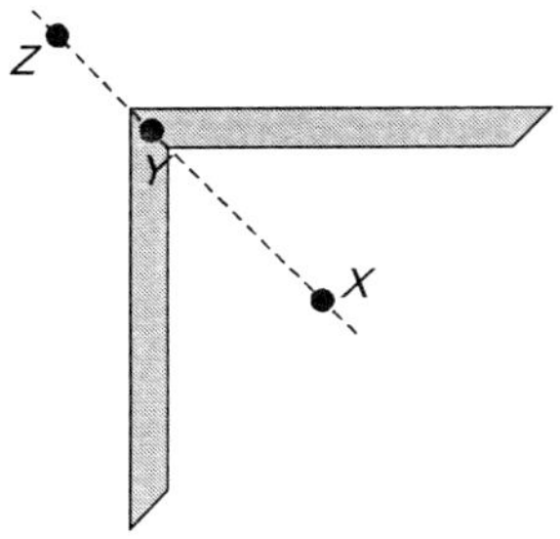

(a) X (b) Y
(c) Z (d) None [1996 : 1 M]

7.2 In a section, shear centre is a point through which, if the resultant load passes, the section will not be subjected to any
(a) Bending (b) Tension
(c) Compression (d) Torsion [1999 : 1 M]

7.3 Four columns of the same material and having identical geometric properties are supported in different ways as shown below:

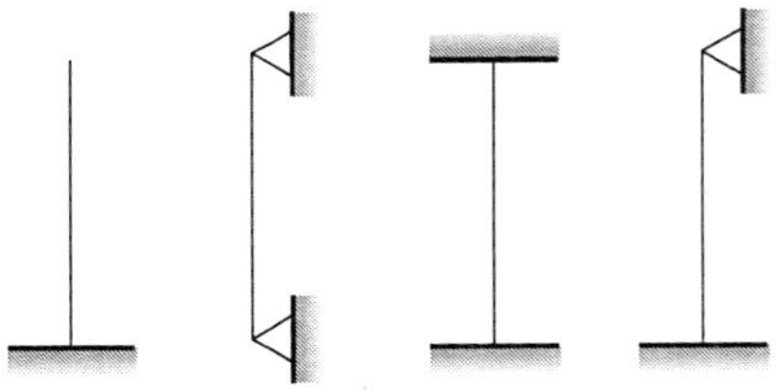

(I) (II) (III) (IV)

It is required to arrange these four columns in the increasing order of their respective first buckling loads. The correct order is given by
(a) I, II, III, IV (b) III, IV, I, II
(c) II, I, IV, III (d) I, II, IV, III
[2000 : 1 M]

7.4 A long structural column (length = L) with both ends hinged is acted upon by an axial compressive load, P. The differential equation governing the bending of column is given by:

$$EI\frac{d^2y}{dx^2} = -Py$$

where y is the structural lateral deflection and EI is the flexural rigidity. The first critical load on column responsible for its buckling is given by

(a) $\frac{\pi^2 EI}{L^2}$ (b) $\frac{\sqrt{2}\pi^2 EI}{L^2}$

(c) $\frac{2\pi^2 EI}{L^2}$ (d) $\frac{4\pi^2 EI}{L^2}$ [2003 : 2 M]

7.5 The buckling load $P = P_{cr}$ for the column AB in fig., as K_T approaches infinity, becomes $\alpha\frac{\pi^2 EI}{L^2}$

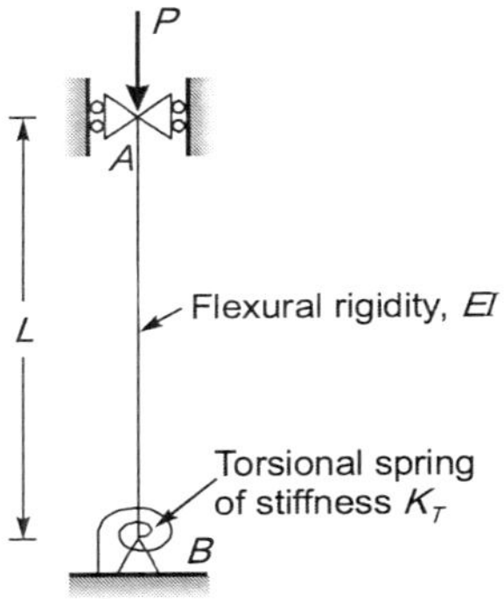

where α is equal to
(a) 0.25 (b) 1.00
(c) 2.05 (d) 4.00 [2006 : 1 M]

7.6 A steel column, pinned at both ends, has a buckling load of 200 kN. If the column is restrained against lateral movement at its mid-height, its buckling load will be
(a) 200 kN (b) 283 kN
(c) 400 kN (d) 800 kN [2007 : 1 M]

7.7 Cross-section of a column consisting of two steel strips, each of thickness t and width b is shown in the figure below. The critical loads of the column with perfect bond and without bond between the strips are P and P_0 respectively. The ratio P/P_0 is

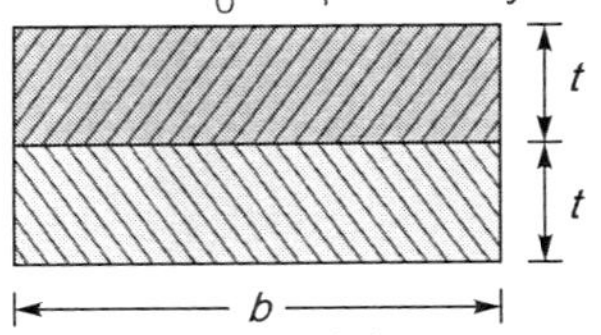

(a) 2 (b) 4
(c) 6 (d) 8 [2008 : 2 M]

7.8 A rigid bar GH of length L is supported by a hinge and a spring of stiffness K as shown in the figure below. The buckling load, P_{cr}, for the bar will be

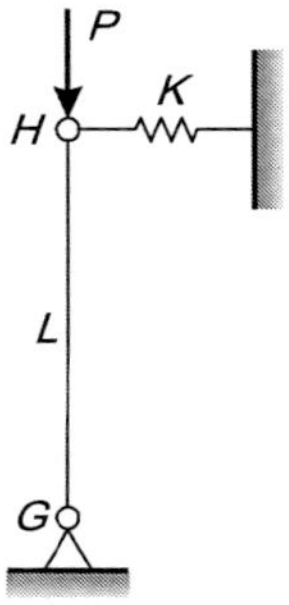

(a) 0.5 KL (b) 0.8 KL
(c) 1.0 KL (d) 1.2 KL [2008 : 2 M]

7.9 The point within the cross sectional plane of a beam through which the resultant of the external loading on the beam has to pass through to ensure pure bending without twisting of the cross-section of the beam is called

(a) moment centre (b) centroid
(c) shear centre (d) elastic center

[2009 : 1 M]

7.10 The effective length of a column of length L fixed against rotation and translation at one end and free at other end is

(a) 0.5 L (b) 0.7 L
(c) 1.414 L (d) 2 L [2010 : 1 M]

7.11 The ratio of the theoretical critical buckling load for a column with fixed ends to that of another column with the same dimensions and material, but with pinned ends, is equal to

(a) 0.5 (b) 1.0
(c) 2.0 (d) 4.0 [2012 : 1 M]

7.12 The sketch shows a column with a pin at the base and rollers at the top.

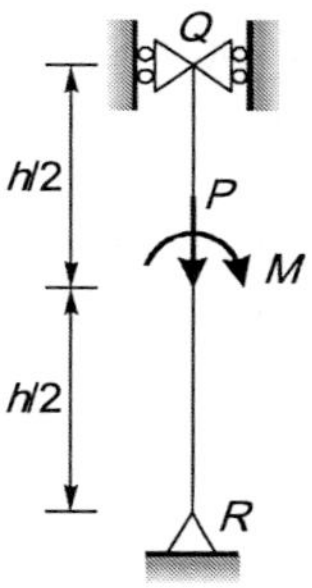

It is subjected to an axial force P and a moment M at mid-height. The reaction(s) at R is/are

(a) a-vertical force equal to P
(b) a vertical force equal to $P/2$
(c) a vertical force equal to P and a horizontal force equal to M/h
(d) a vertical force equal to $P/2$ and a horizontal force equal to M/h [2012 : 2 M]

7.13 Two steel column P (length L and yield strength $f_y = 250$ MPa) and Q (length $2L$ and yield strength $f_y = 500$ MPa) have the same cross-section and end condition the ratio of buckling load of column P to that of column Q is

(a) 0.5 (b) 1.0
(c) 2.0 (d) 4.0 [2013 : 1 M]

7.14 The possible location of shear centre of the channel section, shown below is,

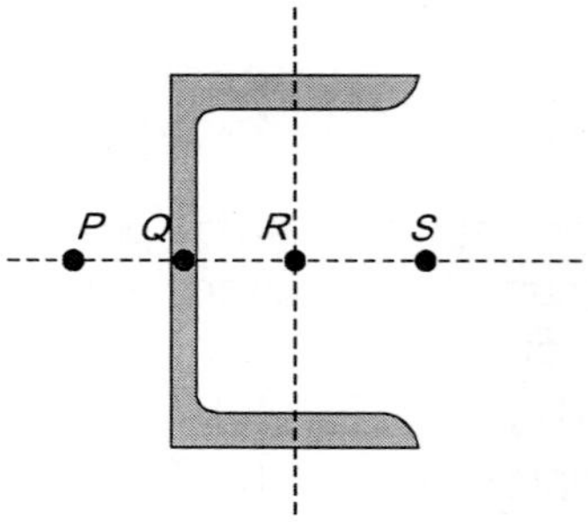

(a) P (b) Q
(c) R (d) S [2014 : 1 M, Set-I]

7.15 If the following equation establishes equilibrium in slightly bent position, the mid-span deflection of a member shown in the figure is

$$\frac{d^2y}{dx^2} + \frac{P}{EI}y = 0$$

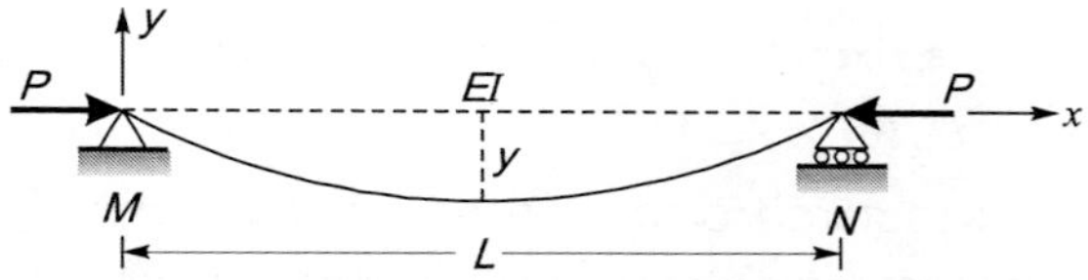

If a is amplitude constant for y, then

(a) $y = \frac{1}{P}\left(1 - a\cos\frac{2\pi x}{L}\right)$

(b) $y = \frac{1}{P}\left(1 - a\sin\frac{2\pi x}{L}\right)$

(c) $y = a\sin\frac{n\pi x}{L}$

(d) $y = a\cos\frac{n\pi x}{L}$ [2014 : 2 M, Set-I]

7.16 In a system two connected rigid bars AC and BC are of identical length, L with pin supports at A and B. The bars are interconnected at C by a frictionless hinge. The rotation of the hinge is restrained by a rotational spring of stiffness, k.

The system initially assumes a straight line configuration, ACB. Assuming both the bars as weightless, the rotation at supports, A and B, due to a transverse load, P applied at C is

(a) $\frac{PL}{4k}$ (b) $\frac{PL}{2k}$

(c) $\frac{P}{4k}$ (d) $\frac{Pk}{4L}$

[2015 : 2 M, Set-II]

7.17 Consider two axially loaded columns, namely, 1 and 2, made of a linear elastic material with Young's modulus 2×10^5 MPa, square cross-section with side 10 mm, and length 1 m. For Column 1, one end is fixed and the other end is free. For Column 2, one end is fixed and the other end is pinned. Based on the Euler's theory, the ratio (up to one decimal place) of the buckling load of Column 2 to the buckling load of Column 1 is ________.

[2017 : 2 M, Set-I]

7.18 A column of height h with a rectangular cross-section of size $a \times 2a$ has a buckling load of P. If the cross-section is changed to $0.5a \times 3a$ and its height changed to $1.5h$, the buckling load of the redesigned column will be

(a) $\frac{P}{12}$ (b) $\frac{P}{4}$

(c) $\frac{P}{2}$ (d) $\frac{3P}{4}$

[2018 : 1 M, Set-I]

7.19 An elevated cylindrical water storage tank is shown in the figure. The tank has inner diameter of 1.5 m. It is supported on a solid steel circular column of diameter 75 mm and total height (L) of 4 m.

Take, water density = 1000 kg/m^3 and acceleration due to gravity = 10 m/s^2.

If elastic modulus (E) of steel is 200 GPa, ignoring self-weight of the tank, for the supporting steel column to remain unbuckled, the maximum depth (h) of the water permissible (in m, round off to one decimal place) is ________.

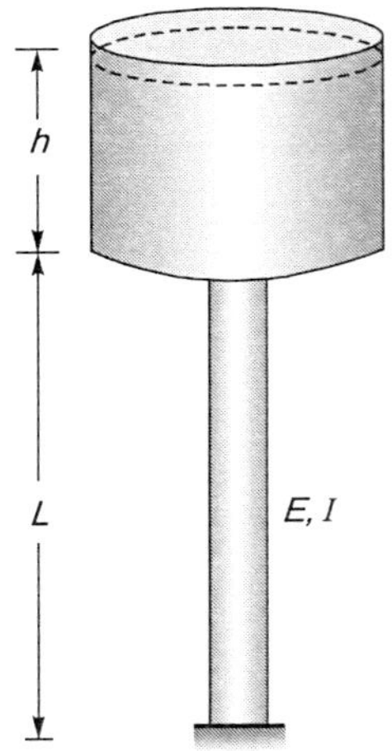

[2021 : 2 M, Set-II]

7.20 Consider the cross-section of a beam made up of thin uniform elements having thickness $t(t << a)$ shown in the figure. The (x, y) coordinates of the points along the center-line of the cross-section are given in the figure.

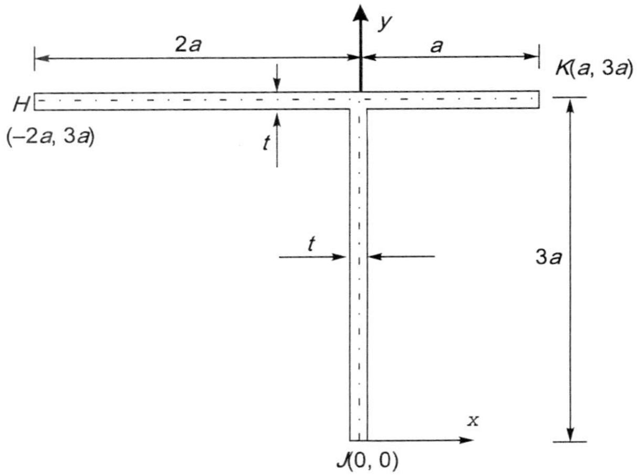

The coordinates of the shear center of this cross-section are:

(a) $x = 0, y = 3a$ (b) $x = 2a, y = 2a$

(c) $x = -a, y = 2a$ (d) $x = -2a, y = a$

[2022 : 1 M, Set-I]

7.21 For a thin-walled section shown in the figure, points P, Q, and R are located on the major bending axis $X-X$ of the section. Point Q is located on the web whereas point S is located at the intersection of the web and the top flange of the section.

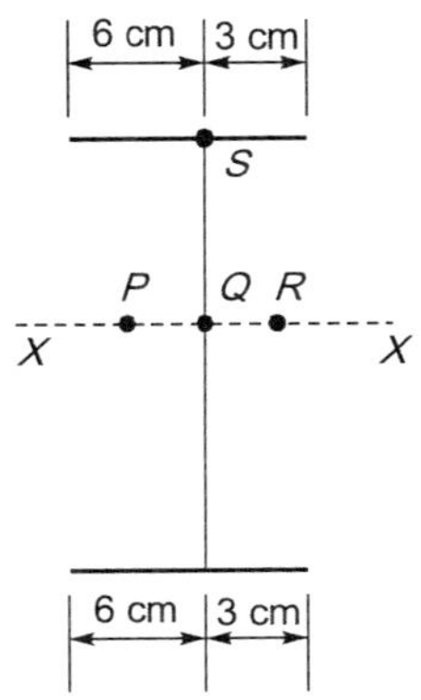

(Figure NOT to scale)

Qualitatively, the shear center of the section lies at

(a) P (b) S

(c) R (d) Q

[2024 : 1 M, Set-II]

Answers Theory of Columns & Shear Centre

7.1 (b)	7.2 (d)	7.3 (d)	7.4 (a)	7.5 (d)	7.6 (d)	7.7 (b)
7.8 (c)	7.9 (c)	7.10 (d)	7.11 (d)	7.12 (c)	7.13 (d)	7.14 (a)
7.15 (c)	7.16 (a)	7.17 (8)	7.18 (a)	7.19 (2.7)	7.20 (a)	7.21 (c)

Explanations Theory of Columns & Shear Centre

7.1 (b)

Shear center will be at Y.

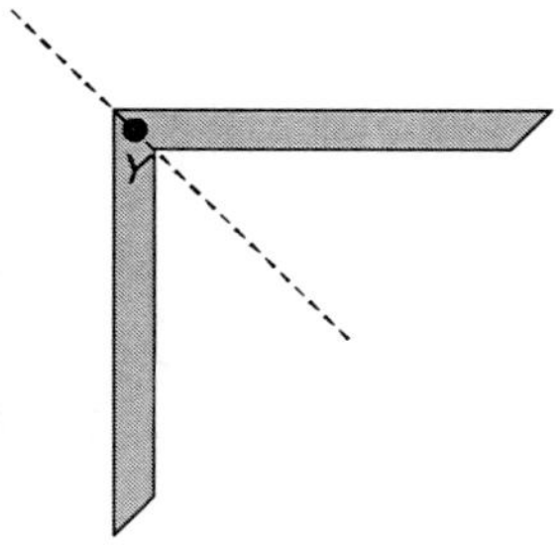

7.2 (d)

If the resultant of shear stress, i.e. shear force does not pass through shear centre, then section will be subjected to a torsional moment in addition to shear force. It can be said that a lateral load acting on a beam will produce bending without twisting only when it passes through shear centre.

7.3 (d)

In case-I, $L_{eff} = 2L$
In case-II, $L_{eff} = L$
In case-III, $L_{eff} = 0.5L$
In case-IV, $L_{eff} = 0.7L$

Now, Buckling load, $P_e = \dfrac{\pi^2 EI}{L^2}$

$$\Rightarrow \quad P_e \propto \frac{1}{L^2}$$

Hence, if other conditions remain same then an increasing the effective length, buckling load will decreae.

Therefore, in the order of increasing buckling load we will be I, II, IV, III.

7.4 (a)

Differential equation governing the bending of column,

$$\frac{EId^2y}{dx^2} = -Py$$

$$\Rightarrow \frac{d^2y}{dx^2} + \frac{P}{EI}y = 0$$

Solution of differential equation,

$$y = A\cos\left(x\sqrt{\frac{P}{EI}}\right) + B\sin\left(x\sqrt{\frac{P}{EI}}\right)$$

when $x = 0$. $y = 0$

$$\Rightarrow \quad 0 = B\sin\left(L\sqrt{\frac{P}{EI}}\right)$$

$$\therefore \quad L\sqrt{\frac{P}{EI}} = \pi \qquad \text{(First value)}$$

$$\Rightarrow \quad P = \frac{\pi^2 EI}{L^2}$$

7.5 (d)

As the k_T approaches infinity, support at B becomes fixed end and support at A also fixed end.

So, $P_{cr} = \dfrac{\pi^2 EI}{(l_e)^2} = \dfrac{4\pi^2 EI}{l^2}$ $[\because l_e = 0.5\,l]$

7.6 (d)

When both ends are hinged, the buckling load is given by,

$$P_{cr} = \frac{\pi^2 EI}{L^2} \quad \Rightarrow \quad 200 = \frac{\pi^2 EI}{L^2}$$

When the lateral movement at the mid-height is not available, then buckling load

$$= \frac{\pi^2 EI}{(L_1)^2} \quad \text{where } L_1 = \frac{L}{2}$$

$$= \frac{4\pi^2 EI}{L^2} = 4 \times 200 = 800 \text{ kN}$$

7.7 (b)

We know that critical load for a column is proportional to moment of inertia irrespective of end conditions of the column i.e.

$$P_{cr} \propto I$$

When the steel strips are perfectly bonded, then,

$$I_{pb} = \frac{b \times (2t)^3}{12} = \frac{8bt^3}{12}$$

When the steel strips are not bonded, then,

$$I_{wb} = 2 \times \frac{bt^3}{12} = \frac{2bt^3}{12}$$

$$\therefore \quad \frac{P}{P_0} = \frac{8bt^3/12}{2bt^3/12} = 4$$

7.8 (c)

Let the deflection in the spring be δ and force in the spring be F.

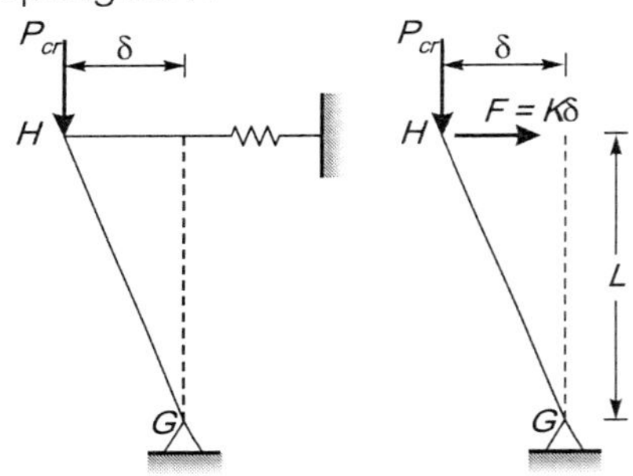

Taking moments about G, we get,

$$P_{cr} \times \delta = F \times L$$

$$\Rightarrow \quad P_{cr} = \frac{K\delta \times L}{\delta} \qquad [\because F = K\delta]$$

$$\Rightarrow \quad P_{cr} = KL$$

7.9 (c)

If the resultant of shear stress, i.e. shear force does not pass through shear centre, then section will be subjected to a torsional moment in addition to shear force. It can be said that a lateral load acting on a beam will produce bending without twisting only when it passes through shear centre.

7.10 (d)

End conditon	Eff. length (l_e)
Both ends hinged	L
One end free and one fixed	2L
Both ends fixed	0.5L
One end fixed and one hinged	$L/\sqrt{2}$

7.11 (d)

Eulars critical load,

$$P_e = \frac{\pi^2 EI}{l_{eft}^2}$$

$$\frac{P_{e_1}}{P_{e_2}} = \left[\frac{(l_{eft})_2}{(l_{eft})_1}\right]^2 = \left[\frac{l}{l/2}\right]^2 = 4$$

7.12 (c)

$$\Sigma F_y = 0$$

$$\Rightarrow \quad V_R - P = 0$$

$$\Rightarrow \quad V_R = P$$

$$\Sigma M_Q = 0$$

$$\Rightarrow \quad H_R \cdot h - M = 0$$

$$\Rightarrow \quad H_R = \frac{M}{h}$$

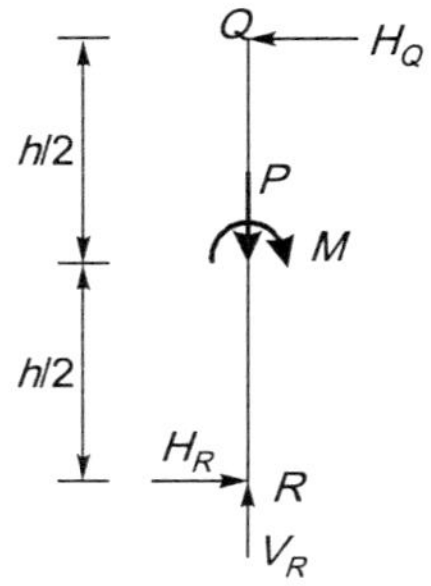

7.13 (d)

Buckling load,

$$P = \frac{\pi^2 EI}{(l_{eff})^2}$$

$$\therefore \quad P_P = \frac{\pi^2 EI}{(L)^2}$$

$$P_Q = \frac{\pi^2 EI}{(2L)^2}$$

$$\therefore \quad \frac{P_P}{P_Q} = 4$$

7.14 (a)

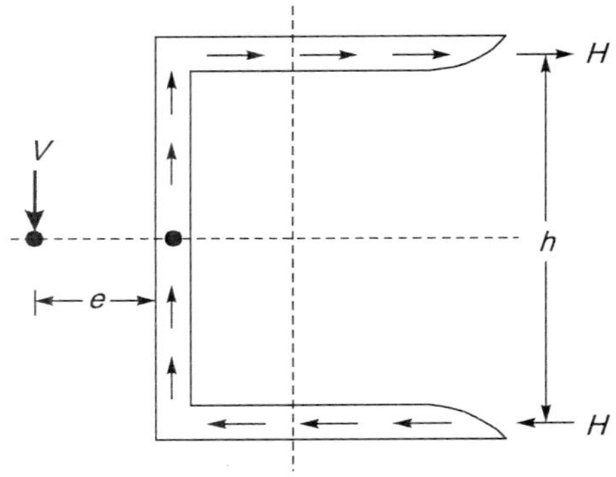

For no twisting

$$V \times e = H \times h$$

$$\Rightarrow \quad e = \frac{Hh}{V}$$

Hence possible location of shear centre is P.

7.15 (c)

$$\frac{d^2y}{dx^2} = -\frac{P}{EI} \times y = -m^2 y$$

∴ Solution of above differential equation is,

$$y = a\sin mx + b\cos mx$$

at $x = 0,\ y = 0$

$\Rightarrow$ $b = 0$

at $x = L,\ y = 0$

$\Rightarrow$ $0 = \sin mL$

$\Rightarrow$ $mL = n\pi$

$\Rightarrow$ $m = \dfrac{n\pi}{L}$

∴ $y = a\sin\dfrac{n\pi x}{L}$

7.16 (a)

Deflection under load $P = \theta L$

Work done by force $P = \dfrac{1}{2}P(\theta L)$

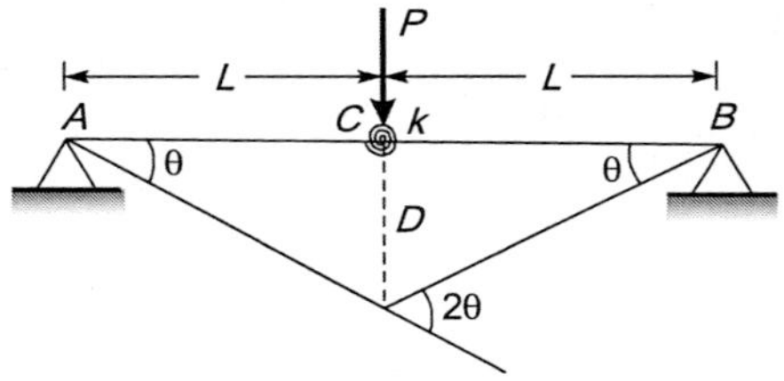

Strain energy stored in spring

$$= \frac{1}{2}[k(2\theta)](2\theta)$$

$$\frac{1}{2}P(\theta L) = \frac{4k\theta^2}{2}$$

$$\Rightarrow \theta = \frac{PL}{4k}$$

7.17 Sol.

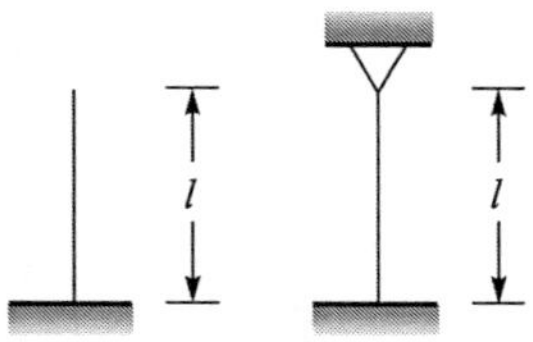

$$P_{cr1} = \frac{\pi^2 E.I}{(2l)^2} = \frac{\pi^2 EI}{4l^2}$$

$$P_{cr2} = \frac{\pi^2 E.I}{\left(\dfrac{l}{\sqrt{2}}\right)^2} = \frac{2\pi^2 EI}{l^2}$$

$$\therefore \text{Required ratio} = \frac{P_{cr2}}{P_{cr1}} = \frac{\dfrac{2\pi^2 EI}{l^2}}{\dfrac{\pi^2 EI}{4l^2}} = 8$$

7.18 (a)

For column, $P = \dfrac{\pi^2 EI_{min}}{L^2}$

$$= \frac{\pi^2 E\left(\dfrac{2a\times a^3}{12}\right)}{h^2} = \frac{\pi^2 Ea^4}{6h^2}$$

For new column, $P' = \dfrac{\pi^2 E\left[\dfrac{3a\times(0.5a)^3}{12}\right]}{(1.5h)^2}$

$$= \frac{1}{12}\times\frac{\pi^2 Ea^4}{6h^2} = \frac{P}{12}$$

7.19 Sol.

Given: $L = 4$ m, $\rho = 1000$ kg/m^3, $g = 10$ m/s^2, $E = 200$ GPa, $D_i = 1.5$ m, $d = 75$ mm

$$\frac{\pi^2 EI}{L_e^2} = \frac{\pi}{4}D_i^2\times h\times\rho g$$

$$\frac{\pi^2 E\times\dfrac{\pi}{64}d^4}{(2L)^2} = \frac{\pi}{4}D_i^2\times h\times\rho g$$

$$h = \frac{4\pi^2 Ed^4}{4\times 64L^2\times D_i^2\times\rho g}$$

$$= \frac{4\pi^2\times 200\times 10^9\times(0.075)^4}{4\times 64\times 4^2\times 1.5^2\times 10^4}$$

$$h = 2.71 \text{ m} \simeq 2.7 \text{ m}$$

7.20 (a)

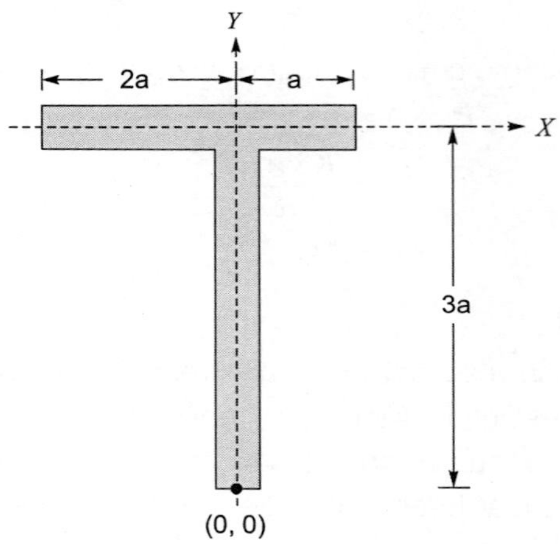

Coordinates of shear centre = (0, 3*a*)

7.21 (c)

Position of shear centre will be at *R*.

Structural Analysis

UNIT II

CONTENTS

Structural Analysis

Syllabus

Statically determinate and indeterminate structures by force/ energy methods; Method of superposition; Analysis of trusses, arches, beams, cables and frames; Displacement methods: Slope deflection and moment distribution methods; Influence lines; Stiffness and flexibility methods of structural analysis.

Analysis of Previous GATE Papers

Exam Year	1 Mark Ques.	2 Marks Ques.	Total Marks
1995	4	1	6
1996	3	1	5
1997	4	–	4
1998	1	2	5
1999	–	–	–
2000	2	–	2
2001	4	3	10
2002	–	–	–
2003	2	5	12
2004	3	3	9
2005	1	5	11
2006	–	2	4
2007	1	4	9
2008	–	3	6
2009	–	1	2
2010	1	1	3
2011	–	–	–
2012	–	–	–
2013	1	3	7
2014 Set-1	2	2	6
2014 Set-2	1	1	3

Exam Year	1 Mark Ques.	2 Marks Ques.	Total Marks
2015 Set-1	2	1	4
2015 Set-2	1	1	3
2016 Set-1	1	1	3
2016 Set-2	1	2	5
2017 Set-1	1	3	7
2017 Set-2	1	1	3
2018 Set-1	–	2	4
2018 Set-2	2	1	4
2019 Set-1	3	2	7
2019 Set-2	1	2	5
2020 Set-1	2	1	4
2020 Set-2	–	2	4
2021 Set-1	2	2	6
2021 Set-2	1	2	5
2022 Set-1	–	4	8
2022 Set-2	1	3	7
2023 Set-1	1	2	5
2023 Set-2	2	1	4
2024 Set-1	1	1	3
2024 Set-2	–	2	4

1 Determinacy and Indeterminacy

1.1 The degree of kinematic indeterminacy of the rigid frame with clamped ends at *A* and *D* as shown in the figure is

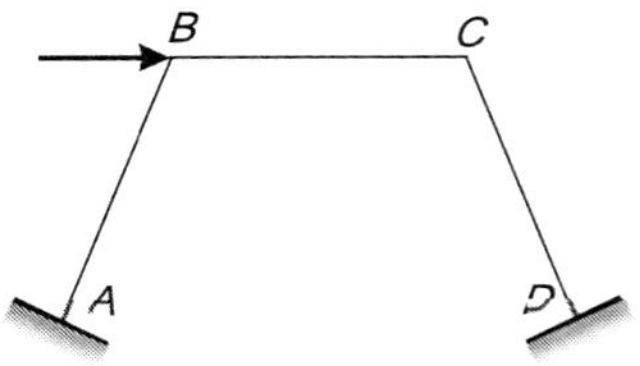

(a) 4 (b) 3
(c) 2 (d) Zero [1997 : 1 M]

1.2 The following two statements are made with reference to the plane truss shown below

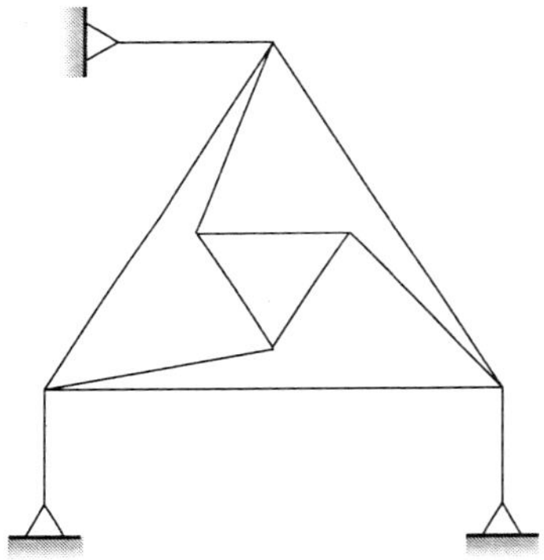

I. The truss is statically determinate
II. The truss is kinematically determinate

With reference to the above statements, which of the following applies?
(a) Both statements are true
(b) Both statements are false
(c) II is true but I is false
(d) I is true but II is false [2000 : 1 M]

1.3 The degree of static indeterminacy, D_s, and the degree of kinematic indeterminacy, D_k, for the plane frame shown below, assuming axial deformations to be negligible, are given by

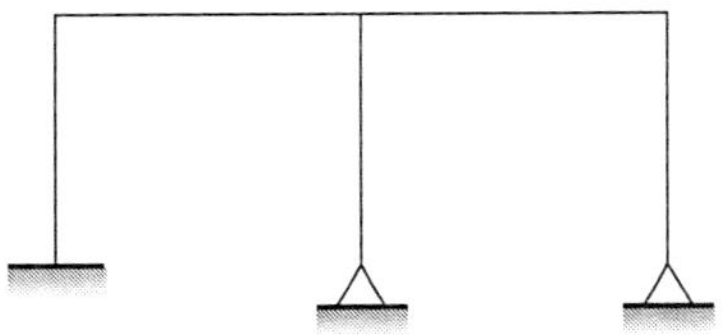

(a) $D_s = 6$ and $D_k = 11$
(b) $D_s = 6$ and $D_k = 6$
(c) $D_s = 4$ and $D_k = 6$
(d) $D_s = 4$ and $D_k = 4$ [2001 : 1 M]

1.4 For the plane frame with an overhang as shown below, assuming negligible axial deformation, the degree of static indeterminacy, *d*, and the degree of kinematic indeterminacy, *k*, are

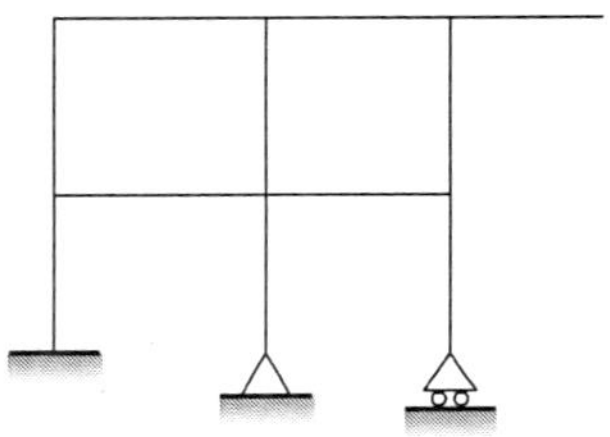

(a) $d = 3$ and $k = 10$
(b) $d = 3$ and $k = 13$
(c) $d = 9$ and $k = 10$
(d) $d = 9$ and $k = 13$ [2004 : 2 M]

1.5 Considering beam as axially rigid, the degree of freedom of a plane frame shown below is

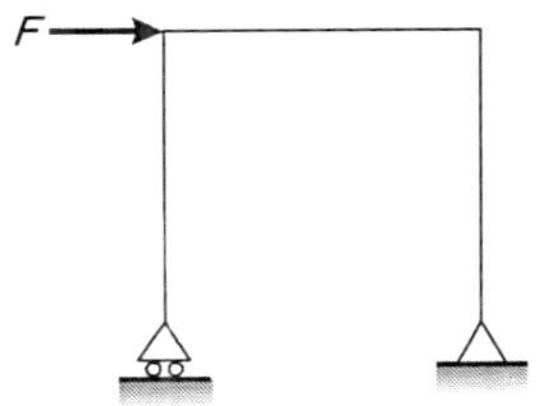

(a) 9 (b) 8
(c) 7 (d) 6 [2005 : 2 M]

1.6 The degree of static indeterminacy of the rigid frame having two internal hinges as shown in the figure below, is

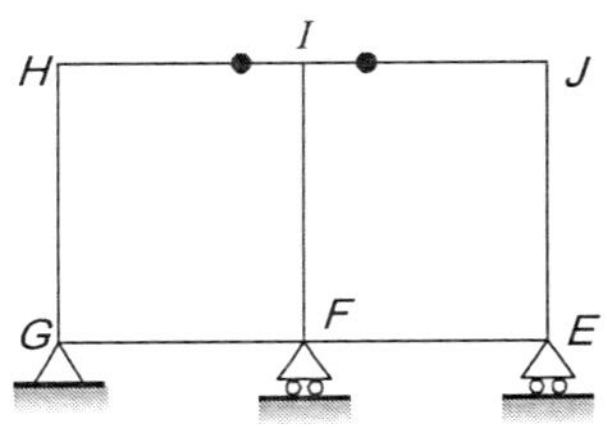

(a) 8 (b) 7
(c) 6 (d) 5 [2008 : 2 M]

1.7 The degree of static indeterminacy of a rigidly jointed frame in a horizontal plane and subjected to vertical load only, as shown in figure below, is

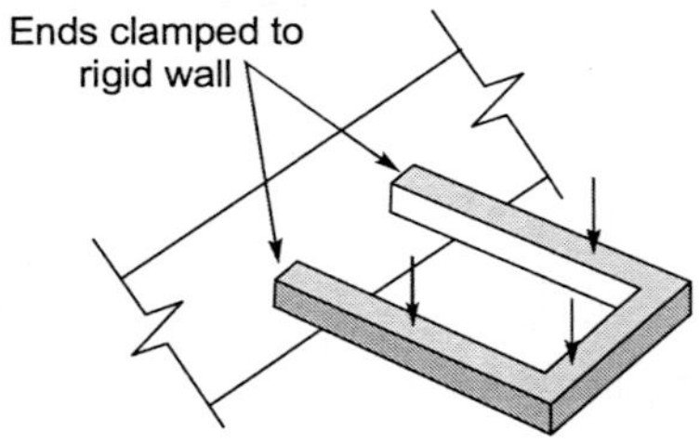

(a) 6 (b) 4
(c) 3 (d) 1 [2009 : 2 M]

1.8 The degree of static indeterminacy of a rigid jointed frame *PQR* supported as shown in figure is

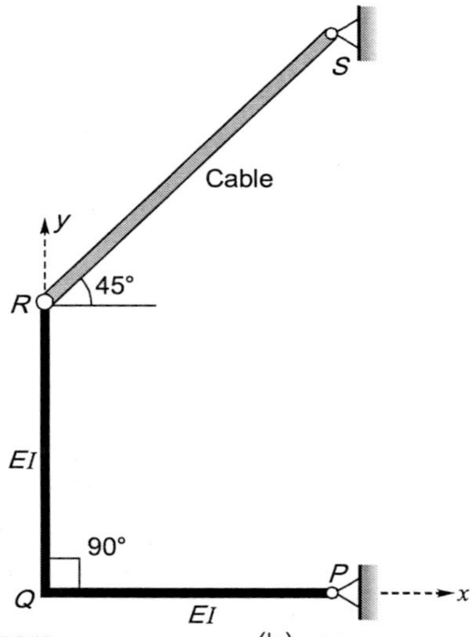

(a) zero (b) one
(c) two (d) unstable

[2014 : 1 M, Set-I]

1.9 The static indeterminacy of the two-span continuous beam with an internal hinge, shown below, is ______________.

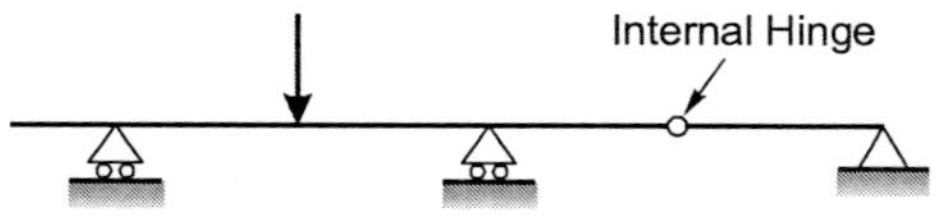

[2014 : 1 M, Set-II]

1.10 A guided support as shown in the figure below is represented by three springs (horizontal, vertical and rotational) with stiffness k_x, k_y and k_θ respectively. The limiting values of k_x, k_y and k_θ are

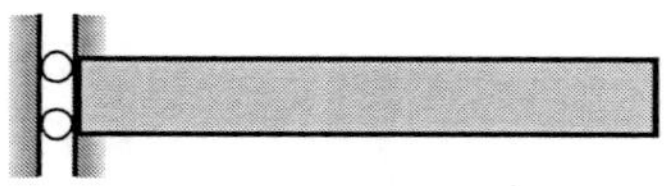

(a) ∞, 0, ∞ (b) ∞, ∞, ∞
(c) 0, ∞, ∞ (d) ∞, ∞, 0

[2015 : 1 M, Set-II]

1.11 The kinematic indeterminacy of the plane truss shown in the figure is

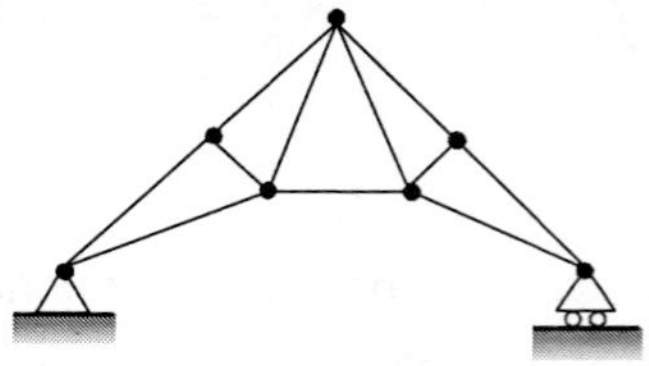

(a) 11 (b) 8
(c) 3 (d) 0 [2016 : 1 M, Set-II]

1.12 A planar truss tower structure is shown in the figure.

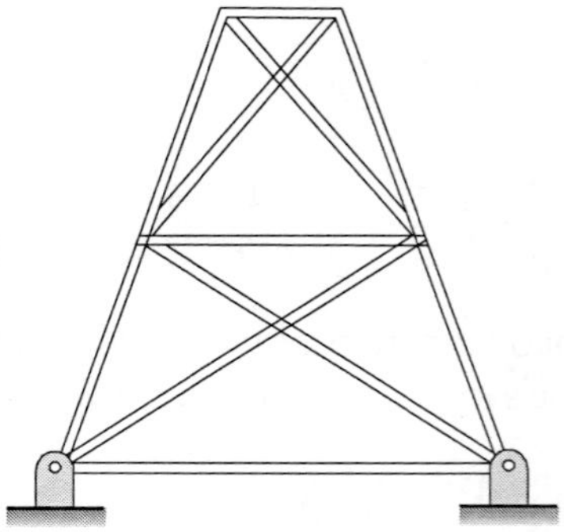

Consider the following statements about the external and internal determinacies of the truss.

P. Externally Determinate
Q. External Static Indeterminacy = 1
R. External Static Indeterminacy = 2
S. Internally Determinate
T. Internal Static Indeterminacy = 1
U. Internal Static Indeterminacy = 2

Which one of the following options is correct?

(a) P-False; Q-True; R-False; S-False; T-False; U-True
(b) P-False; Q-True; R-False; S-False; T-True; U-False
(c) P-False; Q-False; R-True; S-False; T-False; U-True
(d) P-True; Q-True; R-False; S-True; T-False; U-True

[2017 : 2 M, Set-I]

1.13 Consider the frame shown in the figure

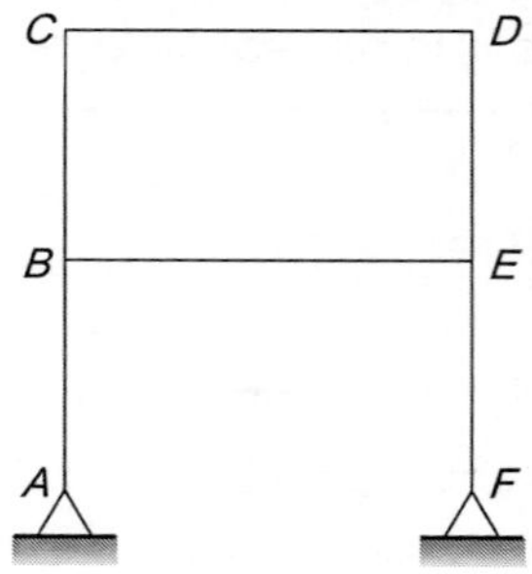

If the axial and shear deformations in different members of the frame are assumed to be negligible, the reduction in the degree of kinematic indeterminacy would be equal to

(a) 5 (b) 6
(c) 7 (d) 8

[2017 : 1 M, Set-II]

1.14 The degree of static indeterminacy of the plane frame as shown in the figure is_______

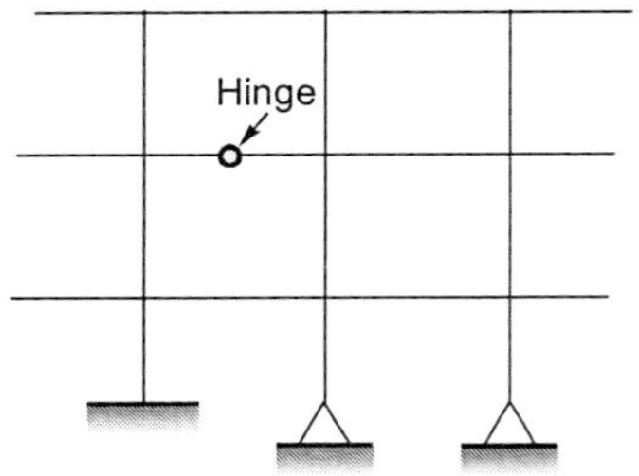

[2019 : 1 M, Set-II]

1.15 Consider a beam PQ fixed at P, hinged at Q, and subjected to a load F as shown in figure (not drawn to scale). The static and kinematic degrees of indeterminacy, respectively, are

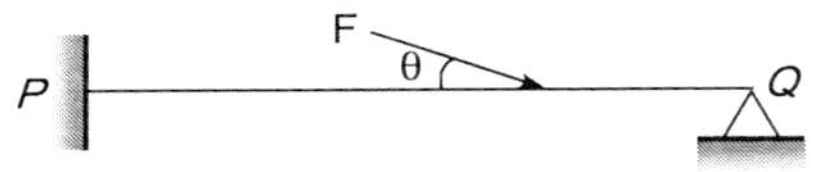

(a) 2 and 0 (b) 2 and 2
(c) 1 and 2 (d) 2 and 1

[2022 : 2 M, Set-II]

1.16 Consider the following three structures:

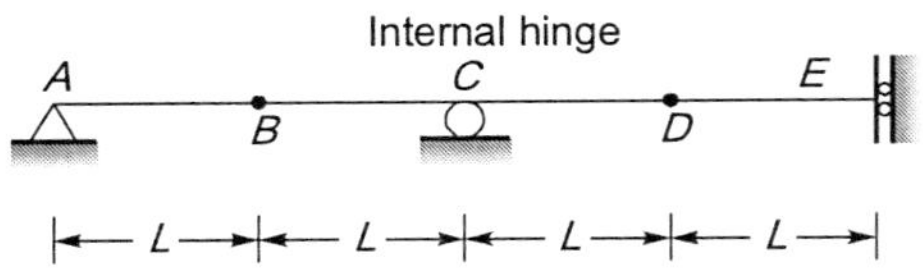

I

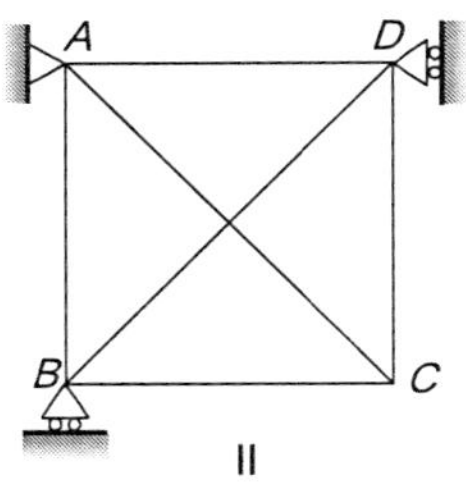

II

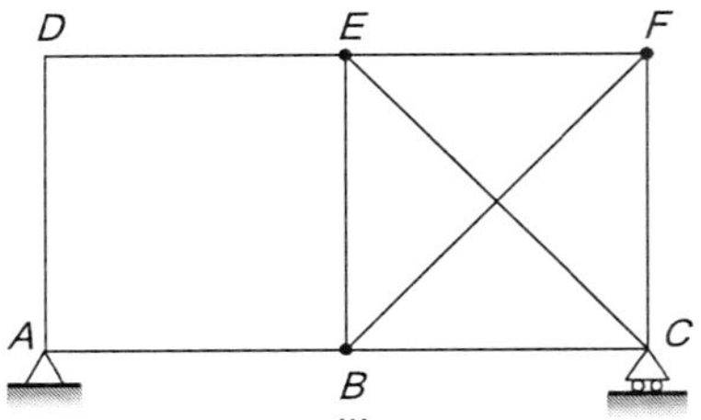

III

Structure I: Beam with hinge support at A, roller at C, guided roller at E, and internal hinges at B and D

Structure II: Pin-jointed truss, with hinge support at A, and rollers at B and D

Structure III: Pin-jointed truss, with hinge support at A and roller at C

Which of the following statements is/are TRUE?

(a) Structure I is unstable
(b) Structure II is unstable
(c) Structure III is unstable
(d) All three structures are stable

[2023 : 1 M, Set-I]

1.17 The plane frame shown in the figure has fixed support at joint A, hinge support at joint F, and roller support at joint I. In the figure, A to I indicate joints of the frame.

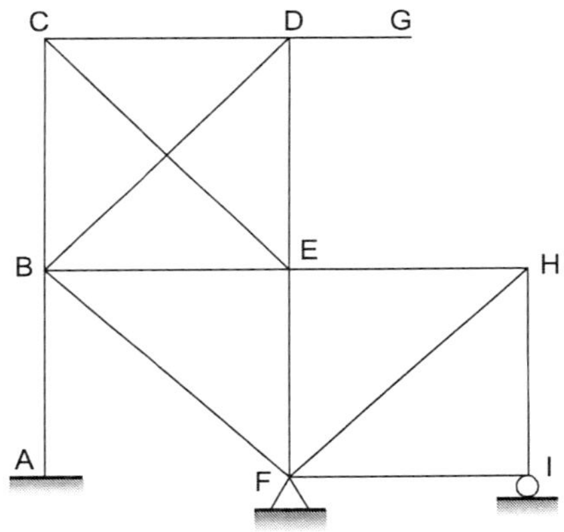

If the axial deformations are neglected, the degree of kinematic indeterminacy is _________(integer).

[2024 : 1 M, Set-I]

Answers Determinacy and Indeterminacy

1.1 (b)	1.2 (d)	1.3 (c)	1.4 (d)	1.5 (b)	1.6 (d)	1.7 (c)
1.8 (a)	1.9 (0)	1.10 (a)	1.11 (a)	1.12 (a)	1.13 (b)	1.14 (15)
1.15 (d)	1.16 (a, b, c)	1.17 (9)				

Explanations Determinacy and Indeterminacy

1.1 (b)

Degree of **kinematic indeterminacy**, D_k is given by,

$$D_k = 3j - r_e - m''$$

Where, j = Number of joints = 4

r_e = No. of reactions = 6

m'' = No. of rigid members = 3

$\therefore$ $D_k = 3 \times 4 - 6 - 3 = 3$

Note :

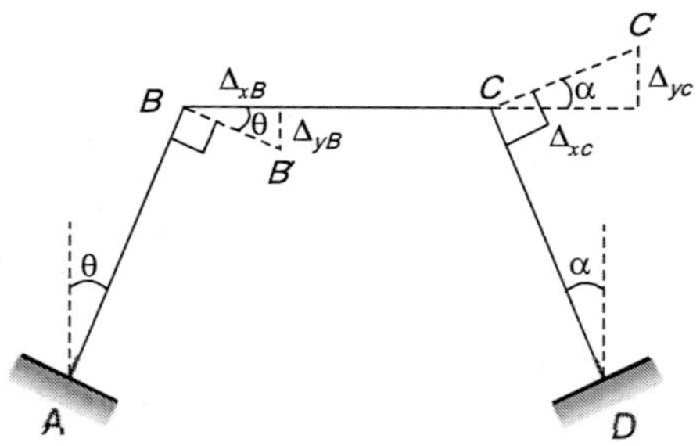

KI = 3 (Δ_{xB}, θ_B, θ_C)

$\Delta_{yB} = \Delta_{xB} \tan\theta$

$\Delta_{xC} = \Delta_{xB}$

$\Delta_{yC} = \Delta_{xC} \tan\alpha = \Delta_{xB} \tan\alpha$

It means, Δ_{yB}, Δ_{xC} and Δ_{yC} are dependent on Δ_{xB}.

1.2 (d)

Degree of **static indeterminacy,**

$$D_s = m + r_e - 2j$$

Here, $m = 12$, $j = 9$ and $r_e = 6$

$\therefore$ $D_s = 12 + 6 - 2 \times 9 = 0$

Degree of **kinematic indeterminacy,**

$$D_k = 2j - r_e - m''$$

Where m'' is no. of axially rigid members

$= 2 \times 9 - 6 - 0 = 12$

$\therefore$ Thus, truss is **statically determinate** and **kinematically indeterminate**.

1.3 (c)

Degree of **static indeterminacy,**

$D_s = 3m + r_e - 3j - r_r$

$= 3 \times 5 + (3 + 2 + 2) - 3 \times 6 - 0 = 4$

Degree of **kinematic indeterminacy,**

$D_k = 3j - r_e - m''$

Where m'' is no. of axially rigid members

$= 3 \times 6 - (3 + 2 + 2) - 5 = 6$

$D_s = 4$ and $D_k = 6$

1.4 (d)

Method-I (D_s by Formula)

$D_s = 3m + r_e - 3j$

m = total members = 11

r_e = total external reactions

$= 3 + 2 + 1 = 6$

j = total number of joints = 10

$\therefore$ $D_s = (3 \times 11) + 6 - (3 \times 10) = 9$

Method-II (Loop method)

For plane frame,

$D_{si} = 3 \times$ No. of closed loops

$D_{se} = R - 3$

$D_{si} = 3 \times 2 = 6$

$D_{se} = 6 - 3 = 3$

$\therefore$ $D_s = 6 + 3 = 9$

Kinematic indeterminacy, $D_k = (3j - r_e) - m''$

where m'' = total number of axially rigid members.

$\Rightarrow$ $k = (3 \times 10 - 6) - 11 = 13$

1.5 (b)

Degree of **kinematic indeterminacy** or **total degree of freedom of joints.**

$D_k = 3j - r_e - m''$

$= 3 \times 4 - 3 - 1$

$= 8$

Note : Only beam is rigid, (not the columns) ($\therefore$ $m' = 1$)

1.6 (d)

Method-I (By Formula)

The degree of **static indeterminacy** for a rigid hybrid frame is given by ,

$D_s = 3m + r_e - r_r - 3(j + j')$

Where, m = total number of members = 9

r_e = total number of external reactions

$= 2 + 1 + 1 = 4$

r_r = total number of released reactions at hybrid joint

$= \Sigma(m_j - 1) = (2 - 1) + (2 - 1) = 2$

j = total number of rigid joints = 6

j' = total number of hybrid joints = 2

$\therefore \quad D_S = (3 \times 9) + 4 - 2 - 3\,(6 + 2)$

$= 27 + 4 - 2 - 24 = 31 - 26 = 5$

Method-II (By Loop Method)

$D_{si} = 3C - r_r$

where C = no. of closed loops

$= 3 \times 2 - 2 = 4$

$D_{se} = r_e - 3 = 1$

$D_s = D_{si} + D_{se} = 4 + 1 = 5$

1.7 (c)

In general, total number of equilibrium equations for a 3-D rigidly jointed frame are 6. i.e.,

$\Sigma F_x = 0,\ \Sigma F_y = 0,\ \Sigma F_z = 0,\ \Sigma M_x = 0,\ \Sigma M_y = 0,\ \Sigma M_z = 0$

But for vertical loading, these equilibrium equations reduces to 3.

i.e. $\Sigma F_y = 0,\ \Sigma M_x = 0$ and $\Sigma M_z = 0$

Also, at fixed support in a 3-D rigidly jointed frame total number of external reactions are 6 i.e.,

$$R_x, R_y, R_z, M_x, M_y, M_z$$

But for vertical loading, these external reactions reduce to 3 i.e.,

$$R_y, M_x, M_z$$

Thus for two fixed supports total external reactions will be 6

Note : If a horizontal grid frame is subjected to vertical loading, then there will be only 3 member forces/support reactions.

$D_s = 6 - 3 = 3$

1.8 (a)

$D_s = D_{s_e} + D_{s_i}$

$= (r_e - 3) + 3C - r_r$

$= (4 - 3) + 3 \times 0 - 1 = 0$

Hence, the structure is stable.

Note : Stable for vertical loading unstable for horizontal loading.

1.9 Sol.

Number of member, $m = 4$

Number of external reaction, $r_e = 4$

Number of joint, $j = 5$

Number of reaction released, $r_r = 1$

Degree of static indeterminacy,

$D_s = 3\,m + r_e - 3j - r_r$

$= 3 \times 4 + 4 - 3 \times 5 - 1 = 0$

1.10 (a)

Guided roller support has 2 reactions:

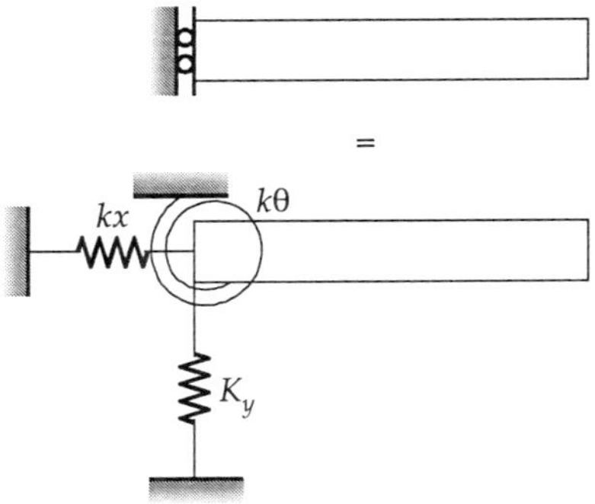

For given support, $k_x = \infty,\ k_y = 0,\ k_\theta = \infty$

1.11 (a)

Kinematic indeterminacy,

$D_k = 2j - r_e = 2 \times 7 - 3 = 11$

1.12 (a)

$D_{se} = r_e - 3 - s = 4 - 3 = 1$

$D_{si} = m - (2j - 3) = 15 - (2 \times 8 - 3)$

$= 2$

Trick : For a truss formed due to combination of simple triangles.

D_{si} = No. of double diagonal panels = 2

$D_{se} = R - 3 = 4 - 3 = 1$

1.13 (b)

DOF of Joints:

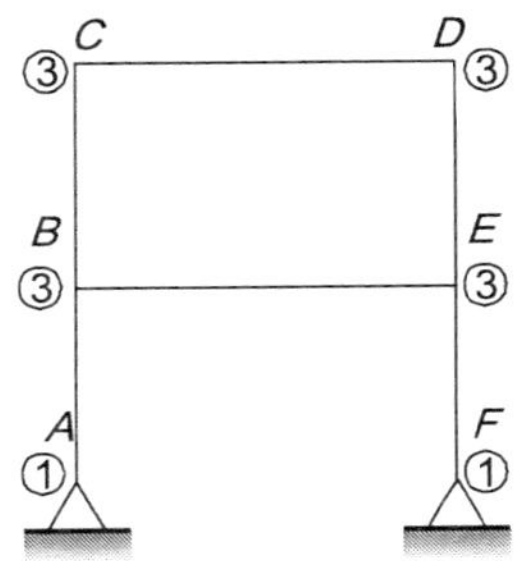

When all members are extensible,

$D_k = 3j - r_e - m'' = 3 \times 6 - 4 - 0 = 14$

D_k (when **extensible**) = 14

D_k(when **inextensible**) = D_k(when **extensible**) – No. of axially rigid member

$= 14 - 6 = 8$

So, reduction in D_k is 6 ($\theta_A, \theta_B, \theta_C, \theta_D, \theta_E, \theta_F$)

Note : Shear deformation is not considered in calculation of D_K.

1.14 Sol.

$$D_{se} = 7 - 3 = 4$$
$$D_{si} = (3 \times 4) - (2 - 1) = 11$$
$$D_s = 15$$

1.15 (d)

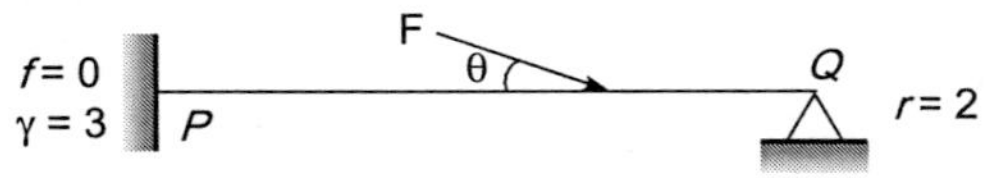

Total static,

$$D_s = D_{sc} + D_{si}$$
$$= (r_e - 3) + (3c - r_r) = (5 - 3) + 0 = 2$$

or
$$D_s = 3m + r_e - 3j$$
$$= 3 \times 1 + 5 - 3 \times 2 = 2$$

Kinematic indeterminacy is given by,

$$D_k = 3j - r_e - m$$
$$= (3 \times 2) - 5 - 0 = 1$$

1.16 (a, b, c)

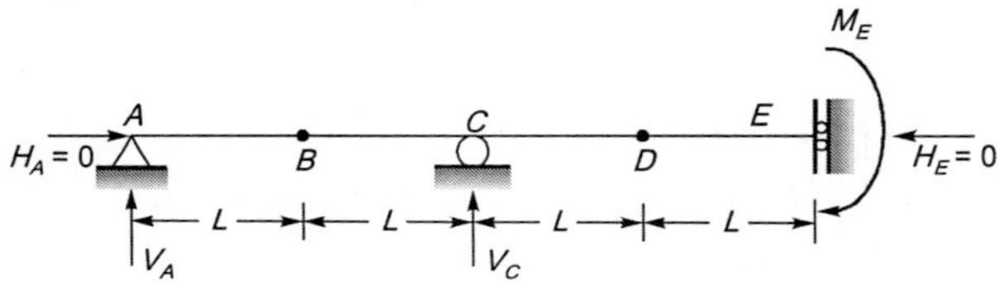

$r = 3$ (V_A, V_C, M_E)

$s = 4$ ($\Sigma F_y = 0$, $\Sigma M = 0$, $BM_B = 0$, $BM_D = 0$)

$D_s = r - s = 3 - 4 = -1$

⇒ Negative sign implies unstable

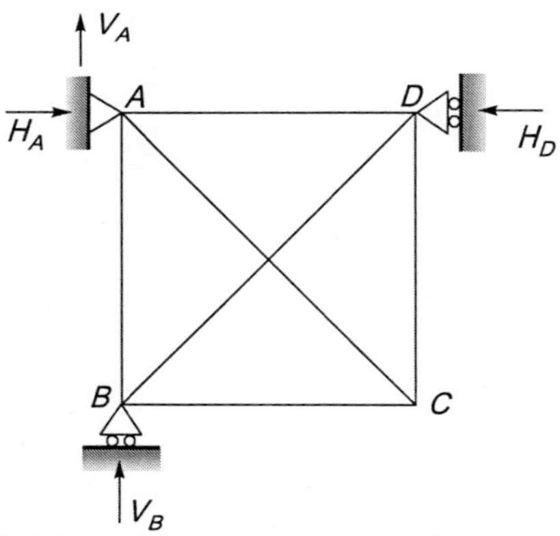

Rigid body rotation is possible since all the four reactions are concurrent at *A*. So the structure is unstable.

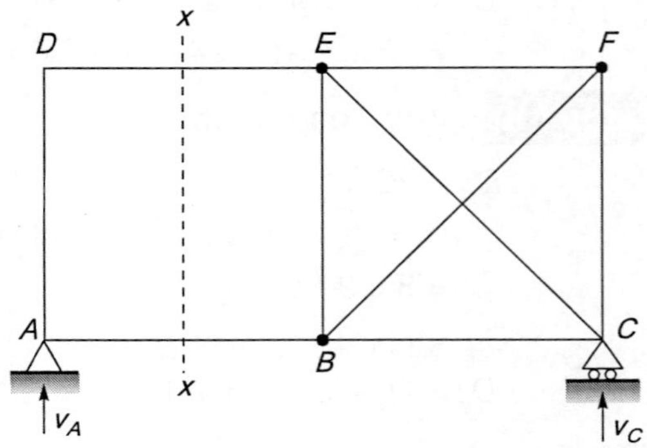

Shear force at x-x = Panel shear force = V_A

There is no member to resist panel shear force. So the structure is unstable.

1.17 Sol.

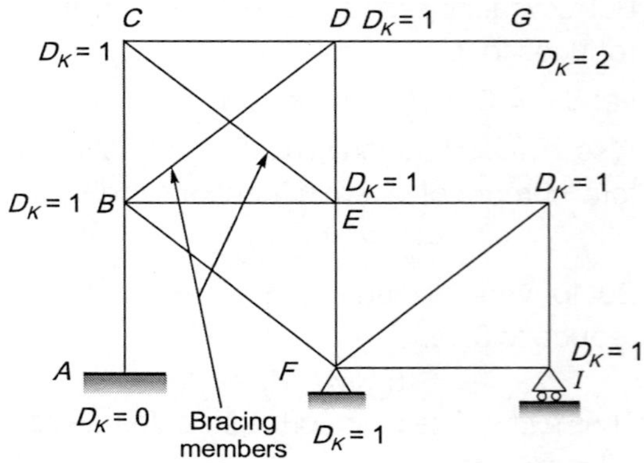

$$\Rightarrow \qquad D_K = 9$$

NOTE: A rigid jointed structure is kept on hinge support at F and on a roller support at I.

If axial deformations are neglected and bracing is done, only rotations are possible at the joints.

2 Influence Line Diagram and Rolling Loads

2.1 A simply supported beam with an overhang is traversed by a unit concentrated moment from the left to the right as shown below

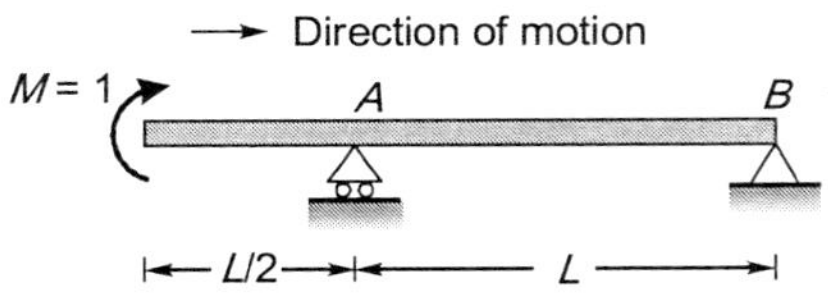

The influence line for reaction at B is given by

(a) 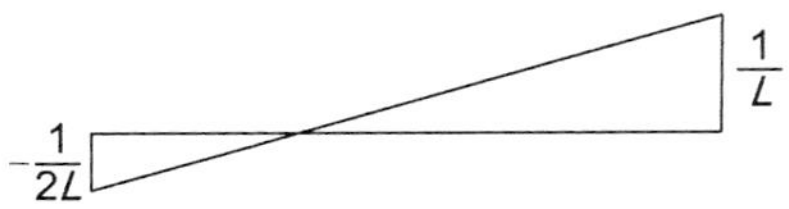

(b) 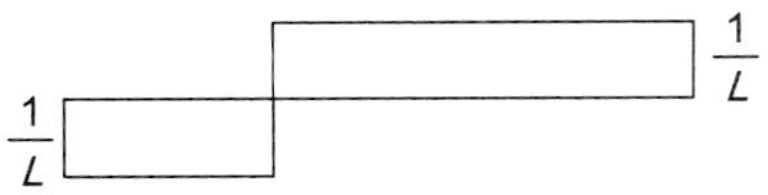

(c) 

(d) Zero everywhere [2000 : 1 M]

2.2 Identify from the following, the correct values of the bending moment M_A (in kNm unit) at the fixed end A in the statically determinate beam shown below (with internal hinges at B and D), when a uniformly distributed load of 10 kN/m is placed on all spans.

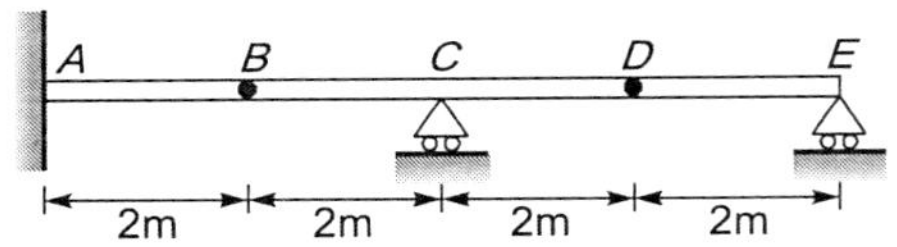

(a) –80 (b) –40
(c) 0 (d) +40 [2001 : 2 M]

2.3 Muller Breslau's principle in structural analysis is used for

(a) drawing influence line diagram for any force function
(b) writing virtual work equation
(c) superposition of load effects
(d) None of these [2003 : 1 M]

Common Data for (2.4, 2.5 and 2.6)

A beam $PQRS$ is 18 m long and is simply supported at points Q and R 10 m apart. Overhangs PQ and RS are 3 m and 5 m respectively. A train of two point loads of 150 kN and 100 kN, 5 m apart, crosses this beam from left to right with 100 kN load leading.

2.4 The maximum sagging moment under the 150 kN load anywhere is

(a) 500 kN-m (b) 450 kN-m
(c) 400 kN-m (d) 375 kN-m [2003 : 2 M]

2.5 During the passage of the loads, the maximum and the minimum reactions at support R, in kN, are respectively

(a) 300 and –30 (b) 300 and –25
(c) 225 and –30 (d) 225 and –25 [2003 : 2 M]

2.6 The maximum hogging moment in the beam anywhere is

(a) 300 kN-m (b) 450 kN-m
(c) 500 kN-m (d) 750 kN-m [2003 : 2 M]

2.7 Consider the beam $ABCD$ and the influence line as shown below. The influence line pertains to

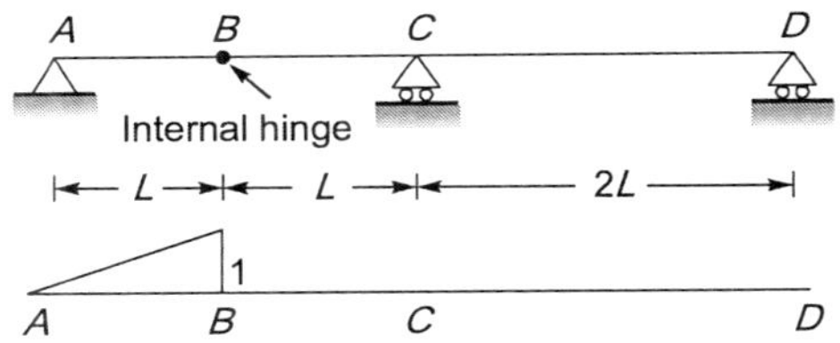

(a) reaction at A, R_A
(b) shear force at B, V_B
(c) shear force on the left of C, V_C^-
(d) shear force on the right of C, V_C^+ [2006 : 2 M]

2.8 The influence line diagram (ILD) shown is for the member

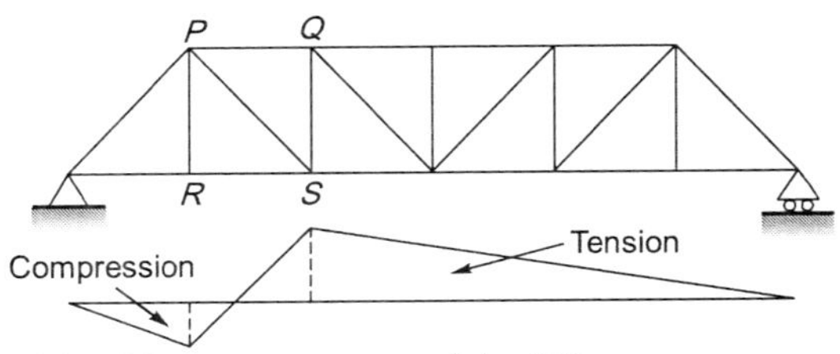

(a) PS (b) RS
(c) PQ (d) QS [2007 : 2 M]

2.9 The span(s) to be loaded uniformly for maximum positive (upward) reaction at support P, as shown in the figure below, is(are)

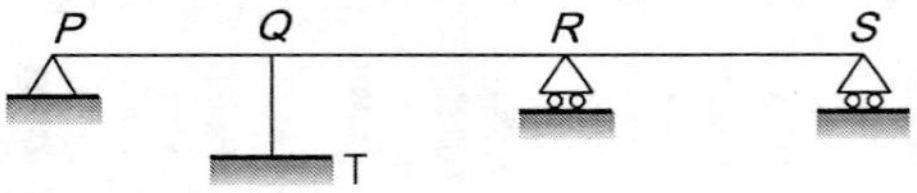

(a) *PQ* only
(b) *PQ* and *QR*
(c) *QR* and *RS*
(d) *PQ* and *RS*

[2008 : 2 M]

2.10 Beam *PQRS* has internal hinges in spans in spans *PQ* and *RS* as shown. The beam may be subjected to a moving distributed vertical load of maximum intensity 4 kN/m of any length anywhere on the beam. The maximum absolute value of the shear force (in kN) that can occur due to this loading just to the right of support *Q* shall be

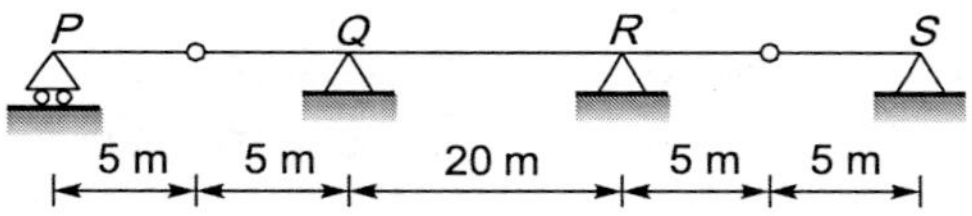

(a) 30 kN
(b) 40 kN
(c) 45 kN
(d) 55 kN [2013 : 2 M]

2.11 In a beam of length *L*, four possible influence line diagrams for shear force at a section located at a distance of *L*/4 from the left end support (marked as *P*, *Q*, *R* and *S*) are shown below. The correct influence line diagram is

P.

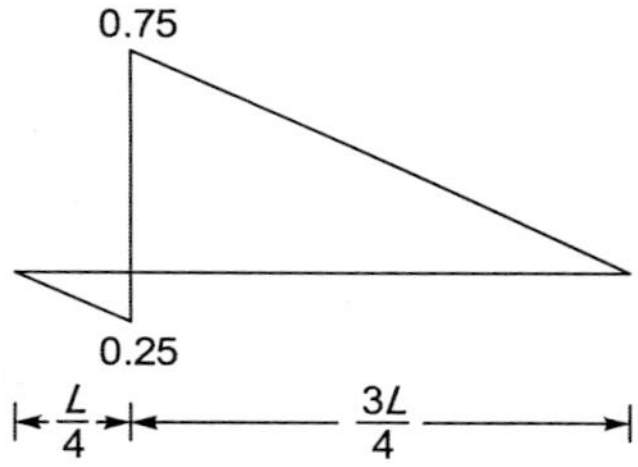

Q.

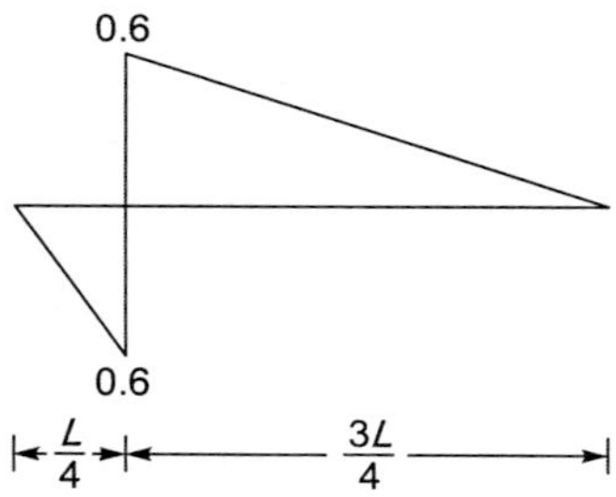

R.

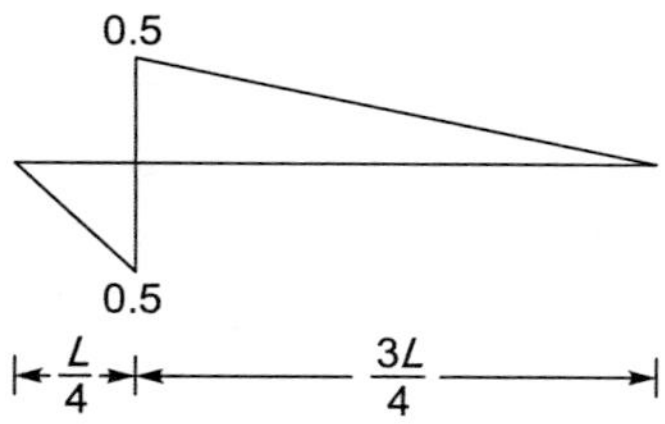

S.

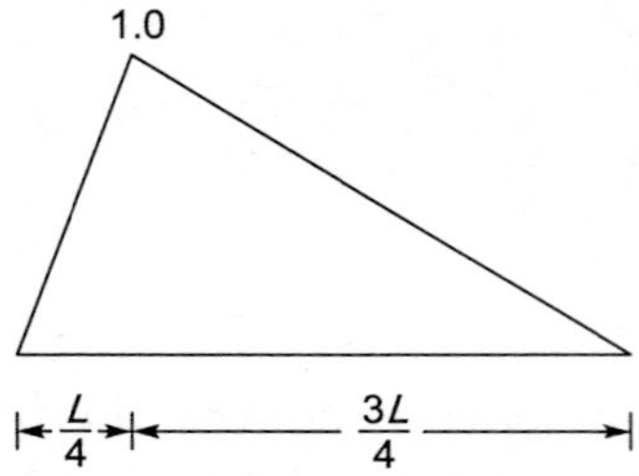

(a) *P*
(b) *Q*
(c) *R*
(d) *S* [2014 : 1 M, Set-I]

2.12 A simply supported beam *AB* of span, $L = 24$ m is subjected to two wheel loads acting at a distance, $d = 5$ m apart as shown in the figure below. Each wheel transmits a load, $P = 3$ kN and may occupy any position along the beam. If the beam is an I-section having section modulus, $S = 16.2$ cm^3, the maximum bending stress (in GPa) due to the wheel loads is ________.

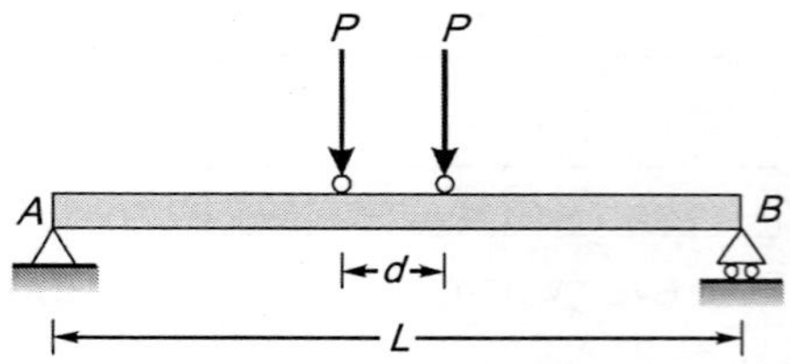

[2015 : 2 M, Set-II]

2.13 Consider the beam *ABCD* shown in the figure

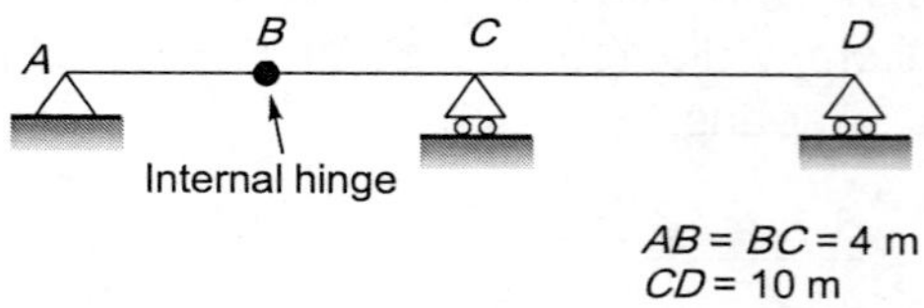

For a moving concentrated load of 50 kN on the beam, the magnitude of the maximum bending moment (in kN-m) obtained at the support *C* will be equal to ________.

[2017 : 2 M, Set-I]

2.14 A long uniformly distributed load of 10 kN/m and a concentrated load of 60 kN are moving together on the beam ABCD shows in the figure (not drawn to scale). The relative positions of the two loads are not fixed. The maximum shear force (in kN, round off to the nearest integer) caused at the internal hinge B due to the two loads is______

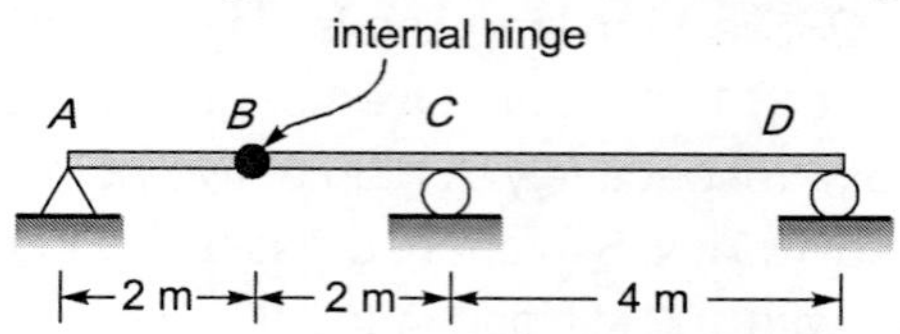

[2019 : 2 M, Set-II]

2.15 Distributed load(s) of 50 kN/m may occupy any position(s) (either continuously or in patches) on the girder PQRST as shown in the figure (not drawn to the scale)

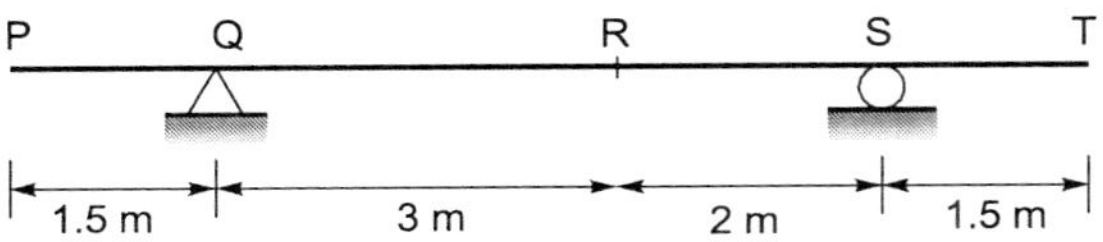

The maximum negative (hogging) bending moment (in kNm) that occurs at point R is

(a) 56.25 (b) 22.50
(c) 150.00 (d) 93.75

[2020 : 2 M, Set-I]

2.16 Refer the truss as shown in the figure (not to scale)

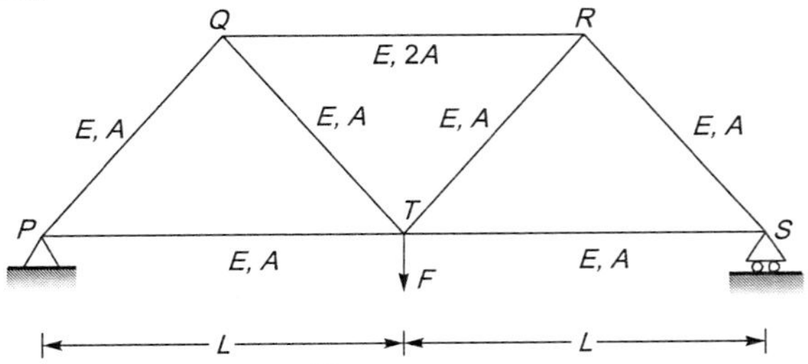

If load, $F = 10\sqrt{3}$ kN, moment of inertia, $I = 8.33 \times 10^6$ mm^4, area of cross section, $A = 10^4$ mm^2 and length, $L = 2$ m for all the members of the truss, the compressive stress (in kN/m^2), in integer) carried by the member Q-R is ______

[2021 : 2 M, Set-I]

2.17 A propped cantilever beam EF is subjected to a unit moving load as shown in the figure (not to scale). The sign convention for positive shear force at the left and right sides of any section is also shown.

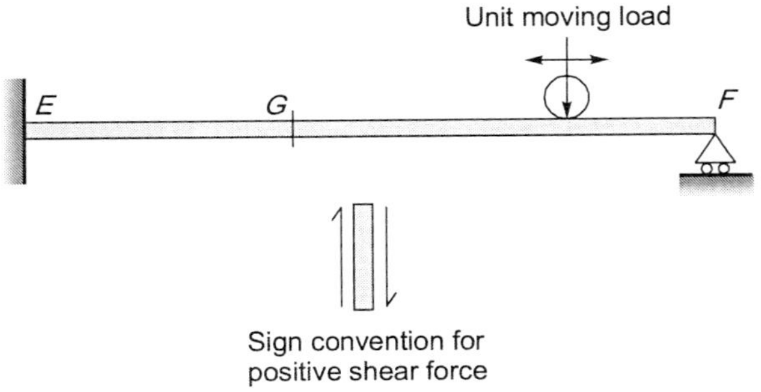

The correct qualitative nature of the influence line diagram for shear force at G is

(a)

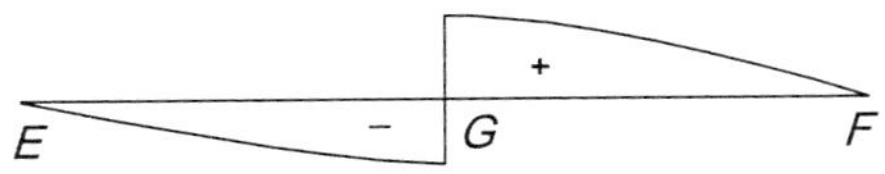

(b)

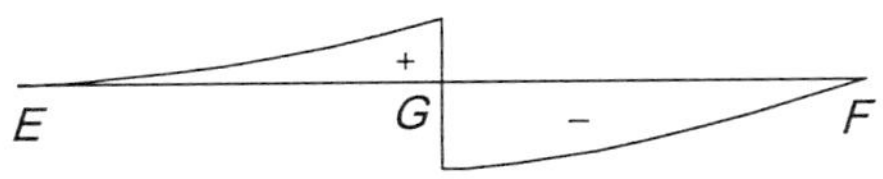

(c)

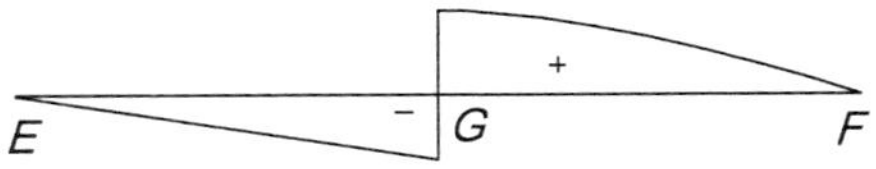

(d)

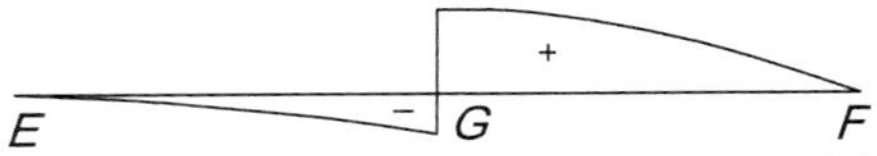

[2021 : 1 M, Set-I]

2.18 A perfectly flexible and inextensible cable is shown in the figure (not to scale), The external loads at F and G are acting vertically.

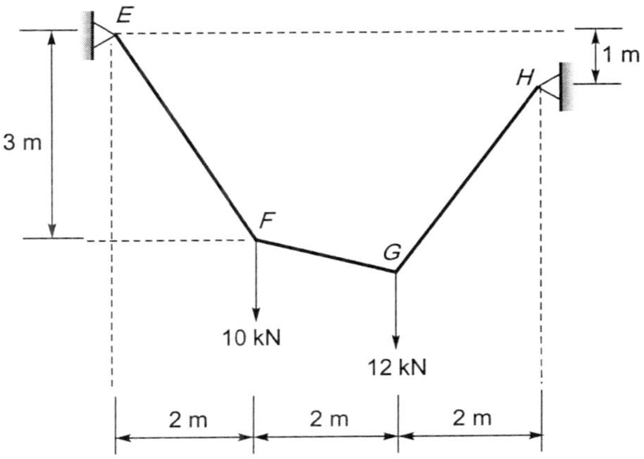

The magnitude of tension in the cable segment FG (in kN, round off to two decimal places) is ______.

[2021 : 2 M, Set-II]

2.19 Consider a simply supported beam *PQ* as shown in the figure. A truck having 100 kN on the front axle and 200 kN on the rear axle, moves from left to right. The spacing between the axles is 3 m. The maximum bending moment at point *R* is ______ kNm. (in integer)

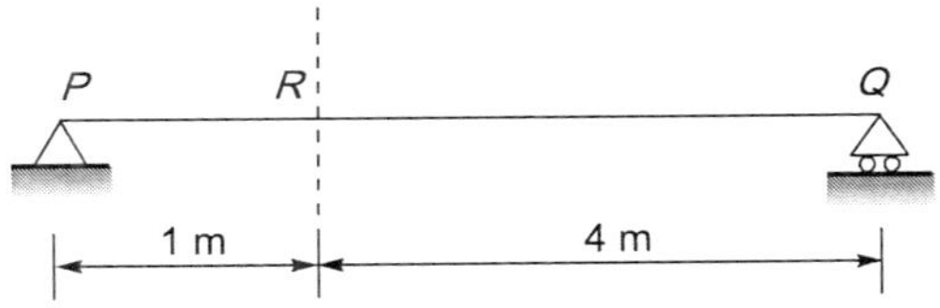

[2022 : 2 M, Set-I]

2.20 Muller-Breslau principle is used in analysis of structures for ______ .

(a) drawing an influence line diagram for any force response in the structure
(b) writing the virtual work expression to get the equilibrium equation
(c) superposing the load effects to get the total force response in the structure
(d) relating the deflection between two points in a member with the curvature diagram in-between

[2023 : 1 M, Set-II]

2.21 The horizontal beam $PQRS$ shown in the figure has a fixed support at point P, an internal hinge at point Q, and a pin support at point R. A concentrated vertically downward load (V) of 10 kN can act at any point over the entire length of the beam.

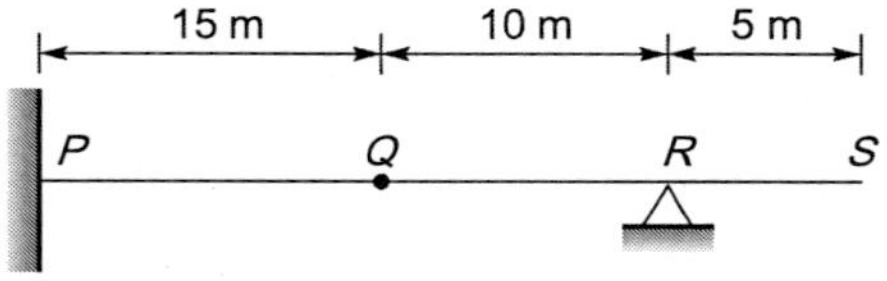

(Figure NOT to scale)

The maximum magnitude of the moment reaction (in kN.m) that can act at the support P due to V is ________ (in integer).

[2024 : 2 M, Set-II]

Answers Influence Line Diagram and Rolling Loads

2.1 (c)	2.2 (c)	2.3 (a)	2.4 (c)	2.5 (a)	2.6 (d)	2.7 (b)
2.8 (a)	2.9 (d)	2.10 (c)	2.11 (a)	2.12 (1.78)	2.13 (200)	2.14 (70)
2.15 (a)	2.16 (500)	2.17 (d)	2.18 (8.24)	2.19 (180)	2.20 (a)	2.21 (150)

Explanations Influence Line Diagram and Rolling Loads

2.1 (c)

Taking $\Sigma M_A = 0$

$\Rightarrow \quad 1 - V_B \times L = 0$

$\Rightarrow \quad V_B = 1/L$

So, ILD for vertical reaction at B is independent of x.

POINTS TO REMEMBER

(i)

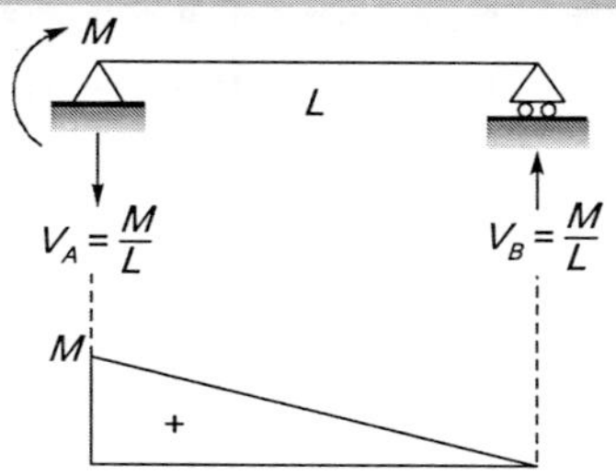

(ii)

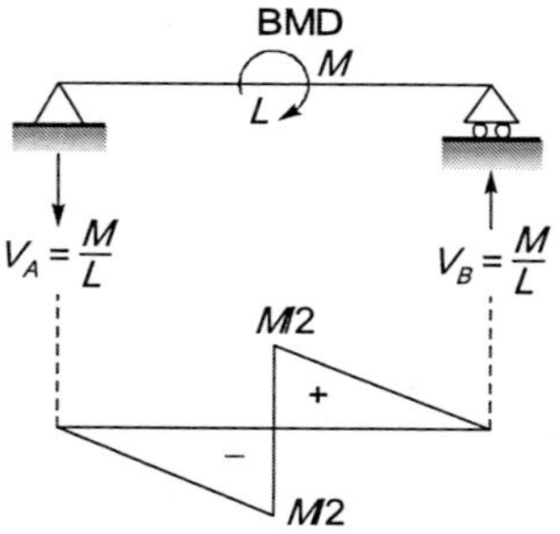

- The external effect of a couple remain same when it is replaced anywhere in its plane (i.e., support reactions donot change) but internal effect will be different (i.e. BMD changes).

2.2 (c)

Method-I (By ILD)

ILD for BM at A by **using Muller Breslau principle** will be,

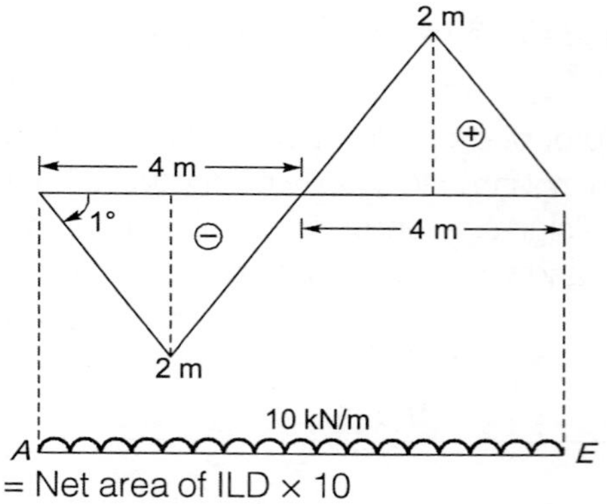

M_A = Net area of ILD × 10

$$M_A = \left[-\left(\frac{1}{2}\times 2\times 4\right)+\left(\frac{1}{2}\times 2\times 4\right)\right]\times 10 = 0 \text{ kNm}$$

Method-II (By Statics)

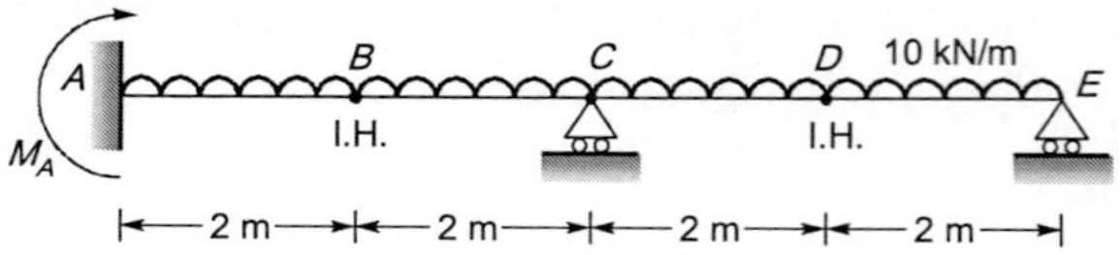

$\Sigma F_V = 0,$

$R_A + R_C + R_E = 10 \times 8 = 80$ kN

$BM_D = 0,$

$2R_E - 10 \times 2 \times 1 = 0$

$R_E = 10$ kN ↑

$BM_B = 0,$

$6R_E - 10 \times 6 \times 3 + 2R_C = 0$

$R_C = 90 - 3R_E = 60$ kN↑

$R_A = 80 - R_C - R_E = 10$ kN↑

$BM_B = 0,$

$M_A - 10 \times 2 \times 1 + 10 \times 2 = 0$

$M_A = 0$ kNm

2.3 (a)

Muller Breslau's principle is a principle on the basis of which it is possible to draw influence lines for various quantities pertaining to a structure. It states that:

"The ordinates of the influence lines for any stress element (such as axial stress, moment or reaction) of any structure are proportional to those of deflection curve, which is obtained by removing the restraint corresponding to that element from the structure."

2.4 (c)

Maximum sagging bending moment under 150 kN load occurs when 150 kN load and resultant of 150 kN and 100 kN are located equidistant from the centre of span. In other words, when the centre of beam lies midway between the resultant load and the load under consideration, maximum sagging moment occurs under the load under consideration.

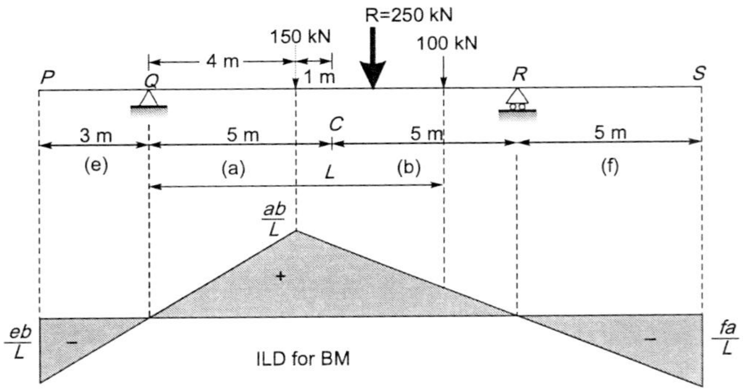

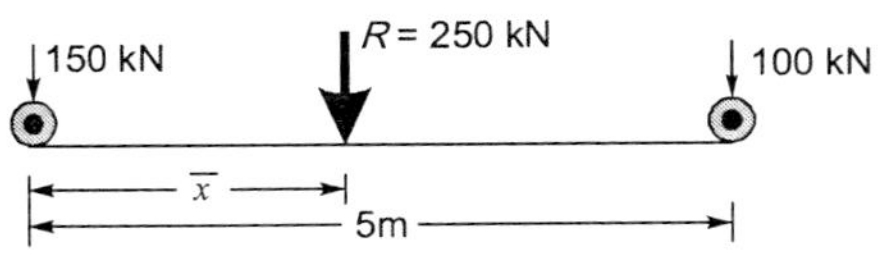

$$(150 + 100)\bar{x} = 150 \times 0 + 100 \times 5$$

$$\bar{x} = 2 \text{ m}$$

Ordinate under 150 kN load

$$\therefore \quad \frac{ab}{L} = \frac{4 \times 6}{10} = 2.4$$

Ordinate under 100 kN load $= \frac{2.4}{6} \times 1 = 0.4$

∴ Sagging BM under 150 kN load

$= 2.4 \times 150 + 0.4 \times 100 = 400$ kN-m

POINTS TO REMEMBER

- Understand the difference between maximum BM at a particular point and maximum BM under a particular load.
- For the first case, you've to draw the ILD of BM at that point and then check for the arrangement, which produces maximum BM at that point.
- For the second case, you've to first assign the location of loads as per criteria explained above in the solution, and then draw ILD of BM at the point at which considered load is located.

2.5 (a)

ILD for reaction at R is given as,

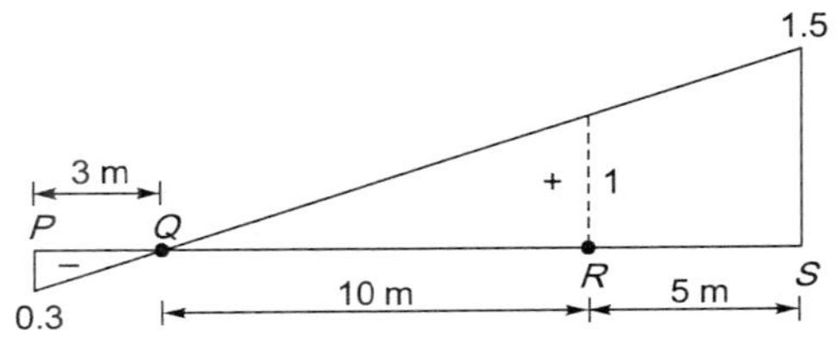

Maximum reaction at R will occur when 100 kN load is at S.

$$V_{R\,(max)} = 100 \times 1.5 + 150 \times 1$$
$$= 150 + 150 = 300 \text{ kN}$$

Minimum reaction at

Case-I: 100 kN is at P

$V_{R\,(min)} = -100 \times 0.3 = -30$ kN ...(i)

Case-II: 150 kN is at P and 100 kN is at 2 m from Q.

$V_{R\,(min)} = -150 \times 0.3 + 100 \times 0.2$

$= -25$ kN ...(ii)

From (i) and (ii),

$V_{R\,(min)} = -30$ kN (Mathematical Minimum)

2.6 (d)

Maximum hogging BM will be at support R because overhang is maximum on that side. ILD for hogging BM at R is,

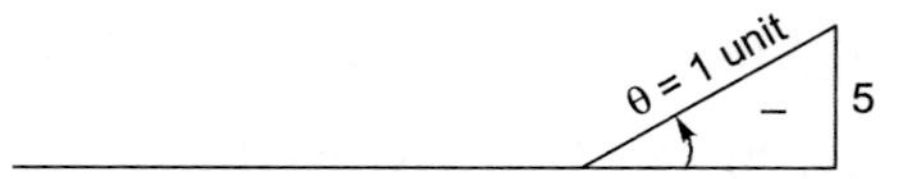

Maximum hogging moment in the beam occurs when 150 kN load is at S.

$$\text{BM at } S = -150 \times 5 = -750 \text{ kNm}$$

2.7 (b)

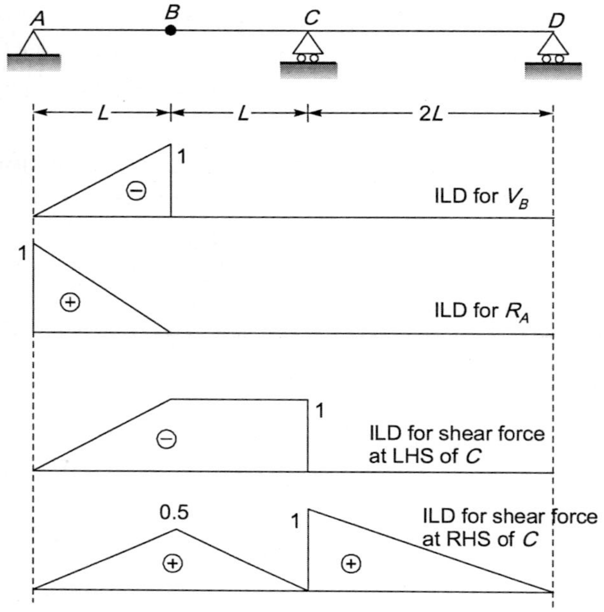

2.8 (a)

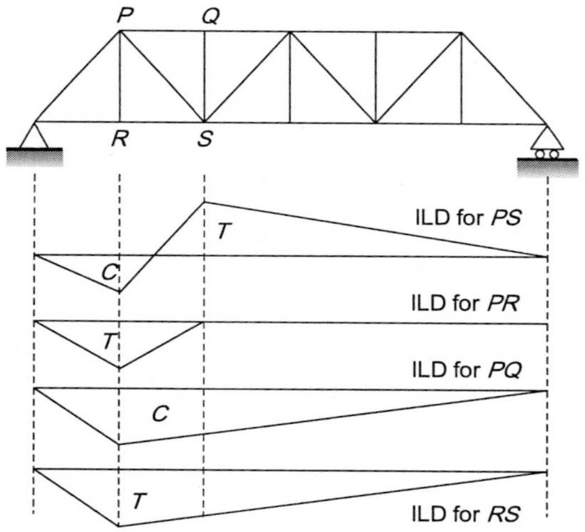

Note: Diagonal member subjected to reversal of stresses.

2.9 (d)

With the help of **Muller Breslau principle**, we can draw the ILD for reaction at P.

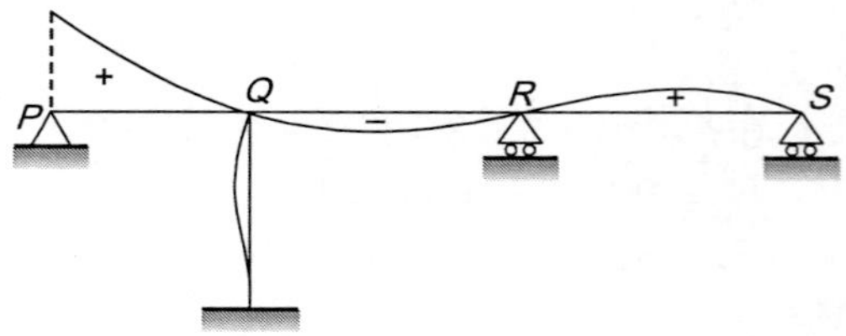

ILD for vertical reaction at P

The reaction is positive in spans PQ and RS. Hence, the spans PQ and RS should be loaded uniformly for maximum positive reaction at P.

2.10 (c)

Drawing the ILD of shear force just to right of Q by using **Muller Breslau's principle**. A cut is made just to the right of Q, since cut is very close to support Q, therefore displacement of left portion is almost zero and that to the right portion will be 1.

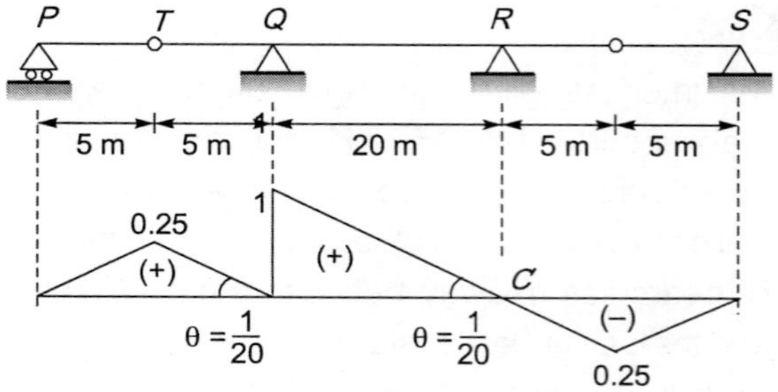

ILD for SF the right of Q

If moving distributed load is present over span PR then we get maximum shear force just to the right of Q.

$$\Rightarrow \quad \text{SF} = \left[\left(\frac{1}{2} \times 0.25 \times 10\right) + \left(\frac{1}{2} \times 1 \times 20\right)\right] \times 4$$

$$\text{SF} = 45 \text{ kN}$$

2.11 (a)

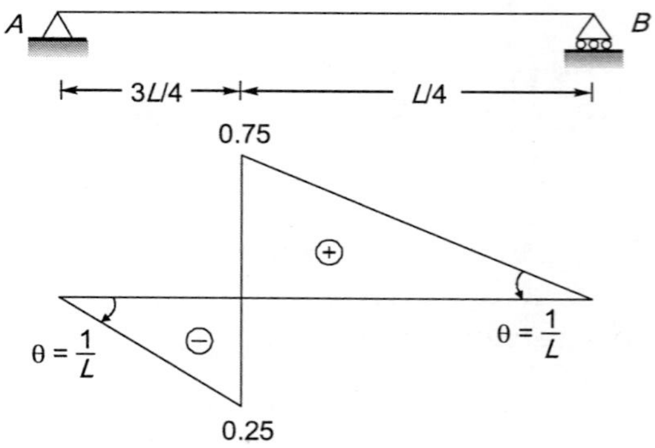

ILD for SF at $L/4$ from left support

2.12 Sol.

Method-I (By ILD)

C.G. of system = 2.5 m from any load

For maximum bending moment, system of load should be,

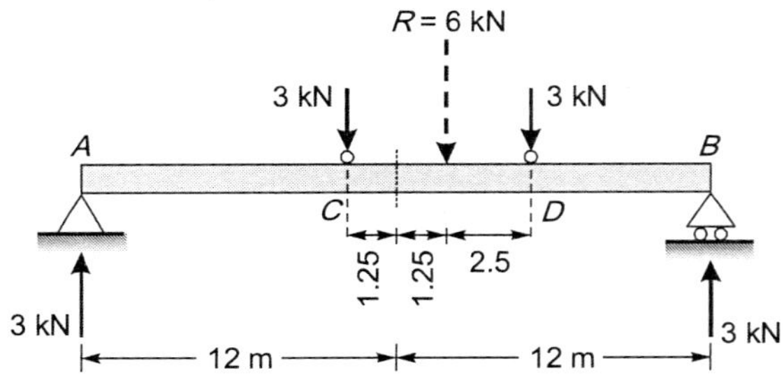

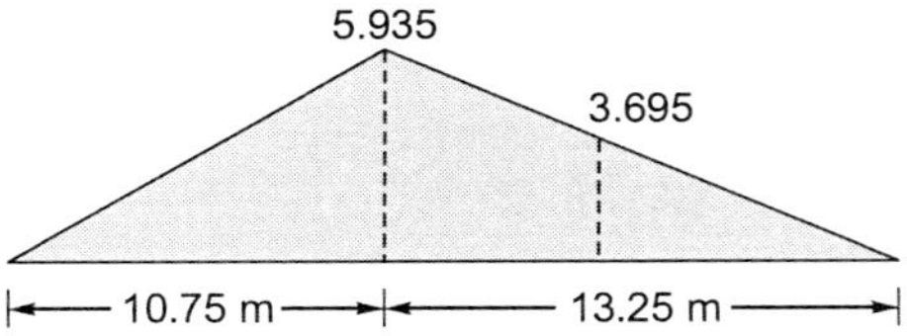

Maximum BM will occur below load at C.

Ordinate of ILD at C

$$= \frac{(12-1.25)\times(12+1.25)}{24}$$

$$= 5.935$$

Ordinate of ILD at D

$$= \frac{5.935\times 8.25}{13.25} = 3.695$$

Maximum BM = 5.935 × 3 + 3.695 × 3

= 28.89 kN-m

Maximum bending stress,

$$\frac{M}{Z} = \frac{28.89\times 10^3}{16.2\times 10^{-6}\times 10^9}$$

$$= 1.783 \text{ GPa}$$

Method-II (By Statics)

Let the section at which absolute max. BM ocurs be at x distance from left support.

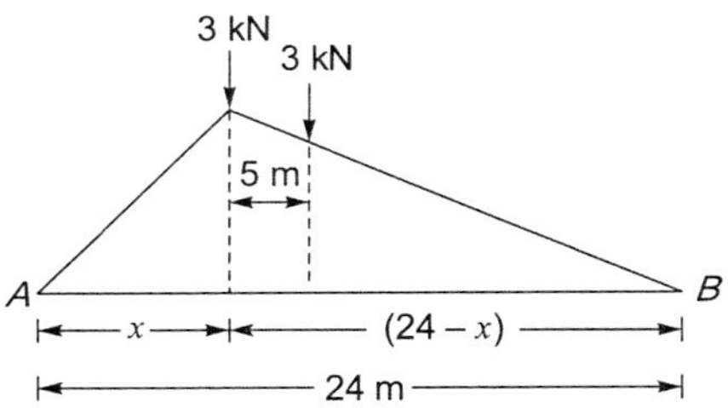

Max. BM at section

$$C = 3\times\left[\frac{x(24-x)}{24}\right]+3\times\left[\frac{x(24-x)}{24}\times\frac{(24-x-5)}{(24-x)}\right]$$

$$= 3\left[\frac{x(24-x)}{24}\right]+3\times\left[\frac{x(19-x)}{24}\right]$$

$$= \frac{43x-2x^2}{8}$$

For absolute max. BM,

$$\frac{dM_c}{dx} = 0$$

$$\Rightarrow \quad \frac{d}{dx}\left(\frac{43x-2x^2}{8}\right) = 0$$

$$\Rightarrow \quad x = 10.75 \text{ m}$$

Absolute max BM

$$= \frac{43\times 10.75-2\times 10.75^2}{8} = 28.89 \text{ kNm}$$

Absolute max bending stress

$$= \frac{28.89\times 10^3}{16.2\times 10^{-6}\times 10^9} = 1.783 \text{ GPa}$$

AVOID MISTAKE

(i)

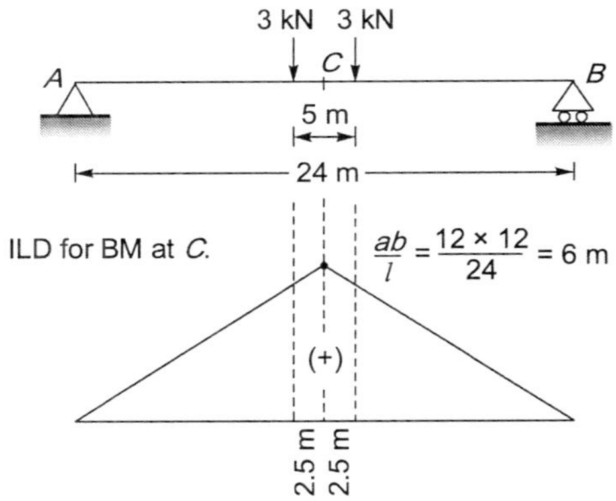

$$\text{BM at C} = 2\times\left[3\times\frac{6}{12}\times(12-2.5)\right]$$

$$= 28.5 \text{ kNm}$$

Max. bending stress

$$\frac{M}{Z} = \frac{28.5\times 10^3}{16.2\times 10^{-6}\times 10^9} = 1.759 \text{ GPa}$$

(ii)

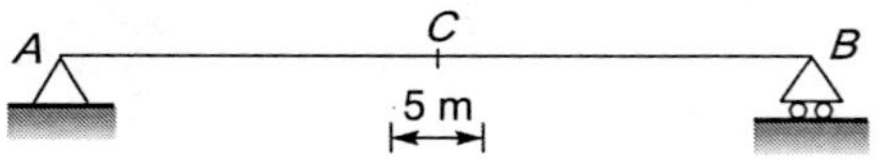

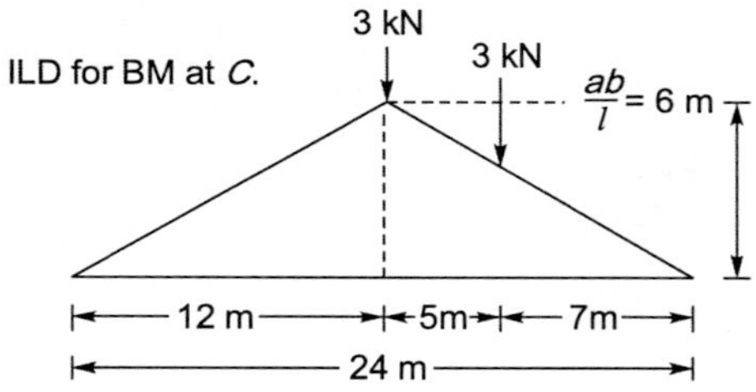

$$\text{BM at C} = 3\times6+3\times\frac{6}{12}\times7 = 28.5 \text{ kNm}$$

Max. bending stress

$$\frac{M}{Z} = \frac{28.5\times10^3}{16.2\times10^{-6}\times10^9}$$

$$= 1.759 \text{ GPa}$$

This is the maximum bending stress at mid= span and not the absolute max. bending stress in the beam.

2.13 Sol.

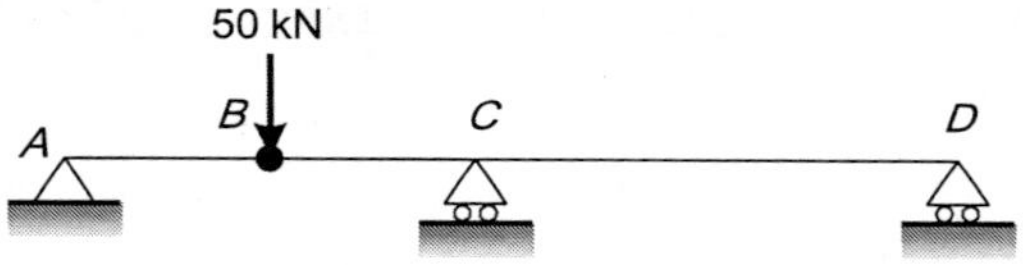

ILD for BM at *C*

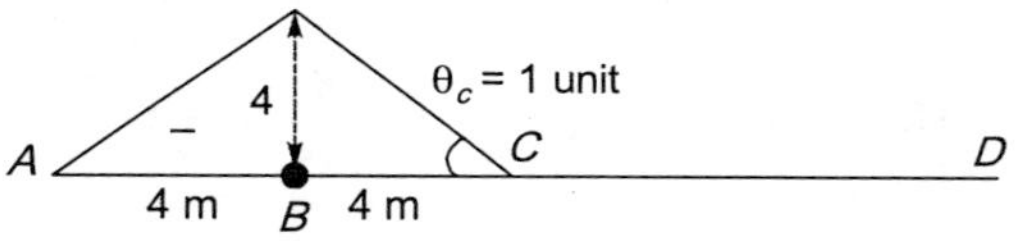

Max. BM occurs at *C*, when 50 kN load is at *B*.

Max. BM at *C* = 50 × (–4) = –200 kNm

2.14 Sol.

ILD for V_B

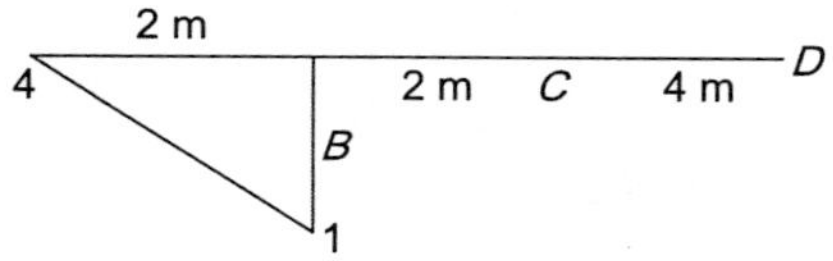

$$\text{Max. shear } V_B = \left[\left(\frac{1}{2}\times2\times1\times10\right)+(60\times1)\right]$$

$$= 70 \text{ kN}$$

Note: The maximum shear force at hinge B may be +70 kN or –70 kN. So, student can challenge this questions.

2.15 (a)

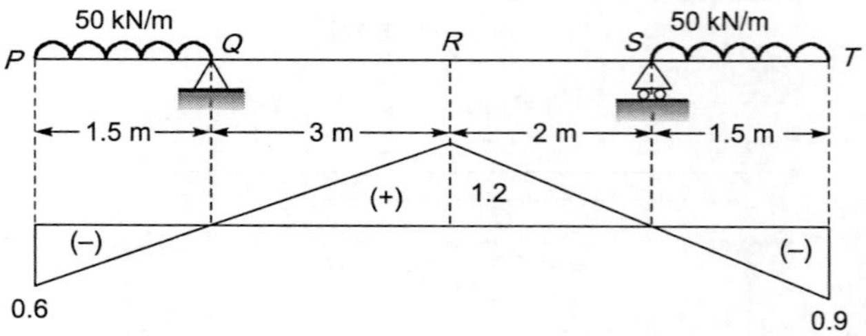

ILD for BM at R:

To get maximum hogging, BM at R, keep UDL over PQ and ST.

Max. –ve BM at R

$$= 50\left[-\frac{1}{2}\times1.5\times0.6\right] +50\left[-\frac{1}{2}\times1.5\times0.9\right]$$

$$= 56.25 \text{ kNm}$$

2.16 Sol.

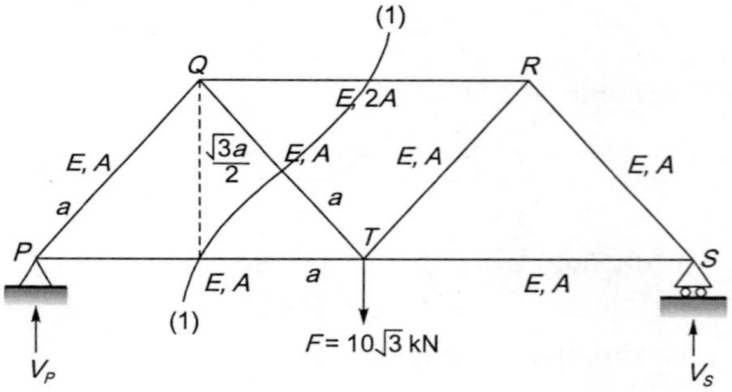

$$V_P = V_S = 5\sqrt{3} \text{ kN}$$

Considering equilibrium of LHS of section (1)-(1)

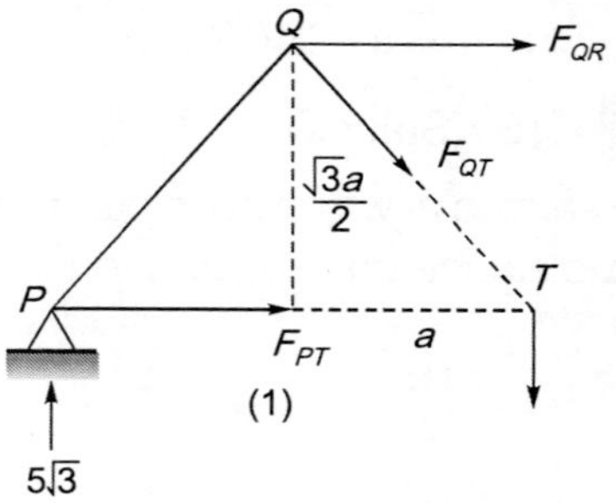

Taking moment about '*T*'

$$\Sigma M_T (\circlearrowright) = 0$$

$$\left(5\sqrt{3}\times a\right)+F_{QR}\left(\frac{\sqrt{3}a}{2}\right) = 0$$

$$\Rightarrow \quad F_{QR} = -10 \text{ kN or } 10 \text{ kN (C)}$$

Compressive stress in member QR (σ_C)

$$\sigma_C = \frac{F_{QR}}{2A} = \frac{10\text{ kN}}{2\times\left(10^4\times10^{-6}\right)\text{m}^2} = 500\text{ kN/m}^2$$

2.17 (d)

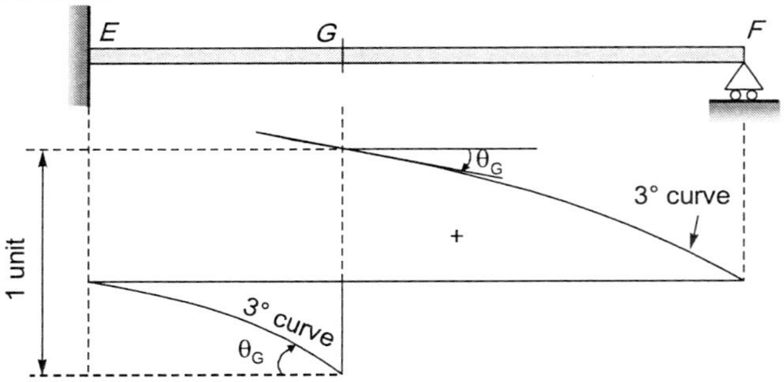

As per Muller Breslau principle ILD for stress function (shear -V_G) will be a combination of curves (3° curves).

2.18 Sol.

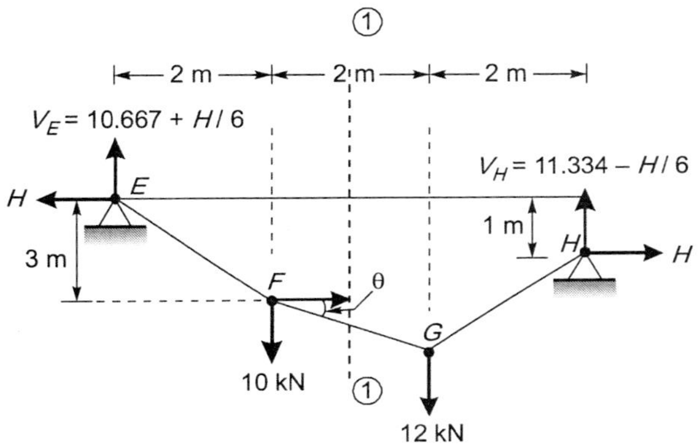

$$\Sigma M_F = 0$$

$$\left(10.667+\frac{H}{6}\right)\times 2-3H = 0$$

$$\Rightarrow (10.667\times 2)+\frac{H}{3}-3H = 0$$

$$\Rightarrow \quad H = \frac{(10.667\times 2)}{2.667} = 8.00\text{ kN}$$

$$\Rightarrow \quad V_E = 10.667+\frac{8}{6} = 12.00\text{ kN}$$

$$\Rightarrow \quad V_H = 11.334-\frac{8}{6} = 10.00\text{ kN}$$

Consider LHS of section (1)-(1)

$$T\cos\theta = 8.0$$

$$T\sin\theta = 12.00 - 10 = 2.0$$

$$\therefore T^2\cos^2\theta + T^2\sin^2\theta = (8)^2 + (2)^2$$

$$\Rightarrow \quad T = 8.246\text{ kN}$$

2.19 Sol.

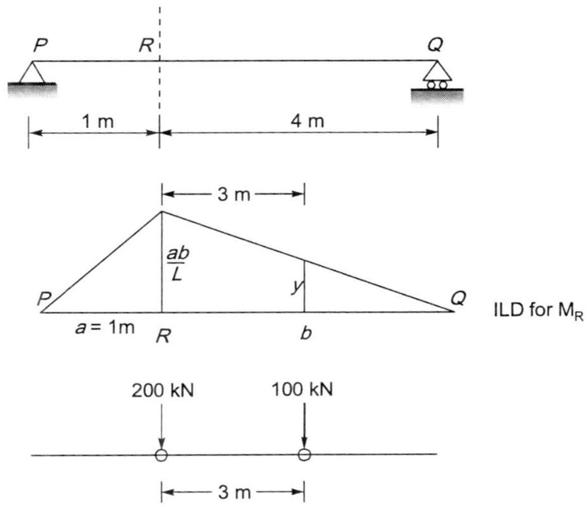

$$\frac{ab}{L} = \frac{1\times 4}{5} = 0.8\text{ m}$$

$$\frac{ab/L}{b} = \frac{y}{4-3} \Rightarrow \frac{0.8}{4} = \frac{y}{1} = 0.2\text{ m}$$

To get maximum BM at R,

$$\text{BM}_{max} = 200\times\frac{ab}{L}+100\times y$$

$$\text{BM}_{max} = 200 \times 0.8 + 100 \times 0.2$$

$$= 180\text{ kNm}$$

2.20 (a)

As per Muller-Breslau principle, The ordinates of influence lines for any stress elements (such as axial force, moment or reaction) of any structure are proportional to those of deflection curve, which is obtained by removing the restraint corresponding to that element from the structure and introducing in its place a deformation into the primary structure which remains.

2.21 Sol.

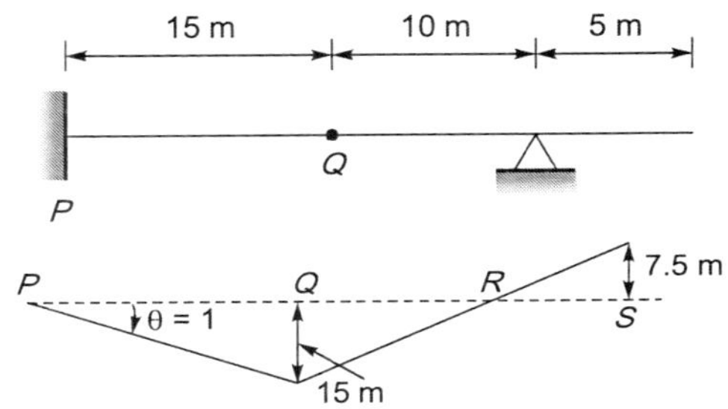

ILD for moment at P,

$\therefore$ Maximum moment at P will be when load is at Q,

$\Rightarrow (\text{BM})_{max} = 10\text{ kN} \times 15\text{ m} = 150\text{ kN-m}$

3 Arches

3.1 Reaction at support *B* of the structure is

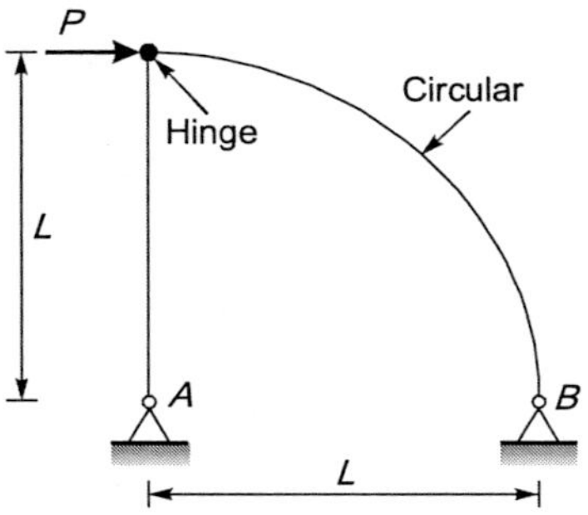

(a) P (b) $P\sqrt{2}$

(c) $\frac{P}{\sqrt{2}}$ (d) $\frac{P}{2}$ **[1996 : 1 M]**

3.2 A three hinged arch shown in the figure is quarter of a circle. If the vertical and horizontal components of reaction at *A* are equal, the value of θ is

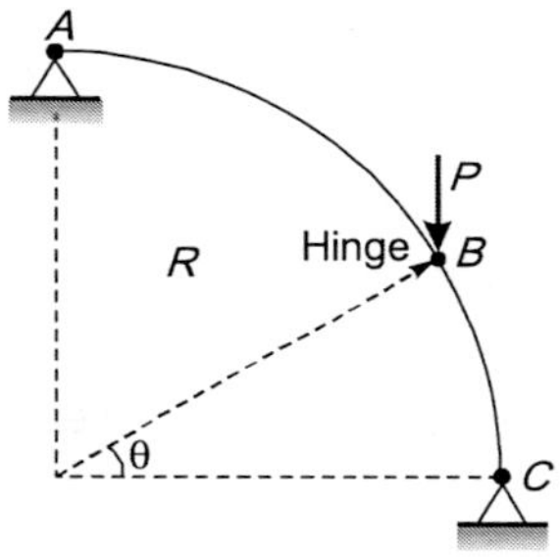

(a) 60° (b) 45°

(c) 30° (d) None in (0°, 90°)

[1998 : 1 M]

3.3 A three hinged parabolic arch *ABC* has a span of 20 m and a central rise of 4 m. The arch has hinges at the ends and at the centre. A train of two point loads of 20 kN and 10 kN, 5 m apart, crosses this arch from left to right, with 20 kN load leading. The maximum thrust induced at the supports is

(a) 25.00 kN (b) 28.13 kN

(c) 31.25 kN (d) 32.81 kN

[2004 : 2 M]

3.4 A three hinged parabolic arch having a span of 20 m and a rise of 5 m carries a point load of 10 kN at quarter span from the left end as shown in the figure. The resultant reaction at the left support and its inclination with the horizontal are respectively

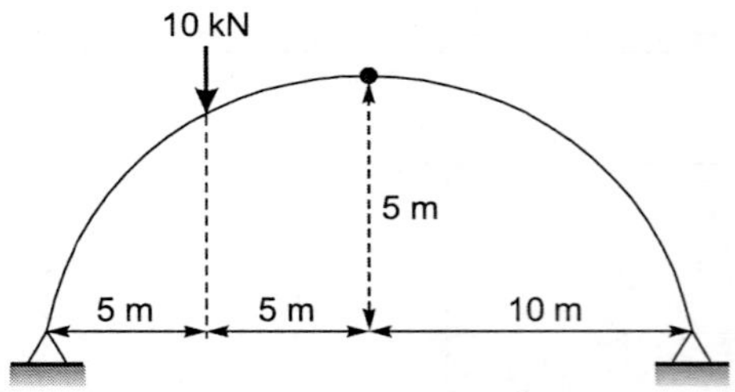

(a) 9.01 kN and 56.31°

(b) 9.01 kN and 33.69°

(c) 7.50 kN and 56.31°

(d) 2.50 kN and 33.69° **[2010 : 2 M]**

3.5 The figure shows a two-hinged parabolic arch span *L* subjected to a uniformly distributed load of intensity *q* per unit length.

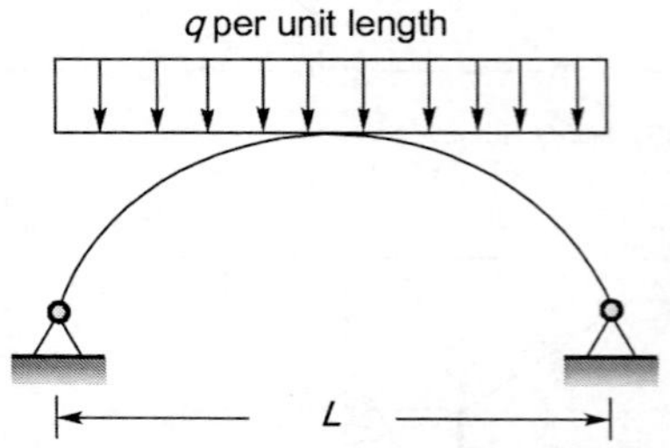

The maximum bending moment in the arch is equal to

(a) $\frac{qL^2}{8}$ (b) $\frac{qL^2}{12}$

(c) zero (d) $\frac{qL^2}{10}$

[2017 : 1 M, Set-I]

3.6 A planar elastic structure is subjected to uniformly distributed load, as shown in the figure (not drawn to the scale)

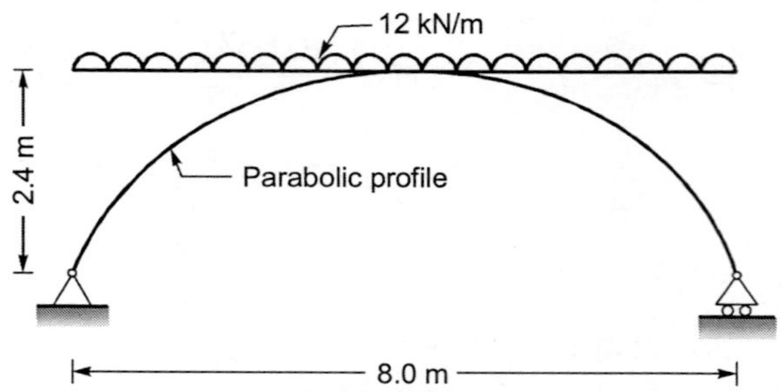

Neglecting self-weight, the maximum bending moment generated in the structure (in kNm, round off to the nearest integer), is ________ .

[2020 : 1 M, Set-I]

3.7 A semi-circular bar of radius R m, in a vertical plane, is fixed at the end G, as shown in the figure. A horizontal load of magnitude P kN is applied at the end H. The magnitude of the axial force, shear force, and bending moment at point Q for $\theta = 45°$, respectively, are

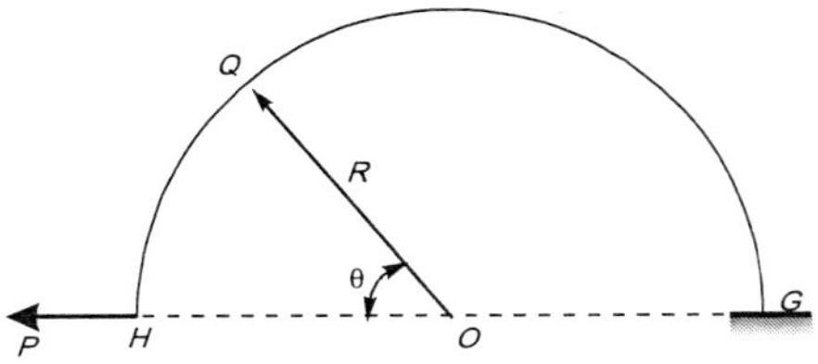

(a) $\frac{P}{\sqrt{2}}$ kN, $\frac{P}{\sqrt{2}}$ kN, and $\frac{PR}{\sqrt{2}}$ kNm

(b) $\frac{P}{\sqrt{2}}$ kN, $\frac{P}{\sqrt{2}}$ kN, and 0 kNm

(c) 0 kNm, $\frac{P}{\sqrt{2}}$ kN, and $\frac{PR}{\sqrt{2}}$ kNm

(d) $\frac{P}{\sqrt{2}}$ kN, 0 kNm, and $\frac{PR}{\sqrt{2}}$ kNm

[2022 : 2 M, Set-I]

Answers Arches

3.1	(b)	3.2	(d)	3.3	(c)	3.4	(a)	3.5	(c)	3.6	(96)	3.7	(a)

Explanations Arches

3.1 (b)

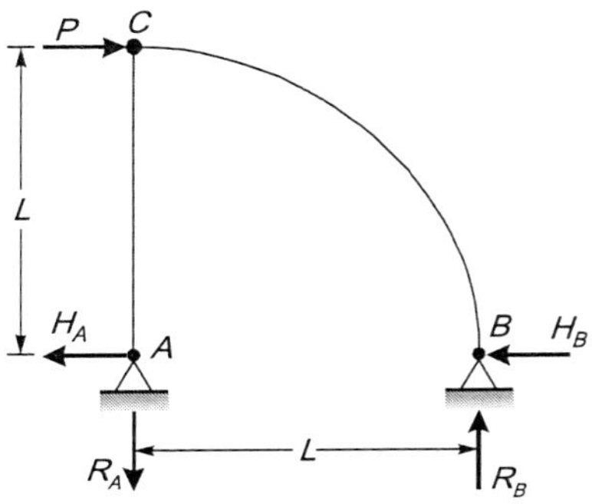

$\Sigma M_A = 0$

$\Rightarrow \quad P \times L = R_B \times L$

$\therefore \quad R_B = P$

$BM_C = 0 \quad$ ($\because$ C is internal hinge)

$\Rightarrow \quad R_B L - H_B L = 0$

$\Rightarrow \quad H_B = R_B = P$

$\therefore$ Reaction at support B,

$$= \sqrt{R_B^2 + H_B^2} = \sqrt{P^2 + P^2} = \sqrt{2}P$$

Note : As 'AC' is a link

$\therefore \quad H_A = 0$

3.2 (d)

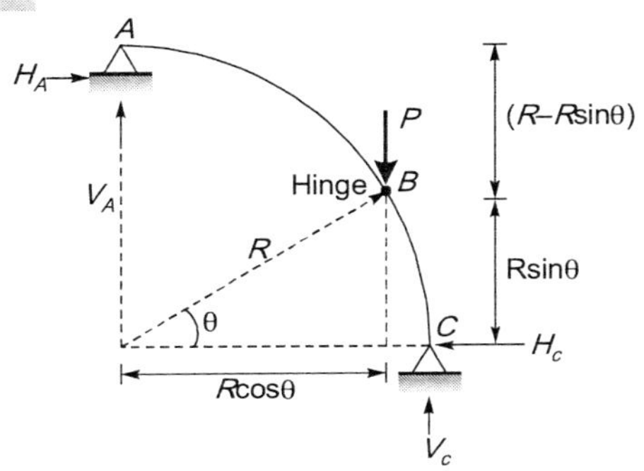

Given: $H_A = V_A$

Now, $BM_B = 0 \quad$ ($\because$ B is internal hinge)

$\Rightarrow \quad V_A R\cos\theta + H_A R(1 - \sin\theta) = 0$

$\Rightarrow \quad H_A R\cos\theta + H_A R(1 - \sin\theta) = 0$

$\Rightarrow \quad \cos\theta + 1 - \sin\theta = 0$

$\Rightarrow \quad \sin\theta - \cos\theta = 1$

$\Rightarrow \quad \sin^2\theta + \cos^2\theta - 2\sin\theta\cos\theta = 1$

$\Rightarrow \quad 1 - \sin 2\theta = 1$

$\Rightarrow \quad \sin 2\theta = 0$

$\Rightarrow \quad 2\theta = 0; 180°$

$\Rightarrow \quad \theta = 0°; 90°$

$\therefore \quad \theta$ is none in (0°, 90°)

3.3 (c)

In case of a three hinged parabolic arch, the influence line diagram for horizontal thrust is linear.

Maximum thrust will be induced at the supports when 20 kN load is at the crown.

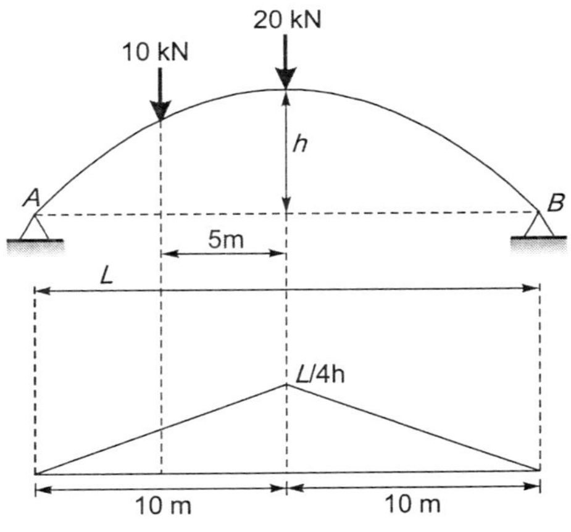

ILD for horizontal thrust

Ordinate of the ILD at a distance of 5 m from A

$$= \frac{5}{10} \times \frac{L}{4h} = \frac{1}{2} \times \left(\frac{20}{4 \times 4}\right) = 0.625$$

Also, $\frac{L}{4h} = \frac{20}{4 \times 4} = 1.25$

Thus, **horizontal thrust,**

$$H = 10 \times 0.625 + 20 \times 1.25 = 31.25 \text{ kN}$$

3.4 (a)

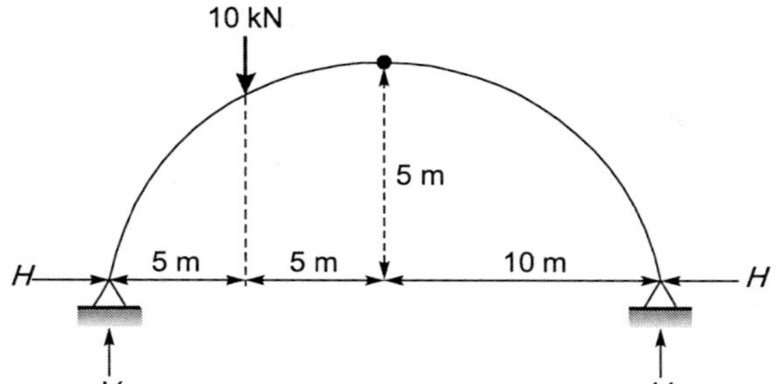

Label the diagram

Let the vertical reactions at left and right support be V_L and V_R upwards respectively.

Taking moments about right support, we get,

$V_L \times 20 - 10 \times 15 = 0$

$\Rightarrow \quad V_L = 7.5$ kN

$\therefore \quad V_R = 10 - 7.5 = 2.5$ kN

Let the horizontal reaction at left support be H from left to right.

Taking moments about the crown from left, we get,

$7.5 \times 10 - 10 \times 5 - H \times 5 = 0$

$\Rightarrow \quad H = 5$ kN

Resultant reaction,

$$R = \sqrt{H^2 + V_L^2}$$

$$= \sqrt{5^2 + (7.5)^2} = 9.01 \text{ kN}$$

Let the resultant reaction at the left support makes an angle θ with the horizontal.

$$\tan\theta = \frac{7.5}{5}$$

$\Rightarrow \quad \theta = \tan^{-1}(1.5)$

$\Rightarrow \quad \theta = 56.31°$

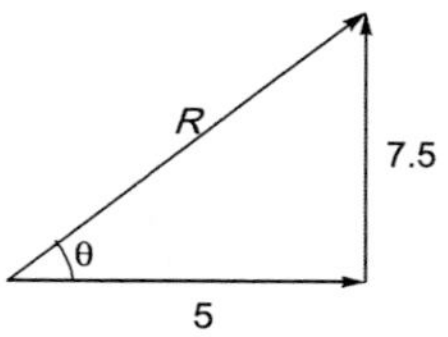

3.5 (c)

If a two hinged or three hinged parabolic arch is subjected to UDL throughout its length, bending moment is zero everywhere.

POINTS TO REMEMBER

For a given loading if the free BMD corresponds to the actual shape of the structure, then such structures are called funicular polygon.

For a funicular profile the BM and SF at any section is zero.

3.6 Sol.

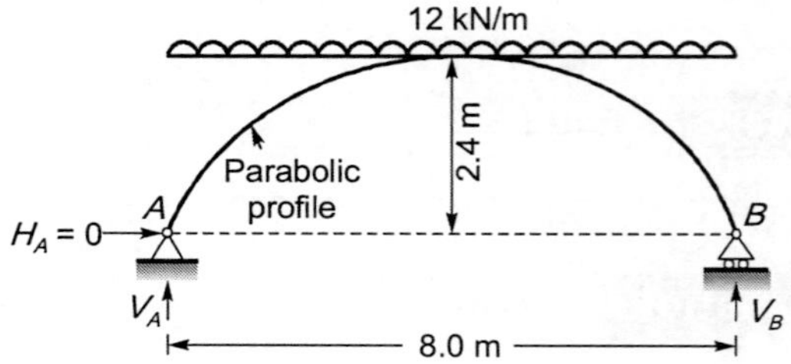

$$\Sigma F_y = 0$$

$\Rightarrow V_A + V_B - wL = 0$ (Also $V_A = V_B$

$\therefore \quad V_A = V_B = \frac{wL}{2} = \frac{12 \times 8}{2} = 48$ kN

As horizontal thrust is zero so it behaves like a beam (curved beam)

$$M_{max} = \frac{wL^2}{8} \text{(At crown)}; = \frac{12 \times 8^2}{8} = 96 \text{ kNm}$$

3.7 (a)

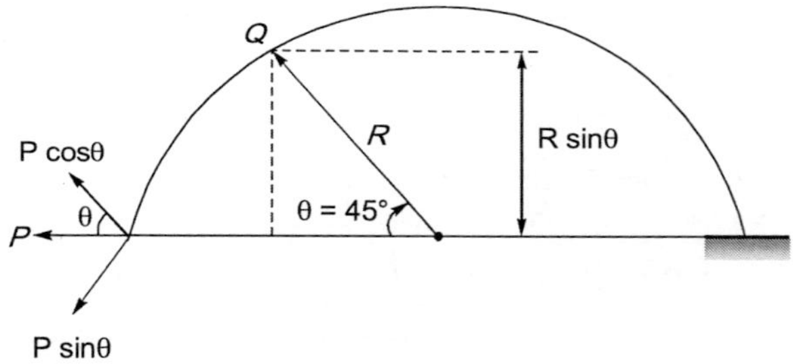

Axial force at Q,

$$F_Q = P\sin\theta = \frac{P}{\sqrt{2}} \quad (\because \theta = 45°)$$

Shear force at Q,

$$S_Q = P\cos\theta = \frac{P}{\sqrt{2}} \quad (\because \theta = 45°)$$

Bending moment at Q,

$$M_Q = PR \sin\theta = \frac{PR}{\sqrt{2}} (\because \theta = 45°)$$

■■■■

4 Methods of Structural Analysis

4.1 The magnitude of the bending moment at the fixed support of the beam is equal to

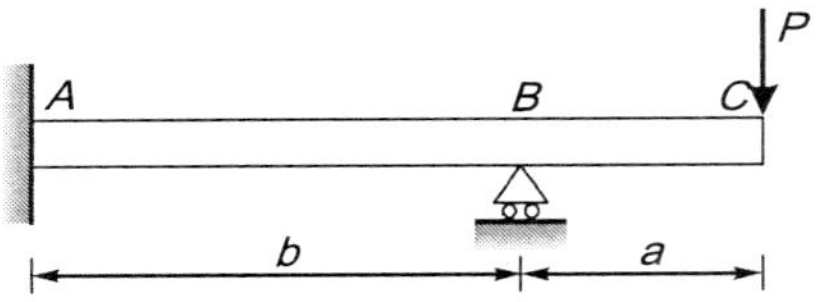

(a) $P.a$ (b) $\frac{P.a}{2}$
(c) $P.b$ (d) $P.(a+b)$ [1995 : 1 M]

4.2 The number of simultaneous equations to be solved in the slope deflection method is equal to
(a) the degree of static indeterminacy
(b) the degree of kinematic indeterminacy
(c) the number of joints in the structure
(d) None of the above [1995 : 1 M]

4.3 A single bay single storey portal frame has hinged left support and a fixed right support. It is loaded with uniformly distributed load on the beam. Which one of the following statements is true with regard to the deformation of the frame?
(a) It would sway to the left side
(b) It would sway to the right side
(c) It would not sway at all
(d) None of the above [1995 : 1 M]

4.4 Bending moments at joint b and c of the portal frame are respectively

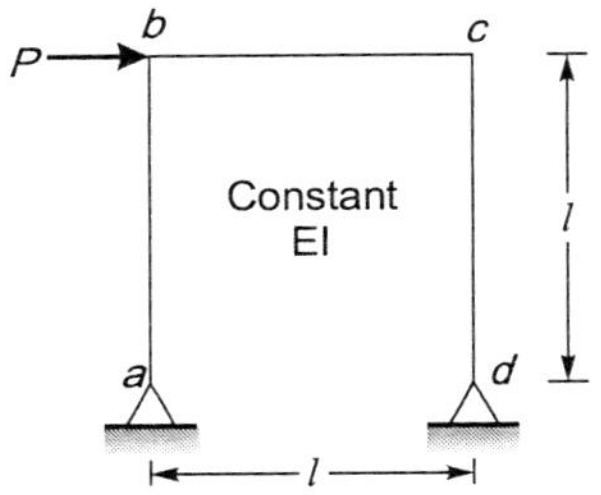

(a) $+\frac{pl}{2}, -\frac{pl}{2}$ (b) $+\frac{pl}{2}, +\frac{pl}{2}$
(c) $-\frac{pl}{4}, -\frac{pl}{4}$ (d) $+\frac{pl}{4}, +\frac{pl}{4}$
[1996 : 2 M]

4.5 For the frame shown in the figure, the maximum bending moment in the column is

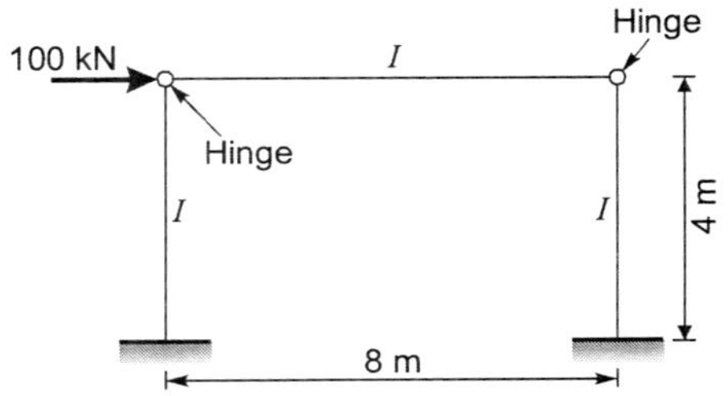

(a) Zero (b) 400 kNm
(c) 100 kNm (d) 200 kNm
[1997 : 1 M]

4.6 The order of the flexibility matrix for a structure is
(a) equal to the number of redundant forces
(b) more than the number of redundant forces
(c) less than the number of redundant forces
(d) equal to the number of redundant forces plus three [1997 : 1 M]

4.7 Identify the correct deflection diagram corresponding to the loading in the plane frame shown below:

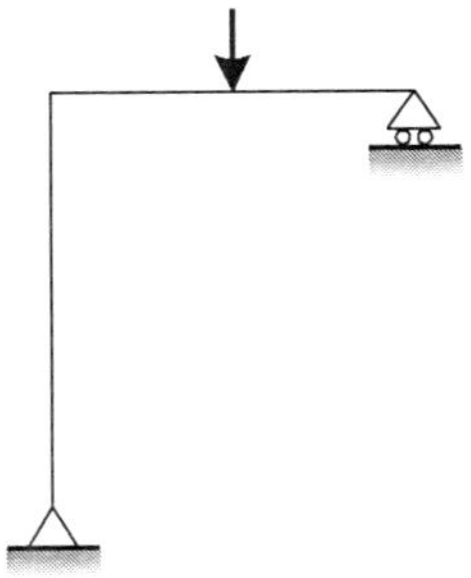

(a)
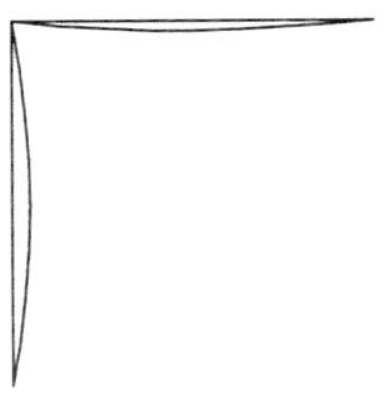

(b)
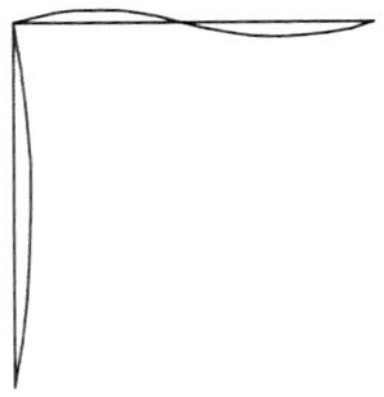

(c)
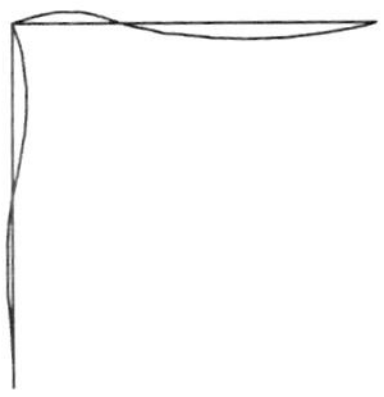

(d)
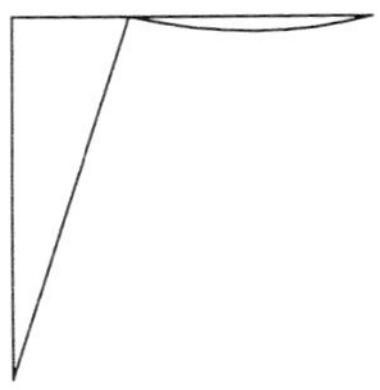

[2001 : 1 M]

4.8 Identify the false statement from the following, pertaining to the methods of structural analysis
(a) Influence lines for stress resultants in beams can be drawn using Muller Breslau's Principle
(b) The Moment Distribution Method is a force method of analysis, not a displacement method
(c) The Principle of Virtual Displacements can be used to establish a condition of equilibrium
(d) The Substitute Frame Method is not applicable to frames subject to significant sideway

[2001 : 1 M]

4.9 The frame below shows three beam elements *OA*, *OB* an *OC*, with identical length *L* and flexural rigidity *EI*, subject to an external moment *M* applied at the rigid joint *O*. The correct set of bending moments $\{M_{OA}, M_{OB}, M_{OC}\}$ that develop at *O* in the three beam elements *OA*, *OB* and *OC* respectively, is

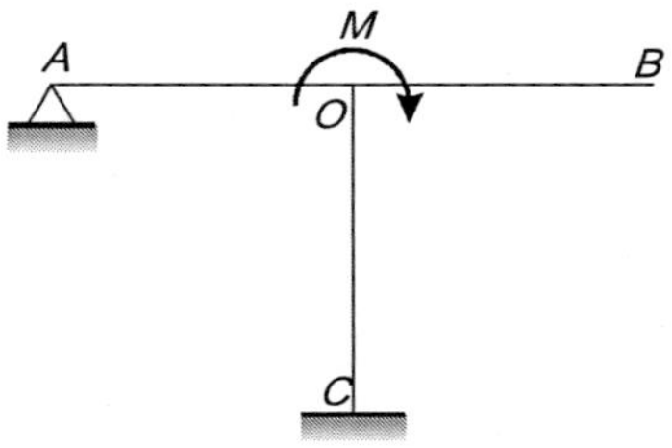

(a) {3*M*/8, *M*/8, 4*M*/8}
(b) {3*M*/11, 4*M*/11, 4*M*/11}
(c) {*M*/3, *M*/3, *M*/3}
(d) {3*M*/7, 0, 4*M*/7} [2001 : 2 M]

4.10 The end moment (in kNm) developed in the roof level beams in the laterally loaded frame shown below (with all columns having identical cross-sections), according to the Cantilever Method of simplified analysis is

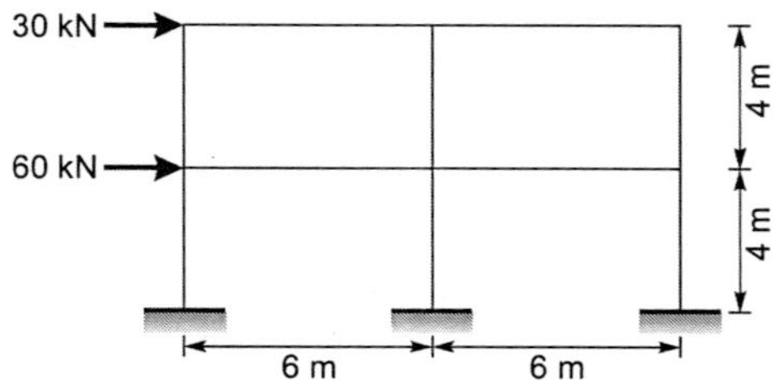

(a) 7.5 (b) 15
(c) 20 (d) 30 [2001 : 2 M]

4.11 The unit load method used in structural analysis is
(a) applicable only to statically indeterminate structures
(b) another name for stiffness method
(c) an extension of Maxwell's reciprocal theorem
(d) derived from Castigliano's theorem

[2004 : 1 M]

4.12 For a linear elastic structural system, minimization of potential energy yields
(a) compatibility conditions
(b) constitutive relations
(c) equilibrium equations
(d) strain-displacement relations [2004 : 1 M]

4.13 The plane frame below is analyzed by neglecting axial deformations. Following statements are made with respect to the analysis:

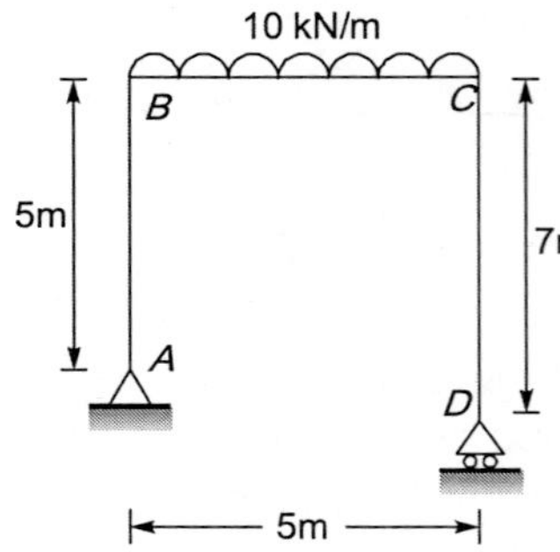

$EI = 81380$ kN-m^2

1. Column *AB* carries axial force only.
2. Vertical deflection at the center of beam *BC* is 1 mm.

With reference to the above statements, which of the following applies?
(a) Both the statements are true
(b) Statement 1 is true but 2 is false
(c) Statement 2 is true but 1 is false
(d) Both the statements are false [2004 : 2 M]

4.14 Match **List-I** with **List-II** and select the correct answer using the codes given below the lists:

List-I
A. Slope deflection method
B. Moment distribution method
C. Method of three moments
D. Castigliano's second theorem

List-II
1. Force method
2. Displacement method

Codes:

	A	B	C	D
(a)	1	2	1	2
(b)	1	1	2	2
(c)	2	2	1	1
(d)	2	1	2	1

[2005 : 2 M]

4.15 All members of the frame shown below have the same flexural rigidity *EI* and length *L*. If a moment *M* is applied at joint *B*, the rotation of the joint is

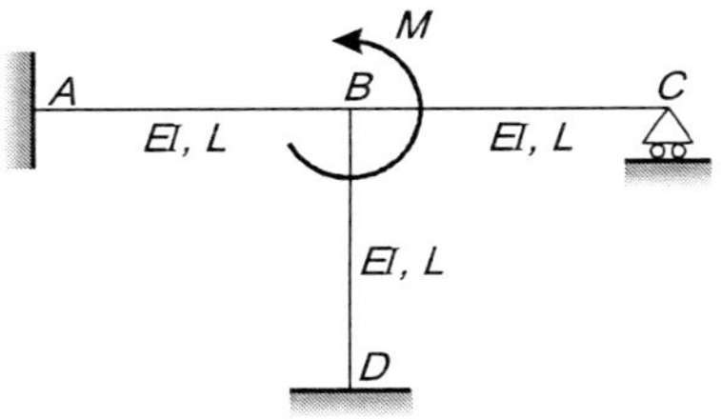

(a) $\frac{ML}{12EI}$ (b) $\frac{ML}{11EI}$

(c) $\frac{ML}{8EI}$ (d) $\frac{ML}{7EI}$ [2005 : 2 M]

4.16 Carry-over factor C_{AB} for the beam shown in the figure below is

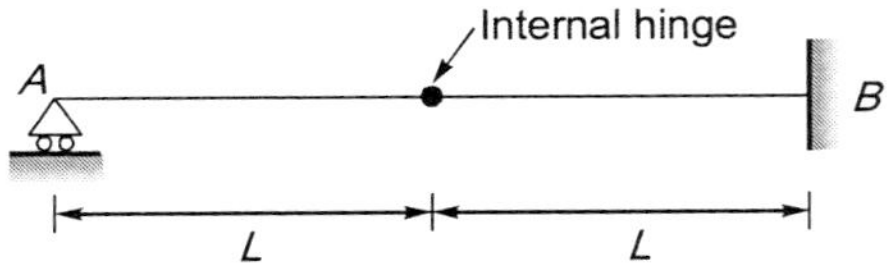

(a) 1/4 (b) 1/2
(c) 3/4 (d) 1 [2006 : 2 M]

Common Data for Questions 4.20 and 4.21:

A two span continuous beam having equal spans each of length L is subjected to a uniformly distributed load w per unit length. The beam has constant flexural rigidity.

4.17 The reaction at the middle support is

(a) wL (b) $\frac{5wL}{2}$

(c) $\frac{5wL}{4}$ (d) $\frac{5wL}{8}$ [2007 : 2 M]

4.18 The bending moment at the middle support is

(a) $\frac{wL^2}{4}$ (b) $\frac{wL^2}{8}$

(c) $\frac{wL^2}{12}$ (d) $\frac{wL^2}{16}$ [2007 : 2 M]

4.19 All members in the rigid-jointed frame shown are prismatic and have the same flexural stiffness EI. Find the magnitude of BM at Q (in kN-m) due to given loading.

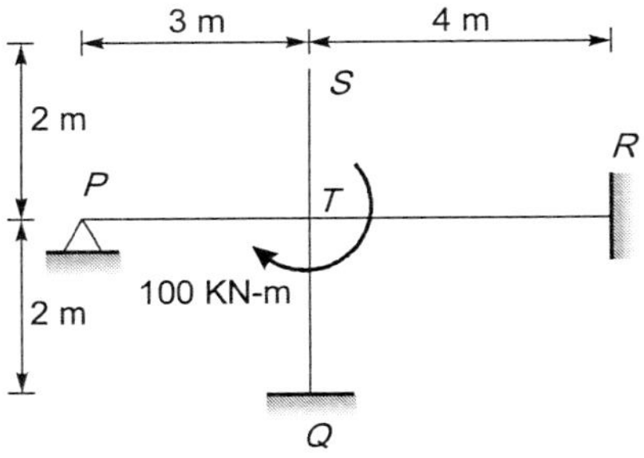

[2013 : 2 M]

4.20 Considering the symmetry of a rigid frame as shown below, the magnitude of the bending moment (in kNm) at P (preferably using the moment distribution method) is

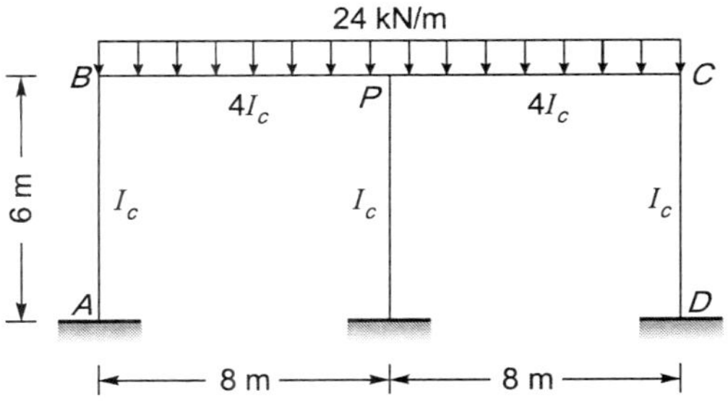

(a) 170 (b) 172
(c) 176 (d) 178

[2014 : 2 M, Set-II]

4.21 For the beam shown below, the value of the support moment M is _____kN-m.

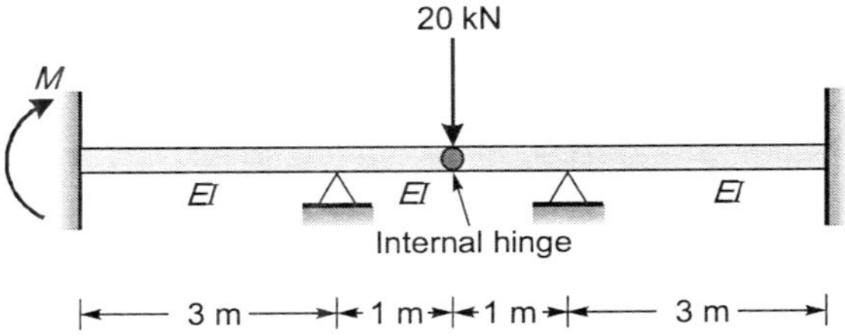

[2015 : 1 M, Set-I]

4.22 The portal frame shown in the figure is subjected to a uniformly distributed vertical load w (per unit length).

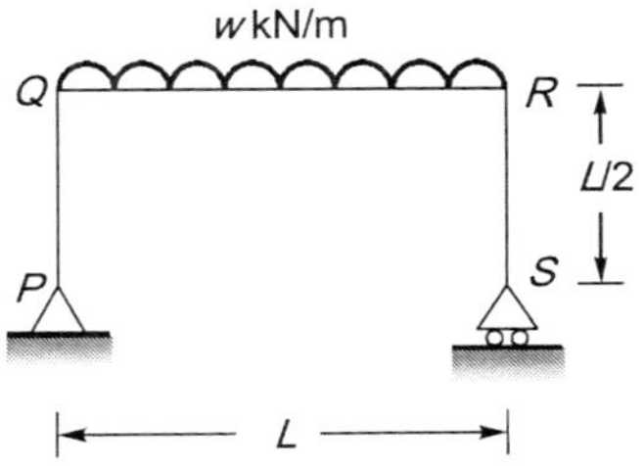

The bending moment in the beam at the join 'Q' is

(a) zero (b) $\frac{wL^2}{24}$ (hogging)

(c) $\frac{wL^2}{12}$ (hogging) (d) $\frac{wL^2}{8}$ (sagging)

[2016 : 2 M, Set-II]

4.23 The value of M in the beam ABC shown in the figure is such that the joint B does not rotate.

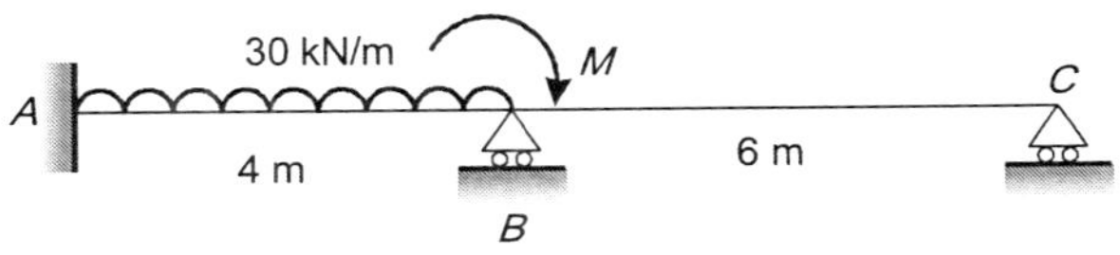

The value of support reaction (in kN) at B should be equal to ________.

[2017 : 2 M, Set-I]

4.24 Consider the portal frame shown in the figure and assume the modulus of elasticity, $E = 2.5 \times 10^4$ MPa and the moment of inertia, $I = 8 \times 10^8$ mm^4 for all the members of the frame.

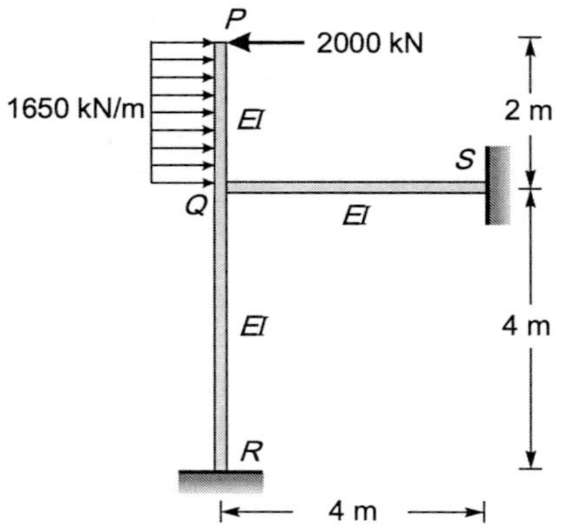

The rotation (in degrees, up to one decimal place) at the rigid joint Q would be __________.

[2017 : 2 M, Set-II]

4.25 A vertical load of 10 kN acts on a hinge located at a distance of $L/4$ from the roller support Q of a beam of length L (see figure).

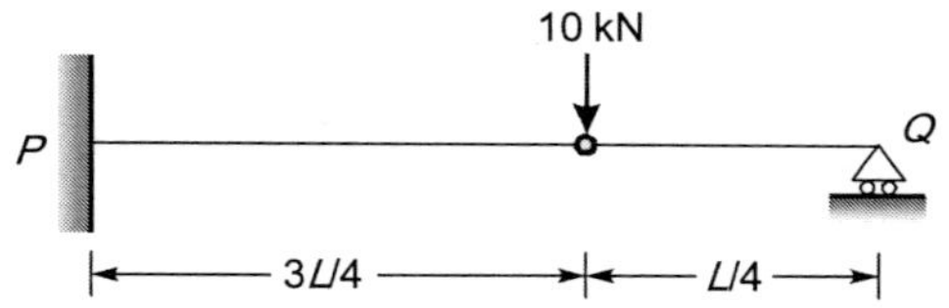

The vertical reaction at support Q is

(a) 0.0 kN (b) 2.5 kN
(c) 7.5 kN (d) 10.0 kN

[2018 : 1 M, Set-II]

4.26 A prismatic beam P-Q-R of flexural rigidity $EI = 1 \times 10^4$ kNm^2 is subjected to a moment of 180 kNm at Q as shown in the figure.

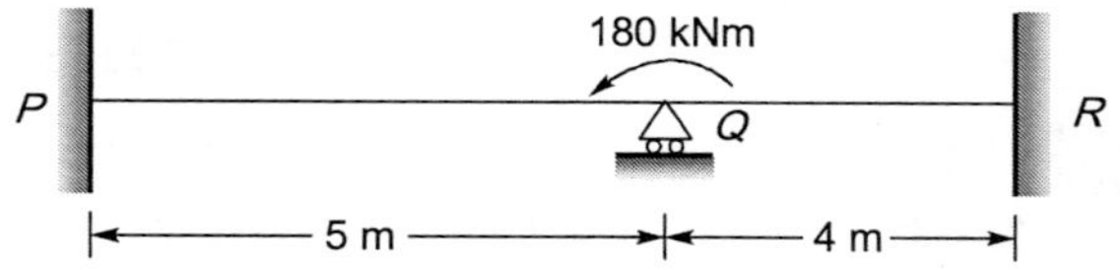

The rotation at Q (in rad, up to two decimal places) is ________.

[2018 : 2 M, Set-II]

4.27 The rigid-jointed plane frame QRS shown in figure is subjected to load P at the joint R. Let the axial deformations in the frame be neglected. If the support S undergoes a settlement of $\Delta = \dfrac{PL^3}{\beta EI}$, the vertical reaction at the support S will be become zero when b is equal to

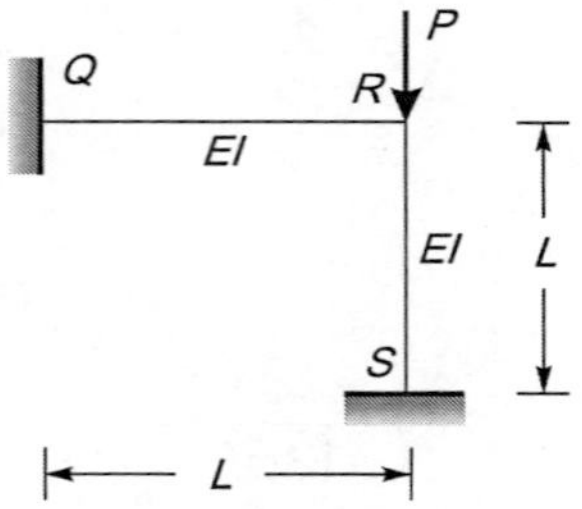

(a) 7.5 (b) 3.0
(c) 48.0 (d) 0.1

[2019 : 2 M, Set-I]

4.28 A portal frame shown in figure (not drawn to scale) has a hinge support at joint P and a roller support at joint R. A point load of 50 kN is acting at joint R in the horizontal direction. The flexural rigidity. EI, of each member is 10^6 kNm^2. Under the applied load, the horizontal displacement (in mm, round off to 1 decimal place) of joint R would be______.

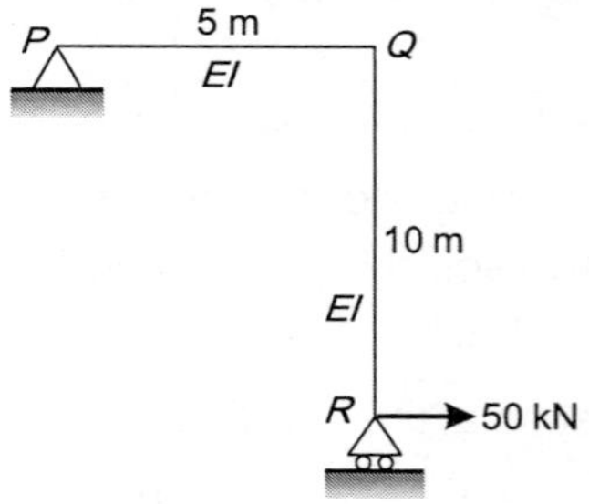

[2019 : 2 M, Set-I]

4.29 The plane truss has hinge supports at P and W and is subjected to the horizontal forces as shown in the figure (not drawn to the scale).

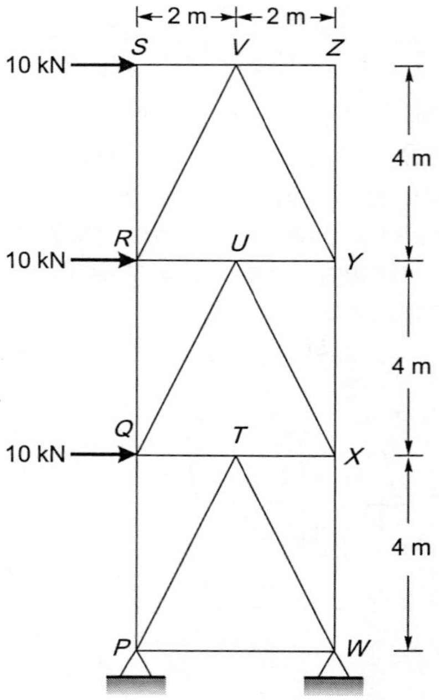

Representing the tensile force with '+' sign and the compressive force with '–' sign, the force in member XW (in kN, round off to the nearest integer), is ________.

[2020 : 2 M, Set-II]

4.30 A frame EFG is shown in the figure. All members are prismatic and have equal flexural rigidity. The member FG carries a uniformly distributed load w per unit length. Axial deformation of any member is neglected.

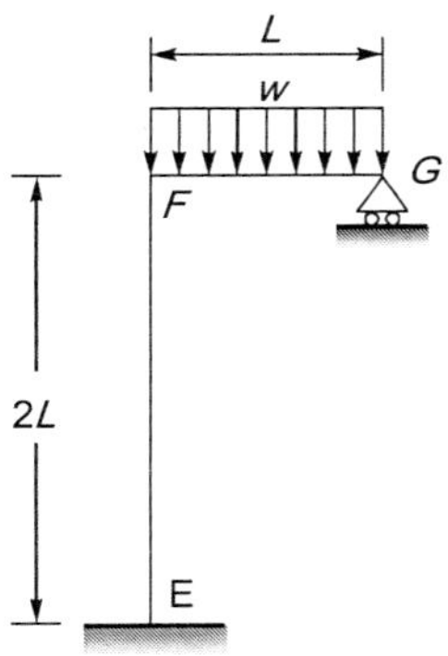

Considering the joint F being rigid, the support reaction at G is

(a) 0.375 wL
(b) 0.453 wL
(c) 0.482 wL
(d) 0.500 wL

[2021 : 2 M, Set-II]

4.31 Consider the linearly elastic plane frame shown in the figure. Members HF, FK and FG are welded together at joint F. Joints K, G and H are fixed supports. A counter-clockwise moment M is applied at joint F. Consider flexural rigidity $EI = 10^5$ kN-m^2 for each member and neglect axial deformations.

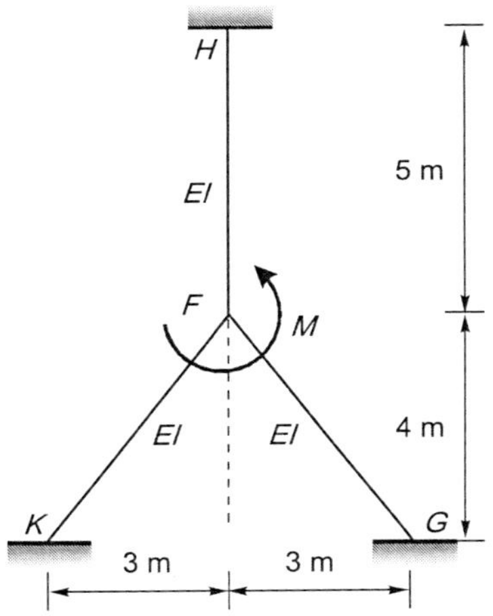

If the magnitude (absolute value) of the support moment at H is 10 kN-m, the magnitude (absolute value) of the applied moment M (in kN-m) to maintain static equilibrium is ________. (round off to the nearest integer)

[2022 : 2 M, Set-I]

4.32 Which of the following statement(s) is/are correct?

(a) If a structure is acted upon by two force system P_a and P_b, in equilibrium separately, the external virtual work done by a system forces P_b during the deformations caused by another system of forces P_a is equal to the external virtual work done by the P_a system during the deformation caused by the P_b system.

(b) If a linearly elastic structure is subjected to a set of loads, the partial derivative of the total strain energy with respect to the load at any point is equal to the deflection at that point.

(c) The shear force in a conjugate beam loaded by the M/EI diagram of the real beam is equal to the corresponding deflection of the real beam.

(d) If a linearly elastic structure is subjected to a set of loads the partial derivative of the total strain energy with respect to the deflection at any point is equal to the load applied at that point,

[2022 : 2 M, Set-II]

4.33 The linearly elastic planar structure shown in the figure is acted upon by two vertical concentrated forces. The horizontal beams UV and WX are connected with the help of the vertical linear spring with spring constant $k = 20$ kN/m. The fixed supports are provided at U and X. It is given that flexural rigidity $EI = 10^5$ kN-m^2, $P = 100$ kN. and $a = 5$ m. Force Q is applied at the center of beam WX such that the force in the spring VW becomes zero.

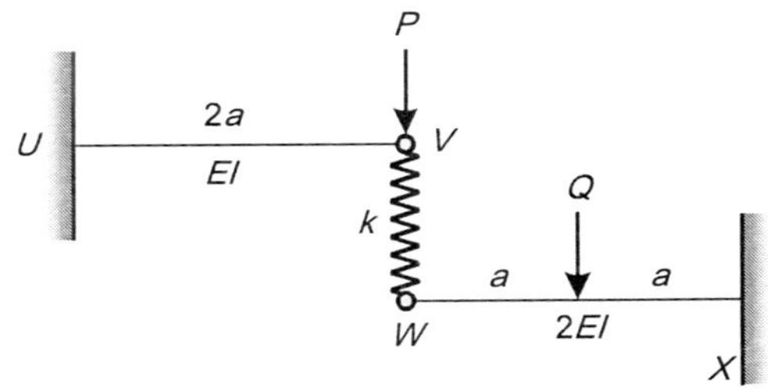

The magnitude of force Q (in kN) is ______ . (round off to the nearest integer)

[2022 : 2 M, Set-II]

4.34 In the frame shown in the figure (not to scale), all four members (AB, BC, CD, and AD) have the same length and same constant flexural rigidity. All the joints A, B, C, and D are rigid joints. The midpoints of AB, BC, CD, and AD, are denoted by E, F, G, and H, respectively. The frame is in unstable equilibrium under the shown forces of magnitude P acting at E and G. Which of the following statements is/are TRUE?

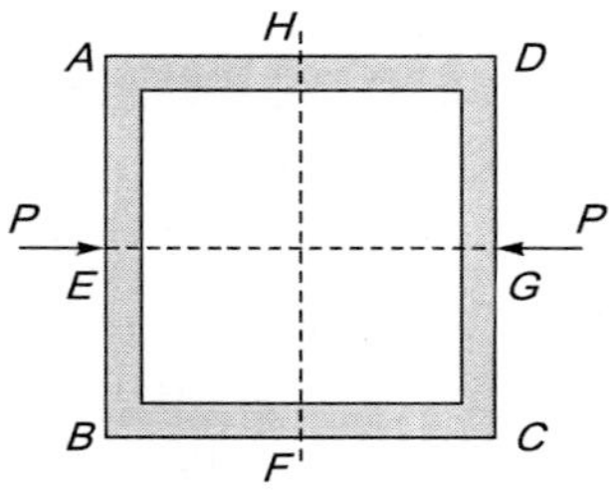

(a) Shear forces at H and F are zero
(b) Horizontal displacements at H and F are zero
(c) Vertical displacements at H and F are zero
(d) Slopes at E, F, G, and H are zero

[2023 : 1 M, Set-II]

4.35 For the frame shown in the figure (not to scale), all members (AB, BC, CD, GB, and CH) have the same length, L and flexural rigidity, EI The joints at B and C are rigid joints, and the supports A and D are fixed supports. Beams GB and CH carry uniformly distributed loads of w per unit length. The magnitude of the moment reaction at A is wL^2/k. What is the value of k (in integer)?

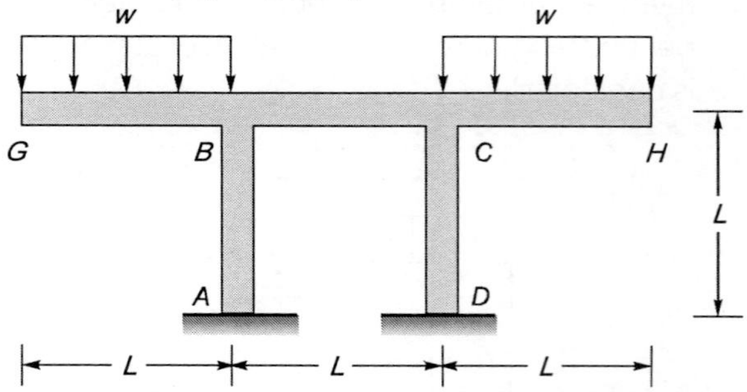

[2023 : 2 M, Set-II]

Answers Methods of Structural Analysis

4.1 (b)	4.2 (b)	4.3 (a)	4.4 (a)	4.5 (d)	4.6 (a)	4.7 (d)
4.8 (b)	4.9 (d)	4.10 (b)	4.11 (d)	4.12 (a)	4.13 (a)	4.14 (c)
4.15 (b)	4.16 (d)	4.17 (c)	4.18 (b)	4.19 (25)	4.20 (c)	4.21 (5)
4.22 (a)	4.23 (40)	4.24 (1.00)	4.25 (a)	4.26 (0.01)	4.27 (a)	4.28 (25)
4.29 (30)	4.30 (c)	4.31 (60)	4.32 (a, b, d)	4.33 (640)	4.34 (a, b, d)	4.35 (6)

Explanations Methods of Structural Analysis

4.1 (b)

Method-I

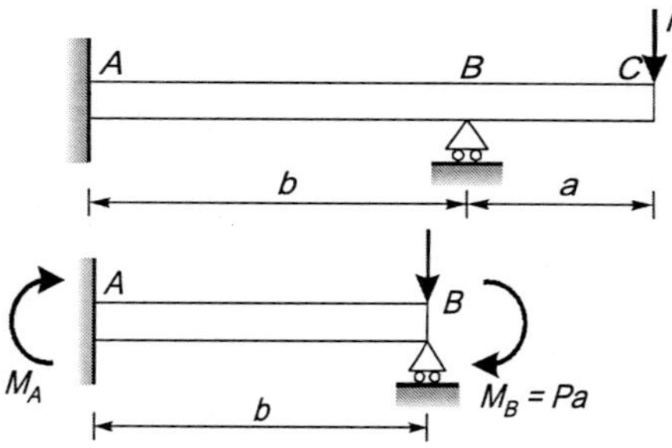

∴ The bending moment at A will be,

$$M_A = \frac{M_B}{2} = \frac{Pa}{2} \text{ (Carry-over moment)}$$

Method-II

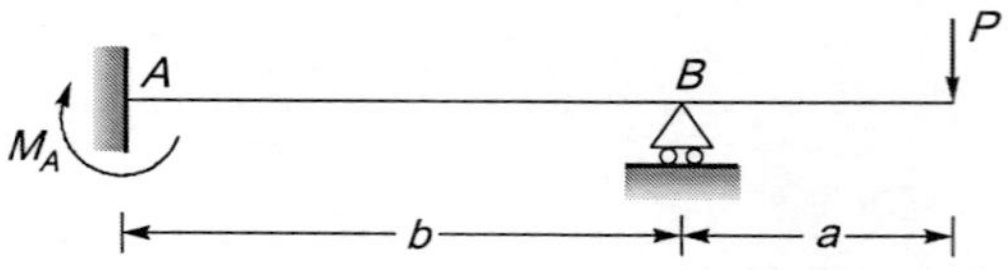

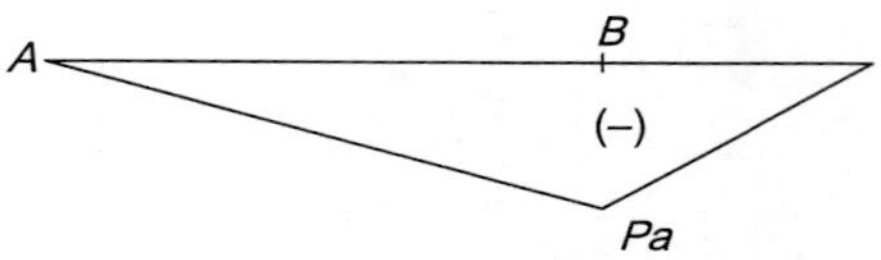

BMD for simply supported beam

+

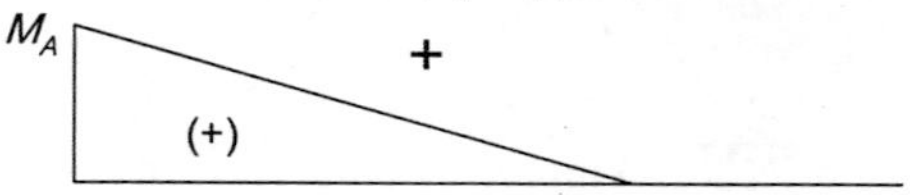

BMD due to fixed end moment

Using moment area method,

$\delta_{B/A}$ = deflection of point B w.r.t. tangent at $A = 0$

$$\Rightarrow \quad \left(-\frac{1}{2}\times Pa\times b\times\frac{b}{3}\right)+\left(\frac{1}{2}\times M_A\times b\times\frac{2b}{3}\right)=0$$

$$\Rightarrow \quad M_A = \frac{Pa}{2}$$

4.2 (b)

Slope deflection method is a displacement method, so number of equations required is equal to number of kinematic indeterminacy.

4.3 (a)

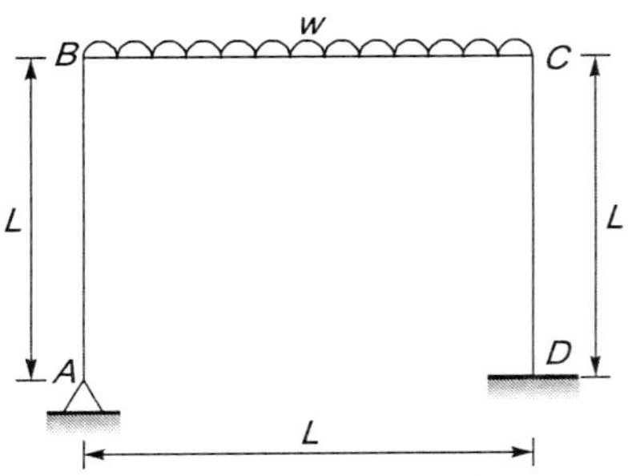

Stiffness of $BA = \frac{3EI}{L}$

Stiffness of $CD = \frac{4EI}{L}$

The frame will sway on that side where stiffness is less.

$\therefore$ Frame would sway to the left side.

4.4 (a)

Method-I

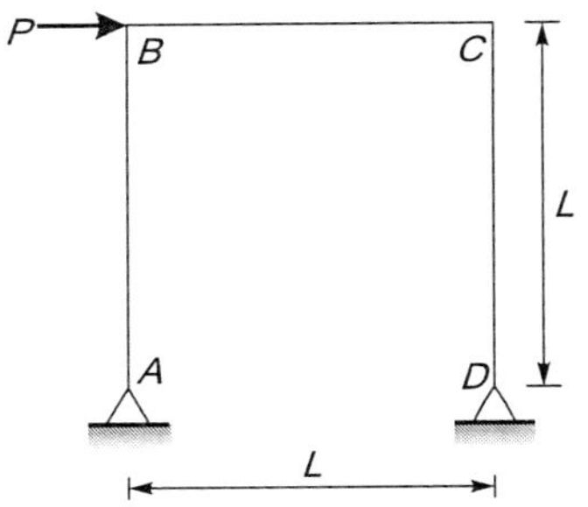

=

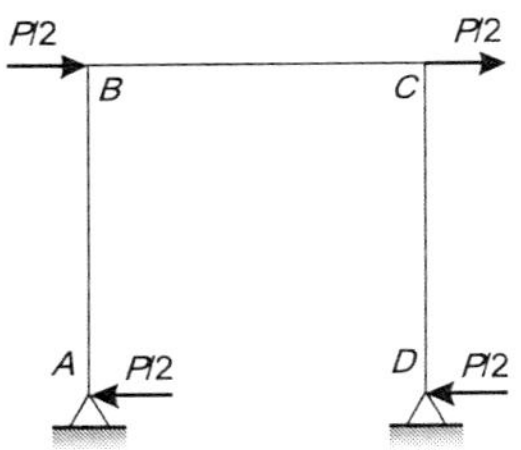

Asymmetric structure

\+

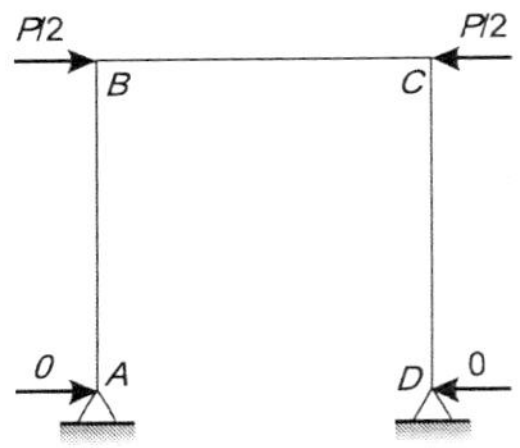

Symmetric structure

Horizontal reaction at '*a*' or '*d*' is equal to horizontal reaction at '*a*' or '*d*' in asymmetric structure plus horizontal reaction at '*a*' or '*d*' in symmetric structure.

$$H_A = \frac{P}{2}+0$$

$\therefore$ Bending moment at '*b*' $= \frac{Pl}{2}$

Similarly bending moment at '*c*' $= -\frac{Pl}{2}$

Method-II

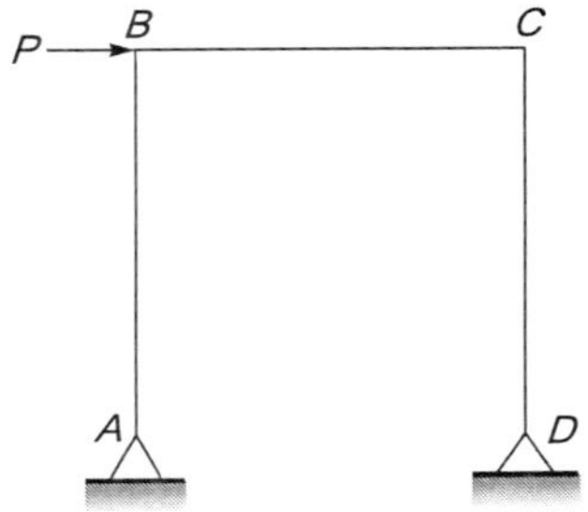

As the two columns *AB* and *CD* are identical, the load *P* will be distributed equally on the two columns.

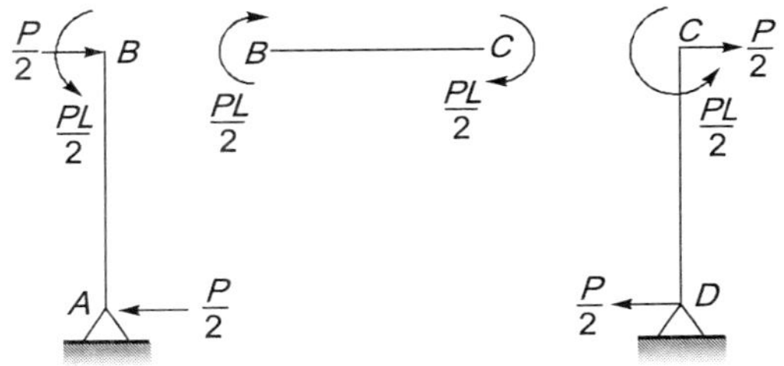

$$\text{BM at } B = +\frac{P}{2}\times L = +\frac{PL}{2}$$

$$\text{BM at } C = -\frac{P}{2}\times L = -\frac{PL}{2}$$

4.5 (d)

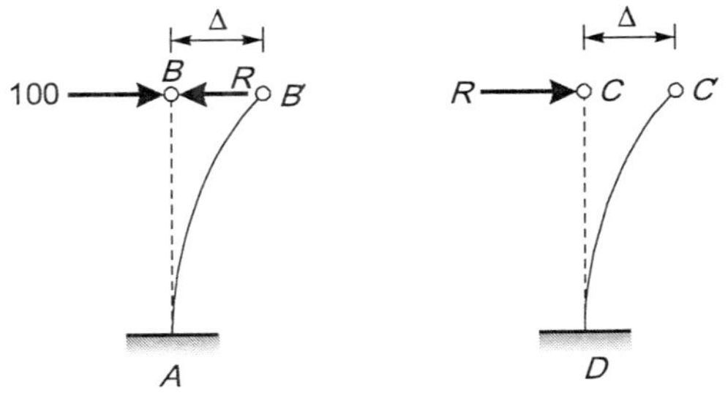

$\because \quad BB' = CC'$

$$\Rightarrow \quad \frac{(100-R)(4)^3}{3EI} = \frac{R(4)^3}{3EI}$$

$\therefore \quad R = 50$ kN

Maximum bending moment in column

$$= 50 \times 4 = 200 \text{ kNm}$$

AVOID MISTAKE

Note that BM is asked at B and C in this question, so sign convention of BM will be used.
Same directionof moments at B and C doesn't mean nature of bending moment.

i.e.,

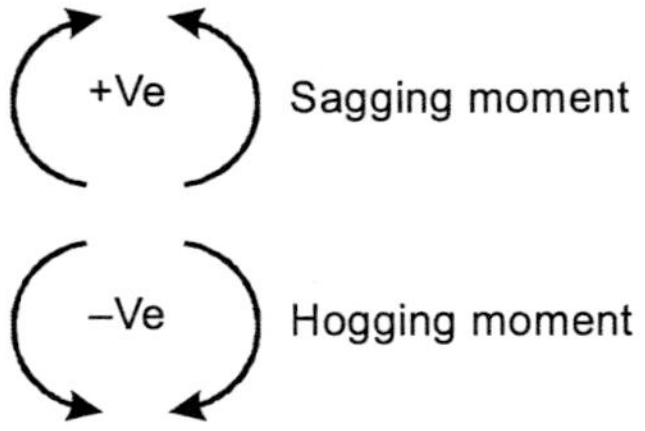

4.6 (a)

Flexibility matrix is force method and order of matrix is equal to D_s(static indeterminacy). In force method, forces are unknown so order is equal to number of redundant forces.

4.7 (d)

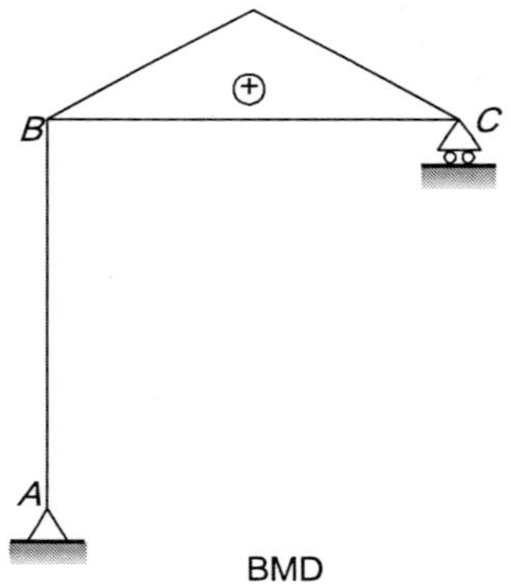

BMD

Since, there is no bending moment in vertical leg so there will be no bending in member *AB*. And to maintain angle between *AB* and *BC* at joint *B*, frame will sway as given below.

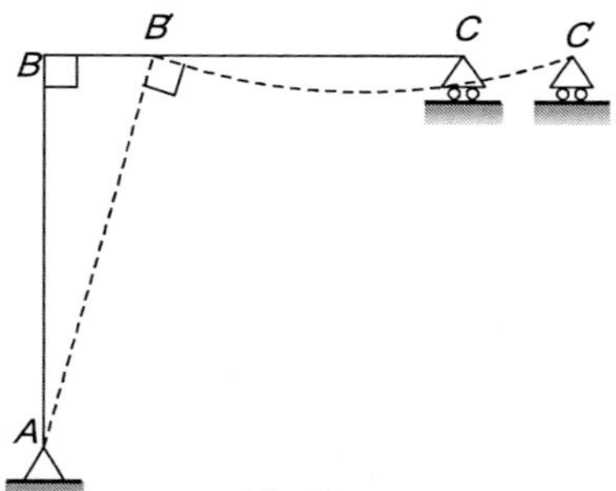

Elastic curve

4.8 (b)

MDM is a displacement method in which we calculate joint rotation by method of successive approximation.

4.9 (d)

Joint	Member	RS	TS	DF
	OA	$\frac{3}{4}\frac{I}{L}$		$\frac{3}{7}$
O	*OB*	0	$\frac{7}{4}\frac{I}{L}$	0
	OC	$\frac{I}{L}$		$\frac{4}{7}$

$$M_{OA} = M \times DF \text{ for } OA = M \times \frac{3}{7} = \frac{3}{7}M$$

$$M_{OB} = 0$$

$$M_{OC} = M \times DF \text{ for } OC = M \times \frac{4}{7} = \frac{4}{7}M$$

4.10 (b)

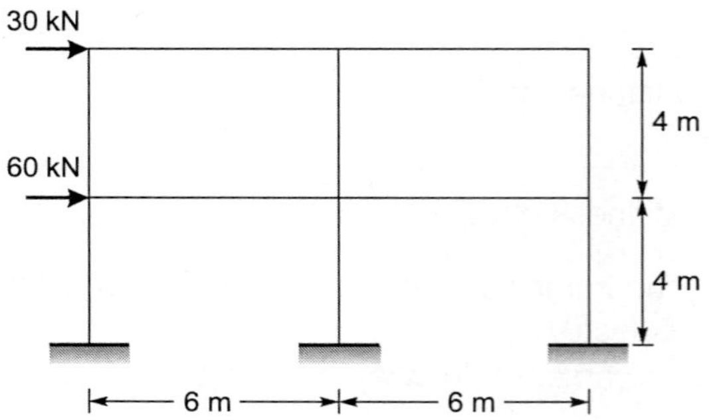

According to **cantilever method**, point of contraflexure lies at mid point of each span and direct stresses (axial) in column due to horizontal forces are directly proportional to their distance from the centroidal vertical axis of frame.
$\bar{h} = CG$ of frame will be at mid column due to symmetry.

$$\bar{h} = \frac{1 \times 0 + 1 \times 6 + 1 \times 12}{3} = 6 \text{ m}$$

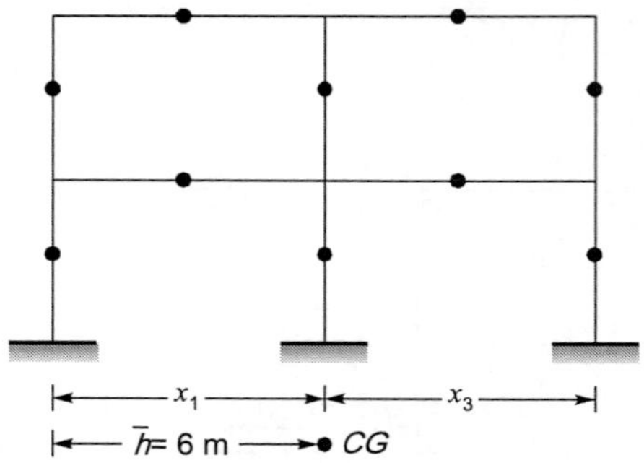

Since, area of each column are equal so axial force in each column will be proportional to their distance from CG.

$$\frac{V_1}{x_1} = \frac{V_3}{x_3}$$

As distance of V_2 from $CG = 0$,

$$\therefore \quad V_2 = 0$$

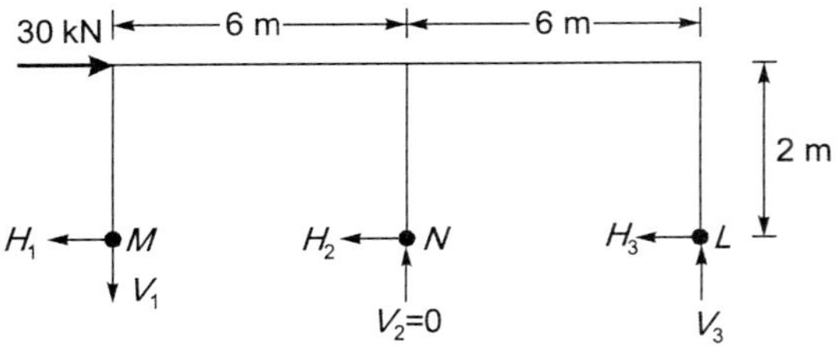

$$\frac{V_1}{6} = \frac{V_3}{6}$$

$$\Rightarrow \quad V_1 = V_3$$

Take moment about L,

$$30 \times 2 = V_1 \times 12$$

$$V_1 = 5 \text{ kN}$$

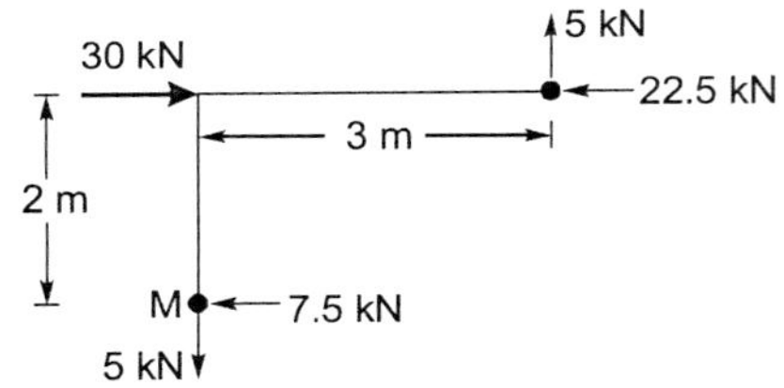

and from equilibrium of forces, end moment at roof level beam = 7.5 × 2 = 15 kNm

REFERENCE MATERIAL

1. *Read cantilever method from R.C. Hibbeler "Structural Analysis" Chapter 7*

4.11 (d)

Unit load method is used to find deflection at any point of structure. It is derived from Castigliano's theorem.

$$\Delta = \frac{\partial U}{\partial Q} = \int M \frac{\partial M}{\partial Q} \frac{dx}{EI} = \int \frac{Mm\,dx}{EI}$$

4.12 (a)

In a linear elastic structural system, for potential energy (U) to be minimum:

$$\frac{\partial U}{\partial P} = 0$$

The above equation represents the compatibility condition at the point where load P is acting.

4.13 (a)

$$H_A = H_D = 0 \; [\because D \text{ is a roller support}]$$

Let vertical reaction at A and D be R_A and R_D respectively.

Taking moments about D, we get,

$$R_A \times 5 - 10 \times 5 \times 2.5 = 0$$

$$\Rightarrow \quad R_A = 25 \text{ kN}$$

$$\therefore \quad R_D = (10 \times 5) - 25 = 25 \text{ kN}$$

As there is no horizontal reaction, beam BC will behave as a simply supported beam. Deflected shape is as shown below.

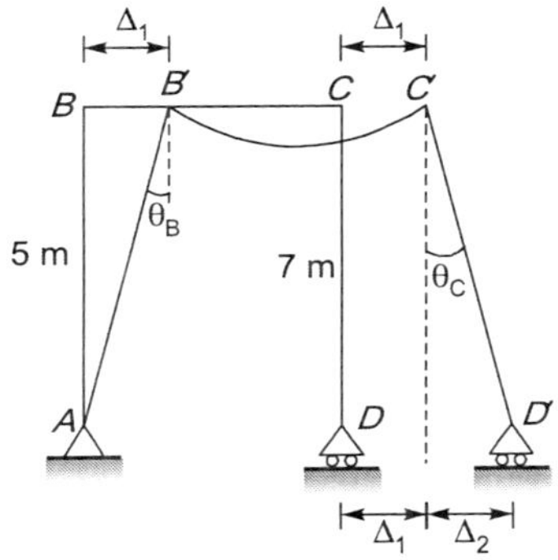

Deflected Shape

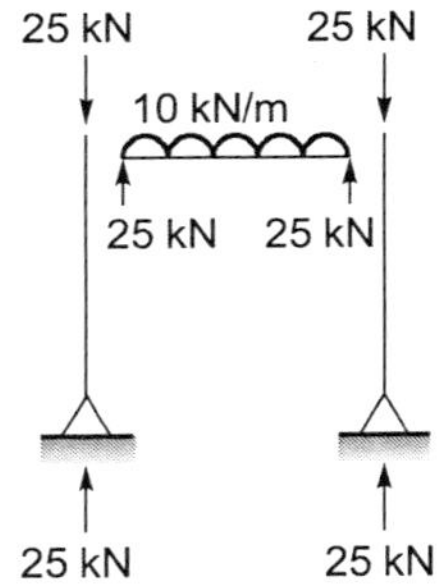

Full body diagram

Thus, deflection at the centre of beam

$$= \frac{5}{384} \frac{wL^4}{EI}$$

$$= \frac{5}{384} \times \frac{10 \times (5)^4}{81380} \times 10^3$$

$$= 1.00 \text{ mm}$$

Column AB will carry only axial force equal to R_A i.e. 25 kN.

4.14 (b)

Method-I

$\Rightarrow$ Stiffness of member $\quad AB = \dfrac{4EI}{L}$

$\Rightarrow$ Stiffness of membner $\quad BC = \dfrac{3EI}{L}$

$\Rightarrow$ Stiffness of member $\quad BD = \dfrac{4EI}{L}$

Total stiffness of the joint

$$= \frac{4EI}{L} + \frac{3EI}{L} + \frac{4EI}{L} = \frac{11EI}{L}$$

$\therefore$ Rotation of joint $\quad B = \dfrac{ML}{11EI}$

Method-II

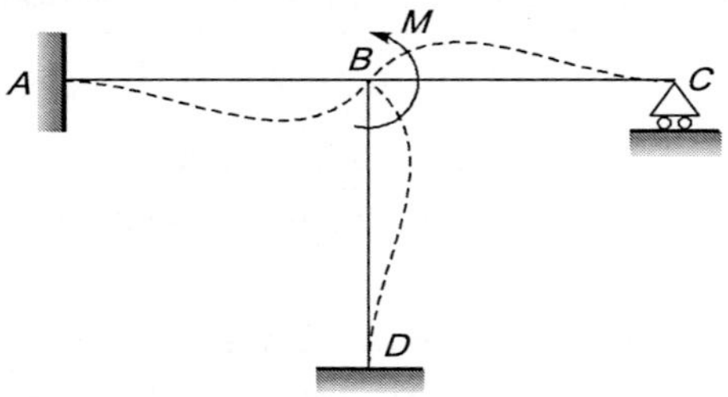

At Joint B

$$M_{BA} + M_{BD} + M_{BC} = M$$

$$\frac{4EI}{L}\theta_B + \frac{4EI}{L}\theta_B + \frac{3EI}{L}\theta_B = M$$

$$\Rightarrow \quad \frac{11EI}{L}\theta_B = M$$

$$\Rightarrow \quad \theta_B = \frac{ML}{11EI}$$

Method-III

Joint B:

Member	Stiffness	Distribution factor	Moments
BA	$\frac{4EI}{L}$	$\frac{4}{11}$	$\frac{4M}{11}$
BC	$\frac{3EI}{L}$	$\frac{3}{11}$	$\frac{3M}{11}$
BD	$\frac{4EI}{L}$	$\frac{4}{11}$	$\frac{4M}{11}$
	$\Sigma = \frac{11EI}{L}$		

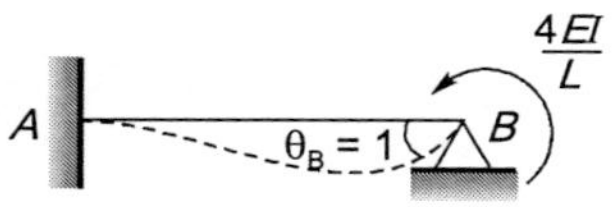

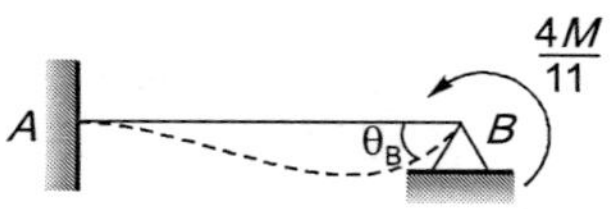

$$k_{BA} = \frac{4EI}{L}$$

$$M_{BA} = \frac{4M}{11}$$

Again $\quad M_{BA} = k_{BA} \cdot \theta_B$

$$\Rightarrow \quad \theta_B = \frac{\frac{4M}{11}}{\frac{4EI}{L}} = \frac{ML}{11EI}$$

4.15 (b)

Stiffness of BA at B = $\frac{4EI}{L}$

Stiffness of BC at B = $\frac{3EI}{L}$

Stiffness of BD at B = $\frac{4EI}{L}$

Total stiffness of joint B = $\frac{4EI}{L} + \frac{4EI}{L} + \frac{3EI}{L}$

$= \frac{11EI}{L}$

Rotation of joint B = $\frac{M}{\frac{11EI}{L}} = \frac{ML}{11EI}$

4.16 (d)

Method-I

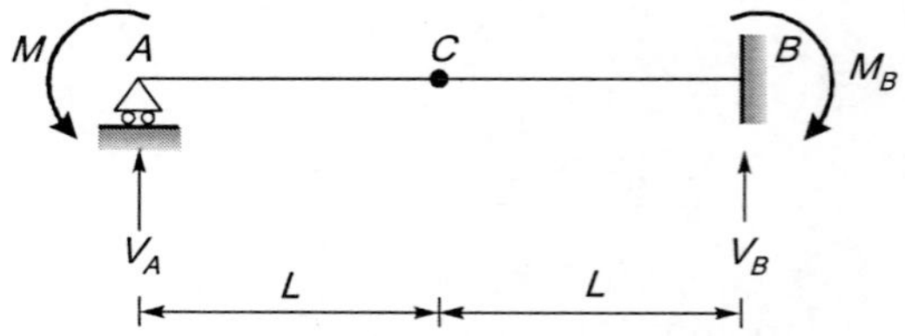

Applying a moment M at A.

Let reaction at A and B is V_A and V_B respectively, and carryover moment be M_B.

Taking $\quad M_C = 0 \quad$ (from left)

$$V_A \times L - M = 0$$

$$\Rightarrow \quad V_A = \frac{M}{L}$$

$$\therefore \quad V_B = -\frac{M}{L}$$

Now, taking $M_C = 0$ from right,

$$-M_B + V_B \times L = 0$$

$$\Rightarrow \quad M_B = \left(\frac{-M}{L}\right) \times L$$

$$\Rightarrow \quad M_B = -M$$

Carryover factor = $\frac{\text{Carryover moment}}{\text{Applied moment}}$

$= \frac{M}{M} = 1$

Method-II By FBD

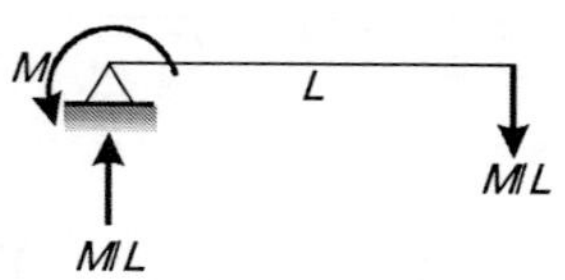

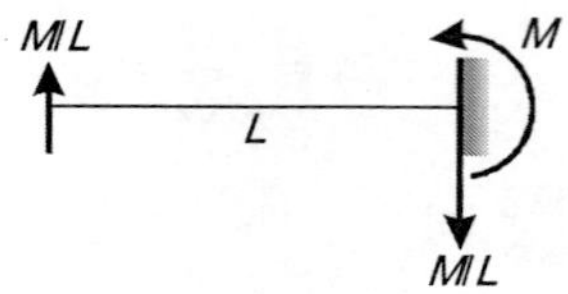

4.17 (c)

Method-I

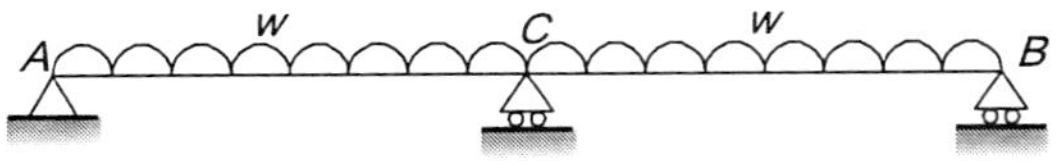

Removing support C, we get deflection at C as

$$\delta_c = \frac{5}{384} \times \frac{w(2L)^4}{EI}(\downarrow)$$

Let reaction at support C is R, then considering only R, we get deflection at C as,

$$\delta'_c = \frac{R(2L)^3}{48EI}(\uparrow)$$

For no deflection at C,

$$\delta_c - \delta'_c = 0$$

$$\Rightarrow \quad \frac{5}{384} \times \frac{w(2L)^4}{EI} = \frac{R \times (2L)^3}{48EI}$$

$$\Rightarrow R = \frac{5}{384} \times 48w \times 2L = \frac{5wL}{4}$$

Method-II

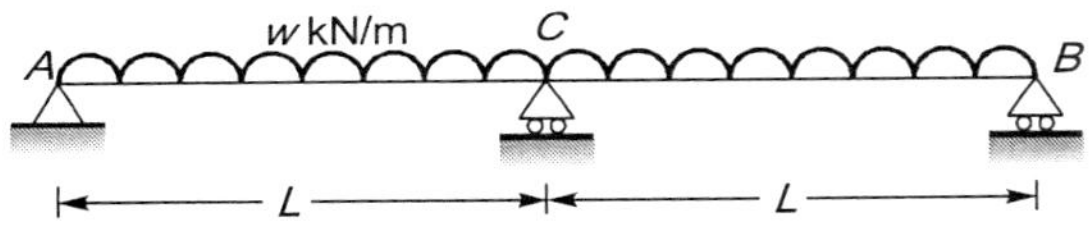

Fixed end moments:

$$M_{FAC} = -\frac{wL^2}{12}, \quad M_{FCB} = -\frac{wL^2}{12}$$

$$M_{FCA} = +\frac{wL^2}{12}, \quad M_{FBC} = +\frac{wL^2}{12}$$

Distribution factors:

$$k_{BA} = k_{BC} = 0.5$$

($\because$ Identical members)

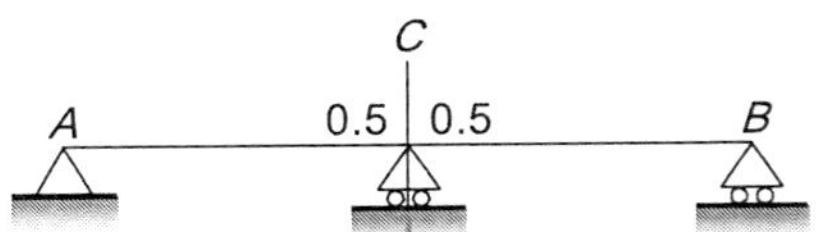

	A	C	C	B
DF		0.5	0.5	
FEMs	$\frac{-wL^2}{12}$	$\frac{+wL^2}{12}$	$\frac{-wL^2}{12}$	$\frac{+wL^2}{12}$
Unlocking θ_A & θ_B	$\frac{+wL^2}{12}$			$\frac{-wL^2}{12}$
Carry over		$\frac{+wL^2}{24}$	$\frac{-wL^2}{24}$	
	0	$\frac{+wL^2}{8}$	$\frac{-wL^2}{8}$	0

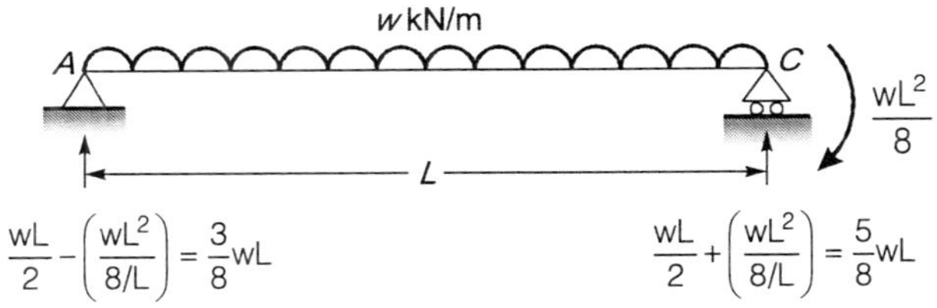

$$R_B = \frac{5}{8}wL \times 2 = \frac{5}{4}wL$$

[$\because$ Reaction at B from both the spans AB and BC wil be same]

$$BM_B = -\frac{wL^2}{8}$$

4.18 (b)

Bending moment at the support $C = \frac{wL^2}{8}$ from distribution table method-II.

Or, from method-I

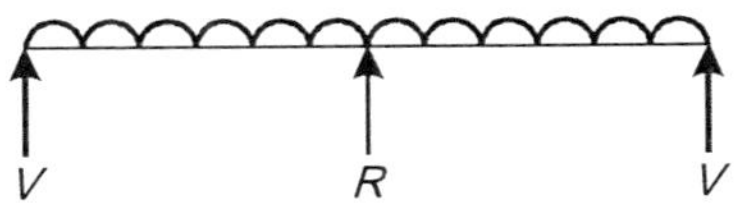

$$\Sigma F_y = 0$$

$$\Rightarrow 2V + R - w(2L) = 0$$

$$\Rightarrow \quad V = \frac{2wL - \frac{5wL}{4}}{2} = \frac{3wL}{8}$$

Now, $\quad BM_C = V \times L - \frac{wL^2}{2} = -\frac{wL^2}{8}$

4.19 Sol.

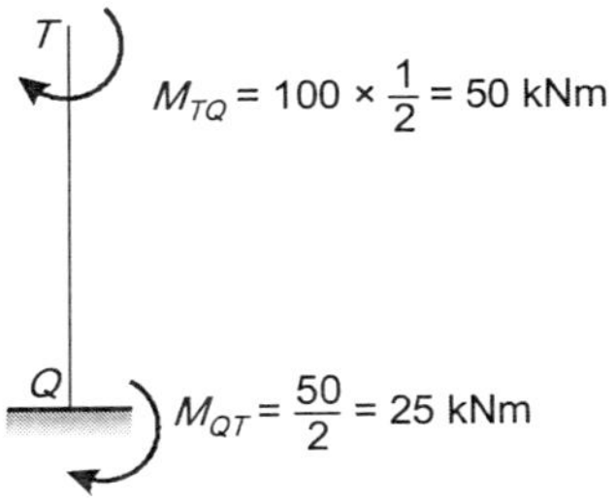

Joint	Member	RS	TRS	DF
T	TP	$\frac{3}{4}\frac{I}{3} = \frac{I}{4}$	I	$\frac{1}{4}$
	TS	0		0
	TR	$\frac{I}{4}$		$\frac{1}{4}$
	TQ	$\frac{I}{2}$		$\frac{1}{2}$

$\therefore \; M_Q = 25$ kNm

4.20 (c)

As the frame and loading is symmetrical, $\theta_P = 0$. So, joint P can be replaced by a fixed support as shown.

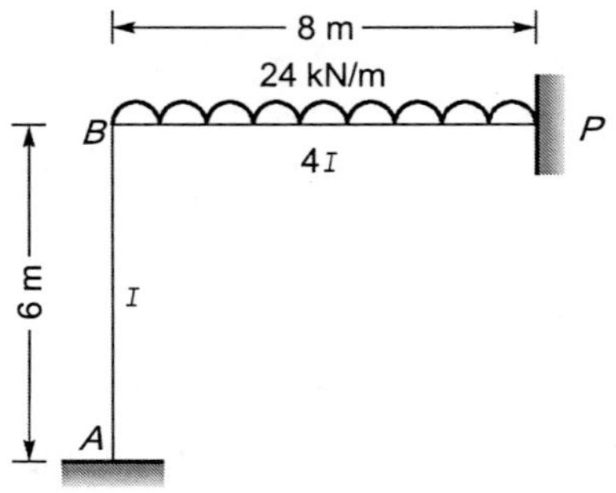

Fixed end moment:

$M_{fAB} = 0, \quad M_{fBA} = 0$

$$M_{fBP} = -\frac{wl^2}{12} = -\frac{24 \times 8^2}{12} = -128 \text{ kNm}$$

$$M_{fPB} = +\frac{wl^2}{12} = +128 \text{ kNm}$$

Distribution factor:

Joint B	Member	Stiffness	Distribution factor
	BP	4I/8	3/4
	BA	1/6	1/4
		$\Sigma = \frac{2}{3}I$	

	A	B		P
	AB	BA	BP	PB
D.F.	0	0.25	0.75	0
Fem	0	0	–128	+128
Balance		32	96	
COM	16			48
Final moment	+16	+32	–32	+176

$BM_P = -176$ kNm

By symmetry, the BMD is as shown

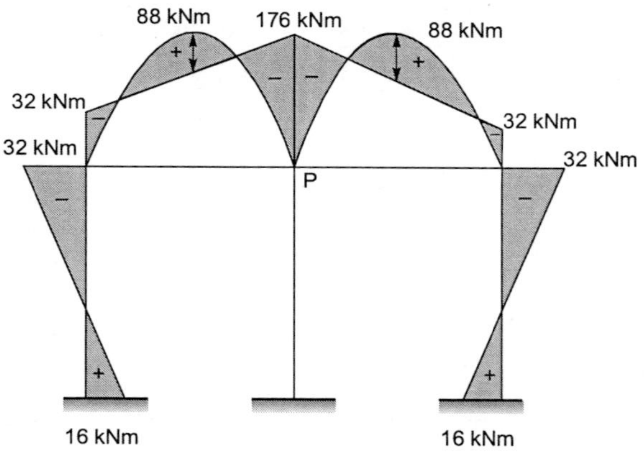

4.21 Sol.

By symmetry we can consider two parts of beam separately with half the load.

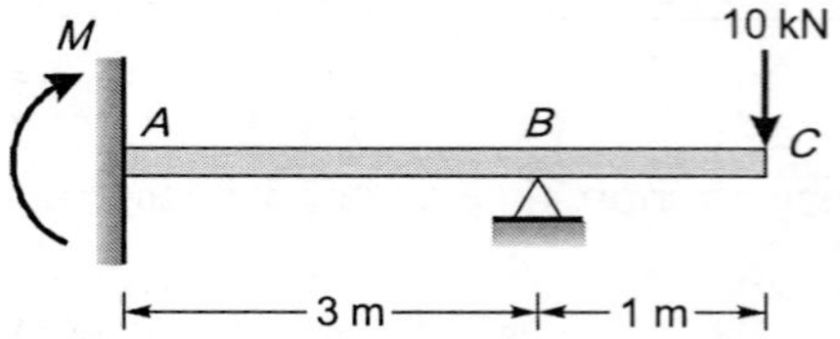

Moment at joint B = 10 kNm

Carry over moment at joint A due to 10 kNm

moment at propped end $= \frac{10}{2} = 5$ kN-m

4.22 (a)

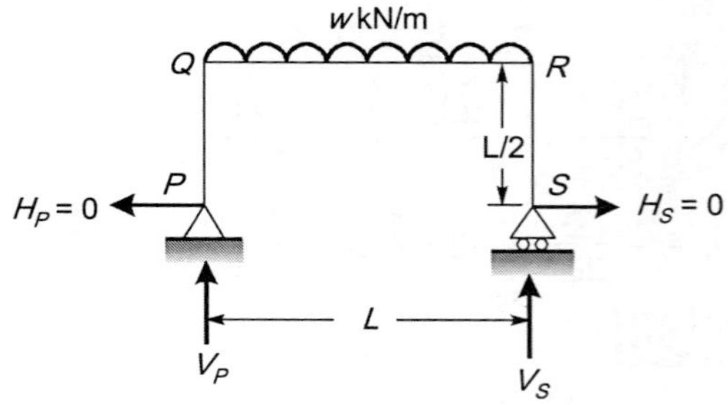

As there is no horizontal force,

Hence, $\quad H_P = H_S = 0$

$$\Sigma M_Q = H_P \times \frac{L}{2} = 0$$

$\therefore \quad$ BM at Q = 0

4.23 Sol.

Fixed end moment:

$$M_{fAB} = \frac{-30 \times 4^2}{12} = -40 \text{ kNm}$$

$$M_{fBA} = +40 \text{ kNm}$$

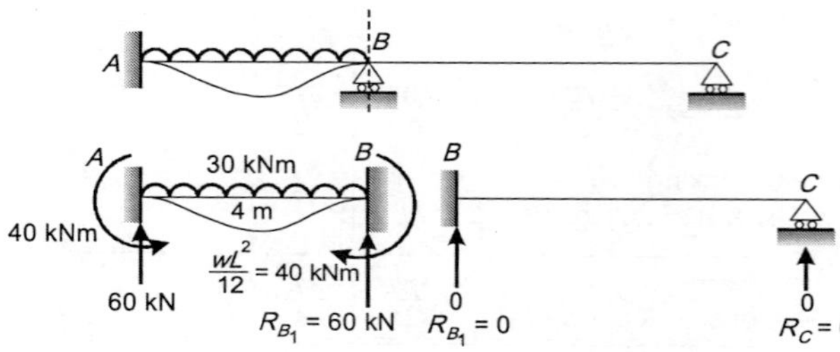

By super position,

$R_B = R_{B_1} + R_{B_2} = 60$ kN

Note : The value of 'M' shown in the figure,

$M = M_{BA} + M_{BC} = 40$ kNm

POINTS TO REMEMBER

There is no need to balance moments at *B* as the final slope at *B* is to be zero, which means there is an imaginary fixed support at *B*.

4.24 Sol.

Method-I

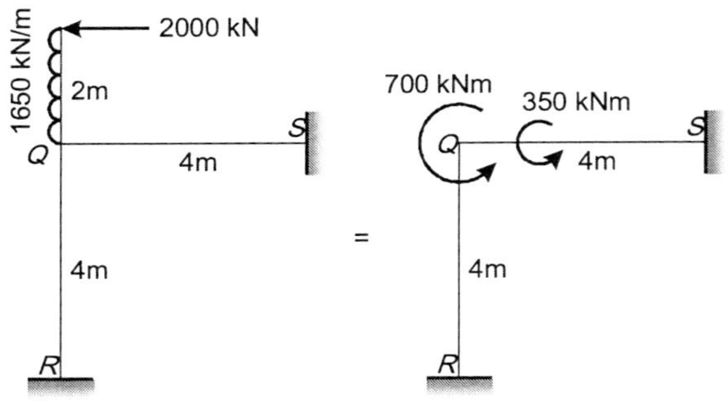

$$\text{Moment at } Q = \frac{1650\times 2^2}{2} - 2000\times 2$$

$$= -700 \text{ kNm (ACW)}$$

$$EI = 2.5 \times 10^4 \times 8 \times 10^8$$

$$= 20 \times 10^{12} \text{ Nmm}^2$$

$$= 20000 \text{ kNm}^2$$

As the two members *QR* and *QS* are identical, moment will be equally divided i.e., 350 kNm each.

$$\theta_Q = \frac{Ml}{4EI} = \frac{350\times 4}{4\times 20000}$$

$$= 0.0175 \text{ rad}$$

$$= \frac{0.0175\times 180°}{\pi} = 1.0027°$$

Method-II

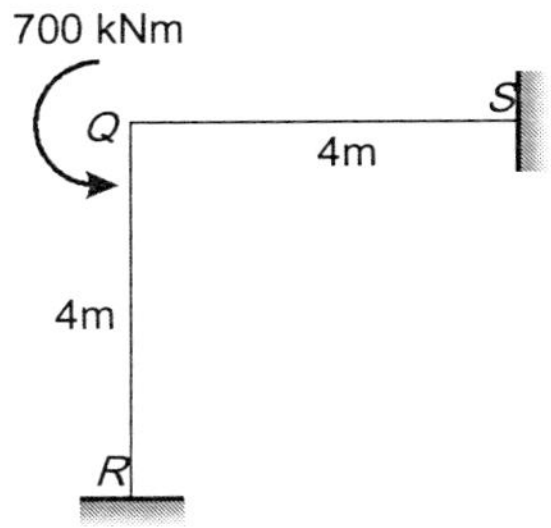

Fixed end moment:

$$M_{fRQ} = M_{fQR} = M_{fQS} = M_{fSQ} = 0 \text{ kNm}$$

Also, $\theta_R = \theta_S = 0$ (Fixed support at R & S)

Slope deflection equation:

$$M_{QR} = 0 + \frac{2EI}{4}(2\theta_Q + \theta_R) = EI\theta_Q$$

$$M_{QS} = 0 + \frac{2EI}{4}(2\theta_Q + \theta_S) = EI\theta_Q$$

Joint equilibrium equation at joint Q,

$$M_{QR} + M_{QS} = -700$$

$$\Rightarrow \quad 2EIQ_Q = -700$$

$$\Rightarrow \quad Q_Q = -\frac{350}{EI} = -\frac{350}{20000}$$

$$= 0.0175 \text{ rad.} = 1.0027°$$

4.25 (a)

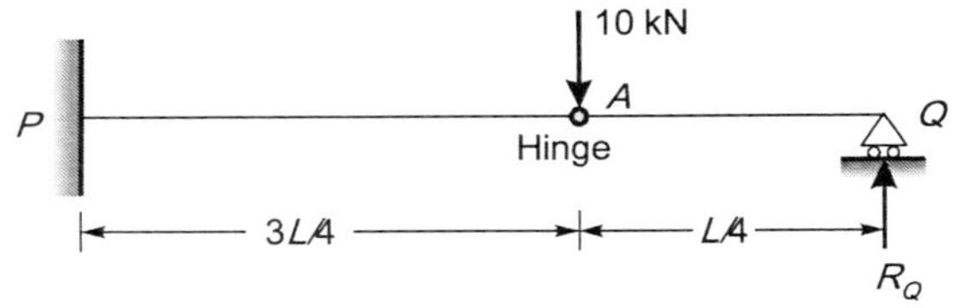

Bending moment about hinge point *A* = 0 (consider the right hand side of *A*)

$$R_Q \times \frac{L}{4} = 0$$

$$R_Q = 0 \text{ kN}$$

4.26 Sol.

Method-I

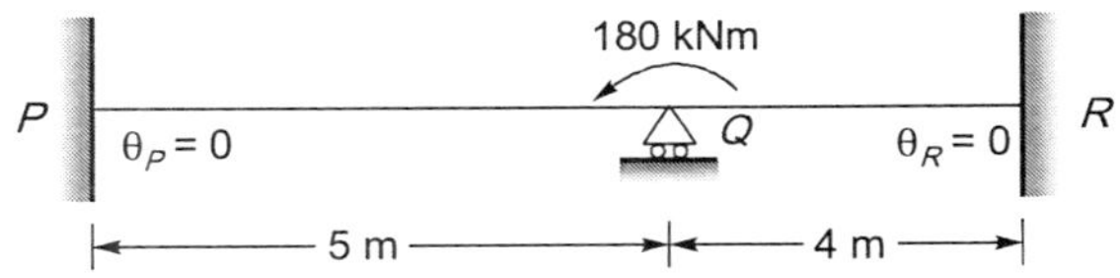

$$K_Q = K_{QP} + K_{QR}$$

$$= \frac{4EI}{5} + \frac{4EI}{4} = 1.8\ EI$$

$$\theta_Q = \frac{M_Q}{K_Q} = \frac{180}{1.8\times 10^4} = 0.01 \text{ rad}$$

Method-II

$$M_{QP} = M_{fQP} + \frac{2EI}{l}[2\theta_Q + \theta_P]$$

$$\{M_{fQP} = 0,\ \theta_P = 0\}$$

$$= \frac{4\times 10^4}{5}\times\theta_Q = 8000\theta_Q$$

$$M_{QR} = M_{fQR} + \frac{2EI}{l}[2\theta_Q + \theta_R]$$

$$\{M_{fQR} = 0,\ \theta_R = 0\}$$

$$= \frac{2\times10^4}{4}[2\theta_Q] = 10000Q$$

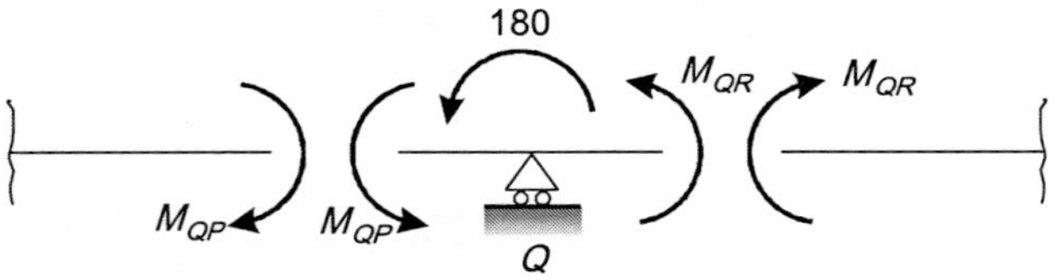

$$\Sigma M_Q = 0$$

$$\Rightarrow -M_{QP} - M_{QR} - 180 = 0$$

$$\Rightarrow 18000\cdot\theta_Q = -180$$

$$\Rightarrow \theta_Q = 0.01 \text{ radian (anticlockwise)}$$

4.27 (a)

Assume 'R' sinks by Δ.

Sway analysis:

$$M_{F_{QR}} = \frac{-6EI\delta}{l^2}$$

$$M_{F_{RQ}} = \frac{-6EI\delta}{l^2}$$

$$M_{F_{RS}} = M_{F_{SR}} = 0$$

D.F. at R | 0.5 | 0.5 |

End moment distribution,

	Q	R (0.5)	R (0.5)	S
FEM	$\frac{-6EI\delta}{l^2}$	$\frac{-6EI\delta}{l^2}$	0	0
Balance		$\frac{3EI\delta}{l^2}$	$\frac{3EI\delta}{l^2}$	
COM	$\frac{1.5EI\delta}{l^2}$			$\frac{1.5EI\delta}{l^2}$
Final end moments	$\frac{-4.5EI\delta}{l^2}$	$\frac{-3EI\delta}{l^2}$	$\frac{3EI\delta}{l^2}$	$\frac{1.5EI\delta}{l^2}$

Sway force

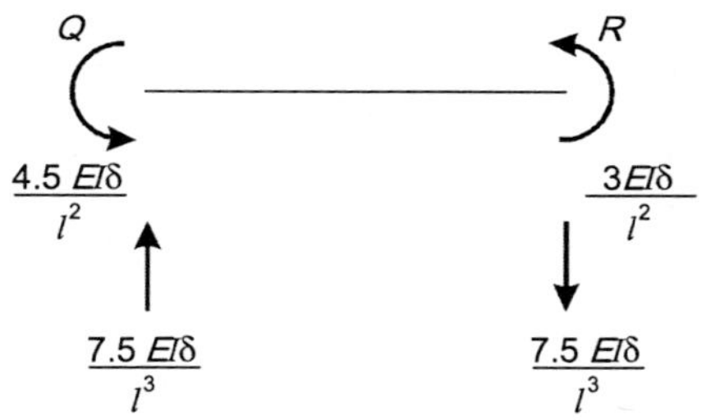

So, Sway force,

$$P = \frac{7.5EI\delta}{l^3}$$

So, $\delta = \dfrac{Pl^3}{7.5EI}$

$\therefore$ $\beta = 7.5$

Alternate method:

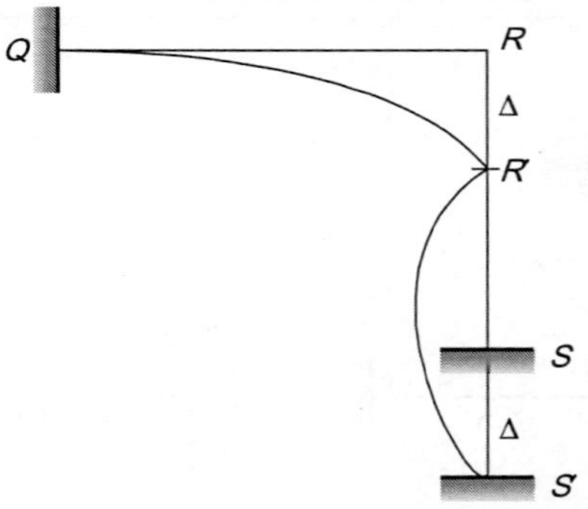

Elastic Curve

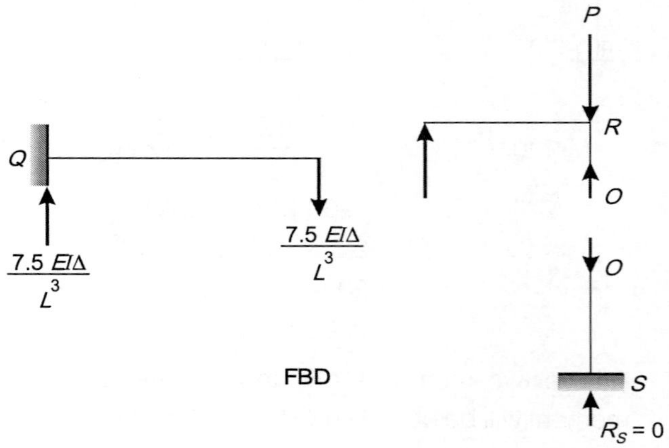

FBD

$$\Sigma F_v = 0 \text{ (At joint R)}$$

$$\therefore \quad P = \frac{7.5EI\Delta}{L^3}$$

$$\therefore \quad \Delta = \frac{PL^3}{7.5EI}$$

4.28 Sol.

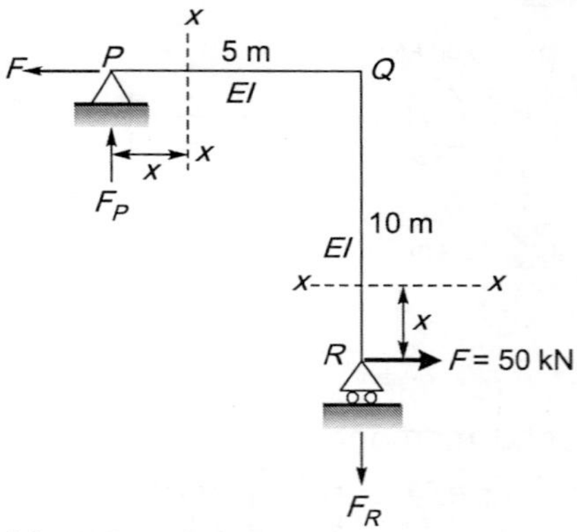

$$\Sigma M_P = 0$$

$$F \times 10 - F_R \times 5 = 0$$

$$F_R = 2F = F_P$$

$$\Delta_R = \frac{\partial U}{\partial F} = \frac{\partial U_{PQ}}{\partial F} + \frac{\partial U_{QR}}{\partial F}$$

$$= \frac{1}{EI}\int M_{x_{PQ}} \times \frac{\partial M_{x_{PQ}}}{\partial F}dx + \frac{1}{EI}\int M_{x_{QR}} \times \frac{\partial M_{x_{QR}}}{\partial F}dx$$

$$M_{x_{PQ}} = F_P \times x = 2F\cdot x$$

$$\frac{\partial M_{x_{PQ}}}{\partial F} = 2x;\quad M_{x_{QR}} = Fx;\quad \frac{\partial M_{x_{QR}}}{\partial F} = x$$

$$\therefore\quad \Delta R = \frac{1}{EI}\int_0^5 4Fx^2dx + \frac{1}{EI}\int_0^{10} Fx^2dx$$

$$= \frac{4F}{EI}\left[\frac{x^3}{3}\right]_0^5 + \frac{F}{EI}\left[\frac{x^3}{3}\right]_0^{10}$$

$$= \frac{200}{EI}\times\frac{125}{3} + \frac{50}{EI}\times\frac{1000}{3} = \frac{25000}{EI}$$

$$= \frac{25000}{10^6}\text{ m} = 25\text{mm}$$

4.29 Sol.

Force in PQ

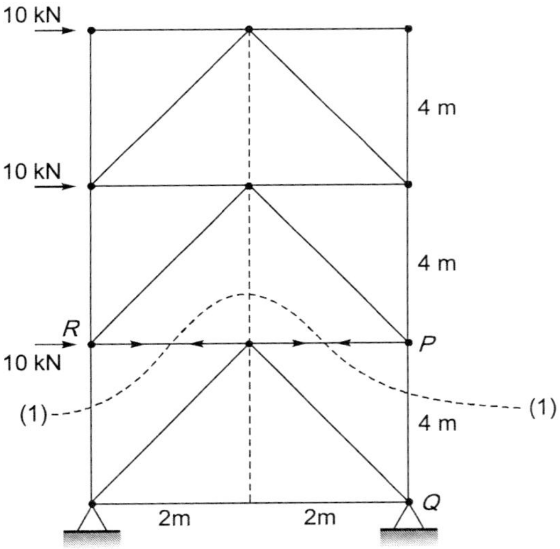

Considering the section above (1) – (1)

Taking moment about 'R'

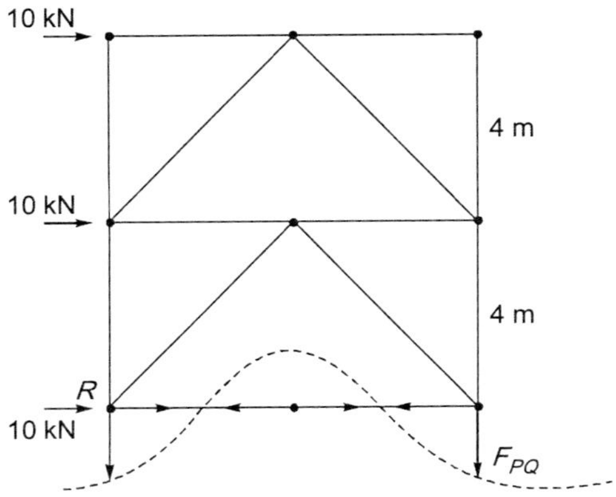

$$\Sigma M_{RL} = 0$$

$$(10 \times 4) + (10 \times 8) + F_{PQ} \times 4 = 0$$

$$F_{PQ} = -\frac{120}{4} = -30\text{ kN}$$

$$= 30\text{ kN (Comp.)}$$

4.30 (c)

Compatibility condition

$$\frac{\partial U}{\partial R} = 0$$

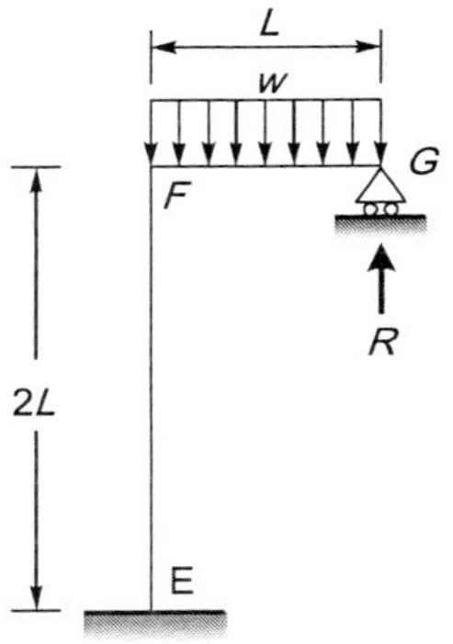

$$\Rightarrow\quad \frac{\partial}{\partial R}\left[\int \frac{M^2}{2EI}dx\right] = 0$$

$$\Rightarrow\quad \frac{\partial}{\partial R}\left[\int_0^L \frac{\left(Rx - \frac{wx^2}{2}\right)^2}{2EI}dx + \int_0^{2L} \frac{\left(RL - \frac{wL^2}{2}\right)^2}{2EI}dx\right] = 0$$

$$\Rightarrow \int_0^L \frac{2\left(Rx - \frac{wx^2}{2}\right)}{2EI}(x)dx + \int_0^{2L}\frac{\left(RL - \frac{wL^2}{2}\right)}{EI}(L)dx = 0$$

$$\Rightarrow \int_0^L \frac{\left(Rx^2 - \frac{wx^3}{2}\right)dx}{EI} + \int_0^{2L}\frac{\left(RL^2 - \frac{wL^3}{2}\right)dx}{EI} = 0$$

$$\Rightarrow \frac{RL^3}{3} - \frac{w}{2}\times\frac{L^4}{4} + RL^2(2L) - \frac{wL^3}{2}(2L) = 0$$

$$\Rightarrow \frac{RL^3}{3} - \frac{wL^4}{8} + 2RL^3 - wL^4 = 0$$

$$\Rightarrow 2RL^3 + \frac{RL^3}{3} = \frac{wL^4}{8} + wL^4 = \frac{9}{8}wL^4$$

$$\Rightarrow \frac{7RL^3}{3} = \frac{9}{8}wL^4$$

$$\Rightarrow R = \frac{27}{56}wL = 0.482wL$$

Alternative

$$\frac{wL^4}{8EI} + \frac{wL^2(2L)}{2EI}\times L = \frac{RL^3}{3EI} + \frac{RL\times 2L\times L}{EI}$$

$$\frac{wL^4}{8EI} + \frac{wL^4}{EI} = \frac{RL^3}{3EI} + \frac{2RL^3}{EI}$$

$$\frac{9wL^4}{8EI} = \frac{7RL^3}{3EI}$$

$$R = \frac{27wL}{56} = 0.482wL$$

4.31 Sol.

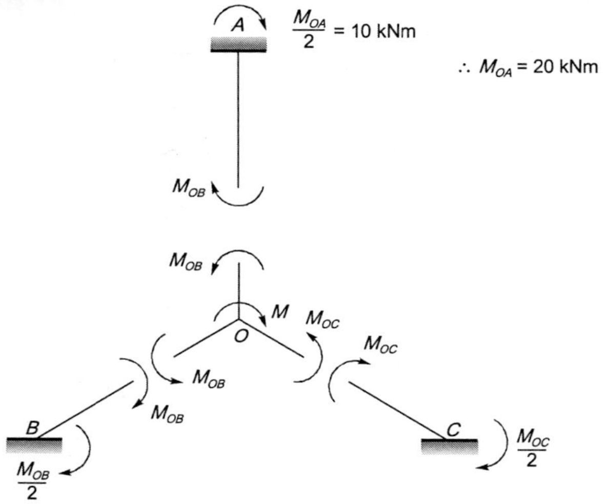

Joint	Member	RS	TRS	DF
O	OB	1/5	35/5	1/3
	OB	1/5		1/3
	OC	1/5		1/3

$$M_{OB} = M \text{ DF for } OA$$

$$\Rightarrow \quad M_{OB} = M \times \frac{1}{3}$$

$$\Rightarrow \quad M = 20 \times 3 = 60 \text{ mm}$$

4.32 (a, b, d)

(a) As per Bett's theorem,

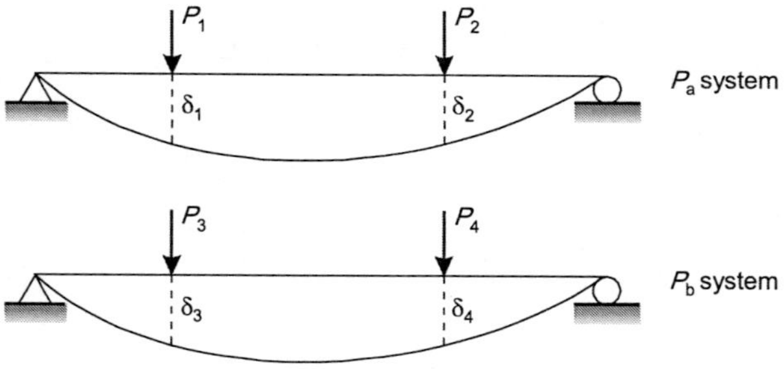

$$P_1\delta_3 + P_2\delta_4 = P_3\delta_1 + P_4\delta_2$$

(b) As per Castigliano Theorem-II,

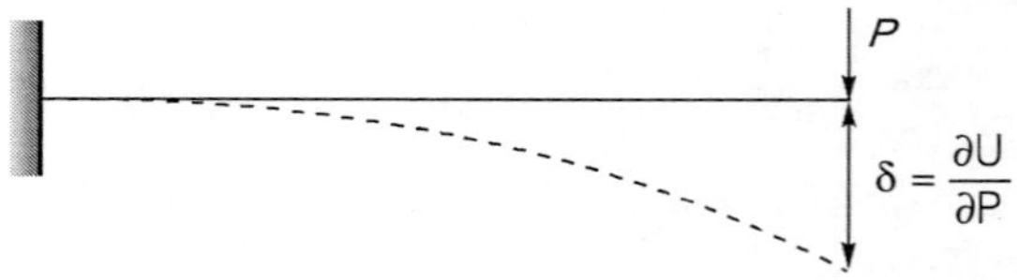

U = total strain energy, P = Load applied

(c) The shear force in conjugate beam loaded by M/EI diagram of real beam equal to corresponding slope of real beam. Therefore, given statement is incorrect.

(d) As per Castigliano Theorem-I,

$$\frac{\partial U}{\partial \delta} = P$$

4.33 Sol.

Given that there is no force in spring, so deformation in spring will be zero.

From here it can be concluded that distance travelled by V and W are equal.

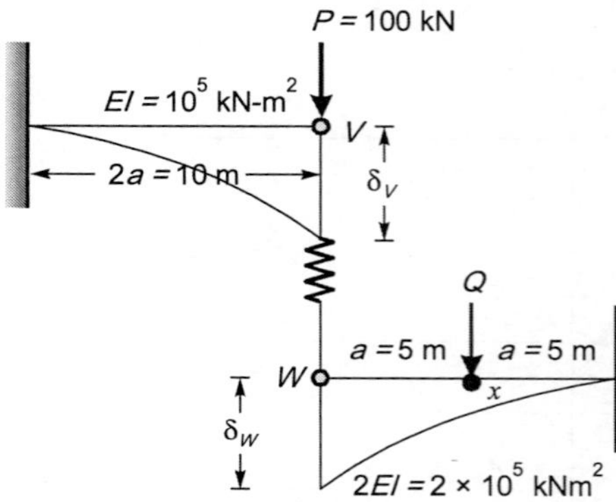

$$\delta_V = \delta_W$$

$$\Rightarrow \quad \delta_V = \delta_x + Q_x L_{wx}$$

$$\Rightarrow \quad \frac{P(2a)^3}{3EI} = \frac{Qa^3}{3(2EI)} + \frac{Qa^2}{2(2EI)}(a)$$

$$\Rightarrow \quad \frac{100(10)^3}{3(10^5)} = \frac{Q(5)^3}{3(2\times10^5)} + \frac{Q(5)^2(5)}{2(2\times10^5)}$$

$$\therefore \quad Q = 640 \text{ kN}$$

4.34 (a, b, d)

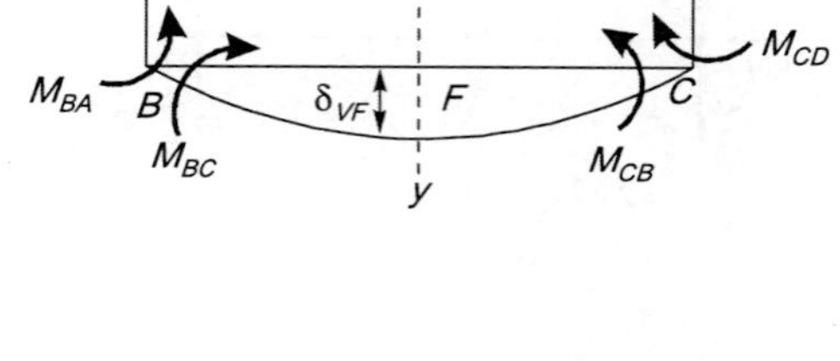

Shear forces at H and F are zero since it is subjected to pure bending.

Horizontal displacements at H and F are zero since axial deformations are neglected.

At the axis of symmetry, slopes must be zero for the deflection profile to be symmetrical.

$\therefore$ Slopes at E, F, G, H are zero.

Note: In actual question unstable equilibrium is given but in the figure equal force P is acting so there should be stable equilibrium as forces are balancing each other.

4.35 Sol.

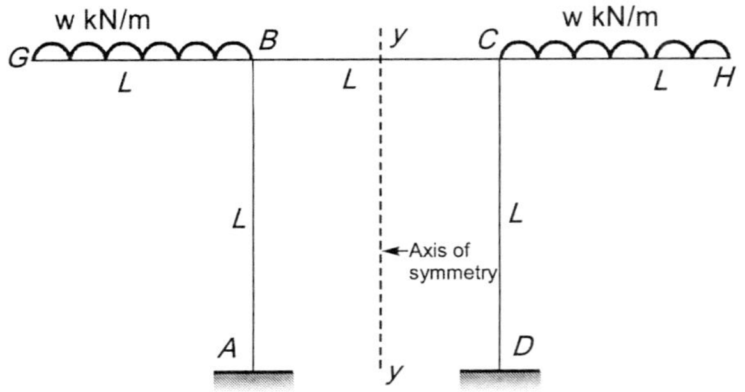

Distribution factors

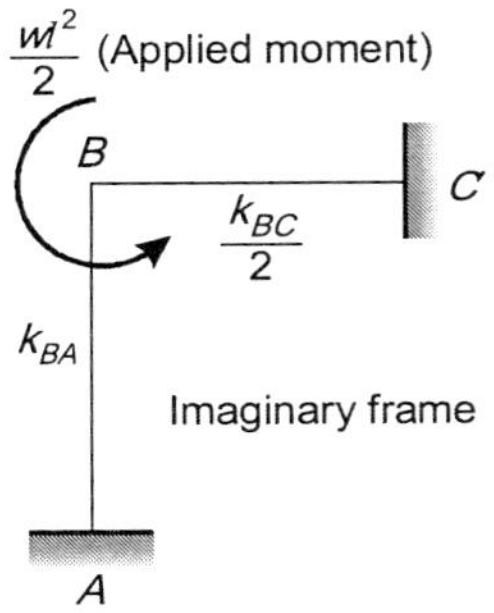

Joint	Member	k	Σk	D.F = $k/\Sigma k$
B	BA	$\frac{1}{L}$	$\frac{3I}{2L}$	$\frac{2}{3}$
	BC	$\frac{1}{2}\left(\frac{1}{L}\right)$		$\frac{1}{3}$

End moment distribution

Joint	A	B		C
D.F.	0	$\frac{2}{3}$	$\frac{1}{3}$	—
Balance		$\frac{-wl^2}{3}$	$\frac{-wl^2}{6}$	
C.O.M.	$\frac{-wl^2}{6}$			—
Final end moments	$\frac{-wl^2}{6}$	$\frac{-wl^2}{3}$	$\frac{-wl^2}{6}$	—

$$\therefore \quad \text{Moment at } A = \frac{wl^2}{6}$$

$$\therefore \quad \alpha = 6$$

Alternatively by stiffness matrix

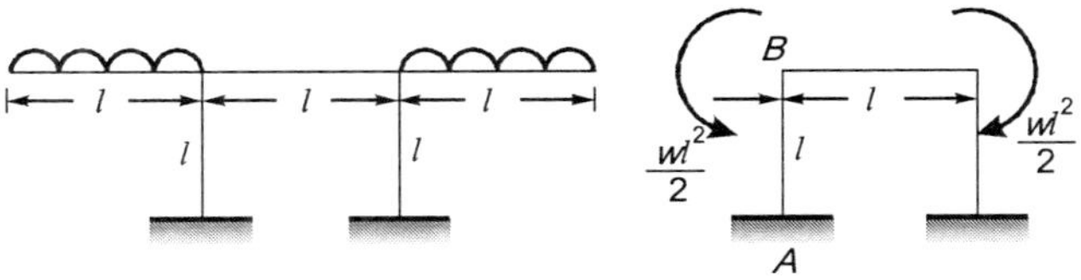

$$D_k = 1 \text{ (i.e. } \theta_3)$$
$$\theta_c = f(\theta_B)$$

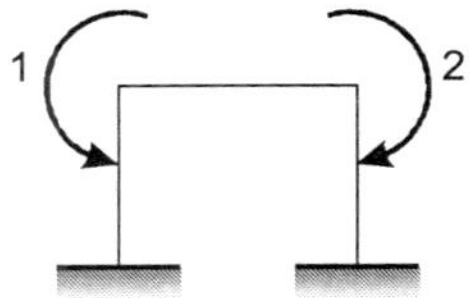

Stiffness matrix

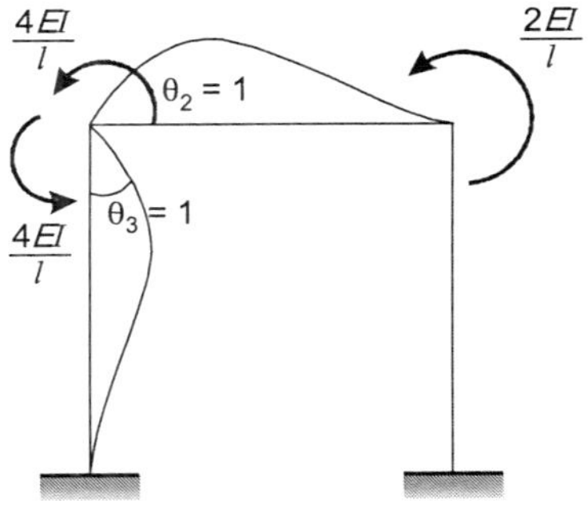

$$[k] = \begin{bmatrix} \frac{8EI}{l} & \frac{2EI}{l} \\ \frac{2EI}{l} & \frac{8EI}{l} \end{bmatrix} = \frac{2EI}{l}\begin{bmatrix} 4 & 1 \\ 1 & 4 \end{bmatrix}$$

Now, $\quad [P] = [P_L] + [k][\Delta]$

$$\Rightarrow \quad 0 = \begin{bmatrix} \frac{-wl^2}{2} \\ \frac{wl^2}{2} \end{bmatrix} + \frac{2EI}{l}\begin{bmatrix} 4 & 1 \\ 1 & 4 \end{bmatrix}\begin{bmatrix} \theta_B \\ \theta_C \end{bmatrix}$$

$$\frac{wl^2}{2} = \frac{8EI}{l}\theta_B + \frac{2EI}{l}\theta_C$$

$$\frac{wl^2}{2} = \frac{6EI}{l}\theta_B \qquad (\because \theta_B = -\theta_C)$$

$$\therefore \quad \theta_B = \frac{wl^2}{12EI}$$

$$\therefore \quad M_{BA} = \frac{4EI}{l}\theta_B = \frac{4EI}{l}\frac{wl^2}{12EI} = \frac{wl^2}{3}$$

$$\therefore \quad M_{BA} = \frac{M_{BA}}{2} = \frac{wl^2}{6}$$

Alternatively by

Slope deflection method

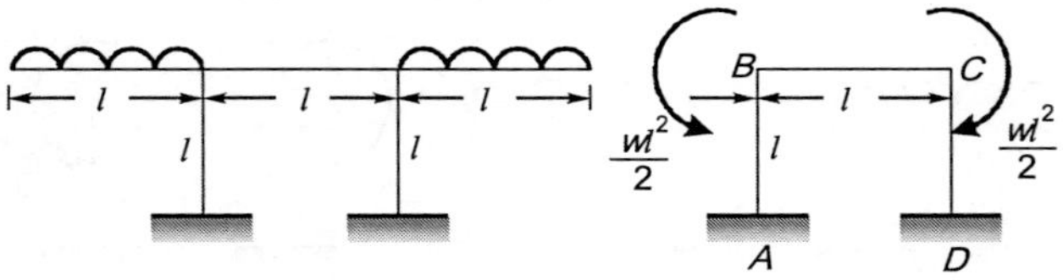

$$M_{BA} = \frac{2EI}{l}(2\theta_B + \theta_A) = \frac{4EI}{l}\theta_B$$

$$M_{BC} = \frac{2EI}{l}(2\theta_B + \theta_C)$$

$\theta_B = -\theta_C$ due to symmetry

$$\therefore \quad M_{BC} = \frac{2EI}{l}\theta_B + \frac{4EI}{l}\theta_B$$

Joint B equilibrium equation, comliting

$$M_{BA} + M_{BC} = \frac{wl^2}{2}$$

$$\Rightarrow \quad \frac{6EI}{l}\theta_B = \frac{wl^2}{2}$$

$$\Rightarrow \quad \theta_B = \frac{+wl^3}{12EI}$$

$$M_{BA} = \frac{4EI}{l} \times \frac{wl^3}{12EI} = \frac{+wl^2}{3EI}$$

$$\therefore \quad M_B = \frac{M_{BA}}{2} = \frac{M_{BA}}{2} = \frac{wl^2}{6EI}$$

■■■■

5 Trusses

5.1 The forces in members '*a, b, c*' in the truss shown are, respectively

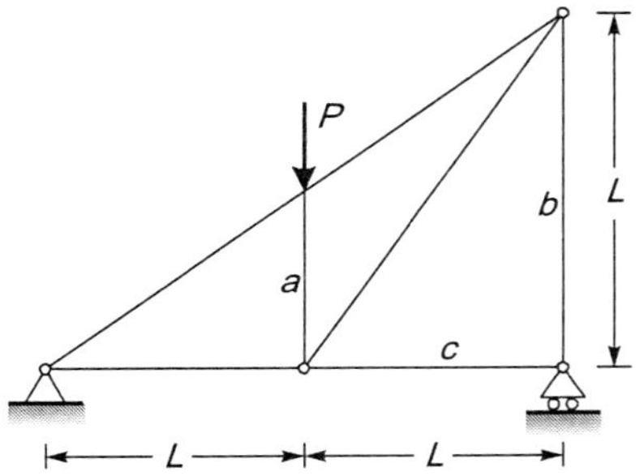

(a) $P, \frac{P}{2}, 0$ (b) $\frac{P}{2}, P, 0$

(c) P, P, P (d) $\frac{P}{2}, \frac{P}{2}, 0$ [1995 : 2 M]

5.2 If the deformation of the truss members are as shown in parantheses, the rotation of the member *bd* is

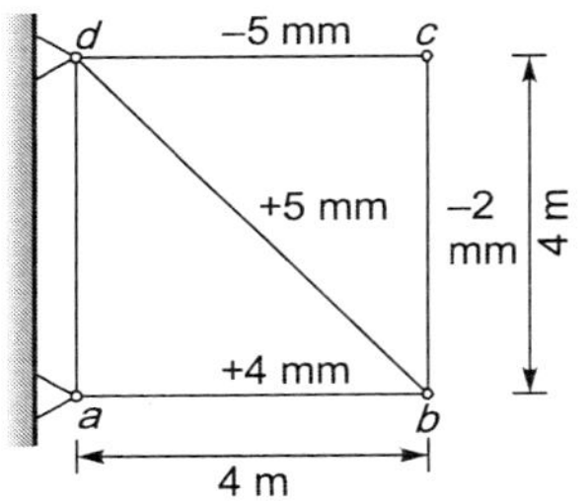

(a) 0.5×10^{-3} radian (b) 1.0×10^{-2} radian

(c) 1.5×10^{-2} radian (d) 2.0×10^{-2} radian

[1996 : 1 M]

5.3 The force in the member *DE* of the truss shown in the figure is

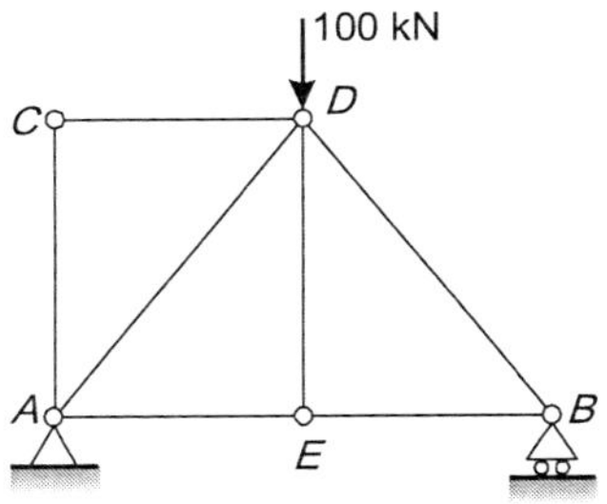

(a) 100.0 kN (b) Zero

(c) 35.5 kN (d) 25.0 kN

[1997 : 1 M]

5.4 The strain energy stored in member *AB* of the pin-jointed truss as shown in figure where *E* and *A* are same for all members, is

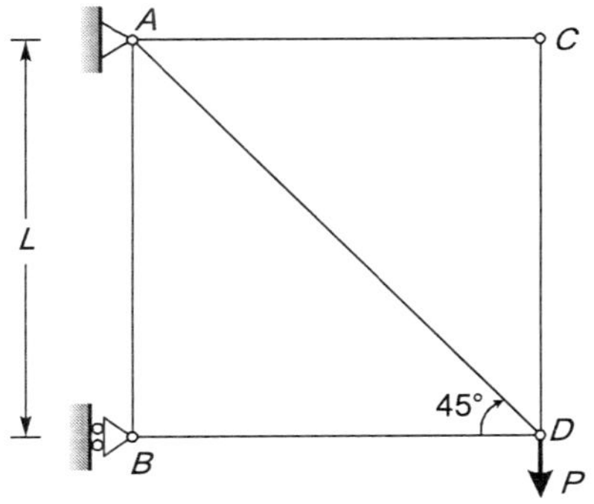

(a) $\frac{2P^2L}{AE}$ (b) $\frac{P^2L}{AE}$

(c) $\frac{P^2L}{2AE}$ (d) Zero

[1998 : 2 M]

5.5 Identify the false statement from the following, pertaining to the effects due to a temperature rise ΔT in the bar *BD* alone in the plane truss shown below:

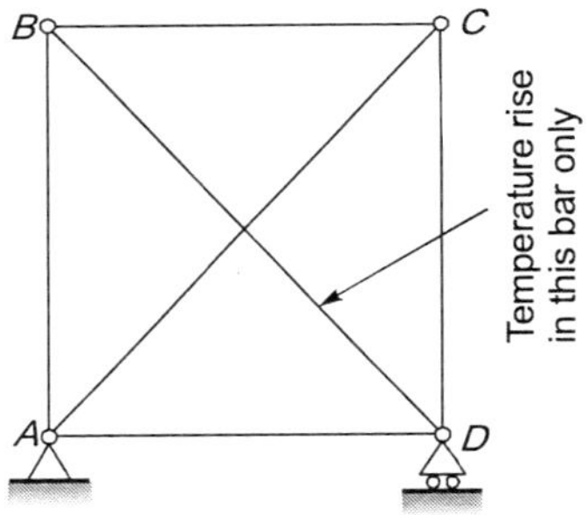

(a) No reactions develop at supports *A* and *D*

(b) The bar *BD* will be subject to a tensile force

(c) The bar *AC* will be subject to a compressive force

(d) The bar *BC* will be subject to a tensile force

[2001 : 1 M]

5.6 In a redundant joint model, three bar members are pin connected at *Q* as shown in the figure. Under some load placed at *Q*, the elongation of the members *MQ* and *OQ* are found to be 48 mm and 35 mm respectively. Then the horizontal displacement '*u*' and the vertical displacement '*v*' of the node *Q*, in mm, will be respectively,

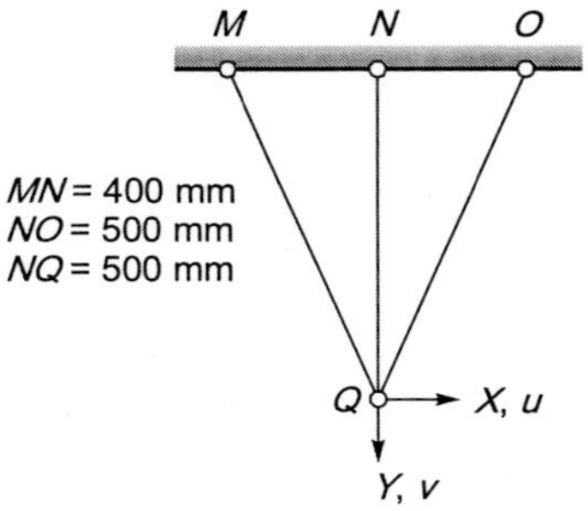

(a) –6.64 and 56.14 (b) 6.64 and 56.14
(c) 0.0 and 59.41 (d) 59.41 and 0.0

[2003 : 2 M]

5.7 A truss, as shown in the figure, is carrying 180 kN load at node L_2. The force in the diagonal member M_2U_4 will be

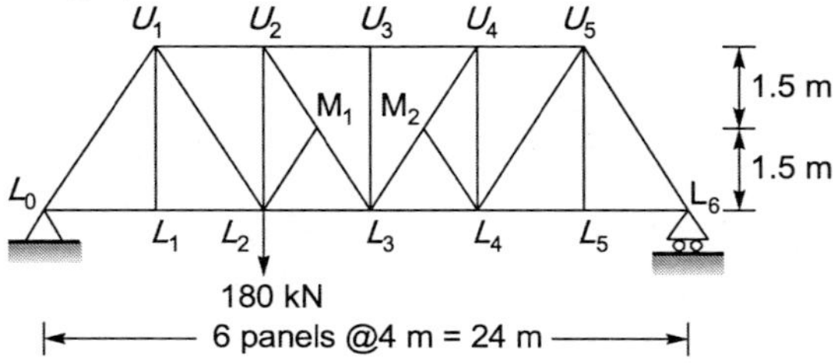

(a) 100 kN tension
(b) 100 kN compression
(c) 80 kN tension
(d) 80 kN compression [2003 : 2 M]

5.8 For the plane truss shown in the figure, the number of zero force members for the given loading is

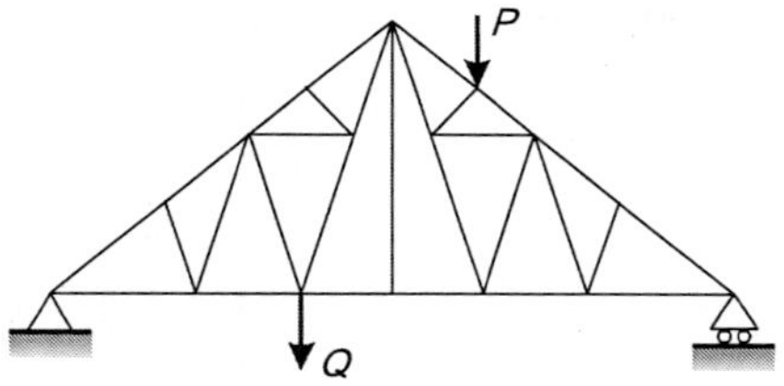

(a) 4 (b) 8
(c) 11 (d) 13 [2004 : 1 M]

Common Data for (5.9 and 5.10)

A truss is shown in the figure. Members are of equal cross section A and same modulus of elasticity E. A vertical force P is applied at point C.

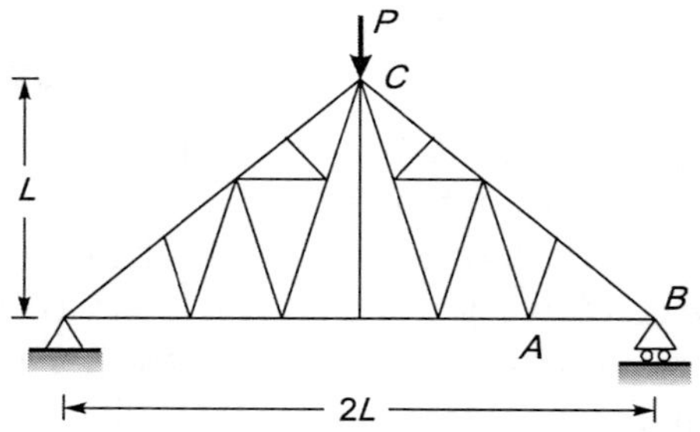

5.9 Force in the member AB of the truss is

(a) $\frac{P}{\sqrt{2}}$ (b) $\frac{P}{\sqrt{3}}$
(c) $\frac{P}{2}$ (d) P [2005 : 2 M]

5.10 Deflection of the point C is

(a) $\left(\frac{2\sqrt{2}+1}{2}\right)\frac{PL}{EA}$ (b) $\sqrt{2}\frac{PL}{EA}$
(c) $(2\sqrt{2}+1)\frac{PL}{EA}$ (d) $(\sqrt{2}+1)\frac{PL}{EA}$

[2005 : 2 M]

5.11 The right triangular truss is made of members having equal cross sectional area of 1550 mm^2 and Young's modulus of 2×10^5 MPa. The horizontal deflection of the joint Q is

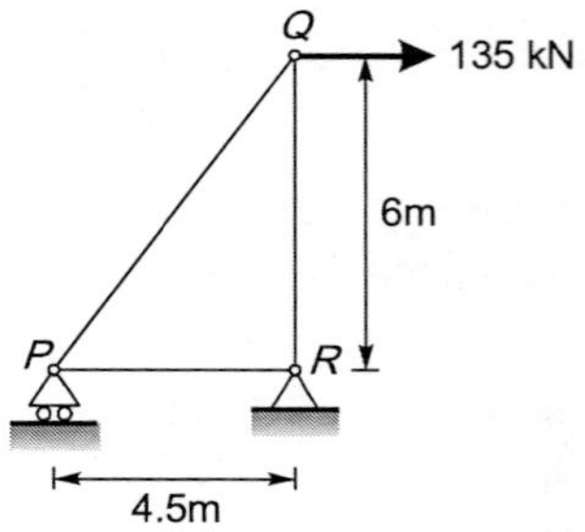

(a) 2.47 mm (b) 10.25 mm
(c) 14.31 mm (d) 15.68 mm

[2007 : 2 M]

5.12 The members EJ and IJ of a steel truss shown in the figure below are subjected to a temperature rise of 30°C. The coefficient of thermal expansion of steel is 0.000012 per °C per unit length. The displacement (mm) of joint E relative to joint H along the direction HE of the truss, is

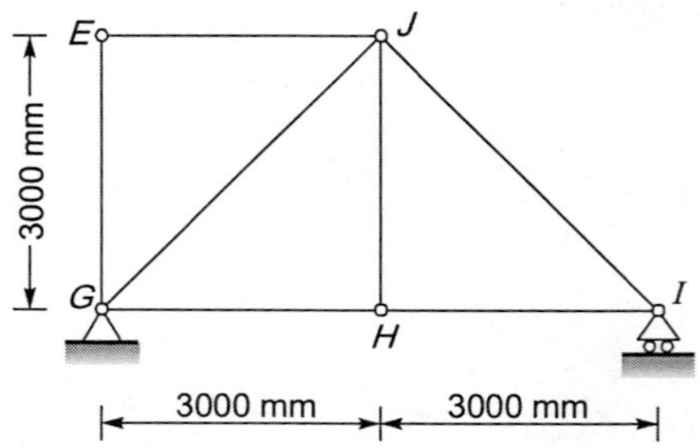

(a) 0.255 (b) 0.589
(c) 0.764 (d) 1.026 [2008 : 2 M]

5.13 For the truss shown in the figure, the force in the member QR is

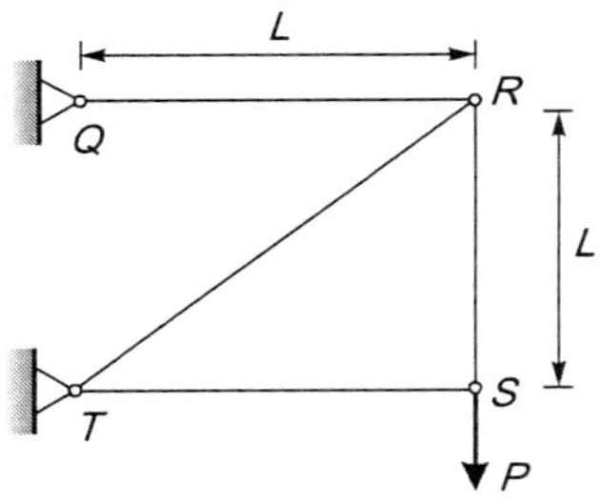

(a) zero (b) $\frac{P}{\sqrt{2}}$

(c) P (d) $\sqrt{2}P$ [2010 : 1 M]

5.14 The pin-jointed 2-D truss is loaded with a horizontal force of 15 kN at joint S and another 15 kN vertical force at joint U as shown in figure. Find the force in member RS (in kN) and report your answer taking tension as +ve and compression as –ve.

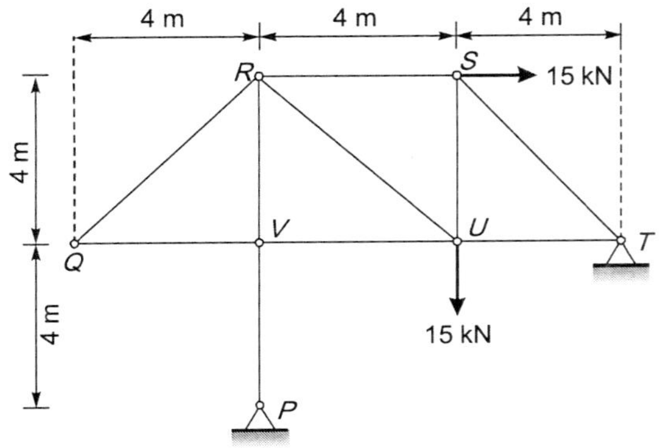

[2013 : 1 M]

5.15 A uniform beam weighing 1800 N is supported at E & F by cable $ABCD$. Determine the tension force in segment AB at this cable (correct to 1 decimal place). Assume the cable $ABCD$, BE and CF are weightless

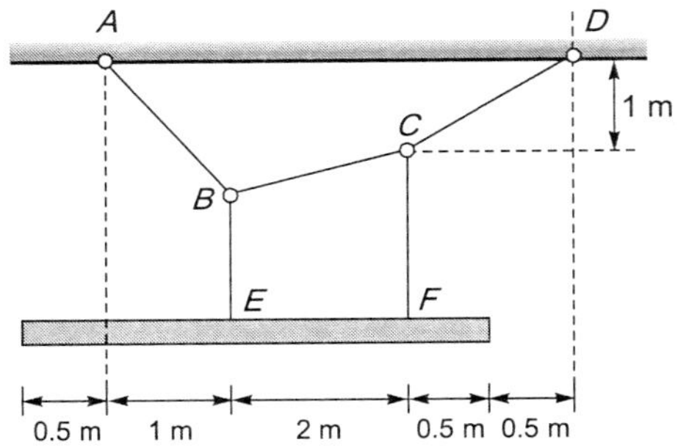

[2013 : 2 M]

5.16 Mathematical idealization of a crane has three bars with their vertices arranged as shown in the figure with a load of 80 kN hanging vertically. The coordinates of the vertices are given in parentheses. The force in the member QR, F_{QR} will be

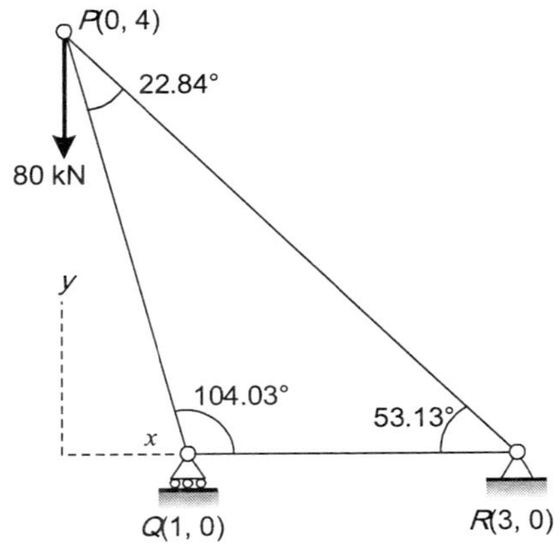

(a) 30 kN Compressive
(b) 30 kN Tensile
(c) 50 kN Compressive
(d) 50 kN Tensile [2014 : 2 M, Set-I]

5.17 For the truss shown below, the member PQ is short by 3 mm. The magnitude of vertical displacement of joint R (in mm) is ________.

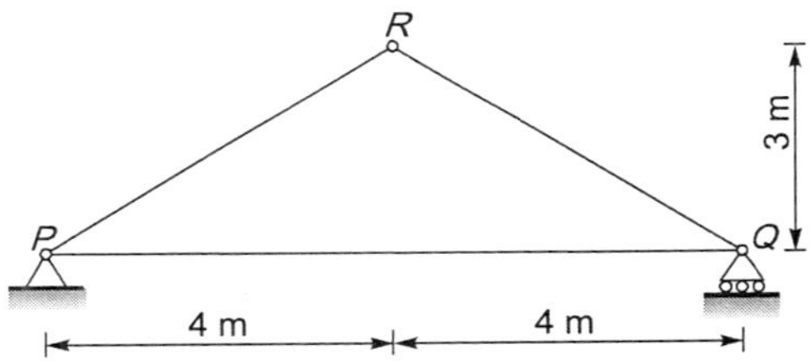

[2014 : 2 M, Set-I]

5.18 For the 2D truss with the applied loads shown below, the strain energy in the member XY is ____________ kN-m. For member XY, assume AE = 30 kN, where A is cross-section area and E is the modulus of elasticity.

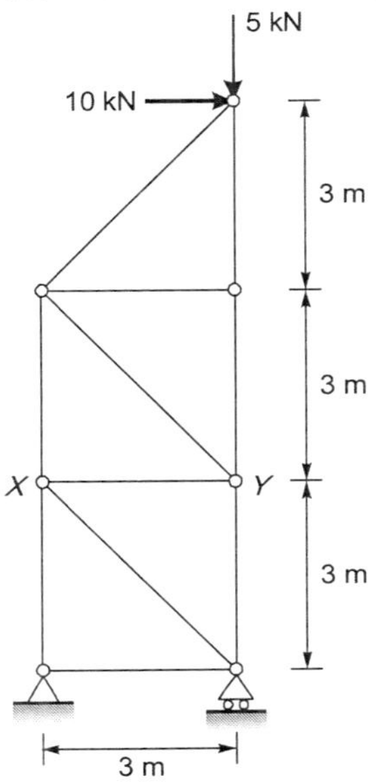

[2015 : 2 M, Set-I]

5.19 Consider the plane truss with load P as shown in the figure. Let the horizontal and vertical reactions at the joint B be H_B and V_B, respectively and V_C be vertical reaction at the joint C.

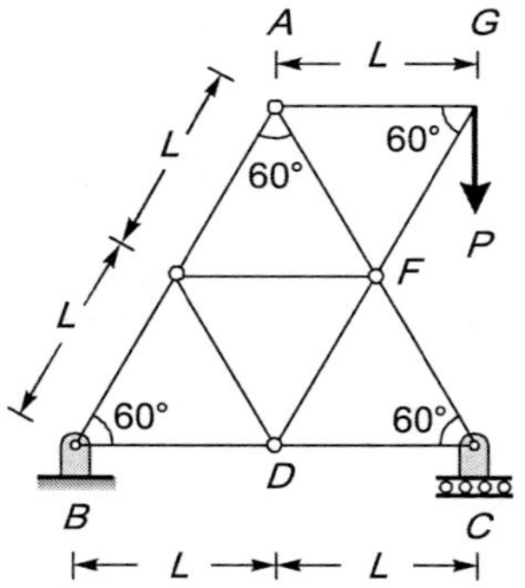

Which one of the following sets gives the correct values of V_B, H_B and V_C?

(a) $V_B = 0;\ H_B = 0;\ V_C = P$
(b) $V_B = P/2;\ H_B = 0;\ V_C = P/2$
(c) $V_B = P/2;\ H_B = P\ (\sin 60°);\ V_C = P/2$
(d) $V_B = P;\ H_B = P\ (\cos(60°));\ V_C = 0$

[2016 : 1 M, Set-I]

5.20 A plane truss with applied loads is shown in the figure.

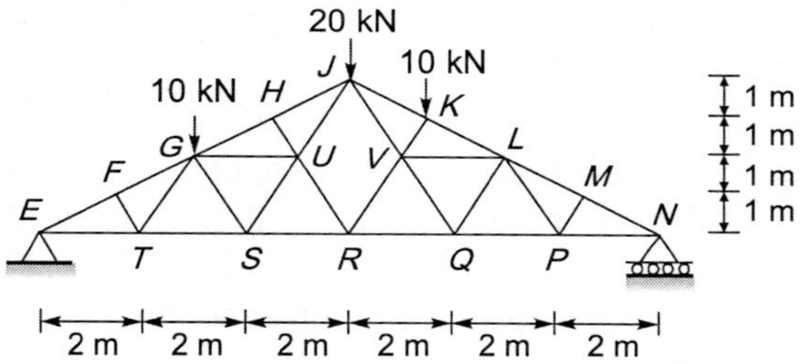

The members which do not carry any force are

(a) FT, TG, HU, MP, PL
(b) ET, GS, UR, VR, QL
(c) FT, GS, HU, MP, QL
(d) MP, PL, HU, FT, UR

[2016 : 2 M, Set-I]

5.21 Consider the structural system shown in the figure under the action of weight W. All the joints are hinged. The properties of the members in terms of length (L), are (A) and the modulus of elasticity (E) are also given in the figure. Let L, A and E be 1 m, 0.05 m^2 and 30×10^6 N/m^2, respectively, and W be 100 kN.

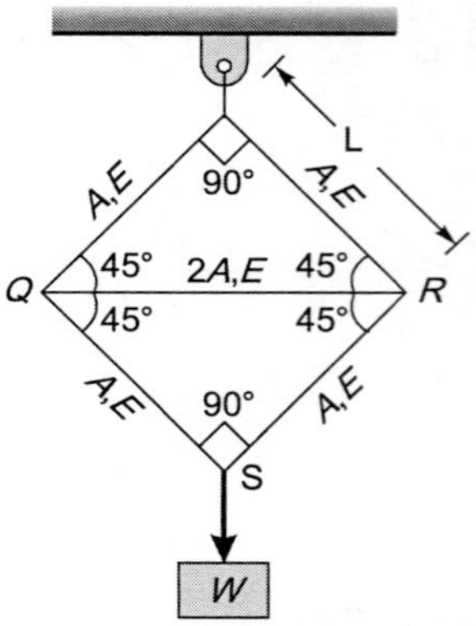

Which one of the following sets gives the correct values of the force, stress and change in length of the horizontal member QR?

(a) Compressive force = 25 kN;
Stress = 250 kN/m^2 ; Shortening = 0.0118 m
(b) Compressive force = 14.14 kN;
Stress = 141.4 kN/m^2; Extension = 0.0118 m
(c) Compressive force = 100 kN;
Stress = 1000 kN/m^2; Shortening = 0.0417 m
(d) Compressive force = 100 kN;
Stress = 1000 kN/m^2; Extension = 0.0417 m

[2016 : 2 M, Set-II]

5.22 Consider the deformable pin-jointed truss with loading, geometry and section properties as shown in figure.

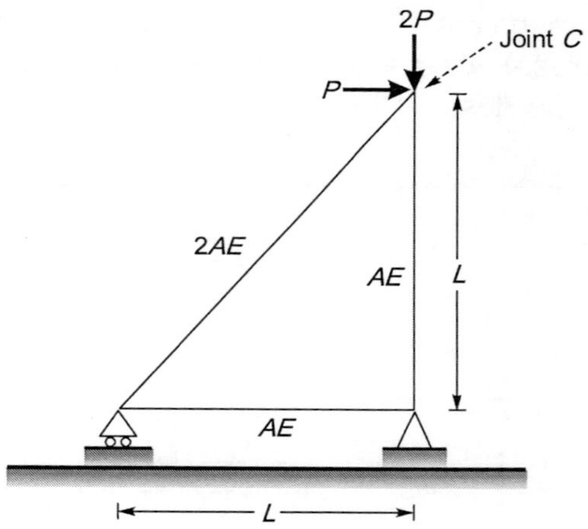

Given that $E = 2 \times 10^{11}$ N/m^2, $A = 10$ mm^2, $L = 1$ m and $P = 1$ kN. the horizontal displacement of Joint C (in mm, up to one decimal place) is ________

[2018 : 2 M, Set-I]

5.23 All the members of the planar truss (see figure), have the same properties in terms of area of cross-section (A) and modulus of elasticity (E),

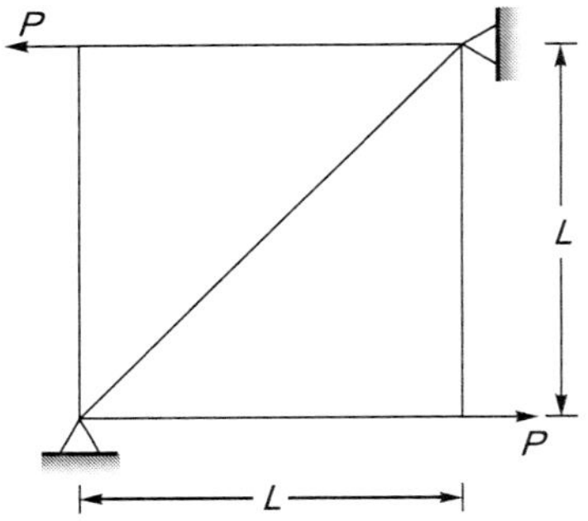

For the loads shown on the truss, the statement that correctly represents the nature of forces in the members of the truss is:

(a) There are 3 members in tension, and 2 members in compression

(b) There are 2 members in tension, 2 members in compression, and 1 zero-force member

(c) There are 2 members in tension, 1 member in compression, and 2 zero-force members

(d) There are 2 members in tension, and 3 zero-force Members

[2018 : 1 M, Set-II]

5.24 A plane truss is shown in the figure (not drawn to scale).

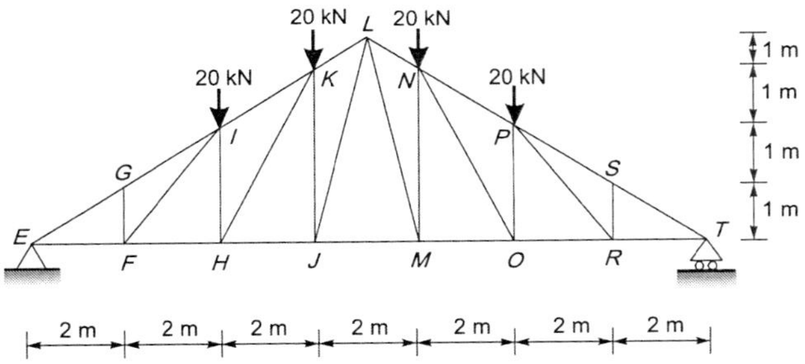

Which one of the following options contains ONLY zero force members in the truss?

(a) FG, FI, HI, RS (b) FI, FG, RS, PR

(c) FG, FH, HI, RS (d) FI, HI, PR, RS

[2019 : 1 M, Set-I]

5.25 Consider the pin-jointed plane truss shown in figure (not drawn to scale). Let R_P, R_Q, and R_R denote the vertical reactions (upward positive) applied by the supports at P, Q and R, respectively, on the truss. The correct combination of (R_P, R_Q, R_R) is represented by

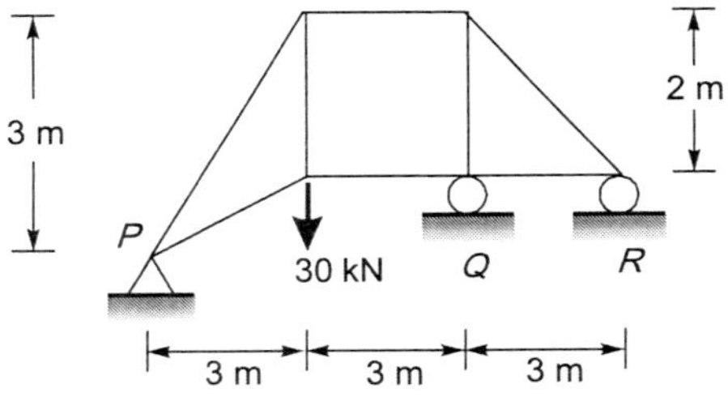

(a) (10, 30, –10) kN (b) (30, –30, 30) kN

(c) (20, 0, 10) kN (d) (0, 60, –30) kN

[2019 : 1 M, Set-I]

5.26 Consider the planar truss shown in the figure (not drawn to the scale)

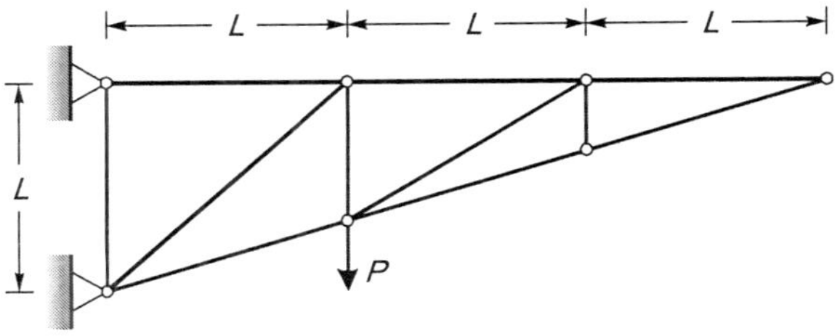

Neglecting self-weight of the members, the number of zero-force members in the truss under the action of the load P, is

(a) 6 (b) 9

(c) 7 (d) 8

[2020 : 1 M, Set-I]

5.27 Joints I, J, K, L, Q and M of the frame shown in the figure (not drawn to the scale) are pins. Continuous members IQ and JL are connected through a pin at N. Continuous members JM and KQ are connected through a pin at P. The frame has hinge supports at joints R and S. The loads acting at joints I, J and K are along the negative Y direction and the loads acting at joints L, M are along the positive X direction.

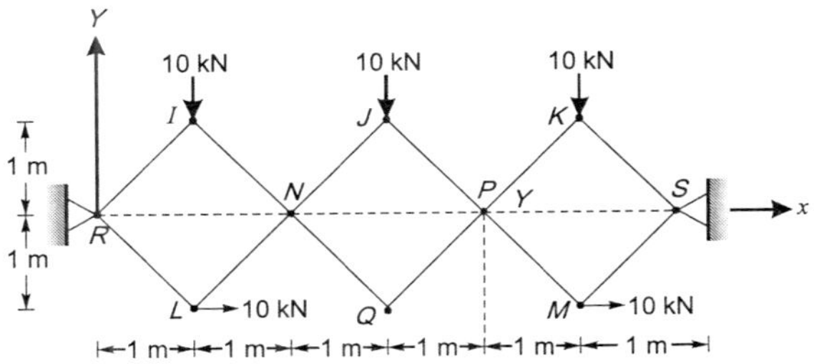

The magnitude of the horizontal component of reaction (in kN) at S, is

(a) 15 (b) 10

(c) 5 (d) 20

[2020 : 2 M, Set-II]

5.28 A truss *EFGH* is shown in the figure, in which all the members have the same axial rigidity *R*. In the figure, *P* is the magnitude of external horizontal forces acting at joints *F* and *G*.

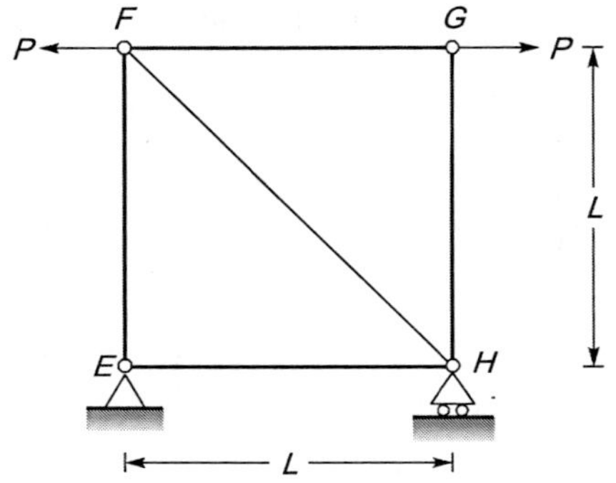

If $R = 500 \times 10^3$ kN, $P = 150$ kN and $L = 3$ m, the magnitude of the horizontal displacement of joint *G* (in mm, round off to one decimal place) is ______

[2021 : 1 M, Set-I]

5.29 The plane truss shown in the figure is subjected to an external force *P*. It is given that $P = 70$ kN, $a = 2$ m, and $b = 3$ m.

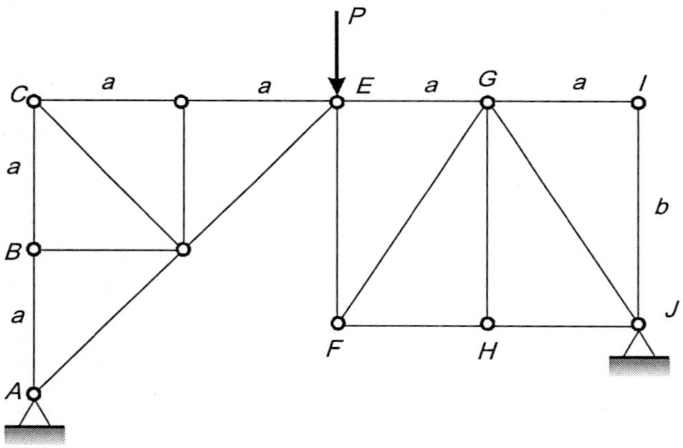

The magnitude (absolute value) of force (in kN) in member *EF* is _______. (round off to the nearest integer)

[2022 : 2 M, Set-I]

5.30 Consider the pin-jointed truss shown in the figure (not to scale). All members have the same axial rigidity, AE. Members QR, RS, and ST have the same length *L*. Angles QBT, RCT, SDT are all 90°. Angles BQT, CRT, DST are all 30°. The joint T carries a vertical load P. The vertical deflection of joint T is $k\dfrac{PL}{AE}$. What is the value of *k*?

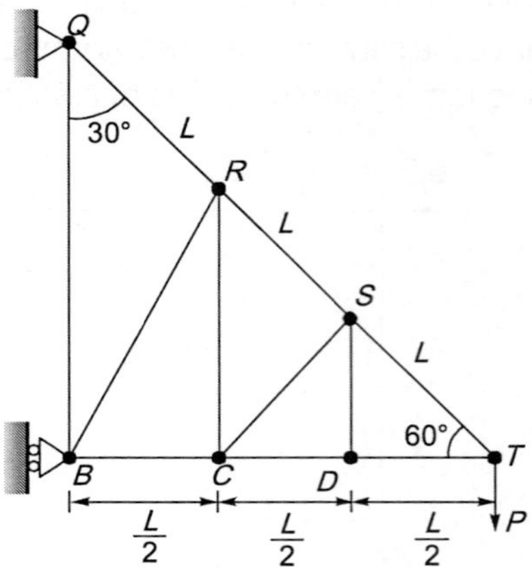

(a) 1.5 (b) 4.5
(c) 3.0 (d) 9.0

[2023 : 2 M, Set-I]

5.31 An idealised bridge truss is shown in the figure. The force in Member U_2L_3 is _________ kN (round off to one decimal place).

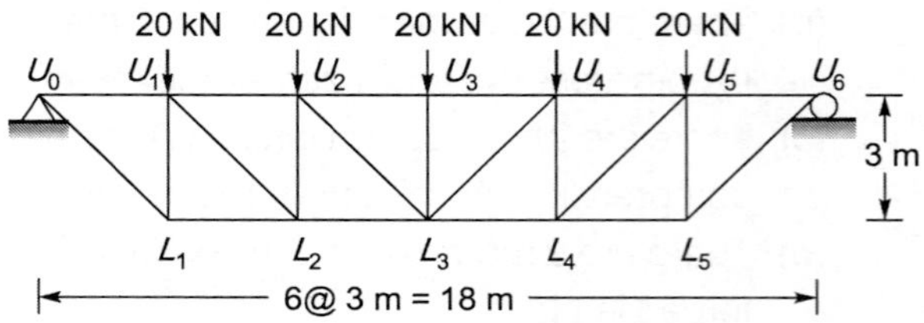

[2023 : 2 M, Set-I]

5.32 The plane truss shown in the figure has 13 joints and 22 members. The truss is made of a homogeneous, prismatic, linearly elastic material. All members have identical axial rigidity. A to M indicate the joints of the truss. The truss has pin supports at joints A and L and roller support at joint K. The truss is subjected to a 10 kN vertically downward force at joint H and a 10 kN horizontal force in the rightward direction at joint B as shown

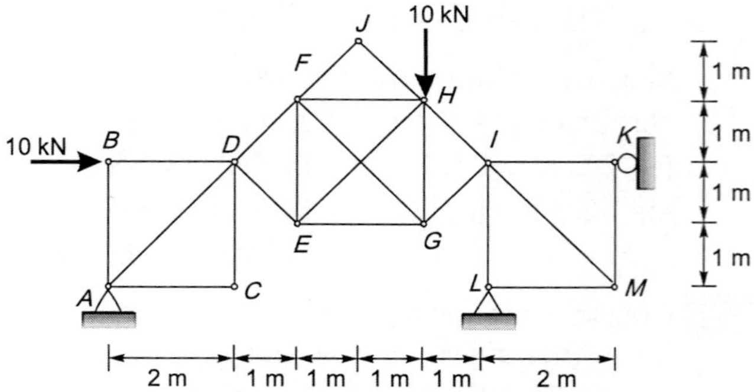

(Figure NOT to scale)

The magnitude of the reaction (in kN) at the pin support L is _____. (rounded off to 1 decimal place).

[2024 : 2 M, Set-I]

Answers Trusses

5.1 (a)	5.2 (*)	5.3 (b)	5.4 (d)	5.5 (b)	5.6 (b)	5.7 (a)
5.8 (b)	5.9 (c)	5.10 (a)	5.11 (d)	5.12 (c)	5.13 (c)	5.14 (0)
5.15 (1311.96)	5.16 (a)	5.17 (2)	5.18 (5)	5.19 (a)	5.20 (a)	5.21 (c)
5.22 (Sol.)	5.23 (d)	5.24 (b)	5.25 (b)	5.26 (d)	5.27 (a)	5.28 (0.90)
5.29 (30)	5.30 (b)	5.31 (14.1)	5.32 (7.5)			

Explanations Trusses

5.1 (a)

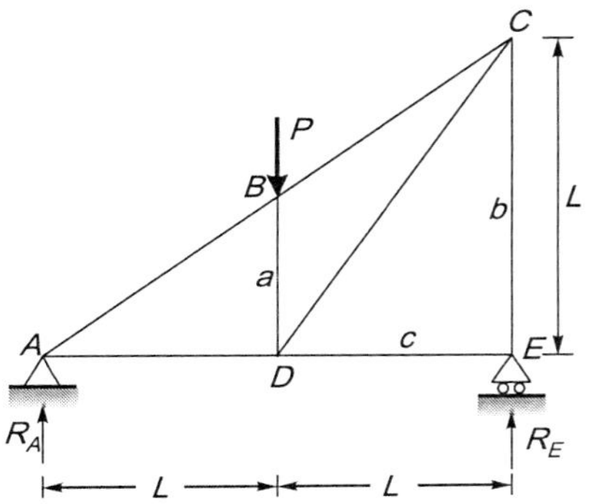

From equilibrium conditions,

$$R_A + R_E = P$$

$$\Sigma M_E = 0$$

$$\therefore \quad R_A \times 2L - P \times L = 0$$

$$\Rightarrow \quad R_A = \frac{P}{2}$$

and $$R_E = \frac{P}{2}$$

Consider Joint B, member 'a' and load P are along same line.

$$\Rightarrow \quad F_{BD} = P$$

Now consider Joint E,

$$\Sigma V_E = 0$$

$$\Rightarrow \quad R_E = P_{CE} = P/2$$

Consider equilibrium of horizontal force at E,

$$\Rightarrow \quad F_{ED} = 0$$

POINTS TO REMEMBER

For two pairs of colinear forces in equilibrium

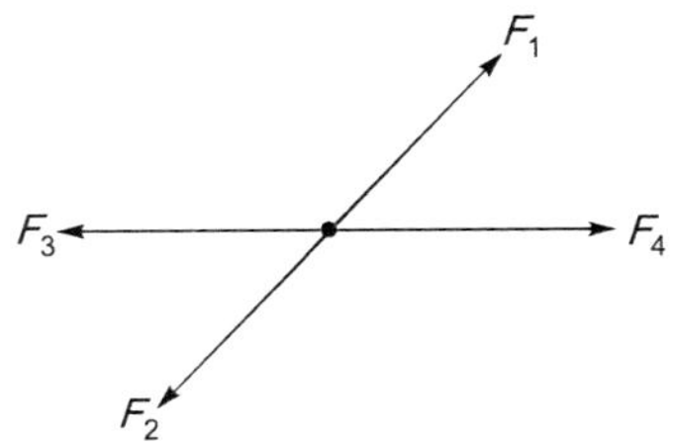

For equation, $F_1 = F_2$

$F_3 = F_4$

5.2 (*)

$$F_{cb} = F_{cd} = 0$$

($\because$ Two non-collinear member meeting at a joint)

To calculate the rotation of member bd, we will apply a unit couple as shown below,

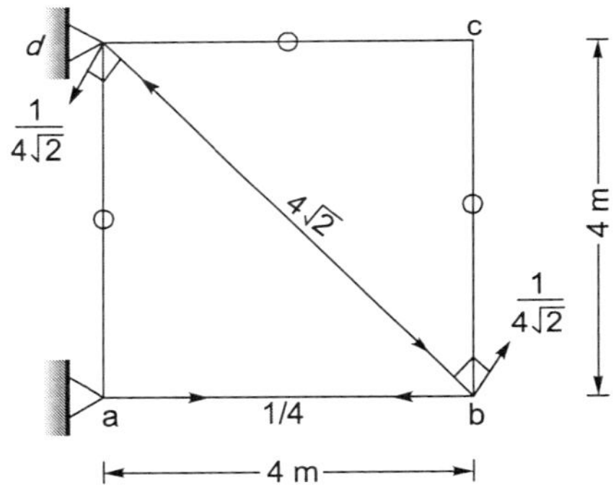

At joint 'b':

$\Sigma F_y = 0$

$$\Rightarrow \quad F_{db} \sin 45° = \frac{1}{4\sqrt{2}} \times \cos 45°$$

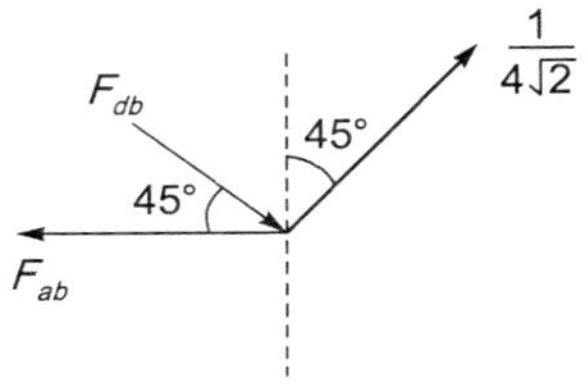

$$\therefore \quad F_{db} = \frac{1}{4\sqrt{2}} \text{ kN} \quad \text{(Compressive)}$$

$\Sigma F_x = 0$

$$\Rightarrow \quad F_{db} \cos 45° + \frac{1}{4\sqrt{2}} \sin 45° = F_{ab}$$

$\therefore \quad F_{ab} = \frac{1}{4}\text{kN} \quad$ (Tensile)

Hence, rotation of member *bd* will be given as,

$$\left(-\frac{1}{4\sqrt{2}} \times 0.005 + \frac{1}{4} \times 0.004\right) = 1.16 \times 10^{-4} \text{ radian.}$$

(Anticlock)

POINTS TO REMEMBER

- Due to hinge supports A and D, member AD can not elongate or contrract due to applied loads. Since, deformation is zero, force in the member AD is also zero.
  However, had the member AD been subjected to temperature changes force would have developed in the member AD due to restraint offered by supports against deformation.

5.3 (b)

Consider joint *E*,

Taking equilibrium of vertical forces, we get

$$F_{DE} = 0$$

5.4 (d)

Method-I

Degree of static indeterminacy

$$D_s = m + r_e - 2j$$
$$= 5 + 3 - 2 \times 4$$
$$= 8 - 8 = 0$$

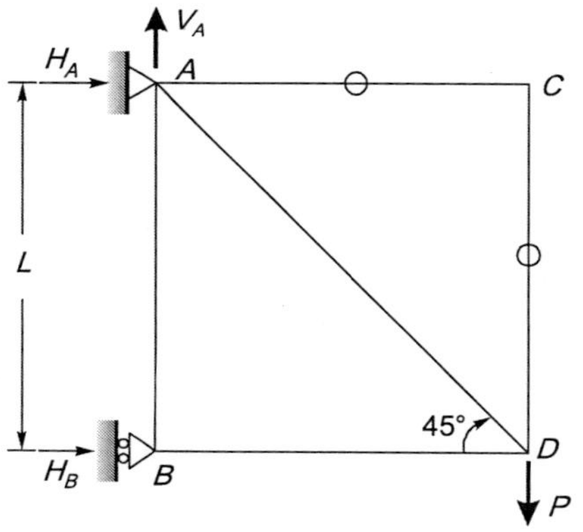

$\therefore$ Truss is determinate.

Now, from equation of equilibrium,

$$\Sigma F_y = 0$$
$$\Rightarrow \quad V_A = P$$
also, $$\Sigma M_B = 0$$
$$\Rightarrow \quad P \times L + H_A \times L = 0$$
$$\Rightarrow \quad H_A = -P$$
$$\Sigma F_x = 0$$
$$\Rightarrow \quad H_A + H_B = 0$$
$$\Rightarrow \quad H_B = P$$

Consider joint *B*,

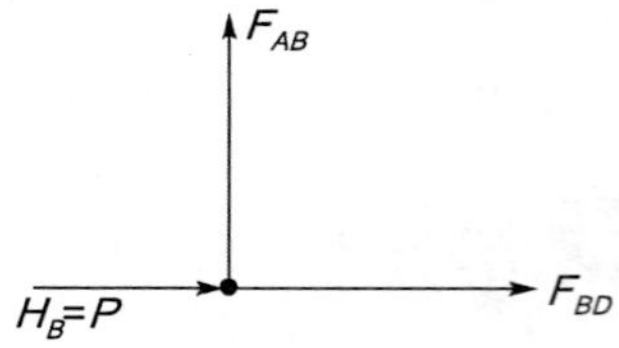

$$\Sigma F_y = 0$$
$$\Rightarrow \quad F_{AB} = 0$$

$\therefore$ strain energy stored in members *AB*,

$$U_{AB} = \frac{F_{AB}^2 \cdot L_{AB}}{2AE} = 0$$

5.5 (b)

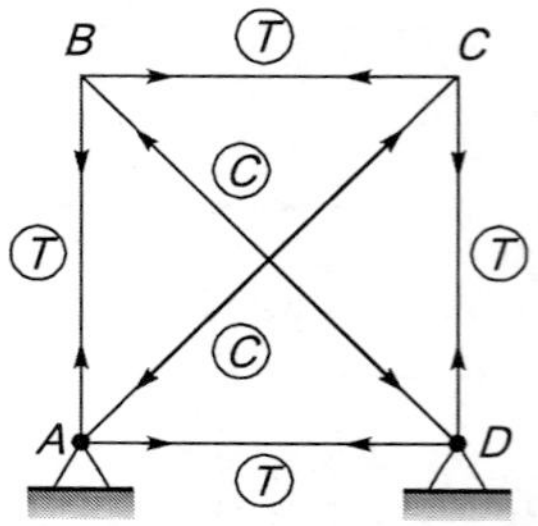

As temperature increases the bar *AD* will tend to elongate but joint *B* and *D* will offer resistance. Hence bar *BD* will be in compression. Once nature of force in BD is known, forces in other members can be determined by joint equilibrium.

5.6 (b)

Let $\angle MQN = \theta_1$ and $\angle OQN = \theta_2$

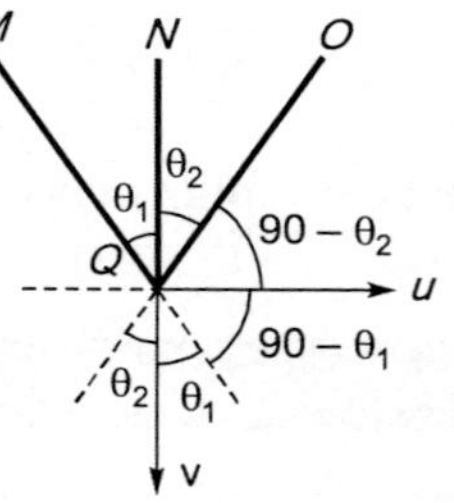

$$\therefore \quad \tan\theta_1 = \frac{MN}{NQ} = \frac{400}{500} = 0.8$$
$$\Rightarrow \quad \sin\theta_1 = 0.625 \text{ and } \cos\theta_1 = 0.781$$
and $$\tan\theta_2 = \frac{NO}{NQ} = \frac{500}{500} = 1$$
$$\Rightarrow \quad \sin\theta_2 = 0.707 \text{ and } \cos\theta_2 = 0.707$$

Elongation of member MQ in terms of u and v will be,

$u \sin \theta_1 + v \cos \theta_1 = 48$

or, $0.625u + 0.781v = 48$ ...(i)

Elongation of member OQ in terms of u and v will be,

or, $-u \sin \theta_2 + v \cos \theta_2 = 35$

$-0.707u + 0.707v = 35$ ...(ii)

Solving (i) and (ii) we get,

$u = 6.64$ mm and $v = 56.14$ mm

5.7 (a)

Let reactions at L_0 and L_6 be R_{L_0} and R_{L_6} respectively.

$\therefore \; R_{L_0} + R_{L_6} = 180$ ...(i)

Taking moments about L_0, we get,

$R_{L_6} \times 24 - 180 \times 8 = 0$

$\Rightarrow \; R_{L_6} = 60$ kN

$\therefore$ From (i), we get,

$R_{L_0} = 120$ kN

Assuming a section passing through members U_3U_4, M_2L_3 and L_3L_4 and considering left portion

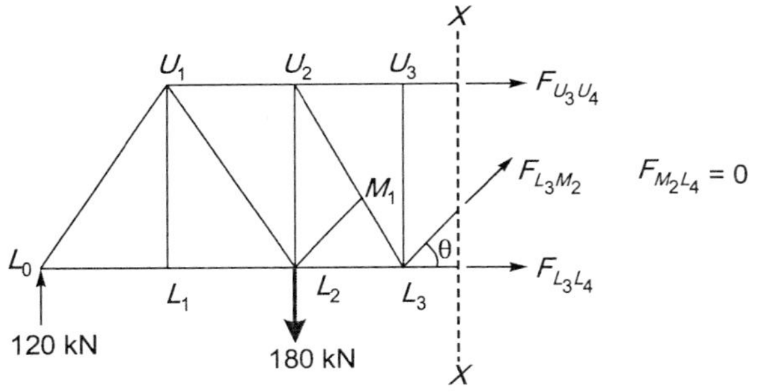

Considering equation of LHS of X-X

$\Sigma F_V = 0$

$\Rightarrow \; 120 - 180 + F_{L_3M_2} \sin \theta = 0$

$\Rightarrow \; F_{L_3M_2} = 60 \times \frac{5}{3} = 100 \text{ kN (T)}$

5.8 (b)

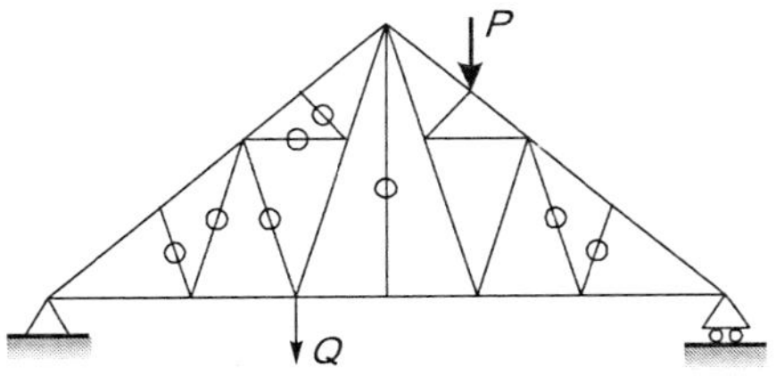

If three members meet at a joint and out of them two are collinear, then non-collinear member will carry zero force, provided that there is no external load at the joint.

Thus using above statement, we arrive at 8 zero force members which are highlighted by 'o' sign.

5.9 (c)

Since the truss is loaded symmetrically, the reactions will be equal.

Thus reaction at $B = R_B = \frac{P}{2}$

Now considering joint equilibrium at B,

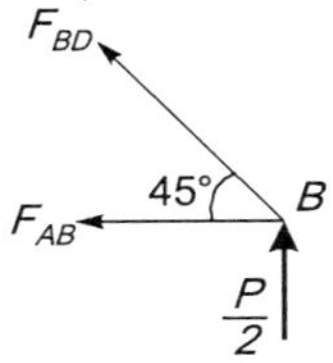

$\because \; \Sigma F_y = 0$

$F_{BD} \sin \theta + \frac{P}{2} = 0$ ...(i)

$\because \; \Sigma F_x = 0$

$F_{BD} \cos \theta + F_{AB} = 0$ ...(ii)

From (i), we have

$$F_{BD} = -\frac{P}{2 \sin 45^\circ} = \frac{-P}{\sqrt{2}}$$

Now, from equation (ii)

$\therefore \; F_{AB} = -F_{BD} \cos \theta$

$$= -\left(\frac{-P}{\sqrt{2}}\right) \times \cos 45^\circ = \frac{P}{2}$$

5.10 (a)

It is evident from the diagram that all the interior members will carry zero force.

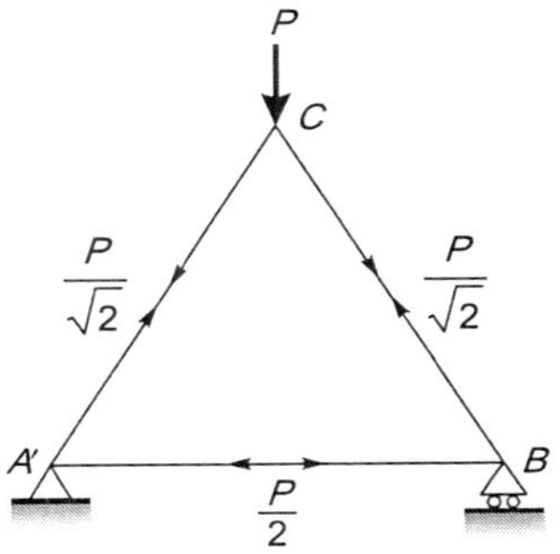

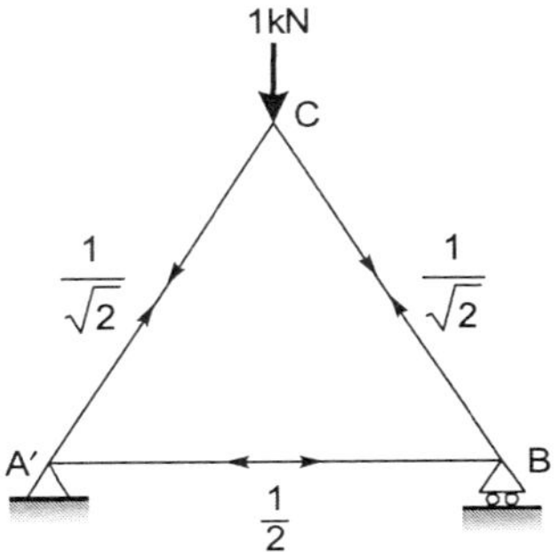

Assuming tensile forces as positive and compressive forces as negative.

Member	P	K	L	$\frac{PKL}{AE}$
$A'C$	$\frac{-P}{\sqrt{2}}$	$\frac{-1}{\sqrt{2}}$	$\sqrt{2}L$	$\frac{PL}{\sqrt{2}AE}$
BC	$\frac{-P}{\sqrt{2}}$	$\frac{-1}{\sqrt{2}}$	$\sqrt{2}L$	$\frac{PL}{\sqrt{2}AE}$
$A'B$	$\frac{P}{2}$	$\frac{1}{2}$	$2L$	$\frac{PL}{2AE}$

Deflection at C, $\delta_c = \Sigma \frac{PKL}{AE}$

Where K is the force in member due to unit load applied at C.

$$= \frac{PL}{\sqrt{2}AE} + \frac{PL}{\sqrt{2}AE} + \frac{PL}{2AE}$$

$$= \frac{\sqrt{2}PL + \sqrt{2}PL + PL}{2AE} = \left(\frac{2\sqrt{2}+1}{2}\right)\frac{PL}{AE}$$

5.11 (d)

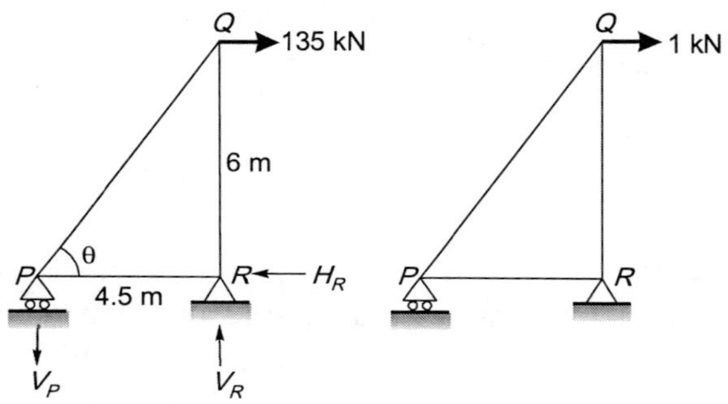

$$\Sigma F_x = 0$$

$$\Rightarrow \quad -H_R + 135 = 0$$

$$\Rightarrow \quad H_R = 135 \text{ kN}$$

Now, $\tan\theta = \frac{6}{4.5}$

$$\sin\theta = \frac{6}{7.5}; \cos\theta = \frac{4.5}{7.5}$$

Taking $\Sigma M_R = 0$

$$\Rightarrow -V_P \times 4.5 + 135 \times 6 = 0$$

$$\Rightarrow \quad V_P = 180 \text{ kN}$$

$$\therefore \quad V_R = V_P = 180 \text{ kN}$$

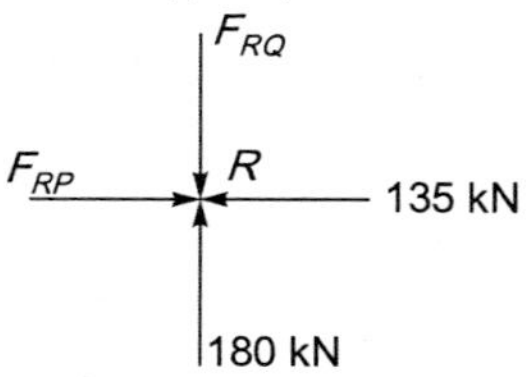

FBD of joint R

$$\Sigma F_y = 0$$

$$-F_{RQ} + 180 = 0$$

$$F_{RQ} = 180 \text{ kN (Comp.)}$$

and $\Sigma F_x = 0$

$$F_{RP} - 135 = 0$$

$$F_{RP} = 135 \text{ kN (Comp.)}$$

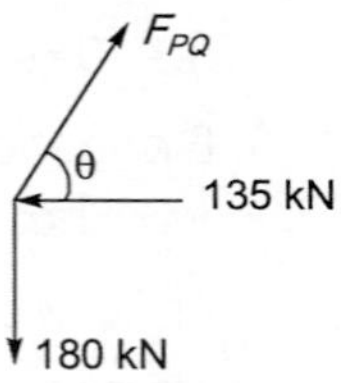

FBD of joint P

$$\Sigma F_x = 0$$

$$F_{PQ}\cos\theta - 135 = 0$$

$$F_{PQ} = \frac{135 \times 7.5}{4.5}$$

$$F_{PQ} = 225 \text{ kN (Tension)}$$

Tension (+) compression (–) and $K = \frac{P}{135}$ for each member,

Member	P	K	L	$\frac{PKL}{AE}$
PQ	225	1.67	7.5	$\frac{2818.125}{AE}$
QR	–180	–1.33	6	$\frac{1436.4}{AE}$
RP	–135	–1	4.5	$\frac{607.5}{AE}$

$$\delta_Q = \Sigma \frac{PKL}{AE}$$

$$= \frac{2818.125 + 1436.4 + 607.5}{AE}$$

$$= \frac{(4862.025 \times 10^3) \times 10^3}{(1550 \times 10^{-6}) \times (2 \times 10^5 \times 10^{-6})}$$

$$= 15.68 \text{ mm}$$

5.12 (c)

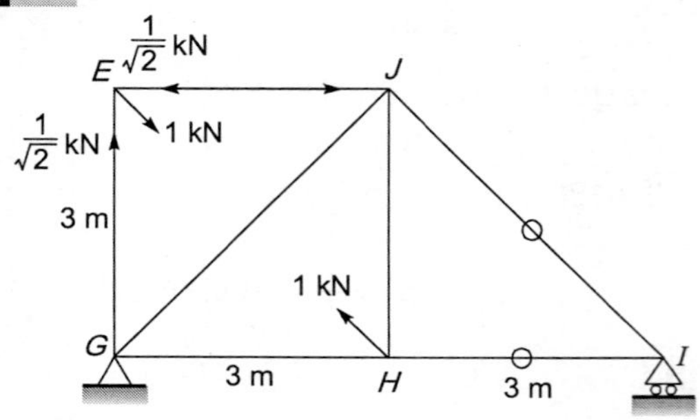

As only EI and IJ are subjected to temperature rise, so we will calculate forces in only these members due to unit load.

$$F_{EJ} = \frac{-1}{\sqrt{2}} \text{ kN (Comp.)}$$

Relative displacement of joint E and H

$$= \left(\frac{-1}{\sqrt{2}}\right) \times 3000 \times 0.000012 \times 30$$

$= 0.764$ mm

'–ve' sign means displacement is in opposite direction of unit load.

5.13 (c)

Using **method of joints** and considering joint S, we get,

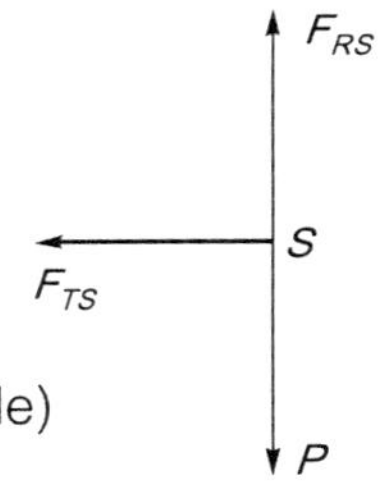

$$\Sigma F_x = 0$$
$$\Rightarrow \quad F_{TS} = 0$$
and
$$\Sigma F_y = 0$$
$$\Rightarrow \quad F_{RS} - P = 0$$
$$\Rightarrow \quad F_{RS} = P \text{ (tensile)}$$

Considering joint R, we get

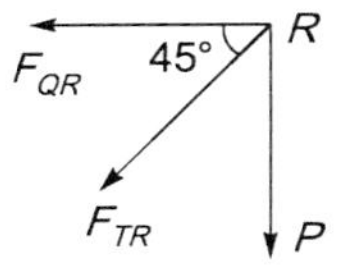

$$\Sigma F_y = 0$$
$$\Rightarrow \quad -F_{TR} \sin 45° - P = 0$$
$$\Rightarrow \quad F_{TR} = -P\sqrt{2} \text{ (compressive)}$$
$$\Sigma F_x = 0$$
$$\Rightarrow \quad -F_{TR} \cos 45° - F_{QR} = 0$$
$$\Rightarrow \quad P\sqrt{2} \times \frac{1}{\sqrt{2}} - F_{QR} = 0$$
$$\Rightarrow \quad F_{QR} = P \text{ (tension)}$$

5.14 Sol.

Take moment about $V, M_V = 0$,

$$\Rightarrow \quad R_H \times 4 = 0$$
$$\therefore \quad R_H = 0$$
$$\Sigma M_T = 0$$
$$\Rightarrow R_V \times 8 - 15 \times 4 + 15 \times 4 = 0$$
$$\therefore \quad R_V = 0$$

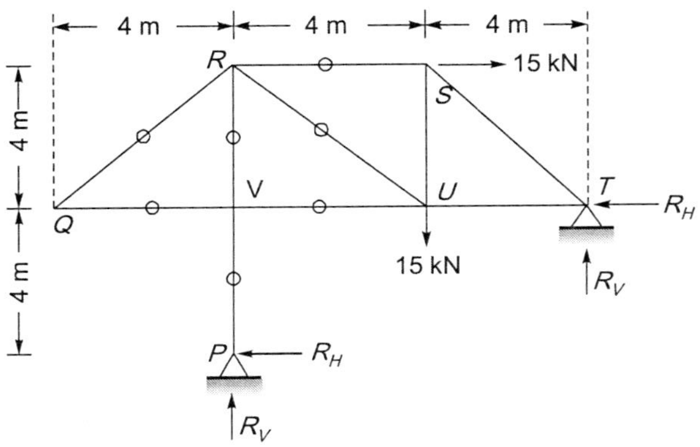

We know that if two members meet at a joint which are not collinear and also there is no external forces acting on that joint, then both members will carry zero forces.

$$\therefore \quad F_{QV} = F_{QR} = 0$$

Now, consider joint P,

$$\Sigma H = 0$$
$$\Rightarrow \quad R_H = 0,$$
and
$$\Sigma F_V = 0,$$
$$F_{PV} = R_V$$

Taking moment about T,

$$\Sigma M_T = 0$$
$$\Rightarrow \quad 8R_V + 15 \times 4 - 15 \times 4 = 0$$
$$\Rightarrow \quad R_V = 0$$

Joint V,

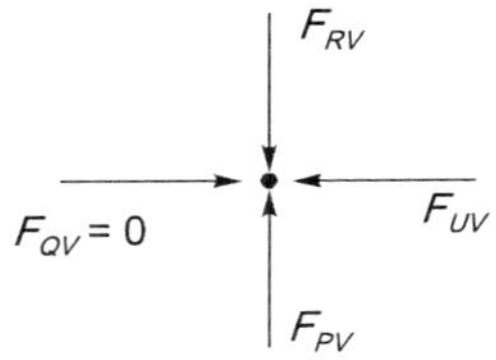

$\Sigma H = 0$: $\quad F_{UV} = F_{QV} = 0$

$\Sigma V = 0$; $\quad F_{RV} = F_{PV} = R_V = 0$

Now, member RS and RU area non-collinear members meeting at a joint with no external force acting on the joint.

$$\Rightarrow \quad F_{RS} = R_{RU} = 0$$

5.15 Sol.

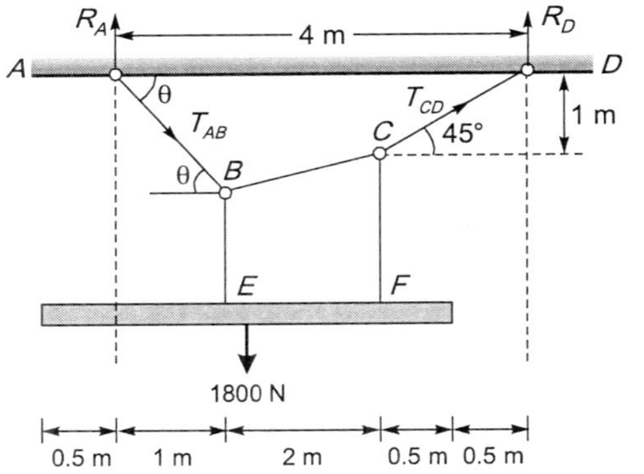

Taking moment about A,

$$R_D \times 4 = 1800 \times 1.5$$
$$\Rightarrow \quad R_D = 675 \text{ N}$$
$$R_A = 1800 - R_D = 1125 \text{ N}$$

At joint D,

$$\Sigma F_y = 0$$
$$\Rightarrow \quad T_{CD} = \frac{R_D}{\sin 45} = 675\sqrt{2} \text{ N}$$

Taking free body diagram as shown in figure,

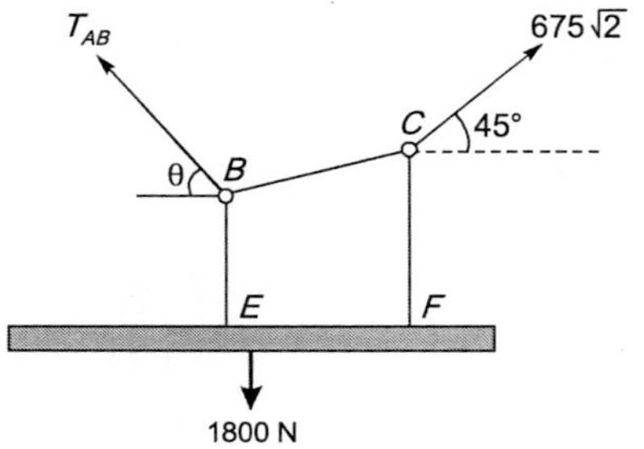

$$\Sigma F_y = 0$$

$$\Rightarrow \quad T_{AB}\sin\theta + 675\sqrt{2}\sin 45° = 1800$$

$$\Rightarrow \quad T_{AB}\sin\theta = 1125 \qquad \text{...(i)}$$

$$\Sigma H = 0$$

$$\Rightarrow \quad T_{AB}\cos\theta = 675\sqrt{2}\cos 45°$$

$$\Rightarrow \quad T_{AB}\cos\theta = 675 \qquad \text{...(ii)}$$

From eq. (i) and (ii),

$$\tan\theta = \frac{5}{3}$$

$$\Rightarrow \quad \sin\theta = 0.8575$$

Putting value of sinθ in eq. (i),

$$T_{AB} \times 0.8575 = 1125$$

$$\therefore \quad T_{AB} = 1311.96 \text{ N}$$

5.16 (a)

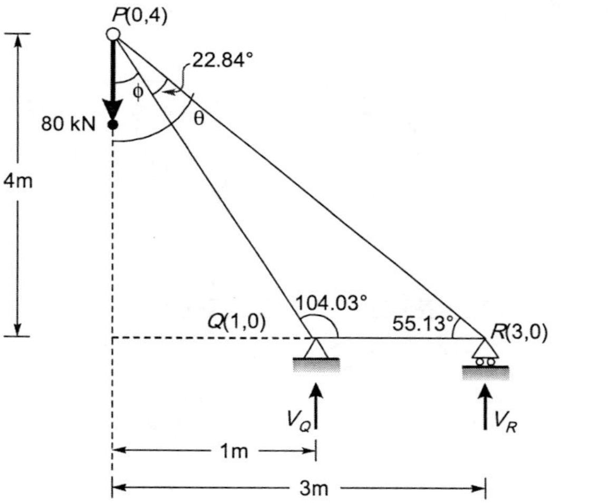

$$\Sigma F_y = 0$$

$$\Rightarrow \quad V_Q + V_R = 80 \qquad \text{...(i)}$$

$$\Sigma M_R = 0$$

$$\Rightarrow \quad 80 \times 3 = V_Q \times 2$$

$$\therefore \quad V_Q = 120 \text{ kN}$$

$$\Rightarrow \quad V_R = -40 \text{ kN} \quad \text{From eq. (i)}$$

$$\tan\phi = \frac{1}{4}$$

$$\Rightarrow \quad \sin\phi = \frac{1}{\sqrt{17}}, \cos\phi = \frac{4}{\sqrt{17}}$$

$$\therefore \quad \tan\theta = \frac{3}{4}$$

$$\Rightarrow \quad \cos\theta = \frac{4}{5}$$

Consider joint Q,

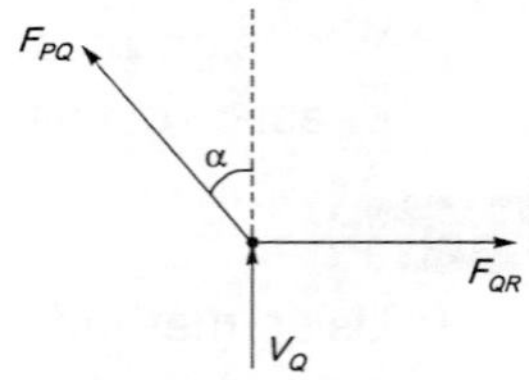

$$\alpha = 104.03° - 90° = 14.03°$$

$$\Sigma F_y = 0$$

$$\Rightarrow \quad F_{PQ}\cos\alpha + V_Q = 0$$

$$\Rightarrow \quad F_{PQ}\cos 14.03° + 120 = 0$$

$$\Rightarrow \quad F_{PQ} = -123.6897 \text{ kN}$$

$$\Sigma F_x = 0$$

$$\Rightarrow \quad F_{PQ}\sin\alpha = F_{QR}$$

$$\therefore \quad F_{QR} = -29.986 \text{ kN}$$

$$\simeq 30 \text{ kNm (comp.)}$$

5.17 Sol.

PQ is short by 3 mm

We have to find out vertical displacement of joint R in mm,

$$\Delta_R = \Sigma u(\lambda)$$

Apply unit load at R as shown below

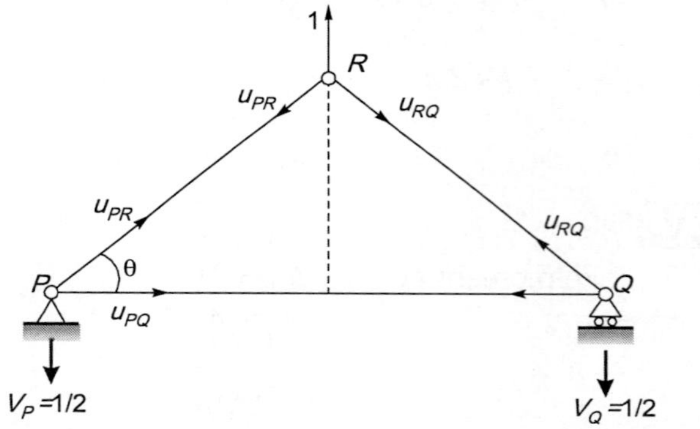

Due to symmetric loading, $V_P = V_Q = \frac{1}{2}$.

Consider P:

$\Sigma V = 0$,

$$u_{PR}\sin\theta = \frac{1}{2}$$

$\Sigma H = 0$,

$$u_{PQ} + u_{PR}\cos\theta = 0$$

$$u_{PQ} = -u_{PR}\cos\theta = -\frac{1}{2\sin\theta}.\cos\theta$$

$$u_{PQ} = -\frac{1}{2}\cot\theta = \frac{-1\times 4/3}{2} = \frac{-2}{3}$$

$$\Delta_R = u_{PQ} \times \lambda_{PQ} = \frac{-2}{3}(-3)$$

$$= 2 \text{ mm } \textbf{upwards}$$

POINTS TO REMEMBER

We need not calcualte u_{PR} and u_{RQ}, as $\lambda_{PR} = \lambda_{RQ} = 0$

5.18 Sol.

First calculating reactions,

$$\Sigma M_A = 0$$

$$\Rightarrow \quad R_B \times 3 - 10 \times 9 - 5 \times 3 = 0$$

$$R_B = 35 \text{ kN}$$

$$H_A = 10 \text{ kN}$$

$$\Sigma M_B = 0$$

$$\Rightarrow \quad R_A \times 3 + 10 \times 9 = 0$$

$$R_A = -30 \text{ kN}$$

Let us cut a section 1-1 as shown in figure and consider the lower part.

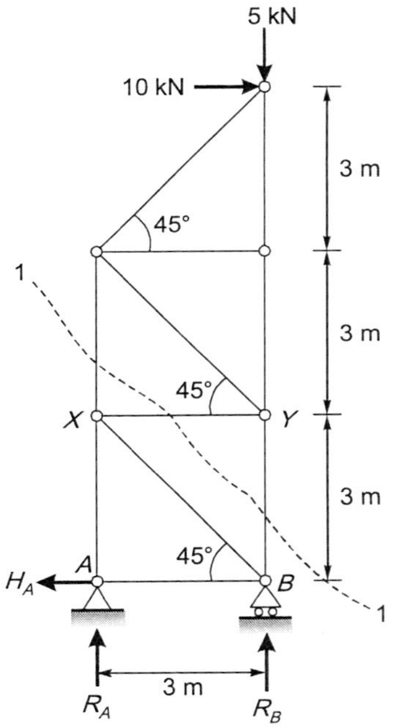

Given AE = 30 kN and L = 3 m,

Calculation for force P in XY member.

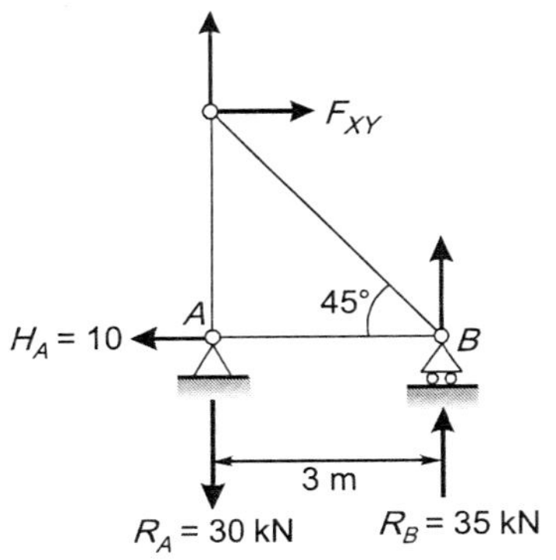

$$\Sigma F_{horizontal} = 0$$

$$\Rightarrow \quad F_{XY} = 10 \text{ kN}$$

Strain energy,

$$U = \frac{P^2 L}{2AE} = \frac{10^2 \times 3}{2 \times 30} = 5 \text{ kNm}$$

5.19 (a)

$$\Sigma F_y = 0$$

$$\Rightarrow \quad V_B + V_C = P \quad \text{... (i)}$$

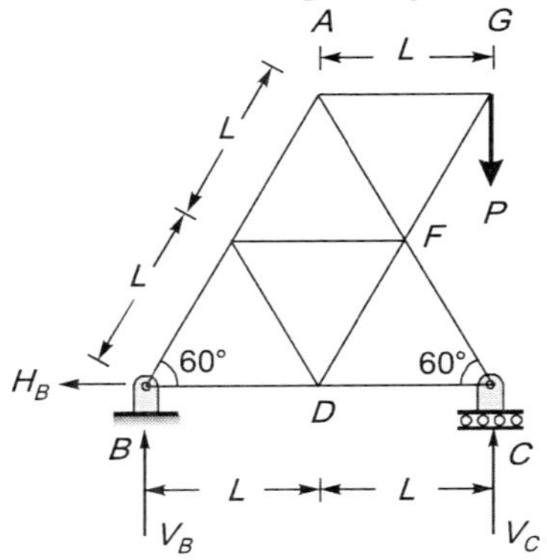

$$\Sigma M_B = 0;$$

$$\Rightarrow -V_C \times 2L + P \times 2L = 0$$

$$\Rightarrow \quad V_C = P$$

From (i), $\quad V_B = 0$

$$\Sigma F_x = 0$$

$$\Rightarrow \quad H_B = 0$$

5.20 (a)

If there members meet at a joint and out of them are collinear, then non collinear member will carry zero force provided that there is no external load at the joint. Use this statement to check the members with zero force.

5.21 (c)

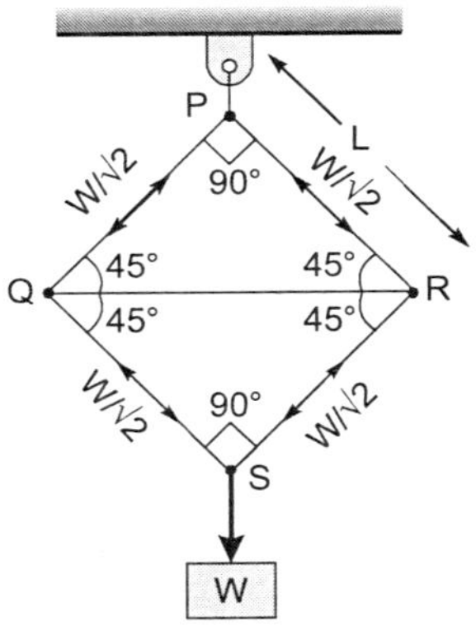

Given data:

$$L = 1\text{m}, \ A = 0.05 \text{ m}^2,$$

$$E = 30 \times 10^6 \text{ N/m}^2$$

Consider joint 'S',

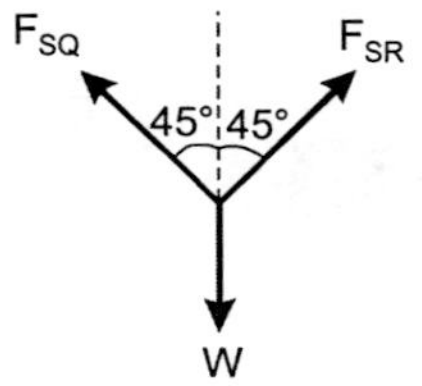

$$\Sigma F_x = 0$$

$$\Rightarrow \quad F_{SQ} = F_{SR}$$

$$\Sigma F_y = 0$$

$$\Rightarrow \quad 2F_{SQ} \cos 45° = W$$

$$\Rightarrow \quad F_{SQ} = \frac{W}{2} \times \sqrt{2} = \frac{W}{\sqrt{2}}$$

As the truss is symmetrical,

$$\therefore \quad F_{QP} = F_{PR} = \frac{W}{\sqrt{2}} \quad \text{(Tensile)}$$

Now consider joint 'Q',

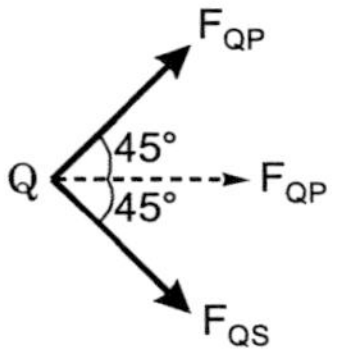

$$\Sigma F_y = 0$$

$$\Rightarrow \quad F_{QP} = F_{QS} = \frac{W}{\sqrt{2}}$$

$$\Sigma F_x = 0$$

$$\Rightarrow \frac{F_{QP}}{\sqrt{2}} + \frac{F_{QS}}{\sqrt{2}} + F_{QR} = 0$$

$$\Rightarrow \quad F_{QR} = -W$$

$$\therefore \quad F_{QR} = -100 \text{ kN} \quad \text{(Compressive)}$$

Stress in member QR,

$$\sigma_{QR} = \frac{F_{QR}}{(2A)} \quad (\because \text{Area of } QR = 2A)$$

$$\Rightarrow \quad \sigma_{QR} = \frac{100}{2 \times 0.05}$$

$$= \frac{100 \times 100}{2 \times 5} = 1000 \text{ kN/m}^2$$

$$\therefore \quad \sigma_{QR} = 1000 \text{ kN/m}^2$$

As the member QR is subjecterd to compression, it will go under shortening.

$\therefore$ Shortening,

$$\Delta = \frac{F_{QR} \cdot L_{QR}}{2AE}$$

$$L_{QR} = \sqrt{2}L$$

$$\therefore \quad \Delta = \frac{100 \times 10^3 \times \sqrt{2} \times 1}{2 \times 0.05 \times 30 \times 10^6}$$

$$= \frac{100 \times 10^3 \times \sqrt{2}}{0.1 \times 30 \times 10^6} = \frac{\sqrt{2}}{30}$$

$$\therefore \quad \Delta = 0.0471 \text{ m}$$

5.22 Sol.

Force is each member due to applied loading.

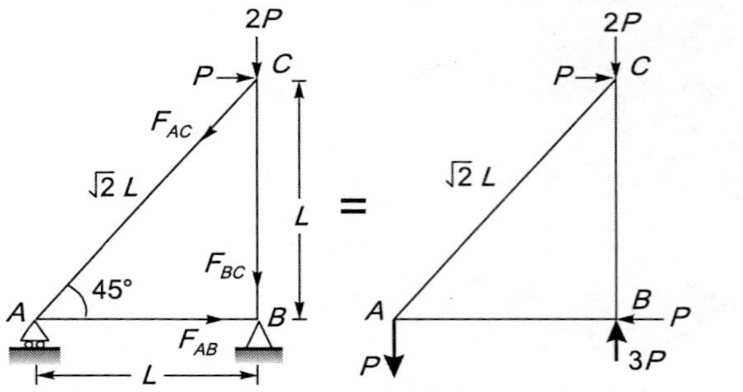

$$F_{AB} = P \text{(Comp.)}$$

$$F_{BC} = 3P \text{(Comp.)}$$

$$F_{AC} = \sqrt{2}P \text{ (Tension)}$$

Force in each member due to unit load.

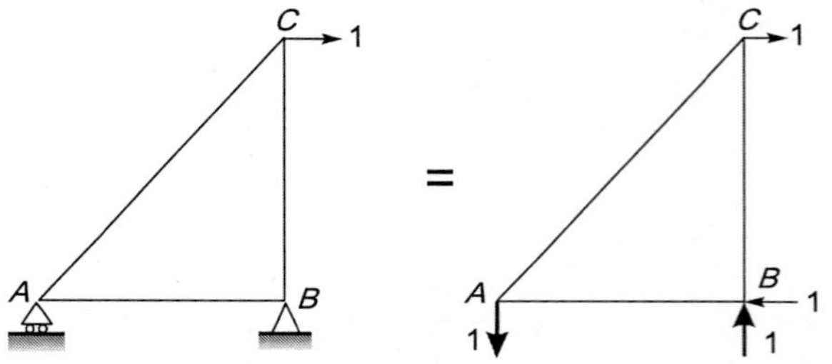

$$F_{AB} = 1 \text{ (Comp.)}$$

$$F_{BC} = 1 \text{ (Comp.)}$$

$$F_{AC} = \sqrt{2} \text{ (Tension)}$$

Member	P	k	L	AE	$\frac{PkL}{AE}$
AB	$-P$	-1	L	AE	$\frac{PL}{AE}$
BC	$-3P$	-1	L	AE	$\frac{3PL}{AE}$
CA	$\sqrt{2}P$	$\sqrt{2}$	$\sqrt{2}L$	$2AE$	$\frac{2\sqrt{2}PL}{2AE}$
					$\Sigma = \frac{5.414PL}{AE}$

$$\therefore \quad \text{Total deflection} = \delta_{H_C} = \Sigma \frac{PkL}{AE}$$

$$= \frac{5.414 \times (1000\text{N}) \times (1000\text{mm})}{(10\text{mm}^2) \times (2 \times 10^5 \text{ N/mm}^2)} = 2.7 \text{ mm}$$

5.23 (d)

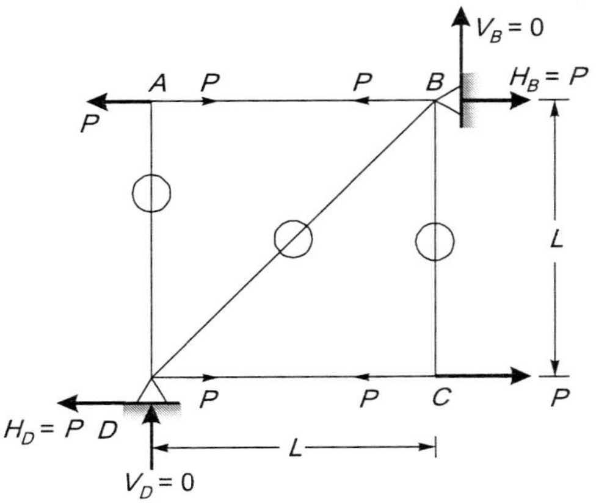

Since member BD neither elongate nor contract.

Hence, $F_{BD} = 0$

So there are 2 tension members (AB and DC) and 3 zero force members (AD, BD, BC).

5.24 (b)

- At joint G, GI and EG are collinear and GF is non-collinear, so force in GF will be zero as there is no external load at G.
- At joint F, EF and FH are collinear and FI is non-collinear, so force in FI will be zero as there is no external load at F.

Similarly in RS and RP, forces will be zero.

5.25 (b)

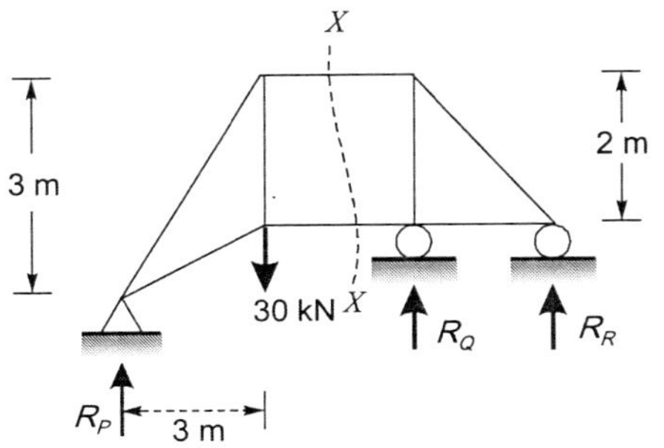

Using, $\Sigma F_V = 0$

$$R_P + R_R + R_Q = 30 \text{ kN} \quad \text{... (i)}$$

Taking moment about P,

$$\Sigma M_P = 0$$

$$\Rightarrow (30 \times 3) - R_Q \times 6 - R_R \times 9 = 0$$

$$\Rightarrow 2R_Q + 3R_R = 30 \quad \text{... (ii)}$$

Again considering the equilibrium of RHS of section X-X

$\Sigma F_V = 0$

$$\therefore \quad R_R = -R_Q \quad \text{... (iii)}$$

Using (i), (ii) and (iii)

$$R_P = 30 \text{ kN}$$
$$R_Q = -30 \text{ kN}$$
$$R_R = +30 \text{ kN}$$

5.26 (d)

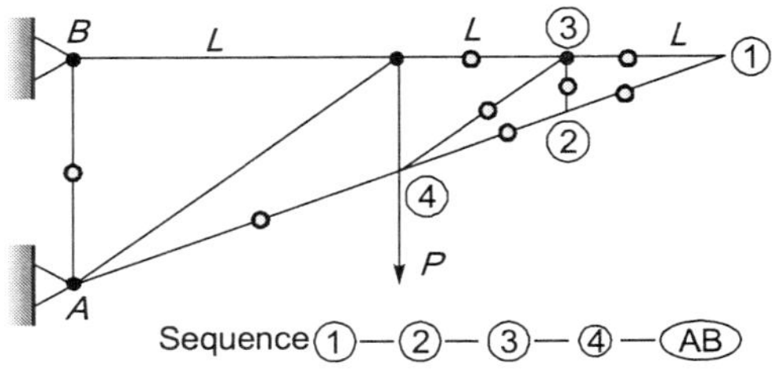

As $\Delta_{AB} = 0$, hence $F_{AB} = 0$

Total number of zero force member = 8

5.27 (a)

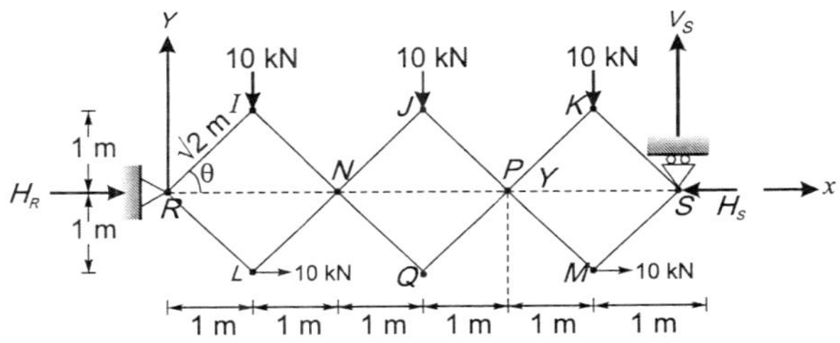

Remove hinge at support S and replace it with roller support as shown in the figure.

Ist **Step :** Find coordinates of all the points where forces are acting.

$$y_I = \sqrt{2}\sin\theta,\ y_j = \sqrt{2}\sin\theta,\ y_k = \sqrt{2}\sin\theta$$

$$x_L = \sqrt{2}\cos\theta,\ x_m = 5\sqrt{2}\cos\theta,\ x_s = 6\sqrt{2}\cos\theta$$

IInd **Step :** Find virtual displacements of all the points.

$$\delta y_I = \sqrt{2}\cos\theta d\theta,\ \delta y_j = \sqrt{2}\cos\theta d\theta,$$

$$\delta y_k = \sqrt{2}\cos\theta d\theta$$

$$\delta x_L = -\sqrt{2}\sin\theta d\theta,\ \delta x_m = -5\sqrt{2}\sin\theta d\theta,$$

$$\delta x_s = -6\sqrt{2}\sin\theta d\theta$$

IIIrd **Step :** Use principle of virtual work to find unknown horizontal force H_S

$$\Rightarrow \delta U = 0$$

$$= \left[-10 \times \sqrt{2}\cos\theta d\theta\right] \times 3 + \left[10 \times -\sqrt{2}\sin\theta d\theta\right]$$

$$+ \left[10 \times -5\sqrt{2}\sin\theta d\theta\right] - \left[H_s \times -6\sqrt{2}\sin\theta d\theta\right]$$

$$H_s = \frac{30\sqrt{2}\cos\theta + 10\sqrt{2}\sin\theta + 50\sqrt{2}\sin\theta}{6\sqrt{2}\sin\theta}$$

Substituting, $\theta = 45°$, $H_s = \frac{90}{6} = 15$ kN

Note : *Sign conventions*

If a force acts along positive x-axis or positive y-axis, take it as positive.

If a force acts along negative x-axis or negative y-axis, take it as negative.

5.28 Sol.

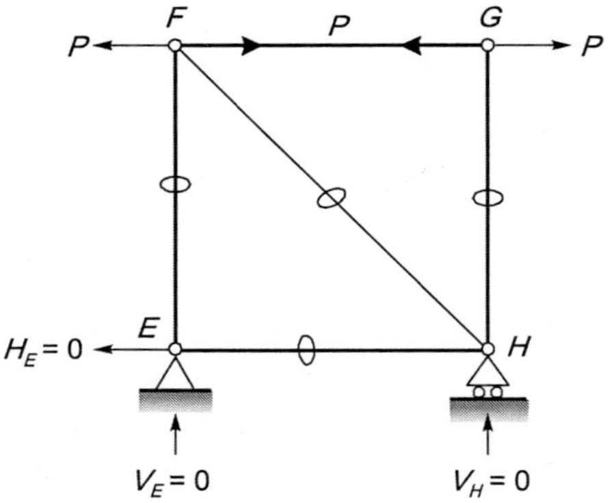

(P force system)

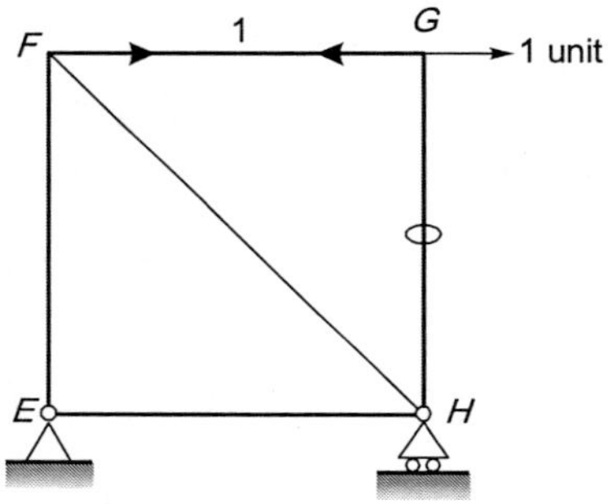

(k force system)

Note: No need to calculate 'k' force in all members because 'P' force is zero fore all members except FG

By unit load method

1. $$\Delta_{HG} = \sum_{i=1}^{n} \frac{P_i k_i L_i}{A_i E_i}$$

Δ_{HG} = Horizontal deflection at joint G

$$\therefore \quad \Delta_{HG} = \underbrace{\frac{P \times 1 \times L}{AE}}_{\text{FG-member}} + \underbrace{\text{'O'}}_{\text{(For all other members)}}$$

$$\Delta_{HG} = \frac{PL}{AE} = \left(\frac{150 \times 3}{500 \times 10^3} \times 10^3\right) \text{ mm}$$

$= 0.90$ mm

5.29 Sol.

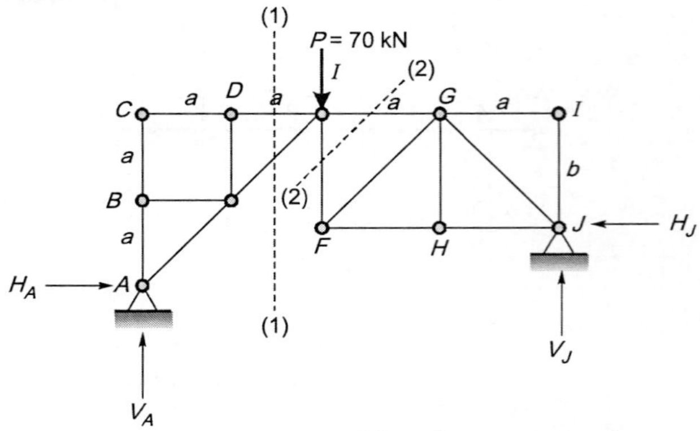

$$\Sigma M_J = 0$$

$\Rightarrow V_A \times 8 - H_A \times 1 - 70 \times 4 = 0$

$\Rightarrow \quad 8V_A - H_A = 280 \quad$ ...(i)

To find H_A: (Cut the truss by 1-1)

Consider left hand side,

$$\Sigma M_E = 0$$

$\Rightarrow \quad V_A \times 4 - H_A \times 4 = 0$

$\Rightarrow \quad V_A = H_A \quad$ ... (ii)

Using (i) in (ii), $8V_A - V_A = 280$

$\Rightarrow \quad 7V_A = 280$

$\Rightarrow \quad V_A = 40$ kN

$\therefore \quad H_A = 40$ kN

$\therefore \quad H_J = 40$ kN

$\therefore \quad V_J = 70 - 40 = 30$ kN

To find force in member EF: (Cut the truss by 2-2)

Consider right hand side,

Force in the member EF,

$$F_{EF} = V_J$$

$\Rightarrow \quad F_{EF} = 30$ kN

5.30 (b)

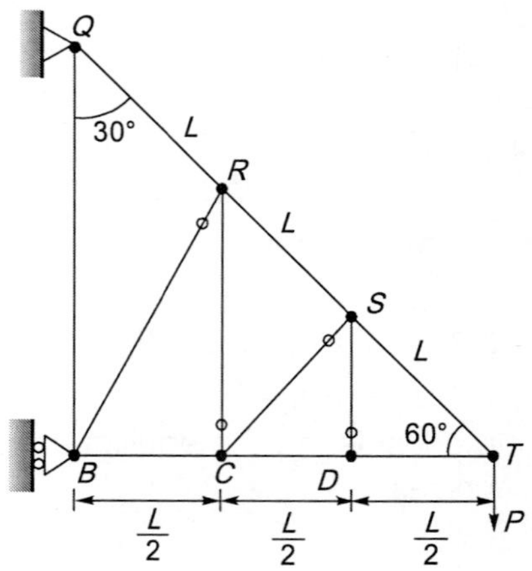

At joint T,

$$\Sigma F_y = 0$$

$$\Rightarrow \quad F_{TS}\sin 60^\circ = P$$

$$F_{TS} = \frac{2P}{\sqrt{3}}(T)$$

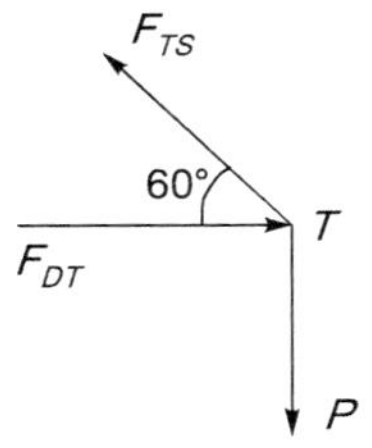

$$\Sigma F_x = 0$$

$$F_{TS}\cos 60^\circ = F_{DT}$$

$$F_{DT} = \frac{2P}{\sqrt{3}}\left(\frac{1}{2}\right) = \frac{P}{\sqrt{3}}(C)$$

$$F_{TS} = F_{SR} = F_{RQ} = \frac{2P}{\sqrt{3}}(T)$$

$$F_{DT} = F_{CD} = F_{BL} = \frac{P}{\sqrt{3}}(C)$$

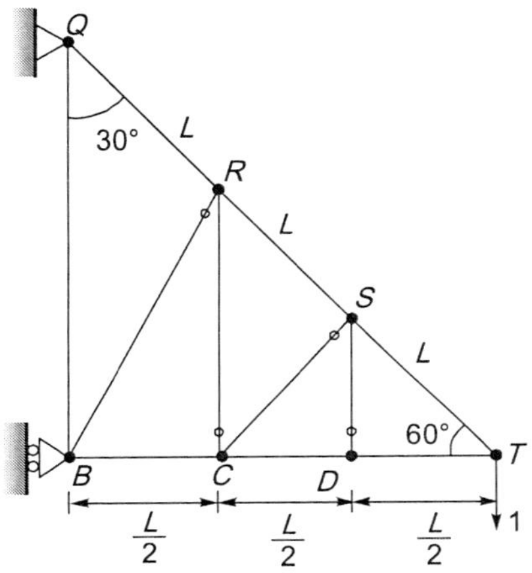

Similarly,

$$k_{ST} = k_{SR} = k_{RQ} = \frac{2}{\sqrt{3}}(T)$$

$$k_{DT} = k_{CD} = k_{BC} = \frac{1}{\sqrt{3}}(T)$$

$$\therefore \delta_{VT} = \Sigma\frac{PkL}{AE} = \frac{\left(\frac{2P}{\sqrt{3}}\right)\left(\frac{2}{\sqrt{3}}\right)L\times 3}{AE} + \frac{\left(\frac{P}{\sqrt{3}}\right)\left(\frac{1}{\sqrt{3}}\right)\frac{L}{2}\times 3}{AE}$$

$$\delta_{VT} = \frac{4PL}{AE} + \frac{PL}{2AE} = \frac{4.5PL}{AE}$$

$$\Rightarrow \quad k = 4.5$$

5.31 Sol.

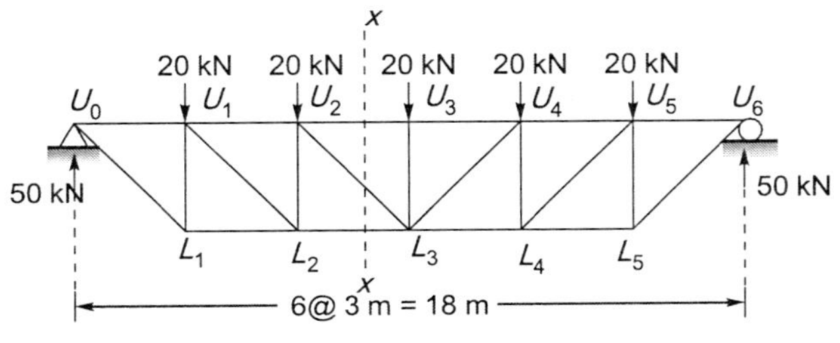

Reactions at supports:

Because of symmetry, vertical reaction at supports

$$= \frac{100}{2} = 50 \text{ kN}$$

Applying method of sections and taking the section as shown in the figure above,

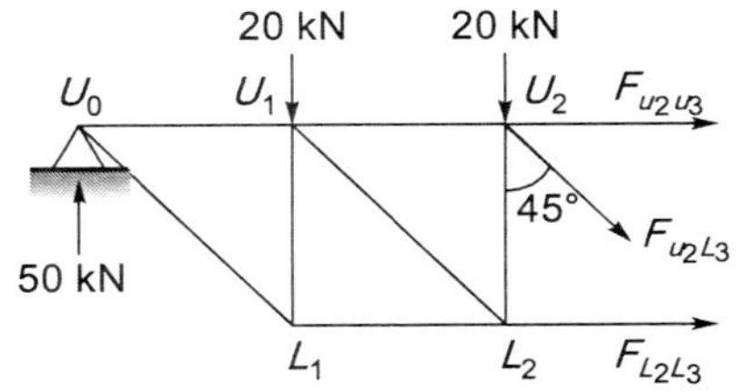

$$\Sigma F_y = 0$$

$$\Rightarrow F_{u2}L_3\cos 45^\circ + 20 + 20 - 50 = 0$$

$$F_{u2L3}\cos 45^\circ = 10$$

$$F_{u2L3} = 10\sqrt{2} \text{ kN} = 14.1 \text{ kN} \quad \text{(Tension)}$$

5.32 Sol.

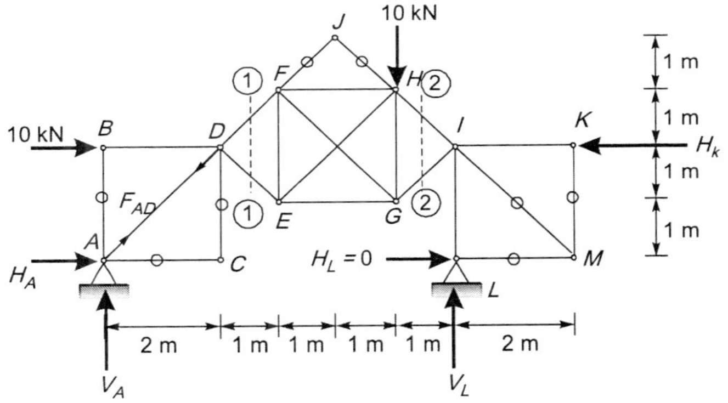

Considering left portion of section 1-1

$$\Sigma M_D = 0$$

$$\Rightarrow V_A \times 2 - H_A \times 2 = 0$$

$$\Rightarrow \quad V_A = H_A \quad \text{...(i)}$$

Considering right portion of section 2 - 2

$$\Sigma M_I = 0 \Rightarrow H_L \times 2 = 0$$

$$\Rightarrow \quad H_L \times 2 = 0$$

Considering left portion of section 2 - 2

$$\Sigma M_I = 0 \Rightarrow V_A \times 6 - H_A \times 2 - 10 \times 1 = 0$$

$$\Rightarrow 6V_A - 2H_A = 10$$

$$\Rightarrow \quad V_A = 2.5 \text{ kN} \quad [\because H_A = V_A]$$

Now, $\quad \Sigma F_y = 0$

$$\Rightarrow V_A + V_L - 10 = 0$$

$$\Rightarrow \quad V_L = 7.5 \text{ kN}$$

6 Matrix Method of Structural Analysis

6.1 Which one of the following statements is true with regard to the flexibility method of analysis?
(a) The method is used to analyze determinate structures
(b) The method is used only for manual analysis of indeterminate structures
(c) The method is used for analysis of flexible structures
(d) The method is used for analysis of indeterminate structure with lesser degree of static indeterminacy **[1995 : 1 M]**

6.2 Horizontal stiffness coefficient, K_{11} of bar AB is given by

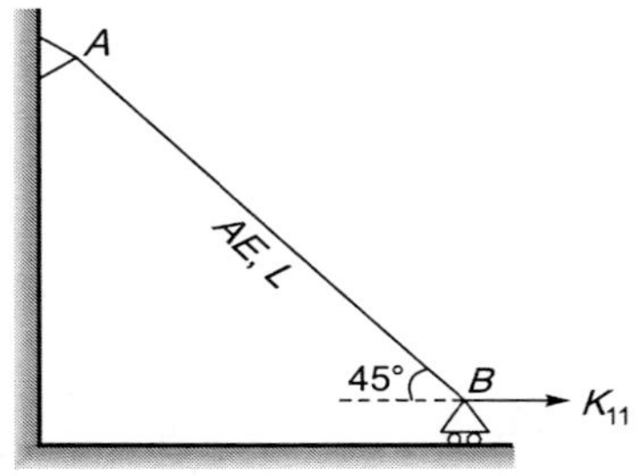

(a) $\frac{AE}{L\sqrt{2}}$ (b) $\frac{AE}{2L}$
(c) $\frac{AE}{L}$ (d) $\frac{2AE}{L}$ **[1996 : 1 M]**

6.3 The stiffness matrix of a beam element is given as $\left(\frac{2EI}{L}\right)\begin{bmatrix} 2 & 1 \\ 1 & 2 \end{bmatrix}$. Then the flexibility matrix is

(a) $\left(\frac{L}{2EI}\right)\begin{bmatrix} 2 & 1 \\ 1 & 2 \end{bmatrix}$ (b) $\left(\frac{L}{6EI}\right)\begin{bmatrix} 2 & -1 \\ -1 & 2 \end{bmatrix}$
(c) $\left(\frac{L}{3EI}\right)\begin{bmatrix} 2 & -1 \\ -1 & 2 \end{bmatrix}$ (d) $\left(\frac{L}{5EI}\right)\begin{bmatrix} 2 & -1 \\ -1 & 2 \end{bmatrix}$
[1998 : 2 M]

6.4 The stiffness K of a beam deflecting in a symmetric mode, as shown in the figure, is

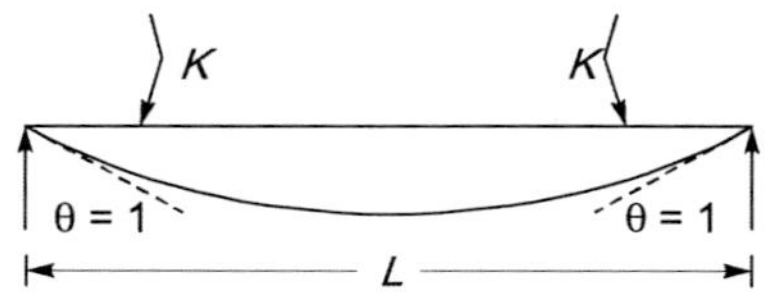

EI uniform

(a) $\frac{EI}{L}$ (b) $\frac{2EI}{L}$
(c) $\frac{4EI}{L}$ (d) $\frac{6EI}{L}$ **[2003 : 1 M]**

6.5 For a linear elastic frame, if stiffness matrix is doubled with respect to the existing stiffness matrix, the deflection of the resulting frame will be
(a) twice the existing value
(b) half the existing value
(c) the same as existing value
(d) indeterminate value **[2005 : 1 M]**

6.6 The stiffness coefficient k_{ij} indicates
(a) force at i due to a unit deformation at j
(b) deformation at j due to a unit force at i
(c) deformation at i due to a unit force at j
(d) force at j due to a unit deformation at i
[2007 : 1 M]

6.7 For the beam shown below, the stiffness coefficient K_{22} can be written as

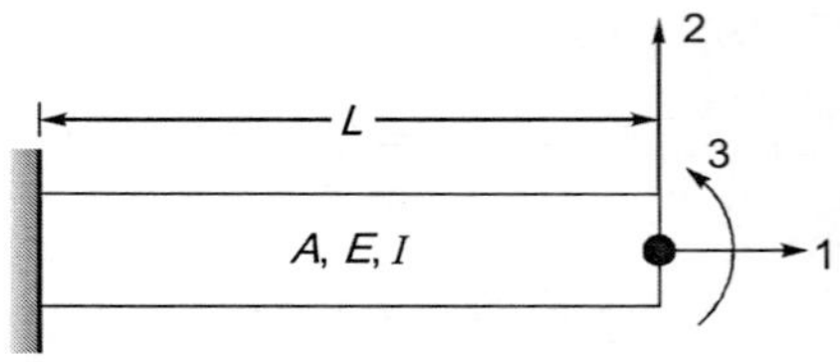

Note : 1, 2 and 3 are the d.o.f

(a) $\frac{6EI}{L^2}$ (b) $\frac{12EI}{L^3}$
(c) $\frac{3EI}{L}$ (d) $\frac{EI}{6L^2}$ **[2015:1 M, Set-I]**

6.8 Employ stiffness matrix approach for the simply supported beam as shown in the figure to calculate unknown displacements/rotations. Take length, $L = 8$ m; modulus of elasticity, $E = 3 \times 10^4$ N/mm^2; moment of inertia, $I = 225 \times 10^6$ mm^4.

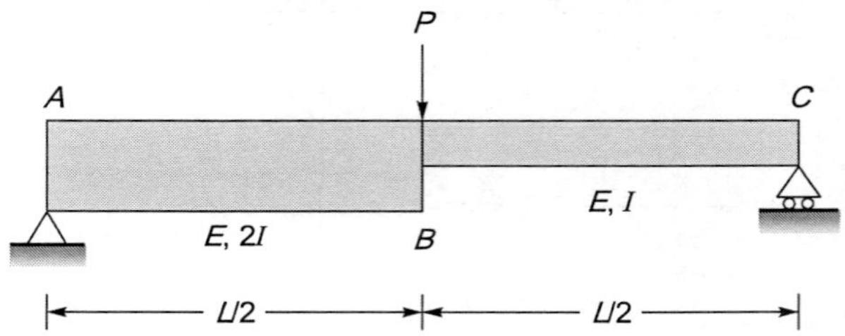

The mid-span deflection of the beam (in mm, round off to integer) under $P = 100$ kN in downward direction will be ________.

[2021 : 2 M, Set-I]

■■■■

Answers Matrix Method of Structural Analysis

6.1 (d) 6.2 (b) 6.3 (b) 6.4 (b) 6.5 (b) 6.6 (a) 6.7 (b)

6.8 (119)

Explanations Matrix Method of Structural Analysis

6.1 (d)

Flexibility method is used when $D_k > D_s$.

In flexibility method, the number of compatibility equations needed is equal to degree of static determinacy so lesser static indeterminacy facilitates easy analysis.

6.2 (b)

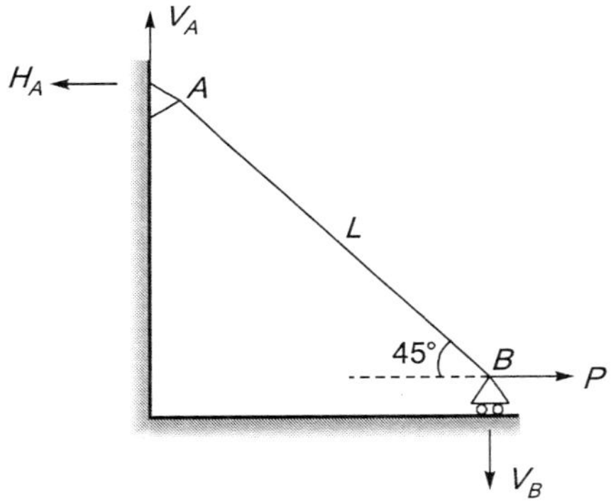

k_{11} means force required at 1 to produce unit deflection at 1.

$$\Sigma F_x = 0$$

$$\Rightarrow \quad H_A = P$$

$$\Sigma M_B = 0$$

$$\Rightarrow \quad V_A \times \frac{L}{\sqrt{2}} - H_A \times \frac{L}{\sqrt{2}} = 0$$

$$\Rightarrow \quad V_A \times \frac{L}{\sqrt{2}} - P \times \frac{L}{\sqrt{2}} = 0$$

$$\Rightarrow \quad V_A = P$$

$$\Sigma F_y = 0$$

$$\Rightarrow \quad V_A = V_B$$

$$\therefore \quad V_B = P$$

Consider joint B,

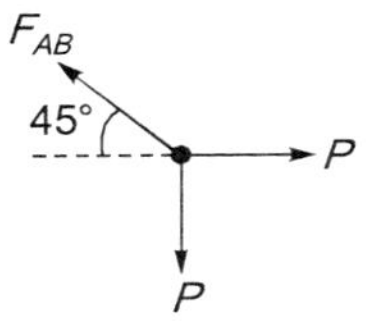

$$\Sigma F_x = 0$$

$$\Rightarrow \quad F_{AB} \cos 45° = P$$

$$\Rightarrow \quad F_{AB} = P\sqrt{2}$$

Deflection of joint B in horizontal direction is,

$$\delta_{11} = \frac{\delta_{AB}}{\cos 45°}$$

$$= \frac{F_{AB} \times L}{AE} \times \sqrt{2} = \frac{\sqrt{2}PL}{AE} \times \sqrt{2} = \frac{2PL}{AE}$$

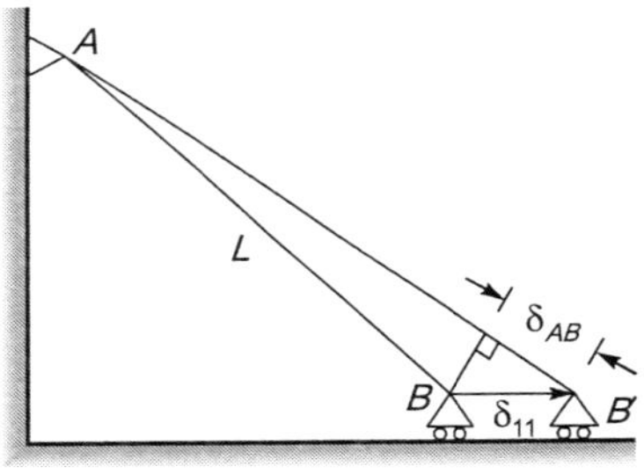

$$\therefore \quad k_{11} = \frac{P}{\delta_{11}} = \frac{AE}{2L}$$

Method-II (Direct Formula)

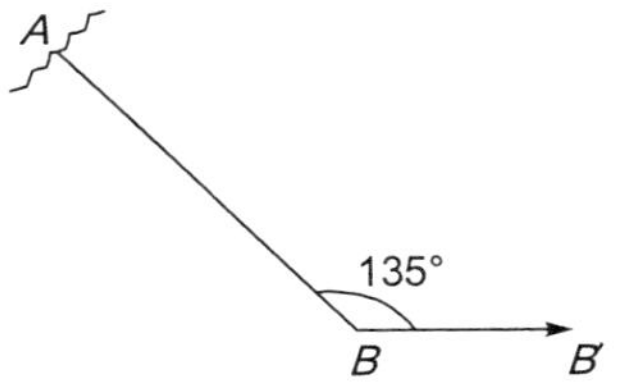

$$k_{11} = \frac{AE}{L} \cos^2 135° = \frac{AE}{2L}$$

6.3 (b)

Method-I

Given: $$[k] = \frac{2EI}{L}\begin{bmatrix} 2 & 1 \\ 1 & 2 \end{bmatrix} = \begin{bmatrix} \frac{4EI}{L} & \frac{2EI}{L} \\ \frac{2EI}{L} & \frac{4EI}{L} \end{bmatrix}$$

$\therefore \quad [f] = [k]^{-1}$

and $\quad [k]^{-1} = \dfrac{\text{adj}[k]}{|k|}$

$$\text{adj}[k] = \begin{bmatrix} \dfrac{4EI}{L} & -\dfrac{2EI}{l} \\ -\dfrac{2EI}{L} & \dfrac{4EI}{L} \end{bmatrix}$$

and $\quad |k| = \dfrac{12E^2I^2}{L^2}$

$$[k]^{-1} = \frac{\text{adj}[k]}{|k|} = \frac{L}{6EI}\begin{bmatrix} 2 & -1 \\ -1 & 2 \end{bmatrix}$$

Method-II

$$[f] = [k]^{-1}$$

$\Rightarrow \quad [f]\,[k] = I$

Proceed by option approach and check which of the option satisfy above criteria.

6.4 (b)

The beam will deflect in symmetric mode when same moment M is applied at both ends. Slope at either ends will be given as $\dfrac{ML}{2EI}$. We know that stiffness is the moment required to produce unit slope.

$\therefore \quad \theta = \dfrac{ML}{2EI}$

$\Rightarrow \quad M = \dfrac{2EI}{L}\theta \qquad [\because \theta = 1]$

$\Rightarrow \quad M = \dfrac{2EI}{L} = K$

POINTS TO REMEMBER

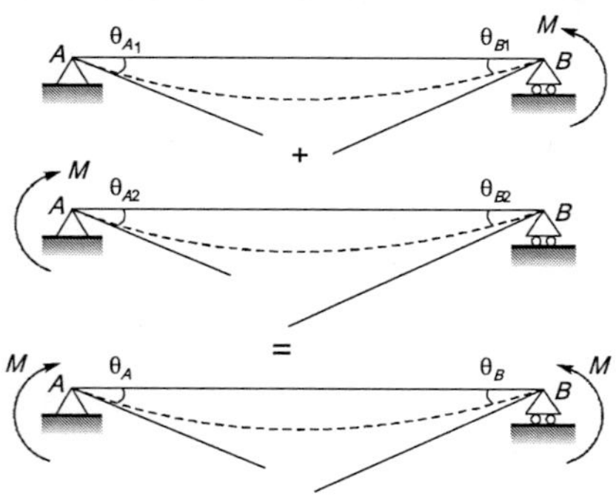

$\theta_{A_1} = \dfrac{ML}{6EI} \qquad \theta_{B_1} = \dfrac{ML}{3EI}$

$\theta_{A_2} = \dfrac{ML}{3EI} \qquad \theta_{B_2} = \dfrac{ML}{6EI}$

By superposition method,

$$\theta_A = \theta_{A_1} + \theta_{A_2} = \frac{ML}{2EI}$$

$$\theta_B = \theta_{B_1} + \theta_{B_2} = \frac{ML}{2EI}$$

6.5 (b)

Stiffness matrix and deflection are related as,

$$P = k\delta$$

Thus, when stiffness matrix is doubled, then deflection will reduce to half of the existing value.

6.6 (a)

Stiffness(k) is the force required to produce unit deformation.

Thus, k_{ij} denotes force required in direction i due to unit deformation (displacement) in direction j.

6.7 (b)

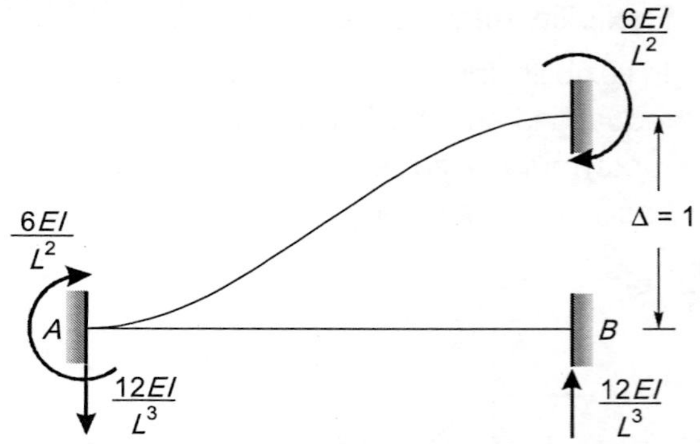

By giving unit displacement in 2nd direction without giving displacement in any other direction, the force developed R_B, is,

$$R_B = K_{22} = \frac{12EI}{L^3} \qquad (\because \Delta = 1)$$

6.8 Sol.

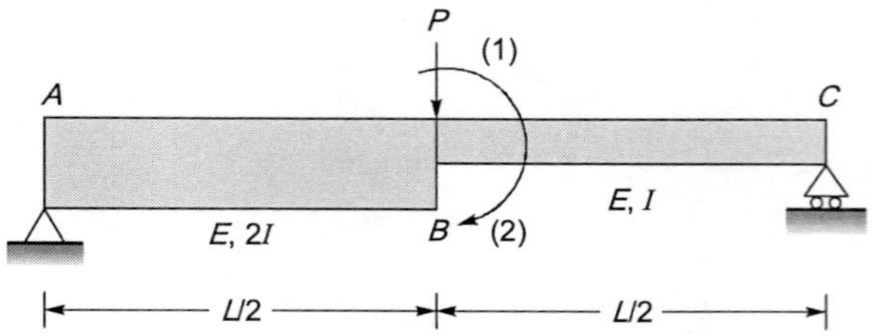

By stiffness matrix method

Step-1: Generation of stiffness matrix

Column-1

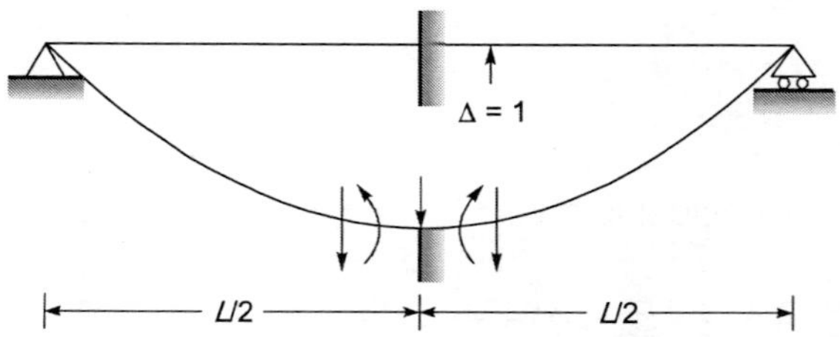

$$k_{11} = \frac{3E(2I)}{(L/2)^3} + \frac{3E(I)}{(L/2)^3} = \frac{72EI}{L^3}$$

$$k_{21} = -\frac{3E(2I)}{(L/2)^2} + \frac{3E(I)}{(L/2)^2} = -\frac{12EI}{L^2}$$

Column-2

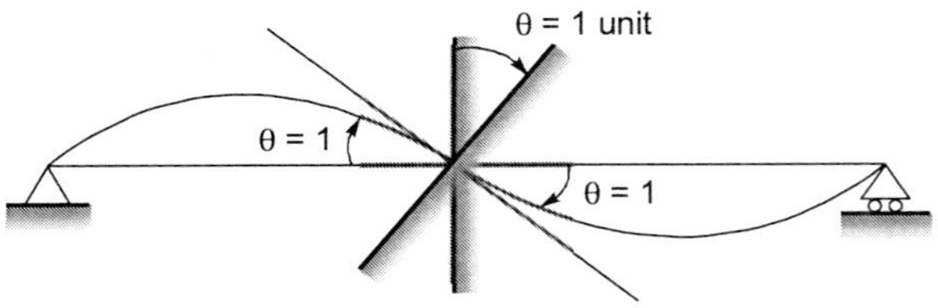

$$k_{22} = \frac{3E(2I)}{(L/2)} + \frac{3E(I)}{(L/2)} = 18\frac{EI}{L}$$

Stiffness matrix $[k]$

$$= \begin{bmatrix} 72\frac{EI}{L^3} & -\frac{12EI}{L^2} \\ -\frac{12EI}{L^2} & \frac{18EI}{L} \end{bmatrix}$$

Step-2: Calculation of unknown Nodal displacements (Δ_B, θ_B)

Using $[P] = [k]\,[\Delta]$

$$\Rightarrow \quad \begin{bmatrix} P \\ O \end{bmatrix} = \begin{bmatrix} \frac{72EI}{L^2} & -\frac{12EI}{L^2} \\ -\frac{12EI}{L^2} & \frac{18EI}{L} \end{bmatrix} \begin{bmatrix} \Delta_B \\ \theta_B \end{bmatrix}$$

On solving

$$\Delta_B = \frac{PL^3}{64EI}(\downarrow)$$

$$\theta_B = \frac{PL^2}{96EI}(\text{CW})$$

$$\therefore \quad \Delta_B = \frac{\left(100\times10^3\right)\times(8000)^3}{64\times3\times10^4\times225\times10^6}$$

$$= 118.519 \text{ mm} \simeq 119 \text{ mm}$$

■■■■

7 Structural Dynamics

7.1 A simple mass-spring oscillatory system consists of a mass m, suspended from a spring of stiffness k. Considering z as the displacement of the system at any time t, the equation of motion for the free vibration of the system is $m\ddot{z} + kz = 0$. The natural frequency of the system is

(a) $\frac{m}{k}$ (b) $\frac{k}{m}$

(c) $\sqrt{\frac{m}{k}}$ (d) $\sqrt{\frac{k}{m}}$

[2019 : 1 M, Set-I]

7.2 A plane frame shown in the figure (not to scale) has linear elastic springs at node H. The spring constants are $k_x = k_y = 5 \times 10^5$ kN/m and $k_\theta = 3 \times 10^5$ kNm/rad.

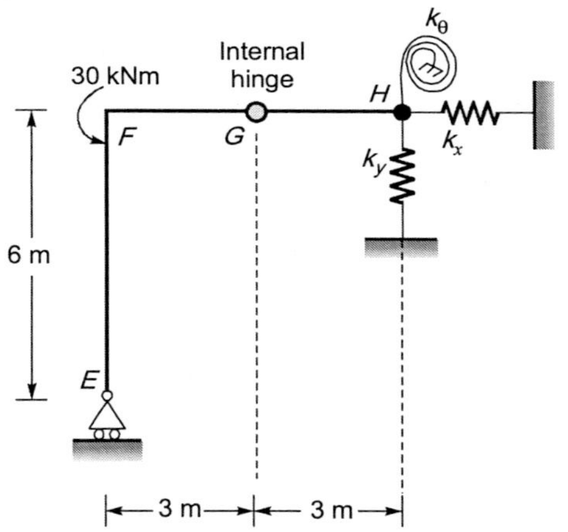

For the externally applied moment of 30 kNm at node F, the rotation (in degrees, round off to 3 decimals) observed in the rotational spring at node H is__________

[2019 : 2 M, Set-II]

7.3 A single story building model is shown in the figure. The rigid bar of mass 'm' is supported by three massless elastic columns whose ends are fixed against rotation. For each of the columns, the applied lateral force (P) and corresponding moment (M) are also shown in the figure. The lateral deflection (δ) of the bar is given by $\delta = \frac{PL^3}{12EI}$, where L is the effective length of the column, E is the Young's modulus of elasticity and I is the area moment of inertia of the column cross-section with respect to its neutral axis.

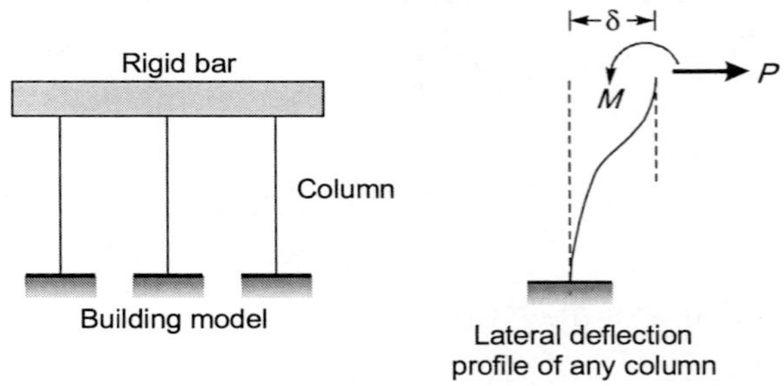

For the lateral deflection profile of the columns as shown in the figure, the natural frequency of the system for horizontal oscillation is

(a) $6\sqrt{\frac{EI}{mL^3}}$ rad/s (b) $\frac{2}{L}\sqrt{\frac{EI}{m}}$ rad/s

(c) $\frac{1}{L}\sqrt{\frac{2EI}{m}}$ rad/s (d) $2\sqrt{\frac{6EI}{mL^3}}$ rad/s

[2021 : 1 M, Set-II]

7.4 An undamped spring-mass system with mass m and spring stiffness k is shown in the figure. The natural frequency and natural period of this system are ω rad/s and T s. respectively. If the stiffness of the spring is doubled and the mass is halved, then the natural frequency and the natural period of the modified system, respectively, are

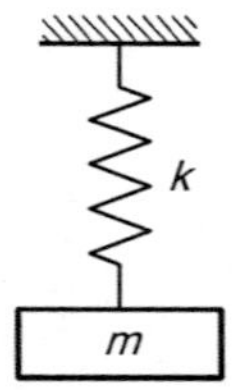

(a) ω rad/s and T s
(b) $\omega/2$ rad/s and $2T$ s
(c) 4 ω rad/s and $T/4$ s
(d) 2 ω rad/s and $T/2$ s

[2022 : 1 M, Set-II]

7.5 A linearly elastic beam of length $2l$ with flexural rigidity EI has negligible mass. A massless spring with a spring constant k and a rigid block of mass m are attached to the beam as shown in the figure.

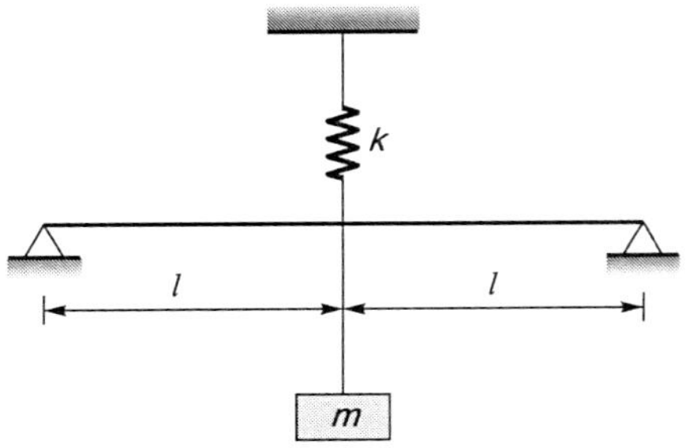

The natural frequency of this system is

(a) $\sqrt{\dfrac{6EIk}{(kl^3+6EI)m}}$

(b) $\sqrt{\dfrac{48EIk}{(kl^3+48EI)m}}$

(c) $\sqrt{\dfrac{kl^3+48EI}{ml^3}}$

(d) $\sqrt{\dfrac{kl^3+6EI}{ml^3}}$

[2024 : 2 M, Set-II]

Answers Structural Dynamics

7.1 (d) 7.2 (0.0057) 7.3 (a) 7.4 (d) 7.5 (d)

Explanations Structural Dynamics

7.1 (d)

$$m\ddot{z} + kz = 0 \Rightarrow \ddot{z} + \frac{k}{m}z = 0$$

Comparing with $\ddot{z} + \omega_n^2 z = 0$

We get $\omega_n = \sqrt{\dfrac{k}{m}}$.

7.2 Sol.

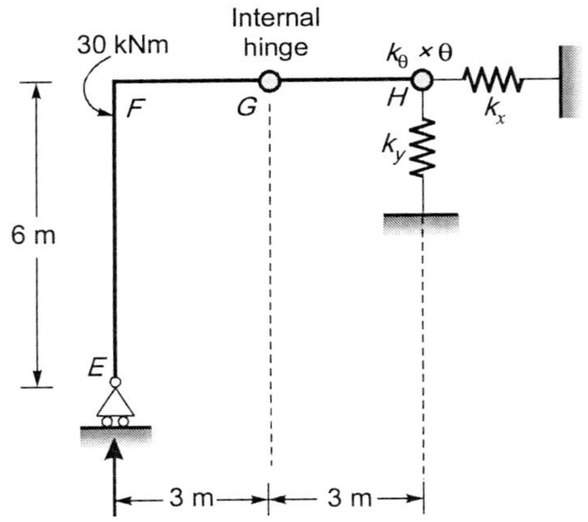

No moment can be taken by segment FE

$\therefore \quad M_{FE} = 0$

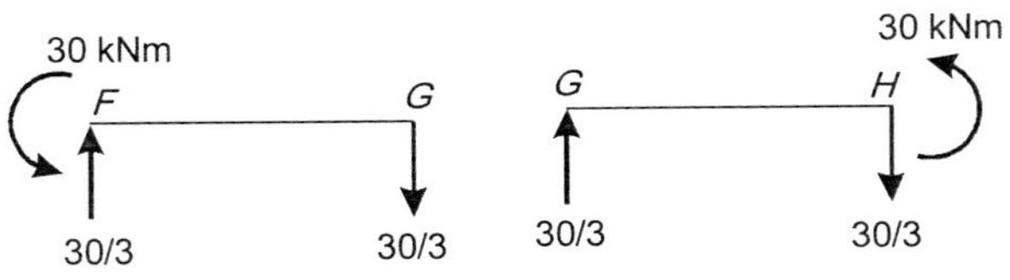

$$M_\theta = k_\theta \times \theta$$

$$30 \text{ kNm} = 3 \times 10^5 \text{ kNm/rad} \times \theta$$

$$\Rightarrow \quad \theta = 1 \times 10^{-4} \text{ radians} = 0.0057°$$

7.3 (a)

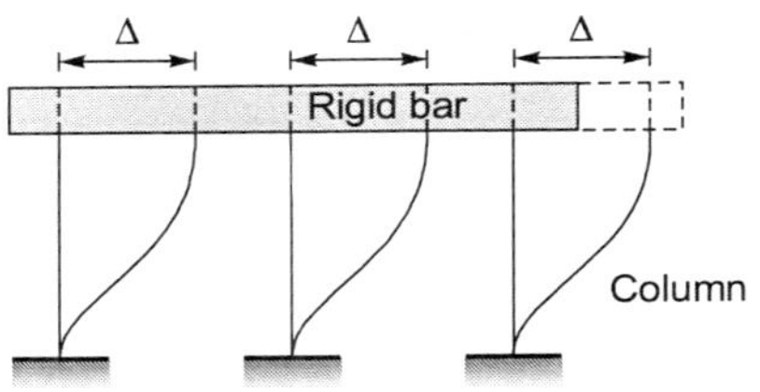

As the deflection will be same in all the 3 columns, so it represents a parallel connection.

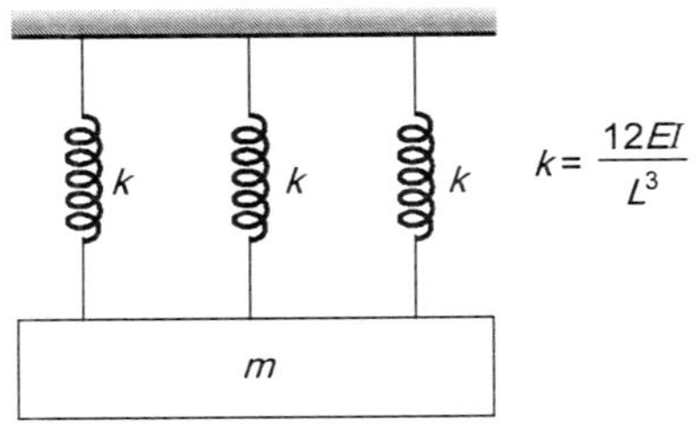

$$k_{eq} = 3k = \frac{36EI}{L^3}$$

Natural frequency,

$$\omega = \sqrt{\frac{k}{m}} = \sqrt{\frac{36EI}{mL^3}} = 6\sqrt{\frac{EI}{mL^3}} \text{ rad/s}$$

7.4 (d)

Natural frequency, $\omega_n = \sqrt{\frac{k}{m}}$ rad/s

Time period, $T_n = \frac{2\pi}{\omega_n}$

For new system, $m' = \frac{m}{2}$

$k' = 2k$

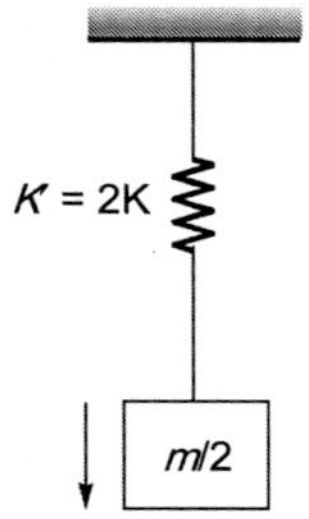

∴ New natural frequency,

$$\omega_n' = \sqrt{\frac{k'}{m'}} = \sqrt{\frac{2k}{m/2}} = \sqrt{\frac{4k}{m}}$$

$$\Rightarrow \quad \omega_n' = 2\sqrt{k/m} \quad \because \sqrt{k/m} = \omega_n$$

$$\Rightarrow \quad \omega_n' = 2\omega_n$$

∴ Frequency will be double.

Now, new time period,

$$T_n' = \frac{2\pi}{\omega_n'} = \frac{2\pi}{2\omega_n} \quad \left(\because \frac{2\pi}{\omega_n} = T_n\right)$$

$$\Rightarrow \quad T_n' = \frac{T_n}{2}$$

∴ Time period will become half.

7.5 (d)

Here, weight (mg) is shared by the beam and top spring and the deflection in beam and top spring will be same.

∴ Beam and top spring will be in parallel.

Beam stiffness:

$$K_{beam} = \frac{48EI}{(2l)^3} = \frac{6EI}{l^3}$$

Equivalent stiffness:

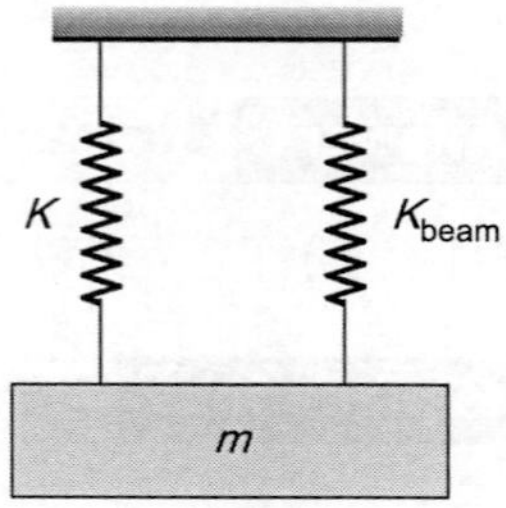

$$K_{eq} = k + K_{beam}$$

$$= k + \frac{6EI}{l^3}$$

∴ Natural frequency,

$$\omega_n = \sqrt{\frac{K_{eq}}{m}} = \sqrt{\frac{k + \frac{6EI}{l^3}}{m}}$$

$$= \sqrt{\frac{kl^3 + 6EI}{ml^3}} \text{ rad/s}$$

■■■■

RCC Structures and Prestressed Concrete

UNIT

III

CONTENTS

RCC Structures and Prestressed Concrete

Syllabus

Working stress and Limit state design concepts; Design of beams, slabs, columns; Bond and development length; Prestressed concrete beams.

Analysis of Previous GATE Papers

Exam Year	1 Mark Ques.	2 Marks Ques.	Total Marks
1995	7	–	7
1996	2	–	2
1997	6	–	6
1998	2	1	4
1999	4	–	4
2000	1	–	1
2001	–	1	2
2002	2	1	4
2003	3	5	13
2004	2	5	12
2005	2	6	14
2006	1	5	11
2007	–	6	12
2008	2	5	12
2009	2	0	2
2010	1	4	5
2011	3	2	7
2012	2	3	8
2013	2	1	4
2014 Set-1	2	3	8
2014 Set-2	4	–	4

Exam Year	1 Mark Ques.	2 Marks Ques.	Total Marks
2015 Set-1	4	1	6
2015 Set-2	1	2	5
2016 Set-1	1	2	5
2016 Set-2	1	1	3
2017 Set-1	1	1	3
2017 Set-2	1	1	3
2018 Set-1	3	2	7
2018 Set-2	3	2	7
2019 Set-1	1	0	1
2019 Set-2	1	2	5
2020 Set-1	1	2	5
2020 Set-2	1	2	5
2021 Set-1	–	2	4
2021 Set-2	–	1	2
2022 Set-1	1	1	3
2022 Set-2	–	2	4
2023 Set-1	3	1	5
2023 Set-2	2	1	4
2024 Set-1	3	1	5
2024 Set-2	1	1	3

1 Working Stress & Limit State Method

1.1 The basic assumption of plane sections normal to the neutral axis before bending remaining plane and normal to theneutral axis after bending, leads to
(a) uniform strain over the beam cross-section
(b) uniform stress over the beam cross-section
(c) linearly varying strain over the cross-section
(d) stresses which are proportional to strains at the cross-section

[1995 : 1 M]

1.2 The span to depth ratio limit is specified in **IS 456 : 1978** for the reinforced concrete beams, in order to ensure that the
(a) tensile crack width is below a limit
(b) shear failure is avoided
(c) stress in the tension reinforcement is less than the allowable value
(d) deflection of the beam is below a limiting value

[1995 : 1 M]

1.3 The modulus of rupture of concrete gives
(a) the direct tensile strength of the concrete
(b) the direct compressive strength of the concrete
(c) the tensile strength of the concrete under bending
(d) the characteristic strength of the concrete

[1995 : 1 M]

1.4 The effective width of a reinforced concrete T-beam flange under compression, according to **IS 456 : 1978**, given l_0 is the distance between the adjacent zero moment points, b is the breadth of the rib and D is the thickness of the flange, is

(a) $\frac{l_0}{6} + B + 6D$ (b) $l_0 + 6D$

(c) $\frac{l_0}{6} + 6D$ (d) $\frac{l_0}{6} + b$ [1995 : 1 M]

1.5 The cylinder strength of the concrete is less than the cube strength because of
(a) the difference in the shape of the cross-section of the specimens
(b) the difference in the slenderness ratio of the specimens
(c) the friction between the concrete specimens and the steel plate of the testing machine
(d) the cubes are tested without capping but the cylinders are tested with capping

[1997 : 1 M]

1.6 The permissible bending tensile stress in concrete for the vertical wall of an RC water tank made of M 25 concrete is
(a) 8.5 N/mm^2 (b) 6.0 N/mm^2
(c) 2.5 N/mm^2 (d) 1.8 N/mm^2

[1997 : 1 M]

1.7 Permissible bending tensile stress in high yield strength deformed bars of grade 415 in a beam is
(a) 190 N/mm^2 (b) 230 N/mm^2
(c) 140 N/mm^2 (d) None of the above

[1997 : 1 M]

1.8 The characteristic strength of concrete is defined as that compressive strength below which not more than
(a) 10% of results fail
(b) 5% of results fail
(c) 2% of results fail
(d) None of the above [1999 : 1 M]

1.9 Maximum strain at the level of compression steel for a rectangular section having effective cover to compression steel as d' and neutral axis depth from compression face as x_u is

(a) $0.0035\left(1-\frac{d'}{x_u}\right)$ (b) $0.002\left(1-\frac{d'}{x_u}\right)$

(c) $0.0035\left(1-\frac{x_u}{d'}\right)$ (d) $0.002\left(1-\frac{x_u}{d'}\right)$

[1999 : 1 M]

1.10 The following two statements are made with reference to a simply supported under reinforced RCC beam
I. Failure takes place by crushing of concrete before the steel has yielded.
II. The neutral axis moves up as the load is increased.

With reference to the above statements, which of the following applies?

(a) Both the statements are false
(b) I is true but II is false
(c) Both the statements are true
(d) I is false but II is true [2000 : 1 M]

1.11 Read the following two statements:
I. Maximum strain in concrete at the outermost compression fibre is taken to be 0.0035 in bending
II. The maximum compressive strain in concrete in axial compression is taken as 0.002

Keeping the provisions of **IS 456 : 2000** on limit state design in mind, which of the following is true:
(a) Statement I is True but II is False
(b) Statement I is False but II is True
(c) Both statement I and II are True
(d) Both statement I and II are False
[2002 : 1 M]

1.12 As per the provisions of **IS 456 : 2000**, the (short term) modulus of elasticity of M 25 grade concrete (in N/mm^2) can be assumed to be
(a) 25000 (b) 28500
(c) 30000 (d) 36000 [2002 : 1 M]

1.13 Maximum strains in an extreme fibre in concrete and in the tension reinforcement (Fe-415 grade and E_s = 200 kN/mm^2) in a balanced section at limit state of flexure are respectively
(a) 0.0035 and 0.0038
(b) 0.002 and 0.0018
(c) 0.0035 and 0.0041
(d) 0.002 and 0.0031 [2003 : 1 M]

1.14 The working stress method of design specifies the value of modular ratio, $m = 280/(3\sigma_{cbc})$, where σ_{cbc} is the allowable stress in bending compression in concrete. To what extent does the above value of 'm' make any allowance for the creep of concrete?
(a) No compensation
(b) Full compensation
(c) Partial compensation
(d) The two are unrelated [2003 : 1 M]

1.15 **List-I** contains some properties of concrete/cement and **List-II** contains list of some tests on concrete/cement. Match the property with the corresponding test.

List-I
A. Workability of concrete
B. Direct tensile strength of concrete
C. Bond between concrete and steel
D. Fineness of cement

List-II
1. Cylinder splitting test
2. Vee-Bee test
3. Surface area test
4. Fineness modulus test
5. Pull out test

Codes:

	A	B	C	D
(a)	2	1	5	3
(b)	4	5	1	3
(c)	2	1	5	4
(d)	2	5	1	4

[2003 : 2 M]

1.16 In the limit state design method of concrete structures, the recommended partial material safety factor (γ_m) for steel according to **IS 456 : 2000** is
(a) 1.5 (b) 1.15
(c) 1.00 (d) 0.87 [2004 : 1 M]

1.17 For avoiding the limit state of collapse, the safety of RC structures is checked for appropriate combinations of Dead Load (DL), Imposed Load or Live Load (IL), Wind Load (WL) and Earthquake Load (EL). Which of the following load combinations is NOT considered?
(a) 0.9 DL + 1.5 WL
(b) 1.5 DL + 1.5 WL
(c) 1.5 DL + 1.5 WL + 1.5 EL
(d) 1.2 DL + 1.2 IL + 1.2 WL [2004 : 1 M]

1.18 The partial factor of safety for concrete as per **IS 456 : 2000** is
(a) 1.50 (b) 1.15
(c) 0.87 (d) 0.446 [2005 : 1 M]

1.19 The flexural strength of M30 concrete as per **IS 456 : 2000** is
(a) 3.83 MPa (b) 5.47 MPa
(c) 21.23 MPa (d) 30.0 MPa
[2005 : 2 M]

1.20 In a random sampling procedure for cube strength of concrete, one sample consists of X number of specimens. These specimens are tested at 28 days and average strength of these X specimens is considered as test result of the sample, provided the individual variation in the strength of specimens is not more than $\pm$ Y per cent of the average strength. The values of X and Y as per **IS 456 : 2000** are
(a) 4 and 10 respectively
(b) 3 and 10 respectively

(c) 4 and 15 respectively
(d) 3 and 15 respectively [2005 : 2 M]

Linked Answer Questions 1.21 and 1.22

Assume straight line instead of parabola for stress-strain curve of concrete as given below and partial factor of safety as 1.0.

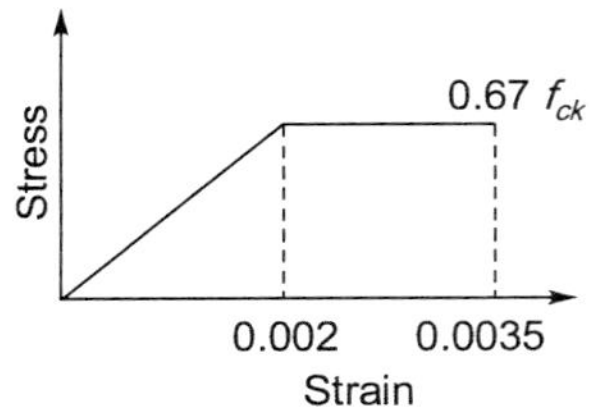

A rectangular under-reinforced concrete section of 300 mm width and 500 mm effective depth is reinforced with 3 bars of grade Fe-415, each of 16 mm diameter. Concrete mix is M-20.

1.21 The depth of the neutral axis from the compression fibre is
(a) 76 mm (b) 81 mm
(c) 87 mm (d) 100 mm [2005 : 2 M]

1.22 The depth of the neutral axis obtained as per **IS 456 : 2000** differs from the depth of neutral axis obtained in above question by
(a) 15 mm (b) 20 mm
(c) 25 mm (d) 32 mm
[2005 : 2 M]

1.23 If the characteristic strength of concrete f_{ck} is defined as the strength below which not more than 50% of the test results are expected to fall, the expression for f_{ck} in terms of mean strength f_m and standard deviation S would be
(a) $f_m - 0.1645\ S$ (b) $f_m - 1.645\ S$
(c) f_m (d) $f_m + 1.645\ S$
[2006 : 1 M]

1.24 Assuming concrete below the neutral axis to be cracked, the shear stress across the depth of a singly-reinforced rectangular beam section
(a) increases parabolically to the neutral axis and then drops suddenly to zero value.
(b) increases parabolically to the neutral axis and then remains constant over the remaining depth.
(c) increases linearly to the neutral axis and then remains constant up to the tension steel.
(d) increases parabolically to the neutral axis and then remains constant up to the tension steel.
[2006 : 2 M]

1.25 As per **IS 456 : 2000**, consider the following statements:
1. The modular ratio considered in the working stress method depends on the type of steel used.
2. There is an upper limit on the nominal shear stress in beams (even with shear reinforcement) due to the possibility of crushing of concrete in diagonal compression.
3. A rectangular slab whose length is equal to its width may not be a two-way slab for some support conditions.

The TRUE statements are
(a) 1 and 2 (b) 2 and 3
(c) 1 and 3 (d) 1, 2 and 3
[2006 : 2 M]

1.26 Consider the following statements:
1. The width-to-thickness ratio limitations on the plate elements under compression in steel members are imposed by **IS 800 : 1984** in order to avoid fabrication difficulties.
2. In a doubly reinforced concrete beam, the strain in compressive reinforcement is higher than the strain in the adjoining concrete.
3. If a cantilever I-section supports slab construction all along its length with sufficient friction between them, the permissible bending stress in compression will be the same as that in tension.

The TRUE statements are
(a) 1 and 2 (b) 2 and 3
(c) 1 and 3 (d) 1, 2, and 3
[2006 : 2 M]

1.27 Consider the following statements:
1. The compressive strength of concrete decreases with increase in water-cement ratio of the concrete mix.
2. Water is added to the concrete mix for hydration of cement and workability.
3. Creep and shrinkage of concrete are independent of the water-cement ratio in the concrete mix.

The TRUE statements are
(a) 1 and 2 (b) 1, 2 and 3
(c) 2 and 3 (d) only 2
[2007 : 2 M]

1.28 Consider the following statements:

1. Modulus of elasticity of concrete increases with increase in compressive strength of concrete.
2. Brittleness of concrete increases with decrease in compressive strength of concrete.
3. Shear strength of concrete increase with increase in compressive strength of concrete.

The TRUE statements are

(a) 2 and 3 (b) 1, 2 and 3
(c) 1 and 2 (d) 1 and 3

[2007 : 2 M]

1.29 A reinforced concrete structure has to be constructed along a sea coast. The minimum grade of concrete to be used as per **IS 456 : 2000** is

(a) M 15 (b) M 20
(c) M 25 (d) M 30 [2008 : 1 M]

1.30 Un-factored maximum bending moments at a section of a reinforced concrete beam resulting from a frame analysis are 50, 80, 120 and 180 kNm under dead, live, wind and earthquake loads respectively. The design moment (kNm) as per **IS 456: 2000** for the limit state of collapse (flexure) is

(a) 195 (b) 250
(c) 345 (d) 372 [2008 : 2 M]

1.31 The modulus of rupture of concrete in terms of its characteristic cube compressive strength (f_{ck}) in MPa according to **IS 456: 2000** is

(a) $5000\ f_{ck}$ (b) $0.7\ f_{ck}$
(c) $5000\sqrt{f_{ck}}$ (d) $0.7\sqrt{f_{ck}}$

[2009 : 1 M]

1.32 For limit state of collapse, the partial safety factors recommended by **IS 456 : 2000** for estimating the design strength of concrete and reinforcing steel are respectively

(a) 1.15 and 1.5 (b) 1.0 and 1.0
(c) 1.5 and 1.15 (d) 1.5 and 1.0

[2009 : 1 M]

1.33 Match **List-I** (List of test methods for evaluating properties of concrete) with **List-II** (List of properties) and select the correct answer using the codes given below the lists:

List-I

A. Resonant frequency test
B. Rebound hammer test
C. Split cylinder test
D. Compacting factor test

List-II

1. Tensile strength
2. Dynamic modulus of elasticity
3. Workability
4. Compressive strength

Codes:

	A	B	C	D
(a)	2	4	1	3
(b)	2	1	4	3
(c)	2	4	3	1
(d)	4	3	1	2

[2010 : 2 M]

1.34 The cross-section of a thermo-mechanically treated (TMT) reinforcing bar has

(a) soft ferrite-pearlite throughout.
(b) hard martensite throughout.
(c) a soft ferrite-pearlite core with a hard martensitic rim.
(d) a hard martensitic core with a soft pearlite-bainitic rim. [2011 : 1 M]

1.35 As per **IS 456: 2000**, in the Limit State Design of a flexural member, the strain in reinforcing bars under tension at ultimate state should not be less than

(a) $\dfrac{f_y}{E_s}$ (b) $\dfrac{f_y}{E_s}+0.002$

(c) $\dfrac{f_y}{1.15E_s}$ (d) $\dfrac{f_y}{1.15E_s}+0.002$

[2012 : 1 M]

1.36 Maximum possible value of compaction factor for fresh (green) concrete is

(a) 0.5 (b) 1.0
(c) 1.5 (d) 2.0 [2013 : 1 M]

1.37 Match the information given in **Group-I** with those in **Group-II**.

Group-I

P. Factor to decrease ultimate strength to design strength
Q. Factor to increase working load to ultimate load for design
R. Statical method of ultimate load analysis
S. Kinematical mechanism method of ultimate load analysis

Group-II

1. Upper bound on ultimate load
2. Lower bound on ultimate load
3. Material partial safety factor
4. Load factor

	P	Q	R	S
(a)	1	2	3	4
(b)	2	1	4	3
(c)	3	4	2	1
(d)	4	3	2	1

[2014 : 1 M, Set-I]

1.38 While designing, for a steel column of Fe250 grade, a base plate resting on a concrete pedestal of M20 grade, the bearing strength of concrete (in N/mm^2) in limit state method of design as per **IS 456 : 2000** is ________.

[2014 : 1 M, Set-I]

1.39 **Group-I** contains representative stress-strain curves as shown in the figure, while **Group-II** gives the list of materials. Match the stress-strain curves with the corresponding materials.

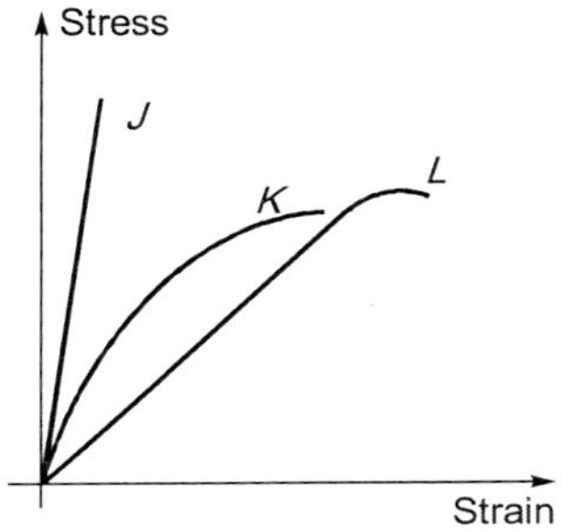

Group-I	Group-II
P. Curve J	1. Cement paste
Q. Curve K	2. Coarse aggregate
R. Curve L	3. Concrete

	P	Q	R
(a)	1	3	2
(b)	2	3	1
(c)	3	1	2
(d)	3	2	1

[2014 : 1 M, Set-II]

1.40 The target mean strength f_{cm} for concrete mix design obtained from the characteristic strength f_{ck} and standard deviation σ, as defined in **IS 456 : 2000**, is

(a) $f_{ck} + 1.35\,\sigma$ (b) $f_{ck} + 1.45\,\sigma$
(c) $f_{ck} + 1.55\,\sigma$ (d) $f_{ck} + 1.65\,\sigma$

[2014 : 1 M, Set-II]

1.41 The flexural tensile strength of M25 grade of concrete, in N/mm^2, as per **IS 456 : 2000** is ________.

[2014 : 1 M, Set-II]

1.42 The modulus of elasticity, $E = 5000\sqrt{f_{ck}}$ where f_{ck} is the characteristic compressive strength of concrete, specified in **IS 456 : 2000** is based on

(a) tangent modulus
(b) initial tangent modulus
(c) secant modulus
(d) chord modulus

[2014 : 1 M, Set-II]

1.43 In a reinforced concrete section, the stress at the extreme fibre in compression is 5.80 MPa. The depth of neutral axis in the section is 58 mm and the grade of concrete is M25. Assuming linear elastic behavior of the concrete, the effective curvature of the section (in per mm) is

(a) 2.0×10^{-6} (b) 3.0×10^{-6}
(c) 4.0×10^{-6} (d) 5.0×10^{-6}

[2014 : 2 M, Set-I]

1.44 Workability of concrete can be measured using slump, compaction factor and Vee-bee time. Consider the following statements for workability of concrete

(i) As the slump increases, the Vee-bee time increases.
(ii) As the slump increases, the compaction factor increases.

Which of the following is TRUE?

(a) Both (i) and (ii) are True
(b) Both (i) and (ii) are False
(c) (i) is True and (ii) is False
(d) (i) is False and (ii) is True

[2015 : 1 M, Set-I]

1.45 Consider the singly reinforced beam shown in the figure below:

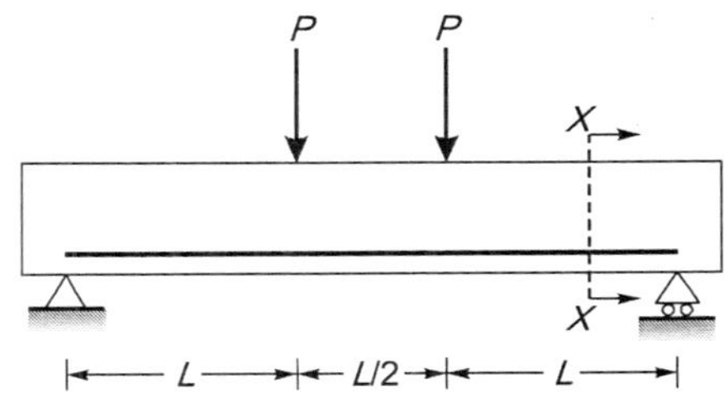

At cross-section XX, which of the following statements is TRUE at the limit state?

(a) The variation of stress is linear and that of strain is non-linear
(b) The variation of strain is linear and that of stress is non-linear
(c) The variation of both stress and strain is linear
(d) The variation of both stress and strain is non-linear

[2015 : 1 M, Set-I]

1.46 Consider the following statements for air-entrained concrete:

(i) Air-entrainment reduces the water demand for a given level of workability.

(ii) Use of air-entrained concrete is required in environments where cyclic freezing and thawing is expected.

Which of the following is TRUE?

(a) Both (i) and (ii) are True
(b) Both (i) and (ii) are False
(c) (i) is True and (ii) is False
(d) (i) is False and (ii) is True [2015:1 M, Set-I]

1.47 Consider the singly reinforced beam section given below (left figure). The stress block parameters for the cross-section from **IS 456 : 2000** are also given below (right figure). The moment of resistance for the given section by the limit state method is __________ kN-m.

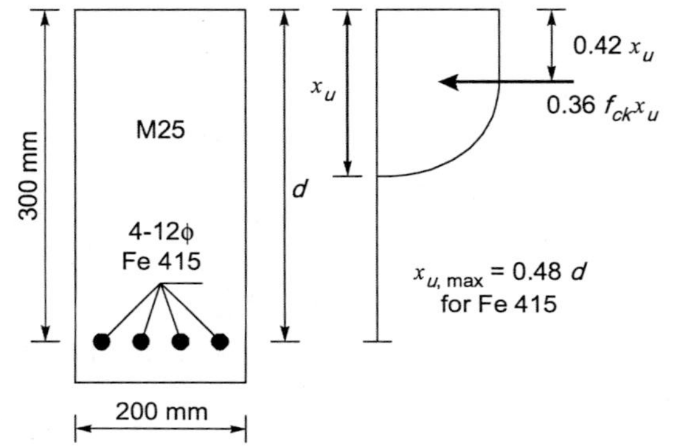

[2015 : 2 M, Set-I]

1.48 According to the concept of Limit State Design as per **IS 456 : 2000**, the probability of failure of a structure is _________.

[2015 : 2 M, Set-II]

1.49 For M25 concrete with creep coefficient of 1.5, the long-term static modulus of elasticity (expressed in MPa) as per the provisions of IS:456-2000 is ______ [2016 : 2 M, Set-I]

1.50 According to **IS 456-2000**, which one of the following statements about the depth of neutral axis $x_{u,bal}$ for a balanced reinforced concrete section is correct?

(a) $x_{u,bal}$ depends on the grade of concrete only
(b) $x_{u,bal}$ depends on the grade of steel only
(c) $x_{u,bal}$ depends on both the grade of concrete and grade of steel
(d) $x_{u,bal}$ does not depend on the grade of concrete and grade of steel [2017 : 1 M, Set-I]

1.51 Let the characteristic strength be defined as that value, below which not more than 50% of the results are expected to fall. Assuming a standard deviation of 4 MPa, the target mean strength (in MPa) to be considered in the mix design of a M25 concrete would be

(a) 18.42 (b) 21.00
(c) 25.00 (d) 31.58

[2017 : 1 M, Set-II]

1.52 The deformation in concrete due to sustained loading is

(a) creep (b) hydration
(c) segregation (d) shrinkage

[2018 : 1 M, Set-I]

1.53 The frequency distribution of the compressive strength of 20 concrete cube specimen is given in the table.

f (MPa)	Number of specimens with compressive strength equal to f
23	4
28	2
22.5	5
31	5
29	4

If μ is the mean strength of the specimens and σ is the standard deviation, the number of specimens (out of 20) with compressive strength less than $\mu - 3\sigma$ is _______.

[2018 : 1 M, Set-I]

1.54 As per IS 456 : 2000, the minimum percentage of tension reinforcement (up to two decimal places) required in reinforced-concrete beams of rectangular cross-section (considering effective depth in the calculation of area) using Fe500 grade steel is _______.

[2018 : 1 M, Set-II]

1.55 In the reinforced beam section shown in the figure (not drawn to scale), the nominal cover provided at the bottom of the beam as per IS 456-2000, is

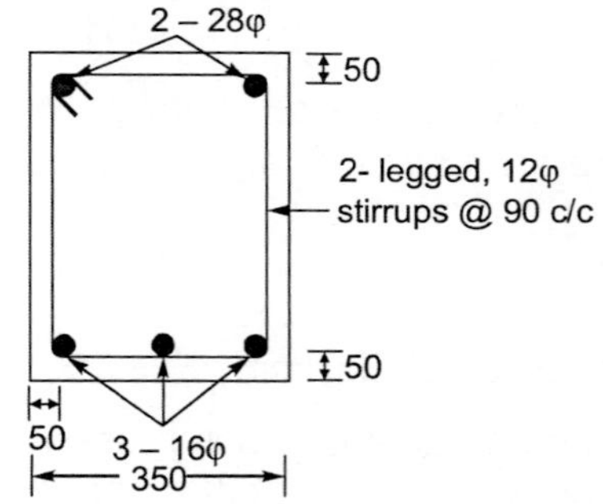

All dimensions are in mm

(a) 30 mm (b) 50 mm
(c) 42 mm (d) 36 mm

[2019 : 1 M, Set-I]

1.56 The characteristic compressive strength of concrete required in a project is 25 MPa and the standard deviation in the observed compressive strength expected at site is 4 MPa. The average compressive strength of cubes tested at different water-cement (w/c) ratios using the same material as is used for the project is given in the table.

w/c(%)	45	50	55	60
Average compressive strength of cubes (MPa)	35	25	20	15

The water-cement ratio (in percent, round off to the lower integer) to be used in the mix is ________.

[2019 : 1 M, Set-II]

1.57 In the context of provisions relating to durability of concrete, consider the following assertions:

Assertion (1): As per IS 456-2000, air entrainment to the extent of 3% to 6% is required for concrete exposed to marine environment.

Assertion (2): The equivalent alkali content (in terms of Na_2O equivalent) for a cement containing 1% and 0.6% of Na_2O and K_2O, respectively, is approximately 1.4% (rounded to 1 decimal place).

Which one of the following statements is CORRECT?

(a) Both Assertion (1) and Assertion (2) are FLASE
(b) Assertion (1) is TRUE and Assertion (2) is FLASE
(c) Both Assertion (1) and Assertion (2) are TRUE
(d) Assertion (1) is FLASE and Assertion (2) is TRUE

[2019 : 2 M, Set-II]

1.58 When a specimen of M25 concrete is loaded to a stress level of 12.5 MPa, a strain of 500×10^{-6} is recorded. If this load is allowed to stand for a long time, the strain increases to 1000×10^{-6}. In accordance with the provisions of IS:456-2000, considering the long-term effects, the effective modulus of elasticity of the concrete (in MPa) is________.

[2019 : 2 M, Set-II]

1.59 The singly reinforced concrete beam section shown in the figure (not drawn to the scale) is made of M25 grade concrete and Fe500 grade reinforcing steel. The total cross-sectional area of the tension steel is 942 mm^2.

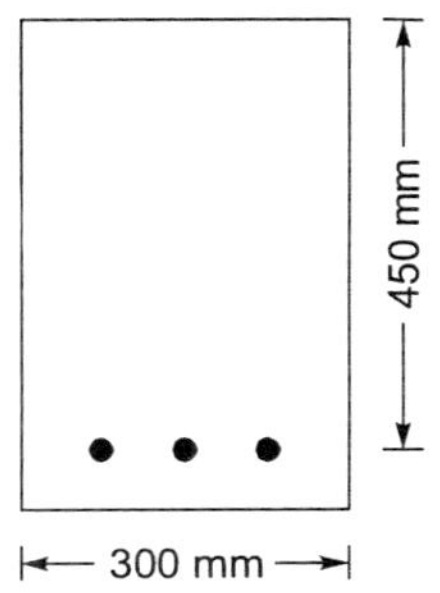

As per Limit State Design of IS 456 : 2000, the design moment capacity (in kNm round off to two decimal places) of the beam section, is __________.

[2020 : 2 M, Set-I]

1.60 As per **IS 456:2000**, the pH value of water for concrete mix shall NOT be less than

(a) 6.0 (b) 5.0
(c) 4.5 (d) 5.5

[2020 : 1 M, Set-II]

1.61 A rectangular cross-section of a reinforced concrete beam is shown in the figure. The diameter of each reinforcing bar is 16 mm. The values of modulus of elasticity of concrete and steel are 2.0×10^4 MPa and 2.1×10^5 MPa, respectively.

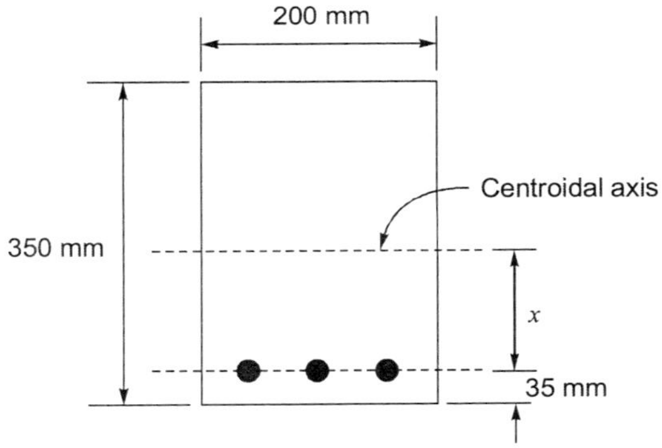

The distance of the centroidal axis from the centerline of the reinforcement (x) for the uncracked section (in mm, found off to one decimal place) is ________.

[2021 : 2 M, Set-II]

1.62 In the context of elastic theory of reinforced concrete, the modular ratio is defined as the ratio of

(a) Young's modulus of elasticity of reinforcement material to Young's modulus of elasticity of concrete.
(b) Young's modulus of elasticity of concrete to Young's modulus of elasticity of reinforcement material.

(c) shear modulus of reinforcement material to the shear modulus of concrete.

(d) Young's modulus of elasticity of reinforcement material to the shear modulus of concrete.

[2022 : 1 M, Set-I]

1.63 Read the following statements relating to flexure of reinforced concrete beams:

I. In over-reinforced sections, the failure strain in concrete reaches earlier than the yield strain in steel.

II. In under-reinforced sections, steel reaches yielding at a load lower than the load at which the concrete reaches failure strata.

III. Over-reinforced beams are recommended in practice as compared to the under-reinforced beams.

IV. In balanced sections, the concrete reaches failure strain earlier than the yield strain in tensile steel.

Each of the above statements is either True or False.

Which one of the following combinations is correct?

(a) (I)-True, (II)-True, (III)-False, (IV)-False
(b) (I)-False, (II)-True, (III)-True, (IV)-False
(c) (I)-False, (II)-False, (III)-True, (IV)-False
(d) (I)-True, (II)-True, (III)-False, (IV)-True

[2022 : 2 M, Set-II]

1.64 Creep of concrete under compression is defined as the ____.

(a) increase in the magnitude of strain under constant stress
(b) increase in the magnitude of stress under constant strain
(c) decrease in the magnitude of strain under constant stress
(d) decrease in the magnitude of stress under constant strain

[2023 : 1 M, Set-I]

1.65 A singly reinforced concrete beam of balanced section is made of M20 grade concrete and Fe415 grade steel bars. The magnitudes of the maximum compressive strain in concrete and the tensile strain in the bars at ultimate state under flexure, as per IS 456: 2000 are ________, respectively. (round off to four decimal places)

(a) 0.0035 and 0.0038
(b) 0.0020 and 0.0018
(c) 0.0035 and 0.0041
(d) 0.0020 and 0.0031

[2023 : 1 M, Set-I]

1.66 In cement concrete mix design, with the increase in water-cement ratio, which one of the following statements is TRUE?

(a) Compressive strength decreases but workability increases
(b) Compressive strength increases but workability decreases
(c) Both compressive strength and workability decrease
(d) Both compressive strength and workability increase

[2023 : 1 M, Set-I]

1.67 Consider a doubly reinforced RCC beam with the option of using either Fe250 plain bars or Fe500 deformed bars in the compression zone. The modulus of elasticity of steel is 2×10^5 N/mm^2. As per IS 456 : 2000, in which type(s) of the bars, the stress in the compression steel (f_{sc}) can reach the design strength ($0.87f_y$) at the limit state of collapse?

(a) Fe250 plain bars only
(b) Fe500 deformed bars only
(c) Both Fe250 plain bars and Fe500 deformed bars
(d) Neither Fe250 plain bars nor Fe500 deformed bars

[2023 : 2 M, Set-I]

1.68 M20 concrete as per IS 456: 2000 refers to concrete with a design mix having ______

(a) an average cube strength of 20 MPa
(b) an average cylinder strength of 20 MPa
(c) a 5-percentile cube strength of 20 MPa
(d) a 5-percentile cylinder strength of 20 MPa

[2023 : 1 M, Set-II]

1.69 Consider the singly reinforced section of a cantilever concrete beam under bending, as shown in the figure (M25 grade concrete, Fe415 grade steel). The stress block parameters for the section at ultimate limit state, as per IS 456 : 2000 notations, are given. The ultimate moment of resistance for the section by the Limit State Method is ________ kN.m (round off to one decimal place).

[Note: Here, As is the total area of tension steel bars, b is the width of the section, d is the effective depth of the bars, f_{ck} is the characteristic compressive cube strength of concrete, f_y is the yield stress of steel, and x_u is the depth of neutral axis.]

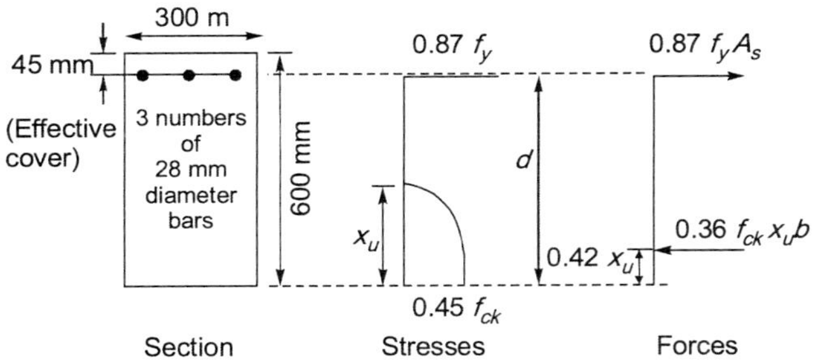

[2023 : 2 M, Set-II]

1.70 Concrete of characteristic strength 30 MPa is required. If 40 specimens of concrete cubes are to be tested, the minimum number of specimens having at least 30 MPa strength should be

(a) 35 (b) 37
(c) 38 (d) 39

[2024 : 1 M, Set-I]

1.71 The following figure shows the arrangement of formwork for casting a cantilever RC beam.

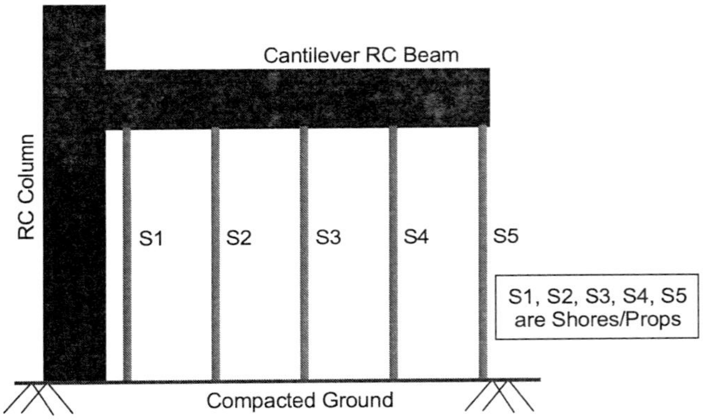

(a) S3 → S4 → S2 → S5 → S1
(b) S5 → S4 → S3 → S2 → S1
(c) S1 → S2 → S3 → S4 → S4
(d) S3 → S2 → S4 → S1 → S5

[2024 : 1 M, Set-I]

1.72 Consider a balanced doubly reinforced concrete section. If the material and other sectional properties remain unchanged, for which of the following cases will the section becomes under-reinforced?

(a) Area of compression reinforcement is increased.
(b) Area of tension reinforcement is increased.
(c) Area of compression reinforcement is decreased.
(d) Area of tension reinforcement is decreased.

[2024 : 1 M, Set-I]

■■■■

Answers Working Stress & Limit State Method

1.1	(c)	1.2	(d)	1.3	(c)	1.4	(a)	1.5	(c)	1.6	(d)	1.7	(b)
1.8	(b)	1.9	(a)	1.10	(d)	1.11	(c)	1.12	(a)	1.13	(a)	1.14	(c)
1.15	(a)	1.16	(b)	1.17	(c)	1.18	(a)	1.19	(a)	1.20	(d)	1.21	(a)
1.22	(c)	1.23	(c)	1.24	(d)	1.25	(b)	1.26	(b)	1.27	(a)	1.28	(d)
1.29	(d)	1.30	(d)	1.31	(d)	1.32	(c)	1.33	(a)	1.34	(c)	1.35	(d)
1.36	(b)	1.37	(c)	1.38	(9)	1.39	(b)	1.40	(d)	1.41	(3.5)	1.42	(c)
1.43	(c)	1.44	(d)	1.45	(b)	1.46	(a)	1.47	(42.8)	1.48	(0.0975)	1.49	(10,000)
1.50	(b)	1.51	(c)	1.52	(a)	1.53	(0)	1.54	(0.17)	1.55	(a)	1.56	(46)
1.57	(d)	1.58	(12500)	1.59	(158.28)	1.60	(a)	1.61	(129.41)	1.62	(a)	1.63	(a)
1.64	(a)	1.65	(a)	1.66	(a)	1.67	(a)	1.68	(c)	1.69	(301.0)	1.70	(c)
1.71	(b)	1.72	(a, d)										

Explanations Working Stress & Limit State Method

1.1 (c)

In WSM as well as LSM, it is assumed that plane section before bending remains plane after bending and normal to neutral axis after bending. It leads to the fact that strain varies linearly as the depth from neutral axis increases.

1.2 (d)

The vertical deflection limits may generally be assumed to be satisfied provided the span to depth ratios are satisfied as specified in **IS 456 Cl 23.2.1.**

(a) Basic values of span to effective depth ratios for span upto 10 m.

Cantilever	7
Simply supported	20
Continuous	26

(b) For spans > 10 m, the values in (a) may be multiplied by $\left(\frac{10}{\text{span}}\right)$ in metres, except for cantilever in which case deflection calculations should be made.

1.3 (c)

Tensile strength of concrete in flexure is known as modulus of rupture.

Modulus of rupture, $f_{cr} = 0.7\sqrt{f_{ck}}$

1.4 (a)

Effective width (b_f) for T-beams $= \frac{L_0}{6} + b_W + 6D_f$

Effective width (b_f) for for L-beams

$$= \frac{L_0}{12} + b_W + 3D_f$$

1.5 (c)

Middle portion of cylinder is having negligible effect of friction between top and bottom face of cylinder to machine because of the height difference between cube and cylinder. Hence it gives lesser compressive strength compared to cube.

1.6 (d)

The permissible tensile stress in concrete on water face (MPa).

Stress	Grade of Concrete			
	M 20	M 25	M 30	M 35
Direct tension	1.2	1.3	1.5	1.6
Bending tension	1.7	1.8	2.0	2.2

1.7 (b)

1.8 (b)

According to IS 456 : 2000,Clause 36.1. The term characteristic strength means that value of the strength of material below which not more than 5% of test result are expected to fall.

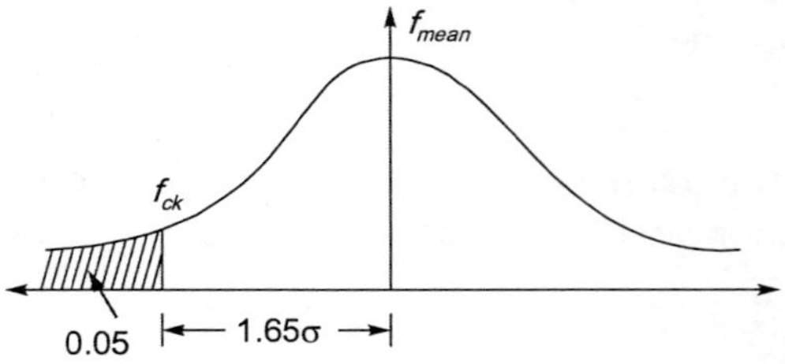

$$f_{mean} = f_{ck} + 1.65\,\sigma$$

$$\therefore \quad f_{ck} = f_{mean} - 1.65\,\sigma$$

1.9 (a)

Ultimate strength in compression,

$$\epsilon_{cu} = 0.0035$$

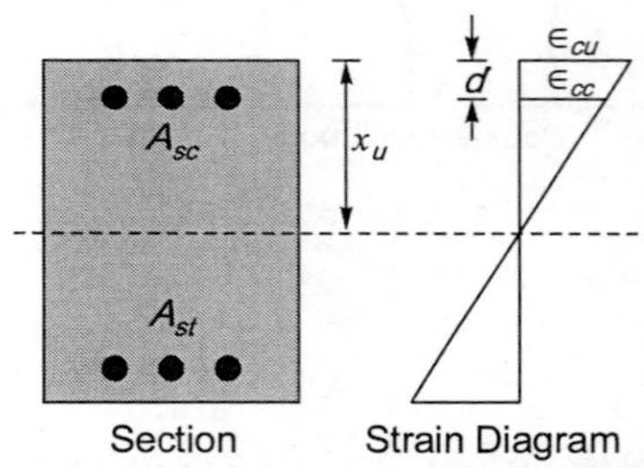

Let maximum strain at level of compression steel is ϵ_{cc}

∴ From geometry,

$$\frac{\epsilon_{cu}}{x_u} = \frac{\epsilon_{cc}}{x_u - d'}$$

$$\Rightarrow \quad \epsilon_{cc} = 0.0035\left(1 - \frac{d'}{x_u}\right)$$

1.10 (d)

In under reinforced section, steel yields first hence it does not take any additional stress and force taken by steel become constant. However, compression stress in concrete do increases with additional strain. Thus neutral axis and centre of gravity of compressive force further shift upward to maintain equilibrium.

1.11 (c)

Refer **IS 456 : 2000, Cl. 38.1.**

1.12 (a)

As per clause 6.2.3.1 of IS 456 : 2000, short term static modulus of elasticity in N/mm², E_c is given by

$$E_c = 5000\sqrt{f_{ck}} = 5000\sqrt{25} = 25000 \text{ N/mm}^2$$

1.13 (a)

Maximum strain in extreme fibres for concrete = 0.0035

Maximum strain in extreme fibres for steel

$$= 0.002 + \frac{0.87 f_y}{E_s}$$

$$= 0.002 + \frac{0.87 \times 415}{200 \times 10^3} = 0.0038$$

1.14 (c)

As per Annex B, clause B-2.1.2.d of IS 456 : 2000

The expression given for $m = \dfrac{280}{3\sigma_{cbc}}$, partially takes into account long term effects such as creep. Therefore this m in not the same as modular ratio derived based on value of E_c given in 6.2.3.1.

1.15 (a)

- Fineness of cement is measured by Blaine's air permeability test in which surface air is passed through cement and head loss is measured which is caliberated with fineness of cement.
- Workability of concrete is measured withe Vee-bee consistometer test.
- Bond between concrete and steel is tested with pull out test.

1.16 (b)

As per IS 456 : 2000, Partial safety factor for
Concrete : 1.50
Steel : 1.15

1.17 (c)

Refer **IS 456 : 2000, Table 18.**

1.18 (a)

Refer **IS 456 : 2000, Cl. 36.4.2.1.**

1.19 (a)

As per **IS 456 : 2000,** the flexural strength of concrete is given as,

$$f_{cr} = 0.7\sqrt{f_{ck}} = 0.7\sqrt{30} = 3.83 \text{ MPa}$$

1.20 (d)

As per **IS 456 : 2000, clause 15.3** three test specimens shall be made for each sample for testing at 28 days. As per **clause 15.4** the individual variation should not be more than ± 15 per cent of the average.

1.21 (a)

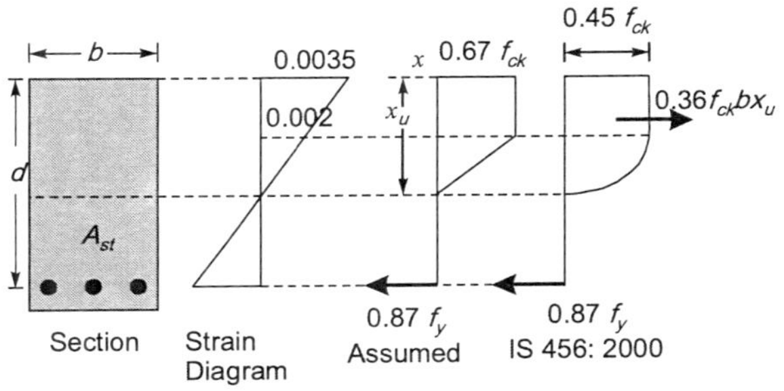

$$\frac{AB}{AC} = \frac{0.002}{0.0035}$$
[from assumed stress diagram]

$$\Rightarrow \quad \frac{AB}{AC} = \frac{4}{7}$$

$$\Rightarrow \quad AB = \frac{4}{7}x_u$$

But, $BC = AC - AB$

$$= x_u - \frac{4}{7}x_u = \frac{3}{7}x_u$$

Total compressive force

$$= 0.67 f_{ck} \times \frac{3}{7} x_u \times b + \frac{0.67 f_{ck}}{2} \times \frac{4}{7} x_u \times b$$

$= 0.48 f_{ck}\, b\, x_u$

Total tensile force = $0.87 f_y A_{st}$

For actual depth of neutral axis,

Total compressive force = Total tensile force

$$\Rightarrow \quad 0.48 f_{ck}\, b\, x_u = 0.87 f_y A_{st}$$

$$\Rightarrow \quad 0.48 \times 20 \times 300 \times x_u = 0.87 \times 415 \times 3 \times \frac{\pi}{4} \times (16)^2$$

$$\Rightarrow \quad x_u = 75.84 \simeq 76 \text{ mm}$$

1.22 (c)

As per **IS 456 : 2000**,

total compressive force = $0.36\ f_{ck}\ b\ x_u$

and total tensile force = $0.87\ f_y A_{st}$

∴ For actual depth of neutral axis

$$0.36\ f_{ck}\ b\ x_u = 0.87\ f_y A_{st}$$

$$\Rightarrow\ 0.36 \times 20 \times 300 \times x_u = 0.87 \times 415 \times 3 \times \frac{\pi}{4} \times (16)^2$$

$$\Rightarrow\ x_u = 100.82 \simeq 101 \text{ mm}$$

∴ Difference between the depth of neutral axis calculated as per given stress strain curve and the depth of neutral axis calculated as per **IS 456 : 2000** is = 101 − 76 = 25 mm

1.23 (c)

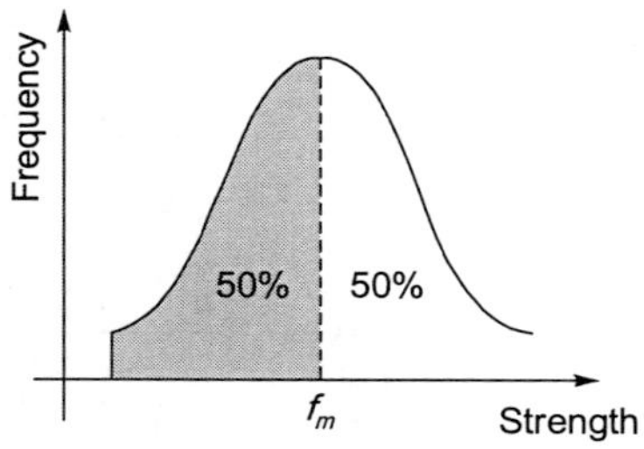

According to question, f_{ck} is defined as the strength below which 50% of test results are expected to fall according to curve.

∴ Here $f_{ck} = f_{mean}$

1.24 (d)

Shear stress distribution for reinforced concrete beam of rectangular section is

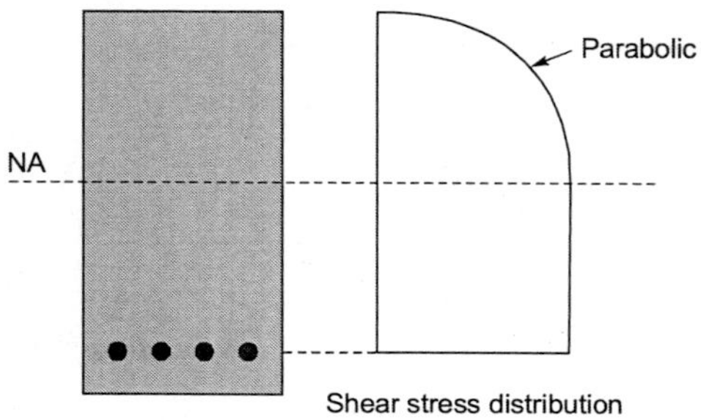

Shear stress distribution

1.25 (b)

Statement-1.

Modular ratio, $m = \dfrac{280}{3\sigma_{cbc}}$

σ_{cbc} = permissible compressive stress due to bending

It clearly show that modular ratio (m) is independent of type of steel.

Statement-2.

Upper limit of nominal shear $\tau_{c\,max}$ depends upon the grade of concrete. Upper limit is to avoid the failure of concrete in diagonal compression.

Statement-3.

For a slab to be designed as two-way slab, it must be supported on all four edges and $\dfrac{L}{B} \le 2$.

1.26 (b)

Statement-1

The width to thickness ratio limitations on the plate elements under compression in steel members are imposed by **IS 800 : 1984** to avoid buckling, not the fabrication difficulties.

Statement-2

Concrete undergoes creep strains in the compression zone due to continued compressive stress. Hence the strain in concrete goes on increasing with time.

This leads to increase in total strain in steel because of the flexural and compressive strain in steel.

Statement-3

Since friction is sufficient between the I-section and the slab, buckling will not occur and hence the permissible bending stress in compression will be same as that in tension.

1.27 (a)

1. With increase in water cement ratio of concrete mix, void ratio increases, hence compressive strength of concrete will decreases.
2. Main function of water in concrete is hydration of cement and workability of concrete mixture.
3. Creep and shrinkage of concrete depend upon water cement ratio of concrete mix.
   (So statement 3 is wrong)

1.28 (d)

1. Modulus of elasticity of concrete,
   $$E = 5000\sqrt{f_{ck}}$$
   ∴ modulus of elasticity of concrete increases with increase in compressive strength of concrete.
2. Brittleness of high strength concrete is more as compared to low strength concrete.

3. Shear strength of concrete is directly a function of compressive strength of concrete,

$$\tau_c \propto f_{ck}$$

(So statement 2 is wrong)

1.29 (d)

As per clause **8.2.8** of **IS 456 : 2000** concrete in sea water or exposed directly along the sea coast shall be atleast M 20 grade in the case of plain concrete and M 30 in case of reinforced concrete.

1.30 (d)

Note: Wind and earthquake loads are not considered simultaneously.

As explained in solution 1.4, design moment will be maximumof the three case:

Case-1: 1.5 DL + 1.5 LL

$= 1.5 \times 50 + 1.5 \times 80 = 195$ kNm

Case-2: 1.5 DL + 1.5 (max of EL or WL)

$= 1.5 \times 50 + 1.5 \times 180 = 345$ kNm

Case-3: 1.2 DL + 1.2 LL + 1.2 (max of EL or WL)

$= 1.2 \times 50 + 1.2 \times 80 + 1.2 \times 180$

$= 372$ kNm

$\therefore$ Design moment = 372 kNm

AVOID MISTAKE

Design moment with EL/WL = 1.5 (DL + LL + EL/WL) which is wrong coefficient is 1.2, not 1.5

1.31 (d)

Refer **IS 456 : 2000, Cl. 6.2.3.2.**

1.32 (c)

Refer **IS 456 : 2000, Cl. 36.4.2.1.**

1.33 (a)

A. **Resonant frequency test** is used to determine dynamic modulus of elasticity whereas destructive test like cube test etc measures the static modulus of elasticity.

B. **Rebound hammer test** is a non-destructive test used for compressive strength of concrete.

C. **Split cylinder test** is used to determine the direct tensile strength of concrete.

D. **Compaction factor test** is used to measure workability of concrete.

1.34 (c)

TMT or thermo mechanicaly treated bars are high strength reinforcement bars having a tough outer core and a soft inner core.

They provide strength is well as enough ductility to the structure.

1.35 (d)

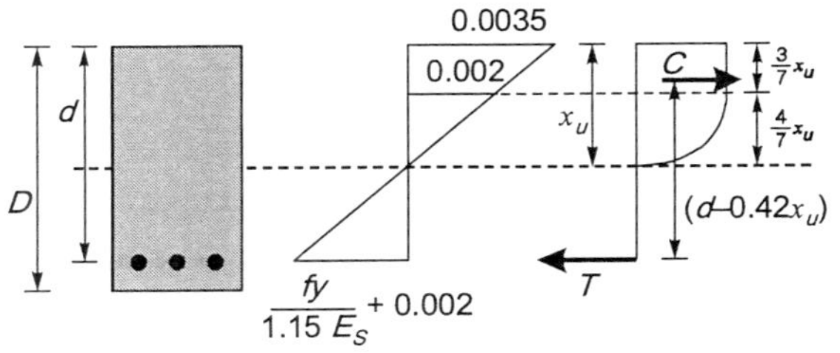

1.36 (b)

Compaction factor

$$= \frac{\text{Weight of partially compacted concrete}}{\text{Weight of fully compacted concrete}}$$

Hence, its maximum value for fresh concrete can be 1.

1.37 (c)

- Partial safety factor = $\frac{\text{Ultimate strength}}{\text{Design strength}}$
- Load factor = $\frac{\text{Ultimate load}}{\text{Working load}}$
- Statical method (Lower bound theorem) satisfies equilibrium and yield condition.
- Kinematic method (Upper bound theorem) satisfies equilibrium and mechanism condition.

1.38 Sol.

Permissible bearing stress = $0.45\, f_{ck}$

$= 0.45 \times 20 = 9$ N/mm^2

1.39 (b)

The leaner or less stiller the material, lesser is the slope of stress-strain curve, which is a measure of modulus of elasticity. Hence the diagram is self explanatory.

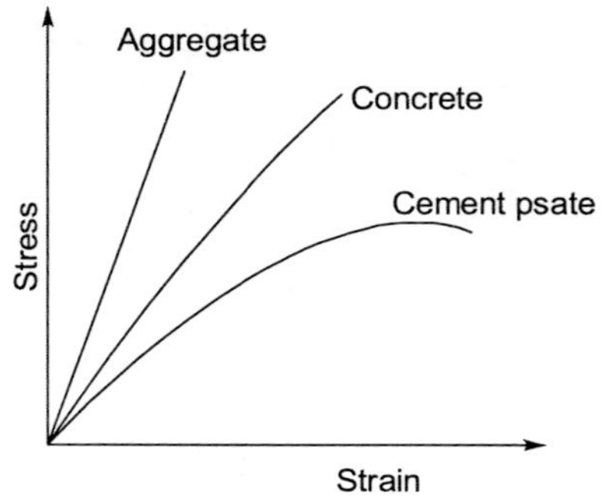

1.40 (d)

Refer **IS 456 : 2000, Cl. 9.2.2.**

1.41 Sol.

Flexural strength = $0.7\sqrt{f_{ck}}$ = 3.5 N/mm^2

1.42 (c)

Refer **IS 456 : 2000, Cl. 6.2.3.1**

1.43 (c)

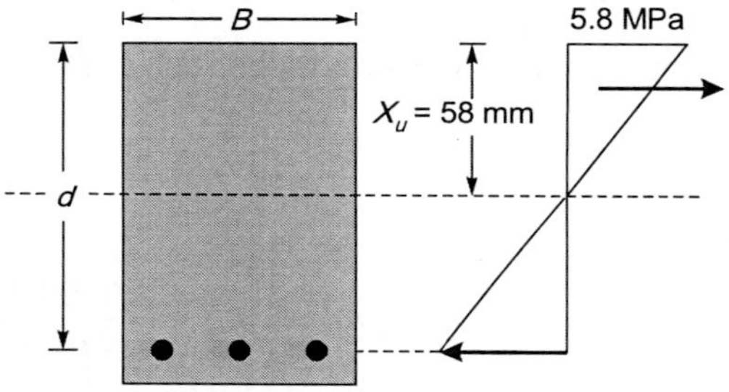

Modulus of elasticity of concrete,

$$E = 5000\sqrt{25}$$
$$= 25000 \text{ N/mm}^2$$

$$\therefore \quad \frac{M}{I} = \frac{\sigma}{y} = \frac{E}{R}$$

$$\Rightarrow \quad \frac{1}{R} = \frac{\sigma}{Ey} = \frac{5.8}{EX_u}$$

$$\text{Curvature} = \frac{5.8}{58 \times 25000}$$
$$= 4 \times 10^{-6} \text{ Per mm}$$

$$\Rightarrow \quad \text{Curvature} = 4 \times 10^{-6} \text{ per mm}$$

1.44 (d)

As slump increase, water content increases means workability increases.

Compaction factor

$$= \frac{\text{Weight of partially compacted concrete}}{\text{Weight of fully compacted concrete}}$$

∴ Compaction factor will also increase

The time required for complete remoulding in seconds is considered as a measure of workability and is expressed as the number of Vee-Bee seconds.

∴ As workability increases, Vee-Bee time decreases.

1.45 (b)

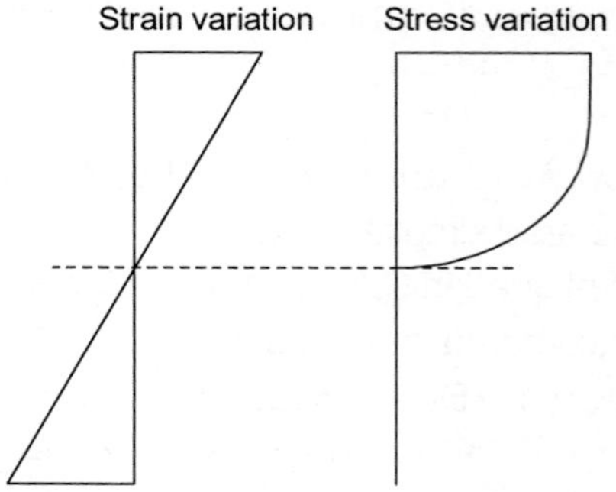

1.46 (a)

(i) An air-entraining agent introduces air in the form of bubbles that occupy upto 5% of the volume of concrete distributed uniformly throughout the cement paste. Thus, for the same amount of water/cement ratio, we get higher workability.

(ii) Since water takes up the air voids in the concrete and its freezing and melting causes cracks in concrete. hence using air entraiment agents improves the freezing and throwing resistance.

1.47 Sol.

$$A_{st} = 4 \times \frac{\pi}{4} \times 12^2 = 452.38 \text{ mm}^2$$

$$x_{u,\text{lim}} = 0.48 \times 300 = 144 \text{ mm}$$

$$\because \quad C = T$$

$$\Rightarrow \quad 0.36 f_{ck} b x_u = 0.87 f_y A_{st}$$

$$\Rightarrow \quad x_u = \frac{0.87 f_y A_{st}}{0.36 f_{ck} b} = 90.73 \text{ mm}$$

$\because$ $x_u < x_{u,\text{lim}}$, it is **under reinforced section**

$$\Rightarrow \quad \text{MOR} = 0.36 f_{ck} b x_u (d - 0.42 x_u)$$
$$= 42.7708 \times 10^6 \text{ N-mm} = 42.7708 \text{ kN-m}$$

1.48 Sol.

The structure may fail when,

(i) the load exceeds the design load

(ii) the strength is less than the characteristic strength

(iii) both the load exceeds the design load and strength is less than the characteristic strength. The probability of load exceeding design load = 5%

(iv) The probability of strength less than characteristic strength = 5%.

The probability of failure is

= 0.05 × 0.95 + 0.05 × 0.95 + 0.05 × 0.05

= 0.0975

1.49 Sol.

Long term static modulus of elasticity,

$$E_\theta = \frac{E_C}{1+\theta} = \frac{5000\sqrt{f_{ck}}}{1+\theta} = \frac{5000\sqrt{25}}{1+1.5} = 10{,}000$$

1.50 (b)

$$x_{u,\text{ bal}} = \frac{700}{1100+0.87f_y}\times d$$

So it depends upon grade of steel only.

1.51 (c)

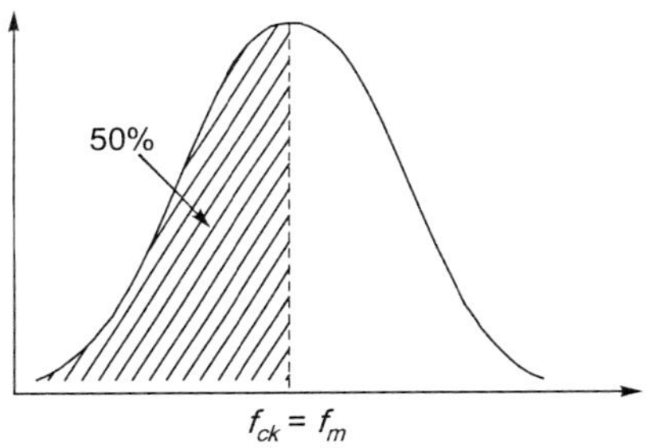

If f_{ck} is the value below which not more than 50% of test results are expected then,

$$f_m = f_{ck}$$

[Mean strength = Characteristics strength]

So target mean strength of concrete to be considered in design mix.

= Mean strength

$= f_m = f_{ck} = 25$ MPa

1.52 (a)

Creep is inelastic deformation with time due to sustained loading.

1.53 Sol.

Average strength,

$$\mu = \frac{(4\times 23)+(2\times 28)+(5\times 22.5)+(5\times 31)+(4\times 29)}{20}$$

= 26.575 MPa

$$\sigma = \sqrt{\frac{\Sigma(\mu-f)^2}{n-1}}$$

$$= \sqrt{\frac{(26.575-23)^2\times 4+(26.575-28)^2\times 2+(26.575-22.5)^2\times 5+(26.575-31)^2\times 5+(26.575-29)^2\times 4}{(20-1)}}$$

= 3.7

Now, $\mu - 3\sigma = 26.575 - 3 \times 3.7 = 15.48$

Thus, no specimen is having compressive strength less than $\mu - 3\sigma$.

No. of specimen = 0.

1.54 Sol.

From clause 26.5.1.1 of IS 456 : 2000, we know minimum reinforcement is given by

$$\frac{A_{st}}{bd} = \frac{0.85}{f_y}$$

$$\therefore \text{ Min. \% of steel} = \frac{A_{st}}{bd}\times 100 = \frac{0.85}{f_y}\times 100$$

For Fe500, min. % of steel

$$= \frac{0.85\times 100}{500} = 0.17\%$$

1.55 (a)

$$\text{Nominal cover} = \text{Effective cover} - \frac{\phi_m}{2} - \phi_{st}$$

$$= 50 - \frac{16}{2} - 12 = 30 \text{ mm}$$

1.56 Sol.

Target mean strength

$= f_{ck} + 1.65\sigma$

$= 25 + 1.65 \times 4.0 = 31.6$

Water content required

$$= 50 - \frac{(50-45)}{(35-25)}\times(31.6-25)$$

= 46.7%

Say 46%(round off to the lower integer)

1.57 (d)

Refer **IS 456 : 2000,**

1.58 Sol.

Initial strain = 500×10^{-6}

stress = 12.5 N/mm²

$$E_c = \frac{\text{stress}}{\text{strain}} = \frac{12.5}{500\times 10^{-6}}$$

= 25000 N/mm²

$$E_c = 5000\sqrt{25} = 25000 \text{ N/mm}^2$$

Total strain after long time = 1000×10^{-6}

$$E_{ce} = \frac{E_c}{1+\theta}$$

$$\theta = \frac{\text{Ultimate strain due to creep}}{\text{Elastic strain}}$$

$$= \frac{(1000-500)\times 10^{-6}}{500\times 10^{-6}} = 1.0$$

Effective modulus of elasticity

$$E_{ce} = \frac{E_C}{1+\theta} = \frac{25000}{1+1.0} = 12500 \text{ N/mm}^2$$

1.59 Sol.

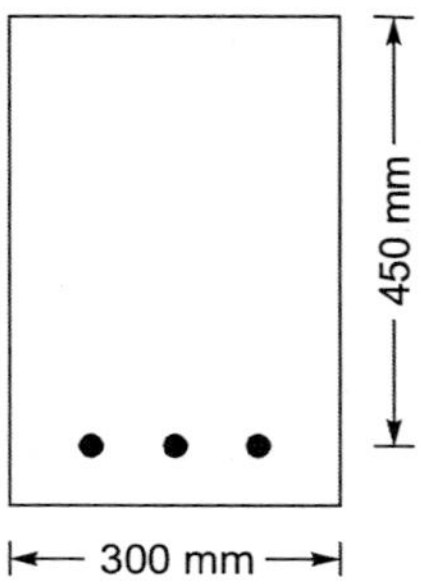

Given: M25 concrete,

Fe500 steel

Width, B = 300 mm

Effective depth,

d = 450 mm

A_{st} = 942 mm²

M_u = ?

(i) $x_{ulim} = 0.46 \times d = 0.46 \times 450 = 207$ mm

(ii) $x_u = \dfrac{0.87 \cdot f_y \cdot A_{st}}{0.36 \cdot f_{ck} \cdot B} = \dfrac{0.87\times 500\times 942}{0.36\times 25\times 300}$

= 151.77 M

(iii) $x_u < x_{ulim}$

It is an under reinforcement section.

(iv) $M_u = 0.36 \cdot f_{ck} \cdot B \cdot x_u \cdot (d - 0.42x_u)$

$= 0.36 \times 25 \times 300 \times 151.77 \times (450 - 0.42 \times 151.77)/10^6 = 158.28$ kN-m

1.60 (a)

Minimum pH value of water for concrete = 6.0

As per IS code provision no. 5.4.2, the pH value of water shall not less than 6.0.

1.61 Sol.

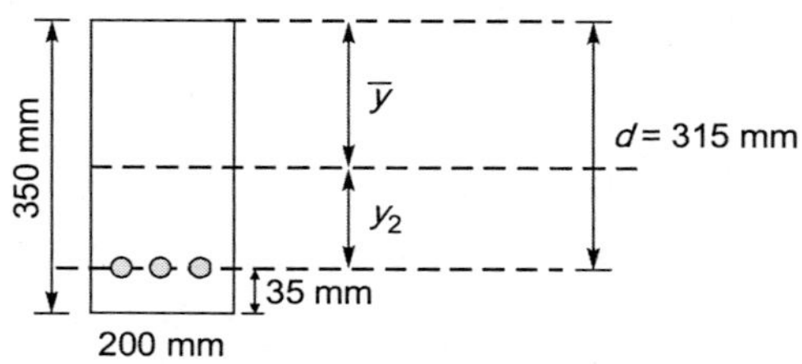

$$m = \frac{E_S}{E_C} = \frac{2.1\times 10^5}{2\times 10^4} = 10.5$$

$$A_{st} = 3\times\frac{\pi}{4}(16)^2 = 603.20 \text{ mm}^2$$

$$\bar{y} = \frac{\left(B\cdot D\cdot\frac{D}{2}+(m-1)\times A_{st}\times d\right)}{B\cdot D+(m-1)\cdot A_{st}}$$

$$= \frac{\left(200\times\frac{350^2}{2}+(10.5-1)\times 603.2\times 315\right)}{200\times 350+(10.5-1)\times 603.2}$$

= 185.59 mm

Distance of N-A from reinforcement

$$y_2 = d - \bar{y}$$

$= 315 - 185.59 = 129.41$ mm

1.62 (a)

The modular ratio, $m = \dfrac{E_s}{E_c} = \dfrac{280}{3\sigma_{cbc}}$

1.63 (a)

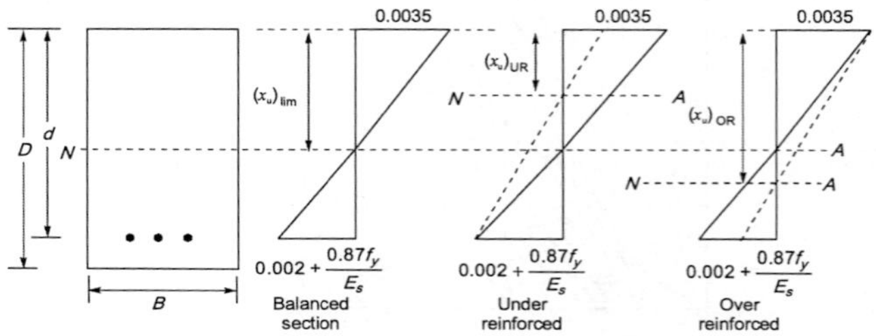

1.64 (a)

Refer **IS 456 : 2000, Cl. 6.2.5.**

1.65 (a)

In a balanced section, strain diagram under flexure is

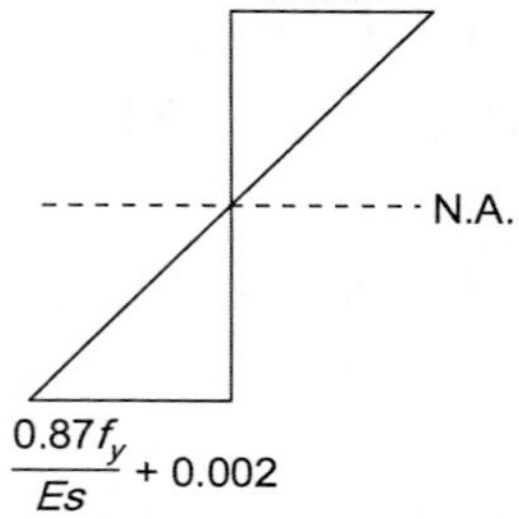

So, strain in concrete = 0.0035

(irrespective of grade of concrete)

Strain in steel $= \dfrac{0.87\times 415}{2\times 10^5}+0.002 = 0.0038$

1.66 (a)

As the water cement ratio increases, workability of concrete increases and strength decreases.

1.67 (a)

As per **IS 456 : 2000,** only in case of Fe250, stress (f_{sc}) can reach the design strength.

1.68 (c)

Characteristic strength is defined as that strength of the material below which not more than 5% of test results are expected to fail. So M20 concrete is the concrete having a 5 percentile cube strength of 20 MPa.

1.69 Sol.

Given: B = 300 mm, d = 600 – 45 = 555 mm

$$\text{Area of steel} = 3 \times \frac{\pi}{4} \times 28^2 = 1847.256 \text{ mm}^2$$

Depth of N.A.:

$0.36 f_{ck}\, b x_u = 0.87 f_y A_{st}$

$\Rightarrow 0.36 \times 25 \times 300 \times x_u = 0.87 \times 415 \times 1847.256$

$$\Rightarrow x_u = \frac{0.87 \times 415 \times 1847.256}{0.36 \times 25 \times 300} = 247.02 \text{ mm}$$

Now,

$x_{u,\text{lim}} = 0.48d = 0.48 \times 555 = 266.4 \text{ mm} > x_u$

$\therefore$ Under reinforced section.

M.R. $= 0.36\, f_{ck} b x_u\, (d - 0.42 x_u)$

$= 0.36 \times 25 \times 300 \times 247.02\,(555 - 0.42 \times 247.02)$

$= 300964059.6$ N-mm $\simeq 301.0$ kNm

1.70 (c)

As per IS456:2000, atleast 95% of the specimens should have strength equal to or more than the characteristic strength.

$\therefore$Minimum number of specimens = 0.95×40

= 38

1.71 (b)

Form work should be removed in such a way that the cantilever beam should remain cantilever.

1.72 (a, d)

For URS, $x_u < x_{u,\,\text{lim}}$

For doubly reinforced beam,

$$x_v = \frac{0.87 f_y A_{st} - (fsc - 0.45fck) A_{sc}}{0.36 fck\, B}$$

$x_u(\downarrow)$ if $A_{st}(\downarrow)$ and $A_{sc}(\uparrow)$

■■■■

2 Shear, Torsion, Bond, Anchorage & Development Length

2.1 If ϕ=nominal diameter of reinforcing bar, f_s = compressive stress in the bar and f_{bd} = design bond stress of concrete, the anchorage length, L_a of straight bar in compression is equal to

(a) $L_a = \frac{\phi f_s}{f_{bd}}$ (b) $L_a = \frac{\phi f_s}{2f_{bd}}$

(c) $L_a = \frac{\phi f_s}{\pi f_{bd}}$ (d) $L_a = \frac{\phi f_s}{4f_{bd}}$

[1996 : 1 M]

2.2 Consider the following two statements related to reinforced concrete design, and identify whether they are True or False.

I. Curtailment of bars in the flexural tension zone in beams reduces the shear strength at the cut-off locations.

II. When a rectangular column section is subject to biaxially eccentric compression, the neutral axis will be parallel to the resultant axis of bending.

(a) Both statements I and II are True
(b) Statement I is True, and statement II is False
(c) Statement I is False, and statement II is True
(d) Both statements I and II are False

[2001 : 2 M]

Common Data for Questions 2.3 and 2.4:

At the limit state of collapse, an RC beam is subjected to flexural moment 200 kN-m, shear force 20 kN and torque 9 kN-m. The beam is 300 mm wide and has a gross depth of 425 mm, with an effective cover of 25 mm. The equivalent nominal shear stress (τ_{ve}) as calculated by using the design code turns out to be lesser than the design shear strength (τ_c) of the concrete.

2.3 The equivalent shear force (V_c) is

(a) 20 kN (b) 54 kN
(c) 56 kN (d) 68 kN [2004 : 2 M]

2.4 The equivalent flexural moment (M_{eq}) for designing the longitudinal tension steel is

(a) 187 kN-m (b) 200 kN-m
(c) 209 kN-m (d) 213 kN-m

[2004 : 2 M]

2.5 In the design of a reinforced concrete beam the requirement for bond is not getting satisfied. The economical option to satisfy the requirement for bond is by

(a) bundling of bars
(b) providing smaller diameter bars more in number
(c) providing larger diameter bars less in number
(d) providing same diameter bars more in number

[2008 : 1 M]

Common Data for Questions 2.6 and 2.7:

A reinforced concrete beam of rectangular cross section of breadth 230 mm and effective depth 400 mm is subjected to a maximum factored shear force of 120 kN. The grades of concrete, main steel and stirrup steel are M 20, Fe 415 and Fe 250 respectively. For the area of main steel provided, the design shear strength τ_c as per **IS 456: 2000** is 0.48 N/mm^2. The beam is designed for collapse limit state.

2.6 The spacing (mm) of 2-legged 8 mm stirrups to be provided is

(a) 40 (b) 115
(c) 250 (d) 400 [2008 : 2 M]

2.7 In addition, the beam is subjected to a torque whose factored value is 10.90 kN-m. The stirrups have to be provided to carry a shear (kN) equal to

(a) 50.42 (b) 130.56
(c) 151.67 (d) 200.23 [2008 : 2 M]

2.8 Consider a reinforcing bar embedded in concrete. In a marine environment this bar undergoes uniform corrosion, which leads to the deposition of corrosion products on its surface and an increase in the apparent volume of the bar. This subjects the surrounding concrete to expansive pressure. As a result, corrosion induced cracks appear at the surface of concrete. Which of the following statements is **TRUE**?

(a) Corrosion causes circumferential tensile stresses in concrete and the cracks will be parallel to the corroded reinforcing bar.
(b) Corrosion causes radial tensile stresses in concrete and the cracks will be parallel to the corroded reinforcing bar.
(c) Corrosion causes circumferential tensile stresses in concrete and the cracks will be

perpendicular to the direction of the corroded reinforcing bar.

(d) Corrosion causes radial tensile stresses in concrete and the cracks will be perpendicular to the direction of the corroded reinforcing bar.

[2011 : 1 M]

2.9 Consider two RCC beams, *P* and *Q*, each having the section 400 mm × 750 mm (effective depth, d = 750 mm) made with concrete having τ_{cmax} = 2.1 N/mm^2. For the reinforcement provided and the grade of concrete used, it may be assumed that the τ_c = 0.75 N/mm^2. The design shear in beam *P* is 400 kN and in beam *Q* is 750 kN. Considering the provisions of **IS 456 : 2000**, which of the following statements is **TRUE**?

(a) Shear reinforcement should be designed for 175 kN for beam *P* and the section for beam Q should be revised.

(b) Nominal shear reinforcement is required for beam *P* and the shear reinforcement should be designed for 120 kN for beam *Q*.

(c) Shear reinforcement should be designed for 175 kN for beam *P* and the shear reinforcement should be designed for 525 kN for beam *Q*.

(d) The sections for both beams *P* and *Q* need to be revised.

[2011 : 2 M]

2.10 Consider a bar of diameter '*D*' embedded in a large concrete block as shown in the adjoining figure, with a pull out force *P* being applied. Let σ_b and σ_{st} be the bond strength (between the bar and concrete) and the tensile strength of the bar, respectively. If the block is held in position and it is assumed that the material of the block does not fail, which of the following options represents the maximum value of *P*?

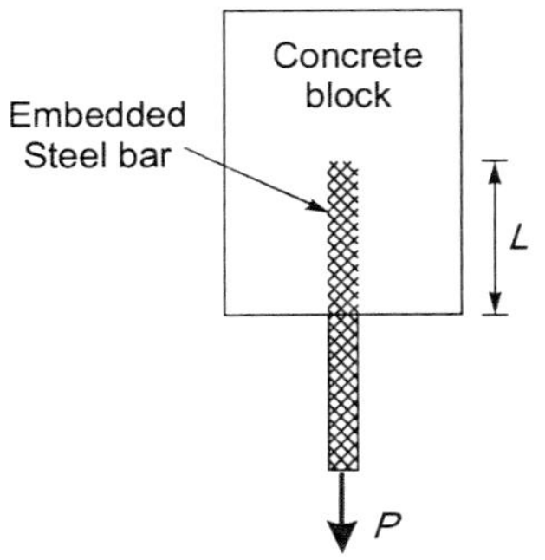

(a) Maximum of $\left(\frac{\pi}{4}D^2\sigma_b\right)$ and $(\pi DL\sigma_{st})$

(b) Maximum of $\left(\frac{\pi}{4}D^2\sigma_{st}\right)$ and $(\pi DL\sigma_b)$

(c) Minimum of $\left(\frac{\pi}{4}D^2\sigma_{st}\right)$ and $(\pi DL\sigma_b)$

(d) Minimum of $\left(\frac{\pi}{4}D^2\sigma_b\right)$ and $(\pi DL\sigma_{st})$

[2011 : 2 M]

2.11 As per **IS 456 : 2000** for M20 grade concrete and plain bars in tension the design bond stress τ_{bd} = 1.2 MPa. Further, **IS 456 : 2000** permits this design bond stress value to be increased by 60% for HYSD bars. The stress in the HYSD reinforcing steel bars in tension, σ_s = 360 MPa. Find the required development length, L_d, for HYSD bars in terms of the bar diameter, ϕ. ______

[2013 : 1 M]

2.12 A rectangular beam of width (b) 230mm and effective depth (d) 450 mm is reinforced with four bars of 12 mm diameter. The grade of concrete is M20 and grade of steel is Fe500. Given that for M20 grade of concrete the ultimate shear strength, τ_{uc} = 0.36 N/mm^2 for steel percentage, p = 0.25, and τ_{uc}= 0.48 N/mm^2 for p = 0.50. For a factored shear force of 45 kN, the diameter (in mm) of Fe500 steel two legged stirrups to be used at spacing of 375 mm, should be

(a) 8 (b) 10
(c) 12 (d) 16

[2014 : 2 M, Set-I]

2.13 The development length of a deformed reinforcement bar can be expressed as $\left(\frac{1}{k}\right)\left(\frac{\phi\sigma_s}{\tau_{bd}}\right)$. From the IS:456-2000, the value of k can be calculated as ________.

[2015 : 1 M, Set-I]

2.14 In shear design of an RC beam, other than allowable shear strength of concrete (τ_c), there is also an additional check suggested in **IS 456 : 2000** with respect to the maximum permissible shear stress (τ_{cmax}). The check for τ_{cmax} is required to take care of

(a) additional shear resistance from reinforcing steel

(b) additional shear stress that comes from accidental loading

(c) possibility of failure of concrete by diagonal tension

(d) possibility of crushing of concrete by diagonal compression

[2016 : 1 M, Set-I]

2.15 As per **IS 456 : 2000** for the design of reinforced concrete beam, the maximum allowable shear stress (τ_{cmax}) depends on the

(a) grade of concrete and grade of steel
(b) grade of concrete only
(c) grade of steel only
(d) grade of concrete and percentage of reinforcement

[2016 : 1 M, Set-II]

2.16 An RCC beam of rectangular cross-section has factored shear of 200 kN at its critical section. Its width b is 250 mm and effective depth d is 350 mm. Assume design shear strength τ_c of concrete as 0.62 N/mm^2 and maximum allowable shear stress $\tau_{c\ max}$ in concrete as 2.8 N/mm^2. If two legged 10 mm diameter vertical stirrups of Fe250 grade steel are used, then the required spacing (in cm, up to one decimal place) as per limit state method will be _______

[2018 : 2 M, Set-I]

2.17 A reinforcing steel bar, partially embedded in concrete, is subjected to a tensile force P. The figure that appropriately represents the distribution of the magnitude of bond stress (represented as hatched region), along the embedded length of the bar, is

(a)
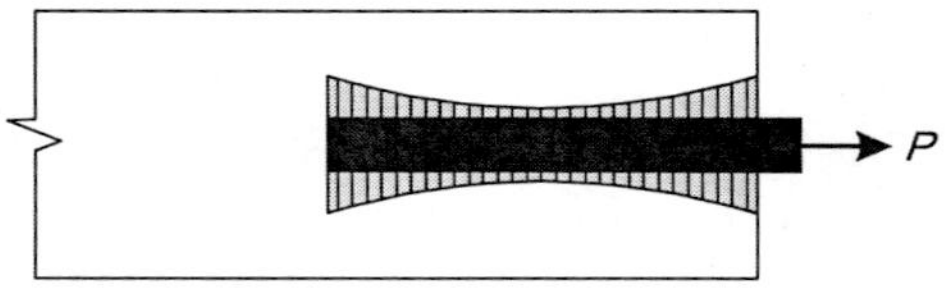

(b)
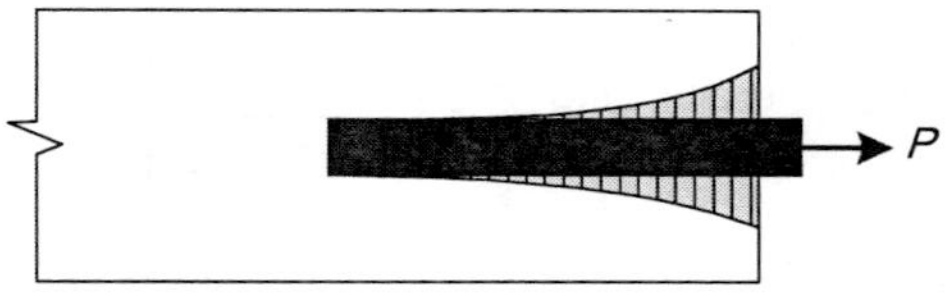

(c)
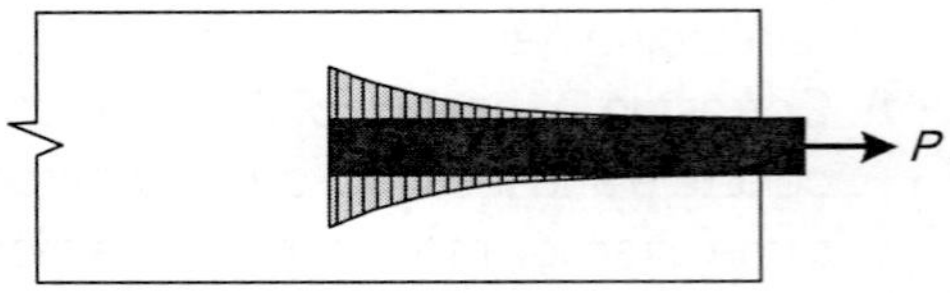

(d)
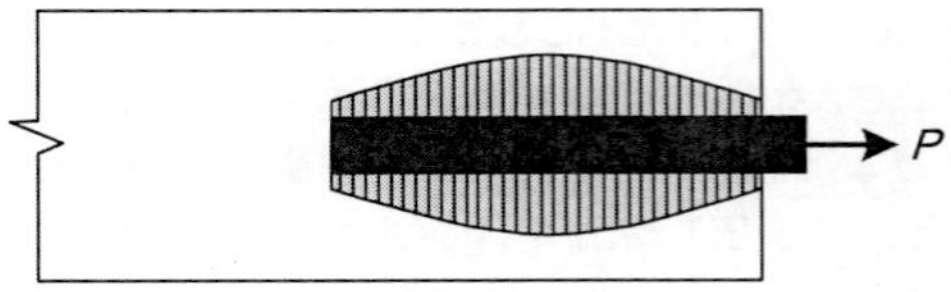

[2020 : 1 M, Set-I]

2.18 A reinforced concrete beam with rectangular cross section (width = 300 mm, effective depth = 580 mm) is made of M30 grade concrete. It has 1% longitudinal tension reinforcement of Fe 415 grade steel. The design shear strength for this beam is 0.66 N/mm^2. The beam has to resist a factored shear force of 440 kN. The spacing of two-legged, 10 mm diameter vertical stirrups of Fe 415 grade steel is _______mm. (round off to the nearest integer)

[2022 : 2 M, Set-I]

2.19 With regard to the shear design of RCC beams, which of the following statements is/are TRUE?

(a) Excessive shear reinforcement can lead to compression failure in concrete
(b) Beams without shear reinforcement, even if adequately designed for flexure, can have brittle failure
(c) The main (longitudinal) reinforcement plays no role in the shear resistance of beam
(d) As per IS456 : 2000, the nominal shear stress in the beams of varying depth depends on both the design shear force as well as the design bending moment

[2023 : 1 M, Set-II]

■■■■

Answers **Shear, Torsion, Bond, Anchorage & Development Length**

2.1	(*)	2.2	(a)	2.3	(d)	2.4	(b)	2.5	(b)	2.6	(b)	2.7	(c)
2.8	(a)	2.9	(a)	2.10	(c)	2.11	(46.8)	2.12	(a)	2.13	(6.4)	2.14	(d)
2.15	(b)	2.16	(8.2)	2.17	(b)	2.18	(101.1)	2.19	(a, b, d)				

Explanations Shear, Torsion, Bond, Anchorage & Development Length

2.1 (*)

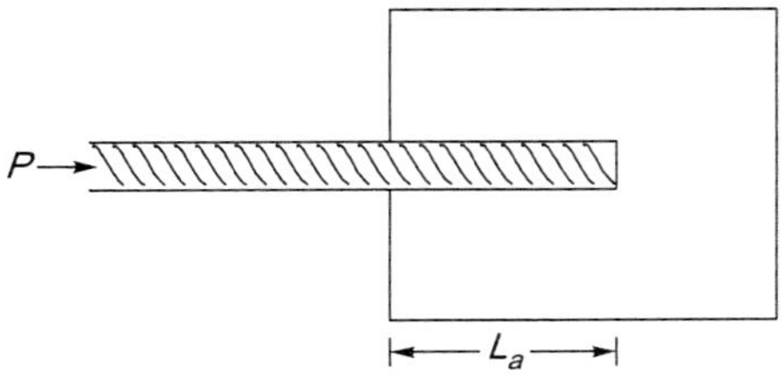

A_s = Area of cross-section

f_s = Compressive stress in the bar

$$P = f_s\,A_s$$
$$= f_s \frac{\pi}{4}\phi^2$$

This compressive force P, should be transferred from steel to concrete through bond.

$$P = \text{Perimeter} \times L_a \times f_{bd} = \pi\phi\, L_a\, f_{bd}$$
$$f_s \frac{\pi}{4}\phi^2 = \pi\phi L_A f_{bd}$$
$$L_a = \frac{f_s\phi}{4 f_{bd}}$$

f_{bd} is taken 25% more according to clause 26.2.1 of IS 456 : 2000 in case of compression

$$\therefore \quad L_a = \frac{f_s\phi}{4\times 1.25 f_{bd}} = \frac{f_s\phi}{5 f_{bd}}$$

2.2 (a)

(i) Curtailment of bars in flexure tension zone reduces the shear strength because shear strength is dependent on the % of steel, grade of steel and τ_c at that section.

(ii) Resultant axis of bending and neutral axis will always be perpendicular.

2.3 (d)

The equivalent shear force,

$$V_c = V + 1.6\,\frac{T}{B} = 20 + \frac{1.6\times 9}{300\times 10^{-3}} = 68 \text{ kN}$$

2.4 (b)

Since $\tau_{ve} < \tau_c$ so section will be designed for M_u only. (As per 41.3.2 and 41.3.3 of IS code, section should be design for M_{ue} only if $\tau_{ve} < \tau_c$.

AVOID MISTAKE

$$M_{eq} = M + \frac{T}{1.7}\left(1+\frac{D}{B}\right)$$
$$= 200 + \frac{9}{1.7}\left(1+\frac{0.425}{0.300}\right)$$
$$= 213 \text{ kNm}$$

Since $\tau_{ve} < \tau_c$ so section will be designed for M_u only i.e., 200 kNm.

2.5 (b)

$$\text{Bond stress } (\tau_{bd}) = \frac{\text{Tensile force}}{(n\pi\phi)\sigma_{st}}$$

τ_{bd} should be less than permissible value, if it is greater than $(\tau_{bd})_{\text{permissible}}$ then best economical solution is to reduce the diameter of bar and increase its number.

Also bond strength is dependent on the surface area of bar. Hence for providing the same A_{st} we can choose smaller diameter bars and increase their number and hence surface area will increase.

2.6 (b)

Ultimate design shear force,

$$V_u = 120 \text{ kN}$$

Ultimate shear capacity of concrete,

$$V_{uc} = \tau_c bd$$
$$= 0.48 \times 230 \times 400 = 44.16 \text{ kN}$$

Ultimate shear to be resisted by shear reinforcement,

$$V_{us} = V_u - V_{uc} = 120 - 44.16 = 75.84 \text{ kN}$$

For vertical stirrups, spacing is given as

$$S_v = \frac{0.87\, f_y A_{sv}\, d}{V_{us}}$$
$$= \frac{0.87\times 250\times 2\times \pi\times (8)^2\times 400}{75.84\times 10^3\times 4}$$
$$= 115.32 \text{ mm}$$

2.7 (c)

Equivalent shear,

$$V_{ue} = V_u + \frac{1.6 T_u}{b}$$
$$= 120 + \frac{1.6\times 10.90\times 1000}{230}$$
$$= 120 + 75.83 = 195.83 \text{ kN}$$
$$V_{uc} = 44.16 \text{ kN}$$

$\therefore$ Ultimate shear to be resisted by stirrups,

$$V_{us} = V_{ue} - V_{uc} = 195.83 - 44.16 = 151.67 \text{ kN}$$

2.8 (a)

Corrosion in steel occupies more volume than volume of reinforcing steel, hence exerts radial pressure on adjoining concrete layer. As a consequence, cracking results along reinforcing bars and concrete starts spalling.

2.9 (a)

Shear strength of the concrete = $\tau_c \times Bd$

$= 0.75 \times 400 \times 750 \times 10^{-3} = 225$ kN

Maximum shear strength of concrete

$= \tau_{cmax} \times Bd$

$= 2.1 \times 400 \times 750 \times 10^{-3} = 630$ kN

Design shear in beam *P* is 400 kN which is less than the maximum shear that the concrete can sustain. But the design shear is more than the nominal shear of 225 kN.

∴ Shear reinforcement is designed for

$= 400 - 225 = 175$ kN

Design shear in beam *Q* is 750 kN which is more than the maximum shear that the concrete can sustain. Hence the beam *Q* should be redesigned.

2.10 (c)

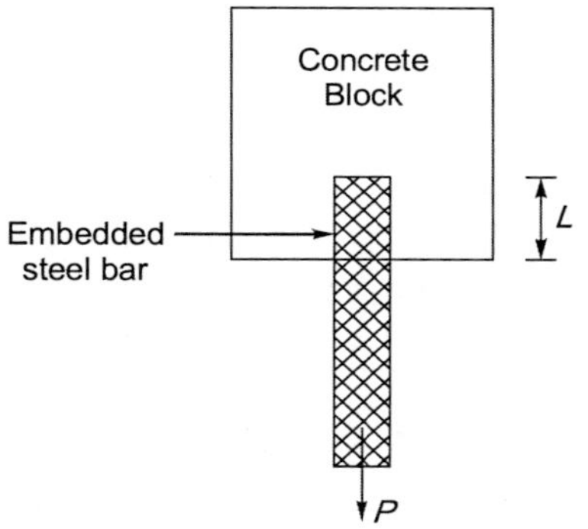

Maximum force resisted by bond strength $= \sigma_b(\pi DL)$

Maximum tensile force resisted by bar

$$= \sigma_{st}\left(\frac{\pi}{4}D^2\right)$$

∴ Maximum value of P = Minimum of

$$\left\{\sigma_b(\pi DL) \text{ or} \left(\sigma_{st}\frac{\pi}{4}D^2\right)\right\}$$

2.11 Sol.

$$L_d = \frac{\phi\sigma_{st}}{4\tau_{bd}} = \frac{\phi \times 360}{4 \times 1.2 \times 1.60} = 46.875\,\phi$$

POINTS TOREMEMBER

1. As per clause 26.2.1 of IS 456 : 2000 for deformed bars conforming to IS 1786, Design bond stress values τ_{bd}, should be increased by 60 percent.
2. For bars in compression, these values of τ_{bd} shall be increased by 25 percent.

   Note: As per latest addition to IS 456 : 2000.
3. For fusion bonded epoxy coated deformed bars, design bond stress values shall be taken as 80 percent of τ_{bd} given in 26.2.1.1.m

2.12 (a)

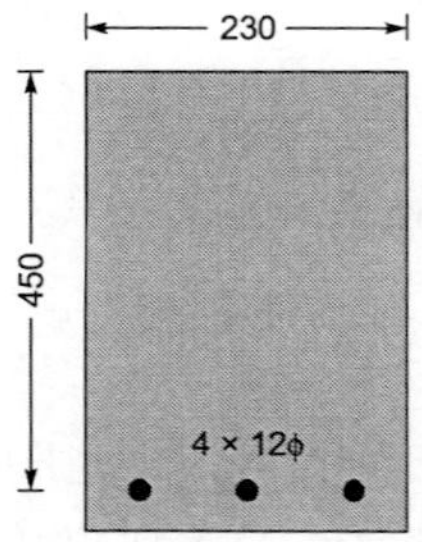

$\tau_{uc} = 0.36$ N/mm² [For M20]

for $\frac{A_{st}}{bd} \times 100 = 0.25$

$\tau_{uc} = 0.48$ N/mm² [For M20]

for $\frac{A_{st}}{bd} \times 100 = 0.5$

Factored $SF = 45$ kN $= V_u$

We have to calculate the dia of Fe 500 2-Legged stirrup to be used at a spacing of 325 mm c/c

$$\tau_v = \frac{V_u}{bd} = \frac{45 \times 1000}{230 \times 450} = 0.4348 \text{ N/mm}^2$$

∴ %tensile steel

$$= \frac{4 \times \frac{\pi}{4}(12)^2}{230 \times 450} \times 100 = 0.437\%$$

$$\tau_c = 0.36 + \frac{0.12}{0.25} \times (0.437 - 0.25)$$

$= 0.45$ N/mm²

Since $\tau_v - \tau_c < 0$

⇒ Min shear reinforcement is required

⇒ Min shear reinforcement is given by

$$\frac{A_{sv}}{bS_v} = \frac{0.4}{0.87f_y}$$

$$A_{sv} = \frac{0.4\times(S_v)(b)}{0.87f_y}$$

Since we limit f_y to 415 N/mm² hence,

$$A_{sv} = 2\times\frac{\pi}{4}(\phi)^2 = \frac{0.4\times325\times230}{0.87\times415}$$

$$= 82.814 \text{ mm}^2$$

$$\phi = 7.26 \text{ mm}$$

adopt ϕ = 8 mm

POINTS TOREMEMBER

As per clause 40.4 of IS 456 : 2000, the characteristic strength of the stirrup or bent up reinforcement shall nto be taken greater than 415 N/mm².

2.13 Sol.

Bond strength of concrete

= Tensile force in steel

$$\tau_{bd}\times(L_d\pi\phi) = 0.87f_y\times\left(\frac{\pi}{4}\phi^2\right)$$

$$\Rightarrow \quad L_d = \frac{0.87f_y\phi}{4\tau_{bd}}$$

For deformed bars τ_{bd} value is increased by 60%,

$$\therefore \quad L_d = \frac{\phi\sigma_s}{4\times(1.6\tau_{bd})} = \frac{\phi\sigma_s}{6.4\tau_{bd}}$$

$\therefore \quad k = 6.4$

2.14 (d)

The check for $\tau_{c\,max}$ is an additional check to take care of possibility of crushing of concrete by design of compression.

2.15 (b)

From table 20 of IS 456 : 2000 it is evident that $\tau_{c\,max}$ values are dependent only on concrfete grade.

Concrete grade	M-15	M-20	M-25	M-30	M-35	M-40
$\tau_{c,max}$ N/mm²	2.5	2.8	3.1	3.5	3.7	4.0

2.16 Sol.

Nominal shear stress,

$$\tau_v = \frac{V_u}{bd} = \frac{200\times10^3}{250\times350} = 2.286 \text{ N/mm}^2 < \tau_{c,max}$$

(OK)

SF taken by stirrups

$= (\tau_v - \tau_c)\, bd$

$= (2.286 - 0.62) \times 250 \times 350 = 145.75$ kN

$$\text{Now, } V_{us} = \frac{0.87f_yA_{sv}d}{S_v}$$

$$\Rightarrow \quad S_v = \frac{0.87\times250\times350\times2\times\frac{\pi}{4}\times10^2}{145.75\times10^3}$$

$= 82$ mm $= 8.2$ cm

AVOID MISTAKE

We tend to answer the spacing in millimeters. But here don't forget to answer in the desired units, i.e, centimeters.

2.17 (b)

As the axial tensile force P is applied, bond stress will be developed between concrete and steel which will resist the movement of steel. The magnitude of this stress wil be highest at begining of embedded length and least at other end.

2.18 Sol.

B = 300 mm, d = 580 mm, V_u = 440 kN

Concrete used is M30 and steel is Fe415

1. $V_{cu} = \tau_cBd = 0.66\times300\times\frac{580}{1000} = 114.84$ kN
2. $V_{su} = V_u - V_{cu} = 440 - 114.84$ kN $= 325.16$ kN
3. Spacing of 2-legged shear reinforcement,

$$S_v = \frac{A_{sv}\times0.87f_y\times d}{V_{su}}$$

$$\Rightarrow S_v = \frac{2\times\frac{\pi}{4}\times(10)^2\times0.87\times415\times580}{325.16\times1000}$$

$\Rightarrow S_v = 101.16$ mm (say 101 mm)

2.19 (a, b, d)

- Check of '$\tau_{c,max}$' is applied in order to prevent excessive shear reinforcement causing compression failure in concrete.
- Minimum shear reinforcement is providded to provide ductility to beam.
- $$\tau_v = \frac{V_u \pm \frac{M_u}{d}\tan\beta}{bd}$$

■■■■

3 Footing, Columns, Beams and Slabs

3.1 The lateral ties in a reinforced concrete rectangular column under axial compression are used to
(a) avoid the buckling of the longitudinal steel under compression
(b) provide adequate shear capacity
(c) provide adequate confinement to concrete
(d) reduce the axial deformation of the column

[1995 : 1 M]

3.2 Which one of the following set of values give the minimum clear cover (in mm) for the main reinforcements in the slab, beam, column and footing respectively, according to **IS 456: 1978**?
(a) 20, 25, 30, 75 (b) 5, 15, 25, 50
(c) 15, 25, 40, 75 (d) None of these

[1995 : 1 M]

3.3 In a reinforced concreted beam column, the increase in the flexural strength along with the increase in the axial strength occurs
(a) beyond the elastic limit of the material
(b) when the yielding of the tension reinforcement governs the strength
(c) when the crushing of the concrete in the compression zone governs the strength
(d) never

[1995 : 1 M]

3.4 The correct reinforcement details at the corner of a rectangular water tank, in horizontal plane is shown by

(a)
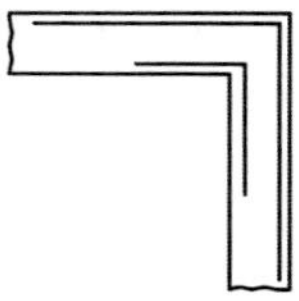

(b)
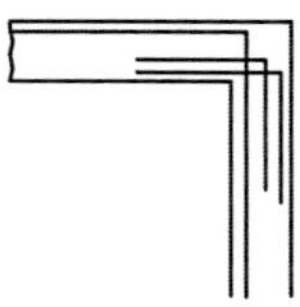

(c)
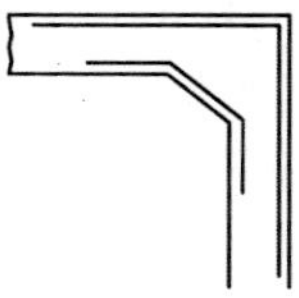

(d)
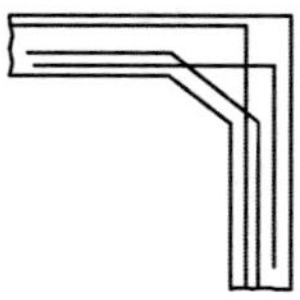

[1996 : 1 M]

3.5 **IS 456: 1978** recommends to provide certain minimum steel in a RCC beam
(a) to ensure compression failure
(b) to avoid rupture of steel in case a flexural failure occurs
(c) to hold the stirrup steel in position
(d) to provide enough ductility to the beam

[1997 : 1 M]

3.6 An isolated T-beam is used as a walkway. The beam is simply supported with an effective span of 6 m. The effective width of flange, for the cross-section shown in figure is

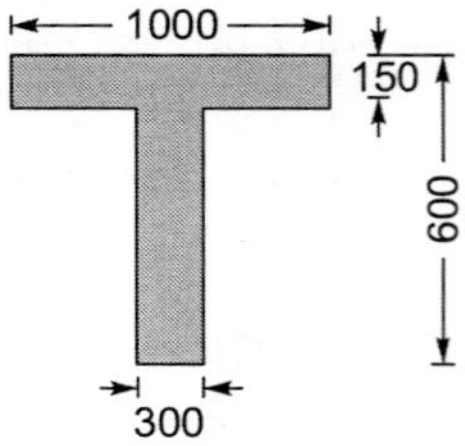

(a) 900 mm (b) 1000 mm
(c) 1259 mm (d) 2200 mm

[1998 : 1 M]

3.7 The plan of stairs supported at each end by landing spanning parallel with risers is shown in figure. The effective span of staircase slab is

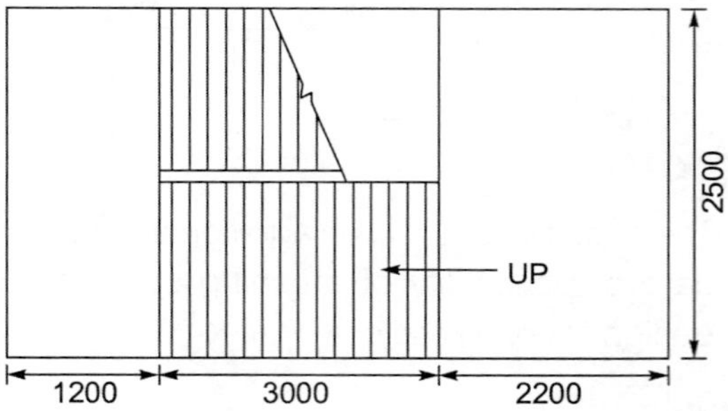

(a) 3000 mm (b) 4600 mm
(c) 4750 mm (d) 6400 mm

[1998 : 1 M]

3.8 A reinforced concrete wall carrying vertical loads is generally designed as per recommendations given for columns. The ratio of minimum reinforcements in the vertical and horizontal directions is
(a) 2 : 1 (b) 5 : 3
(c) 1 : 1 (d) 3 : 5

[1998 : 2 M]

3.9 In reinforced concrete, pedestal is defined as compression member, whose effective length does not exceed its dimension by

(a) 12 times (b) 3 times
(c) 16 times (d) 8 times

[1999 : 1 M]

3.10 The minimum area of tension reinforcement in a beam shall be greater than

(a) $\frac{0.85bd}{f_y}$ (b) $\frac{0.87f_y}{bd}$
(c) $0.04\ bd$ (d) $\frac{0.4bd}{y}$

[1999 : 1 M]

3.11 As per the provisions of **IS 456: 2000**, in the limit state method for, design of beams, the limiting value of the depth of neutral axis in a reinforced concrete beam of effective depth 'd' is given as

(a) 0.53 d
(b) 0.48 d
(c) 0.46 d
(d) any of the above depending on the different grades of steel.

[2002 : 2 M]

3.12 The effective length of a column in a reinforced concrete building frame, as per **IS 456: 1978**, is independent of the

(a) frame type i.e., braced (no sway) or unbraced (with sway)
(b) span of the beam
(c) height of the column
(d) loads acting on the frame

[2003 : 1 M]

3.13 Top ring beam of an Intze tank carries a hoop tension of 120 kN. The beam cross-section is 250 mm wide and 400 mm deep and it is reinforced with 4 bars of 20 mm diameter of Fe-415 grade. Modular ratio of the concrete is 10. The tensile stress in N/mm^2 in the concrete is

(a) 1.02 (b) 1.07
(c) 1.20 (d) 1.32

[2003 : 2 M]

3.14 A concrete column carries an axial load of 450 kN and a bending moment of 60 kN-m at its base. An isolated footing of size 2 m × 3 m side along the plane of the bending moment, is provided under the column. Centres of gravity of column and footing coincide. The net maximum and the minimum pressures in kN/m^2 on soil under the footing are respectively,

(a) 95 and 55 (b) 95 and 75
(c) 75 and 55 (d) 75 and 75

[2003 : 2 M]

Common Data for Questions 3.15 and 3.16:

A reinforced concrete beam, size 200 mm wide and 300 mm deep overall is simply supported over a span of 3 m. It is subjected to two point loads P of equal magnitude placed at middle third points. The two loads are gradually increased simultaneously. Beam is reinforced with 2 HYSD bars of 16 mm diameter placed at an effective cover of 40 mm on bottom face and nominal shear reinforcement. The characteristic compressive strength and the bending tensile strength of the concrete are 20.0 N/mm^2 and 2.2 N/mm^2 respectively.

3.15 Ignoring the presence of tension reinforce-ment, the value of load *P* in kN when the first flexure crack will develop in the beam is

(a) 4.5 (b) 5.0
(c) 6.6 (d) 7.5

[2003 : 2 M]

3.16 The theoretical failure load of the beam for attainment of limit state of collapse in flexure is

(a) 23.7 kN (b) 25.6 kN
(c) 28.7 kN (d) 31.6 kN

[2003 : 2 M]

3.17 An RC short column with 300 mm × 300 mm square cross-section is made of M 20 grade concrete and has 4 numbers, 20 mm diameter longitudinal bars of Fe-415 steel. It is under the action of a concentric axial compressive load. Ignoring the reduction in the area of concrete due to steel bars, the ultimate axial load carrying capacity of the column is

(a) 1659 kN (b) 1548 kN
(c) 1198 kN (d) 1069 kN

[2004 : 2 M]

3.18 An RC square footing of side length 2 m and uniform effective depth 200 mm is provided for a 300 mm × 300 mm column. The line of action of the vertical compressive load passes through the centroid of the footing as well as of the column. If the magnitude of the load is 320 kN, the nominal transverse (one way) shear stress in the footing is

(a) 0.26 N/mm^2 (b) 0.30 N/mm^2
(c) 0.34 N/mm^2 (d) 0.75 N/mm^2

[2004 : 2 M]

3.19 A rectangular column section of 250 mm × 400 mm is reinforced with five steel bars of grade Fe-500, each of 20 mm diameter. Concrete mix is M 30. Axial load on the column section with minimum eccentricity as per **IS 456: 2000** using limit state method can be applied upto

(a) 1707.37 (b) 1805.30
(c) 1806.40 (d) 1903.7 **[2005 : 2 M]**

Linked Answer Questions 3.20 and 3.21:

In the design of beams for the limit state of collapse in flexure as per **IS 456 : 2000**, let the maximum strain in concrete be limited to 0.0025 (in place of 0.0035). For this situation, consider a rectangular beam section with breadth as 250 mm, effective depth as 350 mm, area of tension steel as 1500 mm^2, and characteristic strengths of concrete and steel as 30 MPa and 250 MPa respectively.

3.20 The depth of neutral axis for the balanced failure is

(a) 140 mm (b) 156 mm
(c) 168 mm (d) 185 mm **[2006 : 2 M]**

3.21 At the limiting state of collapse in flexure, the force acting on the compression zone of the section is

(a) 326 kN (b) 389 kN
(c) 424 kN (d) 542 kN **[2006 : 2 M]**

Common Data for Questions 3.22 and 3.23:

A single reinforced rectangular concrete beam has a width of 150 mm and an effective depth of 330 mm. The characteristic compressive strength of concrete is 20 MPa and the characteristic tensile strength of steel is 415 MPa. Adopt the stress block for concrete as given in **IS 456 : 2000** and take limiting value of depth of neutral axis as 0.48 times the effective depth of the beam.

3.22 The limiting value of the moment of resistance of the beam in kN-m is

(a) 14.52 (b) 49.65
(c) 45.08 (d) 156.82 **[2007 : 2 M]**

3.23 The limiting area of tension steel in mm^2 is

(a) 473.9 (b) 412.3
(c) 373.9 (d) 312.3 **[2007 : 2 M]**

3.24 A reinforced concrete column contains longitudinal steel equal to 1 per cent of net cross-sectional area of the column. Assume modular ratio as 10. The loads carried (using the elastic theory) by the longitudinal steel and the net area of concrete, are P_s and P_c respectively. The ratio P_s/P_c expressed as per cent is

(a) 0.1 (b) 1
(c) 1.1 (d) 10 **[2008 : 2 M]**

Linked Answer Questions 3.25 and 3.26:

A doubly reinforced rectangular concrete beam has a width of 300 mm and an effective depth of 500 mm. The beam is reinforced with 2200 mm^2 of steel in tension and 628 mm^2 of steel in compression. The effective cover for compression steel is 50 mm. Assume that both tension and compression steel yield. The grades of concrete and steel used are M20 and Fe250, respectively. The stress block parameters (rounded off to first two decimal places) for concrete shall be as per **IS 456 : 2000**.

3.25 The depth of neutral axis is

(a) 205.30 mm (b) 184.56 mm
(c) 160.91 mm (d) 145.30 mm
[2010 : 2 M]

3.26 The moment of resistance of the section is

(a) 206.00 kN-m (b) 209.20 kN-m
(c) 237.80 kN-m (d) 251.90 kN-m
[2010 : 2 M]

3.27 A 16 mm thick plate measuring 650 mm × 420 mm is used as a base plate for an ISHB 300 column subjected to a factored axial compressive load of 2000 kN. As per **IS 456 : 2000**, the minimum grade of concrete that should be used below the base plate for safely carrying the load is

(a) M15 (b) M20
(c) M30 (d) M40 **[2011 : 1 M]**

Linked Answer for (Q.3.28 and Q.3.29)

The cross-section at mid-span of a beam at the edge of a slab is shown in the sketch. A portion of the slab is considered as the effective flange width for the beam. The grades of concrete and reinforcing steel are M25 and Fe415 respectively. The total area of reinforcing bars (A_s) is 4000 mm^2. At the ultimate limit sate, x_u denotes the depth of the neutral axis from the top fibre. Treat the section as under-reinforced and flanged (x_u > 100 mm).

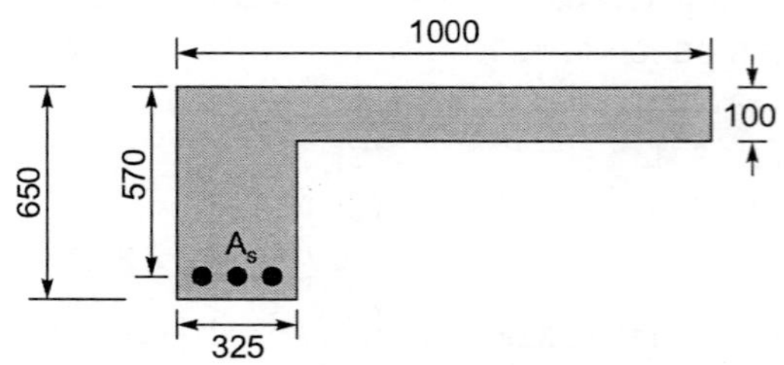

3.28 The value of x_u (in mm) computed as per the Limit State Method of **IS 456 : 2000** is

(a) 200.0 (b) 223.3
(c) 236.3 (d) 273.6 **[2012 : 2 M]**

3.29 The ultimate moment capacity (in kNm) of the sections as per the Limit State Method of **IS 456 : 2000** is
(a) 475.2 (b) 717.0
(c) 756.4 (d) 762.5 [2012 : 2 M]

3.30 For a beam of cross-section, width = 230 mm and effective depth = 500 mm, the number of rebars of 12 mm diameter required to satisfy minimum tension reinforcement requirement specified by **IS 456 : 2000** (assuming grade of steel reinforcement as Fe500) is ________.
[2014 : 2 M, Set-I]

3.31 A column of size 450 mm × 600 mm has unsupported length of 3.0 m and is braced against side sway in both directions. According to **IS 456 : 2000**, the minimum eccentricities (in mm) with respect to major and minor principle axes are
(a) 20.0 and 20.0 (b) 26.0 and 21.0
(c) 26.0 and 20.0 (d) 21.0 and 15.0
[2015 : 1 M, Set-II]

3.32 A haunched (varying depth) reinforced concrete beam is simply supported at both ends, as shown in the figure. The beam is subjected to a uniformly distributed factored load of intensity 10 kN/m. The design shear force (expressed in kN) at the section X-X of the beam is________. (Neglect variation of load due to change in section)

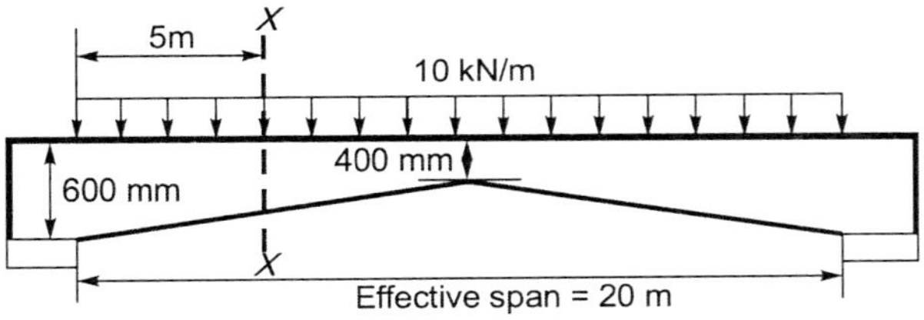

[2016 : 2 M, Set-II]

3.33 A reinforced concrete (RC) beam with width of 250 mm and effective depth of 400 mm is reinforced with Fe415 steel. Consider 50 mm effective cover. As per the provisions of **IS 456 : 2000**, the minimum and maximum amount of tensile reinforcement (expressed in mm^2) for the section are, respectively
(a) 250 and 3500 (b) 205 and 4500
(c) 270 and 2000 (d) 300 and 2500
[2016 : 2 M, Set-I]

3.34 Two rectangular under-reinforced concrete beam sections *X* and *Y* are similar in all aspects except that the longitudinal compression reinforcement in section *Y* is 10% more. Which one of the following is the correct statement?
(a) Section *X* has less flexural strength and is less ductile than section *Y*
(b) Section *X* has less flexural strength but is more ductile than section *Y*
(c) Section *X* and *Y* have equal flexural strength but different ductility
(d) Sections *X* and *Y* have equal flexural strength and ductility.
[2018 : 1 M, Set-I]

3.35 An RCC short column (with lateral ties) of rectangular cross-section of 250 mm × 300 mm is reinforced with four members of 16 mm diameter longitudinal bars. The grades of steel and concrete are Fe415 and M20, respectively. Neglect eccentricity effect. Considering limit state of collapse in compression (IS 456 : 2000), the axial load carrying capacity of the column (in kN, up to one decimal place) is ______
[2018 : 2 M, Set-I]

3.36 A reinforced-concrete slab with effective depth of 80 mm is simply supported at two opposite ends on 230 mm thick masonry walls. The centre-to-centre distance between the walls is 3.3 m. As per IS 456 : 2000, the effective span of the slab (in m, up to two decimal places) is __________.
[2018 : 1 M, Set-II]

3.37 A structural member subjected to compression, has both translation and rotation restrained at one end, while only translation is restrained at the other end. As per IS 456 : 2000, the effective length factor recommended for design is
(a) 0.50 (b) 0.65
(c) 0.70 (d) 0.80
[2018 : 1 M, Set-II]

3.38 A singly-reinforced rectangular concrete beam of width 300 mm and effective depth 400 mm is to be designed using M25 grade concrete and Fe500 grade reinforcing steel. For the beam to be under-reinforced, the maximum number of 16 mm diameter reinforcing bars that can be provided is
(a) 3 (b) 4
(c) 5 (d) 6
[2018 : 2 M, Set-II]

3.39 The cross-section of the reinforced concrete beam having an effective depth of 500 mm is shown in the figure (not drawn to the scale). The grades of concrete and steel used are M35 and Fe550, respectively. The area of tension reinforcement is

400 mm². It is given that corresponding to 0.2% proof stress, the material safety factor is 1.15 and the yield strain of Fe550 steel is 0.0044.

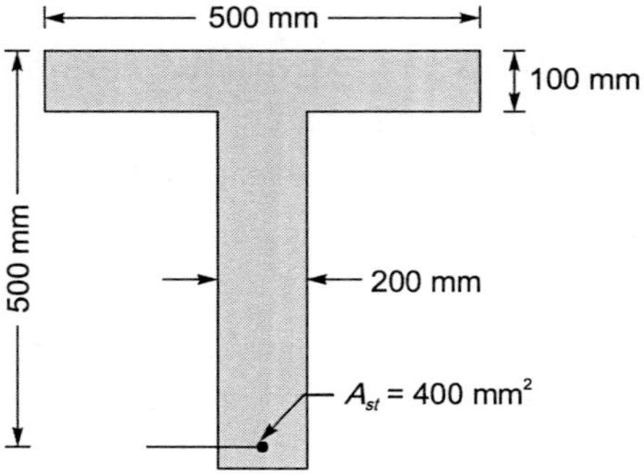

As per **IS 456:2000**, the limiting depth (in mm, round off to the nearest integer) of the neutral axis measured from the extreme compression fiber, is ________.

[2020 : 2 M, Set-II]

3.40 A combined trapezoidal footing of length L supports two identical square columns (P_1 and P_2) of size 0.5 m × 0.5 m, as shown in the figure. The columns P_1 and P_2 carry loads of 2000 kN and 1500 kN, respectively.

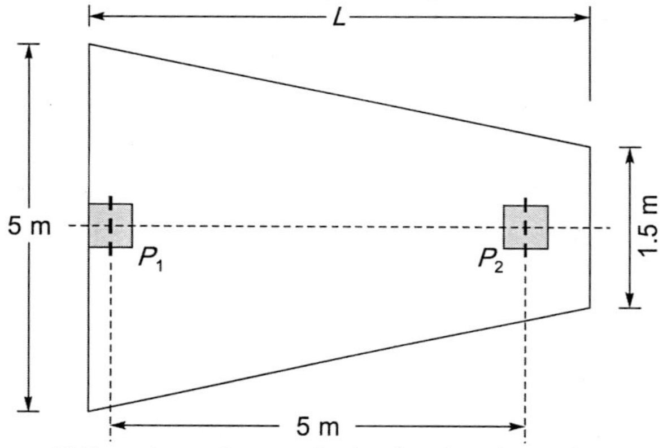

If the stress beneath the footing is uniform, the length of the combined footing L (in m, round off to two decimal places) is ______.

[2021 : 2 M, Set-I]

3.41 A slab panel with an effective depth of 250 mm is reinforced with 0.2% main reinforcement using 8 mm diameter steel bars. The uniform center-to-center spacing (in mm) at which the 8 mm diameter bars are placed in the slab panel is ________ (rounded off to the nearest integer).

[2024 : 2 M, Set-I]

3.42 A simply supported, uniformly loaded, two-way slab panel is torsionally unrestrained. The effective span lengths along the short span (x) and long span (y) directions of the panel are l_x and l_y, respectively. The design moments for the reinforcements along the x and y directions are M_{ux} and M_{uy}, respectively. By using Rankine-Grashoff method, the ratio M_{ux}/M_{uy} is proportional to

(a) $(l_y/l_x)^2$
(b) l_y/l_x
(c) l_x/l_y
(d) $(l_x/l_y)^2$

[2024 : 1 M, Set-II]

3.43 A concrete column section of size 300 mm × 500 mm as shown in the figure is subjected to both axial compression and bending along the major axis. The depth of the neutral axis (x_u) is 1.1 times the depth of the column, as shown.

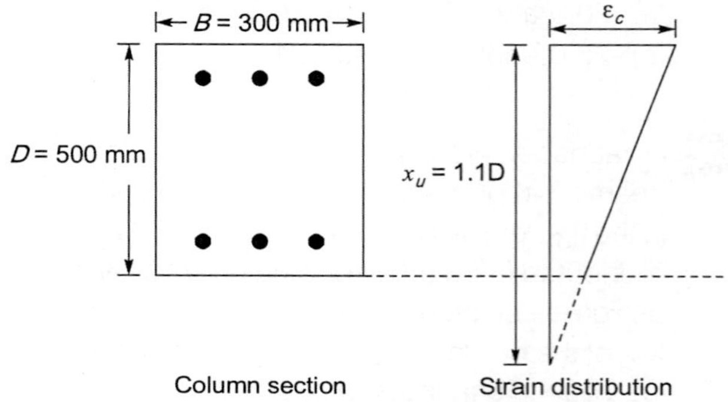

(Figure NOT to scale)

The maximum compressive strain (ε_c) at highly compressive extreme fiber in concrete, where there is no tension in the section, is ________ × 10^{-3} (rounded off to 2 decimal places).

[2024 : 2 M, Set-II]

Answers Footing, Columns, Beams and Slabs

3.1 (a)	3.2 (c)	3.3 (b)	3.4 (d)	3.5 (d)	3.6 (a)	3.7 (b)
3.8 (d)	3.9 (b)	3.10 (a)	3.11 (d)	3.12 (d)	3.13 (b)	3.14 (a)
3.15 (c)	3.16 (d)	3.17 (c)	3.18 (a)	3.19 (a)	3.20 (b)	3.21 (a)
3.22 (c)	3.23 (a)	3.24 (d)	3.25 (c)	3.26 (b)	3.27 (b)	3.28 (c)
3.29 (b)	3.30 (2)	3.31 (b)	3.32 (65)	3.33 (b)	3.34 (a)	3.35 (918.1)
3.36 (3.15)	3.37 (d)	3.38 (c)	3.39 (221.52)	3.40 (5.83)	3.41 (100)	3.42 (a)
3.43 (3.28)						

Explanations Footing, Columns, Beams and Slabs

3.1 (a)

There are two kinds of reinforcement in a column, longitudinal reinforcement and transverse reinforcement. The purpose of transverse reinforcement is to hold the vertical bars in position so that individual bars cannot buckle outwards and split the concrete. Transverse reinforcement does not contribute to the strength of a column. Transverse reinforcement may be in the form of lateral ties or spirals.

3.2 (c)

The recommended value of clear covers as per **IS code 456 : 1978** for slab, beams, columns and footings are respectively 15, 25, 40 and 75 mm.

Note : This is as per **IS 456 : 1978**, not **IS 456 : 2000**.

As per new code it will be 20, 25, 40, 75.

For slab, if diameter of bar is upto 12 mm, clear cover may be reduced by 5 mm.

3.3 (b)

When the strength of column is governed by yield of tension steel then some amount of tension is nullified in tension steel, if we apply additional axial compressive force. However tension bars are able to take more tension due to which moment carrying capacity of beam column is increased.

Refer to sol. 3.2. In the region III, which is called as tension controlled region, when we move along the curve from A to B, P_u as well as M_u increases.

Hence, in a reinforced concrete beam column the increase in the flexural strength along with the increase in the axial strength occurs when the yielding of the tension reinforcement governs the strength.

3.4 (d)

For proper interlocking and anchorage, figure (d) suggest the best alternative . Hauches are alsoa provided in figure (d), and (c) which further increase the shear strength.

3.5 (d)

Minimum shear reinforcement is provided for the following reasons:

1. A sudden failure of beam is prevented in case the concrete cover bursts and the bond between steel and concrete is lost.
2. Possibility of brittle shear failure is prevented which would have occurred without any shear reinforcement.
3. During the concreting process at site, it holds the reinforcing bars in position while pouring concrete.
4. Tie effect of compression steel makes the section more effective.
5. It improves dowel action of longitudinal tension bars.

3.6 (a)

For an **isolated T-beam,**

For simply supported beam, $L_0 = L_{eff}$

Effective width of flange,

$$b_f = \frac{l_0}{\left(\frac{l_0}{B}+4\right)} + b_w = \frac{6\times1000}{\frac{6000}{1000}+4} + 300 = 900 \text{ mm}$$

3.7 (b)

$$x = \frac{1200}{2} = 600 \text{ mm}$$

$$c = 3000 \text{ mm}$$

$$y = \frac{2200}{2} = 1100 \text{ mm} \not> 1000 \text{ mm}$$

If $x < 1000$ mm and $y \geq 1000$ mm then

Effective span of stair case slab

$= (x + c + 1000) = (600 + 3000 + 1000) = 4600$ mm

AVOID MISTAKE

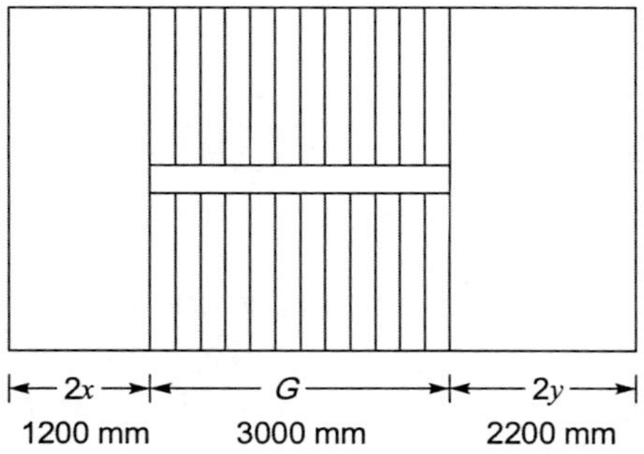

As per IS 4456 : 2000

Effective span of stair case slab

$= \text{Going} + x + y$

We must take care, maximum value of x and y can be 1 m.

$$\therefore \text{ From } x = \frac{1200}{2} = 600 \text{ mm}$$

$$y = \frac{2200}{2} = 1100 \text{mm}$$

$$\therefore \quad l_{eff} = G + x + y$$

$$= 3000 + 600 + 1000 = 4600 \text{ mm}$$

Common error:

$$l = G + x + y$$

$$= 3000 + 600 + 1100 = 4700 \text{ mm}$$

3.8 (d)

As per IS 456 : 2000, Clause 32.5

Minimum value	Deformed bars ($\not>$ 16mm)	Other bars	Welded wire fabric $\not>$ (16 mm)
$\left(\frac{\text{Vertical steel}}{\text{Gross concrete}}\right) \times 100$	0.12%	0.15%	0.12%
$\left(\frac{\text{Horizontal steel}}{\text{Gross concrete}}\right) \times 100$	0.20%	0.25%	0.20%

$$\therefore \frac{\text{Minimum vertical reinforcement}}{\text{Minimum horizontal reinforcement}} = \left(\frac{0.12}{0.20}\right)$$

$$\text{or } \left(\frac{0.15}{0.25}\right)$$

Both equal to 0.6 i.e.m, 3 : 5

3.9 (b)

As per **IS 456 : 2000** pedestal is a vertical compression member whose effective length is less than 3 times the least lateral dimension.

3.10 (a)

According to **clause 26.5.1.1** of **IS 456 : 2000**, minimum area of tension reinforcement,

$$\frac{A_{st\ min.}}{bd} = \frac{0.85}{f_y}$$

$$\Rightarrow \quad A_{st\ min.} = \frac{0.85\ bd}{f_y}$$

3.11 (d)

$$x_{u\ lim} = \frac{700}{1100 + 0.87 f_y}$$

Values of $x_{u\ lim}$ for various grades of steel are tabulated below for ready reference of students.

	Fe250	Fe415	Fe500
x_{ulim}	$0.5313 \approx 0.53$	$0.4791 \approx 0.48$	$0.456 \approx 0.46$

3.12 (d)

Effective length of a column depends upon:

(i) Flexural stiffness (EI/L) of beams joining at a point.

(ii) Flexural stiffness (EI/L) of columns joining at a point.

(iii) Type of frame (sway or non-sway).

3.13 (b)

The tensile stress in the concrete is given by,

$$\sigma_t = \frac{\text{Hoop tension}}{(b \times d) + (m - 1) A_{st}}$$

$$= \frac{120 \times 10^3}{(250 \times 400) + (10 - 1) \times 4 \times \frac{\pi}{4} \times (20)^2}$$

$$= 1.07 \text{ N/mm}^2$$

3.14 (a)

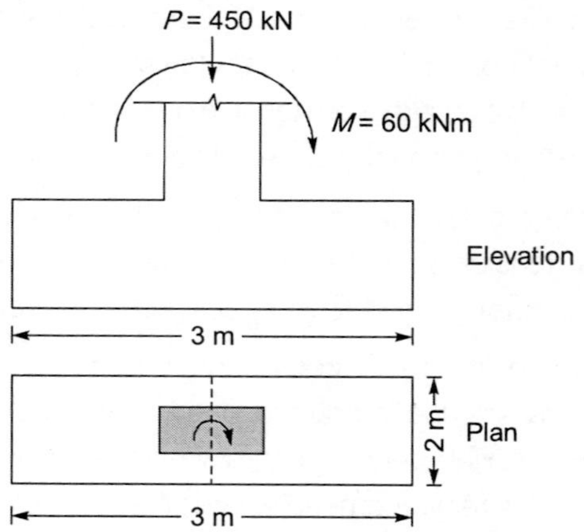

Area of footing,

$$A = 2 \times 3 = 6 \text{ m}^2$$

Section modulus,

$$Z = \frac{bd^2}{6} = \frac{2 \times 3^2}{6} = 3 \text{ m}^3$$

$$\therefore \quad \sigma = \frac{P}{A} \pm \frac{M}{Z} = \frac{450}{6} \pm \frac{60}{3} = 75 \pm 20$$

$$\therefore \quad \sigma_{max} = 95 \text{ kN/m}^2$$

and $\sigma_{min} = 55 \text{ kN/m}^2$

3.15 (c)

Method-I

Bending tensile strength, $\sigma_{bt} = 2.2 \text{ N/mm}^2$

Characteristic compressive strength,

$$f_{ck} = 20 \text{ N/mm}^2$$

∴ Maximum stress in concrete,

$$\sigma_{cbc} = 0.45 f_{ck} = 0.45 \times 20 = 9 \text{ N/mm}^2$$

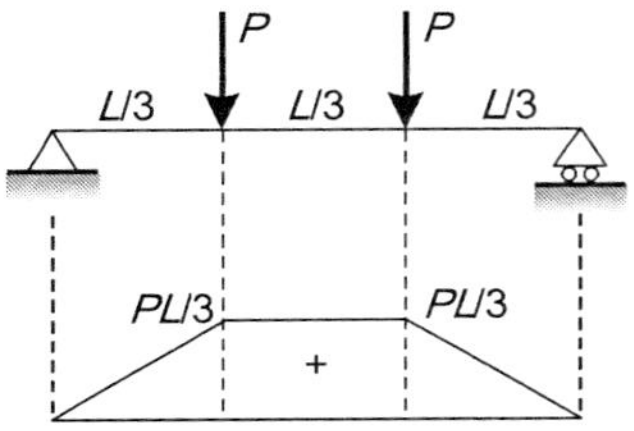

BMD

Section modulus,

$$Z = \frac{bD^2}{6} = \frac{200 \times (300)^2}{6}$$

$$= 3 \times 10^6 \text{ mm}^3$$

∴ Maximum moment,

$$M_{max} = Z\sigma_{bt} = 3 \times 10^6 \times 2.2$$

$$= 6.6 \times 10^6 = 6.6 \text{ kN-m}$$

$$M_{max} = \frac{PL}{3}$$

$$\Rightarrow \quad \frac{P \times 3}{3} = 6.6$$

$$\Rightarrow \quad P = 6.6 \text{ kN}$$

Method-II

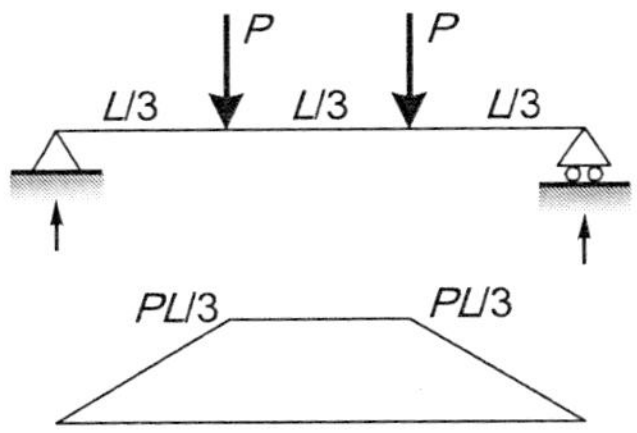

Maximum bending moment, $M = \frac{PL}{3}$

From standard flexure equation, $\sigma_t = \frac{My}{I}$

σ_t = Bending tensile stress at bottom of beam

$$\therefore \quad \sigma_t = \frac{My}{I}$$

$$\frac{2.2\text{N}}{\text{mm}^2} = \frac{\frac{PL}{3} \times \left(\frac{300}{2}\right)}{\frac{200 \times 300^3}{12}}$$

$$2.2 = \frac{P \times \frac{3000}{3} \times \frac{300}{2}}{\frac{200 \times 300^3}{12}}$$

$$\therefore \quad P = 6600 \text{ N} = 6.6 \text{ kN}$$

3.16 (d)

$$A_{st} = 2 \times \frac{\pi}{4} \times (16)^2 = 402.12 \text{ mm}^2$$

For limit state of collapse,

$$x_u = \frac{0.87 f_y A_{st}}{0.36 f_{ck} b}$$

$$= \frac{0.87 \times 415 \times 402.12}{0.36 \times 20 \times 200} = 100.8 \text{ mm}$$

But $x_{u,\,lim} = 0.48\, d = 0.48 \times (300 - 40)$

$$= 124.8 \text{ mm}$$

Since, $x_u < x_{u,\,lim}$

the section is under reinforced.

$$\therefore \quad M_u = 0.87 f_y A_{st} (d - 0.42\, x_u)$$

$$= 0.87 \times 415 \times 402.12 \times (260 - 0.42 \times 100.8)$$

$$= 31.6 \text{ kN-m}$$

But $M_u = \frac{PL}{3}$

$$\Rightarrow \quad \frac{P \times 3}{3} = 31.6$$

$$\Rightarrow \quad P = 31.6 \text{ kN}$$

3.17 (c)

The ultimate load carrying capacity of the column will be,

$$P_{uz} = 0.45 f_{ck} A_c + 0.75 f_y A_{sc}$$

...[clause 39.6 IS 456 : 2000]

$$= 0.45 \times 20 \times (300)^2 + 0.75 \times 415 \times 4 \times \frac{\pi}{4} \times (20)^2 = 1201.13 \text{ kN}$$

It may be noted that the equation,

$$P_u = 0.4 f_{ck} A_c + 0.67 f_y A_{sc}$$

cannot be used here because it is applicable only when minimum eccentricy does not exceed 0.05D. Also, it is given in the problem that the column is under the action of a concentric axial load.

3.18 (a)

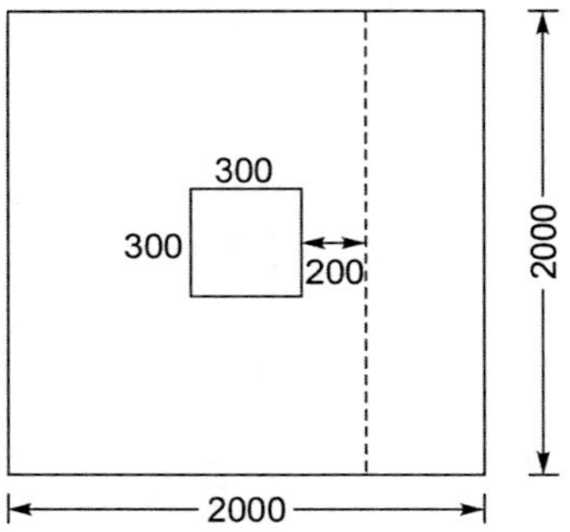

Plan

Uniform pressure,

$$p_0 = \frac{P}{A} = \frac{320}{2 \times 2} = 80 \text{ kN/m}^2$$

The critical section for shear is at a distance of d = 200 mm from the face of the column

$$V_u = p_0 B\left[\frac{1}{2}(B-b) - d\right]$$

$$= 80 \times 2 \times \left[\frac{1}{2}(2-0.3) - 0.2\right] = 104 \text{ kN}$$

Shear stress,

$$\tau_u = \frac{V_u}{Bd} = \frac{104 \times 10^3}{2 \times 10^3 \times 200} = 0.26 \text{ N/mm}^2$$

3.19 (a)

The axial load on the column can be given as

$$P_u = 0.4 f_{ck} A_c + 0.67 f_y A_{sc}$$

$$A_c = A_g - A_{sc}$$

$$= (250 \times 400) - \left[5 \times \frac{\pi}{4} \times (20)^2\right]$$

$$= 98429.20 \text{ mm}^2$$

$$f_y = 500 \text{ N/mm}^2$$

$$A_{sc} = 5 \times \frac{\pi}{4} \times (20)^2 = 1570.8 \text{ mm}^2$$

$$\therefore \quad P_u = (0.4 \times 30 \times 98429.20) + (0.67 \times 500 \times 1570.8)$$

$$= 1707.37 \text{ kN}$$

It may be noted that as per **IS 456: 2000** the above formula is not applicable for $e_{min} < 0.05$ D.

3.20 (b)

As per **IS 456 : 2000**, maximum strain in tension reinforcement at the time of collapse should notbe less than $\frac{f_y}{1.15 E_s} + 0.002$.

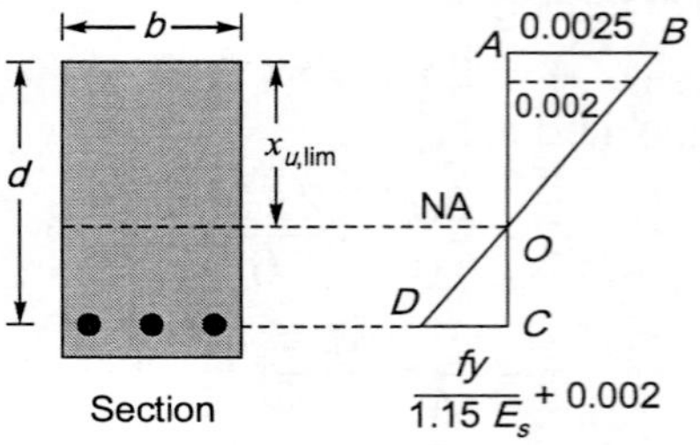

Strain Diagram

From similar triangles ΔOAB and ΔOCD, we have,

$$\frac{x_{u,lim}}{0.0025} = \frac{d - x_{u,lim}}{\frac{f_y}{1.15 E_s} + 0.002}$$

$$\Rightarrow \quad \frac{d - x_{u,lim}}{x_{u,lim}} = \frac{\frac{f_y}{1.15 E_s} + 0.002}{0.0025}$$

$$\Rightarrow \quad \frac{d}{x_{u,lim}} = 1 + \frac{\frac{0.87 f_y}{E_s} + 0.002}{0.0025}$$

$$\Rightarrow \quad \frac{d}{x_{u,lim}} = \frac{0.0045 + \frac{0.87 f_y}{E_s}}{0.0025}$$

$$\Rightarrow \quad \frac{d}{x_{u,lim}} = \frac{0.0045 E_s + 0.87 f_y}{0.0025 E_s}$$

$[\because E_s = 2 \times 10^5 \text{ MPa}]$

$$\Rightarrow \quad \frac{x_{u,lim}}{d} = \frac{500}{900 + 0.87 f_y}$$

$$\Rightarrow \quad x_{u,lim} = \left[\frac{500}{900 + 0.87 \times 250}\right] \times 350$$

$$\Rightarrow \quad x_{u,lim} = 156.6 \text{ mm} \approx 156 \text{ mm}$$

3.21 (a)

In strain diagram,

$$\frac{OE}{OA} = \frac{0.002}{0.0025}$$

$$\Rightarrow \quad OE = \frac{4}{5} x_u$$

$$\Rightarrow \quad OE = 0.8 x_u$$

$$\therefore \quad AE = x_u - 0.8 x_u = 0.2 x_u$$

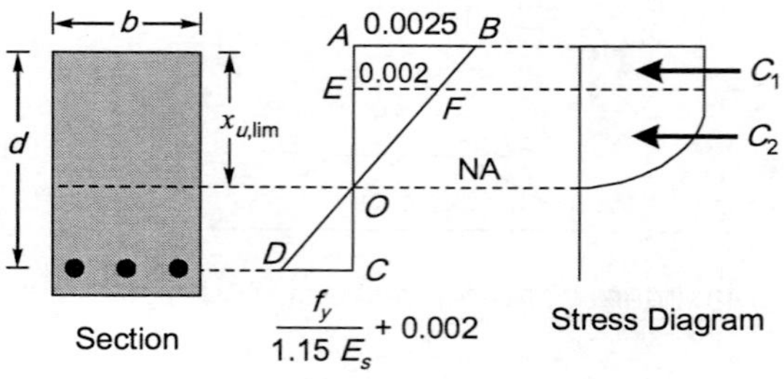

Strain Diagram

$\therefore$ Total compressive force,

$$= 0.45 f_{ck} \times 0.2 x_u \times b + \frac{2}{3} \times 0.45 f_{ck} \times 0.8 x_u \times b$$

$$= 0.33 f_{ck} bx_u$$

Actual NA depth,

$$C = T$$

$$\Rightarrow 0.33 f_{ck} bx_u = 0.87 f_y A_{st}$$

$$\Rightarrow x_u = \frac{0.87 \times 250 \times 1500}{0.33 \times 30 \times 250} = 131.8 \text{ mm}$$

$\therefore$ Total compressive force,

$$= 0.33 \times 30 \times 250 \times 131.8 = 326 \text{ kN}$$

3.22 (c)

$$x_{u,\lim} = 0.48 d$$

$$= 0.48 \times 330 = 158.4 \text{ mm}$$

$\therefore$ Moment of resistance

$$= 0.36 f_{ck} b x_{u,\lim} (d - 0.42 x_{u,\lim})$$

$$= 0.36 \times 20 \times 150 \times 158.4(330 - 0.42 \times 158.4)$$

$$= 45.07 \text{ kN-m}$$

3.23 (a)

Moment of resistance

$$= 0.87 f_y A_{st} (d - 0.42 x_{u,\lim})$$

$$\Rightarrow 45.07 \times 10^6 = 0.87 \times 415 \times A_{st} (330 - 0.42 \times 158.4)$$

$$\Rightarrow A_{st} = 473.82 \text{ mm}^2$$

3.24 (d)

If A is the gross cross-sectional area of the column, A_{st} is the area of steel, A_c is net area of concrete, m is modular ratio and f_c is stress in concrete, then, as per question

$$A_{st} = \frac{1}{100} \times A_c = 0.01 A_c$$

From elastic theory, we know that the strain in concrete will beequal to the strain in steel.

$$\therefore \quad e_c = e_s$$

$$\frac{\sigma_c}{A_c E_c} = \frac{P_s}{A_s E_s}$$

Percentage of $\left(\frac{P_s}{P_c}\right)$

$$= \frac{P_s}{P_c} \times 100 = \left(\frac{E_s}{E_c}\right)\left(\frac{A_s}{A_c}\right) \times 100$$

$$= m \times \frac{0.01 A_c}{A_c} \times 100 = 10 \times 1 \times 100 = 10\%$$

3.25 (c)

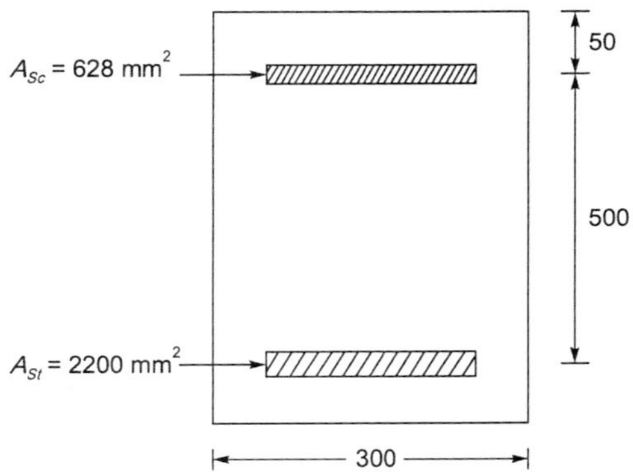

The resultant compressive force in the concrete is given by,

$$C_{uc} = 0.36 f_{ck} bx_u$$

The resultant compressive force in the compression steel is given by,

$$C_{us} = (f_{sc} - 0.45 f_{ck}) A_{sc}$$

The resultant tensile force in the tensile steel is given by,

$$T_u = 0.87 f_y A_{st}$$

we know that,

$$C_{uc} + C_{us} = T_u$$

$$\Rightarrow 0.36 f_{ck} bx_u + (f_{sc} - 0.45 f_{ck}) A_{sc} = 0.87 f_y A_{st}$$

$$\Rightarrow x_u = \frac{0.87 f_y A_{st} - (f_{sc} - 0.45 f_{ck}) A_{sc}}{0.36 f_{ck} b}$$

Consider, $f_{sc} = 0.87 f_y$ (as both steel yields)

$$\Rightarrow x_u = \frac{0.87 \times 250 \times 2200 - (0.87 \times 250 - 0.45 \times 20) \times 628}{0.36 \times 20 \times 300}$$

$$\Rightarrow x_u = 160.91 \text{ mm}$$

But $x_{u,\max} = 0.53 d$ for Fe 250

$$\therefore \quad x_{u,\max} = 0.53 \times 500 = 265 \text{ mm}$$

$x_u < x_{u,\max}$, hence the assumption $f_{sc} = 0.87 f_y$ is justified.

3.26 (b)

Moment of resistance M_u is given by,

$$M_u = C_{uc}(d - 0.42 x_u) + C_{us}(d - d')$$

$$= 0.36 f_{ck} bx_u(d - 0.42 x_u) + (f_{sc} - 0.45 f_{ck}) A_{sc}(d - d')$$

$$= 0.36 \times 20 \times 300 \times 160.91(500 - 0.42 \times 160.91) + (0.87 \times 250 - 0.45 \times 20) \times 628 (500 - 50)$$

$$= 209215652.1 \times 10^{-6} \text{ kN-m} = 209.21 \text{ kN-m}$$

3.27 (b)

Working axial load

$$= \frac{\text{Factored axial load}}{1.5}$$

$$= \frac{2000}{1.5} = 1333.33 \text{ kN}$$

The developed bearing pressure on concrete is given as,

$$\sigma_{all} = \frac{\text{Direct load}}{\text{Area of base plate}}$$

$$= \frac{1333.33 \times 10^3}{650 \times 420} = 4.88 \text{ N/mm}^2$$

The developed stress in direct compression in various grades of concrete as per **IS 456 : 2000** are tabulated below = 0.25 f_{ck} (WSM)

Grades	M10	M15	M20	M25	M30	M35	M40	M45	M50
Stress	2.5	4.0	5.0	6.0	8.0	9.0	10.0	11.0	12.0

The permissible stress in concrete should be more than the developed bearing pressure. Thus the minimum grade of concrete which should be used is **M 20**.

3.28 (c)

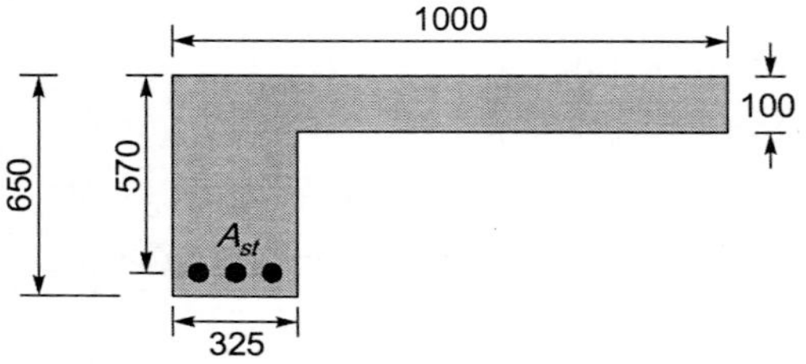

Given, X_u > 100 mm and section as under reinforced as per **IS 456 : 2000**

Effective width of flange of L-beam:

$$b_f = \frac{l_0}{12} + b_w + 3D_f$$

Where l_0 = Distance between point of zero moment in beam

D_f = Depth of flange

b_w = Width of web

Taking b_f = 1000 mm

∴ Length of beam is not given

Assuming $D_f < \frac{3}{7} x_u$

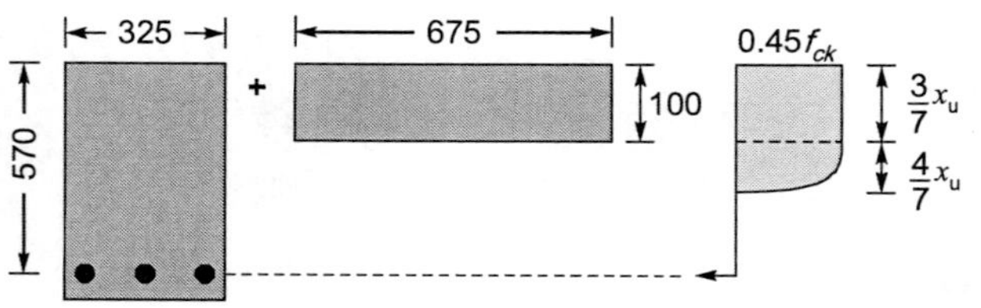

∴ Equating compressive force and tensile force

$C = T$

$$0.36 f_{ck} x_u b_w + (100 - b_w) D_f \times 0.45 f_{ck} = 0.87 f_y A_{st}$$

$$0.36 \times 25 \times x_u \times 325 + 675 \times 100 \times 0.45 \times 25$$

$$= 0.87 \times 415 \times 4000$$

∴ x_u = 234.13 mm

Check $\frac{3}{7} x_u$ = 100.34 > D_f

It shows our assumption was correct.

So, x_u = 234.13 mm

(Nearest match = 236 mm)

3.29 (b)

$M_u = 0.36 f_{ck} (325) X_u(d - 0.42 X_u) + 0.45$

or, $f_{ck}(675)\, 100\, (d - 50)$

$= 0.36 \times 25 \times 325 \times 234.13\,(570 - 0.42 \times 234.13) + 0.45 \times 25 \times 675 \times 100\,(570 - 50) = 717$ kN-m

Take 717.00 kNm

3.30 Sol.

$$\frac{A_{st\,min}}{bd} = \frac{0.85}{f_y}$$

or, $$A_{st} = \frac{0.85}{500} \times 230 \times 500 \text{ mm}^2$$

$$n = \frac{A_{st}}{\frac{\pi d^2}{4}} = \frac{0.85 \times 230}{\frac{\pi}{4}(12)^2} = 1.729 = 2 \text{ bars}$$

3.31 (b)

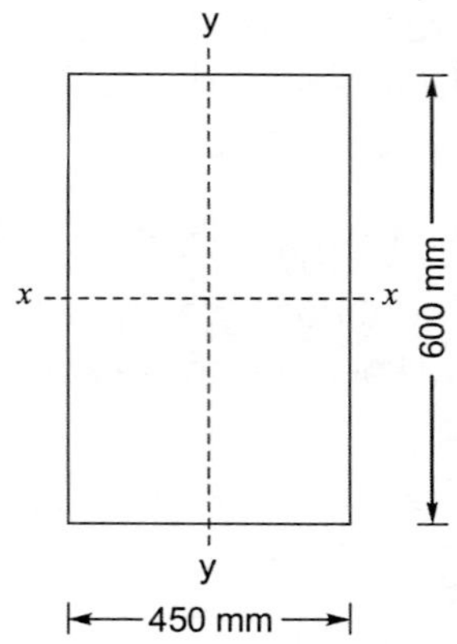

$$e_{min} = \text{maximum}\begin{cases} \frac{L}{500} + \frac{D}{30} \\ 20 \text{ mm} \end{cases}$$

x-x will be major axis and y-y will be minor axis.

and $$e_{min\,xx} = \frac{3000}{500} + \frac{600}{30} = 26 \text{ mm}$$

∴ $$e_{min\,yy} = \frac{3000}{500} + \frac{450}{30} = 21 \text{ mm}$$

3.32 Sol.

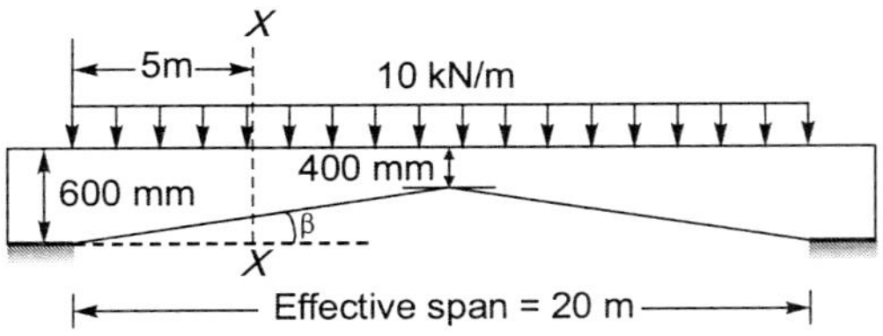

Shear force at section X-X,

$$V_u = 100 - 5 \times 10 = 50 \text{ kN}$$

Depth at section X-X,

$$d = 400 + \frac{200}{10} \times 5 = 500 \text{ mm} = 0.5 \text{ m}$$

Moment at section X-X,

$$M_u = 100 \times 5 - 10 \times 2.5 \times 5 = 375 \text{ kNm}$$

Design shear force at section X-X,

$$V_{u,\text{design}} = V_u + \frac{M_u}{d} \tan\beta$$

$$= 50 + \frac{375}{0.5} \times \frac{200}{10000} = 65 \text{ kN}$$

POINTS TOREMEMBER

- As per clause 40.1.1 of IS 456 : 2000, nominal shear stress in beams of varying depth is given by $\tau_v = \left(\dfrac{V_u \pm \dfrac{M_u}{d} t_{a\beta}}{bd} \right)$.
- The negative sign in the formula applies when the bending moment M_u increase numerically in the same direction as the effective depth increases and positive sign when the moment decreases numerically in this direction.

3.33 (b)

Minimum tension reinforcement is given by,

$$\frac{A_s}{bd} = \frac{0.85}{f_y}$$

$$A_s = \frac{0.85 \times 250 \times 400}{415} = 205 \text{ mm}^2$$

Maximum compression reinforcement = $0.04\, bD$

$$= 0.04 \times 250 \times 450$$

$$= 4500 \text{ mm}^2$$

3.34 (a)

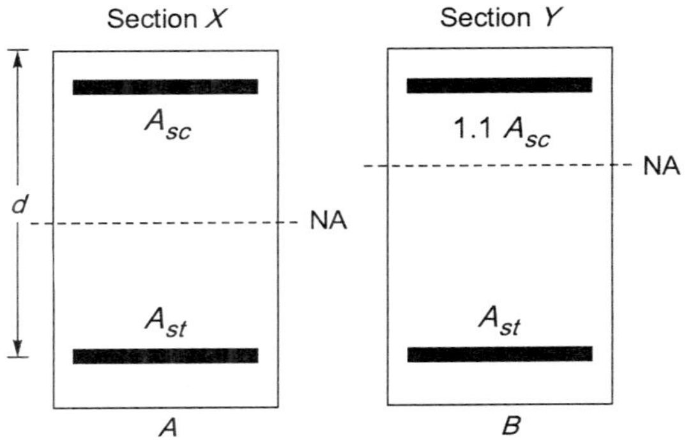

Due to presence of more compression steel in section *Y*, NA of section of *Y* is above than as of *X*. It means *Y* is more under-reinforced than *X* so ductility of *Y* is more.

Since compression steel of *Y* is more so flexure resistance of *X* is less than as of *Y*.

3.35 Sol.

Since eccentricity effect is being neglected so column can be considered as concentrically loaded.

Ultimate axial load carrying capacity of column.

$$P_u = 0.45\, f_{ck} A_g + (0.75\, f_y - 0.45\, f_{ck})\, A_{sc}$$

$$= 0.45 \times 20 \times 250 \times 300 + (0.75 \times 415 - 0.45 \times 20)\; 4 \times \frac{\pi}{4} \times 16^2 = 918.1 \text{ kN}$$

3.36 Sol.

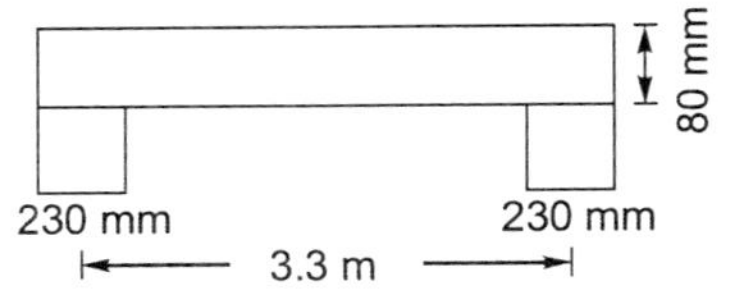

Here support width = 230 mm

Effective depth of slab = 80 mm

Clear span = 3.3 − 0.23 = 3.07 m

$$\text{Effective span} = \text{Min. of} \begin{cases} \text{Clear span} + d \\ \text{Clear span} + w \\ = \text{c/c spacing} \end{cases}$$

$$l_{\text{eff}} = \text{Min. of} \begin{cases} 3.07 + 0.80 = 3.15 \text{ m} \\ 3.3 \text{ m} \end{cases}$$

$$\therefore\; l_{\text{eff}} = 3.15 \text{ m}$$

3.37 (d)

One end is fixed

Other end is pin jointed

Effective length of column

(as per IS:456–2000) = 0.80 L

REFERENCE MATERIAL

Table 28, Pg 94 IS 456 : 2000

3.38 (c)

$$B = 300 \text{ mm}$$

$$d = 400 \text{ mm (effective depth)}$$

M25 and Fe500

$$0.87 f_y A_{st} = 0.36 f_y x_{u\,lim} b$$

$$\therefore \quad A_{st\ lim} = \frac{0.36}{0.87}\left(\frac{f_{ck}}{f_y}\right) x_{ulim} b$$

$$A_{st,\ lim} = 0.414\left(\frac{f_{ck}}{f_y}\right) x_{u,lim} b$$

$$= 0.414\left(\frac{25}{500}\right) 0.46 \times 400 \times 300 = 1142.64 \text{ mm}^2$$

$$\text{Number of 16 mm } \phi = \frac{1142.64}{\frac{\pi}{4}(16)^2} = 5.68$$

For $A_{st} < A_{st\ lim}$, maximum number of bars to be provided is 5 to keep the beam as under-reinforced section. If the reinforcement provided exceed 5.68 No. (say 6 and more) the section will become over-reinforced.

3.39 Sol.

For a RCC T-Beam

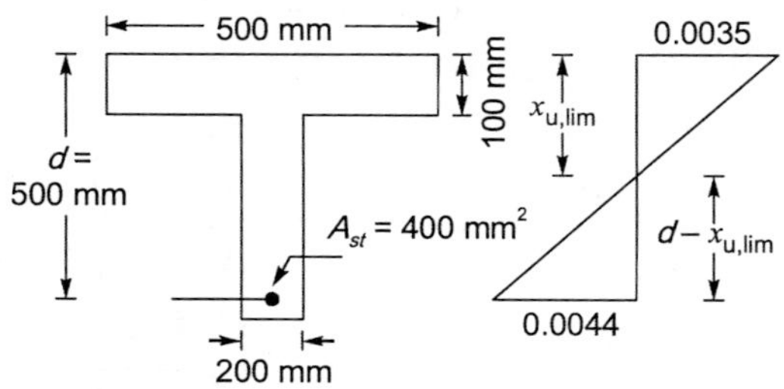

(For limiting depth of neutral axis)

Considering $d = 500$ mm

$$\frac{0.0035}{x_{u,\,lim}} = \frac{0.0044}{d - x_{u,\,lim}}$$

$$d - x_{u,lim} = \frac{0.0044}{0.0035} \times x_{u,lim}$$

$$35 \times 500 = 35x_{u,lim} + 44x_{u,lim} = 79x_{u,lim}$$

$$x_{u,lim} = \frac{35 \times 500}{79} = 221.52 \text{ mm}$$

Limiting depth of neutral axis

$$x_{u,lim} = 221.52 \text{ mm}$$

3.40 Sol.

C.G. of load from P_1

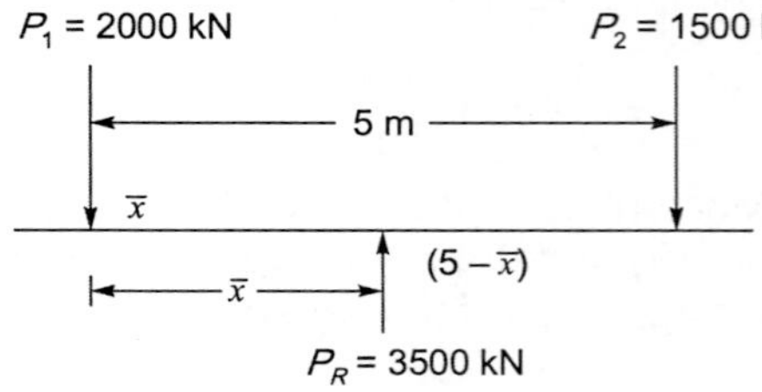

$$P_R \bar{x} = P_1 \times 0 + P_2 \times 5$$

$$\bar{x} = \frac{1500 \times 5}{3500} = 2.143 \text{ m}$$

Distance of C.G. of footing from face of P_1

$$\bar{y} = \bar{x} + 0.25 = 2.393 \text{ m}$$

C.G. of footing,

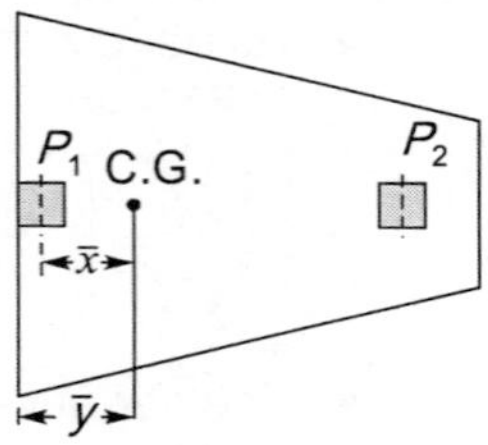

$$\bar{y} = \left(\frac{B_1 + 2B_2}{B_1 + B_2}\right) \times \frac{L}{3}$$

$$2.393 = \left(\frac{5 + 2 \times 1.5}{5 + 1.5}\right) \times \frac{L}{3}$$

$$L = 5.833 \text{ m say } 5.83 \text{ m}$$

3.41 Sol.

Effective depth, d = 250 mm

Reinforcement, A_{st} = 0.2%

Diameter of bar, ϕ = 8 mm

S = ?

$$A_{st}(\text{in } \%) = \frac{A_{st}}{B.d} \times 100$$

$$\Rightarrow 0.2 = \frac{\left(n \times \frac{\pi}{4}\phi^2\right) \times 100}{B.d}$$

$$\Rightarrow \frac{0.2}{100} = \frac{\left(\frac{B}{S}\right) \times \frac{\pi}{4} \times 8^2}{B \times 250}$$ where S is spacing

$$\Rightarrow S = \frac{\frac{\pi}{4} \times 8^2 \times 100}{0.2 \times 250}$$

$$\Rightarrow S = 100.53 \text{ mm}$$

Provide spacing of 100 mm center-to-center.

3.42 (a)

By Rankine Grashoff method,

$$M_{ux} = \frac{r^4}{1+r^4} w_u \times \frac{l_x^2}{8}$$

$$M_{uy} = \frac{1}{1+r^4} w_u \times \frac{l_y^2}{8}$$

$$\text{Now, } \frac{M_{ux}}{M_{uy}} = r^4 \cdot \frac{l_x^2}{l_y^2}$$

$$= \left(\frac{l_y}{l_x}\right)^4 \times \left(\frac{l_x}{l_y}\right)^2 = \left(\frac{l_y}{l_x}\right)^2$$

3.43 Sol.

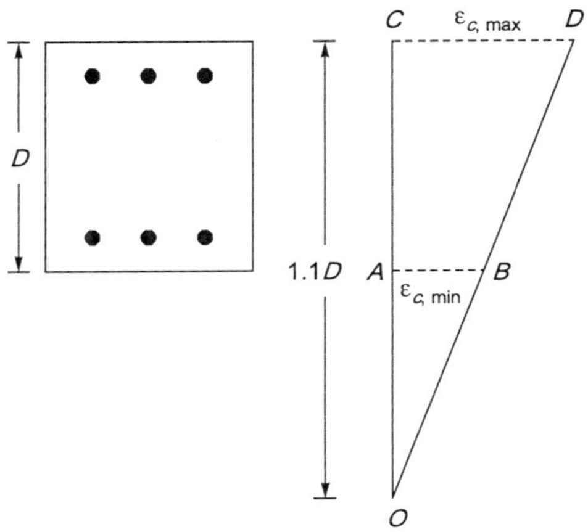

$$\varepsilon_{c,\max} = 0.0035 - 0.75\, \varepsilon_{c,\min}$$

From similar ΔOAB and ΔOCD,

$$\frac{\varepsilon_{c,\min}}{0.1D} = \frac{\varepsilon_{c,\max}}{1.1D}$$

$$\Rightarrow \quad \varepsilon_{c,\min} = \frac{\varepsilon_{c,\max}}{11}$$

$$\therefore \quad \varepsilon_{c,\max} = 0.0035 - 0.75\left(\frac{\varepsilon_{c,\max}}{11}\right)$$

$$\Rightarrow \quad \varepsilon_{c,\max} = 3.28 \times 10^{-3}$$

■■■■

4 Prestressed Concrete

4.1 A prestressed concrete rectangular beam of size 300 mm × 9 00 mm is prestressed with an initial prestressing force of 700 kN at an eccentricity of 350 mm at midspan. Stress at top beam section due to prestress alone, in N/mm^2 is

(a) –3.46 (T) (b) 2.59 (C)
(c) zero (d) 8.64 (C) **[1997 : 1 M]**

4.2 A simply supported prestressed concrete beam is 6 m long and 300 mm wide. Its gross depth is 600 mm. It is prestressed by horizontal cable tendons at a uniform eccentricity of 100 mm. The prestressing tensile force in the cable tendons is 1000 kN. Neglect the self weight of beam. The maximum normal compressive stress in the beam at transfer is (in N/mm^2)

(a) zero (b) 5.55
(c) 11.11 (d) 15.68 **[2004 : 2 M]**

4.3 **IS 1343 : 1980** limits the minimum charac-teristic strength of pre-stressed concrete for post tensioned work and pretension work as

(a) 25 MPa, 30 MPa respectively
(b) 25 MPa, 35 MPa respectively
(c) 30 MPa, 35 MPa respectively
(d) 30 MPa, 40 MPa respectively **[2005 : 1 M]**

4.4 A concrete beam of rectangular cross section of 200 mm × 400 mm is prestressed with a force 400 kN at eccentricity 100 mm. The maximum compressive stress in the concrete is

(a) 12.5 N/mm^2 (b) 7.5 N/mm^2
(c) 5.0 N/mm^2 (d) 2.5 N/mm^2
[2005 : 2 M]

4.5 The percentage loss of prestress due to anchorage slip of 3 mm in a concrete beam of length 30 m which is post-tensioned by a tendon with an initial stress of 1200 N/mm^2 and modulus of elasticity equal to 2.1×10^5 N/mm^2 is

(a) 0.0175 (b) 0.175
(c) 1.75 (d) 17.5 **[2007 : 2 M]**

4.6 A concrete beam of rectangular cross-section of size 120 mm (width) and 200 mm (depth) is prestressed by a straight tendon to an effective force of 150 kN at an eccentricity of 20 mm (below the centroidal axis in the depth direction). The stresses at the top and bottom fibres of the section are

(a) 2.5 N/mm^2 (compression), 10 N/mm^2 (C)
(b) 10 N/mm^2 (tension), 2.5 N/mm^2 (C)
(c) 3.75 N/mm^2 (tension), 3.75 N/mm^2 (C)
(d) 2.75 N/mm^2 (compression), 3.75 N/mm^2 (C)
[2007 : 2 M]

4.7 A pre-tensioned concrete member of section 200 mm × 250 mm contains tendons of area 500 mm^2 at centre of gravity of the section. The prestress in the tendons is 1000 N/mm^2. Assuming modular ratio as 10, the stress (N/mm^2) in concrete is

(a) 11 (b) 9
(c) 7 (d) 5 **[2008 : 2 M]**

4.8 A rectangular concrete beam of width 120 mm and depth 200 mm is prestressed by pretensioning to a force of 150 kN at an eccentricity of 20 mm. The cross-sectional area of the prestressing steel is 187.5 mm^2. Take modulus of elasticity of steel and concrete as 2.1×10^5 MPa and 3.0×10^4 MPa respectively. The percentage loss of stress in the prestressing steel due to elastic deformation of concrete is

(a) 8.75 (b) 6.125
(c) 4.81 (d) 2.19 **[2010 : 2 M]**

4.9 As per Indian standard code of practice for prestressed concrete (**IS 1343 : 1980**) the minimum grades of concrete to be used for post-tensioned and pre-tensioned structural elements are respectively

(a) M20 for both (b) M40 and M30
(c) M15 and M20 (d) M30 and M40
[2010 : 1 M]

4.10 Which one of the following is categorised as a long-term loss of prestresss in a pre-stressed concrete member?

(a) Loss due to elastic shortening
(b) Loss due to friction
(c) Loss due to relaxation of strands
(d) Loss due to anchorage slip [2012 : 1 M]

4.11 A concrete beam pre-stressed with a parabolic tendon is shown in the sketch. The eccentricity of the tendon is measured from the centroid of the cross-section. The applied pre-stressing force at service is 1620 kN. The uniformly distributed load of 45 kN/m includes the self-weight.

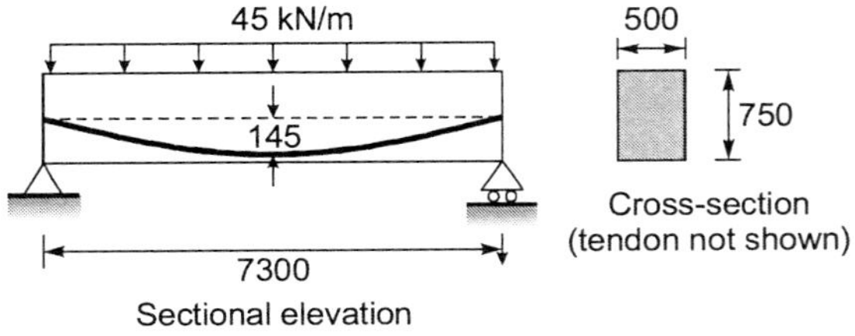

(All dimensions are in mm)

The stress (in N/mm^2) in the bottom fibre at mid-span is
(a) tensile 2.90
(b) compressive 2.90
(c) tensile 4.32
(d) compressive 4.32 [2012 : 2 M]

4.12 A rectangular concrete beam 250 mm wide and 600 mm deep is prestressed by means of 16 wire of high tensile steel wires, each of 7 mm dia. located at 200 mm from the bottom face of the beam at a given section. If the effective prestress in the wires is 700 MPa. What is the maximum sagging BM (in kN-m) due to live load which this section of the beam can withstand without causing tensile stresses at the bottom face of the beam. Neglect the self weight of the beam.
[2013 : 2 M]

4.13 In a pre-stressed concrete beam section shown in the figure, the net loss is 10% and the final pre-stressing force applied at X is 750 kN. The initial fiber stresses (in N/mm^2) at the top and bottom of the beam were:

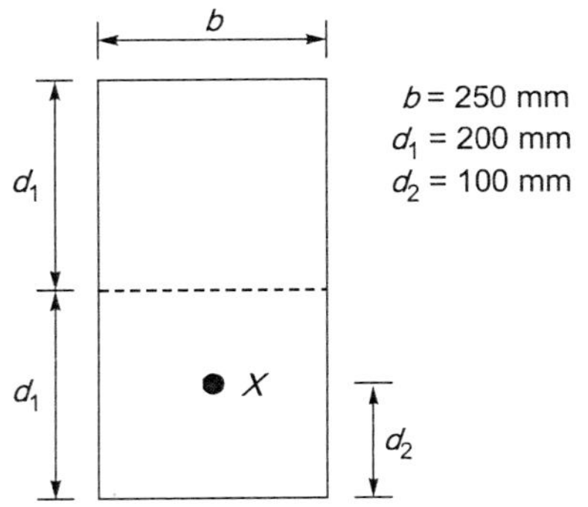

(a) 4.166 and 20.833
(b) –4.166 and –20.833
(c) 4.166 and –20.833
(d) –4.166 and 20.833 [2015 : 2 M, Set-II]

4.14 A pre-tensioned rectangular concrete beam 150 mm wide and 300 depth is prestressed with three straight tendons, each having a cross-sectional area of 50 mm^2, to an initial stress of 1200 N/mm^2. The tendons are located at 100 mm from the soffit of the beam. If the modular ratio is 6, the loss of prestressing force (in kN, up to one decimal place) due to the elastic deformation of concrete only is ________.
[2017 : 2 M, Set-I]

4.15 A simply supported rectangular concrete beam of span 8 m has to be prestressed with a force of 1600 kN. The tendon is of parabolic having zero eccentricity at the supports. The beam has to carry an external uniformly distributed load of intensity 30 kN/m. Neglecting the self-weight of the beam, the maximum dip (in meters, up to two decimal places) of the tendon at the mid-span to balance the external load should be ______.
[2017 : 2 M, Set-II]

4.16 A 6 m long simply-supported beam is prestressed as shown in the figure.

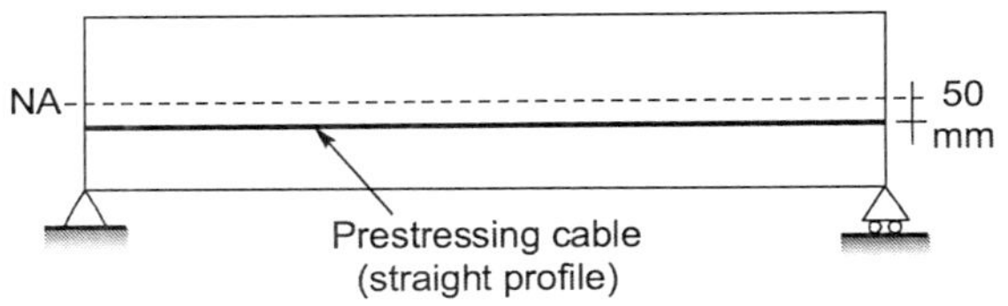

The beam carries a uniformly distributed load of 6 kN/m over its entire span. If the effective flexural rigidity $EI = 2 \times 10^4$ kNm^2 and the effective prestressing force is 200 kN, the net increase in length of the prestressing cable (in mm, up to two decimal places) is ________.
[2018 : 2 M, Set-II]

4.17 A simply supported prismatic concrete beam of rectangular cross-section, having a span of 8 m, is prestressed with an effective prestressing force of 600 kN. The eccentricity of the prestressing tendon is zero at supports and varies linearly to a value of e at the mid-span. In order to balance an external concentrated load of 12 kN applied at the mid-span, the required value of e (in mm, round off to the nearest integer) of the tendon, is _______.
[2020 : 2 M, Set-I]

4.18 A concrete beam of span 15 m, 150 mm wide and 350 mm deep is prestressed with a parabolic cable as shown in the figure (not drawn to the scale). Coefficient of friction for the cable is 0.35, and coefficient of wave effect is 0.0015 per metre.

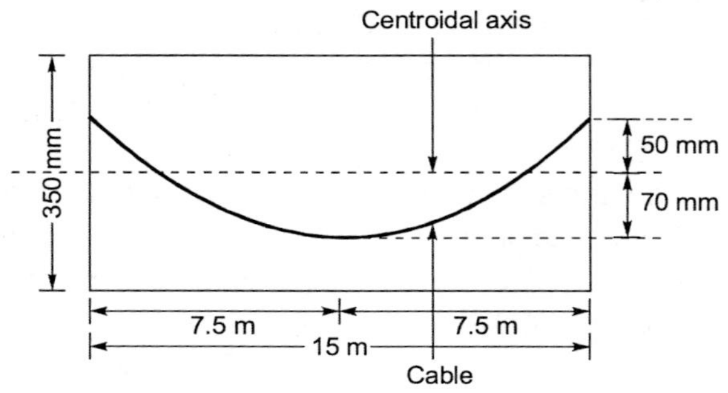

If the cable is tensioned from one end only, the percentage loss (round off to one decimal place) in the cable force due to friction, is ________.

[2020 : 2 M, Set-II]

4.19 A prismatic cantilever prestressed concrete beam of span length, $L = 1.5$ m has one straight tendon placed in the cross-section as shown in the following figure (not to scale). The total prestressed force of 50 kN in the tendon is applied at $d_c = 50$ mm from the top in the cross-section of width, $b = 200$ mm and depth, $d = 300$ mm.

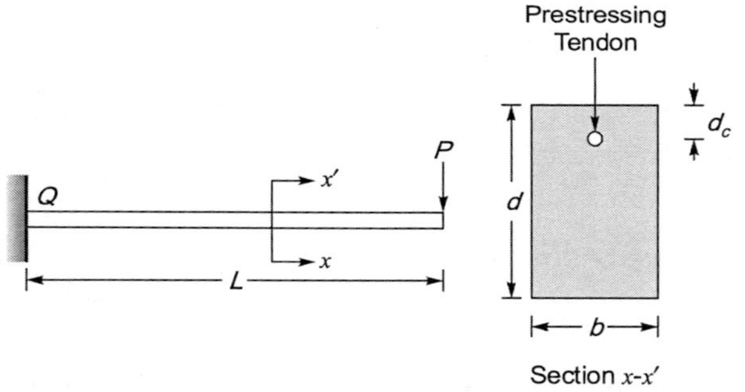

If the concentrated load, P = 5 kN, the resultant stress (in MPa, in integer) experienced at point 'Q' will be________ [2021 : 2 M, Set-I]

4.20 A post-tensioned concrete member of span 15 m and cross-section of 450 mm × 450 mm is prestressed with three steel tendons, each of cross-sectional area 200 mm^2. The tendons are tensioned one after another to a stress of 1500 MPa. All the tendons are straight and located at 125 mm from the bottom of the member. Assume the prestress to be the same in all tendons and the modular ratio to be 6. The average loss of prestress, due to elastic deformation of concrete, considering all three tendons is

(a) 28.32 MPa (b) 7.08 MPa
(c) 14.16 MPa (d) 42.48 MPa

[2022 : 2 M, Set-II]

4.21 An inverted T-shaped concrete beam (B1) in the figure, with centroidal axis X-X is subjected to an effective prestressing force of 1000 kN acting at the bottom kern point of the beam cross-section. Also consider an identical concrete beam (B2) with the same grade of concrete but without any prestressing force.

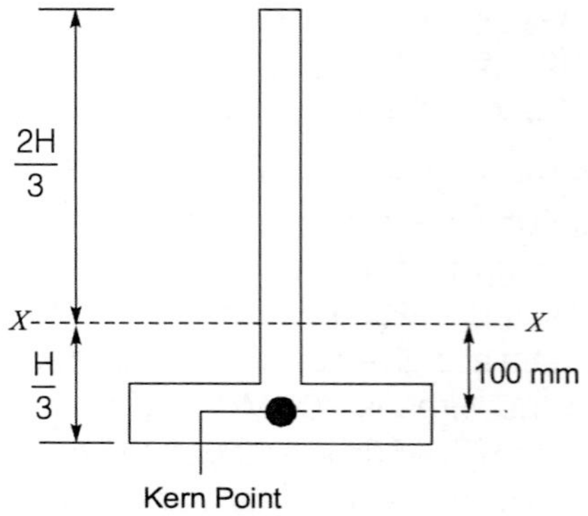

(Figure NOT to scale)

The additional cracking moment (in kNm) that can be carried by beam B1 in comparison to beam B2 is ______. (rounded off to the nearest integer).

[2024 : 2 M, Set-I]

Answers Prestressed Concrete

4.1	(a)	4.2	(c)	4.3	(d)	4.4	(a)	4.5	(c)	4.6	(a)	4.7	(b)
4.8	(b)	4.9	(d)	4.10	(c)	4.11	(b)	4.12	(86.205)	4.13	(d)	4.14	(4.8)
4.15	(0.15)	4.16	(0.12)	4.17	(40)	4.18	(4.49)	4.19	(0.56)	4.20	(c)	4.21	(300)

Explanations Prestressed Concrete

4.1 (a)

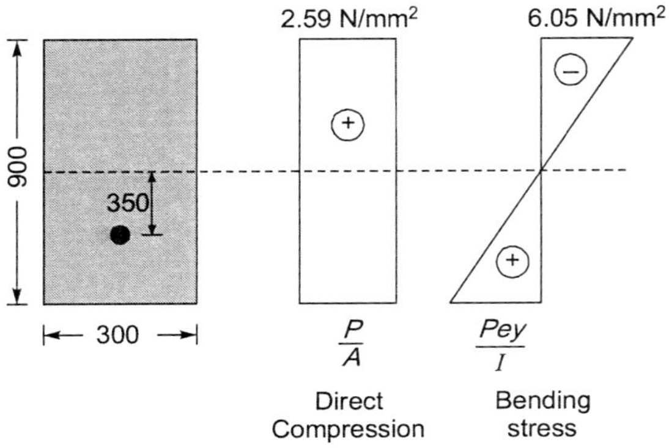

$$\frac{P}{A} = \frac{700 \times 10^3}{300 \times 900} \text{N/mm}^2 = 2.59 \text{ N/mm}^2$$

$$\frac{Pey}{I} = \frac{700 \times 10^3 \times 350 \times 6}{300 \times 900^2} = 6.05 \text{ N/mm}^2$$

∴ Stress at top fibre due to prestress force

$$= \frac{P}{A} - \frac{Pey}{I} = 2.59 - 6.05$$

$$= -3.46 \text{ N/mm}^2 \text{ (tensile)}$$

4.2 (c)

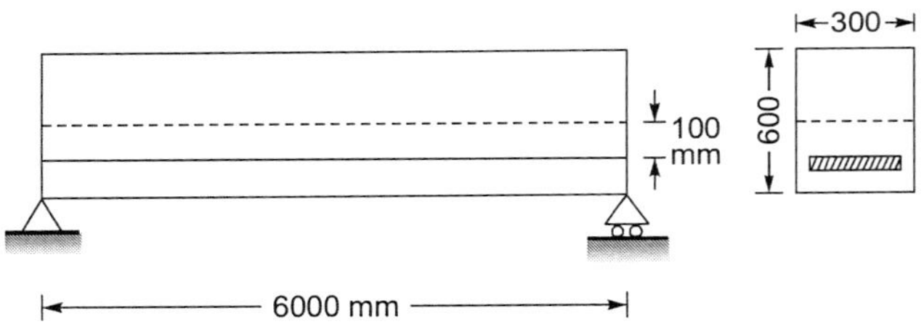

$$P = 1000 \text{ kN}$$

$$Z = \frac{bd^2}{6} = \frac{300 \times 600^2}{6}$$

$$= 18 \times 10^6 \text{ mm}^3$$

$$e = 100 \text{ mm}$$

$$A = 300 \times 600 = 180000 \text{ mm}^2$$

Maximum compressive stress,

$$\sigma = \frac{P}{A} + \frac{Pe}{Z}$$

$$= \frac{1000 \times 10^3}{180000} + \frac{1000 \times 10^3 \times 100}{18 \times 10^6}$$

$$= 11.11 \text{ N/mm}^2$$

4.3 (d)

4.4 (a)

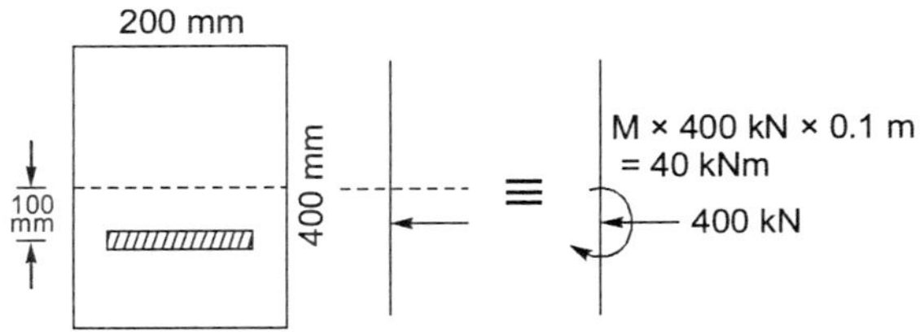

$$\sigma_{max} = \frac{P}{A} + \frac{Pe}{Z}$$

$$= \frac{400 \times 10^3}{200 \times 400} + \frac{400 \times 10^3 \times 100 \times 6}{200 \times (400)^2}$$

$$= 5 + 7.5 = 12.5 \text{ N/mm}^2$$

4.5 (c)

Loss of prestress due to anchorage slip

$$= \frac{\Delta L}{L} \times E_s = \frac{3 \times 10^{-3}}{30} \times 2.1 \times 10^5$$

$$= 21 \text{ N/mm}^2$$

% loss of prestress

$$= \frac{21}{1200} \times 100 = 1.75 \%$$

4.6 (a)

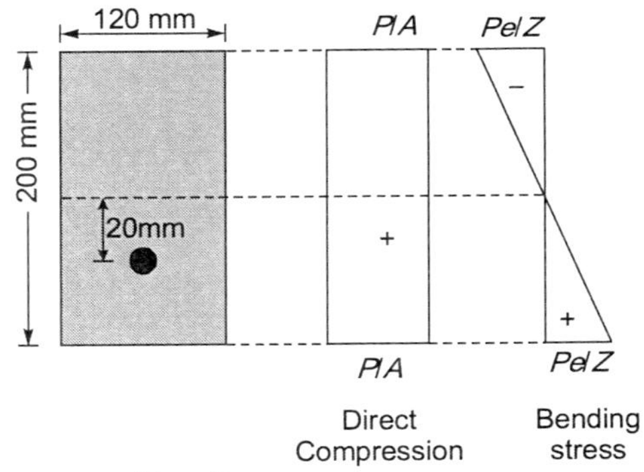

$$\sigma_{top} = \frac{P}{A} - \frac{Pe}{Z}$$

$$= \frac{150 \times 10^3}{120 \times 200} - \frac{150 \times 10^3 \times 20 \times 6}{120 \times (200)^2}$$

$$= 6.25 - 3.75$$

$$= 2.5 \text{ N/mm}^2 \text{ (compression)}$$

$$\sigma_{bottom} = \frac{P}{A} + \frac{Pe}{Z} = 6.25 + 3.75$$

$$= 10 \text{ N/mm}^2 \text{ (compression)}$$

4.7 (b)

Given

Modular ratio, m = 10

Prestress in tendon = 1000 N/mm^2

Area of tendon = 500 mm^2

$\therefore$Prestressing force = 1000 × 500 N = 500 kN

There will be loss in prestress due to elastic shortening of concrete = mf_c

f_c = Stress in concrete at the level of steel

$$f_c = \frac{P}{A}+\frac{Me}{I}=\frac{P}{A}+\frac{Pe^2}{I}=\frac{5\times10^5}{200\times250}+0=10\text{ N/mm}^2$$

$\therefore$ Loss = mf_c = 10 × 10 = 100 N/mm^2

$$\text{Now stress in concrete} = \frac{P'}{A}=\frac{900\times500}{200\times250}$$

$$= 9\text{ N/mm}^2$$

4.8 (b)

Stress in concrete at the level of tendon,

$$f_c = \frac{P}{A}+\frac{Pe^2}{I}$$

$$= \frac{150\times10^3}{120\times200}+\frac{150\times20^2\times12\times10^3}{120\times200^3}$$

$$= 7\text{ MPa}$$

Loss of prestress due to elastic deformation,

$$\Delta f_s = \frac{E_s}{E_c}\times f_c = \frac{2.1\times10^5}{3.0\times10^4}\times7 = 49\text{ MPa}$$

Total stress in steel,

$$f_s = \frac{P}{A_s} = \frac{150\times10^3}{187.5} = 800\text{ MPa}$$

$\therefore$Percentage loss of stress

$$= \frac{\Delta f_s}{f_s}\times100 = \frac{49}{800}\times100 = 6.125\%$$

4.9 (d)

4.10 (c)

Loss due to relaxation, shrinkage of concrete and due to creep are long term losses.

4.11 (b)

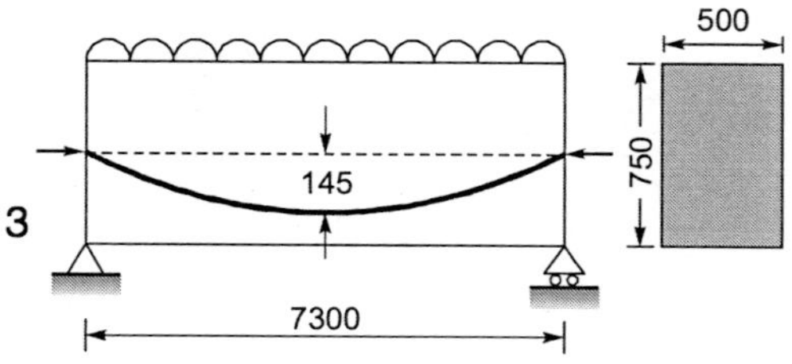

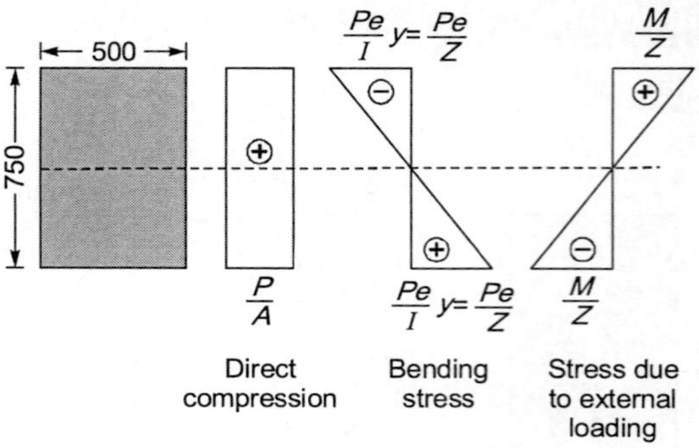

$$\text{Total stress at Bottom} = \frac{P}{A}+\frac{Pe}{I}y-\frac{M}{Z}$$

$$\text{Here, } Z = \frac{bd^2}{6} = \frac{500\times750^2}{6}$$

$$= 46.875\times10^6\text{ mm}^3.$$

e = 145 mm; P = 1620 kN

$$M = \frac{wl^2}{8} = \frac{45\times7.3^2}{8} = 299.76\text{ kNm}$$

$$\therefore \frac{P}{A} = \frac{1620\times10^3}{500\times750} = 4.32\text{ N/mm}^2$$

$$\frac{Pe}{Z} = \frac{1620\times10^3\times145}{46.875\times10^6} = 5.01\text{ N/mm}^2$$

$$\frac{M}{Z} = \frac{299.76\times10^6}{46.875\times10^6} = 6.39\text{ N/mm}^2$$

$\therefore$ Total stress at bottom = 4.32 + 5.01 − 6.39

= 2.94 N/mm^2. **(Comp.)**

4.12 Sol.

Since the tensile stress at bottom face of the beam is zero.

$$\frac{P}{A}+\frac{Pe}{I}y-\frac{M}{I}y = 0$$

$$P = 700\times\frac{\pi}{4}(7)^2\times16 = 431.026\times10^3\text{ N}$$

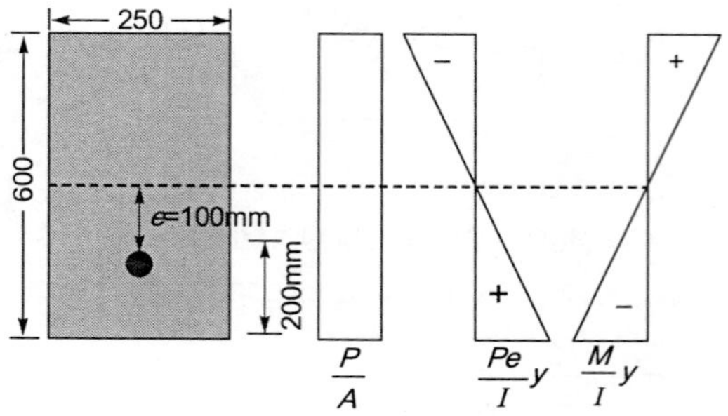

Since the prestressing force is located at 200 mm from the bottom face of the beam.

$\therefore$ Eccentricity = 300 − 200 = 100 mm

$$\frac{431.03\times10^3}{250\times600}+\frac{431.03\times10^3\times100}{250\times600^2}\times6$$

$$= \frac{M \times 6}{250 \times 600^2}$$

$\Rightarrow 2.87 + 2.87 = 6.66 \times 10^{-8}\, M$

$\Rightarrow M = 86.205$ kNm

4.13 (d)

Given, net loss = 10%

Final prestressing force after loss = 750 kN

Hence, initial prestressing force

$$= \frac{750}{0.9} = 833.33 \text{ kN}$$

Initial stresses at top and bottom

$$= \frac{P}{A} \pm \frac{Pe}{Z} = \frac{833.33 \times 10^3}{250 \times 400} \pm \frac{833.33 \times 10^3 \times 100 \times 6}{250 \times 400^2}$$

$= 8.33 \pm 12.50$

Stress at top = 8.33 – 12.50 = –4.17 (tensile)

Stress at bottom = 8.33 + 12.50 = 20.83 (Comp.)

4.14 Sol.

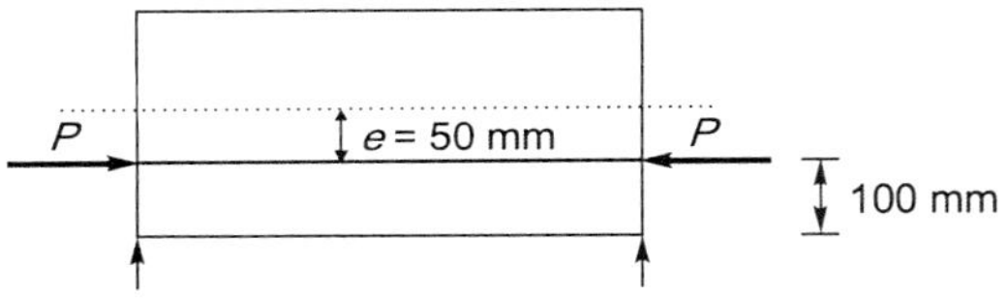

Prestressing force,

$P = 3 \times 50 \times 1200 = 180000$ N.

$$e = \left(\frac{D}{2} - 100\right) = \left(\frac{300}{2} - 100\right) = 50 \text{ mm}$$

Stress in concrete at the location of steel

$$= \frac{P}{A} + \frac{P \cdot e}{I} e$$

$$= \frac{180000}{150 \times 300} + \frac{180000 \times 50^2}{\left(\frac{150 \times 300^3}{12}\right)}$$

$= 4.0 + 1.33 = 5.333$

Loss of stress

$= m \times f_c = 6 \times 5.3333 = 32$ N/mm²

Loss of prestressing force

$$= \frac{3 \times 50 \times 32}{1000} \text{ kN} = 4.8 \text{ kN}$$

4.15 Sol.

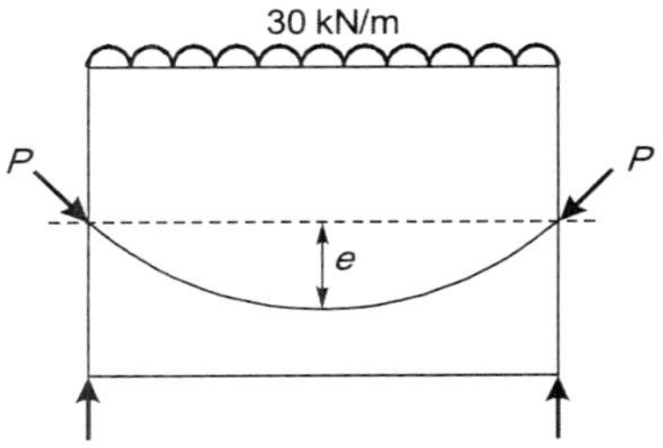

We know, $BM_{max} = \frac{wl^2}{8} = \frac{30 \times 8^2}{8} = 240$ kNm

Note: Tendon profile shall follow the shape of bending moment diagram.

Also $BM_{max} = Pe$

$$\therefore \quad e = \left(\frac{BM}{P}\right) = \frac{240 \times 10^6 \text{ Nmm}}{1600 \times 10^3}$$

$e = 150$ mm

$e = 0.15$ m

4.16 Sol.

Span of PSC beam = 6 m

$EI = 2 \times 10^4$ kNm² $= 2 \times 10^{13}$ N-mm²

$P = 200$ kN

Total UDL = 6 kN/m

eccentricity = e = 50 mm

(a) Slope of beam due to P-force

$$\theta_1 = \frac{P.e.L}{2EI}$$

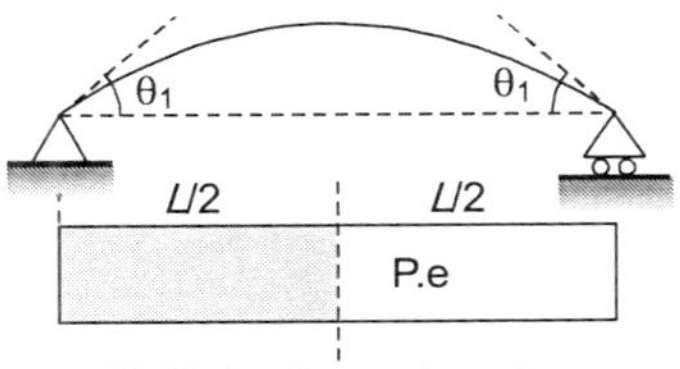

BMD due to prestressing

$$= \frac{200 \times 10^3 \times 50 \times 6000}{2 \times 2 \times 10^{13}} = 1.5 \times 10^{-3} \text{ (upward)}$$

(b) Slope of beam due to UDL

$$\theta_2 = \frac{wL^3}{24EI}$$

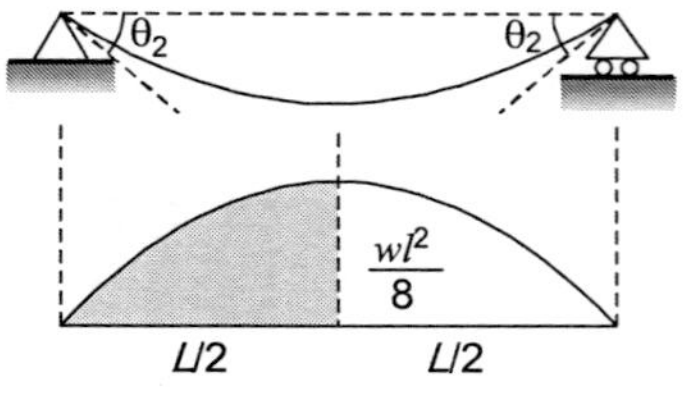

BMD due to loading

$$= \frac{6\times(6000)^3}{24\times2\times10^{13}} = (+)\ 2.7\times10^{-3}\ \text{(downward)}$$

(c) Net slope of beam

$$\theta = \theta_1 + \theta_2$$

$$= (-)1.5\times10^{-3} + (+)2.7\times10^{-3} = 1.2\times10^{-3}$$

(d)

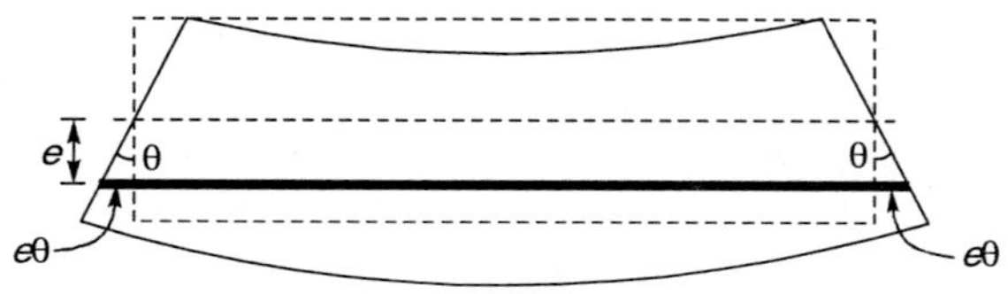

Total net increase in length

$= 2e\theta = 2\times50\times1.2\times10^{-3} = 0.12$ mm

4.17 Sol.

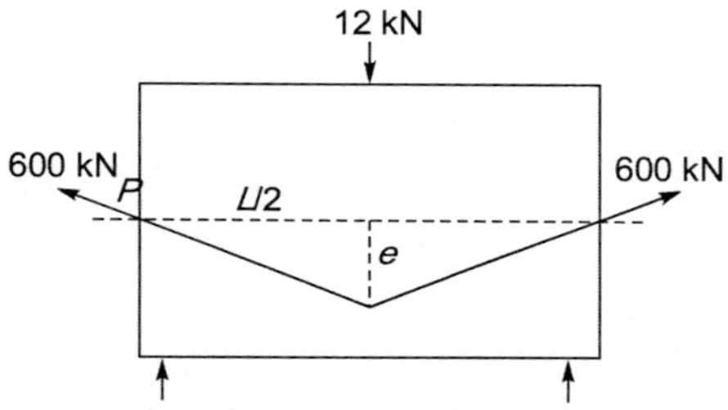

Prestressing force, $P = 600$ kN

Simply supported span $= L = 8$ m

To support a point load applied at mid span,

$$W = 12\text{ kN}$$

Balancing load = Point load

$$2P\sin\theta = W$$

$$2P\left(\frac{e}{L/2}\right) = 2$$

$$\frac{2Pe\times2}{L} = W,\quad \frac{4Pe}{L} = W$$

$$e = \frac{WL}{4P} = \frac{12000\text{ N}\times8000\text{ mm}}{4\times600\times1000\text{ N}} = 40\text{ mm}$$

4.18 Sol.

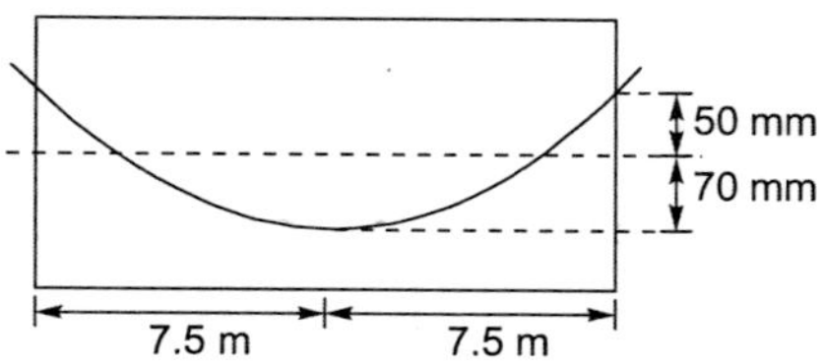

Jacking from one end

$$x = L = 15\text{ m}$$

Wobble correction factor, $K = 0.0015$

Coefficient of friction $= 0.35 = \mu$

Prestressing force, P = Not given Prestressing stress, p_0 = Unknown

Change of gradient,

$$\alpha = \tan\alpha = \frac{8h}{L} = \frac{8\times120}{15000} = 0.064$$

Loss due to friction $= p_0\,(kx + \mu\alpha)$

% loss of stress in steel due to friction

$$= \frac{p_0(Kx+\mu\alpha)}{p_0}\times100$$

$= (0.0015\times15 + 0.35\times0.064)\times100 = 4.49\%$

4.19 Sol.

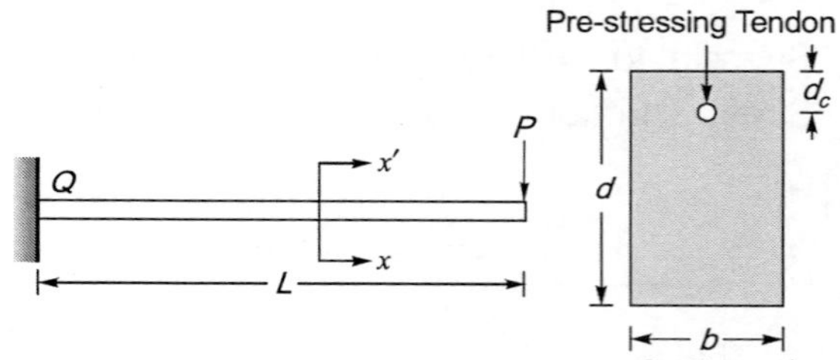

$$e = \frac{D}{2} - 50 = \frac{300}{2} - 50 = 100\text{ mm}$$

$$DL = 0.2\times0.3\times1.0\times25 = 1.50\text{ kN/m}$$

$$P = 50\text{ kN} = 50000\text{ N}$$

$$W = 5\text{ kN}$$

$$\text{Maximum BM} = \frac{wL^2}{2} + WL$$

$$= \frac{1.5\times1.5^2}{2} + 5\times1.50 = 9.1875\text{ kNm}$$

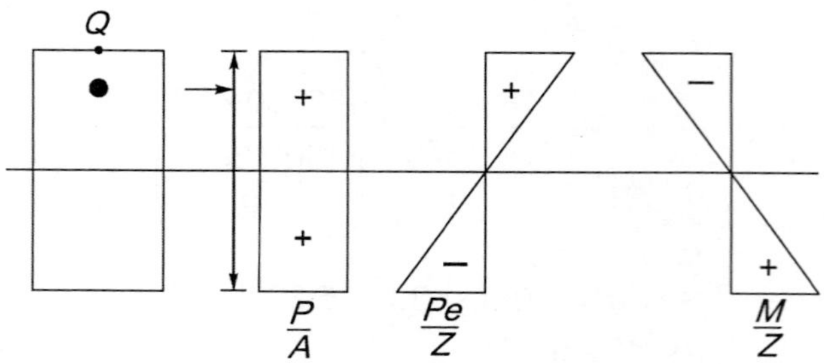

$$\frac{P}{A} = \frac{50000}{200\times300} = 0.833\text{ N/mm}^2$$

$$\frac{Pe}{Z} = \frac{50000\times100}{200\times\frac{300^2}{6}} = 1.67\text{ N/mm}^2$$

$$\frac{M}{Z} = \frac{9.1875\times10^6}{200\times\frac{300^2}{6}} = 3.0625\ \text{N/mm}^2$$

Stress at Q,

$$\text{Stress at } Q = \frac{P}{A}+\frac{Pe}{Z}-\frac{M}{Z}$$

$= 0.833 + 1.67 - 3.0625 = -0.56\ \text{N/mm}^2$ (Tensile)

4.20 (c)

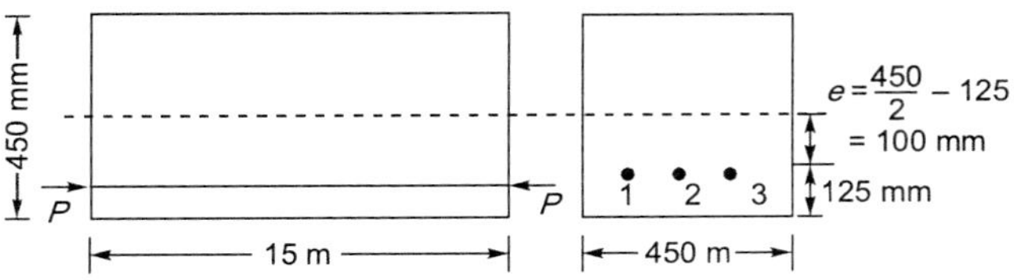

Given: Area of steel of each bar = 200 mm²

Initial prestress = 1500 N/mm²

∴ Prestress force in each tendon,

$$P = p_o A_s$$

$$\Rightarrow\ P = 1500\times\frac{200}{1000} = 300\ \text{kN in each wire}$$

As per given condition,

Loss of stress in 3rd wire = 0

Loss in wire-2, is due to P_3

Loss in wire-1, is due to P_2 and P_3 both (Neglecting effect of self weight, because self weight will result in gain of stress, not loss). Now, Loss of stress due to elastic shortening in wire-2,

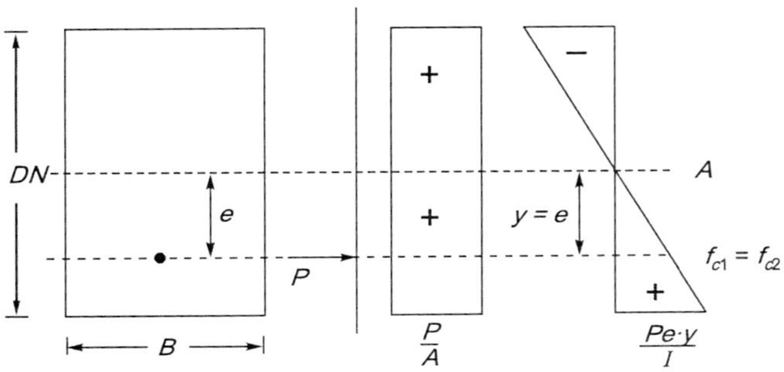

$$\therefore\quad f_{c1} = f_{c2} = \frac{P_3}{A}+\frac{P_3 e.e}{I} = \frac{P_3}{A}+\frac{P_3e^2}{I}$$

$$\Rightarrow\quad f_{c1} = f_{c2}$$

$$= \frac{300\times1000}{450\times450}+\frac{300\times1000\times100^2}{\left(450\times\frac{450^3}{12}\right)}$$

$$\Rightarrow\quad f_{c1} = f_{c2} = 2.36\ \text{N/mm}^2$$

∴Loss of stress in wire-2

$= m(f_c)_{avg.} = 6\times2.36 = 14.16\ \text{N/mm}^2$

Loss of stress due to elastic shortening in wire-1, due to $P_2 + P_3 = 2P = 600$ kN

$$\therefore\quad f_{c_1}{'} = f_{c_2}{'} = \frac{P_2+P_3}{A}+\frac{P_2+P_3}{I}(e^2)$$

$$\Rightarrow\quad f_{c_1}{'} = f_{c_2}{'} = \frac{2P}{A}+\frac{2P}{I}(e^2) = 2\left(\frac{P}{A}+\frac{P}{I}e^2\right)$$

$$\Rightarrow\quad f_{c_1}{'} = f_{c_2}{'} = 2\times(f_{c1}\text{ for wire no. 2})$$

$$\Rightarrow\quad f_{c_1}{'} = f_{c_2}{'} = 2\times2.36 = 4.72\ \text{N/mm}^2$$

∴Loss in stress in wire-1

$= m\,(f_c)_{avg.} = 6\times4.72 = 28.32\ \text{N/mm}^2$

Thus, average loss of stress in all three wires

$$= \frac{0+14.16+28.32}{3} = 14.16\ \text{N/mm}^2$$

4.21 Sol.

At bottom fibre,

$$\frac{P}{A}+\frac{Pey_b}{I}-\frac{M_{cr}}{I}y_b = 0 \quad\text{...(i)}$$

where P = Prestressing force

e = Eccentricity of P-force

M_{cr} = cracking moment for beam B1

y_t and y_b are distance of top and bottom fibre from neutral axis respectively,

At top fibre

$$\frac{P}{A}-\frac{Pey_t}{I} = 0 \text{ [P-force is at kern point]}$$

$$\Rightarrow\frac{I}{A} = ey_t \quad\text{...(ii)}$$

Putting (ii) in (i)

$$\frac{M_{cr}}{I}y_b = \frac{P}{A}+\frac{Pey_b}{I}$$

$$\Rightarrow M_{cr} = \frac{PI}{Ay_b}+Pe$$

$$= \frac{Pey_t}{y_b}+Pe$$

$$= \frac{Pe(2H/3)}{(H/3)}+Pe$$

$$= 3Pe = 3\times1000\times\frac{100}{1000} = 300\ \text{kNm}$$

■■■■

Design of Steel Structures

UNIT IV

CONTENTS

Design of Steel Structures

Syllabus

Working stress and Limit state design concepts; Design of tension and compression members, beams and beam- columns, column bases; Connections - simple and eccentric, beam-column connections, plate girders and trusses; Concept of plastic analysis -beams and frames.

Analysis of Previous GATE Papers

Exam Year	1 Mark Ques.	2 Marks Ques.	Total Marks
1995	5	–	5
1996	5	–	5
1997	4	–	4
1998	2	1	4
1999	5	–	5
2000	2	–	2
2001	1	1	3
2002	2	3	8
2003	2	4	10
2004	2	5	12
2005	1	4	9
2006	–	3	6
2007	–	3	6
2008	1	2	5
2009	2	2	6
2010	–	2	4
2011	1	2	5
2012	1	1	3
2013	1	2	5
2014 Set-1	2	1	4
2014 Set-2	–	1	2

Exam Year	1 Mark Ques.	2 Marks Ques.	Total Marks
2015 Set-1	–	2	4
2015 Set-2	2	1	4
2016 Set-1	1	2	5
2016 Set-2	–	2	4
2017 Set-1	–	1	2
2017 Set-2	–	1	2
2018 Set-1	2	1	4
2018 Set-2	1	2	5
2019 Set-1	1	2	5
2019 Set-2	1	2	5
2020 Set-1	–	1	2
2020 Set-2	1	1	3
2021 Set-1	–	1	2
2021 Set-2	–	1	2
2022 Set-1	–	1	2
2022 Set-2	–	1	2
2023 Set-1	1	1	3
2023 Set-2	–	1	2
2024 Set-1	–	–	0
2024 Set-2	2	1	4

1 Structural Fasteners

1.1 As per **IS 800 : 1984**, the minimum pitch of rivets in a row is recommended as the diameter of the rivet times
(a) 2.0
(b) 2.5
(c) 3.0
(d) 4.0
[1995 : 1 M]

1.2 Generally, fatigue life of welded steel structure to fatigue life of riveted steel structure ratio is
(a) smaller than 1
(b) equal to 1
(c) greater than 1
(d) greater than 2.1
[1996 : 1 M]

1.3 Maximum size of fillet weld for a plate of square edge is
(a) 1.5 mm less than the thickness of the plate
(b) one half of the thickness of the plate
(c) thickness of the plate itself
(d) 1.5 mm more than the thickness of the plate
[1997 : 1 M]

1.4 Factor of safety adopted by **IS 800 : 1984** while arriving at the permissible stress in axial compression is
(a) 2.00
(b) 1.00
(c) 1.67
(d) 1.50
[1997 : 1 M]

1.5 Two steel plates each of width 150 mm and thickness 10 mm are connected with three 20 mm diameter rivets placed in a zig-zag pattern. The pitch of the rivets is 75 mm and gauge is 60 mm. If the allowable tensile stress is 150 MPa, the maximum tensile force that the joint can withstand is
(a) 195.66 kN
(b) 195.00 kN
(c) 192.75 kN
(d) 225.00 kN
[1999 : 1 M]

1.6 Identify the most effective butt joint (with double cover plates) for a plate in tension from the patterns (plan view) shown below, each comprising 6 identical bolts with the same pitch and gauge:

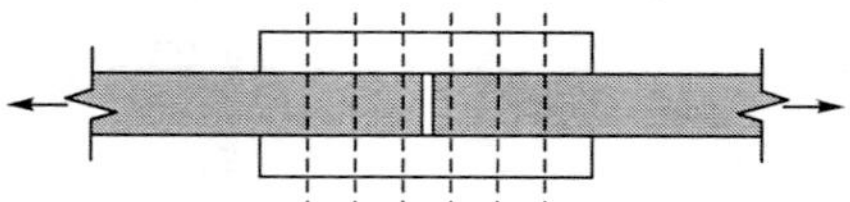

Common Elevation
(all plates have same thickness)

(a) 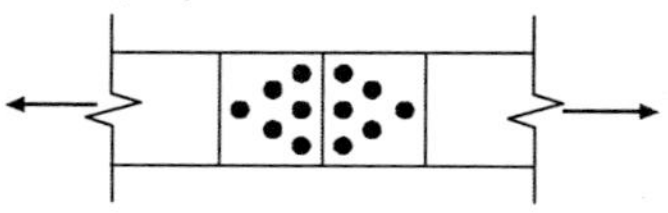

(b) 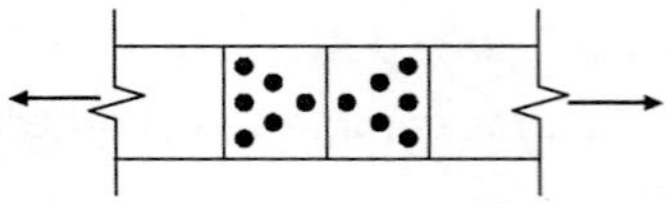

(c) 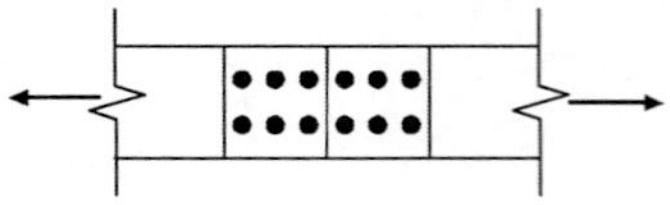

(d) 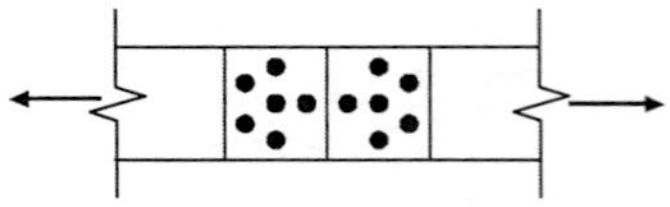

[2001 : 1 M]

1.7 ISA 100 × 100 × 10 mm (cross sectional area = 1908 mm^2) is welded along *A* and *B* as shown in figure, such that the length of the weld along *A* and *B* are l_1 and l_2, respectively. Which of the following is a possibly acceptable combination of l_1 and l_2.

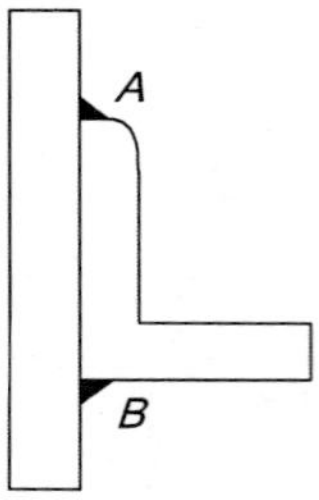

(a) l_1 = 60 mm and l_2 = 150 mm
(b) l_1 = 150 mm and l_2 = 60 mm
(c) l_1 = 150 mm and l_2 = 150 mm
(d) Any of the above depending on the size of the weld
[2002 : 2 M]

1.8 Rivet value is defined as
(a) lesser of the bearing strength of rivet and the shearing strength of the rivet
(b) lesser of the bearing strength of rivet and the tearing strength of thinner plate
(c) greater of the bearing strength of rivet and the shearing of the rivet
(d) lesser of the shearing strength of the rivet and the tearing strength of thinner plate
[2004 : 1 M]

1.9 A moment M of magnitude 50 kN-m is transmitted to a column flange through a bracket by using four 20 mm diameter rivets as shown in the figure:

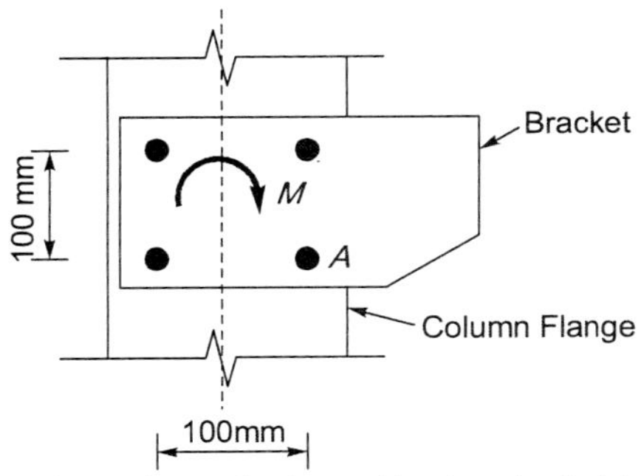

The shear force induced in the rivet *A* is

(a) 250 kN (b) 176.8 kN
(c) 125 kN (d) 88.4 kN **[2004 : 2 M]**

1.10 A fillet-welded joint of 6 mm size is shown in the figure. The welded surfaces meet at 60-90 degree and permissible stress in the fillet weld is 108 MPa. The safe load that can be transmitted by the joint is

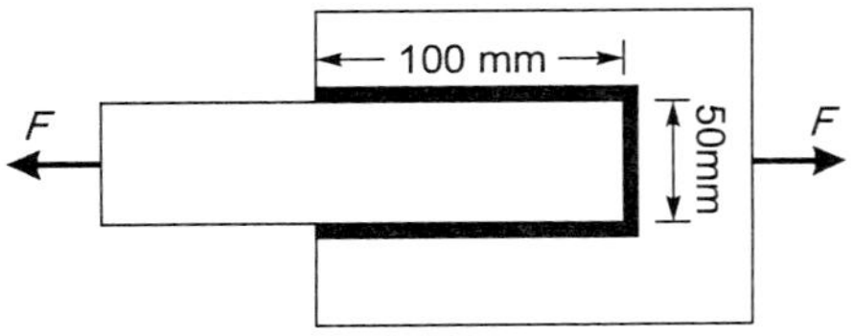

(a) 162.7 kN (b) 151.6 kN
(c) 113.4 kN (d) 109.5 kN

[2005 : 2 M]

1.11 Which one of the following is NOT correct for steel sections as per **IS 800 : 1984**?

(a) The maximum bending stress in tension or in compression in extreme fibre calculated on the effective section of a beam shall not exceed 0.66 f_y.
(b) The bearing stress in any part of a beam when calculated on the net area shall not exceed 0.75 f_y.
(c) The direct stress in compression on the gross sectional area of axially loaded compression member shall not exceed 0.6 f_y.
(d) None of the above [2005 : 2 M]

1.12 A bracket connection is made with four bolts of 10 mm diameter and supports a load of 10 kN at an eccentricity of 100 mm. The maximum force to be resisted by any bolt will be

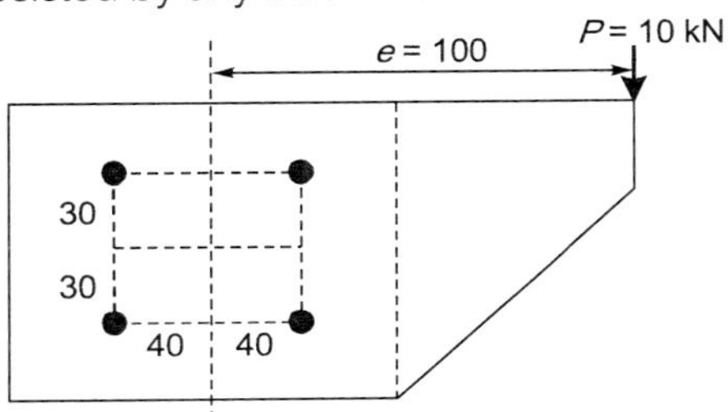

(all distance are in mm)

(a) 5 kN (b) 6.5 kN
(c) 6.8 kN (d) 7.16 kN **[2007 : 2 M]**

1.13 Rivets and bolts subjected to both shear stress ($\tau_{vf,\ cal}$) and axial tensile stress ($\sigma_{tf,\ cal}$) shall be so proportioned that the stresses do not exceed the respective allowable stresses τ_{vf} and σ_{tf} and the value of $\left(\dfrac{\tau_{vf,cal}}{\tau_{vf}} + \dfrac{\sigma_{tf,cal}}{\sigma_{tf}}\right)$ does not exceed

(a) 1.0 (b) 1.2
(c) 1.4 (d) 1.8 [2008 : 2 M]

1.14 A 12 mm thick plate is connected to two 8 mm thick plates, on either side through a 16 mm diameter power driven field rivet as shown in the figure below. Assuming permissible shear stress as 90 MPa and permissible bearing stress as 270 MPa in the rivet, the rivet value of the joint is

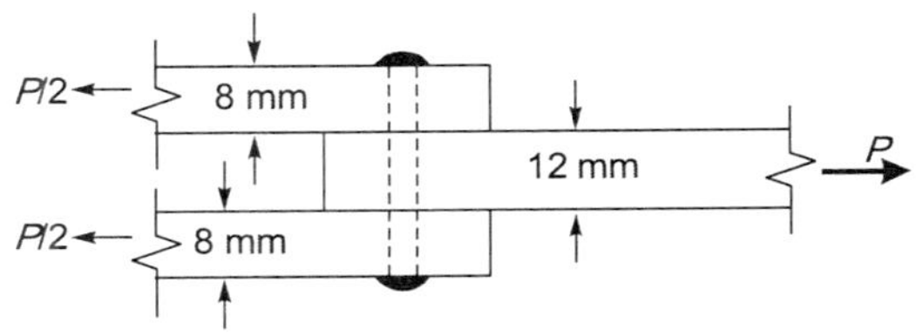

(a) 56.70 kN (b) 43.29 kN
(c) 36.19 kN (d) 21.65 kN

[2009 : 2 M]

1.15 A double cover butt riveted joint is used to connect two flat plates of 200 mm width and 14 mm thickness as shown in the figure. There are twelve power driven rivets of 20 mm diameter at a pitch of 50 mm in both directions on either side of the plate. Two cover plates of 10 mm thickness are used. The capacity of the joint in tension considering bearing and shear ONLY, with permissible bearing and shear stresses as 300 MPa and 100 MPa respectively is

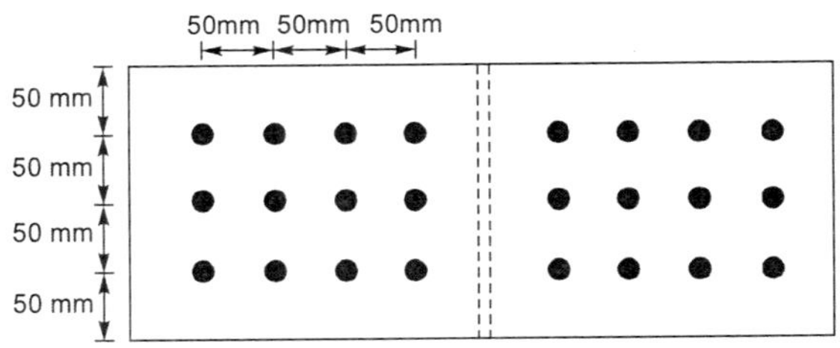

(a) 1083.6 kN (b) 871.32 kN
(c) 541.8 kN (d) 433.7 kN

[2010 : 2 M]

1.16 Two plates, subjected to direct tension, each of 10 mm thickness and having widths of 100 mm and 175 mm, respectively are to be fillet welded with an overlap of 200 mm. Given that the permissible weld stress is 110 MPa and the permissible stress in steel is 150 MPa, the length of the weld required using the maximum permissible weld size as per **IS 800 : 1984** is

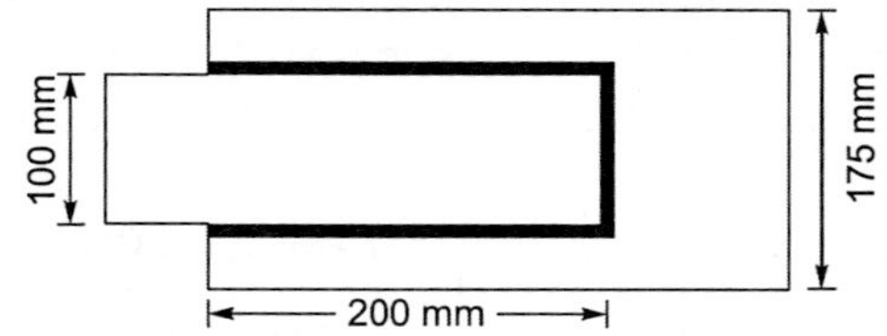

(a) 245. 3 mm (b) 229.2 mm
(c) 205.5 mm (d) 194.8 mm

[2010 : 2 M]

1.17 For the fillet weld of size '*s*' shown in the adjoining figure the effective throat thickness is

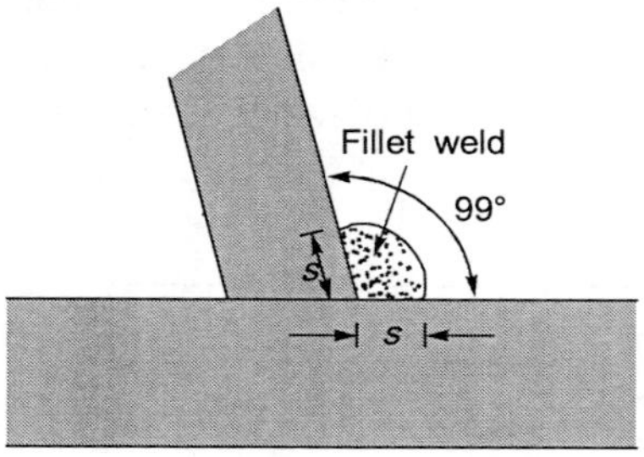

(a) 0.61s (b) 0.65s
(c) 0.70s (d) 0.75s [2011 : 1 M]

1.18 In a steel plate with bolted connection, the rupture of the net section is a mode of failure under
(a) tension (b) compression
(c) flexure (d) shear [2012 : 1 M]

1.19 Two plates are connected by fillet welds of size 10 mm and subjected to tension, as shown in the sketch. The thickness of each plate is 12 mm. The yield stress and the ultimate tensile stress of steel are 250 MP and 410 MPa respectively. The welding is done in the workshop (γ_{mm} = 1.25). As per the Limit State Method of **IS 800:2007**, the minimum length (rounded off to the nearest higher multiple of 5 mm) of each weld to transmit a force *P* equal to

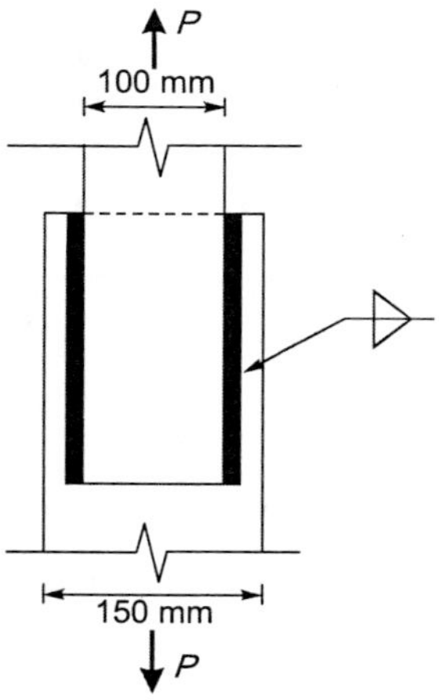

(a) 100 mm (b) 105 mm
(c) 110 mm (d) 115 mm

[2012 : 2 M]

1.20 The tension and shear force (both in kN) in each bolt of the joint, as shown below, respectively are

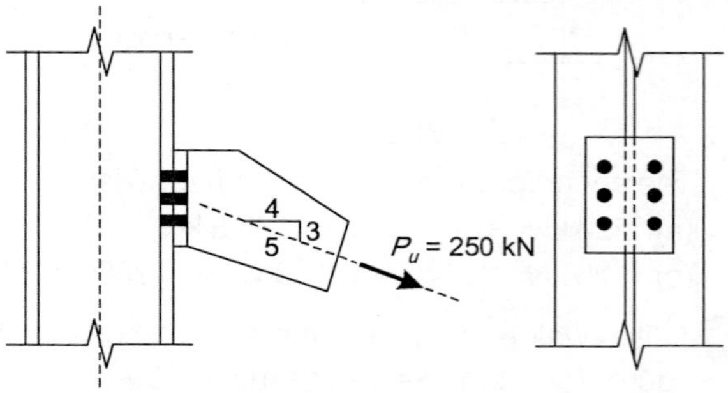

(a) 30.33 and 20.00 (b) 30.33 and 25.00
(c) 33.33 and 20.00 (d) 33.33 and 25.00

[2014 : 2 M, Set-I]

1.21 A bracket plate connected to a column flange transmits a load of 100 kN as shown in the following figure. The maximum force for which the bolts should be designed is _________ kN.

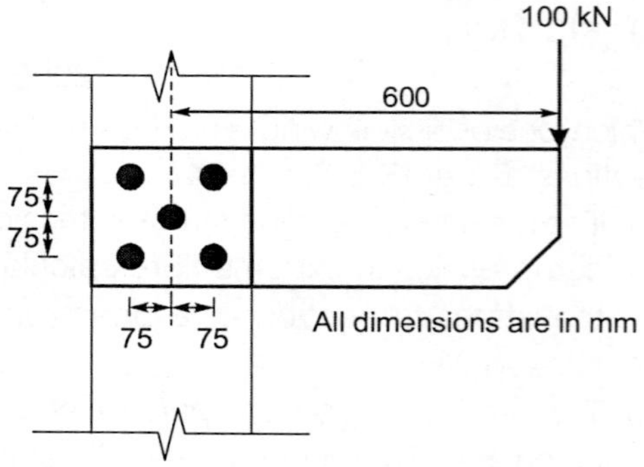

[2015 : 2 M, Set-I]

1.22 Prying forces are
(a) shearing forces on the bolts because of the joints
(b) tensile forces due to the flexibility of connected parts
(c) bending forces on the bolts because of the joints
(d) forces due the friction between connected parts

[2015 : 1 M, Set-II]

1.23 Two plates are connected by fillet welds of size 10 mm and subjected to tension, as shown in the figure. The thickness of each plate is 12 mm. The yield stress and the ultimate tensile stress of steel are 250 MPa and 410 MPa, respectively. The welding is done in the workshop (γ_{mw} = 1.25).

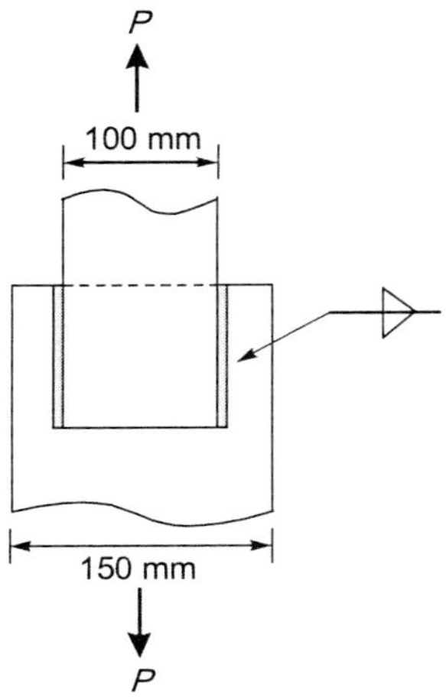

As per the Limit State Method of IS 800 : 2007, the minimum length (rounded off to the nearest higher multiple of 5 mm) of each weld to transmit a force P equal to 270 kN (factored) is

(a) 90 mm (b) 105 mm
(c) 110 mm (d) 115 mm

[2016 : 2 M, Set-I]

1.24 Two bolted plates under tension with alternative arrangement of bolt holes are shown in figures 1 and 2. The hole diameter, pitch, and gauge length are d, p and g, respectively

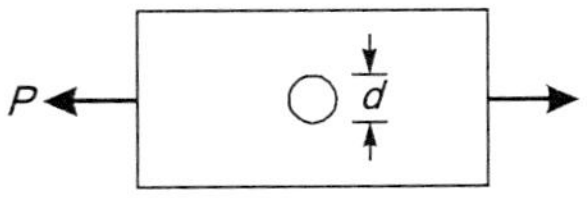

Figure 1

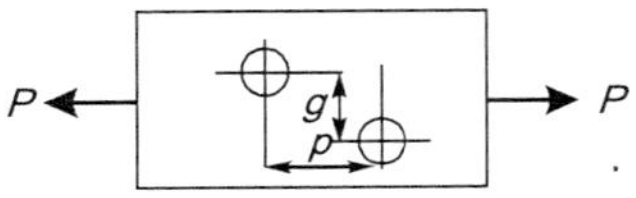

Figure 2

Which one of the following conditions must be ensured to have higher net tensile capacity of configuration shown in figure 2 than that shown in figure 1?

(a) $p^2 > 2gd$ (b) $p^2 < \sqrt{4gd}$
(c) $p^2 > 4gd$ (d) $p > 4gd$

[2016 : 2 M, Set-II]

1.25 A column is subjected to a load through a bracket as shown in the figure

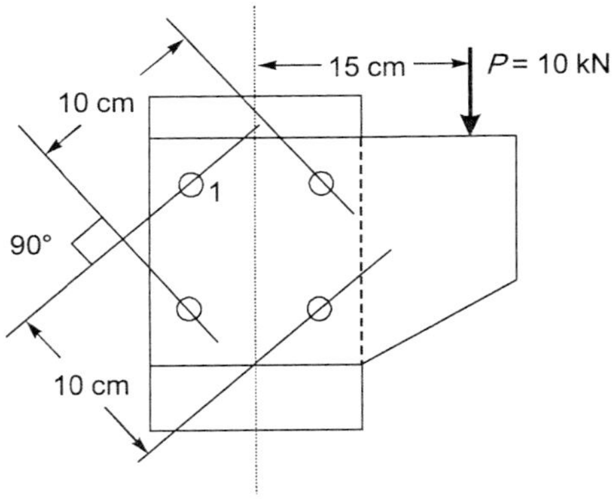

The resultant force (in kN, up to one decimal place) in the bolt 1 is ________.

[2017 : 2 M, Set-I]

1.26 Two plates of 8 mm thickness each are connected by a fillet weld of 6 mm thickness as shown in the figure.

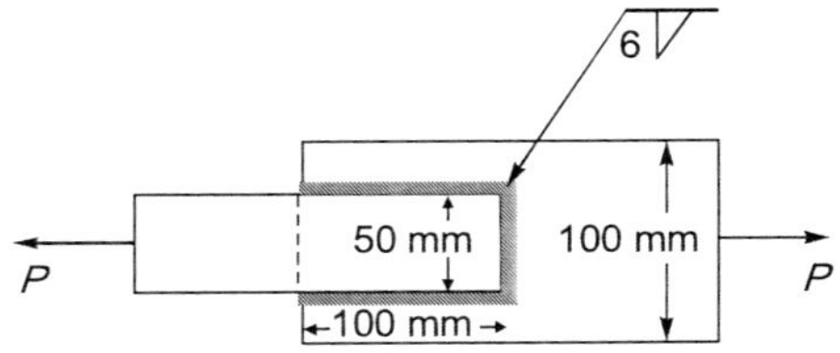

The permissible stresses in the plate and the weld are 150 MPa and 110 MPa, respectively. Assuming the length of the weld shown in the figure to be the effective length, the permissible load P (in kN) is ________.

[2017 : 2 M, Set-II]

1.27 In a fillet weld, the direct shear stress and bending tensile stress are 50 MPa and 150 MPa, respectively. As per IS 800: 2007, the equivalent stress (in MPa, up to two decimal places) will be ________.

[2018 : 1 M, Set-I]

1.28 A fillet weld is simultaneously subjected to factored normal and shear stresses of 120 MPa and 50 MPa, respectively. As per IS 800 : 2007, the equivalent stress (in MPa. up to two decimal places) is ________.

[2018 : 1 M, Set-II]

1.29 Four bolts P, Q, R and S of equal diameter are used for a bracket subjected to a load of 130 kN as shown in the figure.

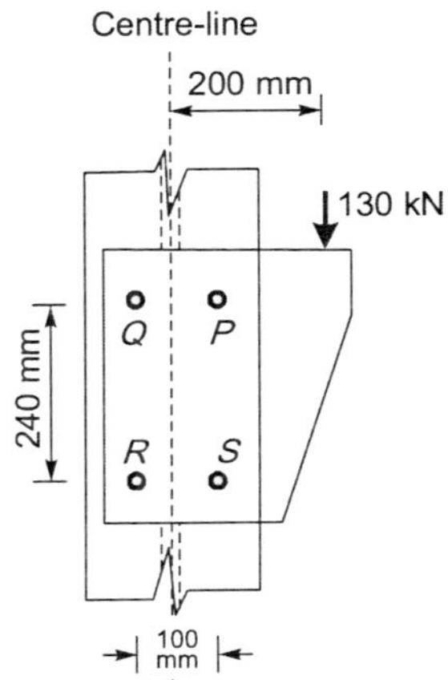

The force in bolt P is

(a) 32.50 kN (b) 69.32 kN
(c) 82.50 kN (d) 119.32 kN

[2018 : 2 M, Set-II]

1.30 A 16 mm thick gusset plate is connected to the 12 mm thick flange plate of an I-section using fillet welds on both sides as shown in figure (not drawn to scale). The gusset plate is subjected to point of 350 kN acting at a distance of 100 mm from the flange plate. Size of fillet weld is 10 mm.

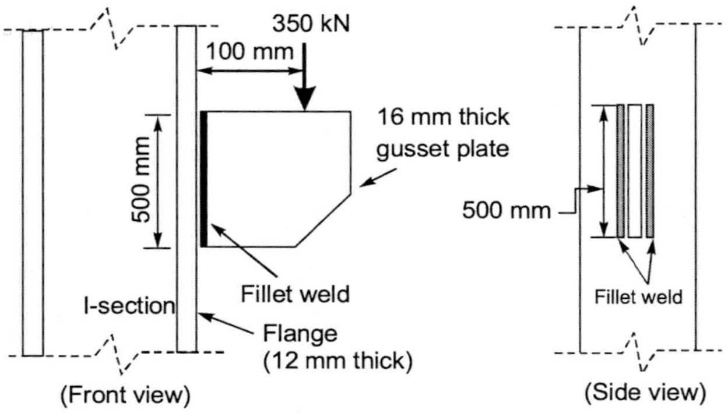

The maximum resultant stress (in MPa, round off to 1 decimal place) on the fillet weld along the vertical plane would be_______.

[2019 : 2 M, Set-I]

1.31 The flange and web plates of the doubly symmetric built-up section are connected by continuous 10 mm thick fillet welds as shown in the figure (not drawn to the scale). The moment of inertia of the section about its principal axis X-X is 7.73×10^6 mm^4. The permissible shear stress in the fillet welds is 100 N/mm^2. The design shear strength of the section is governed by the capacity of the fillet welds.

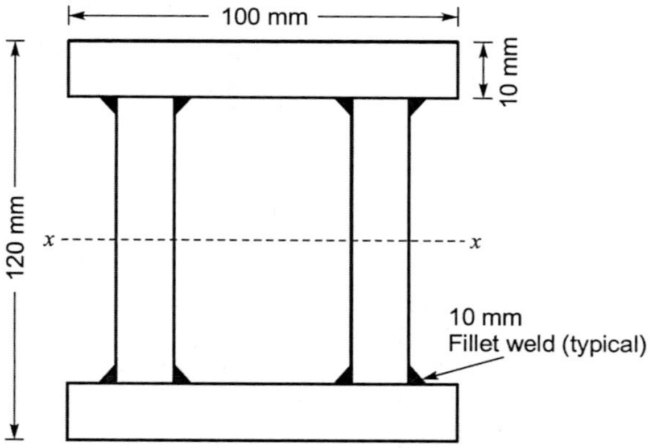

The maximum shear force (in kN, round off to one decimal place) that can be carried by the section, is _______.

[2020 : 2 M, Set-I]

1.32 Two steel plates are lap jointed in a workshop using 6 mm thick fillet weld as shown in the figure (not drawn to the scale). The ultimate strength of the weld is 410 MPa.

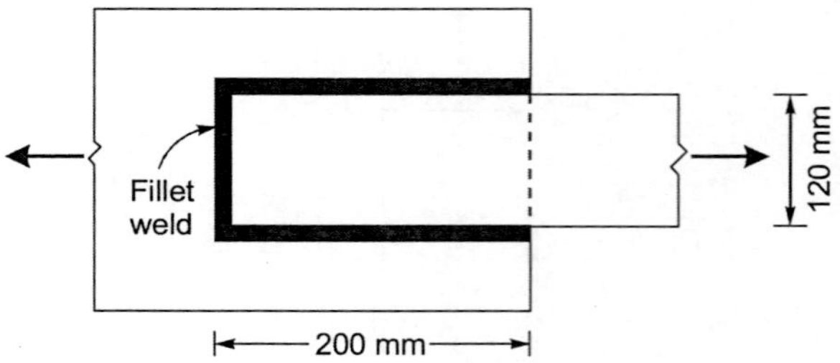

As per Limit State Design is **IS 800 : 2007**, the design capacity (in kN, round off to three decimal places) of the welded connection, is ______.

[2020 : 2 M, Set-II]

1.33 A column is subjected to total load (P) of 60 kN supported through a bracket connection, as shown in the figure (not to scale).

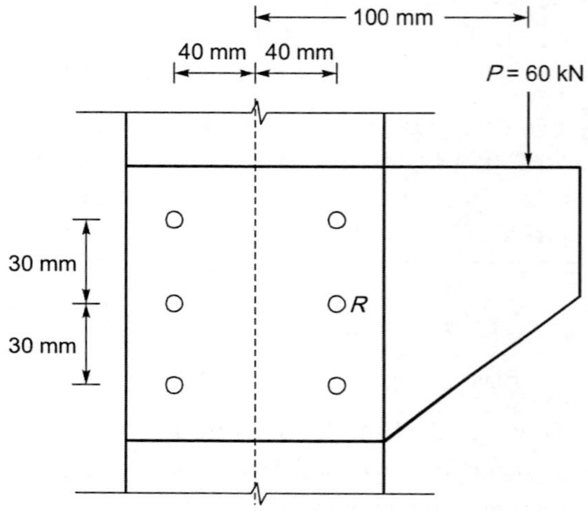

The resultant force in bolt R (in kN, round off to one decimal place) is ______

[2021 : 2 M, Set-I]

1.34 A weld is used for joining an angle section ISA 100 mm × 100 mm × 10 mm to a gusset plate of thickness 15 mm to transmit a tensile load. The permissible stress in the angle is 150 MPa and the permissible shear stress on the section through the throat of the fillet weld is 108 MPa. The location of the centroid of the angle is represented by C_{yy} in the figure, where C_{yy} = 28.4 mm. The area of cross-section of the angle is 1903 mm^2. Assuming the effective throat thickness of the weld to be 0.7 times the given weld size, the lengths L_1 and L_2 (rounded off to the nearest integer) of the weld required to transmit a load equal to the full strength of the tension member are, respectively

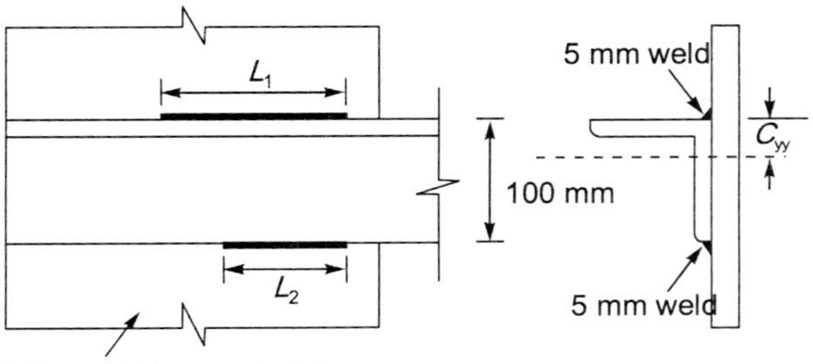

15 mm thick gusset plate

(a) 541 mm and 214 mm
(b) 214 mm and 541 mm
(c) 380 mm and 151 mm
(d) 151 mm and 380 mm [2022 : 2 M, Set-I]

1.35 Consider the fillet-welded lap joint shown in the figure (not to scale). The length of the weld shown is the effective length. The welded surfaces meet at right angle. The weld size is 8 mm, and the permissible stress in the weld is 120 MPa. What is the safe load P (in kN, rounded off to one decimal place) that can be transmitted by this welded joint? ________.

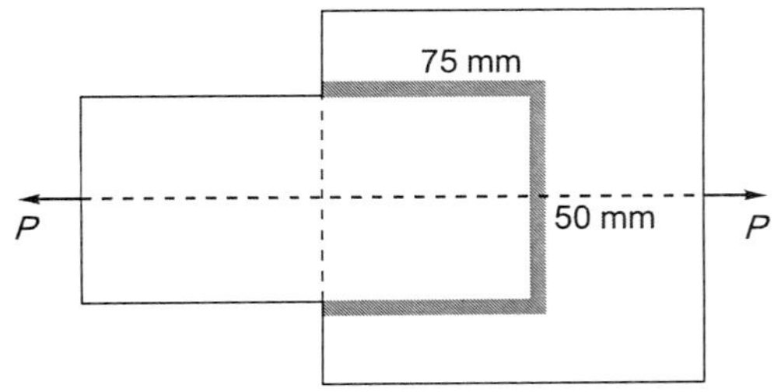

[2023 : 1 M, Set-I]

1.36 Two plates are connected by fillet welds of size 10 mm and subjected to tension, as shown in the figure. The thickness of each plate is 12 mm. The yield stress and the ultimate stress of steel under tension are 250 MPa and 410 MPa, respectively. The welding is done in the workshop (partial safety factor, y_{mw} = 1.25). As per the Limit State Method of IS 800: 2007, what is the minimum length (in mm, rounded off to the nearest higher multiple of 5 mm) required of each weld to transmit a factored force P equal to 275 kN?

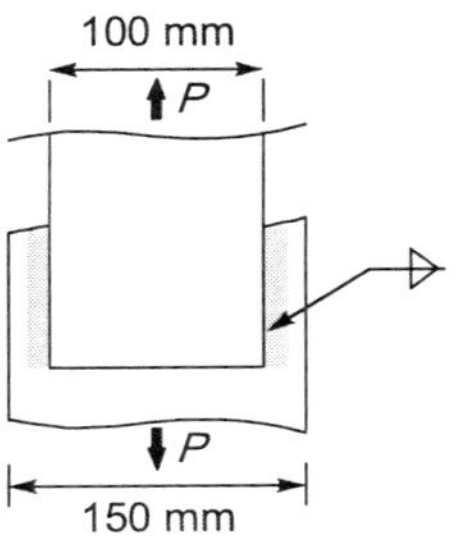

(a) 100
(b) 105
(c) 110
(d) 115

[2023 : 2 M, Set-II]

1.37 The structural design method that DOES NOT take into account the safety factors on the design loads is

(a) limit state method.
(b) ultimate load method.
(c) load factor method.
(d) working stress method.

[2024 : 1 M, Set-II]

Answers Structural Fasteners

1.1	(b)	1.2	(a)	1.3	(a)	1.4	(c)	1.5	(c)	1.6	(a)	1.7	(a)
1.8	(a)	1.9	(b)	1.10	(c)	1.11	(b)	1.12	(d)	1.13	(c)	1.14	(b)
1.15	(b)	1.16	(b)	1.17	(b)	1.18	(a)	1.19	(b)	1.20	(d)	1.21	(156.2)
1.22	(b)	1.23	(b)	1.24	(c)	1.25	(6)	1.26	(60)	1.27	(173.2)	1.28	(147.9)
1.29	(b)	1.30	(105.3)	1.31	(393.5)	1.32	(413.58)	1.33	(28.2)	1.34	(a)	1.35	(134.4)
1.36	(b)	1.37	(d)										

Explanations Structural Fasteners

1.1 (b)

As per IS 800 : 2007, clause 10.2.2, the minimum spacing between centres of fasteners shall not be less than 2.5 time the nominal diameter of the fastener.

1.2 (a)

Welded connection are weak under fatigue loading and prone to cracking.

∴ Fatigue life of welded structure is always less than the fatigue life of riveted structure.

$$\Rightarrow \frac{\text{Fatigue life of welded structure}}{\text{Fatigue life of riveted structure}} < 1$$

1.3 (a)

According to **clause 10.5.8.1 of IS 800 : 2007**, where a fillet weld is applied to the square edge of a part, the specified size of the weld should generally be at least 1.5 mm less than the edge thickness in order to avoid washing down of the exposed arris.

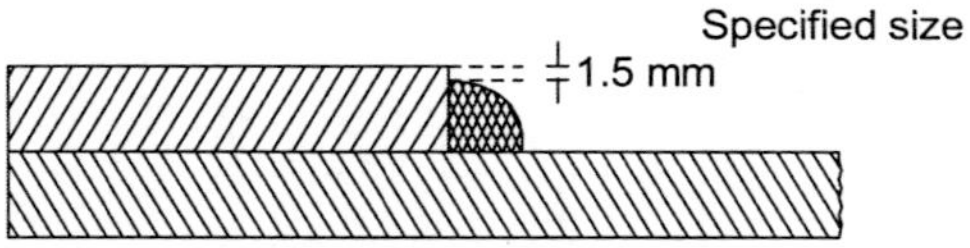

Fillet weld on square edge of plate

POINTS TO REMEMBER

1.

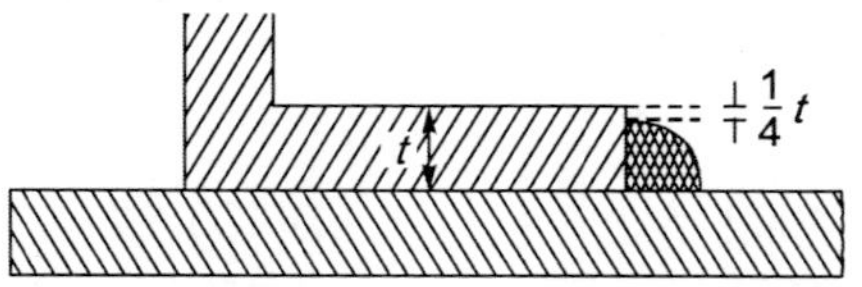

Fillet weld on round toe of rolled section

2. Where the size specified for fillet weld is such that the parent metal will not project beyond the weld no menting of the outer cover or covers shall be allowed to occur to such an extent as to reduce the throat thickness.

1.4 (c)

Permissible stress in axial compression as per IS 800

$$1984 = 0.6\sigma_y$$

$$\therefore \quad \text{Factor of safety} = \frac{1}{0.6} = 1.67$$

1.5 (c)

Arrangement-1

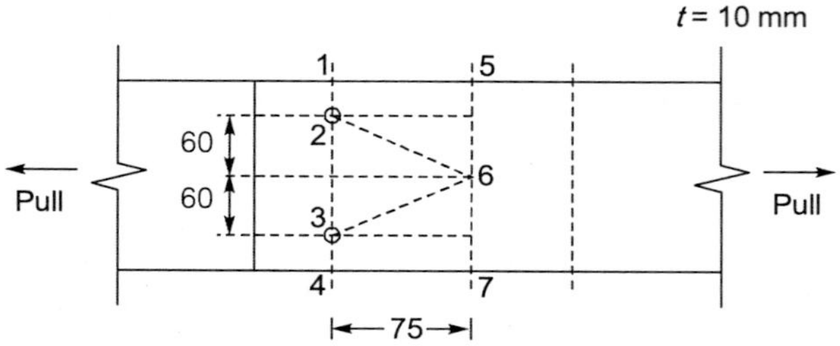

Arrangement -1

Given: Diameter of rivet = 20 mm

∴Gross diameter of hole = 20 + 1.5 = 21.5 mm

Note: According to IS 800 : 2007 hole diameter for rivet or bolts (+16-22 mm) = (nominal diameter + 2 mm)

Since the question was of 1991, solving by previous code regulations.

For most critical sectional are, consider different sections as follows:

Case-1: Section 1-2-3-4

$$A_{net1} = t \times (B - nd)$$
$$= 10 \times (150 - 2 \times 21.5) = 1070 \text{ mm}^2$$

Case-2: Section 5-6-7

$$A_{net2} = t \times (B - nd)$$
$$= 10 \times (150 - 1 \times 21.5) = 1285 \text{ mm}^2$$

Case-3: Section 1-2-6-3-4

$$A_{net3} = t \times \left[B - bd + \frac{\Sigma st^2}{4g}\right]$$
$$= 10 \times \left[150 - 3 \times 21.5 + \frac{2 \times 75^2}{4 \times 60}\right]$$
$$= 1323 \text{ mm}^2$$

$\therefore A_{net}$ minimum $= A_{net\,1}$

i.e., $A_{net} = 1070 \text{ mm}^2$

∴ Maximum tensile force permitted at that joint

$$= 150 \times 1070 = 160.5 \text{ kN}$$

Arrangement-2

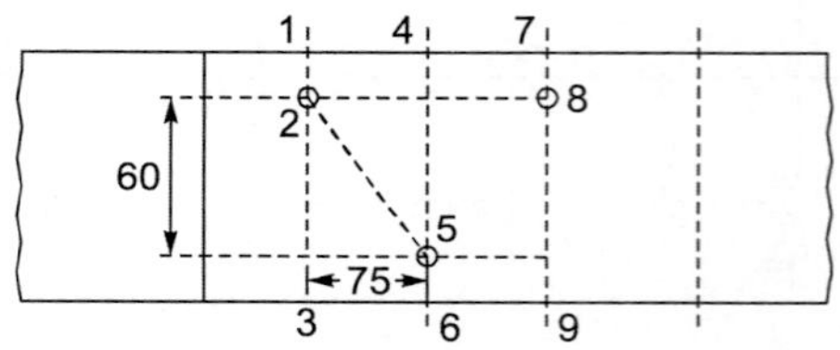

Arrangement -2

Case-1: Section 1-2-3

$$A_{net1} = t \times (B - nd)$$
$$= 10 \times (150 - 21.5) = 1285 \text{ mm}^2$$

Case-2: Section 1-2-5-6

$$A_{net2} = t \times \left[(B - nd) + \frac{\Sigma s_i^2}{4g}\right]$$
$$= 10 \times \left[150 - 2 \times 21.5 + \frac{75^2}{4 \times 60}\right] = 1304.4 \text{ mm}^2$$

$\therefore A_{net}$ minimum $= A_{net1}$

i.e., $A_{net} = 1285 \text{ mm}^2$

∴ Maximum tensile foce permitted at that joint

$$= 150 \times 1285 = 192.75 \text{ kN}$$

$$\therefore \quad P_{max} = 192.75 \text{ kN}$$

We can attain this strength by arranging the riverts as per second arrangement.

1.6 (a)

POINTS TO REMEMBER

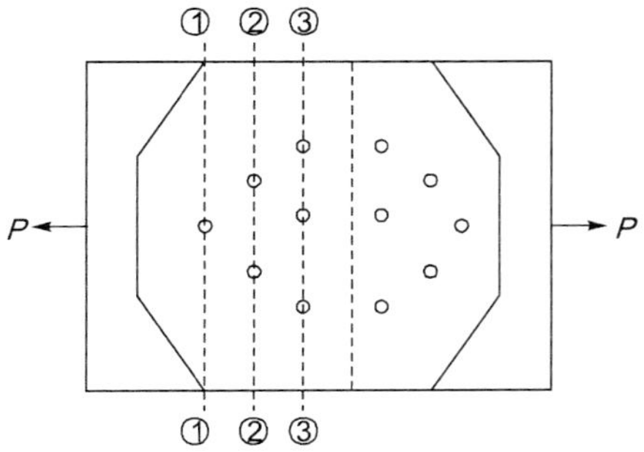

The plates resists the total load P at section (1)-(1) and successively lesser load at (2)-(2) and (3)-(3). Force in main plate just before (1)-(1) in equal to P. As soon as the flow of force crosses (1)-(1), the bolt absorb force equal to their strength and transfer it to the cover plate. Thus after (1)-(1) force to be resisted by main plate is lesser than the force at or before section (1)-(1). Further reduction takes place after (2)-(2) again.

Suggestions:

1. Provide lesser no. of bolts at section (1)-(1) to increase A_{net}.
2. It can be increased gradually inwards and the lesser net area can also take up that lesser load.

Also note it will be reverse in the case of cover plate, where the force is maximum at the section (3)-(3). Thus for the main plate, critical section is (1)-(1) and for cover plate is section (3)-(3).

1.7 (a)

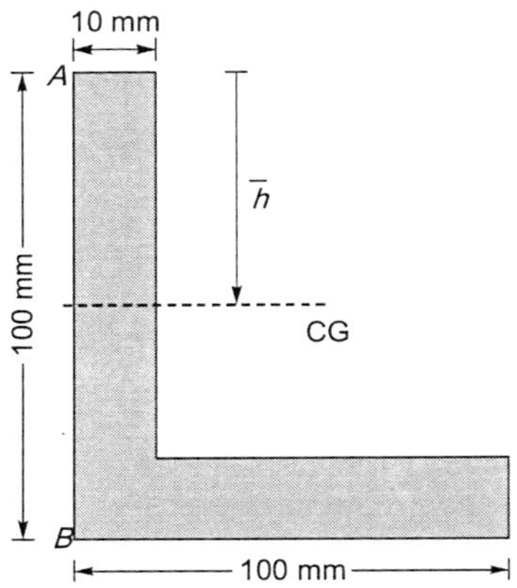

CG of angle from A will be,

$$\bar{h} = \frac{(100\times 10)50 + 90\times 10\times(90+5)}{100\times 10 + 90\times 100}$$

$= 71.31$ mm joint A

Let P_1 and P_2 force is resisted by weld at A and B respectively then by taking moment about CG

$$P_1 \times 71.33 = P_2 \times (100 - 71.33)$$

$$P_1 = 0.4P_2$$

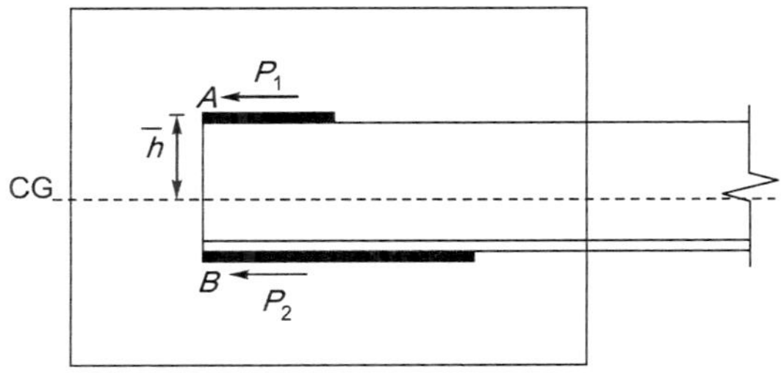

As $P_2 > P_1$ (about 2.5 times) weld required on 'B' side will be more than A by 2.5 times.

If l_1 = length of weld at A = 60 mm

l_2 = length of weld at B = 60 × 2.5 = 150 mm

1.8 (a)

Rivet values and bolt value is defined as the overall strength of fastner for design purpose. It is taken as minimum of shearing and bearing strength.

REFERENCE MATERIAL

You may visit http://nptel.ac.in/courses/105106 112/ 3_connections/9_Examples.pdf for much better understanding.

1.9 (b)

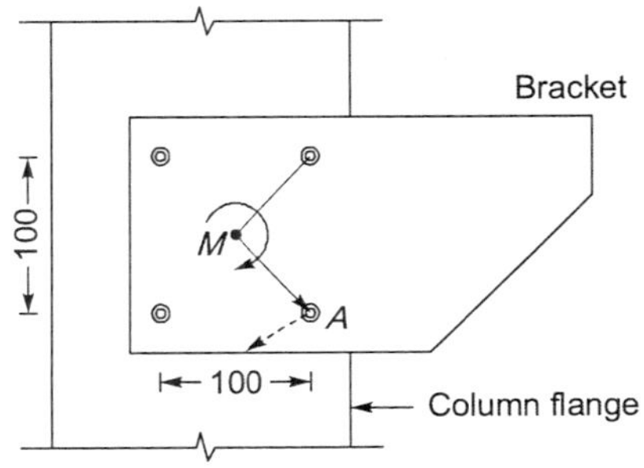

The centroid of rivets will be on the symmetric lines and each rivet will be at,

$$r = \sqrt{(50)^2 + (50)^2} = 50\sqrt{2} \text{ mm}$$

Shear force in rivet A is given by,

$$F_s = \frac{Mr}{\Sigma r^2} = \frac{50\times 10^3 \times 50\sqrt{2}}{4\times(50\sqrt{2})^2} = 176.8 \text{ kN}$$

1.10 (c)

Method-I

Total weld length = (100 × 2) + 50 = 250 mm

Strength of weld per mm length

$= 0.70 \times 6 \times 108 = 453.6$ N

Maximum load taken by joint

$$= \frac{453.6 \times 250}{1000} = 113.4 \text{ kN}$$

Method-II

Size of weld, s = 6 mm

Thickness at throat = 0.7s = 4.2 mm

Length of the weld = 250 mm

∴ Safe load = Allowable stress × Area resisting shear

$= 108 \times 4.2 \times 250 = 113.4$ kN

1.11 (b)

The bearing stress is any part of a beam when calculated on the gross area shall not exceed $0.75 f_y$

1.12 (d)

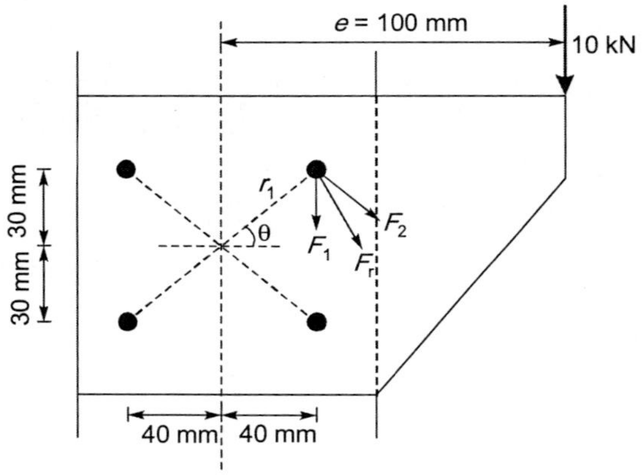

Bolt will be subjected to direct shear and shear force due to twisting moment.

F_1 = Force on bolt due to direct shear

F_2 = Force on bolt due to twisting moment

$$F_1 = \frac{P}{4} = \frac{10}{4} = 2.5 \text{kN}$$

F_2 = stress due to moment

$$F_2 = \frac{(Pe) \times r_1}{\Sigma r_i^2}$$

$$r_1 = r_2 = r_4 = \sqrt{30^2 + 40^2} = 50 \text{ mm}$$

$$F_2 = \frac{10 \times 100 \times 50}{2500 \times 4} = 5 \text{ kN}$$

$$\tan\theta = \frac{30}{40}$$

$$\theta = 37°$$

Resultant force

$$= \sqrt{F_1^2 + F_2^2 + 2F_1F_2\cos\theta}$$

$$= \sqrt{(2.5)^2 + (5)^2 + 2 \times 2.5 \times 5 \times \cos 37°}$$

$= 7.16$ kN

AVOID MISTAKE

If you notice, θ is the angle between F_1 and F_2. It is not the angle between horizontal and line joining centre to bolt.

Here both angles are same, but when the loading is inclined both angles are different.

Students must take care of it.

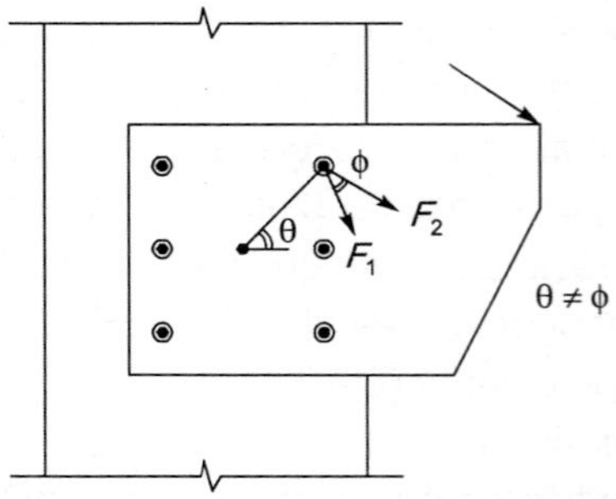

1.13 (c)

As per clause **8.9.4.5** of **IS 800 : 1984** rivets and bolts subject to both shear and axial tension shall be so proportioned that the shear and axial stresses calculated do not exceed the respective allowable stresses τ_{vf} and σ_{tf} and the expression

$$\left(\frac{\tau_{vf,cal}}{\tau_{vf}} + \frac{\sigma_{tf,cal}}{\sigma_{tf}}\right) \text{ does not exceed 1.4.}$$

Also note, According the **IS 800 : 2007, clause no. 10.3.6,** A bolt required to resist both design shear force (V_{sd}) and design tensile force (T_b) at the same time shall satisfy:

$$\left(\frac{V_{sb}}{V_{db}}\right)^2 + \left(\frac{T_b}{T_{db}}\right)^2 \leq 1.0$$

V_{sb} = Factored SF acting on bolt.

V_{db} = Design shear capacity

T_b = Factored tensile force acting on bolt.

T_{db} = Design tension capacity

1.14 (b)

Rivet value is the lesser of shearing strength or bearing strength of rivet.

Shearing strength of rivet

$= 2 \times \frac{\pi}{4} \times (16 + 1.5)^2 \times 90 = 43.29$ kN

Bearing strength of rivet

$= (16 + 1.5) \times 12 \times 270 = 56.7$ kN

$\therefore$ Rivet value of the joint = 43.29 kN

1.15 (b)

Strength of one rivet in double shear

$$= 2 \times \frac{\pi (d')^2}{4} \times \sigma_s = 2 \times \frac{(20 + 1.5)^2 \pi}{4} \times 100 \times 10^{-3}$$

$= 72.61$ kN

Strength of the riveted joint in double shear

$= 12 \times 72.61 = 871.32$ kN

Strength of one rivet in bearing $= d' t \times \sigma_b$

$= (20 + 1.5) \times 14 \times 300 \times 10^{-3} = 90.3$ kN

Strength of the riveted joint in bearing

$= 12 \times 90.3 = 1083.6$ kN

Thus the strength of joint will be governed by shearing and it will be equal to 871.32 kN

1.16 (b)

The maximum size of a fillet weld is obtained by subtracting 1.5 mm from the thickness of the thinner member to be jointed.

$\therefore \quad S = 10 - 1.5 = 8.5$ mm

Strength of the thinner plate

$= 100 \times 10 \times 150 \times 10^{-3} = 150$ kN

Strength of the weld

$= L \times S \times 0.7 \times \sigma_{ws}$

$\Rightarrow \quad 150 \times 10^3 = L \times 8.5 \times 0.7 \times 110$

$\Rightarrow \quad L = 229.18 \text{ mm} \simeq 229.2 \text{ mm}$

1.17 (b)

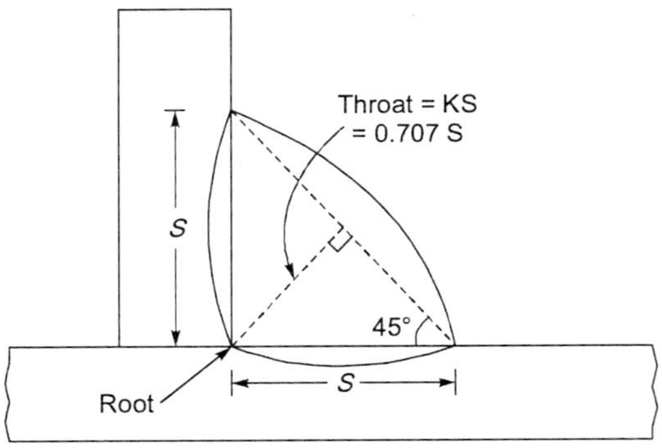

(a) Throat of weld

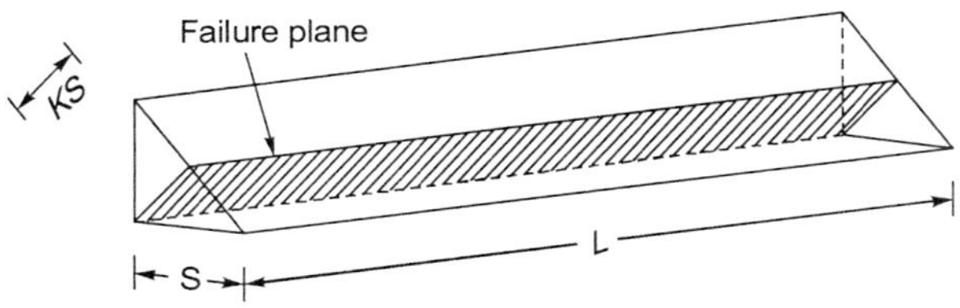

(b) Effective length of weld

Effective throat thickness is the shortest distance from the root of fillet weld to the face of the line joining the toes. The effective throat thickness should not be less than 3 mm.

Effective throat thickness = Ks.

where s is the size of weld in mm and K is a constant. The value of K depends upon the angle between the fusion faces.

Angle	60°-90°	91°-100°	101°-106°	107°-113°	114°-120°
K	0.70	0.65	0.60	0.55	0.50

1.18 (a)

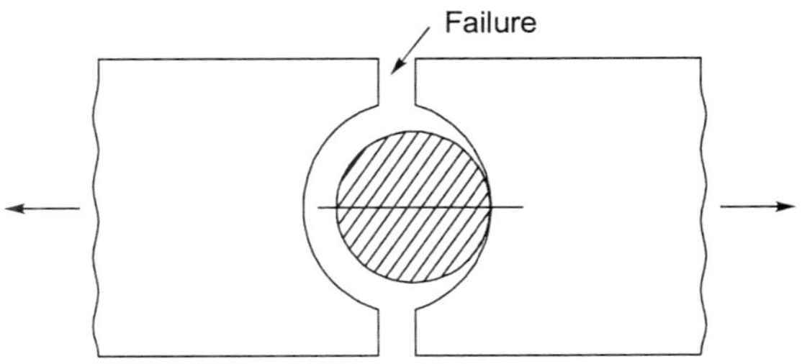

Tension failure of plate

1.19 (b)

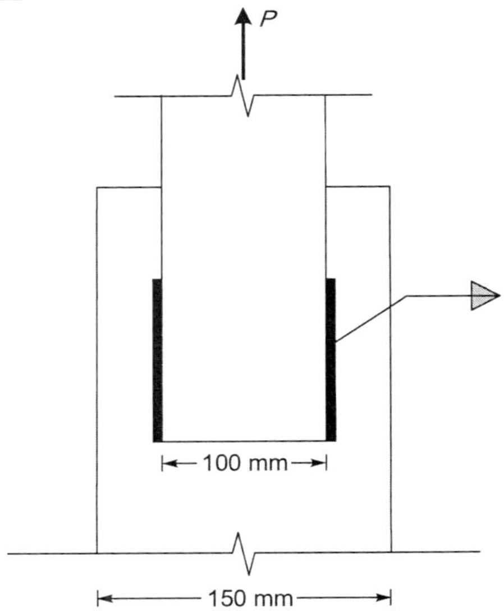

Maximum force transmitted by plate = P

$$P = A_g \times \frac{f_y}{1.1} = (12 \times 100) \times \frac{250}{1.1}$$

$= 272.72$ kN

Force resisted by weld,

$P = 272.72$ kN ...(i)

Let length of weld on either side = l

Size of weld (S) = 10 mm

$$\text{force resisted by weld} = \frac{0.7 \times S}{1.25} \times \left(\frac{f_u}{\sqrt{3}}\right) \times l$$

Size = 10 mm

Throat thickness = 0.7 × 10 = 7 mm

$$= \frac{7 \times 410}{1.25 \times \sqrt{3}} \times l \qquad \text{...(ii)}$$

From equations (i) and (ii),

$$272.72 \times 10^3 = \frac{7 \times 410}{1.25 \times \sqrt{3}} \times l$$

$$l = 205.8 \text{ mm}$$

$$\text{Length eq. of each side} = \frac{l}{2} = 103 \text{ mm} \simeq 105 \text{ mm}$$

AVOID MISTAKE

103 mm should be scaled to 105 mm to safer side and not as 100 mm; which would be unsafe.

1.20 (d)

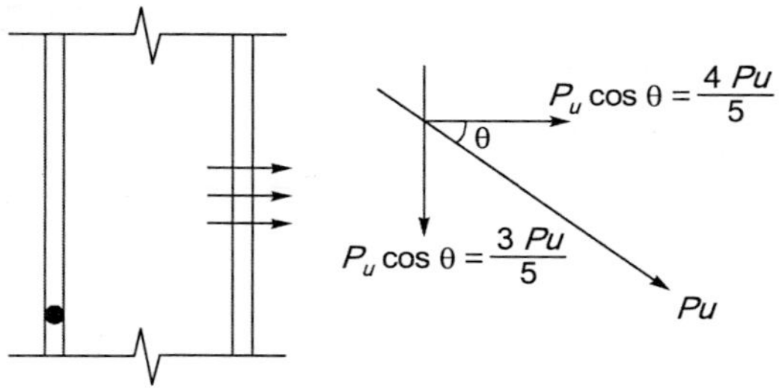

$$\tan\theta = \frac{3}{4},\ \cos\theta = \frac{4}{5},\ \sin\theta = \frac{3}{5}$$

$$\text{Tension in each bolt} = \frac{4P_u}{5 \times 6}$$

$$\frac{4 \times 250}{5 \times 6} = 33.33 \text{ kN}$$

$$\text{Shear in each bolt} = \frac{3P_u}{5 \times 6} = \frac{3 \times 250}{5 \times 6} = 25 \text{ kN}$$

1.21 Sol.

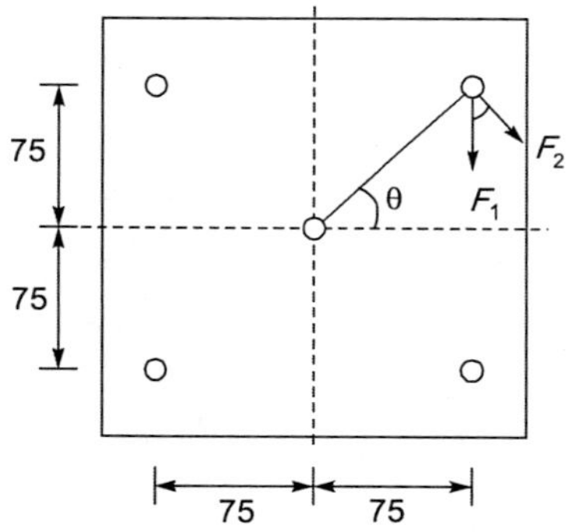

$$F_1 = \frac{100}{5} = 20 \text{ kN}$$

$$F_2 = \frac{Per}{\sum r^2} = \frac{(100 \times 0.6) \times (75\sqrt{2} \times 10^{-3})}{(75\sqrt{2})^2 \times 4 \times 10^{-6}}$$

$$= \frac{\sqrt{2}}{10} \times 1000 = 141.42 \text{ kN}$$

$$\theta = 45°$$

$$\Rightarrow \quad F_R = \sqrt{(20)^2 + (141.42)^2 + 2 \times 20 \times 141.42 \cos 45°} = 156.2 \text{ kN}$$

1.22 (b)

In the connections subjected to tension, due to flexibility of connected parts, can lead to deformations that increase the tension applied to bolts. This additional tension force is called as pyring force.

1.23 (b)

Maximum force carried by plates,

$$P = \frac{A_g f_y}{\gamma_{m0}} = \frac{(100 \times 12) \times 250}{1.1} = 272.73 \text{ kN}$$

Load carried by each weld

$$= \frac{P}{2} = 136.36 \text{ kN}$$

For minimum length of weld,
Strength of weld = Load carried by weld

$$l_w t_t \frac{f_u}{\sqrt{3}\gamma_{mw}} = 136.36 \times 10^3$$

$$\text{or,} \quad l_w \times (10 \times 0.7) \times \frac{410}{\sqrt{3} \times 1.2} = 136.36 \times 10^3$$

$\Rightarrow$ $l_w = 102.9$ mm
= 105 mm (rounded off to nearest higher multiple of 5 mm)

1.24 (c)

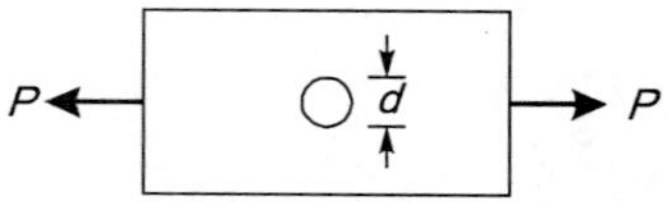

Figure 1

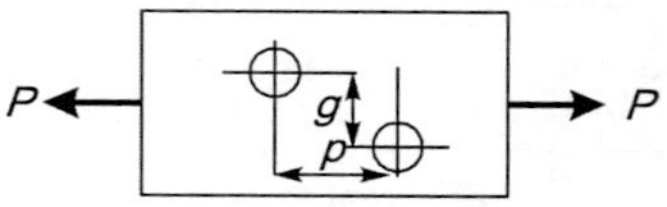

Figure 2

Tensile strength of plate in arrangement (2) will be greater than in arrangement (1),

* As per IS code 800:2007 close 6.3

$$\left(0.9A_{net}\frac{f_{up}}{\gamma_{m1}}\right)_2 > \left(0.9A_{net}\frac{f_{up}}{\gamma_{m1}}\right)_1$$

$$(A_{net})_2 > (A_{net})_1$$

$$\left[\left(B-2d+\frac{p^2}{4g}\right)t\right]_2 > [(B-d)t]_1$$

$$B-2d+\frac{p^2}{4g} > B-d$$

$$\frac{p^2}{4g} > d$$

$$p^2 > 4gd$$

1.25 Sol.

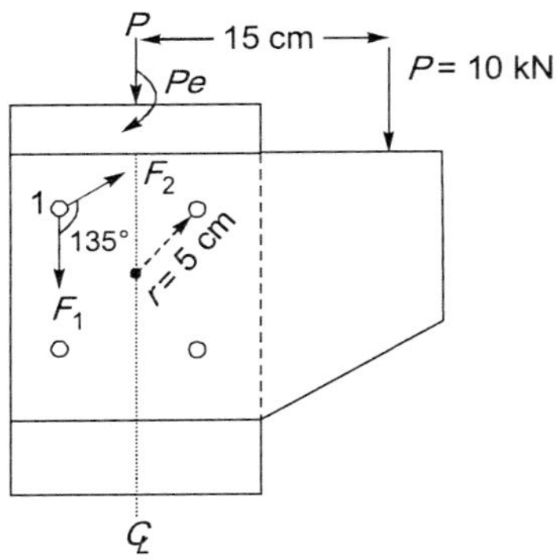

$$F_1 = \frac{P}{n} = \frac{10}{4} = 2.5 \text{ kN}$$

$$F_2 = \frac{Pe}{\Sigma r_i^2}\times r_1 = \frac{10\times 15}{4\times 5^2}\times 5 = 7.5 \text{ kN}$$

$$F_R = \sqrt{F_1^2+F_2^2+2F_1F_2\cos\theta}$$

$$= \sqrt{2.5^2+7.5^2+2\times 2.5\times 7.5\times \cos 135°} = 6 \text{ kN}$$

1.26 Sol.

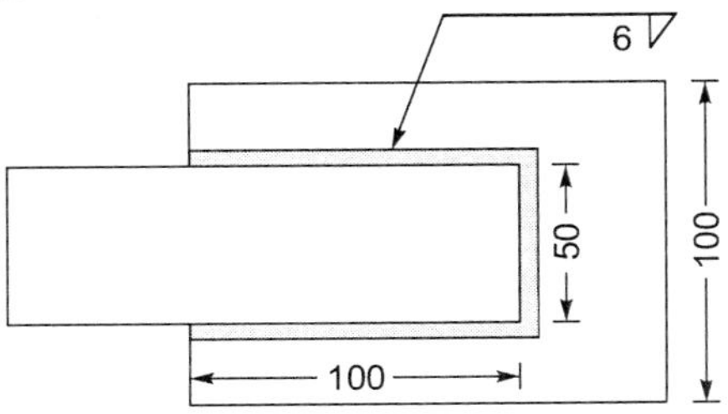

$$P_w = f_b \times l_{eff} \times t_t$$

$$= 110\ (100 + 100 + 50) \times 0.7 \times 6$$

$$= 115.5 \text{ kN}$$

$$P_{plate} = \sigma_{at} \times A_{net} = 150 \times (50 \times 8) = 60 \text{ kN}$$

So permissible load = 60 kN

POINTS TO REMEMBER

Representation of type of weld

Types of weld	Representation
1. Fillet weld	◺
2. Groove weld	
• Square	‖
• V	∨
• Bevel	V
• U	Y
• U	Y
• V with broad root face	Y
• Bevel with broad root face	Y
• Weld with raised edge	八
3. Seam weld	⌀
4. Spot weld	○
5. Plug weld	⊓
6. Field weld	●

1.27 Sol.

Direct bending tensile stress,

$$f_a = 150 \text{ MPa}$$

Direct shear stress,

$$q = 50 \text{ MPa}$$

According to IS 800 : 2007, clause 10.5.10.1.1

The equivalent stress,

$$f_e = \sqrt{f_a^2+3q^2} \le \frac{f_u}{\sqrt{3}\gamma_{mw}}$$

$$f_e = \sqrt{150^2+3\times 50^2} = 173.21 \text{ MPa}$$

$$\Rightarrow \quad f_e \le \frac{f_u}{\sqrt{3}\gamma_{mw}}$$

$$\left(= \frac{400}{\sqrt{3}\times 1.25} = 184.75 \text{ MPa}\right) \text{ (Hence OK)}$$

So, equivalent stress of 173.21 MPa.

1.28 Sol.

Factored normal stress, f_a = 120 MPa

Factored shear stress, q = 50 MPa

According to IS-800 : 2007, clause 10.5.10.1.1

The equivalent stress,

$$f_e = \sqrt{f_a^2+3q^2} \le \frac{f_u}{\sqrt{3}\,y_{mw}}$$

$$f_e = \sqrt{120^2+3\times 50^2} = 147.99$$

$$\Rightarrow \quad f_e \le \frac{f_u}{\sqrt{3}\gamma_{mw}} = \frac{400}{\sqrt{3}\times 1.25}$$

$= (184.75 \text{ MPa})$

So, equivalent stress 147.99 MPa.

1.29 (b)

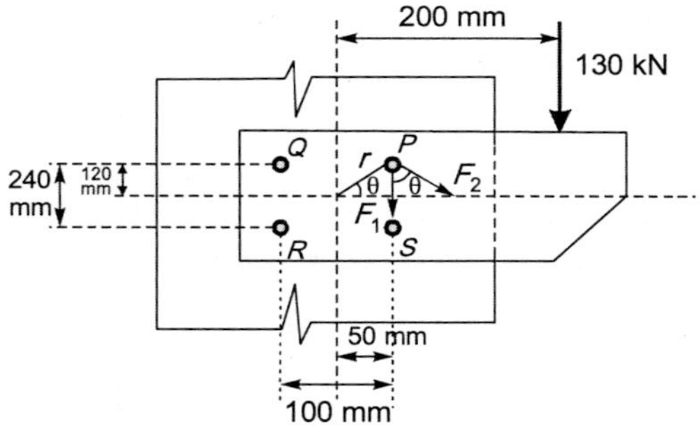

$$F_1 = \frac{P}{n} = \frac{130}{4} = 32.5 \text{ kN}$$

$$F_2 = \frac{Pe}{\Sigma r_i^2}\times r_p$$

$$r_p = r_Q = r_R = r_S = \sqrt{50^2+120^2} = 130 \text{ mm}$$

$$F_2 = \frac{(130\times 200)\times 130}{4\times 130^2} = \frac{200}{4} = 50 \text{ kN}$$

$$\cos\theta = \frac{50}{130}$$

$$\Rightarrow \quad F_R = \sqrt{\left[(32.5)^2+(50)^2+(2\times 32.5\times 50)\times\left(\frac{50}{130}\right)\right]}$$

$= 69.33 \text{ kN}$

1.30 Sol.

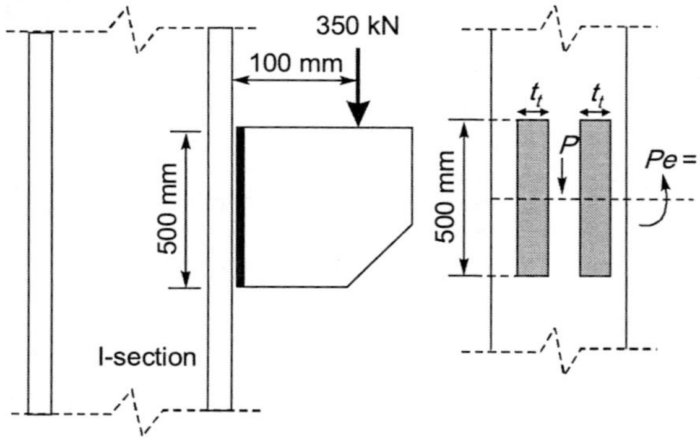

$P = 150$ kN, $e = 100$ mm

t_t = Effective throat thickness

$t_t = 0.7S$

S = Size of weld = 10 mm

$\therefore \quad t_t = 0.7 \times 10 = 7$ mm

Shear stress due to direct force,

$$q = \frac{350\times 10^3}{2\times 500\times 7}\text{ MPa} = 50 \text{ MPa}$$

Normal stress due to bending,

$$f_b = \frac{My}{I}$$

$$f_b = \frac{Pe\times d/2}{2\times\frac{t_t d^3}{12}}$$

$$= \frac{250\times 10^3\times 100\times 500/2}{2\times 7\times 500^3/12}\text{ MPa} = 60 \text{ MPa}$$

Resultant stress,

$$f_e = \sqrt{f_b^2+3\times q^2} = \sqrt{60^2+3\times 50^2}$$

$= 105.35 \text{ MPa}$

1.31 Sol.

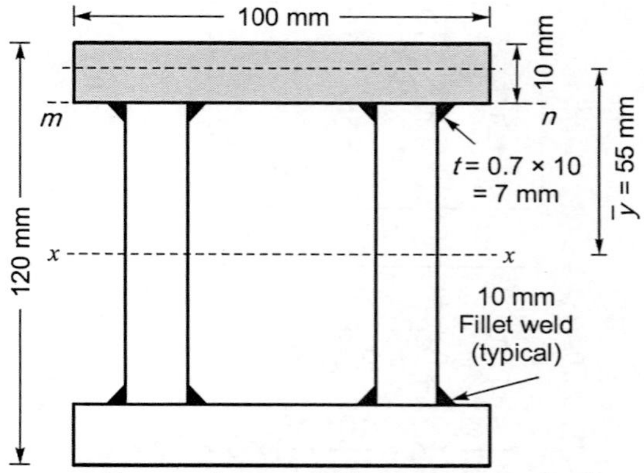

q = Shear stress at the level mn in the weld

$$= 100 \text{ MPa} = \frac{FA\bar{y}}{Ib}$$

F = Shear force at the given section

A = Area of the cross-section above the level mn = 100 × 10 mm^2

$\bar{y}$ = C.G. of shaded area above the level mn = 60 − 5 = 55 m

$I = 7.73 \times 10^6$ mm^4

b = Width of weld at mn (4 welds) = 4 × t = 4 × 7 = 28 mm

Throat thickness,

$t = 0.7 \times s = 0.7 \times 10 \times 4 = 28$ mm

$$\therefore \quad 100 = \frac{F\times(100\times 10)\times 55}{7.73\times 10^6\times 28}$$

$$F = \frac{100\times 7.73\times 10^6\times 28}{1000\times 55}$$

$= 393.527 \text{ kN} = 393.5 \text{ kN}$

1.32 Sol.

Design capacity of welded connection

$$P_s = f_b \times l_{eff} \times t_t$$

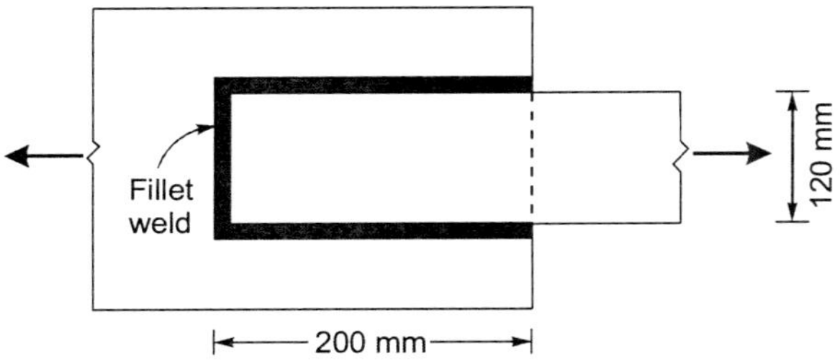

$$P = \frac{410}{\sqrt{3} \times 1.25} \times 520 \times 0.7 \times 6$$

$$= \frac{716352}{\sqrt{3}} = 413586 \text{ N}$$

Design capacity,

$$P = 413.586 \text{ kN}$$

1.33 Sol.

$$F_1 = \frac{P}{n} = \frac{60}{6} = 10 \text{ kN}$$

$$F_2 = \frac{P \cdot e}{\Sigma r_i^2} \times r_R = \frac{60 \times 100 \text{ mm}}{4 \times 50^2 + 2 \times 40^2} \times 40$$

$$F_2 = \frac{60 \times 100 \times 40}{10000 + 3200}$$

$$= \frac{60 \times 100 \times 40}{13200} = \frac{200}{11} \text{ kN}$$

$$F_R = F_1 + F_2 = 10 + \frac{200}{11} = \frac{310}{11} = 28.2 \text{ kN}$$

1.34 (a)

As per IS : 800-2007 Clause no. 11.2.1

Full strength of tension member,

$$P = \sigma_{at} \times A_g$$

$$\Rightarrow \quad P = 150 \times 1903$$

$$\Rightarrow \quad P = 285450 \text{ N}$$

For weld strength of L_1,

$$P_1 = L_1 \times (0.7 \times s) \times 108$$

$$= L_1 \times 0.7 \times 5 \times 108$$

$$\Rightarrow \quad P_1 = 378L_1 \quad \text{...(i)}$$

For weld strength of L_2,

$$P_2 = L_2 \times (0.7 \times s) \times 108$$

$$= L_2 \times 0.7 \times 5 \times 108$$

$$\Rightarrow \quad P_2 = 378L_2 \quad \text{...(ii)}$$

Applying moment about CG equal to zero,

$$P_1 \times 28.4 = P_2 \times (100 - 28.4)$$

$$\Rightarrow \quad \frac{P_1}{P_2} = \frac{71.6}{28.4}$$

$$\frac{L_1 \times 378}{L_2 \times 378} = \frac{71.6}{28.4} = 2.521$$

$$\Rightarrow \quad L_1 = 2.521 L_2 \quad \text{...(iii)}$$

$\therefore$ Total weld strength = Tensile strength

$\Rightarrow (L_1 + L_2) \times 3.5 \times 108 = 285450$

$\Rightarrow L_1 + L_2 = 755.158$ mm

Using equation (iii),

$2.521 L_2 + L_2 = 755.158$ mm

$$\Rightarrow \quad L_2 = 214.472 \text{ mm}$$

$$\therefore \quad L_1 = 540.685 \text{ mm}$$

1.35 Sol.

Given: $s = 8$ mm, $\tau_{vw} = 120$ MPa,

$L_w = 2 \times 75 + 50 = 200$ mm

The safe load is calculated as per WSM

$$\therefore \quad P_{safe} = \tau_{vw} \times A_w$$

$$= \tau_{vw} \times L_w \times (0.7 \times s) \quad [\because t_t = 0.7s]$$

$$= 120 \times [2 \times 75 + 50] \times 0.7 \times 8 \times 10^{-3} \text{ kN}$$

$$= 134.4 \text{ kN}$$

1.36 (b)

Since the given weld is fillet weld. Thus, strength of fillet weld,

$$P = L_w (ks) \frac{f_u}{\sqrt{3} \gamma_{mw}}$$

$$275 \times 10^3 = L_w \times 0.7 \times 10 \times \frac{410}{\sqrt{3} \times 1.25}$$

$$L_w = 207.45 \text{ mm}$$

$$\text{Length of weld on each side} = \frac{207.45}{2}$$

$$= 103.73 \text{ mm} \simeq 105 \text{ mm}$$

1.37 (d)

■■■■

2 Tension Member

2.1 The net effective cross sectional area calculated in the steel angle tension member design, accounts for
(a) the tensile force and bolt holes
(b) the eccentricity of the end connections and the bolt holes
(c) the effectiveness of the tack connection along the length
(d) the effectiveness of the end connection

[1995 : 1 M]

2.2 ISA 100 × 100 × 10 mm (cross-sectional area = 1908 mm²) serves as tensile member. This angle is welded to a gusset plate along *A* and *B* appropriately as shown in figure. Assuming the yield strength of the steel to be 260 N/mm² the tensile strength of this member can be taken to be approximately

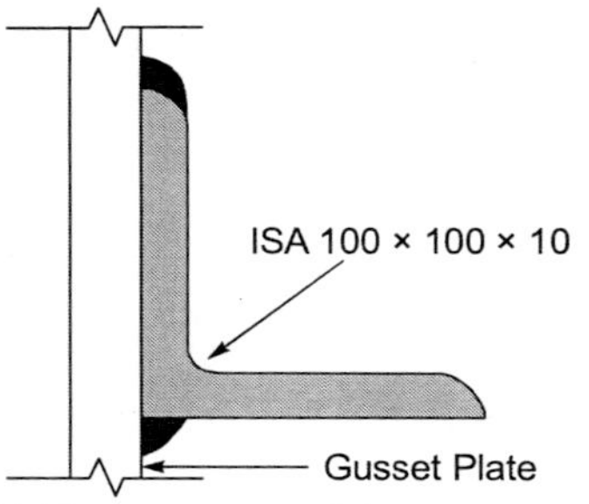

(a) 500 kN (b) 300 kN
(c) 225 kN (d) 375 kN **[2002 : 2 M]**

Common Data for Questions 2.3 and 2.4:
A truss tie consisting of 2 ISA 75 × 75 × 8 mm carries a pull of 150 kN. At ends the two angles are connected, one each on either side of a 10 mm thick gusset plate, by 18 mm diameter rivets arranged in one row. The allowable stresses in rivet are f_s = 90.0 N/mm² and f_{br} = 250 N/mm².

2.4 Minimum number of rivets required at each end is
(a) 2 (b) 3
(c) 4 (d) 5 **[2003 : 2 M]**

2.3 Maximum tensile stress in the tie in N/mm² is
(a) 93.6 (b) 87.5
(c) 77.2 (d) 66.0 **[2003 : 2 M]**

2.5 Two equal angles ISA 100 mm × 100 mm of thickness 10 mm are placed back-to-back and connected to the either side of a gusset plate through a single row of 16 mm diameter rivets in double shear. The effective areas of the connected and unconnected legs of each of these angles are 775 mm² and 950 mm², respectively. If these angles are NOT tack riveted, the net effective area of this pair of angles is
(a) 3650 mm² (b) 3450 mm²
(c) 3076 mm² (d) 2899 mm²

[2004 : 2 M]

2.6 The permissible stress in axial tension σ_{st} in steel member on the net effective area of the section shall not exceed (f_y is the yield stress)
(a) 0.80 f_y (b) 0.75 f_y
(c) 0.60 f_y (d) 0.50 f_y **[2005 : 1 M]**

2.7 In the design of welded tension members, consider the following statements:
1. The entire cross-sectional area of the connected leg is assumed to contribute to the effective area in case of angles.
2. Two angles back-to-back and tack-welded as per the codal requirements may be assumed to behave as a tee section.
3. A check on slenderness ratio may be necessary in some cases.

The TRUE statements are
(a) only 1 and 2 (b) only 2 and 3
(c) only 1 and 3 (d) 1, 2 and 3

[2006 : 2 M]

2.8 A steel flat of rectangular section of size 70 × 6 mm is connected to a gusset plate by three bolts each having a shear capacity of 15 kN in holes having diameter 11.5 mm. If the allowable tensile stress in the flat is 150 MPa, the maximum tension that can be applied to the flat is

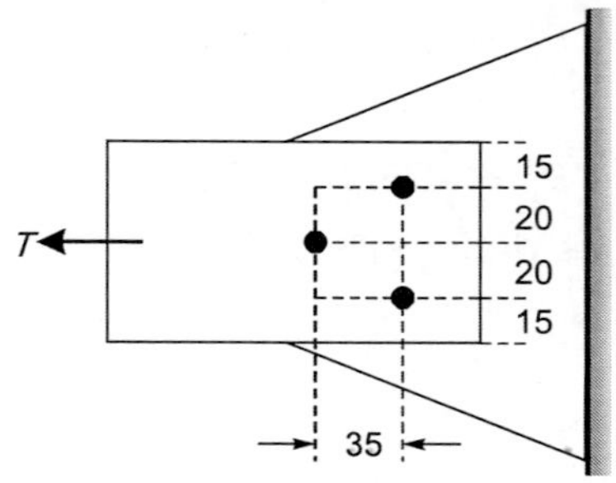

(a) 42.3 kN (b) 52.65 kN
(c) 59.5 kN (d) 63.0 kN **[2007 : 2 M]**

2.9 A steel member '*M*' has reversal of stress due to live loads, whereas another member '*N*' has reversal of stress due to wind load. As per IS 800 : 2007, the maximum slenderness ratio permitted is

(a) less for member '*M*' than that of member '*N*'
(b) more for member '*M*' than for member '*N*'
(c) same for both the members
(d) not specified in the Code

[2015 : 1 M, Set-II]

Answers Tension Member

2.1 (b) 2.2 (c) 2.3 (c) 2.4 (a) 2.5 (d) 2.6 (c) 2.7 (d) 2.8 (a) 2.9 (a)

Explanations Tension Member

2.1 (b)

When both the legs of angles are connected, i.e. shear lag effect is tno considered, then net effective cross-sectional area (A_n) is taken as minimum of A_{n1} and An_2

where A_{n1} = Gross area minus bolt hole projected area

and A_{n2} = Net area considering eccentricity of conection

2.2 (c)

A_1 = Net area of connected leg
$= (100 - 5) \times 10$
$= 950\ mm^2$

A_2 = Net area of outstand leg
$= (100 - 5) \times 10$
$= 950\ mm^2$

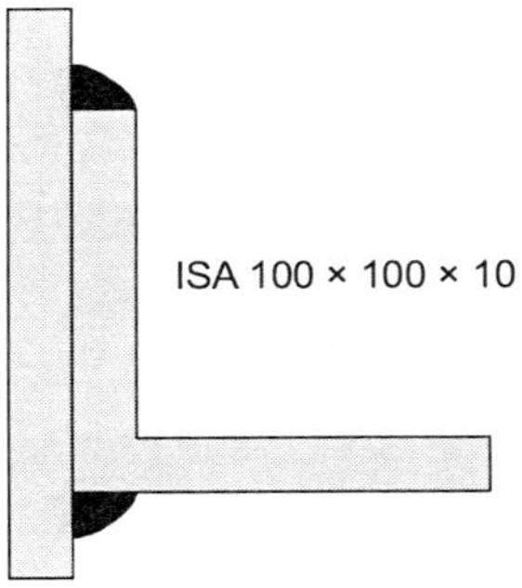

$$k = \frac{3A_1}{3A_1 + A_2} = 0.75$$

Hence $A_{net} = A_1 + kA_2$
$= 950 + 0.75 \times 950$
$= 1662.5\ mm^2$

Tensile strength of member
$= 0.6 \times f_y \times A_{net}$
$= 0.6 \times 260 \times 1662.5 = 259.4$ kN

So, on the safe side, tensile strength can be taken as 225 kN.

2.4 (c)

Strength of rivets in shearing,

$$= \frac{2 \times \pi(18 + 1.5)^2 \times 90}{4} = 53.76 \text{ kN}$$

Strength of rivets in bearing,

$= (18 + 1.5) \times 10 \times 250 = 48.75$ kN

$\therefore$ Rivet value = 48.75 kN

$$\text{Number of rivets} = \frac{150}{48.75} = 3.08 \approx 4$$

POINTS TO REMEMBER

Clearances for Fastener Holes
(Clause 10.2.1)

S.No.	Nominal size of Fastener, *d* mm	Size of the Hole = Nominal Diameter of the Fastener + Clearances mm			
		Standard Clearance in Diameter and Width of slot	Over size Clearance in Diameter	Clearance in the Length of the Slot	
				Short Slot	Long Slot
(1)	(2)	(3)	(4)	(5)	(6)
(i)	12-14	1.0	3.0	4.0	2.5 *d*
(ii)	16-22	2.0	4.0	6.0	2.5 *d*
(iii)	24	2.0	6.0	8.0	2.5 *d*
(iv)	Larger than 24	3.0	8.0	10.0	2.5 *d*

2.3 (a)

Note: Maximum tensile stress is obtained when angles are not tack bolted.

Assuming both the angle sections are not tack riveted,

$$\therefore\ k_1 = \frac{3A_1}{3A_1 + A_2} = \frac{3 \times 412}{3 \times 412 + 568} = 0.685$$

$\therefore\ A_e = (A_1 + k_1A_2) \times 2$
$= (412 + 0.685 \times 568) \times 2 = 1602.2\ mm^2$

$\therefore$ Maximum tensile stress

$$= \frac{150 \times 10^3}{1602.3} = 93.6 \text{ N/mm}^2$$

Thus the most close option is (a)

Note: As per IS : 800 2007, shear lag factor $\beta \nless 0.7$.

2.5 (d)

When angles are not tack riveted, they will be considered as single angles connected on one side of gusset plate,

$$A_e = (A_1 + kA_2) \times 2$$

$$k = \frac{3A_2}{3A_1 + A_2} = \frac{3 \times 775}{3 \times 775 + 950} = 0.71$$

$$A_e = (775 + 0.71 \times 950) \times 2$$
$$= 2899 \text{ mm}^2$$

Note: As per IS : 800-2007, $\beta \nless 0.7$. So take $\beta = k = 0.7$.

2.6 (c)

S.No.	Types of stresses	Notation	Allowable Stress	FOS
1.	Axial tensile	σ_{at}	$0.6f_y$	1.67
2.	Max. axial compressive	σ_{ac}	$0.6f_y$	1.67
3.	Bending tensile	σ_{bt}	$0.66f_y$	1.15
4.	Max. bending compressive	σ_{bc}	$0.66f_y$	1.15
5.	Average shear	τ_{va}	$0.4f_y$	2.5
6.	Max. shear	τ_{vm}	$0.45f_y$	2.22
7.	Bending stress	σ_p	$0.75f_y$	1.33
8.	Stress in slab base	σ_{bs}	185	—

2.7 (d)

Statement-1: Correct because correction factor k, is only applied to A_2 i.e., area of outstanding leg.

Statement-2: Correct because correction factor $k = 1$, when angles are connected back to back and tacked. Hence $A_{net} = A_1 + A_2$ as in case of T-section.

Statement-3: Correct because when slenderness ratio increases beyond 400, it is called as cable.

2.8 (a)

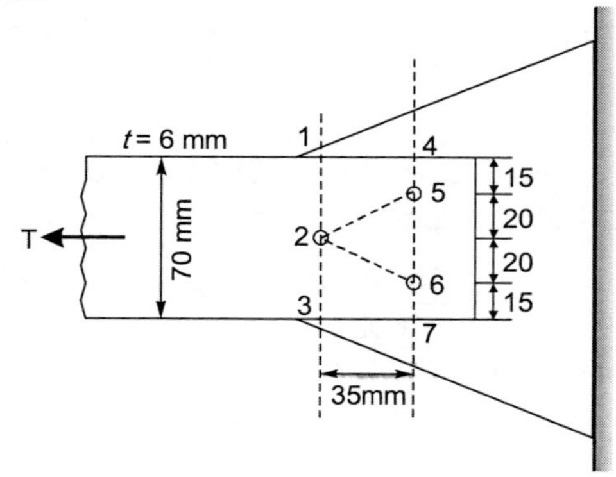

Given:

$$d_{hole} = 11.5 \text{ m}$$
$$\sigma_{at} = 150 \text{ N/mm}^2$$

Shear capacity of each bolt = 15 mm

Case-I: Let us suppose all bolt fail in shear

$\therefore$ Max T allowed = 3 × 15 = 45 kN

Hence only option below 45 kN = 42.3 kN i.e., (a)

Case-II: By failure along section (1)-(2)-(3)

Max. T allowed $= (B - d_{hole})\, t\sigma_{at}$

$$= (70 - 11.5) \times 6 \times 150$$
$$= 52.65 \text{ kN}$$

Case-III: By failure along section (4)-(5)-(2)-(6)-(7)

$$A_{net} = \left[(B - 3 \times d_{hole}) + \frac{S^2}{4g} \times 2\right] t$$

$$= \left[(70 - 3 \times 11.5) + \frac{2 \times 35^2}{4 \times 20}\right] \times 6$$

$$= 396.75 \text{ mm}^2$$

$\therefore$ Maximum T allowed

$$= A_{net} \times \sigma_{at} = 396.75 \times 150 = 59.5 \text{ kN}$$

Case-IV: By failure along (4)-(5)-(6)-(7)

$$A_{net} = (B - 2 \times d_{hole}) \times t$$
$$= (70 - 2 \times 11.5) \times 6 = 282 \text{ mm}^2$$

Max. T allowed

$$= A_{net}\, \sigma_{at} = 282 \times 150$$

Min. value among all 4 cases = 42.3 kN

Hence option (a) is correct.

2.9 (a)

As per IS 800 : 2007

S.No.	Member	Maximum Slenderness Ratio
1.	A tension member in which a reversal of direct stress due to loads other than wind or seismic forces occurs.	180
2.	A member normally acting as a tie in a roof truss or a bracing system but subjected to possible reversal of stresses resulting from the action of the wind or earthquake forces.	350
3.	Members always under tension (other than pre-tensioned members).	400

■■■■

3 Compression Member

3.1 The maximum allowable compressive stress corresponding to lateral buckling in a discretely laterally supported symmetrical I-beam does not depend upon

(a) the modulus of elasticity
(b) the radius of gyration about the minor axis
(c) the span/length of the beam
(d) the ratio of overall depth to thickness of the flange [1995 : 1 M]

3.2 As per **IS 800 : 1984**, the maximum allowable slenderness ratio of compression members carrying forces resulting from dead load and superimposed load is

(a) 180 (b) 250
(c) 300 (d) 400 [1995 : 1 M]

3.3 The effective length of a circular electric pole of length L and constant diameters erected on ground is

(a) 0.80 L (b) 1.20 L
(c) 1.50 L (d) 2.00 L [1996 : 1 M]

3.4 In **IS 800 : 1984** the permissible compressive stress in column is based on

(a) Euler formula
(b) Secant formula
(c) Rankine Gordan formula
(d) Rankine Merchant formula [1996 : 1 M]

3.5 The problem of lateral buckling can arise only in those steel beams which have

(a) moment of inertia about the bending axis larger than the other
(b) moment of inertia about the bending axis smaller than the other
(c) fully supported compression flange
(d) None of the above [1999 : 1 M]

3.6 Consider the following two statements related to structural steel design, and identify whether they are True or False

I. The Euler buckling load of a slender steel column depends on the yield strength of steel.
II. In the design of laced column, the maximum spacing of the lacing does not depend on the slenderness of column as a whole.

(a) Both statement I and II are True
(b) Statement I is True, and statement is False
(c) Statement I is False, and statement II is True
(d) Both statements I and II are False

[2001 : 2 M]

3.7 In the design of lacing system for a built-up steel column, the maximum allowable slenderness ratio of a lacing bar is

(a) 120 (b) 145
(c) 180 (d) 250 [2003 : 1 M]

3.8 A strut in a steel truss is composed of two equal angles ISA 150 mm × 150 mm of thickness 10 mm connected back-to-back to the same side of a gusset plate. The cross sectional area of each angle is 2921 mm^2 and moment of inertia ($I_{xx} = I_{yy}$) is 6335000 mm^4. The distance of the centroid of the angle from its surface ($C_x = C_y$) is 40.8 mm. The minimum radius of gyration of the strut is

(a) 93.2 mm (b) 62.7 mm
(c) 46.6 mm (d) 29.8 mm

[2004 : 2 M]

3.9 Consider the following statements:

1. Effective length of a battened column is usually increased to account for the additional load on battens due to the lateral expansion of columns.
2. As per **IS 800 : 1984**, permissible stress in bending compression depends on both Euler buckling stress and the yield stress of steel.
3. As per **IS 800 : 1984**, the effective length of a column effectively held in position at both ends but not restrained against rotation, is taken to be greater than that in the ideal end conditions.

The TRUE statements are

(a) 1 and 2 (b) 2 and 3
(c) 1 and 3 (d) 1, 2 and 3

[2006 : 2 M]

3.10 Consider the following statements for a compression member:

1. The elastic critical stress in compression increases with decrease in slenderness ratio.
2. The effective length depends on the boundary conditions at its ends.
3. The elastic critical stress in compression is independent of the slenderness ratio
4. The ratio of the effective length to its radius of gyration is called as slenderness ratio.

Which of the above statements is/are correct?

(a) 2 and 3 (b) 3 and 4
(c) 2, 3 and 4 (d) 1, 2 and 4

[2009 : 2 M]

3.11 The square root of the ratio of moment of inertia of the cross-section to its cross-sectional area is called

(a) second moment of area
(b) slenderness ratio
(c) section modulus
(d) radius of gyration [2009 : 1 M]

3.12 A steel column is restrained against both translation and rotation at one end and is restrained only against rotation but free to translate at the other end. Theoretical and design (IS:800-2007) values, respectively, of effective length factor of the column are

(a) 1.0 and 1.2 (b) 1.2 and 1.0
(c) 1.0 and 1.0 (d) 1.2 and 1.2

[2019 : 1 M, Set-II]

3.13 The critical bending compressive stress in the extreme fibre of a structural steel section is 1000 MPa. It is given that the yield strength of the steel is 250 MPa, width of flange is 250 mm and thickness of flange is 15 mm. As per the provisions of IS: 800-2007. the non-dimensional slenderness ratio of the steel cross-section is

(a) 0.50 (b) 0.75
(c) 0.25 (d) 2.00

[2019 : 2 M, Set-II]

3.14 A homogeneous, prismatic, linearly elastic steel bar fixed at both the ends has a slenderness ratio (l/r) of 105, where l is the bar length and r is the radius of gyration. The coefficient of thermal expansion of steel is $12 \times 10^{-6}/°$ C. Consider the effective length of the steel bar as $0.5l$ and neglect the self-weight of the bar.

The differential increase in temperature (rounded off to the nearest integer) at which the bar buckles is

(a) 85 °C (b) 250 °C
(c) 400 °C (d) 298 °C

[2024 : 2 M, Set-II]

■■■■

Answers Compression Member

3.1 (a)	3.2 (a)	3.3 (d)	3.4 (d)	3.5 (a)	3.6 (d)	3.7 (b)
3.8 (c)	3.9 (a)	3.10 (d)	3.11 (d)	3.12 (a)	3.13 (a)	3.14 (d)

Explanations Compression Member

3.1 (a)

The allowable compressive stress depends upon D/T ratio and l/r_{min} ratio as per **IS code 800 : 1984.** It does not depend upon the modulus of elasticity.

3.2 (a)

Type of member	Maximum slenderness ratio
A member carrying compressive loads resulting from dead load and superimposed loads	180
A member subjected to compressive loads resulting from wind/earthquake force provided the deformation of such members does not adversely affect the stress in any part of the structure	250
A member normally carrying tension but subjected to reversal of stress due to wind or earthquake force	350

3.3 (d)

When the pole sways perpendicular to electric wire that it is carrying, the effective length L_{eff} becomes 2.0 L.

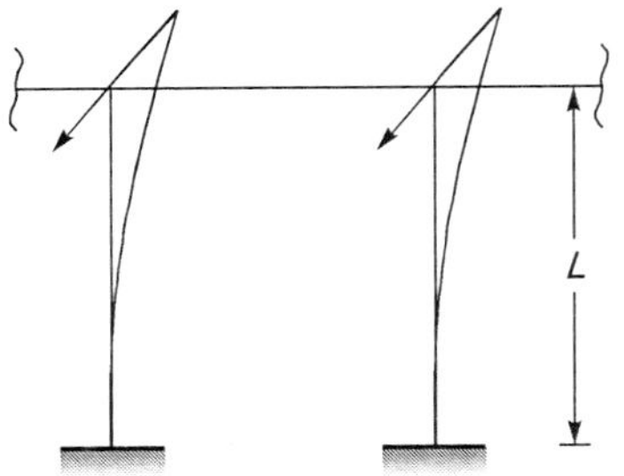

When it sways along the electric wire, $L_{eff} = 0.8L$.

$\therefore$ Answer = 2.0 L.

3.4 (d)

Permissible stress in axial compression (σ_{ac}) is based on **Rankine-merchant formula** and its magnitude is,

$$\sigma_{ac} = 0.6 \frac{f_{cc} \cdot f_y}{(f_{cc}^n + f_y^n)^{1/n}}$$

where, f_{cc} = elastic critical stress in compression = $\frac{\pi^2 E}{\lambda^2}$

λ = slenderness ratio = l/r

l = effective length of the compression member

r = appropriate radius of gyration of the member (minimum value)

E = modulus of elasticity of steel = 2×10^5 MPa

n = a factor assumed as 1.4

3.5 (a)

- Lateral buckling or flexural buckling means bending or warping of steel element out of its axis or plane under the action of lateral force.
- It is a deflection caused by bending about the axis corresponding to the largest slenderness ratio.
- This is usually the minor prinicpal axis–The one with minimum radius of gyration.

3.6 (d)

I. Euler's buckling load = $\frac{\pi^2 EI}{l^2}$

$\therefore$ Euler's buckling load is independent of yield strength of steel.

II. Maximum spacing of lacing bars shall be such that the maximum slendernes of the main member between adjacent lacing connection should not be greater than 50 or $0.7 \times \lambda_{whole}$.

3.7 (b)

As per **Clause 7.6.6.3 of IS 800 : 2007,** the slenderness ratio of lacing bars $\frac{kL}{r}$, shall not exceed 145.

POINTS TO REMEMBER

Here, kL = Effective length of lacing bar

L = Distance between inner end bolts/rivets/weld

For single lacing k = 1.0

For double lacing k = 0.7

For welding k = 0.7

3.8 (c)

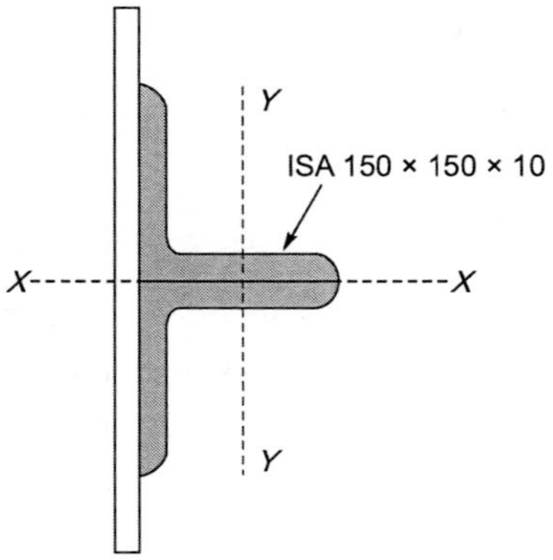

Total cross sectional area

$= 2A = 2 \times 2921 = 5842 \text{ mm}^2$

The minimum moment of inertia will be about $Y - Y$ axis,

$I_{yy} = 2\,I_{yi} = 2 \times 6335000 = 12670000 \text{ mm}^4$

Minimum radius of gyration,

$$r_{min} = \sqrt{\frac{I_{yy}}{2A}} = \sqrt{\frac{12670000}{5842}} = 46.6 \text{ mm}$$

POINTS TO REMEMBER

Students get confused with C_x and C_y

C_x = distance of CG from x-axis

C_y = distance of CG from y-axis

3.9 (a)

Statement 1: Correct

Because effective length of battened column is increased by 10%. Increased length means, more slenderness ratio and hence, the column is assumed toslightly weaker than it actually is to account for more safety.

Statement-2: Correct

Statement-3: Wrong

Because, the ideal condition .... same.

The ideal condition is that column is effectively held in position at both ends but not restrained against rotation. **IS 800 : 1984** prescribes the same value of effective length as taken for ideal end condition. Hence 3 is false.

3.10 (d)

Statement-1: Correct

Because, $f_{CD} \propto \dfrac{1}{\lambda}$

Statement-2: Correct

Because

Boundary conditions	Effective length
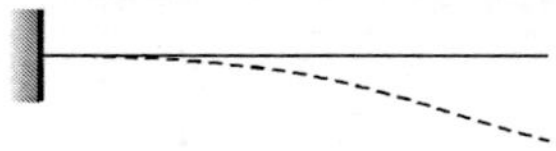	2.0 L
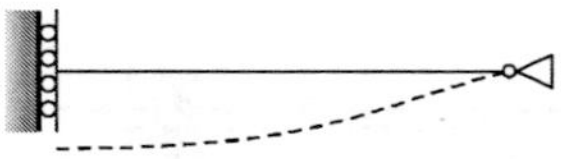	2.0 L
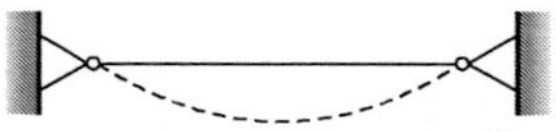	1.0 L
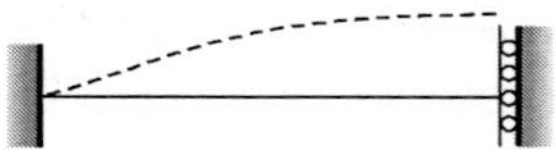	1.2 L
	0.8 L
	0.65 L

Statement-3: Incorrect

Because, $f_{cr} = \dfrac{\pi^2 E}{\lambda^2}$, where λ = slenderness ratio of compression member.

Statement-4: Correct

Because, $\lambda = \dfrac{kL}{r}$

kL = effective length

r = radius of gyration

3.11 (d)

Radius of gyration, $r = \sqrt{\dfrac{\text{Moment of inertia}}{\text{Cross-sectional area}}}$

3.12 (a)

3.13 (a)

$$\lambda = \sqrt{\frac{f_y}{f_{cr}}} = \sqrt{\frac{250}{1000}} = 0.5$$

3.14 (d)

$$\frac{Pe}{A} = \alpha \Delta T E$$

$$\Rightarrow \quad \frac{\pi^2 E}{\lambda^2} = \alpha \Delta T E$$

$$\Rightarrow \quad \Delta T =$$

$$\frac{\pi^2}{\alpha \cdot \lambda^2} = \frac{\pi^2}{\alpha\left[\frac{0.5L}{R}\right]^2} \qquad \left[\because \lambda = \frac{L}{R}\right]$$

$$= \frac{\pi^2}{12 \times 10^{-6} \times [0.5 \times 105]^2} = 298^\circ\text{C}$$

4 Beams

4.1 Generally the maximum deflection/span ratio of a steel member should not exceed

(a) $\frac{1}{750}$ (b) $\frac{1}{500}$

(c) $\frac{1}{325}$ (d) $\frac{1}{250}$ [1996 : 1 M]

4.2 Allowable average shear stress in an unstiffened web for beams made of steel of grade 250 N/mm^2 is

(a) 250 N/mm^2 (b) 165 N/mm^2

(c) 150 N/mm^2 (d) 100 N/mm^2

[1997 : 1 M]

4.3 A steel beam supporting loads from the floor slab as well as from wall is termed as

(a) Stringer beam (b) Lintel beam

(c) Spandrel beam (d) Header beam

[1999 : 1 M]

4.4 An ISMB 500 is used as a beam in a multi-storey construction. From the viewpoint of structural design, it can be considered to be 'laterally restrained' when

(a) the tension flange is 'laterally restrained'

(b) the compression flange is 'laterally restained'

(c) the web is adequately stiffened

(d) the conditions in (A) and (C) are met

[2002 : 1 M]

4.5 When designing steel structures, one must ensure that local buckling in webs dos not take place. This check may not be very critical when using rolled steel sections because

(a) Quality control at the time of manufacture of rolled sections is very good

(b) Web depths available are small

(c) Web stiffeners are in built in rolled sections

(d) Depth to thickness ratio (of the web) are appropriately adjusted [2002 : 1 M]

4.6 **List-I** contains some elements in design of a simply supported plate girder and **List-II** gives some qualitative locations on the girder. Match the items of two lists as per good design practice and relevant codal provisions.

List-I

A. Flange splice

B. Web splice

C. Bearing stiffeners

D. Horizontal stiffener

List-II

1. At supports (minimum)
2. Away from centre of span
3. Away from support
4. In the middle of span
5. Longitudinally somewhere in the compression flange

Codes:

	A	B	C	D
(a)	2	3	1	5
(b)	4	2	1	3
(c)	3	4	2	1
(d)	1	5	2	3

[2003 : 2 M]

4.7 A square steel slab base of area 1 m^2 is provided for a column made of two rolled channel sections. The 300 mm × 300 mm column carries an axial compressive load of 2000 kN. The line of action of the load passes through the centroid of the column section as well as of the slab base. The permissible bending stress in the slab base is 185 MPa. The required minimum thickness of the slab base is

(a) 110 mm (b) 89 mm

(c) 63 mm (d) 55 mm [2004 : 2 M]

4.8 An unstiffened web I-section is fabricated from a 10 mm thick plate by fillet welding as shown in the figure. If yield stress of steel is 250 MPa, the maximum shear load that section can take is

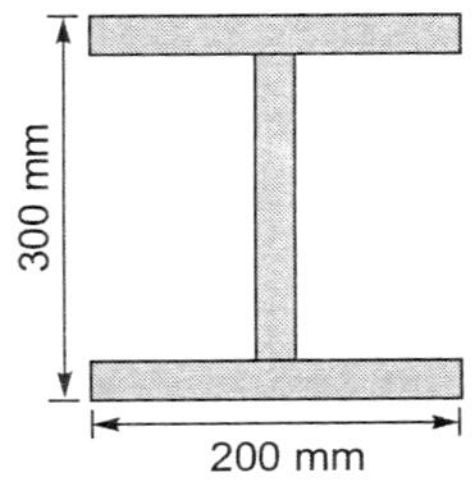

(a) 750 kN (b) 350 kN

(c) 337.5 kN (d) 300 kN [2005 : 2 M]

4.9 The adjoining figure shows a schematic representation of a steel plate girder to be used as a simply supported beam with a concentrated load. For stiffeners, *PQ* (running along the beam axis) and *RS* (running between the top and bottom flanges) which of the following pairs of statements will be TRUE?

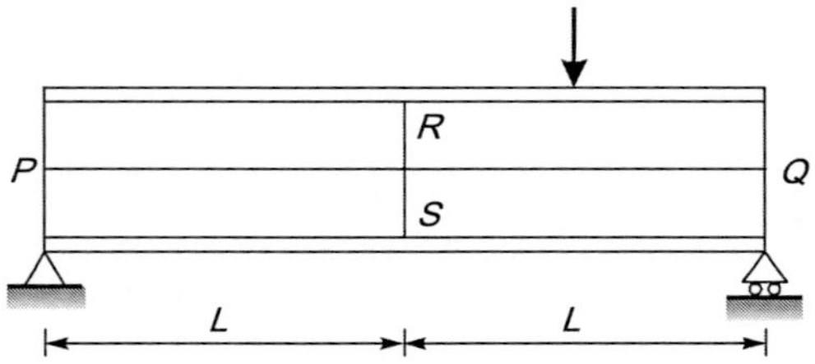

(a) (i) *RS* should be provided under the concentrated load only
  (ii) *PQ* should be placed in the tension side of the flange.
(b) (i) *RS* helps to prevent local buckling of the web.
  (ii) *PQ* should be placed in the compression side of the flange.
(c) (i) *RS* should be provided at supports.
  (ii) *PQ* should be placed along the neutral axis
(d) (i) *RS* should be provided away from points of action of concentrated loads.
  (ii) *PQ* should be provided on the compression side of the flange.

[2011 : 2 M]

4.10 A steel section is subjected to a combination of shear and bending actions. The applied shear force is V and the shear capacity of the section is V_s. For such a section, high shear force (as per **IS 800 : 2007**) is defined as

(a) $V > 0.6\ V_s$ (b) $V > 0.7\ V_s$
(c) $V > 0.8\ V_s$ (d) $V > 0.9\ V_s$

[2014 : 1 M, Set-I]

4.11 The semi-compact section of a laterally unsupported steel beam has an elastic section modulus, plastic section modulus and design bending compressive stress of 500 cm^3, 650 cm^3 and 200 MPa respectively. The design flexural capacity (expressed in kNm) of the section is_______. [2016 : 1 M, Set-I]

4.12 A steel column of ISHB 350 @ 72.4 kg/m is subjected to a factored axial compressive load of 2000 kN. The load is transferred to a concrete pedestal of grade M20 through a square base plate. Consider bearing of concrete as 0.45 f_{ck}, where f_{ck} is the characteristic strength of concrete. Using limit state method and neglecting the self weight of base plate and steel column, the length of a side of the base plate to be provided is

(a) 39 cm (b) 42 cm
(c) 45 cm (d) 48 cm

[2018 : 1 M, Set-I]

4.13 Assuming that there is no possibility of shear buckling in the web, the maximum reduction permitted by IS 800-2007 in the (low-shear) design bending strength of a semi-compact steel section due to high shear is

(a) Zero
(b) governed by the area of the flange
(c) 50%
(d) 25%

[2019 : 1 M, Set-I]

4.14 A rolled I-section beam is supported on a 75 mm wide bearing plate as shown in the figure. Thicknesses of flange and web of the I-section are 20 mm and 8 mm, respectively. Root radius of the I-section is 10 mm. Assume: material yield stress, f_y = 250 MPa and partial safety factor for material. γ_{mo} = 1.10.

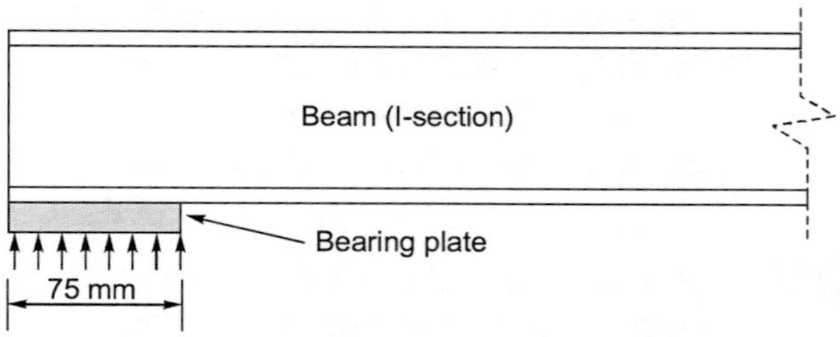

As per IS: 800-2007, the web bearing strength (in kN, round off to 2 decimal places) of the beam is_________.

[2019 : 2 M, Set-II]

Answers Beams

4.1	(c)	4.2	(d)	4.3	(a)	4.4	(b)	4.5	(d)	4.6	(a)	4.7	(d)
4.8	(d)	4.9	(b)	4.10	(a)	4.11	(100)	4.12	(d)	4.13	(a)	4.14	(272.73)

Explanations Beams

4.1 (c)

Generally, the maximum deflection for a beam shall not exceed deflection for a beam shall not exceed 1/325 of span. This limit may be exceeded in cases where greater deflection would not reduce the efficiency of structure.

4.2 (d)

According to **IS 800 : 1984**, allowable shear stress for unstiffened web of steel

$= 0.4\, f_y = 0.4 \times 250 = 100$ N/mm^2

4.3 (a)

- Spandral beam supports load from exterior wall and slab and spans from column to column.
- Stringer beams are secondary beams used to carry the load from the slab till the cross beam are located at truss nodes in a truss bridge.

4.4 (b)

A beam in which compression flange is restrained laterally is called a laterally restrained beam.

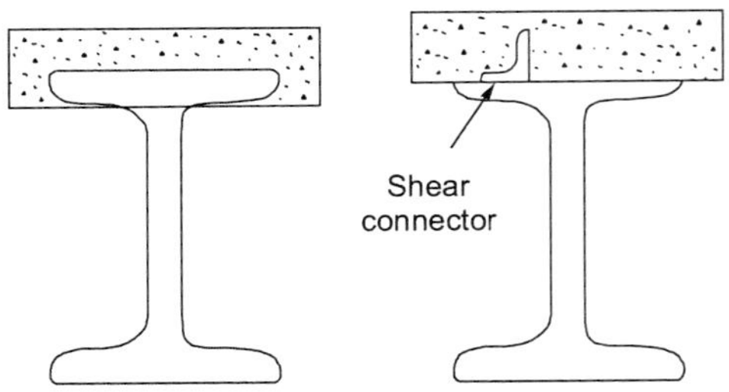

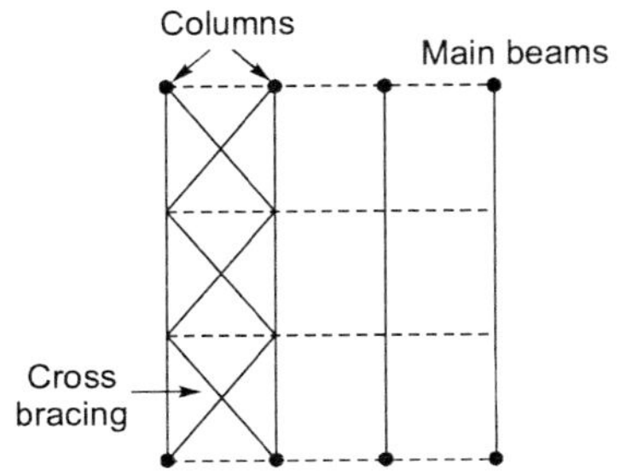

Beam with lateral support

4.5 (d)

Local buckling can be prevented by adopting higher thickness of element i.e., by controlling width to thickness ratio of component plates of steel sections.

IS 800 : 2007 places limit on width to thickness ratio of component plates of steel section and given in table 2 of IS 800 : 2007.

4.6 (a)

Flangesplice: Away from the centre of span because flanges bear the bending stresses and bending moment in lesser at the places away from centre.

Web splice: Web takes care of shearing stress and it is maximum at supports.

Bearing stiffners: Bearing stiffners are to be placed at the places where concentrated loads or reaciton acts i.e., at supports.

Horizontal stiffners: The task of horizontal stiffners is to prevent buckling of web.

4.7 (d)

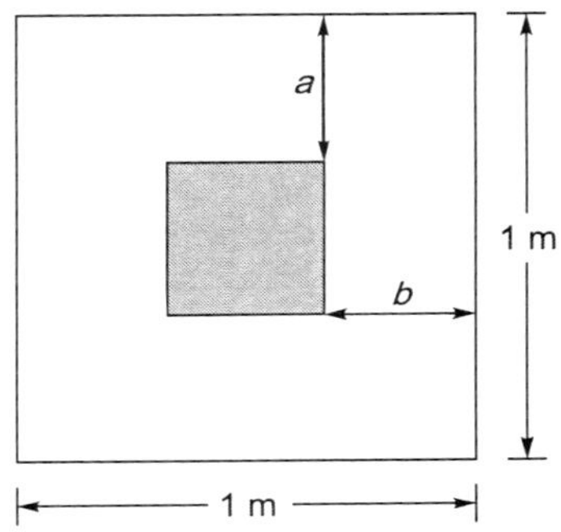

300 mm × 300 mm column leaves projections as,

$$a = b = \frac{1000 - 300}{2} = 350 \text{ mm}$$

Pressure on the underside of base slab

$$w = \frac{2000}{1} = 2000 \text{ kN/m}^2 = 2 \text{ N/mm}^2$$

As per IS : 800-1984,
Thickness of slab base,

$$t = \sqrt{\frac{3w}{F_b}\left(a^2 - \frac{b^2}{4}\right)} = \sqrt{\frac{3 \times 2}{185}\left(350^2 - \frac{350^2}{4}\right)}$$

$$t \simeq 55 \text{ mm}$$

POINTS TO REMEMBER

In LSM $\quad t = \sqrt{\dfrac{2.5w\left(a^2 - 0.3b^2\right)\gamma_{mo}}{f_y}} > t_f$

w = pressure from below the slab
a = larger projection
b = smaller projection
t_f = flange thickness of compression member

4.8 (d)

Shear force will be resisted by shaded portion

The web depth to thickness ratio

$$= \frac{300 - 20}{10} = 28 < 85$$

$\therefore$ $V = f_s t_w d$

$d = 300$ mm, $f_s = 0.4 f_y$

$f_y = 0.4 \times 250 = 100$ N/mm^2

$t_w = 10$ mm

$\Rightarrow$ $V = 100 \times 10 \times 300 = 300$ kN

4.9 (b)

PQ is a horizontal stiffener in the given plate girder. Horizontal stiffeners are also called longitudinal stiffeners. The horizontal stiffener are provided in the compression zone of the web. The first horizontal stiffener is provided at one-fifth of the distance from the compression flange to the tension flange. If required another stiffener is provided at the neutral axis. Horizontal stiffeners are not continuous and are provided between vertical stiffeners.

RS is a vertical stiffener in the given plate girder. Vertical stiffeners are also called transverse stiffeners. It is assumed that the vertical stiffener is not subjected to any load and is selected to provide necessary lateral stiffness only and can therefore, be crimped or joggled for tight fittings. Such stiffeners increase the buckling resistance of the web caused by shear.

4.10 (a)

According to IS 800 : 2007, clause 9.2.1 No reduction in moment capacity of the section is necessary as long as the cross-section is not subjected to high shear force (factored value of shear force in less than or equal to 60 percent of the shear strength of the section).

4.11 Sol.

Design flexural capacity,

$$M_d = \beta_b Z_p \frac{f_y}{\gamma_{m0}}$$

$\beta_b = 1.0$ (for plastic and compact sections)

$= \frac{Z_e}{Z_p}$ (for semi-compact sections)

Z_p = plastic section modulus

Z_e = elastic section modulus

f_y = yeild stress of steel provided

γ_{m0} = material factor of safety of steel against yeilding

As, design bending stress is directly given,

$$M_d = \beta_b Z_p f_d$$

$$= \frac{Z_e}{Z_p} \times Z_p \times f_d = Z_e \times f_d$$

$= 500 \times 10^3 \times 200 \times 10^{-6} = 100$ kN-m

4.12 (d)

Area required for base plate

$$= \frac{\text{Factored load}}{\text{Bearing capacity of concrete}}$$

$$= \frac{2000 \times 10^3}{0.45 \times 20} = 222222.2 \text{ mm}^2$$

So, side of base plate $= \sqrt{\text{Area}}$

$= 471.405$ mm $= 40.8$ cm $\simeq 47.14$ cm

Since, provided area must be more than required so answer should be 48 cm.

4.13 (a)

As per IS 800 : 2007

For semi compact section

(i) In low shear case ($V \leq 0.6\ V_d$)

$$M_d = Z_e f_y / \gamma_{mo}$$

(ii) In high shear case ($V > 0.6\ V_d$)

$$M_d = Z_e f_y / \gamma_{mo}$$

So reduction is zero.

4.14 Sol.

Web bearing strength

$$= \left[b + 2.5(t_f + R)\right] \times t_w \times \frac{f_y}{\gamma_{mo}}$$

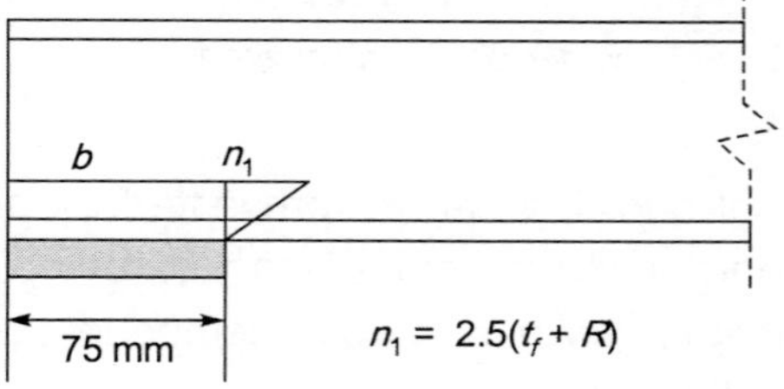

$$= \left[75 + 2.5(20 + 10)\right] \times 8 \times \frac{250}{1.1} = 272.73 \text{ kN}$$

5 Plate Girders and Industrial Roofs

5.1 Effective length of a rafter member between two nodes at a distance L, perpendicular to the plane of the truss is

(a) 2.00 L (b) 0.85 L

(c) 1.50 L (d) 1.00 L **[1997 : 1 M]**

5.2 The maximum permissible deflection for a gantry gride, spanning over 6 m, on which an EOT (electric overhead travelling) crane of capacity 200 kN is operating, is

(a) 8 mm (b) 10 mm

(c) 12 mm (d) 18 mm **[1998 : 1 M]**

5.3 Which of the following elements of a pitched roof industrial steel building primarily resists lateral load parallel to the ridge?

(a) bracings (b) purlins

(c) truss (d) columns **[2003 : 1 M]**

5.4 In a plate girder, the web plate is connected to the flange plates by fillet welding. The size of the fillet welds is designed to safely resist

(a) the bending stresses in the flanges

(b) the vertical shear force at the section

(c) the horizontal shear force between the flanges and the web plate

(d) the forces causing buckling in the web

[2004 : 1 M]

Answers Plate Girders and Industrial Roofs

5.1 (d) 5.2 (a) 5.3 (a) 5.4 (c)

Explanations Plate Girders and Industrial Roofs

5.1 (d)

5.2 (a)

The deflection of gantry girders under dead and imposed loads should not exceed the following values as per **IS 800 : 1984**

(a) Where cranes are manually operated $L/500$

(b) Where electric overhead travelling $L/750$ crane are operated upto 50t or 500 kN

(c) Where electric overhead travelling $L/1000$ cranes are operated over 50t or 500 kN

(d) Other moving loads such as charging cars etc. $L/600$

Where L = span of crane runway girder.

$\therefore$ For EOT crane with load capacity of 200 kN, maximum deflection

$$= \frac{L}{750} = \frac{6000}{750} = 8 \text{ mm}$$

5.3 (a)

Truss are quite stable in their own plane. However in the out of plane, these are very weak. To make the trusses stable to carry out of plane loads, along the ridge and lateral loads (wind and earthquake) the truss must be braced.

5.4 (c)

The size of the fillet welds in design to safely resist the horizontal shear force between the flanges and the web.

6 Plastic Analysis

6.1 Which one of the following conditions, both elastic and plastic methods of analysis of indeterminate structures have to satisfy?
(a) yield condition
(b) mechanism condition
(c) equilibrium
(d) compatibility of deformation [1995 : 1 M]

6.2 If yy is the centroidal axis of a T-beam section subjected to plastic moment M_p, the neutral axis lies

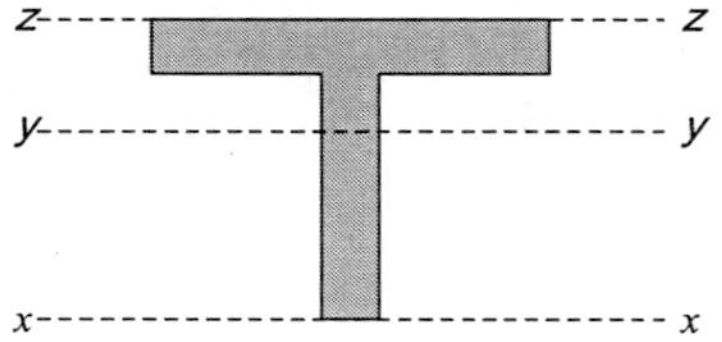

(a) above the line zz
(b) between the lines yy and zz
(c) between the lines xx and yy
(d) below the line xx [1996 : 1 M]

6.3 A propped cantilever beam is shown in the figure. The plastic moment capacity of the beam is M_p. The collapse load P is

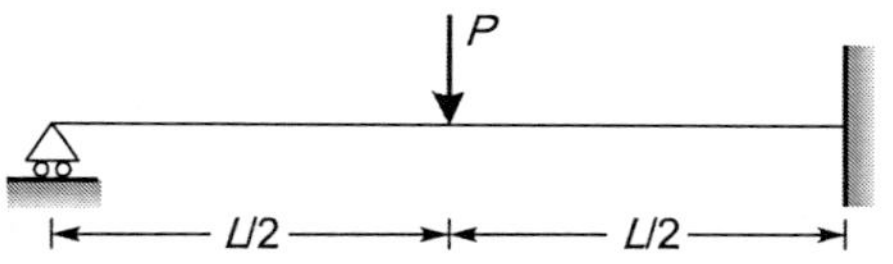

(a) $4\,M_p/L$ (b) $6\,M_p/L$
(c) $8\,M_p/L$ (d) $12\,M_p/L$ [1998 : 1 M]

6.4 The plastic modulus of a section is 4.8×10^{-4} m^2. The shape factor is 1.2. The plastic moment capacity of the section is 120 kNm. The yield stress of the material is
(a) 100 MPa (b) 240 MPa
(c) 250 MPa (d) 300 MPa
[1998 : 2 M]

6.5 For a fixed beam with span L, having plastic moment capacity of M_p, the ultimate central concentrated load will be
(a) $\frac{4M_p}{L}$ (b) $\frac{M_p}{8L}$
(c) $\frac{6M_p}{L}$ (d) $\frac{8M_p}{L}$ [1999 : 1 M]

6.6 The shape factor of the section shown in figure is

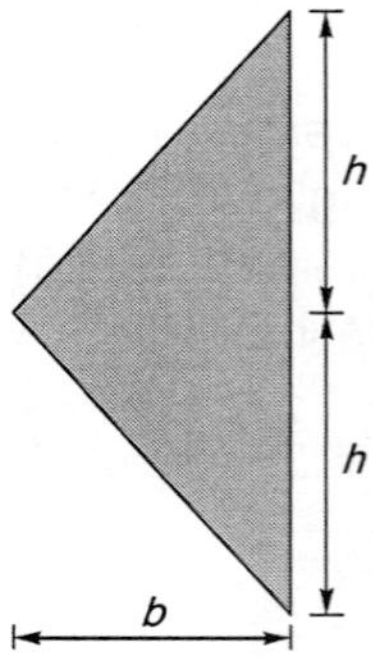

(a) 1.5 (b) 1.12
(c) 2 (d) 1.7 [1999 : 1 M]

6.7 The four cross-sections shown below are required to be ordered in the increasing order of their respective shape factor

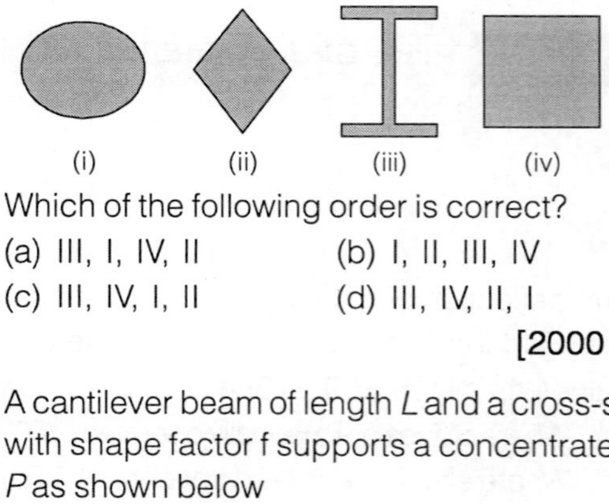

(i) (ii) (iii) (iv)

Which of the following order is correct?
(a) III, I, IV, II (b) I, II, III, IV
(c) III, IV, I, II (d) III, IV, II, I
[2000 : 1 M]

6.8 A cantilever beam of length L and a cross-section with shape factor f supports a concentrated load P as shown below

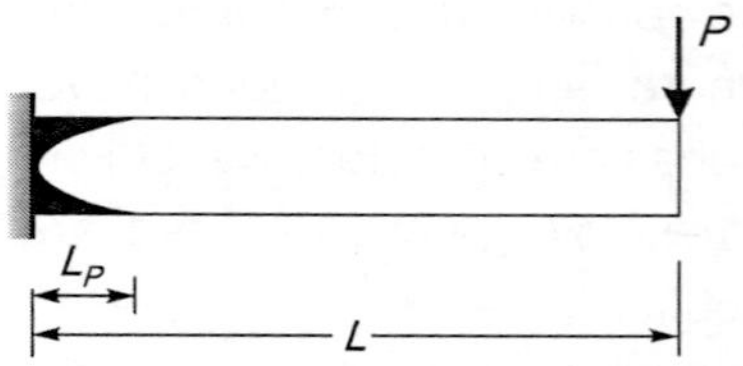

The length L_p of the plastic zone, when the maximum bending moment, equals the plastic moment M_p, given by
(a) $\frac{L_p}{L} = \frac{1}{f}$ (b) $\frac{L_p}{L} = L(1-f)$

(c) $\frac{L_p}{L} = 1 - \frac{1}{\sqrt{f}}$ (d) $\frac{L_p}{L} = 1 - \frac{1}{f}$

[2000 : 1 M]

6.9 A steel beam (with a constant EI and span L) is fixed at both ends and carries a uniformly distributed load (w kN/m), which is gradually increased till the beam reaches the stage of plastic collapse (refer to the following figure). Assuming 'B' to be at mid-span, which of the following is true

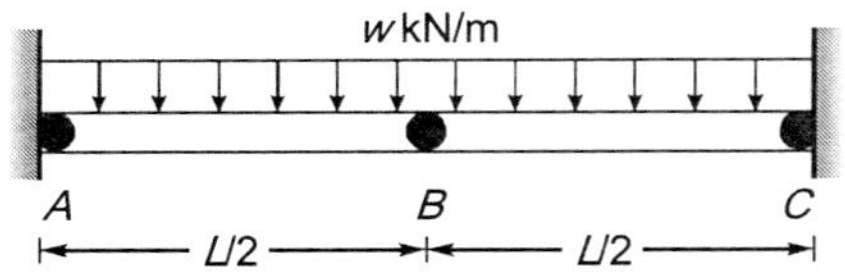

(a) Hinges are formed at A, B and C together
(b) Hinges are formed at B and then at A and C together
(c) Hinges are formed at A and C together and then at B
(d) Hinges are formed at A and C only

[2002 : 2 M]

6.10 A steel portal frame has dimensions, plastic moment capacities and applied loads as shown in the figure. The vertical load is always twice of the horizontal load. The collapse load P required for the development of a beam mechanism is

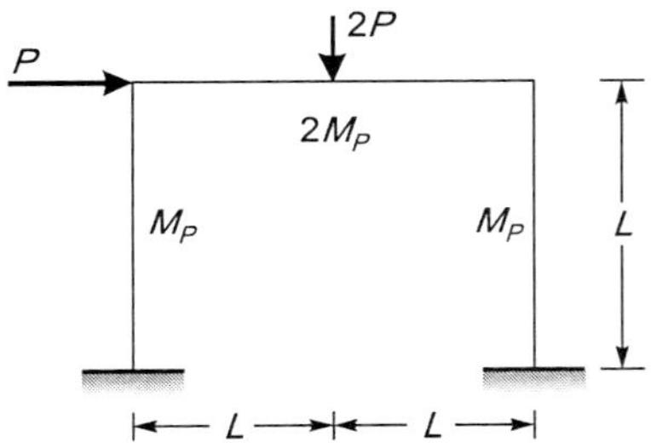

(a) $\frac{3M_P}{L}$ (b) $\frac{4M_P}{L}$
(c) $\frac{6M_P}{L}$ (d) $\frac{8M_P}{L}$ [2003 : 2 M]

6.11 A propped cantilever of span L is carrying a vertical concentrated load acting at midspan. The plastic moment of the section is M_P. The magnitude of the collapse load is

(a) $\frac{8M_P}{L}$ (b) $\frac{6M_P}{L}$
(c) $\frac{4M_P}{L}$ (d) $\frac{2M_P}{L}$ [2004 : 2 M]

6.12 A cantilever beam of length L, width b and depth d is loaded with a concentrated vertical load at the tip. If yielding starts at a load P, the collapse load shall be

(a) 2.0 P (b) 1.5 P
(c) 1.2 P (d) P [2005 : 2 M]

6.13 When the triangular section of a beam as shown below becomes a plastic hinge, the compressive force acting on the section (with σ_y denoting the yield stress) becomes

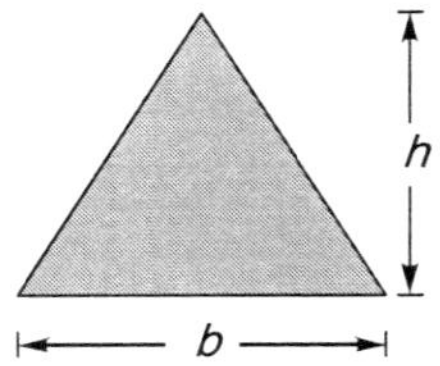

(a) $\frac{bh\sigma_y}{4}$ (b) $\frac{2bh\sigma_y}{9}$
(c) $\frac{bh\sigma_y}{2}$ (d) $\frac{bh\sigma_y}{3}$ [2006 : 2 M]

6.14 The plastic collapse load W_P for the propped cantilever supporting two point loads as shown in figure in terms of plastic moment capacity, M_P is given by

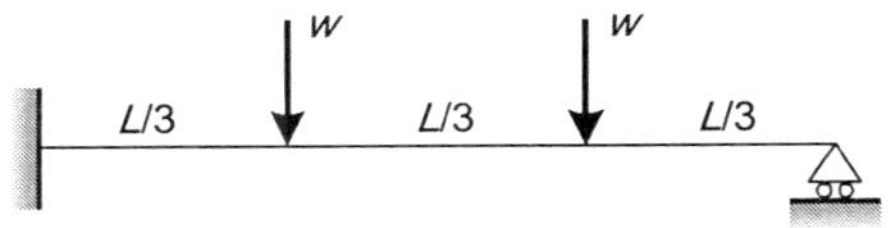

(a) $\frac{3M_P}{L}$ (b) $\frac{4M_P}{L}$
(c) $\frac{5M_P}{L}$ (d) $\frac{6M_P}{L}$ [2007 : 2 M]

6.15 The shape of the cross-section, which has the largest shape factor, is

(a) rectangular (b) I-section
(c) diamond (d) solid circular

[2008 : 1 M]

6.16 A continuous beam is loaded as shown in the figure below. Assuming a plastic moment capacity equal to M_P, the minimum load at which the beam would collapse is

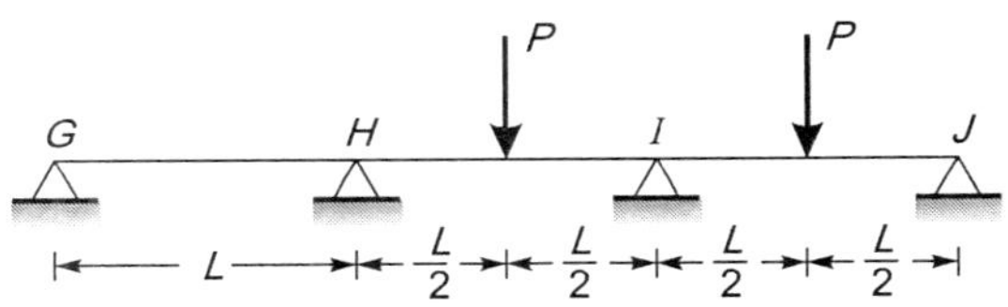

(a) $\frac{4M_P}{L}$ (b) $\frac{6M_P}{L}$
(c) $\frac{8M_P}{L}$ (d) $\frac{10M_P}{L}$ [2008 : 2 M]

6.17 In the theory of plastic bending of beams, the ratio of plastic moment to yield moment is called
(a) shape factor
(b) plastic section modulus
(c) modulus of resilience
(d) rigidity modulus [2009 : 1 M]

6.18 The value of W that results in the collapse of the beam shown in the adjoining figure and having a plastic moment capacity of M_p is

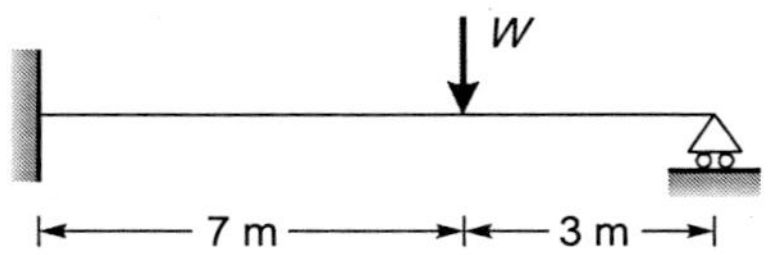

(a) $(4/21)M_p$ (b) $(3/10)M_p$
(c) $(7/21)M_p$ (d) $(13/21)M_p$
[2011 : 2 M]

6.19 As per IS 800 : 2007 the cross-section in which extreme fibre can reach the yield stress but can not develop the plastic moment of resistance due to local buckling is classified as
(a) plastic section
(b) compact section
(c) semi compact section
(d) shear section [2013 : 1 M]

Linked Answer Questions (6.20 and 6.21):
A propped cantilever made of a prismatic steel beam is subjected to a concentrated load P at mid span as shown.

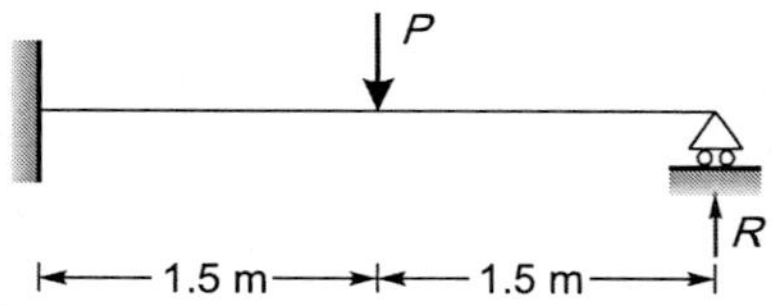

6.20 If load P = 80 kN, find the reaction R (in kN) (correct to 1 decimal place) using elastic analysis ________.
[2013 : 2 M]

6.21 If the magnitude of load P is increased till collapse and the plastic moment carrying capacity of steel beam section is 90 kNm, determine reaction R (in kN) (correct to 1-decimal place) using plastic analysis ________.
[2013 : 2 M]

6.22 The ultimate collapse load (P) in terms of plastic moment M_p by kinematic approach for a propped cantilever of length L with P acting at its mid-span as shown in the figure, would be

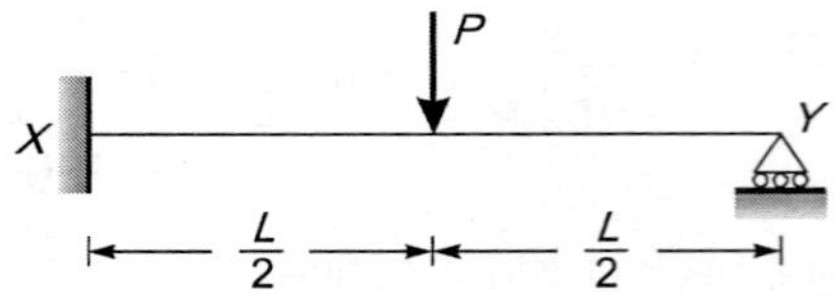

(a) $P = \frac{2M_p}{L}$ (b) $P = \frac{4M_p}{L}$
(c) $P = \frac{6M_p}{L}$ (d) $P = \frac{8M_p}{L}$
[2014 : 1 M, Set-I]

6.23 A prismatic beam (as shown below) has plastic moment capacity of M_p, then the collapse load P of the beam is

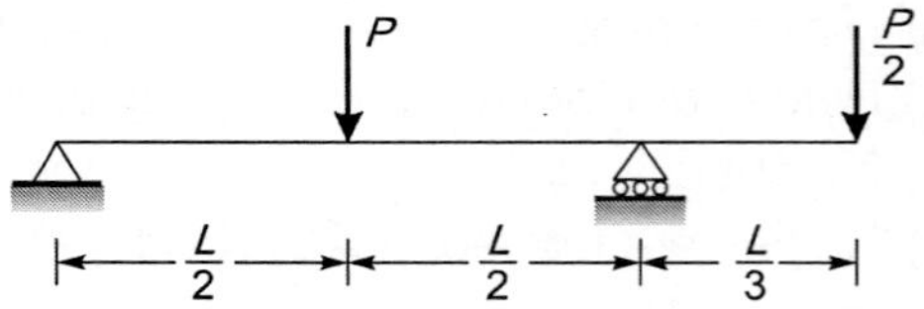

(a) $\frac{2M_p}{L}$ (b) $\frac{4M_p}{L}$
(c) $\frac{6M_p}{L}$ (d) $\frac{8M_p}{L}$
[2014 : 2 M, Set-II]

6.24 For formation of *collapse mechanism* in the following figure, the minimum value of P_u is $\frac{cM_p}{L}$. M_p and $3M_p$ denote the plastic moment capacities of beam sections as shown in this figure. The value of c is ________.

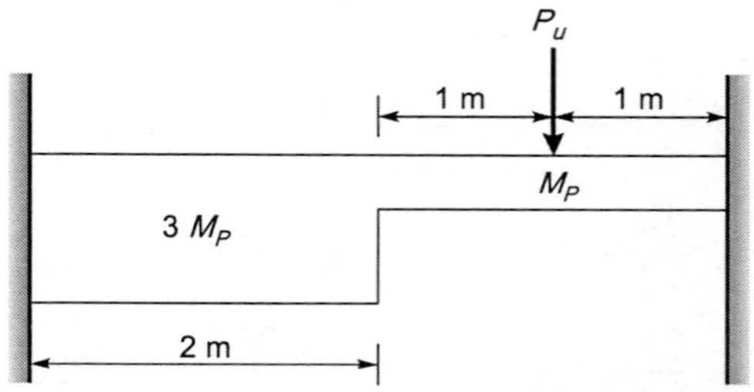

[2015 : 2 M, Set-I]

6.25 A fixed end beam is subjected to a load, W at 1/3rd span from the left support as shown in the figure. The collapse load of the beam is

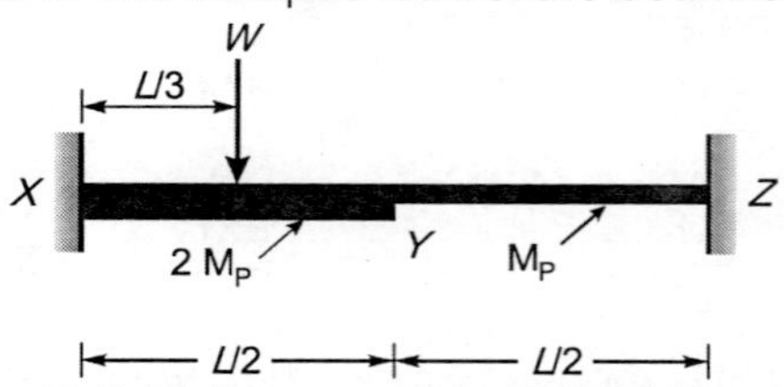

(a) 16.5 M_p/L (b) 15.5 M_p/L
(c) 15.0 M_p/L (d) 16.0 M_p/L
[2015 : 2 M, Set-II]

6.26 A propped cantilever of span L carries a vertical concentrated load at the mid-span. If the plastic moment capacity of the section is M_p, the magnitude of the collapse load is

(a) $\frac{8M_P}{L}$ (b) $\frac{6M_P}{L}$

(c) $\frac{4M_P}{L}$ (d) $\frac{2M_P}{L}$

[2016 : 2 M, Set-I]

6.27 A fixed-end beam is subjected to a concentrated load (P) as shown in the figure. The beam has two different segments having different plastic moment capacities (M_p, $2M_p$) as shown.

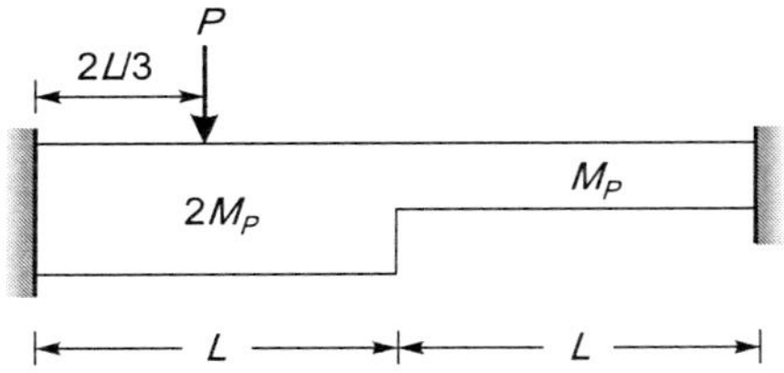

The minimum value of load (P) at which the beam would collapse (ultimate load is)

(a) $\frac{7.5M_P}{L}$ (b) $\frac{5.0M_P}{L}$

(c) $\frac{4.5M_P}{L}$ (d) $\frac{2.5M_P}{L}$

[2016 : 2 M, Set-II]

6.28 The dimension of a symmetrical welded I-section are shown in the figure.

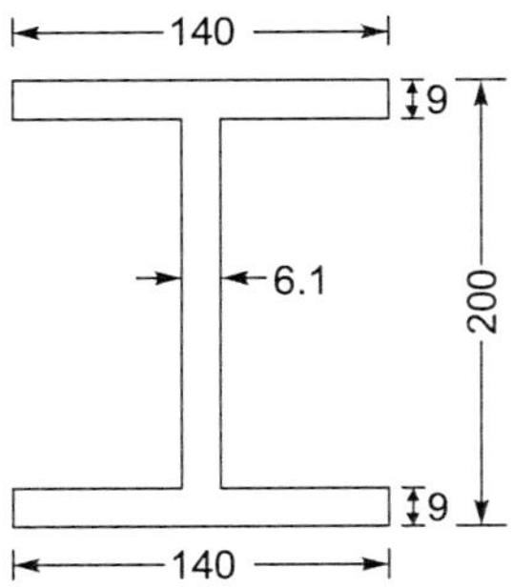

(All dimension are in mm)

The plastic section modulus about the weaker axis (in cm^3, up to one decimal place) is ______

[2018 : 2 M, Set-I]

6.29 A prismatic propped cantilever beam of span L and plastic moment capacity M, is subjected to a concentrated load at its mid-span. If the collapse load of the beam is $\alpha\frac{M_p}{L}$, the value of α is ____.

[2018 : 2 M, Set-II]

6.30 If the section shown in figure turns from fully elastic to fully plastic, the depth of neutral axis (NA), $\bar{y}$, decreases by

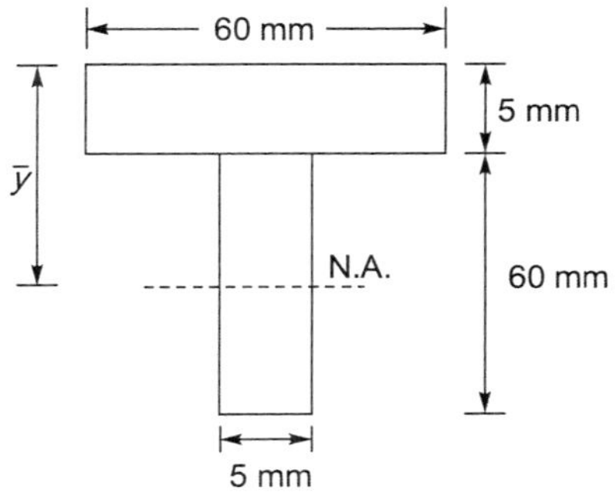

(a) 12.25 mm (b) 15.25 mm

(c) 10.75 mm (d) 13.75 mm

[2019 : 2 M, Set-I]

6.31 The ratio of the plastic moment capacity of a beam section to its yield moment capacity is termed as

(a) aspect ratio (b) load factor

(c) shape factor (d) slenderness ratio

[2020 : 1 M, Set-II]

6.32 A prismatic steel beam is shown in the figure.

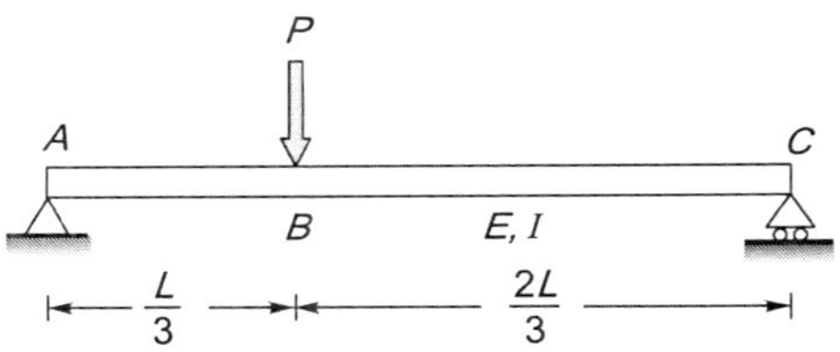

The plastic moment, M_p calculated for the collapse mechanism using static method and kinematic method is

(a) $M_{p,\text{static}} < \frac{2PL}{9} = M_{p,\text{kinematic}}$

(b) $M_{p,\text{static}} = \frac{2PL}{9} \neq M_{p,\text{kinematic}}$

(c) $M_{p,\text{static}} > \frac{2PL}{9} = M_{p,\text{kinematic}}$

(d) $M_{p,\text{static}} = \frac{2PL}{9} = M_{p,\text{kinematic}}$

[2021 : 2 M, Set-II]

6.33 For the square steel beam cross-section shown hi the figure, the shape factor about Z - Z axis is S and the plastic moment capacity is M_p. Consider yield stress f_y = 250 MPa and a = 100 mm.

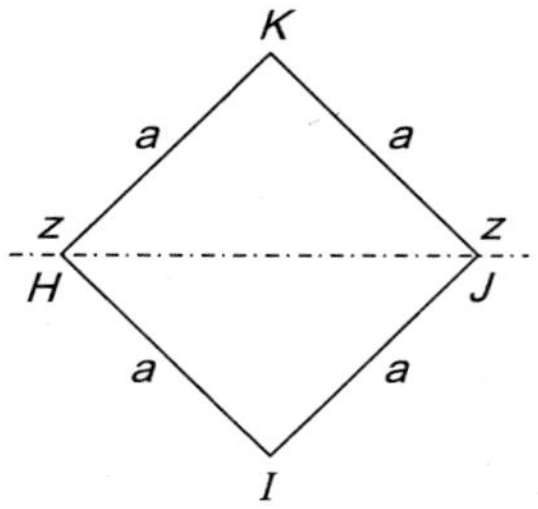

The values of S and M_P (rounded off to one decimal place) are

(a) $S = 2.0$, $M_P = 100.2$ kN-m
(b) $S = 1.5$, $M_P = 58.9$ kN-m
(c) $S = 2.0$, $M_P = 58.9$ kN-m
(d) $S = 1.5$, $M_P = 100.2$ kN-m

[2022 : 2 M, Set-II]

6.34 Consider the horizontal axis passing through the centroid of the steel beam cross-section shown in the figure. What is the shape factor (rounded off to one decimal place) for the cross-section?

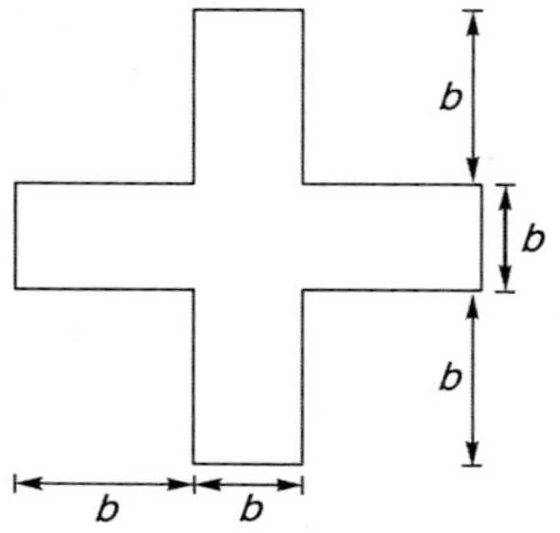

(a) 1.5 (b) 1.7
(c) 1.3 (d) 2.0

[2023 : 2 M, Set-I]

6.35 The steel angle section shown in the figure has elastic section modulus of 150.92 cm^3 about the horizontal $X-X$ axis, which passes through the centroid of the section.

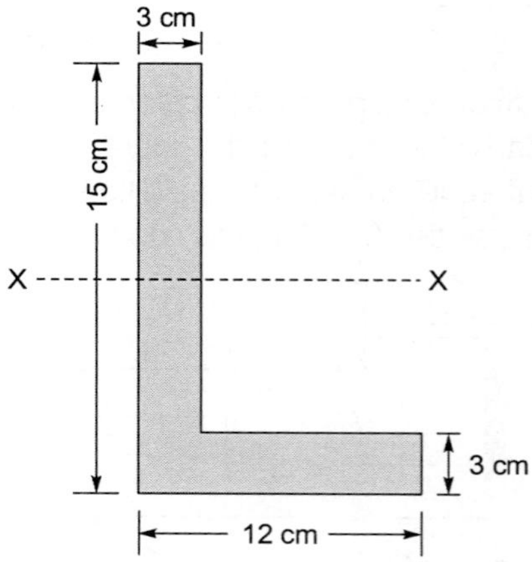

(Figure NOT to scale)

The shape factor of the section is ________ (rounded off to 2 decimal places).

[2024 : 1 M, Set-II]

Answers Plastic Analysis

6.1	(c, d)	6.2	(b)	6.3	(b)	6.4	(c)	6.5	(d)	6.6	(c)	6.7	(c)
6.8	(d)	6.9	(c)	6.10	(a)	6.11	(b)	6.12	(b)	6.13	(a)	6.14	(b)
6.15	(c)	6.16	(b)	6.17	(a)	6.18	(d)	6.19	(c)	6.20	(25)	6.21	(60)
6.22	(c)	6.23	(c)	6.24	(3.33)	6.25	(c)	6.26	(b)	6.27	(a)	6.28	(89.9)
6.29	(6)	6.30	(d)	3.31	(c)	6.32	(d)	6.33	(c)	6.34	(b)	6.35	(1.79)

Explanations Plastic Analysis

6.1 (c, d)

Plastic and elastic method of analysis of indeterminate structures must satisfy equilibrium and compatibility conditions.

6.2 (b)

In case of plastic moment M_p, neutral axis passes through the point which divides the given section into two equal parts having equal area. Thus neutral axis must be between YY and ZZ.

6.3 (b)

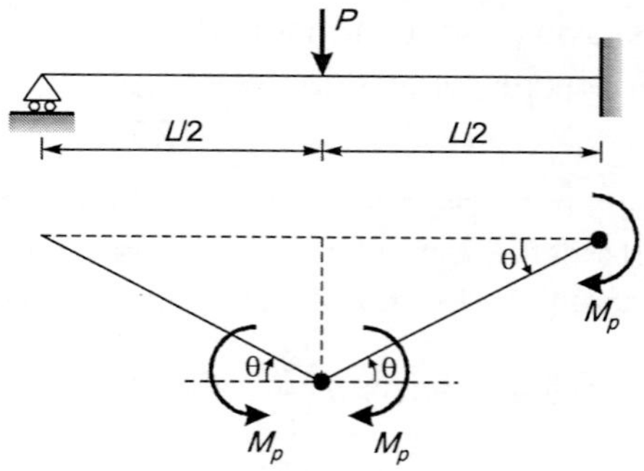

By principle of virtual work,

$$M_p\theta + M_p\theta + M_p\theta = P\left(\frac{L}{2}\theta\right)$$

$$\Rightarrow 3M_p = P\frac{L}{2} \quad \Rightarrow \quad P = \frac{6M_p}{L}$$

$$\therefore \text{ Collapse load, } P = \frac{6M_p}{L}$$

AVOID MISTAKE

Student tend to form a hinge at roller support. Be aware of it.

6.4 (c)

Given data,

Plastic section modulus,

$$Z_p = 4.8 \times 10^{-4} \text{ m}^3$$

Shape factor = 1.2

Plastic moment capacity

$$M_p = 120 \text{ kNm}$$

Yield stress, f_y = ?

$$\because \text{Shape factor} = \frac{Z_p}{Z_e}$$

$$\Rightarrow \quad Z_e = \frac{4.8\times10^{-4}}{1.2} = 4\times10^{-4}\text{m}^3$$

$$\text{Also shape factor} = \frac{M_p}{M_y}$$

$$\Rightarrow \quad 1.2 = \frac{120}{M_y}$$

$$\Rightarrow \quad M_y = \frac{120}{1.2} = 100 \text{ kNm}$$

$$\because \quad M_y = f_y Z_e$$

$$\Rightarrow \quad f_y = \frac{M_y}{Z_e} = \frac{100}{4\times10^{-4}}$$

$= 25 \times 10^4$ kN/m^2 $= 250 \times 10^6$ N/m^2

$=$ 250 N/m^2 = 250 MPa

$\therefore$ Yield stress of material = 250 MPa

6.5 (d)

Method-I

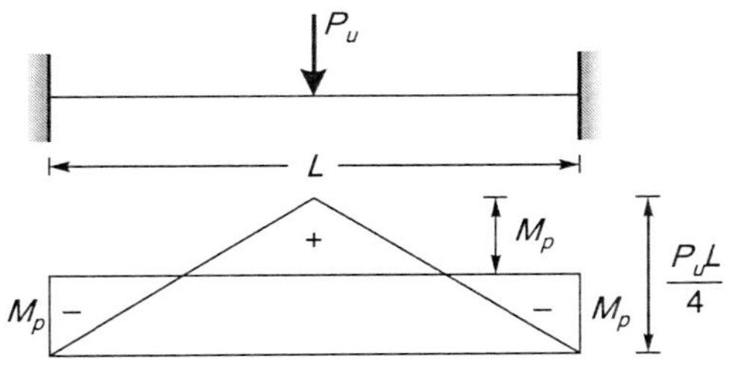

Consider a beam with span L having plastic moment capacity M_p. Let P_u be the Collapse load. Considering the bending moment diagram, from Geometry of Collapse we have,

$$M_p + M_p = \frac{P_uL}{4}$$

$$\Rightarrow \text{Collapse load, } P_u = \frac{8M_p}{L}$$

Method-II

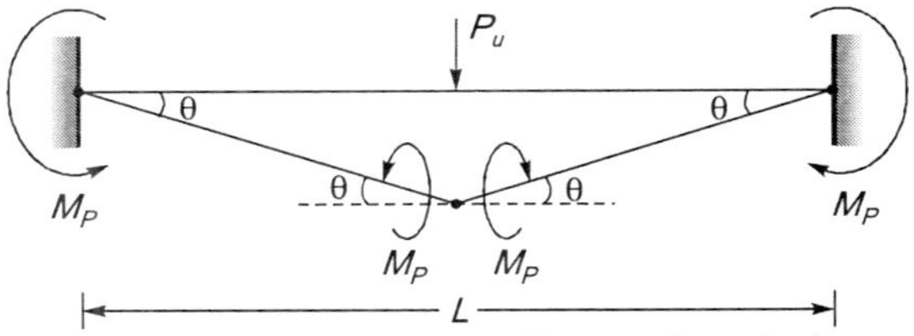

Equating external and internsl work done

$$P_u \times \frac{L}{2}\theta = M_P\theta + M_P\theta + M_P\theta + M_P\theta$$

$$P_u = \frac{8M_P}{L}$$

6.6 (c)

Method-I

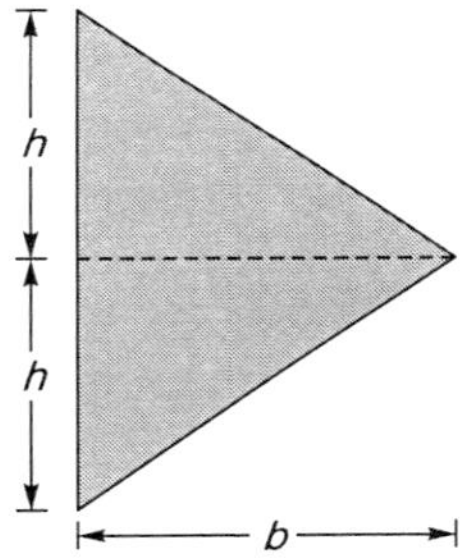

$$\text{Shape factor, } S = \frac{Z_p}{Z_y}$$

$$\text{where,} \quad Z_y = \frac{I_{xx}}{y} = \frac{I_{xx}}{h}$$

$$\text{and} \quad I_{xx} = 2\times\frac{bh^3}{12} = \frac{bh^3}{6}$$

$$\therefore \quad Z_y = \frac{bh^2}{6}$$

$$\text{and} \quad Z_p = A_1\bar{y}_1 + A_2\bar{y}_2$$

$$= 2\times\frac{1}{2}\times bh\times\frac{h}{3} = \frac{bh^2}{3}$$

$\therefore$ shape factor,

$$S = \frac{bh^2/3}{bh^2/6} = 2$$

Method-II

Alternate method to calculate Z_y

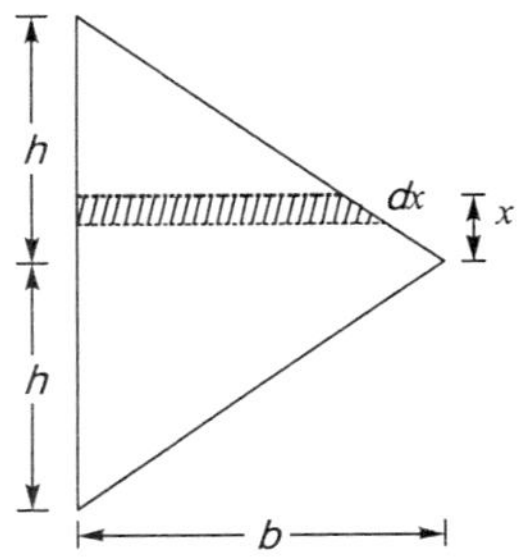

$$Z_y = \frac{I_y}{h}$$

$$I_y = 2\int_0^h \frac{b}{h}(h-x)x^2 dx$$

$$I_y = \frac{2b}{h}\int_0^h \left(hx^2 - x^2\right)dx$$

$$I_y = \frac{2}{h}b\left[h^4\left(\frac{1}{3}-\frac{1}{4}\right)\right] = \frac{bh^3}{6}$$

$$Z_y = \frac{bh^3}{bh} = \frac{bh^2}{6}$$

$$\frac{Z_p}{Z_y} = \frac{bh^3/3}{bh^2/6} = 2.0$$

6.7 (c)

S.No.	Shape	Shape factor
1.	Circle	1.67
2.	Diamond	2.0
3.	I-section	As per dimensions but less than 1.5
4.	Rectangle	1.5

6.8 (d)

$$\frac{M_p}{M} = f = \frac{L}{L-L_p}$$

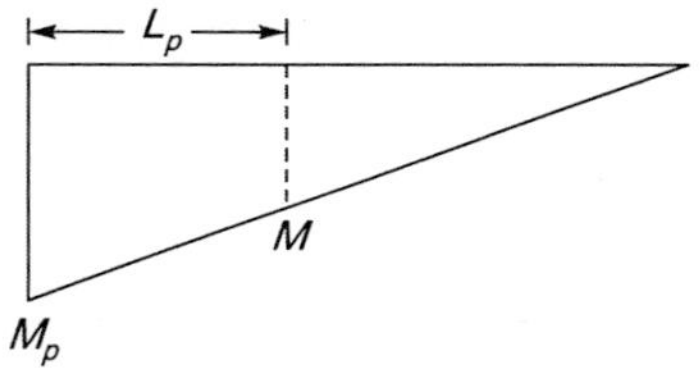

$$\Rightarrow \quad \frac{1}{f} = \frac{L-L_p}{L} = 1-\frac{L_p}{L}$$

or $$\frac{L_p}{L} = 1-\frac{I}{f}$$

POINTS TO REMEMBER

The value of the moment of sections adjacent of the yield zone for a certain length is more than the yield moment. This length is known as hinge length and is dependent on

1. Loading
2. Geometry of section

6.9 (c)

Initially when load is increased maximum –ve BM occurs at end $\frac{Wl^2}{12}$ and plastic hinges are formed at ends and when load is further increased it behaves like simply supported beam and third hinge is formed at mid span when load increases to

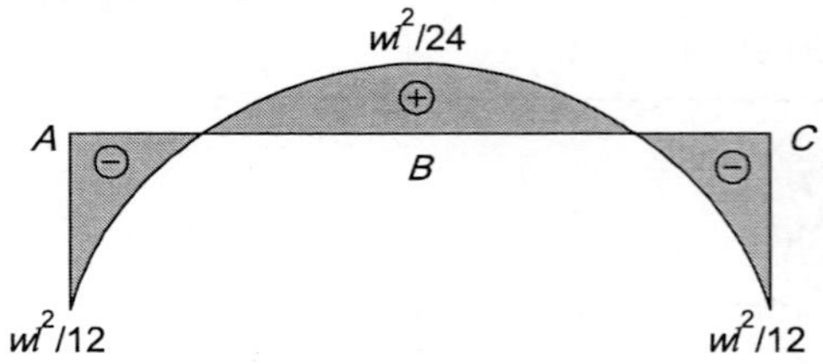

$$W_u = \frac{16M_p}{l^2}$$

Ist two hinges are formed at (A) and (C) together at load,

$$W_u = \frac{12M_p}{l^2}$$

and then third hinge is formed at load,

$$W_u = \frac{16M_p}{l^2}$$

6.10 (a)

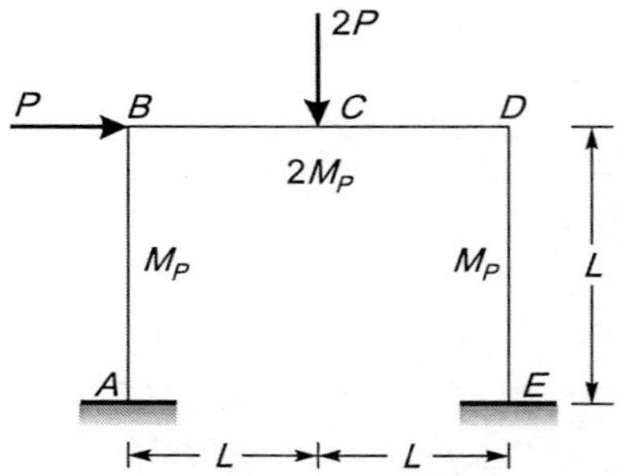

The possible locations of the plastic hinges are at *A*, *B*, *C*, *D* and *E*.

Number of plastic hinges, $N = 5$

Degree of redundancy, $r = 6 - 3 = 3$

Number of possible independent mechanisms,

$$n = N - r = 5 - 3 = 2$$

Thus, the two independent mechanisms are

1. Beam mechanism
2. Sway mechanism

Beam mechanism: The ends *B* and *D* of the frame *ABDE* are fixed joints. Therefore, the beam *BD* acts as a fixed end beam.

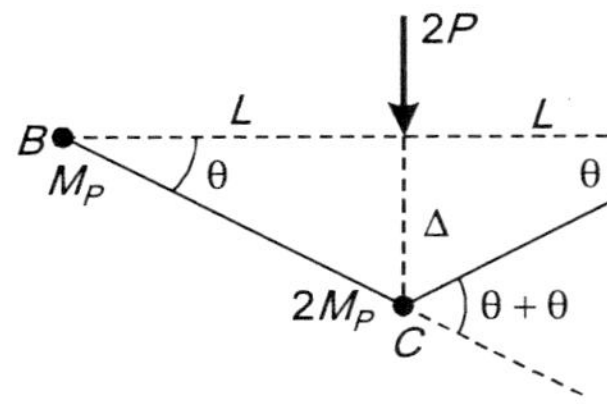

External work done

$= \text{load} \times \text{deflection}$

$= 2P \times \Delta = 2P \times L \times \theta = 2PL\theta$

Internal work done

$= \text{moment} \times \text{rotation}$

$= M_P\theta + 2M_P(\theta + \theta) + M_P\theta = 6\ M_P\theta$

By principal of virtual work,

External work done

= Internal work done

$$2PL\theta = 6\ M_P\theta$$

$$\Rightarrow \quad P = \frac{3M_P}{L}$$

6.11 (b)

External work done

$= \text{load} \times \text{deflection}$

$$= W_u \times \Delta = W_u \times \frac{L}{2} \times \theta$$

$$= \frac{W_u L\theta}{2}$$

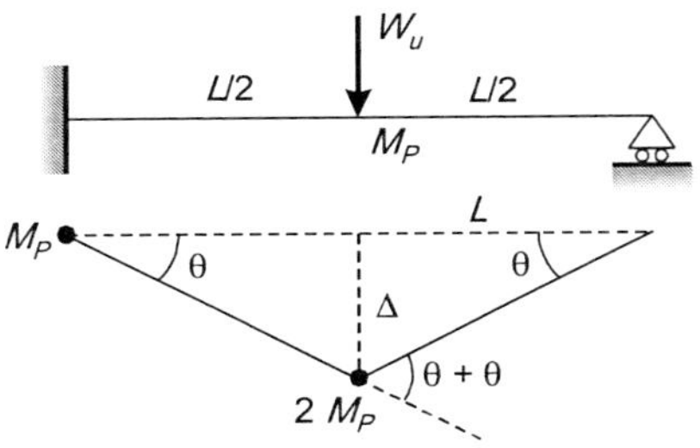

Internal work done

$= \text{moment} \times \text{rotation}$

$= M_P\theta + M_P(\theta + \theta) = 3\ M_P\theta$

By principle of virtual work,

External work done = Internal work done

$$\frac{W_u L\theta}{2} = 3\ M_P\theta$$

$$W_u = \frac{6M_P}{L}$$

6.12 (b)

Yielding starts at P, therefore yielding moment at the fixed end is given by,

$$M_y = PL \quad \text{...(i)}$$

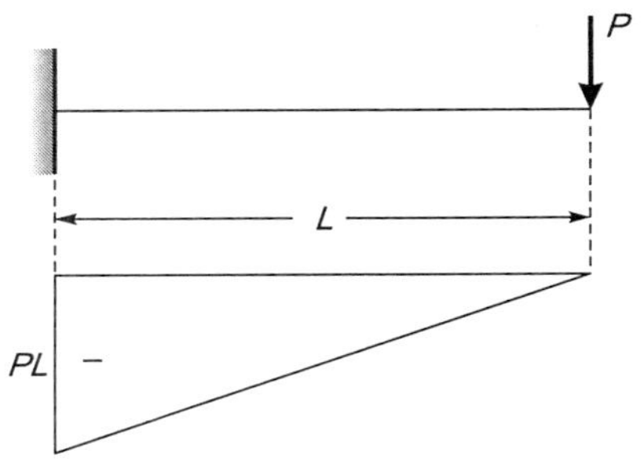

Now, if W_u is the collapse load, then collapse moment at the fixed end is given by,

$$M_p = W_u L \quad \text{...(ii)}$$

But, for rectangular beam,

$$\frac{M_p}{M_y} = 1.5 \qquad \left[\because \text{shape factor} = \frac{M_p}{M_y}\right]$$

$$\Rightarrow \quad \frac{W_u L}{PL} = 1.5$$

$$\Rightarrow \quad W_u = 1.5\,P$$

6.13 (a)

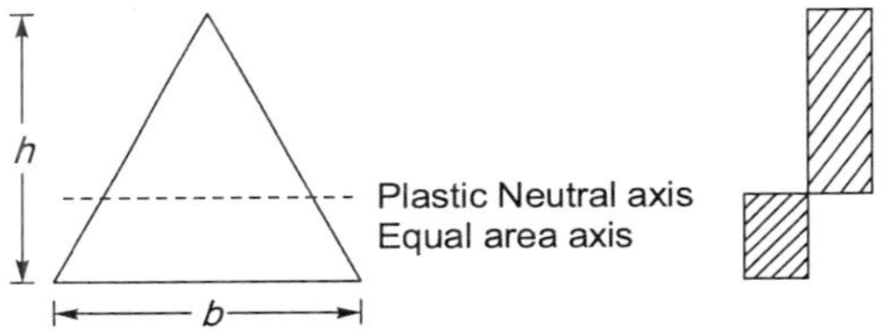

We know that the neutral axis of the plastified section is the equal area axis.

∴Compressive force,

$$= \sigma_y \times \frac{\text{Area of triangular section}}{2}$$

$$= \sigma_y \times \frac{bh}{2 \times 2} = \frac{bh\sigma_y}{4}$$

6.14 (b)

Method-I

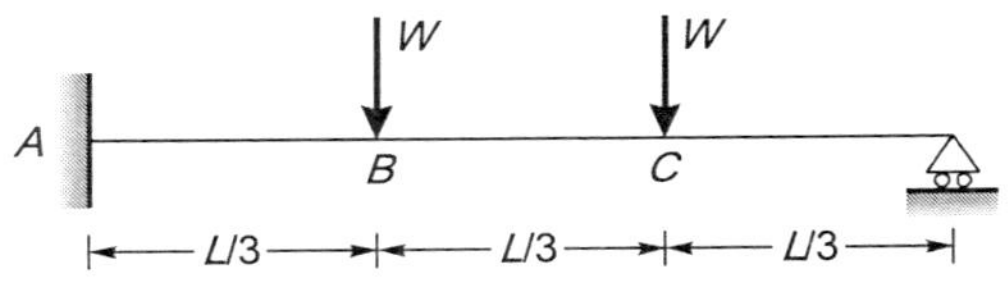

$$D_s = (3 + 1) - 3 = 1$$

No. of plastic hinges formed $= D_s + 1 = 2$

I[st] case: Plastic hinge formed at A and B

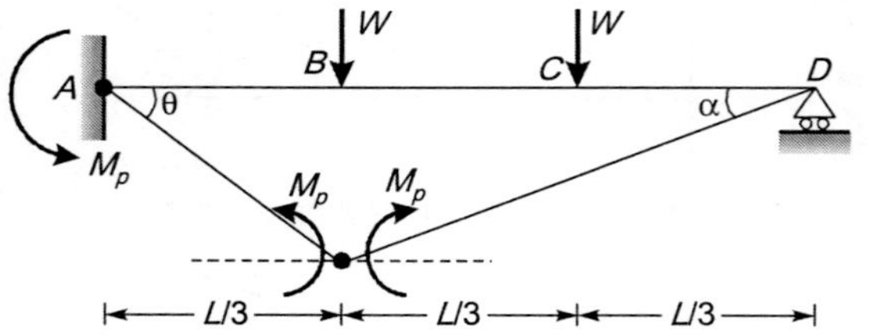

$$\frac{L}{3}\theta = \frac{2L}{3}\alpha$$

$$\theta = 2\alpha$$

By virtual work method

External work = Internal work

$$W\left(\frac{L}{3}\theta\right) + W\left(\frac{L}{3}\alpha\right) = M_p\theta + M_p(\theta + \alpha)$$

Put $\theta = 2\alpha$

$$W_u = \frac{5M_p}{L} \qquad \text{...(i)}$$

IInd case: Plastic hinge at (A) and (C)

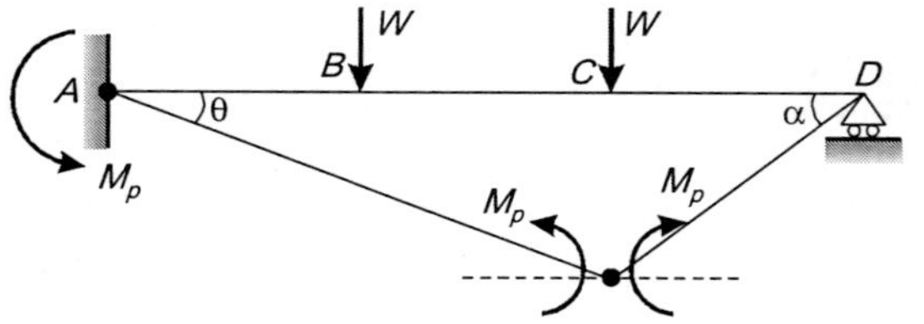

$$\frac{2L}{3}\theta = \frac{L}{3}\alpha$$

$$\alpha = 2\theta$$

External work = Internal work

$$\frac{WL}{3}\theta + \frac{2WL}{3}\theta = M_p\theta + M_p(\theta + \alpha)$$

Put $\alpha = 2\theta$

$$W_u = \frac{4M_p}{L}$$

Hence, collapse load

$$W_u = \frac{4M_p}{L}$$

Method-II

Before rushing onto the solutionby drucwing possible mechanisms, we can analysis the expected rough BMD of the beam.

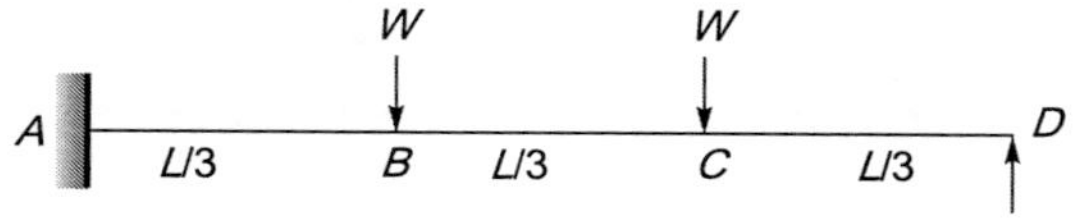

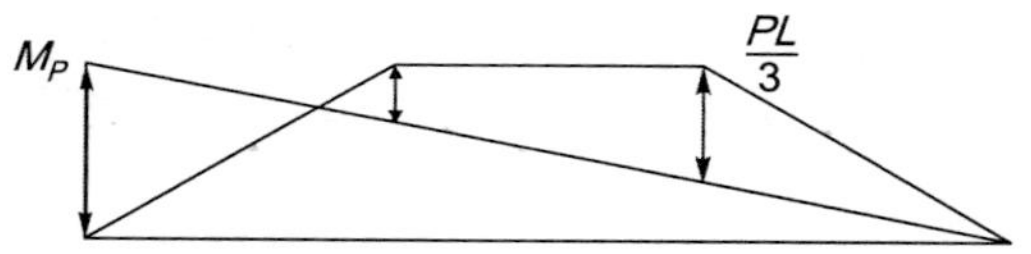

If you notice largest ordinate occurs at *A* and *C*.

Hence mechanism 2 will be our obvious and only choice.

6.15 (c)

Shape factors of some cross-section are as follows:

(i) Rectangle – 1.5
(ii) I-Section – 1.14
(iii) Diamond – 2
(iv) Triangle – 2.34
(v) Circle – 1.7

6.16 (b)

Method-I

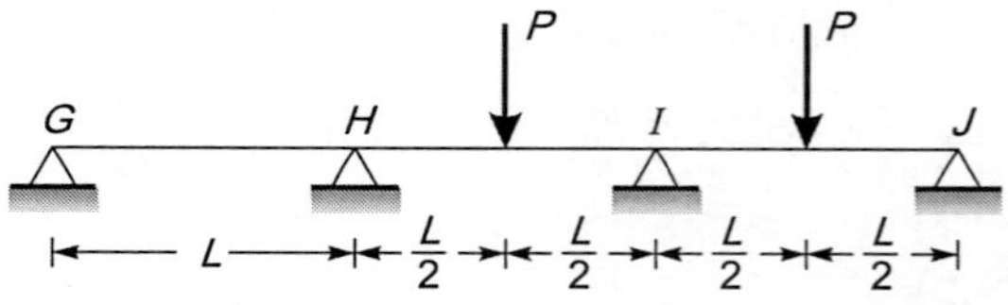

For collapse in IJ

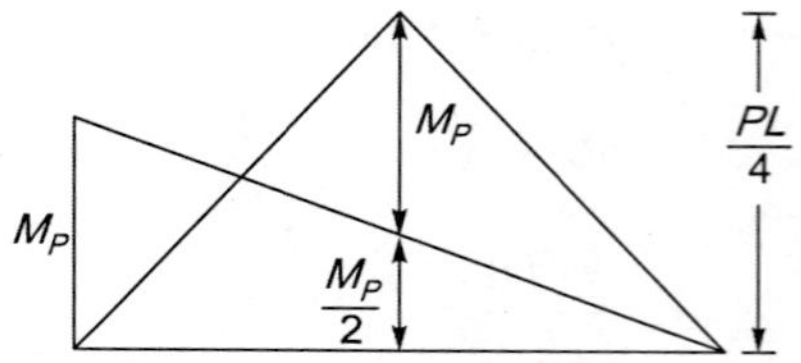

$$M_P + \frac{M_P}{2} = \frac{PL}{4}$$

$$\Rightarrow \quad \frac{3M_P}{2} = \frac{PL}{4}$$

$$\Rightarrow \quad P = \frac{6M_P}{L}$$

For collapse in HI

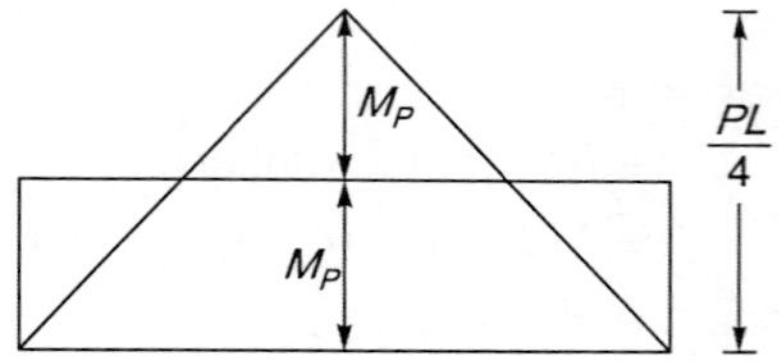

$$M_P + M_P = \frac{PL}{4}$$

$$\Rightarrow \quad P = \frac{8M_P}{L}$$

∴ Minimum load for collapse $= \dfrac{6M_P}{L}$

Method-II

In continuous span beam, collapse of individual span govern the collapse load.

(a) ∴ Collapse in HI in similar to collapse in fixed beam with central point load

$$\therefore \quad P_u = \frac{8M_P}{L} \qquad \text{(Standard Result)}$$

(b) Also collapse in IJ is similar to collapse in propped cantilever with centra point load.

$$\therefore \quad P_u = \frac{6M_P}{L} \qquad \text{(Standard Result)}$$

6.17 (a)

Slope factor may be defined as the ratio of the plastic moment and the yield moment of the section.

It is a function of the cross-section from or shape and is represented by S.

$$S = \frac{M_P}{M_y} = \frac{f_y Z_p}{f_y Z_e} = \frac{Z_p}{Z_e}$$

6.18 (d)

Method-I

Number of possible hinges, $n = 2$

Statical indeterminacy, $r = 1$

Number of independent mechanisms,

$$i = n - r = 2 - 1 = 1$$

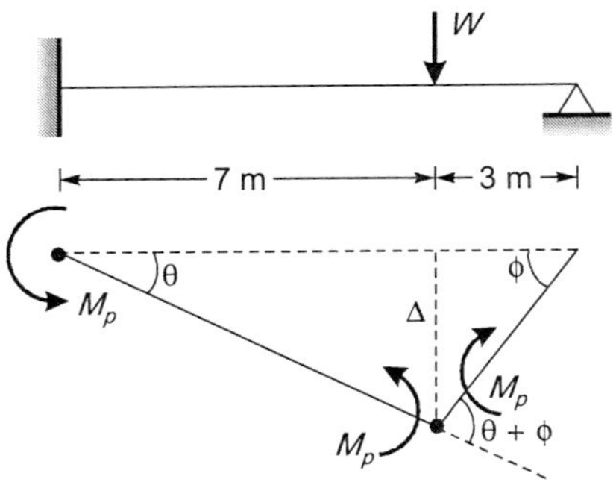

$$\Delta = 7\theta = 3\phi$$

By principle of virtual work, we get,

$$W \times \Delta - M_p\theta - M_p(\theta + \phi) = 0$$

$$\Rightarrow \quad W \times 7\theta = 2M_p\theta + M_p\phi$$

$$\Rightarrow \quad W \times 7\theta = 2M_p\theta + M_p \times \frac{7\theta}{3}$$

$$\Rightarrow \quad 7W = \frac{6M_p + 7M_p}{3}$$

$$\Rightarrow \quad W = \frac{13M_p}{21}$$

Method-II

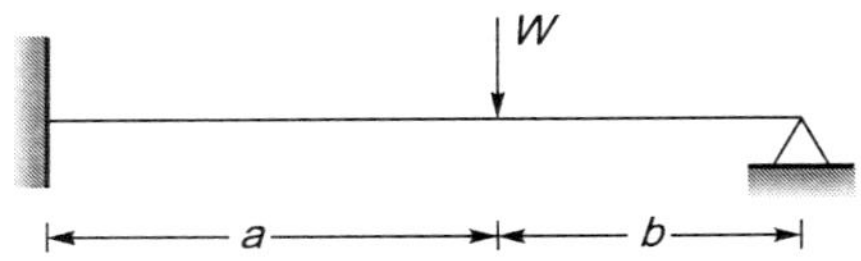

Where $(a + b) = L$

For standard cases

$$W_u = M_p\frac{(L+b)}{ab} = M_P \times \frac{(10+3)}{7\times 3}$$

Here $a = 7$ m, $b = 3$ m

$$W_u = \frac{13}{21}M_P$$

6.19 (c)

Plastic section: Cross-section which can develop plastic hinges and have the rotation capacity required for the failure of the structure by formation of plastic mechanism.

Compact section: Cross-section which can develop plastic moment of resistance but have inadequate plastic hinge rotation capacity for formation of plastic mechanism before buckling.

Semi compact: Cross-sections, in which the expense fibre in compression can reach yield stress, but cannot develop plastic moment of resistance due to location buckling.

Slender: Cross-sections in which the elements buckle locally even before attainment of yield stress are closed as sledner sections.

6.20 Sol.

Equating deflection at end B,

Let, $\quad l = 3$ m

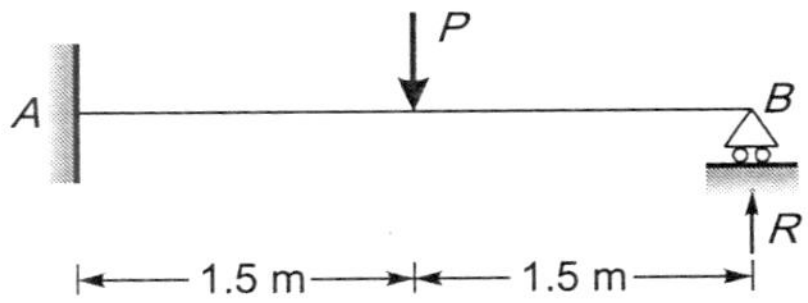

$$\therefore \quad \frac{R(l)^3}{3EI} = \frac{P(1.5)^3}{3EI} + \frac{P(1.5)^2}{2EI} \times 1.5$$

$$\Rightarrow \quad \frac{R(3)^3}{3EI} = \frac{1.125P}{EI} + \frac{1.6875P}{EI}$$

$$\frac{9R}{EI} = \frac{1.125P}{EI} + \frac{1.6875P}{EI}$$

$$\Rightarrow \quad R = \frac{5}{16}P$$

$$R = \frac{5}{16} \times 80 = 25 \text{ kN}$$

6.21 Sol.

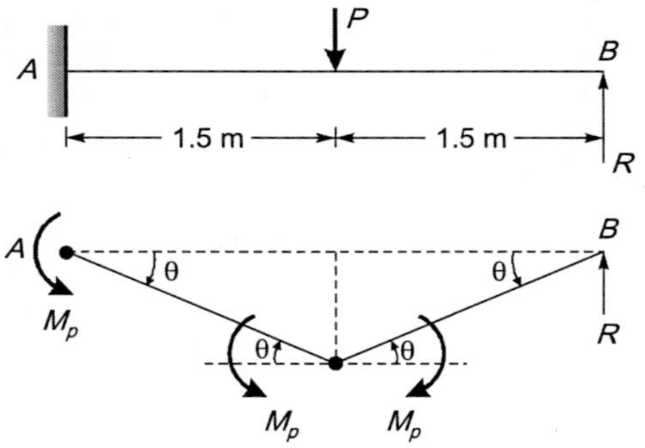

$$-3M_p\theta + P\frac{l}{2}\theta = 0$$

$$\therefore \quad P_u = \frac{6M_p}{l} = \frac{6\times 90}{3} = 180 \text{ kN}$$

At collapse condition $\Sigma M_A = 0$,

$R \times 3 + M_p - P \times 1.5 - M_p + M_p = 0$

$\Rightarrow \quad R = 60$ kN

6.22 (c)

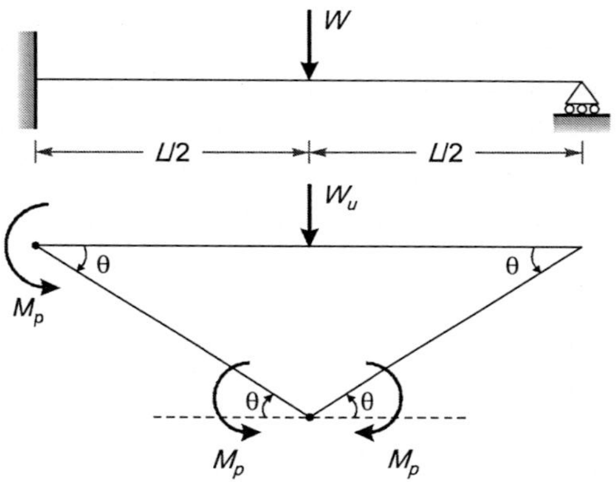

From principal of virtual work

$$-M_p\theta - M_p\theta - M_p\theta + W_u\frac{L}{2}\theta = 0$$

$$\Rightarrow \quad W_u = \frac{6M_p}{L}$$

6.23 (c)

Here degree of static indeterminacy = 0

$\therefore$ Number of plastic hinges required for mechanical equlibrium

$$= D_s + 1 = 0 + 1 = 1$$

Mechanism 1:

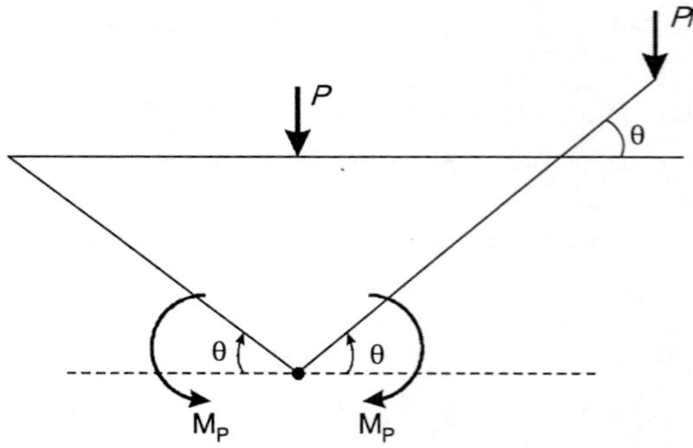

From **principal of virtual work**

$$-M_p\theta - M_p\theta + P\frac{L}{2}\theta - \frac{P}{2}\times\frac{L}{3}\theta = 0$$

$$\Rightarrow \quad -2M_p\theta + \frac{PL}{2}\theta - \frac{PL}{6}\theta = 0$$

$$\Rightarrow \quad 2M_p = \frac{PL}{2} - \frac{PL}{6} = \frac{(3-1)PL}{6} = \frac{1}{3}PL$$

$$\Rightarrow \quad P = \frac{6Mp}{L}$$

Mechanism 2:

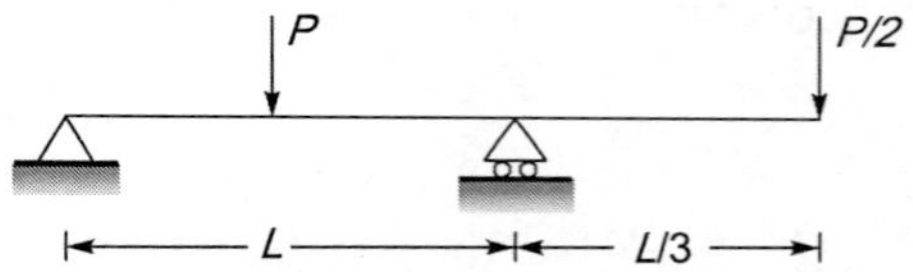

From principal of virtual work, equating external work donw to internal work done

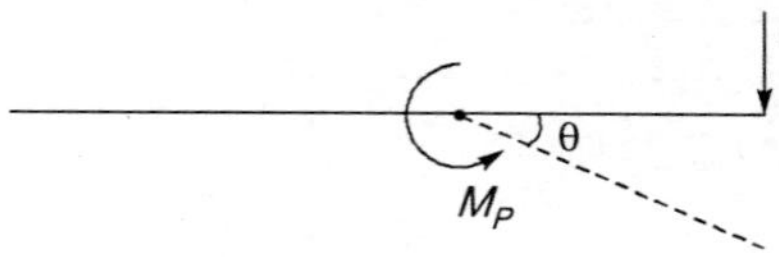

$$M_p\theta = \frac{P_u}{2}\times\frac{L}{3}\theta$$

$$\therefore \quad P_u = \left[\frac{6M_P}{L}\right]$$

AVOID MISTAKE

In mechanism-1, students sometimes make hinge at roller support. This can't be any case because it will cause fialure in overhang portion.

6.24 Sol.

As L is not defined in question. So, by assuming $L = 1$ m

Mechanism 1:

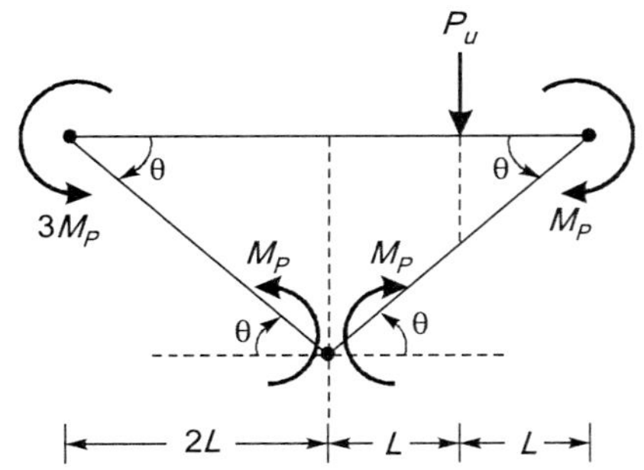

$$3M_P\theta + M_P\theta + M_P\theta + M_P\theta = P \times \theta \times L$$

$$\Rightarrow \quad L \times P_u = 6\,M_P$$

$$\Rightarrow \quad P_u = \frac{6\,M_P}{L}$$

$$\Rightarrow \quad c = 6$$

Mechanism 2:

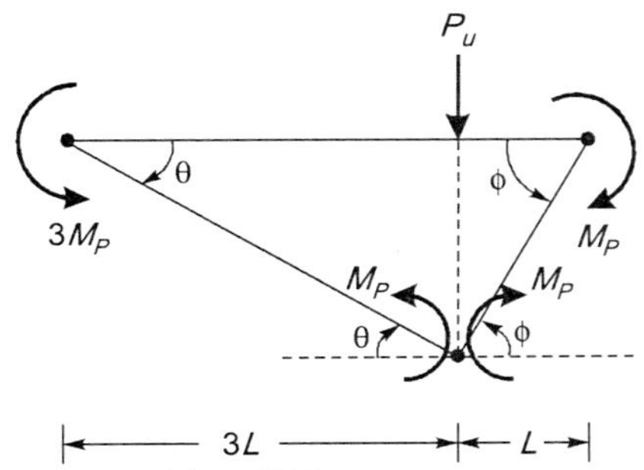

$$\Rightarrow \quad L\phi = 3L\theta$$

$$\phi = 3\theta$$

$$4M_P\theta + 2M_P\phi = PL\phi$$

$$\frac{4M_P\phi}{3} + 2M_P\phi = PL\phi$$

$$\frac{10\,M_P\phi}{3} = PL\phi$$

$$\Rightarrow \quad P = \frac{10\,M_P}{3L} \Rightarrow c = \frac{10}{3}$$

Failure load obtained out of two mechanisms

$$= \frac{10\,M_P}{3L}$$

$$\therefore \quad c = \frac{10}{3} = 3.33$$

Mechanism 3:

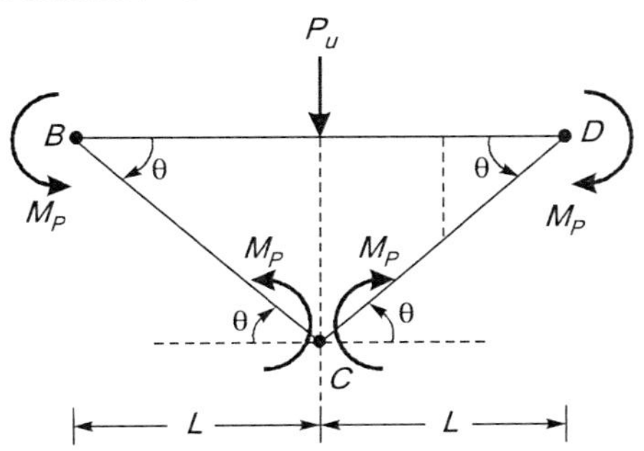

$$M_P\theta + 2M_P\theta + M_P\theta = P_u \times L\theta$$

$$\Rightarrow \quad 4M_P\theta = P_u L\theta$$

$$\Rightarrow \quad P_u = \frac{4\,M_P}{L}$$

$$c = 4$$

So the minimum value of $c = 3.33$

6.25 (c)

No. of plastic hinges formed at collapse

$$= r + 1 = 3$$

There can be two collapse mechanisms

Case (I)

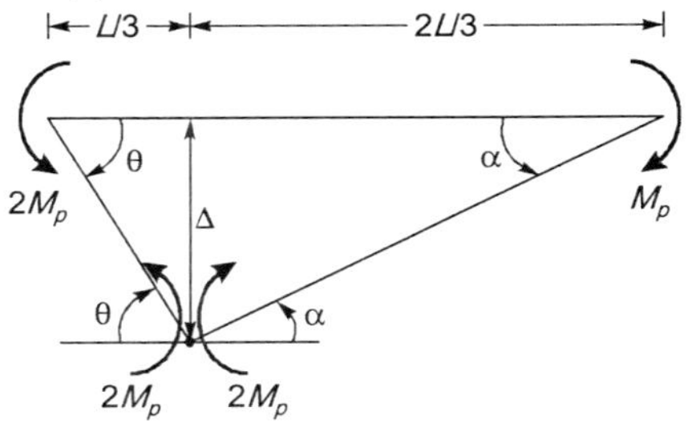

$$\Delta = \frac{L}{3} \times \theta = \frac{2L}{3} \times \alpha$$

$$\Rightarrow \quad \theta = 2\alpha$$

Internal work done = External work done

$$\Rightarrow \quad 2M_p\theta + 2M_p\theta + 2M_p\alpha + M_p\alpha$$

$$= W \times \frac{L}{3}\theta$$

$$\Rightarrow 4M_p\theta + 3M_p\frac{\theta}{2} = \frac{WL}{3}\theta$$

$$\Rightarrow \quad \frac{11}{2}M_p\theta = \frac{WL}{3}\theta$$

$$\Rightarrow \quad W = \frac{16.5}{L}M_p$$

Case (II)

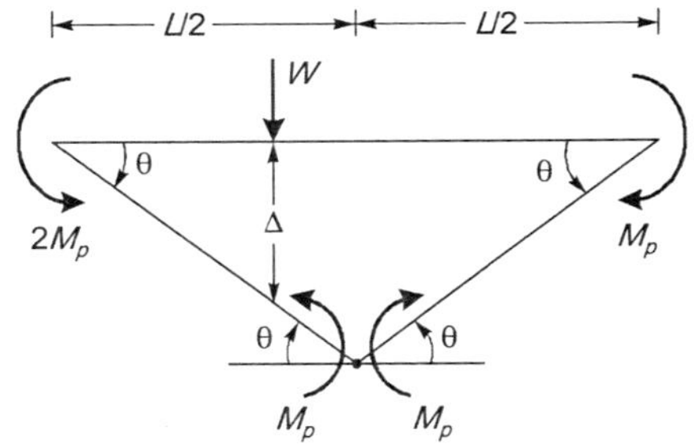

$$2M_p\theta + M_p\theta + M_p\theta + M_p\theta = \frac{WL}{3} \times \theta$$

$$W = \frac{15\,M_p}{L}$$

Hence, minimum collapse load is $\frac{15\,M_p}{L}$.

6.26 (b)

6.27 (a)

$D_S = 2$

$\therefore$ Number of plastic hinge required for complete collapse $= D_S + 1$

$= 2 + 1 = 3$

Mechanism 1:

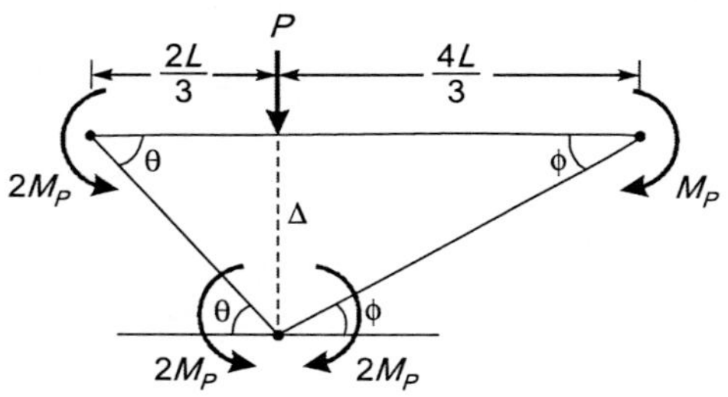

$$\Delta = \frac{2L}{3\theta} = \frac{4L}{3}$$

$$\Rightarrow \quad \theta = 2\theta$$

For principal of virtual work done,

$$-2M_P\theta - 2M_P\theta - 2M_P\phi - M_P\phi + P\left(\frac{2L}{3}\theta\right) = 0$$

$$\Rightarrow \quad 4M_P\theta + 3M_P\phi = \frac{2PL}{3}\theta$$

$$\Rightarrow \quad 8M_P\phi + 3M_P\phi = \frac{4PL}{3}\phi$$

$$\Rightarrow \quad 11M_P = \frac{4P_uL}{3}$$

$$\Rightarrow \quad P_u = \frac{33}{4}\frac{M_P}{L}$$

$$\therefore \quad P_u = 8.25\frac{M_P}{L}$$

Mechanism 2:

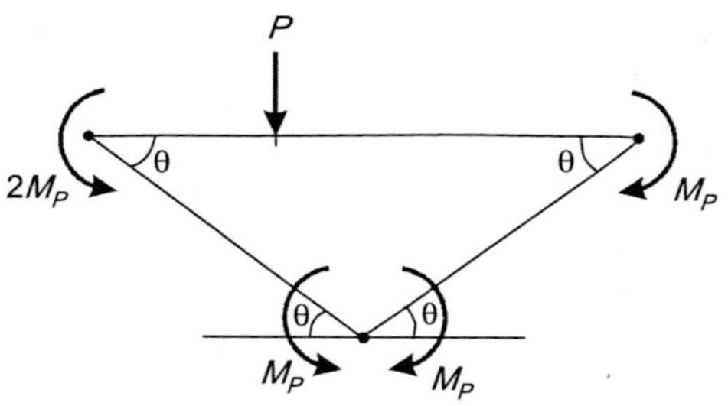

$$2M_P\theta + M_P\theta + M_P\theta + M_P\theta = P\left(\frac{2L}{3}\right)$$

$$\Rightarrow \quad 5M_P = \frac{2PL}{3}$$

$$\Rightarrow \quad P_u = \frac{15M_P}{2L}$$

$$\therefore \quad P_u = 7.5\frac{M_P}{L}$$

6.28 Sol.

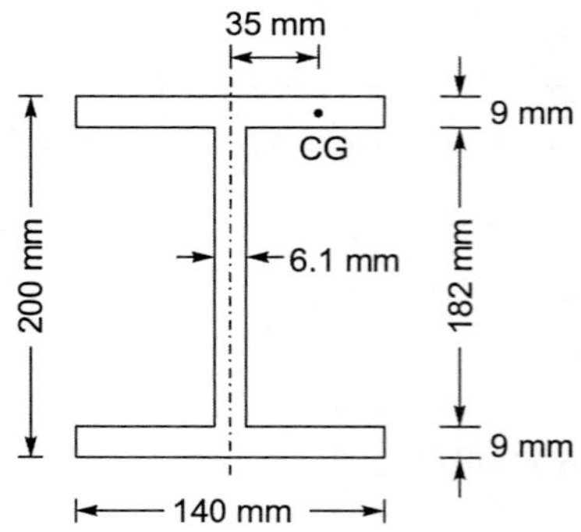

$$Z_p = \frac{A}{2}(\bar{y}_1 + \bar{y}_2)$$

$$= 2\times\frac{140\times 9}{2}(35+35) + \frac{182\times 6.1}{2}\left(\frac{6.1}{4}+\frac{6.1}{4}\right)$$

$$= 88200 + 1693.055 = 88893.055\text{ mm}^3 = 89.9\text{ cm}^3$$

Note: Keep in mind the units and decimal places.

6.29 Sol.

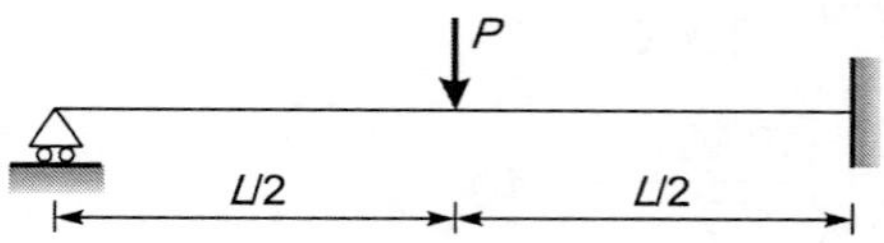

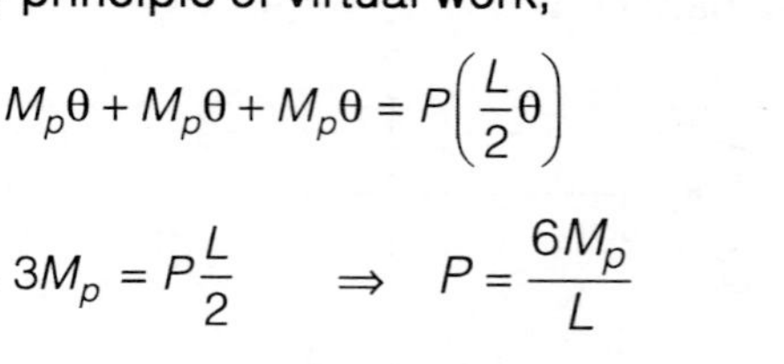

By principle of virtual work,

$$M_p\theta + M_p\theta + M_p\theta = P\left(\frac{L}{2}\theta\right)$$

$$\Rightarrow 3M_p = P\frac{L}{2} \quad \Rightarrow \quad P = \frac{6M_p}{L}$$

$\therefore$ Collapse load, $P = \dfrac{6M_p}{L}$

Hence α is 6.

6.30 (d)

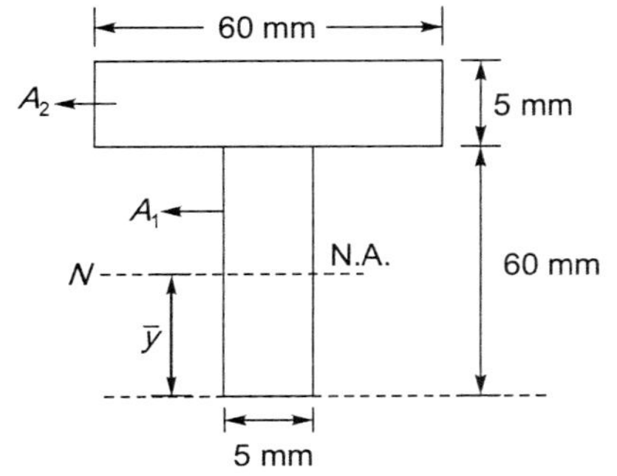

$$\bar{y} = \frac{A_1\bar{y}_1 + A_2\bar{y}_2}{A_1 + A_2}$$

$$= \frac{60\times5\times\frac{60}{2} + 60\times5\times\left(60+\frac{5}{2}\right)}{60\times5+60\times5}$$

$= 46.25$ mm

NA - Neutral axis

The section is unsymmetrical about the NA and hence the equal area axis (EA) has to be located

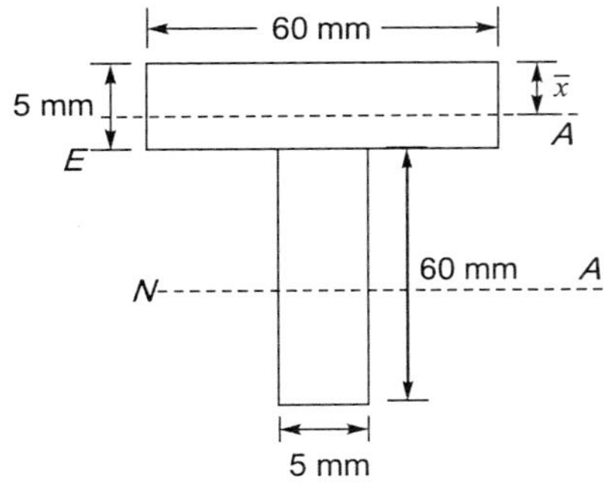

Let, $\bar{x} \le 5$ mm

$$60\times\bar{x} = \frac{(60\times5)+60\times5}{2}$$

$\bar{x} = 5$ mm

∴NA shifts by,

$60 - \bar{y} = 60 - 46.25$

$= 13.75$ mm

6.31 (c)

Ratio of $\frac{M_p}{M_y}$ = Shape factor

6.32 (d)

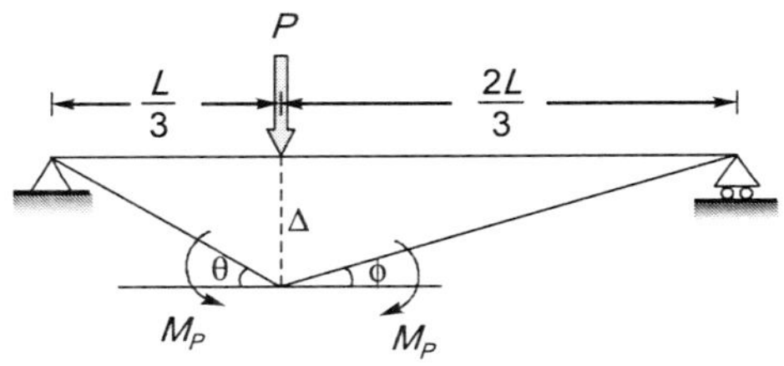

At collapse,

$$M_p\theta + M_p\phi = P\Delta$$

$$\Rightarrow \quad 3M_P\frac{\Delta}{l} + \frac{3M_P\Delta}{2l} = P\Delta$$

$$M_P = \frac{2Pl}{9}$$

Also, $\quad M_{P,\text{ static}} = M_{p,\text{ kinematic}}$

6.33 (c)

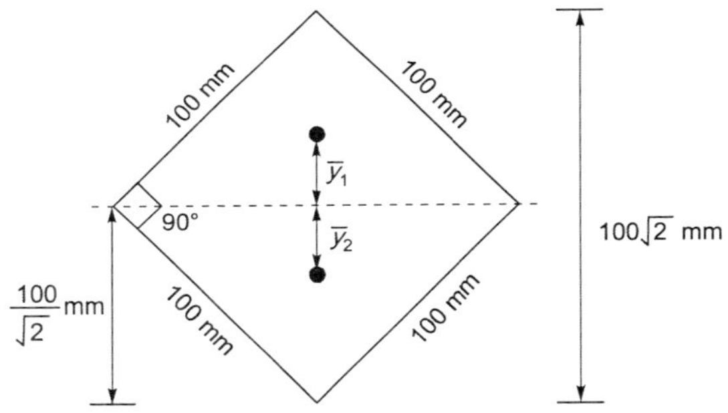

Plastic moment capacity,

$$M_P = f_y Z_P$$

where, $\quad Z_P = \frac{A}{2}(\bar{y}_1 + \bar{y}_2)$

$$\Rightarrow \quad Z_P = \frac{100\times100}{2}\left[\frac{100}{3\sqrt{2}} + \frac{100}{3\sqrt{2}}\right]$$

$\Rightarrow \quad Z_P = 235.702 \times 10^3$ mm^3

$\therefore \quad M_p = f_y Z_P = 250 \times 235.702 \times 10^3 \times 10^{-6}$

$\Rightarrow \quad M_p = 58.925$ kN-m

Now, yield moment capacity,

$$M_y = f_y Z_e$$

$$\Rightarrow M_y = 250\times\frac{a^3}{6\sqrt{2}} = 250\times\frac{(100)^3}{6\sqrt{2}} = 29.46 \text{ kNm}$$

∴ Shape factor,

$$SF = \frac{M_p}{M_y} = \frac{58.925}{29.46} = 2$$

6.34 (b)

As we know,

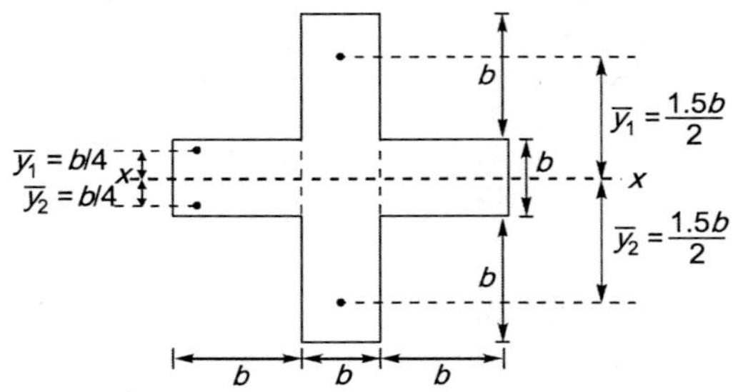

$$Z_p = \frac{A}{2}[\bar{y}_1 + \bar{y}_2]$$

$$= (b \times 1.5b)\left[\frac{1.5b}{2} + \frac{1.5b}{2}\right] + \left[2b \times \frac{b}{2}\right] \times \left[\frac{b}{4} + \frac{b}{4}\right]$$

$$= 2.75b^3$$

$$Z = \frac{I_{xx}}{y} = \frac{\left[\frac{b \times (3b)^3}{12} + \frac{2b \times b^3}{12}\right]}{1.5b} = \frac{29}{18}b^3$$

$$\text{Shape factor, S.F.} = \frac{Z_p}{Z} = \frac{2.75b^3}{\frac{29}{18}b^3} = 1.7$$

6.35 Sol.

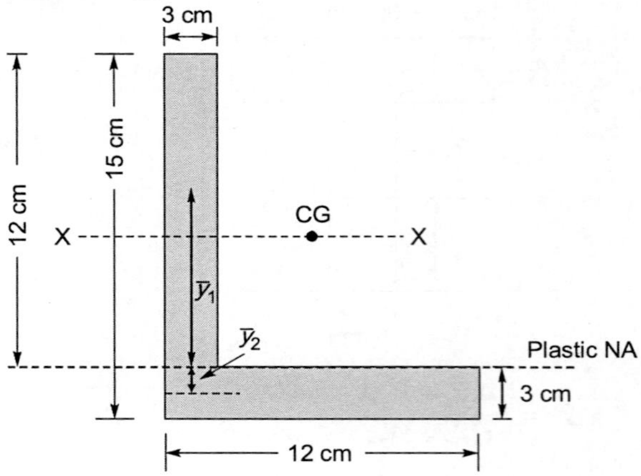

$$\text{Gross area of section} = b_e t$$

$$= (15 + 12 - 3) \times 3$$

$$= 72 \text{ cm}^2$$

Location of plastic neutral axis is as shown in figure.

$$\therefore \quad A_1 = A_2 = 36 \text{ cm}^2$$

where A_1 is the area above plastic N.A. and A_2 is the area below plastic N.A.

$$\therefore \quad Z_{Px} = \frac{A}{2}(\bar{y}_1 + \bar{y}_2) = \frac{72}{2}\left(6 + \frac{3}{2}\right)$$

$$= 270 \text{ cm}^3$$

$$\therefore \text{Shape factor, S.F.} = \frac{Z_P}{Z_e} = \frac{270}{150.92} = 1.79$$

■■■■

Geotechnical Engineering

UNIT

V

CONTENTS

Geotechnical Engineering

Syllabus

Soil Mechanics: Three-phase system and phase relationships, index properties; Unified and Indian standard soil classification system; Permeability - one dimensional flow, Seepage through soils – two - dimensional flow, flow nets, uplift pressure, piping, capillarity, seepage force; Principle of effective stress and quicksand condition; Compaction of soils; One- dimensional consolidation, time rate of consolidation; Shear Strength, Mohr's circle, effective and total shear strength parameters, Stress-Strain characteristics of clays and sand; Stress paths.

Foundation Engineering: Sub-surface investigations - Drilling bore holes, sampling, plate load test, standard penetration and cone penetration tests; Earth pressure theories - Rankine and Coulomb; Stability of slopes – Finite and infinite slopes, Bishop's method; Stress distribution in soils – Boussinesq's theory; Pressure bulbs, Shallow foundations – Terzaghi's and Meyerhoff's bearing capacity theories, effect of water table; Combined footing and raft foundation; Contact pressure; Settlement analysis in sands and clays; Deep foundations – dynamic and static formulae, Axial load capacity of piles in sands and clays, pile load test, pile under lateral loading, pile group efficiency, negative skin friction.

Analysis of Previous GATE Papers

Exam Year	1 Mark Ques.	2 Marks Ques.	Total Marks
1995	14	–	14
1996	15	–	15
1997	13	–	13
1998	6	6	18
1999	9	–	9
2000	9	–	9
2001	5	5	15
2002	5	4	13
2003	5	11	27
2004	6	10	26
2005	4	9	22
2006	4	8	20
2007	3	10	23
2008	3	10	23
2009	3	6	15
2010	5	5	15
2011	6	5	16
2012	5	5	15
2013	4	6	16
2014 Set-1	4	4	12
2014 Set-2	4	5	14

Exam Year	1 Mark Ques.	2 Marks Ques.	Total Marks
2015 Set-1	5	5	15
2015 Set-2	4	5	14
2016 Set-1	3	6	15
2016 Set-2	3	4	11
2017 Set-1	6	4	14
2017 Set-2	4	4	12
2018 Set-1	4	5	14
2018 Set-2	4	5	14
2019 Set-1	5	5	15
2019 Set-2	4	6	16
2020 Set-1	5	5	15
2020 Set-2	4	5	14
2021 Set-1	3	4	11
2021 Set-2	3	5	13
2022 Set-1	6	4	14
2022 Set-2	5	3	11
2023 Set-1	4	6	16
2023 Set-2	4	6	16
2024 Set-1	3	5	13
2024 Set-2	3	5	13

1 Properties of Soils

1.1 The consistency of a saturated cohesive soil is affected by
(a) water content
(b) particle size distribution
(c) density index
(d) coefficient of permeability [1995 : 1 M]

1.2 Write True/False in the answer book.
The void ratio of soil can exceed unity.
[1995 : 1 M]

1.3 Which one of the following relations is not correct?
(a) $e = \frac{n}{1-n}$ (b) $\gamma_{sat} = \frac{G+e}{1+e}\gamma_w$
(c) $n = \frac{e}{1-e}$ (d) $e = \frac{wG}{S}$
Where e = void ratio n = porosity, w = water content, S =degree of saturation, γ_{sat} = saturated unit weight of soil, γ_w = unit weight of water
[1996 : 1 M]

1.4 Consistency Index for a clayey soil is [LL = Liquid Limit, PL = Plastic Limit, PI = Plasticity Index, w = Natural moisture content]
(a) $\frac{LL-w}{PI}$ (b) $\frac{w-LL}{PI}$
(c) $LL - PL$ (d) 0.5 w [1997 : 1 M]

1.5 If the porosity of a soil sample is 20%, the void ratio is
(a) 0.20 (b) 0.80
(c) 1.00 (d) 0.25 [1997 : 1 M]

1.6 The ratio of unconfined compressive strength of an undisturbed sample of soil to that of a remoulded sample, at the same water content, is known as
(a) activity (b) damping
(c) plasticity (d) sensitivity
[1997 : 1 M]

1.7 If soil is dried beyond its shrinkage limit, it will show
(a) Large volume change
(b) Moderate volume change
(c) Low volume change
(d) No volume change [1998 : 1 M]

1.8 Principle involved in the relationship between submerged unit weight and saturated weight of a soil is based on
(a) Equilibrium of floating bodies
(b) Archimedes' principle
(c) Stokes' law
(d) Darcy's law [1999 : 1 M]

1.9 A soil sample in its natural state has mass of 2.290 kg and a volume of 1.15×10^{-3} m^3. After being oven dried, the mass of the sample is 2.035 kg. G_s for soil is 2.68. The void ratio of the natural soil is
(a) 0.40 (b) 0.45
(c) 0.55 (d) 0.53 [1999 : 1 M]

1.10 The toughness index of clayey soils is given by
(a) Plasticity index/Flow index
(b) Liquid limit/Plastic limit
(c) Liquidity index/Plastic limit
(d) Plastic limit/Liquidity index [1999 : 1 M]

1.11 A soil sample has a void ratio of 0.5 and its porosity will be close to
(a) 50% (b) 66%
(c) 100% (d) 33% [2000 : 1 M]

1.12 A borrow pit soil has a dry density of 17 kN/m^3. How many cubic metres of this soil will be required to construct an embankment of 100 m^3 volume with a dry density of 16 kN/m^3.
(a) 94 m^3 (b) 106 m^3
(c) 100 m^3 (d) 90 m^3 [2000 : 1 M]

1.13 The void ratio and specific gravity of a soil are 0.65 and 2.72 respectively. The degree of saturation (in percent) corresponding to water content of 20% is
(a) 65.3 (b) 20.9
(c) 83.7 (d) 54.4 [2001 : 1 M]

1.14 The void ratios at the densest, loosest and the natural states of a sand deposit are 0.2, 0.6 and 0.4, respectively. The relative density of the deposit is
(a) 100% (b) 75%
(c) 50% (d) 25% [2002 : 1 M]

1.15 The following data was obtained from a liquid test conducted on a soil sample

Number of blows	17	22	25	28	34
Water content (%)	63.8	63.1	61.9	60.6	60.5

The liquid limit of the soil is
(a) 63.1% (b) 62.8%
(c) 61.9% (d) 60.6% [2002 : 1 M]

1.16 The undrained cohesion of a remoulded clay soil is 10 kN/m^2. If the sensitivity of the clay is 20, the corresponding remoulded compressive strength is
(a) 5 kN/m^2 (b) 10 kN/m^2
(c) 20 kN/m^2 (d) 200 kN/m^2
[2004 : 1 M]

1.17 The ratio of saturated unit weight to dry unit weight of a soil is 1.25. If the specific gravity of solids (G_s) is 2.65, the void ratio of the soil is
(a) 0.625 (b) 0.663
(c) 0.944 (d) 1.325 [2004 : 1 M]

1.18 A saturated soil mass has a total density 22 kN/m^3 and a water content of 10%. The bulk density and dry density of this soil are
(a) 12 kN/m^3 and 20 kN/m^3 respectively
(b) 22 kN/m^3 and 20 kN/m^3 respectively
(c) 19.8 kN/m^3 and 19.8 kN/m^3 respectively
(d) 23.2 kN/m^3 and 19.8 kN/m^3 respectively
[2005 : 2 M]

1.19 The water content of a saturated soil and the specific gravity of soil solids were found to be 30% and 2.70, respectively. Assuming the unit weight of water to be 10 kN/m^3, the saturated unit weight (kN/m^3) and the void ratio of the soil are
(a) 19.4, 0.81 (b) 18.5, 0.30
(c) 19.4, 0.45 (d) 18.5, 0.45
[2007 : 2 M]

1.20 The liquid limit (*LL*), plastic limit (*PL*) and shrinkage limit (*SL*) of a cohesive soil satisfy the relation
(a) $LL > PL < SL$ (b) $LL > PL > SL$
(c) $LL < PL < SL$ (d) $LL < PL > SL$
[2008 : 2 M]

1.21 A soil is composed of solid spherical grains of identical specific gravity and diameter between 0.075 mm and 0.0075 mm. If the terminal velocity of the largest particle falling through water without flocculation is 0.5 mm/s, that for the smallest particle would be (in mm/s)
(a) 0.005 (b) 0.05
(c) 5 (d) 50 [2011 : 1 M]

1.22 In its natural condition a soil sample has a mass of 1.980 kg and a volume of 0.001 m^3. After being completely dried in an oven; the mass of the sample is 1.800 kg. Specific gravity is 2.7. Unit weight of water is 10 kN/m^3. The degree of saturation of soil is
(a) 0.65 (b) 0.7
(c) 0.54 (d) 0.61
[2013 : 1 M]

1.23 A given cohesionless soil has e_{max} = 0.85 and e_{min} = 0.50. In the field, the soil is compacted to a mass density of 1800 kg/m^3 at a water content of 8%. Take the mass density of water as 1000 kg/m^3 and G_s as 2.7. The relative density (in %) of the soil is
(a) 56.43 (b) 60.25
(c) 62.87 (d) 65.71
[2014 : 2 M, Set-I]

1.24 A certain soil has the following properties: G_s = 2.71, n = 40% and w = 20%. The degree of saturation of the soil (rounded off to the nearest percent) is ________.
[2014 : 1 M, Set-II]

1.25 A fine-grained soil has 60% (by weight) silt content. The soil behaves as *semi-solid* when water content is between 15% and 28%. The soil behaves *fluid-like* when the water content is more than 40%. The '*Activity*' of the soil is
(a) 3.33 (b) 0.42
(c) 0.30 (d) 0.20
[2015 : 1 M, Set-I]

1.26 An earth embankment is to be constructed with compacted cohesionless soil. The volume of the embankment is 5000 m^3 and the target dry unit weight is 16.2 kN/m^3. Three nearby sites (see figure below) have been identified from where the required soil can be transported to the construction site. The void ratios (e) of different sites are shown in the figure. Assume the specific gravity of soil to be 2.7 for all three sites. If the cost of transportation per km is twice the cost of excavation per m^3 of borrow pits, which site would you choose as the most economic solution? (Use unit weight of water = 10 kN/m^3)

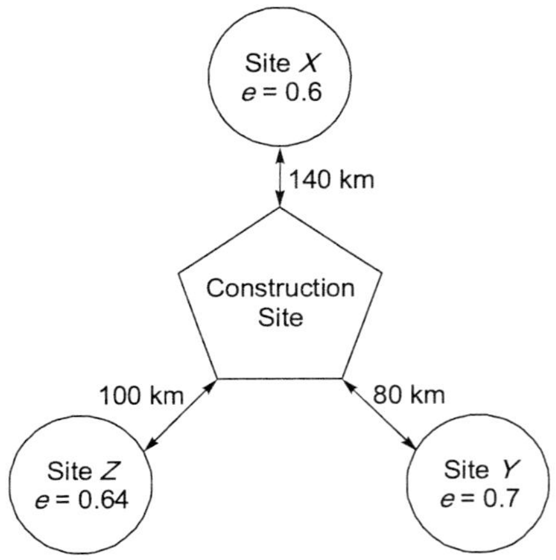

(a) Site X (b) Site Y
(c) Site Z (d) Any of the sites

[2015 : 2 M, Set-I]

1.27 If the water content of a fully saturated soil mass is 100%, the void ratio of the sample is
(a) less than specific gravity of soil
(b) equal to specific gravity of soil
(c) greater than specific gravity of soil
(d) independent of specific gravity of soil

[2015 : 1 M, Set-II]

1.28 A 588 cm^3 volume of moist sand weighs 1010 gm. Its dry weight is 918 gm and specific gravity of solids, G is 2.67. Assuming density of water as 1 gm/cm^3, the void ratio is ___.

[2015 : 2 M, Set-II]

1.29 The porosity (n) and the degree of saturation (S) of a soil sample are 0.7 and 40%, respectively. In a 100 m^3 volume of the soil, the volume (expressed in m^3) of air is ______

[2016 : 2 M, Set-I]

1.30 The laboratory tests on a soil sample yield the following results; natural moisture content = 18%, liquid limit = 60%, plastic limit = 25%, percentage of clay sized fraction = 25%. The liquidity index and activity (as per the expression proposed by Skempton) of the soil, respectively, are
(a) –0.2 and 1.4 (b) 0.2 and 1.4
(c) –1.2 and 0.714 (d) 1.2 and 0.714

[2017 : 2 M, Set-I]

1.31 Let G be the specific gravity of soil solids, w is the water content in the soil sample, γ_w the unit weight of water, and γ_d the dry unit weight of the soil. The equation for the zero air voids line in a compaction test plot is

(a) $\gamma_d = \dfrac{G\gamma_w}{1+Gw}$ (b) $\gamma_d = \dfrac{G\gamma_w}{Gw}$

(c) $\gamma_d = \dfrac{Gw}{1+\gamma_w}$ (d) $\gamma_d = \dfrac{Gw}{1-\gamma_w}$

[2017 : 1 M, Set-II]

1.32 In a shrinkage limit test, the volume and mass of a dry soil pat are found to be 50 cm^3 and 88 g. respectively. The specific gravity of the soil solids is 2.71 and the density of water is 1 g/cc. The shrinkage limit (in % up to two decimal places) is ________.

[2018 : 1 M, Set-I]

1.33 The clay mineral, whose structural units are held together by potassium bond is
(a) Halloysite (b) Illite
(c) Kaolinite (d) Smectite

[2018 : 1 M, Set-II]

1.34 A soil has specific gravity of its solids equal to 2.65. The mass density of water is 1000 kg/m^3. Considering zero air voids and 10% moisture content of the soil sample, the dry density (in kg/m^3, round off to 1 decimal place) would be ______.

[2019 : 1 M, Set-I]

1.35 The notation "SC" as per Indian Standard Soil Classification System refers to
(a) Silty clay (b) Clayey sand
(c) Sandy clay (d) Clayey silt

[2019 : 1 M, Set-II]

1.36 Which one of the following statements is NOT correct?
(a) the cohesion of normally consolidated clay is zero when tri-axial test is conducted under consolidated undrained condition.
(b) In case of a point load, Boussinesq's equation predicts higher value of vertical stress at a point directly beneath the load as compared to Westergaard's equation.
(c) The ultimate bearing capacity of a strip foundation supported on the surface of sandy soil increase in direct proportion to the width of footing.
(d) A clay deposit with a liquidity index greater than unity is in a state of plastic consistency.

[2020 : 1 M, Set-I]

1.37 Soil deposit formed due to transportation by wind is termed as
(a) lacustrine deposit
(b) alluvial deposit
(c) estuarine deposit
(d) aeolian deposit

[2020 : 1 M, Set-II]

1.38 A soil has dry weight of 15.5 kN/m^3, specific gravity of 2.65 and degree of saturation of 72%. Considering the unit weight of water as 10 kN/m^3, the water content of the soil (in %, round off to two decimal places) is ____. [2020 : 1 M, Set-II]

1.39 A sample of 500 g dry sand, when poured into a 2 litre capacity cylinder which is partially filled with water, displaces 188 cm^3 of water. The density of water is 1 g/cm^3. The specific gravity of the sand is
(a) 2.55 (b) 2.72
(c) 2.66 (d) 2.52
[2020 : 1 M, Set-II]

1.40 A partially-saturated soil sample has natural moisture content of 25% and bulk unit weight of 18.5 kN/m^3. The specific gravity of soil solids is 2.65 and unit weight of water is 9.81 kN/m^3. The unit weight of the soil sample on full saturation is
(a) 21.12 kN/m^3 (b) 20.12 kN/m^3
(c) 18.50 kN/m^3 (d) 19.03 kN/m^3
[2021 : 1 M, Set-I]

1.41 From laboratory investigations, the liquid limit, plastic limit, natural moisture content and flow index, of a soil specimen are obtained as 60%, 27%, 32% and 27 respectively. The corresponding toughness index and liquidity index of the soil specimen, respectively, are
(a) 0.15 and 1.22 (b) 0.19 and 6.60
(c) 1.22 and 0.15 (d) 6.60 and 0.19
[2021 : 2 M, Set-II]

1.42 An uncompacted heap of soil has a volume of 10000 m^3 and void ratio of 1. If the soil is compacted to a volume of 7500 m^3, then the corresponding void ratio of the compacted soil is _______. (round off to one decimal place)
[2022 : 1 M, Set-I]

1.43 The specific gravity of a soil is 2.60. The soil is at 50% degree of saturation with a water content of 15%. The void ratio of the soil is _____.
(a) 0.35 (b) 0.78
(c) 0.87 (d) 1.28
[2023 : 1 M, Set-I]

1.44 Consider the statements P and Q.
P: Soil particles formed by mechanical weathering, and close to their origin are generally subrounded.
Q: Activity of the clay physically signifies its swell potential.
Which one of the following options is CORRECT?.
(a) Both P and Q are FALSE
(b) Both P and Q are TRUE
(c) P is TRUE and Q is FALSE
(d) P is FALSE and Q is TRUE
[2024 : 1 M, Set-I]

1.45 An embankment is constructed with soil by maintaining the degree of saturation as 75% during compaction. The specific gravity of soil is 2.68 and the moisture content is 17% during compaction. Consider the unit weight of water as 10 kN/m^3, the dry unit weight (in kN/m^3) of the compacted soil is ______. (rounded off to 2 decimal places)
[2024 : 1 M, Set-I]

1.46 The in-situ percentage of voids of a sand deposit is 50%. The maximum and minimum densities of sand determined from the laboratory tests are 1.8 g/cm^3 and 1.3 g/cm^3, respectively, Assume the specific gravity of sand as 2.7.

The relative density index of the in-situ sand is _________

(rounded off to 2 decimal places).
[2024 : 2 M, Set-II]

Answers Properties of Soils

1.1	(a)	1.2	(*)	1.3	(c)	1.4	(a)	1.5	(d)	1.6	(d)	1.7	(d)
1.8	(b)	1.9	(d)	1.10	(a)	1.11	(d)	1.12	(a)	1.13	(c)	1.14	(c)
1.15	(c)	1.16	(c)	1.17	(b)	1.18	(b)	1.19	(a)	1.20	(b)	1.21	(a)
1.22	(c)	1.23	(d)	1.24	(81.3)	1.25	(c)	1.26	(a)	1.27	(b)	1.28	(0.71)
1.29	(42)	1.30	(a)	1.31	(a)	1.32	(0.19)	1.33	(b)	1.34	(2094.8)	1.35	(b)
1.36	(d)	1.37	(d)	1.38	(19.28)	1.39	(c)	1.40	(d)	1.41	(c)	1.42	(0.5)
1.43	(b)	1.44	(d)	1.45	(16.75)	1.46	(0.13 or 13%)						

Explanations Properties of Soils

1.1 (a)

Consistency is a term which is used to describe the degree or firmness of a soil in a qualitative manner by using descriptions such as soft, medium firm, stiff or hard. It indicates the relative ease with which a soil can be deformed.

The **consistency** of clays are considerably influenced by the amount of water present in them.

1.2 Sol.

TRUE

$$e = \frac{V_v}{V_s}$$

There cannot be an upper limit to the volume of void.

1.3 (c)

$$n = \frac{e}{1+e}$$

$$\Rightarrow \quad n + ne = e$$

$$\Rightarrow \quad e(1-n) = n$$

$$\Rightarrow \quad e = \frac{n}{1-n}$$

1.4 (a)

Relative Consistency, I_C or **Consistency index** is defined as the ratio of the difference between the liquid limit and natural water content of a soil to its plasticity index.

$$I_C = \frac{w_L - w_N}{I_p}$$

$$\Rightarrow \quad I_C = \frac{LL - w}{PI}$$

1.5 (d)

As we know,

$$n = \frac{e}{1+e}$$

$$\therefore \quad 0.2 = \frac{e}{1+e}$$

Solving we get,

$$e = 0.25$$

1.6 (d)

$$\text{Sensitivity } (S_t) = \frac{\begin{pmatrix}\text{Unconfined}\\ \text{compressive}\\ \text{strength}\end{pmatrix}\text{undistrubed}}{\begin{pmatrix}\text{Unconfined}\\ \text{compressive}\\ \text{strength}\end{pmatrix}\text{Demoulded}}$$

it represents degree of disturbance achieved on moulding.

1.7 (d)

At **shrinkage limit**, saturation = 100%; any reduction in volume leads to decrease in decrease of volume of water which is further occupied by air. Thus there will be no volume change.

Note: Shrinkage limit is minimum water content at which soil is completely saturated.

1.8 (b)

When a soil solid particles is submerged water, an upward force which is given by Archimedis principle acts on the particles which reduces the weight of particles. So submerged unit weight = Saturated unit weight – Unit weight of water.

1.9 (d)

$$G = \frac{\gamma_s}{\gamma_w}$$

$$\therefore \quad \gamma_s = 2.68 \times 9.81 = 26.291 \text{ kN/m}^3$$

$$\gamma_s = \frac{W_s}{V_s}$$

$$\Rightarrow V_s = \frac{W_s}{\gamma_s} = \frac{2.035 \times 9.81}{26.291 \times 10^3} = 0.75 \times 10^{-3} \text{ m}^3$$

$$V_v = V - V_s$$

$$= 1.15 \times 10^{-3} - 0.75 \times 10^{-3} = 0.4 \times 10^{-3}$$

$$\therefore \quad e = \frac{V_v}{V_s} = \frac{0.4 \times 10^{-3}}{0.75 \times 10^{-3}} = 0.53$$

1.10 (a)

Toughness index, I_T is defined as the ratio of plasticity index to the flow index.

$$I_T = I_P / I_F$$

1.11 (d)

$$n = \frac{e}{1+e} = \frac{0.5}{1.5} = 33\%$$

1.12 (a)

$V \propto (1 + e),\ \gamma_d \propto \frac{1}{V}$

Dry densities of soil in pit and embankment respectively are:

$$\gamma_2 = 16 \text{ kN/m}^3$$
$$\gamma_1 = 17 \text{ kN/m}^3$$
$$V_1 = \text{Volume of soil req from pit}$$
$$V_2 = \text{Volume of embankment}$$
$$V_2 = 100$$
$$V_1 = \frac{16}{17} \times 100 = 94 \text{ m}^3$$

1.13 (c)

Given $e = 0.65$

$G = 2.72$

$w = 20\%$

From $e = \frac{wG}{S}$

$$\therefore \quad S = \frac{wG}{e} = \frac{0.2 \times 2.72}{0.65}$$
$$= 0.8369 \approx 83.7\%$$

1.14 (c)

Relative density

$$= \frac{e_{\max} - e_{natural}}{e_{\max} - e_{\min}} \times 100$$
$$= \frac{0.6 - 0.4}{0.6 - 0.2} \times 100 = 50\%$$

1.15 (c)

The **liquid limit** is the water content corresponding to **25 number of blows**

$\therefore \quad w_L = 61.9\%$

1.16 (c)

We know that,

$$c_u = \frac{q_u}{2}$$
$$\Rightarrow \quad q_u = 2\,c_u = 2 \times 10$$
$$= 20 \text{ kN/m}^2$$

Note: Sensitivity of the clay is of no use.

1.17 (b)

$$\gamma_{sat} = \left(\frac{G+e}{1+e}\right)\gamma_w \quad \text{...(i)}$$
$$\gamma_d = \frac{G\gamma_w}{1+e} \quad \text{...(ii)}$$
$$\Rightarrow \quad \frac{\gamma_{sat}}{\gamma_d} = \frac{G+e}{G}$$
$$\Rightarrow \quad 1.25 = \frac{2.65 + e}{2.65}$$
$$\Rightarrow \quad e = 0.663$$

1.18 (b)

Total density is bulk density, $\gamma_t = 22$ kN/m^3

$$\text{Dry density, } \gamma_d = \frac{\gamma_t}{1+w} = \frac{22}{1+\frac{10}{100}} = \frac{22}{1.1}$$
$$= 20 \text{ kN/m}^3$$

1.19 (a)

We know that,

$$S \times e = G \times w$$
$$\Rightarrow \quad 1 \times e = 2.70 \times 0.30$$
$$\Rightarrow \quad e = 0.81$$

and

$$\gamma_{sat} = \left(\frac{G+e}{1+e}\right)\gamma_w$$
$$= \left(\frac{2.7 + 0.81}{1 + 0.81}\right) \times 10$$
$$= 19.4 \text{ kN/m}^3$$

1.20 (b)

As we increase the water content in cohesive soil, soil state changes from solid to semi-solid to plastic to liquid. Therefore

Liquid limit > Plastic limit > Shrinkage limit

1.21 (a)

According to **Stoke's law,**

$$v \propto D^2$$
$$\Rightarrow \quad \frac{v_1}{v_2} = \left(\frac{D_1}{D_2}\right)^2 \quad \Rightarrow \quad \frac{0.5}{v_2} = \left(\frac{0.075}{0.0075}\right)^2$$
$$\Rightarrow \quad v_2 = \frac{0.5}{100} \quad \Rightarrow \quad v_2 = 0.005 \text{ mm/s}$$

1.22 (c)

$$\rho = \frac{M}{V} = \frac{1.980}{0.001} = 1980\ \text{kg/m}^3$$

$$\rho_d = \frac{M_d}{V} = \frac{1.800}{0.001} = 1800\ \text{kg/m}^3$$

$$\rho_d = \frac{\rho}{1+w}$$

$$\Rightarrow \quad 1800 = \frac{1980}{1+w}$$

$$\Rightarrow \quad w = 0.1$$

$$\because \quad \rho_d = \frac{G\cdot\rho_w}{1+e}$$

$$\Rightarrow \quad 1800 = \frac{2.7\times1000}{1+e}$$

$$\Rightarrow \quad e = 0.5$$

$$\because \quad wG = e\cdot S$$

$$\Rightarrow 0.1 \times 2.7 = 0.5S$$

$$\Rightarrow \quad S = 0.54$$

1.23 (d)

$$e_{max} = 0.85$$

$$e_{min} = 0.5$$

$$\rho_{field} = 1800\ \text{kg/m}^3$$

at water content = 8%

$$\rho_w = 1000\ \text{kg/m}^3$$

$$G_S = 2.7$$

Relative density, I_D = ?

$$\rho = \frac{G\rho_w(1+w)}{1+e} = \frac{2.7\times1000(1.08)}{1+e}$$

$$\Rightarrow \quad 1+e = \frac{2.7\times1000\times1.08}{1800}$$

$$\Rightarrow \quad e = 0.62$$

$$\Rightarrow \quad I_D = \frac{e_{max}-e}{e_{max}-e_{min}}\times100$$

$$= \frac{0.85-0.62}{0.85-0.5}\times100$$

$$= 65.714\%$$

1.24 Sol.

$$Se = wG$$

and

$$e = \frac{n}{1-n} = 0.67$$

$$\Rightarrow \quad S = \frac{2.71\times20}{0.67} = 81.3\%$$

1.25 (c)

$$I_p = W_L - W_P = 40 - 28 = 12\%$$

$$\text{Activity} = \frac{I_P}{C(\%\text{clay fraction finer than } 2\,\mu)}$$

$$= \frac{12}{40} = 0.3$$

1.26 (a)

Volume of solids at embankment,

$$V_s = \frac{W_s}{G_s\gamma_w}$$

$$V_s = \frac{16.2\times5000}{2.7\times10}\text{m}^3$$

$$= 3000\ \text{m}^3$$

Volume of solids will remain constant,

also,
$$V = V_V + V_s$$

$$= eV_s + V_s = V_s(1+e)$$

$$V = 3000\,(1+e)$$

$$V_X = 3000\times1.6 = 4800\ \text{m}^3$$

$$V_Y = 3000\times1.7 = 5100\ \text{m}^3$$

$$V_Z = 3000\times1.64 = 4920\ \text{m}^3$$

Let cost of excavation be α/m^3,

$$P_X = 4800\alpha + 140\times\alpha\times2$$

$$= 5080\alpha$$

$$P_Y = 5100\alpha + 80\times\alpha\times2$$

$$= 5260\alpha$$

$$P_Z = 4920\alpha + 100\times\alpha\times2$$

$$= 5120\alpha$$

Hence P_X is least.

$\therefore$ Hence total cost for site X is least.

1.27 (b)

$$S.e = w.G_s$$

Given $S = 1$ (Fully saturated soil)

$$w = 100\% = 1$$

$$e = G_s$$

1.28 Sol.

$$\gamma_d = \frac{918}{588} = 1.561\ \text{g/cc}$$

$$\gamma_d = \frac{G\gamma_w}{1+e}$$

$$1.561 = \frac{2.67\times1}{1+e}$$

$$e = 0.71$$

1.29 Sol.

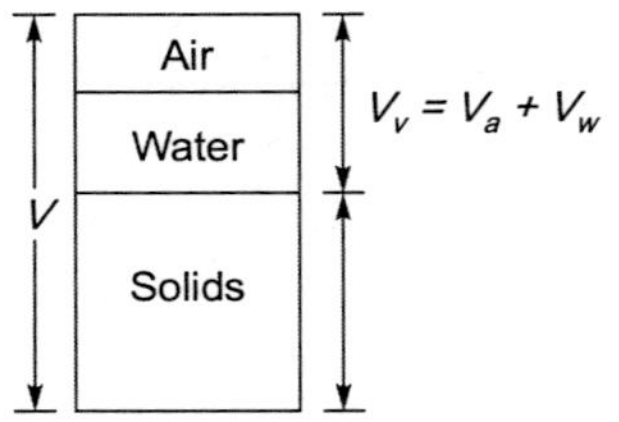

$n = 0.7$, $S = 40\%$, $V = 100\ m^3$, $V_a = ?$

$$n = \frac{V_v}{V}$$

$$\Rightarrow \quad 0.7 = \frac{V_v}{100}$$

$$\Rightarrow \quad V_v = 70\ m^3$$

Now, $$S = \frac{V_w}{V_v} = \frac{V_v - V_a}{V_v} = 1 - \frac{V_a}{V_v}$$

$$\Rightarrow \quad 0.40 = 1 - \frac{V_a}{V_v}$$

$$\Rightarrow \quad \frac{V_a}{V_v} = 0.60$$

$$\Rightarrow \quad V_a = 0.6 \times 70 = 42\ m^3$$

1.30 (a)

Given data: $w_n = 18\%$, $w_L = 60\%$, $w_p = 25\%$

% of clay size particle = 25%

Liquidity index,

$$I_L = \frac{w_n - w_p}{w_L - w_p} = \frac{0.18 - 0.25}{0.60 - 0.25} = -0.2$$

$$\text{Activity} = \frac{I_p}{\%C} = \frac{w_L - w_p}{\%C} = \frac{60 - 25}{25} = 1.4$$

1.31 (a)

Percentage air void line is relation between dry unit weight and water content at constant air void. Hence equation of **zero air void line** is

$$\gamma_d = (1 - n_a)\frac{G\gamma_w}{1 + wG} \qquad (n_a = 0)$$

$$\gamma_d = \frac{G\gamma_w}{1 + wG}$$

1.32 Sol.

Dry soil mass = 88 gm

Volume of dry soil = 50 cc

Dry density of soil mass,

$$\rho_d = \frac{M_d}{V_d} = \frac{88}{50} \text{ gm/cc}$$

$$\because \text{ Shrinkage limit, } w_s = \frac{1}{R} - \frac{1}{G} = \frac{1}{(\rho_d/\rho_w)} - \frac{1}{G}$$

$$w_s = \frac{\rho_w}{\rho_d} - \frac{1}{G}$$

$$= \frac{1}{1.76} - \frac{1}{2.71} = 0.1990$$

1.33 (b)

Kaolinite → Hydrogen bond

Illite → Ionic Bond/Potassium Bond

Montmorillonite → Water Bond

1.34 Sol.

$w = 10\%$ and has zero air voids

$\gamma_d = ?$

$\gamma_w = 1000\ kg/m^3$

$G = 2.65$

Zero air voids

$\therefore \quad \eta_a = 0$ and $S = 1$

$\Rightarrow \quad Se = wG$

$$e = \frac{0.1 \times 2.65}{1} = 0.265$$

$$\gamma_d = \frac{G\gamma_w}{1+e} = \frac{2.65 \times 1000}{1 + 0.265}$$

$$= 2094.86\ kg/m^3$$

1.35 (b)

SC → Clayey sand

1.36 (d)

A clay deposit with liquidty index greater then 1, will be in liquid stage of consistency.

$$\because \quad I_L = \frac{W_n - W_p}{W_L - W_p} > 1$$

$$\therefore \quad W_n > W_L$$

1.37 (d)

Soil deposited by wind is Aeolian soil.

1.38 Sol.

$\gamma_d = 15.5\ kN/m^3$, $G = 2.65$, $S = 72\%$

$$\gamma_d = \frac{G\gamma_w}{1+e} = \frac{2.65 \times 10}{1+e} = 15.5$$

$$e = 0.7096$$

$$w = \frac{Se}{G} = \frac{0.72 \times 0.7096}{2.65} = 0.1928$$

$$w = 19.28\%$$

1.39 (c)

Weight of dry sand,

$$(W_s) = 500 \text{ gm}$$

Volume of water displaced,

$$(V_s) = 188 \text{ cm}^3$$

$$\gamma_s = \frac{W_s}{V_s}$$

$$G_s = \frac{\gamma_s}{\gamma_w} = \frac{500/188}{1} = 2.66$$

1.40 (d)

$$w = 0.25,\ \gamma_t = 18.5 \text{ kN/m}^3$$

$$G_s = 2.65,\ \gamma_w = 9.81$$

$$\gamma_t = \frac{G_s\gamma_w(1+w)}{1+e}$$

$$\Rightarrow \quad e = \frac{2.65 \times 9.81 \times 1.25}{18.5} - 1$$

$$\Rightarrow \quad e = 0.756$$

At full saturation,

$$S = 1$$

$$\Rightarrow \quad \gamma_{sat} = \frac{(G_s + e)\gamma_w}{1+e}$$

$$\gamma_{sat} = \frac{(2.65 + 0.756) \times 9.81}{1.756}$$

$$= 19.03 \text{ kN/m}^3$$

1.41 (c)

Flow index = 27

Plasticity index,

$$I_P = W_L - W_P = 33$$

Toughness index,

$$I_T = \frac{I_P}{I_f} = \frac{33}{27} = 1.22$$

Liquidity index,

$$I_L = \frac{W_n - W_P}{W_L - W_P} = 0.151$$

1.42 Sol.

$$(V_s)_{heap} = (V_s)_{compacted\ soil}$$

$$\Rightarrow \left(\frac{V_T}{1+e}\right)_{heap} = \left(\frac{V_T}{1+e}\right)_{compacted\ soil}$$

$$\Rightarrow \quad \frac{10000}{1+1} = \frac{7500}{1+e}$$

$$\Rightarrow \quad e = 0.5$$

1.43 (b)

Given: Specific gravity, $G = 2.6$

Degree of Saturation, $S = 0.5$

Water content, $w = 0.15$

As we know

$$Se = Gw$$

$$\Rightarrow \quad 0.5 \times e = 2.6 \times 0.15$$

$$\Rightarrow \quad e = \frac{2.6 \times 0.15}{0.5} = 0.78$$

1.44 (d)

1.45 Sol.

$$\text{Dry density, } \gamma_d = \frac{G\gamma_w}{1+e}$$

$$e = \frac{wG}{S} = \frac{0.17 \times 2.68}{0.75} = 0.6$$

$$\therefore \quad \gamma_d = \frac{2.68 \times 10}{1+0.6} = 16.75 \text{ kN/m}^3$$

1.46 Sol.

Minimum dry density, $\gamma_{dmin} = 1.3 \text{ g/cm}^3$

Maximum dry density, $\gamma_{dmax} = 1.8 \text{ g/cm}^3$

Specific gravity of soil solids, $G = 2.7$

Porosity of soil, $n = 0.5$

$$\text{Void ratio, } e = \frac{n}{1-n} = \frac{0.5}{0.5} = 1$$

In-situ dry density,

$$\gamma_d = \frac{G\rho_w}{1+e} = \frac{2.7 \times 1}{1+1} = 1.35 \text{ g/cm}^3$$

Now, Relative density,

$$RD = \frac{\frac{1}{\gamma_{dmin}} - \frac{1}{\gamma_d}}{\frac{1}{\gamma_{dmin}} - \frac{1}{\gamma_{dmax}}} = \frac{\frac{1}{1.3} - \frac{1}{1.35}}{\frac{1}{1.3} - \frac{1}{1.8}} = 0.13$$

So, relative density = 0.13 or 13%

2 Classification of Soils and Clay Minerals

2.1 A soil having particles of nearly the same size is known as
(a) well graded (b) uniformly graded
(c) poorly graded (d) gap graded
[1995 : 1 M]

2.2 The equation of the A-line in the Plasticity Chart is ...
[1995 : 1 M]

2.3 Soil transported by wind are known as .... soils.
[1995 : 1 M]

2.4 The particle size distribution curves are extremely useful for the classification of
(a) fine grained soils
(b) coarse grained soils
(c) both coarse grained and fine grained soils
(d) silts and clays [1996 : 1 M]

2.5 The shape of clay particle is usually
(a) angular (b) flaky
(c) tubular (d) rounded
[1997 : 1 M]

2.6 Some of the structural strength of a clayey material that is lost by remoulding is slowly recovered with time. This property of soils to undergo an isothermal gel-to-sol-to-gel transformation upon agitation and subsequent rest is termed
(a) Isotropy (b) Anisotropy
(c) Thixotropy (d) Allotropy
[1998 : 1 M]

2.7 The values of liquid limit and plasticity index for soils having common geological origin in a restricted locality usually define
(a) a zone above A-line
(b) a straight line parallel to A-line
(c) a straight line perpendicular to A-line
(d) points may be anywhere in the plasticity chart
[1999 : 1 M]

2.8 Data from a sieve analysis conducted on a given sample of soil showed that 67% of the particles passed through 75 micron IS sieve. The liquid limit and plastic limit of the finer fraction was found to be 45 and 33 percent respectively. The group symbol of the given soil as per **IS : 1498 - 1970** is
(a) *SC* (b) *MI*
(c) *CH* (d) *MH* [2002 : 1 M]

2.9 A soil mass contains 40% gravel, 50% sand and 10% silt. This soil can be classified as
(a) silty sandy gravel having coefficient of uniformity less than 60.
(b) silty gravelly sand having coefficient of uniformity equal to 10.
(c) gravelly silty sand having coefficient of uniformity greater than 60.
(d) gravelly silty sand and its coefficient of uniformity cannot be determined.
[2005 : 2 M]

Linked Answer Questions 2.10 and 2.11

Laboratory sieve analysis was carried out on a soil sample using a complete set of standard IS sieves. Out of 500 g of soil used in the test, 200 g was retained on IS 600μ sieve, 250 g was retained on IS 500 μ sieve and the remaining 50 g was retained on IS 425 μ sieve.

2.10 The coefficient of uniformity of the soil is
(a) 0.9 (b) 1.0
(c) 1.1 (d) 1.2 [2006 : 2 M]

2.11 The classification of the soil is
(a) *SP* (b) *SW*
(c) *GP* (d) *GW* [2006 : 2 M]

2.12 Sieve analysis on a dry soil sample of mass 1000 g showed that 980 g and 270 g of soil pass through 4.75 mm and 0.075 mm sieve, respectively. The liquid limit and plastic limits of the soil fraction passing through 425 μ sieves are 40% and 18%, respectively. The soil may be classified as
(a) *SC* (b) *MI*
(c) *CI* (d) *SM*
[2007 : 2 M]

2.13 Group symbols assigned to silty sand and clayey sand are respectively
(a) *SS* and *CS* (b) *SM* and *CS*
(c) *SM* and *SC* (d) *MS* and *CS*
[2008 : 1 M]

2.14 The laboratory test results of a soil sample are given below:
Percentage finer than 4.75 mm = 60
Percentage finer than 0.075 mm = 30
Liquid Limit = 35%
Plastic Limit = 27%
The soil classification is
(a) *GM* (B) *SM*
(c) *GC* (D) *ML-MI* [2009 : 2 M]

2.15 A fine grained soil has liquid limit of 60 and plastic limit of 20. As per the plasticity chart, according to IS classification, the soil is represented by the letter symbols
(a) *CL* (b) *CI*
(c) *CH* (d) *CL-ML* [2010 : 1 M]

2.16 As per the Indian Standard soil classification system, a sample of silty clay with liquid limit of 40% and plasticity index of 28% is classified as
(a) *CH* (b) *CI*
(c) *CL* (d) *CL-ML* [2012 : 1 M]

2.17 As per Indian Standard Soil Classification System **(IS 1498 : 1970)**, an expression for A-line is
(a) $I_p = 0.73\,(w_L - 20)$
(b) $I_p = 0.70\,(w_L - 20)$
(c) $I_p = 0.73\,(w_L - 10)$
(d) $I_p = 0.70\,(w_L - 10)$ [2014 : 1 M, Set-II]

2.18 The clay mineral primarily governing the swelling behavior of Black Cotton soil is
(a) Halloysite (b) Illite
(c) Kaolinite (d) Montmorillonite
[2014 : 1 M, Set-II]

2.19 A fine grained soil is found to be plastic in the water content range of 26-48%. As per **Indian Standard Classification System,** the soil is classified as
(a) *CL* (b) *CH*
(c) *CL-ML* (d) *CI*
[2016 : 1 M, Set-I]

2.20 As per the Unified Soil Classification System (USCS), the type of soil represented by 'MH' is
(a) Inorganic silts of high plasticity with liquid limit more than 50%.
(b) Inorganic silts of low plasticity with liquid limit less than 50%.
(c) Inorganic clays of high plasticity with liquid limit less than 50%.
(d) Inorganic clays of low plasticity with liquid limit more than 50%
[2021 : 1 M, Set-II]

Answers Classification of Soils and Clay Minerals

2.1	(b)	2.2	(Sol.)	2.3	(Sol.)	2.4	(b)	2.5	(b)	2.6	(c)	2.7	(b)
2.8	(b)	2.9	(c)	2.10	(d)	2.11	(a)	2.12	(a)	2.13	(c)	2.14	(a)
2.15	(c)	2.16	(b)	2.17	(a)	2.18	(d)	2.19	(d)	2.20	(a)		

Explanations Classification of Soils and Clay Minerals

2.1 (b)

- A soil having particles of all the sizes is known as well graded soil.
- A soil having particles of nearly the same size is known as uniformly graded soil.

2.2 Sol.

$$I_p = 0.73\,(w_L - 20)$$

2.3 Sol.

Water transported soil are of three types:
1. Alluvial deposit.
2. Lacustrine deposit.
3. Marine deposit.

Wind transported soil are 'Aeolin deposits'.

2.4 (b)

Properties of coarse grained soil majority depends upon grain size distribution while the properties of fine grained soil little bit depends on grain size distribution. Fine grained soil's properties depend upon structure, shape, origin etc. particle size distribution curve extremely helps in determining gradation and uniformity of coarse grained soils.

2.5 (b)

The majority of soils may be regarded as either sands or clays. Sand grains generally have a rounded shape. Clay particles have a flaky shape.

2.6 (c)

Thixotropy is the property of soil due to which loss in shear strength due to moulding at unaltered moisture content may be regained.

2.7 (b)

- Soil having common geological origin i.e., soil is formed by same weathering agent.
- No soil exist above U-line is plasticity chat.
- Soil can not be perpendicular for a plasticity chart.

2.8 (b)

% passing 75 μ sieve > 50%

So the soil might be, silt (*M*) or clay (*C*).

$$I_P \text{ of soil} = w_L - w_P$$
$$= 45 - 33 = 12$$

Equation of A line is given by

$$I_P = 0.73(w_L - 20)$$
$$= 0.73(45 - 20) = 18.25$$

$\because$ I_P of soil < I_P of A-line

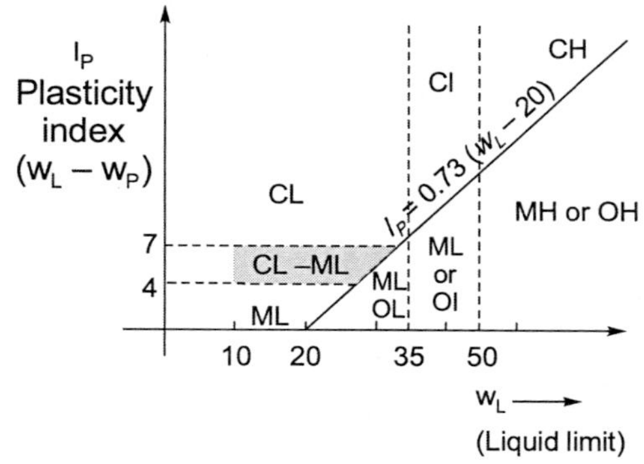

Plasticity Chart

$\therefore$ It will lie below A-line and also $35 < w_L < 50$

So it is *MI*.

2.9 (c)

Soil		Sieve	Weight retained (%)	Cumulative weight retained (%)	Cumulative Weight Passing (N%)
Gravel	40%	4.75 mm	40%	40%	60%
Sand	50%	0.075 mm	50%	90%	10%
Silt	10%	0.002 mm	10%	100%	0

2.10 (d)

S.No.	Sieve	Weight size	Cum. weight retained	% retained	% passed
1.	600μ	200	200	40	60
2.	500μ	250	450	90	10
3.	425μ	50	500	100	0

$$\therefore \quad D_{60} = 600\,\mu$$
$$D_{10} = 500\,\mu$$

Now uniformity coefficient,

$$C_u = \frac{D_{60}}{D_{10}} = \frac{600\mu}{500\mu} = 1.2$$

2.11 (a)

More than 50% of the soil pass through 600 μ sieve which means that even a greater percentage of the soil passes through 4.75 mm sieve.

Hence, the soil is definitely **sandy soil**. Also, we know that for a sand to be considered well graded, C_u should be greater than 6. Here, in this case C_u = 1.2. Thus, the soil is **poorly graded sand**.

2.12 (a)

$$\% \text{ retained on 4.75 mm} = \frac{20}{1000} \times 100 = 2\%$$

$$\% \text{ passing 75 } \mu \text{ sieve} = \frac{270}{1000} \times 100 = 27\%$$

$\therefore$ > 50% of coarse fraction is between 4.75 mm and 75 μ

Hence, sandy soil and % fines > 12%

Soil is clayey sand (*SC*).

2.13 (c)

First part of the dual symbol indicates gradation whereas second part indicates the nature of fineness.

$\Rightarrow$ Silty sand = *SM*

$\Rightarrow$ Clayey sand = *SC*

2.14 (a)

% soil retained on 4.75 mm sieve = 40%

% soil retained on 75 μ sieve = 70 – 40 = 30%

Gravel > Sand

So, it is gravel.

Also % fines is 30%

$$I_P = w_L - w_P = 35 - 27 = 8\%$$

$$(I_P)_{A\,line} = 0.73\,(w_L - 20) = 10.95\%$$

$$(I_P)_{soil} < (I_P)_{A\,line}$$

∴ Silt > Clay

Hence the given soil is **GM**.

2.15 (c)

Plasticity index,

$$I_P = w_L - w_P = 60 - 20 = 40$$

Equation of A line is given by,

$$I_P = 0.73(w_L - 20) = 0.73(60 - 20) = 29.2$$

Thus the given soil lies above the A-Line. The liquid limit of the soil is more than 50, hence the soil will be **CH**.

2.16 (b)

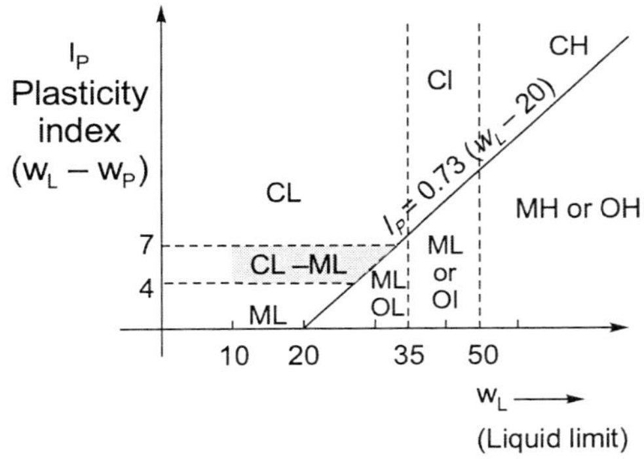

Plasticity Chart

I_P of soil = 28%

I_P of A-line = $0.73(w_L - 20)$

$= 0.73(40 - 20) = 14.6$

∵ I_P of soil > I_P of A-line

∴ It will lie above A-line and also $35 < w_L < 50$

So it is CI.

2.17 (b)

In case of fine grained soils, classification is done using plasticity chart.

As per plasticity chart,

$$I_p = 0.73(W_L - 20)$$

2.18 (a)

Montmorrilouite mineral has significant affinity for water, and results in expansion and seveling. Black cotton soil contains montmoriltonite mineral in large proportion.

2.19 (d)

Soil is plastic in the range of 26-48%.

Hence plastic limit = 26% and liquid limit = 48%. Because liquid limit is in the range of 35% to 50%. Hence intermediate compressible soil.

2.20 (a)

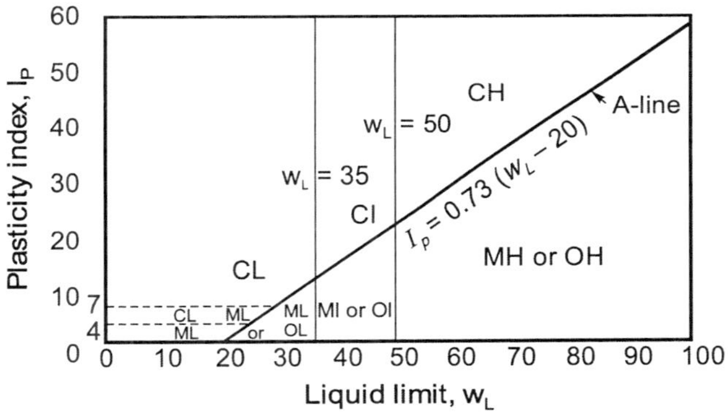

3 Effective Stress and Permeability

3.1 According to Darcy's law for flow through porous media, the velocity is proportional to
(a) effective stress
(b) hydraulic gradient
(c) cohesion
(d) stability number [1997 : 1 M]

3.2 A soil mass has coefficients of horizontal and vertical permeability as 9×10^{-7} cm/s and 4×10^{-7} cm/s, respectively. The transformed coefficient of permeability of an equivalent isotropic soil mass is
(a) 9×10^{-7} cm/s (b) 4×10^{-7} cm/s
(c) 13×10^{-7} cm/s (d) 6×10^{-7} cm/s
[1997 : 1 M]

3.3 A river 5 m deep consists of a sand bed with saturated unit weight of 20 kN/m^3. γ_w = 9.81 kN/m^3. The effective vertical stress at 5 m from the top of sand bed is
(a) 41 kN/m^2 (b) 51 kN/m^2
(c) 55 kN/m^2 (d) 53 kN/m^2
[1999 : 1 M]

3.4 For an anisotropic soil, permeabilities in x and y directions are k_x and k_y respectively in a two dimensional flow. The effective permeability k_{eff} for the soil is given by
(a) $k_x + k_y$
(b) $\frac{k_x}{k_y}$
(c) $\left(k_x^2 + k_y^2\right)^{1/2}$
(d) $\left(k_x k_y\right)^{1/2}$
[1999 : 1 M]

3.5 For the soil strata shown in figure, the water table is lowered by drainage by 2 m and if the top 2 m thick silty sand stratum remains saturated by capillary action even after lowering of water table, the increase in effective vertical pressure in kPa at mid-height of clay layer will be

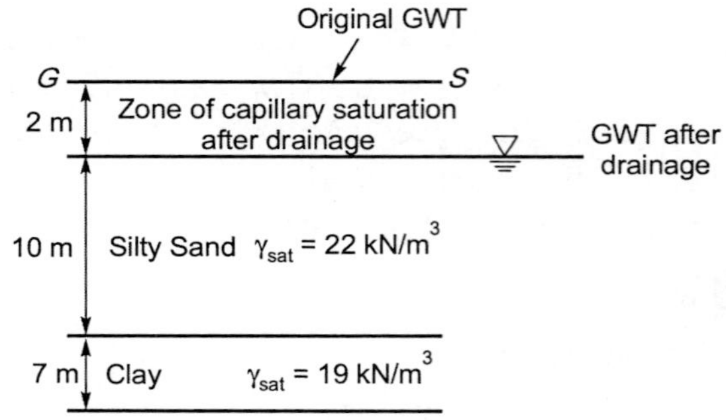

(a) 0.2 (b) 2
(c) 20 (d) 200 [2003 : 2 M]

3.6 A 10 m thick clay layer is underlain by a sand layer of 20 m depth (see figure below). The water table is 5 m below the surface of clay layer. The soil above the water table is capillary saturated. The value of γ_{sat} is 19 kN/m^3. The unit weight of water is γ_w. If now the water table rises to the surface, the effective stress at a point P on the interface will

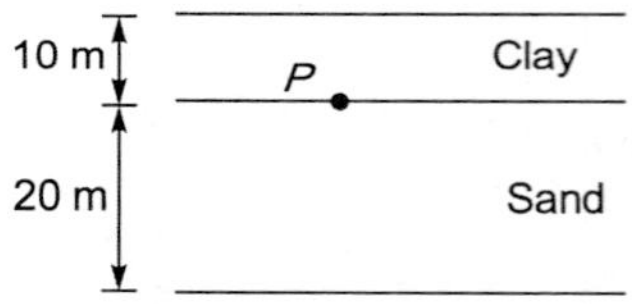

(a) increase by $5\gamma_w$ (b) remain unchanged
(c) decrease by $5\gamma_w$ (d) decrease by $10\gamma_w$
[2004 : 1 M]

3.7 In a constant head permeameter with cross section area of 10 cm^2, when the flow was taking place under a hydraulic gradient of 0.5, the amount of water collected in 60 seconds is 600 cc. The permeability of the soil is
(a) 0.002 cm/s (b) 0.02 cm/s
(c) 0.2 cm/s (d) 2.0 cm/s
[2005 : 2 M]

3.8 Assuming that a river bed level does not change and the depth of water in river was 10 m, 15 m and 8 m during the months of February, July and December respectively of a particular year. The average bulk density of the soil is 20 kN/m^3. The density of water is 10 kN/m^3. The effective stress at a depth of 10 m below the river bed during these months would be

(a) 300 kN/m^2 in February, 350 kN/m^2 July and 320 kN/m^2 in December
(b) 100 kN/m^2 in February, 100 kN/m^2 July and 100 kN/m^2 in December
(c) 200 kN/m^2 in February, 250 kN/m^2 July and 180 kN/m^2 in December
(d) 300 kN/m^2 in February, 350 kN/m^2 July and 280 kN/m^2 in December

[2005 : 2 M]

3.9 Which of the following statement is NOT true in the context of capillary pressure in soils?
(a) Water is under tension in capillary zone
(b) Pore water pressure is negative in capillary zone
(c) Effective stress increases due to capillary zone
(d) Capillary pressure is more in coarse grained soils

[2006 : 1 M]

3.10 Match **List-I** with **List-II** and select the correct answer using the codes given below the lists:

List-I
A. Constant head permeability test
B. Consolidation test
C. Pycnometer test
D. Negative skin friction

List-II
1. Pile foundation
2. Specific gravity
3. Clay soil
4. Sand

Codes:

	A	B	C	D
(a)	4	3	2	1
(b)	4	2	3	1
(c)	3	4	2	1
(d)	4	1	2	3

[2007 : 2 M]

Linked Answer Questions 3.11 and 3.12:

The ground conditions at a site are shown in the figure below.

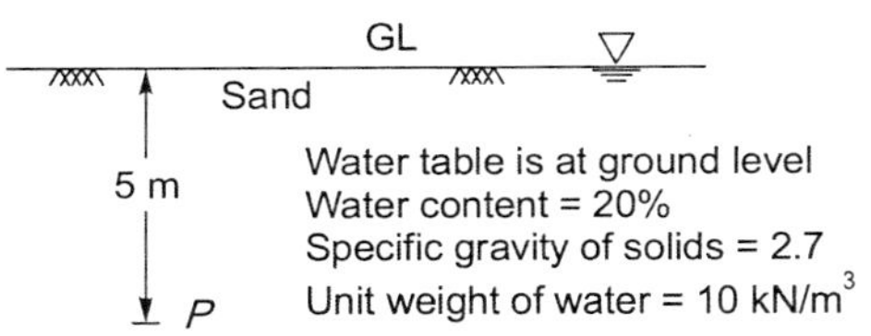

3.11 The saturated unit weight of the sand (kN/m^3) is
(a) 15 (b) 18
(c) 21 (d) 24 [2008 : 2 M]

3.12 The total stress, pore water pressure and effective stress (kN/m^2) at the point *P* are, respectively
(a) 75, 50 and 25 (b) 90, 50 and 40
(c) 105, 50 and 55 (d) 120, 50 and 70

[2008 : 2 M]

3.13 An open ended steel barrel of 1 m height and 1 m diameter is filled with saturated fine sand having coefficient of permeability of 10^{-2} m/s. The barrel stands on a saturated bed of gravel. The time required for the water level in the barrel to drop by 0.75 m is
(a) 58.9 s (b) 75 s
(c) 100 s (d) 150 s [2010 : 2 M]

Common Data for Questions 3.14 and 3.15:

A sand layer found at sea floor under 20 m water depth is characterised with relative density = 40%, maximum void ratio = 1.0, minimum void ratio = 0.5, and specific gravity of soil solids = 2.67. Assume the specific gravity of sea water to be 1.03 and the unit weight of fresh water to be 9.81 kN/m^3.

3.14 What would be the effective stress (rounded off to the nearest integer value of kPa) at 30 m depth into the sand layer?
(a) 77 kPa (b) 273 kPa
(c) 268 kPa (d) 281 kPa

[2011 : 2 M]

3.15 What would be the change in the effective stress (rounded off to the nearest integer value of kPa) at 30 m depth into the sand layer if the sea water level permanently rises by 2 m?
(a) 19 kPa (b) 0 kPa
(c) 21 kPa (d) 22 kPa [2011 : 2 M]

3.16 Two soil specimens with identical geometric dimensions were subjected to falling head permeability tests is the laboratory under identical conditions. The fall of water head was measured after an identical time interval. The ratio of initial to final water heads for the test involving the first specimen was 1.25. If the coefficient of permeability of the second specimen is 5 times that of the first, the ratio of initial to final water heads in the test involving the second specimen is
(a) 3.05 (b) 3.80
(c) 4.00 (d) 6.25 [2012 : 2 M]

3.17 The soil profile below a lake with water level at elevation = 0 m and lake bottom at elevation = −10 m is shown in the figure, where *k* is the permeability coefficient. A piezometer (stand pipe) installed in the sand layer shows a reading of +10 m elevation. Assume that the piezometric head is uniform in the sand layer. The quantity of water (in m^3/s) flowing into the lake from the sand layer through the silt layer per unit area of the lake bed is

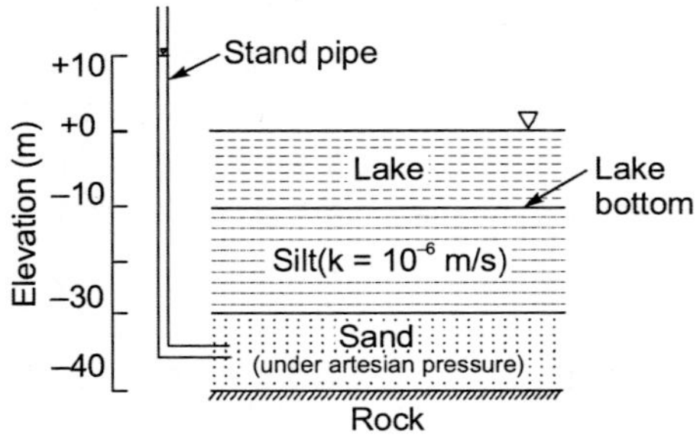

(a) 1.5×10^{-6} (b) 2.0×10^{-6}
(c) 1.0×10^{-6} (d) 0.5×10^{-6}

[2013 : 2 M]

3.18 A non-homogeneous soil deposit consists of a silt layer sandwiched between a fine-sand layer at top and a clay layer below. Permeability of the silt layer is 10 times the permeability of the clay layer and one-tenth of the permeability of the sand layer. Thickness of the silt layer is 2 times the thickness of the sand layer and two-third of the thickness of the clay layer. The ratio of equivalent horizontal and equivalent vertical permeability of the deposit is ________.

[2015 : 2 M, Set-I]

3.19 Water flows from P to Q through two soil samples, Soil 1 and Soil 2, having cross sectional area of 80 cm^2 as shown in the figure. Over a period of 15 minutes, 200 ml of water was observed to pass through any cross section. The flow conditions can be assumed to be steady state. If the coefficient of permeability of Soil 1 is 0.02 mm/s, the coefficient of permeability of Soil 2 (expressed in mm/s) would be________.

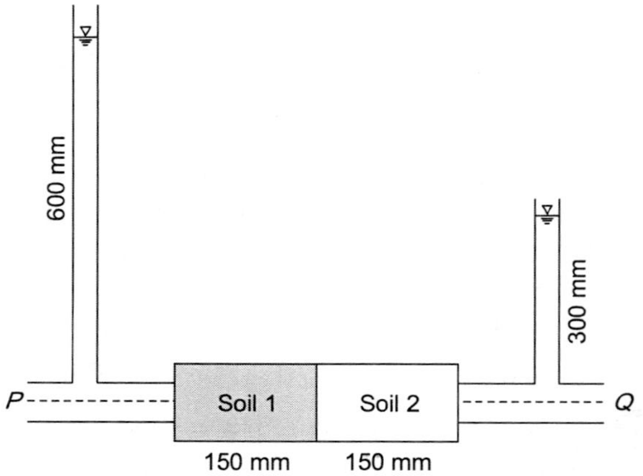

[2016 : 2 M, Set-II]

3.20 Three soil specimens (Soil 1, Soil 2 and Soil 3), each 150 mm long and 100 mm diameter, are placed in series in a constant head flow set-up as shown in the figure. Suitable screens are provided at the boundaries of the specimens to keep them intact. The values of coefficient of permeability of Soil 1, Soil 2 and Soil 3 are 0.01, 0.003 and 0.03 cm/s, respectively.

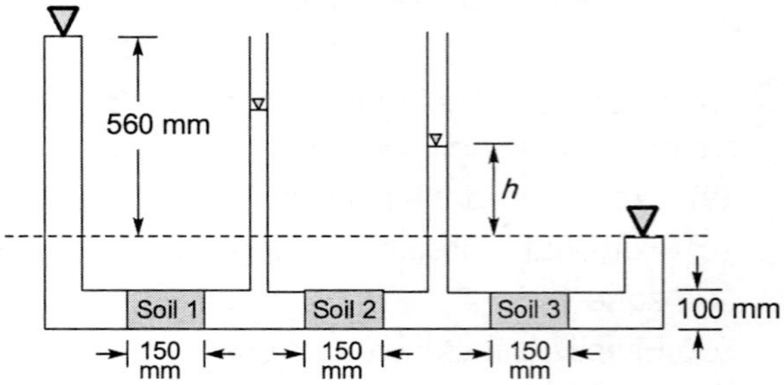

The value of h in the set-up is

(a) 0 mm (b) 40 mm
(c) 255 mm (d) 560 mm

[2018 : 2 M, Set-II]

3.21 An anisotropic soil deposit has coefficient of permeability in vertical and horizontal directions as k_z and k_x, respectively. For constructing a flow net, the horizontal dimension of the problem's geometry is transformed by a multiplying factor of

(a) $\frac{k_x}{k_z}$ (b) $\sqrt{\frac{k_z}{k_x}}$

(c) $\sqrt{\frac{k_x}{k_z}}$ (d) $\frac{k_z}{k_x}$

[2019 : 1 M, Set-II]

3.22 Constant head permeability tests were performed on two soil specimens, S1 and S2. The ratio of height of the two specimens ($L_{s1}:L_{s2}$) is 1.5, the ratio of the diameter of specimens ($D_{s1}:D_{s2}$) is 0.5, and the ratio of the constant head ($h_{s1}:h_{s2}$) applied on the specimens is 2.0. If the discharge from both the specimens is equal, the ratio of the permeability of the soil specimens ($k_{s1}:k_{s2}$) is__________.

[2019 : 2 M, Set-II]

3.23 A confined aquifer of 15 m constant thickness is sandwiched between two aquicludes as shown in the figure (not drawn to scale).

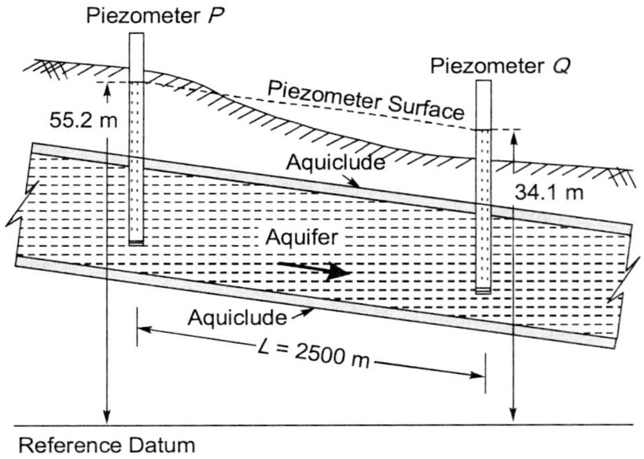

The heads indicated by two piezometers P and Q are 55.2 m and 34.1 m, respectively. The aquifer has a hydraulic conductivity of 80 m/day and its effective porosity is 0.25. If the distance between the piezometers is 2500 m, the time taken by the water to travel through the aquifer from piezometer location P to Q (in days, round off to 1 decimal place) is________.

[2019 : 2 M, Set-II]

3.24 Velocity of flow is proportional to the first power of hydraulic gradient in Darcy's law. The law is applicable to

(a) transitional flow in porous media
(b) turbulent flow in porous media
(c) laminar as well as turbulent flow in porous media
(d) laminar flow in porous media

[2020 : 1 M, Set-I]

3.25 The soil profile at a site up to a depth of 10 m is shown in the figure (not drawn to the scale). The soil is preloaded with a uniform surcharge (q) of 70 kN/m^2 at the ground level. The water table is at a depth of 3 m below ground level. The soil unit weight of the respective layers is shown in the figure. Consider unit weight of water as 9.81 kN/m^3 and assume that the surcharge (q) is applied instantaneously.

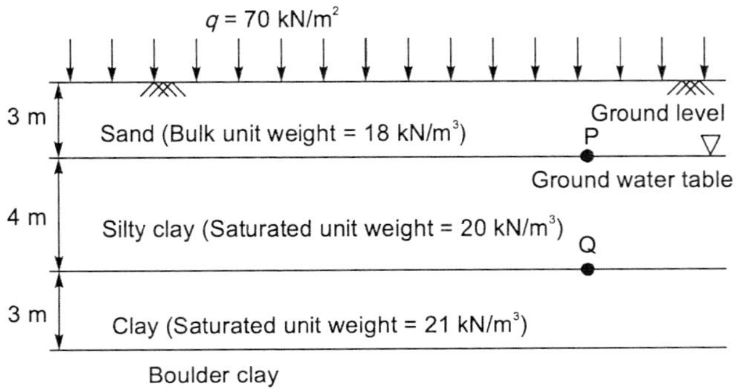

Immediately after preloading, the effective stresses (in kPa) at points P and Q respectively, are

(a) 36 and 126
(b) 36 and 90
(c) 54 and 95
(d) 124 and 204

[2020 : 2 M, Set-I]

3.26 Water flows in the upward direction in a tank through 2.5 m thick sand layer as shown in the figure. The void ratio and specific gravity of sand are 0.58 and 2.7, respectively. The sand is fully saturated. Unit weight of water is 10 kN/m^3.

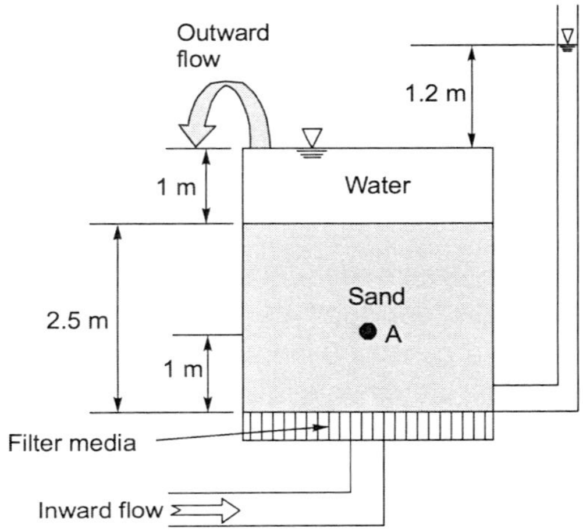

The effective stress (in kPa, round off to two decimal places) at point A, located 1 m above the base of tank, is _________.

[2020 : 2 M, Set-I]

3.27 Permeability tests were carried out on the samples collected from two different layers as shown in the figure (not drawn to the scale). The relevant horizontal (k_h) and vertical (k_v) coefficients of permeability are indicated for each layer.

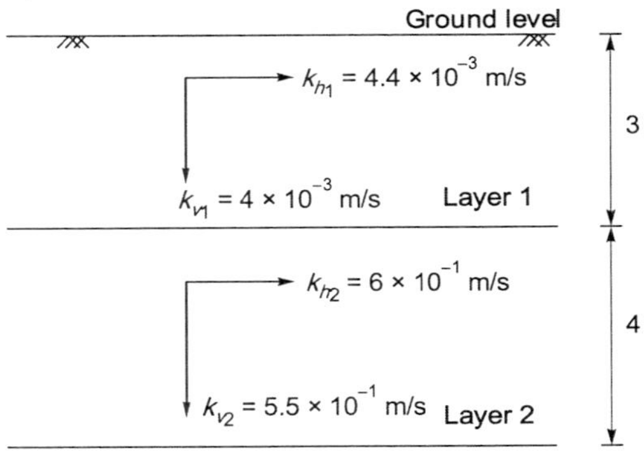

The ratio of the equivalent horizontal to vertical coefficients of permeability, is

(a) 37.29
(b) 80.20
(c) 0.03
(d) 68.25

[2020 : 2 M, Set-II]

3.28 A constant head permeability test was conducted on a soil specimen under a hydraulic gradient of 2.5. The soil specimen has specific gravity of 2.65 and saturated water content of 20%. If the coefficient of permeability of the soil is 0.1 cm/s, the seepage velocity (in cm/s, round off to two decimal places) through the soil specimen is ______. [2020 : 2 M, Set-II]

3.29 Four different soils are classified as CH, ML, SP, and SW, as per the Unified Soil Classification System. Which one of the following options correctly represents their arrangement in the decreasing order of hydraulic conductivity?
(a) SW, SP, ML, CH (b) CH, ML, SP, SW
(c) SP, SW, CH, ML (d) ML, SP, CH, SW
[2022 : 1 M, Set-I]

3.30 Let ψ represent soil suction head and K represent hydraulic conductivity of the soil. If the soil moisture content θ increases, which one of the following statements is TRUE?
(a) ψ decreases and K increases
(b) ψ increases and K decreases.
(c) Both ψ and K decrease.
(d) Both ψ and K increase [2022 : 1 M, Set-I]

3.31 A soil sample is underlying a water column of height h_1, as shown in the figure. The vertical effective stresses at point *A*, *B* and *C* are σ_A', s_B', and s_C' respectively. Let γ_{sat} and γ' be the saturated and submerged unit weights of the soil sample, respectively and γ_w be the unit weight of water. Which one of the following expressions correctly represents the sum $(\sigma_A' + \sigma_B' + \sigma_C')$?

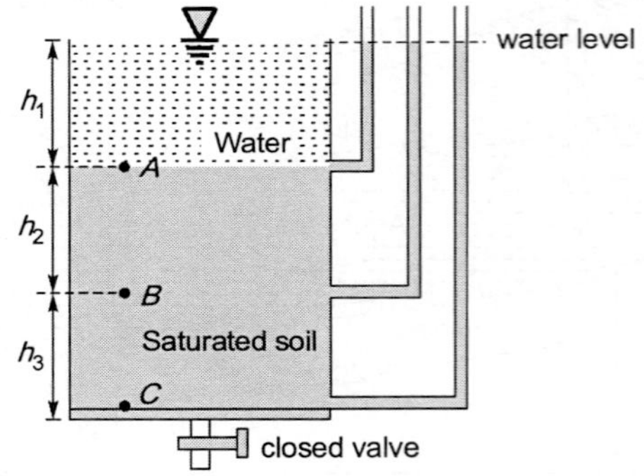

(a) $(h_1 + h_2 + h_3)\gamma_{sat}$
(b) $(2h_2 + h_3)\gamma'$
(c) $(h_2 + h_3)(\gamma_{sat} - \gamma_w)$
(d) $(h_1 + h_2 + h_3)g'$ [2022 : 1 M, Set-II]

3.32 A vertical sheet pile wall is installed in an anisotropic soil having coefficient of horizontal permeability, k_H and coefficient of vertical permeability, k_V. In order to draw the flow net for the isotropic condition, the embedment depth of the wall should be scaled by a factor of ______, without changing the horizontal scale.

(a) $\sqrt{\frac{k_H}{k_V}}$ (b) $\sqrt{\frac{k_V}{k_H}}$

(c) 1.0 (d) $\frac{k_H}{k_V}$

[2023 : 1 M, Set-II]

Answers Effective Stress and Permeability

3.1	(b)	3.2	(d)	3.3	(b)	3.4	(d)	3.5	(c)	3.6	(c)	3.7	(d)
3.8	(b)	3.9	(d)	3.10	(a)	3.11	(c)	3.12	(c)	3.13	(d)	3.14	(c)
3.15	(b)	3.16	(a)	3.17	(d)	3.18	(10.9)	3.19	(0.045)	3.20	(b)	3.21	(b)
3.22	(3)	3.23	(925.7)	3.24	(d)	3.25	(c)	3.26	(8.939)	3.27	(a)	3.28	(0.72)
3.29	(a)	3.30	(a)	3.31	(b)	3.32	(a)						

Explanations Effective Stress and Permeability

3.1 (b)

Darcy established that the flow occurring per unit time is directly proportional to the head causing flow and the area of cross-section of the soil sample but is inversely proportional to the length of the soil sample.

$\therefore \quad q \propto \frac{\Delta h}{L} A \quad \Rightarrow \quad q = kiA$

Dividing both sides by *A*,

$$\frac{q}{A} = v = ki$$

$\therefore \quad v \propto i$ i.e., hydraulic gradient

3.2 (d)

$$k_{xy} = \sqrt{k_H k_V} = \sqrt{9\times10^{-7}\times4\times10^{-7}}$$
$$= 6\times10^{-7} \text{ cm/s}$$

3.3 (b)

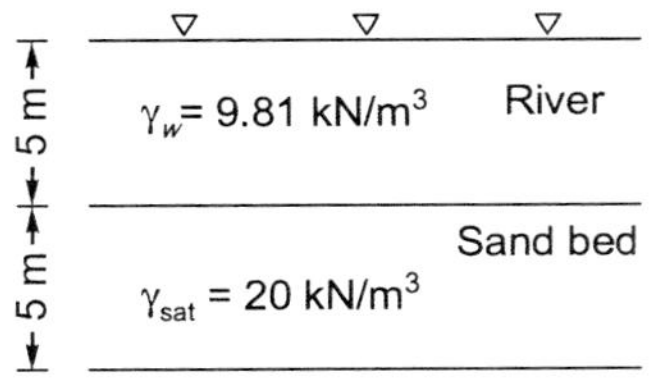

Total vertical stress at 5 m from top of sand bed

$$= 5\times9.81 + 5\times20 = 149.05 \text{ kN/m}^2$$

Pore water pressure at 10 m

$$= 9.81\times10 = 98.1 \text{ kN/m}^2$$

∴ Effective vertical stress

$$= 50.95 \text{ kN/m}^2 \approx 51 \text{ kN/m}^2$$

3.4 (d)

For an anisotropic medium, dimensions in horizontal direction are transformed and equivalent permeability (k_e) for medium is given as

$$k_e = \sqrt{k_x k_y}$$

3.5 (c)

In the capillary zone, pore pressure is negative. The total stress will not change due to lowering of water table because the stratum is still saturated. Thus, the increase in effective vertical pressure will be only due to pore pressure.

Thus increase in effective stress,

$$= h\gamma_w = 2\times10 = 20 \text{ kN/m}^2$$

3.6 (c)

In first case, when the water table is 5 m below the surface and the clay over water table is capillary saturated, the effective stress is given by,

$$\bar{\sigma}_1 = \gamma_{sat}\times5 + \gamma'\times5$$
$$= \gamma_{sat}\times5 + (\gamma_{sat}-\gamma_w)\times5$$
$$\Rightarrow \quad \bar{\sigma}_1 = 10\gamma_{sat} - 5\gamma_w \qquad \ldots(i)$$

In the second case, when water table rises to the surface, the effective stress is given by,

$$\bar{\sigma}_2 = \gamma'\times10 = (\gamma_{sat}-\gamma_w)\times10 = 10\gamma_{sat} - 10\gamma_w$$
$$= (10\gamma_{sat} - 5\gamma_w) - 5\gamma_w \quad [\because \text{ from (i)}]$$
$$\Rightarrow \quad \bar{\sigma}_2 = \bar{\sigma}_1 - 5\gamma_w$$

Thus in second case, the effective stress at the interface will decrease by $5\gamma_w$.

3.7 (d)

$$Q = kiA \Rightarrow k = \frac{Q}{iA} = \frac{600/60}{0.5\times10}$$
$$= 2.0 \text{ cm/s}$$

3.8 (b)

Since the river bed level does not change and the depth of water is changing only above the river bed, the effective stress below the river bed will not change.

Total stress at a depth of 10 m below river bed,

$$\sigma = 20\times10 = 200 \text{ kN/m}^2$$

Pore water pressure,

$$u = 10\times10 = 100 \text{ kN/m}^2$$

∴ Effective stress,

$$\bar{\sigma} = \sigma - u = 200 - 100 = 100 \text{ kN/m}^2$$

Thus effective stress will be 100 kN/m² each in the month of February, July and December.

3.9 (d)

In coarse grained soil the size of voids is more than that of fine grained soil.

∴ Capillary rise is more in fine grained soils and as a result of this, the capillary pressure is also more than the coarse grained soils.

3.10 (a)

- Pyrometer test is used to find out specific gravity of soil solids.
- Constant head permeability test is used to find out permeability of sandy soil.
- Consolidation test is used in case of clayey soils.
- Negative skin friction is found in pile foundation.

3.11 (c)

$$\therefore \quad e = \frac{Gw}{S}$$
$$\Rightarrow \quad e = \frac{2.7\times0.2}{1} \Rightarrow e = 0.54$$
$$\therefore \quad \gamma_{sat} = \left(\frac{G+e}{1+e}\right)\times\gamma_w = \left(\frac{2.7+0.54}{1+0.54}\right)\times10$$
$$= 21.04 \text{ kN/m}^3$$

3.12 (c)

Total stress,

$$\sigma = \gamma_{sat} \times 5 = 21 \times 5 = 105 \text{ kN/m}^2$$

Pore water pressure,

$$u = \gamma_w \times 5 = 10 \times 5 = 50 \text{ kN/m}^2$$

Effective stress,

$$\bar{\sigma} = \sigma - u = 105 - 50 = 55 \text{ kN/m}^2$$

3.13 (d)

$$k = 2.303\frac{aL}{At}\log 10\left[\frac{h1}{h2}\right]$$

$$10^{-2} \text{ m/sec} = 2.303\frac{A}{A}\frac{1}{t}\log\left[\frac{1}{0.25}\right]$$

$$t = 138.65 \text{ sec}$$

$$\simeq 150 \text{ sec}$$

3.14 (c)

Relative density is given by,

$$I_D(\%) = \left(\frac{e_{max} - e_{nat}}{e_{max} - e_{min}}\right) \times 100$$

$$\Rightarrow \quad 40 = \left(\frac{1 - e_{nat}}{1 - 0.5}\right) \times 100$$

$$\Rightarrow \quad 0.4 \times 0.5 = 1 - e_{nat}$$

$$\Rightarrow \quad e_{nat} = 0.8$$

Also, $G_{sea} = \dfrac{\gamma_{sea}}{\gamma_w}$

$$\Rightarrow \quad \gamma_{sea} = 1.03 \times 9.81$$

$$\Rightarrow \quad \gamma_{sea} = 10.1043 \text{ kN/m}^3$$

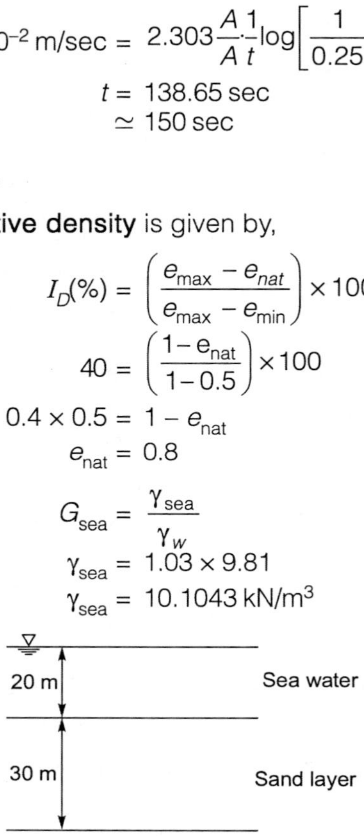

$$\gamma_{sat} = \left(\frac{G\gamma_w + Se_{nat} \times \gamma_{sea}}{1 + e_{nat}}\right)$$

$$= \left(\frac{2.67 \times 9.81 + 1 \times 0.8 \times 10.1043}{1 + 0.8}\right)$$

$$= 19.0423 \text{ kN/m}^3$$

Total stress at the bottom of sand layer,

$$\sigma = 20 \times \gamma_{sea} + 30 \times \gamma_{sat}$$

$$= 20 \times 10.1043 + 30 \times 19.0423$$

$$= 773.355 \text{ kN/m}^2$$

Pore water pressure at the bottom of sand layer,

$$u = (20 + 30) \times 10.1043$$

$$= 505.215 \text{ kN/m}^2$$

Effective stress at the bottom of sand layer,

$$\bar{\sigma} = \sigma - u = 773.355 - 505.215$$

$$= 268.14 \text{ kN/m}^2$$

3.15 (b)

The effective stress will not alter due to change in sea water level. The total stress will increase due to increase in sea water level. The pore water pressure will also increase in the same proportion, thus nullifying the effect of each other.

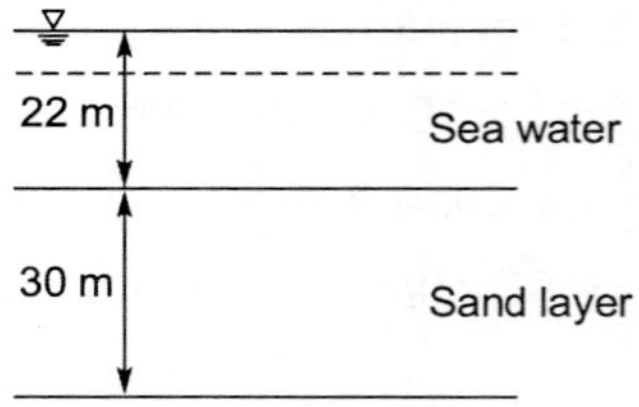

3.16 (a)

Given, For **first specimen**,

$$\frac{h_1}{h_2} = 1.25 \ \& \ k_2 = 5k_1$$

then for **second specimen**, $\dfrac{h'_1}{h'_2} = ?$

We know that; for falling head permeability test,

$$k = \frac{2.303aL}{A \cdot t}\log_{10}\left(\frac{h_1}{h_2}\right)$$

$$\therefore \quad \frac{k_2}{k_1} = \frac{\log_{10}(h'_1/h'_2)}{\log_{10}(h_1/h_2)}$$

$$\Rightarrow \quad 5\log_{10}1.25 = \log_{10}\left(\frac{h'_1}{h'_2}\right)$$

$$\Rightarrow \quad \frac{h'_1}{h'_2} = 3.05$$

3.17 (d)

From the diagram;

10 m of head is lost in 20 m of silt layer before reaching lake,

$$i = \frac{h}{z} = \frac{10}{20} = 0.5$$

$$Q = kiA = 10^{-6} \times 0.5 \times 1 = 0.5 \times 10^{-6} \text{ m}^3\text{/s}$$

3.18 Sol.

①	Fine sand	H_1
②	Silt	$H_2 = 2H_1$
③	Clay	$H_3 = \frac{3}{2}H_2 = 3H_1$

and $k_1 = 10\,k_2 = 100\,k_3$

$$k_2 = \frac{1}{10}k_1$$

$$\Rightarrow \quad k_H = \frac{k_1H_1 + k_2H_2 + k_3H_3}{H_1 + H_2 + H_3}$$

$$= \frac{100k_3H_1 + 10k_3 \times 2H_1 + k_3 \times 3H_1}{H_1 + 2H_1 + 3H_1} = 20.5\,k_3$$

$$k_v = \frac{H_1 + H_2 + H_3}{\frac{H_1}{k_1} + \frac{H_2}{k_2} + \frac{H_3}{k_3}}$$

$$= \frac{H_1 + 2H_1 + 3H_1}{\frac{H_1}{100k_3} + \frac{2H_1}{10k_3} + \frac{3H_1}{k_3}} = 1.869\,k_3$$

$$= \frac{k_H}{k_v} = \frac{2.05k_3}{1.869k_3} = 10.9684$$

3.19 Sol.

$$\text{Discharge} = \frac{200\text{ m}l}{15\text{ min}} = \frac{200\text{ cm}^3}{15 \times 60\text{ sec}}$$

$$= \frac{200 \times 10^3}{900}\ \frac{\text{mm}^3}{\text{sec}}$$

As per Darcy's Law,

$$q = k_{avg}iA$$

$$k_{avg} = \frac{\Sigma Z_i}{\Sigma \frac{Z_i}{k_i}}$$

($\because$ flow is normal to bedding plane)

$$k_{avg} = \frac{150 + 150}{\frac{150}{0.02} + \frac{150}{k}}$$

$$i = \frac{\text{Head difference}}{\text{length}} = \frac{600 - 300}{300} = 1$$

$$A = 80\text{ cm}^2 = 80 \times 10^2\text{ mm}^2$$

$$q = \left[\frac{150 + 150}{\frac{150}{0.02} + \frac{150}{k}}\right] \times 1 \times 80 \times 10^2$$

$$= \frac{200 \times 10^3}{900}$$

$$k = 0.045\text{ mm/sec}$$

3.20 (b)

In normal to bedding plane flow (series arrangement), Discharge will be same and Head loss and Hydraulic gradient will be different.

$$q = K_1i_1A = K_2i_2A$$

$$= K_3i_3A = K_{avg} \cdot \left(\frac{H_L}{L}\right)A$$

$$K_{avg_1} = \frac{\Sigma Z_i}{\Sigma \frac{Z_i}{K_i}} = \frac{150 + 150 + 150}{\frac{150}{0.01} + \frac{150}{0.003} + \frac{150}{0.03}}$$

$$= 0.0064$$

Total head loss $= H_L = 560$ mm

$$\therefore K_3 \cdot \frac{h}{150} \times A = K_{avg} \cdot \frac{560}{(150 + 150 + 150)}A$$

$$0.03 \cdot \left(\frac{h}{150}\right) = 0.0064\left(\frac{560}{450}\right)$$

$$h = 40\text{ mm}$$

3.21 (b)

$$X = X_T\sqrt{\frac{k_x}{k_z}}$$

Transformed horizontal dimension, $X_T = X\sqrt{\frac{k_z}{k_x}}$

3.22 Sol.

$$\frac{L_{s_1}}{L_{s_2}} = 1.5; \quad \frac{D_{s_1}}{D_{s_2}} = 0.5$$

$$\frac{h_{s_1}}{h_{s_2}} = 2; \quad \frac{k_{s_1}}{k_{s_2}} = ?$$

Discharge is same.

$$k_1i_1A_1 = k_2i_2A_2$$

$$k_1\frac{h_{s1}}{L_{s1}} \times \frac{\pi}{4} \times D_{s_1}^2 = k_2\frac{h_{s2}}{L_{s2}} \times \frac{\pi}{4} \times D_{s_2}^2$$

$$\frac{k_{s1}}{k_{s2}} = \frac{L_{s1}}{L_{s2}} \times \frac{h_{s2}}{h_{s1}} \times \frac{D_{s_2}^2}{D_{s_1}^2}$$

$$= 1.5\times\frac{1}{2}\times\left(\frac{1}{0.5}\right)^2 = 3$$

3.23 Sol.

$$v = ki,\ v_s = \frac{ki}{n}$$

$$V_s = 80\times\left(\frac{55.2-34.1}{2500}\times\frac{1}{0.25}\right) = \frac{2500}{\text{time}}$$

Time = 925.651 day

= 925.7 days

3.24 (d)

Darcy's law is valid for laminar flow condition in porous media.

3.25 (c)

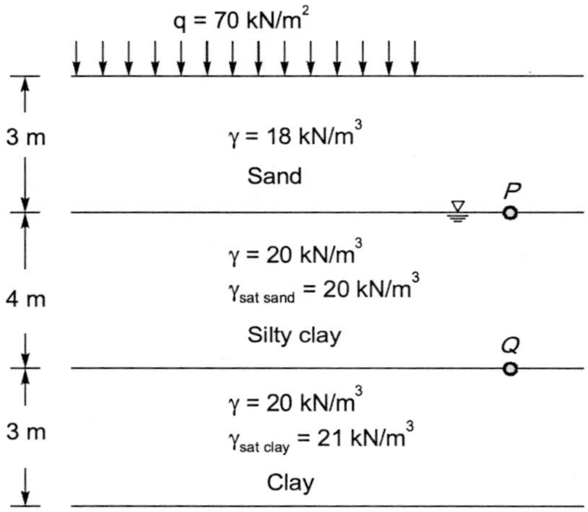

Since, surcharge (q = 70 kN/m²) is applied instantaneously, hence excess pore pressure (u_i = 70 kPa) is developed at point P and Q [GWT level is at level P]

At point P:

Total stress $\sigma = q + 3\gamma = 70 + 3 \times 18$

Pore water pressure

= Hydrostatics pore pressure + Excess pore pressure

$= 0 + u_i$

= 0 + 70 = 70 kN/m²

Effective stress,

$\bar{\sigma} = \sigma - u = 54$ kPa

At point Q:

Total stress, $\sigma = q + 3\gamma + 4\gamma_{sat}$

$= 70 + 3 \times 18 + 4 \times 20$

Pore pressure,

u = Hydrostatics pore pressure + Excess pore pressure

$= 4\gamma_w + u_i = 4 \times 9.81 + 70$

Effective stress,

$\bar{\sigma} = \sigma - u = 94.76$ kPa

3.26 Sol.

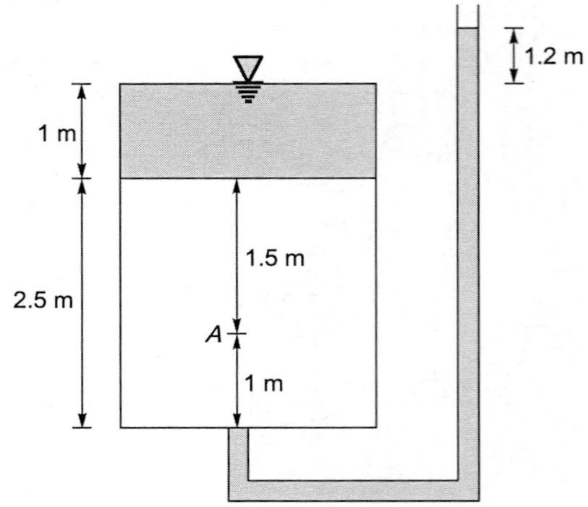

$$\gamma_{sub} = \left(\frac{G-1}{1+e}\right)\gamma_\omega = \left(\frac{2.7-1}{1+0.58}\right)\times 10$$

$= 10.759$

$$\bar{\sigma} = z\gamma_{sub} - iz\gamma_\omega$$

$$= 1.5(\gamma_{sub}) - \left(\frac{1.2}{2.5}\right)\times 1.5\times\gamma_\omega$$

= 8.939 kN/m²

3.27 (a)

$$\frac{k_{eq.H}}{k_{eq.V}} = \frac{\dfrac{\Sigma K_i Z_i}{\Sigma Z_i}}{\dfrac{\Sigma Z_i}{\Sigma \dfrac{Z_i}{K_i}}}$$

$$= \frac{\dfrac{4.4\times10^{-3}\times3+6\times10^{-1}\times4}{7}}{\dfrac{7}{\dfrac{3}{4\times10^{-3}}+\dfrac{4}{5.5\times10^{-1}}}}$$

$$= \frac{0.3447}{9.24\times10^{-3}} = 37.29$$

3.28 Sol.

$$\text{Void ratio, } e = \frac{wG}{s} = \frac{0.2\times2.65}{1} = 0.53$$

$$\text{Porosity, } n = \frac{e}{1+e} = 0.3464$$

Seepage velocity,

$$V_s = \frac{v}{n} = \frac{Ki}{n}$$

$$= \frac{0.1 \times 2.5}{0.3464}$$
$$= 0.72 \text{ cm/sec}$$

3.29 (a)

$K \propto (\text{dia.})^2$

Diameter sizer of soil:

Sand > Silt > Clay

Permeability hydraulic conductivity,

Sand > ML > CH

SW > SP > ML > CH

3.30 (a)

As we know, $h_c \propto \frac{1}{R}$ and $K \propto S$

∴ As the soil moisture content increases, radius of meniscus (R) increases which result decrease in height of capillary rise (h_c) and thus, suction head (ψ) decreases.

∴ As the soil moisture, content increases, degree of saturation (S) increases resulting in increase in hydraulic conductivity (K).

3.31 (b)

$$\sigma_A' = \gamma_w h_1 - \gamma_w h_1 = 0$$
$$\sigma_B' = \gamma_w h_1 + \gamma_{sat} h_2 - \gamma_w (h_1 + h_2) = \gamma' h_2$$
$$\sigma_C' = \gamma_w h_1 + \gamma_{sat} (h_2 + h_3) - \gamma_w (h_1 + h_2 + h_3)$$
$$= \gamma' (h_2 + h_3)$$

$$\therefore \sigma_A' + \sigma_B' + \sigma_C'$$
$$= 0 + \gamma' h_2 + \gamma' (h_2 + h_3) = \gamma' [2h_2 + h_3]$$

3.32 (a)

A steady-state, homogeneous, anisotropic system can be mathematically transformed into an isotropic system by coordinate transformation, creating what is sometimes called a transformed section.

The coordinates in the true anisotropic sysem are x and z. In the transformed isotropic system, the coordinates are X and Z, where

$$X = x$$

and,
$$Z = z\sqrt{\frac{k_x}{k_z}}$$

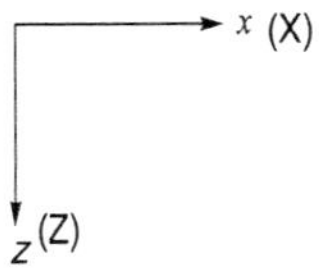

$$Z = z\sqrt{\frac{k_H}{k_V}}$$

So, the embedment depth of wall should be scaled by factor of $\sqrt{\frac{k_H}{k_V}}$.

■■■■

4 Seepage Analysis

4.1 The piezometric head at point *C*, in the experimental set up shown in figure when the flow takes place under a constant head through the soils *A* and *B* is

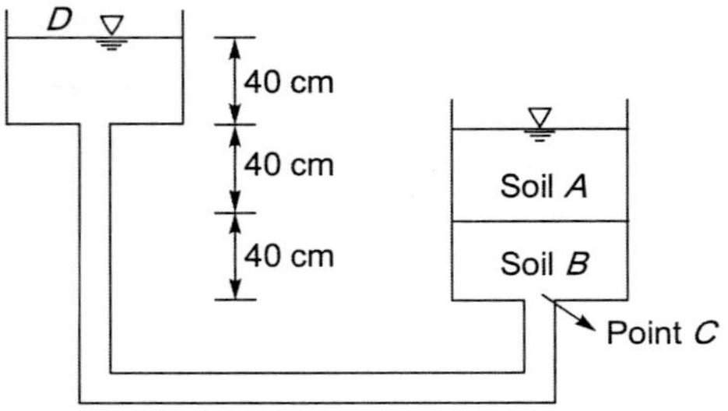

(a) 0 cm (b) 40 cm
(c) 80 cm (d) 120 cm **[1996 : 1 M]**

4.2 Seepage force per unit volume can be expressed as
(a) $i\gamma_w L$ (b) iL
(c) $\gamma_w h$ (d) $i\gamma_w$ **[1996 : 1 M]**

4.3 Piping in soil occurs when
(a) the soil is highly porous
(b) sudden change in permeability occurs
(c) effective pressure becomes zero
(d) the soil is highly stratified **[1996 : 1 M]**

4.4 The soils most susceptible to liquefaction are
(a) saturated dense sands
(b) saturated fine and medium sands of uniform particle size
(c) saturated clays of uniform size
(d) saturated gravels and cobbles
[1996 : 1 M]

4.5 The proposed dam shown in figure is 90 m long and the coefficient of permeability of the soil is 0.0013 mm/second. The quantity of water (m^3) that will be lost per day by seepage is (rounded to the nearest number)

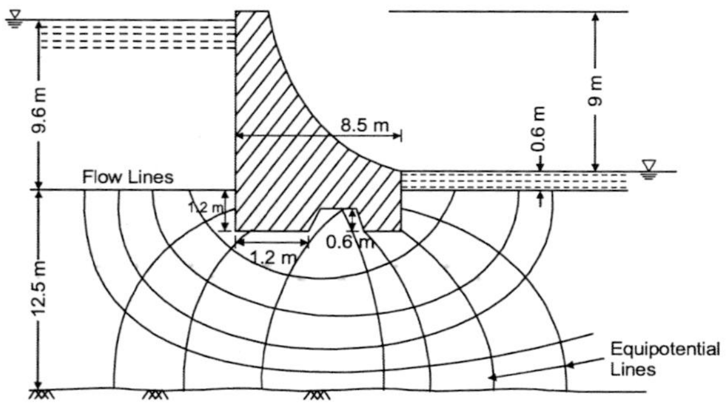

(a) 55 (b) 57
(c) 59 (d) 61 **[1998 : 2 M]**

4.6 The two tubes shown in figure (i) may be considered to be permeameters. Dimensions of the sample in figure (ii) are alike, and the elevations of head water and tail water are the same for both the figure *A*, *B*,... etc. indicate points and *AB*, *AE*,... etc. indicated heads. Head loss through these samples are

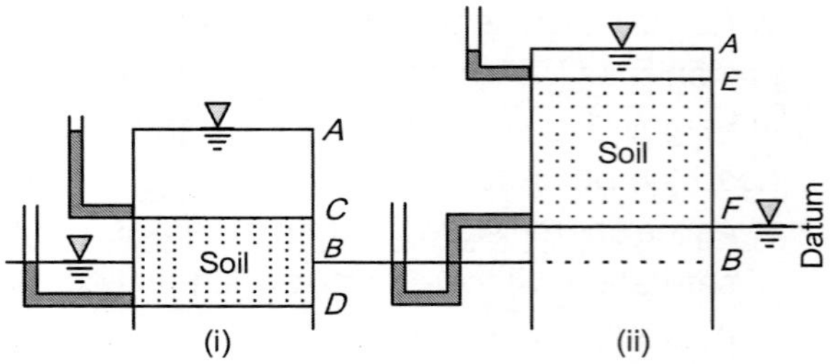

(a) (i) *BD*, (ii) *FB* (b) (i) *AC*, (ii) *AE*
(c) (i) *AD*, (ii) *AF* (d) (i) *AB*, (ii) *AB*
[1999 : 1 M]

4.7 The coefficients of permeability of a soil in horizontal and vertical directions are 3.46 and 1.5 m/day respectively. The base length of a concrete dam resting in this soil is 100 m. When the flow net is developed for this soil with 1 : 25 scale factor in the vertical direction, the reduced base length of the dam will be
(a) 2.63 m (b) 4.00 m
(c) 6.08 m (d) 5.43 m **[2001 : 2 M]**

4.8 The specific gravity and insitu void ratio of a soil deposit are 2.71 and 0.85 respectively. The value of the critical hydraulic gradient is
(a) 0.82 (b) 0.85
(c) 0.92 (d) 0.95
[2002 : 1 M]

4.9 A masonry dam is founded on pervious sand having porosity equal to 45% and specific gravity of sand particles is 2.65. For a desired factor of safety of 3 against sand boiling, the maximum permissible upward gradient will be
(a) 0.225 (b) 0.302
(c) 1.0 (d) None of these
[2003 : 1 M]

4.10 A unit volume of a mass of saturated soil is subjected to horizontal seepage. The saturated unit weight is 22 kN/m^3 and the hydraulic gradient is 0.3. The resultant body force on the soil mass is

(a) 1.98 kN (b) 6.6 kN
(c) 11.49 kN (d) 22.97 kN

[2004 : 1 M]

4.11 The figure below shows two flow lines for seepage across an interface between two soil media of different coefficients of permeability. If entrance angle α_1 = 30°, the exit angle α_2 will be

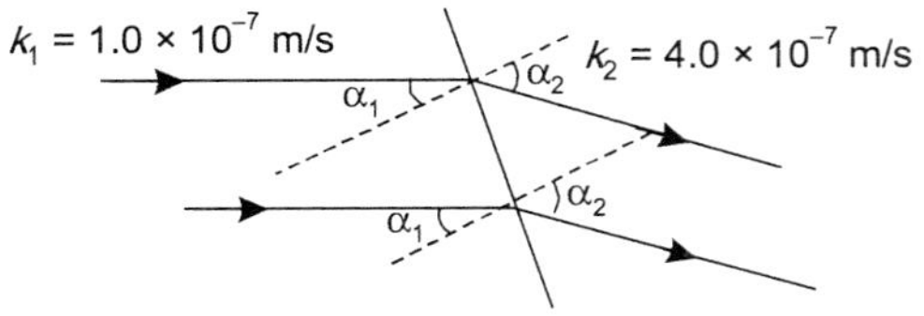

(a) 7.50° (b) 14.03°
(c) 66.59° (d) 75.96° [2004 : 2 M]

4.12 The range of void ratio between which quick sand condition occurs in cohesionless granular soil deposits is

(a) 0.4 – 0.5 (b) 0.6 – 0.7
(c) 0.8 – 0.9 (d) 1.0 – 1.1

[2006 : 1 M]

4.13 To provide safety against piping failure, with a factor of safety of 5, what should be the maximum permissible exit gradient for soil with specific gravity of 2.5 and porosity of 0.35?

(a) 0.155 (b) 0.167
(c) 0.195 (d) 0.213 [2006 : 1 M]

Common Data for Questions 4.14 and 4.15:

Water is flowing through the permeability apparatus as shown in the figure. The coefficient of permeability of the soil is 'k' m/s and the porosity of the soil sample is 0.50.

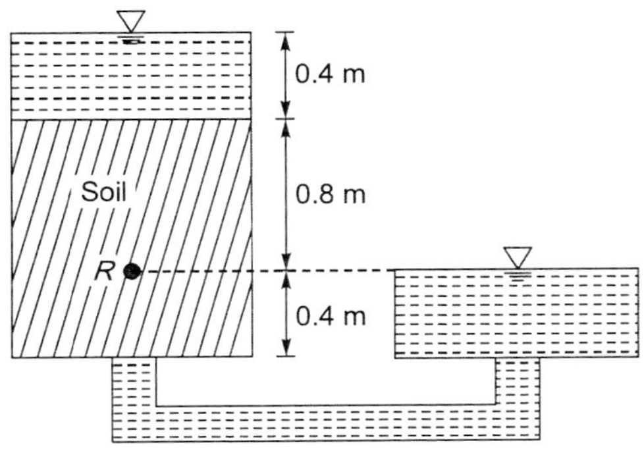

4.14 The total head, elevation head and pressure head in metres of water at the point R shown in the figure are

(a) 0.8, 0.4, 0.4 (b) 1.2, 0.4, 0.8
(c) 0.4, 0, 0.4 (d) 1.6, 0.4, 1.2

[2007 : 2 M]

4.15 What are the discharge velocity and seepage velocity through the soil sample?

(a) k, $2k$ (b) $\frac{2}{3}k, \frac{4}{3}k$
(c) $2k$, k (d) $\frac{4}{3}k, \frac{2}{3}k$ [2007 : 2 M]

4.16 Steady state seepage is taking place through a soil element at Q, 2 m below the ground surface immediately downstream of the toe of an earthen dam as shown in the sketch. The water level in a piezometer installed at P, 500 mm above Q, is at the ground surface. The water level in a piezometer installed at R, 500 mm below Q, is 100 mm above the ground surface. The bulk saturated unit weight of the soil is 18 kN/m^3 and the unit weight of water is 9.81 kN/m^3. The vertical effective stress (in kPa) at Q is

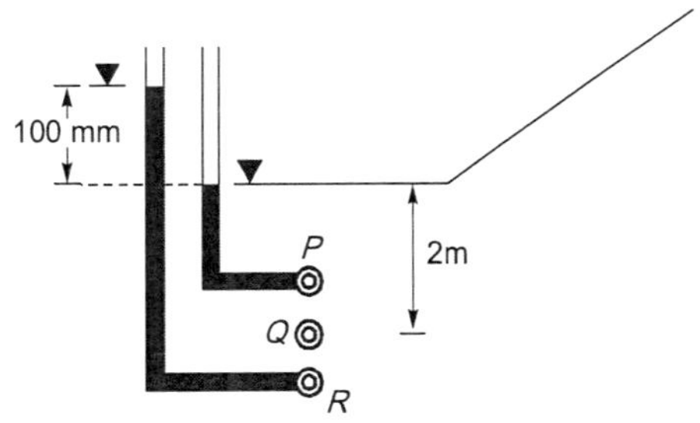

(a) 14.42 (b) 15.89
(c) 16.38 (d) 18.34 [2012 : 2 M]

4.17 Quick sand condition occurs when

(a) the void ratio of the soil becomes 1.0
(b) the upward seepage pressure in soil becomes zero
(c) the upward seepage pressure in soil becomes equal to the saturated unit weight of the soil
(d) the upward seepage pressure in soil becomes equal to the submerged unit weight of the soil

[2010 : 1 M]

4.18 For a saturated sand deposit, the void ratio and the specific gravity of solids are 0.70 and 2.67, respectively. The critical (upward) hydraulic gradient for the deposit would be

(a) 0.54 (b) 0.98
(c) 1.02 (d) 1.87 [2011 : 1 M]

Common Data for Questions 4.19 and 4.20

The flow net around a sheet pile wall is shown in the sketch. The properties of the soil are permeability coefficient = 0.09 m/day (isotropic), specific gravity = 2.70 and void ratio = 0.85. The sheet pile wall and the bottom of the soil are impermeable.

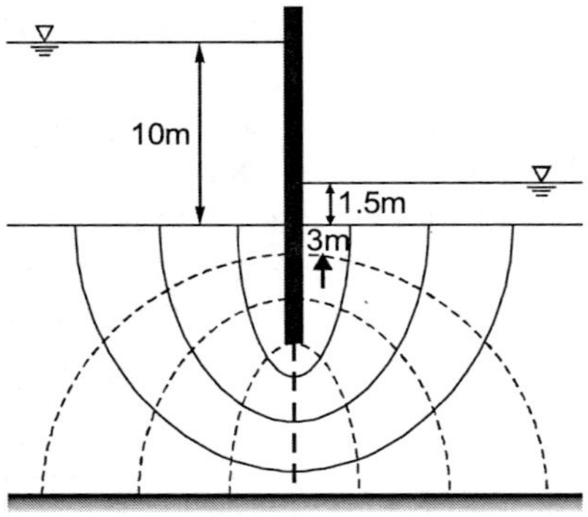

4.19 The seepage loss (in m^3 per day per unit length of the wall) of water is

(a) 0.33 (b) 0.38
(c) 0.43 (d) 0.54 [2012 : 2 M]

4.20 The factor of safety against the occurrence of piping failure is

(a) 3.55 (b) 2.93
(c) 2.60 (d) 0.39 [2012 : 2 M]

4.21 The ratio N_f/N_d is known as shape-factor where N_f is the number of flow channels and N_d is the equipotential drop. The flownet always drawn with a constant b/a ratio where b and a are distance between two consecutive flow lines and equipotential lines respectively. Assuming that b/a ratio remains same, the shape factor of a flow net will change if the

(a) upstream and downstream heads are interchanged.
(b) soil in the flow space is changed.
(c) dimension of the flow space are changed
(d) head difference causing the flow is changed

[2013 : 1 M]

4.22 Laplace equation for water flow in soil is givenby

$$\frac{\partial^2 H}{\partial x^2}+\frac{\partial^2 H}{\partial y^2}+\frac{\partial^2 H}{\partial z^2}=0$$

Head H does not vary in y and z-direction.

Boundary condition $x = 0$, $H = 5$ m, $\frac{dH}{dx} = -1$.

What is the value of H at $x = 1.2$?

[2013 : 2 M]

4.23 The flow net constructed for the dam is shown in the figure below. Taking the coefficient of permeability as 3.8×10^{-6} m/s, the quantity of flow (in cm^3/s) under the dam per meter of dam is ______.

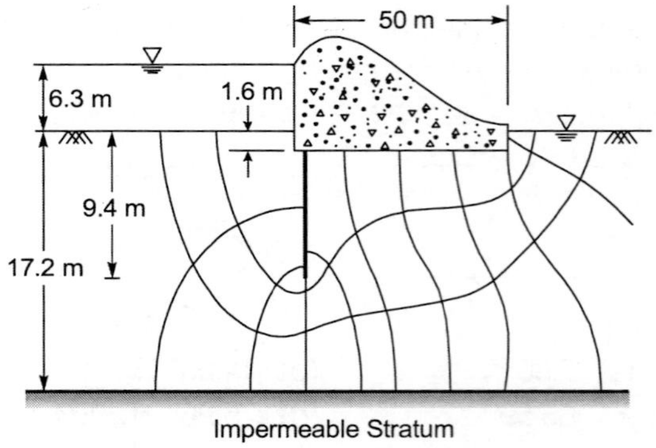

[2014 : 2 M, Set-I]

4.24 Water is flowing at a steady rate through a homogeneous and saturated horizontal soil strip of 10 m length. The strip is being subjected to a constant water head (H) of 5 m at the beginning and 1 m at the end. If the governing equation of flow in the soil strip is $\frac{d^2H}{dx^2} = 0$ (where x is the distance along the soil strip), the value of H (in m) at the middle of the strip is ________.

[2014 : 2 M, Set-II]

4.25 Which of the following statements is TRUE for the relation between discharge velocity and seepage velocity?

(a) Seepage velocity is always smaller than discharge velocity
(b) Seepage velocity can never be smaller than discharge velocity
(c) Seepage velocity is equal to the discharge velocity
(d) No relation between seepage velocity and discharge velocity can be established

[2015 : 1 M, Set-I]

4.26 A seepage flow condition is shown in the figure. The saturated unit weight of the soil $\gamma_{sat} = 18$ kN/m^3. Using unit weight of water, $\gamma_w = 9.81$ kN/m^3, the effective vertical stress (expressed in kN/m^2) on plane X-X is ______

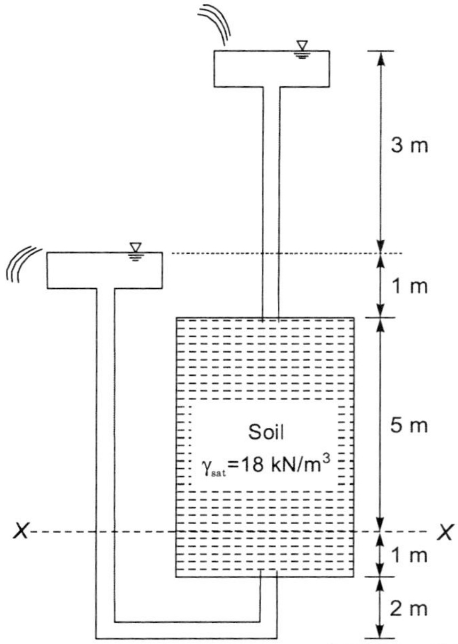

[2016 : 2 M, Set-I]

4.27 Seepage is occurring through a porous media shown in the figure. The hydraulic conductivity values (k_1, k_2, k_3) are in m/day.

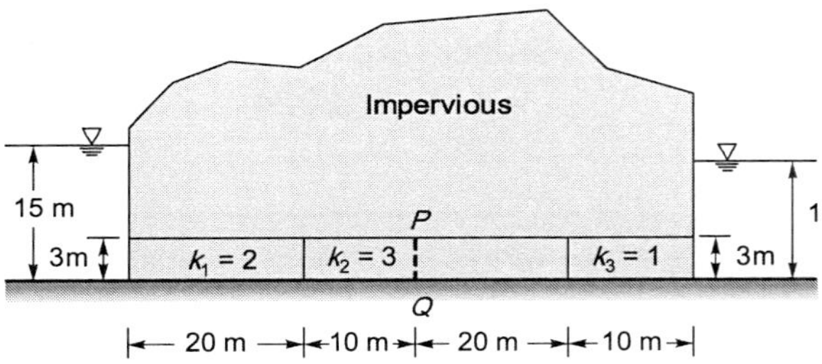

The seepage discharge (m^3/day per m) through the porous media at section PQ is

(a) 7/12 (b) 1/2
(c) 9/16 (d) 3/4

[2016 : 2 M, Set-I]

4.28 The relationship between the specific gravity of sand (G) and the hydraulic gradient (i) to initiate quick condition in the sand layer having porosity of 30% is

(a) $G = 0.7i + 1$ (b) $G = 1.43i - 1$
(c) $G = 1.43i + 1$ (d) $G = 0.7i - 1$

[2016 : 1 M, Set-II]

4.29 The seepage occurring through an earthen dam is represented by a flownet comprising of 10 equipotential drops and 20 flow channels. The coefficient of permeability of the soil is 3 mm/min and the head loss is 5 m. The rate of seepage (expressed in cm^3/s per *m* length of the dam) through the earthen dam is______

[2016 : 2 M, Set-II]

4.30 A sheet pile has an embedment depth of 12 m in a homogeneous soil stratum. The coefficient of permeability of soil is 10^{-6} m/s. Difference in the water levels between the two sides of the sheet pile is 4 m. The flow net is constructed with five number of flow lines and eleven number of equipotential lines. The quantity of seepage (in cm^3/s per m), up to one decimal place) under the sheet pile is ________.

[2017 : 1 M, Set-II]

4.31 A flownet below a dam consists of 24 equipotential drops and 7 flow channels. The difference between the upstream and downstream water levels is 6 m. The Length of the flow line adjacent to the toe of the dam at exit is 1 m. The specific gravity and void ratio of the soil below the dam are 2.70 and 0.70. respectively. The factor of safety against piping is

(a) 1.67 (b) 2.5
(c) 3.4 (d) 4

[2018 : 1 M, Set-II]

4.32 At a construction site, a contractor plans to make an excavation as shown in the figure.

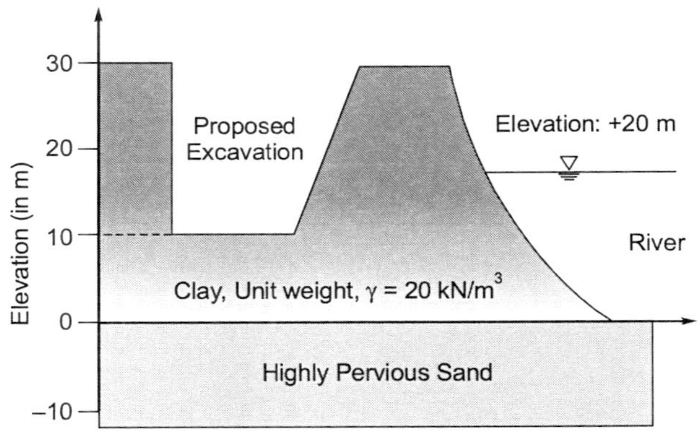

The water level in the adjacent river is at an elevation of +20.0 m. Unit weight of water is 10 kN/m^3. The factor of safety (up to two decimal places) against sand boiling for the proposed excavation is _______ [2018 : 2 M, Set-I]

4.33 Consider a two-dimensional flow through isotropic soil along x-direction and z-direction. If h is the hydraulic head, the Laplace's equation of continuity is expressed as

(a) $\frac{\partial h}{\partial x} + \frac{\partial h}{\partial x}\frac{\partial h}{\partial z} + \frac{\partial h}{\partial z} = 0$

(b) $\frac{\partial h}{\partial x} + \frac{\partial h}{\partial z} = 0$

(c) $\dfrac{\partial^2 h}{\partial x^2}+\dfrac{\partial^2 h}{\partial x \partial z}+\dfrac{\partial^2 h}{\partial z^2}=0$

(d) $\dfrac{\partial^2 h}{\partial x^2}+\dfrac{\partial^2 h}{\partial z^2}=0$ [2019 : 1 M, Set-I]

4.34 For the flow setup shown in the figure (not to scale), the hydraulic conductivities of the two soil samples, Soil 1 and Soil 2, are 10 mm/s and 1 mm/s, respectively. Assume the unit weight of water as 10 kN/m^3 and ignore the velocity head. At steady state, what is the total head (in m, rounded off to two decimal places) at any point located at the junction of the two samples?

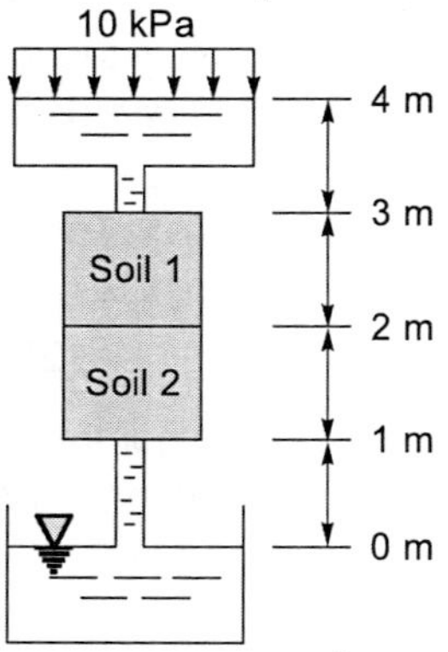

[2023 : 2 M, Set-II]

4.35 Consider the statements P, Q, and R.

P : Compacted fine-grained soils with flocculated structure have isotropic permeability.

Q : Phreatic surface/line is the line along which the pore water pressure is always maximum.

R : The piping phenomenon occurring below the dam foundation is typically known as blowout piping.

Which of the following option(s) is/are CORRECT?

(a) P is FALSE and Q is TRUE

(b) Both Q and R are FALSE

(c) Both P and R are TRUE

(d) P is TRUE and R is FALSE

[2024 : 2 M, Set-II]

4.36 A homogeneous earth dam has a maximum water head difference of 15 m between the upstream and downstream sides. A flownet was drawn with the number of potential drops as 10 and the average length of the element as 3 m. Specific gravity of the soil is 2.65. For a factor of safety of 2.0 against piping failure, void ratio of the soil is (rounded off to 2 decimal places).

[2024 : 2 M, Set-II]

■■■■

Answers Seepage Analysis

4.1	(b)	4.2	(d)	4.3	(c)	4.4	(b)	4.5	(b)	4.6	(d)	4.7	(a)
4.8	(c)	4.9	(b)	4.10	(c)	4.11	(c)	4.12	(b)	4.13	(c)	4.14	(c)
4.15	(a)	4.16	(b)	4.17	(d)	4.18	(b)	4.19	(b)	4.20	(c)	4.21	(c)
4.22	(3.8)	4.23	(7.182)	4.24	(3)	4.25	(b)	4.26	(65.4)	4.27	(b)	4.28	(c)
4.29	(500)	4.30	(1.6)	4.31	(d)	4.32	(1)	4.33	(d)	4.34	(4.54)	4.35	(b, d)
4.36	(0.66)												

Explanations Seepage Analysis

4.1 (b)

Assume datum at tail water level

$$DH = -80$$
$$PH = 120$$
$$TH = 40$$

4.2 (d)

Seepage pressure is exerted by the water on the soil due to friction drag. This drag force/seepage force always acts in the direction of flow.

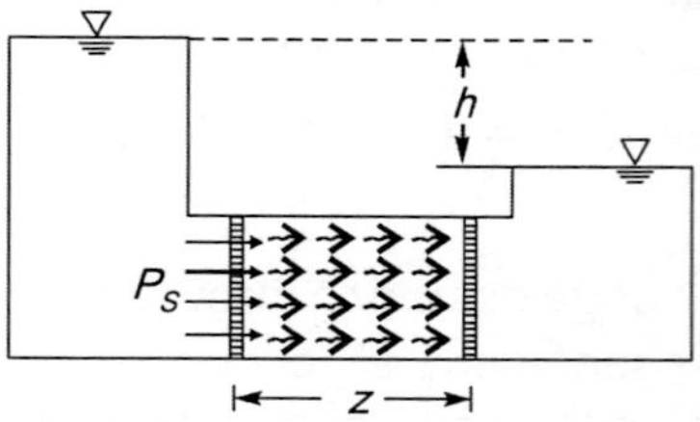

The seepage pressure is given by,

$$P_s = h\gamma_w$$

where, P_s = Seepage pressure

γ_w = 9.81 kN/m^3.

Here, h = head loss

and z = length

$F_s = hA\gamma_w$

where, F_s = Seepage force.

$f_s = i\gamma_w$

where, f_s = Seepage force per unit vol.

$i = h/z$

where, i = Hydraulic gradient.

4.3 (c)

When upward exit gradient approaches critical hydraulic gradient, boiling condition occur, leading to erosion of soil such erosion gradually cause a pipe shaped discharrge channel. This phenomenon is known as piping.

4.4 (b)

When out soil is saturated and a larger rate of loading is applied then positive pore water pressure will develop. Due to which effective stress reduces which results in decrease in the strength of soil. When effective stress reduces to zero then soil will lose all its shear strength and this phenomena is known as liquefaction of soil.

This phenomena majority occurs in the saturated fine and medium sands of uniform particle size.

4.5 (b)

Given data:

Coefficient of permeability,

$k = 0.0013$ mm/sec

$= 0.0013 \times 10^{-3}$ m/sec

Length of dam,

$L = 90$ m

$$\because \quad q = kH\frac{N_f}{N_d}$$

Here, N_f = Number of flow channels

$= 4 + 1 = 5$

N_d = Number of equipotential drops

$= 7 + 1 = 8$

$H = 9.6 - 0.6 = 9$ m

$$\therefore q = 0.0013 \times 10^{-3} \times 9 \times \frac{5}{8} \times 9 \times 3600 \times 24$$

$= 57$ m^3/day

4.6 (d)

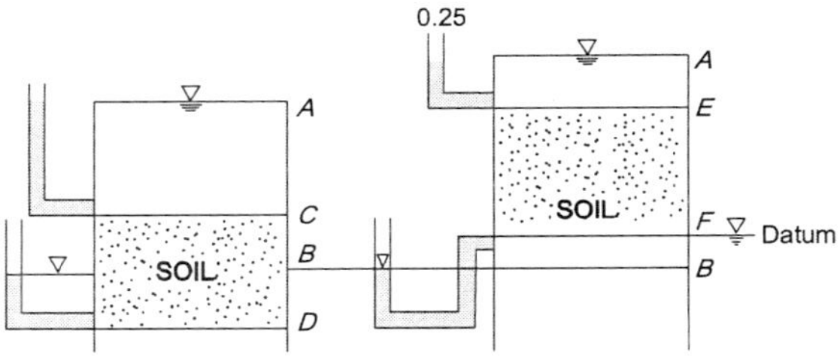

Head loss is change of height in permeameters,

$\therefore$ In sample (1),

$h = AB$

In sample (2),

$h = AB$

4.7 (a)

$K_x = 3.46$ m/day

$K_y = 1.5$ m/day

$l_b = 100$ m (base length)

scale factor in vertical direction = 1 : 25

scale factor in horizontal direction

$$= \frac{1}{25}\sqrt{\frac{K_y}{K_x}} = \frac{1}{37.969}$$

$\therefore$ Reduced base length = scale $\times l_b$

$$= \frac{100}{37.969} = 2.63 \text{ m}$$

4.8 (c)

$$i_c = \frac{G-1}{1+e} = \frac{2.71-1}{0.85+1} = \frac{1.71}{0.85} = 0.924$$

4.9 (b)

Critical hydraulic gradient,

$$i_c = \frac{G-1}{1+e} = (G-1)(1-n)$$

$$\therefore \quad i_c = (2.65-1)(1-0.45) = 0.9075$$

Factor of safety

$$= \frac{\text{Critical hydraulic gradient}}{\text{Maximum permissible gradient}}$$

$\Rightarrow$ Maximum permissible gradient

$$= \frac{0.9075}{3} = 0.3025$$

4.10 (c)

Volume of soil = 1 m^3

$\gamma_{sat} = 22$ kN/m^3

Hydraulic gradient = i = 0.3

$$F_{horizontal} = i\gamma_w \cdot \text{volume}$$
$$= 0.3 \times 9.81 \times 1$$
$$= 2.943 \text{ kN}$$
$$F_{vertical} = \gamma' \cdot \text{volume}$$
$$= (22 - 9.81) \times 1 = 12.19 \text{ kN}$$
$$F_{resultant} = \sqrt{F_h^2 + F_v^2} = 12.54 \text{ kN}$$

Note: Nearby answer is (c)

4.11 (c)

$$\frac{k_1}{k_2} = \frac{\tan\alpha_1}{\tan\alpha_2}$$
$$\Rightarrow \frac{1.0\times10^{-7}}{4.0\times10^{-7}} = \frac{\tan 30°}{\tan\alpha_2}$$
$$\Rightarrow \tan\alpha_2 = 2.309$$
$$\Rightarrow \alpha_2 = 66.59°$$

4.12 (b)

The specific gravity of cohesionless granular soils (sands) does not vary much and for all practical purposes it is taken to be 2.65. Critical hydraulic gradient should be nearly 1 for quick sand condition.

i.e. $i_{cr} = \dfrac{G-1}{1+e}$

From the above equation, the void ratio range is found to be between 0.6 and 0.7.

4.13 (c)

$$i_{cr} = \frac{G-1}{1+e} = (G-1)(1-n)$$
$$= (2.5-1)(1-0.35) = 0.975$$

∴ Maximum permissible exit gradient,

$$= \frac{i_{cr}}{\text{Factor of safety}} = \frac{0.975}{5} = 0.195$$

4.14 (c)

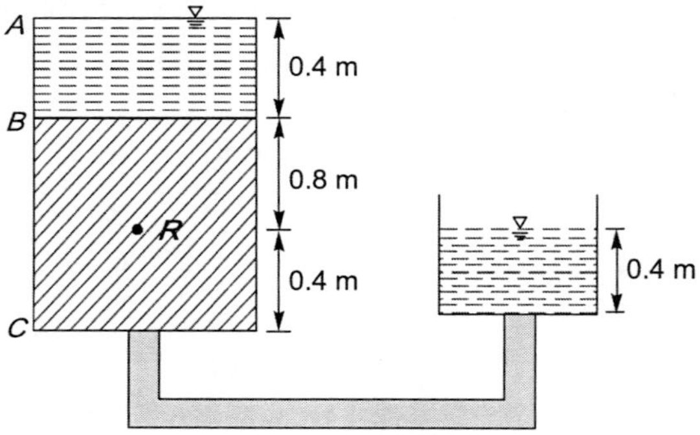

Points	DH	PH	TH
A	1.6 m	0	1.6 m
B	1.2 m	0.4 m	1.6 m
C	0	0.4 m	0.4 m

where,

DH = Datum head

PH = Pressure head

TH = Total head

Loss of head in soil,
in soil length between B and C,

$$(h) = 1.6 - 0.4 = 1.2 \text{ m}$$

$$\text{Hydraulic gradient (i)} = \frac{h}{z} = \frac{1.2}{1.2} = 1$$
$$(\text{Head loss})_{R\rightarrow exit} = i \times \text{length} = 1 \times 0.4 = 0.4 \text{ m}$$
$$(TH)_R = (TH)_{exit} + 0.4$$
$$= 0 + 0.4 = 0.4$$
$$(DH)_R = 0.4$$

4.15 (a)

$$\text{Seepage velocity, } (v_s) = \frac{v}{n} = \frac{v}{0.5} = 2v$$

[porosity n = 0.5]

as per **Darcy's law**, $v = ki$

as $i = 1$

$v = k$ m/s

Discharge velocity $= k$

Seepage velocity $= 2k$

4.16 (b)

$$\text{Hydraulic gradient } i = \frac{100 \text{ mm}}{1000 \text{ mm}} = 0.1$$

Head loss between Q & $P = iz = 0.1 \times 0.5 = 0.05$

Pressure head a $Q = 2 + 0.05 = 2.05$ m

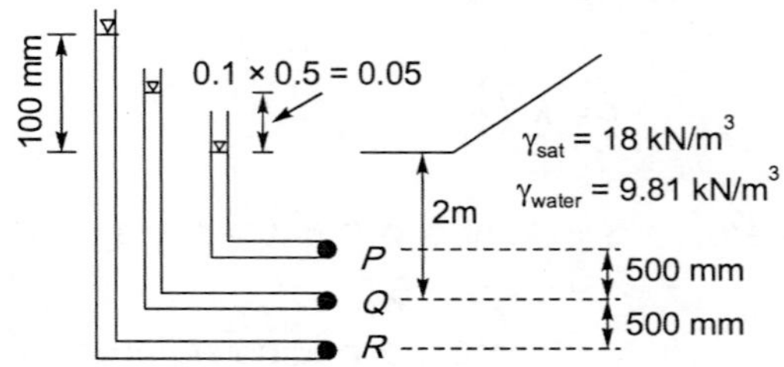

$$\sigma = (1.5 + 0.5)\gamma_{sat}$$
$$u = 2.05\gamma_w$$
$$\sigma' = \sigma - u$$
$$= 2\gamma_{sat} - 2.05\gamma_w = 15.89 \text{ kPa}$$

4.17 (d)

In upward flow condition,

Effective stress, $\bar{\sigma} = \gamma_{sub} z - iz\gamma_w$

Quick sand occurs when $\bar{\sigma}$ is zero.

4.18 (b)

Critical hydraulic gradient is given by,

$$i_{cr} = \frac{G-1}{1+e} = \frac{2.67-1}{1+0.70} = 0.98$$

4.19 (b)

$$k = 0.09 \text{ m/day}$$
$$G = 2.70$$
$$e = 0.85$$
$$q = k H_L \frac{N_f}{N_d}$$
$$H_L = H_1 - H_2 = 10 - 1.5 = 8.5 \text{ m}$$
N_f = No. of flow lines $= 3 + 1 = 4$

N_d = No of equipotential drops $= 7 + 1 = 8$

$$\therefore \quad q = 0.09 \times 8.5 \times 4/8 = 0.38 \text{ m}^2\text{/day}$$

4.20 (c)

$$H_L = H_1 - H_2 = 10 - 1.5 = 8.5 \text{ m}$$

Loss of head per drop

$$= \frac{8.5}{8} = 1.0625$$

∴ Hydraulic gradient,

$$i = \frac{1.0625}{3} = 0.354$$

$$\therefore \quad \text{FOS} = \frac{i_c}{i} = \frac{0.9189}{0.354} = 2.60$$

4.21 (c)

Flow net depends on given boundary conditions and if the boundary conditions does not change then $\frac{N_f}{N_d}$ will not change.

4.22 Sol.

Since H does not vary in y and z direction.

$$\therefore \quad \frac{\partial^2 H}{\partial x^2} = 0$$

Integrating both sides, we get,

$$\frac{\partial H}{\partial x} = C_1 \quad \text{...(i)}$$

Integrating again,

$$H = C_1 x + C_2$$

At $x = 0,\ H = 5$

$\therefore \quad 5 = C_2$

At $x = 0,\ \frac{dH}{dx} = -1$

From eq. (i),

$$C_1 = -1$$
$$\therefore \quad H = -x + 5$$

At $x = 1.2$ m,

$H = 5 - 1.2 = 3.8$ m

4.23 Sol.

Quantity of flow,

$$Q = kH\frac{N_f}{N_d}$$

Here, N_f = No. of flow channels = 3

N_d = No. of equipotential drops = 10

Given, $k = 3.8 \times 10^{-6}$ m/s

and $H = 6.3$ m

$$\therefore \quad Q = 3.8 \times 10^{-6} \times 6.3 \times \frac{3}{10} \times 1$$
$$= 7.182 \times 10^{-6} \text{ m}^3\text{/s}$$
$$= 7.182 \times 10^{-6} \times 10^6 \text{cm}^3\text{/s}$$
$$Q = 7.182 \text{ cm}^3\text{/s}$$

4.24 Sol.

Given, equation of the flow of soil strip is,

$$\frac{d^2H}{dx^2} = 0$$

$$\Rightarrow \quad \frac{dH}{dx} = C_1$$

$$\Rightarrow \quad H = C_1 x + C_2$$

at $x = 0,\ H = 5$ m

$$\Rightarrow \quad C_2 = 5$$
$$\Rightarrow \quad H = C_1 x + 5$$

at $x = 10,\ H = 1$ m

$$\Rightarrow \quad 1 = C_1 \times 10 + 5$$
$$\Rightarrow \quad 10\,C_1 = -4$$

$$\Rightarrow \quad C_1 = -\frac{2}{5}$$

$$\Rightarrow \quad H = -\frac{2}{5}x + 5$$

at $x = 5$ m

$$H = -\frac{2}{5}\times 5 + 5$$

$$H = 3 \text{ m}$$

4.25 (b)

Seepage velocity is given by,

$$V_s = \frac{V}{n}$$

where, V = discharge velocity,

$$n = \frac{V_V}{V} < 1$$

Hence, $V_s > V$

4.26 Sol.

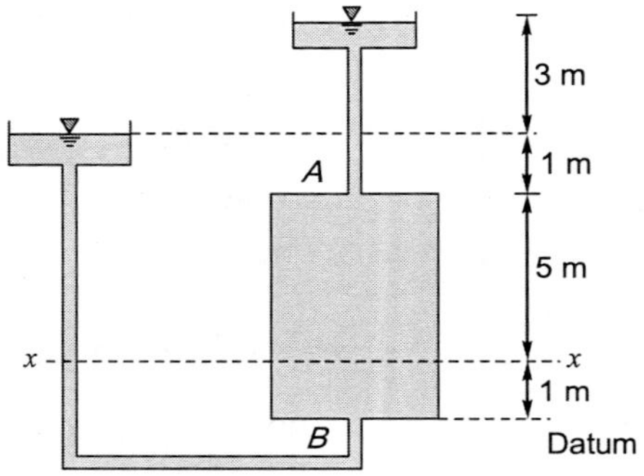

	Datum head	Pressure head	Total head
A	6	4	10
B	0	7	7
x-x	1	6.5	7.5

Hydraulic gradient,

$$i = \frac{\text{Hydraulic head difference}}{\text{Length}} = \frac{H_A - H_B}{6}$$

$$i = \frac{10-7}{6} = \frac{3}{6} = \frac{1}{2}$$

$$\text{Force} = h\gamma_w$$

Head loss at x-x

$$= i \times 5 = 2.5\,\gamma_w$$

Effective stress at x-x

$$= 5\,\gamma_{sub} + 2.5\,\gamma_w = 5 \times (18 - 9.81) + 2.5 \times 9.81$$

$$= 65.475 \text{ kN/m}^2$$

4.27 (b)

Flow is normal to bedding plane

$$\therefore k_{avg} = \frac{\Sigma z_i}{\Sigma \frac{z_i}{k_i}} = \frac{20+30+10}{\frac{20}{2}+\frac{30}{3}+\frac{10}{1}} = 2 \text{ m/day}$$

Hydraulic gradient,

$$i = \frac{\text{Head difference}}{\text{Length}} = \frac{15-10}{60} = \frac{5}{60}$$

Seepage discharge,

$$= k_{avg}\, i A = 2 \text{ m/day} \times \frac{5}{60} \times 3 \text{ m}$$

$$= \frac{1}{2} \text{ m}^3\text{day/m}$$

4.28 (c)

$$i_c = \frac{G-1}{1+e} = (G-1)(1-n) \quad \because \frac{1}{1+e} = 1-n$$

$$i_c = (G-1)(1-0.3) = (G-1)\times 0.7$$

$$G = \frac{i_c}{0.7} + 1 = 1.43 i_c + 1$$

4.29 Sol.

$$q = \frac{KH \cdot N_f}{N_d}$$

$$K = \frac{3 \text{ mm}}{\text{min}} = \frac{3\times 10^{-3}\text{m}}{60}\text{m/s}$$

$H = 5$ m, $N_f = 20$, $N_d = 10$

$$q = \frac{3\times 10^{-3}}{60}\times 5 \times \frac{20}{10}\text{m}^3\text{/sec per m length of dam}$$

$$q = \frac{3\times 10^{-3}\times 5 \times 20}{60 \times 10}\times 10^6 \text{ cm}^3\text{/sec per m length of dam}$$

$q = 500$ cm^3/sec per m length of dam

4.30 Sol.

N_f = No. of flow lines – 1 = 5 – 1 = 4

N_d = No. of equipotential lines – 1 = 11 – 1 = 10

$H = 4$ m

Discharge,

$$q = \frac{KHN_f}{N_d} = 10^{-6}\times 4 \times \frac{4}{10} = 16 \times 10^{-7} \text{ m}^3\text{/sec/m}$$

$q = 16 \times 10^{-7} \times 10^6$ cm^3/sec/m

$q = 1.6$ cm^3/s per meter length of dam

4.31 (d)

$N_f = 7 \quad N_d = 24 \quad H = 6$ m

Critical Hydraulic Gradient,

$$i_c = \frac{G-1}{1+e} = \frac{2.7-1}{1+0.7} = 1$$

Exit Gradient (i_{exit})

$$= \frac{\Delta h}{l} = \frac{\left(\frac{H}{N_d}\right)}{l} = \frac{\left(\frac{6}{24}\right)}{1\text{m}} = \frac{1}{4}$$

$$\text{F.O.S.} = \frac{i_c}{i_{exit}} = \frac{1}{\left(\frac{1}{4}\right)} = 4$$

4.32 Sol.

$$\text{FOS} = \frac{\text{Submerged wt.}}{\text{Seepage pressure}}$$

$$= \frac{\text{Submerged wt.}}{\text{Seepage head} \times \gamma_w}$$

$$\text{FOS} = \frac{10\gamma'}{(20-10)\gamma_w} = 1$$

4.33 (d)

$$\frac{\partial^2 h}{\partial x^2} + \frac{\partial^2 h}{\partial z^2} = 0$$

4.34 Sol.

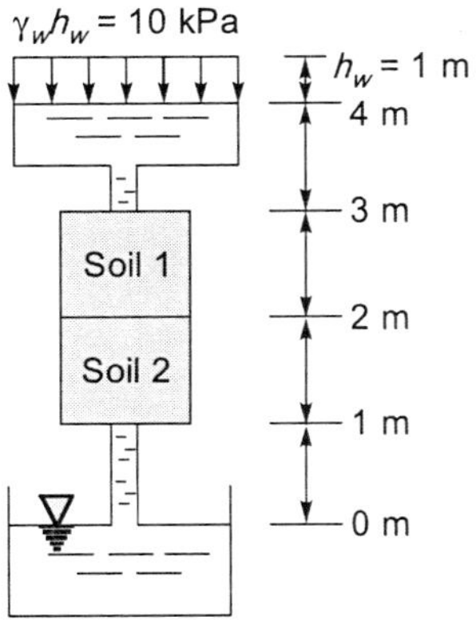

$$\gamma_w h_w = 10 \text{ kPa}$$

$$\Rightarrow \quad h_w = 1 \text{ m}$$

$$(\text{TH})_{entry} = \text{D.H} + \text{P.H} = 3 + 2$$

$= 5$ m

$$(\text{T·H})_{middle} = (\text{TH})_{entry} - h_{L1}$$

Now, soil is in series,

$$\text{So, } \frac{\Sigma z_i}{\Sigma \frac{z_i}{k_i}} \times \left(\frac{H_L}{L}\right)_{Total} \times A = k_1 \times \frac{h_{L_1}}{L_1} \times A$$

$$\Rightarrow \quad \frac{1+1}{\frac{1}{10}+\frac{1}{1}} \times \left(\frac{5}{2}\right) \times A = 10 \times \frac{h_{L_1}}{1} \times A$$

$$\Rightarrow \quad h_{L_1} = 0.4545$$

$$\therefore \quad (\text{T.H})_{middle} = 5 - 0.4545$$

$$= 4.54 \text{ m}$$

4.35 (b, d)

4.36 Sol.

Total head loss, $h_L = 15$ m

Number of equipotential drops, $N_d = 10$

Average length, L = 3 m

Specific gravity, $G = 2.65$

Factor of safety, FOS = 2

Hydraulic gradient, $i = \frac{h_L}{L} = \frac{15}{3 \times 10} = 0.5$

Now, $\text{FOS} = \frac{i_C}{i}$, where i_C is critical hydraulic gradient

$$\Rightarrow \quad 2 = \frac{i_C}{0.5}$$

$$\Rightarrow \quad i_C = 1$$

$$\text{But} \quad i_C = \frac{G-1}{1+e}$$

$$\Rightarrow \quad 1 = \frac{2.65-1}{1+e}$$

$$\Rightarrow \quad e = 0.65$$

■■■■

5 Stress Distribution in the Soil

5.1 The vertical stress at depth, z directly below the point load P is (k is a constant)

(a) $k\frac{P}{z}$ (b) $k\frac{P}{z^3}$

(c) $k\frac{P}{z^2}$ (d) $k\frac{P}{\sqrt{z}}$ [1997 : 1 M]

5.2 A point load of 700 kN is applied on the surface of thick layer of saturated clay. Using Boussinesq's elastic analysis, the estimated vertical stress (σ_v) at a depth of 2 m and a radial distance of 1.0 m from the point of application of the load is

(a) 47.5 kPa (b) 47.6 kPa

(c) 47.7 kPa (d) 47.8 kPa

[1998 : 2 M]

5.3 A 25 kN point load acts on the surface of an infinite elastic medium. The vertical pressure intensity in kN/m^2 at a point 6.0 m below and 4.0 m away from the load will be

(a) 132 (b) 13.2

(c) 1.32 (d) 0.132 [2003 : 1 M]

5.4 The vertical stress at some depth below the corner of a 2 m × 3 m rectangular footing due to a certain load intensity is 100 kN/m^2. What will be the vertical stress in kN/m^2 below the centre of a 4 m × 6 m rectangular footing at the same depth and same load intensity?

(a) 25 (b) 100

(c) 200 (d) 400 [2007 : 1 M]

5.5 A footing 2 m × 1 m exerts a uniform pressure of 150 kN/m^2 on the soil. Assuming a load dispersion of 2 vertical to 1 horizontal, the average vertical stress (kN/m^2) at 1.0 m below the footing is

(a) 50 (b) 75

(c) 80 (d) 100 [2008 : 2 M]

5.6 The vertical stress at point P_1 due to the point load Q on the ground surface as shown in figure is σ_z. According to Boussinesq's equation, the vertical stress at point P_2 shown in figure will be

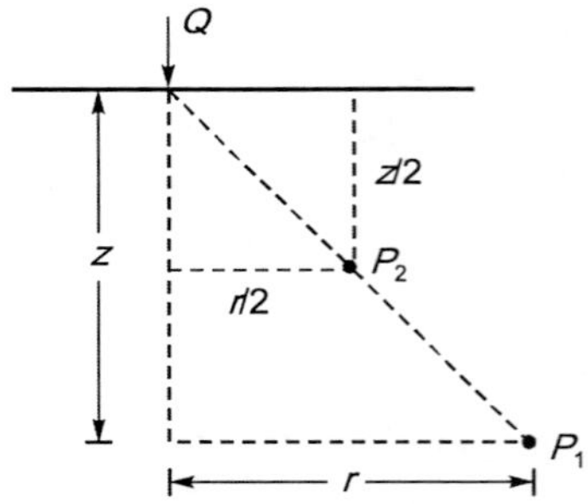

(a) $\sigma_z/2$ (b) σ_z

(c) $2\sigma_z$ (d) $4\sigma_z$ [2010 : 2 M]

5.7 A uniformly distributed line load of 500 kN/m is acting on the ground surface. Based on Boussinesq's theory, the ratio of vertical stress at depth 2 m to that at 4 m, right below the line of loading, is

(a) 0.25 (b) 0.5

(c) 2.0 (d) 4.0

[2017 : 1 M, Set-I]

5.8 Consider a square-shaped area $ABCD$ on the ground with its centre at M as shown in the figure. Four concentrated vertical loads of P = 5000 kN are applied on this area, one at each corner.

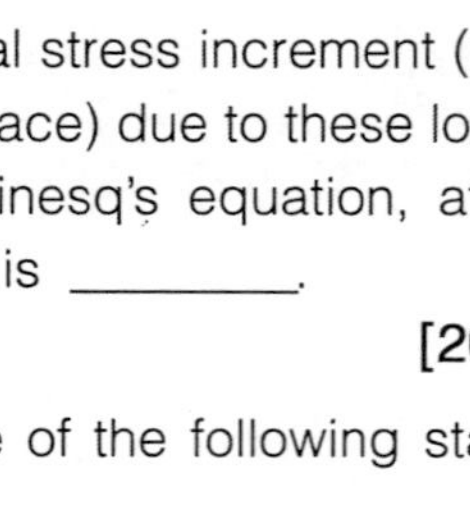

The vertical stress increment (in kPa, up to one decimal place) due to these loads according to the Boussinesq's equation, at a point 5 right below M, is ________.

[2017 : 2 M, Set-II]

5.9 Which one of the following statements is NOT correct?

(a) When the water content of soil lies between its liquid limit and plastic limit, the soil is said to be in plastic state.

(b) Boussinesq's theory is used for the analysis of stratified soil.

(c) The inclination of stable slope in cohesive soil can be greater than its angle of internal friction.

(d) For saturated dense fine sand, after applying overburden correction, if the Standard Penetration Test value exceeds 15, dilatancy correction is to be applied.

[2018 : 1 M, Set-II]

5.10 In a soil specimen, the total stress, effective stress, hydraulic gradient and critical hydraulic gradient are σ, σ', i and i_c respectively. For initiation of quicksand condition, which one of the following statements is TRUE?

(a) $\sigma' = 0$ and $i = i_c$ (b) $\sigma' \neq 0$ and $i = i_c$

(c) $\sigma' \neq 0$ and $i \neq i_c$ (d) $\sigma = 0$ and $i = i_c$

[2019 : 1 M, Set-I]

5.11 A concentrated load of 500 kN is applied on an elastic half space. The ratio of the increase in vertical normal stress at depths of 2 m and 4 m along the point of the loading, as per Boussinesq's would be_______.

[2019 : 1 M, Set-I]

5.12 A 2 m × 4 m rectangular footing has to carry a uniformly distributed load of 120 kPa. As per the 2:1 dispersion method of stress distribution, the increment in vertical stress (in kPa) at a depth of 2 m below the footing is _________.

[2019 : 2 M, Set-II]

5.13 A concentrated vertical load of 3000 kN is applied on a horizontal ground surface. Points P and Q are at depths 1 m and 2 m below the ground, respectively, along the line of application of the load. Considering the ground to be a linearly elastic, isotropic, semi-infinite medium, the ratio of the increase in vertical stress at P to the increase in vertical stress at Q is ________. (in integer)

[2022 : 1 M, Set-I]

5.14 A concentrically loaded isolated square footing of size 2 m x 2 m carries a concentrated vertical load of 1000 kN. Considering Boussinesq's theory of stress distribution, the maximum depth (in m) of the pressure bulb corresponding to 10 % of the vertical load intensity will be ___________. (round off to two decimal all places)

[2022 : 1 M, Set-II]

5.15 Consider that a force P is acting on the surface of a half-space (Boussinesq's problem). The expression for the vertical stress (σ_z) at any point (r, z), within the half-space is given as,

$$\sigma_z = \frac{3P}{2\pi}\frac{z^3}{(r^2+z^2)^{\frac{5}{2}}}$$

where, r is the radial distance, and z is the depth with downward direction taken as positive. At any given r, there is a variation of σ_z along z, and at a specific z, the value of σ_z will be maximum. What is the locus of the maximum σ_z?

(a) $z^2 = \frac{3}{2}r^2$ (b) $z^3 = \frac{3}{2}r^2$

(c) $z^5 = \frac{5}{2}r^2$ (d) $z^3 = \frac{5}{2}r^2$

[2023 : 2 M, Set-I]

■■■■

Answers Stress Distribution in the Soil

5.1	(c)	5.2	(d)	5.3	(d)	5.4	(d)	5.5	(a)	5.6	(d)	5.7	(c)
5.8	(190.8)	5.9	(b)	5.10	(a)	5.11	(4)	5.12	(40)	5.13	(4)	5.14	(4.37)
5.15	(a)												

Explanations Stress Distribution in the Soil

5.1 (c)

As per **Boussinesq equation**, for a point load the vertical stress at depth z, directly below the point load P is given by,

$$\sigma_z = k \times \frac{Q}{z^2}$$

where, $k = \frac{3}{2\pi}\left[\frac{1}{1+\left(\frac{r}{z}\right)^2}\right]^{5/2}$

= Constant known as **Boussinesq influence factor**.

5.2 (d)

Given Data, Q= 700 kN, z = 2 m, r = 1.0 m

Vertical stress,

$$\sigma_v = \frac{3Q}{2\pi z^2}\left[\frac{1}{1+\frac{r^2}{z^2}}\right]^{5/2} = \frac{3\times 700}{2\pi(2)^2}\left[\frac{1}{1+\left(\frac{1}{2}\right)^2}\right]^{5/2}$$

$= 47.8$ kPa

5.3 (d)

As per **Boussinesq's equation**, the vertical stress σ_z at a point located at a depth 'z' and a horizontal distance 'r' from the point of application of point load Q is,

$$\sigma_z = \frac{3Q}{2\pi z^2}\left[\frac{1}{1+(r/z)^2}\right]^{5/2}$$

$$\Rightarrow \quad \sigma_z = \frac{3\times 25}{2\pi\times(6)^2}\left[\frac{1}{1+(4/6)^2}\right]^{5/2}$$

$= 0.132$ kN/m^2

5.4 (d)

For same load intensity and same depth,

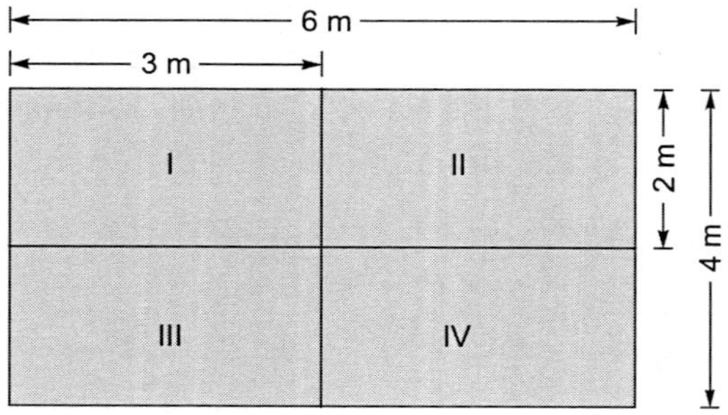

All part I, II, II and IV are of equal area

Vertical stress at the centre of rectangular footing (4m × 6m)

= [Vertical stress at the centre of part-I (2m × 3m)] × 4

= 100 × 4 = 400 kN/m^2

5.5 (a)

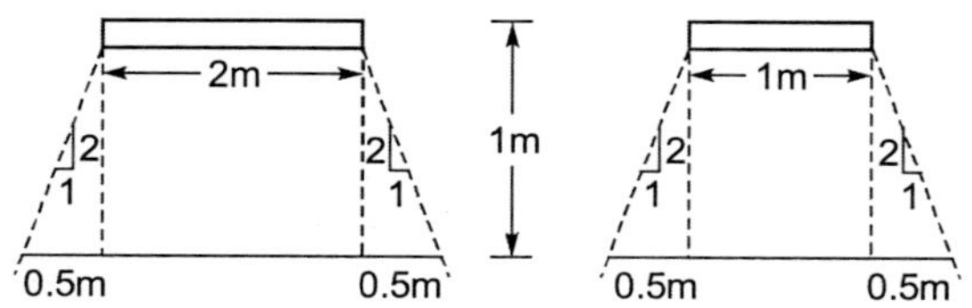

Total dispersion area

= (2 + 0.5 + 0.5) × (1 + 0.5 + 0.5) = 3 m × 2 m

5.6 (d)

As per Boussinesq equation, the vertical stress σ_z at a point located at a depth 'z' and a horizontal distance 'r' from the point of application of point load Q is

$$\sigma_z = \frac{3Q}{2\pi z^2}\left[\frac{1}{1+\left(\frac{r}{z}\right)^2}\right]^{5/2}$$

For a horizontal distance $\frac{r}{2}$ and a depth $\frac{z}{2}$, the stress will be

$$= \frac{3Q}{2\pi\left(\frac{z}{2}\right)^2}\left[\frac{1}{1+\left(\frac{r/2}{z/2}\right)^2}\right]^{5/2}$$

$$= 4\times\frac{3Q}{2\pi z^2}\left(\frac{1}{1+\left(\frac{r}{z}\right)^2}\right)^{5/2} = 4\sigma_z$$

5.7 (c)

$q' = 500$ kN/m

$x = 0$ (Just below the line load)

$$\sigma_z = \frac{2q'}{\pi z}\left(\frac{1}{1+\frac{x^2}{z^2}}\right)^2 = \frac{2q'}{\pi z}$$

$$\text{Ratio} = \frac{(\sigma_z)_{at,2m}}{(\sigma_z)_{at,4m}} = \frac{\frac{2q'}{\pi\times 2}}{\frac{2q'}{\pi\times 4}} = \frac{4}{2} = 2$$

5.8 Sol.

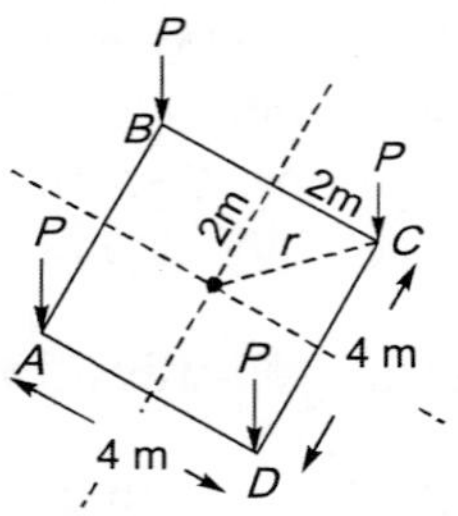

$P = 5000$ kN

$$r = \sqrt{2^2+2^2} = 2\sqrt{2}\text{ m}$$

According to **Boussinesq's equation,**

$$\sigma_z = k\frac{Q}{z^2} = \frac{3}{2\pi}\left(\frac{1}{1+\frac{r^2}{z^2}}\right)^{5/2}\times\frac{Q}{z^2}$$

$\Rightarrow$ Vertical stress at 5 m below point M,

$$\sigma_z = 4\times\frac{3}{2\pi}\times\left\{\frac{1}{1+\left(\frac{2\sqrt{2}}{5}\right)^2}\right\}^{5/2}\times\frac{5000}{5^2}$$

$= 190.8$ kPa

5.9 (b)

Boussinesq's assumed soil as isotropic hence not applicable for stratified soil.

5.10 (a)

For quick sand condition

(i) Effective stress ($\bar{\sigma}$) = 0

(ii) $i = i_c$

i - Hydraulic gradient

i_c - Critical hydraulic gradient

$$i_c = \frac{\gamma_{sub}}{\gamma_w} = \frac{G_s - 1}{1+e}$$

5.11 Sol.

Q = 500 kN

$$\sigma_z = \frac{3}{2\pi}\frac{Q}{z^2} \text{ (Just below the loading)}$$

$$\frac{\sigma_{z_1}}{\sigma_{z_2}} = \frac{\frac{3}{2\pi}\times\frac{500}{2^2}}{\frac{3}{2\pi}\times\frac{500}{4^2}} = \frac{16}{4} = 4.$$

5.12 Sol.

The area of rectangular footing = 2 m × 4 m

q = 120 kPa

as 2:1 dispersion method of stress distribution

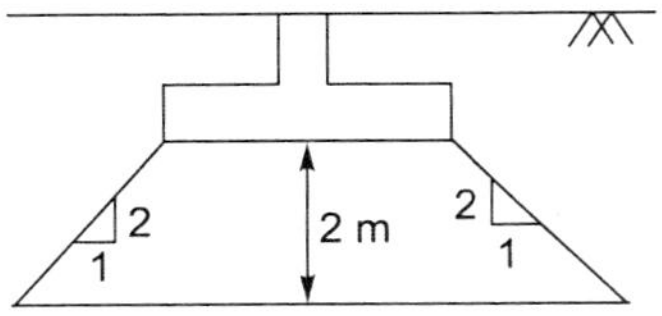

$$\Delta\bar{\sigma} = \frac{q(B\times L)}{(B+2nZ)(L+2nZ)}$$

$$= \frac{120\times2\times4}{\left(2+2\times\frac{1}{2}\times2\right)\left(4+2\times\frac{1}{2}\times2\right)}$$

$$\Delta\bar{\sigma} = 40 \text{ kPa}$$

5.13 Sol.

$$\sigma_z = \frac{3}{2\pi}\left(\frac{1}{1+\frac{r^2}{Z^2}}\right)^{5/2}\frac{Q}{Z^2}$$

$\because$ $r = 0$

$$\Rightarrow \sigma_z = \frac{3}{2\pi}\frac{Q}{Z^2}$$

$$\therefore (\sigma_z)_p = \frac{3}{2\pi}\frac{Q}{(1)^2}$$

$$\therefore (\sigma_z)_Q = \frac{3}{2\pi}\frac{Q}{(2)^2}$$

$$\Rightarrow \frac{(\sigma_z)_p}{(\sigma_z)_q} = \frac{\frac{3}{2\pi}\frac{Q}{(1)^2}}{\frac{3}{2\pi}\frac{Q}{(2)^2}} = 4$$

5.14 Sol.

Given, $\sigma_z = 0.1\times\frac{Q}{2\times2}$ ...(i)

As per Boussinesq's theorye,

$$\sigma_z = k_B\frac{Q}{Z^2} \quad ...(ii)$$

where, $k_B = \frac{3}{2\pi}\left[\frac{1}{1+\left(\frac{r}{Z}\right)^2}\right]^{5/2}$

For maximum depth,

$r = 0$

$$\Rightarrow k_B = \frac{3}{2\pi} \quad ...(iii)$$

From eq. (i), (ii) and (iii), we get

$$\therefore 0.1\times\frac{Q}{4} = \frac{3}{2\pi}\times\frac{Q}{Z^2}$$

$$\Rightarrow Z = 4.37 \text{ m}$$

5.15 (a)

At given r for maximum σ_z,

$$\frac{d\sigma_z}{dz} = 0$$

(taking r constant)

$$\sigma_z = \frac{3P}{2\pi}\frac{z^3}{(r^2+z^2)^{5/2}}$$

$$\frac{d\sigma_z}{dz} = \frac{3P}{2\pi}\left[\frac{(r^2+z^2)^{5/2}3z^2 - z^3\cdot\frac{5}{2}(r^2+z^2)^{3/2}\cdot2z}{(r^2+z^2)^5}\right]$$

$= 0$

$$\Rightarrow (r^2+z^2)^{3/2}z^2\left[3(r^2+z^2)-5z^2\right] = 0$$

$$\Rightarrow 3r^2 + 3z^2 - 5z^2 = 0$$

$$\Rightarrow 3r^2 = 2z^2$$

$$\Rightarrow z^2 = \frac{3r^2}{2}$$

■■■■

6 Compaction of Soil

6.1 The unit weight of a soil at zero air voids depends on
(a) specific gravity (b) water content
(c) unit weight of water (d) All of the above
[1995 : 1 M]

6.2 Write True/False in the answer book.
The measure of soil compaction is its wet density
[1995 : 1 M]

6.3 In a compaction test, as the compaction effort is increased, the optimum moisture content
(a) decreases
(b) remains same
(c) increases first and then decreases
(d) increases [1997 : 1 M]

6.4 Compaction of an embankment is carried out in 500 mm thick layers. The rammer used for compaction has a foot area of 0.05 m^2 and the energy imparted in every drop of rammer is 400 N-m. Assuming 50% more energy in each pass over the compacted area due to overlap, the number of passes required to develop compactive energy equivalent to Indian Standard light compaction for each layer would be
(a) 10 (b) 16
(c) 20 (d) 26 [2003 : 2 M]

6.5 A clayey soil has a maximum dry density of 16 kN/m^3 and optimum moisture content of 12%. A contractor during the construction of core of an earth dam obtained the dry density 15.2 kN/m^3 and water content 11%. This construction is acceptable because
(a) the density is less than the maximum dry density and water content is on dry side of optimum
(b) the compaction density is very low and water content is less than 12%
(c) the compaction is done on the dry side of the optimum
(d) both the dry density and water content of the compacted soil are within the desirable limits
[2005 : 1 M]

6.6 In a standard proctor test, 1.8 kg of moist soil was filling the mould (volume = 944 cc) after compaction. A soil sample weighing 23 g was taken from the mould and ovendried for 24 hours at a temperature of 110°C. Weight of the dry sample was found to be 20 g. Specific gravity of soil solids is G = 2.7. The theoretical maximum value of the dry unit weight of the soil at that water content is equal to
(a) 4.67 kN/m^3 (b) 11.5 kN/m^3
(c) 16.26 kN/m^3 (d) 18.85 kN/m^3
[2006 : 2 M]

6.7 Compaction by vibratory roller is the best method of compaction in case of
(a) moist silty sand
(b) well graded dry sand
(c) clay of medium compressibility
(d) silt of high compressibility
[2008 : 1 M]

6.8 Deposit with flocculated structure is formed when
(a) clay particles settle on sea bed
(b) clay particles settle on fresh water lake bed
(c) sand particles settle on river bed
(d) sand particles settle on sea bed
[2009 : 1 M]

6.9 In a compaction test, *G*, *w*, *S* and *e* represent the specific gravity, water content, degree of saturation and void ratio of the soil sample, respectively. If γ_w represents the unit weight of water and γ_d represents the dry unit weight of the soil, the equation for zero air voids line is

(a) $\gamma_d = \dfrac{G\gamma_w}{1+Se}$ (b) $\gamma_d = \dfrac{G\gamma_w}{1+Gw}$

(c) $\gamma_d = \dfrac{Gw}{1+\gamma_w S}$ (d) $\gamma_d = \dfrac{Gw}{1+Se}$

[2010 : 1 M]

6.10 Two series of compaction tests were performed in the laboratory on an inorganic clayey soil employing two different level of compaction energy per unit volume of soil. With regard to the above tests, the following two statements are made
I. The optimum moisture content is expected to be more for the tests with higher energy.
II. The maximum dry density is expected to be more for the tests with higher energy.

The CORRECT option evaluating the above statement is
(a) Only I is TRUE
(b) Only II is TRUE
(c) Both I and II are TRUE
(d) Neither I nor II is TRUE [2012 : 1 M]

6.11 Following statement are made on compacted soil, where DS stands for soil compaction on Dry Side of OMC and WS stands for soil compacted on Wet Side of OMC. Identify incorrect statement
(a) Soil structure is flocculated on DS and dispersed on WS.
(b) Construction of pore water pressure is low on DS and High on WS.
(c) Soil on drying, shrinkage is high on DS and Low on WS.
(d) On addition to water, swelling is high on DS and low on WS. [2013 : 1 M]

6.12 OMC-SP and MDD-SP denote the optimum moisture content and maximum dry density obtained from standard Proctor compaction test, respectively. OMC-MP and MDD-MP denote the optimum moisture content and maximum dry density obtained from the modified Proctor compaction test, respectively. Which one of the following is correct?
(a) OMC-SP < OMC-MP and MDD-SP < MDD-MP
(b) OMC-SP > OMC-MP and MDD-SP < MDD-MP
(c) OMC-SP < OMC-MP and MDD - SP >MDD-MP
(d) OMC-SP > OMC-MP and MDD-SP > MDD-MP
[2016 : 2 M, Set-II]

6.13 With reference to the compaction test conducted on soils, which of the following is INCORRECT?
(a) Peak point of the compaction curve gives the maximum dry unit weight and optimum moisture content
(b) With increase in the compaction effort, the maximum dry unit weight increases
(c) With increase in the compaction effort, the optimum moisture content decreases
(d) Compaction curve crosses the zero-air-voids curve
[2023 : 2 M, Set-I]

Answers Compaction of Soil

6.1	(d)	6.2	(False)	6.3	(a)	6.4	(d)	6.5	(a)	6.6	(d)	6.7	(b)
6.8	(a)	6.9	(b)	6.10	(b)	6.11	(c)	6.12	(b)	6.13	(d)		

Explanations Compaction of Soil

6.1 (d)

Unit weight of soil,

$$\gamma_t = \frac{G_s(1+w)\gamma_w}{1+e} \quad \text{Since, } e = 0$$

$\therefore$ γ_t depends on G_s, γ_w and w.

6.2 Sol.

FALSE, The measure of soil compaction is its dry density.

6.3 (a)

For a given type of compaction, the higher the compactive effort, the higher the maximum dry unit weight and lower the optimum moisture content.

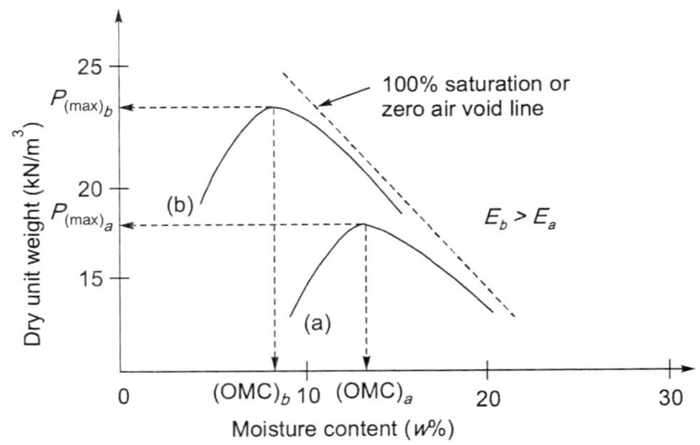

Effect of compactive effort

6.4 (d)

As per **Indian Standard light compaction test**, a hammer of 2.6 kg is allowed to fall from a height of 310 mm and 3 layers are tamped 25 times in a mould of volume 1000 cc.

∴ Energy imparted

$= 2.6 \times 25 \times 3 \times 9.81 \times \frac{310}{1000} = 593.01$ N-m

For embankment compaction, the volume of soil covered by rammer

$$= 0.05 \times 10^4 \times \frac{500}{10} = 25000 \text{ cc}$$

∴ Energy imparted in every drop of hammer

$$= \left(1+\frac{50}{100}\right) \times 400 = 600 \text{ N-m}$$

If 'n' number of passes are required to develop equivalent energy to **Indian Standard light compaction test**, then,

$$n \times \frac{600}{25000} = \frac{593.01}{1000}$$

$$\Rightarrow \quad n = 24.71 \simeq 25 \text{ (approx.)}$$

However, If the data of **Standard Proctor Test** are taken i.e. weight of hammer = 2.495 kg; Volume of mould = 944 cc, then the value of '*n*' will be close to 26.

6.5 (a)

6.6 (d)

Bulk density of the soil

$$= \frac{1.8 \times 1000}{944} = 1.907 \text{ gm/cc}$$

Water content of soil

$$= \left(\frac{23-20}{20}\right) \times 100 = 15\%$$

For, theoretical maximum dry unit weight, degree of saturation should be 100%.

$$Se = Gw$$

$$\Rightarrow \quad e = \frac{2.7 \times 0.15}{1} \quad \Rightarrow \quad e = 0.405$$

The theoretical maximum dry unit weight may be given as,

$$\gamma_{d,\max} = \frac{G\gamma_w}{1+e} = \frac{2.7 \times 9.81}{1+0.405} = 18.85 \text{kN/m}^3$$

6.7 (b)

Compaction by vibratory roller is suitable for sands. This method is used in embankments for oil storage tanks etc.

6.8 (a)

Flocculated structure is formed when clay particles settle on sea bed, since sea bed is polar in nature.

6.9 (b)

The bulk unit weight of soil is given by,

$$\gamma = \frac{(G+Se)\gamma_w}{1+e}$$

For dry unit weight, degree of saturation,

$$S = 0 \qquad \therefore \gamma_d = \frac{G\gamma_w}{1+e}$$

But we know that,

$$Gw = Se \qquad \therefore \gamma_d = \frac{G\gamma_w}{1+\frac{Gw}{S}}$$

For zero air voids, degree of saturation,

$$S = 100\% \qquad \therefore \gamma_d = \frac{G\gamma_w}{1+Gw}$$

6.10 (b)

Optimum moisture content of the soil decreases with increasing compactive efforts.

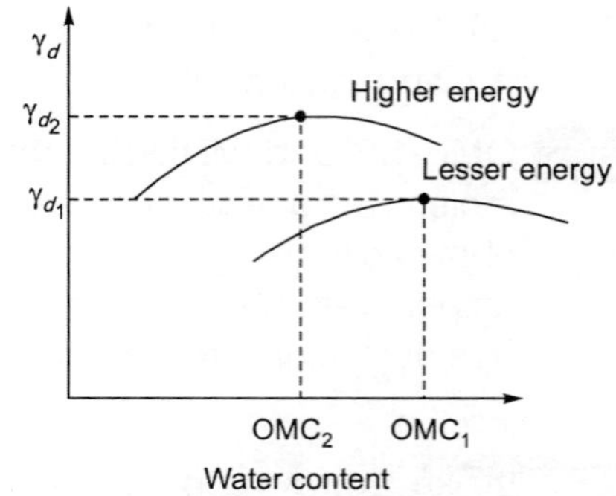

6.11 (c)

Soils compacted on the wet of optimum have nearly parallel orientations of particles which allows the particles to pack more efficiently as compacted to the randomly oriented particles on the dry side of optimum. Therefore, soil on WS tend to exibit more shrinkage.

6.12 (b)

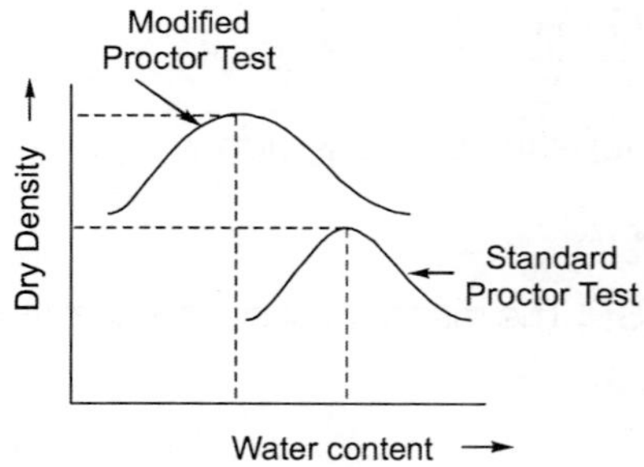

OMC – SP > OMC – MP

MDD – SP < MDD – MP

6.13 (d)

Zero air void line do not cross compaction curve.

7 Consolidation of Soils

7.1 The slope of the e-log $\bar{\sigma}$ curve for a soil mass gives
(a) coefficient of permeability, k
(b) coefficient of consolidation C_v
(c) compression index, C_c
(d) coefficient of volume compressibility, m_v

[1997 : 1 M]

7.2 The stress-strain behaviour of soils as shown in the figure corresponds to

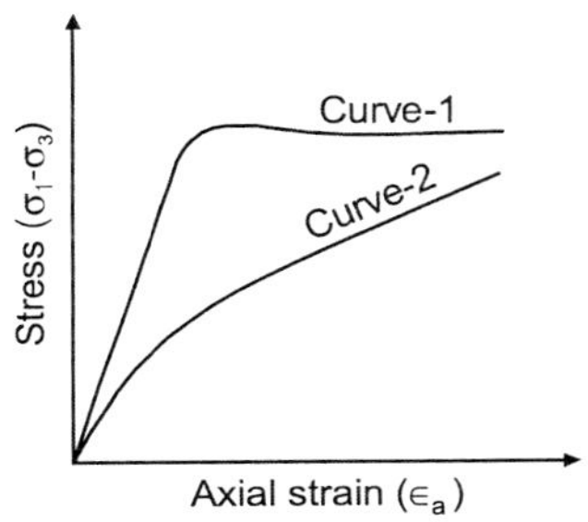

(a) Curve-1: Loose sand and normally consolidated clay
Curve-2: Loose sand and over consolidated clay
(b) Curve-1: Dense sand and normally consolidated clay
Curve-2: Loose sand and over consolidated clay
(c) Curve-1: Dense sand and over consolidated clay
Curve-2: Loose sand and normally consolidated clay
(d) Curve-1: Loose sand and over consolidated clay
Curve-2: Dense sand and normally consolidated clay

[1998 : 1 M]

7.3 The time for a clay layer to achieve 90% consolidation is 15 years. The time required to achieve 90% consolidation, if the layer were twice as thick, 3 times more permeable and 4 times more compressible would be
(a) 70 years (b) 75 years
(c) 80 years (d) 85 years

[1998 : 2 M]

7.4 Consolidation in soils
(a) is a function of the effective stress
(b) does not depend on the present stress
(c) is a function of the pore water pressure
(d) is a function of the total stress

[1999 : 1 M]

7.5 Identify the two False statements from the following four statements
I. The consolidation of soil happens due to the change in total stress.
II. When Standard Penetration Tests are performed in fine sands below the water table, the dilation correction is applied after the overburden correction is applied.
III. Over consolidated clays will have predominantly cohesive strength as compared to the frictional strength.
IV. Compaction of soils is due to expulsion of water.
(a) II and III (b) I and IV
(c) I and III (d) II and IV

[2001 : 2 M]

7.6 The time for a clay layer to achieve 85% consolidation is 10 years. If the layer was half as thick, 10 times more permeable and 4 times more compressible then the time that would be required to achieve the same degree of consolidation is
(a) 1 year (b) 5 year
(c) 12 years (d) 16 years

[2002 : 2 M]

7.7 A double draining clay layer, 6 m thick, settles by 30 mm in three years under the influence of certain loads. Its final consolidation settlement has been estimated to be 120 mm. If a thin layer of sand having negligible thickness is introduced at a depth of 1.5 m below the top surface, the final consolidation settlement of clay layer will be
(a) 60 mm (b) 120 mm
(c) 240 mm (d) None of these

[2003 : 1 M]

7.8 At a reclamation site for which the soil strata is shown in figure, a 3 m thick layer of a fill material is to be laid instantaneously on the top surface. If the coefficient of volume compressibility, m_v for clay is 2.2×10^{-4} m²/kN, the consolidation settlement of the clay layer due to placing of fill material will be

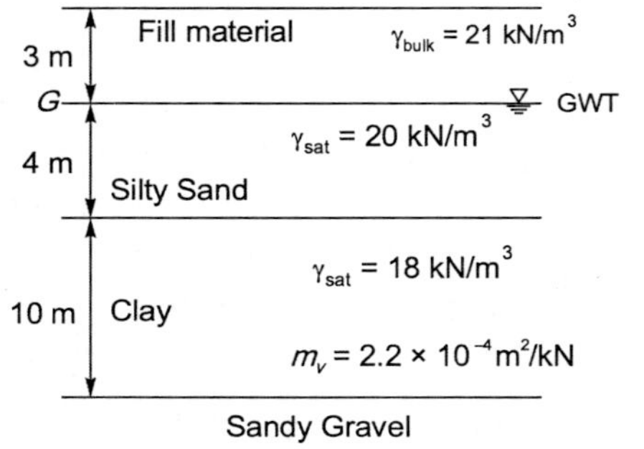

(a) 69.5 mm (b) 139 mm
(c) 228 mm (d) 278 mm [2003 : 2 M]

Common Data for Questions 7.10 and 7.11:

Figure shows the geometry of a strip footing supporting the load bearing walls of a three storied building and the properties of clay layer.

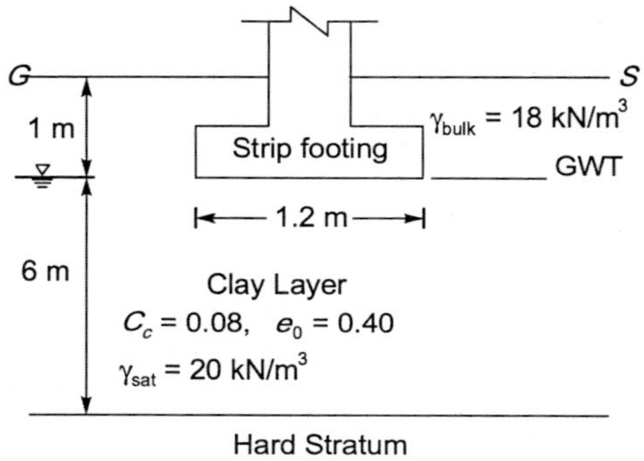

7.9 If the pressure acting on the footing is 40 kPa, the consolidation settlement of the footing will be
(a) 0.89 mm (b) 8.9 mm
(c) 89.0 mm (d) None of these
[2003 : 2 M]

7.10 If the elastic modulus and the Poisson's ratio of the clay layer are respectively 50×10^3 kPa and 0.4 and if the influence factor for the strip footing is 1.75, the elastic settlement of the footing will be
(a) 0.41 mm (b) 1.41 mm
(c) 14.1 mm (d) None of these
[2003 : 2 M]

7.11 A 6 m thick clay layer undergoes 90% consolidation four times faster under two-way drainage as compared to one-way drainage. In an identical clay layer of 15 m thickness, two-way drainage will be faster as compared to one-way drainage by
(a) 8 times (b) 4 times
(c) 2.5 times (d) 2 times [2004 : 2 M]

7.12 Root time method is used to determine
(a) T, time factor
(b) C_v, coefficient of consolidation
(c) a_v, coefficient of compressibility
(d) m_v, coefficient of volume compressibility
[2005 : 1 M]

Linked Answer Questions 7.13 and 7.14:

The average effective overburden pressure on 10 m thick homogeneous saturated clay layer is 150 kPa. Consolidation test on an undisturbed soil sample taken from the clay layer showed that the void ratio decreased from 0.6 to 0.5 by increasing the stress intensity from 100 kPa to 300 kPa (G = 2.65).

7.13 The initial void ratio of the clay layer is
(a) 0.209 (b) 0.563
(c) 0.746 (d) 1.000 [2006 : 2 M]

7.14 The total consolidation settlement of the clay layer due to the construction of a structure imposing an additional stress intensity of 200 kPa is
(a) 0.10 m (b) 0.25 m
(c) 0.35 m (d) 0.50 m [2006 : 2 M]

Linked Answer Questions 7.15 and 7.16:

The ground conditions at a site are as shown in the figure. The water table at the site which was initially at a depth of 5 m below the ground level got permanently lowered to a depth of 15 m below the ground level due to pumping of water over a few years. Assume the following data:
(i) unit weight of water = 10 kN/m³
(ii) unit weight of sand above water table = 18 kN/m³
(iii) unit weight of sand and clay below the water table = 20 kN/m³
(iv) coefficient of volume compressibility = 0.25 m²/MN

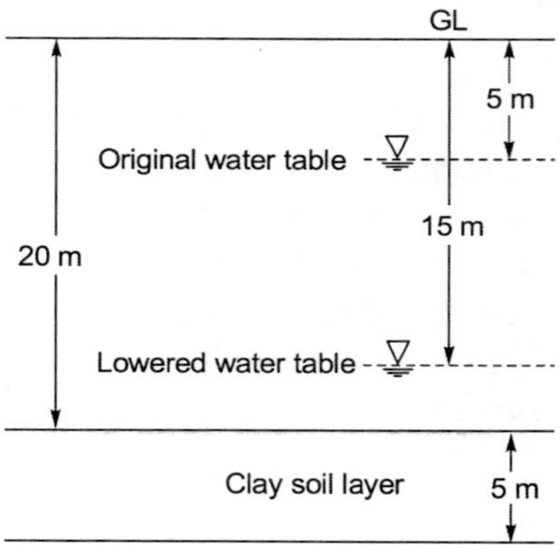

7.15 What is the change in the effective stress in kN/m^2 at mid-depth of the clay layer due to the lowering of the water table?

(a) 0 (b) 20
(c) 80 (d) 100 [2007 : 2 M]

7.16 What is the compression of the clay layer in mm due to the lowering of the water table?

(a) 125 (b) 100
(c) 25 (d) 0 [2007 : 2 M]

7.17 A saturated clay stratum draining both at the top and bottom undergoes 50 per cent consolidation in 16 years under an applied load. If an additional drainage layer were present at the middle of the clay stratum, 50 per cent consolidation would occur in

(a) 2 years (b) 4 years
(c) 8 years (d) 16 years

[2008 : 2 M]

Linked Answers Questions 7.18 and 7.19:

7.18 A saturated undisturbed sample from a clay strata has moisture content of 22.22% and specific weight of 2.7. Assuming $\gamma_w = 10$ kN/m^3, the void ratio and the saturated unit weight of the clay, respectively are

(a) 0.6 and 16.875 kN/m^3
(b) 0.3 and 20.625 kN/m^3
(c) 0.6 and 20.625 kN/m^3
(d) 0.3 and 16.975 kN/m^3 [2009 : 2 M]

7.19 Using the properties of the clay layer derived from the above question, the consolidation settlement of the same clay layer under a square footing (neglecting its self weight) with additional data shown in the figure below (assume the stress distribution as 1 H : 2V from the edge of the footing and γ_w = 10 kN/m^3) is

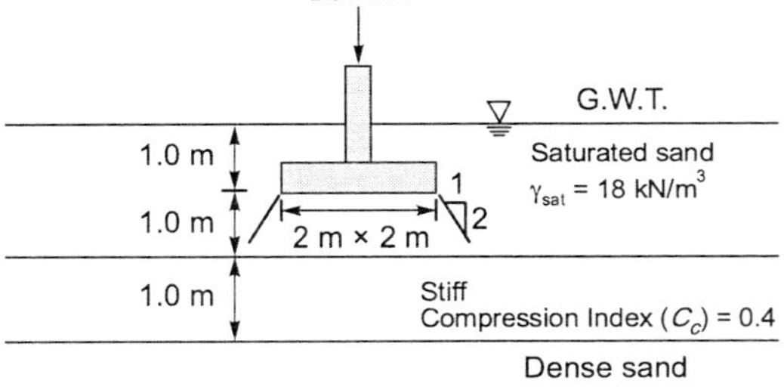

(a) 32.78 mm (b) 61.75 mm
(c) 79.5 mm (d) 131.13 mm

[2009 : 2 M]

7.20 The *e*-log *p* curve shown in the figure is representative of

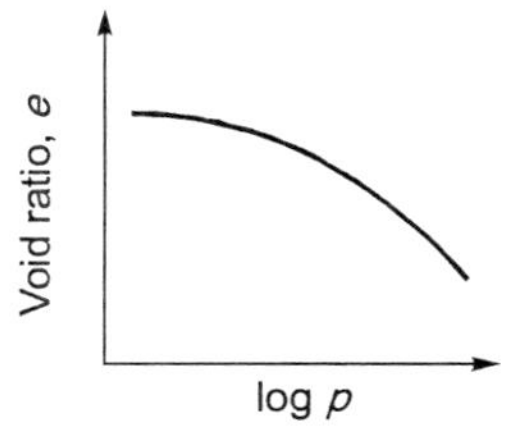

(a) Normally consolidated clay
(b) Over consolidated clay
(c) Under consolidated clay
(d) Normally consolidated clayey sand

[2010 : 1 M]

7.21 Identical surcharges are placed at ground surface at sites *X* and *Y*, with soil conditions shown alongside and water table at ground surface. The silty clay layers at *X* and *Y* are identical. The thin sand layer at *Y* is continuous and free-draining with a very large discharge capacity. If primary consolidation at *X* is estimated to complete in 36 months, what would be the corresponding time for completion of primary consolidation at *Y*?

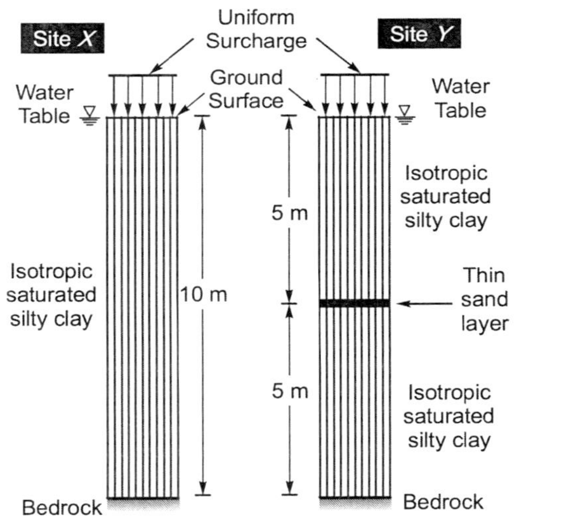

(a) 2.25 months (b) 4.5 months
(c) 9 months (d) 36 months

[2011 : 2 M]

7.22 A layer of normally consolidated, saturated silty clay of 1 m thickness is subjected to one dimensional consolidation under a pressure increment of 20 kPa. The properties of the soil are: specific gravity = 2.7, natural moisture content = 45%, compression index = 0.45, and recompression index = 0.05. The initial average effective stress within the layer is 100 kPa. Assuming Terzaghi's theory to be applicable, the primary consolidation settlement (rounded off to the nearest mm) is

(a) 2 mm (b) 9 mm
(c) 14 mm (d) 16 mm [2012 : 2 M]

7.23 The following data are given for the laboratory sample.

$$\sigma_o' = 175 \text{ kPa}, e_o' = 1.1,$$
$$\sigma_o' + \Delta\sigma_o' = 300 \text{ kPa}, e = 0.9$$

If thickness of the clay specimen is 25 mm, the value of coefficient of volume compressibility is_____ $\times 10^{-4}$ m^2/kN.

[2014 : 2 M, Set-I]

7.24 A water tank is to be constructed on the soil deposit shown in the figure below. A circular footing of diameter 3 m and depth of embedment 1 m has been designed to support the tank. The total vertical load to be taken by the footing is 1500 kN. Assume the unit weight of water as 10 kN/m^3 and the load dispersion pattern as 2 V : 1 H. The expected settlement of the tank due to primary consolidation of the clay layer is______ mm.

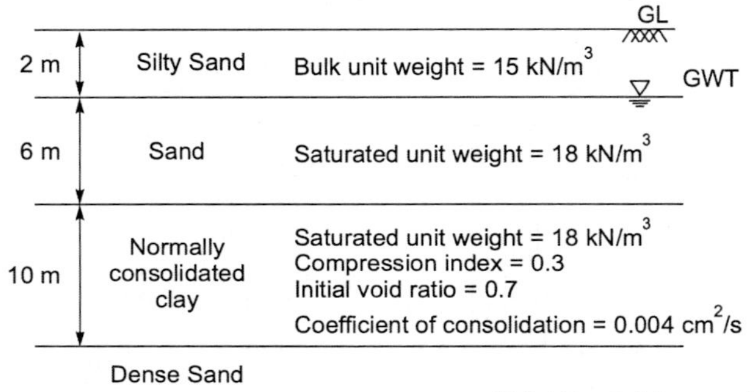

[2015 : 2 M, Set-I]

7.25 A 20 m thick clay layer is sandwiched between a silty sand layer and a gravelly sand layer. The layer experiences 30 mm settlement in 2 years. Given:

$$T_v = \begin{cases} \dfrac{\pi}{4}\left(\dfrac{U}{100}\right)^2 & \textit{for } U \le 60\% \\ 1.781 - 0.933\log_{10}(100-U) & \textit{for } U > 60\% \end{cases}$$

Where T_v is the time factor and U is the degree of consolidation in %.

If the coefficient of consolidation of the layer is 0.003 cm^2/s, the deposit will experience a total of 50 mm settlement in the next ________ years.

[2015 : 2 M, Set-I]

7.26 A 4 m thick layer of normally consolidated clay has an average void ratio of 1.30. Its compression index is 0.6 and coefficient of consolidation is 1 m^2/yr. If the increase in vertical pressure due to foundation load on the clay layer is equal to the existing effective overburden pressure, the change in the thickness of the clay layer is _____ mm.

[2015 : 2 M, Set-II]

7.27 An undisturbed soil sample was taken from the middle of a clay layer (i.e., 1.5 m below GL), as shown in figure. The water table was at the top of clay layer. Laboratory test results are as follows:

Natural water content of clay	:	25%
Preconsolidation pressure of clay	:	60 kPa
Compression index of clay	:	0.50
Recompression index of clay	:	0.05
Specific gravity of clay	:	2.70
Bulk unit weight of sand	:	17 kN/m^3

A compacted fill of 2.5 m height with unit weight of 20 kN/m^3 is placed at the ground level.

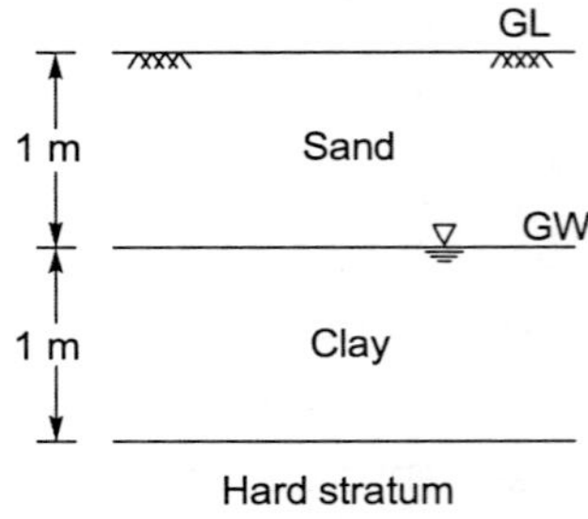

Assuming unit weight of water as10 kN/m^3, the ultimate consolidation settlement (expressed in mm) of the clay layer is______

[2016 : 2 M, Set-I]

7.28 The results of a consolidation test on an undisturbed soil, sampled at a depth of 10 m below the ground level are as follows :

Saturated unit weight : 16 kN/m^3
Pre-consolidation pressure : 90 kPa

The water table was encountered at the ground level. Assuming the unit weight of water as 10 kN/m^3, the over-consolidation ratio of the soil is

(a) 0.67 (b) 1.50
(c) 1.77 (d) 2.00

[2016 : 1 M, Set-II]

7.29 A 3 m thick clay layer is subjected to an initial uniform pore pressure of 145 kPa as shown in the figure.

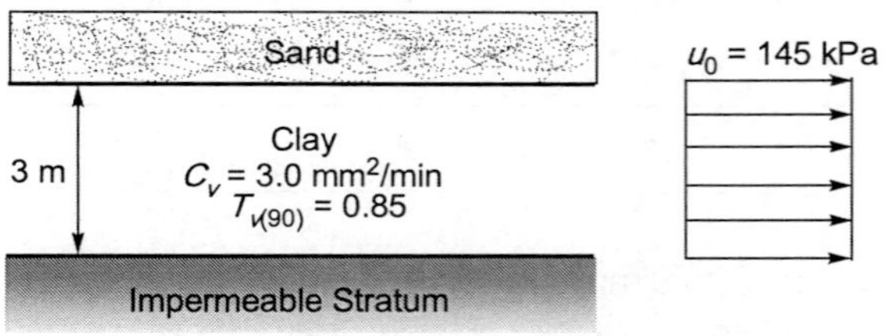

For the given ground conditions, the time (in days, rounded to the nearest integer) required for 90% consolidation would be ________.

[2017 : 1 M, Set-I]

7.30 Consider the following statements related to the pore pressure parameters, *A* and *B*

P. A always lies between 0 and 1.0

Q. A can be less than 0 or greater than 1.0

R. B always lies between 0 and 1.0

S. B can be less than 0 or greater than 1.0

For these statements, which one of the following options is correct?

(a) *P* and *R* (b) *P* and *S*

(c) *Q* and *R* (d) *Q* and *S*

[2017 : 1 M, Set-II]

7.31 The void ratio of a soil is 0.55 at an effective normal stress of 140 kPa. The compression index of the soil is 0.25. In order to reduce the void ratio to 0.4, an increase in the magnitude of effective normal stress (in kPa, up to one decimal place) should be ________. [2018 : 2 M, Set-I]

7.32 The compression curve (void ratio, *e* vs. effective stress, σ'_v) for a certain clayey soil is a straight line in a semi-logarithmic plot and it passes through the points (e = 1.2; σ'_v = 50 kPa) and (e = 0.6; σ'_v = 800 kPa). The compression index (up to two decimal places) of the soil is ________.

[2018 : 2 M, Set-II]

7.33 Construction of a new building founded on a clayey soil was completed in January 2010. In January 2014, the average consolidation settlement of the foundation in clay was recorded as 10 mm. The ultimate consolidation settlement was estimated in design as 40 mm. Considering double drainage to occur at the clayey soil site, the expected consolidation settlement in January 2019 (in mm, roundoff to the nearest integer) will be ________.

[2019 : 1 M, Set-II]

7.34 A 10 m thick clay layer is resting over a 3 m thick sand layer and is submerged. A fill of 2 m thick sand with unit weight of 20 kN/m^3 is placed above the clay layer to accelerate the rate of consolidation of the clay layer. Coefficient of consolidation of clay is 9×10^{-2} m^2/year and coefficient of volume compressibility of clay is 2.2×10^{-4} m^2/kN. Assume Taylor's relation between time factor and average degree of consolidation.

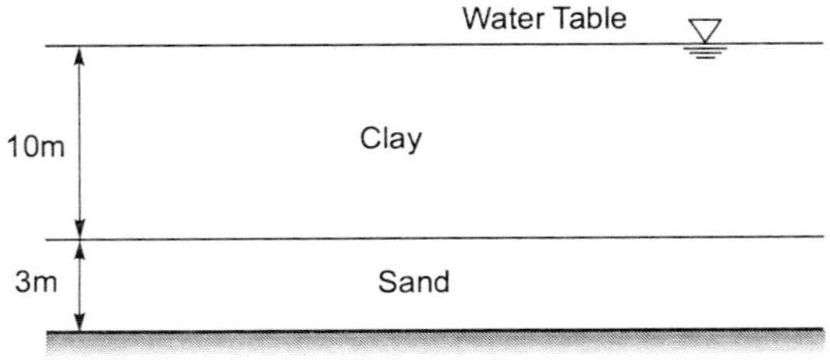

The settlement (in mm, *round off to two decimal places*) of the clay layer, 10 years after the construction of the fill, is ________.

[2020 : 2 M, Set-I]

7.35 A one-dimensional consolidation test is carried out on a standard 19 mm thick clay sample. The oedometer's deflection gauge indicates a reading of 2.1 mm, just before removal of the load, without allowing any swelling. The void ratio is 0.62 at this stage. The initial void ratio (round off to two decimal places) of the standard specimen is ________.

[2020 : 1 M, Set-II]

7.36 A footing of size 2 m × 2 m transferring a pressure of 200 kN/m^2, is placed at a depth of 1.5 m below the ground as shown in the figure (not drawn to the scale). The clay stratum is normally consolidated. The clay has specific gravity of 2.65 and compression index of 0.3.

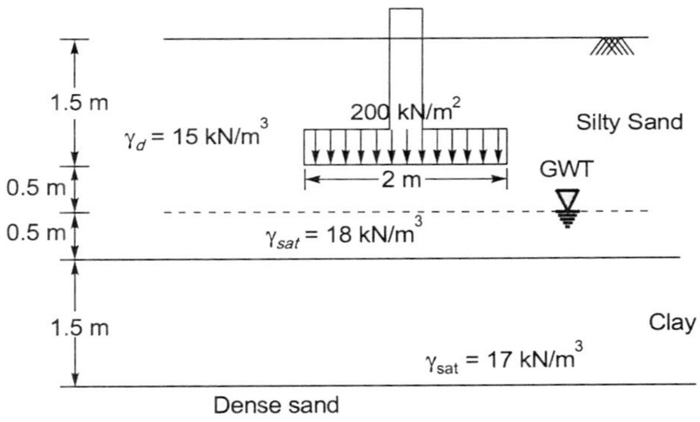

Considering 2 : 1 (vertical to horizontal) method of load distribution and γ_w = 10 kN/m^3, the primary consolidation settlement (in mm, round off to two decimal places) of the clay stratum is ________.

[2020 : 2 M, Set-II]

7.37 In an Oedometer apparatus, a specimen of fully saturated clay has been consolidated under a vertical pressure of 50 kN/m^3 and is presently at equilibrium. The effective stress and pore water pressure immediately on increasing the vertical stress to 150 kN/m^2, respectively are

(a) 50 kN/m^2 and 100 kN/m^2

(b) 100 kN/m^2 and 50 kN/m^2

(c) 150 kN/m^2 and 0

(d) 0 and 150 kN/m^2 [2021 : 1 M, Set-I]

7.38 The soil profile at a construction site is shown in the figure (not to scale). Ground water table (GWT) is at 5 m below the ground level at present. An old well data shows that the ground water table was as low as 10 m below the ground level in the past. Take unit weight of water. γ_w = 9.81 kN/m^3.

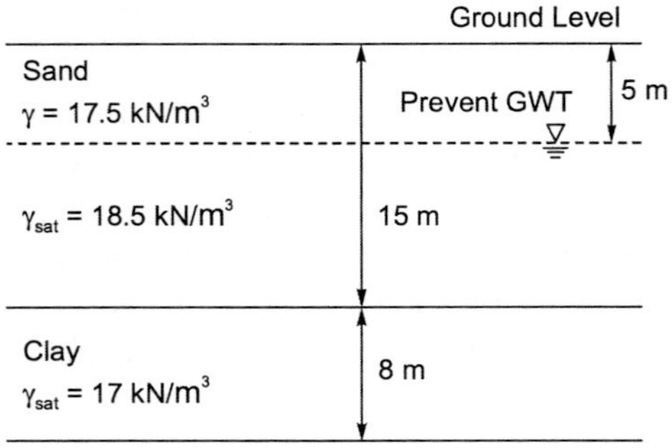

The over consolidation ratio (OCR) (round off to two decimal places) at the mid-point of the clay layer is _______. [2021 : 2 M, Set-I]

7.39 The soil profile at a road construction site is as shown in figure (not to scale). A large embankment is to be constructed at the site. The ground water table (GWT) is located at the surface of the clay layer, and the capillary rise in the sandy soil is negligible. The effective stress at the middle of the clay layer after the application of the embankment loading is 180 kN/m^2. Take unit weight of water, γ_w = 9.81 kN/m^3.

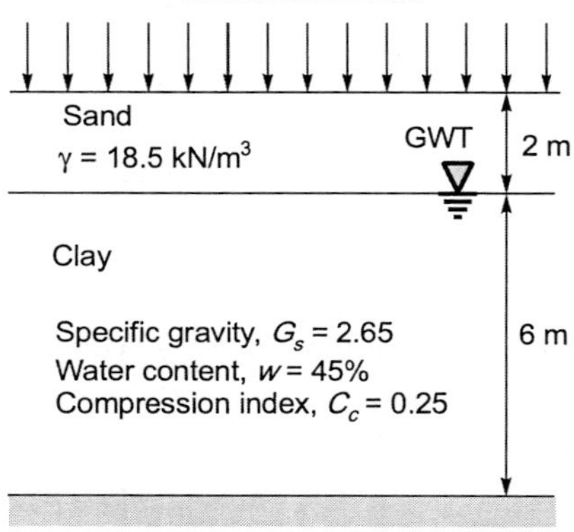

The primary consolidation settlement (in m, round off to two decimal places) of the clay layer resulting from this loading will be _______.

[2021 : 2 M, Set-II]

7.40 The void ratio of a clay soil sample M decreased from 0.575 to 0.510 when the applied pressure is increased from 120 kPa to 180 kPa. For the same increment in pressure, the void ratio of another clay soil sample N decreases from 0.600 to 0.550, If the ratio of hydraulic conductivity of sample M to sample N is 0.125, then the ratio of coefficient of consolidation of sample M to sample N (round off to three decimal places) is_______.

[2021 : 2 M, Set-II]

7.41 A clay layer of thickness H has a preconsolidation pressure p_c and an initial void ratio e_0. The initial effective overburden stress at the mid-height of the layer is p_0. At the same location, the increment in effective stress due to applied external load is Δp. The compression and swelling indices of the clay are C_c and C_s respectively. If $p_0 < p_c < (p_0 + \Delta p)$, then the correct expression to estimate the consolidation settlement (S_c) of the clay layer is

(a) $S_c = \frac{H}{1+e_0}\left[C_c \log\frac{p_c}{p_0} + C_s \log\frac{p_0+\Delta p}{p_c}\right]$

(b) $S_c = \frac{H}{1+e_0}\left[C_s \log\frac{p_c}{p_0} + C_c \log\frac{p_0+\Delta p}{p_c}\right]$

(c) $S_c = \frac{H}{1+e_0}\left[C_c \log\frac{p_0}{p_c} + C_s \log\frac{p_0+\Delta p}{p_c}\right]$

(d) $S_c = \frac{H}{1+e_0}\left[C_s \log\frac{p_0}{p_c} + C_c \log\frac{p_0+\Delta p}{p_c}\right]$

[2021 : 2 M, Set-II]

7.42 The correct match between the physical states of the soils given in **Group-I** and the governing conditions given in **Group-II** is

Group-I

1. normally consolidated soil.
2. quick clay
3. sand in critical state.
4. clay of high plasticity.

Group-II

P. sensitivity > 16
Q. dilation angle = 0
R. liquid limit > 50
S. over consolidation ratio = 1

(a) 1-S, 2-P, 3-Q, 4-R
(b) 1-Q, 2-S, 3-P, 4-R
(c) 1-Q, 2-P, 3-R, 4-S
(d) 1-S, 2-Q, 3-P, 4-R [2022 : 2 M, Set-I]

7.43 A raft foundation of 30 m × 25 m is proposed to be constructed at a depth of 8 m in a sand layer. A 25 m thick saturated clay layer exists 2 m below the base of the raft foundation. Below the clay layer, a dense sand layer exists at the site. A 25 mm thick undisturbed sample was collected from the mid-depth of the clay layer and tested in a laboratory oedometer under double drainage condition. It was found that the soil sample had undergone 50% consolidation settlement in 10 minutes.

The time (in days) required for 25% consolidation settlement of the raft foundation will be ________. (round off to the nearest integer)

[2022 : 2 M, Set-I]

7.44 A saturated compressible clay layer of thickness h is sandwiched between two sand layers, as shown in the figure. Initially, the total vertical stress and pore water pressure at point P. which is located at the mid-depth of the clay layer were 150 kPa and 25 kPa. respectively. Construction of a building caused an additional total vertical stress of 100 kPa at P. When the vertical effective stress at P is 175 kPa. the percentage of consolidation in The clay layer at P is __________. (in integer)

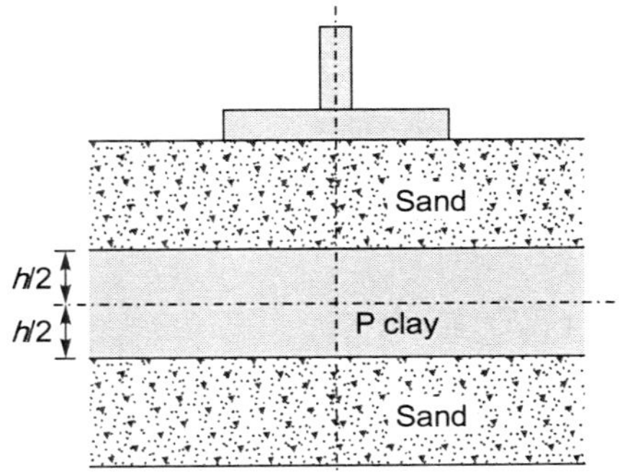

[2022 : 1 M, Set-II]

7.45 The total primary consolidation settlement (S_c) of a building constructed on a 10 m thick saturated clay layer is estimated to be 50 mm. After 300 days of the construction of the building. primary consolidation settlement was reported as 10 mm. The additional time (in days) required to achieve 50% of S_c will be ______. (rounded off to the nearest integer).

[2024 : 2 M, Set-I]

Answers Consolidation of Soils

7.1	(c)	7.2	(c)	7.3	(c)	7.4	(a)	7.5	(b)	7.6	(a)	7.7	(b)
7.8	(b)	7.9	(d)	7.10	(b)	7.11	(b)	7.12	(b)	7.13	(b)	7.14	(d)
7.15	(c)	7.16	(b)	7.17	(b)	7.18	(c)	7.19	(b)	7.20	(b)	7.21	(c)
7.22	(d)	7.23	(7.61)	7.24	(53.2)	7.25	(4.39)	7.26	(314)	7.27	(36.7)	7.28	(b)
7.29	(1771)	7.30	(c)	7.31	(417.3)	7.32	(0.498)	7.33	(15)	7.34	(18.8)	7.35	(0.82)
7.36	(74.27)	7.37	(a)	7.38	(1.22)	7.39	(0.33)	7.40	(0.095)	7.41	(b)	7.42	(a)
7.43	(1736)	7.44	(50)	7.45	(1575)								

Explanations Consolidation of Soils

7.1 (c)

Compression Index, C_c is given as,

$$C_c = \frac{\Delta e}{\log_{10}\left(\frac{\bar{\sigma}_2}{\bar{\sigma}_1}\right)} = \frac{e_1 - e_2}{\log_{10}\left(\frac{\bar{\sigma}_2}{\bar{\sigma}_1}\right)}$$

7.2 (c)

Stress strain curve for dense sands and loose sands are similar in shape to the stress-strain curves for over consolidated and normally consolidated clays respectively.

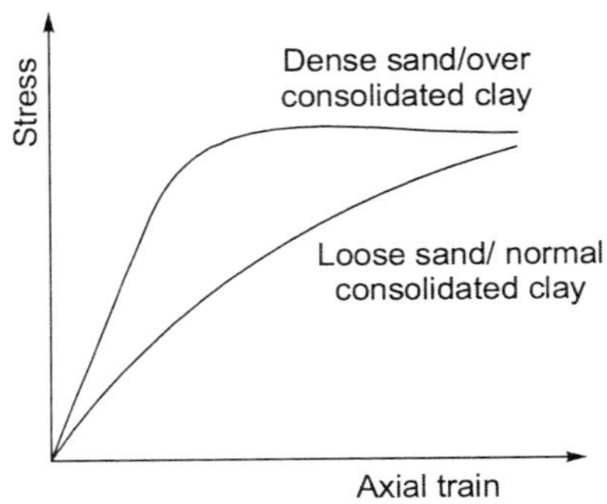

7.3 (c)

$$\because \quad T_v = C_v \frac{t}{d^2}$$

$$\Rightarrow \quad t \propto d^2$$

$$\Rightarrow \quad \frac{t_1}{t_2} = \left(\frac{d_1}{d_2}\right)^2$$

$$\Rightarrow \quad \frac{15}{t_2} = \left(\frac{H}{2H}\right)^2$$

$\Rightarrow \quad t_2 = 60$ years

As the clay layer is 3 times more permeable and 4 times more compressible, therefore time required for 90% consolidation

$$= 60 \times \frac{4}{3} = 80 \text{ years}$$

7.4 (a)

Terzaghi stated that effective stress controls the beahviour of soil like compression and strength.

$\Rightarrow$ Compression = f(effective stress)

By using the principle of effective stress, **Terzaghi** derived the consolidation theory for saturated soil.

7.5 (b)

(I) Consolidation of soil takes place due to the change in effective stress.

(IV) Compaction of soil is due to expulsion of air.

7.6 (a)

$$k = C_v m_v \gamma_w$$

$$T_v = \frac{C_v t}{h^2}$$

$$\therefore \quad t_1 = \frac{h_1^2 T_v}{k_1} m_v \gamma_w$$

$$\text{and} \quad t_2 = \frac{h_2^2 T_v}{k_2} m_v \gamma_w$$

$$\therefore \quad \frac{t_1}{t_2} = \left(\frac{h_1}{h_2}\right)^2 \times \frac{T_v}{T_v} \times \frac{k_2}{k_1} \times \frac{m_{v1}}{m_{v2}}$$

$$\frac{t_2}{t_1} = \left(\frac{h_2}{h_1}\right)^2 \times \frac{k_1}{k_2} \times \frac{m_{v2}}{m_{v1}}$$

$$= \frac{1}{4} \times 1 \times \frac{1}{10} \times 4 = \frac{1}{10}$$

$$\therefore \quad t_2 = \frac{t_1}{10} = \frac{10}{10} = 1 \text{ year}$$

7.7 (b)

Final settlement will not be altered. However, the rate of settlement of clay layer will increase.

7.8 (b)

Increase in effective stress,

$$\Delta\bar{\sigma} = 3 \times 21 = 63 \text{ kN/m}^2$$

Consolidation settlement of the clay layer

$$= H_0 m_v \Delta\bar{\sigma}$$
$$= 10 \times 2.2 \times 10^{-4} \times 63$$
$$= 0.1386 \text{ m} = 138.6 \text{ mm} \approx 139 \text{ mm}$$

7.9 (d)

Initial effective stress at the centre of compressible clay layer,

$$\bar{\sigma} = 18 \times 1 + 3 \times (20 - 9.81)$$
$$= 48.57 \text{ kN/m}^2$$

(Assuming load dispersion as 2V : 1H)

Width of strip at mid of clay layer i.e., 4 m from ground surface,

$$B' = 1.2 + 2 \times 0.5 \times 3 = 4.2 \text{ m}$$

Increase in pressure at the centre of clay layer,

$$\bar{\sigma}_0 = \frac{40 \times 1.2}{4.2} = 11.43 \text{ kN/m}^2$$

$\Rightarrow$ Settlement of footing,

$$\Delta H = \frac{C_c H_o}{1 + e_o} \log_{10}\left(\frac{\bar{\sigma} + \bar{\sigma}_0}{\bar{\sigma}}\right)$$

$$= \frac{0.08 \times 6000}{1 + 0.4} \log_{10}\left(\frac{48 + 11.43}{48}\right) = 31.8 \text{ mm}$$

7.10 (b)

Elastic settlement is given by

$$S_i = qB\left(\frac{1 - \mu^2}{E}\right) I_f$$

where,

$$\therefore \quad S_i = 40 \times 1.2 \times \left[\frac{1 - (0.4)^2}{50 \times 10^3}\right] \times 1.75$$

$= 1.41$ mm

7.11 (b)

The time required for any degree of consolidation, can be obtained from the **Taylor's formula**,

$$T_v = \frac{C_v t}{d^2}$$

Now, C_v is constant for same type of soils, T_v will also remain constant for same degree of consolidation. Thus, $t \propto d^2$ where d is length of drainage path.

Hence, in the clay layer of 15 m thickness, two way drainage will be four times faster in comparison to one way drainage.

7.12 (b)

Square root of tme fitting method and casagrande's logarithm time fitting method is used to find out coefficient of consolidation, C_v.

7.13 (b)

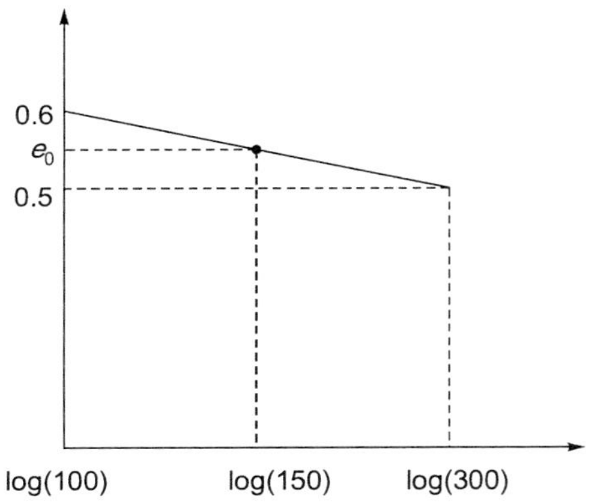

$$\frac{0.6 - e_0}{\log 150 - \log 100} = \frac{e_0 - 0.5}{\log 300 - \log 150}$$

$$e_0 = 0.563$$

7.14 (d)

$$\Delta H = \frac{C_c H_0}{1+e_0}\log_{10}\left[\frac{\bar{\sigma}_0 + \Delta\bar{\sigma}}{\bar{\sigma}_0}\right]$$

$$= \left(\frac{0.2096 \times 10}{1+0.563}\right)\log_{10}\left[\frac{150+200}{150}\right]$$

$$= 0.493 \text{ m} \approx 0.5 \text{ m}$$

7.15 (c)

Initial total, pore and effective stresses at central layer are,

$\sigma_{c-c} = 5 \times 18 + 15 \times 20 + 2.5 \times 20 = 440$ kN/m^2

$u_{c-c} = (15 + 2 - 5) \times 10 = 175$ kN/m^2

$(\sigma')_1 = 265$ kN/m^2

After the drawdown of GWT, changes in total and pore pressure at central layer are as follows,

$\sigma_{c-c} = 15 \times 18 + 5 \times 20 + 2.5 \times 20$

$= 420$ kN/m^2

$u_{c-c} = 7.5 \times 10 = 75$ kN/m^2

$(\sigma')_2 = 345$ kN/m^2

$\Rightarrow \quad \Delta\sigma' = 345 - 265$

$\Delta\sigma' = 80$ kN/m^2

7.16 (b)

$$\Delta H = H_0 m_v \Delta\sigma'$$

$$= \frac{5 \times 0.25}{10^3} \times 80 = 100 \text{ mm}$$

7.17 (b)

From Taylor's formula, we have,

$$T_v = \frac{C_v t}{d^2}$$

Here T_v and C_v are constant

[For same soil and same degree of consolidation]

$\therefore \quad t \propto d^2$

$$\Rightarrow \quad \frac{t_1}{t_2} = \left(\frac{d_1}{d_2}\right)^2$$

$$\Rightarrow \quad \frac{16}{t_2} = \left(\frac{H}{H/2}\right)^2$$

$$\Rightarrow \quad t_2 = \frac{16}{4}$$

$$\Rightarrow \quad t_2 = 4 \text{ years}$$

7.18 (c)

$$S \times e = G \times w$$

$$\Rightarrow \quad 1 \times e = \frac{2.7 \times 22.22}{100} \quad [\because \text{ soil is saturated}]$$

$$\Rightarrow \quad e = 0.5999$$

$$\Rightarrow \quad e \approx 0.6$$

$$\therefore \quad \gamma_{sat} = \left(\frac{G+e}{1+e}\right)\gamma_w$$

$$= \left(\frac{2.7+0.6}{1+0.6}\right) \times 10$$

$$= 20.625 \text{ kN/m}^3$$

7.19 (b)

The consolidation settlement at the centre of clay layer is given by,

$$\Delta H = \frac{C_c H_0}{1+e_0}\log_{10}\left[\frac{\bar{\sigma}_0 + \Delta\bar{\sigma}}{\bar{\sigma}_0}\right]$$

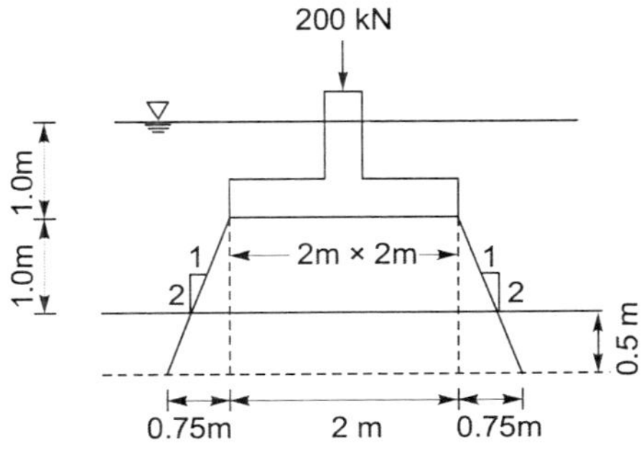

Load distribution dimensions at the centre of clay layer

$= 2 + 0.75 + 0.75 = 3.5$ m

$\therefore$ Increase in stress due to load

$$= \frac{200}{3.5 \times 3.5} = 16.33 \text{ kN/m}^2$$

Effective stress at the centre of clay layer

$= (18 - 10) \times 2 + (20.625 - 10) \times 0.5$

$= 16 + 5.31 = 21.31 \text{ kN/m}^2$

$$\therefore \Delta H = \frac{0.4 \times 1}{1 + 0.6} \times \log_{10}\left[\frac{21.31 + 16.33}{21.31}\right]$$

$= 0.06176 \text{ m} = 61.67 \text{ mm}$

7.20 (b)

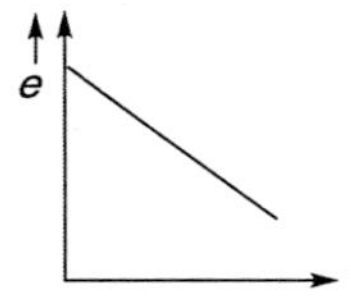

Log p→ Normally consolidated clay

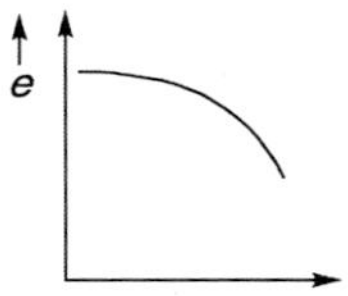

Log p→ over consolidated clay

7.21 (c)

- Site y will consolidate fully when lower soil will consolidated fully.
- Upper layer $d = 5/2$ m $= 2.5$ m
- Lower layer $= d = 5$ m

The time required for any degree of consolidation can be obtained from the **Taylor's formula,**

$$T_v = \frac{C_v t}{d^2}$$

Since C_v is constant for same type of soil and T_v is constant for same degree of consolidation, $t \propto d^2$ where d is the length of drainage path.

$$\frac{t_1}{t_2} = \left(\frac{d_1}{d_2}\right)^2$$

$$\Rightarrow \quad \frac{36}{t_2} = \left(\frac{10}{10/2}\right)^2$$

$$\Rightarrow \quad t_2 = \frac{36}{4}$$

$$\Rightarrow \quad t_2 = 9 \text{ months}$$

7.22 (d)

Given : $\Delta\bar{\sigma} = 20$ kPa, $\bar{\sigma}_0 = 100$ kPa

$G = 2.7$

$w = 45\%$

$C_c = 0.45$

$C_r = 0.05$

$H_0 = 1$ m

$$\because \quad \Delta H = \frac{C_c H_0}{1 + e_0} \log_{10}\left[\frac{\bar{\sigma}_0 + \Delta\bar{\sigma}}{\bar{\sigma}_0}\right]$$

Also, $wG = eS$

$\Rightarrow \quad 0.45 \times 2.7 = e.1$

$\Rightarrow \quad e = 1.215$

$$\therefore \quad \Delta H = \frac{0.45 \times 1}{1 + 1.215} \log_{10}\left(\frac{(100 + 20)}{100}\right)$$

$$= \frac{0.45}{2.215} \log_{10}\left(\frac{120}{100}\right)$$

$= 0.016 \text{ m} = 16 \text{ mm}$

7.23 Sol.

$$m_v = \frac{a_v}{1 + e_0} = \frac{\frac{\Delta e}{\Delta\bar{\sigma}}}{1 + e_0} = \frac{(1.1 - 0.9)}{125 \times 2.1}$$

$= 7.619 \times 10^{-4} \text{ m}^2/\text{kN}$

7.24 Sol.

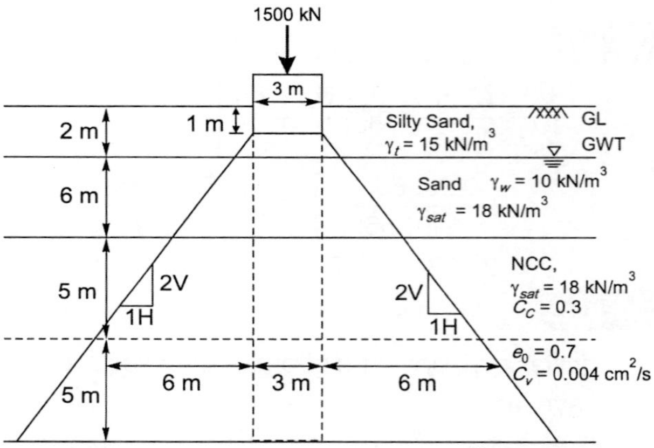

$$\Delta H = \frac{C_c H_0}{1 + e_0} \log_{10}\left[\frac{\bar{\sigma}_0 + \Delta\bar{\sigma}}{\bar{\sigma}_0}\right]$$

H_0 = Thickness of clay layer

$= 10$ m

$\bar{\sigma}_0$ = Effective stress before application of load at the centre of clay layer

$\bar{\sigma}_0 = (15 \times 2) + (18 - 10) \times 6 + (18 - 10) \times 5$

$= 118 \text{ kN/m}^2$

$\Delta\bar{\sigma}$ = Increase in effective stress due to the structural weight at the centre of clay layer

$$\Delta\bar{\sigma} = \frac{1500}{\frac{\pi}{4}\times(3+12)^2} = 8.488\,\text{kN/m}^2$$

$$\Rightarrow \quad \Delta H = \frac{0.3\times 10}{1+0.7}\log_{10}\left(\frac{118+8.488}{118}\right)$$

$$= 0.053236\text{ m} = 53.236\text{ mm}$$

7.25 Sol.

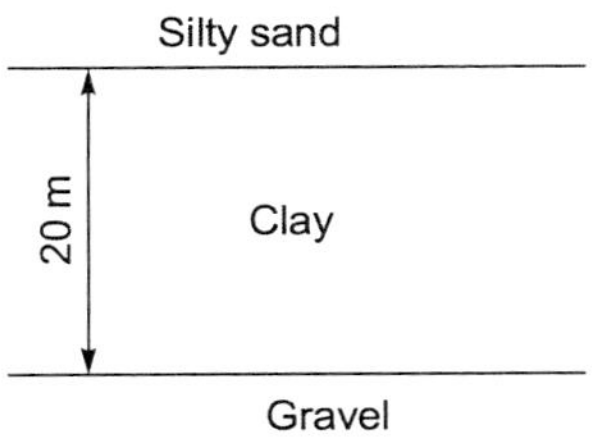

$$T_v = \frac{C_v t}{d^2}$$

$$\Rightarrow \quad T_v = \frac{0.003\times 2\times 86400\times 365}{\left(\frac{20}{2}\times 100\right)^2}$$

$$T_v = 0.1892 < 0.02827$$

$$T_v = \frac{\pi}{4}U^2 = 0.1892 \qquad (\therefore U < 60\%)$$

$$\Rightarrow \quad U = 0.4908$$

$\Rightarrow$ Total consolidation

$$= \frac{30}{0.4908} = 61.12\text{ mm}$$

Degree of consolidation for 50 mm settlement,

$$U = \frac{50}{61.12} = 0.818 = 81.8\%$$

$$\Rightarrow \quad T_v = 0.60535 = \frac{C_v t}{H^2}$$

$$\Rightarrow \quad t = \frac{0.60535\times(10)^2}{0.003\times 10^{-4}} = 6.398\text{ years}$$

$\Rightarrow$ Additional number of years = 4.398

7.26 Sol.

$$e = 1.3$$

$$C_C = 0.6$$

$$C_V = 1\text{ m}^2\text{/year}$$

$$\Delta H = \frac{C_c H_0}{1+e_0}\log_{10}\left[\frac{\bar{\sigma}_0+\Delta\bar{\sigma}}{\bar{\sigma}_0}\right]$$

As increase in vertical pressure ($\Delta\sigma$) is equal to existing effective over burden pressure $(\bar{\sigma}_0)$,

$$= \frac{0.6\times 4}{1+1.3}\log_{10}\left(\frac{\bar{\sigma}_0+\bar{\sigma}_0}{\bar{\sigma}_0}\right)$$

$$= \frac{0.6\times 4}{2.3}\log_{10}2$$

$$= 0.314\text{ m} = 314\text{ mm}$$

7.27 Sol.

Unit weight of clay,

$$\gamma_{sat} = \frac{G+e}{1+e}\times\gamma_w$$

$$\because \quad Se = wG$$

$$e = \frac{wG}{S} = \frac{wG}{1} = 0.25\times 2.7 = 0.675$$

$$\therefore \quad \gamma_{sat} = \frac{2.7+0.675}{1+0.675}\times 10 = 120.15\text{ kN/m}^3$$

Present effective stress at the middle of clay layer,

$$\bar{\sigma}_0 = \sigma - U = 1\times\gamma_{sand} + 0.5(\gamma_{sat}-\gamma_w)$$

$$= 1\times 17 + 0.5(20.15-10)$$

$$= 22.075\text{ kN/m}^2$$

Change in effective stress due to compacted fill

$$\Delta\bar{\sigma}_0 = 2.5\times 20 = 50\text{ kN/m}^2$$

Since pre-consolidation pressure is 60 kPa, hence from 22.075 to 60 kPa

Over consolidation stage

from 60 kPa to (20.075 + 60 = 72.075)

Normal consolidated stage,

Hence consolidation settlement,

$$\Delta H = \frac{H_o c_r}{1+e_0}\log\left(\frac{\bar{\sigma}_1}{\bar{\sigma}_0}\right) + \frac{H_o c_c}{1+e_0}\log\left(\frac{\bar{\sigma}_2}{\bar{\sigma}_1}\right)$$

$$\Delta H = \frac{1\times 0.05}{1+0.675}\log\left(\frac{60}{22.075}\right) + \frac{1\times 0.50}{1+0.675}\log\left(\frac{72.075}{60}\right)$$

$$= 0.0367\text{ m} = 36.7\text{ mm}$$

7.28 (b)

$$\text{OCR} = \frac{\text{Maximum effective stress in past}}{\text{Maximum effective stress in present}}$$

Maximum effective stress in present

$$= 10\,\gamma_{sat} - 10\,\gamma_w$$

$$= 10\times 16 - 10\times 10 = 60\text{ kN/m}^2$$

$\therefore \quad OCR = \frac{90}{60} = 1.5$

7.29 Sol.

As per **Terzaghi's 1–D consolidation theory.**

$$T_v = C_v \frac{t}{d^2}$$

One way drainage condition d = 3 m

For 90% consolidation,

$$(T_v)_{90} = C_v \cdot \frac{t_{90}}{d^2}$$

$$0.85 = 3 \times 10^{-6} \times \frac{t_{90}}{(3)^2}$$

$$t_{90} = \frac{0.85 \times 9}{3 \times 10^{-6}} \text{ min}$$

$$t_{90} = 1770.83 \simeq 1771 \text{ days}$$

7.30 (c)

Pore pressure parameter B lies in between 0 to 1 and **pore pressure parameter A** may be as low as –0.5 and may be as high as 3.

7.31 Sol.

$$\Delta H = \frac{H_0 \Delta e}{1+e_0} = \frac{H_0 C_c}{1+e_0} \log\left(\frac{\sigma_0 + \Delta\sigma}{\sigma_0}\right)$$

$$\Delta H = H_0\left(\frac{0.55-0.40}{1+0.50}\right) = \frac{H_0 \times 0.25}{1+0.50} \log\left(\frac{140+\Delta\sigma}{140}\right)$$

$$\frac{0.15}{0.25} = \frac{3}{5} = \log\left(\frac{140+\Delta\sigma}{140}\right)$$

$$\Delta\sigma = 417.35 \text{ kPa}$$

7.32 Sol.

Compression Index

$$C_c = \frac{\Delta e}{\log\left(\frac{\bar{\sigma}_1}{\bar{\sigma}_0}\right)} = \frac{1.2-0.6}{\log\left(\frac{800}{50}\right)}$$

$$C_c = \frac{0.6}{\log(16)} = \frac{0.6}{1.204} = 0.4982$$

7.33 Sol.

Jan. 2010 – Jan. 2014 (4 years)

⇒ Settlement is 10 mm

ΔH = 40 mm

2 – way drainage

Settlement in Jan. 2019 (in 9 years) = ?

In 4 years,

$$\%U = \frac{\Delta h}{\Delta H} \times 100 = \frac{10}{40} \times 100$$

$$= 25\%$$

$$T_v = C_v \frac{t}{d^2} \qquad (\because \%U < 60\%)$$

$$\Rightarrow \quad T_v = \frac{\pi}{4} U^2$$

$$\frac{\pi}{4}\left(\frac{25}{100}\right)^2 = \frac{C_v}{d^2} \times 4 \text{years} \qquad \text{...(i)}$$

In 9 years

$$T_v = \frac{C_v t}{d^2} = \left\{\frac{\pi}{4} \times \frac{(0.25)^2}{4}\right\} \times 9 \text{years}$$

$$= 0.11044$$

$\because \quad (T_v)_{60} = 0.283$

$\therefore \quad \%U < 60\%$

$$\therefore \quad T_v = 0.11044 = \frac{\pi}{4} U^2$$

$$\%U = 0.3749 = 37.499\%$$

$$\therefore \quad \%U = \frac{\Delta h}{\Delta H} \times 100$$

$$= \frac{\Delta h}{40 \text{mm}} \times 100 = 37.499$$

$$\Delta h = 14.99 \text{ mm} = 15 \text{ mm}$$

7.34 Sol.

$$\Delta\bar{\sigma} = 2 \times 20 = 40 \text{ kN/m}^2$$

$$\Delta H = m_v \Delta\bar{\sigma} H$$

$$= 2.2 \times 10^{-4} \times 40 \times 10 \times 10^3 \text{ mm}$$

$$= 88 \text{ mm}$$

$$T_v = \frac{C \times t}{H^2} = \frac{9 \times 10^{-2} \times 10}{5^2} = 0.036$$

$$T_v = \frac{\pi}{4} U^2$$

$$U = \sqrt{\frac{0.036 \times 4}{\pi}} = 0.214$$

Δh after 10 years = 0.214 × 88 = 18.832 mm

7.35 Sol.

Oedometer reading = 2.1 mm

16 mm thick, $e_f = 0.62$

$$\frac{\Delta H}{H_0} = \frac{\Delta e}{1+e_0} = \frac{e_0 - e_f}{1+e_0}$$

$$\frac{2.1\text{ mm}}{19\text{ mm}} = \frac{e_0 - 0.62}{1+e_0}$$

$$e_0 = 0.82$$

7.36 Sol.

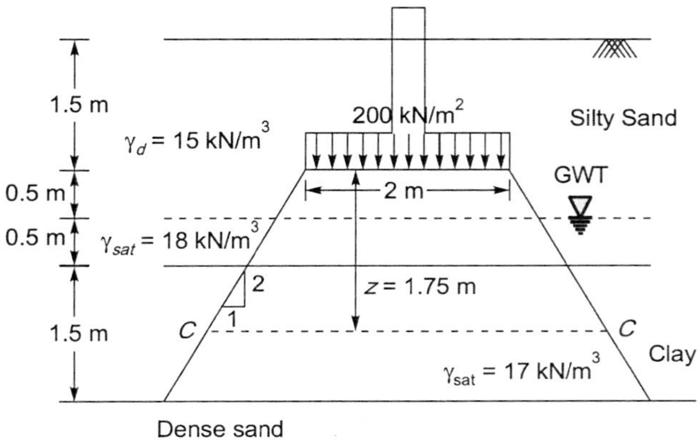

For clay layer,

$$\gamma_{sat} = \left(\frac{G+e}{1+e}\right)\gamma_w$$

$$\Rightarrow \left(\frac{2.65+e}{1+e}\right)10 = 17 \text{ kN/m}^3$$

$$e_0 = 1.357$$

1. $H_0 = 1.5$ m
2. $(\bar{\sigma}_0)_{c-c} = [2\gamma_d + 0.5\gamma_{sat} + 0.75\gamma_{sat}] - 1.25\gamma_w$
   $= (2 \times 15 + 0.5 \times 18 + 0.75 \times 17) - 1.25 \times 10$
   $= 39.25$ kN/m^2
3. $$\Delta\bar{\sigma} = \frac{\text{Force}}{\text{Area}} = \frac{qB^2}{(B+2nZ)^2}$$
   $$= \frac{200 \times 2 \times 2}{\left(2 + 2 \times \frac{1}{2} \times 1.75\right)^2} = 56.88 \text{ kN/m}^2$$
4. $$\Delta H = \frac{H_0 C_C}{1+e_0}\log\left(\frac{\bar{\sigma}_0 + \Delta\bar{\sigma}}{\bar{\sigma}_0}\right)$$
   $$= \frac{1.5 \times 0.3}{1+1.357}\log\left(\frac{39.25+56.88}{39.25}\right)$$
   $$= 74.27 \text{ mm}$$

7.37 (a)

Stress is increased suddenly, hence entire change will be taken by water $\Delta\bar{\sigma} = \Delta U = 100$ kPa. There will be no change in effective stress

$\therefore \quad \bar{\sigma} = 50$ kPa

7.38 Sol.

$$\text{OCR} = \frac{\bar{\sigma}_c}{\bar{\sigma}_0}$$

$\bar{\sigma}_c$ = Preconsolidation stress

$\bar{\sigma}_0$ = Present effective stress

$\bar{\sigma}_c = 17.5 \times 10 + (18.5 - 9.81) \times 5 + (17 - 9.81) \times 4$
$= 247.21$ kN/m^2

$\bar{\sigma}_0 = 17.5 \times 5 + (18.5 - 9.81) \times 10 + (17 - 9.81) \times 4$
$= 203.16$ kN/m^2

$$\text{OCR} = \frac{247.21}{203.16} = 1.22$$

7.39 Sol.

Primary consolidation settlement

$$\Delta H = \frac{C_c H}{1+e_0}\log_{10}\left(\frac{\bar{\sigma}_0 + \Delta\bar{\sigma}}{\bar{\sigma}_0}\right)$$

$\bar{\sigma}_0 + \Delta\bar{\sigma}$ = Effective stress at the center of clay layer after embankment loading
= 180 kN/m^2

$\bar{\sigma}_0$ = Effective stress at the centre of clay layer before embankment loading

γ_{sub} of clay layer

$$= \frac{G-1}{1+e}\gamma_w = \frac{(2.65-1)}{1+\frac{wG}{1}} \times 9.81$$

$$= \frac{1.65 \times 9.81}{1+0.45 \times 2.65}$$

$= 7.383$ kN/m^3

$\bar{\sigma}_0 = (18.5 \times 2) + (7.383) \times 3$
$= 59.149$ kNm/^{2}

$$e_0 = \frac{wG}{1} = 0.45 \times 2.65 = 1.1925$$

$$\Rightarrow \quad \Delta H = \frac{0.25 \times 6}{1+1.1925} \log_{10}\left(\frac{180}{59.149}\right)$$

$$= 0.33 \text{ m}$$

7.40 Sol.

$$m_v = \frac{a_v}{1+e_0} = \frac{\Delta e}{(1+e_0)\times(\Delta\bar{\sigma})}$$

($\Delta\bar{\sigma}$ is same for both M and N)

$$m_{v1} = \frac{0.575-0.510}{(1+0.575)\times\Delta\bar{\sigma}}$$

$$m_{v2} = \frac{0.600-0.550}{(1+0.600)\times\Delta\bar{\sigma}}$$

$$\frac{C_{v_1}}{C_{v_2}} = \frac{\dfrac{k_1}{m_{v_1}\gamma_w}}{\dfrac{k_2}{m_{v_2}\gamma_w}} = \frac{k_1}{k_2}\times\frac{m_{v_2}}{m_{v_1}}$$

$$= 0.125\left(\frac{1.575}{1.6}\right)\times\left(\frac{0.60-0.55}{0.575-0.510}\right)$$

$$= 0.0947 \simeq 0.095$$

7.41 (b)

If preconsolidation stress, P_c is greater than p_0 but less than $(p_0 + \Delta p)$, then settlement is computed in two parts:

1. Settlement for pressure, p_0 to p_c.
2. Settlement for pressure, p_c to $(p_0 + \Delta p)$.

7.42 (a)

For normally consolidated soil, OCR = 1
For quick clay, Sensitivity > 16
For sand in critical state, Dilation angle is zero
For clay of high plasticity, $w_L > 50$

7.43 Sol.

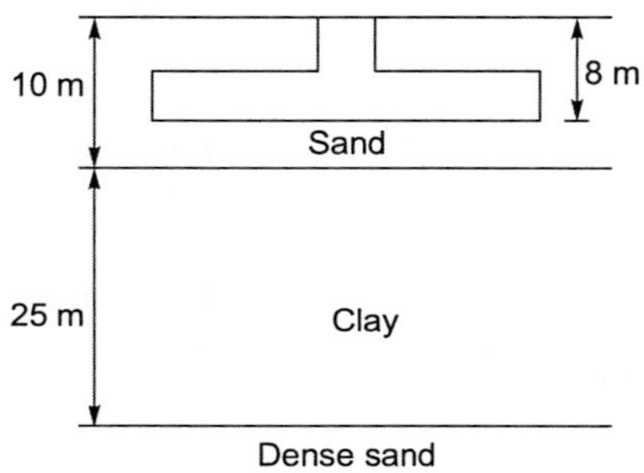

Lab	Field
2 way	1 way
$d = \frac{H_0}{2} = \frac{25}{2}$ mm	
50% → 10 min	25% → t

$$(T_v)_{50\%} = C_v \frac{t}{d^2} \quad \text{(from lab)}$$

$$\Rightarrow \quad \frac{\pi}{4}(0.5)^2 = C_v \times \frac{10 \text{ min}}{\left(\frac{25}{2}\times 10^{-3}\right)^2}$$

$$\Rightarrow \quad C_v = \frac{\frac{\pi}{4}(0.5)^2 \times (12.5\times10^{-3})^2}{10} \quad \text{...(i)}$$

$$(T_v)_{25\%} = C_v \frac{t}{d^2} \quad \text{(from field)}$$

$$\Rightarrow \quad \frac{\pi}{4}(0.25)^2 = C_v \frac{t}{(12.5)^2} \quad \text{...(ii)}$$

Since the soil is same, therefore C_v is same, From eq. (i) and (ii),

$$\Rightarrow \quad \frac{\pi}{4}(0.25)^2 = \frac{\frac{\pi}{4}(0.5)^2 \times (12.5\times10^{-3})^2}{10} \times \frac{t}{(12.5)^2}$$

$$\Rightarrow \quad t = 2500000 \text{ min} = 1736 \text{ days}$$

7.44 Sol.

We know degree of consolidation,

$$\%U = \frac{u_i - u}{u_i}\times 100$$

Initial excess pore water pressure,

$$u_i = 100 \text{ kPa}$$

Initial total stress,

$$\sigma = 150 \text{ kPa}$$

Initial effective stress,

$$\sigma_0' = \sigma - u = 150 - 25 = 125 \text{ kPa}$$

Total stress after construction of building

$$= 150 + 100 = 250 \text{ kPa}$$

When the vertical effective stress at P is 175 kPa,

the change in effective stress is 175 – 125 = 50 kPa, so during this, the change in pore water pressure is

$$\therefore \quad \Delta\sigma = \Delta\sigma' + \Delta u$$

$$\Rightarrow (250 - 150) = (175 - 125) + \Delta u$$

$$\Rightarrow \quad \Delta u = 50 \text{ kPa}$$

$$\therefore \quad \%U = \frac{100 - 50}{100} \times 100 = 50\%$$

7.45 Sol.

As we know,

$(T_v)_{50} = \frac{C_v t}{d^2}$ where, C_v is coefficient of consolidation

$$\frac{\pi}{4}(0.5)^2 = \frac{C_v}{d^2} \times t_2 \quad \text{...(1)}$$

For 10 mm settlement,

$$\%U = \frac{\Delta h}{\Delta H} \times 100 = \frac{10}{50} \times 100\% = 20\%$$

$$(T_v)_{20} = \frac{C_v}{d^2} \times t$$

$$\Rightarrow \quad \frac{\pi}{4}(0.2)^2 = \frac{C_v}{d^2} \times 300 \quad \text{...(2)}$$

From (1) and (2)

$$\frac{\frac{\pi}{4}(0.5)^2}{\frac{\pi}{4}(0.2)^2} = \frac{\frac{C_v}{d^2} \times t_2}{\frac{C_v}{d^2} \times 300}$$

$$\Rightarrow \quad t_2 = \left(\frac{0.5}{0.2}\right)^2 \times 300 = 1875 \text{ days}$$

Additional number of days = (1875 – 300) days = 1575 days

■■■■

8 Shear Strength of Soil

8.1 The unconfined compressive strength of a 'stiff clay' falls in the range
(a) less than 50 kN/m^2
(b) 50 to 100 kN/m^2
(c) 100 to 200 kN/m^2
(d) Above 200 kN/m^2 [1995 : 1 M]

8.2 The appropriate field test to determine the undrained shear strength of soft clay is
(a) plate load test
(b) static cone penetration test
(c) standard penetration test
(d) vane shear test [1995 : 1 M]

8.3 Vane shear test is normally used for determining in situ shear strength of
(a) soft clays (b) sand
(c) stiff clays (d) gravel [1997 : 1 M]

8.4 What is the shear strength in terms of effective stress on a plane within a saturated soil mass at a point where the total normal stress is 295 kPa and the pore water pressure is 120 kPa? The effective stress and shear strength parameters are c' = 12 kPa and ϕ' = 30°. [1998 : 2 M]

8.5 In a drained triaxial compression test, a saturated specimen of a cohesionless sand fails under a deviatoric stress of 3 kgf/cm^2 when the cell pressure is 1 kgf/cm^2. The effective angle of shearing resistance of sand is about
(a) 37° (b) 45°
(c) 53° (d) 20° [2000 : 1 M]

8.6 The following two statements are made with respect to different sand samples having the same relative density. Identify if they are true or false
I. Poorly graded sands will have lower friction angle than the well graded sands.
II. The particle size has no influence on the friction angle of sand.
(a) II is true but I is false
(b) Both are false statements
(c) Both are true statements
(d) I is true but II is false [2001 : 1 M]

8.7 If the effective shear stress strength parameters of a soil are c' = 10 kPa and ϕ' = 30°, the shear strength on a plane within the saturated soil mass at a point where the total normal stress is 300 kPa and pore water pressure is 150 kPa will be
(a) 90.5 kPa (b) 96.6 kPa
(c) 101.5 kPa (d) 105.5 kPa
[2002 : 2 M]

8.8 In a tri-axial test carried out on a cohesionless soil sample with a cell pressure of 20 kPa, the observed value of applied stress at the point of failure was 40 kPa. The angle of internal friction of the soil is
(a) 10° (b) 15°
(c) 25° (d) 30° [2002 : 2 M]

8.9 In an undrained triaxial test on a saturated clay, the Poisson's ratio is
(a) $\frac{\sigma_3}{\sigma_1+\sigma_3}$ (b) $\frac{\sigma_3}{\sigma_1-\sigma_3}$
(c) $\frac{\sigma_1-\sigma_3}{\sigma_3}$ (d) $\frac{\sigma_1+\sigma_3}{\sigma_3}$
[2004 : 1 M]

8.10 For a triaxial shear test conducted on a sand specimen at a confining pressure of 100 kN/m^2 under drained conditions, resulted in a deviator stress ($\sigma_1 - \sigma_3$) at failure of 100 kN/m^2. The angle of shearing resistance of the soil would be
(a) 18.43° (b) 19.47°
(c) 26.56° (d) 30° [2005 : 2 M]

8.11 A sample of saturated cohesionless soil tested in a drained triaxial compression test showed an angle of internal friction of 30°. The deviatoric stress at failure for the sample at a confining pressure of 200 kPa is equal to
(a) 200 kPa (b) 400 kPa
(c) 600 kPa (d) 800 kPa [2006 : 2 M]

8.12 A clay soil sample is tested in a triaxial apparatus in consolidated-drained conditions at a cell pressure of 100 kN/m^2. What will be the pore water pressure at a deviator stress of 40 kN/m^2?
(a) 0 (b) 20 kN/m^2
(c) 40 kN/m^2 (d) 60 kN/m^2
[2007 : 1 M]

8.13 A direct shear test was conducted on a cohesionless soil ($c = 0$) specimen under a normal stress of 200 kN/m^2. The specimen failed at a shear stress of 100 kN/m^2. The angle of internal friction of the soil (degrees) is

(a) 26.6 (b) 29.5
(c) 30.0 (d) 32.6 [2008 : 2 M]

8.14 For a sample of dry, cohesionless soil with friction angle, ϕ, the failure plane will be inclined to the major principal plane by an angle equal to

(a) ϕ (b) 45°
(c) $45° - \phi/2$ (d) $45° + \phi/2$
[2011 : 1 M]

8.15 A field vane shear testing instrument (shown alongside) was inserted completely into a deposit of soft, saturated silty clay with the vane rod vertical such that the top of the blades were 500 mm below the ground surface. Upon application of a rapidly increasing torque about the vane rod, the soil was found to fail when the torque reached 4.6 N-m. Assuming mobilization of undrained shear strength on all failure surfaces to be uniform and the resistance mobilized on the surface of the vane rod to be negligible, what would be the peak undrained shear strength (rounded off to the nearest integer value of kPa) of the soil?

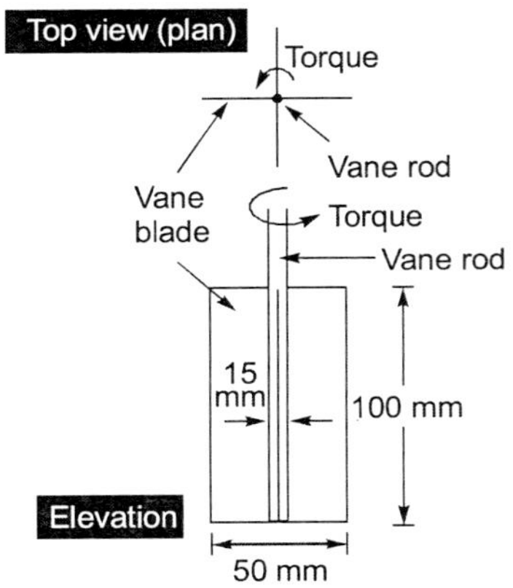

(a) 5 kPa (b) 10 kPa
(c) 15 kPa (d) 20 kPa [2011 : 2 M]

8.16 The effective stress friction angle of a saturated, cohesionless soil is 38°. The ratio of shear stress to normal effective stress on the failure plane is

(a) 0.781 (b) 0.616
(c) 0.488 (d) 0.438 [2012 : 1 M]

8.17 For a saturated cohesive soil, a triaxial test yields the angle of internal friction (ϕ) as zero. The conducted test is

(a) Consolidated Drained (CD) test
(b) Consolidated Undrained (CU) test
(c) Unconfined Compression (UC) test
(d) Unconsolidated Undrained (UU) test
[2014 : 1 M, Set-I]

8.18 Stress path equation for tri-axial test upon application of deviatoric stress is, $q = 10\sqrt{3} + 0.5p$. The respective values of cohesion, c (in kPa) and angle of internal friction, ϕ are:

(a) 20 and 20° (b) 20 and 30°
(c) 30 and 30° (d) 30 and 20°
[2015 : 2 M, Set-II]

8.19 Which of the following statements is NOT correct?

(a) Loose sand exhibits contractive behavior upon shearing
(b) Dense sand when sheared under undrained condition, may lead to generation of negative pore pressure
(c) Black cotton soil exhibits expansive behavior
(d) Liquefaction is the phenomenon where cohesionless soil near the downstream side of dams or sheet-piles loses its shear strength due to high upward hydraulic gradient
[2015 : 1 M, Set-I]

8.20 In an unconsolidated undrained triaxial test, it is observed that an increase in cell pressure from 150 kPa to 250 kPa leads to a pore pressure increase of 80 kPa. It is further observed that, an increase of 50 kPa in deviatoric stress results in an increase of 25 kPa in the pore pressure. The value of *Skempton's pore pressure parameter B* is

(a) 0.5 (b) 0.625
(c) 0.8 (d) 1.0
[2015 : 1 M, Set-I]

8.21 In the consolidated undrained triaxial test on a saturated soil sample, the pore water pressure is zero

(a) during shearing stage only
(b) at the end of consolidation stage only
(c) both at the end of consolidation and during shearing stages
(d) under none of the above conditions
[2016 : 1 M, Set-I]

8.22 A drained triaxial compression test on a saturated clay yielded the effective shear strength parameters as $c' = 15$ kPa and $\phi' = 22°$. Consolidated Undrained triaxial test on an identical sample of this clay at a cell pressure of 200 kPa developed a pore water pressure of 150 kPa at failure. The deviator stress (expressed in kPa) at failure is______. [2016 : 2 M, Set-I]

8.23 A soil sample is subjected to a hydrostatic pressure, σ. The Mohr circle for any point in the soil sample would be

(a) a circle of radius σ and center at the origin
(b) a circle of radius σ and center at a distance σ from the origin
(c) a point at a distance σ from the origin
(d) a circle of diameter σ and center at the origin

[2017 : 1 M, Set-I]

8.24 **Group I** lists the type of gain or loss of strength in soils. **Group II** lists the property or process responsible for the loss or gain of strength in soils.

Group I

P. Regain of strength with time
Q. Loss of strength due to cyclic loading
R. Loss of strength due to upward seepage
S. Loss of strength due to remoulding

Group II

1. Boiling
2. Liquefaction
3. Thixotropy
4. Sensitivity

The correct match between **Group I** and **Group II** is

(a) P-4, Q-1, R-2, S-3 (b) P-3, Q-1, R-2, S-4
(c) P-3, Q-2, R-1, S-4 (d) P-4, Q-2, R-1, S-3

[2017 : 1 M, Set-I]

8.25 A consolidated undrained $(\overline{CU})$ triaxial compression test is conducted on a normally consolidated clay at a confining pressure of 100 kPa. The deviator stress at failure is 80 kPa, and the pore-water pressure measured at failure is 50 kPa. The effective angle of internal friction (in degrees, up to one decimal place) of the soil is

[2017 : 2 M, Set-I]

8.26 Following are the statements related to the stress path in a triaxial testing of soils:

P. If $\sigma_1 = \sigma_3$, the stress point lies at the origin of the *p*-*q* plot.
Q. If $\sigma_1 = \sigma_3$, the stress point lies on the *p*-axis of the *p*-*q* plot.
R. If $\sigma_1 > \sigma_3$, both the stress points *p* and *q* are positive.

for the above statements, the correct combination is

(a) P-False; Q-True; R-True
(b) P-True; Q-False; R-True
(c) P-False; Q-True; R-False
(d) P-True; Q-False; R-False [2017 : 2 M, Set-II]

8.27 A conventional drained triaxial compression test was conducted on a normally consolidated clay sample under an effective confining pressure of 200 KPa. The deviator stress at failure was found to be 400 kPa. An identical specimen of the same clay sample is isotropically consolidated to a confining pressure of 200 kPa and subjected to standard undrained triaxial compression test. If the deviator stress at failure is 150 kPa, the pore pressure developed (in kPa, up to one decimal place) is ________ [2018 : 2 M, Set-I]

8.28 The total horizontal and vertical stresses at a point *X* in a saturated sandy medium are 170 kPa and 300 kPa, respectively. The static pore-water pressure is 30 kPa. At failure, the excess pore-water pressure is measured to be 94.50 kPa. and the shear stresses on the vertical arid horizontal planes passing through the point *X* are zero. Effective cohesion is 0 kPa and effective angle of internal friction is 36°. The shear strength (in kPa, up to two decimal places) at point *X* is ________.

[2018 : 2 M, Set-II]

8.29 In a drained tri-axial compression test, a sample of sand fails at deviator stress of 150 kPa under confining pressure of 50 kPa. The angle of internal friction (in degree, round off to the nearest integer) of the sample, is ________.

[2020 : 1 M, Set-I]

8.30 The total stress paths corresponding to different loading conditions, for a soil specimen under the isotropically consolidated stress state (O), are shown below:

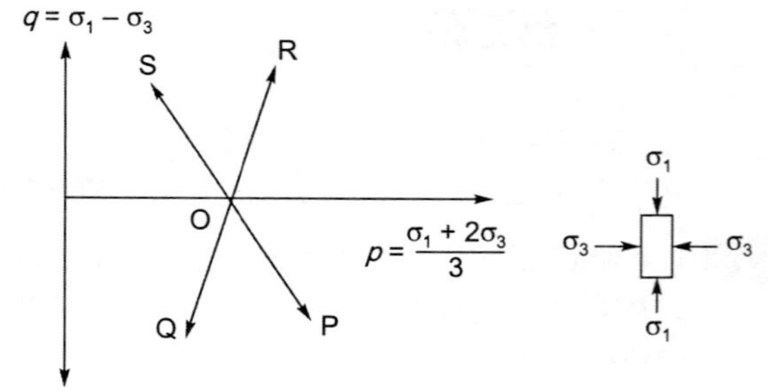

Stress Path	Loading Condition
OP	I. Compression loading (σ_1 – increasing; σ_3 – constant)
OQ	II. Compression unloading (σ_1 – constant; σ_3 – decreasing)
OR	III. Extension unloading (σ_1 – decreasing; σ_3 – constant)
OS	IV. Extension loading (σ_1 – constant; σ_3 – increasing)

The correct match between the stress paths and the listed loading conditions, is

(a) OP–I, OQ–II, OR–IV, OS–III
(b) OP–III, OQ–II, OR–I, OS–IV
(c) OP–IV, OQ–III, OR–I, OS–II
(d) OP–I, OQ–III, OR–II, OS–IV

[2020 : 2 M, Set-I]

8.31 Based on drained triaxial shear tests on sands and clays, the representative variations of volumetric strain (ΔV/V) with the shear strain (γ) is shown in the figure.

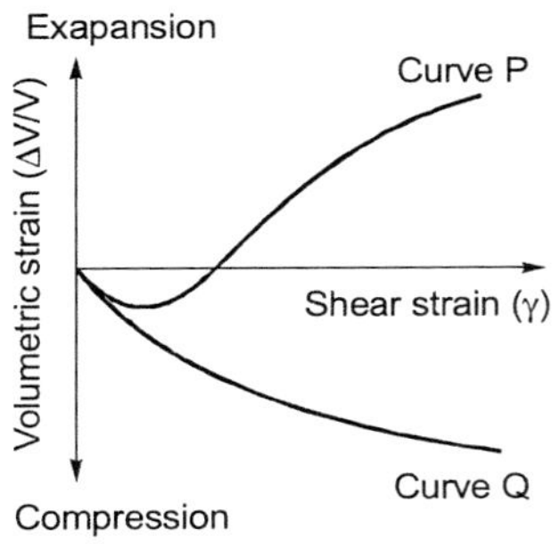

Choose the correct option regarding the representative behaviour exhibited by Curve P and Curve Q.

(a) Curve P represents loose sand and normally consolidated clay, while Curve Q represents dense sand and overc onsolidated clay.
(b) Curve P represents loose sand and over consolidated clay, while Curve Q represents dense sand and normally consolidated clay.
(c) Curve P represents dense sand and over consolidated clay while Curve Q represents loose sand and normally consolidated clay.
(d) Curve P represents dense sand and normally consolidated clay, while Curve Q represents loose sand and over consolidated clay.

[2021 : 2 M, Set-I]

8.32 The most appropriate triaxial test to assess the long-term stability of an excavated clay slope is

(a) consolidated undrained test
(b) unconsolidated undrained test
(c) unconfined compression test
(d) consolidated drained test

[2021 : 1 M, Set-II]

8.33 In a triaxial unconsolidated undrained (UU) test on a saturated clay sample, the cell pressure was 100 kPa, If the deviatoric stress at failure was 150 kPa, then the undrained shear strength of the soil is__________kPa. (in integer)

[2022 : 2 M, Set-II]

8.34 Match the following in Column X with Column Y:

Column X

P. In a triaxial compression test, with increase of axial strain in loose sand under drained shear condition, the volumetric strain.
Q. In a triaxial compression test, with increase of axial strain in loose sand under undrained shear condition, the excess pore water pressure.
R. In a triaxial compression test, the pore pressure parameter "B" for a saturated soil.
S. For shallow strip footing in pure saturated clay, Terzaghi's bearing capacity factor (N_q) due to surcharge.

Column Y

I. decreases
II. increase
III. remains same
IV. is always 0.0
V. is always 1.0
VI. is always 0.5

Which one of the following combinations is correct?

(a) (P)-(I), (Q)-(II), (R)-(V), (S)-(VI)
(b) (P)-(I), (Q)-(III), (R)-(VI), (S)-(IV)
(c) (P)-(I), (Q)-(I), (R)-(IV), (S)-(V)
(d) (P)-(I), (Q)-(II), (R)-(V), (S)-(V)

[2022 : 1 M, Set-II]

8.35 A drained direct shear test was carried out on a sandy soil. Under a normal stress of 50 kPa, the test specimen failed at a shear stress of 35 kPa. The angle of internal friction of the sample is ________degree (round off to the nearest integer).

[2023 : 1 M, Set-I]

8.36 An unconfined compression strength test was conducted on a cohesive soil. The test specimen failed at an axial stress of 76 kPa. The undrained cohesion (in kPa, *in integer*) of the soil is ______ .

[2023 : 1 M, Set-II]

8.37 In the given figure, Point *O* indicates the stress point of a soil element at initial non-hydrostatic stress condition. For the stress path (OP), which of the following loading conditions is correct?

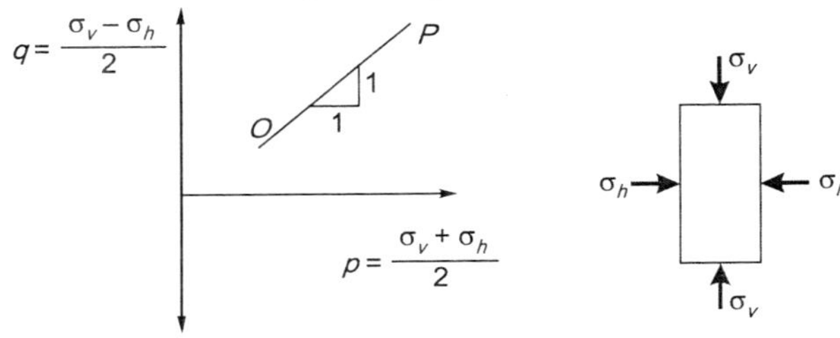

(a) σ_v is increasing and σ_h is constant
(b) σ_v is constant and σ_h is increasing
(c) σ_v is increasing and σ_h is decreasing
(d) σ_v is decreasing and σ_h is increasing

[2023 : 2 M, Set-II]

8.38 A consolidated drained (CD) triaxial test was carried out on a sand sample with the known effective shear strength parameters, $c' = 0$ and $\phi' = 30$. In the test, prior to the failure, when the sample was undergoing axial compression under constant cell pressure, the drainage valve was accidentally closed. At the failure, 360 kPa deviatoric stress was recorded along with 70 kPa pore water pressure. If the test is repeated without such error, and no back pressure is applied in either of the tests, what is the deviatoric stress (in kPa, in integer) at the failure? _______.

[2023 : 2 M, Set-II]

8.39 A soil sample was consolidated at a cell pressure of 20 kPa and a back pressure of 10 kPa for 24 hours during a consolidated undrained (CU) triaxial test. The cell pressure was increased to 30 kPa on the next day and it resulted in the development of pore water pressure of 1 kPa. The soil sample failed when the axial stress was gradually increased to 50 kPa. The pore water pressure at failure was recorded as 21 kPa. The value of skempton's pore pressure parameter *B* for the soil sample is ______ (round off to 2 decimal places).

[2024 : 2 M, Set-I]

8.40 Which one of the following saturated fine-grained soils can attain a negative Skempton's pore pressure coefficient (A)?
(a) Over-consolidated clays
(b) Lightly-consolidated clays
(c) Quick clays
(d) Normally-consolidated clays

[2024 : 1 M, Set-II]

8.41 A drained triaxial test was conducted on a saturated sand specimen using a stress-path triaxial testing system. The specimen failed when the axial stress reached a value of 100kN/m^2 from an initial confining pressure of 300kN/m^2.

The angle of shearing plane (in degrees) with respect to horizontal is ________ (rounded off to the nearest integer).

[2024 : 2 M, Set-II]

Answers Shear Strength of Soil

8.1	(c)	8.2	(d)	8.3	(a)	8.4	(113.04)	8.5	(a)	8.6	(d)	8.7	(b)
8.8	(d)	8.9	(a)	8.10	(b)	8.11	(b)	8.12	(a)	8.13	(a)	8.14	(d)
8.15	(b)	8.16	(a)	8.17	(d)	8.18	(b)	8.19	(d)	8.20	(c)	8.21	(b)
8.22	(104.37)	8.23	(c)	8.24	(c)	8.25	(26.4)	8.26	(a)	8.27	(125)	8.28	(52.52)
8.29	(37)	8.30	(c)	8.31	(c)	8.32	(d)	8.33	(75)	8.34	(d)	8.35	(35)
8.36	(38)	8.37	(a)	8.38	(500)	8.39	(0.1)	8.40	(a)	8.41	(49)		

Explanations Shear Strength of Soil

8.1 (c)

Consistency	q_u (kPa)
Very soft	0-24
Soft	24-48
Medium	48-96
Stiff	96-192
Very stiff	192-383
Hard	>383

8.2 (d)

Generally soft clays are very sensitive and obtaining undisturbed sample is very difficult. So, Shear strength of such soil is highly affected during sampling and handling. For such type of soils, **Vane shear test** can be done in the field.

8.3 (a)

A difficulty often encountered in determining the shearing resistance of **soft, saturated clay** deposits in the field is in obtaining undisturbed samples. The shear strength of such significant clays may be significantly altered during the process of sampling and handling. **Vane shear test** offers a method of overcoming this problem.

8.4 Sol.

Given data:

Total normal stress,

$$\sigma_n = 295 \text{ kPa}$$

Pore water pressure,

$$u = 120 \text{ kPa}$$

$$c' = 12 \text{ kPa and } \phi' = 30°$$

Effective stress,

$$\bar{\sigma}_n = \text{Total stress} - \text{Pore water pressure}$$

$$= \sigma - u = 295 - 120 = 175 \text{ kN/m}^2$$

$\therefore$ Shear strength,

$$\tau = c + \bar{\sigma}_n \tan\phi'$$

$$= 12 + 175 \tan 30° = \mathbf{113.04 \text{ kN/m}^2}$$

8.5 (a)

$$\sigma_1 - \sigma_3 = 3 \text{ kgf/cm}^2$$

$$\phi' = ?$$

Then $\sigma_1 = (\sigma_1 - \sigma_2) - \text{cell pressure}$

$$= 3 + 1 = 4$$

Also $\sigma_1 = \sigma_3 \tan^2\alpha$

$\Rightarrow$ $1 \times \tan^2\alpha = 4$

$\therefore$ $\tan\alpha = 2$

or $\alpha = 63.43°$

Now, $\alpha = 45° + \dfrac{\phi'}{2} = 63.43°$

$\Rightarrow$ $\phi' = 36.86 \simeq 37°$

8.6 (d)

A well graded soil with high coefficient of uniformity has a higher angle of friction as compared to poorly graded soil.

8.7 (b)

$$\bar{\sigma}_n = \sigma - u$$

$$= 300 - 150 = 150 \text{ kN/m}^2$$

$$\tau = c + \bar{\sigma}_n \tan\phi'$$

$$= 10 + (150) \tan 30° = 96.6 \text{ kPa}$$

8.8 (d)

Given, $\sigma_1 = \sigma_d + \sigma_3$

$$= 40 + 20 = 60 \text{ kPa}$$

$$\sigma_1 = \sigma_3 \tan^2\left(45° + \frac{\phi}{2}\right) + 2c\tan\left(45° + \frac{\phi}{2}\right)$$

Where $c = 0$

$$60 = 20\tan^2\left(45° + \frac{\phi}{2}\right)$$

$$\phi = 30°$$

8.9 (a)

In undrained triaxial test on saturated clay, the volume change due to confining pressure is zero.

$\Rightarrow$ $\varepsilon_3 = 0$

$$\frac{\sigma_3}{E} - \frac{\mu\sigma_1}{E} - \frac{\mu\sigma_3}{E} = 0 \qquad [\sigma_2 = \sigma_3]$$

$$\mu = \frac{\sigma_3}{\sigma_1 + \sigma_3}$$

8.10 (b)

We know that,

$$\sigma_3 = 100 \text{ kN/m}^2$$

$$\sigma_d = 100 \text{ kN/m}^2$$

$$\sigma_1 = \sigma_3 + \sigma_d$$

$$= 100 + 100 = 200 \text{ kN/m}^2$$

$$\sigma_1 = \sigma_3 \tan^2\left(45° + \frac{\phi}{2}\right) + 2c\tan\left(45° + \frac{\phi}{2}\right)$$

Where $c = 0$

$$200 = 100\tan^2\left(45° + \frac{\phi}{2}\right)$$

$$\phi = 19.47°$$

8.11 (b)

$$\sigma_1 = \sigma_3 \tan^2\left(45 + \frac{\phi}{2}\right) + 2c\tan\left(45 + \frac{\phi}{2}\right)$$

$$\Rightarrow \sigma_1 = 200\tan^2\left(45 + \frac{30}{2}\right) + 2 \times 0 \times \tan\left(45 + \frac{30}{2}\right)$$

$$\Rightarrow \sigma_1 = 600 \text{ kN/m}^2$$

$\therefore$ Deviatoric stress at failure

$$= \sigma_1 - \sigma_3 = 600 - 200 = 400 \text{ kPa}$$

8.12 (a)

In the consolidated drained test, the soil sample is first consolidated under an appropriate cell pressure. With the cell pressure kept at the same value, the soil sample is then sheared by applying the deviator stress so slowly that excess pore water pressure does not develop during the test. Thus, at any stage of the test, the total stresses are the effective stresses.

8.13 (a)

$$\tau = c + \bar{\sigma} \tan\phi \qquad [\because c = 0]$$

$\therefore$ $100 = 0 + 200 \tan\phi$

$\Rightarrow$ $\phi = \tan^{-1}(0.5)$

$\Rightarrow$ $\phi = 26.6°$

8.14 (d)

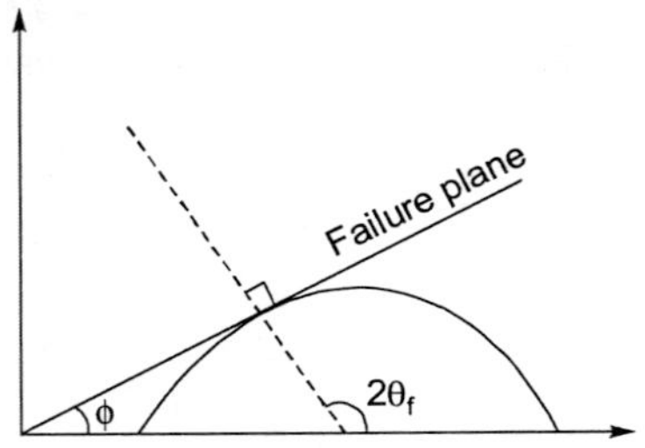

Angle between major principal plane and failure plane is θ_f from Mohr's circle

$$2\theta_f = 90° + \phi$$

$$\theta_f = 45° + \frac{\phi}{2}$$

8.15 (b)

When the top end of the vane completely shear the soil, Then the torque at failure is given by

$$\tau = \frac{T}{\pi D^2\left(\frac{H}{2}+\frac{D}{6}\right)} = c_u + \bar{\sigma}_n \tan\phi$$

$$(\phi = 0°)$$

$$T = c_u \pi\left(\frac{D^2 H}{2}+\frac{D^3}{6}\right)$$

$$\Rightarrow \frac{4.6}{1000} = c_u \times \pi \times \left(\frac{(0.05)^2 \times 0.1}{2}+\frac{(0.05)^3}{6}\right)$$

$$\Rightarrow \quad c_u = 10 \text{ kPa}$$

8.16 (a)

$$\because \quad \tau = c + \bar{\sigma}_n \tan\phi$$

For sands, $c = 0$

$$\Rightarrow \quad \tau = \bar{\sigma}_n \tan\phi$$

$$\therefore \quad \frac{\tau}{\sigma_n} = \tan\phi = \tan 38° = 0.781\text{s}$$

8.17 (d)

8.18 (b)

$$q = \frac{\sigma_1 - \sigma_3}{2}$$

$$p = \frac{\sigma_1 + \sigma_3}{2}$$

$$q = 10\sqrt{3} + 0.5p$$

$$\frac{\sigma_1 - \sigma_3}{2} = 10\sqrt{3} + 0.5\left(\frac{\sigma_1 + \sigma_3}{2}\right)$$

$$\sigma_1 = 3\sigma_3 + 40\sqrt{3} \qquad \text{...(i)}$$

From stress relationship

$$\sigma_1 = \sigma_3 \tan^2\left(45° + \frac{\phi}{2}\right) + 2c\tan\left(45° + \frac{\phi}{2}\right) \qquad \text{...(ii)}$$

Comparing eq. (i) and (ii)

$$\tan^2\left(45° + \frac{\phi}{2}\right) = 3$$

$$\phi = 30°$$

$$2c\tan\left(45° + \frac{\phi}{2}\right) = 40\sqrt{3}$$

$$c = 20 \text{ kPa}$$

8.19 (d)

The phenomenon where cohesionless soil near the downstream side of dams or sheet piles loses its shear strength due to high upward hydraulic gradient is called **piping**.

8.20 (c)

$$B = \frac{\Delta u}{\Delta\sigma_3} = \frac{80}{250 - 150} = 0.80$$

8.21 (b)

In CU test, drainage is allowed for only Ist stage i.e. consolidation stage.

So, pore water pressure does not develop during this stage.

8.22 Sol.

Cell pressure,

$$\sigma_3 = 200 \text{ kPa}$$

Pore water pressure,

$$u = 150 \text{ kPa}$$

$$\therefore \bar{\sigma}_3 = \sigma_3 - U = 200 - 150 = 50 \text{ kPa}$$

Effective shear parameters (c' = 150 kPa and $\phi°$ = 22°) is given,

$$\therefore \bar{\sigma}_1 = \bar{\sigma}_3 \tan^2\left(45 + \frac{\phi'}{2}\right) + 2c'\tan\left(45 + \frac{\phi'}{2}\right)$$

$$\bar{\sigma}_1 = 50 \times \tan^2\left(45 + \frac{22}{2}\right) + 2 \times 15\tan\left(45 + \frac{22}{2}\right)$$

$$\bar{\sigma}_1 = 154.376 \text{ kPa}$$

Now,

$$\bar{\sigma}_1 = \bar{\sigma}_3 + \sigma_d$$

$$\sigma_d = \bar{\sigma}_1 - \bar{\sigma}_3 = 154.376 - 50$$

Deviator stress, σ_d = 104.376 kPa

8.23 (c)

Method-I

Hydrostatic pressure acts equally in all directions.

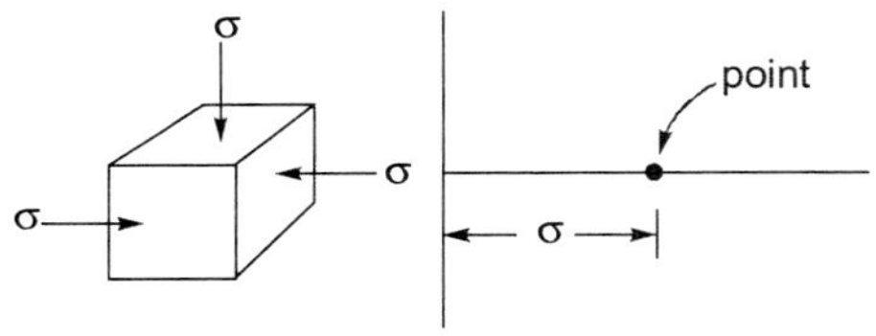

Method-II

$$\text{Radius of Mohr circle} = \sqrt{\left(\frac{\sigma_x - \sigma_y}{2}\right)^2 + \tau^2}$$

Given: $\sigma_x = \sigma_y = \sigma_z = \sigma, \tau = 0$

$$R = \sqrt{\left(\frac{\sigma - \sigma}{2}\right)^2 + 0^2} = 0$$

$$\text{Centre} = \frac{\sigma_x - \sigma_y}{2} = \frac{\sigma + \sigma}{2} = \sigma$$

Mohr circle is a point at a distance of σ from origin.

8.24 (c)

- Loss in strength of soil due to remoulding at same water content is termed as **sensitivity**.
- Over a period of time soil regain a part of its lost strength is termed as **thixotropy**.
- When **seepage** takes place in upward direction, seepage pressure acts in upward direction and effective stress is reduced, consequently shear strength is reduced.
- In **liquefaction**, due to dynamic/cyclic loading in loose saturated sand, effective stress decreases and decrease in shear strength is recorded.

8.25 Sol.

CU test is conducted.

$\sigma_3 = 100$ kPa, $\sigma_d = 80$ kPa, $u = 50$ kPa

$$\sigma_1 = \sigma_3 + \sigma_d = 100 + 80 = 180 \text{ kPa}$$

$$\bar{\sigma}_1 = \sigma_1 - u = 180 - 50 = 130 \text{ kPa}$$

$$\bar{\sigma}_3 = \sigma_3 - u = 100 - 50 = 50 \text{ kPa}$$

$$\bar{\sigma}_1 = \bar{\sigma}_3 \tan^2\left(45 + \frac{\phi'}{2}\right) + 2C'\tan\left(45 + \frac{\phi'}{2}\right)$$

Assume $C' = 0$,

$$\bar{\sigma}_1 = \bar{\sigma}_3 \tan^2\left(45 + \frac{\phi'}{2}\right)$$

$$130 = 50\tan^2\left(45 + \frac{\phi'}{2}\right)$$

$$\phi' = 26.4°$$

8.26 (a)

On stress path in terms of *p-q* plot,

$$p = \frac{\sigma_1 + \sigma_3}{2}$$

$$q = \frac{\sigma_1 - \sigma_3}{2}$$

If $\sigma_1 = \sigma_3$ then $q = 0$ and $p \neq 0$ hence point does not lie at origin, point will lies on the p-axis.

If $\sigma_1 > \sigma_3$, both the stress point will be positive.

8.27 Sol.

I[st] **Specimen** : Drained condition

$\bar{\sigma}_3 = 200$ kPa : $\sigma_d = 400$ kPa : $\bar{\sigma}_1 = 600$ kPa

IInd **Specimen** : Undrained condition

$$\sigma_3 = 200 \text{ kPa}$$

$$\sigma_d = 150 \text{ kPa}$$

$$\sigma_1 = \sigma_3 + \sigma_d = 350 \text{ kPa}$$

Let pore pressure developed is u

$$\therefore \quad \bar{\sigma}_3 = (200 - u)$$

$$\bar{\sigma}_1 = (350 - u)$$

From stress relationship (For I[st] specimen)

$$\bar{\sigma}_1 = \bar{\sigma}_3 \tan^2\left(45° + \frac{\phi}{2}\right) + 2c\tan\left(45° + \frac{\phi}{2}\right)$$

For clay under drained condition $c = 0$

$$\therefore \quad \bar{\sigma}_1 = \bar{\sigma}_3 \tan^2\left(45° + \frac{\phi}{2}\right)$$

$$600 = 200\tan^2\left(45° + \frac{\phi}{2}\right)$$

$$\phi = 30°$$

For second specimen,

$$\bar{\sigma}_1 = \bar{\sigma}_3 \tan^2\left(45° + \frac{\phi}{2}\right)$$

$$(350 - u) = (200 - u)\tan^2\left(45° + \frac{30°}{2}\right)$$

$$u = 125 \text{ kPa}$$

8.28 Sol.

Normal stress

$$\because \quad \sigma_n = \frac{300 + 170}{2} + \frac{300 - 170}{2}\cos 2\alpha$$

$$\alpha = 45° + \frac{36°}{2} = 63°$$

$$\sigma_n = 196.79 \text{ kPa}$$

$$\Rightarrow \quad u = \text{Hydrostatic pore pressure} + \text{Excess pore pressure}$$

$$= 30 + 94.5 = 124.5 \text{ kPa}$$

$$\sigma_n' = \sigma_n - u = 196.79 - 124.5$$

$$\sigma_n' = 72.29 \text{ kPa}$$

Shear strength

$$\tau_f = c + \sigma_n' \tan\phi = 72.29 \tan 36 = 52.52 \text{ kPa}$$

8.29 Sol.

Given: $\sigma_d = 150$; $\sigma_3 = 50$; $\sigma_1 = 200$

For sand ($C = 0$);

$$\sigma_1 = \sigma_3 \tan^2\left(45 + \frac{\phi}{2}\right) + 2c\tan\left(45 + \frac{\phi}{2}\right)$$

$$200 = 50\tan^2\left(45 + \frac{\phi}{2}\right)$$

$$\phi = 36.87°$$

So, the angle of internal friction to the nearest integer is 37°.

8.30 (c)

I. Compression loading

OR

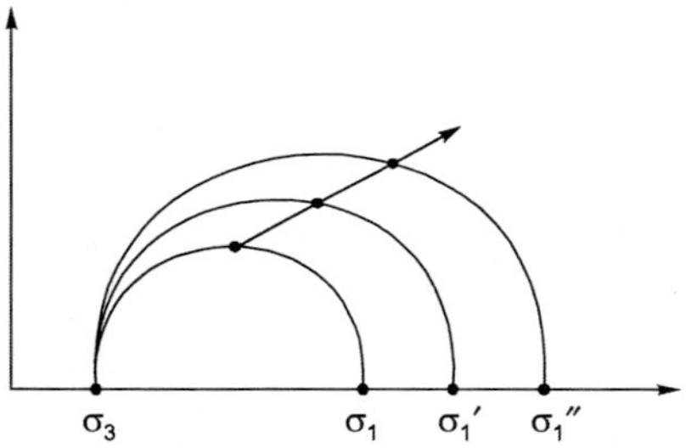

II. Compression unloading

OS

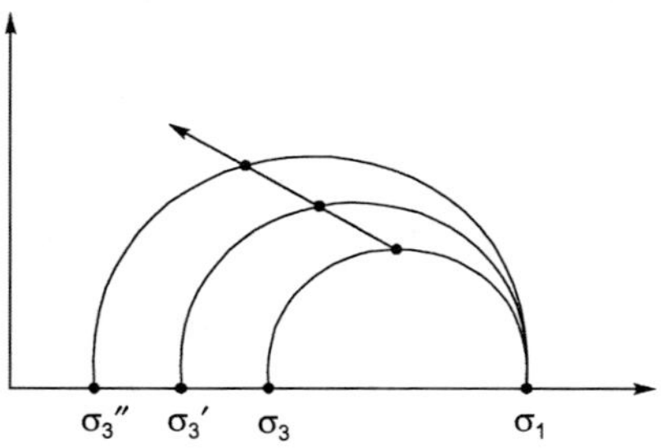

III. Extension unloading

OQ

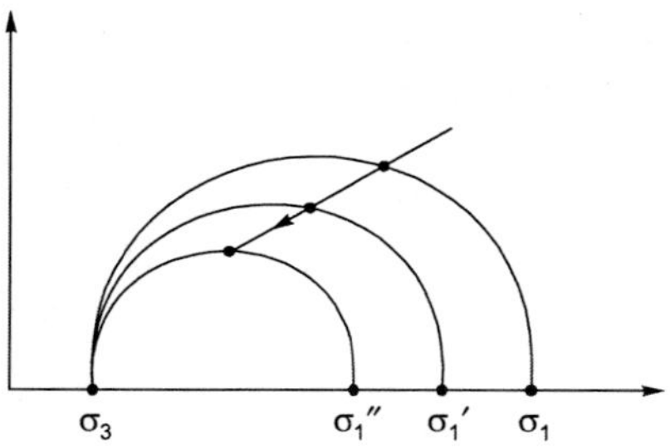

IV. Extension loading

OP

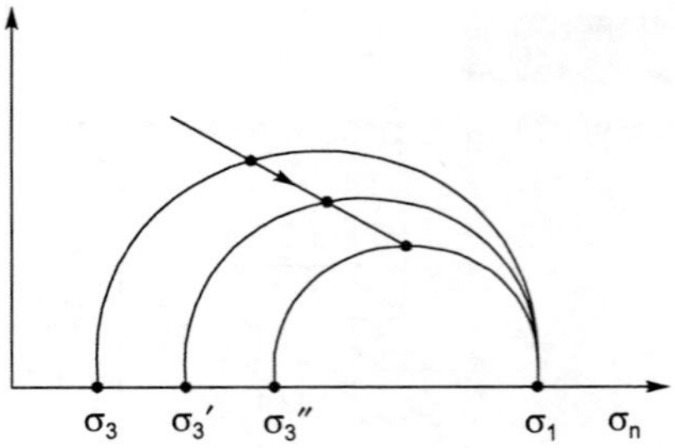

8.31 (c)

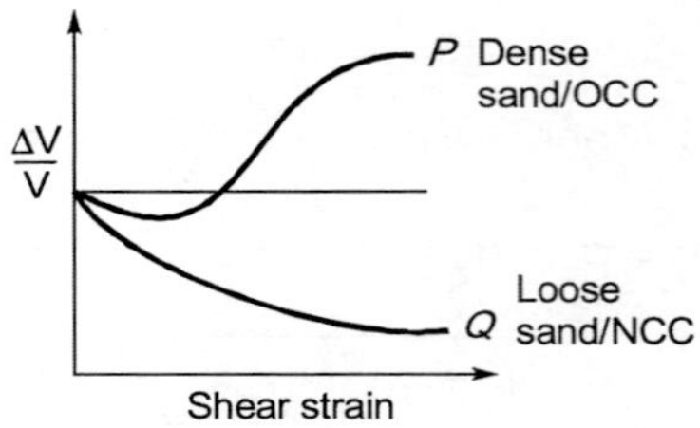

8.32 (d)

- CU test is performed for short term analysis such as of earthen dam during sudden drawdonw.
- CD test is s slow test. So, it is used for long term stability of excavated clay slope.

8.33 Sol.

Given: UU test for clay ($\phi = 0$),

$$\sigma_d = 150 \text{ kPa}$$

$$\sigma_3 = 100 \text{ kPa}$$

$$\sigma_1 = \sigma_3 + \sigma_d = 250 \text{ kPa}$$

$$\therefore \quad \sigma_1 = \sigma_3\tan^2\left(45 + \frac{\phi}{2}\right) + 2C\tan\left(45 + \frac{\phi}{2}\right)$$

$$\Rightarrow \quad 250 = 100 + 2C$$

$$\Rightarrow \quad C = 75 \text{ kPa}$$

∴ Shear strength,

$$\tau = C + \bar{\sigma}_n \tan\phi = 75 \text{kPa}$$

8.34 (d)

(P) Change in volume with axial stran

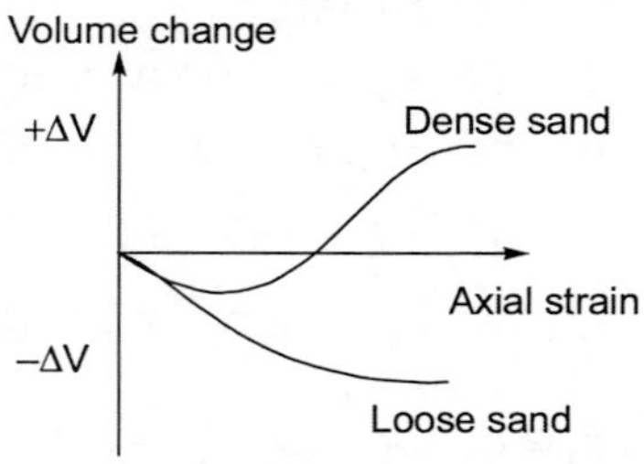

(Q) Change in pore water pressure with axial strain:

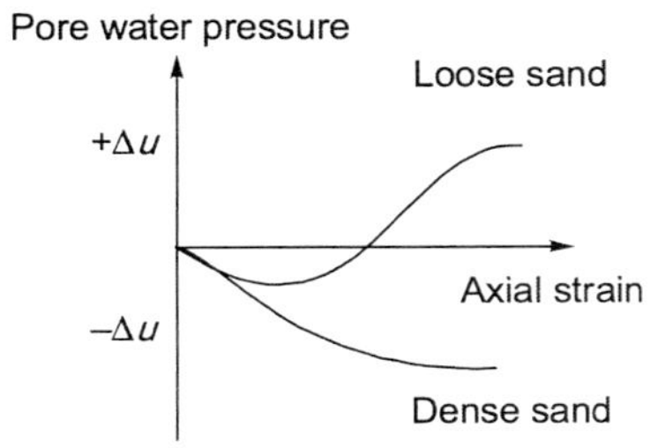

(R) Skempton parameter B value for saturated cohesive soil is 1.

(S) For clay, $N_q = 1$, $N_c = 5.7$ and $N_\gamma = 0$ $(\because \phi = 0)$

8.35 Sol.

Given, Normal stress, $\sigma_n = 50$ kPa

Shear stress, $\tau = 35$ kPa

$$\tau = c + \sigma_n \tan\phi$$

$\Rightarrow \quad 35 = 50\tan\phi$

$(\because c = 0$, sandy soil)

$\Rightarrow \quad \tan\phi = 0.7$

$\Rightarrow \quad \phi = 35°$

8.36 Sol.

$\because \quad \sigma_3 = 0 \quad$ (UCS test)

So, $\sigma_1 = \sigma_3 \tan^2\left(45 + \frac{\phi}{2}\right) + 2c\tan\left(45 + \frac{\phi}{2}\right)$

$\Rightarrow 76 = 2c\tan\left(45 + \frac{0}{2}\right) \Rightarrow c = \frac{76}{2} = 38$ MPa

8.37 (a)

σ_v = Major principal stress

σ_h = Minor principal stress

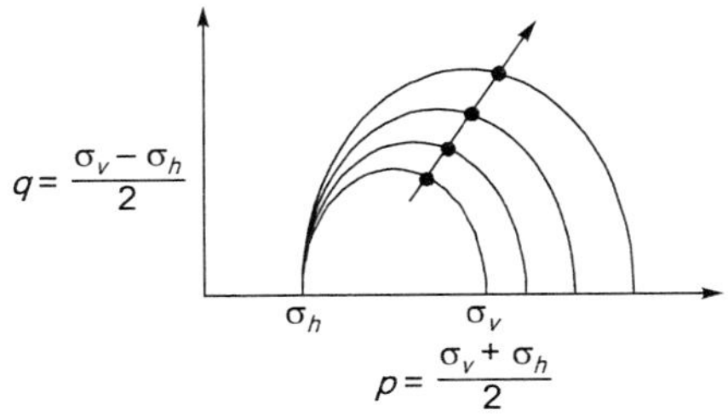

So, σ_h is constant and σ_v is increasing.

8.38 Sol.

$$\bar{\sigma}_1 = \bar{\sigma}_3 \tan^2\left(45 + \frac{\phi}{2}\right) + 2c'\tan\left(45 + \frac{\phi}{2}\right)$$

$\because \quad \bar{\sigma}_3 = \sigma_3 - 70$

$\bar{\sigma}_1 = \sigma_1 - 70 = \sigma_3 + \sigma_d$

$- 70 = (\sigma_3 + 360 - 70)$

$\Rightarrow \quad (\sigma_3 + 360 - 70) = (\sigma_3 - 70)\tan(45 +$ $30/2) \quad [\because c = 0]$

$\Rightarrow \quad \sigma_3 = 250$ kPa

Now, $\quad \sigma_1 = \sigma_3\tan^2(45 + \phi/2)$

$[\because c = 0]$

$\Rightarrow \quad (\sigma_3 + \sigma_d) = \sigma_3\tan^2(45 + 30/2)$

$\Rightarrow \quad (250 + \sigma_d) = 250 \times 3$

$\Rightarrow \quad \sigma_d = 500$ kPa

8.39 Sol.

Given: $\quad \sigma_3 = 20$ kN/m^2

Back pressure = 10 kN/m^2

$\Delta u_c = 1$ kPa

$\sigma_3' = 30$ kN/m^3

$\Delta\sigma_3 = (30 - 20)$ kN/m^2 = 10 kN/m^2

$$B = \frac{\Delta u_c}{\Delta\sigma_3} = \frac{1}{10} = 0.1$$

8.40 (a)

8.41 Sol.

Given: $\sigma_d = 100$ kN/m^2, $\sigma_3 = 300$ kN/m^2

Now, $\quad \sigma_1 = \sigma_3 \tan^2\alpha + \cancel{2c}^{\,0}\tan\alpha$

$\Rightarrow \quad (\sigma_d + \sigma_3) = \sigma_3 \times \tan^2\alpha \left[\therefore \alpha = 45° + \frac{\phi}{2}\right]$

$\Rightarrow (100 + 300) = 300\tan^2\alpha$

$\Rightarrow \quad 400 = 300\tan^2\alpha$

$\Rightarrow \quad \tan^2\alpha = \frac{4}{3}$,

$\tan\alpha = 1.1547$

$= \tan^{-1}(1.1547) = 49.1° \simeq 49°$

■■■■

9 Retaining Wall/Earth Pressure Theories

9.1 For the determination of earth pressure, Coulomb's wedge theory assumes that
(a) the back of wall is smooth and vertical
(b) the soil is non homogeneous and anisotropic
(c) the slip surface is circular
(d) the wall surface is rough [1996 : 1 M]

9.2 The wall shown in figure has failed. The cause of failure or the error made in the design of the failed wall is

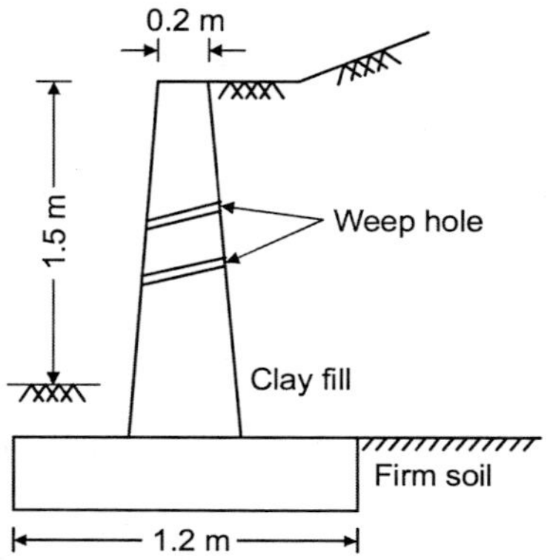

(a) deep slip surface failure
(b) overturning
(c) no proper drainage in the clay backfill
(d) translational failure [1996 : 1 M]

9.3 Coulomb's theory of earth pressure is based on
(a) the theory of elasticity
(b) the theory of plasticity
(c) empirical rules
(d) wedge theory [1997 : 1 M]

9.4 The depth of tension cracks in a soft clay ($\phi_u = 0°$) is
(a) $\frac{4C_u}{\gamma}$ (b) $\frac{2C_u}{\gamma}$
(c) $\frac{C_u}{\gamma}$ (d) $\frac{C_u}{2\gamma}$ [1997 : 1 M]

9.5 In cohesive soils the depth of tension crack (Z_{cr}) is likely to be

(a) $Z_{cr} \geq \frac{2c}{\gamma}\tan\left(45° - \frac{\phi}{2}\right)$

(b) $Z_{cr} \geq \frac{2c}{\gamma}\tan\left(45° + \frac{\phi}{2}\right)$

(c) $Z_{cr} \geq \frac{4c}{\gamma}\sqrt{\tan\left(45° - \frac{\phi}{2}\right)}$

(d) $Z_{cr} \geq \frac{4c}{\gamma}\sqrt{\tan\left(45° + \frac{\phi}{2}\right)}$ [1998 : 1 M]

9.6 The total active thrust on a vertical wall 3 m high retaining a horizontal sand backfill (unit weight γ_t = 20 kN/m^3, angle of shearing resistance ϕ'= 30°) when the water table is at the bottom of the wall, will be
(a) 30 kN/m (b) 35 kN/m
(c) 40 kN/m (d) 45 kN/m [1998 : 2 M]

9.7 Cohesion in soil
(a) decreases active pressure and increases passive resistance
(b) decreases both active pressure and passive resistance
(c) increases the active pressure and decreases the passive resistance
(d) increases both active pressure and passive resistance [1999 : 1 M]

9.8 To have zero active pressure intensity at the tip of a wall in cohesive soil, one should apply a uniform surcharge intensity of
(a) 2c tan α (b) 2c cot α
(c) –2c tan α (d) –2c tan α
[2000 : 1 M]

9.9 A braced cut, 5 m wide and 7.5 m deep is proposed in a cohesionless soil deposit having effective cohesion c' = 0 and effective friction angle, ϕ' = 36°. The first row of struts is to be installed at a depth of 0.5 m below ground surface and spacing between the struts should be 1.5 m. If the horizontal spacing of struts is 3 m and unit weight of the deposit (in kN) is 20 kN/m^3, the maximum strut load will be
(a) 70.87 (b) 98.72
(c) 113.90 (d) 151.86 [2003 : 2 M]

9.10 An unsupported excavation is made to the maximum possible depth in a clay soil having γ_t = 18 kN/m^3, c = 100 kN/m^2, ϕ = 30°. The active earth pressure, according to Rankine's theory, at the base level of the excavation is

(a) 115.47 kN/m^2 (b) 54.36 kN/m^2
(c) 27.18 kN/m^2 (d) 13.0 kN/m^2

[2004 : 2 M]

9.11 A retaining wall of height 8 m retains dry sand. In the initial state, the soil is loose and has a void ratio of 0.5, γ_d = 17.8 kN/m^3 and ϕ = 30°. Subsequently, the backfill is compacted to a state where void ratio is 0.4, γ_d = 18.8 kN/m^3 and ϕ = 35°. The ratio of initial passive thrust to the final passive thrust, according to Rankine's earth pressure theory, is

(a) 0.38 (b) 0.64
(c) 0.77 (d) 1.55 [2004 : 2 M]

9.12 Using ϕ_u = 0 analysis and assuming planar failure as shown, the minimum factor of safety against shear failure of a vertical cut of height 4 m in a pure clay having c_u = 120 kN/m^2 and γ_{sat} = 20 kN/m^3 is

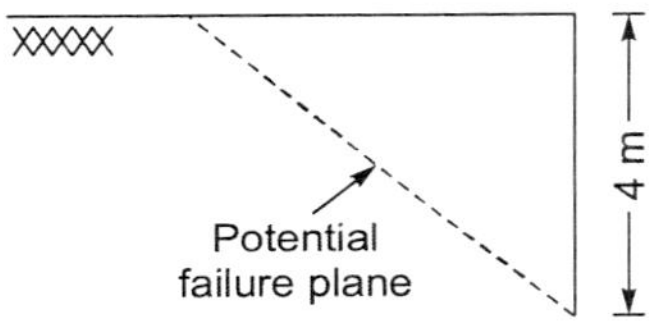

(a) 1 (b) 6
(c) 10 (d) 20 [2004 : 2 M]

9.13 A 3 m high retaining wall is supporting a saturated sand (saturated due to capillary action) of bulk density 18 kN/m^3 and angle of shearing resistance 30°. The change in magnitude of active earth pressure at the base due to rise in ground water table from the base of the footing to the ground surface shall (γ_w = 10 kN/m^3)

(a) increase by 20 kN/m^2
(b) decrease by 20 kN/m^2
(c) increase by 30 kN/m^2
(d) decrease by 30 kN/m^2 [2005 : 2 M]

9.14 Figure given below shows a smooth vertical gravity retaining wall with cohesionless soil backfill having an angle of internal friction ϕ.

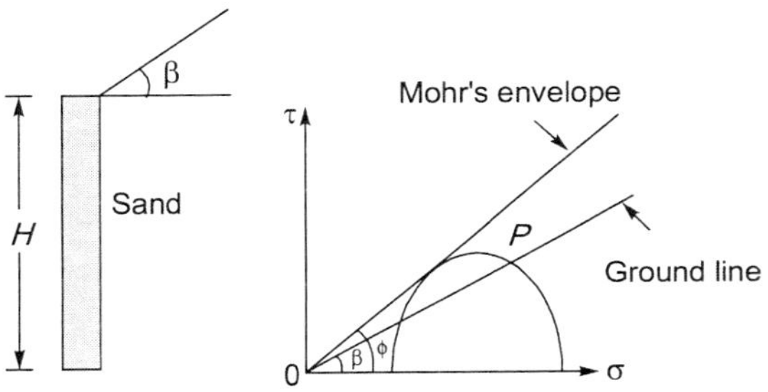

In the graphical representation of Rankine's active earth pressure for the retaining wall shown in figure, length *OP* represents

(a) vertical stress at the base
(b) vertical stress at a height *H*/3 from the base
(c) lateral earth pressure at the base
(d) lateral earth pressure at a height *H*/3 from the base [2006 : 1 M]

9.15 When a retaining wall moves away from the backfill, the pressure exerted on the wall is termed as

(a) passive earth pressure
(b) swelling pressure
(c) pore pressure
(d) active earth pressure [2008 : 1 M]

9.16 If σ_h, σ_v, σ'_h and σ'_v represent the total horizontal stress, total vertical stress, effective horizontal stress and effective vertical stress on a soil element, respectively, the coefficient of earth pressure at rest is given by

(a) $\frac{\sigma_h}{\sigma_v}$ (b) $\frac{\sigma'_h}{\sigma'_v}$

(c) $\frac{\sigma_v}{\sigma_h}$ (d) $\frac{\sigma'_v}{\sigma'_h}$ [2010 : 1 M]

9.17 A smooth rigid retaining wall moves as shown in the sketch causing the backfill material to fail. The backfill material is homogenous and isotropic and obeys the Mohr-Coulomb failure criterion. The major principal stress is

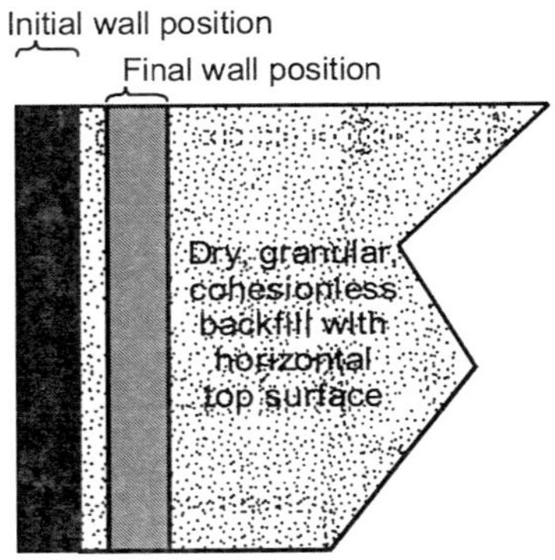

(a) parallel to the wall face and acting downwards
(b) normal to the wall face
(c) oblique to the wall face acting downwards
(d) oblique to the wall face acting upwards

[2012 : 1 M]

9.18 Two soil profile are used as backfill behind a retaining wall as shown in the figure, where γ_t is total unit weight and c' & ϕ' are effective shear parameters. Find the resultant active earth pressure in kN/m

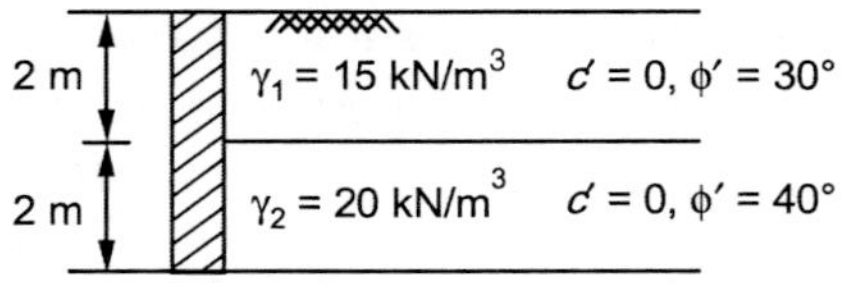

(a) 31.7 (b) 35.2
(c) 57.8 (d) 57

[2013 : 2 M]

9.19 Surcharge loading required to be placed on the horizontal backfill of a smooth retaining vertical wall so as to completely eliminate tensile crack is

(a) $2c$ (b) $2cK_a$
(c) $2c\sqrt{K_a}$ (d) $\dfrac{2c}{\sqrt{K_a}}$

[2015 : 1 M, Set-II]

9.20 A 6 m high retaining wall having a smooth vertical back face retains a layered horizontal backfill. Top 3 m thick layer of the backfill is sand having an angle of internal friction, $\phi = 30°$ while the bottom layer is 3 m thick clay with cohesion, $c = 20$ kPa. Assume unit weight for both sand and clay as 18 kN/m^3. The total active earth pressure per unit length of the wall (in kN/m) is:

(a) 150 (b) 216
(c) 156 (d) 196

[2015 : 2 M, Set-II]

9.21 A vertical cut is to be made in a soil mass having cohesion c, angle of internal friction ϕ, and unit weight γ. Considering K_a as the K_p as the coefficients of active and passive earth pressure, respectively, the maximum depth of unsupported excavation is

(a) $\dfrac{4c}{\gamma\sqrt{K_p}}$ (b) $\dfrac{2c\sqrt{K_p}}{\gamma}$
(c) $4c\sqrt{K_p}$ (d) $\dfrac{4c}{\gamma\sqrt{K_a}}$

[2016 : 1 M, Set-I]

9.22 A homogeneous gravity retaining wall supporting a cohesionless backfill is shown in the figure. The lateral active earth pressure at the bottom of the wall is 40 kPa.

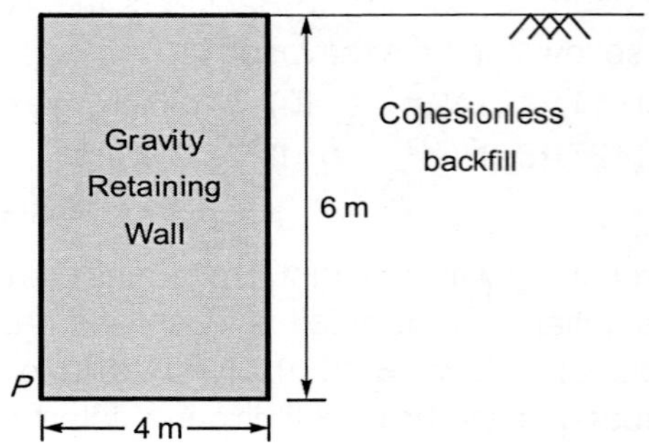

The minimum weight of the wall (expressed in kN per m length) required to prevent it from overturning about its toe (Point P) is

(a) 120 (b) 180
(c) 240 (d) 360

[2016 : 2 M, Set-I]

9.23 The soil profile at a site consists of a 5 m thick sand layer underlain by a c-ϕ soil a shown in figure. The water table is found 1 m below the ground level. The entire soil mass is retained by a concrete retaining wall and is in the active state. The back of the wall is smooth and vertical. The total active earth pressure (expressed in kN/m^2) at point A as per Rankine's theory is______.

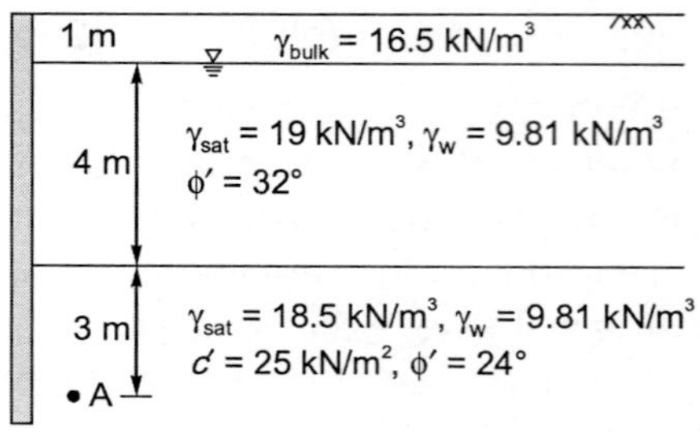

[2016 : 2 M, Set-II]

9.24 Consider a rigid retaining wall with partially submerged cohesionless backfill with a surcharge. Which one of the following diagrams closely represents the Rankine's active earth pressure distribution against this wall?

(a) 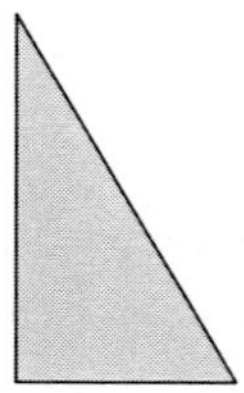 (b) 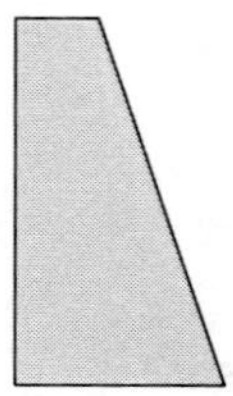

(c) 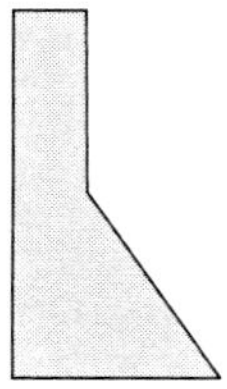 (d) 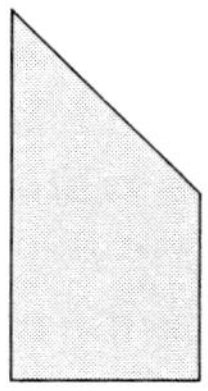

[2017 : 1 M, Set-II]

9.25 A rigid smooth retaining wall of height 7 m with vertical backface retains saturated clay as backfill. The saturated unit weight and undrained cohesion of the backfill are 17.2 kN/m^3 and 20 kPa, respectively. The difference in the active lateral forces on the wall (in kN per meter length of wall, up to two decimal places), before and after the occurrence of tension cracks is ________

[2018 : 2 M, Set-I]

9.26 A 3 m high vertical earth retaining wall retains a dry granular backfill with angle of internal friction of 30° and unit weight of 20 kN/m^3. If the wall is prevented from yielding (no movement), the total horizontal thrust (in kN per unit length) on the wall is

(a) 0
(b) 30
(c) 45
(d) 270

[2018 : 2 M, Set-II]

9.27 A retaining wall of height H with smooth vertical backface supports a backfill inclined at an angle β with the horizontal. The backfill consists of cohesionless soil having angle of internal friction ϕ. If the active lateral thrust acting on the wall is P_a, which one of the following statements is TRUE?

(a) P_a acts at a height $\frac{H}{3}$ from the base of the wall and at an angle ϕ with the horizontal

(b) P_a acts at a height $\frac{H}{2}$ from the base of the wall and at an angle ϕ with the horizontal

(c) P_a acts at a height $\frac{H}{3}$ from the base of the wall and at an angle β with the horizontal

(d) P_a acts at a height $\frac{H}{2}$ from the base of the wall and at an angle β with the horizontal

[2019 : 1 M, Set-I]

9.28 A vertical retaining wall of 5 m height has to support soil having unit weight of 18 kN/m^3, effective cohesion of 12 kN/m^2, and effective friction angle of 30°. As per Rankine's earth pressure theory and assuming that a tension crack has occurred, the lateral active thrust on the wall per meter length (in kN/m, round off to two decimal places), is ________.

[2020 : 2 M, Set-I]

9.29 A 5 m high vertical wall has a saturated clay backfill. The saturation unit weight and cohesion of clay are 18 kN/m^3 and 20 kPa, respectively. The angle of internal friction of clay is zero. In order to prevent development of tension zone, the height of the wall is required to be increased. Dry sand is used as backfill above the clay for the increased portion of the wall. The unit weight and angle of internal friction of sand are 16 kN/m^3 and 30°, respectively. Assume that the back of the wall is smooth and top of the backfill is horizontal. To prevent the development of tension zone, the minimum height (in m, round off to one decimal place) by which the wall has to be raised, is __________.

[2020 : 2 M, Set-II]

9.30 The cohesion (c), angle of internal friction (ϕ) and unit weight (γ) of a soil are 15 kPa. 20° and 17.5 kN/m^3, respectively. The maximum depth of unsupported excavation in the soil (in m, round off to two decimal places) is _______

[2021 : 1 M, Set-I]

9.31 A retaining wall of height 10 m with clay backfill is shown in the figure (not to scale). Weight of the retaining wall is 5000 kN per m acting at 3.3 from the toe of the retaining wall. The interface friction angle between base of the retaining wall and the base soil is 20°. The depth of clay in front of the retaining wall is 2.0 m. The properties of the clay backfill and the clay placed in front of the retaining wall are the same. Assume that the tension crack is filled with water. Use Rankine's earth pressure theory. Take unit weight of water. γ_w = 9.81 kN/m^3.

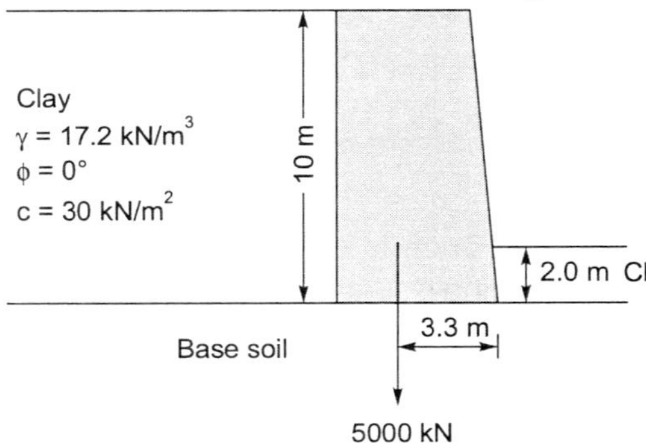

The factor of safety (round off to two decimal places) against sliding failure of the retaining wall after ignoring the passive earth pressure will be ____________

[2021 : 2 M, Set-I]

9.32 Let σ_v' and σ_h' denote the effective vertical stress and effective horizontal stress, respectively. Which one of the following conditions must be satisfied for a soil element to reach the failure state under Rankine's passive earth pressure condition?

(a) $\sigma_v' < \sigma_h'$
(b) $\sigma_v' > \sigma_h'$
(c) $\sigma_v' = \sigma_h'$
(d) $\sigma_v' + \sigma_h' = 0$

[2022 : 1 M, Set-I]

9.33 As per Rankine's theory of earth pressure, the inclination of failure planes is $\left(45+\frac{\phi}{2}\right)^\circ$ with respect to the direction of the minor principal stress.

The above statement is correct for which one of the following options?

(a) Only the active state and not the passive state
(b) Only the passive state and not the active state
(c) Both active as well as passive states
(d) Neither active nor passive state

[2022 : 2 M, Set-I]

9.34 A smooth vertical retaining wall supporting layered soils is shown in figure. According to Rankine's earth pressure theory, the lateral active earth pressure acting at the base of the wall is __________ kPa (round off to one decimal place).

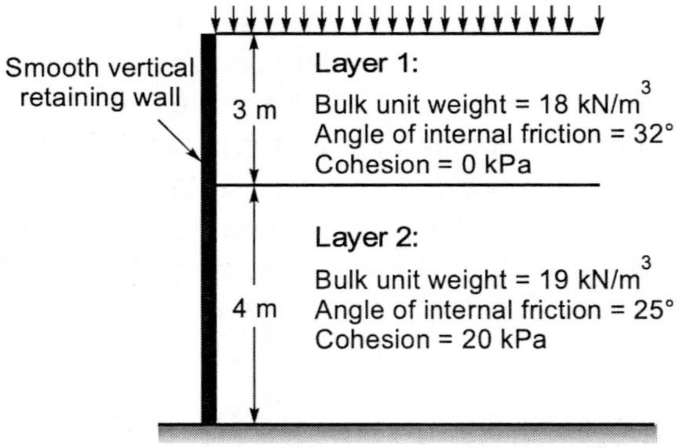

[2023 : 2 M, Set-I]

9.35 A vertical trench is excavated in a clayey soil deposit having a surcharge load of 30 kPa. A fluid of unit weight 12 kN/m^3 is poured in the trench to prevent collapse as the excavation proceeds. Assume that the fluid is not seeping through the soil deposit. If the undrained cohesion of the clay deposit is 20 kPa and saturated unit weight is 18 kN/m^3, what is the maximum depth of unsupported excavation (in m, rounded off to two decimal places)?

[2023 : 2 M, Set-I]

9.36 The figure shows a vertical retaining wall with backfill consisting of cohesive-frictional soil and a failure plane developed due to passive earth pressure. The forces acting on the failure wedge are: P as the reaction force between the wall and the soil, R as the reaction force on the failure plane, C as the cohesive force along the failure plane and W as the weight of the failure wedge. Assuming that there is no adhesion between the wall and the wedge, identify the most appropriate force polygon for the wedge.

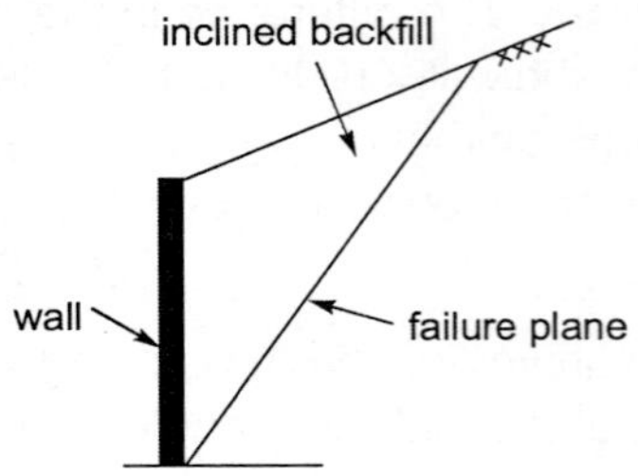

(a)

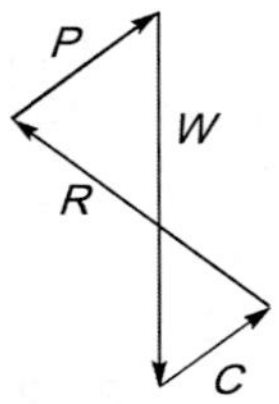

(b)

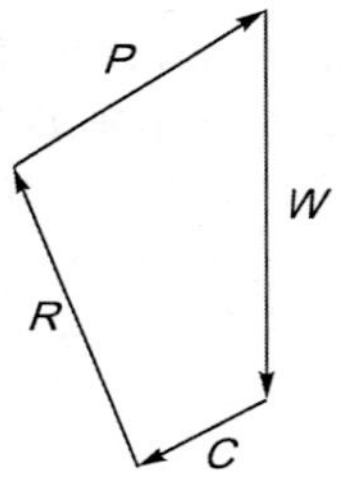

(c)

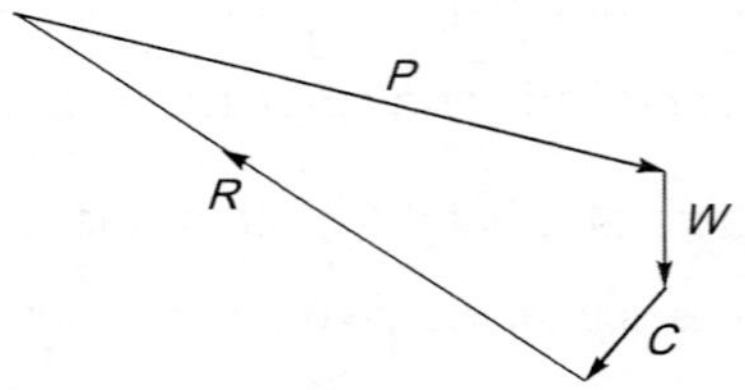

(d)

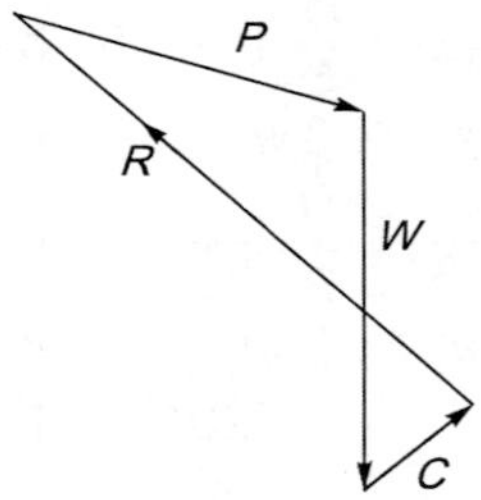

[2023 : 2 M, Set-II]

9.37 A vertical smooth rigid retaining wall is supporting horizontal ground with dry cohesionless backfill having a friction angle of 30°. The inclinations of failure planes with respect to the major principal plane for Rankine's active and passive earth pressure condition, respectively are ____.

(a) 30° and 30° (b) 30° and 60°
(c) 60° and 60° (d) 60° and 30°

[2024 : 2 M, Set-I]

9.38 Consider the statements P and Q related to the analysis/design of retaining walls.

P : When a rough retaining wall moves toward the backfill, the wall friction force/resistance mobilizes in upward direction along the wall.

Q : Most of the earth pressure theories calculate the earth pressure due to surcharge by neglecting the actual distribution of stresses due to surcharge.

Which one of the following options is CORRECT?

(a) P is FALSE and Q is TRUE
(b) Both P and Q are TRUE
(c) P is TRUE and Q is FALSE
(d) Both P and Q are FALSE

[2024 : 2 M, Set-II]

■■■■

Answers Retaining Wall/Earth Pressure Theories

9.1	(d)	9.2	(b)	9.3	(d)	9.4	(b)	9.5	(b)	9.6	(a)	9.7	(a)
9.8	(a)	9.9	(c)	9.10	(a)	9.11	(c)	9.12	(b)	9.13	(a)	9.14	(a)
9.15	(d)	9.16	(b)	9.17	(b)	9.18	(a)	9.19	(d)	9.20	(a)	9.21	(d)
9.22	(a)	9.23	(69.65)	9.24	(b)	9.25	(46.7)	9.26	(c)	9.27	(c)	9.28	(21.7)
9.29	(2.5)	9.30	(4.9)	9.31	(4.28)	9.32	(a)	9.33	(c)	9.34	(35.4)	9.35	(3.33)
9.36	(c)	9.37	(c)	9.38	(a)								

Explanations Retaining Wall/Earth Pressure Theories

9.1 (d)

Rankine's theory assumes the wall surface to be smooth while coulomb's wedge theory assumed it to be rough.

9.2 (b)

The outward motion is resisted by sliding resistance along the base of the wall and by the passive resistance of the soil lying above the toe of the wall. The active thrust also tends to overturn the wall around the toe.

9.3 (d)

Coulomb considered equilibrium of entire wedge. He assumed that the wedge acts as a rigid body and friction is distributed uniformly on the rupture surface.

9.4 (b)

$$p_a = k_a \gamma z - 2c_u\sqrt{k_a}$$

Let z_c = Depth of tension crack

At $z = z_c,\ p_a = 0$

$$\therefore \quad 0 = k_a \gamma z_c - 2c_u\sqrt{k_a}$$

$$\therefore \quad z_c = \frac{2c_u}{\gamma\sqrt{k_a}}$$

For $\phi_u = 0,\ k_a = 1$

∴ Depth of tension crack,

$$z_c = \frac{2c_u}{\gamma}$$

9.5 (b)

Depth of tension crack,

$$Z_{cr} = \frac{2c}{\gamma}\tan\left(45° + \frac{\phi}{2}\right)$$

9.6 (a)

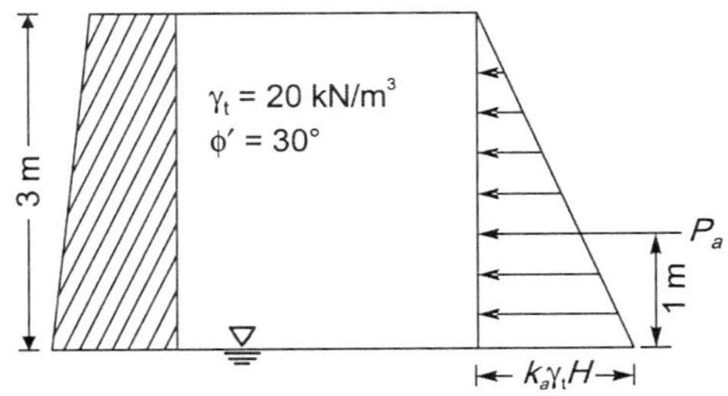

$$k_a = \frac{1-\sin 30°}{1+\sin 30°} = \frac{1}{3}$$

Total active thrust,

$$P_a = \frac{1}{2}k_a\gamma_t H^2 = \frac{1}{2}\times\frac{1}{3}\times 20\times 3^2 = 30 \text{ kN/m}$$

9.7 (a)

$$p_a = k_a\gamma Z - 2c\sqrt{k_a}$$
$$p_p = k_p\gamma Z + 2c\sqrt{k_p}$$

9.8 (a)

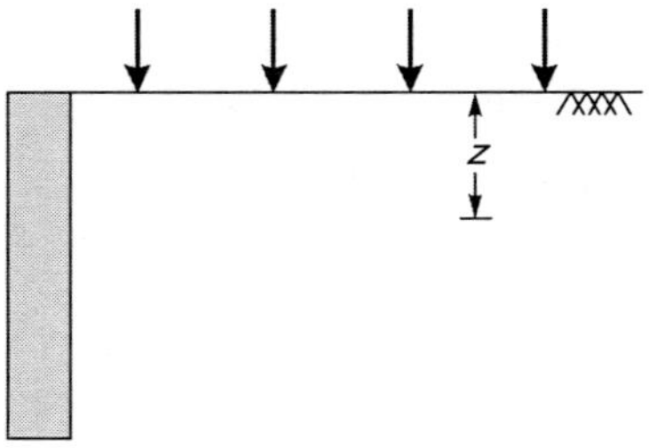

Active earth pressure

$$P_a = K_a\sigma_z - 2\sqrt{K_a}$$
$$P_a = K_a(q+\gamma z) - 2c\sqrt{K_a}$$

Given, at $z = 0$, $P_a = 0$

$$0 = K_a q - 2c\sqrt{K_a}$$
$$q = \frac{2c}{\sqrt{K_a}} = \frac{2c}{\sqrt{\frac{1}{\tan^2\left(45+\frac{\phi}{2}\right)}}}$$
$$= 2c\tan\left(45+\frac{\phi}{2}\right) = 2c\tan\alpha$$

α = angle of failure plane with major principal plane.

$$\alpha = 45+\frac{\phi}{2}$$

9.9 (c)

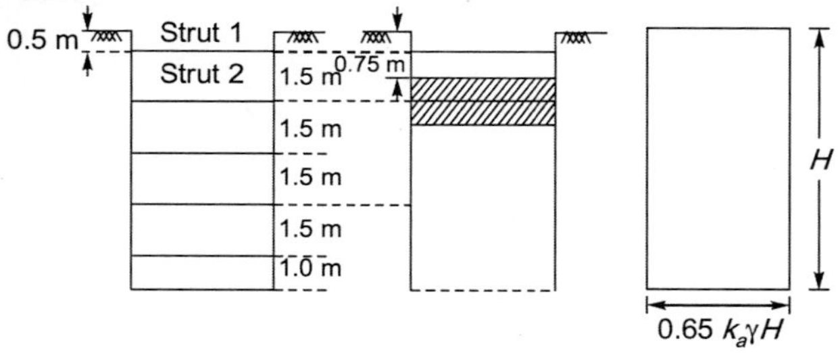

Pressure distribution

$$K_a = \frac{1-\sin\phi'}{1+\sin\phi'} = \frac{1-\sin 36°}{1+\sin 36°}$$
$$= 0.2596$$

$\therefore$ Uniform pressure on side of cut

$$= 0.65\, K_a\gamma H$$
$$= 0.65 \times 0.2596 \times 20 \times 7.5$$
$$= 25.31 \text{ kN/m}^2$$

Area covered by strut 1,

$$= 3 \times (0.5 + 0.75)$$
$$= 3.75 \text{ m}^2$$

Area covered by strut 2 and downward struts

$$= 3 \times (0.75 + 0.75) = 4.5 \text{ m}^2$$

$\therefore$ Maximum force will be carried by strut 2 and downward struts

$$= 25.31 \times 4.5 = 113.89 \text{ kN} \simeq 113.90 \text{ kN}$$

9.10 (a)

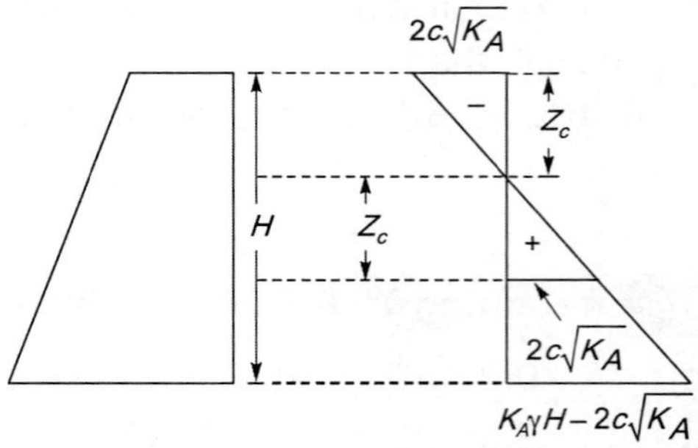

The excavation is unsupported which means the maximum possible depth upto which the clay soil can be excavated is $2Z_c$. The active earth pressure at this depth is,

$$P_A = 2c\sqrt{K_A} \qquad \left[\because K_A = \frac{1-\sin\phi}{1+\sin\phi}\right]$$
$$= 2 \times 100 \times \sqrt{\frac{1-\sin 30°}{1+\sin 30°}} = 115.47 \text{ kN/m}^2$$

Alternatively:

$$Z_c = \frac{2c}{\gamma\sqrt{K_A}}$$
$$\Rightarrow \quad 2Z_c = \frac{4c}{\gamma\sqrt{K_A}}$$

$\therefore$ Active pressure at a depth, $2Z_c$

$$= \gamma_t \times 2\,Z_c \times K_A - 2c\sqrt{K_A}$$
$$= 18 \times \frac{4\times 100\times\sqrt{3}}{18} \times \frac{1}{3} - 2\times 100\times\frac{1}{\sqrt{3}}$$
$$= 230.94 - 115.47 = 115.47 \text{ kN/m}^2$$

9.11 (c)

Passive earth pressure for loose sand is,

$$K_P = \frac{1+\sin\phi}{1-\sin\phi} = \frac{1+\sin 30°}{1-\sin 30°} = 3$$

Now, $P_{P_1} = \frac{1}{2}K_P\gamma H^2 = \frac{1}{2}\times 3\times 17.8\times 8^2$

$$= 1708.8 \text{ kN/m}$$

Passive earth pressure for compacted sand is,

$$K_P = \frac{1+\sin\phi}{1-\sin\phi} = \frac{1+\sin 35°}{1-\sin 35°} = 3.69$$

Now, $P_{P_2} = \frac{1}{2} K_P \gamma H^2$

$= \frac{1}{2} \times 3.69 \times 18.8 \times 8^2 = 2219.90$ kN/m

Ratio of P_{P_1} and $P_{P_2} = \frac{1708.8}{2219.90} = 0.77$

9.12 (b)

The maximum height of unsupported cut

$$= \frac{4c}{\gamma\sqrt{K_A}}$$

For $\phi_u = 0$; $K_A = \frac{1-\sin 0°}{1+\sin 0°} = 1.0$

∴Maximum height of unsupported cut

$$= \frac{4 \times 120}{20} = 24 \text{ m}$$

Actual height of cut = 4 m

∴Factor of safety $= \frac{24}{4} = 6$

9.13 (a)

When the water table is at the base of footing, the active earth pressure will be,

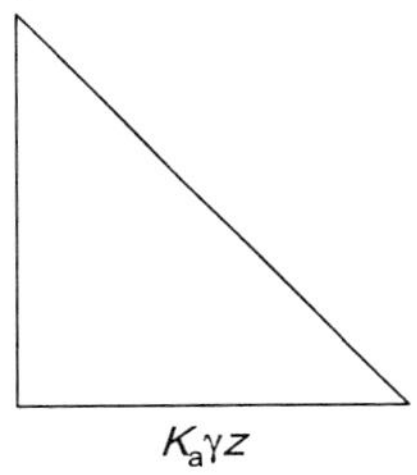

$$P_1 = K_a\gamma z = \left(\frac{1-\sin\phi}{1+\sin\phi}\right) \times \gamma \times z$$

$$= \frac{1-\sin 30°}{1+\sin 30°} \times 18 \times 3$$

$$= 1/3 \times 18 \times 3 = 18 \text{ kN/m}^2$$

When the water table rises upto the surface, the thrust due to water will also be added to the active earth pressure.

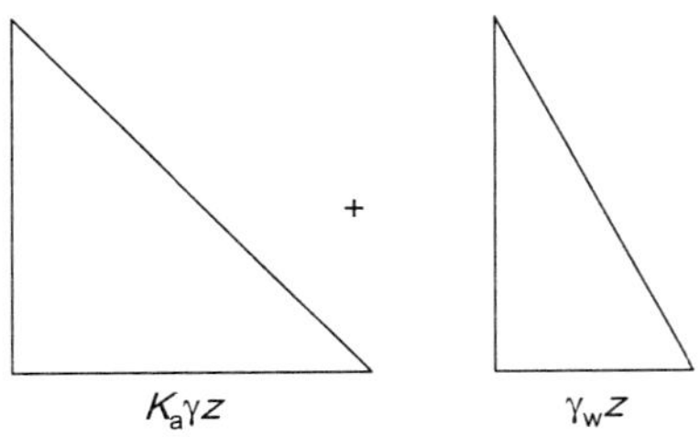

$$P_2 = K_a\gamma' z + \gamma_w z$$

$$= \frac{1}{3} \times (18-10) \times 3 + 10 \times 3$$

$$= 8 + 30 = 38 \text{ kN/m}^2$$

∴ change in active earth pressure

$$= P_2 - P_1 = 38 - 18$$

$$= 20 \text{ kN/m}^2 \text{ (increase)}$$

Note: In strict terms γ_{sat} should be taken

9.14 (a)

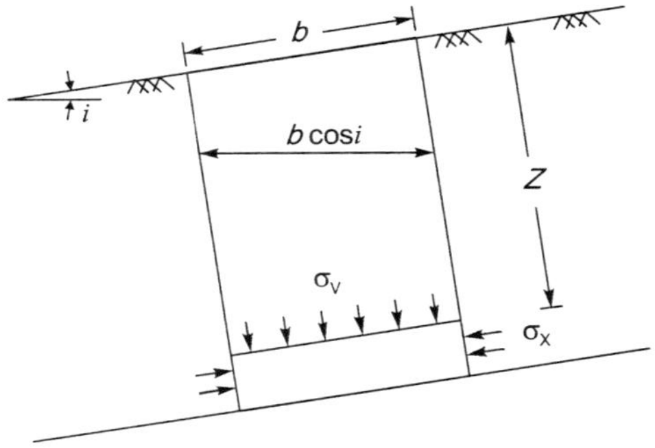

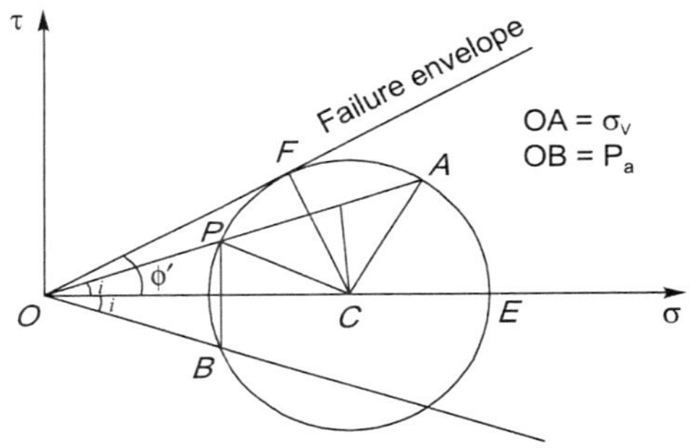

The Mohr circle corresponding to the active limiting conditions. The vertical stress σ_v is represented by the line *OA* making an angle *i* with the horizontal.

9.15 (d)

When the wall moves away from the backfill, pressure in horizontal direction decreases as the soil expands outward. When this pressure becomes minimum, then the state is known as Rankine's active state and pressure is known as active earth pressure.

9.16 (b)

Coefficient of earth pressure at rest is the ratio of intensity of earth pressure at rest to the vertical stress at a specified depth.

$$K_0 = \frac{\sigma_h}{\sigma_v}$$

But for a submerged soil, $K_0 = \frac{\sigma'_h}{\sigma'_v}$

9.17 (b)

In case of active state, stress in vertical direction is maximum. So, major principal stress is normal to wall face.

9.18 (a)

$$k_{a_1} = \frac{1-\sin 30°}{1+\sin 30°} = \frac{1}{3} = 0.33$$

$$k_{a_2} = \frac{1-\sin 40°}{1+\sin 40°} = 0.217$$

$$P_{a_1} = \frac{1}{2} k_{a_1} \gamma_1 H_1^2 = \frac{1}{2} \times \frac{1}{3} \times 15 \times 2^2 = 10 \text{ kN/m}$$

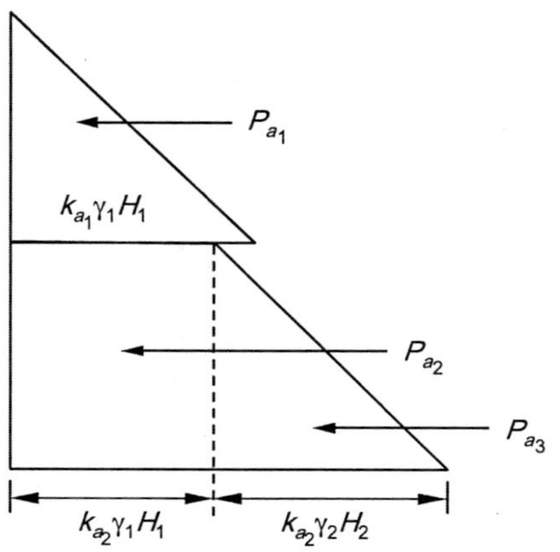

$$P_{a_2} = k_{a_2} \gamma_1 H_1 \cdot H_2$$
$$= 0.217 \times 15 \times 2 \times 2 = 13.02 \text{ kN/m}$$

$$P_{a_3} = \frac{1}{2} k_{a_2} \gamma_2 H_2^2$$

$$= \frac{1}{2} \times 0.217 \times 20 \times 2^2 = 8.68 \text{ kN/m}$$

$$P_a = P_{a_1} + P_{a_2} + P_{a_3}$$
$$= 10 + 13.02 + 8.68$$
$$P_a = 31.7 \text{ kN/m}$$

9.19 (d)

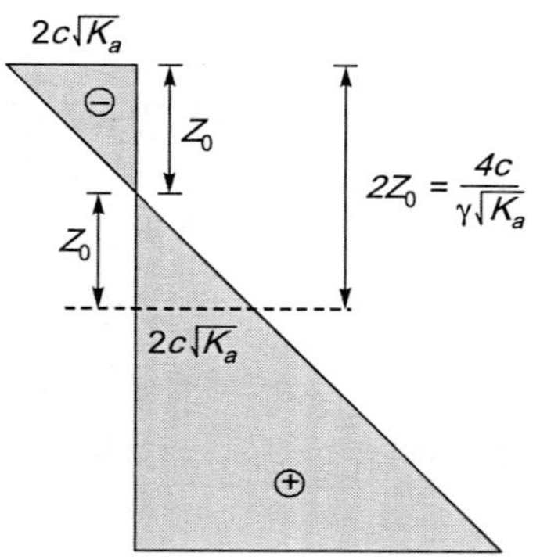

$$K_a q = 2c\sqrt{K_a}$$

$$q = \frac{2c}{\sqrt{K_a}}$$

9.20 (a)

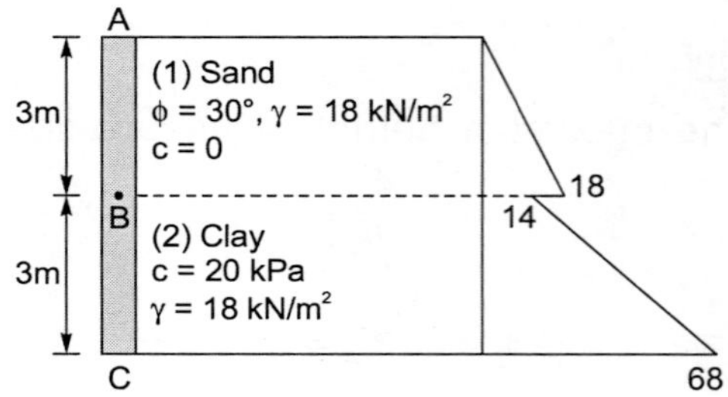

$$\phi_{clay} = 0$$

$$K_{a_1} = \frac{1-\sin\phi}{1+\sin\phi} = \frac{1}{3}$$

$$K_{a_2} = 1$$

$$p_B = K_{a_1} \gamma z = \frac{1}{3} \times 18 \times 3 = 18 \text{ kN/m}^2$$

$$p_{B'} = K_{a_2} \gamma z - 2c\sqrt{K_{a_2}}$$
$$= 1 \times 18 \times 3 - 2 \times 20\sqrt{1} = 14$$

$$p_C = K_{a_2} \gamma z - 2c\sqrt{K_{a_2}}$$
$$= 18 \times 6 - 2 \times 20\sqrt{1} = 68 \text{ kN/m}^2$$

Total active earth pressure per unit length of wall,

$$P_a = \frac{1}{2} \times 18 \times 3 + \frac{1}{2}(68 + 54) \times 3 = 150 \text{ kN/m}$$

9.21 (d)

In case of active earth pressure in cohesive soil,

$$p_a = k_a \gamma z - 2C\sqrt{k_a}$$

At, $z = z_0$, $p_a = 0$

$$\Rightarrow \quad Z_0 = 2z_0$$
$$= \frac{4C}{\gamma\sqrt{k_a}}$$

9.22 (a)

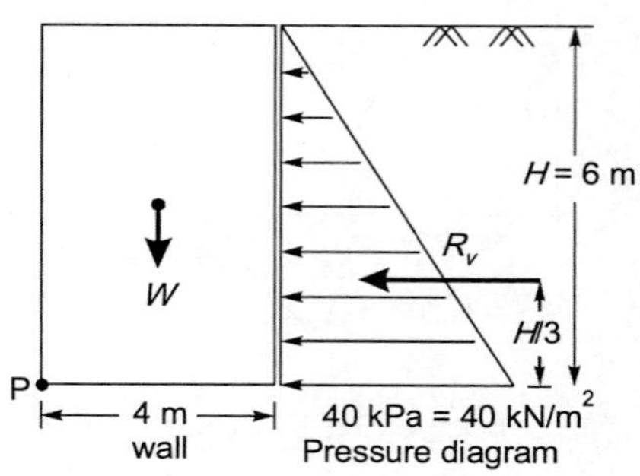

W = weight of wall

R_v = Reaction from cohesion less backfill

$= \frac{1}{2} \times 40 \times H = \frac{1}{2} \times 40 \times 6 = 120$ kN

To prevent overturning about point P,

$$M_0 \le M_R$$

$$R_v \times \frac{H}{3} \le W \times 2$$

$$120 \times \frac{6}{3} \le W \times 2$$

$$\Rightarrow \quad W \ge 120 \text{ kN}$$

∴ Minimum weight of wall = 120 kN

9.23 Sol.

In c-ϕ soil,

Earth pressure,

$$p_a = k_a\sigma_v - 2c\sqrt{k_a}$$

$$k_a = \frac{1-\sin\phi}{1+\sin\phi} = \frac{1-\sin 24}{1+\sin 24} = 0.4217$$

Note: Below water table, water pressure will not be multiplied by k_a,

at point A,

$\sigma_v = 1 \times \gamma_b + 4\gamma_{sat} + 3\gamma_{sat}$

$\sigma_v = 1 \times \gamma_b + (4 \times \gamma' + 4 \times \gamma_w) + (3\gamma' + 3\gamma_w)$

$\sigma_v = (1 \times \gamma_b + 4\gamma' + 3\gamma') + (4\gamma_w + 3\gamma_w)$

Earth Pressure at 'A',

$$p_a = k_a\sigma_v - 2c\sqrt{k_a}$$

$$= k_a[\gamma_b + 4\gamma' + 3\gamma'] + 4\gamma_w + 3\gamma_w - 2 \times c'\sqrt{k_a}$$

$$= 0.4217[16.5 + 4(19-9.81) + 3(18.5-9.81)]$$

$$+7 \times 9.81 - 2 \times 25 \times \sqrt{0.4217}$$

$$p_a = 69.65 \text{ kN/m}^2$$

9.24 (b)

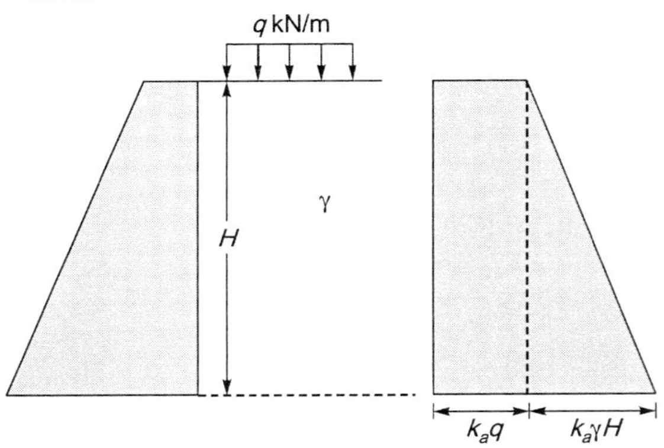

9.25 Sol.

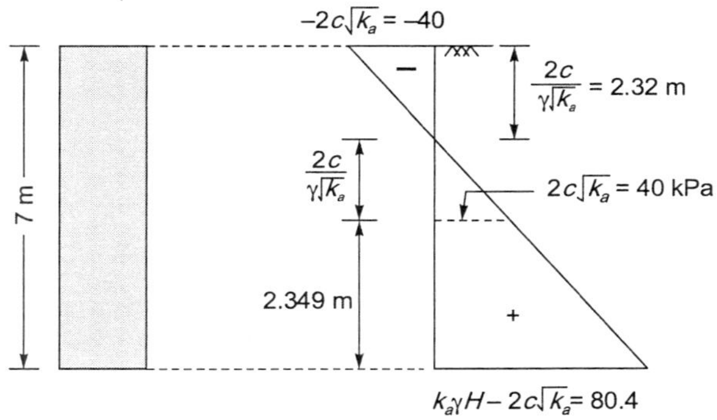

For clay $\phi = 0$

$$\therefore \quad k_a = \frac{1-\sin 0}{1+\sin 0} = 1$$

Earth pressure when tension cracks are not developed.

$$P_a = \frac{1}{2}(40 + 80.4) \times 2.349 = 141.4098$$

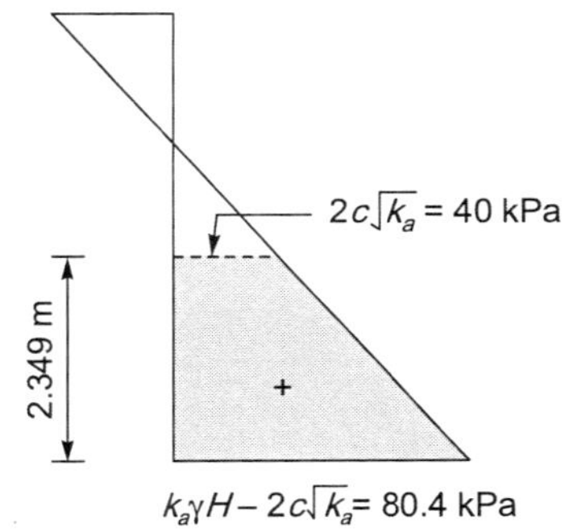

Earth pressure when tension cracks are developed

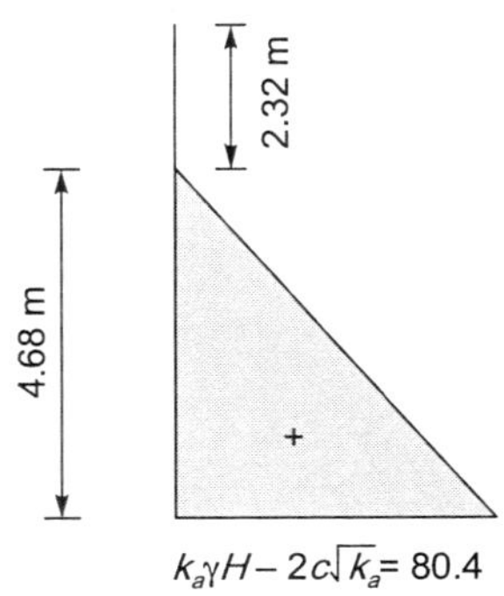

$$P_a = \frac{1}{2} \times 80.4 \times 4.68 = 188.136$$

Difference = 141.4098 – 188.136

= –46.7262 kN/m^2

9.26 (c)

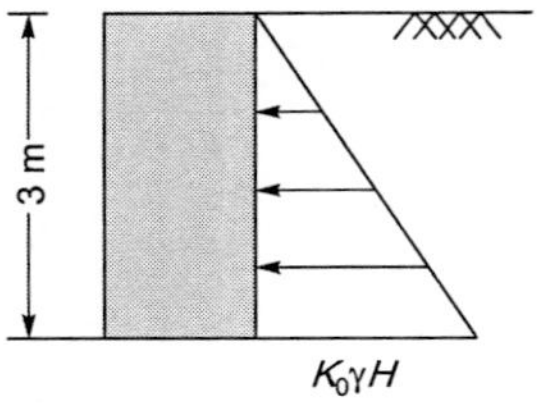

Soil is dry sand

$\therefore\ K_0 = 1 - \sin\phi = 1 - \sin 30° = 0.5$

Total horizontal thrust

$$P_0 = \frac{1}{2}K_0\gamma H \cdot H = \frac{1}{2}\times\frac{1}{2}\times 20\times 3^2 = 45 \text{ kN/m}$$

9.27 (c)

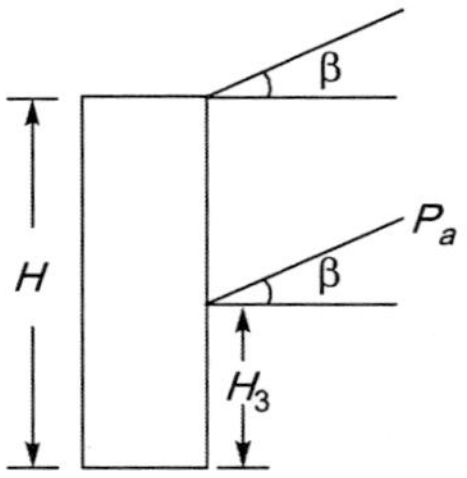

P_a will act at $\frac{H}{3}$ from base, at an angle of 'β' with horizontal.

9.28 Sol.

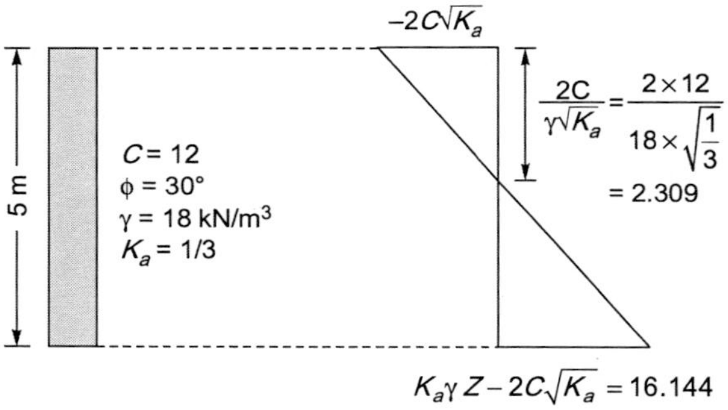

After tension crack

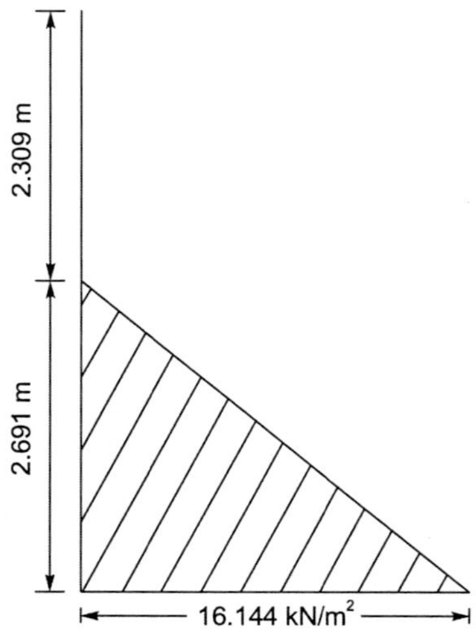

$$P_a = \frac{1}{2}\times 16.144(5 - 2.309) = 21.714 \text{ kN/m}$$

9.29 Sol.

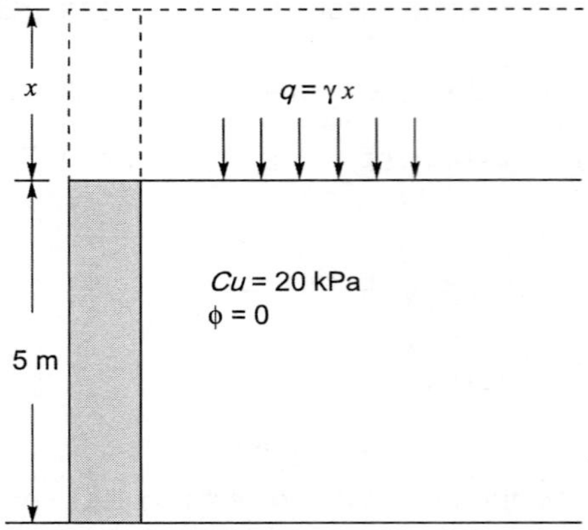

To prevent tension crack,

$$q = \frac{2c}{\sqrt{k_a}} = \frac{2\times 20}{1} = 40$$

$$q = \gamma_d x = 40 \qquad (x = \text{min. height of raise})$$

$$x = \frac{40}{16} = 2.5 \text{ m}$$

9.30 Sol.

Maximum depth of unsupported excavation,

$$H = \frac{4C}{\gamma_t\sqrt{K_a}}$$

$$k_a = \frac{1-\sin\phi}{1+\sin\phi} = \frac{1-\sin 20°}{1+\sin 20°} = 0.49$$

$$\Rightarrow \qquad H = \frac{4\times 15}{17.5\sqrt{0.49}} = 4.9 \text{ m}$$

9.31 Sol.

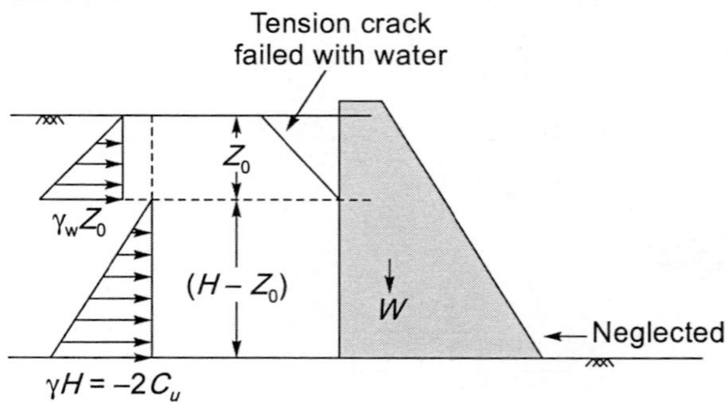

$$k_a = \frac{1-\sin\phi}{1+\sin\phi} = \frac{1-\sin 0}{1+\sin 0} = 1$$

Depth of tension crack is 'Z_0'

$$\Rightarrow\ Z_0 = \frac{2C_u}{\gamma\sqrt{k_a}} = \frac{2\times 30}{17.2\sqrt{1}} = 3.488 \text{ m}$$

Total active thrust due to soil and water

$$\frac{1}{2}(\gamma H - 2C_u)\times(H - Z_0)\times 1 + \frac{1}{2}\gamma_w Z_0\times Z_0\times 1$$

$$= \frac{1}{2}(17.2\times10-2\times30)\times(10-3.488)$$

$$\times1+\frac{1}{2}\times9.81\times3.488\times3.488 \text{ kNm}$$

$$= 424.347 \text{ kN/m}$$

(acting horizontally)

Frictional resistance

$$= N\tan\delta = W\tan\delta$$

$$= 5000 \times \tan 20° = 1819.851$$

$$\text{FOS} = \frac{\text{Resisting force}}{\text{Actuating force}} = \frac{1819.581}{424.347} = 4.289$$

9.32 (a)

For passive earth pressure conditions,

$K_p > 1$ (Generally)

$$K_p = \frac{\sigma_h}{\sigma_v} > 1$$

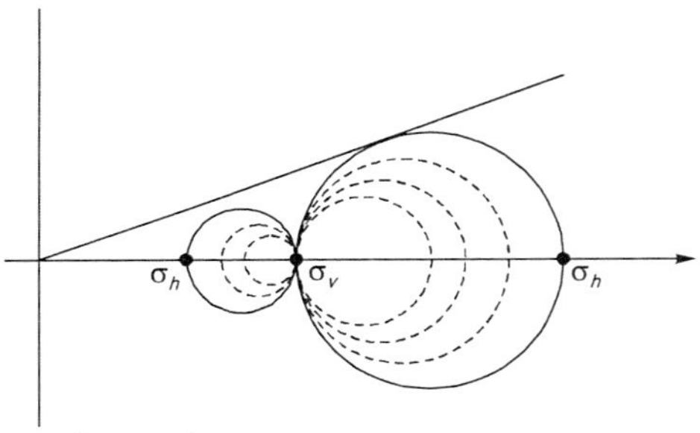

$$\sigma'_h > \sigma'_v$$

9.33 (c)

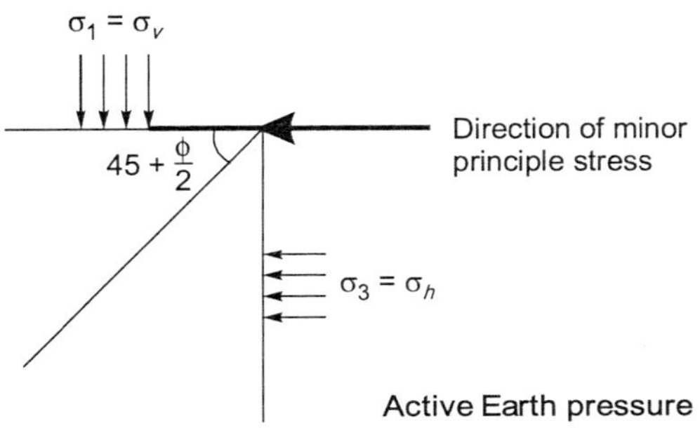

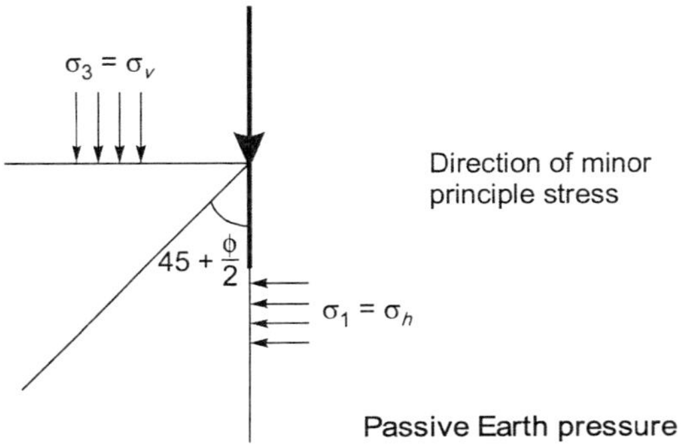

9.34 Sol.

Active earth pressure at base,

$$\therefore \quad \sigma = k_a(\gamma z + q) - 2C\sqrt{ka}$$

where, $$k_a = \frac{1-\sin\phi}{1+\sin\phi} = \frac{1-\sin25°}{1+\sin25°} = 0.406$$

$$q = 20 + 18 \times 3 = 74 \text{ kN/m}^2$$

$$\gamma z = 19 \times 4 = 76 \text{ kN/m}^2$$

$$C = 20 \text{ kN/m}^2 \qquad [\because \text{Given}]$$

So, $\sigma = 0.406\times(76+74) - 2\times20\times\sqrt{0.406}$

$$= 35.413 \text{ kN/m}^2 \simeq 35.4 \text{ kN/m}^2$$

9.35 Sol.

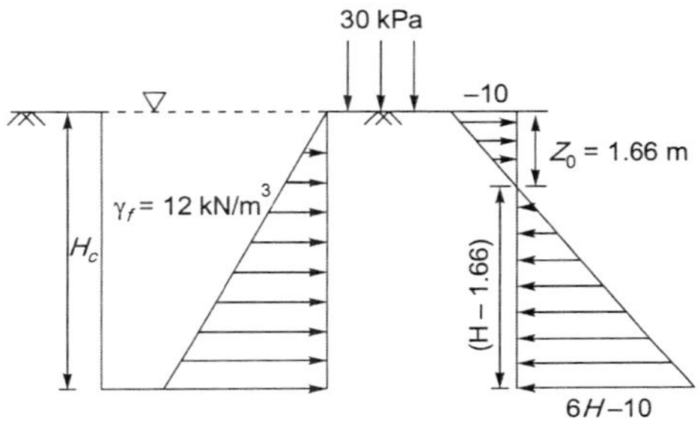

Now, active pressure at depth z,

$$\sigma_z = k_a\gamma z - 2C\sqrt{k_a} - 12z + kq$$

At $z = 0$

[$\because$ $k = 1$ for $\phi = 0°$, $q = 30$ kN/m^2; $C = 20$ kN/m^2]

$$\sigma_z = -2 \times C + q = 2 \times 20 + 30$$

$$= -40 + 30 = -10 \text{ kN/m}^2$$

At z_0, $\quad \sigma_z = 0$

$$18z_0 - 40 - 12z_0 + 30 = 0$$

$$\Rightarrow \quad 6z_0 - 10 = 0$$

$$\Rightarrow \quad z_0 = 1.66 \text{ m}$$

At depth $H, \sigma_H = k_a\gamma H - 2C\sqrt{k_a} - 12H + k_q$

So, $$\sigma_H = 18H - 40 - 12H + 30$$

$$= 6H - 10$$

For unsupported depth of excavation, total active thrust must be zero.

So, $$\frac{1}{2}\times10\times1.66 = \frac{1}{2}\times(H-1.66)\times(6H-10)$$

$$\Rightarrow \quad H = \frac{10}{3} = 3.33 \text{ m}$$

Alternatively:

$$\sigma_v = q + \gamma z$$
$$= 30 + 18z$$
$$p_a = k_a\sigma_v - 2c\sqrt{k_a}$$
$$p_a = 30 + 18z - 2 \times 20 = 18z - 10$$

For critical depth, p_a = 0

$$18z_c - 10 = 0$$
$$z_c = \frac{10}{18}\text{m} = 0.556 \text{ m}$$

For the maximum depth of unsupported excavation, the active earth pressure force at that depth should be zero.

$$P_a = \frac{1}{2}(18H_c - 10)(H_c - 0.56) - \frac{1}{2}10 \times 0.56 - \frac{1}{2}\gamma_f H_c^2$$
$$0 = (9H_c - 5)(H_c - 0.56) - 5 \times 0.56 - 6H_c^2$$
$$0 = 3H_c^2 - 10.04H_c + 5 \times 0.56 - 5 \times 0.56$$
$$3H_c^2 - 10.04H_c = 0$$
$$H_c = 0,\ 3.33 \text{ m}$$

Therefore, the maximum depth of unsupported excavation is 3.33 m.

9.36 (c)

9.37 (c)

For Rankine's active pressure condition

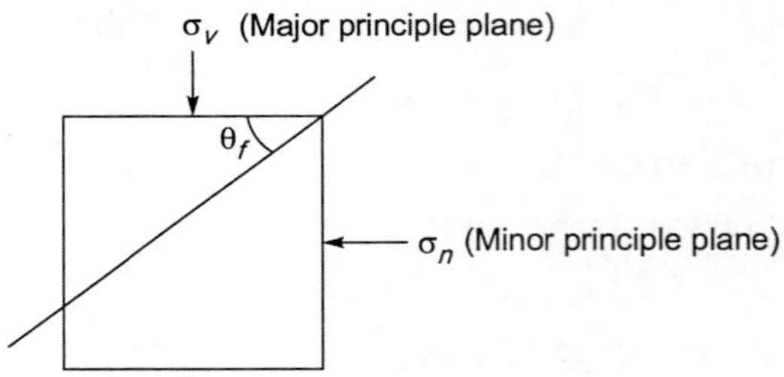

where, $\theta_f = 45° + \frac{\phi}{2} = 45° + \frac{30°}{2} = 60°$

For Rankine's passive pressure condition

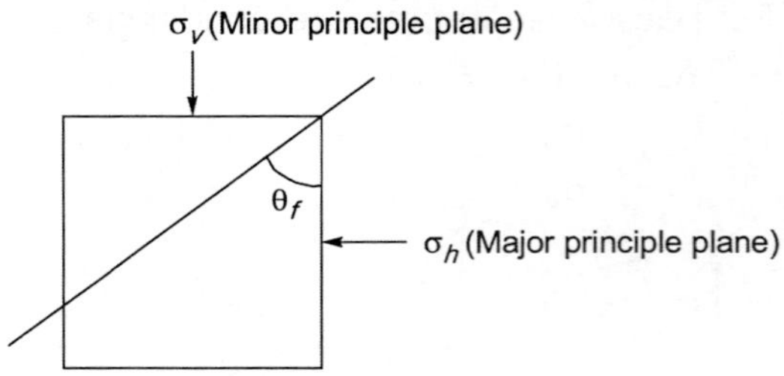

where, $\theta_f = 45° + \frac{\phi}{2} = 45° + \frac{30°}{2} = 60°$

9.38 (b)

10 Stability Analysis of Slopes

10.1 A 40° slope is excavated to a depth of 10 m is a deep layer of saturated clay of unit weight 20 kN/m^3; the relevant shear strength parameters are c_u = 72 kN/m^2 and ϕ_u = 0. The rock ledge is at a great depth. The Taylor's stability coefficient for ϕ_u = 0 and 40° slope angle is 0.18. The factor of safety of the load is

(a) 2.0 (b) 2.1
(c) 2.2 (d) 2.3 [1998 : 2 M]

10.2 A deep cut of 7 m has to be made in a clay with unit weight 16 kN/m^3 and a cohesion of 25 kN/m^2. What will be the factor of safety is one has to have a slope angle of 30°? Stability number is given to be 0.178 (from Taylor's chart) for a depth factor of 3.

(a) 0.80 (b) 1.1
(c) 1.25 (d) 1.0 [2000 : 1 M]

10.3 With respect to a c-ϕ soil in an infinite slope, identify if the following two statements are True or False.

I. The stable slope angle can be grater than ϕ
II. The factor of safety of the slope does not depend on the height of soil in the slope.

(a) Both statements are False
(b) I is True but II is False
(c) I is False but II is True
(d) Both statements are True [2001 : 1 M]

10.4 The critical slip circle for a slope is shown below along with the soil properties

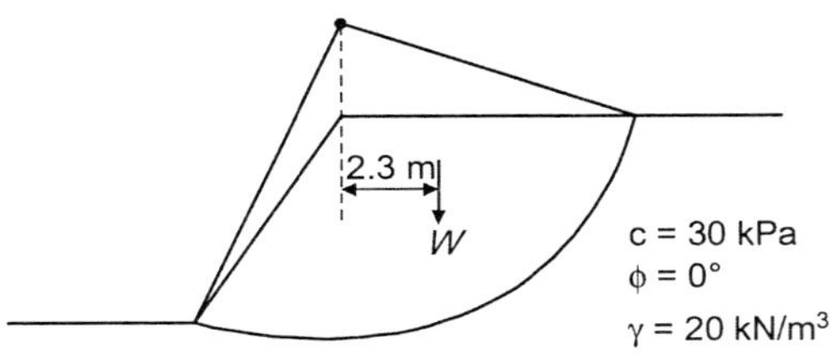

The length of the arc of the slip circle is 15.6 m and the area of soil within the slip circle is 82 m^2. The radius of the slip circles is 10.3 m. The factor of safety against the slip circle failure is nearly equal to

(a) 1.05 (b) 1.22
(c) 0.78 (d) 1.28 [2001 : 2 M]

10.5 An infinite slope is to be constructed in a soil. The effective stress strength parameters of the soil are c' = 0 and ϕ' = 30°. The saturated unit weight of the slope is 20 kN/m^3 and the unit weight of water is 10 kN/m^3. Assuming that seepage is occurring parallel to the slope, the maximum slope angle for a factor of safety of 1.5 would be

(a) 10.89° (b) 11.30°
(c) 12.48° (d) 14.73° [2002 : 2 M]

10.6 A granular soil possesses saturated density of 20 kN/m^3. Its effective angle of internal friction is 35 degrees. If the desired factor of safety is 1.5, the safe angle of slope for this soil, when seepage occurs at and parallel to the slope surface, will be

(a) 25° (b) 23°
(c) 20° (d) 13° [2003 : 1 M]

Common Data for Questions 10.7 and 10.8:

A canal having side slopes 1:1 is proposed to be constructed in a cohesive soil to a depth of 10 m below the ground surface. The soil properties are ϕ_u = 15°, c_u = 12 kPa, e = 1.0, G_s = 2.65.

10.7 If Taylor's Stability Number, S_n is 0.08 and if the canal flows full, the factor of safety with respect to cohesion against failure of the canal bank slopes will be

(a) 3.7 (b) 1.85
(c) 1.0 (d) None of these
[2003 : 2 M]

10.8 If there is a sudden drawdown of water in the canal and if Taylor's Stability Number for the reduced value of ϕ_w is 0.126, the factor of safety with respect to cohesion against the failure of bank slopes will be

(a) 1.85 (b) 1.18
(c) 0.84 (d) 0.53 [2003 : 2 M]

10.9 An infinite soil slope with an inclination of 35° is subjected to seepage parallel to its surface. The soil has c' = 100 kN/m^2 and ϕ' = 30°. Using the concept of mobilized cohesion and friction, at a factor of safety of 1.5 with respect to shear strength, the mobilized friction angle is

(a) 20.02° (b) 21.05°
(c) 23.33° (d) 30.00° [2004 : 2 M]

10.10 For two infinite slopes (one in dry condition and other in submerged condition) in a sand deposit having the angle of shearing resistance 30°, factor of safety was determined as 1.5 (for both slopes). The slope angles would have been
(a) 21.05° for dry slope and 21.05° for submerged slope
(b) 19.47° for dry slope and 18.40° for submerged slope
(c) 18.4° for dry slope and 21.05° for submerged slope
(d) 22.6° for dry slope and 19.47° for submerged slope [2005 : 2 M]

10.11 List-I below gives the possible types of failure for a finite soil slope and List-II gives the reasons for these different types of failure. Match the items in List-I with the items in List-II and select the correct answer from the codes given below the lists:

List-I
A. Base failure
B. Face failure
C. Toe failure

List-II
1. Soils above and below the toe have same strength
2. Soil above the toe is comparatively weaker
3. Soil above the toe is comparatively stronger

Codes:

	A	B	C
(a)	1	2	3
(b)	2	3	1
(c)	2	1	3
(d)	3	2	1

[2006 : 2 M]

10.12 The factor of safety of an infinite soil slope shown in the figure having the properties $c = 0$, $\phi = 35°$, $\gamma_{dry} = 16$ kN/m^3 and $\gamma_{sat} = 20$ kN/m^3 is approximately equal to

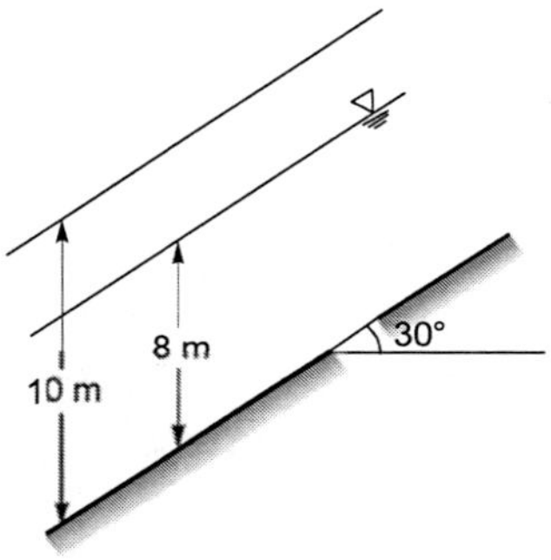

(a) 0.70 (b) 0.80
(c) 1.00 (d) 1.20 [2007 : 2 M]

10.13 The soil profile above the rock surface at 25° infinite slope is shown in figure where S_u is undrain shear stress and γ_t is total unit weight. The slip will occur at a depth of

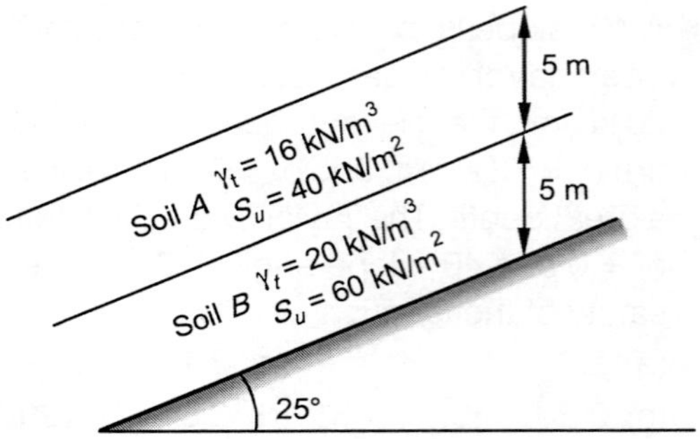

(a) 8.83 m (b) 9.79 m
(c) 7.83 m (d) 6.53 m [2013 : 2 M]

10.14 A long slope is formed in a soil with shear strength parameters: $c' = 0$ and $\phi' = 34°$. A firm stratum lies below the slope and it is assumed that the water table may occasionally rise to the surface, with seepage taking place parallel to the slope. Use $\gamma_{sat} = 18$ kN/m^3 and $\gamma_w = 10$ kN/m^3. The maximum slope angle (in degrees) to ensure a factor of safety of 1.5, assuming a potential failure surface parallel to the slope, would be
(a) 45.3 (b) 44.7
(c) 12.3 (d) 11.3

[2014 : 1 M, Set-I]

10.15 An infinitely long slope is made up of a c-φ soil having the properties: cohesion (c) = 20 kPa, and dry unit weight (γ_d) = 16 kN/m^3. The angle of inclination and critical height of the slope are 40° and 5 m, respectively. To maintain the limiting equilibrium, the angle of internal friction of the soil (in degrees) is ________. [2014 : 2 M, Set-II]

10.16 In friction circle method of slope analysis, if r defines the radius of the slip circle, the radius of friction circle is
(a) $r \sin \phi$ (b) r
(c) $r \cos \phi$ (d) $r \tan \phi$

[2015 : 1 M, Set-II]

10.17 The infinite sand slope shown in the figure is on the verge of sliding failure. The ground water table coincides with the ground surface. Unit weight of water $\gamma_w = 9.81$ kN/m^3.

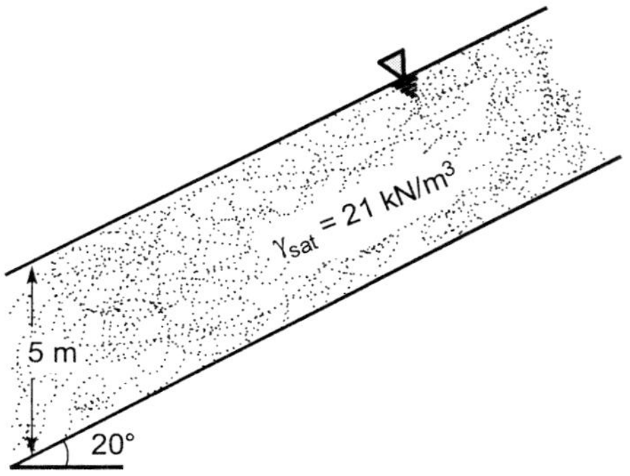

The value of the effective angle of internal friction (in degrees, up to one decimal place) of the sand is ________. [2017 : 2 M, Set-I]

10.18 For the construction of a highway, a cut is to be made as shown in the figure.

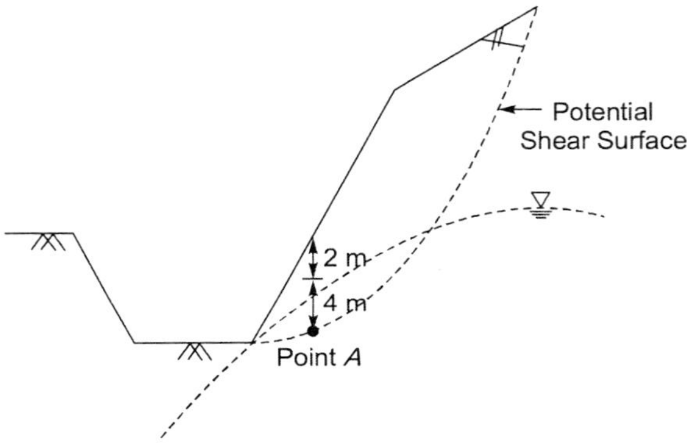

The soil exhibits c' = 20 kPA, ϕ' = 18°, and the undrained shear strength = 80 kPa. The unit weight of water is 9.81 kN/m³. The unit weights of the soil above and below the ground water table are 18 and 20 kN/m³, respectively. If the shear stress at point A is 50 kPa, the factors of safety against the shear failure at this point, considering the undrained and drained conditions, respectively, would be

(a) 1.6 and 0.9
(b) 0.9 and 1.6
(c) 0.6 and 1.2
(d) 1.2 and 0.6

[2017 : 2 M, Set-II]

10.19 A granular soil has a saturated unit weight of 20 kN/m³ and an effective angle of shearing resistance of 30°. The unit weight of water is 9.81 kN/m³. A slope is to be made on this soil deposit in which the seepage occurs parallel to the slope up to the free surface. Under this seepage condition for a factor of safety of 1.5, the safe slope angle (in degree. round off to 1 decimal place) would be________.

[2019 : 2 M, Set-I]

10.20 A 3 m × 3 m square precast reinforced concrete segments to be installed by pushing them through an existing railway embankment for making an underpass as shown in the figure. A reaction arrangement using precast PCC blocks placed on the ground is to be made for the jacks.

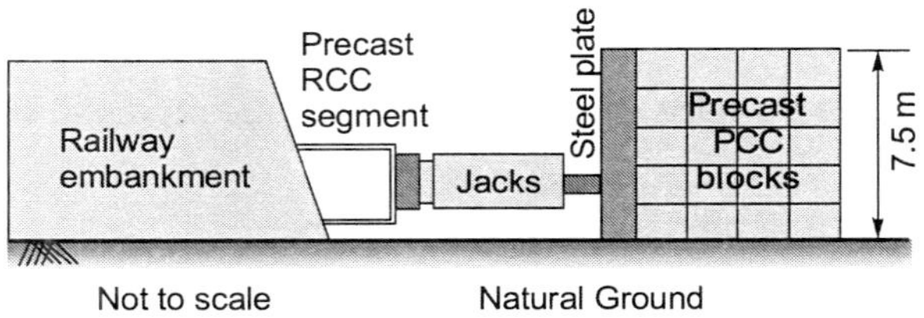

At each stage, the jacks are required to apply a force of 1875 kN to push the segment. The jacks will react against the rigid steel plate placed against the reaction arrangement. The footprint area of reaction arrangement on natural ground is 37.5 m². The unit weight of PCC block is 24 kN/m³. The properties of the natural ground are: c = 17 kPa. ϕ = 25° and γ = 18 kN/m³. Assuming that the reaction arrangement has rough interface and has the same properties that of soil, the factor of safety (round off to 1 decimal place) against shear failure is________.

[2019 : 2 M, Set-I]

10.21 An earthen dam of height H is made of cohesive soil whose cohesion and unit weight are c and γ, respectively. If the factor of safety against cohesion is F_c, the Taylor's stability number (S_n) is

(a) $\dfrac{c}{F_c \gamma H}$
(b) $\dfrac{cF_c}{\gamma H}$
(c) $\dfrac{\gamma H}{cF_c}$
(d) $\dfrac{F_c \gamma H}{c}$

[2019 : 1 M, Set-II]

10.22 A fully submerged infinite sandy slope has an inclination of 30° with the horizontal. The saturated unit weight and effective angle of internal friction of sand are 18 kN/m³ and 38°, respectively. The unit weight of water is 10 kN/m³. Assume that the seepage is parallel to the slope. Against shear failure of the slope, the factor of safety (round off to two decimal places) is ________.

[2020 : 1 M, Set-I]

10.23 A 10 m high slope of dry clay soil (unit weight = 20 kN/m³), with a slope angle of 45° and the circular slip surface, is shown in the figure (not drawn to the scale). The weight of the slip wedge is denoted by W. The undrained unit cohesion (c_u) is 60 kPa.

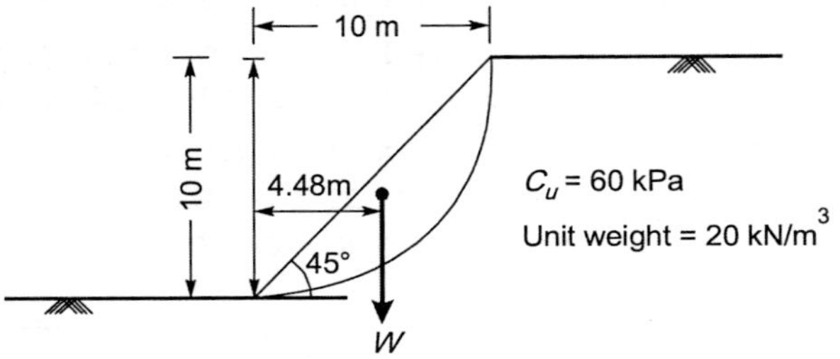

The factor of safety of the slope against slip failure, is

(a) 0.58 (b) 1.84
(c) 1.57 (d) 1.67

[2020 : 2 M, Set-II]

10.24 An unsupported slope of height 15 m is shown in the figure (not to scale), in which the slope face makes an angle 50° with the horizontal. The slope material comprises purely cohesive soil having undrained cohesion 75 kPa. A trial slip circle KLM with a radius 25 m, passes through the crest and toe of the slope and it subtends an angle 60° at its centre O. The weight of the active soil mass (W, bounded by KLMN) is 2500 kN/m, which is acting at a horizontal distance of 10 m from toe of the slope. Consider the water table to be present at a very large depth from the ground surface.

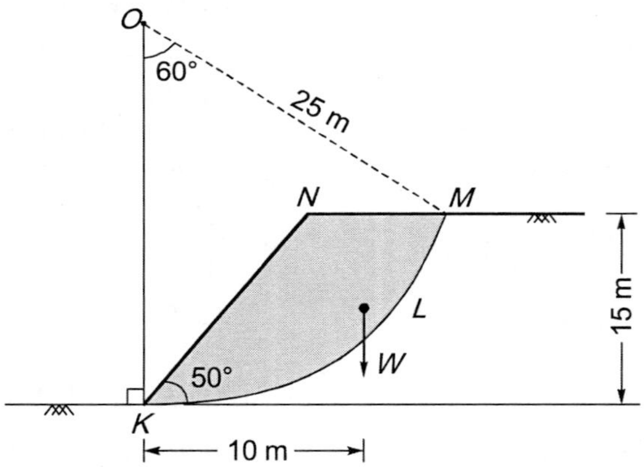

Considering the trial slip circle KLM, the factor of safety against the failure of slope under undrained condition (round off to two decimal places) is ___________

[2021 : 2 M, Set-I]

10.25 Three soil samples are taken from different locations (I, II and III) of the potential failure plane. Which is the most appropriate shear strength test for each of the sample to identify the failure mechanism? Identify the correct combination from the following options:

P: Triaxial compression test
Q: Triaxial extension test
R: Direct shear or shear box test
S: Vane shear test

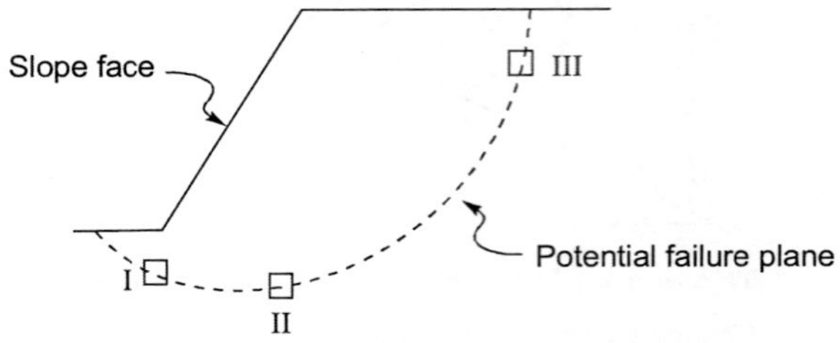

(a) I-Q, II-R, III-P (b) I-R, II-P, III-Q
(c) I-S, II-Q, III-R (d) I-P, II-R, III-Q

[2023 : 1 M, Set-I]

10.26 A soil having the average properties, bulk unit weight = 19 kN/m^3; angle of internal friction = 25° and cohesion = 15 kPa, is being formed on a rock slope existing at an inclination of 35° with the horizontal. The critical height (in m) of the soil formation up to which it would be stable without any failure is ___________ (round off to one decimal place).

[Assume the soil is being formed parallel to the rock bedding plane and there is no ground water effect.]

[2023 : 2 M, Set-I]

10.27 An infinite slope is made up of cohesionless soil with seepage parallel to and upto the sloping surface. The angle of slope is 30° with respect to horizontal ground surface. The unit weight of the saturated soil and water are 20 kN/m^3 and 10 kN/m^3 respectively.

The minimum angle of shearing resistance of soil (in degrees) for the critically stable condition of the slope is _____. (rounded off to the nearest integer).

[2024 : 2 M, Set-I]

Answers Stability Analysis of Slopes

10.1 (a)	10.2 (c)	10.3 (b)	10.4 (d)	10.5 (a)	10.6 (d)	10.7 (b)
10.8 (d)	10.9 (b)	10.10 (a)	10.11 (d)	10.12 (a)	10.13 (a)	10.14 (d)
10.15 (22.44)	10.16 (a)	10.17 (34.3)	10.18 (a)	10.19 (11.1)	10.20 (2.0)	10.21 (a)
10.22 (0.601)	10.23 (*)	10.24 (1.96)	10.25 (a)	10.26 (5)	10.27 (49.10)	

Explanations Stability Analysis of Slopes

10.1 (a)

Taylors stability coefficient,

$$S_n = \frac{C}{\gamma F_c H}$$

$$\Rightarrow \quad F_c = \frac{C}{\gamma H S_n} = \frac{72}{20 \times 10 \times 0.18} = 2$$

10.2 (c)

$H = 7$ m; $\gamma = 16$ kN/m^3, $C = 25$ kN/m^2

$s = 0.178$ for (D.F. = 3 and $i = 30°$), $F_c = ?$

$$s = \frac{C}{F_c \gamma H}, \quad F_c = \frac{25}{0.178 \times 16 \times 7} = 1.25$$

Note: *In this question, there is no use of depth factor = 3*

Depth factor, $n_d = \dfrac{D+H}{H}$

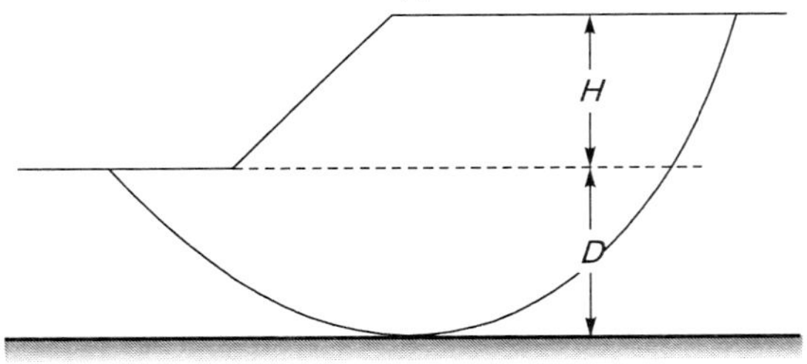

L = Depth of hard stratum below toe

H = Height of slope above toe

10.3 (b)

- In case of cohesionless soil, slope angle cannot be greater than ϕ.
- FOS depend on height of soil in the slope.

10.4 (d)

$$\text{FOS} = \frac{\text{Resisting moment}}{\text{Actuating moment}}$$

$$= \frac{CLR}{W \times 2.3} = \frac{30 \times 15.6 \times 10.3}{20 \times 82 \times 2.3} = 1.28$$

10.5 (a)

$$\tan \phi_w = \frac{\gamma'}{\gamma_{sat}} \left(\frac{1}{F_d} \tan \phi' \right)$$

$$= \frac{10}{20} \left(\frac{1}{1.5} \tan 30° \right) \text{ or } \phi_w = 10.89°$$

10.6 (d)

Given $\phi' = 35°$; $\gamma_{sat} = 20$ kN/m^3;

$F = 1.5$; $\beta = ?$; $\gamma' = \gamma_{sat} - \gamma_w$

Factor of safety,

$$F = \frac{\gamma' \tan \phi'}{\gamma_{sat} \tan \beta}$$

$$\Rightarrow \quad 1.5 = \frac{(\gamma_{sat} - \gamma_w) \tan 35°}{\gamma_{sat} \tan \beta}$$

$$\Rightarrow \quad \tan \beta = \frac{(20 - 10) \tan 35°}{1.5 \times 20}$$

$$\Rightarrow \quad \beta = 13.14° \simeq 13°$$

10.7 (b)

When the canal flows full, the canal bank slopes will be submerged under water.

$$\therefore \quad \gamma' = \left(\frac{G_s - 1}{1 + e} \right) \gamma_w = \left(\frac{2.65 - 1}{1 + 1} \right) \times 9.81$$

$$= 8.09 \text{ kN/m}^3$$

Taylor's stability number,

$$S_n = \frac{c_u}{F_c \, \gamma_{sub} \, H}$$

$$\Rightarrow \quad F_c = \frac{c_u}{S_n \times \gamma_{sub} \times H} = \frac{12}{0.08 \times 8.09 \times 10}$$

$$= 1.85$$

10.8 (d)

Even if there is a sudden drawdown of water in the canal, the soil remains saturated for sometime. Hence, unit weight of the soil will be the saturated unit weight i.e.

$$\gamma_{sat} = \gamma' + \gamma_w = 8.09 + 9.81$$

$$= 17.9 \text{ kN/m}^3$$

$\therefore$ Taylor's stability number,

$$S_n = \frac{c_u}{F_c \, \gamma_{sat} \, H}$$

$$\Rightarrow \quad F_c = \frac{c_u}{S_n \times \gamma_{sat} \times H}$$

$$= \frac{12}{0.126 \times 17.9 \times 10} = 0.53$$

10.9 (b)

The factor of safety with respect to shear strength is given by,

$$F_s = \frac{\tau_f}{\tau} = \frac{c' + \sigma_n' \tan\phi'}{\tau}$$

The shear strength mobilised, thus may be written as,

$$\tau = \frac{c'}{F_s} + \sigma_n' \frac{\tan\phi'}{F_s}$$

$$= c_m + \sigma_n' \tan \phi_m'$$

$$\therefore \quad \tan\phi_m' = \frac{\tan\phi'}{F_s}$$

$$\Rightarrow \quad \phi_m' = \tan^{-1}\left(\frac{\tan 30^\circ}{1.5}\right) = 21.05^\circ$$

10.10 (a)

For dry or submerged slopes, Factor of safety

$$= \frac{\tan\phi}{\tan\beta}$$

$$\Rightarrow \quad \tan\beta = \frac{\tan 30^\circ}{1.5}$$

$$\Rightarrow \quad \beta = 21.05^\circ$$

Note:

- *There is difference between submerged condition and saturated condition. In submerged condition seepage does not take place through the soil parallel to slope, so in calculation of FOS, both in numerator and denominator we use γ_{sub}.*
- *When soil is in saturated condition, seepage takes place and water table is assumed to be parallel to the slope.*

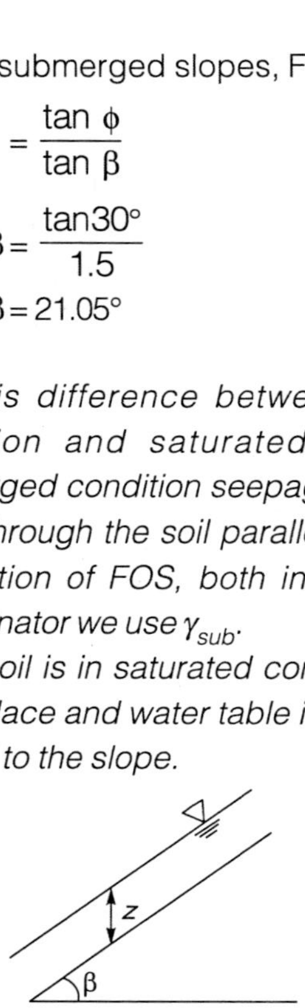

$$FOS = \frac{\gamma_{sub} z \cos^2\beta \tan\phi}{\gamma_{sat} z \cos\beta \sin\beta}$$

- *FOS in submerged condition is more than that in saturated condition.*

10.11 (d)

Face failure or slope failure can occur when the slope angle β is very high and the soil close to the toe is quite strong or the soil in the upper part of slope is relatively weak.

Base failure can occur when the soil below the toe is relatively weak and soft and the slope is flat.

Toe failure occurs in steep slopes when the soil mass above the base and below the base is homogeneous.

10.12 (a)

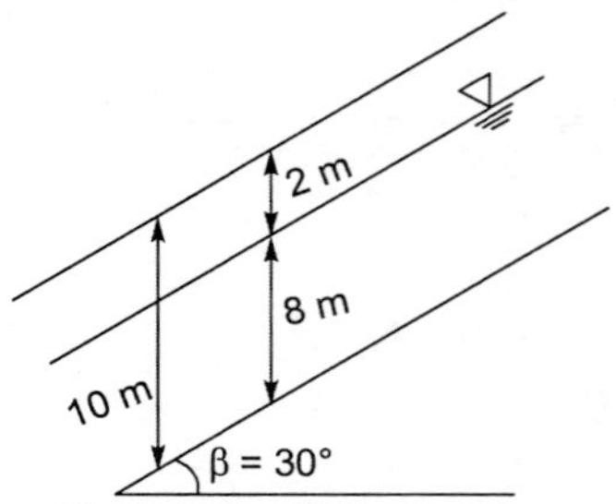

Given, $C = 0, \quad \phi = 35^\circ$

$$\gamma_{dry} = 16 \text{ kN/m}^3$$

$$\gamma_{sat} = 20 \text{ kN/m}^3$$

$$\gamma_{sub} = \gamma_{sat} - \gamma_w = 20 - 10 = 10 \text{ kN/m}^3$$

Factor of safety,

$$F = \left[1 - \frac{\gamma_w h}{\gamma_{avg} z}\right] \frac{\tan\phi}{\tan\beta}$$

Assuming,

$$\gamma_w = 10 \text{ kN/m}^3$$

$$\therefore \quad F = \left[1 - \frac{10 \times 8}{20 \times 10}\right] \frac{\tan 35^\circ}{\tan 30^\circ} = 0.73$$

10.13 (a)

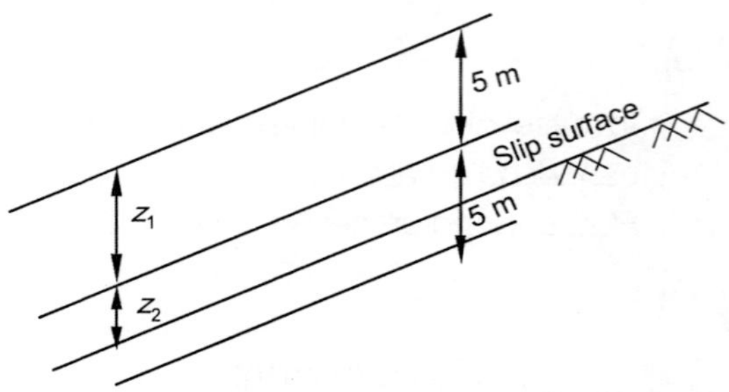

The slip will occur when shear stress is greater then or equal to shear strength.

$$\tau \geq S_u$$

or, $\quad (\gamma_1 z_1 + \gamma_2 z_2)\sin\beta \cos\beta \geq S_u$

$$\Rightarrow \quad \frac{(16 \times 5 + 20 \times z_2)\sin 2\beta}{2} \geq S_u$$

$$\Rightarrow \quad \frac{(80+20z_2)\sin 50°}{2} \geq 60$$

$\therefore \quad z_2 = 3.83$ m

$\therefore$ Depth of slip = 5 + 3.83 = 8.83 m

10.14 (d)

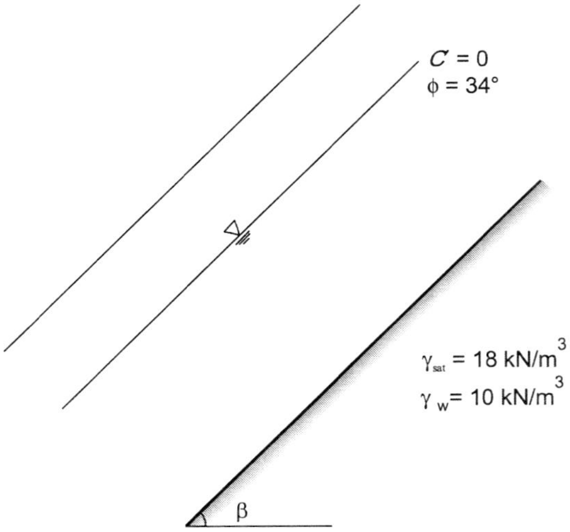

$$\text{FOS} = \frac{\gamma_{sub}}{\gamma_{sat}} \cdot \frac{\tan\phi}{\tan\beta}$$

$$\Rightarrow \quad \tan\beta = \left(\frac{\gamma_{sub}}{\gamma_{sat}} \cdot \frac{\tan\phi}{FOS}\right)$$

$$\Rightarrow \quad \tan\beta = \frac{\gamma_{sub}}{\gamma_{sat}} \cdot \frac{\tan\phi}{1.5}$$

$$\tan\beta = \frac{(18-10)}{18\times 1.5} \times \tan 34°$$

$$\Rightarrow \quad \beta = 11.30°$$

10.15 Sol.

$$F_s = \frac{c}{\gamma H \sin\beta.\cos\beta} + \frac{\tan\phi}{\tan\beta} \geq 1$$

c = 20 kPa, γ = 16 kN/m^3, β = 40° and H = 5 m $\Rightarrow$ $\phi \geq 22.44°$

10.16 (a)

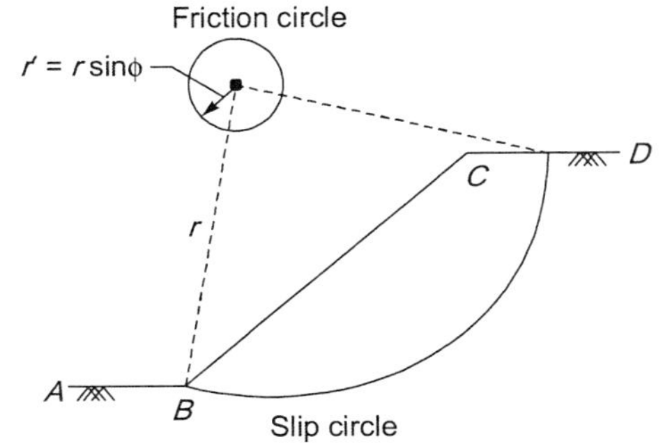

10.17 Sol.

Infinite sand slope:

GWT coincides with the ground surface, therefore

$$\text{F.O.S.} = \frac{\gamma'}{\gamma_{sat}} \times \frac{\tan\phi'}{\tan\beta}$$

As slope is on the verge of sliding failure hence,

$$\text{F.O.S.} = \frac{\gamma' \tan\phi'}{\gamma_{sat} \tan\beta} = 1$$

$$(21 - 9.81)\tan\phi' = 21 \tan 20°$$

$$\phi' = 34.3°$$

10.18 (a)

Drained parameters:

$c' = 20$ kPa, ϕ' = 18°

Undrained shear strength,

$S = 80$ kPa

Normal stress at point A,

$$\bar{\sigma}_n = \sigma_n - u = 2 \times \gamma_b + 4\gamma'$$

$$\bar{\sigma}_n = 2 \times 18 + 4\,(20 - 9.81)$$

$$= 76.76 \text{ kN/m}^2$$

Shear stress at point A = 50 kPa

FOS under **undrained condition**

$$= \frac{\text{Shear strength}}{\text{Sthear stress}} = \frac{80}{50} = 1.6$$

FOS in **drained condition**

$$= \frac{c + \bar{\sigma}_n \tan\phi'}{\tau}$$

$$= \frac{20 + 76.76\tan 18°}{50} = 0.9788$$

10.19 Sol.

γ_{sat} = 20 kN/m^3 ϕ' = 30 γ_w = 9.81 kN/m^3

FOS = 1.5

Seepage at GL

$$\text{FOS} = \frac{\gamma'}{\gamma_{sat}} \times \frac{\tan\phi}{\tan\beta}$$

$$1.5 = \frac{(20-9.81)}{20} \times \frac{\tan 30}{\tan\beta}$$

$$\tan\beta = 0.196$$

$$\beta = 11.089° \simeq 11.1°$$

10.20 Sol.

$$\text{FOS} = \frac{S}{\tau} = \frac{c + \bar{\sigma}_n \tan\phi}{\tau}$$

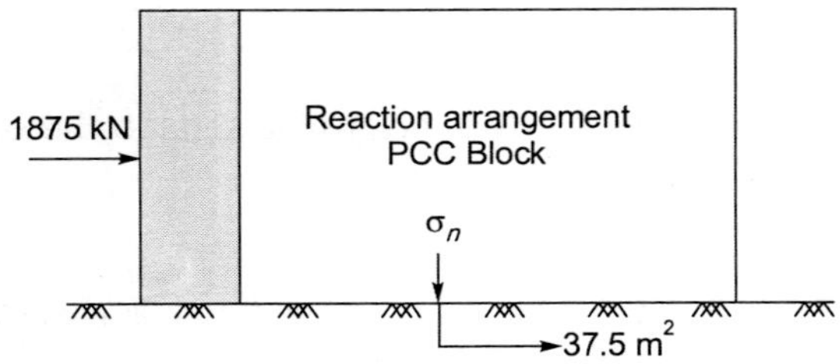

$$\tau = \frac{\text{Jack force}}{\text{Area}} = \frac{1875\text{kN}}{37.5 \text{ m}^2} = 50 \text{ kN/m}^2$$

Shear strength of soil

$$S = c + \bar{\sigma}_n \tan\phi$$

$$\sigma_n = \frac{7.5 \times 37.5 \times \gamma_{concrete}}{\text{Area}}$$

$$\sigma_n = \frac{7.5 \times 37.5 \times 24}{37.5} = 180 \text{ kN/m}^2$$

$$S = c + \bar{\sigma}_n \tan\phi = 17 + 180 \times \tan 25°$$

$$S = 100.935 \text{ kPa}$$

$$\text{FOS} = \frac{S}{\tau} = \frac{100.935}{50} = 2.018 \simeq 2.0$$

10.21 (a)

$$S_n = \frac{c}{\gamma H_c} = \frac{c}{\gamma F_c H} \qquad \left\{\because F_c = \frac{H_c}{H}\right\}$$

10.22 Sol.

$$\text{F.O.S.} = \frac{\gamma'}{\gamma_{sat}} \cdot \frac{\tan\phi}{\tan\beta}$$

$$= \left(\frac{18-10}{18}\right)\frac{\tan 38°}{\tan 30°} = 0.601$$

10.23 (*)

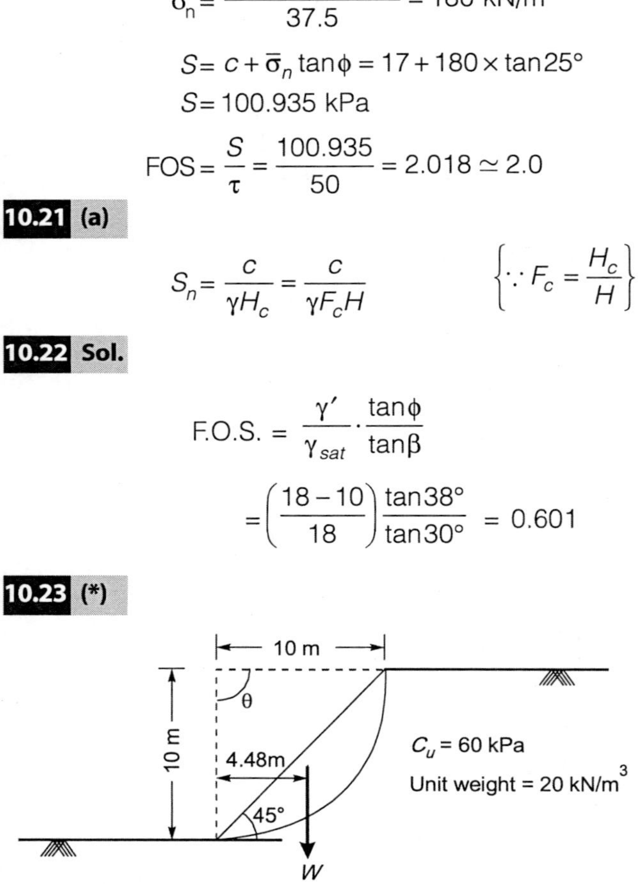

Consider unit length of slope

Area of circular arc

$$= \frac{\theta}{360} \times \pi r^2 - \text{Area of } \Delta$$

$$= \frac{90}{360} \times \pi \times 10^2 - \frac{1}{2} \times 10 \times 10 = 28.54 \text{ m}^2$$

Height of wedge = Volume × γ = (Area × 1) × γ

= 28.54 × 1 × 20 = 570.8 kN

$$\text{FOS} = \frac{M_R}{M_0} = \frac{[c \times (r\theta)] \times r}{W \times x}$$

$$= \frac{60 \times 10 \times \frac{\pi}{2} \times 10}{570.8 \times 4.48} = 3.68$$

10.24 Sol.

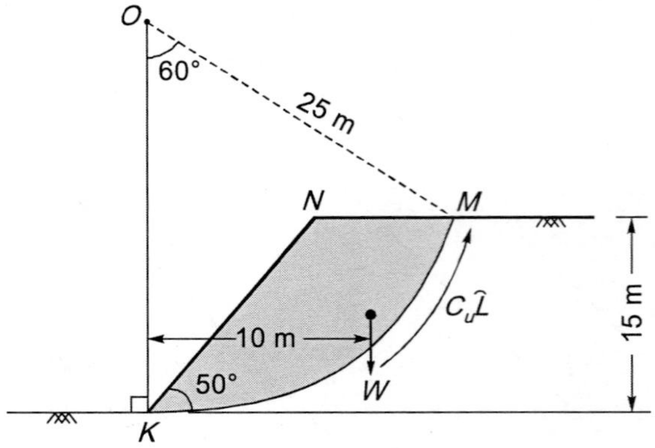

$$\text{FOS} = \frac{\text{Resisting moment}}{\text{Actuating moment}}$$

$$\text{FOS} = \frac{C_u l R}{w\bar{x}}$$

l = Length of ac KLM

$\bar{x}$ = Distance of 'w' from toe

$$\Rightarrow \quad \text{FOS} = \frac{75 \times 2\pi \times 25 \times \frac{60}{360} \times 25}{2500 \times 10}$$

$$\Rightarrow \quad \text{FOS} = 1.96$$

10.25 (a)

10.26 Sol.

For critical height of slope

$$C + \gamma H_C \cos^2\beta \tan\phi = \gamma H_C \sin\beta\cos\beta$$

$$\Rightarrow 15 + 19 \times H_C \times \cos^2 35° \tan 25° = 19 \times H_C \times \sin 35° \cos 35°$$

$$\Rightarrow \quad H_C = 5.03 \text{ m}$$

10.27 Sol.

$$\text{FOS} = \frac{\gamma_{sub}}{\gamma_{sat}} \times \frac{\tan\phi}{\tan\beta}$$

Here, $\beta = 30°$

$\gamma_{sat} = 20 \text{ kN/m}^3$

$\gamma_w = 10 \text{ kN/m}^3$

$\therefore \gamma_{sub} = \gamma_{sat} - \gamma_w = 20 - 10 = 10 \text{ kN/m}^3$

For critically stable condition, FOS = 1

$$\because \quad 1 = \frac{20-10}{20} \times \frac{\tan\phi}{\tan 30°}$$

$$\Rightarrow \quad \phi = 49.10°$$

11 Shallow Foundation and Bearing Capacity

11.1 Between bearing capacity and settlement, the proportioning of a footing in sand is more often governed by settlement. [1995 : 1 M]

11.2 Write True/False in the answer book.
The bulb of pressure under a strip footing forms in the direction of its length. [1995 : 1 M]

11.3 Use of N-values not corrected for overburden pressure leads to ... design of footings at shallow depths

[1995 : 1 M]

11.4 The nature of contact pressure distribution under a rigid footing resting on a sandy soil and subjected to uniformly distributed load is as shown in

(a)
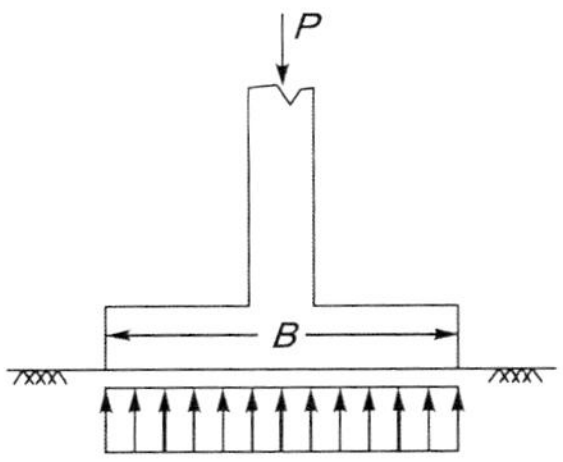

(b)
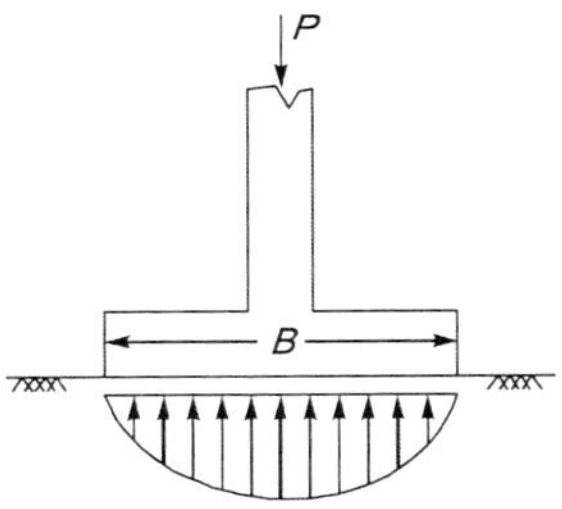

(c)
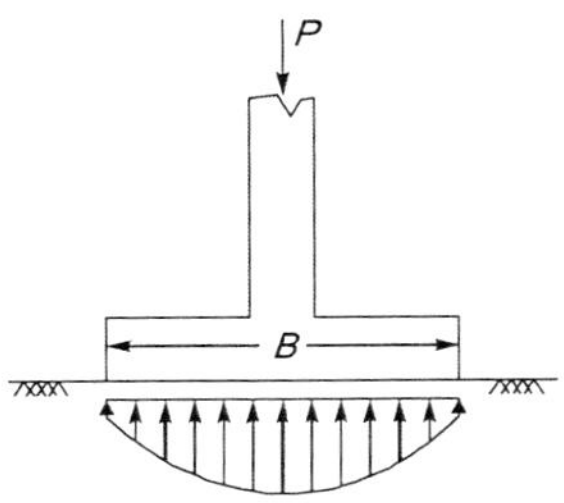

(d)
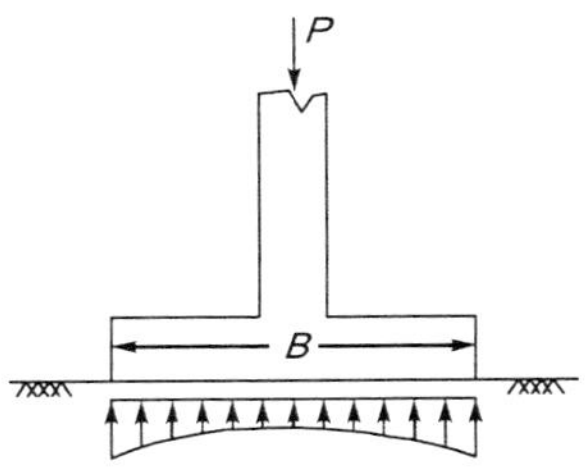

[1996 : 1 M]

11.5 A foundation is considered as shallow if its depth is

(a) less than 1 meter
(b) greater than its width
(c) equal to or less than its width
(d) greater than 1 meter [1996 : 1 M]

11.6 For the strip footing on a saturated clay, for the given failure surface, the bearing capacity equation takes the form

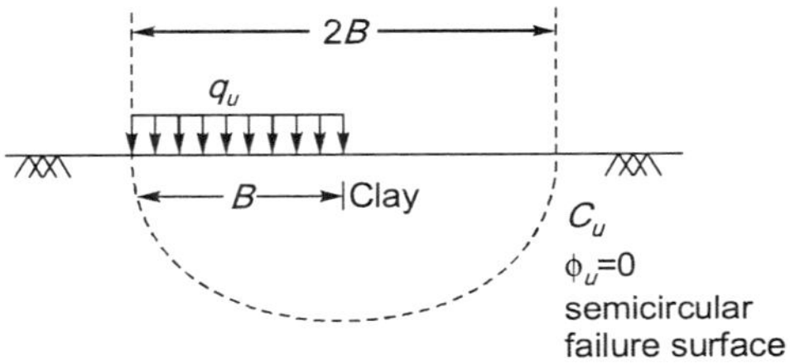

(a) $5.7C_u$ (b) $5.14C_u$
(c) $4\pi C_u$ (d) $2\pi C_u$

Where

C_u = undrained shear strength
ϕ_u = angle of internal friction
B = width of strip footing
q_c = ultimate bearing capacity of soil

[1996 : 1 M]

11.7 The settlement of prototype in clayey soil may be estimated using plate load test data from the following expression

(a) $S_{prototype} = S_{plate} \times \left(\frac{B_{prototype}}{B_{plate}}\right)$

(b) $S_{prototype} = S_{plate} \times \left(\frac{B_{plate}}{B_{prototype}}\right)$

(c) $S_{prototype} = S_{plate} \times \left(\frac{2B_{prototype}}{B_{prototype} + B_{plate}}\right)^2$

(d) $S_{prototype} = S_{plate} \times \left(\frac{B_{prototype} + B_{plate}}{2B_{prototype}}\right)^2$

[1998 : 1 M]

11.8 The two criteria for the determination of allowable bearing capacity of a foundation are
(a) tensile failure and compression failure
(b) tensile failure and settlement
(c) bond failure and shear failure
(d) shear failure and settlement [2000 : 1 M]

11.9 The ultimate bearing capacity of a soil is 300 kN/m^2. The depth of foundation is 1 m and unit weight of soil is 20 kN/m^3. Choosing a factor of safety of 2.5, the net safe bearing capacity is
(a) 100 kN/m^3 (b) 112 kN/m^3
(c) 80 kN/m^3 (d) 100.5 kN/m^3
[2000 : 1 M]

11.10 Two footings, one circular and the other square, are founded on the surface of a purely cohensionless soil. The diameter of the circular footing is same as that of the side of the square footing. The ratio of their ultimate bearing capacities is
(a) 3/4 (b) 4/3
(c) 1.0 (d) 1.3 [2000 : 1 M]

11.11 The following two statements are made with reference to the calculation of net bearing capacity of a footing in pure clay soil ($\phi = 0$) using Terzaghi's bearing capacity theory. Identify if they are True or False.
I. Increase in footing width will result in increase in bearing capacity.
II. Increase in depth of foundation will result in higher bearing capacity
(a) Both statements are True
(b) Both statements are False
(c) I is True but II is False
(d) I is False but II is True [2001 : 1 M]

11.12 The width and depth of a footing are 2 and 1.5 m respectively. The water table at the site is at a depth of 3 m below the ground level. The water table correction factor for the calculation of the bearing capacity of soil is
(a) 0.875 (b) 1.000
(c) 0.925 (d) 0.500 [2001 : 1 M]

11.13 A plate load test was conducted in sand on a 300 mm diameter plate. If the plate settlement was 5 mm at a pressure of 100 kPa, the settlement (in mm) of a 5 m × 8 m rectangular footing at the same pressure will be
(a) 9.4 (b) 18.6
(c) 12.7 (d) 17.8 [2001 : 2 M]

11.14 The observed value of the standard penetration number (N) at 10 m depth of silty sand deposit is 13. The unit weight of the soil is 16 kN/m^3. The N value after correcting for the presence of fines will be
(a) 12 (b) 13
(c) 14 (d) 15 [2002 : 1 M]

11.15 In a plate load test conducted on cohesionless soil, a 600 mm square test plate settles by 15 mm under a load intensity of 0.2 N/mm^2. All conditions remaining the same, settlement of a 1m square footing will be
(a) less than 15 mm
(b) greater than 25 mm
(c) 15.60 mm
(d) 20.50 mm [2003 : 1 M]

11.16 Two circular footings of diameters D_1 and D_2 are resting on the surface of the same purely cohesive soil. The ratio of their gross ultimate bearing capacities is
(a) $\frac{D_1}{D_2}$ (b) 1.0
(c) $\left(\frac{D_1}{D_2}\right)^2$ (d) $\frac{D_2}{D_1}$ [2004 : 1 M]

11.17 The figure given below represents the contact pressure distribution underneath a

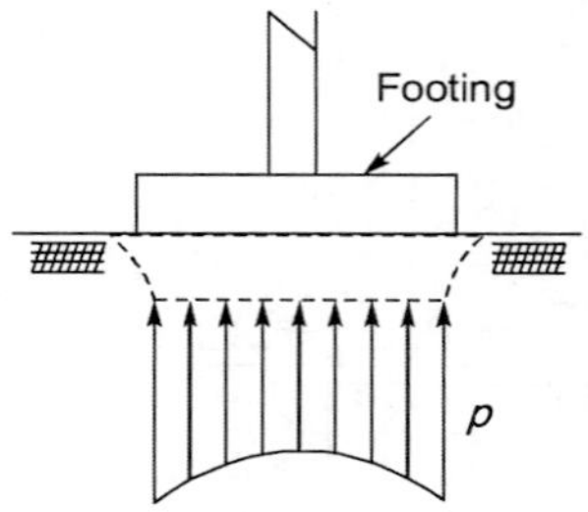

(a) rigid footing on saturated clay
(b) rigid footing on sand
(c) flexible footing on saturated clay
(d) flexible footing on sand
[2004 : 2 M]

11.18 There are two footings resting on the ground surface. One footing is square of dimension 'B'. The other is strip footing of width 'B'. Both of them are subjected to a loading intensity of q. The pressure intensity at any depth below the base of the footing along the centre line would be
(a) equal in both footings
(b) large for square footing and small for strip footing

(c) large for strip footing and small for square footing
(d) more for strip footing at shallow depth ($\leq B$) and more for square footing at large depth ($>B$)

[2005 : 1 M]

11.19 A strip footing (8 m wide) is designed for a total settlement of 40 mm. The safe bearing capacity (shear) was 150 kN/m^2 and safe allowable soil pressure was 100 kN/m^2. Due to importance o f the structure, now the footing is to be redesigned for total settlement of 25 mm. The new width of the footing will be

(a) 5 m (b) 8 m
(c) 12 m (d) 12.8 m [2005 : 2 M]

11.20 During the subsurface investigations for design of foundations, a standard penetration test was conducted at 4.5 m below the ground surface. The record of number of blows is given below:

Penetration depth (cm)	No. of blows
0 – 7.5	3
7.5 – 15	3
15 – 22.5	6
22.5 – 30	6
30 – 37.5	8
37.5 – 45	7

Assuming the water table at ground level, soil as fine sand and correction factor for overburden as 1.0, the corrected 'N' value for the soil would be

(a) 18 (b) 19
(c) 21 (d) 33 [2005 : 2 M]

11.21 The number of blows observed in a Standard Penetration Test (SPT) for different penetration depths are given as follows:

Penetration of sampler	No. of blows
0 – 150 mm	6
150 – 300 mm	8
300 – 450 mm	10

The observed N value is

(a) 8 (b) 14
(c) 18 (d) 24 [2007 : 1 M]

11.22 The bearing capacity of a rectangular footing of plan dimensions 1.5 m × 3 m resting on the surface of a sand deposit was estimated as 600 kN/m^2 when the water table is far below the base of the footing. The bearing capacities in kN/m^2 when the water level rises to depths of 3 m, 1.5 m and 0.5 m below the base of the footing are

(a) 600, 600, 400 (b) 600, 450, 350
(c) 600, 500, 250 (d) 600, 400, 250

[2007 : 2 M]

11.23 A test plate 30 cm × 30 cm resting on a sand deposit settles by 10 mm under a certain loading intensity. A footing 150 cm × 200 cm resting on the same sand deposit and loaded to the same load intensity settles by

(a) 2.0 mm (b) 27.8 mm
(c) 30.2 mm (d) 50.0 mm

[2008 : 2 M]

Linked Answer Questions 11.24 and 11.25:

A column is supported on a footing as shown in the figure below. The water table is at a depth of 10 m below the base of the footing.

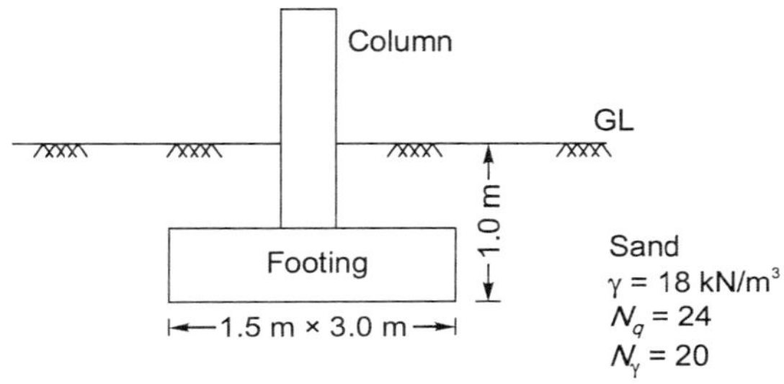

11.24 The net ultimate bearing capacity (kN/m^2) of the footing based on Terzaghi's bearing capacity equation is

(a) 216 (b) 432
(c) 630 (d) 846 [2008 : 2 M]

11.25 The safe load (kN) that the footing can carry with a factor of safety 3 is

(a) 282 (b) 648
(c) 945 (d) 1269 [2008 : 2 M]

11.26 Dilatancy correction is required when a strata is

(a) cohesive and saturated and also has N value of SPT > 15
(b) saturated silt/fine sand and N value of SPT < 10 after the overburden correction
(c) saturated silt/fine sand and N value of SPT > 15 after the overburden correction
(d) coarse sand under dry condition and N value of SPT < 10 after the overburden correction

[2009 : 1 M]

11.27 A plate load test is carried out on a 300 mm × 300 mm plate placed at 2 m below the ground level to determine the bearing capacity of a 2 m × 2 m footing placed at same depth of 2 m on a homogeneous sand deposit extending 10 m below ground. The ground water table is 3 m below the ground level. Which of the following factors does not require a correction to the bearing capacity determined based on the load test?

(a) Absence of the overburden pressure during the test
(b) Size of the plate is much smaller than the footing size
(c) Influence of the ground water table
(d) Settlement is recorded only over a limited period of one or two days [2009 : 2 M]

Common Data for Questions 11.28 and 11.29:
Examine the test arrangement and the soil properties given below:

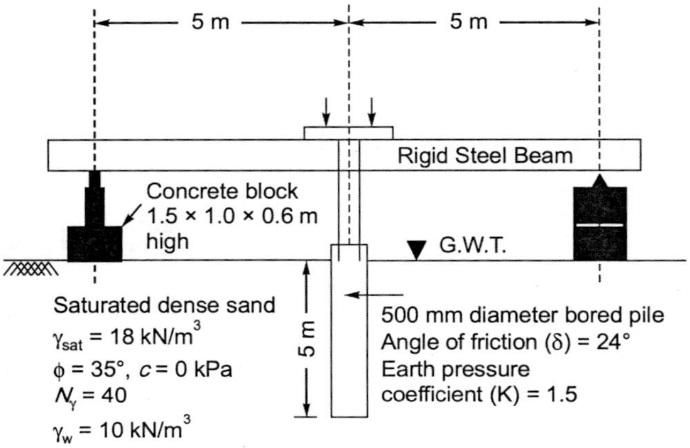

11.28 The maximum pressure that can be applied with a factor of safety of 3 through the concrete block, ensuring no bearing capacity failure in soil using Terzaghi's bearing capacity equation without considering the shape factor, depth factor and inclination factor is

(a) 26.67 kPa (b) 60 kPa
(c) 90 kPa (c) 120 kPa [2009 : 2 M]

11.29 The maximum resistance offered by the soil through skin friction while pulling out the pile from the ground is

(a) 104.9 kN (b) 209.8 kN
(c) 236 kN (d) 472 kN

[2009 : 2 M]

Linked Answer Questions 11.30 and 11.31:
The unconfined compressive strength of a saturated clay sample is 54 kPa.

11.30 The value of cohesion for the clay is

(a) zero (b) 13.5 kPa
(c) 27 kPa (d) 54 kPa [2010 : 2 M]

11.31 If a square footing of size 4 m × 4 m is resting on the surface of a deposit of the above clay, the ultimate bearing capacity of the footing (as per Terzaghi's equation) is

(a) 1600 kPa (b) 316 kPa
(c) 200 kPa (d) 100 kPa

[2010 : 2 M]

11.32 Likelihood of general shear failure for an isolated footing in sand decreases with

(a) decreasing footing depth
(b) decreasing inter-granular packing of the sand
(c) increasing footing width
(d) decreasing soil grain compressibility

[2011 : 1 M]

11.33 Two geometrically identical isolated footings, *X* (linear elastic) and *Y* (rigid), are loaded identically (shown alongside). The soil reactions will

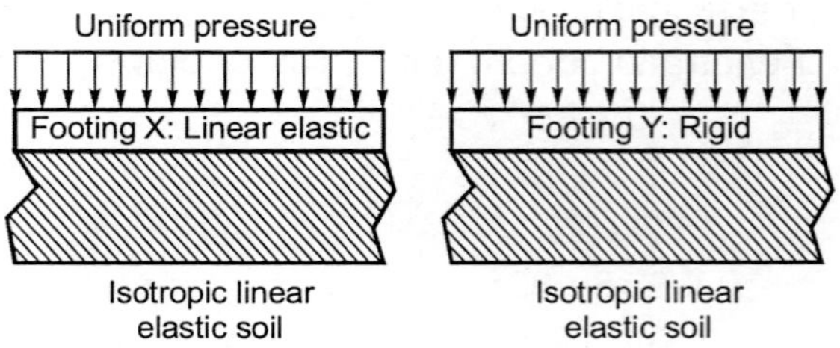

(a) be uniformly distributed for *Y* but not for *X*
(b) be uniformly distributed for *X* but not for *Y*
(c) be uniformly distributed for both *X* and *Y*
(d) not be uniformly distributed for both *X* and *Y*

[2011 : 1 M]

11.34 An embankment is to be constructed with a granular soil (bulk unit weight = 20 kN/m^3) on a saturated clayey silt deposit (undrained shear strength = 25 kPa). Assuming undrained general shear failure and bearing capacity factor of 5.7. The maximum height (in m) of the embankment at the point of failure is

(a) 7.1 (b) 5.0
(c) 4.5 (d) 2.5 [2012 : 1 M]

Linked Answer Questions (11.35 and 11.36):
A multistorey building with a basement is to be constructed, the top 4 m contains loose silt below which dense sand layer is present upto a great depth. Ground water table is at the ground surface. The foundation consists of the basement slab of 6 m width which will rest on the top of dense sand as shown in figure. For dense sand saturated unit weight is 20 kN/m^3 and bearing capacity factor N_q = 40, N_γ = 45, for loose silt

saturated unit weight = 18 kN/m^3, N_q = 15, N_γ = 20. Effective cohesion C' is zero for both soils. Neglect shape factor and depth factor. Average elastic modulus E, and Poisson ratio μ of dense sand is 60 × 10^3 kN/m^2 and 0.3 respectively.

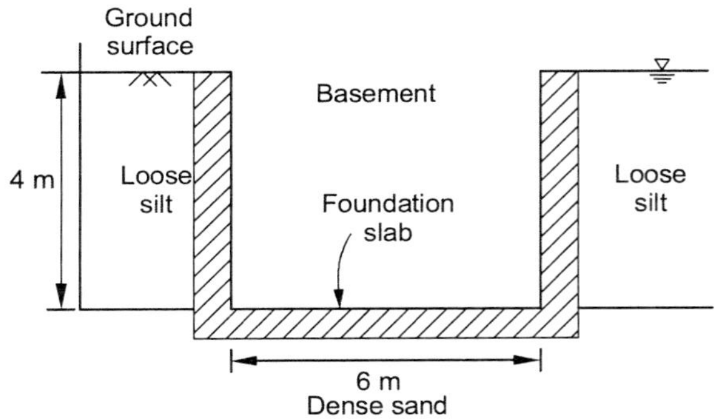

11.35 Using factor of safety = 3, the net safe bearing capacity (in kN/m^2) of the foundation is
(a) 610 (b) 320
(c) 983 (d) 693 [2013 : 2 M]

11.36 The foundation slab is subjected to vertical downward stresses equal to net safe bearing capacity derived in the above question. Using influence factor I_f = 2.0, and neglecting embedment depth and rigidity corrections, the immediate settlement of the dense sand layer will be
(a) 58 mm (b) 111 mm
(c) 179 mm (d) 126 mm [2013 : 2 M]

11.37 The contact pressure for a rigid footing resting on clay at the centre and the edges are respectively
(a) maximum and zero
(b) maximum and minimum
(c) zero and maximum
(d) minimum and maximum
[2014 : 1 M, Set-II]

11.38 **Group-I** contains representative load-settlement curves for different modes of bearing capacity failures of sandy soil. **Group-II** enlists the various failure characteristics. Match the load-settlement curves with the corresponding failure characteristics.

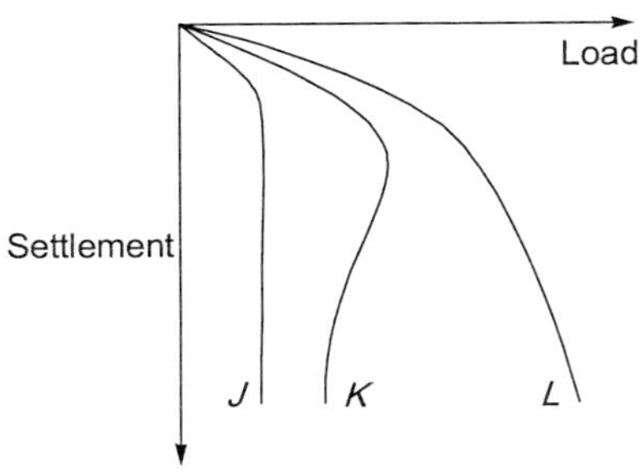

	Group-I		Group II
P.	Curve J	1.	No apparent heaving of soil around the footing
Q.	Curve K	2.	Rankine's passive zone develops imperfectly
R.	Curve L	3.	Well defined slip surface extends to ground surface

	P	Q	R
(a)	1	3	2
(b)	3	2	1
(c)	3	1	2
(d)	1	2	3

[2014 : 2 M, Set-I]

11.39 **Group-I** enlists in-situ field tests carried out for soil exploration, while **Group-II** provides a list of parameters for sub-soil strength characterization. Match the type of tests with the characterization parameters.

Group-I
P. Pressuremeter Test (PMT)
Q. Static Cone Penetration Test (SCPT)
R. Standard Penetration Test (SPT)
S. Vane Shear Test (VST)

Group-II
1. Menard's modulus (E_m)
2. Number of blows (N)
3. Skin resistance (f_c)
4. Undrained cohesion (c_u)

	P	Q	R	S
(a)	1	3	2	4
(b)	1	2	3	4
(c)	2	3	4	1
(d)	4	1	2	3

[2014 : 2 M, Set-II]

11.40 A circular raft foundation of 20 m diameter and 1.6 m thick is provided for a tank that applies a bearing pressure of 110 kPa on sandy soil with Young's modulus, E_s' = 30 MPa and Poisson's ratio, ν_s = 0.3. The raft is made of concrete (E_c = 30 GPa and ν_c = 0.15). Considering the raft as rigid, the elastic settlement (in mm) is
(a) 50.96 (b) 53.36
(c) 63.72 (d) 66.71
[2014 : 2 M, Set-II]

11.41 A square footing (2 m × 2 m) is subjected to an inclined point load, P as shown in the figure below. The water table is located well below the base of the footing. Considering one-way eccentricity, the net safe load carrying capacity of the footing for a factor of safety of 3.0 is ________ kN.

The following factors may be used:

Bearing capacity factors:

N_q = 33.3, N_γ = 37.16;

Shape factors: $F_{qs} = F_{\gamma s}$ = 1.314;

Depth factors: $F_{qd} = F_{\gamma d}$ = 1.113;

Inclination factors: F_{qi} = 0.444, $F_{\gamma i}$ = 0.02.

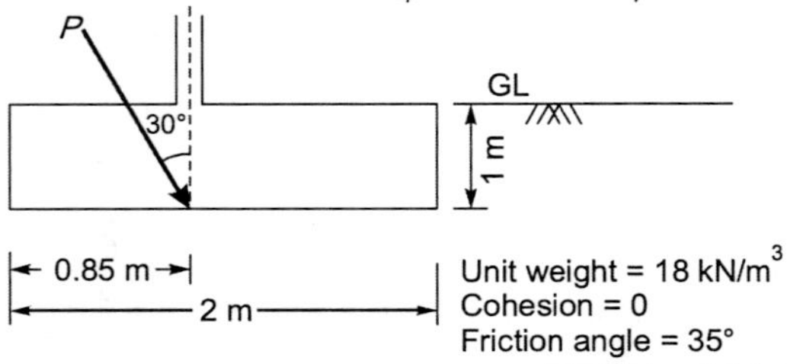

[2015 : 2 M, Set-I]

11.42 Net ultimate bearing capacity of a footing embedded in a clay stratum

(a) increases with depth of footing only
(b) increases with size of footing only
(c) increases with depth and size of footing
(d) is independent of depth and size of footing

[2015 : 1 M, Set-II]

11.43 A strip footing is resting on the surface of a purely clayey soil deposit. If the width of the footing is doubled, the ultimate bearing capacity of the soil

(a) becomes double
(b) becomes half
(c) becomes four-times
(d) remains the same

[2016 : 1 M, Set-II]

11.44 A 4 m wide strip footing is founded at a depth of 1.5 m below the ground surface in a c-ϕ soil as shown in the figure. The water table is at a depth of 5.5 m below ground surface. The soil properties are : c' = 35 kN/m^2, ϕ' = 28.63°, γ_{sat} = 19 kN/m^3, γ_{bulk} = 17 kN/m^3 and γ_w = 9.81 kN/m^3. The values of bearing capacity factors for different ϕ' are given.

ϕ'	N_c	N_q	N_γ
15°	12.9	4.4	2.5
20°	17.7	7.4	5.0
25°	25.1	12.7	9.7
30°	37.2	22.5	19.7

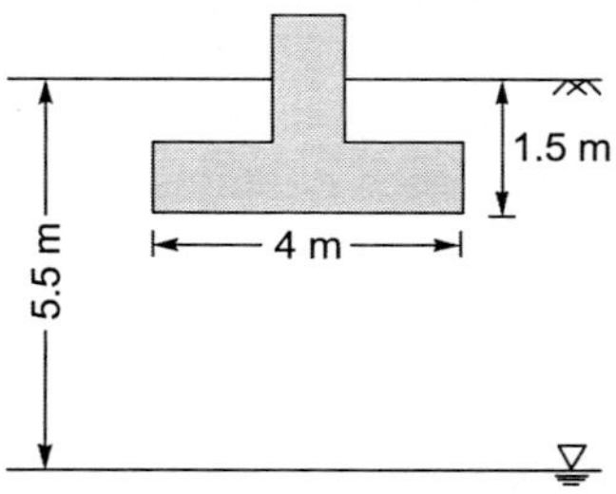

Using Terzaghi's bearing capacity equation and a factor of safety F_s = 2.5, the net safe bearing capacity (expressed in kN/m^2) for local shear failure of the soil is_______

[2016 : 2 M, Set-II]

11.45 The plate load test was conducted on a clayey strata by using a plate of 0.3 m × 0.3 m dimensions, and the ultimate load per unit area for the plate was found to be 180 kPa. The ultimate bearing capacity (in kPa) of a 2 m wide square footing would be

(a) 27
(b) 180
(c) 1200
(d) 2000

[2017 : 1 M, Set-II]

11.46 A strip footing is resting on the ground surface of a pure clay bed having an undrained cohesion c_u. The ultimate bearing capacity of the footing is equal to

(a) $2\pi c_u$
(b) πc_u
(c) $(\pi + 1)\, c_u$
(d) $(\pi + 2)\, c_u$

[2017 : 1 M, Set-I]

11.47 The percent reduction in the bearing capacity of a strip footing resting on sand under flooding condition (water level at the base of the footing) when compared to the situation where the water level is at a depth much greater than the width of footing, is approximately

(a) 0
(b) 25
(c) 50
(d) 100

[2018 : 1 M, Set-I]

11.48 The width of a square footing and the diameter of a circular footing are equal. If both the footings are placed on the surface of sandy soil, the ratio of the ultimate bearing capacity of circular footing to that of square footing will be

(a) $\frac{4}{3}$
(b) 1
(c) $\frac{3}{4}$
(d) $\frac{2}{3}$

[2018 : 1 M, Set-I]

11.49 The contact pressure and settlement distribution for a footing are shown in the figure.

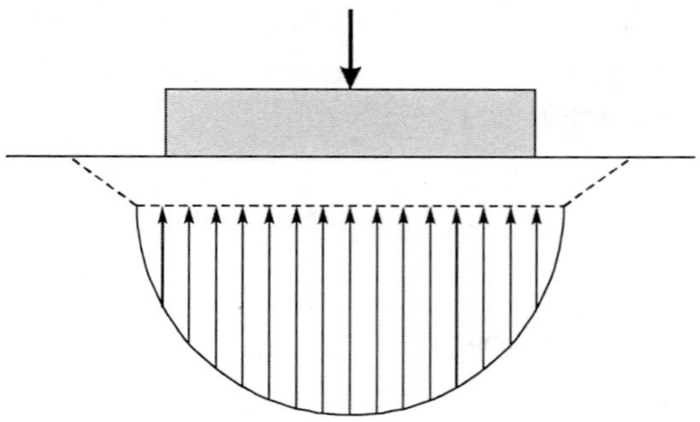

The figure corresponds to a

(a) rigid footing on granular soil

(b) flexible footing on granular soil
(c) flexible footing on saturated clay
(d) rigid footing on cohesive soil

[2018 : 1 M, Set-II]

11.50 For the following statements:

P: The lateral stress in the soil while being tested in an oedometer is always at-rest.

Q: For a perfectly rigid strip footing at deeper depths in a sand deposit, the vertical normal contact stress at the footing edge is greater than that at its centre.

R: The corrections for overburden pressure and dilatancy are not applied to measured SPT-N values in case of clay deposits.

The correct combination of the statements is

	P	Q	R
(a)	True	True	False
(b)	False	False	True
(c)	False	False	False
(d)	True	True	True

[2019 : 2 M, Set-I]

11.51 A square footing of 4 m side is placed at 1 m depth in a sand deposit. The dry unit weight (γ) of sand is 15 kN/m^3. This footing has an ultimate bearing capacity of 600 kPa. Consider the depth factors: $d_q = d_\gamma = 1.0$ and the bearing capacity factor: $N_\gamma = 18.75$. This footing is placed at a depth of 2 m in the same soil deposit. For a factor of safety of 3.0 as per Terzaghi's theory, the safe bearing capacity (in kPa) of this footing would be_______.

[2019 : 2 M, Set-I]

11.52 A square footing of 2 m sides rests on the surface of a homogeneous soil bed having the properties: cohesion c = 24 kPa, angle of internal friction ϕ = 25°, and unit weight γ = 18 kN/m^3. Terzaghi's bearing capacity factors for ϕ = 25° are N_c = 25.1, N_q = 12.7, N_γ = 9.7, N_c' = 14.8., N_q' = 5.6, and N_γ' = 3.2. The ultimate bearing capacity of the foundation (in kPa, round off to 2 decimal places) is___________

[2019 : 2 M, Set-II]

11.53 In a soil investigation work at a site, Standard Penetration Test (SPT) was conducted at every 1.5 m interval up to 30 m depth. At 3 m depth, the observed number of hammer blows for three successive 150 mm penetrations were 8, 6 and 9, respectively. The SPT- N-value at 3 m depth, is

(a) 14 (b) 17
(c) 23 (d) 15

[2020 : 1 M, Set-I]

11.54 A rectangular footing of size 2.8 m × 3.5 m is embedded in a clay layer and a vertical load is placed with an eccentricity of 0.8 m as shown in the figure (not to scale). Take Bearing capacity factors: N_c = 5.14, N_q = 1.0, and N_γ = 0.0: Shape factors: s_c = 1.16, s_q = 1.0 and s_γ = 1.0; Depth factors: d_c = 1.1. d_q = 1.0 and d_γ = 1.0; and Inclination factors: i_c =1,0 and i_q = 1.0 and i_γ = 1.0.

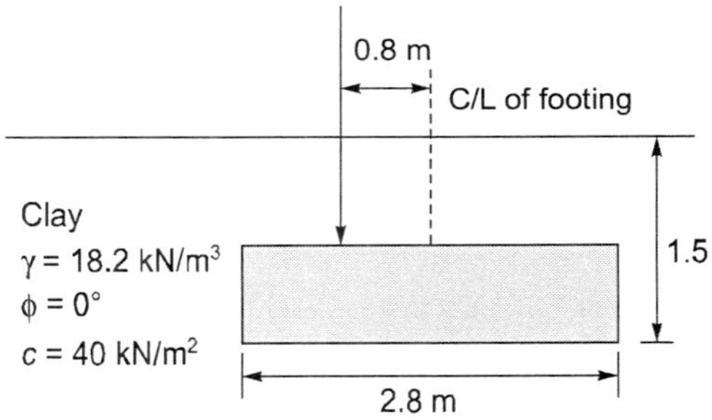

Using Meyerhoff's method, the load (in kN, round off to two decimal places) that can be applied on the footing with a factor of safety of 2.5 is_____.

[2021 : 2 M, Set-II]

11.55 Read the following statements:

(P) While designing a shallow footing in sandy soil, monsoon season is considered for critical design in terms of bearing capacity,

(Q) For slope stability of an earthen dam, sudden drawdown is never a critical condition.

(R) In a sandy sea beach, quicksand condition can arise only if the critical hydraulic gradient exceeds the existing hydraulic gradient.

(S) The active earth thrust on a rigid retaining wall supporting homogeneous cohesionless backfill will reduce with the lowering of water table in the backfill.

Which one of the following combinations is correct?

(a) (P)-False, (Q)-True, (R)-False, (S)-False
(b) (P)-False, (Q)-True, (R)-True, (S)-True
(c) (P)-True, (Q)-False, (R)-False, (S)-False
(d) (P)-True, (Q)-False, (R)-True, (S)-True

[2022 : 1 M, Set-II]

11.56 A square footing of size 2.5 m × 2.5 m is placed 1.0 m below the ground surface on a cohesionless homogeneous soil stratum. Considering that the groundwater table is located at the base of the footing, the unit weights of soil above and below the groundwater table are 18 kN/m^3 and 20 kN/m^3, respectively, and the bearing capacity factor N_q is 58, the net ultimate bearing capacity of the soil is estimated as 1706 kPa (unit weight of water = 10 kN/m^3).

Earlier, a plate load test was carried out with a circular plate of 30 cm diameter in the same foundation pit during a dry season, when the water table was located beyond the plate influence zone. Using Terzaghi's bearing capacity formulation, what is the ultimate bearing capacity (in kPa) of the plate?

(a) 110.16 (b) 61.20
(c) 204.00 (d) 163.20

[2023 : 2 M, Set-I]

11.57 A standard penetration test (SPT) was carried out at a location by using a manually operated hammer dropping system with 50% efficiency. The recorded SPT value at a particular depth is 28. If an automatic hammer dropping system with 70% efficiency is used at the same location, the recorded SPT value will be

(a) 28 (b) 20
(c) 40 (d) 25

[2023 : 1 M, Set-II]

11.58 The reason(s) of the nonuniform elastic settlement profile below a flexible footing, resting on a cohesionless soil while subjected to uniform loading, is/are:

(a) Variation of friction angle along the width of the footing
(b) Variation of soil stiffness along the width of the footing
(c) Variation of friction angle along the depth of the footing
(d) Variation of soil stiffness along the depth of the footing

[2023 : 1 M, Set-II]

11.59 A square footing is to be designed to carry a column load of 500 kN which is resting on a soil stratum having the following average properties: bulk unit weight = 19 kN/m^3; angle of internal friction = 0° and cohesion = 25 kPa. Considering the depth of the footing as 1 m and adopting Meyerhoff's bearing capacity theory with a factor of safety of 3, the width of the footing (in m) is ________ (round off to one decimal place)

[Assume the applicable shape and depth factor values as unity; ground water level at greater depth.]

[2023 : 2 M, Set-II]

11.60 A 2 m wide strip footing is founded at a depth of 1.5 m below the ground level in a homogenous, pure clay bed. The clay bed has unit cohesion 40 kPa. Due to seasonal fluctuations of water table from peak summer to peak monsoon period, the net ultimate bearing capacity of the footing as per Terzaghi's theory

(a) decrease (b) become zero
(c) remain the same (d) increase

[2024 : 1 M, Set-I]

11.61 The contact pressure distribution shown in the figure belongs to a

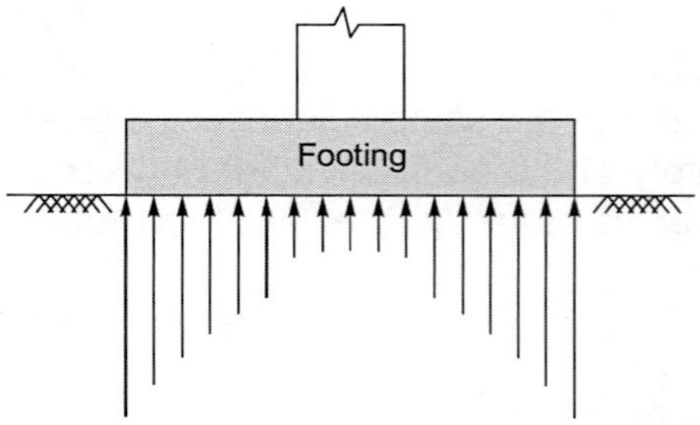

(a) flexible footing resting on a cohesionless soil.
(b) rigid footing resting on a cohesive soil.
(c) flexible footing resting on a cohesive soil.
(d) rigid footing resting on a cohesionless soil.

[2024 : 1 M, Set-II]

■■■■

Answers Shallow Foundation and Bearing Capacity

11.1	(Sol.)	11.2	(Sol.)	11.3	(Sol.)	11.4	(b)	11.5	(c)	11.6	(d)	11.7	(a)
11.8	(d)	11.9	(b)	11.10	(a)	11.11	(b)	11.12	(a)	11.13	(d)	11.14	(b)
11.15	(d)	11.16	(b)	11.17	(a)	11.18	(c)	11.19	(d)	11.20	(c)	11.21	(c)
11.22	(a)	11.23	(b)	11.24	(c)	11.25	(c)	11.26	(c)	11.27	(d)	11.28	(b)
11.29	(a)	11.30	(c)	11.31	(c)	11.32	(b)	11.33	(b)	11.34	(a)	11.35	(*)
11.36	(*)	11.37	(d)	11.38	(a)	11.39	(a)	11.40	(b)	11.41	(439.5)	11.42	(d)
11.43	(d)	11.44	(298.48)	11.45	(b)	11.46	(d)	11.47	(c)	11.48	(c)	11.49	(a)
11.50	(d)	11.51	(270)	11.52	(353.92)	11.53	(d)	11.54	(440.7)	11.55	(c)	11.56	(a)
11.57	(b)	11.58	(b)	11.59	(3.4)	11.60	(c)	11.61	(b)				

Explanations Shallow Foundation and Bearing Capacity

11.1 Sol.

TRUE

Generally bearing capacity of sands based on shear criteria is greater than that of clays and sands do not fail in shear.

11.2 Sol.

TRUE

11.3 Sol.

UNDER ESTIMATED

Penetration resistance or the N=value is influenced by the overburden pressure of two granular soils processing the same relative density but having different confining pressures, the one with a higher confining pressure gives a higher N-value. Since the confining pressure increases with depth, the N-values at shallow depths are underestimated and hence lead to underestimated design of footing.

11.4 (b)

11.5 (c)

In a shallow foundation, depth of foundation is less than or equal to its width whereas in a deep foundation, depth of foundation is greater than is width.

11.6 (d)

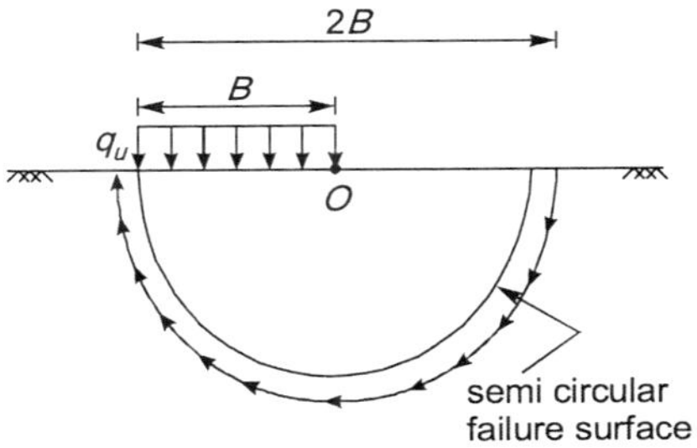

In this case, failure will cause a rotation about point 0.

Moment causing rotation

$= \text{Load} \times \text{Leore arm}$

$= q_u \times B \times B/2$

Moment resisting rotation

$= \text{shear strength} \times \text{length of arc} \times \text{lower arm}$

$$= s \times \left(\frac{2\pi B}{2}\right) \times B$$

At failure these are equal,

$$q_u \times B \times \frac{B}{2} = C_u \times \pi B^2$$

$$q_u = 2\pi C_u$$

***Note:** It is fellinius theory which is not in syllabus now.*

11.7 (a)

For dense sand,

$$\frac{S_F}{S_P} = \left[\frac{B_f(B_p+0.3)}{B_p(B_f+0.3)}\right]^2$$

For clays, $\frac{S_F}{S_p} = \frac{B_F}{B_p}$

11.8 (d)

Allowable bearing capacity is the maximum intensity of loading that can be imposed on the soil with no possibility of shear failure or excessive settlement.

11.9 (b)

$$q_u = 300 \text{ kN/m}^2$$
$$D = 1 \text{ m}, \gamma = 20 \text{ kN/m}^3, F = 2.5$$

Then, $q_{nu} = q_u - \gamma D$

$$= 300 - 20 \times 1 = 280 \text{ kN/m}^2$$

$$q_{ns} = \frac{q_{nu}}{F} = \frac{280}{2.5} = 112 \text{ kN/m}^2$$

11.10 (a)

Since $c = 0$; $D_f = 0$, ($\therefore$ founded on surface)

$$\frac{(q_f)_{cir}}{(q_f)_{sq}} = \frac{0.3\gamma\beta N_\gamma}{0.4\gamma\beta N_\gamma} = \frac{3}{4}$$

11.11 (b)

$$q_u = cN_c + \gamma D_f N_q + 0.5\gamma B N_\gamma$$

For pure clay,

$$\phi = 0, N_c = 5.7, N_q = 1, N_\gamma = 0$$

$\therefore$ $q_u = cN_c + \gamma D_f$

$$q_{nu} = cN_c + \gamma D_f - \gamma D_f = cN_c$$

Thus, net ultimate bearing capcity for pure clay soil ($\phi = 0$) is independent of width and depth of footing.

11.12 (a)

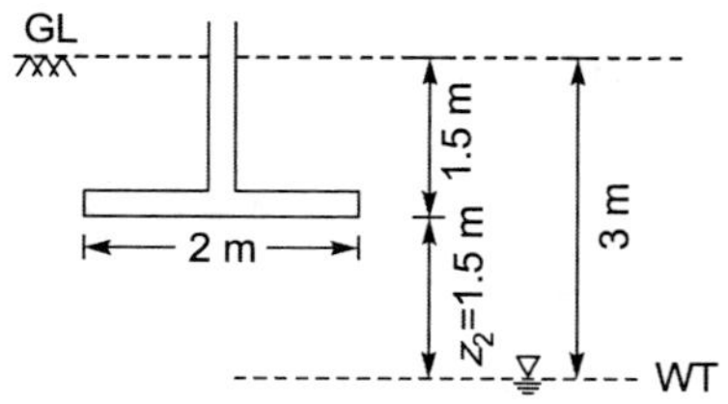

$$W_\gamma = \frac{1}{2}\left(1+\frac{z_2}{B}\right) = 0.5 + 0.5 \times \frac{1.5}{2} = 0.875$$

11.13 (d)

For cohesionless soil,

$$S_f = S_p\left[\frac{B_f(B_p+0.3)}{B_p(B_f+0.3)}\right]^2$$

$$= 5\left[\frac{5(0.3+0.3)}{0.3(5+0.3)}\right]^2 = 17.8 \text{ mm}$$

***Note:** For clayey soil* $\frac{S_F}{S_P} = \frac{B_F}{B_P}$

11.14 (b)

Given SPT number $N = 13$ (<15)

Correction of fines is not required as N is not greater than 15

$$N_{corrected} = 13$$

11.15 (d)

Settlement of footing in a cohesionless soil is given by,

$$S_f = S_p\left[\frac{B_f(B_p+30)}{B_p(B_f+30)}\right]^2$$

$$\Rightarrow \quad S_f = 15\left[\frac{100(60+30)}{60(100+30)}\right]^2$$

$$\Rightarrow \quad S_f = 19.97 \text{ mm}$$

11.16 (b)

We know that $q_u = 1.3\,cN_c + qN_q + 0.3\,\gamma D N_\gamma$

where D is the diameter of the **circular footing**

Now, for purely cohesive soils, $\phi = 0$

$N_c = 5.7$, $N_q = 1$, $N_\gamma = 0$

Therefore, the gross ultimate bearing capacity of a footing (of any shape) resting on the surface of a purely cohesive clay is constant irrespective of its size. Thus, the ratio of gross ultimate bearing capacities of the two circular footings will be 1.0.

11.17 (a)

11.18 (c)

$$(\sigma_z)_{square} = \frac{qB^2}{(B+2nz)^2}$$

$$(\sigma_z)_{strip} = \frac{qB}{(B+2nz)}$$

11.19 (d)

Load per metre length of footing,

$P = B \times$ {Minimum of bearing capacity and allowable soil pressure}

$= 8 \times$ Minimum of [150, 100]

$= 8 \times 100 = 800$ kN/m

Allowable pressure for 25 mm settlement

$$= \frac{100 \times 25}{40} = 62.5 \text{ kN/m}^2$$

Load resisted = Width × Allowable pressure

$$800 = B \times 62.5$$

$$\Rightarrow \quad B = 12.8 \text{ m}$$

11.20 (c)

The number of blows required for the first 150 mm of penetration is disregarded, and only the number of blows required for the last 300 mm of penetration is added together.

∴ Number of blows,

$$N' = 6 + 6 + 8 + 7 = 27$$

(i) Correction for overburden pressure

Corrected value of observed

$$N' = C_N N' = 1.0 \times 27 = 27$$

(ii) Correction for dilatancy

It is to be applied when number of blows obtained after overburden correction exceeds 15 in saturated fine sands and silts.

$$\therefore \quad N = 15 + 0.5\,(27 - 15)$$

$$= 15 + 6 = 21$$

11.21 (c)

The number of blows for the first 150 mm penetration of the sampler are disregarded. The number of blows for the next 300 mm penetration are recorded as the observed N value.

∴ Observed N value = 8 + 10 = 18

11.22 (a)

The bearing capacity of the sand deposit will remain same until the water table rises to a depth less than the width of the footing.

Thus, when water table is 3.0 below footing, the bearing capacity is 600 kN/m^2; similarly when water table is 1.5 m below footing, the bearing capacity remaining same i.e. 600 kN/m^2. But when the water table rises to a depth of 0.5 m below the footing, the sand deposit will be saturated for a depth of 1.5 – 0.5 = 1 m. Thus, the bearing capacity below the base of footing will decrease due to submergence of soil (1 m).

11.23 (b)

$$S_p = S_f\left[\frac{B_p(B_f + 0.3)}{B_f(B_p + 0.3)}\right]^2 \text{ for sandy soils}$$

$$\Rightarrow \quad 10 = S_f\left[\frac{0.3(1.5 + 0.3)}{1.5(0.3 + 0.3)}\right]^2$$

$$\Rightarrow \quad S_f = \frac{10}{0.36} \Rightarrow S_f = 27.8 \text{ mm}$$

11.24 (c)

Ultimate bearing capacity of the footing as per **Terzaghi's equation** for rectangular footing is given as

$$q_u = \left(1 + 0.3\frac{B}{L}\right) cN_c + \gamma D_f N_q + \frac{1}{2}\gamma B\left(1 - 0.2\frac{B}{L}\right) N_\gamma$$

But for sand, $c = 0$

$$\therefore \quad q_u = (18 \times 1 \times 24) + \frac{1}{2} \times 18 \times 1.5 \times \left(1 - 0.2 \times \frac{1.5}{3.0}\right) \times 20$$

$$\Rightarrow \quad q_u = 432 + 243$$

$$\Rightarrow \quad q_u = 675 \text{ kN/m}^2$$

The net ultimate bearing capacity,

$$q_{nu} = q_u - \gamma D_f$$

$$= 675 - 18 \times 1 = 657 \text{ kN/m}^2$$

Note: None of the options given is correct, but option (c) is most close.

11.25 (c)

Using net ultimate bearing capacity as 630 kN/m^2

$$\text{Safe load} = \frac{630 \times 1.5 \times 3.0}{3} = 945 \text{ kN}$$

11.26 (c)

In case of fine sand or silt below water, apparently high values may be noted for SPT number. No value of SPT, corrected for overburden pressure i.e. N_1 is further correctd for dialatancy and given

as $N_2 = 15 + \frac{1}{2}(N_1 - 15)$

11.27 (d)

- Overburden pressure correction should be applied.
- Settlement of footing depends on size of plate in sandy soil.
- Ground water table correction factor should also be applied.

11.28 (b)

$$c = 0$$
$$D_f = 0 \quad (\therefore \text{ Footing is at surface})$$
$$\gamma' = \gamma_{sat} - \gamma_w = 8 \text{ kN/m}^3$$

without considering shape and depth factor, safe bearing capacity,

$$q_u = cN_c + \gamma' D_f N_q + 0.5B\gamma' N_\gamma$$

$$q_s = \frac{q_u - \bar{\sigma}}{\text{FOS}} + \bar{\sigma}$$

$$= \frac{q_u - \gamma' D_f}{\text{FOS}} + \gamma' D_f$$

$$= \frac{1}{3}[0.5 \times 1 \times 8 \times 40]$$

$$= 53.34 \text{ kN/m}^2$$

Nearest option is 60 kN/m².

11.29 (a)

Ultimate (maximum) skin friction resistance, Q_f is given by,

$$Q_f = K\bar{\sigma}_{av} A_s \tan\delta$$

where, K = lateral earth pressure coefficient

$= 1.5$

$\bar{\sigma}_{av}$ = average effective overburden pressure over the embedded length of pile

$$= \frac{(\gamma_{sat} - \gamma_w) \times L}{2} = \frac{(18-10) \times 5}{2}$$

$= 20 \text{ kN/m}^2$

A_s = Surface area of pile in contact with soil

$$\therefore \quad Q_f = 1.5 \times 20 \times \pi \times 0.5 \times 5 \times \tan 24^\circ$$

$= 104.9$ kN

11.30 (c)

Cohesion of clay, $c = \frac{q_u}{2} = \frac{54}{2} = 27$ kPa

11.31 (c)

The ultimate bearing capacity of a square footing in clay as per **Terzaghi's** bearing capacity equation is given by,

$$q_{ult} = 1.3\,cN_c \qquad [\because N_c \text{ for clay} = 5.7]$$
$$= 1.3 \times 27 \times 5.7$$
$$= 200 \text{ kPa}$$

11.32 (b)

General shear failure is generally associated with dense sand. So if intergranular packing of sand decreases then sand will behave as loose sand. Hence likelihood of general shear failure with decrease.

11.33 (b)

For rigid footing, soil reaction will not be uniformly distributed and for flexible footing, it will be uniformly distributed.

11.34 (a)

Given, $\gamma = 20 \text{ kN/m}^3$

Shear strength,

$$s = 25 \text{ kN/m}^2$$

Bearing capacity factor, $N_c = 5.7$

$$\because \quad S = c \Rightarrow c = 25 \text{ kN/m}^2$$
$$q_u = cN_c = 25 \times 5.7$$
$$\Rightarrow \quad \gamma D_f = 25 \times 5.7$$
$$\Rightarrow \quad D_f = 7.1 \text{ m}$$

11.35 (*)

$$q_u = 1.3cN_c + (\gamma')_{silt}(N_q)_{silt} D_f + 0.4B(\gamma')_{sand}(N_\gamma)_{sand}$$

$\because c = 0$

$= 8.19 \times 4 \times 15 + 0.5 \times 10.19 \times 6 \times 45$

$= 1867.05 \text{ kN/m}^2$

$Q_{nu} = q_u - \gamma' D_f = 1867.05 - (8.19 \times 4)$

$= 1834.29 \text{ kN/m}^2$

Safe bearing capacity $= Q_{ns} = \frac{Q_{nu}}{F}$

$$= \frac{1867.05 - 8.19 \times 4}{3} = 611.43 \text{ kN/m}^2$$

* (Marks to All)

Comment: But above solution is not correct because N_c, N_q and N_γ depends upon ϕ-soil below the footing and answer is matching by taking N_q of soil above the footing. Hence in GATE solution marks to all.

11.36 (*)

Immediate settlement,

$$S_i = \frac{qB(1-\mu^2)I_t}{E_s}$$

$$S_i = \frac{610\times6\times(1-0.3^2)\times2}{60\times10^3} = 111 \text{ mm}$$

∗ (Marks to All)

11.37 (d)

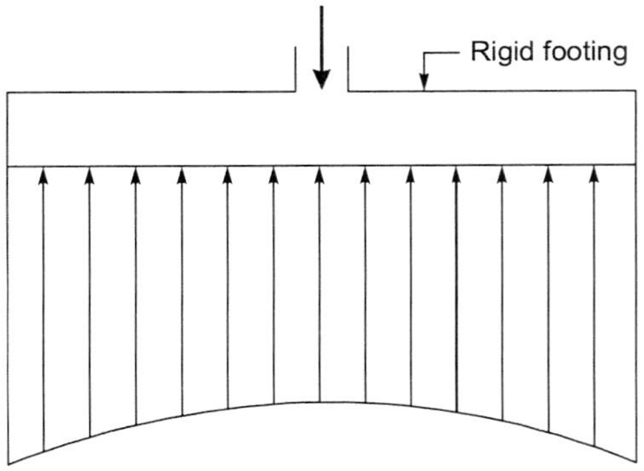

Stress distribution for cohesive soil

Minimum at centre and maximum at edge.

11.38 (a)

- In general shear failure shown by curve L, well defined slip surface extends to ground surface.
- In local shear failure shown by curve k, Rankine's passive zone develops imperfectly.
- For punching shear faiilure shown by curve J, no appearant takes place around the footing.

11.39 (a)

- SCPT is used to find skin restistance of footing.
- IN SPT, numbers of blows required to find the stated penetration i.e. 300 mm are noted.
- Vane shear test is used to find out undrained cohesion of soil.

11.40 (b)

Elastic settlement of rigid footing

$$= \frac{qB(1-\mu^2)}{E_s}\times I_f$$

Shape factor for circular rigid footing,

$$I_t = 0.8$$

$$\Rightarrow \quad S_f = 0.8\left[\frac{110\times20(1-0.3^2)}{30}\right]$$

$$= 53.386 \text{ mm}$$

11.41 Sol.

$$B' = B - 2e_x = 2 - 2\times0.15 = 1.7 \text{ m}$$

As per Meyerhof theory

$$q_u = CN_cS_cd_c\,i_c + \gamma D_fN_qS_qd_qi_q + 0.5\,B'\gamma N_\gamma S_\gamma d_r i_r$$

$$q_u = 18\times1\,(1.314\times1.113\times0.444)\times33.3 + 10.5\times1.7\times18\times37.16 \times(1.314\times1.113\times0.02)$$

$$= 405.84 \text{ kN/m}^2$$

$$q_{ns} = \frac{q_u-\sigma}{F} = \frac{q_u-\gamma D_f}{F}$$

$$= \frac{405.84-18\times1}{3} = 129.28 \text{ kN/m}^2$$

Net safe load = $q_{ns}\times$ Area

$= q_{ns}\times(B'\times L) = 129.28\times1.7\times2 = 439.55$ kN

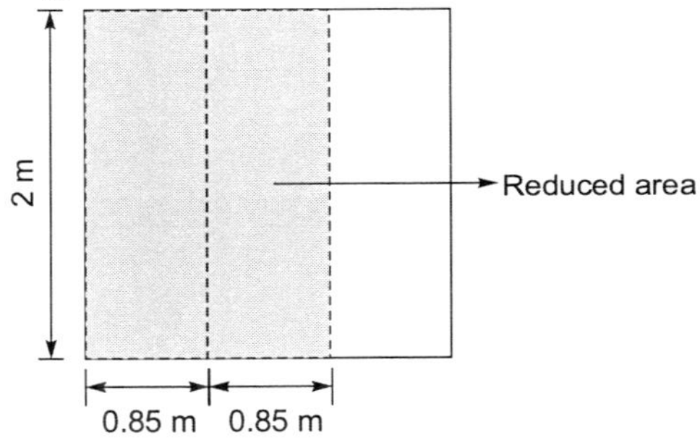

11.42 (d)

$q_{nu} = cN_c$ (as per Terzaghi) independent of depth and size of footing.

11.43 (d)

In case of clay ultimate bearing capacity is independent of width of footing.

11.44 Sol.

Local shear failure is occuring hence modified c and ϕ should be used,

$$c_m = \frac{2}{3}c = \frac{2}{3}\times35 = 23.33 \text{ kN/m}^2$$

$$\tan\phi_m = \frac{2}{3}\tan\phi = \frac{2}{3}\tan28.63$$

$$\phi_m = 20°$$

for $\phi_m = 20°$, $N_c = 17.7$, $N_q = 7.4$, $N_\gamma = 5.0$

⇒ Water table is at

$D_f + B = 1.5 + 4 = 5.5$ m hence no effect on bearing capacity

⇒ As per **Terzaghi** for strip footing

$$q_u = cN_c + \gamma D_f N_q + 0.5 B\gamma N_\gamma$$

$$q_u = \frac{2}{3}\times 35\times 17.7 + 17 + 1.5\times 7.4 + 0.5 \times 4 \times 17 \times 5$$

$$q_u = 771.7 \text{ kN/m}^2$$

$$q_{nu} = q_u - \bar{\sigma} = q_u - \gamma D_f = 771.7 - 17 \times 1.5$$

$$q_{nu} = 746.2 \text{ kN/m}^2$$

Net safe bearing capacity,

$$q_{ns} = \frac{q_{nu}}{F} = \frac{746.2}{2.5} = 298.48 \text{ kN/m}^2$$

11.45 (b)

- **Plate load test** was conducted on a clay strata.
- Bearing capacity for the plate is 180 kPa.
- Ultimate bearing capacity in case of clay is independent of width of footing, hence ultimate bearing capacity for footing will be 180 kPa.

11.46 (d)

Footing is at surface,

Hence, $D_f = 0$

$$q_u = CN_c + \gamma D_f N_q + 0.5 B\gamma N_\gamma$$

⇒ For clay

$$N_\gamma = 0,\ N_q = 1$$

$\therefore \quad q_u = CN_c$

As per **Terzaghi,**

$$N_c = 5.7$$

and as per **Meyerhoff** and **Prandtl**,

$$N_c = 5.14$$

$\therefore \quad q_u = (\pi + 2)C_u = (3.14 + 2)C_u = 5.14\ C_u$

11.47 (c)

For strip footing resting on sand ($c = 0$)

$$q_u = 0.5B\gamma N_\gamma$$

$\therefore \quad q_u = 0.5B\gamma' N_\gamma$

$\because \quad \gamma' \simeq \frac{1}{2}\gamma_{sat}$

***Note:** If water table rises to ground level then both γ will reduce to γ′ hence percentage reduction would be approximately 50%.*

11.48 (c)

Footing placed on surface

$\therefore \quad D_f = 0$

For square footing,

$$q_u = CN_c + \gamma D_f N_q + 0.4B\gamma N_\gamma$$

For circular footing,

$$q_u = CN_c + \gamma D_f N_q + 0.3B\gamma N_\gamma$$

For sandy soil,

$$C = 0$$

$$\text{Ratio } \frac{(q_u)_{circular}}{(q_u)_{square}} = \frac{3}{4}$$

11.49 (a)

Rigid footing on granular soil.

11.50 (d)

For rigid footings resting on a surface of cohesionless soil the settlement must be uniform and the pressure distribution is zero at the edges and maximum at the centre but for very deep rigid footing on sand the contact pressure distribution may be more like that of rigid footing on cohesive soil i.e., more at edges than at center.

Hence statement Q is true.

11.51 Sol.

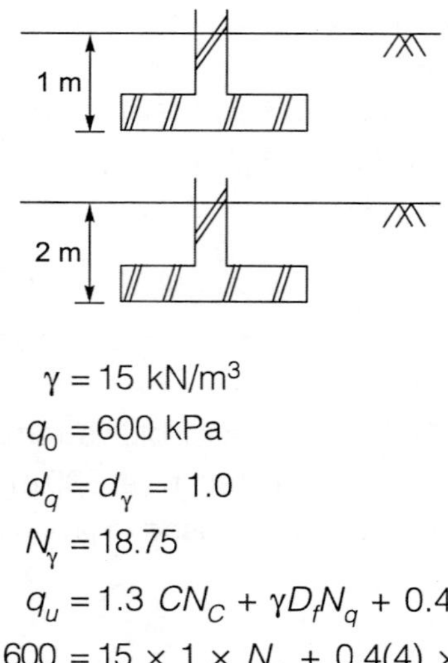

$$\gamma = 15 \text{ kN/m}^3$$

$$q_0 = 600 \text{ kPa}$$

$$d_q = d_\gamma = 1.0$$

$$N_\gamma = 18.75$$

$$q_u = 1.3\ CN_C + \gamma D_f N_q + 0.4B\gamma N_\gamma$$

$$600 = 15 \times 1 \times N_q + 0.4(4) \times 15\times 18.75$$

$$N_q = 10$$

$$\text{FOS} = 3$$

$$q_{safe} = ??$$

$$q_u = 1.3CN_C + \gamma D_f N_q + 0.4B\gamma N_\gamma$$

$= 0 + 15 \times 2 \times 10 + 0.4 \times 4 \times 15 \times 18.75$

$= 750 \text{ kN/m}^2$

$$q_s = \frac{q_u - \bar{\sigma}}{FOS} + \bar{\sigma} = \frac{q_u - \gamma D_f}{FOS}$$

$$= \frac{750 - 15 \times 2}{3} + 15 \times 2 = 270 \text{ kN/m}^2$$

11.52 Sol.

$c = 24 \text{ kN/m}^2$, $\phi = 25°$, $\gamma = 18 \text{ kN/m}^3$

$N_c = 25.1$, $N_q = 12.7$, $N_\gamma = 9.7$

$N_c' = 14.8$, $N_q' = 5.6$, $N_\gamma' = 3.2$

Square footing $B = 2$ m, $L = 2$ m, $D_f = 0$ (Surface footing)

Since,

$\phi = 25° < 28°$

$\rightarrow$ Assume local shear failure.

Hence, $c_m = \frac{2}{3}c = \frac{2}{3} \times 24 = 16 \text{ kN/m}^2$

$N_c' = 14.8$, $N_q' = 5.6$, $N_\gamma' = 3.2$

Ultimate bearing capacity of square footing,

$q_u = 1.3c_m N_c' + \gamma D_f N_q' + 0.4\gamma B N_\gamma'$

$q_u = 1.3 \times 16 \times 14.8 + 0 + 0.4 \times 18 \times 2 \times 3.2$

$q_u = 307.84 + 46.08$

$q_u = 353.92 \text{ kN/m}^2$

11.53 (d)

No. of blows for each 150 mm penetration 8, 6 and 9.

We will not consider first 150 mm number of blows.

Hence, for last 300 mm, number of blows are 15.

Hence, observed SPT number = 15.

11.54 Sol.

Given data: 2.8×3.56, $e = 0.8$, $D_f = 1.5$ m

$N_c = 5.14$, $N_q = 1, N_r = 0$

$S_c = 1.16$, $S_q = 1, S_r = 1.0$

$d_c = 1.1$, $d_q = 1, d_r = 1.0$

$i_c = 1$, $i_q = 1$, $i_r = 1.0$

$\gamma = 18.2$, $c = 40 \text{ kN/m}^2$

$B' = B - 2e_x = 2.8 - 2(0.8) = 1.2$ m

$L' = L = 3.5$

Ultimate bearing capacity

$q_u = CN_cS_cd_ci_c + \gamma D_f N_q S_q d_q i_q + 0.5B'\gamma N_r S_r d_r$

$q_u = 40 \times 5.14 \times 1.16 \times 1.1 + 18.2 \times 1.5 \times 1 + 0$

$q_{nu} = q_u - \gamma D_f = 262.346 \text{ kN/m}^2$

$$q_{ns} = \frac{q_{nu}}{F} = 104.938 \text{ kN/m}^2$$

$$\text{Net safe load} = q_{ns} \times A = \frac{q_{nu}}{F} \times (B' \times L)$$

$= 104.938 \times 1.2 \times 3.5$ kN

Load applied on the footing = 440.740 kN

11.55 (c)

- Stability of slope will dfinitely be affected by sudden drawdown condition. Even, it is one of the reason of failure of slope.
- Active thrust decrease, when water table is lowered.

  For eg:. When water table is at ground level, then active pressure at base is $k_a\gamma' H + \gamma_w H$ and when there is no water table than active thrust at base is $k_a\gamma H$.
- Quick sand condition will not occur if $i_c > i$.

11.56 (a)

Since soil is cohesionless, $c = 0$

$$q_{nu} = q_u - \bar{\sigma} = q_u - \gamma D_f$$

$$\Rightarrow 1706 = \{1.3CN_C + \gamma D_f N_q + 0.4B\gamma N_\gamma\} - \gamma D_f$$

$\Rightarrow$ $1706 = 18 \times 1 \times 58 + 0.4 \times 2.5 \times (20 - 10) N_\gamma - 18 \times 1$

$\Rightarrow$ $N_\gamma = 68$

For Plate: [Surcharge at the plate level is zero]

$q_u = 1.3CN_c + \gamma D_f N_q + 0.3B\gamma N_\gamma$

$\Rightarrow$ $q_u = 0 + 0.3 \times 0.3 \times 18 \times 68$

$\Rightarrow$ $q_u = 110.16$ kPa

11.57 (b)

$$\text{Efficiency} \propto \frac{1}{\text{Number of blows}} \propto \frac{1}{SPT \text{ value}}$$

$\Rightarrow$ $\eta_1 N_1 = \eta_2 N_2$

$\Rightarrow$ $0.5 \times 28 = 0.7 \times N_2$

$$\Rightarrow \quad N_2 = \frac{0.5 \times 28}{0.7} = 20$$

11.58 (b)

11.59 Sol.

As per Meyerhoff's theory,

$$q_u = CN_cS_cd_ci_c + qN_qS_qd_qi_q + 0.5\gamma BN_\gamma S_\gamma d_\gamma i_\gamma$$

For $\phi = 0$, $N_c = 5.14$, $N_q = 1$ and $N_\gamma = 0$

Also, considering shape, depth and inclination factor as 1, we get

$\Rightarrow \quad q_u = 5.14 \times 25 + \gamma D_f$

$\Rightarrow \quad q_{nu} = q_u - \gamma D_f = 5.14 \times 25$

$$\Rightarrow q_n = \frac{q_{nu}}{\text{FOS}} = \frac{5.14 \times 25}{3} = 42.83 \text{ kN/m}^2$$

Also, $\quad Q_{ns} = 500 \text{ kN} = q_{ns} \times B^2$

$\Rightarrow 42.83 \times B^2 = 500$

$\Rightarrow \quad B = 3.4167 \text{ m} \simeq 3.4 \text{ m}$

11.60 (c)

Ultimate bearing capacity,

$$q_u = C\overset{5.7}{N_c} + \gamma Df \overset{1}{N_q} + 0.5 B\gamma \overset{0}{N_\gamma}$$

$$q_u = 5.7C + \gamma D_f$$

$$q_{nu} = q_u - \gamma D_f \; 5.7C$$

Net ultimate bearing capacity,

So, q_{nu} does not depend on presence of water table.

11.61 (b)

■■■■

12 Deep Foundation

12.1 Write True/False in the answer book.
Friction piles are also called 'floating piles'.
[1995 : 1 M]

12.2 Negative skin friction occurs when
(a) an upward drag exists in the pile
(b) the surrounding soil settles more than the pile
(c) the pile passes continuously through a firm soil
(d) the driving operation begins [1996 : 1 M]

12.3 The value of bearing capacity factor for cohesion, N_c, for piles as per Meyerhoff, is taken as
(a) 6.2 (b) 12.0
(c) 9.0 (d) 5.14 [1996 : 1 M]

12.4 Sand drains are used to
(a) reduce the settlement
(b) accelerate the consolidation
(c) increase the permeability
(d) transfer the load [1997 : 1 M]

12.5 Well foundation are commonly used as foundation for the following structures:
(a) Water tanks
(b) Bridge
(c) Buildings
(d) Reciprocating machines [1997 : 1 M]

12.6 The group efficiency of pile group
(a) will be always less than 100%
(b) will be always greater than 100%
(c) may be less than 100% or more than 100%
(d) will be more than 100% for pile group in cohesionless soils and less than 100% for those in cohesive soils [2000 : 1 M]

12.7 Identify the two True statements from the following four statements.
I. Negative skin friction is higher on floating piles than on end bearing piles
II. All other things being the same is footings on sand, the footing with smaller width will have lower settlement at the same net pressure.
III. The void ratio of soils in always less than 1.0
IV. For determining the depth of embedment of anchored sheet piles, net moment at the anchor elevation is set to zero.
(a) I and IV (b) I and III
(c) II and IV (d) II and III
[2001 : 2 M]

12.8 For the (3 × 3) pile group shown in the figure, the settlement of pile group, in a normally consolidated clay stratum having properties as shown in the figure, will be

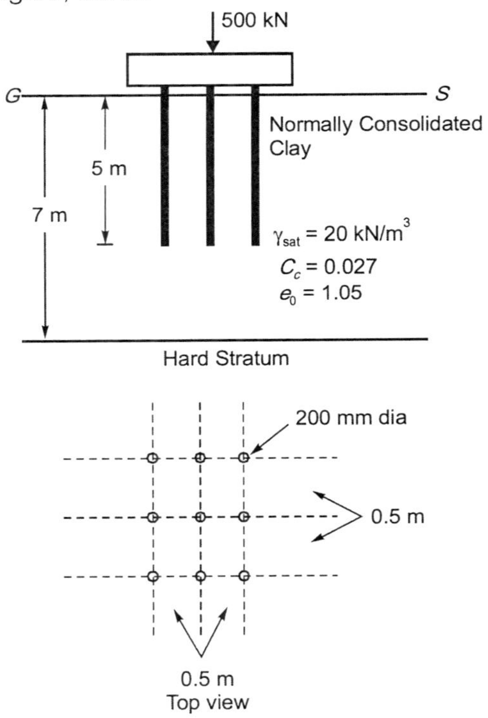

(a) 13.2 mm (b) 12.775 mm
(c) 7.345 mm (d) None of these
[2003 : 2 M]

12.9 Match the items of **List-I** with **List-II** and select the correct answer using the codes given below the lists:
List-I
A. Modulus of subgrade reaction
B. Relative density and strength
C. Skin friction and point bearing resistance
D. Elastic constants

List-II
1. Cyclic pile load test
2. Pressure meter test
3. Plate load test
4. Standard penetration tests
5. Dynamic cone penetration test

Codes:

	A	B	C	D
(a)	1	3	2	5
(b)	1	2	4	3
(c)	2	5	1	3
(d)	3	4	1	2

[2003 : 2 M]

Common Data for Questions 12.10 and 12.11:

A group of 16 piles of 10 m length and 0.5 m diameter is installed in a 10 m thick stiff clay layer underlain by rock. The pile-soil adhesion factor is 0.4; average shear strength of soil on the sides is 100 kPa; undrained shear strength of the soil at the base is also 100 kPa.

12.10 The base resistance of a single pile is (in kN)
(a) 40.00 (b) 88.35
(c) 100.00 (d) 176.71 [2004 : 2 M]

12.11 Assuming 100% efficiency, the group side resistance is
(a) 5026.5 kN (b) 10000.0 kN
(c) 10053.1 kN (d) 20106.0 kN
[2004 : 2 M]

12.12 Negative skin friction in a soil is considered when the pile is constructed through a
(a) fill material
(b) dense coarse sand
(c) over consolidated stiff clay
(d) dense fine sand [2005 : 1 M]

12.13 For the soil profile shown in figure below, the minimum number of precast concrete piles of 300 mm diameter required to safely carry the load for a given factor of safety of 2.5 (assuming 100% efficiency for the pile group) is equal to

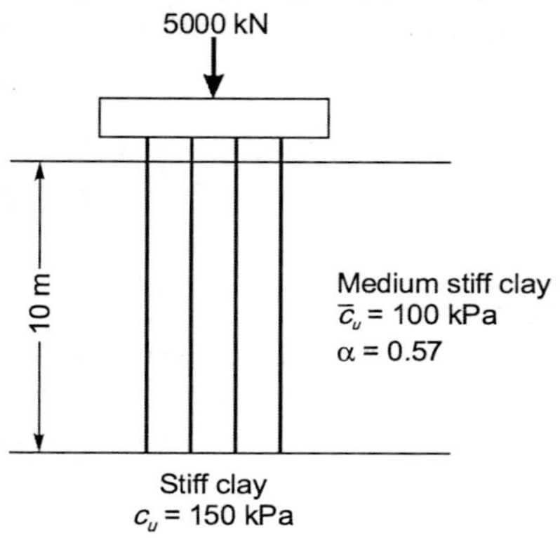

(a) 10 (b) 15
(c) 20 (d) 25 [2006 : 2 M]

12.14 What is the ultimate capacity in kN of the pile group shown in the figure assuming the group to fail as a single block?

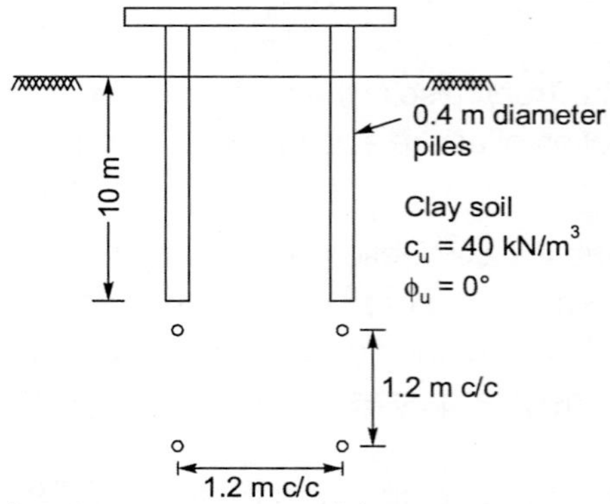

(a) 921.6 (b) 1177.6
(c) 2438.6 (d) 3481.6 [2007 : 2 M]

12.15 A pile of 0.50 m diameter and of length 10 m is embedded in a deposit of clay. The undrained strength parameters of the clay are cohesion = 60 kN/m^2 and the angle of internal friction = 0. The skin friction capacity (kN) of the pile for an adhesion factor of 0.6, is
(a) 671 (b) 565
(c) 283 (d) 106 [2008 : 2 M]

12.16 A precast concrete pile is driven with a 50 kN hammer ralling through a height of 1.0 m with an efficiency of 0.6. The set value observed is 4 mm per blow and the combined temporary compression of the pile, cushion and the ground is 6 mm. As per Modified Hiley Formula, the ultimate resistance of the pile is
(a) 3000 kN (b) 4285.7 kN
(c) 8333 kN (d) 11905 kN
[2009 : 1 M]

12.17 The ultimate load capacity of a 10 m long concrete pile of square cross section 500 mm × 500 mm driven into a homogeneous clay layer having undrained cohesion value of 40 kPa is 700 kN. If the cross section of the pile is reduced to 250 mm × 250 mm and the length of the pile is increased to 20 m, the ultimate load capacity will be
(a) 350 kN (b) 632.5 kN
(c) 722.5 kN (d) 1400 kN [2010 : 2 M]

12.18 A singly under-reamed, 8-m long, RCC pile (shown in the adjoining figure) weighing 20 kN with 350 mm shaft diameter and 750 mm under-ream diameter is installed within stiff, saturated silty clay (undrained shear strength is 50 kPa, adhesion factor is 0.3, and the applicable bearing capacity factor is 9) to counteract the impact of soil swelling on a structure constructed above. Neglecting suction and the contribution of the under-ream to the adhesive shaft capacity, what would be the estimated ultimate tensile capacity (rounded off to the nearest integer value of kN) of the pile?

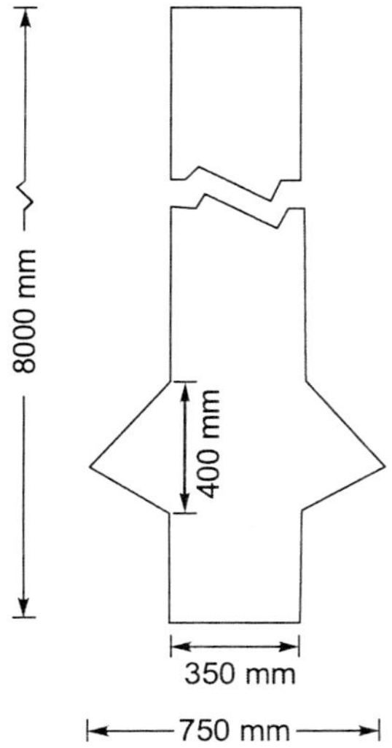

(a) 132 kN (b) 156 kN
(c) 287 kN (d) 301 kN [2011 : 2 M]

12.19 Four columns of building are to be located within a plot size of 10 m × 10 m. The expected load on each column is 4000 kN. Allowable bearing capacity of soil deposit is 100 kN/m^2. The type of foundation to be used is

(a) Isolated foundation
(b) Raft foundation
(c) Pile foundation
(d) Combined foundation [2013 : 1 M]

12.20 The action of negative skin friction on the pile is to

(a) increase the ultimate load on the pile
(b) reduce the allowable load on the pile
(c) maintain the working load on the pile
(d) reduce the settlement of the pile

[2014 : 1 M, Set-I]

12.21 A single vertical friction pile of diameter 500 mm and length 20 m is subjected to a vertical compressive load. The pile is embedded in a homogeneous sandy stratum where: angle of internal friction (φ) = 30°, dry unit weight (γ_d) = 20 kN/m^3 and angle of wall friction (δ) = $2\varphi/3$. Considering the coefficient of lateral earth pressure (K) = 2.7 and the bearing capacity factor (N_q) = 25, the ultimate bearing capacity of the pile (in kN) is _______. [2014 : 2 M, Set-II]

12.22 A pile of diameter 0.4 m is fully embedded in a clay stratum having 5 layers, each 5 m thick as shown in the figure below. Assume a constant unit weight of soil as 18 kN/m^3 for all the layers. Using λ-method (λ = 0.15 for 25 m embedment length) and neglecting the end bearing component, the ultimate pile capacity (in kN) is____.

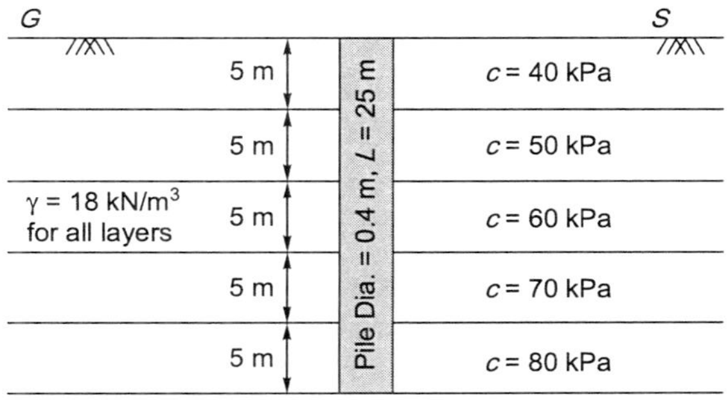

[2015 : 2 M, Set-II]

12.23 It is proposed to drive H-piles up to a depth of 7 m at a construction site. The average surface area of the H-pile is 3 m^2 per meter length. The soil at the site is homogeneous sand, having an effective friction angle of 32°. The ground water table (GWT) is at a depth of 2 m below the ground surface. The unit weights of the soil above and below the GWT are 16 kN/m^3 and 19 kN/m^3, respectively. Assume the earth pressure coefficient, K = 1.0, and the angle of wall friction, δ = 23°. The total axial frictional resistance (in kN, up to one decimal place) mobilized on the pile against the driving is __________.

[2017 : 2 M, Set-I]

12.24 Two identical concrete piles having the plan dimensions 50 cm × 50 cm are driven into a homogeneous sandy layer as shown in the figures. Consider the bearing capacity factor N_q for ϕ = 30° as 24.

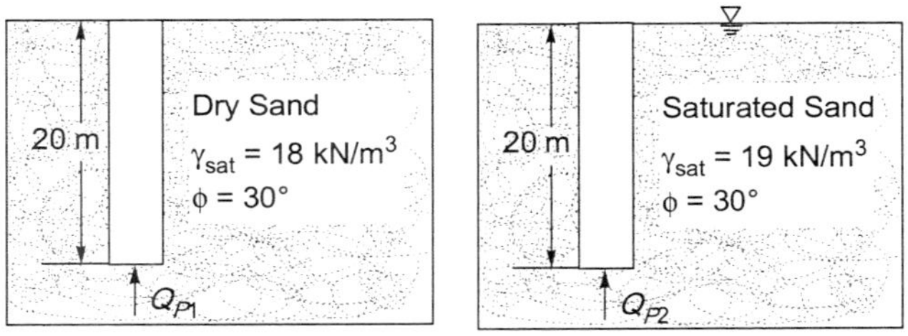

If Q_{P1} and Q_{P2} represent the ultimate point bearing resistances of the piles under dry and submerged conditions, respectively, which one of the following statements is correct?

(a) $Q_{P1} > Q_{P2}$ by about 100%
(b) $Q_{P1} < Q_{P2}$ by about 100%

(c) $Q_{P1} > Q_{P2}$ by about 5%
(d) $Q_{P1} < Q_{P2}$ by about 5% [2017 : 2 M, Set-II]

12.25 A 0.5 m × 0.5 m square concrete pile is to be driven in a homogeneous clayey soil having undrained shear strength, c_u = 50 kPa and unit weight, γ = 18.0 kN/m^3. The design capacity of the pile is 500 kN. The adhesion factor α is given 0.75. The length of the pile required for the above design load with a factor of safety of 2.0 is

(a) 5.2 m (b) 5.8 m
(c) 11.8 m (d) 12.5 m

[2018 : 2 M, Set-I]

12.26 A group of nine piles in a 3 × 3 square pattern is embedded in a soil strata comprising dense sand underlying recently filled clay layer, as shown in the figure. The perimeter of an individual pile is 126 cm. The size of pile group is 240 cm × 240 cm. The recently filled clay has undrained shear strength of 15 kPa and unit weight of 16 kN/m^3.

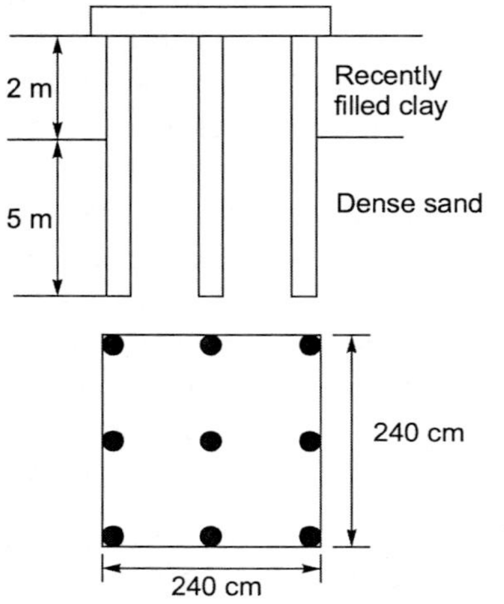

The negative frictional load (in kN, up to two decimal places) acting on the pile group is ________. [2018 : 2 M, Set-II]

12.27 A reinforced concrete circular pile of 12 m length and 0.6 m diameter is embedded in stiff clay which has an undrained unit cohesion of 110 kN/m^2. The adhesion factor is 0.5. The Net Ultimate Pullout (uplift) Load for the pile (in kN, round off to 1 decimal place) is________.

[2019 : 2 M, Set-I]

12.28 A timber pile of length 8 m and diameter 0.2 m is driven with a 20 kN drop hammer, falling freely from a height of 1.5 m. The total penetration of the pile in the last 5 blows is 40 mm. Use the Engineering News Record expression. Assume a factor of safety of 6 and empirical factor (allowing reduction in the theoretical set, due to energy losses) of 2.5 cm. The safe load carrying capacity of the pile (in kN, round off to 2 decimal places) is ________.

[2019 : 2 M, Set-II]

12.29 At a site, Static Cone Penetration Test was carried out. The measured point (tip) resistance q_c was 1000 kPa at a certain depth. The friction ratio (f_r) was estimated as 1% at the same depth.

The value of sleeve (side) friction (in kPa) at that depth was ________ . (in integer)

[2022 : 1 M, Set-I]

12.30 A square concrete pile of 10 m length is driven into a deep layer of uniform homogeneous clay. Average unconfined compressive strength of the clay, determined through laboratory tests on undisturbed samples extracted from the clay layer, is 100 kPa. If the ultimate compressive load capacity of the driven pile is 632 kN, the required width of the pile is ______ mm. (in integer)

(Bearing capacity factor N_c = 9; adhesion factor α = 0.7)

[2022 : 2 M, Set-I]

12.31 A group of total 16 piles are arranged in a square grid format, The center-to-center spacing (s) between adjacent piles is 3 m. The diameter (d) and length of embedment of each pile are 1 m and 20 m, respectively, The design capacity of each pile is 1000 kN in the vertical downward direction. The pile group efficiency η_g is given by

$$\eta_g = 1 - \frac{\theta}{90}\left[\frac{(n-1)m + (m-1)n}{mn}\right]$$

where m and n are number of rows and columns in the plan grid of pile arrangement, and $\theta = \tan^{-1}\left(\frac{d}{s}\right)$.

The design value of the pile group capacity (in kN) in the vertical downward direction is ______. (round off to the nearest integer)

[2022 : 2 M, Set-II]

12.32 A group of 9 friction piles are arranged in a square grid maintaining equal spacing in all directions. Each pile is of diameter 300 mm and length 7 m. Assume that the soil is cohesionless with effective friction angle ϕ' = 32°. What is the center-to-center spacing of the piles (in m) for the pile group efficiency of 60%?

(a) 0.582 (b) 0.486
(c) 0.391 (d) 0.677

[2023 : 1 M, Set-I]

12.33 A circular pile of diameter 0.6 m and length 8 m was constructed in a cohesive soil stratum having the following properties: bulk unit weight = 19 kN/m^3; angle of internal friction = 0° and cohesion = 25 kPa.

The allowable load the pile can carry with a factor of safety of 3 is _______ kN (round off to one decimal place).

[Adopt: Adhesion factor, α = 1.0 and Bearing capacity factor, N_c = 9.0]

[2023 : 2 M, Set-II]

12.34 A reinforced concrete pile of 10 m length and 0.7 m diameter is embedded in a saturated pure clay with unit cohesion of 50 kPa. If the adhesion factor is 0.5, the net ultimate uplift pullout capacity (in kN) of the pile is _______ (rounded off to the nearest integer).

[2024 : 1 M, Set-II]

■■■■

Answers Deep Foundation

12.1 (True)	12.2 (b)	12.3 (c)	12.4 (b)	12.5 (b)	12.6 (c)
12.7 (c)	12.8 (d)	12.9 (d)	12.10 (d)	12.11 (c)	12.12 (a)
12.13 (c)	12.14 (d)	12.15 (b)	12.16 (b)	12.17 (b)	12.18 (d)
12.19 (b)	12.20 (b)	12.21 (6174.49)	12.22 (1625.77)	12.23 (390.8)	12.24 (a)
12.25 (c)	12.26 (472.3)	12.27 (1244.1)	12.28 (151.51)	12.29 (10)	12.30 (400)
12.31 (11085)	12.32 (*)	12.33 (146.9)	12.34 (550)		

Explanations Deep Foundation

12.1 Sol.

True

Friction piles transfer their load to the ground through skin friction. These types of pile foundations are commonly known as floating pile foundations.

12.2 (b)

'**Negative skin friction**' or '**downward drag**' is a phenomenon which occurs when a soil layer surrounding a portion of the pile settles more than the pile. Such relative motion may occur when the clay stratum undergoes consolidation due to,

1. A fill recently placed over the clay stratum.
2. Lowering of the ground water table.
3. Reconsolidation occurring due to disturbance caused by pile driving in sensitive clay stratum, etc.

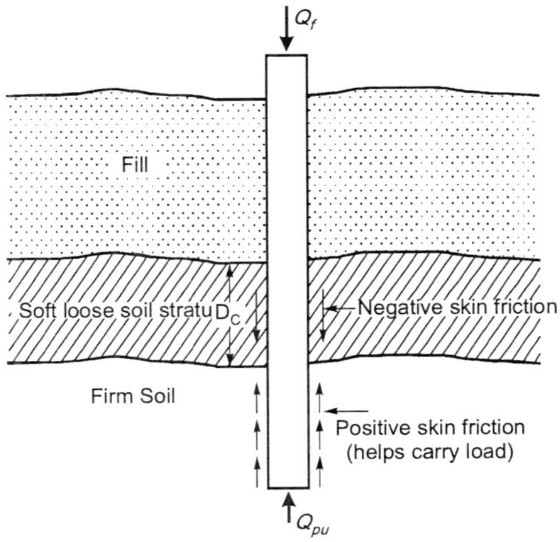

Negative skin friction on a pile

Negative skin friction increases gradually as the consolidation of the clay layer proceeds since the effective overburden pressure gradually increases due to dissipation of excess pore pressure.

12.3 (c)

The value of N_c in **Mayerhoff's** bearing capacity equation for shallow and deep foundation is same as were given by Prandtl which were user than the **Terzaghi's** N_c value.

For cohesive soils N_c value as per **Mayerhoff's** is **9**.

12.4 (b)

The slow rate of consolidation of saturated clays of low permeability may be accelerated by means of vertical sand drains which provide for radial drainage, resulting in shortening of drainage path. Consolidation is then mainly due to horizontal radial drainage which results in a faster dissipation of excess pore water pressure.

12.5 (b)

Well foundations are also, like the pile foundations deep foundations. Well foundations provide a solid and massive foundation for heavy loads and are useful in situations where the foundation is provided in running water.

12.6 (c)

Group efficiency, n_g of ple group is defined as

$$n_g = \frac{Q_{ug}}{nQ_{up}}$$

n_g can be greater than or less than 1.

12.7 (c)

The void ratio can be greater than 1.

12.8 (d)

Assume equivalent raft at a distance of $\frac{2L}{3} = \frac{2}{3} \times 5$ from the top ground level and load dispersion at 2 H : 1 V.

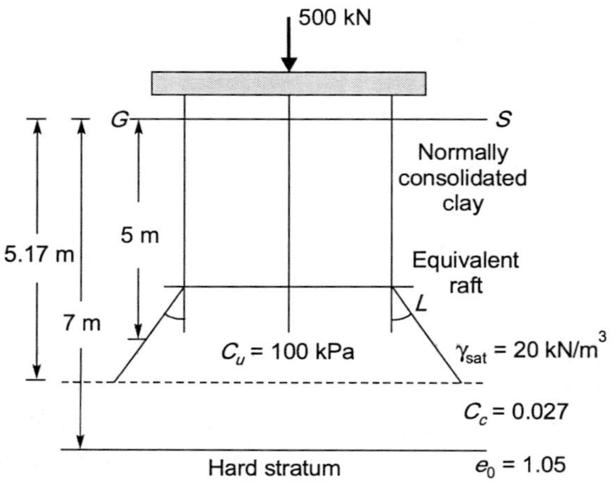

Compressible layer thickness

$$= 7 - 5 + \frac{5}{3} = 3.667 \text{ m}$$

Average stress at the center of layer,

$$\bar{\sigma}_0 = \gamma \times 5.17$$

$$= 20 \times 5.17 = 103.4 \text{ kN/m}^2$$

Increased width of equivalent raft at depth

$$= 1.2 + \frac{\left(\frac{5}{3} + 2\right)}{2} = 3.03 \text{ m}$$

Pressure increase due to load on pile,

$$\Delta\sigma = \frac{500}{3.03^2} = 54.34 \text{ kN/m}^2$$

$$\Delta H = \frac{C_c H_0}{1 + e_0} \log\left(\frac{\bar{\sigma}_0 + \Delta\sigma}{\bar{\sigma}_0}\right)$$

$$= \frac{0.027 \times \frac{11}{3}}{1 + 1.05} \log\left(\frac{103.4 + 54.34}{103.4}\right) = 8.8 \text{ mm}$$

12.9 (d)

- Relative density and strength of soil can be related to SPT value.
- Modulus of subgrade reaction of soil is found out using plate load test.
- Cyclic pile test can be used to find out separately the end bearing and friction strength of a pile.

12.10 (d)

Base resistance of a single pile = $9c_u A_b$

where, $c_u = 100 \text{ kN/m}^2$

$$A_b = \frac{\pi}{4} \times (0.5)^2$$

∴ Base resistance

$$= 9 \times 100 \times \frac{\pi}{4} \times (0.5)^2 = 176.71 \text{ kN}$$

12.11 (c)

$$\eta = \frac{Q_{ug}}{nQ_u} \qquad [\because \eta = 100\%]$$

$$Q_{ug} = nQ_u = n \times \alpha\, c_u\, PL_c$$

$$= 16 \times 0.4 \times 100 \times \pi \times 0.5 \times 10 = 10053.1 \text{ kN}$$

12.12 (a)

Negative skin friction is experienced when the soil around the pile settles at a faster rate than pile. Thus, piles installed in freshly prepared fills of soft compressive deposits are subject to a downward drag. This downward drag on the pile surface, where the soil moves down relative to the pile, adds to the structural loads and is called **Negative skin friction**.

12.13 (c)

The ultimate load capacity of piles

$= 2.5 \times 5000 = 12{,}500$ kN

The load capacity of single pile

$= c_u N_c A_b + \alpha \bar{c}_u A_s$

$= 150 \times 9 \times \frac{\pi}{4} \times (0.3)^2 + 0.57 \times 100 \times 10 \times 0.3\pi$

$= 632.64$ kN

$\therefore$ Number of piles $= \frac{12500}{632.64} = 19.75 \approx 20$

12.14 (d)

Pile groupis falling as a single block

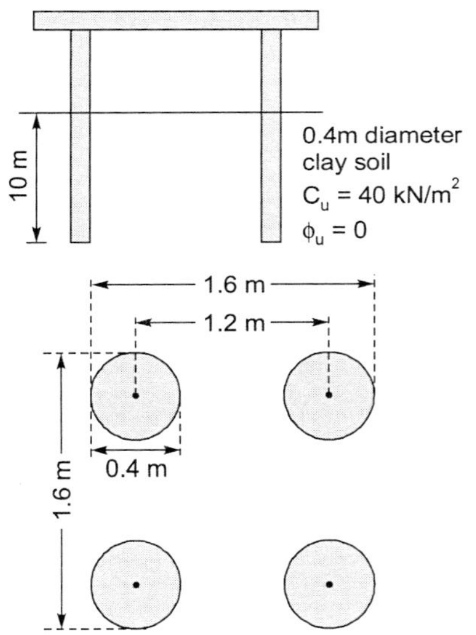

Size of pile group 1.6 m × 1.6 m

The ultimate load capacity of the pile group by block failure,

$$Q_{ug} = c_u N_c A_b + P_b L c_u$$

$$= 40 \times 9 \times (1.2 + 0.4)^2 + 4 \times (1.2 + 0.4) \times 10 \times 40$$

$$= 3481.6 \text{ kN}$$

12.15 (b)

The **skin friction capacity** of a pile is given by

$$Q_f = \alpha c_u A_s$$

$$= 0.6 \times 60 \times \pi \times 0.50 \times 10$$

$$= 565.48 \text{ kN} = 565 \text{ kN}$$

12.16 (b)

Ultimate driving resistance as per modified **Hiley's formula** is given by,

$$R = \frac{wh\eta}{S + 0.5C} = \frac{50 \times 1 \times 0.6}{4 \times 10^{-3} + 0.5 \times 6 \times 10^{-3}}$$

$$= 4285.7 \text{ kN}$$

12.17 (b)

The ultimate load of a pile in clay is given by,

$$Q_u = cNA_b + \alpha c A_s$$

$\Rightarrow$ $700 = 40 \times 9 \times 0.5 \times 0.5 + \alpha \times 40 \times 0.5 \times 4 \times 10$

$\Rightarrow$ $\alpha = 0.7625$

Again, $Q_u = 40 \times 9 \times 0.25 \times 0.25 + 0.7625 \times 40 \times 0.25 \times 4 \times 20$

$Q_u = 632.5$ kN

12.18 (d)

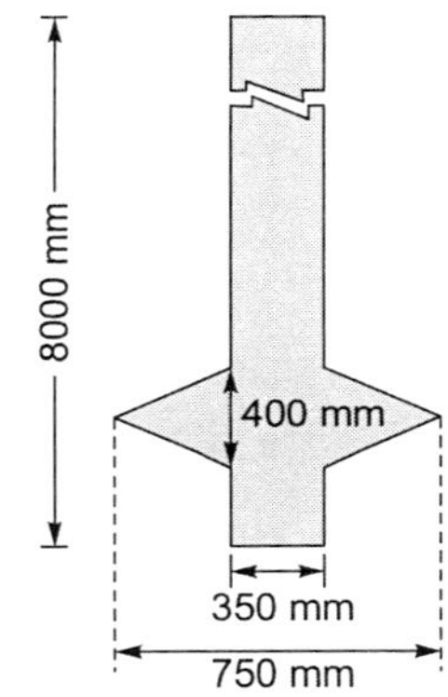

Weight of pile

$P_s = 20$ kN

Shaft dia, $D_0 = 350$ mm

Under ream dia,

$D_u = 750$ mm

Undrained shear strength,

$C = 50$ kPa

Adhesion factor

$\alpha = 0.3$

$N_c = 9$

Ultimate tensile capacity will be due to

1. Fricrtion along the length of pile (P_1)
2. Bearing action caused by underreamed portion (P_2)
3. Dead weight of pile (P_3)

Tensile capcity due to friction

$$P_1 = f_s \times A_s = aC\pi D_0 \times L$$

$$= 0.3 \times 50 \times \pi \times 0.35 \times 7.6 = 125.286 \text{ kN}$$

Tensile capacity due to bearing action

$$P_2 = CN_cA$$

$$= 50 \times 9 \frac{\pi \times (D_u^2 - D_0^2)}{4}$$

$$= 50 \times 9 \frac{\pi \times (0.75^2 - 0.35^2)}{4} = 155.51 \text{ kN}$$

$$P = P_1 + P_2 + P_3$$

$$= 125.286 + 155.51 + 20 = 300.796 \text{ kN}$$

12.19 (b)

Load on each column = 400 kN

Bearing capacity of soil = 100 kN/m^2

$\therefore$ Area of each footing required

$$= \frac{4000}{100} = 40 \text{ m}^2$$

$\therefore$ Total area required = 4 × 4 = 16 m^2

$$\therefore \quad \frac{\text{Foundation area}}{\text{Plot area}} = \frac{40}{100} = 40\%$$

If the foundation area is more than 40% of plot area, then it is suitable otherwise provide raft foundation or combined footing.

12.20 (b)

12.21 Sol.

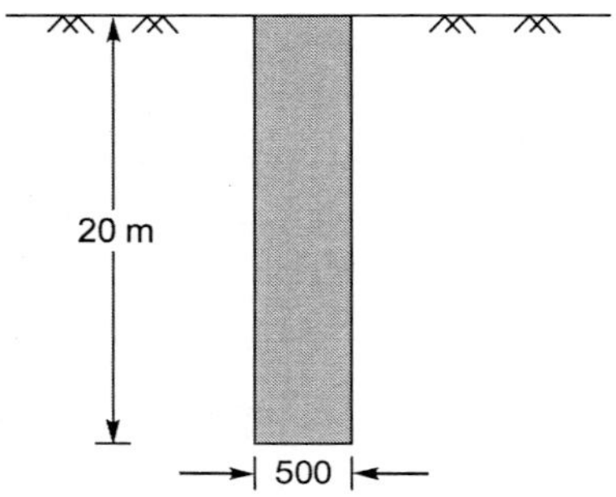

Homogeneous sandy stratum

$$\phi = 30^\circ$$

$$\gamma_d = 20 \text{ kN/m}^3$$

Wall friction angle,

$$\delta = \frac{2}{3}\phi = \frac{2}{3} \times 30 = 20^\circ$$

Lateral earth pressure coefficient,

$$K = 2.7$$

Bearing capacity factor,

$$N_q = 25$$

Ultimate load capacity,

$$Q_u = ?$$

Vertical effective stress at 20 m,

$$\overline{\sigma_v} = 20 \times 20 = 400 \text{ kN/m}^2$$

From 0 to 20 m, unit point bearing resistance and skin friction resistance remain constant at,

$$\overline{\sigma_v} = 400 \text{ kN/m}^2$$

The ultimate load capacity is given by,

$Q_u = q_{up} \cdot A_p + q_s \cdot A_s = f_s$ ($Q_{up} = 0$ for friction pile)

where $\quad q_s = \frac{1}{2} \cdot \overline{\sigma_v} \cdot K \cdot \tan\delta$

$$= \frac{1}{2} \times 400 \times 2.7 \tan 20^\circ = 196.54 \text{ kN/m}^2$$

Skin friction resistance,

$$Q_u = f_s = 196.54 \times \pi \times 0.5 \times 20 = 6174.49 \text{ kN}$$

12.22 Sol.

Ultimate bearing capacity

$= \lambda(\sigma_{v\,av} + 2C_u) A_s$

$= 0.15[18 \times 12.5 + 2C_u] \times (\pi \times 0.4 \times 25)$

C_u is the average value of all layers i.e., 60 kPa.

$= 0.15[18 \times 12.5 + 2 \times 60] \times (\pi \times 0.4 \times 25)$

$= 1625.77$ kN

12.23 Sol.

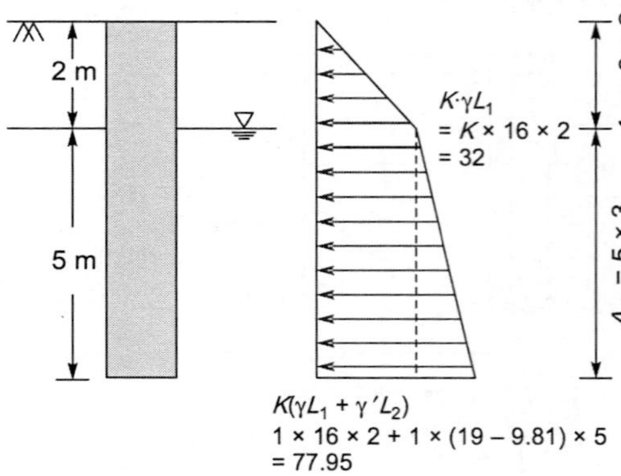

$$Q_{sf} = \frac{1}{2}k\gamma L_1 \tan\delta\, A_{s_1} + \frac{1}{2}K\left[\gamma L_1 + \gamma' L_2\right]\tan\delta A_{s_2}$$

$$= \frac{1}{2} \times 1 \times 16 \times 2 \tan 23^\circ (3 \times 2)$$

$$+ \frac{1}{2} \times 1\left[16 \times 2 + 77.95\right] \times \tan 23^\circ (5 \times 3)$$

$$= 390.8 \text{ kN}$$

12.24 (a)

End bearing resistance

$= \text{Effective stress at base} \times N_q = \gamma_l \times N_q$

$$Q_{P_1} = qN_q = \gamma D_f N_q$$

$$Q_{P_1} = 18 \times 20 \times 24 = 8640 \text{ kN}$$

$$Q_{P_2} = \gamma' \, LN_q = (19 - 9.81)\, 20 \times 24 = 4411.2$$

12.25 (c)

$$C_u = 50 \text{ kPa}, \gamma = 18 \text{ kN/m}^3, \text{FOS} = 2$$

$$Q_{up} = 9C \times B^2 + \alpha\bar{C}(4BL) = 1000 \text{ kN}$$

$1000 = 9 \times 50 \times 0.5 \times 0.5 + 0.75 \times 50\,(4 \times 0.5\,L)$

$L = 11.83$ m

12.26 Sol.

3 × 3 pile group

Negative skin friction for group.

$= \alpha\bar{c}(4BL)$ + weight of soil in negative zone.

$= \alpha\bar{c}(4BL) + \gamma[\text{Area} \times \text{length}]$

$= 1 \times 15[4 \times 2.4 \times 2] + 16[2.4^2 \times 2]$

$= 472.32$ kN

Negative skin friction in individual action.

$= n[\alpha\bar{c}\ (\text{perimeter} \times l]$

$= 9[0.5 \times 15 \times 1.26 \times 2]$ [Assume $\alpha = 0.5$]

$= 170.1$ kN

Negative skin friction is maximum of above (two)

$Q_{nf} = 472.32$ kN

12.27 Sol.

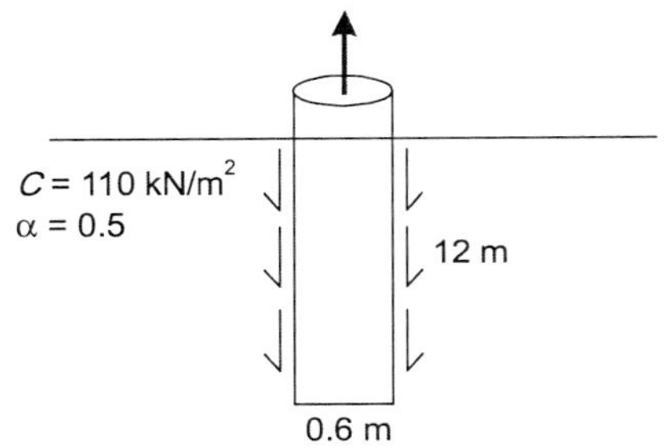

Net ultimate pullout

$= \alpha\bar{C}A_s = 0.5 \times 110(\pi dL)$

$= 0.5 \times 110\,(\pi \times 0.6 \times 12) = 1244.1$ kN

12.28 Sol.

$L = 8$ m; $d = 0.2$ m, 20 kN = $W \rightarrow$ drop hammer

$H = 1.5$ m

Penetration in 5 blows = 40 mm

∴ in 1 blow

$= \frac{40}{5} = 8 \text{ mm} = 0.8 \text{ cm}$

$$Q_{safe} = \left(\frac{WH}{s+c}\right) \times \frac{1}{FOS}$$

$$= \frac{1}{6}\left[\frac{20\text{kN} \times (1.5 \times 100)}{0.8\text{cm} + 2.5\text{cm}}\right]$$

$Q_{safe} = 151.51$ kN

12.29 Sol.

Friction ratio,

$$f_r = \frac{q_s}{q_t} = 1\%$$

where, q_s = Sleeve friction

q_t = Tip friction

$$\therefore \quad \frac{q_s}{q_t} = \frac{1}{100}\ (q_t = 1000 \text{ kPa})$$

$$\Rightarrow \quad q_s = \frac{1000}{100} = 10 \text{ kPa}$$

12.30 Sol.

$$C_u = \frac{q_u}{2} = \frac{100}{2} = 50 \text{ kN/m}^2$$

$$\therefore \quad Q_{up} = q_b A_b + q_s A_s$$

$$\Rightarrow \quad Q_{up} = 9 \times CB^2 + \alpha\bar{C}(4BL)$$

$\Rightarrow 632 \text{ kN} = 9 \times 50 \times B^2 + 0.7 \times 50\,(4B \times 10)$

$\Rightarrow \quad B = 0.4$ m

$\Rightarrow \quad B = 400$ mm

12.31 Sol.

Pile group efficiency,

$$\eta_g = 1 - \frac{\theta}{90}\left\{\frac{m(n-1)+n(m-1)}{mn}\right\}$$

where, $\theta = \tan^{-1}\left(\frac{d}{s}\right) = \tan^{-1}\left(\frac{1}{3}\right) = 18.43°$

Pile group efficiency is also given by

$$\eta_g = \frac{Q_{ug}}{nQ_{up}}$$

$$\Rightarrow \quad Q_{ug} = \eta_g n\, Q_{up}$$

$$\Rightarrow \quad Q_{ug} = \left[1 - \frac{18.43}{90}\left\{\frac{4(4-1)+4(4-1)}{4 \times 4}\right\}\right]$$

$$(16 \times 1000) = 11085 \text{kN}$$

12.32 (*)

By Converse – Labarre formula

$$\eta_g = \left[1 - \frac{\theta}{90}\left(\frac{m(n-1)+n(m-1)}{mn}\right)\right] \times 100$$

$$\Rightarrow \quad 0.6 = 1 - \frac{\theta}{90}\left(\frac{3 \times 2 + 3 \times 2}{3 \times 3}\right)$$

$$\Rightarrow \frac{\theta}{90}\left(\frac{6+6}{9}\right) = 0.4$$

$\Rightarrow \quad \theta = 27°$

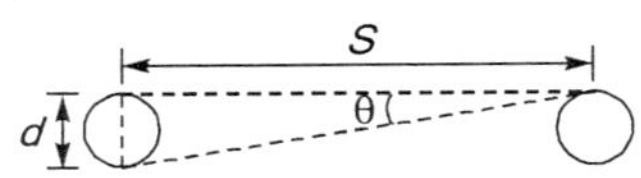

Now, $\tan\theta = \frac{d}{S}$

where d is diameter and S is spacing.

So, Spacing, $S = \frac{d}{\tan\theta} = \frac{0.3}{\tan 27°} = 0.589$

Alternatively,

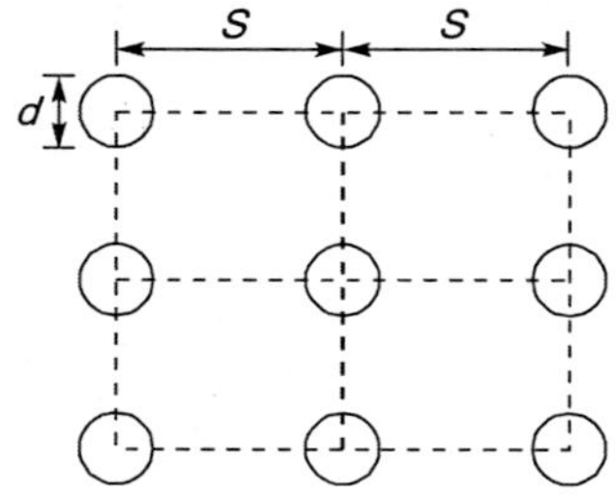

Group efficiency $= \frac{Q_{ug}}{nQ_{us}}$

$$\Rightarrow \quad 0.6 = \frac{\frac{1}{2}(\gamma l)\tan\delta(2S+d)\times l\times 4}{9\times\frac{1}{2}(\gamma l)\tan\delta(pdl)}$$

$$\Rightarrow \quad 0.6 = \frac{(2S+d)\times 4}{9\times\pi d}$$

$$\Rightarrow \quad S = \frac{1}{2}\left[\frac{0.6\times 9\times\pi\times 0.3}{4} - 0.3\right] = 0.486$$

12.33 Sol.

$$Q_{up} = 9c\left(\frac{\pi}{4}D^2\right) + \alpha\bar{c}(\pi DL)$$

$$= 9\times 25\left(\frac{\pi}{4}\times(0.6)^2\right) + 1\times 25(\pi\times 0.6\times 8)$$

$\therefore \quad Q_{up} = 440.608$ kN

$\therefore$ Allowable load $= Q_{ap} = \frac{Q_{up}}{FOS} = \frac{440.608}{3}$

$= 146.869$ kN $\simeq 146.9$ kN

12.34 Sol.

Length of pile, $L = 10$ m, Diameter of pile, $d = 0.7$ m

Cohesion, $C = 50$ kPa

Adhesion factor, $\alpha = 0.5$

Net ultimate uplift pullout capacity,

$T = \alpha C\pi dL$

$= 0.5 \times 50 \times \pi \times 0.7 \times 10 = 549.78$ kN

$\simeq 550$ kN

13 Soil Stabilization and Soil Exploration

13.1 The most commonly used sample for obtaining a disturbed sample of the soil is
(a) split spoon sampler
(b) open drive sampler
(c) piston sampler
(d) thin wall shell by tube sampler [1996 : 1 M]

13.2 Mechanical stabilization requires
(a) mixing of two or more types of natural soils
(b) addition of chemical to soils
(c) addition of lime to soils
(d) addition of cementing material to soils
[1996 : 1 M]

13.3 Match the **List-I** (Boring methods) with **List-II** (Field conditions) and select the correct answer using the codes given below the lists:

List-I
A. Auger Boring
B. Wash Boring
C. Percussion Drilling
D. Rotary Drilling

List-II
1. Below water table in all soil types except hard soils and rocks
2. Large diameter boreholes over 150 mm in size
3. Explorations for shallow foundations and highways
4. Bouldery and gravelly strata

Codes:

	A	B	C	D
(a)	3	1	4	2
(b)	1	2	4	3
(c)	2	3	4	1
(d)	3	1	2	4

[2003 : 2 M]

13.4 In the context of collecting undisturbed soil samples of high quality using a spoon sampler, following statements are made:
1. Area ratio should be less than 10%.
2. Clearance ratio should be less than 1%.

With reference to above statements, which of the following applies?
(a) Both the statements are true
(b) Statements 2 is true but 1 is false
(c) Statement 1 is true but 2 is false
(d) Both the statements are false
[2004 : 2 M]

13.5 The degree of disturbance of the sample collected by the sampler is expressed by a term called the "area ratio". If the outer diameter and inner diameter of the sampler are D_o and D_i respectively, the area ratio is given by

(a) $\dfrac{D_o^2 - D_i^2}{D_i^2}$ (b) $\dfrac{D_i^2 - D_o^2}{D_i^2}$

(c) $\dfrac{D_o^2 - D_i^2}{D_o^2}$ (d) $\dfrac{D_i^2 - D_o^2}{D_o^2}$

[2014 : 1 M, Set-I]

13.6 Which of the following statements is TRUE for degree of disturbance of collected soil sample?
(a) Thinner the sampler wall, lower the degree of disturbance of collected soil sample
(b) Thicker the sampler wall, lower the degree of disturbance of collected soil sample
(c) Thickness of the sampler wall and the degree of disturbance of collected soil sample are unrelated
(d) The degree of disturbance of collected soil sample is proportional to the inner diameter of he sampling tube [2015 : 1 M, Set-I]

13.7 A core cutter of 130 mm height has inner and outer diameters of 100 mm and 106 mm. respectively. The area ratio of the core cutter (in %, up to two decimal places) is ______.
[2018 : 1 M, Set-I]

13.8 The dimensions of a soil sampler are given in the table.

Parameter	Cutting edge	Sampling tube
Inside diameter (mm)	80	86
Outside diameter (mm)	100	90

For this sampler, the outside, clearance ratio (in percent, round off to 2 decimal places) is ______.
[2019 : 2 M, Set-II]

13.9 The internal (d_i) and external (d_o) diameters of a Shelby sampler are 48 mm and 52 mm, respectively. The area ratio (A_r) of the sampler (in %, round off to two decimal places) is ______.

[2021 : 1 M, Set-II]

13.10 The inside diameter of a sampler tube is 50 mm, The inside diameter of the cutting edge is kept such that the inside Clearance Ratio (ICR) is 1.0 % to minimize the friction on the sample as the sampler tube enters into the soil. The inside diameter (in mm) of the cutting edge is ________ . (round off to two decimal places)

[2022 : 2 M, Set-II]

13.11 Which of the following statement(s) is/are CORRECT?

(a) Swell potential of soil decreases with an increase in the shrinkage limit.

(b) In electrical resistivity tomography, the depth of current penetration is half of the spacing between the electrodes.

(c) Among the several corrections to be applied to the SPT-N value, the dilatancy correction is applied before all other corrections.

(d) Both loose and dense sands with different initial void ratios can attain similar void ratio at large strain during shearing.

[2024 : 2 M, Set-I]

Answers Soil Stabilization and Soil Exploration

13.1 (a) 13.2 (a) 13.3 (a) 13.4 (c) 13.5 (a) 13.6 (a) 13.7 (12.36)

13.8 (11.11) 13.9 (17.36) 13.10 (49.50) 13.11 (a, d)

Explanations Soil Stabilization and Soil Exploration

13.1 (a)

Split spoon sampler is used in SPT to find out disturbed sample of soil.

13.2 (a)

Addition of chemical, lime and cementing materials are chemical stabilization.

Mixing of two or more type of natural soil and compaction are mechanical stabilization technique.

13.3 (a)

In percussion boring, soil and rocks formation are broken by repeated blows of heavy chisel or bit suspended by a cable or drill rod. Suitable for advancing a hole in all types of soil, boulders and rock. Rotary drilling in very fast method of advancing hole in both rocks and soil. A drill bit, fixed to the lower end of drill rods, is rotated by a suitable chuck.

13.4 (c)

Area ratio should be less than 10% for a undisturbed sample. But in case of collecting a sample using split spoon sampler the area ratio is as high as 30%. So it is impossible to get a undisturbed sample with the help of a split spoon sampler. Also, the clearance ratio should be from 1 to 3 per cent.

13.5 (a)

Area ratio, A_r is expressed as

$$\%A_r = \frac{D_0^2 - D_i^2}{D_1^2} \times 100$$

Area ratio should be kept as low as possible.

13.6 (a)

As thickness of sampler increases, disturbance increases.

13.7 Sol.

$$\text{Area ratio} = \frac{A_{outer} - A_{inner}}{A_{inner}} \times 100$$

$$= \frac{\frac{\pi}{4}D_0^2 - \frac{\pi}{4}D_i^2}{\frac{\pi}{4}D_i^2} \times 100 = \frac{\frac{\pi}{4}(106)^2 - \frac{\pi}{4}(100)^2}{\frac{\pi}{4}(100)^2} \times 100$$

$= 12.36\%$

13.8 Sol.

Outside clearance.

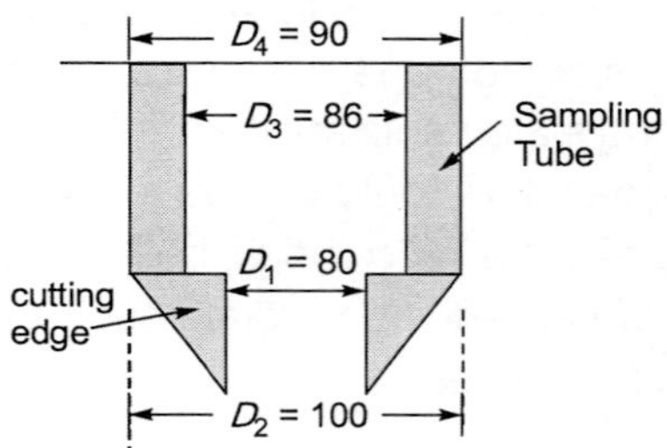

Outside clearance,

$$= \left[\frac{D_2 - D_4}{D_4}\right] \times 100 = \left[\frac{100 - 90}{90}\right] \times 100$$

$= 11.11\%$

13.9 Sol.

Outside diameter= 52 mm

Inside diameter = 48 mm

$$A_r = \frac{\frac{\pi}{4}(D_2)^2 - \frac{\pi}{4}(D_1)^2}{\frac{\pi}{4}(D_1)^2} \times 100$$

$$A_r = \frac{(52)^2 - (48)^2}{(48)^2} \times 100 = 17.36\%$$

13.10 Sol.

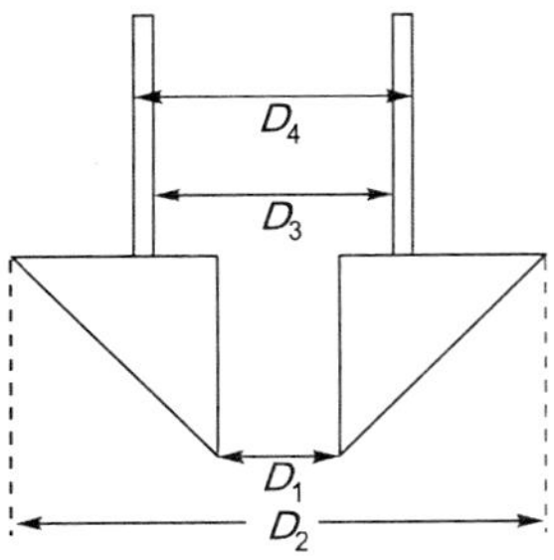

∴ Inside clearance ratio,

$$C_i = \frac{D_3 - D_1}{D_1} \times 100$$

$$\Rightarrow \quad 1 = \frac{50 - D_1}{D_1} \times 100$$

$$\Rightarrow \quad D_1 = 49.50 \text{ mm}$$

13.11 (a, d)

(i) Lower is the shrinkage → Higher is the swelling and shrinkage

(ii)

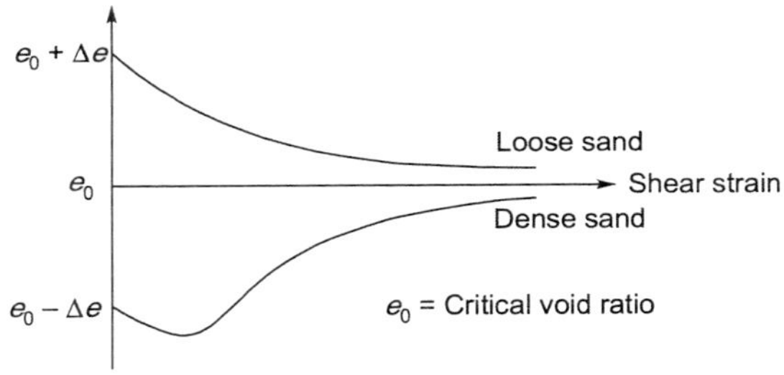

Fig. Shear strain vs Void ratio curve

(iii) Over burden correction is applied before dilatancy correction.

■■■■

Fluid Mechanics & Fluid Machines

UNIT VI

CONTENTS

Fluid Mechanics & Fluid Machines

Syllabus

Fluid Mechanics: Properties of fluids, fluid statics; Continuity, momentum and energy equations and their applications; Potential flow, Laminar and turbulent flow; Flow in pipes, pipe networks; Concept of boundary layer and its growth; Concept of lift and drag.

Hydraulics: Forces on immersed bodies; Flow measurement in channels and pipes; Dimensional analysis and hydraulic similitude; Channel Hydraulics - Energy-depth relationships, specific energy, critical flow, hydraulic jump, uniform flow, gradually varied flow and water surface profiles.

Analysis of Previous GATE Papers

Exam Year	1 Mark Ques.	2 Marks Ques.	Total Marks
1995	7	–	7
1996	17	–	17
1997	12	–	12
1998	3	2	7
1999	3	–	3
2000	8	–	8
2001	1	3	7
2002	2	2	6
2003	3	8	19
2004	6	10	26
2005	3	7	17
2006	2	8	18
2007	2	7	16
2008	1	7	15
2009	1	3	7
2010	2	2	6
2011	3	2	7
2012	3	–	3
2013	3	2	7
2014 Set-1	2	7	16
2014 Set-2	2	4	10
2015 Set-1	3	4	11
2015 Set-2	3	3	9
2016 Set-1	1	3	7
2016 Set-2	1	3	7
2017 Set-1	2	3	8
2017 Set-2	1	3	7
2018 Set-1	4	3	10
2018 Set-2	1	3	7
2019 Set-1	2	3	8
2019 Set-2	2	2	6
2020 Set-1	3	3	9
2020 Set-2	2	2	6
2021 Set-1	2	2	6
2021 Set-2	2	2	6
2022 Set-1	2	2	6
2022 Set-2	–	4	8
2023 Set-1	1	3	7
2023 Set-2	1	3	7
2024 Set-1	1	3	7
2024 Set-2	2	3	8

Fluid Properties and Manometry

1.1 The dimensions of a pressure gradient in a fluid flow are
(a) $ML^{-1}T^{2}$
(b) $ML^{-3}T^{-2}$
(c) $ML^{-2}T^{-2}$
(d) $M^{-1}L^{-3}T^{-2}$
[1995 : 1 M]

1.2 A fluid is one which can be defined as a substance that
(a) has some shear stress at all points
(b) can deform indefinitely under the action of the smallest shear force
(c) has the small shear stress in all directions
(d) is practically incompressible
[1996 : 1 M]

1.3 Shear stress in the newtonian fluid is proportional to
(a) Pressure
(b) Strain
(c) Strain rate
(d) The inverse of the viscosity [1996 : 1 M]

1.4 A liquid of density ρ and dynamic viscosity μ flows steadily down an inclined plane in a thin sheet of constant thickness t. Neglecting air friction the shear stress on the bottom surface due to the liquid flow is (where θ is the angle the plane makes with the horizontal)
(a) $\rho g t \sin\theta$
(b) $\rho g t \cos\theta$
(c) $\mu\sqrt{g/t}$
(d) ρg [1996 : 1 M]

1.5 A U-tube with both limbs open to atmosphere contains two immiscible liquids of densities ρ_1 & ρ_2 (figure shown below). Under equilibrium, the distance h is given by

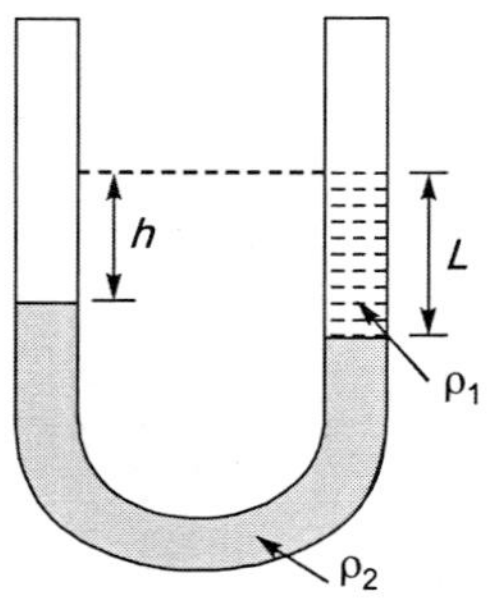

(a) $h = L\left(1-\frac{\rho_1}{\rho_2}\right)$
(b) $h = L\left(\frac{\rho_1}{\rho_2}-1\right)$
(c) $h = L\left(\frac{\rho_2}{\rho_1}\right)$
(d) $h = 0$ [1996 : 1 M]

1.6 The unit of dynamic viscosity of a fluid is
(a) m^2/s
(b) Ns/m^2
(c) $Pa\ s/m^2$
(d) $kg\ s^2/m^2$
[1997 : 1 M]

1.7 With increase of temperature, viscosity of a fluid
(a) Does not change
(b) Always increases
(c) Always decrease
(d) Increase, if fluid is a gas and decrease if it is a liquid [1997 : 1 M]

1.8 Two pipelines, one carrying oil (mass density 900 kg/m^3) and the other water, are connected to a manometer as shown in figure. By what amount the pressure in the water pipe should be increased so that the mercury levels in both the limbs of the manometer become equal? (Mass density of mercury = 13550 kg/m^3 and $g = 9.81$m/s^2)

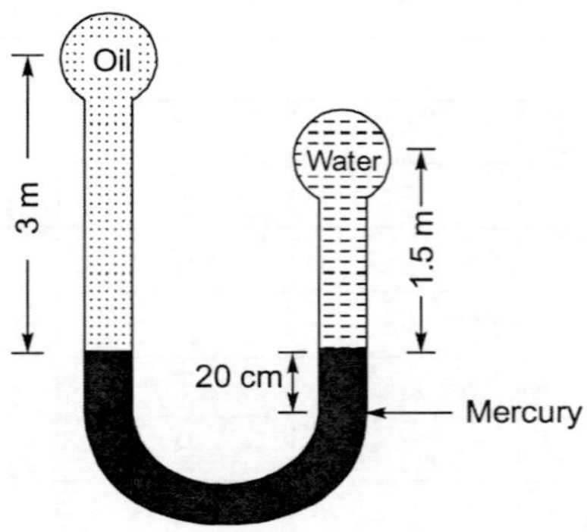

(a) 24.7 kPa
(b) 26.5 kPa
(c) 26.7 kPa
(d) 28.9 kPa
[2003 : 2 M]

1.9 In the inclined manometer shown in the figure below, the reservoir is large. Its surface may be assumed to remain at a fixed elevation. A is connected to a gas pipeline and the deflection noted on the inclined glass tube is

100 mm. Assuming $\theta = 30°$ and the manometric fluid as oil with specific gravity of 0.86, the pressure at *A* is

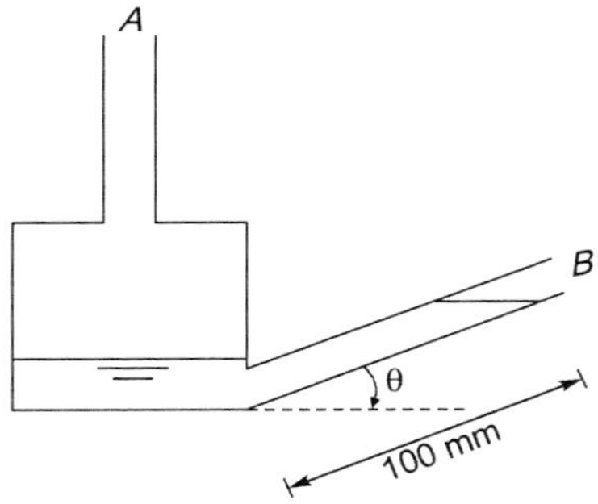

(a) 43 mm water (vacuum)
(b) 43 mm water
(c) 86 mm water
(d) 100 mm water [2004 : 1 M]

1.10 The necessary and sufficient condition for a surface to be called as a 'free surface' is
(a) no stress should be acting on it
(b) tensile stress acting on it must be zero
(c) shear stress acting on it must be zero
(d) no point on it should be under any stress [2006 : 1 M]

1.11 The dimension for kinematic viscosity is

(a) $\frac{L}{MT}$ (b) $\frac{L}{T^2}$

(c) $\frac{L^2}{T}$ (d) $\frac{ML}{T}$ [2014 : 1 M, Set-I]

1.12 Group I contains the types of fluids while Group II contains the shear stress-rate of shear relationship of different types of fluids, as shown in the figure

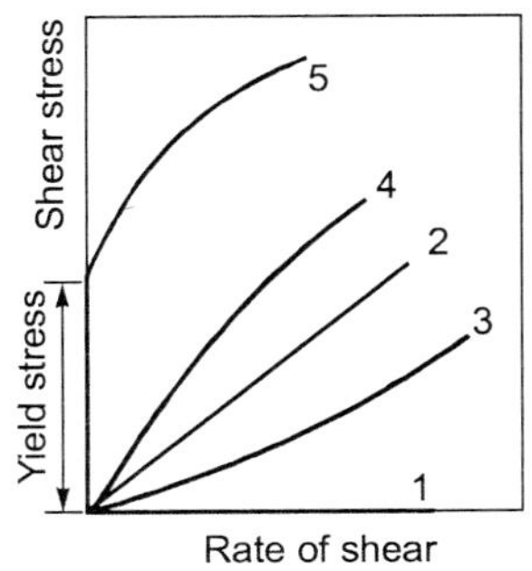

	Group I	Group II
P.	Newtonian fluid	1. Curve 1
Q.	Pseudo plastic fluid	2. Curve 2
R.	Plastic fluid	3. Curve 3
S.	Dilatant fluid	4. Curve 4
		5. Curve 5

The correct match between Group I and Group II is
(a) P-2, Q-4, R-1, S-5
(b) P-2, Q-5, R-4, S-1
(c) P-2, Q-4, R-5, S-3
(d) P-2, Q-1, R-3, S-4 [2016 : 1 M, Set-II]

1.13 The figure shows a U-tube having a 5 mm × 5 mm square cross-section filled with mercury (specific gravity = 13.6) up to a height of 20 cm in each limb (open to the atmosphere).

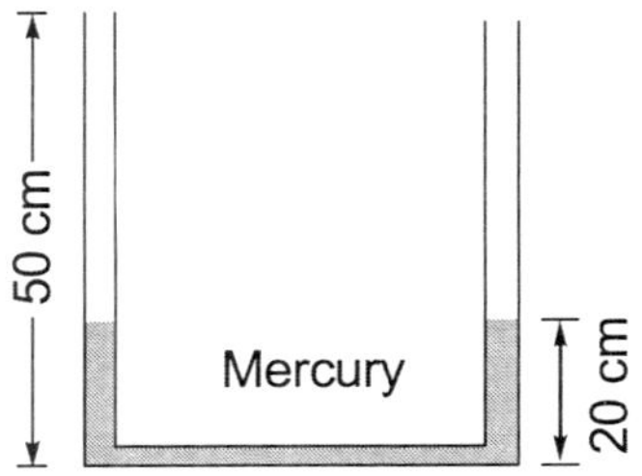

If 5 cm³ of water is added to the right limb, the new height (in cm, up to two decimal places) of mercury in the Left limb will be ________. [2017 : 2 M, Set-II]

1.14 A closed tank contains 0.5 m thick layer of mercury (specific gravity = 13.6) at the bottom. A 2.0 m thick layer of water lies above the mercury layer. A 3.0 m thick layer of oil (specific gravity = 0.6) lies above the water layer. The space above the oil layer contains air under pressure. The gauge pressure at the bottom of the tank is 196.2 kN/m². The density of water is 1000 kg/m³ and the acceleration due to gravity is 9.81 m/s². The value of pressure in the air space is
(a) 92.214 kN/m² (b) 95.644 kN/m²
(c) 98.922 kN/m² (d) 99.321 kN/m² [2018 : 2 M, Set-I]

1.15 A three-fluid system (immiscible) is connected to a vacuum pump. The specific gravity values of the fluids (S_1, S_2) are given in the figure.

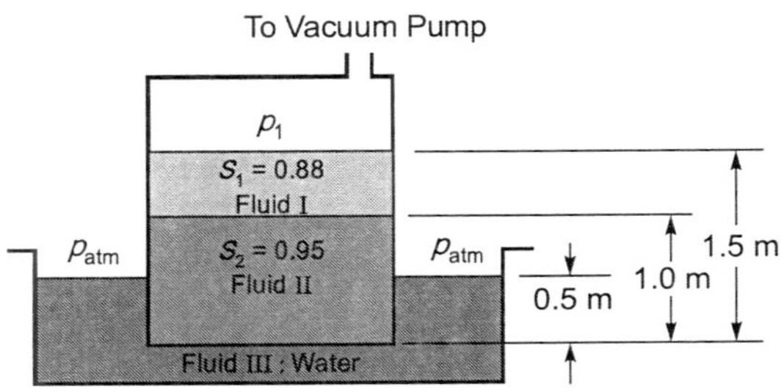

Unit weight of water, $\gamma_w = 9.81$ kN/m³
Atmospheric Pressure, $p_{atm} = 95.43$ kPa

The gauge pressure value (in kNm², up to two decimal places) of p_1 is ________. [2018 : 2 M, Set-II]

1.16 'Kinematic viscosity' is dimensionally represented as

(a) $\frac{M}{L^2T}$ (b) $\frac{L^2}{T}$

(c) $\frac{M}{LT}$ (d) $\frac{T^2}{L}$

[2021 : 1 M, Set-I]

1.17 A lake has a maximum depth of 60 m. If the mean atmospheric pressure in the lake region is 91 kPa and the unit weight of the lake water is 9790 N/m^3, the absolute pressure (in kPa, round off to two decimal places) at the maximum depth of the lake is _______.

[2021 : 1 M, Set-II]

1.18 The dimension of dynamic viscosity is:

(a) $M L^{-2} T^{-2}$ (b) $M L^0 T^{-1}$

(c) $M L^{-1} T^{-2}$ (d) $M L^{-1} T^{-1}$

[2022 : 2 M, Set-II]

1.19 The pressure in a pipe at X is to be measured by an open manometer as shown in figure. Fluid A is oil with a specific gravity of 0.8 and Fluid B is mercury with a specific gravity of 13.6. The absolute pressure at X is _______ kN/m^2 (round off to one decimal place).

[Assume density of water as 1000 kg/m^3 and acceleration due to gravity as 9.81 m/s^2 and atmospheric pressure as 101.3 kN/m^2]

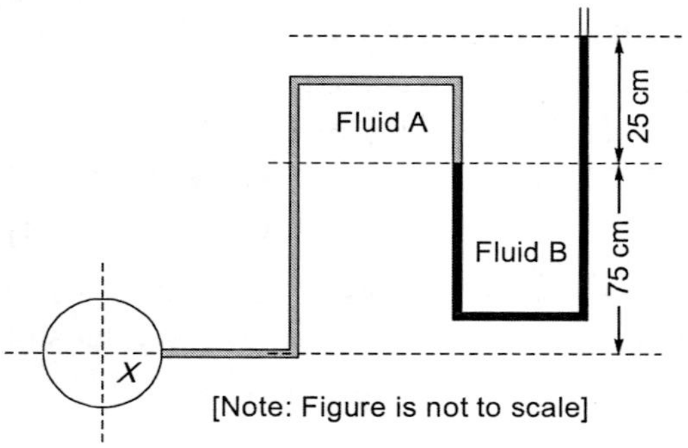

[Note: Figure is not to scale]

[2023 : 1 M, Set-II]

1.20 A 5 m × 5 m closed tank of 10 m height contains water and oil, is connected to an overhead water reservoir as shown in the figure.

Use γ_w = 10 kN/m^3 and specific gravity of oil = 0.8.

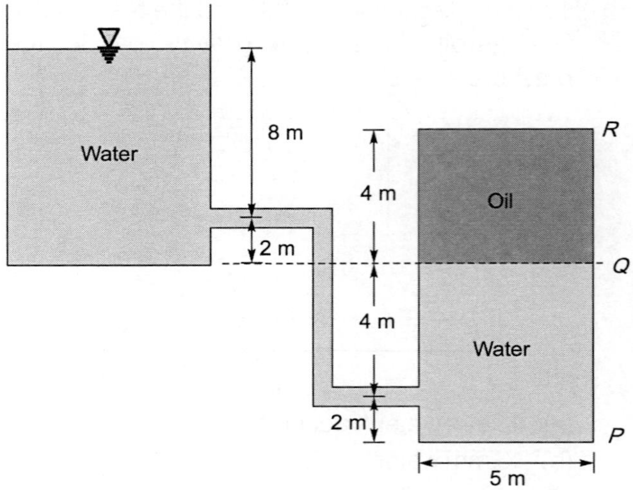

(Figure NOT to scale)

The total force (in kN) due to pressure on the side PQR of the tank is __________.
(rounded off to the nearest integer).

[2024 : 2 M, Set-I]

1.21 The following figure shows a plot between shear stress and velocity gradient for materials/fluids P.Q, R. S, and T.

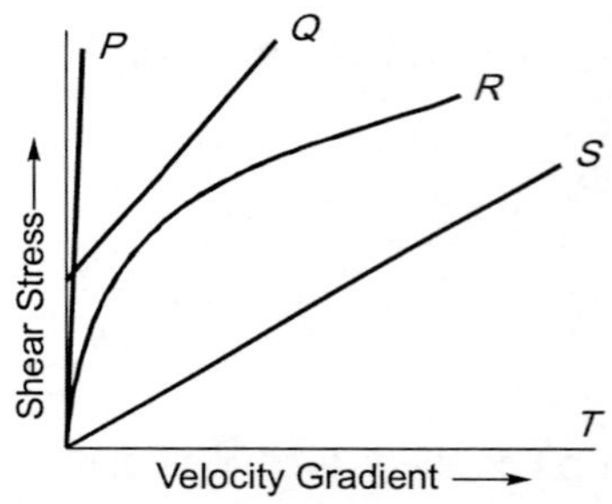

Which one of the following options is CORRECT?

(a) P → Ideal Fluid; Q → Ideal Bingham plastic
R → Non-Newtomian fluid; T → Real solid

(b) P → Ideal Fluid; Q → Ideal Bingham plastic
R → Non-Newtonian fluid; S → Newtonian fluid

(c) P → Real solid; Q → Ideal Bingham plastic
S → Newtonian fluid; T → Ideal Fluid

(d) P → Real solid; Q → Newtonian fluid
R → Ideal Bingham plastic; T → Ideal Fluid

[2024 : 1 M, Set-II]

■■■■

Answers Fluid Properties and Manometry

1.1	(c)	1.2	(b)	1.3	(c)	1.4	(a)	1.5	(a)	1.6	(b)	1.7	(d)
1.8	(a)	1.9	(b)	1.10	(c)	1.11	(c)	1.12	(c)	1.13	(20.74)	1.14	(a)
1.15	(Sol.)	1.16	(b)	1.17	(678.4)	1.18	(d)	1.19	(140.5)	1.20	(5580)	1.21	(c)

Explanations Fluid Properties and Manometry

1.1 (c)

Pressure gradient = pressure intensity per unit length

$$= \frac{\text{Dimension of pressure intensity}}{\text{Dimension of length}}$$

$$= \frac{ML^{-1}T^{-2}}{L^1} = ML^{-2}T^{-2}$$

Note: SI unit of force is Newton (N) but in dimensional analysis, everything has to be converted into basis dimensions and that is why N is converted into kgm/s² giving us the dimension as [MLT⁻²]

1.2 (b)

A fluid is a substance which deforms continuously under the influence of shearing forces

1.3 (c)

- According to Newton's law of viscosity, shear stress is directly proportional to the rate of angular deformation (strains rate) or velocity gradient across the flow.

$$\tau \propto \frac{d\theta}{dt}$$

$$\tau = \mu\frac{du}{dy}$$

where μ = absolute or dynamic viscosity

1.4 (a)

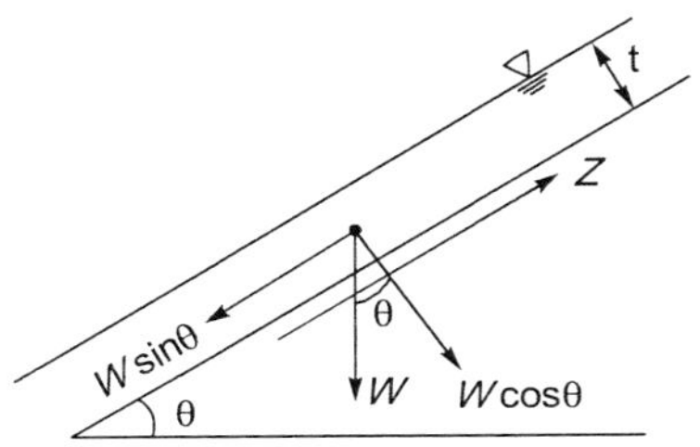

$$W\sin\theta = \text{Resistive force}$$

$$\rho(At)g\sin\theta = \tau A$$

$$\rho g t\sin\theta = \tau$$

1.5 (a)

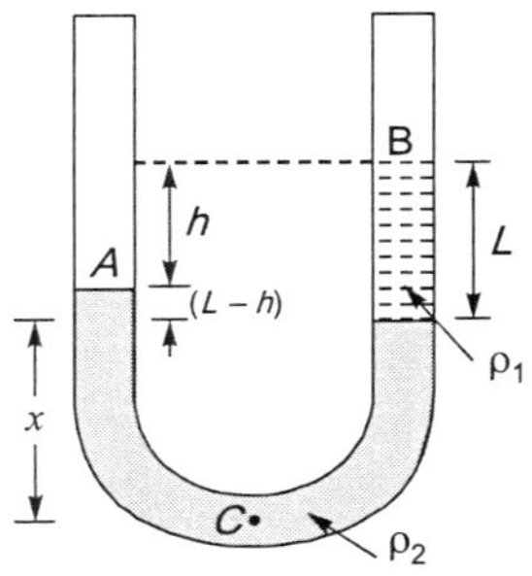

$$P_A + (L-h)\rho_2 g - L\rho_1 g = P_B$$

$$(L-h)\rho_2 g = L\rho_1 g \qquad (P_A = P_B = P_{atm})$$

$$h = L\left(1-\frac{\rho_1}{\rho_2}\right)$$

1.6 (b)

Dynamic viscosity,

$$\mu = \frac{\text{Shear stress}}{\frac{\text{Change of velocity}}{\text{Change of distance}}}$$

$$= \frac{\text{Force}/(\text{Length})^2}{1/\text{Time}}$$

$$= \frac{\text{Force}\times\text{Time}}{(\text{Length})^2} = \frac{\text{Ns}}{\text{m}^2}$$

Note:

	Unit	Dimension
Dynamic viscosity (μ)	N-s/m² or Pas.	$[ML^{-1}T^{-1}]$
Kinematic viscosity (ν)	m²/s or cm²/s	$[L^2T^{-1}]$

1 poise = 0.1 NS/m²; 1 stokes = 10^{-4} m²/s

1.7 (d)

In case of liquid, cohesive force between molecules cause viscosity. hence at higher temperature, when molecules posses more energy, they can oppose large intermolecular forces more strongly.

As a result, energized molecule can move more freely, thus viscosity of liquid decrease with increase in temp.

In case of gases, viscosity is caused by molecular collision. At high temp. more molecular collision per unit volume per unit time occurs thus increasing viscosity.

1.8 (a)

Initially, the pressure difference between water and oil can be given as,

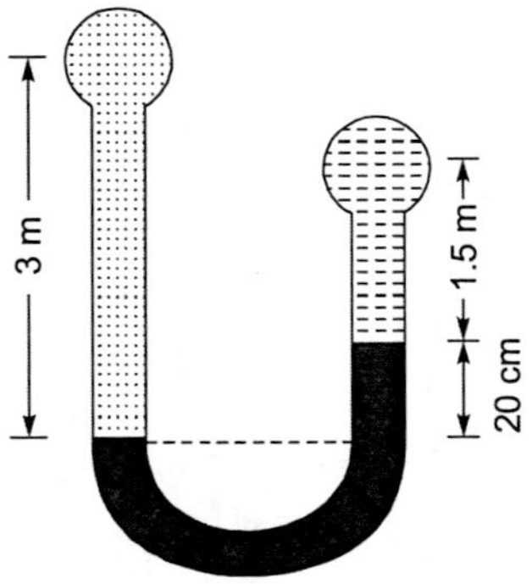

p_{wi} + (1000 × 9.81 × 1.5) + (13550 × 9.81 × 20 × 10^{-2}) – (900 × 9.81 × 3) = p_0

$\Rightarrow \quad p_{wi} + 14715 + 26585.1 - 26487 = p_0$

$\Rightarrow \quad p_0 - p_{wi} = 14813.1$

$\Rightarrow \quad p_0 - p_{wi} = 14.8 \text{ kPa} \quad \ldots\text{(i)}$

To make the same level of mercury in both the limbs, the level of mercury should be decreased by 10 cm in right limb and increased by 10 cm in left limb. So the final level will be

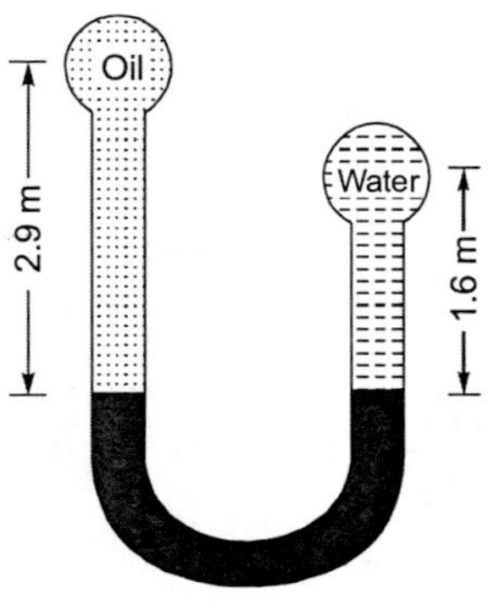

$p_{wf} + (1000 \times 9.81 \times 1.6) - (900 \times 9.81 \times 2.9) = p_0$

$\Rightarrow \quad p_0 - p_{wf} = -9.9 \text{ kPa} \quad \ldots\text{(ii)}$

Subtracting (ii) from (i) we get,

$p_{wf} - p_{wi} = 14.8 + 9.9$

$p_{wf} - p_{wi} = 24.7 \text{ kPa (increase)}$

1.9 (b)

Total pressure at *A* will be equal to the pressure at the surface of manometric fluid because above the surface, the gas pipeline is connected.

$\therefore$ Pressure at A = 100 sin 30° mm of oil
= 50 mm of oil

We know that,

$$S_1h_1 = S_2h_2$$

$\Rightarrow \quad 0.86 \times 50 = 1 \times h_2$

$\Rightarrow \quad h_2 = 43 \text{ mm of water}$

Here, we have done analysis on the basis of gauge pressure and not absolte pressure by neglecting atmospheric pressure.

AVOID MISTAKE

- Sometimes it is confusing to analyse inclined limbs and students tend to take the length of the limb to calculate the pressure.
- For E.g. in this problem someone may simply write pressure at A = 100 mm of oil, but that would be wrong. It is because pressure at any point in fluid depends only on the vertical height of fluid over it (for static fluid).

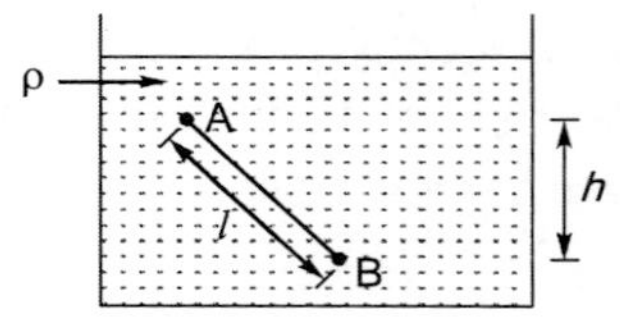

$$P_A - P_B = \rho g h$$

1.10 (c)

A free surface is the surface that is subjected to zero parallel shear force. This condition applies to both fluids and solids and hence is universal.

1.11 (c)

The SI unit of **kinematic viscosity** is m^2/s

$\therefore$ Dimension of **kinematic viscosity** $[\nu] = L^2/T$.

$$\text{Kinematic viscosity } (\nu) = \frac{\text{Dynamic viscosity } (\mu)}{\text{Mass density } (\rho)}$$

$$= \frac{kg/ms}{kg/m^3} = \frac{kg}{ms} \times \frac{m^3}{kg} = \frac{m^2}{s}$$

$$[\nu] = [L^2 T^{-1}]$$

1.12 (c)

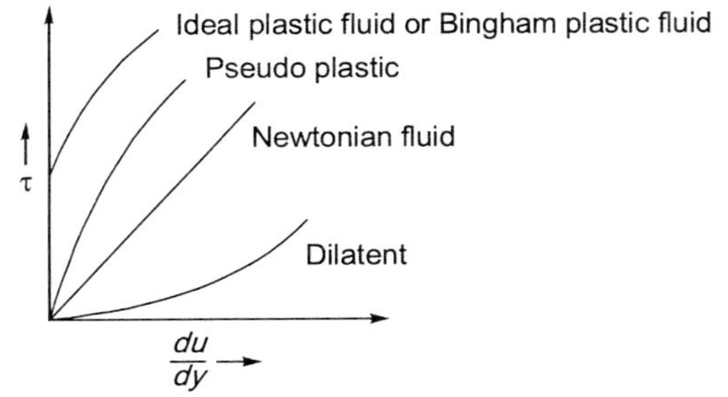

1.13 Sol.

Volume of water added = 5 cm^3

Cross-section of U tube

= 5 mm × 5 mm = 0.25 cm^2

Height of water column in U tube

$$= \frac{\text{Volume of water added}}{\text{Area of U tube}}$$

$$= \frac{5\ cm^3}{0.25\ cm^2} = 20\ cm$$

Now due to conservation of volume, rise in left limb will be equal to fall in right limb.

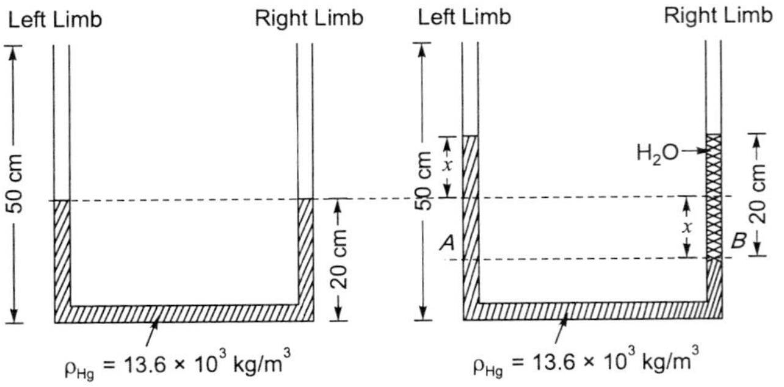

$$P_A = P_B$$

$$(13.6 \times 10^3)g \times (2x) = (10^3)\, g\, (20)$$

$$x = \frac{1.0}{3.6} = 0.735\ cm$$

So, the new height (in cm to two decimal place of mercury in the left limb will be)

$$= 20 + 0.74 = 20.74\ cm$$

1.14 (a)

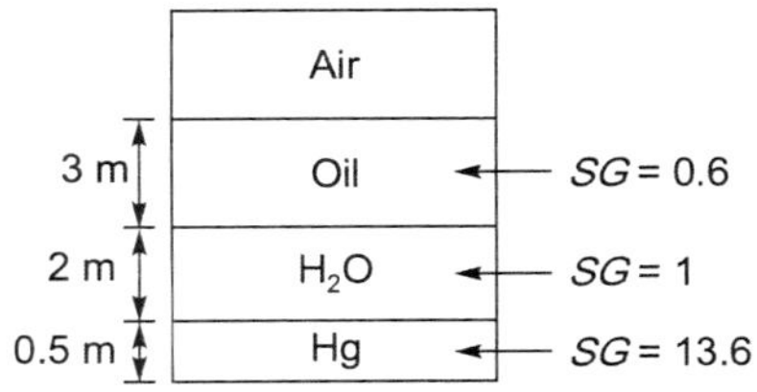

P_{air} is in gauge pressure.

$$P_{air} + (0.6 \times 10^3)(9.81)(3) + (10^3)(9.81)(2) + (13.6 \times 10^3)(9.81)(0.5) = 196.2 \times 10^3$$

$$P_{air} = 92.214\ kN/m^2$$

Note: It is a closed chamber, hence concept of absolute pressure cannot be applied. Calculations have to be done in the form of gauge pressure.

1.15 Sol.

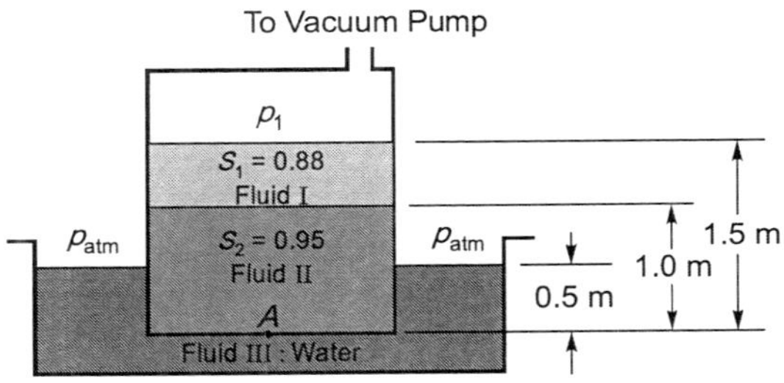

Taking P_1 is in gauge pressure.

$$P_A = P_1 + (0.88 \times 10^3)\cdot(9.81)(0.5) + (0.95 \times 10^3)(9.81)(1)$$

$$(10^3)(9.81)(0.5) = P_1 + (0.88 \times 10^3)\cdot(9.81)(0.5) + (0.95 \times 10^3)(9.81)(1)$$

$$P_1 = -8.73\ kN/m^2$$

1.16 (b)

Kinematic viscosity of fluid, (ν) is defined as dynamic viscosity (μ) divided by mass density of fluid (ρ).

i.e. $\nu = \frac{\mu}{\rho}$

Its unit is L^3T^{-1}.

1.17 Sol.

Absolute pressure at maximum depth of the lake = $P_{atm} + \rho g h$

$$= 91 + \frac{9790(60)}{1000} = 678.4 \text{ kPa}$$

1.18 (d)

Unit of dynamic viscosity

$$= \frac{kg}{m.s} \text{ or } \frac{Ns}{m^2} = ML^{-1}T^{-1}$$

1.19 Sol.

$P_x - (800 \times 9.81 \times 0.75) - (13600 \times 9.81 \times 0.25)$
$= P_{atm}$

$P_x = (101.3 \times 10^3) + (800 \times 9.81 \times 0.75) + (13600 \times 9.81 \times 0.25)$

$P_x = 140540 \text{ N/m}^2$

$P_x = 140.54 \text{ kN/m}^2$

1.20 Sol.

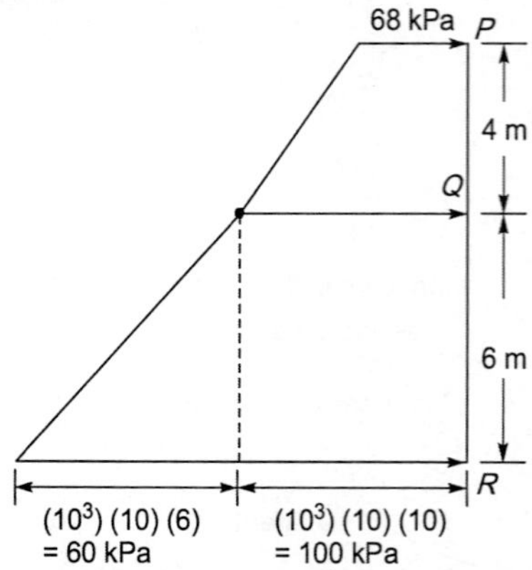

$P_P + (800)(10)(4) = (10^3)(10)(10)$, where P_p is pressure at P

$\Rightarrow P_P + (32 \times 10^3) = (100 \times 10^3)$

$\Rightarrow \quad P_P = 68 \times 10^3$ Pa

$\Rightarrow \quad P_P = 68$ kPa

Now, force on PQR,

$$F = \left[\frac{1}{2}(68 + 100)4 + \frac{1}{2}(100 + 160)6\right]5$$

$= (336 + 780)5$

$= 5580$ kN

1.21 (c)

■■■■

2 Hydrostatic Forces

2.1 The force per unit width on the rectangular door submerged in a liquid of density ρ (figure shown below) is

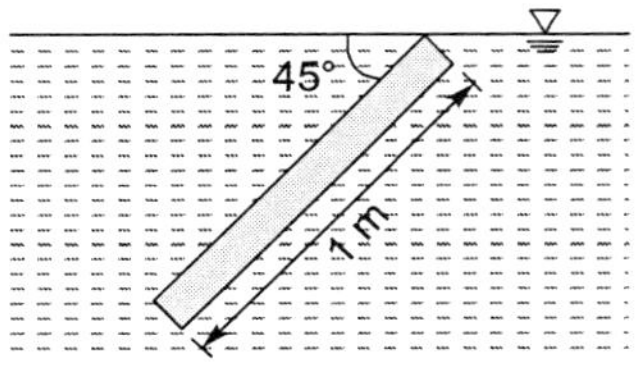

(a) $\sqrt{2}\rho g$ (b) $\frac{\rho g}{2}$

(c) $\frac{\rho g}{2\sqrt{2}}$ (d) $\frac{2}{\rho g}$ [1996 : 1 M]

2.2 A vertical triangular plane area, submerged in water, with one side in the free surface, vertex downward and latitude 'h' has the pressure centre below the free surface by

(a) $h/4$ (b) $h/3$

(c) $2h/3$ (d) $h/2$ [1996 : 1 M]

2.3 The centre of pressure of a liquid on a plane surface immersed vertically in a static body of liquid, always lies below the centroid of the surface area, because

(a) in liquids the pressure acting is same in all directions

(b) there is no shear stress in liquids at rest

(c) the liquid pressure is constant over depth

(d) the liquid pressure increases linearly with depth

[1997 : 1 M]

2.4 In which one of the following arrangement would the vertical force on the cylinder due to water be the maximum?

(a)

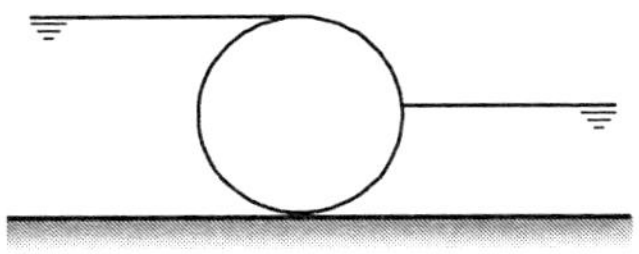

(b)

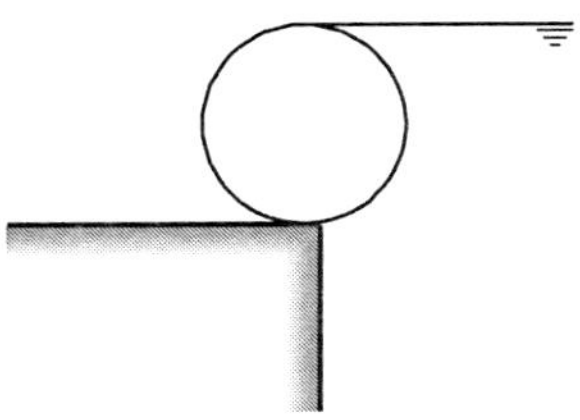

(c)

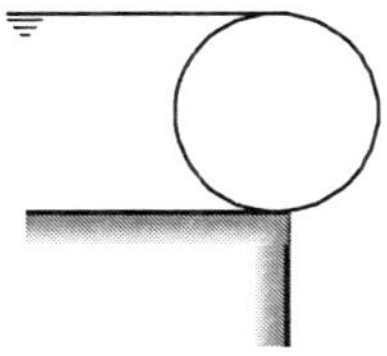

(d)

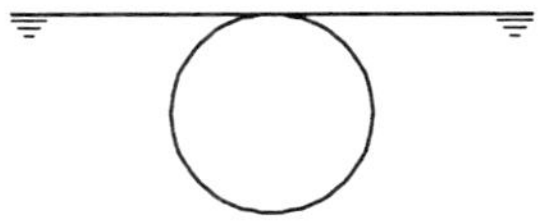

[1998 : 1 M]

2.5 The force 'F' required at equilibrium on the semi-cylindrical gate shown below is

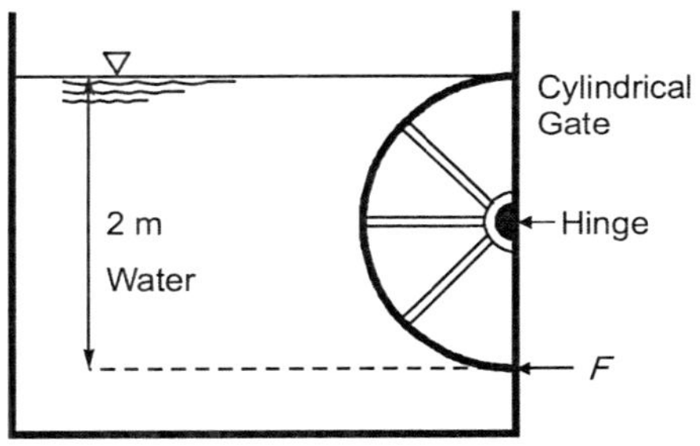

(a) 9.81 kN (b) 0.0 kN

(c) 19.62 kN (d) None of these

[2002 : 1 M]

2.6 Cross-section of an object (having same section normal to the paper) submerged into a fluid consists of a square of sides 2 m and triangle as shown in the figure. The object is hinged at point P that is one meter below the fluid free surface. If the object is to be kept in the position as shown in the figure, the value of 'x' should be

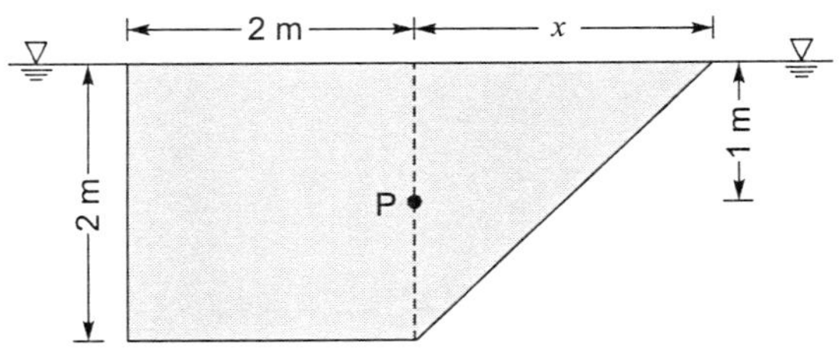

(a) $2\sqrt{3}$ m (b) $4\sqrt{3}$ m

(c) 4 m (d) 8 m [2005 : 2 M]

2.7 Three rigid buckets, shown as in the figures (1), (2) and (3), are of identical heights and base areas. Further, assume that each of these buckets have negligible mass and are full of water. The weights of water in these buckets are denoted as W_1, W_2, and W_3 respectively. Also, let the force of water on the base of the bucket be denoted as F_1, F_2, and F_3 respectively. The option giving an accurate description of the system physics is

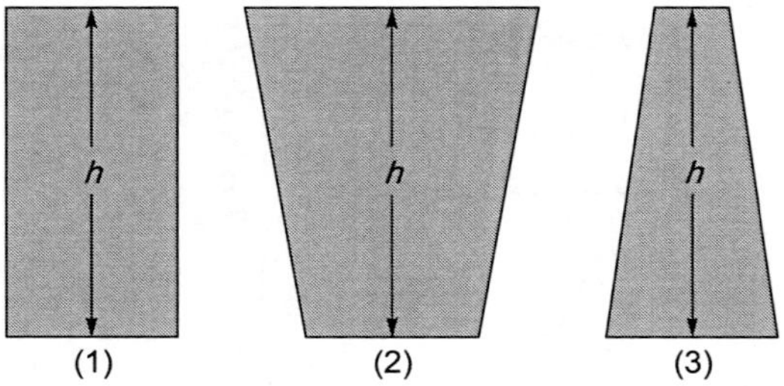

All three buckets have the same base area

(a) $W_2 = W_1 = W_3$ and $F_2 > F_1 > F_3$
(b) $W_2 > W_1 > W_3$ and $F_2 > F_1 > F_3$
(c) $W_2 = W_1 = W_3$ and $F_1 = F_2 = F_3$
(d) $W_2 > W_1 > W_3$ and $F_1 = F_2 = F_3$

[2014 : 2 M, Set-I]

2.8 A triangular gate with a base width of 2 m and a height of 1.5 m lies in a vertical plane. The top vertex of the gate is 1.5 m below the surface of a tank which contains oil of specific gravity 0.8. Considering the density of water and acceleration due to gravity to be 1000 kg/m³ and 9.81 m/s², respectively, the hydrostatic force (in kN) exerted by the oil on the gate is ________.

[2015 : 2 M, Set-II]

2.9 A sector gate is provided on a spillway as shown in the figure. Assuming $g = 10$ m/s², the resultant force per meter length (expressed in kN/m) on the gate will be________.

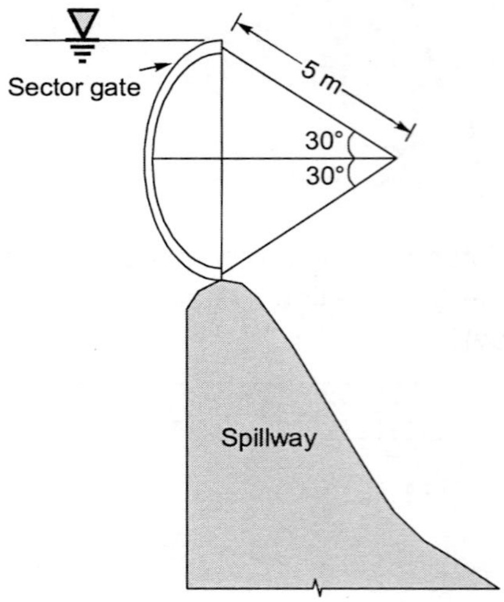

[2016 : 2 M, Set-I]

2.10 A body floating in a liquid is in a stable state of equilibrium if its
(a) metacentre lies below its centre of gravity
(b) metacentre lies above its centre of gravity
(c) metacentre coincides with its centre of gravity
(d) centre of gravity is below its centre of buoyancy

[2020 : 1 M, Set-I]

2.11 A cylinder (2.0 m diameter, 3.0 m long and 25 kN weight) is acted upon by water on one side and oil (specific gravity = 0.8) on other side as shown in the figure.

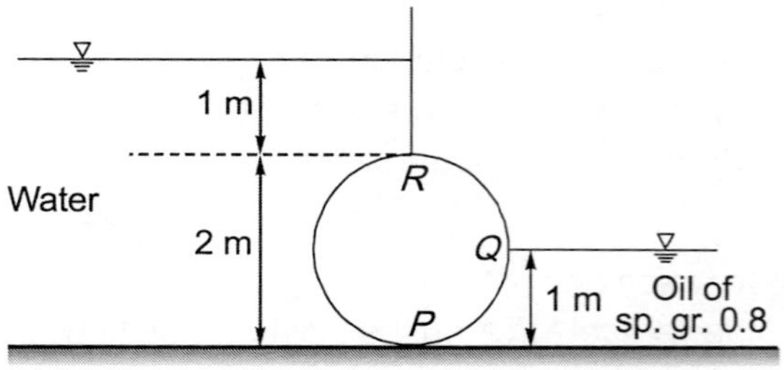

The absolute ratio of the net magnitude of vertical forces to the net magnitude of horizontal forces (round off to two decimal places) is ________

[2021 : 2 M, Set-I]

Answers Hydrostatic Forces

2.1	(c)	2.2	(d)	2.3	(d)	2.4	(d)	2.5	(b)	2.6	(a)	2.7	(d)
2.8	(Sol.)	2.9	(Sol.)	2.10	(b)	2.11	(Sol.)						

Explanations Hydrostatic Forces

2.1 (c)

Force per unit width is given as $F = \rho g \bar{h} A$

Where $\bar{h}$ = vertical depth of the C.G. of plate from free surface.

Hence,
$$F = \rho g\left(\frac{1}{2}\sin 45^\circ\right)(1\times 1)$$
$$= \rho g \frac{1}{2}\times\frac{1}{\sqrt{2}} = \frac{\rho g}{2\sqrt{2}}$$

2.2 (d)

Centre of pressure is a point on a surface through which the resultant force due to pressure passes.

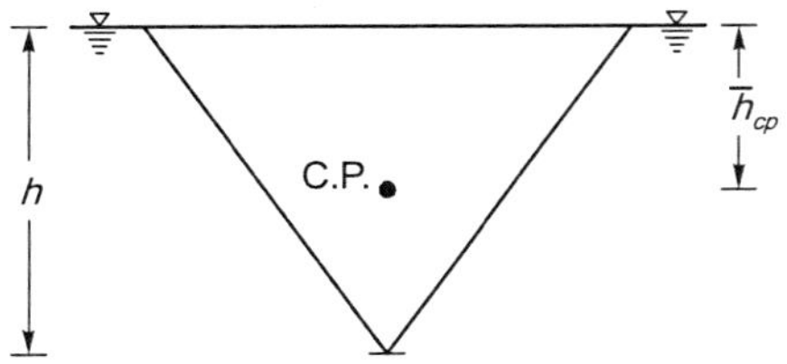

Centre of pressure,

$$\bar{h}_{cp} = \bar{h} + \frac{I_{Gxx}\sin^2\theta}{\bar{h}A}$$

$$= \frac{h}{3} + \frac{bh^3/36}{\frac{bh}{2}\times\frac{h}{3}} = \frac{h}{3}+\frac{h}{6} = \frac{h}{2}$$

2.3 (d)

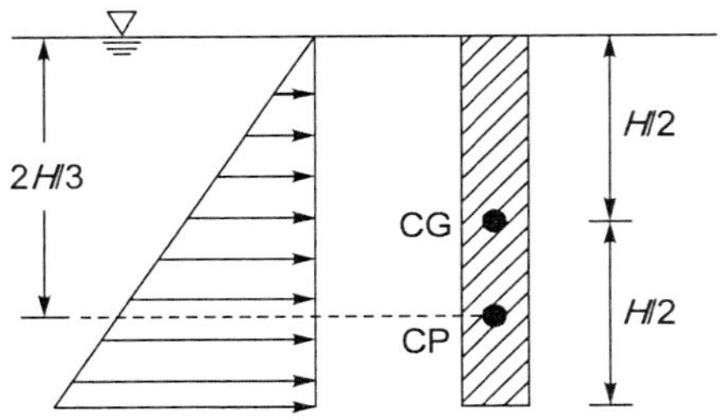

Analysing Physically:

Centre of pressure (CP) always lies below centre of gravity (CG) of the plane surface because pressure distribution is triangular and weight distribution is linear.

2.4 (d)

Vertical component of hydrostatic force is equal to the weight of liquid of shaded portion in figure given below:

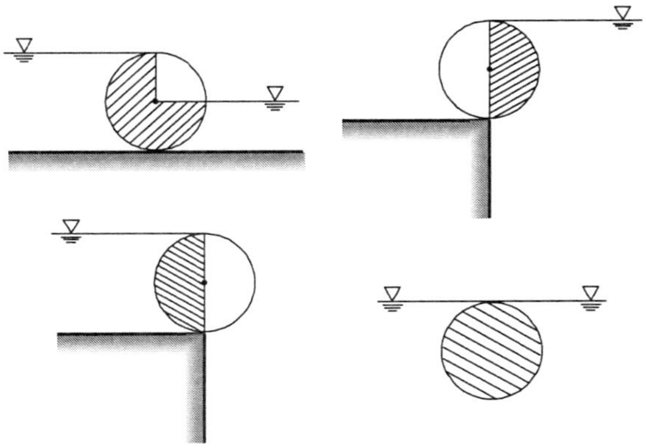

So it is clear that maximum force is experienced in (d).

2.5 (b)

For the gate to be in equilibrium and not have any rotation, summation of moment of all the forces about the hinge must be zero.

Depth of water (H) = 2 m.

Consider unit width, of cylinder,

$$F_H = \rho g \bar{h} A_v$$

$$= 1000\times 9.81\times\left(\frac{2}{2}\right)\times(2\times 1) = 19.62 \text{ kN/m width}$$

Vertical component,

$$F_v = \rho g\,(\text{volume})$$

$$= 1000\times 9.81\left(\frac{\pi R^2}{2}\times 1\right) = 15.41 \text{ kN/m width}$$

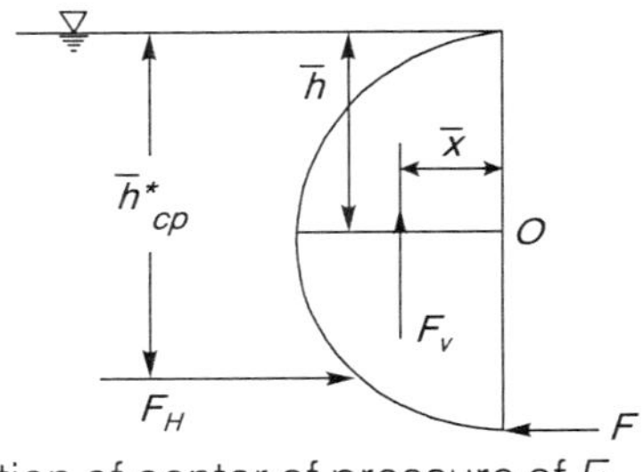

location of center of pressure of F_H,

$$\bar{h}^*_{cp} = \frac{2}{3}H = \frac{4}{3}\text{ m}$$

location of center of pressure of F_V,

$$\therefore \quad \bar{x} = \frac{4R}{3\pi} = \frac{4\times 1}{3\pi} = 0.424 \text{ m}$$

Moment about hinge,

$$\therefore \quad F_H\times(\bar{h}^*_{cp} - \bar{h}) = (F_v\times\bar{x}) + (F\times 1)$$

$$\therefore \quad 19.62\times\left(\frac{4}{3}-1\right) = 15.41\times 0.424 + F\times 1$$

$$\therefore \quad F = 0 \text{ kN}$$

2.6 (a)

Since the body is in equilibrium, hence net moment about P must be zero.

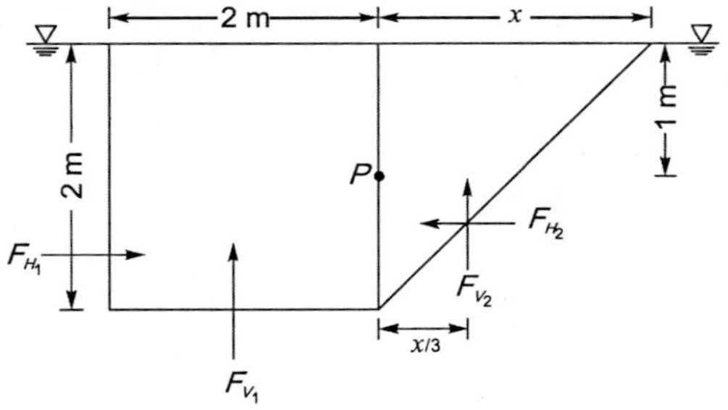

F_{H_1} and F_{H_2} act at 2/3m from water surface, hence cancel each other.Taking moments of both the forces about P, we get,

$$F_{V_1} \times 1 = F_{V_2} \times \frac{x}{3} \quad \text{(Assume width = 1)}$$

$$\rho(2\times 2\times 1)g\times 1 = \rho\left[\frac{1}{2}\times x\times x\times 2\times 1\right]y\times\frac{x}{3}$$

$$x = 2\sqrt{3} \text{ m}$$

2.7 (d)

Bucket → identical height
→ identical base area

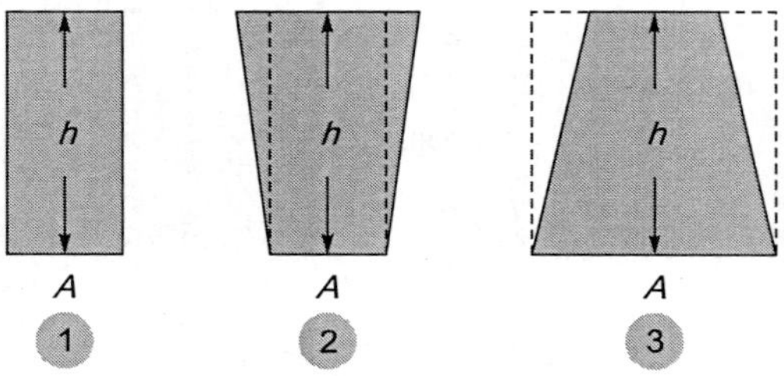

$\Rightarrow \quad W_2 > W_1 > W_3$

Force on the base in each case will be equal to $= \gamma_w h A$

Hence, $\quad F_1 = F_2 = F_3$

2.8 Sol.

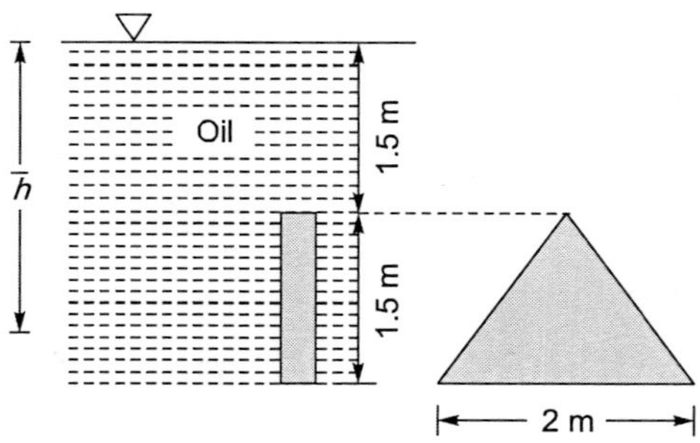

$$\bar{h} = 1.5 + 1.5\left(\frac{2}{3}\right) = 2.5 \text{ m}$$

$$F_R = \rho g\bar{h}A$$

$$= (800)(9.81)(2.5)\left(\frac{1}{2}\times 2\times 1.5\right)$$

$$= 29430 \text{ N} = 29.43 \text{ kN}$$

2.9 Sol.

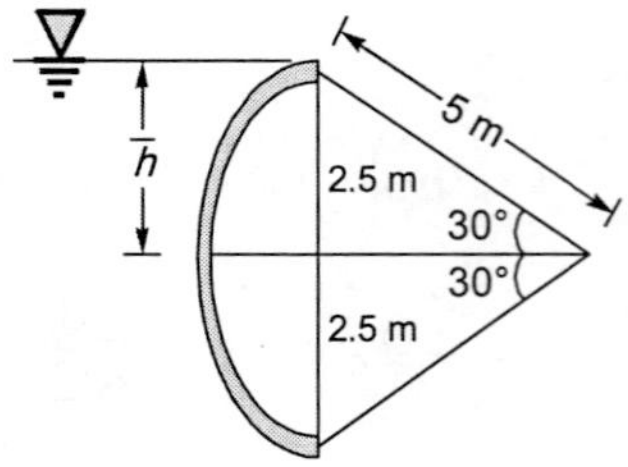

$$F_x = \rho g\bar{h}A_V = (10^3)(10)\left(\frac{5}{2}\right)(5\times 1)$$

$$= 125 \text{ kN per unit width}$$

$$F_y = \rho g \forall$$

$$\text{where } \forall = \frac{\pi(5)^2}{6} - \left(\frac{1}{2}\times 5\times 5\cos 30°\right) = 2.264 \text{m}^3$$

$$F_y = (10)^3(10)(2.264) = 22.64 \text{ kN}$$

Resultant force per unit width,

$$F_R = \sqrt{F_x^2 + F_y^2} = \sqrt{125^2 + 22.64^2} = 127.03 \text{ kN}$$

2.10 (b)

For stability of floating body, M lies above G.

$$GM > 0$$

2.11 Sol.

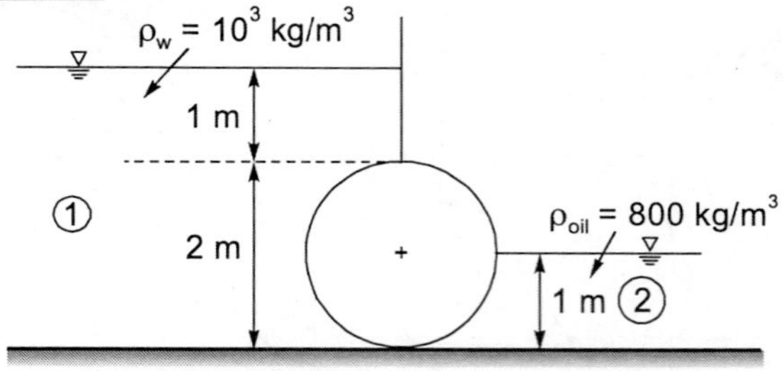

Net horizontal force (F_H) due to liquids

$$F_H = F_{H1} - F_{H2}$$

$$= \rho_w g\bar{h}_1 A_{V_1} - \rho_{oil} g\bar{h}_2 A_{V_2}$$

$$= (10^3)(9.81)\left(1+\frac{2}{2}\right)(2\times 3) - (800)(9.81)\left(\frac{1}{2}\right)(1\times 3)$$

$$F_H = 105.948 \text{ kN } (\rightarrow)$$

Net vertical force (F_V) due to liquids

$$F_V = F_{V_1} + F_{V_2} - W = \rho_w g\forall_1 + \rho_{oil} g\forall_2 - 25\times 10^3$$

$$= (10^3)(9.81)\left(\frac{\pi(1)^2\times 3}{2}\right) + (800)(9.81)\left(\frac{\pi}{4}(1)^2\times 3\right) - 25\times 10^3$$

$$= 39.719 \text{ kN } (\uparrow)$$

$$\frac{F_V}{F_H} = \frac{39.719}{105.948} = 0.374$$

■■■■

3 Buoyancy and Floatation

3.1 One of the following statements is true with regards to bodies that float or are submerged in liquids
(a) For a body wholly submerged in a liquid the stability is ensured if the centre of buoyancy is below the centre of gravity of the body
(b) For a body floating in a liquid the stability is ensured if the centre of buoyancy is below the centre of gravity of the body
(c) For a body floating in a liquid the stability is ensured if the centre of buoyancy and the centre of gravity, regardless of the relative positions of the centre of buoyancy and gravity
(d) For a body floating in a liquid the stability is ensured if the centre of buoyancy is below the centre of gravity and the metacentre is above both the centres of gravity and buoyancy

[1997 : 1 M]

3.2 In an iceberg, 15% of the volume projects above the sea surface. If the specific weight of sea water is 10.5 kN/m^3, the specific weight of iceberg in kN/m^3 is
(a) 12.52 (b) 9.81
(c) 8.93 (d) 7.83

[1999 : 1 M]

3.3 A 15 cm length of steel rod with relative density of 7.4 is submerged in a two layer fluid. The bottom layer is mercury and the top layer is water. The height of top surface of the rod above the liquid interface in 'cm' is
(a) 8.24 (b) 7.82
(c) 7.64 (d) 7.38

[2001 : 2 M]

3.4 For a body completely submerged in a fluid, the centre of gravity (G) and centre of Buoyancy (O) are known. The body is considered to be in stable equilibrium if
(a) O does not coincide with the centre of mass of the displaced fluid
(b) G coincides with the centre of mass of the displaced fluid
(c) O lies below G
(d) O lies above G

[2011 : 1 M]

■■■■

Answers Buoyancy and Floatation

3.1 (d) 3.2 (c) 3.3 (d) 3.4 (d)

Explanations Buoyancy and Floatation

3.1 (d)

The stability of a floating body is determined from the position of Metacenter (M). If the point M is above G (Centre of gravity), the floating body will be in stable equilibrium.

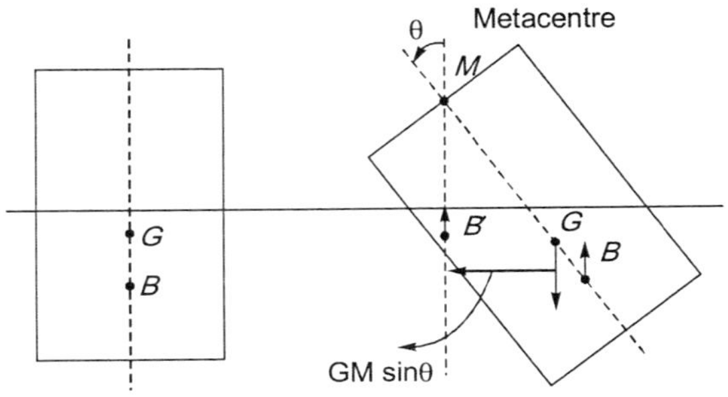

Restoring couple = W ($\overline{GM}$) sinθ

Metacentric height

Case-(a): If M below G
BM < BG: Unstable equilibrium

Case-(b): If M above G
BM > BG: Stable equilibrium

Case-(c): If M at G
BM = BG: Neutral equilibrium

3.2 (c)

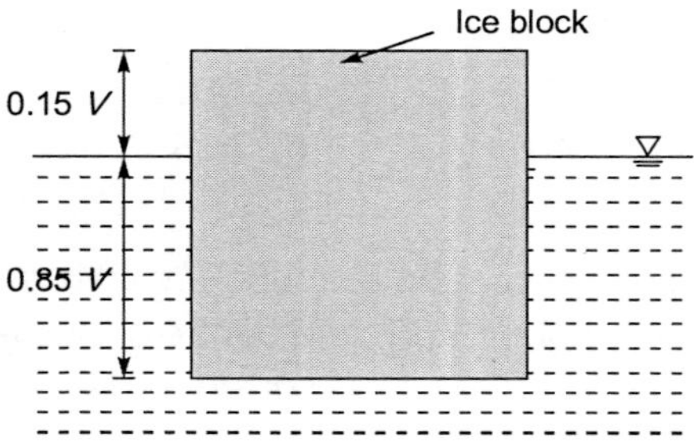

Let the total volume of ice cube be V

Given volume projected above sea surface = 0.15V

FBD of ice cube

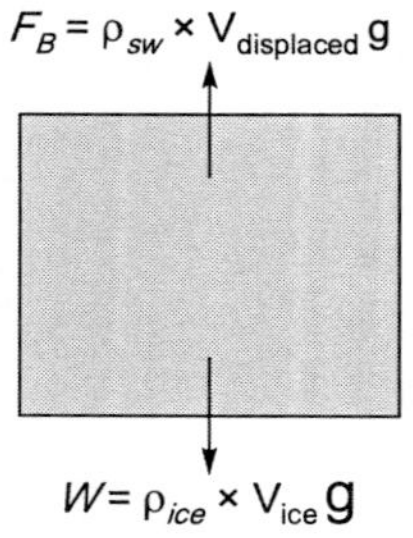

For equlibirum

$$W = F_B$$
$$\rho_{ice} V_{ice}\, g = \rho_{sw} V_{displaced}\, g$$
$$(\rho_{ice}\, g)(V) = (\rho_{sw}\, g)(0.85\, V)$$
$$(\rho_{ice}\, g) = 10.5 \times 0.85$$
$$= 8.93 \text{ kN/m}^2$$

3.3 (d)

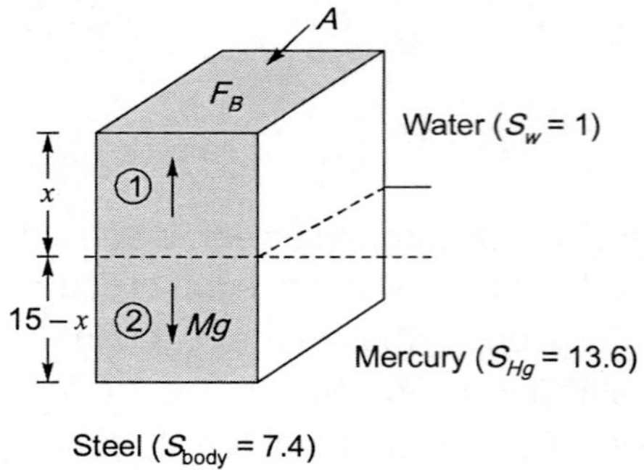

For equilibrium, weight

$$Mg = F_B$$
$$Mg = F_{B_1} + F_{B_2}$$
$$\rho_{body} V_{body}\, g = \rho_w V_1 g + \rho_{Hg} V_2 g$$

Divide by ρ_w

$$\frac{\rho_{body}}{\rho_w} V_{body} = V_1 + \frac{\rho_{Hg}}{\rho_w} V_2$$
$$7.4\,(15 \times A) = (A \times x) + 13.6\,(A \times 15 - x)$$
$$x = 7.38 \text{ cm}$$

3.4 (d)

A completely submerged body will be in stable equilibrium when CG lies below the centre of buoyancy. It will be in unstable equilibrium when CG lies above the centre of buoyancy. And when the CG coincides with the centre of buoyancy, the body will be in neutral equilibrium.

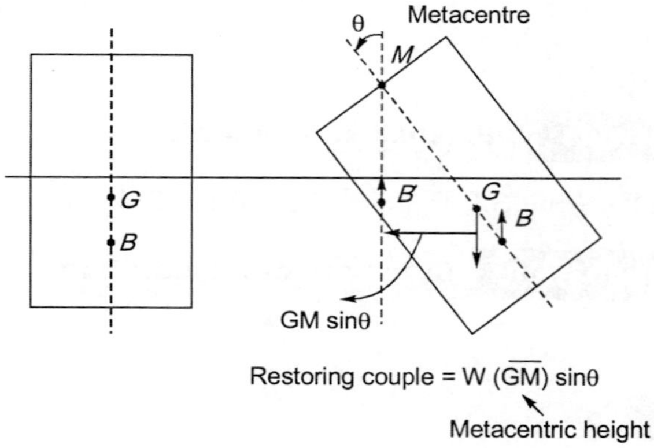

4 Fluid Kinematics

4.1 X-component of velocity in a 2-D incompressible flow is given by $u = y^2 + 4xy$. If Y-component of velocity 'v' equals zero at $y = 0$, the expression for 'v' is given by

(a) $4y$ (b) $2y^2$

(c) $-2y^2$ (d) $2xy$ [1996 : 1 M]

4.2 Two flow patterns are represented by their stream functions ψ_1 and ψ_2 as given below

$\psi_1 = x^2 + y^2$

$\psi_2 = 2xy$

If these two patterns are superimposed on one another, the resulting streamline pattern can be represented by one of the following

(a) A family of parallel straight lines

(b) A family of circles

(c) A family of parabolas

(d) A family of hyperbolas [1997 : 1 M]

4.3 The relation that must hold for the flow to be irrotational is

(a) $\frac{\partial u}{\partial y} - \frac{\partial v}{\partial x} = 0$ (b) $\frac{\partial u}{\partial x} = \frac{\partial u}{\partial y}$

(c) $\frac{\partial^2 u}{\partial x^2} + \frac{\partial^2 v}{\partial y^2} = 0$ (d) $\frac{\partial u}{\partial y} = -\frac{\partial v}{\partial x}$

[2000 : 1 M]

4.4 For a two-dimensional irrotational flow, the velocity potential is defined as $\phi = \log_e(x^2 + y^2)$. Which of the following is a possible stream function, ψ, for this flow?

(a) $\frac{1}{2}\tan^{-1}(y/x)$ (b) $\tan^{-1}(y/x)$

(c) $2\tan^{-1}(y/x)$ (d) $2\tan^{-1}(x/y)$

[2003 : 1 M]

4.5 A velocity field is given as $\overline{V} = 2y\hat{i} + 3x\hat{j}$ where x and y are in metres. The acceleration of a fluid particle at $(x, y) = (1, 1)$ in the x direction is

(a) 0 (b) 5.00 m/s^2

(c) 6.00 m/s^2 (d) 8.48 m/s^2

[2004 : 2 M]

4.6 The velocity in m/s at a point in a two-dimensional flow is given as $\overline{V} = 2\hat{i} + 3\hat{j}$. The equation of the stream line passing through the point is

(a) $3dx - 2dy = 0$ (b) $2x + 3y = 0$

(c) $3dx + 2dy = 0$ (d) $xy = 6$

[2004 : 2 M]

4.7 The x component of velocity in a two dimensional incompressible flow is given by $u = 1.5x$. At the point $(x, y) = (1, 0)$, the y component of velocity $v = 0$. The equation for the y component of velocity is

(a) $v = 0$ (b) $v = 1.5y$

(c) $v = -1.5x$ (d) $v = -1.5y$

[2004 : 1 M]

Common Data for Questions 4.8 and 4.9:

The laminar flow takes place between closely spaced parallel plates as shown in figure below. The velocity profile is given by $u = V\frac{y}{h}$. The gap height, h, is 5 mm and the space is filled with oil (specific gravity = 0.86, viscosity $\mu = 2 \times 10^{-4}$ Ns/m^2). The bottom plate is stationary and the top plate moves with a steady velocity of $V = 5$ cm/s. The area of the plate is 0.25 m^2.

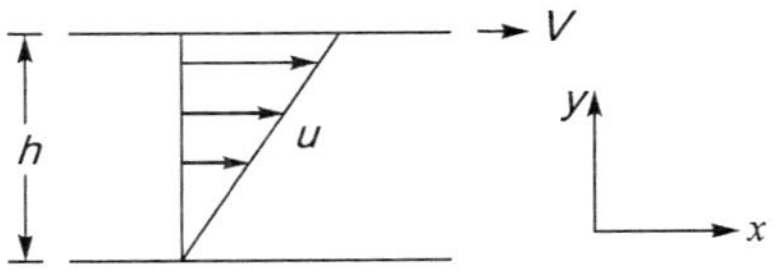

4.8 The rate of rotation of a fluid particle is given by

(a) $\omega_y = 0; \omega_z = -\frac{V}{2h}$

(b) $\omega_y = 0; \omega_z = -\frac{V}{h}$

(c) $\omega_y = \frac{V}{h}; \omega_z = \frac{V}{h}$

(d) $\omega_y = \frac{V}{h}; \omega_z = 0$ [2004 : 2 M]

4.9 The power required to keep the plate in steady motion is

(a) 5×10^{-4} watts (b) 10^{-5} watts

(c) 2.5×10^{-5} watts (d) 5×10^{-5} watts

[2004 : 2 M]

4.10 An inert tracer is injected continuously from a point in an unsteady flow field. The locus of locations of all the tracer particles at an instance of time represents

(a) Streamline (b) Pathline

(c) Streamtube (d) Streakline

[2005 : 1 M]

4.11 A stream function is given by:

$$\psi = 2x^2y + (x+1)y^2$$

The flow rate across a line joining points $A(3, 0)$ and $B(0, 2)$ is

(a) 0.4 units (b) 1.1 units

(c) 4 units (d) 5 units [2005 : 2 M]

4.12 The circulation 'Γ' around a circle of radius 2 units for the velocity field $u = 2x + 3y$ and $v = -2y$ is

(a) -6π units (b) -12π units

(c) -18π units (d) -24π units

[2005 : 2 M]

4.13 The velocity field for a flow is given by: $\vec{V} = (5x + 6y + 7z)\hat{i} + (6x + 5y + 9z)\hat{j} + (3x + 2y + \lambda z)\hat{k}$ and the density varies as $\rho = \rho_0 \exp(-2t)$. In order that the mass is conserved, the value of λ should be

(a) –12 (b) – 10

(c) – 8 (d) 10 [2006 : 2 M]

4.14 For a 2-D flow field, the stream function ψ is given as $\psi = \frac{3}{2}(y^2 - x^2)$.

The magnitude of discharge occurring between the stream line passing through points (0, 3) and (3, 4) is

(a) 6 units (b) 3 units

(c) 1.5 units (d) 2 units [2013 : 1 M]

4.15 An incompressible homogeneous fluid is flowing steadily in a variable diameter pipe having the large and small diameters as 15 cm and 5 cm, respectively. If the velocity at a section at the 15 cm diameter portion of the pipe is 2.5 m/s, the velocity of the fluid (in m/s) at a section falling in 5 cm portion of the pipe is ______.

[2014 : 1 M, Set-I]

4.16 A plane flow has velocity components $u = \frac{x}{T_1}$, $v = -\frac{y}{T_2}$ and $w = 0$ along x, y and z directions respectively, where T_1 ($\neq 0$) and T_2 ($\neq 0$) are constants having the dimension of time. The given flow is incompressible if

(a) $T_1 = -T_2$ (b) $T_1 = -\frac{T_2}{2}$

(c) $T_1 = \frac{T_2}{2}$ (d) $T_1 = T_2$

[2014 : 1 M, Set-II]

4.17 A particle moves along a curve whose parametric equations are: $x = t^3 + 2t$, $y = -3e^{-2t}$ and $z = 2\sin(5t)$, where x, y and z show variations of the distance covered by the particle (in cm) with time t (in s). The magnitude of the acceleration of the particle (in cm/s^2) at $t = 0$ is ______.

[2014 : 2 M, Set-I]

4.18 In a two-dimensional steady flow field, in a certain region of the x-y plane, the velocity component in the x-direction is given by $v_x = x^2$ and the density varies as $\rho = \frac{1}{x}$. Which of the following is a valid expression for the velocity component in the y-direction, v_y?

(a) $v_y = -\frac{x}{y}$ (b) $v_y = \frac{x}{y}$

(c) $v_y = -xy$ (d) $v_y = xy$

[2015 : 1 M, Set-I]

4.19 A nozzle is so shaped that the average flow velocity changes linearly from 1.5 m/s at the beginning to 15 m/s at its end in a distance of 0.375 m. The magnitude of the convective acceleration (in m/s^2) at the end of the nozzle is ________.

[2015 : 1 M, Set-II]

4.20 The velocity components of a two dimensional plane motion of a fluid are: $u = \frac{y^3}{3} + 2x - x^2y$ and $v = xy^2 - 2y - \frac{x^3}{3}$.

The correct statement is:

(a) Fluid is incompressible and flow is irrotational

(b) Fluid is incompressible and flow is rotational

(c) Fluid is compressible and flow is irrotational

(d) Fluid is compressible and flow is rotational

[2015 : 2 M, Set-II]

4.21 A flow field is given by $u = y^2$, $v = -xy$, $w = 0$. Value of the z-component of the angular velocity (in radians per unit time, up to two decimal places) at the point (0, –1,1) is ______.

[2018 : 1 M, Set-I]

4.22 The velocity field in a flow system is given by $v = 2i + (x + y)j + (xyz)k$. The acceleration of the fluid at (1, 1, 2) is
(a) 4i + 12**k**
(b) 2i + 10**k**
(c) j + **k**
(d) 4j + 10**k**
[2019 : 1 M, Set-II]

4.23 Uniform flow with velocity U makes an angle θ with the y-axis, as shown in the figure

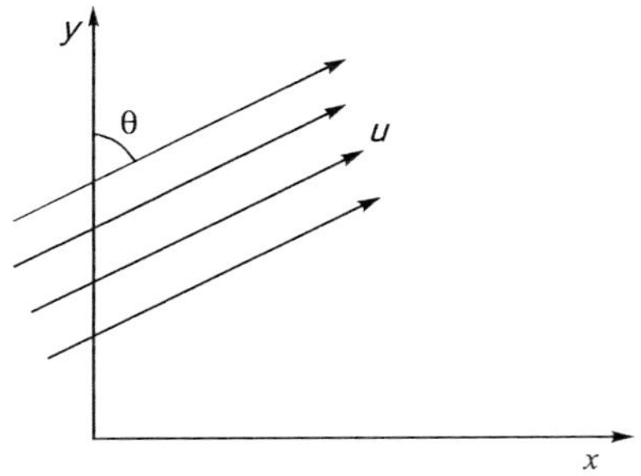

The velocity potential (ϕ), is
(a) ±U (x sinθ − y cosθ)
(b) ±U (y sinθ + x cosθ)
(c) ±U (y sinθ − x cosθ)
(d) ±U (x sinθ + y cosθ) [2020 : 1 M, Set-I]

4.24 The velocity components in the x and y directions for an incompressible flow are given as $u = (-5 + 6x)$ and $v = -(9 + 6y)$, respectively. The equation of the streamline is
(a) $(-5 + 6x)(9 + 6y)$ = constant
(b) $\frac{-5+6x}{9+6y}$ = constant
(c) $\frac{9+6y}{-5+6x}$ = constant
(d) $(-5+6x)-(9+6y)$ = constant
[2020 : 1 M, Set-II]

4.25 A flow velocity field $\vec{V} : \vec{V}(x,y)$ for a fluid is represented by
$\vec{V} = 3\hat{i} + (5x)\hat{j}$.
In the context of the fluid and the flow, which one of the following statements is CORRECT?
(a) The fluid is incompressible and the flow is irrotational.
(b) The fluid is compressible and the flow is rotational.
(c) The fluid is incompressible and the flow is rotational.
(d) The fluid is compressible and the flow is irrotational.
[2024 : 2 M, Set-I]

■■■■

Answers Fluid Kinematics

4.1	(c)	4.2	(a)	4.3	(a)	4.4	(c)	4.5	(c)	4.6	(a)	4.7	(d)
4.8	(a)	4.9	(c)	4.10	(d)	4.11	(c)	4.12	(b)	4.13	(c)	4.14	(b)
4.15	(22.5)	4.16	(d)	4.17	(12)	4.18	(c)	4.19	(540)	4.20	(a)	4.21	(1.50)
4.22	(d)	4.23	(d)	4.24	(a)	4.25	(c)						

Explanations Fluid Kinematics

4.1 (c)

Given, $u = y^2 + 4xy$
$v = 0$ at $y = 0$

For a 2-D incompressible flow the continuty equation is,

$$\frac{\partial u}{\partial x} + \frac{\partial v}{\partial y} = 0$$

$$\Rightarrow \quad 4y + \frac{\partial v}{\partial y} = 0$$

$$\Rightarrow \quad \int \partial v = -\int 4y \cdot \partial y$$

$$\Rightarrow \quad V = \frac{-4y^2}{2} + C$$

$$\Rightarrow \quad V = -2y^2 + C$$

Now at $y = 0$, $V = 0$

$$\Rightarrow \quad 0 = 0 + C$$

$$\Rightarrow \quad C = 0$$

$$\therefore \quad v = -2y^2$$

i.e. y-component of velocity is $-2y^2$.

4.2 (a)

$$\psi_1 = x^2 + y^2$$
$$\psi_2 = 2xy$$

For **superimposed flow,**

$$\psi = \psi_1 + \psi_2 = x^2 + y^2 + 2xy$$

Since $\Delta = 0$ and $h^2 = ab$

So it represents a pair of parallel straight lines.

General 2nd degree equation of conic section

$ax^2 + 2hxy + by^2 + 2gx + 2fy + c = 0$

The above equation represents a pair of parallel straight lines, if

$$\Delta = abc + 2fgh - af^2 - bg^2 - ch^2 = 0$$

and $h^2 = ab$

4.3 (a)

For flow to be irrotational,

$$\omega_z = 0$$

i.e., $$\frac{1}{2}\left(\frac{\partial v}{\partial x} - \frac{\partial u}{\partial y}\right) = 0$$

or $$\frac{\partial u}{\partial y} - \frac{\partial v}{\partial x} = 0$$

4.4 (c)

We know that,

$$\frac{\partial \psi}{\partial y} = \frac{\partial \phi}{\partial x}$$

$$\Rightarrow \frac{\partial \psi}{\partial y} = \frac{\partial}{\partial x} \log_e (x^2 + y^2)$$

$$\Rightarrow \frac{\partial \psi}{\partial y} = \frac{2x}{x^2 + y^2}$$

Integrating both sides, we get,

$$\psi = \int \frac{2x}{x^2 + y^2} dy + C$$

$$\Rightarrow \psi = 2x \times \frac{1}{x} \tan^{-1}\left(\frac{y}{x}\right) + C$$

$$\Rightarrow \psi = 2 \tan^{-1}\left(\frac{y}{x}\right) + C$$

However, it may be noted that 'C' is a function of x.

4.5 (c)

Given $\vec{V} = 2y\hat{i} + 3x\hat{j}$

i.e. $u = 2y$ and $v = 3x$

$$a_x = \frac{\partial u}{\partial t} + u\frac{\partial u}{\partial x} + v\frac{\partial u}{\partial y}$$

$$\Rightarrow a_x = 0 + (2y \times 0) + (3x \times 2)$$

$$\Rightarrow a_x = 6x$$

Now at (1, 1)

$$a_x = 6x = 6 \times 1 = 6 \text{ m/s}^2$$

4.6 (a)

$$\vec{V} = 2\hat{i} + 3\hat{j}$$

The equation of streamlime

$$\frac{dx}{u} = \frac{dy}{v}$$

$$3dx - 2dy = 0$$

4.7 (d)

For a two-dimensional incompressible flow, the continuity equation is given by,

$$\frac{\partial u}{\partial x} + \frac{\partial v}{\partial y} = 0$$

$$\Rightarrow 1.5 + \frac{\partial v}{\partial y} = 0$$

$$\Rightarrow \frac{\partial v}{\partial y} = -1.5$$

$\Rightarrow$ $v = -1.5\,y + C$

But at $y = 0,\ v = 0$

$\therefore$ $0 = -1.5 \times 0 + C$

$\Rightarrow$ $C = 0$

Thus $v = -\,1.5\,y$

4.8 (a)

Given $u = \dfrac{Vy}{h}$

$v = 0$ and $w = 0$

$$\omega_x = \frac{1}{2}\left(\frac{\partial w}{\partial y} - \frac{\partial v}{\partial z}\right) = \frac{1}{2}(0 - 0) = 0$$

$$\omega_y = \frac{1}{2}\left(\frac{\partial u}{\partial z} - \frac{\partial w}{\partial x}\right) = \frac{1}{2}(0 - 0) = 0$$

$$\omega_z = \frac{1}{2}\left(\frac{\partial v}{\partial x} - \frac{\partial u}{\partial y}\right) = \frac{1}{2}\left(0 - \frac{V}{h}\right) = \frac{-V}{2h}$$

4.9 (c)

Shear stress,

$$\tau = \frac{\mu \partial u}{\partial y} = \frac{\mu V}{h}$$

Shear force,

$$F = \tau A$$

Power required to keep the plate in steady motion

$= F \times \text{Velocity}$

$= \tau A \times V = \dfrac{\mu V}{h} \times A \times V$

$= \dfrac{\mu V^2 A}{h} = \dfrac{2\times10^{-4}\times(5\times10^{-2})^2\times0.25}{5\times10^{-3}}$

$= 2.5 \times 10^{-5}$ Watts

4.10 (d)

Streamline: A streamline is an imaginary curve drawn through a flowing fluid in such a way that the tangent to it at any point gives the direction of flow at that point.

Pathline: Actual path travelled by any individual fluid partial over a time period is called pathline.

Streamtube: It is formed by the group of streamlines.

Streakline: It is locus of all fluid particles passing through a fixed point at a particular time.

4.11 (c)

The rate of flow is the difference in value of stream function at points *A* and *B*.

$\therefore \quad Q = |\psi_A - \psi_B|$

$\psi_A = 2 \times (3)^2 \times 0 + (3+1) \times (0)^2 = 0$

$\psi_B = 2 \times (0)^2 \times 2 + (0+1) \times (2)^2 = 4$

$\therefore \quad Q = |0 - 4| = 4$ units

4.12 (b)

Circulation = Vorticity × Area

$= \left(\dfrac{\partial v}{\partial x} - \dfrac{\partial u}{\partial y}\right) \times \text{Area}$

$= (0-3) \times \pi \times (2)^2 = -12\pi$ units

4.13 (c)

The continuity equation is given by

$$\frac{\partial \rho}{\partial t} + \frac{\partial(\rho u)}{\partial x} + \frac{\partial(\rho v)}{\partial y} + \frac{\partial(\rho w)}{\partial z} = 0$$

Now $\quad \rho = \rho_0 e^{-2t}$

$$\frac{\partial \rho}{\partial t} = -2\rho_0 e^{-2t} = -2\rho$$

$$\rho\frac{\partial u}{\partial x} = 5\rho;\ \rho\frac{\partial v}{\partial y} = 5\rho;\ \rho\frac{\partial w}{\partial z} = \lambda\rho$$

$\therefore \quad -2\rho + 5\rho + 5\rho + \lambda\rho = 0$

$\Rightarrow \quad 8\rho + \lambda\rho = 0$

or $\quad 8 + \lambda = 0$

or $\quad \lambda = -8$

4.14 (b)

$Q_{per\ unit\ width} = |\Psi_1 - \Psi_2|$

At (0, 3)

$$\psi_1 = \frac{3}{2}(y^2 - x^2) = \frac{3}{2}(3^2 - 02)$$

$$= \frac{27}{2}$$

At (3, 4)

$$\psi_2 = \frac{3}{2}(y^2 - x^2)$$

$$= \frac{3}{2}(4^2 - 3^2) = 24 - \frac{27}{2}$$

$$Q_{per\ unit\ width} = \left|\frac{27}{2} - 24 + \frac{27}{2}\right| = 3 \text{ units}$$

4.15 Sol.

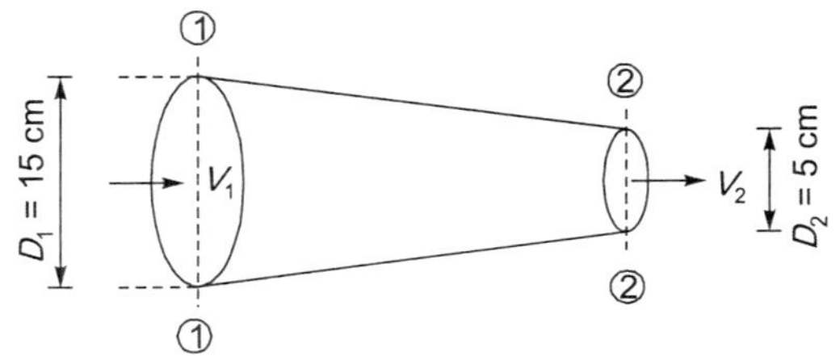

Given: $\quad V_1 = 2.5$ m/s

$D_1 = 15$ cm

Apply continuing equation

$D_2 = 5$ cm

$\rho_1 A_1 V_1 = \rho_2 A_2 V_2$

Also, $\quad \rho_1 = \rho_2$

$A_1 V_1 = A_2 V_2$

$$\frac{\pi}{4}(15)^2 \times 2.5 = \frac{\pi}{4}(5)^2 \times V_2$$

$\Rightarrow \quad V_2 = 2.5 \times 9 = 22.5$ m/s

4.16 (d)

For a flow to exist, continuity should be satisfied.

$$\Rightarrow \quad \frac{\partial u}{\partial x} + \frac{\partial v}{\partial y} = 0$$

$$\Rightarrow \quad \frac{1}{T_1} - \frac{1}{T_2} = 0$$

$$\Rightarrow \quad T_1 = T_2$$

4.17 Sol.

$x = t^3 + 2t$

$y = -3e^{-2t}$

$$z = 2\sin(5t)$$

$$\frac{dx}{dt} = 3t^2 + 2$$

$$\Rightarrow \quad a_x = \frac{d^2x}{dt^2} = 6t$$

$$\frac{dy}{dt} = -3e^{-2t} \times (-2) = 6e^{-2t}$$

$$\Rightarrow \quad a_y = \frac{d^2y}{dt^2} = -12e^{-2t}$$

$$\Rightarrow \quad \frac{dz}{dt} = 2 \times 5\cos(5t) = 10\cos(5t)$$

$$\Rightarrow \quad a_z = \frac{d^2z}{dt^2} = -50\sin 5t$$

$$\vec{a} = a_x\hat{i} + a_y\hat{j} + a_z\hat{k}$$

$$\vec{a} \text{ at } t = 0 = 0\hat{i} - 12\hat{j} + 0\hat{k}$$

$$\vec{a} = -12\hat{j}$$

$\Rightarrow$ Magnitude of acceleration at $t = 0$,

$$= 12\ \text{cm/s}^2$$

$$|\vec{a}| = \sqrt{a_x^2 + a_y^2 + a_z^2}$$

$$= 12\ \text{cm/s}^2$$

4.18 (c)

Continuity equation in steady flow,

$$\frac{\partial(\rho u)}{\partial x} + \frac{\partial(\rho v)}{\partial y} = 0$$

$$\frac{\partial}{\partial x}\left[\frac{1}{x}\cdot x^2\right] + \frac{\partial}{\partial y}\left(\frac{1}{x}\cdot v_y\right) = 0$$

$$\frac{\partial}{\partial x}(x) + \frac{1}{x}\frac{\partial v_y}{\partial y} = 0$$

$$1 + \frac{1}{x}\frac{\partial v_y}{\partial y} = 0$$

$$-x = \frac{\partial v_y}{\partial y}$$

On integrating both sides,

$$v_y = -xy$$

4.19 Sol.

$$\text{Convective acceleration} = \frac{u\partial u}{\partial x}$$

$$= 15 \times \frac{15 - 1.5}{0.375} = 540\ \text{m/s}^2$$

4.20 (a)

Apply continuity equation for 2D in compressible flow,

$$\frac{\partial u}{\partial x} + \frac{\partial v}{\partial y} = 0$$

$$\Rightarrow \quad (2 - 2xy) + (2xy - 2) = 0$$

$$0 = 0$$

Hence, **incompressible flow**.

$$\omega_z = \frac{1}{2}\left[\frac{\partial v}{\partial x} - \frac{\partial u}{\partial y}\right]$$

$$= \frac{1}{2}\left[y^2 - \frac{3x^2}{3} - \left(\frac{3y^2}{3} - x^2\right)\right]$$

$$= \frac{1}{2}\left[y^2 - x^2 - y^2 + x^2\right] = 0$$

Hence, flow is **irrotational**.

4.21 Sol.

$$\omega_z = \frac{1}{2}\left[\frac{\partial v}{\partial x} - \frac{\partial u}{\partial y}\right]$$

$$= \frac{1}{2}\left[\frac{\partial}{\partial x}(-xy) - \frac{\partial}{\partial y}(y)^2\right]$$

$$= \frac{1}{2}[-y - 2y] = -\frac{3y}{2}$$

At point (0, –1, 1)

$$\omega_z = -\frac{3}{2} \times -1 = 1.50\ \text{rad/s}$$

4.22 (d)

$$\vec{v} = 2\hat{i} + (x + y)\hat{j} + xyz\hat{k}$$

$$u = 2$$

$$v = x + y$$

$$w = xyz$$

$$a_x = u\frac{\partial u}{\partial x} + v\frac{\partial u}{\partial y} + w\frac{\partial u}{\partial z} + \frac{\partial u}{\partial t} = 0$$

$$a_y = u\frac{\partial v}{\partial x} + v\frac{\partial v}{\partial y} + w\frac{\partial v}{\partial z} + \frac{\partial v}{\partial t} = 2 + (x + y)$$

$$a_y = x + y + 2$$

$$a_z = u\frac{\partial w}{\partial x} + v\frac{\partial w}{\partial y} + w\frac{\partial w}{\partial z} + \frac{\partial w}{\partial t}$$

$$= 2(yz) + (x + y)(xz) + xyz(xy)$$

$$a_z = 2yz + x^2z + xyz + x^2y^2z$$

At (1, 1, 2)

$$a_y = 1 + 1 + 2 = 4$$

$$a_z = 2(1)(2) + (1)^2(2) + (1)(1)(2) + (1)^2(1)^2(2)$$

$$= 4 + 2 + 2 + 2 = 10$$

$$\vec{a} = 4\hat{j} + 10\hat{k}$$

4.23 (d)

Velocity in x-depth,

$$u_x = u\sin\theta$$

Velocity in y-depth,

$$u_y = u\cos\theta$$

$$-\frac{\partial\phi}{\partial x} = u_x$$

Integrating it $\phi = -u_x x + f(y) + c$

$$= -(u\sin\theta)x + f(y) + c \quad ...(i)$$

$$-\frac{\partial\phi}{\partial y} = u_y$$

Integrating it $\phi = -u_y\, y + f(x) + c$

$$= -(u\cos\theta)y + f(x) + c \quad ...(ii)$$

By equation (i) and (ii),

$$\phi = -u(x\sin\theta + y\cos\theta)$$

If we take $\frac{\partial\phi}{\partial x} = u_x$ and $\frac{\partial\phi}{\partial y} = u_y$

Then $\phi = u(x\sin\theta + y\cos\theta)$

So, $\phi = \pm u(x\sin\theta + y\cos\theta)$

4.24 (a)

Given: $u = -5 + 6x$; $v = -(9 + 6y)$

Equation of streamline

$$\frac{dx}{u} = \frac{dy}{v}$$

$$\frac{dx}{-5+6x} = \frac{dy}{-(9+6y)}$$

Integrating it,

$$\ln(-5+6x)^{1/6} = -\ln(9+6y)^{1/6} + \ln C^{1/6}$$

$$\frac{1}{6}\ln(-5+6x)\cdot(9+6y) = \frac{1}{6}\ln C$$

Take antilog,

$$(-5 + 6x)(9 + 6y) = \text{constant}$$

$$u.v = \text{constant}$$

4.25 (c)

Continuity equation for incompressible flow:

$$\frac{\partial u}{\partial x} + \frac{\partial v}{\partial y} = 0$$

$$\therefore \frac{\partial}{\partial x}(3) + \frac{\partial}{\partial y}(5x) = 0 + 0$$

$$= 0 \text{ (Incompressible Flow)}$$

$$\omega_z = \frac{1}{2}\left(\frac{\partial v}{\partial x} - \frac{\partial u}{\partial y}\right)$$

$$= \frac{1}{2}\left[\frac{\partial}{\partial x}(5x) - \frac{\partial}{\partial y}(3)\right]$$

$$= \frac{1}{2}(5) = 2.5 \neq 0$$

(Flow is rotational)

■■■■

5 Fluid Dynamics & Flow Measurements

5.1 Shear stress develops on a fluid element, if
(a) the fluid is at rest
(b) the fluid container is subject to uniform linear acceleration
(c) the fluid is inviscid
(d) the fluid is viscous and the flow is nonuniform
[1995 : 1 M]

5.2 If, for a fluid in motion, pressure at a point is same in all directions, then the fluid is
(a) a real fluid
(b) a Newtonian fluid
(c) an Ideal fluid
(d) a Non-Newtonian fluid [1996 : 1 M]

5.3 The equation $gz + \frac{V^2}{2} + \frac{P}{\rho}$ = constant along a stream line holds true for
(a) steady, frictionless, compressible fluid
(b) steady, uniform, incompressible fluid
(c) steady, frictionless, incompressible fluid
(d) unsteady incompressible fluid
[1996 : 1 M]

5.4 A body moving through still water at 6 m/sec produces a water velocity of 4 m/sec at a point 1 m ahead. The difference in pressure between the nose and the point 1 m ahead would be
(a) 2000 N/m^2 (b) 10000 N/m^2
(c) 19620 N/m^2 (d) 98100 N/m^2
[1998 : 2 M]

5.5 The reading of differential manometer of a venturimeter, placed at 45° to the horizontal is 11 cm. If the venturimeter is turned to horizontal position, the manometer reading will be
(a) zero (b) $\frac{11}{\sqrt{2}}$ cm
(c) 11 cm (d) $11\sqrt{2}$ cm
[2000, 2005 : 1 M]

5.6 A frictionless fluid of density ρ flow through a bent pipe as shown below. If A is the cross sectional area and V is the velocity of flow, the forces exerted on segment 1 – 2 of the pipe in the x and y directions are, respectively,

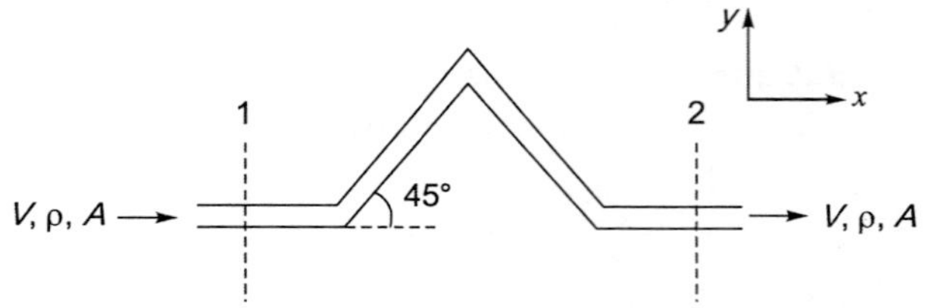

(a) ρAV^2; 0 (b) ρAV^2; $\sqrt{2}\rho AV^2$
(c) 0; 0 (d) 0; $\frac{1}{\sqrt{2}}\rho AV^2$
[2004 : 1 M]

5.7 At two points 1 and 2 in a pipeline the velocities are V and $2V$, respectively. Both the points are at the same elevation. The fluid density is ρ. The flow can be assumed to be incompressible, inviscid, steady and irrotational. The difference in pressures P_1 and P_2 at points 1 and 2 is
(a) $0.5\rho V^2$ (b) $1.5\rho V^2$
(c) $2\rho V^2$ (d) $3\rho V^2$ [2007 : 1 M]

5.8 Match **List-I** (Devices) with **List-II** (Uses) and select the answer using the codes given below the lists:

List-I
A. Pitot tube
B. Manometer
C. Venturimeter
D. Anemometer

List-II
1. Measuring pressure in a pipe
2. Measuring velocity of flow in a pipe
3. Measuring air and gas velocity
4. Measuring discharge in a pipe

Codes:

	A	B	C	D
(a)	1	2	4	3
(b)	2	1	3	4
(c)	2	1	4	3
(d)	4	1	3	2

[2010 : 1 M]

5.9 **Group-I** lists a few devices while **Group-II** provides information about their uses. Match the devices with their corresponding use.

Group-I
P. Anemometer

Q. Hygrometer
R. Pitot Tube
S. Tensiometer

Group-II
1. Capillary potential of soil water
2. Fluid velocity at a specific point in the flow stream
3. Water vapour content of air
4. Wind speed

	P	Q	R	S
(a)	1	2	3	4
(b)	2	1	4	3
(c)	4	2	1	3
(d)	4	3	2	1

[2014 : 1 M, Set-II]

5.10 A venturimeter, having a diameter of 7.5 cm at the throat and 15 cm at the enlarged end, is installed in a horizontal pipeline of 15 cm diameter. The pipe carries an incompressible fluid at a steady rate of 30 litres per second. The difference of pressure head measured in terms of the moving fluid in between the enlarged and the throat of the venturimeter is observed to be 2.45 m. Taking the acceleration due to gravity as 9.81 m/s^2, the coefficient of discharge of the venturimeter (correct up to two places of decimal) is ______.

[2014 : 2 M, Set-I]

5.11 A venturimeter having a throat diameter of 0.1 m is used to estimate the flow rate of a horizontal pipe having a diameter of 0.2 m. For an observed pressure difference of 2 m of water head and coefficient of discharge equal to unity, assuming that the energy losses are negligible, the flow rate (in m^3/s) through the pipe is approximately equal to

(a) 0.500
(b) 0.150
(c) 0.050
(d) 0.015

[2014 : 2 M, Set-II]

5.12 Water flows through a 90° bend in a horizontal plane as depicted in the figure.

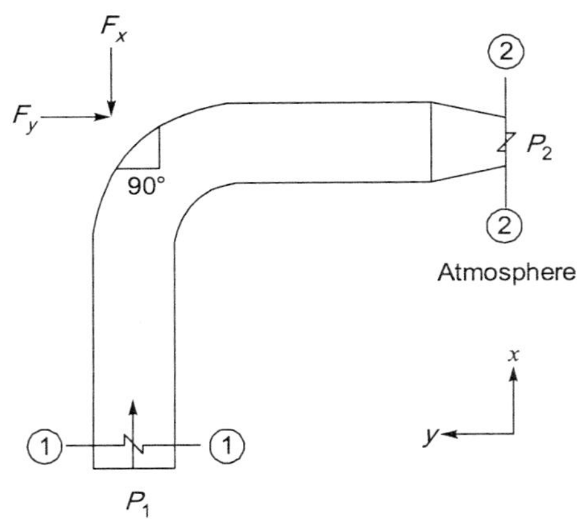

A pressure of 140 kPa is measured at section 1-1. The inlet diameter marked at section 1-1 is $\frac{27}{\sqrt{\pi}}$ cm, while the nozzle diameter marked at section 2-2 is $\frac{14}{\sqrt{\pi}}$ cm. Assume the following:

(i) Acceleration due to gravity = 10 m/s^2.
(ii) Weights of both the bent pipe segment as well as water are negligible.
(iii) Friction across the bend is negligible.

The magnitude of the force (in kN, up to two decimal places) that would be required to hold the pipe section is ________.

[2017 : 2 M, Set-I]

5.13 Bernoulli's equation is applicable for

(a) viscous and compressible fluid flow
(b) inviscid and compressible fluid flow
(c) inviscid and incompressible fluid flow
(d) viscous and incompressible fluid flow

[2018 : 1 M, Set-I]

5.14 A circular duct carrying water gradually contracts from a diameter of 30 cm to 15 cm. The figure (not drawn to scale) shows the arrangement of differential manometer attached to the duct.

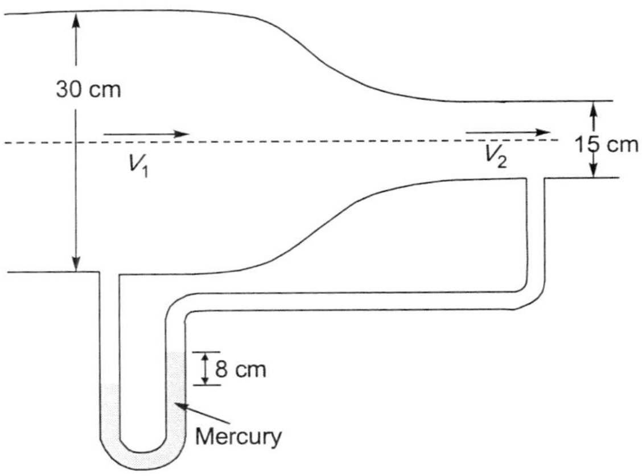

When the water flows, the differential manometer shows a deflection of 8 cm of mercury (Hg). The values of specific gravity of mercury and water are 13.6 and 1.0, respectively. Consider the acceleration due to gravity, g = 9.81 m/s^2. Assuming frictionless flow, the flow rate (in m^3/s, round off to 3 decimal places) through the duct is ________.

[2019 : 1 M, Set-I]

5.15 A circular water tank of 2 m diameter has a circular orifice of diameter 0.1 m at the bottom. Water enters the tank steadily at a flow rate of 20 litre/s and escapes through the orifice. The coefficient of discharge of the orifice is 0.8. Consider the acceleration due to gravity as 9.81 m/s^2 and neglect frictional loses. The height of the water level (in m, round off to two decimal places) in the tank at the steady state, is ________.

[2020 : 2 M, Set-I]

5.16 A venturimeter as shown in the figure (not to scale) is connected to measure the flow of water in a vertical pipe of 20 cm diameter.

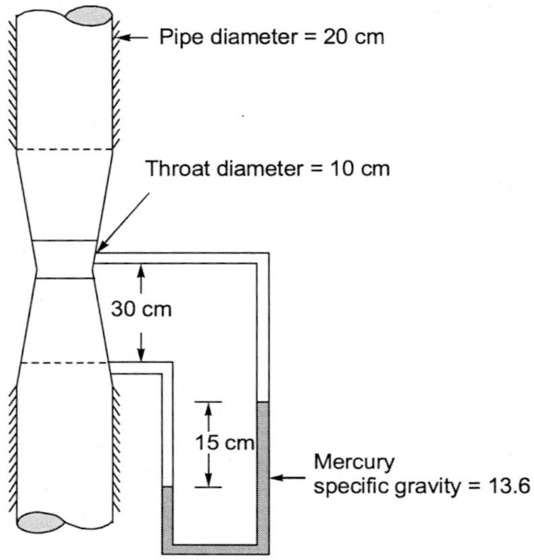

Assume g = 9.8 m/s^2. When the deflection in the mercury manometer is 15 cm, the flow rate (in lps, round off to two decimal places) considering no loss in the venturimeter is ________.

[2021 : 2 M, Set-II]

5.17 A 2 m × 1.5 m tank of 6 m height is provided with a 100 mm diameter orifice at the center of its base. The orifice is plugged and the tank is filled up to 5 m height. Consider the average value of discharge coefficient as 0.6 and acceleration due to gravity (g) as 10 m/s^2. After unplugging the orifice, the time (in seconds) taken for the water level to drop from 5 m to 3.5 m under free discharge condition is ____________ (rounded off to 2 decimal places).

[2024 : 2 M, Set-II]

Answers Fluid Dynamics & Flow Measurements

5.1	(d)	5.2	(c)	5.3	(c)	5.4	(b)	5.5	(c)	5.6	(c)	5.7	(b)
5.8	(c)	5.9	(d)	5.10	(0.95)	5.11	(c)	5.12	(2.55)	5.13	(c)	5.14	(0.081)
5.15	(0.5164)	5.16	(49.40)	5.17	(103.08)								

Explanations Fluid Dynamics & Flow Measurements

5.1 (d)

Shear stress acting on a fluid element is given as

$$\tau = \mu \frac{du}{dy}$$

For $\tau \neq 0$

$\mu \neq 0$ i.e., fluid should be viscous

and $\frac{du}{dy} \neq 0$ i.e., flow should be non-uniform.

5.2 (c)

For an ideal fluid, no shear stresses exist and pressure is same in all direction.

In reality however, ideal fluid does not exist. We study newtonian fluid which is a practical fluid.

5.3 (c)

Assumption in Bernoulli's theorem along a streamline are as follows:

1. Flow should be steady.
2. Flow should be frictionless (inviscous).
3. Flow should be incompressible.

5.4 (b)

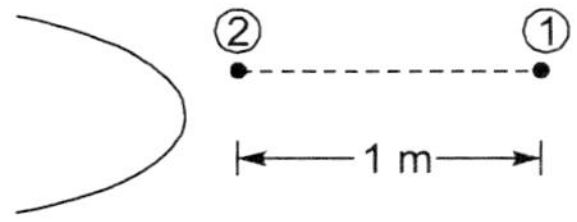

Apply **Bernoullie's equation** for section-1 (still water 1 metre ahead of nose) and section-2 (moving water at the nose),

$$\frac{P_1}{\rho g}+\frac{V_1^2}{2g}+Z_1 = \frac{P_2}{\rho g}+\frac{V_2^2}{2g}+Z_2$$

Assume $Z_1 = Z_2$

$$\Rightarrow \quad \frac{P_1-P_2}{\rho g} = \frac{(V_2^2-V_1^2)}{2g}$$

$$\Rightarrow \quad (P_1-P_2) = \frac{(V_2^2-V_1^2)}{2g}\times\rho g$$

$$= \frac{16-36}{2}\times 1000 = -10000 \text{ N/m}^2$$

$\therefore \quad (P_1-P_2) = 10000 \text{ N/m}^2$

$\therefore$ Difference in pressure between the nose and the point 1 m ahead = 10000 N/m^2.

5.5 (c)

The manometer reading of a venturimeter does not depend upon the inclination of venturimeter with the horizontal.

5.6 (c)

Force exerted on segment 1-2 in x-direction,

$F_x = \rho Q\,[V_F - V_i] = \rho Q\,[V - V] = 0$

Force exerted on segment 1-2 in y-direction,

$F_y = \rho Q[0-0] = 0$

5.7 (b)

Applying **Bernoulli's equation**, we get,

$$\frac{P_1}{\rho g}+\frac{(V)^2}{2g}+z = \frac{P_2}{\rho g}+\frac{(2V)^2}{2g}+z$$

$$\Rightarrow \quad \frac{P_1-P_2}{\rho g} = \frac{4V^2-V^2}{2g}$$

$$\Rightarrow \quad P_1-P_2 = \frac{3}{2}\rho V^2$$

$$\Rightarrow \quad P_1-P_2 = 1.5\,\rho V^2$$

Since the flow is assumed to be inviscid, steady and irrotational, Bernoulli's equation has been applied directly.

5.8 (c)

Pitot tube: It is used to measure velocity of fluid. Difference in the readings of pitot tube and piezometer (h), indicates the velocity head. This is with the assumption that pierometric head at A and C is same.

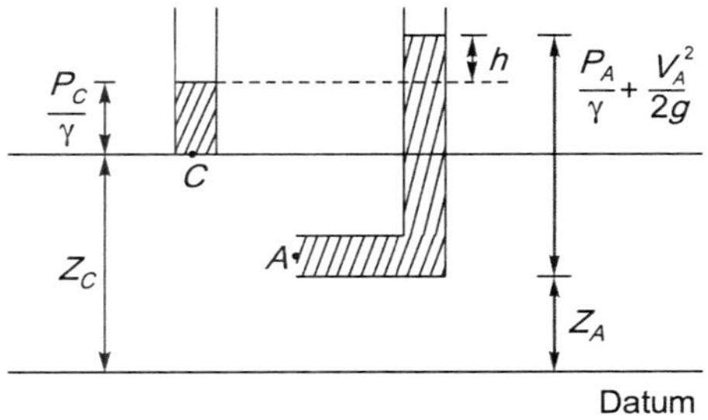

$$\text{i.e.,} \quad \frac{P_A}{\gamma}+Z_A = \frac{P_C}{\gamma}+Z_C$$

$$\left(\frac{P_A}{\gamma}+Z_A+\frac{V_A^2}{2g}\right)-\left(\frac{P_C}{\gamma}+Z_C\right) = h$$

$$\frac{V_A^2}{2g} = h$$

$$V_A = \sqrt{2gh}$$

Manometer: Measurely pressure in a pipe.

Venturimeter: Measuring discharge in a pipe.

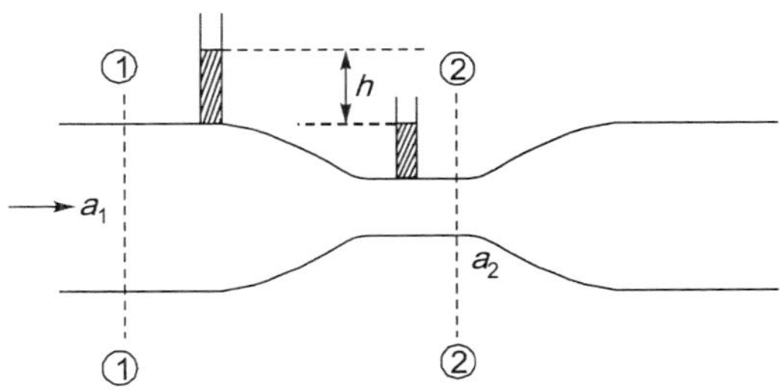

$$Q = C_d \times \frac{a_1 a_2}{\sqrt{a_1^2-a_2^2}}\sqrt{2gh}$$

Anemometer: Used to measure air velocity.

5.9 (d)

Hygrometer: A hygrometer is an instrument used for measuring the amount of humidity and water vapour in the atmosphere, in soil or in confined spaces.

Tensiometer: It is used to measure capillary potential of soil water.

5.10 Sol.

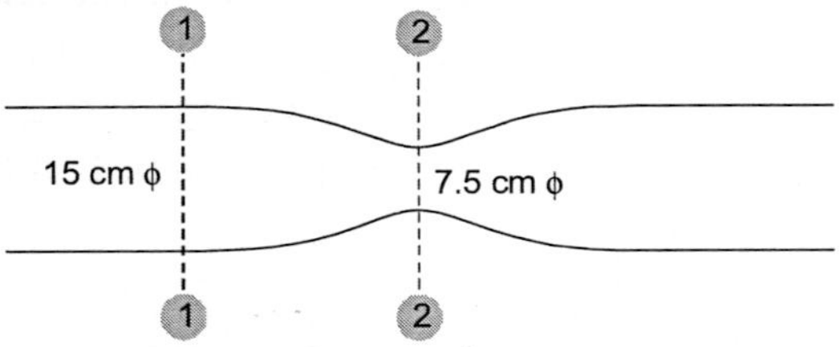

$$Q = 30\ l/s$$

$$\left(\frac{P_1}{w}+Z_1\right)-\left(\frac{P_2}{w}+Z_2\right) = 2.45\,\text{m} = h$$

$$g = 9.81$$

$$Q = C_d \frac{A_1A_2}{\sqrt{A_1^2 - A_2^2}}\sqrt{2gh}$$

$$C_d = \frac{Q}{\frac{A_1A_2}{\sqrt{A_1^2 - A_2^2}}\sqrt{2gh}}$$

$$= \frac{30\times10^{-3}\,\text{m}^3/\text{s}}{\frac{\frac{\pi}{4}(0.15)^2\times(0.075)^2}{\sqrt{(0.15)^2-(0.075)^2}}\times\sqrt{2\times9.81\times2.45}}$$

$$C_d = 0.95$$

5.11 (c)

Diameter of throat, $D_2 = 0.1$ m
Diameter of pipe, $D_1 = 0.2$ m
Pressure difference,

$$\frac{P_1 - P_2}{w} = 2\text{ m} = h\text{ (say)}$$

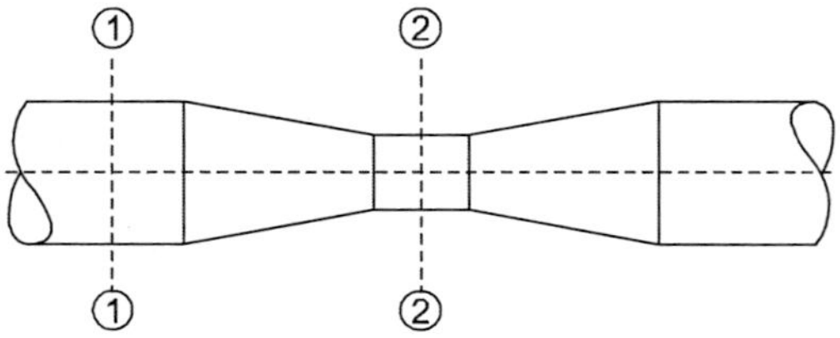

Since it is given that losses are negligible. Take coefficient of discharge as,

$$C_D = 1$$

$$\text{Discharge, } Q = \frac{C_D.A_1A_2\sqrt{2gh}}{\sqrt{A_1^2 - A_2^2}}$$

$$A_1 = \frac{\pi}{4}(0.2)^2$$

and $A_2 = \frac{\pi}{4}(0.1)^2$

$$\Rightarrow \quad Q = \frac{1\times\frac{\pi^2}{4}\times(0.2)^2\frac{\pi}{4}(0.1)^2\sqrt{2\times9.81\times2}}{\left(\frac{\pi}{4}\right)\sqrt{(0.2)^4-(0.1)^4}}$$

$$= 0.0508 \approx 0.05\ \text{m}^3/\text{sec}$$

5.12 Sol.

$$P_1A_1 - F_x = \dot{m}\left(V_{2_x} - V_{1_x}\right) \qquad (\dot{m} = 0)$$

$$P_1A_1 = F_x$$

or

$$F_x = \left(140\times10^3\right)\left[\frac{\pi}{4}\left(\frac{0.27}{\sqrt{\pi}}\right)^2\right]$$

$$= 2551.5\ \text{N}$$

$$-P_2A_2 + F_y = \dot{m}\left(V_{2_y} - V_{1_y}\right) \qquad (\dot{m} = 0)$$

$$F_y = P_2A_2 \qquad (P_2 = 0)$$

$$F_y = 0$$

$$\text{Resultant force} = \sqrt{F_x^2 + F_y^2}$$

$$= \sqrt{(2551.5)^2+0} = 2551.5\ \text{N} \approx 2.552\ \text{kN}$$

5.13 (c)

Bernoulli's equation is application for:

1. Flow along a stream line
2. When the effect of viscosity is negligible i.e., invisid flow
3. Incompressible flow condition
4. Steady flow condition

Note: *Even if flow is rotational, Bernoulli's equation can be applied between two points on the same stream line.*

5.14 Sol.

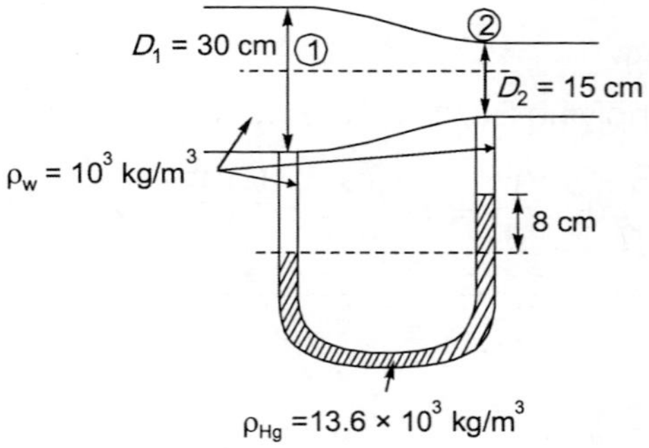

$$Q = \frac{A_1A_2}{\sqrt{A_1^2 - A_2^2}}\sqrt{2gh}$$

$$h = x\left[\frac{\rho_{Hg}}{\rho_w} - 1\right] = 0.08(13.6-1)$$

$$= 1.008\ \text{m} \qquad (A_1 = 4A_2)$$

$$Q = \frac{4A_2 \times A_2}{\sqrt{16A_2^2 - A_2^2}} \times \sqrt{2 \times 9.81 \times 1.008}$$

$$= \frac{4}{\sqrt{15}} \times A_2 \times \sqrt{2 \times 9.81 \times 1.008}$$

$$= \frac{4}{\sqrt{15}} \times \frac{\pi}{4}(0.15)^2 \times \sqrt{2 \times 9.81 \times 1.008}$$

$Q = 0.081$ m³/s (round off to 3 decimal place)

5.15 Sol.

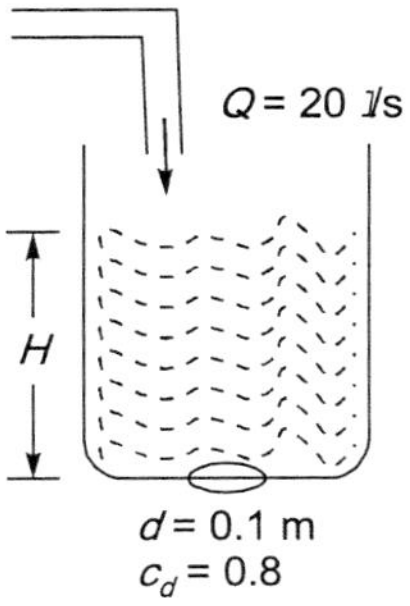

Assume H is the level of water in the tank in steady condition.

For steady water level in the tank

Discharge through orifice

= Water enters in the tank

$$c_d \cdot a \cdot \sqrt{2gH} = 20 \times 10^{-3}$$

$$0.8 \times \frac{\pi}{4}(0.1)^2 \sqrt{2gH} = 0.02$$

$$H = 0.5164 \text{ m}$$

5.16 Sol.

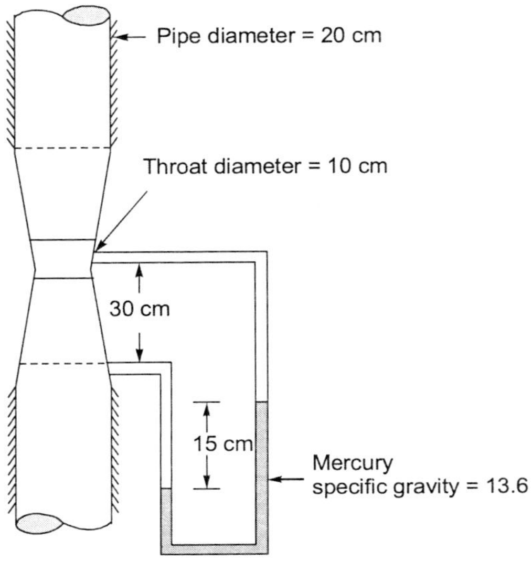

Discharge,

$$Q = C_d \frac{A_1 A_2}{\sqrt{A_1^2 - A_2^2}} \sqrt{2gh} \; [C_d = 1]$$

$$h = X\left(\frac{\rho_m}{\rho} - 1\right)$$

$$= 0.15\left(\frac{13.6 \times 10^3}{10^3} - 1\right) = 1.89 \text{ m}$$

$$Q = \frac{A_1 A_2}{A_2\sqrt{\left(\frac{A_1}{A_2}\right)^2 - 1}} \times \sqrt{2 \times 9.8 \times 1.89}$$

$$= \frac{\frac{\pi}{4}(0.2)^2}{\sqrt{(2)^4 - 1}} \times \sqrt{2 \times 9.8 \times 1.89}$$

$= 49.395$ l/s $\simeq 49.40$ l/s

5.17 Sol.

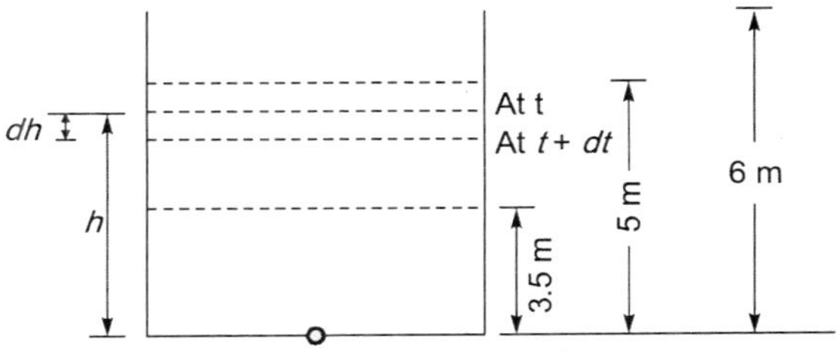

Coefficient of discharge, $C_d = 0.6$

Acceleration due to gravity, $g = 10$ m/s²

Area, $A = (2 \times 1.5)$ m² $= 3$ m²

Let, in dt time, the water level decreases by 'dh'.

$$\therefore \quad -Adh = C_d \times a \times \sqrt{2gh} \cdot dt$$

$$\Rightarrow \frac{-A}{aC_d\sqrt{2g}} \cdot \frac{dh}{\sqrt{h}} = dt$$

Integrating it,

$$\Rightarrow \frac{-A}{aC_d\sqrt{2g}} \int_{H_1}^{H_2} \frac{dh}{\sqrt{h}} = \int_0^T dt$$

$$\Rightarrow \frac{A}{aC_d\sqrt{2g}} \int_{H_2}^{H_1} \frac{dh}{\sqrt{h}} = T$$

$$\frac{A}{C_d a\sqrt{2g}} \times \left[2\sqrt{H}\right]_{H_2}^{H_1} = T$$

$$\Rightarrow T = \frac{2A}{C_d a\sqrt{2g}}(\sqrt{H_1} - \sqrt{H_2})$$

$$= \frac{2 \times (2 \times 1.5)}{0.6 \times \frac{\pi}{4} \times 0.1^2 \times \sqrt{2(10)}} \times \left(\sqrt{5} - \sqrt{3.5}\right)$$

$= 103.99$ sec $\simeq 104$ sec

6 Viscous Flow of Incompressible Fluid

6.1 The friction factor of laminar required flow in a circular pipe is proportional to
(a) Reynold's number
(b) Inverse to the Reynold's number
(c) Square of the Reynold's number
(d) Square root of the Reynold's number
[1996 : 1 M]

6.2 The kinetic energy correction factor for a fully developed laminar flow through a circular pipe is
(a) 1.00 (b) 1.33
(c) 2.00 (d) 1.50 [1997 : 1 M]

Common Data for Questions 6.3 and 6.4:
An upward flow of oil (mass density 800 kg/m^3, dynamic viscosity 0.8 kg/m-s) takes place under laminar conditions in an inclined pipe of 0.1 m diameter as shown in the figure. The pressures at sections 1 and 2 are measured as p_1 = 435 kN/m^2 and p_2 = 200 kN/m^2.

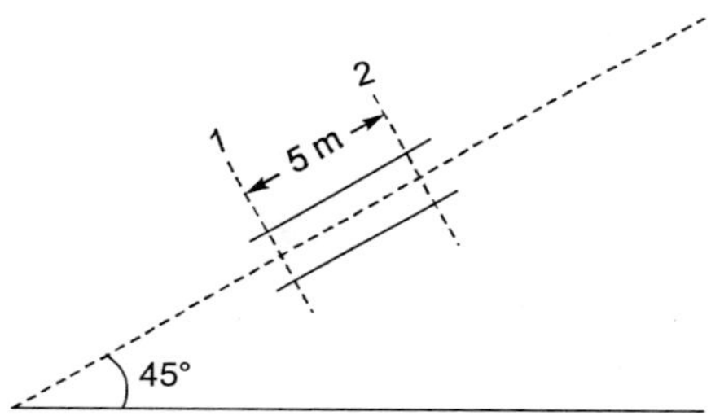

6.3 The discharge in the pipe is equal to
(a) 0.100 m^3/s (b) 0.127 m^3/s
(c) 0.144 m^3/s (d) 0.161 m^3/s
[2006 : 2 M]

6.4 If the flow is reversed, keeping the same discharge, and the pressure at section 1 is maintained as 435 kN/m^2, the pressure at section 2 is equal to
(a) 488 kN/m^2 (b) 549 kN/m^2
(c) 586 kN/m^2 (d) 614 kN/m^2
[2006 : 2 M]

6.5 Flow rate of a fluid (density = 1000 kg/m^3) in a small diameter tube is 800 mm^3/s. The length and the diameter of the tube are 2 m and 0.5 mm, respectively. The pressure drop in 2 m length is equal to 2.0 MPa. The viscosity of the fluid is
(a) 0.025 N-s/m^2 (b) 0.012 N-s/m^2
(c) 0.00192 N-s/m^2 (d) 0.00102 N-s/m^2
[2007 : 2 M]

6.6 With reference to a standard Cartesian (x, y) plane, the parabolic velocity distribution profile of fully developed laminar flow in x-direction between two parallel, stationary and identical plates that are separated by distance, h, is given by the expression

$$u = -\frac{h^2}{8\mu}\frac{dp}{dx}\left[1-4\left(\frac{y}{h}\right)^2\right]$$

In this equation, the y = 0 axis lies equidistant between the plates at a distance h/2 from the two plates, p is the pressure variable and μ is the dynamic viscosity term. The maximum and average velocities are, respectively

(a) $u_{max} = -\frac{h^2}{8\mu}\frac{dp}{dx}$ and $u_{average} = \frac{2}{3}u_{max}$

(b) $u_{max} = \frac{h^2}{8\mu}\frac{dp}{dx}$ and $u_{average} = \frac{2}{3}u_{max}$

(c) $u_{max} = -\frac{h^2}{8\mu}\frac{dp}{dx}$ and $u_{average} = \frac{3}{8}u_{max}$

(d) $u_{max} = \frac{h^2}{8\mu}\frac{dp}{dx}$ and $u_{average} = \frac{3}{8}u_{max}$

[2014 : 2 M, Set-II]

6.7 For a steady incompressible laminar flow between two infinite parallel stationary plates, the shear stress variation is
(a) linear with zero value at the plates
(b) linear with zero value at the center
(c) quadratic with zero value at the plates
(d) quadratic with zero value at the center
[2017 : 1 M, Set-I]

6.8 Consider a laminar flow in the x-direction between two infinite parallel plates (Couette flow). The lower plate is stationary and the upper plate is moving with a velocity of 1 cm/s in the x-direction. The distance between the plates is 5 mm and the dynamic viscosity of the fluid is 0.01 N-s/m^2. If the shear stress on the lower plate is zero, the pressure gradient, $\frac{\partial p}{\partial x}$, (in N/m^2 per m, round off to 1 decimal place) is_______.

[2019 : 2 M, Set-I]

6.9 A solid sphere of radius, r and made of material with density, ρ_s is moving through the atmosphere (constant pressure, p) with a velocity, v. The net force ONLY due to atmospheric pressure (F_p) acting on the sphere at any time t, is

(a) $4\pi r^2 p$
(b) $\frac{4}{3}\pi r^3 \rho_s \frac{dv}{dt}$
(c) $\pi r^2 p$
(d) zero

[2019 : 1 M, Set-II]

6.10 A fluid flowing steadily in a circular pipe of radius R has a velocity that is everywhere parallel to the axis (centreline) on the pipe. The velocity distribution along the radial direction is $V_r = U\left(1-\frac{r^2}{R^2}\right)$, where r is the radial distance as measured from the pipe axis and U is the maximum velocity at $r = 0$. The average velocity of the fluid in the pipe is

(a) $\left(\frac{5}{6}\right)U$
(b) $\frac{U}{4}$
(c) $\frac{U}{2}$
(d) $\frac{U}{3}$

[2021 : 2 M, Set-I]

6.11 The ratio of the momentum correction factor to the energy correction factor for a laminar flow in a pipe is

(a) $\frac{1}{2}$
(b) $\frac{3}{2}$
(c) $\frac{2}{3}$
(d) 1

[2021 : 1 M, Set-II]

■■■■

Answers Viscous Flow of Incompressible Fluid

6.1 (b)	6.2 (c)	6.3 (b)	6.4 (d)	6.5 (c)	6.6 (a)	6.7 (b)
6.9 (d)	6.10 (c)	6.11 (c)				

Explanations Viscous Flow of Incompressible Fluid

6.1 (b)

Friction factor for lqminar flow through pipe is

$$f = \frac{64}{Re}$$

$$f \propto \frac{1}{Re}$$

6.2 (c)

The actual value of α depends on the velocity distribution at the flow section. The value of α for turbulent flow in pipes lies between 1.03 to 1.06, which is approximately equal to 1, because in turbulent flow the velocity distribution is very close to uniform velocity distribution. However for laminar flow in pipes the value of α is 2.

6.3 (b)

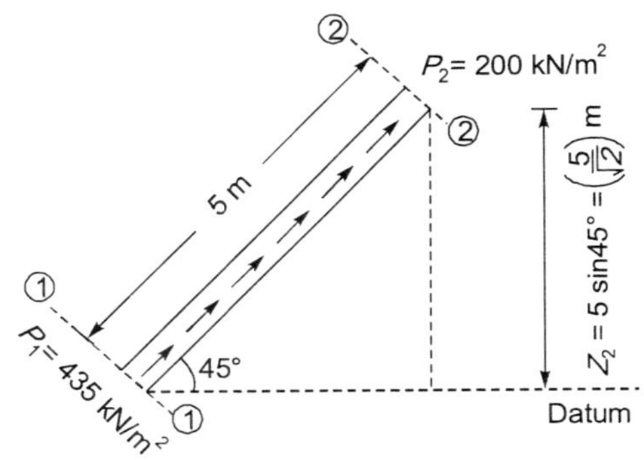

Apply energy equation between (1) and (2)

$$\frac{P_1}{\rho g} + \frac{V_1^2}{2g} + Z_1 = \frac{P_2}{\rho g} + \frac{V_2^2}{2g} + Z_2 + h_f$$

Where $$h_f = \frac{32\mu VL}{\rho g D^2}$$

$$\frac{435\times10^3}{800\times9.81}+\frac{V^2}{2g}+0 = \frac{200\times10^3}{800\times9.81}+\frac{V^2}{2g}+\frac{5}{\sqrt{2}}+\frac{32(0.8)V(5)}{(800)(9.81)(0.1)^2}$$

$$\Rightarrow \frac{435\times10^3-200\times10^3}{800\times9.81}-\frac{5}{\sqrt{2}} = \frac{32\times0.8\times V\times5}{800\times9.81\times(0.1)^2}$$

$$\Rightarrow V = 16.19 \text{ m/s}$$

$$\therefore Q = AV = \frac{\pi\times(0.1)^2}{4}\times16.19 = 0.127 \text{ m}^3\text{/s}$$

6.4 (d)

For reversal of flow

Total energy at section (2) – loses = Total energy at section (1)

When the flow is reversed, then,

$$\frac{p_1}{\rho g}+\frac{32\,\mu VL}{\rho g D^2} = \frac{p_2}{\rho g}+5\sin45°$$

$$\Rightarrow \frac{435\times10^3}{800\times9.81}+\frac{32\times0.8\times16.19\times5}{800\times9.81\times(0.1)^2} = \frac{p_2}{800\times9.81}+\frac{5}{\sqrt{2}}$$

$$\Rightarrow p_2 = 614.48 \text{ kN/m}^2 = 614 \text{ kN/m}^2$$

6.5 (c)

Applying **Hagen-Poiseuille equation**, the drop in pressure is given by,

$$p_1-p_2 = \frac{32\,\mu VL}{D^2}$$

$$Re = \frac{\rho VD}{\mu} \qquad \text{But } Q = AV$$

$$\Rightarrow V = \frac{Q}{A} = \frac{800\times4}{\pi\times(0.5)^2} = 4074.37 \text{ mm/s} \simeq 4.07 \text{ m/s}$$

$$\therefore 2\times10^6 = \frac{32\times\mu\times4.07\times2}{(0.5\times10^{-3})^2}$$

$$\Rightarrow \mu = 1.92\times10^{-3} \text{ Ns/m}^2 = 0.00192 \text{ Ns/m}^2$$

***Note:** We have assumed laminar flow in this problem*

6.6 (a)

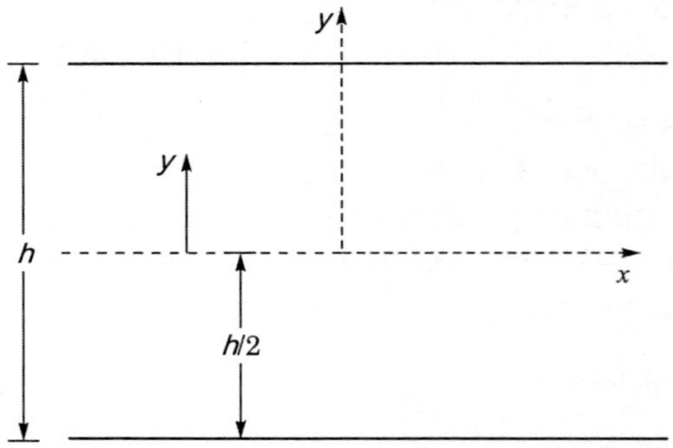

Assume width = 1

Velocity expression for a laminar flow between two parallel plates is,

$$U = -\frac{h^2}{8\mu}\left(\frac{dp}{dx}\right)\left[1-4\left(\frac{y}{h}\right)^2\right]$$

At $y = 0$,

$$U = U_{max}$$

$$\Rightarrow U_{max} = -\frac{h^2}{8\mu}\left(\frac{dp}{dx}\right)$$

Discharge, dQ = Area × Velocity

$$\Rightarrow dQ = \left[-\frac{h^2}{8\mu}\left(\frac{dp}{dx}\right)\left(1-4\left(\frac{y}{h}\right)^2\right)\right](dy\times1)$$

$$\Rightarrow Q = -\frac{h^2}{8\mu}\left(\frac{dp}{dx}\right)\int_{-h/2}^{h/2}\left(1-\frac{4y^2}{h^2}\right)dy$$

$$= -\frac{h^2}{8\mu}\left(\frac{dp}{dx}\right)\left[y-\frac{4y^3}{3h^2}\right]_{-h/2}^{h/2}$$

$$= -\frac{h^2}{8\mu}\left(\frac{dp}{dx}\right)\left[\left\{\frac{h}{2}-\left(-\frac{h}{2}\right)\right\}-\left\{\frac{4}{3h^2}\left(\frac{h^3}{8}-\left(-\frac{h^3}{8}\right)\right)\right\}\right]$$

$$= -\frac{h^3}{12\mu}\left(\frac{dp}{dx}\right)$$

$$\because Q = AU_{avg}$$

$$-\frac{h^3}{12\mu}\left(\frac{dp}{dx}\right) = (h\times1)\times U_{avg}$$

$$\Rightarrow U_{avg} = -\frac{h^2}{12\mu}\left(\frac{dp}{dx}\right)$$

$$\frac{U_{avg}}{U_{max}} = \frac{-\frac{h^2}{12\mu}\left(\frac{dp}{dx}\right)}{-\frac{h^2}{8\mu}\left(\frac{dp}{dx}\right)} = \frac{8}{12} = \frac{2}{3}$$

$$\therefore U_{avg} = \frac{2}{3}U_{max}$$

6.7 (b)

Shear stress variation between parallel stationary plats.

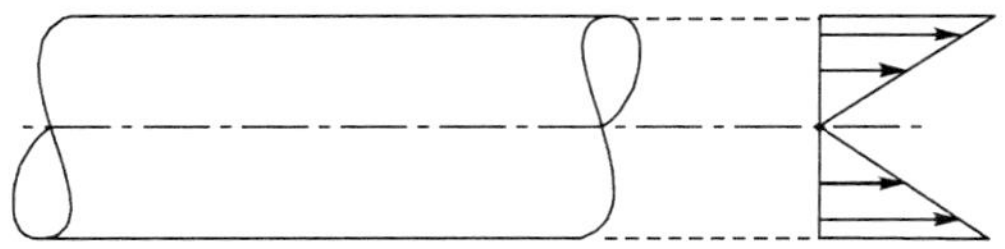

6.8 Sol.

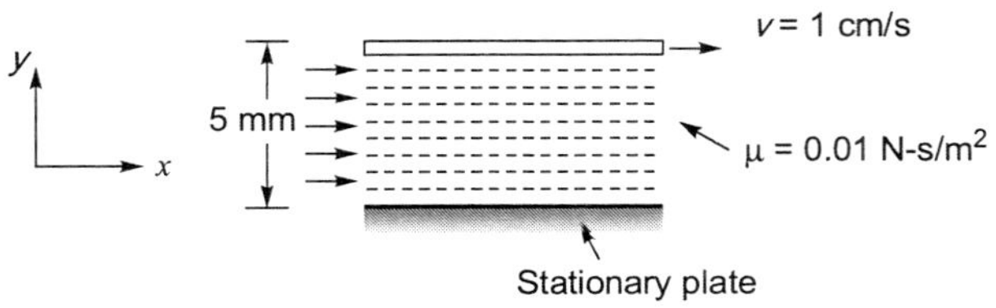

Given, $\tau_{lower\ plate} = 0$

Velocity distribution equation for Couette flow

$$u = -\frac{1}{2\mu}\left(\frac{\partial P}{\partial x}\right)(Hy - y^2) + \frac{V}{H} \times y$$

Now shear distribution at wall

$$\tau_0 = \mu \cdot \left.\frac{du}{dy}\right|_{y=0}$$

$$0 = \mu \cdot \frac{d}{dy}\left[-\frac{1}{2\mu}\left(\frac{\partial P}{\partial x}\right)(Hy - y^2) + \frac{V}{H} \cdot y\right]_{y=0}$$

$$0 = -\frac{1}{2\mu}\left(\frac{\partial P}{\partial x}\right)H + \frac{V}{H}$$

$$\frac{H}{2\mu}\left(\frac{\partial P}{\partial x}\right) = \frac{V}{H}$$

$$\left(\frac{\partial P}{\partial x}\right) = \frac{2\mu V}{H^2} = \frac{2(0.01)(0.01)}{(0.005)^2}$$

$$= 8.0 \text{ N/m}^2\text{/m}$$

6.9 (d)

As the pressure is constant and body is spherical, so net force on body due to pressure will be zero.

6.10 (c)

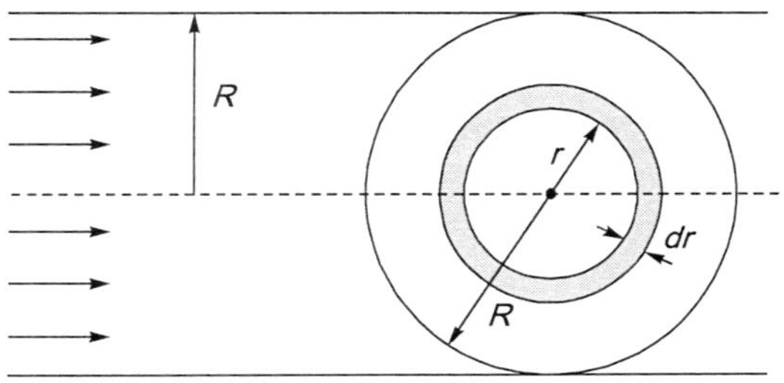

$$u = U\left(1 - \frac{r^2}{R^2}\right)$$

$$\dot{m} = \int_0^R \rho(2\pi r dr)u$$

$$= 2\pi\rho U \int_0^R \left(1 - \frac{r^2}{R^2}\right) r dr$$

$$\rho(\pi R^2)\bar{u} = 2\pi\rho U\left[\frac{R^2}{2} - \frac{R^4}{R^2 \times 4}\right]$$

$$\bar{u} = \frac{U}{2}$$

$\bar{u}$ = Mean velocity

U = Max velocity

6.11 (c)

For laminar flow through pipe.

$$\frac{\text{Momentum correction factor}}{\text{Kinetic energy correction factor}}$$

$$= \frac{4/3}{2} = \frac{4}{6} = \frac{2}{3}$$

■■■■

7 Flow Through Pipes

7.1 Due to aging of a pipeline, its carrying capacity has decreased by 25%. The corresponding increase in the Darcy Weisbach friction factor, f is ..... %

[1995 : 1 M]

7.2 In network of pipes
(a) the algebraic sum of discharges around each circuit is zero
(b) the algebraic sum of discharges around each circuit should not be zero
(c) the elevation of hydraulic grade line is assumed for each junction point
(d) elementary circuits are replaced by equivalent pipes

[1996 : 1 M]

7.3 While deriving an expression for loss of head due to a sudden expansion in a pipe, in addition to the continuity and impulse momentum equations, one of the following assumptions is made
(a) Head loss due to friction is equal to the head loss is eddying motion
(b) The mean pressure in eddying fluid is equal to the downstream pressure
(c) The mean pressure in eddying fluid is equal to the upstream pressure
(d) Head lost in eddies is neglected

[1997 : 1 M]

7.4 If a single pipe of length L and diameter D is to be replaced by three pipes of same material, same length and equal diameter $d(d < D)$, to convey the same total discharge under the same head loss, then d and D are related by

(a) $D = \frac{d}{3^{1/5}}$ (b) $D = \frac{d}{2^{5/3}}$

(c) $D = \frac{d}{3^{2/3}}$ (d) $D = \frac{d}{2^{3/2}}$

[1997 : 1 M]

7.5 Three reservoirs A, B and C are interconnected by pipes as shown in the figure. Water surface elevations in the reservoirs and the Piezometric head at the junction J are indicated in the figure.

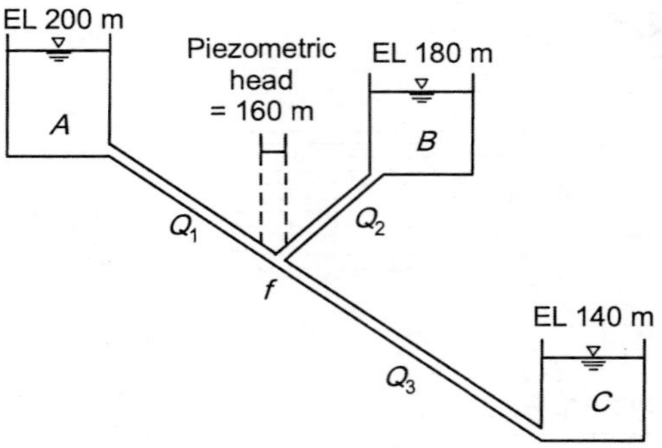

Discharges Q_1, Q_2 and Q_3 are related as
(a) $Q_1 + Q_2 = Q_3$ (b) $Q_1 = Q_2 + Q_3$
(c) $Q_2 = Q_1 + Q_3$ (d) $Q_1 + Q_2 + Q_3 = 0$

[1998 : 1 M]

7.6 Cavitation is caused by
(a) High velocity (b) Low pressure
(c) High pressure (d) High temperature

[2000 : 1 M]

7.7 In the Bernoulli's equation, used in pipe flow, each term represents
(a) Energy per unit weight
(b) Energy per unit mass
(c) Energy per unit volume
(d) Energy per unit flow length [2001 : 1 M]

Common Data for Questions 7.8 and 7.9:

A pipeline (diameter 0.3 m, length 3 km) carries water from point P to point R (see figure). The piezometric heads at P and R are to be maintained at 100 m and 80 m, respectively. To increase the discharge, a second pipe is added in parallel to the existing pipe from Q to R. The length of the additional pipe is also 2 km. Assume the friction factor, $f = 0.04$ for all pipes and ignore minor losses.

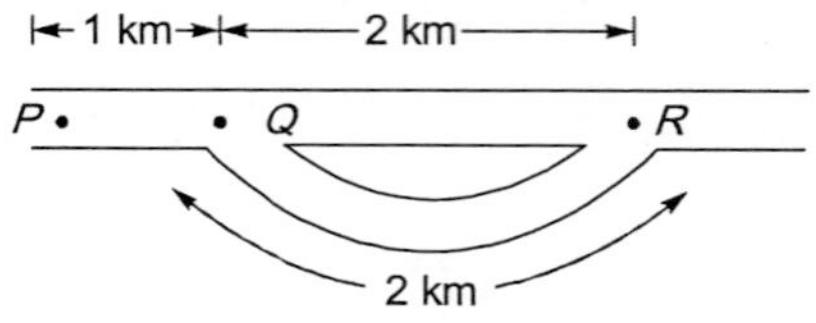

7.8 What is the increase in discharge if the additional pipe has same diameter (0.3 m)?
(a) 0% (b) 33%
(c) 41% (d) 67% [2003 : 2 M]

7.9 If there is no restriction on the diameter of the additional pipe, what would be the maximum increase in discharge theoretically possible from this arrangement?
(a) 0% (b) 50%
(c) 67% (d) 73% [2003 : 2 M]

7.10 A fire protection system is supplied from a water tower with a bent pipe as shown in the figure. The pipe friction f is 0.03. Ignoring all minor losses, the maximum discharge, Q, in the pipe is

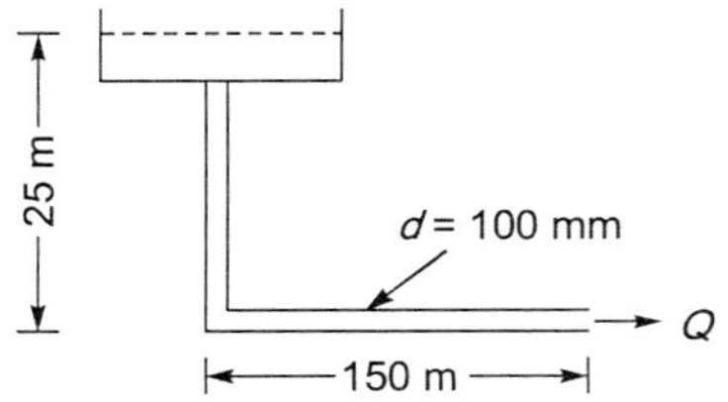

(a) 31.7 lit/sec (b) 24.0 lit/sec
(c) 15.9 lit/sec (d) 12.0 lit/sec
[2004 : 2 M]

7.11 Water flows through a 100 mm diameter pipe with a velocity of 0.015 m/sec. If the kinematic viscosity of water is 1.13×10^{-6} m^2/sec, the friction factor of the pipe material is
(a) 0.0015 (b) 0.032
(c) 0.037 (d) 0.048 [2009 : 2 M]

7.12 Water (γ_w = 9.879 kN/m^3) flows with a flow rate of 0.3 m^3/sec through a pipe AB of 10 m length and of uniform cross-section. The end B is above end A and the pipe makes an angle of 30° to the horizontal. For a pressure of 12 kN/m^2 at the end B, the corresponding pressure at the end A is
(a) 12.0 kN/m^2 (b) 17.0 kN/m^2
(c) 56.4 kN/m^2 (d) 61.4 kN/m^2
[2009 : 2 M]

7.13 A single pipe of length 1500 m and diameter 60 cm connects two reservoirs having a difference of 20 m in their water levels. The pipe is to be replaced by two pipes of the same length and equal diameter 'd' to convey 25% more discharge under the same head loss. If the friction factor is assumed to be the same for all the pipes, the value of 'd' is approximately equal to which of the following options?
(a) 37.5 cm (b) 40.0 cm
(c) 45.0 cm (d) 50.0 cm [2011 : 2 M]

7.14 The circular water pipes shown in the sketch are flowing full. The velocity of flow (in m/s) in the branch pipe "R" is

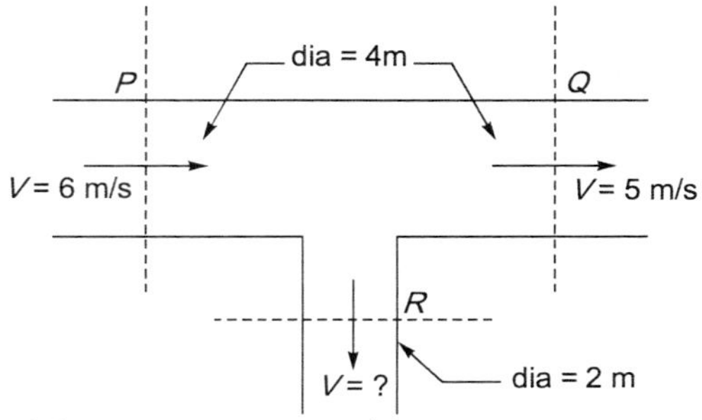

(a) 3 (b) 4
(c) 5 (d) 6 [2012 : 1 M]

7.15 A 2 km pipe of 0.2 m diameter connects two reservoirs. The difference between the water levels in the reservoir is 8 m. The Darcy Weisbach friction factor of the pipe is 0.04. Accounting for frictional entry and exit losses. The velocity in the pipe in (m/sec) is
(a) 0.63 (b) 0.35
(c) 2.52 (d) 1.25 [2013 : 2 M]

7.16 An incompressible fluid is flowing at a steady rate in a horizontal pipe. From a section, the pipe divides into two horizontal parallel pipes of diameters d_1 and d_2 (where $d_1 = 4d_2$) that run for a distance of L each and then again join back to a pipe of the original size. For both the parallel pipes, assume the head loss due to friction only and the Darcy-Weisbach friction factor to be the same. The velocity ratio between the bigger and the smaller branched pipes is _____.
[2014 : 2 M, Set-I]

7.17 A straight 100 m long raw water gravity main is to carry water from an intake structure to the jack well of a water treatment plant. The required flow through this water main is 0.21 m^3/s. Allowable velocity through the main is 0.75 m/s. Assume f = 0.01, g = 9.81 m/s^2. The minimum gradient (in cm/100 m length) to be given to this gravity main so that the required amount of water flows without any difficulty is _______. [2014 : 2 M, Set-I]

7.18 For steady incompressible flow through a closed-conduit of uniform cross-section, the direction of flow will always be
(a) from higher to lower elevation
(b) from higher to lower pressure
(c) from higher to lower velocity
(d) from higher to lower piezometric head
[2015 : 1 M, Set-I]

7.19 Two reservoirs are connected through a 930 m long, 0.3 m diameter pipe, which has a gate valve.

The pipe entrance is sharp (loss coefficient = 0.5) and the valve is half-open (loss coefficient = 5.5). The head difference between the two reservoirs is 20 m. Assume the friction factor for the pipe as 0.03 and g = 10 m/s^2. The discharge in the pipe accounting for all minor and major losses is ______ m^3/s.

[2015 : 2 M, Set-I]

7.20 A pipe of 0.7 m diameter has a length of 6 km and connects two reservoirs A and B. The water level in reservoir A is at an elevation 30 m above the water level in reservoir B. Halfway along the pipe line, there is a branch through which water can be supplied to a third reservoir C. The friction factor of the pipe is 0.024. The quantity of water discharged into reservoir C is 0.15 m^3/s. Considering the acceleration due to gravity as 9.81 m/s^2 and neglecting minor losses, the discharge (in m^3/s) into the reservoir B is _________.

[2015 : 2 M, Set-II]

7.21 A triangular pipe network is shown in the figure

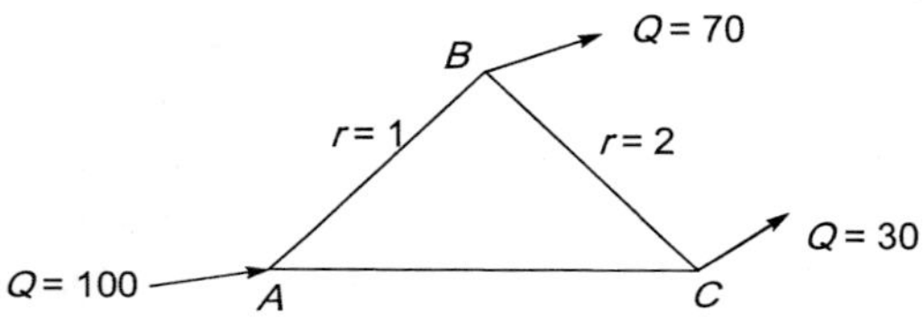

The head loss in each pipe is given by $h_f = rQ^{1.8}$, with the variables expressed in a consistent set of units. The value of r for the pipe AB is 1 and for the pipe BC is 2. If the discharge supplied at the point A (i.e., 100) is equally divided between the pipes AB and AC, the value of r (up to 2 decimal places) for the pipe AC should be _________.

[2017 : 1 M, Set-I]

7.22 Water is pumped at a steady uniform flow rate of 0.01 m^3/s through a horizontal smooth circular pipe of 100 mm diameter. Given that the Reynolds number is 800 and g is 9.81 m/s^2, the head loss (in meters, up to one decimal place) per km length due to friction would be _________.

[2017 : 2 M, Set-II]

7.23 Two water reservoirs are connected by a siphon (running full) of total length 5000 m and diameter of 0.10 m, as shown below (figure not drawn to scale).

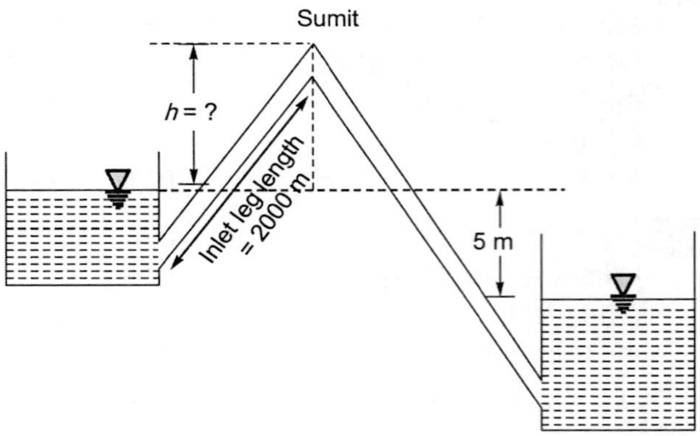

The inlet leg length of the siphon to its summit is 2000 m. The difference in the water surface levels of the two reservoirs is 5 m. Assume the permissible minimum absolute pressure at the summit of siphon to be 2.5 m of water when running full. Given: friction factor, f = 0.02 throughout, atmospheric pressure = 10.3 m of water and acceleration due to gravity g = 9.81 m/s^2. Considering only major loss using Darcy-Weisback equation, the maximum height of the summit of siphon from the water level of upper reservoir, h (in m, round off to 1 decimal place) is_______.

[2019 : 2 M, Set-I]

7.24 Two identical pipes (i.e., having the same length, same diameter, and same roughness) are used to withdraw water from a reservoir. In the first case, they are attached in series and discharge freely into the atmosphere. In the second case, they are attached in parallel and also discharge freely into the atmosphere. Neglecting all minor losses, and assuming that the friction factor is same in both the cases, the ratio of the discharge in the parallel arrangement to that in the series arrangement (round off to 2 decimal places) is _______.

[2019 : 2 M, Set-II]

7.25 Three reservoir P, Q and R are interconnected by pipes as shown in the figure (not drawn to the scale). Piezometric head at the junction S of the pipes is 100 m. Assume acceleration due to gravity as 9.81 m/s^2 and density of water as 1000 kg/m$^{`3}$. the length of the pipe from junction S to the inlet of reservoir R is 180 m.

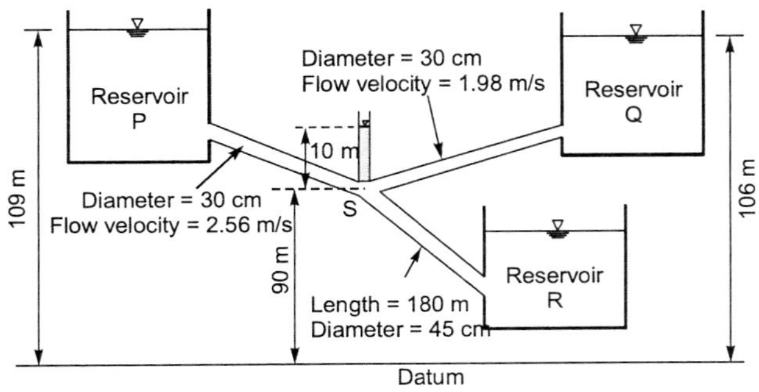

Considering head loss only due to friction (with friction factor of 0.03 for all the pipes), the height of water level in the lowermost reservoir R (in m, round off to one decimal places) with respect to the datum, is ________.

[2020 : 2 M, Set-I]

7.26 A cast iron pipe of diameter 600 mm and length 400 m carries water from a tank and discharges freely into air at a point 4.5 m below the water surface in the tank. The friction factor of the pipe is 0.018. Consider acceleration due to gravity as 9.81 m/s^2. The velocity of the flow in pipe (in m/s, round off to two decimal places) is ________.

[2020 : 2 M, Set-II]

7.27 Two reservoirs are connected by two parallel pipes of equal length and of diameters 20 cm and 10 cm, as shown in the figure (not drawn to scale). When the difference in the water levels of the reservoirs is 5 m, the ratio of discharge in the larger diameter pipe to the discharge in the smaller diameter pipe is ________. (round off to two decimal places)

(Consider only loss due to friction and neglect all other losses. Assume the friction factor to be the same for both the pipes)

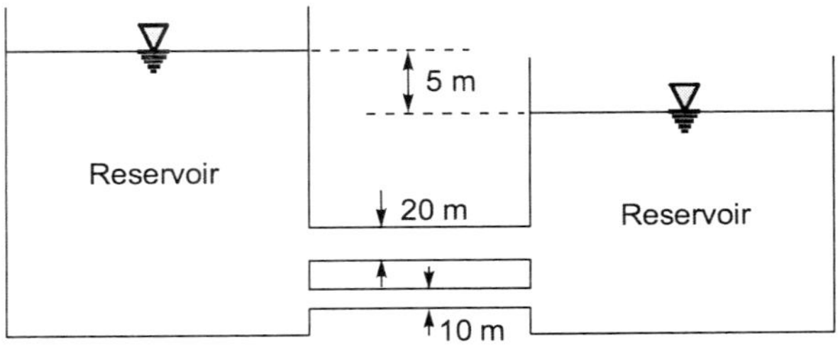

[2022 : 2 M, Set-I]

7.28 A 500 m long water distribution pipeline P with diameter 1.0 m, is used to convey 0.1m^3/s of flow. A new pipeline Q, with the same length and flow rate, is to replace P. The friction factors for P and Q are 0.04 and 0.01, respectively. The diameter of the pipeline Q (in meters) is (rounded off to 2 decimal places).

[2024 : 2 M, Set-II]

Answers Flow Through Pipes

7.1	(1.77)	7.2	(b)	7.3	(c)	7.4	(a)	7.5	(a)	7.6	(b)	7.7	(a)
7.8	(c)	7.9	(d)	7.10	(b)	7.11	(d)	7.12	(d)	7.13	(d)	7.14	(b)
7.15	(a)	7.16	(2)	7.17	(4.8)	7.18	(d)	7.19	(0.141)	7.20	(0.572)	7.21	(0.62)
7.22	(66.11)	7.23	(5.8)	7.24	(2.83)	7.25	(97.51)	7.26	(2.56)	7.27	(5.66)	7.28	(0.75)

Explanations Flow Through Pipes

7.1 Sol.

As we know,

$$h_f = \frac{8Q^2}{\pi^2 g} \times \frac{fL}{D^5}$$

$$f \propto \frac{1}{Q^2}$$

If Q is decreased by 25%, then,

$$f' = \frac{k}{(0.75Q)^2} = 1.77f$$

∴ Value of **Darcy's Weisbach friction factor** is increased by 77%.

7.2 (b)

7.3 (c)

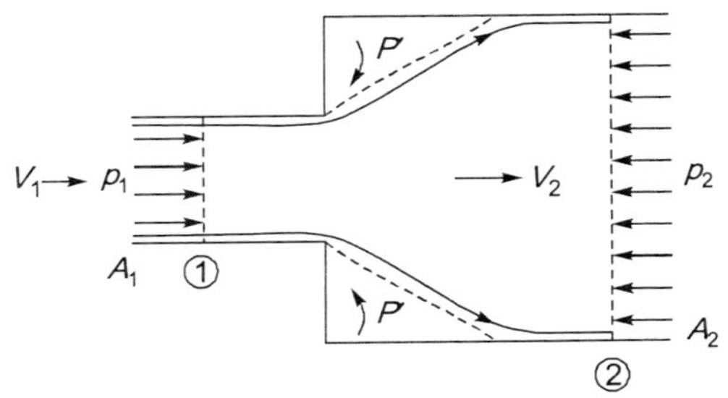

From impulse momentum equation,

$$P_1A_1 + P'(A_2 - A_1) - P_2A_2 = \rho Q(V_2 - V_1)$$

Assume $\quad P' = P_1$

$$(P_1 - P_2)A_2 = \rho A_2V_2(V_2 - V_1)$$

From Bernoulli's eq.

$$\frac{P_1 - P_2}{\rho g} = \frac{V_2(V_2 - V_1)}{g} \quad ...(i)$$

$$\frac{P_1}{\rho g} + \frac{V_1^2}{2g} + z_1 = \frac{P_2}{\rho g} + \frac{V_2^2}{2g} + z_2 + h_L \quad (z_1 = z_2)$$

$$h_L = \frac{P_1 - P_2}{\rho g} + \frac{V_1^2 - V_2^2}{2g} \quad ...(ii)$$

By eq. (i) and (ii)

$$h_L = \frac{V_2^2 - V_1V_2}{g} + \frac{V_1^2 - V_2^2}{2g}$$

$$h_L = \frac{2V_2^2 - 2V_1V_2 + V_1^2 - V_2^2}{2g}$$

$$h_L = \frac{V_1^2 + V_2^2 - 2V_1V_2}{2g}$$

$$h_L = \frac{(V_1 - V_2)^2}{2g}$$

7.4 (a)

For series connection

$$\frac{L}{D^5} = \frac{L_1}{D_1^5} + \frac{L_2}{D_2^5} + \frac{L_3}{D_3^5}$$

Since $\quad L_1 = L_2 = L_3 = L$

and $\quad D_1 = D_2 = D_3 = d$

$$\therefore \quad \frac{L}{D^5} = \frac{L}{d^5} + \frac{L}{d^5} + \frac{L}{d^5}$$

$$\Rightarrow \quad \frac{1}{D^5} = \frac{3}{d^5}$$

$$\Rightarrow \quad D = d/(3)^{1/5}$$

7.5 (a)

7.6 (b)

When a liquid flows into a region where its pressure is reduced from vapour pressure, it starts vaporizing (or boiling), and vapour pockets or bubbles are formed in the liquid.

When these bubbles start to travel in a region where pressure increases, bubble start to burst and form cavity. This process is known as cavitation.

7.7 (a)

Bernoulli's equation, used in pipe flow

$$\frac{P}{\rho g} + \frac{V^2}{2g} + Z = \text{Constant}$$

Here, all terms in energy per unit weight.

7.8 (c)

The difference between piezometric head between P and R is 100 – 80 = 20 m. Now, if the minor losses are neglected, then the only loss is the head loss.

Let the initial discharge be Q_1,

$$h_f = \frac{8Q^2}{\pi^2 g} \times \frac{fL}{D^5}$$

or $$h_f = \frac{fLQ_2^2}{12.1D^5}$$

$$\Rightarrow \quad 20 = \frac{0.04 \times 3000 \times Q_1^2}{12 \times (0.3)^5}$$

$$\Rightarrow \quad Q_1 = 0.0697 \text{ m}^3/\text{s}$$

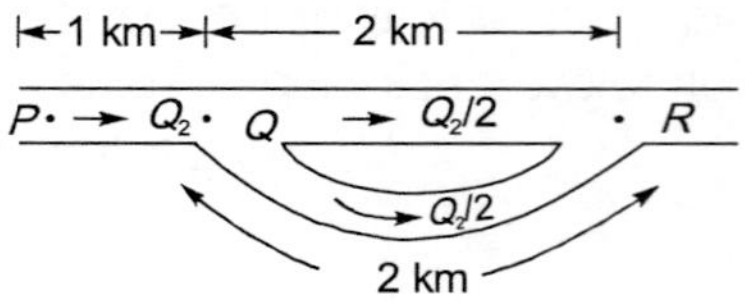

$$h_L = h_{PQ} + h_{QR}$$

$$\therefore \quad 20 = \frac{fL_{PQ}Q_2^2}{12 \times (0.3)^5} + \frac{fL_{QR}\left(\frac{Q_2}{2}\right)^2}{12 \times (0.3)^5}$$

$$\Rightarrow \quad 20 = \frac{0.04 \times 1000 \times Q_2^2}{12 \times (0.3)^5} + \frac{0.04 \times 2000 \times Q_2^2}{4 \times 12 \times (0.3)^5}$$

$$\Rightarrow \quad Q_2 = 0.0986 \text{ m}^3/\text{s}$$

∴ % increase in discharge

$$= \left(\frac{Q_2 - Q_1}{Q_1}\right) \times 100$$

$$= \left(\frac{0.0986 - 0.0697}{0.0697}\right) \times 100 = 41.4\%$$

Note: In parallel combination of pipes, head loss between the junction remains same for different pipes whereas for series combination, it gets added up i.e.,

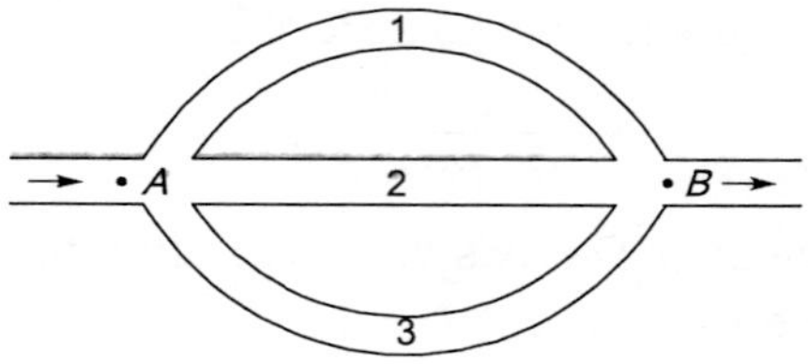

$$h_{AB} = h_1 = h_2 = h_3$$

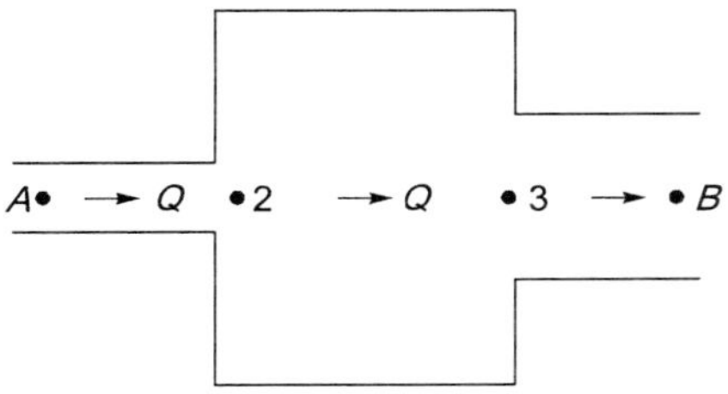

$$h_{AB} = h_{A2} + h_{23} + h_{3B}$$

7.9 (d)

If the diameter of additional pipe is very large, the head loss will be zero in it theoretically. Therefore the total head loss will occur between *P* and *Q* only.

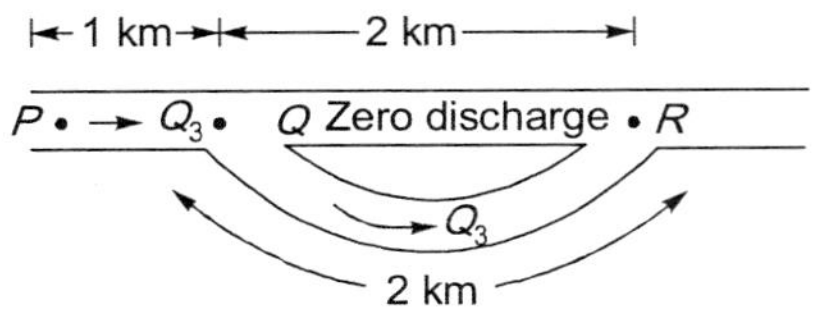

$$h_L = \frac{fL_{PQ}Q_3^2}{12d^5}$$

$$\Rightarrow \quad 20 = \frac{0.04 \times 1000 \times Q_3^2}{12 \times (0.3)^5}$$

$$\Rightarrow \quad Q_3 = 0.1207 \text{ m}^3/\text{s}$$

% maximum increase in discharge

$$= \left(\frac{Q_3 - Q_1}{Q_1}\right) \times 100$$

$$= \left(\frac{0.1207 - 0.0697}{0.0697}\right) \times 100 = 73.17\% \approx 73\%$$

7.10 (b)

$$h_f = \frac{8Q^2}{\pi^2 g}\frac{fL}{D^5}$$

$$\Rightarrow \quad 25 = \frac{8Q^2}{\pi^2 g} \times \frac{(0.03)(175)}{(0.1)^5}$$

$$Q = 24 \text{ litre/sec}$$

Note: *In this problem, all minor losses have been neglected. Hence the total lossof 25 m head will be in the form of friction only.*

7.11 (d)

Reynolds number,

$$Re = \frac{VD}{\nu} = \frac{0.015 \times 100 \times 10^{-3}}{1.13 \times 10^{-6}}$$

$$= 1327.43 < 2000$$

Hence flow is laminar

$$\therefore \quad f = \frac{64}{Re} = \frac{64}{1327.43} = 0.048$$

Note: *The equation* $f = \frac{64}{Re}$ *is valid only in case of laminar flow. It should not be used for cases where Re > 2000.*

7.12 (d)

Applying **Bernoulli's equation** between *A* and *B*, we get,

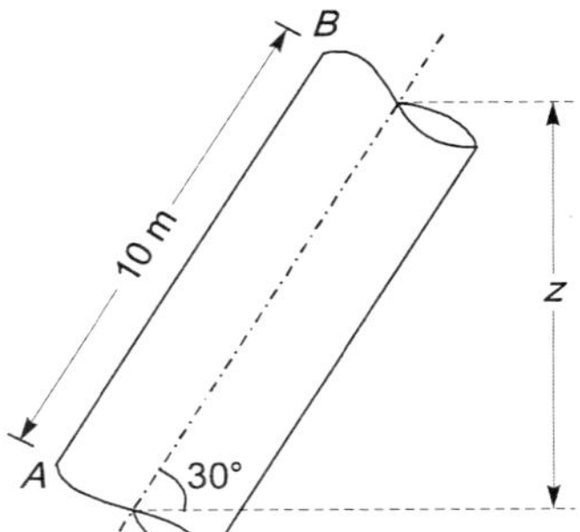

$z = 10 \sin 30°$
$= 5$ cm

$$\frac{p_A}{\gamma_w} + \frac{V^2}{2g} + 0 = \frac{p_B}{\gamma_w} + \frac{V^2}{2g} + z$$

$$\Rightarrow \quad p_A = p_B + \gamma_w z$$

$$\Rightarrow \quad p_A = 12 + 9.879 \times 5$$

$$\Rightarrow \quad p_A = 12 + 49.4$$

$$= 61.4 \text{ kN/m}^2$$

POINTS TO REMEMBER

- In this question, we have assumed datum at A. As stated earlier choosing of datum will have no change in final answer.
- Also we have assumed the losses to be zero in this case.

7.13 (d)

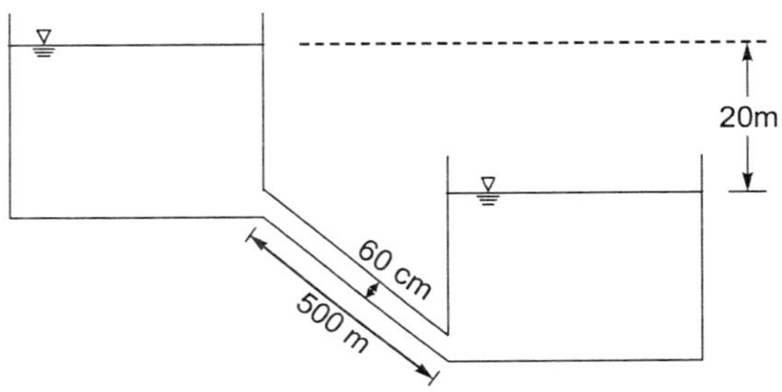

For **single pipe,**

$$h_f = \frac{8fLQ_1^2}{\pi^2 g D^5}$$

Given, $L = 1500$ m, $D = 0.6$ m, $h_f = 20$ m

$$Q_1 = \frac{0.112}{\sqrt{f}} \quad ...(i)$$

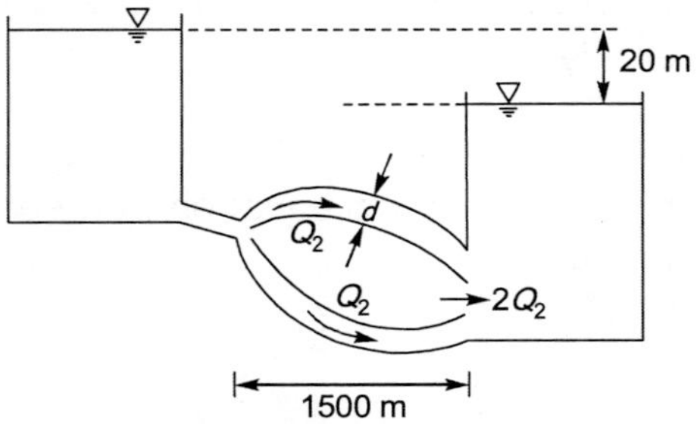

In case of **double pipe** discharge is increased by 25% than that of previous case hence,

$$2Q_2 = 1.25\, Q_1$$
$$Q_2 = 0.625\, Q_1$$

$$20 = \frac{8F \times 500 \times (0.625Q_1)^2}{9.81 \times \pi^2 \times d^5}$$

$$Q_1 = \frac{0.402 \times d^{5/2}}{\sqrt{f}}$$

$$\frac{fQ_1^2}{d^5} = 0.413$$

$$Q_1^2 = \frac{0.413d^5}{f}$$

$$Q_1 = \frac{0.642d^{5/2}}{\sqrt{f}} \quad ...(ii)$$

From eq. (i) and (ii), we get,

$$d = 49.7 \text{ cm} \approx 50 \text{ cm}$$

7.14 (b)

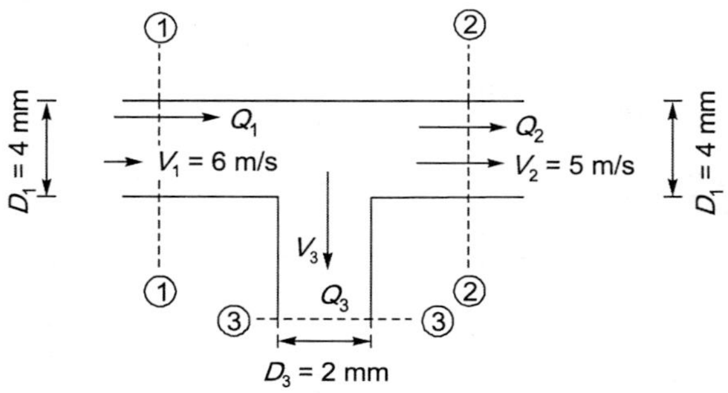

$$Q_1 = Q_2 + Q_3$$

$$\Rightarrow \quad A_1V_1 = A_2V_2 + A_3V_3$$

$$\Rightarrow \quad \frac{\pi}{4}(4)^2 \times 6 = \frac{\pi}{4}(4)^2 \times 5 + \frac{\pi}{4}(2)^2 \times V_3$$

$$\Rightarrow \quad 16 \times 6 = 16 \times 5 + 4V_3$$

$$\Rightarrow \quad 16 = 4\,V_3$$

$$\Rightarrow \quad V_3 = 4 \text{ m/s}$$

7.15 (a)

Frictional loss,

$$h_f = \frac{fLV^2}{D \times 2g}$$

Loss of head at entrance,

$$h_L = \frac{0.5V^2}{2g}$$

Loss at exit of pipe,

$$h_c = \frac{V^2}{2g}$$

$$\therefore \quad 8 = \frac{fLV^2}{D \times 2g} + \frac{0.5V^2}{2g} + \frac{V^2}{2g}$$

$$\Rightarrow \quad 8 = \frac{0.04 \times 2000 \times V^2}{2 \times g \times 0.2} + \frac{V^2}{2g} + \frac{0.5\,V^2}{2g}$$

$$\Rightarrow \quad 8 = 20.3873\,V^2 + 0.07645\,V^2$$

$$\Rightarrow \quad V = 0.625 \text{ m/s}$$
$$\cong 0.63 \text{ m/s}$$

7.16 Sol.

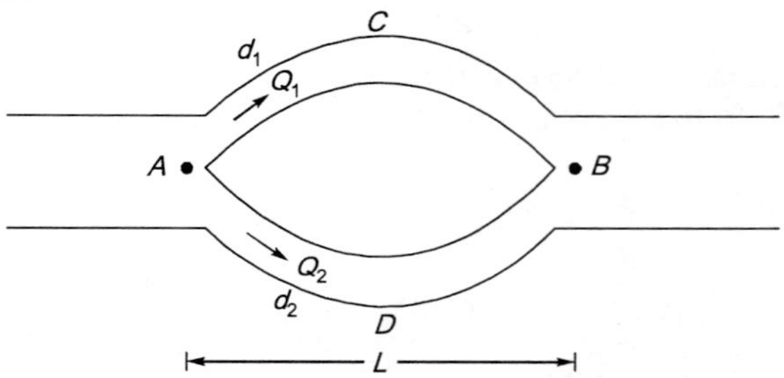

$$d_1 = 4d_2$$

$$\frac{flV_1^2}{2gd_1} = \frac{flV_2^2}{2gd_2}$$

Given $\quad f$ = Same for both pipe

$$\Rightarrow \quad \frac{V_1^2}{d_1} = \frac{V_2^2}{d_2}$$

or $$\frac{V_1^2}{V_2^2} = \frac{d_1}{d_2} = 4$$

or $$\frac{V_1}{V_2} = 2$$

Note: *In this problem, Hagen-Poiseuille equation can't be applied because in the problem laminar flow is not mentioned.*

7.17 Sol.

$$Q = 0.21 \text{ m}^3/\text{s}$$

Allowable velocity = 0.75 m/s

$$f = 0.01$$

$$g = 9.81 \text{ m/s}^2$$

$$Q = A\,V$$

$$0.21 = \left(\frac{\pi}{4}d^2\right)(0.75)$$

$$\Rightarrow \quad d = 0.597 \text{ m}$$

$$\Rightarrow \quad h_f = \frac{flV^2}{2gd} = \frac{0.01 \times 100 \times (0.75)^2}{2 \times 9.81 \times 0.597} \text{m}$$

$$= 4.8 \text{ cm}$$

$$= 0.048 \text{ m}$$

$$\Rightarrow \text{Min. gradient} = \frac{h_f}{l} = \frac{4.8\text{cm}}{100\text{m}}$$

Hence, **answer is 4.8.**

7.18 (d)

In closed uniform conduit, velocity head remains constant, thus flow will be from higher to lower Piezometric head.

$$\text{Piezometric head} = \frac{p}{\rho g} + z.$$

7.19 Sol.

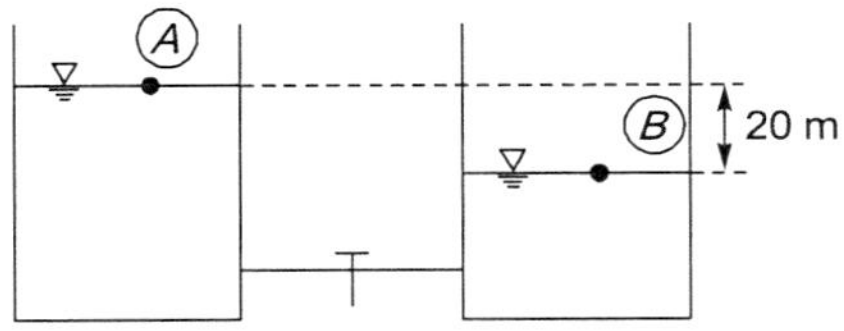

Given:

$f = 0.03$, $L = 930$ m, $D = 0.3$ m, $g = 10$ m/s^2

Apply **energy equation** between A and B,

$$20 = \frac{0.5V^2}{2g} + \frac{(0.03)(930)\cdot V^2}{2g(0.3)} + \frac{5.5V^2}{2g} + \frac{V^2}{2g}$$

$$\text{or,} \quad 20 = \frac{(0.03)(930)}{(0.3)} \cdot \frac{V^2}{2g} + \frac{7.0V^2}{2g}$$

$$\text{or,} \quad 20 = \frac{100V^2}{2g}$$

or, $V = 2$ m/sec

$$\text{But} \quad Q = \frac{\pi}{4}(0.3)^2(2)$$

$$\therefore \quad Q = 0.1414 \text{ m}^3/\text{sec}$$

7.20 Sol.

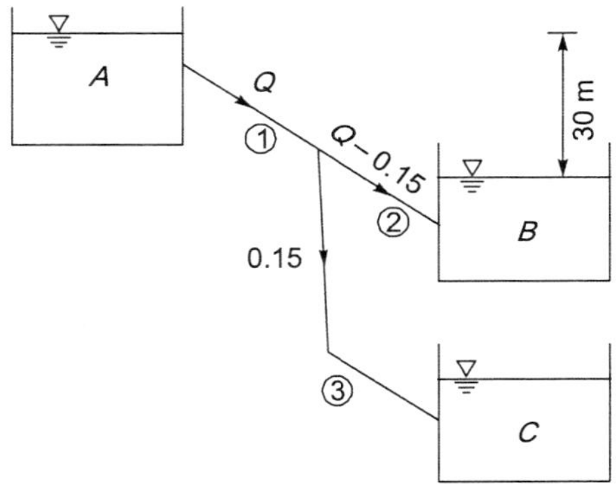

$f = 0.024$, $d = 700$ mm, $l = 6000$ m

$$h_f = \frac{flQ^2}{12.1d^5}$$

$$30 = h_{f_1} + h_{f_2}$$

$$\text{or,} \quad 30 = \frac{0.024 \times 3000 \times Q^2}{12.1 \times 0.7^5} + \frac{0.024 \times 3000 \times (Q - 0.15)^2}{12.1 \times (0.7)^5}$$

$$= 35.40\,Q^2 + 35.40\,(Q - 0.15)^2$$

$$\therefore \; Q = 0.722 \text{ m}^3/\text{s}$$

$$\therefore \; Q_B = Q - 0.15 = 0.572 \text{ m}^3/\text{s}$$

7.21 Sol.

Given $\qquad h_f = r \cdot Q^{1.8}$

Because of the given condition of equal discharge distribution in pipe AB & AC, the discharge in AB and AC will be 50 and 50. Now satisfy continuity at point B and C, discharge from C to B will be 20.

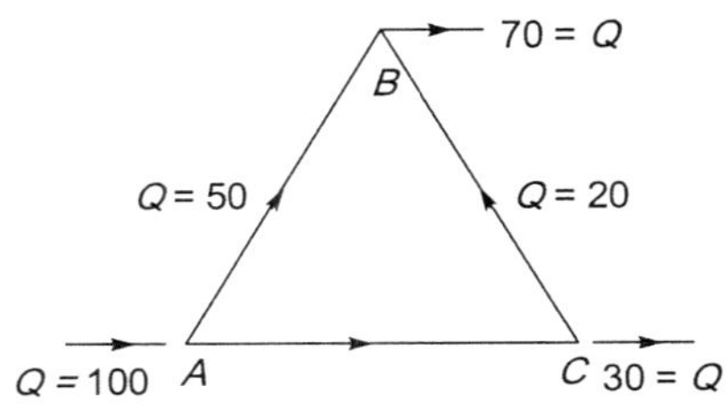

For close Loop $ABCA$,

$$\Sigma h_f = 0$$

$$\Rightarrow 1(50)^{1.8} - 2(20)^{1.8} - r(50)^{1.8} = 0$$

$$\Rightarrow 1143.26 - 439.42 = r(50)^{1.8}$$

$$r = 0.62$$

7.22 Sol.

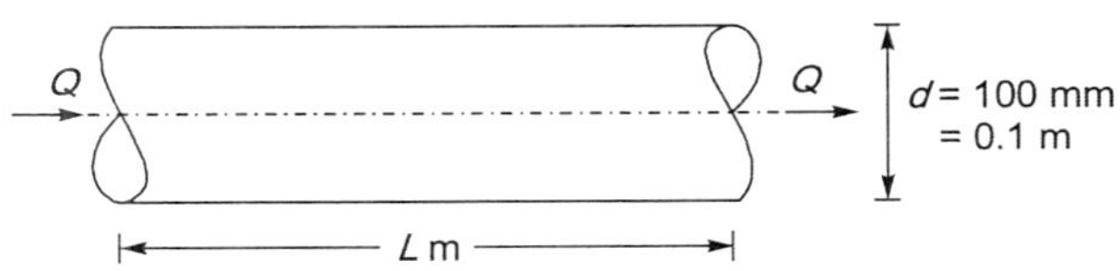

Head loss due to frictions,

$$h_L = \frac{fLQ^2}{12.1d^5}$$

Here, f = friction factor

$$= \frac{64}{\text{Re}} = \frac{64}{800} = 0.08$$

∴ Head loss per metre length,

$$\frac{h_L}{L} = \frac{fQ^2}{12.1d^5} = \frac{0.08\times(0.01)^3}{12.1\times(0.1)^5}$$

$$\frac{h_L}{L} = \frac{0.8}{12.1} = 0.06611 \text{ meter/meter}$$

$$= 66.11 \text{ meter/km}$$

7.23 Sol.

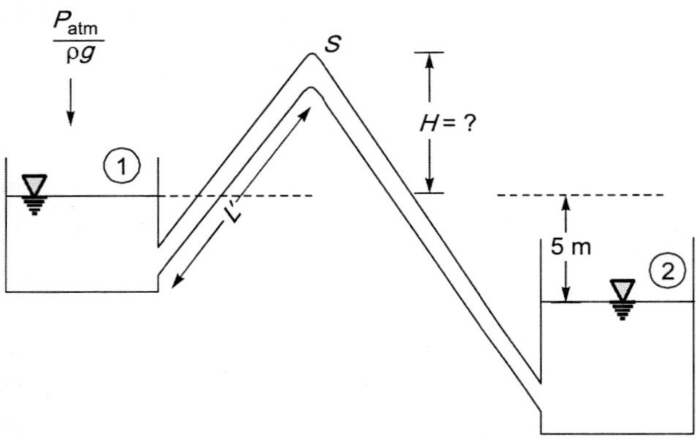

Given: L = 5000 m, D = 0.1 m, L' = 2000 m,

$$\frac{P_s}{\rho g} = 2.5 \text{ m of } H_2O,$$

$$f = 0.02,\ \frac{P_{atm}}{\rho g} = 10.3 \text{ m of } H_2O,\ g = 9.81 \text{ m/s}^2$$

Consider only major losses

Apply energy equation between (1) and (2)

$$\frac{P_1}{\rho g} + \frac{V_1^2}{2g} + z_1 = \frac{P_2}{\rho g} + \frac{V_2^2}{2g} + z_2 + h_{f(1-2)}$$

$$5 = h_{f\,(1-2)}$$

$$5 = \frac{fLV^2}{2gD} = \frac{(0.02)\times(5000)}{0.1}\frac{V^2}{2g}$$

$$\frac{V^2}{2g} = 0.005 \text{ m}$$

Apply energy equation between (1) and (S)

$$\frac{P_1}{\rho g} + \frac{V_1^2}{2g} + z_1 = \frac{P_S}{\rho g} + \frac{V_S^2}{2g} + z_S + h_{f(1-S)}$$

$$10.3 + 0 + 0 = 2.5 + 0.005 + H + \frac{fL'}{D}\frac{V_S^2}{2g}$$

$$10.3 = 2.505 + H + \frac{(0.02)(2000)}{0.1}\times(0.005)$$

$$10.3 = 2.505 + H + 2$$

$$H = 5.795$$

$$H = 5.8 \text{ m (upto 1 decimal place)}$$

7.24 Sol.

Given: Two identical pipes of same length (L), diameter (D) and roughness (k_s).

1st case (Series)

Assume height of reservoir = H

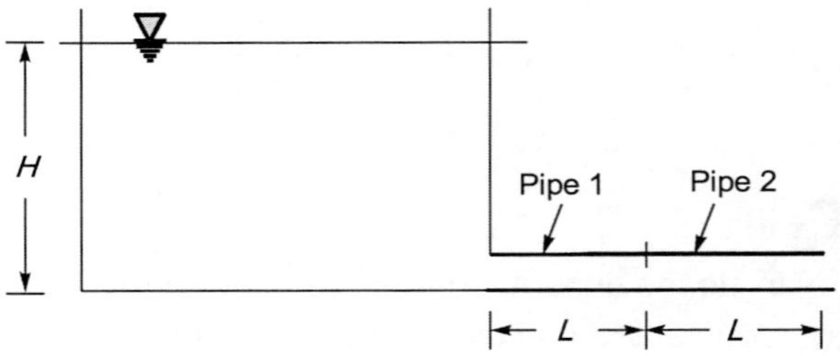

$$H = h_{f1} + h_{f2} \quad \text{...(i)}$$

$$H = 2h_{f1} \quad (h_{f_1} = h_{f_2})$$

$$H = 2\left[\frac{8Q_1^2}{\pi^2 g}\times\frac{f\cdot L}{D^5}\right]$$

2nd case (Parallel)

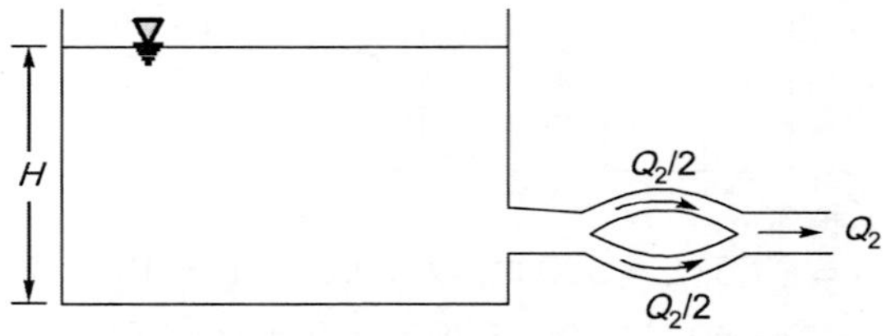

$$H = \frac{8}{\pi^2 g}\left(\frac{Q_2}{2}\right)^2\times\frac{fL}{D^5}$$

$$H = \frac{8}{\pi^2 g}\times\frac{fL}{D^5}\times\frac{Q_2^2}{4} \quad \text{..(ii)}$$

By eq. (i) and (ii)

$$2\left[\frac{8Q_1^2}{\pi^2 g}\times\frac{f\cdot L}{D^5}\right] = \frac{8}{\pi^2 g}\times\frac{fL}{D^5}\times\frac{Q_2^2}{4}$$

$$\frac{8}{\pi^2 g}\times\frac{fL}{D^5}2Q_1^2 = \frac{8}{\pi^2 g}\times\frac{fL}{D^5}\frac{Q_2^2}{4}$$

$$\frac{Q_2}{Q_1} = \sqrt{8}$$

$$= 2.83 \text{ (round off to 2 decimal place)}$$

7.25 Sol.

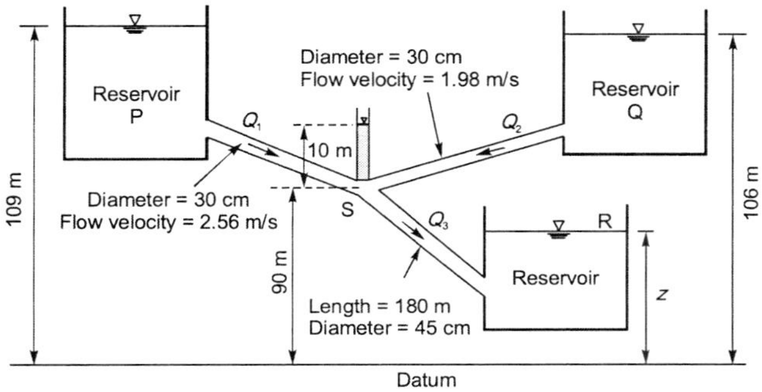

Apply conutinuity

$$Q_3 = Q_1 + Q_2 = A_1V_1 + A_2V_2$$

$$= \frac{\pi}{4}(0.3)^2(2.56) + \frac{\pi}{4}(0.3)^2(1.98) = 0.3209 \text{ m}^3/\text{s}$$

Apply energy eq. between (S) and (R)

$$H_s = H_r + h_f$$

$$100 = z + \frac{8Q_3^2}{\pi^2 g} \times \frac{fL_3}{D_3^5}$$

$$100 = z + \frac{8(0.3209)^2}{\pi^2 g} \times \frac{(0.03)(180)}{(0.45)^5}$$

$$z = 97.51 \text{ m}$$

7.26 Sol.

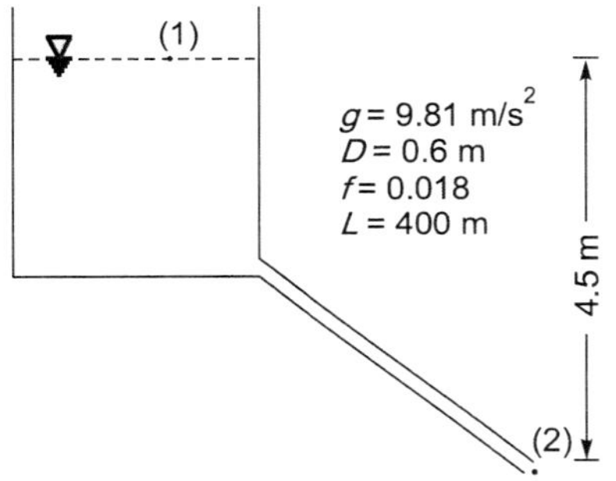

Apply energy equation between (1) and (2)

$$\frac{P_1}{\rho g} + \frac{V_1^2}{2g} + z_1 = \frac{P_2}{\rho g} + \frac{V_2^2}{2g} + z_2 + h_f$$

$$4.5 = \frac{f \cdot L \cdot V^2}{2gD} + 0.5\frac{V^2}{2g} + \frac{V^2}{2g}$$

$$4.5 = \frac{(0.018)(400) \cdot V^2}{2(9.81)(0.6)} + \frac{1.5V^2}{2g} = \frac{12V^2}{2g} + \frac{1.5V^2}{2g}$$

$$V^2 = 6.54$$

Velocity of flow, V = 2.557 m/s $\simeq$ 2.56 m/s

7.27 Sol.

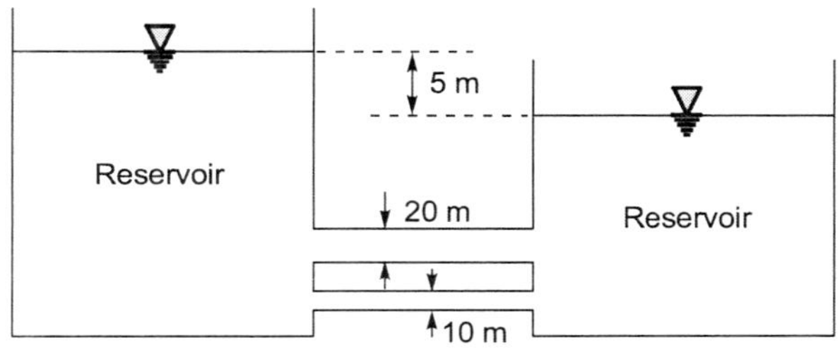

For parallel pipes,

$$h_{f1} = h_{f2}$$

$$\Rightarrow \quad \frac{8Q_1^2}{\pi^2 g} \times \frac{fL}{D_1^5} = \frac{8Q_2^2}{\pi^2 g} \times \frac{fL}{D_2^5}$$

$$\Rightarrow \quad \left(\frac{Q_1}{Q_2}\right)^2 = \left(\frac{D_1}{D_2}\right)^5$$

$$\Rightarrow \quad \frac{Q_1}{Q_2} = \left(\frac{D_1}{D_2}\right)^{5/2} = \left(\frac{0.2}{0.1}\right)^{5/2} = 5.66$$

7.28 Sol.

Here, $h_{fP} = h_{fQ}$

$$\Rightarrow \frac{8q^2}{\pi^2 g} \times \frac{0.04 \times L}{(1)^5} = \frac{8q^2}{\pi^2 g} \times \frac{0.01 \times L}{D^5}$$

where q is discharge and L is length of pipeline.

$$\therefore \quad D = \left(\frac{1}{4}\right)^{1/5} = 0.75 \text{ m}$$

■■■■

8 Boundary Layer Theory, Drag and Lift

8.1 On an immersed body in a flowing fluid, the lift force is
(a) due to buoyant force
(b) always in the opposite direction to gravity
(c) due to wake phenomenon
(d) the dynamic fluid force component normal to approach velocity
[1996 : 1 M]

8.2 Velocity distribution in a boundary layer flow over a plate is given by

$$(u/u_m) = 1.5\,\eta$$

where, $\eta = y/\delta$; y is the distance measured normal to the plate; δ is the boundary layer thickness; and u_m is the maximum velocity at $y = \delta$. If the shear stress τ, acting on the plate is given by

$$\tau = K(\mu\, u_m)/\delta$$

Where, μ is the dynamic viscosity of the fluid, K takes the value of
(a) 0 (b) 1
(c) 1.5 (d) None of these
[2002 : 1 M]

8.3 A flat plate is kept in an infinite fluid medium. The fluid has a uniform free-stream velocity parallel to the plate. For the laminar boundary layer formed on the plate, pick the correct option matching List-I and List-II.

List-I
A. Boundary layer thickness
B. Shear stress at the plate
C. Pressure gradient along the plate

List-II
1. Decreases in the flow direction
2. Increases in the flow direction
3. Remains unchanged

Codes:

	A	B	C
(a)	1	2	3
(b)	2	2	2
(c)	1	1	2
(d)	2	1	3

[2003 : 1 M]

8.4 A solid sphere (diameter 6 mm) is rising through oil (mass density 900 kg/m^3, dynamic viscosity 0.7 kg/m-s) at a constant velocity of 1 cm/s. What is the specific weight of the material from which the sphere is made? (Take $g = 9.81$m/s^2)
(a) 4.3 kN/m^3 (b) 5.3 kN/m^3
(c) 8.7 kN/m^3 (d) 12.3 kN/m^3
[2003 : 2 M]

8.5 An aircraft is flying in level flight at a speed of 200 km/hr through air (density, ρ = 1.2 kg/m^3, and viscosity $\mu = 1.6 \times 10^{-5}$ Ns/m^2). The lift coefficient at this speed is 0.4 and the drag coefficient is 0.0065. The mass of the aircraft is 800 kg. The effective lift area of the aircraft is
(a) 21.2 m^2 (b) 10.6 m^2
(c) 2.2 m^2 (d) 1.1 m^2 [2004 : 1 M]

8.6 A thin flat plate 0.5 m × 0.7 m in size settles in a large tank of water with a terminal velocity of 0.12 m/s. The coefficients of drag $C_D = \dfrac{1.328}{\sqrt{R_L}}$ for a laminar boundary layer and $C_D = \dfrac{0.072}{(R_L)^{1/5}}$ for a turbulent boundary layer, where R_L, is the plate Reynolds number.
Assume $\mu = 10^{-3}$ Ns/m^2 and $\rho = 1000$ kg/m^3

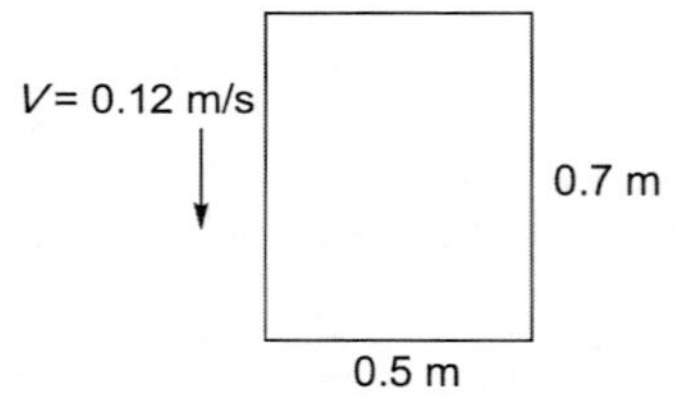

The submerged weight of the plate is
(a) 0.0115 N (b) 0.0118 N
(c) 0.0231 N (d) 0.0376 N
[2004 : 2 M]

8.7 The thickness of the laminar boundary layer on a flat plate at a point A is 2 cm and at a point B, 1 m downstream of A, is 3 cm. What is the distance of A from the leading edge of the plate?
(a) 0.50 m (b) 0.80 m
(c) 1.00 m (d) 1.25 m [2006 : 2 M]

Linked Answer Questions 8.8 and 8.9 :
An automobile with projected area 2.6 m^2 is running on

a road with a speed of 120 km per hour. The mass density and the kinematic viscosity of air are 1.2 kg/m^3 and 1.5×10^{-5} m^2/s, respectively. The drag coefficient is 0.30.

8.8 The drag force on the automobile is
(a) 620 N (b) 600 N
(c) 580 N (d) 520 N [2008 : 2 M]

8.9 The metric horse power required to overcome the drag force is
(a) 33.23 (b) 31.23
(c) 23.23 (d) 20.23 [2008 : 2 M]

8.10 Velocity distribution in a boundary layer is given by $\frac{u}{U_\infty} = \sin\left(\frac{\pi}{2}\frac{y}{\delta}\right)$, where u is the velocity at vertical coordinate y, U_∞ is the free stream velocity and δ is the boundary layer thickness. The values of U_∞ and δ are 0.3 m/s and 1.0 m, respectively. The velocity gradient $\left(\frac{\partial u}{\partial y}\right)$ (in s^{-1}, round off to two decimal places) at y= 0, is ________.
[2020 : 1 M, Set-II]

8.11 Two discrete spherical particles (P and Q) of equal mass density are independently released in water. Particle P and particle Q have diameters of 0.5 mm and 1.0 mm. respectively. Assume Stoke's law is valid.
The drag force on particle Q will be ________ times the drag force on particle P. (round off to the nearest integer).
[2022 : 2 M, Set-II]

8.12 Which of the following statements is/are TRUE?
(a) The thickness of a turbulent boundary layer on a flat plate kept parallel to the flow direction is proportional to the square root of the distance from the leading edge
(b) If the streamlines and equipotential lines of a source are interchanged with each other, the resulting flow will be a sink
(c) For a curved surface immersed in a stationary liquid, the vertical component of the force on the curved surface is equal to the weight of the liquid above it
(d) For flow through circular pipes, the momentum correction factor for laminar flow is larger than that for turbulent flow
[2023 : 2 M, Set-II]

■■■■

Answers Boundary Layer Theory, Drag and Lift

8.1 (d)	8.2 (c)	8.3 (d)	8.4 (b)	8.5 (b)	8.6 (c)	8.7 (b)
8.8 (d)	8.9 (c)	8.10 (Sol.)	8.11 (Sol.)	8.15 (c, d)		

Explanations Boundary Layer Theory, Drag and Lift

8.1 (d)
Drag force is the component of dynamic fluid force in the direction of flow whereas the lift force is normal to the direction of flow.

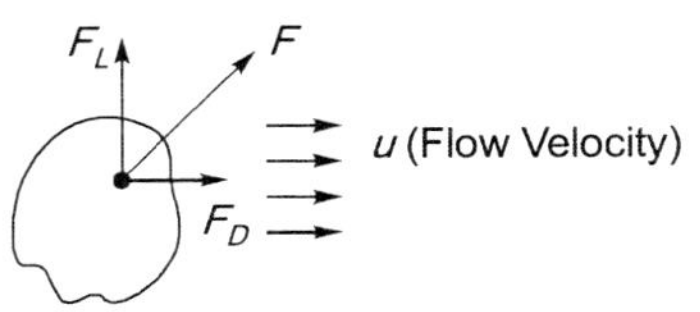

8.2 (c)

Given, $\frac{u}{u_m} = 1.5\eta;\ \eta = \frac{y}{\delta}$

$$\therefore \quad u = 1.5\, u_m \times \frac{y}{\delta}$$

$$\frac{du}{dy} = 1.5\frac{u_m}{\delta}$$

Now, $$\tau = \mu\frac{du}{dy} = 1.5\,\mu\frac{u_m}{\delta}$$

$$\therefore \quad k = 1.5$$

8.3 (d)

1. Laminar boundary layer thickness

$$\delta = \frac{5x}{\sqrt{Re_x}} \qquad \text{i.e., } \delta \propto \sqrt{x}$$

δ increases in the direction of flow.

2. Shear stress at the plate

$$\tau_{o,x} = \frac{0.332}{\sqrt{Re_x}}\rho U_\infty^2$$

$$\tau_{o,x} \propto \frac{1}{\sqrt{x}}$$

Shear stress decreases in the direction of flow.

3. Pressure gradient $\frac{dP}{dx} = 0$ in the direction of flow.

8.4 (b)

Method-I

The solid sphere will be acted upon by three forces viz. (1) drag force (F_D), (2) force of buoyancy (F_B) and (3) self weight of sphere (W)

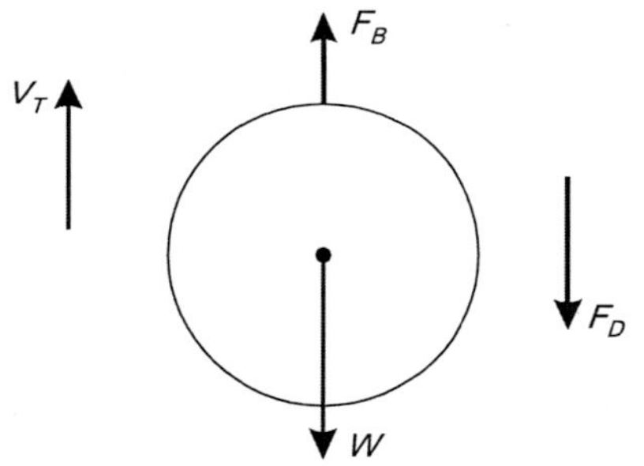

$$W = \rho_s Vg, \quad F_B = \rho_0 Vg$$

$$F_D = C_D A \frac{\rho_0 V^2}{2}$$

Reynolds Number,

$$Re = \frac{\rho_0 V_T D}{\mu} = \frac{900 \times 1 \times 10^{-2} \times 6 \times 10^{-3}}{0.7}$$

$$= 0.07714 < 0.2 \text{ (ok)}$$

$$\therefore \quad C_D = \frac{24}{Re} = \frac{24}{0.07714} = 311.12$$

Now, from the free body diagram, we have,

$$W + F_D = F_B$$

$$\Rightarrow \quad W = F_B - F_D$$

$$\Rightarrow \quad \rho_s V_g = \rho_0 V_g - C_D A \frac{\rho_0 V_T^2}{2}$$

$$\Rightarrow \quad w_s V = \rho_0 V_g - C_D \frac{A\rho_0 V_T^2}{2}$$

$$\Rightarrow \quad w_s = \rho_0 g - \frac{C_D A \rho_0 V_T^2}{2V}$$

$$\Rightarrow \quad w_s = 900 \times 9.81 - \frac{\left[\begin{array}{l} 6 \times 311.12 \times \pi \times (6 \times 10^{-3})^2 \\ \times 900 \times (1 \times 10^{-2})^2 \end{array}\right]}{4 \times \pi \times (6 \times 10^{-3})^3}$$

$$\Rightarrow \quad w_s = 5328.9 \text{ N/m}^3 = 5.3 \text{ kN/m}^3$$

Method-II

$$D = 6 \text{ mm} = 6 \times 10^{-3} \text{ m}$$

$$f = 900 \text{ kg/m}^3, \quad \mu = 0.7 \text{ kg/ms}$$

Terminal velocity,

$$V_T = 1 \text{ cm/s} = 0.01 \text{ m/s}$$

$$V_T = \frac{gd^2}{18\mu}(\rho_{air} - \rho_{material})$$

$$0.01 = \frac{9.81 \times \left(6 \times 10^{-3}\right)^2}{18 \times 0.7}(900 - \rho_m)$$

$$356.77 = (900 - \rho_m)$$

$$\rho_m = 543.22 \text{ kg/m}^3$$

$$\rho_m g = 543.22 \times 9.81 \text{ N/m}^3$$

$$W_s = 5.32 \text{ kN/m}^3$$

8.5 (b)

The lift force will be equal to self weight of the aircraft for equilibrium,

$$F_L = W$$

$$\Rightarrow \quad C_L A_L \frac{\rho V^2}{2} = mg$$

$$\Rightarrow \quad \frac{0.4 \times A_L \times 1.2 \times \left(\frac{200 \times 10^3}{60 \times 60}\right)^2}{2} = 800 \times 9.81$$

$$\Rightarrow \quad A_L = 10.6 \text{ m}^2$$

8.6 (c)

Submerged weight of the plate = Drag force on the plate

$$R_L = \frac{\rho VL}{\mu} = \frac{1000 \times 0.12 \times 0.7}{10^{-3}}$$

$$= 84 \times 10^3$$

The critical Reynolds number for flat plate is 5×10^5.

As R_L < critical Reynolds number thus laminar boundary layer will grow through out the length of plate.

$$\therefore \quad C_D = \frac{1.328}{\sqrt{R_L}} = \frac{1.328}{\sqrt{84 \times 10^3}} = 4.58 \times 10^{-3}$$

$\therefore$ Drag force (for both the sides of plate)

$$F_D = 2C_D A\left(\frac{\rho V^2}{2}\right)$$
$$= 4.58 \times 10^{-3} \times 0.5 \times 0.7 \times 1000 \times (0.12)^2$$
$$= 0.0231 \text{ N}$$

Note:

- *Since the boundary layer will form on both sides of plate so total drag force is twice of one side.*
- *Under equilibrium condition (when plate is settling with terminal velocity) drag force is equal to submerged weight of the plate.*

8.7 (b)

Thickness (δ) of laminar boundary layer at a distance 'x' from theleading edge is given by

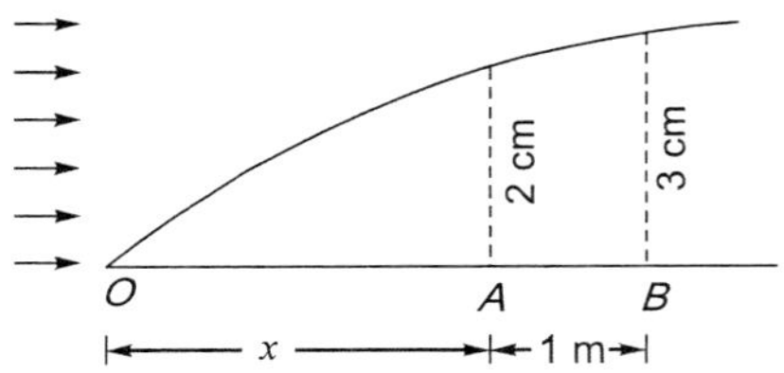

$$\delta = \frac{5x}{\sqrt{\text{Re}_x}}$$

$\Rightarrow \quad \delta_A \propto \sqrt{x} \qquad \left[\because \text{Re}_x = \frac{\rho V x}{\mu}\right]$

$\therefore \quad \delta_A \propto \sqrt{x}\,; \qquad \delta_B \propto \sqrt{x+1}$

$$\frac{\delta_A}{\delta_B} = \sqrt{\frac{x}{x+1}}$$

$$\sqrt{\frac{x}{x+1}} = \frac{2}{3}; \qquad \frac{x}{x+1} = \frac{4}{9}$$

$$9x = 4x + 4; \quad x = 0.8 \text{ m}$$

8.8 (d)

The drag force on the automobile may be given as,

$$F_D = C_D A \times \frac{\rho V^2}{2}$$

Here, $C_D = 0.30$, $A = 2.6 \text{ m}^2$, $\rho = 1.2 \text{ kg/m}^3$, $V = 120$ kmph

$$\therefore \quad F_D = 0.30 \times 2.6 \times \frac{1.2 \times (120 \times 1000)^2}{2 \times (60 \times 60)^2}$$
$$= 520 \text{ N}$$

8.9 (c)

Power required = Drag force × Velocity

$$= 520 \times \frac{120 \times 1000}{60 \times 60} = 17333.33 \text{ watt}$$

(1 mhp = 735.5 watt)

$$= \frac{17333.33}{735.5} \text{ metric horse power}$$
$$= 23.567 \text{ metric horse power}$$

Note:

1 Metric horse power = 735.5 Watt

1 Horse Power = 745.7 Watt

8.10 Sol.

Given : $\quad \frac{u}{u_\infty} = \sin\left(\frac{\pi}{2}\cdot\frac{y}{\delta}\right)$

$u_\infty = 0.3$ m/s

$\delta = 1$ m

$$\frac{du}{dy} = \frac{d}{dy}\left[u_\infty \cdot \sin\left(\frac{\pi}{2}\cdot\frac{y}{\delta}\right)\right]$$
$$= \frac{u_\infty \cdot \pi}{2\delta}\cos\left(\frac{\pi}{2}\cdot\frac{y}{\delta}\right)$$

At $y = 0$ and $\delta = 1$ m

$$\left.\frac{du}{dy}\right|_{y=0} = \frac{0.3\pi}{2(1)}\cos\left(\frac{\pi}{2}\cdot\frac{0}{1}\right) = 0.47 \text{ s}^{-1}$$

8.11 Sol.

ρ = density of water

ρ_b = density of particle

Diameter of P, D_1 = 0.5 mm

Diameter of Q, D_2 = 1 mm

$$\therefore \quad \frac{F_{D2}}{F_{D1}} = \frac{\text{Drag force for particle } Q}{\text{Drag force for particle } P} = ?$$

Let's take basic case of Stoke's law i.e. a particle of diameter, d moving downward in fluid.

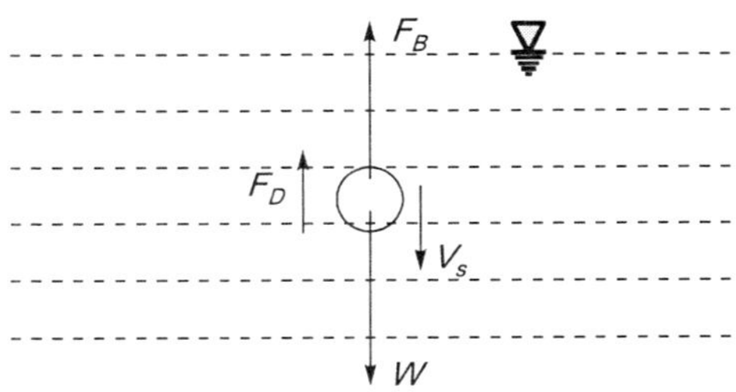

F_B = Buoyancy force, F_D = Drag force

V_s = Setting velocity, W = Weight of particle

$$\therefore \quad W = F_B + F_D$$

$$\Rightarrow \quad mg - F_B = 3\pi\mu D V_s$$

$$\Rightarrow (\rho_b - \rho)V_{sphere} = 3\pi\mu D V_s$$

$$\left(\because \text{Volume of sphere} = \frac{4}{3}\pi r^3\right)$$

$$\Rightarrow \quad V_s = \frac{(\rho_b - \rho)D^2 g}{18\mu}$$

$$\Rightarrow \quad V_s \propto D^2$$

Now,
$$\frac{F_{D_2}}{F_{D_1}} = \frac{3\pi\mu D_2 V_{s_2}}{3\pi\mu D_1 V_{s_1}}$$

$$\Rightarrow \quad \frac{F_{D_2}}{F_{D_1}} = \frac{D_2}{D_1} \times \frac{D_2^2}{D_1^2}$$

$$\Rightarrow \quad \frac{F_{D_2}}{F_{D_1}} = \left(\frac{D_2}{D_1}\right)^3 = \left(\frac{1}{0.5}\right)^3$$

$$\therefore \quad \frac{F_{D_2}}{F_{D_1}} = \frac{\text{Drag force for particle } Q}{\text{Drag force for particle } P} = 8$$

8.12 (c, d)

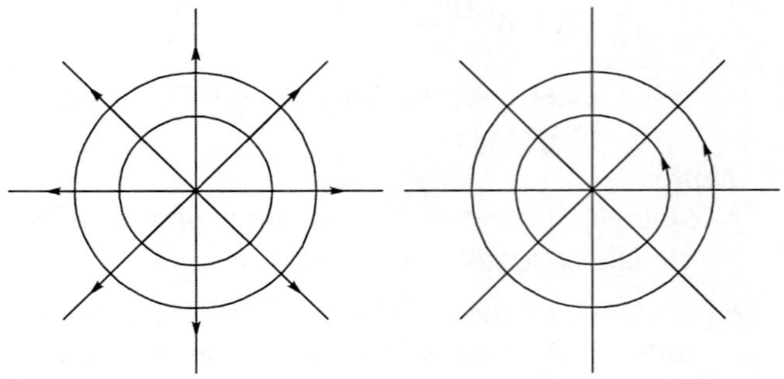

Thickness of boundary layer in turbulent flow:

$$\frac{\delta}{x} = \frac{0.376}{Re_x^{1/5}} = \frac{0.376}{\left(\frac{\rho u_\infty x}{\mu}\right)^{1/5}}$$

$$\Rightarrow \quad \delta = \frac{0.376 x^{4/5}}{\left(\frac{\rho u_\infty}{\mu}\right)^{1/5}}$$

$$\Rightarrow \quad \delta \propto x^{4/5}$$

■■■■

9 Turbulent Flow

9.1 The Prandtl mixing length for turbulent flow through pipes is
(a) dependent on shear stress at the wall
(b) a universal constant
(c) zero at the pipe wall
(d) independent of radial distance from pipe axis
[1996 : 1 M]

9.2 The flow of water (mass density = 1000 kg/m^3 and kinematic viscosity = 10^{-6} m^2/s) in a commercial pipe, having equivalent roughness k_s as 0.12 mm, yields an average shear stress at the pipe boundary = 600 N/m^2. The value of k_s/δ' (δ' being the thickness of laminar sub-layer) for this pipe is
(a) 0.25 (b) 0.50
(c) 6.0 (d) 8.0 [2008 : 2 M]

9.3 A rough pipe of 0.5 m diameter, 300 m length and roughness height of 0.25 mm, carries water (kinematic viscosity = 0.9×10^{-6} m^2/s) with velocity of 3 m/s. Friction factor (f) for laminar flow is given by f = 64/Re, and for turbulent flow it is given by $\frac{1}{\sqrt{f}} = 2\log_{10}\left(\frac{r}{k}\right) + 1.74$, where, Re = Reynolds number, r = radius of pipe, k = roughness height and g = 9.81 m/s^2. The head loss (in m, up to three decimal places) in the pipe due to friction is ________.
[2018 : 2 M, Set-II]

■■■■

Answers Turbulent Flow

9.1 (c) 9.2 (d) 9.3 (4.594)

Explanations Turbulent Flow

9.1 (c)

As per Prandtl, $l = ky$
where, l is mixing length
y is the distance from the pipe wall; k is karman constant = 0.4.

9.2 (d)

We know that,

$$\delta' = \frac{11.6\,\nu}{V^*} \quad \text{But} \quad V^* = \sqrt{\frac{\tau_0}{\rho}}$$

$$\therefore \quad \delta' = \frac{11.6\,\nu}{\sqrt{\tau_0/\rho}}$$

$$\Rightarrow \quad \delta' = 11.6 \times 10^{-6} \times \sqrt{\frac{1000}{600}} \times 1000$$

$$\Rightarrow \quad \delta' = 0.015 \text{ mm}$$

$$\therefore \quad \frac{k_s}{\delta'} = \frac{0.12}{0.015} = 8.0$$

POINTS TO REMEMBER

- For $\frac{k_s}{\delta'} < 0.25$ pipe is hydrodynamically smooth

$0.25 < \frac{k_s}{\delta'} < 6$ in transition state

$\frac{k_s}{\delta'} > 6$ pipe is hydrodynamically rough

9.3 Sol.

$$Re = \frac{\rho \cdot V.D}{\mu} = \frac{V \cdot D}{\nu}$$

$$= \frac{3 \times (0.5)}{0.9 \times 10^{-6}} = 1.67 \times 10^6$$

Means Re > 2000, turbulent flow.

So, $$\frac{1}{\sqrt{f}} = 2\log_{10}\frac{D}{2k_s} + 1.74$$

$$\frac{1}{\sqrt{f}} = 2\log_{10}\frac{0.5}{2 \times 0.25 \times 10^{-3}} + 1.74$$

$$f = 0.01669$$

$$h_f = \frac{f \cdot L \cdot V^2}{2gD} = \frac{(0.01669)(300)(3)^2}{2 \times 9.81 \times 0.5}$$

$$= 4.594 \text{ m}$$

■■■■

10 Dimensional Analysis

10.1 A 1 : 30 model of an ogee spillway crest records an acceleration of 1.3 m/s^2 at a certain location. The homologous value of the acceleration in the prototype in m/s^2, is

(a) 0.043 (b) 0.237
(c) 1.300 (d) 7.120 [1995 : 1 M]

10.2 The repeating variables in dimensional analysis should

(a) include the dependent variable
(b) have amongst themselves all the basic dimensions
(c) be derivable from one another
(d) exclude one of the basic dimensions

[1996 : 1 M]

10.3 Both Reynold's and Froude numbers assume significance in one of the following examples:

(a) Motion of submarine at large depths
(b) Motion of ship in deep seas
(c) Cruising of a missile in air
(d) Flow over spillways [1997 : 1 M]

10.4 In a 1/50 model of a spillway, the discharge was measured to be 0.3 m^3/sec. The corresponding prototype discharge in m^3/sec is

(a) 2.0 (b) 15.0
(c) 106.0 (d) 5303.0 [2001 : 2 M]

10.5 A laboratory model of a river is built to a geometric scale of 1 : 100. The fluid used in the model is oil of mass density 900 kg/m^3. The highest flood in the river is 10,000 m^3/s. The corresponding discharge in the model shall be

(a) 0.095 m^3/s (b) 0.100 m^3/s
(c) 0.105 m^3/s (d) 10.5 m^3/s

[2003 : 1 M]

10.6 The height of a hydraulic jump in the stilling pool of 1 : 25 scale model was observed to be 10 cm. The corresponding prototype height of the jump is

(a) not determinable from the data given
(b) 2.5 m
(c) 0.5 m
(d) 0.1 m [2004 : 2 M]

10.7 The flow of glycerin (kinematic viscosity $\nu = 5 \times 10^{-4}$ m^2/s) in an open channel is to be modeled in a laboratory flume using water ($\nu = 10^{-6}$ m^2/s) as the flowing fluid. If both gravity and viscosity are important, what should be the length scale (i.e., ratio of prototype to model dimensions) for maintaining dynamic similarity?

(a) 1 (b) 22
(c) 63 (d) 500 [2006 : 2 M]

10.8 A 1 : 50 scale model of a spillway is to be tested in the laboratory. The discharge in the prototype is 1000 m^3/s. The discharge to be maintained in the model test is

(a) 0.057 m^3/s (b) 0.08 m^3/s
(c) 0.57 m^3/s (d) 5.7 m^3/s

[2007 : 2 M]

10.9 A river reach of 2.0 km long with maximum flood discharge of 10000 m^3/s is to be physically modeled in the laboratory where maximum available discharge is 0.20 m^3/s. For a geometrically similar model based on equality of Froude number, the length of the river reach (m) in the model is

(a) 26.4 (b) 25.0
(c) 20.5 (d) 18.0 [2008 : 2 M]

10.10 **Group-I** contains dimensionless parameter and **Group-II** contains ratio.

Group-I
P. Mach number
Q. Reynold number
R. Weber number
S. Froude number

Group-II
1. Ratio of inertial force and gravity force.
2. Ratio of fluid velocity and velocity of sound.
3. Ratio of inertial force and viscous force.
4. Ratio of inertial force and surface tension force.

Correct match of the dimensionless parameter in Group-I with Group-II is

	P	Q	R	S
(a)	3	2	4	1
(b)	3	4	2	1
(c)	2	3	4	1
(d)	1	3	2	4

[2013 : 1 M]

10.11 The drag force, F_D, on a sphere kept in a uniform flow field depends on the diameter of the sphere, D; flow velocity, V; fluid density, ρ; and dynamic viscosity, μ. Which of the following options represents the non-dimensional parameters which could be used to analyze this problem?

(a) $\frac{F_D}{VD}$ and $\frac{\mu}{\rho VD}$

(b) $\frac{F_D}{\rho VD^2}$ and $\frac{\rho VD}{\mu}$

(c) $\frac{F_D}{\rho V^2 D^2}$ and $\frac{\rho VD}{\mu}$

(d) $\frac{F_D}{\rho V^3 D^3}$ and $\frac{\mu}{\rho VD}$ [2015 : 2 M, Set-I]

10.12 The relationship between the length scale ratio (L_r) and the velocity scale ratio (V_r) in hydraulic models, in which Froude dynamic similarity is maintained, is

(a) $V_r = L_r$ (b) $L_r = \sqrt{V_r}$

(c) $V_r = L_r^{1.5}$ (d) $V_r = \sqrt{L_r}$

[2015 : 1 M, Set-II]

10.13 A 1:50 model of a spillway is to be tested in the laboratory. The discharge in the prototype spillway is 1000 m^3/s. The corresponding discharge (in m^3/s up to two decimal places) to be maintained in the model, neglecting variation in acceleration due to gravity, is ________. [2018 : 1 M, Set-I]

10.14 In a laboratory, a flow experiment is performed over a hydraulic structure. The measured values of discharge and velocity are 0.05 m^3/s and 0.25 m/s, respectively. If the full scale structure (30 times bigger) is subjected to a discharge of 270 m^3/s, then the time scale (model to full scale) value (up to two decimal places) is ______

[2018 : 2 M, Set-I]

10.15 With respect to fluid flow, match the following in Column X with Column Y:

Column X	Column Y
P. Viscosity	I. Mach number
Q. Gravity	II. Reynolds number
R. Compressibility	III. Euler number
S. Pressure	IV. Froude number

Which one of the following combinations is correct?

(a) P – II, Q – IV, R – I, S – III
(b) P – III, Q – IV, R – I, S – II
(c) P – IV, Q – II, R – I, S – III
(d) P – II, Q – IV, R – III, S – I

[2022 : 1 M, Set-I]

Answers Dimensional Analysis

10.1	(c)	10.2	(b)	10.3	(b)	10.4	(d)	10.5	(b)	10.6	(b)	10.7	(c)
10.8	(a)	10.9	(a)	10.10	(c)	10.11	(c)	10.12	(d)	10.13	(0.06)	10.14	(0.182)
10.15	(a)												

Explanations Dimensional Analysis

10.1 (c)

This problem is similar to problem 10.1

For a spillway, we use **Froude Model Law,**

$$\therefore \quad \frac{a_m}{a_p} = 1, \ \frac{1.3}{a_p} = 1, \ a_p = 1.3 \text{ m/s}^2$$

10.2 (b)

Folloiwng points are needed in selecting the repeating variables.

(i) They must have amongst themselves all the basic dimensions involved in the problem.

(ii) The dependent variable must not be chosen as a repeating variable.

(iii) Usually a length parameter (such as a diameter D or head over a weir, H); a typical velocity V and the fluid density ρ are convenient set of repeating variables.

10.3 (b)

Ship is a partially submerged body. Submerged part is influenced by viscous forces and hence by Reynold's law. Open portion exposed to atmosphere is influenced by gravitational force and hence by **Froude's law**.

Note: *For motion of submarine and cruising of a missile, Reynolds model is applicable.*

For flow over spillway, Froude model is applicable.

10.4 (d)

$$Q_r = L_r^{2.5}$$

$$\frac{Q_m}{Q_p} = \left(\frac{1}{50}\right)^{2.5}$$

$$\frac{0.3}{Q_p} = \left(\frac{1}{50}\right)^{2.5}$$

$$\therefore \quad Q_p = 0.3 \times (50)^{2.5} = 5303.3 \text{ m}^3/\text{s}$$

10.5 (b)

$$Q_r = L_r^{2.5}$$

$$\frac{Q_m}{Q_p} = \left(\frac{1}{100}\right)^{2.5}$$

$$\frac{Q_m}{10000} = \left(\frac{1}{100}\right)^{2.5}$$

$$Q_m = 10000 \times \left(\frac{1}{100}\right)^{2.5} = 0.1 \text{ m}^3/\text{s}$$

10.6 (b)

Froude law is also used in case of free surface flow such as flow over weir, spillway, open channel flow, hydraulic jump, flow of liquid through orifice etc.

$$L_r = \frac{L_m}{L_p} = \frac{1}{25}$$

$$\Rightarrow \quad L_p = L_m \times 25$$

$$\Rightarrow \quad L_p = 10 \times 25$$

$$= 250 \text{ cm} = 2.5 \text{ m}$$

10.7 (c)

Given that both gravity and viscosity are important, we will have to satisfy both Froude & Reynold's model simultaneously.

From Froude's law,

$$(F_r)_m = (F_r)_p$$

$$\left(\frac{V}{\sqrt{Lg}}\right)_m = \left(\frac{V}{\sqrt{Lg}}\right)_p \qquad (g_m = g_p)$$

$$\frac{V_r}{\sqrt{L_r}} = 1 \qquad \text{...(i)}$$

From Reynold's law

$$\frac{V_r L_r}{V_r} = 1 \qquad \text{...(ii)}$$

Divide eq. (ii) and (i)

$$\frac{\sqrt{L_r}\, L_r}{V_r} = 1$$

$$V_r = L_r^{3/2}$$

$$\text{or} \quad L_r = (V_r)^{2/3}$$

$$= \left(\frac{V_p}{V_m}\right)^{2/3} = \left(\frac{5 \times 10^{-4}}{10^{-6}}\right)^{2/3} = (500)^{2/3}$$

$$L_r = 63$$

10.8 (a)

Froude model law will be applicable in this case.

$$Q_r = (L_r)^{2.5}$$

$$\frac{Q_m}{Q_p} = \left(\frac{L_m}{L_p}\right)^{5/2}$$

$$\Rightarrow \quad Q_m = 1000 \times \left(\frac{1}{50}\right)^{2.5} = 0.057 \text{ m}^3/\text{sec}$$

10.9 (a)

$$Q_r = (L_r)^{2.5}$$

$$\Rightarrow \quad \frac{Q_m}{Q_p} = (L_r)^{5/2}$$

$$\Rightarrow \quad \frac{0.20}{10000} = \left(\frac{L_m}{2 \times 1000}\right)^{5/2}$$

$$\Rightarrow \quad L_m = 26.4 \text{ m}$$

10.10 (c)

1. Mach number (M)

$$M = \sqrt{\frac{\text{Inertial force}}{\text{Elasticity force or compressibility force}}}$$

$$= \sqrt{\frac{\rho L^2 V^2}{k L^2}} = \frac{V}{\sqrt{k/\rho}} = \frac{V}{C}$$

C = Velocity of sound in the medium.

$$= \left\{\frac{V}{k/\rho}\right\}$$

2. Reynold's number (Re)

$$\text{Re} = \frac{\text{Inertial force}}{\text{Viscous force}} = \frac{\rho L^2 V^2}{\mu VL} = \frac{\rho VL}{\mu}$$

3. Weber number (W)

$$\text{W} = \sqrt{\frac{\text{Inertial force}}{\text{Surface tension force}}} = \sqrt{\frac{\rho L^2 V^2}{\sigma L}} = \frac{V}{\sqrt{\sigma/\rho L}}$$

4. Froude number (Fr)

$$\text{Fr} = \sqrt{\frac{\text{Inertial force}}{\text{Gravity force}}} = \sqrt{\frac{\rho V^2 L^2}{\rho L^3 g}} = \frac{V}{\sqrt{gL}}$$

10.11 (c)

For sphere, $A = \frac{\pi}{4}D^2$

$$F_D = C_D \times \frac{1}{2} \times \rho \times A \times V^2$$

or,
$$F_D = C_D \times \frac{1}{2} \times \rho \times \left(\frac{\pi}{4}D^2\right) \times V^2$$

or,
$$\frac{F_D}{\rho D^2 V^2} = \frac{1}{2} \times \left(\frac{\pi}{4}\right) \times C_D$$

Hence, right hand side is dimensionless, so

$\frac{F_D}{\rho D^2 V^2}$ should also be a dimensionless parameter.

Option (c) $\frac{F_D}{\rho V^2 D^2}, \frac{\rho V D}{\mu}$

10.12 (d)

As per **Froude Law**

$$(F_r)_m = (F_r)_p$$

$$\left(\frac{V}{\sqrt{Lg}}\right)_m = \left(\frac{V}{\sqrt{Lg}}\right)_p \qquad (g_m = g_p)$$

$$\Rightarrow \quad V_r = \sqrt{L_r}$$

10.13 Sol.

Froude law is valid

$$Q_r = L_r^{2.5}$$

$$\frac{Q_m}{Q_p} = \left(\frac{1}{50}\right)^{2.5}$$

$$\frac{Q_m}{1000} = \left(\frac{1}{50}\right)^{2.5}$$

$$Q_m = 0.0566 \text{ m}^3/\text{s}$$

So,
$$Q_m \simeq 0.06 \text{ m}^3/\text{s}$$

10.14 Sol.

Froude Law $\quad (Fr)_m = (Fr)_p$

$$\left(\frac{V}{\sqrt{Lg}}\right)_m = \left(\frac{V}{\sqrt{Lg}}\right)_p \qquad (g_m = g_p)$$

$$V_r = \sqrt{L_r}$$

or
$$\frac{L_r}{T_r} = \sqrt{L_r}$$

$$T_r = \sqrt{L_r}$$

$$T_r = \sqrt{\frac{1}{30}} = 0.1826$$

10.15 (a)

Reynolds number,

$$\text{R}_\text{e} = \frac{\text{Intertial forces}}{\text{Viscous forces}}$$

Euler number, $\text{E}_\text{u} = \sqrt{\frac{\text{Intertial forces}}{\text{Pressure forces}}}$

Froude number, $\text{F}_\text{r} = \sqrt{\frac{\text{Intertial forces}}{\text{Gravity forces}}}$

Mach number, $\text{M}_\text{a} = \sqrt{\frac{\text{Intertial forces}}{\text{Elastic forces}}}$

■■■■

11 Impact of Jets and Turbines

11.1 A hydraulic turbine develops a power of 10^4 metric horse power while running at a speed of 100 revolutions per minute, under a head of 40 m. Its specific speed is nearest to one of the following

(a) 100 (b) 628
(c) 523 (d) 314

[1997 : 1 M]

11.2 A nozzle discharging water under head H has an outlet area "a" and discharge coefficient $C_d = 1.0$. A vertical plate is acted upon by the fluid force F_j when held across the free jet and by the fluid force F_n when held against the nozzle to stop the flow.

The ratio $\frac{F_j}{F_n}$ is

(a) $\frac{1}{2}$ (b) 1
(c) $\sqrt{2}$ (d) 2 [1998 : 2 M]

11.3 A hydraulic turbine has a discharge of 5 m^3/s, when operating under a head of 20 m with a speed of 500 rpm. If it is to operate under a head of 15 m, for the same discharge, the rotational speed in rpm will approximately be

(a) 433 (b) 403
(c) 627 (d) 388 [1999 : 1 M]

11.4 A horizontal jet strikes a frictionless vertical plate (the plan view is shown in the figure). It is then divided into two parts, as shown in the figure. If the impact loss is neglected, what is the value of θ?

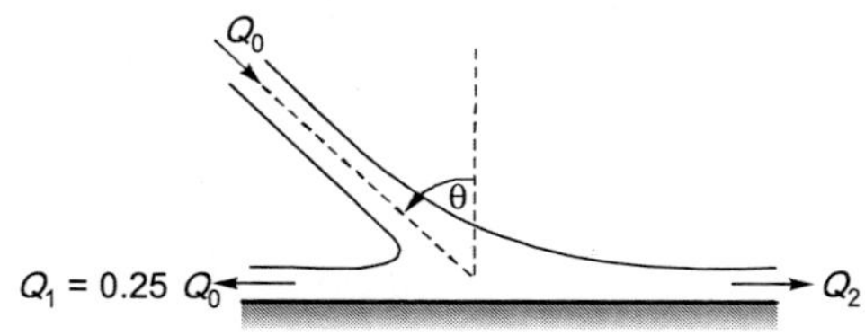

(a) 15° (b) 30°
(c) 45° (d) 60° [2003 : 2 M]

11.5 A tank and a deflector are placed on a frictionless trolley. The tank issues water jet (mass density of water = 1000 kg/m^3), which strikes the deflector and turns by 45°. If the velocity of jet leaving the deflector is 4 m/s and discharge is 0.1 m^3/s, the force recorded by the spring will be

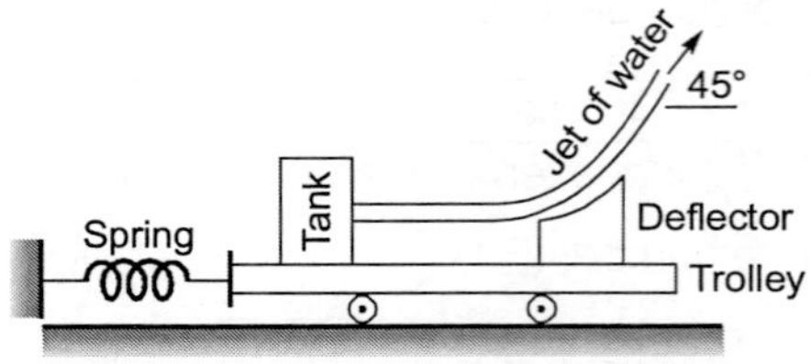

(a) 100 N (b) $100\sqrt{2}$ N
(c) 200 N (d) $200\sqrt{2}$ N

[2005 : 2 M]

11.6 A horizontal water jet with a velocity of 10 m/s and cross sectional area of 10 mm^2 strikes a flat plate held normal to the flow direction. The density of water is 1000 kg/m^3. The total force on the plate due to the jet is

(a) 100 N (b) 10 N
(c) 1 N (d) 0.1 N [2007 : 2 M]

11.7 A horizontal jet of water with its cross-sectional area of 0.0028 m^2 hits a fixed vertical plate with a velocity of 5 m/s. After impact the jet splits symmetrically in a plane parallel to the plane of the plate. The force of impact (in N) of the jet on the plate is

(a) 90 (b) 80
(c) 70 (d) 60

[2014 : 2 M, Set-I]

11.8 A horizontal nozzle of 30 mm diameter discharges a steady jet of water into the atmosphere at a rate of 15 litres per second. The diameter of inlet to the nozzle is 100 mm. The jet impinges normal to a flat stationary plate held close to the nozzle end. Neglecting air friction and considering the density of water as 1000 kg/m^3, the force exerted by the jet (in N) on the plate is ____.

[2014 : 2 M, Set-II]

11.9 A square plate is suspended vertically from one of its edges using a hinge support as shown in figure. A water jet of 20 mm diameter having a velocity of 10 m/s strikes the plate at its mid-point, at an angle of 30° with the vertical. Consider g as 9.81 m/s^2 and neglect the self-weight of the plate. The force F (expressed in N) required to keep the plate in its vertical position is_____.

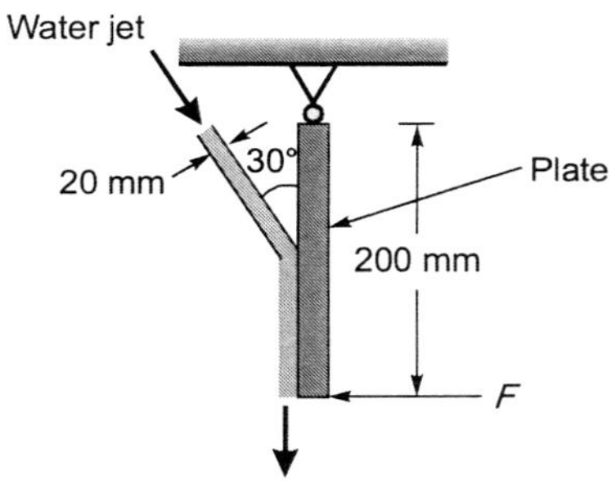

[2016 : 2 M, Set-II]

11.10 A penstock of 1 m diameter and 5 km length is used to supply water from a reservoir to an impulse turbine. A nozzle of 15 cm diameter is fixed at the end of the penstock. The elevation difference between the turbine and water level in the reservoir is 500 m. Consider the head loss due to friction as 5% of the velocity head available at the jet. Assume unit weight of water = 10 kN/m^3 and acceleration due to gravity (g) = 10 m/s^2. If the overall efficiency is 80%, power generated (expressed in kW and rounded to nearest integer) is________.

[2016 : 2 M, Set-II]

11.11 A jet of water having a velocity of 20 m/s strikes a series of plates fixed radially on a wheel revolving in the same direction as the jet at 15 m/s. What is the percentage efficiency of the plates? (round off to one decimal place)

(a) 37.5 (b) 66.7
(c) 50.0 (d) 88.9

[2023 : 2 M, Set-I]

Answers Impact of Jets and Turbines

11.1 (a)	11.2 (d)	11.3 (b)	11.4 (b)	11.5 (d)	11.6 (c)	11.7 (c)
11.8 (Sol.)	11.9 (Sol.)	11.10 (Sol.)	11.11 (a)			

Explanations Impact of Jets and Turbines

11.1 (a)

Specific speed is given by,

$$N_s = \frac{N\sqrt{P}}{H^{5/4}}$$

N = 100 revolutions per minute
H = 40 m
P = 10^4 metric horse power

$$\therefore \quad N_s = \frac{100 \times \sqrt{10^4}}{(40)^{5/4}}$$

$$= \frac{100 \times 100}{(40)^{1.25}} = 99.40 \cong 100$$

Note: Specific speed is not a unit less quantity. The dimensionless specific speed is given as

$$N_s = \frac{N\sqrt{P/\rho}}{g^{5/4}H^{5/4}}$$

11.2 (d)

For jet strikes at stationary vertical plate

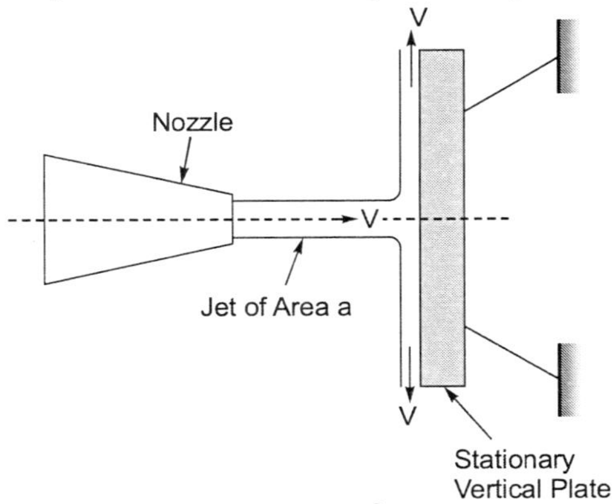

$$F_j = \rho \cdot a \cdot V^2$$

where, a = outlet area of jet, V = velocity of flow

For plate held against the nozzle to stop the flow

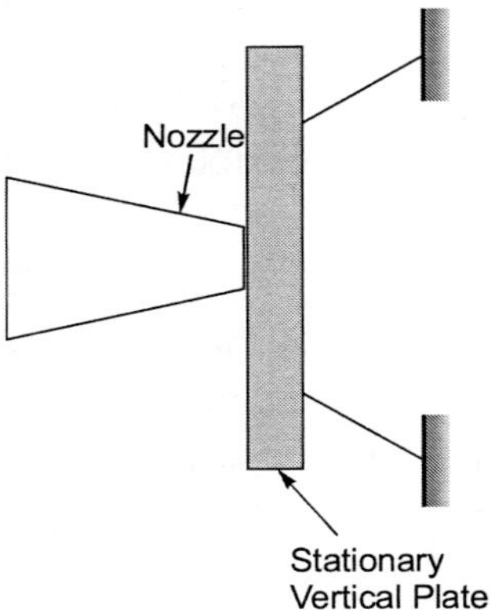

$$F_n = P \cdot a$$

$$= (\rho \cdot g \cdot H) \cdot a \qquad \left[H = \frac{V^2}{2g}\right]$$

$$= \rho \cdot g \cdot \frac{V^2}{2g} \cdot a = \frac{\rho \cdot a \cdot V^2}{2}$$

So, $$\frac{F_j}{F_n} = \frac{\rho \cdot a \cdot V^2}{\rho \cdot a \cdot V^2/2} = 2$$

11.3 (b)

$$Q = A_{flow} \times V_{flow} \qquad \left[A_{flow} \propto D^2, V_{flow} \propto \sqrt{H}\right]$$

$$Q \propto D^2 \cdot \sqrt{H}$$

$$Q \propto \frac{H}{N^2} \cdot \sqrt{H}$$

$$Q \propto \frac{H^{3/2}}{N^2}$$

$$\frac{QN^2}{H^{3/2}} = \text{constant}$$

Here Q is constant

$$\frac{N^2}{H^{3/2}} = \text{constant}$$

$$\frac{(500)^2}{20^{3/2}} = \frac{N^2}{(15)^{3/2}}$$

$$N = 402.96 \approx 403 \text{ rpm}$$

11.4 (b)

From continuity equation, we get,

$$Q_0 = Q_1 + Q_2$$

$$\Rightarrow \quad Q_0 = 0.25\,Q_0 + Q_2$$

$$\Rightarrow \quad Q_2 = 0.75\,Q_0$$

Since, the impact losses are neglected, the velocity will remain unchanged in the direction of Q_1 and Q_2.

i.e., $$V_0 = V_1 = V_2$$

Applying **impulse momentum equation**, we get,

$$\rho Q_0 V_0 \sin\theta = \rho Q_2 V_2 - \rho Q_1 V_1$$

$$\Rightarrow \quad Q_0 \sin\theta = 0.75\,Q_0 - 0.25\,Q_0$$

$$\Rightarrow \quad \sin\theta = 1/2$$

$$\Rightarrow \quad \theta = 30°$$

11.5 (d)

In such type of questions choose control volume in such a way that the element in which force is required is just outside the control volume.

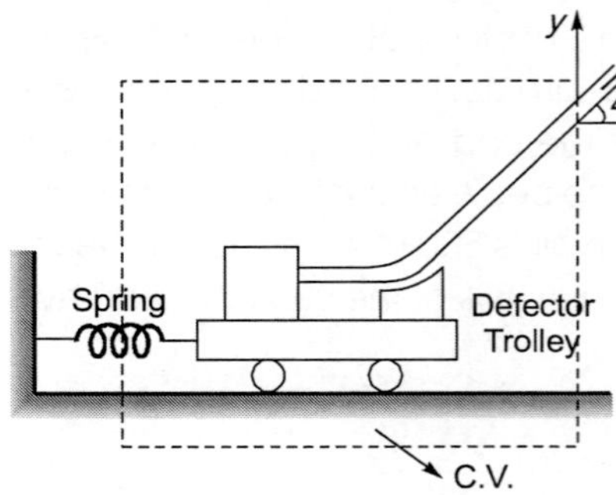

Now, apply impulse momentum thorem in x direction.

Net force in x-direction = (change in momentum)$_x$

$$F_{spring} = \rho Q(V\cos 45° - 0)$$

$$= 1000 \times 0.1 \times 4 \cos 45°$$

$$F_{spring} = 200\sqrt{2} \text{ N}$$

11.6 (c)

Method-I

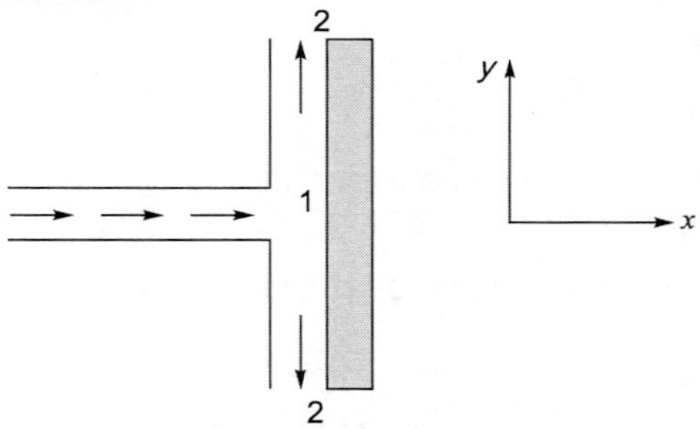

Velocity of jet, $V = 10$ m/s

Cross sectional area,

$$a = 10 \text{ mm}^2$$

Dicharge, $$Q = a \times V$$

$$= 10 \times 10^{-6} \times 10 = 10^{-4} \text{ m}^3/\text{s}$$

Force on the plate in x-direction

= Change in momentum in x-direction

$$= \rho Q(V_1 - V_2)$$

$$= 100 \times 10^{-4}(10 - 0) = 1 \text{ N}$$

Method-II

$$F = \rho.a.V^2$$
$$F = 1000 \times 10 \times 10^{-6} \times (10)^2$$
$$F = 1\text{ N}$$

11.7 (c)

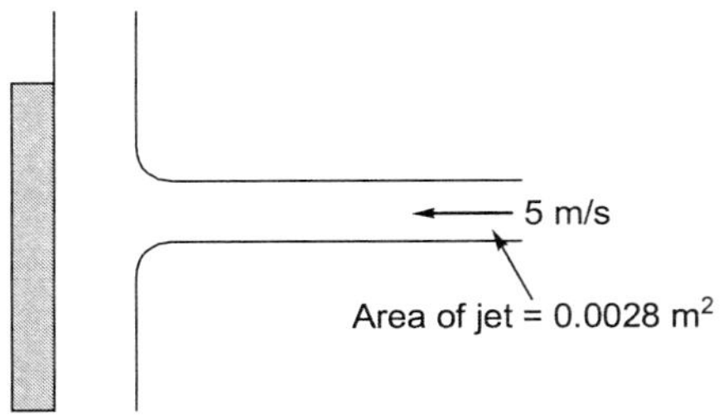

$$\text{Force on plate} = (\rho_w.a.V)V = \rho_w.a.V^2 = 1000 \times 0.0028 \times (5)^2 = 70\text{ N}$$

Note: *This problem can also be solved by impulse momentum theorem as done in previous problem.*

11.8 Sol.

$$\text{Force} = \rho a V^2$$
$$Q = aV = 15\text{ litre/sec (given)}$$
$$\Rightarrow \quad \text{Force} = \rho\frac{Q^2}{a} = 1000 \times \frac{15^2 \times 10^{-6}}{\frac{\pi}{4}\times 30^2} \times \frac{1}{10^{-6}}$$
$$= 1000 \times \frac{1}{\pi}\text{N} = 318.3\text{ N}$$

11.9 Sol.

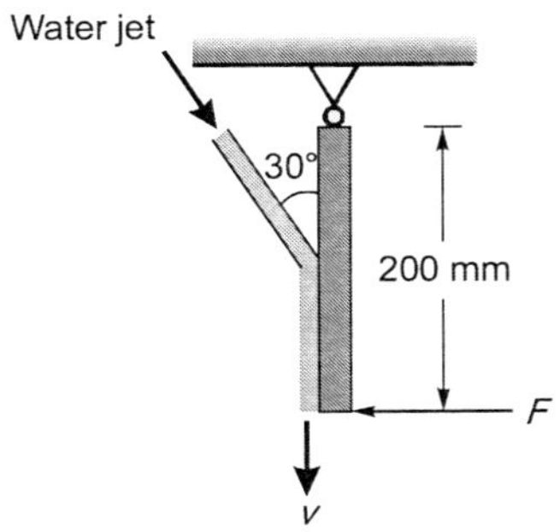

Force exerted by jet in x-direction,

$$F_x = \dot{m}[V\sin\theta - 0]$$
$$= \rho Q \times V\sin\theta = \rho AV \times V\sin\theta$$
$$= 1000 \times \frac{\pi}{4}(0.02)^2 \times (10)^2 \sin 30° = 15.7079\text{ N}$$

Taking moment about hinge,

$$F_x \times \frac{0.2}{2} = F \times 0.2$$
$$F = 7.85\text{ N}$$

11.10 Sol.

Apply **energy equaiton** at the free surface of reservior and exit of nozzle,

$$500 = \text{Head loss} + \frac{V_1^2}{2g}$$
$$500 = 0.05\frac{V_1^2}{2g} + \frac{V_1^2}{2g}$$
$$\sqrt{\frac{2\times 10\times 500}{1.05}} = V_1$$
$$V_1 = 97.59\text{ m/sec}$$

Water Power (W.P)

$$= \frac{1}{2}\dot{m}V_1^2$$
$$= \frac{1}{2}(10^3)\times\frac{\pi}{4}(0.15)^2(97.59)$$
$$= 8212.178\text{ kW}$$

Now, $$\eta_o = \frac{\text{Shaft power (S.P)}}{\text{W.P}}$$
$$0.8 = \frac{S.P}{8212.178}$$
$$S.P = 6569.74\text{ kW} \simeq 6570\text{ kW}$$

Note: *In this problem C_v is considered as 1.*

11.11 (a)

Given:

$v_1 = 20$ m/s

$u = 15$m/s

$[u_1 = u_2 = u]$

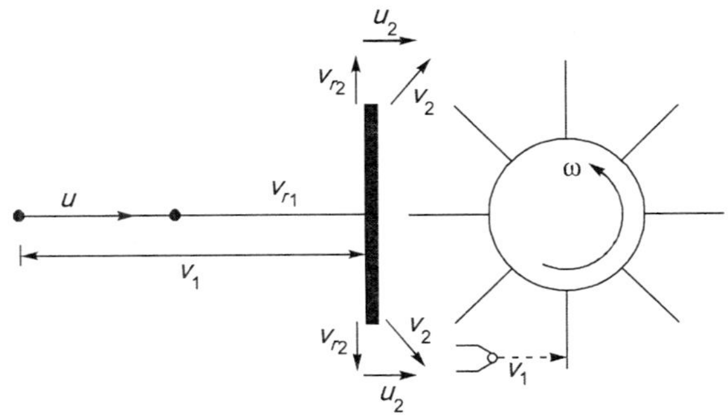

$$\text{Efficiency, } \eta = \frac{\text{Work done per second}}{\text{Imput water power}}$$
$$= \frac{F_x \cdot u}{\frac{1}{2}\dot{m}v_1^2} = \frac{\dot{m}[v_1 - u]\times u}{\frac{1}{2}\dot{m}v_1^2}$$
$$= \frac{[20-15]\times 15}{\frac{1}{2}\times(20)^2} = 0.375 = 37.5\%$$

12 Hydraulic Pumps

12.1 The expression for the specific speed of a pump
(a) does not include the diameter of the impeller
(b) yield larger values for radial pumps than for axial flow pump
(c) is necessarily nondimensional
(d) includes power as one of the variables

[1995 : 1 M]

12.2 At a rated capacity of 44 cumecs, a centrifugal pump develops 36 m of head when operating at 1450 rpm. Its specific speed is
(a) 654
(b) 509
(c) 700
(d) 90

[1996 : 1 M]

12.3 The comparison between pumps operating in series and in parallel is
(a) Pumps operating in series boost the discharge, whereas pumps operating in parallel boost the head
(b) Pumps operating in parallel boost the discharge, whereas pumps operating in series boost the head
(c) In both cases there would be a boost in discharge only
(d) In both case there would be a boost in head only

[1998 : 1 M]

12.4 If the pump head is 75 m, discharge is 0.464 m^3/s and the motor speed is 1440 rpm at rated condition, the specific speed of the pump is about
(a) 4
(b) 26
(c) 38
(d) 1440

[2000 : 1 M]

12.5 A pump can lift water at a discharge of 0.15 m^3/s to a head of 25 m. The critical cavitation number (σ_c) for the pump is found to be 0.144. The pump is to be installed at a location where the barometric pressure is 9.8 m of water and the vapour pressure of water is 0.30 m of water. The intake pipe friction loss is 0.40 m. Using the minimum value of NPSH (Net Positive Suction Head), the maximum allowable elevation above the sump water surface at which the pump can be located is
(a) 9.80 m
(b) 6.20 m
(c) 5.50 m
(d) None of these

[2002 : 2 M]

12.6 The allowable Net Positive Suction Head (NPSH) for a pump provided by the manufacturer for a flow of 0.05 m^3/s is 3.3 m. The temperature of water is 30°C (vapour pressure head absolute = 0.44 m), atmospheric pressure is 100 kPa absolute and the head loss from the reservoir to pump is 0.3 N-m/N. The maximum height of the pump above the suction reservoir is
(a) 10.19 m
(b) 6.89 m
(c) 6.15 m
(d) 2.86 m

[2004 : 2 M]

12.7 Identify the FALSE statement from the following. The specific speed of the pump increases with
(a) increase in shaft speed
(b) increase in discharge
(c) decrease in gravitational acceleration
(d) increase in head

[2006 : 1 M]

12.8 If a centrifugal pump has an impeller speed of N (in rpm), discharge Q (in m^3/s) and the total head H (in m), the expression for the specific speed N_s of the pump is given by

(a) $N_s = \dfrac{NQ^{0.5}}{H^{0.5}}$
(b) $N_s = \dfrac{NQ^{0.5}}{H}$
(c) $N_s = \dfrac{NQ^{0.5}}{H^{0.75}}$
(d) $N_s = \dfrac{NQ}{H^{0.75}}$

[2017 : 1 M, Set-II]

■■■■

Answers Hydraulic Pumps

12.1 (a) 12.2 (a) 12.3 (b) 12.4 (c) 12.5 (c) 12.6 (c) 12.7 (d) 12.8 (c)

Explanations Hydraulic Pumps

12.1 (a)

$$N_s = \frac{N\sqrt{Q}}{H^{3/4}}$$

12.2 (a)

Specific speed of a centrifugal pump,

$$N_s = \frac{N\sqrt{Q}}{(H)^{3/4}} = \frac{1450\sqrt{44}}{(36)^{3/4}} = 654$$

12.3 (b)

Pumps operating in series carries same discharge but different heads. Whereas pumps operating in parallel carries same head but different discharge.

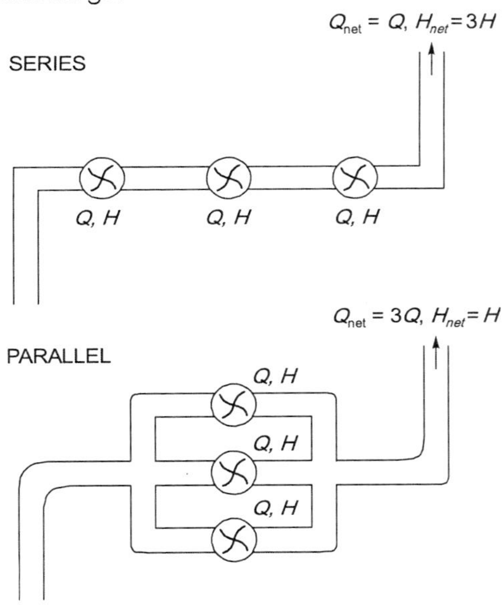

12.4 (c)

$H = 75$ m, $Q = 0.464$ m³/sec, $N = 1440$ rpm

Then $N_s = \frac{N\sqrt{Q}}{H^{3/4}} = \frac{1440\sqrt{0.464}}{(75)^{3/4}} \simeq 38$

12.5 (c)

$$\sigma_c = \frac{NPSH}{H_m}$$

$$\therefore \quad 0.144 = \frac{NPSH}{25}$$

$$\therefore \quad \text{NPSH} = 3.6 \text{ m}$$

$$\text{NPSH} = H_{atm} - H_v - h_s - h_{fs}$$

$$3.6 = 9.8 - 0.3 - h_s - 0.4$$

$$\therefore \quad h_s = 5.5 \text{ m}$$

12.6 (c)

The maximum suction lift,

$$h_s = \left(\frac{p_a - p_v}{\rho g}\right) - NPSH - h_{fs}$$

$$p_a = 100 \text{ kPa};$$

$$\rho g = 1000 \times 9.81 = 9810 \text{ N/m}^3$$

$$\frac{p_v}{\rho g} = 0.44 \text{ m}$$

$$\frac{p_a}{\rho g} = \frac{100 \times 10^3}{9810} = 10.19 \text{ m}$$

$$\therefore \quad h_s = 10.19 - 0.44 - 3.3 - 0.3 = 6.15 \text{ m}$$

12.7 (d)

The specific speed of a pump is given by,

$$N_S = \frac{N\sqrt{Q}}{(H_m)^{3/4}}$$

As the H_m increases, N_s decreases so the statement is false.

12.8 (c)

Specific speed of pump,

$$N_s = \frac{N\sqrt{Q}}{H^{3/4}} = \frac{NQ^{0.5}}{H^{0.75}}$$

■■■■

13 Open Channel Flow

13.1 A steady discharge of 1 cumec flows uniformly in a rectangular channel 1 m wide at a depth of 250 mm. The slope of the channel bed is
(a) adverse
(b) steep
(c) critical
(d) mild
[1995 : 1 M]

13.2 In deriving the equation for the hydraulic jump in the rectangular channel in terms of the conjugate depths and the initial Froude Number,
(a) continuity equation and energy equation are used
(b) continuity equation and momentum equation are used
(c) equations of continuity, momentum and energy are used
(d) gradually varied flow equation is used
[1995 : 1 M]

13.3 A section in open channel at which a fixed relationship exists between the discharge and the depth of flow is called as a ...... section.
[1995 : 1 M]

13.4 The percentage error in the computed discharge over a triangular notch corresponding to an error of 1% in the measurement of the head over the notch would be
(a) 1.0
(b) 1.5
(c) 2.0
(d) 2.5
[1995 : 1 M]

13.5 Flow at critical depth takes place in an open channel when
(a) for a given specific energy, discharge is maximum
(b) for a given discharge, specific energy is minimum
(c) discharge is minimum for a given specific force
(d) discharge is maximum for a given specific force
[1996 : 1 M]

13.6 The water surface profile resulting from flow underneath the gate in figure is

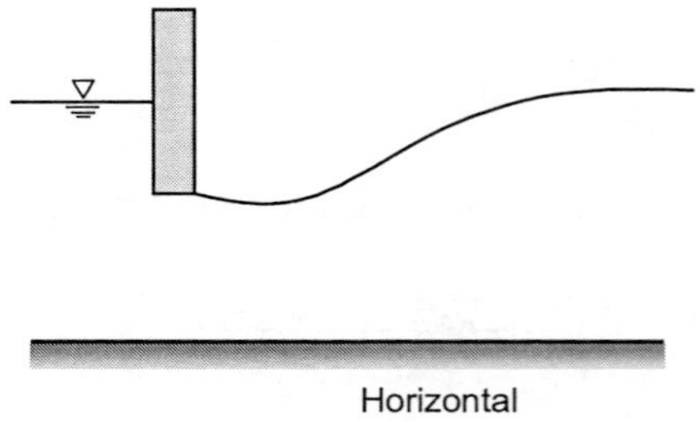

(a) H_3
(b) H_1
(c) H_2
(d) S_2
[1996 : 1 M]

13.7 For a 'best' symmetrical trapezoidal section of an open channel with a given area of section and side slopes, one of the following statements holds true
(a) Half the top width is equal to one of the side slope
(b) Half the top width plus the bottom width is equal to both the side slopes put together
(c) Water depth is equal to half bottom width
(d) Hydraulic mean depth is equal to half the top width
[1997 : 1 M]

13.8 The downstream end of long prismatic channel of mild slope ends in a pool created by a dam. The resulting non-uniform water surface profile can be described as one of the following
(a) M_3 profile ending in a hydraulic jump
(b) M_1 profile that lies above normal depth line
(c) M_2 profile that lies between critical and normal depth lines
(d) M_3 profile that lies between critical and normal depth lines
[1997 : 1 M]

13.9 The sequent depth ratio of a hydraulic jump in a rectangular horizontal channel is 10.30. The Froude Number at the beginning of the jump is
(a) 5.64
(b) 7.63
(c) 8.05
(d) 13.61
[1999 : 1 M]

13.10 Water flows at a depth of 0.1 m with a velocity of 6 m/s in a rectangular channel. The alternate depth is
(a) 0.30 m
(b) 0.40 m
(c) 0.86 m
(d) 0.81 m
[2000 : 1 M]

13.11 Water flows at a rate of 10 m^3/s in a rectangular channel 3 m wide. The critical depth of flow is
(a) 1.13 m (b) 2 m
(c) 1.45 m (d) 1.04 m **[2000 : 1 M]**

13.12 A circular sewer 2 m diameter has to carry a discharge of 2 m^3/s when flowing nearly full. What is the minimum required slope to initiate the flow? Assume manning's $n = 0.015$.
(a) 0.00023 (b) 0.000036
(c) 0.000091 (d) 0.000014
[2000 : 1 M]

13.13 A trapezoidal channel with bottom width of 3 m and side slope of 1V : 1.5 H carries a discharge of 8.0 m^3/sec with the flow depth of 1.5 m. The Froude Number of the flow is
(a) 0.066 (b) 0.132
(c) 0.265 (d) 0.528 **[2001 : 2 M]**

13.14 In a lined rectangular canal, the Froude number of incoming flow is 3.0. A hydraulic jump forms when it meets the pool of water. The depth of flow after the jump formation is 1.51 m. Froude number of flow after the hydraulic jump is
(a) 0.30 (b) 0.71
(c) 0.41 (d) None of these
[2002 : 2 M]

Common Data for Questions 13.15 and 13.16:
A very wide rectangular channel carries a discharge of 8 m^3/s per m width. The channel has a bed slope of 0.004 and Manning's roughness coefficient, $n = 0.015$. At a certain section of the channel, the flow depth is 1 m.

13.15 What Gradually Varied Flow profile exists at this section?
(a) M_2 (b) M_3
(c) S_2 (d) S_3 **[2003 : 2 M]**

13.16 At what distance from this section the flow depth will be 0.9 m? (Use the direct step method employing a single step.)
(a) 65 m downstream
(b) 50 m downstream
(c) 50 m upstream
(d) 65 m upstream **[2003 : 2 M]**

13.17 A hydraulic jump takes place in a triangular channel of vertex angle 90°, as shown in figure. The discharge is 1 m^3/s and the pre-jump depth is 0.5 m. What will be the post-jump depth? (Take $g = 9.81\ m/s^2$)

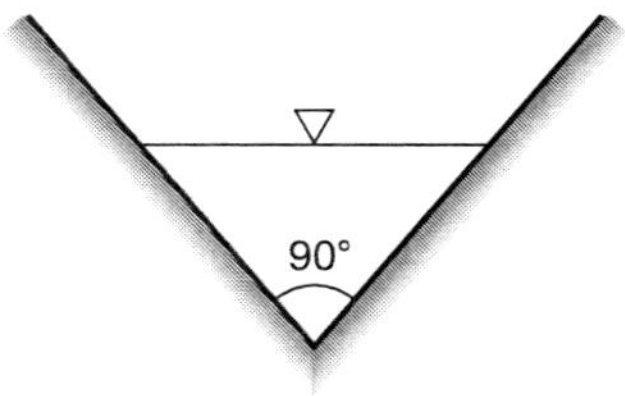

(a) 0.57 m (b) 0.91 m
(c) 1.02 m (d) 1.57 m **[2003 : 2 M]**

13.18 For a pipe of radius, r, flowing half full under the action of gravity, the hydraulic depth is
(a) r (b) $\frac{\pi r}{4}$
(c) $\frac{r}{2}$ (d) $0.379r$ **[2004 : 1 M]**

13.19 A wide channel is 1m deep and has a velocity of flow, V, as 2.13 m/s. If a disturbance is caused, an elementary wave can travel upstream with a velocity of
(a) 1.00 m/s (b) 2.13 m/s
(c) 3.13 m/s (d) 5.26 m/s
[2004 : 1 M]

13.20 A steady flow occurs in an open channel with lateral inflow of q m^3/s per unit width as shown in the figure. The mass conservation equation is

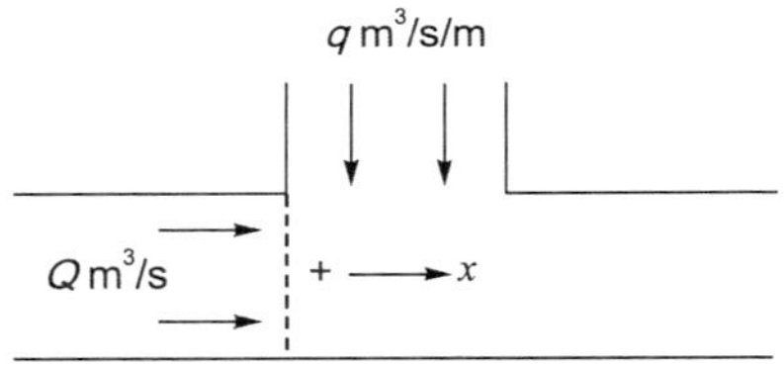

(a) $\frac{\partial q}{\partial x} = 0$ (b) $\frac{\partial Q}{\partial x} = 0$
(c) $\frac{\partial Q}{\partial x} - q = 0$ (d) $\frac{\partial Q}{\partial x} + q = 0$
[2004 : 2 M]

13.21 A steep wide rectangular channel takes off from a reservoir having an elevation of 101.2 m. At the entrance, the bottom elevation of the channel is 100 m. If the slope of the channel is increased by 4%, the discharge per unit length in the channel will be
(a) 2.24 m^2/s
(b) higher than 2.24 m^2/s by 4%
(c) higher than 2.24 m^2/s by 2%
(d) choked **[2004 : 2 M]**

13.22 A horizontal bed channel is followed by a steep bed channel as shown in the figure. The gradually-varied profiles over the horizontal and steep beds are

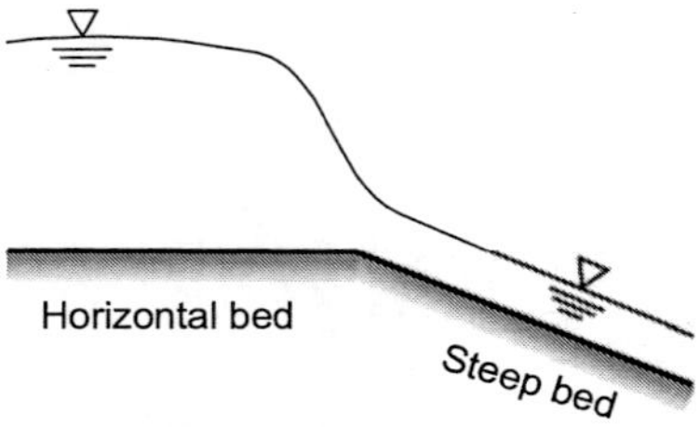

(a) H_2 and S_2 respectively
(b) H_2 and S_1 respectively
(c) H_3 and S_2 respectively
(d) H_3 and S_1 respectively [2005 : 1 M]

13.23 Critical depth at a section of a rectangular channel is 1.5 m. The specific energy at that section is
(a) 0.75 m (b) 1.0 m
(c) 1.5 m (d) 2.25 m [2005 : 2 M]

13.24 A partially open sluice gate discharges water into a rectangular channel. The tail water depth in the channel is 3 m and Froude number is $\frac{1}{2\sqrt{2}}$. If a free hydraulic jump is to be formed at downstream of the sluice gate after the vena contracta of the jet coming out from the sluice gate, the sluice gate opening should be (coefficient of contraction $C_c = 0.9$)
(a) 0.3 m (b) 0.4 m
(c) 0.69 m (d) 0.9 m [2005 : 2 M]

13.25 A channel with a mild slope is followed by a horizontal channel and then by a steep channel. What gradually varied flow profiles will occur?
(a) M_1, H_1, S_1 (b) M_2, H_2, S_2
(c) M_1, H_2, S_3 (d) M_1, H_2, S_2
[2006 : 1 M]

13.26 A hydraulic jump occurs in a rectangular, horizontal, frictionless channel. What would be the pre-jump depth if the discharge per unit width is 2 m^3/s/m and the energy loss is 1 m?
(a) 0.2 m (b) 0.3 m
(c) 0.8 m (d) 0.9 m [2006 : 2 M]

13.27 A very wide rectangular channel is designed to carry a discharge of 5 m^3/s per meter width. The design is based on the Manning's equation with the roughness coefficient obtained from the grain size using Strickler's equation and results in a normal depth of 1.0 m. By mistake, however, the engineer used the grain diameter in mm in the Strickler's equation instead of in meter. What should be the correct normal depth?
(a) 0.32 m (b) 0.50 m
(c) 2.00 m (d) 3.20 m [2006 : 2 M]

13.28 There is a free overfall at the end of a long open channel. For a given flow rate, the critical depth is less than the normal depth. What gradually varied flow profile will occur in the channel for this flow rate?
(a) M_1 (b) M_2
(c) M_3 (d) S_1 [2007 : 1 M]

13.29 A triangular open channel has a vertex angle of 90° and carries flow at a critical depth of 0.30 m. The discharge in the channel is
(a) 0.08 m^3/s (b) 0.11 m^3/s
(c) 0.15 m^3/s (d) 0.2 m^3/s
[2007 : 2 M]

13.30 The flow rate in a wide rectangular open channel is 2.0 m^3/s per metre width. The channel bed slope is 0.002. The Manning's roughness coefficient is 0.012. The slope of the channel is classified as
(a) Critical (b) Horizontal
(c) Mild (d) Steep [2007 : 2 M]

Linked Answer Questions 13.31 and 13.32:

A rectangular open channel needs to be designed to carry a flow of 2.0 m^3/s under uniform flow conditions. The Manning's roughness coefficient is 0.018. The channel should be such that the flow depth is equal to half the width, and the Froude number is equal to 0.5.

13.31 The bed slope of the channel to be provided is
(a) 0.0012 (b) 0.0021
(c) 0.0025 (d) 0.0052 [2007 : 2 M]

13.32 Keeping the width, flow depth and roughness the same, if the bed slope of the above channel is doubled, the average boundary shear stress under uniform flow conditions is
(a) 5.6 N/m^2 (b) 10.8 N/m^2
(c) 12.3 N/m^2 (d) 17.2 N/m^2
[2007 : 2 M]

13.33 A person standing on the bank of a canal drops a stone on the water surface. He notices that the disturbance on the water surface is not traveling upstream. This is because the flow in the canal is
(a) sub-critical (b) super-critical
(c) steady (d) uniform [2008 : 1 M]

Common Data Q.13.34, Q.13.35 and Q.13.36

A rectangular channel 6.0 m wide carries a discharge of 16.0 m^3/sunder uniform flow condition with normal depth of 1.60 m. Manning's 'n' is 0.015.

13.34 The longitudinal slope of the channel is
(a) 0.000585 (b) 0.000485
(c) 0.000385 (d) 0.000285
[2008 : 2 M]

13.35 A hump is to be provided on the channel bed. The maximum height of the hump without affecting the upstream flow condition is
(a) 0.50 m (b) 0.40 m
(c) 0.30 m (d) 0.20 m [2008 : 2 M]

13.36 The channel width is to be contracted. The minimum width to which the channel can be contracted without affecting the upstream flow condition is
(a) 3.0 m (b) 3.8 m
(c) 4.1 m (d) 4.5 m [2008 : 2 M]

13.37 Direct step method of computation for gradually varied flow is
(a) applicable to non-prismatic channels
(b) applicable to prismatic channels
(c) applicable to both prismatic and non-prismatic channels
(d) not applicable to both prismatic and non-prismatic channels [2009 : 1 M]

13.38 A rectangular open channel of width 4.5 m is carrying a discharge of 100 m^3/sec. The critical depth of the channel is
(a) 7.09 m (b) 3.69 m
(c) 2.16 m (d) 1.31 m [2009 : 2 M]

13.39 A mild-sloped channel is followed by a steep-sloped channel. The profiles of gradually varied flow in the channel are
(a) M_3, S_2 (b) M_3, S_3
(c) M_2, S_1 (d) M_2, S_2 [2010 : 1 M]

13.40 The flow in a rectangular channel is subcritical. If width of the channel is reduced at a certain section, the water surface under no-choke condition will
(a) drop at a downstream section
(b) rise at a downstream section
(c) rise at an upstream section
(d) not undergo any change [2010 : 1 M]

13.41 For a rectangular channel section, Match **List-I** (Geometrical elements) with **List-II** (Proportions for hydraulically efficient section) and select the correct answer using the codes given below the lists:

List-I	List-II
A. Top width	1. $y_e/2$
B. Perimeter	2. y_e
C. Hydraulic Radius	3. $2\,y_e$
D. Hydraulic Depth	4. $4y_e$

y_e is the flow depth corresponding to hydraulically efficient section.

Codes:

	A	B	C	D
(a)	2	4	1	3
(b)	3	1	4	2
(c)	3	4	1	2
(d)	3	4	2	1

[2010 : 2 M]

13.42 The Froude number of flow in a rectangular channel is 0.8. If the depth of flow is 1.5 m, the critical depth is
(a) 1.80 m (b) 1.56 m
(c) 1.36 m (d) 1.29 m [2010 : 2 M]

13.43 For a given discharge, the critical flow depth in an open channel depends on
(a) channel geometry only
(b) channel geometry and bed slope
(c) channel geometry, bed slope and roughness
(d) channel geometry, bed slope, roughness and Reynolds number [2011 : 1 M]

13.44 A spillway discharges flood flow at a rate of 9 m^3/s per metre width. If the depth of flow on the horizontal apron at the toe of the spillway is 46 cm, the tail water depth needed to form a hydraulic jump is approximately given by which of the following options?
(a) 2.54 m (b) 4.90 m
(c) 5.77 m (d) 6.23 m [2011 : 2 M]

13.45 The flow in a horizontal, frictionless rectangular open channel is supercritical. A smooth hump is built on the channel floor. As the height of hump is increased, choked condition is attained. With further increase in the height of the hump, the water surface will
(a) rise at a section upstream of the hump
(b) drop at a section upstream of the hump
(c) drop at the hump
(d) rise at the hump [2011 : 1 M]

13.46 A trapezoidal channel is 10.0 m wide at the base and has a side slope of 4 horizontal to 3 vertical. The bed slope is 0.002. The channel is lined with smooth concrete (Manning's n = 0.012). The hydraulic radius (in m) for a depth of flow of 3.0 m is
(a) 20.0 (b) 3.5
(c) 3.0 (d) 2.1 [2012 : 1 M]

13.47 A rectangular open channel of width 5.0 m is carrying a discharge of 100 m^3/s. The Froude number of the flow is 0.8. The depth of flow (in m) in the channel is

(a) 4 (b) 5
(c) 16 (d) 20 [2012 : 1 M]

13.48 For subcritical flow in an open channel, the control section for gradually varied flow profile is

(a) at the downstream end
(b) at the upstream end
(c) at both ends
(d) at any intermediate section [2013 : 1 M]

13.49 The normal depth in a wide rectangular channel is increased by 10%. The % increase in discharge in the channel is

(a) 20.1 (b) 15.4
(c) 10.5 (d) 17.2 [2013 : 2 M]

13.50 A rectangular channel having a bed slope of 0.0001, width 3.0 m and Manning's coefficient 'n' 0.015, carries a discharge of 1.0 m^3/s. Given that the normal depth of flow ranges between 0.76 m and 0.8 m. The minimum width of a throat (in m) that is possible at a given section, while ensuring that the prevailing normal depth is not exceeded along the reach upstream of the contraction, is approximately equal to (assume negligible losses)

(a) 0.64 (b) 0.84
(c) 1.04 (d) 1.24

[2014 : 2 M, Set-I]

13.51 A rectangular channel of 2.5 m width is carrying a discharge of 4 m^3/s. Considering that acceleration due to gravity as 9.81 m/s^2, the velocity of flow (in m/s) corresponding to the critical depth (at which the specific energy is minimum) is _______.

[2014 : 2 M, Set-II]

13.52 A circular pipe has a diameter of 1 m, bed slope of 1 in 1000, and Manning's roughness coefficient equal to 0.01. It may be treated as an open channel flow when it is flowing just full, i.e., the water level just touches the crest. The discharge in this condition is denoted by Q_{full}. Similarly, the discharge when the pipe is flowing half-full, i.e., with a flow depth of 0.5 m, is denoted by Q_{half}. The ratio $\dfrac{Q_{full}}{Q_{half}}$ is

(a) 1 (b) $\sqrt{2}$
(c) 2 (d) 4 [2015 : 1 M, Set-I]

13.53 A hydraulic jump is formed in a 2 m wide rectangular channel which is horizontal and frictionless. The post-jump depth and velocity are 0.8 m and 1 m/s, respectively. The pre-jump velocity is ________ m/s. (use g = 10 m/s^2)

[2015 : 2 M, Set-I]

13.54 A short reach of a 2 m wide rectangular open channel has its bed level rising in the direction of flow at a slope of 1 in 10000. It carries a discharge of 4 m^3/s and its Manning's roughness coefficient is 0.01. The flow in this reach is gradually varying. At a certain section in this reach, the depth of flow was measured as 0.5 m. The rate of change of the water depth with distance, $\dfrac{dy}{dx}$, at this section is ________ (use g = 10 m/s^2).

[2015 : 2 M, Set-I]

13.55 A hydraulic jump takes place in a frictionless rectangular channel. The pre-jump depth is y_p. The alternate and sequent depths corresponding to y_p are y_a and y_s respectively. The correct relationship among y_p, y_a and y_s is

(a) $y_a < y_s < y_p$ (b) $y_p < y_s < y_a$
(c) $y_p < y_s = y_a$ (d) $y_a = y_s = y_p$

[2015 : 1 M, Set-II]

13.56 The pre-jump Froude Number for a particular flow in a horizontal rectangular channel is 10. The ratio of sequent depths (i.e., post-jump depth to pre-jump depth) is_______

[2016 : 1 M, Set-I]

13.57 A 4 m wide rectangular channel, having bed slope of 0.001 carries a discharge of 16 m^3/s. Considering Manning's roughness coefficient = 0.012 and g = 10 m/s^2, the category of the channel slope is

(a) horizontal (b) mild
(c) critical (d) steep

[2016 : 2 M, Set-I]

13.58 A hydraulically efficient trapezoidal channel section has a uniform flow depth of 2 m. The bed width (expressed in m) of the channel is _________

[2016 : 2 M, Set-I]

13.59 A 3 m wide rectangular channel carries a flow of 6 m^3/s. The depth of flow at a section P is 0.5 m. A flat-topped hump is to be placed at the downstream of the section P. Assume negligible energy loss between section P and hump, and consider g as 9.81 m/s^2. The maximum height of the hump (expressed in m) which will not change the depth of flow at section P is________.

[2016 : 2 M, Set-II]

13.60 A 1 m wide rectangular channel has a bed slope of 0.0016 and the Manning's roughness coefficient is 0.04. Uniform flow takes place in the channel at a flow depth of 0.5 m. At a particular section, gradually varied flow (GVF) is observed and the flow depth is measured as 0.6 m. The GVF profile at that section is classified as

(a) S_1 (b) S_2
(c) M_1 (d) M_2

[2017 : 2 M, Set-I]

13.61 A sluice gate used to control the flow in a horizontal channel of unit width is shown in the figure.

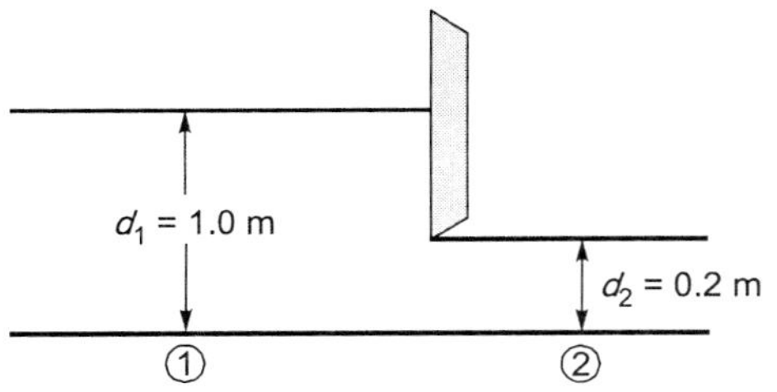

It is obeserved that the depth of flow is 1.0 m upstream of the gate, while the depth is 0.2 m downstream of the gate. Assuming a smooth flow transition across the sluice gate, i.e., without any energy loss, and the acceleration due to gravity as 10 m/s^2, the discharge (in m^3/s, up to two decimal places) passing under the sluice gate is __________.

[2017 : 2 M, Set-I]

13.62 A 1 m wide rectangular channel carries a discharge of 2 m^3/s. The specific energy-depth diagram is prepared for the channel. It is observed in this diagram that corresponding to a particular specific energy, the subcritical depth is twice the supercritical depth. The subcritical depth (in meters, up to two decimal places) is equal to __________.

[2017 : 2 M, Set-II]

13.63 A 10 m wide rectangular channel carries a discharge of 20 m^3/s under critical condition. Using g = 9.81 m/s^2, the specific energy (in m, up to two decimal places) is _________

[2018 : 1 M, Set-I]

13.64 Variation of water depth (y) in a gradually varied open channel flow is given by the first order differential equation

$$\frac{dy}{dx} = \frac{1 - e^{-\frac{10}{3}\ln(y)}}{250 - 45e^{-3\ln(y)}}$$

Given initial conditions: $y(x = 0) = 0.8$ m. The depth (in m, up to three decimal places) of flow at a downstream section at x = 1 from one calculation step of Single Step Euler Method is _______.

[2018 : 2 M, Set-I]

13.65 For a given discharge in an open channel, there are two depths which have the same specific energy. These two depths are known as

(a) alternate depths (b) critical depths
(c) normal depths (d) sequent depths

[2018 : 1 M, Set-II]

13.66 In a 5 m wide rectangular channel, the velocity u distribution in the vertical direction y is given by $u = 1.25y^{\frac{1}{6}}$. The distance y is measured from the channel bed. If the flow depth is 2 m, the discharge per unit width of the channel is

(a) 2.40 m^3/s/m (b) 2.80 m^3/s/m
(c) 3.27 m^3/s/m (d) 12.02 m^3/s/m

[2018 : 2 M, Set-II]

13.67 In a rectangular channel, the ratio of the velocity head to the flow depth for critical flow condition, is

(a) $\frac{3}{2}$ (b) $\frac{1}{2}$

(c) $\frac{2}{3}$ (d) 2 [2019 : 1 M, Set-I]

13.68 A rectangular open channel has a width of 5 m and a bed slope of 0.001. For a uniform flow of depth 2 m, the velocity is 2 m/s. The Manning's roughness coefficient for the channel is

(a) 0.050 (b) 0.002
(c) 0.033 (d) 0.017

[2019 : 2 M, Set-I]

13.69 At the foot of a spillway, water flows at a depth of 23 cm with a velocity of 8.1 m/s, as shown in the figure.

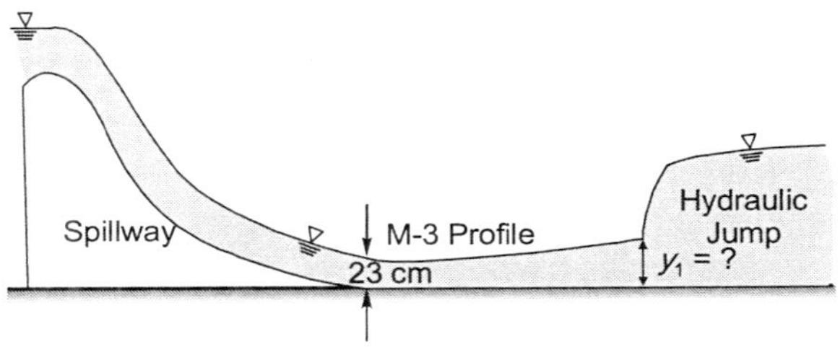

The flow enters as an M-3 profile in the long wide rectangular channel with bed slope = $\frac{1}{1800}$ and Manning's n = 0.015. A hydraulic jump is formed at a certain distance from the foot of the spillway. Assume the acceleration due to gravity, g = 9.81 m/s^2. Just before the hydraulic jump, the depth of flow y_1 (in m, round off to 2 decimal places) is ________.

[2019 : 2 M, Set-II]

13.70 A 4 m wide rectangular channel carries 6 m^3/s of water. The Manning's '*n*' of the open channel is 0.02. Considering g = 9.81 m/s^2, the critical velocity of flow (in m/s, round off to two decimal places) in the channel, is ________.

[2020 : 1 M, Set-I]

13.71 Water flows at the rate of 12 m^3/s in a 6 m wide rectangular channel. A hydraulic jump is formed in the channel at a point where the upstream depth is 30 cm (just before the jump). Considering acceleration due to gravity as 9.81 m/s^2 and density of water as 1000 kg/m^3, the energy loss in the jump is

(a) 114.2 kW
(b) 141.2 J/s
(c) 141.2 h.p.
(d) 114.2 MW

[2020 : 2 M, Set-I]

13.72 A hydraulic jump occurs, in a triangular (V-shaped) channel with side slopes 1:1 (vertical to horizontal). The sequent depths are 0.5 m and 1.5 m. The flow rate (in m^3/s, round off to two decimal places) in the channel is ________.

[2020 : 2 M, Set-II]

13.73 If water is flowing at the same depth in most hydraulically efficient triangular and rectangular channel sections then the ratio of hydraulic radius of triangular section to that of rectangular section is

(a) 2
(b) $\sqrt{2}$
(c) 1
(d) $\frac{1}{\sqrt{2}}$

[2021 : 1 M, Set-I]

13.74 A rectangular open channel of 6 m width is carrying a discharge of 20 m^3/s. Consider the acceleration due to gravity as 9.81 m/s^2 and assume water as incompressible and inviscid. The depth of flow in the channel at which the specific energy of the flowing water is minimum for the given discharge will then be

(a) 3.18 m
(b) 1.04 m
(c) 2.56 m
(d) 0.82 m

[2021 : 2 M, Set-II]

13.75 A rectangular channel with Gradually Varied Flow (GVF) has a changing bed slope. If the change is from a steeper slope to a steep slope, the resulting GVF profile is

(a) S_3
(b) S_1
(c) S_2
(d) either S_1 or S_2, depending on the magnitude of the slopes

[2022 : 1 M, Set-I]

13.76 Depth of water flowing in a 3 m wide rectangular channel is 2 m. The channel carries a discharge of 12 m^3/s. Take g = 9.8 m/s^2.
The bed width (in m) at contraction, which just causes the critical flow, is ________ without changing the upstream water level. (round off to two decimal places)

[2022 : 2 M, Set-I]

13.77 A hydraulic jump takes place in a 6 in wide rectangular channel at a point where the upstream depth is 0.5 m (just before the jump). If the discharge in the channel is 30 m^3/s and the energy loss in the jump is 1.6 m. then the Froude number computed at the end of the jump is ________. (round off to two decimal places)
(Consider the acceleration due to gravity as 10 m/s^2.)

[2022 : 2 M, Set-II]

13.78 Water is flowing in a horizontal, frictionless, rectangular channel. A smooth hump is built on the channel floor at a section and its height is gradually increased to reach choked condition in the channel. The depth of water at this section is y_2 and that at its upstream section is y_1. The correct statement(s) for the choked and unchoked conditions in the channel is/are

(a) In choked condition, y_2 is equal to the critical depth if the flow is supercritical or sub critical.
(b) In choked condition, decreases if the flow is supercritical and increases if the flow is subcritical.
(c) In unchoked condition, y_1 remains unaffected when the flow is supercritical or subcritical.
(d) In chocked conditions, y_1 increases if the flow is supercritical and decreases if the flow is subcritical.

[2022 : 2 M, Set-II]

13.79 When a supercritical stream enters a mild-sloped (M) channel section, the type of flow profile would become _______.

(a) M_1 (b) M_2
(c) M_3 (d) M_1 and M_2

[2023 : 1 M, Set-I]

13.80 A very wide rectangular channel carries a discharge (Q) of 70 m³/s per meter width. Its bed slope changes from 0.0001 to 0.0009 at a point P, as shown in the figure (not to scale). The Manning's roughness coefficient of the channel is 0.01. What water surface profile(s) exist(s) near the point P?

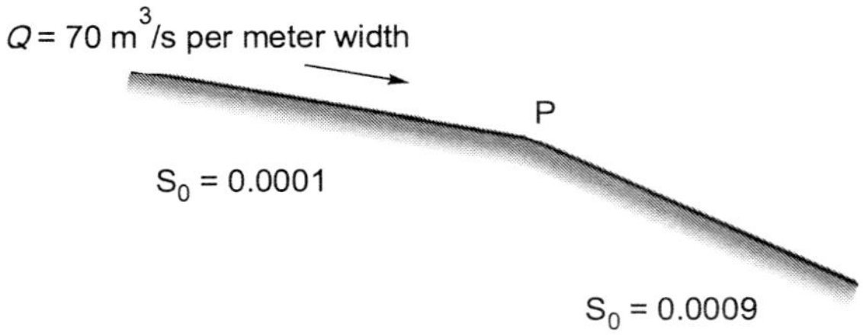

(a) M_2 and S_2 (b) M_2 only
(c) S_2 only (d) S_2 and hydraulic jump

[2023 : 2 M, Set-I]

13.81 A hydraulic jump occurs in a 1.0 m wide horizontal, frictionless, rectangular channel, with a pre-jump depth of 0.2 m and a post-jump depth of 1.0 m. The value of g may be taken as 10 m/s². The values of the specific force at the pre-jump and post-jump sections are same and are equal to (in m³, rounded off to two decimal places) ________.

[2023 : 2 M, Set-I]

13.82 A compound symmetrical open channel section as shown in the figure has a maximum of critical depth(s).

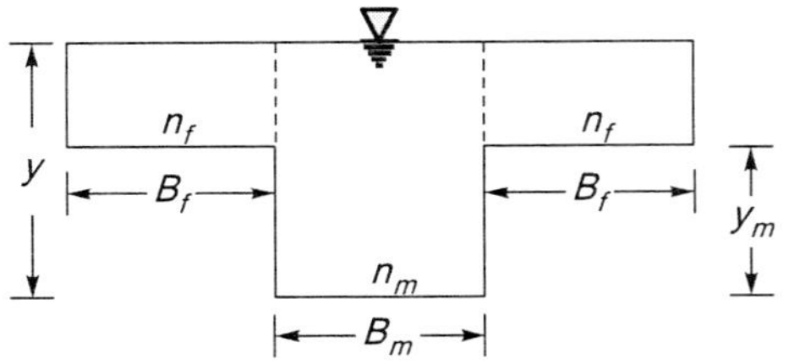

B_m – Bottom width of main channel.
B_f – Bottom width of flood channel.
y_m – Depth of main channel.
y – Total depth of the channel.
n_m – Manning's roughness of the main channel.
n_f – Manning's roughness of the flood channel.

(a) 3 (b) 2
(c) 1 (d) 4

[2023 : 2 M, Set-II]

13.83 The critical flow condition in a channel is given by ______ .

[Note: α – kinetic energy correction factor; Q – discharge; A_c – cross-sectional area of flow at critical flow condition; T_c – top width of flow at critical flow condition; g – acceleration due to gravity]

(a) $\frac{\alpha Q^2}{g} = \frac{A_c^3}{T_c}$ (b) $\frac{\alpha Q}{g} = \frac{A_c^3}{T_c^2}$
(c) $\frac{\alpha Q^2}{g} = \frac{A_c^3}{T_c^2}$ (d) $\frac{\alpha Q}{g} = \frac{A_c^3}{T_c}$

[2023 : 2 M, Set-II]

13.84 The number of degrees of freedom for a natural open channel flow with a mobile bed is

(a) 2 (b) 4
(c) 3 (d) 5

[2024 : 1 M, Set-I]

13.85 A spillway has unit discharge of 7.5 m³/s/m. The flow depth at the downstream horizontal apron is 0.5 m. The tail water depth (in meters) required to form a hydraulic jump is _______. (rounded off to 2 decimal places).

[2024 : 2 M, Set-I]

13.86 A 2 m wide rectangular channel is carrying a discharge of 30 m³/s at a bed slope of 1 in 300. Assuming the energy correction factor as 1.1 and acceleration due to gravity as 10 m/s², the critical depth of flow (in meters) is ________ (rounded off to 2 decimal places).

[2024 : 1 M, Set-II]

13.87 A rectangular channel is 4.0 m wide and carries a discharge of 2.0 m³/s with a depth of 0.4 m. The channel transitions to a maximum width contraction at a downstream location, without influencing the upstream flow conditions. The width (in meters) at the maximum contraction is __________ (rounded off to 2 decimal places).

[2024 : 2 M, Set-II]

■■■■

Answers Open Channel Flow

13.1 (b)	13.2 (b)	13.3 (*)	13.4 (d)	13.5 (*)	13.6 (a)	13.7 (a, b)
13.8 (b)	13.9 (b)	13.10 (d)	13.11 (d)	13.12 (a)	13.13 (*)	13.14 (c)
13.15 (d)	13.16 (c)	13.17 (c)	13.18 (b)	13.19 (a)	13.20 (c)	13.21 (a)
13.22 (a)	13.23 (d)	13.24 (c)	13.25 (d)	13.26 (b)	13.27 (b)	13.28 (b)
13.29 (b)	13.30 (d)	13.31 (b)	13.32 (d)	13.33 (b)	13.34 (a)	13.35 (b)
13.36 (c)	13.37 (b)	13.38 (b)	13.39 (d)	13.40 (a)	13.41 (c)	13.42 (d)
13.43 (a)	13.44 (c)	13.45 (b)	13.46 (d)	13.47 (a)	13.48 (a)	13.49 (d)
13.50 (b)	13.51 (2.504)	13.52 (c)	13.53 (4.83)	13.54 (0.0032)	13.55 (b)	13.56 (13.65)
13.57 (b)	13.58 (2.31)	13.59 (0.20)	13.60 (c)	13.61 (0.816)	13.62 (1.07)	13.63 (1.11)
13.64 (0.793)	13.65 (a)	13.66 (a)	13.67 (b)	13.68 (d)	13.69 (0.42)	13.70 (a)
13.71 (114.0)	13.72 (1.728)	13.73 (d)	13.74 (b)	13.75 (a)	13.76 (2.15)	13.77 (2.43)
13.78 (a, b, c)	13.79 (c)	13.80 (a)	13.81 (0.62)	13.82 (a)	13.83 (a)	13.84 (b)
13.85 (4.55)	13.86 (2.91)	13.87 (3.53)				

Explanations Open Channel Flow

13.1 (b)

$$y_c = \left(\frac{q^2}{g}\right)^{1/3} = \left(\frac{1^2}{9.81}\right)^{1/3} = 0.467 \text{ m}$$

$$F_r^2 = \frac{Q^2T}{gA^3} = \frac{1^2 \times 1}{9.81 \times (0.25)^3}$$

$$F_r = 2.554$$

$$= 0.467 \text{ m}$$

Given, $y_n = 0.25$ m

$y_c > y_n$

$\Rightarrow s_c < s_0 \Rightarrow$ steep slope

For steep slope $y_c > y_n$

For critical slope $y_c = y_n$

For mild slope $y_c < y_n$

13.2 (b)

The derivation for equation of hydraulic jump in terms of conjugate depths and the initial provide number involves continuity and momentum equation. As h_L is unknown hence y_2 cannot be determined by energy equation.

13.3 Sol.

A ***control section*** is defined as a section in which a fixed relationship exits between the discharge and depth of flow. Weirs, spillways, sluice gates are some typical examples of structure which give rise to control sections.

Supercritical flow has upstream control wherea subcritical flow has downstream control.

13.4 (d)

Method-I

For a triangular notch

$$Q = \frac{8}{15} C_d \sqrt{2g} \tan\frac{\theta}{2} H^{5/2}$$

$$dQ = \frac{8}{15} C_d \sqrt{2g} \tan\left(\frac{\theta}{2}\right) \frac{5}{2} H^{3/2} dH$$

$$\frac{dQ}{Q} = \frac{5}{2} \times \frac{dH}{H} = 2.5 \frac{dH}{H}$$

Hence, if error in head measurement is 1%, then the corresponding error in discharge measurement will be 2.5%.

Method-II

For 1% error

$$Q' = \frac{8}{15} C_d \sqrt{2g} \tan\left(\frac{\theta}{2}\right) (x \pm 0.01H)^{5/2}$$

$Q' = 1.025Q$ or $0.975\,Q$

Error = +2.5% or – 2.5%

Note: *For small order of error i.e., <5%, differential form is used.*

13.5 (*)

At minimum specific energy there is only one depth of flow known as critical depth whereas if specific energy is other than minimum energy, there will be two depth of flow which are known as **conjugate depths**.

a, *b*, *d* all correspond to critical condition of flow.

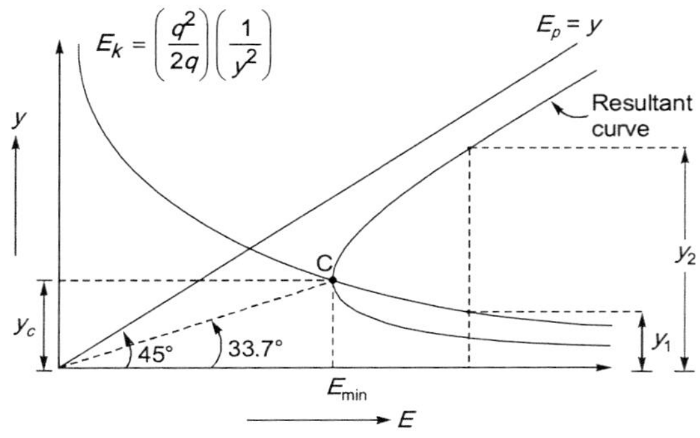

13.6 (a)

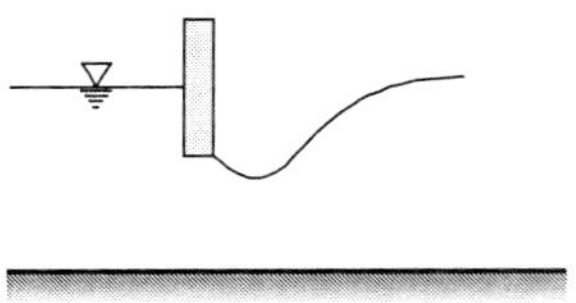

For rectangular channel

$$y_1 + \frac{q^2}{2gy_1^2} = y_2 + \frac{q^2}{2gy_2^2} + h_L$$

H_1 profile does not exist H_2 and S_2 are drawdown profiles and the given surface profile is rising curve. Hence, H_3.

Note:

1. *M_2, S_2, H_2 and A_2 all are falling profiles.*
2. *M_1, M_3, S_1, S_3, H_3, C_1, C_3, A_3 are all rising profile.*

13.7 (a, b)

For a trapezoidal section to be most economical or most efficient, half the top width must be equal to one of the sloping sides of the channel.

It is half of a regular hexagon.

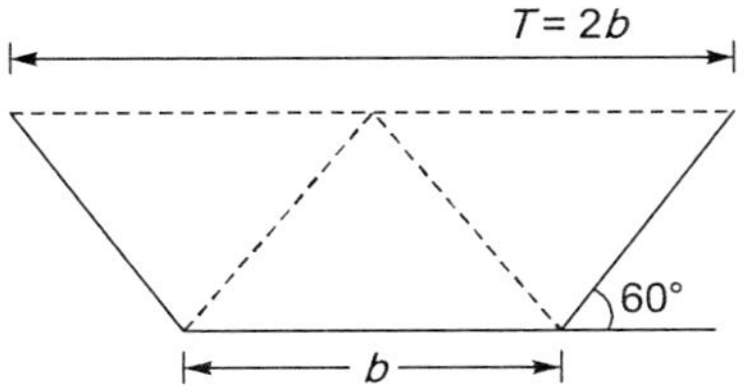

13.8 (b)

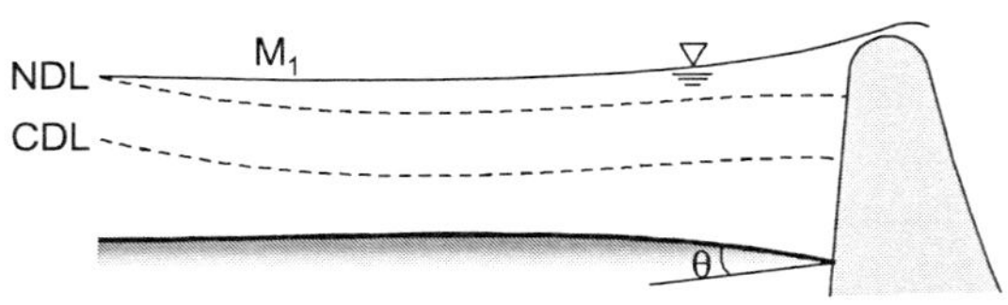

In mild slope, NDL is above CDL, when a subcritical flow gets obstructed by structures such as weirs, dams, control structures and natural factors produce M_1 (back water) curves. M_1 curves extends several kilometers upstream of obstruction.

13.9 (b)

From Belenger's momentum equation

$$\frac{y_2}{y_1} = \frac{1}{2}\left[-1+\sqrt{1+8\,F_{r_1}^{\;2}}\right]$$

$$\Rightarrow \quad 10.3 = \frac{1}{2}\left[-1+\sqrt{1+8\,F_{r_1}^{\;2}}\right]$$

$$\Rightarrow \quad 20.6 = -1+\sqrt{1+8F_{r_1}^{\;2}}$$

$$\Rightarrow \quad (21.6)^2 = 1+8\,F_{r_1}^{\;2}$$

Solving the above equation, we get,

$$F_{r_1} = 7.63$$

13.10 (d)

As we know

$$Q = y_1V_1 = y_2V_2$$

$$\Rightarrow \quad 0.1\times 6 = y_2\times V_2$$

$$\Rightarrow \quad V_2 = \frac{0.6}{y_2}$$

At alternate depths

$$y_1 + \frac{V_1^2}{2g} = y_2 + \frac{V_2^2}{2g}$$

$$0.1+\frac{6^2}{2\times 9.81} = y_2 + \left(\frac{0.6}{y_2}\right)^2 \times \frac{1}{2\times 9.81}$$

$$\Rightarrow \quad 1.937 = y_2 + \frac{0.0183}{y_2^2}$$

13.11 (d)

$$q = \frac{Q}{B} = \frac{10}{3} = 3.333$$

$$\therefore \quad y_c = \left[\frac{q^2}{g}\right]^{1/3} = \left[\frac{(3.333)^2}{9.81}\right]^{1/3} = 1.04\ m$$

13.12 (a)

$$V = \frac{Q}{A} = \frac{2}{\frac{\pi}{4}(2)^2} = 0.637\ \text{m/sec}$$

For circular section, flowing full, the hydraulic radius

$$R = \frac{d}{4} = \frac{2}{4} = 0.5$$

From Manning's formula,

$$V = \frac{1}{n} R^{2/3} S^{1/2}$$

$$\sqrt{S} = \frac{0.637 \times 0.015}{(0.5)^{2/3}} = 0.015$$

$$S = 0.00023$$

13.13 (*)

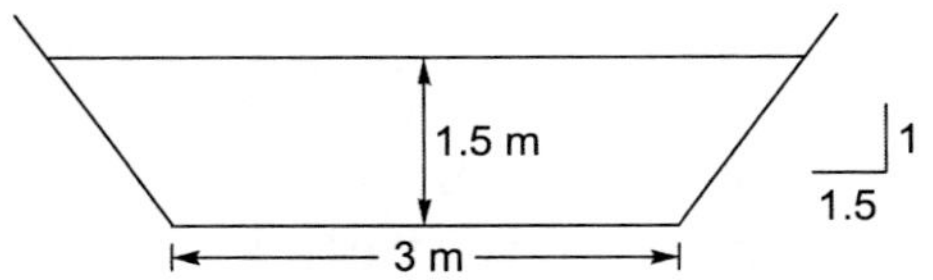

$$Q = 8 \text{ m}^3/\text{s},\ T = 7.5 \text{ m}$$

$$A = \frac{1}{2}(3+7.5)\times 1.5 = 7.875 \text{ m}^2$$

$$V = \frac{Q}{A} = \frac{8}{7.875} = 1.015 \text{ m/s}$$

$$\text{Fr} = \sqrt{\frac{Q^2T}{gA^3}} = \sqrt{\frac{8^2 \times 7.8}{9.81 \times 7.875^3}} = 0.3164$$

Now $\frac{1}{1.5} = \frac{1.5}{x}$

or $x = 2.25$

$$T = 3 + 2 \times 2.25 = 7.5 \text{ m}$$

$$D = \frac{A}{T} = \frac{7.875}{7.5} = 1.05$$

$$F = \frac{V}{\sqrt{gD}} = \frac{1.05}{\sqrt{9.81 \times 1.05}} = 0.316$$

Hence **no** option is **correct**.

13.14 (c)

$$\frac{y_2}{y_1} = \frac{1}{2}\left[-1+\sqrt{1+8F_1^2}\right] \quad \text{...(i)}$$

$$\frac{y_1}{y_2} = \frac{1}{2}\left[-1+\sqrt{1+8F_2^2}\right] \quad \text{...(ii)}$$

Multiply eq. (i) and (ii)

$$\left[-1\sqrt{1+8F_1^2}\right]\left[-1\sqrt{1+8F_2^2}\right] = 4$$

From this, we can find out post jump Froude number from pre jump Froude number or vice verse.

Alternatively

$$F_{r_2}^2 = \frac{q^2}{gy_1^3 \times \frac{y_2^3}{y_1^3}} = \frac{8F_{r_1}^2}{\left(-1+\sqrt{1+8F_{r_1}^2}\right)^3}$$

$$\Rightarrow F_{r_2} = \left[\frac{8 \times 3^2}{\left(-1+\sqrt{1+8\times 3^2}\right)^3}\right]^{1/2} = 0.409$$

13.15 (d)

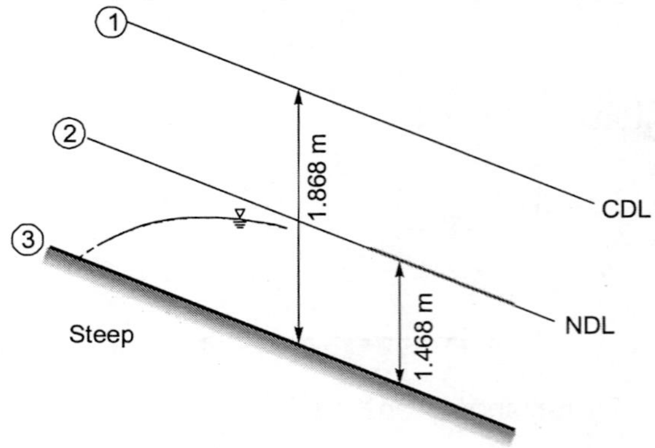

Critical depth,

$$y_c = \left(\frac{q^2}{g}\right)^{1/3} = \left(\frac{8^2}{9.81}\right)^{1/3} = 1.868 \text{ m}$$

From **Manning's equation**, we have,

$$q = \frac{1}{n}(y_n)^{5/3} S^{1/2}$$

$$\Rightarrow 8 = \frac{1}{0.015} \times (y_n)^{5/3}(0.004)^{1/2}$$

$$\Rightarrow y_n = 1.468 \text{ m}$$

Given that,

$$y = 1 \text{ m}$$

$\because \quad y_c > y_n > y$

$\therefore$ The gradually varied profile will be S_3.

13.16 (c)

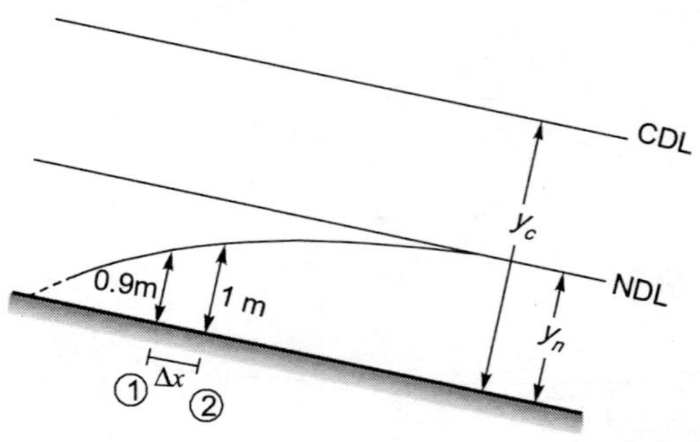

Region of flow (3)

Flow profile (S_3)

As S_3 is a profile hence 0.9 m will occurs upstream of 1 m.

$$\Delta x = \frac{E_1 - E_2}{S_0 - \left(\frac{S_{f_1} - S_{f_2}}{2}\right)}$$

$$E_1 = y_1 + \frac{V_1^2}{2g} = 0.9 + \frac{(8/0.9)^2}{2\times 9.81} = 4.927 \text{ m}$$

$$E_2 = y_2 + \frac{V_2^2}{2g} = 1 + \frac{(8/1)^2}{2\times 9.81} = 4.26 \text{ m}$$

$$S_{f_1} = \frac{Q^2 n^2}{A_1^2 R_1^{4/3}} = \frac{q^2}{y_1^2} \times \frac{x^2}{y_1^{4/3}}$$

$$= \frac{8^2 \times (0.015)^2}{(0.9)^{10/3}} = 0.0204$$

$$S_{f_2} = \frac{q^2 n^2}{(y_2)^{10/3}} = \frac{0.02 \times (0.9)^{10/3}}{1^{10/3}} = 0.0144$$

$$\Delta x = \left|\frac{E_1 - E_2}{S_0 - \left(\frac{S_{f_1} + S_{f_2}}{2}\right)}\right|$$

$$= \frac{4.9527 - 4.26}{0.004 - \left(\frac{0.0204 + 0.0144}{2}\right)}$$

= 49.7 m upstream of 1 m.

Note: *Sometimes $\bar{S}_f$ instead of being calculate as $\left(\frac{S_{f_i} + S_{f_{i+1}}}{2}\right)$ is calculated as S_f for $\left(\frac{y_i + y_{i+1}}{2}\right)$, we should prefer $\bar{S}_f = \left(\frac{S_{f_i} + S_{f_{i+1}}}{2}\right)$.*

AVOID MISTAKE

Depth at midpoint,

$$y = \frac{y_1 + y_2}{2} = \frac{0.9 + 1}{2} = 0.95$$

But, $\Delta x = \dfrac{E_2 - E_1}{S_0 - S_e}$

Now, $q = \dfrac{1}{n}(y)^{5/3}\sqrt{S_e}$

$$\Rightarrow \quad 8 = \frac{1}{0.015} \times (0.95)^{5/3} \times \sqrt{S_e}$$

$$\Rightarrow \quad S_e = 0.0171$$

$$E_2 = y_2 + \frac{V_2^2}{2g} = 1 + \frac{q^2}{2gy_2^2}$$

$$= 1 + \frac{8^2}{2\times 9.81 \times 1^2} = 4.26$$

$$E_1 = y_1 + \frac{V_1^2}{2g} = 0.9 + \frac{q^2}{2gy_1^2}$$

$$= 0.9 + \frac{8^2}{2\times 9.81\times (0.9)^2} = 4.927$$

$$\Delta x = \frac{E_2 - E_1}{S_0 - S_e} = \frac{4.26 - 4.927}{0.004 - 0.017}$$

= 51.31 m (upstream)

13.17 (c)

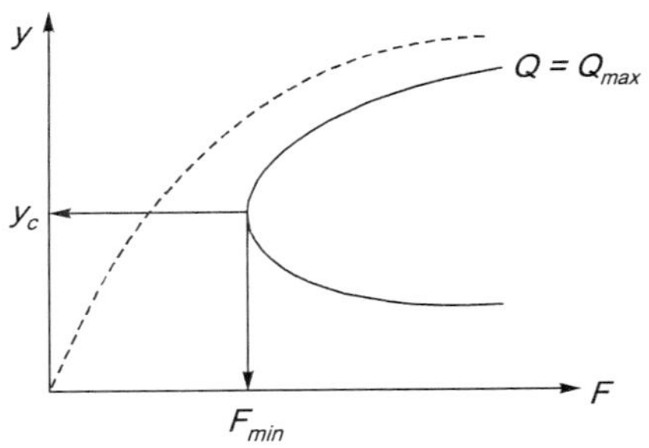

Specific force will remain constant on both sides of hydraulic jump.

$$\therefore \quad \frac{Q^2}{gA_1} + A_1\bar{Z}_1 = \frac{Q^2}{gA_2} + A_2\bar{Z}_2$$

$$A_1 = y_1^2;\ A_2 = y_2^2$$

$$\bar{Z}_1 = \left(\frac{y_1}{3}\right);\ \bar{Z}_2 = \left(\frac{y_2}{3}\right)$$

Given $\quad Q = 1 \text{ m}^3/\text{s};\ y_1 = 0.5 \text{ m}$

$$\therefore \quad \frac{(1)^2}{9.81\times(0.5)^2} + (0.5)^2 \times \left(\frac{0.5}{3}\right) = \frac{(1)^2}{9.81\times y_2^2} + (y_2)^2 \times \left(\frac{y_2}{3}\right)$$

$$\Rightarrow \quad 0.4494 = \frac{0.1019}{y_2^2} + \frac{y_2^3}{3}$$

By hit and trial, we get $y_2 = 1.02$ m

13.18 (b)

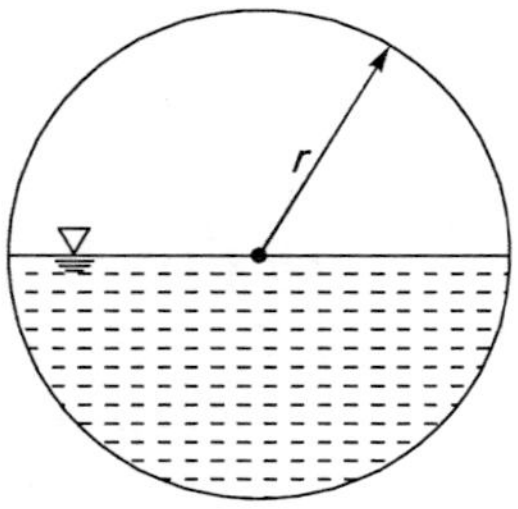

Area, $A = \frac{\pi r^2}{2}$

Top width,

$T = 2r$

$\therefore$ Hydraulic depth

$$= \frac{A}{T} = \frac{\pi r^2}{2 \times 2r} = \frac{\pi r}{4}$$

Note: *Hydraulic depth and hydraulic radius are two different terms:*

- *Hydraulic depth* = $\frac{\text{Area of flow section}}{\text{Top water surface width}}$
- *Hydraulic radius* = $\frac{\text{Area of flow}}{\text{Wetted perimeter}}$
- *Hydraulic radius approaches depth as the channel becomes very wide.*

13.19 (a)

$$F_r = \frac{V}{C_o} = \frac{\text{Flow speed}}{\text{Wave speed}}$$

Note: Control section

At low speed, $F_r < 1$; a small disturbance travels u/s (with a velocity $C_0 - V$ relative to a stationary observer) and affects the u/s conditions.

At high flow velocities, $F_r > 1$, a small disturbance travels u/s at a velocity of $C_0 - V$ relative to a stationary observer. (As $V > C_0$ small disturbance can not travel u/s) and thus the u/s conditions can not be influenced by the d/s disturbance. We can say that the region which remains unaffected controls the flow, so u/s is a control section for supercritical flow.

$$F_r = \frac{2.13}{\sqrt{9.81 \times 1}} = 0.68 < 1$$

(Subcritical flow)

$$C_0 = \sqrt{9.81 \times 1} = 3.13 \text{ m/s}$$

An elementary wave travels u/s with a velocity $C_0 - V$ relative to a stationary observer $= 3.13 - 2.13 = 1$ m/s.

13.20 (c)

Considering an element of width dx.

The mass conservation equation will be,

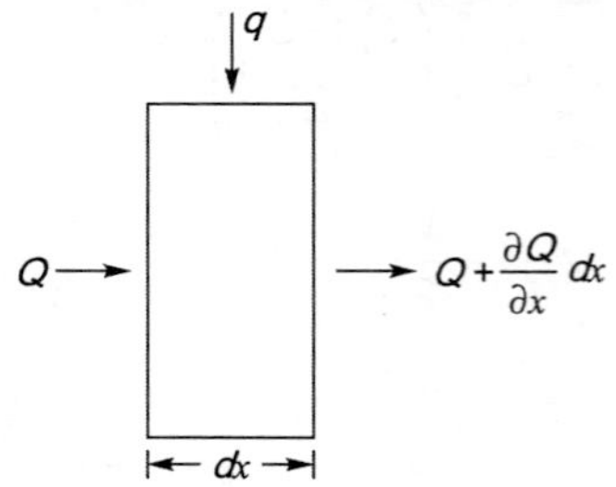

$$Q + qdx - \left(Q + \frac{\partial Q}{\partial x} dx\right) = 0$$

$$\Rightarrow \quad Q + qdx - Q - \frac{\partial Q}{\partial x} dx = 0$$

$$\Rightarrow \quad \frac{\partial Q}{\partial x} - q = 0$$

13.21 (a)

It is a example of super critical canal intake, the reservoir water surface falls to the critical depth at the head of the canal and then towards the water surface follows the S_2 curve. The critical depth occurring at the u/s end of the channel is the control for this flow.

And critical depth only depends upon channel geometry not the slope.

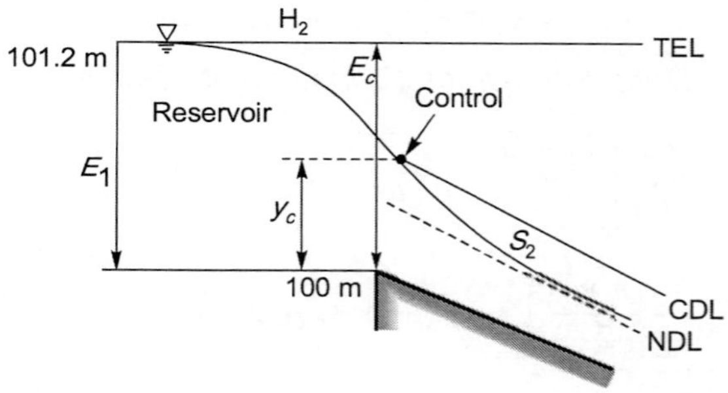

Assuming horizontal bed to be frictionless

$$E_c = 101.2 - 100 = 1.2 \text{ m}$$

$$y_c = \frac{2}{3} E_c = 0.8 \text{ m}$$

$$y_c = \left(\frac{q^2}{g}\right)^{1/3}$$

$$q = 2.24 \text{ m}^2/\text{sec}$$

Note: *On changing the slope discharge will not change; but velocity and depth vary with slope.*

13.22 (a)

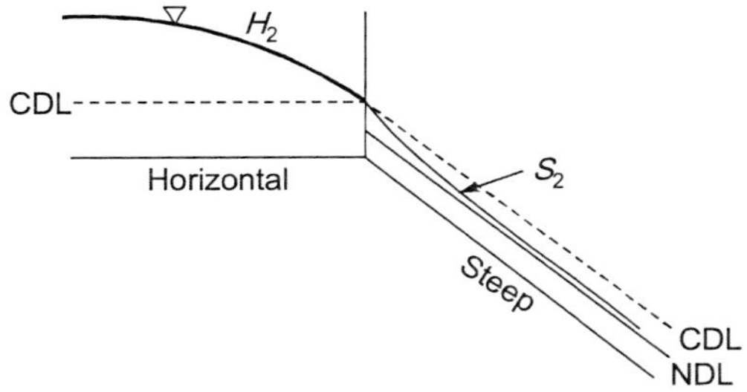

When a flow profile changes from subcritical to supercritical flow, critical depth act as a control point.

POINTS TO REMEMBER

- CDL is independent of the slope.
- NDL is smaller for larger bed slope and vice-versa.
- In region (1) & (3), depth of flow will always increase in the direction of flow and in region (2) depth of flow will always decrease.

13.23 (d)

For rectangular channel,

$$E_c = 1.5\ y_c = 1.5 \times 1.5 = 2.25 \text{ m}$$

point.

POINTS TO REMEMBER

- CDL is independent of the slope.
- For triangular channel
$$E_c = 1.25\ y_c$$
- $$E_c = y_c + \frac{A}{2T}$$

13.24 (c)

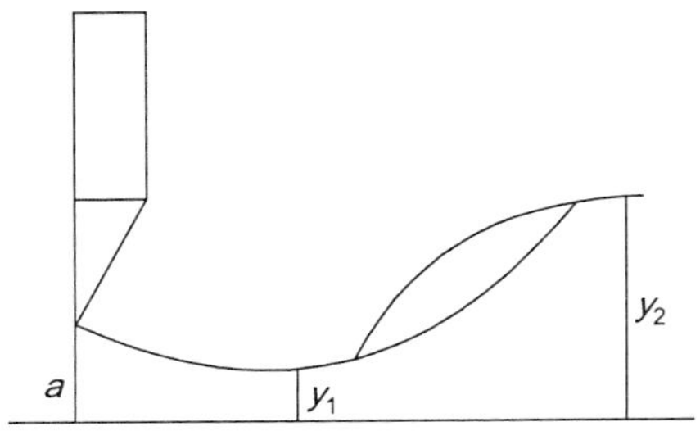

$$y_1 = C_c \times a$$

For hydraulic jump,

$$\frac{y_1}{y_2} = \frac{1}{2}\left[-1+\sqrt{1+8\,F_2^2}\right]$$

Given, $y_2 = 3$ m;

$$F_2 = \frac{1}{2\sqrt{2}}$$

$$\therefore \quad y_1 = \frac{3}{2}\left[-1+\sqrt{1+8\times\frac{1}{4\times 2}}\right]$$

$$\Rightarrow \quad y_1 = 0.62 \text{ m}$$

Sluice gate opening

$$= \frac{y_1}{C_c} = \frac{0.62}{0.9} = 0.69 \text{ m}$$

13.25 (d)

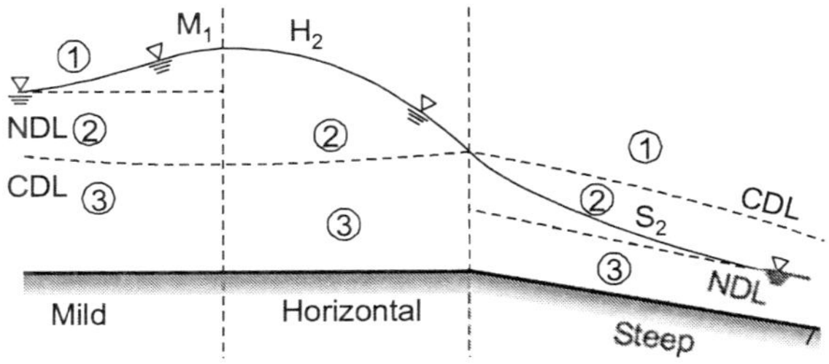

POINTS TO REMEMBER

- Flow is region (1) & (3) always rises in the direction of flow.
- Flow is region (2) always falls in the direction of flow.

13.26 (b)

$$\frac{2q^2}{g} = y_1 y_2 (y_1 + y_2) \quad \text{...(i)}$$

$$H_L = \frac{(y_2-y_1)^3}{4y_1 y_2} \quad \text{...(ii)}$$

Given,

$$H_L = 1 \text{ m}$$

$$(y_2 - y_1)^3 = 4\ y_1 \cdot y_2$$

Put, $y_2 = my_1$ in equation (ii),

$$y_1 = \frac{4\,m}{(m-1)^3}$$

Similarly, put $y_2 = my_1$ in equation (i), we get.

$$m(m+1)\,y_1^3 = \frac{2\times 2^2}{9.81} = 0.8155$$

or, $$m(m+1)\left\{\frac{4\,m}{(m-1)^3}\right\}^3 = 0.8155$$

or, $\frac{m(m+1)(m^3)}{(m-1)^9} = \frac{0.8155}{4^3}$

By trial and error, $m = 5.09$

$$y_1 = 0.3 \text{ m}$$

$$y_2 = 1.527 \text{ m}$$

13.27 (b)

As per Strickler's equation, $n = \frac{d_{50}^{1/6}}{24}$

where, d_{50} should be in 'm'

$$\therefore \quad \frac{n_{correct}}{n_{wrong}} = \left(\frac{1}{1000}\right)^{1/6}$$

Now, as per Mannings equation, in case of wide rectangular channel.

$$q = \frac{1}{n}(y)^{5/3}\sqrt{S}$$

∴ Correct depth of flow,

$$y_{correct} = \left(\frac{n_{correct}}{n_{wrong}}\right)^{3/5} \times y_{wrong}$$

$$= \left[\left(\frac{1}{1000}\right)^{1/6}\right]^{3/5} \times 1 = 0.50 \text{ m}$$

AVOID MISTAKE

$$\frac{n_{correct}}{n_{wrong}} = (1000)^{1/6}$$

To avoid it, proceed from basics.

Actual diameter of grain = d_m = 1000 d mm

Wrong value of n used by engineer

$$n_{wrong} = \frac{(1000d)^{1/6}}{24}$$

and $$n_{correct} = \frac{d^{1/6}}{24}$$

Hence, $$\frac{n_{correct}}{n_{wrong}} = \left(\frac{1}{1000}\right)^{1/6}$$

13.28 (b)

All free fall curves are zone II profiles. Since the critical depth (y_c) is less than the normal depth (y_n), therefore the slope is mild. Thus, the gradually varied profile is M_2.

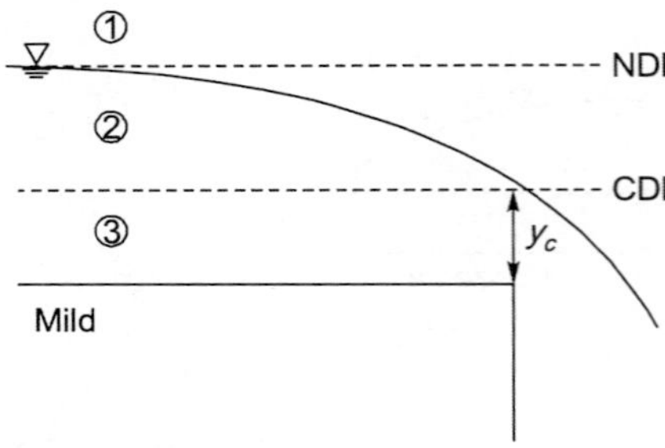

As CDL is below NDL

⇒ Mild slope profile → M_2

13.29 (b)

The side slope of triangular open channel is given as z horizontal to 1 vertical.

When vertex angle is 90°, $z = 1$

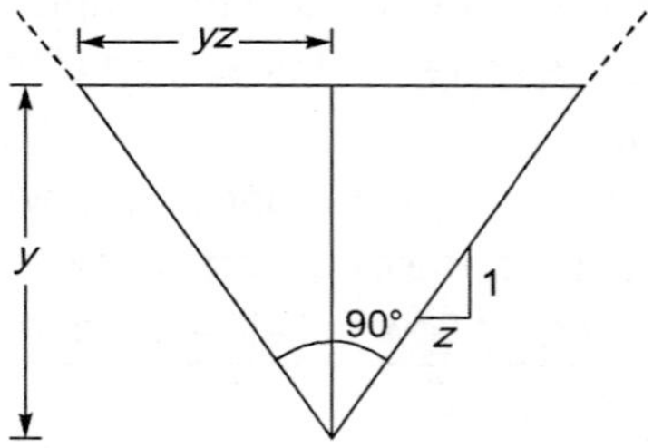

Now, $y_c = \left(\frac{2Q^2}{gz^2}\right)^{1/5}$ for a triangular channel

$$\Rightarrow \quad Q^2 = \frac{gz^2(y_c)^5}{2}$$

$$\Rightarrow \quad Q^2 = \frac{9.81 \times (1)^2 \times (0.30)^5}{2}$$

$$\Rightarrow \quad Q = 0.11 \text{ m}^3/\text{s}$$

Note: Now, as per Mannings equation, Derivation of y_c can be shown as

For critical flow,

$$\frac{Q^2T}{gA^3} = 1 \quad \text{(Always)}$$

Now, $$A = \frac{1}{2} \times y \times 2yz = zy_c^2$$

and $$T = 2yz$$

$$\frac{Q^2 \times 2y_cz}{g \times (y_c^2z)} = 1$$

$$\Rightarrow \quad y_c = \left(\frac{2Q^2}{gz^2}\right)^{1/5}$$

13.30 (d)

The critical depth for a rectangular channel is given by,

$$y_c = \left(\frac{q^2}{g}\right)^{1/3} = \left(\frac{(2)^2}{9.81}\right)^{1/3} = 0.74 \text{ m}$$

Using **Manning's equation,** we get (for a wide rectangular channel)

$$q = \frac{1}{n}(y_n)^{5/3}\sqrt{S}$$

$$\Rightarrow \quad 2 = \frac{1}{0.012} \times (y_n)^{5/3} \times \sqrt{0.002}$$

$$\Rightarrow \quad y_n = 0.69 \text{ m}$$

For steep slopes $y_n < y_c$.

Alternatively

$$S_{cr} = \frac{n^2 g^{10/9}}{q^{2/9}} = \frac{0.01^2 (9.81)^{10/9}}{2^{2/9}}$$

$$= 0.000108$$

13.31 (b)

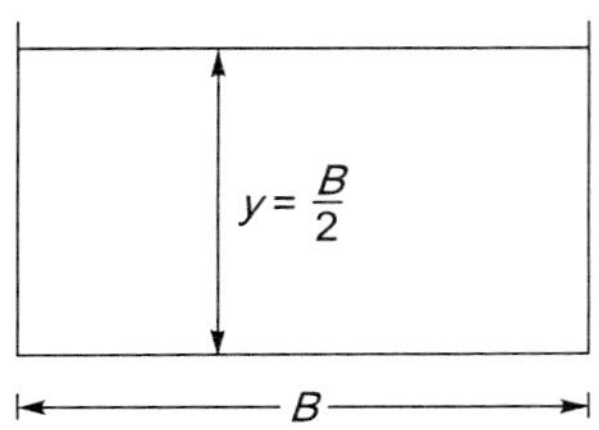

$$A = By = 2y^2$$

$$P = B + 2y = 4y$$

$$R = \frac{A}{P} = \frac{y}{2}$$

Given $Q = 2 \text{ m}^3/\text{s}$, $n = 0.018$, $B = 2y$, $F = 0.5$

We know that,

$$F = \frac{V}{\sqrt{gD}}$$

where V is velocity of flow, D is hydraulic radius,

$$\text{Now} \quad D = \frac{A}{T} = \frac{By}{B} = y$$

$$\therefore \quad F = \frac{Q}{A\sqrt{gy}}$$

$$\Rightarrow \quad 0.5 = \frac{2}{2y^2 \times \sqrt{gy}}$$

$$\Rightarrow \quad 0.5 = \frac{1}{y^2\sqrt{gy}} \quad \text{[Squaring both sides]}$$

$$\Rightarrow 0.25 = \frac{1}{gy^5}$$

$$\Rightarrow \quad y = \left(\frac{1}{0.25 \times 9.81}\right)^{1/5}$$

$$\Rightarrow \quad y = 0.836 \text{ m}$$

Using **Manning's equation**, we get,

$$Q = \frac{1}{n} A(R)^{2/3}\sqrt{S}$$

$$\Rightarrow \quad 2 = \frac{1}{0.018} \times [2 \times (0.836)^2] \times \left(\frac{0.836}{2}\right)^{2/3}\sqrt{S}$$

$$\Rightarrow \sqrt{S} = 0.0461$$

$$\Rightarrow \quad S = 0.0021$$

13.32 (d)

The average shear stress, $\tau_0 = \gamma R S_0$

Bed slope is doubled i.e.

$$S_0 = 2 \times 0.0021 = 0.0042$$

$$\therefore \quad \tau_0 = 9810 \times \frac{0.836}{2} \times 0.0042$$

$$\Rightarrow \quad \tau_0 = 17.22 \text{ N/m}^2$$

13.33 (b)

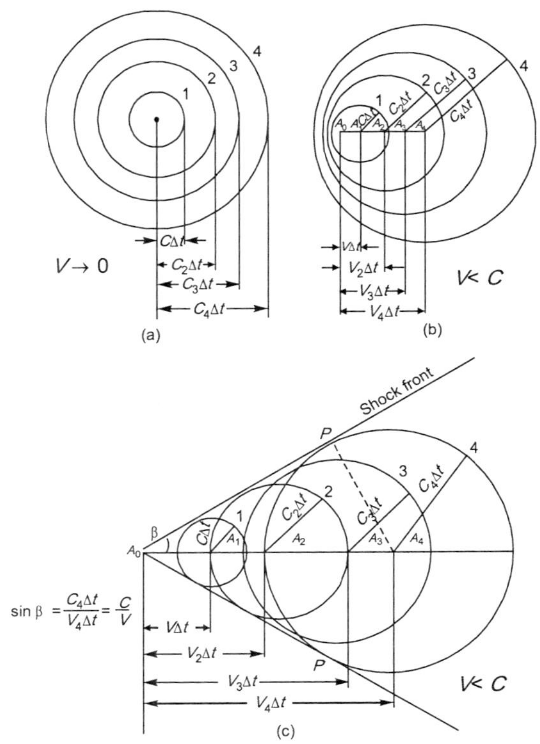

Patterns of disturbance spread;(a) Still water (b) Subcritical flow (c) Supercritical flow

$$\therefore \quad \frac{V}{\sqrt{gy}} > 1$$

or, $$V > \sqrt{gy}$$

$$\therefore \quad V > C$$

Hence, the flow here is super-critical.

13.34 (a)

The velocity as per **Manning's equation** is given by

$$V = \frac{1}{n} R^{2/3} S^{1/2}$$

Here, $Q = 16 \text{ m}^3/\text{s}$, $n = 0.015$, $B = 6.0$ m

$$A = 6.0 \times 1.60 = 9.6 \text{ m}^2$$

$$R = \frac{A}{P} = \frac{9.6}{6.0 + (1.6 \times 2)} = 1.04 \text{ m}$$

But $Q = AV$

$$\therefore \quad Q = A \times \frac{1}{n} R^{2/3} S^{1/2}$$

$$\Rightarrow \quad 16 = 9.6 \times \frac{1}{0.015} \times (1.04)^{2/3} \times S^{1/2}$$

$$\Rightarrow \quad S = 0.00059$$

13.35 (b)

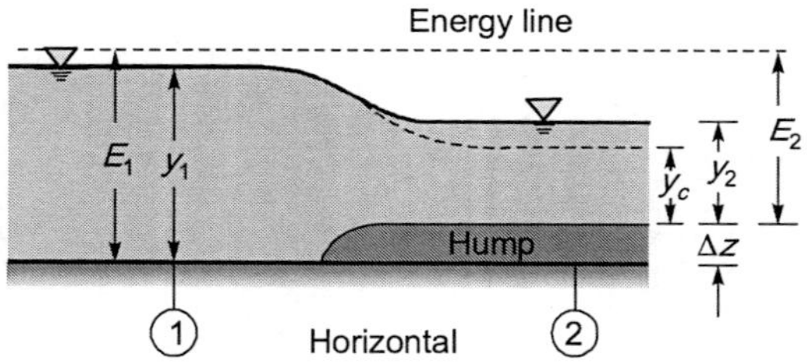

When a hump is introduced on the channel bed the depth of flow at this location will reduce. If ΔZ is the maximum height of hump so that the depth of flow remains same, then,

$$y + \frac{Q^2}{2gA^2} = \Delta Z + E_{min}$$

$$\Rightarrow 1.6 + \frac{(16)^2}{2 \times 9.81 \times (9.6)^2} = \Delta Z + \frac{3}{2} y_c$$

$$\Rightarrow \quad 1.74 = \Delta Z + \frac{3}{2} \times \left[\frac{(16/6)^2}{9.81}\right]^{1/3}$$

$$\Rightarrow \quad \Delta Z = 1.74 - 1.35$$

$$\Rightarrow \quad \Delta Z = 0.39 \approx 0.4 \text{ m}$$

To find ΔZ_{max}, the sectionis assumed to become critical. If ΔZ is raised above this level, the flow will get choked and the upstream depth will change.

13.36 (c)

If the channel is contracted, the depth of flow at this location decreases. Thus width of the channel can be decreased till the specific energy becomes minimum. If the channel is contracted further, the depth of flow starts decreasing.

$$\therefore \quad y + \frac{Q^2}{2gA^2} = E_{min}$$

$$\Rightarrow \quad 1.6 + \frac{(16)^2}{2 \times 9.81 \times (9.6)^2} = \frac{3}{2} y_c$$

$$\Rightarrow \quad 1.74 = \frac{3}{2} \times \left(\frac{Q^2/b^2}{g}\right)^{1/3}$$

$$\Rightarrow \quad 1.74 = \frac{3}{2} \times \left[\frac{(16)^2}{9.81}\right]^{1/3} \times \frac{1}{(b)^{2/3}}$$

$$\Rightarrow \quad b^{2/3} = 2.557$$

$$\Rightarrow \quad b = 4.1 \text{ m}$$

13.37 (b)

The various methods for gradually varied flow computations are:

1. Direct integration method (Chow's method) Used for prismatic channel as area at any section can be calculated while using direct step method.
2. Numerical method
   (a) *Direct step method:* It does not involve any trial and error and is applicable for prismatic channels only.
   (b) *Standard step method:* It involves trial and error and is applicable for prismatic as well as natural channels (non-prismatic).
3. *Graphical method:* Applicable for both prismatic as well as nonprismatic channels.

13.38 (b)

The critical depth for a rectangular channel is given by,

$$y_c^3 = \frac{q^2}{g}$$

But $q = \frac{Q}{B} = \frac{100}{4.5} = 22.22 \text{ m}^2\text{/sec}$

$\therefore \quad y_c^3 = \frac{(22.22)^2}{9.81}$

$\Rightarrow \quad y_c = 3.69 \text{ m}$

13.39 (d)

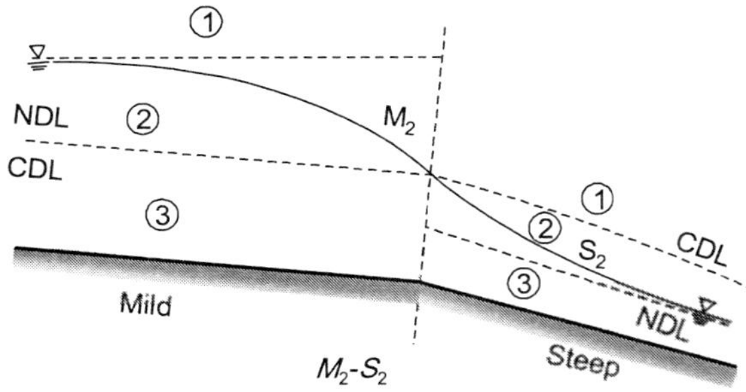

13.40 (a)

Flow is subcritical and if width of flow is reduced discharge intensity will increase. For increased discharge intensity, depth of flow decreases for particular specific energy.

Note: *Upstream condition is not affected before choking condition at downstream section.*

13.41 (c)

Let the width of rectangular channel be B and depth of flow be y.

Area of flow, $A = By$

Wetted perimeter,

$$P = B + 2y = \frac{A}{y} + 2y$$

For hydraulically most efficient section, P should be minimum.

$$\therefore \quad \frac{dP}{dy} = 0$$

$$\Rightarrow \quad \frac{-A}{y^2} + 2 = 0$$

$$\Rightarrow \quad A_e = 2y_e^2$$

$$\Rightarrow \quad B_e y_e = 2y_e^2$$

$$\Rightarrow \quad B_e = 2y_e$$

Hydraulic radius,

$$R = \frac{A}{P} = \frac{2y_e^2}{2y_e + 2y_e} = \frac{y_e}{2}$$

Top width,

$$T = B = 2y_e$$

Perimeter,

$$P = B + 2y_e = 2y_e + 2y_e = 4y_e$$

Hydraulic depth,

$$D = \frac{A}{T} = \frac{2y_e^2}{2y_e} = y_e$$

13.42 (d)

Froude Number for a rectangular channel is given by,

$$(F_r)^{2/3} = \frac{y_c}{y_n}$$

$$y_c = 1.5\,(0.8)^{2/3} = 1.292 \text{ m}$$

13.43 (a)

It may be noted that the **critical flow condition** is governed solely by the channel geometry and discharge. Other channel properties such as bed slope and roughness do not influence the critical flow condition for any given discharge.

13.44 (c)

$$\frac{2q^2}{g} = y_1 y_2 (y_1 + y_2)$$

$$\Rightarrow \quad \frac{2 \times 9^2}{9.81} = 0.46\, y_2\,(0.46 + y_2)$$

$$\Rightarrow \quad y_2^2 + 0.46\, y_2 - 35.899 = 0$$

$$y_2 = 5.765 \text{ mn} = -6.225 \text{ m}$$

13.45 (b)

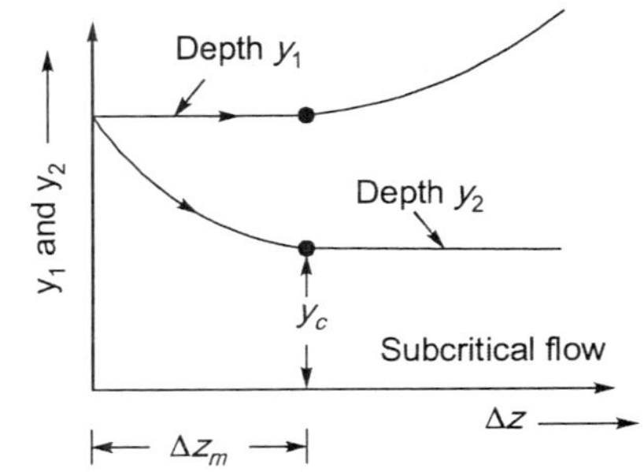

Variation of y_1 and y_2 in subcritical flow over a hump

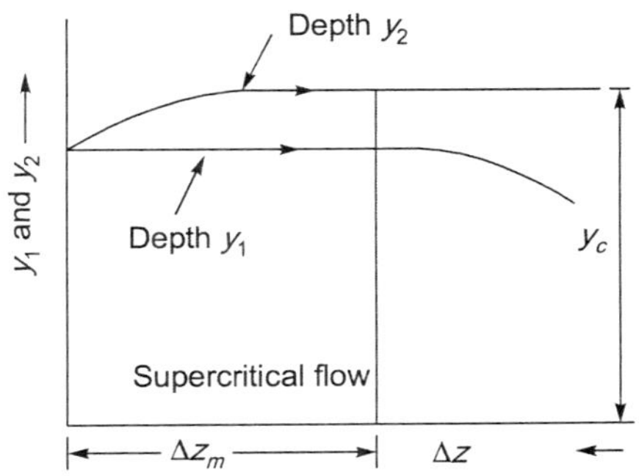

13.46 (d)

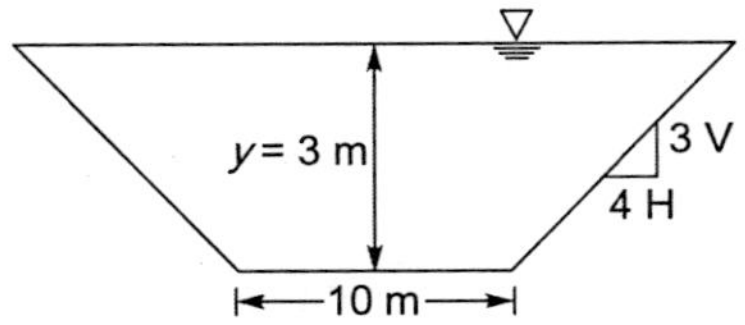

$$\because \quad A = (B + ny)\,y = \left(10+\frac{4}{3}\times 3\right)\times 3$$

$$= (10 + 4)3 = 42 \text{ m}^2$$

$$P = B+2y\sqrt{n^2+1}$$

$$= 10 + 6\sqrt{(4/3)^2+1} = 10+6\times\frac{5}{3} = 20\text{m}$$

$$\therefore \quad R = \frac{A}{P} = \frac{42}{20} = 2.1.$$

13.47 (a)

$$0.8^2 = \frac{100^2\times 5}{10\times(5y)^3}$$

$$\Rightarrow \quad y = 3.96 \text{ m}$$

13.48 (a)

A control section is defined as a section is which a fixed relationship exits between the discharge and depth of flow.

For subcritical flow, the control section for gradually varied flow profiles is at the downstream end while supercritical flows are governed by control sections existing at the upstream end of the channel section.

13.49 (d)

Method-I

The discharge in the channel is given by

$$Q = \frac{1}{n}AR^{2/3}S^{1/2}$$

$$\Rightarrow \quad Q = \frac{1}{n}(By)\left(\frac{By}{B+2y}\right)^{2/3} S^{1/2}$$

For a wide rectangular channel,

$$B \gg y$$

$$\therefore \quad Q = \frac{1}{n}(By)(y)^{2/3}S^{1/2}$$

$$\therefore \quad Q \propto y^{5/3}$$

Let $Q_1 = k(y)^{5/3}$

$\therefore$ $Q_2 = k(1.1y)^{5/3}$

$$\therefore \quad \frac{Q_2-Q_1}{Q_1}\times 100 = \frac{(1.1y)^{5/3}-(y)^{5/3}}{(y)^{5/3}}\times 100$$

$$= 17.21\%$$

Method-II **(Approximate Method)**

$$Q = \frac{1}{n}By\left(\frac{By}{B+2y}\right)^{2/3}\sqrt{S}$$

For $B \gg y$

$$Q \propto y^{5/3}$$

$$\frac{dQ}{Q} = \frac{5}{3}\frac{dy}{y}$$

$$\frac{dQ}{Q} = \frac{5}{3}\times 10\% = 17.33\%$$

13.50 (b)

$$n = 0.015$$

$$Q = 1 \text{ m}^3/\text{s}$$

Normal depth of flow between 0.76 m to 0.8 m.

If prevailing normal depth of flow is not exceeded, there must not be chocking of the section or there must be just chocking.

Thus the width of the section should be such that for the prevailing specific energy there should be critical flow at the contracted section.

i.e. $$\frac{3}{2}\left(\frac{q^2}{g}\right)^{1/3} = E_C = E_{initial}$$

$$\Rightarrow \quad \frac{3}{2}\left[\frac{\left(\frac{Q}{B_{min}}\right)^2}{g}\right]^{1/3} = E_{initial} \qquad \text{...(i)}$$

Let us now calculate $E_{initial}$,

$$Q = \frac{1}{2}AR^{2/3}S_0^{1/2}$$

$$\Rightarrow \quad 1 = \frac{1}{0.015}(3y)\left(\frac{3y}{3+2y}\right)^{2/3}(0.0001)^{1/2}$$

$$\Rightarrow \quad y = 0.78 \text{ m}$$

Note: Which lies between 0.76-0.80 m.

$$\therefore \quad E_{initial} = y+\frac{q^2}{2gy^2}$$

$$= 0.78+\frac{\left(\frac{1}{3}\right)^2}{2\times 9.81\times(0.78)^2} = 0.7893 \text{ m}$$

From eq. (i)

$$\Rightarrow \quad \frac{3}{2}\left[\frac{\left(\frac{Q}{B_{min}}\right)^2}{g}\right]^{1/3} = 0.7893$$

$$\Rightarrow \quad \frac{3}{2}\frac{(1)^{2/3}}{g^{1/3}(B_{min})^{2/3}} = 0.7893$$

$$B_{min} = 0.836 \text{ m}$$

13.51 Sol.

$$Q = 4 \text{ m}^3\text{/s}, \; B = 2.5 \text{ m}$$

$$\Rightarrow \quad q = \frac{4}{2.5} \text{m}^3\text{/s/m} = 1.6 \text{ m}^3\text{/s/m}$$

$$y_c^3 = \frac{q^2}{g} \Rightarrow y_c = 0.639 \text{ m}$$

At critical depth velocity head $= \frac{y_c}{2}$

$$\Rightarrow \quad \frac{V^2}{2g} = \frac{0.639}{2}$$

$$\Rightarrow \quad V = \sqrt{0.639 \times g} = 2.504 \text{ m/s}$$

13.52 (c)

Given diameter, $d = 1$ m

Slope, $S = \frac{1}{1000}$

$n = 0.01$

$$\frac{Q_{full}}{Q_{half}} = \frac{\left[\frac{A}{n}(R)^{2/3}S^{1/2}\right]_{full}}{\left[\frac{A}{n}(R)^{2/3}S^{1/2}\right]_{half}}$$

$$= \frac{\left[AR^{2/3}\right]_{full}}{\left[AR^{2/3}\right]_{half}}$$

$$R_{Full} = R_{Half} = \frac{d}{4} = \frac{\frac{\pi d^2}{4}}{\frac{\pi d^2}{8}} = 2$$

13.53 Sol.

$$y_2 = 0.8 \text{ m}$$

$$v_2 = 1 \text{ m/s}$$

$$F_{r_2} = \frac{v_2}{\sqrt{gD}} = \frac{1}{\sqrt{10 \times \frac{0.8 \times 2}{2}}}$$

$$F_{r_2} = 0.3535$$

$$y_1 = \frac{y_2}{2}\left(-1+\sqrt{1+8F_{r_2}^2}\right)$$

$$= \frac{0.8}{2}\left(-1+\sqrt{1+8\times 0.3535^2}\right) = 0.165643$$

$$\Rightarrow \quad v_1 y_1 = v_2 y_2$$

$$\Rightarrow \quad v_1 = \frac{1 \times 0.8}{0.165643} = 4.83 \text{ m/s}$$

13.54 Sol.

$$\text{Adverse slope} = -\frac{1}{10000}$$

$$Q = 4 \text{ m}^3\text{/s}$$

$$n = 0.01$$

$$y = 0.5 \text{ m}$$

$$\frac{dy}{dx} = \frac{S_0 - S_f}{1 - F_r^2}$$

$$F_r = \frac{V}{\sqrt{gy}} = \frac{Q}{2y\sqrt{gy}}$$

Note: If width of channel is given always use $R = A/P$ and if the width is not given then force a wide rectangular channel use $R = y$.

$$= \frac{4}{2 \times 0.5\sqrt{10 \times 0.5}} = 1.789$$

$$R = \frac{A}{P} = \frac{2 \times 0.5}{2 + 2 \times 0.5} = \frac{1}{3}$$

$$Q = \frac{1}{n}AR^{2/3}S_f^{1/2}$$

On solving for S_f,

$$S_f^{1/2} = 6.923 \times 10^{-3}$$

$$\frac{dy}{dx} = \frac{-\frac{1}{10000} - 6.923 \times 10^{-3}}{1-(1.789)^2}$$

$$= 3.191 \times 10^{-3} \simeq 0.0032$$

1355 (b)

$$y_p < y_s < y_a$$

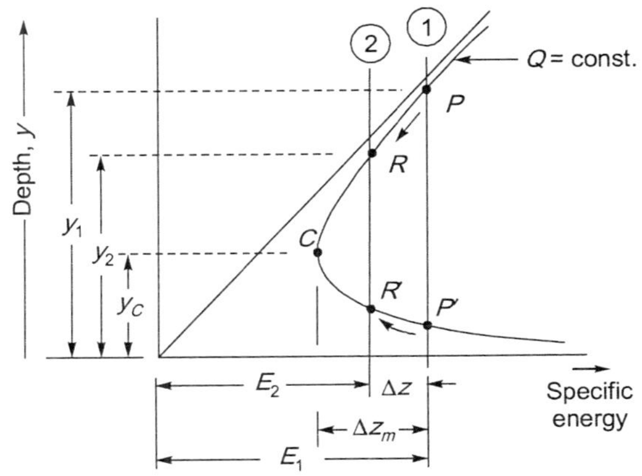

13.56 Sol.

From Belenger's momentum equation.

$$\frac{y_2}{y_1} = \frac{1}{2}\left[-1+\sqrt{1+8F_1^2}\right]$$

$$= \frac{1}{2}\left[-1+\sqrt{1+8\times100}\right]$$

or, $$\frac{y_2}{y_1} = 13.65$$

13.57 (b)

$$Q = \frac{1}{n}AR^{2/3}S^{1/2}$$

$$A = By_n$$

$$R = \frac{By_n}{B+2y_n}$$

$$16 = \frac{1}{0.012}\times By_n\times\left(\frac{By_n}{B+2y_n}\right)^{2/3}\times(0.001)^{1/2}$$

or, $$6.07 = 4\times\frac{y_n^{5/3}}{(4+2y_n)^{2/3}}$$

or, $$1.5178 = \frac{y_n^{5/3}}{(4+2y_n)^{2/3}}$$

or, $$y_n = \left[1.5178\times(4\times 2y)^{2/3}\right]^{3/5}$$

$\therefore$ $$y_n = 3.3 \text{ m}$$

Critical depth, $y_c = \left(\frac{q^2}{g}\right)^{1/3} = \left(\frac{4^2}{10}\right)^{1/3} = 1.169$

$y_n > y_c \Rightarrow$ Flow is subcritical.

Hence channel slope is **mild**.

13.58 Sol.

For hydraulically efficient trapezoidal channel is half of a regular hexagon.

Hence all sides are equal.

From figure,

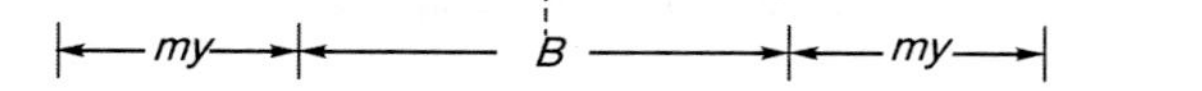

$\Rightarrow$ $B\sin 60° = d$

$\therefore$ $$B = \frac{d}{\sin 60°} = 2.31\text{m}$$

13.59 Sol.

The maximum hump (Δz) that can be provided is given by,

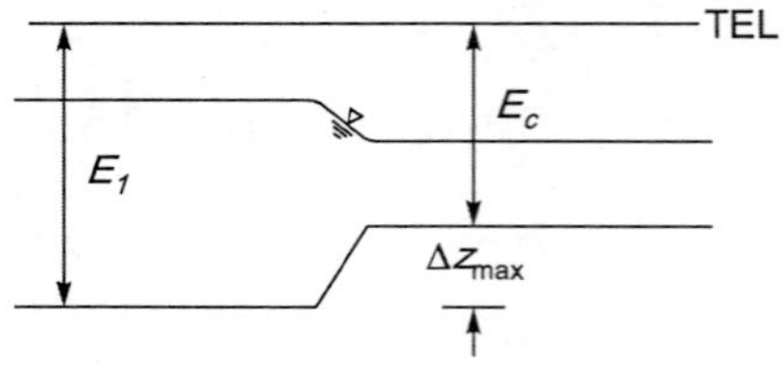

$$y+\frac{Q^2}{2gA^2} = \Delta Z_{hump}+\frac{3}{2}y_c$$

Here, $A = 3\times0.5 = 1.5 \text{ m}^2$

$$y_c = \left(\frac{Q^2}{gB^2}\right)^{1/3} = \left(\frac{6^2}{9.81\times3^2}\right)^{1/3}$$

$$= 0.7415 \text{ m}$$

$$E_1 = EC\times\Delta z_{max}$$

$$\Rightarrow 0.5+\frac{6^2}{2\times9.81\times1.5^2} = \Delta Z_{hump}+\frac{3}{2}\times0.7415$$

$\Rightarrow$ $$\Delta Z_{hump} = 0.20 \text{ m}$$

13.60 (c)

$$Q = \frac{1}{n}\times A\times R^{2/3}\times S^{1/2}$$

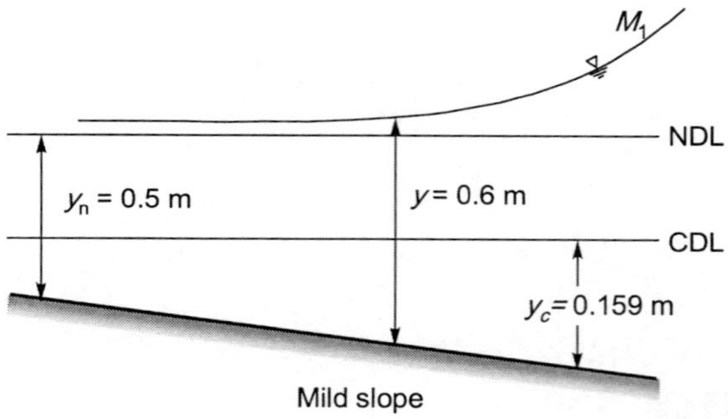

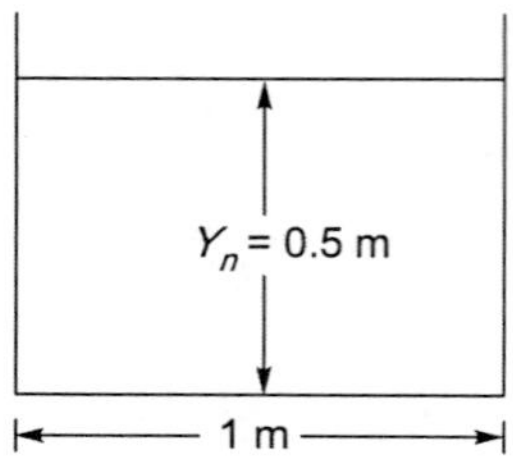

$$A = 1\times0.5 = 0.5 \text{ m}^2$$

$$P = 1+(0.5\times2) = 2 \text{ m}$$

$$Q = \frac{1}{0.04}\times0.5\times\left(\frac{0.5}{2}\right)^{2/3}\times(0.0016)^{1/2}$$

$$= 0.1984 \text{ m}^3/\text{s}$$

$$q = \frac{Q}{1} = 0.1984 \text{ m}^3/\text{s/m}$$

For rectangular channel,

$$y_c = \left(\frac{q^2}{g}\right)^{1/3} = \left(\frac{0.1984^2}{9.81}\right)^{1/3} = 0.159 \text{ m}$$

Since, $y_n > y_c$ (Thus it is Mild Slope)

13.61 Sol.

Applying **Bernoulli's energy balance equation**

$$y_1 + \frac{Q^2}{2gA_1^2} = y_2 + \frac{Q^2}{2gA_2^2}$$

Given, $y_1 = 1$ m, $b = 1$ m, $y_2 = 0.2$

$$\Rightarrow 1 + \frac{q^2 \times 1^2}{2g(1\times 1)^2} = 0.2 + \frac{q^2 \times 1^2}{2g \times (0.2 \times 1)^2}$$

$$\Rightarrow \quad 0.8 = \frac{q^2}{2g}\left(\frac{1}{(0.2)^2} - 1\right)$$

$$0.8 = \frac{q^2}{20} \times 24$$

$$q = 0.816 \text{ m}^2/\text{s}$$

13.62 Sol.

Let Supercritical depth = Y

Subcritical depth = $2Y$

As the specific energy E remains constant for alternate depths,

$$\therefore \quad E_1 = E_2$$

$$Y_1 + \frac{Q^2}{2gB^2Y_1^2} = Y_2 + \frac{Q^2}{2gB^2Y_2^2}$$

$$\Rightarrow Y + \frac{2^2}{2 \times 9.81 \times 1^2 \times Y^2} = 2Y + \frac{2^2}{2 \times 9.81 \times 1 \times (2Y)^2}$$

$$\Rightarrow \quad Y + \frac{0.204}{Y^2} = 2Y + \frac{0.051}{Y^2}$$

$$\Rightarrow \quad \frac{0.204 - 0.051}{Y^2} = Y$$

$$\Rightarrow \quad Y^3 = 0.1530$$

$$\Rightarrow \quad Y = 0.534$$

$\therefore$ Subcritical depth

$= 0.534 \times 2 = 1.0697 \simeq 1.07$ m

13.63 Sol.

$$E_c = \frac{3}{2}y_c$$

$$y_c = \left(\frac{q^2}{g}\right)^{1/3}$$

Here, $$q = \frac{20}{10} = 2$$

$$\Rightarrow \quad E_c = \frac{3}{2}\left(\frac{2^2}{9.81}\right)^{1/3} = 1.11 \text{ m}$$

13.64 Sol.

$$\frac{dy}{dx} = \frac{1 - e^{-\frac{10}{3}\ln y}}{250 - 45e^{-3\ln y}}$$

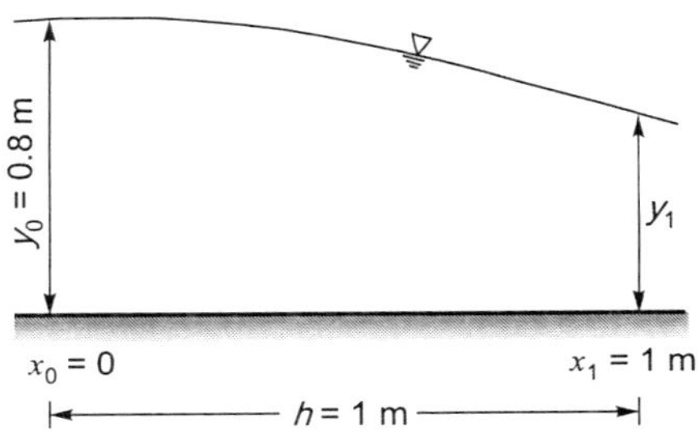

$$y_1 = y_0 + hf(x_0, y_0)$$

$$\Rightarrow \quad y_1 = 0.8 + 1\left[\frac{1 - e^{-\frac{10}{3}\ln 0.8}}{250 - 45e^{-3\ln 0.8}}\right]$$

$$= 0.8 + 1\left(\frac{-1.1039}{162.109}\right) = 0.793 \text{ m}$$

13.65 (a)

Depth with same specific energy are called Alternate depths of flow. It represents a subcritical depth of flow (Y_2) and a supercriitical depth of flow (Y_1).

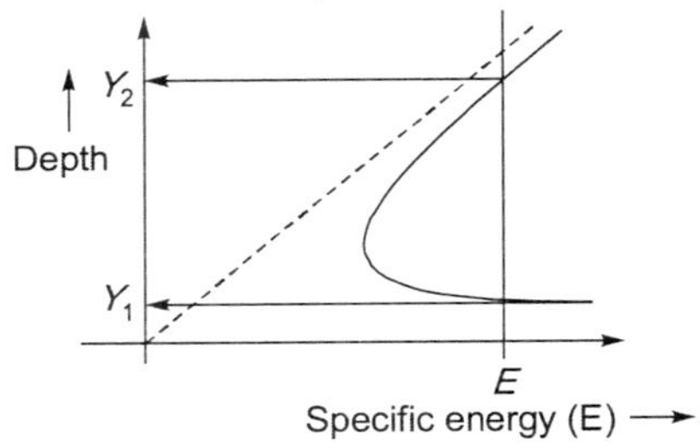

13.66 (a)

$$U_y = 1.25\, y^{1/6}$$

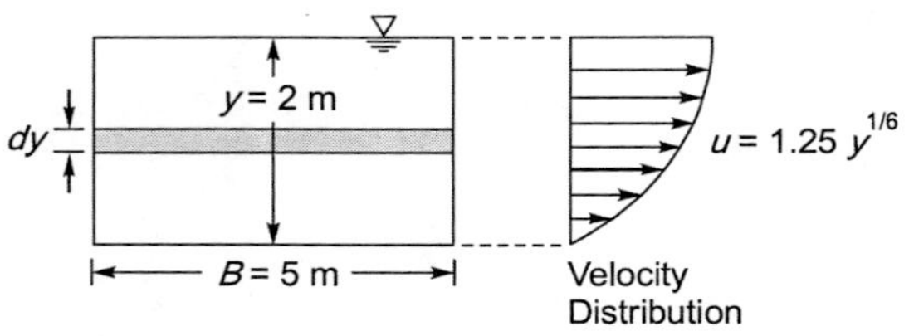

Discharge through elementary strip,

$$dQ = 5dy \times u$$

$\therefore$ Total discharge,

$$Q = \int_0^2 (5dy)\left(1.25y^{1/6}\right)$$

$$\Rightarrow \quad Q = 6.25\int_0^2 y^{1/6}dy = 6.25 \times \left[\frac{y^{1/6+1}}{\frac{1}{6}+1}\right]_0^2$$

$$\because \quad Q = 12.026 \text{ m}^3$$

$\therefore$ Discharge per unit width,

$$q = \frac{12.026}{5} = 2.405 \text{ m}^3\text{/s/m}$$

13.67 (b)

Critical velocity head

$$= \frac{V_C^2}{2g} = \frac{Q_C^2}{2gA^2} \quad \text{...(i)}$$

For a critical flow

$$\frac{Q_c^2 T}{gA^3} = 1$$

$$\therefore \quad \frac{Q_c^2}{gA^2} = \frac{A}{T} \quad \text{...(ii)}$$

From (i) and (ii),

$$\frac{v_c^2}{2g} = \frac{A}{2T}$$

$\therefore$ For Rectangular channel

$$\left(\frac{v_c^2}{2g}\right) = \frac{B \cdot Y_c}{2B} = \frac{Y_c}{2}$$

$$\therefore \quad \frac{v_c^2/2g}{Y_c} = \frac{1}{2}$$

13.68 (d)

Manning's equation is given by

$$V = \frac{1}{n}R^{2/3}S^{1/2}$$

$$\Rightarrow \quad 2.0 = \frac{1}{n}\left(\frac{BY}{B+2Y}\right)^{2/3} \times S^{1/2}$$

$$\Rightarrow \quad 2.0 = \frac{1}{n}\left(\frac{5\times 2}{5+(2\times 2)}\right)^{2/3} \times (0.001)^{1/2}$$

$$\Rightarrow \quad n = 0.01696 = 0.017$$

13.69 Sol.

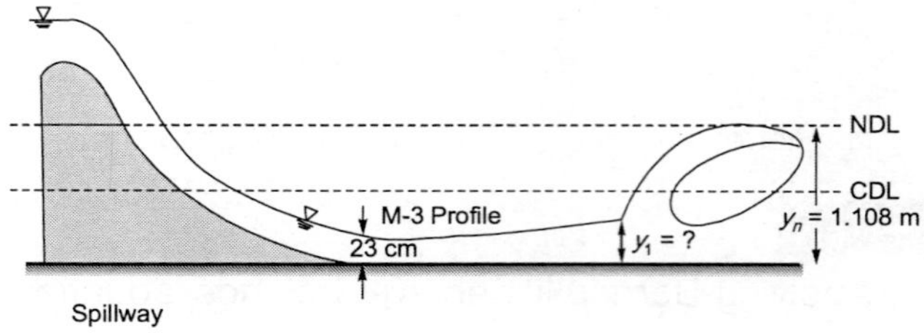

$$y = 0.23\text{m}, \; V = 8.1 \text{ m/s}$$

$$\therefore \quad q = Vy = 0.23 \times 8.1$$

$$= 1.863 \text{ m}^3\text{/s-m}$$

$$S_0 = \frac{1}{800}$$

$$n = 0.015$$

y_n = Normal depth of flow

$R = y$ for wide rectangular channel

By Manning's equation

$$\therefore \quad q = \frac{y_n}{n}R^{2/3}S_0^{1/2}$$

$$\Rightarrow 1.863 = \frac{y_n^{5/3}}{0.015} \times \left(\frac{1}{1800}\right)^{1/2}$$

$$\Rightarrow \quad y_n = 1.108 \text{ m}$$

y_1 is conjugate depth of y_0,

$$\therefore \quad \frac{y_1}{y_n} = \frac{-1+\sqrt{1+8Fr_n^2}}{2} \qquad \left(Fr_n^2 = \frac{q^2}{gy_n^3}\right)$$

$$\therefore \quad y_1 = \frac{-1+\sqrt{1+8\times\frac{1.863^2}{9.81\times 1.108^3}}}{2} \times 1.108$$

$$= 0.418 \text{ m} \simeq 0.42 \text{ m}$$

13.70 Sol.

$$\text{Critical depth } (Y_C) = \left(\frac{q^2}{g}\right)^{1/3}$$

$$= \left(\frac{1.5^2}{9.81}\right)^{1/3} = 0.612 \text{ m}$$

Critical velocity $(V_C) = \sqrt{gY_C} = \sqrt{9.81 \times 0.612}$
$= 2.45$ m/s

13.71 (a)

Assuming channel bed to be horizontal and frictionless.

Discharge per unit width,

$$q = \frac{12}{6} = 2 \text{ m}^3\text{/s/m}$$

Initial Froude No.

$$F_r = \left(\frac{q^2}{gY_1^3}\right)^{1/2}$$

$$= \left(\frac{2^2}{9.81 \times 0.3^3}\right)^{1/2} = 3.88$$

From Belenger's Momentum equation for a rectangular channel

$$\frac{Y_2}{Y_1} = \frac{1}{2}\left(-1 + \sqrt{1 + 8F_1^2}\right)$$

$$= \frac{1}{2}\left(-1 + \sqrt{1 + 8 \times 3.88^2}\right)$$

$$= 5.018$$

$\therefore \quad Y_2 = 5.018 \times 0.3 = 1.505$ m

Head loss in the jump,

$$h_L = \frac{(Y_2 - Y_1)^3}{4Y_1Y_2}$$

$$= \frac{(1.505 - 0.3)^3}{4 \times 1.505 \times 0.3}$$

$$= 0.968 \text{ m}$$

Power lost in the jump

$= \gamma_w Q h_L$
$= (9.81 \times 12 \times 0.968)$ kW
$= 114.04$ kW

13.72 Sol.

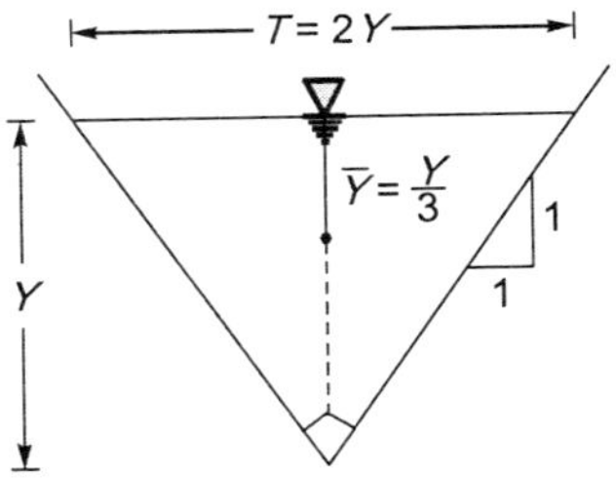

Area of triangular channel,

$$A = \frac{1}{2} \times 2Y \times Y = Y^2$$

$$\bar{Y} = \frac{Y}{3}$$

For a horizontal and frictionless channel

$$\text{Specific Force } (F) = A\bar{Y} + \frac{Q^2}{Ag} = \text{Constant}$$

$$\Rightarrow Y^2\left(\frac{Y}{3}\right) + \frac{Q^2}{(Y^2)g} = \text{Constant}$$

$$\Rightarrow \quad \frac{Y^3}{3} + \frac{Q^2}{gY^2} = \text{Constant}$$

If Y_1 and Y_2 are conjugate depths

$$\frac{Y_1^3}{3} + \frac{Q^2}{gY_1^2} = \frac{Y_2^3}{3} + \frac{Q^2}{gY_2}$$

$$\Rightarrow \frac{0.5^3}{3} + \frac{Q^2}{g \times 0.5^2} = \frac{1.5^3}{3} + \frac{Q^2}{g \times 1.5^2}$$

$$\Rightarrow \quad \frac{1.5^3}{3} - \frac{0.5^3}{3} = \frac{Q^2}{g}\left(\frac{1}{0.5^2} - \frac{1}{1.5^2}\right)$$

$$Q = 1.728 \text{ m}^3\text{/sec}$$

13.73 (d)

Efficient channel section

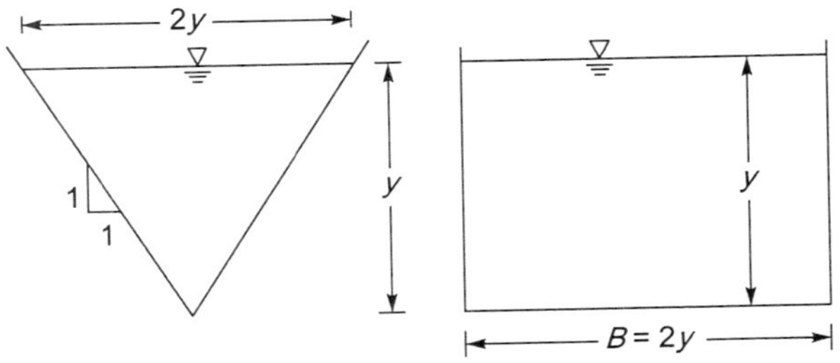

$A = y^2$	$A = 2y^2$
$P = 2\sqrt{2}y$	$P = 4y$
$R_I = \dfrac{y}{2\sqrt{2}}$	$R_{II} = \dfrac{y}{2}$

$$\therefore \quad \frac{R_I}{R_{II}} = \frac{1}{\sqrt{2}}$$

13.74 (b)

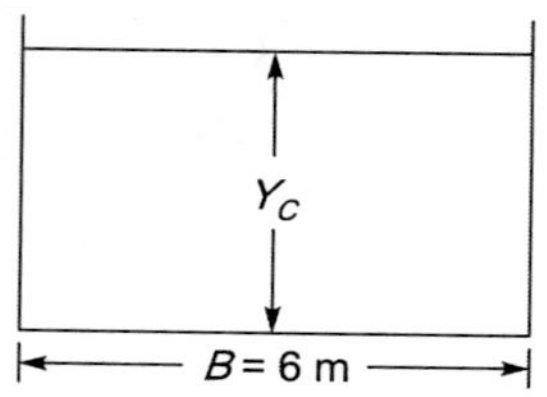

Minimum specific energy will correspond to a critical flow condition.

The critical depth,

$$Y_C = \left[\frac{q^2}{g}\right]^{1/3} = \left[\frac{(20/6)^2}{9.81}\right]^{1/3} = 1.042 \text{ m}$$

13.75 (a)

Steeper slope → More bed slope than steep slope.

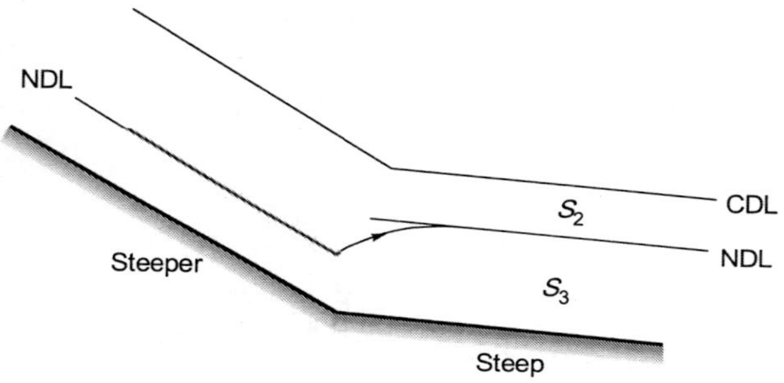

13.76 Sol.

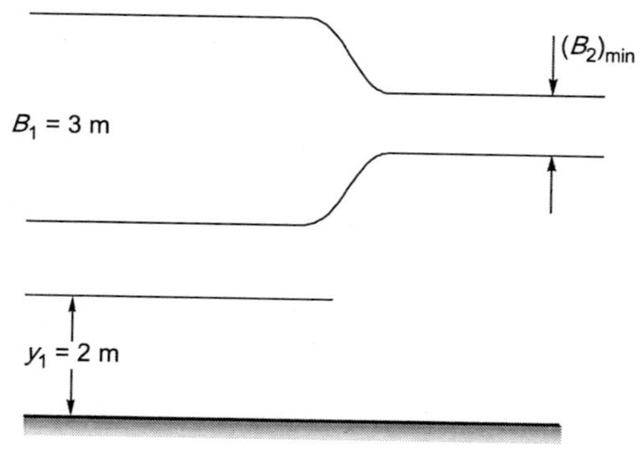

$$Q = A_1 V_1$$

$$\Rightarrow \quad 12 = 3 \times 2 \times V_1$$

$$\Rightarrow \quad V_1 = 2 \text{ m/s}$$

$$\therefore \quad F_{r_1} = \frac{V_1}{\sqrt{gy_1}} = \frac{2}{\sqrt{g \times 2}} = 0.45 < 1$$

(Subcritical flow)

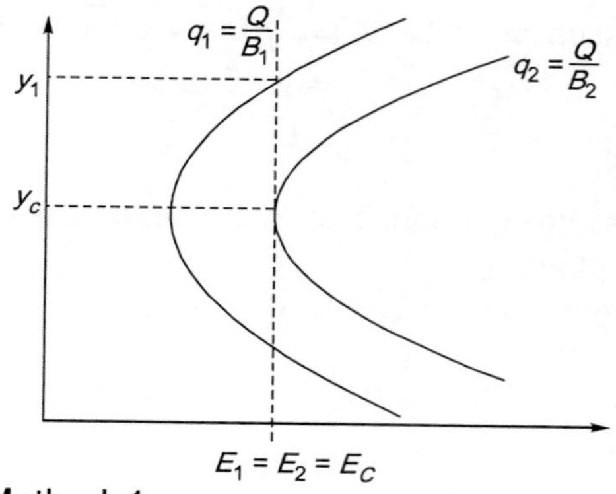

Method 1:

$$(B_2)_{min} = \sqrt{\frac{27Q^2}{8gE_1^3}} \quad \text{... (i)}$$

$$\Rightarrow \quad E_1 = y_1 + \frac{V_1^2}{2g} = 2 + \frac{(2)^2}{2g} = 2.2038 \text{ m}$$

Using (i),

$$(B_2)_{min} = \sqrt{\frac{27 \times 12^2}{8 \times 9.81 \times (2.2038)^3}} = 2.15 \text{ m}$$

Method 2: $E_1 = E_C = \frac{3}{2} y_c$

$$y_c = 1.46 \text{ m}$$

$$\Rightarrow \quad \left(\frac{q_2^2}{g}\right)^{1/3} = 1.46 \text{ m}$$

$$\Rightarrow \quad q_2 = 5.57 \text{ m}^3\text{/s/m}$$

$$\therefore \quad \frac{Q}{(B_2)_{min}} = 5.57$$

$$\Rightarrow \quad (B_2)_{min} = 2.15 \text{ m}$$

13.77 Sol.

Given, Discharge in channel,

$$Q = 30 \text{ m}^3\text{/s}$$

Width of channel,

$$B = 6 \text{ m}$$

Energy loss,

$$E_L = 1.6 \text{ m}$$

Depth before jump,

$$y_1 = 0.5 \text{ m}$$

Froude's number after jump,

$$F_2 = ?$$

Method I:

$$\therefore \quad V_1 = \frac{Q}{By_1} = \frac{30}{6 \times 0.5} = 10 \text{ m/s}$$

$$\therefore \quad F_{r1} = \frac{V_1}{\sqrt{gy_1}} = \frac{10}{\sqrt{10 \times 0.5}} = 4.47$$

We know for conjugatal sequent depths,

$$\frac{y_2}{y_1} = \frac{-1+\sqrt{1+8F_{r1}^2}}{2}$$

$$\Rightarrow \quad \frac{y_2}{0.5} = \frac{-1+\sqrt{1+84.47^2}}{2}$$

$$\Rightarrow \quad y_2 = 2.92 \text{ m}$$

$$\therefore \quad F_{r2}^2 = \frac{Q^2T}{gA_2^3} = \frac{30^2 \times 6}{10 \times (6 \times 2.95)^3} = 0.100$$

$$\Rightarrow \quad F_{r2} = 0.32$$

[This should be the answer]

Method II:

Now as per given energy loss,

$$E_L = \frac{(y_2 - y_1)^3}{4y_1y_2}$$

$$\Rightarrow \quad 1.6 = \frac{(y_2 - y_1)^3}{4y_1y_2}$$

$$\Rightarrow \quad 1.6 \times 4 \times 0.5 \times y_2 = (y_2 - 0.5)^3$$

$$\Rightarrow \quad y_2 = 2.5 \text{ m}, -0.0527 \text{ m}, -0.947 \text{ m}$$

$$\therefore \quad y_2 = 2.5 \text{ m}$$

$$\therefore \quad F_{r_2} = \sqrt{\frac{Q^2T}{gA_2^3}} = \sqrt{\frac{30^2 \times 6}{10 \times (6 \times 2.5)^3}} = 0.4$$

Note: We are getting two different values of Froude number from given data.

$$E_L = \frac{(y_2 - y_1)^3}{4y_1y_2} = \frac{(2.922 - 0.5)^3}{4 \times 2.922 \times 0.5}$$

$$E_L = 2.43 \text{ m}$$

[This is not equal to energy loss given]

Therefore given data is incosistent.

13.78 (a, b, c)

Subcritical flow:

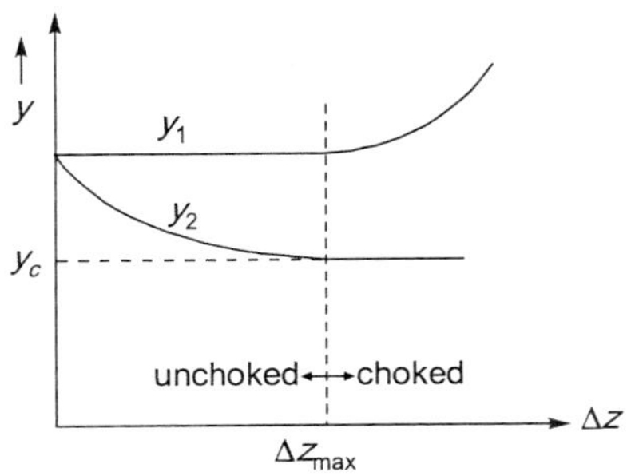

Supercritical flow:

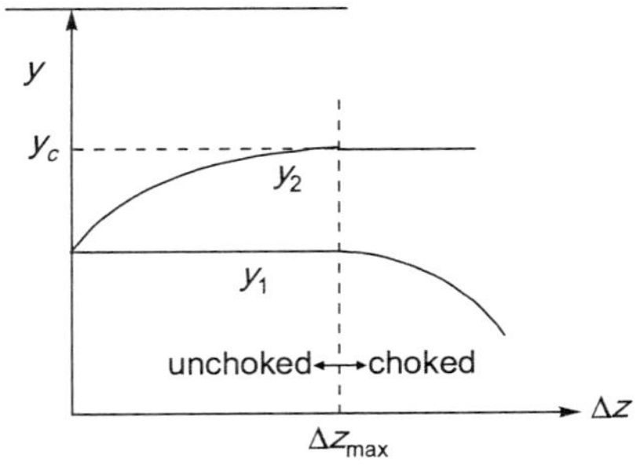

13.79 (c)

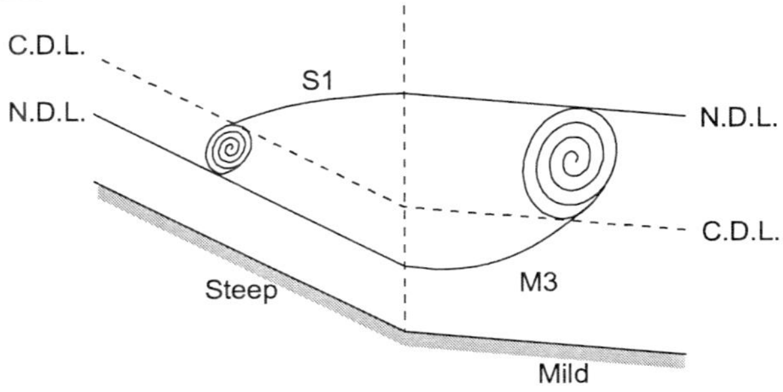

13.80 (a)

For a wide Rectangular channel ($B >>> y$),

$$R = \frac{A}{P} = \frac{By}{B+2y} \approx \frac{By}{B} = y$$

Dischare per unit width, q is given by,

$$\Rightarrow \quad q = \left(\frac{1}{n}y^{2/3}S^{1/2}\right)y$$

$$\Rightarrow \quad q = \frac{1}{n}y^{5/3}S^{1/2}$$

Calculation of S_c,

$$y_c = \left(\frac{q^2}{g}\right)^{1/3}$$

$$\Rightarrow \quad q = \frac{1}{n}y_c^{5/3}S_c^{1/2}$$

$$\Rightarrow q = \frac{1}{n}\left(\frac{q^2}{g}\right)^{5/9} S_c^{1/2} = \frac{1}{n}\frac{q^{10/9}}{g^{5/9}}S_c^{1/2}$$

$$\Rightarrow \quad S_c^{1/2} = \frac{nqg^{5/9}}{q^{10/9}}$$

$$\Rightarrow \quad S_c = \left(\frac{ng^{5/9}}{q^{1/9}}\right)^2 = \left(\frac{0.01 \times 9.8^{5/9}}{70^{1/9}}\right)^2$$

$= 0.00049 \quad (> 0.0001 \text{ and } < 0.0009)$

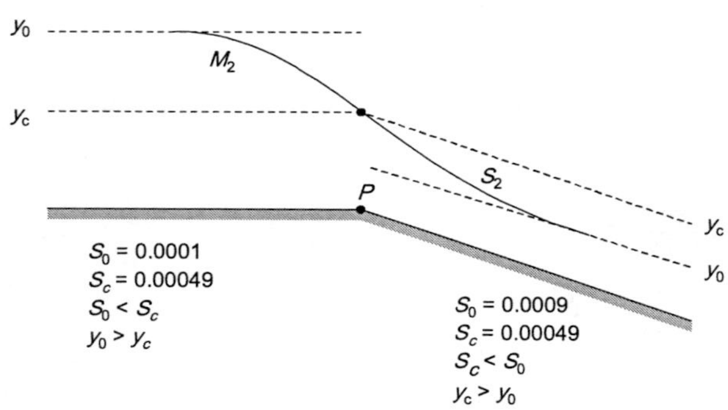

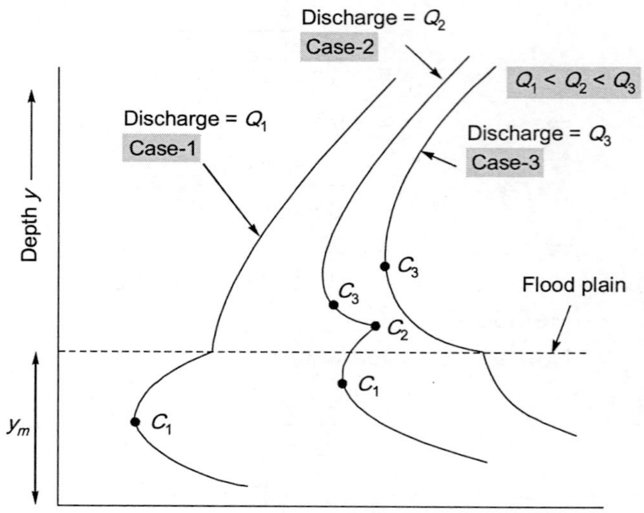

13.81 Sol.

Given, $\quad y_1 = 0.2$ m, $\; y_2 = 1$ m

As we know

$$\frac{2q^2}{g} = y_{1a}y_2(y_1 + y_2)$$

$$\Rightarrow \quad \frac{2q^2}{10} = 0.2 \times 1(0.2 + 1)$$

$$\Rightarrow \quad q^2 = \frac{10}{2} \times 0.2 \times 1.2 = 1.2$$

$$q = 1.095 \text{ m}^3/\text{s/m}$$

$$\text{Now, specific force} = \frac{Q^2}{Ag} + A\bar{y} = \frac{q^2B^2}{By_1g} + \frac{By_1y_1}{2}$$

$$= \left(\frac{1.095^2}{0.2 \times 102} + \frac{0.2^2}{2}\right) \times 1 = 0.6195 \text{ m}^3 \simeq 0.62 \text{ m}^3$$

So, Specific force per unit width = 0.62 m³/m

13.82 (a)

Case 1:

For $Q = Q_1$ and depth of flow $(y) < y_m$ then, y_{c1} is the critical depth.

Case 2:

For $Q = Q_2$ and depth of flow $(y) > y_m$ then, y_{c1} , y_{c2} and y_{c3} be the critical depth.

Case 3:

Only y_{c3} will exist at any discharge Q_3 such that $y_{c3} > y_3$

So, the maximum number of critical depth would be three.

13.83 (a)

Froude number, $F_r = \dfrac{V}{\sqrt{gD}}$; where

D = hydraulic depth $= \dfrac{A}{T}$

$$\Rightarrow \quad F_r^2 = \frac{V^2}{gD}$$

$$\Rightarrow \quad F_r^2 = \frac{Q^2T}{gA^2A} = \frac{Q^2T}{gA^3}$$

For critical condition,

$$F_r = 1$$

$$\Rightarrow \quad \frac{Q^2T_c}{gA^3} = 1$$

$$\Rightarrow \quad \frac{Q^2}{g} = \frac{A_c^3}{T_c}$$

Now, taking into account, kinetic energy correction factor,

$$\frac{\alpha Q^2}{g} = \frac{A_c^3}{T_c}$$

13.84 (b)

Degree of freedom in a mobile boundary channel are bed width, depth, slope and planiform.

13.85 Sol.

Unit discharge(q) = 7.5 m³/s/m

$$\because \quad \frac{q^2}{g} = \frac{y_1y_2(y_1 + y_2)}{2}$$

$$\Rightarrow \quad \frac{7.5^2}{9.81} = 0.5y_2\left(\frac{0.5+y_2}{2}\right)$$

Tail water depth, y_2 = 4.55 m

13.86 Sol.

Discharge, $Q = 30\ m^3/s$

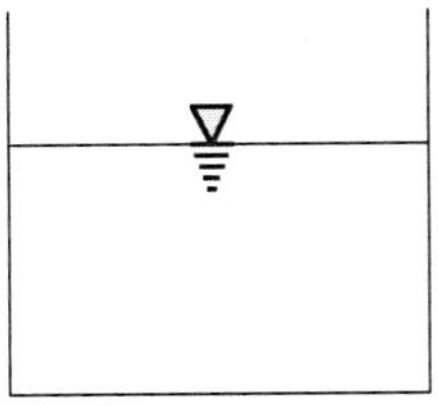

2 m

$g = 10\ m/s^2$

Width of channel, B = 2 m

Now, discharge intensity,

$$q = \frac{Q}{B} = \frac{30}{2} = 15\ m^3/s/m$$

Now critical energy, E_c is given as

$$E_c = y_c + \alpha\cdot\frac{v^2}{2g}$$

$$\Rightarrow \quad E_c = y_c + \alpha\cdot\frac{q^2}{2g\cdot y_c^2}$$

$$\Rightarrow \quad 1.5y_c = y_c + \alpha\cdot\frac{q^2}{2g\cdot y_c^2}$$

$$\Rightarrow \quad 0.5y_c = \alpha\cdot\frac{q^2}{2g\cdot y_c^2}$$

$$\Rightarrow \quad y_c^3 = \alpha\cdot\frac{q^2}{g}$$

$$\Rightarrow \quad y_c = \left(\frac{\alpha\cdot q^2}{g}\right)^{1/3}$$

$$\Rightarrow \quad y_c = \left(\frac{1.1\times 15^2}{10}\right)^{1/3} = 2.914\ m \approx 2.91m$$

Alternatively,

At critical condition,

$$\frac{\alpha Q^2}{g} = \frac{A_c^3}{T_c}$$

Where, A_c = Critical area of flow $= (By_c) = 2y_c$

T_c = Critical top width = B = 2 m

$$\Rightarrow \quad \frac{1.1\times 30^2}{10} = \frac{(2y_c)^3}{2}$$

$$\therefore \quad \frac{1.1\times 30^2}{40} = y_c^3$$

$\Rightarrow y_c = 2.914\ m \simeq 2.91\ m$

13.87 Sol.

Width of channel, B = 4 m

Discharge, $Q = 2\ m^3/s$

Depth of flow, y = 0.4 m

Now, $$E_1 = y_1 + \frac{V^2}{2y} = y_1 + \frac{Q^2}{2gA^2}$$

$$= 0.4 + \frac{(2)^2}{2\times 9.81\times(4\times 0.4)^2} = 0.4796\ m$$

At maximum contraction

$$E_1 = E_{min} = 1.5y_c$$

where y_c is depth at section of critical maximum contraction

$\therefore \quad 0.4796 = 1.5\ y_c$

$\Rightarrow \quad y_c = 0.3197$ m

Now, $$\left(\frac{q^2}{g}\right)^{1/3} = 0.3197$$

$$\Rightarrow \quad \left(\frac{q^2}{9.81}\right)^{1/3} = 0.3197$$

$\Rightarrow \quad q = 0.566\ m^3/s/m$

Now, $$\frac{Q}{B_{min}} = 0.566$$

where, B_{min} is width at maximum contraction section.

$$\Rightarrow \quad B_{min} = \frac{2}{0.566} = 3.534\ m \approx 3.53m$$

Alternatively,

$$B_{min} = \sqrt{\frac{27Q^2}{8gE_1^3}}$$

Where, E_1 = 0.4796 m

$$\text{So, } B_{min} = \sqrt{\frac{27\times 2^2}{8\times 9.81\times 0.4796^3}} = 3.53\ m$$

Environmental Engineering

UNIT VII

CONTENTS

Environmental Engineering

Syllabus

Water and Waste Water Quality and Treatment: Basics of water quality standards – Physical, chemical and biological parameters; Water quality index; Unit processes and operations; Water requirement; Water distribution system; Drinking water treatment.

Sewerage system design, quantity of domestic wastewater, primary and secondary treatment. Effluent discharge standards; Sludge disposal; Reuse of treated sewage for different applications.

Air Pollution: Types of pollutants, their sources and impacts, air pollution control, air quality standards, Air quality Index and limits.

Municipal Solid Wastes: Characteristics, generation, collection and transportation of solid wastes, engineered systems for solid waste management (reuse/ recycle, energy recovery, treatment and disposal).

Analysis of Previous GATE Papers

Exam Year	1 Mark Ques.	2 Marks Ques.	Total Marks
1995	5	–	5
1996	11	–	11
1997	11	–	11
1998	7	2	11
1999	8	–	8
2000	6	3	12
2001	3	3	9
2002	5	3	11
2003	6	9	24
2004	3	9	21
2005	4	9	22
2006	4	6	16
2007	3	6	15
2008	2	7	16
2009	3	6	15
2010	3	4	11
2011	3	5	13
2012	2	4	10
2013	2	3	8
2014 Set-1	2	3	8
2014 Set-2	2	4	10

Exam Year	1 Mark Ques.	2 Marks Ques.	Total Marks
2015 Set-1	2	2	6
2015 Set-2	3	4	11
2016 Set-1	4	3	10
2016 Set-2	5	4	13
2017 Set-1	2	4	10
2017 Set-2	3	4	11
2018 Set-1	2	4	10
2018 Set-2	3	4	11
2019 Set-1	2	4	10
2019 Set-2	2	4	10
2020 Set-1	3	5	13
2020 Set-2	3	4	11
2021 Set-1	4	6	16
2021 Set-2	3	4	11
2022 Set-1	4	5	11
2022 Set-2	5	3	11
2023 Set-1	3	5	13
2023 Set-2	3	5	13
2024 Set-1	3	3	9
2024 Set-2	2	3	8

1 Water Demand

1.1 Water distribution systems are sized to meet the

(a) Maximum hourly demand
(b) Average hourly demand
(c) Maximum daily demand and fire demand
(d) Average daily demand and fire demand

[1998 : 1 M]

1.2 The present population of a community is 28000 with an average water consumption of 4200 m^3/d. The existing water treatment plant has a design capacity of 6000 m^3/d. It is expected that the population will increase to 44000 during the next 20 years. The number of years from now when the plant will reach its design capacity, assuming an arithmetic rate of population growth, will be

(a) 5.5 years (b) 8.6 years
(c) 15.0 years (d) 16.5 years

[2004 : 2 M]

1.3 Match the following attributes of a city with the appropriate scale of measurements.

Attribute	Scale of measurement
(P) Average temperature (°C) of a city	(I) Interval
(Q) Name of city	(II) Ordinal
(R) Population density of a city	(III) Nominal
(S) Ranking of a city based on ease of business	(VI) Ratio

Which one of the following combinations is correct?

(a) (P)-(II), (Q)-(I), (R)-(IV), (S)-(III)
(b) (P)-(I), (Q)-(II), (R)-(III), (S)-(II)
(c) (P)-(I), (Q)-(III), (R)-(IV), (S)-(II)
(d) (P)-(II), (Q)-(III), (R)-(IV), (S)-(I)

[2022 : 1 M, Set-II]

■■■■

Answers Water Demand

1.1 (c) 1.2 (c) 1.3 (c)

Explanations Water Demand

1.1 (c)

Water distribution systems are sized to meet the maximum of the following:

(i) Maximum hourly draft of maximum day
(ii) Coincident draft = Maximum daily demand + Fire demand

POINTS TO REMEMBER

- The total draft is not taken as the sum of maximum hourly demand and fire demand becasue the coincidence that fire may break out at the time of maximum hourly draft is not considered.

Design life (yrs)	50	30	15	30	15	30	15	30	30
Source	Pipe-I	Pumps	Pipe-I	Pumps	Water treatment plant	Pipe-I	Service reservoir	Pipe-II	Distribution system
Well, dam Reservoir, etc.									
Design Discharge	Q_{MD}	Q_{MP}	Q_{Pump}	Q_{MD}	Q_{MD}	Q_{MD}	V_{SR}	Max. of Q_{MH} & Q_{CD}	Max. of Q_{MH} & Q_{CD}

- Here, V_{SR} =
  - Balancing Storage
  - Fire storage
  - Emergency storage for 7 days
- If pumps are not operatd continuously for 24 hours.

Then,

$$Q_{pump} = \left[\frac{24}{\text{No. of hours in the day for which pumps are running}}\right] \times \text{(}Q_{MD}\text{ + Reservoir for breakdown and repairs)}$$

1.2 (c)

Present population, $P_0 = 28000$

Population after 20 years, $P_n = 44000$

∴ Increase in population per year, $\bar{x}$

$$P_n = P_0 + n\bar{x}$$

$$\bar{x} = \frac{P_n - P_0}{n} \qquad (n = 20)$$

$$= \frac{44000 - 28000}{20} = 800$$

Now, for population for 28000, water consumption = 4200 m³/d

Hence, population for water consumption of 6000 m³/d

$$= \frac{28000}{4200} \times 6000$$

$$= 40000 \text{ persons}$$

$$= \text{Population at design capacity}$$

∴ No. of years from now when plant will reach at design capacity

$$P_n = P_0 + n\bar{x}$$

$$n = \frac{40000 - 28000}{800} = 15 \text{ years}$$

1.3 (c)

- Nominal data are a type of categorical data; that is, they are used to represent named quality.
- Ordinal data are a type of categorical data. That is they are used to represent named qualities of things, so that they can be sorted in order in their rank.
- Interval data are a type of categorical data. That is, they represent measured quantities of things.
- Ratio data are a type of categorical data that is, they represent measured quantity of things.

■■■■

2 Quality Characteristics of Water

2.1 The temporary hardness of water is caused by
(a) dissolved carbon dioxide
(b) bicarbonates and carbonates of calcium and magnesium
(c) bicarbonates of sodium and potassium
(d) carbonates of calcium and magnesium
[1995 : 1 M]

2.2 The most important water quality parameter for domestic use of water is
(a) carbonate hardness
(b) non-carbonate hardness
(c) coliform group of organisms
(d) chlorides [1996 : 1 M]

2.3 Presence of fluoride in water greater than permissible level of 1.5 mg/L causes
(a) cardiovascular disease
(b) methemoglobinemia
(c) hepatitis
(d) dental fluorosis [1996 : 1 M]

2.4 Alkalinity of water can be defined correctly in one of the following ways
(a) it is the measure of ability of water to neutralize oxygen
(b) it is the measure of ability of water to neutralize carbonates
(c) it is the presence of ions in water that will neutralize hydrogen ions
(d) it is the measure of ability of water to neutralize hydroxides [1997 : 1 M]

2.5 MPN index is a measure of one of the following
(a) Coliform bacteria
(b) BOD_5
(c) Dissolved Oxygen content
(d) Hardness [1997 : 1 M]

2.6 The microbial quality of treated piped water supplies is monitored by
(a) Microscopic examination
(b) Plate count of heterotrophic bacteria
(c) Coliform MPN test
(d) Identification of all pathogens [1998 : 1 M]

2.7 Excessive fluoride in drinking water causes
(a) Alzheimer's disease
(b) Mottling of teeth and embrittlement of bones
(c) Methamoglobinemia
(d) Skin cancer [1998 : 1 M]

2.8 Temporary hardness in water is caused by the presence of
(a) Bicarbonates of Ca and Mg
(b) Sulphates of Ca and Mg
(c) Chlorides of Ca and Mg
(d) Nitrates of Ca and Mg [1999 : 1 M]

2.9 Blue baby disease (methaemoglobinemia) in children is caused by the presence of excess
(a) Chlorideds (b) Nitrates
(c) Fluoride (d) Lead [1999 : 1 M]

2.10 Two samples of water *A* and *B* have pH values of 4.4 and 6.4 respectively. Then how many times sample *A* is more acidic than sample *B*?
(a) 0 (b) 50
(c) 100 (d) 200 [1999 : 1 M]

2.11 The Ca^{2+} concentration and Mg^{2+} concentration of a water sample are 160 mg/lit and 40 mg/lit as their ions respectively. The total hardness of the water sample in terms of $CaCO_3$ in mg/lit is approximately equal to
(a) 120 (b) 200
(c) 267 (d) 567 [2001 : 2 M]

2.12 In natural water, hardness is mainly caused by
(a) Ca^{++} and Mn^{++} (b) Ca^{++} and Fe^{++}
(c) Na^{+} and K^{+} (d) Ca^{++} and Mg^{++}
[2002 : 1 M]

2.13 The results of analysis of a raw water sample are given below:

Turbidity	:	5 mg/L
pH	:	7.4
Fluorides	:	2.5 mg/L
Total Hardness	:	300 mg/L
Iron	:	3.0 mg/L
MPN	:	50 per 100 mL

From the data given above, it can be inferred that water needs removal of

(a) turbidity followed by disinfection
(b) fluorides and hardness
(c) iron, followed by disinfection
(d) fluorides, hardness and iron followed by disinfection [2003 : 1 M]

2.14 Zero hardness of water is achieved by
(a) lime soda process
(b) excess lime treatment
(c) ion exchange treatment
(d) excess alum and lime treatment
[2003 : 1 M]

2.15 Results of a water sample analysis are as follows:

Cation	Concentration (mg/l)	Equivalent weight
Na^+	40	23
Mg^{+2}	10	12.2
Ca^{+2}	55	20
K^+	2	39

(milliequivalent weight of $CaCO_3$ = 50 mg/meq). Hardness of the water sample in mg/L as $CaCO_3$ is
(a) 44.8 (b) 89.5
(c) 179 (d) 358 [2003 : 2 M]

2.16 Most of the turbidity meters work on the scattering principle. The turbidity value so obtained is expressed in
(a) *CFU* (b) *FTU*
(c) *JTU* (d) *NTU* [2004 : 1 M]

2.17 Hardness of water is directly measured by titration with ethylene-di-amine-tetracetic acid (EDTA) using
(a) eriochrome black T indicator
(b) ferroin indicator
(c) methyl orange indicator
(d) phenolphthalein indicator [2004 : 1 M]

2.18 The organism, which exhibits very nearly the characteristics of an ideal pathogenic indicator is
(a) Entamoeba histolytica
(b) Escherichia coli
(c) Salmonella typhi
(d) Vibrio comma [2004 : 1 M]

2.19 A standard multiple-tube fermentation test was conducted on a sample of water from a surface stream. The results of the analysis for the confirmed test are given below.

Sample size (ml)	No. of positive results out of 5 tubes	No. of negative results out of 5 tubes
1.0	4	1
0.1	3	2
0.01	1	4

MPN index and 95% confidence limits for combination of positive results when five tubes used per dilutions (10 mL, 1.0 mL, 0.1 mL)

Combination of positive	MPN index per 100 mL	95% confidence limit	
		Lower	Upper
4-2-1	26	12	65
4-3-1	33	15	77

Using the above MPN index table, the Most Probable Number (MPN) of the sample is
(a) 26 (b) 33
(c) 260 (d) 330 [2004 : 2 M]

2.20 1 TCU is equivalent to the colour produced by
(a) 1 mg/L of chloroplatinate ion
(b) 1 mg/L of platinum ion
(c) 1 mg/L platinum in form of chloroplatinate ion
(d) 1 mg/L of organo-chloroplatinate ion
[2005 : 1 M]

2.21 If tomato juice is having a pH of 4.1, the hydrogen ion concentration will be
(a) 10.94×10^{-5} mol/L
(b) 9.94×10^{-5} mol/L
(c) 8.94×10^{-5} mol/L
(d) 7.94×10^{-5} mol/L [2005 : 2 M]

2.22 A synthetic sample of water is prepared by adding 100 mg Kaolinite (a clay mineral), 200 mg glucose, 168 mg NaCl, 120 mg $MgSO_4$, and 111 mg $CaCl_2$ to 1 liter of pure water. The concentrations of total solids (TS) and fixed dissolved solids (FDS) respectively in the solution in mg/L are equal to
(a) 699 and 599 (b) 599 and 399
(c) 699 and 199 (d) 699 and 399
[2006 : 1 M]

Linked Answer Questions 2.23 and 2.24:

A water contains the following dissolved ions:

$[Na^+]$ = 56 mg/L; $[Ca^{2+}]$ = 40 mg/L;
$[Mg^{2+}]$ = 30 mg/L; $[Al^{3+}]$ = 3 mg/L;
$[HCO_3^-]$ = 190 mg/L; $[Cl^-]$ = 165 mg/L
Water pH is 7.

Atomic weights: Ca : 40; Mg : 24; Al : 27; H : 1; C : 12; O : 16; Na : 23; Cl : 35.5

2.23 The total hardness of the sample in mg/L as $CaCO_3$ is

(a) 484 (b) 450
(c) 242 (d) 225 [2006 : 2 M]

2.24 The non-carbonate hardness of the sample in mg/L as $CaCO_3$ is

(a) 225 (b) 156
(c) 86 (d) 0 [2006 : 2 M]

2.25 The presence of hardness in excess of permissible limit causes

(a) cardio-vascular problems
(b) skin discolouration
(c) calcium deficiency
(d) increased laundry expenses [2007 : 1 M]

2.26 The alkalinity and the hardness of a water sample are 250 mg/L and 350 mg/L as $CaCO_3$, respectively. The water has

(a) 350 mg/L carbonate hardness and zero non-carbonate hardness
(b) 250 mg/L carbonate hardness and zero non-carbonate hardness
(c) 250 mg/L carbonate hardness and 350 mg/L non-carbonate hardness
(d) 250 mg/L carbonate hardness and 100 mg/L non-carbonate hardness

[2007 : 1 M]

2.27 A wastewater sample contains $10^{-5.6}$ mmol/L of OH^- ions at 25°C. The pH of this sample is

(a) 8.6 (b) 8.4
(c) 5.6 (d) 5.4 [2008 : 2 M]

2.28 Match **List-I** (Estimation method) with **List-II** (Corresponding indicator) and select the correct answer using the codes given below the lists:

List-I

A. Azide modified Winkler method for dissolved oxygen
B. Dichromate method for chemical oxygen demand
C. EDTA titrimetric method for hardness
D. Mohr or Argentometric method for chlorides

List-II

1. Eriochrome Black T
2. Ferroin
3. Potassium chromate
4. Starch

Codes:

	A	B	C	D
(a)	3	2	1	4
(b)	4	2	1	3
(c)	4	1	2	3
(d)	4	2	3	1

[2008 : 2 M]

Common Data for Questions 2.29 and 2.30:

Following chemical species were reported for water sample from a well:

Species	Concentration (mill equivalent/l)
Chloride (Cl^-)	15
Sulphate (SO_4^{2-})	15
Carbonate (CO_3^{2-})	05
Bicarbonate (HCO_3^-)	30
Calcium (Ca^{2+})	12
Magnesium (Mg^{2+})	18
pH	8.5

2.29 Total hardness in mg/L as $CaCO_3$ is

(a) 1500 (b) 2000
(c) 3000 (d) 5000 [2009 : 2 M]

2.30 Alkalinity present in the water in mg/L as $CaCO_3$ is

(a) 250 (b) 1500
(c) 1750 (d) 5000 [2009 : 2 M]

Common Data for Questions 2.31 and 2.32:

Ion concentrations obtained for a groundwater sample (having pH = 8.1) are given below:

Ion	Ca^{2+}	Mg^{2+}	Na^{2+}	HCO_3^-	SO_4^{2-}	Cl^-
Ion Concentration (mg/l)	100	6	15	250	45	39
Atomic weight	Ca = 40	Mg = 24	Na = 23	H = 1 C = 12 O = 16	S = 32 O = 16	Cl = 35.5

2.31 Total hardness (mg/L as $CaCO_3$) present in the above water sample is

(a) 205 (b) 250
(c) 275 (d) 308 [2010 : 2 M]

2.32 Carbonate hardness (mg/L as $CaCO_3$) present in the above water sample is

(a) 205 (b) 250
(c) 275 (d) 289 [2010 : 2 M]

2.33 Anaerobically treated effluent has MPN of total coliform as 10^6/100 mL. After chlorination, the MPN value declines to 10^2/100 mL. The percent removal (%R) and log removal (log R) of total coliform MPN is

(a) $\%R = 99.90$; $\log R = 4$
(b) $\%R = 99.90$; $\log R = 2$
(c) $\% R = 99.99$; $\log R = 4$
(d) $\% R = 99.99$; $\log R = 2$ [2011 : 1 M]

2.34 A water sample has a pH of 9.25. The concentration of hydroxyl ions in the water sample is
(a) $10^{-9.25}$ moles/L
(b) $10^{-4.75}$ mmoles/L
(c) 0.302 mg/L
(d) 3.020 mg/L [2012 : 2 M]

2.35 Match the given water properties in **Group-I** to the given titrants shown in **Group-II**.

	Group-I		Group-II
P.	Alkalinity	1.	N/35.5 $AgNO_3$
Q.	Hardness	2.	N/40 $Na_2S_2O_3$
R.	Chlorine	3.	N/50 H_2SO_4
S.	Dissolved oxygen	4.	N/50 EDTA

(a) P-1, Q-2, R-3, S-4
(b) P-2, Q-1, R-4, S-3
(c) P-3, Q-4, R-1, S-2
(d) P-4, Q-3, R-2, S-1
[2013 : 1 M]

2.36 Some of the nontoxic metals normally found in natural water are
(a) arsenic, lead and mercury
(b) calcium, sodium and silver
(c) cadmium, chromium and copper
(d) iron, manganese and magnesium
[2014 : 1 M, Set-I]

2.37 For a sample of water with the ionic composition shown in the figure below, the carbonate and non-carbonate hardness concentrations (in mg/l as $CaCO_3$), respectively are:

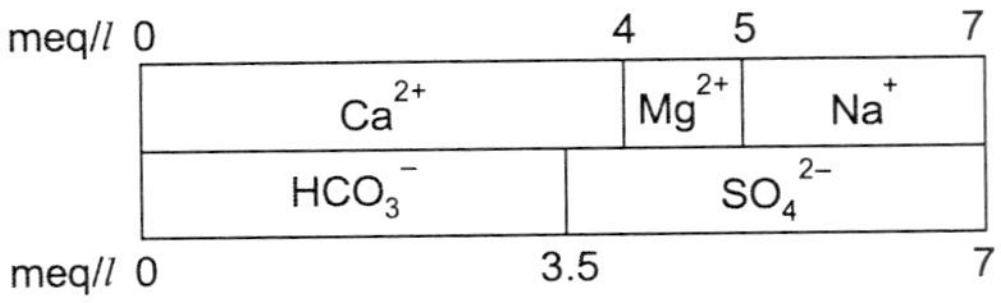

(a) 200 and 50
(b) 175 and 75
(c) 75 and 175
(d) 50 and 200
[2014 : 2 M, Set-I]

2.38 A groundwater sample was found to contain 500 mg/L total dissolved solids (TDS). TDS (in %) present in the sample is ________.
[2015 : 1 M, Set-II]

2.39 A sample of water has been analyzed for common ions and results are presented in the form of a bar diagram as shown.

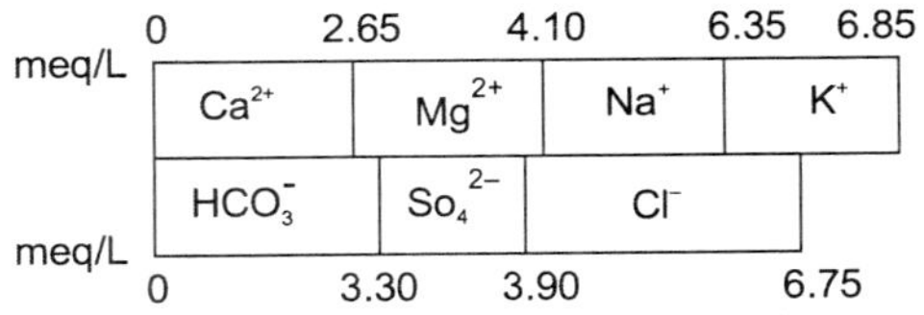

The non-carbonate hardness (expressed in mg/l as $CaCO_3$) of the sample is
(a) 40
(b) 165
(c) 195
(d) 205
[2016 : 2 M, Set-II]

2.40 Effluent from an industry '*A*' has a pH of 4.2. The effluent from another industry '*B*' has double the hydroxyl (OH^-) ion concentration than the effluent from industry '*A*'. pH of effluent from the industry '*B*' will be ________ [2016 : 2 M, Set-I]

2.41 The analysis of a water sample produces the following results:

Ion	Milligram per mill-equivalent of the ion	Concentration (mg/l)
Ca^{2+}	20.0	60
Mg^{2+}	12.2	36.6
Na^+	23.0	92
K^+	39.1	78.2
Cl^-	35.5	71
SO_4^{2-}	48.0	72
HCO_3^-	61.0	122

The total hardness (in mg/l as $CaCO_3$) of the water sample is ________.
[2017 : 2 M, Set-II]

2.42 As per IS 10500:2012, for drinking water in the absence of alternate source of water, the permissible limits for chloride and sulphate, in mg/L, respectively are
(a) 250 and 200
(b) 1000 and 400
(c) 200 and 250
(d) 500 and 1000
[2018 : 1 M, Set-II]

2.43 A water sample analysis data is given below:

Ion	Concentration (mg/l)	Atomic weight
Ca^{2+}	60	40
Mg^{2+}	30	24.31
HCO_3^-	400	61

The carbonate hardness (expressed as mg/L of $CaCO_3$, up to one decimal place) for the water sample is _______ [2018 : 2 M, Set-I]

2.44 Analysis of a water sample revealed that the sample contains the following species.

$CO_3^{2-}, Na^+, PO_4^{3-}, Al^{3+}, H_2CO_3, Cl^-, Ca^{2+}, Mg^{2+}, HCO_3^-, Fe^{2+}, OH^-$

Concentrations of which of the species will be required to compute alkalinity?

(a) $CO_3^{2-}, H^+, HCO_3^-, OH^-$

(b) $H^+, H_2CO_3, HCO_3^-, OH^-$

(c) $CO_3^{2-}, H_2CO_3, HCO_3^-, OH^-$

(d) $CO_3^{2-}, H^+, H_2CO_3, HCO_3^-$

[2019 : 1 M, Set-II]

2.45 An amount of 35.67 mg HCl is added to distilled water and the total solution volume is made to one litre. The atomic weights of H and Cl are 1 and 35.5, respectively. Neglecting the dissociation of water, the pH of the solution, is

(a) 2.50 (b) 2.01
(c) 3.01 (d) 3.50

[2020 : 1 M, Set-I]

2.46 During chlorination process, aqueous (aq) chlorine reacts rapidly with water to from Cl^-, HOCl, and H^+ as shown below:

$$Cl_2\,(aq) + H_2O \rightleftharpoons HOCL + Cl^- + H^+$$

The most active disinfectant in the chlorination process from amongst the following, is

(a) H_2O (b) H^+
(c) Cl^- (d) HOCl

[2020 : 1 M, Set-I]

2.47 The ion product of water (pK_w) is 14. If a rain water sample has a pH of 5.6, the concentration of OH^- in the sample (in 10^{-9} mol/litre, round off to one decimal place), is _______.

[2020 : 2 M, Set-II]

2.48 Alkalinity of water, in equivalent/litre (eq/litre), is given by

$$\{HCO_3^-\} + 2\{CO_3^{2-}\} + \{OH^-\} - \{H^+\}$$

where, { } represents concentration in mol/litre. For a water sample, the concentration of HCO_3^- $= 2 \times 10^{-3}$ mol/litre, $CO_3^{2-} = 3.04 \times 10^{-4}$ mol/litre and the pH of water = 9.0. The atomic weights are : Ca = 40; C = 12; and O = 16. If the concentration of OH^- and H^+ are NEGLECTED, the alkalinity of the water sample (in mg/litre as $CaCO_3$), is

(a) 65.2 (b) 50.0
(c) 100.0 (d) 130.4

[2020 : 2 M, Set-II]

2.49 A water sample is analysed for coliform organisms by the multiple-tube fermentation method. The results of confirmed test are as follows:

Sample Size (ml)	Number of positive result out of 5 tubes	Number of negative results out of 5 tubes
0.01	5	0
0.001	3	2
0.001	1	4

The most probable number (MPN) of coliform organisms for the above results is to be obtained using the following MPN index.

MPN Index for various Combination of Positive Results when Five Tubes used per dilution of 10.0 ml, 1.0 ml and 0.1 ml

Combination of positive tubes	MPN index per 100 ml
0-2-4	11
1-3-5	19
4-2-0	22
5-3-1	110

The MPN of coliform organisms per 100 ml is

(a) 110 (b) 1100
(c) 1100000 (d) 110000

[2021 : 2 M, Set-I]

2.50 A 50 ml sample of industrial wastewater is taken into a silica crucible. The empty weight of the crucible is 54.352 g. The crucible with the sample is dried in a hot air oven at 104°C till a constant weight of 55.129 g. Thereafter, the crucible with the dried sample is fired at 600°C for 1 h in a muffle furnace, and the weight of the crucible along with residue is determined as 54.783 g. The concentration of total volatile solids is ______.

(a) 15540 mg/*l* (b) 8620 mg/*l*
(c) 6920 mg/*l* (d) 1700 mg/*l*

[2021 : 2 M, Set-I]

2.51 The hardness of a water sample is measured directly by titration with 0.01 M solution of ethylenediamine tetraacetic acid (EDTA) using eriochrome black T (EBT) as an indicator. The EBT reacts and forms complexes with divalent metallic cations present in the water. During titration, the EDTA replaces the EBT in the complex. When the replacement of EBT is complete at the end point of the titration, the colour of the solution changes from

(a) blue to colourless
(b) wine red to blue
(c) blue-green to reddish brown
(d) reddish brown to pinkish yellow

[2021 : 1 M, Set-II]

2.52 The total hardness in raw water is 500 milligram per liter as $CaCO_3$. The total hardness of this raw water, expressed in milligram equivalent per liter, is (Consider the atomic weights of Ca, C, and O as 40 g/mol, 12 g/mol, and 16 g/mol, respectively.)

(a) 10 (b) 100
(c) 1 (d) 5 [2022 : 1 M, Set-I]

2.53 In a water sample, the concentrations of Ca^{2+}, Mg^{2+} and HCO_3^- are 100 mg/L, 36 mg/L and 122 mg/L, respectively. The atomic masses of various elements are:
Ca = 40, Mg = 24, H = 1, C = 12, O = 16.
The total hardness and the temporary hardness in the water sample (in mg/L as $CaCO_3$) will be

(a) 400 and 100, respectively.
(b) 400 and 300, respectively.
(c) 500 and 100, respectively.
(d) 800 and 200, respectively.

[2022 : 2 M, Set-I]

2.54 A 100 mg of HNO_3 (strong acid) is added to water, bringing the final volume to 1.0 liter. Consider the atomic weights of H, N, and O, as 1 g/mol. 14 g mol. and 16 g/mol, respectively. The final pH of this water is (Ignore the dissociation of water.)

(a) 6.5 (b) 2.8
(c) 8.5 (d) 3.8

[2022 : 1 M, Set-II]

2.55 Identify the waterborne diseases caused by viral pathogens:

(a) Acute anterior poliomyelitis
(b) Cholera
(c) Infectious hepatitis
(d) Typhoid fever [2023 : 1 M, Set-I]

■■■■

Answers Quality Characteristics of Water

2.1	(b)	2.2	(c)	2.3	(d)	2.4	(c)	2.5	(a)	2.6	(c)	2.7	(b)
2.8	(a)	2.9	(b)	2.10	(c)	2.11	(d)	2.12	(d)	2.13	(d)	2.14	(c)
2.15	(c)	2.16	(b, d)	2.17	(a)	2.18	(b)	2.19	(d)	2.20	(c)	2.21	(d)
2.22	(d)	2.23	(c)	2.24	(c)	2.25	(d)	2.26	(d)	2.27	(d)	2.28	(b)
2.29	(a)	2.30	(c)	2.31	(c)	2.32	(a)	2.33	(c)	2.34	(c)	2.35	(c)
2.36	(d)	2.37	(b)	2.38	(0.05)	2.39	(a)	2.40	(4.5)	2.41	(300)	2.42	(b)
2.43	(327.8)	2.44	(a)	2.45	(c)	2.46	(d)	2.47	(3.98)	2.48	(d)	2.49	(d)
2.50	(c)	2.51	(b)	2.52	(a)	2.53	(a)	2.54	(b)	2.55	(a, c)		

Explanations Quality Characteristics of Water

2.1 (b)

Temporary hardness of water is caused by the carbonates (CO_3^{2-}) and bicarbonates (HCO_3^-) of calcium and magnesium.
Temporary hardness can be removed by simply boiling the water.

Permanent Hardness (non-carbonate hardness) is calused by the sulphates, chlorides and nitrates of calcium or magnesium present in water.
They cannot be removed by boiling and require special treatment for softening.

2.2 (c)

For domestic water supply, water should be free from any harmful pathogens which can cause diseases. So, coliform test is most important while other water quality parameter like carbonate hardness, non-carbonate hardness and chlorides are tolerable upto a certain limit.

2.3 (d)

Permissible limit:

(i) Acceptable limit is upto 1 mg/*l*.
This concentration is effective in reducing dental cavities.

(ii) Cause for rejection is greater than 1.5 mg/*l*.
Higher concentratiions exceeding this may cause molting of teeth. Excessive consumption over long time or concentration > 5 mg/*l* causes crippling of bones or bone fluorisis.

2.4 (c)

Alkanity is the capcity of water to resist change in pH that would make the water more acidic. It is also defined as the measure of waters' ability to neutralize acidity.

Hence, alkalinity is the presence of ions in water that will neutralize hydrogen ions.

2.5 (a)

MPN (Most Probable Number) index is a measure of coliform bacteria.

The testing and counting of pathogenic bacteria in laboratory is very difficult. Hence, coliform bacteria (harmless aerobic micro-organisms) are used as indicator organism for probable presence of pathogenes. They live longer in water than pathogenic bacteria. Hence, in coliform is absent pathogenes will also be absent.

2.6 (c)

Pathogenic bacteria are tested and counted in laboratory with great difficulty. Hence instead of testing pathogens we tested and counted coliform bacteria which themselves are harmless but their presence or absence indicates the presence or absence of pathogens.

Coliform bacteria survive for a longer period them the pathogens and the method used to count coliform bacteria is Most Probable Number (MPN).

2.7 (b)

Fluoride content upto 1 ppm helps to prevent dental cavities but if its concentration is more than 1.5 ppm then it results in de-colouration of teeth called mottling of teeth and deformation of bones if it is more than 5 ppm.

2.8 (a)

Temporary hardness in water is caused by the presence of carbonates and bicarbonates of Ca and Mg.

2.9 (b)

The nitrates represent the fully oxidized organic matter, and such waters may not be harmful. However, the presence of too much of nitrate in water may adversely affect the health of infants, causing a disease called **methemoglobinemia** called "**blue baby disease**".

2.10 (c)

$$p(H) = \log_{10}[H^+]$$

$$-\log_{10}[H^+] = p(H)_A$$

$$\Rightarrow \quad [H^+]_A = 10^{-4.4}$$

$$\therefore \quad [H^+]_B = 10^{-6.4}$$

$$\therefore \quad \frac{[H^+]_A}{[H^+]_B} = \frac{10^{-4.4}}{10^{-6.4}} = 100$$

Sample A is 100 times more acidic than sample B.

2.11 (d)

For total hardness of water sample in terms of cations, convert the concentration of various cations causing hardness into that of $CaCO_3$ by equating the no. of gram equivalents of a cation wityh $CaCO_3$.

Hence, Total hardness

$$= \frac{[Ca^{2+}]}{\text{Eq. wt. of Ca}} \times \text{Eq. wt. of } CaCO_3 + \frac{[Mg^{2+}]}{\text{Eq. wt. of Mg}} \times \text{Eq. wt. of } CaCO_3$$

$$= 160 \times \frac{50}{20} + 40 \times \frac{50}{12} = 566.67 \text{ mg}/l$$

2.12 (d)

Hardness in water is due to multivalent metallic cations and is mainly due to Ca^{++} and mg^{++} because these are most commonly present in water.

2.13 (d)

Characteristics	Permissible limits	Given value
pH	6.5-8.5	7.4
Turbidity	5 mg/l	5 mg/l
Fluorides	greater than 1 mg/l and less than 1.5 mg/l	2.5 mg/l
Total Hardness	75-115 mg/l	300 mg/l
Iron	0.3 mg/l	3.0 mg/l
MPN	Shall not be detectable in any 100 ml	50 per 100 ml

Here, Fluorides, Total Hardness, Iron and MPN are deviating from their permissible limits.

2.14 (c)

Water of zero hardness can be obtained from ion exchange method only. It is also called **zeolite process or base exchange treatment**.

2.15 (c)

$$\text{Total hardness} = [Mg^{2+}] \times \frac{\text{eq. wt. of } CaCO_3}{\text{eq. wt. of } Mg^{2+}} + [Ca^{2+}] \times \frac{\text{eq. wt. of } CaCO_3}{\text{eq. wt. of } Ca^{2+}}$$

$$= \left(10 \times \frac{50}{12.2}\right) + \left(55 \times \frac{50}{20}\right) = 178.48 \simeq 179 \text{ mg/L}$$

In addition to Ca^{2+} and Mg^{2+}, other divalent metallic ions like Fe^{2+}, Mn^{2+}, Sr^{2+}, etc may also cause hardness but their presence is negligible in most of the waters. Na^+ and K^+ do not contribute in hardness.

2.16 (b, d)

JTU (Jackson Turbidity Unit) is based on absorbing principle.

NTU (Naphelometric Turbidity Unit) is based on scattering principle. It is also referred to as **FTU (Formazin Turbidity Unit)** since formazin polymer is used to prepare the standard sample. More appropriate is FTU since formazine is added to prepare the standard turbid water

2.17 (a)

Hardness of water is determined by titration with EDTA using EBT as an indicator which is blue/black in colour.

2.18 (b)

E-coli act as an ideal pathogenic indicator as its presence or absence can be used to indicate the presence or absence of pathogens.

2.19 (d)

The MPN for the combination 4-3-1 is 33 but the standard sample sizes are 10 mL, 1.0 mL and 0.1 mL. The specimen sample sizes are 1.0 mL, 0.1 mL and 0.01 mL, since the specimen sample is 10 times less diluted than the standard sample, henc, the bacterial density in the specimen sample will be 10 times more than that is the standard sample. Therefore MPN = 33 × 10 = 330.

2.20 (c)

1 TCU is the colour produced by 1 mg of platinum-cobalt in the form of chloroplatinate ions dissolved in 1 litre of distilled water.

2.21 (d)

$$pH = -\log_{10}[H^+]$$
$$\Rightarrow \quad 4.1 = -\log_{10}[H^+]$$
$$\Rightarrow \quad 10^{-4.1} = [H^+]$$
$$\Rightarrow \quad [H^+] = 7.94 \times 10^{-5} \text{ mol/L}$$

2.22 (d)

Total Solids = 100 + 200 + 168 + 120 + 111
= 699 mg/l

Fixed dissolved solids
= 168 (NaCl) + 120 ($MgSO_4$) + 111 ($CaCl_2$)
= 399 mg/l

Note: Glucose is organic dissolved solids.

2.23 (c)

Total hardness (due to multivalent metallic cations)

$$= [Mg^{2+}] \times \frac{\text{eq. wt. of } CaCO_3}{\text{eq. wt. of } Mg^{2+}} + \left[Ca^{2+}\right] \times \frac{\text{eq. wt. of } CaCO_3}{\text{eq. wt. of } Ca^{2+}} + \left[Al^{+3}\right] \times \frac{50}{9}$$

$$= \left(30 \times \frac{50}{12}\right) + \left(40 \times \frac{50}{20}\right) + 3 \times \frac{50}{9}$$

= 242 in mg/L as $CaCO_3$

2.24 (c)

Alkalinity (dur to CO_3^{2} and HCO_3^- of multivalent metallic cations)

$$= \frac{[HCO_3^-]}{\text{eq. wt. of } HCO_3^-} \times \text{eq. wt. of } CaCO_3$$

$$= 190 \times \frac{50}{61} = 155.74 \simeq 156 \text{ mg/L as } CaCO_3$$

Carbonate hardness
= Minimum (Total hardness, Alkalinity)
= Minimum (242, 186)
= 156 mg/L as $CaCO_3$
Non-carbonate hardness
= Total hardness – Carbonate hardness
= 242 – 156 = 86 mg/L as $CaCO_3$

2.25 (d)

Hardness of water is the characteristic which prevents the formation of sufficeint leather or foam, when such hard water is mixed with soap. It is usually caused by the prsence of Ca and Mg salts present in water, which form scum by reaction with soap.
Hence, hard water does not cause any health problem but they may lead to greater soap consumption, thereby increasing the loundry expenses.

2.26 (d)

Carbonate hardness is equal to the total hardness or alkalinity, whichever is lesser.
In this case alkalinity (250 mg/L) is less than total hardness (350 mg/L).
Carbonate hardness = 250 mg/L as $CaCO_3$
Non-corbonate hardness is the total hardness in excess of the alkalinity i.e.
Non-carbonate hardness
= Total hardness – alkalinity = 350 – 250
= 100 mg/L as $CaCO_3$

2.27 (d)

$$pOH = -\log[OH^-]$$
$$\Rightarrow \quad pOH = -\log(10^{-5.6} \times 10^{-3})$$
$$\Rightarrow \quad pOH = -\log(10^{-8.6})$$
$$pOH = 8.6$$
But $pH + pOH = 14$
$$\Rightarrow \quad pH = 14 - 8.6$$
$$\Rightarrow \quad pH = 5.4$$

2.28 (b)

2.29 (a)

Total hardness (due to multivalent metallic cations)
= Milliequivalent of (Ca^{2+} and Mg^{2+})
= 12 + 18
= 30 milliequivalent/l
= (30 × equivalent weight of $CaCO_3$) mg/l
= (30 × 50) mg/l
⇒ Total hardness
= 1500 mg/l as $CaCO_3$

2.30 (c)

Alkalinity (de to CO_3^{2-}, HCO_3^-, OH^-)
= Milliequivalent of (CO_3^{2-} and HCO_3^-)
= 5 + 30
= 35 milliequivalent/l
= (35 × eq. wt. of $CaCO_3$) mg/l = (35 × 50) mg/l
⇒ Alkalinity = 1750 mg/l

2.31 (c)

Total hardness

$$= [Mg^{2+}] \times \frac{\text{eq. wt. of } CaCO_3}{\text{eq. wt. of } Mg^{2+}} + [Ca^{2+}] \times \frac{\text{eq. wt. of } CaCO_3}{\text{eq. wt. of } Ca^{2+}}$$

$$= 6 \times \frac{50}{12} + 100 \times \frac{50}{20} = 275 \text{ mg/L}$$

2.32 (a)

Total alkalinity in water consists of alkalinity caused by CO_3^{2-}, HCO_3^- and OH^-. A little negative alkalinity is also caused by H^+.

$$\therefore \text{ Total Alkalinity} = [HCO_3^-] \times \frac{\text{Eq. wt. of } CaCO_3}{\text{Eq. wt. of } HCO_3^-}$$

$$= 250 \times \frac{50}{61} = 205 \text{ mg/L}$$

Carbonate hardness is equal to the total hardness or alkalinity, whichever is less. The carbonate hardness of the sample will be 205 mg/L
∴ Non-carbonate hardness
= Total hardness – carbonate hardness
= 275 – 205 = 70 mg/L

2.33 (c)

$$\%R = \left(\frac{10^6 - 10^2}{10^6}\right) \times 100 = 0.9999 \times 100$$
$$= 99.99\%$$
$$\log R = \log 10^6 - \log 10^2 = 6 - 2 = 4$$

2.34 (c)

$$pH = 9.25$$
$$\because \quad pH + pOH = 14$$

$\Rightarrow \quad pOH = 14 - 9.25 = 4.75$

$\Rightarrow -\log_{10}[OH^-] = 4.75$

$\therefore \quad [OH^-] = 1.778 \times 10^{-5}\left(\frac{mol}{lit}\right)$

$\therefore \quad \frac{mol}{lit}$ of $OH^- = 1.778 \times 10^{-5}$

$\therefore \quad \frac{gm}{lit}$ of $OH^- = 1.778 \times 10^{-5} \times 17$

$= 30.226 \times 10^{-5} = 02.6 \times 10^{-6}$ gm/lit

$\therefore \quad [OH^-] = 302.6 \times 10^{-3}$ mg/lit

$= 0.302$ mg/lit.

2.35 (c)

2.36 (d)

Although manganese is toxic, but only in very high concentration. And it is also a necessary consituent is water.

Hence, according to the options available, best possible answer is (d)

2.37 (b)

Total hardness

$= 5$ meq/l (due to Ca^{2+} and Mg^{2+})

$= (5 \times 50)$ mg/l as $CaCO_3$

$= 250$ mg/l as $CaCO_3$

Alkalinity (due to HCO_3^-)

$=3.5$ Meq/l $= (3.5 \times 50)$ mg/l as $CaCO_3$

$= 175$ mg/l as $CaCO_3$

Carbonate hardness

= Min. (tota hardness, Alkalinity)

= Minimum (250, 175)

$= 175$ mng/l as $CaCO_3$

Non-carbonate hardness

= Total hardness – Carbonate hardness

$= 250 - 175 = 75$ mg/l as $CaCO_3$

2.38 Sol.

1 kg = 1 litre

TDS (in %) in the sample

$$= \frac{500 \times 10^{-3} g}{1000 g} \times 100 = 0.05\%$$

2.39 (a)

Hardness is due to multivalent metallic cations, i.e. Ca^{2+} and Mg^{2+}

Total hardness (mg/l as $CaCO_3$)

= (Total meq/l) × (eq. weight of $CaCO_3$ in mg)

$= (4.1) \times 50$ mg/l as $CaCO_3$

$= 205$ mg/l as $CaCO_3$

Alkalinity is due to the presence of HCO_3^- in this case,

Alkalinity (mg/l as $CaCO_3$)

$= 3.3 \times 50$ mg/l as $CaCO_3 = 165$ mg/l as $CaCO_3$

Now, non-carbonate hardness

= Total Hardness in excess of alkalinity

$= 205 - 165 = 40$ mg/l as $CaCO_3$

2.40 Sol.

$[pH]_A = 4.2 = -\log [H^+]_A$

$\Rightarrow \quad [H^+]_A = 10^{-4.2}$

As $[H^+] \times [OH^-] = 10^{-14}$

$\therefore \quad [OH^-]_A = 10^{-9.8}$

$\Rightarrow \quad [OH^-]_B = 2 \times 10^{-9.8}$

Now, $p[OH]_B = -\log [OH^-]_B$

$\Rightarrow \quad p[OH]_B = 9.8 - \log 2$

As, $pH + pOH = 14$

$\Rightarrow \quad p[H]_B = 4.5$

2.41 Sol.

Milligrah per mill-equivalent of the ion

= equivalent weight

$\therefore$ Milliequivalent of $Ca^{2+} = \frac{60}{20} = 3$

and, milliequivalent of $Mg^{+2} = \frac{36.6}{12.2} = 3$

Total hardness (due to Mg^{2+} and Ca^{2+})

= Milliequivalent of Ca^{2+} and Mg^{2+}

$= 3 + 3 = 6$ milliequivalent/l

$= (6 \times 50)$ mg/l as $CaCO_3$

$= 300$ mg/l as $CaCO_3$

2.42 (b)

As per IS-10500 : 2012, Table 2

Permissible limit in absence of alternate source.

Chloride (as Cl^-)	Sulphate as $[SO_4]^{-2}$
1000 mg/l	400 mg/l

2.43 Sol.

Carbonate hardness = min. (Total hardness, alkalinity)

Total hardness

$$= \left(\frac{60}{20} \times 50 + \frac{30}{12.155} \times 50\right) \text{ mg/}l \text{ as } CaCO_3$$

$= 150 + 123.406$

$= 273.406$ mg/l as $CaCO_3$

$$\text{Alkalinity} = \left(\frac{400}{61}\times 50\right) \text{ mg/}l \text{ as } CaCO_3$$

$= 327.868$ mg/l as $CaCO_3$

$\therefore$ CH = 273.4 mg/l as $CaCO_3$

2.44 (a)

Alkalinity is defined as ability of water to neutralize the acid or hydronium ion

Alkalinity (A_T) of water = $[HCO_3^-] + [CO_3^{2-}] + [B(OH)_4^-] + [H_3(SiO_4)^-] + [HS^-]$ + [organic anions] + $[OH^-] - [H^+]$

From given options of ions in problem answer is (a).

i.e. CO_3^{2-}, H^+, HCO_3^-, OH^-

2.45 (c)

$$HCl \rightarrow H^+ + Cl^-$$

1 mole of HCL gives 1 mole H^+ ions

36.5 gm of HCl gives 1 gm of H^+ ions

$$35.67 \text{ mg} = \frac{1}{36.5}\times 35.67 = 0.977 \text{ mg of } H^+$$

$$= \frac{0.977\times 10^{-3}}{1} = 9.77 \times 10^{-4} \text{ moles of } H^+$$

$$pH = -\log_{10}[H^+]$$

$$= -\log_{10}[9.77 \times 10^{-4}] = -\log_{10}9.77 + 4\log_{10}10$$

$$= 4 - 0.989 = 3.01$$

2.46 (d)

2.47 Sol.

$$pH + pOH = 14$$

$$pOH = 14 - 5.6 = 8.4$$

$$-\log [OH^-] = 8.4$$

$$[OH^-] = 10^{-8.4} \text{ moles/}lt$$

$$= 10^{-8.4+9} \times 10^{-9} \text{ moles/}lt = 3.98 \times 10^{-9} \text{ moles/}lt$$

2.48 (d)

Alkalinity of water sample is due to presence of $[HCO_3^-]$ and $[CO_3^{2-}]$

Total alkalinity = 1 mole of $[HCO_3^-]$

+ 2 mole of $[CO_3^{2-}]$

in terms of $CaCO_3$

$$= (2 \times 10^{-3} \times 50 + 2 \times 3.04 \times 10^{-4} \times 50) \times 10^3 \text{ mg/}l$$

$$= 130.4 \text{ mg/}l \text{ as } CaCO_3$$

2.49 (d)

The sample size is 0.01 ml, 0.001 ml and 0.0001 ml which is 1000 times lesser than the standard of 10 ml, 1 ml and 0.1 ml.

The positive set is 5-3-1 and w.r.t. the positive combination,

MPN/100 ml will be 110 × 1000 = 110000

2.50 (c)

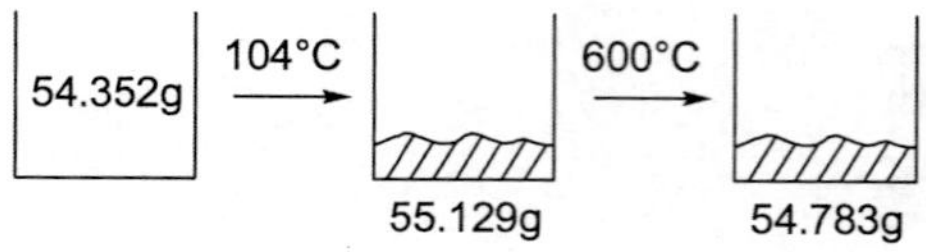

Volatile solids weight

= 55.129 − 54.783 = 0.346 gm (in 50 ml)

$$\text{So, in one litre} = \frac{346}{0.05} = 6920 \text{ mg/litre}$$

2.51 (b)

2.52 (a)

Total Hardness = 500 mg/l as $CaCO_3$

No. of equivalents

$$= \frac{(\text{Weight})_{gram}}{\text{Equivalent weight of } CaCO_3}$$

$\Rightarrow$ No. of equivalent

$$= \frac{500\times 10^{-3} g}{\frac{100}{2}} = \frac{500\times 10^{-3} g}{50 \text{ g/eq}} = 0.01 \text{ eq}$$

$\Rightarrow$ No. of equivalent

$= 0.01 \times 10^3$ m-eq = 10 m-eq

2.53 (a)

Total hardness,

$$TH = \left[Ca^{+2}\right]\times\frac{50}{20} + \left[Mg^{+2}\right]\times\frac{50}{12}$$

$$\Rightarrow \quad TH = 100\times\frac{50}{20} + 36\times\frac{50}{12}$$

$\Rightarrow$ TH = 400 mg/l as $CaCO_3$

Total alkalinity,

$$TA = \left[HCO_3^-\right]\times\frac{50}{61} = 122\times\frac{50}{61}$$

$\Rightarrow$ TA = 100 mg/l as $CaCO_3$

Temporary hardness (carbonate hardness),

$$CH = \min \begin{cases} TA = 100 \text{ mg}/l \text{ as } CaCO_3 \\ TH = 400 \text{ mg}/l \text{ as } CaCO_3 \end{cases}$$

$\Rightarrow$ CH = 100 mg/l as $CaCO_3$

2.54 (b)

$HNO_3 \rightarrow H^+ + NO_3^-$ {Molecular mass HNO_3 = 63g}

1 mole of HNO_3 produces 1 mole of H^+ means

$\Rightarrow$ 63g of HNO_3 produce 1 g of H^+

$\therefore$ 100 mg of HNO_3 will produce $= \frac{1}{63} \times 100$

= 1.587 mg of H^+

= 1.587×10^{-3} g of H^+ in 1 lit

= 1.587×10^{-3} mol/l

$\therefore$ $pH = -\log_{10}[H^+] = -\log_{10}[1.587 \times 10^{-3}]$

$\Rightarrow$ pH = 2.8

2.55 (a, c)

■■■■

3 Water Treatment

3.1 Breakpoint chlorination of water involves addition of chlorine in an amount sufficient to
(a) react with any ammonia and readily oxidizable organic matter
(b) kill giardia cysts
(c) react with inorganic matter
(d) reduce bacterial growth in filters
[1995 : 1 M]

3.2 The treatment that should be given to the water from a deep tube well is
(a) pre-settling only
(b) coagulation and flocculation only
(c) filtration only
(d) disinfection only [1996 : 1 M]

3.3 Flocculation is a process
(a) that removes algae from stabilization pond effluent
(b) that promotes to aggregation of small particles into larger particles to enhance their removal by gravity
(c) that mixes the coagulant in water
(d) None of these [1997 : 1 M]

3.4 Pathogens are usually removed by
(a) chemical precipitation
(b) sedimentation
(c) activated sludge process
(d) chlorination [1997 : 1 M]

3.5 Design parameters for rapid mixing units are
(a) velocity gradient and the volume of mixing basin
(b) viscosity and velocity gradient
(c) viscosity, velocity gradient and the volume of the mixing basin
(d) detention time and viscosity of water
[1997 : 1 M]

3.6 The absorbent most commonly used in water and waste treatment is
(a) sand of grain size from 0.1 to 2 mm
(b) activated carbon granules of size 0.1 to 2 mm
(c) ordinary wood shavings of fine size
(d) coal-tar [1997 : 1 M]

3.7 Among the following disinfectants of waste water, the one that is most commonly used is
(a) Chlorine dioxide
(b) Chlorine
(c) Ozone
(d) UV-radiation [1997 : 1 M]

3.8 Coagulation flocculation with alum is performed
(a) immediately before chlorination
(b) immediately after chlorination
(c) after rapid sand filtration
(d) before rapid sand filtration [1998 : 1 M]

3.9 For a flow of 5.7 MLD (million litres per day) and a detention time of 2 hours, the surface area of a rectangular sedimentation tank to remove all particles having setting velocity of 0.33 mm/s is
(a) 20 m^2 (b) 100 m^2
(c) 200 m^2 (d) 400 m^2 [1998 : 2 M]

3.10 The disinfection efficiency of chlorine in water treatment
(a) is not dependent of pH value
(b) is increase by increased pH value
(c) remains constant at all pH values
(d) is reduced by increased pH values
[2000 : 1 M]

3.11 The following characteristics pertain to the sand filters used in water industry.
I. Filtration rate is 1 to 4 $m^3/m^2/day$
II. Typical duration of operation in one run is 24 to 72 hours
III. Operating cost is low
Which of the above characteristics pertain to slow sand filters?
(a) I, II and III (b) I and II
(c) II and III (d) I and III
[2000 : 1 M]

3.12 Use of coagulants such as alum
(a) results in reduction of pH of the treated water
(b) results in increase in pH of the treated water
(c) results in no change in pH of the treated water
(d) may cause an increase or decrease of pH of the treated water [2000 : 1 M]

3.13 Aeration of water is done to remove
(a) Suspended impurities
(b) Colour
(c) Dissolved Salts
(d) Dissolved gases [2001 : 1 M]

3.14 The following chemical is used for coagulation
(a) Ammonium Chloride
(b) Aluminium Chloride
(c) Aluminium Sulphate
(d) Copper Sulphate [2001 : 1 M]

3.15 A town has an existing horizontal flow sedimentation tank with an overflow rate of 17 $m^3/day/m^2$, and it is desirable to remove particles that have settling velocity of 0.1 mm/sec. Assuming the tank is an ideal sedimentation tank, the percentage of particles removal is approximately equal to
(a) 30% (b) 50%
(c) 70% (d) 90% [2001 : 2 M]

3.16 In disinfection, which of the following forms of chlorine is most effective in killing the pathogenic bacteria?
(a) Cl (b) OCl
(c) NH_2Cl (d) HOCl [2002 : 1 M]

3.17 For a water treatment plant having a flow rate of 432 m^3/hr, what is the required plan area of a Type-I settling tank to remove 90% of the particles having a settling velocity of 0.12 cm/sec is
(a) 120 m^2 (b) 111 m^2
(c) 90 m^2 (d) 100 m^2 [2002 : 2 M]

3.18 Zero hardness of water is achieved by
(a) lime soda process
(b) excess lime treatment
(c) ion exchange mehtod
(d) excess alum and lime treatment
[2003 : 1 M]

3.19 Match **List-I** (Type of water impurity) with **List-II** (Method of treatment) and select the correct answer using the codes given below the lists:

List-I
A. Hardness
B. Brackish water from sea
C. Residual MPN from filters
D. Turbidity

List-II
1. Reverse Osmosis
2. Chlorination
3. Zeolite Treatment
4. Coagulation and Flocculation
5. Coagulation, Flocculation and Filtration

Codes:

	A	B	C	D
(a)	1	2	4	5
(b)	3	2	2	4
(c)	2	1	3	5
(d)	3	1	2	5

[2003 : 2 M]

3.20 An ideal horizontal flow settling basin is 3 m deep having surface area 900 m^2. Water flows at the rate of 8000 m^3/d, at water temperature 20°C ($\mu = 10^{-3}$ kg/m-s and $\rho = 1000$ kg/m^3). Assuming Stoke's law to be valid, the proportion (percentage) of spherical sand particles (0.01 mm in diameter with specific gravity 2.65), that will be removed, is
(a) 32.5 (b) 67
(c) 87.5 (d) 95.5 [2003 : 2 M]

Common Data for Questions 2.21 and 2.22:

A water treatment plant treating 10 MLd of water requires 20 mg/L of filter Alum, $Al_2(SO_4)_3.18\,H_2O$. The water has 6 mg/L of alkalinity as $CaCO_3$. (Al = 26.97, S = 32, O = 16, H = 1, Ca = 40, and C= 12).

3.21 Total alkalinity requirement (10^6 mg per day as $CaCO_3$) matching filter Alum, shall be
(a) 180 (b) 120
(c) 90 (d) 60 [2003 : 2 M]

3.22 Quantity of Quick Lime required (10^6 mg per year as CaO) shall be
(a) 2132 (b) 3000
(c) 4132 (d) 6132 [2003 : 2 M]

3.23 The design parameter for flocculation is given by a dimensionless number Gt, where G is the velocity gradient and t is the detention time. Values of Gt ranging from 10^4 to 10^5 are commonly used, with t ranging from 10 to 30 min. The most preferred combination of G and t to produce smaller and denser flocs is
(a) large G values with short t
(b) large G values with long t
(c) small G values with short t
(d) small G values with long t [2004 : 2 M]

3.24 Chlorine gas used for disinfection combines with water to form hypochlorous acid (HOCL). The HOCL ionizes to form hypochlorite (OCL^-) in a reversible reaction:
HOCL $\Leftrightarrow$ $H^+ + OCL^-$ ($k = 2.7 \times 10^{-8}$ at 20°C), the equilibrium of which is governed by pH. The sum of HOCL and OCL^- is known as free chlorine

residual and HOCL is the more effective disinfectant. The 90% fraction of HOCL in the free chlorine residual is available at a pH value

(a) 4.8 (b) 6.6
(c) 7.5 (d) 9.4 [2004 : 2 M]

3.25 A circular primary clarifier processes an average flow of 5005 m^3/d of municipal waste water. The overflow rate is 35 $m^3/m^2/d$. The diameter of clarifier shall be

(a) 10.5 m (b) 11.5 m
(c) 12.5 m (d) 13.5 m [2005 : 2 M]

Linked Answer Questions 3.26 and 3.27:

A city is going to install the rapid sand filter after the sedimentation tanks. Use the following data.

Design loading rate to the filter – 200 m^3/m^2d
Design flow rate – 0.5 m^3/s
Surface area per filter box – 50 m^2

3.26 The surface area required for the rapid sand filter will be

(a) 210 m^2 (b) 215 m^2
(c) 216 m^2 (d) 218 m^2 [2005 : 2 M]

3.27 The number of filters required shall be

(a) 3 (b) 4
(c) 6 (d) 8 [2005 : 2 M]

Common Data for Questions 3.28 and 3.29:

In a rapid sand filter, the time for reaching particle break through (T_B) is defined as the time elapsed from start of filter run to the time at which the turbidity of the effluent from the filter is greater than 2.5 NTU. The time for reaching terminal head loss (T_H) is defined as the time elapsed from the start of the filter run to the time when head loss across the filter is greater than 3 m.

3.28 The effect of increasing the filter depth (while keeping all other conditions same) on T_B and T_H is

(a) T_B increases and T_H decreases
(b) both T_B and T_H increase
(c) T_B decreases and T_H increases
(d) both T_B and T_H decrease [2006 : 2 M]

3.29 The effect of increasing the filter loading rate (while keeping all other conditions same) on T_B and T_H is

(a) T_B increases and T_H decreases
(b) both T_B and T_H increase
(c) T_B decreases and T_H increases
(d) both T_B and T_H decrease [2006 : 2 M]

Common Data for Questions 3.30 and 3.31:

A plain sedimentation tank with a length of 20 m, width of 10 m, and a depth of 3 m is used in a water treatment plant to treat 4 million litres of water per day (4 MLD). The average temperature of water is 20°C. The dynamic viscosity of water is 1.002×10^{-3} N-s/m^2 at 20°C. Density of water is 998.2 kg/m^3. Average specific gravity of particles is 2.65.

3.30 What is the surface overflow rate in the sedimentation tank?

(a) 20 m^3/m^2/day (b) 40 m^3/m^2/day
(c) 67 m^3/m^2/day (d) 133 m^3/m^2/day
[2007 : 2 M]

3.31 What is the minimum diameter of the particle which can be removed with 100% efficiency in the above sedimentation tank?

(a) 11.8×10^{-3} mm (b) 16.0×10^{-3} mm
(c) 50×10^{-3} mm (d) 160×10^{-3} mm
[2007 : 2 M]

3.32 A water treatment plant is required to process 28800 m^3/d of raw water (density = 1000 kg/m^3, kinematic vicosity = 10^{-6} m^2/s). The rapid mixing tank imparts a velocity gradient of 900 s^{-1} to blend 35 mg/L of alum with the flow for a detention time of 2 minutes. The power input (W) required for rapid mixing is

(a) 32.4 (b) 36
(c) 324 (d) 32400 [2008 : 2 M]

3.33 A horizontal flow primary clarifier treats wastewater in which 10%, 60% and 30% of particles have settling velocities of 0.1 mm/s, 0.2 mm/s and 1.0 mm/s respectively. What would be the total percentage of particles removed if clarifier operates at a Surface Overflow Rate (SOR) of 43.2 m^3/m^2-d?

(a) 43% (b) 56%
(c) 86% (d) 100% [2009 : 2 M]

3.34 Consider the following unit processes commonly used in water treatment; rapid mixing (RM), flocculation (F), primary sedimentation (PS), secondary sedimentation (SS), chlorination (C) and rapid sand filtration (RSF). The order of these unit processes (first to last) in a conventional water treatment plant is

(a) PS – RSF – F – RM – SS – C
(b) PS – F – RM – RSF – SS – C
(c) PS – F – SS – RSF – RM – C
(d) PS – RM – F – SS – RSF – C [2011 : 1 M]

3.35 Chlorine gas (8 mg/L as Cl_2) was added to a drinking water sample. If the free chlorine residual and pH was measured to be 2 mg/L (as Cl_2) and 7.5, respectively, what is the concentration of residual OCl^- ions in the water? Assume that the chlorine gas added to the water is completely converted to HOCl and OCl^-. Atomic Weight of Cl: 35.5

Given: $OCl^- + H^+ \overset{K}{\rightleftharpoons} HOCl$, $K = 10^{7.5}$

(a) 1.408×10^{-5} moles/L
(b) 2.817×10^{-5} moles/L
(c) 5.634×10^{-5} moles/L
(d) 1.127×10^{-4} moles/L [2011 : 2 M]

3.36 A town is required to treat 4.2 m^3/min of raw water for daily domestic supply. Flocculating particles are to be produced by chemical coagulation. A column analysis indicated that an overflow rate of 0.2 mm/s will produce satisfactory particle removal in a settling basin at a depth of 3.5 m. The required surface area (in m^2) for settling is

(a) 210 (b) 350
(c) 1728 (d) 21000 [2012 : 2 M]

3.37 A water treatment plant, having discharge 1 m^3/sec, has 14 filters to treat the water. Each filter is having 50 m^2 area, but due to backwashing activity 2 filters are non operational. Calculate hydraulic loading rate in m^3/day.m^2.

[2013 : 1 M]

3.38 A settling tank is designed for a surface overflow rate of 30 m^3/day.m^2. Assuming specific gravity of sediment particles = 2.65. Density of water, γ_w = 1000 kg/m^3, dynamic viscosity of water m_{water} = 0.001 Ns/m^2 and stokes law is valid. The approximate minimum size of particles which can be completely removed is

(a) 0.01 mm (b) 0.02 mm
(c) 0.03 mm (d) 0.04 mm

[2013 : 2 M]

3.39 16 MLD of water is flowing through a 2.5 km long pipe of diameter 45 cm. The chlorine at the rate of 32 kg/d is applied at the entry of this pipe so that disinfected water is obtained at the exit. There is a proposal to increase the flow through this pipe to 22 MLD from 16 MLD. Assume the dilution coefficient, n = 1. The minimum amount of chlorine (in kg per day) to be applied to achieve the same degree of disinfection for the enhanced flow is

(a) 60.50 (b) 44.00
(c) 38.00 (d) 23.27

[2014 : 2 M, Set-1]

3.40 The potable water is prepared from turbid surface water by adopting the following treatment sequence.

(a) Turbid surface water → Coagulation → Flocculation → Sedimentation → Filtration → Disinfection → Storage & Supply
(b) Turbid surface water → Disinfection → Flocculation → Sedimentation → Filtration → Coagulation → Storage & Supply
(c) Turbid surface water → Filtration → Sedimentation → Disinfection → Flocculation → Coagulation → Storage & Supply
(d) Turbid surface water → Sedimentation → Flocculation → Coagulation → Disinfection → Filtration → Storage & Supply

[2014 : 2 M, Set-I]

3.41 A suspension of sand like particles in water with particles of diameter 0.10 mm and below is flowing into a settling tank at 0.10 m^3/s. Assume g = 9.81 m/s^2, specific gravity of particles = 2.65, and kinematic viscosity of water = 1.0105×10^{-2} cm^2/s. The minimum surface area (in m^2) required for this settling tank to remove particles of size 0.06 mm and above with 100% efficiency is _____.

[2014 : 2 M, Set-II]

3.42 A surface water treatment plant operates round the clock with a flow rate of 35 m^3/min. The water temperature is 15 °C and jar testing indicated an alum dosage of 25 mg/l with flocculation at a Gt value of 4×10^4 producing optimal results. The alum quantity required for 30 days (in kg) of operation of the plant is _______.

[2014 : 2 M, Set-II]

3.43 An effluent at a flow rate of 2670 m^3/d from a sewage treatment plant is to be disinfected. The laboratory data of disinfection studies with a chlorine dosage of 15 mg/l yield the model $N_t = N_o e^{-0.145t}$ where N_t = number of micro-organisms surviving at time t (in min.) and N_o = number of micro-organisms present initially (at t = 0). The volume of disinfection unit (in m^3) required to achieve a 98% kill of micro-organisms is _____.

[2014 : 2 M, Set-II]

3.44 Consider a primary sedimentation tank (PST) in a water treatment plant with Surface Overflow Rate (SOR) of 40 m^3/m^2/d. The diameter of the spherical particle which will have 90 percent theoretical removal efficiency in this tank is _______ μm. Assume that settling velocity of the particles in water is described by Stoke's Law.
Given: Density of water = 1000 kg/m^3; Density

of particle = 2650 kg/m^3; g = 9.81 m/s^2; Kinematic viscosity of water (ν) = 1.10 × 10^{-6} m^2/s.

[2015 : 2 M, Set-I]

3.45 A water treatment plant of capacity, 1 m^3/s has filter boxes of dimensions 6 m × 10 m. Loading rate to the filters is 120 m^3/day/m^2. When two of the filters are out of service for back washing, the loading rate (in m^3/day/m^2) is ______.

[2015 : 2 M, Set-II]

3.46 A water supply board is responsible for treating 1500 m^3/day of water. A settling column analysis indicates that an overflow rate of 20 m/day will produce satisfactory removal for a depth of 3.1 m. It is decided to have two circular settling tanks in parallel. The required diameter (expressed in m) of the settling tanks is_____.

[2016 : 1 M, Set-II]

3.47 The spherical grit particles, having a radius of 0.01 mm and specific gravity of 3.0, need to be separated in a settling chamber. It is given that

- g = 9.81 m/s^2
- the density of the liquid in the settling chamber = 1000 kg/m^3
- the kinematic viscosity of the liquid in the settling chamber = 10^{-6} m^2/s

Assuming laminar conditions, the settling velocity (in mm/s, up to one decimal place) is _________.

[2017 : 2 M, Set-I]

3.48 A municipal corporation is required to treat 1000 m^3/day of water. It is found that an overflow rate of 20 m/day will produce a satisfactory removal of the discrete suspended particles at a depth of 3 m. The diameter (in meters, rounded to the nearest integer) of a circular settling tank designed for the removal of these particles would be ______.

[2017 : 2 M, Set-II]

3.49 A rapid sand filter comprising a number of filter beds is required to produce 99 MLD of potable water. Consider water loss during backwashing as 5%, rate of filtration as 6.0 m/h and length to width ratio of filter bed as 1.35. The width of each filter bed is to be kept equal to 5.2 m. One additional filter bed is to be provided to take care of break-down repair and maintenance. The total number of filter beds required will be

(a) 19 (b) 20
(c) 21 (d) 22

[2018 : 2 M, Set-I]

3.50 A flocculation tank contains 1800 m^3 of water, which is mixed using paddles at an average velocity gradient G of 100/s. The water temperature and the corresponding dynamic viscosity are 30°C and 0.798 × 10^{-3} Ns/m^2, respectively. The theoretical power required to achieve the stated value of G (in kW, up to two decimal places) is____.

[2018 : 2 M, Set-II]

3.51 At a small water treatment plant which has 4 Filters, the rates of filtration and backwashing are 200 m^3/d/m^2 and 1000 m^3/d/m^2, respectively. Backwashing is done for 15 min per day. The maturation, which occurs initially as the filter is put back into service after cleaning, takes 30 min. It is proposed to recover the water being wasted during backwashing and maturation. The percentage increase in the filtered water produced (up to two decimal places) would be ________.

[2018 : 2 M, Set-II]

3.52 A completely mixed dilute suspension of sand particles having diameters 0.25, 0.35, 0.40. 0.45 and 0.50 mm are filled in a transparent glass column of diameter 10 cm and height 2.50 m. The suspension is allowed to settle without any disturbance. It is observed that all particles of diameter 0.35 mm settle to the bottom of the column in 30 s. For the same period of 30 s. The percentage removal (round off to integer value) of particles of diameters 0.45 and 0.50 mm from the suspension is_______.

[2019 : 1 M, Set-I]

3.53 Sedimentation basin in a water treatment plant is designed for a flow rate of 0.2 m^3/s. The basin is rectangular with a length of 32 m. width of 8 m and depth of 4 m. Assume that the settling velocity of these particles is governed by the Stokes' law. Given: density of the particles = 2.5 g/cm^3, density of water = 1 g/cm^3, dynamic viscosity of water = 0.01 g/(cm.s) gravitational acceleration = 980 cm/s^2. If the incoming water contains particles of diameter 25 μm (spherical and uniform), the removal efficiency of these particles is

(a) 100% (b) 51%
(c) 78% (d) 65%

[2019 : 2 M, Set-I]

3.54 A 0.80 m deep bed of sand filter (length 4 m and width 3 m) is made of uniform particles (diameter = 0.40 mm, specific gravity = 2.65, shape factor = 0.85) with bed porosity of 0.4. The bed has to be backwashed at a flow rate of 3.60 m^3/min.

During backwashing, if the terminal settling velocity of sand particles is 0.05 m/s, the expanded bed depth (in m, round off to 2 decimal places) is_______.

[2019 : 2 M, Set-I]

3.55 A wastewater is to be disinfected with 35 mg/L of chlorine to obtain 99% kill of micro-organisms. The number of micro-organisms remaining alive (N_t) at time t, is modelled by $N_t = N_o e^{-kt}$, where N_o is number of micro-organisms at $t = 0$, and k is the rate of kill. The wastewater flow rate is 36 m^3/h and $k = 0.23\ min^{-1}$. If the depth and width of the chlorination tank are 1.5 m and 1.0 m respectively, the length of the tank (in m, round off to 2 decimal places) is_______.

[2019 : 2 M, Set-I]

3.56 A water treatment plant treats 6000 m^3 of water per day. As a part of the treatment process, discrete particles are required to be settled in a clarifier. A column test indicates that an overflow rate of 1.5 m per hour would produce the desired removal of particles through settling in the clarifier having a depth of 3.0 m. The volume of the required clarifier, (in m^3, round off to 1 decimal place) would be_______.

[2019 : 2 M, Set-II]

3.57 Chlorine is used as the disinfectant in a municipal water treatment plant. It achieves 50 percent of disinfection efficiency measured in terms of killing the indicator microorganisms (E-Coli) in 3 minutes. The minimum time required to achieve 99 percent disinfection efficiency would be

(a) 19.93 minutes
(b) 11.93 minutes
(c) 9.93 minutes
(d) 21.93 minutes

[2019 : 2 M, Set-II]

3.58 Surface Overflow Rate (SOR) of a primary settling tank (discrete settling) is 20000 litre/m^2 per day. Kinematic viscosity of water in the tank is 1.01×10^{-2} cm^2/s. Specific gravity of the settling particles is 2.64. Acceleration due to gravity is 9.81 m/s^2. The minimum diameter (in μm, round off to one decimal place) of the particles that will be removed with 80% efficiency in the tank, is _______.

[2020 : 2 M, Set-I]

3.59 A water supply scheme transports 10 MLD (Million Litres per Day) water through a 450 mm diameter pipeline for a distance of 2.5 km. A chlorine dose of 3.50 mg/litre is applied at the starting point of the pipeline to attain a certain level of disinfection at the downward end. It is decided to increase the flow rate from 10 MLD to 13 MLD in the pipeline. Assume exponent for concentration, n = 0.86. With this increased flow, in order to attain the same level of disinfection, the chlorine does (in mg/litre) to be applied at the starting point should be

(a) 5.55
(b) 4.75
(c) 3.95
(d) 4.40

[2020 : 2 M, Set-I]

3.60 Two identically sized primary settling tanks receive water for Type-I settling (discrete particles in dilute suspension) under laminar flow conditions. The surface overflow rate (SOR) maintained in the two tanks are 30 m^3/m^2.d and 15 m^3/m^2.d. The lowest diameters of the particles, which shall be settled out completely under SORs of 30 m^3/m^2.d are designated as d_{30} and d_{15} respectively. The ratio $\frac{d_{30}}{d_{15}}$ (round off to two decimal places), is _______.

[2020 : 1 M, Set-II]

3.61 A water filtration unit is made of uniform-size sand particles of 0.4 mm diameter with a shape factor of 0.84 and specific gravity of 2.55. The depth of the filter bed is 0.70 m and the porosity is 0.35. The filter bed is to be expanded to a porosity of 0.65 by hydraulic backwash. If the terminal settling velocity of sand particles during backwash is 4.5 cm/s, the required backwash velocity is

(a) 6.35×10^{-3} m/s
(b) 5.79×10^{-3} m/s
(c) 0.69 cm/s
(d) 0.75 cm/s

[2021 : 2 M, Set-II]

3.62 Which of the following process(es) can be used for conversion of salt water into fresh water?

(a) Microfiltration
(b) Electrodialysis
(c) Ultrafiltration
(d) Reverse osmosis

[2022 : 1 M, Set-I]

3.63 A flocculator tank has a volume of 2800 m^3. The temperature of water in the tank is 15°C, and the average velocity gradient maintained in the tank is 100/s. The temperature of water is reduced to 5°C, but all other operating conditions including the power input are maintained as the same. The decrease in the average velocity gradient (in %) due to the reduction in water temperature is (round off to nearest integer).

[Consider dynamic viscosity of water at 15°C and 5°C as 1.139×10^{-3} N-s/m^2 and 1.518×10^{-3} N-s/m^2, respectively]

[2023 : 2 M, Set-I]

3.64 Which of the following is/are NOT active disinfectant(s) in water treatment?

(a) •OH (hydroxyl radical)
(b) O_3 (ozone)
(c) OCl^- (hypochlorite ion)
(d) Cl^- (chloride ion)

[2023 : 1 M, Set-II]

3.65 A water treatment plant treats 25 MLD water with a natural alkalinity of 4.0 mg/*l* (as $CaCO_3$). It is estimated that during coagulation of this water, 450 kg/day of calcium bicarbonates ($Ca(HCO_3)_2$) is required based on alum dosage.
Consider the atomic weight as: Ca-40, H-1, C-12, O-16.
The quantity of pure quick lime CaO (in kg) required for this process per day is ______ (round off to 2 decimal places)

[2024 : 2 M, Set-I]

3.66 Activated carbon is used to remove a pollution from wastewater in a mixed batch reactor, which follows first-order reaction kinetics.
At a reaction rate of 0.38/day, the time (in days) required to remove the pollutant by 95% is ______ (rounded off to 1 decimal place).

[2024 : 2 M, Set-I]

3.67 A hypothetical multimedia filter, consisting of anthracite particles (specific gravity: 1.50), silica sand (specific gravity: 2.60), and ilmenite sand (specific gravity: 4.20), is to be designed for treating water/wastewater. After backwashing, the particles should settle forming three layers: coarse anthracite particles at the top of the bed, silica sand in the middle, and small ilmenite sand particles at the bottom of the bed.
Assume
(i) Slow discrete settling (Stoke's law is applicable)
(ii) All particles are spherical
(iii) Diameter of silica sand particles is 0.20 mm
The CORRECT option fulfilling the diameter requirements for this filter media is

(a) diameter of anthracite particles is slightly less than 0.35 mm and diameter of ilmenite particles is slightly greater than 0.141 mm.
(b) diameter of anthracite particles is slightly greater than 0.35 mm and diameter of ilmenite particles is slightly less than 0.141 mm.
(c) diameter of anthracite particles is slightly greater than 0.64 mm and diameter of ilmenite particles is slightly less than 0.10 mm.
(d) diameter of anthracite particles is slightly less than 0.64 mm and diameter of ilmenite particles is slightly less than 0.10 mm.

[2024 : 2 M, Set-II]

3.68 A circular settling tank is to be designed for primary treatment of sewage at a flow rate of 10 million liters/day. Assume a detention period of 2.0 hours and surface loading rate of 40000 liters/m^2/ day. The height (in meters) of the water column in the tank is ______ (rounded off to 2 decimal places).

[2024 : 2 M, Set-II]

■■■■

Answers Water Treatment

3.1	(a)	3.2	(d)	3.3	(b)	3.4	(d)	3.5	(c)	3.6	(b)	3.7	(b)
3.8	(d)	3.9	(c)	3.10	(d)	3.11	(d)	3.12	(a)	3.13	(c)	3.14	(c)
3.15	(b)	3.16	(d)	3.17	(c)	3.18	(c)	3.19	(d)	3.20	(c)	3.21	(c)
3.22	(d)	3.23	(a)	3.24	(b)	3.25	(d)	3.26	(c)	3.27	(c)	3.28	(a)
3.29	(d)	3.30	(a)	3.31	(b)	3.32	(d)	3.33	(b)	3.34	(d)	3.35	(a)
3.36	(b)	3.37	(144)	3.38	(b)	3.39	(a)	3.40	(a)	3.41	(31.2)	3.42	(37,800)
3.43	(50.02)	3.44	(22.576)	3.45	(144)	3.46	(6.91)	3.47	(0.436)	3.48	(8)	3.49	(c)
3.50	(14.36)	3.51	(7.525)	3.52	(Sol.)	3.53	(d)	3.54	(1.21)	3.55	(8.01)	3.56	(500)
3.57	(a)	3.58	(14.46)	3.59	(b)	3.60	(6.35)	3.61	(a)	3.62	(b, d)	3.63	(13)
3.64	(d)	3.65	(99.56 or 155.56)			3.66	(7.9)	3.67	(a)	3.68	(3.33)		

Explanations Water Treatment

3.1 (a)

Break point chlorination represents the dose of chlorination, beyond which any further additional chlornie will appear as free residual chlorine.

Break point chlorination is done to react with any ammonia, reducing compounds like Fe, Mn, readily oxidizable organic matter and micro-organism.

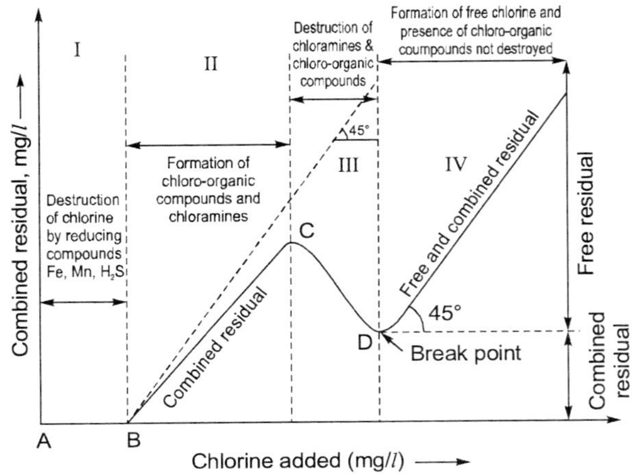

The graph shown above gives the detailed explanation of the steps involved in the process.

3.2 (d)

3.3 (b)

- The surface charge on colloidal particels gives them long-term stability; and hence the particles which would have settled or coalesced are mutually repelled by their like charges.
- Hence, chemical compounds called goagulant are added in the water and mixing is done such that the coagulant completely disperses into the entire mass of water and neutralise the charge on suspended solids, after getting threshold energy.
- In order to increase the opportunity of these particles to form the contact with each other, slow mixing is also induced in the subsequent process called flocculation.

In this process, neutralized particles are brought in intimate contact with each other, so as to promote the agglomeration resulting in bigger size patciles which can get easilty settled in the following sedimentations process.

3.4 (d)

The free chlorine as well as the combined chlorine will cause germicidal action on bacteria and pathogens. The free chlorine will instantaneously kill the pathogens while the combined chlorine will provide long term germicidal effect.

3.5 (c)

Rapid mixing units (or mechanical mixers) are used for mixing in sedimentation coagulation proces.

For a mechanicalmixer, the resultant intensity of mixing is bsed on rate of power inputk P.

and $G = \sqrt{\dfrac{P}{\mu V}}$

$\Rightarrow$ $P = \mu V G^2$

Where, P is power in watts

μ is dynamic viscosity in Ns/m^2

V is volume of tank in m^3

3.6 (b)

Activated carbon is a specially treated carbon, which possesses the property of absorbing and attracting impurities, such as gases, liquids, and finely divided solids. Because of its excellent property of absorbing impurities, which is due to its high surface area to volume ratio. It is widely used for removing tastes and odours from public supplies.

3.7 (b)

Chlorine in its various forms is invariably and almost universally used for disinfecting public water supplies. It is cheap, reliable, easy to handle, easily measurable and above all, it is capable of providing residual disinfecting effects for long periods, thus affording complete protection against future recontamination of water in the distribution system.

POINTS TO REMEMBER

1. Chlorine dioxide
   - It is 2.5 times stonger than free chlorine.
   - Due to high unstability, it has to be used immediately after production.

2. Ozone
   - It is an excellent diinfectant, but very costly.
   - Due to high unstablity, it does not provide any safety against future recontamination.
3. Un-radiation
   - Highly effective in killing bacteria.
   - It is very costly and therefore not used in public supplies.
   - It is used in hospitals and for drinking water.

3.8 (d)

Coagulation flocculation is always done prior to filtration. Filtration can either be by slow sand or rapid sand filters.

POINTS TO REMEMBER

Sequence of various methods used for purifying the water is

Screening → Plain sedimentaiton → Sedimentation aided with coagulation → Filtration → Disinfection → Aeration → Softening → Miscellaneous treatments such as fluoridations, recarbonation, lining, desalination, etc.

3.9 (c)

For 100% efficiency,

$$V_s = V_0$$

Where, V_s = Selting velocity

V_0 = Overflow rate

$$\therefore \quad V_0 = \frac{Q}{S.A}$$

$$\Rightarrow \quad S.A. = \frac{Q}{V_0}$$

$$= \frac{\dfrac{5.7\times10^6\times10^{-3}}{24\times3600}\,\text{m}^3/\text{sec}}{0.33\times10^{-3}\,\text{m/sec}} \simeq 200\ \text{m}^2$$

POINTS TO REMEMBER

- The overflow rate,

$$V_0 = \frac{\text{Flow of water}}{\text{Surface area of settling basin}}$$

- Any particle with settling velocity greater than V_0 will settle out, while other particles will settle out in the ratio V_s / V_0.

3.10 (d)

Chlorination is most effective at a certain pH range. The disinfection efficiency of chlorine in water treatment is reduced by increasing or decreasing the pH value

$$CL_2 + H_2O \xrightarrow{pH>5} HOCl + HCl$$

$$HOCl \underset{pH<7}{\overset{pH>8}{\rightleftharpoons}} H^+ + OCl^-$$

HOCl (Hypochlorous acid) is the most destructive form of chlroine for disinfection (80% more than OCl^-).

Therefore, from above two reactions:

(i) pH should not be lesser than 5, otherwise HOcl will not be formed.

(ii) pH should be kept slightly less than 7, since HOCl is predominant at this pH.

3.11 (d)

For slow sand filters:

1. Filtration rates for design is taken as 100-200 $l/hr/m^2$ which is in the range given in statement 1.
2. Frequency at cleaning is 1-3 months. Hence 2nd statement is wrong.
3. Operation cost is low since it works under the gravity and no other energy source is required.

3.12 (a)

$$\underbrace{Al_2(SO_4)_3.18\,H_2O}_{\text{Alum}} + \underbrace{3\,Ca(HCO_3)_2}_{\text{Alkalinity}} \longrightarrow$$

$$2Al(OH)_3\downarrow + 6CO_2\uparrow + \underbrace{3CaSO_4}_{\text{Permanent hardness}} + 18H_2O$$

From the above equation, the following inferences can be drawn:

1. Alum reacts will alkalinity and hence it decreases the alkalinity, thereby decresing the pH.
2. Precipitate of $Al(OH)_3$ is formed.
3. Permanent Hardness ($CaSO_4$) is induced in water.
4. CO_2 gas is evolved which may reduce the pH due to the formation of carbonic acid.

3.13 (c)

Aeration is the process in which water is brought in intimate contact with air in order to allow the absorption of oxygen in it so as to carryout the removal of following dissolved impurities:

1. Dissolved gases like CO_2 and H_2S.
2. Dissolved volatile liquids like phenol.
3. Dissolved organic matter.
4. Dissolved minerals like Fe and Mn.

3.14 (c)

Aluminium sulphate (or alumn) is used for coagulation. Its chemical formula is $Al_2(SO_4)_3$ $18H_2O$.

POINTS TO REMEMBER

Other coagulants used are:

- Copperas ($FeSO_4 \cdot 7H_2O$) Hydrated ferrous sulphate.
- Chlorinated copperas $[Fe_2(SO_4)_3 + FeCl_3]$
- Sodium aluminate ($Na_2Al_2O_4$)

3.15 (b)

Overflow rate= 17 m³/day/m² = 17 m/day

$$= \frac{17 \times 1000}{24 \times 60 \times 60} = 0.196 \text{ mm/s}$$

Percentage particle removal

$$= \frac{V_s \text{ (Settling velocity)}}{V_0 \text{(Overflow rate)}} \times 100$$

$$= \frac{0.1}{0.196} \times 100 = 50\%$$

3.16 (d)

HOCl, OCl^-, Cl_2 are called freely available chlorine. Out of these form, HOCl in most effective and is 80% more reactive than OCl^-.

3.17 (c)

$$\eta = \frac{V_s}{V_0} \times 100$$

$$\Rightarrow \quad 90 = \frac{0.12}{V_0} \times 100$$

$$\Rightarrow \quad V_0 = 0.133 \text{ cm/sec}$$

Now, $Q = 432 \text{ m}^3/\text{hr} = 0.12 \text{ m}^3/\text{sec}$

Now, $V_0 = \dfrac{Q}{BL}$

Where BL = Plan area

$$\therefore \quad BL = \frac{Q}{V_0} = \frac{0.12}{0.133 \times 10^{-2}} = 90.225 \text{ m}^2$$

$$\Rightarrow \quad BL \simeq 90 \text{ m}^2$$

3.18 (c)

Zero hardness of water is achieved by Zeolite method (ion exchange method) or by Demineralisation method.

3.19 (d)

Metods to remove temporary hardness are:

1. Boiling
2. Lime addition

Method ro remove permanent hardness are:

1. Lime-soad process
2. Zeolite treatment
3. Deminerealisation

3.20 (c)

Settling velocity of sand particles,

$$V_s = \frac{g}{18}(S_s - 1)\frac{d^2}{\nu}$$

$$= \frac{9.81}{18} \times (2.65 - 1) \times \frac{(0.01 \times 10^{-3})^2}{(10^{-3}/1000)}$$

$$= \frac{9.81}{18} \times 1.65 \times 10^{-4}$$

$$V_s = 8.9925 \times 10^{-5} \text{ m/sec}$$

Detention time

$$= \frac{\text{Volume of tank}}{\text{Water flow rate}}$$

$$= \frac{900 \times 3}{8000} \text{ days} = \frac{2700}{8000} \times 24 = 8.1 \text{ hrs}$$

Over flow rate

$$V_0 = \frac{\text{Depth of tank}}{\text{Detention time}} = \frac{3}{8.1 \times 3600}$$

$$= 1.0288 \times 10^{-4} \text{ m/s}$$

$$\therefore \quad \text{Removal efficiency} = \frac{V_s}{V_0} \times 100$$

$$= \frac{8.9925 \times 10^{-5}}{1.0288 \times 10^{-4}} \times 100 = 87.41\% \simeq 87.5\%$$

3.21 (c)

(i) $\underset{666 \text{ gm}}{Al_2(SO_4)_3 \cdot 18H_2O} + \underset{(3 \times 162) \text{ gm}}{3Ca(HCO_3)_2}$
$\rightarrow 3CaSO_4 + 2Al(OH)_3\downarrow + 6CO_2\uparrow + 18H_2O$

(ii) $\underset{100 \text{ gm}}{CaCO_3} + H_2O + CO_2 \rightarrow \underset{162 \text{ gm}}{Ca(HCO_3)_2}$

From (i) it is clear that 666 gm filter Alum can neutralize (3 × 162) gm of alkalinity as $Ca(HCO_3)_2$. Again from (ii) it is clear that this (3 × 162) gm of alkalinity as $Ca(HCO_3)_2$ is equivalent to (3 × 100) gm of alkalinity as $CaCO_3$.

∴ Total Alkalinity of filter Alum as $CaCO_3$, for water requiring 20 mg/L of filter alum

$$= \frac{3 \times 100}{666} \times 20 \text{ mg/L} = 9 \text{ mg/L}$$

Hence total alkalinity requirement

$= 10 \times 10^6 \times 9$ mg/day $= 90 \times 10^6$ mg/day

3.22 (d)

Natural alkalinity available in water = 6 mg/L

Total alkalinity required to be added in the form of quick lime = 9 – 6 = 3 mg/L

$$CaCO_3 \longrightarrow CaO + CO_2$$

100 gm 56 gm

∴ 56 gm of CaO (quick lime) is required to produce 100 gm of $CaCO_3$

∴ Quantity of CaO required

$$= \frac{56}{100} \times 3 \text{ mg/L} = 1.68 \text{ mg/L}$$

Total amount of CaO required

$= 10 \times 10^6 \times 1.68 \times 365 = 6132 \times 10^6$ mg per year

3.23 (a)

Large *G* values with short times tend to produce small, dense flocs, while low *G* values and long times produce larger, lighter flocs. Since large, dense flocs are more easily removed in the settling basin, it may be advantageous to vary the *G* values over the length of the flocculation basin. The small dense flocs produced at high *G* values subsequently combine into larger flocs at the lower *G* values. Reduction in *G* values by a factor of 2 from the influent end to the effluent end of the flocculator has been shown to be effective.

3.24 (b)

Method-I

According to the questions,

- $HOCl + OCl^- =$ Free chlorine residual
- In free chlorine residual, 90% fraction of HOCl is available.

  This means that 10% fraction of OCl^- is available.

  Hence, in the free chlorine residual

  Let $\underset{0.9}{HOCl} \rightleftharpoons H^+ + \underset{0.1}{OCl^{-1}}$

$$k = \frac{[H^+][OCl^-]}{[HOCl^-]}$$

$$\Rightarrow [H^+] = \frac{2.7 \times 10^{-8} \times 0.9}{0.1} = 24.3 \times 10^{-8} \text{ mol/L}$$

Now, $pH = -\log[H^+] = -\log[24.3 \times 10^{-8}]$

$\Rightarrow$ $pH = 6.6$

Method-II

1. $Cl_2 + H_2O \xrightarrow{pH>5} HOCl + HCl$

   Hence, for 90% HOCl, pH can't be less than 5.

2. $HOCl \underset{pH<7}{\overset{pH>8}{\rightleftharpoons}} H^+ + OCl^-$

   Hence pH has to be slightly less than 7, for HOCl to be predominant.

   From above 2 conditions, the only option that is possible is pH = 6.6.

3.25 (d)

$$\text{Area of clarifier} = \frac{\text{Average flow of waste water}}{\text{Overflow rate}}$$

$$= \frac{5005}{35} = 143 \text{ m}^2$$

$$\therefore \quad \frac{\pi}{4} D^2 = 143$$

$$\Rightarrow \quad D^2 = \frac{143 \times 4}{\pi}$$

$$\Rightarrow \quad D = 13.5 \text{ m}$$

3.26 (c)

Total surface area

$$= \frac{\text{Design flow rate}}{\text{Design loading rate}} = \frac{0.5}{200/(24 \times 60 \times 60)}$$

$$= \frac{24 \times 60 \times 60 \times 0.5}{200} = 216 \text{ m}^2$$

3.27 (c)

Number of filters

$$= \frac{\text{Total surface area}}{\text{Surface area of each filter box}}$$

$$= \frac{216}{50} = 4.32 \simeq 5$$

Provide one additional filter as a stand-by filter to be used during cleaning, maintenance, etc.

Hence, number of filters = 5 + 1 = 6.

328 (a) and 3.29 (d)

The effect of increase in certain rapid sand filter and influent particle characteristics on time (T_B) for reaching particle breakthrough (turbidity > 2.5 NTU) in the effluent, and time (T_H) of reaching terminal head-loss (> 3 m) across filter bed are given below. Explain the results based on you understanding of filter operation and particle removal mechanisms operative in rapid sand filters. Assume that when one characteristic is changed, all other characteristic remain fixed.

Parameter	Time to break-through T_B	Time to terminal headloss T_H
Filter dpeth (L)	Increase	Decrease
Superficial velocity (V_s), ($m^3/m^2/h$)	Decrease	Decrease
Influent particle concentration, (C_o), (mg/L)	Decrease	Decrease
Floc strength	Increase	Decrease
Collector diameter (d), m	Decrease	Increase
Porosity, α	Decrease	Increase
Particles not destablized by addtion of coagulants	Decrease	Increase

Effect of increase in filter depth (L)

T_B : T_B will ***increase*** since particles in water will have the opportunity to potentially interact with more collectors, and hence the change for particle attachment to a collector is more.

T_H: Since bed depth is more, water is encounter more resistance is passing through the bed. Hence T_H will ***decrease***.

Effect of increase in superficial velocity (V_s)

T_B : Since pore velocity is more, the shear forces experienced by collected particles will be more. Hence particle detachment will be more, leading to a ***decrease*** in T_B.

T_H: Since particle loading rate on the filter and rate of particle collection in the filter will increase, the porosity of the filter bed will decrease more rapidly, leading to more rapid head-loss buildup and hence a ***decrease*** in T_H.

Effect of increase in influent particle concentration (C_o)

T_B : Particle loading rate on the filter, rate of particle collection and hence porosity of the filter bed will decrease more rapidly. Thus pre velocity in the filter bed will increase more rapidly, leading to more rapid particle detachment. This will lead to ***decrease*** in T_B.

T_H: Since particle loading rate on the filter and rate of particle collection in the filter will increase, the porosity of the filter bed will decrease more rapidly, leading to more rapid head-loss buildup and hence a ***decrease*** in T_H.

Effect of increase in floc strength

T_B : Particle detachment due to shear forces will become less prominent leading to increase in T_B.

T_H: Since particle retention in the filter will be more, the porosity of the filter bed will decline faster leading to faster buildup of head-loss and hence T_H will ***decrease***.

Effect of increase in collector diameter (d)

T_B : Increase in collector diameter will mean the presence of a lesser number of collector in the filter media. This will result is lesser number of potential interaction between a particle and a collector. Thus T_B is expected to ***decrease***.

T_H: Since lesser number of particle will be collected as mentioned above, the rate of head-loss buildup will be lower, leading to ***increase*** in T_H.

Effect of increase in porosity (α)

It must be realized that porosity of the filter media does not depend on particle size, but on grading of particles. Well-graded sand, i.e., with a broader particle size distribution will have less porosity, as compared to poorly graded sand with a narrower particle size distribution.

T_B : In case poorly graded sand (with more porosity) is used as the filter media, the influent particles will potentially encounter lesser number of collectors and will not be collected efficiently. Hence T_B will ***decrease***.

T_H: Increase in porosity of sand and poor particle collection as described above will result in lower rate of head-loss buildup and hence increase in T_H.

Effect of non-addition of coagulant

T_B : If no coagulants are added, the particles will be stable and hence will not attach efficiently to the filter media. Hence T_B will ***decrease***.

T_H: Since rate of particle collection in the filter media will be lower as described above, the rate of head-loss buildup will also be lower, leading to an increase in T_H.

3.29 (d)

3.30 (a)

Surface overflow rate

$$= \frac{Q}{BL} = \frac{4\times10^6\times10^{-3}}{10\times20} = 20 \text{ m}^3/\text{m}^2/\text{day}$$

3.31 (b)

Let the minimum diameter of the particle that can be removed with 100% efficiency be 'd'.
For that particle,
Settling velocity (V_s) ≥ Surface overflow rate

$$\Rightarrow \frac{gd^2}{18\mu}(\rho_s - \rho_w) = \frac{20}{86400}\frac{\text{m}}{\text{sec}}$$

$$\Rightarrow \frac{9.81d^2(2.65\times998.2-998.2)}{18\times1.002\times10^{-3}} = \frac{20}{86400}$$

$$\Rightarrow d^2 = 2.584\times10^{-10}$$

$$\Rightarrow d = 1.607\times10^{-5} \text{ m}$$

$$\Rightarrow d = 16\times10^{-6} \text{ m}$$

$$\Rightarrow d = 16\times10^{-3} \text{ mm}$$

3.32 (d)

The power input (W) required for rapid mixing is expressed in terms of temporal mean velocity gradient, G, expressed by the equation,

$$G = \left[\frac{P}{\mu V}\right]^{1/2} \Rightarrow P = G^2\mu V$$

where μ = dynamic viscosity of raw water

$= $ kinematic viscosity × density

$= 10^{-6}\times10^3 = 10^{-3}$ N-s/m²

$G = 900 \text{ s}^{-1}$

t = 2 minutes = 120 seconds

Discharge of raw water

$$= \frac{28800}{24\times60\times60} = 0.33 \text{ m}^3/\text{s}$$

Volume of water in tank,

$V = 0.33\times120 = 40 \text{ m}^3$

$\therefore$ $P = (900)^2\times10^{-3}\times40 = 32400$ watts

3.33 (b)

Surface overflow rate (V_0)

$= 43.2 \text{ m}^3/\text{m}^2\text{-d} = 43.2$ m/d

$$= \frac{43.2\times10^3}{24\times60\times60} \text{ mm/s} = 0.5 \text{ mm/s}$$

Particles which have setting velocities more than surface overflow rate will be 100% removed. Other particles are removed in the ratio of $\left(\frac{V_s}{V_0}\right)$.

$\therefore$ Total percentage of particles removed

$$= 30 + 60\times\frac{0.2}{0.5} + 10\times\frac{0.1}{0.5}$$

$$= 30 + 24 + 2 = 56\%$$

3.34 (d)

Rapid mixing is the process done to disperse the coagulant into the entire mass of water after that flocculation occurs. Flocculated particles are then removed in secondary sedimentation tank followed by rapid sand filter. Water is than chlorinated and sent for supply.

3.35 (a)

$$OCl^- + H^+ \xrightleftharpoons{K} HOCl;\ K = 10^{7.5}$$

Now,

$$\frac{[HOCl]}{[OCl^-][H^+]} = K$$

$$pH = 7.5$$

$$[H^+] = 10^{7.5}$$

$$\left[\frac{HOCl}{OCl^-}\right] = 10^{7.5}\times10^{-7.5} = 1$$

$$\therefore \quad [HOCl] = [OCl^-]$$

$$\begin{bmatrix}\text{Free chlorine}\\ \text{residual}\end{bmatrix} = [HOCl] + [OCl^-]$$

Free chlorine residual = 2 mg/l

$$[\text{Free chlorine residual}] = \frac{2\times10^{-3}}{\text{Mol. wt of } Cl_2}$$

$$= \frac{2\times10^{-3}}{2\times35.5}$$

$$= 2.817\times10^{-5} \text{ moles}/l$$

$$\therefore \quad 2.817\times10^{-5} = [HOCl] + [OCl^-]$$

$$\therefore \quad [OCl^-] = \frac{2.817\times10^{-5}}{2}$$

$$\because \quad [OCl^-] = [HOCl]$$

$$= 1.4084\times10^{-5} \text{ moles}/l$$

3.36 (b)

Overflow rate, $V_0 = 0.2\times10^{-3}$ m/s

Water required to treat raw water,

$$4.2 \text{ m}^3/\text{min} = \frac{4.2}{60}\text{m}^3/\text{s} = 0.07 \text{ m}^3/\text{s}$$

$\therefore$ Required surface area, A

$$= \frac{0.07}{0.2 \times 10^{-3}} = 350 \text{ m}^2$$

3.37 Sol.

$Q = 1 \text{ m}^3/\text{sec}$
No. of working filters = 12
S.A. of each filter = 50 m^2

$\therefore$ Hydraulic loading rate $= \frac{Q}{A}$

$$= \frac{1}{12 \times 50} \text{m}^3/\text{sec.m}^2$$

$$= \frac{1 \times 24 \times 60 \times 60}{12 \times 50} \text{m}^3/\text{day.m}^2 = 144 \text{ m}^3/\text{day.m}^2$$

3.38 (b)

To calculate minimum size of particles equating settling velocity to overflow rate, we get,

$$V_0 = V_s$$

$$\Rightarrow \quad V_s = \frac{(G_s - 1)\gamma_w d^2}{18\mu}$$

Now $\quad V_0 = \frac{30 \text{m}^3}{\text{day.m}^2} = \frac{30}{24 \times 60 \times 60} \frac{\text{m}}{\text{s}}$

$$\Rightarrow \frac{30}{24 \times 60 \times 60} = \frac{(2.65 - 1) \times 9.81 \times 10^3 \times d^2}{18 \times 0.001}$$

Solving we get,

$$d = 1.965 \times 10^{-5} \text{ m} \approx 0.02 \text{ mm}$$

3.39 (a)

In the disinfection process we have the relationship,

$$tC^n = K$$

where t = time required to kill all organism

c = concentration of disinfectant

n = dilution coefficient

k = constant

$$\Rightarrow \quad t_1 C_1^n = t_2 C_2^n$$

in our case n = 1

$$\Rightarrow \quad t_1 C_1 = t_2 C_2$$

$$t_1 = \frac{L}{V_1}$$

L = length of pipe ; V_1 = velocity of flow

$$t_1 = \frac{L}{Q_1/A} = \frac{L \cdot A}{Q_1}$$

$$C_1 = \frac{W_1}{Q_1},$$

where W_1 = weight of disinfectant per day;
Q_1 = discharge per day,

$$\Rightarrow \quad \frac{L \cdot A}{Q_1} \times \frac{W_1}{Q_1} = \frac{L \cdot A}{Q_2} \times \frac{W_2}{Q_2}$$

$$\Rightarrow \quad W_2 = \frac{Q_2^2}{Q_1^2} \times W_1$$

$$= \left(\frac{22}{16}\right)^2 \times 32 \text{ kg/day} = 60.5 \text{ kg/day}$$

3.40 (a)

3.41 Sol.

For particles of size 0.06 mm,

$$V_s = \frac{(G_s - 1) g d^2}{18\nu} = \frac{(2.65 - 1) \times 9.81 \times (6 \times 10^{-2})^2}{18 \times 1.0105 \times 10^{-2}}$$

$$= \frac{1.65 \times 9.81 \times 36 \times 10^{-4}}{18 \times 1.0105 \times 10^{-2} \times 10^2} = 3.204 \times 10^{-3} \text{ m/s}$$

$$V_0 = \frac{Q}{BL}$$

For 100% efficient removal of 0.06 mm size particles

$$V_s = V_0$$

$$\Rightarrow \quad BL = \frac{0.1}{3.204 \times 10^{-3}} = 31.214 \text{ m}^2$$

3.42 Sol.

Given data

Flow rate, $\quad Q = 35 \text{ m}^3/\text{min}$

$Gt = 4 \times 10^4$

Alum dosage = 25 mg/lit.

Alum quantity (kg) required for 30 days

$= 35 \times 10^3 \times 60 \times 24 \times 30 \times 25 \times 10^{-6}$

$= 37{,}800$ kg

3.43 Sol.

$Q = 2670 \text{ m}^3/\text{d}$; $N_t = \text{No. } e^{-0.145t}$

Let x be the no. of microorganisms (M.O.) present initially.

98% kill of M.O. imlies that at time 't' 2% of M.O. are still surviving

$\therefore$ M.O. surviving at time 't' $= \frac{2}{100}x$

$$\therefore \quad \frac{2}{100}x = x \cdot e^{-0.145t}$$

$\therefore$ t = 26.979 min = 0.018736 days

$\therefore$ Volume = $Q.t$

$= 2670 \times 0.018736 = 50.0244 \text{ m}^3$

3.44 Sol.

$$\% \text{ removal} = \frac{V_s'}{V_s} \times 100$$

$$V_s' = 0.9 \times V_s = 0.9 \times \text{Overflow rate}$$

$$V_s' = \frac{0.9 \times 40}{86400} \text{m/s}$$

$$\frac{(G-1)gd^2}{18\nu} = \frac{0.9 \times 40}{86400}$$

$$d = \sqrt{\frac{0.9 \times 40 \times 18\rho_w \nu}{86400 \times (G_s - 1) \times \rho_w \times g}}$$

$$= \sqrt{\frac{0.9 \times 40 \times 18 \times 1.1 \times 10^{-6}}{86400 \times 1.64 \times 9.81}}$$

$$= 2.2576 \times 10^{-5} \text{ m} = 22.576 \ \mu\text{m}$$

3.45 Sol.

Total water filtered in a day = 60 × 60 × 24

= 86400 m^3/day

Total surface area of filter required

$$= \frac{86400 \text{ m}^3\text{/day}}{120 \text{ m}^3\text{/day/m}^2} = 720\text{m}^2$$

Area of one filter = 6 × 10 = 60 m^2

$$\text{Total no. of filter} = \frac{720}{60} = 12 \text{ filtres}$$

As two filters are out of service, number of filters working = 12 – 2 = 10

Total surface area of filter = 60 × 10 = 600 m^2

The new loading rate

$$= \frac{86400 \text{ m}^3\text{/day}}{600 \text{ m}^2} = 144 \text{ m}^3\text{/day/m}^2$$

3.46 Sol.

Discharge to be treated by one tank,

$$Q = \frac{1500}{2} = 750 \text{ m}^3\text{/day}$$

$$\text{Surface area, } A = \frac{Q}{OFR} = \frac{750}{20} = 37.5 \text{ m}^2$$

$$\Rightarrow \quad 37.5 = \pi \times \frac{d^2}{4}$$

$$\Rightarrow \quad d = 6.91 \text{ m}$$

3.47 Sol.

$$V_s = \frac{(G-1)gd^2}{18\nu} = \frac{(3-1)9.81(0.02 \times 10^{-3})^2}{18 \times 10^{-6}} \times 10^3$$

= 0.436 mm/sec

3.48 Sol.

Q_0 = 1000 m^3/d OFR = 20 m/day

$$SA = \frac{Q_o}{OFR} = \frac{1000}{20} = 50 \text{ m}^2$$

Assuming diameter of tank to be 'd'

$$\frac{\pi}{4} \times d^2 = SA = 50$$

$$\Rightarrow \quad d = 7.98 \text{ m} \simeq 8 \text{ m}$$

3.49 (c)

Total water to be filtered

= 99 × 1.05 MLD = 103.95 MLD

(Addition of 5% to be used for backwashing)

$$\frac{L}{B} = 1.35 \quad \text{where } B = 5.2 \text{ m}$$

$\therefore$ L = 7.02 m

$\therefore$ Surface area of each filter = 36.504 m^2

Total surface area required

$$= \frac{\text{Discharge through filter}}{\text{Rate of filtration}}$$

$$= \frac{103.95 \times 10^3}{6 \times 24} = 721.875 \text{ m}^2$$

Total no. of working units required

$$= \frac{721.875}{36.504} = 19.77 \text{ filters} = 20 \text{ filters}$$

1 unit is to added as standby, thus total no. of units required = 21

3.50 Sol.

Power required

$$P = \mu V G^2$$

$$= 0.798 \times 10^{-3} \text{ Ns/m}^2 \times 1800 \text{ m}^3 \times (1005)^2$$

= 14364 Nm/s or Watt = 14.364 kW = 14.36 kW

3.51 Sol.

Let total area of filters be 1 m^2

Water used for backwashing

$$= 1000 \times \frac{15}{24 \times 60} = 10.4166 \text{ m}^3$$

Water used for maturation

$$= 200 \times \frac{30}{24 \times 60} = 4.166 \text{ m}^3$$

Total water wasted for backwashing and maturation

= 10.4166 + 4.166 = 14.58 m^3

Water to be treated by filtered

$$= 200 \times \frac{23.25}{24} = 193.75 \text{ m}^3\text{/day}$$

% increase in filtered water produced

$$= \frac{14.58}{193.75} \times 100 = 7.525\%$$

3.52 Sol.

Since sand particle of size 0.35 mm settles to the bottom of the column in 30 sec particles having size greater than 0.35 mm i.e. 0.45 and 0.50 mm will also settle in suspension at the bottom of column by 100% in 30 sec, infact these bigger sized particle will settle by 100% in less than 30 sec. So answer is 100%.

3.53 (d)

Flow rate, $Q_0 = 0.2$ m³/sec

Plan area,(PA) = LB = 32 × 8 = 256 m²

(OFR) over flow rate

$$= \frac{Q_0}{PA} = \frac{0.2}{256} = 7.8125 \times 10^{-4} \text{ m/s}$$

Now, settling velocity of particle of size 25 μm be u_s

$$u_s = \frac{(G-1)\gamma_w d^2}{18\mu}$$

$$= \frac{(2.5-1)9.81\times10^{-3}(25\times10^{-6})^2}{18\times0.01\times10^{-3}\times10^2}$$

$= 5.10 \times 10^{-4}$ m/sec

$$\eta_{removal} = \frac{u_s}{OFR}\times100$$

$$= \frac{5.10\times10^{-4}}{7.8125\times10^{-4}}\times100 = 65.28\% \simeq 65\%$$

3.54 Sol.

Head loss of expanded bed = Head loss of unexpanded bed

$$D'(1-\eta')(G-1) = D(1-\eta)(G-1)$$

$$D'(1-\eta') = D(1-\eta)$$

$$D' = \frac{D(1-\eta)}{(1-\eta')}$$

Now, $$\eta' = \left(\frac{v_B}{v_s}\right)^{0.22}$$

Backwash velocity, $$v_B = \frac{Q_B}{P_A} = \frac{Q_B}{LB}$$

$$\eta' = \left(\frac{3.6}{4\times3\times60\times0.05}\right)^{0.22} = 0.602$$

$$D' = 0.8\frac{(1-0.4)}{(1-0.602)} = 1.21\text{m}$$

3.55 Sol.

$$N_t = N_0 e^{-kt}$$

$$\%\text{kill} = \frac{N_0 - N_t}{N_0}\times100$$

$$99 = \frac{N_0 - N_0 e^{-kt}}{N_0}\times100$$

$$e^{-kt} = 0.01$$

$$\Rightarrow \quad -kt = \ln(0.01)$$

$$t = \frac{4.602}{0.23} = 20.022 \text{ min}$$

Now, Volume of tank,

$$V = Q_0 \times t$$

$$V = \frac{36\times20.022}{60} = 12.01\text{m}^3$$

Also, $$V = LBH$$

$$L = \frac{V}{BH} = \frac{12.01}{1\times1.5} = 8.01\text{m}$$

3.56 Sol.

Design discharge, Q_0 = 6000 m³/d

Overflow rate OFR (v_s) = 1.5 m/hr

$$SA = \frac{Q_0}{OFR}$$

Volume, V = SA × Depth (H)

$$= \frac{Q_0}{OFR}\times H = \frac{6000\times3}{1.5\times24}; \qquad V = 500 \text{ m}^3$$

3.57 (a)

During disinfection variations of micro-organism is given by

$$N_t = N_o e^{-kt}$$

N_t = No. of micro-organism at time t

N_o = No. of micro-organism at time 0

So, disinfection efficiency at any time 't',

$$\eta_t = \frac{N_o - N_t}{N_o}\times100$$

For t = 3 min; η_3 = 50%

$$\eta_3 = \frac{N_o - N_o e^{-k\times3}}{N_o}\times100 = 50$$

$$k = 0.231 \text{ min}^{-1}$$

Now for η_t = 99%

$$\eta_t = \frac{N_o - N_t}{N_o}\times100 = 99$$

$$\frac{N_o - N_o e^{-0.231\times t}}{N_o}\times100 = 99; \quad t = 19.93 \text{ min}$$

3.58 Sol.

$$\eta = 80 = \frac{u_s}{v_s}\times100$$

$$u_s = \frac{0.8 \times 20000 \times 10^{-3}}{86400} = 1.85 \times 10^{-4} \text{ m/sec}$$

$$u_s = \frac{(G-1)gd^2}{18\nu}$$

$$1.85 \times 10^{-4} = \frac{(2.64-1) \times 9.81 \times d^2}{18 \times 1.01 \times 10^{-2} \times 10^{-4}}$$

Diameter of particles,

$$d = 1.446 \times 10^{-5} \text{ m} = 14.46 \ \mu\text{m}$$

3.59 (b)

Watson's Law,

$$tc^n = \text{Constant}$$

$$t_1 c_1^n = t_2 c_2^n$$

$$\frac{d_1}{v_1} c_1^n = \frac{d_2}{v_2} c_2^n$$

$$\frac{d_1}{Q_1} A_1 c_1^n = \frac{d_2 A_2 c_2^n}{Q_2}$$

$$d_1 = d_2,\ A_1 = A_2$$

$$\frac{(3.5)^{0.86}}{10} = \frac{(c_2)^{0.86}}{13}$$

$$c_2 = \left(\frac{13}{10}\right)^{1/0.86} \times 3.5 = 4.747 = 4.75 \text{ mg}/l$$

3.60 Sol.

For type-I setting, Stokes law is applicable.

$$V_s \propto d^2$$

$$\frac{d_{30}^2}{d_{15}^2} = \frac{30}{15} = 2; \qquad \frac{d_{30}}{d_{15}} = \sqrt{2} = 1.41$$

3.61 (a)

n' = Porosity of expanded bed

$$n' = \left(\frac{V_B}{V_s}\right)^{0.22}$$

$$0.65 = \left(\frac{V_B}{4.5 \text{ cm/s}}\right)^{0.22}$$

$$V_B = 6.35 \times 10^{-3} \text{ m/s}$$

3.62 (b, d)

3.63 Sol.

$$G_1 = \sqrt{\frac{P}{\mu_1 \times V}} \quad \text{...(i)}$$

$$G_2 = \sqrt{\frac{P}{\mu_2 \times V}} \quad \text{...(ii)}$$

Divide (i) by (ii), we get

$$\frac{G_1^2}{G_2^2} = \frac{\mu_2}{\mu_1}$$

$$G_2^2 = \left[\frac{(100 s^{-1})^2 \times 1.139 \times 10^{-3} \text{ N-s/m}^2}{1.518 \times 10^{-3} \text{ N-s/m}^2}\right]^{1/2}$$

$$= 86.62 \text{ S}^{-1}$$

$$\% \text{ change in } G = \frac{G_1 - G_2}{G_1} \times 100$$

$$= \frac{100 - 86.62}{100} \times 100 = 13.38\%$$

3.64 (d)

Ozone quickly decomposes to generate highly reactive free radicals. The ozone's oxidation potential (–2.7 V) is greater than that of the chlorine (–1.36 V) or hypochlorite ion (–1.49 V), substances widely used in wastewater treatment such as oxidants. Ozone is surpassed only by the hydroxyl radical (•OH) and fluoride in its oxidation capacity.

3.65 Sol.

Alkalinity present in water = 4 mg/L

Discharge of raw water = 25 MLD

So, Alkalinity present in water (as $CaCO_3$)

$= 4 \times 10^{-6} \times 25 \times 10^6$

$= 100$ kg/day

Now, requirement of $Ca(HCO_3)_2$ = 450 kg/day

Requirement of $Ca(HCO_3)_2$ as $CaCO_3$

$$= \frac{450 \times 50}{81} = 277.78 \text{ kg/day}$$

Now, additional alkalinity required as $CaCO_3$ = 277.78 – 100 = 177.78 kg/day

100 gm of $CaCO_3$ is equivalent to 56 gm of CaO.

So, 177.78 kg/day of $CaCO_3$ is equivalent to (56/100) × 177.78 i.e. 99.56 kg/day of CaO.

Alternatively,

Quantity of pure quick lime

$$= \frac{450 \text{ kg/d} \times \text{Eq.mass of CaO}}{\text{Equivalent mass of } Ca(HCO_3)_2}$$

$$= \frac{450 \text{ kg/d}}{81 \text{ g}} \times 28 \text{ g} = 155.56 \text{ kg/d}$$

3.66 Sol.

For first order kinetics,

$$N_t = N_0 e^{-kt}$$

For efficiency of 95%,

$$N_t = \left(1 - \frac{95}{100}\right)^{N_0}$$

$$= 0.05\ N_0$$

$\therefore \quad 0.05 = e^{-0.38t}$

Taking natural log on both sides,

$$\ln 0.05 = -0.38t \times \ln e$$

$\Rightarrow \quad t = 7.9$ days

3.67 (a)

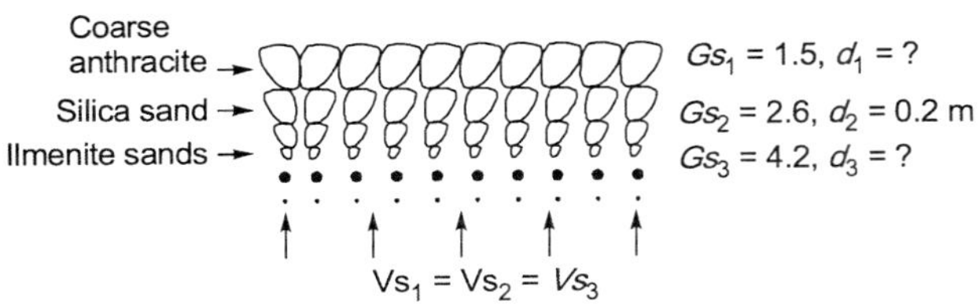

$$\frac{1}{18\mu}(G_{S1} - 1)\rho g d_1^2 = \frac{\rho g}{18\mu}(G_{S_2} - 1)d_2^2$$

$$= \frac{\rho g}{18\mu}(G_{S_3} - 1)d_3^2$$

$$\Rightarrow \frac{\rho g}{18\mu}(G_S - 1)d_1^2 = \frac{\rho g}{18\mu}(G_{S_2} - 1)d_2^2$$

$$= \frac{\rho g}{18\mu}(G_{S_3} - 1)d_3^2$$

$\Rightarrow \quad (G_{S1} - 1)d_1^2 = (G_{S2} - 1).d_2^2$

$\Rightarrow (1.5 - 1)d_1^2 = (2.6 - 1) \times (0.2)^2$

$\Rightarrow \quad d_1 = 0.358$ mm

$(G_{S2} - 1)d_2^2 = (G_{S3} - 1).d_3^2$

Also, $(2.6 - 1) \times (0.2)^2 = (4.2 - 1)d_3^2$

$\Rightarrow \quad d_3 = 0.141$ mm

The logic is that if we are creating 3 layers for better filtration and if we want those layers to be intact, even after backwashing, then we need to ensure that as the backwashing is stopped, each layer settles with the same velocity so that the stratification is maintained as it is.

3.68 Sol.

Discharge, $Q = 10$ MLD

Detention time, $t_d = 2$ hr

Overflow rate, $V_0 = 40000$ litres/m²/day

Now, volume of tank, $V = Q \times t_d$

$$= 10 \times 10^6 \times \left(\frac{2}{24}\right) \times 10^{-3}$$

$$= 833.33\ m^3$$

Also, surface area of tank,

$$SA = \frac{Q}{V_0} = \frac{10 \times 10^6}{40000} = 250\ m^2/day$$

So, height of the tank $= \dfrac{V}{SA} = \dfrac{833.33}{250} = 3.33$ m

■■■■

4 Sources of Water Supply, Distribution System and Well Hydraulics

4.1 Water table drops by 3 m in irrigable land of 50 hectare. If porosity and specific retention are 0.30 and 0.10 respectively, the change in storage in hectare-meter is

(a) 15 (b) 30
(c) 45 (d) 60 [1996 : 1 M]

4.2 Storage coefficient of a compressible confined aquifer is a function of

(a) specific weight of water, thickness of the aquifer, compressibility of the aquifer and that of water
(b) permeability, thickness and compressibility of aquifer and compressibility of water
(c) transmissibility of the aquifer and compressibility of water
(d) transmissibility of aquifer and specific yield of aquifer [1997 : 1 M]

4.3 A soil formation through which only seepage is possible, being partly permeable and capable of giving insignificant yield, is classified as

(a) Aquifer (b) Aquitard
(c) Aquifuge (d) Aquiclude

[1999 : 1 M]

4.4 In a steady radial flow into an intake, the velocity is found to vary as $(1/r^2)$, where r is the radial distance. The acceleration of the flow is proportional to

(a) $\frac{1}{r^5}$ (b) $\frac{1}{r^3}$
(c) $\frac{1}{r^4}$ (d) $\frac{1}{r}$ [1999 : 1 M]

4.5 In an area of 200 hectare, water table drops by 4 m. If the porosity is 0.35 and the specific retention is 0.15, change in volume of storage in the aquifer is

(a) 160 m^3 (b) 1.6×10^6 m^3
(c) 8×10^6 m^3 (d) 1.6×10^3 m^3

[2000 : 1 M]

4.6 When there is an increase in the atmospheric pressure, the water level in a well penetrating in a confined aquifer

(a) increase
(b) decreases
(c) may increase or decrease depending on the nature of the aquifer
(d) does not undergo any change

[2002 : 1 M]

4.7 Water is pumped from a well tapping an unconfined aquifer at a certain discharge rate and the steady state drawdown (X) in an observation well is monitored. Subsequently, the pumping discharge is doubled and the steady state drawdown in the same observation well is found to be more than double (i.e., more than $2X$). This disproportionate drawdown is caused by

(a) well losses
(b) decrease in the saturated thickness of the aquifer
(c) nonlinear flow
(d) delayed gravity yield [2003 : 1 M]

4.8 Two observation wells penetrated into a confined aquifer and located 1.5 km apart in the direction of flow, indicated head of 45 m and 20 m. If the coefficient of permeability of the aquifer is 30 m/day and porosity is 0.25, the time of travel of an inert tracer from one well to another is

(a) 416.7 days (b) 500 days
(c) 750 days (d) 3000 days

[2005 : 2 M]

4.9 atch **List-I** with **List-II** and select the correct answer using the codes given below the lists:

List-I

A. Release valve
B. Check valve
C. Gate valve
D. Pilot valve

List-II

1. Reduce high inlet pressure to lower outlet pressure
2. Limit the flow of water to single direction
3. Remove air from the pipeline
4. Stopping the flow of water in the pipeline

Codes:

	A	B	C	D
(a)	3	2	4	1
(b)	4	2	1	3
(c)	3	4	2	1
(d)	1	2	4	3

[2005 : 2 M]

4.10 For steady flow to a fully penetrating well in a confined aquifer, the drawdowns at radial distances of r_1 and r_2 from the well have been measured as s_1 and s_2 respectively, for a pumping rate of Q. The transmissivity of the aquifer is equal to

(a) $\frac{Q}{2\pi}\left(\frac{\ln\frac{r_2}{r_1}}{s_1 - s_2}\right)$ (b) $\frac{Q\ \ln(r_2 - r_1)}{2\pi\ (s_1 - s_2)}$

(c) $\frac{Q}{2\pi}\ln\left[\frac{r_2 / r_1}{s_2 / s_1}\right]$ (d) $2\pi Q\frac{r_2 - r_1}{\ln\left(\frac{s_2}{s_1}\right)}$

[2006 : 1 M]

4.11 A volume of 3.0×10^6 m^3 of groundwater was pumped out from an unconfined aquifer uniformly from an area of 5 km^2. The pumping lowered the water table from initial level of 102 m to 99 m. The specific yield of the aquifer is

(a) 0.20 (b) 0.30
(c) 0.40 (d) 0.50 2008 : 2 M]

4.12 The relationship among specific yield (S_y), specific retention (S_r) and porosity (η) of an aquifer is

(a) $S_y = S_r + n$ (b) $S_y = S_r - n$
(c) $S_y = n - S_r$ (d) $S_y = S_r + 2n$

[2009 : 1 M]

4.13 A well of diameter 20 cm fully penetrates a confined aquifer. After a long period of pumping at a rate of 2720 litres per minute, the observations of drawdown taken at 10 m and 100 m distances from the centre of the well are found to be 3 m and 0.5 m respectively. The transmissivity of the aquifer is

(a) 676 m^2/day (b) 576 m^2/day
(c) 526 m^2/day (d) 249 m^2/day

[2010 : 2 M]

4.14 In an aquifer extending over 150 hectare, the water table was 20 m below ground level. Over a period of time the water table dropped to 23 m below the ground level. If the porosity of aquifer is 0.40 and the specific retention is 0.15, what is the change in ground water storage of the aquifer?

(a) 67.5 ha-m (b) 112.5 ha-m
(c) 180.0 ha-m (d) 450.0 ha-m

[2011 : 2 M]

4.15 The relationship between porosity (η), specific yield (S_y) and specific retention (S_r) of an unconfined aquifer is

(a) $S_y + S_r = \eta$ (b) $S_y + \eta = S_r$
(c) $S_r + \eta = S_y$ (d) $S_y + S_r + \eta = 1$

[2015 : 1 M, Set-II]

4.16 The hardness of a ground water sample was found to be 420 mg/L as $CaCO_3$. A softener containing ion exchange resins was installed to reduce the total hardness to 75 mg/L as $CaCO_3$ before supplying to 4 households. Each household gets treated water at a rate of 540 L/day. If the efficiency of the softener is 100%, the bypass flow rate (expressed in L/day) is________.

[2016 : 1 M, Set-II]

4.17 A tracer takes 100 days to travel from Well-1 to Well-2 which are 100 m apart. The elevation of water surface in Well-2 is 3 m below that in Well-1. Assuming porosity equal to 15%, the coefficient of permeability (expressed in m/day) is

(a) 0.30 (b) 0.45
(c) 1.00 (d) 5.00

[2016 : 2 M, Set-II]

4.18 Water table of an aquifer drops by 100 cm over an area of 1000 km^2. The porosity and specific retention of the aquifer material are 25% and 5%, respectively. The amount of water (expressed in km^3) drained out from the area is________.

[2016 : 1 M, Set-II]

4.19 Dupuit's assumptions are valid for

(a) artesian aquifer (b) confined aquifer
(c) leaky aquifer (d) unconfined aquifer

[2018 : 1 M, Set-II]

4.20 In a homogeneous unconfined aquifer of area 3.00 km^2, the water table was at an elevation of 102.00 m. After a natural recharge of volume 0.90 million cubic meter (Mm^3), the water table rose to 103.20 m. After this recharge, ground water pumping took place and the water table dropped down to 101.2 m. The volume of ground water pumped after the natural recharge, expressed (in Mm^3 and round off to two decimal places), is ______.

[2020 : 2 M, Set-I]

4.21 Two reservoirs are connected through a homogeneous and isotropic aquifer having hydraulic conductivity (k) of 25 m/day and effective porosity (η) of 0.3 as shown in the figure (not to scale). Ground water is flowing in the aquifer at the steady state.

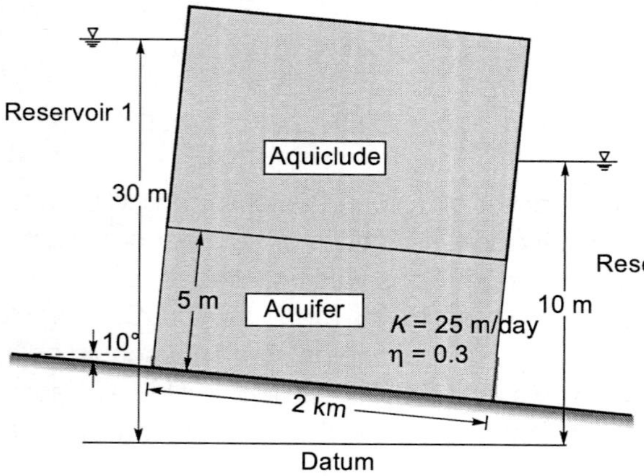

If water is reservoir 1 is contaminated then the time (in days, round off to one decimal place) taken by the contaminated water to reach to Reservoir 2 will be ________

[2021 : 1 M, Set-I]

4.22 A tube-well of 20 cm of diameter fully penetrate a horizontal, homogeneous and isotropic confined aquifer of infinite horizontal extent. The aquifer is of 30 m uniform thickness. A steady pumping at the rate of 40 litres/s from the well for a long time results in a steady drawdown of 4 m at the well face. The subsurface flow to the well due to pumping is steady, horizontal and Darcian and the radius of influence of the well is 245 m. The hydraulic conductivity of the aquifer (in m/day, round off to integer) is ________.

[2021 : 2 M, Set-I]

4.23 A 30 cm diameter well fully penetrates an unconfined aquifer of saturated thickness 20 m with hydraulic conductivity of 10 m day. Under the steady pumping rate for a long time, the drawdowns in two observation wells located at 10 m and 100 m from the pumping well are 5 m and 1 m. respectively. The corresponding pumping rate (in m^3/day) from the well is (rounded off to 2 decimal places).

[2024 : 1 M, Set-I]

■■■■

Answers Sources of Water Supply, Distribution System and Well Hydraulics

4.1	(b)	4.2	(a)	4.3	(b)	4.4	(a)	4.5	(b)	4.6	(b)	4.7	(c)
4.8	(c)	4.9	(a)	4.10	(a)	4.11	(a)	4.12	(c)	4.13	(b)	4.14	(b)
4.15	(a)	4.16	(385.7)	4.17	(d)	4.18	(0.2)	4. 19	(d)	4.20	(1.5)	4.21	(2400)
4.22	(36)	4.23	(1855)										

Explanations Sources of Water Supply, Distribution System and Well Hydraulics

4.1 (b)

Porosity (n) = Specific yield (S_y) + Specific retention (S_r)

$\Rightarrow$ $S_y = 0.3 - 0.1 = 0.2$

And $S_y = \dfrac{\text{Volume of water extraded}}{\text{Volume of aquifer}}$

$\Rightarrow$ $0.2 = \dfrac{\text{Change in storage}}{\text{Volume of aquifer}}$

$\Rightarrow$ Change in storage $= 0.2 \times 50 \times 3 = 30$ ha-m

4.2 (a)

The storage coefficient, S can be expressed as,

$$S = \gamma(n\beta + \alpha)B$$

The storage coefficient S represents the volume of water released by a column of a confined aquifer of unit cross-sectional area under a unit decrease in piezometric head,

where $\gamma = \rho d$ = weight of unit volume of water

β = compressibility of water

α = compressibility of pores

B = thickness of the aquifer

4.3 (b)

Aquitard is that geological formation which does not yield water freely to wells due to its less permeability, although seepage is possible through it. The yield from such formation is thus insignificant.

POINTS TO REMEMBER

1. **Aquifer:** These are the geological formations which are both permeable and porous. Hence, sufficient discharge can be obtained through it. ex. Fine sand and course silt.
2. **Aquiclude:** These are the geological formations which are highly porous but impermeable. Hence, water cannot be extracted through them. Ex. Clay.
3. **Aquitered:** These formations are porous but less permeable. Hence, water does not flow through them but instead it seeps. Ex. Sandy clay.
4. **Aquifuge:** These formation are neither porous nor permeable. Ex. Granite rock.

4.4 (a)

$$v \propto \frac{1}{r^2}; \qquad v = \frac{C}{r^2}$$

Since, the flow is radially inward,

$$v = \frac{dr}{dt} = \frac{-C}{r^2}$$

$$a = \frac{dv}{dt} = \frac{d^2r}{dt^2} = -C\frac{d}{dt}\left(\frac{1}{r^2}\right)$$

$$= \frac{-2C}{r^3}\cdot\frac{dr}{dt} = \frac{-2C}{r^3}\left(\frac{1}{r^2}\right) = \frac{-2C}{r^5}$$

$$\therefore \quad a \propto \frac{1}{r^5}$$

4.5 (b)

$$V = 200 \times 10^4 \times 4 = 800 \times 10^4 \text{ m}^2$$

$$S_r = 0.15;\ n = 0.35$$

The specific yield,

$$S_y = n - S_r = 0.35 - 0.15 = 0.20$$

From $\quad S_y = \dfrac{\Delta V}{V}$

$$\Delta V = 0.2 \times 800 \times 10^4 = 1.6 \times 10^6 \text{ m}^3$$

4.6 (b)

Water level in a well penetrating a confined aquifer will rise up to the piezometric head (or potentiometric surface). With an increase in atmospheric pressure the water surface goes down (phenomenon called as Inverse Barometer effect) and hence water level in the well will decrease.

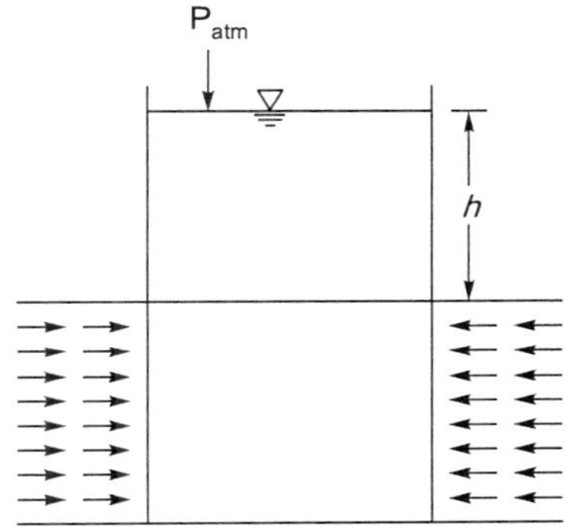

$$P_{aquifer} = P_{atm} - \gamma_w h$$

$$\Rightarrow \quad (P_{aquifer} - P_{atm}) = \gamma_w h$$

Hence, if P_{atm} increase, h will decrease.

4.7 (c)

By **Theim's formula**, the discharge rate for an unconfined aquifer is given as,

$$Q = \frac{\pi k(h_2^2 - h_1^2)}{2.3\log_{10}\left(\frac{r_2}{r_1}\right)}$$

From the above formula it is obvious that pumping rate in a well in an unconfined aquifer follows nonlinear relationship with drawdown.

4.8 (c)

The average flow velocity or discharge velocity,

$$v = ki = 30 \times \left(\frac{45 - 20}{1500}\right) = 0.5 \text{ m/day}$$

Seepage velocity,

$$v_s = \frac{v}{n} = \frac{0.5}{0.25} = 2 \text{ m/day}$$

$$\therefore \text{ Travel time} = \frac{1500}{2} = 750 \text{ days}$$

4.9 (a)

All of these are pipe appurtenances which are required for the proper functioning of the pipeline.

4.10 (a)

Transmittivity, $T = kD$

where, $\quad k = \dfrac{Q\ln\left(\frac{r_2}{r_1}\right)}{2\pi D(h_2 - h_1)}$

$$\Rightarrow \qquad T = \frac{Q \ln\left(\frac{r_2}{r_1}\right)}{2\pi(S_1 - S_2)}$$

where, $(h_2 - h_1)$ = Difference in piezometric heads
= $(S_1 - S_2)$
= Difference in drawdowns.

4.11 (a)

Specific yield

$$= \frac{\text{Volume of water pumped out of the aquifer}}{\text{Total volume of aquifer}}$$

$$= \frac{3.0 \times 10^6}{5 \times 10^6 \times (102 - 99)} = 0.20$$

4.12 (c)

4.13 (b)

$$Q = \frac{2\pi T(s_1 - s_2)}{\ln\left(\frac{r_2}{r_1}\right)}$$

$$T = \frac{2720 \times 10^{-3} \times 24 \times 60 \times ln\left(\frac{100}{10}\right)}{2 \times \pi \times (3 - 0.5)}$$

$= 574.26 \text{ m}^2/\text{day} \simeq 576 \text{ m}^2/\text{day}$

4.14 (b)

We know that,

$$n = S_y + S_r$$

$$\Rightarrow \quad S_y = S_r - n = 0.15 - 0.40 = -0.25$$

The negative sign signifies the decrease in storage.

Change in volume of aquifer

= 150 × (23 – 20) = 450 ha-m

Specific yield is nothing but the actual volume of water that can be extracted by force of gravity from a unit volume of aquifer.

∴ Change in water storage

= 450 × 0.25 = 112.5 ha-m

4.15 (a)

4.16 Sol.

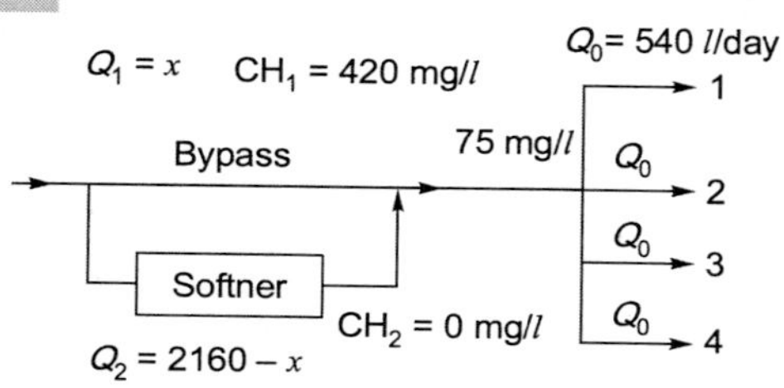

Total water to be treated

= 540 × 4 = 2160 l/day

Resultant hardness required = 75 mg/l.

Let bypass rate be x l/day having hardness of 420 mg/l

Resultant hardness required = 75 mg/l

$$CH_{mix} = \frac{CH_1Q_1 + CH_2Q_2}{Q}$$

$$75 = \frac{x \times 420 + (2160 - x)0}{2160}$$

$$x = \frac{2160 \times 75}{420} = 385.7 \text{ l/day}$$

4.17 (d)

Seepage velocity,

$$V_s = \frac{v}{n} = \frac{\text{distance}}{\text{time}}$$

as per Darcy,

$$v = ki$$

$$\frac{ki}{n} = \frac{100 \text{ m}}{100 \text{ days}}$$

$$i = \frac{\text{head difference}}{\text{length}} = \frac{3}{100}$$

$$\frac{k \times 3}{0.15 \times 100} = \frac{100}{100} \text{ m/day}$$

$$k = 5 \text{ m/day}$$

4.18 Sol.

$$S_r + S_y = n$$

$$\frac{5}{100} + \frac{V_w}{10^3 \times 10^6 \times 1} = \frac{25}{100}$$

$$V_w = 0.2 \times 10^9 \text{ m}^3 = 0.2 \text{ km}^3$$

4.19 (d)

Dupuit's theory assumptions hold that groundwater flows horizontally in an unconfined aquifer and that ground water discharge is proportional to saturated aquifer thickness.

4.20 Sol.

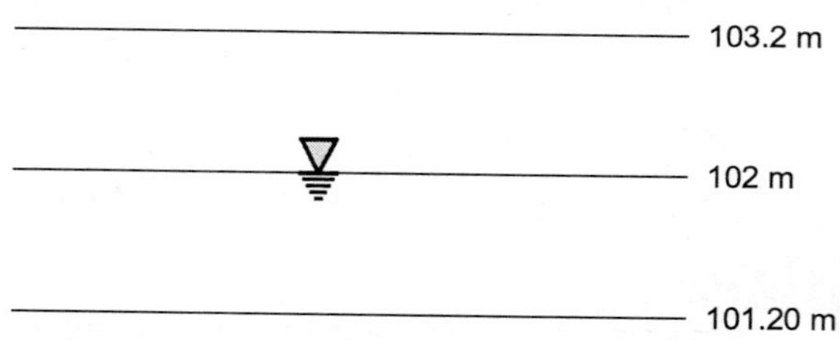

$$V_R = 0.9 \text{ Mm}^3$$

$$V = 3 \times (103.2 - 102) = 3 \times 1.2 = 3.6 \text{ Mm}^3$$

$$y_s \text{ or } y_R = \frac{V_R}{V} = \frac{0.9}{3.6}$$

Now, $$y_s = \frac{V_D}{V}$$

$$V_D = \frac{0.9}{3.6}\left[3 \times (103.2 - 101.2)\right]$$

$$V_D = 1.5 \text{ Mm}^3$$

4.21 Sol.

$$k = 25 \text{ m/d}$$

$$n = 0.3$$

$$i = \frac{20 \text{ m}}{2000 \text{ m}} = 0.01$$

$$V = ki$$

$$V_s = \frac{V}{n}$$

$$t = \frac{l}{V_s}$$

$$t = 2400 \text{ days}$$

4.22 Sol.

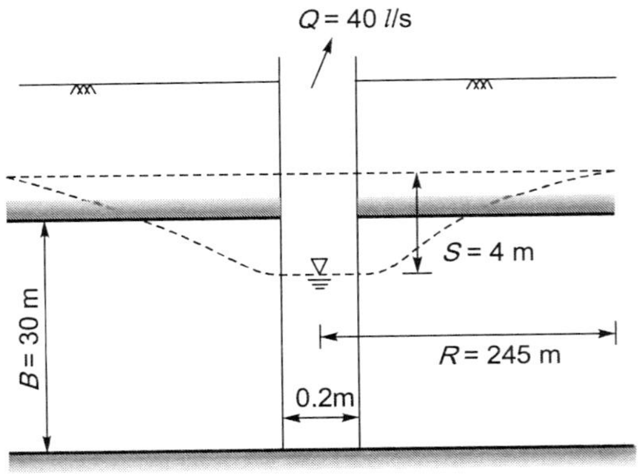

$$40 \times 10^{-3} \text{ m}^3/\text{s} = \frac{2\pi k \times 30 \text{ m} \times 4 \text{ m}}{\ln\left(\frac{245}{0.1}\right)}$$

$$k = 4.14 \times 10^{-4} \text{ m/s}$$

or, $$k = 35.77 \text{ m/d}$$

$$\approx 36 \text{ m/d}$$

(round off to nearest integer)

4.23 Sol.

Thickness of aquifer, $H = 20$ m

$$r_1 = 10 \text{ m}$$

$$r_2 = 100 \text{ m}$$

$$h_2 = H - S_{w1} = 20 - 5 = 15 \text{ m}$$

$$h_1 = H - S_{w2} = 20 - 1 = 19 \text{ m}$$

Now discharge from an unconfined aquifer is given as

$$\Rightarrow \quad Q = \frac{\pi \cdot K\left[h_1^2 - h_2^2\right]}{\ln\left[\frac{r_2}{r_1}\right]}$$

$$= \frac{\pi \times 10 \times \left[19^2 - 15^2\right]}{\ln\left[\frac{100}{10}\right]}$$

$$Q = 1855.55 \text{ m}^3/\text{day}$$

■■■■

5 Quality and Characteristics of Waste Water

5.1 Biochemical oxygen demand (BOD) of waste water is a measure of
(a) total concentration of biochemicals
(b) total concentration of organic matter
(c) concentration of biodegradable organic matter
(d) concentration of chemically oxidizable matter
[1995 : 1 M]

5.2 Chemical Oxygen Demand (COD) of a sample is always greater than Biochemical Oxygen Demand (BOD) since it represents
(a) biodegradable organic matter only
(b) biodegradable and non-biodegradable organic matter
(c) non-biodegradable organic matter
(d) inorganic matter [1996 : 1 M]

5.3 A waste water sample diluted to 100 times with aeration water had an initial dissolved oxygen (DO) of 7.0 mg/L and after 5 days of incubation at 20°C, the DO was zero. The BOD of waste water is
(a) 700 mg/L
(b) 100 mg/L
(c) Cannot be determined
(d) 7 mg/L [1996 : 1 M]

5.4 A single rapid test to determine the pollution status of river water is
(a) biochemical oxygen demand
(b) chemical oxygen demand
(c) total organic solids
(d) dissolved oxygen [1996 : 1 M]

5.5 The BOD_5 of a surface water sample is 200 mg/litre at 20°C. The value of the reaction constant is $K = 0.2$ day^{-1} with base 'e'. The ultimate BOD of the sample is
(a) 126 mg/litre (b) 544 mg/litre
(c) 146 mg/litre (d) 316 mg/litre
[1997 : 1 M]

5.6 In a BOD test using 5% dilution of the sample (15 mL of sample and 285 mL of dilution water), dissolved oxygen values for the sample and dilution water blank bottles after five days incubation at 20°C were 3.8 were and 8.8 mg/L, respectively. Dissolved oxygen originally present in the undiluted sample was 0.8 mg/L. The 5 days 20°C BOD of the sample is
(a) 116 mg/L (b) 108 mg/L
(c) 100 mg/L (d) 92 mg/L
[1998 : 2 M]

5.7 In a BOD test, 5 ml of waste is added to 295 m*l* of aerated pure water. Initial dissolved oxygen (DO) content of the diluted sample is 7.8 mg/*l*. After 5 days of incubation at 20°C, the DO content of the sample is reduced to 4.4 mg/*l*. The BOD of the waste water is
(a) 196 mg/*l* (b) 200 mg/*l*
(c) 204 mg/*l* (d) 208 mg/*l*
[1999 : 1 M]

5.8 Standard 5 day BOD of a waster water sample is nearly x% of the ultimate BOD, where x is
(a) 48 (b) 58
(c) 68 (d) 78 [1999 : 1 M]

5.9 The BOD removal efficiency, in percentage, during primary treatment, under normal conditions is about
(a) 65% (b) 85%
(c) 30% (d) Zero [2000 : 1 M]

5.10 The dissolved oxygen (DO) in an unseeded sample of diluted waste at 20°C having an initial DO of 9.5 mg/L is measured to be 3.5 mg/L after 5 days. The dilution factor is 0.03 and the reaction rate constant, k = 0.22/day (to the base e in the decay curve at 30°C).
Estimate
(i) The 5 day BOD of the waste at 20°C
(ii) Ultimate carbonaceous BOD [2000 : 2 M]

5.11 Consider a glucose solution ($C_6H_{12}O_6$) of molarity 1.75×10^{-3} that is completely oxidized to CO_2 and H_2O. Find the amount of oxygen required for this reaction. The chemical mass balance equation for the above reaction is given as

$C_6H_{12}O_6 + 6O_2 \rightarrow CO_2 + 6H_2O$
(Atomic weight are: (C = 12, H = 1, O = 16))
[2000 : 2 M]

5.12 If the $BOD_{5,}$ 20°C of waste is 150 mg/L and the reaction rate constant (to the base '*e*') at 20°C is 0.35/day, then the ultimate BOD is mg/L is
(a) 97.5 (b) 181.5
(c) 212.9 (d) 230.5 [2001 : 2 M]

5.13 In a domestic wastewater sample, COD and BOD were measured. Generally which of the following statement is true for their relative magnitude?
(a) COD = BOD
(b) COD > BOD
(c) COD < BOD
(d) Nothing can be said [2002 : 1 M]

5.14 The theoretical oxygen demand of a 0.001 mol/L glucose solution is
(a) 180 mg/L (b) 192 mg/L
(c) 90 mg/L (d) 96 mg/L
[2002 : 2 M]

5.15 A wastewater sample has an initial BOD of 222 mg/L. The first order BOD decay coefficient is 0.4/day. The BOD consumed (in mg/L) in 5 days is
(a) 150 (b) 192
(c) 30 (d) 50 [2002 : 2 M]

5.16 A portion of waste water sample was subjected to standard BOD test (5 days, 20°C), yielding a value of 180 mg/L. The reaction rate constant (to the base 'e') at 20°C was taken as 0.18 per day. The reaction rate constant at other temperature may be estimated by $k_T = k_{20}\,(1.047)^{T-20}$. The temperature at which the other portion of the sample should be tested, to exert the same BOD in 2.5 days, is
(a) 4.9°C (b) 24.9°C
(c) 31.7°C (d) 35.0°C [2004 : 2 M]

5.17 Water samples (*X* and *Y*) from two different sources were brought to the laboratory for the measurement of dissolved oxygen (DO) using modified Winkler method. Samples were transferred to 300 mL BOD bottles. 2 mL of $MnSO_4$ solution and 2 mL of alkaliodide-azide reagent were added to the bottles and mixed. Sample *X* developed a brown precipitate, whereas sample *Y* developed a white precipitate. In reference to these observations, the correct statement is
(a) both the samples were devoid of DO
(b) sample *X* was devoid of DO while sample *Y* contained DO
(c) sample *X* contained DO while sample *Y* was devoid of DO
(d) both the samples contained DO
[2004 : 2 M]

5.18 **List-I** contains some properties of water/waste water and **List-II** contains list of some tests on water/ waste water. Match **List-I** with **List-II** and select the correct answer using the codes given below the lists:

List-I
A. Suspended solids concentration
B. Metabolism of biodegradable organics
C. Bacterial concentration
D. Coagulant dose

List-II
1. BOD
2. MPN
3. Jar test
4. Turbidity

Codes:

	A	B	C	D
(a)	2	1	4	3
(b)	4	1	2	3
(c)	2	4	1	3
(d)	4	2	1	3

[2005 : 2 M]

5.19 Total Kjeldahl nitrogen is a measure of
(a) total organic nitrogen
(b) total organic and ammonia nitrogen
(c) total ammonia nitrogen
(d) total inorganic and ammonia nitrogen
[2005 : 1 M]

5.20 To determine the BOD_5 of a waste water sample, 5, 10 and 50 mL aliquots of the waste water were diluted to 300 mL and incubated at 20°C in BOD bottles for 5 days. The results were as follows.

S.No.	Waste water initial volume (m*l*)	DO, DO (mg/*l*)	DO after 5 days (mg/*l*)
1.	5	9.2	6.9
2.	10	9.1	4.4
3.	50	8.4	0.0

Based on the data, the average BOD_5 of the waste water is equal to
(a) 139.5 mg/L (b) 126.5 mg/L
(c) 109.8 mg/L (d) 72.2 mg/L
[2006 : 1 M]

5.21 The 5-day BOD of a wastewater sample is obtained as 190 mg/L (with k = 0.01 h^{-1}). The ultimate oxygen demand (mg/L) of the sample will be

(a) 3800 (b) 475
(c) 271 (d) 190 [2008 : 2 M]

5.22 If the BOD_3 of a wastewater sample is 75 mg/L and reaction rate constant k (base e) is 0.345 per day, the amount of BOD remaining in the given sample after 10 days is

(a) 3.12 mg/L (b) 3.45 mg/L
(c) 3.69 mg/L (d) 3.92 mg/L

[2010 : 2 M]

5.23 A sample of domestic sewage is digested with silver sulphate, sulphuric acid, potassium dichromate and mercuric sulphate in chemical oxygen demand (COD) test. The digested sample is then titrated with standard ferrous ammonium sulphate (FAS) to determine the un-reacted amount of

(a) mercuric sulphate
(b) potassium dichromate
(c) silver sulphate
(d) sulphuric acid [2012 : 1 M]

5.24 A student began an experiment of 5 day 20°C BOD on Monday. Since the 5th day fell on Saturday. The final DO reading was taken on Monday. On calculation BOD (i.e. 7 day 20°C) was found to be 150 mg/L. What would be the 5 day, 20°C BOD (in mg/L). Assume value of BOD rate constant (K) at standard temperature of 20°C as 0.23/day (base e).

[2013 : 2 M]

5.25 Ultimate BOD of a river water sample is 20 mg/L. BOD rate constant (natural log) is 0.15 day^{-1}. The respective values of BOD (in %) exerted and remaining after 7 days are:

(a) 45 and 55 (b) 55 and 45
(c) 65 and 35 (d) 75 and 25

[2015 : 2 M, Set-II]

5.26 The 2-day and 4-day BOD values of a sewage sample are 100 mg/l and 155 mg/l, respectively. The value of BOD rate constant (expressed in per day) is________.

[2016 : 2 M, Set-I]

5.27 For a wastewater sample, the three-day biochemical oxygen demand at incubation temperature of 20°C($BOD_{3day,20°C}$) is estimated as 200 mg/l. Taking the value of the first order BOD reaction rate constant as 0.22 day^{-1}, the five-day BOD (expressed in mg/l) of the wastewater at incubation temperature of 20°C ($BOD_{5day,20°C}$) would be_____.

[2016 : 2 M, Set-II]

5.28 Two wastewater streams A and B, having an indentical ultimate BOD are getting mixed to form the stream C. The temperature of the stream A is 20°C and the temperature of the stream C is 10°C. It is given that

- the 5-day BOD of the stream A measured at 20°C = 50 mg/l
- BOD rate constant (base 10) at 20°C = 0.115 per day
- temperature coefficient = 1.135

The 5-day BOD (in mg/l, up to one decimal place) of the stream C, calculated at 10°C, is __________.

[2017 : 2 M, Set-I]

5.29 For a given water sample, the ratio between $BOD_{5\text{-day, }20°C}$ and the ultimate BOD is 0.68. The value of the reaction rate constant k (on base e) (in day^{-1}, up to two decimal places) is __________.

[2017 : 2 M, Set-II]

5.30 The ultimate BOD (L_0) a wastewater sample is estimated as 87% of COD. The COD of this wastewater is 300 mg/L. Considering first order BOD reaction rate constant k (use natural log) = 0.23 per day and temperature coefficient θ = 1.047, the BOD value (in mg/L, up to one decimal place) after three days of incubation at 27°C for the wastewater will be ______

[2018 : 2 M, Set-I]

5.31 A river has a flow of 1000 million litres per day (MLD), BOD_5 of 5 mg/litre and Dissolved Oxygen (DO) level of 8 mg/litre before receiving the wastewater discharge at a location. For the existing environmental conditions, the saturation DO level is 10 mg/litre in the river. Wastewater discharge of 100 MLD with the BOD_5 of 200 mg/litre and DO level of 2 mg/litre falls at that location. Assuming complete mixing of wastewater and river water, the immediate DO deficit (in mg/litre, round off to two decimal places), is ________.

[2020 : 1 M, Set-I]

5.32 The relationship between oxygen consumption and equivalent biodegradable organic removal (i.e. BOD) in a closed container with respect to time is shown in the figure.

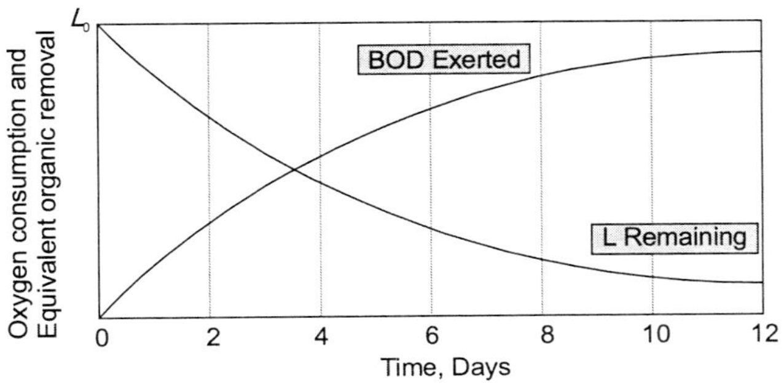

Assume that the rate of oxygen consumption is directly proportional to the amount of degradable organic matter and is expressed as $\frac{dL_t}{dt} = -kL_t$,

where, L_t (in mg/litre) is the oxygen equivalent of the organics remaining at time t and k (in d^{-1}) is the degradation rate constant. L_0 is the oxygen of organic matter at time, $t = 0$

In the above context, the correct expression is

(a) $L_t = L_0(1 - e^{-kt})$ (b) $BOD_t = L_0 - L_t$

(c) $L_0 = L_t e^{-kt}$ (d) $BOD_5 = L_5$

[2020 : 1 M, Set-II]

5.33 A sample of water contain an organic compound $C_8H_{16}O_8$ at a concentration of 10^{-3} mol/litre. Given that the atomic weight of C = 12 g/mol, H = 1 g/mol, and O = 16 g/mol, the theoretical oxygen demand of water (in g of O_2 per litre, round off to two decimal places), is ________.

[2020 : 2 M, Set-II]

5.34 Which one of the following is correct?

(a) The most important type of species involved in the degradation of organic mater in the case of activated sludge process based wastewater treatment is *chemoheterotrophs*.

(b) For an effluent sample of a sewage treatment plant, the ratio $BOD_{5\text{-day-}20°C}$ upon ultimate BOD is more than 1.

(c) The partially treated effluent from a food processing industry, containing high concentration of biodegradable organism is being discharge into a flowing river at a point P. If the rate of degradation of the organics is higher than the rate of aeration, then dissolved oxygen of the river water will be lowest at point P.

(d) A young lake characterised by low nutrient content and low plant productivity is called *eutrophic* lake.

[2021 : 1 M, Set-I]

5.35 Ammonia nitrogen is present in a given wastewater sample as the ammonium ion (NH_4^+) and ammonia (NH_3). If pH is the only deciding factor for the proportion of these two constituents, which of the following is a correct statement?

(a) At pH 7.0, NH_4^+ and NH_3 will be found in equal measures.

(b) At pH below 9.25, NH_3 will be predominant.

(c) At pH 7.0, NH_4^+ will be predominant.

(d) At pH above 9.25, only NH_4^+ will be present

[2021 : 2 M, Set-I]

5.36 For wastewater coming from a wood pulping industry, Chemical Oxygen Demand (COD) and 5-day Biochemical Oxygen Demand (BOD_5) were determined. For this wastewater, which of the following statement(s) is/are correct?

(a) $COD > BOD_5$ (b) $COD \neq BOD_5$

(c) $COD < BOD_5$ (d) $COD = BOD_5$

[2022 : 1 M, Set-I]

5.37 Henry's law constant for transferring O_2 from air into water, at room temperature, is 1.3 m mol/liter-atm. Given that the partial pressure of O_2 in the atmosphere is 0.21 atm, the concentration of dissolved oxygen (mg/liter) in water in equilibrium with the atmosphere at room temperature is

(Consider the molecular weight of O_2 as 32 g/mol)

(a) 8.7 (b) 0.8

(c) 198.1 (d) 0.2

[2022 : 2 M, Set-I]

5.38 A wastewater sample contains two nitrogen species, namely ammonia and nitrate. Consider the atomic weight of N, H, and O as 14 g/mol, 1 g/mol, and 16 g/mol, respectively. In this wastewater, the concentration of ammonia is 34 mg NH_3/liter and that of nitrate is 6.2 mg NO_3^-/liter. The total nitrogen concentration in this wastewater is _______ milligram nitrogen per liter.

(round off to one decimal place)

[2022 : 2 M, Set-I]

5.39 A 2 % sewage sample (in distilled water) was incubated for 3 days at 27 °C temperature. After incubation, a dissolved oxygen depletion of 10 mg/L was recorded. The biochemical oxygen demand

(BOD) rate constant at 27 °C was found to be 0.23 day^{-1} (at base e). The ultimate BOD (in mg/L) of the sewage will be__________. (round off to the nearest integer)

[2022 : 2 M, Set-I]

5.40 A water treatment plant has a sedimentation basin of depth 3 m, width 5 m, and length 40 m. The water inflow rate is 500 m^3/h. The removal fraction of particles having a settling velocity of 1.0 m/h is_______. (round off to one decimal place) (Consider the particle density as 2650 kg/m^3 and liquid density as 991 kg/m^3)

[2022 : 2 M, Set-I]

5.41 In a city, the chemical formula of biodegradable fraction of municipal solid waste (MSW) is $C_{100}H_{250}O_{80}N$. The waste has to be treated by forced-aeration composting process for which air requirement has to be estimated.

Assume oxygen in air (by weight) = 23% and density of air = 1.3 kg/m^3. Atomic mass: C = 12, H = 1, O = 16, N = 14.

C and H are oxidized completely whereas N is converted only into NH_3 during oxidation.

For oxidative degradation of 1 tonne of the waste, the required theoretical volume of air (in m^3/tonne) will be (round off the nearest integer)

(a) 1092
(b) 4749
(c) 1418
(d) 8025

[2022 : 1 M, Set-II]

5.42 The theoretical aerobic oxidation of biomass ($C_5H_7O_2N$) is given below:

$$C_5H_7O_2N + 5O_2 \rightarrow 5CO_2 + NH_3 + 2H_2O$$

The biochemical oxidation of biomass is assumed as a first-order reaction with a rate constant of 0.23/d at 20°C (logarithm to base *e*). Neglecting the second-stage oxygen demand from its biochemical oxidation, the ratio of BOD_5 at 20°C to total organic carbon (TOC) of biomass is _______ (round off to two decimal places). [Consider the atomic weights of C, H, O and N as 12 g/mol, 1 g/mol, 16 g/mol and 14 g/mol, respectively]

[2023 : 2 M, Set-II]

5.43 Which one of the following products is NOT obtained in anaerobic decomposition of glucose?

(a) H_2O
(b) CH_4
(c) H_2S
(d) CO_2

[2024 : 1 M, Set-II]

■■■■

Answers Quality and Characteristics of Waste Water

5.1	(c)	5.2	(b)	5.3	(c)	5.4	(d)	5.5	(d)	5.6	(d)	5.7	(c)
5.8	(c)	5.9	(c)	5.10	(401.2)	5.11	(0.336)	5.12	(b)	5.13	(b)	5.14	(b)
5.15	(b)	5.16	(d)	5.17	(c)	5.18	(b)	5.19	(b)	5.20	(a)	5.21	(c)
5.22	(c)	5.23	(b)	5.24	(128.13)	5.25	(c)	5.26	(0.29)	5.27	(276.19)	5.28	(21.21)
5.29	(0.23)	5.30	(160.2)	5.31	(2.545)	5.32	(b)	5.33	(0.25)	5.34	(a)	5.35	(c)
5.36	(a, b)	5.37	(a)	5.38	(29.4)	5.39	(1003)	5.40	(40)	5.41	(4748.57)	5.42	(1.82)
5.43	(c)												

Explanations Quality and Characteristics of Waste Water

5.1 (c)

The extent of organic matter present in water can be estimated by supplying oxygen to this sample and finding the oxygen consumed by the organic matter present in water. This oxygen demand is known as biochemical oxygen demand (BOD). The BOD of raw water will indicate the extent of biodegradable organic matter present, thus indicating the extent of treatment required for purifying this water to make it safe and wholesome.

5.2 (b)

Chemical oxygen demand (COD) is a measure of total organic matter (bio-degradable as well as non-bio-degradable) present in sewage.

BOD of sewage is the amount of biologically degradable organic matter present in the sewage.

5.3 (c)

$$DO_i = 7 \text{ mg}/l$$
$$DO_f = 0 \text{ mg}/l$$

If $DO_f = 0$ mg/l, BOD of waste water can't be determined as it is not known at which day (time) did DO become zero. Because we need DO at $t = 5$ days to determine BOD_5 and DO could have because zero at some time before $t = 5$ datys also.

5.4 (d)

To determine pollution status of river stream rapid test is DO. DO can be rapidly found by using multimeter.

5.5 (d)

$$BOD_5 = 200 \text{ mg/litre}$$
$$k = 0.2 \text{ day}^{-1} \text{ (base e)}$$
$$BOD_5 = L(1 - e^{-k \times t})$$
$$200 = L(1 - e^{-0.2 \times 5})$$
$$L = 316 \text{ mg}/l$$

5.6 (d)

$$\text{DO of sample} = \frac{(15 \times 0.8) + (285 \times 8.8)}{(15 + 285)}$$
$$= 8.4 \text{ mg/L} = DO_i$$
$$DO_f = 3.8 \text{ mg/L}$$
$$\text{Dilution factor} = \frac{300}{15} = 20$$
$$\therefore \quad BOD_5 = (8.4 - 3.8) \times 20 = 92 \text{ mg/L}$$

5.7 (c)

BOD_5 = (Initial DO – Final DO) × Dilution factor

$$= (7.8 - 4.4) \times \frac{300}{5} = 204 \text{ mg}/l$$

5.8 (c)

BOD of water during 5 days at 20°C is generally taken as the standard demand and is about 68% of the total demand. And 10 day BOD is about 90% of the total.

5.9 (c)

BOD removal efficiency

= 30-35%

Suspended solids removal efficiency

= 60-70%

5.10 Sol.

(i) $$BOD_5 \text{ (at 20°C)} = (DO_i - DO_f) \times \text{Dilution factor}$$
$$= (9.5 - 3.5) \times \frac{1}{0.03}$$

$$\left(\because \text{Dilution factor} = \frac{\text{Final total volume after dilution}}{\text{Initial volume}} > 1\right)$$

$$\Rightarrow \quad BOD_5 = 200 \text{ mg}/l$$

(ii) $$BOD_5 = L_0(1 - e^{-kt})$$
$$k_{30} = k_{20}((1.047)^{(30-20)}$$
$$\Rightarrow \quad 0.22 = k_{20}(1.047)^{10}$$
$$\Rightarrow \quad k_{20} = 0.138$$
$$\text{So,} \quad L_0 = \frac{200}{1 - e^{-0.138 \times 5}}$$
$$\Rightarrow \quad L_0 = 401.26 \text{ mg/L}$$

5.11 Sol.

$$C_6H_{12}O_6 + 6O_2 \rightarrow CO_2 + 6H_2O$$
$$\text{Molarity} = \frac{\text{Moles of solute}}{\text{Volume of solution}}$$

For 1 litres solution,

$$1.75 \times 10^{-3} = \frac{\text{Moles of glucose}}{1}$$

ÞMoles of glucose or $C_6H_{12}O_6 = 1.75 \times 10^{-3}$

From above reaction,

$$\text{Moles of oxygen} = 6 \times (1.75 \times 10^{-3})$$
$$= 10.5 \times 10^{-3}$$

$$\text{And,} \quad \text{Moles} = \frac{\text{Given mass}}{\text{Molecular mass}}$$

Hence, mass of oxygen

$= (10.5 \times 10^{-3}) \times 32 = 0.336$ g

5.12 (b)

Method-I

$$k_D = \frac{0.35}{2.3} = 0.15/\text{day}$$
$$\text{Now} \quad 150 = L[1 - 10^{-0.15 \times 5}] = 0.822\, l$$
$$\therefore \quad L = \frac{150}{0.822} = 181.54 \text{ mg}/l$$

Method-II

$$BOD_5 = L\left(1 - e^{-kt}\right)$$
$$\Rightarrow \quad 150 = L\left(1 - e^{-0.35 \times 5}\right)$$
$$\Rightarrow \quad L = 181.54 \text{ mg}/l$$

5.13 (b)

COD = (Biodegradable + Non-biodegradable) Organic Matter
BOD = Biodegradable Organic Matter
∴ COD > BOD

5.14 (b)

$C_6H_{12}O_6 + 6O_2 \rightarrow 6CO_2 + 6H_2O$

1 mol of glucose → 6 moles of O_2

$0.001\frac{\text{mol}}{l}$ of glucose $\rightarrow 0.006\frac{\text{mol}}{l}$ of O_2

Now, number of moles

$$= \frac{\text{Wt (g)}}{\text{Molecular Wt (g)}}$$

$$\therefore \quad \text{Wt} = 0.006 \times 32$$
$$= 0.192 \text{g}/l = 192 \text{ mg}/l$$

5.15 (b)

Initial BOD = Ultimate BOD
Now, BOD consumed in 5 days,

$$BOD_5 = BOD_u [1 - e^{kt}]$$
$$= 222 [1 - e^{-0.4 \times 5}]$$
$$= 191.95 \text{ mg}/l \simeq 192 \text{ mg}/l$$

5.16 (d)

Given data:

$k_1 = 0.18,\ T_1 = 20°C$
$t_1 = 5$ days, $y_1 = 180$ mg/l
$k_2 = ?,\ T_2 = ?$
$t_2 = 2.5$ days, $y_2 = 180$ mg/l
L_{01} = Ultimate BOD = L_{02}

$$\therefore \quad \frac{y_1}{y_2} = \frac{1 - e^{-k_1 t_1}}{1 - e^{-k_2 t_2}}$$

$$\Rightarrow \quad \frac{180}{180} = \frac{1 - e^{0.18 \times 5}}{1 - e^{-2.5 k_2}}$$

$$\Rightarrow \quad e^{2.5 k_2} = e\, e^{0.9}$$

$$\Rightarrow \quad k_2 = 0.36$$

Now, $k_T = k_{20°C}(1.047)^{T-20}$

$$\Rightarrow \quad 0.36 = 0.18\,(1.047)^{T-20}$$

$$\Rightarrow \quad T = 35.09°C$$

5.17 (c)

$Mn^{2+} + 2OH^- \longrightarrow Mn(OH)_2 \downarrow$
(in water) White ppt.
...in absence of Dissolved Oxygen

$Mn^{2+} + 2OH^- + \frac{1}{2}O_2 \longrightarrow MnO_2^- \downarrow$
Brown ppt.
... in presence of Dissolved Oxygen

5.18 (b)

- Turbidity gives indirect measurement of suspended solid concenrtation.
- MPN (Most Probable No.) is a test used to determine bacterial concentration.
- Jar test is a test used to know optiomum coagulant dose.

5.19 (b)

The sum total of organic nitrogen and ammonia nitrogen is called Kjeldahl nitrogen.

5.20 (a)

For sample number-3:
DO after 5 days = 0 mg/l
Hence this data is to be neglected as nothing can be determined from this test.

Sample-1:

$$BOD = (DO_i - DO_f) \times \text{dilution factor}$$
$$= (9.2 - 6.9) \times \frac{300}{5} = 138 \text{ mg}/l$$

Sample-2:

$$BOD = (9.1 - 4.4) \times \frac{300}{10} = 141 \text{ mg}/l$$

$$\text{Average BOD} = \frac{138 + 141}{2} = 139.5 \text{ mg}/l$$

Note: Sample-3 data is excluded.
As $DO_f = 0$ for sample 3
Hence, BOD cannot be determined for this sample.

5.21 (c)

The BOD of t days may be given as,

$$L_t = L\,[1 - e^{-k \times t}]$$
$$\Rightarrow \quad 190 = L\,[1 - e^{-0.01 \times 24 \times 5}]$$
$$\Rightarrow \quad L = 271 \text{ mg/L}$$

5.22 (c)

The BOD at any instant is given by,

$$L_t = L(1 - e^{-kt})$$
$$\Rightarrow \quad 75 = L(1 - e^{-0.345})$$

$$\Rightarrow \quad L = \frac{75}{1-0.355}$$

$$\Rightarrow \quad L = 116.32 \text{ mg/L}$$

BOD after 10 days,

$$L_{10} = L(1 - e^{-0.345 \times 10})$$
$$= 116.32 \times 0.968 = 112.63 \text{ mg/L}$$

Amount of BOD remaining $= L - L_{10}$

$$= 116.32 - 112.63$$
$$= 3.69 \text{ mg/L}$$

5.23 (b)

REFERENCE MATERIAL

1. *Environmental Engineering, Vol. II, By SK Garg, Ch-7, Quality and Characteristics of Sewage, Topic: Chemical Oxygen Demand.*

5.24 Sol.

$$BOD_7 = L\,[1 - e^{-k \times t}]$$
$$\therefore \quad 150 = L\,[1 - e^{-0.23 \times 7}]$$
$$\Rightarrow \quad L = 187.54 \text{ mg/L}$$
$$\therefore \quad BOD_5 = L\,[1 - e^{-k \times t}]$$
$$= 187.539 \times [1 - e^{-0.23 \times 5}]$$
$$= 128.13 \text{ mg/L}$$

5.25 (c)

BOD remaining after 7 days

$$= BOD_u\, e^{-kt}$$
$$= 20 \times e^{-0.15 \times 7}$$
$$= 20 \times 0.35 = 7$$

$$\% \text{ remaining} = \frac{7 \times 100}{20} = 35\%$$

BOD exerted in 7 days

$$= (100 - 35)\% = 65\%$$

5.26 Sol.

$$Y_2 = 100 \text{ mg}/l$$
$$t_2 = 2 \text{ day}$$
$$Y_4 = 155 \text{ mg}/l$$
$$t_4 = 4 \text{ day}$$
$$K_D = ?$$
$$Y_2 = Y_0(1 - e^{-K_D \cdot t_2})$$
$$Y_4 = Y_0(1 - e^{-K_D \cdot t_4})$$
$$\frac{Y_2}{Y_4} = \frac{1 - e^{-K_D \times 2}}{1 - e^{-K_D \times 4}}$$
$$\frac{100}{155} = \frac{1 - e^{-2K_D}}{1 - e^{-4K_D}}$$
$$1.55 = \frac{1 - e^{-4k}}{1 - e^{-2k}}$$
$$1.55 - 1.55x = 1 - x^2$$
$$x^2 - 1.55x + 0.55 = 0$$
$$x = 1 \text{ or } x = 0.55$$
$$e^{-2k} = 1 \text{ or } e^{-2k} = 0.55$$
$$-2k = \ln 1 \text{ or } -2k = \ln 0.55$$
$$k = 0 \text{ or } k = 0.299$$
$$\therefore \quad k = 0.299$$

5.27 Sol.

$$BOD_5 = L_0\left(1 - e^{-0.22 \times 5}\right) \quad \text{...(i)}$$
$$BOD_3 = L_0\left(1 - e^{-0.22 \times 3}\right) \quad \text{...(ii)}$$

From equation (i) and (ii), we get,

$$\frac{BOD_5}{BOD_3} = \frac{\left(1 - e^{-0.22 \times 5}\right)}{\left(1 - e^{-0.22 \times 3}\right)}$$

$$BOD_5 = 200 \times \frac{0.667}{0.483}$$
$$= 200 \times 1.38 = 276.19 \text{ mg}/l$$

5.28 Sol.

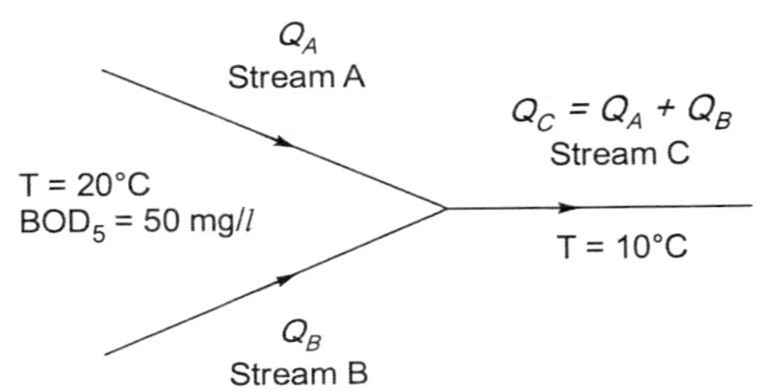

For stream A,

$$BOD_5 = L_0(1 - 10^{-k_D \times 5})$$
$$50 = L_0\,(1 - 10^{-0.115 \times 5})$$
$$L_0 = 68.13 \text{ mg}/l$$

$$L_{0C} = \frac{Q_A L_0 + Q_B L_0}{Q_A + Q_B} = L_0$$

$$k_{D_{10°C}} = k_{D_{20°C}} \times [1.135]^{10-20}$$
$$= 0.115 \times (1.135)^{-10}$$
$$= 0.0324 \text{ day}^{-1}$$

For stream C,

$$(BOD_5)_{10°C} = L_0\left[1 - 10^{-k_{D_{10°C}} \times 5}\right]$$
$$= 68.13\left[1 - 10^{-0.0324 \times 5}\right]$$
$$= 21.21 \text{ mg}/l$$

5.29 Sol.

$$BOD/20°C = L_0\ (1 - e^{-k\ 20°C \times 5})$$

$$\frac{BOD/20°C}{L_0} = 0.68 = (1 - e^{-k\ 20°C \times 5})$$

$$K, 20°C = 0.227 \text{ day}^{-1} \simeq 0.23 \text{ day}^{-1}$$

5.30 Sol.

Ultimate BOD = 0.87 COD

$$= 0.87 \times 300 = 261 \text{ mg}/l$$
$$BOD_3 = L_0\ (1 - e^{-k_{27} \times 3})$$
$$k_{27} = k_{20}\ (1.047)^{T-20}$$

For municipal sewage, at standard temperature, value of k (base e) = 0.23 per day. Thus, value 0.23 per day given is w.r.t. the standard temperature of 20°C.

$$= 0.23\ (1.047)^{27-20} = 0.317 \text{ days}^{-1}$$
$$BOD_3 = 261\ (1 - e^{-0.317 \times 3})$$
$$= 160.226 \text{ mg}/l$$

5.31 Sol.

$$DO_{mix} = \frac{DO_S \cdot Q_S + DO_R \cdot Q_R}{Q_S + Q_R}$$
$$= \frac{2 \times 100 + 8 \times 1000}{100 + 1000}$$
$$= 7.45 \text{ mg}/l$$
$$DO = DO_{sat} - DO_{mix}$$
$$= 10 - 7.45 = 2.545 \text{ mg}/l$$

5.32 (b)

5.33 Sol.

$C_8H_{16}O_8$ of conc. 10^{-3} moles/lt required O_2 (in gm/lt)

$$C_8H_{16}O_8 + 8O_2 \rightarrow 8CO_2 + 8H_2O$$

1 mole 8 mole

1 mole of $C_8H_{16}O_2$ requires 8 moles of O_2 for its decomposition

240 gm = 128 gm

or 10^{-3} moles = 8×10^{-3} moles

$= 8 \times 10^{-3} \times 32 = 0.256$ gm/lt

5.34 (a)

Chemoheterotrophs are the organisms which generate energy with the help of chemical reactions by consuming organic matter during their metabolism. These are the species which help to degrade the organic matter in ASP.

5.35 (c)

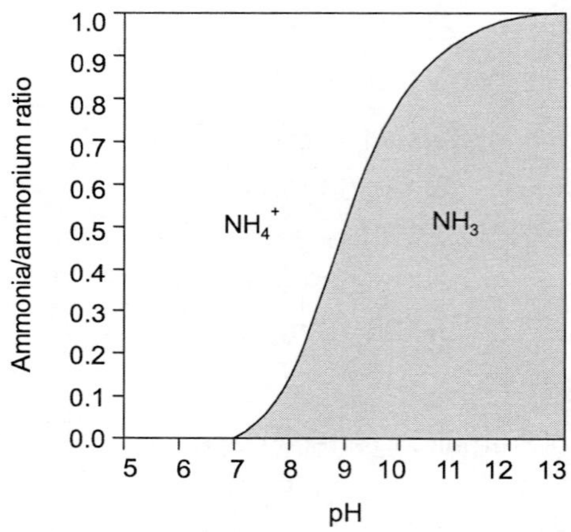

From the above curve, it is evident that at pH 7.0, NH_4^+ will be predominant.

5.36 (a, b)

5.37 (a)

Concentration of O_2 dissolved

= 1.3 mmol/litre/atm × 0.21 atm

= 0.273 mmol/litre

Mass of O_2 dissolved

$= 0.273 \times 10^{-3}$ mol/litre × 32g

$= 8.736 \times 10^{-3}$ g/litre = 8.736 mg/litre

5.38 Sol.

Total nitrogen concentration

$$= \left(\frac{34}{17} + \frac{6.2}{62}\right) \times 14 = 29.4 \text{ mg}/l \text{ as Nitrogen}$$

5.39 Sol.

BOD_3 (27°C)

$$= (DO_i - DO_f) \times \left(\frac{V_1 + V_2}{V_1}\right) = 10 \text{ mg}/l \times 50$$

$$= 500 \text{ mg}/l$$

$$\therefore \quad BOD_3 (27°C) = L_o(1 - e^{-kt})$$

$$\Rightarrow \quad 500 = L_o (1 - e^{-0.23 \times 3})$$

$$\Rightarrow \quad L_o = 1003 \text{ mg}/l$$

5.40 Sol.

Surface over flow rate,

$$V_o = \frac{500}{5 \times 40} = 2.5 \text{ m/h}$$

$$\therefore \quad \eta = \frac{V_s}{V_o} \times 100 = \frac{1}{2.5} = 0.4 \text{ or } 40\%$$

5.41 (b)

$$\underset{2744g}{C_{100}H_{250}O_{80}N} + \underset{3896g}{121.75O_2} \rightarrow 100CO_2 + 123.5H_2O + NH_3$$

$\therefore$ If 1 tonne (or 1000 kg) of waste is present,
Oxygen required

$$= \left(\frac{3896}{2744}\right) \times 1000 \text{ kg} = 1419.825 \text{ kg}$$

Now, 1.3 kg of air contains = 0.23 × 1.3 kg = 0.299 kg of oxygen

$\therefore$ 1419.825 kg of oxygen will be contained by

$$= \left(\frac{1.3}{0.299}\right) \times 1419.825$$

$$= 6173.152 \text{ kg of air}$$

$$\therefore \quad \text{Volume of air} = \frac{6173.152}{1.3} = 4748.57 \text{ m}^3$$

5.42 Sol.

$$C_5H_7O_2N + 5O_2 \rightarrow 5CO_2 + NH_3 + 2H_2O$$

Molar mass of $C_5H_7O_2N$ = 113g

TOC in $C_5H_2O_2$ N = 5 × 12 = 60g

Ultimate BOD = 5 × 32 = 160g

$$BOD_5(20°C) = 160(1 - e^{-0.23 \times 5})$$

$$= 109.34 \text{g}$$

$$\frac{BOD_5(20°C)}{TOC} = \frac{109.34}{60} = 1.82$$

5.43 (c)

Since sulphur is not present in glucose and this H_2S can never be formed be it aerobic or anaerobic decomposition.

■■■■

6 Disposal of Sewage Effluents

6.1 The 'Sag' in the dissolved oxygen curve results because

(a) it is a function of the rate of addition of oxygen to the stream

(b) it is a function of the rate of depletion of oxygen from the stream

(c) it is a function of both addition and depletion of oxygen from the stream

(d) the rate of addition is linear but the rate of depletion is non-linear **[1997 : 1 M]**

6.2 The minimum dissolved oxygen content (ppm) in a river necessary for the survival of aquatic life is

(a) 0 (b) 2

(c) 4 (d) 8 **[1999 : 1 M]**

6.3 Match **List-I** (Characteristics of sewage discharged into inland waters) with **List-II** (Allowable limit, mg/L) and select the correct answer using the codes given below the lists:

	List-I		List-II
A.	BOD_5	1.	250
B.	COD	2.	30
C.	Oil and Grease	3.	20
D.	Total Suspended Solids	4.	10
		5.	5
		6.	3

Codes:

	A	B	C	D
(a)	2	5	4	2
(b)	4	1	6	4
(c)	3	1	4	2
(d)	2	1	6	3

[2003 : 2 M]

6.4 In a certain situation, waste water discharged into a river mixes with the river water instantaneously and completely. Following is the data available:

Waste water DO = 2.00 mg/L
Discharge rate = 1.10 m^3/s

River water DO = 8.3 mg/L
Flow rate = 8.70 m^3/s
Temperature = 20°C

Initial amount of DO in the mixture of waste and river shall be

(a) 5.3 mg/L (b) 6.5 mg/L

(c) 7.6 mg/L (d) 8.4 mg/L

[2005 : 2 M]

6.5 A waste water stream (flow = 2 m^3/s, ultimate BOD = 90 mg/*l*) is joining a small river (flow = 12 m^3/s, ultimate BOD = 5 mg/l). Both water streams get mixed up instantaneously. Cross-sectional area of the river is 50 m^2. Assuming the de-oxygenation rate constant, $K' = 0.25$/day, the BOD (in mg/*l*) of the river water, 10 km downstream of the mixing point is

(a) 1.68 (b) 12.63

(c) 15.46 (d) 1.37

[2014 : 2 M, Set-II]

6.6 The wastewater from a city, containing a high concentration of biodegradable organics, is being steadily discharged into a flowing river at a location S. If the rate of aeration of the river water is lower than the rate of degradation of the organics, then the dissolved oxygen of the river water.

(a) is lowest at the location S.

(b) is lowest at a point upstream of the location S.

(c) remains constant all along the length of the river.

(d) is lowest at a point downstream of the location S. **[2017 : 1 M, Set-I]**

6.7 A stream with a flow rate of 5 m^3/s is having an ultimate BOD of 30 mg/litre. A wastewater discharge of 0.20 m^3/s having BOD_5 of 500 mg/litre joins the stream at a location and instantaneously gets mixed up completely. The cross-sectional area of the stream is 40 m^2 which remains constant. BOD exertion rate constant is 0.3 per day (logarithm base to *e*). The BOD (in mg/litre round off to two decimal places) remaining at 3 km downstream from the mixing location, is ________. **[2020 : 2 M, Set-I]**

6.8 A system of seven river segments is shown in the schematic diagram. The R_i's, Q_i's, and C_i's (i = 1 to 7) are the river segments, their corresponding flow rates, and concentrations of a conservative pollutant, respectively. Assume complete mixing at the intersections, no additional water loss or gain in the system, and steady state condition. Given: Q_1 = 5 m^3/s; Q_2 = 15 m^3/s; Q_4 = 3 m^3/s; Q_6 = 8 m^3/s; C_1 = 8 kg/m^3; C_2 = 12 kg/m^3; C_6 = 10 kg/m^3. What is the steady state concentration (in kg/m^3, rounded off to two decimal place) of the pollutant in the river segment 7 ? _______

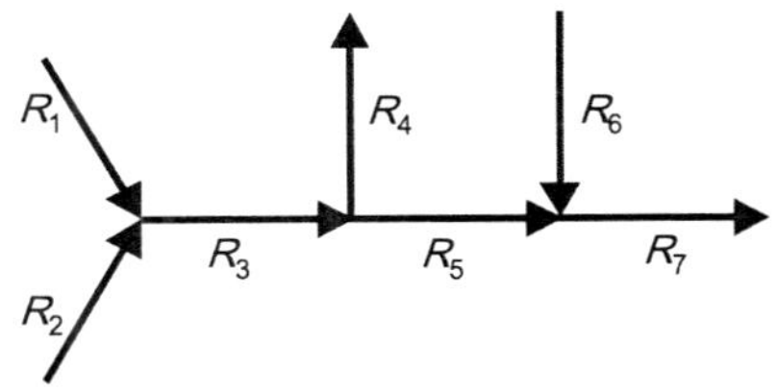

[2023 : 2 M, Set-II]

Answers Disposal of Sewage Effluents

6.1 (c) 6.2 (c) 6.3 (c) 6.4 (c) 6.5 (c) 6.6 (d) 6.7 (49.57) 6.8 (10.68)

Explanations Disposal of Sewage Effluents

6.1 (c)

Oxygen sag curve is a result of reoxygenation and deoxygenation curve. Therefore '**sag**' in the dissolved oxygen curve is a function of both addition and depletion of oxygen from the stream.

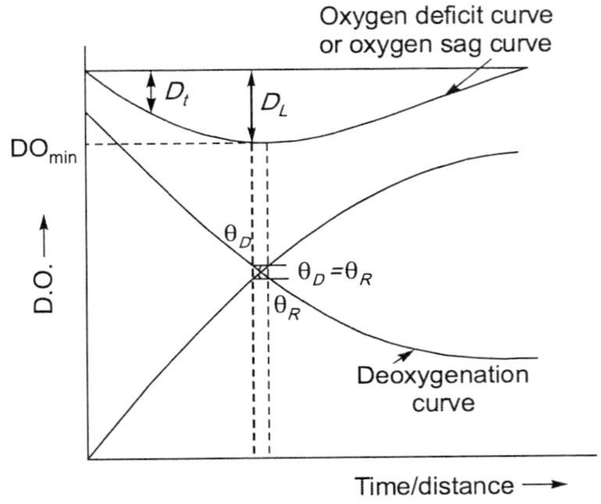

($D_t = DO_{sat} - DO_t$)

6.3 (c)

Characteristics	Permissible limit, mg/l
BOD_5	20
COD	250
Oil and grease	10
Total suspended solids	30

6.4 (c)

$$DO_{mix} = \frac{Q_W DO_W + Q_R DO_R}{Q_W + Q_R}$$

$$= \frac{(1.10 \times 2.00) + (8.70 \times 8.3)}{1.10 + 8.70}$$

$$= 7.6 \text{ mg/L}$$

6.5 (c)

Flow of waste water stream,

$$Q_w = 2 \text{ m}^3\text{/sec}$$

Ultimate BOD,

$$Y_w = 90 \text{ mg}/l = 90 \text{ gm/m}^3$$

Flow of river,

$$Q_R = 12 \text{ m}^3\text{/sec}$$

Ultimate BOD of river,

$$Y_R = 0.5 \text{ mg}/l = 5 \text{ gm/m}^3$$

BOD of mixture,

$$Y_0 = \frac{(2 \times 90) + (12 \times 5)}{2 + 12}$$

$$= \frac{180 + 60}{14} = \frac{240}{14}$$

$$= 17.143 \text{ gm/m}^3$$

$$k_D = 0.434 \text{ K}$$

$$= 0.434 \times 0.25$$

$$= 0.1085$$

$$Y_t = Y_0\left[1 - (10)^{-k_D t}\right] \quad \text{...(i)}$$

Area of river = 50 m^2

Flow of river = 12 m^3/sec

$$\text{Stream velocity} = \frac{14}{50} = 0.28 \text{ m/sec}$$

$$\text{Time taken,} \quad t = \frac{10\text{km}}{0.28\text{m/s}} = 0.4133 \text{ days}$$

∴ From eq. (i),

$$BOD_{left} = Y_0 - Y_t = Y_0 \times 10^{-kt}$$

$$= 17.143 \times 10^{-0.1085 \times 0.4133}$$

$$= 15.46 \text{ mg}/l$$

6.6 (d)

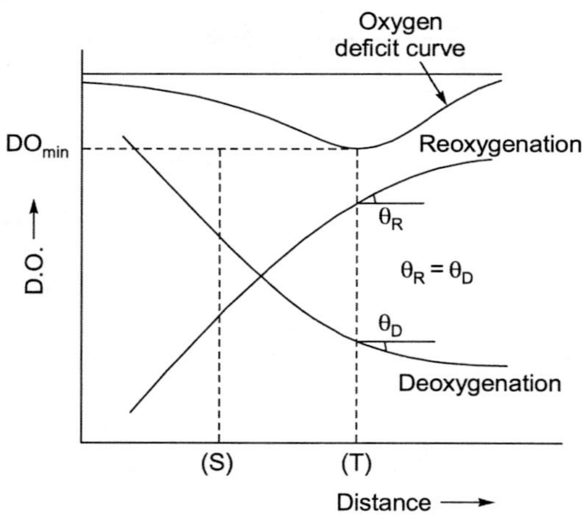

- Before point T, rate of aeration < rate of degradation, so DO continuously decrease from S till T.
- At point T, rate of aeration = Rate of degradation. Hence DO is minimum at T.
- After T, rate of aeration > Rate of degradation, so DO starts increasing.

Hence, DO is minimum at some point downstream of S.

6.7 Sol.

Time, $t = \dfrac{\text{Distance } (d)}{\text{Velocity } (v)}$

where, $v = \dfrac{Q_S + Q_R}{A} = \dfrac{0.2+5}{40} = 0.13 \text{ m/sec}$

$$t = \frac{3 \times 10^3}{0.13 \times 86400} = 0.26 \text{ days}$$

$$BOD_5 = BOD_u(1 - e^{-k \times 5})$$

$$BOD_u = \frac{500}{(1 - e^{-0.3 \times 5})} = 643.66 \text{ mg}/l$$

$$DO_{mix} = \frac{Q_R BOD_u + Q_S \cdot BOD_u}{Q_S + Q_R}$$

$$= \frac{5 \times 30 + 0.2 \times 643.66}{5 + 0.2}$$

$$= 53.6 \text{ mg}/l$$

The BOD remaining at 3 km downstream from the mixing location,

$$L_t = L_0 e^{-k \times t}$$

$$= 53.6 e^{-0.3 \times 0.26}$$

$$= 49.57 \text{ mg}/l$$

6.8 Sol.

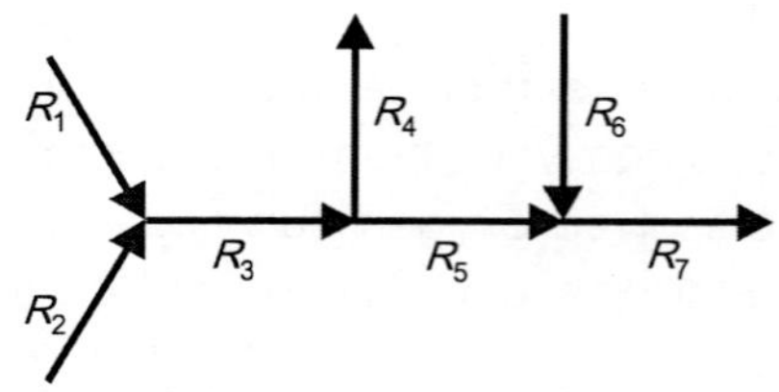

Given, $Q_1 = 5 \text{ m}^3/\text{s}$ $\quad Q_2 = 15 \text{ m}^3/\text{s}$

$Q_4 = 3 \text{ m}^3/\text{s}$ $\quad Q_6 = 8 \text{ m}^3/\text{s}$

$C_1 = 8 \text{ kg/m}^3$ $\quad C_2 = 12 \text{ kg/m}^3$

$C_6 = 10 \text{ kg/m}^3$

$Q_3 = Q_1 + Q_2 = 5 + 15 = 20 \text{ m}^3/\text{s}$

$$C_3 = \frac{(5)(8) + (15)(12)}{5 + 15} = 11 \text{ kg/m}^3$$

$Q_5 = Q_3 - Q_4 = 20 - 3 = 17 \text{ m}^3/\text{s}$

$C_5 = C_3 = 11 \text{ kg/m}^3$

$Q_7 = Q_5 + Q_6 = 17 + 8 = 25 \text{ m}^3/\text{s}$

$$C_7 = \frac{(17)(11) + (8)(10)}{25} = 10.68 \text{ kg/m}^3$$

■■■■

7 Design of Sewers and Sewerage System

7.1 The drop manholes are provided in sewerage system when there is
(a) change in alignment of sewer line
(b) change in size of sewers
(c) change in the elevation of ground level
(d) change from gravity system to pressure system

[1996 : 1 M]

7.2 At the same mean velocity, the ratio of head loss per unit length for a sewer pipe flowing full to that for the same pipe flowing half full would be
(a) 2.0 (b) 1.63
(c) 1.00 (d) 0.61 [1998 : 1 M]

7.3 An inverted siphon is a
(a) device for distributing septic tank effluent to a soil absorption system
(b) device for preventing overflow from elevated water storage tank
(c) derive for preventing crown corrosion of sewer
(d) section of sewer which is dropped below the hydraulic grade line in order to avoid an obstacle. [1998 : 1 M]

7.4 A cicular sewer of 2 m diameter has to carrya discharge of 2 m^2/s when flowing nearly full. What is the mainimum required slope to initiate the flow? Assume Manning's $n = 0.015$.
(a) 0.00023 (b) 0.000036
(c) 0.00091 (d) 0.000014

[2000 : 2 Mark]

7.5 An existing 300 mm diameter circular sewer is laid at a slope of 1 : 280 and carries a peak discharge of 1728 m^3/d. Use the partial flow diagram shown in the given figure and assume Manning's $n = 0.015$.

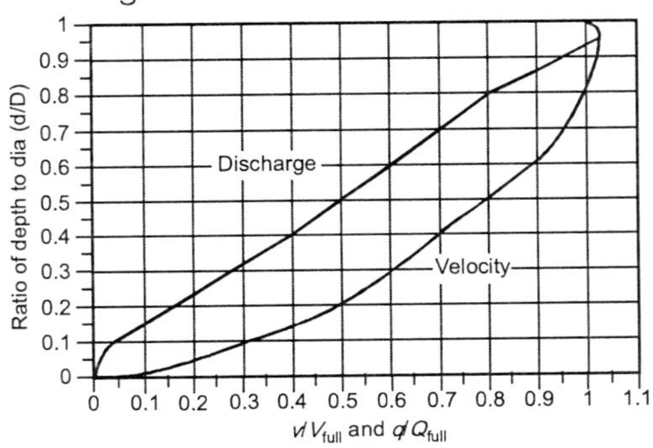

Partial flow diagram for circular sewer

At the peak discharge, the depth of flow and the velocity are, respectively,
(a) 45 mm and 0.28 m/s
(b) 120 mm and 0.50 m/s
(c) 150 mm and 0.57 m/s
(d) 300 mm and 0.71 m/s [2004 : 2 M]

Directions: The following items consists of two statements; one labelled as **'Assertion (A)'** and the other as **'Reason (R)'**. You are to examine these two statements carefully and select the answers to these items using the codes given below:

Codes:
(a) both A and R are true and R is the correct explanation of A
(b) both A and R are true but R is not a correct explanation of A
(c) A is true but R is false
(d) A is false but R is true

7.6 **Assertion [A]:** The crown of the outgoing larger diameter sewer is always matched with the crown of incoming smaller diameter sewer.
Reason [R]: It eliminates backing up of sewage in the incoming smaller diameter sewer.

[2008 : 2 M]

7.7 **Assertion [A]** : At a manhole, the crown of the outgoing sewer should not be higher than the crown of the incoming sewer.
Reason [R] : Transition from a larger diameter incoming sewer to a smaller diameter outgoing sewer at a manhole should not be made.

[2012 : 1 M]

7.8 Crown corrosion in a reinforced concrete sewer is caused by:
(a) H_2S (b) CO_2
(c) CH_4 (d) NH_3

[2016 : 1 M, Set-I]

Answers Design of Sewers and Sewerage System

7.1 (c) 7.2 (c) 7.3 (d) 7.4 (a) 7.5 (b) 7.6 (a) 7.7 (b) 7.8 (a)

Explanations Design of Sewers and Sewerage System

7.1 (c)

Drop manholes provided when there is change in elevation of ground level.

7.2 (c)

Head loss, $h_f = \dfrac{fLV^2}{2gd}$

For same pipe, diameter (d) is same and as velocity is also same for both the cases.

Hence, h_f is also the same.

$\therefore$ For same pipe with equal velocity, ratio of head loss will always be 1.

7.3 (d)

Inverted siphon (or depressed sewer) is used to enable the flow of storm water and waste water under obstructions such as rivers.

7.4 (a)

$$Q = 2 \text{ m}^3/\text{s}$$

Diameter, $d = 2$ m

$$R = \frac{d}{4} = \frac{2}{4} = 0.5 \text{ m},$$

$$n = 0.015$$

$$Q = A \times \frac{1}{n} R^{2/3} S^{1/2}$$

$$\Rightarrow \quad 2 = \frac{\pi}{4} \times 2^2 \times \frac{1}{0.015} \times (0.5)^{2/3} S^{1/2}$$

$$\Rightarrow \quad S = 0.00023$$

7.5 (b)

$$Q = \left(\frac{\pi}{4} D^2\right) \cdot \frac{R^{2/3}\, S^{1/2}}{n} \qquad [R = D/4]$$

$$= \frac{\pi}{4} \times 0.3^2 \times \left(\frac{0.3}{4}\right)^{2/3} \times \frac{\sqrt{1/280}}{0.015}$$

$$= 0.05 \text{ m}^3/\text{s}$$

$$Q = 4327.3 \text{ m}^3/\text{d}$$

$$q = 1728 \text{ m}^3/\text{d}$$

$$\therefore \quad \frac{q}{Q} = 0.4$$

From graph, $\dfrac{d}{D} = 0.4$ $\qquad \therefore d = 120$ mm

From graph, $\dfrac{v}{V} = 0.7$

$$\therefore \quad v = 0.7 \times \frac{1}{n} R^{2/3}\, S^{1/2}$$

$$= 0.496 \text{ m/s} \approx 0.5 \text{ m/s}$$

7.7 (b)

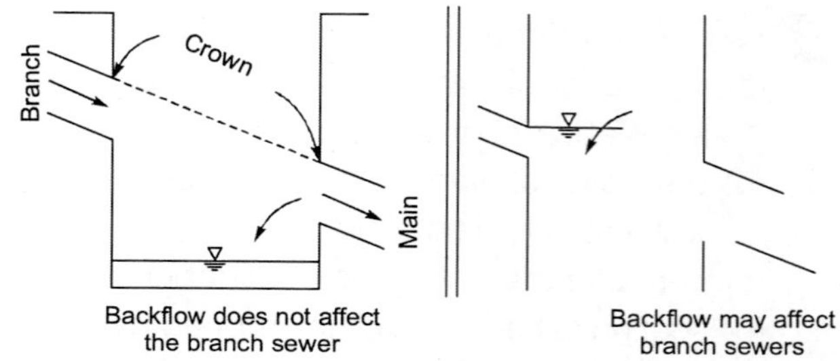

To control the backing up of sewage in the incoming sewer, the crown of the outgoing sewer should not be higher than the crown of incoming sewer.

There may be more than one incoming sewer in a manhole that's why diameter of outgoing sewer should be larger than the incoming sewer.

7.8 (a)

Bacteria in the slime under flowing sewage convert sulphate in the sewage into sulphides. Sulphides in the liquid make their way to the surface of the sewage and reduced into the sewer atmosphere or Hydrogen Sulphide (H_2S) gas. Bacterial action converts atmospheric H_2S gas to Sulphuric Acid which causes corrosion in the crown of the pipe and this corrosion is called crown corrosion.

■■■■

8 Treatment of Waste Water

8.1 A trickling filter is primarily a
(a) straining process to remove suspended solids from sewage
(b) biological oxidation process to remove BOD from sewage
(c) straining process to remove turbidity from water
(d) straining process to remove bacterial from water [1995 : 1 M]

8.2 A septic tank is
(a) an aerobic method on site sewage treatment
(b) an anaerobic method of on site sewage treatment
(c) a physical method of water treatment
(d) a physicochemical method of water treatment [1995 : 1 M]

8.3 The removal of dissolved organic matter occurs in
(a) slow sand filters
(b) trickling filters
(c) rapid sand filters
(d) dual media filters [1996 : 1 M]

8.4 The main constituents of gas generated during the anaerobic digestion of sewage sludge are
(a) carbon dioxide and methane
(b) methane and ethane
(c) carbon dioxide and carbon monoxide
(d) carbon monoxide and nitrogen [1996 : 1 M]

8.5 Presence of excess nitrates in river water indicates
(a) recent pollution of water with sewage
(b) past pollution of water with sewage
(c) immediate pollution of water with sewage
(d) no pollution of water with sewage [1996 : 1 M]

8.6 A typical biological process in treating waste water using aerated lagoon can be described by one of the following schematic diagrams

(a) 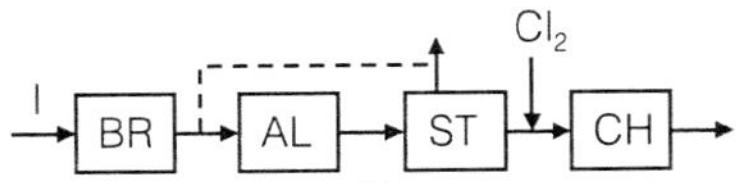

(b) 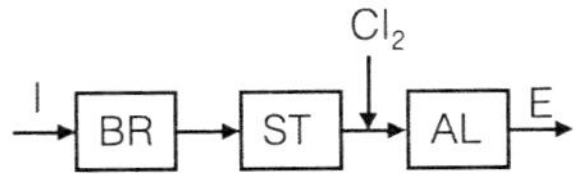

(c) 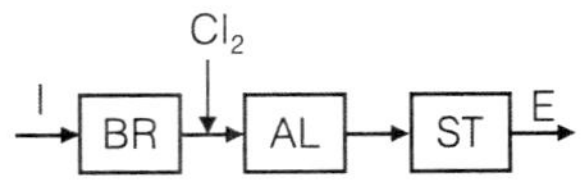

(d) 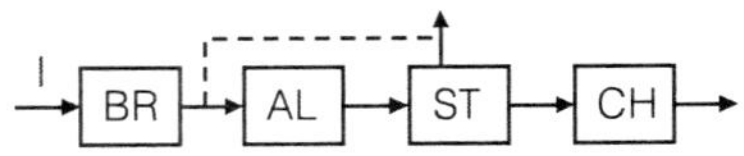

[1997 : 1 M]

8.7 Sewage treatment in an oxidation pond is accomplished primarily by
(a) algal-bacterial symbiosis
(b) algal photosynthesis only
(c) bacterial oxidation only
(d) chemical oxidation only [1998 : 1 M]

8.8 Chlorine is sometimes used in sewage treatment
(a) to avoid flocculation
(b) to increase biological activity of bacteria
(c) to avoid bulking of activated sludge
(d) to help in grease separation [1999 : 1 M]

8.9 Critical factors for the activated sludge treatment process are
(a) maximum hourly flow rate
(b) maximum and minimum flow rate
(c) maximum hourly flow rate and maximum daily organic load
(d) minimum hourly flow rate and minimum daily organic load [2000 : 1 M]

8.10 The unit in which both sedimentation and digestion processes of sludge take place simultaneously is
(a) Skimming Tank (b) Imhoff Tank
(c) Deritus Tank (d) Digestion Tank [2001 : 1 M]

8.11 A Trickling filter is designed to remove
(a) Settleable Solids
(b) Colloidal Solids
(c) Dissolved Organic Matter
(d) None of the above [2002 : 1 M]

8.12 Settling test on a sample drawn from Aeration Tank liquor of ASP (MLSS = 2800 mg/L) was carried out with 1 litre sample. The test yielded a settled volume of 200 mL. The value of Sludge Volume Index shall be

(a) 14.0 (b) 34.2
(c) 71.4 (d) 271 [2003 : 2 M]

8.13 Which of the following sewage treatment methods has inherent problems of odour, ponding and fly nuisance?
(a) UASB system
(b) Activated sludge process
(c) Trickling filters
(d) Stabilization ponds [2003 : 1 M]

8.14 From amongst the following sewage treatment options, largest land requirements for a given discharge will be needed for
(a) trickling filter (b) anaerobic pond
(c) oxidation ditch (d) oxidation pond
[2003 : 1 M]

Common Data for Questions 8.15 and 8.16:

A conventional activated sludge plant treating 1000 m^3/d of municipal waste water disposes off its anaerobically digested sludge on relatively impervious farm land. Use the following data:

1. Raw sewage: SS = 225 mg/L (70% volatile); BOD = 190 mg/L
2. Primary settling: SS = 50% removal; BOD = 30% removal.
3. Excess activated sludge : 0.4g VSS produced per gram BOD applied (80% volatile of total)
4. Anaerobic digester: VSS reduced 50%; Digested sludge concentration = 6% Sludge specific gravity = 1
5. Application on farm land: 2 m^3/ha-d

8.15 Total volatile suspended solids to be anaerobically digested (kg/d VSS) shall be
(a) 144 (b) 168
(c) 233 (d) 245 [2003 : 2 M]

8.16 Area requirements (hectare) for disposal of the sludge on farmland shall be
(a) 2.95 (b) 1.95
(c) 0.95 (d) 0.55 [2003 : 2 M]

8.17 An analysis for determination of solids in the return sludge of Activated Sludge Process was done as follows:
1. A crucible was dried to a constant mass of 62.485 g.
2. 75 mL of a well-mixed sample was taken in the crucible.
3. The crucible with the sample was dried to a constant mass of 65.020 g in a drying oven at 104°C.
4. The crucible with the dried sample was placed in a muffle furnace at 600°C, for an hour. After cooling, the mass of the crucible with residues was 63.145 g.

The concentration of organic fraction of solids present in the return sludge sample is
(a) 8800 mg/L (b) 25000 mg/L
(c) 33800 mg/L (d) 42600 mg/L
[2004 : 2 M]

8.18 The following data are given for a channel-type grit chamber of length 7.5 m.
1. flow-through velocity = 0.3 m/s
2. the depth of waste water at peak flow in the channel = 0.9 m
3. specific gravity of inorganic particles = 2.5
4. $g = 9.80$ m/s^2, $\mu = 1.002 \times 10^{-3}$ N-s/m^2 at 20°C, $\rho_w = 1000$ kg/m^3

Assuming that the Stoke's law is valid, the largest diameter particle that would be removed with 100 per cent efficiency is
(a) 0.04 mm (b) 0.21 mm
(c) 1.92 mm (d) 6.64 mm
[2004 : 2 M]

8.19 Bulking sludge refers to having
(a) F/M < 0.3/*d*
(b) 0.3/*d* < F/M < 0.6/*d*
(c) F/M = zero
(d) F/M > 0.6/d [2005 : 1 M]

8.20 In aerobic environment, nitrosomonas convert
(a) NH_3 to NO_2 (b) NO_2^- to NO_3^-
(c) NH_3 to N_2O (d) NO_2^- to HNO_3
[2005 : 1 M]

8.21 Match **List-I** with **List-II** and select the correct answer using the codes given below the lists:

List-I
A. Thickening of sludge
B. Stabilization of sludge
C. Conditioning of sludge
D. Reduction of sludge

List-II
1. Decrease in volume of sludge by chemical oxidation
2. Separation of water by heat or chemical treatment
3. Digestion of sludge
4. Separation of water by floatation or gravity

Codes:

	A	B	C	D
(a)	4	3	1	2
(b)	3	2	4	1
(c)	4	3	2	1
(d)	2	1	3	4

[2005 : 2 M]

Common Data for Questions 8.22 and 8.23:

A completely mixed activated sludge process is used to treat a wastewater flow of 1 million litres per day (1 MLD) having a BOD_5 of 200 mg/L. The biomass concentration in the aeration tank is 2000 mg/L and the concentration of the net biomass leaving the system is 50 mg/L. The aeration tank has a volume of 200 m^3.

8.22 What is the hydraulic retention time of the wastewater in aeration tank?

(a) 0.2 h (b) 4.8 h
(c) 10 h (d) 24 h [2007 : 2 M]

8.23 What is the average time for which the biomass stays in the system?

(a) 5 h (b) 8 h
(c) 2 days (d) 8 days [2007 : 2 M]

8.24 Two biodegradable components of municipal solid waste are

(a) plastics and wood
(b) cardboard and glass
(c) leather and tin cans
(d) food wastes and garden trimmings

[2008 : 1 M]

8.25 Match **List-I** (Terminology) with **List-II** (Definition/Brief Description) and select the correct answer using the codes given below the lists:

List-I
A. Primary treatment
B. Secondary treatment
C. Unit operation
D. Unit process

List-II
1. Contaminant removal by physical forces
2. Involving biological and/or chemical reaction
3. Conversion of soluble organic matter to biomass
4. Removal of solid materials from incoming wastewater

Codes:

	A	B	C	D
(a)	4	3	1	2
(b)	4	3	2	1
(c)	3	4	2	1
(d)	1	2	3	4

[2008 : 2 M]

8.26 An aerobic reactor receives wastewater at a flow rate of 500 m^3/d having a COD of 2000 mg/L. The effluent COD is 400 mg/L. Assuming that wastewater contains 80% biodegradable waste, the daily volume of methane produced by the reactor is

(a) 0.224 m^3 (b) 0.280 m^3
(c) 224 m^3 (d) 280 m^3

[2009 : 2 M]

8.27 Match **List-I** with **List-II** and select the correct answer by using the codes given below the lists:

List-I
A. Grit chamber
B. Secondary settling tank
C. Activated sludge process
D. Trickling filter

List-II
1. Zone settling
2. Stoke's law
3. Aerobic
4. Contact stabilization

Codes:

	A	B	C	D
(a)	1	2	3	4
(b)	2	1	3	4
(c)	1	2	4	3
(d)	2	1	4	3

[2009 : 2 M]

Linked Answer Questions 8.28 and 8.29:

The sludge from the aeration tank of the activated sludge process (ASP) has solids content (by weight) of 2%. This sludge is put in a sludge thickener, where sludge volume is reduced to half. Assume that the amount of solids in the supernatant from the thickener is negligible, the specific gravity of sludge solids is 2.2 and the density of water is 1000 kg/m^3.

8.28 What is the density of the sludge removed from the aeration tank?

(a) 990 kg/m^3 (b) 1000 kg/m^3
(c) 1011 kg/m^3 (d) 1022 kg/m^3

[2011 : 2 M]

8.29 What is the solids content (by weight) of the thickened sludge?

(a) 3.96% (b) 4.00%
(c) 4.04% (d) 4.10% [2011 : 2 M]

Common Data for Questions 8.30 and 8.31:

An activated sludge system (sketched below) is operating at equilibrium with the following information. Waste water related data : flow rate = 500 m^3/hour, influent BOD = 150 mg/L, effluent BOD = 10 mg/L. Aeration tank related data : hydraulic retention time = 8 hours, mean-cell-residence time = 240 hours, volume = 4000 m^3, mixed liquor suspended solids = 2000 mg/L.

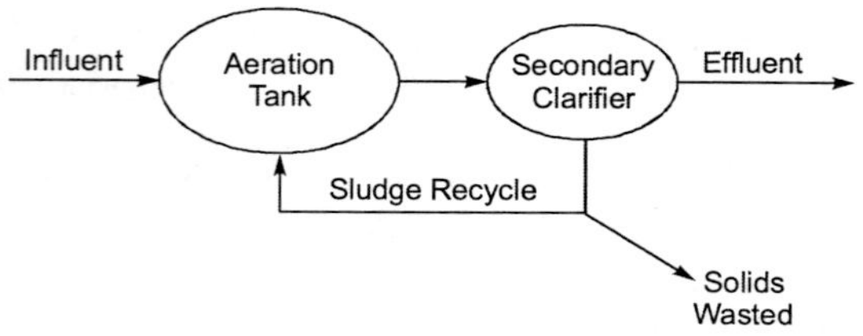

8.30 The food-to-biomass (F/M) ratio (in kg BOD per kg biomass per day) for the aeration tank is

(a) 0.015 (b) 0.210
(c) 0.225 (d) 0.240 [2012 : 2 M]

8.31 The mass (in kg/day) of solids wasted from the system is

(a) 24000 (b) 1000
(c) 800 (d) 33 [2012 : 2 M]

8.32 The dominating microorganisms in an activated sludge process reactor are

(a) aerobic heterotrophs
(b) anaerobic heterotrophs
(d) autotrophs
(d) phototrophs [2014 : 1 M, Set-II]

8.33 The amount of CO_2 generated (in kg) while completely oxidizing one kg of CH_4 to the end products is _______.

[2014 : 1 M, Set-I]

8.34 Total Kjeldahl Nitrogen (TKN) concentration (mg/L as *N*) in domestic sewage is the sum of the concentration of

(a) organic and inorganic nitrogen in sewage
(b) organic nitrogen and nitrate in sewage
(c) organic nitrogen and ammonia in sewage
(d) ammonia and nitrate in sewage

[2015 : 1 M, Set-I]

8.35 In a wastewater treatment plant, Primary Sedimentation Tank (PST) designed at an overflow rate of 32.5 m^3/day/m^2 is 32.5 m long, 8.0 m wide and liquid depth of 2.25 m. If the length of the weir is 75 m, the weir loading rate (in m^3/day/m) is __________.

[2015 : 2 M, Set-II]

8.36 Match the items in **Group-I** with those in **Group-II** and choose the right combination

Group-I
P. Activated sludge process
Q. Rising of sludge
R. Conventional nitrification
S. Biological nitrogen removal

Group-II
1. Nitrifiers and denitrifiers
2. Autotrophic bacteria
3. Heterotrophic bacteria
4. Denitrifiers

(a) P-3, Q-4, R-2, S-1
(b) P-2, Q-3, R-4, S-1
(c) P-2, Q-2, R-4, S-1
(d) P-1, Q-4, R-2, S-3 [2016 : 1 M, Set-I]

8.37 The wastewater having an organic concentration of 54 mg/*l* is flowing at a steady rate of 0.8 m^3/day through a detention tank of dimensions 2 m × 4 m × 2 m. If the contents of the tank are well mixed and the decay contant is 0.1 per day, the outlet concentration (in mg/*l*, up to one decimal place) is ______. [2017 : 2 M, Set-I]

8.38 The most important type of species involved in the degradation of organic matter in the case of activated sludge process is

(a) autotrophs (b) heterotrophs
(c) prototrophs (d) photo-autotrophs

[2017 : 1 M, Set-II]

8.39 A schematic flow diagram of a completely mixed biological reactor with provision for recycling of solids is shown in the figure.

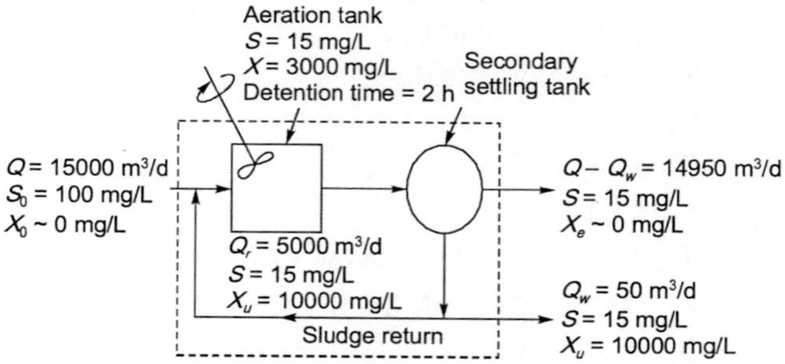

S_0, S = readily biodegradable soluble BOD, mg/L
Q, Q_r, Q_w = flow rates, m^3/d
X_0, X, X_e, X_u = microorganism concentrations (mixed-liquor volatile suspended solids or MLVSS), mg/L

The mean cell residence time (in days, up to one decimal place) is _____. [2018 : 2 M, Set-II]

8.40 Consider the reactor shown in the figure. The flow rate through the reactor is Q m^3/h. The concentrations (in mg/L) of a compound in the influent and effluent are C_o and C, respectively. The compound is degraded in the reactor following the first order reaction. The mixing condition of the reactor can be varied such that the reactor becomes either a completely mixed flow reactor (CMFR) or a plug-flow reactor (PFR). The length of the reactor can be adjusted in these two mixing conditions to L_{CMFR} and L_{PFR} while keeping the cross-section of the reactor constant. Assuming steady state and for C/C_o = 0.8, the value of L_{CMFR}/L_{PFR} (round off to 2 decimal places) is________.

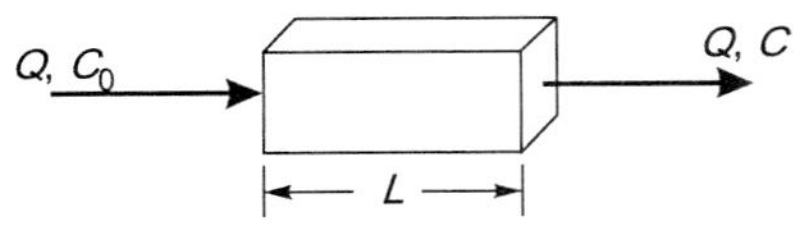

[2019 : 2 M, Set-II]

8.41 A secondary clarifier handles a total flow of 9600 m^3/d from the aeration tank of a conventional activated-sludge treatment system. The concentration of solids in the flow from the aeration tank is 3000 mg/*l*. The clarifier is required to thicken the solids to 12000 mg/*l*, and hence it is to be designed for a solid flux of 3.2 kg/m^2h. The surface area of the designed clarifier for thickening (in m^2, in integer) is _______.

[2021 : 2 M, Set-I]

8.42 A grit chamber of rectangular cross-section is to be designed to remove particles with diameter of 0.25 mm and specific gravity of 2.70. The terminal settling velocity of the particles is estimated as 2.5 cm/s. The chamber is having a width of 0.50 m and has to carry a peak wastewater flow of 9720 m^3/d giving the depth of flow as 0.75 m. If a flow-through velocity of 0.3 m/s has to be maintained using a proportional weir at the outlet end of the chamber, the minimum length of the chamber (in m, in integer) to remove 0.25 mm particles completely is ________.

[2021 : 2 M, Set-II]

8.43 An activated sludge process (ASP) is designed for secondary treatment of 7500 m^3/day of municipal wastewater. After primary clarifier, the ultimate BOD of the influent, which enters into ASP reactor is 200 mg/L. Treated effluent after secondary clarifier is required to have an ultimate BOD of 20 mg/L. Mix liquor volatile suspended solids (MLVSS) concentration in the reactor and the underflow is maintained as 3000 mg/L and 12000 mg/L, respectively. The hydraulic retention time and mean cell residence time are 0.2 day and 10 days, respectively. A representative flow diagram of the ASP is shown below.

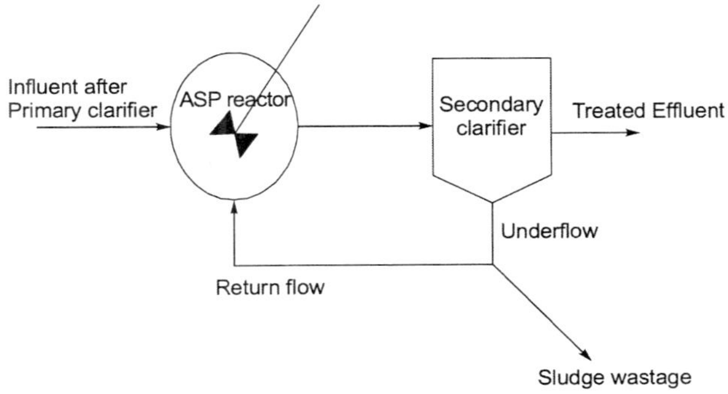

The underflow volume (in m^3/day, round off to one decimal place) of sludge wastage is _______.

[2021 : 2 M, Set-II]

8.44 A sewage treatment plant receives sewage at a flow rate of 5000 m^3/day. The total suspended solids (TSS) concentration in the sewage at the inlet of primary clarifier is 200 mg/L. After the primary treatment, the TSS concentration in sewage is reduced by 60 %. The sludge from the primary clarifier contains 2 % solids concentration. Subsequently, the sludge is subjected to gravity thickening process to achieve a solids concentration of 6%. Assume that the density of sludge, before and after thickening, is 1000 kg/m^3. The daily volume of the thickened sludge (in m^3 /day) will be_______. (round off to the nearest integer)

[2022 : 1 M, Set-II]

8.45 In the following table, identify the correct set of associations between the entries in Column-1 and Column-2.

Column-1
P: Reverse Osmosis
Q: Trickling Filter
R: Coagulation
S: Adsorption

Column-2
I: Ponding
II: Freundlich Isotherm
III: Concentration Polarization
IV: Charge Neutralization

(a) P-II, Q-I, S-III
(b) Q-III, R-II, S-IV
(c) P-IV, R-I, S-II
(d) P-III, Q-I, R-IV

[2023 : 2 M, Set-I]

8.46 The wastewater inflow to an activated sludge plant is 0.5 m^3/s, and the plant is to be operated with a food to microorganism ratio of 0.2 mg/mg-d. The concentration of influent biodegradable organic matter of the wastewater to the plant (after primary settling) is 150 mg/L, and the mixed liquor volatile suspended solids concentration to be maintained in the plant is 2000 mg/L. Assuming that complete removal of biodegradable organic matter in the tank, the volume of aeration tank (in m^3, in integer) required for the plant is _______.

[2023 : 2 M, Set-I]

8.47 In the context of water and wastewater treatments, the correct statements are:

(a) particulate matter may shield microorganisms during disinfection

(b) ammonia decreases chlorine demand

(c) phosphorous stimulates algal and aquatic growth

(d) calcium and magnesium increase hardness and total dissolved solids

[2023 : 2 M, Set-II]

8.48 Which of the following statements is/are TRUE for the aerobic composting of sewage sludge?

(a) Bulking agent is added during the composting process to reduce the porosity of the solid mixture

(b) Leachate can be generated during composting

(c) Actinomycetes are involved in the process

(d) In-vessel composting systems cannot be operated in the plug-flow mode

[2023 : 2 M, Set-II]

Answers Treatment of Waste Water

8.1	(b)	8.2	(b)	8.3	(b)	8.4	(a)	8.5	(b)	8.6	(b)	8.7	(a)
8.8	(c)	8.9	(c)	8.10	(b)	8.11	(c)	8.12	(c)	8.13	(c)	8.14	(d)
8.15	(a)	8.16	(d)	8.17	(b)	8.18	(b)	8.19	(a)	8.20	(a)	8.21	(c)
8.22	(b)	8.23	(d)	8.24	(d)	8.25	(a)	8.26	(c)	8.27	(b)	8.28	(c)
8.29	(a)	8.30	(c)	8.31	(c)	8.32	(a)	8.33	(2.75)	8.34	(c)	8.35	(11.267)
8.36	(a)	8.37	(18)	8.38	(b)	8.39	(7.5)	8.40	(1.12)	8.41	(375)	8.42	(9)
8.43	(37.5)	8.44	(10)	8.45	(d)	8.46	(16200)	8.47	(a, c, d)	8.48	(b, c)		

Explanations Treatment of Waste Water

8.1 (b)

- Trickling filter consist of tanks of coarser filtering media, over which the sewage is allowed to sprinkle or trickle down, by means of spray nozzles or rotary distributors.
- In these, the decomposition of the organic matter and the resultant purification of the sewage is brought about by a population of micro-organism.
- Hence, it is a biological oxidation process to remove BOD from sewage.

REFERENCE MATERIAL

1. *Environmental Engineering, Vol. II, By* SK Garg, Ch-9, Treatment of Sewage, Topic, Trickling filters.

8.2 (b)

A **septic tank** may be defined as a primary sedimentation tank with a longer detention period (12 to 36 hours) and with extra provisions for digestion of the settled sludge. Since the digestion of the settled sludge is carried out by anaerobic decomposition process, the septic tank unit is generally classified under the units which work on the principle of anaerobic decomposition. It is called an onsite treatment because raw sludge is directly fed and no other treatment unit is used in the spetic tank.

8.3 (b)

As the waste water which trickles through the filter media, biomass grows attached to the media surface, normally in the period of 2 weeks making the filter ready for operation. During operation the organic matter present in waste water is metabolised by the biomass attached to the solid medium. The attached biomass is called biological film or slime layer. The slime layer has a thickness of 0.1 to 2 mm and it consists of aerobic and anaerobic layer.

8.4 (a)

Anaerobic digestion is normally adopted for primary sludge because primary sludge contains large amount of readily available organics that would induce rapid growth, of biomass if treated anaerobically.

Anaerobic decomposition process produce less biomass and primary function of anaerobic digester is to convert as much of sludge as possible to end products like liquid and gases while producing little biomass.

Waste-water contains wide varieties of organism. These organism are broadly classified as

(a) Acid formers
(b) Methane formers

Acid formers: Acid formers consists of facultative and anaerobic bacteria and include organisms which solubilize the organic acid through hydrolysis.

The soluble end products are then fermented to acids and alcohol of lower molecular weight.

Methane former: Methane formers are strictly anaerobic and converts acid and alcohol along with hydrogen and CO_2 to methane.

8.5 (b)

The presence of nitrogen in sewage indicates the presence of organic matter and may occur in one or more form:

(a) Free ammonia
(b) Albuminoid nitrogen
(c) Nitrites
(d) Nitrates

Free ammonia indicates very fast stage of decomposition of organic matter (thus indicating recently stated sewage).

Albuminoid nitrogen indicates quantity of nitrogen present in sewage before the decomposition of organic matter is started.

Nitrites indicate the presence of partly decomposed (not fully oxidized) organic matter.

Nitrates indicates the presence of fully oxidized organic matter. Presence of nitrates indicates the most stable form of nitrogenous matter (past pollution) contained in sewage, thus indicating well oxidized treated sewage. Even though 'nitrates' have been defined as the non-objectionable final end product in aerobic treatment of sewage.

8.6 (b)

- Aerated lagoons are in form of earthen basin, where waste water flows through while aerations is done by mechanical aerators. There are no settling tanks and sludge recirculation in these.
- Hence, according to the given options, (b) is the best answer, as it does not include recirculation and settling tank is also present before aerated lagoon.

8.7 (a)

- In oxidation pond, the algae (which are microcopic plants) while growing in the presence of sunlight produce oxygen by the action of photosynthesis, and this oxygen is utilized by bacteria for oxidizing the waste organic matter.
- The end products of organic matter oxidized by bacteria (viz. CO_2, NH_3, PO_4, H_2O) are required by algae to grow and produce oxygen.
- Hence, there exist a beneficial association between algae and bacteria, known as symbiotic relationship.

8.8 (c)

A sludge that exhibits poor settling characteristics is called a buckling sludge. Filamentons micro-organisms are responsible for a buckled sludge. Large surface area to volume ratios of these micro-organisms raetard their settling velocities.

The most successful methods to conrol the buckling of sludge ara:

(i) reduction of the sludge age to less then 6 days, and

(ii) chlorination of return activated sludge.

8.9 (c)

Critical factors are maximum hourly flow rate and maximum daily organic loading for an ASP.

8.10 (b)

Imhoff tank is a modified form of septic tank. It contains two chambers, one is an upper chamber in which sedimentation takes place and another is cover chamber in which the digestion of collected sludge takes place.

POINTS TO REMEMBER

- They combine the advantages of both sedimentation and septic tanks.
- They find use in small treatment plants requireing only primary treatment.
- They are economical and don't require skilled supervision.
- Results obtained are good, with 60-65% removal of solids and 30-40% removal of BOD.

8.11 (c)

As the waste water which trickles through the filter media, biomass grows attached to the media surface, normally in the period of 2 weeks making the filter ready for operation. During operation the organic matter present in waste water is metabolised by the biomass attached to the solid medium. The attached biomass is called biological film or slime layer. The slime layer has a thickness of 0.1 to 2 mm and it consists of aerobic and anaerobic layer.

8.12 (c)

Sludge Volume Index (SVI) is defined as the volume occupied (in mL) by 1 g of solids in themixed liquid after settling for 30 min.

Now, MLSS = 2800 mg/L

$\therefore$ In 1 L sample,

Solids = 2800 mg

$\Rightarrow$ W = 2.8 g

and settled volume = 200 mL

Hence, $SVI = \frac{V}{W} = \frac{200}{2.8} = 71.43$

8.13 (c)

Following operational problems occur in standard rate trickling filters (because of their low hydraulic loading rate)

1. **Fly Nuisance:** Larvas of insects act as food for the flies. As filter medium is open to atmospheres, so insects grow over the surface of the filter.
2. **Odour problem:** Due to anaerobic decomposition of wastewater during intermittent flow.
3. **Ponding problem:** Due to chocking of the voids as sloughing do not take place due to low HLR.

8.14 (d)

In oxidation pond, depth is kept less so that sunlight can penetrate upto full depth to perform aerobic decomposition. If the depth is less than we require more surface area means more land requirements.

8.15 (a)

Waste water discharge,

$Q = 1000\ m^3/d$

In raw sewage,

$SS = 225$ mg/L (70% volatile)

$\therefore\ VSS = 0.7 \times 225$ mg/L

$= 0.7 \times 225 \times 1000 \times 10^3$ mg/d

$= 157.5$ kg/d

BOD/day $= 190 \times 1000 \times 10^3$ mg/d

$= 190$ kg

$\therefore$ BOD after primary settling

$= 0.7 \times 190 = 133$ kg

Now, total volatile SS (kg/d) to be anaerobically digested.

= VSS is sludge proviuced after primary treatment + VSS in sludge produced by secondary treatment

$= (157.5 \times 0.5) + (0.4 \times 133)$

$= 131.95$ kg

8.16 (d)

VSS in digested sludge

$= 0.5 \times 131.95$

$= 65.975$ kg

Digested sludge concentration = 6%

$\Rightarrow$ 6 kg solids gives 100 kg digested sludge

$\Rightarrow$ for 65.975 kg solids

Digested sludge $= 65.975 \times \frac{100}{6}$

$= 1099.583$ kg

Sludge specific graity = 1

$\Rightarrow$ Volume of digested sludge

$= \frac{1099.583}{1\,\rho_w} = \frac{1099.583}{1000} = 1 : 1\ m^3$

Now, area required,

$$ha = \frac{\text{Volume of digested sludeg } (m^3)}{\text{Application of farm land } (m^3/\text{ha-d})}$$

$$= \frac{1.1}{2} = 0.55 \text{ ha}$$

8.17 (b)

Weight of sample of 104°C = 65.02 g
Weight of samle at 600°C = 63.145 g

Note: Organic matter will get converted into carbon dioxide, water vapour and other gases at 600°.

Concentration of organic fraction of solids

$$= \frac{\text{Weight of organic fraction}}{\text{Volume of sample}}$$

$$= \frac{65.02 - 63.145}{75} \times \frac{10^3}{10^{-3}} = 25000 \text{ mg}/l$$

8.18 (b)

Flow through velocity, $V = \dfrac{Q}{BH}$

$$\Rightarrow \quad \frac{Q}{B} = V \times H$$

$$= 0.3 \times 0.9 = 0.27 \text{ m}^2/\text{s}$$

For 100% efficiency settling velocity (V_s) = SOR

$$V = \frac{Q}{BL} = \frac{0.27}{7.5} = 0.036 \text{ m/s}$$

And $$V_s = \frac{\rho_w g}{18}\left(\frac{S_s - 1}{\mu}\right) d^2 = 0.036$$

$$\Rightarrow \quad d^2 = \frac{0.036 \times \mu \times 18}{\rho_w \times g \times (S_s - 1)}$$

$$\Rightarrow \quad d^2 = \frac{0.036 \times 1.002 \times 10^{-3} \times 18}{1000 \times 9.81 \times (2.5 - 1)}$$

$$\Rightarrow \quad d^2 = 4.42 \times 10^{-8}$$

$$\therefore \quad d = 2.10 \times 10^{-4} \text{ m} = 0.21 \text{ mm}$$

8.19 (a)

Buckling sludge (sludge having poor settling characteristics) is formed due to the growth of filamentous micro-organisms (whose growth is not complete) which is due to

1. Low nutrient content
2. High sludge age
3. Low F/M ratio (<3/d)

8.20 (a)

Nitrification:
- Nitrosomonas Bacteria : NH_3 to NO_2^-
- Nitrobacter Bacteria : NO_2^- to NO_3^-

Denitrification:
- Paracoccus Denitrificant : NO_3^- to N_2

8.21 (c)

8.22 (b)

The hydraulic retention time is given by,

$$t = \frac{V}{Q} \times 24$$

where V = Volume of aeration tank
Q = Quantity of waste water flow into the aeration tank excluding the quantity of recycled sludge

$$\therefore \quad t = \frac{200}{1 \times 10^6 \times 10^{-3}} \times 24 = 4.8 \text{ h}$$

8.23 (d)

The average time for which the biomass stays in the system is known as sludge age (θ_c) and it is given by,

$$\theta_c = \frac{\text{Mass of biomass in the system}}{\text{Mass of biomass leaving the system per day}}$$

$$\Rightarrow \quad \theta_c = \frac{200 \times 10^3 \times 2000}{50 \times 10^6}$$

$$\Rightarrow \quad \theta_c = 8 \text{ days}$$

8.24 (d)

Biologically active or biodegradable components of wastes oxidise by bacteria. Food wastes and garden trimmings, both are organic wastes which are biodegradable.

8.25 (a)

In unit operation, only physical changes takes place and not chemical reactions.
Whereas in unit process, only chemical changes like oxidation, reduction, etc, takes place.

8.26 (c)

Organic matter stabilized per day

$= Q[S_i - S_o] \times \eta$

$= 500 \times 10^3 (2000 - 400) \times 0.8 = 640$ kg

$\because$ 1 kg BOD generates 0.35 m^3 methane

$\therefore$ 640 kg BOD generates $0.35 \times 640 = 224$ m^3

8.27 (b)

8.28 (c)

2% of solids + 98% water → 100% of sludge
let x be the density of sludge.

$$\text{Volume} = \frac{\text{Mass}}{\text{Density}}$$

$$\therefore \quad \frac{100}{x} = \frac{98}{1000} + \frac{2}{2.2 \times 1000}$$

...{Harmonic mean}

$$\therefore \quad x = 1011 \text{ kg/ms}$$

8.29 (a)

Now,

$$V_1 = \text{Volume of sludge (initial)}$$
$$= \text{Volume of solids} + \text{Volume of water}$$
$$= \frac{2}{2.2 \times 1000} + \frac{98}{1000}$$
$$V_1 = 0.0989 \text{ m}^3$$

Volume of sludge after thickness

$$= V_2 = \frac{V_1}{2}$$
$$V_2 = 0.04945 \text{ m}^3$$
$$\therefore \quad 0.04945 = \text{Vol. of solids} + \text{Vol. of water}$$
$$\therefore \quad 0.04945 = \frac{2}{2.2 \times 1000} + \frac{x}{1000}$$
$$\therefore \quad x = 48.545\%$$

∴ Solids content (% by Wt.)

$$= \frac{2}{2 + 48.545} \times 100 = 3.957\% \approx 3.96\%$$

8.30 (c)

Given,

Flow rate Q_0 = 500 m^3/hour
Influent BOD S_0 = 150 mg/lit
effluent BOD, = 10 mg/lit.

Hydraulic retention time = 8 hour = $\frac{1}{3}$ day

Mean – cell resistance time,
(θ_C) = 240 hour = 10 day
M · L · S · S, (X) = 2000 mg/lit.

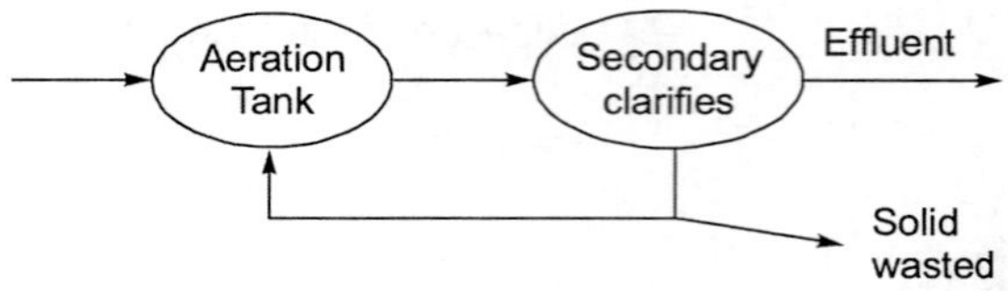

$$\frac{F}{M} = \frac{\text{Mass of BOD applied per day}}{\text{Mass of biomass}}$$

$$\Rightarrow \quad \frac{F}{M} = \frac{Q_0 S_0}{VX} = \frac{500 \times 150}{4000 \times 2000} \times 24$$

$$= 0.225 \text{ kg Biomass per day}$$

8.31 (c)

Mean cell recidence time,

$$\theta_C = \frac{\text{Mass of biomass}}{\text{Mass of biomass wasted/day}}$$

$$\Rightarrow \quad \theta_C = \frac{VX}{Q_w X_u}$$

$$\Rightarrow \quad Q_w X_u = \frac{VX}{\theta_c} = \frac{4000 \times 2000 \times 10^{-3}}{10}$$

$$= 800 \text{ kg/day}$$

8.32 (a)

8.33 Sol.

$$\underset{16g}{CH_4} + 2O_2 \rightarrow \underset{44g}{CO_2} + 2H_2O$$

⇒ 16 g of CH_4 when completely oxidized leads to 44 g of CO_2

⇒ 1 kg of CH_4 when completely oxidized leads to

$$\frac{44}{16} \times 1 = 2.75 \text{kg} CO_2$$

8.34 (c)

Kjehldahl Nitrogen = Organic nitrogen + free ammonia.

8.35 Sol.

$$V_0 = \frac{Q}{BL}$$
$$Q = (32.5 \text{ m}^3/\text{d/m}^2) \times 32.5 \times 8$$
$$= 8450 \text{ m}^3/\text{d}$$

Weir length = 75 m

∴ Weir loading rate,

$$q = \frac{8450}{75} = 112.67 \text{ m}^3/\text{d/m}$$

8.36 (a)

8.37 Sol.

$Q = 0.8 \text{ m}^3/\text{d}$

Detention time,

$$D_t = \frac{V}{Q} = \frac{4 \times 2 \times 2 \text{ m}^3}{0.8 \text{ m}^3/\text{day}} = 20 \text{ days}$$

$$L_t = \frac{L_0}{1 + k_e t} = \frac{54}{1 + 0.1 \times 20} = 18 \text{ mg}/l$$

8.38 (b)

Activated sludge process is designed primarily for satisfaction of carbonaceous BOD which is done by heterotrophs.

8.39 Sol.

$$\theta_C = \frac{VX}{(Q_0 - Q_w)Xe + Q_w X_u}$$

$$= \frac{Q_0 HRT\, X}{(Q_0 - Q_w)X_e + Q_w X_u} \quad (X_e = 0)$$

$$= \frac{15000 \times \frac{2}{24} \times 3000}{50 \times 10000} = 7.5 \text{ days}$$

8.40 Sol.

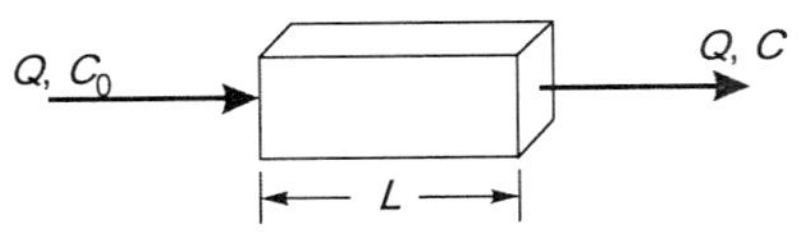

For (CMFR) completely mixed flow reactor

$$c = \frac{c_0}{1 + kt}$$

For (PFR) plug flow reactor

$$c = c_0 e^{-kt}$$

As $c/c_0 = 0.8$

For CMFR $0.8 = \dfrac{1}{1 + kt_{CMFR}}$

$$\Rightarrow \quad t_{CMFR} = \frac{0.25}{k} \quad \text{...(i)}$$

For PFR $0.8 = e^{-kt_{PFR}}$

$$\Rightarrow \quad t_{PFR} = \frac{0.22314}{k} \quad \text{...(ii)}$$

Now for steady state

$v =$ constant

and $L = vt$

So,

$$\frac{L_{CMFR}}{L_{PFR}} = \frac{vt_{CMFR}}{vt_{PFR}} = \frac{0.25}{0.22314} = 1.12$$

8.41 Sol.

$Q = 9600 \text{ m}^3/\text{d}, \quad X = 3000 \text{ mg}/l$

$X_u = 12000 \text{ mg}/l$

Solids loading rate or solid flux = 3.2 kg/m²-h

Surface area,

$$A = \frac{\text{Total quantity of solids entering}}{\text{Solid flux}}$$

$$A = \frac{9600 \text{ m}^3/\text{d} \times 3000 \text{ mg}/l}{3.2 \times 10^6 \text{mg/m}^2\text{-h}}$$

$$= \frac{9600 \times 10^3 \ l/d \times 3000 \text{ mg}/l}{3.2 \times 10^6 \text{ mg/m}^2\text{-h} \times 24 \ h/d} = 375 \text{ m}^2$$

8.42 Sol.

Minimum length of chamber

$L_{min} = V_{flow} \times t_d$

$V_{flow} = 0.30$ m/s

$$t_d = \frac{H}{V_{settling}} = \frac{0.75}{2.5 \times 10^{-2}} = 30 \text{ sec}$$

$\Rightarrow \quad L_{min} = 0.3 \text{ m/sec} \times 30 \text{ sec} = 9 \text{ m}$

8.43 Sol.

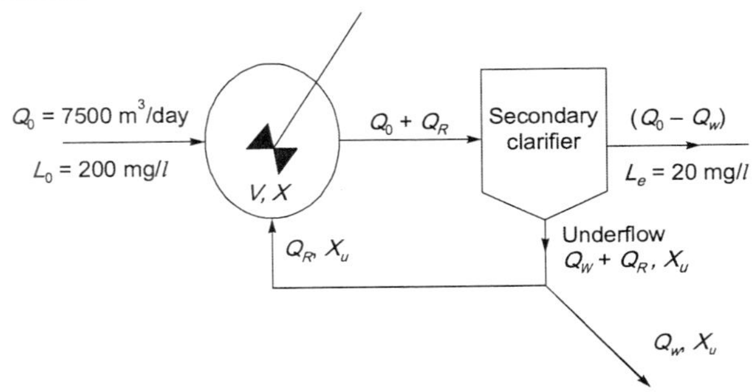

Given, $X = 3000 \text{ mg}/l, \quad X_u = 12000 \text{ mg}/l$

Since $\text{HRT} = \dfrac{V}{Q_0}$

$\Rightarrow$ Volume of reactor,

$V = Q_0 \times \text{HRT}$

$= 7500 \times 0.2 \text{ m}^3 = 1500 \text{ m}^3$

Sludge age,

$$\theta_C = \frac{VX}{(Q_0 - Q_w)X_e + Q_w X_u} \quad (X_e \simeq 0)$$

$$10 = \frac{1500 \times 3000}{Q_w \times 12000}$$

$Q_w = 37.5 \text{ m}^3/\text{day}$

8.44 Sol.

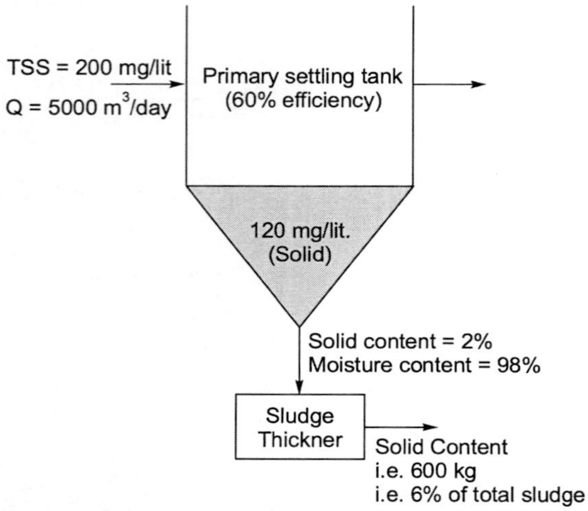

Solid content in thickener

$$= \frac{5000 \times 10^3 l/d \times 120 \text{ mg/lit}}{10^6} = 600 \text{ kg/day}$$

i.e., 6% of total mass of sludge

$\therefore$ Total mass of sludge

$$= \frac{600 \times 100}{6} = 10000 \text{ kg/d}$$

$\therefore$ Volume of sludge

$$= \frac{M_{sludge}}{\rho_{sludge}} = \frac{10000}{1000} = 10 \text{ m}^3\text{/day}$$

8.45 (d)

Reverse osmosis –Concentration polarization
Trickling filter –Ponding (Operational troubb)
Coagulation –Charge Neutralization
Adsorption –Freundlich Isotherm

8.46 Sol.

Given: $Q = 0.5 \text{ m}^3\text{/sec}$

$$\frac{F}{M} = 0.2 \text{ mg/mg–d}$$

$$S_0 = 150 \text{ mg/L}$$

$$X = 2000 \text{ mg/L}$$

Now, $$\frac{F}{M} = \frac{Q_0 S_0}{V \cdot X}$$

$$0.2\, d^{-1} = \frac{0.5 \frac{\text{m}^3}{\text{sec.}} \times 86400 \frac{\text{sec}}{d} \times 150 \frac{\text{mg}}{L}}{V \times 2000 \frac{\text{mg}}{L}}$$

$$V = 16200 \text{ m}^3$$

8.47 (a, c, d)

8.48 (b, c)

■■■■

9 Municipal Solid Waste Management

9.1 The composition of a certain MSW sample and specific weights of its various components are given below:

Component	Percent by weight	Specific weight (kg/m^3)
Food	50	300
Dirt and Ash	30	500
Plastics	10	65
Wood and Yard waste	10	125

Specific weight (kg/m^3) of the MSW sample is

(a) 319 (b) 217

(c) 209 (d) 199 [2006 : 2 M]

9.2 50 g of CO_2 and 25 g of CH_4 are produced from the decomposition of municipal solid waste (MSW) with a formula weight of 120 g. What is the average per capita green house gas production in a city of 1 million people with a MSW production rate of 500 ton/day?

(a) 104 g/day (b) 120 g/day

(c) 208 g/day (d) 313 g/day

[2007 : 2 M]

9.3 A coastal city produces municipal solid waste (MSW) with high moisture content, high organic materials, low calorific value and low inorganic materials. The most effective and sustainable option for MSW management in that city is

(a) Composting (b) Dumping in sea

(c) Incineration (d) Landfill [2010 : 1 M]

9.4 A landfill is to be designed to serve a population of 200000 for a period of 25 years. The solid waste (SW) generation is 2 kg/person/day. The density of the un-compacted SW is 100 kg/m^3 and a compaction ratio of 4 is suggested. The ratio of compacted fill (i.e. SW + Cover) to compacted SW is 1.5. The landfill volume (in million m^3) required is ______.

[2015 : 2 M, Set-II]

9.5 Solid waste generated from an industry contains only two components, X and Y as shown in the table below:

Component	Composition (% weight)	Density (kg/m^3)
X	c_1	ρ_1
Y	c_2	ρ_2

Assuming $(c_1 + c_2) = 100$, the composite density of the solid waste (ρ) is given by

(a) $\dfrac{100}{\left(\dfrac{c_1}{\rho_1}+\dfrac{c_2}{\rho_2}\right)}$

(b) $100\left(\dfrac{\rho_1}{c_1}+\dfrac{\rho_2}{c_2}\right)$

(c) $100\,(c_1\rho_1 + c_2\rho_2)$

(d) $100\left(\dfrac{\rho_1\rho_2}{c_1\rho_1+c_2\rho_2}\right)$ [2015 : 1 M, Set-I]

9.6 The composition of a municipal solid waste sample is given below:

Component	Percent by Mass	Moisture content (%)	Energy content (kJ/kg on as discarded basis)
Food waste	20	70	2500
Paper	10	4	10000
Cardboard	10	4	8000
Plastics	10	1	14000
Garden trimmings	40	60	3500
Wood	5	20	14000
Tin Cans	5	2	100

The difference between the energy content of the waste sample calculated on dry basis and as discarded basis (in kJ/kg) would be ______.

[2017 : 2 M, Set-II]

9.7 There are 20,000 vehicles operating in a city with an average annual travel of 12,000 km per vehicle. The NO_x emission rate is 2.0 g/km per vehicle. The total annual release of NO_x will be

(a) 4,80,000 kg (b) 4,800 kg

(c) 480 kg (d) 48 kg

[2018 : 1 M, Set-I]

9.8 A city generates 40×10^6 kg of municipal solid waste (MSW) per year, out of which only 10% is recovered/recycled and the rest goes to landfill.

The landfill has a single lift of 3 m height and is compacted to a density of 550 kg/m^3. If 80% of the landfill is assumed to be MSW, the landfill area (in m^2, up to one decimal place) required would be ______.

[2018 : 1 M, Set-I]

9.9 A waste activated sludge (WAS) is to be blended with green waste (GW). The carbon (C) and nitrogen (N) contents, per kg of WAS and GW on dry basis are given in the table.

Parameter	WAS	GW
Carbon (g)	54	360
Nitrogen (g)	10	6

The ratio of WAS to GW required (up to two decimal places) to achieved a blended C:N ratio of 20:1 on dry basis is ________.

[2018 : 2 M, Set-I]

9.10 A coal containing 2% sulfur is burned completely to ash in a brick kiln at a rate of 30 kg/min. The sulfur content in the ash was found to be 6% of the initial amount of sulfur present in the coal fed to the brick kiln. The molecular weights of *S*, *H* and *O* are 32, 1 and 16 g/mole, respectively. The annual rate of sulfur dioxide (SO_2) emission from the kiln (in tonnes/year, up to two decimal places) is ________.

[2018 : 2 M, Set-II]

9.11 Raw municipal solid waste (MSW) collected from a city contains 70% decomposable material that can be converted to methane. The water content of the decomposable material is 35%. An elemental analysis of the decomposable material yields the following mass percent.

C : H : O : N : other = 44 : 6 : 43 : 0.8 : 6.2

The methane production of the decomposable material is governed by the following stoichiometric relation

$$C_aH_bO_cN_d + nH_2O \rightarrow mCH_4 + sCO_2 + dNH_3$$

Given atomic weights: C = 12, H = 1, O = 16, N = 14. The mass of methane produced (in grams, round off to 1 decimal place) per kg of raw MSW will be________.

[2019 : 2 M, Set-II]

9.12 A waste to energy plant burns dry solid waste of composition : Carbon = 35%, Oxygen = 26%, Hydrogen = 10%, Sulphur = 6%, Nitrogen = 3% and Inerts = 20%. Burning rate is 1000 tonnes/d. Oxygen in air by weight is 23%. Assume complete conversion of Carbon to CO_2, Hydrogen to H_2O, Sulphur to SO_2 and Nitrogen to NO_2.

Given Atomic weights: H = 1, C = 12, N = 14, O = 16, S = 32.

The stoichiometric (theoretical) amount of air (in tonnes/d, round off to the nearest integer) required for complete burning of this waste, is _________.

[2020 : 2 M, Set-II]

9.13 Which one of the following statements is correct?

(a) Combustion is an exothermic process, which takes place in the absence of oxygen.

(b) Pyrolysis is an exothermic process, which takes place in the absence of oxygen.

(c) Pyrolysis is an endothermic process, which takes place in the absence of oxygen.

(d) Combustion is an endothermic process, which takes place in the abundance of oxygen.

[2021 : 1 M, Set-I]

9.14 Determine the correctness or otherwise of the following Assertion [A] and the Reason [R].

Assertion [A]: One of the best ways to reduce the amount of solid wastes is to reduce the consumption of raw materials,

Reason [R]: Solid wastes are seldom generated when raw materials are converted to goods for consumption.

(a) Both [A] and [R] are true and [R] is the correct reason for [A].

(b) Both [A] and [R] are false.

(c) [A] is true but [R] is false.

(d) Both [A] and [R] are true but [R] is not the correct reason for [A]

[2021 : 1 M, Set-II]

9.15 To design an optimum municipal solid waste collection route, which of the following is/are NOT desired:

(a) Waste collection on congested roads, should not occur during rush hours in morning or evening,

(b) The last collection point on a route should be as close as possible to the waste disposal facility.

(c) Collection should occur in the uphill direction.

(d) Collection vehicle should not travel twice down the same street in a day.

[2022 : 1 M, Set-II]

9.16 In a solid waste handling facility, the moisture contents (MC) of food waste, paper waste, and glass waste were found to be MCf. MCp, and MCg, respectively. Similarly, the energy contents (EC) of plastic waste, food waste, and glass waste were found to be ECpp. ECf, and ECg. respectively. Which of the following statements) is/are correct?

(a) MCf < MCp < MCg

(b) ECpp < ECf < ECg

(c) ECpp > ECf > ECg

(d) MCf > MCp > MCg

[2022 : 2 M, Set-II]

9.17 At a municipal waste handling facility, 30 metric ton mixture of food waste, yard waste, and paper waste was available. The moisture content of this mixture was found to be 10%. The ideal moisture content for composting this mixture is 50%, The amount of water to be added to this mixture to bring its moisture content to the ideal condition is _______ metric ton, (in integer)

[2022 : 2 M, Set-II]

9.18 Which of the following statements is/are TRUE for the Refuse-Derived Fuel (RDF) in the context of Municipal Solid Waste (MSW) management?

(a) Higher Heating Value (HHV) of the unprocessed MSW is higher than the HHV of RDF processed from the same MSW

(b) RDF can be made in the powdered form

(c) Inorganic fraction of MSW is mostly converted to RDF

(d) RDF cannot be used in conjunction with oil

[2023 : 1 M, Set-I]

9.19 The composition and energy content of a representative solid waste sample are given in the table. If the moisture content of the waste is 26%, the energy content of the solid waste on dry-weight basis is MJ/kg (round off to one decimal place).

Components	% by Mass	Energy content as - discarded basis [MJ/kg]
Food waste	20	4.5
Paper	45	16
Card board	5	14
Plastic	10	32
Other	20	8.0

[2023 : 2 M, Set-I]

9.20 In the context of Municipal Solid Waste Management, 'Haul' in 'Hauled Container System operated in conventional mode' includes the ________ .

(a) time spent by the transport truck at the disposal site

(b) time spent by the transport truck in traveling between a pickup point and the disposal site with a loaded container

(c) time spent by the transport truck in picking up a loaded container at a pickup point

(d) time spent by the transport truck in driving from the depot to the first pickup point

[2023 : 1 M, Set-II]

9.21 The following table gives various components of Municipal Solid Waste (MSW) and a list of treatment/separation techniques.

Component of MSW

P Ferrous metals

Q Aluminum and copper

R Food waste

S Cardboard

Treatment/separation technique

i. Incineration

ii. Rapid composting

iii. Eddy current separator

iv. Magnetic separator

(a) P-iv. Q-iii, R-ii, S-i

(b) P-iii, Q-iv, R-ii, S-i

(c) P-iii, Q-iv, R-i, S-ii

(d) P-iv, Q-iii, R-i, S-ii

[2024 : 1 M, Set-I]

9.22 An organic waste is represented as $C_{240}O_{200}H_{180}N_5S$.

(Atomic weights: S-32, H-1, C-12, O-16, N-14).

Assume complete conversion of S to SO_2 while burning.

SO_2 generated (in grams) per kg of this waste is ________

(rounded off to 1 decimal place).

[2024 : 2 M, Set-II]

■■■■

Answers Municipal Solid Waste Management

9.1 (b)	9.2 (d)	9.3 (a)	9.4 (13.68)	9.5 (a)	9.6 (3870)	9.7 (a)
9.8 (27273.7)	9.9 (1.64)	9.10 (592.88)	9.11 (137.6)	9.12 (6965)	9.13 (c)	9.14 (c)
9.15 (c)	9.16 (c, d)	9.17 (24)	9.18 (b)	9.19 (18.4)	9.20 (b)	9.21 (a)
9.22 (10.1)						

Explanations Municipal Solid Waste Management

9.1 (b)

Let specific weight of MSW sample is X kg/m^3

Now, $\rho = \dfrac{\text{Total weight}}{\text{Total volume}}$

$$\frac{100}{X} = \frac{50}{300}+\frac{30}{500}+\frac{10}{65}+\frac{10}{125}$$

$$\therefore \quad \frac{100}{X} = 0.4605$$

$$\therefore \quad X = 217.15 \text{ kg/m}^3 \approx 217 \text{ kg/m}^3$$

9.2 (d)

We know that both CO_2 and CH_4 are green house gases.

$\therefore$ Total green house gases produced by 120 gm of MSW

$= 50 \text{ g } CO_2 + 25 \text{ g of } CH_4 = 75 \text{ gm}$

500 ton of MSW will produce green house gases

$$= 500 \times 10^6 \times \frac{75}{120} = 312.5 \times 10^6 \text{ g/day}$$

Per capita average production of green house gases

$$= \frac{312.5\times10^6}{1\times10^6} \text{ g/day} = 312.5 \text{ g/day} \approx 313 \text{ g/day}$$

9.3 (a)

Incineration can be adopted when the calorific value of the MSW is high. Land fill can be adopted when the density of MSW is high. Since in the given MSW the quantity of inorganic material is low, its density is less. Composting can be adopted when the MSW contains high organic content. Barging the MSW into the sea is now a days generally not used and has become obsolete.

9.4 Sol.

Solid waste generated per day

$= 2 \times 200000 = 400000$ kg

Total solid waste generated in 25 years

$= 400000 \times 365 \times 25 = 3.65 \times 10^9$ kg

Compaction ratio,

$$4 = \frac{\text{Volume before compaction}}{\text{Volume after compaction}}$$

Volume of waste before compaction,

$$V = \frac{3.65\times10^9}{100} = 3.65\times10^7 \text{m}^3$$

Volume of waste after compaction,

$$V' = \frac{3.65\times10^7}{4} = 9.125\times10^6 \text{ m}^3$$

$$\frac{\text{SW + cover}}{\text{SW}} = \frac{\text{SW}}{\text{SW}}+\frac{\text{cover}}{\text{SW}} = 1.5$$

$$\frac{\text{Cover}}{\text{SW}} = 0.5$$

$$\text{Cover} = 0.5 \times 9.125 \times 10^6 = 4.5625 \times 10^6 \text{ m}^3$$

$$\text{Total volume} = \text{SW + Cover} = (9.125 + 4.5625) \times 10^6 \text{ m}^3 = 13.6875 \text{ million m}^3$$

9.5 (a)

Let density of sludge is ρ,

$$\therefore \quad \frac{c_1+c_2}{\rho} = \frac{c_1}{\rho_1}+\frac{c_2}{\rho_2}$$

Assuming $c_1 + c_2 = 100$

$$\therefore \quad \rho = \frac{100}{\frac{c_1}{\rho_1}+\frac{c_2}{\rho_2}}$$

9.6 Sol.

Component	Percent by Mass	Moisture content (%)	Dry content (in %)	Energy as discarded (kJ/kg)	Total dry mass	Total energy as discarded (in kJ/kg)
Food waste	20	70	30	2500	6	500
Paper	10	4	96	10000	9.6	1000
Cardboard	10	4	96	8000	9.6	800
Plastics	10	1	99	14000	9.9	1400
Garden trimmings	40	60	40	3500	16	1400
Wood	5	20	80	14000	4	700
Tin Cans	5	2	98	100	4.9	5
					$\Sigma = 60$	$\Sigma = 5805$

Energy on dry basis: It will be the total energy when whole mass is dry.

Since in this case only 60% mass in dry.

Energy corresponding to 60% drymass = 5805 kJ/kg

$\therefore$ Energy corresponding to 100% dry mass

$$= \frac{100}{60} \times 5805 = 9675 \text{ kJ/kg}$$

So energy as on dry basis = 9675 kJ/kg

Difference = (9675 − 5805) = 3870 kJ/kg

9.7 (a)

Total no. of kms. travelled by all the vehicles

$= 20000 \times 12000$ km $= 24 \times 10^7$ km

Total NO_x emission

$= 2$ g/km $\times 24 \times 10^7$ km

$= 48 \times 10^7$ g $= 48 \times 10^4$ kg

9.8 Sol.

Total weight generated by city

$= 40 \times 10^6$ kg/year

Recovered weith = 10%

$\therefore$ Weight of MSW going into landfill

$= 0.9 \times 40 \times 10^6$ kg/year

$= 36 \times 10^6$ kg/year

Compacted density

$= 550$ kg/m^3

Compacted volume of MSW

$$= \frac{36 \times 10^6 \text{ kg/year}}{550 \text{ kg/m}^3}$$

$= 65454.5454$ m^3/year

Given, Volume of MSW

$= 0.8 \times$ Total landfill volume

$\therefore$ Total landfill volume

$$= \frac{65454.5454}{0.8} \text{ m}^3\text{/year}$$

$= 81818.18175$ m^3/year

Height of landfill = 3 m

$$\therefore \text{Area of landfill} = \frac{81818.18175}{3}$$

$= 27272.7$ m^2/year

9.9 Sol.

Let the ratio of WAS and GW be $x : 1$

According to question

$$\frac{54x + 360 \times 1}{10x + 6 \times 1} = \frac{20}{1}$$

$\Rightarrow \quad 54x + 360 = 200x + 120$

$\Rightarrow \quad 146x = 240$

$\Rightarrow \quad x = 1.64$

9.10 Sol.

Coal burned in one year

$= 30 \times 24 \times 60 \times 365$

$= 1.5768 \times 10^7$ kg

$$\text{Sulfur content} = \frac{2}{100} \times 1.5768 \times 10^7 \times 10^{-6}$$

$= 315.36$ tonnes/year

Sulfur content in ash

$$= \frac{6}{100} \times 315.36$$

$= 18.92$ tonnes/year

Sulfur converted to SO_2 = 315.36 − 18.92

= 296.44 tonnes/year

$$S + O_2 \rightarrow SO_2$$

1 mole of S is present in 1 mole of SO_2

32 gm of S is present in 64 gm of SO_2

$\therefore$ Rate of SO_2 emission

$$= \frac{64}{32} \times 296.44 = 592.88 \text{ tonnes /year}$$

9.11 Sol.

mass of MSW = 1 kg

mass of decomposable material = 0.7×10^3 = 700 gm

mass of decomposable solid = 0.65×700 = 455 gm

% mass C : H : O : N : other = 44 : 6 : 43 : 0.8 : 6.2

$$\text{Moles of } C = \frac{455}{12} \times \frac{44.00}{100} = 16.68 \ (a)$$

$$H = \frac{455}{1} \times \frac{6}{100} = 27.3 (b)$$

$$Q = \frac{455}{16} \times \frac{43}{100} = 12.22 (c)$$

$$N = \frac{455}{14} \times \frac{0.8}{100} = 0.26 (d)$$

$$C_aH_bO_cN_d + \left(a - \frac{b}{4} - \frac{c}{2} + \frac{3}{4}d\right) H_2O$$

$$\rightarrow \left(\frac{a}{2} + \frac{b}{8} - \frac{c}{4} - \frac{3d}{8}\right) CH_4$$

$$+ \left(\frac{a}{2} - \frac{b}{8} + \frac{c}{4} + \frac{3d}{8}\right) CO_2 + dNH_3$$

Moles of CH_4 formed by, 1 mole of decomposable material

$$= \frac{a}{2} + \frac{b}{8} - \frac{c}{4} - \frac{3d}{8}$$

$$= \frac{16.68}{2} + \frac{27.3}{8} - \frac{12.22}{4} - \frac{3}{8} \times 0.26 = 8.6$$

Mass of CH_4 formed

$= 8.6 \times 16 = 137.6$ gm

9.12 Sol.

$$\underbrace{C}_{12} + \underbrace{O_2}_{32} \longrightarrow CO_2$$

Oxygen required for 350 tonne/day

$$\frac{32}{12} \times 350 = 933.33 \text{ tonnes/day}$$

$$\underbrace{4H}_{4} + \underbrace{O_2}_{32} \longrightarrow 2H_2O$$

Oxygen required for 100 tonne/day

$$\frac{32}{4} \times 100 = 800 \text{ tonnes/day}$$

$$\underbrace{S}_{32} + \underbrace{O_2}_{32} \longrightarrow SO_2$$

Oxygen required for 60 tonne/day

$$\frac{32}{32} \times 60 = 60 \text{ tonnes/day}$$

$$\underbrace{N}_{14} + \underbrace{O_2}_{32} \longrightarrow NO_2$$

Oxygen required for 60 tonne/day

$$\frac{32}{14} \times 30 = 68.57 \text{ tonnes/day}$$

Total oxygen required

$= 933.33 + 800 + 60 + 68.57$

$= 1861.9$ tonne/day

Available O_2 in waste

$= 260$ tonne/day

Required $= 1861.9 - 260$

$= 1601.9$ tonne/day

Amount of air required

$$= \frac{1601.9}{0.23} = 6964.78 \text{ tonne/day}$$

$\simeq 6965$ tonne/day

9.13 (c)

Pyrolysis is an endothermic process as there is a substantial heat input required to raise the biomass to the reaction temperature.

9.14 (c)

9.15 (c)

9.16 (c, d)

9.17 Sol.

Total municipal waste = 30 MT

∴Moisture content present

$$= 10\% = \frac{10}{100} \times 30 = 3 \text{ MT}$$

∴ Solid mass = 30 − 3 = 27 MT

Ideally, for composting moisture content of mix = 50% (given)

Let, total mass of waste ideally is x

∴ $x = 2.7 + 0.5;\quad x = 54$

∴ Total mass of waste iideally for composting = 54 MT

Amount of water in ideal situtation

$$= \frac{50}{100} \text{ of } x = 0.5 \times 54 = 27 \text{ MT}$$

∴Amount of water added = 27 − 3 = 24 MT

(∴ 3 MT water is already added)

9.18 (b)

9.19 Sol.

Energy content (as discarded basis):

[0.2 × 4.5 + 0.45 × 16 + 0.05 × 14 + 0.1 × 32 + 0.2 × 8] = 13.6 MJ/kg

Energy content (on dry basis) for moisture content of 26%:

$$= \frac{13.6 \text{ MJ/kg} \times 100\%}{(100-26)\%} = 18.38 \text{MJ/kg}$$

$\simeq 18.4$ MJ/kg

9.20 (b)

'Haul' in hauled container system operated in conventional mode includes the time required to reach the disposal site, starting after a container whose contents are to be emptied has been loaded on the truck, plus the time after leaving the disposal site until the truck arrives at the location where the empty container is to be redeposited.

9.21 (a)

9.22 Sol.

Molecular weight of $C_{240}O_{200}H_{180}N_5$ S is:

12 × 240 + 16 × 200 + 1 × 180 + 14 × 5 + 32 = 6362 gm

$$SO_2 \text{ generated} = \frac{64}{6362} \times 1000 \text{ gm/kg}$$

$= 10.059$ gm/kg $\simeq 10.1$ gm/kg

10 Air and Noise Pollution

10.1 The cumulative noise power distribution curve at a certain location is given below.

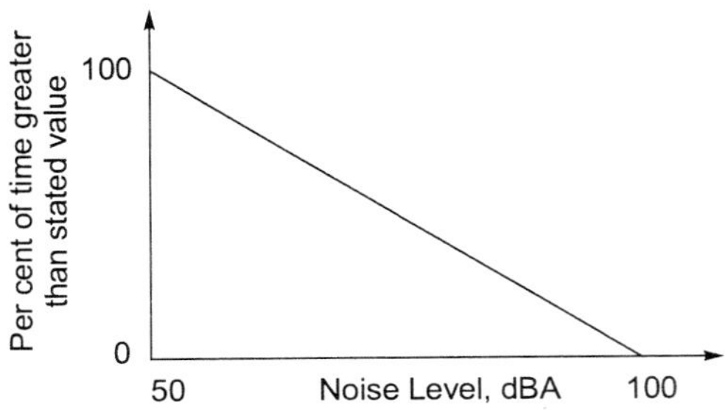

The value of L_{40} is equal to

(a) 90 dBA (b) 80 dBA
(c) 70 dBA (d) 60 dBA [2006 : 1 M]

10.2 The mean indoor airborne Chloroform ($CHCl_3$) concentration in a room was determined to be 0.4 μg/m³.
Use the following data: T = 293K, P = 1 atmosphere, $R = 82.05 \times 10^{-6}$ atm-m³/mol-K.

Atomic weights: C = 12, H = 1, Cl = 35.5. This concentration expressed in parts per billion (volume basis, ppbv) is equal to

(a) 1.00 ppbv (b) 0.20 ppbv
(c) 0.10 ppbv (d) 0.08 ppbv

[2006 : 2 M]

10.3 The dispersion of pollutants in atmosphere is maximum when

(a) environmental lapse rate is greater than adiabatic lapse rate
(b) environmental lapse rate is less than adiabatic lapse rate
(c) environmental lapse rate is equal to adiabatic lapse rate
(d) maximum mixing depth is equal to zero

[2007 : 1 M]

10.4 Two electrostatic precipitators (ESPs) are in series. The fractional efficiencies of the upstream and downstream ESPs for size d_p are 80% and 65%, respectively. What is the overall efficiency of the system for the same d_p?

(a) 100% (b) 93%
(c) 80% (d) 65% [2007 : 2 M]

10.5 Two primary air pollutants are

(a) sulphur oxide and ozone
(b) nitrogen oxide and peroxyacetylnitrate
(c) sulphur oxide and hydrocarbon
(d) ozone and peroxyacetylnitrate

[2008 : 1 M]

10.6 The reference pressure used in the determination of sound pressure level is

(a) 20 μPa (b) 20 dB
(c) 10 μPa (d) 10 dB [2009 : 1 M]

10.7 Particulate matter (fly ash) carried in effluent gases from the furnaces burning fossil fuels are better removed by

(a) Cotton bag house filter
(b) Electrostatic precipitator (ESP)
(c) Cyclone
(d) Wet scrubber [2009 : 1 M]

10.8 Match **List-I** with **List-II** and select the correct answer by using the codes given below the lists:

	List-I		List-II
A.	Coriolis effect	1.	Rotation of earth
B.	Fumigation	2.	Lapse rate and vertical temp. profile
C.	Ozone layer	3.	Inversion
D.	Max. mixing depth (mixing height)	4.	Dobson

Codes:

	A	B	C	D
(a)	2	1	4	3
(b)	2	1	3	4
(c)	1	3	2	4
(d)	1	3	4	2

[2009 : 2 M]

10.9 According to the Noise Pollution (Regulation and Control) Rules, 2000, of the Ministry of Environment and Forests, India, the day time and night time noise level limits in ambient air for residential area expressed in dB(A) L_{eq} are

(a) 50 and 40 (b) 55 and 45
(c) 65 and 55 (d) 75 and 70

[2010 : 1 M]

10.10 An air parcel having 40° temperature moves from ground level to 500 m elevation in dry air following the "adiabatic lapse rate". The resulting temperature of air parcel at 500 m elevation will be

(a) 35°C (b) 38°C
(c) 41°C (d) 44°C [2010 : 1 M]

10.11 Consider four common air pollutants found in urban environments, NO, SO_2, Soot and O_3. Among these which one is the secondary air pollutant?

(a) O_3 (b) NO
(c) SO_2 (d) Soot [2011 : 1 M]

10.12 Total suspended particularte matter (TSP) concentration in ambient air is to be measured using a high volume sampler. The filter used for this purpose had an initial dry weight of 9.787 g. The filter was mounted in the sampler and the initial air flow rate through the filter was set at 1.5 m^3/min. Sampling continued for 24 hours. The airflow after 24 hours was measured to be 1.4 m^3/min. The dry weight of the filter paper after 24 hour sampling was 10.283 g. Assuming a linear decline in the air flow rate during sampling, what is the 24 hour average TSP concentration in the ambient air?

(a) 59.2 $\mu g/m^3$ (b) 118.6 $\mu g/m^3$
(c) 237.5 $\mu g/m^3$ (d) 574.4 $\mu g/m^3$

[2011 : 2 M]

10.13 Elevation and temperature data for places are tabulated below:

Elevation 'm'	Temp. °C
4	21.25
444	15.70

based on data, lapse rate can be referred as

(a) super-adiabatic (b) sub-adiabatic
(c) neutral (d) inversion

[2013 : 2 M]

10.14 The two air pollution control devices that are usually used to remove very fine particles from the flue gas are

(a) Cyclone and Venturi Scrubber
(b) Cyclone and Packed Scrubber
(c) Electrostatic Precipitator and Fabric Filter
(d) Settling Chamber and Tray Scrubber

[2014 : 1 M, Set-II]

10.15 The concentration of Sulfur Dioxide (SO_2) in ambient atmosphere was measured as 30 $\mu g/m^3$. Under the same conditions, the above SO_2 concentration expressed in ppm is ________.

Given: $\frac{P}{(RT)} = 41.6 \text{mol/m}^3$,

where P = Pressure; T = Temperature; R = Universal gas constant; Molecular weight of SO_2 = 64.

[2015 : 2 M, Set-I]

10.16 SO_2 and CO adversely affect

(a) oxygen carrying capacity of blood and functioning of lungs respectively.
(b) functioning of the respiratory system and brain respectively.
(c) functioning of the respiratory system and oxygen carrying capacity of blood respectively.
(d) functioning of air passages and chest respectively.

[2015 : 1 M, Set-II]

10.17 Pre-cursors to photochemical oxidants are

(a) NO_X, VOCs and sunlight
(b) SO_2, CO_2 and sunlight
(c) H_2S, CO and sunlight
(d) SO_2, NH_3 and sunlight

[2016 : 1 M, Set-I]

10.18 It was decided to construct a fabric filter, using bags of 0.45 m diameter and 7.5 m long, for removing industrial stack gas containing particulates. The expected rate of airflow into the fiter is 10 m^3/s. If the filtering velocity edge 2.0 m/min, the minimum number of bags (rounded to nearest higher integer) required for continuous cleaning operation is

(a) 27 (b) 29
(c) 31 (d) 32

[2016 : 1 M, Set-I]

10.19 An electrostatic precipitator (ESP) with 5600 m^2 of collector plate area is 96 percent efficient in treating 185 m^3/s of flue gas from a 200 MW thermal power plant. It was found that in order to achieve 97 percent efficiency, the collector plate area should be 6100 m^2. In order to increase the efficiency to 99 percent, the ESP collector plate area (expressed in m^2) would be ________.

[2016 : 2 M, Set-I]

10.20 The atmospheric layer closest to the earth surface is
(a) the mesosphere (b) the stratosphere
(c) the thermosphere (d) the troposphere
[2016 : 1 M, Set-II]

10.21 The sound pressure (expressed in μPa) of the faintest sound that a normal healthy individual can hear is
(a) 0.2 (b) 2
(c) 20 (d) 55
[2016 : 1 M, Set-II]

10.22 A noise meter located at a distance of 30 m from a point source recorded 74 dB. The reading at a distance of 60 m from the point source would be ______dB.
[2016 : 2 M, Set-II]

10.23 The equivalent sound power level (in dB) of the four sources with the noise levels of 60 dB, 69 dB, 70 dB and 79 dB is ___________.
[2017 : 2 M, Set-I]

10.24 As per Noise Pollution (Regulation and Control) Rules 2000 of India, the day time noise limit for a residential zone, expressed in dB (A) L_{eq}, is
(a) 55 (b) 65
(c) 75 (d) 85
[2017 : 1 M, Set-II]

10.25 Following observations have been made for the elevation and temperature to ascertain the stability of the atmosphere:

Elevation (in m)	Temperature (in °C)
10	15.5
60	15.0
130	14.3

The atmosphere is classified as
(a) Stable (b) Unstable
(c) Neutral (d) Inverse
[2017 : 1 M, Set-II]

10.26 Which one of the following is NOT present in the acid rain?
(a) HNO_3 (b) H_2SO_4
(c) H_2CO_3 (d) CH_3COOH
[2017 : 1 M, Set-I]

10.27 In the figures. Group I represents the atmospheric temperature profiles (*P*, *Q*, *R* and *S*) and Group II represents dispersion of pollutants from a smoke stack (1, 2, 3 and 4). In the figures of Group I, the dashed line represents the dry adiabatic lapse rate, whereas the horizontal axis represents temperature and the vertical axis represents the altitude.

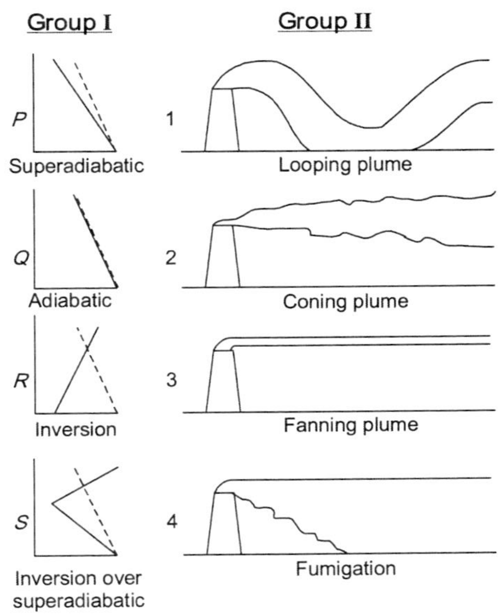

The correct match is
(a) P-1, Q-2, R-3, S-4
(b) P-1, Q-2, R-4, S-3
(c) P-1, Q-4, R-3, S-2
(d) P-3, Q-1, R-2, S-4 [2018 : 1 M, Set-II]

10.28 Which one or the following is a secondary pollutant?
(a) Carbon Monoxide
(b) Hydrocarbon
(c) Ozone
(d) Volatile Organic Carbon (VOC)
[2019 : 1 M, Set-I]

10.29 A sample of air analysed at 0°C and 1 atm pressure is reported to contain 0.02 ppm (parts per million) of NO_2. Assume the gram molecular mass of NO_2 as 46 and its volume at 0°C and 1 atm pressure as 22.4 litres per mole. The equivalent NO_2 concentration (in microgram per cubic meter, round off to 2 decimal places) would be ______.
[2019 : 2 M, Set-I]

10.30 Which one of the following options contains ONLY primary air pollutants?
(a) Ozone and peroxyacetyl nitrate
(b) Hydrocarbons and ozone
(c) Hydrocarbons and nitrogen oxides
(d) Nitrogen oxides and peroxyacetyl nitrate
[2019 : 1 M, Set-II]

10.31 A gaseous chemical has a concentration of 41.6 μmol/m^3 in air at 1 atm pressure and temperature 293 K. The universal gas constant R is 82.05×10^{-6} (m^3 atm)/(mol K). Assuming that ideal gas law is valid, the concentration of the gaseous chemical (in ppm, round off to one decimal place), is _______.

[2020 : 2 M, Set-I]

10.32 A gas contains two types of suspended particle having average sizes of 2 μm and 50 μm. Amongst the options given below, the most suitable pollution control strategy for removal of these particles is

(a) electrostatic precipitator followed by cyclonic separator
(b) bag filter followed by electrostatic precipitator
(c) settling chamber followed by bag filter
(d) electrostatic precipitator followed by venturi scrubber

[2020 : 1 M, Set-II]

10.33 A baghouse filter has to treat 12 m^3/s of waste gas continuously. The baghouse is to be divided into 5 sections of equal cloth area such that one section can be shut down for cleaning and/or repairing, while the other 4 sections continue to operate. An air-to-cloth ratio of 6.0 m^3/min-m^2 cloth will provide sufficient treatment to the gas. The individual bags are of 32 cm in diameter and 5 m in length. The total number of bags (in integer) required in the baghouse is _______.

[2021 : 2 M, Set-I]

10.34 The liquid forms of particulate air pollutants are

(a) fly ash and fumes (b) smoke and spray
(c) dust and mist (d) mist and spray

[2021 : 1 M, Set-I]

10.35 Which of the following statement(s) is/are correct?

(a) Long term exposure to the increased level of photochemical smog becomes a cause of chest constriction and irritation of the mucous membrane.
(b) Volatile organic compounds act as one of the precursors to the formation of photochemical smog in the presence of sunlight.
(c) Increased levels of carbon monoxide in the indoor environment result in the formation of carboxyhemoglobin and the long term exposure becomes a cause of cardiovascular diseases.
(d) Increased levels of volatile organic compounds in the indoor environment will result in the formation of photochemical smog which is a cause of cardiovascular diseases.

[2021 : 1 M, Set-II]

10.36 Read the statements given below:

(i) Value of the wind profile exponent for the 'very unstable' atmosphere is smaller than the wind profile exponent for the 'neutral' atmosphere.
(ii) Downwind concentration of air pollutants due to an elevated point source will be inversely proportional to the wind speed.
(iii) Value of the wind profile exponent for the 'neutral' atmosphere is smaller than the wind profile exponent for the 'very unstable' atmosphere.
(iv) Downwind concentration of air pollutants due to an elevated point source will be directly proportional to the wind speed.

Select the correct option.

(a) (i) is True and (iv) is True
(b) (iii) is False and (iv) is False
(c) (ii) is False and (iii) is False
(d) (i) is False and (iii) is True

[2021 : 2 M, Set-II]

10.37 A sample of air analyzed at 25 °C and 1 atm pressure is reported to contain 0.04 ppm of SO_2. Atomic mass of S = 32. O = 16. The equivalent SO_2 concentration (in μg/m^3) will be ________. (round off to the nearest integer)

[2022 : 1 M, Set-II]

10.38 A process equipment emits, 5 kg/h of volatile organic compounds (VOCs). If a hood placed over the process equipment captures 95% of the VOCs, then the fugitive emission in kg/h is

(a) 0.25 (b) 4.75
(c) 2.50 (d) 0.48

[2022 : 2 M, Set-II]

10.39 Which one of the following statements is TRUE for Greenhouse Gas (GHG) in the atmosphere?

(a) GHG absorbs the incoming short wavelength solar radiation to the earth surface, and allows the long wavelength radiation coming from the earth surface to pass through
(b) GHG allows the incoming long wavelength solar radiation to pass through to the earth surface, and absorbs the short wavelength radiation coming from the earth surface

(c) GHG allows the incoming long wavelength solar radiation to pass through to the earth surface, and allows the short wavelength radiation coming from the earth surface to pass through

(d) GHG allows the incoming short wavelength solar radiation to pass through to the earth surface, and absorbs the long wavelength radiation coming from the earth surface

[2023 : 1 M, Set-I]

10.40 Which of the following statements is/are TRUE in relation to the Maximum Mixing Depth (or Height) 'D_{max}' in the atmosphere?

(a) D_{max} is always equal to the height of the layer of unstable air

(b) Ventilation coefficient depends on D_{max}

(c) A smaller D_{max} will have a smaller air pollution potential if other meteorological conditions remain same

(d) Vertical dispersion of pollutants occurs up to D_{max}

[2023 : 2 M, Set-I]

10.41 For the elevation and temperature data given in the table, the existing lapse rate in the environment is _____ °C/100 m (round off to two decimal places).

Elevation from ground level (m)	Temperature (°C)
5	14.2
325	16.9

[2023 : 1 M, Set-II]

10.42 Match the following air pollutants with the most appropriate adverse health effects:

Air pollutant

(P) Aromatic hydrocarbons

(Q) Carbon monoxide

(R) Sulfur oxides

(S) Ozone

Health effect to human and/or test animal

(I) Reduce the capability of the blood to carry oxygen

(II) Bronchitis and pulmonary emphysema

(III) Damage of chromosomes

(IV) Carcinogenic effect

(a) (P) – (II), (Q) – (I), (R) – (IV), (S) – (III)

(b) (P) – (IV), (Q) – (I), (R) – (III), (S) – (II)

(c) (P) – (III), (Q) – (I), (R) – (II), (S) – (IV)

(d) (P) – (IV), (Q) – (I), (R) – (II), (S) – (III)

[2023 : 2 M, Set-II]

10.43 The primary air pollutant(s) is/are

(a) Sulphur dioxide (b) Sulphuric acid

(c) Lead (d) Ozone

[2024 : 1 M, Set-I]

10.44 For assessing the compliance with the emissions standards of incineration plants, correction needs to be applied to the measured concentrations of air pollutants. The emission standard (based on 11% Oxygen) for HCl is 50 mg/Nm^3 and the measured concentrations of HCl and Oxygen in flue gas are 42 mg/Nm^3 and 13%, respectively. Assuming 21% Oxygen in air, the CORRECT statement is:

(a) Compliance is there, as there is no need to apply the correction since Oxygen is greater than 11% and HCl emission is lesser than the emission standard.

(b) Compliance is there, as the corrected HCl emission is lesser than the emission standard

(c) No compliance, as the Oxygen is greater than 11% in the flue gas.

(d) No compliance, as the corrected HCl emission is greater than the emission standard.

[2024 : 2 M, Set-I]

10.45 What is the CORRECT match between the air pollutants and treatment techniques given in the table?

Air pollutants	Treatment techniques
P - NO_2	i - Flaring
Q - SO_2	ii - Cyclonic seperator
R - CO	iii - Lime scrubbing
S - Particles	iv - NH_3 injection

(a) P-i, Q-ii, R-iii, S-iv (b) P-ii, Q-iii, R-iv, S-i

(c) P-ii, Q-i, R-iv, S-iii (d) P-iv, Q-iii, R-i, S-ii

[2024 : 1 M, Set-II]

■■■■

Answers Air and Noise Pollution

10.1 (b)	10.2 (d)	10.3 (a)	10.4 (b)	10.5 (c)	10.6 (a)	10.7 (b)
10.8 (d)	10.9 (b)	10.10 (a)	10.11 (a)	10.12 (c)	10.13 (a)	10.14 (c)
10.15 (0.0113)	10.16 (c)	10.17 (a)	10.18 (b)	10.19 (8011.78)	10.20 (d)	10.21 (c)
10.22 (67.979)	10.23 (79.928)	10.24 (a)	10.25 (b)	10.26 (d)	10.27 (a)	10.28 (c)
10.29 (41.07)	10.30 (c)	10.31 (1)	10.32 (c)	10.33 (30)	10.34 (d)	10.35 (a, b, c)
10.36 (b)	10.37 (105)	10.38 (a)	10.39 (d)	10.40 (b, d)	10.41 (0.84)	10.42 (d)
10.43 (a, c)	10.44 (d)	10.45 (d)				

Explanations Air and Noise Pollution

10.1 (b)

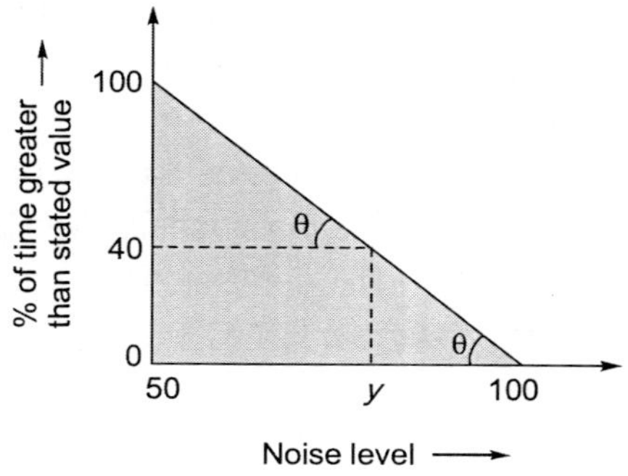

Let, $L_{40} = y$

Now, $\dfrac{100-0}{100-50} = \dfrac{100-40}{y-50}$

$\therefore \quad 2y - 100 = 60$

$\therefore \quad 2y = 160$

$\therefore \quad y = 80 \text{ dB}$

$\therefore \quad L_{40} = 80 \text{ dB}$

The L_N concept: The paameter L_N is a statistical meaure suggesting how frequently a particular sound level is exceeded. The value of L_N will indicate the sound pressure level that will exceed for N% of the gauging time.

AVOID MISTAKE

Equation of line is given by

$$\frac{x}{100} + \frac{y}{100} = 1$$

We need to calculat x and $y = 40$

$$\therefore \quad \frac{x}{100} + \frac{y}{100} = 1$$

$$\therefore \quad x = 60 \text{ dB}$$

Which is wrong becaus at origin (0, 0) is not there. Hence equation of line can not be written like this.

10.2 (d)

Method-I

The relation between mg/m^3 and ppm is,

$$1\,\mu\text{g/m}^3 = \frac{1\,\text{ppm} \times \text{molecular mass of pollutant in g/mol}}{\text{L/mol of pollutant at given temp. and pressure } (V_2)} \times 10^3\ \text{L/m}^3$$

By Avogadro's law, we know that,

$$\frac{P_1V_1}{T_1} = \frac{P_2V_2}{T_2}$$

where P_1 = 1 atmosphere;

T_1 = 0°C = 273 K

V_1 = volume at 0°C at 1 atmosphere = 22.4 L/mol

Given P_2 = 1 atmosphere;

T_2 = 293 K

$$\therefore \quad V_2 = \frac{P_1 V_1 T_2}{P_2 T_1}$$

$$V_2 = \frac{22.4 \times 293}{273}$$

$$V_2 = 24.04 \text{ L/mol}$$

$\because$ 0.4 μg/m^3 of $CHCl_3$

$$= \frac{CHCl_3 \text{ in ppm} \times 119.5 \times 10^3}{24.04}$$

$\Rightarrow$ $CHCl_3$ in ppm = 8.0469×10^{-5}

$CHCl_3$ in ppbv = $8.0469 \times 10^{-5} \times 10^3$

= 0.08 ppbv

Method-II

No. of moles in 0.4 mg

$$= \frac{\text{Weight of chloroform}}{\text{Molecular weight of chloroform}}$$

$$= \frac{0.4 \times 10^{-6} g}{119.5 \ g}$$

$$= 3.35 \times 10^{-9} \text{ moles}$$

3.35×10^{-9} moles of chloroform exist in 1 m^3 of air from ideal gas law

$$PV = nRT$$

Since $P = 1$ atm; $V = nRT$

$$v = 3.35 \times 10^{-9} \times 82.04 \times 10^{-6} \times 293$$

$$= 8.04 \times 10^{-11}$$

$$= \frac{0.08}{10^9} m^3$$

$\therefore$ Volume of chloroform = $0.08 \times 10^{-9} m^3$ in 1 m^3 of indoor air

or 0.08 m^3 in 10^9 m^3 air

$\therefore$ Concentration

$$= 0.08 \ m^3 \text{ in 1 billion} = 0.08 \text{ ppb}$$

10.3 (a)

When the environment lapse rate is more than the adiabatic lapse rate, a rising parcel of warm lighter air (pollutants) will continue to lift up; where as a parcel of heavier cooler air will continue to come down. In such circumstances the environment is unstable and the dispersion of pollutants will be rapid due to marked vertical mixing of the air. The prevailing Environmental Lapse Rate (ELR) is such a condition is known as super Adiabatic Lapse Rate (ALR).

10.4 (b)

Given data: $\eta_1 = 80\%$; $\eta_2 = 65\%$

Since the efficiency of upstream ESP is 80%, only 20% of the particulate are not removed. These remaining particulates will face the downstream ESP whose efficiency is 65%.

$\therefore$ Particulates removed by downstream,

$$\text{ESP} = 20 \times \frac{65}{100} = 13\%.$$

The two ESPs are connected in series, therefore the overall efficiency = 80 +13 =93%.

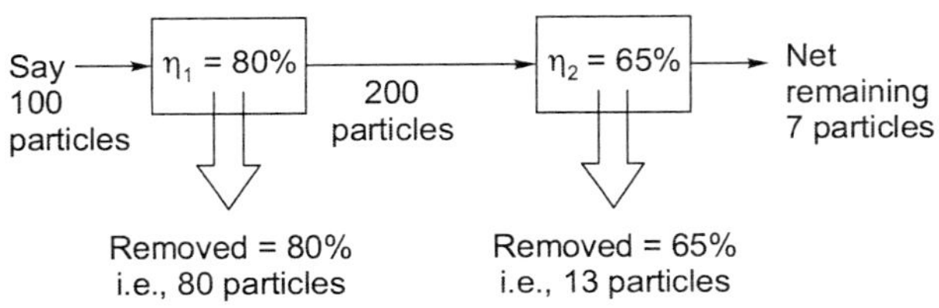

10.5 (c)

Primary pollutant is an air pollutant emitted directly from a source.

Examples: Sulphur dioxide, Carbon monoxide, Nitrogen Dioxide, lead, Hydrocarbons, Allergic Pollens, Spores, particulate matter, Methane, Ammonia, Chlorofluorocarbons, Toxic Metals etc.

Secondary pollutants are not emitted directly. The secondary pollutants form when the primary pollutants react with themselves or other component of the atmosphere.

Examples: Ground level Ozone, Smog, POPs (Persistent Organic Pollutants), Sulphuric acid, Formaldehydes and PAN (Peroxyacyl Nitrates).

10.6 (a)

The sound pressure of the faintest sound that can be heard by a normal healthy individual is about 20 μPa. Hence, this pressure is used as reference pressure in determination of sound pressure level.

10.7 (b)

Electrostatic precipitator is generally used in ***thermal power plants*** and are used for removed of ***fly ash***.

Name of Series	Min. particle size that can be removed, in μm	Efficiency (%)
Gravitational settling chamber	> 50 μm	< 50%
Cyclone collector	5.25 μm	50-90%
Wet-scrubbers (cyclonic scrubbers)	> 2.5 μm	< 80%
ESP	> 1 μm	95-99%
Fabric filters	< 1 μm	> 99%

10.8 (d)

Coriolis Effect: The rotation of earth causes an interesting phenomenon on free moving objects on earth. Objects on Northern Hemisphere are deflected to the right while on Southern Hemisphere are deflected to the left. The coriolis effect thus tries to force winds to shift towards the right or left.

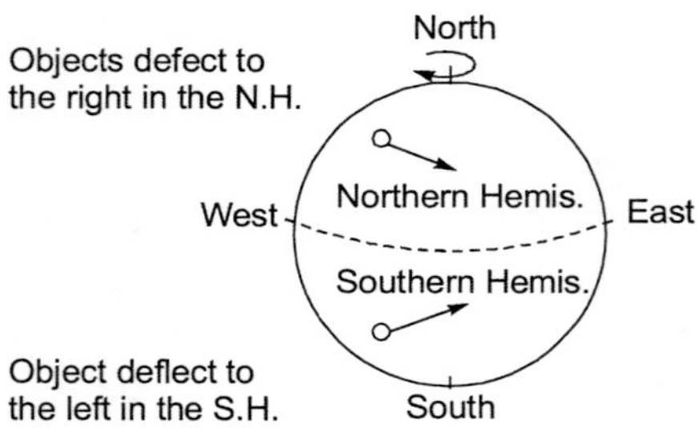

Funigation: When an inversion layer occurs at a short distance above the top of the stack and super adiabatic conditions prevail below the stack, then the plume is said to be fumigating.

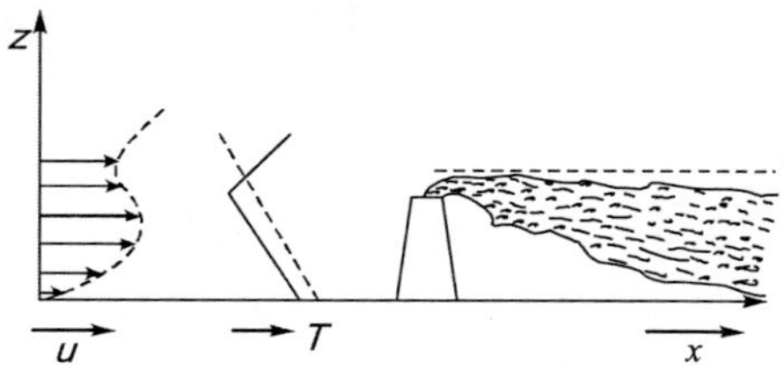

10.9 (b)

Ambient air quality standards in respect of Noise as per Noise Pollution (Regulation and Control) Rules, 2000, of the Ministry of Environment & Forests, India are tabulated below:

Category of Area/Zone	Day time	Night time
Industrial Area	75	70
Commercial Area	65	55
Residential Area	55	45
Silence Zone	50	40

10.10 (a)

Dry air cools at the rate of 9.8°C per km and it is called dry adiabatic lapse rate. In saturated (wet) air, this rate is calculated to be 6°C per km and is known as wet adiabatic lapse rate.

Resulting temperature of air

$$= 40 - 9.8 \times \frac{500}{1000} = 40 - 4.9 = 35.1°\text{C}$$

10.11 (a)

Soot is a general term that refers to impure carbon particles resulting from the incomplete combustion of hydrocarbons. Soot, as an air borne contaminant in the environment has many different sources but they are all the result of some form of **pyrolysis**.

10.12 (c)

Average airflow rate

$$= 1.5 - \left(\frac{1.5 - 1.4}{24 \times 60}\right) \times 12 = 1.45 \text{ m}^3/\text{min}$$

Total airflow through the sampler during 24 hours

$$= 1.45 \times 24 \times 60 = 2088 \text{ m}^3$$

TSP concentration in the ambient air

$$= \left(\frac{10.283 - 9.787}{2088}\right) \times 10^6 = 237.5\ \mu\text{g/m}^3$$

10.13 (a)

Ambient lapse rate

$$= \frac{21.25 - 15.70}{(444 - 4)} \times 1000$$

$$= 12.6\ °\text{C/km} > 9.8°\text{C/km}$$

When the ambient lapse rate exceeds the adiabatic lapse rate, the ambient lapse rate is said to be super adiabatic.

10.14 (c)

10.15 Sol.

1 m^3 of air has 30 μg SO_2

10^6 m^3 of air has 30g SO_2

$$30\text{g } SO_2 = \frac{30}{64} \text{mole } SO_2$$

From $PV = nRT$

$$V = \frac{nRT}{P} = \frac{n}{P/RT} = \frac{(30/64)}{41.6 \text{ mol/m}^3}$$

$$\text{Volume of } SO_2 = \frac{30}{64 \times 41.6}$$

$$= 0.01127 \text{ m}^3$$

⇒ Concentration of SO_2 in PPM = 0.0113 ppm

10.16 (c)

SO_2: The bad effects include pain when taking a deep breath, coughing, throat irritation, and breathing difficulties. High concentration of SO_2 can affect lung function, worsen asthma attacks and worsen existing heart disease in surefire groups.

CO: Carbon monoxide reduces the amount of oxygen carried by haemoglobin around the body is red blood cells. The result in that vital organs, such as the brain, nervous tissues and the heart do not receive enough oxygen to work properly.

10.17 (a)

The atmospheric conditions favourable for formation of photochemical smog are:

1. The presence of high concentration of hydrocarbons and nitrogen oxides in the atmosphere.
2. Air stagnation, as not to cause dispersal of air pollutants.
3. Abundant sunlight.

10.18 (b)

Surface area of one bag

$$= \pi \times d \times l$$
$$= \pi \times 0.45 \times 7.5 = 10.6028 \text{ m}^2$$

The required area of fabric filter

$$= \frac{10}{\left(\frac{2}{60}\right)} = 300 \text{ m}^2$$

$\therefore$Number of bag required$= \frac{300}{10.6028} = 28.294$

So No. of bag required will be 29.

10.19 Sol.

As per **Deutsch-Anderson equation**, to determine the collection efficiency of the precipitator under ideal conditions,

$$\eta = 1 - e^{-w(A/Q)}$$

η = Collection efficiency of the precipitator

w = Migration velocity, cm/s

A = The effective collecting plate area of the precipitator, m^2

Q = Gas flow through the precipitator, m^3/s

$$0.96 = 1 - e^{-\left(\frac{w}{Q}\right)\times 5600} \quad \text{...(i)}$$

$$0.97 = 1 - e^{-\left(\frac{w}{Q}\right)\times 6100} \quad \text{...(ii)}$$

From both equations,

$$\frac{w}{Q} = 5.748 \times 10^{-4}$$

Now for 99% efficiency, plate area required,

$$0.99 = 1 - e^{-5.748\times 10^{-4}\times A}$$

$$\Rightarrow \quad A = 8011.78 \text{ m}^2$$

10.20 (d)

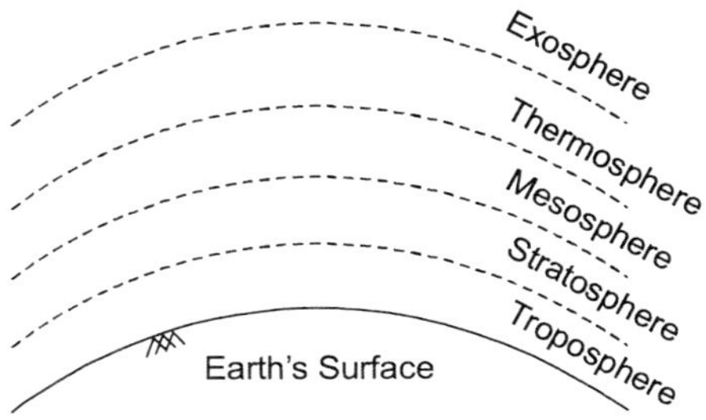

10.21 (c)

The sound pressure of the faintest sound that a normal healthy individual can hear is 20 μPa. It is taken as reference sound pressure level.

A 20 μPa pressure is 0dB on the sound pressure level scale.

10.22 Sol.

Method-I

$$L_2 = L_1 - \left|20\log\left(\frac{R_2}{R_1}\right)\right|$$

$= 74 - |\,20 \log 0.5\,| = 74 - 6.02$

$= 74 - 20 \log 0.5 = 67.98$ dB

Method-II

Since, Intensity $\propto (\text{Pressure})^2_{rms} \propto \frac{1}{r^2}$

$$(\text{Pressure})_{rms} \propto \frac{1}{r}$$

where, r = distance between point source and noise meter

$$\text{Soind in decible} = 20\log_{10}\left(\frac{p_{rms}}{20}\right)$$

$$74 = 20\log_{10}\left(\frac{p_{rms}30}{20}\right)$$

$$\therefore \quad p_{rms\ 30} = 100237.44$$

$$\text{Since} \quad p_{rms} \propto \frac{1}{r}$$

$$\frac{(P_{rms})_{60}}{(P_{rms})_{30}} = \frac{30}{60} = \frac{1}{2}$$

Noise level at 60 m

$$= 20\log_{10}\frac{(P_{rms})_{60}}{20}$$

$$= 20\left[\log_{10}\frac{(P_{rms})_{30}}{20} - \log_{10}^{2}\right] = 67.979 \text{ dB}$$

10.23 Sol.

$$L = 20\log_{10}\left(\frac{p_{rms}}{20\ \mu\text{Pa}}\right) dB$$

$$L_1 = 60 = 20\log_{10}\frac{p_{rms_1}}{20\mu\text{Pa}}$$

$$\frac{p_{rms_1}}{20\mu\text{Pa}} = 1000$$

Similarly,

$$\frac{p_{rms_2}}{20\mu\text{Pa}} = 2818.3829$$

$$\frac{p_{rms_3}}{20\mu\text{Pa}} = 3162.2776$$

$$\frac{p_{rms_4}}{20\mu\text{Pa}} = 8912.5093$$

$$\left(\frac{p_{rms}}{20\mu\text{Pa}}\right)_{eq} = \sqrt{1000^2 + (2818.3829)^2 + (3162.2776)^2 + (8912.5)^2}$$

$$= 9918.4729$$

$$L_{eq} = 20\log_{10}\frac{p_{rms}}{20\mu\text{Pa}}$$

$$= 20\log_{10}(9918.47) = 79.928 \text{ dB}$$

10.24 (a)

10.25 (b)

$$\text{ELR}_{1-2} = \frac{15.5-15}{(60-10)\times10^{-3}} = 10°\text{C/km}$$

$$\text{ELR}_{2-3} = \frac{15-14.3}{(130-60)\times10^{-3}} = 10°\text{C/km}$$

Since ELR > ALR (9.8°C/km)

Atmosphere is **unstable**.

10.26 (d)

Acid rain primarily consist of HNO_3, H_2SO_4 and H_2CO_3.

Acid rain results when SO_2, NO_x, CO_2 react with water, oxygen and other chemicals to form sulphuric acid, nitric acid and cabonic acid. Rain water with pH of about 4.0 is called as acid rain.

10.27 (a)

10.28 (c)

Ozone is considered as secondary air pollutant.

10.29 Sol.

$$1 \text{ ppm of } NO_2 = \frac{1 \text{ part of } NO_2}{10^6 \text{ parts of air}}$$

$$= \frac{1\ m^3 \text{ of } NO_2}{10^6\ m^3 \text{ of air}}$$

At STP i.e. 0°C (273°K) and 1 atm P, volume of 1 mole of NO_2 = 22.4 litres.

at STP i.e. 0°C (273°K) and 1 atm P

22.4 litres of NO_2 has mass of 46 gm

$$1 \text{ litre of } NO_2 \text{ has mass of } = \frac{46}{22.4} = 2.053 \text{ gm}$$

1000 litres or 1 m^3 has mass of = 2053.57 gm

$$\therefore\quad 1 \text{ ppm of } NO_2 = \frac{2053.57 \text{ gm}}{10^6\ m^3}$$

$$= \frac{2053.57\times10^6}{10^6}\ \mu g/m^3$$

$$1 \text{ ppm of } NO_2 = 2053.57\ \mu g/m^3$$

$$\therefore\quad 0.02 \text{ ppm of } NO_2 = 2053.57\times0.02 = 41.07\ \mu g/m^3$$

10.30 (c)

Hydrocarbons and nitrogen oxides are considered primary air pollutants.

10.31 Sol.

$$PV = nRT$$

$$V = \frac{nRT}{P} = \frac{41.6\times10^{-6}\times32.05\times10^{-6}}{1}\times293$$

$$= 10^{-6}\ m^3$$

41.6 μ mole of gas volume of $10^{-6}\ m^3$

$$\text{So, } 1 \text{ ppm} = \frac{1 \text{ part of gas}}{10^6 \text{ parts of air}} = \frac{1\ m^3 \text{ of gas}}{10^6\ m^3 \text{ of air}}$$

$$1 \text{ ppm} = \frac{41.6\times10^6\ \mu\text{moles}}{10^6 m^3}$$

So, 41.6 μ moles/m^3 = 1 ppm

10.32 (c)

10.33 Sol.

Given, discharge to be passed = 12 m^3/s

Out of 5 sections, 4 will operate at a time.

Each bag is

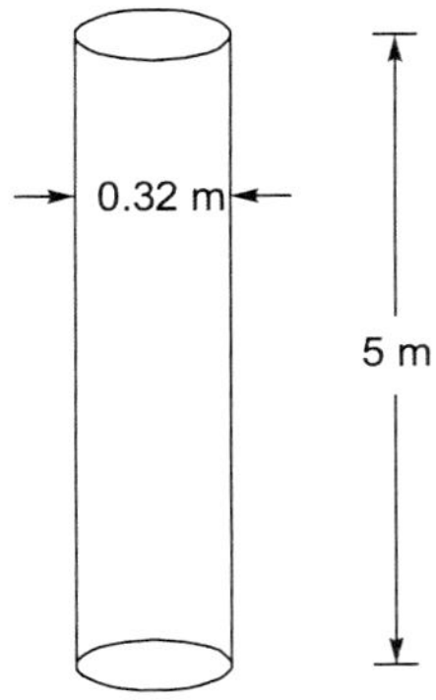

Thus, surface area of each bag

$$= \pi \times 0.32 \times 5 \text{ m}^2$$
$$= 5.0265 \text{ m}^2$$

Total surface area required wr.t. given discharge

$$= \frac{12 \text{ m}^3/\text{s}}{6 \text{ m}^3/\text{min-m}^2}$$

$$= \frac{12 \text{ m}^3/\text{s}}{\frac{6}{60} \text{ m}^3/\text{s-m}^2} = 120 \text{ m}^2$$

$\therefore$ Working bag filters required

$$= \frac{120 \text{ m}^2}{5.0265 \text{ m}^2} = 23.87$$

or 24 bag filters

But, since it is asked total (which includes a standby set also), then,

Total no. of bag filters

$$= 24 \times \frac{5}{4} = 30$$

10.34 (b)

The liquid forms of particulate air pollutants are mist and spray.

Note: Mist is a cloud made of very small drops of water in the air just above the ground which reduces the visibility.

10.35 (a, b, c)

10.36 (b)

10.37 Sol.

$\therefore$ For ideal gas,

$$PV = nRT$$

$$\Rightarrow \quad \frac{P_1V_1}{n_1T_1} = \frac{P_2V_2}{n_2T_2}$$

$\because$ Pressure is same in both case ($P_1 = P_2$)

$$\therefore \quad \frac{V_1}{T_1} = \frac{V_2}{T_2}$$

Now, $T_1 = 273$ K

and $T_2 = 273 + 25 = 298$ K

$$\therefore \quad \frac{V_1}{273} = \frac{V_2}{298}$$

$\because$ 1 mole of gas occupies 22.4 lit of volume at 0°C temperature and 1 atm pressure.

$$\therefore \quad \frac{22.4}{237} = \frac{V_2}{298}$$

$\Rightarrow$ $V_2 = 24.45$ lit

= 1 mole of gas at 25°C

$\therefore 10^6$ m^3 of air= 0.04 m^3 of SO_2

= 40 lit of SO_2

$\therefore 10^6$ m^3 of air= $\frac{40}{24.45}$ = 1.636 mole of SO_2

$\Rightarrow 10^6$ m^3 of air= 1.636 × 64 g of SO_2

$\Rightarrow$ 1 m^3 of air = 104.7 μg of SO_2

$\simeq$ 105 μg of SO_2

10.38 (a)

If hood captures 95% of VOCs then the fugitive emission is 5(1 – 0.95) = 0.25 kg/h.

10.39 (d)

10.40 (b, d)

Maximum Mixing Depth (MMD)

- The dispersion of pollutants in the lower atmosphere is greatly aided by the convective and turbulent mixing that takes place.
- The vertical extent to which this mixing takes place depends on the environmental lapse rate which varies diurnally, from season to season and is also affected by topographical features.

- The depth of the convective mixing layer in which vertical movement of pollutants is possible, is called the maximum mixing depth (MMD).
- The ventilation coefficient is the product of mixing depth and the average wind speed.

10.41 Sol.

The existing lapse rate in the environment,

$$\text{ELR} = -\frac{\Delta T}{\Delta H} = -\frac{16.9 - 14.2}{3.25 - 5}\,°\text{C/m}$$

$$\text{ELR} = 2.7 \times \frac{100}{320}\,°\text{C/100 m}$$

$$\text{ELR} = 0.84°\text{C/100 m}$$

10.42 (d)

10.43 (a, c)

10.44 (d)

As per CPCB, oxygen correction factor is calculated as shown below–

$$E_s = \frac{(21 - 0_s)E_m}{21 - 0_m} \quad ...(i)$$

Where

E_s = Calculated emission concentration in % at the standard percentage oxygen concentration in %

E_m = Measured emission concentration in %

0_s = Standard oxygen concentration in %

0_m = Measured oxygen concentration in %

Note : This correction is done only when measured % oxygen concentration is higher than standard % oxygen (i.e. 11% here)

So, $E_s = \frac{21 - 11}{21 - 13} \times 42 = 52.5$ Mg/Nm^3

Hence, no compliance and the corrected HCl emission is greater than emission standard i.e. 50 mg/Nm^3.

10.45 (d)

■■■■

Irrigation Engineering

UNIT VIII

CONTENTS

Irrigation Engineering

Syllabus

Types of irrigation systems and methods; Crop water requirements - Duty, delta, evapo-transpiration; Gravity Dams and Spillways; Lined and unlined canals, Design of weirs on permeable foundation; cross drainage structures.

Analysis of Previous GATE Papers

Exam Year	1 Mark Ques.	2 Marks Ques.	Total Marks
1995	1	–	1
1996	1	–	1
1997	–	–	–
1998	–	1	2
1999	–	–	–
2000	2	–	2
2001	–	1	2
2002	–	1	2
2003	2	1	4
2004	1	1	3
2005	2	3	8
2006	–	1	2
2007	2	1	4
2008	2	3	8
2009	1	1	3
2010	–	2	4
2011	–	–	–
2012	–	1	2
2013	–	1	2
2014 Set-1	1	–	1

Exam Year	1 Mark Ques.	2 Marks Ques.	Total Marks
2014 Set-2	–	1	2
2015 Set-1	1	–	1
2015 Set-2	–	1	2
2016 Set-1	–	1	2
2016 Set-2	1	–	1
2017 Set-1	–	1	2
2017 Set-2	–	1	2
2018 Set-1	–	–	–
2018 Set-2	1	–	1
2019 Set-1	1	–	1
2019 Set-2	1	–	1
2020 Set-1	1	–	1
2020 Set-2	1	2	5
2021 Set-1	–	1	2
2021 Set-2	–	1	2
2022 Set-1	1	1	3
2022 Set-2	–	1	2
2023 Set-1	1	–	1
2023 Set-2	1	–	1
2024 Set-1	–	1	2
2024 Set-2	–	1	2

1 Water Requirements of Crops

1.1 Delta (Δ) in cm, Duty (*D*) in hectare/cumec and Base period (*B*) in days are related as

(a) $\Delta = 864\ B/D$ (b) $B = 864\ D/\Delta$
(c) $B = 864\ \Delta/D$ (d) $D = 8.64\ B/\Delta$

[1996 : 1 M]

1.2 If duty (*D*) is 1428 hectares/cumec and base period (*B*) is 120 days for an irrigated crop, then delta (Δ) in meters is given by

(a) 102.8 (b) 0.73
(c) 1.38 (d) 0.01 [2000 : 1 M]

1.3 A tube well having a capacity of 4 m^3/hour operates for 20 hours each day during the irrigation season. How much area can be commanded if the irrigation interval is 20 days and depth of irrigation is 7 cm?

(a) $1.71 \times 10^4\ m^2$ (b) $1.14 \times 10^4\ m^2$
(c) $22.9 \times 10^4\ m^2$ (d) $2.29 \times 10^4\ m^2$

[2000 : 1 M]

1.4 A field was supplied water from an irrigation tank at a rate of 120 lit/sec to irrigate an area of 2.5 ha. The duration of irrigation is 8 hours. It was found that the actual delivery at the field, which is about 4 km from the tank, was 100 lit/sec. The runoff loss in the field was estimated as 800 m^3. the application efficiency in this situation is

(a) 62% (b) 72%
(c) 76% (d) 80% [2001 : 2 M]

1.5 A canal was designed to supply the irrigation needs of 1200 hectares of land growing rice of 140 days base period having a Delta of 134 cms. If this canal water is used to irrigate wheat of base period 120 days having a Delta of 52 cm, the area (in hectares) that can be irrigated is

(a) 2650 (b) 3608
(c) 543 (d) None of these

[2002 : 2 M]

1.6 The total irrigation depth of water, required by a certain crop in its entire growing period (150 days), is 25.92 cm. The culturable command area for a distributary channel is 100,000 hectares. The distributary channel shall be designed for a discharge

(a) less than 2 cumecs
(b) 2 cumecs
(c) 20 cumecs
(d) more than 20 cumecs [2003 : 1 M]

1.7 The moisture content of soil in the root zone of an agricultural crop at certain stage is found to be 0.05. The field capacity of the soil is 0.15. The root zone depth is 1.1 m. The consumptive use of crop at this stage is 2.5 mm/day and there is no precipitation during this period. Irrigation efficiency is 65%. It is intended to raise the moisture content to the field capacity in 8 days through irrigation. The necessary depth of irrigation is ($\gamma_d = 1.5$)

(a) 115 mm (b) 169 mm
(c) 200 mm (d) 285 mm

[2003 : 1 M]

1.8 A sprinkler irrigation system is suitable when

(a) the land gradient is steep and the soil is easily erodible
(b) the soil is having low permeability
(c) the water table is low
(d) the crops to be grown have deep roots

[2004 : 1 M]

1.9 A canal irrigates a portion of a culturable command area to grow sugarcane and wheat. The average discharges required to grow sugarcane and wheat are, respectively, 0.36 and 0.27 cumecs. The time factor is 0.9. The required design capacity of the canal is

(a) 0.36 cumecs (b) 0.40 cumecs
(c) 0.63 cumecs (d) 0.70 cumecs

[2004 : 2 M]

1.10 The culturable commanded area for a distributary is $2 \times 10^8\ m^2$. The intensity of irrigation for a crop is 40%. If kor water depth and kor period for the crop are 14 cm and 4 weeks, respectively, the peak demand discharge is

(a) 2.63 m^3/s (b) 4.63 m^3/s
(c) 8.58 m^3/s (d) 11.58 m^3/s

[2005 : 2 M]

1.11 In a cultivated area, the soil has porosity of 45% and field capacity of 38%. For a particular crop, the root zone depth is 1.0 m, the permanent wilting point is 10% and the consumptive use is 15 mm/d. If the irrigation efficiency is 60%, what should be the frequency of irrigation such that the moisture content does not fall below 50% of the maximum available moisture?

(a) 5d (b) 6d
(c) 9d (d) 15d [2006 : 2 M]

1.12 The consumptive use of water for a crop during a particular stage of growth is 2.0 mm/day. The maximum depth of available water in the root zone is 60 mm. Irrigation is required when the amount of available water is 50% of the maximum available water in the root zone. Frequency of irrigation should be

(a) 10 days (b) 15 days
(c) 20 days (d) 25 days

[2007 : 1 M]

1.13 The culturable command area for a distributed channel is 20,000 hectares. Wheat is grown in the entire area and the intensity of irrigation is 50%. The kor period for wheat is 30 days and the kor water depth is 120 mm. The outlet discharge for the distributary should be

(a) 2.85 m^3/s (b) 3.21 m^3/s
(c) 4.63 m^3/s (d) 5.23 m^3/s

[2007 : 2 M]

1.14 An outlet irrigates an area of 20 ha. The discharge (*l/s*) required at this outlet to meet the evapotranspiration requirement of 20 mm occurring uniformly in 20 days neglecting other field losses is

(a) 2.52 (b) 2.31
(c) 2.01 (d) 1.52 [2008 : 2 M]

1.15 An agricultural land of 437 ha is to be irrigated for a particular crop. The base period of the crop is 90 days and the total depth of water required by the crop is 105 cm. If a rainfall of 15 cm occurs during the base period, the duty of irrigation water is

(a) 437 ha/cumec (b) 486 ha/cumec
(c) 741 ha/cumec (d) 864 ha/cumec

[2009 : 2 M]

Common Data for Questions 1.16 and 1.17:

The moisture holding capacity of the soil in a 100 hectare farm is 18 cm/m. The field is to be irrigated when 50 percent of the available moisture in the root zone is depleted. The irrigation water is to be supplied by a pump working for 10 hours a day, and water application efficiency is 75 per cent. Details of crops planned for cultivation are as follows:

Crop	Root zone depth (m)	Peak rate of moisture use (mm/day)
X	1.0	5.0
Y	0.8	4.0

1.16 The capacity of irrigation system required to irrigate crop '*X*' in 36 hectares is

(a) 83 litres/sec (b) 67 litres/sec
(c) 57 litres/sec (d) 53 litres/sec

[2010 : 2 M]

1.17 The area of crop '*Y*' that can be irrigated when the available capacity of irrigation system is 40 litres/sec is

(a) 40 hectares (b) 36 hectares
(c) 30 hectares (d) 27 hectares

[2010 : 2 M]

1.18 Wheat crop requires 55 cm of water during 120 days of base period. The total rainfall during this period is 100 mm. Assume the irrigation efficiency to be 60%. The area (in ha) of the land which can be irrigated with a canal flow of 0.01 m^3/s is

(a) 13.82 (b) 18.85
(c) 23.04 (d) 230.40

[2012 : 2 M]

1.19 The transplantation of rice requires 10 days and total depth of water required during transplantation is 48 cm. During trans-plantation there is an effective rainfall (useful for irrigation) of 8 cm. The duty of irrigation water in hectare/cumecs is

(a) 612 (b) 216
(c) 300 (d) 108 [2013 : 2 M]

1.20 Irrigation water is to be provided to a crop in a field to bring the moisture content of the soil from the existing 18% to the field capacity of the soil at 28%. The effective root zone of the crop is 70 cm. If the densities of the soil and water are 1.3 g/cm^3 and 1.0 g/cm^3 respectively, the depth of irrigation water (in mm) required for irrigating the crop is _______.

[2014 : 2 M, Set-II]

1.21 The two columns below show some parameters and their possible values.

Parameter

P. Gross Command Area
Q. Permanent Wilting Point
R. Duty of canal water
S. Delta of wheat

Value

I. 100 hectares/cumec

II. 6°C

III. 1000 hectares

IV. 1000 cm

V. 40 cm

VI. 0.12

Which of the following options matches the parameters and the values correctly?

(a) P-I, Q-II, R-III, S-IV

(b) P-III, Q-VI, R-I, S-V

(c) P-I, Q-V, R-VI, S-II

(d) P-III, Q-II, R-V, S-IV [2015 : 1 M, Set-I]

1.22 A field channel has cultivable commanded area of 2000 hectares. The intensities of irrigation for gram and wheat are 30% and 50% respectively. Gram has a kor period of 18 days, kor depth of 12 cm, while wheat has a kor period of 18 days and a kor depth of 15 cm. The discharge (in m^3/s) required in the field channel to supply water to the commanded area during the kor period is ________.

[2015 : 2 M, Set-II]

1.23 The culturable command area of a canal is 10000 ha. The area grows only two crops-rice in the Kharif season and wheat in the Rabi season. The design discharge of the canal is based on the rice requirements, which has an irrigated area of 2500 ha, base period of 150 days and delta of 130 cm. The maximum permissible irrigated area (in ha) for wheat, with a base period of 120 days and delta of 50 cm, is ________.

[2017 : 2 M, Set-II]

1.24 The intensity of irrigation for the Kharif season is 50% for an irrigation project with culturable command area of 50,000 hectares. The duty for the Kharif season is 1000 hectare/cumec. Assuming transmission loss of 10%, the required discharge (in cumec, up to two decimal places) at the head of the canal is ________.

[2018 : 1 M, Set-II]

1.25 The command area of a canal grows only one crop. i.e., wheat. The base period of wheat is 120 days and its total water requirement, Δ, is 40 cm. If the canal discharge is 2 m^3/s, the area, in hectares, rounded off to the nearest integer, which could be irrigated (neglecting all losses) is________

[2019 : 1 M, Set-II]

1.26 The data for an agricultural field for a specific month are given below:

Pan Evaporation = 100 mm

Effective Rainfall = 20 mm (after deducting losses due to runoff and deep percolation)

Crop Coefficient = 0.4

Irrigation Efficiency = 0.5

The amount of irrigation water (in mm) to be applied to the field in that month, is

(a) 80 (b) 40

(c) 20 (d) 0

[2020 : 1 M, Set-I]

1.27 Crops are grown in a field having soil, which has field capacity of 30% and permanent wilting point of 13%. The effective depth of root zone is 80 cm. Irrigation water is supplied when the average soil moisture drops to 20%. Consider density of the soil as 1500 kg/m^3 and density of water as 1000 kg/m^3. If the daily consumptive use of water for the crops is 2 mm, the frequency of irrigating the crops (in days), is

(a) 7 (b) 13

(c) 10 (d) 11

[2020 : 2 M, Set-II]

1.28 A reservoir with a live storage of 300 million cubic metre irrigates 40000 hectares (1 hectare = 10^4 m^2) of a crop with two fillings of the reservoir. If the base period of the crop is 120 days, the duty for this crop (in hectares per cumec, round off to integer) will then be ________.

[2021 : 2 M, Set-II]

1.29 During a particular stage of the growth of a crop, the consumptive use of water is 2.8 mm/day. The amount of water available in the soil is 50 % of the maximum depth of available water in the root zone. Consider the maximum root zone depth of the crop as 80 mm and the irrigation efficiency as 70 %. The interval between irrigation (in days) will be ________. (round off to the nearest integer)

[2022 : 1 M, Set-I]

1.30 A pump with an efficiency of 80% is used to draw groundwater from a well for irrigating a flat field of area 108 hectares. The base period and delta for paddy crop on this field are 120 days and 144 cm, respectively. Water application efficiency in the field is 80%. The lowest level of water in the well is 10 m below the ground. The minimum required horse power (h.p.) of the pump is ________. (round off to two decimal places)

(Consider 1 h.p. = 746 W; unit weight of water = 9810 N/m^3)

[2022 : 2 M, Set-II]

1.31 In a certain month, the reference crop evapotranspiration at a location is 6 mm day, If the crop coefficient and soil coefficient are 1.2 and 0.8, respectively, the actual evapotranspiration in mm/day is

(a) 8.00 (b) 7.20
(c) 6.80 (d) 5.76

[2022 : 2 M, Set-II]

1.32 A canal supplies water to an area growing wheat over 100 hectares. The duration between the first and last watering is 120 days, and the total depth of water required by the crop is 35 cm. The most intense watering is required over a period of 30 days and requires a total depth of water equal to 12 cm. Assuming precipitation to be negligible and neglecting all losses, the minimum discharge (in m^3/s, rounded off to three decimal places) in the canal to satisfy the crop requirement is ____________.

[2023 : 1 M, Set-I]

Answers Water Requirements of Crops

1.1	(a)	1.2	(b)	1.3	(d)	1.4	(b)	1.5	(a)	1.6	(d)	1.7	(d)
1.8	(a)	1.9	(d)	1.10	(b)	1.11	(*)	1.12	(b)	1.13	(c)	1.14	(b)
1.15	(d)	1.16	(b)	1.17	(d)	1.18	(a)	1.19	(b)	1.20	(91)	1.21	(b)
1.22	(1.4275)	1.23	(5200)	1.24	(27.78)	1.25	(5184)	1.26	(b)	1.27	(*)	1.28	(691.2)
1.29	(14)	1.30	(30.82)	1.31	(d)	1.32	(0.046)						

Explanations Water Requirements of Crops

1.1 (a)

$\therefore \quad \Delta = 8.64\ B/D$

Here, Δ is in m B = base period in days, D = duty of crop in ha/cumec

$\Rightarrow \quad \Delta = 864\dfrac{B}{D}$ cm

POINTS TOREMEMBER

- **Duty:** It is defined as the area of land irrigated by a unit discharge of water flowing continuously for the direction of the base period of the crop.
- **Delta:** It is the total depth of water in cm required by a crop to come to maturity.
  It is denoted by Δ.
- **Derivation:**
  Let base period of crop be 'B' days.
  1 cumec of water B applied for 'B' days.
  Then, volume of water,
  $$V = 1 \times (B \times 24 \times 60 \times 60) = 86400\ B \text{ cubic meters}$$
  Let V cumic meter of water meter D ha of land
  $$D \text{ ha of land} = D \times 10^4\ m^2$$

$$\therefore \quad \Delta = \frac{V}{A} = \frac{86400B}{D\times 10^4}$$
$$= \frac{8.64B}{D} \text{ (in meters)}$$
$$= \frac{864B}{D} \text{ (in centimeters)}$$

1.2 (b)

$$D = 1428 \text{ Ha/cum}$$
$$B = 120 \text{ days}$$
$$\Delta = \frac{8.64B}{D} = \frac{8.64\times 120}{1428} = 0.726 \text{ m} \simeq 0.73 \text{ m}$$

1.3 (d)

Method-I

$Q = 4 \times 20 = 80$ m^3/day

or $Q = 9.26 \times 10^{-4}$ m^3/sec

Given $B = 20$ days,

$\Delta = 7$ cm $= 0.07$ m

$\therefore$ Duty at outlet

$$D = \frac{8.64B}{\Delta} = \frac{8.64 \times 20}{0.07}$$
$$= 2468 \text{ Ha/cumec}$$

Area $A = D \times Q$

$= 2468 \times 9.26 \times 10^{-4} = 2.29 \text{ Ha} = 2.29 \times 10^4 \text{ m}^2$

Method-II

$$Q = 4 \text{ m}^3/\text{hr}$$

Total time of operation of tubewell

$$= 20 \times 20 = 400 \text{ hrs}$$

[∵ Irigation interval is 20 days @ 20 hours per day]

Irrigation depth,

$$d = 7 \text{ cm} = 0.07 \text{ m}$$

$$\therefore \text{Commanded area} = \frac{Q \times time}{d} = \frac{4 \times 400}{0.07}$$
$$= 2.29 \times 10^4 \text{ m}^2$$

1.4 (b)

Water application efficiency,

$$\eta_a = \frac{\text{Water stored in the root zone during irrigation}}{\text{Water delivered to the field}} \times 100$$

Water supplied to field during 8 hour @ 100 lit/s

$= 100 \times 8 \times 60 \times 60 = 2880 \text{ m}^3$

Runoff loss in the field = 800 m^3

Water stored in the root zone

$$= 2880 - 800 = 2080 \text{ m}^3$$

$$\therefore \quad \eta_a = \frac{2080}{2880} \times 100 = 72.22\%$$

1.5 (a)

As discharge is same in both cases

$$\text{and, discharge } (Q) = \frac{\text{Area} \times \text{delta}}{\text{base period}}$$

$$\therefore \quad \frac{1200 \times 134}{140} = \frac{A(\text{in ha}) \times 52}{120}$$

$$\Rightarrow \quad A = 2650.55 \text{ hectares}$$

1.6 (d)

Given Delta, $\Delta = 25.92 \text{ cm} = 0.2592 \text{ m}$

Base period, $B = 150$ days

$$\text{Then Duty, } D = \frac{8.64\,B}{\Delta} = \frac{8.64 \times 150}{0.2592}$$

$$\Rightarrow \quad D = 5000 \text{ ha/cumec}$$

$$\therefore \text{Minimum discharge} = \frac{100{,}000}{5000} = 20 \text{ cumecs}$$

But the distributary channel should be designed for a discharge more than 20 cumecs.

It is because $Q = 20$ m^3/s is required if uniform discharge is to be supplied.

But the demand will be more (in peak period) and sometimes the canal will even be closed for maintenance purpose. Hence, design should be done for discharge more than 20 m^3/s.

1.7 (d)

$$d_w = \frac{\gamma_d}{\gamma_w} \times d \times (\text{FC} - \text{Existing moisture content})$$

∴Depth of water required to raise the moisture content to the field capacity (without considering loss due to evaporation).

$$= 1.5 \times 1.1 \times (0.15 - 0.05)$$
$$= 0.165 \text{ m}$$

Adding, consumptive use @ 2.5 mm/day for 8 days.

$$\text{Water required} = 0.165 + \frac{8 \times 2.5}{1000} = 0.185 \text{ m}$$

Irrigation efficiency = 65%

∴ The necessary depth of irrigation

$$= \frac{185}{0.65} \text{ mm} = 284.62 \text{ mm}$$

Note: *This question is different from others as it is mentioned in the question itself that irrigation is done in a period of 8 days. So, to account for consumptive use in the period of 8 days, a depth of 2.5 × 8 mm is to be added to raise it to field capacity as otherwise, if not added, moisture content won't be raised to field capacity by the end of 8 days.*

1.8 (a)

The conditions favouring the use of **sprinkler irrigation method** are:

(i) When the land topography is irregular.

(ii) When the land gradient is steeper and soil is easily erodible.

(iii) When the land soil is excessively permeable.

(iv) When the water table is high.

(v) When the seasonal water requirement is low.

(vi) When the water availability is difficult and scarce.

1.9 (d)

Wheat is a Rabi crop i.e., it requires irrigation

during winter. Whereas, sugarcane requires irrigation throughout the year.
Hence, the discharge required by both the crops has to be added to obtain the design discharge.
Hence, design discharge

$$= \frac{\text{Total discharge}}{\text{Time factor}} = \frac{0.36+0.27}{0.9} = 0.7 \text{ m}^3/\text{s}$$

POINTS TOREMEMBER

1. **Time factor:** It is the ratio of actual operating period of a distributary to the crop period.
   For example, if time factor = 0.8 and crop season is of 180 days.
   Then, distributary receives its full supply for a period of 180 × 0.8 = 144 days
2. **Capacity factor:** It is the ratio of mean supply discharge in a canal during a period to its designed full capacity.

1.10 (b)

Area to be irrigated

$$= (\text{CCA} \times \text{Intensity of irrigation})$$

$$= 2\times10^8 \times \frac{40}{100} = 0.8 \times 10^8 \text{ m}^2$$

$$\text{Volume} = \text{Area} \times \text{Kor depth}$$

$$= 0.8 \times 10^8 \times 0.14 = 1.2 \times 10^7 \text{ m}^3$$

$$\text{Discharge} = \frac{\text{Volume}}{\text{Kor period}}$$

$$= \frac{1.2\times10^7}{4\times7\times24\times60\times60} = 4.63 \text{ m}^3/\text{s}$$

1.11 (*)

Depth of maximum available moisture,

$$d_w = \frac{\gamma d}{\gamma_w}(F-\phi)$$

Field capacity

$$= \frac{\text{weight of water retained by soil}}{\text{weight of same area of soil}}$$

$$= \frac{\gamma_w \times V_v}{\gamma \times V}$$

But $\frac{V_v}{V} = n$

$$\therefore \text{Field capacity} = \frac{\gamma_w}{\gamma} \times n$$

$$\Rightarrow \frac{\gamma_w}{\gamma} = \frac{n}{\text{Field capacity}}$$

$$\Rightarrow \frac{\gamma_w}{\gamma} = \frac{0.45}{0.38}$$

$$\therefore d_w = \frac{0.45}{0.38} \times 1 \times [0.38 - 0.1]$$

$$= 0.331 \text{ m}$$

But the moisture content should not fall below 50%.

∴ Depth of water required

$$= 0.5 \times 0.331 = 0.165 \text{ m}$$

This much depth of water has to be stored in root zone.

∴ Frequency of irrigation

$$= \frac{\text{Depth stored in root zone}}{\text{Consumptive use}}$$

$$= \frac{165}{15} = 11 \text{ days}$$

1.12 (b)

Consumptive use of water = 2.0 mm/day
Maximum depth of available water = 60 mm
If amount of available water is reduced by 50%, then irrigation is required for a depth

$$= \frac{50}{100} \times 60 = 30 \text{ mm}$$

$$\therefore \text{Frequency of irrigation} = \frac{30}{2} = 15 \text{ days}$$

1.13 (c)

Method-I

Area to be irrigated = Culturable command area × Intensity of irrigation

$$= 20{,}000 \times \frac{50}{100}$$

$$= 10{,}000 \text{ ha}$$

Duty, $D = \frac{8.64 \text{ B}}{\Delta} = \frac{8.64\times30}{120\times10^{-3}}$

$$= 2160 \text{ ha/cumec}$$

$$\therefore \text{Outlet discharge} = \frac{\text{Area to be irrigated}}{\text{Duty of crop}}$$

$$= \frac{10{,}000}{2160} = 4.63 \text{ m}^3/\text{s}$$

Method-II

Water supplied in kor period by distributary
= Water required by the crop

$$\Rightarrow Q \times B = A \times D$$

$$\Rightarrow \quad Q = \frac{(10000 \times 10^4) \times 0.12}{30 \times 86400}$$

$$\Rightarrow \quad Q = 4.63 \text{ m}^3/\text{s}$$

1.14 (b)

Volume of water required for evapo-transpiration

$= (\text{Area} \times \text{Depth})$

$= 20 \times 10^4 \times 20 \times 10^{-3} \times 10^3$

$= 4 \times 10^6$ litres

∴ Discharge required at outlet

$$= \left(\frac{\text{Volume}}{\text{Time}}\right)$$

$$= \frac{4 \times 10^6}{20 \times 24 \times 60 \times 60}$$

$= 2.31$ litres/sec

1.15 (d)

Total depth of water required by the crop = 105 cm

Contribution of rainfall during the base period = 15 cm

Water to be provided by irrigation = Delta

$$\Delta = 105 - 15 = 90 \text{ cm}$$

$$\text{Duty of crop} = \frac{864\,B}{\Delta} = \frac{864 \times 90}{90}$$

$= 864$ ha/cumec

POINTS TOREMEMBER

1. Duty $= \dfrac{8.64B}{\Delta}$, if Δ is in meters
2. Duty $= \dfrac{864B}{\Delta}$, if Δ is in cm

   In both ase, B is in days and Duty (D) is in ha/cumec

1.16 (b)

Moisture holding capacity of soil = 18 cm /m

Root zone depth of crop X = 1.0 m

Moisture holding capacity of soil for crop

$X = 18 \times 1 = 18$ cm

Allowable depletion of moisture

$$= \frac{50}{100} \times 18 = 9 \text{ cm}$$

Consumptive use of water for crop

$X = 0.5$ cm/day

$$\therefore \text{ Frequency of irrigation} = \frac{9}{0.5} = 18 \text{ days}$$

Thus 9 cm depth of water is to be applied to the crop X in the next 18 days.

Irrigation requirement

$$= \frac{\text{Net water depth}}{\text{Water application efficiency}}$$

$$= \frac{9}{0.75} = 12 \text{ cm}$$

Quantity of water required

$$= \frac{12}{100} \times 36 \times 10^4 = 43200 \text{ m}^3$$

∴ Discharge through pump (if pump works for 10 hrs per day)

$$= \frac{43200}{18 \times 10 \times 60 \times 60} \times 10^3 \text{ litres/sec}$$

$= 66.67$ litres/sec $\simeq 67$ litres/sec

1.17 (d)

Moisture holding capacity of soil for crop

$Y = 18 \times 0.8 = 14.4$ cm

Allowable depletion of moisture

$$= \frac{50}{100} \times 14.4 = 7.2 \text{ cm}$$

Consumptive use of water for crop,

$Y = 0.4$ cm/day

$$\therefore \text{ Frequency of irrigation} = \frac{7.2}{0.4} = 18 \text{ days}$$

Thus 7.2 cm depth of water is to be applied to the crop Y in the next 18 days.

$$\text{Irrigation requirement} = \frac{7.2}{0.75} = 9.6 \text{ cm}$$

Quantity of water required = Discharge through pump when pump works @ 10 hrs per day

$$\Rightarrow \frac{9.6}{100} \times A \times 10^4 = 40 \times 10^{-3} \times 18 \times 10 \times 60 \times 60$$

$\Rightarrow A = 27$ hectares

1.18 (a)

Base period, B = 120 days

Δ = Net water required by the crop during base period

= Total water requirement – Rainfall

= 55 – 10 = 45 cm = 0.45 m

Now, $D = \frac{8.64B}{\Delta} = \frac{8.64 \times 120}{0.45}$

$\Rightarrow$ $D = 2304$ ha/cumec

and $Q = 0.01$ m³/s

$\therefore$ Area, $A = D \times Q$

$= 2304 \times 0.01$ ha $= 23.04$ ha

Efficiencyof irrigation = 60%

$\therefore$ Area which can be irrigated

$$= 23.04 \times \frac{60}{100} = 13.82 \text{ ha}$$

1.19 (b)

Δ = Total water depth required – Effective rainfall

= 48 – 8 = 40 cm

$$\text{Duty (D)} = \frac{864B}{\Delta(\text{cm})} = \frac{864 \times 10}{40} = 216 \text{ ha/cumec}$$

1.20 Sol.

Given,

Root zone depth,

$d = 70$ cm

Field capacity,

$F_C = 28\%$

Existing moisture content,

$w = 18\%$

Density of soil,

$\gamma = 1.3$ gm/cm³

Density of water,

$\gamma_w = 1.0$ gm/cm³

Depth of irrigation water required,

$$d_w = \frac{\gamma}{\gamma_w} d(F_c - w)$$

$$= \frac{1.3}{1.0} \times (70 \times 10)(28\% - 18\%)$$

$$= 1.3 \times (70 \times 10) \times \frac{10}{100} = 91 \text{ mm}$$

Note: *It is to be understood that* $\gamma_d = \frac{\gamma_t}{1+w}$, *but as it is not mentioned clearly in the question that the density of 1.3 g/cc is computed at which moisture content, hence, assuming the given density of 1.3 g/cc to be the dry density.*

If it would have been given that density of 1.3 g/cc is at moisture content 18%, then we would have computed the dry density from the formula

$$\gamma_d = \frac{\gamma_t}{1+w}.$$

1.21 (b)

Gross Command Area is measured in hectares. (P-III)

Duty has a unit of ha/cumec. (R-I)

Delta of wheat can be practically 40 cm

$(S - V)$

1.22 Sol.

Gram and wheat are Rabi crops.

So both are grown in the same period.

Discharge required for gram,

$$Q_g = \frac{A_g}{D_g} = \frac{2000 \times 0.3}{8.64 \times \frac{18}{0.12}} = 0.4630 \text{ m}^3/\text{s}$$

Discharge required for wheat,

$$Q_w = \frac{A_w}{D_w} = \frac{2000 \times 0.5}{8.64 \times \frac{18}{0.15}} = 0.9645 \text{ m}^3/\text{s}$$

Total discharge = $Q_g + Q_w = 1.4275$ m³/s

POINTS TOREMEMBER

- Important Rabi crops are wheat, barley, gram, pea, mustard.
- Important Kharif crops are Rice, Jowar, Bajra, Groundnut, Jute.

1.23 Sol.

Method-I

CCA = 10000 ha

For rice, $\Delta_r = 130$ cm $= 1.3$ m

$A_r = 2500$ ha

$B_r = 150$ days

Duty, $D_r = \frac{8.64B_r}{\Delta_r} = \frac{8.64 \times 150}{1.30}$

$D_r = 996.923$ ha/cumecs

$$D_r = \frac{A}{Q}$$

$$Q = \frac{2500}{996.923} \text{ m}^3/\text{s} = 2.5077 \text{ m}^3/\text{s}$$

For wheat also, discharge is 2.5077 m³/s,

$B_w = 120$ days

$\Delta = 50$ cm

$$A_w = Q \times D_w$$

$$D_w = \frac{8.64 \times B_w}{\Delta_w}$$

$$= \frac{8.64 \times 120}{0.5} = 2073.6$$

$$\therefore \quad A_w = 2.5077 \times 2073.6$$

$$= 5199.966 \text{ ha} \simeq 5200 \text{ ha}$$

Method-II

Dischage, $Q = \dfrac{\text{Area } (A) \times \text{Depth} (\Delta)}{\text{Base period } (B)}$

Now, equate Q for ricve and wheat

Hence, $Q_r = Q_w$

$$\Rightarrow \quad \frac{A_r \times \Delta_r}{B_r} = \frac{A_w \times \Delta_w}{B_w}$$

$$\Rightarrow \quad \frac{2500 \times 130}{150} = \frac{A_w \times 50}{120}$$

$$\Rightarrow \quad A_w = 5200 \text{ ha}$$

1.24 Sol.

Culturable command area = 50000 ha

Intensity of irrigation for kharif season = 50%

∴Area under kharif = 25000 ha

Duty for kharif season = 1000 ha/cumec

$$\text{Duty} = \frac{\text{Area}}{\text{Discharge}}$$

∴ Discharge at the head of field

$$Q = \frac{25000 \text{ ha}}{1000 \text{ ha/cumec}}$$

$$= 25 \text{ cumec}$$

Transmission/conveyance loss = 10%

$\therefore \quad \eta_{conveyance} = 90\%$

Discharge at the head of canal

$$= \frac{25}{0.9} \text{cumec} = 27.78 \text{ cumec}$$

1.25 Sol.

Given data:

Base period,

B = 120 days

Delta of crop,

Δ = 40 cm

Discharge, Q=2 m^3/s

Area to be irrigated,

A = ?

∴Duty of water,

$$\Delta = \frac{864 \times B}{\Delta} \text{ ha/cumec}$$

and Area to be irrigated;

$$A = Q \times D$$

$$\Rightarrow \quad A = 2 \times m^3/s \times \frac{864 \times 120}{40} \times \frac{ha}{cumec}$$

$$\Rightarrow \quad A = 5184 \text{ ha}$$

1.26 (b)

Water required by crop = 100 × 0.4 mm = 40 mm

Effective rainfall = 20 mm

Additional water requried = 20 mm

Amount of water required after accounting irrigation efficiency = $\dfrac{20}{0.5}$ = 40 mm

1.27 (*)

Field capacity, FC = 30%

Permanent wilting point,

PWP = 13%

Depth of water, $d_w = \dfrac{\gamma_d}{\gamma_w} \cdot d \times (FC - OMC)$

(d = Effective depth of root zone)

$$= \frac{1500}{1000} \times 80(0.3 - 0.2)$$

= 12 cm or 120 mm

Consumptive use = 2 mm/day

So, frequency of irrigation = $\dfrac{120}{2}$ = 60 days

1.28 Sol.

Live storage = 300 Mm^3

Area = 40000 hectare

Since 2 filling so volume of water needed

= 600 Mm^3

B = 120 days

$$\text{Duty} = \frac{8.64B}{\Delta}$$

$$\Delta = \frac{600 \times 10^6}{40000 \times 10^4} = 1.5 \text{ m}$$

$$\text{Duty} = \frac{8.64 \times 120}{1.5} = 691.2 \text{ ha/cumec}$$

1.29 Sol.

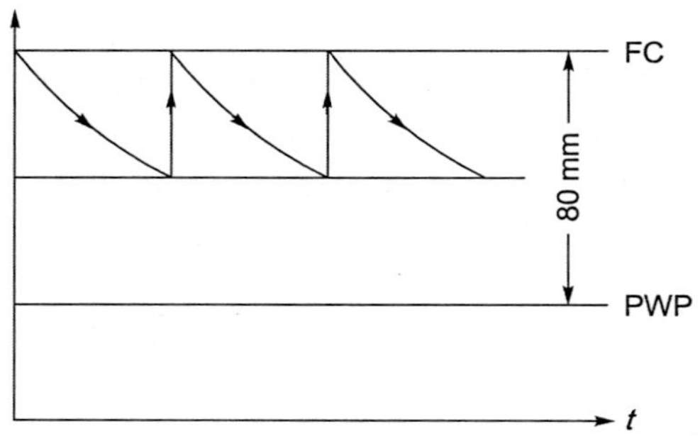

Frequency, $f = \dfrac{\text{Depth of water needed}}{\text{Consumptive use}}$

$\Rightarrow \quad f = \dfrac{40}{2.8} = 14.28$ days

$\Rightarrow \quad f = 14$ days

(Considering depletion amount)

1.30 Sol.

Duty of the crop in the field,

$$D = \frac{A}{Q} = 8.64\frac{B}{\Delta}$$

where, $\quad Q$ = discharge required by crop

Irrigated, $\quad A = 108$ ha

Delta for crop,

$\Delta = 1.44$ m

Base period of crop,

$B = 120$ days

$$\therefore \quad Q = \frac{A\Delta}{8.64B}$$

$$= \frac{108 \times 1.44}{8.64 \times 120}$$

$= 0.15 \text{ m}^3/\text{s}$

Water required to be pumped

$$= \frac{0.15}{0.8} = 0.1875 \text{m}^3/\text{s}$$

Power required by pump $= \dfrac{\gamma QH}{0.8}$

$$= \frac{9810 \text{N/m}^3 \times 0.1875 \text{ m}^3/\text{s} \times 10}{0.8 \times 746}$$

$= 30.82$ h.p.

1.31 (d)

Given: $\quad Et_0 = 6$ mm/day

$K_c = 1.2$

$K_s = 0.8$

We know that based on Basal's equation of reference evapotranspiration,

$$Et_c = K_c \cdot K_s \, Et_0$$

$\Rightarrow \quad Et_c = 1.2 \times 0.8 \times 6$

$\Rightarrow \quad Et_c = 5.76$ mm/day

1.32 Sol.

Given: Area : 100 ha

Total depth of water, $\Delta_2 = 35$ cm

Total time period, $B_2 = 120$ days

Kor depth, $\Delta_1 = 12$ cm

Kor period, $B_1 = 30$ days

(i) Kor duty, Δ_1 =

$$\frac{8.64 \times B_1}{\Delta_1} = \frac{8.64 \times 30}{0.12} = 2160 \text{ ha/m}^3/\text{sec}$$

Corresponding discharge

$$= \frac{\text{Area}}{\text{Duty}} = \frac{100}{2160} = 0.046 \text{m}^3/\text{sec}$$

(ii) Overall duty,

$$\Delta_2 = \frac{8.64 \times B_2}{\Delta_2} = \frac{8.64 \times 120}{0.35}$$

$= 2962.28 \text{ ha/m}^3/\text{sec}$

Corresponding discharge

$$= \frac{\text{Area}}{\text{Duty}} = \frac{100}{2962.28} = 0.034 \text{ m}^3/\text{sec}$$

Minimum discharge required will be maximum of the above two.

$\therefore \quad Q_{req.} = 0.046 \text{ m}^3/\text{sec}$

■■■■

2 Design of Stable Channels

2.1 The medium size of the sediment particles in an alluvial channel is 0.4 mm. For a discharge of 3 m^3/sec, the Lacey regime slope would be........

[1995 : 1 M]

2.2 On which of the canal systems, R.G. Kennedy, executive engineer in the Punjab Irrigation Department made his observations for proposing his theory on stable channels?

(a) Krishna Western Delta canals
(b) Lower Bari Doab canals
(c) Lower Chenab canals
(d) Upper Bari Doab canals [2005 : 1 M]

2.3 A triangular irrigation lined canal carries a discharge of 25 m^3/s at bed slope = $\frac{1}{6000}$. If the side slopes of the canal are 1 : 1 and Manning's coefficient is 0.018, the central depth of flow is equal to

(a) 2.98 m (b) 3.62 m
(c) 4.91 m (d) 5.61 m [2005 : 2 M]

2.4 As per the Lacey's method for design of alluvial channels, identify the TRUE statement from the following:

(a) Wetted perimeter increases with an increase in design discharge.
(b) Hydraulic radius increases with an increase in silt factor.
(c) Wetted perimeter decreases with an increase in design discharge.
(d) Wetted perimeter increases with an increase in silt factor.

[2007 : 1 M]

2.5 A stable channel is to be designed for a discharge of Q m^3/s with silt factor f as per Lacey's method. The mean flow velocity (m/s) in the channel is obtained by

(a) $\left[\frac{Qf^2}{140}\right]^{1/6}$ (b) $\left[\frac{Qf}{140}\right]^{1/3}$

(c) $\left[\frac{Q^2f^2}{140}\right]^{1/6}$ (d) $0.48 \times \left[\frac{Q}{f}\right]^{1/3}$

[2008 : 1 M]

2.6 The depth of flow in a alluvial channel is 1.5 m. If critical velocity ratio is 1.1 and Manning's n is 0.018, the critical velocity of the channel as per Kennedy's method is

(a) 0.713 m/s (b) 0.784 m/s
(c) 0.879 m/s (d) 1.108 m/s

[2009 : 1 M]

2.7 An unlined canal under regime conditions along with a silt factor of 1 has a width of flow 71.25 m. Assuming the unlined canal as a wide channel, the corresponding average depth of flow (in m, round off to two decimal places) in the canal will be ________.

[2021 : 2 M, Set-I]

2.8 A standard round bottom triangular canal section as shown in the figure has a bed slope of 1 in 200. Consider the Chezy's coefficient as 150 $m^{1/2}$/s.

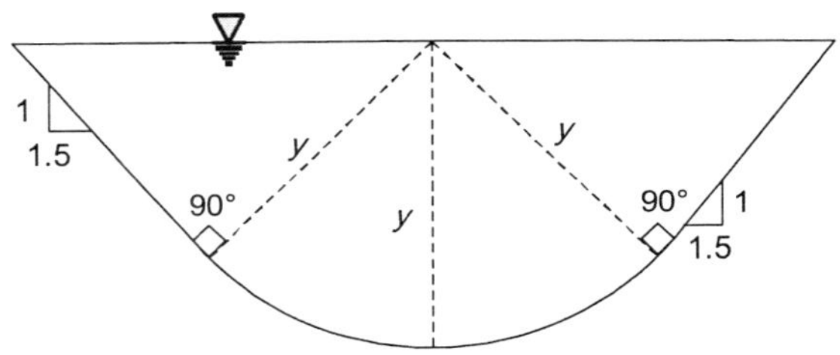

(Figure NOT to scale)

The normal depth of flow, y (in meters) for carrying a discharge of 20 m^3/s is ______(rounded off to 2 decimal places).

[2024 : 2 M, Set-I]

2.9 A round-bottom triangular lined canal is to be laid at a slope of 1 in 1500, to carry a discharge of 25 m^3/s. The side slopes of the canal cross-section are to be kept at 1.25H : 1V. If Manning's roughness coefficient is 0.013, the flow depth (in meters) will be in the range of

(a) 1.94 to 1.97 (b) 2.39 to 2.42
(c) 2.24 to 2.27 (d) 2.61 to 2.64

[2024 : 2 M, Set-II]

■■■■

Answers Design of Stable Channels

2.1 (Sol.) 2.2 (d) 2.3 (b) 2.4 (a) 2.5 (a) 2.6 (b) 2.7 (2.94)
2.8 (1.10) 2.9 (b)

Explanations Design of Stable Channels

2.1 Sol.

Lacey regime slope

$$S = \frac{(f)^{5/3}}{3340(Q)^{1/6}}$$

Now, $f = 1.76\sqrt{d_{mm}} = 1.76\sqrt{0.4} = 1.11$

$Q = 3\ m^3/s$

$$\therefore \quad S = \frac{(1.11)^{5/3}}{3340 \times (3)^{1/6}} = \frac{1}{3370.76} \approx \frac{1}{3371}$$

2.2 (d)

R.G. kennedy made his observations on the upper Bari Doab canal system in Punjab, India, for proposing his theory on stable channels.

2.3 (b)

$A = y^2(\theta + \cot\theta)$ $[\theta = 45°]$

$P = 2y(\theta + \cot\theta)$

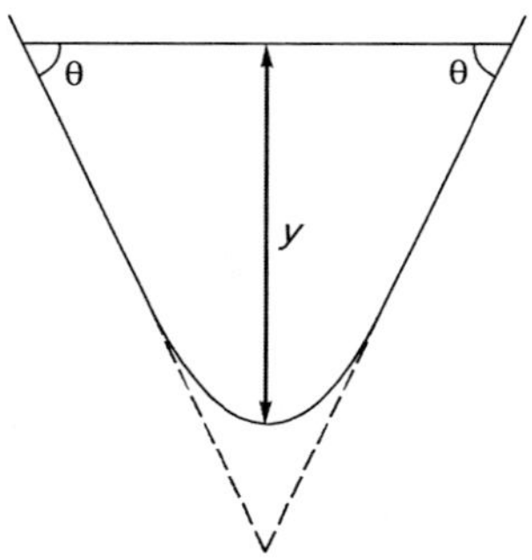

$$A = y^2\left(\frac{\pi}{4}+1\right)$$

$$P = 2y\left(\frac{\pi}{4}+1\right)$$

$$R = \frac{A}{P} = \frac{y}{2}$$

From Manning's equation

$$Q = \frac{1}{n}AR^{2/3}S^{1/2}$$

$$25 = \frac{1}{0.018} \times y^2\left(\frac{\pi}{4}+1\right)\left(\frac{y}{2}\right)^{2/3}\left(\frac{1}{6000}\right)^{1/2}$$

$y = 3.62$ m

Hence option (b) is correct.

2.4 (a)

1. Wetted perimeter,

$$P = 4.75\sqrt{Q}$$

$$\Rightarrow \quad P \propto \sqrt{Q}$$

Hence, P increase with increase in design discharge.

2. Hydraulic radius,

$$R = \frac{5}{2}\frac{V^2}{f}$$

$$R \propto \frac{1}{f}$$

Hence, R decreases with increase in silt factor.

Also consider,

$$V = \left(\frac{Qf^2}{140}\right)^{1/6}$$

or $$V^6 = \frac{Qf^2}{140}$$

or $$Q \propto \frac{1}{f^2}$$

2.5 (a)

According to Lacey's theory,

(i) The mean flow velocity (m/s) in the channel is given as

$$V = \left(\frac{Qf^2}{140}\right)^{1/6}$$

(ii) Hydraulic radius (m) is given as

$$R = 2.5\left(\frac{V^2}{f}\right)$$

(iii) Perimeter, $P = 4.75\sqrt{Q}$

(iv) Bed slope, $S = \dfrac{f^{5/3}}{3340Q^{1/6}}$

where, Q is discharge in m^3/s
V is velocity in m/s
f is a silt factor.

2.6 (b)

The critical velocity as per Kennedy's method is given by,

$V_0 = 0.55\ my^{0.64} = 0.55 \times 1.1 \times (1.5)^{0.64}$
$= 0.784$ m/s

2.7 Sol.

$f = 1$

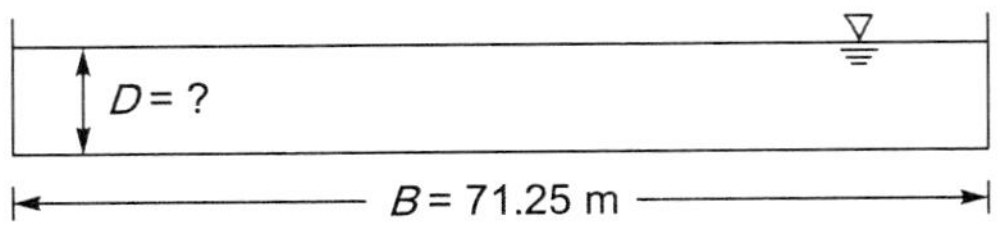

$R = D$ (for wide rectangular channel)

$$Af^2 = 140\left(\frac{2}{5}fR\right)^{5/2}$$

$$(BD)\ f^2 = 140\left(\frac{2}{5}f \times D\right)^{5/2}$$

$$(71.25 \times D) \times 1 = 140\left(\frac{2}{5} \times 1 \times D\right)^{5/2}$$

$$D \times 0.5089 = \left(\frac{2}{5}\right)^{5/2} \times (D)^{5/2}$$

$$D^{3/2} = 5.029 \text{ or } D = 2.94 \text{ m}$$

2.8 Sol.

$$\tan\theta = \frac{1}{1.5} \therefore \theta = 0.588 \text{ rad} = 33.69°$$

Discharge, $Q = A \times V$

Now, Cross-section area, $A = y^2\ (\theta_{radian} + \cot\theta)$
$= y^2\ (0.588 + 1.5)$
$= 2.088\ y^2$

Perimeter, $P = 2y\ (\theta_{radian} + \cot\theta)$

Hydraulic radius, $R = \dfrac{A}{P}$

$$= \frac{y^2(\theta_{radian} + \cot\theta)}{2y(\theta_{radian} + \cot\theta)} = \frac{y}{2}$$

Now, Velocity, $V = C\sqrt{R.S} = 150\sqrt{\left(\dfrac{y}{2}\right) \times \dfrac{1}{200}}$

$$= \frac{150}{20} y^{1/2}$$

Now, $Q = A \times V$

$\Rightarrow$ $20 = 2.088\ y^2 \times \dfrac{150}{20} \cdot y^{1/2}$

$\Rightarrow$ $y = 1.10$ m

2.9 (b)

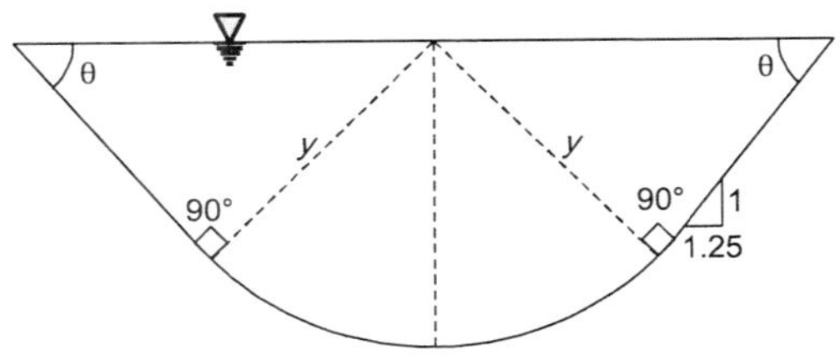

From the figure, $\tan\theta = \dfrac{1}{1.25}$

$\therefore$ $\theta = 38.66°$

$= 38.66° \times \dfrac{\pi}{180°} = 0.675$ radians

Also, $\cot\theta = 1.25$

Cross sectional area, $A = y^2\ (\theta + \cot\theta)$
$= y^2\ [0.675 + 1.25] = 1.925\ y^2$

Perimeter, $P = 2y\ (\theta + \cot\theta)$

Now, Hydraulic radius,

$$R = \frac{A}{P} = \frac{y^2(\theta + \cot\theta)}{2y(\theta + \cot\theta)} = \frac{y}{2}$$

Now, Discharge, $Q = A \times \dfrac{1}{n} R^{2/3} S^{1/2}$

$$\Rightarrow 25 = (1.925y^2) \times \frac{1}{0.013} \times \left(\frac{y}{2}\right)^{2/3} \times \left(\frac{1}{1500}\right)^{1/2}$$

$\Rightarrow$ $y = 2.405 \text{ m} \simeq 2.41 \text{ m}$

■■■■

3 Design & Construction of Gravity Dams

3.1 The base width of an elementary profile of a gravity dam of height H is b. The specific gravity of the material of the dam is G and uplift pressure coefficient is K. The correct relationship for no tension at the heel is given by

(a) $\frac{b}{H} = \frac{1}{\sqrt{G-K}}$ (b) $\frac{b}{H} = \sqrt{G-K}$

(c) $\frac{b}{H} = \frac{1}{G-K}$ (d) $\frac{b}{H} = \frac{1}{K\sqrt{G-K}}$

[2008 : 1 M]

3.2 A concrete gravity dam section is shown in the figure. Assuming unit weight of water as 10 kN/m^3 and unit weight of concrete as 24 kN/m^3, the uplift force per unit length of the dam (expressed in kN/m) at PQ is______

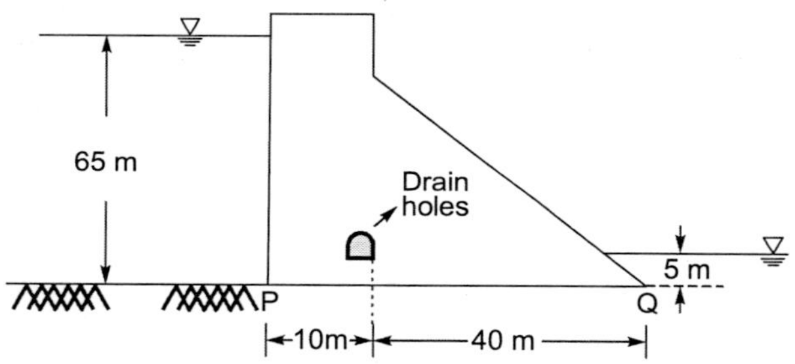

[2016 : 2 M, Set-I]

3.3 A concrete dam holds 10 m of static water as shown in the figure (not drawn to the scale). The uplift assumed to vary linearly from full hydrostatic head at the heel, to zero at the toe of dam. The coefficient of friction between the dam and foundation soil is 0.45. Specific weights of concrete and water are 24 kN/m^3 and 9.81 kN/m^3, respectively.

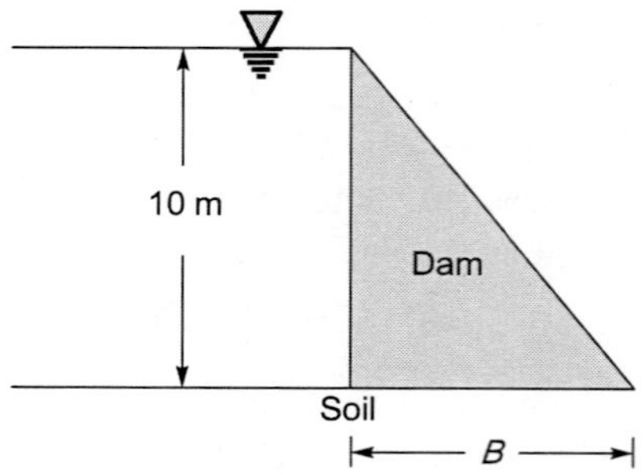

For NO sliding condition, the required minimum base width B (in m, round off to two decimal places) is ________.

[2020 : 2 M, Set-II]

■■■■

Answers Design & Construction of Gravity Dams

3.1 (a) **3.2** (10500) **3.3** (15.873)

Explanations Design & Construction of Gravity Dams

3.1 (a)

Tension will not be developed at the heel with full reservoir, when,

$$b \geq \frac{H}{\sqrt{G-K}}$$

POINTS TOREMEMBER

- For no sliding, $b \leq \dfrac{H}{\mu(G-K)}$

  Where, μ = Cofficient of friction

3.2 Sol.

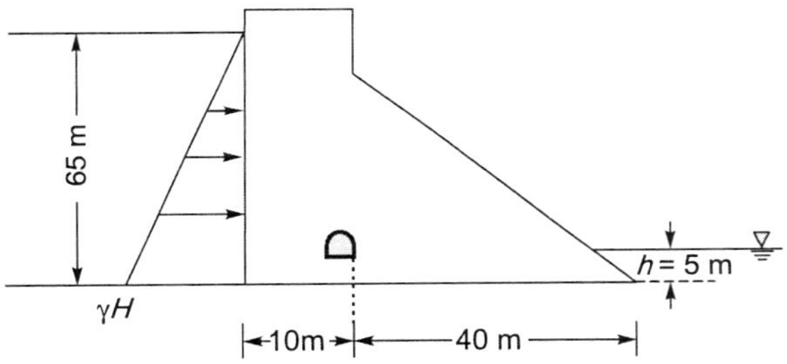

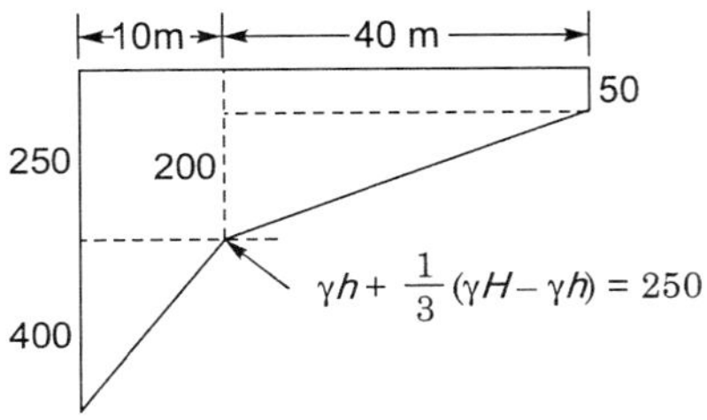

Total uplift pressure

$$= 250 \times 10 + 40 \times 50 + \frac{1}{2} \times 200 \times 40 + \frac{1}{2} \times 400 \times 10$$

$$= 2500 + 2000 + 4000 + 2000$$

$$= 10500 \text{ kN/m}$$

3.3 Sol.

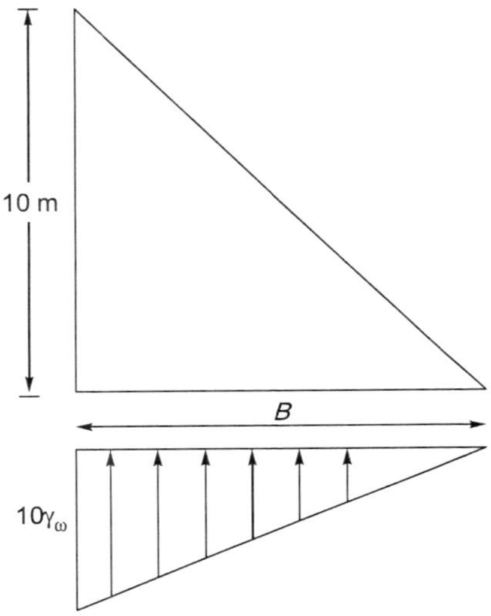

$$\mu = 0.45$$

$$\gamma_{conc.} = 24 \text{ kN/m}^3$$

$$G_{conc.} = 2.4$$

$$B_{min.\ sliding} = \frac{H}{\mu(G-1)}$$

$$B_{min.\ sliding} = \frac{10}{0.45(2.4-1)}$$

$$= 15.873 \text{ m}$$

4 Theories of Seepage, Spillways and Miscellaneous

4.1 In connection with the design of a barrage, identify the correct matching criteria of design with the items of design

Item of design
(i) Width of waterway
(ii) Level and length of downstream floor
(iii) Depth of sheet piles and total length of barrage floor
(iv) Barrage floor thickness

Criteria of Design
A. Scour depth and exit gradient
B. Lacey's formula for wetted perimeter and discharge capacity of the barrage as computed by weir equations
C. Uplift pressure variation
D. Hydraulic jump consideration

Codes:

	(i)	(ii)	(iii)	(iv)
(a)	A	B	C	D
(b)	D	C	B	A
(c)	B	A	D	C
(d)	B	D	A	C

[1998 : 2 M]

4.2 While designing a hydraulic structure, the piezometric head at bottom of the floor is computed as 10 m. The datum is 3 m below floor bottom. The assured standing water depth above the floor is 2 m. The specific gravity of the floor material is 2.5. The floor thickness should be
(a) 2.00 m (b) 3.33 m
(c) 4.40 m (d) 6.00 m [2003 : 2 M]

4.3 Which one of the following equations represents the downstream profile of Ogee spillway with vertical upstream face? (x, y) are the coordinates of the point on the downstream profile with origin at the crest of the spillway and H_d is the design head.

(a) $\frac{y}{H_d} = -0.5\left(\frac{x}{H_d}\right)^{1.85}$

(b) $\frac{y}{H_d} = -0.5\left(\frac{x}{H_d}\right)^{1/1.85}$

(c) $\frac{y}{H_d} = -2.0\left(\frac{x}{H_d}\right)^{1.85}$

(d) $\frac{y}{H_d} = -2.0\left(\frac{x}{H_d}\right)^{1/1.85}$ [2005 : 1 M]

4.4 Uplift pressure at points *E* and *D* (figure-*A*) of a straight horizontal floor of negligible thickness with a sheet pile at downstream end are 28% and 20%, respectively. If the sheet pile is at upstream end of the floor (figure-*B*), the uplift pressures at points D_1 and C_1 are

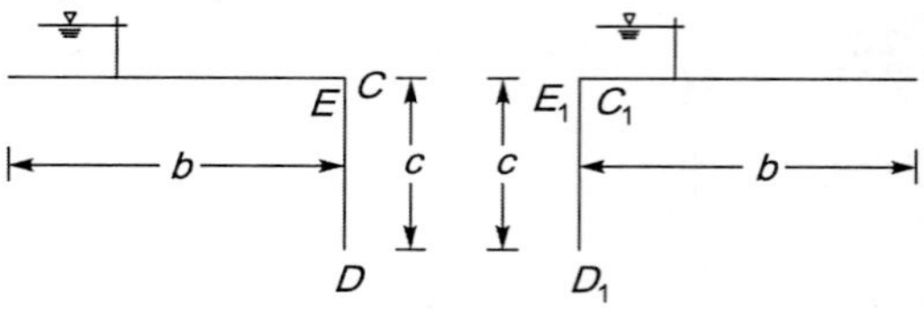

Fig. A Fig. B

(a) 68% and 60% respectively
(b) 80 % and 72% respectively
(c) 88% and 70% respectively
(d) 100% and zero respectively [2005 : 2 M]

4.5 A launching apron is to be designed at downstream of a weir for discharge intensity of 6.5 m³/s/m. For the design of launching aprons the scour depth is taken two times of Lacey scour depth. The silt factor of the bed material is unity. If the tailwater depth is 4.4m, the length of launching apron in the launched position is
(a) $\sqrt{5}$ m (b) 4.7 m
(c) 5 m (d) $5\sqrt{5}$ m [2005 : 2 M]

4.6 A weir on a permeable foundation with downstream sheet pile is shown in the figure below. The exit gradient as per Khosla's method is

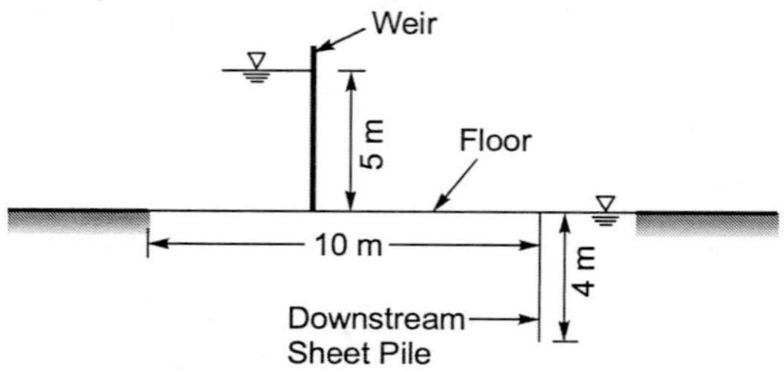

(a) 1 in 6.0 (b) 1 in 5.0
(c) 1 in 3.4 (d) 1 in 2.5 **[2008 : 2 M]**

4.7 Water emerges from an ogee spillway with velocity = 13.72 m/s and depth = 0.3 m at its toe. The tail water depth required to form a hydraulic jump at the toe is
(a) 6.48 m (b) 5.24 m
(c) 3.24 m (d) 2.24 m **[2008 : 2 M]**

4.8 A conventional flow duration curve is a plot between
(a) Flow and percentage time flow is exceeded
(b) Duration of flooding and ground level elevation
(c) Duration of water supply in a city and proportion of area receiving supply exceeding this duration
(d) Flow rate and duration of time taken to empty a reservoir at that flow rate
[2014: 1 M, Set-I]

4.9 Profile of a weir on permeable foundation is shown in figure I and an elementary profile of upstream pile only case' according to Khosla's theory is shown in figure II. The uplift pressure heads at key points *Q*, *R* and *S* are 3.14 m, 2.75 m and 0 m, respectively (refer figure II)

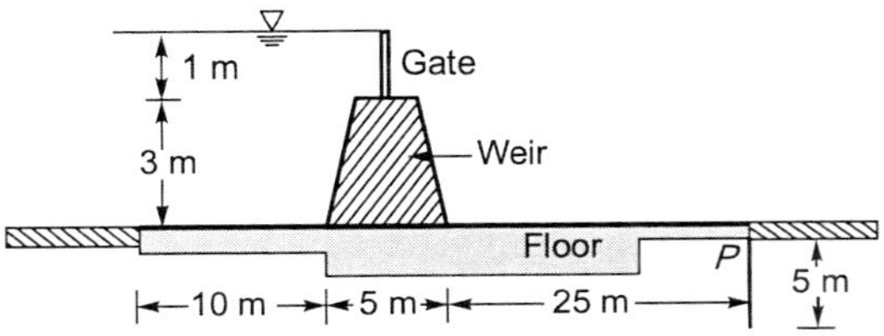

Figure I

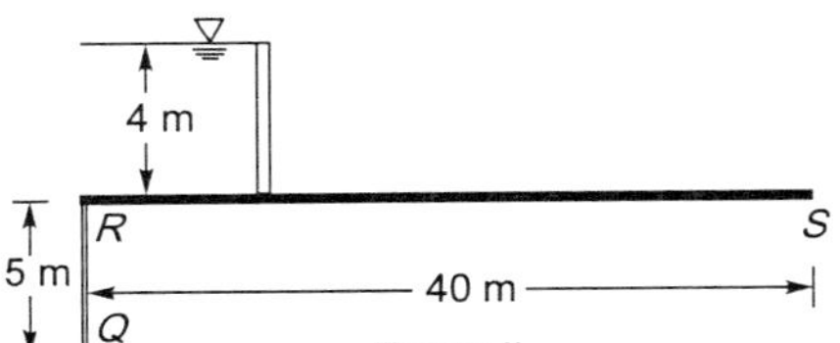

Figure II

What is the uplift pressure head at point *P* downstream of the weir (junction of floor and pile as shown in the figure-I)?
(a) 2.75 m (b) 1.25 m
(c) 0.8 m (d) Data not sufficient
[2016 : 1 M, Set-II]

4.10 **Group I** contains three broad classes of irrigation supply canal outlets. **Group II** presents hydraulic performance attributes.

Group I
P. Non-modular outlet
Q. Semi-modular outlet
R. Modular outlet

Group II
1. Outlet discharge depends on the water levels in both the supply canal as well as the receiving water course
2. Outlet discharge is fixed and is independent of the water levels in both the supply canal as well as the receiving water course
3. Outlet discharge depends only on the water level in the supply canal

The correct match of the items in **Group I** with the items in **Group II** is
(a) P-1; Q-2; R-3 (b) P-3; Q-1; R-2
(c) P-2; Q-3; R-1 (d) P-1; Q-3; R-2
[2017 : 2 M, Set-I]

4.11 If the path of an irrigation canal is below the bed level of a natural stream, the type of cross-drainage structure provided is
(a) Level crossing (b) Super passage
(c) Aqueduct (d) Sluice gate
[2019 : 1 M, Set-I]

4.12 Superpassage is a canal cross-drainage structure in which
(a) canal water flows under pressure below a natural stream
(b) natural stream water flows under pressure below a canal
(c) canal water flows with free surface below a natural stream
(d) natural stream water flows with free surface below a canal **[2020 : 1 M, Set-II]**

4.13 In the context of cross-drainage structures, the correct statement(s) regarding the relative positions of a natural drain (stream/river) and an irrigation canal, is/are
(a) In an aqueduct, natural drain water goes under the irrigation canal, whereas in a super-passage, natural drain water goes over the irrigation canal.
(b) In a level crossing, natural drain water goes through the irrigation canal.
(c) In an aqueduct, natural drain water goes over the irrigation canal, whereas in a super-passage, natural drain water goes under the irrigation canal.
(d) In a canal syphon, natural drain water goes through the irrigation canal.
[2022 : 2 M, Set-I]

4.14 dentify the cross-drainage work in the figure.

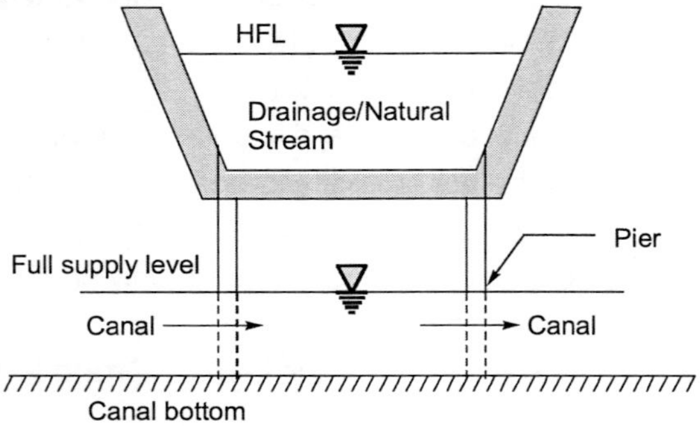

(a) Super passage
(b) Aqueduct
(c) Siphon aqueduct
(d) Level crossing

[2023 : 1 M, Set-II]

Answers Theories of Seepage, Spillways and Miscellaneous

4.1	(d)	4.2	(a)	4.3	(a)	4.4	(b)	4.5	(d)	4.6	(c)	4.7	(c)
4.8	(a)	4.9	(b)	4.10	(d)	4.11	(b)	4.12	(c)	4.13	(a, b)	4.14	(a)

Explanations Theories of Seepage, Spillways and Miscellaneous

4.1 (d)

4.2 (a)

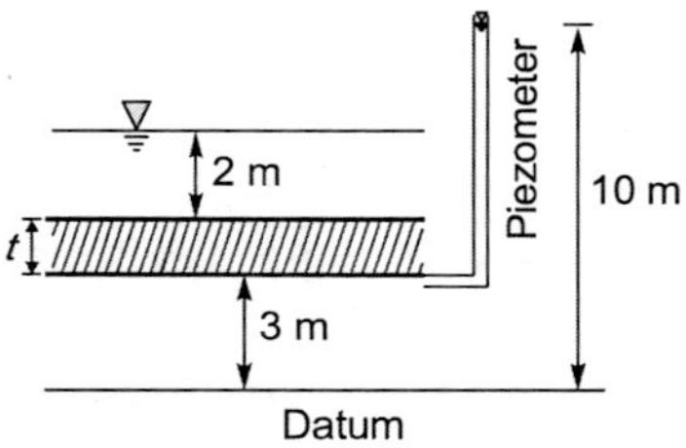

$G_s = 2.5$

Piezometric head = 10 m

Datum head = 3 m

Piezometric head = Pressure head + Datum head

$\Rightarrow$ Pressure head = 10 – 3 = 7 m

At **critical condition,**

$$\text{Uplift} = \text{Downward pressure}$$

$$7 \times \gamma_w = (\gamma_w \times 2) + (\gamma_{floor} \times t)$$

$$\frac{5\gamma_w}{\gamma_{floor}} = t = \frac{5}{G} = \frac{5}{2.5}$$

$\therefore$ $t = 2\text{ m}$

4.3 (a)

For a spillway having a vertical u/s face, the d/s crest is given by the equation,

$$x^{1.85} = -2\,H_d^{0.85} \times y$$

$$\Rightarrow \quad x^{1.85} = -2\,H_d^{0.85}\, y \times \frac{H_d}{H_d}$$

$$\Rightarrow \quad x^{1.85} = -2\frac{H_d^{1.85} y}{H_d}$$

$$\Rightarrow \quad \frac{x^{1.85}}{-2\,H_d^{1.85}} = \frac{y}{H_d}$$

$$\Rightarrow \quad \frac{y}{H_d} = -0.5\left(\frac{x}{H_d}\right)^{1.85}$$

4.4 (b)

Uplift pressure at D_1,

$$\phi_{D_1} = 100 - \phi_D = 100 - 20 = 80\%$$

Uplift pressure at C_1,

$$\phi_{C_1} = 100 - \phi_E = 100 - 28 = 72\%$$

4.5 (d)

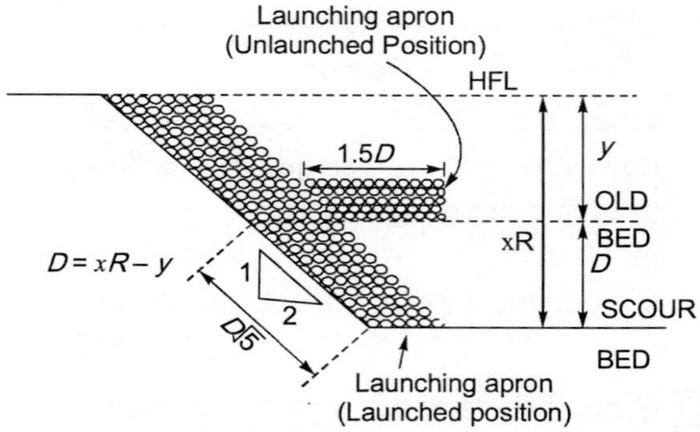

Lacey's scour depth,

$$R = 1.35\left(\frac{q^2}{f}\right)^{1/3} = 1.35 \times \left(\frac{6.5^2}{1}\right)^{1/3} = 4.7\text{ m}$$

Scour depth $= 2R = 9.4$ m

$\therefore \quad D = 2R - 4.4 = 5$ m

Thus, launching apron in launched position $= D\sqrt{5} = 5\sqrt{5}$ m.

4.6 (c)

The exit gradient may be given as,

$$G_E = \frac{H}{d} \times \frac{1}{\pi\sqrt{\lambda}}$$

where $\lambda = \frac{1+\sqrt{1+\alpha^2}}{2}$ and $\alpha = \frac{b}{d}$

Here $b = 10$ m, $d = 4$ m, $H = 5$ m

$$\therefore \quad \alpha = \frac{b}{d} = \frac{10}{4} = 2.5$$

and $$\lambda = \frac{1+\sqrt{1+(2.5)^2}}{2} = 1.85$$

$$\therefore \quad G_E = \frac{5}{4} \times \frac{1}{\pi \times \sqrt{1.85}} = \frac{1}{3.4}$$

4.7 (c)

Water depth required to form a hydraulic jump may be given as,

$$y_2 = \frac{y_1}{2}\left[-1+\sqrt{1+\frac{8V_1^2}{gy_1}}\right]$$

$$= \frac{0.3}{2} \times \left[-1+\sqrt{1+\frac{8\times(13.72)^2}{9.81\times 0.3}}\right] = 3.24 \text{ m}$$

4.8 (a)

A conventional flow durtion curve is a plot between the flow rate (usually expressed in cubic meters per second or cubic feet per second) along the y-axis and the percentage of time the flow is equaled or exceed along the x-axis.

4.9 (b)

$$\phi = \frac{\text{Pressure head at that point}}{\text{Total head}} \times 100$$

$$\Rightarrow \quad \phi_R = \frac{2.75}{4} \times 100 = 68.75\%$$

$$\phi_P = 100 - \phi_R = 31.25\%$$

Now, $$\phi_P = \frac{\text{Pressure head at point P}}{\text{Total head}} \times 100$$

$$31.25 = \frac{h}{4} \times 100$$

$$\therefore \quad h = 1.25 \text{ m}$$

4.10 (d)

Non-modular outlet : These are the outlets whose discharge depends on the difference in water levels in the distributing channel and the water course. The discharge of such outlets, therefore, varies with the variation of the water levels in the distributing channel and the water course.

Semi-modular outlet : These are the outlets whose discharge varies with the variation of the water level in the distribution channel but it is independent of the water level in the water course, so long as the minimum working head required for their working is available.

Modular outlet : These are the outlets whose discharge is independent of the water levels in the distributing channel and the water course, within reasonable working limits. In other words modular outlets maintain a constant discharge irrespective of variation of the water levels in the distributing channel and the water course.

4.11 (b)

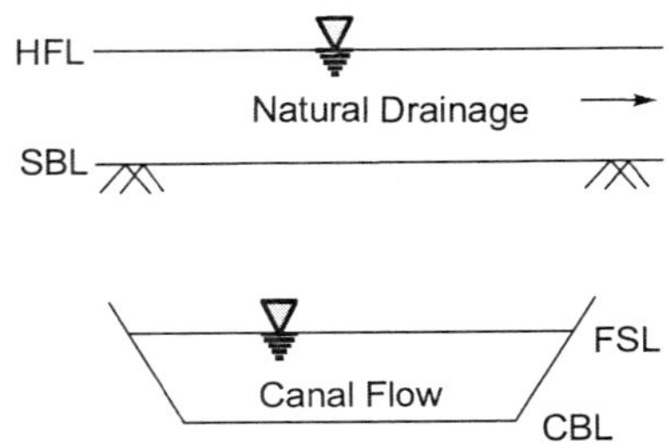

where, HFL: High Flood Level of Drain
SBL: Stream Bed Level
FSL: Full Supply Level of Canal
CBL: Canal Bed Level

Superpassage, as the given elevation condition suits CDW (as shown in diagram).

4.12 (c)

Superpassage is a canal cross-drainage structure in which canal water flows with free surface below a natural stream.

4.13 (a, b)

Aqueduct: Natural drain waer goes under the irrigation canal i.e. canal is above the natural drain.

In a level crossing, natural drain water goes through the irrigation.

4.14 (a)

■■■■

Engineering Hydrology

UNIT IX

CONTENTS

Engineering Hydrology

Syllabus

Hydrologic cycle, precipitation, evaporation, evapo-transpiration, watershed, infiltration, unit hydrographs, hydrograph analysis, reservoir capacity, flood estimation and routing, surface run-off models, ground water hydrology - steady state well hydraulics and aquifers; Application of Darcy's Law.

Analysis of Previous GATE Papers

Exam Year	1 Mark Ques.	2 Marks Ques.	Total Marks
1995	2	1	4
1996	1	–	1
1997	3	–	3
1998	1	1	3
1999	5	–	5
2000	1	–	1
2001	3	1	5
2002	1	2	5
2003	1	5	11
2004	–	2	4
2005	2	2	6
2006	–	3	6
2007	–	3	6
2008	1	–	1
2009	–	2	4
2010	1	–	1
2011	1	2	5
2012	1	4	9
2013	2	2	6
2014 Set-1	1	–	1
2014 Set-2	2	–	2

Exam Year	1 Mark Ques.	2 Marks Ques.	Total Marks
2015 Set-1	–	2	4
2015 Set-2	–	1	2
2016 Set-1	2	–	2
2016 Set-2	–	1	2
2017 Set-1	1	1	3
2017 Set-2	2	1	4
2018 Set-1	1	2	5
2018 Set-2	1	1	3
2019 Set-1	2	1	4
2019 Set-2	1	1	3
2020 Set-1	1	–	1
2020 Set-2	2	–	2
2021 Set-1	–	–	–
2021 Set-2	1	1	3
2022 Set-1	–	1	2
2022 Set-2	1	1	3
2023 Set-1	1	2	5
2023 Set-2	1	2	5
2024 Set-1	–	3	6
2024 Set-2	–	1	2

1 Precipitation and General Aspects of Hydrology

1.1 A hyetograph is a graph representing
(a) rainfall volume with time
(b) rainfall intensity with time
(c) rainfall volume with duration
(d) rainfall intensity over an area [1997 : 1 M]

1.2 An isochrone is a line on the basin map
(a) joining raingauge stations having equal rainfall duration
(b) joining points having equal rainfall depth in a given time interval
(c) joining points having equal time of travel of surface runoff to the catchment outlet
(d) joining points which are at equal distance from the catchment outlet

[1999 : 1 M]

1.3 Match **List-I** with **List-II** and select the correct answers using the codes given below the lists:

	List-I		List-II
A.	Rainfall intensity	1.	Isohyets
B.	Rainfall excess	2.	Cumulative rainfall
C.	Rainfall averaging	3.	Hyetograph
D.	Mass curve	4.	Direct runoff hydrograph

Codes:

	A	B	C	D
(a)	1	3	2	4
(b)	3	4	1	2
(c)	1	2	4	3
(d)	3	4	2	1

[2003 : 2 M]

1.4 The intensity of rainfall and time interval of a typical storm are:

Time interval (minutes)	Intensity of rainfall (mm/minute)
0-10	0.7
10-20	1.1
20-30	2.2
30-40	1.5
40-50	1.2
50-60	1.3
60-70	0.9
70-80	0.4

The maximum intensity of rainfall for 20 minutes duration of the storm is
(a) 1.5 mm/minute (b) 1.85 mm/minute
(c) 2.2 mm/minute (d) 3.7 mm/minute

[2005 : 1 M]

1.5 An isohyet is a line joining points of
(a) equal temperature
(b) equal humidity
(c) equal rainfall depth
(d) equal evaporation [2013 : 1 M]

1.6 In a catchment, there are four rain-gauge stations, *P*, *Q*, *R* and *S*. Normal annual precipitation values at these stations are 780 mm, 850 mm, 920 mm and 980 mm, respectively. In the year 2013, stations *Q*, *R* and *S*, were operative but *P* was not. Using the normal ratio method, the precipitation at station *P* for the year 2013 has been estimated as 860 mm. If the observed precipitation at stations *Q* and *R* for the year 2013 were 930 mm and 1010 mm, respectively; what was the observed precipitation (in mm) at station *S* for that year?

[2015 : 2 M, Set-I]

1.7 A catchment is idealized as a 25 km × 25 km square. It has five rain gauges, one at each corner and one at the center, as shown in the figure.

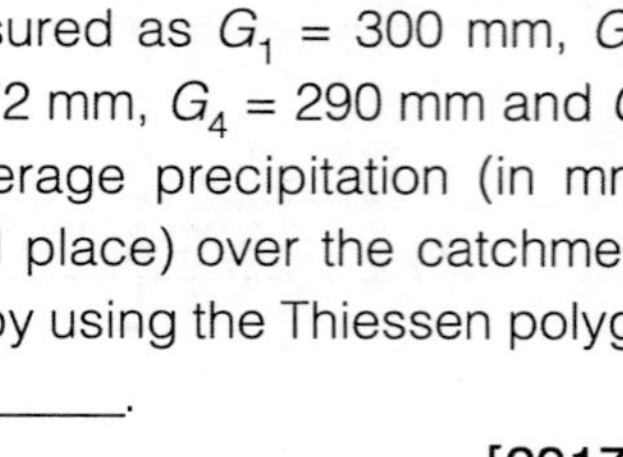

During a month, the precipitation at these gauges is measured as G_1 = 300 mm, G_2 = 285 mm, G_3 = 272 mm, G_4 = 290 mm and G_5 = 288 mm. The average precipitation (in mm, up to one decimal place) over the catchment during this month by using the Thiessen polygon method is ________.

[2017 : 2 M, Set-II]

1.8 Rainfall depth over a watershed is monitored through six number of well distributed rain gauges. Gauged data are given below:

Rain Gauge Number	1	2	3	4	5	6
Rainfall Depth (mm)	470	465	435	525	480	510
Area of Thiessen Polygon ($\times 10^4$ m^2)	95	100	98	80	85	92

The Thiessen mean value (in mm, up to one decimal place) of the rainfall is______

[2018 : 2 M, Set-I]

1.9 catchment may be idealised as a rectangle. There are three rain gauges located inside the catchment at arbitrary locations. The average precipitation over the catchment is estimated by two methods:

(i) Arithmetic mean (P_A), and

(ii) Thiessen polygon (P_T). Which one of the following statements is correct?

(a) P_A is always smaller than P_T

(b) P_A is always equal to P_T

(c) There is no definite relationship between P_A and P_T

(d) P_A is always greater than P_T

[2019 : 1 M, Set-I]

1.10 A catchment may be idealized as a circle of radius 30 km. There are five rain gauges, one at the center of the catchment and four on the boundary (equi-spaced), as shown in the figure (not to scale).

The annual rainfall recorded at these gauges in a particular year are given below.

Gauge	G_1	G_2	G_3	G_4	G_5
Rainfall (mm)	910	930	925	895	905

Using the Thiessen polygon method, what is the average rainfall (in mm, rounded off to two decimal places) over the catchment in that year ________.

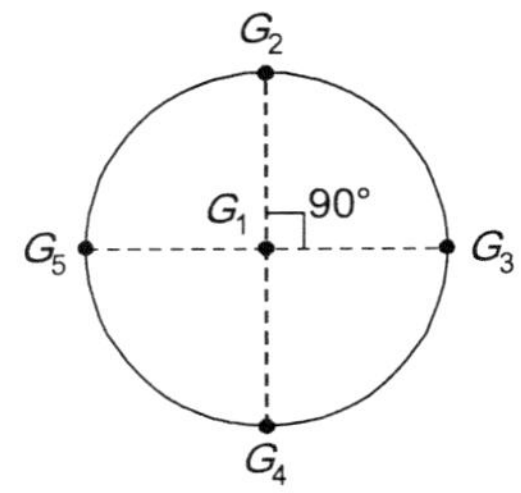

[2023 : 2 M, Set-II]

Answers Precipitation and General Aspects of Hydrology

1.1 (b) 1.2 (c) 1.3 (b) 1.4 (b) 1.5 (c) 1.6 (1093.43) 1.7 (287.37)
1.8 (479.09) 1.9 (c) 1.10 (912.5)

Explanations Precipitation and General Aspects of Hydrology

1.1 (b)

A hyetograph is a graph representing rainfall intensity with time over an area.

On the other hand, a hydrograph is a graph showing the rate of flow (discharge) versus time at a specific point of a river or channel.

1.2 (c)

If points on the area having equal time of travel are considered and located on a map of the catchment, a line joining them is called an isochrone.

POINTS TOREMEMBER

- **Isohyets:** Line joining points of equal rainfall.
- **Isopleths:** Line joining points of equal evapotranspiration.
- **Isochrones:** Line joining points having equal time of travel of runoff to catchment outlet.

1.3 (b)

Rainfall intensity	:	Hyetrograph
Rainfall excess	:	Direct runoff hydrograoh
Rainfall averaging	:	Isohyetls (used in isohyetal method
Mass curve	:	Cumulative rainfall

1.4 (b)

Maximum intensity for 20 minutes is in (20-40) minutes period.

$$\text{Maximum intensity} = \frac{2.2\times 10 + 1.5\times 10}{10+10}$$

$$= 1.85 \text{ mm/minute}$$

1.5 (c)

Isohyet is a line joining points on a map that receive equal amounts of precipitation within a specified time period.

1.6 Sol.

Let P_p = Precipitation in 2013 at station P

N_p = Normal annual precipitation at station P

$$\frac{P_P}{N_P} = \frac{1}{3}\left[\frac{P_Q}{N_Q}+\frac{P_R}{N_R}+\frac{P_S}{N_S}\right]$$

$$\frac{860}{780} = \frac{1}{3}\left[\frac{930}{850}+\frac{1010}{920}+\frac{P_s}{980}\right]$$

$$\Rightarrow \quad P_s = 1093.43 \text{ mm}$$

1.7 Sol.

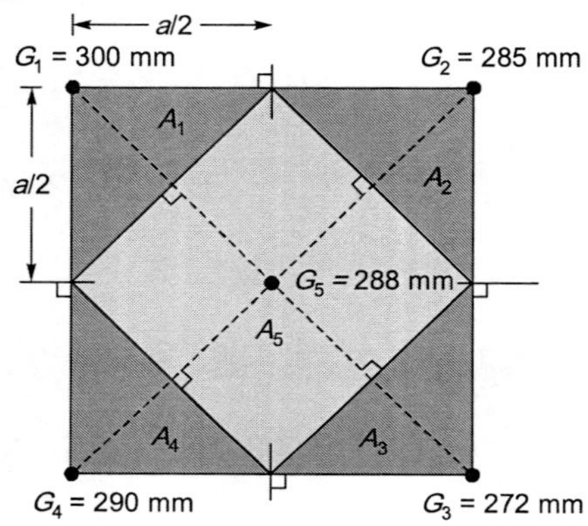

Let sides of square be, a = 25 km

Now area of the polygon is calculated as,

$$A_1 = A_2 = A_3 = A_4$$

$$= \frac{1}{2}\times\frac{a}{2}\times\frac{a}{2} = \frac{a^2}{8} = \frac{25^2}{8} = 78.125$$

$$A_5 = \text{Area of square of sides } \frac{\sqrt{2}a}{2} = \frac{a}{\sqrt{2}}$$

$$= \frac{a^2}{2} = \frac{25^2}{2} = 312.5 \text{ km}^2$$

∴ Mean precipitation,

$$\bar{P} = \frac{G_1A_1+G_2A_2+G_3A_3+G_4A_4+G_5A_5}{A_1+A_2+A_3+A_4+A_5}$$

$$\therefore \bar{P} = \frac{(300+285+272+290)\times 78.125+288\times 312.5}{(4\times 78.125)+312.5}$$

$$= 287.375 \text{ mm}$$

1.8 Sol.

Thiessen mean value of rainfall;

$$P_{avg} = \frac{\sum_{i=1}^{6} P_i A_i}{\sum_{i=1}^{6} A_i}$$

$$\Rightarrow P_{avg} = \frac{470\times 95+465\times 100+435\times 98+525\times 80+480\times 85+510\times 92}{95+100+98+80+85+92}$$

$$= 479.09 \text{ mm}$$

1.9 (c)

The result from Thiessen polygon method is more accurate than arithmetic mean method. But there is no any close relationship between values obtained by Thiessen polygon method and Arithmetic mean method.

∴ There is no any relation between P_A and P_T.

1.10 Sol.

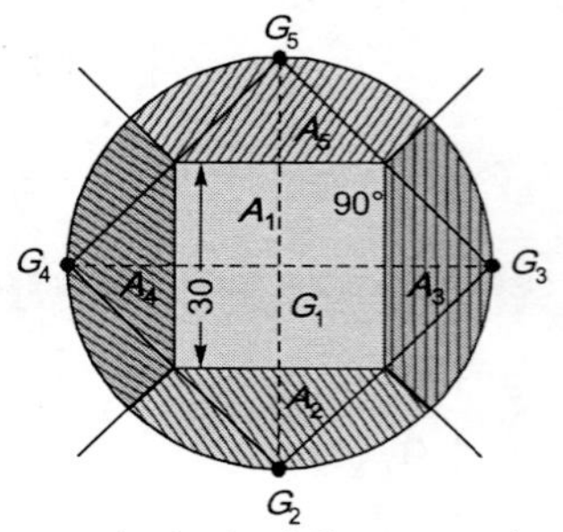

Area of circle (A) = $\pi \times (30)^2$ km^2

A_1 = 30 × 30 = 900 km^2

$A_2 = A_3 = A_4 = \; = A_5$

$$= \frac{\pi\times 30^2 - 900}{4} = 481.858 \text{ km}^2$$

∴ Average rainfall,

$$\bar{P} = \frac{G_1A_1+G_2A_2+G_3A_3+G_4A_4+G_5A_5}{A}$$

$$= \frac{910\times 900+481.858(930+925+895+905)}{\pi\times 30^2}$$

$$= 912.555 \text{ mm}$$

■■■■

2 Evaporation, Transpiration, Evapotranspiration and Stream Flow Measurement

2.1 A lysimeter is used to measure
(a) infiltration (b) evaporation
(c) evapotranspiration (d) radiation
[1995 : 1 M]

2.2 Lysimeter and Tensiometer are used to measure respectively, one of the following groups of quantities
(a) Capillary potential and permeability
(b) Evapotranspiration and capillary potential
(c) Velocity in channels and vapour pressure
(d) Velocity in pipes and pressure head
[1997 : 1 M]

2.3 The Bowen ratio is defined as
(a) Ratio of heat and vapour diffusivities
(b) Proportionality constant between vapour heat flux and sensible heat flux
(c) Ratio of actual evapotranspiration and potential evapotranspiration
(d) Proportionality constant between heat energy used up in evaporation and the bulk radiation from a water body
[1998 : 1 M]

2.4 The number of revolutions of a current meter in 50 seconds were found to be 12 and 30 corresponding to the velocities of 0.25 and 0.46 m/s respectively. What velocity (in m/s) would be indicated by 50 revolutions of that current meter in one minute?
(a) 0.42 (b) 0.50
(c) 0.60 (d) 0.73
[1999 : 1 M]

2.5 In a river, discharge is 173 m^3/s; water surface slope is 1 in 6000; and stage at the gauge station is 10.0 m. If during a flood, the stage at the gauge station is same and the water surface slope is 1 in 2000, the flood discharge in m^3/s, is approximately
(a) 371 (b) 100
(c) 519 (d) 300 [1999 : 1 M]

2.6 Isopleths are lines on a map through points having equal depth of
(a) Rainfall (b) Infiltration
(c) Evapotranspiration (d) Total runoff
[2001 : 1 M]

2.7 The plan area of a reservoir is 1 km^2. The water level in the reservoir is observed to decline by 20 cm in a certain period. During this period the reservoir receives a surface inflow of 10 hectare-meters, and 20 hectare-meters are abstracted from the reservoir for irrigation and power. The pan evaporation and rainfall recorded during the same period at a nearby meteorological station are 12 cm and 3 cm respectively. The calibrated pan factor is 0.7. The seepage loss from the reservoir during this period in hectare-meters is
(a) 0.0 (b) 1.0
(c) 2.4 (d) 4.6
[2003 : 2 M]

2.8 The ratio of actual evapo-transpiration to potential evapo-transpiration is in the range
(a) 0.0 to 0.4 (b) 0.6 to 0.9
(c) 0.0 to 1.0 (d) 100 to 2.0
[2012 : 1 M]

2.9 The top width and the depth of flow in a triangular channel were measured as 4 m and 1 m, respectively. The measured velocities on the centre line at the water surface. 0.2 m and 0.8 m below the surface are 0.7 m/s, 0.6 m/s and 0.4 m/s respectively. Using two-point method of velocity measurement, the discharge (in m^3/s) in the channel is
(a) 1.4 (b) 1.2
(c) 1.0 (d) 0.8 [2012 : 2 M]

2.10 The average surface area of a reservoir in the month of June is 20 km^2. In the same month, the average rate of inflow is 10 m^3/s, outflow rate is 15 m^3/s, monthly rainfall is 10 cm, monthly seepage loss is 1.8 cm and the storage change is 16 million m^3. The evaporation (in cm) in that month is
(a) 46.8 (b) 136.0
(c) 13.6 (d) 23.4
[2015 : 2 M, Set-II]

2.11 The cross-section of a small river is sub-divided into seven segments of width 1.5 m each. The average depth, and velocity at different depths were measured during a field campaign at the middle of each segment width. The discharge computed by the velocity area method for the given data is ________ m^3/s (round off to one decimal place).

Segment	Average depth (D) (m)	Velocity (m/s) at different depths		
		0.2 D	0.6 D	0.8 D
1	0.40	--	0.40	--
2	0.70	0.76	--	0.70
3	1.20	1.19	--	1.13
4	1.40	1.25	--	1.10
5	1.10	1.13	--	1.09
6	0.80	0.69	--	0.65
7	0.45	--	0.42	--

[2023 : 2 M, Set-II]

Answers Evaporation, Transpiration, Evapotranspiration and Stream Flow Measurement

2.1 (c) 2.2 (b) 2.3 (a) 2.4 (c) 2.5 (d) 2.6 (c) 2.7 (d)
2.8 (c) 2.9 (c) 2.10 (d) 2.11 (8.5)

Explanations Evaporation, Transpiration, Evapotranspiration and Stream Flow Measurement

2.1 (c)

Measurement of evapotranspiration is done in two ways either by using **lysimeter** or by the use of **field plots**. Lysimeter studies are time consuming and expansive.

2.2 (b)

An isochrone is a line joining points that have the same travel time from a specific location, often used in transportation planning or mapping services.

2.3 (a)

The sensible heat transfer from water surface to air (H_a) cannot be readily measured. It is estimated by using **Bowen's ratio** (β) given by the expression,

$$\beta = \frac{H_a}{\rho L E_L} = 6.1\times10^{-4}\times P_a\left(\frac{T_w - T_a}{e_w - e_a}\right)$$

Where, P_a = atmospheric pressure in mm of Hg.
e_w = saturated vapour pressure in mm of Hg
e_a = actual vapour pressure in mm of Hg
T_w = temperature of water surface in °C and
T_a = temperature of air in °C

2.4 (c)

For a **current meter,**

$$V = aN_s + b$$

For $V = 0.25$ m/s

$N_s = 12$

$\therefore \quad 0.25 = 12a + b \quad ...(i)$

$N_s = 30$

$0.46 = 30a + b \quad ...(ii)$

Solving eq. (i) and (ii), we get,

$a = 0.01167$

$b = 0.11$

For 50 revolutions in one minute,

∴ Number of revolutions in 50 seconds

$$= \frac{50\times50}{60} = 41.67$$

∴ Velocity indicated,

$$V = aN_s + b$$

$= 0.01167 \times 41.67 + 0.11 = 0.596 \cong 0.60$

2.5 (d)

$$Q = A.V = A\left[\frac{1}{n}R^{2/3}S^{1/2}\right]$$

$$\simeq \frac{1}{n}By^{5/3}S^{1/2}$$

[∵, stage (y) - constant]

$$Q \propto \sqrt{S}$$

$$\frac{Q_2}{Q_1} = \sqrt{\frac{S_2}{S_1}}$$

$$Q_2 = 300 \text{ m}^3/\text{s}$$

2.6 (c)

The mean annual PET (in cm) over various parts of the country is shown on the map. Isopleths are the line on a map through places having equal depth of "evapotransipiration".

2.7 (d)

Inflow to reservoir,

$$I = 10 \text{ ha-m}$$

Outflow from reservoir,

$$Q = 20 \text{ ha-m}$$

Evaporation loss,

$$E = 1 \times 10^6 \times \frac{12}{100} \times 0.7$$

$$= 84000 \text{ m}^3 = 8.4 \text{ ha-m}$$

Rainfall, $P = 1 \times 10^6 \times \frac{3}{100}$

$$= 30000 \text{ m}^3 = 3 \text{ ha-m}$$

Change in storage,

$$\Delta S = -\left(1\times10^6\times\frac{20}{100}\right)$$

$$= -200000 \text{ m}^3 = -20 \text{ ha-m}$$

We know that,

$(I + P) - (E + Q + \text{seepage}) = \Delta S$

$\Rightarrow$ $(10 + 3) - (8.4 + 20 + \text{seepage}) = -20$

$\Rightarrow$ $13 - 28.4 - \text{seepage} = -20$

$\Rightarrow$ Seepage = 4.6 ha-m

2.8 (c)

The ratio of actual evapo-transpiration to potential evapotranspiration can vary, but it is typically expressed as a percentage and falls within a range, often between 0% and 100%.

2.9 (c)

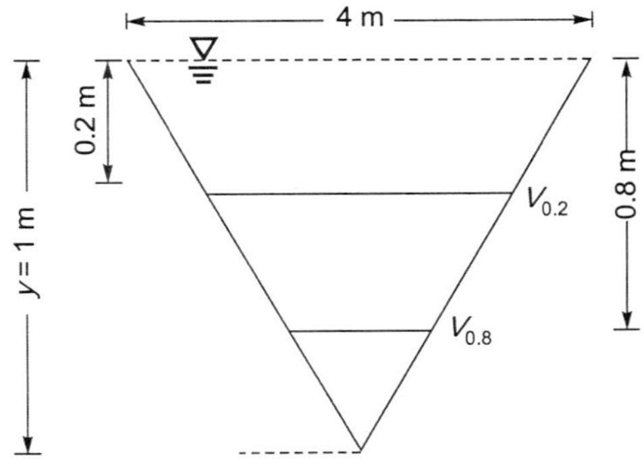

$V_{0.2} = 0.6$ m/s

$V_{0.8} = 0.4$ m/s

$$V = \frac{V_{0.2}+V_{0.8}}{2} = \frac{0.6+0.4}{2} = 0.5 \text{ m/s}$$

$$Q = AV = \frac{1}{2}\times1\times4\times0.5 = 1.0 \text{ m}^3/\text{s}.$$

POINTS TOREMEMBER

Mean velocity of a stream is determined using the following methods.

(i) One point method: ($v_{0.6d}$)

(ii) Two point method: ($v_{0.2d}$ and $v_{0.8d}$) avg.

(iii) Three point method: ($v_{0.2d}$, $v_{0.6d}$ and $v_{0.8d}$)avg.

2.10 (d)

Let 'x' cm evaporation takes place in month of June.

Total inflow = I + P

$$= \left(\frac{10\times30\times24\times60\times60}{20\times10^6}\times100\right)+10$$

$$= 139.6 \text{ cm}$$

Total outflow = Q+ S+ E

$$= \left(\frac{15\times30\times24\times60\times60}{20\times10^6}\times100\right)+1.8+x$$

$$= 196.2 + x \text{ cm}$$

As total outflow is more than total inflow, therefore depression in storage takes place.

Depression in storage

$$= -\frac{16\times10^6}{20\times10^6}\times100 = -80 \text{ cm}$$

$$\Rightarrow \quad 139.6-(196.2+x) = -80$$

$$-x = -80+56.6$$

$$\therefore \quad x = 23.4 \text{ cm}$$

2.11 Sol.

Segment (1)	Average depth (D) m (2)	Velocity of depths 0.2 D (3)	0.6 D (4)	0.8 D (5)	$\frac{0.2D+0.8D}{2}$ (6)	Average Velocity
1	0.40	--	0.40	--	--	0.4
2	0.70	0.76	--	0.70	0.73	0.13
3	1.20	1.19	--	1.13	1.16	1.16
4	1.40	1.25	--	1.10	1.175	1.175
5	1.10	1.13	--	1.09	1.11	1.11
6	0.80	0.69	--	0.65	0.67	0.67
7	0.45	--	0.42	--	--	0.42

Effective width of 1st section = $(\bar{W}_1)$

$$= \frac{\left(W_1+\frac{W_2}{2}\right)^2}{2W_1} = \frac{\left(1.5+\frac{1.5}{2}\right)^2}{2\times1.5} = 1.6875 \text{ m}$$

Effective width of 2nd to 6th section ($W_2 = W_2$ .... $= W_6$)

$$= \left(\frac{1.5+1.5}{2}\right) = 1.5 \text{ m}$$

Effective width of 7th section

$$= \frac{\left(1.5+\frac{1.5}{2}\right)^2}{2\times1.5} = 1.6875\ m$$

$$\therefore \quad \text{Discharge} = \sum_{i=1}^{7}\Delta Q_i \ \sum_{i=1}^{7}\Delta Q_i = \sum_{i=1}^{7} W_i \times y_1 \times \bar{V}_i$$

$= (1.6875 \times 0.4 \times 0.4) + (1.5 \times 0.7 \times 0.73) + (1.5 \times 1.2 \times 1.16) + (1.5 \times 1.4 \times 1.175) + (1.5 \times 1.1 \times 1.11) + (1.5 \times 0.8 \times 0.67) + (1.6875 \times 0.45 \times 0.42)$

$= 8.564 \text{ m}^3\text{/sec}$

3 Infiltration, Runoff and Hydrographs

3.1 The peak discharge of the instantaneous unit hydrograph of a basin, when compared to the peak discharge of a 4 hour unit hydrograph of that basin, would be

(a) greater (b) equal
(c) equal or lesser (d) lesser

[1995 : 1 M]

3.2 A 2-hour unit hydrograph can be approximated as trapezoidal as shown in figure. The unit hydrograph refers to catchment of area

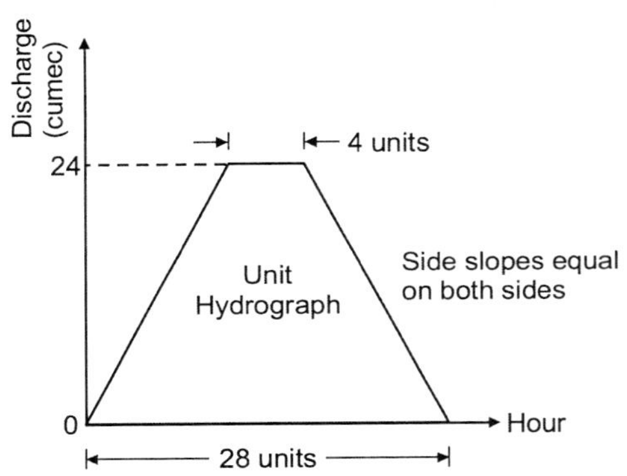

(a) 138.24 km^2 (b) 0.0384 km^2
(c) 384 m^2 (d) 3840 m^2

[1996 : 1 M]

3.3 The ordinate of the Instantaneous Unit Hydrograph (IUH) of a catchment at any time *t*, is

(a) the slope of the 1 hour unit hydrograph at that time
(b) the slope of the direct runoff unit hydrograph at that time
(c) difference in the slope of the *S* curve and 1 hour unit hydrograph
(d) the slope of the *S* curve with effective rainfall intensity of 1 cm/hr

[1999 : 1 M]

3.4 A direct runoff hydrograph due to an isolated storm with an effective rainfall of 2 cm was trapezoidal in shape as shown in figure. The hydrograph corresponds to a catchment area (in sq. km) of

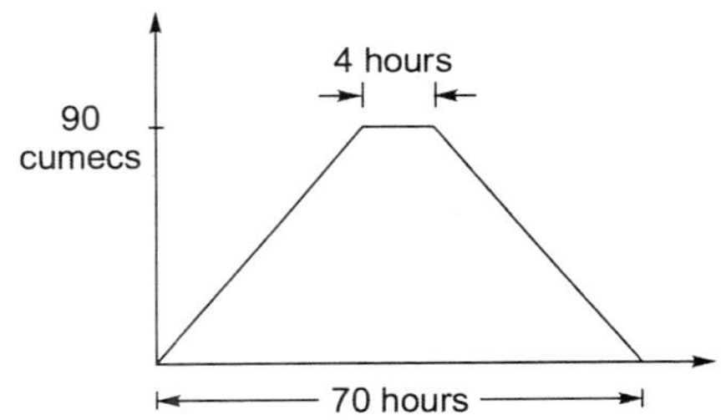

(a) 790.2 (b) 599.4
(c) 689.5 (d) 435.3

[1999 : 1 M]

3.5 The parameters in Horton's infiltration equation $[f(t) = f_c + (f_0 - f_c)e^{-kt}]$ are given as, f_0 = 7.62 cm/hour, f_c = 1.34 cm/hour and k = 4.182/hour. For assumed continuous ponding the cumulative infiltration at the end of 2 hours is

(a) 2.68 cm (b) 1.50 cm
(c) 1.34 cm (d) 4.18 cm

[2000 : 1 M]

3.6 The direct runoff hydrograph of a storm obtained from a catchment is triangular in shape and has a base period of 80 hours. The peak flow rate is 30 m^3/s and catchment area is 86.4 km^2. The rainfall excess that has resulted the above hydrograph is

(a) 5 cm (b) 8 cm
(c) 10 cm (d) 16 cm

[2001 : 2 M]

3.7 A 6-hour unit hydrograph (UH) of a catchment is triangular in shape with a total time base of 36 hours and a peak discharge of 18 m^3/s. The area of the catchment (in sq. km) is

(a) 233
(b) 117
(c) 1.2
(d) Sufficient information not available

[2002 : 1 M]

3.8 During a 6-hour storm, the rainfall intensity was 0.8 cm/hour on a catchment of area 8.6 km^2. The measured runoff volume during this period was 256000 m^3. The total rainfall that was lost due to infiltration, evaporation, and transpiration (in cm/hour) is

(a) 0.80
(b) 0.304
(c) 0.496
(d) sufficient information not available

[2002 : 2 M]

3.9 The rainfall on five successive days in a catchment were measured as 3, 8, 12, 6 and 2 cms. If the total run-off at the outlet from the catchment was 15 cm, the value of the ϕ-index (in mm/hour) is
(a) 0.0
(b) 1.04
(c) 1.33
(d) sufficient information not available

[2002 : 2 M]

3.10 The vertical hydraulic conductivity of the top soil at certain stage is 0.2 cm/hr. A storm of intensity 0.5 cm/hr occurs over the soil for an indefinite period. Assuming the surface drainage to be adequate, the infiltration rate after the storm has lasted for a very long time, shall be
(a) smaller than 0.2 cm/hr
(b) 0.2 cm/hr
(c) between 0.2 & 0.5 cm/hr
(d) 0.5 cm/hr

[2003 : 1 M]

3.11 While applying the Rational formula for computing the design discharge, the rainfall duration is stipulated as the time of concentration because
(a) this leads to the largest possible rainfall intensity
(b) this leads to the smallest possible rainfall intensity
(c) the time of concentration is the smallest rainfall duration for which the Rational formula is applicable
(d) the time of concentration is the largest rainfall duration for which the Rational formula is applicable

[2003 : 2 M]

Common Data for Questions 3.12 and 3.13:

An average rainfall of 16 cm occurs over a catchment during a period of 12 hours with a uniform intensity. The unit hydrograph (unit depth = 1 cm, duration = 6 hours) of the catchment rises linearly from 0 to 30 cumecs in six hours and then falls linearly from 30 to 0 cumecs in the next 12 hours. ϕ index of the catchment is known to be 0.5cm/hr. Base flow in the river is known to be 5 cumecs.

3.12 Peak discharge of the resulting direct runoff hydrograph shall be
(a) 150 cumecs (b) 225 cumecs
(c) 230 cumecs (d) 360 cumecs

[2003 : 2 M]

3.13 Area of the catchment in hectares is
(a) 97.20 (b) 270
(c) 9720 (d) 27000 [2003 : 2 M]

3.14 The average rainfall for a 3 hour duration storm is 2.7 cm and the loss rate is 0.3 cm/hr. The flood hydrograph has a base flow of 20 m^3/s and produces a peak flow of 210 m^3/s. The peak of a 3-h unit hydrograph is
(a) 125.50 m^3/s (b) 105.50 m^3/s
(c) 77.77 m^3/s (d) 70.37 m^3/s

[2004 : 2 M]

3.15 The rainfall during three successive 2 hour periods are 0.5, 2.8 and 1.6 cm. The surface runoff resulting from this storm is 3.2 cm. The ϕ index value of this storm is
(a) 0.20 cm/hr (b) 0.28 cm/hr
(c) 0.30 cm/hr (d) 0.80 cm/hr

[2004 : 2 M]

Linked Answer Questions 3.16 and 3.17:

A four hour unit hydrograph of a catchment is triangular in shape with base of 80 hours. The area of the catchment is 720 km^2. The base flow and ϕ-index are 30 m^3/s and 1 mm/h, respectively. A storm of 4 cm occurs uniformly in 4 hours over the catchment.

3.16 The peak discharge of four hour unit hydrograph is
(a) 40 m^3/s (b) 50 m^3/s
(c) 60 m^3/s (d) 70 m^3/s

[2005 : 2 M]

3.17 The peak flood discharge due to the storm is
(a) 210 m^3/s (b) 230 m^3/s
(c) 260 m^3/s (d) 720 m^3/s

[2005 : 2 M]

3.18 During a 3 hour storm event, it was observed that all abstractions other than infiltration are negligible. The rainfall was idealized as 3 one hour storms of intensity 10 mm/hr, 20 mm/hr and 10 mm/hr respectively and the infiltration was idealized as a Horton curve, $f = 6.8 + 8.7 \exp(-t)$(f in mm/hr and t in hr). What is the effective rainfall?
(a) 10.00 mm (b) 11.33 mm
(c) 12.43 mm (d) 13.63 mm [2006 : 2 M]

Common Data for Questions 3.19 and 3.20:

For a catchment, the *S*-curve (or *S*-hydrograph) due to a rainfall of intensity 1 cm/hr is given by $Q = 1 - (1 + t)\exp(-t)$ (t in hr and Q in m^3/s).

3.19 What is the area of the catchment?

(a) 0.01 km^2 (b) 0.36 km^2
(c) 1.00 km^2 (d) 1.28 km^2

[2006 : 2 M]

3.20 What will be the ordinate of a 2-hour unit hydrograph for this catchment at $t = 3$ hour?

(a) 0.13 m^3/s (b) 0.20 m^3/s
(c) 0.27 m^3/s (d) 0.54 m^3/s

[2006 : 2 M]

3.21 An isolated 4-hour storm occurred over a catchment as follows:

Time	1st hr	2nd hr	3rd hr	4th hr
Rainfall (mm)	9	28	12	7

The ϕ index for the catchment is 10 mm/h. The estimated runoff depth from the catchment due to the above storm is

(a) 10 mm (b) 16 mm
(c) 20 mm (d) 23 mm

[2007 : 2 M]

Common Data for Questions 3.22 and 3.23:

Ordinates of a 1-hour unit hydrograph at 1 hour intervals, starting from time $t = 0$ are 0, 2, 6, 4, 2, 1 and 0 m^3/s.

3.22 Catchment area represented by this unit hydrograph is

(a) 1.0 km^2 (b) 2.0 km^2
(c) 3.2 km^2 (d) 5.4 km^2

[2007 : 2 M]

3.23 Ordinate of a 3-hour unit hydrograph for the catchment at $t = 3$ hours is

(a) 2.0 m^3/s (b) 3.0 m^3/s
(c) 4.0 m^3/s (d) 5.0 m^3/s

[2007 : 2 M]

Common Data for Questions 3.24 and 3.25:

One hour triangular unit hydrograph of a watershed has the peak discharge of 60 m^3/sec at 10 hours and time base of 30 hours. The ϕ index is 0.4 cm per hour and base flow is 15 m^3/sec.

3.24 The catchment area of the watershed is

(a) 3.24 km^2 (b) 32.4 km^2
(c) 324 km^2 (d) 3240 km^2

[2009 : 2 M]

3.25 If there is rainfall of 5.4 cm in 1 hour, the ordinate of the flood hydrograph at 15th hour is

(a) 225 m^3/sec (b) 240 m^3/sec
(c) 249 m^3/sec (d) 258 m^3/sec

[2009 : 2 M]

3.26 A watershed got transformed from rural to urban over a period of time. The effect of urbanization on storm runoff hydrograph from the watershed is to

(a) decrease the volume of runoff
(b) increase the time to peak discharge
(c) decrease the time base
(d) decrease the peak discharge

[2011 : 1 M]

Common Data for Questions 3.27 and 3.28:

The ordinates of a 2-*h* unit hydrograph at 1 hour intervals starting from time $t = 0$, are 0, 3, 8, 6, 3, 2 and 0 m^3/s. Use trapezoidal rule for numerical integration, if required.

3.27 What is the catchment area represented by the unit hydrograph?

(a) 1.00 km^2 (b) 2.00 km^2
(c) 7.92 km^2 (d) 8.64 km^2

[2011 : 2 M]

3.28 A storm of 6.6 cm occurs uniformly over the catchment in 3 hours. If ϕ-index is equal to 2 mm/h and base flow is 5 m^3/s, what is the peak flow due to the storm?

(a) 41.0 m^3/s (b) 43.4 m^3/s
(c) 53.0 m^3/s (d) 56.2 m^3/s

[2011 : 2 M]

3.29 Match **List-I** (contains parameters) with **List-II** (methods/instruments) and select the correct answer using the code given below the Lists:

List-I
A. Streamflow velocity
B. Evapo-transpiration rate
C. Infiltration rate
D. Wind velocity

List-II
1. Anemomenter
2. Penman's method
3. Horton's method
4. Current meter

Codes:

	A	B	C	D
(a)	1	2	3	4
(b)	4	3	2	1
(c)	4	2	3	1
(d)	1	3	2	4

[2012 : 2 M]

Linked Answer Questions 3.30 and 3.31:

The drainage area of a watershed is 50 km^2. The ϕ-index is 0.5 cm/hour and the base flow at the outlet is 10 m^3/s. One hour unit hydrograph (unit depth = 1 cm) of the watershed is triangular in shape with a time base of 15 hours. The peak ordinate occurs at 5 hours.

3.30 The peak ordinate (in m^3/s/cm) of the unit hydrograph is

(a) 10.00 (b) 18.52

(c) 37.03 (d) 185.20

[2012 : 2 M]

3.31 For a storm of depth of 5.5 cm and duration of 1 hour, the peak ordinate (in m^3/s) of the hydrograph is

(a) 55.00 (b) 82.60

(c) 92.60 (d) 102.60

[2012 : 2 M]

Common Data Questions (3.32 and 3.33):

Storm-I of duration 5 hrs. gives a direct run-off of 4 cm and has an average intensity of 2 cm/hr.

Storm-II of 8 hr. duration gives a run-off of 8.4 cm.

(Assume ϕ-index is same for both the storms.)

3.32 The value of ϕ-index (in cm/hr) is

(a) 1.2 (b) 1.6

(c) 1 (d) 1.4

[2013 : 2 M]

3.33 Intensity of **storm-II** in cm/hr is

(a) 2 (b) 1.5

(c) 1.75 (d) 2.25 [2013 : 2 M]

3.34 An isolated 3-h rainfall event on a small catchment produces a hydrograph peak and point of inflection on the falling limb of the hydrograph at 7 hours and 8.5 hours respectively, after the start of the rainfall. Assuming, no losses and no base flow contribution, the time of concentration (in hours) for this catchment is approximately

(a) 8.5 (b) 7.0

(c) 6.5 (d) 5.5

[2014 : 1 M, Set-II]

3.35 The 4-hr unit hydrograph for a catchment is given in the table below. What would be the maximum ordinate of the S-curve (in m^3/s) derived from this hydrograph?

Time (hr)	0	2	4	6	8	10	12	14	16	18	20	22	24
Unit Hydrograph ordinate (m^3/s)	0	0.6	3.1	10	13	9	5	2	0.7	0.3	0.2	0.1	0

[2015 : 2 M, Set-I]

3.36 The direct runoff hydrograph in response to 5 cm rainfall excess in a catchment is shown in the figure. The area of catchment (expressed in hectares) is_______.

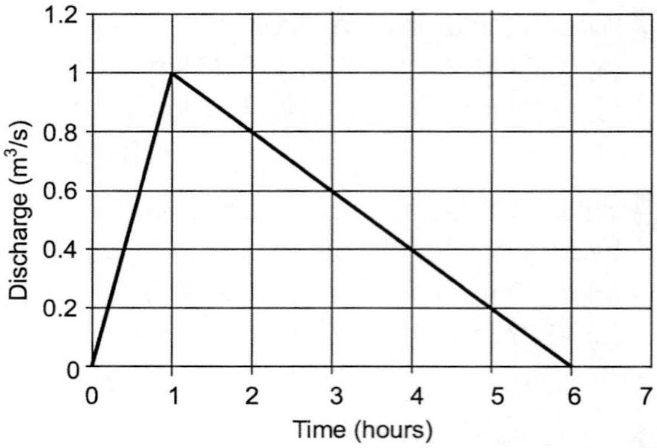

[2016 : 1 M, Set-I]

3.37 The ordinates of a one-hour unit hydrograph at sixty minute interval are 0, 3, 12, 8, 6, 3 and 0 m^3/s. A two-hour storm of 4 cm excess rainfall occurred in the basin from 10 AM. Considering constant base flow of 20 m^3/s, the flow of the river (expressed in m^3/s) at 1 PM is_______.

[2016 : 2 M, Set-II]

3.38 The ordinates of a 2-hour unit hydrograph for a catchment are given as

Time (*h*)	0	1	2	3	4
Ordinate (m^3/s)	0	5	12	25	41

The ordinate (in m^3/s) of a 4-hour unit hydrograph for this catchment at the time of 3h would be_________. [2017 : 1 M, Set-I]

3.39 An effective rainfall of 2-hour duration produced a flood hydrograph peak of 200 m^3/s. The flood hydrograph has a base flow of 20 m^3/s. If the spatial average rainfall in the watershed for the duration of storm is 2 cm and the average loss rate is 0.4 cm/hour, the peak of 2-hour unit hydrograph (in m^3/s-cm, up to one decimal place) is _________.

[2017 : 2 M, Set-I]

3.40 The infiltration capacity of a soil follows the Horton's exponential model, $f = c_1 + c_2e^{-kt}$. During an experiment, the initial infiltration capacity was observed to be 200 mm/h. After a long time, the infiltration capacity was reduced to 25 mm/h. If the infiltration capacity after 1 hour was 90 mm/h, the value of the decay rate constant, k (in h^{-1}, up to two decimal places) is _________.

[2017 : 1 M, Set-II]

3.41 During a storm event in a certain period, the rainfall intensity is 3.5 cm/hour and the ϕ-index is 1.5 cm/hour. The intensity of effective rainfall (in cm/hour, up to one decimal place) for this period is ______________.

[2017 : 1 M, Set-II]

3.42 The infiltration rate f in a basin under ponding condition is given by $f = 30 + 10e^{-2t}$, where, f is in mm/h and t is time in hour. Total depth of infiltration (in mm, up to one decimal place) during the last 20 minutes of a storm of 30 minutes duration is ____

[2018 : 2 M, Set-I]

3.43 The total rainfall in a catchment of area 1000 km^2, during a 6 h storm, is 19 cm. The surface runoff due to this storm computed from triangular direct runoff hydrograph is 1×10^8 m^3. The ϕ-index for this storm (in cm/h, up to one decimal place) is ________.

[2018 : 2 M, Set-II]

3.44 The hyetograph of a storm event of duration 140 minutes is shown in the figure.

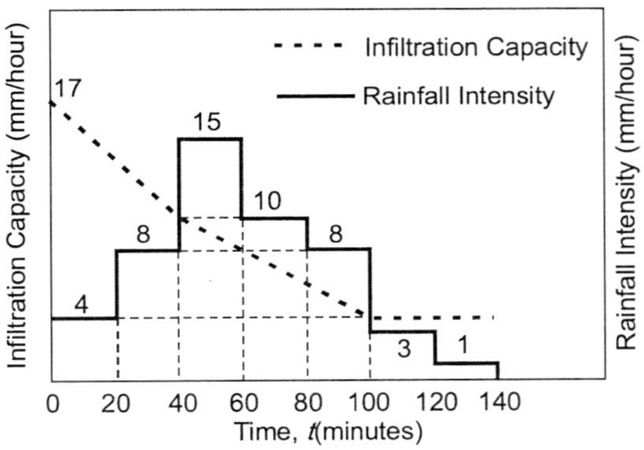

The infiltration capacity at the start of this event (t = 0) is 17 mm/hour, which linearly decreases to 10 mm/hour after 40 minutes duration. As the event progresses, the infiltration rate further drops down linearly to attain a value of 4 mm/hour at t = 100 minutes and remains constant thereafter till the end of the storm event. The value of the infiltration index, φ (in mm/hour, round off to 2 decimal places), is_____.

[2019 : 2 M, Set-I]

3.45 The ordinates, u, of a 2-hour unit hydrograph (i.e., for 1 cm of effective rain), for a catchment are shown in the table.

t (hour)	0	1	2	3	4	5	6	7	8	9	10	11	12
u (m^3/s)	0	2	8	18	32	45	30	19	12	7	3	1	0

A 6-hour storm occurs over the catchment such that the effective rainfall intensity is 1 cm/hour for the first two hours, zero for the next two hours, and 0.5 cm/hour for the last two hours. If the base flow is constant at 5 m^3/s, the peak flow due to this storm (in m^3/s, round off to 1 decimal place) will be________.

[2019 : 2 M, Set-II]

3.46 A triangular direct runoff hydrograph due to a storm has a time base of 90 hours. The peak flow of 60 m^3/s occurs at 20 hours from the start of the storm. The area of catchment is 300 km^2. The rainfall excess of the storm (in cm), is

(a) 5.40 (b) 2.00
(c) 3.24 (d) 6.48

[2020 : 1 M, Set-II]

3.47 The hyetograph in the figure corresponds to a rainfall event of 3 cm.

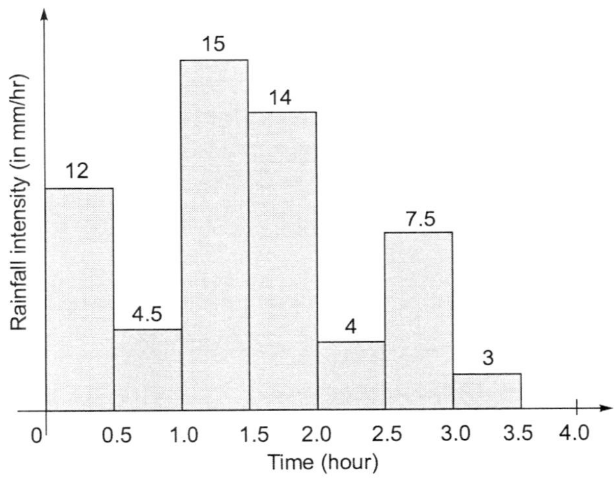

If the rainfall event has produced a direct runoff of 1.6 cm, the ϕ-index of the event (in mm/hour, round off to one decimal place) would be _______.

[2021 : 2 M, Set-II]

3.48 A 12-hour unit hydrograph (of 1 cm excess rainfall) of a catchment is of a triangular shape with a base width of 144 hour and a peak discharge of 23 m^3/s. The area of the catchment (in km^2, round off to the nearest integer) is _________.

[2021 : 1 M, Set-II]

3.49 A two-hour duration storm event with uniform excess rainfall of 3 cm occurred on a watershed. The ordinates of streamflow hydrograph resulting from this event are given in the table.

Time (hours)	0	1	2	3	4	5	6	7
Streamflow (m^3/s)	10	16	34	40	31	25	16	10

Considering a constant base flow of 10 m^3/s, the peak flow ordinate (in m^3/s) of one-hour unit hydrograph for the watershed is ________ . (in integer)

[2022 : 2 M, Set-I]

3.50 The ordinates of a one-hour unit hydrograph for a catchment are given below:

t(hour)	0	1	2	3	4	5	6	7
Q(m^3/s)	0	9	21	18	12	5	2	0

Using the principle of superposition, a D-hour unit hydrograph for the catchment was derived from this one-hour unit hydrograph. The ordinates of the D-hour unit hydrograph were obtained as 3 m^3/s at t = 1 hour and 10 m^3/s at t = 2 hour. The value of D (in integer) is ___________.

[2023 : 1 M, Set-I]

3.51 A 12-hour storm occurs over a catchment and results in a direct runoff depth of 100 mm. The time-distribution of the rainfall intensity is shown in the figure (not to scale). The ϕ-index of the storm is (in mm, rounded off to two decimal places) ___________.

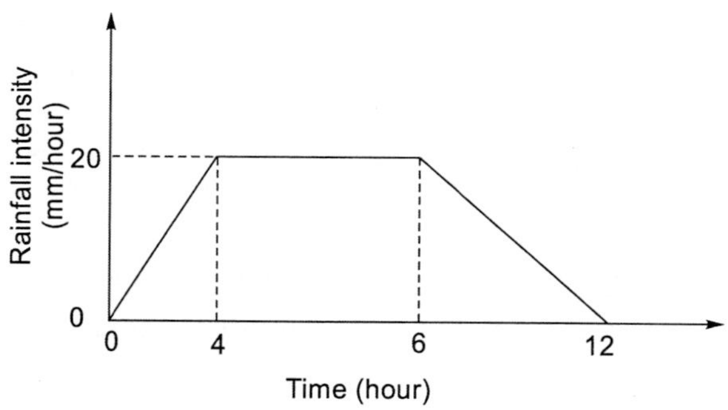

[2023 : 2 M, Set-I]

3.52 In Horton's equation fitted to the infiltration data for a soil, the initial infiltration capacity is 10 mm/h; final infiltration capacity is 5 mm/h; and the exponential decay constant is 0.5 /h. Assuming that the infiltration takes place at capacity rates, the total infiltration depth (in mm) from a uniform storm of duration 12 h is ______________. (round off to one decimal place)

[2023 : 2 M, Set-I]

3.53 The ordinates of a 1-hour unit hydrograph (UH) are given below.

Time(hours)	0	1	2	3	4	5
Ordinates of 1-hour UH(m^3/s)	0	13	50	80	95	85

Time(hours)	6	7	8	9	10	11
Ordinates of 1-hour UH(m^3/s)	55	35	15	10	3	0

These ordinates are used to derive a 3-hour UH. The peak discharge (in m^3/s) for the derived 3-hour UH is _________ (rounded off to the nearest integer).

[2024 : 2 M, Set-I]

3.54 A storm with a recorded precipitation of 11.0 cm, as shown in the table, produced a direct run-off of 6.0 cm.

Time from start (hours)	1	2	3	4	5	6	7	8
Recorded cumulative precipitation (cm)	0.5	1.5	3.1	5.5	7.3	8.9	10.2	11.0

The ϕ-index of this storm is _________ cm/hr (rounded off to 2 decimal places).

[2024 : 2 M, Set-II]

■■■■

Answers Infiltration, Runoff and Hydrographs

3.1	(a)	3.2	(a)	3.3	(d)	3.4	(b)	3.5	(d)	3.6	(a)	3.7	(b)
3.8	(b)	3.9	(*)	3.10	(c)	3.11	(c)	3.12	(b)	3.13	(c)	3.14	(b)
3.15	(c)	3.16	(b)	3.17	(a)	3.18	(d)	3.19	(b)	3.20	(c)	3.21	(c)
3.22	(d)	3.23	(c)	3.24	(c)	3.25	(b)	3.26	(c)	3.27	(c)	3.28	(a)
3.29	(c)	3.30	(b)	3.31	(d)	3.32	(a)	3.33	(d)	3.34	(d)	3.35	(Sol.)
3.36	(21.6)	3.37	(60)	3.38	(15.0)	3.39	(150)	3.40	(0.9904)	3.41	(2.0)	3.42	(11.74)
3.43	(1.5)	3.44	(7.2475)	3.45	(97)	3.46	(c)	3.47	(4.2)	3.48	(59.16)	3.49	(12.66)
3.50	(3)	3.51	(3.60)	3.52	(70.0)	3.53	(87)	3.54	(0.64)				

Explanations Infiltration, Runoff and Hydrographs

3.1 (a)

The limiting case of a unit hydrograph of zero duration is indicated as the duration $D \to 0$. As D is reduced, the intensity of rainfall excess being equal to $1/D$ increases and the unit hydrograph becomes more skewed. Therefore, the peak increases.

3.2 (a)

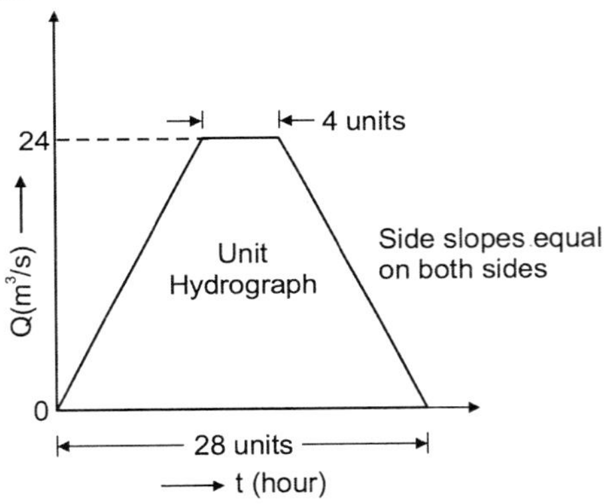

Unit hydrograph is a hydrograph resulted from 1 cm of rainfall excess depth.
Therefore, volume of rainfall excess

$$= \frac{1}{2}(28+4)\times 3600\times 24$$

also, volume of rainfall excess $= A\times\frac{1}{100}$

$$\therefore \quad \frac{1}{2}(28+4)\times 3600\times 24 = A\times\frac{1}{100}$$

$$\Rightarrow \quad A = 138.24\times 10^6 \text{ m}^2$$

$$\Rightarrow \quad A = 138.24 \text{ km}^2$$

3.3 (d)

The limiting case of a **unit hydrograph** of zero duration is known as instantaneous unit hydrograph (IUH). The ordinate of one IUH at any time 't' is the slope of S-curve of intensity 1 cm/hr.

$$\text{IUH} = \frac{1}{i}\frac{dS}{dt}$$

For $\quad i = 1$ cm/hr, $\quad \text{IUH} = \frac{dS}{dt}$

3.4 (b)

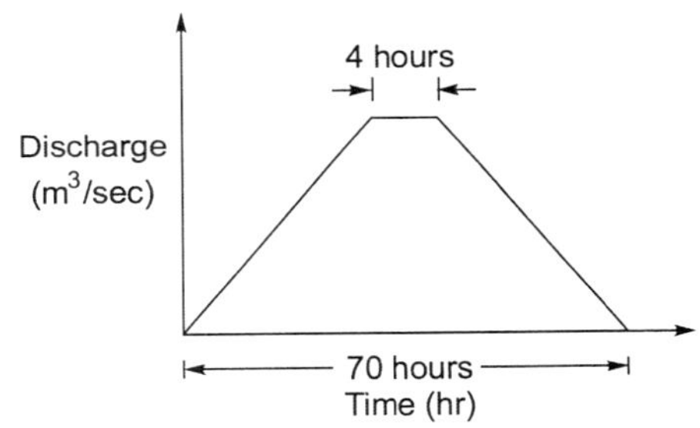

Area of given DRH = Volume of direct runoff

$$= \left[\frac{1}{2}(70+4)\times 60\times 60\times 90\right]\text{m}^3$$

$$= 11{,}988{,}000 \text{ m}^3$$

Effective rainfall = 2 cm

∴ Catchment area

$$= \frac{\text{Volume of DRH}}{\text{Effective rainfall}}$$

$$= \left(\frac{11{,}988{,}000}{2\times 10^{-2}}\right)\text{m}^2$$

$$= 599.4 \text{ km}^2$$

3.5 (d)

Cumulative infiltration at end of 2 hrs,

$$f_p(t = 2\text{ hr}) = \int_0^2 \left(1.34 + 6.28e^{-4.182t}\right)dt$$
$$= 4.18 \text{ cm}$$

3.6 (a)

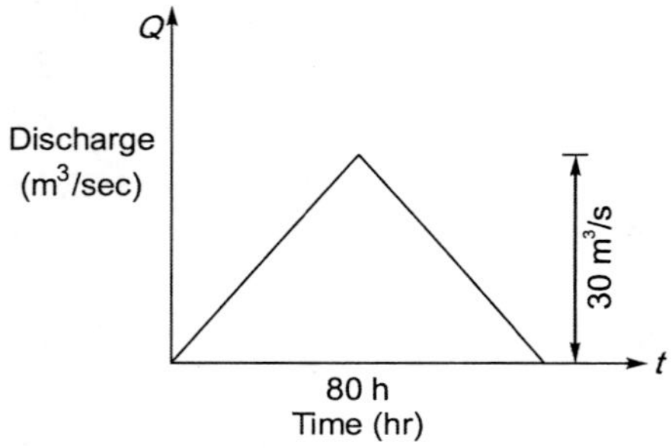

$$A = 86.4 \text{ km}^2$$

Rainfall excess

$$= \frac{\text{Volume of direct runoff}}{\text{Catchment area}}$$

$$= \frac{1}{2} \times \frac{30 \times 80 \times 60 \times 60 \text{m}^3}{86.4 \times 10^6 \text{m}^2} \times 100 = 5 \text{ cm}$$

3.7 (b)

Volume of direct runoff

$$= \frac{1}{2} \times 36 \times 3600 \times 18 = \frac{A \times 1}{100}$$

$$3.6 \times 3.6 \times 9 \times 10^6 = A$$
$$A = 116.64 \text{ km}^2 \simeq 117 \text{ km}^2$$

3.8 (b)

Infiltration (per 6 hr)

$$= \text{Rainfall} - \text{Runoff}$$
$$= 0.8 \times 6 - \frac{256000}{8.6 \times 10^6} \times 100$$
$$= 1.823 \text{ cm}$$

∴ Infiltration (per hr)

$$= \frac{1.823}{6} = 0.304 \text{ cm/hr}$$

3.9 (*)

Method-I

$$\text{W-index} = \frac{P - Q}{t}$$
$$= \frac{(3 + 8 + 12 + 6 + 2) - 15}{5}$$
$$= 3.2 \text{ cm/day}$$

Now, as ϕ-index ≥ w-index

∴ The rainfall of magnitude 3 cm and 2 cm will not participate in rainfall excess depth (i.e., these will completely infiltrate)

$$\therefore \quad \phi\text{-index} = \frac{\text{Infiltration during rainfall excess depth}}{\text{Time period of rainfall excess}}$$

$$= \frac{(8 + 12 + 6) - 15}{3} = 3.67 \text{ cm/day}$$

i.e.,
$$= \frac{3.67 \times 10}{24} \text{ mm/hr} = 1.525 \text{ mm/h}$$

Method-II

Consider time period of rainfall excess,

$$t_e = 5 \text{ hours}$$
$$\phi\text{-index} = \frac{(3 + 8 + 12 + 6 + 2) - 15}{5}$$
$$= 3.2 \text{ cm/hr}$$

But $\phi > (2.3)$

⇒ $t_e = 3$ hours

$$\phi\text{-index} = \frac{(8 + 12 + 6) - 15}{3}$$
$$= 3.67 \text{ cm/hr}$$

For $\phi = 3.67$ cm/hr, $t_e = 3$ hrs. (OK)

Hence, $\phi = 3.67$ cm/hr

3.10 (c)

Infiltration depends upon the permeability of soil, rainfall characteristics, surface cover, temperature etc.

Initially, the soil is partially saturated. In partially saturated soils, entrapped air causes blockage to the flow of water. Hence, the permeability of partially saturated soil is lesser than that of a fully saturated soil.

As the storm continues for an indefinite period, the soil becomes saturated and its permeability increases. Consequently, rate of infiltration also increase (i.e., > 0.2 cm/hr).

Also, it is stated that the surface drainage is adequate. It means either the water infiltrates or flows as surface runoff but doesn't stand at its location.

Hence, infiltration rate $\not>$ 0.5 cm/hr.

⇒Infiltration rate lies between 0.2 and 05 cm/hr.

3.11 (c)

Time of concentration is the time when whole catchment starts contributing to runoff at the outlet.

For an impervious surface, at $t > t_c$ runoff rate becomes equal to rate of rainfall

$$\Rightarrow \quad Q_p = pA$$

But in practice impervious surface doesn't occur.

$$\therefore \quad Q_p = KpA$$

3.12 (b)

The DRH of the storm of 12 hours can be prepared by superimposing another 6-hr UH lagged by 6-hr.

Rainfall excess = 16 – 0.5 × 12 = 10 cm

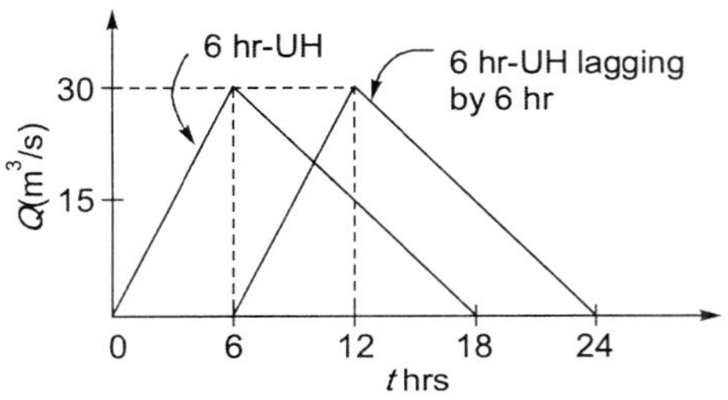

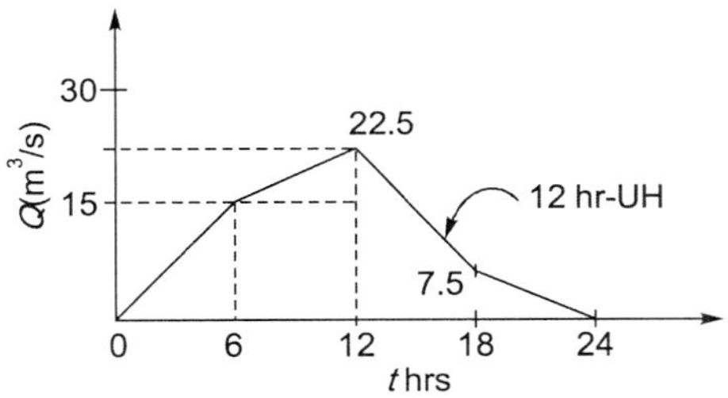

Time	6 hr UH (m^3/s)	6 hr. UH (m^3/s) logged by 6 hr	12hr UH (m^3/s)
0	0	0	0
6	30	30	15
12	15	15	22.5
18	0	0	7.5
24	–	–	0

Peak discharge of resulting direct runoff is given by,

$$Q_P = 22.5 \times \text{rainfall excess}$$
$$= 22.5 \times 10 = 225 \text{ cumecs}$$

3.13 (c)

Catchment area × rainfall excess = Area under UH

$$A \times \frac{1}{100} = \frac{30 \times 18 \times 60 \times 60}{2}$$

$$\Rightarrow \quad A = 9720 \times 10^4 \text{ m}^2$$

$$\Rightarrow \quad A = 9720 \text{ ha}$$

3.14 (b)

Rainfall excess = 2.7 – 0.3 × 3 = 1.8 cm

Peak discharge of DRH = 210 – 20 = 190 m^3/s

$$\text{Peak of 3-h UH} = \frac{190}{1.8} = 105.55 \text{ m}^3/\text{s}$$

3.15 (c)

Total rainfall = 0.5 + 2.8 + 1.6 = 4.9 cm

Excess rainfall = 4.9 – 3.2 = 1.7 cm

$$w\text{-index} = \frac{1.7}{6} = 0.28 \text{ cm/hr}$$

This value being more than $\frac{0.5}{2} = 0.25$ cm/hr, therefore it will not prorduce run-off.

$$\therefore \quad t_e = 4 \text{ hrs}$$

Excess rainfall = 2.8 + 1.6 – 3.2 = 1.2 cm

$$\phi\text{-index} = \frac{1.2}{4} = 0.3 \text{ cm/hr}$$

3.16 (b)

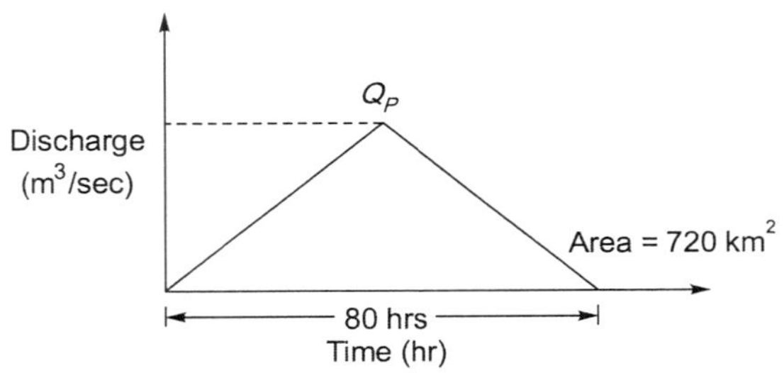

$$\therefore \frac{1}{2} \times Q_P \times 80 \times 60 \times 60 = 720 \times 10^6 \times 1 \times 10^{-2}$$

$$\Rightarrow \quad Q_P = 50 \text{ m}^3/\text{sec}$$

3.17 (a)

Depth of precipitation = 4 cm

Infiltration in 4 hrs

= 4 × 1 = 4 mm = 0.4 cm

Run off depth = 4 – 0.4 = 3.6 cm

Peak of DRH of 4 hrs

= 50 × 3.6 = 180 m^3/sec

Peak of flood discharge,

$$Q = 180 + 30 = 210 \text{ m}^3/\text{sec}$$

3.18 (d)

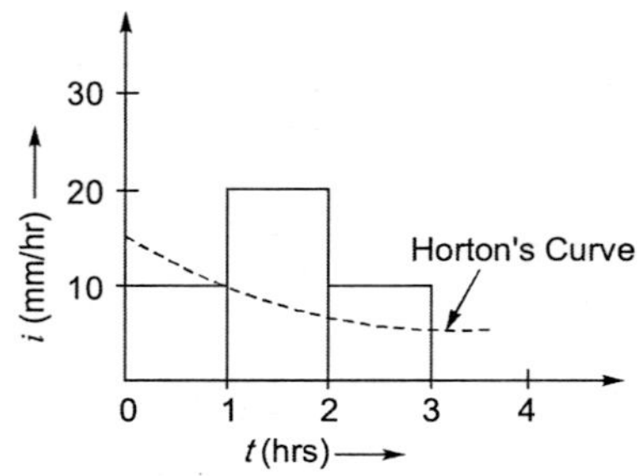

Time	0	1	2	3
f (mm/hr)	15.5	10	8	7.2

From **superposition** of hyetograph and infiltration curve it is clear that first hour does not contribute to effective rainfall (i.e. rainfall excess).

For t = 1 to t = 3, the effective rainfall is,

$$r_e = 20 + 10 - \int_1^3 (6.8 + 8.7\,e^{-t})dt$$

$$= 30 - \left[(6.8t)_1^3 + 8.7\int_1^3 e^{-t}dt\right]$$

$$= 30 - [6.8 \times 2 - 8.7\,(e^{-t})_1^3]$$

$$= 30 - 13.6 + 8.7\left[\frac{1}{e^3} - \frac{1}{e^1}\right]$$

$$= 16.4 - 2.77$$

$$= 13.63 \text{ mm}$$

AVOID MISTAKE

Effective rainfall

$$= (10+20+10) - \int_0^3 6.8 + 8.7e^{-t}.dt$$

$$= 11.33 \text{ mm}$$

Note that first hour does't contribute to effective rainfall as $f > i$.

where f = infiltration capacity

i = rainfall intensity

3.19 (b)

The saturation discharge for S-curve,

$$Q_S = \lim_{t\to\infty} Q = \lim_{t\to\infty} Q\left[1 - \frac{(1+t)}{e^t}\right]$$

$$= \lim_{t\to\infty} Q\left[1 - \frac{1}{e^t}\right] = 1 \text{ m}^3/\text{s}$$

Rainfall intensity,

$$i = 1 \text{ cm/hr} = \frac{1}{360000} \text{ m/s}$$

$\therefore$ Catchment area $\times\, i = Q_S$

$\Rightarrow$ Catchment area $= \dfrac{1}{1/360000}$

$\Rightarrow$ Catchment area = 360000 m^2 = 0.36 km^2

3.20 (c)

The duration of S-curve = 1 hr (∵ i = 1 cm/hr)

The ordinate of 2-hr UH is obtained by the following procedure:

Step 1: The ordinate of S-curve at t = 3 hr

$$S_3 = 1 - (1 + 3)e^{-3} = 0.8 \text{ m}^3/\text{s}$$

Step 2: The ordinate of S-curve at

$$t = 3 - 2 = 1 \text{ hr}$$

$$S_1 = 1 - (1 + 1)e^{-1} = 0.26 \text{ m}^3/\text{s}$$

Step 3: $S_3 - S_1 = 0.8 - 0.26 = 0.54$ m^3/s

Step 4: The ordinate of UH at t = 3 hr is

$$(S_3 - S_1) \times \frac{1}{2} = 0.27 \text{ m}^3/\text{s} \quad \left[\because \begin{array}{l} t_{sc} = 1 \text{ hr} \\ t_{UH} = 2 \text{ hr} \end{array}\right]$$

3.21 (c)

Method-I

$$\phi\text{-index} = \frac{\text{Total Infiltration}}{\text{Time of rainfall excess}}$$

$$\Rightarrow \quad 10 = \frac{(28+12) - \text{Runoff}}{\text{Time of rainfall excess}}$$

But precipitation in the 1st and 4th hours is less than the ϕ-index i.e. the rainfall during these two hours is ineffective which means time of rainfall excess will be 2 hrs.

$$\therefore \quad 10 = \frac{40 - \text{Runoff}}{2}$$

$$\Rightarrow \quad \text{Runoff} = 20 \text{ mm}$$

Method-II

Only 2nd and 3rd hr will contribute to the rainfall excess as $i > \phi$.

Rainfall excess = (28 – 10) + (12 – 10)

= 20 mm

3.22 (d)

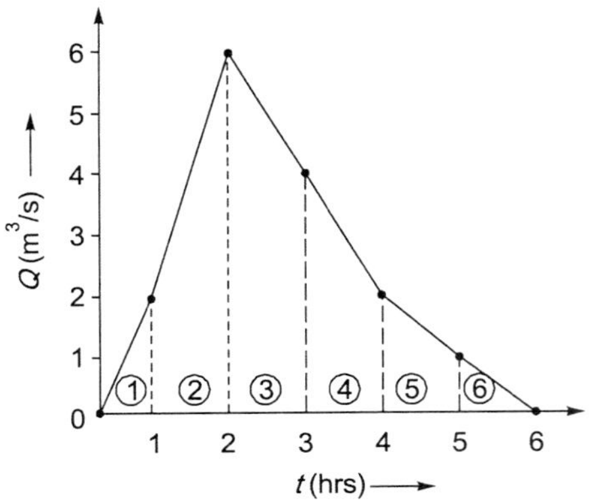

Volume of rainfall excess = Area water U-H

$$= \left[\left(\frac{1}{2}\times 2\times 1\right)+\left(\frac{2+6}{2}\right)\times 1+\left(\frac{6+4}{2}\right)\times 1 + \left(\frac{4+2}{2}\right)\times 1+\left(\frac{2+1}{2}\right)\times 1+\left(\frac{1}{2}\times 1\times 1\right)\right]$$

$\times\, 3600\ \text{m}^3$

$= (15 \times 3600)\ \text{m}^3$

Also, volume of rainfall excess = (Area of catchment in m^2) × 0.01 m

$$\Rightarrow \text{ Area of catchment} = \frac{15\times 36000}{0.01}\ \text{m}^2$$

$$= 5.4 \times 10^6\ \text{m}^2 = 5.4\ \text{km}^2$$

3.23 (c)

Time (h)	Ordinate of 1 h-UH (m³/s)			DRH of 3 cm in 3 h (m³/s)	Ordinate of 3 h. UH (m³/s)
	0 hr	1 hr	2 hr		
0	0	–	–	0	0/3 = 0
1	2	0	–	2	2/3 = 0.67
2	6	2	0	8	8/3 = 2.67
3	4	6	2	12	12/3 = 4
4	2	4	6	12	12/3 = 4
5	1	2	4	7	7/3 = 2.33
6	0	1	2	3	3/3 = 1
7	–	0	1	1	1/3 = 0.33
8	–	–	0	0	0 = 0

Thus at time $t = 3h$, the ordinate of $3h$ – UH is 4 m^3/s.

3.24 (c)

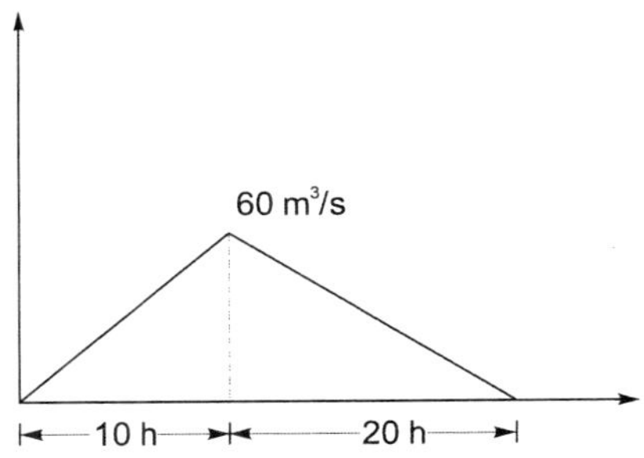

Depth = 1 cm

Volume of runoff = $A \times 1 \times 10^{-2}$ m^2

But, Volume of runoff = Area of the unit hydrograph

$$= \frac{1}{2} \times 30 \times 60 \times 3600$$

$$= 324 \times 10^4\ \text{m}^3$$

$\therefore \quad A \times 10^{-2} = 324 \times 10^4$

$\Rightarrow \quad A = 324 \times 10^6\ \text{m}^2$

$\Rightarrow \quad A = 324\ \text{km}^2$

3.25 (b)

Total rainfall = 5.4 cm

Infiltration in 1 hour = ϕ-index × 1

= 0.4 × 1 = 0.4 cm

Runoff depth = 5.4 – 0.4 = 5 cm

The discharge in the unit hydrograph decreases from 60 m^3/sec to 0 in a period of 20 hours.

$$\therefore \text{Discharge at the 15th hour} = 60 - \frac{60}{20}\times 5$$

$= 45$ m^3/sec

$\therefore$ Ordinate of DRH of 1h = 45 × 5 = 225 m^3/s

Ordinate of flood hydrograph,

Q = 225 + Base flow

= 225 + 15 = 240 m^3/sec

3.26 (c)

The effect of urbanization on storm runoff hydrograph from a watershed typically includes increased peak flows, shortened lag times, and higher volumes of runoff due to decreased infilteration rates and increased impervious surfaces such as roads and buildings. This leads to faster response times to rainfall events and increased risk of flooding. Additionally, changes in land use and drainage patterns can alter the shape and characteristics of the hydrograph.

3.27 (c)

For UH, depth of effective rainrall = 1 cm

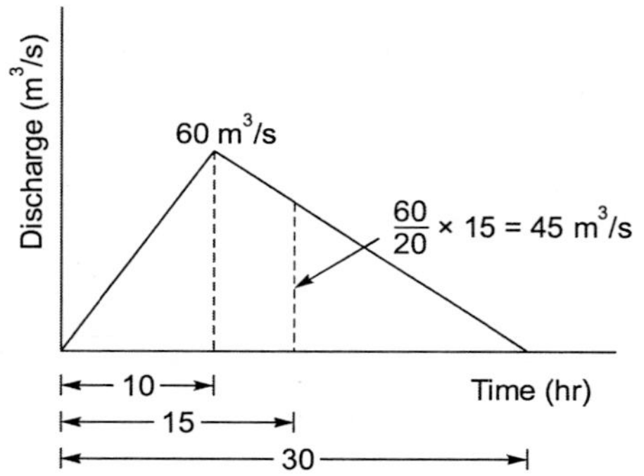

Volume of runoff = $A \times 10^{-2}$ m^3

Volume of runoff = Area of UH

$$A \times 10^{-2} = \frac{1}{2} \times 60 \times (30 \times 3600)$$

$$A = 324 \times 10^6 \text{ m}^2 = 324 \text{ km}^2$$

3.28 (a)

(1) t	(2) 2 h UH	(3) Imaginary off setted S-curve laggeds by 2h	(4) S-curve ordinates	(5) S-curve lagged by 3 h	(6) Col. (4) – Col. (5)	(7) 3 h-UH Col. (6) × 2/3
0	0	–	0	–	0	0
1	3	–	3	–	3	2
2	8	0	8	–	8	5.33
3	6	3	9	0	9	6
4	3	8	11	3	8	5.33
5	2	9	11	8	3	2
6	0	11	11	9	2	1.33
7	0	11	11	11	0	0
8	0	11	11	11	0	0

Peak of 3 h-UH = 6 m^3/s

Runoff due to storm

$$= 6.6\frac{2}{10} \times 3 = 6 \text{ cm}$$

Peak flow due to storm

$$= 6 \times 6 + 5 = 41 \text{ m}^3/\text{s}$$

3.29 (c)

3.30 (b)

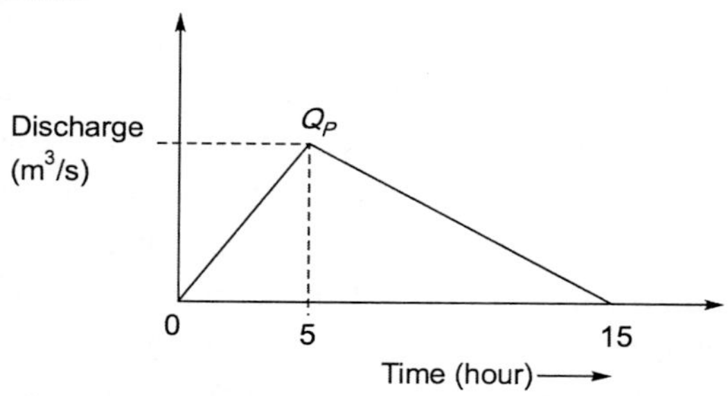

Peak ordinate of unit hydrograph can be found by equating the Area of triangle by volume of the catchment.

$$\Rightarrow \frac{1}{2} \cdot Q_P \cdot (15 \times 3600) = 50 \times 10^6 \times 1 \times 10^{-2}$$

$$\Rightarrow Q_P = 18.52 \text{ m}^3/\text{s}$$

3.31 (d)

Total infiltration = 0.5 × 1 = 0.5 cm

Total runoff = 5.5 – 0.5 = 5.0 cm

∴ Peak of the direct runoff hydrograph due to 5 cm of rainfall excess depth

$$= 18.52 \times 5 = 92.60 \text{ cm}$$

Peak of the flood hydrograph

= Peak of D.R.H + Base flow

= 92.60 + 10

= 102.60 m^3/s

3.32 (a)

$$\phi\text{-index} = \frac{2 \times 5 - 4}{5}$$

$$= \frac{10-4}{5} = \frac{6}{5} = 1.2 \text{ cm/hr}$$

3.33 (d)

Let intensity of storm-II be P cm/hr,

$$\therefore \quad 1.2 = \frac{P \times 8 - 8.4}{8}$$

$$\Rightarrow 1.2 \times 8 + 8.4 = P \times 8$$

$$P = 2.25 \text{ cm/hr}$$

3.34 (d)

In a small catchment,

Time of concentration = Lag time of peak flow

$$= 7 - \left(\frac{3}{2}\right) = 5.5 \text{ hr}$$

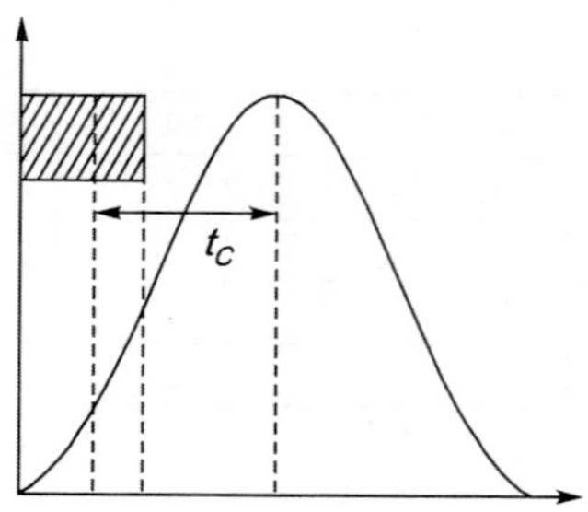

3.35 Sol.

Time (hr)	0	2	4	6	8	10	12	14	16	18	20	22	24
Unit Hydrograph ordinate (m^3/s)	0	0.6	3.1	10	13	9	5	2	0.7	0.3	0.2	0.1	0
4 hr shifted UH			0	0.6	3.1	10	13	9	5	2	0.7	0.3	0.2
8 hr shifted UH					0	0.6	3.1	10	13	9	5	2	0.7
12 hr shifted UH							0	0.6	3.1	10	13	9	5
16 hr shifted UH									0	0.6	3.1	10	13
20 hr shifted UH											0	0.6	3.1
S curve ordinates (m^3/s)	0	0.6	3.1	10.6	16.1	19.6	21.1	21.6	21.8	21.9	22	22	22

.....

Hence, **maximum ordinate** is 22 m^3/s.
Switch the columns and rows of this table.

3.36 Sol.

$$\frac{1}{2}\times 1\times 6\times 3600 = A\times\frac{5}{100}$$

$$\Rightarrow 2160 \times 100 = A$$

$$\Rightarrow \quad A = 216000 \text{ m}^2 = 21.6 \text{ Ha}$$

3.37 Sol.

Flood hydrograph at:

Time (1)	Ordinate of 1 hr UH (m^3/s) (2)	Offset ordinate (m^3/s) (3)	Ordinate of 2 hr DRH (m^3/s) (4) = (2) + (3)	Ordinate of DRH of 4 cm rainfall excess (m^3/s) (5)= (4)×4/2	Ordinate of flood hydrograph (m^3/s) (6) = (5) +20
10 : 00 am	0	–	0	0	20
11 : 00 am	3	0	3	6	26
12 : 00 pm	12	3	15	30	50
01 : 00 pm	8	12	20	40	60
02 : 00 pm	6	8	14	28	48
03 : 00 pm	3	6	9	18	38
05 : 00 pm	0	3	3	6	26
			0	0	20

Ordinate of 01 : 00 pm = 60 m^3/s

3.38 Sol.

Time (hr)	Ordinate of 2 hr unit hydrograph	Ordinate of 2 hr unit hydrograph lag by 2hr	Ordinate of 4 hr DRH	Ordinate of 4hr unit hydrograph = $\frac{\text{Ordinate of given DRH}}{2 \text{ cm}}$
0	0	—	0	0
1	5	—	5	2.5
2	12	0	12	6.0
3	25	5	30	15
4	41	12	53	26.5
		25	25	12.5
		41	41	20.5

∴ Ordinate of 4 hr U.H. at 3 hr duration = 15.0 m^3/sec.

3.39 Sol.

Peak of 2 hr direct runoff hydrograph = Peak of 2 hr flood hydrograph – Base flow

$$= 200 - 20 = 180 \text{ m}^3\text{/s}$$

The rainfall excess depth for a given duration of 2 hr effective storm/rainfall is given by,

Rainfall excess depth

$$= 2\text{ cm} - 0.4\,\frac{\text{cm}}{\text{hr}}\times 2\text{ hr}$$

$$= 1.2 \text{ cm}$$

∴ Peak of 2 hr UH

$$= \frac{180}{1.2} = 150 \text{ m}^3\text{/s}$$

3.40 Sol.

Horton's expression for infiltration capacity of a soil is given as,

$$f = C_1 + C_2 \,.e^{-kt}$$

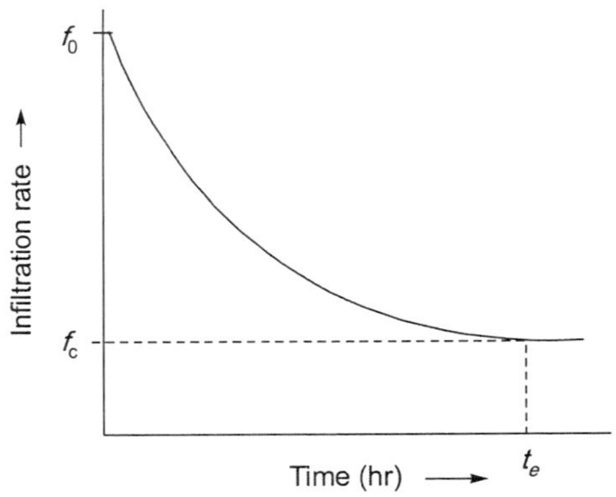

Here, C_1 = Constant or steady stage infiltration capacity

C_2 = Difference between initial and steady stage infiltration capacity = $(f_0 - f_c)$

f = Infiltration capacity of any time 't'

$$\therefore \quad f_p = f_c + (f_0 - f_c)\, e^{-k_h t}$$

$$\Rightarrow \quad 90 = 25 + (200 - 25)\, e^{-k_h 1}$$

$$\Rightarrow \quad k_h = 0.9904 \text{ hr}^{-1}$$

3.41 Sol.

Let the rainfall duration is 't' hours

∴Total rainfall = 3.5 cm/hr × t hr = 3.5 t

Total infiltration = φ-index × t = 1.5 t

∴Intensity of rainfall excess

$$= \frac{\text{Total rainfall excess}}{\text{Time period}}$$

$$= \frac{3.5t - 1.5t}{t} \text{ cm/hr}$$
$$= 2.0 \text{ cm/hr}$$

3.42 Sol.

Infiltration rate $f(t) = 30 + 10e^{-2t}$

Total infiltration depth in time 10 min. to 30 min.

i.e., 0.166 hour to 0.5 hour

$$= \int_{0.166}^{0.5} \left(30 + 10e^{-2t}\right) dt$$
$$= 11.74 \text{ mm}$$

AVOID MISTAKE

Time 't' is to be put in hours.

3.43 Sol.

$$\text{Surface runoff} = \frac{1 \times 10^8 \text{m}^3}{1000 \times 10^6 \text{m}^2}$$
$$= 0.1 \text{ m} = 10 \text{ cm}$$

Total Rainfall = 19 cm

$$\text{Rainfall intensity} = \frac{19}{6} = 3.167 \text{ cm/hr}$$

$$w\text{-index} = \frac{P - Q}{t}$$
$$= \frac{\text{Total infiltration}}{\text{Total duration of storm}}$$

$$\therefore \quad w\text{-index} = \frac{19 - 10}{6} = 1.5 \text{ cm/hr}$$

As intensity of rainfall > w-index and rainfall intensity is uniform therefore ϕ-index = w-index = 1.5 cm/hr.

3.44 Sol.

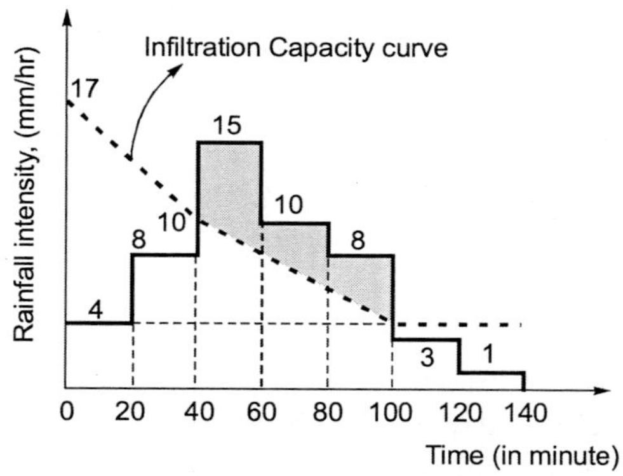

$$P = (4 + 8 + 15 + 10 + 8 + 3 + 1) \times \frac{20}{60}$$
$$= 16.33 \text{ mm}$$

$$Q = (15 - 10) \times \frac{20}{60} + (10 - 8) \times \frac{20}{60}$$
$$+ (8 - 6) \times \frac{20}{60} + \frac{1}{2} \times 2 \times \frac{20}{60} + \frac{1}{2}$$
$$+ \frac{1}{2} \times 2 \times \frac{20}{60} + \frac{1}{2} \times 2 \times \frac{20}{60} = 4 \text{ mm}$$

$$\text{W-index} = \frac{P - Q}{t} = \frac{16.33 - 4}{\frac{140}{60}}$$
$$= 5.28 \text{ mm/hr}$$

Since, $\phi \geq W$

Assume, ϕ = 5.28 mm/hr

$\Rightarrow$ Corrected,

$$\phi = \frac{16.33 - 4 - 4 \times \frac{20}{60} - 3 \times \frac{20}{60} - 1 \times \frac{20}{60}}{\left(\frac{140 - 30 - 20 - 20}{60}\right)}$$
$$= 7.2475 \text{ mm/hr}$$

3.45 Sol.

Rainfall excess in 1st two hours,

$$R_1 = 1 \text{ cm/hr} \times 2 \text{ hr} = 2 \text{ cm}$$

Rainfall excess in 2nd two hours

$$R_2 = 0$$

Rainfall excess in 3rd two hours,

$$R_3 = 0.5 \text{ cm/hr} \times 2\text{hr} = 1 \text{ cm}$$

Time (hr)	ord. of 2 hr UH say u	R_1u = $2u$	R_2u = 0	R_3u = $1 \cdot u$	ord. of 6 hr Complex DRH	ord. of 6 hr Flood Hyd. = 6 hr DRH + base Flow
0	0	0	–	–	0	
1	2	4	–	–	4	
2	8	16	0	–	16	
3	18	36	0	–	36	
4	32	64	0	0	64	Peak flow
5	45	90	0	2	92	92 + 5
6	30	60	0	8	68	= 97 m³/s
7	19	38	0	18	56	
8	12	24	0	32	56	
9	7	14	0	45	59	
10	3	6	0	30	36	
11	1	2	0	19	21	
12	0	0	0	12	12	
				7	7	
				3	3	
				1	1	
				0	0	

Alternate method:

Flood peak = (90 + 2) + 5

= 97 m^3/s

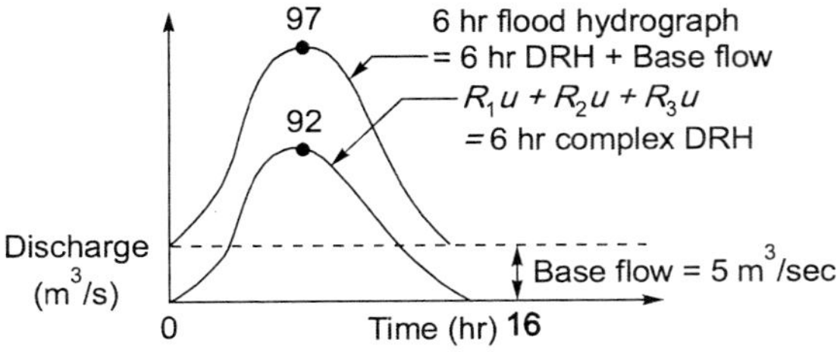

3.46 (c)

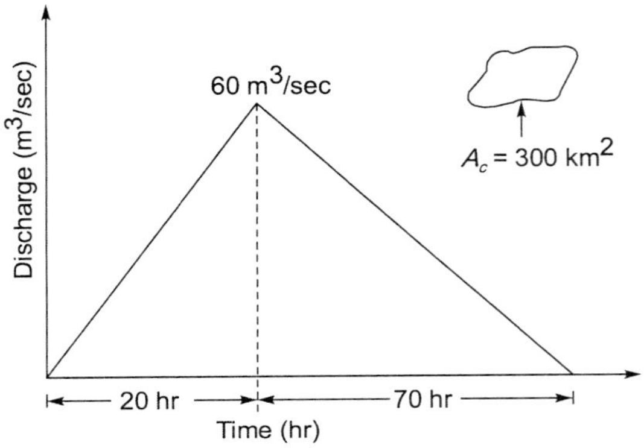

$$\Rightarrow \left[\frac{\frac{1}{2}\times 60\,m^3/s\times 90\times 3600\text{ s}}{300\times 10^6 m^2}\times 100\right]\text{cm}$$

= Rainfall excess

$\therefore$ Rainfall excess = 3.24 cm

3.47 Sol.

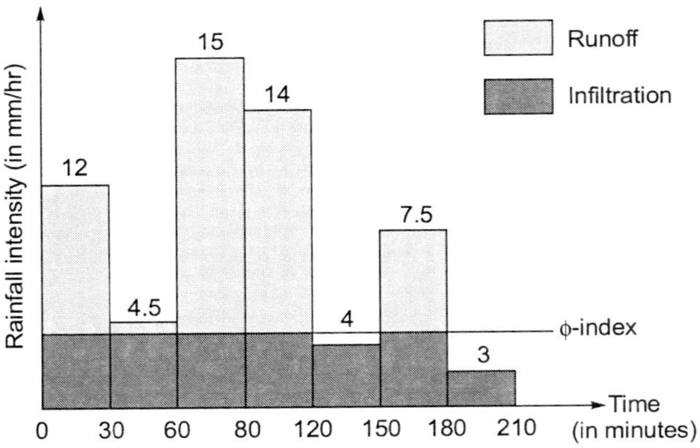

Total rainfall = 3 cm

Total runoff = 1.6 cm

$\therefore$Total infiltration = 3 − 1.6 = 1.4 cm

$$\therefore \quad \text{W-index} = \frac{\text{Total infiltration}}{\text{Total duration of storm}}$$

$$= \frac{1.4}{(210/60)}\text{ cm/hr}$$

= 0.4 cm/hr = 4 mm/hr

As ϕ-index > W-index

Hence storm of intensities 4 mm/hr and 3 mm/hr will not produce rainfall exam.

$$\phi\text{-index} = \frac{\text{Total infiltration in which rainfall excess occur}}{\text{Time period in which rainfall excess occur}}$$

$$= \frac{\text{Total infiltration} - \text{Infiltration in which no rainfall excess occur}}{T_{excess}}$$

$$= \frac{14\text{ mm} - \left(4\times\frac{30}{60}+3\times\frac{30}{60}\right)\text{mm}}{\left(\frac{150}{60}\right)\text{hr}} = 4.2\text{ mm/hr}$$

3.48 Sol.

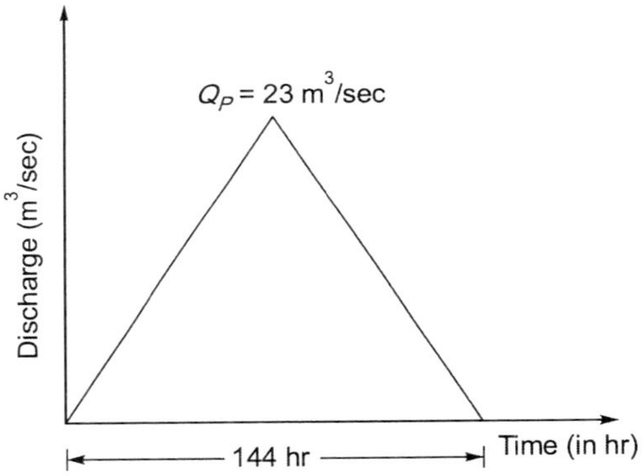

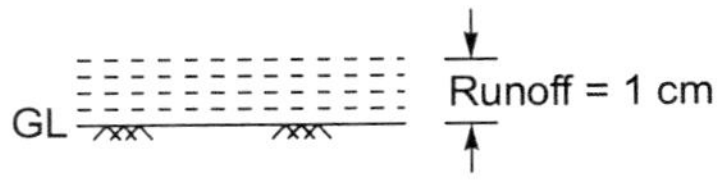

Area of catchment = A km^2

Area of hydrograph = Total direct runoff volume

$$\Rightarrow \frac{1}{2}\times 23\ m^3/\text{sec}\times 144\times 3600\text{ sec}$$

= Area of catchment × Runoff depth

$$\Rightarrow \quad \frac{1}{2}\times 23\times 144\times 3600 m^3$$

$$= A\times\frac{1}{100}\text{ m}$$

$A = 596.16 \times 10^6\ m^2$

$\therefore$Area of catchment

= 596.16 km^2

3.49 Sol.

Time	Ordinate of 2hr DRH	Ordinate of 2hr UH = $\frac{\text{Ord. of 2 hr DRH}}{\text{Rainfall excess of 3cm}}$	S-curve lag by 2hr	S-curve Ordinates (S_2)	S-curve lag by 1 hr	Δy	Ordinate of 1hr UH = $\frac{\Delta y \times 2}{1}$
0	0	0	—	0	—	0	
1	6	2	—	2	0	2	
2	24	8	0	8	2	6	
3	30	10.33	2	12	8	4	
4	21	7	8	18.33	12	6.33	6.33 × 2 = 12.66
5	15	5	12	17	18.33	−1.33	
6	6	2	18.33	20.33	17	3.33	
7	0	0	17	17	20.33		
			20.33	20.33	17		
			17	17	20.33		
					17		

Peak flow ordinate of 1 hr UH = 12.66 m^3/s

3.50 Sol.

Time	1hr-UH	1hr UH lagged by 1 hr	1hr-UH lagged by 1 more hr	3 hr DRH	Ordinate of 3hr UH = $\frac{\text{Ordinate of 3hr DRH}}{3\text{cm}}$
(1)	(2)	(3)	(4)	(5)	(6)
0	0	–	–	0	0
1	9	0	–	9	3
2	21	9	0	30	10
3	18	21	9	48	16
4	12	18	21	51	17
5	5	12	18	35	11.67
6	2	5	12	19	6.33
7	0	2	5	7	2.33
		0	2	2	0.67
			0	0	0

- Firstly lag the UH by 1 hr. By lagging 1 hr. We obtained a 2 hr UH of ordinate 4.5 m^3/sec. at 1 hr and 15 m^3/sec at 2 hr.
- So, further lag the UH by 1 more hr we obtained a 3 hr, UH of ordinates 3 m^3/sec at 1 hr and 10 m^3/sec at 2 hr. Therefore D = 3 hr.

3.51 Sol.

Given: Time = 12 hr

Runoff, R = 100 mm

Total depth of precipitation,

$$P = \frac{1}{2}[20\times4] + 2\times20 + \frac{1}{2}\times20\times6 = 140 \text{ mm}$$

∴ Total infiltration, $I = P - R$

$= 140 - 100 = 40$ mm

No, W-Index $= \frac{I}{t} = \frac{140-100}{12} = 3.33$ mm/hr

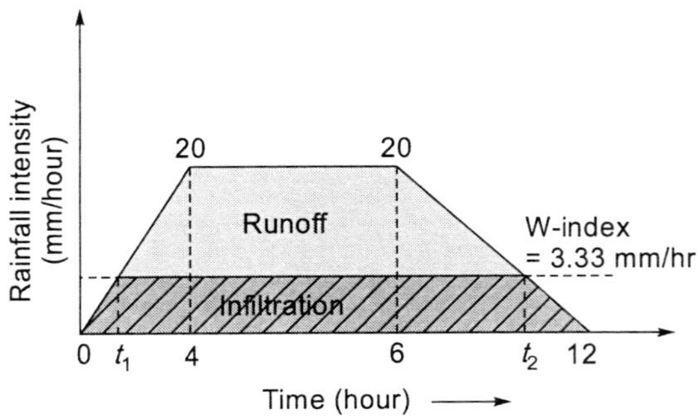

Now, $\frac{20}{4} = \frac{3.33}{t_1}$

$\Rightarrow \quad t_1 = 0.67$ hr

Similarly, $\frac{20}{6} = \frac{3.33}{t_2}$

$\Rightarrow \quad t_2 = 0.99$ hr

Now,

$$\phi\text{-Index} = \frac{\text{Total infiltration in which rainfall excess occur}}{\text{Time period in rainfall which excess occur}}$$

$$= \frac{I - \text{infiltration in which no runoff}}{\text{Time excess}}$$

$$\phi\text{-Index} = \frac{140 - \left[\frac{1}{2}\times 3.33\times 0.67 + \frac{1}{2}\times 3.33\times 0.99\right]}{12 - 0.67 - 0.99}$$

$= 3.6$ mm/hr

Check: Runoff = Area of hatched portion

$$= \frac{1}{2}\times(20-3.6)\times(4-0.67)+(20-3.6)\times 2+\frac{1}{2}\times(20-3.6)\times(6-0.99)$$

$= 100$ mm (Hence, OK)

$\therefore$ ϕ-Index = 3.60 mm/hr

3.52 Sol.

Initial infiltration capacity, f_0 = 10 mm/h
Final infiltration capacity, f_c = 5 mm/h
Horton's Decay constant, k_h = 0.5/h

Now, $F_P = \int_0^T \left[f_c + [f_0 - f_c]e^{-k_h t}\right]dt$

$$= \int_0^{12} \left(5 + (10-5)e^{-0.5t}\right)dt$$

$$= [5t]_0^{12} + \left[\frac{5e^{-0.5t}}{(-0.5)}\right]_0^{12}$$

$$= [5\times 12] + \left[-10e^{-0.5\times 12} - \left[-10e^{-0.5\times 0}\right]\right]$$

$= 60 + [9.975]$

$= 69.97$ mm $\simeq 70$ mm

Alternatively:

For large value of 't' cumulative infiltration capacity is given by,

$$F_P = f_c t + \frac{(f_0 - f_c)}{k_h}$$

$$= (5\times 12) + \frac{(10-5)}{0.5} = 70\text{ mm}$$

3.53 Sol.

	(I)	(II)	(III)	(IV)	(V) = IV/3
Time (hr)	Ordinate (m^3/s)	Lagged by 1 Hrs	Lagged by 2 Hrs	Ordinate of 3 Hrs DRH	Ordinate of 3 H UH
0	0	–	–	0	0
1	13	0	–	13	4.33
2	50	13	0	63	21
3	80	50	13	143	47.67
4	95	80	50	225	75
5	85	95	80	260	86.67
6	55	85	95	235	78.33
7	35	55	85	175	58.33
8	15	35	55	105	35
9	10	15	35	60	20
10	3	10	15	28	9.33
11	0	3	10	13	4.33
12		0	3	3	1
13			0	0	0
14					

Maximum ordinate of 3H UH is 86.67 m^3/s

$\therefore$ Maximum ordinate is 87 m^3/s (to nearest integer)

3.54 Sol.

Time from start (hr)	1	2	3	4	5	6	7	8
Accumulated precipitation (cm)	0.5	1.5	3.1	5.5	7.3	8.9	10.2	11.0
Incremental precipitation (cm)	0.5	1	1.6	2.4	1.8	1.6	1.3	0.8
Rainfall intensity (cm/hr)	0.5	1	1.6	2.4	1.6	1.6	1.3	0.8

Total infiltration, I = Total rainfall – Total runoff
= 11 – 6 = 5 cm

$\therefore$ W-index $= \frac{5}{8} = 0.625$ cm/hr

As the 1 hour rainfall is less than W-index. So,

$$\phi\text{-index} = \frac{5 - 0.5\times 1}{(8-1)} = 0.6429\text{ cm/hr} \approx 0.64\text{ cm/hr}$$

4 Floods, Flood Routing and Flood Control

4.1 Match the following:
(a) Coriolis force (b) Khosla's curve
(c) Gumbel's method (d) Manning's method
(i) Weirs and barrages
(ii) Open channel flow
(iii) River meandering
(iv) Estimation of flood peaks [1995 : 2 M]

4.2 Muskingham method for routing of flood
(a) is used for routing floods through reservoirs
(b) is a method of routing that uses continuity and momentum equations
(c) is a hydrologic method of routing floods through streams
(d) is one in which only energy equation is used
[1997 : 1 M]

4.3 The return period for the annual maximum flood of a given magnitude is 8 years. The probability that this flood magnitude will be exceeded once during the next 5 years is
(a) 0.625 (b) 0.366
(c) 0.487 (d) 0.529 [1998 : 2 M]

4.4 The stage-discharge relation in a river during the passage of flood is measured. If q_f is the discharge at the stage when water surface is falling and q_r is the discharge at the same stage when water surface is rising, then
(a) $q_f = q_r$
(b) $q_f < q_r$
(c) $q_f > q_r$
(d) $\frac{q_f}{q_r}$ = constant for all stages [2001 : 1 M]

4.5 A linear reservoir is one in which
(a) Storage varies linearly with time
(b) Storage varies linearly with outflow rate
(c) Storage varies linearly with inflow rate
(d) Storage varies linearly with elevation
[2001 : 1 M]

4.6 When the outflow from a storage reservoir is uncontrolled as in a freely operating spillway, the peak of outflow hydrograph occurs at
(a) the point of intersection of the inflow and outflow hydrographs
(b) a point, after the intersection of the inflow and outflow hydrographs
(c) the tail of inflow hydrograph
(d) a point, before the intersection of the inflow and outflow hydrographs [2005 : 1 M]

4.7 A flood wave with a known inflow hydrograph is routed through a large reservoir. The outflow hydrograph will have
(a) attenuated peak with reduced time-base
(b) attenuated peak with increased time-base
(c) increased peak with increased time-base
(d) increased peak with reduced time-base
[2008 : 1 M]

4.8 Match **List-I** with **List-II** and select the correct answer using the codes given below the lists:

List-I
A. Evapotranspiration
B. Infiltration
C. Synthetic Unit Hydrograph
D. Channel Routing

List-II
1. Penman method
2. Snyder's method
3. Muskingum method
4. Horton's method

Codes:

	A	B	C	D
(a)	1	3	4	2
(b)	1	4	2	3
(c)	3	4	1	2
(d)	4	2	1	3

[2010 : 1 M]

4.9 A 1 hour rainfall of 10 cm has return period of 50 year. The probability that 1 hour of rainfall 10 cm or more will occur in each of two successive years is
(a) 0.04 (b) 0.2
(c) 0.02 (d) 0.0004 [2013 : 1 M]

4.10 In reservoirs with an uncontrolled spillway, the peak of the plotted outflow hydrograph

(a) lies outside the plotted inflow hydrograph
(b) lies on the recession limb of the plotted inflow hydrograph
(c) lies on the peak of the inflow hydrograph
(d) is higher than the peak of the plotted inflow hydrograph [2014 : 1 M, Set-I]

4.11 The Muskingum model of routing a flood through a stream reach is expressed as $O_2 = K_0I_2 + K_1I_1 + K_2O_1$, where K_0, K_1 and K_2 are the routing coefficients for the concerned reach, I_1 and I_2 are the inflows to the reach, and O_1 and O_2 are the outflows from the reach corresponding to time steps 1 and 2 respectively. The sum of K_0, K_1 and K_2 of the model is

(a) – 1
(b) – 0.5
(c) 0.5
(d) 1 [2014 : 1 M, Set-II]

4.12 The type of flood routing (**Group-I**) and the equation(s) used for the purpose (**Group-II**) are given below.

Group-I
P. Hydrologic flood routing
Q. Hydraulic flood routing

Group-II
1. Continuity equation
2. Momentum equation
3. Energy equation

(a) P-1; Q-1, 2 and 3
(b) P-1; Q-1 and 2
(c) P-1 and 2; Q-1
(d) P-1 and 2; Q-1 and 2 [2016 : 1 M, Set-I]

4.13 For routing of flood in a given channel using the Muskingum method, two of the routing coefficients are estimated as $C_0 = -0.25$ and $C_1 = 0.55$. The value of the third coefficient C_2 would be _______.
[2018 : 1 M, Set-I]

4.14 A culvert is designed for a flood frequency of 100 years and a useful life of 20 years. The risk involved in the design of the culvert (in percentage, up to two decimal places) is _______.
[2018 : 1 M, Set-II]

4.15 The probability that the annual maximum flood discharge will exceed 25000 m^3/s, at least once in next 5 years is found to be 0.25. The return period of this flood event (in years, round off to 1 decimal place) is_______.
[2019 : 1 M, Set-I]

4.16 An inflow hydrograph is routed through a reservoir to produce an outflow hydrograph. The peak flow of the inflow hydrograph is P_I and the time of occurrence of the peak is t_I. The peak flow of the outflow hydrograph is P_o and the time of occurrence of the peak is t_o. Which one of the following statements is correct?

(a) $P_I > P_o$ and $t_I > t_o$
(b) $P_I > P_o$ and $t_I < t_o$
(c) $P_I < P_o$ and $t_I < t_o$
(d) $P_I < P_o$ and $t_I > t_o$ [2019 : 1 M, Set-II]

4.17 The probability that a 50 year flood may NOT occur at all during 25 years life of a project (round off to two decimal places), is _______.
[2020 : 1 M, Set-I]

4.18 Muskingum method is used in
(a) hydrologic channel routing
(b) hydraulic channel routing
(c) hydrologic reservoir routing
(d) hydraulic reservoir routing
[2020 : 1 M, Set-II]

4.19 A flood control structure having an expected life of n years is designed by considering a flood of return period T years. When $T = n$, and $n \to \infty$, the structure's hydrologic risk of failure in percentage is _______ . (rounded off to one decimal place)
[2022 : 1 M, Set-II]

4.20 Match **Column X** with **Column Y**:

Column X	Column Y
(P) Horton eqution	(I) Design of alluvial channel
(Q) Penman method	(II) Maximum flood dischage
(R) Chezy's formula	(III) Evapotranspiration
(S) Lacey's theory	(IV) Infiltration
(T) Dicken's formula	(V) Flow velocity

Which one of the following combinations is correct ?
(a) (P)-(IV), (Q)-(III), (R)-(II), (S)-(I), (T)-(V)
(b) (P)-(III), (Q)-(IV), (R)-(I), (S)-(V), (T)-(II)
(c) (P)-(IV), (Q)-(III), (R)-(V), (S)-(I), (T)-(II)
(d) (P)-(III), (Q)-(IV), (R)-(V), (S)-(I), (T)-(II)
[2022 : 2 M, Set-II]

4.21 Which one of the following options provides the correct match of the terms listed in Column-1 and Column-2?

Column-1
P. Horton equation
Q. Muskingum method
R. Penman method

Column-2
I. Precipitation
II. Flood frequency
III. Evapotranspiration
IV. Infiltration
V. Channel routing

(a) P-IV, Q-V, R-III (b) P-III, Q-IV, R-I
(c) P-IV, Q-III, R-II (d) P-III, Q-I, R-IV

[2023 : 1 M, Set-II]

4.22 A 2 m × 2 m tank of 3 m height has inflow, outflow and stirring mechanisms. Initially, the tank was half filled with fresh water. At $t = 0$, an inflow of salt a solution of concentration 5 g/m^3 at the rate of 2 litre/s and an outflow of the well stirred mixture at the rate of 1 litre/s are initiated. This process can be modelled using the following differential equation.

$$\frac{dm}{dt} + \frac{m}{6000+t} = 0.01$$

where m is the mass (grams) of salt at time t (seconds). The mass of the salt (in grams) in the tank at 75% of its capacity is ________. (rounded off to 2 decimal places).

[2024 : 2 M, Set-I]

4.23 The return period of a large earthquake for a given region is 200 years. Assuming that earthquake occurrence follows Poisson's distribution, the probability that it will be exceeded at least once in 50 years is ________%. (rounded off to the nearest integer).

[2024 : 2 M, Set-I]

Answers Floods, Flood Routing and Flood Control

4.1	(Sol.)	4.2	(c)	4.3	(b)	4.4	(b)	4.5	(b)	4.6	(a)	4.7	(b)
4.8	(b)	4.9	(d)	4.10	(b)	4.11	(d)	4.12	(b)	4.13	(0.7)	4.14	(18.209)
4.15	(17.9)	4.16	(b)	4.17	(0.603)	4.18	(a)	4.19	(63.2)	4.20	(c)	4.21	(a)
4.22	(25.5)	4.23	(22)										

Explanations Floods, Flood Routing and Flood Control

4.1 Sol.

(a) (iii), (b) (i), (c) (iv), (d) (ii)

4.2 (c)

In reservoir routing, the storage is a unique function of outflow discharge, $S = f(Q)$.

However, in channel routing the storage is a function of both outflow and inflow discharge and hence a different routing method is needed. From **Muskingham method** of hydrological channel, routing storage is given by:

$$S = k\left[xI^m + (1-x)Q^m\right]$$

Muskingham method being a hydrologic routing method makes use of education of continuity.

4.3 (b)

$$P = \frac{1}{8}$$

$$q = \frac{7}{8}$$

$$P(x=1) = 5C_1\left(\frac{1}{8}\right)\left(\frac{7}{8}\right)^4 = 0.3664$$

AVOID MISTAKE

Note that probability of exceedance once during the next 5 years has been asked, and not the probability of exceeding atleast once in next 5 years

$$P\,(\text{atleast once}) = 1 - q^n = 1 - \left(\frac{7}{8}\right)^5 = 0.487$$

4.4 (b)

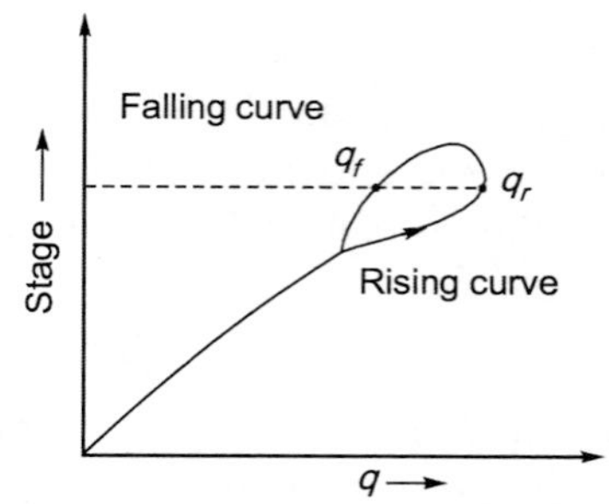

At the same stage, more discharge passes through the river during rising stages than in falling ones.

Note: *Stage discharge relation is also known as rating curve.*

4.5 (b)

4.6 (a)

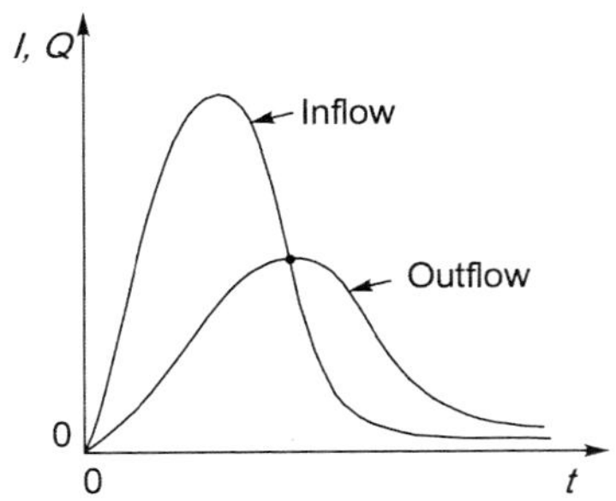

Reservoir with uncontrolled spillway

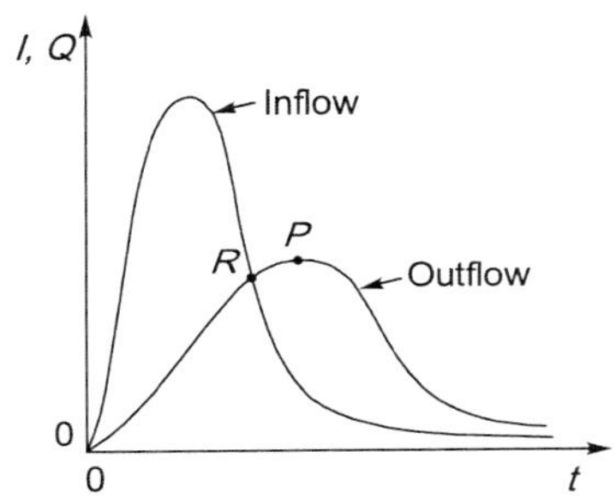

Reservoir with controlled spillway

When outflow from a reservoir is uncontrolled as in freely operating spillway, then the peak of outflow hydrograph will occur at a point of intersection of inflow and outflow curves, whereas if outflow from a reservoir is controlled then peak will occur after the intersection of the curve.

4.7 (b)

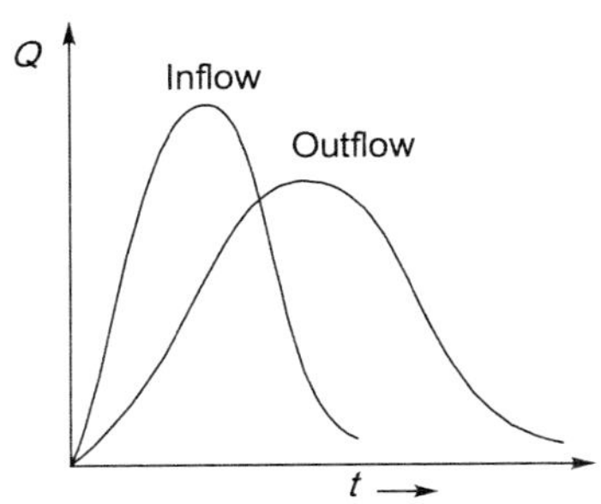

The inflow hydrograph will have attenuated peak with increased time-based.

4.8 (b)

4.9 (d)

Return period of rainfall,

$$T = 50 \text{ years}$$

$\therefore$ Probability of occurrence once in 50 years,

$$p = \frac{1}{50} = 0.02$$

Probability of occurrence in each of 2 successive year = $p^2 = (0.02)^2 = 0.0004$

4.10 (b)

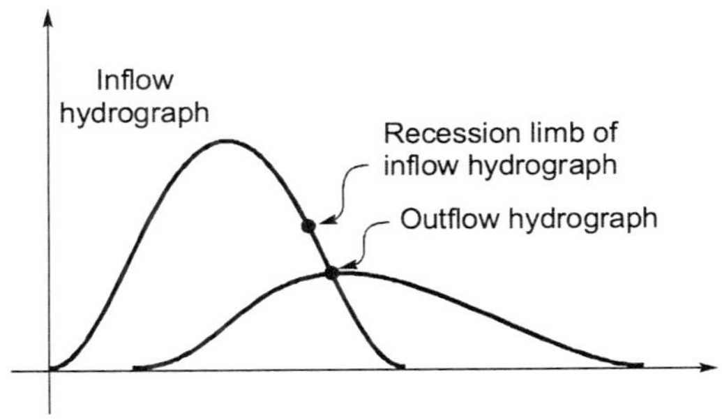

4.11 (d)

In Muskingum method of flood routing,

$k_0 + k_1 + k_2 = 1$

4.12 (b)

Hydrologic flood routing : Continuity equation

Hydraulic flood routing: Continuity and momentum equations

4.13 Sol.

In Muskingum flood routing method

$C_0 + C_1 + C_2 = 1$

$\Rightarrow \quad C_2 = 1 - (-0.25) - 0.55 = 0.7$

4.14 Sol.

Risk = The probability of a flood to occur at least once in n-successive years.

$$\begin{aligned} \therefore \quad \text{Risk} &= 1 - q^n \\ &= 1 - (1 - P)^n \\ &= 1 - \left(1 - \frac{1}{T}\right)^n = 1 - \left(1 - \frac{1}{100}\right)^{20} \\ &= 1 - (0.99)^{20} = 0.18209 = 18.209\% \end{aligned}$$

4.15 Sol.

$$\text{Risk} = 1 - q^n$$

$$\Rightarrow \quad 0.25 = 1 - (q)^5$$

$\Rightarrow \quad q = 0.944087$

$\Rightarrow \quad 1 - p = 0.944087$

$\Rightarrow \quad 1 - \frac{1}{T} = 0.944087$

$\therefore \quad T \simeq 17.9$ years

4.16 (b)

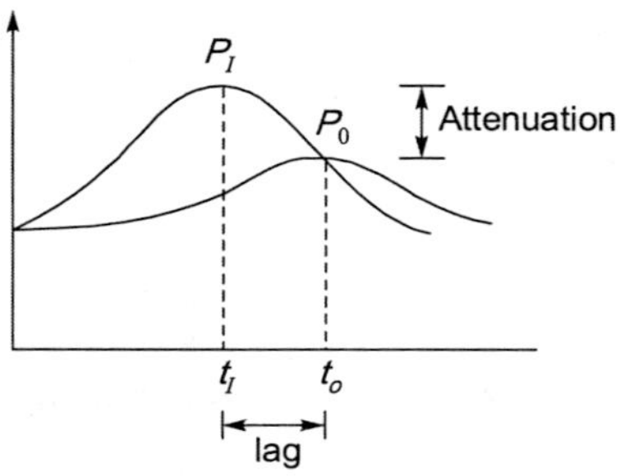

The outflow from the reservoir is uncontrolled therefore peak of outflow hydrograph will occur at the junction of inflow and outflow hydrograph.

$P_I > P_0$

$t_I < t_o$

4.17 Sol.

$$P = \frac{1}{T} = \frac{1}{50} = 0.02$$

$$q = 1 - P = 0.98$$

$\therefore$ Probability of non-occurance of an event is given by,

$$= q^n$$
$$= (0.98)^{25}$$
$$= 0.603$$

4.18 (a)

The muskingum method is a hydrological technique used to model and analyze river flow and flood routing.

4.19 Sol.

$$R = 1 - q^n$$

$$\Rightarrow \quad R = 1 - (1 - p)^n$$

$$R = 1 - \left[1 - \frac{1}{T}\right]^n$$

Given that $\quad T = n$ and $n \to \infty$

$$\Rightarrow \quad R = 1 - \left[\lim_{n\to\infty}\left(1 - \frac{1}{n}\right)^n\right]$$

$$\Rightarrow \quad R = 1 - \left[\lim_{n\to\infty}\left\{1 + \left(-\frac{1}{n}\right)\right\}^n\right] \quad ...(i)$$

We know that

$$\lim_{n\to\infty}(1 + x)^{1/x} = e,$$

So let's take $\frac{-1}{n} = y$

$$\Rightarrow \quad n = \frac{-1}{y}$$

At $n = \infty$; $y = 0$

$$\text{Now, } \lim_{n\to\infty}\left(1 - \frac{1}{n}\right)^n = \lim_{y\to 0}(1 + y)^{-1/y}$$

$$= \lim_{y\to 0}(1 + y)^{\frac{1}{y}(-1)}$$

$$= \left[\lim_{y\to 0}(1 + y)^{\frac{1}{y}}\right]^{-1}$$

$$= e^{-1}$$

Put in eq. (i),

$\therefore \quad R = 1 - e^{-1}$

$\Rightarrow \quad R = 1 - \frac{1}{e}$

$\Rightarrow \quad R = 0.6321$ i.e. 63.2%

4.20 (c)

4.21 (a)

4.22 Sol.

Total capacity of tank = 2 × 2 × 3 = 12 m^3

At 50% capacity,

Volume (V_{50}) = 0.5 × 12 = 6 m^3

At 75% capacity,

Volume (V_{75}) = 0.75 × 12 m^3 = 9 m^3

Now, increase in volume $\Delta V = V_{75} - V_{50}$ = (9 − 6)m^3 = 3 m^3

Also, Inflow (I) = 2 l/s

Outflow (Q) = 1 l/s

Rate of change storage,

$$\frac{\Delta S}{\Delta t} = I - Q = 2 - 1 = 1\ l/s$$

$$= \frac{3000 l}{1 l/s} = 3000 \text{ sec}$$

$$\text{Now, } \frac{dm}{dt} + \frac{m}{6000 + t} = 0.01$$

$$\Rightarrow \frac{dm}{dt} + \frac{m}{6000 + 3000} = 0.01$$

$$\Rightarrow \frac{dm}{dt} + \frac{m}{9000} = 0.01$$

$$\Rightarrow \frac{dm}{0.01 - \frac{m}{9000}} = dt$$

Integrating both sides

$$\Rightarrow \int_{m_o=0}^{m} \frac{dm}{0.01 - \frac{m}{9000}} = \int_{t=0}^{t=3000} dt$$

$$\Rightarrow \frac{\left[\ln\left(0.01 - \frac{m}{9000}\right)\right]_0^m}{-\frac{1}{9000}} = (3000 - 0)$$

$$\Rightarrow \ln\left(0.01 - \frac{m}{9000}\right) - \ln 0.01 = -\frac{1}{9000} \times 3000$$

$$\Rightarrow \ln\left(0.01 - \frac{m}{9000}\right) = -4.939$$

$$\Rightarrow 0.01 - \frac{m}{9000} = 7.162 \times 10^{-3}$$

$$\Rightarrow \quad m = 25.5 \text{ gm}$$

4.23 Sol.

$$\Rightarrow \quad \text{Risk} = 1 - \left[\frac{\lambda^r \cdot e^{-\lambda}}{r!}\right]$$

$$= 1 - \left[\frac{\lambda^0 \cdot e^{-\lambda}}{0!}\right]$$

$$\lambda = np = 50 \times \frac{1}{200} = 0.25$$

Now, $\text{Risk} = 1 - [e^{-0.25}]$

$= 0.2212$ or $22.12\% \simeq 22\%$

■■■■

Transportation Engineering

UNIT X

CONTENTS

Transportation Engineering

Syllabus

Transportation Infrastructure: Geometric design of highways - cross-sectional elements, sight distances, horizontal and vertical alignments. Geometric design of railway Track – Speed and Cant. Concept of airport runway length, calculations and corrections; taxiway and exit taxiway design.

Highway Pavements: Highway materials - desirable properties and tests; Desirable properties of bituminous paving mixes; Design factors for flexible and rigid pavements; Design of flexible and rigid pavement using IRC codes

Traffic Engineering: Traffic studies on flow and speed, peak hour factor, accident study, statistical analysis of traffic data; Microscopic and macroscopic parameters of traffic flow, fundamental relationships; Traffic signs; Signal design by Webster's method; Types of intersections; Highway capacity.

Analysis of Previous GATE Papers

Exam Year	1 Mark Ques.	2 Marks Ques.	Total Marks
1995	2	1	4
1996	10	–	10
1997	11	–	11
1998	4	6	10
1999	3	–	3
2000	6	–	6
2001	3	2	7
2002	3	3	9
2003	2	5	12
2004	3	5	13
2005	4	5	14
2006	3	6	15
2007	3	6	15
2008	4	6	16
2009	2	4	10
2010	2	3	8
2011	3	3	9
2012	3	3	9
2013	3	3	9
2014 Set-1	2	4	10
2014 Set-2	2	5	12

Exam Year	1 Mark Ques.	2 Marks Ques.	Total Marks
2015 Set-1	2	3	8
2015 Set-2	2	3	8
2016 Set-1	2	2	6
2016 Set-2	2	2	6
2017 Set-1	3	3	9
2017 Set-2	2	3	8
2018 Set-1	3	2	7
2018 Set-2	3	3	9
2019 Set-1	3	4	11
2019 Set-2	4	4	12
2020 Set-1	3	4	11
2020 Set-2	3	3	9
2021 Set-1	3	4	11
2021 Set-2	3	3	9
2022 Set-1	1	5	11
2022 Set-2	3	4	11
2023 Set-1	2	2	6
2023 Set-2	2	2	6
2024 Set-1	3	4	11
2024 Set-2	3	4	11

1 Highway Geometric Design & Planning

1.1 At highway stretches where the required overtaking sight distance cannot be provided, it is necessary to incorporate
(a) at least twice the stopping sight distance
(b) half the required overtaking sight distance
(c) one third the required overtaking sight distance
(d) three times the stopping sight distance
[1995 : 1 M]

1.2 The ideal form of curve for the summit curve is
(a) spiral (b) parabola
(c) circle (d) lemniscate
[1995 : 1 M]

1.3 Transition curves are provided on the approach to horizontal curves in order to
(a) increase jerk to allowable levels
(b) minimize the length of the horizontal curve
(c) simplify the laying out and construction of the horizontal curve
(d) reduce jerk to allowable levels
[1996 : 1 M]

1.4 On sag (or valley) curves the available sight distance is determined based on
(a) design speed
(b) height of obstacle
(c) height of driver eye
(d) nighttime driving conditions **[1996 : 1 M]**

1.5 Maximum allowable grades are lower for railways than for highways because
(a) construction costs become prohibitive for railways at high grades
(b) trains are longer than vehicles which use the highways
(c) high grades causes discomfort to passengers
(d) steel wheels on steel rails have lower frictional coefficient than rubber tyres on pavements
[1996 : 1 M]

1.6 Sliding considerations for stopped vehicles on super elevated horizontal curves provide the following bound on the amount of super elevation, e,
(a) $e \geq$ coefficient of rolling friction
(b) $e \geq$ coefficient of side friction
(c) $e \leq$ coefficient of rolling friction
(d) $e \leq$ coefficient of side friction **[1996 : 1 M]**

1.7 The value of the camber recommended for cement concrete roads in areas of heavy rainfall is
(a) 1 in 25 (b) 1 in 33
(c) 1 in 40 (d) 1 in 50 **[1997 : 1 M]**

1.8 The reaction time for calculation of stopping distance may be assumed as
(a) 5 secs (b) 2.5 secs
(c) 0.5 secs (d) 10.0 secs
[1997 : 1 M]

1.9 The superelevation needed for a vehicle travelling at a speed of 60 kmph on a curve of radius 128 m on a surface with a coefficient of friction of 0.15 is
(a) 0.071 (b) 0.015
(c) 0.022 (d) Zero **[1997 : 1 M]**

1.10 If an ascending gradient of 1 in 50 meets another ascending gradient of 1 in 30 then the deviation angle is
(a) $\frac{1}{50}$ (b) $\frac{1}{75}$
(c) $\frac{1}{30}$ (d) $\frac{8}{150}$ **[1997 : 1 M]**

1.11 At highway stretches where the required overtaking sight distance cannot be provided, it is necessary to incorporate in such sections the following
(a) at least twice the stopping sight distance
(b) half of the required overtaking sight distance
(c) one third of the required overtaking sight distance
(d) three times the stopping sight distance
[1998 : 1 M]

1.12 For a highway with design speed of 100 kmph, the safe overtaking sight distance is (assume acceleration as 0.53 m/s^2)
(a) 300 m (b) 750 m
(c) 320 m (d) 470 m **[1998 : 2 M]**

1.13 A parabolic curve is used to connect a 4% upgrade with a 2% downgrade as shown in the figure. The highest point on the summit curve is at a distance of (measured horizontally from the first tangent point FTP)

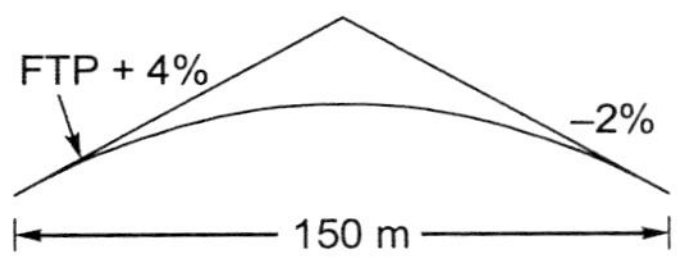

(a) 50 m (b) 60 m
(c) 75 m (d) 100 m [1999 : 1 M]

1.14 Width of carriage way for a single lane is recommended to be
(a) 7.5 m (b) 7.0 m
(c) 3.75 m (d) 5.5 m [2000 : 1 M]

1.15 Stopping sight distance is the minimum distance available on a highway which is the
(a) distance of sufficient length to stop the vehicle without collision
(b) distance visible to a driver during night driving
(c) height of the object above the road surface
(d) distance equal to the height of the driver's eye above the road surface
[2000 : 1 M]

1.16 The ruling minimum radius of horizontal curve of a national highway in plane terrain for a ruling design speed of 100 km/hour with e = 0.07 and f = 0.15 is close to
(a) 250 m (b) 360 m
(c) 36 m (d) 300 m [2000 : 1 M]

1.17 Design rate of super elevation for horizontal highway curve of radius 450 m for a mixed traffic condition, having a speed of 125 km/hour is
(a) 1.0 (b) 0.05
(c) 0.07 (d) 0.154 [2000 : 1 M]

1.18 The design value of lateral friction coefficient of highway is
(a) 1.5 (b) 0.50
(c) 0.35 (d) 0.15 [2001 : 1 M]

1.19 Camber on highway pavement is provided to take care of
(a) Centrifugal force (b) Drainage
(c) Sight Distance (d) Off-tracking
[2001 : 1 M]

1.20 A valley curve has a descending gradient of 1 in 40 followed by a ascending gradient of 1 in 50. The length of the valley curve required for a design speed of 80 km/hour for comfort condition is
(a) 199 m (b) 116 m
(c) 58 m (d) 37 m [2001 : 2 M]

1.21 Stopping sight distance and frictional coefficients are
(a) directly proportional to each other
(b) inversely proportional to each other
(c) unrelated
(d) either directly or inversely proportional to each other depending on the nature of pavement
[2002 : 1 M]

1.22 A car is moving at a speed of 72 km/hr on a road having 2% upward gradient. The driver applies brakes when he sees an obstruction. If his reaction time is 1.5 seconds, assuming that the coefficient of friction between the pavement and tyre as 0.15, calculate the distance traversed before the car finally stops
(a) 24 m (b) 150 m
(c) 1056 m (d) 324 m [2002 : 2 M]

1.23 The design speed for a National Highway is 100 kmph. If the maximum permissible superelevation is 0.10 and the coefficient of lateral friction is 0.15, the ruling minimum radius of horizontal curve on the highway should be
(a) 260 m (b) 315 m
(c) 380 m (d) 410 m [2003 : 2 M]

1.24 The Star and Grid pattern of road network was adopted in
(a) Nagpur Road Plan
(b) Lucknow Road Plan
(c) Bombay Road Plan
(d) Delhi Road Plan [2004 : 1 M]

1.25 Three new roads P, Q and R are planned in a district. The data for these roads are given in the table below.

Road	Length (km)	Number of villages with population		
		Less than 2000	2000-5000	More than 5000
P	20	8	6	1
Q	28	19	8	4
R	12	7	5	2

Based on the principle of maximum utility, the order of priority for these three roads should be
(a) P, Q, R (b) Q, R, P
(c) R, P, Q (d) R, Q, P [2004 : 2 M]

1.26 For a road with camber of 3% and the design speed of 80 km/hr, the minimum radius of the curve beyond which NO superelevation is needed is
(a) 1680 m (b) 948 m
(c) 406 m (d) 280 m
[2004 : 2 M]

1.27 The co-efficient of friction in the longitudinal direction of a highway is estimated as 0.396. The braking distance for a car moving at a speed of 65 km/hr is

(a) 87 m (b) 45 m
(c) 42 m (d) 40 m [2004 : 2 M]

1.28 The length of summit curve on a two lane two way highway depends upon

(a) allowable rate of change of centrifugal acceleration
(b) coefficient of lateral friction
(c) required stopping sight distance
(d) required overtaking sight distance

[2005 : 1 M]

1.29 Pradhan Mantri Gram Sadak Yojna (PMGSY), launched in the year 2000, aims to provide rural connectivity with all-weather roads. It is proposed to connect the habitations in plain areas of population more than 500 persons by the year

(a) 2005 (b) 2007
(c) 2010 (d) 2012 [2005 : 1 M]

1.30 A road is having a horizontal curve of 400 m radius on which a super-elevation of 0.07 is provided. The coefficient of lateral friction mobilized on the curve when a vehicle is travelling at 100 kmph is

(a) 0.07 (b) 0.13
(c) 0.15 (d) 0.4 [2005 : 2 M]

1.31 A vehicle moving at 60 kmph on an ascending gradient of a highway has to come to stop position to avoid collision with a stationary object. The ratio of lag to brake distance is 6:5. Considering total reaction time of the driver as 2.5 seconds and the coefficient of longitudinal friction as 0.36, the value of ascending gradient (%) is

(a) 3.3 (b) 4.8
(c) 5.3 (d) 6.8 [2006 : 2 M]

1.32 At a horizontal curve portion of a 4 lane undivided carriageway, a transition curve is to be introduced to attain required superelevation. The design speed is 60 kmph and radius of the curve is 245 m. Assume length of wheel base of a longest vehicle as 6 m, superelevation rate as 5% and rate of introduction of this superelevation as 1 in 150. The length of the transition curve (m) required, if the pavement is rotated about inner edge is

(a) 81.4 (b) 85.0
(c) 91.5 (d) 110.2 [2006 : 2 M]

1.33 The extra widening required for a two-lane National Highway at a horizontal curve of 300 m radius, considering a wheel base of 8 m and a design speed of 100 kmph is

(a) 0.42 m (b) 0.62 m
(c) 0.82 m (d) 0 .92 m

[2007 : 2 M]

1.34 While designing a hill road with a ruling gradient of 6%, if a sharp horizontal curve of 50 m radius is encountered, the compensated gradient at the curve as per the Indian Roads Congress specifications should be

(a) 4.4% (b) 4.75%
(c) 5.0% (d) 5.25% [2007 : 2 M]

1.35 The design speed on a road is 60 kmph. Assuming the driver reaction time of 2.5 seconds and coefficient of friction of pavement surface as 0.35, the required stopping distance for two-way traffic on a single lane road is

(a) 82.1 m (b) 102.4 m
(c) 164.2 m (d) 186.4 m

[2007 : 2 M]

Linked Answer Questions 1.36 and 1.37:

A horizontal circular curve with a centre line radius of 200 m is provided on a 2-lane, 2-way SH section. The width of the 2-lane road is 7.0 m. Design speed for this section is 80 km per hour. The brake reaction time is 2.4 s, and the coefficients of friction in longitudinal and lateral directions are 0.355 and 0.15, respectively.

1.36 The safe stopping sight distance on the section is

(a) 221 m (b) 195 m
(c) 125 m (d) 65 m [2008 : 2 M]

1.37 The set-back distance from the centre line of the inner lane is

(a) 7.93 m (b) 8.10 m
(c) 9.60 m (d) 9.77 m [2008 : 2 M]

1.38 A road is provided with a horizontal circular curve having deflection angle of 55° and centre line radius of 250 m. A transition curve is to be provided at each end of the circular curve of such a length that the rate of gain of radial acceleration is 0.3 m/s^3 at a speed of 50 km per hour. Length of the transition curve required at each of the ends is

(a) 2.57 m (b) 33.33 m
(c) 35.73 m (d) 1666.67 m

[2008 : 2 M]

1.39 The value of lateral friction or side friction used in the design of horizontal curve as per Indian Roads Congress guidelines is

(a) 0.40 (b) 0.35
(c) 0.24 (d) 0.15 [2009 : 1 M]

1.40 A crest vertical curve joins two gradients of +3% and –2% for a design speed of 80 km/h and the corresponding stopping sight distance of 120 m. The height of driver's eye and the object above the road surface are 1.20 m and 0.15 m respectively. The curve length (which is less than stopping sight distance) to be provided is

(a) 120 m (b) 152 m
(c) 163 m (d) 240 m [2009 : 2 M]

1.41 Consider the following statements in the context of geometric design of roads:

1. A simple parabolic curve is an acceptable shape for summit curves.
2. Comfort to passengers is an important consideration in the design of summit curves.

The correct option evaluating the above statements and their relationship is

(a) 1 is true, 2 is false
(b) 1 is true, 2 is true and 2 is the correct reason for 1
(c) 1 is true, 2 is true and 2 is NOT the correct reason for 1
(d) 1 is false and 2 is true [2010 : 2 M]

1.42 The design speed for a two-lane road is 80 kmph. When a design vehicle with a wheelbase of 6.6 m is negotiating a horizontal curve on that road, the off-tracking is measured as 0.096 m. The required widening of carriageway of the two-lane road on the curve is approximately:

(a) 0.55 m (b) 0.65 m
(c) 0.75 m (d) 0.85 m [2010 : 2 M]

1.43 A vehicle negotiates a transition curve with uniform speed v. If the radius of the horizontal curve and the allowable jerk are R and J, respectively, the minimum length of the transition curve is

(a) $R^3/(vJ)$ (b) $J^3/(Rv)$
(c) v^2R/J (d) $v^3/(RJ)$ [2011 : 1 M]

1.44 If v is the initial speed of a vehicle, g is the gravitational acceleration, G is the upward longitudinal slope of the road and f_r is the coefficient of rolling friction during braking, the braking distance (measured horizontally) for the vehicle to stop is

(a) $\dfrac{v^2}{g(G+f_r)}$ (b) $\dfrac{v^2}{2g(G+f_r)}$
(c) $\dfrac{vg}{(G+f_r)}$ (d) $\dfrac{vf_r}{(G+g)}$

[2011 : 2 M]

1.45 The following data are related to a horizontal curved portion of a two lane highway: length of curve = 200 m, radius of curve = 300 m and width of pavement = 7.5 m. In order to provide a Stopping Sight Distance (SSD) of 80 m, the set back distance (in m) required from the centre line of the inner lane of the pavement is

(a) 2.54 (b) 4.55
(c) 7.10 (d) 7.96 [2012 : 2 M]

Common Data Questions (1.46 and 1.47):

For a portion of highway descending gradient 1 in 25 meets an ascending gradient 1 in 20. A valley curve needs to be design at a velocity of 90 kmph based on

(i) Head light sight distance equal to stopping sight distance of a level terrain. Consider length of curve > SSD.
(ii) Comfort condition if rate of change of acceleration is 0.5 m/s^3.

Reaction time = 2.5 sec, coefficient of longitudinal friction μ = 0.35. Height of head light = 0.75 m, and beam angle = 1°.

1.46 What is the length of valley curve as per headlight sight distance?

[2013 : 2 M]

1.47 What is the length of valley curve (in meter) based on comfort condition?

[2013 : 2 M]

1.48 The perception-reaction time for a vehicle travelling at 90 km/h, given the coefficient of longitudinal friction of 0.35 and the stopping sight distance of 170 m (assume g = 9.81 m/s^2), is ________ seconds.

[2014 : 2 M, Set-I]

1.49 A road is being designed for a speed of 110 km/hr on a horizontal curve with a super elevation of 8%. If the coefficient of side friction is 0.10, the minimum radius of the curve (in m) required for safe vehicular movement is

(a) 115.0 (b) 152.3
(c) 264.3 (d) 528.5

[2014 : 1 M, Set-II]

1.50 A sign is required to be put up asking drivers to slow down to 30 km/h before entering Zone Y (see figure). On this road, vehicles require 174 m to slow down to 30 km/h (the distance of 174 m includes the distance travelled during the perception-reaction time of drivers). The sign can be read by 6/6 vision drivers from a distance of 48 m. The sign is placed at a distance of x m from the start of Zone Y so that even a 6/9 vision driver can slow down to 30 km/h before entering the zone. The minimum value of x is ________m.

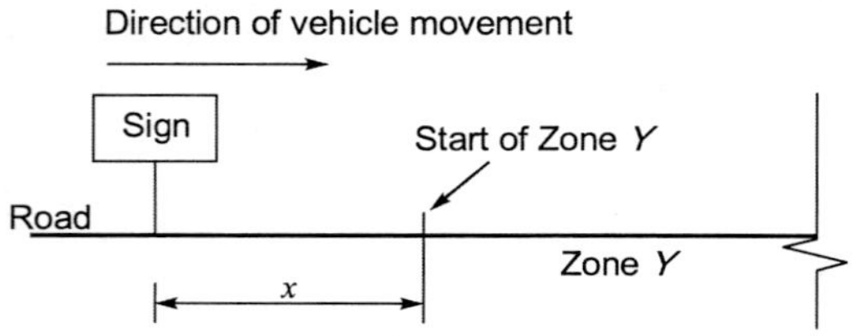

[2015 : 2 M, Set-I]

1.51 The acceleration-time relationship for a vehicle subjected to non-uniform acceleration is,

$$\frac{dv}{dt} = (\alpha - \beta v_0)e^{-\beta t}$$

where, v is the speed in m/s, t is the time in s, α and β are parameters, and v_0 is the initial speed in m/s. If the accelerating behavior of a vehicle, whose driver intends to overtake a slow moving vehicle ahead, is described as,

$$\frac{dv}{dt} = (\alpha - \beta v)$$

Considering α = 2 m/s^2, β = 0.05 s^{-1} and $\frac{dv}{dt} = 1.3 \text{m/s}^2$ at t = 3s, the distance (in m) travelled by the vehicle in 35 s is ________.

[2015 : 2 M, Set-I]

1.52 On a circular curve, the rate of superelevation is e. While negotiating the curve a vehicle comes to a stop. It is seen that the stopped vehicle does not slide inwards (in the radial direction). The coefficient of side friction is f. Which of the following is true?

(a) $e \le f$
(b) $f < e \le 2f$
(c) $e \ge 2f$
(d) none of these

[2015 : 2 M, Set-I]

1.53 A superspeedway in New Delhi has among the highest super-elevation rates of any track on the Indian Grand Prix circuit. The track requires drivers to negotiate turns with a radius of 335 m and 33° banking. Given this information, the coefficient of side friction required in order to allow a vehicle to travel at 320 km/h along the curve is

(a) 1.761
(b) 0.176
(c) 0.253
(d) 2.530

[2015 : 1 M, Set-II]

1.54 A two lane, one-way road with radius of 50 m is predominantly carrying lorries with wheelbase of 5 m. The speed of lorries is restricted to be between 60 kmph and 80 kmph. The mechanical widening and psychological widening required at 60 kmph are designated as $W_{me,60}$ and $W_{ps,60}$, respectively. The mechanical widening and psychological widening required at 80 kmph are designated as $W_{me,80}$ and $W_{ps,80}$, respectively. The correct values of $W_{me,60}$, $W_{ps,60}$, $W_{me,80}$, $W_{ps,80}$, respectively are

(a) 0.89 m, 0.50 m, 1.19 m, and 0.50 m
(b) 0.50 m, 0.89 m, 0.50 m, and 1.19 m
(c) 0.50 m, 1.19 m, 0.50 m, and 0.89 m
(d) 1.19 m, 0.50 m, 0.89 m, and 0.50 m

[2016 : 2 M, Set-I]

1.55 A motorist traveling at 100 km/h on a highway needs to take the next exit, which has a speed limit of 50 km/h. The section of the roadway before the ramp entry has a downgrade of 3% and coefficient of friction (f) is 0.35. In order to enter the ramp at the maximum allowable speed limit, the braking distance (expressed in m) from the exit ramp is_______.

[2016 : 2 M, Set-II]

1.56 A super-elevation e is provided on a circular horizontal curve such that a vehicle can be stopped on the curve without sliding. Assuming a design speed v and maximum coefficient of side friction f_{max}, which one of the following criteria should be satisfied?

(a) $e \le f_{max}$
(b) $e > f_{max}$
(c) no limit on e can be set
(d) $e = \dfrac{1-(f_{max})^2}{f_{max}}$

[2017 : 1 M, Set-I]

1.57 The radius of a horizontal circular curve on a highway is 120 m. The design speed is 60 km/hour, and the design coefficient of lateral friction between the tyre and the road surface is 0.15. The estimated value of superelevation required (if full lateral friction is assumed to develop), and the value of coefficient of friction needed (if no superelevation is provided) will, respectively, be

(a) $\frac{1}{11.6}$ and 0.10 (b) $\frac{1}{10.5}$ and 0.37

(c) $\frac{1}{11.6}$ and 0.24 (d) $\frac{1}{12.9}$ and 0.24

[2017 : 2 M, Set-I]

1.58 While aligning a hill road with a ruling gradient of 6%, a horizontal curve of radius 50 m is encountered. The grade compensation (in percentage, up to two decimal places) to be provided for this case would be ________.

[2017 : 1 M, Set-II]

1.59 A priority intersection has a single-lane one-way traffic road crossing an undivided two-lane two-way traffic road. The traffic stream speed on the single-lane road is 20 kmph and the speed on the two-lane road is 50 kmph. The perception-reaction time is 2.5 s, coefficient of longitudinal friction is 0.38 and acceleration due to gravity is 9.81 m/s^2. A clear sight triangle has to be ensured at this intersection. The minimum lengths of the sides of the sight triangle along the two-lane road and the single-lane road, respectively will be

(a) 50 m and 20 m (b) 61 m and 18 m

(c) 111 m and 15 m (d) 122 m and 36 m

[2018 : 2 M, Set-I]

1.60 A 7.5 m wide two-lane road on a plain terrain is to be laid along a horizontal curve of radius 510 m. For a design speed of 100 kmph, super-elevation is provided as per IRC : 73-1980. Consider acceleration due to gravity as 9.81 m/s^2. The level difference between the inner and outer edges of the road (in m, up to three decimal places) is ________.

[2018 : 2 M, Set-II]

1.61 A car follows a slow moving truck (travelling at a speed of 10 m/s) on a two-lane two-way highway. The car reduces its speed to 10 m/s and follows the truck maintaining a distance of 16 m from the truck. On finding a clear gap in the opposing traffic stream, the car accelerates at an average rate of 4 m/s^2, overtakes the truck and returns to its original lane. When it returns to its original lane, the distance between the car and the truck is 16 m. The total distance covered by the car during this period (from the time it leaves its lane and subsequently returns to its lane after overtaking) is

(a) 64 m (b) 72 m

(c) 128 m (d) 144 m

[2018 : 2 M, Set-II]

1.62 The coefficient of average rolling friction of a road is f_r, and its grade is +G%. If the grade of this road is doubled, what will be the percentage change in the braking distance (for the design vehicle to come to a stop) measured along the horizontal (assume all other parameters are kept unchanged)?

(a) $\frac{0.02G}{f_r + 0.01G} \times 100$ (b) $\frac{2f_r}{f_r + 0.01G} \times 100$

(c) $\frac{0.01G}{f_r + 0.02G} \times 100$ (d) $\frac{f_r}{f_r + 0.02G} \times 100$

[2019 : 1 M, Set-I]

1.63 A parabolic vertical curve is being designed to join a road of grade +5% with a road of grade –3%. The length of the vertical curve is 400 m measured along the horizontal, The vertical point of curvature (VPC) is located an the road of grade +5%. The difference in height between VPC and vertical point of intersection (VPI) (in m, round off to the nearest integer) is_______.

[2019 : 2 M, Set-I]

1.64 In an urban area, a median is provided to separate the opposing streams of traffic. As per IRC : 86-1983, the desirable minimum width (in m, expressed as integer) of the median, is _______.

[2020 : 1 M, Set-I]

1.65 A road in a hilly terrain is to be laid at a gradient of 4.5%. A horizontal curve of radius 100 m is laid at a location on this road. Gradient needs to be eased due to combination of curved horizontal and vertical profiles of the road. As per IRC, the compensated gradient (in %, round off to one decimal place), is ______.

[2020 : 1 M, Set-I]

1.66 The design speed of a two-lane two-way road is 60 km/h and the longitudinal coefficient of friction is 0.36. The reaction time of a driver is 2.5 seconds. Consider acceleration due to gravity as 9.8 m/s^2. The intermediate sight distance (in m, round off to the nearest integer) required for the load is ________.

[2020 : 2 M, Set-II]

1.67 The shape of the most commonly designed highway vertical curve is

(a) circular (multiple radii)

(b) circular (single radius)

(c) parabolic

(d) spiral

[2021 : 1 M, Set-I]

1.68 A highway designed for 80 km/h speed has a horizontal curve section with radius 250 m. If the design lateral friction is assumed to develop fully, the required superelevation is

(a) 0.02 (b) 0.05
(c) 0.07 (d) 0.09

[2021 : 1 M, Set-I]

1.69 The stopping sight distance (SSD) for a level highway is 140 m for the design speed of 90 km/h. The acceleration due to gravity and deceleration rate are 9.81 m/s^2 and 3.5 m/s^2 respectively. The perception/reaction time (in sec, round off to two decimal places) used in the SSD calculation is ________.

[2021 : 2 M, Set-II]

1.70 A horizontal curve is to be designed in a region with limited space. Which of the following measure(s) can be used to decrease the radius of curvature?

(a) Decrease the design speed.
(b) Increase the superelevation.
(c) Increase the design speed.
(d) Restrict vehicles with higher weight from using the facility.

[2022 : 1 M, Set-I]

1.71 A parabolic vertical crest curve connects two road segments with grades +1.0% and –2.0%, If a 200 m stopping sight distance is needed for a driver at a height of 1.2 m to avoid an obstacle of height 0.15 m, then the minimum curve length should be _____m, (round off to the nearest integer)

[2022 : 2 M, Set-II]

1.72 For a horizontal curve, the radius of a circular curve is obtained as 300 m with the design speed as 15 m/s. If the allowable jerk is 0.75 m/s^3, what is the minimum length (in m, in integer) of the transition curve?

[2023 : 1 M, Set-I]

1.73 G_1 and G_1 are the slopes of the approach and departure grades of a vertical curve, respectively. Given $|G_1| < |G_2|$ and $|G_2| \neq |G_2| \neq 0$

Statement 1: $+G_1$ followed by $+G_2$ results in a sag vertical curve.

Statement 2: $-G_1$ followed by $-G_2$ results in a sag vertical curve.

Statement 3: $+G_1$ followed by $-G_2$ results in a crest vertical curve.

Which option amongst the following is true?

(a) Statement 1 and Statement 3 are correct; Statement 2 is wrong
(b) Statement 1 and Statement 2 are correct; Statement 3 is wrong
(c) Statement 1 is correct; Statement 2 and Statement 3 are wrong
(d) Statement 2 is correct; Statement 1 and Statement 3 are wrong

[2023 : 1 M, Set-I]

1.74 Which of the following is equal to the stopping sight distance?

(a) (braking distance required to come to stop) + (distance travelled during the perception-reaction time)
(b) (braking distance required to come to stop) – (distance travelled during the perception-reaction time)
(c) (braking distance required to come to stop)
(d) (distance travelled during the perception-reaction time)

[2023 : 1 M, Set-II]

1.75 As per the Indian Roads Congress guidelines (IRC 86 : 2018), extra widening depends on which of the following parameters?

(a) Horizontal curve radius
(b) Superelevation
(c) Number of lanes
(d) Longitudinal gradient

[2023 : 1 M, Set-II]

1.76 A car is travelling at a speed of 60 km/hr on a section of a National Highway having a downward gradient of 2%. The driver of the car suddenly observes a stopped vehicle on the car path at a distance 130 m ahead, and applies brake. If the brake efficiency is 60%, coefficient of friction is 0.7, driver's reaction time is 2.5 s. and acceleration due to gravity is 9.81 m/s^2, the distance (in meters) required by the driver to bring the car to a safe stop lies in the range

(a) 41 to 45 (b) 126 to 130
(c) 33 to 37 (d) 75 to 79

[2024 : 1 M, Set-I]

1.77 A vertical summit curve on a freight corridor is formed at the intersection of two gradients, +3.0% and -5.0%.

Assume the following:

Only large-sized trucks are allowed on this corridor

Design speed =80kmph

Eye height of truck drivers above the road surface = 2.30 m

Height of object above the road surface for which trucks need to stop =0.35 m

Total reaction time of the truck drivers =2.0 s

Coefficient of longitudinal friction of the road =0.36

Stopping sight distance gets compensated on the gradient

The design length of the summit curve (in meters) to accommodate the stopping sight distance is _______ (rounded off to 2 decimal places).

[2024 : 2 M, Set-II]

Answers Highway Geometric Design & Planning

1.1	(a)	1.2	(c)	1.3	(d)	1.4	(d)	1.5	(d)	1.6	(d)	1.7	(d)
1.8	(b)	1.9	(a)	1.10	(b)	1.11	(a)	1.12	(b)	1.13	(d)	1.14	(c)
1.15	(a)	1.16	(b)	1.17	(c)	1.18	(d)	1.19	(b)	1.20	(c)	1.21	(b)
1.22	(b)	1.23	(b)	1.24	(a)	1.25	(d)	1.26	(b)	1.27	(c)	1.28	(d)
1.29	(b)	1.30	(b)	1.31	(b)	1.32	(d)	1.33	(c)	1.34	(a)	1.35	(c)
1.36	(c)	1.37	(d)	1.38	(c)	1.39	(d)	1.40	(*)	1.41	(a)	1.42	(c)
1.43	(d)	1.44	(b)	1.45	(*)	1.46	(308.9)	1.47	(106.066)	1.48	(3.15)	1.49	(d)
1.50	(142)	1.51	(900.8)	1.52	(a)	1.53	(a)	1.54	(b)	1.55	(92.3)	1.56	(a)
1.57	(c)	1.58	(1)	1.59	(c)	1.60	(0.525)	1.61	(b)	1.62	(c)	1.63	(10)
1.64	(1.2)	1.65	(4)	1.66	(162)	1.67	(c)	1.68	(b)	1.69	(2.028)	1.70	(a, b)
1.73	(a)	1.74	(a)	1.75	(a, c)	1.76	(d)	1.77	(117.93)				

Explanations Highway Geometric Design & Planning

1.1 (a)

The highway stretches at which required overtaking distance can not be provided, intermediate sight distance = 2 × SSD is provided.

1.2 (c)

Ideal shape of summit curve is circular due to constant sight distance throughout the curve.

In practice we use square parabolic curve because it is easy to compute and also give good riding comfort.

1.3 (d)

Transition curve has radius which decreases from infinity at the tangent point to the designed radius of the circular curve. If a curve of radius 'r' takes off from straight road and a vehicle travels on this road then due to the centrifugal force which suddenly acts on the vehicle just after the tangent point, a sudden jerk is felt on the vehicle. This not only causes discomfort to the passengers but also makes it difficult to steer the vehicle safely. If a transition curve is introduced between the straight and circular curve of radius 'R', the centrifugal force will be introduced gradually.

The rate at which this force is introduced can be controlled by adopting suitable shape of the transition curve and by designing its length so that vehicle can have a smooth entry from the straight to the circular curve at the design speed.

1.4 (d)

There is no restriction to sight distance of valley curves during the day time. At right time the only source of visibility is headlight of the vehicles. So, valley curve is designed based on night time driving conditions.

1.5 (d)

Steel wheels on steel rails have lower frictional coefficient than rubber tyres on pavements.

POINTS TOREMEMBER

- The coefficient of side friction prevents sliding of stopped vehicles on superelevated horizontal curves along the lateral direction, whereas longitudinal coefficient of friction prevents sliding of stopped vehicles downfill on the gradients.

1.6 (d)

1.7 (d)

Recommended values of camber for different types of road surfaces,

Sl. No.	Types of road surface	Range of camber in areas of rainfall range Heavy to Light
1.	Cement concrete and high type bituminous surface	1 in 50(2.0%) to 1 in 60(1.7%)
2.	Thin bituminous surface	1 in 40(2.5%) to 1 in 50(2.0%)
3.	Water bound mecadam, and gravel pavement	1 in 33(3.0%) to 1 in 40(2.5%)
4.	Earth	1 in 25(4.0%) to 1 in 33(3.0%)

1.8 (b)

As per IRC, the reaction time for SSD and OSD calculations are taken as 2.5 seconds and 2 seconds respectively.

1.9 (a)

Superelevation is given by the relation,

$$e + f = \frac{V^2}{gR}$$

if V is in km/hr, then,

$$e + f = \frac{V^2}{127R}$$

$$\Rightarrow \quad e + 0.15 = \frac{(60)^2}{127 \times 128}$$

$$\Rightarrow \quad e = 0.071$$

1.10 (b)

$$\frac{1}{N} = \left|\frac{1}{50} - \frac{1}{30}\right|$$

$$\Rightarrow \quad \frac{1}{N} = \left|\frac{-20}{50 \times 30}\right|$$

$$\Rightarrow \quad N = \frac{1}{75}$$

1.11 (a)

Intermediate sight distance (= 2 × SSD) is provided at highway stretches where the required overtaking sight distance can not be provided.

1.12 (b)

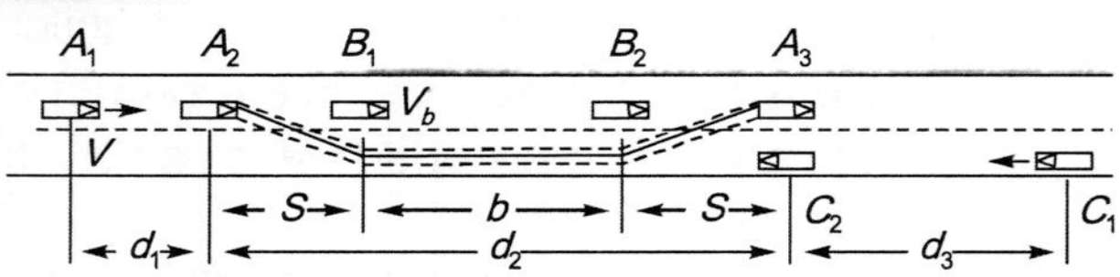

Overtaking Manoeuvre

Overtaking Sight Distance (OSD)

$$= d_1 + d_2 + d_3$$

$$V_b = (V - 16) = (100 - 16) = 84 \text{ km/hr}$$

$$d_1 = 0.278\, V_b t \qquad (\because t = 2 \text{ sec for OSD})$$

$$= 0.278 \times 84 \times 2 = 46.704 \text{ m}$$

$$d_2 = 0.278 V_b T + \frac{1}{2} a T^2 = 0.278 V_b T + 2S$$

and $\quad T = \sqrt{\frac{4S}{a}}$

also, $\quad S = 0.2\, V_b + 6$

$$= 0.2 \times 84 + 6 = 22.8 \text{ m}$$

$$\Rightarrow \quad T = \sqrt{\frac{4 \times 22.8}{0.53}} = 13.118 \text{ sec}$$

$$\therefore \quad d_2 = (0.278 \times 84 \times 13.118) + \frac{1}{2} \times 0.53 \times (13.118)^2$$

$$= 306.331 + 45.60 = 351.931 \text{ m}$$

$$d_3 = 0.278\, V_c T$$

$$= 0.278 \times 100 \times 13.118$$

$$= 364.68 \text{ m}$$

$$\therefore \quad OSD = d_1 + d_2 + d_3$$

$$= 46.704 + 351.931 + 364.68$$

$$= 763.315 \text{ m}$$

POINTS TOREMEMBER

- In OSD calculations, the speed of the vehicle coming from the opposite direction is taken as equal to the design speed of the road i.e., 100 kmph in this case.

1.13 (d)

$$n_1 = +4\%, \; n = -2\%$$

$$N = n_1 - n_2 = 4 - 2(-2) = 6\%$$

The highest point on the summit curve is at a distance given by,

$$D = \frac{n_1}{N} L_s$$

$$D = 150 \times \frac{4}{4 - (-2)} = 100 \text{m}$$

AVOID MISTAKE

$$D = \frac{L_s}{2} = \frac{150}{2} = 75 \text{ m}$$

This is the location of vertical point of intersection (VPI) and not the highest point of summit curve.

1.14 (c)

1.15 (a)

1.16 (b)

$$e + f = \frac{V^2}{127R}$$

$$\Rightarrow \quad R = \frac{(100)^2}{127 \times (0.07 + 0.15)}$$

$$\Rightarrow \quad R = 358 \text{ m} \approx 360 \text{ m}$$

1.17 (c)

$$e = \frac{V^2}{225R} = \frac{(125)^2}{225 \times 450} = 0.154$$

But $0.154 > (e_{max} = 0.07)$

$e > 0.07$ is not provided

Hence design rate of superelevation is,

$$e = 0.07$$

1.18 (d)

The design value of lateral friction is 0.15.

1.19 (b)

Camber is the transverse slope provided on the road surface to drain off the water from the road surface.

1.20 (c)

$$L = 2\left[\frac{Nv^3}{C}\right]^{1/2}$$

$$= 0.38\,[NV^3]^{1/2} \qquad [\text{since } C = 0.6 \text{ m/s}^3]$$

$$N = \left|\frac{1}{40} - \left(-\frac{1}{50}\right)\right| = \frac{1}{40} + \frac{1}{50} = 0.045$$

$$\therefore \quad L = 0.38[0.045 \times 80^3]^{1/2}$$

$$= 57.67 \text{ m} \simeq 58 \text{ m}$$

AVOID MISTAKE

As per IRC, rate of change of radial acceleration is to be taken as 0.6 m³/s for valley curves.

Do not consider relation, $C = \frac{80}{70+V} \text{m}^3/\text{s}$ as it is valid only for spinal nature transition curve, while the valley curves are designed as a combination of two transitional curve of cubic parabola shape.

1.21 (b)

$$SSD = Vt + \frac{V^2}{2gf}$$

As f↑ SSD↓

Hence, SSD and frictional coefficients are inversely proportional.

1.22 (b)

$$V = 72 \text{ kmph} = 20 \text{ m/s}$$

$$S = Vt + \frac{V^2}{2g(f + 0.02)}$$

$$= 30 + \frac{20^2}{2 \times 9.81(0.15 + 0.02)}$$

$$= 150 \text{ m}$$

AVOID MISTAKE

$$SSD = Vt + \frac{V^2}{2g(f \pm n\%)}$$

Take note of +/– sign,

for upward gradient → (+)

for downward gradient → (–)

1.23 (b)

The ruling minimum radius of horizontal curve is given by,

$$R_{min} = \frac{V^2}{127(e+f)} = \frac{(100)^2}{127(0.10 + 0.15)}$$

$$= 314.96 \text{ m} \simeq 315 \text{ m}$$

1.24 (a)

The Star and Grid pattern of road network was adopted in Nagpur Road Plan while Rectangular and Block pattern was adopted in Lucknow Road Plan.

1.25 (d)

Assuming utility units as given below

Population	< 2000	2000-5000	> 5000
Unit	0.5	1	2

Road	Length (km)	Total utility units (served by the road)	Utility per unit length	Priority
P	20	8 × 0.5 + 6 × 1 + 1 × 2	12/20 = 0.6	III
Q	28	19 × 0.5 + 8 × 1 + 4 × 2	25.5/28 = 0.91	II
R	12	7 × 0.5 + 5 × 1 + 2 × 2	12.5/12 = 1.04	I

Thus, the order of priority will be *R*, *Q*, *P*.

1.26 (b)

Taking camber as super elevation,

$$\therefore \quad e = \frac{3}{100} = 0.03$$

$$0.03 = \frac{V^2}{225R}$$

$$\Rightarrow \quad 0.03 = \frac{(80)^2}{225R}$$

$$\therefore \quad R = 948.14 \text{ m}$$

POINTS TO REMEMBER

- *Radius beyond which no superelevation is needed:* It is the radius of curve designed for 75% of design speed, considering coefficient of friction as zero and camber as superelevation.

$$e + f = \frac{V^2}{225R}$$

$$R = \frac{V^2}{225e_{camber}} \qquad [\because f = 0]$$

1.27 (c)

$$\text{Braking distance} = \frac{V^2}{254\,f} = \frac{(65)^2}{254 \times 0.396} = 42 \text{ m}$$

POINTS TO REMEMBER

$$SSD = \text{Lag distance} + \text{Braking distance}$$

$$= Vt + \frac{V^2}{2gf}$$

1.28 (d)

Length of summit curve depends upon:

(i) SSD for single lane two way highway

(ii) OSD for two lane two way highway

1.29 (b)

1.30 (b)

$$e + f = \frac{V^2}{127\,R}$$

$$\Rightarrow \quad f = \frac{V^2}{127\text{ R}} - e = \frac{(100)^2}{127 \times 400} - 0.07$$

$$= 0.13$$

POINTS TO REMEMBER

- Friction is a self-adjusting force.
- Full effect of superelvation (e = 0.07) is to be taken into account.

1.31 (b)

$$\text{Lag distance} = 0.278\,Vt$$

$$= 0.278 \times 60 \times 2.5$$

$$= 41.7 \text{ m}$$

$$\text{Brake distance} = \frac{5}{6} \times 41.7 = 34.75 \text{ m}$$

$$\left[\because \frac{\text{Lag distance}}{\text{Braking distance}} = \frac{6}{5}\right]$$

$$\text{Braking distance} = \frac{V^2}{254(f + 0.01n)}$$

$$\Rightarrow \quad 34.75 = \frac{(60)^2}{254(0.36 + 0.01\text{n})}$$

$$\Rightarrow \quad 0.01\,n = \frac{(60)^2}{254 \times 34.75} - 0.36$$

$$\Rightarrow \quad n = 4.8\%$$

1.32 (d)

Length of transition curve based on:

(i) Allowable rate of centrifugal acceleration

$$C = \frac{80}{75 + V}\text{m}^3/\text{s} = \frac{80}{75 + 60} = 0.59 \text{ m}^3/\text{s}$$

$$L_T = \frac{V^3}{CR} = \frac{(0.278 \times 60)^3}{0.59 \times 245} = 32.10 \text{ m}$$

(ii) As per introduction of superelevation

Extra-widening,

$$\text{W}_\text{m} = \frac{V}{9.5\sqrt{R}} + \frac{nl^2}{2R} = \frac{60}{9.5\sqrt{245}} + \frac{4 \times 6^2}{2 \times 245} = 0.697 \text{ m}$$

The road is rotated about inner edge,

Superlevation,

$$e = (\text{W} + \text{W}_\text{m})$$

$$= 0.05\,(4 \times 3.5 + 0.697)$$

$$= 0.735 \text{ m}$$

Rate of introduction of superelevation = 1 in 150

$$\Rightarrow \quad \text{L}_\text{T} = 150 \times \text{E}$$

$$= 150 \times 0.735$$

$$= 110.25 \text{ m}$$

(iii) IRC minimum length

$$L_T = 2.7\frac{V^2}{R} = \frac{2.7\times 60^2}{245} = 39.67 \text{ m}$$

Length of transition

= max. {32.10, 110.25, 39.67}

= 110.25 m

AVOID MISTAKE

Do not take C = 0.6 m^3/s as in design of valley curves. Also remember to incorporate the effect of extra-widening.

1.33 (c)

The extra widening is given by,

$$E_w = \frac{nL^2}{2R} + \frac{V}{9.5\sqrt{R}}$$

$$= \frac{2\times(8)^2}{2\times 300} + \frac{100}{9.5\sqrt{300}} = 0.82 \text{ m}$$

Special case: Psychological widening is not to be considered for single lane road.

1.34 (a)

Grade compensation is given by,

$\frac{30+R}{R}$ or $\frac{75}{R}$ whichever is minimum

$$\Rightarrow \frac{30+50}{50} = \frac{80}{50} = 1.6$$

or $\frac{75}{50} = 1.5$

$\therefore$ G.C. = 1.5

Ruling gradient = 6%

Compensated gradient = 6 – 1.5 = 4.5% > 4%

So most close option is (a).

POINTS TO REMEMBER

- Grade comensation is not required for a gradient flatter than 4%.
- Compensated gradient should not be less than 4%.

1.35 (c)

$$\text{Stopping Distance} = 0.278\,Vt + \frac{V^2}{254\,f}$$

$$= 0.278\times 60\times 2.5 + \frac{(60)^2}{254\times 0.35} = 82.1 \text{ m}$$

But the traffic is two way, therefore the stopping distance = 2 × 82.1 = 164.2 m

POINTS TO REMEMBER

- For one-way traffic on a single lane road: SSD.
- For two-way traffic on a single lane road: 2× SSD
- For two-way traffic on an undivided two lane road: 2 × SSD
- For two-way traffic on an divided two lane road: SSD

1.36 (c)

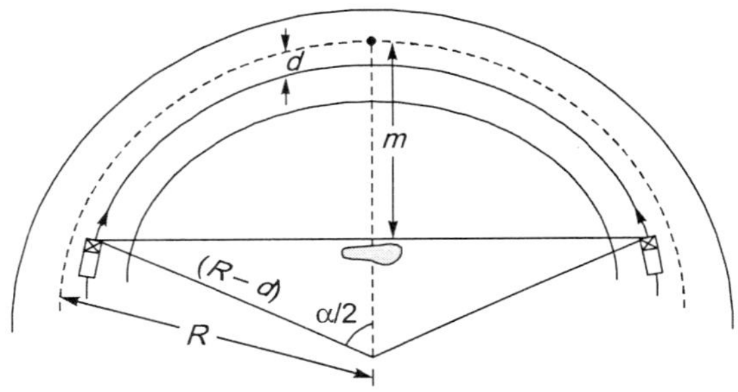

Safe stopping sight distance

$$= 0.278\,Vt + \frac{V^2}{254f}$$

$$= (0.278\times 80\times 2.4) + \frac{(80)^2}{254\times 0.355}$$

$$= 124.35 \text{ m} \approx 125 \text{ m}$$

AVOID MISTAKE

Always check from which line set back distacne has been asked.

POINTS TO REMEMBER

Set back distance from:

- Centerline of inner lane = (m – d)
- Centerline of outer lane = (m + d)
- Inneredge of pavement = (m – 2d)
- Outeredge of pavement = (m + 2d)

1.37 (d)

As no information is given about length of the curve, assume

$L_c >$ SSD

$$m = (R) - (R-d)\cos\frac{\alpha}{2}$$

$$\frac{\alpha}{2} = \frac{180S}{2\pi(R-d)}$$

$$= \frac{180 \times 125}{2 \times \pi\left(200 - \frac{7}{4}\right)} = 18.063$$

$$\therefore \quad m = 200 - \left(200 - \frac{7}{4}\right)\cos(18.063)$$

$$= 11.52 \text{ m}$$

Thus, set back distance from centre line of inner lane = $m - d$

$$= 11.52 - \left(\frac{7}{4}\right) = 9.77 \text{ m}$$

1.38 (c)

If α is the rate of change of radial acceleration, the radial acceleration (a) attained during the time the vehicle passes over the transition curve is given by,

$$a = \alpha t = \alpha \times \frac{L}{v}$$

Radial acceleration,

$$a = \frac{v^2}{R}$$

$$\therefore \quad \alpha \times \frac{L}{v} = \frac{v^2}{R}$$

$$\Rightarrow \quad L = \frac{v^3}{\alpha R}$$

$$\therefore \quad L = \frac{\left(\frac{50 \times 1000}{60 \times 60}\right)^3}{0.3 \times 250}$$

$$\Rightarrow \quad L = 35.73 \text{ m}$$

1.39 (d)

1.40 (*)

When the curve length is less than stopping sight distance, then it is given by,

$$L = 2S - \frac{4.4}{N}$$

$$= 2 \times 120 - \frac{4.4}{0.03 - (-0.02)}$$

$$= 240 - 88 = 152 \text{ m} > S \text{ (120 m)}$$

× Assuming wrong

Now assume,

$L > S$,

$$L = \frac{NS^2}{4.4} = 0.05 \times \frac{120^2}{4.4} = 163.64 \text{m} \quad \text{(OK)}$$

1.41 (a)

When fast moving vehicle travels along a summit curve the centrifugal force will act upwards against gravity and hence a part of the pressure on the tyres and spring of the vehicle suspensions is relieved. So there is no problem of discomfort to passengers on summit curves. The only problem is designing summit curves is to provide adequate sight distances.

Circular summit curve is ideal curve as the sight distance available throughout the length of circular curve is constant. The deviation angles in vertical curves of highways are very small and so between the same tangent points, a simple parabola is nearly congruent with a circular arc; also a parabola is very easy for arithmetical manipulation for computing ordinates. Thus, in actual practice a simple parabolic curve is used as a summit curve instead of the circular arc.

1.42 (c)

Off-tracking, $\frac{l^2}{2R} = 0.096$ m

$$\Rightarrow \quad R = \frac{(6.6)^2}{2 \times 0.096} = 226.875 \text{ m}$$

Extra widening, $E_w = n \times \frac{l^2}{2R} + \frac{V}{9.5\sqrt{R}}$

$$= 2 \times 0.096 + \frac{80}{9.5\sqrt{226.875}}$$

$$= 0.192 + 0.56 = 0.752 \text{ m}$$

1.43 (d)

1.44 (b)

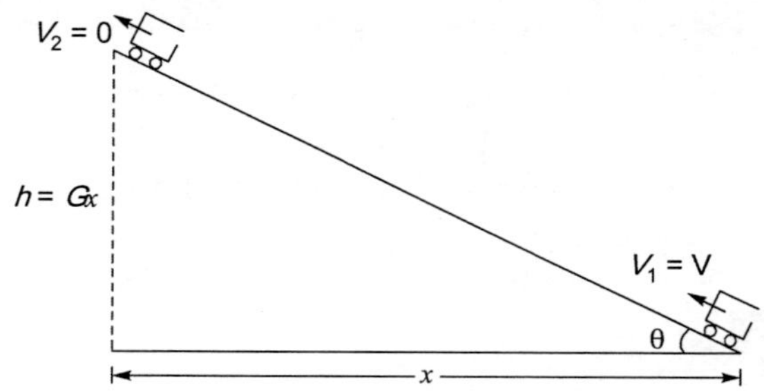

Friction force, $F = f_r(\text{mg} \cos\theta)$

Using law of conservation of energy,

$$\frac{1}{2}mV^2 = mgh + \text{Workdone against friction}$$

$$\Rightarrow \quad \frac{1}{2}mV^2 = mg(Gx) + f_r(mg\cos\theta)(x\sec\theta)$$

$$\Rightarrow \quad \frac{1}{2}mV^2 = mgx(G + f_r)$$

$$\Rightarrow \quad x = \frac{V^2}{2g(G + f_r)}$$

1.45 (*)

Length of curve, $L_c = 200$ m
Radius, $R = 300$ m
Width, $W = 7.5$ m
SSD, $S = 80$ m

$\because \quad L_c > S,$

Set back distance, from centre of the pavement,

$$m = (R) - (R - d)\cos\frac{\alpha}{2}$$

$$\frac{\alpha}{2} = \frac{180 \cdot S}{2\pi(R - d)} = \frac{180 \times 80}{2\pi \times \left(300 - \frac{7.5}{4}\right)} = 7.687$$

$$m = 300 - \left(300 - \frac{7.5}{4}\right)\cos(7.687) = 4.53$$

The setback distance (in m) required from the centre line of the inner lane of the pavement $= (m - d)$

$$= 4.55 - \frac{7.5}{4} = 2.675 \text{ m (not given)}$$

1.46 Sol.

$$N = \left| -\frac{1}{25} - \frac{1}{20} \right| = 0.09$$

$$S = 0.278 \times 90 \times 2.5 + \frac{(90)^2}{254 \times 0.35}$$

$$= 62.55 + 91.1136 = 153.6636$$

Assume $L >$ SSD,

$$L = \frac{NS^2}{(1.50 + 0.035S)}$$

$$\therefore \quad L = \frac{0.09(153.6636)^2}{(1.50 + 0.035 \times 153.6636)}$$

$$= 308.9641 \text{ m} > \text{SSD} \qquad \text{(OK)}$$

1.47 Sol.

$$L_v = 2\left[\frac{NV^3}{C}\right]^{1/2}$$

$$= 2\left[\frac{0.09 \times (0.278 \times 90)^3}{0.5}\right]^{1/2}$$

$$= 106.066 \text{ m}$$

1.48 Sol.

$$\text{SSD} = 0.278vt_r + \frac{(0.278v)^2}{2gf}$$

$$\Rightarrow \quad 170 = 0.278 \times 90 \times t_r + \frac{(0.278 \times 90)^2}{2 \times 9.81 \times 0.35}$$

$$\Rightarrow \quad t_r = 3.1510 \text{ sec.}$$

1.49 (d)

$$e + f = \frac{V^2}{127R}$$

$$\Rightarrow \quad R = \frac{V^2}{127(e + f)}$$

$$= \frac{(110)^2}{127(0.08 + 0.1)} = 528.5 \text{ m}$$

1.50 Sol.

For a driver having 6/6 vision can see the board from distance 48 m.

For a driver having 6/9 vision can see the board from distance $48 \times \frac{6}{9} = 32$ m

Vehicle requires 174 m to slow down to 30 km/hr. So minimum distance of X from the start of Y zone = 174 − 32 = 142 m.

POINTS TO REMEMBER

- A person with 6/9 vision can recognise or see somethings placed at a distance of 6 m, what a normal person can recognise or see when placed at a distance of 9 m.

1.51 Sol.

$$\frac{dv}{dt} = \alpha - \beta v$$

$$\Rightarrow \quad \int \frac{dv}{(\alpha - \beta v)} = \int dt$$

$$\Rightarrow \quad \frac{\ln(\alpha - \beta v)}{-\beta} = t + C$$

at $t = 3$ sec, $\frac{dv}{dt} = \alpha - \beta v = 1.3$ m/s^2

$\Rightarrow \quad \dfrac{\ln(1.3)}{-0.05} = 3 + c \Rightarrow c = -8.247$

$\therefore \quad \ln(2 - 0.05 \times v) = (t - 8.247)(-0.05)$

$\Rightarrow \quad 2 - 0.05 \times v = 1.51\, e^{-0.05t}$

$\Rightarrow \quad v = \dfrac{ds}{dt} = 40 - 30.21 e^{-0.05t}$

$\Rightarrow \quad s = 40t + 604.15\, e^{-0.05t} + c'$

$\therefore$ at $t = 0$, $s = 0 \Rightarrow c' = -604.15$

$\Rightarrow \quad s = 40t + 604.15\, e^{-0.05t} - 604.15$

at $t = 35$ sec

$\Rightarrow \quad s = 40 \times 35 + 604.15\, e^{-0.05 \times 35} - 604.15$

$\Rightarrow \quad s = 900.84$ m

1.52 (a)

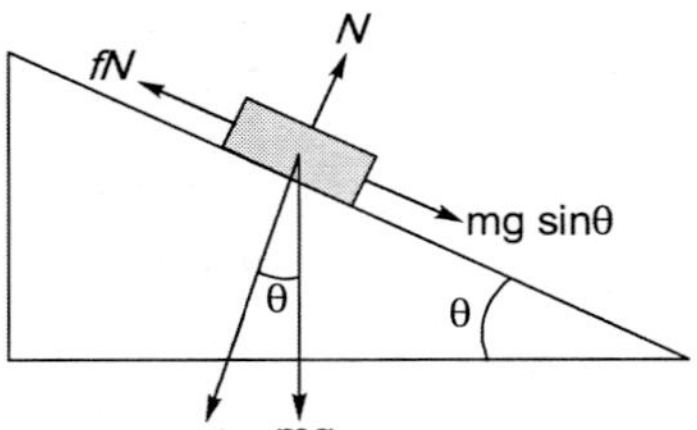

For equilibrium,

$$fN \geq mg \sin\theta$$

$$\Rightarrow \quad f\, mg \cos\theta \geq mg \sin\theta$$

$$\Rightarrow \quad f \cos\theta \geq \sin\theta$$

$$\Rightarrow \quad f \geq \tan\theta$$

$$\Rightarrow \quad f \geq e$$

or $\quad e \leq f$

1.53 (a)

$$R = 335 \text{ m}$$

$$V = 320 \text{ km/hr}$$

$$e + f = \frac{V^2}{127R}$$

$$\Rightarrow \quad f = \frac{V^2}{127R} - e$$

$$= \frac{320^2}{127 \times 335} - \tan 33^\circ$$

$$= 1.76$$

Note: *While deriving the above equation, we consider '1 – ef ≃ 1' but this is not valid in this case because banking provided is very high (tan 33° = 0.65). So formula used should be*

$\dfrac{e+f}{1-ef} = \dfrac{V^2}{127R}$ *which will result in f = 0.685.*

1.54 (b)

Method-I

$$W_{me'60} = \frac{nl^2}{2R} = \frac{2 \times 5^2}{2 \times 50} = 0.5 \text{ m}$$

$$W_{ps'60} = \frac{V}{9.5\sqrt{R}} = \frac{60}{9.5\sqrt{60}} = 0.89 \text{ m}$$

$$W_{me'80} = \frac{nl^2}{2R} = \frac{2 \times 5^2}{2 \times 50} = 0.5 \text{ m}$$

$$W_{ps'80} = \frac{V}{9.5\sqrt{R}} = \frac{80}{9.5\sqrt{50}} = 1.19 \text{ m}$$

Method-II

As mechanical widening is independent of speed of the vehicle,

$$W_{mc',60} = W_{me,80}$$

Hence, opetions (a) and (b) are eliminated.

Now, as psychological widening is directly proportional to the speed of the vehicle.

$$W_{ps,60} < W_{ps,80}$$

Hence, option (b) is the correct answer.

1.55 Sol.

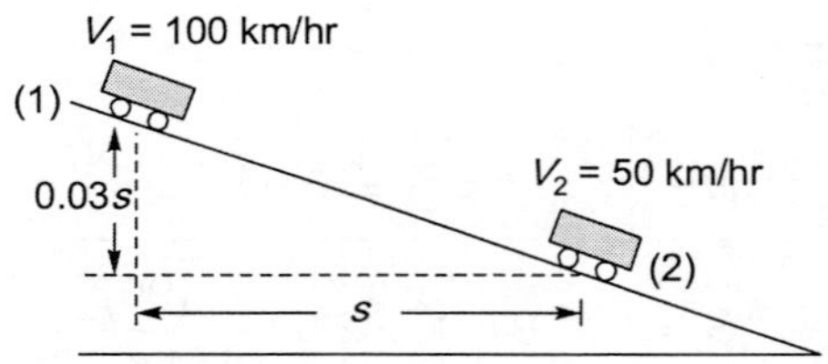

Total energy lost between point (1) and (2)

= Work done by frictional force

$$\frac{1}{2} m v_1^2 - \frac{1}{2} m v_2^2 + mg \times 0.03s = f \times (mg) \times s$$

$$\frac{(0.278 \times 100)^2}{2} - \frac{(0.278 \times 50)^2}{2} + 9.81 \times 0.03 \times s$$

$= 0.35 \times 9.81 \times s$

$$289.815 = 3.1392s$$

$$\Rightarrow \quad s = 92.32 \text{ m}$$

1.56 (a)

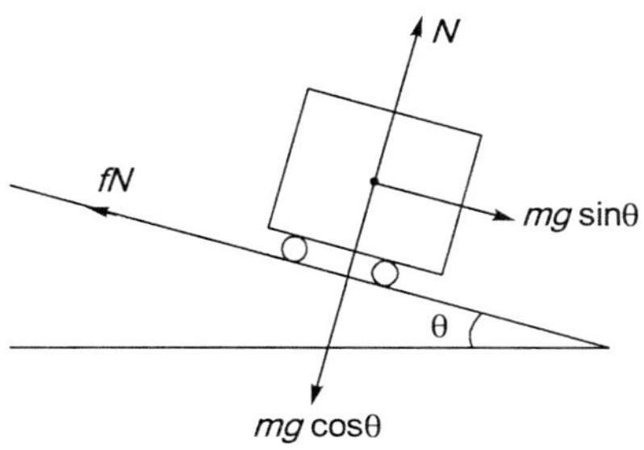

Sliding of stopped vehicle occurs if,

$mg \sin\theta > fn$

$mg \sin\theta > f(mg \cos\theta)$

$\tan\theta > f$

$e > f$

Hence, to prevent sliding $e \leq f_{max.}$

1.57 (c)

Given, R = 120 m, V = 60 kmph and f = 0.15

$$\Rightarrow \quad e + f = \frac{V^2}{127R}$$

$$\Rightarrow \quad e + 0.15 = \frac{60^2}{127 \times 120}$$

$$\Rightarrow \quad e = \frac{1}{11.6}$$

Also, if $e = 0$

$$\Rightarrow \quad f = \frac{V^2}{127R} = \frac{60^2}{127 \times 120} \simeq 0.24$$

1.58 Sol.

$$\text{G.C.} = \text{Minimum of} \left(\frac{75}{R} \text{ or } \frac{30+R}{R}\right)$$

$$= \text{Minimum of} \left(\frac{75}{50} \text{ or } \frac{30+50}{50}\right)$$

$$= \text{Minimum of } (1.5 \text{ or } 1.6)$$

$\therefore$Grade compensation = 1.5%

POINTS TO REMEMBER

- Grade compensation is not required for a gradient flatter then 4%.
- Compensated gradient should not be less than 4%.

  Had ruling gradient be 5%?

  $$\text{Grade compensation} = \text{Min.} \left\{\frac{75}{50} \text{ or } \frac{30+50}{50}\right\}$$

  $$= 1.5\%$$

  Compressed gradient = (5 – 1.5) = 3.5% < 4%

  Hence, compensated gradient = 4%.

  and grade compensation = (5 – 4) = 1%

1.59 (c)

1.60 Sol.

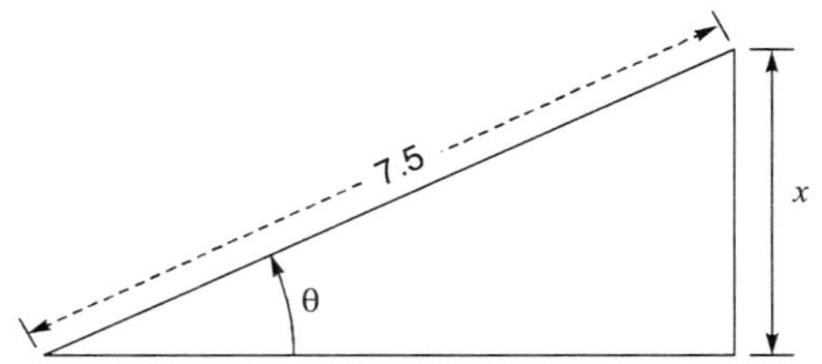

$$e = \frac{V^2}{225R} = \frac{100^2}{225 \times 510} = 0.0871 \not> 0.07$$

Provide, $e = 0.07$

$\sin\theta \simeq 0.07$

$x = eB = 0.07 \times 7.5 = 0.525$ m

1.61 (b)

Overtaking time,

$$T = \sqrt{\frac{4s}{a}} = \sqrt{\frac{4 \times 16}{4}} = 4 \text{ sec}$$

S = Space heady way = 16 m

a = Acceleration = 4 m/s^2

Distance travelled by vehicle = S_2

$$S_2 = uT + \frac{1}{2}aT^2$$

$$= 10 \times 4 + \frac{1}{2} \times 4 \times 4^2 = 72 \text{ m}$$

1.62 (c)

Percentage change in Braking Distance

$$= \left(\frac{L_{B_1} - L_{B_2}}{L_{B_1}}\right) \times 100$$

$$\%\Delta L_B = \frac{\dfrac{V^2}{254[f_r + 0.01G]} - \dfrac{V^2}{254[f_r + 0.02G]}}{\dfrac{V^2}{254[f_r + 0.01G]}}$$

$$= \frac{f_r + 0.02G - f_r - 0.01G}{(f_r + 0.01G)(f_r + 0.02G)} \times (f_r + 0.01G)$$

$$= \left(\frac{0.01G}{f_r + 0.02G} \times 100\right)$$

$$\%\Delta L_B = \left(\frac{0.01G}{f_r + 0.02G} \times 100\right)$$

1.63 Sol.

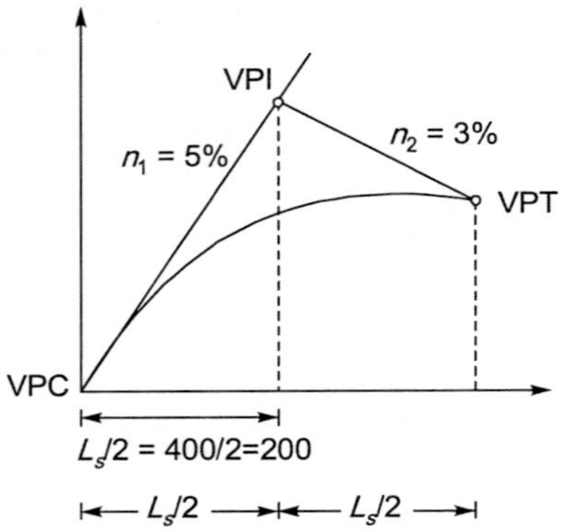

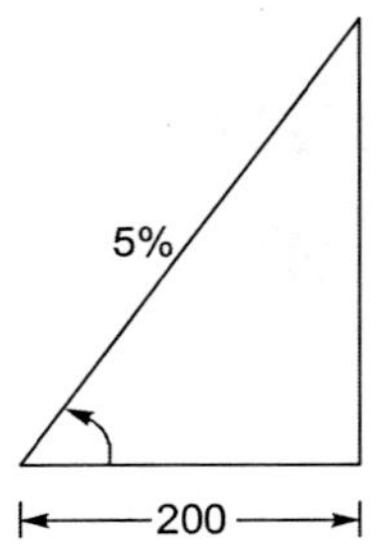

$$\tan\theta = 5\% = \frac{5}{100} = \frac{x}{200}$$

$$\frac{5}{100} = \frac{x}{200}$$

$$\Rightarrow \quad x = 10 \text{ m}$$

1.64 Sol.

As per IRC : 86-1983

Desirable minimum width of median in urban roads = 5 m

And minimum width = 1.2 m

1.65 Sol.

Gradient = 4.5%, R = 100 m

Grade compensation

$$= \left(\frac{30+R}{R}\right) \not> \left(\frac{75}{R}\right)\% = \frac{30+100}{100} \not> \frac{75}{100}$$

$$= 1.3\% \not> 0.75\} \; G.C = 0.75$$

Compansated Gradient

$$= 4.5\% - 0.75 = 3.75 \not< 4\%$$

Hence C.G = 4%

1.66 Sol.

Given : f = 0.36; v = 60 km;

g = 9.8 m/s^2; t_R = 2.5s

Stopping sight distance

$$\text{SSD} = \left(0.278Vt_R + \frac{V^2}{254f}\right)$$

$$= 0.278 \times 60 \times 2.5 + \frac{60^2}{254 \times 0.36}$$

$$= 41.7 + 39.37 = 81 \text{ m}$$

Intermediate sight distance

$$\text{ISD} = 2 \times \text{SSD} = 81 \times 2 = 162 \text{ m}$$

1.67 (c)

1.68 (b)

$$V = 80 \text{ kmph}, \; R = 250 \text{ m}$$

$$e + f = \frac{V^2}{127R}$$

$$e + 0.15 = \frac{80^2}{127 \times 250}$$

$$e = 0.051$$

1.69 Sol.

$$\text{SSD} = 140 \text{ m}$$

$$V = 90 \text{ kmph}$$

$$a = 3.5 \text{ m/s}^2$$

$$\text{SSD} = Vt_R + \frac{V^2}{2gf}$$

$$a = gf$$

$$140 = \left(\frac{5}{18} \times 90 \times t_R\right) + \frac{\left(\frac{5}{18} \times 90\right)^2}{2 \times 3.5}$$

$$t_R = 2.028 \text{ seconds}$$

1.70 (a, b)

On a horizontal curve,

$$e + f = \frac{V^2}{Rg}$$

$$\Rightarrow \quad R = \frac{V^2}{g(e+f)}$$

∴ To decrease radius of curvature, we can

(i) Decrease the design speed.

(ii) Increase the superelevation.

1.71 Sol.

Assume, L_s > SSD

∴ Length of summit curve,

$$L_s = \frac{NS^2}{2[\sqrt{H} + \sqrt{h}]^2}$$

$$\Rightarrow \quad L_s = \frac{[0.01 - (-0.02)] \times 200^2}{2[\sqrt{1.2} + \sqrt{0.15}]^2}$$

$$\Rightarrow \quad L_s = 272.90 \text{ m} > \text{SSD}$$

(∴ Assumption was correct)

1.72 Sol.

Given: $V = 15$ m/s

$R = 300$ m

Jerk, $C = 0.75$ m³/s

So, height of transition curve,

$$L_t = \frac{V^3}{RC} = \frac{15^3}{300 \times 0.75} = 15 \text{ m}$$

For plain and rolling terrain,

$$L = 2.7\frac{V^2}{R} = 2.7\frac{\left(15 \times \frac{18}{5}\right)^2}{300} = 26.24 \text{ m}$$

Minimum length of transition curve = 26.24 m

NOTE: In this question GATE has given two set of ranges (15 – 15) and (26 – 27) since the minimum length of transition curve depends on the type of terrain but the terrain is not given.

1.73 (a)

Condition for summit curve (crest vertical curve),

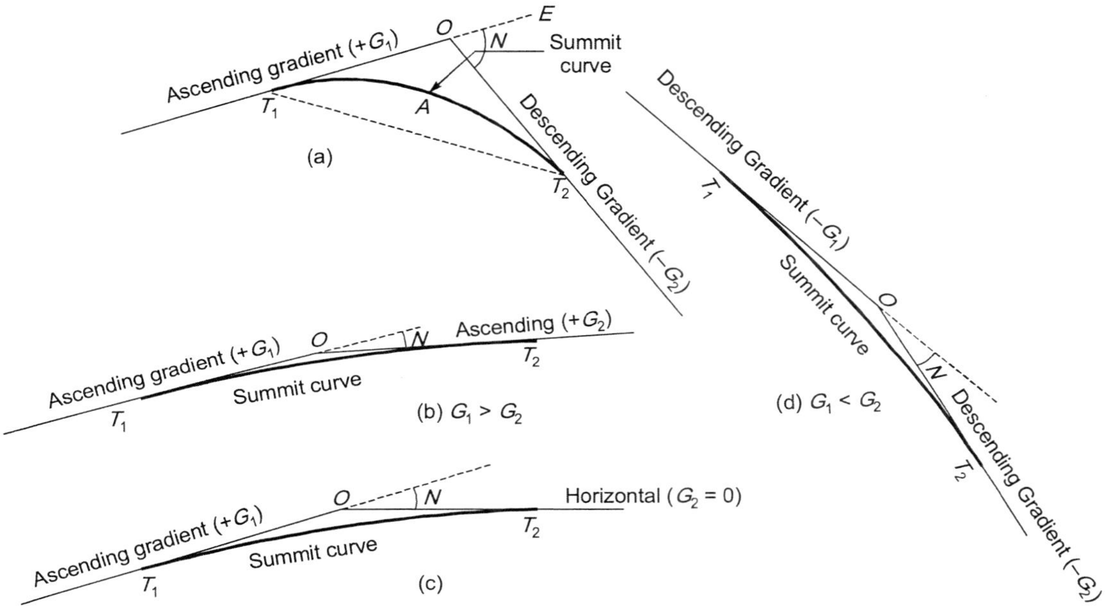

Condition for valley curve (sag vertical curve),

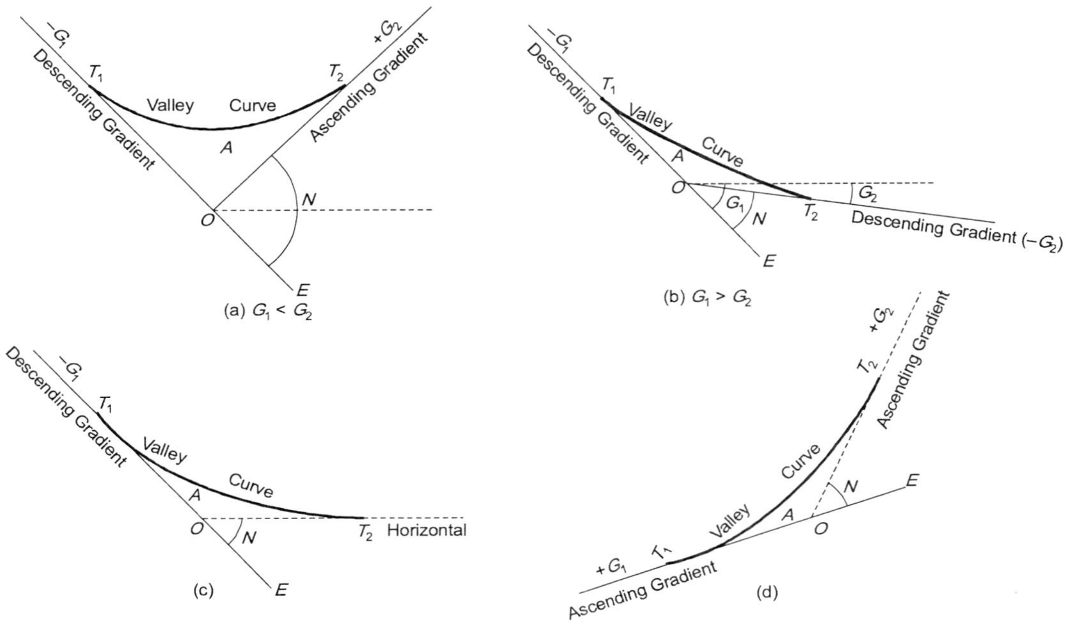

1.74 (a)

Stopping sight distance, SSD

$$= \underbrace{0.278Vt_R}_{\text{Lag distance}} + \underbrace{\frac{V^2}{254(f \pm s)}}_{\text{Braking distance}}$$

1.75 (a, c)

$$\text{Extra widening}, W_E = \frac{nl^2}{2R} + \frac{V}{9.5\sqrt{R}}$$

$$\therefore \quad W_E \propto n$$

$$W_E \propto \frac{1}{R}$$

1.76 (d)

Given, $V = 60$ km/hr, $t_R = 2.5$ sec

Braking efficiency, $\eta = 60\%$

Coefficient of friction, $\mu = 0.7\%$

Down grade, $s = 2\%$

Stopping sight distance = Lag distance + Braking distance

$$= 0.278 \times V \times t_r + \frac{V^2}{254(\eta f - s)}$$

$$= 0.278 \times 60 \times 2.5 + \frac{60^2}{254(0.6\,(0.7) - 0.02)}$$

$$= 41.7 + 35.4 = 77.1 \text{ m}$$

1.77 Sol.

$n_1 = +3\%$
$n_2 = -5\%$
$v = 80$ km/h
$h_1 = 2.30$ m
$h_2 = 0.35$ m
$t_r = 2$ sec.
$f = 0.36$

Deflection angle, $N = |n_1 - n_2| = 8\%$

Now stopping sight distance, S is given as

$$S = 0.278Vt + \frac{v^2}{254f}$$

$$= 0.278 \times 80 \times 2 + \frac{80^2}{254 \times 0.36} = 114.47 \text{ m}$$

Now assume, length of summit curve (L_s) is greater than stopping sight distance.

$$\therefore \quad L_s = \frac{NS^2}{2\left[\sqrt{h_1} + \sqrt{h_2}\right]^2} = \frac{0.08 \times (114.47)^2}{2\left[\sqrt{2.3} + \sqrt{0.35}\right]^2}$$

$$= 117.93 \text{ m} > 114.47 \text{ m} \quad \text{(OK)}$$

■■■■

2 Traffic Engineering

2.1 The speed and delay studies on a defined section of highway are conducted by
(a) radar gun
(b) traffic counters
(c) moving car method
(d) enoscope [2003 : 1 M]

2.2 A traffic stream in a particular direction of a two lane road is moving with a constant speed of 50 kmph, with an average headway of 2.52 seconds. The longitudinal distance between two consecutive vehicles is
(a) 30 m (b) 35 m
(c) 38 m (d) 42 m [2003 : 2 M]

2.3 The road geometrics in India are designed for the
(a) 98th highest hourly traffic volume
(b) 85th highest hourly traffic volume
(c) 50th highest hourly traffic volume
(d) 30th highest hourly traffic volume
[2004 : 1 M]

2.4 A single lane unidirectional highway has a design speed of 65 kmph. The perception-brake-reaction time of drivers is 2.5 seconds and the average length of vehicles is 5m. The coefficient of longitudinal friction of the pavement is 0.4. The capacity of this road in terms of 'vehicles per hour per lane' is
(a) 1440 (b) 750
(c) 710 (d) 680 [2005 : 2 M]

2.5 A transport company operates a scheduled daily truck service between city *P* and city *Q*. One-way journey time between these two cities is 85 hours. A minimum layover time of 5 hours is to be provided at each city. How many trucks are required to provide this service?
(a) 4 (b) 6
(c) 7 (d) 8 [2005 : 2 M]

2.6 Name the traffic survey data which is plotted by means of **Desire lines**.
(a) Accident
(b) Classified volume
(c) Origin and Destination
(d) Speed and Delay [2006 : 1 M]

2.7 For designing a 2-phase fixed type signal at an intersection having North-South and East-West road where only straight ahead traffic is permitted, the following data is available.

Parameter	North	South	East	West
Design Hour Flow (PCU/hr)	1000	700	900	550
Saturation Flow (PCU/hr)	2500	2500	3000	3000

Total time lost per cycle is 12 seconds. The cycle length (seconds) as per Webster's approach is
(a) 67 (b) 77
(c) 87 (d) 91 [2006 : 2 M]

2.8 On an urban road, the free mean speed was measured as 70 kmph and the average spacing between the vehicles under jam condition as 7.0 m. The speed-flow-density equation is given by:

$$U = U_{sf} = \left[1 - \frac{k}{k_j}\right]$$

where,
U = space-mean speed (kmph);
U_{sf} = free mean speed (kmph);
k = density (veh/km);
k_j = jam density (veh/km);
q = flow (veh/hr)

The maximum flow (veh/hr) per lane for this condition is equal to
(a) 2000 (b) 2500
(c) 3000 (d) None of these
[2006 : 2 M]

2.9 If a two-lane National Highway and a two-lane State Highway intersect at right angles, the number of potential conflict points at the intersection, assuming that both the roads are two-way is
(a) 11 (b) 17
(c) 24 (d) 32 [2007 : 1 M]

2.10 In signal design as per Indian Roads Congress specifications, if the sum of the ratios of normal flows to saturation flow of two directional traffic flow is 0.50 and the total lost time per cycle is 10 sec, the optimum cycle length in seconds is

(a) 100 (b) 80
(c) 60 (d) 40 [2007 : 1 M]

2.11 The capacities of "One-way 1.5 m wide sidewalk (persons per hour)" and "One-way 2-lane urban road (PCU per hour, with no frontage access, no standing vehicles and very little cross traffic)" are respectively
(a) 1200 and 2400 (b) 1800 and 2000
(c) 1200 and 1500 (d) 2000 and 1200
[2008 : 1 M]

2.12 The shape of the STOP sign according to IRC : 67-2001 is
(a) circular (b) triangular
(c) octagonal (d) rectangular
[2008 : 1 M]

2.13 A roundabout is provided with an average entry width of 8.4 m, width of weaving section as 14 m, and length of the weaving section between channelizing islands as 35 m. The crossing traffic and total traffic on the weaving section are 1000 and 2000 PCU per hour respectively. The nearest rounded capacity of the roundabout (in PCU per hour) is
(a) 3300 (b) 3700
(c) 4500 (d) 5200 [2008 : 2 M]

2.14 Design parameters for a signalized intersection are shown in the figure below. The green time calculated for major and minor roads are 34 and 18 s, respectively.

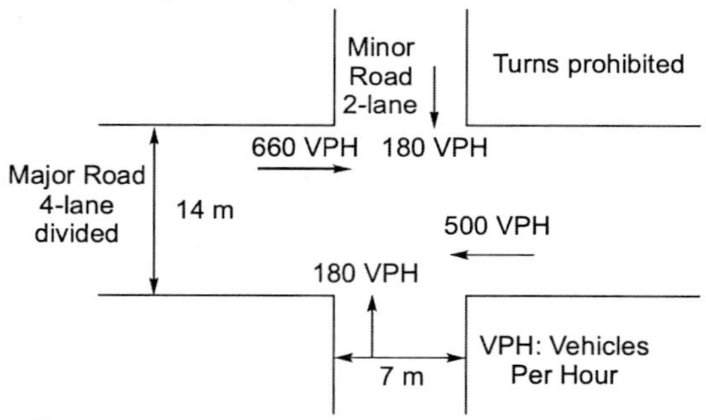

The critical lane volume on the major road changes to 440 vehicles per hour per lane and the critical lane volume on the minor road remains unchanged. The green time will
(a) increase for the major road and remain same for the minor road.
(b) increase for the major road and decrease for the minor road.
(c) decrease for both the roads.
(d) remain unchanged for both the roads.
[2008 : 2 M]

2.15 A linear relationship is observed between speed and density on a certain section of a highway. The free flow speed is observed to be 80 km per hour and the jam density is estimated as 100 vehicles per km length. Based on the above relationship, the maximum flow expected on this section and the speed at the maximum flow will respectively be
(a) 8000 vehicles per hour and 80 km per hour
(b) 8000 vehicles per hour and 25 km per hour
(c) 2000 vehicles per hour and 80 km per hour
(d) 2000 vehicles per hour and 40 km per hour
[2008 : 2 M]

2.16 On a specific highway, the speed-density relationship follows the Greenberg's model [$v = v_f \log_e(k_j/k)$], where v_f and k_j are the free flow speed and jam density respectively. When the highway is operating at capacity, the density obtained as per this model is
(a) ek_j (b) k_j
(c) $\frac{k_j}{2}$ (d) $\frac{k_j}{e}$ [2009 : 2 M]

2.17 A three-phase traffic signal at an intersection is designed for flows shown in the figure below. There are six groups of flows identified by the numbers 1 through 6. Among these 1, 3, 4 and 6 are through flows and 2 and 5 are right turning. Which phasing scheme is **not feasible**?

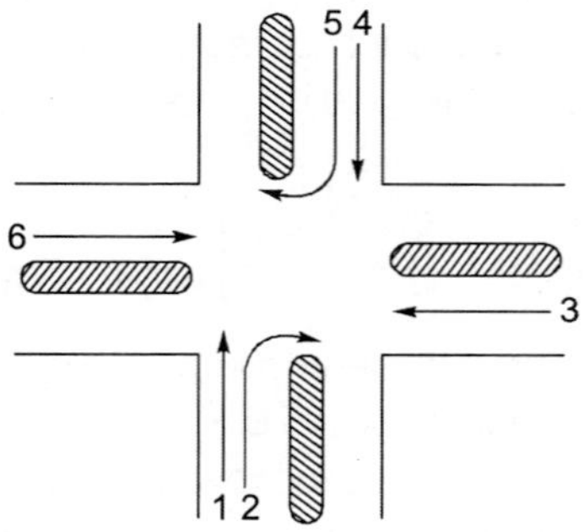

Combination choice	Phase-I	Phase-II	Phase-III
P	1, 4	2, 5	3, 6
Q	1, 2	4, 5	3, 6
R	2, 5	1, 3	4, 6
S	1, 4	2, 6	3, 5

(a) *P* (b) *Q*
(c) *R* (d) *S*
[2009 : 2 M]

2.18 As per **IRC: 67-2001**, a traffic sign indicating the Speed Limit on a road should be of
(a) circular shape with white background and red border
(b) triangular shape with white background and red border
(c) triangular shape with red background and white border
(d) circular shape with red background and white border

[2010 : 1 M]

2.19 The probability that k number of vehicles arrive (i.e. cross a predefined line) in time t is given as $(\lambda t)^k e^{-\lambda t}/k!$, where λ is the average vehicle arrival rate. What is the probability that the time headway is greater than or equal to time t_1?
(a) $\lambda e^{\lambda t_1}$ (b) λe^{-t_1}
(c) $e^{\lambda t_1}$ (d) $e^{-\lambda t_1}$ [2011 : 1 M]

2.20 If the jam density is given as k_j and the free flow speed is given as u_f, the maximum flow for a linear traffic speed-density model is given by which of the following options?
(a) $\frac{1}{4}k_j \times u_f$ (b) $\frac{1}{3}k_j \times u_f$
(c) $\frac{3}{5}k_j \times u_f$ (d) $\frac{2}{3}k_j \times u_f$

[2011 : 2 M]

2.21 The cumulative arrival and departure curve of one cycle of an approach lane of a signalized intersection is shown in the adjoining figure. The cycle time is 50 s and the effective red time is 30s and the effective green time is 20 s. What is the average delay?

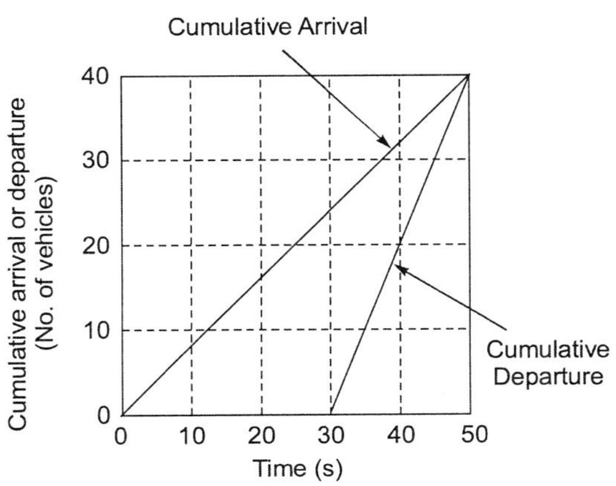

(a) 15 s (b) 25 s
(c) 35 s (d) 45 s [2011 : 2 M]

2.22 Two major roads with two lanes each are crossing in an urban areas to form an un-controlled intersection. The number of conflict points when both roads are one way is "X" and when both roads are two-way is "Y". The ratio of X to Y is
(a) 0.25 (b) 0.33
(c) 0.50 (d) 0.75 [2012 : 1 M]

2.23 A two-lane urban road with one-way traffic has a maximum capacity of 1800 vehicles/hour. Under the jam condition, the average length occupied by the vehicles is 5.0 m. The speed versus density relationship is linear. For a traffic volume of 1000 vehicles/hour, the density (in vehicles/km) is
(a) 52 (b) 58
(c) 67 (d) 75 [2012 : 2 M]

2.24 It was observed that 150 vehicle crossed a particular location of highway in 30 minutes. Assume that vehicle arrival follow a negative exponential distribution. The number of time headways greater than 5 sec. in above observation is

[2013 : 1 M]

2.25 For two major-roads with divided carriage way crossing at right angle, a full clover leaf interchange with four indirect ramps is provided. Following statements are made on turning movement of vehicles to all direction from both road, identity the correct statement.
(a) Merging from left is not possible, but diverging to left not possible.
(b) Merging from left and diverging to left is possible.
(c) Merging from left is possible but diverging is not possible.
(d) Neither merging from left nor diverging to left is possible.

[2013 : 1 M]

2.26 The minimum value of 15 minute peak hour factor on a section of a road is
(a) 0.10 (b) 0.20
(c) 0.25 (d) 0.33

[2014 : 1 M, Set-I]

2.27 The speed-density (*u-k*) relationship on a single lane road with unidirectional flow is u = 70 − 0.7 *k*, where *u* is in km/hr and *k* is in veh/km. The capacity of the road (in veh/hr) is __________.

[2014 : 2 M, Set-I]

2.28 An isolated three-phase traffic signal is designed by Webster's method. The critical flow ratios for three phases are 0.20, 0.30, and 0.25 respectively, and lost time per phase is 4 seconds. The optimum cycle length (in seconds) is ________.

[2014 : 2 M, Set-I]

2.29 The average spacing between vehicles in a traffic stream is 50 m, then the density (in veh/km) of the stream is ________.

[2014 : 1 M, Set-II]

2.30 An observer counts 240 veh/h at a specific highway location. Assume that the vehicle arrival at the location is Poisson distributed, the probability of having one vehicle arriving over a 30-second time interval is ______.

[2014 : 2 M, Set-II]

2.31 A student riding a bicycle on a 5 km one-way street takes 40 minutes to reach home. The student stopped for 15 minutes during this ride. 60 vehicles overtook the student (assume the number of vehicles overtaken by the student is zero) during the ride and 45 vehicles while the student stopped. The speed of vehicle stream on that road (in km/hr) is

(a) 7.5 (b) 12
(c) 40 (d) 60

[2014 : 2 M, Set-II]

2.32 On a section of a highway the speed-density relationship is linear and is given by $v = [80 - (2/3)\,k]$; where v is in km/h and k is in veh/km. The capacity (in veh/h) of this section of the highway would be

(a) 1200 (b) 2400
(c) 4800 (d) 9600

[2014 : 2 M, Set-II]

2.33 A pre-timed four phase signal has critical lane flow rate for the first three phases as 200, 187 and 210 veh/hr with saturation flow rate of 1800 veh/hr/lane for all phases. The lost time is given as 4 seconds for each phase. If the cycle length is 60 seconds, the effective green time (in seconds) of the fourth phase is _______.

[2014 : 2 M, Set-II]

2.34 Which of the following statements CANNOT be used to describe free flow speed (u_f) of a traffic stream?

(a) u_f is the speed when flow is negligible
(b) u_f is the speed when density is negligible
(c) u_f is affected by geometry and surface conditions of the road
(d) u_f is the speed at which flow is maximum and density is optimum

[2015 : 1 M, Set-I]

2.35 The following statements are made related to the lengths of turning lanes at signalised intersections

(i) 1.5 times the average number of vehicles (by vehicle type) that would store in turning lane per cycle during the peak hour.
(ii) 2 times the average number of vehicles (by vehicle type) that would store in turning lane per cycle during the peak hour.
(iii) Average number of vehicles (by vehicle type) that would store in the adjacent through lane per cycle during the peak hour.
(iv) Average number number of vehicles (by vehicle type) that would store in all lanes per cycle during the peak hour.

As per the IRC recommendations, the correct choice for design length of storage lanes is

(a) Maximum of (ii) and (iii)
(b) Maximum of (i) and (iii)
(c) Average of (i) and (iii)
(d) Only (iv)

[2015 : 1 M, Set-II]

2.36 The relation between speed u (in km/h) and density k (number of vehicles/km) for a traffic stream on a road is $u = 70 - 0.7k$. The capacity on this road is ________ vph (vehicles/hour).

[2015 : 2 M, Set-II]

2.37 In a one-lane one-way homogeneous traffic stream, the observed average headway is 3.0s. The flow (expression vehicles/hr) in this traffic stream is ________.

[2016 : 1 M, Set-I]

2.38 While traveling along and against the traffic stream, a moving observer measured the relative flows as 50 vehicles/hr and 200 vehicles/hr, respectively. The average speeds of the moving observer while traveling along and against the stream are 20 km/hr and 30 km/hr, respectively. The density of the traffic stream (expressed in vehicles/km) is________.

[2016 : 2 M, Set-I]

2.39 If the total number of commercial vehicles per day ranges from 3000 to 6000, the minimum percentage of commercial traffic to be surveyed for axle load is

(a) 15 (b) 20
(c) 25 (d) 30

[2016 : 1 M, Set-II]

2.40 The critical flow ratios for a three-phase signal are found to be 0.30, 0.25, and 0.25. The total time lost in the cycle is 10 s. Pedestrian crossings at the junction are not significant. The respective Green times (expressed in seconds and rounded off to the nearest integer) for the three phases are

(a) 34, 28 and 28 (b) 40, 25, and 25
(c) 40, 30 and 30 (d) 50, 25, and 25

[2016 : 2 M, Set-II]

2.41 Vehicles arriving at an intersection from one of the approach roads follow the Poisson distribution. The mean rate of arrival is 900 vehicles per hour. If a gap is defined as the time difference between two successive vehicle arrivals (with vehicles assumed to be points), the probability (up to four decimal places) that the gap is greater than 8 seconds is ________.

[2017 : 1 M, Set-I]

2.42 The queue length (in number of vehicles) versus time (in seconds) plot for an approach to a signalized intersection with the cycle length of 96 seconds is shown in the figure (not drawn to scale).

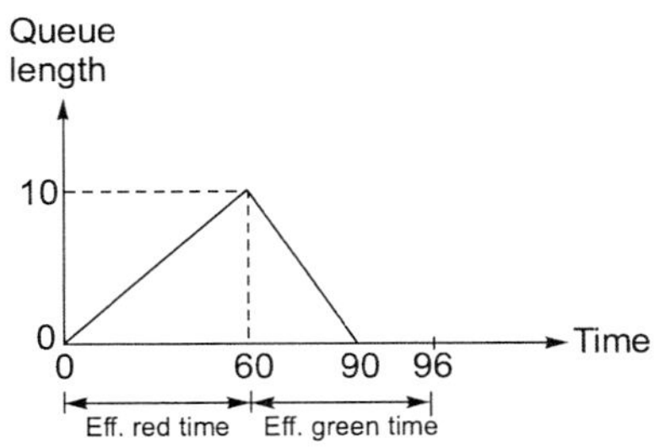

At time, $t = 0$, the light has just turned red. The effective green time is 36 seconds, during which vehicles discharge at the saturation flow rate, s (in vph). Vehicles arrive at a uniform rate, v (in vph), throughout the cycle. Which one of the following statements is TRUE?

(a) $v = 600$ vph, and for this cycle, the average stopped delay per vehicle = 30 seconds.
(b) $s = 1800$ vph, and for this cycle, the average stopped delay per vehicle = 28.125 seconds.
(c) $v = 600$ vph, and for this cycle, the average stopped delay per vehicle = 45 seconds.
(d) $s = 1200$ vph, and for this cycle, the average stopped delay per vehicle = 28.125 seconds.

[2017 : 2 M, Set-I]

2.43 The safety within a roundabout and the efficiency of a roundabout can be increased, respectively by

(a) increasing the entry radius and increasing the exit radius
(b) increasing the entry radius and decreasing the exit radius
(c) decreasing the entry radius and increasing the exit radius
(d) decreasing the entry radius and decreasing the exit radius

[2017 : 2 M, Set-II]

2.44 Two cars P and Q are moving in a racing track continuously for two hours. Assume that no other vehicles are using the track during this time. The expressions relating the distance travelled d (in km) and time t (in hours) for both the vehicles are given as

$P : d = 60t$

$Q : d = 60t^2$

Within the first one hour, the maximum space headway would be

(a) 15 km at 30 minutes
(b) 15 km at 15 minutes
(c) 30 km at 30 minutes
(d) 30 km at 15 minutes

[2017 : 2 M, Set-II]

2.45 The speed-density relationship for a road section is shown in the figure.

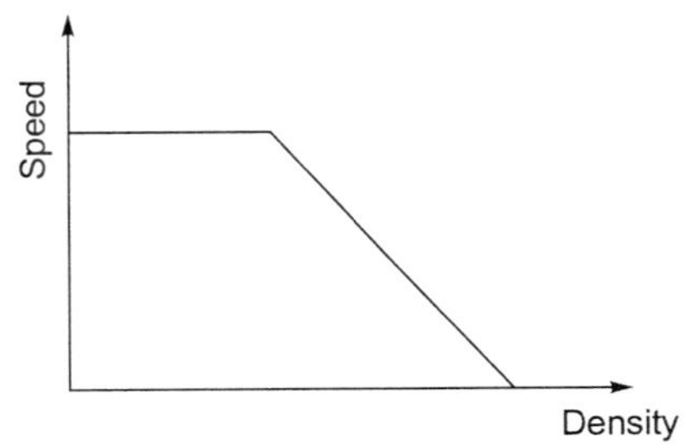

The shape of the flow-density relationship is

(a) piecewise linear
(b) parabolic
(c) initially linear then parabolic
(d) initially parabolic then linear

[2018 : 1 M, Set-I]

2.46 A well-designed signalized intersection is one in which the
(a) crossing conflicts are increased
(b) total delay is minimized
(c) cycle time is equal to the sum of red and green times in all phases
(d) cycle time is equal to the sum of red and yellow times in all phases

[2018 : 1 M, Set-I]

2.47 Peak Hour Factor (PHF) is used to represent the proportion of peak sub-hourly traffic flow within the peak hour. If 15-minute sub-hours are considered, the theoretically possible range of PHF will be
(a) 0 to 1.0 (b) 0.25 to 0.75
(c) 0.25 to 1.0 (d) 0.5 to 1.0

[2018 : 1 M, Set-II]

2.48 The space mean speed (kmph) and density (vehicles/km) of a traffic stream are linearly related. The free flow speed and jam density are 80 kmph and 100 vehicles/km respectively. The traffic flow (in vehicles/h, up to one decimal place) corresponding to a speed of 40 kmph is __________.

[2018 : 2 M, Set-II]

2.49 The maximum number of vehicles observed in any five minute period during the peak hour is 160. If the total flow in the peak hour is 1000 vehicles, the five minute peak hour factor (round off to 2 decimal places) is________.

[2019 : 1 M, Set-I]

2.50 Average free flow speed and the jam density observed on a road stretch are 60 km/h and 120 vehicles/km, respectively. For a linear speed-density relationship, the maximum flow on the road stretch (in vehicles/h) is________.

[2019 : 2 M, Set-I]

2.51 Traffic on a highway is moving at a rate of 360 vehicles per hour at a location. If the number of vehicles arriving on this highway follows Poisson distribution, the probability (round off to 2 decimal places) that the headway between successive vehicles lies between 6 and 10 seconds is________.

[2019 : 2 M, Set-I]

2.52 The speed-density relationship in a mid-block section of a highway follows the Greenshield's model. If the free flow speed is v_f and the jam density is k_j, the maximum flow observed on this section is
(a) $\frac{v_f k_j}{4}$ (b) $\frac{v_f k_j}{2}$
(c) $v_f k_j$ (d) $\frac{v_f k_j}{8}$

[2019 : 1 M, Set-II]

2.53 A vehicle is moving on a road of grade +4% at a speed of 20 m/s. Consider the coefficient of rolling friction as 0.46 and acceleration due to gravity as 10 m/s^2. On applying brakes to reach a speed of 10 m/s, the required braking distance (in m, round off to nearest integer) along the horizontal, is ____________.

[2019 : 1 M, Set-II]

2.54 The uniform arrival and uniform service rates observed on an approach road to a signalized intersection are 20 and 50 vehicles/minute, respectively. For this signal, the red time is 30 s, the effective green time is 30 s, and the cycle length is 60 s. Assuming that initially there are no vehicles in the queue, the average delay per vehicle using the approach road during a cycle length (in *s*, round off to 2 decimal places) is________. [2019 : 2 M, Set-II]

2.55 The speed-density relationship of a highway is given as

$$u = 100 - 0.5k$$

where, u = speed in km per hour, k = density in vehicles per km. The maximum flow (in vehicles per hour, round off to the nearest integer) is ________.

[2019 : 2 M, Set-II]

2.56 The relationship between traffic flow rate (q) and density (D) is shown in the figure.

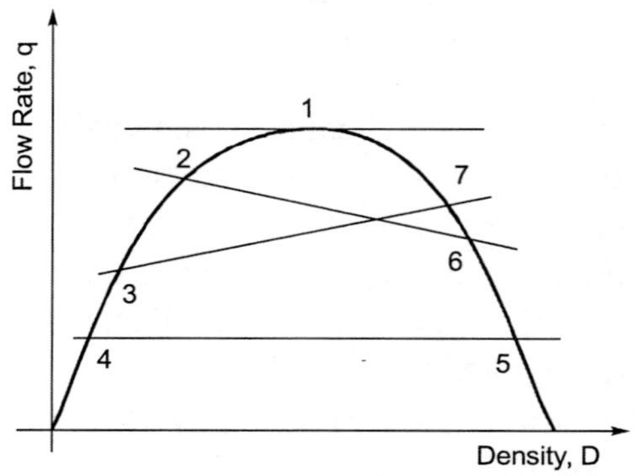

The shock wave condition is depicted by
(a) flow with respect to point 4 and point 5($q_4 = q_5$)
(b) flow changing from point 3 to point 7 ($q_3 < q_7$)

(c) flow changing from point 2 to point 6 ($q_2 > q_6$)
(d) flow with respect to point 1 ($q_1 = q_{max}$)
[2020 : 2 M, Set-I]

2.57 Traffic volume count has been collected on a 2 lane road section which needs upgradation due to severe traffic flow condition. Maximum service flow rate per lane is observed as 1280 veh/h at level of service 'C'. The Peak Hour Factor is reported as 0.78125. Historical traffic volume count provides Annual Average Daily Traffic as 12270 veh/day. Directional split of the traffic flow is observed to be 60 : 40. Assuming that traffic stream consists of 'All Cars' and all drivers are 'Regular Commuters', the number of extra lane(s) (round off to the next higher integer) to be provide, is ________. [2020 : 2 M, Set-I]

2.58 The traffic starts discharging from an approach at an intersection with the signal turning green. The constant headway considered from the fourth or fifth headway position is referred to as
(a) saturation headway
(b) effective headway
(c) discharge headway
(d) intersection headway [2020 : 1 M, Set-II]

2.59 24-h traffic count at a road section was observed to be 1000 vehicles on a Tuesday in the month of July. If daily adjustment factor for Tuesday is 1.121 and monthly adjustment factor for July is 0.913, the Annual Average Daily Traffic (in veh/day, round off to the nearest integer) is _______.
[2020 : 1 M, Set-II]

2.60 The flow-density relationship of traffic on a headway is shown in the figure

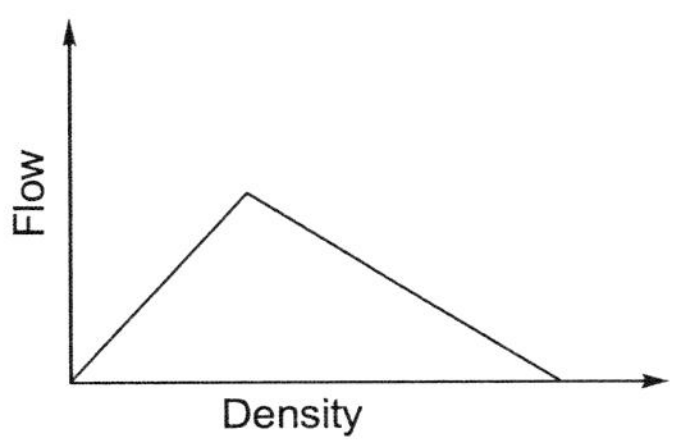

The correct representation of speed-density relationship of the traffic on this highway is

(a)
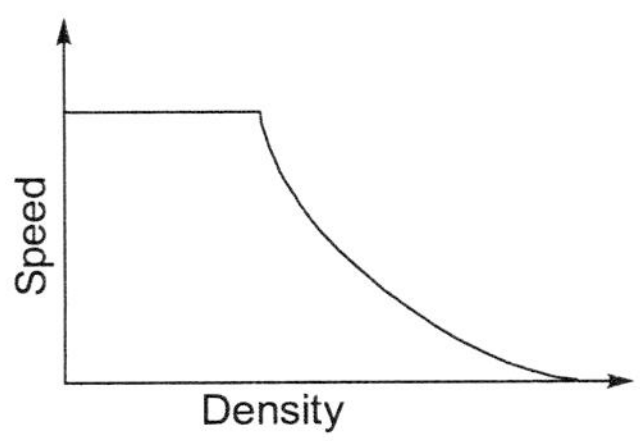

(b)
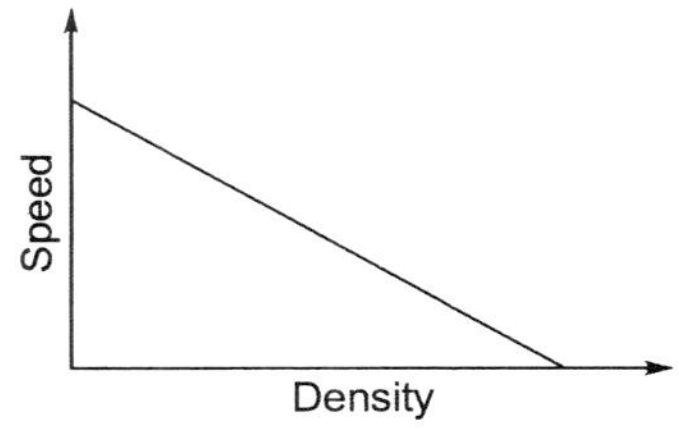

(c)
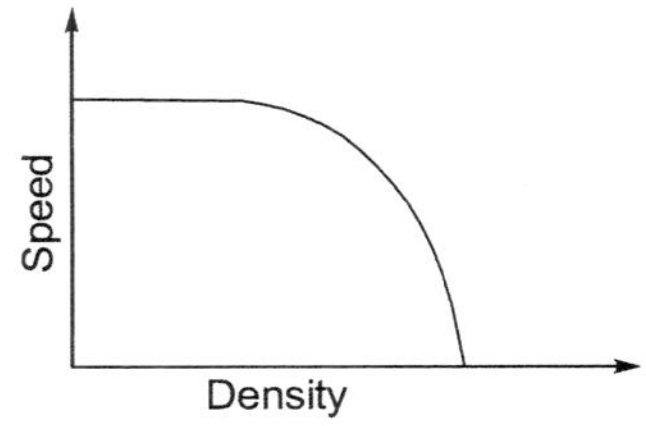

(d)
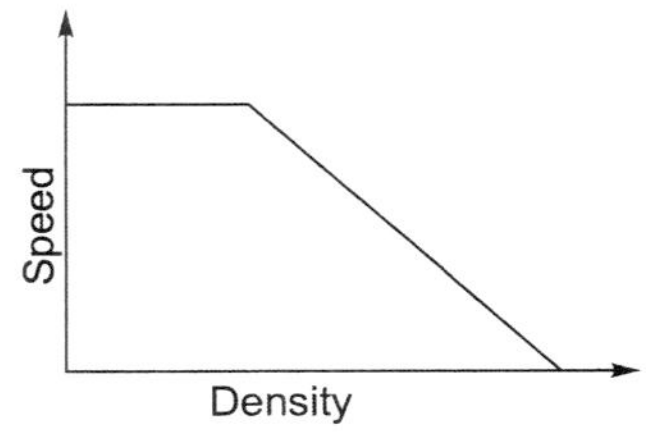

[2020 : 2 M, Set-II]

2.61 A signalized intersection operates in two phases. The lost time is 3 seconds per phase. The maximum ratios of approach flow to saturation flow for the two phases are 0.37 and 0.40. The optimum cycle length using the Webster's method (in seconds, round off to one decimal place) is
[2021 : 1 M, Set-I]

2.62 Spot speeds of vehicles observed at a point on a highway are 40, 55, 60, 65 and 80 km/h. The space-mean speed (in km/h, round off to two decimal places) of the observed vehicles is
[2021 : 2 M, Set-I]

2.63 Vehicular arrival at an isolated intersection follows the Poisson distribution. The mean vehicular arrival rate is 2 vehicle per minute. The probability (round off to two decimal places) that at least 2 vehicles will arrive in any given 1-minute interval is ______
[2021 : 2 M, Set-I]

2.64 On a road, the speed-density relationship of a traffic stream is given by u = 70 – 0.7 k (where speed, u, is in km/h and density, k, is in veh/km). At the capacity condition, the average time headway will be
(a) 1.0 s (b) 2.1 s
(c) 0.5 s (d) 1.6 s
[2021 : 2 M, Set-I]

2.65 In a three-phase signal system design for a four-leg intersection, the critical flow ratios for each phase are 0.18. 0.32 and 0.22. The total loss time in each of the phases is 2s. As per Webster's formula, the optimal cycle length (in *s*, round off to the nearest integer) is ________.

[2021 : 1 M, Set-II]

2.66 Relationship between traffic speed and density is described using a negatively sloped straight line. If v_f is the free-flow speed then the speed at which the maximum flow occurs is

(a) $\frac{v_f}{4}$ (b) 0

(c) $\frac{v_f}{2}$ (d) v_f

[2021 : 1 M, Set-II]

2.67 Consider the four points P, Q, R, and S shown in the Greenshields fundamental speed-flow diagram. Denote their corresponding traffic densities by k_P, k_Q, k_R, and k_S, respectively. The correct order of these densities is

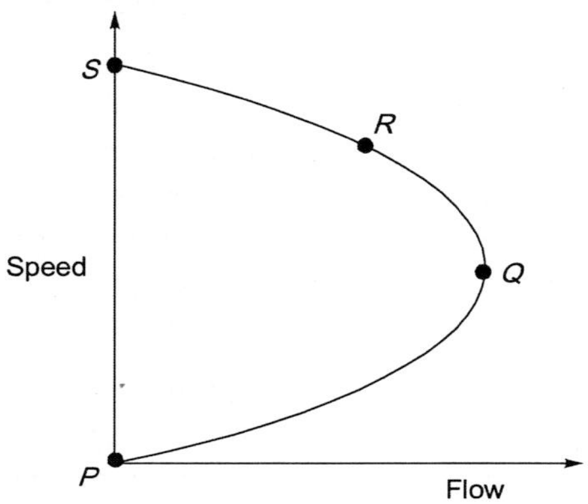

(a) $k_P > k_Q > k_R > k_S$
(b) $k_S > k_R > k_Q > k_P$
(c) $k_Q > k_R > k_S > k_P$
(d) $k_Q > k_R > k_P > k_S$

[2022 : 2 M, Set-I]

2.68 A two-phase signalized intersection is designed with a cycle time of 100 s. The amber and red times for each phase are 4 s and 50 s, respectively. If the total lost time per phase due to start-up and clearance is 2 s, the effective green time of each phase is ______s. (in integer)

[2022 : 2 M, Set-I]

2.69 At a traffic intersection, cars and buses arrive randomly according to independent Poisson processes at an average rate of 4 vehicles per hour and 2 vehicles per hour, respectively. The probability of observing at least 2 vehicles in 30 minutes is ______. (round off to two decimal places)

[2022 : 2 M, Set-I]

2.70 The vehicle count obtained in every 10 minute interval of a traffic volume survey done in peak one hour is given below.

Time Interval (in minutes)	Vehicle Count
0 – 10	10
10 – 20	11
20 – 30	12
30 – 40	15
40 – 50	13
50 – 60	11

The peak hour factor (PHF) for 10 minute sub-interval is ________. (round off to one decimal place)

[2022 : 2 M, Set-I]

2.71 A single-lane highway has a traffic density of 40 vehicles km If the time-mean speed and space-mean speed are 40 kmph and 30 kmph, respectively, the average headway (in seconds) between the vehicles is

(a) 2.25 (b) 6.25×10^{-4}
(c) 3.00 (d) 8.33×10^{-4}

[2022 : 1 M, Set-II]

2.72 The lane configuration with lane volumes in vehicles per hour of a four-arm signalized intersection is shown in the figure. There are only two phases: the first phase is for the East-West and the West-East through movements, and the second phase is for the North-South and the South-North through movements. There are no turning movements. Assume that the saturation flow is 1800 vehicles per hour per lane for each lane and the total lost time for the first and the second phases together is 9 seconds.

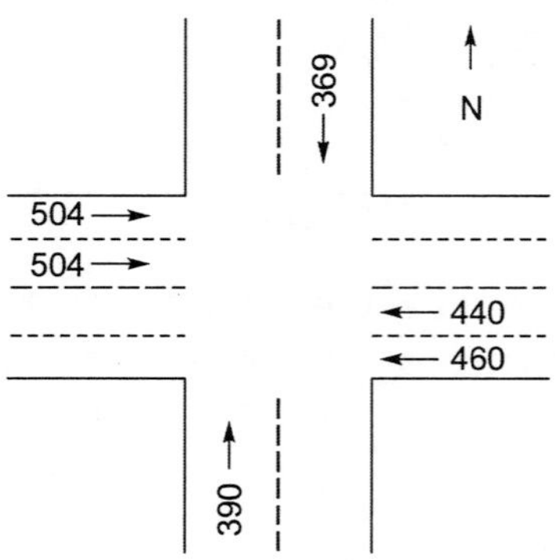

The optimum cycle length (in second), as per the Webster's method, is ______ . (round off to the nearest integer).

[2022 : 1 M, Set-II]

2.73 Assuming that traffic on a highway obeys the Greenshields model, the speed of a shockwave between two traffic streams (P) and (Q) as shown in the schematic is ________ kmph. (in integer)

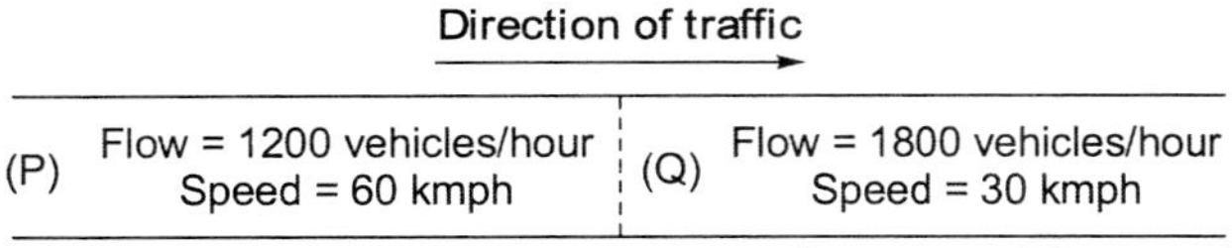

[2022 : 1 M, Set-II]

2.74 For a traffic stream, V is the space mean speed, k is the density, q is the flow. V_f is the free flow speed, and k_j is the jam density. Assume that the speed decreases linearly with density,

Which of the following relation(s) is/are correct?

(a) $q = k_j k - \left(\frac{k_j}{V_f}\right)k^2$

(b) $q = k_j V - \left(\frac{k_j}{V_f}\right)V^2$

(c) $q = V_f V - \left(\frac{V_f}{k_j}\right)V^2$

(d) $q = V_f k - \left(\frac{V_f}{k_j}\right)k^2$

[2022 : 2 M, Set-II]

2.75 A plot of speed-density relationship (linear) of two roads (Road A and Road B) is shown in the figure.

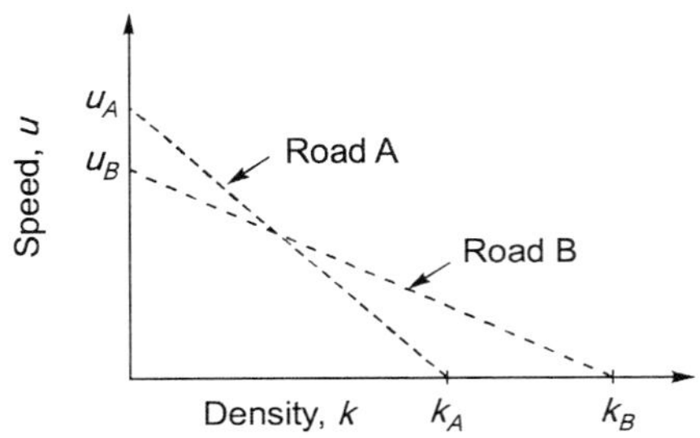

If the capacity of Road A is C_A and the capacity of Road B is C_B, what is $\frac{C_A}{C_B}$?

(a) $\frac{k_A}{k_B}$ (b) $\frac{u_A}{u_B}$

(c) $\frac{k_A u_A}{k_B u_B}$ (d) $\frac{k_A u_B}{k_B u_A}$

[2023 : 2 M, Set-I]

2.76 The figure presents the time-space diagram for when the traffic on a highway is suddenly stopped for a certain time and then released. Which of the following statements are true?

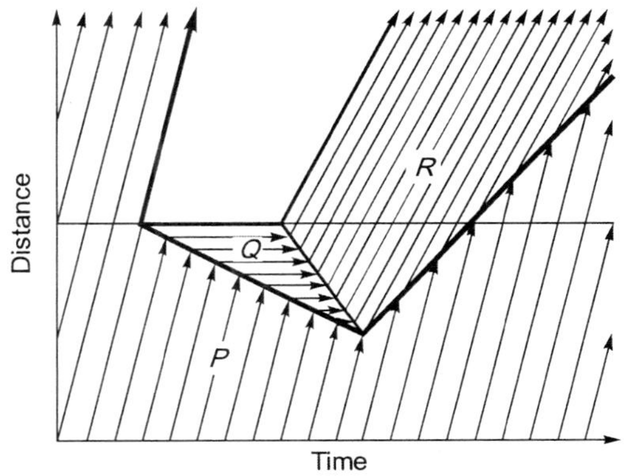

(a) Speed is higher in Region R than in Region P
(b) Volume is lower in Region Q than in Region P
(c) Volume is higher in Region R than in Region P
(d) Density is higher in Region Q than in Region R

[2023 : 2 M, Set-II]

2.77 The figure presents the trajectories of six vehicles within a time-space domain. The number in the parentheses represents unique identification of each vehicle

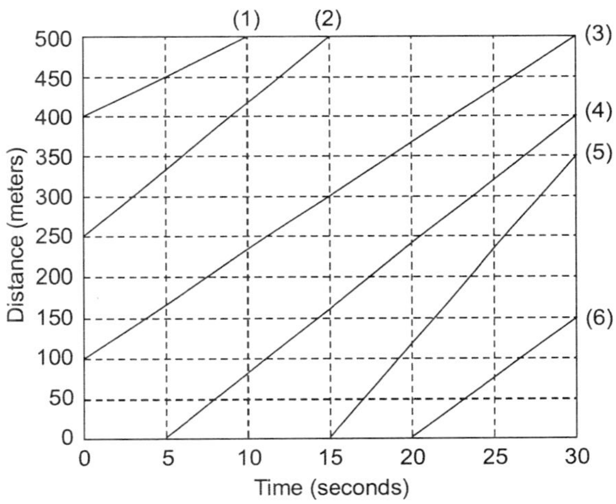

(Figure NOT to scale)

The mean speed (in km/hr) of the vehicles in the entire time-space domain is _____. (rounded off to the nearest integer)

[2024 : 2 M, Set-I]

2.78 The free mean speed is 60 km/hr on a given road. The average space headway at jam density on the road is 8 m. For a linear speed density relationship the maximum flow (in veh/hr/lane) expected on the road is

(a) 2075 (b) 1875
(c) 938 (d) 1038

[2024 : 2 M, Set-I]

2.79 The consolidated data of a spot speed study for a certain stretch of a highway is given in the table.

Speed range (kmph)	Number of observations
0-10	7
10-20	31
20-30	76
30-40	129
40-50	104
50-60	78
60-70	29
70-80	24
80-90	13
90-100	9

The "upper speed limit" (in kmph) for the traffic sign is

(a) 50 (b) 55
(c) 70 (d) 65

[2024 : 2 M, Set-II]

Answers Traffic Engineering

2.1	(c)	2.2	(b)	2.3	(d)	2.4	(c)	2.5	(d)	2.6	(c)	2.7	(b)
2.8	(b)	2.9	(c)	2.10	(d)	2.11	(a)	2.12	(c)	2.13	(b)	2.14	(a)
2.15	(d)	2.16	(d)	2.17	(c)	2.18	(a)	2.19	(d)	2.20	(a)	2.21	(a)
2.22	(a)	2.23	(c)	2.24	(98.88)	2.25	(b)	2.26	(c)	2.27	(1750)	2.28	(92)
2.29	(20)	2.30	(0.270)	2.31	(d)	2.32	(b)	2.33	(15.74)	2.34	(d)	2.35	(c)
2.36	(1750)	2.37	(1200)	2.38	(3)	2.39	(a)	2.40	(a)	2.41	(0.135)	2.42	(b)
2.43	(c)	2.44	(a)	2.45	(c)	2.46	(b)	2.47	(c)	2.48	(2000)	2.49	(0.52)
2.50	(1800)	2.51	(0.18)	2.52	(a)	2.53	(30)	2.54	(12.50)	2.55	(5000)	2.56	(a, b, c)
2.57	(6)	2.58	(a)	2.59	(1023)	2.60	(a)	2.61	(60.9)	2.62	(56.99)	2.63	(0.59)
2.64	(b)	2.65	(50)	2.66	(c)	2.67	(a)	2.68	(48)	269	(0.8)	2.70	(0.8)
2.71	(c)	2.72	(37)	2.73	(15)	2.74	(b, d)	2.75	(c)	2.76	(b, c, d)		
2.77	(57 or 53)			2.78	(b)	2.79	(b)						

Explanations Traffic Engineering

2.1 (c)

Traffic counters are used for traffic volume study and enoscope is used for spot speed study. Radar gun is used to measure he speed of vehicles for low-enforemenet. Floating car method or riding check method or moving car method is used for speed and delay studies.

2.2 (b)

The longitudinal distance = $0.278\ Vt$

$= 0.278 \times 50 \times 2.52$

$= 35.03 \text{ m} \approx 35 \text{ m}$

2.3 (d)

The road geometrics in India are designed for:

(a) 30th highest hourly traffic volume

(b) 98th percentile speed

2.4 (c)

Space headway,

$$S = 0.278\,Vt + \frac{V^2}{254\,f} + L$$

$$= 0.278 \times 65 \times 2.5 + \frac{(65)^2}{254 \times 0.4} + 5$$

$$= 91.8 \text{ m}$$

Capacity,

$$C = \frac{1000\,V}{S} = \frac{1000 \times 65}{91.8}$$

$$= 708.37 \approx 710 \text{ vehicles/hr.}$$

AVOID MISTAKE

Do not take S = 0.7 V + L as taken in overtaking sight distance calculation.

POINTS TO REMEMBER

$$q = VK = V\left(\frac{1000}{5}\right) = \frac{100V}{S}$$

2.5 (d)

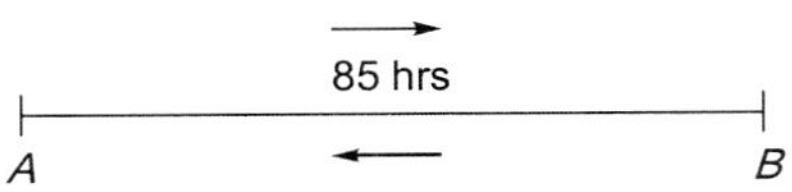

Layover time = 5 hrs.

Total journey time for one round trip

$$= \underset{\text{(A to B)}}{85} + \underset{\text{(Layover)}}{5} + \underset{\text{(B to A)}}{85}$$

$$= 175 \text{ hrs.} = 7.3 \text{ days } (\simeq 8 \text{ days})$$

8 trucks are required to ensure one truck leaves from city P each day.

Service from city Q to city P can be provided by the trucks which have already reached city Q.

2.6 (c)

2.7 (b)

For N-S road and E-W road the higher traffic volume will be taken i.e.

$$q_1 = 1000 \text{ and } q_2 = 900$$

$$S_1 = 2500 \text{ and } S_2 = 3000$$

$$\therefore \quad y_1 = \frac{q_1}{S_1} = \frac{1000}{2500} = 0.4$$

and $$y_2 = \frac{q_2}{S_2} = \frac{900}{3000} = 0.3$$

$$\therefore \quad Y = y_1 + y_2 = 0.4 + 0.3 = 0.7$$

Optimum cycle time,

$$C_0 = \frac{1.5\,L + 5}{1 - Y} = \frac{1.5 \times 12 + 5}{1 - 0.7}$$

$$= 76.67 \approx 77 \text{ seconds}$$

2.8 (b)

Traffic volume = Density × Speed

$$q = Uk$$

$$\Rightarrow \quad q = U_{sf}\left(1 - \frac{k}{k_j}\right)k$$

$$\Rightarrow \quad q = U_{sf}\left(k - \frac{k^2}{k_j}\right)$$

Jam density,

$$k_j = \frac{1000}{\text{average spacing between vehicles}}$$

$$k_j = \frac{1000}{7} \text{ and } U_{sf} = 70 \text{ kmph}$$

For maximum traffic volume, $\frac{dq}{dk} = 0$

$$\Rightarrow \quad U_{sf}\left(1 - \frac{2k}{k_j}\right) = 0 \qquad [\because U_{sf} \neq 0]$$

$$\Rightarrow \quad 1 - \frac{2k}{k_j} = 0$$

$$\Rightarrow \quad k = \frac{k_j}{2}$$

∴ Maximum traffic volume,

$$q_{max} = U_{sf}\left(\frac{k_j}{2} - \frac{(k_j/2)^2}{k_j}\right)$$

$$= U_{sf}\left(\frac{k_j}{2} - \frac{k_j}{4}\right) = U_{sf}\left(\frac{k_j}{4}\right)$$

$$= 70 \times \frac{1000}{7 \times 4} = 2500 \text{ veh/hr}$$

2.9 (c)

On a right angled road intersection with two way traffic the total number of conflict points are 24. This consists of 4 crossing and 12 weaving conflicts which are the major conflict points. The merging and diverging conflicts are considered as minor conflicts, numbering 4 in each case.

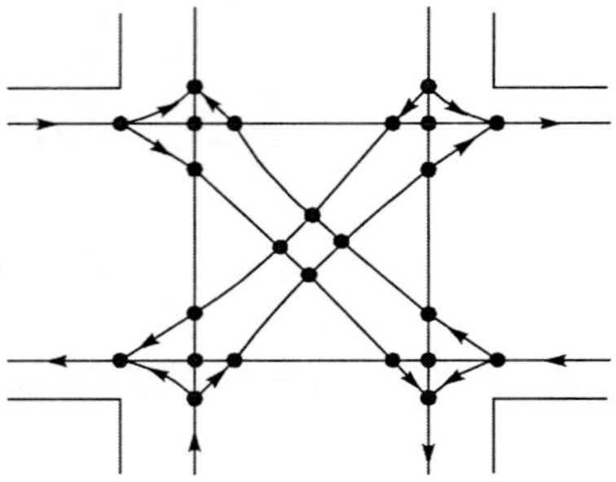

Also remember for 2 lane -1 way roads meeting at an intersection, there are 4 crossing conflicts and 2 merging conflicts.

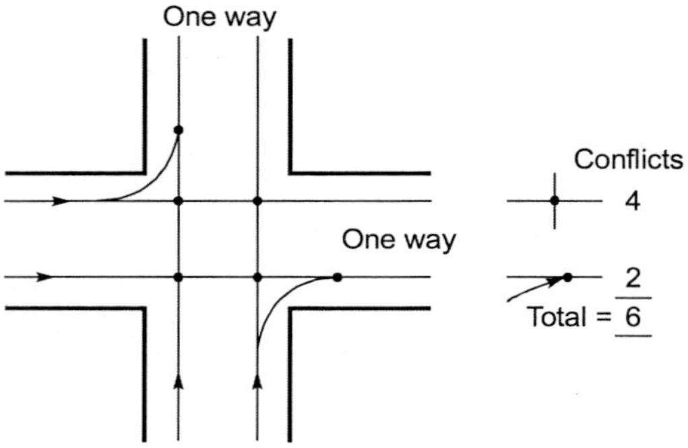

Vehicular conflicts with one-way regulation on both roads.

2.10 (d)

Given $Y = 0.50$, $L = 10$ sec

$\therefore$ Optimum cycle length,

$$C_0 = \frac{1.5L+5}{1-Y} = \frac{1.5\times10+5}{1-0.50} = 40 \text{ sec}$$

2.11 (a)

Tentative capacity values of urban roads suggested by IRC are:

(i) One way two lane road with no frontage access, no standing vehicles and very little cross traffic – 2400 PCU per hour.

(ii) One way two lane road with frontage access, but no standing vehicle and high capacity intersections – 1500 PCU per hour.

(iii) One way two lane road with free frontage access, parked vehicles and heavy cross traffic – 1200 PCU per hour.

2.12 (c)

The stop sign is intended to stop the vehicle on a road way. It is octagonal in shape and red in colour with a white border. Stop sign is a regulatory or mandatory sign. Also note that stop sign and give way sign are the two regulatory signs which are not circular in shape.

2.13 (b)

Width of the weaving section, $w = 14$ m

Average entry width, $e = 8.4$ m

Length of weaving section between channelizing islands, $L = 35$ m

Proportion of weaving traffic is given by,

$$p = \frac{\text{crossing traffic on the weaving section}}{\text{total traffic on the weaving section}}$$

$$\Rightarrow \quad p = \frac{1000}{2000}$$

$$\Rightarrow \quad p = 0.5$$

The capacity of the roundabout may be given as,

$$Q = \frac{280\,w\left(1+\frac{e}{w}\right)\left(1-\frac{p}{3}\right)}{1+\frac{w}{L}}$$

$$Q = \frac{280\times14\times\left(1+\frac{8.4}{14}\right)\times\left(1-\frac{0.5}{3}\right)}{1+\frac{14}{35}}$$

$$= 3733.33 \approx 3700 \text{ PCU per hour}$$

POINTS TO REMEMBER

Wearing ratio

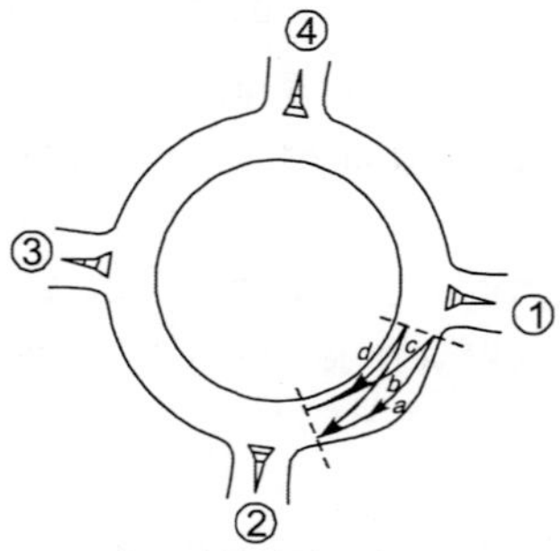

Weaving section (1)-(2)

$a = V_{12}$

$b = V_{13} + V_{14}$

$c = V_{32} + V_{42}$

$d = V_{43}$

$$p = \frac{\text{Weaving Traffic}}{\text{Total traffic}} = \frac{b+c}{a+b+c+d}$$

2.14 (a)

Critical lane volume on major road is increased to 440 veh./hr./lane i.e. 880 veh./hr.

Therefore, green time should be increased for major road and as there is no change in lane volume for minor road, it remains same.

2.15 (d)

The maximum flow may be calculated as,

$$q_{max} = \frac{\text{free flow speed} \times \text{jam density}}{4}$$

$$= \frac{80 \times 100}{4} = 2000 \text{ vehicles per hour}$$

Maximum flow occurs when the speed becomes half of the free flow speed i.e.

$$\text{Speed at } q_{max} = \frac{\text{free flow speed}}{2}$$

$$= \frac{80}{2} = 40 \text{ km per hour}$$

POINTS TO REMEMBER

At max flow (traffic capacity)

$$q_{max} = \frac{1}{4} V_f k_j$$

$$\text{Speed} = \frac{V_f}{2}$$

$$\text{Traffic density} = \frac{k_j}{2}$$

Where V_f = Fre flow speed

k_j = Jam density

2.16 (d)

Given, $v = v_f \log_e\left(\frac{k_j}{k}\right)$

But traffic volume,

$$q = \text{traffic density} \times \text{speed}$$

$$= k \times v$$

$$\Rightarrow \quad q = v_f k \log_e\left(\frac{k_j}{k}\right)$$

For maximum traffic volume i.e., traffic capacity

$$\frac{dq}{dk} = 0$$

$$\Rightarrow \quad v_f \times \frac{d}{dk}\left[k \log_e\left(\frac{k_j}{k}\right)\right] = 0$$

$$\Rightarrow \quad k\frac{d}{dk}\log_e\left(\frac{k_j}{k}\right) + \log_e\left(\frac{k_j}{k}\right)\frac{d}{dk}k = 0$$

$$\Rightarrow \quad k \times \frac{k}{k_j} \times k_j \times \left(-\frac{1}{k^2}\right) + \log_e\frac{k_j}{k} = 0$$

$$\Rightarrow \quad \log_e\frac{k_j}{k} = 1$$

$$\Rightarrow \quad e = \frac{k_j}{k}$$

$$\Rightarrow \quad k = \frac{k_j}{e}$$

POINTS TO REMEMBER

For linear relationship between speed and traffic density, k = $\frac{kj}{2}$ for max traffic flow condition.

For other relations, differentiate q to obtain result.

2.17 (c)

Under phase-II, 1 and 3 cannot move simultaneously. It is also the case under phase-III for 4 and 6.

2.18 (a)

Speed limit sign is a regulatory or mandatory sign and should be of circular shape with white background and red border.

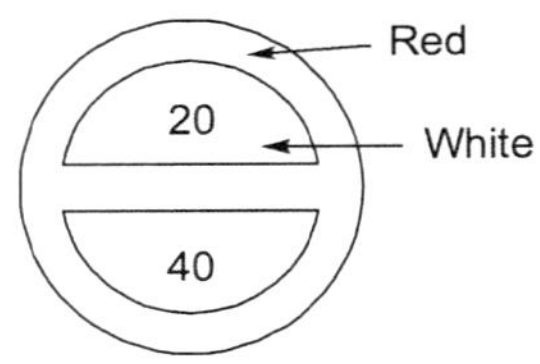

2.19 (d)

Probability of arriving 'n' number of vehicles in time t,

$$P(n) = \frac{(\lambda t)^n e^{-\lambda t}}{n!}$$

If time headway is greater than or equal to time t, then number of vehicles arriving is zero,

$$\Rightarrow \quad P(o) = \frac{(\lambda t_1)^o e^{-\lambda t_1}}{o!} = e^{-\lambda t_1}$$

POINTS TO REMEMBER

(i) Probability for first arrival after time t

$$P(>t) = e^{-\lambda t}$$

(ii) Probability for first arrival within time t

$$P(< t) = 1 - e^{-\lambda t}$$

(iii) Probability for 'n' number of vehicles arriave in time t.

$$P(n, t_0) = \frac{(\lambda t_0)^n e^{-\lambda t_0}}{n!}$$

Where λ = vehicle arrival rate
t = time headway

2.20 (a)

2.21 (a)

Method-I

Average delay

$$= \frac{\begin{pmatrix}\text{Area of triangle formed} \\ \text{by arrival curve}\end{pmatrix} - \begin{pmatrix}\text{Area of triangle formed} \\ \text{by departure curve}\end{pmatrix}}{\text{No. of vehicles}}$$

$$= \frac{\frac{1}{2} \times 40 \times 50 - \frac{1}{2} \times 40 \times 20}{40}$$

= 15 sec

Method-II

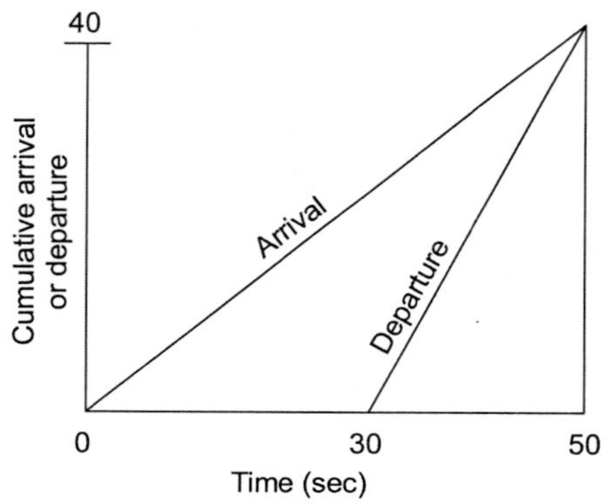

Vehicle 1 arrives at 0 second and departs at 30 seconds.

Vehicle 40 arrivers at 50 seconds and departs at the same time.

Delay for vehicle 1 = 30 second

Delay for vehicle 40 = 0 second

$$\text{Average delay} = \frac{0 + 30}{2} = 15 \text{ seconds}$$

2.22 (a)

When both roads are one-way then total no. of conflict points, $x = 6$

When both roads are two-way then total no . of conflict points, $y = 24$

$$\therefore \quad \frac{x}{y} = \frac{6}{24} = \frac{1}{4} = 0.25$$

Note :

Number of lanes*		Number of potential conflicts		
Road A	Road B	Both roads two-way	A-One-way B-Two-way	Both-roads One-way
2	2	24	11	6
2	3	24	11	8
2	4	32	17	10
3	3	24	13	11
4	4	44	25	18

2.23 (c)

Method-I

Capacity per lane

$$= \frac{1800}{2} = 900 \text{ Vehicle/hour}$$

$$k_j = \frac{1000}{5} = 200 \text{ Veh./km}$$

$$q_{max} = \frac{V_{sf} \times k_j}{4}$$

$$\Rightarrow \quad 900 = \frac{V_{sf} \times 200}{4}$$

$$\therefore \quad V_{sf} = 18 \text{ kmph}$$

$$\therefore \quad V = V_{sf}\left[1 - \frac{k}{k_j}\right]$$

$$\because \quad q = Vk$$

Since traffic volume = 1000 vehicle/hour for two lane

So, for single lane,

$$q = 500 \text{ vehicle/hour}$$

$$q = V_{sf}\left[1-\frac{k}{k_j}\right]k$$

$$\Rightarrow \quad 500 = 18\left[1-\frac{k}{200}\right]k$$

$$\Rightarrow \quad 27.77 = k-\frac{k^2}{200}$$

$$\Rightarrow \quad 5554 = 200\,k - k^2$$

$\Rightarrow$ $k^2 - 200k + 5554 = 0$

$\therefore$ $k = 33.32$ Veh./km (for single lane)

So, for two lane urban road,

$$k = 33.32 \times 2 = 66.64 \approx 67$$

Method-II

$$q_{max} = \frac{V_{sf} \times k_j}{4} = 1800 \text{ vehicles/hr}$$

$$q = V_{sf}\left[1-\frac{k}{k_j}\right]k = 100 \text{ vehicle/hr}$$

$$k_j = \frac{1000}{l} = \frac{1000}{5} = 200 \text{ vehicle/hr}$$

(For single lane)

For two lanes,

$$k_j = 2 \times 200 = 400 \text{ vehicles/hr}$$

$$\frac{q}{q_{max}} = \frac{\left[1-\frac{k}{k_j}\right]l}{\frac{k_j}{4}} = \frac{1000}{1800}$$

$$\Rightarrow \quad \frac{(400-k)k}{\frac{400^2}{4}} = \frac{5}{9}$$

$\Rightarrow$ $400\,k - k^2 = 22222.222$

$\Rightarrow$ $k = 66.67$ vehicles

2.24 Sol.

$$P(x) = \frac{\lambda^n e^{-\lambda}}{x!}$$

Where, $P(x)$ is the probability of x events (vehicle arrivals) in some time interval (t).

λ is the mean arrival rate in that interval.

$$\lambda = \frac{150}{30 \times 60} \times 5 = \frac{5}{12}$$

Now, the probability that zero vehicle arrive in an interval t, denoted as $P(0)$, will be same as the probability that the ***headway*** (into arrival time) ***greater than or equal to t.***

$$\therefore \quad P(x=0) = \frac{\lambda^0 e^{-\lambda}}{0!} = e^{-5/12} = 0.6592$$

$\therefore$ $P(x = 0) = 0.6592$

Number of time headway greater than 5 secs

= (Total no. of headways)$\times P(x = 0)$

= 150×0.6592

= 98.88

2.25 (b)

Merging from left is done using a clover left and diverging to left is done using indirect ramp.

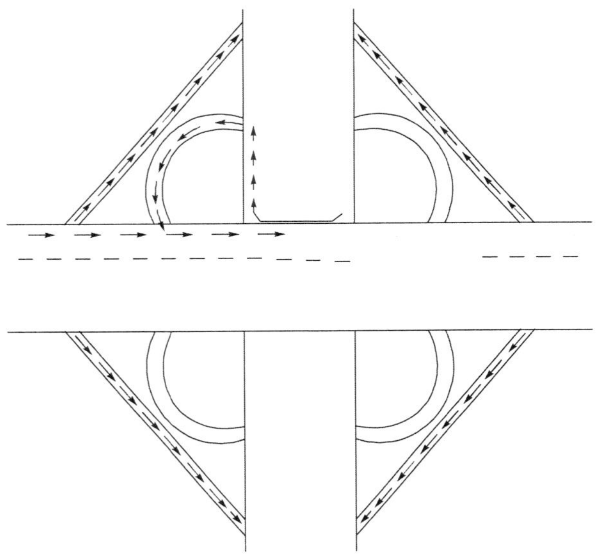

2.26 (c)

Method-I

15 min. peak hr factor is used for traffic intersection design,

$$\text{PHF} = \frac{(V/4)}{V_{15}}$$

V = Peak hourly volume $\left(\text{in}\frac{\text{veh.}}{\text{hr.}}\right)$

V_{15} = Maximum 15 minute volume within the peak hr. (veh.)

Maximum value is 1.0 and minimum value is 0.25

Normal range is 0.7 – 0.98

Method-II

Consider a peak hour with total traffic V.
Let the traffic be divided as such,

Time	0 – 15	15 – 30	30 – 45	45 – 60
No. of Vehicles	0	V	0	0

$$PHF_{15} = \frac{\text{Total traffic (in one hour)}}{\text{Traffic based on peak 15 min.}}$$

$$= \frac{V}{\frac{V}{15}\times 60} = \frac{1}{4}$$

POINTS TO REMEMBER

In general, minimum value of PHF_t (where t is in minutes) $= \frac{t}{60}$.

2.27 Sol.

Method-I

Capacity = Velocity × Density

$\Rightarrow$ $C = V \times K$

$= 70\,K - 0.7\,K^2$

Now, $\frac{dC}{dK} = 70 - 1.4\,K = 0$

$\Rightarrow$ $K = 50$

$\Rightarrow$ Capacity, $C = 70 \times 50 - 0.7(50)^2$

$= 1750$ veh/hr

Method-II

$$u = 70 - 0.7\,k = 70\left[1 - \frac{k}{100}\right]$$

Comparing it with

$$u = u_{sf}\left[1 - \frac{k}{k_j}\right]$$

$u_{sf} = 70$ kmph
$k_j = 100$ vehicles/km

$$\Rightarrow \quad q_{max} = \frac{u_{sf} \times k_j}{4} = \frac{70 \times 100}{4}$$

$= 1750$ vehicles/hr

2.28 Sol.

Sum of the flow,

$$y = y_1 + y_2 + y_3$$
$$= 0.2 + 0.3 + 0.25 = 0.75$$

Total lost time in a cycle $L = 4 \times 3 = 12$ sec.

Optimum cycle length,

$$C_0 = \frac{1.5L + 5}{1 - y}$$

$$\frac{1.5 \times 12 + 5}{1 - 0.75} = 92 \text{ sec.}$$

2.29 Sol.

$$\Rightarrow \quad \text{Density} = \frac{1000}{S} = \frac{1000}{50} = 20 \text{ veh/km}$$

2.30 Sol.

$$P(n, t) = \frac{e^{-\lambda t}.(\lambda t)^n}{n!}$$

Here, λ = no. of vehicles
$= 240$ vehicle/km

$$P(1, 30) = \frac{e^{\frac{-240\times30}{3600}}\left(\frac{240}{3600}\times 30\right)^1}{1!}$$

$$= 2.e^{-2} = 0.2707$$

2.31 (d)

Method-I

$$\text{Velocity of student} = \frac{5\,km}{(40-15)\,min}$$
$$= 12 \text{ km/hr}$$

$$\left(\frac{\text{Vehicle/min}}{\text{Relative speed of vehicle w.r.t. student}}\right)_{moving}$$

$$= \left(\frac{\text{Vehicle/min}}{\text{Relative speed of vehicle w.r.t. student}}\right)_{standing}$$

$$\Rightarrow \quad \frac{60/25}{x - 12} = \frac{45/15}{x - 0}$$

$\Rightarrow$ $x = 60$ km/hr

Method-II

45 vehicles overtook the student when he stopped for 15 minutes.

$$\Rightarrow \quad q = \frac{45}{15} \times 60 = 180 \text{ vehicles/hr}$$

Also, journey time

$$\bar{t} = t_w - \frac{n_y}{q} = \left(\frac{40}{60}\right) - \left(\frac{60+45}{180}\right) = \frac{1}{12}\text{hr}$$

$$\text{Speed of the vehicle stream} = \frac{L}{\bar{t}} = \frac{5}{1/12} = 60 \text{ kmph}$$

2.32 (b)

Method-I

$$v = 80 - \frac{2k}{3}$$

Capacity, $q = v \times k$

$$= 80k - \frac{2k^2}{3}$$

For q to be maximum

$$\frac{dq}{dk} = 0$$

$$\Rightarrow \quad \frac{dq}{dk} = 80 - \frac{4k}{3} = 0$$

$$\Rightarrow \quad k = 60$$

Maximum capacity,

$$q = 80 \times 60 - \frac{2}{3}(60)^2$$

$$= 4800 - \frac{2}{3} \times 3600$$

$$= 2400$$

Method-II

$$V = 80 - \frac{2}{3}k = 80\left[1 - \frac{k}{120}\right]$$

$$\Rightarrow \quad V_{sf} = 80 \text{ kmph}$$

$$k_j = 120 \text{ vehicles/km}$$

$$\therefore \quad q_{max} = \frac{V_{sf} \times k_j}{4} = \frac{80 \times 120}{4}$$

$$= 2400 \text{ vehicles/hr}$$

2.33 Sol.

Flow rates for the first three phase are given as

$$q_1 = 200 \text{ veh/hr}$$

$$q_2 = 187 \text{ veh/hr}$$

and $q_3 = 210$ veh/hr

Saturation flow rate is 1800 veh/hr/lane

Lost time,

$$L = 4 \times 4 = 16 \text{ sec}$$

Length of the cycle,

$$C_0 = 60 \text{ sec}$$

Now, $y_1 = \frac{q_1}{s_1} = \frac{200}{1800}$

$$y_2 = \frac{q_2}{s_2} = \frac{187}{1800}$$

$$y_3 = \frac{q_3}{s_3} = \frac{210}{1800}$$

$$C_0 = \frac{1.5L + 5}{1 - y}$$

$$\Rightarrow \quad 60 = \frac{1.5 \times 16 + 5}{1 - y}$$

$$\Rightarrow \quad 60 = \frac{24 + 5}{1 - y}$$

$$\Rightarrow \quad 1 - y = \frac{29}{60}$$

$$\Rightarrow \quad y = \frac{1 - 29}{60} = \frac{60 - 29}{60}$$

$$= 0.517$$

And $y = y_1 + y_2 + y_3 + y_4$

$$\Rightarrow \quad 0.517 = \frac{597}{1800} + y_4$$

$$\Rightarrow \quad y_4 = 0.185$$

$$G = \frac{y_4}{y}(C_0 - L)$$

$$= \frac{0.185}{0.517}(60 - 16)$$

$$= 15.745 \text{ sec}$$

AVOID MISTAKE

$$G = \frac{y_4}{y} \times C_0 = \frac{0.185}{0.517} \times 60 = 21.47 \text{seconds}$$

2.34 (d)

Free flow speed (V_f) is that speed at which flow and density is negligible.

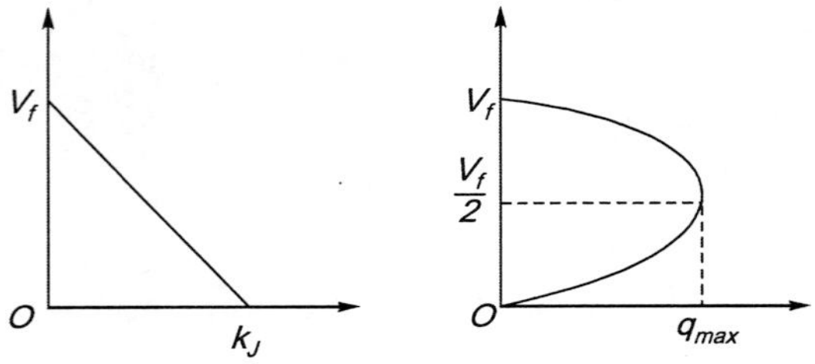

2.35 (c)

Design length is the average of (i) and (iii) as per Indian Roads Congress (IRC) recommendations for storage lanes.

2.36 Sol.

Traffic volume,

$$q = uk$$

$$q = (70 - 0.7k)k$$

$$= 70k - 0.7k^2$$

Maximum possible volume is capacity of road

For maximum volume,

$$\frac{dq}{dk} = 0$$

$$\Rightarrow \quad 70 - 1.4k = 0$$

$$\Rightarrow \quad k = 50 \text{ Veh/km}$$

Now, capacity,

$$C = (70 - 0.7 \times 50) \times 50$$

$$C = 1750 \text{ Veh/hr}$$

2.37 Sol.

Given: Average headway

$$= 3.0 \text{ sec}$$

Flow of traffic stream

$$= \frac{3600}{\text{headway}} \text{ vehicles/hr}$$

$$= \frac{3600}{3} \text{ veh/h} = 1200 \text{ veh/hr}$$

2.38 Sol.

Method-I

Let us assume,

q = traffic flow

k = traffic density

V_w = velocity of test vehicle along the stream

V_a = velocity of test vehicle against the stream

$$\Rightarrow \quad q - kV_w = 50$$

$$\Rightarrow q - k \times 20 = 50 \quad \quad ...(i)$$

$$q + kV_a = 200$$

$$\Rightarrow q + k \times 30 = 200 \quad \quad ...(ii)$$

Solving equation (*i*) and (*ii*), we get

$$k = 3 \text{ Veh/km}$$

Method-II

Consider a 'L' km long stretch.

$$t_w = \frac{L}{V_w} = \frac{L}{20} \text{hr}$$

$$t_a = \frac{L}{V_a} = \frac{L}{30} \text{ hr}$$

$$q = \left(\frac{n_y + n_a}{t_w + t_a}\right) = \frac{\left(50 \times \frac{L}{20} + 200 \times \frac{L}{30}\right)}{\left(\frac{L}{20} + \frac{L}{30}\right)}$$

$$= 110 \text{ vehicles/hr}$$

Journey time,

$$\bar{t} = t_w - \frac{n_y}{q} = \frac{L}{20} - \left(\frac{50 \times \frac{L}{20}}{110}\right)$$

$$= \frac{3}{110} L \text{ hr}$$

$$V = \frac{L}{t} = \frac{L}{\frac{3}{110}L} = 36.67 \text{ kmph}$$

Also, $\quad q = VK$

$$\Rightarrow \quad K = \frac{q}{v} = \frac{110}{36.67} = 3 \text{ vehicles/km}$$

2.39 (a)

2.40 (a)

$$Y = y_1 + y_2 + y_3$$

$$= 0.30 + 0.25 + 0.25$$

$$= 0.80$$

Now Total lost time,

$$L = 10 \text{ sec (given)}$$

$\therefore$ Optimum cycle time,

$$C_0 = \frac{1.5L + 5}{1 - Y}$$

$$\Rightarrow \quad C_0 = \frac{(1.5 \times 10) + 5}{1 - 0.80} = \frac{15 + 5}{0.20} = \frac{20}{0.2}$$

$= 100$ sec

Now green times are calculated by,

$$G_1 = \frac{y_1}{y}(C_0 - L) = \frac{0.30}{0.80}(100-10)$$

$= 33.75 \simeq 34$ sec

$$G_2 = \frac{y_2}{y}(C_0 - L) = \frac{0.25}{0.80}(100-10)$$

$= 28.11 \simeq 28$ sec

$$G_3 = \frac{y_3}{y}(C_0 - L) = \frac{0.25}{0.80}(100-10)$$

$= 28.11 \simeq 28$ sec

2.41 Sol.

$$\lambda = 900 \text{ veh/hour}$$

$$= \frac{900}{3600} = \frac{1}{4} \text{veh/sec}$$

$$P(n, t_0) = \frac{(\lambda t_0)^n e^{-\lambda t_0}}{\eta!}$$

Probability that time gap (time headway) betwen successive vehicles is greater than 8 seconds,

$$P(0,\ 8 \text{ sec}) = e^{-\frac{900}{3600}\times 8} = 0.1353$$

2.42 (b)

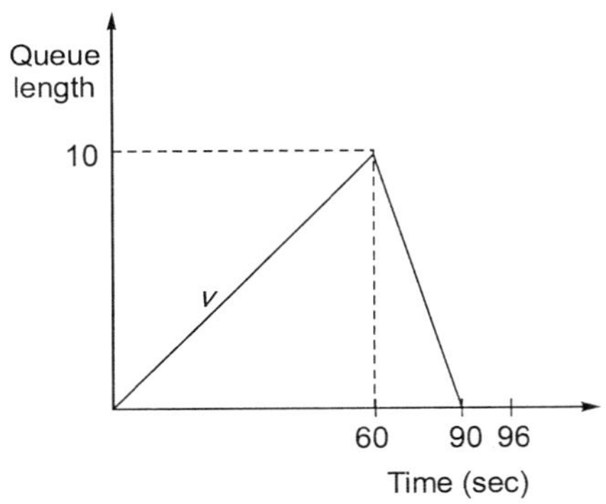

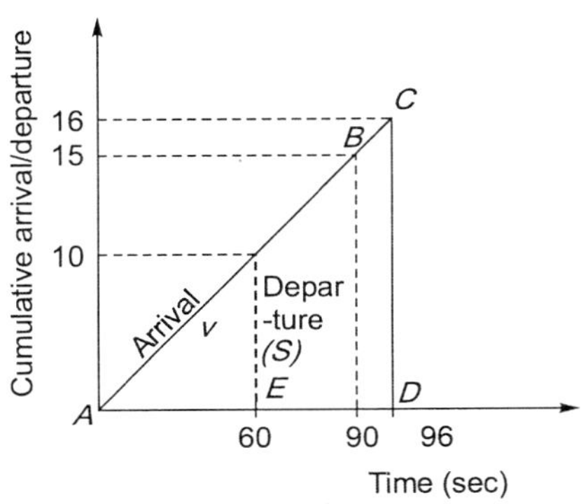

Arrival curve : ABC
Departure curve : EBC

$$v = \frac{11}{60} \times 3600 = 600 \text{ vehicles/hr}$$

$$s = \frac{15}{30} \times 3600 = 1800 \text{ vehicles/hr}$$

Average delay

$$= \frac{\begin{pmatrix}\text{Area of} \\ \text{arrival curve}\end{pmatrix} - \begin{pmatrix}\text{Area of depar-} \\ \text{ture curve}\end{pmatrix}}{\text{No. of vehicles}}$$

$$= \frac{\left(\frac{1}{2}\times 16\times 96\right) - \left[\frac{1}{2}\times 15\times 30 + \left(\frac{15+16}{2}\right)\times 6\right]}{16}$$

$= 28.125$ seconds

2.43 (c)

2.44 (a)

Space headway,

$$S = 60\,t - 60\,t^2$$

$$\frac{dS}{dt} = 60 - 120t = 0$$

$$t = 0.5 \text{ hr} = 30 \text{ minutes}$$

$$\frac{d^2S}{dt^2} = -120 \times 0 \text{ (Maxima)}$$

$\therefore$ Maximum space head,

$$S_{max} = 60 \times 0.5 - 60 \times (0.5)^2$$

$$= 15 \text{ km}$$

2.45 (c)

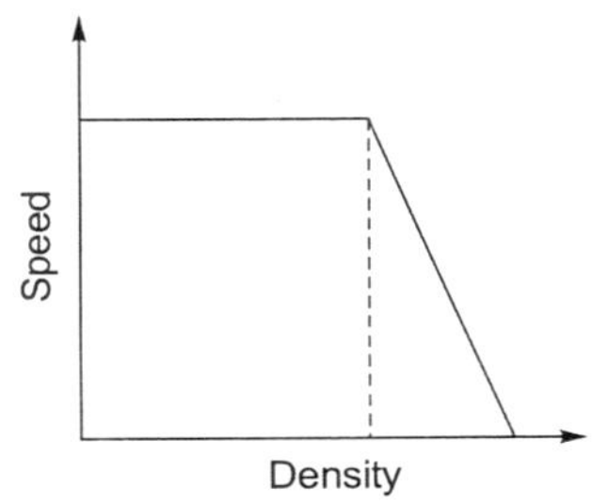

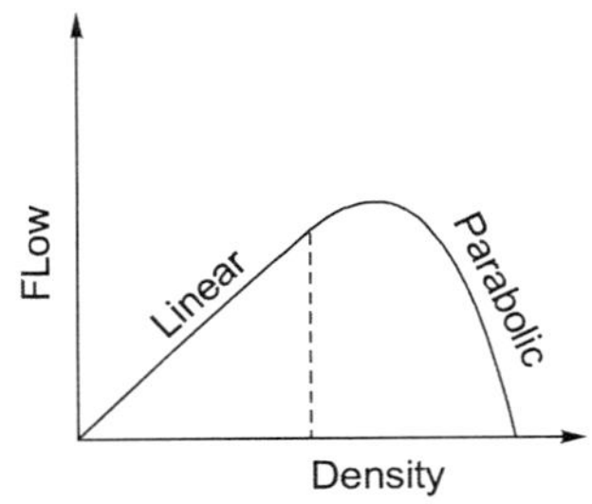

$$q = vk$$

For initial values of k, v is constant

$\therefore$ q will very linearly with k

Later on, v varies linearly with k.

$\therefore$ q will vary parabolically with k.

2.46 (b)

2.47 (c)

$$PHF_{15\text{ min}} = \frac{t}{60} = \frac{15}{60} = 0.25$$

$$0.25 \le PHF_{15} \le 1$$

2.48 Sol.

V_f = Free mean speed = 50 kmph
k_j = Jam density = 100 veh/km
As per linear model (green-shield)

$$V = V_f\left(1-\frac{k}{k_j}\right)$$

⇒ Density @ speed = 40 kmph

$$40 = 80\left(1-\frac{k}{100}\right) = 50 \text{ veh/km}$$

Traffic flow @ density 50 veh/km @ speed 40 kmph

$$q = Vk$$

$$q = V_f\left(k-\frac{k^2}{k_j}\right)$$

$$\Rightarrow \quad q = 80\left(50-\frac{50^2}{100}\right) = 2000 \text{ veh/hr}$$

2.49 Sol.

$$\text{PHF}_5 = \frac{\text{Peak hour flow}}{\left(\frac{60}{5}\right)\times V_5(\text{max.})} = \frac{1000}{\frac{60}{5}\times 160} = 0.52$$

2.50 Sol.

V_f = 60 kmph
k_j = 120 veh/km

$$\text{Max flow} = \frac{1}{4}V_s\times k_s = \frac{1}{4}\times 60\times 120$$
$$= 1800 \text{ veh/hr}$$

2.51 Sol.

$$\lambda = 360 \text{ veh/hr}$$
$$= \frac{360}{3600} = \frac{1}{10} \text{ veh/sec}$$
$$P(6 \text{ to } 10) = e^{-\lambda t_1} - e^{-\lambda t_2}$$
$$= e^{-\left[\frac{1}{10}\times 6\right]} - e^{-\left[\frac{1}{10}\times 10\right]} = 0.18$$

2.52 (a)

$$\text{Maximum flow} = \frac{1}{4}v_f k_j$$

2.53 Sol.

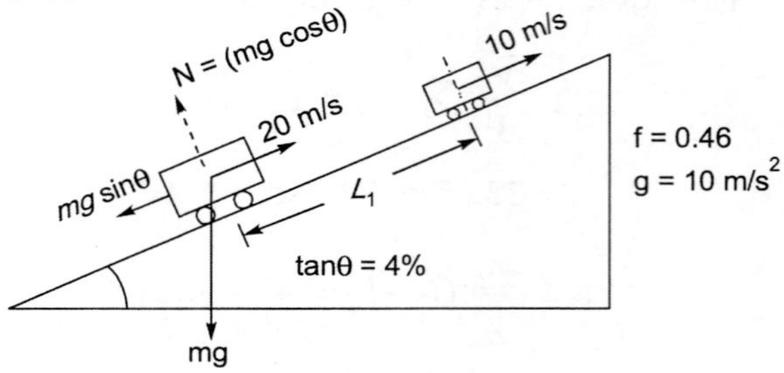

$$\tan\theta = 4\% = 0.04$$
$$\theta = 2.29°$$
$$\cos\theta = 0.999$$
$$\Delta K.E = \text{work done}$$

$$\frac{1}{2}mv_2^2 - \frac{1}{2}mv_1^2 = -L[mg\sin\theta + fmg\cos\theta]$$

$$\frac{m}{2}\left[v_2^2 - v_1^2\right] = -Lmg[\tan\theta + f]$$

$$\Rightarrow \quad \frac{v_1^2 - v_2^2}{2g} = L[\tan\theta + f]$$

$$\frac{20^2 - 10^2}{2\times 10} = L[0.04 + 0.46]$$

$$L = 30 \text{ m}$$

Now braking distance along horizontal

$$= [L\cos\theta] = 30 \times 0.999$$
$$= 29.97 \text{ m} \simeq 30 \text{ m}$$

(Near integer)

2.54 Sol.

Arrival rate = 20 veh/min
Service rate = 50 veh/min
R = 30 sec, G_i = 30 sec, Cycle length = 60 sec

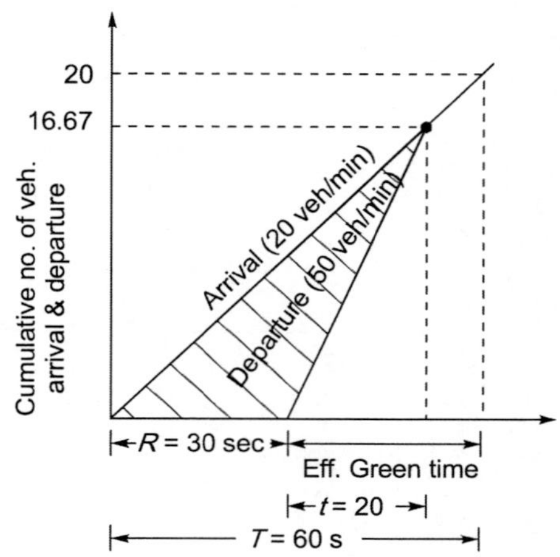

Time corresponding to which no. of arrival becomes same as no. of departure.

$$\Rightarrow \quad 20 \times (R + t) = 50t$$

$$20 \times (30 + t) = 50t$$
$$600 + 20t = 50t$$
$$t = \frac{600}{30} = 20 \text{ sec}$$

Avg. Delay

$$= \left(\frac{\text{Area under arrival line} - \text{Area under departure line}}{\text{Cumulative number of vehicle arrival}}\right)$$

$$= \left(\frac{\frac{1}{2}\times 16.67\times 50 - \frac{1}{2}\times 16.67\times 25}{20}\right)$$

$= 12.50$ sec

2.55 Sol.

$$u = 100 - 0.5k$$
$$u = 100\left[1 - \frac{k}{(100/0.5)}\right]$$

Green shield model,

$$u = V_f\left(1 - \frac{k}{k_J}\right)$$

V_f = free mean speed = 100 kmph

$$k_J = \text{Jam density} = \frac{100}{0.5} = 200 \text{ veh./km}$$

Max flow:

$$q_{max} = \frac{1}{4}V_f k_j = \frac{1}{4}\times 100\times 200$$
$$= 5000 \text{ veh/hr}$$

2.56 (a, b, c)

There can be three types of shock wave:

(i) The forward moving shock wave [fig. (a)]
(ii) The stationary shock wave [fig. (b)]
(iii) The backward moving shock wave [Fig. (c)]

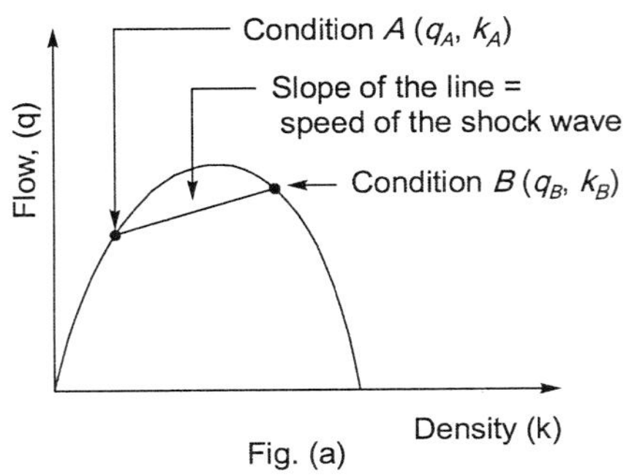

Fig. (a)

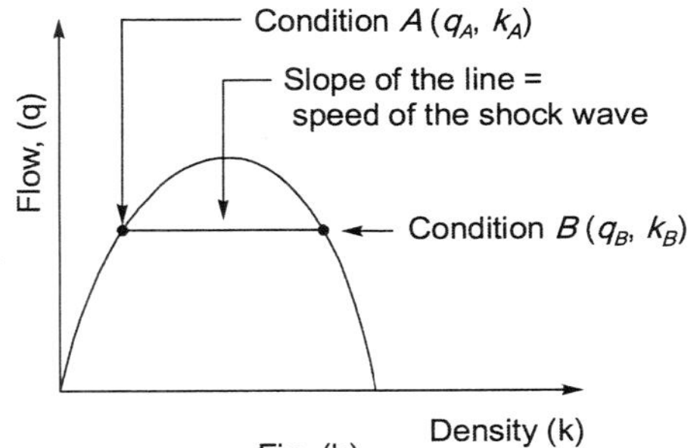

Fig. (b)

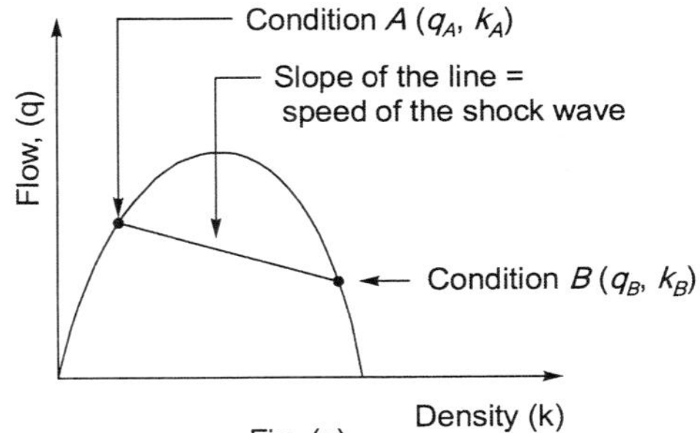

Fig. (c)

2.57 Sol.

Directional design hourly volume (DDHV)

DDHV = AADT × K × D

where D = Volume proportion in major direction

K = The proportion of AADT occuring in peak hour.

DDHV = 12270 × 0.6 × K

[Assumed, K = 1]

= 7362

f_{HV} = Heavy veh. adjustment factor = 1 for car

f_P = Road user familarity adjustment factor

= 1 for regular commuters

As per HCM,

Number of lanes required,

$$N = \frac{DDHV}{PHF \times MSF \times F_{HV} \times f_p}$$
$$= \frac{7362}{0.78125 \times 1280 \times 1 \times 1}$$
$$= 7.362 = 8 \text{ lanes}$$

Number of extra lanes = 8 − 2 = 6 lanes

2.58 (a)

Saturation headway: The constant headway achieved once a stable moving queue is established is called saturation headway.

Discharge headway: Discharge headway is the headway between successive vehicles negotiating an intersection during green time of signal operation.

2.59 Sol.

$$T_{24} = 1000 \text{ veh (Tuesday)}$$
$$\text{DAF} = 1.121$$
$$\text{AADT} = ?$$
$$\text{MAF} = 0.913$$
$$T_{week} = T_{24} \times \text{DAF}$$
$$= 1000 \times 1.121 = 1121$$
$$\text{AADT} = MAF \times ADT = (0.913 \times 1121)$$
$$= 1023.473 = 1023 \text{ VPD}$$

2.60 (a)

2.61 Sol.

$$n = 2,\ L = 3 \times 2 = 6 \text{ sec}$$
$$y_1 = 0.37,\ y_2 = 0.40$$
$$Y = y_1 + y_2$$
$$= 0.37 + 0.40$$
$$= 0.77$$
$$C_0 = \left(\frac{1.5L+5}{1-Y}\right) = \frac{1.5\times6+5}{1-0.77}$$
$$= 60.87 \text{ sec} \simeq 60.9 \text{ sec}$$

2.62 Sol.

$$V_1 = 40,\ V_2 = 55,\ V_3 = 60,\ V_4 = 65,\ V_5 = 80$$
$$V_S = \left(\frac{n}{\frac{1}{V_1}+\frac{1}{V_2}+\ldots+\frac{1}{V_n}}\right) = \frac{5}{\frac{1}{40}+\frac{1}{55}+\frac{1}{60}+\frac{1}{65}+\frac{1}{80}}$$
$$= 56.99 \text{ kmph}$$

2.63 Sol.

$$\lambda = 2 \text{ mean} = 2 \text{ veh/min}$$

Probability that at least 2 vehicle will arrive in any given 1 minute

$$= P(x \geq 2)$$
$$= 1 - [P(x=0) + P(x=1)]$$
$$\because P(x=0) + P(x=1)$$
$$= \frac{e^{-2}(2)^0}{0!} + \frac{e^{-2}(2)^1}{1!} = 0.406$$

Hence required probability

$$= 1 - 0.406 = 0.593 \simeq 0.59$$

2.64 (b)

$$u = 70 - 0.07k$$
$$u = 70\left[1-\frac{k}{\frac{70}{0.7}}\right]$$
$$V_f = 70 \text{ kmph}$$
$$k_j = \frac{70}{0.7} = 100 \text{ veh/km}$$
$$q_{max} = \frac{1}{4}V_f k_j = \left(\frac{1}{4}\times70\times100\right)$$
$$= 1750 \text{ veh/hr}$$
$$q_{max} = \frac{3600}{h_i}$$
$$1750 = \frac{300}{h_i}$$

Average time headway

$$h_i = \frac{3600}{1750} = 2.057 = 2.1 \text{ sec}$$

2.65 Sol.

$$L = 2 \times 3 = 6 \text{ seconds}$$
$$n = 3$$
$$y = (0.18 + 0.32 + 0.22)$$
$$C_0 = \left(\frac{1.5L+5}{1-y}\right)$$
$$= \left(\frac{1.5\times6+5}{1-(0.18+0.32+0.22)}\right)$$
$$= 50 \text{ seconds}$$

2.66 (c)

Speed at maximum flow $= \frac{V_f}{2}$

2.67 (a)

As per Greenshield model,

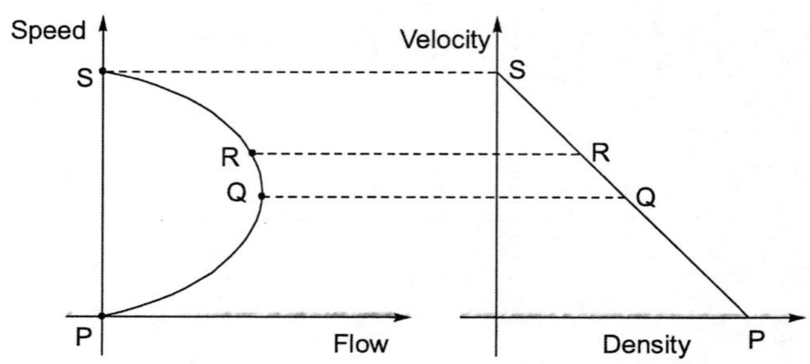

$$k_P > k_Q > k_R > k_S$$

2.68 Sol.

Cycle time, $T = 100$ sec

Amber time = 4 sec

Red time = 50 sec

Lost time per phase = 2 sec

$t_{SL} + t_{CL} = 2$ sec

Now, $T = G + A + R$

$\Rightarrow 100 = G + 4 + 50$

$\Rightarrow G = 46$ sec

Effective green, $G_i = G + A -$ Lost time

$\Rightarrow G_i = 46 + 4 - 2 = 48$ sec

2.69 Sol.

$$\lambda_C = 4 \text{ Vehicle/hr}$$
$$= 2 \text{ Vehicle/30 min}$$
$$\lambda_B = 2 \text{ Vehicle/hr}$$
$$= 1 \text{ Vehicle/30 min}$$
$$\lambda_{Vehicle} = 3 \text{ Vehicle/30 min}$$

Now, $P(x < 2) = 1 - P(x \leq 1)$

$$= 1 - [P(x = 0) + P(x = 1)]$$
$$= 1 - \left[\frac{e^{-\lambda}\lambda^0}{0!} + \frac{e^{-\lambda}\lambda^1}{1!}\right]$$
$$= 1 - e^{-3}(1+3)$$
$$= 1 - \frac{4}{e^3} = 0.8$$

2.70 Sol.

$$\text{PHF}_{10} = \frac{V}{\left(\frac{60}{n}\right)(V_n)_{max}}$$

Now, $V = 10 \times 10 + 11 \times 10 + 12 \times 10 + 15 \times 10 + 13 \times 10 + 11 \times 10$

$\Rightarrow V = 720$ veh

Now, $(V_n)_{max} = 15 \times 10$ (t = 30 to 40 min)

$= 150$ veh

$$\therefore \quad \text{PHF}_{10} = \frac{720}{\frac{60}{10} \times 150} = 0.8$$

2.71 (c)

$$q = kV$$

where, q = Flow of traffic (veh/hr)

V = Space mean speed of vehicle (kmph) = 30 kmph

k = Density of traffic (veh/km) = 40 vehicle/km

$\therefore \quad q = 40 \times 30 = 1200$ veh/hr

Now, $q = \frac{3600}{h_t} = \frac{3600}{1200}$ sec

$\Rightarrow h_t = 3$ sec (time headway)

2.72 Sol.

As per Webster's method,

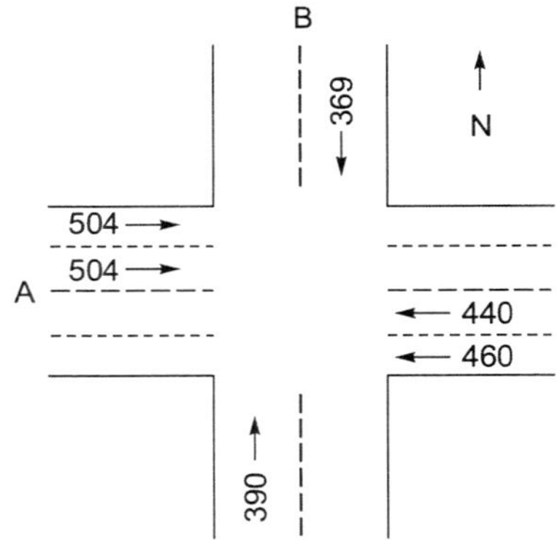

Optimum cycle length,

$$C_0 = \frac{1.5L + 5}{1 - Y}$$

Average flow over lane A and B,

$$q_A = \text{max.}(504, 460) = 504 \text{ veh.}$$
$$q_B = \text{max.}(360, 396) = 396 \text{ veh.}$$

Saturation flow over lane A and B,

$$S_A = S_B = 1800 \text{ veh/hr}$$

$\therefore \quad Y = y_A + y_B$

$$\Rightarrow \quad Y = \frac{q_A}{S_A} + \frac{q_B}{S_B}$$

$$\Rightarrow \quad Y = \frac{504}{1800} + \frac{396}{1800} = \frac{1}{2}$$

Los time = 9s (given)

$\therefore$Optimum cycle length,

$$C_o = \frac{1.5(9) + 5}{1 - \frac{1}{2}} = 37s$$

2.73 Sol.

$$q_P = 1200 \text{ veh/hr}$$
$$V_P = 60 \text{ kmph}$$
$$\therefore \quad K_P = \frac{q_p}{V_p} = \frac{1200}{60} = 20 \text{ veh/km}$$

Similarly, $q_Q = 1800$ veh/hr

$$V_Q = 30 \text{ kmph}$$
$$\therefore \quad K_Q = \frac{q_Q}{V_Q} = \frac{1800}{30} = 60 \text{ veh/km}$$

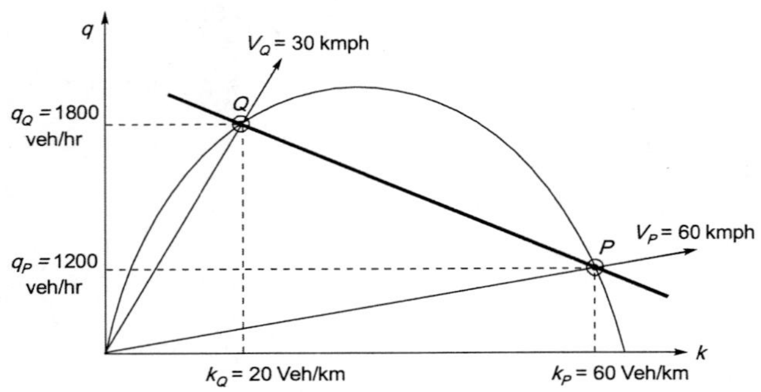

Speed of shock wave (PQ) is, slope of line PQ.

∴Speed of shock wave

$$= \frac{q_Q - q_P}{k_P - k_Q} = \frac{1800 - 1200}{60 - 20}$$

$$= 15 \text{ kmph}$$

2.74 (b, d)

We know as per Greenshield traffic model,

$$V = V_f\left[1 - \frac{k}{k_j}\right] = V_f - \left(\frac{V_f}{k_j}\right)k$$

Where, V_f = Free and low speed
k_j = jam density
k = density
V = speed of vehicle
q = flow of traffic

We know that,

$$q = Vk$$

$$\Rightarrow \quad q = \left[V_f - \left(\frac{V_f}{k_j}\right)k\right]k$$

$$\Rightarrow \quad q = V_f k - \frac{V_f}{k_j}k^2 \quad \text{...(i)}$$

Now, we know that,

$$q = Vk$$

$$\Rightarrow \quad k = \frac{q}{V}$$

Put in equation (i),

$$q = V_f\frac{q}{v} - \frac{V_f}{k_j}\left(\frac{q^2}{v^2}\right)$$

$$\Rightarrow \quad 1 = \frac{V_f}{V} - \frac{V_f}{V^2}\frac{q}{k_j}$$

$$\Rightarrow \quad \frac{V_f q}{V^2 k_j} = \frac{V_f}{V} - 1$$

$$\Rightarrow \quad q = \frac{V^2 k_j}{V_f}\left[\frac{V_f}{V} - 1\right]$$

$$\therefore \quad q = k_j V - \left(\frac{k_j}{V_f}\right)V^2$$

2.75 (c)

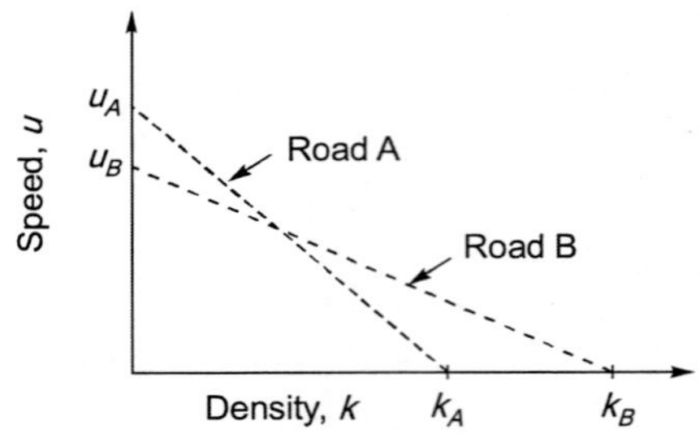

Now, capacity of road A, C_A

$$C_A = q_{\max A} = \frac{1}{4}k_A u_A$$

Similarly, capacity of road B, C_B

$$C_B = q_{\max B} = \frac{1}{4}k_B u_B$$

$$\text{Now, required ratio} = \frac{C_A}{C_B}$$

$$= \frac{\frac{1}{4} \times k_A \times u_A}{\frac{1}{4} \times k_B \times u_B} = \frac{k_A u_A}{k_B u_B}$$

2.76 (b, c, d)

- Option (a) is incorrect because, slope of distance time graph is more in region P than in region R.
- Volume in region Q is lower than in region P because in region Q, vehicles are at halted conditions. So, in region Q, velocity of vehicle is zero and hence, volume is zero. Option (b) is correct.
- Option (c) is correct because slope of shock wave is higher from Q to R in comparision from P to Q.
- Density is higher in region Q than in region R and region P because in region Q, they are practically at Jam density that is maximum. So, option (d) is correct.

2.77 Sol.

Time mean speed, $V_t = \dfrac{\text{Total distance}}{\text{Total time}}$

$$= \frac{D_1 + D_2 + D_3 + D_4 + D_5 + D_6}{t_1 + t_2 + t_3 + t_4 + t_5 + t_6}$$

$$= \frac{(500-400)+(500-250)+(500-100)+(400-0)+(350-0)+(150-0)}{(10-0)+(15-0)+(30-0)+(30-5)+(30-15)+(30-20)}$$

$$= 15.71 \text{ m/s}$$

$$= 15.71 \times \frac{18}{5} \text{km/h}$$

$$= 56.556 \text{ kmph} \simeq 57 \text{ kmph}$$

Alternatively,

Space mean speed, $V_s = \dfrac{n}{\Sigma \dfrac{1}{V_i}}$

$$= \frac{1+1+1+1+1+1}{\dfrac{1}{V_1}+\dfrac{1}{V_2}+\dfrac{1}{V_3}+\dfrac{1}{V_4}+\dfrac{1}{V_5}+\dfrac{1}{V_6}}$$

$$= \frac{6}{\dfrac{1}{\left(\frac{D}{T}\right)_1}+\dfrac{1}{\left(\frac{D}{T}\right)_2}+\dfrac{1}{\left(\frac{D}{T}\right)_3}+\dfrac{1}{\left(\frac{D}{T}\right)_4}+\dfrac{1}{\left(\frac{D}{T}\right)_5}+\dfrac{1}{\left(\frac{D}{T}\right)_6}}$$

$$= \frac{6}{\left(\frac{T_1}{D_1}\right)+\left(\frac{T_2}{D_2}\right)+\left(\frac{T_3}{D_3}\right)+\left(\frac{T_4}{D_4}\right)+\left(\frac{T_5}{D_5}\right)+\left(\frac{T_6}{D_6}\right)}$$

$$= \frac{6}{\left\{\left(\frac{10-0}{500-400}\right)+\left(\frac{15-0}{500-250}\right)+\left(\frac{30-0}{500-100}\right)+\left(\frac{30-5}{400-0}\right)+\left(\frac{30-15}{350-0}\right)+\left(\frac{30-20}{150-0}\right)\right\}}$$

$$= 14.74 \text{ m/s} = 53.064 \text{ km/h}$$

2.78 (b)

Given, Free mean speed,

$$V_f = 60 \text{ kmph}$$

Speed headway, $s = 8$ m

At jam density, $k = k_j$

$$= \frac{1000}{s} = \frac{1000}{8} = 125 \text{ veh/km}$$

For linear speed density relationship,

Maximum flow, $q_{max} = \dfrac{V_f k_j}{4}$

$$= \frac{60 \times 125}{4} = 1875 \text{ veh/hr}$$

2.79 (b)

Speed range (kmph)	Number of observations	Cumulative number of Vehicles	Cumulative %age Passed
0 - 10	7	7	1.4%
10 - 20	31	38	7.6%
20 - 30	76	114	22.8%
30 - 40	129	243	48.6%
40 - 50	104	347	69.4%
50 - 60	78	425	85%
60 - 70	29	454	90.8%
70 - 80	24	478	95.6%
80 - 90	13	491	98.2%
90 - 100	9	500	100%
	$\Sigma N = 500$		

Upper speed limit is corresponding to 85% of cumulative vehicles passed.

$$\therefore \text{ Upper speed limit} = \frac{50+60}{2} = 55 \text{ kmph}$$

■■■■

3 Highway Materials

3.1 The following general statement may be made about the penetration value and softening point of bitumen
(a) higher the penetration value, higher is the softening point
(b) higher the penetration value, lower is the softening point
(c) for very high and very low penetration value, the softening point is very low
(d) absolutely no correlation can be drawn between penetration value and softening point of bitumen
[1996 : 1 M]

3.2 Bituminous materials are used in highway construction primarily because of their
(a) cementing and waterproofing properties
(b) load bearing capacity
(c) high specific gravity
(d) black colour which facilitates road markings
[1996 : 1 M]

3.3 The penetration test for bitumen is conducted at a temperature of
(a) 60°C (b) 37°C
(c) 25°C (d) 50°C [1997 : 1 M]

3.4 The group index of a soil subgrade is 7. The subgrade soil is rated as
(a) poor (b) very poor
(c) good (d) fair [1997 : 1 M]

3.5 Rapid curing cutback bitumen is produced by blending bitumen with
(a) Kerosene (b) Benzene
(c) Diesel (d) Petrol [1999 : 1 M]

3.6 The standard plate size in a plate bearing test for finding modulus of subgrade reaction (k) value is
(a) 100 cm diameter (b) 50 cm diameter
(c) 75 cm diameter (d) 25 cm diameter
[2000 : 1 M]

3.7 Bituminous materials are commonly use in highway construction because of their good
(a) tensile and compression properties
(b) binding and water proofing properties
(c) shear strength and tensile properties
(d) bond and tensile properties
[2000 : 1 M]

3.8 The minimum value of CBR (%) required for granular subbase as per Ministry of Surface Transport (MOST) specification is
(a) 5 (b) 10
(c) 15 (d) 20 [2001 : 1 M]

3.9 Bitumen is derived from
(a) destructive distillation of coal tar
(b) destructive distillation of petroleum
(c) fractional distillation of petroleum
(d) naturally occurring ores [2002 : 1 M]

3.10 In the Marshall method of mix design, the coarse aggregates, fine aggregates, filler and bitumen, having respective specific gravities of 2.62, 2.72, 2.70 and 1.02 are mixed in the ratio of 55,34.6, 4.8 and 5.6 per cent, respectively. The theoretical specific gravity of the mix would be
(a) 2.36 (b) 2.40
(c) 2.44 (d) 2.50 [2003 : 2 M]

3.11 List-I below gives a list of physical properties of aggregates which should be determined to judge their suitability in road construction. **List-II** gives a list of laboratory tests which are conducted to determine these properties. Match **List-I** with **List-II** and select the correct answer from the codes given below the lists:

List-I	List-II
A. Hardness	**1.** Water absorption
B. Porosity	**2.** Impact test
C. Toughness	**3.** Soundness test
D. Durability	**4.** Abrasion test

Codes:

	A	B	C	D
(a)	1	2	3	4
(b)	4	1	2	3
(c)	3	4	1	2
(d)	2	3	4	1

[2003 : 2 M]

3.12 A Marshall specimen is prepared for bituminous concrete with a bitumen content of 5 per cent by weight of total mix. The theoretical and the measured unit weights of the mix are 2.442 g/cm^3 and 2.345 g/cm^3, respectively. The bitumen has a specific gravity of 1.02. The per cent voids in mineral aggregate filled with bitumen (VFB) are

(a) 34.55 (b) 35.9
(c) 73.55 (d) 74.3 [2004 : 2 M]

3.13 List-I contains some properties of bitumen. List-II gives a list of Laboratory Tests conducted on bitumen to determine the properties. Match the property with the corresponding test and select the correct answer using the codes given below the lists:

List-I
A. Resistance to flow
B. Ability to deform under load
C. Safety

List-II
1. Ductility test
2. Penetration test
3. Flash and fire point test

Codes:

	A	B	C
(a)	2	1	3
(b)	2	3	1
(c)	1	2	3
(d)	3	1	2

[2005 : 1 M]

3.14 Bituminous concrete is a mix comprising of
(a) fine aggregate, filler and bitumen
(b) fine aggregate and bitumen
(c) coarse aggregate, fine aggregate, filler and bitumen
(d) coarse aggregate, filler and bitumen

[2005 : 1 M]

3.15 If aggregate size of 50–40 mm is to be tested for finding out the portion of elongated aggregates using length gauge, the slot length of the gauge should be

(a) 81 mm (b) 45 mm
(c) 53 mm (d) 90 mm [2006 : 1 M]

3.16 The consistency and flow resistance of bitumen can be determined from which of the following?
(a) Ductility test
(b) Penetration test
(c) Softening point test
(d) Viscosity test [2007 : 1 M]

3.17 Match List-I (Test) with List-II (Property) and select the correct answers using the codes given below the lists:

List-I
A. Crushing test
B. Los Angeles abrasion test
C. Soundness test
D. Angularity test

List-II
1. Hardness
2. Weathering
3. Shape
4. Strength

Codes:

	A	B	C	D
(a)	2	1	4	3
(b)	4	2	3	1
(c)	3	2	1	4
(d)	4	1	2	3

[2007 : 2 M]

3.18 The specific gravity of paving bitumen as per IS 73 : 1992 lies between

(a) 1.10 and 1.06 (b) 1.06 and 1.02
(c) 1.02 and 0.97 (d) 0.97 and 0.92

[2008 : 1 M]

3.19 A combined value of flakiness and elongation index is to be determined for a sample of aggregates. The sequence in which the two tests are conducted is
(a) elongation index test followed by flakiness index test on the whole sample
(b) flakiness index test followed by elongation index test on the whole sample
(c) flakiness index test followed by elongation index test on non-flaky aggregates
(d) elongation index test followed by flakiness index test on non-elongated aggregates

[2008 : 1 M]

3.20 Aggregate impact value indicates which one of the following property of aggregates?

(a) Durability (b) Toughness
(c) Hardness (d) Strength

[2010 : 1 M]

3.21 In Marshall testing of bituminous mixes, as the bitumen content increases the flow value
(a) remains constant
(b) decreases first and then increases
(c) increases monotonically
(d) increases first and then decreases

[2011 : 1 M]

3.22 Two bitumen samples "*X*" and "*Y*" have softening points 45°C and 60°C, respectively. Consider the following statement:

I. Viscosity of "*X*" will be higher than that of "*Y*" at the same temperature.

II. Penetration value of "*X*" will be lesser than that of "*Y*" under standard conditions.

The CORRECT option evaluating the above statements is

(a) Both I and II are TRUE
(b) I is FALSE and II is TRUE
(c) Both I and II are FALSE
(d) I is TRUE and II is FALSE [2012 : 1 M]

3.23 In compacted cylindrical bituminous mix
VMA = 15% (void mineral aggregate)
V_V = 4.5% (air void)
The magnitude of VFB (void filled bituminous) is

(a) 24 (b) 30
(c) 54 (d) 70 [2013 : 2 M]

3.24 In a Marshall sample, the bulk specific gravity of mix and aggregates are 2.324 and 2.546 respectively. The sample includes 5% of bitumen (by total weight of mix) of specific gravity 1.10. The theoretical maximum specific gravity of mix is 2.441. The void filled with bitumen (VFB) in the Marshall sample (in %) is _______.

[2014 : 2 M, Set-II]

3.25 The penetration value of a bitumen sample tested at 25°C is 80. When this sample is heated to 60°C and tested again, the needle of the penetration test apparatus penetrates the bitumen sample by *d* mm. The value of *d* CANNOT be less than _________ mm.

[2015 : 1 M, Set-I]

3.26 In Marshall method of mix design, the coarse aggregate, fine aggregate, fines and bitumen having respective values of specific gravity 2.60, 2.70, 2.65 and 1.01, are mixed in the relative proportions (% by weight) of 55.0, 35.8, 3.7 and 5.5 respectively. The theoretical specific gravity of the mix and the effective specific gravity of the aggregates in the mix respectively are

(a) 2.42 and 2.63
(b) 2.42 and 2.78
(c) 2.42 and 2.93
(d) 2.64 and 2.78 [2015 : 2 M, Set-II]

3.27 Match the information related to test on aggregates given in **Group-I** with that in **Group-II**.

Group-I
P. Resistance to impact
Q. Resistance to wear
R. Resistance to weathering action
S. Resistance to crushing

Group-II
1. Hardness
2. Strength
3. Toughness
4. Soundness

(a) P-1, Q-3, R-4, S-2
(b) P-3, Q-1, R-4, S-2
(c) P-4, Q-1, R-3, S-2
(d) P-3, Q-4, R-2, S-1 [2015 : 2 M, Set-II]

3.28 During a forensic investigation of pavement failure, an engineer reconstructed the graphs *P*, *Q*, *R* and S, using partial and damaged old reports.

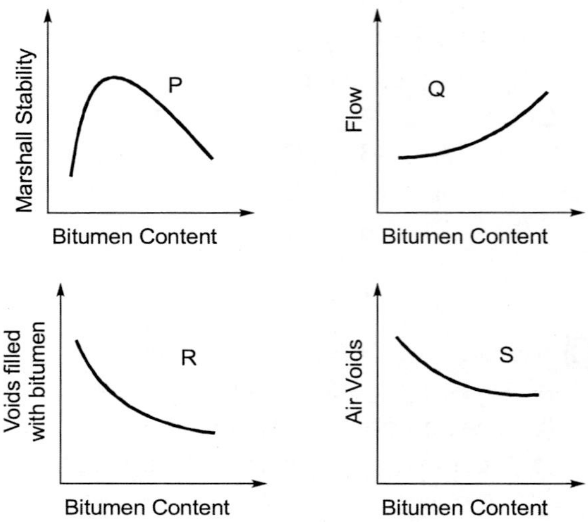

Theoretically plausible correct graphs according to the 'Marshall mixture design output' are

(a) *P*, *Q*, *R* (b) *P*, *Q*, *S*
(c) *Q*, *R*, *S* (d) *R*, *S*, *P*

[2016 : 1 M, Set-I]

3.29 The following observations are made while testing aggregate for its suitability in pavement construction:

(i) Mass of oven-dry aggregate in air = 1000 g
(ii) Mass of saturated surface-dry aggregate in air = 1025 g
(iii) Mass of saturated surface-dry aggregate under water = 625 g

Based on the above observations, the correct statement is

(a) bulk specific gravity of aggregate = 2.5 and water absorption = 2.5%
(b) bulk specific gravity of aggregate = 2.5 and water absorption = 2.4%
(c) apparent specific gravity of aggregate = 2.5 and water absorption = 2.5%
(d) apparent specific gravity of aggregate = 2.5 and water absorption = 2.4%

[2017 : 2 M, Set-I]

3.30 A bitumen sample has been graded as VG30 as per IS : 73-2013. The '30' in the grade means that
(a) penetration of bitumen at 25°C is between 20 and 40
(b) viscosity of bitumen at 60°C is between 2400 and 3600 Poise
(c) ductility of bitumen at 27°C is more than 30 cm
(d) elastic recovery of bitumen at 15°C is more than 30%

[2018 : 1 M, Set-I]

3.31 The initial concavity in the Load-penetration curve of a CBR test is NOT due to
(a) uneven top surface
(b) high impact at start of loading
(c) inclined penetration plunger
(d) soft top layer of soaked soil

[2018 : 1 M, Set-II]

3.32 If the fineness modulus of a sample of fine aggregates is 4.3, the mean size of the particles in the sample is between
(a) 300 μm and 600 μm
(b) 2.36 mm and 4.75 mm
(c) 1.18 mm and 2.36 mm
(d) 150 μm and 300 μm

[2019 : 1 M, Set-II]

3.33 The Los Angeles test for stone aggregates is used to examine
(a) specific gravity (b) abrasion resistance
(c) soundness (d) crushing strength

[2020 : 1 M, Set-I]

3.34 The softening point of bitumen has the same unit as that of
(a) viscosity (b) temperature
(c) distance (d) time

[2021 : 1 M, Set-II]

3.35 In an aggregate mix, the proportions of coarse aggregate, fine aggregate and mineral filler are 55%, 40% and 5%, respectively. The values of bulk specific gravity of the coarse aggregate, fine aggregate and mineral filler are 2.55. 2.65 and 2.70, respectively. The bulk specific gravity of the aggregate mix (round off to two decimal places) is ________.

[2021 : 2 M, Set-II]

3.36 It is given that an aggregate mix has 260 grams of coarse aggregates and 240 grams of fine aggregates. The specific gravities of the coarse and fine aggregates are 2.6 and 2.4. respectively. The bulk specific gravity of the mix is 2.3. The percentage air voids in the mix is ________. (round off to the nearest integer)

[2022 : 2 M, Set-II]

3.37 Which of the following options match the test reporting conventions with the given material tests in the table?

Test reporting convention
(P) Reported as ratio
(Q) Reported as percentage
(R) Reported in temperature
(S) Reported in length

Material test
(I) Solubility of bitumen
(II) Softening point of bitumen
(III) Los Angeles abrasion test
(IV) Flash point of bitumen
(V) Ductility of bitumen
(VI) Specific gravity of bitumen
(VII) Thin film oven test

(a) (P) - (VI); (Q) - (I); (R) - (II); (S) - (VII)
(b) (P) - (VI); (Q) - (III); (R) - (IV); (S) - (V)
(c) (P) - (VI); (Q) - (I); (R) - (II); (S) - (V)
(d) (P) - (VI); (Q) - (III); (R) - (IV); (S) - (VII)

[2023 : 2 M, Set-I]

3.38 Consider the Marshall method of mix design for bituminous mix. With the increase in bitumen content, which of the following statements is/are TRUE?
(a) the Stability decreases initially and then increases
(b) the Flow increases monotonically
(c) the air voids (*VA*) increases initially and then decreases
(d) the voids filled with bitumen (*VFB*) increases monotonically

[2023 : 2 M, Set-II]

3.39 Which one of the following statements related to bitumen is FALSE?

(a) Softer grade bitumen possesses higher softening point than hard grade bitumen.

(b) Kinematic Viscosity is a measure of resistance to the flow of molten bitumen under gravity

(c) Flash point of bitumen is the lowest temperature at which application of a test flame causes vapours of the bitumen to catch an instant fire in the form of flash under specified test conditions.

(d) Ductility test is carried out on bitumen to test its adhesive property and ability to stretch

[2024 : 1 M, Set-I]

3.40 In the context of pavement material characterization, the CORRECT statement(s) is/are

(a) The toughness and hardness of road aggregates are determined by Los Angeles abrasion test and aggregate impact test, respectively.

(b) The load penetration curve of CBR test may need origin correction due to the non-vertical penetrating plunger of the loading machine.

(c) In compacted bituminous mix, Voids in the Mineral Aggregate (VMA) is equal to the sum of total volume of air voids (V_v) and total volume of bitumen (V_b).

(d) Grading of normal (unmodified) bitumen binders is done based on viscosity test results.

[2024 : 2 M, Set-II]

Answers Highway Materials

3.1	(b)	3.2	(a)	3.3	(c)	3.4	(a)	3.5	(d)	3.6	(c)	3.7	(b)
3.8	(d)	3.9	(c)	3.10	(c)	3.11	(b)	3.12	(d)	3.13	(a)	3.14	(c)
3.15	(a)	3.16	(d)	3.17	(d)	3.18	(c)	3.19	(c)	3.20	(b)	3.21	(c)
3.22	(c)	3.23	(d)	3.24	(68.7)	3.25	(8)	3.26	(a)	3.27	(b)	3.28	(b)
3.29	(a)	3.30	(b)	3.31	(b)	3.32	(c)	3.33	(b)	3.34	(b)	3.35	(2.59)
3.36	(8)	3.37	(b, c)	3.38	(b, d)	3.39	(a)	3.40	(b, c, d)				

Explanations Highway Materials

3.1 (b)

Softening point of a sample is lower so at the same temperature viscosity of sample will be lower, hence penetration value of sample will be more.

3.2 (a)

Bitumen is a complex organic material and occurs either naturally or may be obtained artificially during the distillation of petroleum. Bituminous material are very commonly used in highway construction because of their binding/cementing and their water proofing properties. When the bitumen contains some inert material or minerals, it is some times called asphalt. Asphalt is found as deposit in the form of natural asphalt or rock asphalt.

3.3 (c)

3.4 (a)

Higher the value of group index, poorer is the soil as subgrade material.

Type of subgrade soil	Group index range of subgrade
Good	0-1
Fair	2-4
Poor	5-9
Very Poor	10-20

3.5 (d)

Rapid curing cutback bitumen – Petrol, naptha or gasoline

Medium curing cutback bitumen – Kerosene

Slow curing cutback bitumen – High boiling point diluent

3.6 (c)

3.7 (b)

3.8 (d)

For granular sub-base, minimum CBR value should be 20% for cumulative traffic upto 2 msa and 30% for cumulative traffic exceeding 2 msa. For subgrade, minimum CBR value = 10%

3.9 (c)

Bitumen is obtained from fractional distillation of petroleum.
Also note that tar is produced by the destructive distillation of coal or wood.

3.10 (c)

The theoretical specific gravity of the mix is given by,

$$G_t = \frac{100}{\frac{55}{2.62} + \frac{34.6}{2.72} + \frac{4.8}{2.70} + \frac{5.6}{1.02}} = 2.44$$

3.11 (b)

Abnrasion test	:	Hardness
Water absorption	:	Porosity
Impact test	:	Toughness
Soundness test	:	Durability
Crushing test	:	Strength
Flakiness index, Elongation index and angularity number	:	Shape

3.12 (d)

Percentage air voids,

$$V_v = \left[\frac{G_t - G_m}{G_t}\right] \times 100$$

$$= \frac{2.442 - 2.345}{2.442} \times 100 = 3.97\%$$

Percentage volume of bitumen

$$V_b = G_m \times \frac{W_b}{G_b}$$

$$= 2.345 \times \frac{5}{1.02} = 11.5\%$$

$$\text{VMA} = V_v + V_b = 3.97 + 11.5 = 15.461\%$$

$$\therefore \quad \text{VFB} = \frac{V_b\%}{\text{VMA}\%} \times 100$$

$$= \frac{11.5\%}{15.47\%} \times 100\% = 74.33\% = 74.3\%$$

3.13 (a)

Resistance to flow	:	Viscosity test
Hardness	:	Penetration test
Ability to deform under load without breaking	:	Ductility test
Safety	:	Flash and fire point test

3.14 (c)

3.15 (a)

Slot length for elongated aggregate

$$= 1.8 \times \text{mean dimension}$$

$$= 1.8 \times \left(\frac{50 + 40}{2}\right) = 81 \text{ mm}$$

For flakiness index, the slot size

$$= 0.6 \times \left(\frac{50 + 40}{2}\right) = 27 \text{ mm}$$

3.16 (d)

3.17 (d)

3.18 (c)

3.19 (c)

Flakiness index test followed by elongation index test on non-flaky aggregates.

REFERENCE MATERIAL

1. Refer IS : 2386 Part-I for test procedure.

3.20 (b)

3.21 (c)

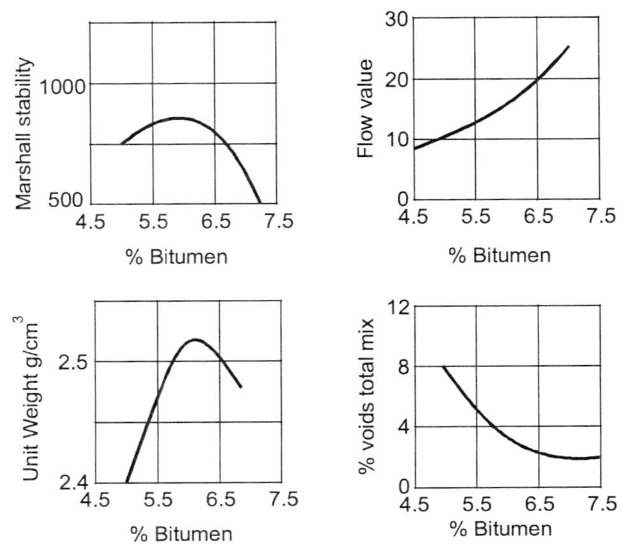

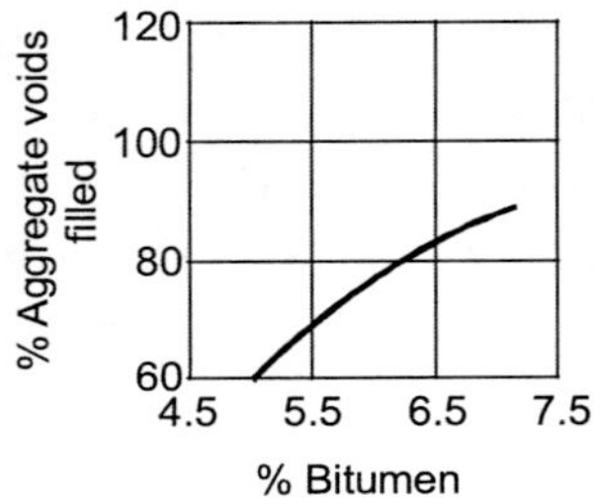

3.22 (c)

Since softening point for Y is higher than X.

$\therefore$ Viscosity of Y is high as compared to X and penetration will be more in X, because it will offer less resistance to penetrate.

$\therefore$ Both statement I & II are false.

3.23 (d)

$$\text{VFB} = \frac{V_B}{V_{VMA}} = \left(\frac{15-4.5}{15}\right) \times 100\% = 70\%$$

3.24 Sol.

$$V_v = \frac{G_t - G_m}{G_t} \times 100 = \frac{2.441 - 2.324}{2.441} \times 100$$

$= 4.79\%$

Percentage volume of birumen,

$$V_b = G_m \times \frac{W_b(\%)}{G_b} = 2.324 \times \frac{5}{1.1} = 10.56\%$$

$\text{VMA} = V_v + V_b = 4.79 + 10.56 = 15.35\%$

$$\therefore \text{VFB} = \frac{V_b(\%)}{VMA} \times 100 = \frac{10.56}{15.35} \times 100 = 68.79\%$$

3.25 Sol.

Relationship between penetration and temperature is given by,

$P = AT + K$

Where P = Penetration

A = Temperature susceptibility

K = Constant

With increase in temperature, penetration increases.

Hence, we will simply take,

$$d \not< \frac{80}{10} = 8 \text{ mm}$$

3.26 (a)

Theoretical specific gravity,

$$G_t = \frac{w_1 + w_2 + w_3 + w_4}{\frac{w_1}{G_1} + \frac{w_2}{G_2} + \frac{w_3}{G_3} + \frac{w_4}{G_b}}$$

$$= \frac{55 + 35.8 + 3.7 + 5.5}{\frac{55}{2.6} + \frac{35.8}{2.7} + \frac{3.7}{2.65} + \frac{5.5}{1.01}} = 2.42$$

Effective specific gravity of aggregates (coarse + fine) is given by,

$$G' = \frac{(55 \times 2.6) + (35.8 \times 2.7)}{55 + 35.8} = 2.639 = 2.64$$

3.27 (b)

3.28 (b)

The graph between VFB and Bitumen content is,

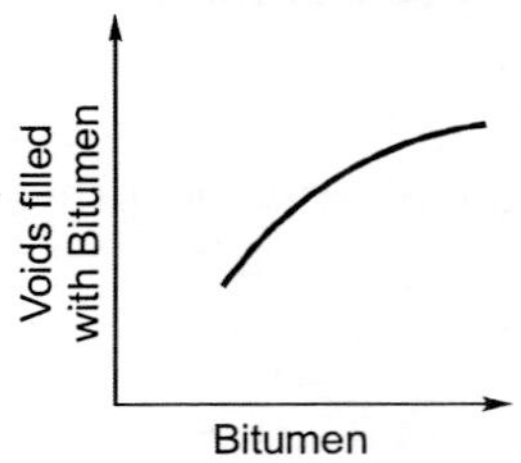

3.29 (a)

$$\text{Bulk specific gravity} = \frac{W_a}{V_{mix}} = \frac{A}{B-C}$$

$$\text{Apparent specific gravity} = \frac{W_a}{V_a} = \frac{A}{A-C}$$

Where, A = Mass of oven dry aggregate in air

B = Mass of Saturated surface dry aggregate in air

C = Mass of Saturated surface dry aggregate under water

$\therefore$ Bulk specific gravity

$$= \frac{1000}{1025 - 625} = \frac{1000}{400} = 2.5$$

Water absorption

$$= \frac{W_w}{W_a} \times 100 = \left(\frac{B-A}{A}\right) \times 100$$

$$= \frac{1025 - 1000}{1000} \times 100 = 2.5\%$$

3.30 (b)

3.31 (b)

Initial concavity in CBR test is due to :

- Improper compaction
- Soft top layer
- Inclined plunger
- Uneven top serface

3.32 (c)

The sieves that are to be used for the sieve analysis of the aggregate (coarse, fine, or all-in-aggregate) for concrete as per IS:2386 (Part-I) -1963 are, 80 mm, 40 mm, 20 mm, 10 mm, 4.75 mm, 2.36 mm, 1.18 mm, 600 μm, 300 μm and 150 μm.

The fineness modulus can be regarded as a weighted average size of a sieve on which material is retained and the sieves being counted from the first sieve.

Fineness modulus of 4.3 indicates size between 4th and 5th sieve i.e., between 1.18 mm and 2.36 mm.

3.33 (b)

The Los Angeles test for stone aggregates is used to examine abrasion resistance.

3.34 (b)

3.35 Sol.

$$G_m = \frac{55+40+5}{\frac{55}{2.55}+\frac{40}{2.65}+\frac{5}{2.70}} = 2.596$$

3.36 Sol.

Theoretical specific gravity,

$$G_t = \frac{\text{Total weight}}{V_{CA}+V_{FA}}, \quad C = \frac{260+240}{\frac{260}{2.6}+\frac{240}{2.4}} = 2.5$$

$$\therefore \quad V_{av} = \frac{G_t - G_m}{G_t}\times 100 \quad \{G_m = 2.3\}$$

$$= \frac{2.5-2.3}{2.5}\times 100 = 8\%$$

3.37 (b, c)

3.38 (b, d)

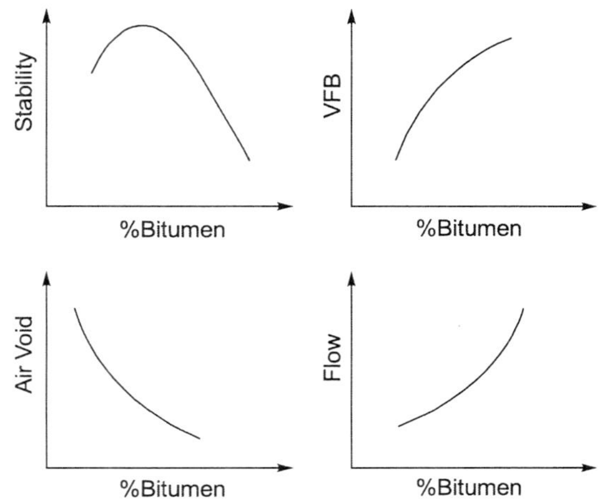

3.39 (a)

Softer grade bitumen has a lower softening point, while harder grade bitumen has a higher softening point. Hence statement A is false.

3.40 (b,c,d)

4 Pavement Design

4.1 In using the data from a plate bearing test for determining the modulus of sub-grade reaction, the value of settlement to be used is
(a) 1.25 mm (b) 2.50 mm
(c) 3.75 mm (d) 1.75 mm
[1995 : 1 M]

4.2 Flexible pavements derive stability primarily from
(a) aggregate interlock, particle friction and cohesion
(b) cohesion alone
(c) the binding power of bituminous materials
(d) the flexural strength of the surface course
[1996 : 1 M]

4.3 Since the moisture content at the bottom of a rigid pavement slab is generally more than at the top
(a) the bottom of the slab is generally in tension
(b) the top of the slab is generally in compression
(c) the bottom of the slab is generally in compression
(d) the bottom of the slab is neither is compression nor in tension
[1996 : 1 M]

4.4 The unit for coefficient of subgrade modulus is
(a) kN/m^3 (b) kN/m^2
(c) kN/m (d) kNm [1997 : 1 M]

4.5 The total thickness of pavement by CBR method depends on the CBR value of
(a) base course (b) surface course
(c) subgrade (d) All layers
[1997 : 1 M]

4.6 The width of expansion joint gap is 2.5 cm in a cement concrete pavement. The spacing between expansion joint for a maximum rise in temperature of 25°C is, (Assuming coefficient of thermal expansion of concrete as 10×10^{-6} per degree C)
(a) 5 m (b) 50 m
(c) 100 m (d) 25 m [1997 : 1 M]

4.7 The modulus of subgrade reaction is obtained from the plate bearing test in the form of load deformation curve. The pressure corresponding to the following settlement value should be used for computing modulus of subgrade reaction
(a) 0.375 cm (b) 0.175 cm
(c) 0.125 cm (d) 0.250 cm
[1998 : 1 M]

4.8 In the plate bearing test, if the load applied is in the form of an inflated type of wheel, then this mechanism corresponds to
(a) rigid plate
(b) flexible plate
(c) semi-rigid plate
(d) semi-elastic plate [1998 : 1 M]

4.9 Base course is used in rigid pavements for
(a) prevention of subgrade settlement
(b) prevention of slab cracking
(c) prevention of pumping
(d) prevention of thermal expansion
[1998 : 1 M]

4.10 What is the equivalent single wheel load of a dual wheel assembly carrying 20,440 N each for pavement thickness of 20 cm? Centre to centre spacing of tyres is 27 cm and the distance between the walls of tyres is 11 cm
(a) 27,600 N (b) 32,300 N
(c) 40,880 N (d) 30,190 N
[1998 : 2 M]

4.11 Plate bearing test with 20 cm diameter plate on soil subgrade yielded a pressure of $1.25\times 10^5\ N/m^2$ at 0.5 cm deflection. What is the elastic modulus of subgrade?
(a) $56.18\times 10^5\ N/m^2$ (b) $29.50\times 10^5\ N/m^2$
(c) $44.25\times 10^5\ N/m^2$ (d) $50.19\times 10^5\ N/m^2$
[1998 : 2 M]

4.12 A two lane single carriage-way is to be designed for a design life period of 15 years. Total two way traffic intensity in the year of completion of construction is expected to be 2000 commercial vehicles per day. Vehicle damage factor = 3.0, Lane distribution factor = 0.75. Assuming an annual rate of traffic growth as 7.5%, the design traffic expressed as cumulative number of standard axles, is

(a) 42.9×16^6 (b) 22.6×16^6
(c) 10.1×10^6 (d) 5.3×16^6
[1999 : 1 M]

4.13 The radius of relative stiffness for a 20 cm thick slab with $E = 3 \times 10^5$ kg/cm^2 and Poisson's ratio = 0.15, resting on a subgrade having modulus of 5 kg/cm^3 is
(a) 10 cm (b) 80 cm
(c) 120 cm (d) 320 cm [2001 : 2 M]

4.14 Dowel bars in concrete pavement are placed
(a) along the direction of traffic
(b) perpendicular to the direction of traffic
(c) along 45° to the direction of traffic
(d) can be placed along any direction
[2002 : 1 M]

4.15 The average daily traffic on a stretch of road is 300 commercial vehicles per lane per day. Design traffic repetition for 10 years when vehicle damage factor is 2.5 and traffic growth rate is 7%, is
(a) 3.8 msa (b) 23.5 msa
(c) 45.4 msa (d) 16 msa
[2002 : 2 M]

4.16 In a concrete pavement
I. Temperature stress is tensile at bottom during day time
II. Load stress is compressive at bottom
(a) Both the statement I and II are correct
(b) Statement I is correct and II is wrong
(c) Statement II is wrong and I is correct
(d) Both statement I and II are incorrect
[2002 : 2 M]

4.17 Temperature stresses in concrete pavements may cause the slab to crack. If a slab cools uniformly then the crack will develop at the which of the following locations of the slab?
(a) At centre
(b) Near edges
(c) At corners
(d) Near edges and at corners [2003 : 1 M]

4.18 The plate load test conducted with a 75 cm diameter plate on soil subgrade yielded a deflection of 2.5 mm under a stress of 800 N/cm^2. The modulus of elasticity of the subgrade soil, in kN/cm^2 is
(a) 141.6 (b) 154.6
(c) 160.0 (d) 185.4 [2003 : 2 M]

4.19 In the context of flexible pavement design, the ratio of contact pressure to tyre pressure is called the Rigidity Factor. This factor is less than unity when the tyre pressure is
(a) less than 0.56 N/mm^2
(b) equal to 0.56 N/mm^2
(c) equal to 0.7 N/mm^2
(d) more than 0.7 N/mm^2 [2004 : 1 M]

4.20 The data given below pertain to the design of a flexible pavement:

Initial traffic = 1213 cvpd
Traffic growth rate = 8% per annum
Design life = 12 years
Vehicle damage factor = 2.5
Distribution factor = 1.0

The design traffic in terms of million standard axles to be catered would be
(a) 0.06 msa (b) 8.40 msa
(c) 21.00 msa (d) 32.26 msa
[2004 : 2 M]

4.21 For a 25 cm thick cement concrete pavement, analysis of stresses gives the following values:

Wheel load stress due to corner loading................ 30 kg/cm^2
Wheel load stress due to edge loading................. 32 kg/cm^2
Warping stress at corner region during summer...... 9 kg/cm^2
Warping stress at corner region during winter......... 7 kg/cm^2
Warping stress at edge region during summer........ 8 kg/cm^2
Warping stress at edge region during winter........... 6 kg/cm^2
Frictional stress during summer............................ 5 kg/cm^2
Frictional stress during winter............................... 4 kg/cm^2

The most critical stress value for this pavement is
(a) 40 kg/cm^2 (b) 42 kg/cm^2
(c) 44 kg/cm^2 (d) 45 kg/cm^2
[2005 : 2 M]

4.22 The following observations were made of an axle-load survey on a road:

Axle load (kN)	Repetitions per day
35-45	800
75-85	400

The standard axle-load is 80 kN. Equivalent daily number of repetitions for the standard axle-load are

(a) 450 (b) 480
(c) 800 (d) 1200 [2005 : 2 M]

4.23 In case of governing equations for calculating wheel load stresses using Westergaard's approach, the following statements are made:

1. Load stresses are inversely proportional to wheel load.
2. Modulus of subgrade reaction is useful for load stress calculation.

(a) Both statements are TRUE
(b) 1 is TRUE and 2 is FALSE
(c) Both statements are FALSE
(d) 1 is FALSE and 2 is TRUE [2006 : 1 M]

4.24 Using **IRC : 37-1984** "Guidelines for the Design of Flexible Pavements" and the following data, choose the total thickness of the pavement.

Number of commercial vehicles when construction is completed.	2723 veh/day
Annual growth rate of the traffic.	5.0%
Design life of the pavement.	10 years
Vehicle damage factor.	2.4
CBR value of the subgrade soil.	5%

Data for 5% CBR value	
Number of Standard Axles, msa	Total Thickness, mm
20	620
25	640
30	670
40	700

(a) 620 mm (b) 640 mm
(c) 670 mm (d) 700 mm [2006 : 2 M]

4.25 A subgrade soil sample was tested using standard CBR apparatus and the observations are given below.

Load (kg)	Penetration (mm)
60.5	2.5
80.5	5.0

Assuming that the load-penetration curve is convex throughout, the CBR value (%) of the sample is

(a) 6.5 (b) 5.5
(c) 4.4 (d) 3.9 [2006 : 2 M]

4.26 The width of the expansion joint is 20 mm in a cement concrete pavement. The laying temperature is 20°C and the maximum slab temperature in summer is 60°C. The coefficient of thermal expansion of concrete is 10×10^{-6} mm/mm/°C and the joint filler compresses up to 50% of the thickness. The spacing between expansion joints should be

(a) 20 m (b) 25 m
(c) 30 m (d) 40 m [2007 : 2 M]

4.27 The following data pertains to the number of commercial vehicles per day for the design of a flexible pavement for a national highway as per **IRC:37-1984**:

Type of commercial vehicles	Number of vehicle per day considering the number of lanes	Vehicle Damage Factor
Two axle trucks	2000	5
Tandem axle trucks	200	6

Assuming a traffic growth factor of 7.5% per annum for both the types of vehicles, the cumulative number of standard axle load repetitions (in million) for a design life of ten years is

(a) 44.6 (b) 57.8
(c) 62.4 (d) 78.7 [2007 : 2 M]

4.28 It is proposed to widen and strengthen an existing 2-lane NH section as a divided highway. The existing traffic in one direction is 2500 commercial vehicles (CV) per day. The construction will take 1 year. The design CBR of soil subgrade is found to be 5 per cent. Given: traffic growth rate for CV = 8 per cent, vehicle damage factor = 3.5 (standard axles per CV), design life = 10 years and traffic distribution factor = 0.75. The cumulative standard axles (msa) computed are

(a) 35 (b) 37
(c) 65 (d) 70 [2008 : 2 M]

4.29 During a CBR test, the load sustained by a remolded soil specimen at 5.0 mm penetration is 50 kg. The CBR value of the soil will be

(a) 10.0% (b) 5.0%
(c) 3.6% (d) 2.4% [2009 : 1 M]

4.30 Which of the following stress combinations are appropriate in identifying the critical condition for the design of concrete pavements?

Type of stress	Location
A. Load	1. Corner
B. Temperature	2. Edge
	3. Interior

Select the correct answer using the codes given below:

(a) A-2, B-3 (b) A-1, B-3
(c) A-3, B-1 (d) A-2, B-2

[2009 : 2 M]

4.31 Consider the following statements in the context of cement concrete pavements:

1. Warping stresses in cement concrete pavements are caused by the seasonal variation in temperature.
2. Tie bars are generally provided across transverse joints of cement concrete pavements.

The correct option evaluating the above statements is

(a) 1 true and 2 false (b) 1 false and 2 true
(c) 1 true and 2 true (d) 1 false and 2 false

[2010 : 2 M]

4.32 A pavement designer has arrived at a design traffic of 100 million standard axles for a newly developing national highway as per IRC:37 guidelines using the following data : life = 15 years, commercial vehicle count before pavement construction = 4500 vehicles/day, annual traffic growth rate = 8%. The vehicle damage factor used in the calculation was

(a) 1.53 (b) 2.24
(c) 3.66 (d) 4.14 [2012 : 2 M]

4.33 Select the strength parameter of concrete used in design of plain jointed cement pavement from the following choices:

(a) Tensile strength (b) Compressive strength
(c) Flexural strength (d) Shear strength

[2013 : 1 M]

4.34 The following statements are related to temperature stresses developed in concrete pavement slabs with free edges (without any restraint):

P. The temperature stresses will be zero during both day and night times if the pavement slab is considered weightless.

Q. The temperature stresses will be compressive at the bottom of the slab during night time if the self-weight of the pavement slab is considered.

R. The temperature stresses will be compressive at the bottom of the slab during day time if the self-weight of the pavement slab is considered

The TRUE statement(s) is(are)

(a) *P* only (b) *Q* only
(c) *P* and *Q* (d) *P* and *R* only

[2014 : 1 M, Set-I]

4.35 A traffic survey conducted on a road yields an average daily traffic count of 5000 vehicles. The axle load distribution on the same road is given in the following table:

Axle load (tonnes)	18	14	10	8	6
Frequency of traffic (%)	10	20	35	15	20

The design period of the road is 15 years, the yearly traffic growth rate is 7.5% and the load safety factor (LSF) is 1.3. If the vehicle damage factor (VDF) is calculated from the above data, the design traffic (in million standard axle load, MSA) is ________.

[2014 : 2 M, Set-I]

4.36 In the context of the **IRC 58-2011** guidelines for rigid pavement design, consider the following pair of statements.

I. Radius of relative stiffness is directly related to modulus of elasticity of concrete and inversely related to Poisson's ratio.

II. Radius of relative stiffness is directly related to thickness of slab and modulus of subgrade reaction.

Which one of the following combinations is correct?

(a) I. True; II. True (b) I. False; II. False
(c) I. True; II. False (d) I. False; II. True

[2016 : 1 M, Set-II]

4.37 The radii of relative stiffness of the rigid pavements *P* and *Q* are denoted by l_P and l_Q, respectively. The geometric and material properties of the concrete slab and underlying soil are given below:

	Concrete					Soil
Pavement	Length of Slab	Breadth of Slab	Thickness of Slab	Modulus of Elasticity	Poisson's Ratio	Subgrade Reaction Modulus
P	*L*	*B*	*h*	*E*	μ	*K*
Q	*L*	*B*	0.5 *h*	*E*	μ	2 *K*

The ratio (up to one decimal place) of $\frac{l_P}{l_Q}$ is ________.

[2017 : 2 M, Set-II]

4.38 Given the following data: design life n = 15 years, lane distribution factor D = 0.75, annual rate of growth of commercial vehicles r = 6%, vehicle damage factor F = 4 and initial traffic in the year of completion of construction = 3000 Commercial Vehicles Per Day (CVPD). As per IRC : 37-2012, the design traffic in terms of cumulative number of standard axles (in million standard axles, up to two decimal places) is ________

[2018 : 2 M, Set-I]

4.39 An isolated concrete pavement slab of length L is resting on a frictionless base. The temperature of the top and bottom fibre of the slab are T_t and T_b, respectively. Given: the coefficient of thermal expansion = α and the elastic modulus = E. Assuming $T_t > T_b$ and the unit weight of concrete as zero, the maximum thermal stress is calculated as

(a) $\dfrac{E\alpha(T_t - T_b)}{2}$ (b) $L\alpha(Tt_b - T_b)$

(c) $E\alpha(T_t - T_b)$ (d) Zero

[2019 : 1 M, Set-I]

4.40 Tie bars of 12 mm diameter are to be provided in a concrete pavement slab. The working tensile stress of the tie bars is 230 MPa, the average bond strength between a tie bar and concrete is 2 MPa. and the joint gap between the slabs is 10 mm. Ignoring the loss of bond and the tolerance factor, the design length of the tie bars (in mm, round off to the nearest integer) is________.

[2019 : 2 M, Set-I]

4.41 Structural failures considered in the mechanistic method of bituminous pavement design are

(a) Fatigue and Rutting
(b) Fatigue and Shear
(c) Shear and Slippage
(d) Rutting and Shear [2019 : 1 M, Set-II]

4.42 A flexible pavement has the following class of loads during a particular hour of the day.

(i) 80 buses with 2-axles (each axle load of 40 kN);
(ii) 160 trucks with 2-axles (front and rear axle loads of 40 kN and 80 kN, respectively)

The equivalent standard axle load repetitions for this vehicle combination as per IRC:37-2012 would be

(a) 250 (b) 320
(c) 180 (d) 240

[2019 : 2 M, Set-II]

4.43 A dowel bar is placed at a contraction joint. When contraction occurs, the concrete slab cracks at predetermined location(s). Identify the arrangement, which shows the correct placement of dowel bar and the place of occurrence of the contraction crack(s).

(a)

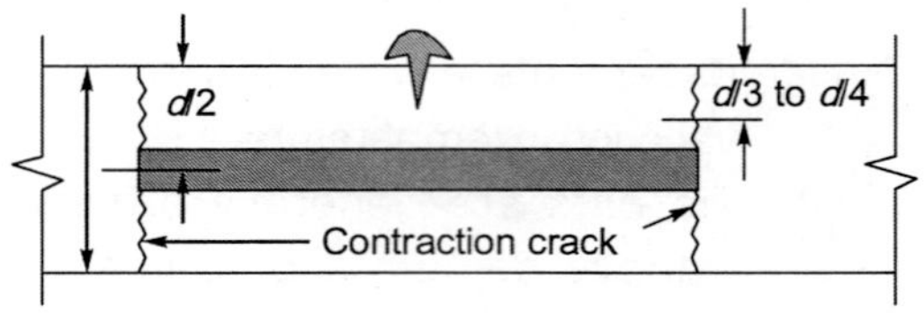

(b)

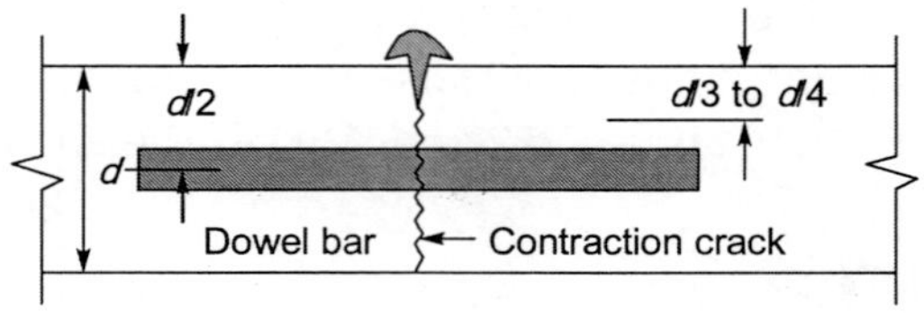

(c)

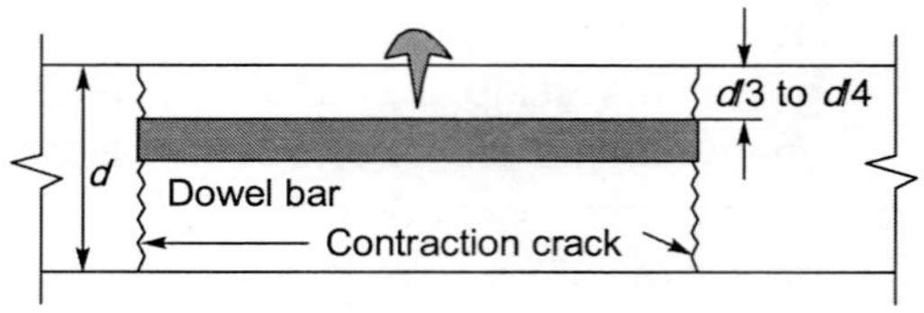

(d)

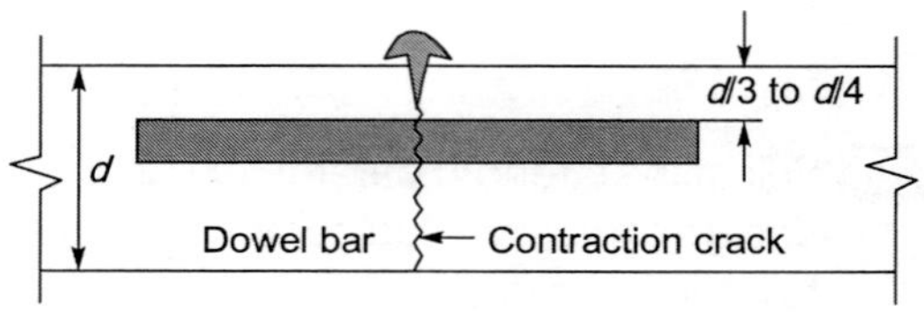

[2020 : 2 M, Set-I]

4.44 For an axle load of 15 tonne on a road, the Vehicle Damage Factor (round off to two decimal places), in terms of the standard axle load of 8 tonne, is ________.

[2020 : 1 M, Set-II]

4.45 For the dual-wheel carrying assembly shown in the figure, P is the load on each wheel, a is the radius of the contact area of the wheel, s is the spacing between the wheels, and d is the clear distance between the wheels. Assuming that the ground is an elastic, homogeneous, and isotropic half space, the ratio of Equivalent Single Wheel Load (ESWL) at depth $z = d/2$ to the ESWL at depth $z = 2s$ is _______. (round off to one decimal place). (Consider the influence angle to be 45° for the linear dispersion of stress with depth)

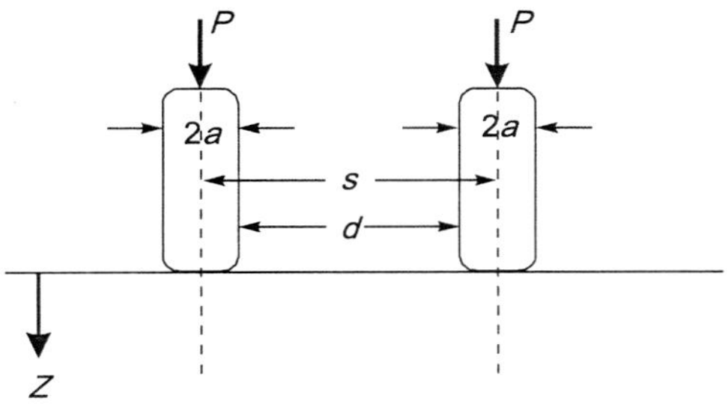

[2022 : 2 M, Set-I]

4.46 The following data is obtained from an axle load survey at a site:

Average rear axle load - 12000 kg.

Number of commercial vehicles = 800 per day

The pavement at this site would be reconstructed over a period of 5 years from the date of survey. The design life of the reconstructed pavement is 15 years. Use the standard axle load as 8160 kg and the annual average vehicle growth rate as 4.0% Assume that Equivalent Wheel Load Factor (EWLF) and Vehicle Damage Factor (VDF) are equal

The cumulative standard axle (in msa) for the pavement design is ________. (rounded off to 2 decimal places)

[2024 : 2 M, Set-I]

4.47 Various stresses in jointed plain concrete pavement with slab size of 3.5 m × 4.5 m are denoted as follows:

Wheel load stress at interior = S_{wl}^{i}

Wheel load stress at edge = S_{wl}^{e}

Wheel load stress at corner = S_{wl}^{c}

Warping stress at interior = S_{t}^{i}

Warping stress at edge = S_{t}^{e}

Warping stress at corner = S_{t}^{c}

Frictional stress between slab and supporting layer = S_f

The critical stress combination in the concrete slab during a summer midnight is

(a) $S_{wl}^{c} + S_{t}^{c} + S_f$ (b) $S_{wl}^{c} + S_{t}^{c}$

(c) $S_{wl}^{e} + S_{t}^{e} - S_f$ (d) $S_{wl}^{e} + S_{t}^{e} + S_f$

[2024 : 1 M, Set-II]

■■■■

Answers Pavement Design

4.1	(a)	4.2	(a)	4.3	(a)	4.4	(a)	4.5	(c)	4.6	(b)	4.7	(c)
4.8	(b)	4.9	(c)	4.10	(d)	4.11	(b)	4.12	(a)	4.13	(b)	4.14	(a)
4.15	(a)	4.16	(b)	4.17	(a)	4.18	(a)	4.19	(d)	4.20	(c)	4.21	(b)
4.22	(a)	4.23	(d)	4.24	(c)	4.25	(c)	4.26	(b)	4.27	(b)	4.28	(b)
4.29	(d)	4.30	(a)	4.31	(d)	4.32	(b)	4.33	(c)	4.34	(c)	4.35	(309.08)
4.36	(b)	4.37	(2)	4.38	(76.45)	4.39	(d)	4.40	(700)	4.41	(a)	4.42	(c)
4.43	(b)	4.44	(12.3)	4.45	(0.5)	4.46	(33.29)	4.47	(b)				

Explanations Pavement Design

4.1 (a)

Plate bearing test is used to evaluate the supporting power of subgrade for use in pavement design by using large diameter plates. After the plate bearing test consolidation specimen is tested as such without soaking by applying increments of pressure and pressure deformation curve is plotted. Pressure corresponding to 75 cm diameter plate is used to report modulus of subgrade reaction required in the plate bearing test to cause a deformation of 0.125 cm.

4.2 (a)

4.3 (a)

As the moisture content at the bottom is more the slab will contract from its natural shape, therefore tensile stress is setup at the bottom of the slab.

4.4 (a)

4.5 (c)

4.6 (b)

$$\delta_1 = \frac{2.5}{2} = 1.25 \text{ cm} = 1.25 \times 10^{-2} \text{ m}$$

$$t_2 - t_1 = 25°C$$

Coefficient of thermal expansion,

$$\alpha = 10 \times 10^{-6}/°C$$

$\therefore \quad \delta_1 = L\alpha(t_2 - t_1)$

$\therefore$ Spacing of expansion joint, L

$$= \frac{\delta_1}{\alpha(t_2 - t_1)} \text{m}$$

$$= \frac{1.25 \times 10^{-2}}{10 \times 10^{-6} \times 25} = 50 \text{ metres}$$

POINTS TO REMEMBER

- In expansion joint, it is ensured that $\Delta/2$ distance is always maintained after expansion.

  Hence $\quad \delta_1 = \frac{\Delta}{2} = L \propto t$

4.7 (c)

The modulus of subgrade reaction 'k' is proportion to the displacement. The displacement level Δ is taken as 0.125 cm in calculating k.

If P is the pressure sustained in kg/cm^2 by the rigid plate of diameter 75 cm at a deflection $\Delta = 0.125$ cm, the modulus of subgrade reaction 'k' is given by,

$$k = \frac{P}{\Delta} = \frac{P}{0.125} \text{kg/cm}^3$$

4.8 (b)

Plate bearing test conducted with a mild steel plate is considered relevant to the condition of rigid plate as in equation $\Delta = \frac{1.18\,pa}{E_s}$, but the wheel loads through inflated rubber tyres may be considered as flexible plate loading or loading with uniformly distributed pressure as in equation

$$\Delta = \frac{1.5\,pa}{E_s}$$

where, p = uniform pressure of loaded plate

a = radius of plate

E_s = modulus of elasticity of the soil

4.9 (c)

Base course and subbase course are used under flexible pavement primarily to improve the load supporting capacity by distributing the load through a finite thickness. Whereas, in rigid pavement base course are used for (i) Preventing mud pumping (ii) Protecting the subgrade against frost action.

4.10 (d)

Given, $\quad S = 27$ cm

$d = 11$ cm

$P = 20440$ kN

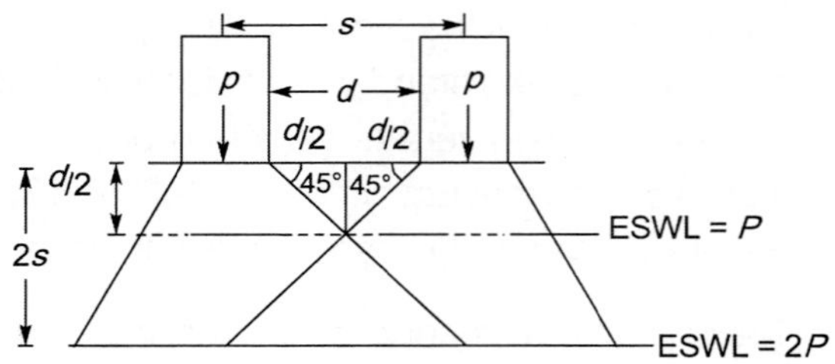

ESWL of $\frac{d}{2}$ depth i.e., at 5.5 cm

$$= 20440 \text{ N}$$

ESWL at $2s$ depth i.e., at 54 cm

$$= 2 \times 20440 = 40880 \text{ N}$$

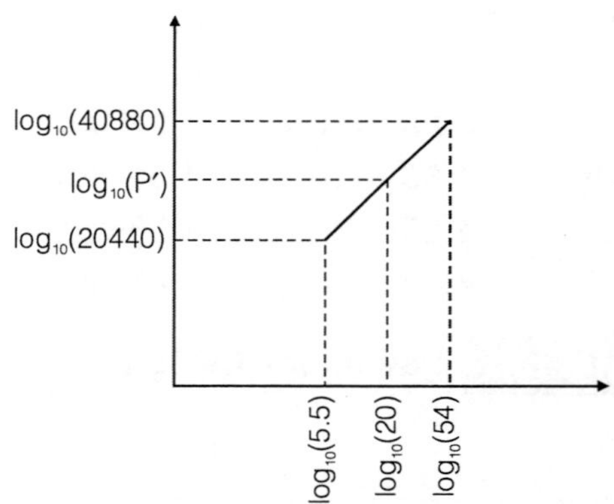

$$\log_{10}(P') = \log_{10}(20440) + \frac{\log_{10}(40880) - \log_{10}(20440)}{\log_{10}(54) - \log_{10}(5.5)} \times [\log_{10}(20) - \log_{10}(5.5)]$$

$\Rightarrow \quad P' = 30242.26$ N

$\therefore$ ESWL = 30242.26 N

4.11 (b)

$$\Delta = 1.18\frac{Pa}{E}$$

Here, a = radius of plate

$$= \frac{20}{2} = 10 \text{ cm} = 0.1\text{m}$$

$$\Rightarrow \frac{0.5}{100} = \frac{1.18 \times 1.25 \times 10^5 \times 0.1}{E}$$

$$\Rightarrow E = \frac{1.18 \times 1.25 \times 10^5 \times 0.1 \times 100}{0.5}$$

$$= 29.5 \times 10^5 \text{ N/m}^2$$

AVOID MISTAKE

Do not use formula, $\Delta = \frac{1.5pa}{E}$ which is actually for flexible plate test.

POINTS TO REMEMBER

- Modulus of subgrade reaction is the reaction pressure sustained by the soil sample under a rigid plate of specified pressure or settlement.
- Elastic modulus of subgrade and modulus of subgrade reaction are two different properties indicating the response of subgrade to the applied loads.

4.12 (a)

The cumulative number of standard axle load, N_s is given by,

$$N_s = \frac{365A\left[(1+r)^n - 1\right]}{r} \times F \times LDF$$

Where A = no. of commercial vehicles per day = 2000

r = annual growth rate = 7.5%

F = vehicle damage factor = 3

Lane distribution factor = 0.75

$$\therefore N_s = \frac{365 \times 2000\left[(1+0.075)^{15} - 1\right]}{0.075} \times 3 \times 0.75$$

$$N_s = 42.899 \times 10^6$$

4.13 (b)

Given, h = 20 cm, $E = 3 \times 10^5$ kg/cm^2

μ = 0.15, k = 5 kg/cm^3

$$L = \left[\frac{Eh^3}{12k(1-\mu^2)}\right]^{1/4}$$

$$= \left[\frac{3 \times 10^5 \times 20^3}{12 \times 5\left[1-(0.15)^2\right]}\right]^{1/4}$$

$$= 79.98 \text{ cm} \approx 80 \text{ cm}$$

POINTS TO REMEMBER

- A certain degree of resistance to slab deflection is offered by the sub-grade and the subgrade deformation is equal to the slab deflection.
- Relative stiffness of slab w.r.t. subgrade is represented by radius of relative stiffness (L).

4.14 (a)

Dowel bars are used in expansion joints and placed along the direction of movement of traffic.

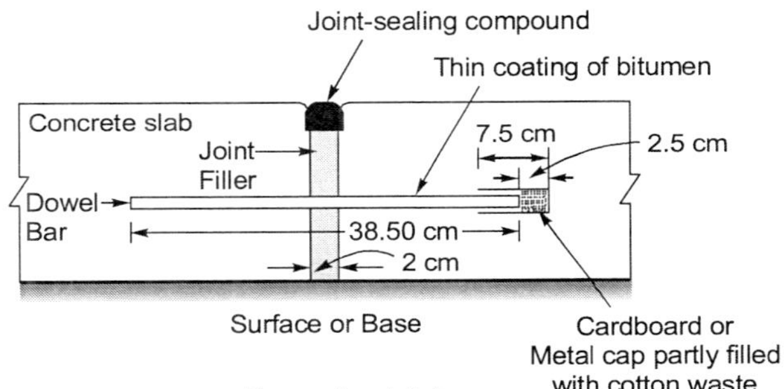

Expansion joint

On the other hand, tie bars are provided across longitudinal joints to transfer the load.

4.15 (a)

$$N_s = 365A\left[\frac{(1+r)^n - 1}{r}\right] \times F$$

$$= 365 \times 300\left[\frac{(1+0.07)^{10} - 1}{0.07}\right] \times 2.5 = 3.8 \text{ msa}$$

4.16 (b)

During the day time, temperature at the top of slab increase and at the bottom decreases. The slab at the bottom tries to move inward, hence tensile stress is developed at the bottom.

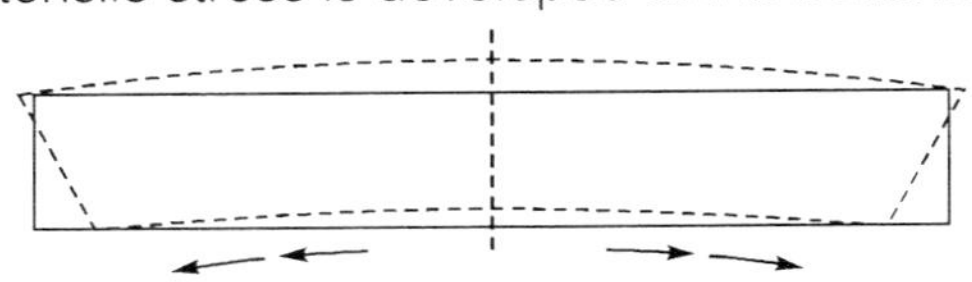

Slab during the day time

Load stress is tensile at bottom at inertia and edge of the slab.

4.17 (a)

The corners and edges are free to contract while cooling. Hence, the cracks due to stresses will develop at the centre of the slab.

POINTS TO REMEMBER

- Uniform cooling of slab will result in built-up of frictional stress. Frictional stress is nearly zero at ends and maximum at centre of slab. hence, cracking will occur at centre.

4.18 (a)

The modulus of elasticity of the subgrade soil is given by,

$$E = \frac{1.18\ pa}{\Delta}$$

$$= \frac{1.18 \times 800 \times 75 \times 10}{2.5 \times 2}$$

$$= 141600\ \text{N/cm}^2$$

$$= 141.6\ \text{kN/cm}^2$$

4.19 (d)

$$\text{Rigidity factor} = \frac{\text{Contact pressure (C.P.)}}{\text{Tyre pressure (T.P.)}}$$

R.F. < 1 when C.P. < 0.7 N/mm^2 and T.P. > 0.7 N/mm^2

4.20 (c)

The cumulative number of standard axle load,

$$N_s = \frac{365A\left[(1+r)^n - 1\right]}{r} \times F$$

$$= \frac{365 \times 1213 \times \left[\left(1+\frac{8}{100}\right)^{12} - 1\right]}{8/100} \times 2.5 \times 10^{-6}$$

$$= 21.00\ \text{msa}$$

4.21 (b)

There are four cases which should be considered:

(i) During summer, critical combination of stresses at edges
=wheel load stress due to edge loading
+ warping stress at edge region during summer
– frictional stress during summer
= 32 + 8 – 5 = 35 kg/cm^2

(ii) During winter, critical combination of stresses at edges
=wheel load stress due to edge loading
+ warping stress at edge region during winter
+ frictional stress during winter
=32 + 6 + 4 = 42 kg/cm^2

(iii) During summer, critical combination of stresses at corner regions
=wheel load stress at corner loading
+ warping stress at corner region during summer
=30 + 9 = 39 kg/cm^2

(iv) During winter, critical combination of stresses at corner region
=wheel load stress at corner loading
+ warping stress at corner region during winter
= 30 + 7 = 37 kg/cm^2

Hence, the critical combination is (ii) i.e. 42 kg/cm^2.

POINTS TO REMEMBER

- No effect of frictional stress has been taken in corner region, as frictional stress is nearly zero at corners. It has to be taken into account only for edge and interior of the slab.

4.22 (a)

From fourth power low

$$\text{Equivalent axle load factor, (EALF)} = \left(\frac{L_0}{L_s}\right)^4$$

Axle load	Avg. axle load	Repetitions per day
35 – 45	40	800
75 – 85	80	400

Total no. of repetitions in terms of standard axle load of 80 kN

$$= \left(\frac{40}{80}\right)^4 \times 800 + \left(\frac{80}{80}\right)^4 \times 400 = 450$$

4.23 (d)

Load stresses are proportional to wheel load. Modulus of subgrade reaction (K) is used to calculate the radius of relative stiffness (L) which eventually is used in load stress calculations.

4.24 (c)

Number of standard axles in msa is given by,

$$N_s = \frac{365\,A\left[(1+r)^n - 1\right]}{r \times 10^6} + F$$

$$= \frac{365 \times 2723 \times \left[\left(1+\frac{5}{100}\right)^{10} - 1\right]}{\frac{5}{100} \times 10^6} \times 2.4$$

$= 30$ msa

Now, for 5% CBR value, the total thickness of the pavement for 30 msa is given as 670 mm.

4.25 (c)

$$CBR_{2.5} = \frac{\text{Load sustained by specimen at 2.5 mm penetration}}{\text{Load sustained by standard aggregatesat 2.5 mm penetration}} \times 100$$

$$= \frac{60.5}{1370} \times 100 = 4.4\%$$

Also,

$$CBR_{5.0} = \frac{\text{Load sustained by specimend at 5.0 mm penetration}}{\text{Load sustained by standard aggregates at 5 mm penetration}} \times 100$$

$$= \frac{80.5}{2055} \times 100 = 3.9\%$$

Thus, the higher value of CBR which is obtained at 2.5 mm penetration is adopted i.e. 4.4%.

POINTS TO REMEMBER

- CBR value = Max $\{CBR_{2.5\,mm}, CBR_{5\,mm}\}$
- Generally, $CBR_{2.5\,mm} < CBR_{5\,mm}$, but if $CBR_{5\,m}$ comes out to be greater than $CBR_{2.5\,mm}$, then test is repeated and higher of the two is adopted as CBR value.

4.26 (b)

The spacing of expansion joint is given by,

$$L_e = \frac{\delta'}{\alpha(T_2 - T_1)}$$

Given $\delta' = \frac{\text{Width of expansion joint}}{2}$

$$= \frac{20}{2} = 10 \text{ mm} = 0.01 \text{ m}$$

$$\therefore \quad L_e = \frac{0.01}{10 \times 10^{-6}(60-20)} = 25 \text{ m}$$

4.27 (b)

$$N_{S_1} = \frac{365A_1\left[(1+r)^n - 1\right]}{r} \times F_1$$

$$= \frac{365 \times 2000\left[\left(1+\frac{7.5}{100}\right)^{10} - 1\right]}{\left(\frac{7.5}{100}\right) \times 10^6} \times 5$$

$= 51.6$ msa

$$N_{S_2} = \frac{365A_2\left[(1+r)^n - 1\right]}{r} \times F_2$$

$$= \frac{365 \times 200\left[\left(1+\frac{7.5}{100}\right)^{10} - 1\right]}{\left(\frac{7.5}{100}\right) \times 10^6} \times 6 = 6.2$$

$$\therefore \quad N_s = N_{S_1} + N_{S_2} = 51.6 + 6.2$$

$= 57.8$ msa

4.28 (b)

Number of commercial vehicles per day, A
= existing traffic × traffic distribution factor
= 2500 × (1.08) × 0.75 = 2025
Annual growth rate of commercial vehicles = 8%
Vehicle damage factor, $F = 3.5$
Design life, $n = 10$ years,
Thus, the cumulative standard axles (msa) may be calculated as

$$N_S = \frac{365\,A\left[(1+r)^n - 1\right]}{r} \times F$$

$$= \frac{365 \times 2025 \times \left[\left(1+\frac{8}{100}\right)^{10} - 1\right]}{\frac{8}{100}} \times 3.5$$

$= 37.47 \approx 37$

POINTS TO REMEMBER

- 'A' in the above formula represents the no. of commercial vehicles AFTER the completion of construction.

4.29 (d)

$$CBR(\%) = \frac{\text{Load sustained by specimen at 5.0 mm penetration}}{\text{Load sustained by standard aggregates at 5.0 mm penetration}} \times 100$$

$$= \frac{50}{2055} \times 100 = 2.4\%$$

4.30 (a)

4.31 (d)

Warping stresses in cement concrete pavements are caused by the daily variation in temperature. Tie bars are generally provided across longitudinal joints of cement concrete pavements.

4.32 (b)

Number of commercial vehicles per day,

$$A = 4500$$

Annual traffic growth rate,

$$r = 8\%$$

Design life,

$$n = 15 \text{ years}$$

The cumulative standard axles,

$$N_s = 100 \times 10^6$$

Vehicle damage factor used

$$= \frac{N_s \times r}{365A\left[(1+r)^n - 1\right]}$$

$$= \frac{100 \times 10^6 \times 0.08}{365 \times 4500 \times \left[(1+0.08)^{15} - 1\right]}$$

$$= 2.24$$

4.33 (c)

Rigid pavement transfers the load by bending action, so flexural strength of concrete used should be high.

4.34 (c)

The temperature stress will be tensile at the bottom of the slab during the day time if the self weight of the pavement slab is considered.

4.35 Sol.

Vehicle damage factor

$$= 0.1\left[\frac{18}{8.2}\right]^4 + 0.2\left[\frac{14}{8.2}\right]^4 + 0.35\left[\frac{10}{8.2}\right]^4$$

$$+0.15\left[\frac{8}{8.2}\right]^4 + 0.20\left[\frac{6}{8.2}\right]^4 = 4.988$$

$$N_s = \frac{365A\left((1+r)^n - 1\right)DF}{r}$$

$$= \frac{365 \times 5000\left((1.075)^{15} - 1\right) \times 1.3 \times 4.988}{0.075}$$

$$N_s = 309.08 \text{ msa}$$

4.36 (b)

Radius of relative stiffness,

$$l = \left[\frac{Eh^3}{12k(1-\mu^2)}\right]^{1/4}$$

As μ increases then $1 - \mu^2$ decreases

∴ Relative stiffness also increases.

∴ Statement 1 is false.

As modulus of subgrade reaction (k) increases, (l) decreases

∴ Statement 2 is false.

4.37 Sol.

Relative stiffness,

$$l = \left[\frac{Eh^3}{12k(1-\mu^2)}\right]^{1/4}$$

$$\Rightarrow \quad \frac{l_P}{l_Q} = \frac{\left(h^3/k\right)^{1/4}}{\left[\frac{(0.5h)^3}{2k}\right]^{1/4}} = 2$$

4.38 Sol.

$$N_s = \frac{365A\left[\left(1+\frac{r}{100}\right)^n - 1\right] \times VDF \times LDF}{\frac{r}{100}}$$

$$= \frac{365 \times 3000\left[1.06^{15} - 1\right] \times 4 \times 0.75}{0.06} \times 10^{-6}$$

$$= 76.45 \text{ msa}$$

4.39 (d)

Frictionless base,

$$\sigma_{thermal} = 0$$

4.40 Sol.

$$L_T = \frac{\phi\sigma_{st}}{2\tau_{bd}}$$

$$= \frac{12\text{mm} \times 230 \text{ MPa}}{2 \times 2 \text{ MPa}} = 690 \text{ mm}$$

The design length of tie bar

$$= 690 + 10 = 700 \text{ mm}$$

4.41 (a)

4.42 (c)

(i) 80 buses with 2 axle with 40 kN each.

$$N_1 = 80 \times 2 = 160$$
$$L_1 = 40 \text{ kN}$$

(ii) 160 trucks: Front axle → 40 kN; Rear axle → 80 kN

$$N_1 = 160,\ L_1 = 40 \text{ kN}$$
$$N_2 = 160,\ L_2 = 80 \text{ kN}$$

Total no. of repetitions: For 40 kN = 160 + 160 = 320; For 80 kN = 160

$$N_1 = 320,\ L_1 = 40 \text{ kN}$$
$$N_2 = 160,\ L_2 = 80 \text{ kN}$$

As per 4th power law

$$N_s = N_1\left(\frac{L_1}{L_5}\right)^4 + N_2\left(\frac{L_2}{L_s}\right)^4$$

$$= 320\left(\frac{40}{80}\right)^4 + 160\left(\frac{80}{80}\right)^4$$

$$= 20 + 160 = 180$$

4.43 (b)

4.44 Sol.

Axle load = 15 T

Standard axle load = 8 T

Velocity damage factor,

$$\text{VDF} = \left[\frac{15}{8}\right]^4 = 12.35$$

4.45 Sol.

$$\text{ESWL @ depth } \frac{d}{2} = P$$

$$\text{ESWL @ depth } 2S = 2P$$

$$\therefore \quad \text{Required ratio} = \frac{P}{2P} = \frac{1}{2} = 0.5$$

4.46 Sol.

Number of commercial vehicles = 800 vehicles per day

For vehicle growth rate of 4%,

Number of commercial vehicles after construction period,

$$A = 800 \times [1 + 0.04]^5$$
$$= 973.32 \text{ veh}$$

Vehicle damage factor,

$$F = \left(\frac{L_0}{L_5}\right)^4 = \left(\frac{12000}{8160}\right)^4 = 4.68$$

$\therefore$ Cumulative standard axle (in msa)

$$= \frac{365\,\text{ADF}\left((1+r)^n - 1\right)}{r \times 10^6} \times \text{LSF}$$

$$= \frac{365 \times 973.32 \times 1 \times 4.68 \times \left[1.04^{15} - 1\right]}{0.04 \times 10^6} = 33.29 \text{ msa}$$

4.47 (b)

During summer at night, maximum stress combination occurs due to load and warping stresses at corner of pavement slab.

■■■■

5 Highway Maintenance

5.1 In highway pavements emulsions are mainly used in
(a) surface dressing
(b) patching and maintenance
(c) bitumen macadam
(d) asphaltic concrete

[1995 : 1 M]

5.2 An important purpose of prime coats is to
(a) promote the bond between the base and the wearing courses
(b) promote the adhesion between an existing wearing surface and a subsequent wearing surface
(c) promote the bond between the subbase course and the subgrade
(d) increases the stability of the subgrade

[1996 : 1 M]

5.3 One of the probable causes of rutting on flexible pavements is
(a) excessive stripping of binder material from the wearing course
(b) use of flaky aggregates in the wearing course
(c) inadequate compaction of pavement layers
(d) high wind speeds

[1996 : 1 M]

5.4 Mud pumping is commonly associated with
(a) bituminous penetration macadam construction
(b) cement concrete pavement on granular subgrade
(c) premixed bituminous construction
(d) cement concrete pavement on clay subgrade

[1997 : 1 M]

5.5 Alligator or map cracking is the common type of failure in
(a) concrete pavements
(b) bituminous surfacing
(c) gravel roads
(d) WBM construction

[1997 : 1 M]

5.6 Road roughness is measured using
(a) Benkelman beam
(b) Bump integrator
(c) Dynamic cone penetrometer
(d) Falling weight deflectormeter

[2012 : 1 M]

5.7 As per IRC : 37-2012, in order to control subgrade rutting in flexible pavements, the parameter to be considered is
(a) horizontal tensile strain at the bottom of bituminous layer
(b) vertical compressive strain on top of subgrade
(c) vertical compressive stress on top of granular layer
(d) vertical deflection at the surface of the pavement

[2018 : 1 M, Set-II]

■■■■

Answers Highway Maintenance

5.1 (b) 5.2 (b) 5.3 (c) 5.4 (d) 5.5 (b) 5.6 (b) 5.7 (b)

Explanations Highway Maintenance

5.1 (b)

Emulsions are used in bituminous road constructions, especially in maintenance and patch repair works. The main advantages of emulsion is that it can be used in wet weather even when it is raining.

5.2 (b)

Bituminous prime coat is the first application of low viscosity liquid bituminous material over an existing porous or absorbent pavement surface like the WBM base course. The main object of priming is to plug in the capillary voids of the loose mineral particles on the existing surface using a binder of low viscosity which can penetrate into capillary voids and promote the adhesion between existing wearing surface and subsequent wearing surface.

5.3 (c)

Probable causes of rutting are heavy channelisd traffic, inadequate compactionof pavement layers, poor stability of pavement material or heavy bullock cart traffic.

5.4 (d)

Mud pumping is recognized when the soil slurry ejects out through the joints and cracks of cement concrete pavement caused during the downward movement of slab under the heavy wheel loads. Pumping is noticed just after the rains in cement concrete pavements that are placed on clayey soil subgrade.

5.5 (b)

Alligator or map cracking is the most common type of failure and occurs due to relative movement of pavement layer materials. This is most common in bituminous surfacing.

5.6 (b)

Bankelmann beam is used to measure the deflection of a flexible pavement under moving wheel loads.

Falling weight deflectometer is used to estimate pavement structural capacity. Bump integrater is used to measure road roughness.

5.7 (b)

As per IRC : 37-2012

$$N = 4.1656 \times 10^{-0.08}\left[\frac{1}{\epsilon_V}\right]^{4.5337}$$

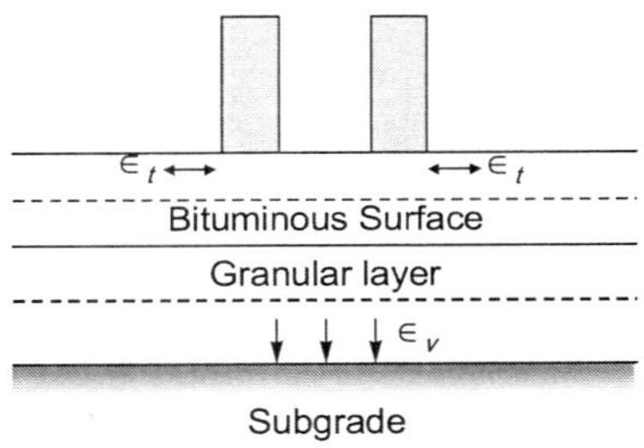

$$N = 1.41 \times 10^{-8}\left[\frac{1}{\epsilon_V}\right]^{4.5337}$$

where N = No. of cumulative standard axle

ϵ_v = Vertical strain in the subgrade

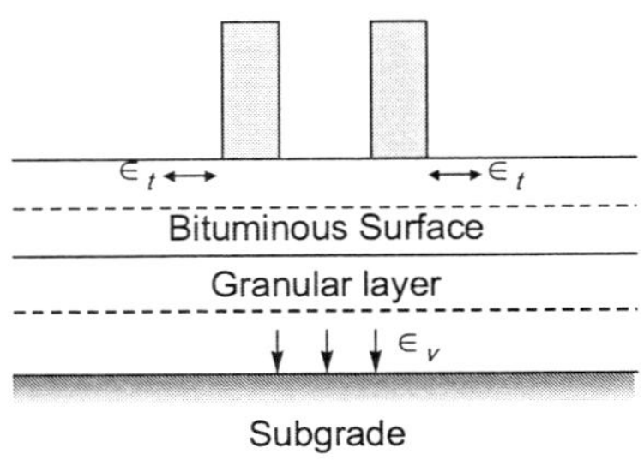

As evident from exprossion, vertical strain in subgrade is the only variable considered for rutting in flexible pavements.

■■■■

6 Railway & Airport Engineering

6.1 A runway is being constructed in a new airport as per the International Civil Aviation Organization (ICAO) recommendations. The elevation and the airport reference temperature of this airport are 535 m above the mean sea level and 22.65°C, respectively. Consider the effective gradient of runway as 1%. The length of runway required for a design-aircraft under the standard conditions is 2000 m. Within the framework of applying sequential corrections as per the ICAO recommendations, the length of runway corrected for the temperature is

(a) 2223 m (b) 2250 m
(c) 2500 m (d) 2750 m

[2017 : 1 M, Set-I]

6.2 For a broad gauge railway track on a horizontal curve of radius R (in m), the equilibrium cant e required for a train moving at a speed of V (in km per hour) is

(a) $e = 1.676\frac{V^2}{R}$ (b) $e = 1.315\frac{V^2}{R}$

(c) $e = 0.80\frac{V^2}{R}$ (d) $e = 0.60\frac{V^2}{R}$

[2017 : 1 M, Set-II]

6.3 A broad gauge railway line passes through a horizontal curved section (radius = 875 m) of length 200 m. The allowable speed on this portion is 100 km/h. For calculating the cant, consider the gauge as centre-to-centre distance between the rail heads, equal to 1750 mm. The maximum permissible cant (in mm, round off to 1 decimal place) with respect to the centre-to-centre distance between the rail heads is________

[2019 : 2 M, Set-II]

6.4 The appropriate design length of a clearway is calculated on the basis of 'Normal Take-off' condition. Which one of the following options correctly depicts the length of the clearway? (Note: None of the option are drawn to scale)

(a) 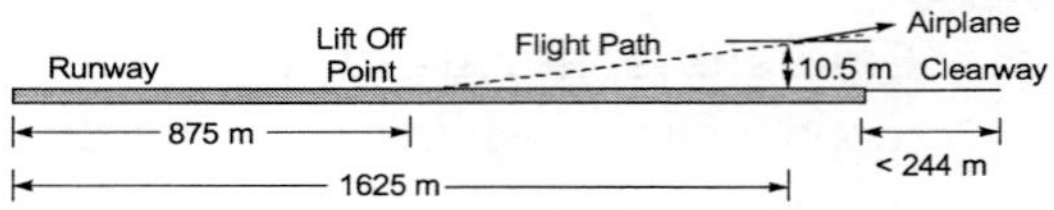

(b) 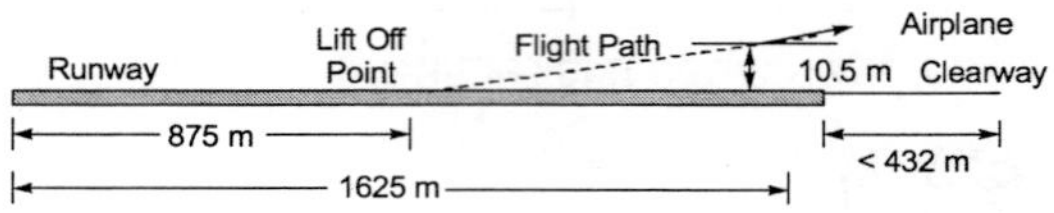

(c) 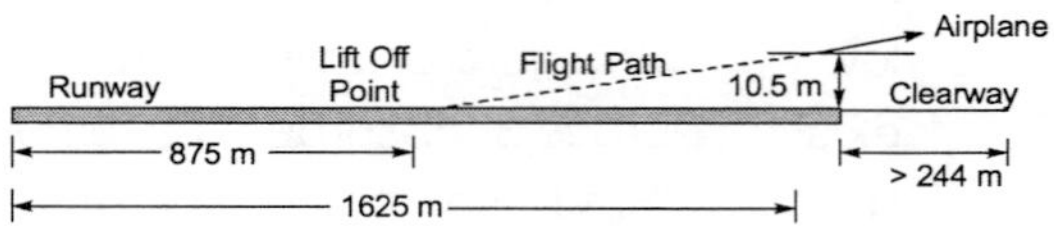

(d) 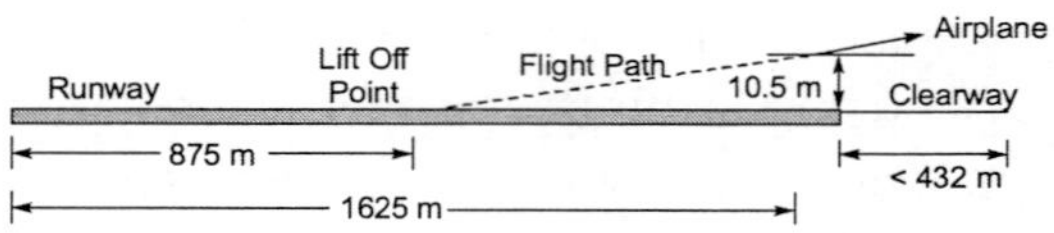

[2020 : 2 M, Set-I]

6.5 For the hottest month of the year at the proposed airport site, the monthly mean of the average daily temperature is 39°C. The monthly mean of the maximum daily temperature is 48°C for the same month of the year. From the given information, the calculated Airport Reference Temperature (in°C), is

(a) 42 (b) 39
(c) 36 (d) 48

[2020 : 2 M, Set-II]

6.6 The longitudinal section of a runway provides the following data:

End-to-end runway (m)	Gradient (%)
0 to 300	+1.2
300 to 600	–0.7
600 to 1100	+0.6
1100 to 1400	–0.8
1400 to 1700	–1.0

The effective gradient of the runway (in %, round off to two decimal places) is ________

[2021 : 2 M, Set-I]

6.7 For a 2° curve on a high speed Broad Gauge (BG) rail section, the maximum sanctioned speed is 100 km/h and the equilibrium speed is 80 km/h. Consider dynamic gauge of BG rail as 1750 mm, The degree of curve is defined as the angle subtended at its center by a 30.5 m arc. The cant deficiency for the curve (in mm. round off to integer) is ________.

[2021 : 2 M, Set-II]

6.8 The base length of the runway at The mean sea level (MSL) is 1500 m. If the runway is located at an altitude of 300 m above the MSL, the actual length (in m) of the runway To be provided is ________. (round off to the nearest integer)

[2022 : 2 M, Set-II]

6.9 As per the International Civil Aviation Organization (ICAO), the basic runway length is increased b x (%) for every y (m) raise in elevation from the Mean Sea Level (MSL) The values of x and y respectively, are

(a) 10% and 1000 m
(b) 4% and 500 m
(c) 5% and 200 m
(d) 7% and 300 m

[2024 : 1 M, Set-I]

6.10 The number of trains and their corresponding speeds for a curved Broad Gauge section with 437 m radius are

- 20 trains at a speed of 40 kmph
- 15 trains at a speed of 50 kmph
- 12 trains at a speed of 60 kmph
- 8 trains at a speed of 70 kmph
- 3 trains at a speed of 80 kmph

If the gauge (centre to centre distance between the rail heads) is taken as 1750 mm, the required equilibrium cant (in mm) will be ________. (rounded off to the nearest integer)

[2024 : 2 M, Set-I]

6.11 In general, the outer edge is raised above the inner edge in horizontal curves for

(a) Railways and Taxiways only
(b) Highways, Railways, and Taxiways
(c) Highways and Railways only
(d) Highways only

[2024 : 1 M, Set-II]

6.12 A horizontal curve of radius 1080 m (with transition curves on either side) in a Broad Gauge railway track is designed and constructed for an equilibrium speed of 70kmph. However, a few years after construction, the Railway Authorities decided to run express trains on this track. The maximum allowable cant deficiency is 10 cm. The maximum restricted speed (in kmph) of the express trains running on this track is ________ (rounded off to the nearest integer).

[2024 : 2 M, Set-II]

6.13 The longitudinal sections of a runway have gradients as shown in the table.

End to end for sections of runway (m)	Gradient (%)
0 to 200	+ 1.0
200 to 600	– 1.0
600 to 1200	+ 0.8
1200 to 1600	+ 0.2
1600 to 2000	– 0.5

Consider the reduced level (RL) at the starting point of the runway as 100 m.

The effective gradient of the runway is

(a) 0.28%
(b) 0.18%
(c) 0.02%
(d) 0.35%

[2024 : 1 M, Set-II]

■■■■

Answers Railway & Airport Engineering

6.1 (c) 6.2 (b) 6.3 (157.5) 6.4 (b, d) 6.5 (a) 6.6 (0.32) 6.7 (57)
6.8 (1605) 6.9 (d) 6.10 (88) 6.11 (c) 6.12 (112) 6.13 (a)

Explanations Railway & Airport Engineering

6.1 (c)

Elevation = 535 m

Airport Reference Temperature = 22.65° C

Effective Gradient = 1%

Runway length = 2000 m under standard conditions.

Runway length correction for T°C.

⇒ Correction for elevation

$$= \frac{7}{100} \times 2000 \times \frac{535}{300} = 249.67 \text{ m}$$

Corrected runway length = 2249.67 m

⇒ Corrected standard T° at 535 m elevation

= 15°C – (0.0065 × 535) = 11.52°C

Correction for T°

$$= \frac{1}{100} \times 2249.67 \times \frac{\Delta T°C}{1°C}$$

$$\Delta T°C = 22.65 - 11.52 = 11.13°C$$

$$= \frac{1}{100} \times 2249.67 \times \frac{11.13°c}{1°c}$$

$$= 250.39 \text{ m}$$

Corrected Runway length

$$= 2249.67 + 250.39$$

$$= 2500.06 \text{ m} \simeq 2500 \text{ m}$$

Combined elevation and temperature correction

$$= 500 \text{ m} < (0.35 \times 2000) \text{ (OK)}$$

POINTS TO REMEMBER

- ICAO recommends combined correction for elevation and temperature should not exceed 35% of basic runway length.
- However, if the combined correction exceeds 35%, correction is checked by specific studies at the site.

6.2 (b)

$$e \text{ (in cm)} = \frac{GV^2}{127R} = \frac{1.676V^2}{127R} \times 100 = 1.315\ \frac{V^2}{R}$$

6.3 Sol.

Allowable speed is given as 100 kmph

$$e_{th} = \frac{GV_{max}^2}{127R}$$

$$e_{th} = \frac{1.750 \times 100^2}{127 \times 875}$$

$$e_{th} = 15.75 \text{ cm}$$

$$= 157.5 \text{ mm}$$

6.4 (b, d)

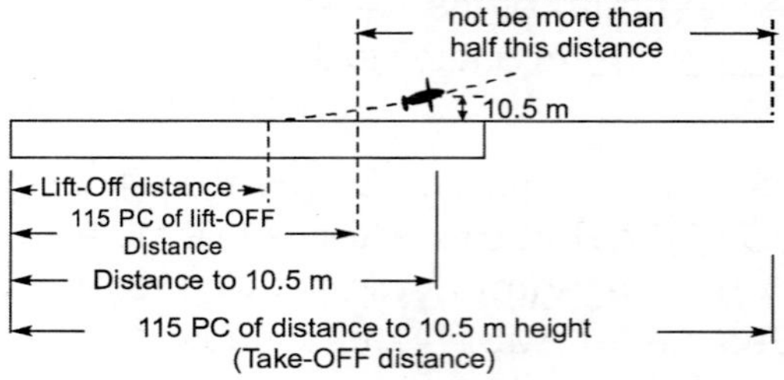

LONGITUDINAL SECTION

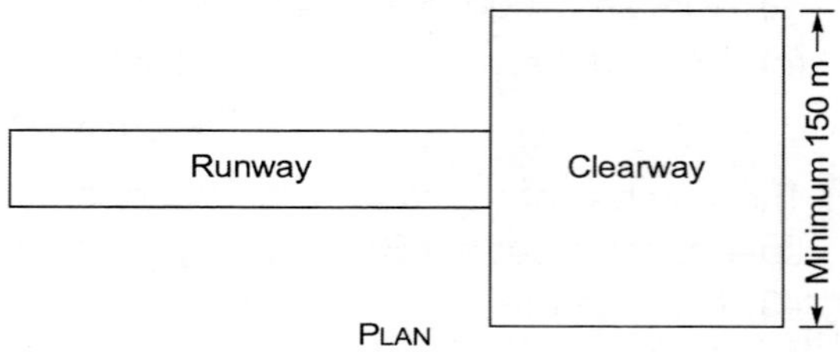

PLAN

$$\text{Clearway} \not> \frac{1}{2}\left[\frac{115}{100} \times 1625 - \frac{115}{100} \times 875\right]$$

$$\not> 431.25 \text{ m}$$

6.5 (a)

Average daily temperature,

$$T_a = 39°C$$

Maximum daily temperature,

$$T_m = 48°C$$

Airport reference temperature,

$$\text{ART} = T_a + \left(\frac{T_m - T_a}{3}\right)$$

$$= 39 + \left(\frac{48 - 39}{3}\right) = 42°C$$

6.6 Sol.

Assuming RL of start of runway as datum i.e., RL = 0 m)

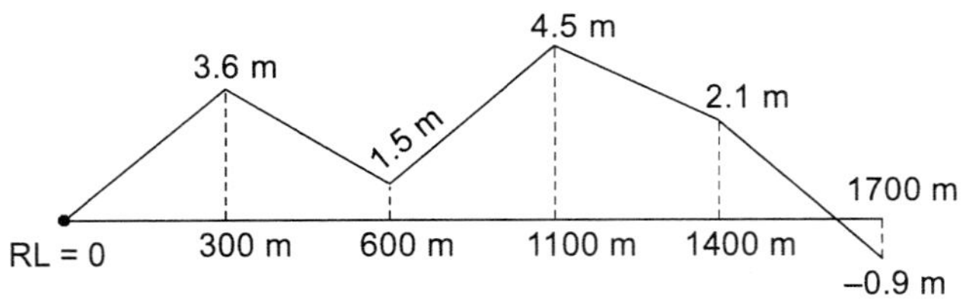

Effective gradient

$$= \left[\frac{\text{Maximum difference in reduced level}}{\text{Total runway length}}\right]$$

$$= \left[\frac{4.5-(-0.9)}{1700}\times 100\right]\%$$

$$= 0.3176\% \simeq 0.32\%$$

6.7 Sol.

Length of curve = Radius × Degree of curve

$$\frac{30.5\times 180}{2°\times\pi} = R$$

$$R = 873.76 \text{ m}$$

$$C_d = C_{th} - C_{act}$$

$$C_{act} = \frac{GV^2}{127R} = \frac{1750\times 100^2}{127\times 873.76}$$

$$= 157.70 \text{ mm}$$

$$C_{ev} = \frac{GV^2}{127R} = \frac{1750\times 80^2}{127\times 873.76}$$

$$= 100.930 \text{ mm}$$

$$C_{def} = C_{act} - C_{th}$$

$$= 157.70 - 100.930$$

$$= 56.77 \text{ mm} \simeq 57 \text{ mm}$$

6.8 Sol.

Basic runway length = 1500 m (at MSL)
Elevation of airport = 300 m above MSL
We know as per elevation correction the length of should be increase by 7% per 300 m elevation above MSL.
Correction due to elevation

$$= \frac{0.07}{300}\times 300\times 1500 = 105 \text{ m}$$

∴Corrected length = 1605 m

6.9 (d)

6.10 Sol.

Average speed of trains,

$$V_{avg} = \frac{20(40)+15(50)+12(60)+8(70)+3(80)}{20+15+12+8+3}$$

$$= \frac{3070}{58} = 52.93 \text{ kmph}$$

Gauge (G) = 1750 mm
∴Equilibrium cant,

$$e_{th} = \frac{GV_{avg}^2}{127\,R} = \frac{1750\times(52.93)^2}{127\times 437}$$

$$= 88.34 \text{ mm}$$

$$= 88 \text{ mm (to nearest integer)}$$

6.11 (c)

In highways and railways, the outer edge is raised above the inner edge in horizontal curve.

6.12 Sol.

$$R = 1080 \text{ m}$$
$$V_{avg} = 70 \text{ km/h}$$
$$CD = 10 \text{ cm}$$
$$G = 1.75 \text{ m}$$

Now, $e_{theoretical} = e_{act.} + \text{CD}$

$$\Rightarrow \quad e_{theoretical} = \frac{GV^2}{127R} + CD$$

$$\Rightarrow \quad \frac{GV_{max}^2}{127R} = \frac{1.75\times 70^2}{127\times 1080} + 0.1 = 0.1625$$

$$\Rightarrow \quad V_{max.} = 112.855 \text{ km/h}$$

6.13 (a)

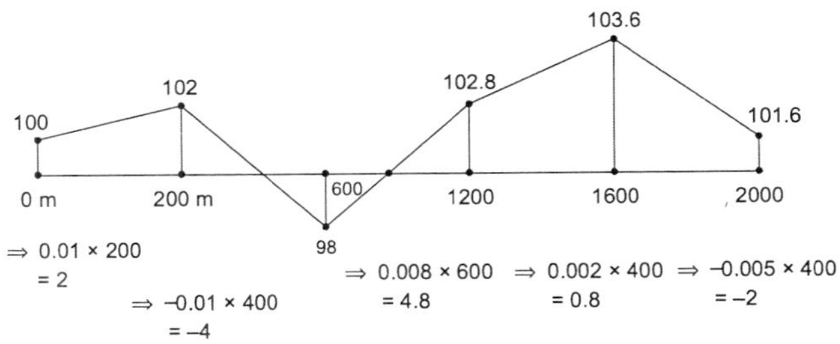

$$\text{Effectie gradient} = \frac{\text{Highest elevation} - \text{Lowest elevation}}{\text{Length}}$$

$$= \frac{103.6-98}{2000}\times 100 = 0.275\% \approx 0.28\%$$

Geometics Engineering

UNIT XI

CONTENTS

Geometics Engineering

Syllabus

Principles of surveying; Errors and their adjustment; Maps - scale, coordinate system; Distance and angle measurement - Levelling and trigonometric levelling; Traversing and triangulation survey; Total station; Horizontal and vertical curves.

Photogrammetry and Remote Sensing - Scale, flying height; Basics of remote sensing and GIS.

Analysis of Previous GATE Papers

Exam Year	1 Mark Ques.	2 Marks Ques.	Total Marks
1995	–	–	–
1996	–	–	–
1997	–	–	–
1998	–	–	–
1999	1	–	1
2000	–	–	–
2001	–	–	–
2002	–	–	–
2003	–	–	–
2004	–	–	–
2005	–	–	–
2006	–	5	10
2007	–	5	10
2008	1	4	9
2009	1	2	5
2010	1	2	5
2011	1	1	3
2012	1	1	3
2013	1	2	5
2014 Set-1	1	2	5
2014 Set-2	1	2	5

Exam Year	1 Mark Ques.	2 Marks Ques.	Total Marks
2015 Set-1	2	2	6
2015 Set-2	2	2	6
2016 Set-1	3	1	5
2016 Set-2	3	1	5
2017 Set-1	2	1	4
2017 Set-2	2	2	6
2018 Set-1	–	2	4
2018 Set-2	–	2	4
2019 Set-1	1	2	5
2019 Set-2	1	2	5
2020 Set-1	–	2	4
2020 Set-2	–	2	4
2021 Set-1	2	1	4
2021 Set-2	2	1	4
2022 Set-1	3	–	3
2022 Set-2	1	1	3
2023 Set-1	2	1	4
2023 Set-2	2	1	4
2024 Set-1	3	2	7
2024 Set-2	3	2	7

1 Fundamental Concepts of Surveying and Linear Measurement

1.1 The plan of a map was photo copied to a reduced size such that a line originally 100 mm, measures 90 mm. The original scale of the plan was 1 : 1000. The revised scale is

(a) 1 : 900 (b) 1 : 1111
(c) 1 : 1121 (d) 1 : 1221

[2007 : 2 M]

1.2 The type of surveying in which the curvature of the earth is taken into account is called

(a) Geodetic surveying
(b) Plane surveying
(c) Preliminary surveying
(d) Topographical surveying

[2008 : 1 M]

1.3 The plan of a survey plotted to a scale of 10 m to 1 cm is reduced in such a way that a line originally 10 cm long now measures 9 cm. The area of the reduced plan is measured as 81 cm^2. The actual area (m^2) of the survey is

(a) 10000 (b) 6561
(c) 1000 (d) 656

[2008 : 2 M]

1.4 The survey carried out to delineate natural features, such as hills, rivers, forests and man-made features, such as towns, villages, buildings, roads, transmission lines and canals is classified as

(a) engineering survey
(b) geological survey
(c) land survey
(d) topographic survey

[2014 : 1 M, Set-II]

1.5 The accuracy of an Electronic Distance Measuring Instrument (EDMI) is specified as ± (a mm + b ppm). Which one of the following statements is correct?

(a) Both a and b remain constant, irrespective of the distance being measured.
(b) a remains constant and b varies in proportion to the distance being measured.
(c) a varies in proportion to the distance being measured and b remains constant.
(d) Both a and b vary in proportion to the distance being measured.

[2017 : 1 M, Set-I]

1.6 A survey line was measured to be 285.5 m with a tape having a nominal length of 30 m. On checking, the true length of the tape was found to be 0.05 m too short. If the line lay on a slope of 1 in 10. the reduced length (horizontal length) of the line for plotting of survey work would be

(a) 285.6 mm (b) 284.5 mm
(c) 283.6 mm (d) 285.0 mm

[2019 : 2 M, Set-I]

1.7 Which of the following is NOT a correct statement?

(a) Basic principle of surveying is to work from whole to parts.
(b) Contours of different elevations may intersect each other in case of an overhanging cliff.
(c) Planimeter is used for measuring 'area'.
(d) The first reading from a level station is a 'Fore Sight'.

[2021 : 1 M, Set-I]

1.8 In general, the CORRECT sequence of surveying operations is

(a) Data analysis → Reconnaissance → Field observations → Map making
(b) Reconnaissance → Field observations → Data analysis → Map making
(c) Reconnaissance → Data analysis → Field observations → Map making
(d) Field observations → Reconnaissance → Data analysis → Map making

[2021 : 1 M, Set-II]

1.9 A line between stations P and Q laid on a slope of 1 in 5 was measured as 350 m using a 50 m tape. The tape is known to be short by 0.1 m.
The corrected horizontal length (in m) of the line PQ will be

(a) 342.52 (b) 349.30
(c) 356.20 (d) 350.70

[2022 : 1 M, Set-I]

■■■■

Answers Fundamental Concepts of Surveying and Linear Measurement

1.1 (b)	1.2 (a)	1.3 (a)	1.4 (d)	1.5 (b)	1.6 (c)	1.7 (d)
1.8 (b)	1.9 (a)					

Explanations Fundamental Concepts of Surveying and Linear Measurement

1.1 (b)

$$\text{Reduction factor} = \frac{90}{100} = 0.9$$

$$\text{Revised scale} = \text{Original scale} \times \text{Reduction factor}$$

$$= \frac{1}{1000} \times 0.9 = \frac{1}{1000/0.9} = \frac{1}{1111}$$

1.2 (a)

Geodetic surveying is that type of surveying in which the shape of the earth is taken into account. **Geodetic surveying** involves spherical trigonometry.

Whereas in plan suveying, the mean surface of earth is assumed tobe flat and not the curved one. This type of survey is used when the area under considerationis of small extent.

1.3 (a)

$$\text{Shrinkage factor} = \frac{9}{10} = 0.9$$

$$\text{Reduced plan area} = (\text{Shrinkage factor})^2 \times \text{Actual plan area}$$

$$\Rightarrow \quad 81 = (0.9)^2 \times \text{Actual plan area}$$

$$\Rightarrow \text{Actual plan area} = 100 \text{ cm}^2$$

$\therefore$ Actual area of survey in m^2

$$= 100 \times (10)^2 = 10000$$

POINTS TO REMEMBER

- $\text{Shrinkage ratio} = \dfrac{\text{Shrink length}}{\text{Actual length}}$
- $(\text{Shrinkage ratio})^2 = \dfrac{\text{Measured area}}{\text{Actual area}}$
- (Correct length) × (Correct scale) = (Wrong scale) × (Measured length)

1.4 (d)

Tropographic survey is done to determine the natural features of a country such as rivers, strains, lakes, wood, hills etc. and artificial features such as roads, railways, canals, towns and villages.

POINTS TO REMEMBER

- **Engineering survey** is done to acquire information for the planning and design of engineering projects like highways, dams, runway line, etc. It also involves topographic survey of the area, earthwork measurement, etc.
- **Geological survey** involves acquiring information about both the surface and sub-surface for assessing the extent of different reserves like minerals, rocks, etc. It helps in determining faults, folds and also the type of foundation, etc.
- **Land survey,** this survey (done on land) involves running of survey lines and determining their length and directions.

1.5 (b)

Accuracy of **EDMI** is generally stated in terms of constant instruments error and a measuring error proportional to the distance being measured: $\pm$ (a mm + b ppm). The first part indicates a constant instrument error that is independent of the length of the line measured. The second component is the distance related error.

1.6 (c)

Given:

Measured length = 285.5 m

Designated tape length = 30 m

Actual tape length = 30 – 0.05 = 29.95 m

Slope along measurement = 1 in 10.

1. **Standardization correction,**

   Actual tape length $l' < l$ (hence negative correction)

Correction per tape length

$= 29.95 - 30 = -0.05$ m

$$\text{Total correction} = \frac{285.5}{30} \times 0.05$$

$= 0.4758$ (negative)

2. **Slope correction:**

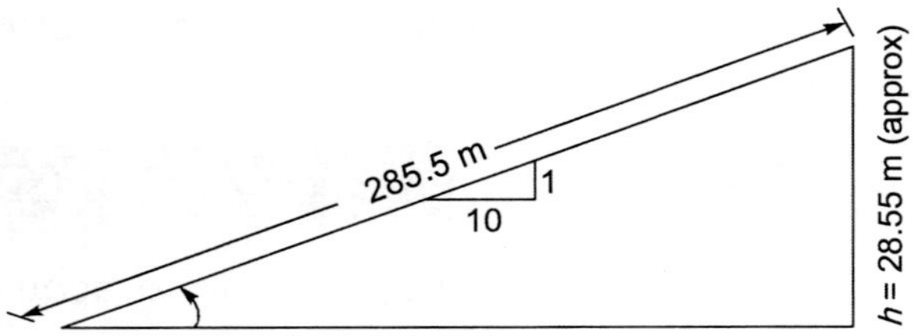

Approximate slope correction

$$= -\frac{h^2}{2l} = \frac{28.55^2}{285.5 \times 2}$$

$= 1.4275$ (negative)

Total correction $= -(1.4275 + 0.4758)$

$= -1.9033$

Correct length = Measured length − Correction

$= 285.5 - 1.9033$

$= 283.596$

$= 283.6$ m

1.7 (d)

First reading from level station is called BS.

1.8 (b)

Reconnaissance → Field observations → Data analysis → Map making

1.9 (a)

Slope = 1 in 5

Measured length,

$l' = 350$ m

Tape length = 50 m

Slope correction (C_{slope}):

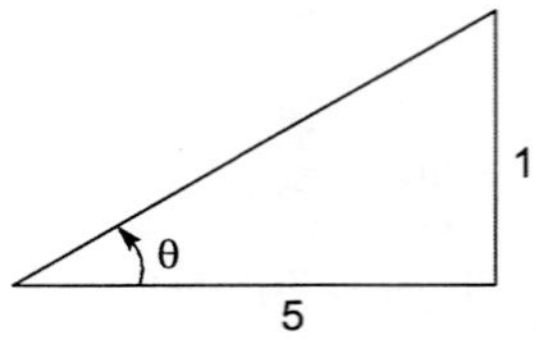

$\therefore \quad \theta = \tan^{-1}\left(\frac{1}{5}\right) = 11.31°$

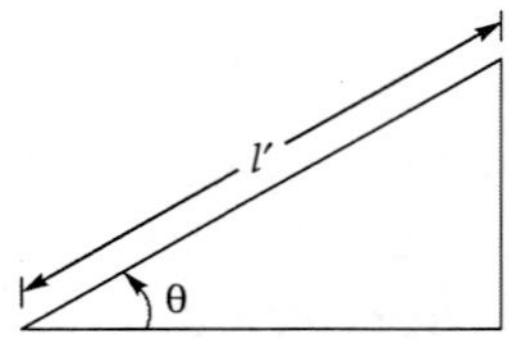

$\therefore \quad C_{slope} = l'\cos\theta - l'$

$= 350(\cos 11.31 - 1) = -6.8$ m

Tape correction, (C_{length}):

True length, $L = 50$ m

$L' = 50 - 0.1 = 49.9$ m

$l' = 350$ m

$$\therefore \quad l = \frac{L'}{L} \times l' = \frac{49.9}{50} \times 350$$

$\Rightarrow \quad l = 349.3$ m

$C_{length} = 349.3 - 350 = -0.7$ m

True length = Measured length + C_{net}

$= 350 + [-6.8 - 0.7] = 342.5$ m

■■■■

2 Compass and Traverse Surveying

2.1 The observed magnetic bearing of a line *OE* was found to be 185°. It was later discovered that station *O* had a local attraction of +1.5°. The true bearing of the line *OE*, considering a magnetic declination of 3.5°E shall be

(a) 180° (b) 187°
(c) 190° (d) 193° [2006 : 2 M]

2.2 In the figure given below, the lengths *PQ* (WCB : 30°) and *QR* (WCB : 45°) respectively up to three places of decimal are

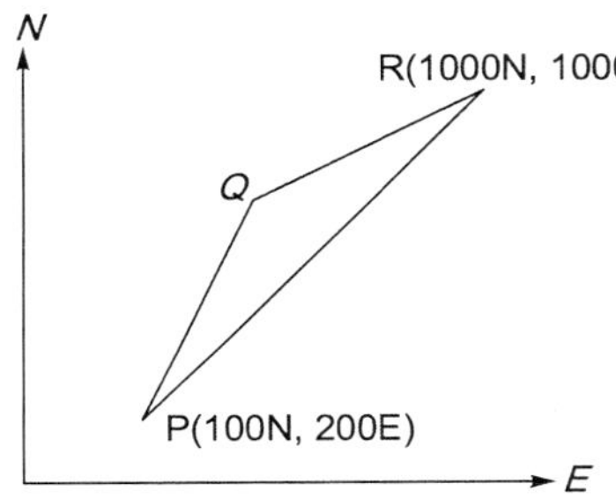

(a) 273.205, 938.186
(b) 273.205, 551.815
(c) 551.815, 551.815
(d) 551.815, 938.186 [2006 : 2 M]

2.3 The following table gives data of consecutive coordinates in respect of a closed theodolite traverse *PQRSP*.

Station	Northing (m)	Southing (m)	Easting (m)	Westing (m)
P	400.75	–	–	300.5
Q	100.25	–	199.25	–
R	–	199.0	299.75	–
S	–	300.0	–	200.5

The magnitude and direction of error of closure in whole circle bearing are

(a) 2.0 m and 45° (b) 2.0 m and 315°
(c) 2.82 m and 315° (d) 3.42 m and 45°
[2007 : 2 M]

2.4 The magnetic bearing of a line *AB* is S 45°E and the declination is 5° West. The true bearing of the line *AB* is

(a) S 45°E (b) S 40°E
(c) S 50°E (d) S 50°W [2007 : 2 M]

2.5 The lengths and bearings of a closed traverse *PQRSP* are given below.

Line	Length (m)	Bearing (WCB)
PQ	200	0°
QR	1000	45°
RS	907	180°
SP	?	?

The missing length and bearing, respectively of the line SP are

(a) 207 m and 270° (b) 707 m and 270°
(c) 707 m and 180° (d) 907 m and 270°
[2008 : 2 M]

2.6 In quadrantal bearing system, bearing of a line varies from

(a) 0° to 360° (b) 0° to 180°
(c) 0° to 90° (d) 0° N to 90°S
[2009 : 1 M]

2.7 The magnetic bearing of a line *AB* was N 59° 30′ W in the year 1967, when the declination was 4° 10′ E. If the present declination is 3° W, the whole circle bearing of the line is

(a) 299° 20′ (b) 307° 40′
(c) 293° 20′ (d) 301° 40′
[2009 : 2 M]

2.8 The observations from a closed loop traverse around an obstacle are

Segment	Observation from station	Length (m)	Azimuth (Clockwise from magnetic north)
PQ	*P*	Missing	33.7500°
AR	*Q*	300.00	86.3847°
RS	*R*	354.524	169.3819°
ST	*S*	450.000	243.9003°
TP	*T*	268.000	317.5000°

What is the value of the missing measurement (rounded off to the nearest 10 mm)?

(a) 396.86 m (b) 396.79 m
(c) 396.05 m (d) 396.94 m
[2011 : 2 M]

2.9 The latitude and departure of a line *AB* are +78 m and –45.1 m respectively. The whole circle bearing of the line *AB* is

(a) 30° (b) 150°
(c) 210° (d) 330° [2013 : 1 M]

2.10 Bearing of the given system is shown below:

Line	Fore bearing	Back bearing
AB	126°45′	308°00′
BC	45° 15′	227° 30′
CD	340° 30′	161° 45′
DE	258° 30′	78° 30′
EA	216° 30′	31° 45′

Applying correction due to local attraction, the correct bearing of line *BC* will be

(a) 48°15° (b) 50°15′
(c) 49°15′ (d) 48°45′ [2013 : 2 M]

2.11 A Theodolite is placed at *A* and a 3 m long vertical staff is held at *B*. The depression angle made at reading of 2.5 m marking on staff is 6°10′. The horizontal distance between *A* and *B* is 2200 m. The height of instrument at *A* is 1.1 m and reduced level of point *A* is 880.88 m using curvature correction and refraction correction determine the R.L. of point B(in m)?

[2013 : 2 M]

2.12 **Group-I** lists tool/instrument while **Group-II** lists the method of surveying. Match the tool/instrument with the corresponding method of surveying.

	Group-I		Group-II
P.	Alidade	1.	Chain surveying
Q.	Arrow	2.	Levelling
R.	Bubble tube	3.	Plain table surveying
S.	Stadia hair	4.	Theodolite surveying

	P	Q	R	S
(a)	3	2	1	4
(b)	2	4	3	1
(c)	1	2	4	3
(d)	3	1	2	4

[2014 : 2 M, Set-I]

2.13 In a closed loop traverse of 1 km total length, the closing errors in departure and latitude are 0.3 m and 0.4 m, respectively. The relative precision of this traverse will be

(a) 1 : 5000 (b) 1 : 4000
(c) 1 : 3000 (d) 1 : 2000

[2015 : 1 M, Set-I]

2.14 In a region with magnetic declination of 2°E, the magnetic Fore Bearing (FB) of a line *AB* was measured as N79°50′E. There was local attraction at *A*. To determine the correct magnetic bearing of the line, a point *O* was selected at which there was no local attraction. The magnetic FB of line *AO* and *OA* were observed to be S52°40′E and N50°20′W, respectively. What is the true FB of line *AB*?

(a) N81°50′E (b) N82°10′E
(c) N84°10′E (d) N77°50′E

[2015 : 2 M, Set-I]

2.15 The bearings of two inaccessible stations, S_1 (Easting 500 m, Northing 500 m) and S_2 (Easting 600 m, Northing 450 m) from a station S_3 were observed as 225° and 153° 26′ respectively. The independent Easting (in m) of station S_3 is:

(a) 450.000 (b) 570.710
(c) 550.000 (d) 650.000

[2015 : 2 M, Set-II]

2.16 The reduced bearing of a 10 m long line is N30°E. The departure of the line is

(a) 10.00 m (b) 8.66 m
(c) 7.52 m (d) 5.00 m

[2016 : 1 M, Set-II]

2.17 The observed bearings of a traverse are given below

Line	Bearing	Line	Bearing
PQ	46°15′	*QP*	226°15′
AR	108° 15′	*RQ*	286° 15′
RS	201° 30′	*SR*	20° 30′
ST	321° 45′	*TS*	141° 45′

The station(s) most likely to be affected by the local attraction is/are

(a) Only *R* (b) Only *S*
(c) *R* and *S* (d) *P* and *Q*

[2017 : 2 M, Set-I]

2.18 The following details refer to a closed traverse:

Line	Consecutive coordinate			
	Northing (m)	Southing (m)	Easting (m)	Westing (m)
PQ	—	437	173	—
QR	101	—	558	—
RS	419	—	—	96
SP	—	83	—	634

The length and direction (whole circle bearing) of closure, respectively are

(a) 1 m and 90° (b) 2 m and 90°
(c) 1 m and 270° (d) 2 m and 270°

[2018 : 2 M, Set-I]

2.19 The interior angles of four triangles are given below:

Triangle	Interior Angles
P	85°, 50°, 45°
Q	100°, 55°, 25°
R	100°, 45°, 35°
S	130°, 30°, 20°

Which of the triangles are ill-conditioned and should be avoided in Triangulation surveys?

(a) Both Q and S (b) Both P and S
(c) Both Q and R (d) Both P and R

[2019 : 1 M, Set-I]

2.20 The data from a closed traverse survey PQRS (run in the clockwise direction) are given in the table

Line	Included angle (in degrees)
PQ	88
QR	92
RS	94
SP	89

The closing error for the traverse PQRS (in degrees) is ________.

[2019 : 1 M, Set-II]

2.21 The length and bearings of a traverse PQRS are:

Segment	Length (m)	Bearing
PQ	40	80°
QR	50	10°
RS	30	210°

The length of line segment SP (in m, round off to two decimal places), is ________.

[2020 : 2 M, Set-I]

2.22 An open traverse PQRST is surveyed using theodolite and the consecutive coordinates obtained are given in the table

Line	Consecutive Coordinates			
	Northing (m)	Southing (m)	Easting (m)	Westing (m)
PQ	110.2	–	45.5	–
QR	80.6	–	–	60.1
RS	–	90.7	–	70.8
ST	–	105.4	55.5	–

If the independent coordinates (Northing, Easting) of station P are (400 m, 200 m) the independent coordinates (in m) of station T, are

(a) 405.3, 229.9 (b) 394.7, 170.1
(c) 194.7, 370.1 (d) 205.3, 429.9

[2020 : 2 M, Set-I]

2.23 Traversing is carried out for a closed traverse PQRS. The internal angles at vertices P, Q, R and S are measured a 92°, 68°, 123° and 77°, respectively. If fore bearing of line PQ is 27°, fore bearing of line Rs (in degrees, in integer) is ________.

[2021 : 2 M, Set-I]

2.24 Which of the following is/are correct statement(s)?

(a) If the whole circle bearing of a line is 270°, its reduced bearing is 90° NW.
(b) Back bearing of a line is equal to Fore bearing ± 180°.
(c) The boundary of water of a calm water pond will represent contour line.
(d) In the case of fixed hair stadia tacheometry, the staff intercept will be larger, when the staff is held nearer to the observation point.

[2021 : 1 M, Set-I]

2.25 For a given traverse, latitudes and departures are calculated and it is found that sum of latitudes is equal to +2.1 m and the sum of departures is equal to –2.8 m. The length and bearing of the closing error, respectively, are

(a) 3.50 m and 53.13° SE
(b) 3.50 m and 53°7′48″ NW
(c) 2.45 m and 53°7′48″ NW
(d) 0.35 m and 53.13° SE

[2021 : 2 M, Set-II]

2.26 The bearing of a survey line is N31°17′. Its azimuth observed from north is ______ deg. (round off to two decimal places)

[2022 : 1 M, Set-I]

2.27 If the magnetic bearing of the Sun at a place at noon is S 2° E. the magnetic declination (in degrees) at that place is

(a) 2° W (b) 4° E
(c) 4° W (d) 2° E

[2022 : 1 M, Set-II]

2.28 The magnetic bearing of the sun for a location at noon is 183° 30′. If the sun is exactly on the geographic meridian at noon, the magnetic declination of the location is ______.

(a) 3° 30′ W (b) 3° 30′ E
(c) 93° 30′ W (d) 93° 30′ E

[2023 : 1 M, Set-II]

2.29 A delivery agent is at a location R. To deliver the order, she is instructed to travel to location P along straight-line paths of RC, CA, AB and BP of 5 km each. The direction of each path is given in the table below as whole circle bearings. Assume that the latitude (L) and departure (D) of R is (0, 0) km. What is the latitude and departure of P (in km, rounded off to one decimal place)?

Paths	RC	CA	AB	BP
Directions (in degrees)	120	0	90	240

(a) L = 2.5; D = 5.0 (B) L = 0.0; D = 5.0
(C) L = 5.0; D = 2.5 (D) L = 0.0; D = 0.0

[2023 : 2 M, Set-II]

2.30 If the number of sides resulting in a closed traverse is increased from three to four, the sum of the interior angles increases by

(a) 360° (b) 270°
(c) 90° (d) 180°

[2024 : 1 M, Set-I]

2.31 A child walks on a level surface from point P to point Q at a bearing of 30°, from point Q to point R at a bearing of 90° and then directly returns to the starting point P at a bearing of 240°. The straight-line paths PQ and QR are 4 m each. Assuming that all bearings are measured from the magnetic north in degrees, the straight-line path length RP (in meters) is (rounded off to the nearest integer).

[2024 : 2 M, Set-II]

Answers Compass and Traverse Surveying

2.1 (b)	2.2 (a)	2.3 (c)	2.4 (c)	2.5 (b)	2.6 (c)	2.7 (b)
2.8 (b)	2.9 (d)	2.10 (d)	2.11 (642.1)	2.12 (d)	2.13 (d)	2.14 (c)
2.15 (c)	2.16 (d)	2.17 (a)	2.18 (a)	2.19 (a)	2.20 (3)	2.21 (44.79)
2.22 (b)	2.23 (218)	2.24 (a, b, c)	2.25 (b)	2.26 (329.7)	2.27 (d)	2.28 (a)
2.29 (b)	2.30 (d)	2.31 (7)				

Explanations Compass and Traverse Surveying

2.1 (b)

Observed magnetic bearing of OE = 185°
Local attraction at O = +1.5°

∴ Corrected magnetic bearing of,

OE = 185° − 1.5° = 183.5°

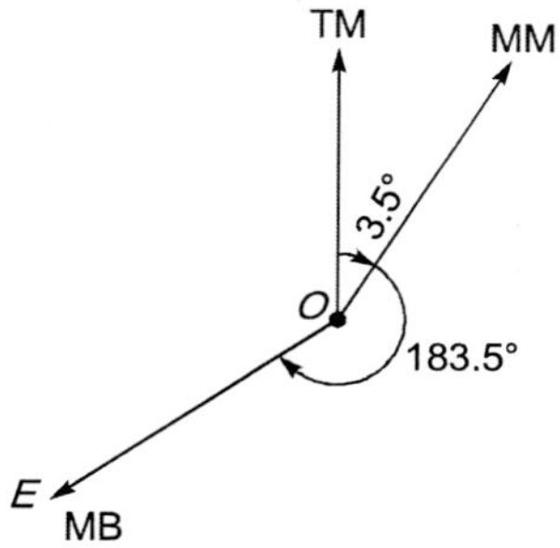

∴ True bearing of OE = Corrected magnetic bearing of OE + Magnetic declination
= 183.5° + 3.5° = 187°

POINTS TO REMEMBER

1. δ = declination toward east
θ = Magnetic bearing
Then, true bearing = $(\theta + \delta)$

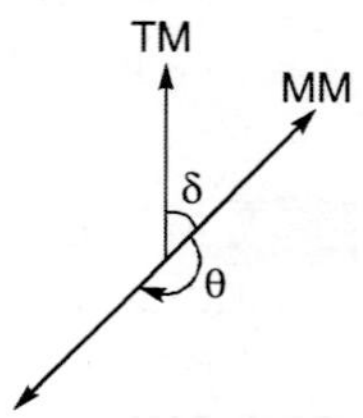

2. δ = declination towards west
θ = magnetic bearing
Then, true bearing = $(\theta - \delta)$

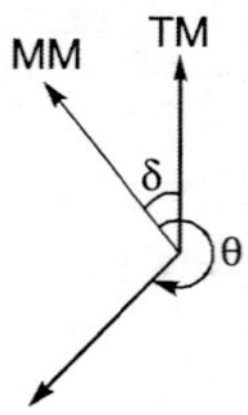

2.2 (a)

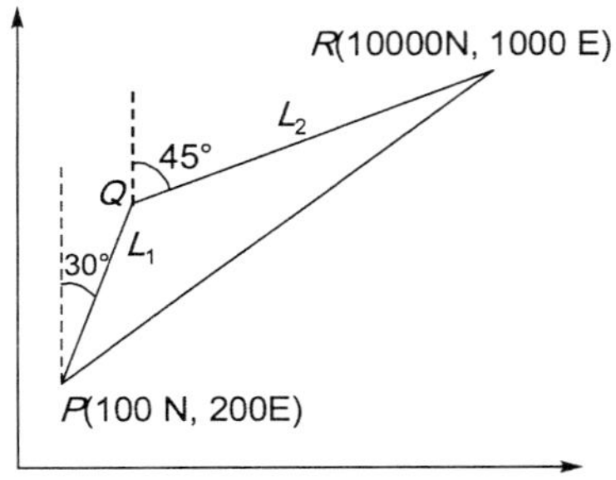

$L_1 \cos 30° + L_2 \cos 45° = (1000 - 100) = 900$ ...(i)

$L_1 \sin 30° + L_2 \sin 45° = (1000 - 200) = 800$ ...(ii)

Solving eq. (i) and (ii), we get

$L_1 = 273.205$ m and $L_2 = 938.186$ m

2.3 (c)

Error in latitude,

$\Delta L = 400.75 + 100.25 - 199.0 - 300.0 = 2$

Error in departure,

$\Delta D = -300.5 + 199.25 + 299.75^{*} - 200.5 = -2$

[* This value is given as 399.75 in the original question paper which is incorrect]

∴Error of closure,

$$e = \sqrt{(\Delta L)^2 + (\Delta D)^2}$$

$$= \sqrt{(2)^2 + (-2)^2} = 2.82 \text{ m}$$

$$\therefore \quad \tan\theta = \frac{\Delta D}{\Delta L}$$

$$\Rightarrow \quad \theta = \tan^{-1}\left(\frac{-2}{2}\right)$$

$$\Rightarrow \quad \theta = -45°$$

$$\text{or} \quad \theta = 360 - 45 = 315°$$

POINTS TO REMEMBER

To determine the direciton of closing error (i.e., θ),

- Note the sign of ΔL and ΔD.
- Consider ΔL in 'y' direction and ΔD is 'x' direction.
- Hence check the quadrant in which θ lies.

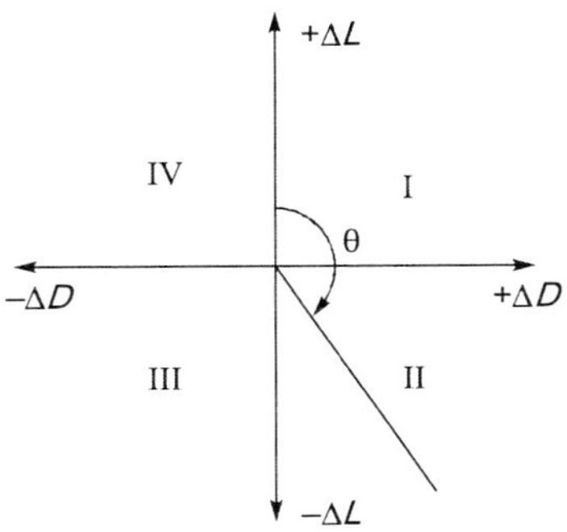

2.4 (c)

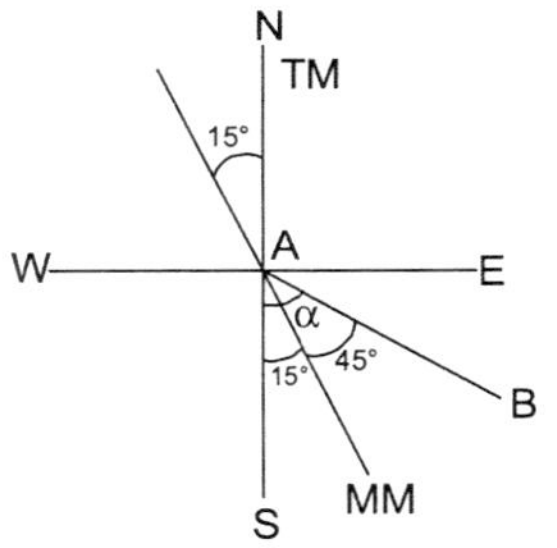

True bearing of AB, (α)

$= MB$ of AB + Magnetic declination

$= 45° + 5° = $ S50° E

2.5 (b)

For a closed traverse sum of latitutes and departures should be zero respectively i.e.

$$\Sigma L = 0$$

$\Rightarrow$ 200 cos0° + 1000 cos 45° + 907 cos 180° + $L \cos\theta = 0$

$\Rightarrow$ $L \cos\theta = -0.10678$ ...(i)

$\Sigma D = 0$

$\Rightarrow$ 200 sin 0° + 1000 sin 45° + 907 sin180° + L sin θ = 0

$\Rightarrow$ $L \sin\theta = -707.10678$ ...(ii)

Dividing (ii) by (i), we get,

$\tan\theta = 6622.09$

$\Rightarrow \theta = 270°$

(Since L cosθ and L sinθ both are −ve. Hence 'θ' lies in 3rd quadrant)

$$\therefore \quad L = \frac{-707.10678}{\sin 270°} = 707.10678 \text{ m}$$

2.6 (c)

2.7 (b)

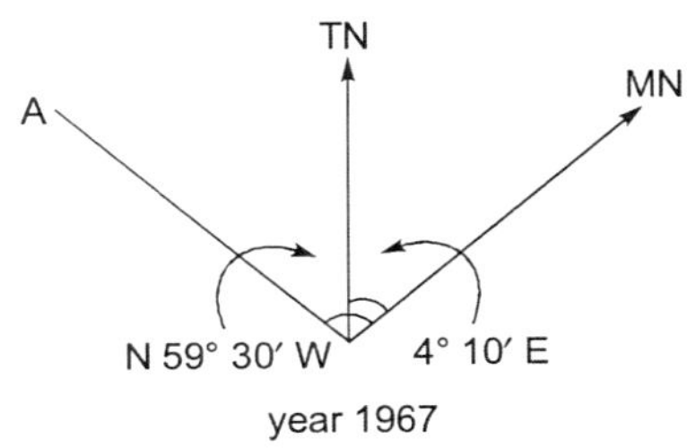

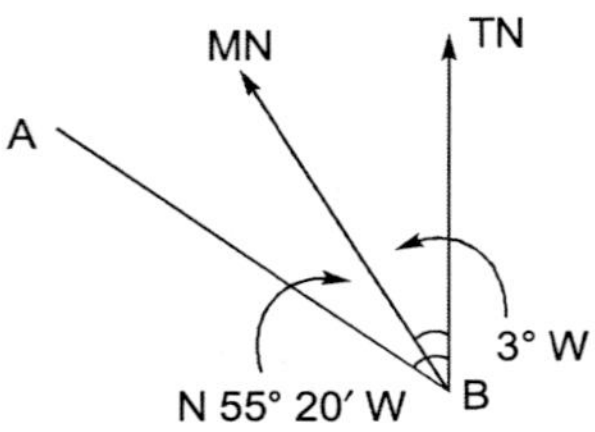

True Bearing of,

AB = Magnetic bearing of AB – Magnetic declination

= N 59°30′W – 4°10′= N 55°20′W

Present Magnetic bearing of,

AB = True bearing AB – Magnetic declination

= N 55°20′W – 3° = N 52°20′W

Whole circle bearing of,

AB = 360° – 52°20′ = 307°40′

2.8 (b)

In a closed loop travers, the algebraic sum of all the latitudes should be equal to zeroi.e., $\Sigma L = 0$

$\Rightarrow$ $L\cos 33.7500° + 300\cos 86.3847° + 354.524\cos 169.3819° + 450\cos 243.900° + 268\cos 317.5° = 0$

$\Rightarrow$ $L\cos 33.75° = 329.9166$

$\Rightarrow$ $L = 396.79$ m

2.9 (d)

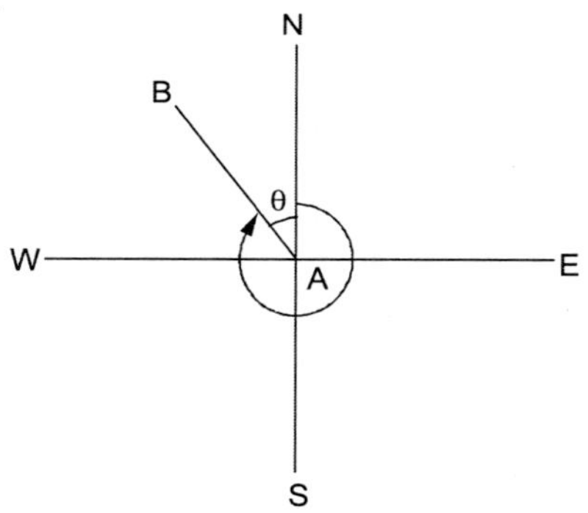

Since the latitude of line is positive and departure is negative, the line lies in the fourth quadrant.

$\therefore$ $l\cos\theta = 78$

$l\sin\theta = -45.1$

$\Rightarrow$ $\tan\theta = -0.578$

$\theta = -30°$

$\therefore$ WCB of AB = 360° – 30° = 330°

2.10 (d)

The difference between back bearing and fore bearing of line DE

= 258° 30' – 78° 30' = 180°

$\therefore$ Station D and E are free from local attraction

Fore bearing of line EA = 216° 30'

$\therefore$ Correct back bearing of line EA

= 216° 30' – 180° = 36° 30'

$\therefore$ Error at A = 31° 45' – 36° 30'

= –4° 45'

Hence correction at,

A = +4° 45'

$\therefore$Correct fore bearing of line AB

= 126° 45' + 4° 45'

= 131° 30'

Correct back bearing of AB

= 131° 30 + 180° = 311° 30'

$\therefore$ Error at B = –3° 30'

Correction at B= +3° 30'

Hence correct bearing of BC

= 45° 15' + 3° 30' = 48° 45'

2.11 Sol.

RL of A = 880.88 m

RL of plane of collimation

= 880.88 + 1.1 = 881.98 m

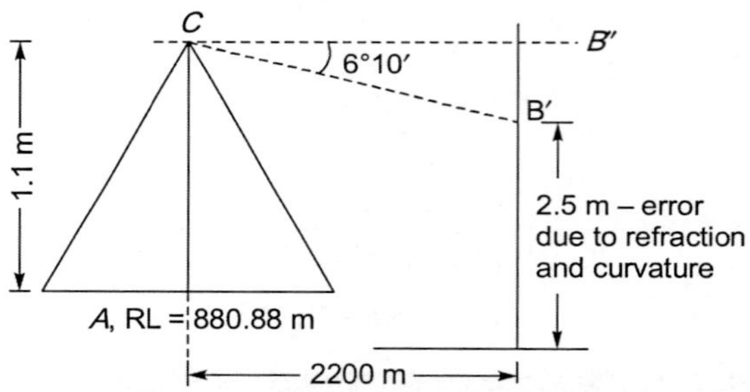

$$\frac{B'B''}{CB''} = \tan 6°10'$$

$B'B''$ = 2200 m × tan 6°10′

= 237.701 m

RL of B = RL ofline of sight – $B'B''$

–(2.5 – Error due to refraction & curvature)

= 881.98 – 237.701 – (2.5 – 0.0673 × P^2)

Here, D = Distance between theodolite and staff in km

$\therefore$ RL of B = 881.98 – 237.701

– (2.5 × 0.0673 × 2.2^2) = 642.105 m

2.12 (d)

1. **Alilade:** It is a straight edge ruler used in plane table surveying whichis used for

sighting the objects and drawing the line son the drawing sheet.

2. **Arrow:** Arrows are used to mark the position of end of the hain or tape on the ground.
3. **Bubble tube:** It is used to check the level of the instrument. When the bubble of the tube comes in the centre, then the instrument in levelled.
4. **Stadia hair:** Lines on diaphragm, the intercept 'S' between 2 stadia hairs of a vertically held rod gives the distance between tacheometer and the road as $D = kS$ (k = stadia constant)

2.13 (d)

$$\text{Relative precision} = \frac{1}{p/e}$$

where, p = Perimeter of traverse
e = Closing error

$$\text{Closing error, } e = \sqrt{0.3^2 + 0.4^2} = 0.5\text{m}$$

$$\text{Relative precision} = \frac{1}{\frac{1000}{0.5}} = \frac{1}{2000}$$

2.14 (c)

Magnetic Declination, $\delta = 2°\text{E}$

Magnetic *FB* of *AB* = N 79°50′ E

To find local attraction at station *A*

As station *O* is free from local attraction

Hence *FB* of *OA* will be correct,

Correct *FB* of *OA* = N 50°20′ W ≅ 309°40′

∴ Correct *BB* of OA = 129°40′

∵ Obserbed *FB* of *AO*

= Observed *BB* of *OA*

= S52°40′E = 127°20′

Error = Observed bearing – Correct bearing

= 127°20′ – 129°40′ = – 2°20′

Correction = +2°20′

Local attraction at station *A*

= +2°20′ ≅ 2°20′ E

∴ Magnetic *FB* of *AB* = N 79°50′ E

δ = 2°E and local attraction = 2°20′E

∴ *TB* of *FB* of *AB* = 79°50′ + 2° 20′ + 2°

= N84°10′E

2.15 (c)

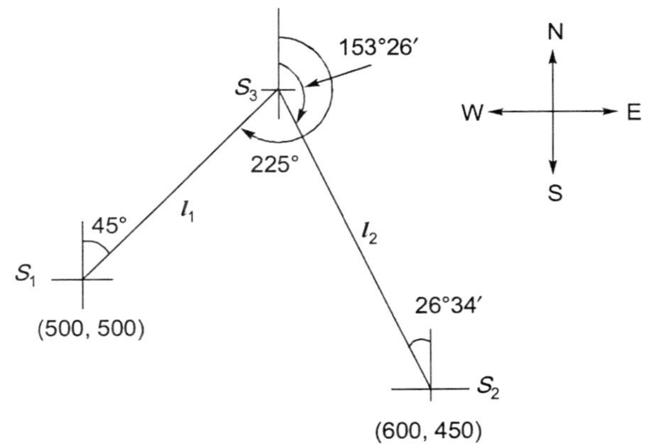

Let $S_1S_3 = l_1$ and $S_2S_3 = l_2$

Northing of S_3:

$500 + l_1 \cos 45° = 450 + l_2 \cos 26°34'$

$l_1 \cos 45° - l_2 \cos 26° 34' = -50$ ...(i)

Easting of S_3:

$500 + l_1 \sin 45° = 600 - l_2 \sin 26°34'$

$l_1 \sin 45° + l_2 \sin 26°34' = 100$ ...(ii)

Solving eq. (i) and eq. (ii),

$l_1 = 70.707$

$l_2 = 111.802$

∴ Easting of $S_3 = 500 + 70.707 \sin 45°$

$= 550$ m

2.16 (d)

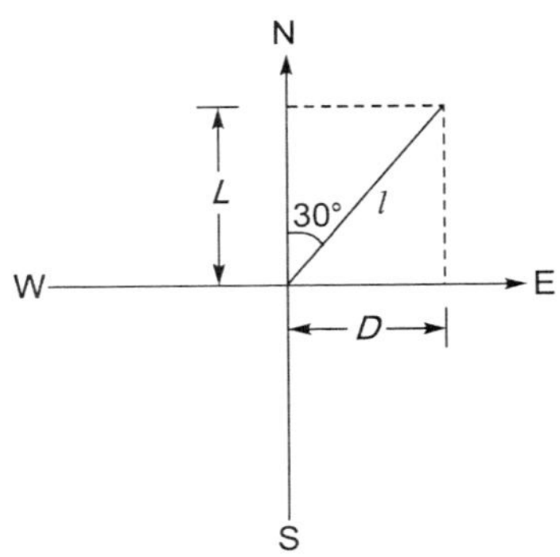

The departure of the line,

$$D = l \sin \theta$$

$$= 10 \sin 30° = \frac{10}{2} = 5 \text{ m}$$

2.17 (a)

Line	FB	BB	Δ = | FB – BB|
PQ	46°15′	226°15′	180°
QR	108° 15′	286° 15′	178°
RS	201° 30′	20° 30′	181°
ST	321° 45′	141° 45′	180°

Station *R* is most likely to be affected.
Since difference of FB and BB for PQ and ST is 180°.
Hence, P, Q, S, T are free from local attraction.

2.18 (a)

ΣL = W northing– E southing
$= (101 + 419) - (437 + 83) = 0$
ΣD = E Easting – E Westing
$= (173 + 558) - (96 + 634) = 1$ m

∴ Length of closure

$$= \sqrt{(\Sigma L)^2 + (\Sigma D)^2} = \sqrt{0^2 + 1^2} = 1\text{m}$$

And, direction of closure,

$$\theta = \tan^{-1}\left(\frac{\Sigma D}{\Sigma L}\right)$$

$$= \tan^{-1}\left(\frac{1}{0}\right) = 90°$$

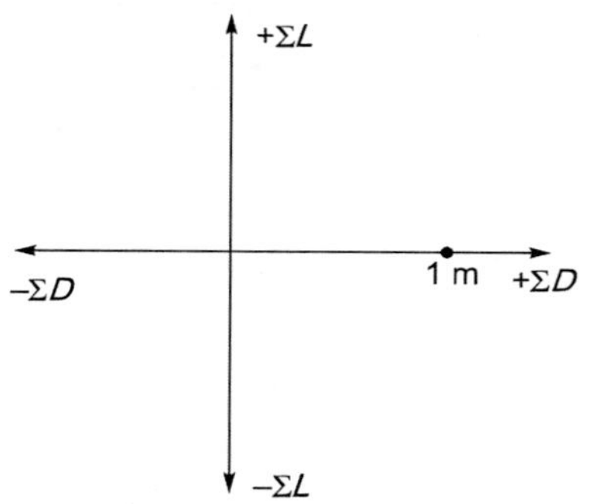

Hence, θ lies in *I* quadrant and θ = 90°

2.19 (a)

For a well conditioned triangle.
The interior angle should not be less than 30°.
In this way, triangle '*Q*' & *S* having less angles (acute angle).
⇒ Q & S are ill-conditioned.

2.20 Sol.

Assuming it as anticlockwise traverse.
Mathematically sum of interior angle for a closed traverse

$= (2n - 4) \times 90 = (2 \times 4 - 4) \times 90$
$= 4 \times 90 = 360°$

Given sum of interior angles,
$= 88 + 92 + 94 + 89 = 363°$

Then error in interior angle = 363 – 360 = 3°

***Note**: In this question as per clockwise traverse included angle should be taken as exterior angle. But if we take exterior angle then we get all interior angles more than 180°.*

2.21 Sol.

$\Delta L = 40\cos 80° + 50\cos 10° + 30\cos 210°$
$= 30.20$
$\Delta D = 40\sin 80° + 50\sin 10° + 30\sin 210°$
$= 33.07$

Length,

$SP = \sqrt{\Delta L^2 + \Delta D^2} = 44.79$ m

2.22 (b)

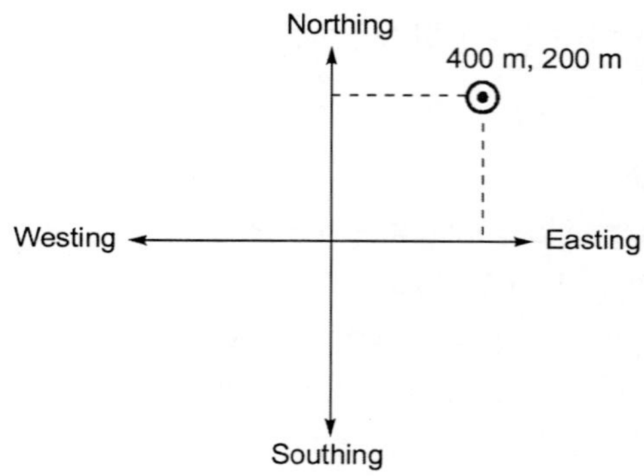

$\Delta L = -5.3$ m
$\Delta D = -29.9$ m

T, Northing{400 + (–5.3) = 394.7 m
T, Easting{200 + (–29.9) = 170.1 m

Independent coordinate of T[394.7m, 170.1m]

2.23 Sol.

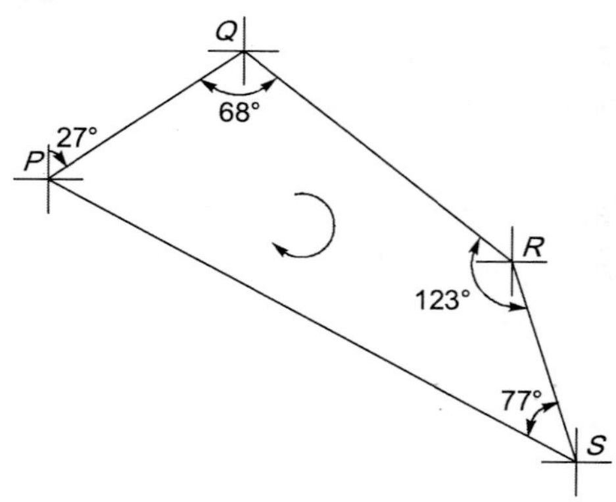

$$Q = \begin{bmatrix} BB \text{ of } PQ = 27° + 180° = 207° \\ FB \text{ of } QR = 207° - 68° = 139° \end{bmatrix}$$

$$R = \begin{bmatrix} BB \text{ of } QR = 139° + 180° = 319° \\ FB \text{ of } RS = 319° - 123° = 196° \end{bmatrix}$$

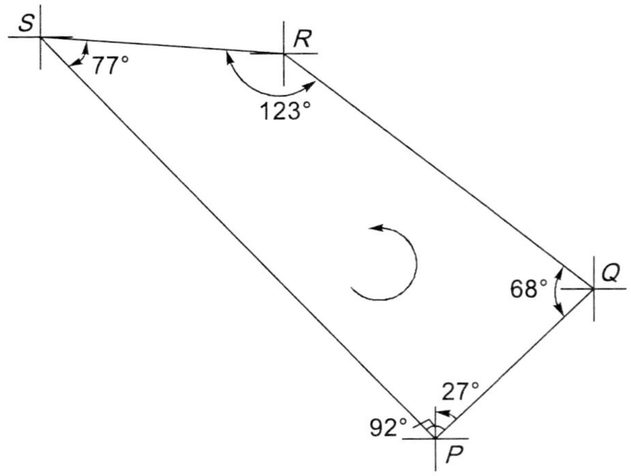

$$Q = \begin{bmatrix} BB \text{ of } PQ = 27° + 180° = 207° \\ FB \text{ of } QR = 207° + 68° = 275° \end{bmatrix}$$

$$R = \begin{bmatrix} BB \text{ of } QR = 275° - 180° = 95° \\ FB \text{ of } RS = 95° + 123° = 218° \end{bmatrix}$$

2.24 (a, b, c)

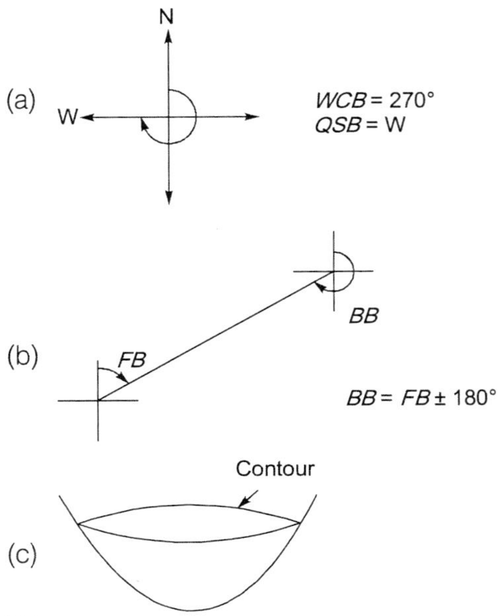

2.25 (b)

$$e_L = +2.1 \text{ m}$$
$$e_D = -2.8 \text{ m}$$
$$e = \sqrt{e_L^2 + e_D^2} = \sqrt{(2.1)^2 + (2.8)^2} = 3.5 \text{ m}$$

Bearing of closing error

$$= \tan^{-1}\left(\frac{e_D}{e_L}\right) = \tan^{-1}\left(\frac{-2.8}{2.1}\right) = -53.13°$$
$$= 53°7'48'' \text{ NW}$$

2.26 Sol.

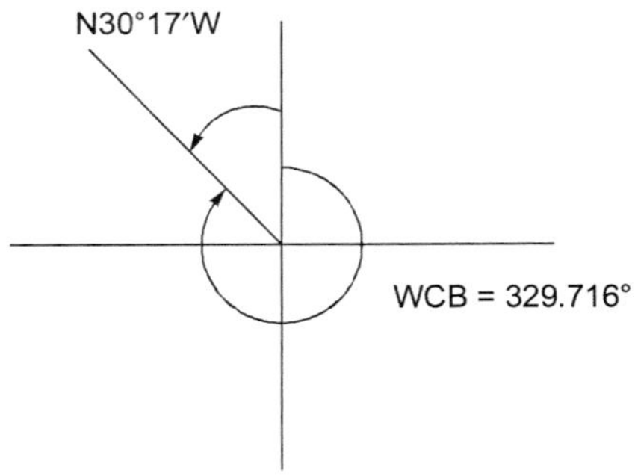

$$\therefore \quad WCB = 360 - \left(31 + \frac{17}{60}\right)°$$
$$\Rightarrow \quad WCB = 329.716°$$

2.27 (d)

Magnetic bearing (MB) of Sun in WCB = 178°. True bearing (TB) of Sun in WCB at noon must be 180°.

$$MB + \delta = TB$$
$$\Rightarrow \quad 178° + \delta = 180°$$
$$\Rightarrow \quad \delta = 2°$$

Since, declination (δ) is found to be positive, so it is East declination of 2° i.e. 2°E.

2.28 (a)

Magnetic declination = True bearing – Magnetic bearing

= 180° – 183°30′ = –3°30′ = 3°30′ W

2.29 (b)

For location R,

Latitude, $L_R = 0$

Departure, $D_R = 0$

For location P,

Latitude, $L_P = L_R + \Sigma\Delta_L$

$= 0 + 5\cos120° + 5\cos0° + 5\cos90° + 5\cos240° = 0$

Departure, $D_P = D_R + \Sigma\Delta_D$

$= 0 + 5\sin120° + 5\sin0° + 5\sin90° + 5\sin240° = 5$

So, $L_P = 0, \quad D_P = 5$

Alternatively,

Path (l)	Direction (θ)	Latitude ($l\cos\theta$)	Departure ($l\sin\theta$)
RC	120°	–2.5	4.33
CA	0°	5	0
AB	90°	0	5
BP	240°	–2.5	–4.33

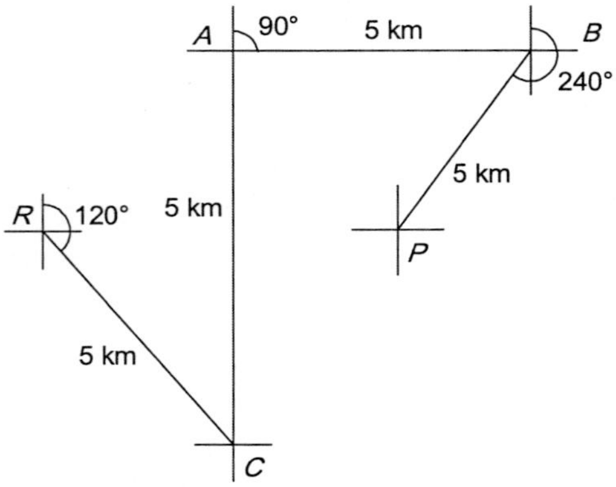

Length of each path (l) = 5 km

Latitude of P in km = Σlatitude = 0

Departure of P in km = Σdeparture = 5

2.30 (d)

Sum of interior angles = $(n - 2) \times 180°$

So, sum of interior angles for 3 sides

$= (3 - 2) \times 180° = 180°$

Now, sum of interior angles for 4 sides

$= (4 - 2) \times 180° = 360°$

$\therefore$ Increase in sum of interior angles

$= 360° - 180° = 180°$

2.31 Sol.

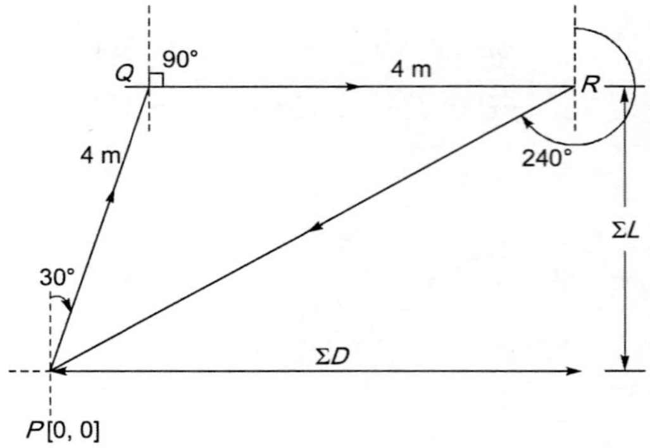

Horizontal distance of traverse from P to R = ΣD

$= PQ \sin 30° + QR \sin 90°$

$= 4 \times \frac{1}{2} + 4 \times 1 = 6 \text{ m}$

Vertical distance of traverse from P to R = ΣL

$= PQ \cos 30° + QR \cos 90°$

$= 4 \times \frac{\sqrt{3}}{2} = 2\sqrt{3}$

$\therefore \quad PR = \sqrt{(\Sigma L)^2 + (\Sigma D)^2}$

$= \sqrt{(6)^2 + \left(2\sqrt{3}\right)^2} = \sqrt{48}$

$= 6.928 \text{ m} \simeq 7 \text{ m}$

■■■■

3 Levelling and Contouring

3.1 A Bench Mark (BM) with Reduced Level (RL) = 155.305 m has been established at the floor of a room. It is required to find out the RL of the underside of the roof (*R*) of the room using Spirit Levelling. The Back Sight (BS) to the BM has been observed as 1.500 m whereas the Fore Sight (FS) to *R* has been observed as 0.575 m (Staff held inverted). The RL(m) of R will be

(a) 155.880 (b) 156.230
(c) 157.380 (d) 157.860

[2006 : 2 M]

3.2 Consider the following figure, which is an extract from a contour map (scale = 1:20, 000) of an area. An alignment of a road at a ruling gradient of 4% is to be fixed from the point *O* and beyond. What should be the radius of the arc with *O* as the center to get the point of alignment of the next contour on the map?

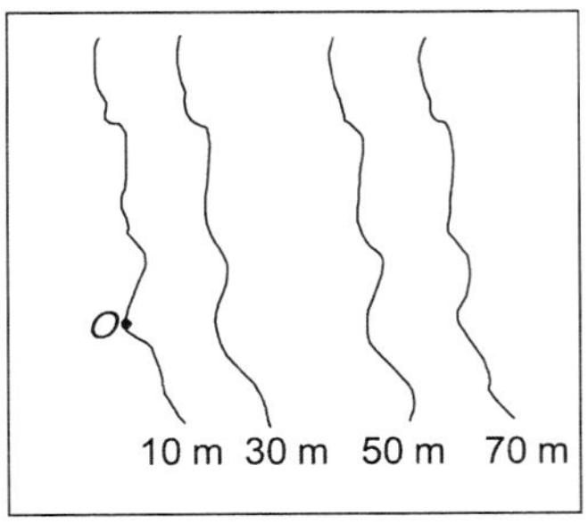

(a) 0.025 cm (b) 0.25 cm
(c) 2.5 cm (d) 5.0 cm

[2006 : 2 M]

3.3 During a levelling work along a falling gradient using a Dumpy Level and a Staff of 3 m length, following successive readings were taken: 1.785, 2.935, 0.360, 1.320. What will be the correct order of booking these four readings in a level book? (BS : Back Sight, IS : Intermediate Sight, FS: Fore Sight)

(a) BS, FS, BS, FS (b) BS, IS, FS, FS
(c) BS, IS, IS, FS (d) BS, IS, BS, FS

[2006 : 2 M]

3.4 The following measurements were made during testing a levelling instrument.

Instrument at	Staff reading at	
	P_1	Q_1
P	2.800 m	1.700 m
Q	2.700 m	1.800 m

P_1 is close to *P* and Q_1 is close to *Q*. If the reduced level of station *P* is 100.000 m, the reduced level of station *Q* is

(a) 99.000 m (b) 100.000 m
(c) 101.000 m (d) 102.000 m

[2007 : 2 M]

3.5 A light house of 120 m height is just visible above the horizon from a ship. The correct distance (m) between the ship and the light house considering combined correction for curvature and refraction, is

(a) 39.098 (b) 42.226
(c) 39098 (d) 42226 [2008 : 2 M]

3.6 Consider the following statements:

Assertion (A): Curvature correction must be applied when the sights are long.

Reason (R): Line of collimation is not a level line but is tangential to the level line.

Of these statements

(a) both A and R are true and R is the correct explanation of A
(b) both A and R are true but R is not a correct explanation of A
(c) A is true but R is false
(d) A is false but R is true [2009 : 2 M]

3.7 A bench mark has been established at the soffit of an ornamental arch at the known elevation of 100.0 m above mean sea level. The back sight used to establish height of instrument is an inverted staff reading of 2.105 m. A forward sight reading with normally held staff of 1.105 m is taken on a recently constructed plinth. The elevation of the plinth is

(a) 103.210 m (b) 101.000 m
(c) 99.000 m (d) 96.790 m

[2010 : 2 M]

3.8 Curvature correction to a staff reading in a differential leveling survey is
(a) always subtractive
(b) always zero
(c) always additive
(d) dependent on latitude [2011 : 1 M]

3.9 Which of the following errors can be eliminated by reciprocal measurements in differential leveling?
I. Error due to earth's curvature
II. Error due to atmospheric refraction
(a) Both I and II (b) I only
(c) II only (d) Neither I nor II
[2012 : 1 M]

3.10 The horizontal distance between two stations P and Q is 100 m. The vertical angles from P and Q to the top of a vertical tower at T are 3° and 5° above horizontal, respectively. The vertical angles from P and Q to the base of the tower are 0.1° and 0.5° below horizontal, respectively. Stations P, Q and the tower are in the same vertical plane with P and Q being on the same side of T. Neglecting earth's curvature and atmospheric refraction, the height (in m) of the tower is
(a) 6.972 (b) 12.387
(c) 12.540 (d) 128.745 [2012 : 2 M]

3.11 The Reduced Levels (RLs) of the points P and Q are +49.600 m and +51.870 m respectively. Distance PQ is 20 m. The distance (in m from P) at which the +51.000 m contour cuts the line PQ is
(a) 15.00 (b) 12.33
(c) 3.52 (d) 2.27
[2014 : 1 M, Set-I]

3.12 A levelling is carried out to establish the Reduced Levels (RL) of point R with respect to the Bench Mark (BM) at P. The staff readings taken are given below:

Staff Station	BS	IS	FS	RL
P	1.655 m			100.000 m
Q	– 0.950 m		–1.500 m	
R			0.750 m	?

If RL of P is +100.000 m, then RL (in m) of R is
(a) 103.355 (b) 103.155
(c) 101.455 (d) 100.355
[2014 : 2 M, Set-I]

3.13 Which of the following statements is FALSE?
(a) Plumb line is along the direction of gravity
(b) Mean Sea Level (MSL) is used as a reference surface for establishing the horizontal control
(c) Mean Sea Level (MSL) is a simplification of the Geoid
(d) Geoid is an equi-potential surface of gravity
[2015 : 1 M, Set-I]

3.14 In a leveling work, sum of the Back Sight (B.S.) and Fore Sight (F.S.) have been found to be 3.085 m and 5.645 m respectively. If the Reduced Level (R.L.) of the starting station is 100.000 m, the R.L. (in m) of the last station is _______.
[2015 : 1 M, Set-II]

3.15 The combined correction due to curvature and refraction (in m) for distance of 1 km on the surface of Earth is
(a) 0.0673 (b) 0.673
(c) 7.63 (d) 0.763
[2015 : 1 M, Set-II]

3.16 Two pegs A and B were fixed on opposite banks of a 50 m wide river. The level was set up at A and the staff readings on Pegs A and B were observed as 1.350 m and 1.550 m, respectively. Thereafter the instrument was shifted and set up at B. The staff readings on Pegs B and A were observed as 0.750 m and 0.550 m, respectively. If the R.L. of Peg A is 100.200 m, the R.L. (in m) of Peg B is _____.
[2015 : 2 M, Set-II]

3.17 The staff reading taken on a workshop floor using a level is 0.645 m. The inverted staff reading taken to the bottom of a beam is 2.960 m. The reduced level of the floor is 40.500 m. The reduced level (expressed in m) of the bottom of the beam is
(a) 44.105 (b) 43.460
(c) 42.815 (d) 41.145
[2016 : 1 M, Set-I]

3.18 The vertical angles subtended by the top of a tower T at two instrument stations set up at P and Q, are shown in the figure. The two stations are in line with the tower and spaced at a distance of 60 m. Readings taken from these two stations on a leveling staff placed at the benchmark (BM = 450.000 m) are also shown in the figure. The reduced level of the top of the tower T (expressed in m) is_________

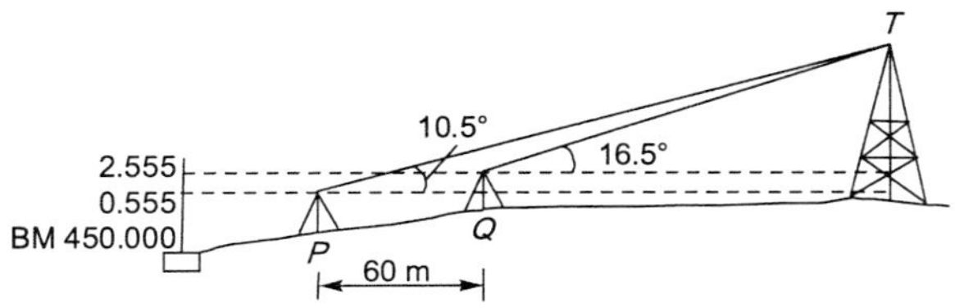

[2016 : 2 M, Set-I]

3.19 An observer standing on the deck of a ship just sees the top of a lighthouse. The top of the lighthouse is 40 m above the sea level and the height of the observer's eye is 5 m above the sea level. The distance (in km, up to one decimal place) of the observer from the lighthouse is ________.

[2017 : 2 M, Set-II]

3.20 A level instrument at a height of 1.320 m has been placed at a station having a Reduced Level (RL) of 112.565 m. The instrument reads –2.835 m on a levelling staff held at the bottom of a bridge deck. The RL (in m) of the bottom of the bridge deck is

(a) 116.720 (b) 116.080
(c) 114.080 (d) 111.050

[2018 : 2 M, Set-II]

3.21 A staff is placed on a benchmark (BM) of reduced level (RL) 100.000 m and a theodolite is placed at a horizontal distance of 50 m from the BM to measure the vertical angles. The measured vertical angles from the horizontal at the staff readings of 0.400 m and 2.400 m are found to be the same. Taking the height of the instrument as 1.400 m, the RL (in m) of the theodolite station is_______.

[2019 : 2 M, Set-I]

3.22 A series of perpendicular offsets taken from a curved boundary wall to a straight survey line at an interval of 6 m are 1.22, 1.67, 2.04, 2.34, 2.14. 1.87, and 1.15 m. The area (in m^2, round off to 2 decimal places) bounded by the survey line, curved boundary wall, the first and the last offsets, determined using Simpson's rule, is__________.

[2019 : 2 M, Set-II]

3.23 A theodolite is set up at station A. The RL of instrument axis is 212.250 m. The angle of elevation to the top of a 4 m long staff, held vertical at station B, is 7°. The horizontal distance between station A and B is 400 m. Neglecting the errors due to curvature of earth and refraction, the RL (in m, round off to three decimal places) of station B is ________.

[2020 : 2 M, Set-II]

3.24 A theodolite was set up at a station P. The angle of depression to a vane 2 m above the foot of a staff held at another station Q was 45°. The horizontal distance between stations P and Q is 20 m. The staff reading at a benchmark S of RL 433.050 m is 2.905 m. Neglecting the errors due to curvature and refraction, the RL of the station Q (in m), is

(a) 431.050 (b) 435.955
(c) 413.050 (d) 413.955

[2020 : 2 M, Set-II]

3.25 Trigonometric levelling was carried out from two stations P and Q to find the reduced level (R. L.) of the top of hillock, as shown in the table. The distance between Stations P and Q is 55 m. Assume Stations P and Q, and the hillock are in the same vertical plane. The R. L. of the top of the hillock (in m) is ____________ (round off to three decimal places).

Station	Vertical angle of top of hilllock	Staff reading on B.M	R.L of BM
P	18° 45′	2.340 m	100.000 m
Q	12° 45′	1.660 m	

[2023 : 2 M, Set-I]

3.26 A bird is resting on a point *P* at a height of 8 m above the Mean Sea Level (MSL). Upon hearing a loud noise, the bird flies parallel to the ground surface and reaches a point *Q* which is located at a height of 3 m above MSL. The ground surface has a falling gradient of 1 in 2. Ignoring the effects of curvature and refraction, the horizontal distance (in meters) between points *P* and *Q* is _______ (in integer).

[2024 : 1 M, Set-I]

3.27 A map is prepared with a scale of 1: 1000 and a contour interval of 1 m. If the distance between two adjacent contours on the map is 10 mm, the slope of the ground between the adjacent contours is

(a) 40% (b) 35%
(c) 10% (d) 30%

[2024 : 2 M, Set-I]

3.28 Differential levelling is carried out from point *P*(BM:+200.000 m) to point *R*. The readings taken are given in the table.

Points	Staff readings (m)		Remarks
	Back Sight	Fore Sight	
P	(–) 2.050		BM: +200.000 m
Q	1.050	0.950	*Q* is a change point
R		(-)1.655	

Reduced Level (in meters) of the point *R* is ________ (rounded off to 3 decimal places).

[2024 : 2 M, Set-II]

Answers Levelling and Contouring

3.1 (c)	3.2 (c)	3.3 (a)	3.4 (c)	3.5 (d)	3.6 (a)	3.7 (d)
3.8 (a)	3.9 (a)	3.10 (b)	3.11 (b)	3.12 (c)	3.13 (b)	3.14 (97.44)
3.15 (a)	3.16 (100.0)	3.17 (a)	3.18 (476.91)	3.19 (33.0)	3.20 (a)	3.21 (100)
3.22 (68.5)	3.23 (257.36)	3.24 (d)	3.25 (137.6)	3.26 (10)	3.27 (c)	3.28 (199.7)

Explanations Levelling and Contouring

3.1 (c)

Height of instrument

$$= \text{RL of bench mark} + \text{BS}$$
$$= 155.305 + 1.500$$
$$= 156.805 \text{ m}$$

Now, the staff is held inverted and the foresight (FS) is 0.575 m,

$$\therefore \quad \text{RL of } R = \text{Height of instrument} + \text{FS}$$
$$= 156.805 + 0.575 = 157.380 \text{ m}$$

3.2 (c)

The contour interval = 20 m

For 4% gradient, the length needed from one contour to another

$$= \frac{20}{0.04} = 500 \text{ m}$$

$$\therefore \text{ Radius of arc} = \frac{500 \times 100}{20000} = 2.5 \text{ cm}$$

3.3 (a)

The first reading is always BS. The second reading is foresight because the instrument has been shifted after this reading which is evident from the low value of third reading. Since the instrument has been shifted the third reading is backlight again. Successively the fourth reading is foresight.

3.4 (c)

$$h = \frac{\text{Difference of staff reading at } P_1 \text{ and } Q_1 \text{ when instrument is at } P + \text{Difference of staff readings at } P_1 \text{ and } Q_1 \text{ when instrument is at } Q}{2}$$

$$\Rightarrow h = \frac{(2.800 - 1.700) + (2.700 - 1.800)}{2}$$

$$\Rightarrow h = 1$$

∴ Reduced level of station,

Q = Reduced level of $P + h$

$= 100.000 + 1 = 101.000$ m

3.5 (d)

Correction due to curvature,

$$C_c = -0.0785 d^2$$

Correction due to refraction,

$$C_r = +0.0112\, d^2$$

∴ Composite correction,

$$C = -0.0785\, d^2 + 0.0112\, d^2$$
$$= -0.0673\, d^2$$

where C is in metres and d is in kilometers

Here, $C = 120$ m; $d = ?$

$$\therefore \quad 120 = 0.0673 \times d^2$$
$$\Rightarrow \quad d = \sqrt{1783.06}$$
$$\Rightarrow \quad d = 42.226 \times 1000$$
$$\Rightarrow \quad d = 42226 \text{ m}$$

3.6 (a)

For long sights, curvature correction must be applied because due to curvature of earth, line of collimation differs from the level line. It is tangential to level line.

3.7 (d)

Height of instrument = BM + Back sight

Since the staff is inverted, the back sight will be negative.

∴ Height of instrument

$=100 + (-2.105) = 97.897$ m

Elevation of plinth = Height of instrument – Fore sight

$= 97.895 - 1.105 = 96.790$ m

3.8 (a)

3.9 (a)

Reciprocal levelling eliminates the errors due to

(i) Error in instrument adjustment.

(ii) Combined effect of earth's curvature and the refraction of the atmosphere.

(iii) Variation in the average refraction.

3.10 (b)

$x (\tan 3 + \tan 0.1) = (x - 100) (\tan 5° + \tan 0.5°)$

$\Rightarrow \quad x = 228.758$

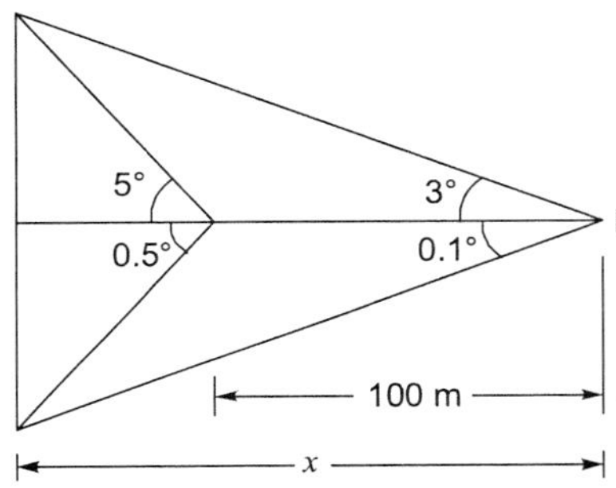

Height of tower= 228.758 (tan 3° + tan0.1°)

$= 12.387$ m

Note: Question has erroreous data, hence many other answers are also possible.

3.11 (b)

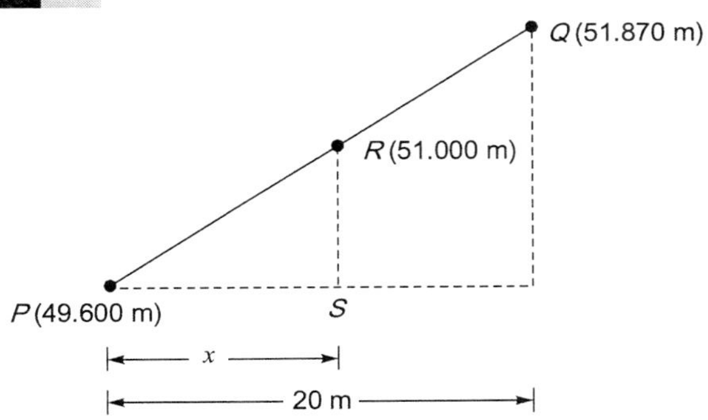

ΔPRS ~ ΔPQR, we get,

$$\Rightarrow \frac{x}{(51-49.6)} = \frac{20}{(51.87-49.6)}$$

$\therefore \quad x = 12.33$ m

3.12 (c)

$HI = RL + BS$

and $RL = HI - FS$

Staff station	BS	IS	FS	RL	HI	RL
P	1.655	–	–	–	101.655	100
Q	–0.95	–	–1.5	–	102.205	103.155
R	–	–	0.75	–	–	101.455

∴ *RL* of *R* = 101.455 m

3.13 (b)

Mean sea level (MSL) is used as a reference surface for establishing the vertical control.

3.14 Sol.

Using Rise and Fall method,

ΣF.S. > ΣB.S.

∴ Fall = ΣF.S. – ΣB.S. = R.L. of first station – R.L. of last station

$= 5.645 - 3.085 = 2.56$ m

R.L. (last station) = R.L. (first station) – Fall

$= 100 - 2.56 = 97.44$ m

3.15 (a)

Correction due to curvature,

$$C_c = -0.0785\, d^2$$

Correction due to refraction,

$$C_r = +0.0112\, d^2$$

∴ Composite correction,

$$C = -0.0785\, d^2 + 0.0112\, d^2$$

$$= -0.0673\, d^2$$

where *d* is in km, C is in m,

$$\Rightarrow \quad C = 0.0673 \times 1^2 = 0.0673 \text{ m}$$

3.16 Sol.

Staff readings shows that station *B* is below station *A*.

$a_1 = 1.350$

$a_2 = 0.550$

$b_1 = 1.550$

$b_2 = 0.750$

$$h_{AB} = \frac{(b_1 - a_1) + (b_2 - a_2)}{2}$$

$$= \frac{(1.550 - 1.350) + (0.750 - 0.550)}{2}$$

$= 0.200$

RL of *B* = RL of *A* – 0.200

$= 100.200 - 0.200$

$= 100.000$ m

3.17 (a)

Height of instrument, HI

= R.L. of floor + staff reading from floor

= 40.500 + 0.645

= 41.145 m

R.L. of bottom of beam

= HI + inverted staff reading taken from bottom of beam

= 41.145 + 2.960

= 44.105 m

3.18 Sol.

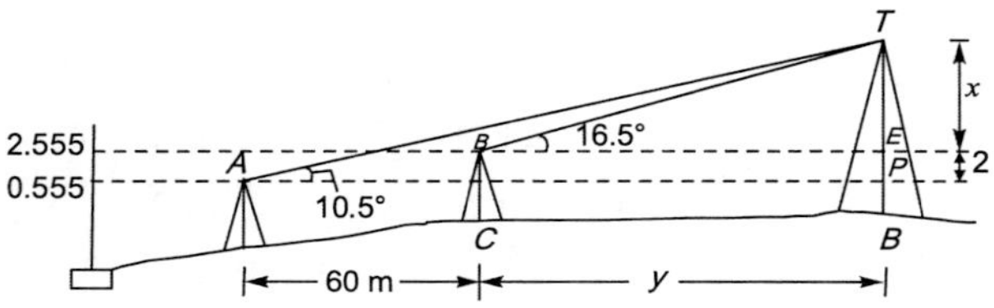

In ΔBET,

$$\tan 16.5° = \frac{x}{y}$$

$$\Rightarrow \quad y = 3.3759x \quad ...(i)$$

In ΔAOT,

$$\tan 10.5 = \frac{x+2}{60+y} \quad ...(ii)$$

Put value of y in eq. (ii),

$$0.1853 = \frac{x+2}{60+3.3759x}$$

$$11.1203 + 0.6255x = x + 2$$

$$x = 24.3568 \text{ m}$$

So, the reduced level of tower,

$$T = 450.000 + 2.555 + 24.356$$

$$= 476.911 \text{ m}$$

3.19 Sol.

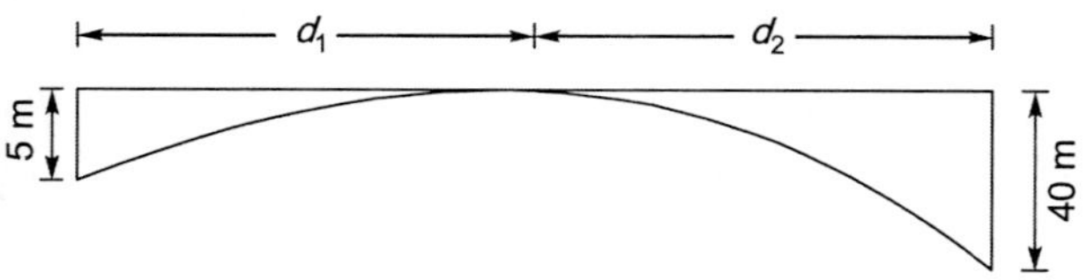

Distance of observer from the lighthouse,

$$d = d_1 + d_2$$

$$= 3.855\left(\sqrt{h_1} + \sqrt{h_2}\right)$$

$$= 3.855\left(\sqrt{5} + \sqrt{40}\right) = 33.0012 \text{ km}$$

3.20 (a)

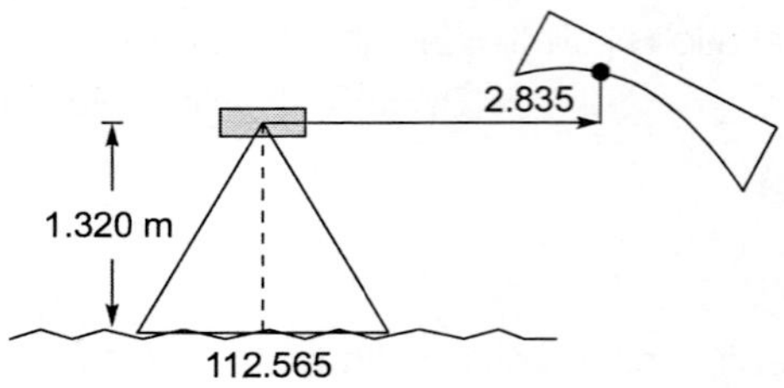

RL of bottom of bridge deck

= 112.565 + 1.320 + (2.835)

= 116.720 m

3.21 Sol.

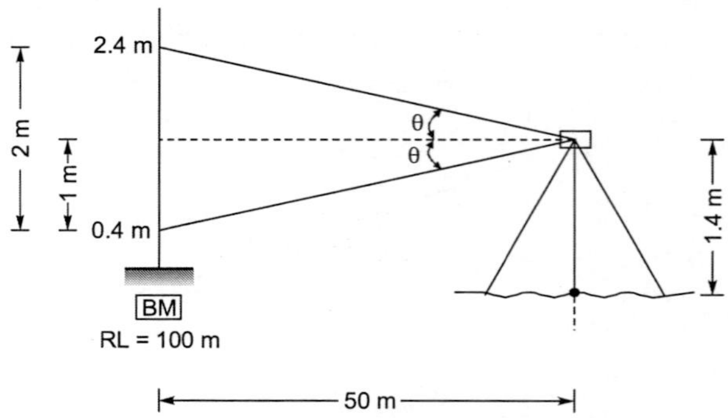

HI = RL of BM + BS

= 100 + (0.4 + 1)

= 101.4

RL of Theodolite station

= HI-height

= 101.4 − 1.4

= 100 m

3.22 Sol.

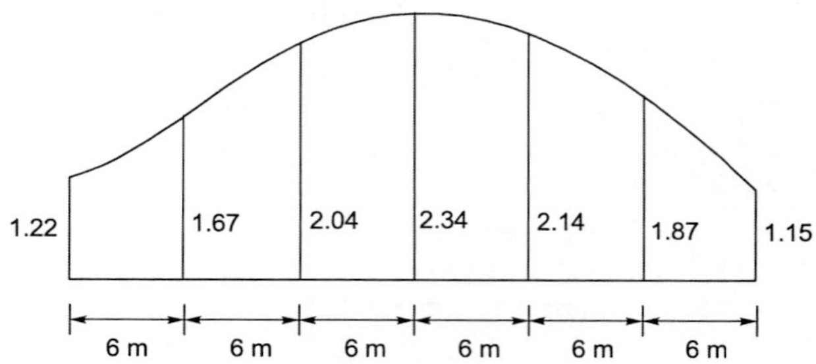

Area by Simpson's rule

$$A = \frac{d}{3}\left[h_0 + h_n + 4(h_1 + h_3 + ...) + 2(h_2 + h_4 + ...)\right]$$

$$= \frac{6}{3}\begin{bmatrix} 1.22 + 1.15 + 4 \times (1.67 + 2.34 + 1.87) \\ +2(2.04 + 2.14) \end{bmatrix}$$

$$= 68.50 \text{ m}^2$$

3.23 Sol.

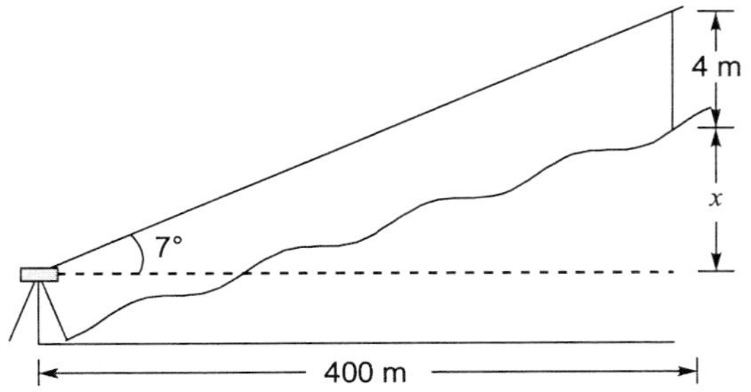

$$V = 400 \tan 7° = 49.113$$

$$x = (49.113 - 4) = 45.113$$

$$RL_B = 212.25 + 45.113 = 257.363 \text{ m}$$

3.24 (d)

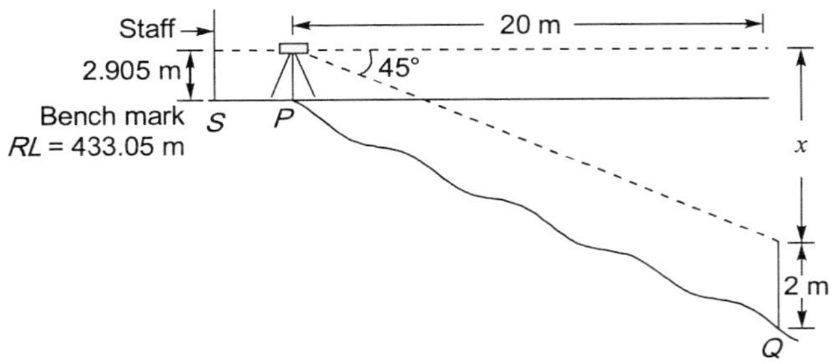

$$\frac{x}{20} = \tan 45°$$

$$x = 20 \text{ m}$$

$$\text{RL of } Q = 433.05 + 2.905 - x - 2$$

$$= 433.05 + 2.905 - 20 - 2$$

$$= 413.955 \text{ m}$$

3.25 Sol.

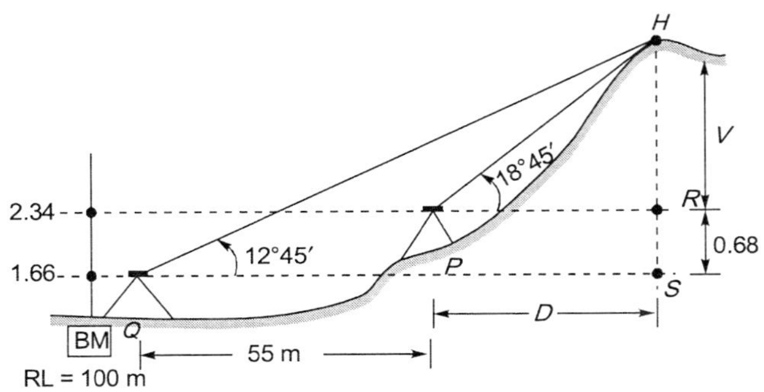

$$\text{Now, } \tan 12°45' = \frac{V + 0.68}{D + 55} \quad \text{...(i)}$$

$$\text{Also, } \tan 18°45' = \frac{V}{D} \quad \text{...(ii)}$$

On solving (i) and (ii), we get

$$V = 35.287 \text{ m}$$

$$D = 103.954 \text{ m}$$

Now, R.L. of hill top = R.L. of BM + 2.34 + V

= 100 + 2.34 + 35.287 = 137.627 m

3.26 Sol.

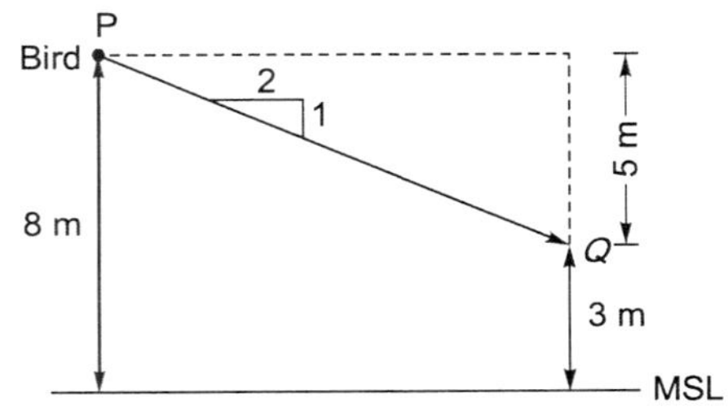

For a falling gradient of 1 in 2,

For vertical distance of 5 m between P and Q,

Horizontal distance between P and Q

= 5 × 2 = 10 m

3.27 (c)

Given, contour Interval = 1 m

Distance between contour in map = 10 mm

$$\text{Map scale} = \frac{1}{1000}$$

∴ Distance between contours = 10 × 1000

= 10,000 mm

$$\therefore \quad \text{Gradient} = \frac{\text{Contour Interval}}{\text{Distance between contour}}$$

$$= \frac{1000 \text{ mm}}{10{,}000 \text{ mm}}$$

$$= \frac{1}{10} = \frac{1}{10} \times 100 = 10\%$$

3.28 Sol.

$\Sigma BS - \Sigma FS$ = Last RL − First RL

$\Rightarrow (1.05 - 2.05) - (0.95 - 1.655) = (RL)_R - 200$

$\Rightarrow \quad RL_R = 199.705 \text{ m}$

■■■■

4 Tacheometric, Curve, Hydrographic and Plane Table Surveying & Theory of Errors

4.1 The relationship between the length (l) and radius (r) of an ideal transition curve is given by

(a) $l \propto r$
(b) $l \propto r^2$
(c) $l \propto \frac{1}{r}$
(d) $l \propto \frac{1}{r^2}$ [1999 : 1 M]

4.2 Two straight lines intersect at an angle of 60°. The radius of a curve joining the two straight lines is 600 m. The length of long chord and mid-ordinates in metres of the curve are

(a) 80.4, 600.0
(b) 600.0, 80.4
(c) 600.0, 39.89
(d) 49.89, 300.0

[2007 : 2 M]

4.3 The focal length of the object glass of a tacheometer is 200 mm, the distance between the vertical axis of the tacheometer and the optical centre of the object glass is 100 mm and the spacing between the upper and lower line of the diaphragm axis is 4 mm. With the line of collimation perfectly horizontal, the staff intercepts are 1 m (top), 2 m (middle), and 3 m (bottom). The horizontal distance (m) between the staff and the instrument station is

(a) 100.3
(b) 103.0
(c) 150.0
(d) 153.0 [2008 : 2 M]

4.4 A tacheometer was placed at point P to estimate the horizontal distances PQ and PR. The corresponding stadia intercepts with the telescope kept horizontal, are 0.320 m and 0.210 m, respectively. The $\angle QPR$ is measured to be 61° 30′ 30″. If the stadia multiplication constant = 100 and stadia addition constant = 0.10 m, the horizontal distance (in m) between the points Q and R is ______.

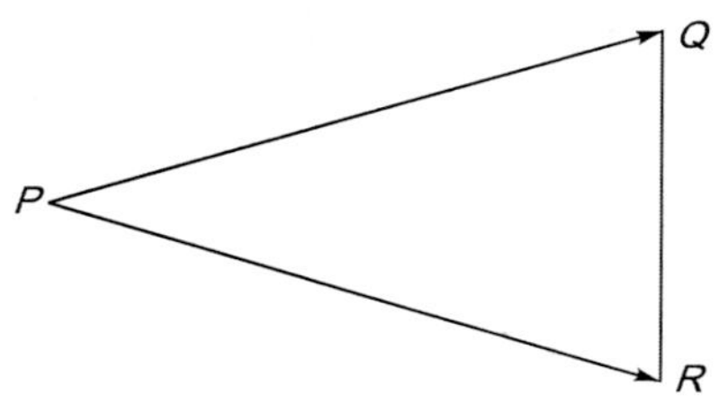

[2014 : 2 M, Set-II]

4.5 The chainage of the intersection point of two straights is 1585.60 m and the angle of intersection is 140°. If the radius of a circular curve is 600.00 m, the tangent distance (in m) and length of the curve (in m), respectively are

(a) 418.88 and 1466.08
(b) 218.38 and 1648.49
(c) 218.38 and 418.88
(d) 418.88 and 218.38 [2014 : 2 M, Set-II]

4.6 In a survey work, three independent angles, X, Y and Z were observed with weights W_X, W_Y, W_Z, respectively. The weight of the sum of angles X, Y and Z is given by

(a) $\dfrac{1}{\left(\dfrac{1}{W_X}+\dfrac{1}{W_Y}+\dfrac{1}{W_Z}\right)}$

(b) $\left(\dfrac{1}{W_X}+\dfrac{1}{W_Y}+\dfrac{1}{W_Z}\right)$

(c) $W_X + W_Y + W_Z$

(d) $W_X^2 + W_Y^2 + W_Z^2$ [2015 : 2 M, Set-I]

4.7 A circular curve of radius R connects two straights with a deflection angle of 60°. The tangent length is

(a) 0.577 R
(b) 1.155 R
(c) 1.732 R
(d) 3.464 R

[2016 : 1 M, Set-II]

4.8 The VPI (vertical point of intersection) is 100 m away (when measured the horizontal) from the VPC (vertical point of curvature). If the vertical curve parabolic, the length of the curve (in meters and measured along the horizontal) is ________.

[2017 : 1 M, Set-II]

4.9 The method of orientation used, when the plane table occupies a position not yet located on the map, is called as

(a) traversing
(b) radiation
(c) levelling
(d) resection

[2017 : 1 M, Set-II]

4.10 A horizontal angle θ is measured by four different surveyors multiple tunes and the values reported are given below.

Surveyor	Angle θ	Number of observations
1	36°30′	4
2	36°00′	3
3	35°30′	8
4	36°30′	4

The most probable value of the angle θ (in degree, round off to two decimal places) is ______.

[2021 : 1 M, Set-II]

4.11 The error in measuring the radius of a 5 cm circular rod was 0.2%, If the cross-sectional area of the rod was calculated using this measurement, then the resulting absolute percentage error in the computed area is _______. (round off to two decimal places)

[2022 : 2 M, Set-II]

4.12 The direct and reversed zenith angles observed by a theodolite are 56° 00′ 00″ and 303° 00′ 00″, respectively. What is the vertical collimation correction?

(a) +1° 00′ 00″
(b) −1° 00′ 00″
(c) −0° 30′ 00″
(d) +0° 30′ 00″

[2023 : 1 M, Set-I]

4.13 Among the following statements relating the fundamental lines of a transit theodolite, which one is CORRECT

(a) The axis of plate level must lie in a plane parallel to the vertical axis.
(b) The axis of altitude level must be perpendicular to the line of collimation
(c) The Vernier of vertical circle must read zero when the line of collimation is vertical.
(d) The line of collimation must be perpendicular to the horizontal axis at its intersection with the vertical axis.

[2024 : 1 M, Set-I]

4.14 What is the CORRECT match between the survey instruments/parts of instruments shown in the table and the operations carried out with them?

Instruments/Parts of instruments	Operations
P - Bubble tube	i - Tacheometry
Q - Plumb bob	ii - Minor movements
R - Tangent screw	iii - Centering
S - Stadia cross-wire	iv - Levelling

(a) P-iv, Q-iii, R-ii, S-i
(b) P-ii, Q-iii, R-iv, S-i
(c) P-i, Q-iii, R-ii, S-iv
(d) P-iii, Q-iv, R-i, S-ii

[2024 : 1 M, Set-II]

4.15 To finalize the direction of a survey, four surveyors set up a theodolite at a station *P* and performed all the temporary adjustments. From the station *P*, each of the surveyors observed the bearing to a tower located at station *Q* with the same instrument without shifting it. The bearings observed by the surveyors are 30° 30′ 00′, 30° 29′ 40″, 30° 30′ 20″ and 30° 31′ 20″. Assuming that each measurement is taken with equal precision, the most probable value of the bearing is

(a) 30° 29′ 40″
(b) 30° 30′ 20″
(c) 30° 31′ 20″
(d) 30° 30′ 00″

[2024 : 1 M, Set-II]

■■■■

Answers **Tacheometric, Curve, Hydrographic and Plane Table Surveying & Theory of Errors**

4.1 (c) 4.2 (b) 4.3 (a) 4.4 (28.8) 4.5 (c) 4.6 (a) 4.7 (a)
4.8 (200) 4.9 (d) 4.10 (36) 4.11 (0.40) 4.12 (d) 4.13 (d) 4.14 (a)
4.15 (b)

Explanations Tacheometric, Curve, Hydrographic and Plane Table Surveying & Theory of Errors

4.1 (c)

For an ideal transition curve,

$$Lr = \text{Constant}$$

$$\Rightarrow \quad L \propto \frac{1}{r}$$

4.2 (b)

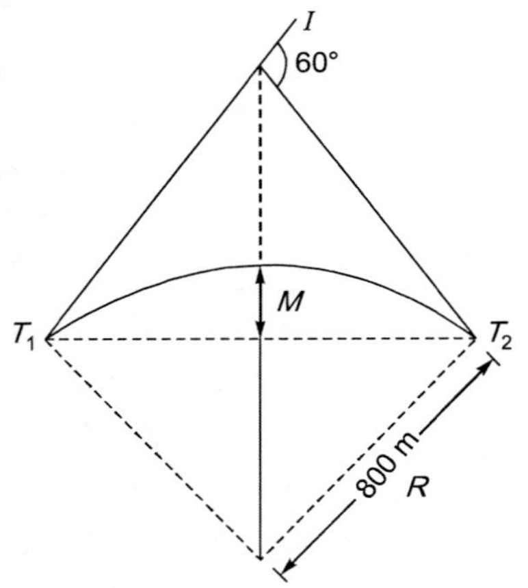

Length of long chord,

$$T_1T_2 = 2R\sin(\Delta/2)$$
$$= 2 \times 600 \times \sin(60/2)$$
$$= 600 \text{ m} \qquad (\because \Delta = 60°)$$

Length of mid-ordinate,

$$M = R[1 - \cos(\Delta/2)]$$
$$= 600[1 - \cos(60/2)]$$
$$= 600 \times 0.134 = 80.4 \text{ m}$$

4.3 (a)

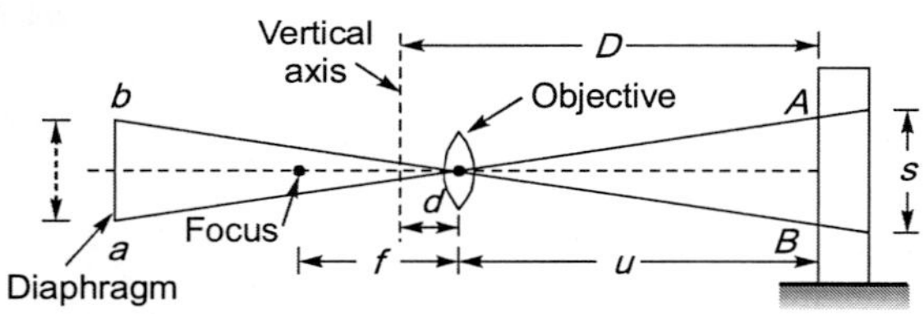

The horizontal distance (D) between the vertical axis and staff may be given as,

$$D = u + d$$

But $u = \left(\frac{f}{i}\right)s + f$

$$D = \left(\frac{f}{i}\right)s + f + d$$

$$= \left(\frac{200}{4}\right) \times (3 - 1) + \frac{200}{1000} + \frac{100}{1000}$$

$$= 100 + 0.3 = 100.3 \text{ m}$$

4.4 Sol.

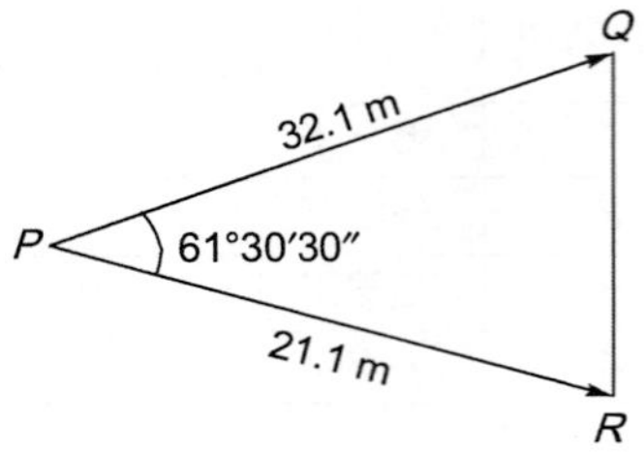

$$PQ = ks + C$$
$$= 100(0.32) + 0.1 = 32.1 \text{ m}$$
$$PR = ks + C$$
$$= 100(0.21) + 0.1 = 21.1 \text{ m}$$

Applying the **cosine rule**

$$QR = \sqrt{PQ^2 + PR^2 - 2(PQ)(PR)\cos\theta}$$

Where, $\theta = 61°30'30'' = 61.508°$

$$= \sqrt{1030.41 + 445.21 - 2(32.1)(21.1)\cos 61.508°}$$

$$= \sqrt{1030.41 + 445.21 - 646.2} = 28.8 \text{ m}$$

4.5 (c)

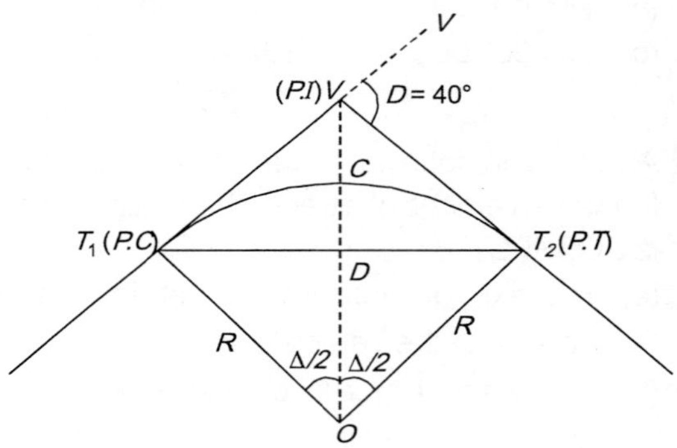

$$\Delta = 180° - 140° = 40°$$

Length of the curve

$$= 2\pi r \times \left(\frac{\Delta}{360°}\right)$$

$$= \frac{\pi \times 600}{180} \times 40 = 418.82 \text{m}$$

Tangent distance (T) is the distance between P-C to $P.I$ (also the distance from $P.I$ to $P.T$)

$$\Rightarrow \quad T = T_1V$$

$$= T_2V = OT_1\tan\frac{\Delta}{2} = R\tan\frac{\Delta}{2}$$

$$= 600 \tan 20° = 218.88 \text{ m}$$

4.6 (a)

Various laws of weight

Let there be quantities A and B with weight W_a and W_b respectively

Operation	Resultant weight
$(A+B)$	$\dfrac{1}{\dfrac{1}{W_a}+\dfrac{1}{W_b}}$
$(A-B)$	$\dfrac{1}{\dfrac{1}{W_a}+\dfrac{1}{W_b}}$
$k.A.$(k=Constant)	$\dfrac{W_a}{k^2}$
$\dfrac{A}{k}$	$W_a k^2$

4.7 (a)

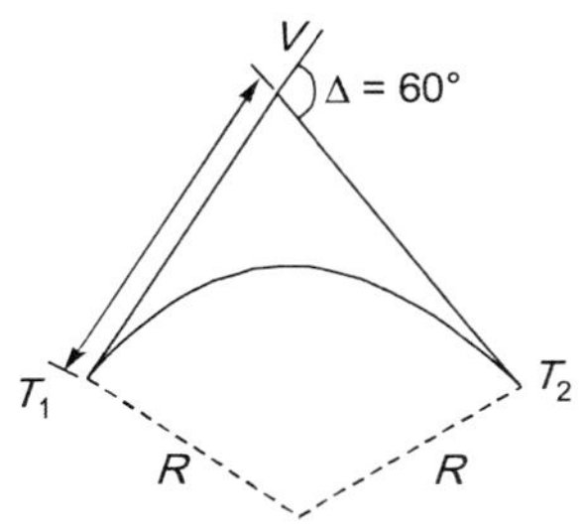

Tangent length,

$$VT_1 = R\tan\frac{\Delta}{2} = R\tan 30° = 0.577\ R$$

4.8 Sol.

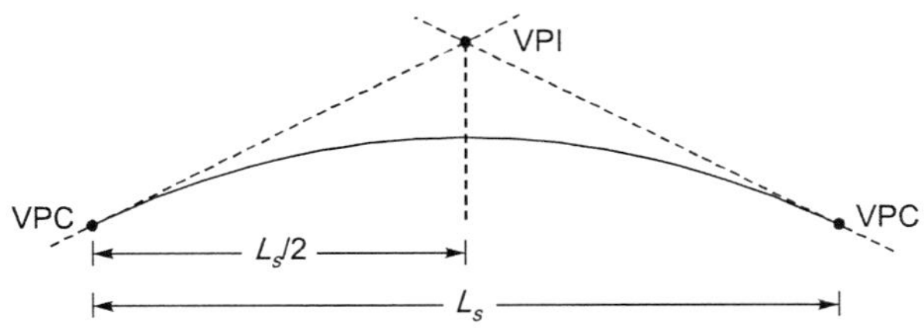

$$L_s = 2 \times 100 = 200 \text{ m}$$

4.9 (d)

1. **Traversing:** 'Traverse' means to pass across. Traversing refer to the framework of series of lines forming an open or closed polygon.
2. **Radiation:** It is a method of locating a point by drawing a radial line from the plane table station to the station under consideration.
3. **Levelling:** It is the method of determining the difference of elevations or levels of different points on the earth's surface.
4. **Resection:** It is the method of locating a station occupied by the plane table when the position of that station has not been plotted earlier on the sheet when the plane table occupied other station.

4.10 Sol.

$$\text{MPV} = \frac{(36°30' \times 4) + (36° \times 3) + (35°30' \times 8) + (36°30' \times 4)}{4+3+8+4}$$

$$= 36°$$

4.11 Sol.

$$A = \pi r^2$$

For percentage error,

$$\partial A = 2\pi r \partial r$$

$$\Rightarrow \quad \frac{\partial A}{A} = \frac{2\pi r \partial r}{A}$$

$$\Rightarrow \quad \frac{\partial A}{A} = \frac{2\pi r \partial r}{\pi r^2}$$

$$\Rightarrow \quad \frac{\partial A}{A} = 2\left[\frac{\partial r}{r}\right] \quad \left\{\frac{\partial r}{r} = 0.2\% \text{ given}\right\}$$

$$\therefore \quad \frac{\partial A}{A} = 2 \times 0.2 = 0.40\%$$

4.12 (d)

ϕ_1 = Observed value of direct zenith angle
= 56° 00′ 00″

ϕ_2 = Observed value of reversed zenith angle
= 303° 00′ 00″

$$\text{Error} = \left[\frac{(\phi_1+\phi_2)-360°}{2}\right]$$

$$= \frac{359°00'00'' - 360°00'00''}{2}$$

$$= -0° 30' 00''$$

So, vertical collimation correction = +0° 30′

4.13 (d)

4.14 (a)

4.15 (b)

Most probable value

$$= \frac{30°30'00'' + 30°29'40'' + 30°30'20'' + 30°31'20''}{4}$$

$$= 30° 30' 20''$$

5 Time Concept, Remote Sensing, GIS, GPS and Photogrammetry

5.1 The local mean time at a place located in Longitude 90° 40′E when the standard time is 6 hours and 30 minutes and the standard meridian is 82° 30′ E is

(a) 5 hours, 2 minutes and 40 seconds
(b) 5 hours, 47 minutes and 20 seconds
(c) 6 hours and 30 minutes
(d) 7 hours, 02 minutes and 40 seconds

[2010 : 1 M]

5.2 The minimum number of satellites needed for a GPS to determine its position precisely is

(a) 2 (b) 3
(c) 4 (d) 24 [2016 : 1 M, Set-I]

5.3 The system that uses the Sun as a source of electromagnetic energy and records the naturally radiated and reflected energy from the object is called

(a) Geographical Information System
(b) Global Positioning System
(c) Passive Remote Sensing
(d) Active Remote Sensing

[2016 : 1 M, Set-I]

5.4 Optimal flight planning for a photogrammetric survey should be carried out considering

(a) only side-lap
(b) only end-lap
(c) either side-lap or end-lap
(d) both side-lap as well as end-lap

[2016 : 1 M, Set-II]

5.5 A tall tower was photographed from an elevation of 700 m above the datum. The radial distances of the top and bottom of the tower from the principal points are 112.50 mm and 82.40 mm, respectively. If the bottom of the tower is at an elevation 250 m above the datum, then the height (expressed in m) of the tower is________

[2016 : 2 M, Set-II]

5.6 The number of spectral bands in the Enhanced Thematic Mapper sensor on the remote sensing satellite Landsat-7 is

(a) 64 (b) 10
(c) 8 (d) 15

[2017 : 1 M, Set-I]

5.7 Two towers, *A* and *B*, standing vertically on a horizontal ground, appear in a vertical aerial photograph as shown in the figure.

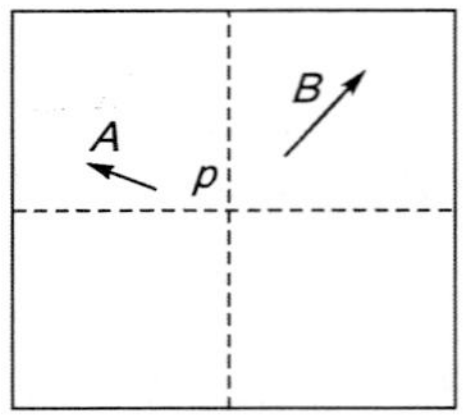

The length of the image of the tower *A* on the photograph is 1.5 cm and of the tower *B* is 2.0 cm. The distance of the top of the tower *A* (as shown by the arrowhead) is 4.0 cm and the distance of the top of the tower *B* is 6.0 cm, as measured from the principal point *p* of the photograph. If the height of the tower *B* is 80 m, the height (in meters) of the tower *A* is ____________.

[2017 : 2 M, Set-II]

5.8 A square area (on the surface of the earth) with side 100 m and uniform height, appears as 1 cm^2 on a vertical aerial photograph. The topographic map shows that a contour of 650 m passes through the area. If focal length of the camera lens is 150 mm, the height from which the aerial photograph was taken, is

(a) 800 m (b) 1500 m
(c) 2150 m (d) 3150 m

[2018 : 2 M, Set-I]

5.9 An aerial photograph of a terrain having an average elevation of 1400 m is taken at a scale of 1:7500. The focal length of the camera is 15 cm. The altitude of the flight above mean sea level (in m, up to one decimal place) is __________.

[2018 : 2 M, Set-II]

5.10 A camera with a focal length of 20 cm fitted in an aircraft is used for taking vertical aerial photographs of a terrain. The average elevation of the terrain is 1200 m above mean sea level (MSL). What is the height above MSL at which an aircraft must fly in order to get the aerial photographs at a scale of 1:8000?

(a) 3200 m (b) 2600 m
(c) 3000 m (d) 2800 m

[2019 : 2 M, Set-II]

5.11 An aerial photograph is taken from a flight at a height of 3.5 km above mean sea level, using a camera of focal length 152 mm. If the average ground elevation is 460 m above mean sea level, then the scale of the photograph is

(a) 1 : 20000 (b) 1 : 20
(c) 1 : 100000 (d) 1 : 2800

[2022 : 1 M, Set-I]

5.12 A student is scanning his 10 inch × 10 inch certificate at 600 dots per inch (dpi) to convert it to raster. What is the percentage reduction in number of pixels if the same certificate is scanned at 300 dpi?

(a) 62 (b) 88
(c) 75 (d) 50

[2023 : 1 M, Set-I]

5.13 If the size of the ground area is 6 km × 3 km and the corresponding photo size in the aerial photograph is 30 cm × 15 cm, then the scale of the photograph is 1 : _______ (in integer)

[2023 : 1 M, Set-II]

5.14 A surveyor observes a zenith angle of 93° 00' 00" during a theodolite survey The corresponding vertical angle is

(a) 87° 00'00" (b) +03° 00'00"
(c) 87° 00'00" (d) –03° 00' 00"

[2024 : 1 M, Set-I]

5.15 For a reconnaissance survey, it is necessary to obtain vertical aerial photographs of a terrain at an average scale of 1:13000 using a camera. If the permissible flying height is assumed as 3000 m above a datum and the average terrain elevation is 1050 m above the datum, the required focal length (in mm) of the camera is

(a) 150 (b) 200
(c) 125 (d) 100

[2024 : 1 M, Set-II]

Answers Time Concept, Remote Sensing, GIS, GPS and Photogrammetry

5.1	(d)	5.2	(c)	5.3	(c)	5.4	(d)	5.5	(120.4)	5.6	(c)	5.7	(240)
5.8	(c)	5.9	(2525)	5.10	(d)	5.11	(a)	5.12	(c)	5.13	(20000)	5.14	(d)
5.15	(a)												

Explanations Time Concept, Remote Sensing, GIS, GPS and Photogrammetry

5.1 (d)

Difference between standard meridian and the longitude of the given place

$$= 90° 40'E - 82°30'E = 8°10'$$

The longitude of the given place is more than the standard meridian. Hence the local mean time of the given place will be ahead of the standard time.

$$8° = \frac{8}{15}\text{hours}$$

= 0 hours and 32 minutes and 0 seconds

$$10' = \frac{10}{15}\text{ minutes}$$

= 0 hours and 0 minutes and 40 seconds

Time for a difference of 8°10′ = 32 min & 40 sec

Local mean time = standard time + 32 min & 40 s

= 6 hrs and 30 min. +32 mins and 40 seconds

=7 hours, 02 minutes and 40 seconds

5.2 (c)

At a minimum, four satellite must be in view of the receiver for it to compute four unknown quantity (three position coordinate and one for clock deviation from satellite time).

5.3 (c)

The sun provide a very convenient source of energy for remote sensing. The Sun's energy is either reflected, as it is for visible wavelengths, or absorbed and then re-emitted, as it is for thermal infrared wavelengths. Remote sensing system which measure energy i.e., naturally available are called as passive remote sensing.

5.4 (d)

5.5 Sol.

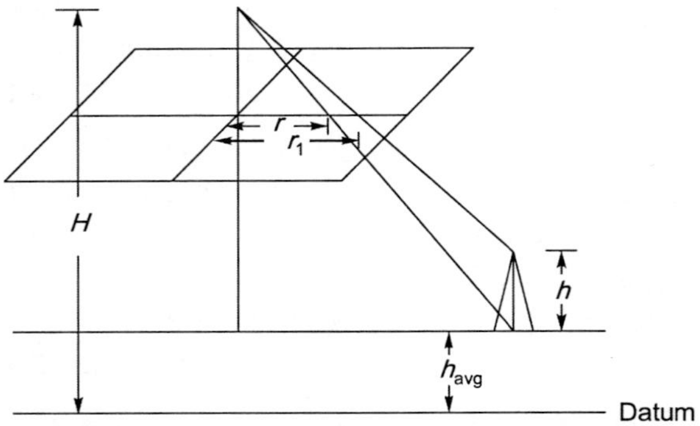

Given: H = 700 m, h_{avg} = 250

Relief distance,

$$d = r_1 - r$$
$$= 112.5 - 82.40 = 30.1 \text{ mm}$$

$$\therefore \quad d = \frac{hr_1}{H - h_{avg}}$$

[where h is height of tower]

$$\Rightarrow \quad 30.1 = \frac{h \times 112.5}{700 - 250}$$

$$\therefore \quad h = 120.4 \text{ mm}$$

5.6 (c)

Total number of spectrum band in Enhanced Thematic Mapper sensor = 8

Band 1 – Blue

Band 2 – Green

Band 3 – Red

Band 4 – Near Infrared (NR)

Band 5 – Shortwave Infrared (SWIR)

Band 6 – Thermal

Band 7 – Shortwave Infrared (SWIR)

Band 8 – Panchromatic

5.7 Sol.

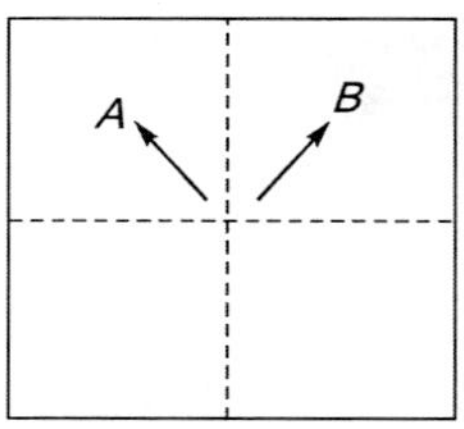

For Tower A:

Relief displacement,

$$d_A = 1.5 \text{ cm}$$
$$r_A = 4 \text{ cm},$$
$$h_A = ?$$

For Tower B:

Relief displacement,

d_B = 2.0 cm, r_B = 6 cm, h_B = 80 m

Relief displacement,

$$d = \frac{h.r}{(H - h_{avg})}$$

For A: $$d_A = \frac{h_A.r_A}{(H - h_{avg})} \Rightarrow 1.5 = \frac{h_A \times 4}{240}$$

$$h_A = 90 \text{ m}$$

For B: $$d_B = \frac{h_B.r_B}{(H - h_{avg})} \Rightarrow 2 = \frac{80 \times 6}{(H - h_{avg})}$$

$$\Rightarrow (H - h_{avg}) = 240 \text{ m}$$

5.8 (c)

$$A = 100 \times 100 \text{ m}^2$$

Area on photo,

$$a = 1 \text{ cm}^2$$

Scale $\quad 1 \text{ cm} = 100 \text{ m}$

$$f = 150 \text{ mm}$$
$$h = 650 \text{ m}$$

$$\text{Scale} = \frac{1}{100} = \frac{1}{100 \times 10^2}$$
$$= \frac{1}{10000} = \frac{f}{H - h}$$

$$\Rightarrow \quad \frac{1}{10000} = \frac{150 \times 10^{-3}}{H - 650}$$

$$\Rightarrow \quad H = 2150 \text{ m}$$

5.9 Sol.

$$h = 1400 \text{ m}$$
$$\text{Scale} = 1 : 7500$$
$$f = 15 \text{ cm}$$

$$\text{Scale} = \frac{f}{H - h}$$

$$\Rightarrow \quad \frac{1}{7500} = \frac{15 \times 10^{-2}}{H - 1400}$$

$$\Rightarrow \quad H = 2525 \text{ m}$$

5.10 (d)

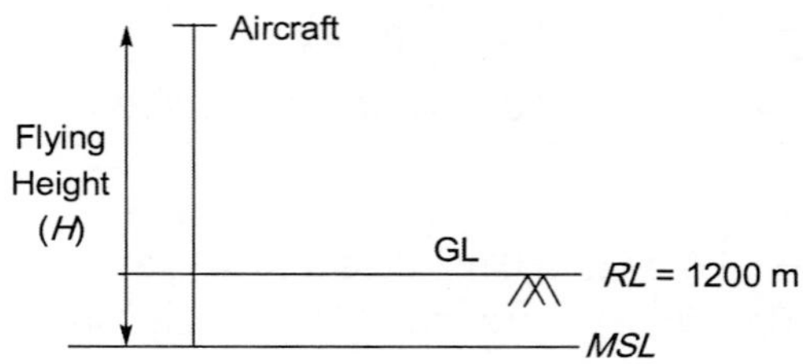

Given focal length = 20 cm

as we know scale of vertical photograph

$$= \frac{f}{H - h_{avg.}}$$

its given as 1 : 8000

Hence, $$\frac{f}{H - h_{avg.}} = \frac{1}{8000}$$

$$\frac{20\text{cm}}{(H-1200)\times 100\text{cm}} = \frac{1}{8000}$$

$\Rightarrow$ $H = 2800$ m

5.11 (a)

Flying height, $H = 3500$ m

$h_{avg} = 460$ m

Focal length, $f = 152$ mm

Scale of photograph,

$$S = \frac{f}{H - h_{avg}}$$

$$\Rightarrow \quad S = \frac{152}{3500 - 460} = \frac{1}{20000}$$

5.12 (c)

Number of dots as per 600 dpi = $(10 \times 600)^2$

= 36000000

Number of dots as per 300 dpi = $(10 \times 300)^2$

= 9000000

Percentage reduction in number of pixels

$$= \frac{36000000 - 9000000}{36000000} \times 100$$

$$= \frac{27}{36} \times 100 = 75\%$$

5.13 Sol.

$$\text{Scale} = \frac{\text{Photo distance}}{\text{Ground distance}}$$

$$= \frac{30\text{cm}}{6\text{km}} = \frac{30\text{cm}}{600000\text{cm}} = \frac{1}{20000}$$

Alternatively,

$$(\text{Scale})^2 = \left(\frac{\text{Area of map}}{\text{Area on ground}}\right)$$

$$= \frac{(30\times 15)}{6\times 3} = \frac{30\times 15}{(6\times 3)\times(100000)^2}$$

$$\Rightarrow \quad \text{Scale} = \sqrt{\frac{25}{(100000)^2}}$$

$$\therefore \quad S = \frac{1}{20000}$$

5.14 (d)

Zenith = 93° 00′ 00″

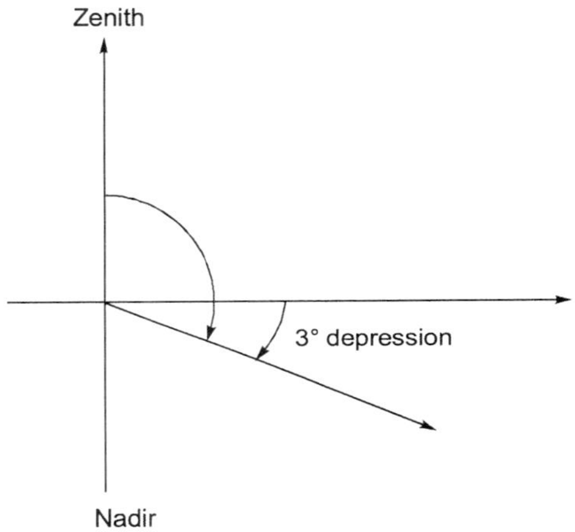

5.15 (a)

$$\text{Scale} = \frac{1}{13000} = \frac{f}{H - h_{avg}}$$

where f is focal length

$$\Rightarrow \quad \frac{1}{13000} = \frac{f}{3000 - 1050}$$

$\Rightarrow$ $f = 0.15$ m $= 150$ mm

■■■■

UNIT

XII

Construction Materials, Construction Management and Engg. Mechanics

CONTENTS

Construction Materials, Construction Management and Engineering Mechanics

Syllabus

Construction Materials: Structural Steel – Composition, material properties and behaviour; Concrete - Constituents, mix design, short-term and long-term properties.

Construction Management: Types of construction projects; Project planning and network analysis - PERT and CPM; Cost estimation.

Engg. Mechanics : System of forces, free-body diagrams, equilibrium equations; Internal forces in structures; Frictions and its applications; Centre of mass; Free Vibrations of undamped SDOF system.

Analysis of Previous GATE Papers

Construction Materials			
Exam Year	1 Mark Ques.	2 Marks Ques.	Total Marks
2015 Set-1	1	–	1
2015 Set-2	–	–	–
2016 Set-1	2	–	2
2016 Set-2	1	–	1
2017 Set-1	–	–	–
2017 Set-2	–	2	4
2018 Set-1	1	–	1
2018 Set-2	1	–	1
2019 Set-1	–	1	2
2019 Set-2	–	–	–
2020 Set-1	1	–	1
2020 Set-2	–	2	4
2021 Set-1	1	–	1
2021 Set-2	1	–	1
2022 Set-1	1	–	1
2022 Set-2	1	–	1
2023 Set-1	–	–	–
2023 Set-2	–	–	–
2024 Set-1	1	–	1
2024 Set-2	–	–	–

Construction Management			
Exam Year	1 Mark Ques.	2 Marks Ques.	Total Marks
2016 Set-1	–	1	2
2016 Set-2	1	1	3
2017 Set-1	–	1	2
2017 Set-2	2	–	2
2018 Set-2	–	–	–
2019 Set-1	–	1	2
2019 Set-2	–	–	–
2021 Set-1	1	2	5
2021 Set-2	1	2	5
2022 Set-1	–	1	2
2022 Set-2	2	2	6
2023 Set-1	–	–	–
2023 Set-2	–	–	–
2024 Set-1	1	1	3
2024 Set-2	2	2	6

Engineering Mechanics			
Exam Year	1 Mark Ques.	2 Marks Ques.	Total Marks
2006	–	1	2
2010	–	1	2
2014 Set-2	1	–	1
2016 Set-1	1	–	1
2016 Set-2	1	1	3
2018 Set-1	–	2	4
2018 Set-2	–	2	4
2019 Set-1	–	–	–
2019 Set-2	–	–	–
2020 Set-1	–	–	–
2020 Set-2	–	1	2
2021 Set-1	–	1	2
2021 Set-2	–	2	4
2022 Set-1	–	1	2
2022 Set-2	–	1	2
2023 Set-1	–	1	2
2023 Set-2	–	–	0
2024 Set-1	–	–	–
2024 Set-2	–	–	–

Construction Materials, Construction Management and Engineering Mechanics

1. Construction Materials

1.1 The results for sieve analysis carried out for three types of sand, *P*, *Q* and *R*, are given in the adjoining figure. If the fineness modulus values of the three sands are given as FM_P, FM_Q and FM_R, it can be stated that

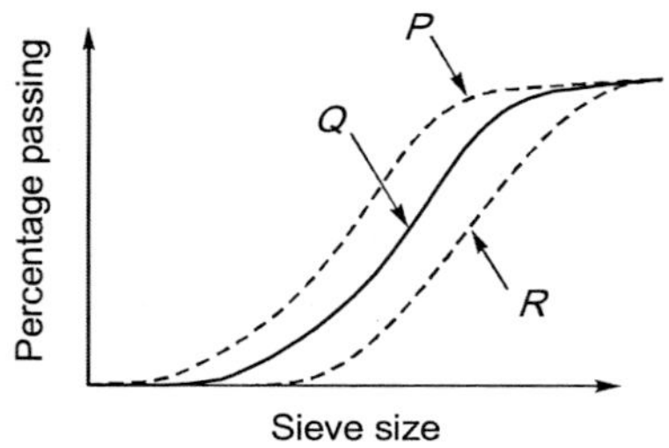

(a) $FM_Q = \sqrt{FM_P \times FM_R}$

(b) $FM_Q = 0.5\,(FM_P + FM_R)$

(c) $FM_P > FM_Q > FM_R$

(d) $FM_P < FM_Q < FM_R$ [2011 : 1 M]

1.2 The composition of an air-entrained concrete is given below:

Water	:	184 kg/m^3
Ordinary Portland Cement (OPC)	:	368 kg/m^3
Sand	:	606 kg/m^3
Coarse aggregate	:	1155 kg/m^3

Assume the specific gravity of OPC, sand and coarse aggregate to be 3.14, 2.67 and 2.74 respectively. The air content is ________ liters/m^3.

[2015 : 2 M, Set-I]

1.3 Bull's trench kiln is used in the manufacturing of

(a) lime (b) cement

(c) bricks (d) none of these

[2016 : 1 M, Set-I]

1.4 The compound which is largely responsible for initial setting and early strength gain of Ordinary Portland Cement is

(a) C_3A (b) C_3S

(c) C_2S (d) C_4AF

[2016 : 1 M, Set-I]

1.5 As per Indian standards for bricks, minimum acceptable compressive strength of any class of burnt clay bricks in dry state is

(a) 10.0 MPa (b) 0.75 MPa

(c) 5.0 MPa (d) 3.5 MPa

[2016 : 1 M, Set-II]

1.6 Group I gives a list of test methods and test apparatus for evaluating some of the properties of Ordinary Portland Cement (OPC) and concrete. Group II gives the list of these properties.

Group-I

P. Le Chattier test

Q. Vee-Bee test

R. Blaine air permeability test

S. The Vicat apparatus

Group-II

1. Soundless of OPC
2. Consistency and setting time of OPC
3. Consistency or workability of concrete
4. Fineness of OPC

The correct match of the items in Group I with the items in Group II is

(a) P-1, Q-3, R-4, S-2 (b) P-2, Q-3, R-1, S-4

(c) P-4, Q-2, R-4, S-1 (d) P-1, Q-4, R-2, S-3

[2017 : 2 M, Set-II]

1.7 Consider the following statements:

P. Walls of one brick thick are measured in square meters.

Q. Walls of one brick thick are measured in cubic meters.

R. No deduction in the brickwork quantity is made for openings in walls up to 0.1 m^2 area.

S. For the measurement of excavation from the borrow pit in a fairly uniform ground, deadman are left at suitable intervals.

For the above statements, the correct option is

(a) P-False; Q-True; R-False; S-True

(b) P-False; Q-True; R-False; S-False

(c) P-True; Q-False; R-True; S-False

(d) P-True; Q-False; R-True; S-True

[2017 : 2 M, Set-II]

1.8 The Le Chatelier apparatus is used to determine
(a) compressive strength of cement
(b) fineness of cement
(c) setting time of cement
(d) soundness of cement [2018 : 1 M, Set-I]

1.9 The setting time of cement is determined using
(a) Le Chatelier apparatus
(b) Briquette testing apparatus
(c) Vicat apparatus
(d) Casagrande's apparatus
[2018 : 1 M, Set-II]

1.10 A box measuring 50 cm x 50 cm x 50 cm is filled to the top with dry coarse aggregate of mass 187.5 kg. The water absorption and specific gravity of the aggregate are 0.5% and 2.5, respectively. The maximum quantity of water (in kg, round off to 2 decimal places) required to fill the box completely is_______.
[2019 : 2 M, Set-I]

1.11 During the process of hydration of cement, due to increase in Dicalcium Silicate (C_2S) content in cement clinker, the heat of hydration
(a) does not change
(b) decreases
(c) initially decreases and then increases
(d) increases
[2020 : 1 M, Set-I]

1.12 The maximum applied load on a cylindrical concrete specimen of diameter 150 mm and length 300 mm tested as per the split tensile strength test guidelines of **IS 5816 : 1999** is 157 kN. The split tensile strength (in MPa, round off to one decimal place) of the specimen is _______.
[2020 : 1 M, Set-II]

1.13 Group-I gives a list of test methods for evaluating properties of aggregates. Group-II gives the list of properties to be evaluated.

Group-I : Test Methods
P. Soundness test
Q. Crushing test
R. Los Angeles abrasion test
S. Stripping value test

Group-II: Properties
1. Strength
2. Resistance to weathering
3. Adhesion
4. Hardness

The correct match of test methods under Group-I to properties under Group-II, is
(a) P-4, Q-1, R-2, S-3
(b) P-2, Q-4, R-3, S-1
(c) P-2, Q-1, R-4, S-3
(d) P-3, Q-4, R-1, S-2
[2020 : 2 M, Set-II]

1.14 Gypsum is typically added in cement to
(a) increase workability
(b) decrease heat of hydration
(c) prevent quick setting
(d) enhance hardening
[2021 : 1 M, Set-I]

1.15 Seasoning of timber for use in construction is done essentially to
(a) smoothen timber surfaces
(b) cut timber in right season and geometry
(c) remove knots from timber logs
(d) increase strength and durability
[2021 : 1 M, Set-II]

1.16 Which of the following equations is correct for the Pozzolanic reaction?
(a) $Ca(OH)_2$ + Reactive Superplasticiser + H_2O $\rightarrow$ C-S-H
(b) $Ca(OH)_2$ + Reactive Silicon dioxide + H_2O $\rightarrow$ C-S-H
(c) $Ca(OH)_2$ + Reactive Sulphates + H_2O $\rightarrow$ C-S-H
(d) $Ca(OH)_2$ + Reactive Sulphur + H_2O $\rightarrow$ C-S-H
[2022 : 1 M, Set-I]

1.17 Match all the possible combinations between Column *X* (Cement compounds) and Column *Y* (Cement properties):

Column X	Column Y
(i) C_3S	(P) Early age strength
(ii) C_2S	(Q) Later age strength
(iii) C_3A	(R) Flash setting
	(S) Highest heat of hydration
	(T) Lowest heat of hydration

Which one of the following combinations is correct?
(a) (i) - (P), (ii) - (Q) and (T), (iii) - (R) and (S)
(b) (i) - (Q) and (T), (ii) - (P) and (S), (iii) - (R)
(c) (i) - (P), (ii) - (Q) and (R), (iii) - (T)
(d) (i) - (T), (ii) - (S), (iii) - (P) and (Q)
[2022 : 2 M, Set-II]

1.18 The elements that DO NOT increase the strength of structural steel are

(a) Carbon (b) Chlorine

(c) Manganese (d) Sulphur

[2024 : 1 M, Set-I]

2. Construction Management

2.1 A construction project consists of twelve activities. The estimated duration (in days) required to complete each of the activities along with the corresponding network diagram is shown below.

Activity		Duration (days)	Activity		Duration (days)
A	Inauguration	1	G	Flooring	25
B	Foundation work	7	H	Electrification	7
C	Structural Construction-1	30	I	Plumbing	7
D	Structural construction-2	30	J	Wood work	7
E	Brick masonry work	25	K	Coloring	3
F	Plastering	7	L	Handing over function	1

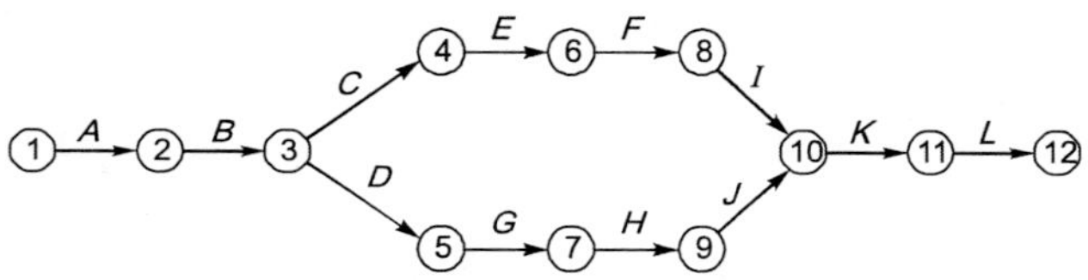

Total floats (in days) for the activities 5-7 and 11-12 for the project are, respectively,

(a) 25 and 1 (b) 1 and 1

(c) 0 and 0 (d) 81 and 0

[2016 : 1 M, Set-II]

2.2 The Optimistic Time (O), Most likely Time (*M*) and Pessimistic Time (*P*) (in days) of the activities in the critical path are given below in the format O-M-P.

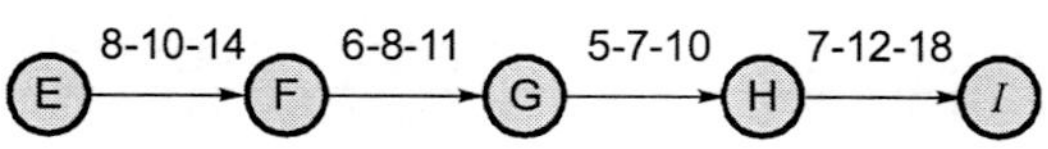

The expected completion time (in days) of the project is ________

[2016 : 2 M, Set-I]

2.3 The activity-on-arrow network of activities for a construction project is shown in the figure. The durations (expressed in days) of the activities are mentioned below the arrows.

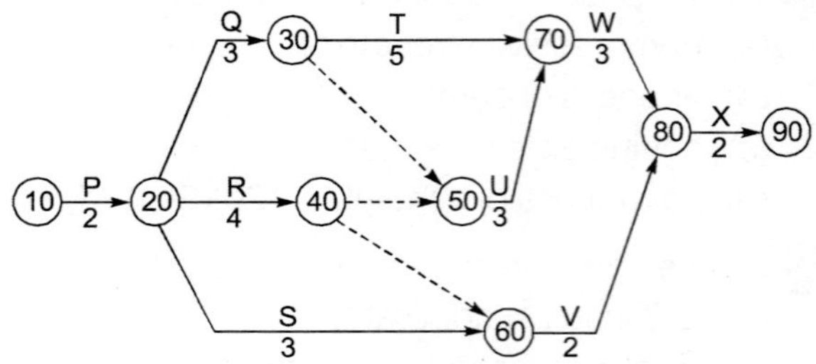

The critical duration for this construction project is

(a) 13 days (b) 14 days

(c) 15 days (d) 16 days

[2016 : 2 M, Set-II]

2.4 The activity details of a project are given below:

Activity	Depends on	Duration (in days)
P	–	6
Q	*P*	15
R	*Q, T*	12
S	*R*	16
T	*P*	10
U	*Q, T*	14
V	*U*	16

The estimated minimum time (in days) for the completion of the project will be ______.

[2017 : 2 M, Set-I]

2.5 Given that the scope of the construction work is well-defined with all its drawings, specifications, quantities and estimates, which one of the following types of contract would be most preferred?

(a) EPC contract

(b) Percentage rate contract

(c) Item rate contract

(d) Lump sum contract

[2017 : 1 M, Set-II]

2.6 For a construction project, the mean and standard deviation of the completion time are 200 days and 6.1 days, respectively. Assume normal distribution and use the value of standard normal deviate, $z = 1.64$ for the 95% confidence level. The maximum time required (in days) for the completion of the project would be ________.

[2017 : 1 M, Set-II]

2.7 The network of a small construction project awarded to a contractors is shown in the following figure. The normal duration, crash duration, normal cost and crash cost of all the activities are shown in the table. The indirect cost incurred by the contractor in INR 5000 per day.

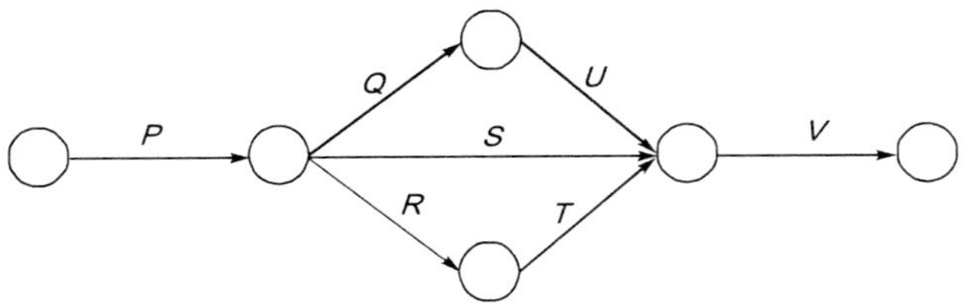

Activity	Normal Duration (days)	Crash Duration (days)	Normal cost (INR)	Crash Cost (INR)
P	6	4	15000	25000
Q	5	2	6000	12000
R	5	3	8000	9500
S	6	3	7000	10000
T	3	2	6000	9000
U	2	1	4000	6000
V	4	2	20000	28000

If the project is target for completion in 16 days, the total cost (in INR) to be incurred by the contractor would be_______.

[2019 : 2 M, Set-I]

2.8 The direct and indirect costs estimated by a contractor for bidding a project is Rs. 160000 and Rs. 20000 respectively. If the mark up applied is 10% of the bid price, the quoted price (in Rs.) of the contractor is

(a) 198000 (b) 196000
(c) 200000 (d) 182000

[2021 : 1 M, Set-I]

2.9 A small project has 12 activities – N, P, Q, R, S, T, U, V, W, X, Y and Z. The relationship among these activities and the duration of these activities are given in the table.

Activity	Duration (in weeks)	Depends upon
N	2	–
P	5	N
Q	3	N
R	4	P
S	5	Q
T	8	R
U	7	R,S
V	2	U
W	3	U
X	5	T,V
Y	1	W
Z	3	X,Y

The total float of the activity "V" (in weeks, in integer) is ______.

[2021 : 2 M, Set-I]

2.10 Contractor X is developing his bidding strategy against Contractor Y. The ratio of Y's bid price to X's cost for the 30 previous bids in which Contractor X has competed against Contractor Y is given in the table.

Ratio of Y's bid price to X's cost	Number of bids
1.02	6
1.04	12
1.06	3
1.10	6
1.12	3

Based on the bidding behaviour of the Contractor Y, the probability of winning against Contractor Y at a mark up of 8% for the next project is

(a) 0%
(b) more than 50% but less than 100%
(c) more than 0% but less than 50%
(d) 100%

[2021 : 2 M, Set-I]

2.11 An equipment has been purchased at an initial cost of ₹ 160000 and has an estimated salvage value of ₹ 10000. The equipment has an estimated life of 5 years. The difference between the book values (in ₹ in integer) obtained at the end of 4th year using straight line method and sum of years digit method of depreciation is _______.

[2021 : 2 M, Set-II]

2.12 In case of bids in Two-Envelop System, the correct option is

(a) Either of the two (Technical and Financial) bids can be opened first
(b) Technical bid is opened first
(c) Both (Technical and Financial) bids are opened simultaneously
(d) Financial bid is opened first

[2021 : 1 M, Set-II]

2.13 The activity details for a small project are given in the Table.

Activity	Duration (days)	Depends on
A	6	-
B	10	A
C	14	A
D	8	B

E	12	C
F	8	C
G	16	D, E
H	8	F, G
K	2	B
L	5	G, K

The total time (in days, in integer) for project completion is ________.

[2021 : 2 M, Set-II]

2.14 The project activities are given in the following table along with the duration and dependency.

Activities	Duration (days)	Depends on
P	10	-
Q	12	-
R	5	P
S	10	Q
T	10	P, Q

Which one of the following combinations is correct?

(a) Total duration of the project = 22 days, Critical path is Q → S

(b) Total duration of the project = 20 days, Critical path is Q → T

(c) Total duration of the project = 22 days, Critical path is P → T

(d) Total duration of the project = 20 days, Critical path is P → R

[2022 : 2 M, Set-I]

2.15 The activities of a project are given in the following table along with their durations and dependency.

Activities	Duration (day)	Depends on
A	10	–
B	12	–
C	5	*A*
D	14	*B*
E	10	*B*, *C*

The total float of the activity E (in day) is ______ . (in integer)

[2022 : 1 M, Set-II]

2.16 Consider the statements P and Q

P: Client's Preliminary Estimate is used for budgeting costs toward the end of planning and design phase.

Q: Client's Detailed Estimate is used for controlling costs during the execution of the project

Which one of the following options is CORRECT?

(a) Both P and Q are TRUE

(b) P is FALSE and Q is TRUE

(c) Both P and Q are FALSE

(d) P is TRUE and Q is FALSE

[2024 : 1 M, Set-I]

2.17 The initial cost of an equipment is Rs. 1,00,000. Its salvage value at the end of accounting life of 5 years is Rs. 10,000. The difference in depreciation (in Rs.) computed using 'double-declining balance method' and 'straight line method' of depreciation in Year-2 is ________ (in positive integer)

[2024 : 2 M, Set-I]

2.18 Consider the following data for a project of 300 days duration.

Budgeted Cost of Work Scheduled (BCWS) = Rs. 200

Budgeted Cost of Work Performed (BCWP) = Rs. 150

Actual Cost of Work Performed (ACWP) = Rs. 190

The 'schedule variance' for the project is

(a) (–)50 days (b) (–)Rs 50

(c) (+)50 days (d) (+)Rs 50

[2024 : 1 M, Set-II]

2.19 Consider the statements *P* and *Q*.

P : In a pure project organization, the project manager maintains complete authority and has maximum control over the project.

Q : A matrix organization structure facilitates quick response to changes, conflicts, and project needs.

Which one of the following options is CORRECT?

(a) Both *P* and *Q* are TRUE

(b) *P* is TRUE and *Q* is FALSE

(c) *P* is FALSE and *Q* is TRUE

(d) Both *P* and *Q* are FALSE

[2024 : 1 M, Set-II]

2.20 The table shows the activities and their durations and dependencies in a project.

Activity	Duration (Days)	Depends on
A	8	-
B	4	A
C	4	B
D	4	C, L
F	4	A
G	4	F
H	6	G, L
K	10	A
L	6	F, K

The total duration (in days) of the project is ________ (in integer).

[2024 : 2 M, Set-II]

2.21 A critical activity in a project is estimated to take 15 days to complete at a cost of Rs. 30,000. The activity can be expedited to complete in 12 days by spending a total amount of Rs. 54,000. Consider the statements *P* and *Q*.

P : It is economically advisable to complete the activity early by crashing, if the indirect cost of the project is Rs. 8,500 per day.

Q : It is economically advisable to complete the activity early by crashing, if the indirect cost of the project is Rs. 10.000 per day.

Which one of the following options is CORRECT?

(a) Both *P* and *Q* are TRUE
(b) *P* is TRUE and *Q* is FALSE
(c) *P* is FALSE and *Q* is TRUE
(d) Both *P* and *Q* are FALSE

[2024 : 2 M, Set-II]

3. Engineering Mechanics

3.1 For the section shown below, second moment of the area about an axis $d/4$ distance above the bottom of the area is

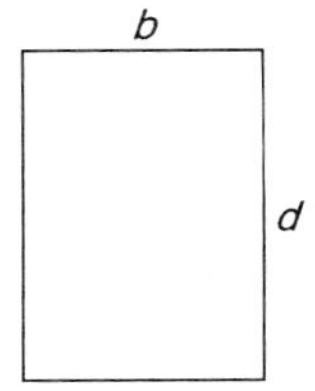

(a) $\frac{bd^3}{48}$ (b) $\frac{bd^3}{12}$
(c) $\frac{7bd^3}{48}$ (d) $\frac{bd^3}{3}$

[2006 : 2 Marks]

3.2 A disc of radius *r* has a hole of radius $r/2$ cut-out as shown. The centroid of the remaining disc (shaded portion) at a radial distance from the centre "*O*" is

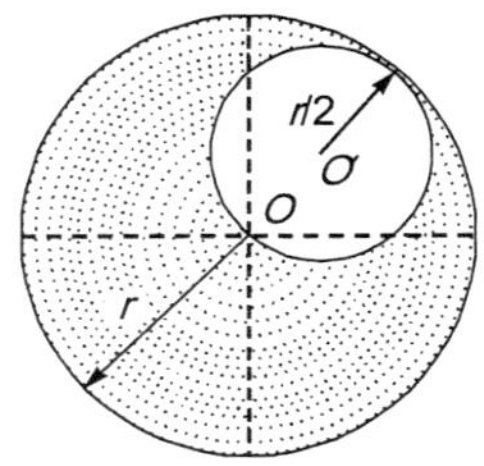

(a) $\frac{r}{2}$ (b) $\frac{r}{3}$
(c) $\frac{r}{6}$ (d) $\frac{r}{8}$ [2010 : 2 Marks]

3.3 Polar moment of inertia (I_p), in cm^4, of a rectangular section having width, b = 2 cm and depth, d = 6 cm is ____________.

[2014 : 1 M, Set-II]

3.4 An assembly made of a rigid arm *A-B-C* hinged at end *A* and supported by an elastic rope *C-D* at end *C* is shown in the figure. The members may be assumed to be weightless and the lengths of the respective members are as shown in the figure

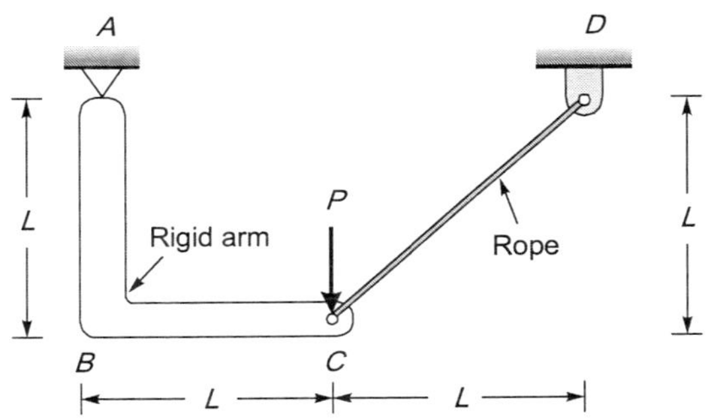

Under the action of a concentrated load *P* at *C* as shown, the magnitude of tension developed in the rope is

(a) $\frac{3P}{\sqrt{2}}$ (b) $\frac{P}{\sqrt{2}}$
(c) $\frac{3P}{8}$ (d) $\sqrt{2}P$

[2016 : 1 M, Set-II]

3.5 The magnitudes of vectors P, Q and R are 100 kN, 250 kN and 150 kN, respectively as shown in the figure

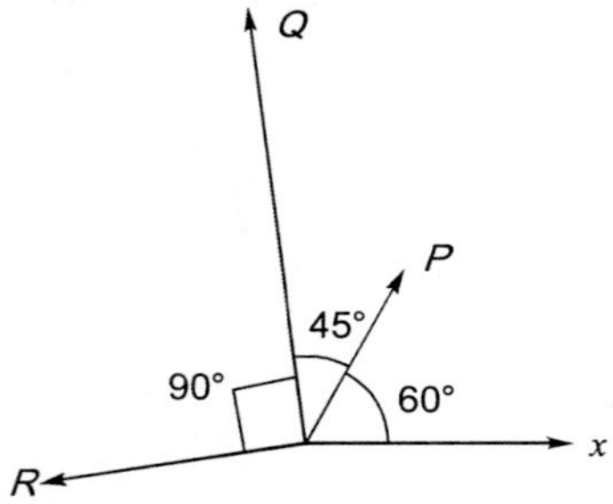

The respective values of the magnitude (in kN) and the direction (with respect to the x-axis) of the resultant vector are

(a) 290.9 and 96.0° (b) 368.1 and 94.7°
(c) 330.4 and 118.9° (d) 400.1 and 113.5°

[2016 : 2 M, Set-I]

3.6 The reaction rate involving reactants A and B is given by $-k[A]^{\alpha}[B]^{\beta}$. Which one of the following statements is valid for the reaction to be a first-order reaction?

(a) $\alpha = 0$ and $\beta = 0$ (b) $\alpha = 1$ and $\beta = 0$
(c) $\alpha = 1$ and $\beta = 1$ (d) $\alpha = 1$ and $\beta = 2$

[2017 : 1 M, Set-I]

3.7 A particle of mass 2 kg is travelling at a velocity of 1.5 m/s. A force $f(t) = 3t^2$ (in N) is applied to it in the direction of motion for a duration of 2 seconds, where t denotes time in seconds. The velocity (in m/s, up to one decimal place) of the particle immediately after the removal of the force is ________. [2017 : 2 M, Set-I]

3.8 An aircraft approaches the threshold of a runway strip at a speed of 200 km/h. The pilot decelerates the aircraft at a rate of 1.697 m/s^2 and takes 18 s to exit the runway strip. If the deceleration after exiting the runway is 1 m/s^2, then the distance (in m, up to one decimal place) of the gate position from the location of exit on the runway is _____

[2018 : 2 M, Set-I]

3.9 A cylinder of radius 250 mm and weight, W = 10 kN is rolled up an obstacle of height 50 mm by applying a horizontal force P at its centre as shown in the figure.

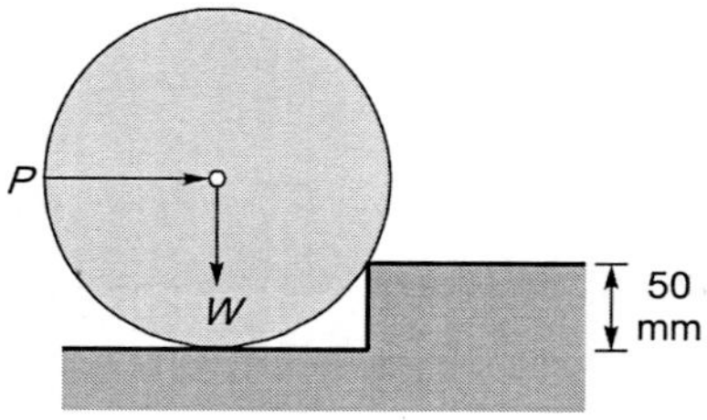

All interfaces are assumed frictionless. The minimum value of P is

(a) 4.5 kN (b) 5.0 kN
(c) 6.0 kN (d) 7.5 kN

[2018 : 2 M, Set-I]

3.10 Two rigid bodies of mass 5 kg and 4 kg are at rest on a frictionless surface until acted upon by a force of 36 N as shown in the Figure. The contact force generated between the two bodies is

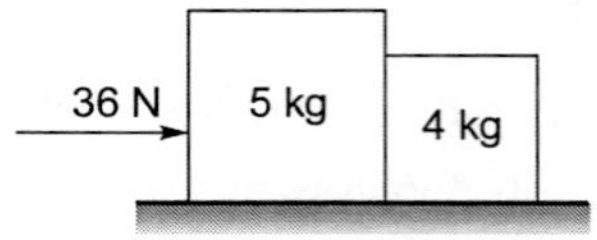

(a) 4.0 N (b) 7.2 N
(c) 9.0 N (d) 16.0 N

[2018 : 2 M, Set-II]

3.11 A cable PQ of length 25 m is supported at two ends at the same level as shown in the figure. The horizontal distance between the supports is 20 m. A point load of 150 kN is applied at point R which divides it into two equal parts.

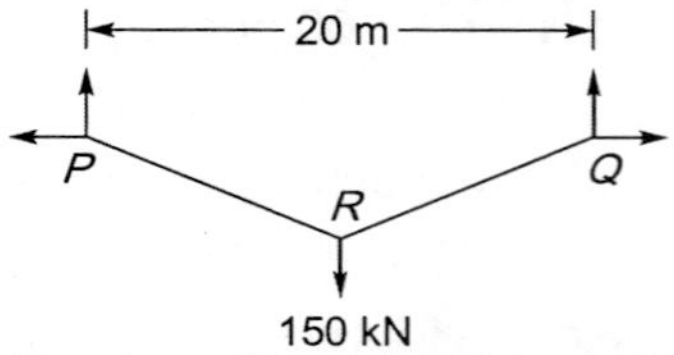

Neglecting the self-weight of the cable, the tension (in kN, integer value) in the cable due to the applied load will be ________.

[2018 : 2 M, Set-II]

3.12 A rigid weightless platform PQRS shown in the figure (not drawn to the scale) can slide freely in the vertical direction. The platform is held in position by the weightless member OJ and four weightless, frictionless rollers. Point O and J are pin connections. A block of 90 kN rests on the platform as shown in the figure.

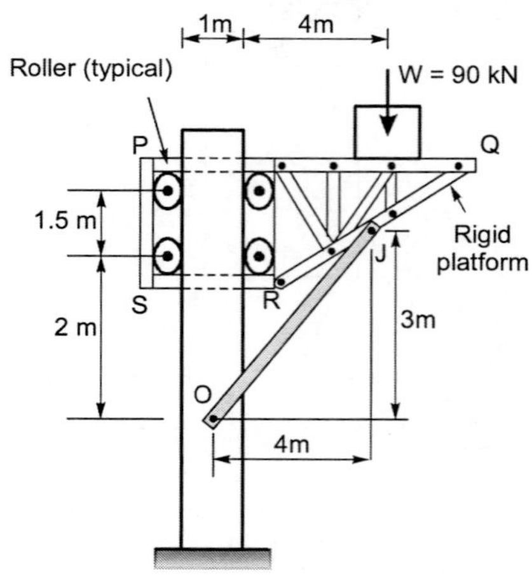

The magnitude of horizontal component of the reaction (in kN) at pin O, is

(a) 180 (b) 150

(c) 90 (d) 120

[2020 : 2 M, Set-I]

3.13 A wedge M and a block N are subjected to force P and Q as shown in the figure. If force P is sufficiently large, then the block N can be raised. The weights of the wedge and the block are negligible compared to the forces P and Q. The coefficient of friction (μ) along the inclined surface between the wedge and the block is 0.2. All other surfaces are frictionless. The wedge angle is 30°.

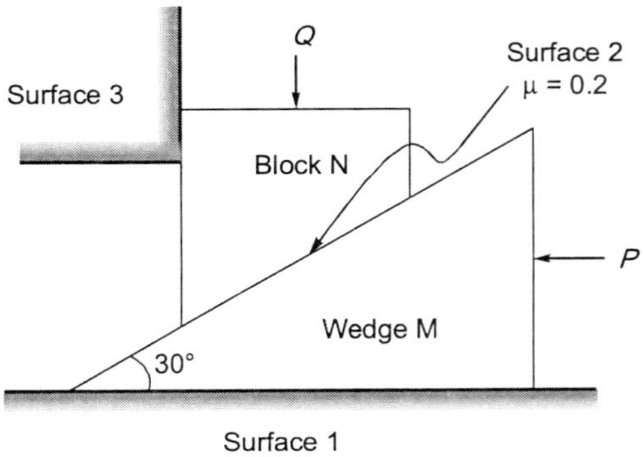

The limiting force P, in terms of Q, required for impending motion of block N to just move it in the upward direction is given as P = αQ. The value of the coefficient 'α' (round off to one decimal place) is

(a) 0.5 (b) 0.9

(c) 2.0 (d) 0.6

[2021 : 2 M, Set-I]

3.14 A prismatic fixed-fixed beam, modelled with a total lumped-mass of 10 kg as a single degree of freedom (SDOF) system is shown in the figure.

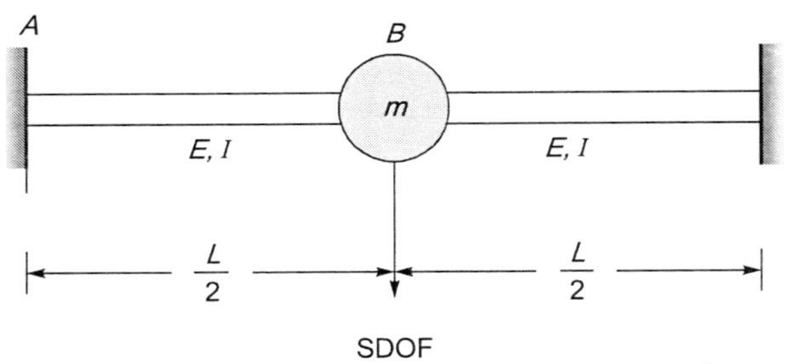

If the flexural stiffness of the beam is $4\pi^2$ kN/m, its natural frequency of vibration (in Hz, in integer) in the flexural mode will be _______.

[2021 : 2 M, Set-II]

3.15 A fire hose nozzle directs a steady stream of water of velocity 50 m/s at an angle of 45° above the horizontal. The stream rises initially but then eventually falls to the ground. Assume water as incompressible and inviscid. Consider the density of air and the air friction as negligible, and assume the acceleration due to gravity as 9.81 m/s^2. The maximum height (in m, round off to two decimal places) reached by the stream above the hose nozzle will then be _________.

[2021 : 2 M, Set-II]

3.16 A horizontal force of P kN is applied to a homogeneous body of weight 25 kN, as shown in the figure. The coefficient of friction between the body and the floor is 0.3. Which of the following statement(s) is/are correct?

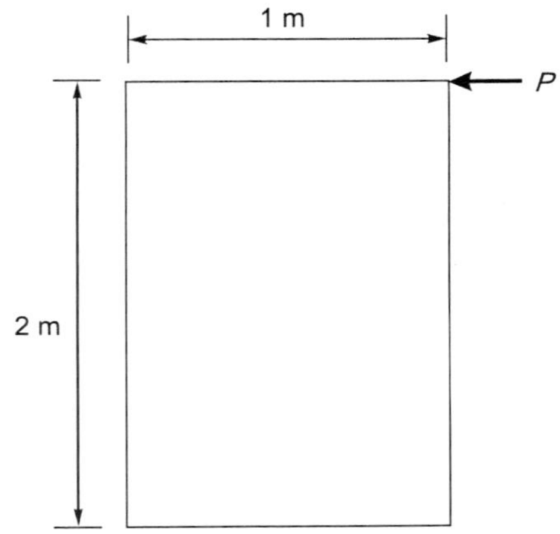

(a) The motion of the body will occur by overturning.

(b) Sliding of the body never occurs.

(c) No motion occurs for $P \leq 6$ kN.

(d) The motion of the body will occur by sliding only.

[2022 : 2 M, Set-I]

3.17 A uniform rod *KJ* of weight *w* shown in the figure rests against a frictionless vertical wall at the point K and a rough horizontal surface at point J, It is given that w = 10 kN, a = 4 in and b = 3 m.

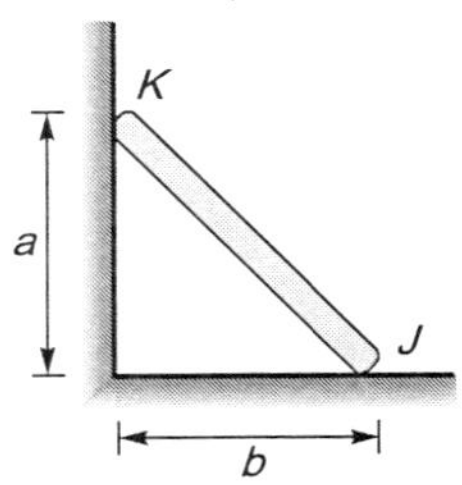

The minimum coefficient of static friction that is required at the point J to hold the rod in equilibrium is_______. (round off to three decimal places)

[2022 : 2 M, Set-II]

3.18 The cross-section of a girder is shown in the figure (not to scale). The section is symmetric about a vertical axis (Y-Y). The moment of inertia of the section about the horizontal axis (X-X) passing through the centroid is ________ cm^4. (round off to nearest integer).

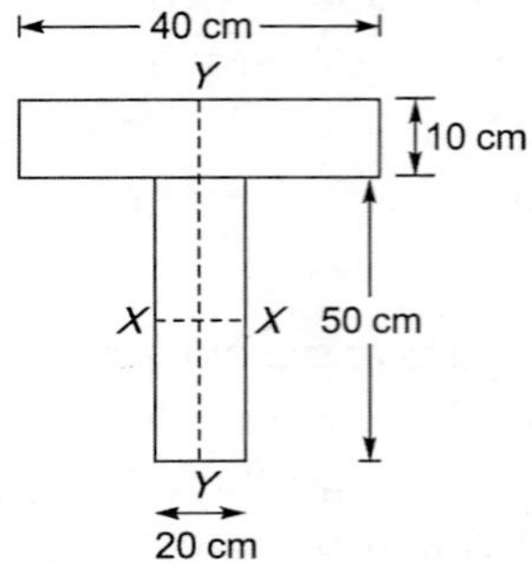

[2023 : 2 M, Set-I]

Answers Construction Materials, Construction Management and Engg. Mechanics

1.1	(a)	1.2	(50.3)	1.3	(c)	1.4	(a)	1.5	(d)	1.6	(a)	1.7	(d)
1.8	(d)	1.9	(c)	1.10	(50.94)	1.11	(b)	1.12	(2.22)	1.13	(c)	1.14	(c)
1.15	(d)	1.16	(b)	1.17	(a)	1.18	(b, d)						
2.1	(c)	2.2	(37.83)	2.3	(c)	2.4	(51)	2.5	(c)	2.6	(210)	2.7	(149500)
2.8	(c)	2.9	(0)	2.10	(c)	2.11	(20000)	2.12	(b)	2.13	(56)		
2.14	(a)	2.15	(1)	2.16	(a)	2.17	(60)	2.18	(b)	2.19	(a)	2.20	(30)
2.21	(*)												
3.1	(c)	3.2	(c)	3.3	(40)	3.4	(b)	3.5	(c)	3.6	(b)	3.7	(5.5)
3.8	(312.8)	3.9	(d)	3.10	(d)	3.11	(125)	3.12	(d)	3.13	(b)	3.14	(10)
3.15	(63.71)	3.16	(a, b, c)	3.17	(0.375)	3.18	(468810)						

Explanations Construction Materials, Construction Management and Engg. Mechanics

1. Construction Materials

1.1 (a)

The term fineness modulus is used to indicate an index number which roughly proportional to average size of particle in entire quantity of aggregate.

$$F_{MQ} = \sqrt{F_{MP} \times F_{MR}} \quad \text{....(Geometric mean)}$$

1.2 Sol.

Let the total volume of air-entrained concrete is unity.

$$\frac{M_C}{\rho_C} + \frac{M_{FA}}{\rho_{FA}} + \frac{M_{CA}}{\rho_{CA}} + V_W + V_{Air} = 1.0$$

$$\frac{368}{3.14 \times 1000} + \frac{606}{2.67 \times 1000} + \frac{1155}{2.74 \times 1000} + \frac{184}{1000} + V_V = 1.0$$

$$0.11720 + 0.22697 + 0.42153$$

$$+ 0.184 + V_v = 1.0$$

$$V_V = 0.0503$$

$$V_V = 0.0503 \times 1000 = 50.3\ l/m^3$$

1.3 (c)

Bull trench kiln is used for brick manufacturing. It is a contnuous kiln has movable chimney. Other than Bull Trench kiln, there are Hoffman kiln and Tunnel kiln are continous type of kiln.

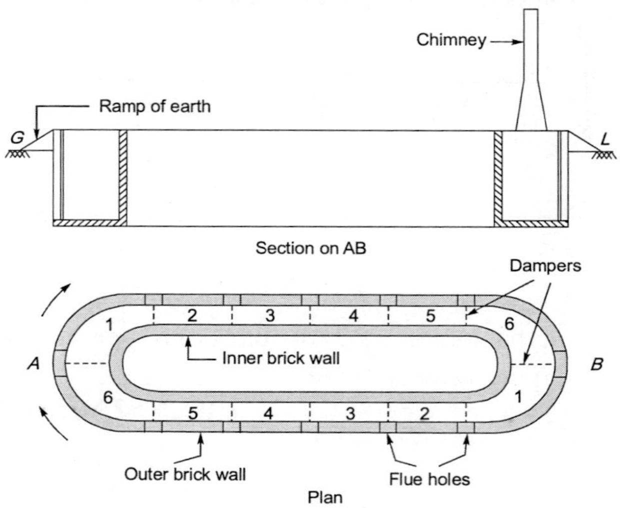

Bull's trench kiln

Section	Arrangement
1	Loading
2	Empty
3	Unloading
4	Cooling
5	Burning
6	Heating

1.4 (a)

C_3S has no role in setting.

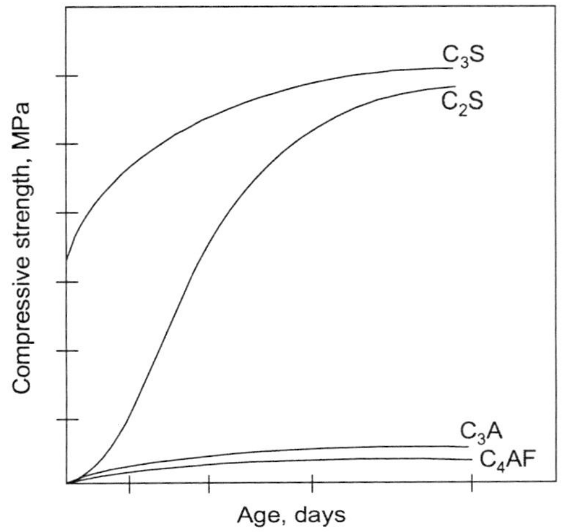

1.5 (d)

As per **IS 1077 : 1992 clause 4.1**, minimum strength of burnt clay bricks is 3.5 MPa. The bricks with crushing strength of 7 to 14 N/mm^2 are graded as A and those having above 14 N/mm^2 are graded as AA.

1.6 (a)

1.7 (d)

1. One brick wall → 20 cm (Nominal thickness)
2. 1.5 brick wall → 30 cm (Nominal thickness)
3. 2 brick wall → 40 cm (Nominal thickness)

where "Nominal thickness" is considered for measuring of wall work.

- Brick wall is measured in m^2 is the brick wall thickness is 10 cm (half brick wall) and 20 cm (full brick wall/one brick wall) and for brick wall greater than one brick wall i.e., 1.5 or 2 brick wall. The brick wall is measured in m^3.
- No deduction for openings and bearings in masonry works is made when
  - (a) opening each upto 1000 sq. cm or 0.1 m^2
  - (b) ends of beam, post, rafters upto 500 sq./m or 0.05 m^2
  - (c) Bed plate, wall, plate, bearing of balcony (chajja) and like upto 10 cm depth, bearing of floor and roof slab are not deducted from masonry.

As per 151200 (Part-I) : 1992, "deadmen and tell tales" in borrow pits is not measured separately and its removal is done after measurement.

1.8 (d)

Le Chatelier Apparatus is used to determine the soundness of cement as per IS code 4031 (part 3); this cement testing procedure is called Le Chatelier test for determining the unsoundness properties of cement due to presence of "free lime".

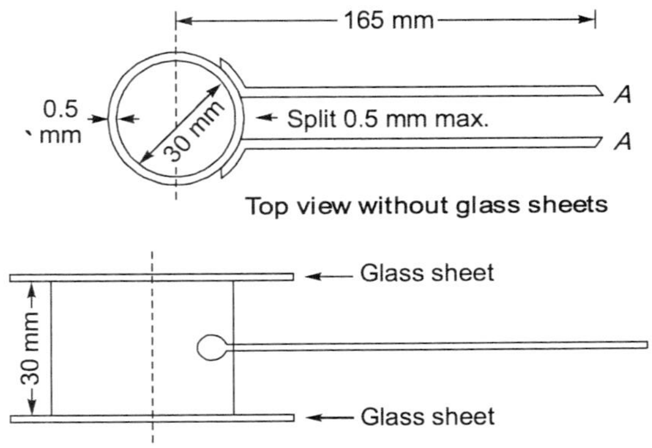

Le Chatelier apparatus

1.9 (c)

Vicat apparatus is used to determine:

(i) Consistency, the percentage of water required to preparing cement pastes for other tests by the help of plunger of length 50 mm and diameter 10 mm.

(ii) Initial setting time, by the help of square needle of cross-section 1 mm × 1 mm.

(iii) Final setting time, by the help of the needle with annular collar.

1.10 Sol.

Given,

Volume of box = 50 × 50 × 50

$= 0.5^3\ m^3 = 0.125\ m^3$

Mass of dry coarse aggregate = 187.5 kg

Water absorption=0.5%

Specific gravity = 2.5

Water required to fill the box = ?

We know,

Volume of particles of Ingredients

= Total volume (air free)

$$\frac{\text{Mass of Coarse Aggregate(kg)}}{G_{CA}} + \frac{\text{Mass of water (kg)}}{G_w} = 0.125 \times 1000$$

$$\frac{187.5}{2.5} + \frac{M_w}{1} = 125$$

$$M_w = 50 \text{ kg}$$

But addition water loss in absorption

$$= \frac{0.5}{100} \times 187.5 = 0.9375 \text{ kg}$$

Total water to be added

$$= 50 + 0.94 = 50.94 \text{ kg}$$

1.11 (b)

1.12 Sol.

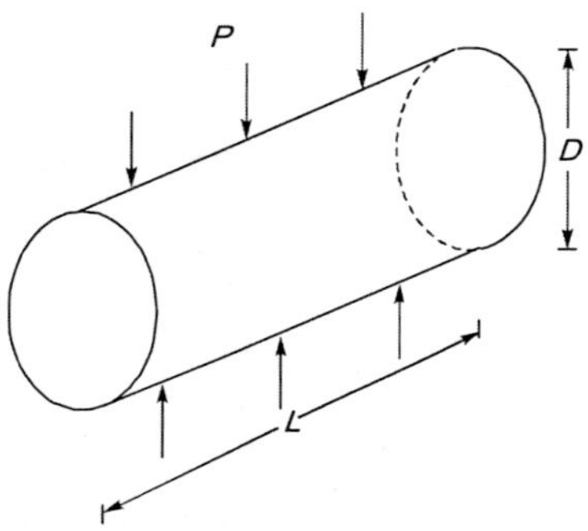

P = 157 kN

D = 150 mm

L = 300 mm

In split tensile strength test, split tensile strength of concrete

$$f_{et} = \frac{2P}{\pi DL} = \frac{2 \times 157000}{\pi \times 150 \times 300} = 2.22 \text{ MPa}$$

1.13 (b)

1.14 (c)

- Gypsum is a retarder added in cement during grinding of clinker in tube mill/ball mill.
- It increases initial setting time.

1.15 (d)

Seasoning provide following advantages:

1. It reduces weight so, transportation become economical.
2. Seasoned timber is less liable to shrinkage or swell.
3. Seasoned timber is durable as it impart resistance against attack of natural agencies like termite.
4. Seasoned timber has more strength, allowable for application of decorative and can burn easily.

1.16 (b)

1.17 (a)

- C_3S is responsible for early age strength i.e. within 7 days.
- C_2S is responsible for progressive strength and releases low heat of hydration i.e. 260 J/g.
- C_3A is responsible for flash setting within 24 hours and also releases maximum heat of hydration i.e. 865 J/g.

1.18 (b, d)

2. Construction Management

2.1 (c)

Total float can be determined once the activity times i.e. EST, EFT, LST and LFT are known.

Total float, $F_T = LST - EST$

$= LFT - EFT$

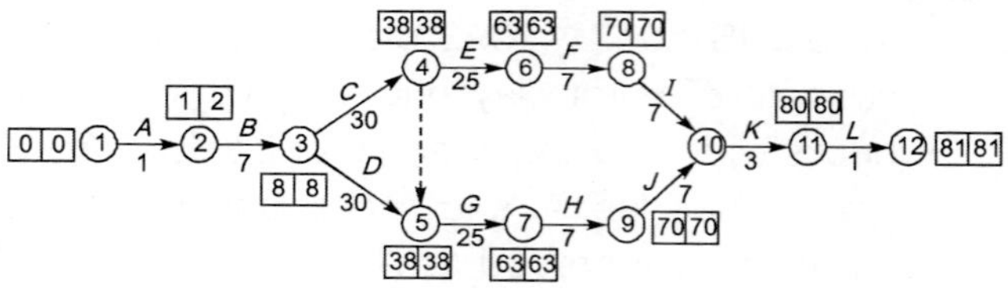

For activity 5-7,

$EST = 38$

$EFT = 63$

$LFT = 63$

$LST = 38$

$F_T = 0$

For activity 11-12,

$EST = 80$

$EFT = 81$

$LFT = 81$

$LST = 80$

$F_T = 0$

Note: *It can be seen directly that since the slack of all events are zero, there is not margin left for the occurence of events and therefore.*

Maximum available line = Time required for completion of activity

$\therefore$ *F_T for all activities is* ***zero.***

2.2 Sol.

$$T_e = \frac{t_0 + 4t_m + t_p}{6}$$

$$T = T_{e_1} + T_{e_2} + T_{e_3} + T_{e_4}$$

$$= \frac{8+4\times10+14}{6} + \frac{6+4\times8+11}{6} + \frac{5+4\times7+10}{6} + \frac{7+4\times12+18}{6}$$

$$= 10.333 + 8.166 + 7.166 + 12.166$$

$$= 37.83 \text{ days}$$

2.3 (c)

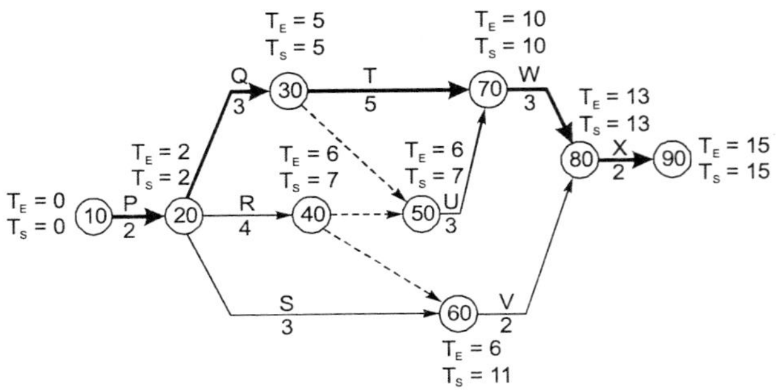

The critical duration is 15 days which is observed along the path *P-Q-T-W-X* which is the critical path of the project.

Note: Critical path is the one which consumes **maximum amount of time.**

2.4 Sol.

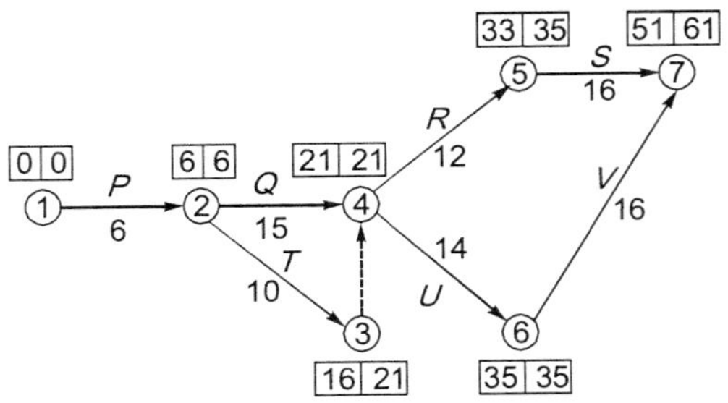

Note: Path of longest duration is defined as the critical path as the activities lying in this path are called critical activities. Any delay in these activity will lead to delay in overall project. The project time given by critical path. Hence the time taken by the critical path is the shortest possible time to finish the project.

Critical Path → P – Q – U – V

$T_{cp} = 6 + 15 + 14 + 16 = 51$ *days*

2.5 (c)

Percentage rate Contract: In this form of contract, the displacemente draws up 'Item rate tender' i.e., bill of quantities with-rate, amount and total amount. The contractors are required to offer to carry out the work as per the rates shown in the specifi price schedule or some percenrage above or some percentage below the rates indicated in the shedule of work attached with the tender.

Item rate contract: In this a contract under takes the execution of work on an item rate basis. He is required to quote rate for individual item of work o the basis of shedule of quantities i.e., BOQ furnished by the department.

Lumpsum concract: In this type, a contractor is required to quote a fixed sum for execution of work complete in all respect in the stipulated time according to the drawing, design and specification supportd to him with the tender.

In item rate contract, contractor only fill its rate against given item and its quantity and contract does not change the item make, drawing, specification during construction without consent of owner/department.

Item rate contract is also called "unit price contract" where contractor bids cost per unit of each item. These works are carried out in accordance with contract documents. The contractor has to perform the quantities of the work actually required in the field at his quoted unit prices, whether they are less or greater than "owner/department" estimates.

2.6 Sol.

Expected completion time of project,

$$t_{E\,project} = 200 \text{ days}$$

Standard deviation of project,

$$\sigma_{project} = 6.1 \text{ days}$$

$$Z = \frac{T_S - T_E}{\sigma_{project}} = \frac{T_S - t_{E\,project}}{\sigma_{project}}$$

$$1.64 = \frac{T_S - 200}{6.1}$$

$$T_S = 200 + 6.1 \times 1.64$$

$$= 200 + 10.004 = 210.004 \simeq 210 \text{ days}$$

2.7 Sol.

Activity	t_n (days)	t_c (days)	C_n (INR)	C_c (INR)	Cost slope $C_s = \frac{C_C - C_n}{t_n - t_c}$
P	6	4	15000	25000	INR 5000/*d*
Q	5	2	6000	12000	INR 2000/*d*
R	5	3	8000	9500	INR 750/*d*
S	6	3	7000	10000	INR 1000/*d*
T	3	2	6000	9000	INR 3000/*d*
U	2	1	4000	6000	INR 2000/*d*
V	4	2	20000	28000	INR 4000/*d*

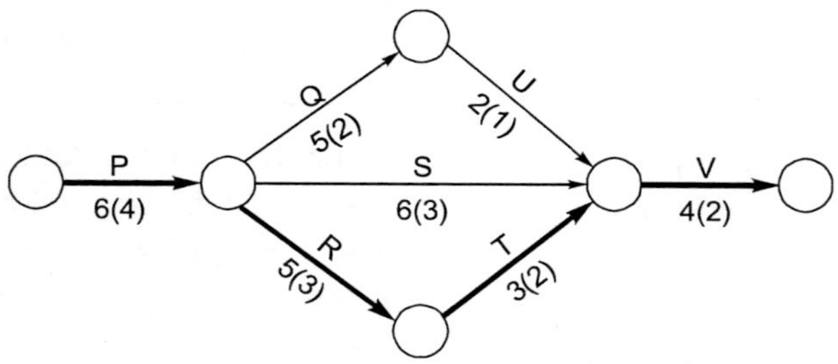

Considering normal durations, the total cost of the project

$$\text{TC} = \text{DC} + \text{IC}$$
$$= \text{INR } 66000 + \text{INR } 5000 \times 18$$
$$= \text{INR } 156000$$

Stage-I: Crash activity R by 1 day

∴ Project becomes

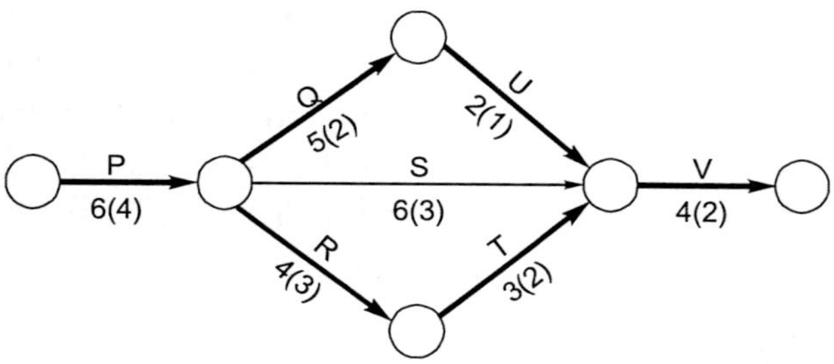

$$\text{TC} = (\text{INR } 6600 + \text{INR } 750) + \text{INR } 5000 \times 17$$
$$= 151750$$

Stage-II: Crash activity Q&R (or U&R) by 1 day

∴ Project becomes

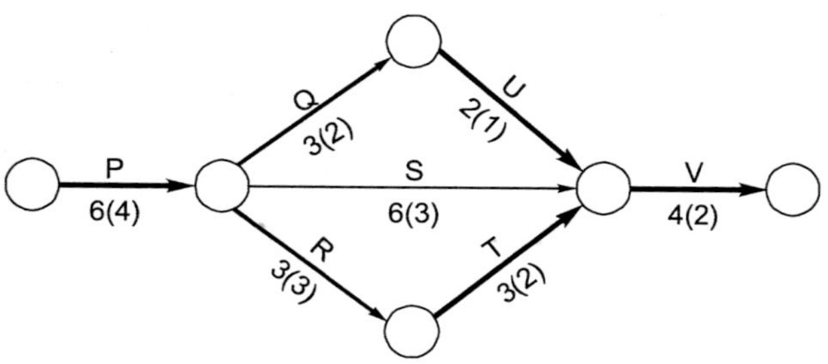

$$\text{TC} = (\text{INR } 66750 + \text{INR } 2000 + \text{INR } 750) + (\text{INR } 5000 \times 16)$$
$$= 149500$$

2.8 (c)

Total cost estimated

= Rs. 160000 + Rs. 20000 = Rs. 180000

Mark up = 10%

0.9 × Quoted price = Total cost estimated

$$\text{Quoted price} = \frac{180000}{0.9} = \text{Rs. } 200000$$

2.9 Sol.

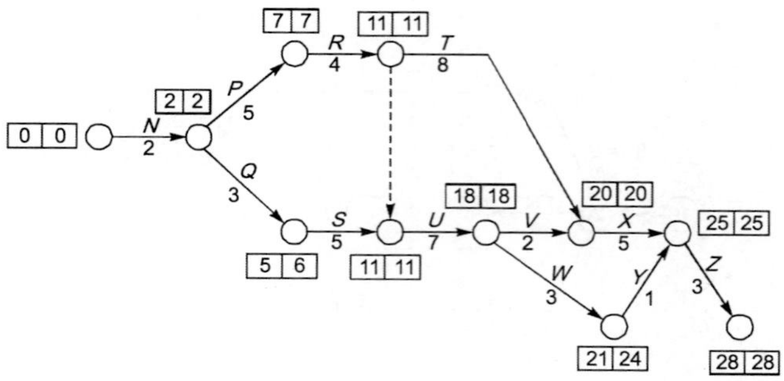

$$F_{T_V} = T_{L_j} - T_{E_i} - t_{ij}$$
$$= 20 - 18 - 2 = 0$$

2.10 (c)

2.11 Sol.

$$C_i = 160000$$
$$C_s = 10000$$
$$n = 5 \text{ years}$$
$$D_m = \frac{C_i - C_s}{n}$$
$$= \frac{160000 - 10000}{5} = 30000$$

Book value of the end of 4 years

$$B_4 = C_i - 4\,D_m$$

$= 160000 - 4 \times 30000 = 40000$

According to sum of years digit method

$$D_m = (C_i - C_s)\left[\frac{(n-m+1)}{\frac{n(n+1)}{2}}\right]$$

$$\frac{n(n+1)}{2} = \frac{5(5+1)}{2} = 15$$

$$\therefore \quad D_1 = (C_i - C_s)\frac{(5-1+1)}{15} = 50000$$

$\therefore \quad B_1 = 160000 - 50000 = 110000$

$$D_1 = 150000 \times \frac{(5-2+1)}{15} = 40000$$

$\therefore \quad B_2 = 110000 - 40000 = 70000$

$$D_3 = 150000 \times \frac{(5-3+1)}{15} = 30000$$

$B_3 = 70000 - 30000 = 40000$

$$D_4 = 150000 \times \frac{(5-4+1)}{15} = 20000$$

$\therefore \quad B_4 = 40000 - 20000 = 20000$

Difference = 40000 − 20000 = 20000

2.12 (b)

2.13 Sol.

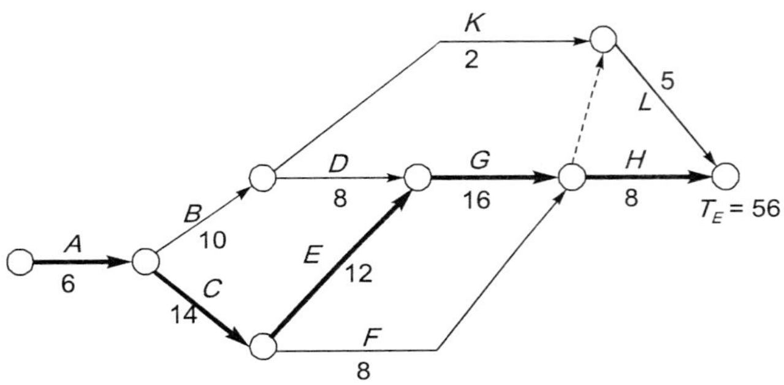

Project duration = 56 days.

2.14 (a)

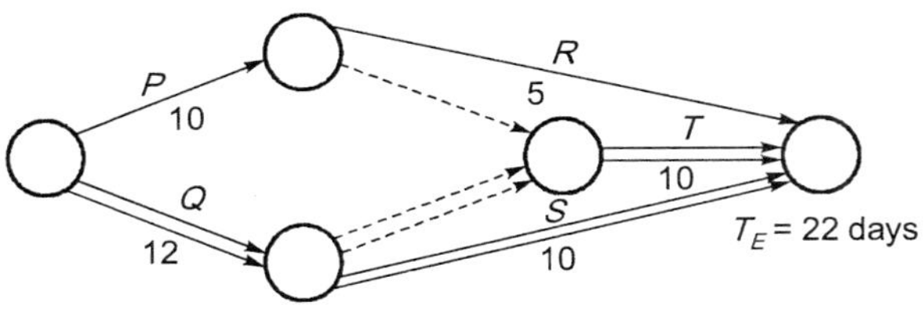

Critical path = Q–S and Q-T also (max. time taken along the path)

= 22 days

2.15 Sol.

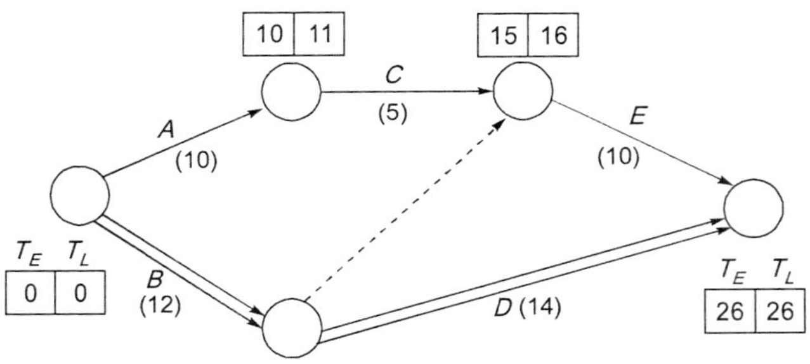

Total float for activity E is given by

$$F_T = T_L^j - T_E^i - t^{ij}$$
$$= 26 - 15 - 10 = 1 \text{ day}$$

Alternatively,

$$F_T = \text{LST} - \text{EST}$$
$$= 16 - 15 = 1 \text{ day}$$
$$F_T = \text{LFT} - \text{EFT}$$
$$= 26 - 25 = 1 \text{ day}$$

2.16 (a)

- During the planning and design phase of a construction project, the client's preliminary estimate is often used as a basis for budgeting costs. Hence statement P is true.
- During the execution phase of a construction project, the clients detailed estimate which is a refined and more detailed version of the preliminary estimate is used for controlling costs. Hence statement Q is true.

2.17 Sol.

Given:

Initial cost, C_i = Rs. 1,00,000

Salvage value, C_s = Rs. 10,000

Useful life, n = 5 years

Straight line method,

$$\text{Depreciation}, D = \frac{C_i - C_s}{n}$$
$$= \frac{10^5 - 10^4}{5}$$
$$= \text{Rs. } 18{,}000$$

As per double declining method,

Fixed factor, F D D B

$$= \frac{2}{n} = \frac{2}{5} = 0.4$$

$\therefore \quad D_1$ = FDDB × C_i = 0.4 × 1,00,000

= 40,000 Rs

Book value, B_1 = Rs. 1,00,000 − Rs. 40,000

= Rs. 60,000

Now, $\quad D_2$ = FDDB × B_1 = 0.4 × 60,000

= Rs. 24,000

Difference in depreciation

= Rs. 24,000 − Rs. 18,000 = Rs. 6000

2.18 (b)

Schedule variance is given by,

$$SV = BCWP - BCWS$$

Here, $BCWP = \text{Rs. } 150$

$BCWS = \text{Rs. } 200$

$\therefore \quad SV = 150 - 200 = (-)\ \text{Rs.} 50$

A negative variance indicates a behind-schedule condition.

2.19 (a)

P: In a pure project organisation, the project manager has the complete authority and control over the project.

Q: In a matrix organisation, employees have more than one boss and work on multiple teams. This leads to quick response to changes, conflicts and project needs.

2.20 Sol.

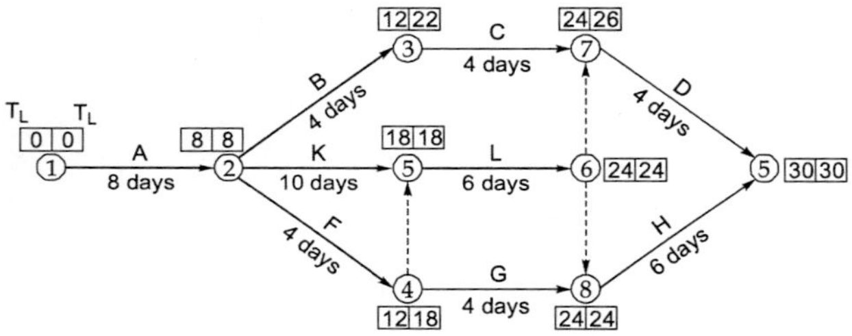

Critical path is along A – K – L – H and project duration is 30 days.

2.21 (*)

$$\text{Cost slope} = \frac{54000 - 30000}{15 - 12} = \text{Rs. } 8000/\text{day}$$

Case-1

Increase in direct cost by crashing of 3 days $= 8000 \times 3 = \text{Rs. } 24000$

Decrease in indirect cost by crashing of 3 days $= 8500 \times 3 = \text{Rs. } 25500$

As the decrease in indirect cost is more than increase and direct cost, so crashing is advantageous.

Case-2

Decrease in indirect cost by crashing of 3 days $= 10000 \times 3 = \text{Rs. } 30000$

As the decrease in indirect cost is more than increase and direct cost, so crashing is advantageous.

3. Engineering Mechanics

3.1 (c)

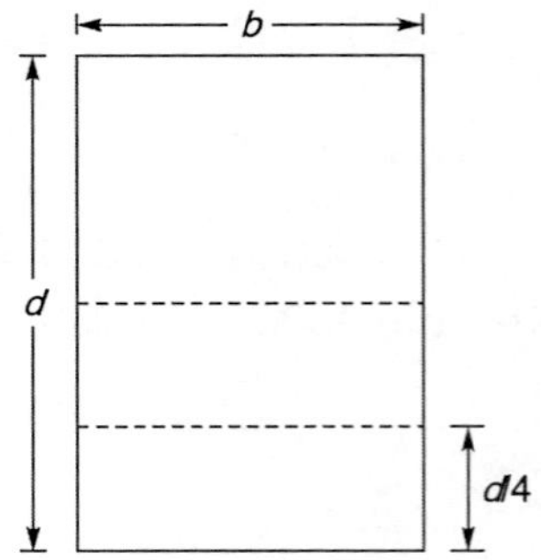

Using **parallel axis theorem**, we get the second moment of inertia as,

$$I = \frac{bd^3}{12} + bd\left(\frac{d}{2} - \frac{d}{4}\right)^2 = \frac{bd^3}{12} + \frac{bd^3}{16} = \frac{7bd^3}{48}$$

3.2 (c)

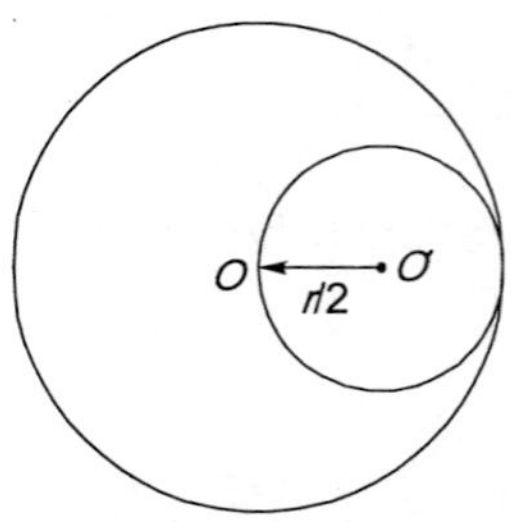

The centroid of the shaded portion of the disc is given by,

$$x = \frac{A_1 x_1 + A_2 x_2}{A_1 + A_2}$$

where x is the radial distance from O.

$$A_1 = \pi r^2;\ x_1 = 0;$$

$$A_2 = -\pi \times \left(\frac{r}{2}\right)^2 = -\frac{\pi r^2}{4}$$

$$x_2 = \frac{r}{2}$$

$$\therefore \quad x = \frac{\pi r^2 \times 0 - \frac{\pi r^2}{4} \times \frac{r}{2}}{\pi r^2 - \frac{\pi r^2}{4}} = -\frac{\pi r^3/2}{3\pi r^2}$$

$$\Rightarrow \quad x = -\frac{r}{6}$$

3.3 Sol.

Polar moment of inertia is measure of an object's ability to resist torsion under specified axis when and torque is being applied. It is also known as "second moment of area, "area moment of inertia", "polar moment of area" or "second area moment".

Polar moment of inertia,

$$I_p = I_x + I_y = \frac{bd^3}{12} + \frac{db^3}{12}$$

$$= \frac{bd}{12}\left(b^2 + d^2\right) = \frac{2\times 6}{12}\left(2^2 + 6^2\right) = 40 \text{ cm}^4$$

3.4 (b)

Method-I

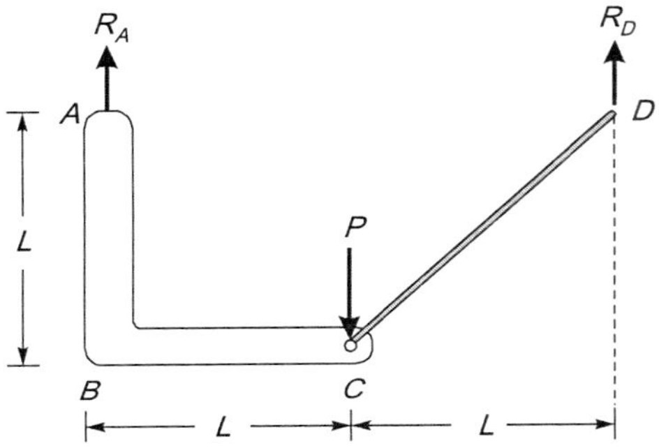

$$\Sigma M_A = 0$$

$$\Rightarrow \quad R_D \times 2L - P \times L = 0$$

$$\Rightarrow \quad R_D = \frac{P}{2}$$

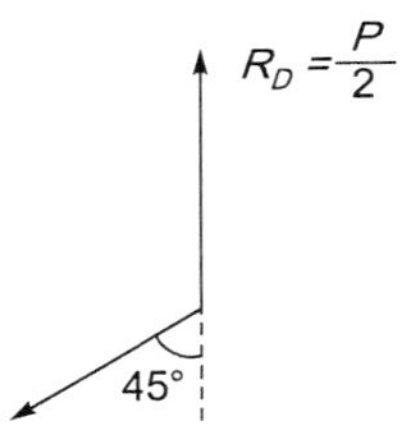

At joint D:

$$\Sigma F_y = 0$$

$$\Rightarrow \quad T\cos 45° = \frac{P}{2}$$

$$\therefore \quad T = \frac{P}{\sqrt{2}}$$

Method-II

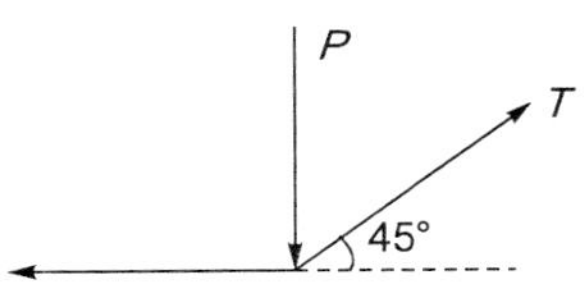

Using sine rule at joint C

$$\frac{-P}{\sin 225°} = \frac{T}{\sin 90°}$$

$$\therefore \quad T = -P\frac{\sin 225°}{\sin 90°}$$

$$T = \frac{+1}{\sqrt{2}}P$$

3.5 (c)

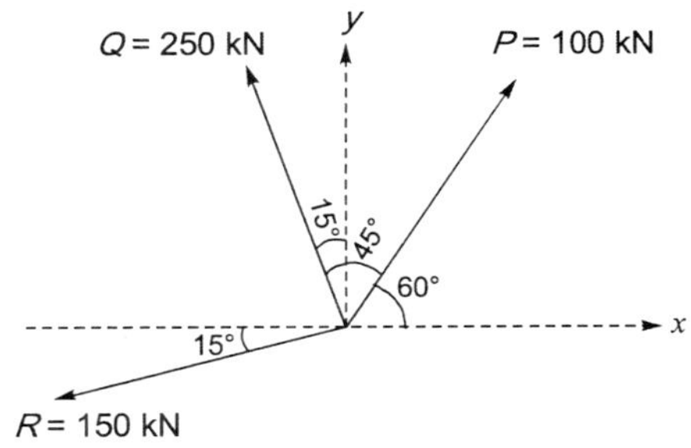

Resultant force in x-direction,

$$R_x = P\cos 60° - Q\sin 15° - R\sin 15°$$

$$= 100\cos 60° - 250\sin 15° - 150\cos 15°$$

$$= -159.59 \text{ kN } (\rightarrow)$$

$$= 159.59 \text{ kN } (\leftarrow)$$

Resultant force in y-direction,

$$R_y = P\sin 60° + Q\cos 15° - R\sin 15°$$

$$= 100\sin 60° + 250\cos 15° - 150\sin 15°$$

$$= 289.26 \text{ kN } (\uparrow)$$

Final resultant force,

$$R = \sqrt{R_x^2 + R_y^2} = 330.4 \text{ kN}$$

$$\theta = \tan^{-1}\left(\frac{159.59}{289.56}\right)$$

$$\theta = 28.9°$$

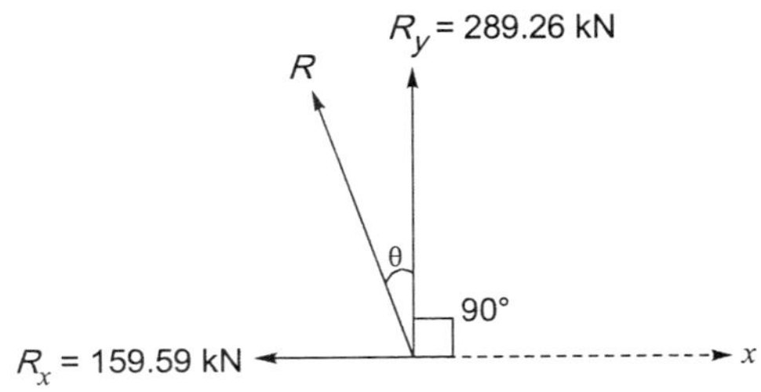

Direction with respect to x-axis

$$= \theta + 90°$$

$$= 28.9° + 90°$$

$$= 118.9°$$

3.6 (b)

The order of reaction is the exponent to which its concentration term in the rate equation is raised.

$$\text{Rate, } r = k[A]^\alpha [B]^\beta$$

For first order reaction,

$$\alpha + \beta = 1$$

3.7 Sol.

$m = 2$ kg, $V_0 = 1.5$ m/sec

$$\int_0^t F(t)dt = m(V - V_0)$$

$$\int_0^2 3t^2 dt = 2(V - 1.5)$$

$$[t^3]_0^2 = 2(V - 1.5)$$

$$\Rightarrow \quad (8 - 0) = 2(V - 1.5)$$

$$V = 5.5 \text{ m/s}$$

3.8 Sol.

Speed of aircraft,

$$u_i = 200 \text{ km/hr}$$

$$= \frac{200 \times 1000}{3600} = 55.56 \text{ m/s}$$

Deceleration of aircraft on the runway

$$= 1.697 \text{ m/s}^2$$

Aircraft takes 18 sec to exit the runway strip.

Speed of the aircraft at the exit of the runway

$$u_f = u_i + at$$

$$= 55.56 - 1.697 \times 18 = 25.014 \text{ m/s}$$

After runway aircraft decelerate with 1 m/s²

Total distance travelled by the aircraft from the location of exit on the runway

$$U^2 = u^2 + 2aS$$

$$0 = 25.014^2 - 2 \times 1 \times S$$

$$S = 312.8 \text{ m}$$

3.9 (d)

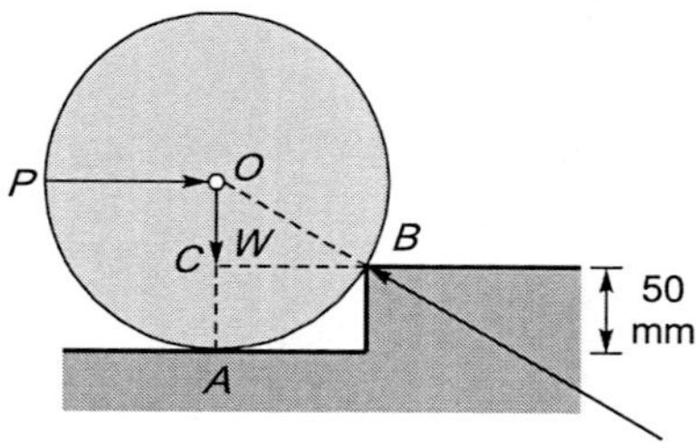

Given: $r = 250$ mm, $W = 10$ kN

Note:

1. When the cylinder will be about to move out of the cylinder, it will loose its contact at point A, only contact will be at point B.
2. Considering equation of cylinder of that instant under P, W and R_B (contact force at B).

$$\Sigma \vec{M}_B = 0$$

$$P \times OC - W \times BC = 0$$

In ΔOCB

$$OC^2 + CB^2 = OB^2$$

$$CB = \sqrt{250^2 - 200^2} = 150$$

$$P = \frac{10 \times 10^3 \times 150}{200} = 2.5 \times 3 \text{ kN} = 7.5 \text{ kN}$$

3.10 (d)

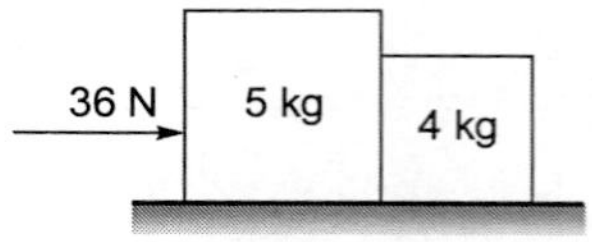

As after action of 36 N, both blocks will move with same acceleration so considering 5 kg and 4 kg together in a system and applying Newton's 2nd law.

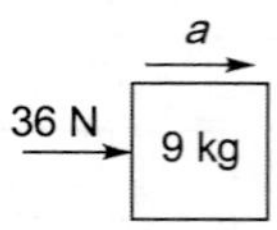

$$a = \frac{36}{9} = 4 \text{ m/s}^2$$

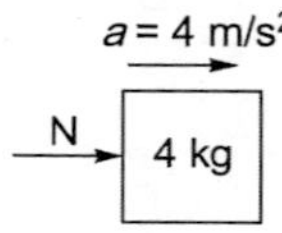

$N = ma$ {Newton's 2nd law}

$N = 4 \times 4 = 16$ N

3.11 Sol.

Method-I

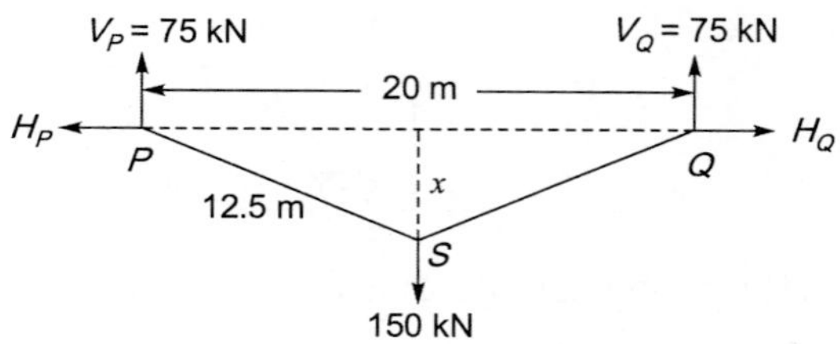

$$12.5^2 = x^2 + 10^2$$

$$\Rightarrow \quad x = 7.5 \text{ m}$$

Bending moment at $S = 0$ {Consider the left part}

$$H_P \times 7.5 = 75 \times 10$$
$$H_P = 100 \text{ kN}$$

Tension in cable

$$= \sqrt{H_p^2 + V_p^2} = \sqrt{75^2 + 100^2}$$
$$= 125 \text{ kN}$$

Method-II

Using sine rule at joint S

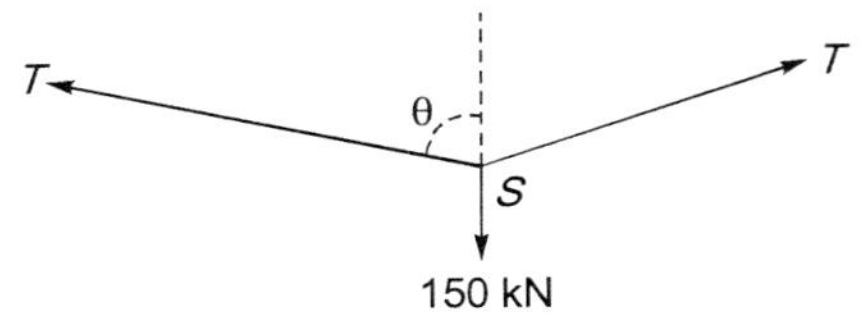

Calculating θ,

$$\theta = \sin^{-1}\left[\frac{10}{12.5}\right] = 53.13°$$

$$\therefore \quad \frac{150}{\sin 106.26} = \frac{T}{\sin 126.87°}$$

$$T = 124.999 \simeq 125 \text{ kN}$$

3.12 (d)

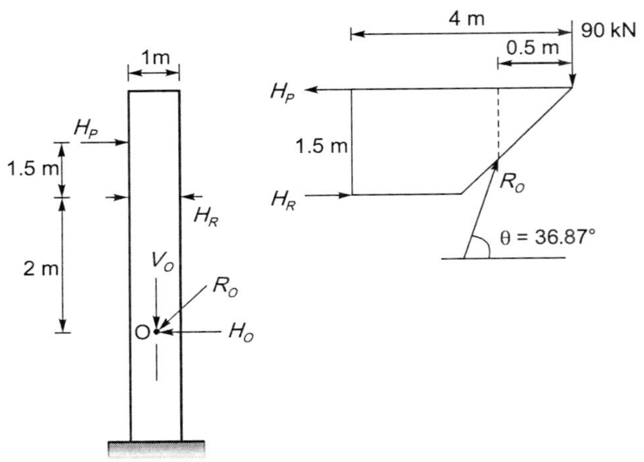

$$\Sigma Y = 0$$
$$\Rightarrow \quad R_o \sin 36.87 - 90 = 0$$
$$R_o = \frac{90}{\sin 3687°} = 150 \text{ kN}$$

Horizontal reaction at $O = H_o$

$= R_o \cos 36.87 = 150 \times \cos 36.87 = 120$ kN

3.13 (b)

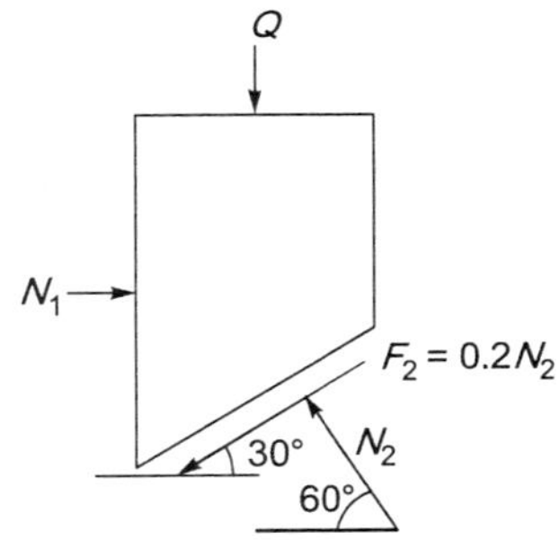

$$\Sigma Y = 0$$
$$\Rightarrow N_2 \sin 60° - 0.2\, N_2 \sin 30° - Q = 0$$
$$Q = 0.766\, N_2 \quad \text{...(i)}$$

F.B.D. of wedge

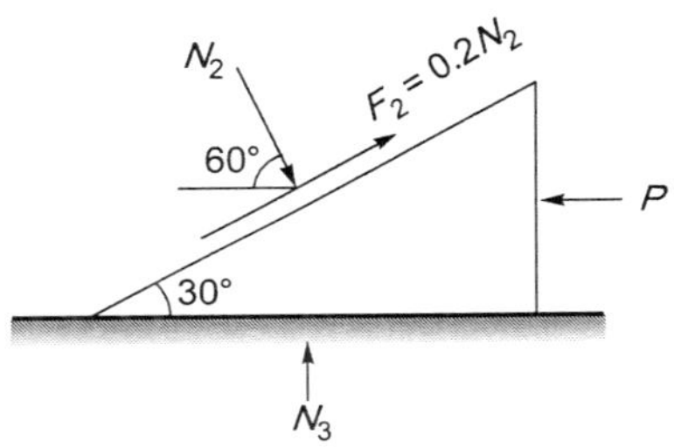

$$\Sigma X = 0$$
$$\Rightarrow 0.2N_2 \cos 30° + N_2 \cos 60° - P = 0$$
$$P = 0.67\, N_2$$
$$\Rightarrow \quad P = 0.67 \times \frac{Q}{0.766}$$
$$\Rightarrow \quad P = 0.875Q \simeq 0.9Q$$
$$P = \alpha Q$$
$$\Rightarrow \quad \alpha = 0.9$$

3.14 Sol.

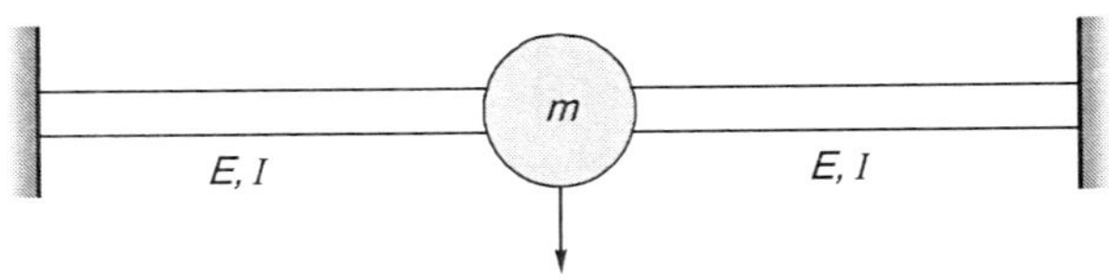

Cyclic frequency,

$$f = \frac{\omega}{2\pi} = \frac{1}{2\pi}\sqrt{\frac{k}{m}} = \frac{1}{2\pi}\sqrt{\frac{(4\pi^2 \times 1000)\text{ N/m}}{10\text{ kg}}} = 10\text{ Hz}$$

3.15 Sol.

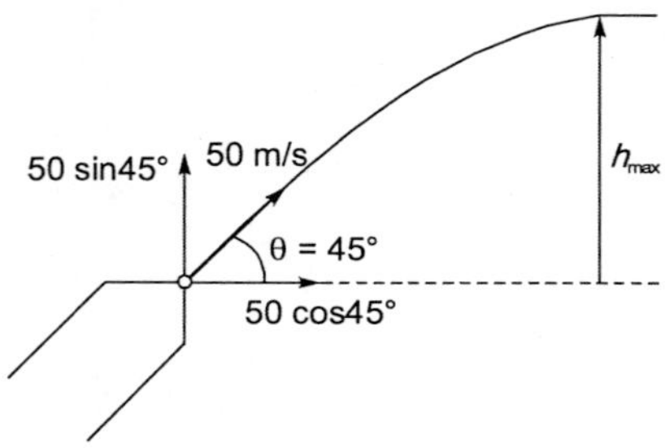

As we know that

$$V^2 - u^2 = 2aS$$

In vertical direction (↑)

$$0^2 - (50 \sin 45°)^2 = 2(-9.81)\, h_{max}$$

$$h_{max} = \frac{(50 \sin 45°)^2}{2(9.81)} = 63.71 \text{ m}$$

3.16 (a, b, c)

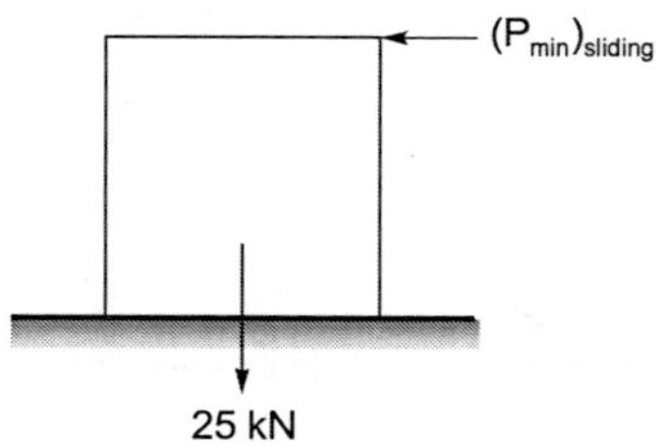

Minimum force for sliding:

Applying equilibrium equation in vertical direction,

Normal reaction = Weight

$$\Rightarrow \quad N = mg = 25 \text{ kN}$$

When the body is on the verge of sliding,

$$(P_{min})_{sliding} = (f_s)_{max} = \mu N$$

$$= 0.3 \times 25 = 7.5 \text{ kN}$$

Minimum force for overturning:

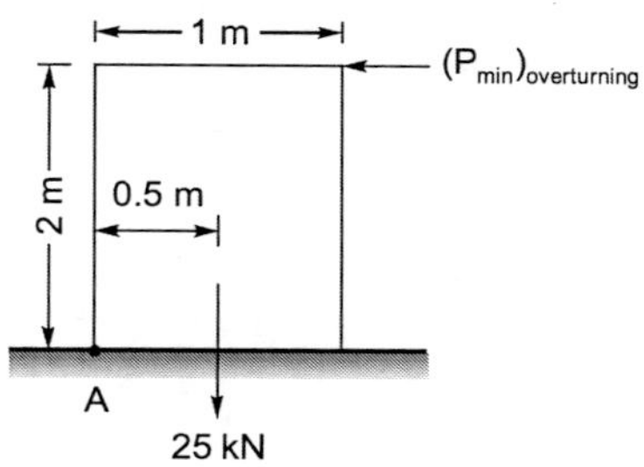

On the verge of overturning,

Overturning moment = Restoring moment

$$\Rightarrow (P_{min})_{overturning} \times 2 = W \times 0.5$$

$$\Rightarrow \quad (P_{min})_{overturning} = \frac{25 \times 0.5}{2} = 6.25 \text{ kN}$$

$\because \quad (P_{min})_{overturning} < (P_{min})_{sliding}$

⇒ Overturning will take place.

∴ Sliding will not take place.

3.17 Sol.

Free body diagram:

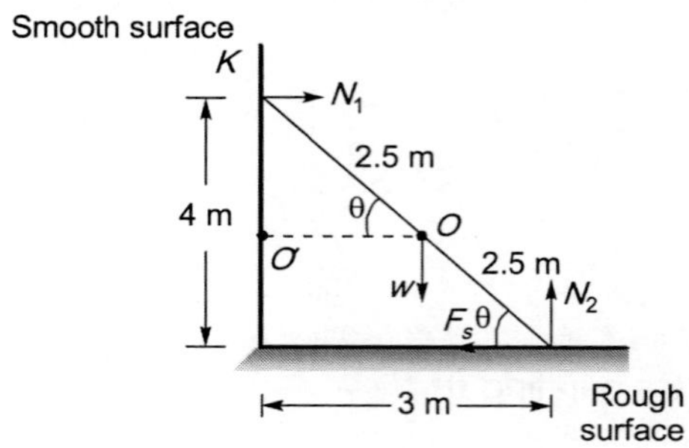

$$\therefore \quad KJ = \sqrt{4^2 + 3^2} = 5 \text{m}$$

$$\therefore \quad KO = OJ = 2.5 \text{ m}$$

$$OO' = 2.5\cos\theta$$

For translational equilibrium,

$$\Sigma \vec{F}_{ext} = 0$$

$$\Rightarrow \quad \Sigma \vec{F}_x = 0 \qquad \text{i.e. } N_1 = F_s$$

$$\Rightarrow \quad \Sigma \vec{F}_y = 0 \qquad \text{i.e. } N_2 = w$$

For rotational equilibrium,

$$(\Sigma \vec{\tau}_{ext})_{\text{about point } K} = 0$$

$$\Rightarrow w \times 2.5\cos\theta + F_s(4) = N_2 \times 3$$

$$\left\{\because \cos\theta \frac{3}{5};\ \text{and } N_2 = w = 10 \text{ kN}\right\}$$

$$\Rightarrow 10\left(2.5 \times \frac{3}{5}\right) + 4F_s = 3(10)$$

$$\Rightarrow \quad F_s = \frac{15}{4} \text{kN}$$

Now, we know that,

$$F_s \le (F_s)_{max}$$

$$\Rightarrow \quad F_s \le \mu_s N_2$$

$$\Rightarrow \quad \frac{15}{4} \le \mu_s \times w$$

$$\Rightarrow \quad \mu_s \ge \frac{15}{4 \times 10}$$

$$\Rightarrow \quad \mu \ge 0.375$$

$$\Rightarrow \quad (\mu)_{min} = 0.375$$

3.18 Sol.

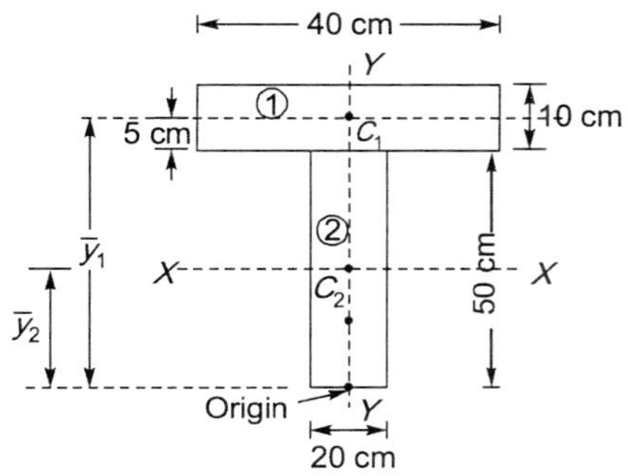

$$A_1 = 40 \times 10 = 400 \text{ cm}^2$$
$$A_2 = 20 \times 50 = 1000 \text{ cm}^2$$
$$\bar{y}_1 = 50 + 5 = 55 \text{ cm}$$
$$\bar{y}_2 = \frac{50}{2} = 25 \text{ cm}$$

Centroid of composite shape, from bottom fibre,

$$\bar{y} = \frac{A_1\bar{y}_1 + A_2\bar{y}_2}{A_1 + A_2}$$

$$\Rightarrow \quad \bar{y} = \frac{(400 \times 55) + (1000 \times 25)}{400 + 1000}$$

$$\Rightarrow \quad \bar{y} = 33.5714 \text{ cm}$$

$$I_{xx} = \left[\frac{40 \times 10^3}{12} + 400(\bar{y}_1 - \bar{y})^2\right] + \left[\frac{20 \times 50^3}{12} + 1000(\bar{y} - \bar{y}_2)^2\right]$$

$$= \left[\frac{40 \times 10^3}{12} + 400(55 - 33.5714)^2\right] + \left[\frac{20 \times 50^3}{12} + 1000(33.5714 - 25)^2\right]$$

$$= 187007.2925 + 281802.2313$$

$$I_{xx} = 468809.5238 \text{ cm}^4 \simeq 468810 \text{ cm}^4$$

■■■■

Engineering Mathematics

UNIT

XIII

CONTENTS

Engineering Mathematics

Syllabus

Linear Algebra: Matrix algebra; Systems of linear equations; Eigen values and Eigen vectors.

Calculus: Functions of single variable; Limit, continuity and differentiability; Mean value theorems, local maxima and minima; Taylor series; Evaluation of definite and indefinite integrals, application of definite integral to obtain area and volume; Partial derivatives; Total derivative; Gradient, Divergence and Curl, Vector identities; Directional derivatives; Line, Surface and Volume integrals.

Ordinary Differential Equation (ODE): First order (linear and non-linear) equations; higher order linear equations with constant coefficients; Euler-Cauchy equations; initial and boundary value problems.

Partial Differential Equation (PDE): Fourier series; separation of variables; solutions of one- dimensional diffusion equation; first and second order one-dimensional wave equation and two-dimensional Laplace equation.

Probability and Statistics: Sampling theorems; Conditional probability; Descriptive statistics – Mean, median, mode and standard deviation; Random Variables – Discrete and Continuous, Poisson and Normal Distribution; Linear regression.

Numerical Methods: Error analysis. Numerical solutions of linear and non-linear algebraic equations; Newton's and Lagrange polynomials; numerical differentiation; Integration by trapezoidal and Simpson's rule; Single and multi-step methods for first order differential equations.

Analysis of Previous GATE Papers

Exam Year	1 Mark Ques.	2 Marks Ques.	Total Marks
1995	12	1	14
1996	10	–	10
1997	9	–	9
1998	5	1	7
1999	10	–	10
2000	9	–	9
2001	3	4	11
2002	3	5	13
2003	3	2	7
2004	3	4	11
2005	4	8	20
2006	2	9	20
2007	3	9	21
2008	2	9	20
2009	3	6	15
2010	3	6	15
2011	3	5	13
2012	3	5	13
2013	2	3	8
2014 Set-1	5	1	7
2014 Set-2	5	2	9

Exam Year	1 Mark Ques.	2 Marks Ques.	Total Marks
2015 Set-1	3	5	13
2015 Set-2	5	4	13
2016 Set-1	5	4	13
2016 Set-2	5	5	15
2017 Set-1	6	5	16
2017 Set-2	4	4	12
2018 Set-1	3	3	9
2018 Set-2	5	4	14
2019 Set-1	2	4	10
2019 Set-2	5	4	13
2020 Set-1	4	5	14
2020 Set-2	5	4	13
2021 Set-1	5	3	11
2021 Set-2	5	4	13
2022 Set-1	4	4	12
2022 Set-2	4	4	12
2023 Set-1	5	4	13
2023 Set-2	5	4	13
2024 Set-1	5	2	9
2024 Set-2	5	3	11

1 Linear Algebra

1.1 The rank of the matrix $A=\begin{bmatrix}1 & -2 & -1\\ -3 & 3 & 0\\ 2 & 2 & 4\end{bmatrix}$ is

(a) One (b) Two
(c) Three (d) None of these

[1994 : 1 M]

1.2 The Eigen vectors of the matrix $\begin{bmatrix}1 & -1\\ -1 & 1\end{bmatrix}$ is/are

(a) (1, 0) (b) (0, 1)
(c) (1, 1) (d) (1, –1)

[1994 : 1 M]

1.3 The inverse of the matrix $A=\begin{bmatrix}-3 & 5\\ 2 & 1\end{bmatrix}$ is

(a) $\begin{bmatrix}\frac{5}{13} & \frac{-1}{13}\\ \frac{2}{13} & \frac{3}{13}\end{bmatrix}$ (b) $\begin{bmatrix}\frac{2}{13} & \frac{5}{13}\\ \frac{-1}{13} & \frac{3}{13}\end{bmatrix}$

(c) $\begin{bmatrix}\frac{-1}{13} & \frac{5}{13}\\ \frac{2}{13} & \frac{3}{13}\end{bmatrix}$ (d) $\begin{bmatrix}\frac{1}{13} & \frac{-5}{13}\\ \frac{2}{13} & \frac{3}{13}\end{bmatrix}$

[1994 : 2 M]

1.4 If A and B are square matrices of size $n \times n$, then which of the following statement is not true

(a) det (AB) = det (A) det (B)
(b) det $(kA) = k^n$ det(A)
(c) det $(A + B)$ = det (A) + det (B)
(d) det $(A^T) = 1/\det(A^{-1})$

[1995 : 1 M]

1.5 If matrix A is $m \times n$ and B is $n \times p$, the number of multiplication operations and addition operations needed to calculate the matrix AB, respectively, are

(a) mn^2p, mpm (b) mpn, $mp(n-1)$
(c) mpn, mpn (d) mn^2p, $(m + p)n$

[1995 : 1 M]

1.6 The rank of the matrix $\begin{bmatrix}1 & 2 & -1\\ 2 & 4 & 6\\ 0 & 0 & -8\end{bmatrix}$ is

(a) 2 (b) 1
(c) 3 (d) None of these

[1995 : 1 M]

1.7 Let A be an invertible matrix and suppose that the inverse of $7A$ is $\begin{bmatrix}-1 & 2\\ 4 & -7\end{bmatrix}$, the matrix A is

(a) $\begin{bmatrix}1 & 2/7\\ 4/7 & 1/7\end{bmatrix}$ (b) $\begin{bmatrix}7 & 2\\ 4 & 1\end{bmatrix}$

(c) $\begin{bmatrix}1 & -4/7\\ -2/7 & 1/7\end{bmatrix}$ (d) $\begin{bmatrix}7 & 4\\ 2 & 1\end{bmatrix}$

[1995 : 1 M]

1.8 The solution(s) to the equations

$2x + 3y = 1$
$x - y = 4$
$4x - y = \alpha$

will exist if α is equal to

(a) –33 (b) 0
(c) 9 (d) None of these

[1995 : 1 M]

1.9 The matrix $B = A^T$, where A is any matrix is

(a) skew symmetric
(b) symmetric about the secondary diagonal
(c) always symmetric
(d) another general matrix

[1996 : 1 M]

1.10 If $A^T = A^{-1}$, where A is a real matrix, then A is

(a) Normal (b) Symmetric
(c) Hermitian (d) Orthogonal

[1996 : 1 M]

1.11 If A and B are non zero square matrices, then $AB = 0$ implies

(a) A and B are orthogonal
(b) either A or B are singular
(c) B is singular
(d) A is singular

[1996 : 1 M]

1.12 The Eigen values of a square symmetric matrix are always
(a) positive (b) real and imaginary
(c) negative (d) real [1996 : 1 M]

1.13 If A and B are two matrices and if AB exists, then BA exists
(a) only if A has as many rows as B has columns
(b) only if both A and B are square matrices
(c) only if A and B are skew matrices
(d) only if both A and B are symmetric
[1997 : 1 M]

1.14 If the determinate of matrix $\begin{bmatrix} 1 & 3 & 2 \\ 0 & 5 & -6 \\ 2 & 7 & 8 \end{bmatrix}$ is 26, then the determinate of the matrix $\begin{bmatrix} 2 & 7 & 8 \\ 0 & 5 & -6 \\ 1 & 3 & 2 \end{bmatrix}$ is
(a) –26 (b) 26
(c) 0 (d) 52 [1997 : 1 M]

1.15 Inverse of matrix $\begin{bmatrix} 0 & 1 & 0 \\ 0 & 0 & 1 \\ 1 & 0 & 0 \end{bmatrix}$ is
(a) $\begin{bmatrix} 0 & 0 & 1 \\ 1 & 0 & 0 \\ 0 & 1 & 0 \end{bmatrix}$ (b) $\begin{bmatrix} 1 & 0 & 0 \\ 0 & 0 & 1 \\ 0 & 1 & 0 \end{bmatrix}$
(c) $\begin{bmatrix} 1 & 0 & 0 \\ 0 & 1 & 0 \\ 0 & 0 & 1 \end{bmatrix}$ (d) $\begin{bmatrix} 0 & 0 & 1 \\ 0 & 1 & 0 \\ 1 & 0 & 0 \end{bmatrix}$
[1997 : 1 M]

1.16 If A is a real square matrix, then AA^T is
(a) unsymmetric
(b) always symmetric
(c) skew symmetric
(d) sometimes symmetric [1998 : 1 M]

1.17 In matrix algebra $AS = AT$(A, S, T, are matrices of appropriate order) implies $S = T$ only if
(a) A is symmetric
(b) A is singular
(c) A is nonsingular
(d) A is skew symmetric [1998 : 1 M]

1.18 If A is any $n \times n$ matrix and k is a scalar, $|kA| = \alpha\,|A|$, where α is
(a) kn (b) n^k
(c) k^n (d) k/n [1999 : 1 M]

1.19 Number of terms in the expansion of general determinant of order n is
(a) n^2 (b) $n!$
(c) n (d) $(n+1)^2$
[1999 : 1 M]

1.20 The equation $\begin{vmatrix} 2 & 1 & 1 \\ 1 & 1 & -1 \\ y & x^2 & x \end{vmatrix} = 0$, represents a parabola passing through the points
(a) (0, 1), (0, 2), (0, – 1)
(b) (0, 0), (–1, 1), (1, 2)
(c) (1, 0), (0, 0), (2, 2)
(d) (1, 2), (2, 1), (0, 0) [1999 : 1 M]

1.21 If A, B, C are square matrices of the same order $(ABC)^{-1}$ is equal to
(a) $C^{-1}A^{-1}B^{-1}$ (b) $C^{-1}B^{-1}A^{-1}$
(c) $A^{-1}B^{-1}C^{-1}$ (d) $A^{-1}C^{-1}B^{-1}$
[2000 : 1 M]

1.22 Consider the following two statements:
I. The maximum number of linearly independent column vectors of a matrix A is called the rank of A.
II. If A is an $n \times n$ square matrix, it will be nonsingular if rank $A = n$.
With reference to the above statements, which of the following applies?
(a) Both the statements are false
(b) Both the statements are true
(c) I si true but II is false
(d) I is false but II is true [2000 : 1 M]

1.23 Determinant of the following matrix is
$$\begin{bmatrix} 5 & 3 & 2 \\ 1 & 2 & 6 \\ 3 & 5 & 10 \end{bmatrix}$$
(a) –76 (b) –28
(c) +28 (d) +72 [2001 : 2 M]

1.24 The product $[P][Q]^T$ of the following two matrices $[P]$ and $[Q]$ is
$$[P] = \begin{bmatrix} 2 & 3 \\ 4 & 5 \end{bmatrix}; \ [Q] = \begin{bmatrix} 4 & 8 \\ 9 & 2 \end{bmatrix}$$

(a) $\begin{bmatrix} 32 & 24 \\ 56 & 46 \end{bmatrix}$ (b) $\begin{bmatrix} 46 & 56 \\ 24 & 32 \end{bmatrix}$

(c) $\begin{bmatrix} 35 & 22 \\ 61 & 42 \end{bmatrix}$ (d) $\begin{bmatrix} 32 & 56 \\ 24 & 46 \end{bmatrix}$

[2001 : 2 M]

1.25 The Eigen value of the matrix $\begin{bmatrix} 5 & 3 \\ 2 & 9 \end{bmatrix}$ are

(a) (5.13, 9.42) (b) (3.85, 2.93)
(c) (9.00, 5.00) (d) (10.16, 3.84)

[2001 : 2 M]

1.26 Eigen values of the following matrix are

$$\begin{bmatrix} -1 & 4 \\ 4 & -1 \end{bmatrix}$$

(a) 3 and – 5 (b) –3 and 5
(c) –3 and – 5 (d) 3 and 5

[2002: 1 M]

1.27 Given Matrix $[A] = \begin{bmatrix} 4 & 2 & 1 & 3 \\ 6 & 3 & 4 & 7 \\ 2 & 1 & 0 & 1 \end{bmatrix}$, the rank of the matrix is

(a) 4 (b) 3
(c) 2 (d) 1 [2003 : 1 M]

1.28 Real matrices $[A]_{3\times1}$, $[B]_{3\times3}$, $[C]_{3\times5}$, $[D]_{5\times3}$, $[E]_{5\times5}$ and $[F]_{5\times1}$ are given. Matrices $[B]$ and $[E]$ are symmetric.
Following statements are made with respect to these matrices.

1. Matrix product $[F]^T [C]^T [B] [C] [F]$ is a scalar.
2. Matrix product $[D]^T [F] [D]$ is always symmetric.

With reference to above statements, which of the following applies?

(a) Statement 1 is true but 2 is false
(b) Statement 1 is false but 2 is true
(c) Both the statements are true
(d) Both the statements are false [2004 : 1 M]

1.29 The eigen values of the matrix $\begin{bmatrix} 4 & -2 \\ -2 & 1 \end{bmatrix}$

(a) are 1 and 4
(b) are –1 and 2
(c) are 0 and 5
(d) cannot be determined [2004 : 2 M]

1.30 Consider the matrices $X_{(4\times3)}$, $Y_{(4\times3)}$ and $P_{(2\times3)}$. The order of $[P(X^TY)^{-1}P^T]^T$ will be

(a) (2×2) (b) (3×3)
(c) (4×3) (d) (3×4)

[2005 : 1 M]

1.31 Consider a non-homogeneous system of linear equations representing mathematically an over-determined system. Such a system will be

(a) consistent having a unique solution
(b) consistent having many solutions
(c) inconsistent having a unique solution
(d) inconsistent having no solution

[2005 : 1 M]

1.32 Consider the system of equations $A_{(n\times n)} X_{(n\times1)} = \lambda_{(n\times1)}$ where, λ is a scalar. Let (λ_i, x_i) be an eigen-pair of an eigen value and its corresponding eigen vector for real matrix A. Let I be a $(n\times n)$ unit matrix. Which one of the following statement is NOT correct?

(a) For a homogeneous $n\times n$ system of linear equations, $(A-\lambda I)x = 0$ having a non-trivial solution, the rank of $(A-\lambda I)$ is less than n.
(b) For matrix A^m, m being a positive integer, (λ_i^m, x_i^m) will be the eigen-pair for all i.
(c) If $A^T = A^{-1}$, then $|\lambda_i| = 1$ for all i.
(d) If $A^T = A$, then λ_i is real for all i.

[2005 : 2 M]

1.33 Solution for the system defined by the set of equations $4y + 3z = 8$; $2x - z = 2$ and $3x + 2y = 5$ is

(a) $x = 0; y = 1; z = \frac{4}{3}$
(b) $x = 0; y = \frac{1}{2}; z = 2$
(c) $x = 1; y = \frac{1}{2}; z = 2$
(d) Non-existent [2006 : 1 M]

1.34 For a given matrix $A = \begin{bmatrix} 2 & -2 & 3 \\ -2 & -1 & 6 \\ 1 & 2 & 0 \end{bmatrix}$, one of the eigenvalues is 3.
The other two eigenvalues are

(a) 2, –5 (b) 3, –5
(c) 2, 5 (d) 3, 5 [2006 : 2 M]

1.35 The minimum and the maximum eigen values of the matrix $\begin{bmatrix} 1 & 1 & 3 \\ 1 & 5 & 1 \\ 3 & 1 & 1 \end{bmatrix}$ are –2 and 6, respectively. What is the other eigen value?

(a) 5 (b) 3
(c) 1 (d) –1 [2007 : 1 M]

1.36 For what values of α and β the following simultaneous equations have an infinite number of solutions?

$x + y + z = 5$
$x + 3y + 3z = 9;$
$x + 2y + \alpha z = \beta$

(a) 2, 7 (b) 3, 8
(c) 8, 3 (d) 7, 2 [2007 : 2 M]

1.37 The inverse of the 2 × 2 matrix $\begin{bmatrix} 1 & 2 \\ 5 & 7 \end{bmatrix}$ is,

(a) $\frac{1}{3}\begin{bmatrix} -7 & 2 \\ 5 & -1 \end{bmatrix}$ (b) $\frac{1}{3}\begin{bmatrix} 7 & 2 \\ 5 & 1 \end{bmatrix}$

(c) $\frac{1}{3}\begin{bmatrix} 7 & -2 \\ -5 & 1 \end{bmatrix}$ (d) $\frac{1}{3}\begin{bmatrix} -7 & -2 \\ -5 & -1 \end{bmatrix}$

[2007 : 2 M]

1.38 The product of matrices $(PQ)^{-1}P$ is

(a) P^{-1} (b) Q^{-1}
(c) $P^{-1}Q^{-1}P$ (d) PQP^{-1} [2008 : 1 M]

1.39 The Eigen values of the matrix $[P] = \begin{bmatrix} 4 & 5 \\ 2 & -5 \end{bmatrix}$ are

(a) –7 and 8 (b) –6 and 5
(c) 3 and 4 (d) 1 and 2

[2008 : 2 M]

1.40 The following simultaneous equations

$$x + y + z = 3$$
$$x + 2y + 3z = 4$$
$$x + 4y + kz = 6$$

will NOT have a unique solution for k equal to

(a) 0 (b) 5
(c) 6 (d) 7 [2008 : 2 M]

1.41 A square matrix B is skew-symmetric if

(a) $B^T = -B$ (b) $B^T = B$
(c) $B^{-1} = B$ (d) $B^{-1} = B^T$

[2009 : 1 M]

1.42 The inverse of the matrix $\begin{bmatrix} 3+2i & i \\ -i & 3-2i \end{bmatrix}$ is

(a) $\frac{1}{12}\begin{bmatrix} 3+2i & -i \\ i & 3-2i \end{bmatrix}$

(b) $\frac{1}{12}\begin{bmatrix} 3-2i & -i \\ i & 3+2i \end{bmatrix}$

(c) $\frac{1}{14}\begin{bmatrix} 3+2i & -i \\ i & 3-2i \end{bmatrix}$

(d) $\frac{1}{14}\begin{bmatrix} 3-2i & -i \\ i & 3+2i \end{bmatrix}$ [2010 : 2 M]

1.43 $[A]$ is square matix which is neither symmetric nor skew-symmetric and $[A]^T$ is its transpose. The sum and difference of these matrices are defined as $[S] = [A] + [A]^T$ and $[D] = [A] - [A]^T$, respectively. Which of the following statements is **TRUE**?

(a) Both $[S]$ and $[D]$ are symmetric
(b) Both $[S]$ and $[D]$ are skew-symmetric
(c) $[S]$ is skew-symmetric and $[D]$ is symmetric
(d) $[S]$ is symmetric and $[D]$ is skew-symmetric

[2011 : 1 M]

1.44 The eigen values of matrix $\begin{bmatrix} 9 & 5 \\ 5 & 8 \end{bmatrix}$ are

(a) – 2.42 and 6.86 (b) 3.48 and 13.53
(c) 4.70 and 6.86 (d) 6.86 and 9.50

[2011 : 2 M]

1.45 There are three matrixes $P(4 \times 2)$, $Q(2 \times 4)$ and $R(4 \times 1)$. The minimum of multiplication required to compute the matrix PQR is

[2013 : 1 M]

1.46 Given the matrices $J = \begin{bmatrix} 3 & 2 & 1 \\ 2 & 4 & 2 \\ 1 & 2 & 6 \end{bmatrix}$ and $K = \begin{bmatrix} 1 \\ 2 \\ -1 \end{bmatrix}$, the product $K^T JK$ is ______.

[2014 : 1 M, Set-I]

1.47 The sum of Eigen values of matrix, $[M]$ is

where $[M] = \begin{bmatrix} 215 & 650 & 795 \\ 655 & 150 & 835 \\ 485 & 355 & 550 \end{bmatrix}$

(a) 915 (b) 1355
(c) 1640 (d) 2180

[2014 : 1 M, Set-I]

1.48 With reference to the conventional Cartesian (x, y) coordinate system, the vertices of a triangle have the following coordinates; $(x_1, y_1) = (1, 0)$; $(x_2, y_2) = (2, 2)$; $(x_3, y_3) = (4, 3)$. The area of the triangle is equal to

(a) $\frac{3}{2}$ (b) $\frac{3}{4}$

(c) $\frac{4}{5}$ (d) $\frac{5}{2}$

[2014 : 1 M, Set-I]

1.49 The determinant of matrix $\begin{bmatrix} 0 & 1 & 2 & 3 \\ 1 & 0 & 3 & 0 \\ 2 & 3 & 0 & 1 \\ 3 & 0 & 1 & 2 \end{bmatrix}$ is ____.

[2014 : 1 M, Set-II]

1.50 The rank of the matrix $\begin{bmatrix} 6 & 0 & 4 & 4 \\ -2 & 14 & 8 & 18 \\ 14 & -14 & 0 & -10 \end{bmatrix}$ is _____.

[2014 : 2 M, Set-II]

1.51 For what value of p the following set of equations will have no solution?

$$2x + 3y = 5$$
$$3x + py = 10$$

[2015 : 1 M, Set-I]

1.52 The smallest and largest Eigen values of the following matrix are

$$\begin{bmatrix} 3 & -2 & 2 \\ 4 & -4 & 6 \\ 2 & -3 & 5 \end{bmatrix}$$

(a) 1.5 and 2.5 (b) 0.5 and 2.5

(c) 1.0 and 3.0 (d) 1.0 and 2.0

[2015 : 2 M, Set-I]

1.53 Let $A = [a_{ij}], 1 \le i, j \le n$ with $n \ge 3$ and $a_{ij} = i.j$. The rank of A is

(a) 0 (b) 1

(c) $n - 1$ (d) n

[2015 : 1 M, Set-II]

1.54 The two Eigen values of the matrix $\begin{bmatrix} 2 & 1 \\ 1 & p \end{bmatrix}$ have a ratio of 3 : 1 for $p = 2$. What is another value of p for which the Eigen values have the same ratio of 3 : 1?

(a) –2 (b) 1

(c) 7/3 (d) 14/3

[2015 : 2 M, Set-II]

1.55 If the entries in each column of a square matrix M add up to 1, then an eigen value of M is

(a) 4 (b) 3

(c) 2 (d) 1

[2016 : 1 M, Set-I]

1.56 Consider the following linear system.

$$x + 2y - 3z = a$$
$$2x + 3y + 3x = b$$
$$5x + 9y - 6z = c$$

This system is consistent if a, b and c satisfy the equation

(a) $7a - b - c = 0$ (b) $3a + b - c = 0$

(c) $3a - b + c = 0$ (d) $7a - b + c = 0$

[2016 : 2 M, Set-II]

1.57 The matrix P is the inverse of a matrix Q. If I denotes the identity matrix, which one of the following options is correct?

(a) $PQ = I$ but $QP \ne I$

(b) $QP = I$ but $PQ \ne I$

(c) $PQ = I$ and $QP = I$

(d) $PQ - QP = I$

[2017 : 1 M, Set-I]

1.58 Consider the matrix $\begin{bmatrix} 5 & -1 \\ 4 & 1 \end{bmatrix}$. Which one of the following statements is TRUE for the eigenvalues and eigen vectors of this matrix?

(a) Eigenvalue 3 has a multiplicity of 2, and only one independent eigenvector exists.

(b) Eigenvalue 3 has a multiplicity of 2, and two independent eigenvector exists.

(c) Eigenvalue 3 has a multiplicity of 2, and no independent eigenvector exists.

(d) Eigenvalue are 3 and –3, and two indepedent eigenvectors exist.

[2017 : 2 M, Set-I]

1.59 Consider the following simultaneous equations (with c_1 and c_2 being constants):

$$3x_1 + 2x_2 = c_1$$
$$4x_1 + x_2 = c_2$$

The characteristics equation for these simultaneous equations is

(a) $\lambda^2 - 4\lambda - 5 = 0$ (b) $\lambda^2 - 4\lambda + 5 = 0$

(c) $\lambda^2 + 4\lambda - 5 = 0$ (d) $\lambda^2 + 4\lambda + 5 = 0$

[2017 : 1 M, Set-II]

1.60 If $A = \begin{bmatrix} 1 & 5 \\ 6 & 2 \end{bmatrix}$ and $B = \begin{bmatrix} 3 & 7 \\ 8 & 4 \end{bmatrix}$, AB^T is equal to

(a) $\begin{bmatrix} 38 & 28 \\ 32 & 56 \end{bmatrix}$ (b) $\begin{bmatrix} 3 & 40 \\ 42 & 8 \end{bmatrix}$

(c) $\begin{bmatrix} 43 & 27 \\ 34 & 50 \end{bmatrix}$ (d) $\begin{bmatrix} 38 & 32 \\ 28 & 56 \end{bmatrix}$

[2017 : 2 M, Set-II]

1.61 For the given orthogonal matrix Q,

$$Q = \begin{bmatrix} \frac{3}{7} & \frac{2}{7} & \frac{6}{7} \\ -\frac{6}{7} & \frac{3}{7} & \frac{2}{7} \\ \frac{2}{7} & \frac{6}{7} & -\frac{3}{7} \end{bmatrix}$$

The inverse is

(a) $\begin{bmatrix} \frac{3}{7} & \frac{2}{7} & \frac{6}{7} \\ -\frac{6}{7} & \frac{3}{7} & \frac{2}{7} \\ \frac{2}{7} & \frac{6}{7} & -\frac{3}{7} \end{bmatrix}$ (b) $\begin{bmatrix} -\frac{3}{7} & -\frac{2}{7} & -\frac{6}{7} \\ \frac{6}{7} & -\frac{3}{7} & -\frac{2}{7} \\ -\frac{2}{7} & -\frac{6}{7} & \frac{3}{7} \end{bmatrix}$

(c) $\begin{bmatrix} \frac{3}{7} & -\frac{6}{7} & \frac{2}{7} \\ \frac{2}{7} & \frac{3}{7} & \frac{6}{7} \\ \frac{6}{7} & \frac{2}{7} & -\frac{3}{7} \end{bmatrix}$ (d) $\begin{bmatrix} -\frac{3}{7} & \frac{6}{7} & -\frac{2}{7} \\ -\frac{2}{7} & -\frac{3}{7} & -\frac{6}{7} \\ -\frac{6}{7} & -\frac{2}{7} & \frac{3}{7} \end{bmatrix}$

[2018 : 1 Marks, Set-I]

1.62 Which one of the following matrices is singular?

(a) $\begin{bmatrix} 2 & 5 \\ 1 & 3 \end{bmatrix}$ (b) $\begin{bmatrix} 3 & 2 \\ 2 & 3 \end{bmatrix}$

(c) $\begin{bmatrix} 2 & 4 \\ 3 & 6 \end{bmatrix}$ (d) $\begin{bmatrix} 4 & 3 \\ 6 & 2 \end{bmatrix}$

[2018 : 1 Marks, Set-I]

1.63 The matrix $\begin{pmatrix} 2 & -4 \\ 4 & -2 \end{pmatrix}$ has

(a) real eigenvalues and eigenvectors
(b) real eigenvalues but complex eigenvectors
(c) complex eigenvalues but real eigenvectors
(d) complex eigenvalues and eigenvectors

[2018 : 2 M, Set-II]

1.64 The rank of the following matrix is $\begin{pmatrix} 1 & 1 & 0 & -2 \\ 2 & 0 & 2 & 2 \\ 4 & 1 & 3 & 1 \end{pmatrix}$

(a) 1 (b) 2
(c) 3 (d) 4

[2018 : 2 M, Set-II]

1.65 The inverse of the matrix $\begin{bmatrix} 2 & 3 & 4 \\ 4 & 3 & 1 \\ 1 & 2 & 4 \end{bmatrix}$ is

(a) $\begin{bmatrix} 10 & -4 & -9 \\ -15 & 4 & 14 \\ 5 & -1 & -6 \end{bmatrix}$ (b) $\begin{bmatrix} -10 & 4 & 9 \\ 15 & -4 & -14 \\ -5 & 1 & 6 \end{bmatrix}$

(c) $\begin{bmatrix} 2 & -\frac{4}{5} & -\frac{9}{5} \\ -3 & \frac{4}{5} & \frac{14}{5} \\ 1 & -\frac{1}{5} & -\frac{6}{5} \end{bmatrix}$ (d) $\begin{bmatrix} -2 & \frac{4}{5} & \frac{9}{5} \\ 3 & -\frac{4}{5} & -\frac{14}{5} \\ -1 & \frac{1}{5} & \frac{6}{5} \end{bmatrix}$

[2019 : 2 M, Set-II]

1.66 Consider the hemi-spherical tank of radius 13 m as shown in the figure (not drawn to scale). What is the volume of water (in m^3) when the depth of water at the centre of the tank is 6 m?

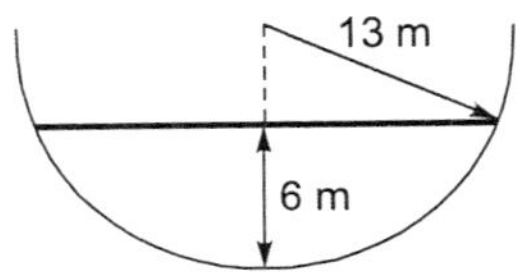

(a) 156π (b) 396π
(c) 468π (d) 78π

[2019 : 2 M, Set-II]

1.67 Consider the system of equations

$$\begin{bmatrix} 1 & 3 & 2 \\ 2 & 2 & -3 \\ 4 & 4 & -6 \\ 2 & 5 & 2 \end{bmatrix} \begin{bmatrix} x_1 \\ x_2 \\ x_3 \end{bmatrix} = \begin{bmatrix} 1 \\ 1 \\ 2 \\ 1 \end{bmatrix}$$

The value of x_3 (round off to the nearest integer), is _______.

[2020 : 2 M, Set-I]

1.68 A 4 × 4 matrix $[P]$ is given below

$$[P]=\begin{bmatrix}0 & 1 & 3 & 0\\ -2 & 3 & 0 & 4\\ 0 & 0 & 6 & 1\\ 0 & 0 & 1 & 6\end{bmatrix}$$

The eigen values of $[P]$ are

(a) 0, 3, 6, 6 (b) 1, 2, 3, 4
(c) 1, 2, 5, 7 (d) 3, 4, 5, 7

[2020 : 2 M, Set-II]

1.69 The rank of matrix $\begin{bmatrix}1 & 2 & 2 & 3\\ 3 & 4 & 2 & 5\\ 5 & 6 & 2 & 7\\ 7 & 8 & 2 & 9\end{bmatrix}$ is

(a) 4 (b) 2
(c) 1 (d) 3

[2021 : 1 M, Set-I]

1.70 If $P=\begin{bmatrix}1 & 2\\ 3 & 4\end{bmatrix}$ and $Q=\begin{bmatrix}0 & 1\\ 1 & 0\end{bmatrix}$ then $Q^T P^T$ is

(a) $\begin{bmatrix}1 & 2\\ 3 & 4\end{bmatrix}$ (b) $\begin{bmatrix}1 & 3\\ 2 & 4\end{bmatrix}$

(c) $\begin{bmatrix}2 & 4\\ 1 & 3\end{bmatrix}$ (d) $\begin{bmatrix}2 & 1\\ 4 & 3\end{bmatrix}$

[2021 : 1 M, Set-I]

1.71 The smallest eigenvalue and the corresponding eigenvector of the matrix $\begin{bmatrix}2 & -2\\ -1 & 6\end{bmatrix}$ respectively, are

(a) 1.55 and $\begin{Bmatrix}2.00\\ -0.45\end{Bmatrix}$

(b) 2.00 and $\begin{Bmatrix}1.00\\ 1.00\end{Bmatrix}$

(c) 1.55 and $\begin{Bmatrix}2.00\\ 0.45\end{Bmatrix}$

(d) 1.55 and $\begin{Bmatrix}-2.55\\ -0.45\end{Bmatrix}$

[2021 : 2 M, Set-II]

1.72 The rank of the matrix $\begin{bmatrix}5 & 0 & -5 & 0\\ 0 & 2 & 0 & 1\\ -5 & 0 & 5 & 0\\ 0 & 1 & 0 & 2\end{bmatrix}$ is

(a) 4 (b) 3
(c) 1 (d) 2

[2021 : 1 M, Set-II]

1.73 If A is a square matrix then orthogonality property mandates

(a) $AA^T = A^{-1}$ (b) $AA^T = 0$
(c) $AA^T = I$ (d) $AA^T = A^2$

[2021 : 1 M, Set-II]

1.74 The matrix M is defined as

$$M=\begin{bmatrix}1 & 3\\ 4 & 2\end{bmatrix}$$

and has eigenvalues 5 and – 2. The matrix Q is formed as

$$Q = M^3 - 4M^2 - 2M$$

Which of the following is/are the eigenvalue(s) of matrix Q ?

(a) 15 (b) 25
(c) –20 (d) –30

[2022 : 1 M, Set-I]

1.75 The Cartesian coordinates of a point P in a right-handed coordinate system are (1, 1, 1). The transformed coordinates of P due to a 45° clockwise rotation of the coordinate system about the positive x-axis are

(a) $(1, 0, \sqrt{2})$ (b) $(1, 0, -\sqrt{2})$

(c) $(-1, 0, \sqrt{2})$ (d) $(-1, 0, -\sqrt{2})$

[2022 : 2 M, Set-I]

1.76 Let y be a non-zero vector of size 2022 × 1. Which of the following statement(s) is/are TRUE?

(a) yy^T is symmetric matrix
(b) yy^T is invertible
(c) yy^T has a rank of 2022
(d) $y^T y$ is an eigen value of yy^T.

[2022 : 1 M, Set-II]

1.77 The components of pure shear strain in a sheared material are given in the matrix form:

$$\varepsilon = \begin{bmatrix}1 & 1\\ 1 & -1\end{bmatrix}$$

Here, Trace (ε) = 0. Given, P = Trace (ε^8) and Q = Trace (ε^{11}).

The numerical value of $(P + Q)$ is _______ . (in integer)

[2022 : 1 M, Set-II]

1.78 P and Q are two square matrices of the same order. Which of the following statement(s) is/are correct?

(a) If P and Q are invertible, then $[PQ]^{-1} = Q^{-1}P^{-1}$.

(b) If P and Q are invertible, then $[PQ]^{-1} = P^{-1}Q^{-1}$.

(c) If P and Q are invertible, then $[QP]^{-1} = P^{-1}Q^{-1}$.

(d) If P and Q are not invertible, then $[PQ]^{-1} = Q^{-1}P^{-1}$.

[2022 : 2 M, Set-II]

1.79 If M is an arbitrary real $n \times n$ matrix, then which of the following matrices will have non-negative eigenvalues?

(a) M^2 (b) MM^T
(c) M^TM (d) $(M^T)^2$

[2023 : 1 M, Set-I]

1.80 For the matrix

$$[A] = \begin{bmatrix} 1 & 2 & 3 \\ 3 & 2 & 1 \\ 3 & 1 & 2 \end{bmatrix}$$

Which of the following statements is/are TRUE?

(a) The eigenvalues of $[A]^T$ are same as the eigenvalues of $[A]$

(b) The eigenvalues of $[A]^{-1}$ are the reciprocals of the eigenvalues of $[A]$

(c) The eigenvectors of $[A]^T$ are same as the eigenvectors of $[A]$

(d) The eigenvectors of $[A]^{-1}$ are same as the eigenvectors of $[A]$

[2023 : 2 M, Set-I]

1.81 For the matrix

$$[A] = \begin{bmatrix} 1 & -1 & 0 \\ -1 & 2 & -1 \\ 0 & -1 & 1 \end{bmatrix}$$

Which of the following statements is/are TRUE?

(a) $[A]\{x\} = \{b\}$ has a unique solution

(b) $[A]\{x\} = \{b\}$ does not have a unique solution

(c) $[A]$ has three linearly independent eigenvectors

(d) $[A]$ is a positive definite matrix

[2023 : 1 M, Set-II]

1.82 Two vectors $[2\ 1\ 0\ 3]^T$ and $[1\ 0\ 1\ 2]^T$ belong to the null space of a 4×4 matrix of rank 2. Which one of the following vectors also belongs to the null space?

(a) $[1\ 1\ -1\ 1]^T$ (b) $[2\ 0\ 1\ 2]^T$
(c) $[0\ -2\ 1\ -1]^T$ (d) $[3\ 1\ 1\ 2]^T$

[2023 : 2 M, Set-II]

1.83 Cholesky decomposition is carried out on the following square matrix $[A]$.

$$[A] = \begin{bmatrix} 8 & -5 \\ -5 & a_{22} \end{bmatrix}$$

Let l_{ij} and a_{ij} be the (i, j)[th] elements of matrices $[L]$ and $[A]$, respectively. If the element l_{22} of the decomposed lower triangular matrix $[L]$ is 1.968, what is the value (rounded off to the nearest integer) of the element a_{22}?

(a) 5 (b) 7
(c) 9 (d) 11

[2023 : 2 M, Set-II]

1.84 What are the eigenvalues of the matrix

$$\begin{bmatrix} 2 & 1 & 1 \\ 1 & 4 & 1 \\ 1 & 1 & 2 \end{bmatrix}?$$

(a) 1, 3, 4 (b) 1, 2, 5
(c) –5, 1, 2 (d) –5, –1, 2

[2024 : 2 M, Set-I]

1.85 The statements P and Q are related to matrices A and B, which are conformable for both addition and multiplication.

$P: (A + B)^T = A^T + B^T$

$Q: (AB)^T = A^T B^T$

Which one of the following options is CORRECT?

(a) P is TRUE and Q is FALSE
(b) P is FALSE and Q is TRUE
(c) Both P and Q are FALSE
(d) Both P and Q are TRUE

[2024 : 1 M, Set-II]

1.86 Consider two matrices $A = \begin{bmatrix} 2 & 1 & 4 \\ 1 & 0 & 3 \end{bmatrix}$ and $B = \begin{bmatrix} -1 & 0 \\ 2 & 3 \\ 1 & 4 \end{bmatrix}$

The determinant of the matrix AB is ________ (in integer).

[2024 : 2 M, Set-II]

Answers Linear Algebra

1.1 (b)	1.2 (c, d)	1.3 (c)	1.4 (c)	1.5 (b)	1.6 (a)	1.7 (a)	1.8 (d)	1.9 (d)
1.10 (d)	1.11 (b)	1.12 (d)	1.13 (b)	1.14 (a)	1.15 (a)	1.16 (b)	1.17 (c)	1.18 (c)
1.19 (b)	1.20 (b)	1.21 (b)	1.22 (b)	1.23 (b)	1.24 (a)	1.25 (d)	1.26 (a)	1.27 (c)
1.28 (a)	1.29 (c)	1.30 (a)	1.31 (a)	1.32 (b)	1.33 (d)	1.34 (b)	1.35 (b)	1.36 (a)
1.37 (a)	1.38 (b)	1.39 (b)	1.40 (d)	1.41 (a)	1.42 (b)	1.43 (d)	1.44 (b)	1.47 (a)
1.48 (a)	1.52 (d)	1.53 (b)	1.54 (d)	1.55 (d)	1.56 (b)	1.57 (c)	1.58 (a)	1.59 (a)
1.60 (a)	1.61 (c)	1.62 (c)	1.63 (d)	1.64 (b)	1.65 (d)	1.66 (b)	1.68 (c)	1.69 (b)
1.70 (c)	1.71 (c)	1.72 (b)	1.73 (c)	1.74 (a, c)	1.75 (a)	1.76 (a, d)	1.78 (a, c)	1.79 (b, c)
1.80 (a, b, d)	1.81 (b, c)		1.82 (a)	1.83 (b)	1.84 (b)	1.85 (a)	1.86 (10)	

Explanations Linear Algebra

1.1 (b)

$\because \quad |A| = 0$

$\Rightarrow \quad \rho(A) < 3$

And as 2 × 2 minor of 3 × 3 matrix is not equal to zero. Hence rank of matrix [A] is two.

1.2 (c, d)

$$\begin{vmatrix} 1-\lambda & -1 \\ -1 & 1-\lambda \end{vmatrix}$$

$\Rightarrow \quad \lambda = 0, 2$

Now, for $\lambda = 0$

$$\begin{bmatrix} 1 & -1 \\ -1 & 1 \end{bmatrix}\begin{bmatrix} x_1 \\ x_2 \end{bmatrix} = \begin{bmatrix} 0 \\ 0 \end{bmatrix}$$

$\Rightarrow \quad x_1 - x_2 = 0$

$\Rightarrow \quad -x_1 + x_2 = 0$

$$\Rightarrow \quad \begin{bmatrix} x_1 \\ x_1 \end{bmatrix} = \begin{bmatrix} 1 \\ 1 \end{bmatrix}$$

$\therefore$ Eigen vector is [1, 1]

For $\lambda = 2$

$$\begin{bmatrix} -1 & -1 \\ -1 & -1 \end{bmatrix}\begin{bmatrix} x_1 \\ x_2 \end{bmatrix} = \begin{bmatrix} 0 \\ 0 \end{bmatrix}$$

$-x_1 - x_2 = 0$

$-x_1 = x_2$

$\therefore$ Eigen vector [–1, 1] or [1, –1]

1.3 (c)

$$A = \begin{bmatrix} -3 & 5 \\ 2 & 1 \end{bmatrix}$$

We know that $A^{-1} = \dfrac{adj[A]}{|A|}$

Co-factors of the matrix [A] is,

$C_{11} = 1 \quad C_{12} = -2$

$C_{22} = -5 \quad C_{22} = -3$

Matrix formed by the co-factor of [A] is,

$$\begin{bmatrix} 1 & -2 \\ -5 & -3 \end{bmatrix} \therefore adj[A] = \begin{bmatrix} 1 & -5 \\ -2 & -3 \end{bmatrix}$$

$\therefore adj[A] = |A| = (-3 \times 1) - (2 \times 5)$

$= -3 - 10 = -13$

$$\therefore \quad A^{-1} = \begin{bmatrix} -1/13 & 5/13 \\ 2/13 & 3/13 \end{bmatrix}$$

1.4 (c)

$det(A + B) \neq det(A) + det(B)$

1.5 (b)

Since n is common in both the matrices, therefore, multiplication operation required will become mpn and addition operations will be $mp(n-1)$,

1.6 (a)

$$det\,A = -8\begin{bmatrix} 1 & 2 \\ 2 & 4 \end{bmatrix} = 0$$

$\therefore \quad$ Rank $\neq 3$

$$det\begin{bmatrix} 2 & -1 \\ 4 & 6 \end{bmatrix} = 16 \neq 0$$

$\therefore \quad$ Rank = 2

1.7 (a)

Inverse of $7A = \begin{bmatrix} -1 & 2 \\ 4 & -7 \end{bmatrix}$

$$\det |\Delta| = \begin{bmatrix} -1 & 2 \\ 4 & -7 \end{bmatrix} = 7 - 8 = -1$$

∴ Matrices of cofactor

$$= \begin{bmatrix} -7 & -4 \\ -2 & -1 \end{bmatrix}$$

∴ It's transpose $A^T = \begin{bmatrix} -7 & -2 \\ -4 & -1 \end{bmatrix}$

∴ Inverse $\dfrac{A^T}{\Delta} = \begin{bmatrix} 7 & 2 \\ 4 & 1 \end{bmatrix}$

$$\therefore \quad A = \frac{1}{7}\left[\frac{A^T}{|\Delta|}\right] = \begin{bmatrix} 1 & 2/7 \\ 4/7 & 1/7 \end{bmatrix}$$

1.8 (d)

Equations are consistent when,

$$\begin{vmatrix} 2 & 3 & -1 \\ 1 & -1 & -4 \\ 4 & -1 & -\alpha \end{vmatrix} = 0$$

$\Rightarrow$ $2(\alpha - 4) - 3(-\alpha + 16) - 1(-1 + 4) = 0$

$\Rightarrow$ $2\alpha - 8 + 3\alpha - 48 + 1 - 4 = 0$

$\Rightarrow$ $5\alpha - 59 = 0$

$\Rightarrow$ $\alpha = \dfrac{59}{5} = 11\dfrac{4}{5}$

1.14 (a)

By interchanging any row or a column, the value of determinant will remain same, but becomes negative. If two rows or columns are interchanged the value remains same with the sign as well.

1.15 (a)

The given matrix,

$$A = \begin{bmatrix} 0 & 1 & 0 \\ 0 & 0 & 1 \\ 1 & 0 & 0 \end{bmatrix}$$

$$\therefore \quad A^{-1} = \frac{\text{Adjoint of matrix } A}{\text{Determinant of } A} = \frac{\text{Adj. A}}{|A|}$$

Where,

$$|A| = \begin{bmatrix} 0 & 1 & 0 \\ 0 & 0 & 1 \\ 1 & 0 & 0 \end{bmatrix} = 0(0-0) - 1(-1) + 0$$

and matrix of cofactors

$$= \begin{bmatrix} 0 & 1 & 0 \\ 0 & 0 & 1 \\ 1 & 0 & 0 \end{bmatrix}$$

∴ Adjoint A = Transpose of cofactors of

$$A = \begin{bmatrix} 0 & 0 & 1 \\ 1 & 0 & 0 \\ 0 & 1 & 0 \end{bmatrix}$$

$$\therefore \quad A^{-1} = \frac{Adj.\ A}{|A|} = \begin{bmatrix} 0 & 0 & 1 \\ 1 & 0 & 0 \\ 0 & 1 & 0 \end{bmatrix}$$

1.18 (c)

$A = n \times n$ matrix

Let A be matrix of 2×2

$$A = \begin{bmatrix} a_{11} & a_{12} \\ a_{21} & a_{22} \end{bmatrix}$$

$$\therefore \quad kA = \begin{bmatrix} ka_{11} & ka_{12} \\ ka_{21} & ka_{22} \end{bmatrix}$$

$\therefore$ $|kA| = ka_{11}(ka_{22}) - ka_{12}\, ka_{21}$

$= k^2(a_{11}a_{12} - a_{12}a_{21})$

$\therefore$ $|kA| = k^n(a_{11}a_{22} - a_{12}a_{21})$

$\Rightarrow$ $|kA| = k^n|A|$

$\therefore$ $\alpha = k^n$

1.20 (b)

$$\begin{vmatrix} 2 & 1 & 1 \\ 1 & 1 & -1 \\ y & x^2 & x \end{vmatrix} = 0$$

$2(x + x^2) - 1(x + y) + 1(x^2 - y) = 0$

$\Rightarrow$ $2x + 2x^2 - x - y + x^2 - y = 0$

$\Rightarrow$ $x + 3x^2 - 2y = 0$

$\Rightarrow$ $2y = 3x^2 + x$

$$x = \frac{3}{2}x^2 + \frac{x}{2}$$

$\Rightarrow$ $3x^2 + x = 2y$

$$x^2 + \frac{x}{3} = \frac{2y}{3}$$

$$\Rightarrow \left(x+\frac{1}{6}\right)^2 - \frac{1}{36} = \frac{2y}{3}$$

$$\Rightarrow \left(x+\frac{1}{6}\right)^2 = \frac{2y}{3} + \frac{1}{36}$$

From the given options we can see that the option (b) safisfies the given equation.

1.23 (b)

$$\begin{vmatrix} 5 & 3 & 2 \\ 1 & 2 & 6 \\ 3 & 5 & 10 \end{vmatrix} = 5(20-30) - 3(10-18) + 2(5-6)$$

$$= -50 + 24 - 2 = -28$$

1.24 (a)

$$[P][Q]^T = \begin{bmatrix} 2 & 3 \\ 4 & 5 \end{bmatrix}\begin{bmatrix} 4 & 9 \\ 8 & 2 \end{bmatrix}$$

$$= \begin{bmatrix} 8+24 & 18+6 \\ 16+40 & 16+10 \end{bmatrix}$$

$$= \begin{bmatrix} 32 & 24 \\ 56 & 46 \end{bmatrix}$$

1.25 (d)

$$\begin{bmatrix} 5-\lambda & 3 \\ 2 & 9-\lambda \end{bmatrix} = 0$$

or $(5-\lambda)(9-\lambda) - 6 = 0$

or $45 - 9\lambda - 5\lambda + \lambda^2 - 6 = 0$

or $\lambda^2 - 14\lambda + 39 = 0$

Solving we get,

$\lambda = 10.1623, 3.837$

Hence eigen value of the matrix are (10.16, 3.837)

1.27 (c)

Consider first 3×3 minors, since maximum possible rank is 3

$$\begin{vmatrix} 4 & 2 & 1 \\ 6 & 3 & 4 \\ 2 & 1 & 0 \end{vmatrix} = 0$$

$$\begin{vmatrix} 2 & 1 & 3 \\ 3 & 4 & 7 \\ 1 & 0 & 1 \end{vmatrix} = 0$$

$$\begin{vmatrix} 4 & 1 & 3 \\ 6 & 4 & 7 \\ 2 & 0 & 1 \end{vmatrix} = 0$$

and $\begin{vmatrix} 4 & 2 & 3 \\ 6 & 3 & 7 \\ 2 & 1 & 1 \end{vmatrix} = 0$

Since all 3×3 minors are zero, now try 2×2 minors.

$$\begin{vmatrix} 4 & 2 \\ 6 & 3 \end{vmatrix} = 0$$

$$\begin{vmatrix} 2 & 1 \\ 3 & 4 \end{vmatrix} = 8 - 3 = 5 \neq 0$$

$\therefore$ rank $= 2$

1.28 (a)

Statement 1 is true.

Consider statement 2

$[D]^T [F] [D]$ is always symmetric.

Now since $[D'FD]' = (FD)'(D')'$

$= (FD)'D$

$= D'F'D$

$= D'FD$

(Since F may or may not be symmetric)

$\therefore$ $D'FD$ is not always symmetric is true.

$\therefore$ 1 is true and 2 is false.

1.29 (c)

Characteristic equation is,

$$\begin{vmatrix} 4-\lambda & -2 \\ -2 & 1-\lambda \end{vmatrix} = 0$$

$$(4-\lambda) \times (1-\lambda) - [(-2) \times (-2)] = 0$$

$$\lambda^2 - 5\lambda = 0$$

$$\Rightarrow \lambda(\lambda - 5) = 0$$

Hence, $\lambda = 0, 5$ are the eigen values

1.30 (a)

With the given order we can say that order of matrices are as follows:

$X^T \rightarrow 3 \times 4$

$Y \rightarrow 4 \times 3$

$X^TY \rightarrow 3 \times 3$

$(X^TY)^{-1} \rightarrow 3 \times 3$

$P \rightarrow 2 \times 3$

$P^T \rightarrow 3 \times 2$

$P(X^TY)^{-1}P^T \rightarrow (2 \times 3)(3 \times 3)(3 \times 2)$

$\rightarrow 2 \times 2$

$\therefore (P(X^TY)^{-1}P^T)^T \rightarrow 2 \times 2$

1.31 (a)

In an over determined system having more equations than variables, it is necessary to have consistent unique solution, by definition.

1.32 (b)

If $A^T = A^{-1}$ i.e. A is orthogonal, we can only say that if λ is an eigen value of A then $1/\lambda$ also will be an eigen value of A, which does not necessarily imply that,

$$|\lambda_i| = 1 \text{ for all } i.$$

1.33 (d)

The augmented matrix for given system is,

$$\left[\begin{array}{ccc|c} 0 & 4 & 3 & 8 \\ 2 & 0 & -1 & 2 \\ 3 & 2 & 0 & 5 \end{array}\right] \xrightarrow{\text{exchange 1st and 2nd row}} \left[\begin{array}{ccc|c} 2 & 0 & -1 & 2 \\ 0 & 4 & 3 & 8 \\ 3 & 2 & 0 & 5 \end{array}\right]$$

then by Gauss elimination procedure

$$\left[\begin{array}{ccc|c} 2 & 0 & -1 & 2 \\ 0 & 4 & 3 & 8 \\ 3 & 2 & 0 & 5 \end{array}\right] \xrightarrow{R_3-\frac{3}{2}R_1} \left[\begin{array}{ccc|c} 2 & 0 & -1 & 2 \\ 0 & 4 & 3 & 8 \\ 0 & 2 & 3/2 & 2 \end{array}\right]$$

$$\xrightarrow{R_3-\frac{2}{4}R_2} \left[\begin{array}{ccc|c} 2 & 0 & -1 & 8 \\ 0 & 4 & 3 & 8 \\ 0 & 0 & 0 & -2 \end{array}\right]$$

For last row we see $0 = -2$ which is inconsistent

$\therefore$ Solution is non-existent for above system.

1.34 (b)

$$\Sigma\lambda_i = \text{Trace}(A)$$

$$\lambda_1 + \lambda_2 + \lambda_3 = \text{Trace}(A)$$

$$= 2 + (-1) + 0 = 1$$

Now $\lambda_1 = 3$

$\therefore 3 + \lambda_2 + \lambda_3 = 1$

$\Rightarrow \lambda_2 + \lambda_3 = -2$

Only choice (b) satisfies this condition.

1.35 (b)

$$\Sigma\lambda_i = \text{Trace}(A)$$

$$\lambda_1 + \lambda_2 + \lambda_3 = 1 + 5 + 1 = 7$$

Now $\lambda_1 = -2, \lambda_2 = 6$

$\therefore -2 + 6 + \lambda_3 = 7$

$$\lambda_3 = 3$$

1.36 (a)

The augmented matrix for this system is,

$$\left[\begin{array}{ccc|c} 1 & 1 & 1 & 5 \\ 1 & 3 & 3 & 9 \\ 1 & 2 & \alpha & \beta \end{array}\right]$$

Using Gauss-elimination method we get

$$\left[\begin{array}{ccc|c} 1 & 1 & 1 & 5 \\ 1 & 3 & 3 & 9 \\ 1 & 2 & \alpha & \beta \end{array}\right] \xrightarrow[R_3-R_1]{R_2-R_1} \left[\begin{array}{ccc|c} 1 & 1 & 1 & 5 \\ 0 & 2 & 2 & 4 \\ 0 & 1 & \alpha-1 & \beta-5 \end{array}\right]$$

$$\xrightarrow{R_3-\frac{1}{2}R_2} \left[\begin{array}{ccc|c} 1 & 1 & 1 & 5 \\ 0 & 2 & 2 & 4 \\ 0 & 0 & \alpha-2 & \beta-7 \end{array}\right]$$

Now, for infinite solution last row must be completely zero

i.e. $\alpha - 2$ and $\beta - 7 = 0$

$\Rightarrow \alpha = 2$ and $\beta = 7$

1.37 (a)

Inverse of $\begin{bmatrix} a & b \\ c & d \end{bmatrix}$ is,

$$\begin{bmatrix} a & b \\ c & d \end{bmatrix}^{-1} = \frac{1}{(ad-bc)}\begin{bmatrix} d & -b \\ -c & a \end{bmatrix}$$

$$\therefore \begin{bmatrix} 1 & 2 \\ 5 & 7 \end{bmatrix} = \frac{1}{(7-10)}\begin{bmatrix} 7 & -2 \\ -5 & 1 \end{bmatrix}$$

$$= \frac{1}{3}\begin{bmatrix} -7 & 2 \\ 5 & -1 \end{bmatrix}$$

1.38 (b)

$$(PQ)^{-1}P = (Q^{-1}P^{-1})P$$

$$= (Q^{-1})(P^{-1}P) = (Q^{-1})(I) = Q^{-1}$$

1.39 (b)

$$A = \begin{bmatrix} 4 & 5 \\ 2 & -5 \end{bmatrix}$$

Characteristic equation of A is,

$$\begin{vmatrix} 4-\lambda & 5 \\ 2 & -5-\lambda \end{vmatrix} = 0$$

$\Rightarrow (4-\lambda)(-5-\lambda) - 2 \times 5 = 0$

$\Rightarrow \lambda^2 + \lambda - 30 = 0$

$\lambda = 5, -6$

1.40 (d)

The augmented matrix for given system is,

$$\left[\begin{array}{ccc|c} 1 & 1 & 1 & 3 \\ 1 & 2 & 3 & 4 \\ 1 & 4 & k & 6 \end{array}\right]$$

Using Gauss elimination we reduce this to an upper triangular matrix to investigate its rank

$$\left[\begin{array}{ccc|c} 1 & 1 & 1 & 3 \\ 1 & 2 & 3 & 4 \\ 1 & 4 & k & 6 \end{array}\right] \xrightarrow[R_3 - R_1]{R_2 - R_1} \left[\begin{array}{ccc|c} 1 & 1 & 1 & 3 \\ 0 & 1 & 2 & 1 \\ 0 & 3 & k-1 & 3 \end{array}\right]$$

$$\xrightarrow[R_3 - 3R_2]{} \left[\begin{array}{ccc|c} 1 & 1 & 1 & 3 \\ 0 & 1 & 2 & 1 \\ 0 & 0 & k-7 & 0 \end{array}\right]$$

Now if $k \neq 7$

rank (A) = rank $(A \mid B) = 3$

$\therefore$ **unique solution.**

If $k = 7$, rank (A) = rank $(A \mid B) = 2$

which is less than number of variables

$\therefore$ when $k = 7$

unique solution is not possible and only infinite solution is possible.

1.41 (a)

A square matrix B is defined as skew-symmetric if and only if $B^T = -B$.

1.42 (b)

$$\begin{bmatrix} a & b \\ c & d \end{bmatrix}^{-1} = \frac{1}{(ad-bc)}\begin{bmatrix} d & -b \\ -c & a \end{bmatrix}$$

$$\therefore \begin{bmatrix} 3+2i & i \\ -i & 3-2i \end{bmatrix}^{-1} = \frac{1}{\left[(3+2i)(3-2i)+i^2\right]}\begin{bmatrix} 3-2i & -i \\ i & 3+2i \end{bmatrix}$$

$$= \frac{1}{12}\begin{bmatrix} 3-2i & -i \\ i & 3+2i \end{bmatrix}$$

1.43 (d)

Since $(A + A^t) = A^t + (A^t)^t$

$= A^t + A$

i.e. $S^t = S$

$\therefore$ S is symmetric

Since $(A - A^t)^t = A^t - (A^t)^t$

$= A^t - A = -(A - A^t)$

i.e. $D^t = -D$

So D is **Skew-Symmetric.**

1.44 (b)

We need eigen values of $A = \begin{bmatrix} 9 & 5 \\ 5 & 8 \end{bmatrix}$

The characteristic equation is,

$$\begin{vmatrix} 9-\lambda & 5 \\ 5 & 8-\lambda \end{vmatrix} = 0$$

$$(9-\lambda)(8-\lambda) - 25 = 0$$

$$\Rightarrow \quad \lambda^2 - 17\lambda + 47 = 0$$

So eigen values are,

$$\lambda = 3.48, 13.53$$

1.45 Sol.

If we multiply QR first then,

$Q_{2\times4} \times R_{(4\times1)}$ having multiplication number 8.

Therefore $P_{(4\times2)}\, QR_{(2\times1)}$ will have minimum number of multiplication = (8 + 8) = 16.

1.46 Sol.

$$J = \begin{bmatrix} 3 & 2 & 1 \\ 2 & 4 & 2 \\ 1 & 2 & 6 \end{bmatrix}$$

$$K = \begin{bmatrix} 1 \\ 2 \\ -1 \end{bmatrix}$$

$$K^T JK = [1\ 2\ -1]\begin{bmatrix} 3 & 2 & 1 \\ 2 & 4 & 2 \\ 1 & 2 & 6 \end{bmatrix}\begin{bmatrix} 1 \\ 2 \\ -1 \end{bmatrix}$$

$$= [6\ 8\ -1]\begin{bmatrix} 1 \\ 2 \\ -1 \end{bmatrix} = 6 + 16 + 1 = 23$$

1.47 (a)

Sum of eigen values = trace of matrix

= 215 + 150 + 550 = 915

1.48 (a)

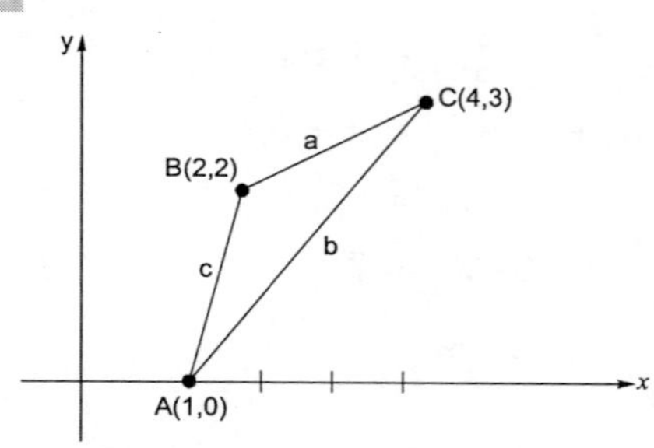

Area of triangle is

$$= \frac{1}{2}\left[x_1(y_2 - y_3) + x_2(y_3 - y_1) + x_2(y_1 - y_2)\right]$$

$$= \frac{1}{2}\left[1(2-3) + 2(3-0) + 4(0-2)\right]$$

$$= \frac{1}{2}[-1+6-8] = \frac{3}{2}$$

1.49 Sol.

$$\Delta = \begin{vmatrix} 0 & 1 & 2 & 3 \\ 1 & 0 & 3 & 0 \\ 2 & 3 & 0 & 1 \\ 3 & 0 & 1 & 2 \end{vmatrix}$$

$R_4 \to R_4 - R_2 - R_3$

$$\Delta = \begin{vmatrix} 0 & 1 & 2 & 3 \\ 1 & 0 & 3 & 0 \\ 2 & 3 & 0 & 1 \\ 0 & -3 & -2 & 1 \end{vmatrix}$$

$R_4 \to R_4 + 3R_1$

$$\Delta = \begin{vmatrix} 0 & 1 & 2 & 3 \\ 1 & 0 & 3 & 0 \\ 2 & 3 & 0 & 1 \\ 0 & 0 & 4 & 10 \end{vmatrix}$$

$R_3 \to R_3 - 3R_1$

$$\Delta = \begin{vmatrix} 0 & 1 & 2 & 3 \\ 1 & 0 & 3 & 0 \\ 2 & 0 & -6 & -8 \\ 0 & 0 & 4 & 10 \end{vmatrix}$$

Interchanging column 1 and column 2 and taking transpose,

$$\Delta = -\begin{vmatrix} 1 & 0 & 0 & 0 \\ 0 & 1 & 2 & 0 \\ 2 & 3 & -6 & 4 \\ 3 & 0 & -8 & 10 \end{vmatrix}$$

$$= -1 \times \begin{vmatrix} 1 & 2 & 0 \\ 3 & -6 & 4 \\ 0 & -8 & 10 \end{vmatrix}$$

$$= -1 \times \{1(-60 + 32) + 2(0 - 30)\}$$

$$= -(-28 - 60) = 88$$

1.50 Sol.

$$\begin{bmatrix} 6 & 0 & 4 & 4 \\ -2 & 14 & 8 & 18 \\ 14 & -14 & 0 & -10 \end{bmatrix}$$

$R_3 \to R_3 - 2R_1 + R_2$

$$\begin{bmatrix} 6 & 0 & 4 & 4 \\ -2 & 14 & 8 & 18 \\ 14-2(6)+(-2) & -14-2(0)+(14) & 0-2(4)+8 & -10-2(4)+(18) \end{bmatrix}$$

$$\begin{bmatrix} 6 & 0 & 4 & 4 \\ -2 & 14 & 8 & 18 \\ 0 & 0 & 0 & 0 \end{bmatrix}$$

Determinant of matrix $\begin{bmatrix} 6 & 0 \\ -2 & 14 \end{bmatrix}$ is not zero.

$\therefore$ Rank is 2.

1.51 Sol.

Given system of equations has no solution if the lines are parallel i.e., their slopes are equal

$$\frac{2}{3} = \frac{3}{p}$$

$$\Rightarrow \qquad p = 4.5$$

1.52 (d)

For eigen values $|A - \lambda I| = 0$

$$\Rightarrow \begin{vmatrix} 3-\lambda & -2 & 2 \\ 4 & -4-\lambda & 6 \\ 2 & -3 & 5-\lambda \end{vmatrix} = 0$$

$\Rightarrow$ $(3 - \lambda)(-20 + 4\lambda - 5\lambda + \lambda^2 + 18) + 2$
$(20 - 4\lambda - 12) + 2(-12 + 8 + 2\lambda) = 0$

$\Rightarrow$ $\lambda^3 - 4\lambda^2 + 5\lambda - 2 = 0$

Only 1 and 2 satisfy this equation.

$\lambda = 1, 1, 2$

Hence,Smallest eigen value = 1 and

Largest eigen value = 2

1.53 (b)

Rank of A = 1

Because each row will be scalar multiple of first row. So we will get only one non-zero row in row Echeleaon form of A.

Alternative:

Rank of A = 1

Because all the minors of order greater than 1 will be zero.

1.54 (d)

$$A = \begin{bmatrix} 2 & 1 \\ 1 & p \end{bmatrix}$$

Let λ_1 and λ_2 be the eigen value of matrix A

$$\therefore \quad \frac{\lambda_1}{\lambda_2} = \frac{3}{1} \text{ for } p = 2$$

Sum of eigen value

$$= \lambda_1 + \lambda_2 = 2 + p \quad \text{...(i)}$$

Product of eigen value

$$= \lambda_1\lambda_2 = 2p - 1 \quad \text{...(ii)}$$

$$\frac{\lambda_1}{\lambda_2} = \frac{3}{1}$$

$$\Rightarrow \quad \lambda_1 = 3\lambda_2$$

From eq. (i).

$$\Rightarrow \quad 3\lambda_2 + \lambda_2 = 2 + p$$

$$4\lambda_2 = 2 + p$$

$$\lambda_2 = \frac{p+2}{4}$$

From eq. (ii)

$$\Rightarrow \quad 3\lambda_2^2 = 2p - 1$$

$$\Rightarrow \quad 3\left(\frac{p+2}{4}\right)^2 = 2p - 1$$

$$\Rightarrow \quad p = 2, \frac{14}{3}$$

OR $\begin{bmatrix} 2 & 1 \\ 1 & p \end{bmatrix}$

$$A - I\lambda = 0$$

$$\begin{bmatrix} 2-\lambda & 1 \\ 1 & p-\lambda \end{bmatrix} = 0$$

$$(2-\lambda)(p-\lambda) - 1 = 0$$

$$\lambda^2 - (p+2)\lambda + (2p-1) = 0$$

By putting values of p from options.

By putting option (d) $\frac{14}{3}$ in above equations gives value $5, \frac{5}{3}$.

Hence ratio of two eigen values $= \frac{5}{5/3} = 3:1$.

1.55 (d)

Consider the '2 × 2' square matrix $M = \begin{bmatrix} a & b \\ c & d \end{bmatrix}$

$$\Rightarrow \quad \lambda^2 - (a+d)\lambda + (ad - bc) = 0 \quad \text{...(i)}$$

Putting $\lambda = 1$, we get

$$1 - (a+d) + ad - bc = 0$$

$$1 - a - d + ad - (1-d)(1-a) = 0$$

$$1 - a - d + ad - 1 + a + d - ad = 0$$

$0 = 0$ which is true.

$\therefore$ $\lambda = 1$ satisfied the eq. (i) but $\lambda = 2, 3, 4$ does not satisfy the eq. (i). For all possible values of a, d.

1.57 (c)

Given that P is inverse of Q.

$$P = Q^{-1} \qquad P = Q^{-1}$$

$$PQ = Q^{-1}Q \qquad QP = QQ^{-1}$$

$$PQ = I \qquad QP = I$$

$$\therefore \quad PQ = QP = I$$

1.58 (a)

$$A = \begin{bmatrix} 5 & -1 \\ 4 & 1 \end{bmatrix}$$

Characteristics equation is $|A - \lambda I| = 0$

$$\begin{vmatrix} 5-\lambda & -1 \\ 4 & 1-\lambda \end{vmatrix} = 0$$

$$5 - 5\lambda - \lambda + \lambda^2 + 4 = 0$$

$$\lambda^2 - 6\lambda + 9 = 0$$

$$\lambda = 3, 3$$

Algebraic multiplicity of eigen value 3 is 2. It has only one independent eigen vector exists.

1.59 (a)

$$[A] = \begin{bmatrix} 3 & 2 \\ 4 & 1 \end{bmatrix}$$

$$[A - \lambda I] = \begin{bmatrix} 3-\lambda & 2 \\ 4 & 1-\lambda \end{bmatrix}$$

$$|A - \lambda I| = 0$$

$$(3-\lambda)(1-\lambda) - 8 = 0$$

$$3 - 4\lambda + \lambda^2 - 8 = 0$$

$$\lambda^2 - 4\lambda - 5 = 0$$

1.60 (a)

$$AB^T = \begin{bmatrix} 1 & 5 \\ 6 & 2 \end{bmatrix}\begin{bmatrix} 7 & 8 \\ 7 & 4 \end{bmatrix} = \begin{bmatrix} 38 & 28 \\ 32 & 56 \end{bmatrix}$$

1.61 (c)

$$|Q| = \frac{3}{7}\left(-\frac{9}{49} - \frac{12}{49}\right) - \frac{2}{7}\left(\frac{18}{49} - \frac{4}{49}\right) + \frac{6}{7}\left(\frac{-36}{49} - \frac{6}{49}\right)$$

$$= -1$$

$$\text{Adj. } Q = \begin{bmatrix} -\frac{21}{49} & \frac{42}{49} & -\frac{14}{49} \\ -\frac{14}{49} & -\frac{21}{49} & -\frac{42}{49} \\ -\frac{42}{49} & -\frac{14}{49} & \frac{21}{42} \end{bmatrix}$$

$$\therefore \quad Q^{-1} = \frac{Adj\,Q}{|Q|} = \begin{bmatrix} \frac{3}{7} & -\frac{6}{7} & \frac{2}{7} \\ \frac{2}{7} & \frac{3}{7} & \frac{6}{7} \\ \frac{6}{7} & \frac{2}{7} & -\frac{3}{7} \end{bmatrix}$$

Or $\because$ Q is orthogonal

$$\therefore \quad Q^{-1} = Q^T$$

1.62 (c)

Option (a): $|A| = 6 - 5 = 1$

Option (b): $|A| = 9 - 4 = 5$

Option (c): $|A| = 12 - 12 = 0$

Option (d): $|A| = 8 - 18 = -10$

Hence matrix (c) is singular.

1.63 (d)

$$A = \begin{bmatrix} 2 & -4 \\ 4 & -2 \end{bmatrix}$$

$$|A - \lambda I| = 0$$

$$\begin{vmatrix} 2-\lambda & -4 \\ 4 & -2-\lambda \end{vmatrix} = 0$$

$$-4 - 2\lambda + 2\lambda + \lambda^2 + 16 = 0$$

$$\lambda^2 + 12 = 0$$

$$\lambda = \pm 2\sqrt{3}i \quad \text{(Complex eigen values)}$$

(1) $\lambda = 2\sqrt{3}i$

Consider $(A - \lambda I)X = 0$

$$\begin{bmatrix} 2-2\sqrt{3}i & -4 \\ 4 & -2-2\sqrt{3}i \end{bmatrix}\begin{bmatrix} x_1 \\ x_2 \end{bmatrix} = \begin{bmatrix} 0 \\ 0 \end{bmatrix}$$

$$2 - 2\sqrt{3}i x_1 = 4x_2$$

$$\frac{x_1}{4} = \frac{x_2}{2-2\sqrt{3}i}$$

$$\begin{bmatrix} x_1 \\ x_2 \end{bmatrix} = \begin{bmatrix} 4 \\ 2-2\sqrt{3}i \end{bmatrix}$$

(2) $\lambda = -2\sqrt{3}i$

Consider $(A - \lambda I)X = 0$

$$\begin{bmatrix} 2+2\sqrt{3}i & -4 \\ 4 & -2+2\sqrt{3}i \end{bmatrix}\begin{bmatrix} x_1 \\ x_2 \end{bmatrix} = \begin{bmatrix} 0 \\ 0 \end{bmatrix}$$

$$2 + 2\sqrt{3}i x_1 = 4x_2$$

$$\frac{x_1}{4} = \frac{x_2}{2+2\sqrt{3}i}$$

$$\begin{bmatrix} x_1 \\ x_2 \end{bmatrix} = \begin{bmatrix} 4 \\ 2+2\sqrt{3}i \end{bmatrix}$$

$\therefore$ Complex Eigen values and complex Eigen vectors.

1.64 (b)

$$A = \begin{pmatrix} 1 & 1 & 0 & -2 \\ 2 & 0 & 2 & 2 \\ 4 & 1 & 3 & 1 \end{pmatrix}$$

$$R_2 \to R_2 - 2R_1,\ R_3 \to R_3 - 4R_1$$

$$\begin{pmatrix} 1 & 1 & 0 & -2 \\ 0 & -2 & 2 & 6 \\ 0 & -3 & 3 & 9 \end{pmatrix}$$

$$R_3 \to R_3 - \frac{3}{2}R_2$$

$$\begin{pmatrix} 1 & 1 & 0 & -2 \\ 0 & -2 & 2 & 6 \\ 0 & 0 & 0 & 0 \end{pmatrix}$$

Number of non zero rows = 2

rank of A = 2

1.66 (b)

Volume of water

$$= \frac{1}{3}\pi h^2(3r - h)$$

$$= \frac{1}{3}\pi \times 6^2 \times (3\times 13 - 6) = 396\pi$$

1.67 Sol.

$$[A : B] = \begin{bmatrix} 1 & 3 & 2 & \vdots & 1 \\ 2 & 2 & -3 & \vdots & 1 \\ 4 & 4 & -6 & \vdots & 2 \\ 3 & 5 & 2 & \vdots & 1 \end{bmatrix}$$

$$\xrightarrow[\text{an Echelon Form}]{\text{Converting into}} \begin{bmatrix} 1 & 3 & 2 & \vdots & 1 \\ 0 & -1 & -2 & \vdots & -1 \\ 0 & 0 & 1 & \vdots & 3 \\ 0 & 0 & 0 & \vdots & 0 \end{bmatrix}$$

$$\Rightarrow \begin{bmatrix} 1 & 3 & 2 \\ 0 & -1 & -2 \\ 0 & 0 & 1 \\ 0 & 0 & 0 \end{bmatrix}\begin{bmatrix} x_1 \\ x_2 \\ x_3 \end{bmatrix} = \begin{bmatrix} 1 \\ -1 \\ 3 \\ 0 \end{bmatrix} \Rightarrow x_3 = 3$$

1.68 (c)

$|P| = 70$ and Trace $(P) = 15$

So, only option, i.e., (c) (1, 2, 5, 7) satisfies.

1.69 (b)

$$\begin{bmatrix} 1 & 2 & 2 & 3 \\ 3 & 4 & 2 & 5 \\ 5 & 6 & 2 & 7 \\ 7 & 8 & 2 & 9 \end{bmatrix}$$

$R_2 \to R_2 - 3R_1$, $R_3 \to R_3 - 5R_1$, $R_4 \to R_4 - 7R_1$

$$A = \begin{bmatrix} 1 & 2 & 2 & 3 \\ 0 & -2 & -4 & -4 \\ 0 & -4 & -8 & -8 \\ 0 & -6 & -12 & -12 \end{bmatrix}$$

$R_3 \to R_3 - 2R_2$, $R_4 \to R_4 - 3R_2$

$$A = \begin{bmatrix} 1 & 2 & 2 & 3 \\ 0 & -2 & -4 & -4 \\ 0 & 0 & 0 & 0 \\ 0 & 0 & 0 & 0 \end{bmatrix}$$

So, $\rho(A)$ = No. of non-zero rows = 2.

1.70 (c)

$$PQ = \begin{bmatrix} 1 & 2 \\ 3 & 4 \end{bmatrix}\begin{bmatrix} 0 & 1 \\ 1 & 0 \end{bmatrix} = \begin{bmatrix} 2 & 1 \\ 4 & 3 \end{bmatrix}$$

$$(PQ)^T = \begin{bmatrix} 2 & 4 \\ 1 & 3 \end{bmatrix}$$

Now using Reversal law

$$Q^T P^T = (PQ)T = \begin{bmatrix} 2 & 4 \\ 1 & 3 \end{bmatrix}$$

1.71 (c)

$$A = \begin{bmatrix} 2 & -2 \\ -1 & 6 \end{bmatrix} \Rightarrow |A - \lambda I| = 0$$

$$\Rightarrow \quad \lambda = (4+\sqrt{6}) \text{ and } (4-\sqrt{6})$$

Eigen vector, for eigen value $(4-\sqrt{6})$

$$AX = \lambda X$$
$$(A - \lambda I)X = 0$$

$$\begin{bmatrix} 2-(4-\sqrt{6}) & -2 \\ -1 & 6-(4-\sqrt{6}) \end{bmatrix}\begin{bmatrix} x_1 \\ x_2 \end{bmatrix} = \begin{bmatrix} 0 \\ 0 \end{bmatrix}$$

$$x_1 = \left(\frac{2}{-2+\sqrt{6}}\right)x_2$$

Let, $x_2 = K$ then $x_1 = \left(\frac{2}{-2+\sqrt{6}}\right)K$

$$\Rightarrow \begin{bmatrix} x_1 \\ x_2 \end{bmatrix} = \begin{bmatrix} \frac{2}{-2+\sqrt{6}}K \\ K \end{bmatrix} \approx \begin{bmatrix} 2 \\ -2+\sqrt{6} \end{bmatrix} = \begin{bmatrix} 2.00 \\ 0.45 \end{bmatrix}$$

1.72 (b)

$$\begin{bmatrix} 5 & 0 & -5 & 0 \\ 0 & 2 & 0 & 1 \\ -5 & 0 & 5 & 0 \\ 0 & 1 & 0 & 2 \end{bmatrix} \xrightarrow{R_3 \longrightarrow R_1 + R_3} \begin{bmatrix} 5 & 0 & -5 & 0 \\ 0 & 2 & 0 & 1 \\ 0 & 0 & 0 & 0 \\ 0 & 1 & 0 & 2 \end{bmatrix}$$

$$\xrightarrow{R_4 \longrightarrow R_4 - \frac{1}{2}R_2} \begin{bmatrix} 5 & 0 & 1 & 0 \\ 0 & 2 & 0 & 0 \\ 0 & 0 & 0 & 0 \\ 0 & 0 & 0 & \frac{3}{2} \end{bmatrix}$$

$$R_3 \longleftrightarrow R_4 \begin{bmatrix} 5 & 0 & 1 & 0 \\ 0 & 2 & 0 & 0 \\ 0 & 0 & 0 & \frac{3}{2} \\ 0 & 0 & 0 & 0 \end{bmatrix}$$

Hence, Rank(A) = 3

1.73 (c)

If, $AA^T = I$ or $A^{-1} = A^T$

Then the matrix is orthogonal.

1.74 (a, c)

Given:

$$\lambda_1 = 5$$
$$\lambda_2 = -2$$
$$Q = M^3 - 4M^2 - 2M$$

By clayey, Haminton's theorem,

$$\therefore \quad Q_{\lambda_1} = (5)^3 - 4(5)^2 - 2(5)$$
$$= 125 - 100 - 10 = 15$$
$$\therefore \quad Q_{\lambda_2} = ((-2)^3 - 4(-2)^2 - 2(-2)$$
$$= -8 - 16 + 4$$
$$= -20$$

1.75 (a)

Using potential matrix about x-axis,

$$\begin{bmatrix} x' \\ y' \\ z' \end{bmatrix} = \begin{bmatrix} 1 & 0 & 0 \\ 0 & \cos\theta & \sin\theta \\ 0 & \sin\theta & \cos\theta \end{bmatrix}\begin{bmatrix} 1 \\ 1 \\ 1 \end{bmatrix}$$

$$= \begin{bmatrix} 1 & 0 & 0 \\ 0 & \frac{1}{\sqrt{2}} & -\frac{1}{\sqrt{2}} \\ 0 & \frac{1}{\sqrt{2}} & \frac{1}{\sqrt{2}} \end{bmatrix} \begin{bmatrix} 1 \\ 1 \\ 1 \end{bmatrix} = \begin{bmatrix} 1 \\ 0 \\ \sqrt{2} \end{bmatrix}$$

1.76 (a, d)

Let, $y = \begin{bmatrix} 1 \\ 1 \\ 1 \end{bmatrix}_{3\times1}$, $y^T = [1 \ 1 \ 1]_{1\times3}$

$$yy^T = \begin{bmatrix} 1 & 1 & 1 \\ 1 & 1 & 1 \\ 1 & 1 & 1 \end{bmatrix}_{3\times3} \text{ and } y^Ty = [3]_{1\times1}$$

Here, Rank of matrix

$$\rho(y) = \rho(y^T) = \rho(yy^T) = \rho(y^Ty) = 1$$

Eigen value of yy^T = 3, 0, 0

$\Rightarrow$ y^Ty is eigen value of yy^T i.e. 3.

- yy^T is symmetrical matrix.
- Since determinant of yy^T is zero, therefore A^{-1} will not exist.
- Rank of yy^T is 1.

1.77 Sol.

$$\varepsilon = \begin{bmatrix} 1 & 1 \\ 1 & -1 \end{bmatrix} \text{ (Given)}$$

Eigen value of ε,

$$\lambda_1 + \lambda_2 = 0$$
$$\lambda_1 \lambda_2 = -2$$
$$\therefore \quad \lambda_1 = \sqrt{2}; \quad \lambda_2 = -\sqrt{2}$$

$\therefore$Eigen value of $\varepsilon^8 = \left(\sqrt{2}\right)^8$ and $\left(-\sqrt{2}\right)^8$

$= 16$ and 16

$\therefore$ Eigen values of $\varepsilon^{11} = \left(\sqrt{2}\right)^{11}$ and $\left(-\sqrt{2}\right)^{11}$

$= 32\sqrt{2}$ and $-32\sqrt{2}$

Now, Trace of $\varepsilon^8 = 16 + 16 = 32$ i.e. P

Trace of $\varepsilon^{11} = 32\sqrt{2} - 32\sqrt{2} = 0$ i.e. Q

$\Rightarrow \quad P + Q = 32$

1.78 (a, c)

$\because$ P and Q are not invertible, so

$$[PQ]^{-1} \neq Q^{-1} P^{-1}$$

If P and Q are invertible then,

$$[QP]^{-1} = P^{-1} Q^{-1}$$

If P and Q are invertible, then

$$[PQ]^{-1} = Q^{-1} P^{-1}$$

1.79 (b, c)

Let us take $M = \begin{bmatrix} 0 & -1 \\ 1 & 0 \end{bmatrix}$ then eigen values are $\lambda = \pm i$

So, eigen values of M^2 are i^2 and $(-i)^2 = -1$ and -1

Eigen values of $(M^T)^2$ are same i.e. -1 and -1

Now, MM^T and M^TM are symmetric matrix, as well as the semi definite matrix and it is the property of +ve semi definite matrix that it's eigen values are non –ve.

Hence correct answer (b) and (c).

1.80 (a, b, d)

Using standard properties of eigen values and eigen vectors,

- Eigen values of A^T and A are same.
- Eigen values of A^{-1} is reciprocal of eigen value of A.
- Eigen vectors of A^{-1} are same as the eigen vectors of A.

  Row space and column space for any matrix are not same. Thus option (c) is false.

1.81 (b, c)

As $|A| = 0$

So, one of the eigen value is zero.

$|A - \lambda I| = 0$

$$\begin{bmatrix} 1-\lambda & -1 & 0 \\ -1 & 2-\lambda & -1 \\ 0 & -1 & 1-\lambda \end{bmatrix} = 0$$

$$(1-\lambda)[(2-\lambda)(1-\lambda)-1]+1[\lambda-1] = 0$$

$\Rightarrow(1 - \lambda)[(2 - \lambda)(1 - \lambda) - 1 - 1] = 0$

$\Rightarrow\lambda(1 - \lambda)(3 - \lambda) = 0$

$\Rightarrow \quad \lambda = 0, 1, 3$

As there are three eigen values so number of linearly independent eigen vector are 3.

Since, $|A| = 0$

So, $[A]\{x\} = \{b\}$ does not have a unique solution

For the positive definite matrix, all the eigen values must be positive.

But, here one eigen value i.e. $\lambda = 0$, so, A is not a positive definite matrix.

1.82 (a)

$\rho(A_{4\times 4}) = 2$

$N(A)$ = Number of column – Rank

$= 4 - 2 = 2$

i.e. Null space of A will consist only two linearly independent vectors which is given as x and y.

Eigen vectors of matrix A,

$$X = \begin{bmatrix} 2 \\ 1 \\ 0 \\ 3 \end{bmatrix} \text{ and } Y = \begin{bmatrix} 1 \\ 0 \\ 1 \\ 2 \end{bmatrix}$$

As these are linearly independent eigen vectors so remaining eigen vectors of null space must be linearly dependent.

$$\text{Hence, } X - Y = \begin{bmatrix} 1 \\ 1 \\ -1 \\ 1 \end{bmatrix}$$

1.83 (b)

$LL^T = A$

$$\begin{bmatrix} L_{11} & 0 \\ L_{21} & L_{22} \end{bmatrix}\begin{bmatrix} L_{11} & L_{21} \\ 0 & L_{22} \end{bmatrix} = \begin{bmatrix} 8 & -5 \\ -5 & a_{22} \end{bmatrix}$$

$L_{11} = 2\sqrt{2}$,

$$L_{21} = -\frac{5}{2\sqrt{2}} \text{ and } L_{21}^2 + L_{22}^2 = a_{22}$$

$$a_{22} = \left(-\frac{5}{2\sqrt{2}}\right)^2 + 1.968 = 6.99 \simeq 7$$

1.84 (b)

$$\lambda_1 + \lambda_2 + \lambda_3 = 8$$

$$\lambda_1 \lambda_2 \lambda_3 = |A| = 10$$

Only option (b) satisfies above condition.

$1 + 2 + 5 = 8$

$1 \times 2 \times 5 = 10$

1.85 (a)

P : It is obviously true because transpose is distributed over addition. i.e. $(A + B)^T = A^T + B^T$

Q : It is false because in matrix multiplication, transpose should follow reversal law

i.e. $(AB)^T = B^T A^T$.

1.86 Sol.

$$A = \begin{bmatrix} 2 & 1 & 4 \\ 1 & 0 & 3 \end{bmatrix}_{2\times 3},$$

$$B = \begin{bmatrix} -1 & 0 \\ 2 & 3 \\ 1 & 4 \end{bmatrix}_{3\times 2}$$

Then, $$AB = \begin{bmatrix} 2 & 1 & 4 \\ 1 & 0 & 3 \end{bmatrix} \times \begin{bmatrix} -1 & 0 \\ 2 & 3 \\ 1 & 4 \end{bmatrix}$$

$$= \begin{bmatrix} 4 & 19 \\ 2 & 12 \end{bmatrix}$$

So, $|AB| = (4)(12) - (2)(19)$

$= 48 - 38 = 10$

■■■■

2 Calculus

2.1 The limit of $\underset{\theta \to 0}{Lt} \dfrac{\sin\theta}{\theta}$ is

(a) 1 (b) 0
(c) ∞ (d) None of these

[1994 : 1 M]

2.2 The function, $y = x^2 + \dfrac{250}{x}$, at $x = 5$

(a) Maximum (b) Minimum
(c) Neither (d) 1 [1994 : 1 M]

2.3 The volume generated by revolving the area bounded by the parabola $y^2 = 8x$ and the line $x = 2$ about y-axis is

(a) $\dfrac{128\pi}{5}$ (b) $\dfrac{5}{128\pi}$
(c) $\dfrac{127}{5\pi}$ (d) None of these

[1994 : 1 M]

2.4 The directional derivative of $f(x, y) = 2x^2 + 3y^2 + z^2$ at point P(2, 1, 3) in the direction of the vector $\vec{a} = \vec{i} - 2\vec{k}$ is

(a) $\dfrac{4}{\sqrt{5}}$ (b) $-\dfrac{4}{\sqrt{5}}$
(c) $\dfrac{\sqrt{5}}{4}$ (d) $-\dfrac{\sqrt{5}}{4}$ [1994 : 1 M]

2.5 The integration of $\int \log x.dx$ has the value

(a) $(x \log x - 1)$ (b) $\log x - x$
(c) $x(\log x - 1)$ (d) None of these

[1994 : 1 M]

2.6 $\lim_{x\to\infty} x\sin\left(\dfrac{1}{x}\right)$ is

(a) ∞ (b) 0
(c) 1 (d) Nonexistent

[1995 : 1 M]

2.7 If $f(0) = 2$ and $f(x) = \dfrac{1}{5-x^2}$, the lower and upper bounds of f(1) estimated by the mean value theorem are

(a) 19, 2.2 (b) 2.2, 2.25
(c) 2.25, 2.5 (d) None of these

[1995 : 1 M]

2.8 The third term in the Taylor series expansion of e^x about a would be

(a) $e^a(x-a)$ (b) $\dfrac{e^a}{2}(x-a)^2$
(c) $\dfrac{e^a}{2}$ (d) $\dfrac{e^a}{6}(x-a)^3$

[1995 : 1 M]

2.9 By reversing the order of integration $\int_0^2 \int_{x^2}^{2x} f(x,y)\,dy\,dx$ may be represented as

(a) $\int_0^2 \int_{x^2}^{2x} f(x,y)\,dy\,dx$ (b) $\int_0^2 \int_y^{\sqrt{y}} f(x,y)\,dx\,dy$
(c) $\int_0^2 \int_{y/2}^{\sqrt{y}} f(x,y)\,dx\,dy$ (d) $\int_{x^2}^{2x} \int_0^2 f(x,y)\,dy\,dx$

[1995 : 1 M]

2.10 The derivative of f(x, y) at point (1, 2) in the direction of vector i + j is $2\sqrt{2}$ and in the direction of the vector –2j is –3. Then the derivative of f(x, y) in direction –i –2j is

(a) $2\sqrt{2} + 3/2$ (b) $-7/\sqrt{5}$
(c) $-2\sqrt{2} - 3/2$ (d) $1/\sqrt{5}$ [1995 : 2 M]

2.11 The area bounded by the parabola $2y = x^2$ and the line $x = y - 4$ is equal to

(a) 6 (b) 18
(c) ∞ (d) Non of these

[1995 : 1 M]

2.12 If a function is continuous at a point its first derivative

(a) may or may not exist
(b) exists always
(c) will not exist
(d) has a unique value [1996 : 1 M]

2.13 The series $f(x) = x + \dfrac{x^3}{3!} + \dfrac{x^5}{5!} + ...$ is the expansion of

(a) cos hx (b) sin x

(c) $\sin hx$ (d) $\cos x$ [1996 : 1 M]

2.14 For real values of x, $\cos(x)$ can be written in one of the forms of a convergent series given below:

(a) $\cos(x) = 1+\frac{x}{1!}+\frac{x^2}{2!}+\frac{x^3}{3!}...\infty$

(b) $\cos(x) = 1-\frac{x^2}{2!}+\frac{x^4}{4!}+\frac{x^5}{5!}...\infty$

(c) $\cos(x) = x-\frac{x^3}{3!}+\frac{x^5}{5!}+\frac{x^7}{7!}...\infty$

(d) $\cos(x) = x-\frac{x^2}{1!}+\frac{x^2}{2!}+\frac{x^3}{3!}...\infty$

[1997 : 1 M]

2.15 Area bounded by the curve $y = x^2$ and lines $x = 4$ and $y = 0$ is given by

(a) 64 (b) $\frac{64}{3}$

(c) $\frac{128}{3}$ (d) $\frac{128}{4}$ [1997 : 1 M]

2.16 $\underset{\theta\to 0}{\text{Lt}} \frac{\sin m\theta}{\theta}$, where m is an integer, is one of the following:

(a) m (b) $m\pi$

(c) $m\theta$ (d) 1 [1997 : 1 M]

2.17 The Taylor's series expansion of $\sin x$ is

(a) $1-\frac{x^2}{2!}+\frac{x^4}{4!}$ (b) $1-\frac{x^2}{4!}+\frac{x^4}{4!}$

(c) $x+\frac{x^3}{3!}+\frac{x^5}{5!}$ (d) $x-\frac{x^3}{3!}+\frac{x^5}{5!}$

[1998 : 1 M]

2.18 Limit of the function $\lim_{n\to\infty} \frac{n}{\sqrt{n^2+n}}$ is

(a) $\frac{1}{2}$ (b) 0

(c) ∞ (d) 1 [1999 : 1 M]

2.19 The function $f(x) = e^x$ is

(a) Even

(b) Odd

(c) Neither even nor odd

(d) None of the above [1999 : 1 M]

2.20 Number of inflection points for the curve $y = x + 2x^4$ is

(a) 3 (b) 1

(c) n (d) $(n + 1)^2$

[1999 : 1 M]

2.21 Value of the function $\lim_{x\to a}(x-a)^{(x-a)}$ is

(a) 1 (b) 0

(c) ∞ (d) a [1999 : 1 M]

2.22 If $f(x, y, z) = (x^2 + y^2 + z^2)^{1/2}$

$\frac{\partial^2 f}{\partial x^2}+\frac{\partial^2 f}{\partial y^2}+\frac{\partial^2 f}{\partial z^2}$ is equal to

(a) Zero

(b) 1

(c) 2

(d) $-3(x^2 + y^2 + z^2)^{-5/2}$ [2000 : 1 M]

2.23 The Taylor expansion of sin x about $x = \pi/6$ is given by

(a) $\frac{1}{2}+\frac{\sqrt{3}}{2}\left(x-\frac{\pi}{6}\right)-\frac{1}{4}\left(x-\frac{\pi}{6}\right)^2-\frac{\sqrt{3}}{12}\left(x-\frac{\pi}{6}\right)^3+...$

(b) $x-\frac{x^3}{3!}+\frac{x^5}{5!}-\frac{x^7}{7!}+...$

(c) $\left(x-\frac{\pi}{6}\right)-\frac{\left(x-\frac{\pi}{6}\right)^3}{3!}+\frac{\left(x-\frac{\pi}{6}\right)^5}{5!}-\frac{\left(x-\frac{\pi}{6}\right)^3}{7!}+...$

(d) $\frac{1}{2}$ [2000 : 1 M]

2.24 The limit of the function $f(x) = [1 - a^4/x^4]$ as $x \to \infty$ is given by

(a) 1 (b) $\exp[-a^4]$

(c) ∞ (d) Zero

[2000 : 1 M]

2.25 The maxima and minima of the function $f(x) = 2x^3 - 15x^2 + 36x + 10$ occurs, respectively, at

(a) x = 3 and x = 2 (b) x = 1 and x = 3

(c) x = 2 and x = 3 (d) x = 3 and x = 4

[2000 : 1 M]

2.26 The value of the integral is $I = \int_0^{\pi/4} \cos^2 x\, dx$

(a) $\frac{\pi}{8}+\frac{1}{4}$ (b) $\frac{\pi}{8}-\frac{1}{4}$

(c) $-\frac{\pi}{8}-\frac{1}{4}$ (d) $-\frac{\pi}{8}+\frac{1}{4}$

[2001 : 1 M]

2.27 The value of the following definite integral is

$$\int_{-\frac{\pi}{2}}^{\frac{\pi}{2}} \frac{\sin 2x}{1+\cos x}dx$$

(a) $-2 \ln 2$
(b) 2
(c) 0
(d) None of these
[2002 : 1 M]

2.28 The following function has a local minima at which value of x

$$f(x) = x\sqrt{5-x^2}$$

(a) $-\frac{\sqrt{5}}{2}$
(b) $\sqrt{5}$
(c) $\frac{\sqrt{5}}{2}$
(d) $-\sqrt{\frac{5}{2}}$
[2002 : 1 M]

2.29 The value of the following improper integral is

$$\int_0^1 x \ln x$$

(a) 1/4
(b) 0
(c) −1/4
(d) 1
[2002 : 2 M]

2.30 The limit of the following sequence as $n \to \infty$ is

$$x_n = n^{1/n}$$

(a) 0
(b) 1
(c) ∞
(d) $-\infty$
[2002 : 2 M]

2.31 If P, Q and R are three points having coordinates (3, −2, −1), (1, 3, 4), (2, 1, −2) in *XYZ* space, then the distance from point *P* to plane *OQR* (*O* being the origin of the coordinate system) is given by
(a) 3
(b) 5
(c) 7
(d) 9
[2003 : 1 M]

2.32 The value of the function

$$f(x) = \lim_{x \to 0} \frac{x^3 + x^2}{2x^3 - 7x^2} \text{ is}$$

(a) 0
(b) $-\frac{1}{7}$
(c) $\frac{1}{7}$
(d) ∞
[2004 : 1 M]

2.33 The function $f(x) = 2x^3 - 3x^2 - 36x + 2$ has its maxima at
(a) $x = -2$ only
(b) $x = 0$ only
(c) $x = 3$ only
(d) both $x = -2$ and $x = 3$
[2004 : 2 M]

2.34 A rail engine accelerates from its stationary position for 8 seconds and travels a distance of 280 m. According to the Mean Value Theorem, the speedometer at a certain time during acceleration must read exactly
(a) 0
(b) 8 kmph
(c) 75 kmph
(d) 126 kmph
[2005 : 2 M]

2.35 What is the area common to the circles $r = a$ and $r = 2a \cos \theta$?
(a) $0.524\, a^2$
(b) $0.614\, a^2$
(c) $1.047\, a^2$
(d) $1.228\, a^2$
[2006 : 2 M]

2.36 Evaluate $\int_0^\infty \frac{\sin t}{t} dt$

(a) π
(b) $\frac{\pi}{2}$
(c) $\frac{\pi}{4}$
(d) $\frac{\pi}{8}$
[2007 : 2 M]

2.37 The value of $\int_0^3 \int_0^x (6 - x - y)\, dx\, dy$ is
(a) 13.5
(b) 27.0
(c) 40.5
(d) 54.0
[2008 : 2 M]

2.38 The $\lim_{x \to 0} \frac{\sin\left[\frac{2}{3}x\right]}{x}$ is

(a) 2/3
(b) 1
(c) 3/2
(d) ∞
[2010 : 1 M]

2.39 Given a function
$f(x, y) = 4x^2 + 6y^2 - 8x - 4y + 8$
The optimal value of $f(x, y)$
(a) is a minimum equal to 10/3
(b) is a maximum equal to 10/3
(c) is a minimum equal to 8/3
(d) is a maximum equal to 8/3
[2010 : 2 M]

2.40 What should be the value of λ such that the function defined below is continuous at $x = \pi/2$?

$$f(x) = \begin{cases} \dfrac{\lambda \cos x}{\frac{\pi}{2} - x} & \text{if } x \neq \pi/2 \\ 1 & \text{if } x = \pi/2 \end{cases}$$

(a) 0
(b) $2/\pi$
(c) 1
(d) $\pi/2$
[2011 : 2 M]

2.41 What is the value of the definite integral,

$$\int_0^a \frac{\sqrt{x}}{\sqrt{x} + \sqrt{a-x}} dx\,?$$

(a) 0
(b) a/2
(c) a
(d) 2a
[2011 : 2 M]

2.42 The infinite series

$$1+x+\frac{x^2}{2!}+\frac{x^3}{3!}+\frac{x^4}{4!}+\ldots \text{ corresponds to}$$

(a) $\sec x$ (b) e^x
(c) $\cos x$ (d) $1+\sin^2 x$

[2011 : 2 M]

2.43 The value of $\int_0^{\pi/6} \cos^4 3\theta \sin^3 6\theta \, d\theta$ is

(a) 0 (b) $\frac{1}{15}$
(c) 1 (d) $\frac{8}{3}$ [2013 : 2 M]

2.44 $\lim_{x\to\infty}\left(\frac{x+\sin x}{x}\right)$ equal to

(a) $-\infty$ (b) 0
(c) 1 (d) ∞ [2014 : 1 M, Set-I]

2.45 The expression $\lim_{\alpha\to 0}\frac{x^\alpha-1}{\alpha}$ is equal to

(a) $\log x$ (b) 0
(c) $x \log x$ (d) ∞

[2014 : 2 M, Set-II]

2.46 $\lim_{x\to\infty}\left(1+\frac{1}{x}\right)^{2x}$ is equal to

(a) e^{-2} (b) e
(c) 1 (d) e^2

[2015 : 1 M, Set-II]

2.47 While minimizing the function $f(x)$, necessary and sufficient conditions for a point x_0 to be a minima are

(a) $f'(x_0)>0$ and $f''(x_0)=0$
(b) $f'(x_0)<0$ and $f''(x_0)=0$
(c) $f'(x_0)=0$ and $f''(x_0)<0$
(d) $f'(x_0)=0$ and $f''(x_0)>0$

[2015 : 1 M, Set-II]

2.48 The optimum value of the function $f(x) = x^2 - 4x + 2$ is

(a) 2 (maximum) (b) 2 (minimum)
(c) –2 (maximum) (d) –2 (minimum)

[2016 : 1 M, Set-II]

2.49 What is the value of $\lim_{\substack{x\to 0\\ y\to 0}}\frac{xy}{x^2+y^2}$?

(a) 1 (b) –1
(c) 0 (d) Limit does not exit

[2016 : 1 M, Set-II]

2.50 The value of $\int_0^\infty \frac{1}{1+x^2}dx+\int_0^\infty \frac{\sin x}{x}dx$ is

(a) $\frac{\pi}{2}$ (b) π
(c) $\frac{3\pi}{2}$ (d) 1

[2016 : 2 M, Set-I]

2.51 The area of the region bounded by the parabola $y = x^2 + 1$ and the straight line $x + y = 3$ is

(a) $\frac{59}{6}$ (b) $\frac{9}{2}$
(c) $\frac{10}{3}$ (d) $\frac{7}{6}$

[2016 : 2 M, Set-I]

2.52 The angle of intersection of the curves $x^2 = 4y$ and $y^2 = 4x$ at point (0, 0) is

(a) 0° (b) 30°
(c) 45° (d) 90°

[2016 : 2 M, Set-II]

2.53 The area between the parabola $x^2 = 8y$ and the straight line $y = 8$ is_______.

[2016 : 2 M, Set-II]

2.54 The quadratic approximation of $f(x) = x^3 - 3x^2 - 5$ a the point $x = 0$ is

(a) $3x^2 - 6x - 5$ (b) $-3x^2 - 5$
(c) $-3x^2 + 6x - 5$ (d) $3x^2 - 5$

[2016 : 2 M, Set-II]

2.55 $\lim_{x\to\infty}\left(\frac{\tan x}{x^2-x}\right)$ is equal to _______.

[2017 : 1 M, Set-I]

2.56 Let x be a continuous variable defined over the interval $(-\infty, \infty)$, and $f(x) = e^{-x-e^{-x}}$. The integral $g(x) = \int f(x)dx$ is equal to

(a) $e^{e^{-x}}$ (b) $e^{-e^{-x}}$
(c) $e^{-e^{x}}$ (d) e^{-x}

[2017 : 1 M, Set-I]

2.57 Let $w = f(x, y)$, where x and y are functions of t. Then, according to the chain rule, $\frac{dw}{dt}$ is equal

(a) $\frac{dw}{dx}\frac{dx}{dt} + \frac{dw}{dy}\frac{dt}{dt}$ (b) $\frac{\partial w}{\partial x}\frac{\partial x}{\partial t} + \frac{\partial w}{\partial y}\frac{\partial y}{\partial t}$

(c) $\frac{\partial w}{\partial x}\frac{dx}{dt} + \frac{\partial w}{\partial y}\frac{dy}{dt}$ (d) $\frac{dw}{dx}\frac{\partial x}{\partial t} + \frac{dw}{dy}\frac{\partial y}{\partial t}$

[2017 : 1 M, Set-II]

2.58 The tangent to the curve represented by $y = x\ln x$ is required to have 45° inclination with the x-axis. The coordinates of the tangent point would be

(a) (1, 0) (b) (0, 1)

(c) (1, 1) (d) $(\sqrt{2}, \sqrt{2})$

[2017 : 2 M, Set-II]

2.59 Consider the following definite integral:

$$I = \int_0^1 \frac{(\sin^{-1} x)^2}{\sqrt{1-x^2}} dx$$

The value of the integral is

(a) $\frac{\pi^3}{24}$ (b) $\frac{\pi^3}{12}$

(c) $\frac{\pi^3}{48}$ (d) $\frac{\pi^3}{64}$

[2017 : 2 M, Set-II]

2.60 At the point $x = 0$, the function $f(x) = x^3$ has

(a) local maximum

(b) local minimum

(c) both local maximum and minimum

(d) neither local maximum nor local minimum

[2018 : 1 Marks, Set-I]

2.61 The value of the integral $\int_0^\pi x\cos^2 x dx$ is

(a) $\frac{\pi^2}{8}$ (b) $\frac{\pi^2}{4}$

(c) $\frac{\pi^2}{2}$ (d) p^2

[2018 : 2 M, Set-I]

2.62 Which of the following is correct?

(a) $\lim_{x\to 0}\left(\frac{\sin 4x}{\sin 2x}\right) = 1$ and $\lim_{x\to 0}\left(\frac{\tan x}{x}\right) = 1$

(b) $\lim_{x\to 0}\left(\frac{\sin 4x}{\sin 2x}\right) = \infty$ and $\lim_{x\to 0}\left(\frac{\tan x}{x}\right) = 1$

(c) $\lim_{x\to 0}\left(\frac{\sin 4x}{\sin 2x}\right) = 2$ and $\lim_{x\to 0}\left(\frac{\tan x}{x}\right) = \infty$

(d) $\lim_{x\to 0}\left(\frac{\sin 4x}{\sin 2x}\right) = 2$ and $\lim_{x\to 0}\left(\frac{\tan x}{x}\right) = 1$

[2019 : 1 M, Set-I]

2.63 For a small value of h, the Taylor series expansion for $f(x+h)$ is

(a) $f(x) + hf'(x) + \frac{h^2}{2}f''(x) + \frac{h^3}{3}f'''(x) + ...\infty$

(b) $f(x) - hf'(x) + \frac{h^2}{2!}f''(x) - \frac{h^3}{3!}f'''(x) + ...\infty$

(c) $f(x) + hf'(x) + \frac{h^2}{2!}f''(x) + \frac{h^3}{3!}f'''(x) + ...\infty$

(d) $f(x) - hf'(x) + \frac{h^2}{2}f''(x) - \frac{h^3}{3}f'''(x) + ...\infty$

[2019 : 1 M, Set-I]

2.64 Which one of the following is NOT a correct statement?

(a) The function $\sqrt[x]{x}, (x > 0)$, has the global minima at $x = e$

(b) The function $\sqrt[x]{x}, (x > 0)$, has the global maxima at $x = e$

(c) The function x^3 has neither global minima nor global maxima

(d) The function $|x|$ has the global minima at $x = 0$

[2019 : 2 M, Set-I]

2.65 Consider the functions: $x = y\ln\phi$ and $y = \phi\ln y$. which one of the following the correct expression for $\frac{\partial\psi}{\partial x}$?

(a) $\frac{\ln\psi}{\ln\phi\ln\psi - 1}$ (b) $\frac{x\ln\phi}{\ln\phi\ln\psi - 1}$

(c) $\frac{\ln\phi}{\ln\phi\ln\psi - 1}$ (d) $\frac{x\ln\phi}{\ln\phi\ln\psi - 1}$

[2019 : 2 M, Set-I]

2.66 The following inequality is true for all x close to 0.

$$2 - \frac{x^2}{3} < \frac{x\sin x}{1-\cos x} < 2$$

What is the value of $\lim_{x\to 0}\frac{x\sin x}{1-\cos x}$?

(a) 1 (b) 0

(c) $\frac{1}{2}$ (d) 2

[2019 : 1 M, Set-II]

2.67 The value of $\lim_{x\to\infty}\frac{x^2-5x+4}{4x^2+2x}$ is

(a) 0 (b) 1

(c) $\frac{1}{4}$ (d) $\frac{1}{2}$

[2020 : 1 M, Set-I]

2.68 The true value of ln(2) is 0.69. If the value of ln(2) is obtained by linear interpolation between ln(1) and ln(6), the percentage of absolute error (round off to the nearest integer), is

(a) 35 (b) 69

(c) 84 (d) 48

[2020 : 1 M, Set-I]

2.69 The area of an ellipse represented by an equation $\frac{x^2}{a^2}+\frac{y^2}{b^2}=1$ is

(a) $\frac{4\pi ab}{3}$ (b) πab

(c) $\frac{\pi ab}{2}$ (d) $\frac{\pi ab}{4}$

[2020 : 1 M, Set-I]

2.70 The value of $\lim_{x\to\infty}\frac{\sqrt{9x^2+2020}}{x+7}$ is

(a) 1 (b) 3

(c) $\frac{7}{9}$ (d) Indeterminable

[2020 : 1 M, Set-II]

2.71 The integral

$$\int_0^1 (5x^3+4x^2+3x+2)\,dx$$

is estimated numerically using three alternative methods namely the rectangular, trapezoidal and Simpson's rules with a common step size. In this context, which one of the following statement is TRUE?

(a) Simpson's rule as well as rectangular rule of estimation will give non zero error.

(b) Only Simpson's rule of estimation will give zero error.

(c) Simpson's rule, rectangular rule as well as trapezoidal rule of estimation will give non-zero error.

(d) Only the rectangular rule of estimation will given zero error.

[2020 : 1 M, Set-II]

2.72 The diameter and height of a right circular cylinder are 3 cm and 4 cm, respectively. The absolute error in each of these two measurements is 0.2 cm. The absolute error in the computed volume (in cm^3, round off to three decimal places), is ______.

[2020 : 2 M, Set-II]

2.73 The volume determined from $\iiint_V 8xyz\,dV$ for $V=[2,3]\times[1,2]\times[0,1]$ will be (in integer) ______.

[2021 : 1 M, Set-I]

2.74 Consider the limit:

$$\lim_{x\to1}\left(\frac{1}{\ln x}-\frac{1}{x-1}\right)$$

The limit (correct up to one decimal place) is ______

[2021 : 1 M, Set-I]

2.75 The value (round off to one decimal place) if $\int_{-1}^{1} xe^{|x|}dx$ is ______.

[2021 : 1 M, Set-II]

2.76 The value of $\lim_{x\to\infty}\frac{x\ln(x)}{1+x^2}$ is

(a) 1.0 (b) 0.5

(c) ∞ (d) 0

[2021 : 1 M, Set-II]

2.77 Consider the following expression:

$z=\sin(y+it)+\cos(y-it)$

where z, y, and t are variables, and $i=\sqrt{-1}$ is a complex number. The partial differential equation derived from the above expression is

(a) $\frac{\partial^2 z}{\partial t^2}+\frac{\partial^2 z}{\partial y^2}=0$ (b) $\frac{\partial^2 z}{\partial t^2}-\frac{\partial^2 z}{\partial y^2}=0$

(c) $\frac{\partial z}{\partial t}-i\frac{\partial z}{\partial y}=0$ (d) $\frac{\partial z}{\partial t}+i\frac{\partial z}{\partial y}=0$

[2022 : 1 M, Set-I]

2.78 Let max {a, b} denote the maximum of two real numbers a and b. Which of the following statement(s) is/are TRUE about the function $f(x) = \max\{3 - x, x - 1\}$?

(a) It is continuous on its domain.
(b) It has a local minimum at $x = 2$.
(c) It has a local maximum at $x = 2$
(d) It is differentiable on its domain.

[2022 : 2 M, Set-I]

2.79 Consider the polynomial $f(x) = x^3 - 6x^2 + 11x - 6$ on the domain S, given by $1 \le x \le 3$. The first and second derivatives are $f'(x)$ and $f''(x)$.

Consider the following statements:

I. The given polynomial is zero at the boundary points $x = 1$ and $x = 3$.
II. There exists one local maxima of $f(x)$ within the domain S.
III. The second derivative $f''(x) > 0$ throughout the domains S.
IV. There exists one local minima $f(x)$ within the domain S.

(a) Only statements II and IV are correct.
(b) Only statements I and IV are correct.
(c) Only statements I, II and III are correct.
(d) Only statements I, II and IV are correct.

[2022 : 1 M, Set-II]

2.80 $\int\left(x - \frac{x^2}{2} + \frac{x^3}{3} - \frac{x^4}{4} + \ldots\right)dx$ is equal to

(a) $-\frac{1}{1-x^2}$ + Constant
(b) $-\frac{1}{1-x}$ + Constant
(c) $\frac{1}{1+x^2}$ + Constant
(d) $\frac{1}{1+x}$ + Constant

[2022 : 2 M, Set-II]

2.81 For the integral

$$I = \int_{-1}^{1} \frac{1}{x^2} dx$$

which of the following statements is TRUE?

(a) $I = 0$
(b) $I = 2$
(c) $I = -2$
(d) The integral does not converge

[2023 : 1 M, Set-I]

2.82 For the function $f(x) = e^x |\sin x|$, $x \in \mathbb{R}$ which of the following statements is/are TRUE?

(a) The function is continuous at all x
(b) The function is differentiable at all x
(c) The function is periodic
(d) The function is bounded

[2023 : 2 M, Set-I]

2.83 A vector field $\vec{p}$ and a scalar field r are given by

$$\vec{p} = (2x^2 - 3xy + z^2)\hat{i} + (2y^2 - 3yz + x^2)\hat{j} + (2z^2 - 3xz + x^2)\hat{k}$$

$$r = 6x^2 + 4y^2 - z^2 - 9xyz - 2xy + 3xz - yz$$

Consider the statement P and Q:

P : Curl of the gradient of the scalar field r is a null vector.

Q : Divergence of curl of the vector field $\vec{p}$ is zero.

Which one of the following options is CORRECT?

(a) Both P and Q are TRUE.
(b) Both P and Q are FALSE
(c) P is TRUE and Q FALSE.
(d) P is FALSE and Q is TRUE.

[2024 : 2 M, Set-I]

2.84 The function $f(x) = x^3 - 27x + 4$, $1 \le x \le 6$ has

(a) Minima point
(b) Maxima point
(c) Saddle point
(d) Inflection point

[2024 : 1 M, Set-II]

2.85 Three vectors $\vec{p}$, $\vec{q}$, and $\vec{r}$ are given as

$$\vec{p} = \hat{i} + \hat{j} + \hat{k}$$
$$\vec{q} = \hat{i} + 2\hat{j} + 3\hat{k}$$
$$\vec{r} = 2\hat{i} + 3\hat{j} + 4\hat{k}$$

Which of the following is/are CORRECT?

(a) $\vec{p} \times (\vec{q} \times \vec{r}) = (\vec{p} \times \vec{q}) \times \vec{r}$
(b) $\vec{r} \cdot (\vec{p} \times \vec{q}) = (\vec{q} \times \vec{p}) \cdot \vec{r}$
(c) $\vec{p} \times (\vec{q} \times \vec{r}) + \vec{q} \times (\vec{r} \times \vec{p}) + \vec{r} \times (\vec{p} \times \vec{q}) = \vec{0}$
(d) $\vec{p} \times (\vec{q} \times \vec{r}) = (\vec{p} \cdot \vec{r})\vec{q} - (\vec{p} \cdot \vec{q})\vec{r}$

[2024 : 2 M, Set-II]

2.86 The expression for computing the effective interest rate (i_{eff}) using continuous compounding for a nominal interest rate of 5% is

$$i_{eff} = \lim_{m \to \infty}\left(1 + \frac{0.05}{m}\right)^m - 1$$

The effective interest rate (in percentage) is ________. (rounded off to 2 decimal places).

[2024 : 2 M, Set-II]

Answers Calculus

2.1	(a)	2.2	(b)	2.3	(d)	2.4	(b)	2.5	(c)	2.6	(c)	2.7	(b)	2.8	(b)	2.9	(*)
2.10	(b)	2.11	(b)	2.12	(a)	2.13	(c)	2.14	(d)	2.15	(b)	2.16	(a)	2.17	(d)	2.18	(d)
2.19	(c)	2.20	(a)	2.21	(a)	2.22	(a)	2.23	(a)	2.24	(a)	2.25	(c)	2.26	(a)	2.27	(c)
2.28	(d)	2.29	(c)	2.30	(b)	2.31	(a)	2.32	(b)	2.33	(a)	2.34	(d)	2.35	(d)	2.36	(b)
2.37	(a)	2.38	(a)	2.39	(a)	2.40	(c)	2.41	(b)	2.42	(b)	2.43	(b)	2.44	(c)	2.45	(a)
2.46	(d)	2.47	(d)	2.48	(d)	2.49	(d)	2.50	(b)	2.51	(b)	2.52	(d)	2.54	(b)	2.56	(b)
2.57	(c)	2.58	(a)	2.59	(a)	2.60	(d)	2.61	(b)	2.62	(d)	2.63	(c)	2.64	(a)	2.65	(a)
2.66	(d)	2.67	(c)	2.68	(d)	2.69	(b)	2.70	(b)	2.71	(b)	2.76	(d)	2.77	(a)	2.78	(a, b)
2.79	(d)	2.80	(*)	2.81	(d)	2.82	(a)	2.83	(a)	2.84	(a)	2.85	(b, c, d)				
2.86	(5.13)																

Explanations Calculus

2.1 (a)

We know that $\lim_{\theta \to 0} \frac{\sin\theta}{\theta} = 1$

$$\lim_{\theta \to 0} \frac{\tan\theta}{\theta} = 1$$

2.2 (b)

$$y = x^2 + \frac{250}{x}$$

$$\Rightarrow \quad \frac{dy}{dx} = 2x + 250\left(\frac{-1}{x^2}\right)$$

if $\quad \frac{dy}{dx} = 0$

they $\quad 2x = \frac{250}{x^2}$

$$\Rightarrow \quad x^3 = 125$$

$$\therefore \quad x = 5$$

Now, $\quad \frac{d^2y}{dx^2} = 2x + 250 \times \frac{2}{x^3}$

of $\quad x = 5$

$$\frac{d^2y}{dx^2} = 2\times 5 + \frac{500}{(5)^3}$$

$$= 14 > 0$$

$\therefore$ the function $y = x^2 + \frac{250}{x}$ attain minimum value at $x = 5$.

2.3 (d)

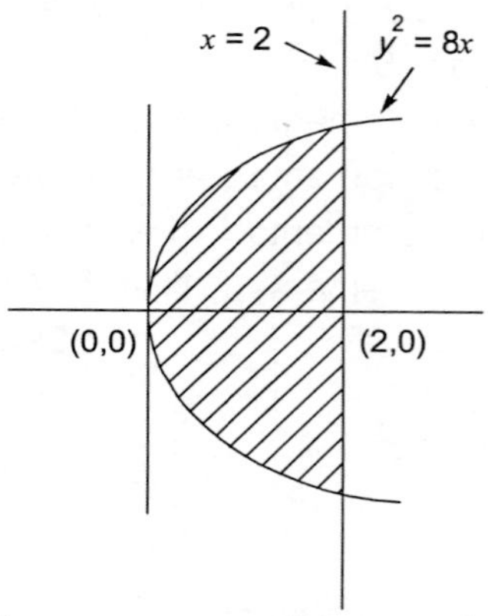

The volume generated by revolving the area bounded by the parabola $y^2 = 8x$ and line $x = 2$ about y-axis is

$$\int_{x=0}^{x=2} \sqrt{8x}\, dx = 2\sqrt{2}\left[\frac{x^{3/2}}{3/2}\right]_0^2$$

$$= 2\sqrt{2} \times \frac{2}{3}[2]^{3/2} = 5.33$$

2.4 (b)

$$\nabla f = 4x\hat{i} + 6y\hat{j} + 2z\hat{k}$$

$$\nabla f|_{P(2,1,3)} = 8\hat{i} + 6\hat{j} + 6\hat{k}$$

Directional derivative in the direction of vector $\vec{a}$

$$= \left(8\hat{i} + 6\hat{j} + 6\hat{k}\right)\frac{\left(\hat{i} - 2\hat{k}\right)}{\sqrt{5}}$$

$$= \frac{(8-12)}{\sqrt{5}} = -\frac{4}{\sqrt{5}}$$

2.5 (c)

$$\int \log x . dx$$

$$= \int \log x . 1 dx$$

$$= \log x \int 1 . dx - \int \left[\left(\frac{d \log x}{dx} \right) \right] \int 1 . dx \, \Big] dx$$

$$= \log x . x - \int \frac{1}{x} . x . dx$$

$$= x \log x - x + c$$

2.6 (c)

$$\lim_{x \to \infty} x \sin\left(\frac{1}{x}\right) = \lim_{x \to \infty} \frac{\sin(1/x)}{(1/x)}$$

Let $\frac{1}{x} = \theta$

$$\Rightarrow \quad \lim_{\theta \to 0} \frac{\sin \theta}{\theta} = 1$$

2.7 (b)

From mean value theorem,

$$\frac{f(1) - f(0)}{1 - 0} = f(x), \; x \in [0, 1]$$

$$\therefore \quad f(1) = f(0) + f(1)$$

$$= 2 + \frac{1}{5 - x^2} \qquad [\because f(0) = 2]$$

$\therefore$ Lower bound of $f(1)$

$$= 2 + \frac{1}{5 - 0} = 2.2$$

Upper bound of $f(1)$

$$= 2 + \frac{1}{5 - 1} = 2.25$$

2.8 (b)

$$f(x) = e^x$$
$$f'(x) = e^x$$

$$f(x) = f(a) + (x - a) f'(a) + \frac{(x - a)^2}{2} f''(a)$$

$$\therefore \text{ Third term} = \frac{(x - a)^2}{2!} f''(a)$$

$$f''(x) = e^x$$
$$\Rightarrow f''(a) = e^a$$

$$\therefore \text{ Third term} = \frac{e^a}{2} (x - a)^2$$

2.9 (*)

$$\int_0^2 \int_{x^2}^{2x} f(x, y) dy . dx$$

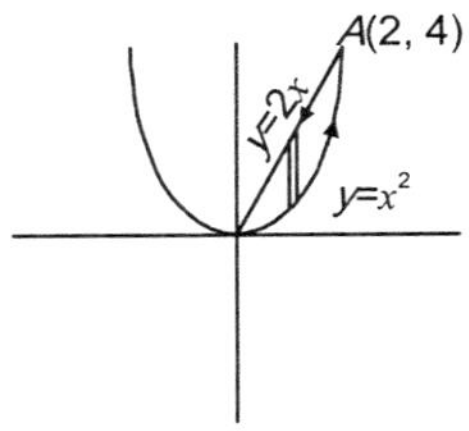

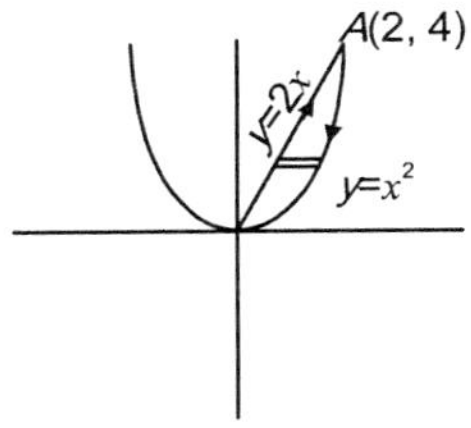

$$= \int_{x=0}^{2} \int_{y=x^2}^{y=2x} f(x, y) dy . dx$$

$$= \int_{y=0}^{y=4} \int_{x=\frac{y}{2}}^{x=\sqrt{y}} f(x, y) dx . dy$$

2.10 (b)

$$\left(\frac{\partial f}{\partial x} i + \frac{\partial f}{\partial y} j \right) \left(\frac{i + j}{\sqrt{2}} \right) = 2\sqrt{2}$$

$$\Rightarrow \quad \frac{\partial f}{\partial x} + \frac{\partial f}{\partial y} = 4 \qquad \text{...(i)}$$

Again,

$$\left(\frac{\partial f}{\partial x} \hat{i} + \frac{\partial f}{\partial y} \hat{j} \right) \frac{2j}{2} = -3$$

$$\Rightarrow \quad \frac{\partial f}{\partial y} = 3 \qquad \text{...(ii)}$$

$\therefore$ From eq. (i) and (ii),

$$\frac{\partial f}{\partial x} = 1$$

$$\therefore \quad \frac{\partial f}{\partial x} i + \frac{\partial f}{\partial y} j = \hat{i} + 3\hat{j}$$

Hence, the directional derivative of $f(x, y)$ in direction $(-i - 2j)$ is

$$= (\hat{i} + 3\hat{j}) \left(\frac{-i - 2j}{\sqrt{5}} \right) = \frac{1 - 6}{\sqrt{5}} = -\frac{7}{\sqrt{5}}$$

2.11 (b)

Given $2y = x^2$, therefore $y = \frac{x^2}{2}$ and $y = x + 4$

$\therefore \quad x^2 - 2x - 8 = 0$

The given curve and the line intersect at $A(-2, 2)$ and $B(4, 8)$

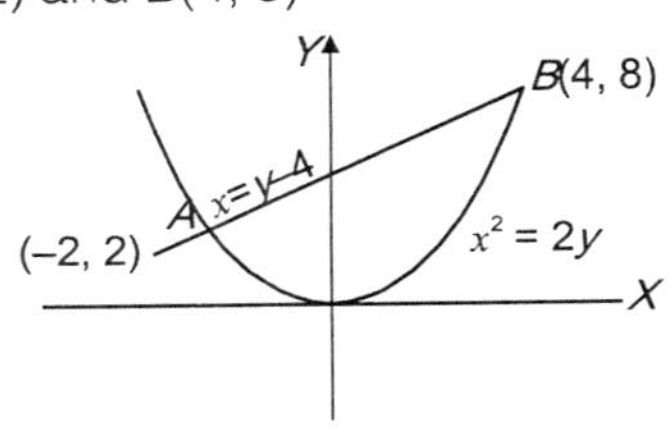

$\therefore$ Area $= \int_{-2}^{4} (y_1 - y_2)\,dx$

$$= \int_{-2}^{4}\left[(x+4)-\frac{x^2}{2}\right]dx = \left[\frac{x^2}{2}+4x-\frac{x^3}{6}\right]_{-2}^{4}$$

$$= \left(8+16-\frac{32}{3}\right)-\left(2-8-\frac{(-2)^3}{6}\right)$$

$$= \left(24-\frac{32}{3}\right)-\left(-6+\frac{4}{3}\right)$$

$$= \frac{40}{3}-\left(-\frac{14}{3}\right) = \frac{54}{3} = 18$$

2.15 (b)

Equation of parabola is, $y = x^2$

Area of parabola $= \frac{1}{3}\times \text{base}\times\text{height}$

$$= \frac{1}{3}\times 4\times 16 = \frac{64}{3}$$

2.16 (a)

$$\lim_{\theta\to 0}\frac{\sin m\theta}{\theta}$$

Dividing and multiplying by 'm' on both sides of numerator and denominator we get

$$\lim_{\theta\to 0} m\left(\frac{\sin m\theta}{m\theta}\right) = m$$

2.18 (d)

$$\lim_{n\to\infty}\frac{n}{\sqrt{n^2+n}}$$

Dividing by 'n', we get

$$= \lim_{n\to\infty}\frac{1}{\sqrt{1+\frac{1}{n}}} = 1$$

2.19 (c)

$$f(x) = e^x$$

For a function to be odd

$$f(-x) = -f(x)$$

For a function to be even

$$f(-x) = f(x)$$
$$f(x) = e^x$$
$$f(-x) = e^{-x}$$

Now, $e^{-x} \neq -f(x)$

and, $\neq f(x)$

$\therefore$ Function is neither even nor odd.

2.20 (a)

$$y = x + 2x^4$$

$$\frac{dy}{dx} = 1 + 8x^3 = 0$$

$$\Rightarrow \quad x^3 = -\frac{1}{8}$$

$$\Rightarrow \quad x = -\frac{1}{2}, -\frac{1}{2}, -\frac{1}{2}$$

Hence number of inflection points= 3.

2.21 (a)

$$\lim_{x\to a}(x-a)^{(x-a)}$$

Let $y = (x-a)$

Taking log on both sides, we get

$$\log y = (x-a)\log(x-a)$$

$$\Rightarrow \log y = (x-a)\left[(x-a)-\frac{(x-a)^2}{2}+\frac{(x-a)^3}{3}-\frac{(x-a)^4}{4}\right]$$

$$\Rightarrow \log y = (x-a)^2-\frac{(x-a)^3}{2}+\frac{(x-a)^4}{3}-\frac{(x-a)^5}{4}+\ldots$$

when $x \to a$

$$\log y = 0$$

$$\Rightarrow y = 1$$

2.22 (a)

$$f = V = (x^2+y^2+z^2)^{-1/2}$$

$$\Rightarrow \quad \frac{\partial V}{\partial x} = -x(x^2+y^2+z^2)^{-3/2}$$

$$\Rightarrow \quad \frac{\partial^2 V}{\partial x^2} = (x^2+y^2+z^2)^{-5/2}(2x^2-y^2-z^2) \quad \ldots\text{(i)}$$

Similarly, $$\frac{\partial^2 V}{\partial y^2} = (x^2+y^2+z^2)^{-5/2}(2y^2-x^2-z^2) \quad \ldots\text{(ii)}$$

and $$\frac{\partial^2 V}{\partial z^2} = (x^2+y^2+z^2)^{-5/2}(2x^2-x^2-y^2) \quad \ldots\text{(iii)}$$

Adding equation (i), (ii) and (iii) we get,

$$\frac{\partial^2 V}{\partial x^2}+\frac{\partial^2 V}{\partial y^2}+\frac{\partial^2 V}{\partial z^2} = 0$$

2.23 (a)

$$y_m = \sin\left(x + \frac{n\pi}{2}\right)$$

$$\Rightarrow \quad f'\left(\frac{\pi}{6}\right) = \sin\left(\frac{\pi}{6} + 0\right) = \frac{1}{2}$$

$$f''\left(\frac{\pi}{6}\right) = \sin\left(\frac{\pi}{6} + \frac{\pi}{2}\right) = -\frac{\sqrt{3}}{2}$$

$$f'\left(\frac{\pi}{6}\right) = \sin\left(\frac{\pi}{6} + \pi\right) = -\frac{1}{4}$$

Hence the required expression is,

$$\sin x = \frac{1}{2} + \frac{\sqrt{3}}{2}\left(x - \frac{\pi}{6}\right) - \frac{1}{4}\left(x - \frac{\pi}{6}\right)^2$$

2.24 (a)

$$f(x) = \lim_{x \to \infty}\left(1 + \frac{a^4}{x^4}\right)$$

$$= \left(1 - \frac{a^4}{\infty}\right) = 1 - 0 = 1$$

2.25 (c)

$$f(x) = 2x^3 - 15x^2 + 36x + 10$$

$$\therefore \quad f'(x) = 6x^2 - 30x + 36 \quad \text{...(i)}$$

and $\quad f''(x) = 12x - 30 \quad$ ...(ii)

For maxima and minima,

$$f'(x) = 0$$

$\Rightarrow \quad 6x^2 - 30x + 36 = 0$

$\Rightarrow \quad x^2 - 5x + 6 = 0$

$\Rightarrow \quad x = 3, 2$

Putting these values of x in eq. (ii) we get,

$$f''(3) = 36 - 30 = +6$$

and $\quad f''(2) = 24 - 30 = -6$

Hence, minimum value occurs at $x = 3$

maximum value occurs at $x = 2$

2.27 (c)

Since the given function is an odd function therefore in the given limits the integral will be zero.

2.28 (d)

$$f(x) = x\sqrt{5 - x^2}$$

$$f'(x) = \sqrt{5 - x^2} + \frac{x \times -2x}{2\sqrt{5 - x^2}}$$

$$5 - x^2 = x^2$$

$$2x^2 = 5$$

$$x = \pm\sqrt{\frac{5}{2}}$$

2.29 (c)

$$\int_0^1 x \log x dx = \log x \int_0^1 x dx - \int_0^1 \left(\frac{1}{x} \int x dx\right) dx$$

$$= \frac{x^2}{2}\log x - \int_0^1 \frac{1}{x}\frac{x^2}{2} dx$$

$$= \left[\frac{x^2}{2}\log x - \frac{x^2}{4}\right]_0^1$$

$$= \left(0 - \frac{1}{4}\right) - (0 - 0) = -\frac{1}{4}$$

2.30 (b)

$$l = \lim_{n \to \infty} n^{1/n}$$

$$\log l = \lim_{n \to \infty} \log n^{1/n}$$

$$= \lim_{n \to \infty} \frac{1}{n} \log n = \frac{\infty}{\infty}$$

by L-Hospital's rules,

$$\lim_{n \to \infty} \frac{1/n}{1} = \frac{1}{\infty} = 0$$

$$\log l = 0$$

$$\Rightarrow \quad l = e^0 = 1$$

2.31 (a)

Given $\quad P(3, -2, -1),\ Q(1, 3, 4)$

$R(2, 1, -2),\quad O(0, 0, 0)$

Equation of plane OQR is,

$$\begin{vmatrix} x - x_1 & y - y_1 & z - z_1 \\ x_2 - x_1 & y_2 - y_1 & z_2 - z_1 \\ x_3 - x_1 & y_3 - y_1 & z_3 - z_1 \end{vmatrix} = 0$$

$$\begin{vmatrix} x - 0 & y - 0 & z - 0 \\ 1 & 3 & 4 \\ 2 & 1 & -2 \end{vmatrix} = 0$$

i.e. $2x - 2y + z = 0$

Now $\perp$ distance of (x_1, y_1, z_1) from $ax + by + cz + d = 0$ is given by,

$$\left|\frac{ax_1 + by_1 + cz_1 + d}{\sqrt{a^2 + b^2 + c^2}}\right|$$

Therefore, $\perp$ distance of $(3, -2, -1)$ from plane

$2x - 2y + z = 0$ is given by,

$$\left| \frac{2\times 3 - 2\times(-2) + (-1)}{\sqrt{2^2 + (-2)^2 + 1^2}} \right| = 3$$

2.32 (b)

$$f(x) = \lim_{x\to 0}\left[\frac{x^3 + x^2}{2x^3 - 7x^2}\right]$$

Since this has $\frac{0}{0}$ form, limit can be found by repeated application of **L' Hospitals rule.**

$$f(x) = \lim_{x\to 0}\left[\frac{3x^2 + 2x}{6x^2 - 14x}\right]$$

$$= \lim_{x\to 0}\left[\frac{6x + 2}{12x - 14}\right]$$

$$= \left[\frac{6\times 0 + 2}{12\times 0 - 14}\right] = -\frac{1}{7}$$

2.33 (a)

Putting $f'(x) = 6x^2 - 6x - 36 = 0$

$\Rightarrow x^2 - x - 6 = 0$

$\Rightarrow x = 3 \text{ or } -2$

Now $f''(x) = 12x - 6$

and $f''(3) = 30 > 0$ (minima)

and $f''(-2) = -30 < 0$ (maxima)

Hence maxima is at $x = -2$ only.

2.34 (d)

Since the position of rail engine $S(t)$ is continuous and differentiable function, according to Lagranges mean value theorem

$\exists t$ where $0 \le t \le 8$ such that

$$S'(t) = v(t) = \frac{S(8) - S(0)}{8 - 0}$$

$$= \frac{(280 - 0)}{(8 - 0)} \text{ m/sec} = \frac{280}{8} \text{ m/sec}$$

$$= \frac{280}{8} \times \frac{3600}{1000} \text{ kmph}$$

$$= 126 \text{ kmph}$$

where $v(t)$ is the velocity of the rail engine.

2.35 (d)

Area common to circles $r = a$ and $r = 2a\cos\theta$ is $1.228a^2$.

2.36 (b)

Let $I(\alpha) = \int_0^\infty e^{-\alpha x}\frac{\sin x}{x}\,dx$ ...(i)

$$\frac{dI}{d\alpha} = \int_0^\infty (-x)e^{-\alpha x}\left(\frac{\sin x}{x}\right)dx$$

$$= -\int_0^\infty e^{-\alpha x}\sin x\,dx$$

$$= -\left[\frac{e^{-\alpha x}}{1+\alpha^2}(-\alpha\sin x - 1.\cos x)\right]_0^\infty$$

$$= -\frac{1}{1+\alpha^2}$$

$$\therefore \frac{dI}{d\alpha} = -\frac{1}{1+\alpha^2}$$

$$dI = \frac{d\alpha}{1+\alpha^2}$$

Intergrating,

$$I = -\tan^{-1}\alpha + C$$

$I(\infty) = -\frac{\pi}{2} + C$ but from (i), $I(\infty) = 0$

$$\therefore -\frac{\pi}{2} + C = 0 \Rightarrow C = \frac{\pi}{2}$$

$$\therefore I(\alpha) = -\tan^{-1}\alpha + \frac{\pi}{2}$$

$$\Rightarrow I(0) = \frac{\pi}{2} - \tan^{-1}0$$

$$\therefore I(0) = \frac{\pi}{2}$$

But from equation (i), $I(0) = \int_0^\infty \frac{\sin x}{x}\,dx$

$$\therefore \int_0^\infty \frac{\sin x}{x}\,dx = \frac{\pi}{2}$$

2.37 (a)

Since the limit is a function of x, we first integrate w.r.t. y and then w.r.t. x.,

$$\int_0^3\int_0^x (6 - x - y)\,dx\,dy = \int_0^3 dx\int_0^x (6 - x - y)\,dy$$

$$= \int_0^3 dx\left[6y - xy - \frac{y^2}{2}\right]_0^x$$

$$= \int_0^3 dx \left(6x - x^2 - \frac{x^2}{2}\right)$$

$$= \int_0^3 \left(6x - \frac{3x^2}{2}\right) dx$$

$$= \left[6\left(\frac{x^2}{2}\right) - \frac{3}{2}\left(\frac{x^3}{3}\right)\right]_0^3$$

$$= \left[3x^2 - \frac{x^3}{2}\right]_0^3$$

$$= 27 - \frac{27}{2} = 13.5$$

2.38 (a)

$$\lim_{x\to 0} \frac{\sin\left[\frac{2}{3}x\right]}{x} = \lim_{\frac{2}{3}x\to 0} \frac{\frac{2}{3}\sin\left[\frac{2}{3}x\right]}{\frac{2}{3}x}$$

Let $\theta = \frac{2}{3}x$

Required limit $= \frac{2}{3}\lim_{\theta\to 0}\frac{\sin\theta}{\theta} = \frac{2}{3}\times 1 = \frac{2}{3}$

2.39 (a)

$$f(x, y) = 4x^2 + 6y^2 - 8x - 4y + 8$$

$$\frac{\partial f}{\partial x} = 8x - 8$$

$$\frac{\partial f}{\partial y} = 12y - 4y$$

Putting, $\frac{\partial f}{\partial x} = 0$ and $\frac{\partial f}{\partial y} = 0$

$8x - 8 = 0$ and $12y - 4y = 0$

Given, $x = 1$ and $y = \frac{1}{3}$

$\left(1, \frac{1}{3}\right)$ is the only stationary point.

$$r = \left[\frac{\partial^2 f}{\partial x^2}\right]_{\left(1, \frac{1}{3}\right)} = 8$$

$$s = \left[\frac{\partial^2 f}{\partial x \partial y}\right]_{\left(1, \frac{1}{3}\right)} = 0$$

$$t = \left[\frac{\partial^2 f}{\partial y^2}\right]_{\left(1, \frac{1}{3}\right)} = 12$$

Since, $rt = 8 \times 12 = 48$

$s^2 = 0$

Since, $rt > s^2$,

we have either a maxima or minima at $\left(1, \frac{1}{3}\right)$

also since, $r = \left[\frac{\partial f}{\partial x^2}\right]_{\left(1, \frac{1}{3}\right)} = 8 > 0$, the point $\left(1, \frac{1}{3}\right)$ is a point of **minima**.

The minimum value is,

$$f\left(1, \frac{1}{3}\right) = 4 \times 1^2 + 6 \times \frac{1}{3^2} - 8 \times 1 - 4 \times \frac{1}{3} + 8$$

$$= \frac{10}{3}$$

So the optimal value of $f(x, y)$ is a minimum equal to $\frac{10}{3}$

2.40 (c)

If $f(x)$ is continuous at $x = \frac{\pi}{2}$

$$\lim_{x\to\frac{\pi}{2}} \frac{\lambda \cos x}{\frac{\pi}{2} - x} = f\left(\frac{\pi}{2}\right) = 1 \quad \text{...(i)}$$

Since the limit is in form of $\frac{0}{0}$, we can use **L'hospitals rule** on LHS of equation (i) and get

$$\lim_{x\to\frac{\pi}{2}} \frac{-\lambda \sin x}{-1} = 1$$

$$\Rightarrow \quad \lambda \sin\frac{\pi}{2} = 1$$

$$\Rightarrow \quad \lambda = 1$$

2.41 (b)

Let $$I = \int_0^a \frac{\sqrt{x}}{\sqrt{x} + \sqrt{a-x}} dx \quad \text{...(i)}$$

Since $\int_0^a f(x)dx = \int_0^a f(a-x)dx$

$$I = \int_0^a \frac{\sqrt{a-x}}{\sqrt{a-x} + \sqrt{x}} dx \quad \text{...(ii)}$$

(i) + (ii) $\Rightarrow$ $2I = \int_o^a \frac{\sqrt{x}+\sqrt{a-x}}{\sqrt{x}+\sqrt{a-x}}dx$

$\Rightarrow$ $2I = \int_o^a dx$

$\Rightarrow$ $2I = a$

$\Rightarrow$ $I = a/2$

2.43 (b)

$$\int_0^{\frac{\pi}{6}} \cos^4 3\theta \sin^3 6\theta \, d\theta$$

$$= 8\int_0^{\frac{\pi}{6}} \cos^6 3\theta \sin^3 3\theta \cos 3\theta \, d\theta$$

Put $\sin 3\theta = t$

$\Rightarrow$ $3\cos 3\theta \, d\theta = dt$

$$\Rightarrow \quad I = \frac{8}{3}\int_0^1 \left(1-t^2\right)^3 t^3.dt$$

$$= \frac{8}{3}\int_0^1 \left(t^6 - 2t^2 + 3t^4\right)t^3 dt$$

$$= \frac{8}{3}\int_0^1 \left(t^3 - t^9 - 3t^5 + 3t^7\right) dt$$

$$= \frac{8}{3}\left[\frac{t^4}{4} - \frac{t^{10}}{10} - \frac{3t^6}{6} + \frac{3t^8}{8}\right]_0^1$$

$$= \frac{8}{3}\left[\frac{1}{4} - \frac{1}{10} - \frac{3}{6} + \frac{3}{8}\right]$$

$$= \frac{8}{3}\left[\frac{30-12-60+45}{120}\right]$$

$$= \frac{1}{15}$$

Alternative Method:

Let $3\theta = t$

$3 \times d\theta = dt$

$$d\theta = \frac{dt}{3}$$

$\theta = \frac{\pi}{6}$ $\quad t = \frac{\pi}{2}$

$\theta = 0$ $\quad t = 0$

$$I = \int_0^{\pi/2} \cos^h t \cdot \sin^3 2t \cdot \frac{dt}{3}$$

$$= \frac{1}{3}\int_0^{\pi/2} \cos^h t \cdot (2\sin t \cos t)^3 \cdot dt$$

$$= \frac{8}{3}\int_0^{\pi/2} \cos^h t \cdot \sin^3 t \cdot \cos^3 t dt$$

$$= \frac{8}{3}\int_0^{\pi/2} \cos^7 t \sin^3 t dt$$

$$= \frac{8}{3}\left[\frac{6\cdot 4\cdot 2\cdot 2}{10\cdot 8\cdot 6\cdot 4\cdot 2}\right] = \frac{1}{15}$$

2.44 (c)

Put $\quad x = \frac{1}{h}$ as $x \to \infty \Rightarrow h \to 0$

$$\lim_{x\to\infty}\left(\frac{x+\sin x}{x}\right) = \lim_{h\to 0}\left(\frac{\frac{1}{h}+\sin\frac{1}{h}}{\frac{1}{h}}\right)$$

$$= \lim_{h\to 0} 1 + \left(\frac{\sin\frac{1}{h}}{\frac{1}{h}}\right) = 1$$

2.45 (a)

$$\lim_{\alpha\to 0}\frac{x^\alpha - 1}{\alpha} \quad \left[\frac{0}{0}\text{form}\right]$$

$$\lim_{\alpha\to 0}\frac{e^{\alpha \ln x} - 1}{\alpha}$$

Use *L*-Hospital Rule

$$\Rightarrow \quad \lim_{\alpha\to 0}\frac{e^{\alpha \ln x} \ln x}{1} = \ln x$$

2.46 (d)

$$y = \lim_{x\to\infty}\left(1+\frac{1}{x}\right)^{2x}$$

$$\Rightarrow \quad \log y = \lim_{x\to\infty} 2x\log\left(1+\frac{1}{x}\right)$$

Which is in the form of $\infty \times 0$.

To convert this into $\frac{0}{0}$ form, we rewrite as,

$$\Rightarrow \quad \log y = \lim_{x\to\infty}\frac{2\log\left(1+\frac{1}{x}\right)}{1/x}$$

Now it is in $\frac{0}{0}$ form.

Using **L'Hospital's rule,**

$$\Rightarrow \quad \log y = \lim_{x\to\infty} \frac{\dfrac{2\times -\frac{1}{x^2}}{1+\frac{1}{x}}}{-\dfrac{1}{x^2}}$$

$$= \lim_{x\to\infty} \frac{2}{1+\frac{1}{x}} = 2$$

$$\therefore \quad y = e^2$$

2.47 (d)

$f(x)$ has a local minimum at $x = x_0$

if $f'(x_0) = 0$

and $f''(x_0) > 0$

2.48 (d)

$$f'(x) = 0$$

$$\Rightarrow \quad 2x - 4 = 0$$

$$\Rightarrow \quad x = 2 \text{ (stationary point)}$$

$$f''(x) = 2 > 0$$

$\Rightarrow$ $f(x)$ is minimum at $x = 2$

i.e., $(2)^2 - 4(2) + 2 = -2$

$\therefore$ The optimum value of $f(x)$ is -2 (minimum)

2.49 (d)

(i) $$\lim_{x\to\infty} \frac{xy}{x^2+y^2} \lim_{y\to\infty}\left(\frac{0}{0^2+y^2}\right) = 0$$

(i.e., put $x = 0$ and then $y = 0$)

(ii) $$\lim_{\substack{x\to 0\\ y\to 0}} \frac{xy}{x^2+y^2} \lim_{x\to 0}\left(\frac{0}{x^2+0}\right) = 0$$

(i.e., put $y = 0$ and then $x = 0$)

(iii) $$\lim_{\substack{x\to 0\\ y\to 0}} \frac{xy}{x^2+y^2} \lim_{x\to 0} \frac{x(mx)}{x^2+m^2x^2}$$

(i.e., put $y = mx$)

$$\lim_{x\to\infty}\left(\frac{m}{1+m^2}\right) = \frac{m}{1+m^2}$$

which depends on m.

2.50 (b)

$$\int_0^\infty \frac{1}{1+x^2}\,dx = \left[\tan^{-1} x\right]_0^\infty$$

$$= \tan^{-1}\infty - \tan^{-1} 0 = \frac{\pi}{2}$$

and $$L(\sin x) = \frac{1}{s^2+1}$$

$$\Rightarrow \quad L\left(\frac{\sin x}{x}\right) = \int_s^\infty \frac{1}{s^2+1}\,dx$$

(Using "division by x")

$$= \left[\tan^{-1} s\right]_s^\infty$$

$$= \tan^{-1}\infty - \tan^{-1}(s) = \cot^{-1}(s)$$

$$\Rightarrow \int_0^\infty e^{-sx}\frac{\sin x}{x}\,dx = \cot^{-1}(s)$$

(Using definition of Laplace transform)

Put $s = 0$, we get

$$\int_0^\infty \frac{\sin x}{x}\,dx = \cot^{-1}(0) = \frac{\pi}{2}$$

$$\frac{1}{1+x^2}\,dx + \int_0^\infty \frac{\sin x}{x}\,dx = \pi$$

2.51 (b)

At the point of intersection of the curves,
$y = x^2 + 1$ and $x + y = 3$ i.e., $y = 3 - x$, we have,

$$x^2 + 1 = 3 - x$$

$$\Rightarrow \quad x^2 + x - 2 = 0$$

$$\Rightarrow \quad x = -2, 1 \text{ and } 3 - x \geq x^2 + 1$$

$\therefore$Required area is $\iint_R dy\,dx$

$$= \int_{x=-2}^{1}\left[\int_{y=x^2+1}^{3-x} dy\right]dx$$

$$= \int_{-2}^{1} \{3-x\} - (x^2+1)\,dx$$

$$= \left(\frac{-x^3}{3} - \frac{x^2}{2} + 2x\right)_{-2}^{1} = \frac{9}{2}$$

2.52 (d)

Given curve,

$$x^2 = 4y \qquad \text{...(i)}$$

and $$y^2 = 4x \qquad \text{...(ii)}$$

$$2x = 4\frac{dy}{dx}$$

$$\Rightarrow \quad \left(\frac{dy}{dx}\right)_{(0,0)} = 0 = m_1 \text{ (say)}$$

$$2y\frac{dy}{dx} = 4$$

$$\Rightarrow \left(\frac{dy}{dx}\right)_{(0,0)} = \infty = m_2$$

Let $m_2 = \frac{1}{m'}$, where $m' = 0$

$$\tan\theta = \left|\frac{m_1 - m_2}{1 + m_1 m_2}\right| = \left|\frac{m_1 m' - 1}{m' + m_1}\right| = \left|\frac{0-1}{0+0}\right| = \infty$$

$$\Rightarrow \theta = \frac{\pi}{2} = 90°$$

2.53 Sol.

Parabola is $x^2 = 8y$

$$y = \frac{x^2}{8} \text{ and straight is } y = 0$$

At the point of intersection, we have,

$$\frac{x^2}{8} = 8$$

$$\Rightarrow \quad x = -8, 8 \text{ and } y = 8 \geq y = \frac{x^2}{8}$$

$\therefore$ Required area is $\int_{x=-8}^{8}\left(8 - \frac{x^2}{8}\right)dx$

$$= 2\int_0^8\left(8 - \frac{x^2}{8}\right)dx \quad \left(\because 8 - \frac{x^2}{8} \text{ is even function}\right)$$

$$= 2\left[8x - \frac{x^3}{24}\right]_0^8 = \frac{256}{3} = 85.33 \text{ sq.units}$$

2.54 (b)

The quadratic approximation of $f(x)$ at the point $x = 0$ is,

$$f(x) = f(0) + \frac{x}{1!}f'(0) + \frac{x^2}{2!}f''(0)$$

$$= (-5) + x\{0\} + \frac{x^2}{2}\{-6\} = -3x^2 - 5$$

2.55 Sol.

$$\underset{x\to 0}{Lt}\frac{\tan x}{x^2 - x} \quad \text{(Applying L'Hospital rule)}$$

$$\underset{x\to 0}{Lt}\frac{\sec^2 x}{2x^2 - 1} = \frac{\sec^2 0}{0-1} = \frac{1}{-1} = -1$$

2.56 (b)

$$f(x) = e^{-x-e^{-x}} = e^{-x}\cdot e^{-e^{-x}}$$

$$y(x) = \int f(x)dx = \int e^{-x}\cdot e^{-e^{-x}}dx$$

Let $\quad e^{-x} = t$

$$-e^{-x}\,dx = dt$$

$$\int f(x)dx = \int e^{-t}\cdot(-dt)$$

$$= \frac{e^{-t}}{-1}\cdot(-d)$$

$$= e^{-t}$$

$$= e^{-\left(e^{-x}\right)} = e^{-e^{-x}}$$

2.57 (c)

$$w = f(x, y)$$

By **Chain rule,**

$$\frac{dw}{dt} = \frac{\partial w}{\partial x}\times\frac{dx}{dt} + \frac{\partial w}{\partial y}\times\frac{dy}{dt}$$

2.58 (a)

$$\tan\theta = \frac{dy}{dx} = \ln x + 1$$

$$\tan 45° = \ln x + 1$$

$$1 = \ln x + 1$$

$$\Rightarrow \quad \ln x = 0$$

$$\therefore \quad x = 1$$

Putting $x = 1$ in the eq. of curve, we get $y = 0$.

2.59 (a)

Let, $\quad \sin^{-1}x = t$

$$\frac{dx}{\sqrt{1-x^2}} = dt$$

$$I = \int_0^{\pi/2} t^2\,dt = \left[\frac{t^3}{3}\right]_0^{\pi/2} = \frac{\pi^3}{24}$$

2.60 (d)

$$f(x) = x^3 \text{ at } x = 0$$

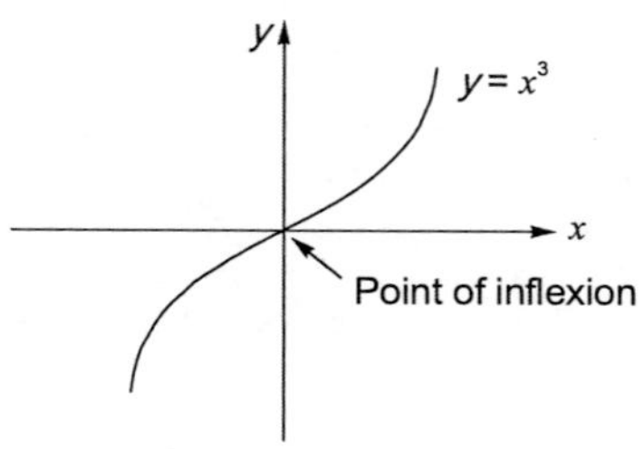

At $x = 0$, the function $y = x^3$ has neither minima nor maxima.

2.61 (b)

The value of $\int_0^{\pi} x\cos^2 x dx$

$$= \int_0^{\pi}\left(\frac{x}{2}+\frac{x\cos 2x}{2}\right)dx$$

$$= \frac{x^2}{4}\bigg|_0^{\pi} + \frac{1}{2}\left(\frac{x\sin 2x}{2}+\frac{\cos 2x}{4}\right)$$

$$= \frac{\pi^2}{4}+\frac{1}{2}\left\{\left(0+\frac{1}{4}\right)-\left(0+\frac{1}{4}\right)\right\}$$

$$= \frac{\pi^2}{4}+\frac{1}{2}\left(\frac{1}{4}-\frac{1}{4}\right)$$

$$= \frac{\pi^2}{4}$$

2.62 (d)

$$\lim_{x\to 0}\frac{\sin 4x}{\sin 2x} = \lim_{x\to 0}\frac{4x}{2x}\frac{\sin 4x}{4x}\times\frac{2x}{\sin 2x} = \frac{4}{2} = 2$$

$$\lim_{x\to 0}\frac{\tan x}{x} = 1$$

2.63 (c)

Taylor series of $f(x + h)$ at x.

$$f(x + h) = f(x) + (x + h - x)$$

$$f'(x) + \frac{(x+h-x)^2}{2!}f''(x)+\ldots$$

$$= f(x)+hf'(x)+\frac{h^2}{2!}f''(x)+\frac{h^3}{3!}f'''(x)+\ldots$$

2.64 (a)

Let $y = x^{1/x}$

$$\log y = \frac{\log x}{x}$$

$$y = e^{\frac{\log x}{x}}$$

y maximum (or) minimum when,

$$f(x) = \frac{\log x}{x} \text{ is maximum (or) minimum}$$

(1) $$f'(x) = \frac{x\left(\frac{1}{x}\right)-\log x}{x^2} = \frac{1-\log x}{x^2}$$

(2) $$f'(x) = 0$$

$$\Rightarrow 1 - \log x = 0$$

$$\log x = 1$$

$$x = e$$

(3) $$f''(x) = \frac{x^2\left(-\frac{1}{x}\right)-2x(1-\log x)}{x^4}$$

$$= \frac{-3x+2x\log x}{x^4}$$

(4) $$f''(e) = -\frac{e}{e^4} < 0 \text{ (max)}$$

$\therefore$ $y = \sqrt[x]{x}$ is maximum at $x = e$

2.65 (a)

$x = \psi \ln\phi,\ y = \phi \ln\psi$

Partially differentiating w.r.t x

$$1 = \psi_x \ln\phi + \frac{\psi}{\phi}\phi_x \quad \ldots (i)$$

$$0 = \phi_x \ln\psi + \frac{\phi}{\psi}\psi_x$$

$$\Rightarrow \frac{\phi}{\psi} = \frac{-\phi_x \ln\psi}{\psi_x}$$

$$\Rightarrow \frac{\phi}{\psi\phi_x} = \frac{-\ln\psi}{\psi_x} \quad \ldots (ii)$$

Put (ii) in (i)

$$1 = \psi_x \ln\phi - \frac{\psi_x}{\ln\psi} = \psi_x\left[\frac{\ln\phi\ln\psi - 1}{\ln\psi}\right]$$

$$\psi_x = \left[\frac{\ln\psi}{\ln\phi\ln\psi - 1}\right]$$

2.66 (d)

$$\lim_{x\to 0}\frac{x\sin x}{1-\cos x}$$

$$\lim_{x\to 0}\frac{\frac{\sin x}{x}}{\frac{1-\cos x}{x}} = \lim_{x\to 0}\frac{\frac{\sin x}{x}}{\frac{1-\cos x}{x^2}} = \frac{1}{1/2} = 2$$

2.67 (c)

It is in $\left(\frac{\infty}{\infty}\right)$ from so by L-Hospital Rule

$$= \lim_{x\to\infty}\left(\frac{2x-5}{8x+2}\right) = \frac{\infty}{\infty} = \lim_{x\to\infty}\left(\frac{2}{8}\right) = \left(\frac{1}{4}\right)$$

2.68 (d)

True value ln2 = 0.69 = T

x	$y = \ln x$
$x_0 = 1$	0
$x_1 = 6$	1.79

Divided differentiation

$$\frac{1.79-0}{6-1} = 0.358 = f[x_0, x_1]$$

Approx: $\ln 2 = f[x_0] + (x - x_0)\, f[x_0, x_1]$

$= 0 + (2 - 1)\, 0.358 = 0.358 = A$

$$\%\text{ error} = \frac{T-A}{T}\times 100 = 48.11\%$$

2.69 (b)

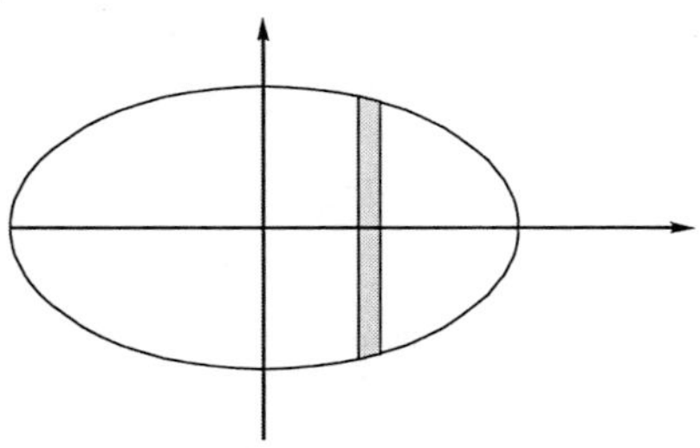

$$\text{Area} = \iint (1)dydx = \int_{x=-a}^{a}\int_{y=-\frac{b}{a}\sqrt{-x^2+a^2}}^{+\frac{b}{a}\sqrt{-x^2+a^2}} (1)\, dydx$$

$$= 4\int_{x=0}^{a}\int_{y=0}^{\frac{b}{a}\sqrt{a^2-x^2}} (1)\, dy\, dx$$

$$= 4\int_{x=0}^{a}\int_{y=0}^{\frac{b}{a}\sqrt{a^2-x^2}} dx = \pi ab$$

2.70 (b)

$$\lim_{x\to\infty}\frac{3x\sqrt{1+\frac{2020}{x^2}}}{x\left(1+\frac{7}{x}\right)} = 3$$

2.71 (b)

Because integral is a polynomial of 3rd degree so Simpson's rule will give error free answer.

2.72 Sol.

Let diameter, $x = 3$ and height $= y = 4$ and error $= \pm 0.2$

$$V = \pi\left(\frac{x}{2}\right)^2 y = \frac{\pi x^2 y}{4}$$

$\therefore \quad V = f(x, y)$

So, $\quad dV = \left(\frac{\partial V}{\partial x}\right)dx + \left(\frac{\partial V}{\partial y}\right)dy$

i.e., $\quad dV = \left(\frac{1}{2}\pi xy\right)dx + \left(\frac{\pi x^2}{4}\right)dy$

$$= \frac{1}{2}\pi\times 3\times 4\times(0.2) + \frac{\pi}{4}\times(3)^2\times(0.2) = 1.65\,\pi$$

$= 1.65 \times 3.14 = 5.18$ (approx)

i.e., absolute error $= |\,5.18\,| = 5.18$

2.73 Sol.

Given, $\quad V = [2, 3,] \times [1, 2] \times [0, 1]$ i.e., $2 < x < 3,\ 1 < y < 2,\ 0 < z < 1$

So, $\quad I = \iiint_V 8xyz\,dV$

$$= 8\iiint_V xyz\,dx\,dy\,dz$$

$$= 8\int_{x=2}^{3} x\,dx \times \int_{y=1}^{2} y\,dy \times \int_{z=0}^{1} z\,dz$$

$$= 8\left[\frac{x^2}{2}\right]_2^3\left[\frac{y^2}{2}\right]_1^2\left[\frac{z^2}{2}\right]_0^1$$

$$= (9-4)(4-1)(1-0)$$

$$= 5 \times 3 \times 1 = 15$$

2.74 Sol.

$$\lim_{x\to 1}\left[\frac{1}{\ln x} - \frac{1}{x-1}\right] = \lim_{x\to 1}\left[\frac{(x-1)-\ln x}{\ln x(x-1)}\right]\left(\frac{0}{0}\text{ form}\right)$$

So using L-Hospital's rule

$$= \lim_{x\to 1}\left[\frac{1-\frac{1}{x}}{\log x + (x-1)\left(\frac{1}{x}\right)}\right] \quad \left(\frac{0}{0}\text{ form}\right)$$

Again using L'Hospital rule

$$\lim_{x\to 1}\left[\frac{1/x^2}{1/x + 1/x^2}\right] = \frac{1}{1+1} = 0.5$$

2.75 Sol.

$$\int_{-1}^{1} xe^{-|x|}dx = 0$$

As $f(x) = xe^{-|x|}$ is an odd function.

2.76 (d)

$$\lim_{x\to\infty}\left(\frac{x\ln x}{x^2+1}\right) \quad \left(\frac{\infty}{\infty}\text{form}\right)$$

Using L'Hospital rule,

$$= \lim_{x\to\infty}\left(\frac{x\left(\frac{1}{x}\right)+\ln x}{2x}\right) \quad \left(\frac{\infty}{\infty}\text{form}\right)$$

$$\lim_{x\to\infty}\left(\frac{0+\frac{1}{x}}{2}\right) = \lim_{x\to\infty}\left(\frac{1}{2x}\right) = \frac{1}{2\times\infty} = 0$$

2.77 (a)

$$z = \sin(y + it) + \cos(y - it)$$

Differentiating w.r.t. y, we get,

$$z_y = \cos(y + it) - \sin(y - it)$$

$$z_{yy} = -\sin(y + it) - \cos(y - it) \quad \text{...(i)}$$

Differentiating With respect to t, we get

$$z_t = \cos(y + it)(i) - \sin(y - it)\,(-i)$$

$$z_{tt} = \sin(y + it) + \cos(y - it) \quad \text{...(ii)}$$

From eq. (i) and (ii)

$$z_{yy} + z_{tt} = 0$$

2.78 (a, b)

For finding intersection point,

$$3 - x = x - 1$$

$$\Rightarrow \quad x = 2$$

So, $3 - x$ and $x - 1$ intersects at $x = 2$.

$$f(x) = \begin{cases} 3-x & x<2 \\ x-1 & x>2 \\ 1 & x=2 \end{cases}$$

Check for continuity:

$$f(2^-) = f(2^+) = f(2)$$

$\therefore$ It is continuous in its domain.

Check for differentiability:

$$f'(x) = \begin{cases} -1 & x<2 \\ 1 & x>2 \end{cases}$$

$\therefore$ It is not differentiable.

For maximum and minima,

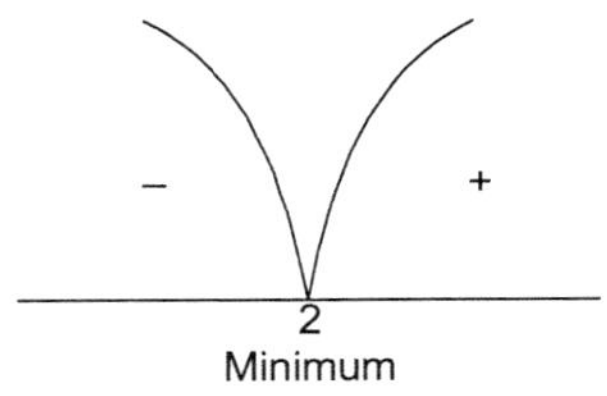

or

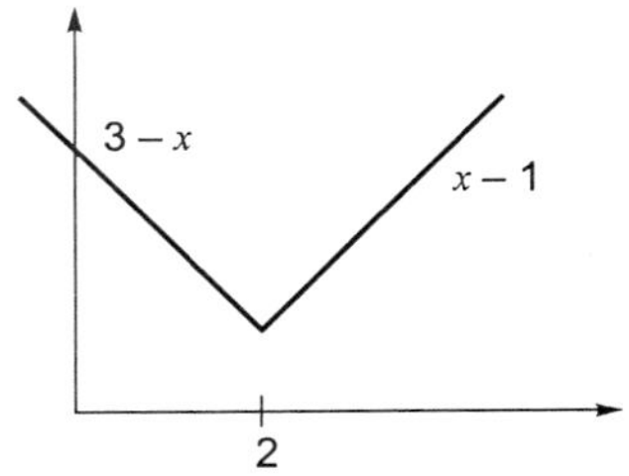

It has local minima at $x = 2$

2.79 (d)

Given: $f(x) = x^3 - 6x^2 + 11x - 6$

$= (x - 1)(x - 2)(x - 3) \quad \text{... (i)}$

- At $x = 1$ and $x = 3$ $f(x) = 0$

$\Rightarrow$ Statement 1 is correct.

Now, $f'(x) = 3x^3 - 12x^2 + 11$

$\therefore \quad f''(x) = 6x - 12 = 6(x - 2)$

- Now, $f''(x) < 0$, if $x < 2$. This is condition for maxima which is, within the domin S.
- Now, $f''(x) > 0$, if $x > 2$. This is condition for minima which is, within the domin S.
- But, $f''(x)$ is not greater than zero throughout the domains S.

2.81 (d)

$f(x) = \frac{1}{x^2}$ is not defined at $x = 0$

$$\text{So,} \quad I = \int_{-1}^{0}\frac{1}{x^2}dx + \int_{0}^{1}\frac{1}{x^2}dx$$

$$= \left(-\frac{1}{x}\right)_{-1}^{0} + \left(-\frac{1}{x}\right)_{0}^{1} = \infty$$

$\frac{1}{x}$ is not defined at $x = 0$

Therefore, the integral does not converge.

2.82 (a)

Given:

$$f(x) = e^x |\sin x|$$

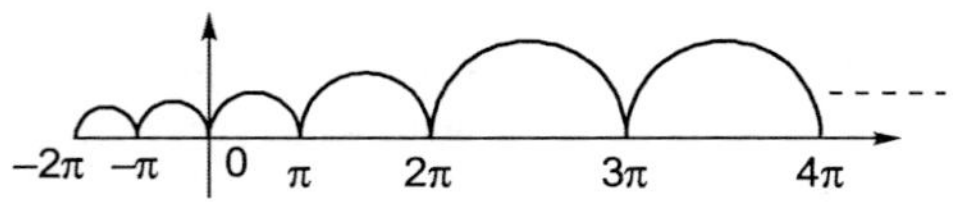

By observing graph we can easily say given function is neither periodic, nor bounded and nor differentiable, but is continuous at all x.

2.83 (a)

$$r = 6x^2 + 4y^2 - z^2 - 9xyz - 2xy + 3xz - yz$$

$$\text{grad } r = \hat{i}\frac{\partial r}{\partial x} + \hat{j}\frac{\partial r}{\partial y} + \hat{k}\frac{\partial r}{\partial z}$$

$$= \begin{matrix} \hat{i}(12x - 9yz - 2y + 3z) \\ +\hat{j}(8y - 9xz - 2x - z) \\ +\hat{k}(-2z - 9xy + 3x - y) \end{matrix}$$

$$\text{So, curl}(\text{grad } \vec{r}) = \vec{\nabla} \times \text{grad } \vec{r}$$

$$= \begin{vmatrix} \hat{i} & \hat{j} & \hat{k} \\ \frac{\partial}{\partial x} & \frac{\partial}{\partial y} & \frac{\partial}{\partial z} \\ (12x - 9yz - 2y + 3z) & (8y - 9xz - 2x - z) & (-2z - 9xy + 3x - y) \end{vmatrix}$$

$$= \begin{matrix} \hat{i}(-9x - 1 + 9x + 1) \\ -\hat{j}(-9y + 3 + 9y - 3) \\ +\hat{k}(-9z - 2 + 9z + 2) \end{matrix} = 0$$

$\therefore$ $\quad P = \vec{0}$ i.e. $P = \text{curl (grad } \vec{r}) = \vec{0}$

$\quad Q = \text{div (curl } \vec{P}) = 0$

(By vector identity)

Hence both are true.

2.84 (a)

$$f(x) = x^3 - 27x + 4$$

$$\therefore \quad f'(x) = 3x^2 - 27$$

$$f'(x) = 0$$

$$\Rightarrow \quad 3x^2 - 27 = 0$$

$$\Rightarrow \quad 3x^2 = 27$$

$$\Rightarrow \quad x^2 = 9$$

$$\Rightarrow \quad x = \pm 3$$

But $\quad 1 \le x \le 6$

$$\therefore \quad x = +3$$

$$f''(x) = 6x$$

$$f''(x)\big|_{x=3} = 6 \times 3 = 18 > 0$$

$$\because \quad f''(x) > 0$$

$\therefore$ At $x = 3$, there will be point of minima.

Hence option (a) is answer.

2.85 (b, c, d)

2.86 Sol.

$$i_{\text{eff}} = \lim_{m \to \infty}\left(1 + \frac{0.05}{m}\right)^m - 1$$

$$y =$$

$$e^{\lim_{x\to\infty} mx\left[1 + \frac{0.05}{m} - 1\right]^m}; \; 1^\infty \text{ form}$$

$$y = e^{\lim_{x\to\infty} mx\left[1 + \frac{0.05}{m} - 1\right]}$$

$$y = e^{\lim_{x\to\infty} mx\left[1 + \frac{0.005}{m} - 1\right]}$$

$$y = e^{\lim_{x\to\infty} 0.05}$$

Finally, $\quad y = e^{0.05} - 1$

$$y = 0.05127 \simeq 5.13\%$$

■■■■

3 Vector Calculus

3.1 The directional derivative of the function f(x, y, x) = x + y at the point P(1, 1, 0) along the direction $\vec{i}+\vec{j}$ is

(a) $\frac{1}{\sqrt{2}}$ (b) $\sqrt{2}$

(c) $-\sqrt{2}$ (d) 2 [1996 : 1 M]

3.2 The directional derivative of the following function at (1, 2) in the direction of (4i + 3j) is

$$f(x, y) = x^2 + y^2$$

(a) 4/5 (b) 4

(c) 2/5 (d) 1 [2002 : 2 M]

3.3 Value of the integral $\oint_c (xydy - y^2dx)$, where, c is the square cut from the first quadrant by the lines x = 1 and y = 1 will be (Use Green's theorem to change the line integral into double integral)

(a) $\frac{1}{2}$ (b) 1

(c) $\frac{3}{2}$ (d) $\frac{5}{3}$ [2005 : 2 M]

3.4 The directional derivative of f(x, y, z) = $2x^2 + 3y^2 + z^2$ at the point P (2, 1, 3) in the direction of the vector a = i – 2 k is

(a) –2.785 (b) –2.145

(c) –1.789 (d) 1.000 [2006 : 2 M]

3.5 A velocity vector is given as

$\vec{V} = 5xy\vec{i} + 2y^2\vec{j} + 3yz^2\vec{k}$.

The divergence of this velocity vector at (1, 1, 1) is

(a) 9 (b) 10

(c) 14 (d) 15 [2007 : 2 M]

3.6 The inner (dot) product of two vectors $\vec{P}$ and $\vec{Q}$ is zero. The angle (degrees) between the two vectors is

(a) 0 (b) 30

(c) 90 (d) 120 [2008 : 2 M]

3.7 For a scalar function f(x, y, z) = $x^2 + 3y^2 + 2z^2$, the gradient at the point P(1, 2, – 1) is

(a) $2\vec{i}+6\vec{j}+4\vec{k}$ (b) $2\vec{i}+12\vec{j}-4\vec{k}$

(c) $2\vec{i}+12\vec{j}+4\vec{k}$ (d) $\sqrt{56}$

[2009 : 1 M]

3.8 For a scalar function f(x, y, z) = $x^2 + 3y^2 + 2z^2$, the directional derivative at the point P(1, 2, – 1) in the direction of a vector $\vec{i}-\vec{j}+2\vec{k}$ is

(a) – 18 (b) $-3\sqrt{6}$

(c) $3\sqrt{6}$ (d) 18 [2009 : 2 M]

3.9 If $\vec{a}$ and $\vec{b}$ are two arbitrary vectors with magnitudes a and b, respectively, $\left|\vec{a}\times\vec{b}\right|^2$ will be equal to

(a) $a^2b^2-\left(\vec{a}\cdot\vec{b}\right)^2$ (b) $ab-\vec{a}\cdot\vec{b}$

(c) $a^2b^2+\left(\vec{a}\cdot\vec{b}\right)^2$ (d) $ab+\vec{a}\cdot\vec{b}$

[2011 : 2 M]

3.10 For the parallelogram OPQR shown in the sketch, $\overline{OP} = a\hat{i}+b\hat{j}$ and $\overline{OR} = c\hat{i}+d\hat{j}$. The area of the parallelogram is

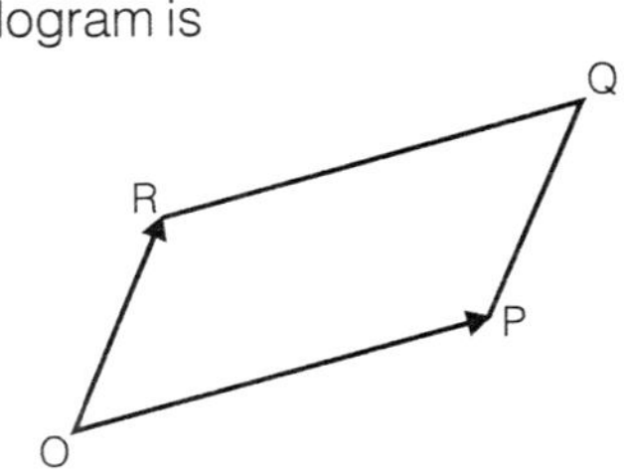

(a) ad – bc (b) ac + bd

(c) ad + bc (d) ab – cd

[2011 : 2 M]

3.11 The directional derivative of the field $u(x, y, z) = x^2 - 3yz$ in the direction of the vector $\left(\vec{i}+\vec{j}-2\vec{k}\right)$ at point (2, – 1, 4) is ________.

[2015 : 2 M, Set-I]

3.12 The divergence of the vector field $V = x^2 i + 2y^3 j + z^4 k$ at $x = 1$, $y = 2$, $z = 3$ is ________.

[2017 : 1 M, Set-II]

3.12 The value (up to two decimal places) of a line integral $\int_C \vec{F}(\vec{r}) \cdot \vec{dr}$, for $\vec{F}(\vec{r}) = x^2\hat{i} + y^2\hat{j}$ along C which is a straight line joining (0, 0) to (1, 1) is ________.

[2018 : 2 M, Set-II]

3.14 Euclidean norm (length) of the vector $[4 \ -2 \ -6]^T$ is

(a) $\sqrt{48}$ (b) $\sqrt{56}$

(c) $\sqrt{24}$ (d) $\sqrt{12}$

[2019 : 1 M, Set-II]

3.15 What is curl of the vector field $2x^2y\mathbf{i} + 5z^2\mathbf{j} - 4uyz\mathbf{k}$?

(a) $-14z\mathbf{i} - 2x^2\mathbf{k}$

(b) $6z\mathbf{i} + 4x^2\mathbf{j} - 2x^2\mathbf{k}$

(c) $-14z\mathbf{i} + 6y\mathbf{j} + 2x^2\mathbf{k}$

(d) $6z\mathbf{i} - 8xy\mathbf{j} + 2x^2y\mathbf{k}$

[2019 : 1 M, Set-II]

3.16 If C represents a line segment between (0, 0, 0) and (1,1, 1) in Cartesian coordinate system, the value (expressed as integer) of the line integral

$$\int_C [(y+z)dx + (x+z)dy + (x+y)dz]$$

is ________.

[2020 : 2 M, Set-I]

3.17 The unit normal vector to the surface $X^2 + Y^2 + Z^2 - 48 = 0$ at the point (4, 4, 4) is

(a) $\frac{1}{\sqrt{3}}, \frac{1}{\sqrt{3}}, \frac{1}{\sqrt{3}}$ (b) $\frac{2}{\sqrt{2}}, \frac{2}{\sqrt{2}}, \frac{2}{\sqrt{2}}$

(c) $\frac{1}{\sqrt{2}}, \frac{1}{\sqrt{2}}, \frac{1}{\sqrt{2}}$ (d) $\frac{1}{\sqrt{5}}, \frac{1}{\sqrt{5}}, \frac{1}{\sqrt{5}}$

[2021 : 1 M, Set-II]

3.18 A function is defined in Cartesian coordinate system as $f(x, y) = xe^y$. The value of the directional derivative of the function (in integer) at the point (2, 0) along the direction of the straight line segment from point (2, 0) to point $\left(\frac{1}{2}, 2\right)$ is ________.

[2021 : 2 M, Set-II]

3.19 For the equation

$$\frac{d^3y}{dx^3} + x\left(\frac{dy}{dx}\right)^{3/2} + x^2y = 0$$

the correct description is

(a) an ordinary differential equation of order 3 and degree 2.

(b) an ordinary differential equation of order 3 and degree 3.

(c) an ordinary differential equation of order 2 and degree 3.

(d) an ordinary differential equation of order 3 and degree 3/2.

[2022 : 1 M, Set-I]

3.20 The concentration $s(x, t)$ of pollutants, in a one-dimensional reservoir at position x and time t satisfies the diffusion equation

$$\frac{\partial s(x,t)}{\partial t} = D\frac{\partial^2 s(x,t)}{\partial x^2}$$

on the domain $0 \le x \le L$, where D is the diffusion coefficient of the pollutants. The initial condition $s(x, 0)$ is defined by the step-function shown in the figure.

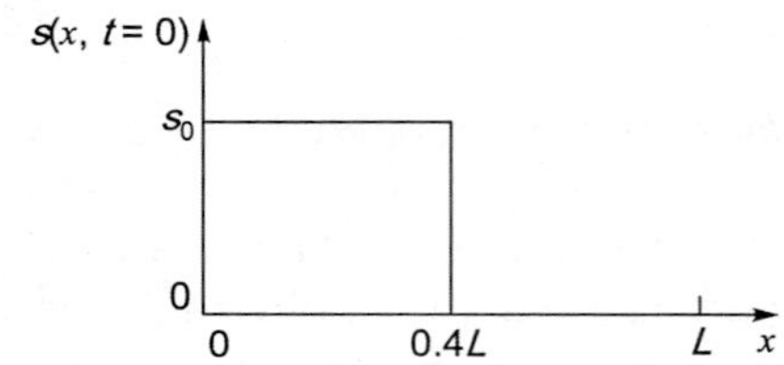

The boundary conditions of the problem are given by $\frac{\partial s(x,t)}{\partial x} = 0$ at the boundary points $x = 0$ and $x = L$ at all times. Consider $D = 0.1$ m²/s, $s_0 = 5$ μmol/m and $L = 10$ m.

The steady-state concentration $\bar{s}\left(\frac{L}{2}\right) = s\left(\frac{L}{2}, \infty\right)$ at the center $x = \frac{L}{2}$ of the reservoir (in μmol/m) is ________. (in integer)

[2022 : 2 M, Set-II]

3.21 Let ϕ be a scalar field, and $\boldsymbol{u}$ be a vector field. Which of the following identities is true for div($\phi\boldsymbol{u}$)?

(a) div($\phi\boldsymbol{u}$) = ϕdiv($\boldsymbol{u}$) + $\boldsymbol{u}$ · grad(ϕ)

(b) div($\phi\boldsymbol{u}$) = ϕdiv($\boldsymbol{u}$) + u × grad(ϕ)

(c) div($\phi\boldsymbol{u}$) = ϕgrad($\boldsymbol{u}$) + $\boldsymbol{u}$ · grad(ϕ)

(d) div($\phi\boldsymbol{u}$) = ϕgrad($\boldsymbol{u}$) + u × grad(ϕ)

[2023 : 1 M, Set-II]

■■■■

Answers Vector Calculus

3.1 (b) 3.2 (b) 3.3 (c) 3.4 (c) 3.5 (d) 3.6 (c) 3.7 (b) 3.8 (b) 3.10 (a)

3.14 (b) 3.15 (a) 3.17 (a) 3.19 (a) 3.21 (a)

Explanations Vector Calculus

3.1 (b)

$$\text{Directional derivative} = \nabla f \cdot \frac{\bar{a}}{|\bar{a}|}$$

$$= (i+j) \cdot \frac{(i+j)}{\sqrt{2}} = \frac{2}{\sqrt{2}} = \sqrt{2}$$

3.2 (b)

$$f(x, y) = x^2 + y^2$$

$$\nabla f = 2xi + 2yj$$

$$\nabla f = 2i + 4j$$

$$e = \frac{4i+3j}{5} \qquad D.D = \nabla f.e$$

$$D.D = \frac{8+12}{5} = \frac{20}{5} = 4$$

3.3 (c)

Green's Theorem is,

$$\oint_c \phi dx + \psi dy = \iint_E \left(\frac{\partial \psi}{\partial x} - \frac{\partial \phi}{\partial y} \right) dxdy$$

Here $\quad I = \oint_c (xydy - y^2dx)$

$$= \oint_c (-y^2)dx + (xy)dy$$

$\therefore \quad \phi = -y^2, \ \psi = xy$

$$\frac{\partial \psi}{\partial x} = y, \ \frac{\partial \phi}{\partial y} = -2y$$

Substituting in **Green's Theorem**, we get,

$$I = \int_{y=0}^{1} \int_{x=0}^{1} [y-(-2y)]dx\,dy$$

$$= \int_{y=0}^{1} \int_{x=0}^{1} 3y\,dx\,dy$$

$$= \int_{y=0}^{1} [3xy]_{x=0}^{1} dy = \int_{y=0}^{1} 3ydy = \frac{3}{2}$$

3.4 (c)

$f = 2x^2 + 3y^2 + z^2$, $P(2, 1, 3)$, $a = i - 2k$

$$\nabla f = i\frac{\partial f}{\partial x} + j\frac{\partial f}{\partial y} + k\frac{\partial f}{\partial z}$$

$$= 4xi + 6yj + 2zk$$

at $P(2, 1, 3) \nabla f = 4 \times 2 \times i + 6 \times 1 \times j + 2 \times 3 \times k$

$$= 8i + 6j + 6k$$

Component of Δf in direction of vector,

$a = i - 2k$ is

$$(8i + 6j + 4k) \times \frac{(i-2k)}{\sqrt{1^2 + (-2)^2}}$$

$$= \frac{8.1 + 6.0 + 6.(-2)}{\sqrt{5}} = \frac{-4}{\sqrt{5}} = -1.789$$

3.5 (d)

$$\vec{V} = 5xyi + 2y^2j + 3yz^2k$$

$$\text{div}(\vec{V}) = \frac{df}{dx} + \frac{d\phi}{dy} + \frac{d\psi}{dz}$$

$$= 5y + 4y + 6yz$$

at $(1, 1, 1) \text{div}(\vec{V}) = 5.1 + 4.1 + 6.1.1 = 15$

3.6 (c)

$$\vec{P} \cdot \vec{Q} = 0$$

$$\vec{P} \cdot \vec{Q} = |P||Q| \cos\theta$$

if $\quad \vec{P} \cdot \vec{Q} = 0$

$\Rightarrow |P||Q| \cos\theta = 0$

since, P and Q are non-zero vectors

$\Rightarrow \quad \cos\theta = 0$

$\Rightarrow \quad \theta = 90°$

3.7 (b)

$$f = x^2 + 3y^2 + 2z^2$$

$$\Delta f = \text{grad } f = i\frac{\partial f}{\partial x} + j\frac{\partial f}{\partial y} + k\frac{\partial f}{\partial z}$$

$$= i(2x) + j(6y) + k(4z)$$

The gradient at $P(1, 2, -1)$ is

$$= i(2 \times 1) + j(6 \times 2) + k(4 \times -1)$$

$$= 2i + 12j - 4k$$

3.8 (b)

$$\Delta f = i\frac{\partial f}{\partial x} + j\frac{\partial f}{\partial y} + k\frac{\partial f}{\partial z}$$

Here $f = x^2 + 3y^2 + 2z^2$

$\therefore \quad \Delta f = i(2x) + j(6y) + k(4z)$

at $p(1, 2, -1)\Delta f = i(2 \times 1) + j(6 \times 2) + k(4x - 1)$

$= 2i + 12j - 4k$

3.10 (a)

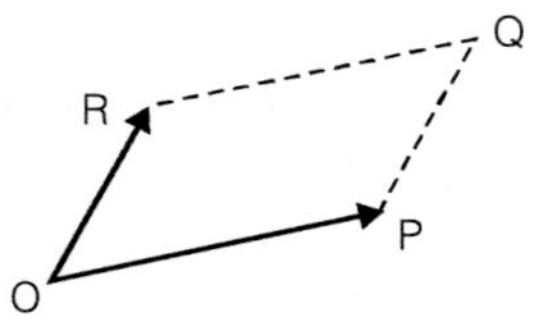

The area of parallelogram *OPQR* in figure shown above, is the magnitude of the vector product

$$= \left|\overrightarrow{OP} \times \overrightarrow{OR}\right|$$

$$\overrightarrow{OP} = a\hat{i} + b\hat{j}$$

$$\overrightarrow{OR} = c\hat{i} + d\hat{j}$$

$$\overrightarrow{OP} \times \overrightarrow{OR} = \begin{vmatrix} i & j & k \\ a & b & 0 \\ c & d & 0 \end{vmatrix}$$

$$= 0\hat{i} + 0\hat{j} + (ad - bc)\hat{k}$$

$$\left|\overrightarrow{OP} \times \overrightarrow{OR}\right| = \sqrt{0^2 + 0^2 + (ad - bc)^2}$$

$$= ad - bc$$

3.11 Sol.

$$u(x, y, z) = x^2 - 3yz$$

$$\nabla u = 2x\hat{i} - 3zj - 3y\hat{k}$$

$$\nabla u|_{\text{at }(2, -1, 4)} = 4\hat{i} + 12\hat{j} - 3\hat{k}$$

Directional derivative

$$= \left(4\hat{i} + 12\hat{j} - 3\hat{k}\right) \cdot \frac{\hat{i} + \hat{j} - \hat{k}}{\sqrt{6}}$$

$$= \frac{4 - 12 - 6}{\sqrt{6}} = -\frac{14}{\sqrt{6}}$$

$$= -\frac{7\sqrt{6}}{3} = -5.715$$

3.12 Sol.

$$\vec{V} = x^2\hat{i} + 2y^3\hat{j} + z^4\hat{k}$$

$$\vec{\nabla}.\vec{V} = \left[\frac{\partial}{\partial x}\hat{i} + \frac{\partial}{\partial y}\hat{j} + \frac{\partial}{\partial z}\hat{k}\right] \cdot \left[x^2\hat{i} + 2y^3\hat{j} + z^4\hat{k}\right]$$

$$= 2x + 6y^2 + 4z^3$$

At (1, 2, 3),

$$\vec{\nabla}.\vec{V} = 2 + 6(2)^2 + 4(3)^3$$

$$= 134$$

3.13 Sol.

$$\bar{F} = x^2\bar{i} + y^2\bar{j}$$

$$\int \bar{F} \cdot d\bar{r} = \int (x^2\bar{i} + y^2\bar{j}) \cdot (dx\,\bar{i} + dy\,\bar{j})$$

$$= \int x^2 dx + y^2 dy$$

(0, 0) to (1, 1) line is $y = x$

$$= \int x^2 dx + x^2 dx = \int_0^1 2x^2 dx$$

$$= 2\left(\frac{x^3}{3}\right)\Bigg|_0^1 = \frac{2}{3} = 0.666$$

3.14 (b)

$$x = \begin{bmatrix} 4 \\ -2 \\ -6 \end{bmatrix}$$

Euclidean norm length

$$= \sqrt{16 + 4 + 36} = \sqrt{56}$$

3.15 (a)

$$\text{curl}\bar{F} = \begin{vmatrix} \bar{i} & \bar{j} & \bar{k} \\ \partial/\partial x & \partial/\partial y & \partial/\partial z \\ 2x^2y & 5z^2 & -4yz \end{vmatrix}$$

$$= \bar{i}(-4z - 10z) - \bar{j}(0 - 0) + \bar{k}(0 - 2x^2)$$

$$= -14z\boldsymbol{i} - 2x^2\boldsymbol{k}$$

3.16 Sol.

$$I = \int_C [(ydx + xdy) + (zdx + xdz) + (zdy + ydz)]$$

$$= \int_C [d(xy) + d(xz) + d(yz)] = (xy + yz + zx)_{(0,0,0)}^{(1,1,1)}$$

$$= (1 + 1 + 1) - (0 + 0 + 0) = 3$$

3.17 (a)

$$\phi = x^2 + y^2 + z^2 - 48, \; P(4, 4, 4)$$

$$\text{grad}\,\phi = \vec{\nabla}\phi = \hat{i}\frac{\partial\phi}{\partial x} + \hat{j}\frac{\partial\phi}{\partial y} + \hat{k}\frac{\partial\phi}{\partial z}$$

$$= (2x)\hat{i} + (2y)\hat{j} + (2z)\hat{k}$$

$$\vec{n} = (\text{grad}\,\phi)_P = 8\hat{i} + 8\hat{j} + 8\hat{k}$$

$$\hat{n} = \frac{\vec{n}}{|\vec{n}|} = \frac{8\hat{i} + 8\hat{j} + 8\hat{k}}{\sqrt{64+64+64}}$$

$$= \frac{\hat{i}+\hat{j}+\hat{k}}{\sqrt{3}} \simeq \left(\frac{1}{\sqrt{3}}, \frac{1}{\sqrt{3}}, \frac{1}{\sqrt{3}},\right)$$

3.18 Sol.

Given : $f(x, y) = xe^y$

P(2, 0) and $Q\left(\frac{1}{2}, 2\right)$

$$\text{grad } f = \hat{i}\left(e^y\right) + \hat{j}\left(xe^y\right) + \hat{k}(0)$$

$$\Rightarrow \quad (\text{grad } f)_p = \hat{i} + 2\hat{j}$$

$$\overrightarrow{PQ} = \left(\frac{1}{2} - 2\right)\hat{i} + (2-0)\hat{j} = -\frac{3}{2}\hat{i} + 2\hat{j}$$

Required directional derivative = $(\text{grad } f)_P \widehat{PQ}$

$$= \left(\hat{i} + 2\hat{j}\right) \times \frac{\left(-\frac{3}{2}\hat{i} + 2\hat{j}\right)}{\sqrt{\frac{9}{4} + 4}} = \frac{\frac{-3}{2} + 4}{\sqrt{\frac{25}{4}}}$$

$$= \frac{\frac{-3+8}{2}}{\left(\frac{5}{2}\right)} = 1$$

3.19 (a)

Given: $\frac{d^3y}{dx^3} + x\left(\frac{dy}{dx}\right)^{3/2} + x^2y = 0$

Since highest order of given equation is 3.
Order = 3

$$(y''' + x^2y) = \left[-x\left(y'\right)^{3/2}\right]$$

Squaring both the sides, we get,

$$\left(\frac{d^3y}{dx^3} + x^2y\right)^2 = \left(-x\frac{dy}{dx}\right)^3$$

Since power of y''' is 2.

Degree of given differential equation = 2.

3.20 Sol.

From figures $s(x, t)$

At $x = 0$

$\Rightarrow \quad s(0, t) = s_0 = 5$

At $x = L$

$\Rightarrow \quad s(L, t) = 0$

The steady state solution of diffusion equation is given by,

$$s(x) = \left(\frac{s(L) - s(0)}{L}\right)x + s(0)$$

$$\Rightarrow \quad s\left(\frac{L}{2}\right) = \left(\frac{0-5}{L}\right)\left(\frac{L}{2}\right) + 5$$

$$= \frac{-5}{2} + 5 = 2.5$$

3.21 (a)

It is special property of vector, $\text{div}(\phi \boldsymbol{u}) = \phi\,\text{div}(\boldsymbol{u}) + \boldsymbol{u}\cdot\text{grad}(\phi)$

■■■■

4 Differential Equations

4.1 The necessary and sufficient condition for the differential equation of the from $M(x, y)\,dx + N(x, y)\,dy = 0$ to be exact is

(a) $M = N$ (b) $\frac{\partial M}{\partial x} = \frac{\partial N}{\partial y}$

(c) $\frac{\partial M}{\partial y} = \frac{\partial N}{\partial x}$ (d) $\frac{\partial^2 M}{\partial x^2} = \frac{\partial^2 N}{\partial y^2}$

[1994 : 1 M]

4.2 The differential equation

$EI\frac{d^4y}{dx^4} + P\frac{d^2y}{dx^2} + ky = 0$ is

(a) Linear of Fourth order
(b) Non-linear of Fourth order
(c) Non-Homogeneous
(d) Linear and Fourth degree [1994 : 1 M]

4.3 The differential equation
$y'' + (s^3 \sin x)^5 y' + y = \cos x^3$ is
(a) homogeneous
(b) nonlinear
(c) second order linear
(d) nonhomogeneous with constant coefficient

[1995 : 1 M]

4.4 For the differential equation,

$f(x, y)\frac{dy}{dx} + g(x, y) = 0$ to be exact,

(a) $\frac{\partial f}{\partial y} = \frac{\partial g}{\partial x}$ (b) $\frac{\partial f}{\partial x} = \frac{\partial g}{\partial y}$

(c) $f = g$ (d) $\frac{\partial^2 f}{\partial x^2} = \frac{\partial^2 g}{\partial y^2}$

[1997 : 1 M]

4.5 The differential equation, $\frac{dy}{dx} + Py = Q$, is a linear equation of first order only if
(a) P is a constant but Q is a function of y
(b) P and Q are functions of y or constants
(c) P is a function of y but Q is a constant
(d) P and Q are functions of x or constants

[1997 : 1 M]

4.6 If c is a constant, solution of the equation $\frac{dy}{dx} = 1 + y^2$ is

(a) $y = \sin(x + c)$ (b) $y = \cos(x + c)$
(c) $y = \tan(x + c)$ (d) $y = e^x + c$

[1999 : 1 M]

4.7 The solution of the following differential equation with boundary conditions $y(0) = 2$ and $y'(1) = -3$ is

$$\frac{d^2y}{dx^2} = 3x - 2$$

(a) $y = \frac{x^3}{3} - \frac{x^2}{2} + 3x - 6$

(b) $y = 3x^3 - \frac{x^2}{2} + 5x + 2$

(c) $y = \frac{x^3}{2} - x^2 - \frac{5x}{2} + 2$

(d) $y = x^3 - \frac{x^2}{2} + 5x + \frac{3}{2}$ [2001 : 2 M]

4.8 Biotransformation of an organic compound having concentration (x) can be modeled using an ordinary differential equation $\frac{dx}{dt} + kx^2 = 0$, where k is the reaction rate constant. If $x = a$ at $t = 0$, the solution of the equation is

(a) $x = ae^{-kt}$ (b) $\frac{1}{x} = \frac{1}{a} + kt$

(c) $x = a(1 - e^{-kt})$ (d) $x = a + kt$

[2004 : 2 M]

4.9 Transformation to linear form by substituting $v = y^{1-n}$ of the equation

$\frac{dy}{dt} + p(t)y = q(t)y^n;\ n > 0$ will be

(a) $\frac{dv}{dt} + (1 - n)pv = (1 - n)q$

(b) $\frac{dv}{dt} + (1 - n)pv = (1 + n)q$

(c) $\frac{dv}{dt} + (1+n)pv = (1-n)q$

(d) $\frac{dv}{dt} + (1+n)pv = (1+n)q$ [2005 : 2 M]

4.10 The solution of $\frac{d^2y}{dx^2} + 2\frac{dy}{dx} + 17y = 0$; $y(0) = 1$, $\frac{dy}{dx}\left(\frac{x}{4}\right) = 0$ in the range $0 < x < \frac{\pi}{4}$ is given by

(a) $e^{-x}\left(\cos 4x + \frac{1}{4}\sin 4x\right)$

(b) $e^{x}\left(\cos 4x - \frac{1}{4}\sin 4x\right)$

(c) $e^{-4x}\left(\cos x - \frac{1}{4}\sin x\right)$

(d) $e^{-4x}\left(\cos 4x - \frac{1}{4}\sin 4x\right)$ [2005 : 2 M]

4.11 The differential equation $\frac{dy}{dx} = 0.25\, y^2$ is to be solved using the backward (implicit) Euler's method with the boundary condition y = 1 at $x = 0$ and with a step size of 1. What would be the value of y at $x = 1$?

(a) 1.33 (b) 1.67
(c) 2.00 (d) 2.33 [2006 : 1 M]

4.12 A spherical nephthalene ball exposed to the atmosphere loses volume at a rate proportional to its instantaneous surface area due to evaporation. If the initial diameter of the ball is 2 cm and the diameter reduces to 1 cm after 3 months, the ball completely evaporates in

(a) 6 months (b) 9 months
(c) 12 months (d) infinite time
[2006 : 2 M]

4.13 The solution of the differential equation, $x^2\frac{dy}{dx} + 2xy - x + 1 = 0$, given that at $x = 1$, $y = 0$ is

(a) $\frac{1}{2} - \frac{1}{x} + \frac{1}{2x^2}$ (b) $\frac{1}{2} - \frac{1}{x} - \frac{1}{2x^2}$

(c) $\frac{1}{2} + \frac{1}{x} + \frac{1}{2x^2}$ (d) $-\frac{1}{2} + \frac{1}{x} + \frac{1}{2x^2}$
[2006 : 2 M]

4.14 The degree of the differential equation $\frac{d^2x}{dt^2} + 2x^3 = 0$ is

(a) 0 (b) 1
(c) 2 (d) 3 [2007 : 1 M]

4.15 The solution for the differential equation $\frac{dy}{dx} = x^2y$ with the condition that $y = 1$ at $x = 0$ is

(a) $y = e^{\frac{1}{2x}}$ (b) $\ln(y) = \frac{x^3}{3} + 4$

(c) $\ln(y) = \frac{x^2}{2}$ (d) $y = e^{\frac{x^3}{3}}$
[2007 : 1 M]

4.16 A body originally at 60°C cools down to 40°C in 15 minutes when kept in air at a temperature of 25°C. What will be the temperature of the body at the end of 30 minutes?

(a) 35.2°C (b) 31.5°C
(c) 28.7°C (d) 15°C [2007 : 2 M]

4.17 The general solution of $\frac{d^2y}{dx^2} + y = 0$ is

(a) $y = P\cos x + Q\sin x$
(b) $y = P\cos x$
(c) $y = P\sin x$
(d) $y = P\sin^2 x$
[2008 : 1 M]

4.18 The equation $k_x\frac{\partial^2 h}{\partial x^2} + k_z\frac{\partial^2 h}{\partial z^2} = 0$ can be transformed to $\frac{\partial^2 h}{\partial x_t^2} + \frac{\partial^2 h}{\partial z^2} = 0$ by substituting

(a) $x_t = x\frac{k_z}{k_x}$ (b) $x_t = x\frac{k_x}{k_z}$

(c) $x_t = x\sqrt{\frac{k_x}{k_z}}$ (d) $x_t = x\sqrt{\frac{k_z}{k_x}}$
[2008 : 2 M]

4.19 Solution of $\frac{dy}{dx} = -\frac{x}{y}$ at $x = 1$ and $y = \sqrt{3}$ is

(a) $x - y^2 = -2$ (b) $x + y^2 = 4$
(c) $x^2 - y^2 = -2$ (d) $x^2 + y^2 = 4$
[2008 : 2 M]

4.21 Solution of the differential equation $3y\frac{dy}{dx} + 2x = 0$ represents a family of

(a) ellipses (b) circles
(c) parabolas (d) hyperbolas

[2009 : 2 M]

4.21 The order and degree of the differential equation

$$\frac{d^3y}{dx^3}+4\sqrt{\left(\frac{dy}{dx}\right)^3+y^2}=0$$ are respectively

(a) 3 and 2 (b) 2 and 3
(c) 3 and 3 (d) 3 and 1

[2010 : 1 M]

4.22 The solution to the ordinary differential equation

$$\frac{d^2y}{dx^2}+\frac{dy}{dx}-6y=0$$ is

(a) $y=c_1e^{3x}+c_2e^{-2x}$

(b) $y=c_1e^{3x}+c_2e^{2x}$

(c) $y=c_1e^{-3x}+c_2e^{2x}$

(d) $y=c_1e^{-3x}+c_2e^{-2x}$

[2010 : 2 M]

4.23 A parabolic cable is held between two supports at the same level. The horizontal span between the supports is L. The sag at the mid-span is h. The equation of the parabola is $y=4h\frac{x^2}{L^2}$, where x is the horizontal coordinate and y is the vertical coordinate with the origin at the centre of the cable. The expression for the total length of the cable is

(a) $\int_0^L\sqrt{1+64\frac{h^2x^2}{L^4}}dx$ (b) $2\int_0^{L/2}\sqrt{1+64\frac{h^3x^2}{L^4}}dx$

(c) $\int_0^{L/2}\sqrt{1+64\frac{h^2x^2}{L^4}}dx$ (d) $2\int_0^{L/2}\sqrt{1+64\frac{h^2x^2}{L^4}}dx$

[2010 : 2 M]

4.24 The solution of the differential equation $\frac{dy}{dx}+\frac{y}{x}=x$, with the condition that $y=1$ at $x=1$, is

(a) $y=\frac{2}{3x^2}+\frac{x}{3}$ (b) $y=\frac{x}{2}+\frac{1}{2x}$

(c) $y=\frac{2}{3}+\frac{x}{3}$ (d) $y=\frac{2}{3x}+\frac{x^2}{3}$

[2011 : 2 M]

4.25 The solution of the ordinary differential equation $\frac{dy}{dx}+2y=0$ for the boundary condition, $y=5$ at $x=1$ is

(a) $y=e^{-2x}$ (b) $y=2e^{-2x}$
(c) $y=10.95\,e^{-2x}$ (d) $y=36.95\,e^{-2x}$

[2011 : 2 M]

4.26 The integrating factor for differential equation

$$\frac{dP}{dt}+k_2P=k_1L_oe^{-k_1t}$$ is

(a) e^{-k_1t} (b) e^{-k_2t}
(c) e^{k_1t} (d) e^{k_2t}

[2014 : 1 M, Set-II]

4.27 Consider the following difference equation

$$x(ydx+xdy)\cos\frac{y}{x}=y(xdy-ydx)\sin\frac{y}{x}$$

Which of the following is the solution of the above equation (c is an arbitrary constant)?

(a) $\frac{x}{y}\cos\frac{y}{x}=c$ (b) $\frac{x}{y}\sin\frac{y}{x}=c$

(c) $xy\cos\frac{y}{x}=c$ (d) $xy\sin\frac{y}{x}=c$

[2015 : 2 M, Set-I]

4.28 Consider the following second order linear differential equation

$$\frac{d^2y}{dx^2}=-12x^2+24x-20$$

The boundary conditions are: at $x=0$, $y=5$ and $x=2$, $y=21$

The value of y at $x=1$ is ________.

[2015 : 2 M, Set-II]

4.29 The respective expressions for complimentary function and particular integral part of the solution of the differential equation

$$\frac{d^4y}{dx^4}+3\frac{d^2y}{dx^2}=108x^2$$ are

(a) $\left[c_1+c_2x+c_3\sin\sqrt{3x}+c_4\cos\sqrt{3x}\right]$ and $\left[3x^4-12x^2+c\right]$

(b) $\left[c_2+c_3\sin\sqrt{3x}+c_4\cos\sqrt{3x}\right]$ and $\left[5x^4-12x^2+c\right]$

(c) $\left[c_1 + c_3 \sin\sqrt{3}x + c_4 \cos\sqrt{3}x\right]$

and $\left[3x^4 - 12x^2 + c\right]$

(d) $\left[c_1 + c_2 x + c_3 \sin\sqrt{3}x + c_4 \cos\sqrt{3}x\right]$

and $\left[5x^4 - 12x^2 + c\right]$

[2016 : 2 M, Set-I]

4.30 The solution of the equation $\frac{dQ}{dt} + Q = 1$ with $Q = 0$ at $t = 0$ is

(a) $Q(t) = e^{-t} - 1$
(b) $Q(t) = 1 + e^{-t}$
(c) $Q(t) = 1 - e^{t}$
(d) $Q(t) = 1 - e^{-t}$

[2017 : 2 M, Set-I]

4.31 Consider the following second-order differential equation:

$y'' - 4y' + 3y = 2t - 3t^2$

The particular solution of the differential equation is

(a) $-2 - 2t - t^2$
(b) $-2t - t^2$
(c) $2t - t^2$
(d) $-2 - 2t - 3t^2$

[2017 : 2 M, Set-II]

4.32 The solution (up to three decimal places) at $x = 1$ of the differential equation $\frac{d^2y}{dx^2} + 2\frac{dy}{dx} + y = 0$ subject to boundary conditions $y(0) = 1$ and $\frac{dy}{dx} = (0) = -1$ is ____.

[2018 : 2 M, Set-I]

4.33 The solution of the equation $x\frac{dy}{dx} + y = 0$ passing through the point (1, 1) is

(a) x
(b) x^2
(c) x^{-1}
(d) x^{-2}

[2018 : 1 M, Set-II]

4.34 Consider the ordinary differential equation $x^2\frac{d^2y}{dx^2} - 2x\frac{dy}{dx} + 2y = 0$. Given the values of $y(1) = 0$ and $y(2) = 2$, the value of $y(3)$ (round off to 1 decimal place), is________.

[2019 : 2 M, Set-I]

4.35 An ordinary differential equation is given below.

$$\left(\frac{dy}{dx}\right)(x\ln x) = y$$

The solution for the above equation is

(**Note:** K denotes a constant in the options)

(a) $y = Kxe^{x}$
(b) $y = Kxe^{-x}$
(c) $y = K\ln x$
(d) $y = Kx\ln x$

[2019 : 2 M, Set-II]

4.36 In the following partial differential equation, θ is a function of t and z, and D and K are functions of θ

$$D(\theta)\frac{\partial^2\theta}{\partial z^2} + \frac{\partial K(\theta)}{\partial z} - \frac{\partial\theta}{\partial t} = 0$$

The above equation is

(a) a second order linear equation
(b) a second order non-linear equation
(c) a second degree non-linear equation
(d) a second degree linear equation

[2020 : 1 M, Set-I]

4.37 The ordinary differential equation $\frac{d^2u}{dx^2} - 2x^2u + \sin x = 0$ is

(a) linear and homogeneous
(b) nonlinear and homogeneous
(c) nonlinear and nonhomogeneous
(d) linear and nonhomogeneous

[2020 : 1 M, Set-II]

4.38 For the Ordinary Differential Equation $\frac{d^2x}{dt^2} - 5\frac{dx}{dt} + 6x = 0$, with initial condition $x(0) = 0$ and $\frac{dx}{dt}(0) = 10$, the solution is

(a) $-10e^{2t} + 10e^{3t}$
(b) $5e^{2t} + 6e^{3t}$
(c) $10e^{2t} + 10e^{3t}$
(d) $-5e^{2t} + 6e^{3t}$

[2020 : 2 M, Set-I]

4.39 The following partial differential equation is defined for $u : u(x, y)$ $\frac{\partial u}{\partial y} = \frac{\partial^2 u}{\partial x^2}$; $y \geq 0$; $x_1 \leq x \leq x_2$

The set of auxiliary conditions necessary to solve the equation uniquely, is

(a) one initial condition and two boundary conditions
(b) three initial conditions
(c) two initial conditions and one boundary condition
(d) three boundary conditions

[2020 : 1 M, Set-II]

4.40 An ordinary differential equation is given below

$$6\frac{d^2y}{dx^2}+\frac{dy}{dx}-y=0$$

The general solution of the above equation (with constants C_1 and C_2), is

(a) $y(x)=C_1e^{-\frac{x}{3}}+C_2xe^{\frac{x}{2}}$

(b) $y(x)=C_1e^{-\frac{x}{3}}+C_2e^{\frac{x}{2}}$

(c) $y(x)=C_1xe^{-\frac{x}{3}}+C_2e^{\frac{x}{2}}$

(d) $y(x)=C_1e^{\frac{x}{3}}+C_2e^{-\frac{x}{2}}$ [2020 : 2 M, Set-II]

4.41 The solution of the second-order differential equation $\frac{d^2y}{dx^2}+2\frac{dy}{dx}+y=0$ with boundary conditions $y(0) = 1$ and $y(1) = 3$ is

(a) $e^{-x}+(3e-1)xe^{-x}$

(b) $e^{-x}+\left[3e\sin\left(\frac{\pi x}{2}\right)-1\right]xe^{-x}$

(c) $e^{-x}-(3e-1)xe^{-x}$

(d) $e^{-x}-\left[3e\sin\left(\frac{\pi x}{2}\right)-1\right]xe^{-x}$

[2021 : 2 M, Set-I]

4.42 If k is a constant, the general solution of $\frac{dy}{dx}-\frac{y}{x}=1$ will be in the form of

(a) $y=x\ln(x)$ (b) $y=xk\ln(k)$

(c) $y=k\ln(kx)$ (d) $y=x\ln(kx)$

[2021 : 2 M, Set-II]

4.43 In the differential equation $\frac{dy}{dx}+\alpha xy=0$, α is a positive constant. If $y = 1.0$ at $x = 0.0$, and $y = 0.8$ at $x = 1.0$, the value of α is ________ (rounded off to three decimal places).

[2023 : 1 M, Set-I]

4.44 The differential equation,

$$\frac{du}{dt}+2tu^2=1,$$

is solved by employing a backward difference scheme within the finite difference framework. The value of u at the $(n-1)^{th}$ time-step, for some n, is 1.75. The corresponding time (t) is 3.14 s. Each time step is 0.01 s long. Then, the value of $(u_n - u_{n-1})$ is ________. (round off to three decimal places).

[2023 : 2 M, Set-I]

4.45 The solution of the differential equation

$\frac{d^3y}{dx^3}-5.5\frac{d^2y}{dx^2}+9.5\frac{dy}{dx}-5y=0$ is expressed as $y=C_1e^{2.5x}+C_2e^{\alpha x}+C_3e^{\beta x}$, where C_1, C_2, C_3, α, and β are constants, with α and β being distinct and not equal to 2.5. Which of the following options is correct for the values of α and β?

(a) 1 and 2 (b) –1 and –2

(c) 2 and 3 (d) –2 and –3

[2023 : 2 M, Set-II]

4.46 The second-order differential equation in an unknown function u: $u(x, y)$ is defined as

$$\frac{\partial^2 u}{\partial x^2}=2$$

Assuming g: g(x), f : $f(y)$, h : h(y), the general solution of the above differential equation

(a) $u = x^2 + f(y) + g(x)$

(b) $u = x^2 + xf(y) + g(x)$

(c) $u = x^2 + xf(y) + h(y)$

(d) $u = x^2 + f(y) + yg(x)$

[2024 : 1 M, Set-I]

4.47 Consider two Ordinary Differential Equations (ODEs):

$$P:\ \frac{dy}{dx}=\frac{x^4+3x^2y^2+2y^4}{x^3y}$$

$$Q:\ \frac{dy}{dx}=\frac{-y^2}{x^2}$$

Which one of the following options is CORRECT?

(a) P is a non-homogeneous ODE and Q is an exact ODE.

(b) P is a nonhomogeneous ODE and Q is not an exact ODE.

(c) P is a homogeneous ODE and Q is not an exact ODE.

(d) P is a homogeneous ODE and Q is an exact ODE.

[2024 : 1 M, Set-II]

■■■■

Answers Differential Equations

4.1 (c)	4.2 (a)	4.3 (c)	4.4 (b)	4.5 (d)	4.6 (c)	4.7 (c)	4.8 (b)	4.9 (a)
4.10 (a)	4.11 (c)	4.12 (a)	4.13 (a)	4.14 (b)	4.15 (d)	4.16 (b)	4.17 (a)	4.18 (d)
4.19 (d)	4.20 (a)	4.21 (a)	4.22 (c)	4.23 (d)	4.24 (d)	4.25 (d)	4.26 (d)	4.27 (c)
4.29 (a)	4.30 (d)	4.31 (a)	4.33 (c)	4.35 (c)	4.36 (b)	4.37 (d)	4.38 (a)	4.39 (a)
4.40 (d)	4.41 (a)	4.42 (d)	4.45 (a)	4.46 (c)	4.47 (c)			

Explanations Differential Equations

4.1 (c)

The differential equation,

$Mdx + Ndy = 0$ is said to be exact when

$$\frac{\partial M}{\partial y} = \frac{\partial N}{\partial x}$$

4.2 (a)

The differential equation

$E.I.\frac{d^4y}{dx^4} + P.\frac{d^2y}{dx^2} + k.y = 0$ has

order 'Four' but degree 'one'

As all derivative consist one degree so we can say the differential equation is linear.

4.3 (c)

y'' indicates that it is of second order.

Degree of y'', y' and y is one.

Hence the given differential equation is second order linear equation.

4.4 (b)

$$f(x,y)\frac{dy}{dx} + g(x,y) = 0$$

$\Rightarrow \quad f(x, y)dy + g(x, y)dx = 0$

$\Rightarrow \quad g(x, y)dx + f(x, y)dy = 0$

Comparing with,

$Mdx + Ndy = 0$

$$M = g(x, y)$$

$$N = f(x, y)$$

Given differential equation will be exact if,

$$\frac{\partial N}{\partial x} = \frac{\partial M}{\partial y}$$

$$\frac{\partial f(x,y)}{\partial x} = \frac{\partial g(x,y)}{\partial y}$$

or $\quad \frac{\partial f}{\partial x} = \frac{\partial g}{\partial y}$

4.6 (c)

$$\frac{dy}{dx} = 1 + y^2$$

$$\int \frac{dy}{1+y^2} = \int dx$$

$$= \tan^{-1} y = x + c$$

$\Rightarrow \quad y = \tan(x + c)$

4.7 (c)

$$\frac{dy}{dx} = \frac{3x^2}{2} - 2x + C_1$$

Integrating, we get,

$$y = \frac{3x^3}{6} - \frac{2x^2}{2} + C_1x + C_2$$

$\therefore \quad y(0) = 2 = 0 + 10 + C_2$

or $\quad C_2 = 2$

Again $\quad -3 = \frac{3}{2}(1)^2 - 2(1) + C_1$

or $\quad -3 = 1.5 - 2 + C_1$

or $\quad C_1 = -3 + 0.5 = -2.5 = -\frac{5}{2}$

$$y = \frac{x^3}{2} - x^2 - \frac{5}{2}x + 2$$

4.8 (b)

$$\frac{dx}{dt} = -kx^2$$

$\Rightarrow \quad \frac{dx}{x^2} = -kdt$

Integrating both sides, $\int \frac{dx}{x^2} = -\int kdt$

$$-\frac{1}{x} = -kt + C$$

$\Rightarrow \quad \frac{1}{x} = kt + C'$

at $t = 0, x = a$

$$\Rightarrow \quad \frac{1}{a} = k \times 0 + C'$$

$$\Rightarrow \quad C' = \frac{1}{a}$$

$$\therefore \quad \frac{1}{x} = kt + \frac{1}{a}$$

4.9 (a)

Given, $\frac{dy}{dt} + p(t)\,y = q(t)\,y^n\,;\; n > 0$

putting $\quad v = y^{1-n}$

$$\frac{dv}{dt} = (1-n)\,y^{-n}\frac{dy}{dt}$$

$$\frac{dy}{dt} = \frac{1}{(1-n)y^{-n}}\frac{dv}{dt}$$

Substituting in the given differential equation, we get,

$$\frac{1}{(1-n)y^{-n}}\frac{dv}{dt} + p(t)y = q(t)y^n$$

Multiplying by $(1-n)\,y^{-n}$, we get,

$$\frac{dv}{dt} + p(t)\,(1-n)y^{1-n} = q(t)\,(1-n)$$

now since $y^{1-n} = v$, we get

$$\frac{dv}{dt} + (1-n)\,pv = (1-n)q$$

where p is $p(t)$ and q is $q(t)$.

4.10 (a)

$$\frac{d^2y}{dx^2} + 2\frac{dy}{dx} + 17y = 0$$

$$y(0) = 1$$

$$\frac{dy}{dx}\left(\frac{\pi}{4}\right) = 0$$

This is a linear differential equation

$D^2 + 2D + 17 = 0$

$$D = -1 \pm 4i$$

$$\therefore \quad y = C_1 e^{(-1+4i)x} + C_2 e^{(-1-4i)x}$$

$$= e^{-x}\, C_1e^{4xi} + C_2e^{-4xi}$$

$$= e^{-x}$$

$$[C_1(\cos 4x + i\sin 4x)] + C_2[\cos(-4x) + i\sin(-4x)]$$

$$= e^{-x}[(C_1 + C_2)\cos 4x + (C_1 - C_2)\, i\sin 4x)]$$

Let $C_1 + C_2 = C_3$ and $(C_1 - C_2)\,i = C_4$

$$y = e^{-x}(C_3\cos 4x + C_4\sin 4x)$$

since $\quad y(0) = 1$

$$\Rightarrow \quad 1 = e^{-0}(C_3\cos 0 + C_4\sin 0)$$

$$\Rightarrow \quad C_3 = 1$$

$$\frac{dy}{dx} = e^{-x}(-4C_3\sin 4x + 4C_4\cos 4x) - e^{-x}[C_3\cos 4x + C_4\sin 4x]$$

$$= e^{-x}[(-4C_3 - C_4)\sin 4x + (4C_4 - C_3)\cos 4x]$$

$\frac{dy}{dx}$ at $x = \frac{\pi}{4}$ is 0

$$\therefore \quad -(-4C_4 - C_3)\,e^{-\pi/4} = 0$$

$$4C_4 = C_3$$

$$C_4 = \frac{C_3}{4} = \frac{1}{4}$$

$$\therefore \quad C_3 = 1 \text{ and } C_4 = \frac{1}{4}$$

$$y = e^{-x}(\cos 4x + \frac{1}{4}\sin 4x)$$

4.11 (c)

$$\frac{dy}{dx} = 0.25y^2 \quad (y = 1 \text{ at } x = 0)$$

$$h = 1$$

Iterative equation for backward (implicit) **Euler methods** for above equation would be,

$$\frac{y_{k+1} - y_k}{h} = 0.25\;y_{k+1}^2$$

$$\Rightarrow y_{k+1} - y_k = 0.25h\;y_{k+1}^2$$

$$\Rightarrow 0.25h\,y_{k+1}^2 - y_{k+1} + y_k = 0$$

putting $k = 0$ in above equation

$$0.25h\;y_1^2 - y_1 + y_0 = 0$$

since $\quad y_0 = 1$ and $h = 1$

$$0.25\;y_1^2 - y_1 + 1 = 0$$

$$y_1 = 2$$

4.12 (a)

$$\frac{dV}{dt} = -kA \qquad \text{...(i)}$$

where $\quad V = \frac{4}{3}\pi r^3$

$$A = 4\pi r^2$$

$$\frac{dV}{dt} = \frac{4}{3}\pi \times 3r^2 \frac{dr}{dt}$$

Substituting in (i) we get,

$$4\pi r^2 \frac{dr}{dt} = -k(4\pi r^2)$$

$$\frac{dr}{dt} = -k$$

$$\Rightarrow \quad dr = -kdt$$

Integrating we get,

$$r = -kt + C$$

at $t = 0,\ r = 1$

$$\Rightarrow \quad 1 = -k \times 0 + C$$

$$\Rightarrow \quad C = 1$$

$$\therefore \quad r = -kt + 1$$

Now at $t = 3$ months $r = 0.5$ cm

$$\therefore \quad 0.5 = -k \times 3 + 1$$

$$\Rightarrow \quad k = \frac{0.5}{3}$$

$$r = -\frac{0.5}{3}t + 1$$

putting $r = 0$

and solving for t gives $0 = -\frac{0.5}{3}t + 1$

$$\Rightarrow \quad t = 6 \text{ months}$$

4.13 (a)

Given, $x^2\frac{dy}{dx} + 2xy - x + 1 = 0 \quad (x = 1,\ y = 0)$

$$x^2\frac{dy}{dx} + 2xy = x - 1$$

Dividing by x^2,

$$\frac{dy}{dx} + \left(\frac{2}{x}\right)y = \left(\frac{x-1}{x^2}\right)$$

which is a linear first order differential equation of the form,

$$\frac{dy}{dx} + Py = Q$$

$$\text{Integrating factor} = \text{I.F} = e^{\int P dx} = e^{\int \frac{2}{x}dx}$$

$$= e^{2\log x}$$

$$= e^{\log_e x^2} = x^2$$

$$y \times \text{I.F} = \int Q(\text{I.F})dx$$

$$y.x^2 = \int\left(\frac{x-1}{x^2}\right).x^2.dx$$

$$= \int(x-1)dx$$

$$yx^2 = \frac{x^2}{2} - x + C$$

Now at $x = 1,\ y = 0$,

i.e. $0 \times 1^2 = \frac{1^2}{2} - 1 + C$

$$0 = \frac{1}{2} - 1 + C$$

$$C = \frac{1}{2}$$

$$\therefore \quad yx^2 = \frac{x^2}{2} - x + \frac{1}{2}$$

$$y = \frac{1}{2} - \frac{1}{x} + \frac{1}{2x^2}$$

4.14 (b)

Degree of a differential equation is the power of its highest order derivative after the differential equation is made free of radicals and fractions if any, in derivative power.

Hence, here the degree is 1, which is power of $\frac{d^2x}{dt^2}$

4.15 (d)

$$\frac{dy}{dx} = x^2y$$

This is variable separable form,

$$\frac{dy}{y} = x^2dx$$

$$\int\frac{dy}{y} = \int x^2dx$$

$$\Rightarrow \quad \log_e y = \frac{x^3}{3} + C_1$$

$$\Rightarrow \quad y = e^{\frac{x^3}{3} + C_1} = e^{C_1} \times e^{\frac{x^3}{3}}$$

$$y = C \times e^{\frac{x^3}{3}}$$

Now at $x = 0,\ y = 1$

$$1 = C \times e^{\frac{0}{3}}$$

$$\Rightarrow \quad C = 1$$

$\therefore \quad y = e^{\frac{x^3}{3}}$ is the solution

4.16 (b)

$$\frac{d\theta}{dt} = -k(\theta - \theta_0)$$

(Newtons law of cooling)

$$\frac{d\theta}{\theta - \theta_0} = -kdt$$

$$\int \frac{d\theta}{\theta - \theta_0} = \int -kdt$$

$\Rightarrow$ ln $(\theta - \theta_0) = -kt + C_1$

$\Rightarrow$ $\theta - \theta_0 = C.e^{-kt}$

$\theta = \theta_0 + C.e^{-kt}$

Given, $\theta_0 = 25°C$

Now at $t = 0, \theta = 60°$

$60 = 25 + C.e^0$

$\Rightarrow$ $C = 35$

$\therefore$ $\theta = 25 + 35\, e^{-kt}$

at $t = 15$ minutes $\theta = 40°C$

$\therefore$ $40 = 25 + 35e^{(-k\times 15)}$

$\Rightarrow$ $e^{-15k} = \frac{3}{7}$

Now at $t = 30$ minutes,

$\theta = 25 + 35\, e^{-30k}$

$= 25 + 35\, (e^{-15k})^2$

$= 25 + 35 \times \left(\frac{3}{7}\right)^2$ $\left(\text{since } e^{-15k} = \frac{3}{7}\right)$

$= 31.428°C \approx 31.5°C$

4.17 (a)

$$\frac{d^2y}{dx^2} + y = 0$$

$D^2 + 1 = 0$

$D = \pm i = 0 \pm 1i$

$\therefore$ General solution is,

$y = e^{0x}[C_1 \cos(1 \times x) + C_2 \sin(1 \times x)]$

$= C_1 \cos x + C_2 \sin x$

$= P \cos x + Q \sin x$

where P and Q are some constants

4.18 (d)

Put $x_t = x\sqrt{\frac{k_z}{k_x}}$

$$\Rightarrow \frac{\partial x_t}{\partial x} = \sqrt{\frac{k_z}{k_x}}$$

$$\Rightarrow \frac{\partial x}{\partial x_t} = \sqrt{\frac{k_x}{k_z}} \quad \text{...(i)}$$

Now given equation is,

$$k_x \frac{\partial^2 h}{\partial x^2} + k_z \frac{\partial^2 h}{\partial z^2} = 0 \quad \text{...(ii)}$$

$$\frac{\partial^2 h}{\partial x_t^2} = \frac{\partial}{\partial x_t}\left(\frac{\partial h}{\partial x_t}\right) = \frac{\partial}{\partial x_t}\left(\frac{\partial h}{\partial x} \times \frac{\partial x}{\partial x_t}\right)$$

$$= \frac{\partial}{\partial x_t}\left(\frac{\partial h}{\partial x} \times \sqrt{\frac{k_x}{k_z}}\right)$$

[from Eqn. (i)]

$$= \sqrt{\frac{k_x}{k_z}} \times \frac{\partial}{\partial x_t}\left(\frac{\partial h}{\partial x}\right)$$

$$= \sqrt{\frac{k_x}{k_z}} \times \left(\frac{\partial}{\partial x}\left(\frac{\partial h}{\partial x}\right) \times \frac{\partial h}{\partial x}\right)$$

$$= \sqrt{\frac{k_x}{k_z}} \times \left(\frac{\partial^2 h}{\partial x^2} \times \sqrt{\frac{k_x}{k_z}}\right)$$

$$= \frac{k_x}{k_z} \times \frac{\partial^2 h}{\partial x^2}$$

$$\therefore \frac{\partial^2 h}{\partial x^2} = \frac{k_z}{k_x} \times \frac{\partial^2 h}{\partial x_t^2}$$

Now substitute in eq. (ii) we get,

$$k_x \times \frac{k_z}{k_x} \times \frac{\partial^2 h}{\partial x_t^2} + k_z \frac{\partial^2 h}{\partial z^2} = 0$$

$$\Rightarrow k_z \frac{\partial^2 h}{\partial x_t^2} + k_z \frac{\partial^2 h}{\partial z^2} = 0$$

$$\Rightarrow \frac{\partial^2 h}{\partial x_t^2} + \frac{\partial^2 h}{\partial z^2} = 0$$

which is the desired form,

$$\therefore x_t = x\sqrt{\frac{k_z}{k_x}}$$

is the correct transformation

4.19 (d)

$$\frac{dy}{dx} = \frac{-x}{y}$$

$\Rightarrow$ $y\,dy = -x\,dx$

$\Rightarrow$ $\int y\,dy = \int -x\,dx$

$$\frac{y^2}{2} = -\frac{x^2}{2} + C$$

at $x = 1, y = \sqrt{3}$

$$\therefore \frac{(\sqrt{3})^2}{2} = \frac{-1^2}{2} + C$$

$\Rightarrow \quad C = 2$

$\therefore$ Solution is,

$$\frac{y^2}{2} = \frac{-x^2}{2} + 2$$

$$\Rightarrow \quad x^2 + y^2 = 4$$

4.20 (a)

$$3y\frac{dy}{dx} + 2x = 0$$

$$\Rightarrow \quad \frac{dy}{dx} = \frac{-2x}{3y}$$

$$\Rightarrow \quad 3y\,dy = -2x\,dx$$

$$\Rightarrow \quad \int 3y\,dy = \int -2x\,dx$$

$$\Rightarrow \quad \frac{3}{2}y^2 = -2 \times \frac{x^2}{2} + C$$

$$\Rightarrow \quad 3y^2 + 2x^2 = C$$

$$\Rightarrow \quad \frac{x^2}{\left(\frac{1}{2}\right)} + \frac{y^2}{\left(\frac{1}{3}\right)} = C$$

$$\Rightarrow \quad \frac{x^2}{\left(\frac{1}{2}C\right)} + \frac{y^2}{\left(\frac{1}{3}C\right)} = 1$$

Which is the equation of a **family of ellipses**.

4.21 (a)

$$\frac{d^3y}{dx^3} + 4\sqrt{\left(\frac{dy}{dx}\right)^3 + y^2} = 0$$

Removing radicals we get,

$$\left(\frac{d^3y}{dx^3}\right)^2 = 16\left[\left(\frac{dy}{dx}\right)^3 + y^2\right]$$

$\therefore$ The order is 3 since highest differential is $\frac{d^3y}{dx^3}$.

The degree is 2 since power of highest differential is 2.

4.22 (c)

$$\frac{d^2y}{dx^2} + \frac{dy}{dx} - 6y = 0$$

$$D^2 + D - 6 = 0$$

$$(D+3)(D-2) = 0$$

$$D = -3 \text{ or } D = 2$$

$\therefore$ Solution is $\quad y = c_1e^{-3x} + c_2e^{2x}$

4.23 (d)

Length of curve $f(x)$ between $x = a$ and $x = b$ is given by,

$$\int_a^b \sqrt{1 + \left(\frac{dy}{dx}\right)^2}\,dx$$

$$y = 4h\frac{x^2}{L^2} \quad \text{...(i)}$$

$$\frac{dy}{dx} = 8h\frac{x}{L^2}$$

since, $\quad y = 0$ at $x = 0$

and $\quad y = h$ at $x = \frac{L}{2}$

(As can be seen from equation (i), by substituting $x = 0$ and $x = L/2$)

$$\therefore \quad \frac{1}{2}(\text{Length of cable}) = \int_0^{L/2} \sqrt{1 + \left(\frac{8hx}{L^2}\right)^2}\,dx$$

$$\text{Length of cable} = 2\int_0^{L/2} \sqrt{1 + 64\frac{h^2x^2}{L^4}}\,dx$$

4.24 (d)

$$\frac{dy}{dx} + \frac{y}{x} = x, \quad y(1) = 1$$

This is a linear differential equation $\frac{dy}{dx} + Py = Q$

with $P = \frac{1}{x}$ and $Q = x$

IF = Integrations factor

$$= e^{\int P dx} = e^{\int \frac{1}{x}dx} = e^{\ln x} = x$$

Solution is,

$$y.(\text{IF}) = \int Q(IF)\,dx + C$$

$$\Rightarrow \quad y.x = \int (x.x)\,dx + C$$

$$\Rightarrow \quad yx = \int x^2 dx + C$$

$$\Rightarrow \quad yx = \frac{x^3}{3} + C$$

$$\Rightarrow \quad y = \frac{x^2}{3} + \frac{C}{x}$$

Now, $y(1) = 1$

$$\Rightarrow \frac{1^2}{3}+\frac{C}{1}=1 \Rightarrow C=\frac{2}{3}$$

So the solution is $y=\frac{x^2}{3}+\frac{2}{3x}$

4.25 (d)

Given, $\frac{dy}{dx}+2y = 0$ and $y(1) = 5$

$$\frac{dy}{dx} = -2y$$

$$\int\frac{dy}{y} = \int -2dx$$

$$\Rightarrow \quad ln\, y = -2x + c$$

$$\Rightarrow \quad y = e^{-2x}\cdot e^{c} = c_1e^{-2x}$$

$$y(1) = c_1e^{-2} = 5 \Rightarrow c_1 = \frac{5}{e^{-2}}$$

So, $y = \frac{5}{e^{-2}}e^{-2x} = 5e^2e^{-2x}$

$= 36.95\, e^{-2x}$

4.27 (c)

$$x(ydx+xdy)\cos\frac{y}{x} = y(xdy-ydx)\sin\frac{y}{x}$$

$$\frac{ydx+xdy}{xdy-ydx} = \frac{y}{x}\tan\frac{y}{x}$$

Let $y = v\cdot x$

$dy = vdx + xdv$

$$\frac{vxdx+vxdx+x^2dv}{vxdx+x^2dv-vxdx} = v\tan v$$

$$\frac{xdv+2vdx}{xdv} = v\tan v$$

$$1+\frac{2v}{x}\frac{dx}{dv} = v\tan v$$

$$\frac{2v}{x}\frac{dx}{dv} = v\tan v - 1$$

$$2\frac{dx}{x} = \left(\tan v-\frac{1}{v}\right)dv$$

Integrating both sides.

$2\log x = \log|\sec v| - \log v + \log c$

$$\Rightarrow \quad x^2 = \frac{c\sec v}{v}$$

$$\Rightarrow \quad x^2\frac{y}{x} = c\sec\frac{y}{x}$$

$$\Rightarrow \quad xy\cos\frac{y}{x} = c$$

4.28 Sol.

$$\frac{d^2y}{dx^2} = -12x^2 + 24x - 20$$

Integrating both sides wrt. x,

$$\frac{dy}{dx} = -4x^3 + 12x^2 - 20x + c_1$$

Integrating both sides wrt. x

$y = -x^4 + 4x^3 - 10x^2 + c_1x + c_2$ ...(i)

At $x = 0$, $y = 5$

$\Rightarrow \quad 5 = c_2$

At $x = 2$, $y = 21$

$\Rightarrow \quad 21 = -16 + 32 - 40 + 2c_1 + c_2$

$2c_1 = 21 + 16 - 2 + 40 - 5$

$2c_1 = 40$

$c_1 = 20$

$\Rightarrow \quad y = -x^4 + 4x^3 - 10x^2 + 20x + 5$

Put $x = 1$

$\Rightarrow \quad y = -1 + 4 - 10 + 20 + 5$

$y = 18$

4.29 (a)

D.E. is $(D^4 + 3D^2)\, y = 108x^2$, $D = \frac{d}{dx}$

A.E. $m^4 + 3m^2 = 0$

$\Rightarrow \quad m^2(m^2 + 3) = 0$

$\Rightarrow \quad m = 0, 0, \pm\pm\sqrt{3}i$

$\therefore$ CF $=(C_1+C_2x)+C_3\sin(\sqrt{3}x)+C_4\cos(\sqrt{3}x)$

and

$$PI = \frac{1}{D^4+3D^2}(108x^2)$$

$$= \frac{1}{3D^2\left[1+\frac{D^2}{3}\right]}(108x^2) = \frac{36}{D^2}\left[1+\frac{D^2}{3}\right]^{-1}(x^2)$$

$$= \frac{36}{D^2}\left[1-\frac{D^2}{3}+\ldots\right](x^2) = \frac{36}{D^2}\left[x^2-\frac{1}{3}(2)+0\right]$$

$$= \iint\left(36x^2-\frac{2}{3}\right)dx\,dx$$

$$= 36\left(\frac{x^4}{(4)(3)}-\frac{2}{3}\frac{x^2}{(2)(1)}\right) = 3x^4-12x^2$$

4.30 (d)

$$\frac{dQ}{dt} + Q = 1$$

Comparing with standard form,

$$I.F. = e^{\int 1 dt} = e^t$$

Solution is

$$Q.e^t = \int 1 \cdot e^t dt = e^t + C$$

$$Q = 1 + Ce^{-t} \quad \text{...(i)}$$

When $t = 0$, $Q = 0$

$\Rightarrow \quad 0 = 1 + C$

$\Rightarrow \quad C = -1$

Therefore, $Q(t) = 1 - e^{-t}$

4.31 (a)

$$y'' - 4y' + 3y = 2t - 3t^2$$

$$(D^2 - 4D + 3)y = 2t - 3t^2$$

$$P.I. = \frac{1}{D^2 - 4D + 3}(2t - 3t^2)$$

$$= \frac{1}{(1-D)(3-D)}(2t - 3t^2)$$

$$= \left(\frac{1/2}{1-D} - \frac{1/2}{3-D}\right)(2t - 3t^2)$$

$$= \frac{1}{2}(1-D)^{-1}(2t - 3t^2) - \frac{1}{6}\left(1 - \frac{D}{3}\right)^{-1}(2t - 3t^2)$$

$$= \frac{1}{2}(1 + D + D^2)(2t - 3t^2)\frac{1}{6}\left(1 + \frac{D}{3} + \frac{D^2}{9}\right)(2t - 3t^2)$$

$$= \frac{1}{2}(2t - 3t^2 + D(2t - 3t^2) + D^2(2t - 3t^2))$$

$$-\frac{1}{6}(2t - 3t^2) + \frac{D(2t - 3t)^2}{3} + \frac{D^2(2t - 3t^2)}{9}$$

$$= \frac{1}{2}\left[2t - 3t^2 + 2 - 6t - 6\right] - \frac{1}{6}\left[2t - 3t^2 + \frac{2 - 6t}{3} + \frac{-6}{9}\right]$$

$$= \frac{1}{2}(-4 - 4t - 3t^2) - \frac{1}{6}(-3t^2)$$

$$= -2 - 2t - t^2$$

4.32 Sol.

$(D^2 + 2D + 1)y = 0 \quad (\therefore \text{ Roots are } -1, -1)$

$$\text{CF} = (C_1 + C_2 x)\, e^{-x}$$

$$y = C_1 e^{-x} + C_2 x\, e^{-x} \quad \text{...(i)}$$

$y(0) = 1 \qquad 1 = C_1 \quad \text{...(ii)}$

$$y' = C_1 e^{-x} + C_2 (e^{-x} - xe^{-x})$$

$y'(0) = -1, \qquad -1 = -C_1 + C_2 \quad \text{...(iii)}$

From eq. (ii) and (iii),

$$C_1 = 1,\ C_2 = 0$$

$\therefore \qquad y = e^{-x}$

At $x = 1$, $y = e^{-1} = \frac{1}{e} = 0.368$

4.33 (c)

$$x\frac{dy}{dx} + y = 0$$

$$x\frac{dy}{dx} = -y$$

$$\frac{dy}{y} = -\frac{dx}{x}$$

$$\int \frac{1}{y} dy = \int \frac{-1}{x} dx$$

$$\ln y = -\ln x + c$$

when $\quad y = 1, x = 1$

$$c = 0$$

$\Rightarrow \qquad y = \frac{1}{x} = x^{-1}$

4.34 Sol.

$$[xD^2 - 2xD + 2]y = 0$$

Put, $\quad x = e^z$

$$xD = D_1$$

$$x^2D^2 = D_1(D_1 - 1)$$

$$[D_1(D_1 - 1) - 2D_1 + 2]y = 0$$

$$[D_1^2 - 3D_1 + 2]y = 0$$

AE: $m^2 - 3m + 2 = 0$

$m = 1, 2$

$$y = C_1 e^z + C_2 e^{2z}$$

$$e^z = x$$

$$y = C_1 x + C_2 x^2$$

$y(1) = 0 \qquad y(2) = 2$

$$y(1) = C_1 + C_2 = 0$$

$$y(2) = 2C_1 + 4C_2 = 2$$

$C_1 = -1,\ C_2 = 1$

$$y = x^2 - x$$

$$y(3) = 6$$

4.35 (c)

$$\frac{dy}{dx}(x \ln x) = y$$

$$\frac{dy}{y} = \frac{dx}{x \ln x}$$

$$\int \frac{dy}{y} = \int \frac{1}{x \ln x} dx + \ln k$$

$$\ln x = t$$

$$\frac{1}{x} dx = dt$$

$$\ln y = \int \frac{dy}{t} + \ln k$$

$$\ln y = \ln t + \ln k$$

$$\ln y = kt$$

$$y = k \ln x$$

4.36 (b)

$\because$ 1st term of given *D*. Equation contains product of dependent variable with it's derivative, so it is non-linear and also we have 2nd order derivative so it's order is two i.e., 2nd order non linear equation.

4.37 (d)

Its solution is of the type $u = f(x)$, i.e., dependent variable is u.

Hence, given equation is Linear & Non-Homogeneous.

4.38 (a)

A.E. is $m^2 - 5m + 6 = 0$

$\Rightarrow$ m = 2, 3 so $C_f = C_1e^{2t} + C_2e^{3t}$.

PI = 0 and G. Solution is x

$$= CF + PI = C_1e^{2t} + C_2e^{3t}$$

and $$\frac{dx}{dt} = 2C_1e^{2t} + 3C_2e^{3t}$$

Now, using initial conditions we get

$$C_1 = -10,\ C_2 = 10.$$

$$x = -10e^{2t} + 10e^{3t}$$

4.39 (a)

Given: DE is $\frac{\partial u}{\partial y} = \frac{\partial^2 u}{\partial x^2}$; $y \geq 0$; $x_1 \leq x \leq x_2$

$\because$ y is given as ≥ 0 so we take it as time. Hence, above equation is nothing but one-*D* heat equation which requires one initial condition and two boundary condition.

4.40 (d)

$$\frac{6d^2y}{dx^2} + \frac{dy}{dx} - y = 0$$

$$(6D^2 + D - 1)y = 0$$

$$6D^2 + 3D - 2D - 1 = 0$$

$$3D(2D + 1) - 1(2D + 1) = 0$$

$$(2D + 1)(3D - 1) = 0$$

$$D = \frac{-1}{2},\ D = \frac{1}{3}$$

$$y = C_1e^{x/3} + C_2e^{-x/2}$$

4.41 (a)

Given $(D^2 + 2D + 1)y = 0$

$$y(0) = 1$$

$$y(1) = 3$$

Auxiliary equation is $m^2 + 2m + 1 = 0$

$\Rightarrow$ $m = -1, -1$

$$CF = (C_1 + C_2x)e^{-x}$$

And PI = 0

Complete solution is

$$y = CF + PI$$

$$y = (C_1 + C_2x)e^{-x} \quad \text{...(i)}$$

Using y(0) = 1, $C_1 = 1$

Using y(1) = 3, $C_2 = 3e - 1$

Hence by eq. (i), $y = [1 + (3e - 1)x]\ e^{-x}$

$$y = e^{-x} + (3e - 1)\ xe^{-x}$$

4.42 (d)

$$\frac{dy}{dx} - \frac{y}{x} = 1$$

$$\frac{dy}{dx} + Py = Q$$

$$P = -\frac{1}{x},\ Q = 1$$

$$IF = e^{\int P dx} = e^{\int \frac{-1}{x} dx} = \frac{1}{x}$$

$$y(IF) = \int Q(IF) dx + c$$

$$y\left(\frac{1}{x}\right) = \int 1 \cdot \frac{1}{x} dx + c$$

$$y\left(\frac{1}{x}\right) = \ln x + \ln k$$

$$y = x \ln (kx)$$

4.43 Sol.

Given, $\frac{dy}{dx} + \alpha xy = 0$

Using variable separable method.

$$\frac{dy}{dx} = \alpha x y$$

$$\Rightarrow \int \frac{dy}{y} = -\alpha \int x dx + \ln C$$

$$\Rightarrow \ln y = -\alpha \frac{x^2}{2} + \ln C$$

$$\Rightarrow \ln\left(\frac{y}{C}\right) = -\frac{\alpha}{2}x^2$$

$$\Rightarrow \quad y = Ce^{\frac{-\alpha}{2}x^2}$$

By putting $y(0) = 1$

we get $C = 1$

By putting $y(1) = 0.8$

$$e^{\frac{-\alpha}{2}} = 0.8$$

$$\Rightarrow \ln e^{-\frac{\alpha}{2}} = \ln 0.8$$

$$\Rightarrow -\frac{\alpha}{2} = \ln(0.8)$$

$$\Rightarrow \quad \alpha = -2 \ln 0.8 = 0.446$$

4.44 Sol.

Given differential equation

$$\frac{du}{dt} + 2tu^2 = 1$$

$$\frac{du}{dt} = f(t, u) = 1 - 2tu^2$$

By Backward Euler Iterative scheme,

$$u_n = u_{n-1} + hf(t_n, u_n)$$

$$u_n = u_{n-1} + h[1 - 2t_n u_n^2] \quad ...(i)$$

Given, $h = 0.01$, $u_{n-1} = 1.75$, $t_{n-1} = 3.14$

So, $t_n = t_{n-1} + h = 3.15$

Putting these values in (i)

$$u_n = 1.75 + 0.01[1 - 2 \times 3.15 \times u_n^2]$$

$$u_n = 1.75 + 0.01 - 0.628u_n^2$$

or $0.628u_n^2 + u_n - 1.75 = 0$

$62.8u_n^2 + 100u_n + 175 = 0$

$\Rightarrow \quad u_n = 1.599 - 17.472$

So, for $u_n = 1.599$ and $u_{n-1} = 1.75$

We have, $u_n - u_{n-1} = 1.599 - 1.75 = -0.151$

4.45 (a)

Auxillary equation is,

$m^3 - 5.5m^2 + 9.5m - 5 = 0$

By solving above equation, we get

$m = 2.5, 1, 2$

So, m_1 and m_2 are 1 and 2.

4.46 (c)

Given: $\dfrac{\partial^2 u}{\partial x^2} = 2$

Integrating with respect to 'x'

$$\int\left(\frac{\partial^2 u}{\partial x^2}\right)dx = 2\int(1)dx + f(y)$$

$$\Rightarrow \quad \frac{\partial u}{\partial x} = 2x + f(y)$$

Again integrating with respect to 'x'

$$u = x^2 + x f(y) + h(y)$$

4.47 (c)

P : $\dfrac{dy}{dx} = \dfrac{x^4 + 3x^2 + 2y^4}{x^3 y}$ is homogeneous DE.

Q : $x^2 dy + y^2 dx = 0$

$$\frac{\partial m}{\partial y} = 2y \neq \frac{\partial M}{\partial x} = 2x$$

$\Rightarrow Q$ is not an exact DE.

■■■■

5 Complex Functions

5.1 e^x is periodic, with a period of

(a) 2π (b) $2i\pi$

(c) π (d) $i\pi$

[1997 : 1 M]

5.2 Which one of the following is NOT true for complex number Z_1 and Z_2?

(a) $\dfrac{Z_1}{Z_2} = \dfrac{Z_1\overline{Z}_2}{|Z_2|^2}$

(b) $|Z_1 + Z_2| \le |Z_1| + |Z_2|$

(c) $|Z_1 - Z_2| \le |Z_1| - |Z_2|$

(d) $|Z_1 + Z_2|^2 + |Z_1 - Z_2|^2 = 2|Z_1|^2 + 2|Z_2|^2$

[2005 : 1 M]

5.3 Consider likely applicability of Cauchy's Integral Theorem to evaluate the following integral counter clockwise around the unit circle c.

$$I = \oint_c \sec z dz,$$

z being a complex variable. The value of I will be

(a) $I = 0$: singularities set $= \phi$

(b) $I = 0$: singularities set

$= \left\{\pm \dfrac{2n+1}{2}\pi;\ n = 0, 1, 2 \ldots\ldots\ldots\right\}$

(c) $I = \pi/2$: singularities set

$= \{\pm n\pi;\ n = 0, 1, 2 \ldots\ldots\ldots\}$

(d) None of above

[2005 : 2 M]

5.4 Using Cauchy's integral theorem, the value of the integral (integration being taken in counterclockwise direction) $\oint_c \dfrac{z^3 - 6}{3z - i} dz$ is

(a) $\dfrac{2\pi}{81} - 4\pi i$ (b) $\dfrac{\pi}{8} - 6\pi i$

(c) $\dfrac{4\pi}{81} - 6\pi i$ (d) 1

[2006 : 2 M]

5.5 The analytic function $f(z) = \dfrac{z-1}{z^2+1}$ has singularities at

(a) 1 and – 1 (b) 1 and i

(c) 1 and – i (d) i and – i

[2009 : 1 M]

5.6 The value of the integral $\int_C \dfrac{\cos(2\pi z)}{(2z-1)(z-3)} dz$ (where C is a closed curve given by $|z| = 1$) is

(a) $-\pi i$ (b) $\dfrac{\pi i}{5}$

(c) $\dfrac{2\pi i}{5}$ (d) πi

[2009 : 2 M]

5.7 For an analytic function, $f(x + iy) = u(x, y) + iv(x, y)$, u is given by $u = 3x^2 - 3y^2$. The expression for v, considering K to be a constant is

(a) $3y^2 - 3x^2 + K$ (b) $6x - 6y + K$

(c) $6y - 6x + K$ (d) $6xy + K$

[2011 : 2 M]

5.8 $z = \dfrac{2-3i}{-5+i}$ can be expressed as

(a) $-0.5 - 0.5i$ (b) $-0.5 + 0.5i$

(c) $0.5 - 0.5i$ (d) $0.5 + 0.5i$

[2014 : 1 M, Set-II]

5.9 Consider the following complex function

$$f(z) = \frac{9}{(z-1)(z+2)^2}$$

Which of the following is one of the residues of the above function?

(a) –1 (b) $\dfrac{9}{16}$

(c) 2 (d) 9

[2015 : 2 M, Set-I]

5.10 Given, $i = \sqrt{-1}$, the value of the definite integral,

$$I = \int_0^{\pi/2} \frac{\cos x + i\sin x}{\cos x - i\sin x}dx \text{ is}$$

(a) 1 (b) −1
(c) i (d) $-i$

[2015 : 1 M, Set-II]

Answers Complex Functions

5.1 (b) 5.2 (c) 5.3 (a) 5.4 (a) 5.5 (d) 5.6 (c) 5.7 (d) 5.8 (b) 5.9 (a)
5.10 (c)

Explanations Complex Functions

5.1 (b)

For a function to be periodic,

$$f(T+x) = f(T)$$

$$e^{x+2\pi i} = e^x e^{2\pi i}$$

$$= e^x[\cos 2\pi i + i\sin 2\pi i]$$

$$= e^x[1+0] = e^x$$

$\therefore$ Period of $e^x = 2\pi i$

5.2 (c)

(a) is true since

$$\frac{Z_1}{Z_2} = \frac{Z_1\,\overline{Z}_2}{Z_2\,\overline{Z}_2} = \frac{Z_1\overline{Z}_2}{|Z_2|^2}$$

(b) is true by triangle inequality of complex number.

(c) is not true since $|Z_1 - Z_2| \geq |Z_1| - |Z_2|$

(d) is true since

$$|Z_1 + Z_2|^2 = (Z_1+Z_2)\,(\overline{Z_1+Z_2})$$

$$= (Z_1+Z_2)\left(\overline{Z}_1+\overline{Z}_2\right)$$

$$= Z_1\overline{Z}_1 + Z_2\overline{Z}_2 + Z_2\overline{Z}_1 + Z_1\overline{Z}_2 \quad \text{...(i)}$$

& $|Z_1 - Z_2|^2 = (Z_1+Z_2)\,(\overline{Z_1-Z_2})$

$$= (Z_1-Z_2)\,(\overline{Z}_1-\overline{Z}_2)$$

$$= Z_1\overline{Z}_1 + Z_2\overline{Z}_2 - Z_2\overline{Z}_1 - Z_1\overline{Z}_2 \quad \text{...(ii)}$$

Adding (i) and (ii) we get,

$$|Z_1+Z_2|^2 + |Z_1-Z_2|^2 = 2Z_1\overline{Z}_1 + 2Z_2\overline{Z}_2$$

$$= 2|Z_1|^2 + 2|Z_2|^2$$

5.3 (a)

$$\int \sec z\,dz = \int \frac{1}{\cos z}dz$$

The poles are at

$$z_0 = (n+1/2)p$$

$$= \ldots -3\pi/2, -\pi/2, \pi/2, +3\pi/2\ldots..$$

None of these poles lie inside the unit circle $|z| = 1$

Hence, sum of residues at poles = 0

$\therefore$ singularities set = ϕ and

$I = 2\pi i$ [sum of residues of $f(z)$ at the poles]

$= 2\pi i \times 0 = 0$

5.4 (a)

Cauchy's integral theorem is,

$$f(a) = \frac{1}{2\pi i}\oint_c \frac{f(z)}{z-a}dz$$

i.e. $$\oint_c \frac{f(z)}{z-a}dz = 2\pi i\, f(a)$$

Now, $$\oint_c \frac{z^3-6}{3z-i}dz = \frac{1}{3}\oint_c \frac{z^3-6}{\left(z-\frac{i}{3}\right)}$$

Applying **Cauchy's integral theorem,**

$$= \frac{1}{3}\left(2\pi i\, f\left(\frac{i}{3}\right)\right) = \frac{1}{3}\left(2\pi i\left[\left(\frac{i}{3}\right)^3 - 6\right]\right)$$

$$= \frac{1}{3}\left[2\pi i\left(\frac{i^3}{27} - 6\right)\right] = \frac{2\pi}{81}i^4 - 4\pi i = \frac{2\pi}{81} - 4\pi i$$

5.5 (d)

$$f(z) = \frac{z-1}{z^2+1} = \frac{z-1}{z^2-i^2} = \frac{z-1}{(z-i)(z+i)}$$

$\therefore$ The singularities arc at $z = i$ and $-i$

5.6 (c)

$$\frac{1}{2}\oint_c \frac{\cos 2\pi z}{\left(z-\frac{1}{2}\right)(z-3)}dz$$

$$= \frac{1}{2}\oint \frac{\frac{\cos 2\pi z}{z-3}}{z-\frac{1}{2}} dz = \frac{1}{2}\times 2\pi i\left[\frac{\cos 2\pi/2}{z-3}\right]_{z=\frac{1}{2}}$$

$$= \pi i\left(\frac{\cos\pi}{\frac{1}{2}-3}\right) = \pi i\left(\frac{-1}{-\frac{5}{2}}\right) = \frac{2\pi i}{5}$$

5.7 (d)

$$f = u + iv$$
$$u = 3x^2 - 3y^2$$

For f to be analysis, we have **Cauchy-Riemann conditions,**

$$\frac{\partial u}{\partial x} = \frac{\partial v}{\partial y} \quad ...(i)$$

$$\frac{\partial u}{\partial y} = \frac{-\partial v}{\partial x} \quad ...(ii)$$

From (i) we have,

$$6x = \frac{\partial v}{\partial y}$$

$$\Rightarrow \int \partial v = \int 6x\partial y$$

$$v = 6\frac{x^2}{2} + f(x)$$

i.e. $v = 3x^2 + f(x)$ ...(iii)

Now applying equation (ii) we get,

$$\frac{\partial u}{\partial y} = \frac{-\partial v}{\partial x}$$

$$\Rightarrow -6y = -\left[6x + \frac{df}{dx}\right]$$

$$\Rightarrow 6x + \frac{df}{dx} = 6y$$

$$\frac{df}{dx} = 6y - 6x$$

By integrating,

$$f(x) = 6yx - 3x^2 + K$$

Substitute in equation (iii),

$$v = 3x^2 + 6yx - 3x^2 + K$$

$$\Rightarrow v = 6yx + K$$

5.8 (b)

$$\frac{(2-3i)}{(-5+i)} = \frac{(2-3i)}{(-5+i)}\times\frac{(-5-i)}{(-5-i)}$$

$$= \frac{-10-2i+15i-3}{25+1} = \frac{-13+13i}{26}$$

$$= -0.5 + 0.5i$$

5.9 (a)

$f(z)$ has poles at $z = 1, -2$

Residue of $f(z)$ at $(z = 1)$

$$= \lim_{z\to 1}(z-1)f(z) = \lim_{z\to 1}\frac{9}{(z+2)^2} = 1$$

Residue of $f(z)$ at $(z = -2)$

$$= \lim_{z\to -2}\frac{d}{dz^{2-1}}\left[(z+2)^2 f(z)\right]$$

$$= \lim_{z\to -2}\frac{d}{dz}\left(\frac{9}{z-1}\right) = \lim_{z\to -2}\frac{-9}{(z-1)^2} = -1$$

5.10 (c)

$$I = \int_0^{\pi/2}\frac{\cos x + i\sin x}{\cos x - i\sin x}dx$$

$$= \int_0^{\pi/2}\frac{e^{ix}}{e^{-ix}}dx = e^{2ix}dx$$

$$= \frac{1}{2i}\left[e^{2ix}\right]_0^{\pi/2} = \frac{1}{2i}\left[e^{i\pi}-1\right] = \frac{1}{2i}(-1-1)$$

$$= -\frac{1}{i} = i$$

■■■■

6 Probability and Statistics

6.1 A box contains 10 screws, 3 of which are defective. Two screws are drawn at random with replacement. The probability that none of the two screws is defective will be

(a) 100% (b) 50%
(c) 49% (d) None of these

[2003 : 1 M]

6.2 A hydraulic structure has four gates which operate independently. The probability of failure of each gate is 0.2. Given that gate 1 has failed, the probability that both gates 2 and 3 will fail is

(a) 0.240 (b) 0.200
(c) 0.040 (d) 0.008 [2004 : 2 M]

6.3 Which one of the following statements is NOT true?

(a) The measure of skewness is dependent upon the amount of dispersion
(b) In a symmetric distribution, the values of mean, mode and median are the same
(c) In a positively skewed distribution: mean > median > mode
(d) In a negatively skewed distribution: mode > mean > median

[2005 : 1 M]

6.4 A class of first year B. Tech. students is composed of four batches *A*, *B*, *C* and *D*, each consisting of 30 students. It is found that the sessional marks of students in Engineering Drawing in batch *C* have a mean of 6.6 and standard deviation of 2.3. The mean and standard deviation of the marks for the entire class are 5.5 and 4.2, respectively. It is decided by the course instructor to normalize the marks of the students of all batches to have the same mean and standard deviation as that of the entire class. Due to this, the marks of a student in batch *C* are changed from 8.5 to

(a) 6.0 (b) 7.0
(c) 8.0 (d) 9.0 [2006 : 2 M]

6.5 There are 25 calculators in a box. Two of them are defective. Suppose 5 calculators are randomly picked for inspection (i.e., each has the same chance of being selected), what is the probability that only one of the defective calculators will be included in the inspection?

(a) $\frac{1}{2}$ (b) $\frac{1}{3}$
(c) $\frac{1}{4}$ (d) $\frac{1}{5}$ [2006 : 2 M]

6.6 A person on a trip has a choice between private car and public transport. The probability of using a private car is 0.45. While using the public transport, further choices available are bus and metro, out of which the probability of commuting by a bus is 0.55. In such a situation, the probability (rounded up to two decimals) of using a car, bus and metro, respectively would be

(a) 0.45, 0.30 and 0.25
(b) 0.45, 0.25 and 0.30
(c) 0.45, 0.55 and 0.00
(d) 0.45, 0.35 and 0.20 [2008 : 2 M]

6.7 The standard normal probability function can be approximated as

$$F(x_N) = \frac{1}{1+\exp(-1.7255\, x_n \,|\, x_n \,|^{0.12})}$$

where x_N = standard normal deviate. If mean and standard deviation of annual precipitation are 102 cm and 27 cm respectively, the probability that the annual precipitation will be between 90 cm and 102 cm is

(a) 66.7% (b) 50.0%
(c) 33.3% (d) 16.7% [2009 : 2 M]

6.8 Two coins are simultaneously tossed. The probability of two heads simultaneously appearing is

(a) 1/8 (b) 1/6
(c) 1/4 (d) 1/2 [2010 : 1 M]

6.9 There are two containers, with one containing 4 red and 3 green balls and the other containing 3 blue and 4 green balls. One ball is drawn at random from each container. The probability that one of the balls is red and the other is blue will be

(a) 1/7 (b) 9/49
(c) 12/49 (d) 3/7 [2011 : 1 M]

6.10 The annual precipitation data of a city is normally distributed with mean and standard deviation as 1000 mm and 200 mm, respectively. The probability that the annual precipitation will be more than 1200 mm is

(a) < 50%
(b) 50%
(c) 75%
(d) 100% [2011 : 2 M]

6.11 In an experiment, positive and negative values are equally likely to occur. The probability of obtaining at most one negative value in five trials is

(a) 1/32
(b) 2/32
(c) 3/32
(d) 6/32 [2011 : 2 M]

6.12 Find the value of λ such that function $f(x)$ is valid probability density function

$f(x) = \lambda(x-1)(2-x)$ for $1 \le x \le 2$
$= 0$ otherwise

[2013 : 2 M]

6.13 The probability density function of evaporation E on any day during a year in a watershed is given by

$$f(E) = \begin{cases} \frac{1}{5} & 0 \le E \le 5 \text{ mm/day} \\ 0 & \text{otherwise} \end{cases}$$

The probability that E lies in between 2 and 4 mm/day in a day in the watershed is (in decimal)______.

[2014 : 1 M, Set-I]

6.14 A fair (unbiased) coin was tossed four times in succession and resulted in the following outcomes: (i) Head, (ii) Head, (iii) Head, (iv) Head. The probability of obtaining a 'Tail' when the coin is tossed again is

(a) 0
(b) $\frac{1}{2}$
(c) $\frac{4}{5}$
(d) $\frac{1}{5}$

[2014 : 1 M, Set-II]

6.15 If $\{x\}$ is a continuous, real valued random variable defined over the interval $(-\infty, +\infty)$ and its occurrence is defined by the density function given as:

$f(x) = \frac{1}{\sqrt{2\pi} * b} e^{-\frac{1}{2}\left(\frac{x-a}{b}\right)^2}$ where 'a' and 'b' are the statistical attributes of the random variable $\{x\}$. The value of the integral

$$\int_{-\infty}^{a} \frac{1}{\sqrt{2\pi} * b} e^{-\frac{1}{2}\left(\frac{x-a}{b}\right)^2} dx$$

(a) 1
(b) 0.5
(c) π
(d) $\pi/2$

[2014 : 1 M, Set-II]

6.16 A traffic office imposes on an average 5 number of penalties daily on traffic violators. Assume that the number of penalties on different days is independent and follows a Poisson distribution. The probability that there will be less than 4 penalties in a day is __________.

[2014 : 2 M, Set-I]

6.17 Consider the following probability mass function (p.m.f.) of a random variable X.

$$p(X,q) = \begin{cases} q & \text{If } X = 0 \\ 1-q & \text{If } X = 1 \\ 0 & \text{otherwise} \end{cases}$$

If q = 0.4, the variance of X is __________.

[2015 : 1 M, Set-I]

6.18 The probability density function of a random variable, x is

$$f(x) = \frac{x}{4}(4-x^2) \text{ for } 0 \le x \le 2 = 0$$

The mean, μ_x of the random variable is ________.

[2015 : 2 M, Set-II]

6.19 Type II error in hypothesis testing is

(a) acceptance of the null hypothesis when it is false and should be rejected
(b) rejection of the null hypothesis when it is true and should be accepted
(c) rejection of the null hypothesis when it is false and should be rejected
(d) acceptance of the null hypothesis when it is true and should be accepted

[2016 : 1 M, Set-I]

6.20 Probability density function of a random variable X is given below

$$f(x) = \begin{bmatrix} 0.25 & \text{if } 1 \le x \ge 5 \\ 0 & \text{otherwise} \end{bmatrix}$$

$P(X \le 4)$

(a) $\frac{3}{4}$
(b) $\frac{1}{2}$
(c) $\frac{1}{4}$
(d) $\frac{1}{8}$

[2016 : 2 M, Set-I]

6.21 The spot speeds (expressed in km/hr) observed at a road section are 66, 62, 45, 79, 32, 51, 56, 60, 53, and 49. The median speed (expressed in km/hr) is ______.
(**Note:** answer with one decimal accuracy)
[2016 : 1 M, Set-II]

6.22 X and Y are two random independent events. It is known that $P(X) = 0.40$ and $P(X \cup Y^C) = 0.7$. Which one of the following is the value of $P(X \cup Y)$?
(a) 0.7 (b) 0.5
(c) 0.4 (d) 0.3
[2016 : 1 M, Set-II]

6.23 If $f(x)$ and $g(x)$ are two probability density functions,

$$f(x) = \begin{cases} \frac{x}{a}+1 & : -a \le x < 0 \\ -\frac{x}{a}+1 & : 0 \le x \le a \\ 0 & : \text{otherwise} \end{cases}$$

$$g(x) = \begin{cases} -\frac{x}{a} & : -a \le x < 0 \\ \frac{x}{a} & : 0 \le x \le a \\ 0 & : \text{otherwise} \end{cases}$$

Which one of the following statements is true?
(a) Mean of $f(x)$ and $g(x)$ are same; Variance of $f(x)$ and $g(x)$ are same
(b) Mean of $f(x)$ and $g(x)$ are same; Variance of $f(x)$ and $g(x)$ are different
(c) Mean of $f(x)$ and $g(x)$ are different; Variance of $f(x)$ and $g(x)$ are same
(d) Mean of $f(x)$ and $g(x)$ are different; Variance of $f(x)$ and $g(x)$ are different
[2016 : 2 M, Set-II]

6.24 A two-faced coin has its faces designated as head (H) and tail (T). This coin is tossed three times in succession to record the following outcomes; H, H, H. If the coin is tossed one more time, the probability (up to one decimal place) of obtaining H again, given the previous realizations of H, H and H, would be ________.
[2017 : 1 M, Set-II]

6.25 The number of parameters in the univariate exponential and Gaussian distributions, respectively, are
(a) 2 and 2 (b) 1 and 2
(c) 2 and 1 (d) 1 and 1
[2017 : 1 M, Set-I]

6.26 For the function $f(x) = a + bx$, $0 \le x \le 1$, to be a valid probability density function, which one of the following statements is correct?
(a) $a = 1, b = 4$ (b) $a = 0.5, b = 1$
(c) $a = 0, b = 1$ (d) $a = 1, b = -1$
[2017 : 2 M, Set-I]

6.27 A probability distribution with right skew is shown in the figure.

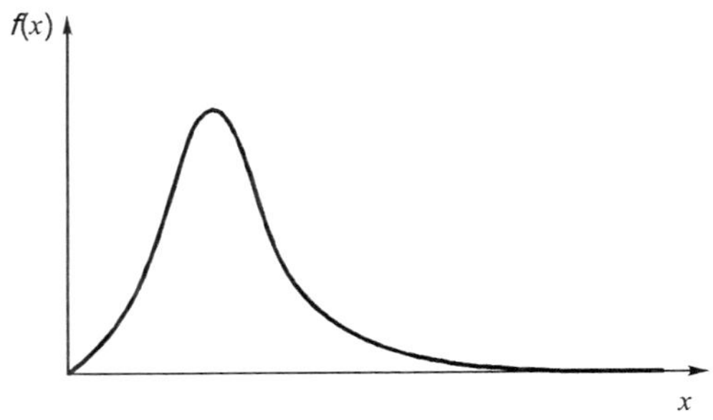

The correct statement for the probability distribution is
(a) Mean is equal to mode
(b) Mean is greater than median but less than mode
(c) Mean is greater than median and mode
(d) Mode is greater than median
[2018 : 1 M, Set-II]

6.28 The graph of a function $f(x)$ is shown in the figure.

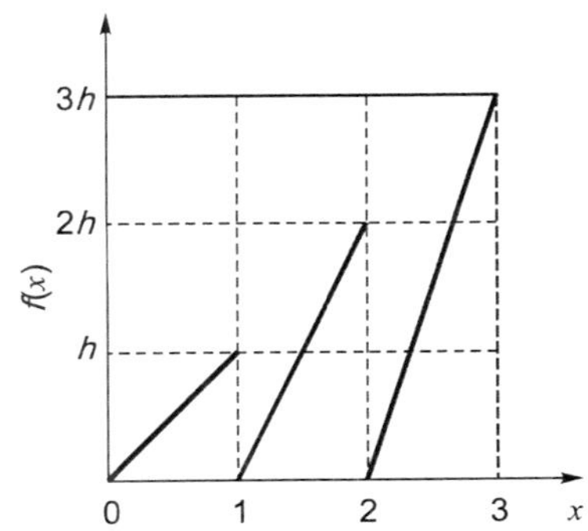

For $f(x)$ to be valid probability density function, the value of h is
(a) $\frac{1}{3}$ (b) $\frac{2}{3}$
(c) 1 (d) 3
[2018 : 1 M, Set-II]

6.29 Probability (up to one decimal place) of consecutively picking 3 red balls without replacement from a box containing 5 red balls and 1 white ball is ________.

[2018 : 1 M, Set-II]

6.30 The probability density function of a continuous random variable distributed uniformly between x and y (for $y > x$) is

(a) $y - x$
(b) $\frac{1}{y-x}$
(c) $x - y$
(d) $\frac{1}{x-y}$

[2019 : 2 M, Set-II]

6.31 A fair (unbiased) coin is tossed 15 times. The probability of getting exactly 8 Heads (round off to three decimal places), is ________.

[2020 : 1 M, Set-II]

6.32 The shape of the cumulative distribution function of Gaussian distribution is

(a) Bell-shaped
(b) S-shaped
(c) Horizontal line
(d) Straight line at 45 degree angle

[2021 : 1 M, Set-I]

6.33 A set of observations of independent variable (x) and the corresponding dependent variable (y) is given below.

x	5	2	4	3
y	16	10	13	12

Based on the data, the coefficient a of the linear regression model

$$y = a + bx$$

is estimated as 6.1. The coefficient b is ____________. (round off to one decimal place)

[2022 : 2 M, Set-I]

6.34 A pair of six-faced dice is rolled twice, The probability that the sum of the outcomes in each roll equals 4 in exactly two of the three attempts is ______ (round off to three decimal places).

[2022 : 1 M, Set-II]

6.35 The probabilities of occurrences of two independent events A and B are 0.5 and 0.8, respectively. What is the probability of occurrence of at least A or B (rounded off to one decimal place)? ____________.

[2023 : 1 M, Set-I]

6.36 Which of the following probability distribution functions (PDFs) has the mean greater than the median?

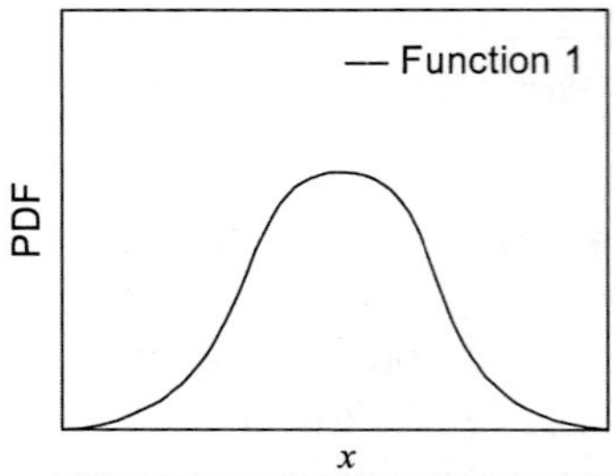

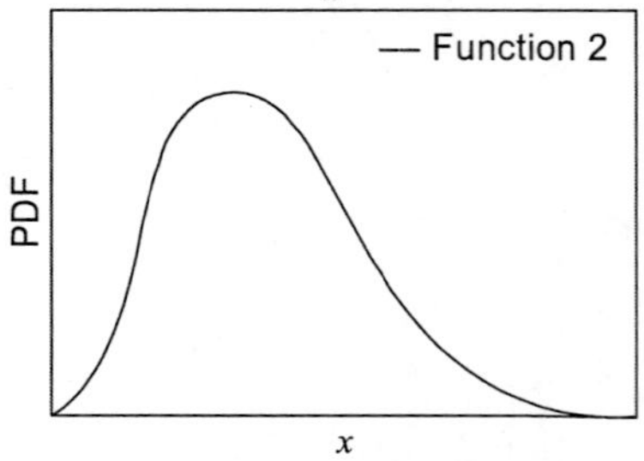

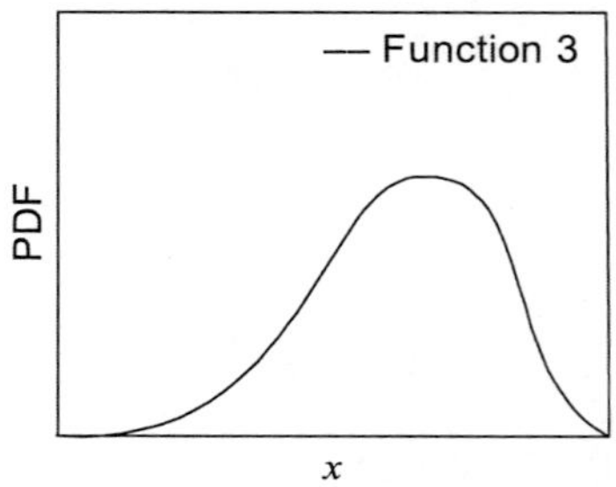

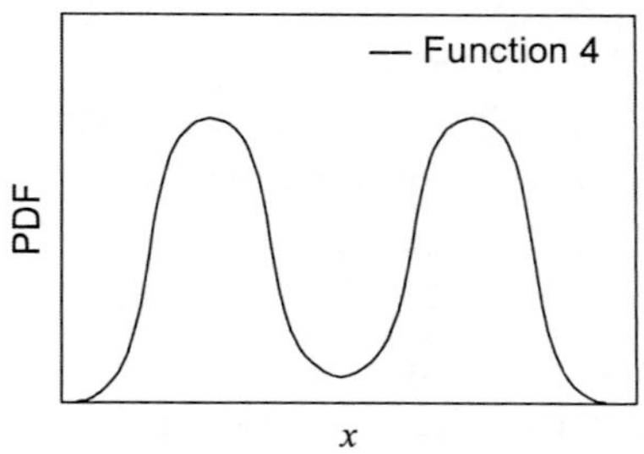

(a) Function 1
(b) Function 2
(c) Function 3
(d) Function 4

[2023 : 1 M, Set-II]

6.37 A remote village has exactly 1000 vehicles with sequential registration numbers starting from 1000. Out of the total vehicles, 30% are without pollution clearance certificate. Further, even- and odd-numbered vehicles are operated on even- and odd-numbered dates, respectively.

If 100 vehicles are chosen at random on an even-numbered date, the number of vehicles expected without pollution clearance certificate is _____ .

(a) 15
(b) 30
(c) 50
(d) 70

[2023 : 1 M, Set-II]

6.38 The probability that a student passes only in Mathematics is $\frac{1}{3}$. The probability that the student passes only in English is $\frac{4}{9}$. The probability that the student passes in both of these subjects is $\frac{1}{6}$. The probability that the student will pass in at least one of these two subjects is

(a) $\frac{17}{18}$ (b) $\frac{1}{18}$

(c) $\frac{11}{18}$ (d) $\frac{14}{18}$

[2024 : 1 M, Set-I]

6.39 In a sample of 100 heart patients, each patient has 80% chance of having a heart attack without medicine *X*. It is clinically known that medicine *X* reduces the probability of having a heart attack by 50%. Medicine *X* is taken by 50 of these 100 patients. The probability that a randomly selected patient, out of the 100 patients, takes medicine *X* and has a heart attack is

(a) 40% (b) 60%

(c) 30% (d) 20%

[2024 : 2 M, Set-II]

Answers Probability and Statistics

6.1 (d)	6.2 (c)	6.3 (d)	6.4 (d)	6.5 (b)	6.6 (a)	6.7 (b)	6.8 (c)	6.9 (c)
6.10 (a)	6.11 (d)	6.14 (b)	6.15 (b)	6.19 (a)	6.20 (a)	6.22 (a)	6.23 (b)	6.25 (b)
6.26 (b)	6.27 (c)	6.28 (a)	6.30 (b)	6.32 (b)	6.36 (b)	6.37 (b)	6.38 (a)	6.39 (d)

Explanations Probability and Statistics

6.1 (d)

This problem is to be solved by hypergeometric distribution, since population is finite.

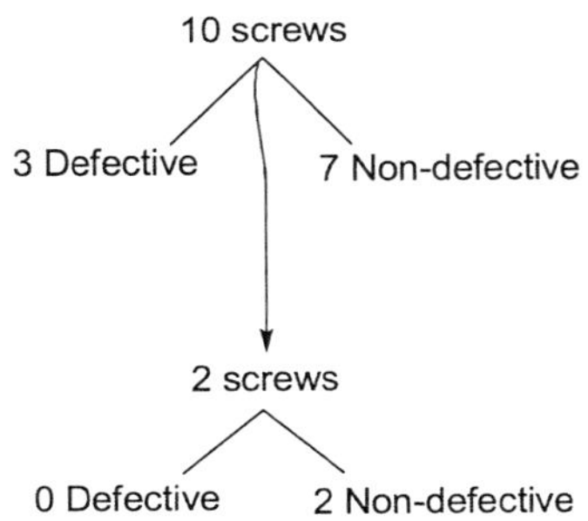

∴ p(none of 2 screws is defective)

$$= \frac{3C_0 \times 7C_2}{10C_2} \times 100\%$$

$$= \frac{7}{15} \times 100\% = 46.6 \simeq 47\%$$

6.2 (c)

p(gate 2 and gate 3 | gate 1 failed)

= p(gate 2 and gate 3)

[all three gates are independent]

= p(gate 2) × p(gate 3)

[gate 2 and 3 fail independently]

= 0.2 × 0.2 = 0.04

6.3 (d)

a, *b*, *c* are true

(d) is not true since in a **negatively skewed distribution**, mode > median > mean

6.4 (d)

Let the mean and standard deviation of the students of batch *C* be μ_c and σ_c respectively, and the mean and standard deviation of entire class of first year students be μ and σ respectively.

Now given,

$$\mu_c = 6.6$$
$$\sigma_c = 2.3$$

and
$$\mu = 5.5$$
$$\sigma = 4.2$$

In order to normalise batch *C* to entire class, the normalised score must be equated.

since $Z = \frac{x-\mu}{\sigma}$

$$Z_c = \frac{x_c - \mu_c}{\sigma_c} = \frac{8.5-6.6}{2.3}$$

Now $Z = \frac{x-\mu}{\sigma} = \frac{x-5.5}{4.2}$

Equating these two and solving, we get,

$$\frac{8.5-6.6}{2.3} = \frac{x-5.5}{4.2}$$

$$\Rightarrow \quad x = 8.969 \simeq 9.0$$

6.5 (b)

Since population is finite, hypergeometric distribution is applicable,

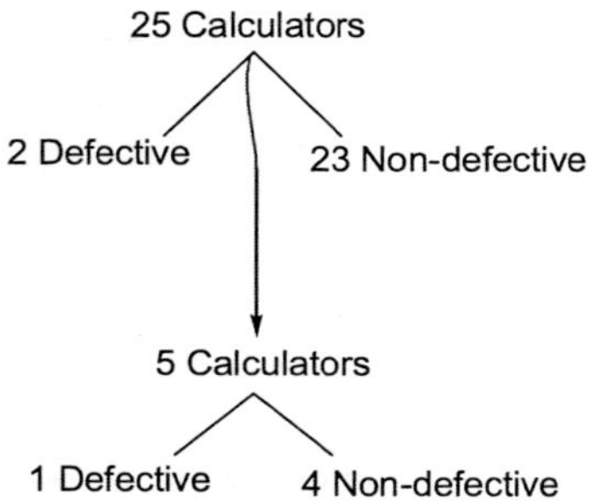

p(1 defective in 5 calculators)

$$= \frac{2C_1 \times 23C_4}{25C_5} = \frac{1}{3}$$

6.6 (a)

Given, p(private car) = 0.45

p(bus | public transport) = 0.55

Since a person has a choice between private car and public transport,

p(public transport) = 1 – p(private car)

$= 1 - 0.45 = 0.55$

p(bus) = p(bus ∩ public transport)

= p(bus | public transport) × p(public transport)

$= 0.55 \times 0.55$

$= 0.3025 \simeq 0.30$

Now p(metro)= 1 – [p(private car) + p(bus)]

$= 1 - (0.45 + 0.30) = 0.25$

∴ p(private car) = 0.45

p(bus) = 0.30

and p(metro) = 0.25

6.7 (b)

Here $\mu = 102$ cm; $\sigma = 27$ cm

$$p(90 \leq x \leq 102) = p\left(\frac{90-102}{27} \leq \geq \leq \frac{102-102}{27}\right)$$

$$= p(-0.44 \leq \geq \leq 0)$$

This area is shown below:

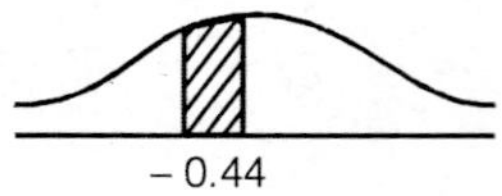

The shades area in above figure is given by $F(0) - F(-0.44)$

$$= \frac{1}{1+\exp(0)} - \frac{1}{1+\exp(-1.7255\,(-0.44)\,(0.44)0.12)}$$

$= 0.5 - 0.3345$

$= 1.1655 \simeq 16.55\%$

closest answer is 16.7%

6.8 (c)

$$p(2\text{ heads}) = \frac{1}{2} \times \frac{1}{2} = \frac{1}{4}$$

6.9 (c)

p(one ball is Red & another is blue)

= p(first is Red and second is Blue)

$$= \frac{4}{7} \times \frac{3}{7} = \frac{12}{49}$$

6.10 (a)

The annual precipitation is normally distributed with $\mu = 1000$ mm and $\sigma = 200$ mm

$$p(x > 1200) = p\left(z > \frac{1200-1000}{200}\right)$$

$$= p(z > 1)$$

Where z is the standard normal variate.

In normal distribution

Now, since $p(-1 < z < 1) \approx 0.68$

($\simeq$ 68% of data is within one standard deviation of mean)

$$p(0 < z < 1) = \frac{0.68}{2} = 0.34$$

So $p(z > 1) = 0.5 - 0.34$

$= 0.16 \simeq 16\%$

Which is <50%

So **choice (a) is correct.**

6.11 (d)

Since negative and positive are equally likely, the distribution of number of negative values is binomial with $n = 5$ and $p = 1/2$

Let X represent number of negative values in 5 trials.

$p(\text{at most 1 negative value})$

$= p(x \le 1)$

$= p(x = 0) + p(x = 1)$

$$= 5C_0\left(\frac{1}{2}\right)^0\left(\frac{1}{2}\right)^5 + 5C_1\left(\frac{1}{2}\right)^1\left(\frac{1}{2}\right)^4 = \frac{6}{32}$$

6.12 Sol.

$$\int_{-\infty}^{\infty} f(x)dx = 1$$

$$f(x) = \begin{cases} \lambda\left(-x^2+3x-2\right) & 1 \le x \le 2 \\ 0 & \text{otherwise} \end{cases}$$

$$\therefore \quad \int_1^2 \lambda\left(-x^2+3x-2\right)dx = 1$$

$$\Rightarrow \quad \lambda\left[-\frac{x^3}{3}+3\frac{x^2}{2}-2x\right]_1^2 = 1$$

$$\Rightarrow \quad \lambda\left[-\left(\frac{8}{3}-\frac{1}{3}\right)+\frac{3}{2}(4-1)-2(2-1)\right] = 1$$

$$\Rightarrow \quad \lambda\left[-\frac{7}{3}+\frac{9}{2}-2\right] = 1$$

$$\Rightarrow \quad \lambda\left[\frac{-14+27-12}{6}\right] = 1$$

$$\Rightarrow \quad \lambda = \frac{6}{1} = 6$$

$$\lambda = 6$$

6.13 Sol.

$$f(E) = \begin{cases} \frac{1}{5} & 0 \le E \le \text{mm/day} \\ 0 & \text{Otherwise} \end{cases}$$

$P(2 < E < 4)$

$$= \int_2^4 f(E)dE = \int_2^4 \frac{1}{5}dE = \frac{1}{5}[E]_2^4$$

$$= \frac{1}{5}(4-2) = \frac{2}{5} = 0.4$$

6.14 (b)

$$P(E) = \frac{n(E)}{n(S)}$$

$n(s) = [\{H\}, \{T\}] = 2$

$n(E) = \{(T)\} = 1$

$$\therefore \quad P(E) = \frac{1}{2}$$

6.16 Sol.

Mean $\lambda = 5$

$P(x < 4) = p(x = 0) + p(x = 1) + p(x = 2) + p(x = 3)$

$$= \frac{e^{-5}s^0}{0!} + \frac{e^{-5}5^1}{1!} + \frac{e^{-5}5^2}{1!} + \frac{e^{-5}5^3}{3!}$$

$$= e^{-5}\left[1+5+\frac{25}{2}+\frac{125}{6}\right] = e^{-5}\left(\frac{118}{3}\right)$$

$= 0.265$

6.17 Sol.

Given, $q = 0.4$

X	0	1
$p(X)$	0.4	0.6

Required value = $V(X) = E(X^2) - [E(X)]^2$

$$E(X) = \sum_i X_i p_i = 0 \times 0.4 + 1 \times 0.6 = 0.6$$

$$E(X^2) = \sum_i X_i^2 p_i = 0^2 \times 0.4 + 1^2 \times 0.6 = 0.6$$

$$\therefore \quad V(X) = E(X^2) - [E(X)]^2 = 0.6 - 0.36 = 0.24$$

6.18 Sol.

$$f(x) = \begin{cases} \frac{x}{4}\left(4-x^2\right), & 0 \le x \le 2 \\ 0 & \text{, Otherwise} \end{cases}$$

$$\mu_x = \int_0^2 xf(x)dx$$

$$\therefore \text{Mean } (\mu_x) = \int_0^2 x\frac{x}{4}\left(4-x^2\right)dx$$

$$= \int_0^2\left(x^2-\frac{x^4}{4}\right)dx$$

$$= \left[\frac{x^3}{3}-\frac{x^5}{20}\right]_0^2 = \frac{8}{3}-\frac{32}{20} = \frac{16}{15} = 1.066$$

6.20 (a)

$$P(x \le 4) = \int_{-\infty}^{4} f(x)dx$$
$$= \int_{-\infty}^{1}(0)dx + \int_{1}^{4}(0.25)dx + \int_{4}^{\infty}(0)dx$$
$$= \frac{1}{4}(x)_1^4 = \frac{1}{4}(4-1) = \frac{3}{4}$$

6.21 Sol.

Median speed is the speed at the middle value in series of spot speeds that are arranged in ascending order. 50% of speed values will be greater than the median 50% will be less than the median.

Ascending order of spot speed studies are
32, 39, 45, 51, 53, 56, 60, 62, 66, 79

Median speed = $\frac{53+56}{2}$ = 54.5 km/hr

6.22 (a)

$P(X \cup Y^c) = 0.7$

$\Rightarrow$ $P(X) + P(Y^c) - P(X)\,P(Y^c) = 0.7$
(Since X, Y are independent events)

$\Rightarrow$ $P(X) + 1 - P(Y) - P(X)\{1 - P(Y)\} = 0$

$\Rightarrow$ $P(Y) - P(X \cap Y) = 0.3$ ...(i)

$P(X \cup Y) = P(X) + P(Y) - P(X \cap Y)$
$= 0.4 + 0.3 = 0.7$

6.23 (b)

Mean of $f(x)$ is $E(x)$

$$= \int_{-a}^{0} x\left(\frac{x}{a}+1\right)dx + \int_{0}^{a} x\left(\frac{-x}{a}+1\right)dx$$

$$= \left(\frac{x^3}{3a}+\frac{x^2}{2}\right)_{-a}^{0} + \left(\frac{-x^3}{3a}+\frac{x^2}{2}\right)_{0}^{a} = 0$$

Variance of $f(x)$ is $E(x)^2 - \{E(x)^2\}$ where,

$$E(x)^2 = \int_{-a}^{0} x^2\left(\frac{x}{a}+1\right)dx + \int_{0}^{a} x^2\left(\frac{-x}{a}+1\right)dx$$

$$= \left(\frac{x^4}{4a}+\frac{x^3}{3}\right)_{-a}^{0} + \left(\frac{-x^4}{4a}+\frac{x^3}{3}\right)_{0}^{a} = \frac{a^3}{6}$$

$\Rightarrow$ Variance is $\frac{a^3}{6}$

Next, mean of $g(x)$ is $E(x)$

$$= \int_{a}^{0} x\left(\frac{-x}{a}\right)dx + \int_{0}^{a} x\left(\frac{x}{a}\right)dx = 0$$

Variance of $g(x)$ is $E(x^2) - \{E(x)\}^2$, where,

$$E(x^2) = \int_{-a}^{0} x^2\left(\frac{-x}{a}\right)dx + \int_{0}^{a} x^2\left(\frac{x}{a}\right)dx = \frac{a^3}{2}$$

$\Rightarrow$ Variance is $\frac{a^3}{2}$

$\therefore$ Mean of $f(x)$ and $g(x)$ are same but variance of $f(x)$ and $g(x)$ are different.

6.25 (b)

In exponential,

$$f(x) = \lambda e^{-\lambda x}; \quad x = 0$$

The parameter is λ.

In Gaussian, $f(x) = \frac{1}{\sigma\sqrt{2\pi}} e^{-\frac{1}{2}\left(\frac{x-\mu}{\sigma}\right)^2}$; $-\infty < x < \infty$

The parameters are μ and σ.
Therefore, **answer is (b).**

6.26 (b)

$$\int_{-\infty}^{\infty} f(x)dx = 1; \quad \int_{0}^{1}(a+bx)dx = 1$$

$$\left(ax + \frac{bx^2}{2}\right)\Bigg|_0^1 = 1; \quad a + \frac{b}{2} = 1$$

Option (b) is satisfying the above equation.

6.27 (c)

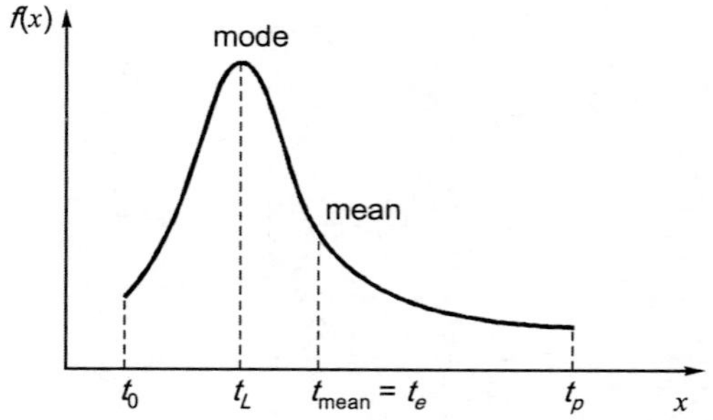

$t_L < t_{mean}$ = Curve is skew to right.
mode < mean
i.e., mean > median and mode

Mean is greater than the mode and the median. This is common for a distribution that is skewed to the right [i.e., bunched up toward the left and a 'tail' stretching toward the right].

6.28 (a)

$$\int_0^3 f(x)dx = 1$$

$$\int_0^1 f(x)dx + \int_1^2 f(x)dx + \int_2^3 f(x)dx = 1$$

$$\frac{h}{2} + \frac{2h}{2} + \frac{3h}{2} = 1$$

$$6h = 2$$

$$\Rightarrow \quad h = \frac{1}{3}$$

6.29 Sol.

Probability, $\bar{P} = \frac{5}{6} \times \frac{4}{5} \times \frac{3}{4} = \frac{1}{2} = 0.5$

6.30 (b)

Probability density function of uniform distribution is

$$f(x) = \frac{1}{y-x}$$

6.31 Sol.

$$P(H) = \frac{1}{2}$$

$$P(T) = \frac{1}{2}$$

Probability of getting exactly 8 heads out of 15 trial

$$= {}^{15}C_8 \left[\frac{1}{2}\right]^8 \times \left[\frac{1}{2}\right]^{15-8} = 0.196$$

6.33 Sol.

x	y	x^2	xy
5	16	25	80
2	10	4	20
4	13	16	52
3	12	9	36
$\Sigma x = 14$	$\Sigma y = 51$	$\Sigma x^2 = 54$	$\Sigma xy = 188$

We know that, normal equation for fitting of straight lines are given by

$$\Sigma y = na + b\,\Sigma x$$

and $\quad \Sigma xy = a\,\Sigma x + b\,\Sigma x^2$

Here, $\quad n = 4$

So, $\quad 51 = 4a + b(14)$

$$188 = a(14) + b(54)$$

$\Rightarrow$ After solving, $a = 6.1$ and $b = 1.9$

6.34 Sol.

$$A = \{(1, 3), (3, 1), (2, 2)\}$$

$$\therefore \quad P(A) = \frac{3}{36} = \frac{1}{12}$$

$$\Rightarrow P(\bar{A}) = \frac{11}{12}$$

$\therefore$ Required probability

$$= P(AA\bar{A}) + P(A\bar{A}A) + P(\bar{A}AA)$$

$$= \left(\frac{1}{12} \times \frac{1}{12} \times \frac{11}{12}\right) + \left(\frac{1}{12} \times \frac{11}{12} \times \frac{1}{12}\right)\left(+\frac{11}{12} \times \frac{1}{12} \times \frac{1}{12}\right)$$

$$= 3 \times \left(\frac{1}{12}\right)^2 \times \frac{11}{12} = 0.019$$

6.35 Sol.

Given, $\quad P(A) = 0.5$ and $P(B) = 0.8$

and A and B are independents.

$$P(A \cup B) = P(A) + P(B) - P(A \cap B)$$

$$= P(A) + P(B) - P(A)P(B)$$

$$= 0.5 + 0.8 - 0.5 \times 0.8 = 0.9$$

6.37 (b)

Probability of selecting even vehicles on even numbered dates = 1

$\therefore$ The no. of vehicles expected without pollution clearance certificate

$$= 100 \times 1 \times 0.3$$

$$= 30$$

6.38 (a)

$$P(M \cup E) = P(M) + P(E) + P(M \cap E)$$

$$= \frac{1}{3} + \frac{4}{9} + \frac{1}{6}$$

$$= \frac{6+8+3}{18} = \frac{17}{18}$$

6.39 (d)

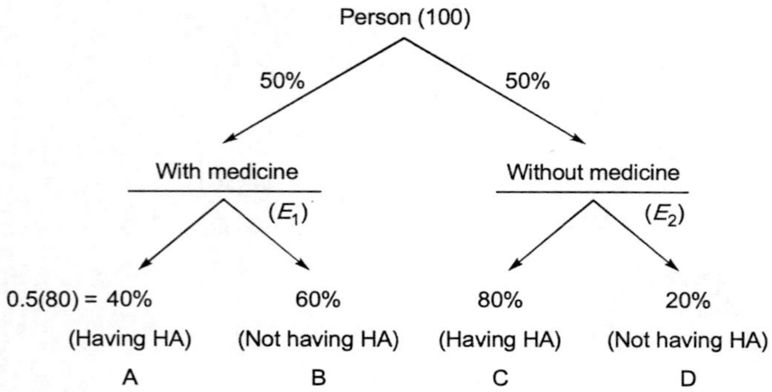

HA: Heat attack

$$\therefore \quad P\left(\frac{A}{E_1}\right) = 50\% \times 40\%$$

$$= \frac{50}{100} \times \frac{40}{100} = \frac{20}{100} = 20\%$$

■■■■

7 Numerical Methods

Linked Answer Questions 7.1 and 7.2:

Give $a > 0$, we wish to calculate its reciprocal value $1/a$ by using Newton Raphson Method for $f(x) = 0$.

7.1 The Newton Raphson algorithm for the function will be

(a) $x_{k+1} = \frac{1}{2}\left(x_k + \frac{a}{x_k}\right)$

(b) $x_{k+1} = \left(x_k + \frac{a}{2}x_k^2\right)$

(c) $x_{k+1} = 2x_k - ax_k^2$

(d) $x_{k+1} = x_k - \frac{a}{2}x_k^2$ [2005 : 2 M]

7.2 For $a = 7$ and starting with $x_0 = 0.2$, the first two iterations will be

(a) 0.11, 0.1299 (b) 0.12, 0.1392

(c) 0.12, 0.1416 (d) 0.13, 0.1428

[2005 : 2 M]

7.3 A 2nd degree polynomial, $f(x)$ has values of 1, 4 and 15 at $x = 0$, 1 and 2, respectively. The integral $\int_0^2 f(x)\,dx$ is to be estimated by applying the trapezoidal rule to this data. What is the error (defined as "true value - approximate value") in the estimate?

(a) $-\frac{4}{3}$ (b) $-\frac{2}{3}$

(c) 0 (d) $\frac{2}{3}$ [2006 : 2 M]

7.4 The following equation needs to be numerically solved using the Newton-Raphson method.

$$x^3 + 4x - 9 = 0$$

The iterative equation for this purpose is (k indicates the iteration level)

(a) $x_{k+1} = \frac{2x_k^3 + 9}{3x_k^2 + 4}$

(b) $x_{k+1} = \frac{3x_k^2 + 4}{2x_k^2 + 9}$

(c) $x_{k+1} = x_k - 3x_k^2 + 4$

(d) $x_{k+1} = \frac{4x_k^2 + 3}{9x_k^2 + 2}$ [2007 : 2 M]

7.5 Given that one root of the equation $x^3 - 10x^2 + 31x - 30 = 0$ is 5, the other two roots are

(a) 2 and 3 (b) 2 and 4

(c) 3 and 4 (d) −2 and −3

[2007 : 2 M]

7.6 Three values of x and y are to be fitted in a straight line in the form $y = a + bx$ by the method of least squares. Given $\Sigma x = 6$, $\Sigma y = 21$, $\Sigma x^2 = 14$ and $\Sigma xy = 46$, the values of a and b are respectively

(a) 2 and 3 (b) 1 and 2

(c) 2 and 1 (d) 3 and 2

[2008 : 2 M]

7.7 If probability density function of a random variable X is

$f(x) = x^2$ for $-1 \le x \le 1$, and

$= 0$ for any other value of x

then, the percentage probability

$P\left(-\frac{1}{3} \le x \le \frac{1}{3}\right)$ is

(a) 0.247 (b) 2.47

(c) 24.7 (d) 247 [2008 : 2 M]

7.8 In the solution of the following set of linear equations by Gauss elimination using partial pivoting $5x + y + 2z = 34$; $4y - 3z = 12$; and $10x - 2y + z = -4$; the pivots for elimination of x and y are

(a) 10 and 4 (b) 10 and 2

(c) 5 and 4 (d) 5 and − 4

[2009 : 2 M]

7.9 The table below gives values of a function $F(x)$ obtained for values of x at intervals of 0.25.

x	0	0.25	0.5	0.75	1.0
$F(x)$	1	0.9412	0.8	0.64	0.50

The value of the integral of the function between the limits 0 to 1 using Simpson's rule is

(a) 0.7854 (b) 2.3562

(c) 3.1416 (d) 7.5000 [2010 : 2 M]

7.10 The square root of a number N is to be obtained by applying the Newton Raphson iterations to the equation $x^2 - N = 0$. If i denotes the iteration index, the correct iterative scheme will be

(a) $x_{i+1} = \frac{1}{2}\left(x_i + \frac{N}{x_i}\right)$ (b) $x_{i+1} = \frac{1}{2}\left(x_i^2 + \frac{N}{x_i^2}\right)$

(c) $x_{i+1} = \frac{1}{2}\left(x_i + \frac{N^2}{x_i}\right)$ (d) $x_{i+1} = \frac{1}{2}\left(x_i - \frac{N}{x_i}\right)$

[2011 : 1 M]

7.11 The estimate of $\int_{0.5}^{1.5} \frac{dx}{x}$ Obtained using Simpson's rule with three-point function evaluation exceeds the exact value by

(a) 0.235 (b) 0.068
(c) 0.024 (d) 0.012 [2011 : 2 M]

7.12 The error in $\left.\frac{d}{dx}f(x)\right|_{x=x_0}$ for a continuous function estimated with h = 0.03 using the central difference formula $\left.\frac{d}{dx}f(x)\right|_{x=x_0} = \frac{f(x_0+h)-f(x_0-h)}{2h}$, is 2×10^{-3}. The values of x_0 and $f(x_0)$ are 19.78 and 500.01, respectively. The corresponding error in the central difference estimate for $h = 0.02$ is approximately

(a) 1.3×10^{-4} (b) 3.0×10^{-4}
(c) 4.5×10^{-4} (d) 9.0×10^{-4}

[2011 : 2 M]

7.13 The magnitude as the error (correct to two decimal places) in the estimation of following integral using simpson 1/3 rule. Take the step length as 1

$$\int_0^4 (x^4+10)dx$$ [2013 : 2 M]

7.14 The integral $\int_{x_1}^{x_2} x^2 dx$ with $x_2 > x_1 > 0$ is evaluated analytically as well as numerically using a single application of the trapezoidal rule. If I is the exact value of the integral obtained analytically and J is the approximate value obtained using the trapezoidal rule, which of the following statements is correct about their relationship?

(a) $J > I$
(b) $J < I$
(c) $J = I$
(d) Insufficient data to determine the relationship

[2015 : 1 M, Set-I]

7.15 The quadratic equation $x^2 - 4x + 4 = 0$ is to be solved numerically, starting with the initial guess $x_0 = 3$. The Newton-Raphson method is applied once to get a new estimate and then the Secant method is applied once using the initial guess and this new estimate. The estimated value of the root after the application of the Secant method is ______.

[2015 : 2 M, Set-I]

7.16 In Newton-Raphson iterative method, the initial guess value (x_{ini}) is considered as zero while finding the roots of the equation:

$$f(x) = -2 + 6x - 4x^2 + 0.5x^3.$$

The correction, Δx, to be added to x_{ini} in the first iteration is __________.

[2015 : 1 M, Set-II]

7.17 For step-size, $\Delta x = 0.4$, the value of following integral using Simpson's 1/3 rule is ________.

$$\int_0^{0.8} \left(0.2 + 25x - 200x^2 + 675x^3 - 900x^4 + 400x^5\right)dx$$

[2015 : 2 M, Set-II]

7.18 Newton-Raphson method is to be used to find foot of equation $3x - e^x + \sin x = 0$. If the initial trail value of the roots is taken as 0.333, the next approximation for the root would be _________.

[2016 : 1 M, Set-I]

7.19 Consider the equation $\frac{du}{dt} = 3t^2 + 1$ with $u = 0$ at $t = 0$. This is numerically solved by using the forward Euler method with a step size. $\Delta t = 2$. The absolute error in the solution in the end of the first time step is ____________.

[2017 : 2 M, Set-I]

7.20 The quadratic equation $2x^2 - 3x + 3 = 0$ is to be solved numerically starting with an initial guess as $x_0 = 2$. The new estimate of x after the first iteration using Newton-Raphson method is _________.

[2018 : 1 M, Set-II]

7.21 The value of the function $f(x)$ is given at n distinct values of x and its value is to be interpolated at the point x^*, using all the n points. The estimate is obtained first by the Lagrange polynomial, denoted by I_L and then by the Newton polynomial, denoted by I_N. Which one of the following statements is correct?

(a) I_L is always greater than I_N
(b) No definite relation exists between I_L and I_N

(c) I_L and I_N are always equal
(d) I_L is always less than I_N

[2019 : 1 M, Set-II]

7.22 A continuous function $f(x)$ is defined. If the third derivative at x_i is to be computed by using the fourth order central finite-divided-difference scheme (the step length = h), the correct formula is

(a) $f'''(x_i) = \dfrac{-f(x_{i+3}) - 8f(x_{i+2}) + 13f(x_{i+1}) + 13f(x_{i-1}) + 8f(x_{i-2}) - f(x_{i-3})}{8h^3}$

(b) $f'''(x_i) = \dfrac{-f(x_{i+3}) + 8f(x_{i+2}) - 13f(x_{i+1}) + 13f(x_{i-1}) - 8f(x_{i-2}) + f(x_{i-3})}{8h^3}$

(c) $f'''(x_i) = \dfrac{f(x_{i+3}) - 8f(x_{i+2}) + 13f(x_{i+1}) + 13f(x_{i-1}) - 8f(x_{i-2}) - f(x_{i-3})}{8h^3}$

(d) $f'''(x_i) = \dfrac{f(x_{i+3}) - 8f(x_{i+2}) - 13f(x_{i+1}) + 13f(x_{i-1}) + 8f(x_{i-2}) + f(x_{i-3})}{8h^3}$

[2020 : 2 M, Set-I]

7.23 The values of abscissa (x) and ordinate (y) of a curve are as follows:

x	y
2.0	5.00
2.5	7.25
3.0	10.00
3.5	13.25
4.0	17.00

By Simpson's 1/3rd rule, the area under the curve (round off to two decimal places) is _______.

[2021 : 2 M, Set-I]

7.24 The value of $\int_0^1 e^x dx$ using the trapezoidal rule with four equal subintervals is

(a) 1.718 (b) 2.192
(c) 1.727 (d) 2.718

[2021 : 2 M, Set-I]

7.25 Numerically integrate. $f(x) = 10x - 20x^2$ from lower limit $a = 0$ to upper limit $b = 0.5$. Use Trapezoidal rule with five equal subdivisions. The value (in units, round off to two decimal places) obtained is _________.

[2021 : 2 M, Set-II]

7.26 Consider the following recursive iteration scheme for different values of variable P with the initial guess $x_1 = 1$:

$$x_{n+1} = \frac{1}{2}\left(x_n + \frac{P}{x_n}\right), n = 1, 2, 3, 4, 5$$

For $P = 2$, x_5 is obtained to be 1.414, rounded-off to three decimal places. For $P = 3$, x_5 is obtained to be 1.732, rounded-off to three decimal places.

If $P = 10$, the numerical value of x_5 is _________ . (round off to three decimal places)

[2022 : 1 M, Set-I]

7.27 Consider the differential equation

$$\frac{dy}{dx} = 4(x+2) - y$$

For the initial condition $y = 3$ at $x = 1$, the value of y at $x = 1.4$ obtained using Euler's method with a step-size of 0.2 is _________. (round off to one decimal place)

[2022 : 2 M, Set-I]

7.28 A function $f(x)$, that is smooth and convex-shaped between interval (x_l, x_u) is shown in the figure. This function is observed at odd number of regularly spaced points. If the area under the function is computed numerically, then .

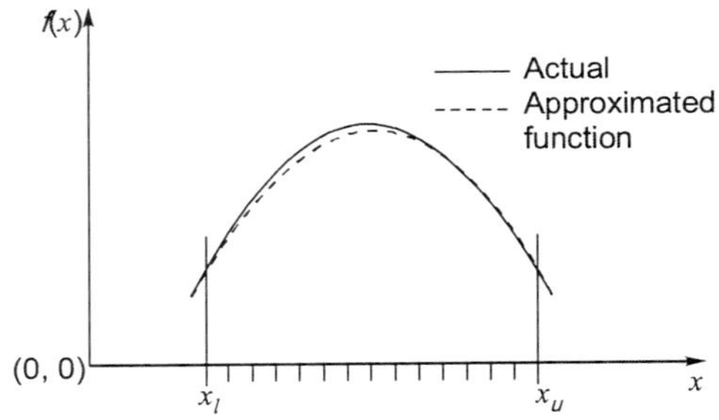

(a) the numerical value of the area obtained using the trapezoidal rule will be less than the actual
(b) the numerical value of the area obtained using the trapezoidal rule will be more than the actual
(c) the numerical value of the area obtained using the trapezoidal rule will be exactly equal to the actual
(d) with the given details, the numerical value of area cannot be obtained using trapezoidal rule

[2023 : 2 M, Set-I]

7.29 The second derivative of a function f is computed using the fourth-order Central Divided Difference method with a step length h.

The CORRECT expression for the second derivative is

(a) $\dfrac{1}{12h^2}\left[f_{i+2} + 16f_{i+1} - 30f_i + 16f_{i-1} - f_{i-2}\right]$

(b) $\frac{1}{12h^2}\left[-f_{i+2}-16f_{i+1}+30f_i-16f_{i-1}-f_{i-2}\right]$

(c) $\frac{1}{12h^2}\left[-f_{i+2}+16f_{i+1}-30f_i+16f_{i-1}+f_{i-2}\right]$

(d) $\frac{1}{12h^2}\left[-f_{i+2}+16f_{i+1}-30f_i+16f_{i-1}-f_{i-2}\right]$

[2024 : 1 M, Set-II]

7.30 Consider the data of f(x) given in the table.

i	0	1	2
x_i	1	2	3
$f(x_i)$	0	0.3010	0.4771

The value of f(1.5) estimated using second-order Newton's interpolation formula is ________ (rounded off to 2 decimal places).

[2024 : 1 M, Set-I]

7.31 The smallest positive root of the equation $x^5-5x^4-10x^3+50x^2+9x-45=0$ lies in the range

(a) $0<x\le 2$
(b) $6\le x\le 8$
(c) $10\le x\le 100$
(d) $2<x\le 4$

[2024 : 1 M, Set-I]

Answers Numerical Methods

7.1 (c) 7.2 (b) 7.3 (a) 7.4 (a) 7.5 (a) 7.6 (d) 7.7 (b) 7.8 (a) 7.9 (a) 7.10 (a) 7.11 (d) 7.12 (d) 7.14 (a) 7.21 (c) 7.22 (b) 7.24 (c) 7.28 (a) 7.29 (d) 7.30 (0.17) 7.31 (a)

Explanations Numerical Methods

7.1 (c)

To calculate $\frac{1}{a}$ using **N-R method,** set up the equation as,

$$x = \frac{1}{a}$$

i.e. $$\frac{1}{x} = a$$

$$\Rightarrow \quad \frac{1}{x_k}-a = 0$$

i.e. $$f(x) = \frac{1}{x_k}-a = 0$$

Now $$f'(x) = -\frac{1}{x^2}$$

$$f(x_k) = \frac{1}{x_k}-a$$

$$f'(x_k) = -\frac{1}{x_k^2}$$

For **N-R method,**

$$x_{k+1} = x_k-\frac{f(x_k)}{f'(x_k)}$$

$$\Rightarrow \quad x_{k+1} = x_k-\frac{(1/x_k-a)}{-\frac{1}{x_k^2}}$$

Simplifying which we get,

$$x_{k+1} = 2x_k-ax_k^2$$

7.2 (b)

For a =7 the iteration equation, becomes,

$$x_{k+1} = 2x_k-7x_k^2$$

with $x_0 = 0.2$

$$x_1 = 2x_0-7x_0^2 = 2\times 0.2-7(0.2)^2 = 0.12$$

and $$x_2 = 2x_1-7x_1^2 = 2\times 0.12-7(0.12)^2 = 0.1392$$

7.3 (a)

$f(x)$ =1, 4, 15 at x = 0, 1 and 2 respectively

$$\int_0^2 f(x)dx = \frac{h}{2}(f_1+2f_2+f_3)$$

(3 point Trapezoidal Rule)

Here, $h = 1$

$$\therefore \int_0^2 f(x)dx = (1+2\times+15) = 12$$

$\therefore$ Approximate value by Trapezoidal Rule = 12

Since $f(x)$ is second degree polynomial, let,

$$f(x) = a_0+a_1x+a_2x^2$$

$$f(0) = 1$$
$$\Rightarrow \quad a_0 + 0 + 0 = 1$$
$$\Rightarrow \quad a_0 = 1$$
$$f(1) = 4$$
$$\Rightarrow \quad a_0 + a_1 + a_2 = 4$$
$$\Rightarrow \quad 1 + a_1 + a_2 = 4$$
$$\Rightarrow \quad a_1 + a_2 = 3 \quad \text{...(i)}$$
$$f(2) = 15$$
$$\Rightarrow \quad a_0 + 2a_1 + 4a_2 = 15$$
$$\Rightarrow \quad 1 + 2a_1 + 4a_2 = 15$$
$$\Rightarrow \quad 2a_1 + 4a_2 = 14 \quad \text{...(ii)}$$

Solving (i) and (ii) $a_1 = -1$ and $a_2 = 4$,

$$\therefore \quad f(x) = 1 - x + 4x^2$$

Now, exact value of $\int_0^2 f(x)dx$

$$= \int_0^2 (1 - x + 4x^2)\,dx$$
$$= \left[x - \frac{x^2}{2} + \frac{4x^3}{3}\right]_0^2 = \frac{32}{3}$$

Error = Exact – Approximate value

$$= \frac{32}{3} - 12 = -\frac{4}{3}$$

7.4 (a)

$$f(x) = x^3 + 4x - 9 = 0$$
$$f'(x) = 3x^2 + 4$$

N–R equation for iteration is,

$$x_{k+1} = x_k - \frac{f(x_k)}{f'(x_k)}$$
$$f(x_k) = x_k^3 + 4x_k - 9$$
$$f'(x_k) = 3x_k^2 + 4$$
$$x_{k+1} = x_k - \frac{(x_k^3 + 4x_k - 9)}{(3x_k^2 + 4)}$$
$$x_{k+1} = \frac{(3x_k^3 + 4x_k) - (x_k^3 + 4x_k - 9)}{3x_k^2 + 4}$$
$$x_{k+1} = \frac{2x_k^3 + 9}{3x_k^2 + 4}$$

7.5 (a)

Sum of roots,

$$\alpha + \beta + \gamma = -\frac{b}{a} = \frac{10}{1} = 10 \quad \text{...(i)}$$

Product of roots, $\alpha\beta\gamma = \dfrac{(-1)^3 d}{a} = 30 \quad \text{...(ii)}$

Given $\gamma = 5$

From eq. (i) $\alpha + \beta = 5$

From eq. (ii) $\alpha\beta = 6$

Solving we get

$$\alpha = 2, \beta = 3$$

7.6 (d)

$$y = a + bx$$

Given

$$n = 3, \Sigma x = 6, \Sigma y = 21, \Sigma x^2 = 14$$

and $\Sigma xy = 46$

$$b = \frac{n\,\Sigma xy - \Sigma x\,\Sigma y}{n\,\Sigma x^2 - (\Sigma x)^2}$$
$$a = \bar{y} - b\bar{x}$$
$$= \frac{\Sigma y}{n} - b\frac{\Sigma x}{n}$$

Substituting, we get

$$b = \frac{(3\times 46) - (6\times 21)}{(3\times 14) - (6)^2} = 2$$
$$a = \frac{21}{3} - 2\times\left(\frac{6}{3}\right) = 3$$
$$\therefore \quad a = 3 \text{ and } b = 2$$

7.7 (b)

Given $f(x) = x^2 \quad -1 \le x \le 1$
$\quad = 0 \quad$ elsewhere

$$p\left(-\frac{1}{3} \le x \le \frac{1}{3}\right) = \int_{-\frac{1}{3}}^{\frac{1}{3}} f(x)\,dx = \int_{-\frac{1}{3}}^{\frac{1}{3}} x^2\,dx$$
$$= \left[\frac{x^3}{3}\right]_{-\frac{1}{3}}^{\frac{1}{3}} = \frac{2}{81}$$

The probability expressed in percentage,

$$p = \frac{2}{81} \times 100 = 2.469\% = 2.47\%$$

7.8 (a)

The equations are,

$$5x + y + 2z = 34$$
$$0x + 4y - 3z = 12$$

and $10x - 2y + z = -4$

The augmented maxtrix for gauss-elimination is,

$$\left[\begin{array}{ccc|c} 5 & 1 & 2 & 34 \\ 0 & 4 & -3 & 12 \\ 10 & -2 & 1 & -4 \end{array}\right]$$

Since in the first column maximum element in absolute value is 10, we need to exchange row 1 with row 3.

$$\left[\begin{array}{ccc|c} 5 & 1 & 2 & 34 \\ 0 & 4 & -3 & 12 \\ 10 & -2 & 1 & -4 \end{array}\right] \xrightarrow{R(1,3)} \left[\begin{array}{ccc|c} 10 & -2 & 1 & -4 \\ 0 & 4 & -3 & 12 \\ 5 & 1 & 2 & 34 \end{array}\right]$$

So the pivot for eliminating x is $a_{11} = 10$.

Now to eliminate y, we need to compass the eliminate in second column at and below the diagonal.

Since $a_{22} = 4$ is already larger in absolute value compares to $a_{32} = 1$

$\therefore$ The pivot element for eliminating y is $a_{22} = 4$ itself.

$\therefore$ The pivots for eliminating x and y are respectively 10 and 4.

7.9 (a)

$$I = \frac{1}{3}h(f_0 + 4f_1 + 2f_2 + 4f_3 + f_4)$$

$$= \frac{1}{3} \times 0.25(1 + 4 \times 0.9412 + 2 \times 0.8 + 4 \times 0.64 + 0.5)$$

$$= 0.7854$$

7.10 (a)

$$x_{i+1} = x_i - \frac{f(x_i)}{f'(x_i)}$$

$$= x_i - \left(\frac{x_i^2 - N}{2x_i}\right)$$

$$= \frac{x_i^2 + N}{2x_i} = \frac{1}{2}\left[x_i + \frac{N}{x_i}\right]$$

7.11 (d)

Exact value of $\int_{0.5}^{1.5} \frac{dx}{x} = [\log x]_{0.5}^{1.5}$

$$= \log(1.5) - \log(0.5) = 1.0986$$

Approximate value by **Simpson's rule** with 3 point is,

$$I = \frac{h}{3}(f(0) + 4f(1) + f(2))$$

$$n_i = n_{pt} - 1 = 3 - 1 = 2$$

(n_{pt} is the number of pts and n_i is the number of intervals)

Here $h = \frac{b-a}{n_i} = \frac{1.5 - 0.5}{2} = 0.5$

The table is

i	x_i	f_i
0	0.5	$\frac{1}{0.5}$
1	1.0	$\frac{1}{1}$
2	1.5	$\frac{1}{1.5}$

$$I = \frac{0.5}{3}\left(\frac{1}{0.5} + 4 \times 1 + \frac{1}{1.5}\right)$$

$$= 1.1111$$

So the estimate exceeds the exact value by,

Approximate value – Exact value

$$= 0.012499$$

$$\approx 0.012$$

7.12 (d)

Error in central difference formula is $0(h^2)$

This means,

$$\text{error} \propto h^2$$

If error for $h = 0.03$ is 2×10^{-3} then,

Error for h = 0.02 is approximately

$$2 \times 10^{-3} \times \frac{(0.02)^2}{(0.03)^2} \simeq 9 \times 10^{-4}$$

7.13 Sol.

Using **Simpson's Rule,**

x	0	1	2	3	4
y	10	11	26	91	266

$$\int_0^4 (x^4 + 10)dx$$

$$= \frac{1}{3}[(10 + 266) + 2(26) + 4(11 + 91)] = 245.33$$

The value of integral,

$$\int_0^4 (x^4 + 10)dx = \left[\frac{x^5}{5} + 10x\right]_0^4$$

$$= \frac{4^5}{5} + 10 \times 4 = 244.8$$

$\therefore$ Magnitude of error = 245.33 – 244.8 = 0.53

7.14 (a)

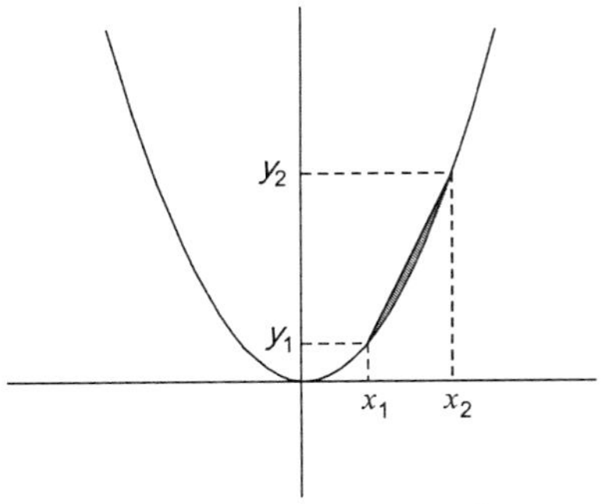

Exact value is computed by integration which follows thee exact shape of graph while computing the area.

Whereas, in **Trapezoidal rule**, the lines joining each points are considered straight lines which is not the exact variation of graph all the time like as shown in figure.

$\therefore \quad J > I$

OR

$$\text{Error} = -\frac{h^3}{12}f''(\xi)\times n_i$$

Here, $f(x) = x^2$

or, $f'(x) = 2x$

or, $f''(x) = 2 > 0$

Since $f''(x)$ is positive, the error is negative.

Since error = exact – approximate.

$= I - J$

and since error is negative in this case $J > I$ is true.

7.15 Sol.

$$f(x) = x^2 - 4x + 4$$
$$f'(x) = 2x - 4$$
$$x_0 = 3$$
$$f(3) = 1,\ f'(3) = 2$$

By **Newton Rapshon method**,

$$x_1 = x_0 - \frac{f(x_0)}{f'(x_0)}$$
$$= 3 - \frac{1}{2} = \frac{5}{2}$$
$$f\left(\frac{5}{2}\right) = \frac{25}{4} - 10 + 4 = \frac{1}{4}$$

By **Secant method**,

$$x_2 = x_1 - \frac{(x_1 - x_0)f(x_1)}{f(x_1) - f(x_0)}$$
$$= \frac{f_1x_0 - f_0x_1}{f_1 - f_0}$$
$$= \frac{\left(\frac{1}{4}\times 3\right) - \left(1\times\frac{5}{2}\right)}{\frac{1}{4} - 1} = \frac{7}{3}$$

7.16 Sol.

$$f(x) = -2 + 6x - 4x^2 + 0.5x^3$$
$$f'(x) = 6 - 8x + 1.5x^2$$
$$x_{ini} = 0$$

By **Newton Raphson Method,**

$$x_1 = x_{ini} - \frac{f(x_{ini})}{f'(x_{ini})} = 0 - \frac{-2}{6}$$

$\Rightarrow \quad x_1 = \frac{1}{3}$

$\therefore \quad \Delta x = x_1 - x_{ini} = \frac{1}{3}$

7.17 Sol.

$a = 0, b = 0.8, \Delta x = 0.4$

$f(x) = 0.2 + 25x - 200x^2 + 675x^3 - 900x^4 + 400x^5$

x	0	0.4	0.8
$f(x)$	0.2	2.456	0.232

By **Simpson's 1/3 Rule**

$$y(x) = \int_0^{0.8} (0.2 + 25x - 200x^2 + 675x^3 - 900x^4 + 400x^5)dx$$
$$= \frac{4}{3}[y_0 + 4y_1 + y_2]$$

$y_0 = y(0) = 0.2$

$y_1 = y(0.4) = 2.456$

$y_2 = y(0.8) = 0.232$

$$y(n) = \frac{0.4}{3}(0.2 + 4\times 2.456 + 0.232) = 1.367$$

7.18 Sol.

According to **Newton-Raphson Method:**

$$X_{N+1} = X_N - \frac{f(X_N)}{f'(X_N)}$$

$$f(x) = 3x - e^x + \sin x$$

or, $f'(x) = 3 - e^x + \cos x$

$\Rightarrow \quad X_1 = X_0 - \frac{f(0.333)}{f'(0.333)}$

$$= 0.333 - \frac{3 \times 0.333 - e^{0.333} + \sin 0.333}{3 - e^{0.333} + \cos 0.333}$$

$$\therefore \quad X_1 = 0.36$$

7.19 Sol.

$$\frac{du}{dt} = 3t^2 + 1$$

$f(u, t) = 3t^2 + 1$

$u_0 = 0$

$t_0 = 0$

$\Delta t = 2$

By Euler's method

$$u_1 = u_0 + hf(u_0 t_0) \qquad t_1 = t_0 + h$$

$$= u_0 + h(3t_0^2 + 1) \qquad = 0 + 2$$

$$= 0 + 2(3(0)^2 + 1) \qquad = 2$$

$$= 2$$

After first iteration $u = 2$ when $t = 2$,

$$\frac{du}{dt} = 3t^2 + 1$$

$$du = (3t^2 + 1)\, dt$$

$$\int du = \int_0^2 (3t^2 + 1) dt$$

$$u = \left(3\frac{t^3}{3} + t\right)\Bigg|_0^2 = 8 + 2 = 10$$

Absolute error = Exact value − approx value

$= 10 - 2 = 8$

7.20 Sol.

Given $f(x) = 2x^2 - 3x + 3$, $x_0 = 2$

$f'(x) = 4x - 3$

By Newton-Rapshon

$$x_1 = x_0 - \frac{f(x_0)}{f'(x_0)}$$

$$= 2 - \frac{2(2)^2 - 3(2) + 3}{4(2) - 3} = 2 - \frac{5}{5} = 1$$

7.22 (b)

$$\left.\frac{\partial^3 u}{\partial x^3}\right|_{x_i} = \frac{-u_{i+3} + 8u_{i+2} - 13u_{i+1} + 13u_{i-1} - 8u_{i-2} + u_{i-3}}{8\Delta h^3}$$

$$f'''(x_i) = \frac{-f(x_{i+3}) + 8f(x_{i+2}) - 13f(x_{i+1}) + 13f(x_{i-1}) - 8f(x_{i-2}) + f(x_{i-3})}{8h^3}$$

7.23 Sol.

$d = 0.5$ unit

- → $y_0 = 5$
- → $y_1 = 7.25$
- → $y_2 = 10$
- → $y_3 = 13.25$
- → $y_4 = 17$

$$A = \frac{d}{3}\left[(y_0 + y_4) + 4(y_1 + y_3) + 2y_2\right]$$

$$= \frac{0.5}{3}\left[(5 + 17) + 4(7.25 + 13.25) + 2 \times 10\right]$$

$= 20.67$ unit^2

7.24 (c)

Given, $n = 4$, $a = 0$, $b = 1$,

Then $h = \frac{b-a}{n} = 0.25$

$\therefore \quad y = f(n) = e^x$

So

$x =$	0	$\frac{1}{4}$	$\frac{1}{2}$	$\frac{3}{4}$	1
$y =$	1	$e^{1/4}$	$e^{1/2}$	$e^{3/4}$	e^1
	y_0	y_1	y_2	y_3	y_4

Using T-rule

$$I = \int_0^1 e^x dx = \frac{h}{2}\left[y_0 + y_4 + 2(y_1 + y_2 + y_3)\right]$$

$$= \frac{0.25}{2}\left[1 + e + 2\left(e^{1/4} + e^{1/2} + e^{3/4}\right)\right] = 1.727$$

7.25 Sol.

$y = 10x - 20x^2$,

$a = 0$, $b = 0.5$, $n = 5$

So, $h = \frac{b-a}{n} = 0.1$

And

$x =$	0	0.1	0.2	0.3	0.4	0.5
$y =$	0	0.8	1.2	1.2	0.8	0
	y_0	y_1	y_2	y_3	y_4	y_5

$$I = \int_0^{0.5} f(x) dx$$

$$= \frac{h}{2}\left[y_0 + y_5 + 2(y_1 + y_2 + y_3 + y_4)\right]$$

$$= \frac{0.1}{2}\left[0 + 0 + 2(0.8 + 1.2 + 1.2 + 0.8)\right]$$

$= 0.40$

7.26 Sol.

When $n \to \infty$

$$x_{n+1} = x_n = x$$

$$\Rightarrow \quad x = \frac{1}{2}\left(x + \frac{P}{x}\right)$$

$$\Rightarrow \quad 2x = \frac{x^2 + P}{x}$$

$$\Rightarrow \quad x^2 - P = 0$$

$$\Rightarrow \quad x = \sqrt{P}$$

For $P = 10$

$$x = x_{n+1} = x_n$$

$$= \sqrt{10} = 3.162$$

7.27 Sol.

	x_0	x_1	x_2
x	1	1.2	1.4
$f(x)$	3	–	?
	y_0	y_1	y_2

For finding y_2, two iteration has to be followed.

$$\frac{dy}{dx} = 4(x+2) - y \text{ (given differential equation)}$$

$\therefore \quad y_1 = y_o + hf(x_o, y_o) = 3 + 0.2\, f(1, 3)$

$\Rightarrow \quad y_1 = 3 + 0.2\,[4(1 + 2) - 3] = 4.8$

$\therefore \quad y_2 = y_1 + hf(x_1, y_1) = 4.8 + 0.2\, f(12, 4.8)$

$\Rightarrow \quad y_2 = 4.8 + 0.2\,[4(1.2 + 2) - 4.8]$

$\Rightarrow \quad y_2 = 4.8 + 1.6 = 6.4$

7.28 (a)

Using the standard property of trapezoidal rule for converse shaped parabola, the numerical value of the area obtained using the trapezoidal rule will be less than the actual.

7.29 (d)

Standard formula for finding 2nd derivative using 4th order divided difference formula is as follows;

$$f''(x) = \frac{1}{12h^2}\left[-f_{i+2} + 16f_{i+1} - 30f_i + 16f_{i-1} - f_{i-2}\right]$$

On comparing the given options with above result, the correction option comes out to be (d).

7.30 (0.17)

It is based on 2nd order interpolation

$$f(x) = y_1 + \frac{y_2 - y_1}{x_2 - x_1}(x - x_1)$$

$f(x)=$

$$y_1 + \frac{y_2 - y_1}{x_2 - x_1}(x - x_1) + \frac{\frac{y_3 - y_2}{x_3 - x_2} - \frac{y_2 - y_1}{x_2 - x_1}}{x_3 - x_1}(x - x_1)(x - x_2)$$

Here,

$y_1 = 0$	$y_2 = 0.3010$	$y_3 = 0.4771$
$x_1 = 1$	$x_2 = 2$	$x_3 = 3$

$$f(x) = 0 + \left[\frac{0.3010 - 0}{2 - 1}\right](x - 1) + \left[\frac{\frac{0.4771 - 0.3010}{3 - 2} - \frac{0.3010 - 0}{2 - 1}}{3 - 1}\right](x - 1)(x - 2)$$

$$\Rightarrow f(x) = 0.3010(x - 1) + \left(\frac{-0.1249}{2}\right)(x - 1)(x - 2)$$

Put $x = 1.5$

$\therefore \quad f(x) =$

$$0.3010(1.5 - 1) + \frac{-0.1249}{2}(1.5 - 1)(1.5 - 2)$$

$$\Rightarrow \quad f(1.5) = 0.1505 + 0.0156$$

$$= 0.1661 \simeq 0.17$$

7.31 (a)

Taking option (a) $0 \le x \le 2$

$$f(0) = 0 - 45 < 0$$

$$f(2) = 2^5 - 5(2)^4 - 10(2)^3 + 50(2)^2 + 9 \times 2 - 45$$

$$= +45 > 0$$

$\left.\begin{array}{l} f(0) < 0 \\ f(2) > 0 \end{array}\right] \Rightarrow$ Hence there will be one root in this interval which will be smallest root as per the given option.

■■■■

8 Laplace Transforms

8.1 The Laplace transform of the function $(t) = e^{at}$ when $t > 0$ and where a is a constant is

(a) $\frac{1}{(s-a)}$ (b) $\frac{1}{(s+a)}$

(c) $\frac{1}{(s-a)^{-1}}$ (d) $\frac{1}{(s+a)^{-1}}$

[1994 : 1 M]

8.2 The inverse Laplace transform of $\frac{s+9}{s^2+6s+13}$ is

(a) $\cos 2t + 9 \sin 2t$

(b) $e^{-3t} \cos 2t + 3e^{-3t} \sin 2t$

(c) $e^{-3t} \sin 2t + 3e^{-3t} \cos 2t$

(d) $e^{-3t} \cos 2t + e^{-3t} \sin 2t$ [1995 : 1 M]

8.3 Laplace transform $L(f)$, where f is the derivative of function, f is given by

(a) $sL(f) - f(0)$ (b) $sL(f) - f(0)$

(c) $s^2L(f) - f(0)$ (d) $L(f)/s - f(0)$

[1996 : 1 M]

8.4 The Laplace Transform of a unit step function $u_a(t)$, defined as

$$u_a(t) = \begin{cases} 0 \text{ for } t < a \\ 1 \text{ for } t > a \end{cases} \text{ is}$$

(a) e^{-as}/s (b) se^{-as}

(c) $s - u(0)$ (d) $se^{-as} - 1$

[1998 : 1 M]

8.5 The Laplace transform of the function

$f(t) = k,\ 0 < t < c$

$= 0,\ c < t < \infty$, is

(a) $\left(\frac{k}{s}\right)e^{-sc}$ (b) $\left(\frac{k}{s}\right)e^{sc}$

(c) ke^{-sc} (d) $\left(\frac{k}{s}\right)\left(1-e^{sc}\right)$

[1999 : 1 M]

8.6 Let $F(s) = \mathcal{L}[f(t)]$ denote the Laplace transform of the function $f(t)$. Which of the following statements is correct?

(a) $\mathcal{L}\left[\frac{df}{dt}\right] = \frac{1}{s}F(s)$ $\mathcal{L}\int_0^t f(\tau)d\tau = sF(s) - f(0)$

(b) $\mathcal{L}\left[\frac{df}{dt}\right] = sF(s) - F(0)$ $\mathcal{L}\left[\int_0^t f(\tau)d\tau\right] = -\frac{dF}{ds}$

(c) $\mathcal{L}\left[\frac{df}{dt}\right] = sF(s) - F(0)$ $\mathcal{L}\left[\int_0^t f(\tau)d\tau\right] = F(s-a)$

(d) $\mathcal{L}\left[\frac{df}{dt}\right] = sF(s) - F(0)$ $\mathcal{L}\left[\int_0^t f(\tau)d\tau\right] = \frac{1}{s}F(s)$

[2000 : 1 M]

8.7 The inverse Laplace Transform of $\frac{1}{(s^2+2s)}$ is

(a) $\left(1-e^{-2t}\right)$ (b) $\frac{\left(1+e^{+2t}\right)}{2}$

(c) $\frac{\left(1-e^{+2t}\right)}{2}$ (d) $\frac{\left(1-e^{-2t}\right)}{2}$

[2001 : 2 M]

8.8 The Laplace transformed of the following function is

$$f(t) = \begin{cases} \sin t & for \quad 0 \le t \le \pi \\ 0 & for \quad 1 > \pi \end{cases}$$

(a) $\frac{1}{1+s^2}$ for all $s > 0$ (b) $\frac{1}{1+s^2}$ for all $s < \pi$

(c) $\frac{1+e^{-\pi s}}{1+s^2}$ for all $s>0$ (d) $\frac{e^{-\pi s}}{1+s^2}$ for all $s > 0$

[2002 : 2 M]

8.9 If L defines the Laplace Transform of a function, $L[\sin(at)]$ will be equal to

(a) $\frac{a}{s^2-a^2}$ (b) $\frac{a}{s^2+a^2}$

(c) $\frac{s}{s^2+a^2}$ (d) $\frac{s}{s^2-a^2}$

[2003 : 2 M]

8.10 Laplace transform for the function $f(x) = \cosh(ax)$ is

(a) $\frac{a}{s^2-a^2}$ (b) $\frac{s}{s^2-a^2}$

(c) $\frac{a}{s^2+a^2}$ (d) $\frac{s}{s^2+a^2}$

[2009 : 2 M]

8.11 The Laplace transform $F(s)$ of the exponential function. $f(t) = e^{at}$ when $t \geq 0$, where a is a constant and $(s - a) > 0$, is

(a) $\frac{1}{s+a}$ (b) $\frac{1}{s-a}$

(c) $\frac{1}{a-s}$ (d) ∞

[2018 : 2 M, Set-II]

8.12 The Laplace transform of $\sinh(at)$ is

(a) $\frac{s}{s^2+a^2}$ (b) $\frac{s}{s^2-a^2}$

(c) $\frac{a}{s^2-a^2}$ (d) $\frac{a}{s^2+a^2}$

[2019 : 1 M, Set-II]

8.13 The function $f(x, y)$ satisfies the Laplace equation

$$\Delta^2 f(x, y) = 0$$

on a circular domain of radius $r = 1$ with its center at point P with coordinates $x = 0$, $y = 0$. The value of this function on the circular boundary of this domain is equal to 3. The numerical value of $f(0, 0)$ is:

(a) 1 (b) 0

(c) 2 (d) 3

[2022 : 2 M, Set-II]

8.14 The steady-state temperature distribution in a square plate ABCD is governed by the 2-dimensional Laplace equation. The side AB is kept at a temperature of 100 °C and the other three sides are kept at a temperature of 0 °C. Ignoring the effect of discontinuities in the boundary conditions at the corners, the steady-state temperature at the center of the plate is obtained as T_0 °C. Due to symmetry, the steady-state temperature at the center will be same (T_0 °C), when any one side of the square is kept at a temperature of 100 °C and the remaining three sides are kept at a temperature of 0 °C. Using the principle of superposition, the value of T_0 is ________ (rounded off to two decimal places).

[2023 : 2 M, Set-II]

Answers Laplace Transforms

8.1 (a) 8.2 (b) 8.3 (a) 8.4 (*) 8.5 (d) 8.6 (d) 8.7 (d) 8.8 (c) 8.9 (b)

8.10 (b) 8.11 (b) 8.12 (c) 8.13 (d)

Explanations Laplace Transforms

8.2 (b)

Taking Laplace transform, we have

$$L^{-1}\left[\frac{s+9}{s^2+6s+13}\right] = L^{-1}\left[\frac{(s+3)+6}{(s+3)^2+2^2}\right]$$

$$= L^{-1}\left[\frac{s+3}{(s+3)^2+2^2}\right] + L^{-1}\left[\frac{6}{(s+3)^2+2^2}\right]$$

$$= e^{-3t}\cos 2t + \frac{6}{2}e^{-3t}\sin 2t = e^{-3t}[\cos 2t + 3\sin 2t]$$

8.5 (d)

$$L[f(t)] = \int_{-\infty}^{\infty} f(t)e^{-st}dt = \int_0^c ke^{-st}dt$$

$$= -\frac{k}{s}e^{-st}\Big|_0^e = -\frac{k}{s}\left(e^{-sc}-1\right) = \frac{k}{s}\left(1-e^{-sc}\right)$$

8.7 (d)

$$\frac{1}{S(S+2)} = \left[\frac{1}{S} - \frac{1}{S+2}\right] = \frac{1}{2}\left[1-e^{-2t}\right]$$

8.9 (b)

$$L[f(t)] = \int_0^{\infty} e^{-st}f(t)\,dt$$

$$\therefore\ L[\sin(at)] = \int_0^{\infty} e^{-st}\sin(at)\,dt = \frac{a}{s^2+a^2}$$

8.11 (b)

$$L(e^{at}) = \frac{1}{S-a}$$

$$L(a^{at}) = \int_0^{\infty} e^{-st}e^{at}dt = \int_0^{\infty} e^{-(s-a)t}dt$$

$$= \frac{e^{-(s-a)t}}{-(s-a)}\Bigg|_0^{\infty} = -\frac{1}{s-a}(0-1) = \frac{1}{s-a}$$

8.12 (c)

$$L(\sinh(at)) = \frac{a}{s^2 - a^2}$$

8.13 (d)

According to given condition given function $f(x, y)$ is nothing but constant function i.e. $f(x, y) = 3$ because this is the only function whose value is 3 at any point on the boundary of unit circle and it is also satisfying Laplace equation, so

$$f(0, 0) = 3$$

8.14 Sol.

Laplace equation is $\frac{\partial^2 u}{\partial x^2} + \frac{\partial^2 u}{\partial y^2} = 0$ ...(i)

Its solution using separation of variables method is

$\therefore u(x, y)$

$= (C_1 \cos\lambda x + C_2 \sin\lambda x)(C_3 e^{\lambda y} + C_4 e^{-\lambda y})$ ...(ii)

Let $AB = CD = 1$

Using $u(x, 0) = 0$

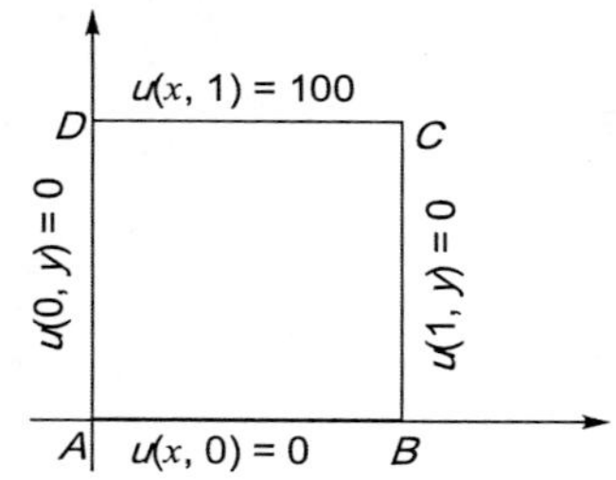

$\Rightarrow C_3 = -C_4$

Using $(0, y) = 0$

$\Rightarrow C_1 = 0$

Substituting constants values in equation (ii)

$u(x, y) = (0 + C_2 \sin\lambda x) \quad (C_3 e^{\lambda y} - C_3 e^{-\lambda y})$

$u(x, y) = a_n \sin\lambda x(e^{\lambda y} - e^{-\lambda y})$

Now, using $u(1, y) = 0$,

$\sin\lambda = 0 \Rightarrow \lambda = n\pi$

Hence

$u(x, y) = \sum a_n \sin(n\pi x)\left[e^{n\pi y} - e^{-n\pi y}\right]$ ...(iii)

Using $u(x, 1) = 100$

$$\Rightarrow a_n = \frac{100}{\sin n\pi x\left[e^{n\pi} - e^{-n\pi}\right]}$$

Using equation (iii),

$$u(x, y) = \sum_{n=1}^{\infty} \frac{100}{e^{n\pi} - e^{-n\pi}}\left(e^{n\pi y} - e^{-n\pi y}\right)$$

At mid point,

$$u\left(\frac{1}{2}, \frac{1}{2}\right) = T_0$$

$$\text{So, } T_0 = \sum_{n=1}^{\infty} \frac{100}{\left(e^{n\pi} - e^{-n\pi}\right)}\left(e^{\frac{n\pi}{2}} - e^{\frac{-n\pi}{2}}\right)$$

$$= 100\left[\left(\frac{e^{\frac{\pi}{2}} - e^{-\frac{\pi}{2}}}{e^{\pi} - e^{-\pi}}\right) + \left(\frac{e^{\pi} - e^{-\pi}}{e^{2\pi} - e^{-2\pi}}\right) + \frac{e^{\frac{3\pi}{2}} - e^{-\frac{3\pi}{2}}}{e^{3\pi} - e^{-3\pi}} + \ldots\right]$$

$$= 100\left[\frac{1}{e^{\frac{\pi}{2}} - e^{-\frac{\pi}{2}}} + \frac{1}{e^{\pi} - e^{-\pi}} + \frac{1}{e^{\frac{3\pi}{2}} - e^{-\frac{3\pi}{2}}} + \ldots\right]$$

$$= 50\left[\frac{1}{\cosh\left(\frac{\pi}{2}\right)} + \frac{1}{\cosh(\pi)} + \frac{1}{\cosh\left(\frac{3\pi}{2}\right)} + \ldots\right]$$

$$= 50\left[\frac{1}{2.509} + \frac{1}{11.59} + \frac{1}{55.66} + \ldots\right]$$

$= 19.928 + 4.314 + 0.903 +$ Neglecting

$= 25.145 \approx 25.1$

■■■■

9 Fourier Series and Transform

9.1 The series $s = 1 - 1 + 1 - 1 \ldots$ is
(a) convergent
(b) monotonous
(c) uniformly convergent
(d) divergent [1996 : 1 M]

9.2 Fourier expansion of the periodic function $f(t) = t^3$ with a period $\left(\frac{2\pi}{\omega}\right)$ has the form

(a) $\frac{\sigma_0}{2} + \sum_{\nu=1}^{\infty}(a_n \cos n\omega t + b_n \sin n\omega t)$

(b) $\frac{\sigma_0}{2} + \sum_{\nu=1}^{\infty} b_n \sin n\omega t$

(c) $\frac{\sigma_0}{2} + \sum_{\nu=1}^{\infty} a_n \cos n\omega t$

(d) $\sum_{\nu=1}^{\infty} b_n \sin n\omega t$ [1996 : 1 M]

9.3 A discontinuous real function can be expressed as
(a) Taylor's series and Fourier's series
(b) Taylor's series and not by Fourier's series
(c) Neither Taylor's series nor Fourier's series
(d) Not by Taylor's series but Fourier's series
[1998 : 1 M]

9.4 The infinite series $1 + \frac{1}{2} + \frac{1}{3} + \ldots$
(a) converges (b) diverges
(c) oscillates (d) unstable
[1998 : 2 M]

9.5 The infinite series $\sum_{n=1}^{\infty} \frac{(n!)^2}{(2n)!}$
(a) Converges (b) Diverges
(c) Is unstable (d) Oscillates
[1999 : 1 M]

9.6 The following integral $\lim_{a\to\infty} \int_1^a x^{-4} dx$
(a) Diverges (b) Converges to $\frac{1}{3}$
(c) Converges to $-\frac{1}{a}$ (d) Converges to 0
[2000 : 1 M]

9.7 A function with a period 2π is shown below

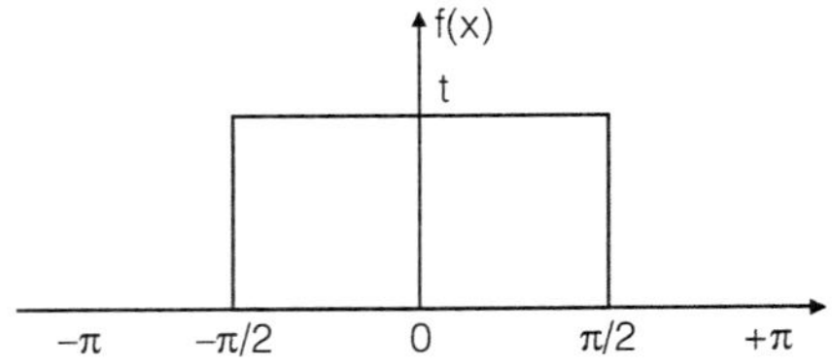

The Fourier series for this function is given by

(a) $f(x) = \frac{1}{2} + \sum_{n=1}^{\infty} \frac{2}{n\pi} \sin\frac{n\pi}{2} \cos nx$

(b) $f(x) = \sum_{n=1}^{\infty} \frac{2}{n\pi} \sin\frac{n\pi}{2} \cos nx$

(c) $f(x) = \frac{1}{2} + \sum_{n=1}^{\infty} \frac{2}{n\pi} \sin\frac{n\pi}{2} \sin nx$

(d) $f(x) = \sum_{n=1}^{\infty} \frac{2}{n\pi} \sin\frac{n\pi}{2} \sin nx$ [2000 : 1 M]

9.8 The limit of the following series as x approaches $\frac{\pi}{2}$ is $f(x) = x - \frac{x^3}{3!} + \frac{x^5}{5!} - \frac{x^7}{7!}$
(a) $\frac{2\pi}{3}$ (b) $\frac{\pi}{2}$
(c) $\frac{\pi}{3}$ (d) 1 [2001 : 1 M]

9.9 The Fourier series expansion of a symmetric and even function, $f(x)$ where
$f(x) = 1 + (2x/\pi), \quad -\pi \le x \le 0$
and $\quad = 1 - (2x/\pi), \quad 0 \le x \le \pi$
will be

(a) $\sum_{n=1}^{\infty} \frac{4}{\pi^2 n^2}(1 + \cos n\pi)$

(b) $\sum_{n=1}^{\infty} \frac{4}{\pi^2 n^2}(1 - \cos n\pi)$

(c) $\sum_{n=1}^{\infty} \frac{4}{\pi^2 n^2}(1 - \sin n\pi)$

(d) $\sum_{n=1}^{\infty} \frac{4}{\pi^2 n^2}(1 + \sin n\pi)$ [2003 : 2 M]

9.10 The summation of series

$S = 2 + \frac{5}{2} + \frac{8}{2^2} + \frac{11}{2^3} + ... \infty$ is

(a) 4.50 (b) 6.0
(c) 6.75 (d) 10.0 [2004 : 1 M]

9.11 The Fourier series of the function,

$$f(x) = 0, \quad -\pi < x \le 0$$
$$= \pi - x, \quad 0 < x < \pi$$

in the interval $[-\pi, \pi]$ is

$$f(x) = \frac{\pi}{4} + \frac{2}{\pi}\left[\frac{\cos x}{1^2} + \frac{\cos 3x}{3^3} + ...\right] + \left[\frac{\sin x}{1} + \frac{\sin 2x}{2} + \frac{\sin 3x}{3} + ...\right]$$

The convergence of the above Fourier series at $x = 0$ gives

(a) $\sum_{n=1}^{\infty} \frac{1}{n^2} = \frac{\pi^2}{6}$

(b) $\sum_{n=1}^{\infty} \frac{(-1)^{n+1}}{n^2} = \frac{\pi^2}{12}$

(c) $\sum_{n=1}^{\infty} \frac{1}{(2n-1)^2} = \frac{\pi^2}{8}$

(d) $\sum_{n=1}^{\infty} \frac{(-1)^{n+1}}{2n-1} = \frac{\pi}{4}$ [2016 : 1 M, Set-II]

9.12 The Fourier series to represent $x - x^2$ for $-\pi \le x \le \pi$ is given by

$$x - x^2 = \frac{a_0}{2} + \sum_{n=1}^{\infty} a_n \cos nx + \sum_{n=1}^{\infty} b_n \sin nx$$

The value of a_0 (round off to two decimal places), is _______.

[2020 : 2 M, Set-II]

9.13 The Fourier cosine series of a function is given by:

$$f(x) = \sum_{n=0}^{\infty} f_n \cos nx$$

For $f(x) = \cos^4 x$, the numerical value of $(f_4 + f_5)$ is ____. (round off to three decimal places)

[2022 : 1 M, Set-I]

9.14 The following function is defined over the interval $[-L, L]$:

$$F(x) = px^4 + qx^5$$

If it is expressed as a Fourier series,

$$F(x) = a_0 + \sum_{n=1}^{\infty}\left\{a_n \sin\left(\frac{\pi x}{L}\right) + b_n \cos\left(\frac{\pi x}{L}\right)\right\}$$

which options amongst the following are true?

(a) a_n, $n = 1, 2, ..., \infty$ depend on p
(b) a_n, $n = 1, 2, ..., \infty$ depend on q
(c) b_n, $n = 1, 2, ..., \infty$ depend on p
(d) b_n, $n = 1, 2, ..., \infty$ depend on q

[2023 : 1 M, Set-I]

9.15 A 5 cm long metal rod AB was initially at a uniform temperature of T_0 °C. Thereafter, temperature at both the ends are maintained at 0 °C. Neglecting the heat transfer from the lateral surface of the rod, the heat transfer in the rod is governed by the one-dimensional diffusion equation $\frac{\partial T}{\partial t} = D\frac{\partial^2 T}{\partial x^2}$, where D is the thermal diffusivity of the metal, given as 1.0 cm²/s.
The temperature distribution in the rod is obtained as

$$T(x, t) = \sum_{n=1,3,5...}^{\infty} C_n \sin\frac{n\pi x}{5} e^{-\beta n^2 t},$$

where x is in cm measured from A to B with $x = 0$ at A, t is in s, C_n are constants in °C, T is in °C, and β is in s^{-1}. The value of β (in s^{-1}, rounded off to three decimal places) is _______.

[2023 : 2 M, Set-II]

Answers Fourier Series and Transform

9.1	(b)	9.2	(d)	9.3	(d)	9.4	(a)	9.5	(a)	9.6	(b)	9.7	(a)	9.8	(d)	9.9	(b)
9.10	(d)	9.11	(c)	9.14	(b, c)												

Explanations Fourier Series and Transform

9.4 (a)

The sum of the series will never tend to zero as n tends to infinity.

9.6 (b)

$$\lim_{a\to\infty}\int_1^a x^{-4}dn = \lim_{a\to\infty}\left[-\frac{1}{3}\left[x^{-3}\right]_1^a\right]$$

$$= \lim_{a\to\infty} -\frac{1}{3}\left[a^{-3}-1\right] = \frac{1}{3}[0-1] = \frac{1}{3}$$

9.7 (a)

$$f(x) = 1 - \frac{\pi}{2} < x < \frac{\pi}{2}$$

$$= 0 \qquad \text{in other region}$$

Then $\dfrac{a_0}{2} = \dfrac{1}{2\pi}\displaystyle\int_{-\pi/2}^{\pi/2} 1dx = \dfrac{1}{2}$

$$a_x = \frac{1}{\pi}\int_{-\pi/2}^{\pi/2} 1\cos x dx = \frac{2}{nx}\sin\frac{nx}{2}$$

$$b_n = 0 \qquad [\because f(x) \text{ is even function}]$$

Hence series is,

$$f(x) = \frac{1}{2} + \sum_{n=1}^{\infty}\frac{2}{n\pi}\sin\frac{nx}{2}\cos nx$$

9.8 (d)

$$\underset{x\to\frac{\pi}{2}}{Lt} f(x) = \underset{x\to\frac{\pi}{2}}{Lt} \sin x = 1$$

9.9 (b)

Since given function is symmetric and even, its Fourier series contain only cosine terms.

$$a_0 = \frac{2}{c}\int_0^c f(x)\,dx$$

and $a_n = \dfrac{2}{c}\displaystyle\int_0^c f(x)\cos\frac{n\pi x}{c}dx$

Using above result and using,

$f(x) = 1 - \dfrac{2x}{\pi}$ we get

$$a_0 = \frac{2}{\pi}\int_0^{\pi}\left(1-\frac{2x}{\pi}\right)dx = \frac{2}{\pi}\left(x-\frac{x^2}{\pi}\right)_0^{\pi} = 0$$

and $a_n = \dfrac{2}{\pi}\displaystyle\int_0^{\pi}\left(1-\frac{2x}{\pi}\right)\cos nx\,dx$

$$= \frac{2}{\pi}\left[\left(1-\frac{2x}{\pi}\right)\frac{\sin nx}{n} - \left(-\frac{2}{\pi}\right)\left(-\frac{\cos nx}{n^2}\right)\right]_0^{\pi}$$

$$= \frac{2}{\pi}\left[-\frac{2\cos n\pi}{\pi n^2} + \frac{2}{\pi n^2}\right]$$

$$= \frac{4}{n^2\pi^2}[1-\cos n\pi]$$

9.10 (d)

S is an arithmetico-geometric series and can be summed up as follows,

$$S = 2 + \left(\frac{5}{2} + \frac{8}{2^2} + \frac{11}{2^3} + \ldots\infty\right)$$

$$= 2 + x \text{ (say)}$$

where $x = \dfrac{5}{2} + \dfrac{8}{2^2} + \dfrac{11}{2^3} + \ldots\infty$ ...(i)

Multipling by 1/2, we get,

$\dfrac{1}{2}x = \dfrac{5}{2^2} + \dfrac{8}{2^3} + \dfrac{11}{2^4} + \ldots\infty$ ...(ii)

Subtracting (ii) from (i), we get,

$$x - \frac{1}{2}x = \frac{5}{2} + \frac{(8-5)}{2^2} + \frac{(11-8)}{2^3} + \frac{(13-11)}{2^4} + \ldots\infty$$

$$\frac{x}{2} = \frac{5}{2} + \left(\frac{3}{2^2} + \frac{3}{2^3} + \frac{3}{2^4} + \ldots\infty\right)$$

$$\frac{x}{2} = \frac{5}{2} + \frac{\frac{3}{2^2}}{\left[1-\frac{1}{2}\right]}$$

$$\frac{x}{2} = \frac{5}{2} + \frac{3}{2} = 4$$

$\therefore$ $x = 8$

and $S = 2 + x = 10$

9.11 (c)

The function is $f(x) = 0,$

$-p < x \le 0$

$= p - x,\ 0 < x < \pi$

And Fourier series is,

$$f(x) = \frac{\pi}{4} + \frac{2}{\pi}\left[\frac{\cos x}{1^2} + \frac{\cos 3x}{3^2} + \frac{\cos 5x}{5^2} + ...\right]$$

$$+\left[\frac{\sin x}{1} + \frac{\sin 2x}{2} + \frac{\sin 3x}{3} + ...\right] \quad ...(i)$$

At $x = 0$, (a point of discontinuity), the fourier series converges to $\frac{1}{2}\left[f(0^{-1}) + f(0^{+})\right]$,

where $f(0^-) = \lim_{x\to 0}(\pi - x) = \pi$

Hence, eq. (i), we get,

$$\frac{\pi}{2} = \frac{\pi}{4} + \frac{2}{\pi}\left[\frac{1}{1^2} + \frac{1}{3^2} + ...\right]$$

$$\Rightarrow \quad \frac{1}{1} + \frac{1}{3^2} + \frac{1}{5^2} + ... \frac{\pi^2}{8}$$

9.12 Sol.

$$a_0 = \frac{1}{\pi}\int_{-\pi}^{\pi} f(x)dx = \frac{1}{\pi}\int_{-\pi}^{\pi}(x - x^2)dx$$

$$= \frac{-1}{\pi}\int_0^{\pi} 2x^2 dx = -\frac{1}{\pi}\left(\frac{2x^3}{3}\right)_0^{\pi}$$

$$= -\frac{2}{3\pi}\left[\pi^3\right] = \frac{-2\pi^2}{3} = -6.58$$

9.13 Sol.

$$f(x) = \cos^4 x = (\cos^2 x)^2 = \left(\frac{1+\cos 2x}{2}\right)^2$$

$$\Rightarrow \quad f(x) = \frac{1}{4}\left[1 + \cos^2 2x + 2\cos 2x\right]$$

$$\Rightarrow \quad f(x) = \frac{1}{4}\left[1 + \frac{1+\cos 4x}{2} + 2\cos 2x\right]$$

$$= \frac{1}{4}\left[\frac{3}{2} + 2\cos 2x + \frac{\cos 4x}{2}\right]$$

$$\Rightarrow \quad f(x) = \frac{3}{8} + 0\cos x + \frac{1}{2}\cos 2x + 0\cos 3x$$

$$+\frac{1}{8}\cos 4x + 0\cos 5x + ...$$

$$\therefore f(4) + f(5) = \frac{1}{8} + 0 = \frac{1}{8}$$

9.14 (b, c)

$$b_n = \frac{1}{l}\int_{-l}^{l} F(x)\cos\frac{(n\pi x)}{l}dx$$

$$= \frac{1}{l}\int_{-l}^{l}(px^4 + qx^5)\cos\left(\frac{n\pi x}{l}\right)dx$$

$$= \frac{1}{l}\int_{-l}^{l} px^4 \cos\frac{(n\pi x)}{l}dx + 0$$ ($\because$ 2nd integral is an odd functions)

Thus, b_n depend on p

Similarly,

$$a_n = \frac{1}{l}\int_{-l}^{l} F(x)\sin\frac{(n\pi x)}{l}dx$$

$$= \frac{1}{l}\int_{-l}^{l}(px^4 + qx^5)\sin\left(\frac{n\pi x}{l}\right)dx$$

$$= 0 + \frac{1}{l}\int_{-l}^{l} qx^5 \sin\frac{(n\pi x)}{l}dx$$ ($\because$ 1st integral is an odd functions)

Thus, a_n depend on q.

9.15 Sol.

$$\frac{\partial T}{\partial t} = \Delta\frac{\partial^2 T}{\partial x^2};\ T(0, t) = 0,$$

$T(5, t) = 0,\ T(x, 0) = T_0,\ D = 1$ ...(1)

Put $D = 1$, it's general solution using separation of variables methods is

$T(x, t) = (C_1 \cos px + C_2 \sin px)C_3 e^{-p^2 t}$ ...(2)

Using $T(0, t) = 0 \Rightarrow C_1 = 0$

Using $T(5, t) = 0 \Rightarrow C_2 C_3 \sin 5p e^{-p^2 t} = 0$

or $\sin 5p = 0 \Rightarrow p = \frac{n\pi}{5}, n \in I$

Hence, by (2),

$T(x, t) = b_n \sin\left(\frac{n\pi x}{5}\right)\cdot e^{\frac{-n^2\pi^2 t}{25}}$ where $b_n = C_2 C_3$

Hence most general solution is

$$T(x, t) = \sum_{n=1}^{\infty} b_n \sin\left(\frac{n\pi x}{5}\right)e^{\frac{-n^2\pi^2 t}{25}} \quad ...(3)$$

Now using $T(x, 0) = T_0$

$$\Rightarrow \quad T_0 = \sum_{n=1}^{\infty} b_n \sin\left(\frac{n\pi x}{5}\right)$$

which is half range fourier sine series for T_0

Hence,

$$b_n = \frac{2}{5}\int_0^5 T_0 \sin\left(\frac{n\pi x}{5}\right) dx = \begin{cases} 0 \ , & n \text{ even} \\ \dfrac{4T_0}{n\pi}, & n \text{ odd} \end{cases}$$

By (3), $T(x, t) = \sum_{n=\text{odd}} \frac{4T_0}{n\pi} \sin\left(\frac{n\pi x}{5}\right) e^{\frac{-n^2\pi^2 t}{25}}$ ...(4)

On comparision with,

$$T(x, t) = \sum_{n=1, 3, 5} C_n \sin\left(\frac{n\pi x}{5}\right) e^{-\beta n^2 t}$$

$$\Rightarrow \quad C_n = \frac{4T_0}{n\pi} \text{ and } \beta$$

$$= \frac{\pi^2}{25} = 0.3947 \approx 0.395$$

■■■■

10 Partial Diffrential Equation

10.1 The number of boundary conditions required to solve the following differential equation is

$$\frac{\partial^2\phi}{\partial x^2}+\frac{\partial^2\phi}{\partial y^2}=0$$

(a) 2 (b) 0
(c) 4 (d) 1

[2001 : 1 M]

10.2 The partial differential equation that can be formed from
$z = ax + by + ab$ has the form (with $p=\frac{\partial z}{\partial x}$ and $q=\frac{\partial z}{\partial y}$)

(a) $z = px + qy$
(b) $z = px + pq$
(c) $z = px + qy + pq$
(d) $z = qy + pq$

[2010 : 2 M]

10.3 The type of partial differential equation

$$\frac{\partial^2 P}{\partial x^2}+\frac{\partial^2 P}{\partial y^2}+3\frac{\partial^2 P}{\partial x\partial y}+2\frac{\partial P}{\partial x}-\frac{\partial P}{\partial y}=0 \text{ is}$$

(a) elliptic (b) parabolic
(c) hyperbolic (d) none of these

[2016 : 1 M, Set-I]

10.4 The solution of the partial differential equation

$\frac{\partial u}{\partial t}=\alpha\frac{\partial^2 u}{\partial x^2}$ is of the form

(a) $C\cos(kt)\left|C_1e^{(\sqrt{k/\alpha})}+C_2e^{-(\sqrt{k/\alpha})x}\right|$

(b) $Ce^{kt}\left|C_1e^{(\sqrt{k/\alpha})x}+C_2e^{-(\sqrt{k/\alpha})x}\right|$

(c) $Ce^{kt}\left|C_1\cos(\sqrt{k/\alpha})x+C_2\sin(-\sqrt{k/\alpha})x\right|$

(d) $C\sin(kt)\left|C_1\cos(\sqrt{k/\alpha})x+C_2\sin-(\sqrt{k/\alpha})x\right|$

[2016 : 1 M, Set-I]

10.5 Consider the following partial differential equation:

$$3\frac{\partial^2\phi}{\partial x^2}+B\frac{\partial^2\phi}{\partial x\partial y}+3\frac{\partial^2\phi}{\partial y^2}+4\phi=0$$

For this equation to be classified as parabolic, the value of B^2 must be ________.

[2017 : 1 M, Set-I]

10.6 The solution at $x = 1$, $t = 1$ of the partial differential equation $\frac{\partial^2 u}{\partial x^2}=25\frac{\partial^2 u}{\partial t^2}$ subject to initial conditions of $u(0) = 3x$ and $\frac{\partial u}{\partial t}(0)=3$ is ________.

(a) 1 (b) 2
(c) 4 (d) 6

[2018 : 2 M, Set-I]

10.7 A one-dimensional domain is discretized into N sub-domains of width Δx with node numbers $i = 0, 1, 2, 3, \ldots, N$. If the time scale is discretized in steps of Δt, the forward-time and centered-space finite difference approximation at i^{th} node and n^{th} time step, for the partial differential equation $\frac{\partial v}{\partial t}=\beta\frac{\partial^2 v}{\partial x^2}$ is

(a) $\frac{v_{i+1}^{(n+1)}-v_i^{(n)}}{\Delta t}=\beta\left[\frac{v_{i+1}^{(n)}-2v_i^{(n)}+v_{i-1}^{(n)}}{2\Delta x}\right]$

(b) $\frac{v_i^{(n+1)}-v_i^{(n)}}{\Delta t}=\beta\left[\frac{v_{i+1}^{(n)}-2v_i^{(n)}+v_{i-1}^{(n)}}{(\Delta x)^2}\right]$

(c) $\frac{v_i^{(n)}-v_i^{(n-1)}}{\Delta t}=\beta\left[\frac{v_{i+1}^{(n)}-2v_i^{(n)}+v_{i-1}^{(n)}}{(\Delta x)^2}\right]$

(d) $\frac{v_i^{(n)}-v_i^{(n-1)}}{2\Delta t}=\beta\left[\frac{v_{i+1}^{(n)}-2v_i^{(n)}+v_{i-1}^{(n)}}{2\Delta x}\right]$

[2019 : 2 M, Set-I]

10.8 For the following partial differential equation.

$$x\frac{\partial^2 f}{\partial x^2}+y\frac{\partial^2 f}{\partial y^2}=\frac{x^2+y^2}{2}$$

Which of the following option(s) is/are CORRECT?

(a) hyperbolic for $x < 0$ and $y > 0$ (b) parabolic for $x > 0$ and $y > 0$

(c) elliptic for $x = 0$ and $y > 0$ (d) elliptic for $x > 0$ and $y > 0$

[2024 : 1 M, Set-I]

10.9 A partial differential equation

$$\frac{\partial^2 T}{\partial x^2}+\frac{\partial^2 T}{\partial y^2}=0$$

is defined for the two-dimensional field T: $T(x, y)$, inside a planar square domain of size 2 m × 2 m. Three boundary edges of the square domain are maintained at value $T = 50$, whereas the fourth boundary edge is maintained at $T = 100$.

The value of T at the center of the domain is

(a) 75.0 (b) 62.5

(c) 87.5 (d) 50.0

[2024 : 1 M, Set-II]

Answers Partial Diffrential Equation

10.1 (a) 10.2 (c) 10.3 (c) 10.4 (b) 10.6 (d) 10.7 (b) 10.8 (a, d) 10.9 (b)

Explanations Partial Diffrential Equation

10.2 (c)

$$z = ax + by + ab \quad \text{...(i)}$$

$$p = \frac{\partial z}{\partial x} = a$$

$$q = \frac{\partial z}{\partial y} = b$$

Substituting a and b in (i) in terms of p and q, we get,

$$z = px + qy + pq$$

10.3 (c)

Comparing the given equation with the general form of second order partial differential equation, we have $A = 1$, $B = 3$, $C = 1 \Rightarrow B^2 - 4AC = 5 > 0$

∴ *PDE* is Hyperbola.

10.4 (b)

The *PDE* $$\frac{\partial u}{\partial t} = \alpha\frac{\partial^2 u}{\partial x^2} \quad \text{...(i)}$$

Solution of (i) is,

$$u(x, t) = (A\cos px + B\sin px)Ce^{-p^2\alpha t}$$

Put $-p^2\alpha = k$

$$\Rightarrow \quad p = \sqrt{-\frac{k}{\alpha}} = \sqrt{\frac{k}{\alpha}}\,i$$

Putting value of p in eq. (i),

$$u(x, t) = \left(A\cos\sqrt{\frac{k}{\alpha}}x + b\sin h\sqrt{\frac{k}{\alpha}}x\right)Ce^{kt}$$

$$= Ce^{kt}\left[A\left\{\frac{e^{\sqrt{\frac{k}{\alpha}}x}+e^{-\sqrt{\frac{k}{\alpha}}x}}{2}\right\}+B\left\{\frac{e^{\sqrt{\frac{k}{\alpha}}x}-e^{-\sqrt{\frac{k}{\alpha}}x}}{2}\right\}\right]$$

$$= Ce^{kt}\left[e^{\sqrt{\frac{k}{\alpha}}x}\left\{\frac{A+B}{2}\right\}+e^{-\sqrt{\frac{k}{\alpha}}x}\left\{\frac{A-B}{2}\right\}\right]$$

$$= Ce^{kt}\left[c_1e^{\sqrt{\frac{k}{\alpha}}x}+c_2e^{-\sqrt{\frac{k}{\alpha}}x}\right]$$

10.5 Sol.

Given that the partial differential equation is parabolic.

$$\therefore \quad B^2 - 4AC = 0 \qquad \text{Here } A = 3$$

$$\therefore B^2 - 4(3)(3) = 0 \qquad C = 3$$

$$B^2 - 36 = 0$$

$$B^2 = 36$$

10.6 (d)

$$\frac{c^2\partial^2 u}{\partial x^2} = \frac{\partial^2 u}{\partial t^2}$$

$$u(x, 0) = f(x)$$
$$u_t(x, 0) = g(x)$$
$$\frac{\partial^2 u}{\partial x^2} = 25\frac{\partial^2 u}{\partial t^2}$$
$$f(x) = 3x$$
$$g(x) = 3$$
$$c^2 = \frac{1}{25}$$

D'Alembert's formula,

$$u(x, t) = \frac{1}{2}\left[f(x+ct)+f(x-ct)\right]+\frac{1}{2c}\int_{x-ct}^{x+ct} g(y)dy$$

$$= \frac{1}{2}\left[3\left(x+\frac{t}{5}\right)+3\left(x-\frac{t}{5}\right)+\frac{1}{2(1/5)}\int_{x-\frac{t}{5}}^{x+\frac{t}{5}} 3dy\right]$$

$$= \frac{1}{2}[6x]+\frac{3}{2}(5)\left[x+\frac{t}{5}-x+\frac{t}{5}\right]$$

$$u(x, t) = 3x + 3t$$

At $x = 1$, $t = 1$

$$u(x, t) = 6$$

10.7 (b)

$$\frac{\partial v}{\partial t} = \beta\frac{\partial^2 v}{\partial x^2} \quad ...(i)$$

Approximate time derivative in equation (i) with forward difference,

$$\frac{\partial v}{\partial t} = \frac{v_i^{n+1}-v_i^n}{\Delta t}+O(\Delta t) \quad ...(ii)$$

Note that the right hand side only in value v at $x = x_i$

Use the central difference approximation to $\left(\frac{\partial^2 v}{\partial x^2}\right)_{x_i}$ and evaluate all the terms at time n.

$$\left.\frac{\partial^2 v}{\partial x^2}\right|_{x_i} = \frac{v_{i-1}^n-2v_i^n+v_{i+1}^n}{\Delta x^2}+0\left(\Delta x^2\right) \quad ...(iii)$$

Substituting equation (ii) in the left hand side of equation (i), substitute the equation (iii) into the right hand side of equation (i), and collect the truncation error terms to get

$$\frac{v_i^{(n+1)}-v_i^{(n)}}{\Delta t} = \beta\left[\frac{v_{i+1}^{(n)}-2v_i^{(n)}+v_{i-1}^{(n)}}{(\Delta x)^2}\right]+0(\Delta t)+0\left(\Delta x^2\right)$$

$$\therefore \frac{v_i^{(n+1)}-v_i^{(n)}}{\Delta t} = \beta\left[\frac{v_{i+1}^{(n)}-2v_i^{(n)}+v_{i-1}^{(n)}}{(\Delta x)^2}\right]$$

10.8 (a, d)

For elliptical

$$B^2 - 4AC < 0$$
$$0 - 4xy < 0$$

i.e. for $x > 0$ and $y > 0$

Hence option (d) is correct.

For hyperbolic

$$B^2 - 4AC > 0$$
$$0 - 4xy > 0$$

i.e. for $x < 0$ and $y > 0$

Hence option (a) is correct.

10.9 (b)

Temperature at centre = Average of temperature at sides

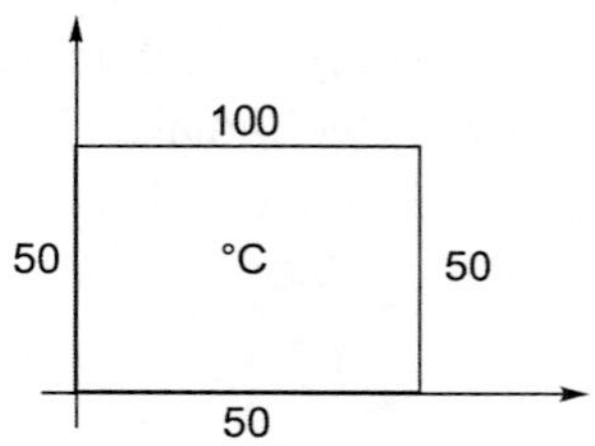

$$= \frac{50+50+50+100}{4} = 62.5$$

■■■■

UNIT XIV

General Aptitude

Verbal Aptitude : Basic English grammar: tenses, articles, adjectives, prepositions, conjunctions, verb-noun agreement, and other parts of speech Basic vocabulary: words, idioms, and phrases in context Reading and comprehension Narrative sequencing.

Quantitative Aptitude : Data interpretation: data graphs (bar graphs, pie charts, and other graphs representing data), 2 and 3-dimensional plots, maps, and tables Numerical computation and estimation: ratios, percentages, powers, exponents and logarithms, permutations and combinations, and series Mensuration and geometry Elementary statistics and probability.

Analytical Aptitude : Logic: deduction and induction ; Analogy : Numerical relations and reasoning.

Spatial Aptitude : Transformation of shapes: translation, rotation, scaling, mirroring, assembling, and grouping Paper folding, cutting, and patterns in 2 and 3 dimensions.

Analysis of Previous GATE Papers

Exam Year	1 Mark Ques.	2 Marks Ques.	Total Marks
2010	5	5	15
2011	5	5	15
2012	5	5	15
2013	5	5	15
2014 (Set-1)	5	5	15
2014 (Set-2)	5	5	15
2015 (Set-1)	5	5	15
2015 (Set-2)	5	5	15
2016 (Set-1)	5	5	15
2016 (Set-2)	5	5	15
2017 (Set-1)	5	5	15
2017 (Set-2)	5	5	15
2018 (Set-1)	5	5	15

Exam Year	1 Mark Ques.	2 Marks Ques.	Total Marks
2018 (Set-2)	5	5	15
2019 (Set-1)	5	5	15
2019 (Set-2)	5	5	15
2020 (Set-1)	5	5	15
2020 (Set-2)	5	5	15
2021 (Set-1)	5	5	15
2021 (Set-2)	5	5	15
2022 (Set-1)	5	5	15
2022 (Set-2)	5	5	15
2023 (Set-1)	5	5	15
2023 (Set-2)	5	5	15
2024 (Set-1)	5	5	15
2024 (Set-2)	5	5	15

General Aptitude

1.1 Which of the following options is the closest in meaning to the word below?

Circuitous

(a) Cyclic (b) Indirect
(c) Confusing (d) Crooked

[2010 : 1 M]

1.2 The question below consists of a pair of related words followed by four pairs of words. Select the pair that best expresses the relation in the original pair.

Unemployed: Worker

(a) Fallow : Land
(b) Unaware : Sleeper
(c) Wit : Jester
(d) Renovated : House [2010 : 1 M]

1.3 Choose the most appropriate word from the options given below to complete the following sentence:

If we manage to .... our natural resources, we would leave a better planet for our children.

(a) uphold (b) restrain
(c) cherish (d) conserve

[2010 : 1 M]

1.4 Choose the most appropriate word from the options given below to complete the following sentence:

His rather casual remarks on politics .... his lack of seriousness about the subject.

(a) masked (b) belied
(c) betrayed (d) suppressed

[2010 : 1 M]

1.5 25 persons are in a room. 15 of them play hockey, 17 of them play football and 10 of them play both hockey and football. Then the number of persons playing neither hockey nor football is

(a) 2 (b) 17
(c) 13 (d) 3 [2010 : 1 M]

1.6 Modern warfare has changed from large scale clashes of armies to suppression of civilian populations. Chemical agents that do their work silently appear to be suited to such warfare; and regretfully, there exist people in militrary establishments who think that chemical agents are usefull fools for their cause.

Which of the following statements best sums up the meaning of the above passage?

(a) Modern warfare has resulted in civil strife.
(b) Chemical agents are useful in modern warfare.
(c) Use of chemical agents in warfare would be undesirable.
(d) People in military establishments like to use chemical agents in war.

[2010 : 2 M]

1.7 If 137 + 276 = 435 how much is 731 + 672?

(a) 534 (b) 1403
(c) 1623 (d) 1531

[2010 : 2 M]

1.8 5 skilled workers can build a wall in 20 days; 8 semi-skilled workers can build a wall in 25 days; 10 unskilled workers can build a wall in 30 days. If a team has 2 skilled, 6 semi-skilled and 5 unskilled workers, how long will it take to build the wall?

(a) 20 days (b) 18 days
(c) 16 days (d) 15 days

[2010 : 2 M]

1.9 Given digits 2, 2, 3, 3, 3, 4, 4, 4, 4 how many distinct 4 digit numbers greater than 3000 can be formed?

(a) 50 (b) 51
(c) 52 (d) 54 [2010 : 2 M]

1.10 Hari (H), Gita (G), Irfan (I) and Saira (S) are siblings (i.e. brothers and sisters). All were born on Ist January. The age difference between any two successive siblings (that is born one after another) is less than 3 years. Given the following facts:

1. Hari's age + Gita's age > Irfan's age + Saira's age.
2. The age difference between Gita and Saira is 1 year. However, Gita is not the oldest and Saira is not the youngest.
3. There are no twins.

In what order were they born (oldest first)?

(a) HSIG (b) SGHI
(c) IGSH (d) IHSG

[2010 : 2 M]

1.11 If log(P) = (1/2)log(Q) = (1/3) log(R), then which of the following options is TRUE?

(a) $P^2 = Q^3R^2$ (b) $Q^2 = PR$
(c) $Q^2 = R^3P$ (d) $R = P^2 Q^2$

[2011 : 1 M]

1.12 Which of the following options is the closest in the meaning to the word below:

Inexplicable

(a) Incomprehensible
(b) Indelible
(c) Inextricable
(d) Infallible [2011 : 1 M]

1.13 Choose the word from the options given below that is most nearly opposite in meaning to the given word:

Amalgamate

(a) merge (b) split
(c) collect (d) separate

[2011 : 1 M]

1.14 Chose the most appropriate word from the options given below to complete the following sentence.

If you are trying to make a strong impression on your audience, you cannot do so by being understated, tentative or______________.

(a) hyperbolic (b) restrained
(c) argumentative (d) indifferent

[2011 : 1 M]

1.15 Choose the most appropriate word(s) from the options given below to complete the following sentence.

I contemplated________ Singapore for my vacation but decided against it.

(a) to visit (b) having to visit
(c) visiting (d) for a visit

[2011 : 1 M]

1.16 P, Q, R and S are four types of dangerous microbes recently found in a human habitat. The area of each circle with its diameter printed in brackets represents the growth of a single microbe surviving human immunity system within 24 hours of entering the body. The danger to human beings varies proportionately with the toxicity, potency and growth attributed to a microbe shown in the figure below:

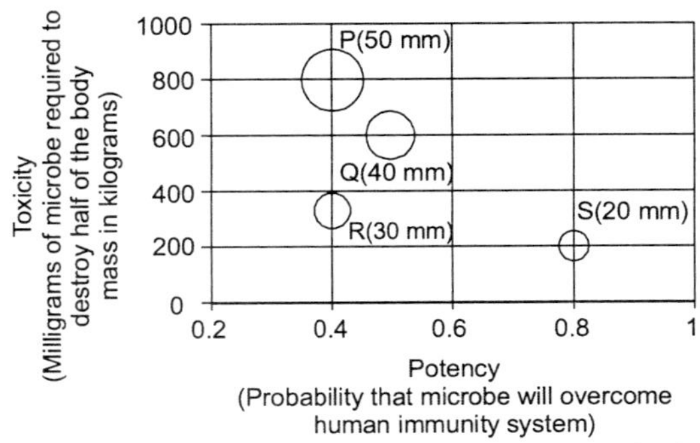

A pharmaceutical company is contemplating the development of a vaccine against the most dangerous microbe. Which microbe should the company target in its first attempt?

(a) P (b) Q
(c) R (d) S

[2011 : 2 M]

1.17 **Few school curricula include a unit on how to deal with bereavement and grief, and yet all students at some point in their lives suffer from losses through death and parting.**

Based on the above passage which topic would not be included in a unit on bereavement?

(a) how to write a letter of condolence
(b) what emotional stages are passed through in the healing process
(c) what the leading causes of death are
(d) how to give support to a grieving friend

[2011 : 2 M]

1.18 A container originally contains 10 litres of pure spirit. From this container 1 litre of spirit is replaced with 1 litre of water. Subsequently, 1 litre of the mixture is again replaced with 1 litre of water and this process is repeated one more time. How much spirit is now left in the container?

(a) 7.58 litres (b) 7.84 litres
(c) 7 litres (d) 7.29 litres

[2011 : 2 M]

1.19 A transporter receives the same number of orders each day. Currently, he has some pending orders (backlog) to be shipped. If he uses 7 trucks, then at the end of the 4th day he can clear all the orders. Alternatively, if he uses only 3 tucks, then all the orders are cleared at the end of the 10th day. What is the minimum number of trucks required so that there will be no pending order at the end of the 5th day?

(a) 4 (b) 5
(c) 6 (d) 7 [2011 : 2 M]

1.20 The variable cost (V) of manufacturing a product varies according to the equation $V = 4q$, where q is the quantity produced. The fixed cost (F) of production of same product reduces with q according to the equation $F = 100/q$. How many units should be produced to minimize the total cost ($V + F$)?

(a) 5 (b) 4
(c) 7 (d) 6 [2011 : 2 M]

1.21 Choose the most appropriate alternative from the options given below to complete the following sentence:

Despite several ____ the mission succeeded in its attempt to resolve the conflict.

(a) attempts (b) setbacks
(c) meetings (d) delegations

[2012 : 1 M]

1.22 The cost function for a product in a firm is given by $5q^2$, where q is the amount of production. The firm can sell the product at a market price of Rs. 50 per unit. The number of units to be produced by the firm such that the profit is maximized is

(a) 5 (b) 10
(c) 15 (d) 25 [2012 : 1 M]

1.23 Choose the most appropriate alternative from the options given below to complete the following sentence.

Suresh's dog is the one _____ was hurt in the stampede.

(a) that (b) which
(c) who (d) whom

[2012 : 1 M]

1.24 Choose the grammatically INCORRECT sentence:

(a) They gave us the money back less the service charges to Three Hundred rupees.
(b) This country's expenditure is not less than that of Bangladesh.
(c) The committee initially asked for a funding of Fifty Lakh rupees, but later settled for a lesser sum.
(d) This country's expenditure on educational reforms is very less.

[2012 : 1 M]

1.25 Which one of the following options is the closest in meaning to the word given below?

Mitigate

(a) Diminish (b) Divulge
(c) Dedicate (d) Denote [2012 : 1 M]

1.26 A political party orders an arch for the entrance to the ground in which the annual conventions is being held. The profile of the arch follows the equation $y = 2x - 0.1x^2$ where y is the height of the arch in meters. The maximum possible height of the arch is

(a) 8 meters (b) 10 meters
(c) 12 meters (d) 14 meters

[2012 : 2 M]

1.27 Wanted Temporary, Part-time persons for the post of Field Interviewer to conduct personal interviews to collect and collate economic data. Requirements: High School-pass, must be available for Day, Evening and Saturday work. Transportation paid, expenses reimbursed.

Which one of the following is the best inference from the above advertisement?

(a) Gender-discriminatory
(b) Xenophobic
(c) Not designed to make the post attractive
(d) Not gender-discriminatory [2012 : 2 M]

1.28 Given the sequence of terms, *AD CG FK JP*, the next term is

(a) *OV* (b) *OW*
(c) *PV* (d) *PW* [2012 : 2 M]

1.29 Which of the following assertions are CORRECT?

P: Adding 7 to each entry in a list adds 7 to the mean of the list
Q: Adding 7 to each entry in a list adds 7 to the standard deviation of the list
R: Doubling each entry n a list doubles the mean of the list
S: Doubling each entry in a list leaves the standard deviation of the list unchanged

(a) P, Q (b) Q, R
(c) P, R (d) R, S [2012 : 2 M]

1.30 An automobile plant contracted to buy shock absorbers from two suppliers X and Y. X supplies 60% and Y supplies 40% of the shock absorbers. All shock absorbers are subjected to a quality test. The ones that pass the quality test are considered reliable. Of X's shock absorbers, 96% are reliable. Of Y's shock absorbers, 72% are reliable.

The probability that a randomly chosen shock absorber, which is found to be reliable, is made by Y is

(a) 0.288 (b) 0.334
(c) 0.667 (d) 0.720 [2012 : 2 M]

1.31 Friendship, No matter how _____ it is, has its limitation
(a) cordial (b) intimate
(c) secret (d) pleasant
[2013 : 1 M]

1.32 A number much greater than 75 it is smaller than 117 is
(a) 91 (b) 93
(c) 89 (d) 96 [2013 : 1 M]

1.33 The pair that best express a relationship similar to that expression in the pair:
Medicine : Health
(a) Science : Experiment
(b) Wealth : Peace
(c) Education : Knowledge
(d) Money : Happiness [2013 : 1 M]

1.34 Which of the following options is closest in meaning to the word given below:
"Primeval"
(a) Modern (b) Historic
(c) Primitive (d) Antique
[2013 : 1 M]

1.35 The professor (I) ordered to (II) the student to go (III) out of the class (IV)
The incorrect one is
(a) (I) (b) II
(c) (III) (d) (IV) [2013 : 1 M]

1.36 A firm is selling it product at ₹ 60/unit. The total cost of production is ₹ 100 and firm is earning total profit of ₹ 500, Later the total cost increased by 30%. By what percentage the price should be increased to maintain the same profit level.
(a) 5 (b) 15
(c) 10 (d) 30 [2013 : 2 M]

1.37 Abhishek is elder to Savan, Savan is younger to Anshul. The correct relations is
(a) Abhishek is elder to Anshul
(b) Anshul is elder to Abhishek
(c) Abhishek and Anshul are of same age
(d) No conclusion can be drawn
[2013 : 2 M]

1.38 Following table provides figures (in rupees) on annual expenditure of a firm for two years 2010 and 2011

Category	2010	2011
Raw material	5200	6240
Power & fuel	7000	9450
Salary & wages	9000	12600
Plant & machinery	20000	25000
Advertising	15000	19500
Research & Development	22000	26400

In 2011, which of the following two categories have registered increase by same percentage?
(a) Raw material and salary and wages
(b) Salary and wages and advertising
(c) Power and fuel and advertising
(d) Raw material and research and development
[2013 : 2 M]

1.39 If $|4x-7| = 5$ then the value of $2|x| - |-x|$ is
(a) $2, \frac{1}{3}$ (b) $\frac{1}{2}, 3$
(c) $\frac{2}{3}, \frac{1}{3}$ (d) $\frac{2}{9}, 3$ [2013 : 2 M]

1.40 x and y are two positive real numbers, such that equation
$2x + y \leq 6$; $x + 2y \leq 8$
For which values of (x, y), the function $f(x, y) = 3x + 6y$ will give maximum value
(a) 4/3, 10/3 (b) 8/3, 20/3
(c) 8/3, 10/3 (d) 4/3, 20/3
[2013 : 2 M]

1.41 A student is required to demonstrate a high level of <u>comprehension</u> of the subject, especially in the social sciences.
The word closest in meaning to <u>comprehension</u> is
(a) understanding (b) meaning
(c) concentration (d) stability
[2014 : 1 M, Set-I]

1.42 Choose the most appropriate word from the options given below to complete the following sentence.
One of his biggest _____ was his ability to forgive.
(a) vice (b) virtues
(c) choices (d) strength
[2014 : 1 M, Set-I]

1.43 Rajan was not happy that Sajan decided to do the project on his own. On observing his unhappiness, Sajan explained to Rajan that he preferred to work independently.
Which one of the statements below is logically

valid and can be inferred from the above sentences?

(a) Rajan has decided to work only in a group.
(b) Rajan and Sajan were formed into a group against their wishes.
(c) Sajan had decided to give in to Rajan's request to work with him.
(d) Rajan had believed that Sajan and he would be working together.

[2014 : 1 M, Set-I]

1.44 If $y = 5x^2 + 3$, then the tangent at $x = 0$, $y = 3$

(a) passes through $x = 0$, $y = 0$
(b) has a slope of +1
(c) is parallel to the x-axis
(d) has a slope of −1

[2014 : 1 M, Set-I]

1.45 A foundry has a fixed daily cost of Rs. 50,000 whenever it operates and a variable cost of Rs. 800Q, where Q is the daily production in tonnes. What is the cost of production in Rs. per tonne for a daily production of 100 tonnes?

[2014 : 1 M, Set-I]

1.46 Choose the most appropriate word from the options given below to complete the following sentence.

A person suffering from Alzheimer's disease __________ short-term memory loss.

(a) experienced
(b) has experienced
(c) is experiencing
(d) experiences

[2014 : 1 M, Set-II]

1.47 Choose the most appropriate word from the options given below to complete the following sentence.

___________ is the key to their happiness; they are satisfied with what they have.

(a) Contentment
(b) Ambition
(c) Perseverance
(d) Hunger

[2014 : 1 M, Set-II]

1.48 Which of the following options is the closest in meaning to the sentence below?

"As a woman, I have no country."

(a) Women have no country.
(b) Women are not citizens of any country.
(c) Women's solidarity knows no national boundaries.
(d) Women of all countries have equal legal rights.

[2014 : 1 M, Set-II]

1.49 In any given year, the probability of an earthquake greater than Magnitude 6 occurring in the Garhwal Himalayas is 0.04. The average time between successive occurrences of such earthquakes is ____ years.

[2014 : 1 M, Set-II]

1.50 The population of a new city is 5 million and is growing at 20% annually. How many years would it take to double at this growth rate?

(a) 3-4 years
(b) 4-5 years
(c) 5-6 years
(d) 6-7 years

[2014 : 1 M, Set-II]

1.51 Find the odd one in the following group: ALRVX, EPVZB, ITZDF, OYEIK

(a) ALRVX
(b) EPVZB
(c) ITZDF
(d) OYEIK

[2014 : 2 M, Set-I]

1.52 Anuj, Bhola, Chandan, Dilip, Eswar and Faisal live on different floors in a six-storeyed building (the ground floor is numbered 1, the floor above it 2, and so on). Anuj lives on an even-numbered floor. Bhola does not live on an odd numbered floor. Chandan does not live on any of the floors below Faisal's floor. Dilip does not live on floor number 2. Eswar does not live on a floor immediately above or immediately below Bhola. Faisal lives three floors above Dilip. Which of the following floor-person combinations is correct?

	Anuj	Bhola	Chandan	Dilip	Eswar	Faisal
(a)	6	2	5	1	3	4
(b)	2	6	5	1	3	4
(c)	4	2	6	3	1	5
(d)	2	4	6	1	3	5

[2014 : 2 M, Set-I]

1.53 The smallest angle of a triangle is equal to two thirds of the smallest angle of a quadrilateral. The ratio between the angles of the quadrilateral is 3:4:5:6. The largest angle of the triangle is twice its smallest angle. What is the sum, in degrees, of the second largest angle of the triangle and the largest angle of the quadrilateral?

[2014 : 2 M, Set-I]

1.54 One percent of the people of country X are taller than 6 ft. Two percent of the people of country Y are taller than 6 ft. There are thrice as many people in country X as in country Y. Taking both countries together, what is the percentage of people taller than 6 ft?

(a) 3.0
(b) 2.5
(c) 1.5
(d) 1.25

[2014 : 2 M, Set-I]

1.55 The monthly rainfall chart based on 50 years of rainfall in Agra is shown in the following figure. Which of the following are true? (k percentile is the value such that k percent of the data fall below that value)

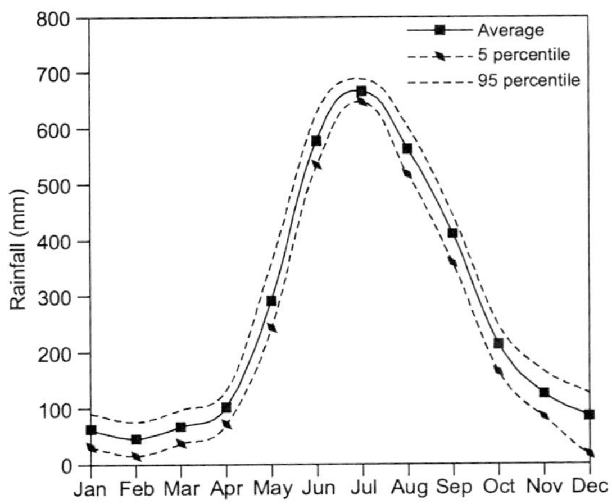

(i) On average, it rains more in July than in December
(ii) Every year, the amount of rainfall in August is more than that in January
(iii) July rainfall can be estimated with better confidence than February rainfall
(iv) In August, there is at least 500 mm of rainfall

(a) (i) and (ii)
(b) (i) and (iii)
(c) (ii) and (iii)
(d) (iii) and (iv)

[2014 : 2 M, Set-I]

1.56 In a group of four children, Som is younger to Riaz. Shiv is elder to Ansu. Ansu is youngest in the group. Which of the following statements is/are required to find the eldest child in the group?

Statements

1. Shiv is younger to Riaz.
2. Shiv is elder to Som.

(a) Statement 1 by itself determines the eldest child.
(b) Statement 2 by itself determines the eldest child.
(c) Statements 1 and 2 are both required to determine the eldest child.
(d) Statements 1 and 2 are not sufficient to determine the eldest child.

[2014 : 2 M, Set-II]

1.57 Moving into a world of big data will require us to change our thinking about the merits of exactitude. To apply the conventional mindset of measurement to the digital, connected world of the twenty-first century is to miss a crucial point. As mentioned earlier, the obsession with exactness is an artefact of the information-deprived analog era. When data was sparse, every data point was critical, and thus great care was taken to avoid letting any point bias the analysis. From "BIG DATA" Viktor Mayer-Schonberger and Kenneth Cukier

The main point of the paragraph is:

(a) The twenty-first century is a digital world
(b) Big data is obsessed with exactness
(c) Exactitude is not critical in dealing with big data
(d) Sparse data leads to a bias in the analysis

[2014 : 2 M, Set-II]

1.58 The total exports and revenues from the exports of a country are given in the two pie charts below. The pie chart for exports shows the quantity of each item as a percentage of the total quantity of exports. The pie chart for the revenues shows the percentage of the total revenue generated through export of each item. The total quantity of exports of all the items is 5 lakh tonnes and the total revenues are 250 crore rupees. What is the ratio of the revenue generated through export of Item 1 per kilogram to the revenue generated through export of Item 4 per kilogram?

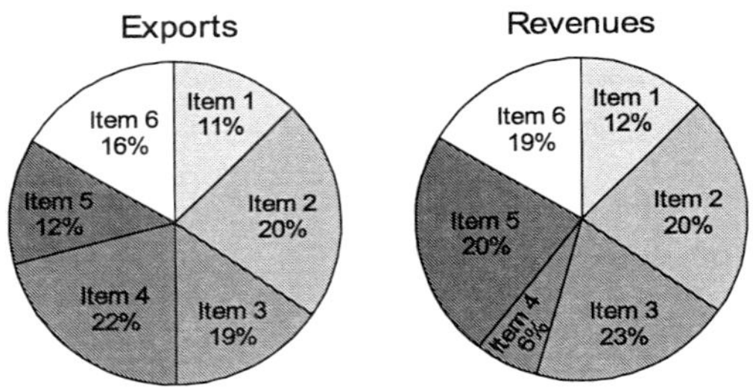

(a) 1 : 2
(b) 2 : 1
(c) 1 : 4
(d) 4 : 1

[2014 : 2 M, Set-II]

1.59 X is 1 km northeast of Y. Y is 1 km southeast of Z. W is 1 km west of Z. P is 1 km south of W. Q is 1 km east of P. What is the distance between X and Q in km?

(A) 1
(b) $\sqrt{2}$
(c) $\sqrt{3}$
(d) 2

[2014 : 2 M, Set-II]

1.60 10% of the population in a town is HIV^+. A new diagnostic kit for HIV detection is available; this kit correctly identifies HIV^+ individuals 95% of the time, and HIV^- individuals 89% of the time. A particular patient is tested using this kit and is found to be positive. The probability that the individual is actually positive is ______.

[2014 : 2 M, Set-II]

1.61 Select the pair that best expresses a relationship similar to that expressed in the pair
Children: Pediatrician
(a) Adult: Orthopaedist
(b) Females: Gynaecologist
(c) Kidney: Nephrologist
(d) Skin: Dermatologist

[2015 : 1 M, Set-I]

1.62 If ROAD is written as URDG, then SWAN should be written as:
(a) VXDQ (b) VZDQ
(c) VZDP (d) UXDQ

[2015 : 1 M, Set-I]

1.63 A function $f(x)$ is linear and has a value of 29 at $x = -2$ and 39 at $x = 3$. Find its value at $x = 5$.
(a) 59 (b) 45
(c) 43 (d) 35

[2015 : 1 M, Set-I]

1.64 Extreme focus on syllabus and studying for tests has become such a dominant concern of Indian students that they close their minds to anything ________ to the requirements of the exam.
(a) related (b) extraneous
(c) outside (d) useful

[2015 : 1 M, Set-I]

1.65 The Tamil version of ___________ John Abraham-starrer *Madras Cafe* _________ cleared by the Censor Board with no cuts last week, but the film's distributors __________ no takers among the exhibitors for a release in Tamil Nadu ________ this Friday.
(a) Mr., was, found, on
(b) a, was, found, at
(c) the, was, found, on
(d) a, being, find at

[2015 : 1 M, Set-I]

1.66 Four cards are randomly selected from a pack of 52 cards. If the first two cards are kings, what is the probability that the third card is a king?
(a) $\frac{4}{52}$ (b) $\frac{2}{50}$
(c) $\left(\frac{1}{52}\right)\times\left(\frac{1}{52}\right)$ (d) $\left(\frac{1}{52}\right)\times\left(\frac{1}{51}\right)\times\left(\frac{1}{50}\right)$

[2015 : 1 M, Set-II]

1.67 Choose the statement where underlined word is used correctly
(a) The minister insured the victims that everything would be all right.
(b) He ensured that the company will not have to bear any loss.
(c) The actor got himself ensured against any accident.
(d) The teacher insured students of good results.

[2015 : 1 M, Set-II]

1.68 Choose the most appropriate word from the options given below to complete the following sentence
The official answered _________ that the complaints of the citizen would be looked into.
(a) respectably (b) respectfully
(c) reputably (d) respectively

[2015 : 1 M, Set-II]

1.69 Mr. Vivek walks 6 meters North-East, then turns and walks 6 meters South-East, both at 60 degrees to East. He further moves 2 meters South and 4 meters West. What is the straight distance in meters between the point he started from and the point he finally reached?
(a) $2\sqrt{2}$ (b) 2
(c) $\sqrt{2}$ (d) $\frac{1}{\sqrt{2}}$

[2015 : 1 M, Set-II]

1.70 Which word is not a synonym for the word *vernacular*?
(a) regional (b) indigeneous
(c) indigent (d) colloquial

[2015 : 1 M, Set-II]

1.71 The exports and imports (in crores of Rs.) of a country from the year 2000 to 2007 are given in the following bar chart. In which year is the combined percentage increase in imports and exports the highest?

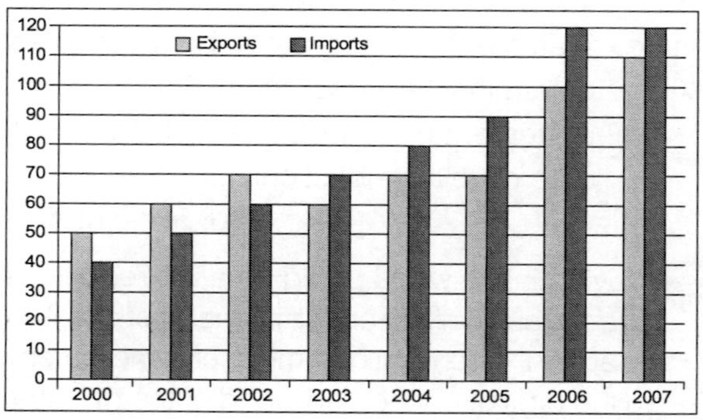

[2015 : 2 M, Set-I]

1.72 Most experts feel that in spite of possessing all the technical skills required to be a batsman of the highest order, he is unlikely to be so due to lack of requisite temperament. He was guilty of throwing away his wicket several times after working hard to lay a strong foundation. His critics pointed out that until he addressed this problem, success at the highest level will continue to elude him.

Which of the statement(s) below is/are logically valid and can be inferred from the above passage?

(i) He was already a successful batsman at the highest level.
(ii) He has to improve his temperamenet in order to become a great batsman.
(iii) He failed to make many of his good starts count.
(iv) Improving his technical skills will guarantee success.

(a) (iii) and (iv) (b) (ii) and (iii)
(c) (i), (ii) and (iii) (d) (ii) only

[2015 : 2 M, Set-I]

1.73 Choose the most appropriate equation for the function drawn as a thick line, in the plot below:

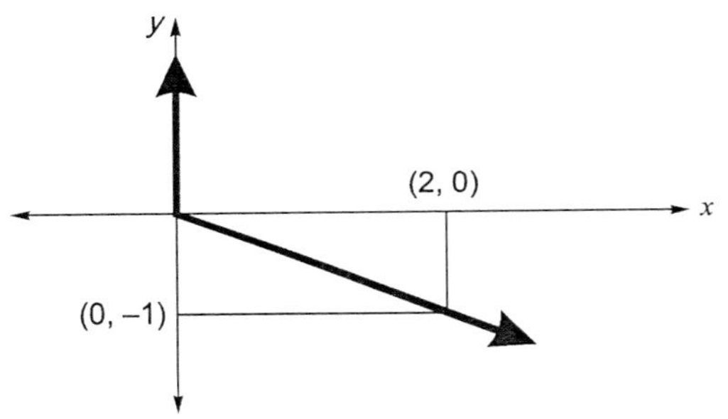

(a) $x = y - |y|$ (b) $x = -(y - |y|)$
(c) $x = y + |y|$ (d) $x = -(y + |y|)$

[2015 : 2 M, Set-I]

1.74 Alexander turned his attention towards India, since he had conquered Persia.

Which one of the statements below is logically valid and can be inferred from the above sentence?

(a) Alexander would not have turned his attention towards India had he not conquered Persia.
(b) Alexander was not ready to rest on his laurels, and wanted to march to India.
(c) Alexander was completely in control of his army and could command it to move towards India.
(d) Since Alexander's kingdom extended to Indian borders after the conquest of Persia, he was keen to move further.

[2015 : 2 M, Set-I]

1.75 The head of a newly formed government desires to appoint five of the six selected members *P*, *Q*, *R*, *S*, *T* and *U* to portfolios of Home, Power, Defense, Telecom, and Finance. *U* does not want any portfolio if *S* gets one of the five. *R* wants either Home or Finance or no portfolio. *Q* says that if *S* gets either Power or Telecom, then she must get the other one. *T* insists on a portfolio if *P* gets one.

Which is the valid distribution of portfolios?

(a) *P*-Home, *Q*-Power, *R*-Defence, *S*-Telecom, *T*-Finance
(b) *R*-Home, *S*-Power, *P*-Defence, *Q*-Telecom, *T*-Finance
(c) *P*-Home, *Q*-Power, *T*-Defence, *S*-Telecom, *U*-Finance
(d) *Q*-Home, *U*-Power, *T*-Defence, *R*-Telecom, *P*-Finance

[2015 : 2 M, Set-I]

1.76 Read the following table giving sales data of five types of batteries for years 2006 to 2012

Year	Type-I	Type-II	Type-III	Type-IV	Type-V
2006	75	144	114	102	108
2007	90	126	102	84	126
2008	96	114	75	105	135
2009	105	90	150	90	75
2010	90	75	135	75	90
2011	105	60	165	45	120
2012	115	85	160	100	145

Out of the following, which type of battery achieved highest growth between the years 2006 and 2012?

(a) Type V (b) Type III
(c) Type II (d) Type I

[2015 : 2 M, Set-II]

1.77 How many four digit numbers can be formed with the 10 digits 0, 1, 2, ... 9 if no number can start with 0 and if repetitions are not allowed?

[2015 : 2 M, Set-II]

1.78 The given question is followed by two statements; select the most appropriate option that solves the question

Capacity of a solution tank *A* is 70% of the

capacity of tank *B*. How many gallons of solution are in tank *A* and tank *B*?

Statements:

I. Tank A is 80% full and tank B is 40% full.

II. Tank A if full contains 14,000 gallons of solution.

(a) Statement I alone is sufficient.
(b) Statement II alone is sufficient.
(c) Either statement I or II alone is sufficient.
(d) Both the statements I and II together are sufficient.

[2015 : 2 M, Set-II]

1.79 The word similar in meaning to 'dreary' is

(a) cheerful (b) dreamy
(c) hard (d) dismal

[2015 : 2 M, Set-II]

1.80 There are 16 teachers who can teach Thermodynamics (TD), 11 who can teach Electrical Sciences (ES), and 5 who can teach both TD and Engineering Mechanics (EM). There are a total of 40 teachers. 6 cannot teach any of the three subjects, i.e. EM, ES or TD. 6 can teach only ES. 4 can teach all three subjects, i.e. EM, ES and TD. 4 can teach ES and TD. How many can teach both ES and EM but not TD?

(a) 1 (b) 2
(c) 3 (d) 4

[2015 : 2 M, Set-II]

1.81 Out of the following four sentences, select the most suitable sentence with respect to grammar and usage.

(a) I will not leave the place until the minister does not meet me.
(b) I will not leave the place until the minister doesn't meet me.
(c) I will not leave the place until the minister meet me.
(d) I will not leave the place until the minister meets me.

[2016 : 1 M, Set-I]

1.82 A rewording of something written or spoken is a ______________.

(a) paraphrase (b) paradox
(c) paradigm (d) paraffin

[2016 : 1 M, Set-I]

1.83 Archimedes said, "Give me a lever long enough and a fulcrum on which to place it, and I will move the world." The sentence above is an example of a __________ statement.

(a) figurative (b) collateral
(c) literal (d) figurine

[2016 : 1 M, Set-I]

1.84 If 'relftaga' means carefree, 'otaga' means careful and 'fertaga' means careless, which of the following could mean 'aftercare'?

(a) zentaga (b) tagafer
(c) tagazen (d) relffer

[2016 : 1 M, Set-I]

1.85 A cube is built using 64 cubic blocks of side one unit. After it is built, one cubic block is removed from every corner of the cube. The resulting surface area of the body (in square units) after the removal is __________.

(a) 56 (b) 64
(c) 72 (d) 96

[2016 : 1 M, Set-I]

1.86 If I were you, I _____ that laptop. It's much too expensive.

(a) won't buy (b) shan't buy
(c) wouldn't buy (d) would by

[2016 : 1 M, Set-II]

1.87 He turned at deaf ear to my request.
What does the underlined pharasal verb mean?

(a) ignored (b) appreciated
(c) twisted (d) returned

[2016 : 1 M, Set-II]

1.88 Choose the most appropriate set of words from the options given below to complete the following sentence.

______ _____ is a will, ______ is a way.

(a) wear, there, their
(b) were, their, there
(c) where, there, there
(d) were, their, their

[2016 : 1 M, Set-II]

1.89 (x% of y) + (y% of x) is equivalent to ____.

(a) 2% of xy (b) 2% of (xy/100)
(c) xy% of 100 (d) 100% of xy

[2016 : 1 M, Set-II]

1.90 The sum of the digits of a two digit number is 12. If the new number formed by reversing the digits is greater than the original number by 54, find the original number.

(a) 39 (b) 57
(c) 66 (d) 93

[2016 : 1 M, Set-II]

1.91 A shaving set company sells 4 different types of razors, Elegance, Smooth, Soft and Executive. Elegance sells at Rs. 48, Smooth at Rs. 63, Soft at Rs. 78 and Executive at Rs. 173 per piece. The table below shows the numbers of each razor sold in each quarter of a year.

Quarter/ Product	Elegance	Smooth	Soft	Executive
Q1	27300	20009	17602	9999
Q2	25222	19392	18445	8942
Q3	28976	22429	19544	10234
Q4	21012	18229	16595	10109

Which product contributes the greatest fraction to the revenue of the company in that year?

(a) Elegance (b) Executive
(c) Smooth (d) Soft

[2016 : 2 M, Set-I]

1.92 Indian currency notes show the denomination indicated in at least seventeen languages. If this is not an indication of the nation's diversity, nothing else is. Which of the following can be logically inferred from the above sentences?

(a) India is a country of exactly seventeen languages.
(b) Linguistic pluralism is the only indicator of a nation's diversity.
(c) Indian currency notes have sufficient space for all the Indian languages.
(d) Linguistic pluralism is strong evidence of India's diversity.

[2016 : 2 M, Set-I]

1.93 Consider the following statements relating to the level of poker play of four players *P*, *Q*, *R* and S. I. *P* always beats *Q* II. *R* always beats *S* III. *S* loses to *P* only sometimes IV. R always loses to *Q*. Which of the following can be logically inferred from the above statements? (i) *P* is likely to beat all the three other players (ii) S is the absolute worst player in the set

(a) (i) only (b) (ii) only
(c) (i) and (ii) (d) neither (i) nor (ii)

[2016 : 2 M, Set-I]

1.94 If $f(x) = 2x^7 + 3x - 5$, which of the following is a factor $f(x)$?

(a) $(x^3 + 8)$ (b) $(x - 1)$
(c) $(2x - 5)$ (d) $(x + 1)$

[2016 : 2 M, Set-I]

1.95 In a process, the number of cycles to failure decreases exponentially with an increase in load. At a load of 80 units, it takes 100 cycles for failure. When the load is halved, it takes 10000 cycles for failure. The load for which the failure will happen in 5000 cycles is ________.

(a) 40.00 (b) 46.02
(c) 60.01 (d) 92.02

[2016 : 2 M, Set-I]

1.96 Two finance companies, *P* and *Q* declared fixed annual rates of interest on the amounts invested with them. The rates of interest offered by these companies may differ from year to year. Year-wise annual rates of interest offered by these companies are shown by the line graph provided below.

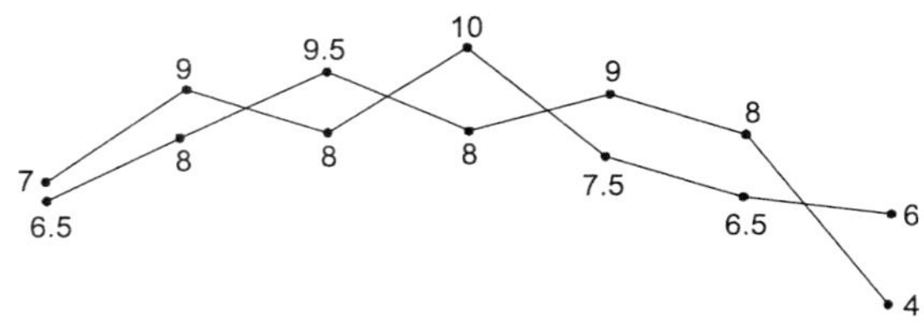

If the amounts invested in the companies, P and *Q* in 2006 are in the ratio 8 : 9, then the amounts received after one year as interest from companies *P* and *Q* would be the ratio

(a) 2 : 3 (b) 3 : 4
(c) 6 : 7 (d) 4 : 3

[2016 : 2 M, Set-II]

1.97 Today we consider, Ashoka as a great ruler because of the copious evidence he left behind in the form of stone carved edicts. Historians tend to correlate greatness of a king at his time with the availability of evidence today.

Which of the following can be logically inferred from the above sentences?

(a) Emperors who do not leave significant sculpted evidence are completely forgotten.
(b) Ashoka produced stone carved edicts to ensure that later historians will respect him.
(c) Statues of kings are a reminder of their greatness.
(d) A king's greatness, as we know him today, is interpreted by historians.

[2016 : 2 M, Set-II]

1.98 Fact 1 : Humans are mammals.

Fact 2 : Some humans are engineers.

Fact 3 : Engineers build house.

If the above statements are facts, which of the following can be logically inferred?

I. All mammals build houses.

II. Engineers are mammals.

III. Some humans are not engineers.

(a) II only (b) III only

(c) I, II and III (d) I only

[2016 : 2 M, Set-II]

1.99 A square pyramid has a base perimeter x and the slant height is half of the perimeter. What is the lateral surface area of the pyramid?

(a) x^2 (b) $0.75\,x^2$

(c) $0.50\,x^2$ (d) $0.25\,x^2$

[2016 : 2 M, Set-II]

1.100 Ananth takes 6 hours and Bharath takes 4 hours to read a book. Both started reading copies of the book at the same time. After how many hours is the number of pages to be rad by Ananth, twice that to be read by Bharath? Assume Ananth and Bharath read all the pages with constant pace.

(a) 1 (b) 2

(c) 3 (d) 4 [2016 : 2 M, Set-II]

1.101 The following sequence of numbers is arranged in increasing order: 1, x, x, x, y, y, 9, 16, 18. Given that the mean and median are equal, and are also equal to twice the mode, the value of y is

(a) 5 (b) 6

(c) 7 (d) 8

[2017 : 1 M, Set-I]

1.102 The bacteria in milk are destroyed when it ___________ heated to 80 degree Celsius.

(a) would be (b) will be

(c) is (d) was

[2017 : 1 M, Set-I]

1.103 Consider the following sentences:

All benches are beds. No bed is a bulb. Some bulbs are lamps.

Which of the following can be inferred?

(i) Some beds are lamps.

(ii) Some lamps are beds.

(a) Only (i) (b) Only (ii)

(c) Both (i) and (ii) (d) Neither (i) nor (ii)

[2017 : 1 M, Set-I]

1.104 If the radius of a right circular cone is increased by 50%, its volume increases by

(a) 75% (b) 100%

(c) 125% (d) 237.5%

[2017 : 1 M, Set-I]

1.105 _________ with someone else's email account is now a very serious offense.

(a) Involving (b) Assisting

(c) Tampering (d) Incubating

[2017 : 1 M, Set-I]

1.106 Two dice are thrown simultaneously. The probability that the product of the numbers appearing on the top faces of the dice is a perfect square is

(a) $\frac{1}{9}$ (b) $\frac{2}{9}$

(c) $\frac{1}{3}$ (d) $\frac{4}{9}$

[2017 : 1 M, Set-II]

1.107 Four cards lie on a table. Each card has a number printed on one side and a colour on the other. The faces visible on the cards are 2, 3, red and blue.

Proposition: If a card has an even value on one side, then its opposite face is red.

The cards which MUST be turned over to verify the above proposition are

(a) 2, red (b) 2, 3, red

(c) 2, blue (d) 2, red, blue

[2017 : 1 M, Set-II]

1.108 The event would have been successful if you _________ able to come.

(a) are (b) had been

(c) have been (d) would have been

[2017 : 1 M, Set-II]

1.109 There was no doubt that their work was thorough.

Which of the words below is closest in meaning to the underlined word above?

(a) pretty (b) complete

(c) sloppy (d) haphazard

[2017 : 1 M, Set-II]

1.110 What is the value of x when

$$81\times\left(\frac{16}{25}\right)^{x+2}\div\left(\frac{3}{5}\right)^{2x+4}=144?$$

(a) 1 (b) –1

(c) –2 (d) Cannot be determined

[2017 : 1 M, Set-II]

1.111 Two machines $M1$ and $M2$ are able to execute any of four jobs, P, Q, R and S. The machines can perform one job on one object at a time, Jobs P, Q, R and S take 30 minutes, 20 minutes, 60 minutes and 15 minutes each respectively. There are 10 objects each requiring exactly 1 job. Job P is to be performed on 2 objects, Job *Q* on 3 objects, Job *R* on 1 object and Job *S* on 4 objects. What is the minimum time needed to complete all the jobs?

(a) 2 hours (b) 2.5 hours
(c) 3 hours (d) 3.5 hours

[2017 : 2 M, Set-I]

1.112 The bar graph below shows the output of five carpenters over one month, each of whom made different items of furniture; chairs, tables and beds.

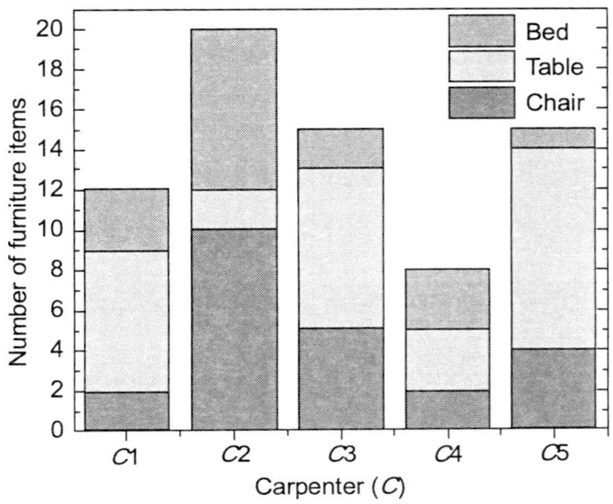

Consider the following statements:

(i) The number of beds made by carpenter *C*2 is exactly the same as the number of tables made by carpenter *C*3.

(ii) The total number of chairs made by all carpenters is less than the total number of tables.

Which one of the following is true?

(a) Only (i) (b) Only (ii)
(c) Both (i) and (ii) (d) Neither (i) nor (ii)

[2017 : 2 M, Set-I]

1.113 Students applying for hostel rooms are alloted rooms in order of seniority. Students already staying in a room will move if they get a room in their preferred list. Preferences of lower ranked applicants are ignored during allocation.

Given the data below, which room will Ajit stay in?

Names	Student Seniority	Current room	Room preference list
Amar	1	P	R, S, Q
Akbar	2	None	R, S
Anthony	3	Q	P
Ajit	4	S	Q, P, R

(a) *P* (b) *Q*
(c) *R* (d) *S*

[2017 : 2 M, Set-I]

1.114 The old concert hall was demolished because of fears that the foundation would be affected by the construction of the new metro line in the area. Modern technology for underground metro construction tried to mitigate the impact of pressurized air pockets created by the excavation of large amounts of soil. But even with these safeguards, it was feared that the soil below the concert hall would not be stable.

From this, one can infer that

(a) the foundations of old buildings create pressurized air pockets underground, which are different to handle during metro construction.

(b) metro construction has to be done carefully considering its impact on the foundations of existing buildings.

(c) old buildings in an area form an impossible hurdle to metro construction in that area.

(d) pressurized air can be used to excavate large amount of soil from underground areas.

[2017 : 2 M, Set-I]

1.115 The last digit of $(2171)^7 + (2172)^9 + (2173)^{11} + (2174)^{13}$ is

(a) 2 (b) 4
(c) 6 (d) 8

[2017 : 2 M, Set-I]

1.116 Bhaichung was observing the pattern of people entering and leaving a car service centre. There was a single window where customers were being served. He saw that people inevitably came out of the centre in the order that they went in. However, the time they spent inside seemed to vary a lot: Some people came out in a matter of minutes while for others it took much longer.

From this, what can one conclude?

(a) The centre operates on a first-come-first-served basis, but with variable service times, depending on specific customer needs.

(b) Customers were served in an arbitrary order, since they took varying amounts of time for service completion in the centre.

(c) Since some people came out within a few minutes of entering the centre, the system is likely to operate on a last-come-first-served basis.

(d) Entering the centre early ensured that one would have shorter service times and most people attempted to do this.

[2017 : 2 M, Set-II]

1.117 The points in the graph below represent the halts of a lift for durations of 1 minute, over a period of 1 hour.

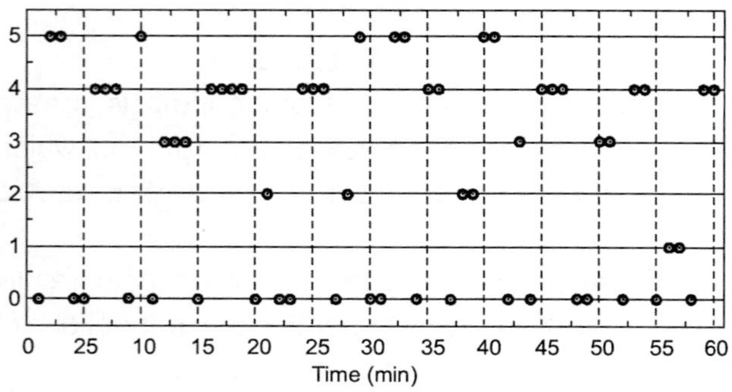

Which of the following statements are correct?

(i) The elevator never moves directly from any non-ground floor to another non-ground floor over the one hour period.

(ii) The elevator stays on the fourth floor for the longest duration over the one hour period.

(a) Only (i) (b) Only (ii)
(c) Both (i) and (ii) (d) Neither (i) nor (ii)

[2017 : 2 M, Set-II]

1.118 Budhan covers a distance of 19 km in 2 hours by cycling one fourth of the time and walking the rest. The next day he cycles (at the same speed as before) for half the time and walks the rest (at the same speed as before) and covers 26 km in 2 hours. The speed in km/h at which Budhan walks is

(a) 1 (b) 4
(c) 5 (d) 6

[2017 : 2 M, Set-II]

1.119 A map shows the elevations of Darjeeling. Gangtok, Kalimpong, Pelling and Siliguri. Kalimpong is at a lower elevation than Gangtok. Pelling is at a lower elevation than Gangtok. Pelling is at a higher elevation than Siliguri. Darjeeling is at a higher elevation than Gangtok.

Which of the following statements can be inferred from the paragraph above?

(i) Pelling is at a higher elevation that Kalimpong.

(ii) Kalimpong is at a lower elevation than Darjeeling.

(iii) Kalimpong is at a higher elevation than Siliguri.

(iv) Siliguri is at a lower elevation than Gangtok.

(a) Only (ii) (b) Only (ii) and (iii)
(c) Only (ii) and (iv) (d) Only (iii) and (iv)

[2017 : 2 M, Set-II]

1.120 *P*, *Q*, *R*, *S*, *T* and *U* are seated around a circular table. R is seated two places to the right of *Q*, *P* is seated three places to the left *R*. *S* is seated opposite *U*. If *P* and *U* now switch seats, which of the following must necessarily be true?

(a) *P* is immediately to the right of *R*

(b) *T* is immediately to the left of *P*

(c) *T* is immediately to the left of *P* or *P* is immediately to the right of *Q*

(d) *U* is immediately to the right of *R* or *P* is immediately to the left of *T*

[2017 : 2 M, Set-II]

1.121 The temperature *T* in a room varies as a function of the outside temperature T_0 and the number of persons in the room *p*, according to the relation $T = K(\theta p + T_0)$, where θ is *K* are constants. What would be the value of θ given the following data?

T_0	p	T
25	2	32.4
30	5	42.0

(a) 0.8 (b) 1.0
(c) 2.0 (d) 10.0

[2018 : 1 M, Set-I]

1.122 "The driver applied the _______ as soon as she approached the hotel where she wanted to take a _________."

The words that best fill the blanks in the above sentence are

(a) brake, break (b) break, break
(c) brake, brake (d) break, brake

[2018 : 1 M, Set-I]

1.123 Tower *A* is 90 m tall and tower *B* is 140 m tall. They are 100 m apart. A horizontal skywalk connects the floors at 70 m in both the towers. If a taut rope connects the top of tower *A* to the bottom of tower *B*, at what distance (in meters) from tower *A* will the rope intersect the skywalk?

[2018 : 1 M, Set-I]

1.124 "It is no surprise that every society has had codes of behaviour; however, the nature of these codes is often ________."

The word that best fills the blank in the above sentence is

(a) unpredictable (b) simple
(c) expected (d) strict

[2018 : 1 M, Set-I]

1.125 Hema's age is 5 years more than twice Hari's age. Suresh's age is 13 years less than 10 times Hari's age. If Suresh is 3 times as old as Hema. How old is Hema?

(a) 14 (b) 17
(c) 18 (d) 19

[2018 : 1 M, Set-I]

1.126 A three-member committee has to be formed from a group of 9 people. How many such distinct committees can be formed?

(a) 27 (b) 72
(c) 81 (d) 84

[2018 : 1 M, Set-II]

1.127 "His face ________ with joy when the solution of the puzzle was ____________ to him." The words that best fill the blanks in the above sentence are

(a) shone, shown (b) shone, shone
(c) shown, shone (d) shown, shown

[2018 : 1 M, Set-II]

1.128 For non-negative integers, a, b, c, what would be the value of $a + b + c$ if $\log a + \log b + \log c = 0$?

(a) 3 (b) 1
(c) 0 (d) –1

[2018 : 1 M, Set-II]

1.129 "Although it does contain some pioneering ideas, one would hardly characterize the work as ________."

The word that best fills the blank in the above sentence is

(a) innovative (b) simple
(c) dull (d) boring

[2018 : 1 M, Set-II]

1.130 $\underbrace{a+a+a+\ldots+a}_{n\text{ times}} = a^2b$ and $\underbrace{b+b+b+\ldots+b}_{m\text{ times}} = ab^2$, where a, b, n and m are natural numbers. What is the value of

$$\left(\underbrace{m+m+m+\ldots+m}_{n\text{ times}}\right)\left(\underbrace{n+n+n+\ldots+n}_{m\text{ times}}\right)?$$

(a) $2a^2b^2$ (b) a^4b^4
(c) $ab(a+b)$ (d) a^2+b^2

[2018 : 1 M, Set-II]

1.131 Consider a sequence of number $a_1, a_2, a_3, \ldots, a_n$ where $a_n = \frac{1}{n} - \frac{1}{n+2}$, for each integer $n > 0$. What is the sum of the first 50 terms?

(a) $\left(1+\frac{1}{2}\right) - \frac{1}{50}$

(b) $\left(1+\frac{1}{2}\right) + \frac{1}{50}$

(c) $\left(1+\frac{1}{2}\right) - \left(\frac{1}{51}+\frac{1}{52}\right)$

(d) $1 - \left(\frac{1}{51}+\frac{1}{52}\right)$

[2018 : 2 M, Set-I]

1.132 A fruit seller sold a basket of fruits at 12.5% loss. Had he sold it for Rs. 108 more, he would have made a 10% gain. What is the loss in Rupees incurred by the fruit seller?

(a) 48 (b) 52
(c) 60 (d) 108

[2018 : 2 M, Set-I]

1.133 Each of the letters arranged as below represents a unique from 1 to 9. The letters are positioned in the figure such that $(A \times B \times C)$, $(B \times G \times E)$ and $(D \times E \times F)$ are equal. Which integer among the following choices cannot be represented by the letters A, B, C, D, E, F or G?

A		D
B	G	E
C		F

(a) 4 (b) 5
(c) 6 (d) 9

[2018 : 2 M, Set-I]

1.134 The price of a wire made of a superalloy material is proportional to the square of its length. The price of 10 m length of the wire is Rs. 1600. What would be the total price (in Rs.) of two wires of lengths 4 m and 6 m?

(a) 768 (b) 832
(c) 1440 (d) 1600

[2018 : 2 M, Set-I]

1.135 What of the following function(s) in an accurate description of the graph for the range(s) indicated?

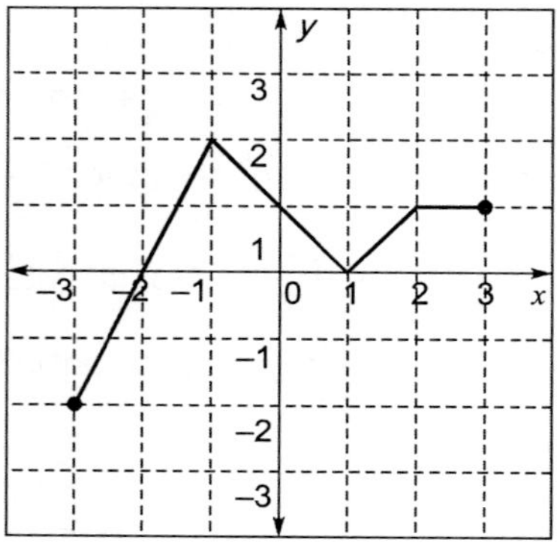

(a) $y = 2x + 4$ for $-3 \leq x \leq -1$
(b) $y = |x - 1|$ for $-1 \leq x \leq 2$
(c) $y = ||x| - 1|$ for $-1 \leq x \leq 2$
(d) $y = 1$ for $2 \leq x \leq 3$
(a) (i), (ii) and (iii) only
(b) (i), (ii) and (iv) only
(c) (i) and (iv) only
(d) (ii) and (iv) only

[2018 : 2 M, Set-I]

1.136 In manufacturing industries, loss is usually taken to be proportional to the square of the deviation from a target. If the loss is Rs. 4900 for a deviation of 7 units, what would be the loss in Rupees for a deviation of 4 units from the target?

(a) 400 (b) 1200
(c) 1600 (d) 2800

[2018 : 2 M, Set-II]

1.137 Each of the letters in the figure below represents a unique integer from 1 to 9. The letters are positioned in the figure such that each of (*A*+*B*+*C*), (*C*+*D*+*E*), (*E*+*F*+*G*) and (*G*+*H*+*K*) is equal to 13. Which integer does *E* represent?

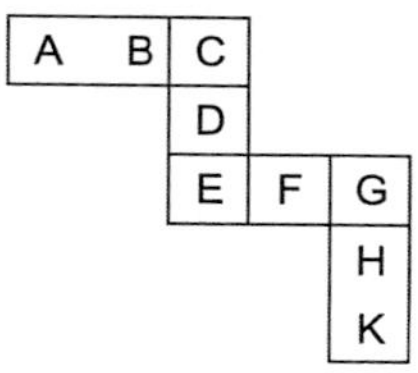

(a) 1 (b) 4
(c) 6 (d) 7

[2018 : 2 M, Set-II]

1.138 The annual average rainfall in a tropical city is 1000 mm. On a particular rainy day (24-hour period), the cumulative rainfall experienced by the city is shown in the graph. Over the 24-hour period. 50% of the rainfall falling on a rooftop, which had an obstruction-free area of 50 m^2, was harvested into a tank. What is the total volume of water collected in the tank in liters?

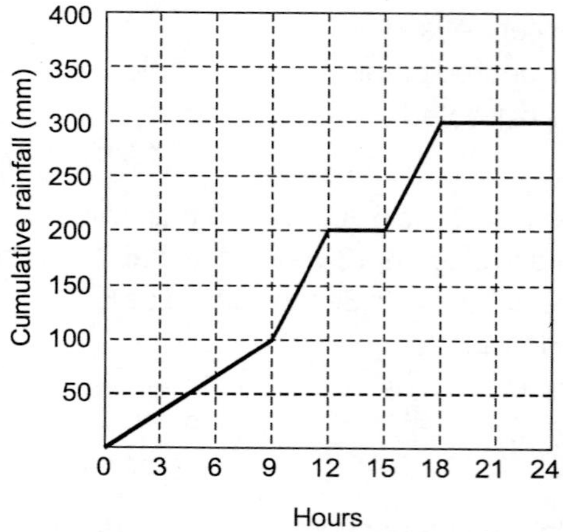

(a) 25,000 (b) 18,750
(c) 7,500 (d) 3,125

[2018 : 2 M, Set-II]

1.139 Given that $\frac{\log P}{y-z} = \frac{\log Q}{z-x} = \frac{\log R}{x-y} = 10$ for $x \neq y \neq z$, what is the value of the product *PQR*?

(a) 0 (b) 1
(c) *xyz* (d) 10^{xyz}

[2018 : 2 M, Set-II]

1.140 A faulty wall clock is known to gain 15 minutes every 24 hours. It is synchronized to the correct time at 9 AM on 11th July. What will be the correct time to the nearest minute when the clock shows 2 PM on 15th July of the same year?

(a) 12:45 PM (b) 12:58 PM
(c) 1:00 PM (d) 2:00 PM

[2018 : 2 M, Set-II]

1.141 If *E* = 10; *J* = 20; *O* = 30; and *T* = 40, what will be *P* + *E* + *S* + *T*?

(a) 51 (b) 120
(c) 82 (d) 164

[2019 : 1 M, Set-I]

1.142 On a horizontal ground, the base of a straight ladder is 6 m away from the base of a vertical pole. The ladder makes an angle of 45° to the horizontal. If the ladder is resting at a point located at one-fifth of the height of the pole from the bottom, the height of the pole is_______meters.

(a) 15 (b) 25
(c) 35 (d) 30

[2019 : 1 M, Set-I]

1.143 They have come a long way in _______ trust among the users.

(a) Creating (b) Created
(c) Creation (d) Create

[2019 : 1 M, Set-I]

1.144 The lecture was attended by quite _______ students, so the hall was not very _____.

(a) few, quite (b) a few, quite
(c) few, quiet (d) a few, quiet

[2019 : 1 M, Set-I]

1.145 The CEO's decision to quit was as shocking to Board as it was to __________.

(a) myself (b) I
(c) me (d) my

[2019 : 1 M, Set-I]

1.146 Suresh wanted to lay a new carpet in his new mansion with an area of 70 x 55 sq. mts. However an area of 550 sq. mts. had to be left out for flower pots. If the cost of carpet is Rs. 50 per sq. mts., how much money (in Rs.) will be spent by Suresh for the carpet now?

(a) Rs. 1,65, 000 (b) Rs. 2,75, 000
(c) Rs. 1,92, 500 (d) Rs. 1,27, 500

[2019 : 1 M, Set-II]

1.147 Daytime temperatures in Delhi can______40°C.

(a) get (b) peak
(c) reach (d) stand

[2019 : 1 M, Set-II]

1.148 A retaining wall with measurements 30 m x 12 m x 6 m was constructed with bricks of dimensions 8 cm x 6 cm x 6 cm If 60% of the wall consists of bricks, the number of bricks used for the construction is________lakhs.

(a) 30 (b) 40
(c) 75 (d) 45

[2019 : 1 M, Set-II]

1.149 Hima Das was_____only Indian athlete to win_______gold for India.

(a) the, a (b) an, a
(c) an, the (d) the, many

[2019 : 1 M, Set-II]

1.150 The growth rate of ABC Motors in 2017 was the same__________XYZ Motors in 2016.

(a) as that off (b) as those of
(c) as off (d) as that of

[2019 : 1 M, Set-II]

1.151 A square has sides 5 cm smaller than the sides of a second square. The area of the larger square is four times the area of the smaller square. The side of the larger square is________cm.

(a) 15.10 (b) 8.50
(c) 18.50 (d) 10.00

[2019 : 2 M, Set-I]

1.152 "The increasing interest in tribal characters might be a mere coincidence, but the timing is of interest. None of this, though, is to say that the tribal hero has arrived in Hindi cinema, or that the new crop of characters represents the acceptance of the tribal character in the industry. The films and characters are too few to be described as a pattern."

What does the word 'arrived' mean in the paragraph above?

(a) came to a conclusion
(b) reached a terminus
(c) went to a place
(d) attained a status

[2019 : 2 M, Set-I]

1.153 The new cotton technology, Bollgard-II, with herbicide tolerant traits has developed into a thriving business in India. However, the commercial use of this technology is not legal in India. Notwithstanding that, reports indicate that the herbicide tolerant Bt cotton had been purchased by farmers at an average of Rs 200 more than the control price of ordinary cotton, and planted in 15% of the cotton growing area in the 2017 Kharif season.

Which one of the following statements can be inferred from the given passage?

(a) Farmers want to access the new technology for experimental purposes.
(b) Farmers want to access the new technology if India benefits from it.
(c) Farmers want to access the new technology even if it is not legal.
(d) Farmers want to access the new technology by paying high price.

[2019 : 2 M, Set-I]

1.154 P, Q, R, S and T are related and belong to the same family. P is the brother of S. Q is the wife of P. R and T are the children of the siblings P and S respectively. Which one of the following statements is necessarily FALSE?

(a) S is the sister-in-law of Q
(b) S is the aunt of R
(c) S is the brother of P
(d) S is the aunt of T [2019 : 2 M, Set-I]

1.155 In a sports academy of 300 people, 105 play only cricket, 70 play only hockey, 50 play only football, 25 play both cricket and hockey, 15 play both hockey and football and 30 play both cricket and football. The rest of them play all three sports. What is the percentage of people who play at least two sports?

(a) 25.00 (b) 28.00
(c) 23.30 (d) 50.00

[2019 : 2 M, Set-I]

1.156 Mohan, the manager, wants his four workers to work in pairs. No pair should work for more than 5 hours. Ram and John have worked together for 5 hours. Krishna and Amir have worked as a team for 2 hours. Krishna does not want to work with Ram. Whom should Mohan allot to work with John, if he wants all the workers to continue working?

(a) Amir (b) Krishna
(c) Ram (d) None of the three

[2019 : 2 M, Set-II]

1.157 "Popular Hindi fiction, despite - or perhaps because of - its wide reach, often does not appear in our cinema. As ideals that viewers are meant to look up to rather than identify with, Hindi film protagonists usually read books of aspirational value: textbooks, English books, or high value literature."

Which one of the following CANNOT be inferred from the paragraph above?

(a) Protagonists in Hindi movies, being ideals for viewers, read only books of aspirational value
(b) People do not look up to writers of textbooks, English books or high value literature
(c) Though popular Hindi fiction has wide reach: it often does not appear in the movies
(d) Textbooks, English books or high literature have aspirational value, but not popular Hindi Fiction

[2019 : 2 M, Set-II]

1.158 Population of state X increased by x% and the population of state Y increased by y% from 2001 to 2011. Assume that x is greater than y. Let P be the ratio of the population of state X to state Y in a given year. The percentage increase in P from 2001 to 2011 is_________.

(a) $\frac{x}{y}$ (b) $\frac{100(x-y)}{100+y}$
(c) $x - y$ (d) $\frac{100(x-y)}{100+x}$

[2019 : 2 M, Set-II]

1.159 An oil tank can be filled by pipe X in 5 hours and pipe Y in 4 hours, each pump working on its own. When the oil tank is full and the drainage hole is open, the oil is drained in 20 hours. If initially the tank was empty and someone started the two pumps together but left the drainage hole open, how many hours will it take for the tank to be filled? (Assume that the rate of drainage is independent of the Head)

(a) 4.00 (b) 1.50
(c) 2.00 (d) 2.50

[2019 : 2 M, Set-II]

1.160 The Newspaper reports that over 500 hectares of tribal land spread across 28 tribal settlements in Mohinitampuram forest division have already been "alienated". A top forest official said, "First the tribals are duped out of their land holdings. Second, the families thus rendered landless are often forced to encroach further into the forests".

On the basis of the information available in the paragraph,_________is/are responsible for duping the tribals.

(a) The Newspaper
(b) it cannot be inferred who
(c) forest officials
(d) landless families

[2019 : 2 M, Set-II]

1.161 It is a common criticism that most of the academicians live in their _____, so, they are not aware of the real life challenges.

(a) glas palaces (b) big flats
(c) ivory towers (d) homes

[2020 : 1 M, Set-I]

1.162 Select the work that fits the analogy:

Fuse : Fusion :: Use : ______

(a) Usage (b) Uses
(c) Usion (d) User

[2020 : 1 M, Set-I]

1.163 His hunger for reading is insatiable. He reads indiscriminately. He is most certainly a/an ________ reader.

(a) all-round (b) voracious
(c) precocious (d) wise

[2020 : 1 M, Set-I]

1.164 If 0, 1, 2, ..., 7, 8, 9 are coded as O, P, Q, ..., V, W, X, then 45 will be coded as _______.

(a) ST (b) SS
(c) SU (d) TS

[2020 : 1 M, Set-I]

1.165 The sum of two positive numebrs is 100. After subtracting 5 from each number, the product of the resulting numbers is 0. One of the original numbers is ______.

(a) 95 (b) 90
(c) 85 (d) 80

[2020 : 1 M, Set-I]

1.166 Select the word that fits the analogy:

Partial : Impartial : : Popular : ________

(a) Impopular
(b) Dispopular
(c) Mispopular
(d) Unpopular

[2020 : 1 M, Set-II]

1.167 After the inauguration of the new building, the head of department (HOD) collated faculty preferences for office space. P wanted a room adjacent to the lab. Q wanted to be close to the lift. R wanted a view of the playground and S wanted a corner office.

Assuming that everyone was satisfied, which among the following shows a possible allocation?

(a)
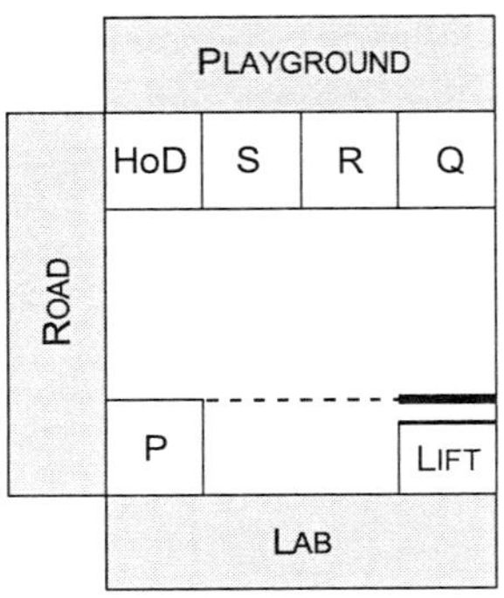

(b)
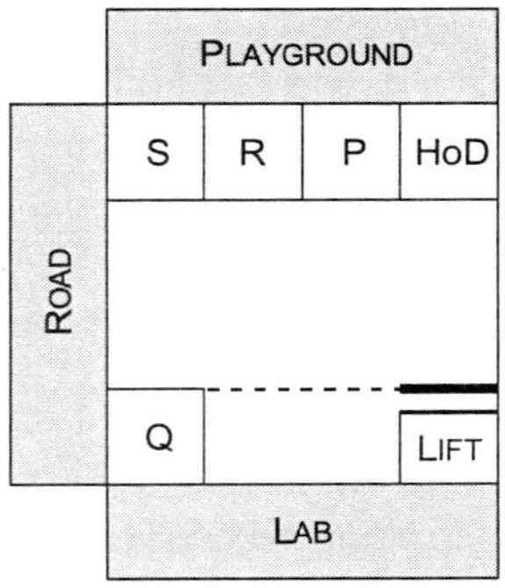

(c)
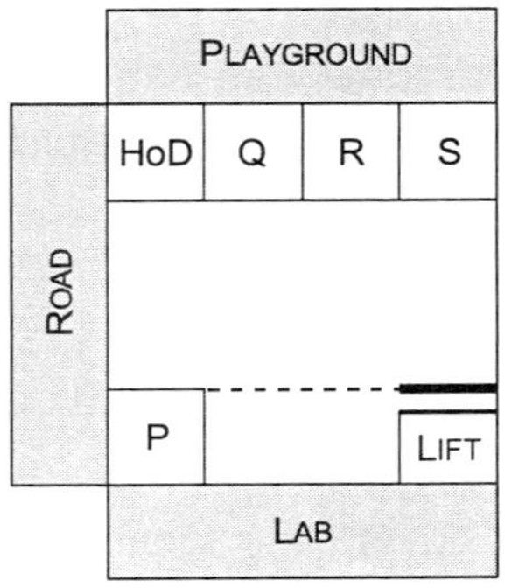

(d)
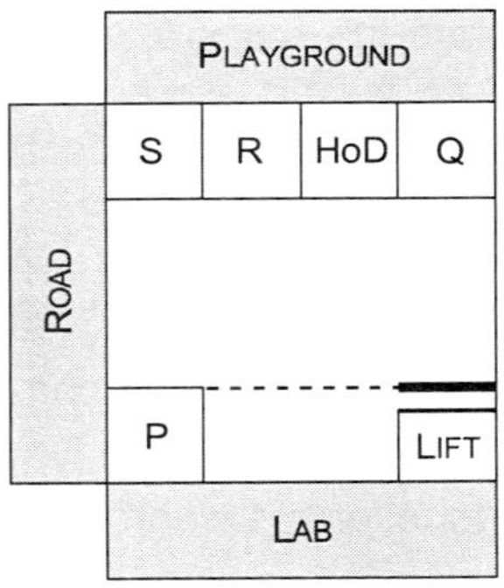

[2020 : 1 M, Set-II]

1.168 Select the most appropriate word that can replace the underlined word without changing the meaning of the sentence:

Now-a-days, most children have a tendency to <u>belittle</u> the legitimate concerns of their parents.

(a) Applaud (b) Begrudge
(c) Disparage (d) Reduce

[2020 : 1 M, Set-II]

1.169 If $f(x) = x^2$ for each $x \in (-\infty, \infty)$, then $\dfrac{f(f(f(x)))}{f(x)}$ is equal to ________.

(a) $f(x)$ (b) $(f(x))^4$
(c) $(f(x))^2$ (d) $(f(x))^3$

[2020 : 1 M, Set-II]

1.170 Rescue teams deployed ________ disaster hit areas combat ________ a lot of difficulties to save the people.

(a) with, at
(b) to, to
(c) with, with
(d) in, with

[2020 : 1 M, Set-II]

1.171 Five friends P, Q, R, S and T went camping. At night, they had to sleep in a row inside the tent. P, Q and T refused to sleep next to R since he snored loudly. P and S wanted to avoid Q as he usually hugged people in sleep.

Assuming everyone was satisfied with the sleeping arrangements, what is the order in which they slept?

(a) RSPTQ
(b) QRSPT
(c) QTSPR
(d) SPRTQ

[2020 : 2 M, Set-I]

1.172 The american psychologist Howard Gardner expounds that human intelligence can be sub-categorised into multiple kinds, in such a way that individuals differ with respect to their relative competence in each kind. Based on this theory, modern educationists insist on prescribing multi-dimensional curriculum and evluation parameters that enable development and assessment of multiple intelligences.

Which of the following statements can be inferred from the given text?

(a) Modern educationists want to develop and asses the theory of multiple intelligences.
(b) Modern educationists insist that the teaching curriculum and evaluation needs to be multi-dimensional.
(c) Howard Gardner wants to develop and assess the theory of multiple intelligences.
(d) Howard Gardner insits that the teaching curriculum and evaluation needs to be multi-dimensional.

[2020 : 2 M, Set-I]

1.173 Insert seven numbers between 2 and 34, such that the resulting sequence including 2 abnd 34 is an arithmetic progression. The sum of these inserted seven numbers is ________.

(a) 124
(b) 120
(c) 126
(d) 130

[2020 : 2 M, Set-I]

1.174 The total expenditure of a family, on different activities in a month, is shown in the pie-chart. the extra money spent on education as compared to transport (in percent) is ____.

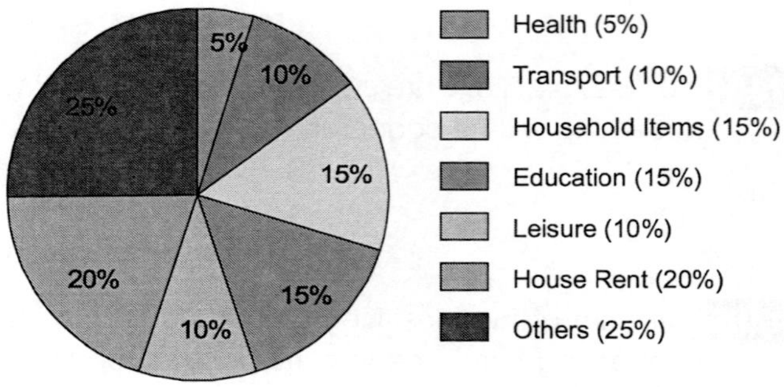

(a) 33.3
(b) 50
(c) 100
(d) 5

[2020 : 2 M, Set-I]

1.175 The unit's place in 26591749^{110016} is ________.

(a) 6
(b) 1
(c) 3
(d) 9

[2020 : 2 M, Set-I]

1.176 Nominal interest rate is defined as the amount paid by the borrower to the lender for using the borrowed amount for a specific period of time. Real interest rate calculated on the basis of actual value (inflation-adjusted), is approximately equal to the difference between nominal rate and expected rate of inflation in the economy.

Which of the following assertions is best supported by the above information?

(a) Under low inflation, real interest rate is low and borrowers get benefited.
(b) Under high inflation, real interest rate is low and borrowers get benefited.
(c) Under low inflation, real interest rate is high and borrowers get benefited.
(d) Under high inflation, real interest rate is low and lenders get benefited.

[2020 : 2 M, Set-II]

1.177 The ratio of 'the sum of the odd positive integers from 1 to 100' to 'the sum of the even positive integers from 150 to 200' is ________.

(a) 50 : 91
(b) 1 : 1
(c) 1 : 2
(d) 45 : 95

[2020 : 2 M, Set-II]

1.178 The monthly distribution of 9 Watt LED bulbs sold by two firms X and Y from January to June 2018 is shown in the pie-chart and the corresponding table. If the total number of LED bulbs sold by two firms during April-June 2018 is 50000, then the number of LED bulbs sold by the firm Y during April-June 2018 is ________.

Months	Ratio of LED bulbs sold by two firms X : Y
January	7 : 8
February	2 : 3
March	2 : 1
April	3 : 2
May	1 : 4
June	9 : 11

January	15%
February	20%
March	30%
April	15%
May	10%
June	10%

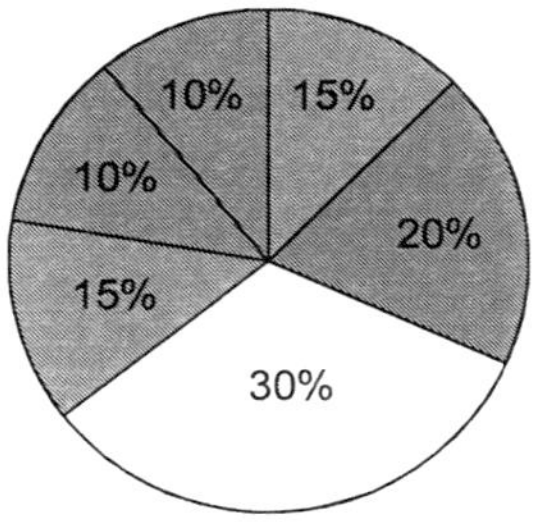

(a) 11250 (b) 9750
(c) 8250 (d) 8750

[2020 : 2 M, Set-II]

1.179 In a school of 1000 students, 300 students play chess and 600 students play football. If 50 students play both chess and football, the number of students who play neither is ________.

(a) 150 (b) 50
(c) 100 (d) 200

[2020 : 2 M, Set-II]

1.180 For the year 2019, which of the previous year's calendar can be used?

(a) 2011 (b) 2013
(c) 2012 (d) 2014

[2020 : 2 M, Set-II]

1.181 The mirror image of the about text about the x-axis is

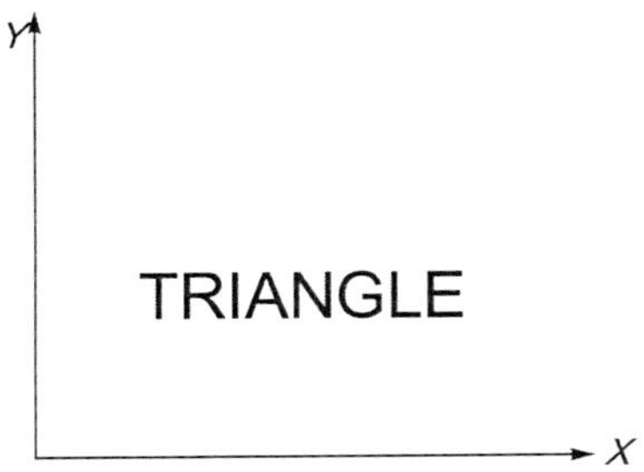

(a) 
(b) 
(c) 
(d) 

[2021 : 1 M, Set-I]

1.182 In a company, 35% of the employees drink coffee, 40% of the employees drink tea and 10% of the employees drink both tea and coffee. What percentage of employees drink neither tea nor coffee?

(a) 15 (b) 25
(c) 35 (d) 40

[2021 : 1 M, Set-I]

1.183 Four persons P, Q, R and S are to be seated in a row. All facing the same direction, but not necessarily in the same order. P and R cannot sit adjacent to each other. S should be seated to the right of Q. The number of distinct seating arrangements possible is:

(a) 4 (b) 2
(c) 8 (d) 6

[2021 : 1 M, Set-I]

1.184 $\oplus$ and $\odot$ are two operators on numbers p and q such that

$$p \oplus Q = \frac{p^2 + q^2}{pq} \text{ and } p \odot q = \frac{p^2}{q};$$

If $x \oplus y = 2 \odot 2$, then $x =$

(a) y (b) $\frac{y}{2}$
(c) $\frac{3y}{2}$ (d) $2y$

[2021 : 1 M, Set-I]

1.185 Getting to the top is ________ than staying on top.

(a) more easy (b) easier
(c) much easy (d) easiest

[2021 : 1 M, Set-I]

1.186 $\oplus$ and $\odot$ are two operators on numbers p and q such that

$p \odot q = p - q$, and $p \oplus q = p \times q$

Then, $(9 \odot (6 \oplus 7)) \odot (7 \oplus (6 \odot 5)) =$

(a) –33 (b) 40
(c) –40 (d) –26

[2021 : 1 M, Set-II]

1.187

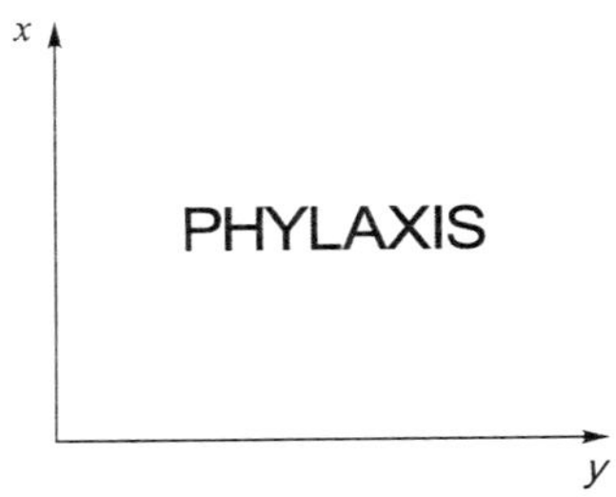

The mirror image of the above text about the x-axis is

(a) dHYΓ∀XIƧ

(b) PHYΓ∀XIƧ

(c) ЬHYΓ∀XIƧ

(d) ЬHYΓ∀XIS

[2021 : 1 M, Set-II]

1.188 (i) Arun and Aparna are here.
(ii) Arun and Aparna is here.
(iii) Arun's families is here.
(iv) Arun's family is here.
Which of the above sentences are grammatically CORRECT?

(a) (i) and (iv) (b) (iii) and (iv)
(c) (i) and (ii) (d) (ii) and (iv)

[2021 : 1 M, Set-II]

1.189 Four persons P, Q, R and S are to be seated in a row. R should not be seated at the second position from the left end of the row. The number of distinct seating arrangements possible is:

(a) 24 (b) 9
(c) 6 (d) 18

[2021 : 1 M, Set-II]

1.190 Two identical cube shaped dice each with faces numbered 1 to 6 are rolled simultaneously. The probability that an even number is rolled out on each dice is:

(a) $\frac{1}{8}$ (b) $\frac{1}{36}$
(c) $\frac{1}{4}$ (d) $\frac{1}{12}$

[2021 : 1 M, Set-II]

1.191 Statement: Either P marries Q or X marries Y
Among the option below, the logical NEGATION of the above statement is:

(a) Neither P marries Q nor X marries Y
(b) X does not marry Y and P marries Q
(c) P marries Q and X marries Y
(d) P does not marry Q and X marries Y

[2021 : 2 M, Set-I]

1.192 A function λ, is defined by

$$\lambda(p,q) = \begin{cases} (p-q)^2, & \text{if } p \geq q \\ p+q, & \text{if } p < q \end{cases}$$

The value of expression $\frac{\lambda[-(-3+2),(-2,+3)]}{[-(-2+1)]}$ is:

(a) -1 (b) 0
(c) $\frac{16}{3}$ (d) 16

[2021 : 2 M, Set-I]

1.193 Five line segments of equal lengths, *PR*, *PS*, *QS*, *QT* and *RT* as used to form a star shown in the figure above.

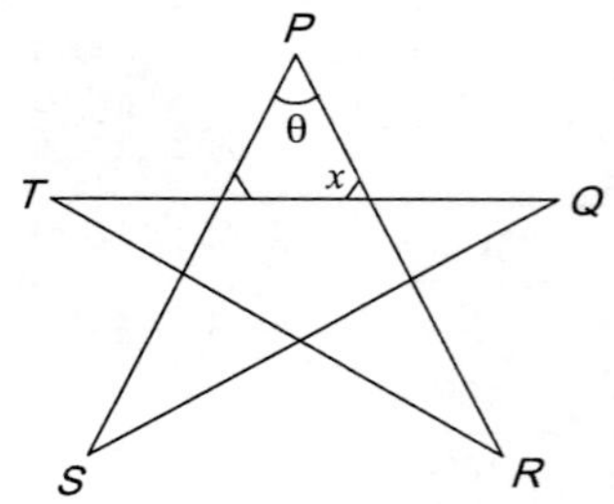

The value of θ, in degrees, is______.

(a) 72 (b) 108
(c) 36 (d) 45

[2021 : 2 M, Set-I]

1.194 Human have the ability to construct worlds entirely in their minds, which don't exist in the physical world. So far as we know, no other species possesses this ability. This skill is so important that we have different worlds to refer to its different flavors, such as imagination invention and innovation.
Based on the above passage, which one of the following is TRUE?

(a) We do not know of any species other than humans who posses the ability to construct mental worlds.
(b) Imagination, invention and innovation are unrelated to the ability to construct mental worlds.
(c) No spices posses the ability to construct worlds in their minds.
(d) The terms imagination, invention and innovation refer to unrelated skill.

[2021 : 2 M, Set-I]

1.195 Consider two rectangular sheets, Sheet *M* and Sheet *N* of dimensions 6 cm × 4 cm each.
Folding operation 1: The sheet is folded into half by joining the short edges of the current shape,

Folding operation 2: The sheet is folded into half by joining the long edges of the current shape.

Folding operation 1 is carried out on Sheet *M* three times,

Folding operation 2 is carried out on Sheet *N* three times,

The ratio of perimeters of the final folded shape of Sheet *N* to the final folded shape of Sheet *M* is_____.

(a) 13 : 7 (b) 7 : 5
(c) 5 : 13 (d) 3 : 2

[2021 : 2 M, Set-I]

1.196 On a planar field, you travelled 3 units East from a point O. Next you travelled 4 units South to arrive at point P. Then you travelled from P in the North-East direction such that you arrive at a point that is 6 units East of point O. Next, you travelled in the North-West direction, so that you arrive at point Q that is 8 units North of point P.

The distance of point Q to point O, in the same units, should be _______.

(a) 6 (b) 5
(c) 4 (d) 3

[2021 : 2 M, Set-II]

1.197 The author said, "Musicians rehearse before their concerts. Actors rehearse their roles before the opening of a new play. On the other hand, I find it strange that many public speakers think they can just walk on to the stage and start speaking. In my opinion, it is no less important for public speakers to rehearse their talks."

Based on the above passage, which one of the following is TRUE?

(a) The author is of the opinion that rehearsing is less important for public speakers than for musicians and actors.
(b) The author is of the opinion that rehearsing is more important only for musicians than public speakers.
(c) The author is of the opinion that rehearsal is more important for actors than musicians.
(d) The author is of the opinion that rehearsing is important for musicians, actors and public speakers.

[2021 : 2 M, Set-II]

1.198 In an equilateral triangle PQR, side PQ is divided into four equal parts, side QR is divided into six equal parts and side PR is divided into eight equal parts. The length of each subdivided part in cm is an integer.

The minimum area of the triangle PQR possible, in cm^2, is

(a) 24 (b) $144\sqrt{3}$
(c) $48\sqrt{3}$ (d) 18

[2021 : 2 M, Set-II]

1.199 1. Some football players play cricket.
2. All cricket players play hockey.

Among the options given below, the statement that logically follows from the two statements 1 and 2 above, is:

(a) All hockey players play football.
(b) Some football players play hockey.
(c) All football players play hockey.
(d) No football player plays hockey.

[2021 : 2 M, Set-II]

1.200

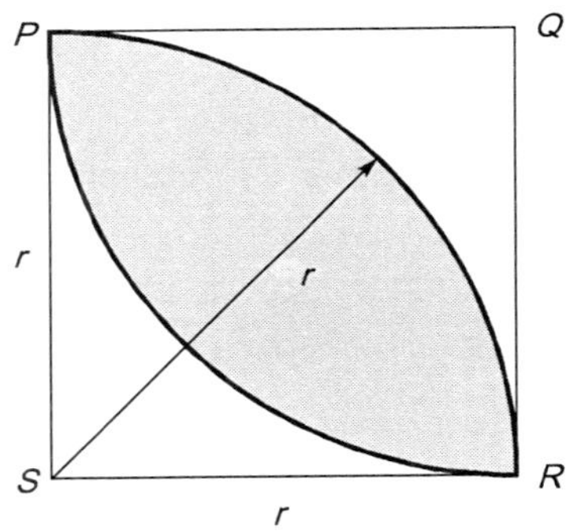

In the figure shown above, PQRS is a square. The shaded portion is formed by the intersection of sectors of circles with radius equal to the side of the square and centers at S and Q.

The probability that any point picked randomly within the square falls in the shaded area is _______.

(a) $\frac{\pi}{2}-1$ (b) $4-\frac{\pi}{2}$
(c) $\frac{\pi}{4}$ (d) $\frac{1}{2}$

[2021 : 2 M, Set-II]

1.201 You should _______ when to say _______.

(a) no/no (b) no/know
(c) know/know (d) know/no

[2022 : 1 M, Set-I]

1.202 Two straight lines pass through the origin $(x_0, y_0) = (0, 0)$. One of them passes through the point $(x_1, y_1) = (1, 3)$ and the other passes through the point $(x_2, y_2) = (1, 2)$.

What is the area enclosed between the straight lines in the interval [0, 1] on the x-axis?

(a) 0.5 (b) 1.5
(c) 1.5 (d) 2.0

[2022 : 1 M, Set-I]

1.203 If

$p : q = 1 : 2$
$q : r = 4 : 3$
$r : s = 4 : 5$

and u is 50% more than s, what is the ratio $p : u$?

(a) 2 : 15 (b) 16 : 15
(c) 1 : 5 (d) 16 : 45

[2022 : 1 M, Set-I]

1.204 Given the statements:

- P is the sister of Q.
- Q is the husband of R.
- R is the mother of S.
- T is the husband of P.

Based on the above information, T is ______ of S.

(a) the grandfather (b) an uncle
(c) the father (d) a brother

[2022: 1 M, Set-I]

1.205 In the following diagram, the point R is the center of the circle. The lines PQ and ZV are tangential to the circle. The relation among the areas of the squares, PXWR, RUVZ and SPQT is

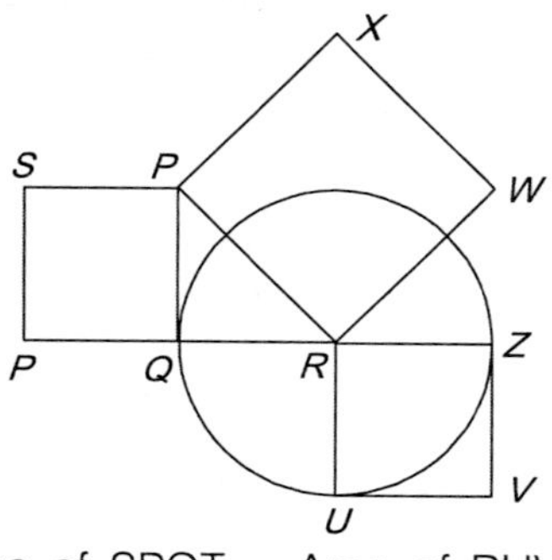

(a) Area of SPQT = Area of RUVZ = Area of PXWR

(b) Area of SPQT = Area of PXWR – Area of RUVZ

(c) Area of PXWR = Area of SPQT – Area of RUVZ

(d) Area of PXWR = Area of RUVZ – Area of SPQT

[2022 : 1 M, Set-I]

1.206 The movie was funny and I ________ .

(a) could helped laughed
(b) couldn't help laughed
(c) couldn't help laughing
(d) could help laughing [2022 : 1 M, Set-II]

1.207

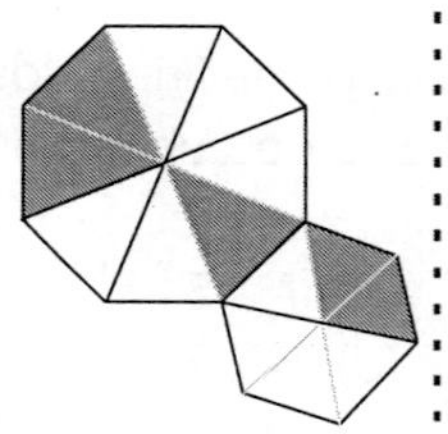

For the picture shown above, which one of the following is the correct picture representing reflection with respect to the mirror shown as the dotted line?

(a) 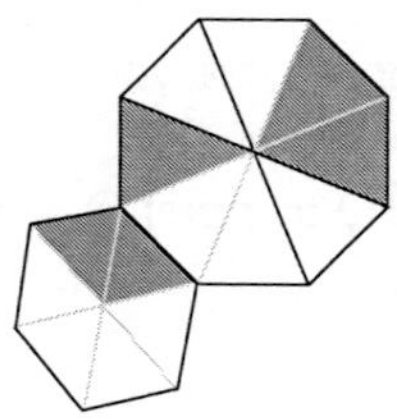 (b) 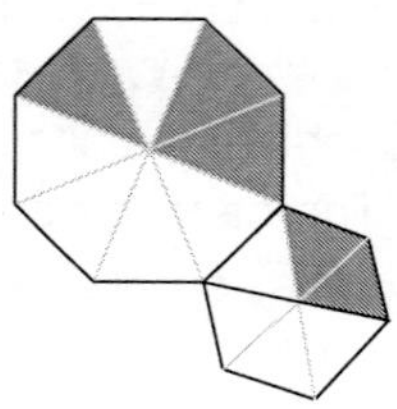

(c) 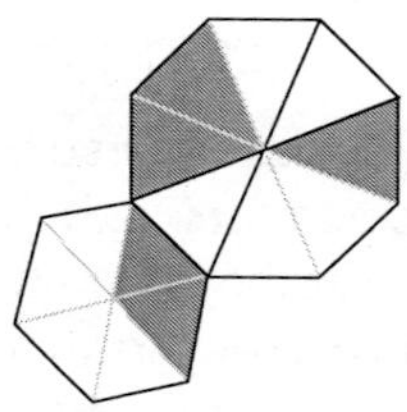 (d) 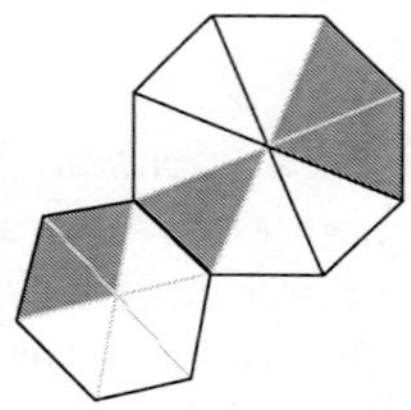

[2022 : 1 M, Set-II]

1.208 An ant walks in a straight line on a plane leaving behind a trace of its movement. The initial position of the ant is at point P facing East.

The ant first turns 72° anticlockwise at P, and then does the following two steps in sequence exactly FIVE times before halting.

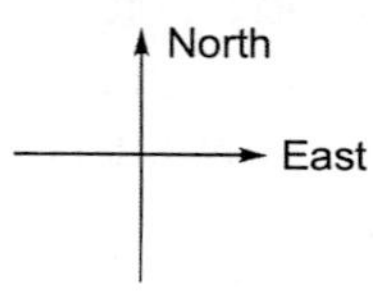

1. moves forward for 10 cm.
2. turns 144° clockwise

The pattern made by the trace left behind by the ant is

(a) 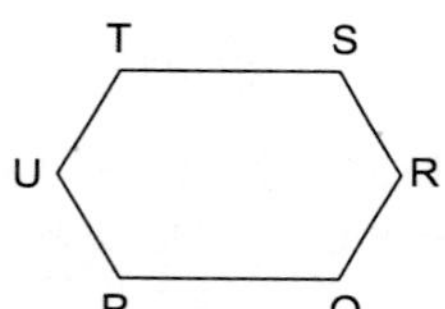

PQ = QR = RS = ST = TP = 10 cm

(b) 

SQ = QT = TR = RP = PS = 10 cm

(c) 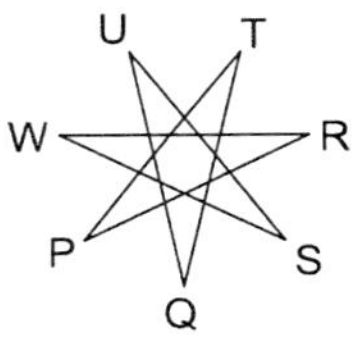

SW = WR = RP = PT = TQ = QU = US = 10 cm

(d) 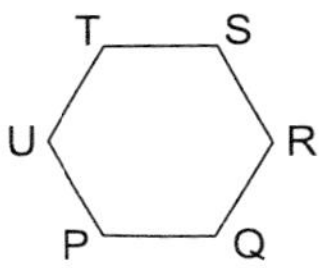

PQ = QR = RS = ST = TU = UP = 10 cm

[2022 : 1 M, Set-II]

1209 In a partnership business the monthly investment by three friends for the first six months is m the ratio 3 :4 : 5. After six months, they had to increase their monthly investments by 10%, 15% and 20%, respectively, of then initial monthly investment. The new investment ratio was kept constant for the next six months.

What is the ratio of their shares in the total profit (m the same order) at the end of the year such that the share is proportional to their individual total investment over the year?

(a) 63 : 86 : 110 (b) 33 : 46 : 60
(c) 22 : 23 : 24 (d) 22 : 33 : 50

[2022 : 1 M, Set-II]

1210 A survey of 450 students about their subjects of interest resulted in the following outcome.

- 150 students are interested m Mathematics.
- 200 students are interested in Physics.
- 175 students are interested in Chemistry.
- 50 students are interested in Mathematics and Physics.
- 60 students are interested in Physics and Chemistry
- 40 students are interested in Mathematics and Chemistry.
- 30 students are interested in Mathematics, Physics and Chemistry.
- Remaining students are interested in Humanities.

Based on the above information, the number of students interested in Humanities is

(a) 45 (b) 10
(c) 40 (d) 30

[2022 : 1 M, Set-II]

1211 Healthy eating is a critical component of healthy aging. When should one start eating healthy? It turns out that it is never too early. For example, babies who start eating healthy in the first year are more likely to have better overall health as they get older.

Which one of the following is the CORRECT logical inference based on the information in the above passage?

(a) Healthy eating is important for those with good health conditions, but not for others
(b) Eating healthy can be started at any age, earlier the better
(c) Eating healthy and better overall health are more correlated at a young age, but not older age
(d) Healthy eating is more important for adults than kids

[2022 : 2 M, Set-I]

1212 P invested ₹ 5000 per month for 6 months of a year and Q invested ₹ x per month for 8 months of the year in a partnership business. The profit is shared in proportion to the total investment made in that year.

If at the end of that investment year, Q receives $\frac{4}{9}$ of the total profit, what is the value of (in ₹)?

(a) 2500 (b) 3000
(c) 4687 (d) 8437

[2022 : 2 M, Set-I]

1213

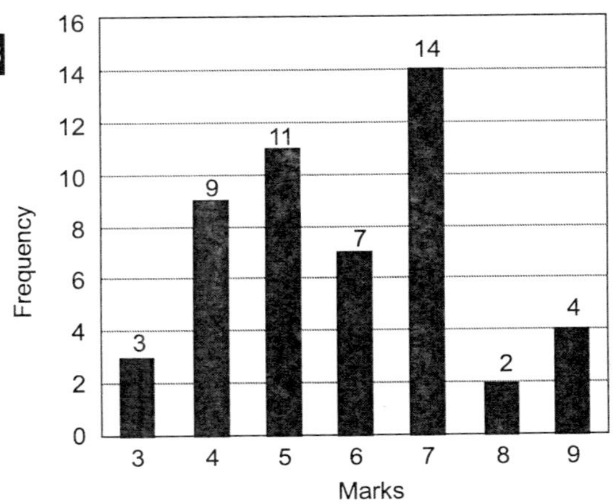

The above frequency chart shows the frequency distribution of marks obtained by a set of students in an exam.

From the data presented above, which one of the following is CORRECT?

(a) mean > mode > median
(b) mode > median > mean
(c) mode > mean > median
(d) median > mode > mean

[2022 : 2 M, Set-I]

1.214 In the square grid shown on the left, a person standing at P2 position is required to move to P5 position.

The only movement allowed for a step involves, "two moves along one direction followed by one move in a perpendicular direction". The permissible directions for movement are shown as dotted arrows in the right.

For example, a person at a given position Y can move only to the positions marked X on the right.

Without occupying any of the shaded squares at the end of each step, the minimum number of steps required to go from P2 to P5 is

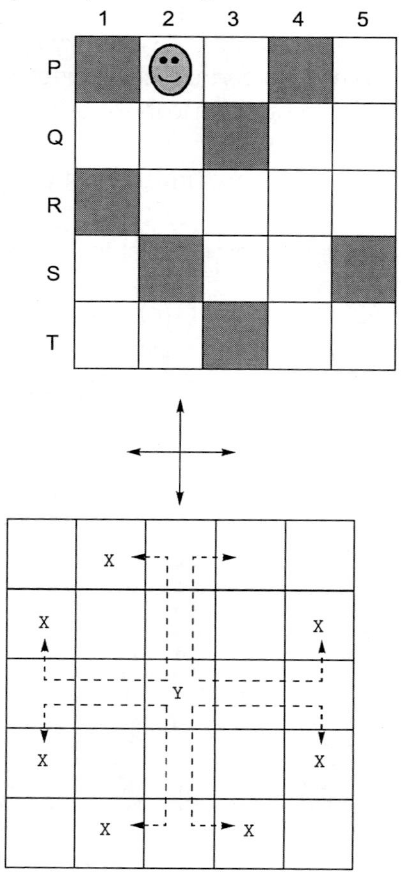

(a) 4
(b) 5
(c) 6
(d) 7

[2022 : 2 M, Set-I]

1.215

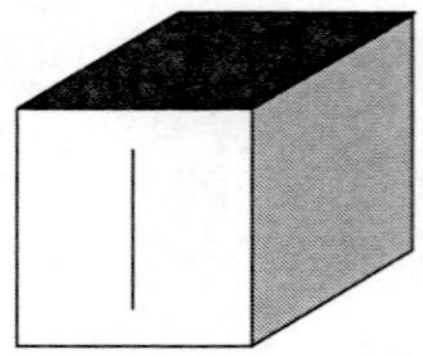

Consider a cube made by folding a single sheet of paper of appropriate shape.

The interior faces of the cube are all blank. However, the exterior faces that are not visible in the above view may not be blank.

Which one of the following represents a possible unfolding of the cube?

(a)
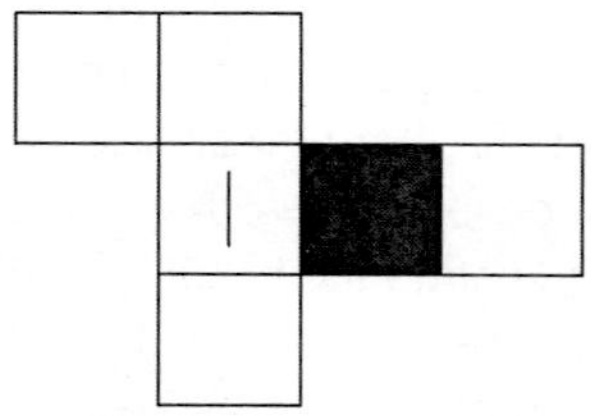

(b)
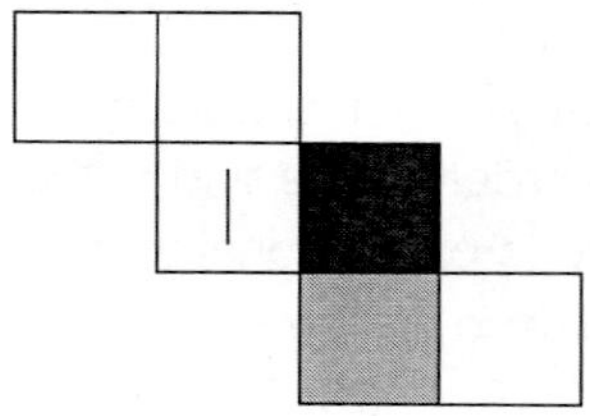

(c)
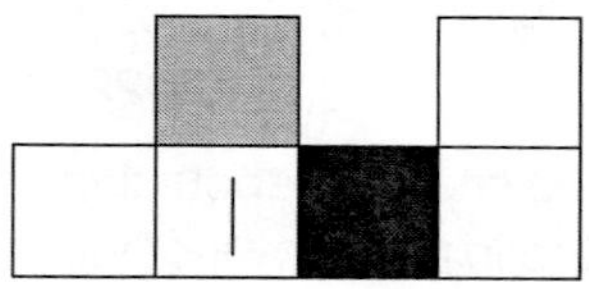

(d)
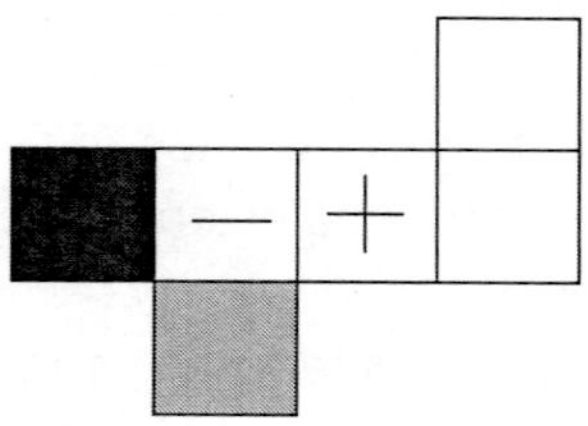

[2022 : 2 M, Set-I]

1.216 $x : y : z = \frac{1}{2} : \frac{1}{3} : \frac{1}{4}$

What is the value of $\frac{x + z - y}{y} = ?$

(a) 1.25
(b) 0.75
(c) 3.25
(d) 2.25

[2022 : 2 M, Set-II]

1.217 Both the numerator and the denominator of $\frac{3}{4}$ are increased by a positive $\frac{15}{17}$ integer, x, and those of — are decreased by the same integer. This operation results m the same value for both the fractions. What is the value of x?

(a) 3 (b) 4
(c) 1 (d) 2

[2022 : 2 M, Set-II]

1.218 Consider the following equations of straight lines:

Line L1 : $2x - 3y = 5$
Line L2 : $3x + 2y = 8$
Line L3 : $4x - 6y = 5$
Line L4 : $6x - 9y = 6$

Which one of the following is the correct statement?

(a) L1 is parallel to L2 and L1 is perpendicular to L3
(b) L2 is parallel to L4 and L2 is perpendicular to L1
(c) L4 is perpendicular to L2 and L4 is parallel to L3
(d) L3 is perpendicular to L4 and L3 is parallel to L2

[2022 : 2 M, Set-II]

1.219 In the last few years, several new shopping malls were opened in the city. The total number of visitors in the malls is impressive. However, the total revenue generated through sales in the shops in these malls is generally low.

Which one of the following is the CORRECT logical inference based on the information in the above passage?

(a) More people are visiting the malls and spending more
(b) Fewer people are visiting the malls but spending more
(c) More people are visiting the malls but not spending enough
(d) Fewer people are visiting the malls and not spending enough

[2022 : 2 M, Set-II]

1.220 Given below are two statements and four conclusions drawn based on the statements.

Statement 1: Some soaps are clean.
Statement 2: All clean objects are wet.

Conclusion I: Some clean objects are soaps.
Conclusion II: No clean object is a soap.
Conclusion III: Some wet objects are soaps.
Conclusion IV: All wet objects are soaps.

Which one of the following options can be logically inferred?

(a) Either conclusion I or conclusion II is correct
(b) Either conclusion III or conclusion IV is correct
(c) Only conclusion I and conclusion III are correct
(d) Only conclusion I is correct

[2022 : 2 M, Set-II]

1.221 "I have not yet decided what I will do this evening; I ______ visit a friend."

(a) mite (b) would
(c) might (d) didn't

[2023 : 1 M, Set-I]

1.222 Eject : Insert : : Advance : ______

(By word meaning)

(a) Advent (b) Progress
(c) Retreat (d) Loan

[2023 : 1 M, Set-I]

1.223 A duck named Donald Duck says "All ducks always lie."

Based only on the information above, which one of the following statements can be logically inferred with certainty?

(a) Donald Duck always lies.
(b) Donald Duck always tells the truth.
(c) Donald Duck's statement is true.
(d) Donald Duck's statement is false.

[2023 : 1 M, Set-I]

1.224 The James Webb telescope, recently launched in space, is giving humankind unprecedented access to the depths of time by imaging very old stars formed almost 13 billion years ago. Astrophysicists and cosmologists believe that this odyssey in space may even shed light on the existence of dark matter. Dark matter is supposed to interact only via the gravitational interaction and not through the electromagnetic-, the weak- or the strong-interaction. This may justify the epithet "dark" in dark matter.

Based on the above paragraph, which one of the following statements is FALSE?

(a) No other telescope has captured images of stars older than those captured by the James Webb telescope.
(b) People other than astrophysicists and cosmologists may also believe in the existence of dark matter.
(c) The James Webb telescope could be of use in the research on dark matter.
(d) If dark matter was known to interact via the strong-interaction, then the epithet "dark" would be justified.

[2023 : 2 M, Set-I]

1.225 In the given figure, PQRSTV is a regular hexagon with each side of length 5 cm. A circle is drawn with its centre at V such that it passes through P. What is the area (in cm^2) of the shaded region? (The diagram is representative)

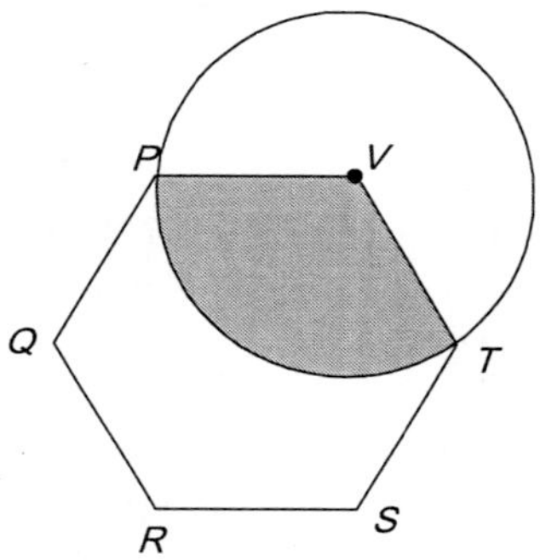

(a) $\frac{25\pi}{3}$ (b) $\frac{20\pi}{3}$
(c) 6π (d) 7π

[2023 : 1 M, Set-I]

1.226 A line of symmetry is defined as a line that divides a figure into two parts in a way such that each part is a mirror image of the other part about that line.

The figure below consists of 20 unit squares arranged as shown. In addition to the given black squares, upto 5 more may be coloured black. Which one among the following options depicts the minimum number of boxes that must be coloured black to achieve two lines of symmetry? (The figure is representative)

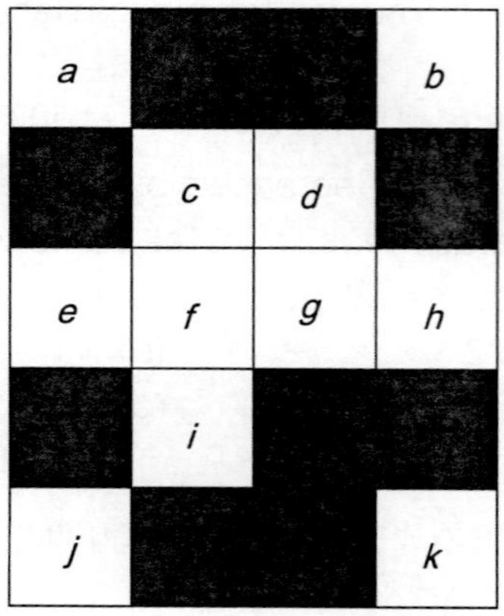

(a) d (b) c, d, i
(c) c, i (d) c, d, i, f, g

[2023 : 1 M, Set-I]

1.227 Based only on the truth of the statement 'Some humans are intelligent', which one of the following options can be logically inferred with certainty?
(a) No human is intelligent.
(b) All humans are intelligent.
(c) Some non-humans are intelligent.
(d) Some intelligent beings are humans.

[2023 : 2 M, Set-I]

1.228 Which one of the options can be inferred about the mean, median, and mode for the given probability distribution (i.e. probability mass function), $P(x)$, of a variable x?

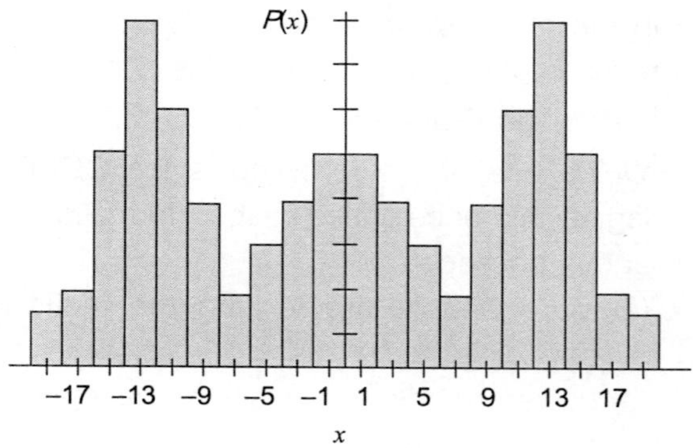

(a) mean = median ≠ mode
(b) mean = median = mode
(c) mean ≠ median = mode
(d) mean ≠ mode = median

[2023 : 2 M, Set-I]

1.229 Let $a = 30!$, $b = 50!$ and $c = 100!$. Consider the following numbers:

$\log_a c, \log_c a, \log_b a, \log_a b$

Which one of the following inequalities is CORRECT?

(a) $\log_c a < \log_b a < \log_a c < \log_a c$

(b) $\log_c a < \log_a b < \log_b a < \log_b c$

(c) $\log_c a < \log_b a < \log_a c < \log_a b$

(d) $\log_b a < \log_c a < \log_a b < \log_a c$

[2023 : 2 M, Set-I]

1.230 A square of side length 4 cm is given. The boundary of the shaded region is defined by one semi-circle on the top and two circular arcs at the bottom, each of radius 2 cm, as shown.

The area of the shaded region is _______ cm^2.

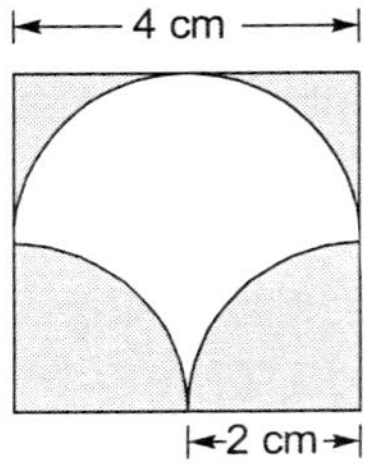

(a) 8 (b) 4
(c) 12 (d) 10

[2023 : 2 M, Set-I]

1.231 The line ran _______ the page, right through the centre, and divided the page into two.

(a) across (b) of
(c) between (d) about

[2023 : 1 M, Set-II]

1.232 Kind : _______ : : Often : Seldom
(By word meaning)

(a) Cruel (b) Variety
(c) Type (d) Kindred

[2023 : 1 M, Set-II]

1.233 Elvesland is a country that has peculiar beliefs and practices. They express almost all their emotions by gifting flowers. For instance, if anyone gifts a white flower to someone, then it is always taken to be a declaration of one's love for that person. In a similar manner, the gifting of a yellow flower to someone often means that one is angry with that person.

Based only on the information provided above, which one of the following sets of statement(s) can be logically inferred with certainty?

(i) In Elvesland, one always declares one's love by gifting a white flower.

(ii) In Elvesland, all emotions are declared by gifting flowers.

(iii) In Elvesland, sometimes one expresses one's anger by gifting a flower that is not yellow.

(iv) In Elvesland, sometimes one expresses one's love by gifting a white flower.

(a) only (ii) (b) (i), (ii) and (iii)
(c) (i), (iii) and (iv) (d) only (iv)

[2023 : 2 M, Set-II]

1.234 Based only on the following passage, which one of the options can be inferred with certainty?

When the congregation sang together, Apenyo would also join, though her little screams were not quite audible because of the group singing. But whenever there was a special number, trouble would begin; Apenyo would try singing along, much to the embarrassment of her mother. After two or three such mortifying Sunday evenings, the mother stopped going to church altogether until Apenyo became older and learnt to behave.

At home too, Apenyo never kept quiet; she hummed or made up silly songs to sing by herself, which annoyed her mother at times but most often made her become pensive. She was by now convinced that her daughter had inherited her love of singing from her father who had died unexpectedly away from home.

[Excerpt from These Hills Called Home by Temsula Ao]

(a) The mother was embarrassed about her daughter's singing at home.

(b) The mother's feelings about her daughter's singing at home were only of annoyance.

(c) The mother was not sure if Apenyo had inherited her love of singing from her father.

(d) When Apenyo hummed at home, her mother tended to become thoughtful.

[2023 : 2 M, Set-II]

1.235 In how many ways can cells in a 3 × 3 grid be shaded, such that each row and each column have exactly one shaded cell? An example of one valid shading is shown.

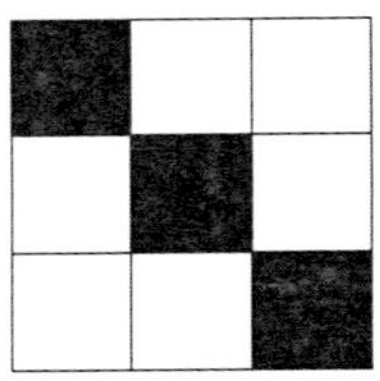

(a) 2 (b) 9
(d) 3 (d) 6

[2023 : 1 M, Set-II]

1.236 There are 4 red, 5 green, and 6 blue balls inside a box. If N number of balls are picked simultaneously, what is the smallest value of N that guarantees there will be at least two balls of the same colour?
One cannot see the colour of the balls until they are picked.

(a) 4 (b) 15
(c) 5 (d) 2

[2023 : 1 M, Set-II]

1.237 Consider a circle with its centre at the origin (O), as shown. Two operations are allowed on the circle.
Operation 1: Scale independently along the x and y axes.
Operation 2: Rotation in any direction about the origin.
Which figure among the options can be achieved through a combination of these two operations on the given circle?

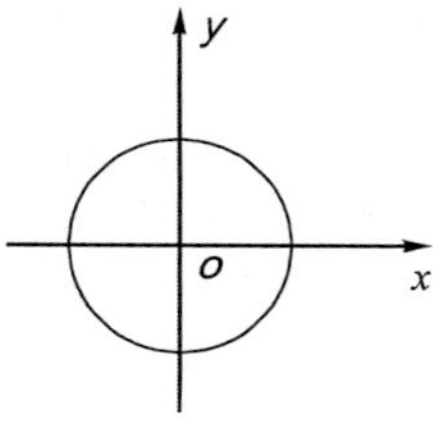

(a)
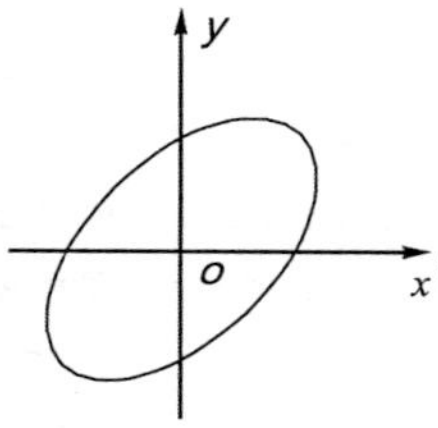

(b)
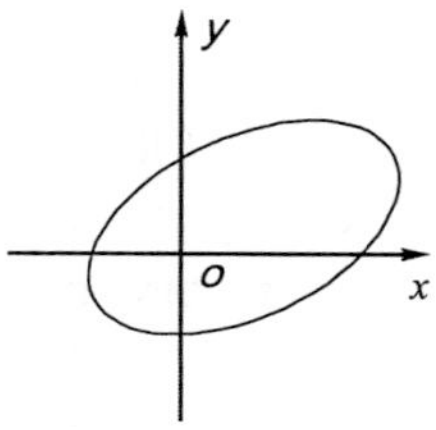

(c)
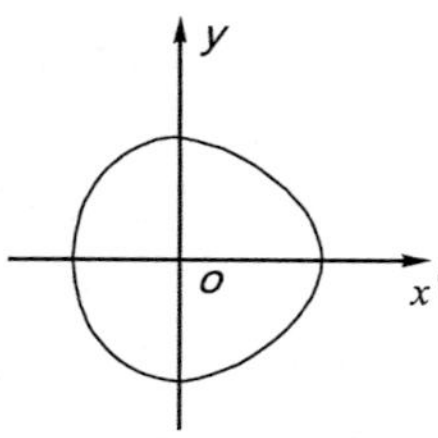

(d)
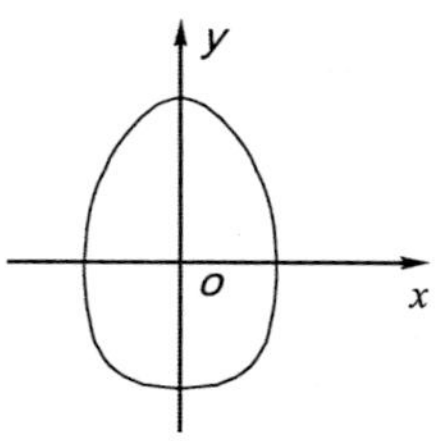

[2023 : 1 M, Set-II]

1.238 Three husband-wife pairs are to be seated at a circular table that has six identical chairs. Seating arrangements are defined only by the relative position of the people. How many seating arrangements are possible such that every husband sits next to his wife?

(a) 16 (b) 4
(c) 120 (d) 720

[2023 : 2 M, Set-II]

1.239 If x satisfies the equation $4^{8^x} = 256$, then x is equal to ______.

(a) $\frac{1}{2}$ (b) $\log_{16} 8$
(c) $\frac{2}{3}$ (d) $\log_4 8$

[2023 : 2 M, Set-II]

1.240 Consider a spherical globe rotating about an axis passing through its poles. There are three points P, Q and R situated respectively on the equator, the north pole, and midway between the equator and the north pole in the northern hemisphere. Let P, Q and R move with speeds v_P, v_Q, and v_R, respectively.
Which one of the following options is CORRECT?

(a) $v_P < v_R < v_Q$ (b) $v_P < v_Q < v_R$
(c) $v_P > v_R > v_Q$ (d) $v_P = v_R \neq v_Q$

[2023 : 2 M, Set-II]

1.241 The chart given below compares the Installed Capacity (MW) of four power generation technologies, T1, T2, T3 and T4 and their Electricity Generation (MWh) in a time of 1000 hours (h).

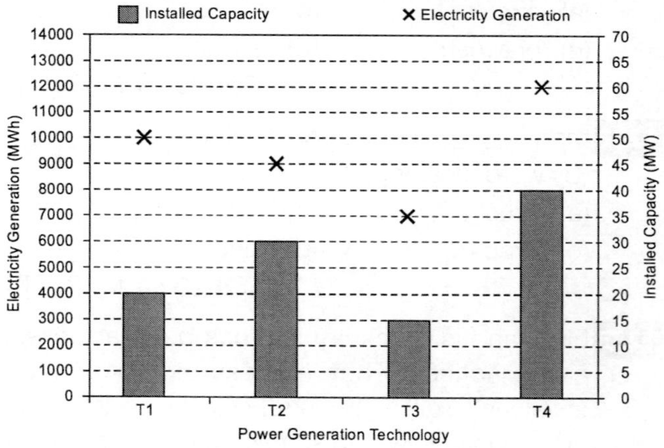

The capacity factor of a power generation technology is:
Capacity Factor =

$$\frac{\text{Electricity Generation (MWh)}}{\text{Installed Capacity (MW)} \times 1000\text{(h)}}$$

Which one of the given technologies has the highest Capacity Factor?

(a) T2 (b) T1
(c) T3 (d) T4

[2024 : 2 M, Set-I]

1.242 In the 4 × 4 array shown below, each cell of the first three columns has either a cross (X) or a number, as per the given rule.

1	1	2	
2	X	3	
2	X	4	
1	2	X	

Rule: The number in a cell represents the count of crosses around its immediate neighboring cells (left, right, top, bottom, diagonals).
As per this rule, the maximum number of crosses possible in the empty column is

(a) 3 (b) 0
(c) 1 (d) 2

[2024 : 2 M, Set-I]

1.243 Three distinct sets of indistinguishable twins are to be seated at a circular table that has 8 identical chairs.. Unique seating arrangements are defined by the relative positions of the people.

How many unique seating arrangements are possible such that each person is sitting next to their twin?

(a) 12 (b) 10
(c) 14 (d) 28

[2024 : 2 M, Set-I]

1.244 For positive integers p and q, with $\frac{p}{q} \neq 1, \left(\frac{p}{q}\right)^{\frac{p}{q}} = p^{\left(\frac{p}{q}-1\right)}$. Then,

(a) $\sqrt[p]{q} = \sqrt[q]{p}$ (b) $\sqrt{q} = \sqrt{p}$
(c) $q^p = p^{2q}$ (d) $q^p = p^q$

[2024 : 1 M, Set-I]

1.245 On a given day, how many times will the second-hand and the minute-hand of a clock cross each other during the clock time 12:05:00 hours to 12:55:00 hours?

(a) 51 (b) 49
(c) 50 (d) 55

[2024 : 1 M, Set-I]

1.246 In the given text, the blanks are numbered (1)-(iv). Select the best match for all the blanks.

From the ancient Athenian arena to the modern Olympic stadiums, athletics ___(i)___ the potential for a spectacle. The crowd ___(ii)___ with bated breath as the Olympian artist twists his body, stretching the javelin behind him. Twelve strides in, he begins to cross-step. Six cross-steps ___(iii)___ in an abrupt stop on his left foot. As his body ___(iv)___ like a door turning on a hinge, the javelin is launched skyward at a precise angle.

(a) (i) holds (ii) wait (iii) culminates (iv) pivot
(b) (i) hold (ii) wait (iii) culminate (iv) pivots
(c) (i) holds (ii) waits (iii) culminate (iv) pivots
(d) (i) hold (ii) Waits (iii) culminates (iv) pivot

[2024 : 2 M, Set-I]

1.247 If '→' denotes increasing order of intensity, then the meaning of the words [simmer → seethe → smolder] is analogous to [break → raze → ______]. Which one of the given options is appropriate to fill the blank?

(a) obliterate (b) fissure
(c) fracture (d) obfuscate

[2024 : 1 M, Set-I]

1.248 Which one of the given options is a possible value of x in the following sequence?
3, 7, 15, x, 63, 127, 255

(a) 35 (b) 31
(c) 45 (d) 40

[2024 : 1 M, Set-I]

1.249 During a half-moon phase, the Earth-Moon-Sun form a right triangle. If the Moon-Earth-Sun angle at this half-moon phase is measured to be 89.85", the ratio of the Earth-Sun and Earth-Moon distances is closest to

(a) 283 (b) 382
(c) 328 (d) 238

[2024 : 2 M, Set-I]

1.250 In a locality, the houses are numbered in the following way:
The house-numbers on one side of a road are consecutive odd integers starting from 301, while the house-numbers on the other side of the road are consecutive even numbers starting from 302. The total number of houses is the same on both sides of the road.

If the difference of the sum of the house-numbers between the two sides of the road is 27, then the number of houses on each side of the rood is

(a) 52 (b) 54
(c) 26 (d) 27

[2024 : 1 M, Set-I]

1.251 Visualize a cube that is held with one of the four body diagonals aligned to the vertical axis. Rotate the cube about this axis such that its view remains unchanged. The magnitude of the minimum angle of rotation is

(a) 120° (b) 90°
(c) 180° (d) 60°

[2024 : 2 M, Set-II]

1.252 Statements:

1. All heroes are winners.
2. All winners are lucky people.

Inferences:

I. All lucky people are heroes.
II. Some lucky people are heroes.
III. Some winners are heroes.

Which of the above inferences can be logically deduced from statements 1 and 2?

(a) Only I and II (b) Only I and III
(c) Only II and III (d) Only III

[2024 : 1 M, Set-II]

1.253 A student was supposed to **multiply** a positive real number p with another positive real number q, Instead, the student **divided** p by q. If the percentage error in the student's answer is 80%, the value of q is

(a) 2 (b) $\sqrt{5}$
(c) 5 (d) $\sqrt{2}$

[2024 : 1 M, Set-II]

1.254 Seven identical cylindrical chalk-sticks are fitted tightly in a cylindrical container. The figure below shows the arrangement of the chalk-sticks inside the cylinder.

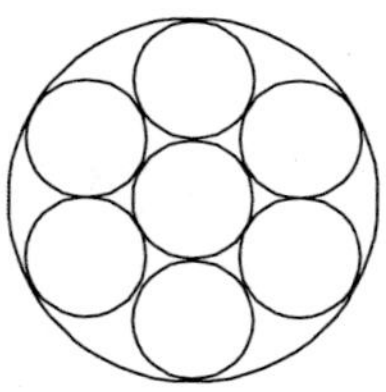

The length of the container is equal to the length of the chalk-sticks. The ratio of the occupied space to the empty space of the container is

(a) 9/2 (b) 5/2
(c) 3 (d) 7/2

[2024 : 2 M, Set-II]

1.255 The plot below shows the relationship between the mortality risk of cardiovascular disease and the number of steps a person walks per day. Based on the data, which one of the following options is true?

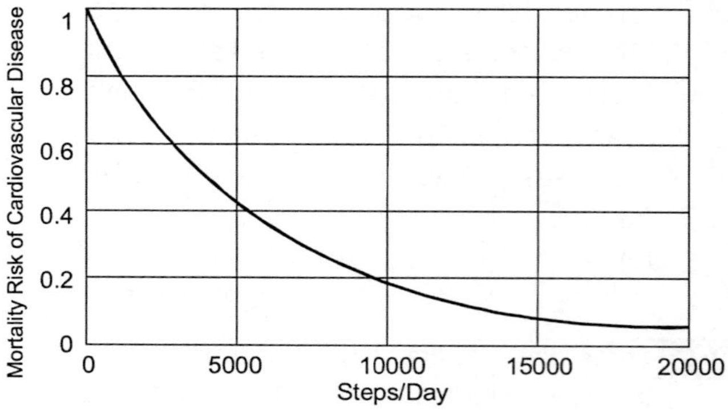

(a) For any 5000 increment in steps/day the largest risk reduction occurs on going from 15000 to 20000.
(b) The risk reduction on increasing the steps/day from 0 to 5000 is less than the risk reduction on increasing the steps/day from 15000 to 20000.
(c) The risk reduction on increasing the steps/day from 0 to 10000 is less than the risk reduction on increasing the steps/day from 10000 to 20000.
(d) For any 5000 increment in steps/day the largest risk reduction occurs on going from 0 to 5000.

[2024 : 2 M, Set-II]

1.256 In the given text, the blanks are numbered (i)-(iv). Select the best match for all the blanks.

Yoko Roi stands __(i)__ as an author for standing __(ii)__ as an honorary fellow, after she stood __(iii)__ her writings that stand __(iv)__ the freedom of speech.

(a) (i) down (ii) out (iii) by (iv) in
(b) (i) down (ii) out (iii) for (iv) in
(c) (i) out (ii) down (iii) in (iv) for
(d) (i) out (ii) down (iii) by (iv) for

[2024 : 2 M, Set-II]

1.257 If the sum of the first 20 consecutive positive odd numbers is divided by 20^2, the result is

(a) 1 (b) 20
(c) 1/2 (d) 2

[2024 : 1 M, Set-II]

1.258 The ratio of the number of girls to boys in class VIII is the same as the ratio of the number of boys to girls in class IX. The total number of students (boys and girls) in classes VIII and IX is 450 and 360, respectively. If the number of girls in classes VIII and IX is the same, then the number of girls in each class is

(a) 200 (b) 175
(c) 250 (d) 150

[2024 : 1 M, Set-II]

1.259 Five cubes of identical size and another smaller cube are assembled as shown in Figure *A*. If viewed from direction *X*, the planar image of the assembly appears as Figure *B*.

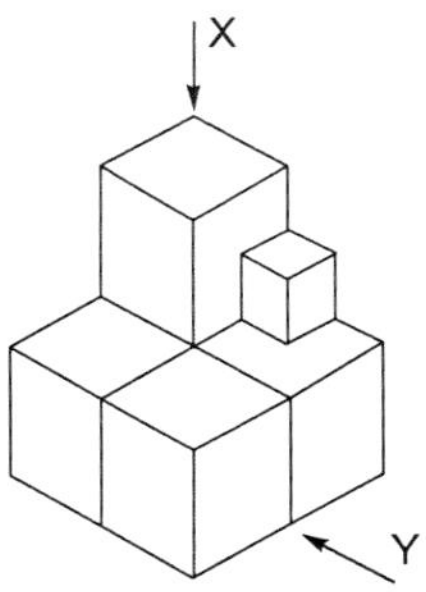

Figure A

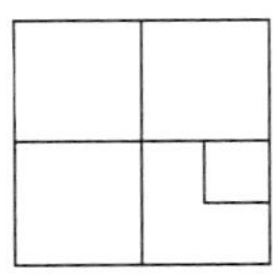
Figure B

(a) 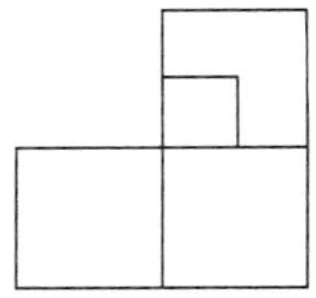
(b) 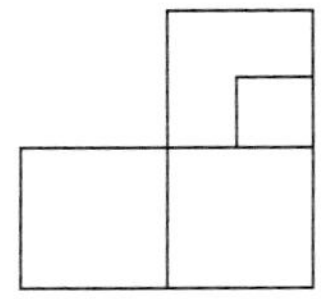
(c) 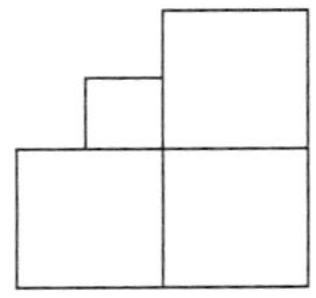
(d) 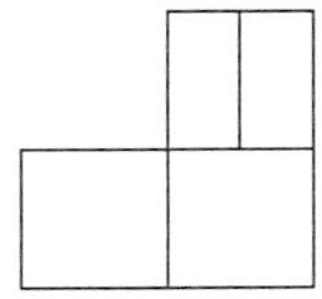

[2024 : 2 M, Set-II]

1.260 If '→' denotes increasing order of intensity, then the meaning of the words

[drizzle → rain → downpour] is analogous to [________ → quarrel → feud].

Which one of the given options is appropriate to fill the blank?

(a) bog (b) dither
(c) bicker (d) dodge

[2024 : 1 M, Set-II]

■■■■

Answers General Aptitude

1.1	(b)	1.2	(a)	1.3	(d)	1.4	(c)	1.5	(d)	1.6	(d)	1.7	(c)
1.8	(d)	1.9	(b)	1.10	(b)	1.11	(b)	1.12	(a)	1.13	(d)	1.14	(b)
1.15	(c)	1.16	(d)	1.17	(c)	1.18	(d)	1.19	(c)	1.20	(a)	1.21	(b)
1.22	(a)	1.23	(a)	1.24	(b)	1.25	(a)	1.26	(b)	1.27	(d)	1.28	(a)
1.29	(c)	1.30	(b)	1.31	(b)	1.32	(d)	1.33	(c)	1.34	(c)	1.35	(b)
1.36	(a)	1.37	(d)	1.38	(d)	1.39	(b)	1.40	(a)	1.41	(a)	1.42	(b)
1.43	(d)	1.44	(c)	1.45	(1300)	1.46	(d)	1.47	(a)	1.48	(c)	1.49	(25)
1.50	(a)	1.51	(d)	1.52	(b)	1.53	(*)	1.54	(d)	1.55	(b)	1.56	(a)
1.57	(c)	1.58	(d)	1.59	(c)	1.60	(0.49)	1.61	(b)	1.62	(b)	1.63	(c)
1.64	(b)	1.65	(c)	1.66	(b)	1.67	(b)	1.68	(b)	1.69	(a)	1.70	(c)
1.71	(Sol.)	1.72	(b)	1.73	(b)	1.74	(d)	1.75	(b)	1.76	(d)	1.77	(4536)
1.78	(d)	1.79	(d)	1.80	(a)	1.81	(d)	1.82	(a)	1.83	(a)	1.84	(c)
1.85	(d)	1.86	(c)	1.87	(a)	1.88	(c)	1.89	(a)	1.90	(a)	1.91	(b)
1.92	(d)	1.93	(d)	1.94	(b)	1.95	(b)	1.96	(d)	1.97	(d)	1.98	(b)
1.99	(d)	1.100	(c)	1.101	(d)	1.102	(c)	1.103	(d)	1.104	(c)	1.105	(c)
1.106	(b)	1.107	(c)	1.108	(b)	1.109	(b)	1.110	(b)	1.111	(a)	1.112	(c)
1.113	(b)	1.114	(b)	1.115	(b)	1.116	(a)	1.117	(d)	1.118	(d)	1.119	(c)
1.120	(c)	1.121	(b)	1.122	(a)	1.123	(22.22)	1.124	(a)	1.125	(d)	1.126	(d)
1.127	(a)	1.128	(a)	1.129	(a)	1.130	(b)	1.131	(c)	1.132	(c)	1.133	(b)
1.134	(b)	1.135	(b)	1.136	(c)	1.137	(b)	1.138	(c)	1.139	(b)	1.140	(b)
1.141	(b)	1.142	(d)	1.143	(a)	1.144	(d)	1.145	(c)	1.146	(a)	1.147	(c)
1.148	(d)	1.149	(a)	1.150	(d)	1.151	(d)	1.152	(d)	1.153	(c)	1.154	(d)
1.155	(a)	1.156	(b)	1.157	(b)	1.158	(b)	1.159	(d)	1.160	(b)	1.161	(c)
1.162	(a)	1.163	(b)	1.164	(a)	1.165	(a)	1.166	(d)	1.167	(d)	1.168	(c)
1.169	(d)	1.170	(d)	1.171	(a)	1.172	(b)	1.173	(c)	1.174	(b)	1.175	(b)
1.176	(b)	1.177	(a)	1.178	(*)	1.179	(a)	1.180	(b)	1.181	(b)	1.182	(c)
1.183	(d)	1.184	(a)	1.185	(b)	1.186	(c)	1.187	(c)	1.188	(a)	1.189	(d)
1.190	(c)	1.191	(a)	1.192	(b)	1.193	(c)	1.194	(a)	1.195	(a)	1.196	(b)
1.197	(d)	1.198	(b)	1.199	(b)	1.200	(a)	1.201	(d)	1.202	(a)	1.203	(d)
1.204	(b)	1.205	(b)	1.206	(c)	1.207	(d)	1.208	(b)	1.209	(a)	1.210	(a)
1.211	(b)	1.212	(b)	1.213	(b)	1.214	(b)	1.215	(d)	1.216	(a)	1.217	(a)
1.218	(c)	1.219	(c)	1.220	(c)	1.221	(c)	1.222	(c)	1.223	(d)	1.224	(d)
1.225	(a)	1.226	(b)	1.227	(d)	1.228	(a)	1.229	(a)	1.230	(a)	1.231	(a)
1.232	(a)	1.233	(d)	1.234	(d)	1.235	(d)	1.236	(a)	1.237	(a)	1.238	(a)
1.239	(c)	1.240	(c)	1.241	(b)	1.242	(d)	1.243	(a)	1.244	(d)	1.245	(c)
1.246	(c)	1.247	(a)	1.248	(b)	1.249	(b)	1.250	(d)	1.251	(a)	1.252	(c)
1.253	(b)	1.254	(d)	1.255	(d)	1.256	(d)	1.257	(a)	1.258	(a)	1.259	(b)
1.260	(c)												

Explanations General Aptitude

1.1 (b)

Circuitous: Deviating from a straight course ⇒ Indirect.

(a) **Cyclic:** Recurring in cycle
(b) **Indirect:** Not leading by straight line
(c) **Confusing:** Lacking clarity
(d) **Crooked:** For shapes (irregular in shape)

1.2 (a)

Unemployed: Worker ⇒ Here one is opposite to other.

(a) Fallow: Land ⇒ Fallow means undeveloped land.
(b) Unaware: sleeper ⇒ Both are same unaware or asleep.
(c) Wit: Jester ⇒ Wit means ability to make jokes and jester is a joker.
(d) Renovated : House ⇒ Renovate means to make better and house can be renovated.

1.3 (d)

(a) Uphold: cause to remain ⇒ not appropriate
(b) Restrain: keep under control ⇒ not appropriate
(c) Cherish: be fond of ⇒ not related
(d) Conserve: Keep in safety and protect from harm, decay, loss, or destruction ⇒ most appropriate.

1.4 (c)

(a) Masked: Hide under a false appearance ⇒ opposite
(b) Belied: Be in contradiction with ⇒ not appropriate
(c) Betrayed: Reveal unintentionally ⇒ most appropriate
(d) Suppressed: To put down by force or authority ⇒ irrelevant

1.5 (d)

Using the set theory formula,

$n(A)$: Number of people who play hockey = 15
$n(B)$: Number of people who play football = 17
$n(A \cap B)$: Persons who play both hockey and football = 10
$n(A \cup B)$: Persons who play either hockey or football or both.

Using the formula,

$$n(A \cup B) = n(A) + n(B) - n(A \cap B)$$
$$n(A \cup B) = 15 + 17 - 10 = 22$$

Thus people who play neither hockey nor football = 25 − 22 = 3

Alternative Method:

Refer to Venn diagram given below:

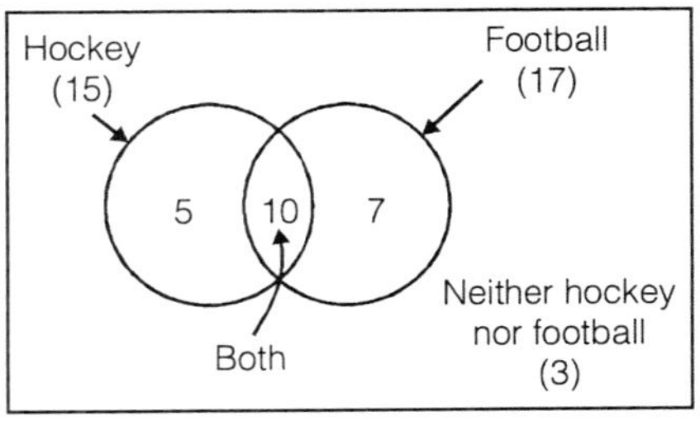

1.6 (d)

(a) Modern warfare has resulted in civil strife: There is no direct consequence of warfare given, so it is not appropriate.
(b) Chemical agents are useful in modern warfare: Passage does not say whether chemical agents are useful or not, so not appropriate.
(c) Use of chemical agents in warfare would be undesirable: Given that people in military think these are useful, undesirable is wrong
(d) People in military establishments like to use chemical agents in war; Correct choice as last statement tells that military people think that chemical agents are useful tools for their cause (work silently in warfare).

1.7 (c)

137 + 276 = 435

This is a octahedral addition thus,

731 + 672(8) = 1623

Alternative Method

7 and 6 added is becoming five means the given two numbers are added on base 8.

$$\begin{array}{r} (137)_8 \\ +(276)_8 \\ \hline (435)_8 \end{array}$$

Hence we have to add the another two given set of numbers also on base 8.

$$\begin{array}{r} (731)_8 \\ +(672)_8 \\ \hline (1623)_8 \end{array}$$

Hence the overall problem was based on identyfying base which was 8 and adding number on base 8.

1.8 (d)

Per day work or rate of 5 skilled workers = $\frac{1}{20}$

⇒ Per day work or rate of one skill worker

$= \frac{1}{5 \times 20} = \frac{1}{100}$

Similarly Per day work or rate of 8 semiskilled

workers = $\frac{1}{25}$

⇒ Per day work or rate of one semi-skill worker

$$= \frac{1}{8 \times 25} = \frac{1}{200}$$

And per day work or rate of 10 unskilled workers

$= \frac{1}{30}$

⇒ Per day work or rate of one semi-skill worker

$$= \frac{1}{10 \times 30} = \frac{1}{300}$$

Thus total per day work of 2 skilled, 6 semiskilled and 5 unskilled workers

$$= \frac{2}{100} + \frac{6}{200} + \frac{5}{300} = \frac{12 + 18 + 10}{600}$$

$$= \frac{40}{600} = \frac{1}{15}$$

Thus time to complete the work is 15 days.

Alternative Method:

Let one day work of skilled semi-skilled and unskilled worker be *a*, *b*, *c* units respectively.

$5a \times 20 = 8b + 25 = 10c \times 30$ = Total unit of work,

$$100a = 200b = 300c$$
$$a = 2b = 3c$$

$$\Rightarrow \quad b = \frac{a}{2} \text{ and } c = \frac{a}{3}$$

Given that 2 skilled, 6 semi-skilled and 5 unskilled workers are working. Let they finish the work in 'x' days.

$(2a + 6b + 5c)x = 5a + 20$
= Total units of work

$$\left(2a + 3a + \frac{5}{3}a\right)x = 5a \times 20$$

$$\frac{20a}{3}x = 5a \times 20$$

$$x = 15 \text{ days}$$

1.9 (b)

We have to make 4 digit numbers, so the number should be start with 3 or 4, two cases possible;

Case (1) thousands digit is 3

Now other three digits may be any of 2, 2, 3, 3, 4, 4, 4, 4.

(a) Using 2, 2, 3
⇒ 223, 232, 322 -------

$\left(\frac{3!}{2!} = 3 \text{ numbers are possible}\right)$

(b) Using 2, 2, 4 ⇒ 224, 242, 422

$\left(\frac{3!}{2!} = 3 \text{ numbers are possible}\right)$

(c) Using 2, 3, 3 ⇒ 233, 323, 332

$\left(\frac{3!}{2!} = 3 \text{ numbers are possible}\right)$

(d) Using 2, 3, 4 ⇒ 234, 243, 324, 342, 423, 432
(3! = 6 numbers are possible)

(e) Using 2, 4, 4 ⇒ 244, 424, 442

$\left(\frac{3!}{2!} = 3 \text{ numbers are possible}\right)$

(f) Using 3, 3, 4 ⇒ 334, 343, 433

$\left(\frac{3!}{2!} = 3 \text{ numbers are possible}\right)$

(g) Using 3, 4, 4 ⇒ 344, 434, 443

$\left(\frac{3!}{2!} = 3 \text{ numbers are possible}\right)$

(h) Using 4, 4, 4 ⇒ 444

$\left(\frac{3!}{3!} = 1 \text{ numbers are possible}\right)$

Total 4 digit numbers in case 1 = 3 + 3 + 3 + 6 + 3 + 3 + 3 + 1 = 25

Case (2) thousands digit is 4 ; Now other three digits may be any of 2, 2, 3, 3, 3, 4, 4, 4.

(a) Using 2, 2, 3 $\Rightarrow$ 223, 232, 322

$\left(\frac{3!}{2!} = 3 \text{ numbers are possible}\right)$

(b) Using 2, 2, 4 $\Rightarrow$ 224, 242, 422

$\left(\frac{3!}{2!} = 3 \text{ numbers are possible}\right)$

(c) Using 2, 3, 3 $\Rightarrow$ 233, 323, 332

$\left(\frac{3!}{2!} = 3 \text{ numbers are possible}\right)$

(d) Using 2, 3, 4 $\Rightarrow$ 234, 243, 324, 342, 423, 432

$\left(\frac{3!}{2!} = 3 \text{ numbers are possible}\right)$

(e) Using 2, 4, 4 $\Rightarrow$ 244, 424, 442

$\left(\frac{3!}{2!} = 3 \text{ numbers are possible}\right)$

(f) Using 3, 3, 3 $\Rightarrow$ 333

$\left(\frac{3!}{3!} = 1 \text{ numbers are possible}\right)$

(g) Using 3, 3, 4 $\Rightarrow$ 334, 343, 433

$\left(\frac{3!}{2!} = 3 \text{ numbers are possible}\right)$

(h) Using 3, 4, 4 $\Rightarrow$ 344, 434, 443

$\left(\frac{3!}{2!} = 3 \text{ numbers are possible}\right)$

(i) Using 4, 4, 4 $\Rightarrow$ 444

$\left(\frac{3!}{3!} = 1 \text{ numbers are possible}\right)$

Total 4 digit numbers in case 2 = 3 + 3 + 3 + 6 + 3 + 3
= 1 + 3 + 1 = 26

Thus total 4 digits numbers using case (1) and case (2)
= 25 + 26 = 51

***Alternative Method / Shortcut method**

As the number is greater than 3000. So thousand's place can be tiehr 3 or 4.

Let's consider the following two cases

Case (I) When thousand's place is 3.

3 a b c

If there is no restricton on number of two's, three's and four's. Then each of *a, b, c* can be filled with 2 or 3 or 4 each in 3 ways.

So 3 × 3 × 3 = 27 numbers are there. Out of which 3222, 3333 are invalid as 2 can be used twice & three thrice only so number of such valid numbers begining with 3 are 27 – 2 = 25.

Case (II) When thousand's place is 4

4 a b c

Without restriction on number of 2's, 3's and 4's *a, b, c* (as explained in case I) can be filled in 27 ways.

Out of these 27 numbers, 4 2 2 2 is only invalid as two have to be used twice only.

So valid numbers are 27 – 1 = 26.

Total numbers from Case (I) & Case (II) 25 + 26 = 51.

1.10 (b)

Suppose: Hari's age : H, Gita's age : G, Saira's age : S, Irfan's age : I

- $H + G > I + S$
- Using Statement (2) both $G - S = 1$ or $S - G = 1$; G can't be oldest and S can't be youngest.
- There are no twins thus using statement (2) either *GS* or *SG* possible.

(A) **HSIG:** not possible as there is I between S and G which is not possible using statement (3)

(B) **SGHI:** *SG* order is possible, $S > G > H > I$ and $G + H > S + I$ (possible)
Because if $\{S = G + 1$; and $G = H + 1$ and $H = I + 2$ then $G + (I + 2) > (G + 1) + I\}$

(C) **IGSH:** according to this $I > G$ and $S > H$ thus adding these both inequalities we get $I + S > G + H$ which is opposite of statement (2) thus not possible.

(D) **IHSG:** according to this $I > H$ and $S > G$ thus adding both inequalities $I + S > H + G$ which is opposite of statement (2) thus not possible.

1.11 (b)

$$\log(P) = \frac{1}{2}\log(Q) = \frac{1}{3}\log(R)$$

$$\Rightarrow \log(P) = \log(Q)^{1/2} = \log(R)^{1/3} = K$$

$$\Rightarrow \quad P = (Q)^{1/2} = (R)^{1/3} = K$$

$$\Rightarrow \quad P = K, Q = K^2, R = K^3 \quad \text{...(i)}$$

Now, only option (B) $Q^2 = PR$ satisfies

$$Q^2 = (K^2)^2 = K^4 \quad \text{From ...(i)}$$

$$P.R = K * K^3 = K^4 \quad \text{From...(i)}$$

Here, $Q^2 = PR$ holds true

1.16 (d)

As we can understand that the danger of a microbe to human being will be directly proportional to potency and growth. At the same time, it will be inversely proportional to toxicity defined (more dangerous will a microbe be if lesser of its milligram is required). So level of dangerous

$\alpha \dfrac{P\uparrow \times G\uparrow}{T\downarrow}$ where P, G and T are the potency, growth and toxicity as defined in question.

So, $D_i = \dfrac{KPG}{T}$ ...(i)

Where K is constant of proportionality.

So level of dangerous of S will be maximum given by $D_S = \dfrac{0.8\times\pi(10\text{mm})^2}{200}$.

Similar calculation for D_P, D_Q, D_R can be done based on (i) to find out that D_S is maximum and so most dangerous among them.

1.18 (d)

Shortcut Method:

Every time if we take 1 litre of mixture out and replace with water, content of pure spirit will keep on reducing by 10%.

So, final quantity of spirit after 3 such operations are,

$$10 \times 0.9 \times 0.9 \times 0.9 = 7.29 \text{ litres}$$

Alternate Solution:

$$\frac{\text{Quantity of spirit left after n}^{\text{th}}\text{ operation}}{\text{Intial quantity of spirit}} = \left(\frac{a-b}{a}\right)^n = \left(1-\frac{b}{a}\right)^n$$

where 'a' is initial quantity of pure spirit and 'b' is quantity taken out and replaced every time.

Hence, quantity of spirit left after 3rd operation

$$= \text{initial quantity} \times \left(1-\frac{1}{10}\right)^3$$

$$= 10 \times 0.9 \times 0.9 \times 0.9$$

$$= 7.29 \text{ litres}$$

1.19 (c)

Let 'y' be the backlog with transporter and 'x' be the number of orders each day. So, as per conditions given in question,

$$4x + y = 28 \quad \text{...(i)}$$

$$10x + y = 30 \quad \text{...(ii)}$$

Solving (i) and (ii),

$$x = \frac{1}{3} \text{ and } y = \frac{80}{3}$$

Since, we need to find out number of trucks so that no pending order will be there at the end of 5th day.

$$5x + y = n \times 5$$

So we need to find 'n', where 'n' is the number of trucks required.

$$n = \frac{5x+y}{5} = \frac{5\times\frac{1}{3}+\frac{80}{3}}{5} = \frac{\frac{85}{3}}{5}$$

Hence, 5.66 truck will be required. As number of trucks have to be natural number. Hence, 6 trucks will be required.

1.20 (a)

(T.C.) Total cost = $V + f$

$$\boxed{T.C = 4q + \frac{100}{q}}$$

As we have to minimize total cost.

Using options,

(a) $q = 5$, T.C. $= 4\times 5 + \dfrac{100}{5} = 40$

(b) $q = 4$, T.C. $= 4\times 4 + \dfrac{100}{4} = 41$

(c) $q = 7$, T.C. $= 4 \times 7 + \frac{100}{7} = 42.285$

(d) $q = 6$, T.C. $= 4 \times 6 + \frac{100}{6} = 40.\overline{66}$

Hence, T.C. is minimum at $q = 5$, **(a) ans.**

∗ always put options to get answer fast.

Alternate Solution:

or T.C. $= 4q + \frac{100}{q}$

for minimum $\frac{d}{dq}\left(4q + \frac{100}{q}\right) = 0$

$\Rightarrow \quad q^2 = 25$ or $q = 5$

$$\frac{d}{dq}\left(4q + \frac{100}{q}\right)_{q=5} > 0$$

Hence, TC (Total cost) is minimum at $q = 5$ **(a) ans.**

1.21 (b)

Setbacks

Despite several setbacks the mission succeeded in its attempt to resolve the conflict.

The word 'Despite' indicates that there has to be a contrast in the sentence, use of the word 'Setbacks' in the blank indicates that despite many problems the mission was successful.

1.22 (a)

Total Cost $C = 5q^2$

Total Sales Revenue,

$S = 50q$

Profit $(P) = (S - C) = 50q - 5q^2$

As the profit have to be maximized,

$P = 50q - 5q^2$

Using options,

(a) $q = 5$

$P = 50 \times 5 - 5 \times 5^2 = 125$

(b) $q = 10$

$P = 50 \times 10 - 5 \times 10^2 = 0$

(c) $q = 15$

$P = 50 \times 15 - 5 \times 15^2 = $ –ive

(d) $q = 25$

$P = 50 \times 25 - 5 \times 25^2 = $ –ive

Hence the maximum profit will happen at $q = 5$

Alternative solution:

$$P = 50q - 5q^2$$

For max/min $\frac{dp}{dq} = 0,$

$\Rightarrow \quad 50 - 10q = 0$

or, $\quad q = 10$

$$\left.\frac{d^2p}{dq^2}\right]_{q=5} = -10 \text{ which is negative}$$

Hence maximum profit will happen at,

$$q = 5$$

1.23 (a)

Suresh's dog is the one that was hurt in the stampede.

That is used with restrictive clauses.

1.24 (b)

The country's expenditure is not less than that of Bangladesh.

The correct statement should be: The country's expenditure is not less than that of Bangladesh's. Because the country's expenditure cannot be compared with Bangladesh. It should be compared with Bangladesh's expenditure.

1.25 (a)

Diminish

Mitigate means to reduce, to lessen etc. So only the word Diminish is close. Rest all choices have no link with the given word.

Divulge means to disclose or reveal which has no link with the given word.

1.26 (b)

$$y = 2x - 0.1x^2$$

For y (height) to be maximum $\frac{dy}{dx} = 0$

$$\frac{dy}{dx} = 2x - 2x = 0$$

$\Rightarrow \quad x = 10$

$$\left.\frac{d^2y}{dx^2}\right]_{x=10} = -0.2 \text{ (–ive)}, \quad \frac{d^2y}{dx^2} < 0$$

Hence maximum height will be at $x = 10$

$$y = 2 \times 10 - 0.1 \times (10)^2 = 10$$

Maximum height = 10 meters.

1.27 (d)

Not gender-discriminatory.

Choice (*A*) cannot be considered since there is no gender discrimination mentioned in the argument.

Choice (*B*) Xenophobic is one who has fear of foreigners, no link with the given argument.

Choice (*C*) It is wrong to say that the profile has not been designed to make the post attractive, since there are certain features which have been added to make the profile lucrative (which are given towards the end of the advertisement, like Transportation paid, expenses reimbursed).

1.28 (a)

AD, CG, FK, JP are given terms. The first letter in each term *A, C, F, J* are respectively 1, 3, 6, 10 letter of alphabetical sequence.

A = 1; 1st letter of alphbetical squence

C = 1 + 2 = 3 ; 3rd letter of alphbetical squence

F = 1 + 2 + 3 = 6 ; 6th letter of alphbetical squence

G = 1 + 2 + 3 + 4 = 10; 10th letter of alphbetical squence

The next term should begin with

1 + 2 + 3 + 4 + 5 = 15th letter of alphabetical sequence. So next term should begin with *O*.

Also in the series *AD, CG, FK, JP*, two letters in each term are separated by 2, 3, 4, 5 letters in between them respectively. Hence next term should be *OV* as there should be a gap of 6 alphabets between *O* and *V*.

Hence answer is *OV*.

1.29 (c)

Adding each and every observation by constant increases the arithmetic mean by same constant. Also multiplying each and every observation by constant mutiplies the arithmetic mean by same constant. While the standard deviation will not show similar effect.

Alternative Solution:

$$\frac{x_1 + x_2 + x_3 \cdots x_N}{N} = \bar{x}$$

$$\frac{(x_1+7)+(x_2+7)+\cdots+(x_n+7)}{N} = \bar{x}+7$$

and
$$\frac{2(x_1 + x_2 \cdots x_N)}{N} = 2\bar{x}$$

Hence only *P*, *R* are correct.

1.30 (b)

Let total 100 shock absorbes are supplied

So *X* supplies = 60 (40% of 100)

Y supplies = 40 (60% of 100)

Reliable supply by *X* = 96% of 60 = 57.6

Reliable supply by *Y* = 72% of 40 = 28.8

So the required probability that random choosen reliable shock absorber is made by *Y* is

Required probability

$$= \frac{28.8}{57.6+28.8} = 0.333 \approx 0.334$$

1.31 (b)

Friendship, No matter how intimate it is, has its limitation. Intimate refers to close personal relations, ex. an intimate friend. It is also characterized by or involving warm friendship or a personally close or familiar association or feeling.

1.33 (c)

Medicine leads to good health. Similarly, Education leads to Knowledge. Science does not lead to experiment. Wealth may not necessarily lead to peace. Also money may not also lead to happiness all the time.

1.34 (c)

Primeval pertains to the first age or ages, especially of the world. Ex: primeval forms of life.

1.38 (d)

% of raw material are salary and wages

$$= \frac{6240-5200}{5200} \times 100 = 20\%$$

% of raw material and research and development

$$= \frac{26400-22000}{22000} \times 100 = 20\%$$

1.39 (b)

$$|4x - 7| = 5$$

$$4x - 7 = 5$$

and $4x - 7 = -5$

$\Rightarrow$ $x = 3$

and $x = 0.5$

$\therefore$ $2|x| - |-x| = 2 \times 3 - 3 = 3$

and $2|x| - |-x| = 2\times\frac{1}{2}-\frac{1}{2}=\frac{1}{2}$

$\Rightarrow \quad = \frac{1}{2}, 3$

1.40 (a)

$$2x + y \le 6$$
$$x + 2y \le 8$$
$$x \le 1.33$$
$$y \le 3.33$$

$\therefore$ option (b) and (c) is rejected because $x \not> 1.33$.

Option (d) is rejected $x \not> 10/3$.

1.45 Sol.

1300

1.49 Sol.

Average time interval $= T = \frac{1}{P}$ = 25 years

1.50 (a)

$$P_{new} = P_{old}\left(1+\frac{r}{100}\right)^n$$

$$\Rightarrow \quad 10 = 5\left(1+\frac{20}{100}\right)^n$$

$$\Rightarrow \quad 2 = (1.2)^n$$

$$\Rightarrow \quad \frac{\log 2}{\log 1.2} = n$$

$$\Rightarrow \quad n = 3.8 \text{ year}$$

1.54 (d)

Let the population of county Y is P

$\therefore$ Population of country X is $3P$

% of people taken than 6 ft

$$= \frac{\frac{3P\times 1}{100}+\frac{P\times 2}{100}}{3P+P} = 1.25$$

1.58 (d)

Total quantity of export = 5 lakh tonnes

Quantity of export of item-1
= 0.11 × 5 = 0.55 lakh tonnes

Quantity of export of item-4 = 0.22 × 5
= 1.1 lakh tonnes

Total revenues generated from item-1
= 250 × 0.12 = 30 crore

Total revenues generated from item-4
= 0.06 × 250 = 15 crore

Revenue generated through export of item-1 per kg $= \frac{30\times 1000000}{0.55\times 1000000}$

Rs. = 545.45/kg

Revenue generated through export of item-4 per kg $= \frac{15\times 10^6}{101\times 10^5}$

Rs. = 136.36/kg

$$\frac{\text{R.G. item 1}}{\text{R.G. item 4}} = \frac{4}{1}$$

1.59 (c)

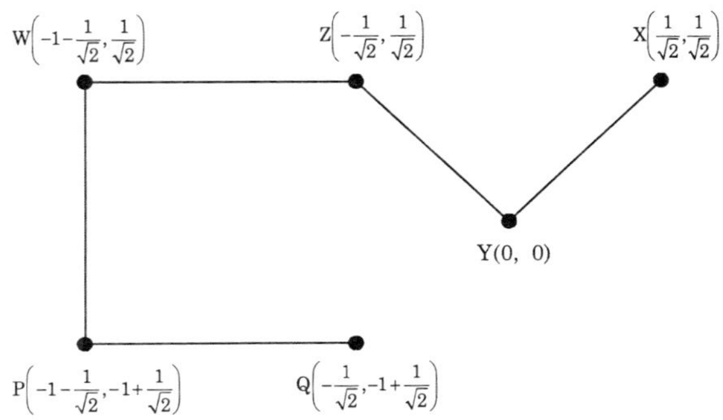

$$XQ = \sqrt{\left(\frac{1}{\sqrt2}+\frac{1}{\sqrt2}\right)^2+(-1)^2} = \sqrt3$$

1.60 Sol.

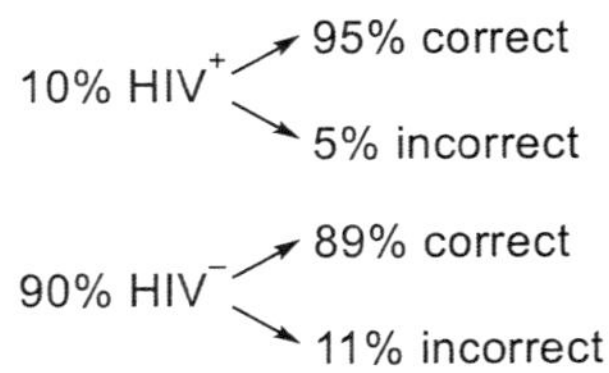

N(declared HIV+) tree diagram is as shown above = 0.1 × 0.95 + 0.9 × 0.11 = 0.194

N(actual HIV+) = 0.1 × 0.95 = 0.095

P(actual HIV+) $= \frac{0.095}{0.194}$ = 0.4896

1.62 (b)

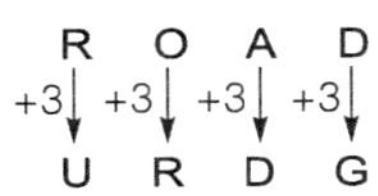

Add 3 to respective letters

So,

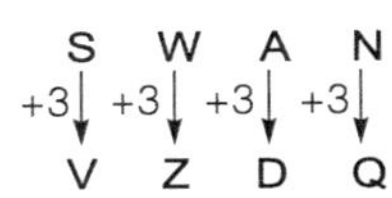

1.63 (c)

Let $f(x) = ax + b$ (linear function)

at $x = -2$, $f(x) = 29$

$\Rightarrow \quad -2a + b = 29 \quad ...(i)$

$x = 3$, $f(x) = 39$

$\Rightarrow \quad 3a + b = 39 \quad ...(ii)(a = 2, b = 33)$

Giving $f(x) = 2x + 33$

$\Rightarrow \quad f(5) = 43$

1.66 (b)

Before drawing third card, only two kings remain among the remaining 50 cards.

$$\text{Probability} = \frac{^2C_1}{^{50}C_1} = \frac{2}{50}$$

1.69 (a)

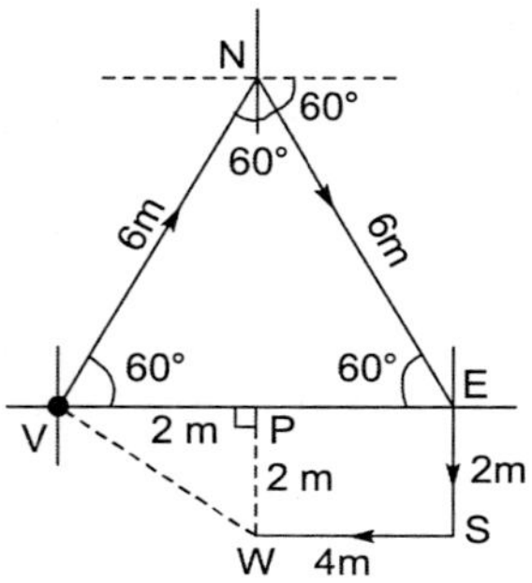

ΔVNE is equilateral Δ,

So $VE = 6$

Also $VP = VE - PE = 6 - 4 = 2$

$VP = 2, PW = 2$

So, $VW = \sqrt{VP^2 + PW^2} = 2\sqrt{2}$

1.71 Sol.

Combined percentage increase in imports and exports for year 2001

$$= \frac{(60+50)-(50+40)}{50+40} \times 100 = 22.22\%$$

Combined percentage increase in imports and exports for year 2002

$$= \frac{(70+60)-(60+50)}{60+50} \times 100 = 18.18\%$$

Combined percentage increase in imports and exports for year 2003

$$= \frac{(70+60)-(60+70)}{70+60} \times 100 = 0$$

Combined percentage increase in imports and exports for year 2004

$$= \frac{(70+80)-(60+70)}{60+70} \times 100 = 15.38\%$$

Combined percentage increase in imports and exports for year 2005

$$= \frac{(70+90)-(70-80)}{70+80} \times 100 = 6.67\%$$

Combined percentage increase in imports and exports for year 2006

$$= \frac{(100+120)-(70+90)}{70+90} \times 100 = 37.5\%$$

Combined percentage increase in imports and exports for year 2007

$$= \left|\frac{(100+120)-(110+120)}{100+120} \times 100\right| = 4.54\%$$

$\therefore$ year 2006 has shown in highest percentage combined growth in imports and exports.

1.73 (b)

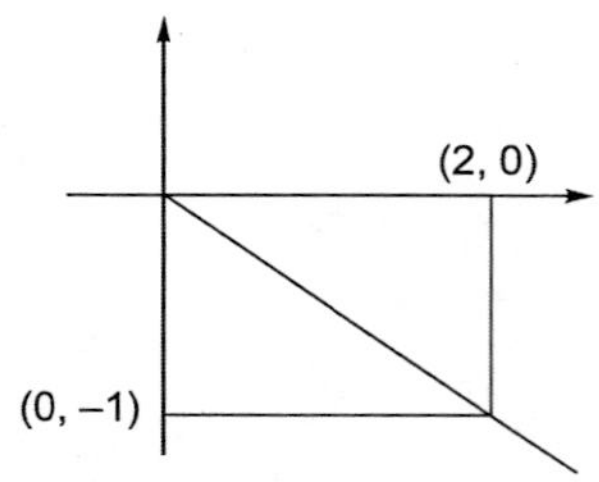

Option (c) is not possible.

$\because \quad x = 0$ for +ve value of y in graph.

Option (a), through valid for +ve values of y, is not correct for –ve values of 'y'.

Option (d) is also not true since x is –ve for +ve values of y which is contradictory to the given graph.

1.75 (b)

Option (a) is not correct as R is given defence. As given in question R wants only Finance or Home.

Option (d) is not corect as R is alloted Telecom (same reason as above).

Option (c) is not possible because S & U cannot be together as per question.

So only option (b) is possible.

1.76 (d)

$$\textbf{Type I: } = \left(\frac{115-75}{75}\right) \times 100 = 53.3\%$$

Type II: $= \left(\frac{85-144}{144}\right) \times 100 = -40.97\%$

Type III: $= \left(\frac{160-114}{114}\right) \times 100 = 40.3\%$

Type IV: $= \left(\frac{145-108}{108}\right) \times 100 = 14.25\%$

Type V: $= \left(\frac{100-102}{102}\right) \times 100 = 1.9\%$

So maximum for Type I.

1.77 Sol.

$A\ B\ C\ D$, left most digit cannot be zero.

$\therefore$ A can have 9 district values B, C, D can have 9, 8, 7 values.

$\therefore \quad 9 \times 9 \times 8 \times 7 = 4536.$

1.78 (d)

Both statements will be required as capacity is required and what portion of capacity is filled is also required.

From (ii) we get full capacity of tank A = 14000

So capacity of B is 20000 $\left(= \frac{14000}{7}\right)$

But for getting quality of solution we also need how much it is full.

Statement I gives A is 80% full and B is 40% full.

i.e., Vol. in A occupied = 80% of 14000

= 11,200

Vol. in B occupied = 40% of 20000 = 8000

Hence both statements will be required for finding out gallons of solution in A and B respectively.

1.80 (a)

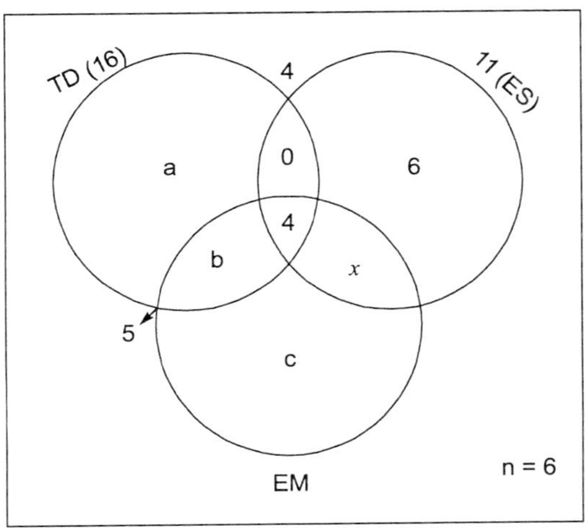

40

See set ES

$0 + 4 + x + 6 = 11$

$x = 1$

No. of teachers ES & EM but not TD is $x = 1$

1.85 (d)

Original surface area = $6\,(4)^2 = 96$

If corner cubes are removed, three exposed surfaces are removed which will create 3 new surfaces in original large cube. So surface area will remain unchanged, i.e. 96.

1.91 (b)

Elegance

$(27300 + 25222 + 28976 + 21012) \times 48 = A$

Executive

$(999 + 8942 + 10234 + 10234 + 10109) \times 173 = B$

Smooth

$(20009 + 9392 + 22429 + 18229) \times 63 = C$

Soft

$(17602 + 18445 + 19544 + 16595) \times 78 = D$

Which is highest for B (executive).

1.93 (d)

From conditions I, II and IV, following inference can be drawn,

P always beats Q

Q always beats R

and R always beats S

Thus, arranging players from best performing to worst performing is,

$$P - Q - R - S$$

Now, condition III states that S loses to P only sometimes. That means S most of the times beats the otherwise best performing player P.

Thus, statement (i) P is likely to beat all the other three players is FALSE, As P is likely to lose from S.

Statement (ii) S is absolutely the worst player in the set is also FALSE as S beats P most of the times.

1.94 (b)

$$f(x) = 2x^7 + 3x - 5$$

Use option (b), if $(x - 1)$ will be a factor then on putting $x - 1 = 0$ i.e. $(x = 1)$ in $f(x)$.

$$f(1) = 2(1)^7 + 3 - 5 = 5 - 5 = 0$$

1.101 (d)

1, x, x, x, y, y, 9, 16, 18

$$\text{Mean} = \frac{1 + 3x + 2y + 9 + 16 + 18}{9}$$

$$= \frac{3x + 2y + 44}{9} \quad \text{...(i)}$$

Median ⇒ As number of terms are odd i.e. 9, so median will be middle i.e. y.

Mode ⇒ Number showing maximum frequency of repetition. So mode is x.

Given ⇒ Mean = Median = 2 Mode

$$\Rightarrow \quad y = 2x \quad \text{...(ii)}$$

From equation (i) and (ii), we get,

$$x = 4, \ y = 8$$

Value of y is 8

1.102 (c)

The bacteria is in milk are destroyed when it is heated to 80° celcius. Use of simple present tense is apt for indication factual statements / scientifically proven statements / universal truth.

1.103 (d)

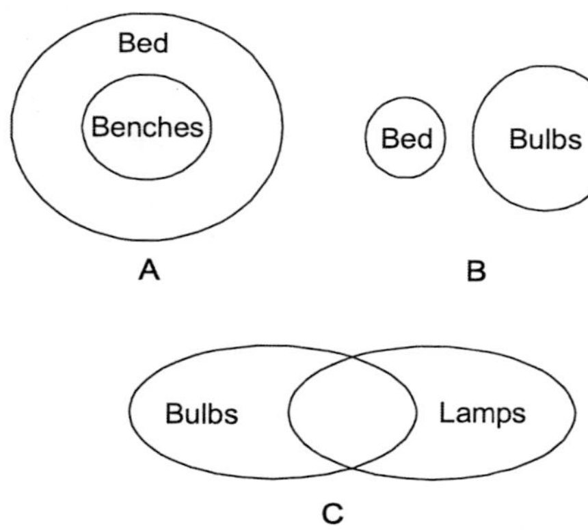

Since there is no direct relation given between **lamps** and **beds**. So, neither will be correct.

1.104 (c)

$$V = \frac{1}{3}\pi(R)^2 H$$

$$V' = \frac{1}{3}\pi(1.5R)^2 H$$

$$= \frac{1}{3}\pi\, 2.25 R^2 H = 2.25\text{ V}$$

Hence increase 125% as,

$$\left[\frac{2.25V - V}{V}\right] \times 100 = 125\%$$

1.105 (c)

Tampering with someone else's email account is a very serious offence. Tampering refers to interfering in a harmful or disruptive manner.

1.106 (b)

Sample space =6 × 6 = 36 events = Total chances

Product of numbers on 2 dice have to perfect square = Favourable chances

(1, 1) (2, 2) (3, 3) (4, 4) (5, 5) (6, 6) (1, 4) (4, 1) ⇒ 8 events

$$\text{Probability} = \frac{\text{Favourable chances}}{\text{Total chances}} = \frac{8}{36} = \frac{2}{9}$$

1.107 (c)

In order to verify proposition we have to turn to card 2 and blue from given four cards as proposition says it has even an one side opposite is red. Vice-verse might or might not be true so, ans. (c) as all other options are eliminated.

1.108 (b)

Use of conditional sentence based on past perfect structure. Conditional clause uses (had + V_3), main clause is based on would / should / could + have + V_3.

1.109 (b)

Thorough refers to somethings that is exhaustively complete, accurate, careful or absolute.

1.110 (b)

$$\text{Put } (x = -1)\ \frac{81\left[\frac{16}{25}\right]^1}{\left[\frac{3}{5}\right]^2} = \frac{\frac{3^4 \times 2^4}{5^2}}{\frac{3^2}{5^2}} = 3^2 \times 2^2 = 144$$

On putting $(x = -1)$

$\Rightarrow$ LHS = RHS

Hence option (b) is correct.

1.111 (a)

P	30 min ×	2 = 60 min
Q	20 min ×	3 = 60 min
R	60 min ×	1 = 60 min
S	15 min ×	4 = 60 min

M_1	P	Q	= 2 hrs
M_2	R	S	= 2 hrs

1.112 (c)

Clearly we can see from graph both (i) and (ii) are correct.

1.113 (b)

Amar ⟶ R
Akbar ⟶ S
Antony ⟶ P
Ajit ⟶ Q

As per their preferences given.

1.115 (b)

Unit digit:

$$(2171)^7 + (2172)^9 + (2173)^{11} + (2174)^{13}$$
$$(1)^{any} + (2)^{4n+1} + (3)^{4n+3} + (4)^{2n+1}$$
$$1 + 2 + 7 + 4 = 14$$

1.116 (a)

People coming out in the same order in which they enter indicates that the centre operates on a first-come-first-serve basis.

1.117 (d)

(i) is incorrect as it has move directly.

(ii) is incorrect as it stayed for maximum duration on ground floor.

1.118 (d)

Let cycling speed = C

and walking speed = W

$$C\left(\frac{1}{2}\right) + W\left(\frac{3}{2}\right) = 19 \quad \text{...(i)}$$

$$C + W = 26 \quad \text{...(ii)}$$

On Solving eq. (i) and (ii) We get,

$$W = 6 \text{ km/hr}$$

1.119 (c)

Following information is given

$K < G$, $P < G$, $S < P$ and $G < D$

From these it can be inferred

$S < P < G < D$ and also $(K < G)$

So, $K < D$ as well but no comparison between K & S or K & P is given

So (ii) and (iv) follows.

1.120 (c)

Following two possibilities can be drawn

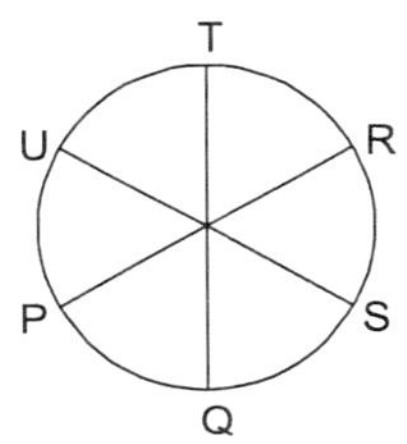

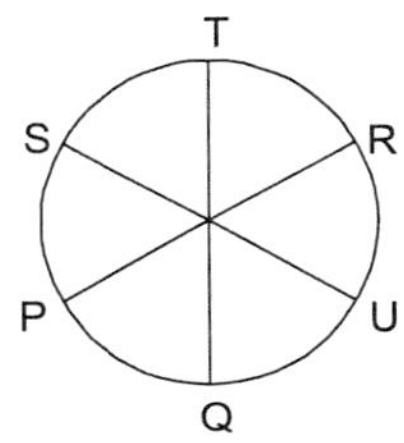

Given in question P & U interchange then new diagram can be drawn.

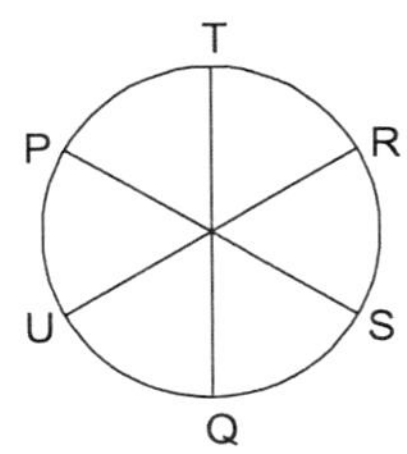

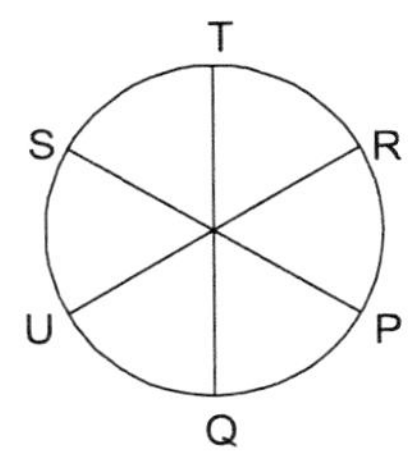

Now verify (c) only follows.

1.121 (b)

$$32.4 = K(2\theta + 25) \quad \text{...(i)}$$

$$42 = K(5\theta + 30) \quad \text{...(ii)}$$

$$\theta = 1$$

1.122 (a)

- Brake is a device which is used for stopping or moving a vehicle.
- Break refers to a pause in work or during an activity.

1.122 Sol.

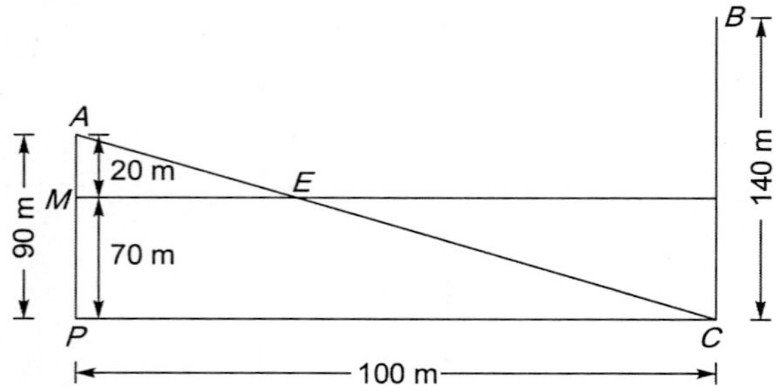

$$\Delta AME \approx \Delta APC$$

$$\frac{AM}{AP} = \frac{ME}{PC}$$

$$\Rightarrow \quad \frac{20}{90} = \frac{ME}{100}$$

$$\Rightarrow \quad ME = 22.22$$

1.124 (a)

Unpredictable - Contrary word required

1.125 (d)

Using options (a) and (c) cannot be answer.

As Hema's age = $2n + 5$, where n is a natural number.

$$\left.\begin{matrix}(2n+5 & \neq 14)\\(2n+5 & \neq 18)\end{matrix}\right\} \text{for any natural no. } n$$

(b) and (d) remains

Let us take Hema's age as option (d) which is 19.

So, Hari's age = 7 and Suresh age = 57

Verifies all condition "answer (d)"

Alternate Method

Hema = 2 Hari + 5

Suresh = 10 Hari − 13 = 3 Hema

Solving equations.

Hari = 7

Hema = 19

Suresh = 57

1.126 (d)

$$^{9}C_3 = \frac{9!}{6! \times 3!} = \frac{9 \times 8 \times 7}{6} = 84$$

1.127 (a)

Shone - It is past - participle and past form of shine.

Shown - To show means to reveal and point out something.

1.128 (a)

As a, b, c are non-negative integers and given

$\log a + \log b + \log c = 0$

$$\log(a \times b \times c) = \log 1$$

$$\Rightarrow \quad a \times b \times c = 1$$

Which can be possible for simple values.

$$a = b = c = 1$$

Hence $\quad a + b + c = 1 + 1 + 1 = 3$

Alternate Method:

Given,

$\log a + \log b + \log c = 0$

As we know $\log 1 = 0$, so each one of them can be zero if $a = b = c = 1$

$\log 1 + \log 1 + \log 1 = 0$

By putting $a = b = 1$ equation satisfies

$$a + b + c = 1 + 1 + 1 = 3$$

1.129 (a)

Innovative is similar to poineer.

1.130 (b)

$$\underbrace{a + a + a + \ldots + a}_{n \text{ times}} = na = a^2 b$$

$$\Rightarrow \quad n = ab \quad \ldots\text{(i)}$$

$$\underbrace{b + b + b + \ldots + b}_{m \text{ times}} = mb = b^2 a$$

$$\Rightarrow \quad m = ab \quad \ldots\text{(ii)}$$

$$\text{So,} \quad \underbrace{[m + m + \ldots\ldots + m]}_{n \text{ times}} \times \underbrace{[n + n + \ldots\ldots + n]}_{m \text{ times}}$$

i.e., $\quad mn \times mn = (mn)^2$

from (i) and (ii) $\quad mn = a^2 b^2$

So, result, $\quad (mn)^2 = (a^2 b^2) = a^4 b^4$

1.131 (c)

Sum of series will be

$$\left(1 - \frac{1}{3}\right) + \left(\frac{1}{2} - \frac{1}{4}\right) + \left(\frac{1}{3} - \frac{1}{5}\right) \ldots\ldots$$

$$\left(\frac{1}{48} - \frac{1}{50}\right) + \left(\frac{1}{49} - \frac{1}{51}\right) + \left(\frac{1}{50} - \frac{1}{52}\right)$$

All like terms will cancel out and we will be left with

$$\left(1 + \frac{1}{2}\right) - \left(\frac{1}{51} + \frac{1}{52}\right)$$

1.132 (c)

$$12.5\%x + 10\%x = 108$$

$$x = \frac{108}{22.5}$$

So loss $108 \times \frac{12.5}{22.5} = 60$

1.133 (b)

$A \times B \times C = B \times G \times E = D \times E \times F = 72$

$8 \times 9 \times 1 = 9 \times 2 \times 4 = 3 \times 4 \times 6 = 72$

Any of A, B, C, D, E, F, G cannot be 5.

1.134 (b)

$$C \propto W^2$$

$$C = kW^2$$

$$\Rightarrow \quad C = k(10)^2 = 100k = 1600$$

$$\Rightarrow \quad k = 16$$

$$C_1 = k(4)^2 = 16k$$

$$C_2 = k(6)^2 = 36k$$

Now total cost $= 52k = 52 \times 16 = 832$

1.135 (b)

Put value and verify

(i) $y = 2x + 4$ is true in $-3 \le x \le -1$
On putting $x = -3$, $y = -2$ and $x = -2$, $y = 0$ and $x = -1$, $y = 2$

(ii) $y = |x - 1|$ is also true $(x = -1, y = 2)$, $(x = 0, y = 1)$ and $(x = 1, y = 0)$

(iv) $y = 1$ in $(2 \le x \le 3)$ always true
(i), (ii) and (iv) are true.

1.136 (c)

Loss $= kd^2$ For duration of 7 units

$4900 = k(7)^2 \Rightarrow k = 100$

Loss $= kd^2$ For duration of 4 units

$= k(4)^2 \Rightarrow 16k = 1600$

1.137 (b)

$A + B + C = 13$ ...(i)

$C + D + E = 13$ ...(ii)

$E + F + G = 13$ ...(iii)

$G + H + K = 13$ ...(iv)

Adding [(i) + (ii) + (iii) + (iv)]

$A + B + C + D + E + F + G + H + K + (C + E + G) = 13 \times 4 = 52$ ...(v)

Also A, B, C, D, E, F, G, H & K represents natural numbers from (1 to 9)

There sum will be given by $\frac{n(n+1)}{2} = 45$

Substituting (iv) $C + E + G = 7$ ...(vi)

Only possibly for sum 7 will be (1, 2, 4)

Now, $C + E$ cannot be (1 and 2)

As eq. (ii) is $C + D + E = 13$

Now, D will become equal to 10 (which is not possible because digits 1 to 9 given)

C & E can be either $\left.\begin{matrix}(1, 4)\\(2, 4)\end{matrix}\right]$ ...(vii)

If $C = H$ from eq. (vi) $C + E + G = 7$

Now, $E + G = 3$

(Not possible in eq. (iii) $E + F + G = 10$, $F = 10$ which is not possible)

So from eq. (vii) only possibility remains is $E = H$.

1.138 (c)

Cumulative rainfall = 300 mm

50% of rainfall $= 300 \times \frac{50}{100} = 150$ mm

Area $= 50\ m^2$

$\Rightarrow$ Volume stored in tank $= 150 \times 10^{-3} \times 50\ m^3$

or $= 7500\ l$

1.139 (b)

$$\log P = 10(y - z)$$

$$\log Q = 10(z - x)$$

$$\log R = 10(x - y)$$

$$\log P + \log Q + \log R = 0$$

$$\log(PQR) = \log 1$$

$$PQR = 1$$

1.140 (b)

9 Am of 11 July of 2 PM on 15th July = 101 hours

$\left(24 + \frac{15}{60}\right)$ hours of incorrect clock = 24 hours of correct clock

$\left(24 + \frac{15}{60}\right)$ hours of IC = 24 hours of correct clock

1 hour of IC $= \frac{96}{97}$ hours of correct clock

$$101 \text{ hour of IC} = \frac{96}{97} \times 101 \text{ hours of correct clock}$$

$= 99.958$ hours of correct clock

$= 99$ hours $+ 0.95876 \times 60$ minutes of correct clock

$= 99$ hours $+ 57.525$ min.

$= 99$ hours and approx 58 min.

So, correct time will be

2 PM, 11th july + (99 hours and 58 minutes) = 12 : 58 PM on 15th July

1.141 (b)

According to given coding

$P = 32,\ E = 10,\ S = 38,\ T = 40$

$P + E + S + T = 32 + 10 + 38 + 40 = 120$

1.142 (d)

Here tan 45° = 1

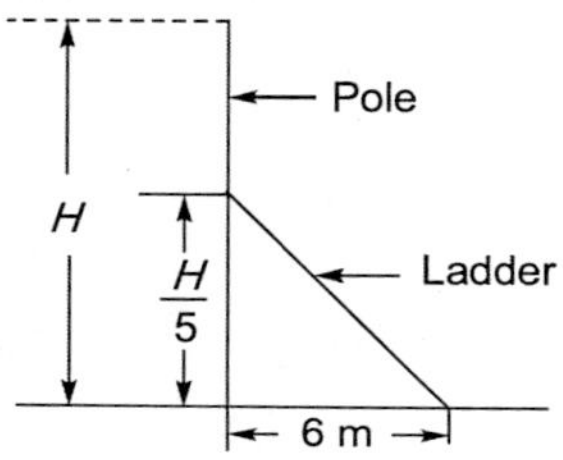

$$\frac{H}{5} = 6$$

$$H = 30 \text{ m}$$

Height of wall = 30 m

1.143 (a)

They have came long way in <u>creating</u> trust among the users.

1.144 (d)

The lecture was attended by quite <u>a few</u> students so tha hall was not very <u>quiet</u>.

- quite a few indicates a fairly large number of units.
- quiet refer to making little or no noise.

1.145 (c)

CEO's decision to quit was a shocking to the board as it was me.

- Board and me are receives of the action hence objective case of pronoun (me) is to be used.

1.146 (a)

Cost of carpet = [70 × 55 − 550] × 50 = Rs. 1,65,000

1.147 (c)

Daytime temperatures in Delhi can **<u>reach</u>** 40°.

1.148 (d)

No. of bricks = x

$\Rightarrow 30 \times 12 \times 6 \times 10^6 \times 0.6 = 8 \times 6 \times 6 \times x$

$x = 4.5 \times 10^6 = 45 \times 10^5 = 45$ lakhs bricks

1.149 (a)

Hima Das was <u>the</u> only Indian athlete to win <u>a</u> gold for India.

1.150 (d)

The growth of ABC motors was same in 2017 <u>as that of</u> *XYZ* motors in 2016.

Subject of the sentence is 'The growth of *ABC* motors'. It is to be compared with the growth of *XYZ* motors hence use of 'that of' is appropriate.

1.151 (d)

Let side of larger square = x

Side of smaller square = $(x - 5)$

Area of larger square = x^2

Area of smaller square = $(x - 5)^2$

Given,Area of larger square

= 4 times of area of smaller square

$$x^2 = 4(x - 5)^2$$

$$\therefore \quad x = 10$$

Side of larger square = 10 cm

1.152 (d)

The paragraph deals with the increasing interest in tribal characters and further goes on to describe that tribal characters haven't attained any special status. There have been very few stories woven around these characters.

1.153 (c)

Farmers want to access the new technology even if it is not legal.

The argument emphasizes on the widespread use of the new cotton technology despite it being illegal. Detail about price is just an additional detail.

1.154 (d)

P is brother of S

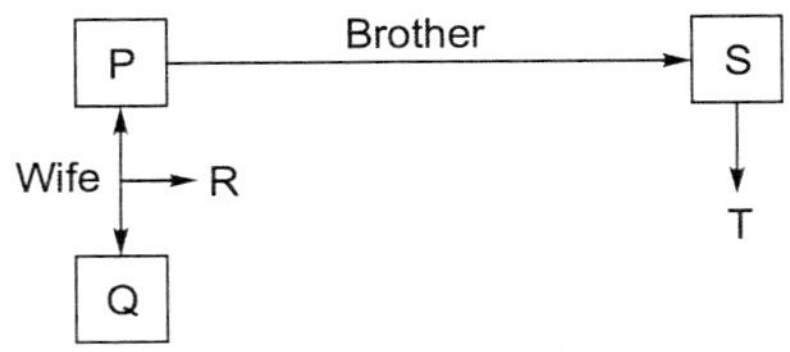

P is male,
S is male or Female
S is sister-in-law of Q may be possible
S is brother of P may be possible
S is aunt of R may be possible
S is the aunt of T is impossible in this case because given T is child of S.

1.155 (a)

All the three sports = 300 − 295

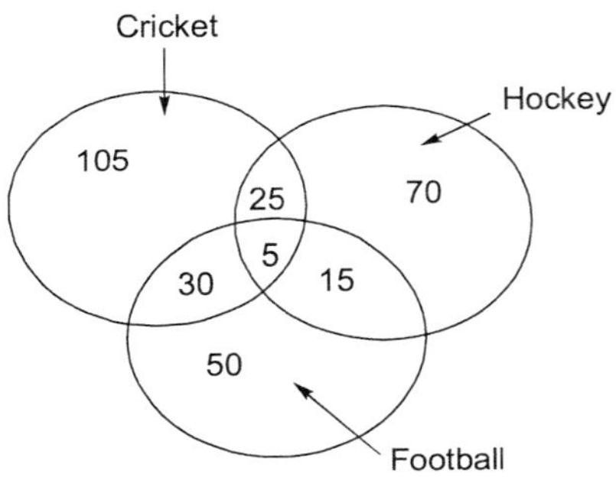

= 5

No. of players playing at least 2 sports

= 25 + 5 + 30 + 15 = 75

Percentage of players playing at least 2 sports

$$= \frac{75}{300} \times 100 = 25$$

1.156 (b)

Ram and John = 5 hours
Krishna and Amir = 2 hours
Krishna does not work with Ram.
Now pair will be
John and Krishna
Ram and Amir
Given all the workers to continue working.
John and Krishna will be one pair.

1.158 (b)

Let a, b be initial population

Given $\frac{a}{b} = p$ (ratio earlier)

$$\frac{a\left(1+\frac{x}{100}\right)}{b\left(1+\frac{y}{100}\right)} = p' \text{ (new ratio)}$$

So, required % change,

$$\frac{p'-p}{p} \times 100 = \left[\frac{\frac{a}{b}\left(\frac{100+x}{100+y}\right) - \frac{a}{b}}{\frac{a}{b}}\right] \times 100$$

$$= \frac{100(x-y)}{100+y}$$

1.159 (d)

Letinlet pipes, A=4 hours
B = 5 hours
outlet pipe, C = 20 hours
In 1 hour (A + B + C) can fill

$$= \frac{1}{4} + \frac{1}{5} - \frac{1}{20} = \frac{8}{20} = \frac{2}{5}$$

(A + B + C) together

$$= \frac{5}{2} = 2.5 \text{ hours}$$

1.160 (b)

The Newspaper is just reporting the matter, it cannot be responsible for duping the tribals.

A top forest official made statement about tribals being duped hence officials cannot be responsible for duping the tribals.

1.164 (a)

0,	1,	2,	3,	4,	5,	6,	7,	8,	9
↓	↓	↓	↓	↓	↓	↓	↓	↓	↓
O	P	Q	R	S	T	U	V	W	X

∴ 45 is coded as 'ST'.

1.165 (a)

If the product of two positive numbers should be zero, one of the number must be zero.

After subtracting 5 if a positive number should become zero, that number should be 5. If one number is 5 and the sum is 100 then the other number must be 95.

Let the two positive numbers be x and y.

∴ $x + y = 100$ ...(i)

$(x-5)(y-5) = 0$ ...(ii)

⇒ $x = 5$ or $y = 5$

If $x = 5$ then $y = 95$

If $y = 5$ then $x = 95$

∴ One of the number is 95 since 5 is not in any of the options.

1.166 (d)

Partial and Impartial are opposite, in the same way popular and unpopular are opposite words.

1.168 (c)

Belittle means to undervalue/ underestimate some as unimportant. Disparage fits in the most appropriate manner.

1.169 (d)

$$f(x) = x^2$$

$$f(f(f(x))) = f(f(x^2)) = f(x^4) = (x^4)^2 = x^8$$

So, $$\frac{f(f(f(x)))}{f(x)} = \frac{x^8}{x^2} = x^6 = (x^2)^3 = (f(x))^3$$

1.171 (a)

Option (a) satisfies the given conditions in the paragraph.

1.173 (c)

2, a, $(a + d)$, $(a + 2d)$, ... $(a + 6d)$, 34

∴ Total number of terms of $AP(n) = 9$

Let sum of seven inserted numbers = S

∴ $$S = \frac{7}{2}[a + (a + 6d)] = 7[a + 3d]$$

$$T_n = 34$$

Also, $a - 2 = (a + d - a)$

$\Rightarrow$ $a - d = 2$

Similarly $a - 2 = 34 - (a + 6d)$

$\Rightarrow$ $a - 2 = 34 - a - 6d$

$\Rightarrow$ $2a = 36 - 6d = 36 - 6(a - 2)$

$\Rightarrow$ $2a = 36 - 6a + 12$

$\Rightarrow$ $8a = 48$

$\Rightarrow$ $a = 6$

∴ $d = a - 2 = 6 - 2 = 4$

∴ $S = 7(a + 3d)$

$= 7(6 + 3 \times 4) = 126$

1.174 (b)

Let total monhly earining

= Rs. 100

Monthly spent on education

$$= \frac{15}{100} \times 100 = \text{Rs.}15$$

Monthly spent on transport

$$= \frac{10}{100} \times 10 = \text{Rs. } 10$$

% money extra spent on education as compard to transportation

$$= \frac{15 - 10}{10} \times 100 = 50\%$$

1.175 (b)

$\simeq 26591749^{(110016)}$

$\simeq$ Unit place of $9^{\text{even}} = 1$

∴ Cyclicity of 9 is (9, 1) (9, 1), (9, 1)

So answer will be 1.

1.177 (a)

$$\frac{\text{Sum of old numbers from 1 to 100}}{\text{Sum of even numbers from 150 to 200}}$$

$$= \frac{50^2}{175 \times 26} = \frac{50 \times 50}{175 \times 26} = \frac{2500}{4550}$$

Ratio = 50 : 91

From 1 to 100 = 50 odd number

From 150 to 200 = 26 even number

1.178 (*)

LED bulbs sold

By $$Y_{\text{April}} = \frac{2}{5} \times 15 = 6\% \quad \text{...(A)}$$

$$Y_{\text{May}} = \frac{4}{5} \times 10 = 8\%$$

$$Y_{\text{June}} = \frac{10}{20} \times 11 = 5.5\%$$

LED bulb sold by Y during April-June

A+ B + C = (6 + 8 + 5.5)% = 19.5%

Total LED bulb sold by X and Y 35% this value = 50000

So, LED bulb sold by Y (April to June)

$$\frac{50000}{35} \times 19.5 = 27857.142$$

[As 35% of total = 5000

$$\text{Total} = \frac{50000}{0.35} = 142857.142$$

Y (April- June) = 19.5% of total 0.195 × 1442857.14 = 27857.142]

No option is matching.

Mistake in paper was that examiner intended to give total as 50000 and the options were place accordingly as 19.5% of 50000 = 9750 which is (d).

But this will be wrong as total is NOT 50000 as per language. So correct answer is 27857.142. Which matches with none of options.

1.179 (a)

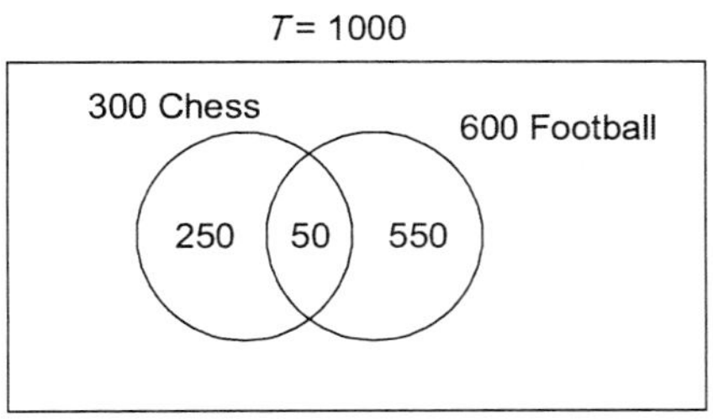

Total number of students playing sports = 850

Total number of students not playing sports = 1000 – 850 = 150

1.180 (b)

Year	Number of odd days
2013	1
2014	1
2015	1
2016	2
2017	1
2018	1
Total	7

Number of odd days in 2019 = 1, so 2013 calendar is same as 2019.

1.181 (b)

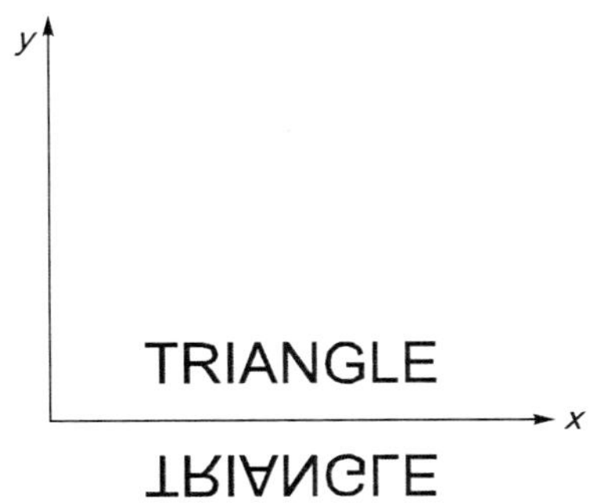

1.182 (c)

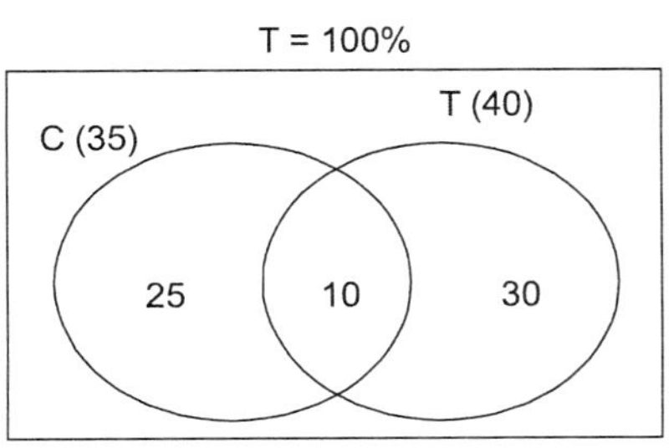

Percentage of employees who drink neither tea nor coffee = 100 – 25 – 10 – 35 = 35

1.183 (d)

Following cases can be PQSR, RQSP, QPSR, QRSP, RQPS and PQRS

and $\begin{bmatrix} P & R \\ R & P \end{bmatrix}$ not possible.

1.184 (a)

$$x \oplus y = 2 \odot 2$$

$$\frac{x^2 + y^2}{xy} = \frac{2^2}{2}$$

$$x^2 + y^2 = 2xy$$

$$(x - y)^2 = 0$$

$$x = y$$

1.185 (b)

When the comparison is between two things we use the second degree of the adjective. The degree form of easy are: (easy - easier - easiest).

1.186 (c)

$$[9 - (6 \times 7)] - [7 \times 1] = -33 - 7$$

$$= -40$$

1.187 (c)

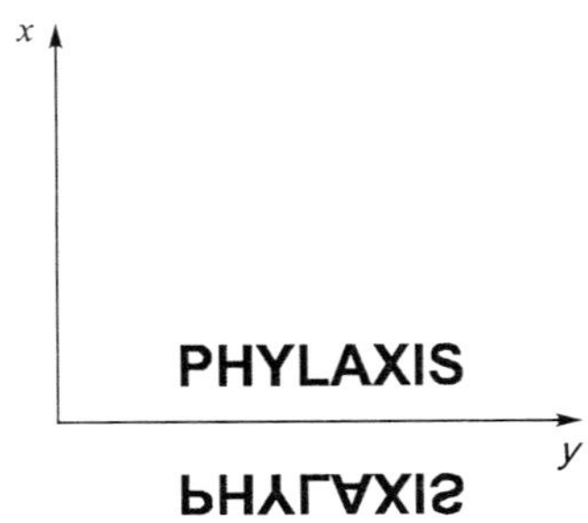

1.188 (a)

Two subject joined with 'and' become plural and hence plural verb is there in first statement, in fourth sentence the subject is family which is singular and takes singular verb.

1.189 (d)

Number of arrangements = 3 × 3! = 18

1.190 (c)

Probability of getting even number on a dice

$$= \frac{3}{6} = \frac{1}{2}$$

$\because$ Two dice are rolled simultaneously,
Hence, probability of an even number rolled out on each dice $= \frac{1}{2} \times \frac{1}{2} = \frac{1}{4}$

1.192 (b)

$$\frac{\lambda(-(-3+2),(-2+3))}{(-(-2+1))} = \lambda\frac{(1,1)}{1} = \lambda(1,1)$$

So, 1st definition will be applicable as $p = q$.
Hence, $\lambda(1, 1) = (1-1)^2 = 0$

1.193 (c)

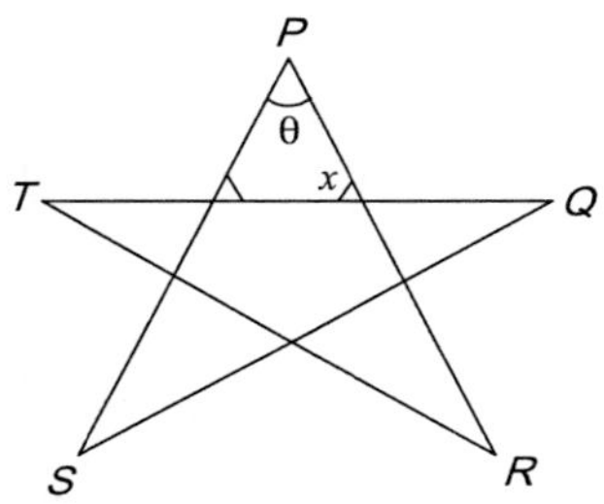

Sum of angle formed at the pentagon = 540°

Each angle of pentagon $= \frac{540}{5} = 108°$

$$\angle x = 180 - 108 = 72°$$

Sum of angle of triangle

$$= 180°$$

$$72° + 72° + \theta = 180°$$

$$\theta = 36°$$

1.194 (a)

Option (b) and (d) are weekend by the word 'UNRELATED SKILLS'. Option (c) is weekend by the expression, no species posses the ability.
Hence answer is option (a) which reflects the information given in the passage.

1.195 (a)

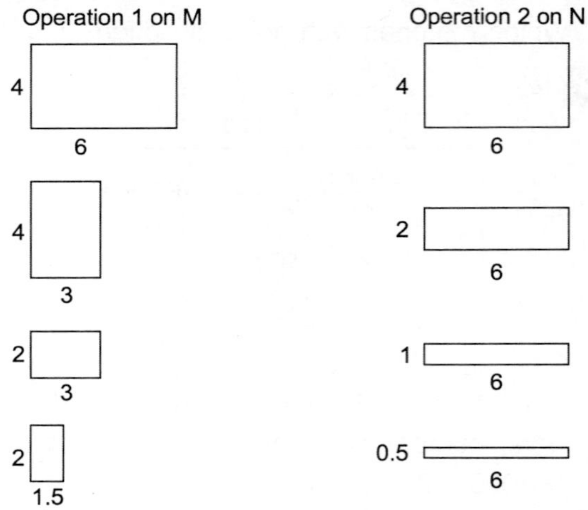

$$(\text{Perimeter})_M = 2\,(2 + 1.5) = 7$$

$$(\text{Perimeter})_N = 2\,(0.5 + 6) = 13$$

$$\text{Required ratio} = \frac{13}{7}$$

1.196 (b)

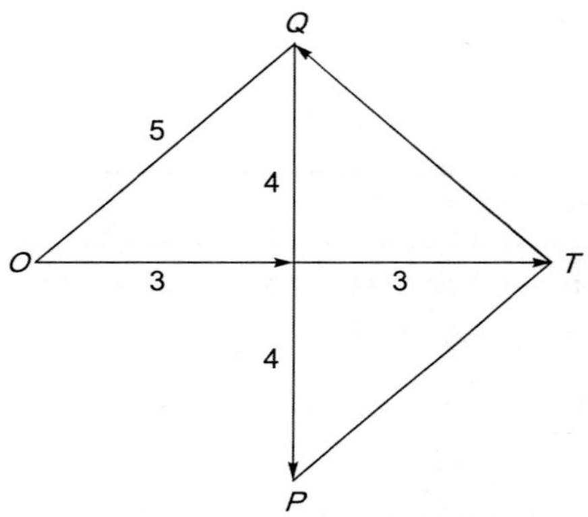

$$OQ = \sqrt{3^2 + 4^2} = 5$$

1.197 (d)

The last sentence of the passage decides the answer with the key words "No Less Important".

1.198 (b)

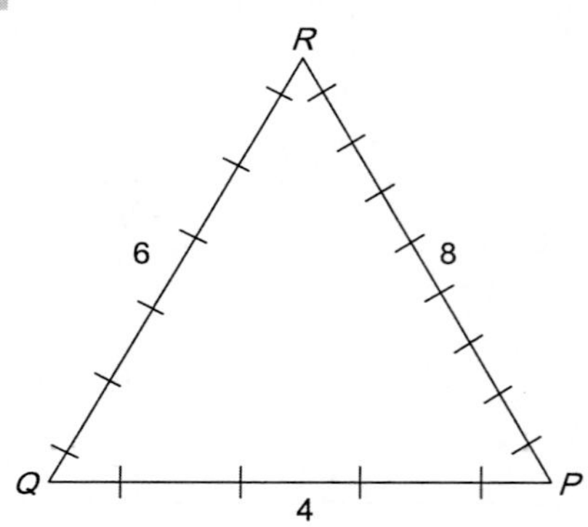

For $\left(\frac{a}{4}, \frac{a}{6}, \frac{a}{8}\right)$ to be integer, *'a'* must be LCM of 4, 6 and 8.

So $a = 24$

$$\text{Area} = \frac{\sqrt{3}}{4}a^2 = \frac{\sqrt{3}}{4} \times 24^2 = 144\sqrt{3}$$

1.199 (b)

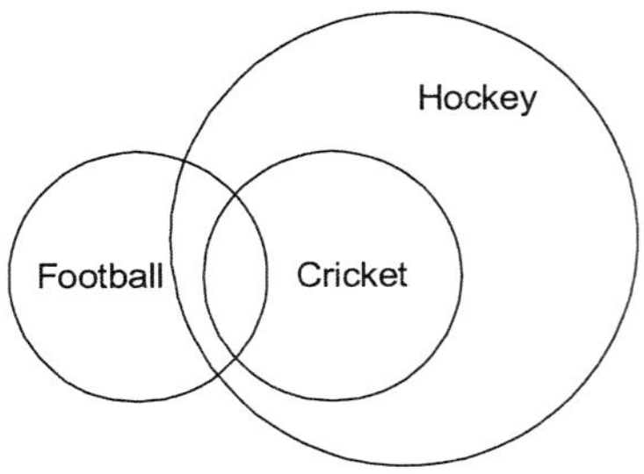

1.200 (a)

Let feasible area be *fA*, then total area is *TA*

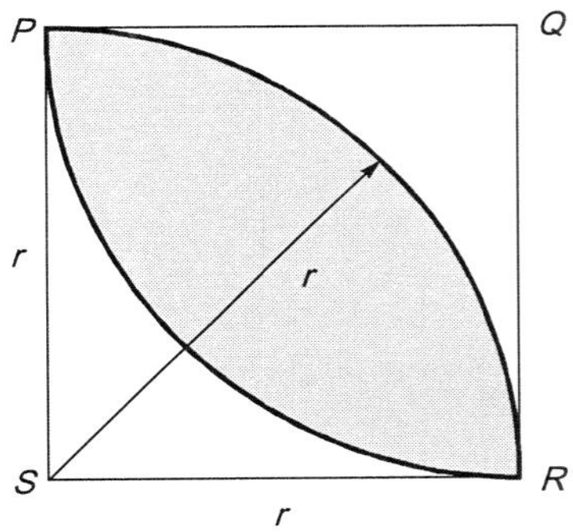

$$\text{So, Probability} = \frac{fA}{TA}$$

$$fA = \left(\frac{\pi r^2}{4} - \frac{r^2}{2}\right) \times 2$$

$$\frac{fA}{TA} = \frac{\left(\frac{\pi r^2}{4} - \frac{r^2}{2}\right) \times 2}{r^2} = \left(\frac{\pi}{2} - 1\right)$$

1.202 (a)

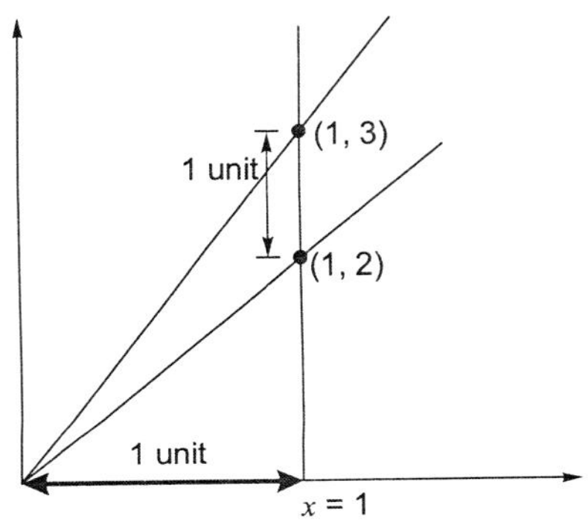

Method 1: Area =

$$\frac{1}{2} \times b \times h = \frac{1}{2} \times 1 \times 1 = 0.5 \text{ unit}^2$$

Method 2:

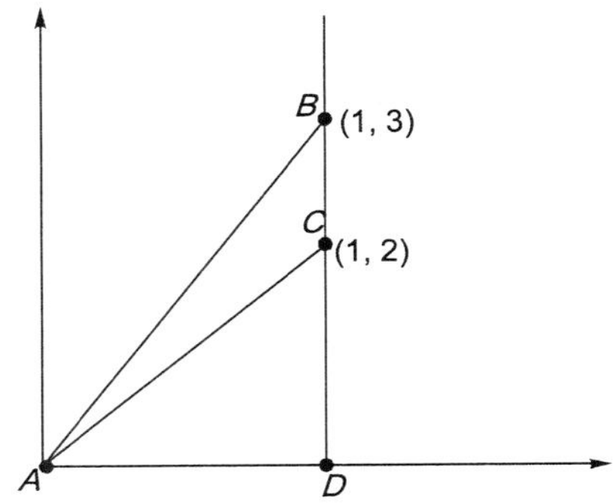

Area *ABC*,

A = Area (ABD) – Area (ACD)

$$= \frac{1}{2} \times 3 \times 1 - \frac{1}{2} \times 2 \times 1 = \frac{1}{2} = 0.5 \text{ unit}^2$$

1.203 (d)

$$\text{Given: } \frac{P}{Q} = \frac{1}{2}, \frac{Q}{R} = \frac{4}{3}$$

$$\frac{R}{S} = \frac{4}{5}$$

$$\frac{R}{S} = \frac{4 \times 3}{5 \times 3}$$

$$\frac{Q}{R} = \frac{4 \times 4}{3 \times 4}$$

$$\frac{P}{Q} = \frac{1 \times 8}{2 \times 8}$$

P	:	Q	:	R	:	S
8	:	16	:	12	:	15

$$U = 15 \times 1.5 = 22.5$$

$$\frac{P}{U} = \frac{8}{22.5} = \frac{8 \times 2}{22.5 \times 2} = \frac{16}{45}$$

1.204 (d)

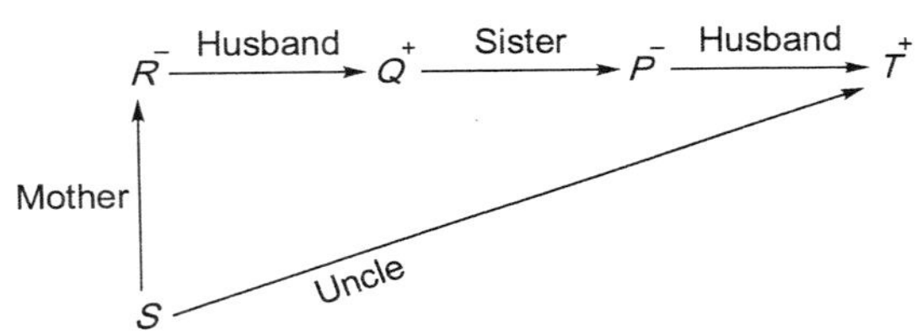

1.205 (b)

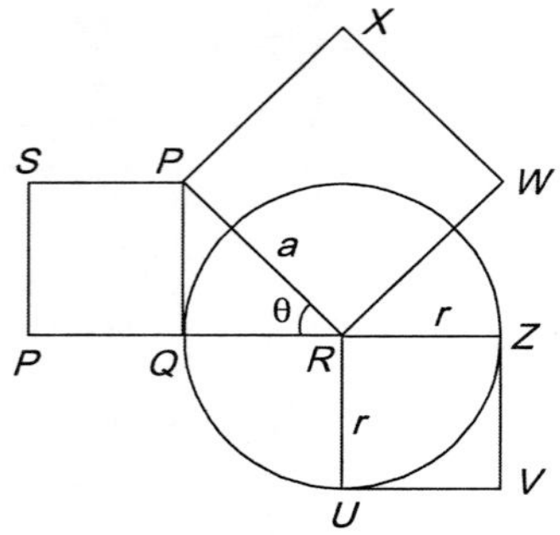

$$R = a\cos\theta$$

$$A_{RUVZ} = r^2 = a^2\cos^2\theta$$

$$A_{PXWR} = a^2$$

$$A_{SPQT} = (a\sin\theta)^2 = a^2\sin^2\theta$$

$$A_{SPQT} = A_{PXWR} - A_{RUVZ}$$

$$= a^2 - a^2\cos^2\theta$$

$$= a^2(1-\cos^2\theta)$$

$$a^2\sin^2\theta = a^2\sin^2\theta$$

1.206 (c)

The movies was funny and I couldn't help laughing.

i.e. movie was so funny that the person couldn't control his/her laugh.

1.208 (b)

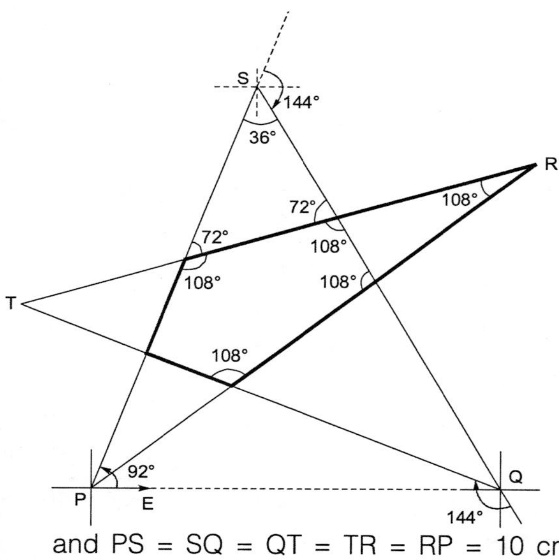

and PS = SQ = QT = TR = RP = 10 cm

1.209 (a)

$$\frac{(\text{Profit})_A}{(\text{Profit})_B} = \frac{(\text{Investment})_A}{(\text{Investment})_B} \times \frac{t_A}{t_B}$$

Given, ratio of initial investment of friends *P*, *Q*, *R* is 3 : 4 : 5 respectively.

Ratio, after increment in investment of P, Q, R.

P increase the investment by 5%.

Q increase the investment by 10%

R increase the investment by 20%

i.e. 3 + (3 × 1.1) : 4 + (4 × 1.5) : 5 + (5 × 1.2)

= 6.3 : 8.6 : 11

= 63 : 86 : 110

1.210 (a)

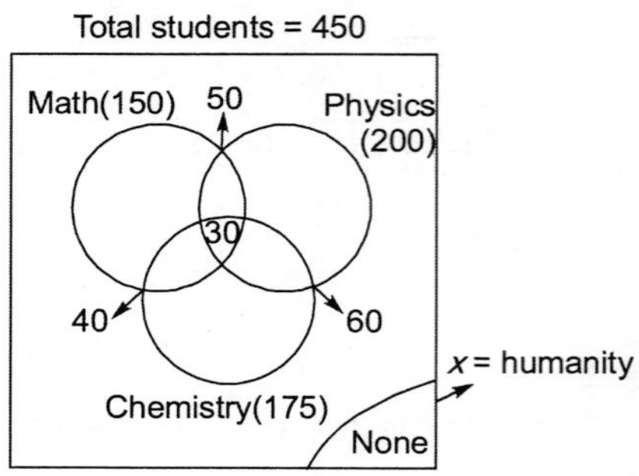

$$T = n(M \cup P \cup C) + x$$

$$450 = [(150 + 200 + 175) - (50 + 60 + 40) + 30] + x$$

$$x = 45$$

1.212 (b)

$$\frac{(P_R)_A}{(P_R)_B} = \frac{(I_A)\times T_A}{(I_B)\times T_B}$$

$$\frac{P_B}{P_{A+B}} = \frac{4}{9}$$

$$\frac{P_B}{P_A} = \frac{4}{5}$$

$$\frac{4}{5} = \frac{5000 \times 6 \text{ months}}{I_B \times 8 \text{ months}}$$

$$I_B = \text{Rs. } 3000$$

1.216 (a)

Given: $x : y : z = \frac{1}{2} : \frac{1}{3} : \frac{1}{4} = 6 : 4 : 3$

As we need to find ratio of $(x + y + z)$ to y, so directly x, y, z values can be taken as

$$x = 6,\ y = 4,\ z = 3$$

$$\therefore \text{The value of } = \frac{x + z - y}{y}$$

$$= \frac{6 + 3 - 4}{4} = \frac{5}{4} = 1.25$$

1.217 (a)

Let's take the intger as x.

Now as per given statement,

$$\frac{3+x}{4+x} = \frac{15-x}{17-x}$$

By solving for x = 3

1.218 (c)

Line L1 : $2x - 3y = 5$Line L2 : $3x + 2y = 8$

$$y = \frac{2x-5}{3}$$

$$y = \frac{8-3x}{2}$$

$$\frac{dy}{dx} = \frac{2}{3} \quad \frac{dy}{dx} = \frac{-3}{2}$$

Line L3 : $4x - 6y = 5$Line L4 : $6x - 9y = 6$

$$y = \frac{4x-5}{6}$$

$$y = \frac{6x-6}{9}$$

$$\frac{dy}{dx} = \frac{2}{3}$$

$$\frac{dy}{dx} = \frac{2}{3}$$

$$\therefore \quad \left(\frac{dy}{dx}\right)_{L1} = \left(\frac{dy}{dx}\right)_{L3} = \left(\frac{dy}{dx}\right)_{L4} = \frac{2}{3}$$

and $\left(\frac{dy}{dx}\right)_{L2} \times \left(\frac{dy}{dx}\right)_{L1}$

$$= \left(\frac{dy}{dx}\right)_{L2} \times \left(\frac{dy}{dx}\right)_{L3} = \left(\frac{dy}{dx}\right)_{L2} \times \left(\frac{dy}{dx}\right)_{L4} = -1$$

∴ L1, L3, L4 are parallel to each other and perpendicular to L2.

1.220 (c)

Statement 1: Vane diagram

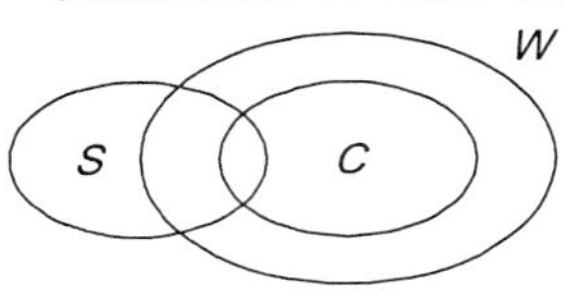

S = Soaps
C = Clean object
W = Wet

Statement 2: Vane diagram

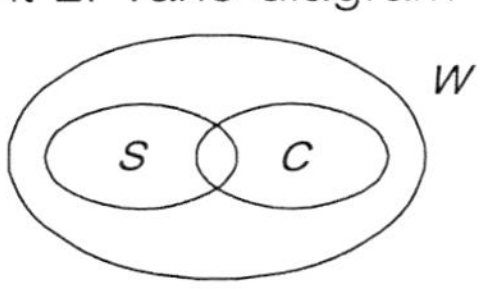

From the above vane diagram we can conculate that conclusion I and conclusion III are correct and conclusion II and IV are in correct.

1.225 (a)

For a polygon with n sides,

Interior angle sum = $(n - 2) \times 180°$

Each angle of regular Hexagon.

$$\frac{(n-2)\times 180°}{n} = \frac{(6-2)\times 180°}{6} = 120°$$

Area of shaded region

$$= \frac{\theta}{360°}\times\pi\times r^2 = \frac{120°}{360°}\times\pi\times 5^2 = \frac{25\pi}{3}$$

1.226 (b)

If box d or i is coloured black, we get one line of symmetry, i.e., horizontal or vertical, respectively. But for two line of symmetry, we need to atleast colour 3 boxes, i.e., c, d and i.

1.227 (d)

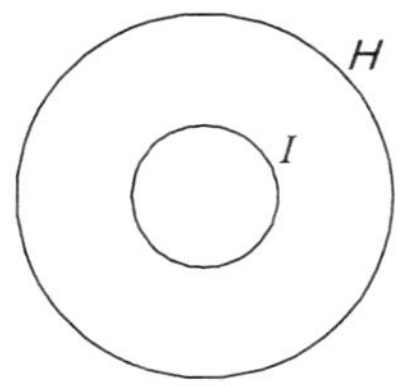

or

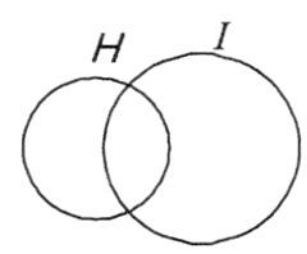

1.229 (a)

Given,

a =30!, b = 50! and c = 100!

We know that $\log_n m = \frac{\log m}{\log n}$

$\log_a c, \log_c a, \log_a b, \log_b a$

$$\frac{\log c}{\log a}, \frac{\log a}{\log c}, \frac{\log b}{\log a}, \frac{\log a}{\log b}$$

$$\Rightarrow \frac{\log 100!}{\log 30!}, \frac{\log 30!}{\log 100!}, \frac{\log 50!}{\log 30!}, \frac{\log 30!}{\log 50!}$$

$$\therefore \frac{\log 30!}{\log 100!} < \frac{\log 30!}{\log 50!} < \frac{\log 50!}{\log 30!} < \frac{\log 100!}{\log 30!}$$

$$\log_c a < \log_b a < \log_a b < \log_a c$$

1.230 (a)

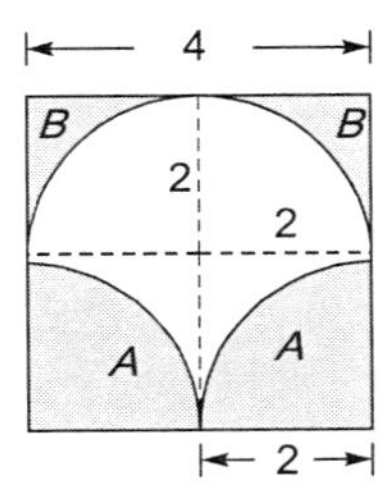

Area of region, $A = \frac{1}{4}\pi(2)^2 = \pi$

Area of region, B = Area of small square – Area of quadrant of circle

$= 4 - \pi$

So, total shaded area $= 2A + 2B$

$= 2 \times \pi + 2 \times (4 - \pi)$

$= 2\pi + 8 - 2\pi = 8$ unit^2

1.235 (d)

Since only one cell is to be shaded in each row and each column. So if one cell is shaded so the size of original square (3 × 3) reduces to (2 × 2) grid square. Thus, number of possible ways = 3 × 2 × 1 = 6

1.236 (a)

Given: 4 Red, 5 Green and 6 Blue balls,

So, in worst case, we can select 3 balls each with different colour, i.e., 1 Red, 1 Green and 1 Blue ball. Now if we select the forth ball, then we found two balls are of same colour. Thus, smallest number of balls that guarantees we have at least two balls of same colour is 4.

1.237 (a)

Operation 1:

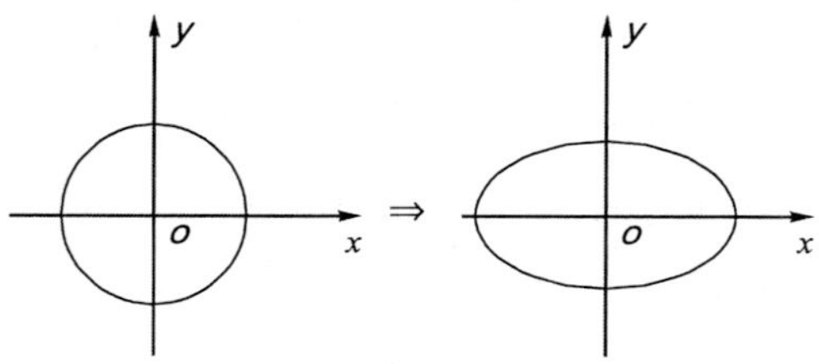

Suppose, $S_{x\text{-axis}} > S_{y\text{-axis}}$

Operation 2:

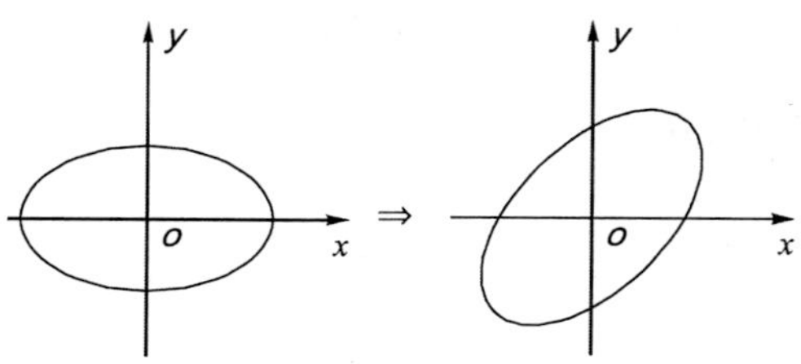

1.238 (a)

Let the three husband – wife pairs be $H_1 W_1$, $H_2 W_2$ and $H_3 W_3$

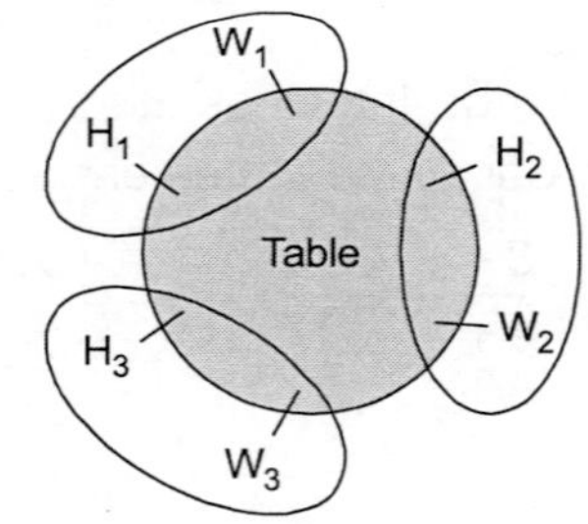

Treat every husband-wife pairs as one block. So, there are three pairs and all three pairs can be arranged by (3 – 1)! ways and in all pairs husband and wife can be arranged by 2! ways.

So, required number of ways are $(3-1)! \times 2! \times 2! \times 2! = 2! \times 2^3 = 16$

1.239 (c)

Given: $4^{8^x} = 256$

$\Rightarrow \quad 4^{8^x} = 4^4$

$\Rightarrow \quad 8^x = 4$

$\Rightarrow \quad \left(2^3\right)^x = 2^2$

So, $\quad x = \dfrac{2}{3}$

1.244 (d)

Given, $p > 0$, $q > 0$ and $\dfrac{p}{q} \neq 1$

$$\left(\frac{p}{q}\right)^{p/q} = p^{\left(\frac{p}{q}-1\right)}$$

$$\left(\frac{p}{q}\right)^{p/q} = p^{\left(\frac{p-q}{q}\right)}$$

$$\Rightarrow \quad \left(\frac{p}{q}\right)^{p} = p^{p-q}$$

$$\Rightarrow \quad \left(\frac{p}{q}\right)^{p} = \frac{p^p}{p^q}$$

$$\Rightarrow \quad q^p = p^q$$

1.245 (c)

Between 12:05:00 to 12:55:00

There are 50 intervals of 1 minute in which a second hand and minute hand crosses one another time. So during 12:05:00 to 12:55:00, minute and second hand crosses 50 times.

1.248 (b)

$$3 \times 2 + 1 = 7$$
$$7 \times 2 + 1 = 15$$
$$15 \times 2 + 1 = \boxed{31}$$
$$31 \times 2 + 1 = 63$$
$$63 \times 2 + 1 = 127$$
$$127 \times 2 + 1 = 255$$

1.250 (d)

301 303 305 ------

302 304 306 ------

Let sum of 301 + 303 + 305 + ... = A
and 302 + 304 + 306 + ... = B

Given, A − B = 27
and 302 − 301 = 1
304 − 303 = 1

As difference of one pair is 1 and total difference is 27.

∴ Number of house in one side is 27.

1.252 (c)

Inference I ×
Inference II ✓
Inference III ✓

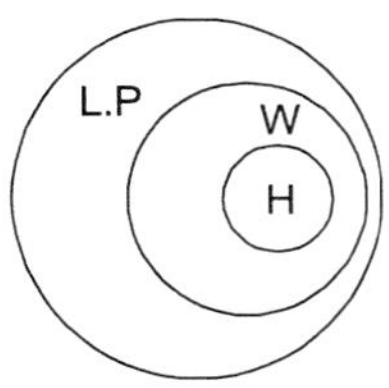

1.253 (b)

$$\frac{p \times q - \frac{p}{q}}{p \times q} = \frac{80}{100} = \frac{4}{5}$$

$$\frac{pq - \frac{p}{q}}{pq} = \frac{4}{5}$$

$$\Rightarrow 1 - \frac{1}{q^2} = \frac{4}{5}$$

$$\Rightarrow \frac{1}{q^2} = \frac{1}{5}$$

$$\Rightarrow q = \sqrt{5}$$

1.254 (d)

Volume of outer cylinder = $\pi R^2 h$

Volume of smaller cylinder = $\pi r^2 h$

$$R = 3r$$

Ratio of occupied space to empty space

$$= \frac{7 \times \pi r^2 h}{9\pi r^2 h - 7 \times \pi r^2 h} = \frac{7}{2}$$

1.257 (a)

The sum of first n odd natural number = n^2

$$\therefore \text{ Result is } \frac{20^2}{20^2} = 1$$

1.258 (a)

	Class VIII	Class IX
Total students	450	360
Number of girls	G	G
Number of boys	450 – G	360 – G

According to question,

$$\frac{G}{450 - G} = \frac{360 - G}{G}$$

$$\Rightarrow G^2 = 360 \times 450 - 450\,G - 360\,G + G^2$$

$$\Rightarrow 810\,G = 360 \times 450$$

$$\Rightarrow G = \frac{360 \times 450}{810} = 200$$

1.260 (c)

- Bicker → To argue about un-important thing or issue
- Quarrel → Angry, argument or disagreement
- Feud → Angry or serious argument between two people that continues for long period of time.

■■■■